Participants at the International Symposium on Sulfur in the Atmosphere, Dubrovnik, September 1977

SULFUR IN THE ATMOSPHERE

PROCEEDINGS OF THE INTERNATIONAL
SYMPOSIUM HELD IN DUBROVNIK
YUGOSLAVIA, 7–14 SEPTEMBER 1977

Other Titles of Interest

BURTON:	Computers and Operations Research Environmental Applications
FUKUSHIMA:	Science for Better Environment
HUGHES-EVANS:	Environmental Education
KATZ:	Air Pollution Problems of the Inorganic Chemistry Industry
ROGERS:	A Short Course in Cloud Physics, 2nd Edition
UNITED NATIONS ECONOMIC COMMISSION FOR EUROPE:	The Gas Industry and the Environment

NOTICE TO READERS

Dear Reader

If your library is not already a standing order customer or subscriber, may we recommend that you place a standing subscription order to receive immediately upon publication all new issues.

The Editors and the Publisher will be glad to receive suggestions or outlines of suitable titles, reviews or symposia for consideration for rapid publication.

Robert Maxwell

Publisher at Pergamon Press

Husar *et al.* Sulfur in the Atmosphere

ERRATA

The publishers and editors regret the following errata:

p. 498: column 2, Table 2: $g\,m^2$ should read $g\,m^{-2}$;

p. 442: Fig. 14 should be replaced by the following:

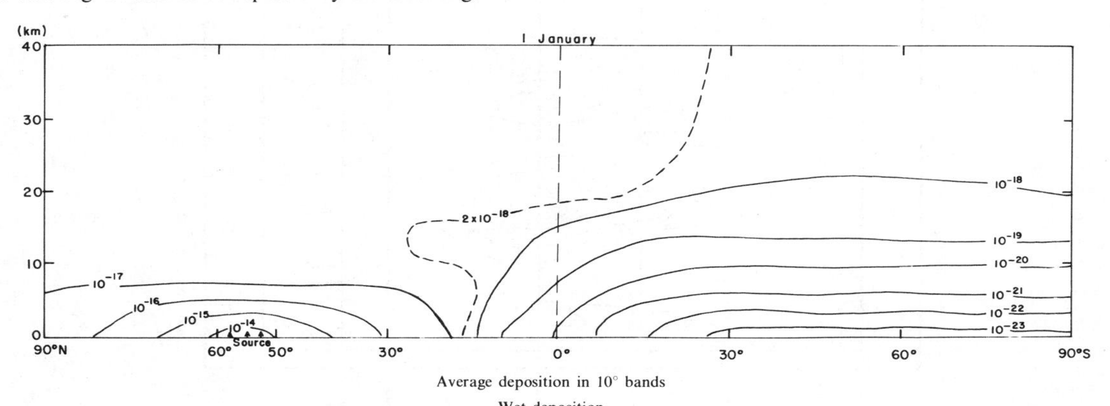

Average deposition in 10° bands

Wet deposition

3.4E-16 5.9E-15 5.4E-14 3.4E-13 5.0E-14 2.9E-15 1.8E-16 3.1E-17 3.0E-18 3.1E-19 2.9E-20 2.2E-21 1.1E-21 1.3E-21 2.1E-21 2.0E-21 8.0E-22 3.8E-22

Dry deposition

3.6E-16 4.5E-15 3.3E-14 1.8E-13 2.3E-14 1.8E-15 1.5E-16 9.1E-18 3.2E-19 1.2E-20 6.7E-22 1.2E-22 4.3E-23 2.4E-23 2.0E-23 1.9E-23 2.5E-23 4.3E-23

Wet & dry deposition

7.0E-16 1.0E-14 8.7E-14 5.2E-13 7.3E-14 4.7E-15 3.3E-16 4.0E-17 3.2E-19 3.2E-19 2.9E-20 2.3E-21 1.2E-21 1.3E-21 2.1E-21 2.0E-21 8.3E-22 4.2E-22

Units of concentration grams of sulfur/grams of air, units of deposition grams of sulfur $m^{-2}\,s^{-1}$, E = power of 10, e.g., E-20 = 10^{-20}, source: 16 grams of sulfur s^{-1} in 50–60°N at ground level.

p. 744: Figs. 8 and 9 should be replaced by the following:

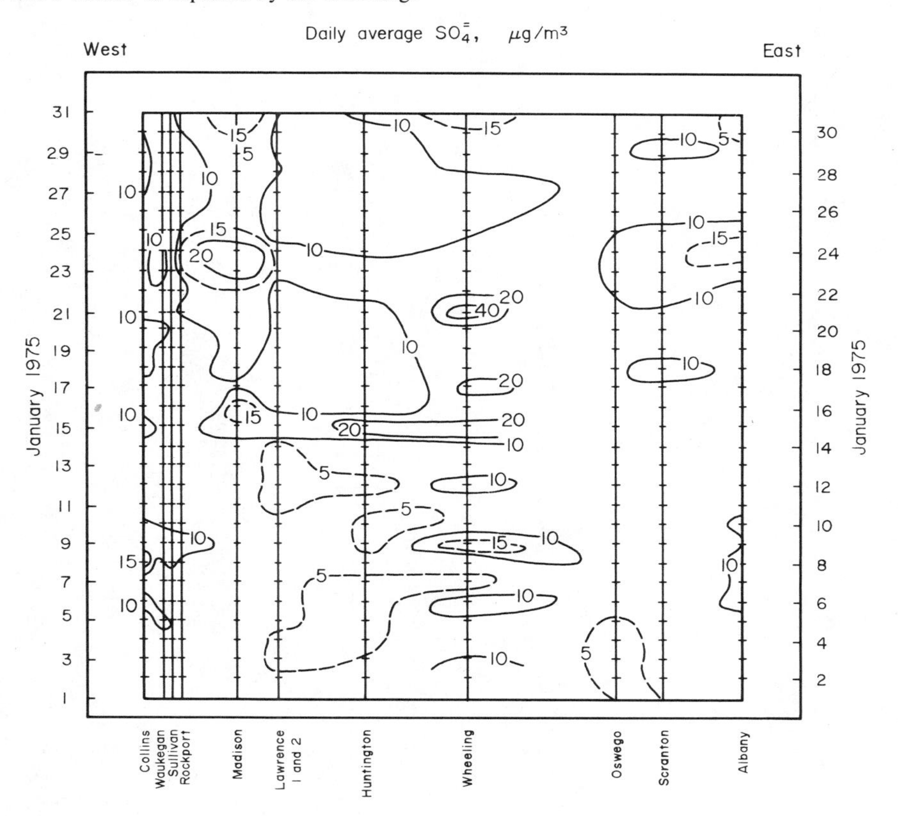

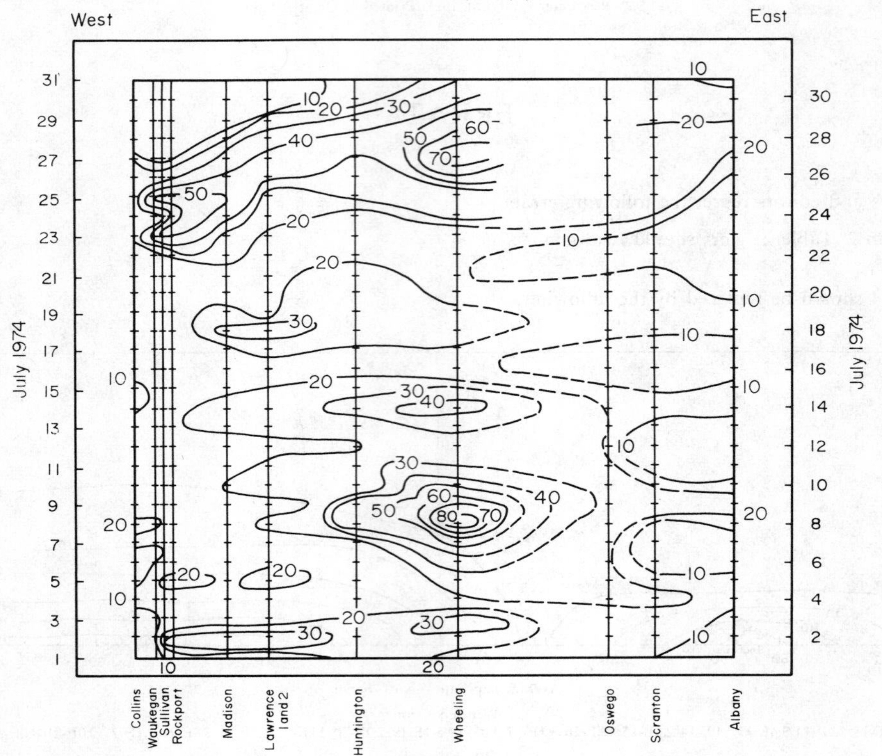

Daily average $SO_4^=$, $\mu g/m^3$
West
East
July 1974
Collins
Waukegan
Sullivan
Rockport
Madison
Lawrence 1 and 2
Huntington
Wheeling
Oswego
Scranton
Albany

SULFUR IN THE ATMOSPHERE

PROCEEDINGS OF THE INTERNATIONAL SYMPOSIUM

HELD IN

DUBROVNIK, YUGOSLAVIA

7–14 SEPTEMBER 1977

Editors:

R. B. HUSAR

J. P. LODGE, Jr.

and D. J. MOORE

Sponsored
by
UNITED NATIONS ENVIRONMENT PROGRAMME
Electric Power Research Institute
U.S. Environmental Protection Agency
U.S. Energy Research and Development Administration

PERGAMON PRESS

OXFORD · NEW YORK · TORONTO · SYDNEY · PARIS · FRANKFURT

U.K.	Pergamon Press Ltd., Headington Hill Hall, Oxford OX3 0BW, England
U.S.A.	Pergamon Press Inc., Maxwell House, Fairview Park, Elmsford, New York 10523, U.S.A.
CANADA	Pergamon of Canada Ltd., 75 The East Mall, Toronto, Ontario, Canada
AUSTRALIA	Pergamon Press (Aust.) Pty. Ltd., 19a Boundary Street, Rushcutters Bay, N.S.W. 2011, Australia
FRANCE	Pergamon Press SARL, 24 rue des Ecoles, 75240 Paris, Cedex 05, France
FEDERAL REPUBLIC OF GERMANY	Pergamon Press GmbH, 6242 Kronberg-Taunus, Pferdstrasse 1, Federal Republic of Germany

First edition 1978

British Library Cataloguing in Publication Data

International Symposium on Sulfur in the Atmosphere, *Dubrovnik, 1977*
Sulfur in the atmosphere.
1. Sulphur compounds – Environmental aspects – Congresses 2. Air – Pollution – Congresses
I. Title II. Husar R B III. Lodge, J P
IV. Moore, D J V. Atmospheric environment
614.7'1 TD885.5.S/ 78-40442

ISBN 0-08-022932-8

Published as Volume 12, Number 1/3, of the journal *Atmospheric Environment* and supplied to subscribers as part of their subscription. Also available to non-subscribers.

Printed in Great Britain by A. Wheaton & Co Ltd, Exeter, Devon

CONTENTS

LIST OF PARTICIPANTS

AUSTRIA

Chalupa, K.

BRAZIL

de Agostinho, J.

CANADA

Barrie, L.A.
Berlie, E.M.
Gormley, F.
Whelpdale, D.M.

CZECHOSLOVAKIA

Kulhanek,
Moldan, B.
Novák, J.
Prusik, B.

DENMARK

Berkowitz, R.
Fenger, J.
Flyger, H.
Prahm, L.P.

FINLAND

Haapala, K.
Lattila, H.
Rantakrans, E.

FRANCE

Boulad, D.
Bricard, J.
Cochrane, C.A.
Delmas, R.
Déspres, A.
Detrie, J.P.
Garnier, A.
Le Parmentier, L.
Madelaine, G.J.
Petit, C.
Reilhac, B.
Zettwoog, P.

GERMANY D.R.

Dietze, G.

GERMANY F.R.

Beilke, S.
Bingemer, H.
Eickel, K.H.
Georgii, H.W.
Gestrich, W.
Gravenhorst, G.
Güsten, H.
Haury, G.
Heits, B.
Israel, G.
Jaenicke, R.

GERMANY F.R. (continued)

Jaeschke, W.
Kümmerer, F.
Lahmann, E.
Platt, U.
Roedel, W.
Roos, M.
Toro, P.
Weber, E.
Zierock, K.H.

HUNGARY

Mészáros, A.
Mészáros, E.
Szepesi, D.
Várhelyi, G.

INDIA

Dave, J.M.
Kelkar, D.N.

IRAQ

Akrawi, A.A.

ITALY

Briattore, L.
Cojutti, A.
Corsi, P.
Fassina, V.
Lazzarini, L.
Liberti, A.
Morelli, F.
Santomauro, L.

KENYA

Mukolwe, E.A.A.

THE NETHERLANDS

Blokker, P.C.
Elshout, A.J.
Slanina, J.
Van Aalst, R.M.
Van der Kooij, J.

NORWAY

Dovland H.
Eliassen, A.
Isaksen, I.
Ottar, B.
Semb, A.

POLAND

Werner, J.W.

SWEDEN

Ahlberg, M.S.
Brosset, C.

SWEDEN (continued)

Granat, L.
Gruvberg, C.
Lindquist, O.
Omsted, G.
Persson, C.
Rodhe, H.

USSR

Maslenikov, V.M.
Rozanov, A.G.
Salov, A.

UNITED KINGDOM

Apsimon, H.M.
Barnes, R.A.
Chamberlain, A.C.
Eggleton, A.
Fisher, B.E.A.
Fowler, D.
Garland, J.
Marsh, A.R.W.
Martin, A.
Moore, D.J.
Reed, L.E.
Scriven, R.A.
Sharkey, J.
Smith, F.B.

UNITED STATES

Beadle, R.W.
Blumenthal, D.L.
Beyerly, R.
Calvert, J.C.
Charlson, R.J.
Cantrell, B.K.
Cobourn, W.G.
Cukor, P.
Friedlander, S.K.
Freiberg, J.
Friend, J.P.
Gage, S.J.
Gillani, N.V.
Eatough, D.J.
Hakkarinen, C.
Hales, J.M.
Hidy, G.M.
Hobbs, P.V.
Hudson, F.P.
Husar, J.
Husar, R.B.
Huntzicker, J.J.
Johnson, W.B.
Kittelson, D.B.
Krey, P.
Krupa, S.V.
Lewellen, W.S.
Lodge, J.P.
Lyons, W.A.
Loo, B.W.
MacCracken, M.C.
McKenzie, R.L.
Middleton, P.

UNITED STATES (continued)

Miller, D.F.
Newman, L.
Olson, T.L.
Pack, D.
de Pena, J.
de Pena, R.
Perhac, R.
Robinson, E.
Schwartz, S.E.
Smith, L.
Stevens, R.K.
Van Valin, C.C.
Whitby, K.T.
White, W.H.
Wilson, W.E., Jr.

YUGOSLAVIA

Bošković, T.
Ćirić,
Fugaš, M.
Gavrilović, M.
Gentilizza, M.
Gburčik, P.
Spasova, D.
Tomić, T.
Vukmirović, Z.
Zupančić, T.

UNEP

Wallén, C.C.

UNESCO

Teller, H.L.

UNECE

Janczak, J.

WMO

Munn, R.E.

Atmospheric Environment Vol. 12, p. 1. Pergamon Press 1978. Printed in Great Britain.

EDITORIAL

'Atmospheric Environment' welcomes the opportunity to publish the Proceedings of the International Symposium on Sulfur in the Atmosphere.

Each of the published papers has been subject to the normal review procedure of this Journal and most have been modified as a result of the referees' comments and/or the discussion at Dubrovnik. In order to achieve publication early in 1978, it has been necessary to exclude some papers where a final version could not be agreed in time.

Editing so many papers on such a short time scale has presented many problems and was possible only because of the excellent co-operation received from the majority of referees and authors, to whom we should like to express our appreciation.

The last three days of the Symposium were given over to workshop sessions. Each of the four working groups produced a report on their deliberations. Abridged versions of their reports, with internal inconsistencies and overlap eliminated as far as was possible without changing the agreed content, are presented in semi-telegraphic style at the front of this issue of the Journal. Discussion of this material and of the Symposium Papers, will, as far as possible, be presented in a single issue of this Journal later this year.

A. E. 12—1/3—A2

Atmospheric Environment Vol. 12, pp. 3-5. Pergamon Press 1978. Printed in Great Britain.

FOREWORD

Where does all the sulfur go?

The daily per capita emission of SO_2 in the U.S., for instance, is about half a kilogram. All that sulfur returns to the ground following its atmospheric residence, but how long will it reside in the atmosphere and what happens during its transport? A particular concern today is what fraction of the SO_2 passes through the environmentally more harmful aerosol phase before being removed.

In the past, the transfer of anthropogenic sulfur through the atmospheric environment has been assessed through the use of "global sulfur cycles" (Eriksson, 1960; Junge, 1963; Robinson and Robbins, 1970; Kellogg et al., 1972; Friend, 1973) with the primary objective to determine the fraction contributed to the global sulfur budget by man. Most recently, Granat et al. (1976) have attributed about 60% of the atmospheric sulfur to man-made sources. While the assessment of the global sulfur budget is necessary and desirable, it is recognized that the bulk of man-made sulfur emissions and their environmental impacts are concentrated in two regions: North-Central Europe and the northeastern United States and adjacent portions of Canada. This necessitates the development of regional budgets (Rodhe, 1972).

The sulfur budget of a region, or for that matter of a single source, is determined by the source configuration and strength as well as by the interaction of transport, transformation and removal (i.e. transmission) processes. Following "Sweden's Case Study for the United Nations Conference on the Human Environment" (1971), a major research project was conducted under the OECD to understand the long-range transmission of sulfur compounds over NW Europe (LRTAP) (Ottar, 1977). Extensive research projects have also been initiated in North America. Reports from the Canadian sulfate transport study (Whelpdale, 1977), the Sulfate Regional Experiment (SURE, Perhac, 1977), the Multi-State Atmospheric Power Production Pollutant Study (MAP3S, MacCracken, 1977), the Midwest Interstate Sulfur Transformation and Transport Study (MISTT, Wilson, 1977) and the Integrated Technology Assessment Study of EPA (Gage et al. 1977) constitute a major component of ISSA.

In the OECD Long Range Transport of Air Pollutants Study, an emission inventory over the region and meteorological transport parameters were used as input into trajectory models for the assessment of transformation and removal rates and for overall budget estimates. The model rate constants for transformation and removal were tuned to provide the best fit between calculated and measured concentration values from monitoring stations. The great utility of this approach is that it incorporates the major processes influencing the sulfur budget, and the rate constants are inherently averages over all sources and over the temporal and spatial scales of interest. It is recognized, however, that the "rate constants" for transformation and removal are actually variables. Past field and laboratory research on SO_2 oxidation rates and mechanisms have implicated temperature, humidity, solar radiation, in-cloud residence time, catalytic heavy metals, soot, ozone, ammonia, hydrocarbons and numerous factors as rate-controlling in the conversion kinetics. The discussion of these processes constitute the scientifically most intriguing component of ISSA.

The burden of establishing the actual mechanisms and rates falls upon the field experiment. The gap between regional scale studies with spatial extent of 1000 km or more and controlled laboratory simulation experiments may be bridged by mesoscale studies of sulfur transmission, also known as plume studies (Wilson, 1977).

Visibility reduction, acid rain, possibly effects on human health, and weather and climate modification are evidently associated with the reaction products rather than with SO_2 itself. Since the sulfate formation typically continues over a period of a day or more, the near field dispersion and maximum ground level concentrations of primary pollutants were not considered in detail. The physical-chemical properties of the aerosol play key roles in determining the adverse effects. Several invited and contributed papers focus on the size distribution and chemical composition of aerosol sulfur compounds, as well as on the associated measurement and monitoring techniques.

The scope of ISSA is restricted to the transport, transformation and removal of sulfur emissions, and to the properties of sulfur compounds in the atmosphere. For the sake of scientific rigor and depth, the important questions related to effects are not considered. Source control techniques have also been omitted from consideration.

The ultimate objective of the symposium is to aid the development of optimal strategies for the control of man-made sulfur compounds. However, control decisions must also incorporate social, economic and political considerations, as well as a value judgement regarding the relative importance of various effects. Given the effect to be minimized (e.g. acid rain, visibility reduction, etc.), the questions that arise are: What compounds are responsible? What are their sources? What are the design or environmental parameters that promote their formation? What changes in source design or emission pattern would minimize a specific effect? While the explicit discussion of the control strategies was beyond the scope of ISSA, the following points, as perceived by us, have implications to the development of control strategies:

1. On global scale, man-made sources contribute about 60% of the total sulfur emissions. In the hot-spot regions of man-made S emissions (NW Europe and NE U.S. and adjacent Canada) accounting for about 1% of the earth surface area, natural sources are not important for the sulfur budget.
2. About 20-50% of the SO_2 oxidizes in the atmosphere to sulfate, while the rest is removed by the vegetation, ground or rain (snow) within about a day or about 500 km. On the average, sulfate resides in the atmosphere for 3-5 days and it is transported to about 1000 to 3000 km from the source.
3. The sulfate formation rate is evidently enhanced by photochemical decay products of hydrocarbon-NO_x emissions.

It is fortunate for ISSA that many of the key researchers in the area of atmospheric sulfur agreed to participate as members of the steering committee, to prepare invited lectures, or to contribute current research results. In Dubrovnik, 165 participants from 22 countries (five continents) presented 85 papers during the first four days. The last three days were devoted to workshop and general discussion. Publication of the proceedings in this issue of Atmospheric Environment a) assured a formal review procedure; b) permits a wider distribution of papers; c) enables referencing of papers to be made to the authors and the journal. This swift publication was made possible by the interest, intensive labor and the close, active involvement of the Executive Editors, Drs. D.J. Moore and J.P. Lodge, Jr. in the preparation of ISSA.

The symposium was co-sponsored by the United Nations Environmental Program, Electric Power Research Institute, U.S. Environmental Protection Agency (Federal Interagency Energy/Environmental Research and Development Program), U.S. Energy and Development Administration, and the American Meteorological Society, and it was under the auspices of Yugoslav Academy of Sciences and Arts. Their support is gratefully appreciated.

To us, organizing the symposium was personally and professionally a most touching and rewarding (and demanding) experience. The organizational effort was shared by our Yugoslav host, Ms. Mirka Fugas, and her co-workers, by Dr. Ottar, as well as by the members of the steering committee: Drs. A. Chamberlain, R.J. Charlson, M. Fugas, H.W. Georgii, G.M. Hidy, MC. MacCracken, E. Meszaros, R.E. Munn, B. Ottar, R.M. Perhac, H. Rodhe, C.C. Wallen, K.T. Whitby and W.E. Wilson, Jr. Their contributions are herewith gratefully acknowledged.

St. Louis, Missouri USA October 1977 Rudolf and Janja Husar

Eriksson E. (1960) The yearly circulation of chloride and sulfur in nature: meteorological, geochemical and pedalogical implications. Part II. Tellus 12, 63-109.

Friend J.P. (1973) The global sulfur cycle. In Chemistry of the Lower Atmosphere. Rasool, S.I. (ed.), pp. 177-201. Plenum Press, New York.

Gage S.J., Smith L.F., Cukor P., Nieman B.L. (1977) Long Range transport of SO_x/MSO_4 from the U.S. EPA/Teknekron Integrated Technology Assessment of Electric Utility Energy Systems. ISSA.

Granat L., Rodhe H. and Hallberg R.O. (1976) The global sulfur cycle. In Svensson, B.H. and Soderlund, R. (eds.), Nitrogen, Phosphorus and Sulphur - global cycles. SCOPE Report 7, Ecol. Bull. (Stockholm) 22, 89-134.

Junge C.E. (1963) Sulfur in the atmosphere. J. Geophys. Res. 68, 3975-3976.

Kellogg W.W., Cadle R.D., Allen E.R., Lazrus A.L. and Martell E.A. (1972) The sulfur cycle. Science 175, 587-596.

MacCracken M.C. (1977) MAP3S: An investigation of atmospheric, energy related pollutants in the northeastern United States. ISSA.

Ottar B. (1977) An assessment of the OECD study of Long Range Transport of Air Pollutants. ISSA.

Perhac R.M. (1977) Sulfate Regional Experiment in Northeastern United States: The 'SURE' program. ISSA.

Robinson E. and Robbins R.C. (1970) Gaseous sulfur pollutants from urban and natural sources. J. Air. Pollut. Control Ass. 20, 303-306.

Rodhe H. (1972) A study of the sulphur budget for the atmosphere over Northern Europe. Tellus 24, 128-138.

Sweden's Case Study (1971) Air pollution across national boundaries: The impact on the environment of sulfur in air and precipitation. Stockholm: Royal Ministry for Foreign Affairs and Royal Ministry of Agriculture, 96 pp.

Whelpdale D.M. (1977) Large scale atmospheric sulfur studies in Canada. ISSA.

Wilson W.E. Jr. (1977) Sulfates in the atmosphere: A progress report on Project MISTT (Midwest Interstate Sulfur Transformation and Transport). ISSA.

Atmospheric Environment Vol. 12, pp. 7-23. Pergamon Press 1978. Printed in Great Britain.

WORKSHOP 1a: PROPERTIES AND METHODS

Atmospheric sulfur is distributed in space and time in compounds in gas, liquid, and solid phases.

Gaseous S compounds are primarily SO_2 and H_2S, secondarily other inorganic and organic species.

Liquid phase S is found in hydrometeors, liquid particles, or liquid films on solid particles. Sulfur is present in the systems

H_2O (ℓ) - H_2SO_4 (ℓ)

H_2O (ℓ) - SO_2 (aq) - SO_2 (g).

Equilibria in these systems are usually coupled to equilibria in the systems

H_2O (ℓ) - NH_3 (aq) - NH_3 (g)

H_2O (ℓ) - HNO_3 (aq) - HNO_3 (g)

Solid phases containing S are numerous, but proof of their existence in aerosols (i.e., identification of crystalline lattices) is limited. The principal compounds seem to include $(NH_4)_2SO_4$, $(NH_4)_3H(SO_4)_2$, NH_4HSO_4, $CaSO_4$, $MgSO_4$, Na_2SO_4, sulfites of transition metals, S(0), and organosulfur compounds. The role of S compounds in nucleation seems proven for H_2SO_4, probable but not established for others.

PROPERTIES

Gaseous Sulfur in Air: The most important S species in air are SO_2 and H_2S. There is evidence also for smaller amounts of inorganic compounds such as SF_6 and of compounds of carbon and sulfur, including mercaptans, $(CH_3)_2S$, CS_2 and COS.

There are also transient species at very low concentrations. These will include RS, HS, SO, HSO_3, SO_3, SO_4 and other reaction intermediates.

Nearly all S-containing gases are in part removed without transformation by contact with surfaces such as soils and plants (dry deposition) and with hydrometeors. The balance of the removal processes involve transformations, which finally lead to the formation of S(VI). SF_6 is an exception, removed from the atmosphere only by photolysis in the stratosphere.

Sulfur in Hydrometeors: S species are introduced into hydrometeors by three processes: (a) sorption from the gas phase, (b) capture of particles, and (c) nucleation of cloud droplets.

The major sulfur species in hydrometeors exist as SO_4^{2-} (10^{-6} to 10^{-2}M) and HSO_3^- (10^{-7} to 10^{-4}M). Minor species include H_2S, H_2SO_3, SO_3^{2-} and HSO_4^- and complexes of metals and organic species.

Solution concentrations are determined by ionization equilibria with other ions, e.g., H^+ and NH_4^+. Chemical transformation can depend upon the presence of strong oxidizing agents, e.g., H_2O_2 and O_3, or upon catalysis involving metallic ions, e.g. Mn^{2+}. The details of these transformations are not yet clear.

Sulfur in Aerosols: Significant properties of aerosols containing sulfur are physical size, shape, molecular composition and optical properties. Sulfur-containing particles exist in two distinct size ranges usually classified as fine and coarse; i.e., smaller or larger than about 1 μm in diameter. The majority of fine particulate S appears in the accumulation region (0.1 - 1 μm); there is evidence for S also in the nucleation region (< 0.1 μm).

Nearly everywhere, degradation of visibility correlates with sulfate in the fine particle region. Since these particles are hygroscopic, their light scattering increases with relative humidity above 70%.

S-containing fine and coarse particles are chemically different. The latter are generally produced by mechanical processes and are neutral or basic on dissolution. The fine particles are neutral or acidic, characterized by a balance among H^+, NH_4^+ and SO_4^{2-} ions. Reduced forms of S have been observed at or near primary sources.

The phase composition of S-containing aerosols may consist of internal systems in phase equilibrium, or of external mixtures. The phase composition of sulfur components is not well characterized. The dynamic coagulation and equilibrium processes (involving H_2O) tend to obscure the real phase equilibirum.

MEASUREMENT METHODS

Gases: Analytical techniques for gaseous S species may be characterized as continuous or discontinuous. Continuous methods usually measure at the concentrations at which they occur; discontinuous methods involve discrete sampling and analysis steps which concentrate the species to be measured.

SO_2 is the only S species that can be measured by commercially available continuous analyzers. These include the correlation spectrometer, the laser absorption spectrometer, the flame-photometric analyzer, and various liquid analyzers based upon conductivity, coulometry or colorimetry. Some are relatively non-specific. Those responding to several S-containing compounds can be made more specific by using selective filters; they are rarely sensitive enough to measure typical non-urban concentrations.

H_2S is typically sampled by concentration in liquid or, preferably, on a chemically impregnated filter. Similar techniques are used for the lowest concentrations of SO_2. Organic S gases are usually collected by cryogenic techniques, either in cold traps or in short chromatographic columns. Recently work has been done involving sampling H_2S and organic S compounds together on gold beads. The technique deserves further evaluation.

Collected samples of mixed S compounds may be separated by gas chromatography or, in the case of sampling on gold beads, by programmed heating. Detection in both cases is usually by flame photometry. Collection in liquid or on impregnated filter media is generally specific for one species, which is then analyzed by colorimetry, fluorometry, chemiluminesence or X-ray fluorescence. SF_6 is invariably measured by electron capture chromatography.

The labile intermediates have not yet been measured, although laser fluorescence and other spectral methods show promise for their measurement.

Hydrometeors: Several methods available for analysis of sulfate have sufficient specificity and sensitivity for the analysis of hydrometeors, e.g., colorimetry, radiochemical methods, XRF and ion chromatography. Similarly sulfite analysis can be by the West & Gaeke method and its modifications, e.g., chemiluminescence.

Sulfate and sulfite need to be related to other components, and analyses should be made to achieve ionic balance. The major ions in precipitation are H^+, NH_4^+, Na^+, Ca^{2+}, Mg^{2+}, K^+, SO_4^{2-}, NO_3^- and Cl^-.

The parameters involved in the oxidation of S(IV) in hydrometeors have not been resolved, hence analyses of likely important components (e.g. heavy metals like Mn and strong oxidizing agents) are required.

Major problems are not with analyses of the above species in water, but with collection and storage of samples. Normally collections are made at ground level but there is increasing emphasis on the analysis of cloud water; collection of these samples needs further investigation, as does investigation of chemical differences from warm or cold precipitation.

Contamination of samples can occur by dustfall. Transformations can result from biological activity associated with insoluble material. The first can be avoided by using a collector open only during precipitation. The second can be reduced by keeping sampling periods short and by removing insoluble material before storage. Samples should be stored for a minimum time and at reduced temperature.

Special techniques and early analysis are required to measure S(IV) species to avoid gain or loss of SO_2 or rapid oxidation to S(VI). Similarly, special precautions, (e.g. the acidification of separate samples) must be adopted for analysis of heavy metals.

Aerosols: Accurate measurements of S in aerosols require attention to both sampling and analysis. Sampling is usually on surfaces such as filters or impactor plates. Sampling artifacts can result from the use of reactive filter surfaces, including the oxidation of SO_2 to sulfate, neutralization of acids and hydration changes. Possibly particle-particle interactions can also induce alterations during sampling. Procedures have been developed for minimizing these effects. Best present methods for determination of atmospheric sulfate permits time resolution of 1/2 hour at 1 $\mu g/m^3$. Some in situ techniques that avoid sampling on surfaces are now being developed. If proven practicable, these techniques should avoid the above problems.

Analysis of collected samples can be either directly on the collection matrix or by sample dissolution. Total S can be analyzed directly without sample preparation by means of X-ray fluorescence. Dissolution permits water soluble sulfate to be determined by wet chemical analysis. The two approaches are generally equivalent for analysis of the fine particle fraction.

The specific stoichiometry of sulfur-containing compounds can be determined by techniques specific for a given compound or by deducing average stoichiometry of the aerosol from the sum of all major components.

Detailed elemental composition of size-fractionated aerosol is valuable in understanding sulfur chemistry. X-ray fluorescence methods are particularly useful for large scale monitoring applications. However, we cannot effectively analyze for some elements of potential importance (e.g. carbon).

Techniques are available for identifying H_2SO_4-ammonium sulfate phases in aerosols. In addition some techniques are available for determining S species other than sulfate.

RECOMMENDATIONS

General: 1. New and existing analytical methods, whether for the determination of ions, molecules, or properties of particles, must be evaluated by interlaboratory comparison with properly calibrated standards and replicate ambient samples, leading to statements of precision and accuracy.

2. Those responsible for promulgating "recommended" or "reference" analytical methods should recognize their burden of proof of utility and adequacy for the intended uses.

3. Further emphasis is needed on development and evaluation of in situ techniques to avoid problems of sampling and sample storage, and to provide more rapid access to the resulting data.

4. Relatively simple and inexpensive methods for measuring gross inputs of sulfur compounds to rural or urban systems should be developed for areas which have neither the need nor the means to carry out highly sophisticated measurements.

Gases: 1. Methods are needed for measurement of those gases present in concentrations well below 1 ppb, and for reaction intermediates such as free radicals to permit evaluation of the relative roles of different reaction paths.

2. There is serious need for techniques for routine measurement of NH_3 and HNO_3 at anticipated ambient levels.

Hydrometeors: 1. Methods for collection and analysis of strong oxidizing agents in hydrometeors, such as H_2O_2 and O_3, should be developed.

2. The stability and reaction rate constants for aqueous S(IV) systems and relevant metal complexes need to be evaluated.

3. Methods of collection of cloud water should be developed.

4. To ensure maximum future use of data and to demonstrate an ionic balance, all of the major ions in hydrometeors should be measured.

5. Hydrometeor collectors should be protected from dustfall, and minimum sampling intervals and storage times must be determined for each network.

Aerosols: 1. Basic research to determine the phase dynamics of S components is needed, also development of measurement techniques to determine phase composition.

2. Aerosol sulfate analysis should be made on size fractionated samples. Collection procedures should use matrices which minimize oxidation, neutralization or hydration changes during sampling.

3. Aerosol standards characterized with respect to physical and chemical parameters should be developed for evaluation of analytical techniques.

4. Standardized procedures for sampling and analysis should be developed for large scale programs.

5. Sampling and measurement techniques for determination of major S-related ambient aerosol components not now readily measured (e.g. carbon) should be developed.

6. Basic research is needed to identify and characterize non-sulfate compounds (e.g. organic or reduced species).

7. Studies are required for aerosols in clouds but not involved in cloud droplets.

WORKSHOP 1b: TRANSFORMATIONS

The conversion of SO_2 in the atmosphere to particulate sulfate is a complex process involving both the chemical conversion of SO_2 to SO_4^{2-} (e.g., H_2SO_4, $(NH_4)_2$, SO_4, etc.) and the incorporation of the sulfate into particles. The principal aspects of these two processes can be considered independently, but a complete treatment must include both. An understanding of these processes is fundamental to the formulation of air quality-emission sources relationships for sulfur-containing aerosols. In this discussion the important mechanisms involved in both processes are discussed and relationships between the two noted. Major areas of uncertainty are identified and recommendations for further research given.

CHEMICAL TRANSFORMATIONS

The oxidation of SO_2 in the atmosphere occurs by gas phase reactions, in aqueous droplets, and on surfaces. The rates of these reactions depend on the environment being considered. Of primary interest are stack plumes, urban air (including urban plumes), and rural air. Both homogeneous gas phase and heterogeneous reactions are important and deserve further study.

Homogeneous Gas Phase Reactions: In the last decade strong laboratory and field evidence for the photochemical gas phase oxidation of SO_2 has been obtained, and this reaction class is now believed to be an important, if not the most important, oxidation route for SO_2 in the atmosphere. Among these the most important are the reactions of SO_2 with HO, HO_2, and CH_3O_2 free radicals. Reactions involving RCHOO, CH_3O, and O (^3P) are of lesser importance. The direct photooxidation of SO_2 is not important. The radicals and atoms are produced in the NO_x-O_3-hydrocarbon-sunlight photochemical cycle and are found in polluted urban air, "clean" rural air, and in stack gas plumes. Rate constants for most of the important reactions have been measured, with general agreement among reported values. Of fundamental importance to the confirmation and improvement of present homogeneous reaction mechanisms of SO_2 oxidation is direct observation of the reactive transients (HO, HO_2, CH_3O_2, etc.) within the various atmospheres.

Computer simulations for typical urban mixtures of pollutants indicate an SO_2 oxidation rate of about 2-4% hr^{-1} for a sunny summer day, in good agreement with atmospheric measurements. However, somewhat higher rates are sometimes observed in smog chamber measurements. In these cases the initial rate of oxidation depended strongly on the ratio of NO_x to non-methane hydrocarbon, but the total integrated conversion to sulfate was relatively insensitive to this ratio. Computer simulations suggest that in clean air, the SO_2 oxidation rates are somewhat slower than in the urban environment. The present level of understanding of homogeneous SO_2 oxidation justifies incorporation in model calculations of rate expressions beyond simple first order, linear dependence of SO_2 concentration.

Mechanism studies are required to establish the nature of the intermediates and the chemical pathways involved in the important reactions. Compounds such as: HO_2SO_2OH, $HOSO_2ONO_2$, and $HOSO_2O_2NO_2$ may be precursors of H_2SO_4 formed in the atmosphere following HO-radical addition to SO_2. The possibility exists that these species might be the active forms with respect to human health effects.

Aqueous Phase Reactions: The aqueous phase oxidation of SO_2 has been studied in both bulk and disperse systems for a number of different reaction types and for a variety of environmental conditions. Systems of interest range from dilute cloud droplets to highly concentrated solutions in the aerosol phase. Although the literature is characterized by considerable disagreement on the rates of the reactions, a general picture as to which of the reactions are likely to be important is emerging.

Rate constants for uncatalyzed oxidation by dissolved O_2 as measured by six independent groups show that this reaction is unimportant. However, data from two other groups give rate constants about two orders of magnitude larger, which, if correct, would make this reaction important in the atmosphere. This discrepancy should be resolved.

The catalyzed oxidation of SO_2 in solution by transition metals (e.g., Fe, Mn) is believed to be important in situations in which relatively high ($>10^{-5}$M) concentrations of catalyst are present in the droplet and in which the total atmospheric concentrations of the catalytic elements are also high. Such conditions can exist in urban and stack plumes and perhaps in urban fogs. In cleaner rural air this reaction would occur only in clouds. However, unless the pH and metal concentrations in cloud water are substantially different from those in rain water, this process is unlikely to be of significance. Both laboratory and field studies of such reactions are necessary. Laboratory simulations of reactions in cloud water should replicate as closely as possible atmospheric conditions to ensure meaningful results.

Recent measurements of the rate of oxidation of S(IV) by dissolved O_3 differ by a factor of 100. If the higher rate constant is correct, then oxidation by O_3 dissolved in water droplets would be significant at atmospheric ozone concentrations of 50 ppb (i.e., background levels). This reaction would not be significant in plumes in which the ozone had been removed by excess NO. Oxidation by dissolved H_2O_2 appears fast and could be an important reaction in the atmosphere at H_2O_2 concentrations of the order of 1 ppb. This level is currently supported by photochemical model calculations but has not yet been reliably measured in the atmosphere. The occurrence of this reaction would establish an important link between the gaseous and aqueous chemistry of SO_2. Measurements of H_2O_2 in the clean and polluted atmospheres are necessary.

The role of NH_3 in the aqueous phase of oxidation of SO_2 has been over-emphasized in the past. The diffusion of NH_3 to a reacting droplet retards the decrease of pH which occurs as sulfate is formed. This permits further dissolution and oxidation of SO_2. However, the chemical analysis of rain water indicates that there is insufficient atmospheric NH_3 to raise the pH to the point where uncatalyzed oxidation by dissolved O_2 can become important. Obviously NH_3 is important in the final transformation of H_2SO_4 to $(NH_4)_2SO_4$. There appears to be a disagreement between measured ambient levels of NH_3 and the concentrations anticipated on the basis of known chemical equilibria. Further studies of the ambient NH_3 levels are necessary to account properly for the conversion of H_2SO_4 to $(NH_4)_2SO_4$.

Although there is no doubt that the particle phase oxidation of SO_2 does occur in the atmosphere (e.g., the London smog of 1952) the relative importance of these reactions in the atmosphere has not been adequately assessed. Field measurements should be undertaken to determine which of the reactions are important in the atmosphere and the rate at which they occur. Specific attention should be given to plumes, clouds, fogs, and high relative humidity situations in general.

Surface Reactions: A variety of measurements indicate that SO_2 is both adsorbed and oxidized on the surfaces of solid particles. Carbon seems a particularly effective surface in this regard. Without rate data, the importance of these reactions cannot be assessed. This mechanism could be significant in stack plumes.

Dynamics of Plume Chemistry: In plumes the relative importance of gas phase, aqueous phase, and surface phase oxidation of SO_2 is uncertain. Considerable variation has been observed in the SO_2 oxidation rate in stack plumes. An initial rapid oxidation rate (occasionally up to 20% hr^{-1}) sometimes occurs during the early history of stack plumes. After longer transport times (30 minutes to several hours depending on meteorological conditions) the admixture of ambient air into the plume can play an important role in the SO_2 oxidation. Measurements by several investigators on such an aged plume from a coal-fired power plant near St. Louis, Missouri, gave oxidation rates varying between 0.5 and 4% hr^{-1} under daytime summer conditions. Nighttime oxidation rates were less than 0.5% hr^{-1}. These variations depend on still undetermined chemical and meteorological factors. Because of the possibility of particle phase oxidation of SO_2 in the plume, a better characterization of primary particulate emissions from stacks is important.

These chemical and dispersion processes cannot be considered independently. Under certain atmospheric conditions plume reaction rates may be controlled by the rate of mixing of ambient air into the plume. A recent model predicts that for all SO_2 oxidation reactions, except the first order, homogeneous gas phase oxidation, the conversion proceeds to a fractional asymptotic limit, the magnitude of which depends on the meteorological and chemical parameters involved. After long transit times of the plume, the chemical processes probably become the dominant rate limiting factor. It is important to carry out further investigations, both field and modeling, that will elucidate the roles of plume dilution and of mixing of the ambient atmosphere into the plume in governing the SO_2 oxidation rate.

Oxidation of Reduced Sulfur Compounds: The rates of the homogeneous reactions of H_2S, CH_3SH, $(CH_3)_2S$, etc., in the atmosphere appear to be adequately defined. However, the mechanisms and products of these reactions are poorly understood and must be investigated further.

AEROSOL DYNAMICS

Aerosol dynamics is the temporal and spatial evolution of both the total particle size distribution and of the individual particle size distributions for particular chemical substances. The particle size distribution at a given moment is the sum of processes involving the chemical production of condensable or condensed species, coagulation, and depositional loss from the atmosphere. Details of the evolution depend critically on the mechanism of gas-to-particle conversion and the particle size distribution of the preexisting aerosol. Other factors such as temperature, relative humidity, and the presence of liquid water in the particles are also important. Knowledge of the size distribution of particulate S and how it evolves is important because of the role of particle size in wet and dry removal from the atmosphere, the amount and location of deposition in the lung, and visibility reduction. Ambient aerosols can be described by three modes as noted in the discussion of Workshop 1a. The boundaries between modes are not fixed but depend on mode concentrations and growth history of the aerosol.

Gas-to-particle conversion processes result in production of particles in the nuclei mode and growth of preexisting particles in the accumulation mode. The former can occur only by homogeneous gas phase reactions. There is qualitative agreement among a number of experimental and theoretical studies of homogeneous heteromolecular nucleation of H_2SO_4-H_2O mixtures. Quantitative comparisons are difficult, however, because of the lack of adequate instrumentation for measurement of particles smaller than 6 nm diameter and the absence of thermodynamic data for the system. In these studies the effect of preexisting particles in the atmosphere was not considered, and hence their relevance to actual atmospheric processes cannot be determined. Newly formed particles coagulate rapidly with each other and with particles in the accumulation mode. Thus the existence of these small particles depends on continuously operating gas-to-particle conversion processes or addition from primary sources. The relative magnitudes of nucleation and condensation are a function of the surface area of the preexisting aerosol. The balance between nucleation and condensation has recently been investigated in a smog chamber study of the SO_2-NO_x-propylene system. A satisfactory theory was developed to describe the particle evolution.

Nucleation has a profound effect on the size distribution of particulate sulfate. In a study of a coal-fired power plant plume, about 5% of the new aerosol volume (which was presumably sulfate) was found to be in the nuclei mode. The coagulation of newly formed particles into the accumulation mode produced a volume mean diameter for this mode less than 0.2 μm, about a factor of two smaller than usually found in urban aerosols where nucleation is of lesser importance. Despite the variability of the size distribution of S in the accumulation mode the major fraction of particulate S is found in this mode. Moreover, results of studies in locations ranging from clean oceanic background to urban areas show that sulfate is always present in this mode in amounts ranging from 5 to 60% of the total mass of this mode.

If gas-to-particle conversion of SO_2 occurs in aqueous droplets, then the sulfate mass distribution would reflect the volume distribution of the particles in which the reaction occurs. (This assumes that the rate controlling step is oxidation of SO_2 within the droplet rather than diffusion of SO_2 to the droplet.) There are no definitive data on this subject. Measurements of the sulfate particle size distribution in Pasadena, California, however, often show a peak in the mass distribution at about 0.8 μm. Such a particle size is too high to have resulted from condensation and suggests the possibility of aqueous phase conversion of SO_2. Further study of this phenomenon is required.

In summary, the dynamics of sulfate aerosols are qualitatively well-understood, but quantitative understanding is less satisfactory. Most of the theoretical developments have been applied to laboratory studies, and quantitative application to the atmosphere is uncertain. Sulfate produced in the homogeneous gas phase oxidation of SO_2 will ultimately reside in the accumulation mode but the exact distribution within this range depends on the balance between nucleation and condensation.

RECOMMENDATIONS

1. Mechanism studies are required to establish the nature of intermediates formed in the gas phase oxidation of SO_2 and the chemical pathways involved.

2. The discrepancy concerning the uncatalyzed oxidation of SO_2 by dissolved O_2 should be resolved.

3. Measurements of atmospheric H_2O_2 are necessary to assess the importance of H_2O_2 in the aqueous phase oxidation of SO_2.

4. Field measurements of atmospheric NH_3 are necessary to account properly for the conversion of H_2SO_4 to $(NH_4)_2SO_4$.

5. Further field and theoretical studies of plumes are necessary to identify the rate controlling chemical and meteorological processes in the oxidation of SO_2. An important aspect of this research should be the characterization of primary particulate emissions from stacks.

6. High priority should be given to the development of techniques for the detection and measurement of the contribution of particle-phase oxidation of SO_2 in real atmospheres as manifested in both the chemical and aerosol dynamics of the S.

7. Continued experimental and theoretical work is needed on aerosol dynamics in the size range below 6 nm and on the interaction of this range with larger particles. Both field and laboratory studies are necessary.

8. More data on the distribution of S with respect to particle size are needed, especially in the size range below 0.1 μm. Improved instrumentation for this purpose should be developed.

WORKSHOP 2: DRY AND WET DEPOSITION

Without removal processes, sulfur concentrations in the atmosphere would increase steadily at the rate of about 70 μg-S $m^{-3}y^{-1}$. Comparing this with the current United States annual standard for SO_2 (40 μg-S m^{-3}) indicates the importance of removal processes to the quality of our atmosphere.

DRY DEPOSITION

Deposition velocities (V_g) may be used (a) to estimate deposition rates from observed concentrations, (b) in atmospheric models. These are usually measured one or two meters above the surface and for the first purpose are used with concentrations measured in the same height range. Occasionally, as in forests, it is difficult to obtain concentration measurements at the appropriate height above the surface. Prediction of concentrations near the surface in modeling calculations is not straight-forward since surface roughness, wind speed and temperature stratification determine transport towards the surface.

Deposition of particles is strongly dependent on particle size. Most sulfate mass lies in the range 0.1 to 1 μm diameter for which the deposition velocity is not expected to exceed 0.1 cm s^{-1}.

Table 1a gives estimates of deposition velocities of SO_2 over various types of vegetation, Table 1b similar information for calcareous and acidic soils.

Table 1a: DEPOSITION VELOCITIES OVER VEGETATION

Vegetation Height	Vegetation example	height (m)	range of V_g (cm s^{-1})	typical V_g [x] (cm s^{-1})
Short	grass	0.1	0.1 to 0.8	0.5
Medium	crops	1.0	0.2 to 1.5	0.7
Tall	forest	10.0	0.2 to 2.0	uncertain

[x] These values were obtained in a humid climate. Much smaller values are likely in arid climates.

Table 1b: DEPOSITION VELOCITIES OVER SOIL

	pH	State	range of V_g (cm s^{-1})	typical (cm s^{-1})
Calcareous	≥7	dry	0.3 to 1.0	0.8
Calcareous	≥7	wet	0.3 to 1.0	0.8
Acid	~4	dry	0.1 to 0.5	0.4
Acid	~4	wet	0.1 to 0.8	0.6

N.B. As yet no information is available to assess V_g on desert sand or lateritic soils.

A mid-day maximum in V_g is frequently observed, reflecting diurnal maxima in stomatal opening and in turbulent transport. A seasonal variation, with minimum values less than 0.3 cm s^{-1} when the vegetation is dry or senescent, is also observed. The presence of water on foliage leads to larger values (1 cm s^{-1} or greater) providing the pH of the water is greater than about 4.

Four other types of surface should be mentioned:

a. Dry snow, $V_g \approx 0.1$ cm s^{-1}. Sparse experimental evidence suggests that snow is a very poor sink for SO_2. Additionally, the surface is naturally smooth so that atmospheric mixing is small. Wet snow behaves like water.

b. Water, V_g, range 0.2 to 1.5 cm s^{-1}, typically 0.7 cm s^{-1}. Natural water surfaces are efficient sinks for SO_2 unless very acid and rates of deposition are determined by atmospheric mixing.

c. Countryside, V_g, range 0.2 to 2.0 cm s^{-1}, typically 0.8 cm s^{-1}. A very few studies of the dry deposition from plumes or of atmospheric budgets provide estimates of V_g over large areas of countryside, generally consistent with estimates over homogeneous surfaces.

d. Cities. Laboratory measurements show that many building materials absorb SO_2 at significant rates, but it is difficult to use the results to estimate deposition to built-up areas. Observations in London during a prolonged stagnant period suggest a deposition velocity of 0.7 cm s^{-1}, but this may be atypical.

Research Requirements: Measurements of dry deposition of SO_2 have mostly been made in selected areas and intermittently. For regional studies, it is necessary to generalize these results.

1. More measurements are needed of dry deposition to coniferous and broad-leaved forests, cold deserts, hot deserts, ice and snow. As gradient and eddy correlation methods require a long fetch over uniform terrain, a method is needed for estimating deposition to composite terrain. Possibly the analogy between deposition of SO_2 and transpiration of water can be exploited further to do this.

2. With few exceptions, measurements of dry deposition have been intermittent. More continuous measurements are needed, or alternatively, more effective parameterization of the results. This includes the effects of wind speed and stability on the aerodynamic resistance, which are particularly important for plume models. Important seasonal effects may be crop growth and senescence, soil moisture, and frost.

3. The eddy correlation method of obtaining vertical fluxes is tending to displace the gradient method in studies of water and heat transfer from crops. Its output is suitable for data acquisition systems. For application to SO_2 or total sulfur, improved fast-response direct analytical methods are needed. Also, more experience is needed to define the necessary conditions of fetch and absence of local sources of SO_2.

4. The catchment balance method of estimating evaporation may also be applied to S compounds, particularly where the sub-soil is impervious. Caution should be exercised in using this method where the sulfur content of soil is large. Catchment studies in arid regions, where resuspension and advection of blown soil occurs, may have to be designed to allow for this contribution to the sulfur budget.

Another possibility is to expose trays of lysimeters containing low-S soil and measure net total S deposition, or dry deposition by seasonal or annual analyses. As an elaboration of this method ^{35}S may be added to the soil. The uptake of atmospheric S is then estimated by analyzing the ratio of ^{35}S to total S in the soil and in any vegetation growing on it, as a function of time.

5. Only very few simultaneous measurements of dry deposition by two or more methods have been used. More estimates of velocity of deposition from plume or regional studies would be particularly useful, as they give mean values over composite terrain.

6. Present estimates of the velocity of deposition of sulfate vary more than do estimates for SO_2. Many measurements of deposition of sulfate particles in the 0.1 to 1 μm diameter size range give velocities of deposition of order 0.1 cm s^{-1}, about an order of magnitude less than for SO_2. However, some experiments give higher results. Also, sulfate may be present in larger particles near emitting sources, in sea spray droplets and in resuspended soil. Estimates of deposition in these circumstances are needed. For sulfate in the submicron range, wind tunnel or field work using ^{35}S or ^{34}S is desirable to confirm the expected deposition rate.

7. At high relative humidity, sulfate particles grow by deliquescence and may become large enough to be impacted on surfaces, particularly trees. Information is needed on the contribution of fog drip to deposition of sulfate.

8. Stomata have been shown to be important for uptake of SO_2. Stomatal aperture depends on the plant's age, health and genetic makeup as well as on CO_2 concentration and meteorological variables. Tropical and desert species differ from temperate species. The effect of these variables on uptake of SO_2 is particularly important to ecologists.

The effect of surface properties on uptake of sulfate particles is unknown. Pubescence may be important, and stomatal behavior may also have some effect.

Monitoring Requirements: 1. Since dry deposition rates are often calculated by multiplying appropriate deposition velocities by surface ambient SO_2 concentrations, there is a need for routine network monitoring of this parameter in rural areas. The monitoring station density will be determined by the meteorological situation and source-receptor distances.

2. Dry deposition is strongly dependent on atmospheric stability and wind speed in the boundary layer, especially over wet surfaces. Meteorological measurements of these parameters on a routine network basis should be made near important receptors.

WET DEPOSITION

Deposition Rates: Table 2 summarizes current knowledge of wet and dry deposition rates. The range given represent spatial variations for each area.

Table 2. REPRESENTATIVE ANNUAL AVERAGE WET AND DRY DEPOSITION RATES

Location		Excess sulfate in precipitation (mg S ℓ^{-1})	Wet deposition rate (g S $m^{-2}y^{-1}$)	Dry deposition rate (g S $m^{-2}y^{-1}$)
heavily industrialized areas	North America	3 - ?	0.1* - 3	?
	Europe	3 - 20	2 - 4	3 - 15
Rural	North America	0.5 - 2	0.1 - 2	0.2 - 2.6
	Europe	0.5 - 3	0.2 - 2	0.5 - 5.0
Remote	North Atlantic	0.2 - 0.6	0.1 - 0.3	0.04 - 0.4
	Other Oceans	0.04	0.01*- 0.2	< 0.1
	Continents	0.1	0.01*- 0.5	0.4

*Low deposition rates result from low precipitation.

In constructing global and regional sulfur budgets, estimates of the ratio of wet to dry deposition should not be used. Given the complexity of precipitation scavenging and dry deposition processes, it is beyond our capability to predict or measure on a global scale this ratio with an accuracy better than a factor of two. On a regional basis, better estimates can be made (e.g., for North America and Europe).

Research Needs: Research is needed in the following areas to better our understanding of precipitation scavenging, so that process may be more realistically included and parameterized in models:

1. Quantitative understanding of how SO_4^{2-} is incorporated into precipitation, including measurements of vertical profiles and in-cloud concentrations of important reactants, correlation of precipitation sulfate with atmospheric SO_2 and SO_4^{2-} concentrations, and measurements of dissolved SO_2 in precipitation.

2. Measurements of other precipitation constituents which affect S chemistry and acidity.

3. Determination of the importance of organic S compounds in precipitation.

4. Additional high-quality measurements of sulfate particle-size distributions and particle size - humidity relationships.

A better understanding of certain meteorological phenomena can lead to improved modeling of wet deposition during long-range transport, and to the diagnosis of such events:

5. Precipitation scavenging and deposition under conditions of orographic precipitation and in other situations of strong topographic influence, such as trapping of pollutants in a valley.

6. Determination of the history of air from which pollutants are scavenged by convective and frontal storms.

To provide more reliable information on the spatial variability and trends of wet deposition, improvements are required in the following areas:

7. Precipitation chemistry measurements over the ocean, including avoidance of sea spray contamination, accurate measurement of precipitation amounts, and pilot studies on the use of various sampling sites, such as drilling platforms, ships, and islands.

8. Establishment in both Europe and North America, and other areas, where desirable and feasible, of long-term monitoring networks (with collectors which exclude dry deposition) to supplement the WMO background network with stations on a grid of approximately 200 km (smaller spacing in areas of strong gradients), with samples collected for no longer than a month, with monthly analysis of as many parameters as possible, and the possibility of short-term, daily-sampling programs or careful intercalibration with shorter-term intensive regional programs.

9. Pilot studies to optimize network density and siting.

Many improvements in the measurement of wet deposition will arise from new requirements of users, whether for modeling or effects research. Often, the research is required to clarify the need, as well as to satisfy the need. These requirements include:

10. The need for greater time resolution in precipitation chemistry sampling, i.e., intrastorm sampling.

11. Statements by ecologists as to chemical parameters and time resolution required for their effects studies.

WORKSHOP 3: BUDGETS, SPATIAL AND TEMPORAL VARIATIONS

MAN-MADE EMISSIONS - Best estimates from fuel burning and other industrial emissions approach ± 15% accuracy for national annual emissions. In some cases accuracy is only within a factor 2. Governments should be encouraged

(i) to give precise data on S content of fuels

(ii) to relate fuel consumption figures to S categories and type of consumer

(iii) to give information on spatial and temporal variation of emissions.

Estimates of primary SO_4^{2-} and H_2S emissions (paper mills, geothermal drilling etc.) are also needed.

NATURAL EMISSIONS - Biogenic H_2S - small fraction of man-made emissions N of 45°N may be dominant in tropics. Sea-salt aerosol sulfate is best estimated from Na^+ measurements. Volcanic emissions about 5% of global man-made SO_2. Important in remote areas and in some industrial regions near sources (e.g. Etna - estimated emission about 2% of total European emission). Vegetation fires - S content largely unknown. Other biogenic S emissions - Unlikely to be important in industrial areas. Further studies of natural emissions are required.

TRENDS AND DISTRIBUTIONS - Observations of sulfate in precipitation extending over 10 years or more are mainly confined to Europe. In N. America, measurements focused on sulfur dioxide and particulate sulfur. SO_2 and sulfate particulate concentrations in air were until recently made independently of the precipitation chemistry measurements. Annual concentration maxima of airborne sulfate (summer) and sulfate in precipitation (spring) do not correspond to man-made emission maxima (winter). Long term trends in SO_2, sulfate concentrations and sulfate content and pH of precipitation do not correlate unambiguously with man-made emissions. Ice core analysis in Greenland shows a three fold increase in sulfate between late 19th century and 1960's which is probably due to man-made sources but volcanic influence may have contributed. Data on the regional distribution of SO_2 is good only north of about 40°N. The latitudinal distribution of SO_2 in N. Atlantic is consistent with transport eastwards from N. America but SO_4 distribution is surprisingly less emphatic. Over NW Europe budget estimates indicate about 2/3 of emission is deposited within 1000 km of source areas. Dry and wet depositions are roughly equal over the area as a whole, but there is considerable spatial variation in the ratio. A similar situation is likely in N. America. Average residence times for mid-latitudes are estimated as; SO_2 about 1 day, SO_4^{2-} 3-5 days, S about 2-3 days. An average of 20-50% of SO_2 will be converted to particulate sulfate before deposition. These figures are subject to seasonal variation and may be quite different in other climates.

Effect of emission characteristics - Low level sources may lose an average of 10-20% of the emissions in the first few tens of km, while elevated sources may not come into contact with the ground in this distance. At greater distances there is little difference in deposition. Most plumes, even those from low level urban sources, have some buoyancy and will tend to become isolated from the surface in stable conditions. This effect will increase long range transport. Occasionally, plumes from tall chimneys may remain above the mixing layer even in convective conditions. Plume budget measurements, including the background outside the plume, can help evaluate the relative importance of deposition and transformation of SO_2.

<u>Burdens and vertical distributions</u> - Sulfur budgets require knowledge of the mass of sulfur species in the atmosphere and their spatial distributions. Column burdens can be estimated if the vertical profile is known, but profiles are so variable that at present a general form cannot be assumed. SO_4^{2-}-S burdens appear to be about evenly divided between remote continental and maritime regions, with the maritime sulfate showing an excess over that due to sea salt. The SO_2-S burden is about evenly divided between these two regions, and is probably somewhat smaller than the SO_4^{2-}-S burden. In industrialized regions the SO_2-S burden is substantially larger than the SO_4^{2-}-S burden. The H_2S and reduced species burdens are largely undetermined.

The upper troposphere, together with the lower troposphere in remote oceanic areas, have been postulated as one entity, with a characteristic composition of the non-sea salt part of the background aerosol. The lower troposphere and the cloud forming zone over industrial and non-industrial continental areas, are also recognized as having possibly characteristic aerosol parameters. The burdens over these identified regions are almost unknown, but are needed for the details of the sulfur cycle, especially to resolve the uncertainties in turn-over times and fluxes. Of these, the lower tropospheric, industrial region SO_2-S and SO_4^{2-}-S are both large and the easiest to estimate. The background aerosol burden has been estimated from Aitken counts and optical measurements, but has not been completely identified chemically. Preliminary measurements indicate sulfate is a major component. Observations of cloud sulfate and SO_2 are required for a full understanding of the sulfur cycle.

<u>High priority should be given to the following</u>

1. Concurrent surface data for atmospheric and precipitation concentrations of SO_2 and SO_4^{2-}. Baseline monitoring stations are well suited for such measurements. In addition, measurements should be done by regional networks in polluted areas and in areas that are likely to become polluted in the future.
2. Vertical profile data at a number of selected stations considered typical of larger regions. Clouds and precipitation should be included.
3. Exploratory measurements on a broad geographic scale in remote oceanic areas, arctic areas, and the upper troposphere, to confirm the role of sulfur compounds in the troposphere background aerosol.
4. Reliable measurements, including vertical profiles, of H_2S, particularly from subtropical and tropical latitudes. The source function of the oceans for organic sulfur compounds, e.g. $(CH_3)_2S$, has to be established.

WORKSHOP 4: MODELS AND LARGE PROJECTS

Models

Modelling in this context includes all efforts to construct a framework for understanding the very complex problems of sulfur emission, transport, transformation and deposition and should explain observed air concentrations, visibility, total pollutant deposition and its characteristics. <u>Long range plume models</u> are used to describe the dispersion from large point sources, integrated area sources, or combinations of these where the concept of a plume contributes to the understanding of the transport and transformation of pollutants over distances up to several 100 km. <u>Regional area models</u>, generally based on numerical analyses on a suitable time and space resolution, may be used for areas and over extended periods of time. Chemical transformations of the pollutants and removal processes (dry and wet deposition) must be included in the models.

Spatial and temporal grid size of a model should be consistent with each other and the questions the model is designed to answer. Availability of meteorological or emissions data may restrict spatial or temporal resolution below the optimum. Most regional transport models currently operate with only one layer. Improved parameterisation in the vertical direction utilising information on wind profiles, mixing height and vertical eddy diffusivity is required e.g. for simulation of night-time storage and transport of sulfur in stable layers aloft, and its subsequent day-time re-entrainment into the mixing domain. The extent of horizontal dispersion and its diurnal variability resulting from wind shear and differences in the boundary layer behaviour and deposition due to variations in the underlying surface (e.g. transport over cold water) require further study. The major portion of annual wet deposition in some areas (e.g. Southern Norway) occurs during a small number of days. Such episodes and those of decreased air quality are simulated most accurately by models with time resolution on the order of hours. Prediction of yearly average concentrations and depositions may be possible without such time resolution.

Emissions must be described on temporal and spatial scales, which are appropriate to the effects being studied and these scales determine the grid size of the models. Time resolution in sufficient detail for calculation of transport during episodes is usually available only for major power plants and not e.g. for area sources. Effective emission height is also important.

The horizontal movement of air masses may be described in terms of trajectories and dispersion from such trajectories based on derived wind fields and eddy motions. Non-horizontal trajectory paths should be used in some situations, such as frontal zones. Additional wind, temperature and humidity data in the vertical would be valuable e.g. the extraction of data for the 925 mb pressure surface. The World Meteorological Organization has recommended reporting of these data by the meteorological services. Special meteorological situations can play a significant role in horizontal transport, e.g. a pronounced maximum in the wind speed at a few hundred metres above ground during stable conditions, (low-level jet) can accentuate long range transport. Orographic features: Large terrain irregularities require special modelling treatment particularly in the treatment of plume dispersion. Stability variations may be exaggerated, direction of transport may be markedly channelled, etc. The horizontal flux of material across the external boundaries (of the model) can be significant. Measurements in remote areas are necessary to determine the input from sources outside the emission area considered.

Vertical motion within air masses may allow some fraction of the emission to ascend out of the boundary layer especially along fronts. Such material will eventually return to the surface, often in precipitation, after very long range transport.

Transformation and removal processes are now often estimated by average rate constants and deposition factors. Improvements in modelling will require time and space variations in these processes. This may create severe complications in the numerical computer procedures and suitable compromises may be necessary. No suitable method for routine measurements of dry deposition is available or foreseen in the near future. Deposition rates are established by special studies of removal processes over various surfaces. Complicated by the need to calculate vertical dispersion in the lowest few hundred meters, particularly at night. Initial deposition and transformation in the grid element where the emission takes place is of particular importance. Further studies under different conditions are needed. Wet deposition is routinely measured, but shows large local variations. Examination of the representativity of these measurements and a better understanding of the behaviour of the pollutants in connection with cloud droplets formation and precipitation scavenging is needed to improve the parameterizations of these processes. Also there is a need to include rainfall events in their actual space and time occurrence during the long range transport, instead of estimating the wet deposition on the basis of average rainfall data for a season. In this relation routine reporting of hourly rainfall data over Europe would be of great value. In addition greater detail on rainfall chemistry patterns in the US is essential. Atmospheric chemical transformations of the sulfur components are affected by other pollutants, substances of natural origin (particularly NH_3) and meteorological factors (e.g. temperature and humidity) Improved representation of these interactions should take account of the space and time variations of these factors including (i) diurnal and seasonal variations, (ii) mapping of sources for important components, (iii) identification of rate controlling processes and (iv) the role of plume mixing with background air. A careful examination of the relative benefits of detailed chemical modelling versus the use of overall parameters, should be made.

Model verification is essential to determine accuracy and to develop confidence in model use. Adequate verification requires a combination of data from regular monitoring and from more intensive measurements. It is important to evaluate simulation in the models of individual atmospheric processes and to select realistic parameters to represent them. Additional studies to verify the generality of such selection in other areas and at other times is essential. Some work has already been done to investigate individual processes. e.g. on large point source and urban plumes. Special field expeditions to measure and relate pollutant concentrations in the air and at the surface in areas of precipitation has been carried out over the North Sea. Similar studies are planned in the U.S. field studies with controlled releases of inert traces such as perfluorocarbons and with tracers of opportunity, e.g. fluorocarbons, can add important additional information on air-trajectories and dispersion especially in complex terrain. By comparing such results with concentrations of active species, information on transformation and removal may be gained. Visibility measurements have recently been used to track horizontal movement of pollution blobs. Indications are that these blobs are heavily burdened with sulfate aerosols. Turbidity measurements together with meteorological data(e.g. relative humidity) could be used to quantify these results and perhaps allow evaluations of transformation rates. Detection of these areas from satellite platforms might also be investigated. Taken together, these efforts could provide specific study periods for model verification. Verification of the models could also be carried out using routine data from monitoring networks or data from intensive monitoring periods. Calibration of the model is then an iterative process between measurements and calculation. Examples of the data which can be used to test models include the following:

From the European OECD Long Range Transport Project a comprehensive data base is available. At about 70 stations, 24 hour ground level SO_2 and SO_4^{2-} concentration and precipitation measurements were collected over a three year period. In addition, about 100 aircraft flights with SO_2 and SO_4^{2-} sampling are reported. Trajectory computations and wind-fields are stored for each day. These data have already been used to evaluate long range transport models. In North America, the surface and airborne measurements, of the SURE and MAP3S programs, particularly during the intensive periods, will offer a substantial data base for model verification. In addition to planned field measurements, the detailed source emission inventory being planned by EPRI and the potential augmentation of meteorological data by ERDA/NOAA could add significantly to the value of the effort. With the methods and data used in the OECD project, an estimated accuracy of ± 50% was obtained for the annual amounts of sulfur deposited in one country as the result of the emissions in any other European country. This uncertainty is mainly governed by the limited precision of the emission data and uncertainties concerning the average deposition rates used. The models did reproduce the main features of the transport of the pollutants. Calculated concentrations during episodes usually agreed with measured values within a factor of 2. The long term time correlation between measured and calculated values at single stations ranged from insignificant to 0.8. The spatial correlation between model calculations and the observations for all stations was 0.9 on an annual basis. Preliminary studies on sulfur with models in the United States have not yet reached the same stage. To improve the day to day agreement between the models and observations a more detailed description of the chemical transformation, deposition processes and vertical layering is likely to be required. For the annual averages the first step would be to improve the emission data and to consider their time variations in some detail. A more engineering approach is required in models for operational use compared with models that are developed to advance the scientific understanding of physical and chemical atmospheric processes. Operational models must be developed to suit the specific needs of the relevant decision policy maker, and are generally more simplified in their attention to scientific detail. Of particular concern in this regard is the development of operational models for developing nations, which should be appropriately matched with their resources. These might take the form of nomographs or tables of dimensionless parameters.

Large scale studies - Results from several large scale scientific projects (OECD/LRTAP, COST, MISTT, SURE, MAP3S, Canada,LRTAP,integrated assessment) were reported during the meeting. In northwest Europe and Canada these studies have responded to the impetus of ecological problems resulting from acid precipitation. In the USA emphasis has been placed more on health effects and air pollutant concentrations. There are, however, no reasons to believe that the atmospheric part of the sulfur cycle is substantially different in these areas. All the studies have confirmed that sulfur compounds do travel long distances (500 km or more) in the atmosphere and have shown that air quality and deposition in any one country is measurably affected by emissions from neighbouring countries. If countries or U.S. Air Quality Control Regions find it desirable to improve air quality away from emission areas and to reduce substantially the total deposition of sulfur within their borders individual national or regional control programmes may achieve only a limited improvement.

The studies have shown that the more important transport of sulfur compounds over long distance is associated with certain combinations of wind conditions which give persistent flow, and with air mass trajectories which may or may not be associated with large area stagnation. In the industrialized areas of north eastern North America multi-day accumulations occur several times a month during the summer and fall. Alignment of point or area SO_2 sources along a path where these conditions occur can therefore result in intensification of sulfur concentrations in downwind areas. Under some meteorological conditions plumes from large point and area sources were identified for several hundred km. Such phenomena need to be considered when selecting sites for new sources. In addition,in the U.S., rate parameters have been determined from measurements in individual power plant and urban plumes out to 200 km. Rain chemistry measurements in conjunction with large scale programmes have shown acid rain (annual average pH as low as 4.2 - 4.3) over large areas in Europe. Similar evidence of acid rain is now being obtained in the U.S. and Canada.

Recommendations: 1. In order to increase the spatial coverage and to better quantify the transformation, transport, and removal of sulfur compounds the existing and newly planned (ECE and STATE) large scale studies need to continue. Such studies could be designed with greater precision if the relationship between pollutant concentrations and ecological and health effects were more precisely understood. A close, iterative exchange is needed between those defining physico-chemical phenomena and those studying environmental effects.

2. It was noted that as part of GEMS/WMO, regular measurements, on a monthly basis, are made of some sulfur compounds in precipitation. While these data are useful for determining long term regional trends the participants felt that further measurements are necessary:

(a) on a short-term basis (3 to 5 years) to provide data for scientific studies of sulfur transport, transformation and removal. As modelling will certainly play a large part in such studies, the data are required with a time resolution no longer than 24 hours.

(b) on a longer-term basis (approximately 20 years) to determine trends of sulfur (and other) compounds relevant to acidity both in air and in precipitation, in areas away from local sources (but not so remote as most WMO regional sites). Here less frequent measurements would suffice.

Significant attention is being given to (a), but (b) should receive further attention.

3. Frequent intercalibration of both sampling techniques and analytical methods. Continuous analysis of data is necesssary to ensure data reliability.

4. Existing large scale studies do not adequately cover intercontinental transport of sulfur compounds or possible weather and climate effects.

5. Since both nitrate and sulfate contribute to acidity in rain, measurements of NO_3^- and NH_4^+ need to be made in addition to SO_4^{2-} and pH measurements.

6. It is recommended that participants in large programs meet as needed to exchange results, to examine program coverage, and to conduct intercalibration activities.

Atmospheric Environment Vol. 12, pp. 25–38. Pergamon Press 1978. Printed in Great Britain.

WATER-SOLUBLE SULPHUR COMPOUNDS IN AEROSOLS

CYRILL BROSSET

Swedish Water and Air Pollution Research Laboratory, P.O. Box 5207, S-402 24 Gothenburg, Sweden

(*Received for publication* 20 *September* 1977)

Abstract—Sampling of particles and analysis of their water-soluble part for H^+, NH_4^+, SO_4^{2-} and NO_3^- were carried out in a remote and an urban area on the Swedish west coast. For the determination of H^+ a special procedure based on Gran's plot was used. High concentrations of sulphate were found in two types of particles of apparently different genetic origin. In fine particles the following solid sulphate phases were identified: $(NH_4)_2SO_4$, $(NH_4)_3H(SO_4)_2$ and NH_4HSO_4. Conditions for the formation and range of existence of these phases are discussed. The observed sulphate maxima are divided according to genetic and to some extent geographic origin of the particles into black and white episodes. Black episodes are associated with Continental air. They are accompanied by increased concentrations of nitrate, dark components and metals, e.g. Mn. Particles characteristic of white episodes seem to occur in air masses of varying origin. The sulphate maxima in this category are usually accompanied by hydrogen ion maxima, and minima for nitrate, dark components and metals. The formation of particle-borne ions in an urban area (Gothenburg, population 500,000) was also studied. For sulphate, this local contribution was found to be 16% of the mean concentration. Processes which might lead to black and white episodes are discussed.

INTRODUCTION

A study of the particles occurring on the Swedish west coast has been carried out, among other things to determine such particle properties as might be of interest from the acidification point of view. The intention has also been to find out if there is a relationship between the ion composition and the genetic and geographic origin of the particles.

The study was initiated in 1972 and is still in progress. So far the following measurement series and part investigations have been carried out: (1) The initial, introductory part comprised primarily the development of sampling and analysis methods. It was concluded with a field test in December 1972 on the Onsala peninsula (clean-air area about 40 km S Gothenburg). (2) Utilising the experience gained from the first field test, another series of field measurements was carried out in the period March–August 1973 at Onsala by means of a special High Volume Sampler. (3) During January–June 1973, i.e. partly in parallel with the last-mentioned series, tests were carried out with fractionated sampling. (4) During August 1975–April 1976 determinations were made of number of particles as well as concentration of certain gas components and particle-borne ions at Onsala and in Gothenburg. The particles were sampled with a conventional (volumetric) sampler.

METHODS

Sampling

For our purposes, a particle sample must satisfy the following requirements: (1) The sample must be sufficiently large for the analyses to be made. (2) The sample must be collected on a filter (or impactor plate) made of a material that does not react with the sample components or otherwise contaminate the sample. The greatest difficulty is finding a filter material which combines other necessary properties with that of absolute inertness to hydrogen ion. (3) The sample must be representative of specific particle populations. The sampling time must therefore be short enough for the sample to represent fairly constant transport conditions and the sampling procedure must permit fractionation with respect to particle size.

In practice, it is very difficult to achieve a sampling in which all these demands are simultaneously met. Normally, therefore, one is forced to resort to compromises.

In the first field measurements the principal aim was to collect, in a relatively short sampling time (6 h), as large a sample as possible. The reason was that the analysis method available at the time was not particularly sensitive. This need led to the development of a special directional high volume sampler (Brosset, 1973). Using a Gelman Acropor AN-5000 filter of 142 mm dia, samples of total suspended matter corresponding to approx $100\ m^3\ air\ h^{-1}$ could be obtained. The Acropor filter was chosen because of its low pressure drop and relatively good chemical properties.

As fractionated sampling was expected to yield better information, both Anderson High Volume Cascade Impactor (ACI) and the BGI Impactor were tested. The plates were covered with Teflon film and an Acropor filter was used as the final stage. The air flow was only $25–35\ m^3\ m^{-1}$, however, which required a sampling time of several days to collect a sufficient amount of particles on all four impactor steps. A simplification was introduced when two-stage fractionation (particles > and < 2–3 μm) was found

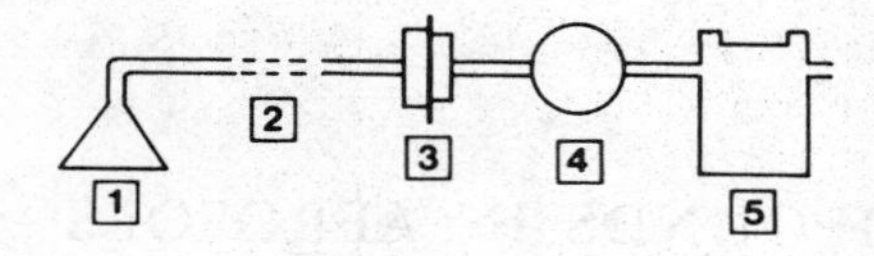

1 POLYTHENE FUNNEL Ø 50 MM

2 POLYPROPYLENE TUBING 1/4"

3 FILTERHOLDER : MILLIPORE 125 x 0002500

FILTER : MILLIPORE FALP 02500

4 PUMP: THOMAS, TYPE 107 CD

5 GASMETER

Fig. 1. Conventional (volumetric) sampler.

to be sufficient for our purposes. However, a cumbersome factor in all fractionated sampling is the large number of samples to be analysed. Since we were primarily interested in H^+, NH_4^+ and SO_4^{2-} ions which, according to recent findings (Brosset, 1976), are present for the most part in fine particles, sampling could in many cases be limited to this fraction.

A further simplification of the sampling procedure was thus possible. The presence of coarse particles in a sample should not affect the analytical result with respect to the above-mentioned ions except possibly for SO_4^{2-} in the case of coarse particles containing sea spray. To test this, comparative measurements were made with a conventional (volumetric) sampler (Fig. 1) and a dichotomous sampler made available by Dr. R. Stevens at the U.S. Environmental Protection Agency. The conventional sampler was dimensioned for samples corresponding to approx 15 m^3 air/24 h which gives sufficient sample volumes also in the Onsala area with the refined analysis technique now available. The filter material used was Millipore FALP 02500 which, in our experience, is the only filter material which does not interfere with H^+. The test results are shown in Table 1. As will be seen, the agreement between the results obtained at Onsala with the conventional sampler and the fine particle fraction of the dichotomous sampler is acceptable.

Analysis

Immediately after sampling, each filter is sealed up in a polyethene box. The Acropor filters are leached in the laboratory in 100 ml distilled H_2O for 30 min. Nowadays we use only the Millipore FALP filters; they are leached with 5 ml *ca.* 5.10^{-5} mol/l. $HClO_4$ to equilibrium, which is checked by EMF-measurement. In this leaching solution H^+ is determined through potentiometric micro-titration. The sample is then diluted to 15 ml. Of these, 5 ml are used for NO_3^- determination (Wood *et al.*, 1967) and 10 ml for NH_4^+ determination. This is done through the addition of alkali followed by determination of NH_3aq with an ion-specific Orion electrode. The solution is then cation-exchanged and of the resulting solution, 5 ml is used for SO_4^{2-} determination (Brosset and Ferm, 1978a).

Determination of hydrogen ion concentration is carried out by most laboratories through ordinary pH-measurements. However, at low concentrations this measurement will give unreliable values. Furthermore, no information is obtained concerning what acids the measured pH represents. In view of our aims, such information was not unimportant: We intended to study the particle-borne, mainly man-made acid composed of the part of H_2SO_4 and HNO_3 which had remained unneutralised by NH_3. These acids, as is known, are generated in the atmosphere through the oxidation of SO_2 and NO_2.

Guided by theoretical considerations, we have arrived at a procedure based on Gran's plot (Gran, 1952) which in most cases seems to give the relevant quantity (Brosset and Ferm, 1978b). As we regard the H^+ determination to be one of the more vital parts of these investigations, a brief description of the method will be given below.

Determination of H^+

In a system with the components H_2O, H_2SO_4 and NH_3 the method permits exact determination of the concentration sum of free hydrogen ion and hydrogen sulphate ion ($c_{H^+} + c_{HSO_4^-}$).

The procedure was proposed by Askne (Askne and Brosset, 1972) for use in investigations of precipitation

Table 1. Comparative test of the conventional volumetric sampler and the dichotomous sampler. Mean concentrations in nmol m^{-3}. Number of observations 11–13

Station	Species	Sampler: Conventional	Dichotomous: Fine part. fraction	Dichotomous: Fine + coarse part. fraction
Onsala	H^+	26	32	28
	NH_4^+	139	150	153
	SO_4^{2-}	78	75	83
	NO_3^-	23	21	24
Gothenburg	H^+	3	2	2
	NH_4^+	62	56	58
	SO_4^{2-}	37	30	32
	NO_3^-	6	5	20

and leaching solutions from particle samples. Titration according to Gran of an acid solution with alkali involves successive additions of alkali followed by (potentiometric) measurement of the hydrogen ion remaining in the sample after each addition. The remaining H^+ is then plotted against the added hydroxyl ion. The slope thus obtained is often linear or has a linear interval. Extrapolation of the linear part to the abscissa gives the amount of OH^- which reduces the stoichiometric hydrogen ion concentration of the sample to zero which, by definition, is the point of equivalence.

In practice, it will be observed that the slope of the plot varies with the total sulphate concentration of the sample. This previously disregarded fact does provide very essential information, which will be seen from the mathematical expression for the slope of Gran's function (Brosset and Ferm, 1978).

If a mixture of strong and weak acids is titrated according to Gran, the slope of the plot can be expressed as follows:

$$\frac{dH^+}{dOH^-} = -\frac{1}{1 + C_1\alpha_1 + C_2\alpha_2 + \ldots} \qquad (1)$$

where $C_1, C_2 \ldots$ are total concentrations of the weak acids and $\alpha_1, \alpha_2 \ldots$ are functions of the dissociation constants of each weak acid and of the H^+ concentration of the solution.

In the system H_2O, H_2SO_4, HNO_3 and NH_3 in which the weak acids HSO_4^- and NH_4^+ are present, the expression of the slope will be

$$\frac{dH^+}{dOH^-} = -\frac{1}{1 + (C_1)_t k_1^{-1}\cdot[1 + c_{H^+}\cdot k_1^{-1}]^{-2} + (C_2)_t\cdot k_2^{-1}\cdot[1 + c_{H^+}\cdot k_2^{-1}]^{-2}} \qquad (2)$$

where

$$(C_1)_t = (C_{HSO_4^-})_t = c_{HSO_4^-} + c_{SO_4^{2-}}$$
$$(C_2)_t = (C_{NH_4^+})_t = c_{NH_4^+} + c_{NH_3}$$
$$k_1 = k_{HSO_4^-}$$
$$k_2 = k_{NH_4^+}$$

By inserting numerical values for the dissociation constants, the slope can be calculated as a function of the hydrogen ion concentration for different concentration ranges of total sulphate and total ammonium. The result of such a calculation is shown in Table 2 (Brosset and Ferm, 1978b). This table gives the following information:

In the acid range, part of the total sulphate is present as HSO_4^-. The dissociation of this medium-weak acid during the first part of the titration will successively alter the slope of the plot. When dissociation is nearly complete, a (pseudo) linear range will be obtained ($\sim 10^{-4} > c_{H^+} > \sim 10^{-5}$). In this range, only free hydrogen ions are titrated. After further decrease of the hydrogen ion concentration, dissociation of the very weak acid NH_4^+ starts, which again alters the slope.

Table 2. $-dH^+/dOH^-$ within the c_{H^+}-interval 10^{-2}–10^{-7} for varying total concentrations of the acids HSO_4^- and NH_4^+. The figures within the dotted line are outside of the (pseudo) linear interval

$(C_{HSO_4^-})_t$	$(C_{NH_4^+})_t$	c_{H^+}: 10^{-2}	10^{-3}	10^{-4}	10^{-5}	10^{-6}	10^{-7}
0	0	1	1	1	1	1	1
10^{-5}	0	1	1	1	1	1	1
10^{-4}	0	1	0.99	0.99	0.99	0.99	0.99
10^{-3}	0	0.97	0.92	0.91	0.91	0.91	0.91
10^{-2}	0	0.80	0.55	0.51	0.50	0.50	0.50
0	10^{-5}	1	1	1	1	0.99	0.63
10^{-5}	10^{-5}	1	1	1	1	0.99	0.63
10^{-4}	10^{-5}	1	0.99	0.99	0.99	0.98	0.63
10^{-3}	10^{-5}	0.97	0.92	0.91	0.91	0.90	0.59
10^{-2}	10^{-5}	0.80	0.55	0.51	0.50	0.50	0.39
0	10^{-4}	1	1	1	1	0.95	0.15
10^{-5}	10^{-4}	1	1	1	1	0.95	0.15
10^{-4}	10^{-4}	1	0.99	0.99	0.99	0.94	0.15
10^{-3}	10^{-4}	0.97	0.92	0.91	0.91	0.86	0.14
10^{-2}	10^{-4}	0.80	0.55	0.51	0.50	0.49	0.13
0	10^{-3}	1	1	1	0.99	0.63	0.017
10^{-5}	10^{-3}	1	1	1	0.99	0.63	0.017
10^{-4}	10^{-3}	1	0.99	0.99	0.98	0.63	0.017
10^{-3}	10^{-3}	0.97	0.92	0.91	0.90	0.59	0.017
10^{-2}	10^{-3}	0.80	0.55	0.51	0.50	0.39	0.017
0	10^{-2}	1	1	1	0.95	0.15	0.002
10^{-5}	10^{-2}	1	1	1	0.95	0.15	0.002
10^{-4}	10^{-2}	1	0.99	0.99	0.94	0.15	0.002
10^{-3}	10^{-2}	0.97	0.92	0.91	0.86	0.14	0.002
10^{-2}	10^{-2}	0.80	0.55	0.51	0.49	0.13	0.002

$K_{HSO_4^-}^{-1} = 100$; $K_{NH_4^+}^{-1} = 1.7\cdot 10^9$.

Had there been no NH_4^+ ions in the system, the slope would have maintained its linearity practically down to the abscissa. This means that extrapolation to the abscissa in the linear range gives the amount of OH^- required to reduce the stoichiometric H^+ concentration in the solution to zero at a practically total absence of dissociation of the NH_4^+ ion. Thus one has determined that part of H_2SO_4 and HNO_3 which has not been neutralised by NH_3.

Equation (2) and Table 2 show that if the system lacks weak acids other than HSO_4^- and NH_4^+, the slope is determined in its (pseudo) linear part of the total concentration of sulphate. If this is known, it is possible under certain conditions (titration at constant ionic strength) to exactly calculate the slope. Disagreement between the calculated and the experimental value indicates the presence of other weak acids interfering with the titration. In that case, exact titration of the kind intended here is often not feasible. Such cases occur sometimes in the titration of leaching solutions of particles, but in samples from remote areas, not very often.

The practical procedure applied by us for the determination of H^+ is described elsewhere (Brosset and Ferm, 1978b).

RESULTS

1. *Period: December, 1972. Station: Onsala (remote). Sampling: high volume sampler (Brosset, 1972): Acropor filter*

As early as 1969 we observed that there was a direct proportionality between the concentration of particle-borne sulphur and the content of dark components (soot) in the particles (Brosset and Nyberg, 1971). It was later shown that increased concentrations of sulphur and soot which occur mostly in winter episodes are related to upper winds from the Continent (Rodhe *et al.*, 1972). The first more detailed investigation of the water-soluble part of such dark particles was carried out in December, 1972 when a sharp particle maximum was present for three days (22–24 December) (Brosset, 1973). Sampling for this period was carried out in two series of eight 6 h samples each. These were analysed primarily for H^+ and SO_4^{2-}. One of the samples (22 December, 18–24.00 h) was studied more closely. The result (Table 3) showed that black particles coming from the Continent could contain a water-soluble part that made up nearly half of the total mass. This water-soluble part consisted largely of ammonium sulphate. It also contained a significant amount of nitrate and was weakly acid. This preliminary test stimulated the initiation of a more comprehensive measurement series.

Table 3. Particle sample from Onsala, 22 December 1972, 18.00–24.00. Mass and ion concentration

Species	Concentration $\mu g\ m^{-3}$	Concentration nmol m^{-3}
Total mass	78.3	—
H^+	0.022	22
NH_4^+	7.0	389
SO_4^{2-}	19.7	205
NO_3^-	10.6	171

Table 4. Observed concentration range and interval of variation for H^+/NH_4^+, Onsala. March–August 1973

Concentration in nmole m^{-3} H^+	NH_4^+	SO_4^{2-}	H^+/NH_4^+
1–219	4–406	6–241	0.02–1.24

2. *Period: March–August, 1973. Station: Onsala (remote). Sampling: high volume sampler (Brosset, 1972): Acropor filter*

In this series, a total of 117 samples were collected and analysed. Sampling time was on the average 24 h, on a few occasions 6 h.

Evaluation of the data (Brosset *et al.*, 1975) showed that individual concentrations could vary within a rather broad range. Table 4 shows that the ratio H^+/NH_4^+ was also subject to considerable variation. As this was of special interest, a more detailed comparison was made: It is shown in diagram form in Fig. 2 and comprises most of the values, excluding those of an exceptional case which are given in Table 5. As will be seen from Fig. 2, the hydrogen ion concentration increased more slowly with the ammonium concentration in the first half of the period than in the latter half.

Table 5 shows that between 21 and 31 May considerably higher concentrations were recorded than for the rest of the period, with a peak between the 28 and 29 May. The ion concentration ratio corresponded approximately to the compound NH_4HSO_4.

It should be pointed out that previously observed high sulphate concentrations have always been associated with high concentrations of dark components. The particles sampled on 28–29 May, on the other hand, were practically colourless. This led to the introduction of the terms black and white episode.

The indication through chemical analysis that these colourless particles may consist largely of ammonium sulphate phases led to an attempt at phase identification by means of X-ray diffraction (Brosset *et al.*, 1975). Powder photographs were taken with a Guinier camera of a number of samples directly on the filter. Clearly discernible lines, but also high background, were obtained. The background was practically eliminated by recrystallisation of each sample. The recrystallised samples showed the same strong lines as the original samples. Through comparison with powder photographs taken of pure synthetic phases it was established that one or two of the following phases dominated in all samples: $(NH_4)_2SO_4$, $(NH_4)_3H(SO_4)_2$ and NH_4HSO_4.

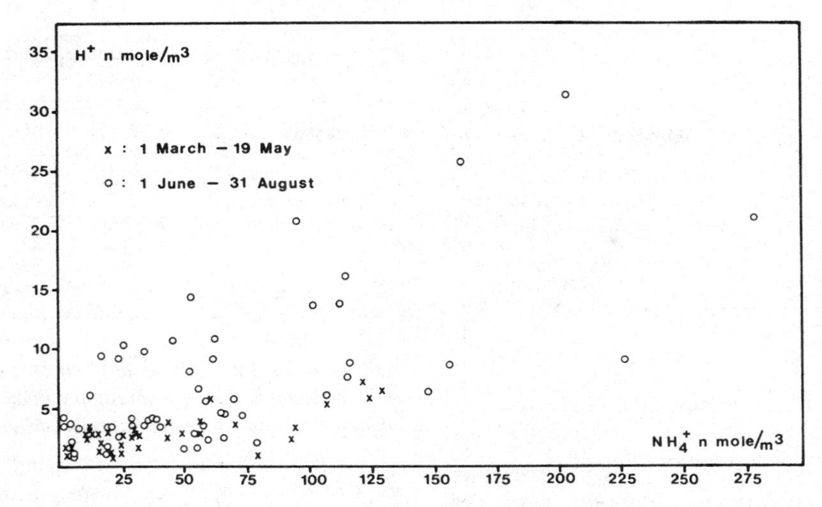

Fig. 2. Content of hydrogen ion and ammonium ion to particulate samples taken at Onsala, 1 March–19 May and 1 June–31 August 1973.

This naturally stimulated interest to understand the conditions under which these phases could be crystallised, which might be achieved with a phase diagram for the system H_2O–H_2SO_4–NH_3. As far as we know, no such complete diagram has as yet been made. However, part of it is available in the form of solubility isotherms at 25°C based on measurements by D'Ans (1913). The diagram is presented in Fig. 3. In his diagram, D'Ans has chosen the components H_2O, SO_3 and $(NH_4)_2O$. It will be seen that five solid phases would be possible in this case, viz. $(NH_4)_2SO_4$, $(NH_4)_3H(SO_4)_2$, NH_4HSO_4, $(NH_4)H_3(SO_4)_2$ and $NH_4HS_2O_7$. The three first are the same as those identified by X-ray diffraction in the particles studied by us.

Table 6 shows part of D'Ans' solubility data expressed as in the original work as mole % for the components $(NH_4)_2O$, SO_3 and H_2O. These were then converted to mole H^+ and NH_4^+ per mole SO_4^{2-}. This made it possible to calculate the ratio H^+/NH_4^+

Table 5. Observations at Onsala during the white episode in May 1973

Sampling				Particle concentration	Ion concentration in nmol m^{-3}				Phases*	
Start date	Time	Stop date	Time	$\mu g\ m^{-3}$	SO_4^{2-}	NH_4^+	H^+	H^+/NH_4^+	Found (X-ray)	Corresp. to equiv. condition
22	12.45	23	14.00	35.3	153	209	35.0	0.17	31	20
23	14.30	23	21.00	59.9	241	402	26.6	0.07	20	20
23	21.00	24	09.10	43.5	166	292	22.7	0.08	20	20
24	09.15	25	14.00	20.2	103	135	45.0	0.33	31	20 + 31
25	15.30	28	15.20	27.0	124	175	20.4	0.12	20 + 31	20
28	15.20	28	24.00	53.7	230	177	219	1.24	11	11
29	00.00	29	06.00	55.8	234	198	212	1.07	11	11
29	06.00	29	12.00	58.9	224	214	214	1.00	31 + 11	31 + 11
29	12.00	30	12.00	35.9	149	217	28.1	0.13	20 + 31	20

* Phase notation: $(NH_4)_2SO_4$:20, $(NH_4)_3H(SO_4)_2$:31, NH_4HSO_4:11.

Table 6. Calculation of H^+/NH_4^+ and p_{NH_3} from solubility data by D'Ans (1913)

	According to D'Ans			Recalculated by the author				
	Mole %			Moles/mole SO_4^{2-}				p_{NH_3}
Solid phase	$(NH_4)_2O$	SO_3	H_2O	NH_3aq	H^+	NH_4^+	H^+/NH_4^+	ppb
$(NH_4)_2SO_4$	17.62	1.73	80.65	18.4	—	2.00	~0	
$(NH_4)_2SO_4$	8.63	8.63	82.74		0	2.00	~0	>0.03
$(NH_4)_2SO_4$	8.91	11.66	79.23		0.47	1.53	0.31	0.029
$(NH_4)_2SO_4$ + $(NH_4)_3H(SO_4)_2$	10.88	15.33	73.79		0.58	1.42	0.41	0.022
$(NH_4)_3H(SO_4)_2$	9.67	18.44	72.19		0.95	1.05	0.90	0.010
$(NH_4)_3H(SO_4)_2$ + NH_4HSO_4	17.47	35.60	46.93		1.02	0.98	1.04	0.009
NH_4HSO_4	4.12	23.29	72 59		1.65	0.35	4.7	0.002
NH_4HSO_4	4.48	31.78	63.74		1.72	0.28	6.1	0.001

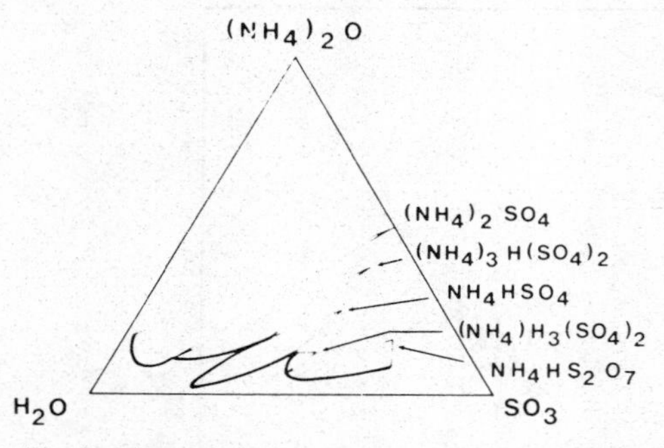

Fig. 3. Phase diagram according to D'Ans (1913).

for each solubility determination, and from this obtain the partial pressure of NH_3 in the gas phase in equilibrium with the solution. For

$$NH_3(g) + H_2O \rightleftarrows NH_3aq(l)$$

$$NH_3aq(l) + H^+ \rightleftarrows NH_4^+ + aq$$

but

$$\frac{c_{NH_3aq}}{p_{NH_3}} = k'$$

and

$$\frac{c_{H^+} \cdot c_{NH_3aq}}{c_{NH_4^+}} = k_1$$

$$\therefore\ p_{NH_3} = \frac{k'}{k_1} \cdot \left(\frac{c_{H^+}}{c_{NH_4^+}}\right)^{-1}.$$

Using approximate values for the constants k' and k_1 at 25°C we get

$$p_{NH_3} = 9 \cdot 10^{-3} \left(\frac{c_{H^+}}{c_{NH_4^+}}\right)^{-1} \text{ ppb.}$$

These calculations have given a rough idea about the levels of atmospheric NH_3 at which the different ammonium sulphate phases may be formed. For crystallisation, however, the water vapour pressure must drop below certain values. These values are not known to us. The determination of the water vapour pressure in connection with the solubility measurement at different temperatures, i.e. preparation of a more complete phase diagram for the system in question, is therefore an important requirement. The availability of such a diagram will not give all the information needed to calculate the conditions under which the water-soluble part of particles is formed, as natural systems of course contain considerably more components than those discussed here. But in some cases it will probably be possible to arrive at acceptable approximations.

The phases identified by X-ray diffraction in the samples taken are shown in Table 5. The table also indicates the phases which according to Table 6 correspond to the experimentally found H^+/NH_4^+ ratios. Considering the sampling time, the agreement is satisfactory.

Table 7. Aerodynamic particle diameter in μm for the respective impactor steps

	ACI	BGI
Impactor		
plate	>7	>13.3
plate	7.0–3.3	13.3–5.4
plate	3.3–2.0	5.4–3.5
plate	2.0–1.1	3.5–2.5
Filter	<1.1	<2.5

3. Period: 26 Jan–4 June 1973 and 25 Feb–16 April 1974. Station: Onsala (remote). Sampling: ACI and BGI impactors. Teflon film and Acropor filter

This measurement series (sampling and analysis) was carried out by Dr. G. Musold, at that time collaborating with us. The aim was to determine the chemical composition of the water-soluble part of different size fractions of particles of different origin. In total, the material comprises seventeen 72 h samples, each in five size fractions defined as shown in Table 7.

The analytical result showed (Brosset, 1976), as was expected, that we were dealing with two populations corresponding to fine and coarse particles (> and <2–3 μm). The analysis data obtained were therefore converted so as to conform to these two fractions.

As we were interested in the component concentrations primarily as a particle property, the concentrations were expressed as equivalent fractions (%). For this purpose, we assumed that the molar concentration sum $2SO_4^{2-} + NO_3^- + Cl^-$ is almost equal to the total equivalent sum. The result is shown in Table 8. In addition to the equivalent composition, it includes the mass and sulphate concentration of fine particles as well as the soot value.

The three latter quantities were used to classify the particles: high content of soot and sulphate relative to total mass is our characterisation of black episodes. Low soot content but high sulphate concentration relative to total mass is characteristic of white episodes. These are of course no rigid criteria but were found to be useful as they seem to correspond to particles of essentially different genetic origin. Transition stages exist. Four samples in this material were considered to represent these stages. On the basis of this classification, 12 of the samples could be grouped as shown in Table 9. This table also includes the main direction of the upper winds during sampling. Guided by these data, the type 'No episode' was divided into a NE and a NW section. The entire material, finally, was condensed in the form of mean values in Table 10.

Even if the values in Table 10 are based on very limited material, they do show a certain consistency that may permit a cautious interpretation to be made.

It is obvious that fine particles are mostly acid and low in chloride, whereas coarse particles are alkaline and contain much chloride.

Table 8. Data obtained at Onsala with fractionated sampling

Sample No.	Date	Main trajectory direct	Fine part. fraction: Mass $\mu g\ m^{-3}$	Soot $\mu g\ m^{-3}$	SO_4^{2-} nmol m^3	Type	Ion concentration in equiv. fractions (%): Fine part. fraction: SO_4^{2-}	NO_3^-	Cl^-	OH^-	H^+	NH_4^+	Coarse part. fraction: SO_4^{2-}	NO_3^-	Cl^-	OH^-	H^+	NH_4^+
	1973:																	
1	Jan 26–Feb 2	W	7	4	28	no ep.	98	2	0	0	9	—	63	11	21	5	0	—
2	Feb 5–12	NW-W	2	2	20	no ep.	70	14	16	0	0	21	38	9	50	3	0	73
3	Feb 16–23	E-NW-N	3	2	28	no ep.	90	5	5	0	5	45	59	6	34	1	0	6
4	Feb 26–Mar 5	NW-W	7	3	41	no ep.	80	17	4	0	4	65	47	24	29	0	0	66
5	Mar 3–12	W-N	1	2	22	no ep.	86	12	2	0	6	61	37	22	35	6	0	9
6	Mar 23–26	W-SW	—	18	221	Black	74	24	2	0	5	54	37	46	13	4	0	27
7	Apr 16–24	N-E	18	3	55	no ep.	61	3	36	0	2	57	5	2	93	0	0	6
8	May 14–21	N	8	1	8	no ep.	52	3	45	0	0	35	8	2	89	0	0	3
9	May 21–23	S	56	5	173	Trans.	56	10	34	0	7	45	16	17	70	0	0	18
10	May 23–25	SW	27	3	105	White	82	2	16	0	16	47	35	9	56	0	0	20
11	May 25–28	SW	(24)	2	135	White	91	2	7	0	18	77	29	25	45	1	0	6
12	May 28–30	W	44	4	169	White	92	1	7	0	14	72	11	21	55	13?	0	20
13	May 30–Jun 4	S-SW	(14)	3	58	Trans.	87	4	9	0	9	53	15	25	59	1	0	2
	1974:																	
14	Feb 25–27	W	20	7	108	Trans.	66	21	13	0	9	65	46	35	18	1	0	28
15	Feb 27–Mar 1	E-SE	20	10	67	Black	75	21	4	0	13	71	29	29	29	13	0	21
16	Mar 1–4	SE-S	37	22	101	Black	75	21	5	0	7	81	21	33	39	7	0	16
17	Apr 8–16	NW-N	12	3	56	Trans.	85	14	2	0	0	69	28	22	50	0	0	3?

Table 9. Ion concentrations in equivalent fractions (%) of the water-soluble part of 'fine' and 'coarse' particles in samples taken at Onsala during different episode situations

Sample No.	Main trajectory direction	SO_4^{2-}	NO_3^-	Cl^-	OH^-	H^+	NH_4^+	SO_4^{2-}	NO_3^-	Cl^-	OH^-	H^+	NH_4^+
		← Fine particles →						← Coarse particles →					
		No episodes—air masses from NE-E											
7	N-E	61	3	36	0	2	57	5	2	93	0	0	6
8	N	52	3	45	0	0	35	8	2	89	0	0	3
		No episodes—air masses from N-W											
2	NW-W	70	14	16	0	0	21	38	9	50	3	0	73
3	E-NW-N	90	5	5	0	5	45	59	6	34	1	0	6
4	NW-W	80	17	4	0	4	65	47	24	29	0	0	66
5	W-N	86	12	2	0	6	61	37	22	35	6	0	9
		Black episodes											
6	W-SW	74	24	2	0	5	54	37	46	13	4	0	27
15	E-SE	75	21	4	0	13	71	29	29	29	13	0	21
16	SE-S	75	21	5	0	7	81	21	33	39	7	0	16
		White episodes											
10	SW	82	2	16	0	16	47	35	9	56	0	0	20
11	SW	91	2	7	0	18	77	29	25	45	1	0	6
12	W	92	1	7	0	14	72	11	21	52	13?	0	20

Black episodes are characterised by moderate acid concentration in the fine particle fraction. Both fractions contain much nitrate. White episodes are characterised by high concentration of acid and low concentration of nitrate in the fine particles. Non-episodic particles introduced from the northern sector are almost neutral. The coarse particles seem to consist of almost only chloride whereas the fine particles contain ammonium sulphate as well.

As established by Rhode *et al.* (1972), black episodes occur in connection with long range transport from the Continent. On the other hand, white episodes observed so far were associated with weak westerly winds. However, as will be shown below, westerly winds are not specific to a white episode.

4. *Period: Aug 1975–April 1976, occasional data from Feb 1977. Station: Onsala (remote) and Gothenburg (urban). Sampling: conventional (volumetric) sampler, Millipore FALP filter*

This is the most comprehensive of the measurement series conducted so far. It was started in 1975. As of August, 24 h samples were collected with conventional (volumetric) samplers (Fig. 2) both at Onsala and at an identically equipped station in central Gothenburg. These stations also monitor particle number (Royco, eight channels) and gas concentrations (O_3, SO_2, NO, NO_x) on a continuous basis. The measurement programme includes meteorological parameters, and trajectories (72 h, 850 mb, arriving at Onsala 06.00 Z and 18.00 Z) are obtained from the Norwegian Institute for Air Research. In addition, sampling with Whatman filter (2 m^3 air/day) is carried out for determination of soot and metals. These measurements are intended to continue.

In that part of the data evaluation reported here the purpose was to elucidate further the conditions during so called episodes and to determine to what extent the long range transported particles affect the air situation in a community. Since we have come across several phenomena of importance in the course of this work, the primary data for the most interesting months are presented in diagram form for soot, H^+, NO_3^- and SO_4^{2-} concentrations (Fig. 4a and b). Even from a first look at these diagrams it is clear that the term episode can be very adequately applied. The concentration increase may occur very rapidly and remain for one or a few days after which, as a rule, the drop is also very rapid. With very few exceptions, there is an extremely good correlation between the respective concentrations at Onsala and in Gothenburg.

Typical episode situations are combined in Table 11. As data from Onsala are directly related to long range transport, these are discussed first. Table 12 presents these data reduced to episode mean values.

Table 10. Mean ion concentrations in equivalent fractions (%) in the water-soluble part of 'fine' and 'coarse' particles sampled at Onsala during different episode situations

Episode, type	SO_4^{2-}	NO_3^-	Cl^-	OH^-	H^+	NH_4^+	SO_4^{2-}	NO_3^-	Cl^-	OH^-	H^+	NH_4^+
	← Fine particles →						← Coarse particles →					
None, N-E	57	3	41	0	1	46	7	2	91	0	0	5
None, N-W	82	12	7	0	4	48	45	15	37	3	0	39
Black	75	22	4	0	8	69	29	36	27	8	0	21
White	88	2	10	0	16	65	25	18	52	5?	0	15

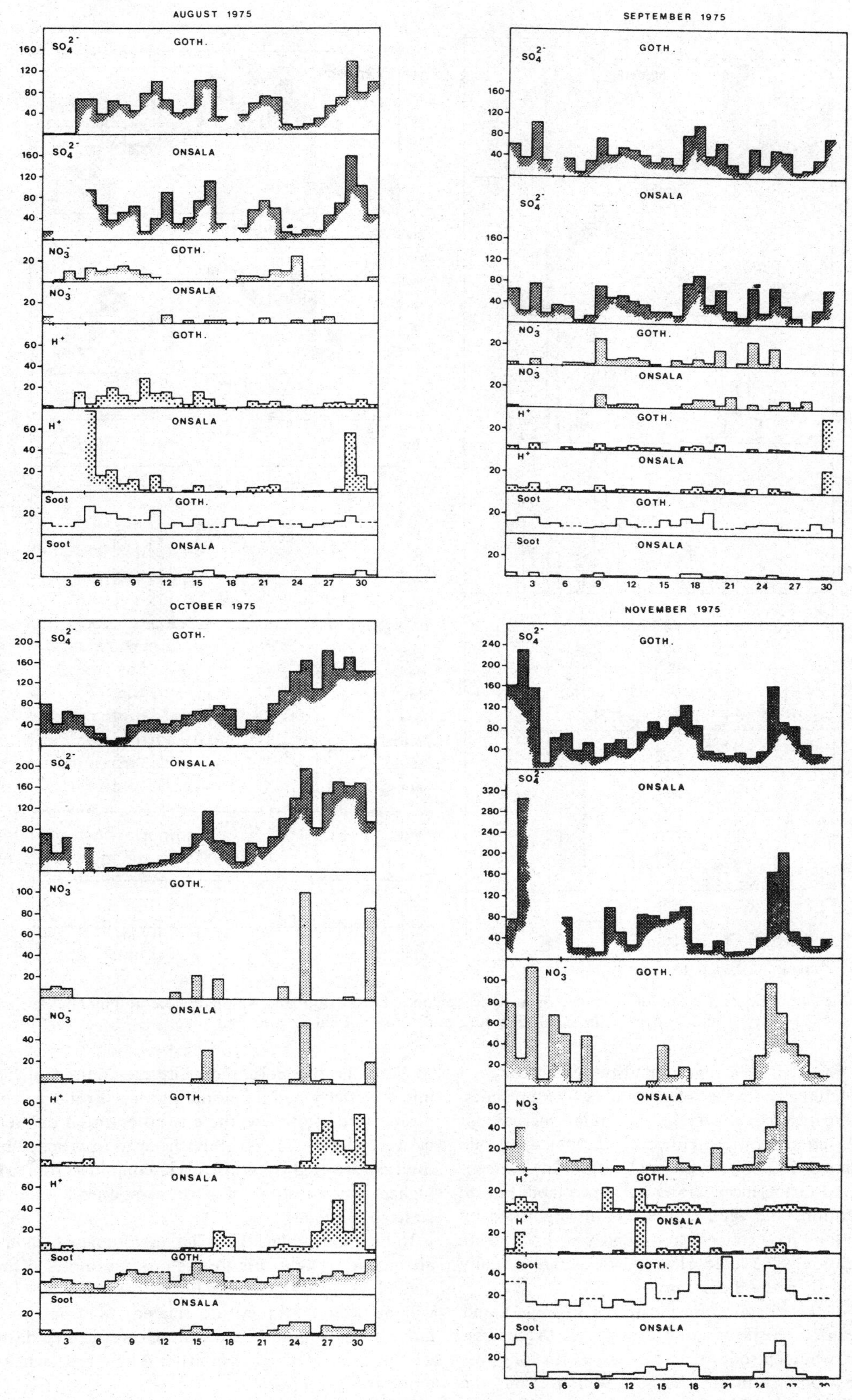

Fig. 4a. Concentrations of particle-borne hydrogen ion, nitrate ion and sulphate ion in nmol m^{-3} and soot in μg m^{-3}. Onsala and Gothenburg, August–November 1975.

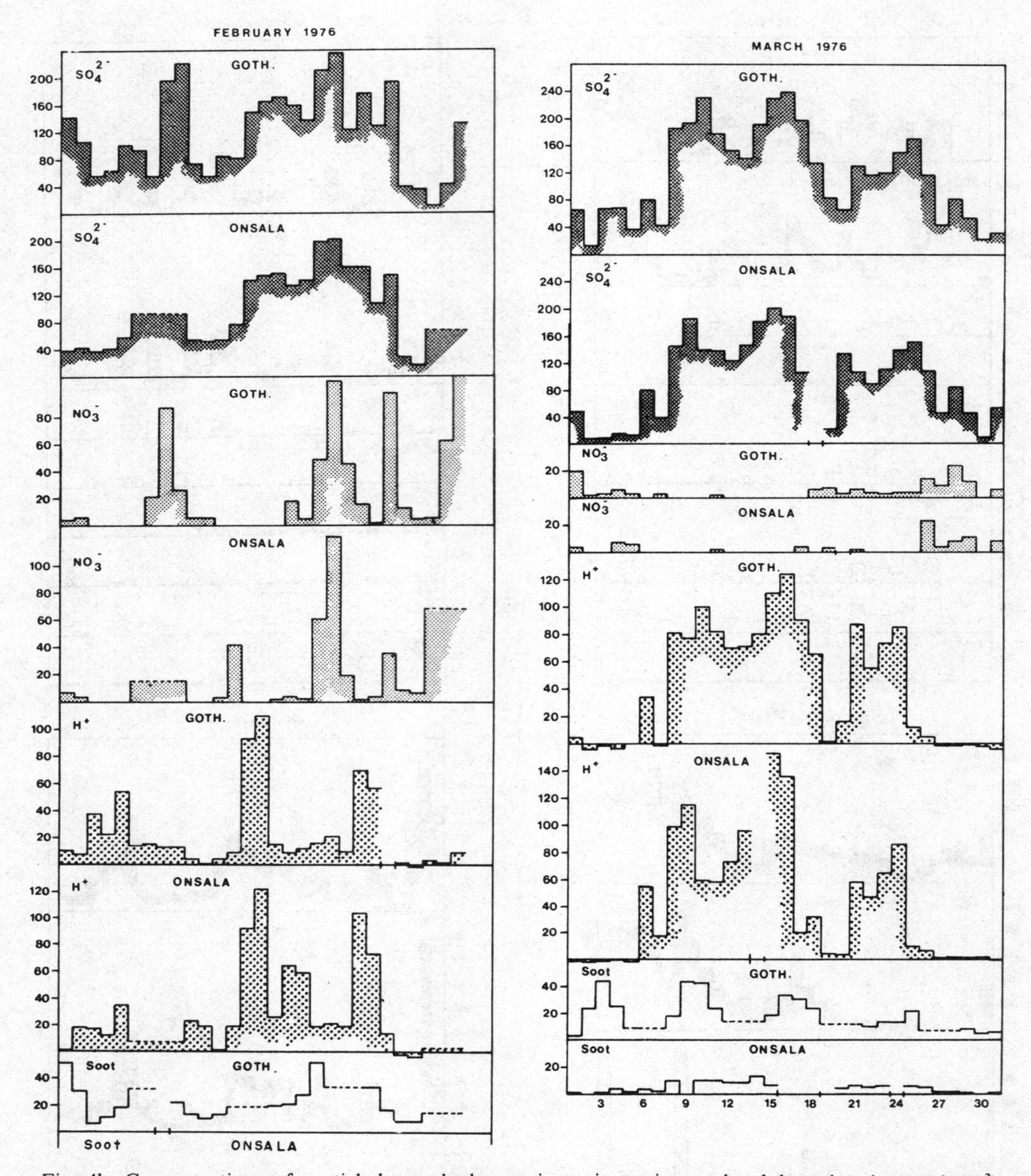

Fig. 4b. Concentrations of particle-borne hydrogen ion, nitrate ion and sulphate ion in nmol m^{-3} and soot in $\mu g\ m^{-3}$. Gothenburg and Onsala, February–March 1976.

As will be seen, this material also demonstrates a clear difference between black and white episodes. Both indicate, on an average, a similar increase in the SO_4^{2-} concentration. However, the black episode is associated with increased soot (and Mn) concentration and furthermore shows an appreciable nitrate concentration. The white episode is distinguished by low soot (and Mn) concentration and very low nitrate concentration. The colourless particles are usually more acid than the dark ones.

Some of the episodes are illustrated in Figs. 5 and 6 by means of corresponding trajectories. According to Fig. 5, white episodes seen to occur with long distance transport from the west as well as from the east. This is new. Previously, white episodes were observed only in air masses moving in from the west. In Table 11, therefore, the white episodes are divided into a westerly and an easterly group. From the composition point of view, there is no essential difference between these two; possibly the white particles from the east are slightly greyish, i.e. contain a somewhat higher concentration of dark components than the westerly particles.

In the case of black particles, certain new facts have also come to light. Figure 6 shows examples of normal black episodes (solid lines) (from Table 11). It will be seen that they are related, as usual, to air movements from the Continent. However, two dashed westerly trajectories are inserted. These represent two unusual black episodes.

In black as well as white episodes, the concentration of SO_2 and NO_2 do not normally exceed 5

Table 11. Daily means at different episodes

Type	Date	Gothenburg					Onsala					
		nmol m^{-3}				$\mu g\ m^{-3}$	nmol m^{-3}				$\mu g\ m^{-3}$	
		H^+	NH_4^+	SO_4^{2-}	NO_3^-	Soot	H^+	NH_4^+	SO_4^{2-}	NO_3^-	Soot	Mn
Black	1975											
	Nov											
	1	7	376	161	78	33	4	164	76	22	32	0.035
	2	14	475	230	26	33	20	604	314	40	39	0.045
	1976											
	Feb											
	8	13	359	197	87	—	7 (8–9)	198 (8–9)	94 (8–9)	15 (8–9)	29	—
	9	13	516	222	26	22					27	—
	19	16	387	212	49	52	19	398	199	61	42	0.040
	20	21	469	235	107	34	21	493	203	122	42	0.044
	21	10	251	124	46		19	333	162	20	29	0.019
	24	—	436	195	98	17	14	339	150	36	29	0.018
Western White	1975											
	Aug											
	5	4	125	67	13	27	77	103	95	<2	1	—
	29	4	200	138	2	18	56	202	159	<2	2	—
	Oct											
	28	26	202	149	2	22	49	276	172	<2	8	0.008
	29	17	246	173	3	18	20	285	165	<2	8	0.009
	30	51	178	146	2	20	66	348	170	<2	5	0.011
Eastern White	1976											
	Feb											
	14	93	165	150	2	19 (14–15)	92	166	142	<2	11	0.016
	15	110	150	166	2		122	148	149	<2	11	0.020
	17	9	286	160	18	21	64	189	135	4	8	—
	18	12	245	138	5	28	59	205	142	2	14	0.011
	22	70	212	178	16	34	106	190	162	2	14	0.010
	23	57	157	129	2	34	73	162	109	4	11	0.008
	March											
	13	71	141	141	2	14 (13–14)	96	166	147	<2	8	—
	14	80	207	187	2		—	219	182	<2	13	—
	15	110	255	230	2	19	153	205	202	<2	6	0.004
	16	124	227	239	2	34	136	190	190	<2	—	—
	21	87	124	130	6	12	58	139	108	2	6	—
	22	55	129	117	4	10	47	121	91	<2	5	—
	23	73	116	120	3	14	65	129	112	<2	6	0.011
	24	85	164	151	4	14	86	160	141	<2	—	—

and 10 ppb, respectively but this did in fact happen in the unusual black episodes that occurred on 25–26 November 1975 (Brosset, 1976) and 2–3 February 1977. The concentrations measured are shown in Table 13 and Fig. 7. Interesting meteorological conditions seem to be connected with these two cases, the only ones of this type so far known to us. According to Dr. K. A. Rahn at the University of Rhode Island, USA, these episodes may be a consequence of special transatlantic transport. This phenomenon will be the subject of further studies.

So far, episodes caused by long distance transport have, for natural reasons, been studied in a remote area. It was later of interest to investigate to what extent they also make themselves felt in a community. In addition to Onsala, therefore, measurements were carried out in Gothenburg, as indicated by Fig. 4a and b. This city has a population of about 500,000 and contains both shipbuilding yards and a number of large industries.

The concentrations of H^+, NH_4^+, SO_4^{2-} and NO_3^- found in the water-soluble part of particles in Onsala

Table 12. Average daily means at different episodes (from Table 11). Onsala. August 1975–April 1976

Episode type	Number of observations	Concentration in nmol m^{-3}				Conc. in $\mu g\ m^{-3}$	
		H^+	NH_4^+	SO_4^{2-}	NO_3^-	Soot	Mn
Black	8	14	341	162	41	34	0.034(6)*
W. White	5	54	243	152	<2	5	0.009(3)
E. White	13	89	167	141	<2	9	0.011(7)

* Number of daily means of Mn.

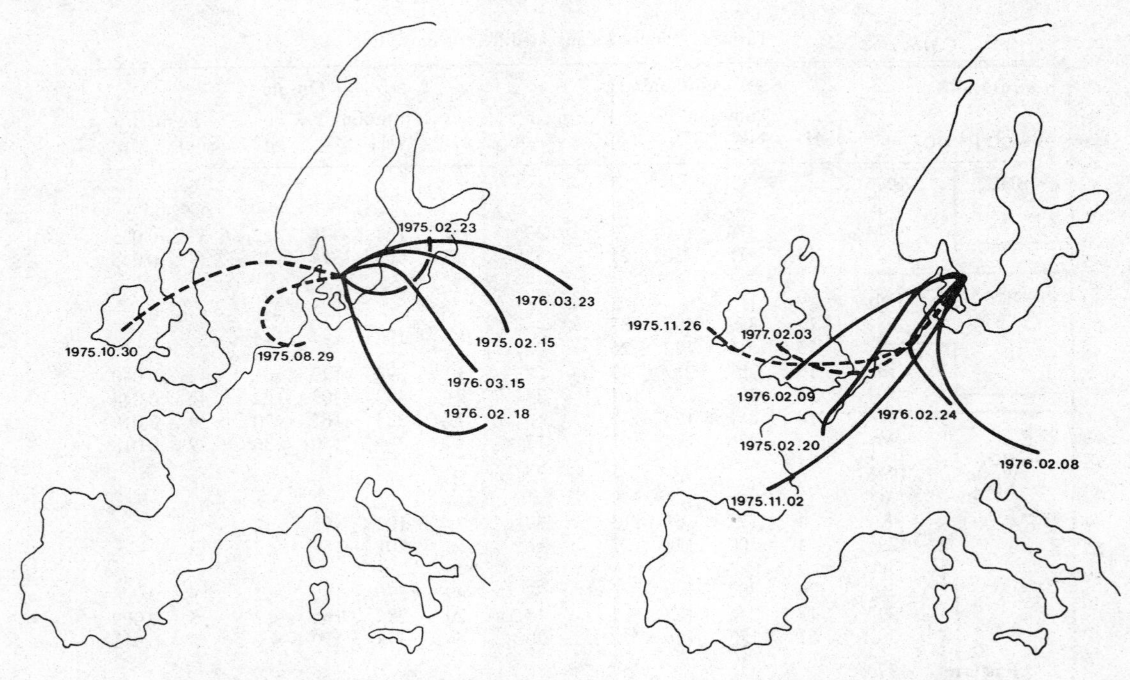

Fig. 5. Trajectories for some western white (-----) and eastern white (——) episodes.
Fig. 6. Trajectories for some normal black (——) and unusual black (-----) episodes.

are characteristic of large areas of the Swedish west coast. It was therefore assumed that the share of long range transported pollutants in Gothenburg is approximately equal to the concentrations observed at about the same time at Onsala. The local contribution to pollutant concentrations measured in Gothenburg would thus be the difference between the Gothenburg and the Onsala values.

These differences were calculated for the entire present material. They are presented in Table 14 in the form of median values for eight different wind sectors. Both the values obtained directly in Gothen-

Table 13. Unusual black episodes. Daily mean concentrations. Onsala

	nmol m^{-3}				μg m^{-3}		ppb			
Date	H^+	NH_4^+	SO_4^{2-}	NO_3^-	Soot	Mn	O_3	SO_2	NO_2	NO_x
1975										
Nov										
24	1	79	40	17	6	0.008	—	0	2	2
25	5	452	163	40	24	0.050	—	10	28	34
26	9	458	201	64	36	0.062	—	8	26	31
27	3	183	52	2	10	0.033	—	0	17	25
28	1	122	43	5	4	—	—	0	2	3
1977										
Feb										
1	2	187	65	95	11	—	19	0	6	10
2	16	186	118	4	18	—	34	5	9	6
3	16	447	270	51	60	—	20	20	15	18
4	4	110	78	19	22	—	41	1	6	6
5	6	79	61	2	4	—	41	0	3	4

Table 14. Median values for SO_4^{2-}, NO_3^- and H^+ in nmol m^{-3} for eight ground level wind directions. Gothenburg, August 1975–April 1976

Wind sector		N-NE	NE-E	E-SE	SE-E	S-SW	SW-W	W-NW	NW-N
SO_4^{2-}	uncorrected	36.0	72.0	71.2	92.8	75.2	49.2	26.5	30.0
	corrected	11.3	25.2	10.0	17.4	7.8	7.4	10.0	10.7
NO_3^-	uncorrected	4.8	4.4	2.7	10.0	11.4	10.8	5.3	4.9
	corrected	2.1	1.8	2.4	6.0	7.8	6.2	2.1	3.6
H^+	uncorrected	3.4	8.6	14.6	7.7	6.0	4.8	2.8	3.2
	corrected	0.4	2.1	3.5	0.9	2.7	2.0	1.0	1.4

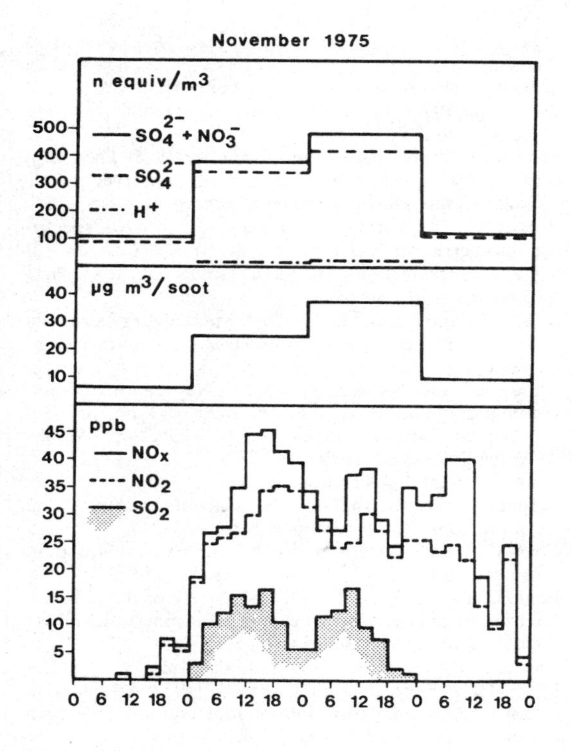

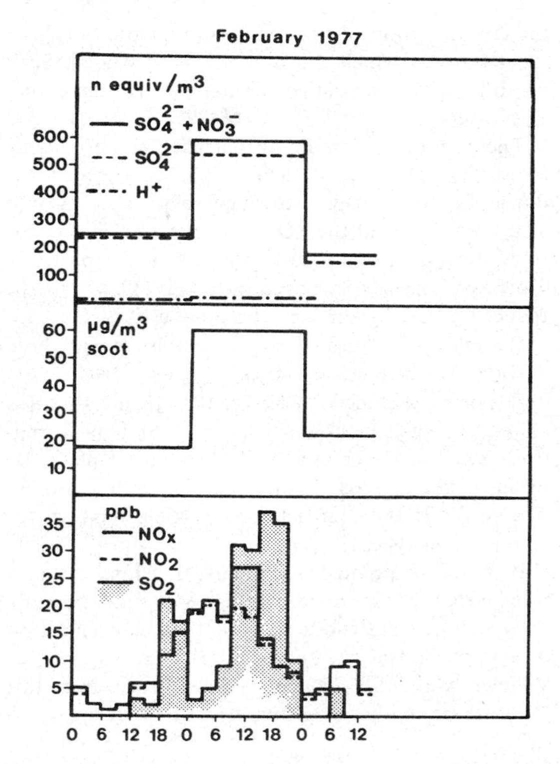

Fig. 7. Unusual black episodes.

burg (uncorrected values) and the differences (corrected values) are given. The percentage distribution of each concentration on local and long distance transport are shown in Table 15. It will be seen that, with the exception of nitrate, the local contribution is small.

POSSIBLE FORMATION PROCESSES FOR BLACK AND WHITE EPISODES

From a large number of samples it had been shown that long range transported sulphate occurs concurrently with two different types of particles, and a plausible explanation was naturally sought. Of course the material is too incomplete to permit any firm conclusions. But we can outline processes which, if they did take place, would lead to occurrences approximately equivalent to black and white episodes.

Assume that we have an air mass which is nearly particle-free but contains *ca.* 200 nmol m^{-3} SO_2 (*ca.* 5 ppb) and has a (probably low) water vapour pressure. Assume further that SO_2 is oxidised in gas phase

Table 15. Percentage of locally (I) produced and long distance-transported (II) ions. Gothenburg, August 1975–April 1976

Ion	I	II
SO_4^{2-}	16%	84%
NO_3^-	41%	59%
H^+	~0%	~100%

through some (e.g. photochemical) process. At complete oxidation (in our assumed case) 200 nmol m^{-3} H_2SO_4 will be formed, i.e. 400 nmol m^{-3} H^+. Assume that the air mass contains or takes up, both during and after oxidation, 200–300 nmol m^{-3} NH_3, most of which reacts with the sulphuric acid droplets. Assume finally that with time, (pseudo)equilibrium will be attained at *ca.* 0.01–0.02 ppb NH_3 and a RH of *ca.* 40%.

Through the outlined process we have arrived at an air mass containing particles which in terms of their nature and concentration are characteristic of a white episode.

Now assume that we have a stack emitting smoke from a combustion process. The flue gas is some hundred degrees warm and contains fly ash, soot, CO_2, H_2O(g), SO_2, Mn(g), etc. In the cooling process, first Mn (Natusch, pers. comm.), then H_2O will be condensed on the particles. SO_2 will dissolve in the water film, oxidise to H_2SO_4 and dissolve the Mn film. A similar process has been simulated in our laboratory (Brosset, 1976). There are now excellent conditions for a catalytic oxidation of SO_2: high SO_2 concentration, high humidity, and a good catalyst distributed over a large area.

The continued oxidation increases the acid concentration of the particles. At the same time, the particles, moving away from the source, approach lower and lower SO_2 concentrations. This necessarily leads to the oxidation process coming to a stop (Junge and Ryan, 1958). The result is particles with a sulphate concentration which stands in a certain relation to the concentration of catalyst (Mn) and the SO_2 gradient through which the particles have passed. Such conditions may be rather similar for a large group of emitters.

These sulphate-containing particles are not likely to catalytically change their sulphate concentration during further transport through relatively clean air. The reasons is that the SO_2 concentration is too low in relation to the acid concentration of the particles. However, the acid concentration is likely to be reduced gradually through the uptake of NH_3.

Particles thus formed are probably of the kind encountered in black episodes.

It can be assumed, however, that in an air mass containing such black particles, the sulphate formation may under certain conditions take place by means of the process characteristic of white episodes. The result is then particles or particle mixtures of a transistion type.

It should be pointed out, finally, that long distance transported particles containing Mn may possibly resume their catalytic activity if their concentration of acid is neutralised and/or if they are exposed to sufficiently high SO_2 concentrations. Processes of this kind are presently being investigated by us.

REFERENCES

Askne C. and Brosset C. (1972) Determination of strong acid in precipitation, lake-water and air-borne matter. Letter to the Editors. *Atmospheric Environment* **6,** 695–696.

Brosset C. (1973) Air-borne acid. *Ambio* **II,** 2–9.

Brosset C. (1976) Air-borne particles: black and white episodes. *Ambio* **V,** 157–163.

Brosset C., Andréasson K. and Ferm M. (1975) The nature and possible origin of acid particles observed at the Swedish west coast. *Atmospheric Environment* **9,** 631–642.

Brosset C. and Ferm M. (1978a) An improved spectrophotometric method for the determination of low sulphate concentrations in aqueous solutions. *Atmospheric Environment* (in press).

Brosset C. and Ferm M. (1978b) Man-made air-borne acidity and its determination. *Atmospheric Environment* (in press).

Brosset C. and Nyberg A. (1971) Investigations of soot and particle-borne sulphur in Sweden. Proc. Sec. Int. Clean Air Congress Washington, 6–11 Dec., pp. 481–489. Academic Press, New York.

D'Ans J. (1913) Zur Kenntnis der Sauren Sulfate VII. Saure Sulfate und Pyrosulfate des Natrium, Kalium und Ammonium. *Z. allg. anorg. Ch.* **80,** 235–245.

Gran G. (1952) Determination of the equivalence point in potentiometric titrations. *Analyst* **77,** 661–671.

Junge C. E. and Ryan T. G. (1958) Study of the SO_2 oxidation in solution and its role in atmospheric chemistry. *Quart. J. Roy. met. Soc.* **84,** 46–55.

Natusch D. F. S. Personal communication.

Rodhe H., Persson C. and Åkesson O. (1972) An investigation into regional transport of soot and sulphate aerosols. *Atmospheric Environment* **6,** 675–693.

Wood E. D., Armstrong F. A. J. and Richards F. A. (1967) Determination of nitrate in sea water by cadmium–copper reduction to nitrate. *J. mar. biol. Ass. U.K.* **47,** 23–31.

Atmospheric Environment Vol. 12. pp. 39–53. Pergamon Press 1978. Printed in Great Britain.

CHEMICAL PROPERTIES OF TROPOSPHERIC SULFUR AEROSOLS

ROBERT J. CHARLSON[1,4], DAVID S. COVERT[2], TIMOTHY V. LARSON[2,3], and ALAN P. WAGGONER[1]

[1]Water and Air Resources Division, Department of Civil Engineering, [2]Department of Environmental Health, [3]Department of Atmospheric Sciences and [4]Institute for Environmental Studies, University of Washington, Seattle, WA 98195, U.S.A.

(*First received* 13 *June* 1977 *and in final form* 10 *August* 1977)

Abstract—Sulfur is widely recognized as an element present in atmospheric aerosol; however, only recently have data been acquired showing the dominance in industrial regions of submicrometer tropospheric aerosol by a family of sulfate compounds ranging from H_2SO_4 to $(NH_4)_2SO_4$. It is possible to infer the presence of other molecular forms and oxidation states. The overall picture is as yet qualitative, with semiquantitative evidence showing that both the urban and rural aerosols in the eastern third of the United States consist mainly of impure sulfate compounds containing substantial amounts of water, with metal and organic compounds as trace inclusions. Among chief physico-chemical consequences of the dominance by sulfates are the fundamental nature of hygroscopic growth and predictable variations in refractive index. It is important to emphasize the role of omnipresent impurities and their possible effects.

1. ROLE AND IMPORTANCE OF CHEMICAL COMPOSITION

Sulfur has been recognized for at least two decades as an important constituent of atmospheric aerosol particles. Junge (1954; 1963) established the ubiquitous occurrence of sulfate compounds in both 'polluted' and 'unpolluted' air at sites near Frankfurt and Boston. In his studies, sulfur was found predominantly in particles below 0.8 μm radius and NH_4^+ coexisting with SO_4^{2-} gave empirical formulas between $(NH_4)_2SO_4$ and NH_4HSO_4. The existence and importance of SO_4^{2-} in aerosol particles is reflected in its inclusion in the lists of substances monitored in sampling programs for both particulate matter and precipitation chemistry.

Possible reasons for the preponderent interest in sulfate compounds as opposed to other forms would seem to include the simple fact that SO_4^{2-} analyses on collected samples are easier to do than more complete molecular analyses or than analyses of other sulfur forms such as sulfite, pyrosulfite or dithionate. Another reason for the interest in sulfate compounds is their ubiquitous presence in one form or another due in part to the stability of SO_4^{2-} as the fully oxidized, +6 oxidation state of sulfur. Some measurements of 'SO_4^{2-}' may in fact be a sulfate compounds formed on a filter from some other atmospheric form.

Table 1 is a list of commonly recognized sulfur compounds which exist or are expected to exist in aerosol particles under atmospheric conditions. Many of the oxy acids and their salts *could* occur, as well as sulfide, polysulfide and organic sulfur compounds and possibly others. Figure 1 depicts a flow diagram for the cycling of sulfur compounds through the atmosphere. Of particular importance in this system is the likelihood of simultaneous co-existence of several sulfur-containing species in atmospheric particles, possibly in different oxidation states.

It is important to recognize that sulfur in particulate matter often begins its atmospheric existence in a molecular form and oxidation state different from that at the time of sampling and analysis. While the most obvious case is the oxidation of SO_2 (gas) to SO_4^{2-} (particles), many other less-studied cases undoubtedly exist. For example, it is suspected that H_2S, RHS or RSR oxidize to SO_2 and then SO_4^{2-} such that during the cycle the oxidation state makes the complete transition from the fully reduced (−2) to the fully oxidized (+6) state.

Only in recent years has an interest been rekindled in defining the molecular/crystalline forms of sulfate compounds and of other forms or oxidation states of sulfur in aerosol particles. It has become clear that there are important reasons for understanding these details of sulfur chemistry. (1) Chemically, SO_4^{2-} is practically inert at normal temperatures in either a condensed form or in an aqueous medium. Nonetheless, the cation associated with the SO_4^{2-} may be strongly reactive. (2) SO_4^{2-} is practically non-toxic although some sulfate compounds are toxic. (3) Optical properties of aerosol particles such as refractive index are governed by the molecular form rather than the atomic or ionic composition. (4) The deliquescent/hygroscopic nature of aerosol particles is governed by molecular form. (5) The molecular form offers clues as to the origin and cycling of the sulfur-bearing particles in the air. (6) Particle shape and state (which are of possible significance to optics, aerodynamics and morphological studies) are at times governed by molecular form.

Table 1a. Oxy acids, their salts and ionized forms which could exist in atmospheric aerosols

Oxy acid	Formula	Salt/ionized Form
sulfuric	H_2SO_4	HSO_4^-, SO_4^{2-} NH_4HSO_4 $(NH_4)_3H(SO_4)_2$ $(NH_4)_2SO_4$ $MgSO_4$ $CaSO_4$ Na_2SO_4 $ZnSO_4$ $(NH_4)_2SO_4 \cdot ZnSO_4$ $R\text{-}O\text{-}SO_3^-$
sulfurous	$SO_2 \cdot xH_2O$	HSO_3^-, SO_3^{2-}
sulfonic	$R\text{-}SO_3\text{-}H$	$R\text{-}SO_3^-$
dithionic	$H_2S_2O_6$	$S_2O_6^{2-}$
hydroxy-sulfonic	$R—C(OH)(SO_3H)$	$R—C(OH)(SO_3^-)$
thiosulfuric	$H_2S_2O_3$	$S_2O_3^{2-}$
polythionic	$H_2S_nO_6$	$S_nO_6^{2-}$
pyrosulfurous	$H_2S_2O_5$	$S_2O_5^{2-}$

Table 1b. Other sulfur compounds

Name	Formula	Ionized form
sulfides	MeS (Me = metal)	most are insoluble
	H_2S	HS^-, S^{2-}
	R-SH	RS^-
	R-S-R	
polysulfides		S_n S^{2-}
chemi-sorbed species	CS_2	
	S	
	SO_2	

The main purposes of this paper are to review our knowledge of the molecular forms and properties of particulate atmospheric sulfur and to outline areas where information is missing.

2. MOLECULAR FORMS OF SULFUR IN TROPOSPHERIC AEROSOLS

There are two fundamentally different approaches for deducing the existence of various molecular or ionic species. Direct analysis suffices for some of the more stable compounds (e.g. sulfate compounds) while the presence of others (such as dissolved SO_2 and its dissociation products) can be deduced by calculation through equilibrium theory or production mechanisms. In some cases, both approaches can be used. In this section we will list the relevant species, segregated into categories by oxidation state, the occurrence of each species and the evidence for it.

It is important to point out that, although these sulfur compounds found in the atmosphere are never present in a pure state, several of the compounds (notably sulfate compounds) are sometimes found in relatively pure form. As a result, *some* of the physical properties of these particles are close to what is expected from the molecular or crystalline composition.

(a) Oxidation state +6

(1) *Sulfuric acid and its neutralization products with ammonia.* (a) Sulfuric acid, H_2SO_4, is the most commonly found atmospheric strong acid. It contributes to acidic precipitation and has been identified as a constituent of submicrometer aerosol particles. Atmospheric H_2SO_4 particles (and their neutralization products with NH_3) are produced both at the source and in the atmosphere by the oxidation of SO_2, and are found preferentially in the accumulation mode, i.e. in particles, between 0.1 and 1.0 μm dia.

The combination of usual tropospheric temperatures $(-20 \leq T \leq 40C)$ and humidities $(20 \leq \text{r.h.} \leq 100\%)$ along with the hygroscopic properties of H_2SO_4 dictate that H_2SO_4 is highly hydrated and usually liquid (Tang, 1976). The composition of bulk H_2SO_4 as a function of the relative humidity with which it is in equilibrium is included in Fig. 2. As a consequence of the large $[H_2O]/[H_2SO_4]$ concentration ratios above about 30% r.h., H_2SO_4 aerosol becomes highly dissociated

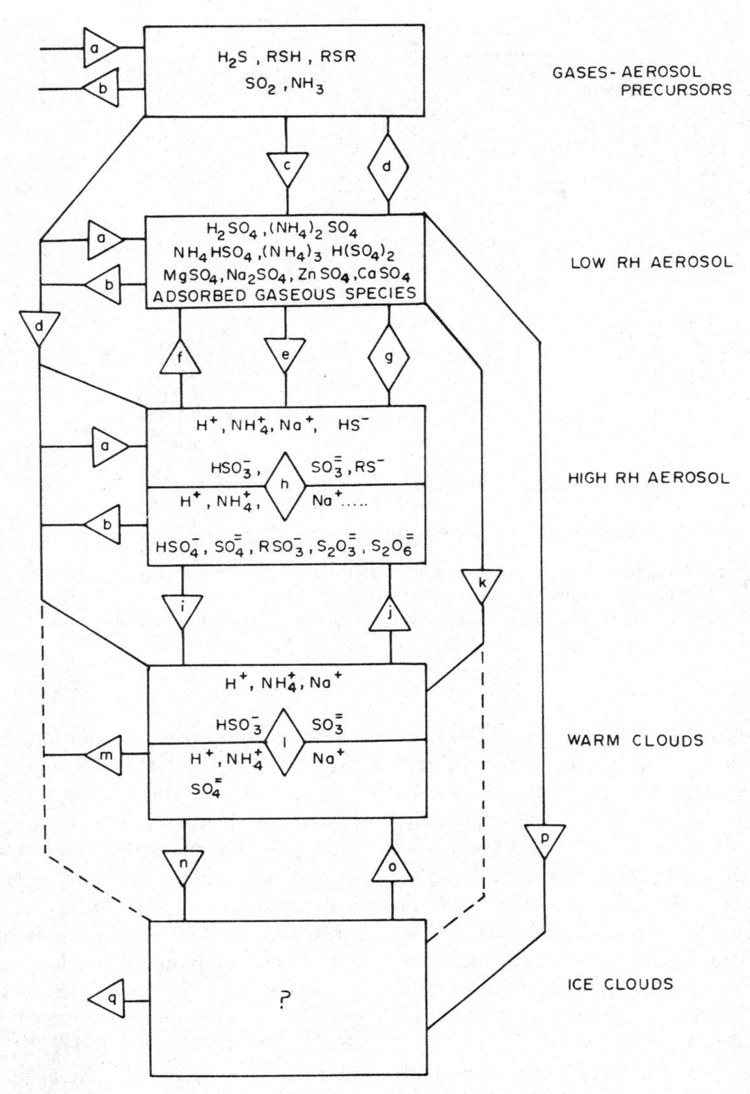

Fig. 1. The tropospheric sulfur cycle. □ Recognizable entities in the atmosphere. △ Processes having single direction of material flow. ◇ Reversible processes. (a) Sources. (b) Sinks. (c) Gas-to-particle conversions. (d) Sorption. (e) Deliquescence. (f) Efflorescence. (g) Raoult's equilibrium. (h) Reaction in concentrated solution droplet. (i) Nucleation and condensation of water. (j) Evaporation. (k) Capture of aerosol by cloud drops. (l) Reaction in dilute solution. (m) Rain. (n) Freezing of supercooled drop by ice nucleus. (o) Melting. (p) Direct sublimation of ice on ice nucleus. (q) Precipitation. R = organic radical.

and thus exhibits strong acid characteristics. Since the second dissociation constant in water is only about 10^{-2} M*, substantial bisulfate (HSO_4^-) ion concentrations can exist in the particles in air. As a result of this acidity and the fundamental nature of modestly concentrated H_2SO_4, a wide variety of inorganic and organic reactions are made possible within or on such aerosol particles (e.g. dissolution of metal oxides or organic dehydration reactions).

Based on observations of fast, *in situ* reactions of atmospheric aerosols with NH_3(g) (Charlson *et al.*, 1974) it can be concluded that acidic sulfate compounds can only exist in situations where the NH_3(g) concentration is very low. Lau and Charlson (1977) estimate that [NH_3(g)] may be as low as 0.01 to 0.1 ppb in some regions, notably those areas with acidic soils as suggested by Junge (1963). Recently, David and co-workers (1977) have measured concentrations of NH_3 as low as 0.1 ppbv in rural Ohio. Consistent with this picture, Vanderpol *et al.* (1975), using the light scattering technique discussed in Section 3. have suggested that acid sulfate particles measured near St. Louis, Missouri, were associated with air mass trajectories arriving from NH_3-poor regions, whereas neutral ammonium sulfate particles were associated with trajectories arriving from NH_3-rich regions.

H_2SO_4 aerosols may also exist briefly near the source as a direct product of oxidation (e.g. in chim-

* M ≡ molar ≡ gram-mole per liter.

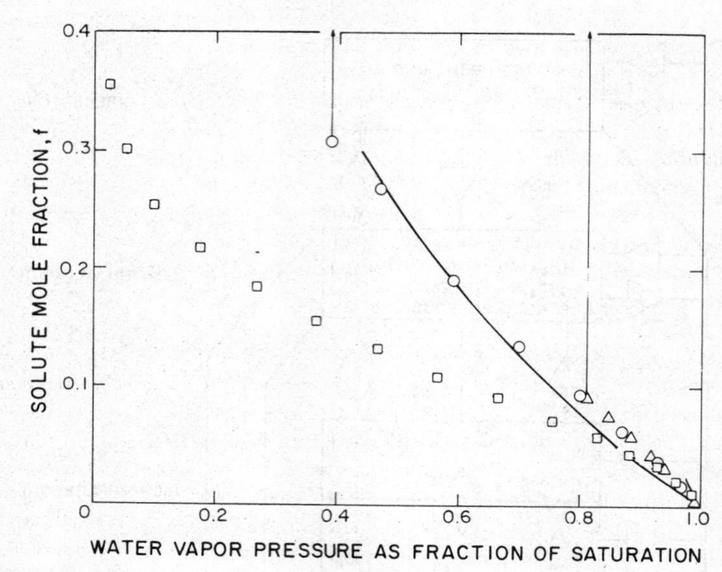

Fig. 2. Solute mole fraction versus water vapor pressure as a fraction of saturation for three sulfate compounds. □ H_2SO_4 (Coordinating Research Council, 1968) ○ NH_4HSO_4 (Tang, in press) △ $(NH_4)_2SO_4$ (Low, 1971)—Raoult's Law assuming complete dissociation.

ney plumes or automobile exhaust where NH_3 is almost completely scavanged by an excess of free acid). There may be other alkaline gases of concern besides NH_3 (such as amines) which would limit the occurrence of free H_2SO_4. Alkaline dusts such as $CaCO_3$ usually occur in particle sizes too large to effectively coagulate with and neutralize submicrometer H_2SO_4. The fate of H_2SO_4 as indicated in Fig. 1 includes removal by precipitation or the reaction with other atmospheric substances, particularly NH_3. The reaction pathway gives rise to a series of compounds as noted below.

(b) Ammonium acid sulfate, NH_4HSO_4, is another acidic sulfate compound known to exist in air. It has been detected and is perhaps an even more common form of acid sulfate than H_2SO_4. It is expected to occur whenever sufficient NH_3 is present to partially neutralize H_2SO_4. Alternatively, it may be produced directly by the NH_3 enhanced oxidation of SO_2 in hydrometers. Unlike H_2SO_4 it is deliquescent but forms a crystalline salt only at very low relative humidities (less than 30–40%). Thus, it will usually exist as a hydrated, aqueous liquid droplet aerosol. As the half-neutralized strong acid, it contributes to low pH in rain and aerosol particles.

(c) Letovicite, $(NH_4)_3H(SO_4)_2$ has been suspected from measurements of the effect of humidity on the light scattering by aerosols (Charlson *et al.*, 1974; Weiss *et al.*, 1977). Other evidence, including wet chemical analysis and X-ray diffraction confirms its presence (Brosset *et al.*, 1974). Its occurrence is apparently less frequent than the more acidic sulfate compounds, or the completely neutralized form $(NH_4)_2SO_4$. According to Tang and Munkelwitz (in press) letovicite is deliquescent at *ca.* 68% r.h., and thus will often exist as solution droplets as in the case of H_2SO_4 or NH_4HSO_4.

(d) Ammonium sulfate, $(NH_4)_2SO_4$, is the fully neutralized ammonium salt of H_2SO_4. It is soluble in water (70 g/100 ml at 0°C) and deliquescent at 80% r.h. It is only weakly acidic in aqueous solutions due to the hydrolysis of the NH_4^+ ion. It is an almost unreactive material and hence is often the last and most stable compound formed prior to removal from the atmosphere. $(NH_4)_2SO_4$ as a secondary aerosol is found in well aged air masses, notably those in which NH_3 is abundant. $(NH_4)_2SO_4$ is found naturally in rare cases as the mineral mascagnite, but due to its H_2O solubility little of it is ever exposed to the earth surface. Hence, naturally produced $(NH_4)_2SO_4$ dust to wind erosion is probably nonexistant. Some $(NH_4)_2SO_4$ may be mechanically released due to its use as a fertilizer. It remains that most $(NH_4)_2SO_4$ in air is a secondary aerosol from oxidation of SO_2.

Atmospheric measurements supporting the existence of H_2SO_4 and its neutralization products with NH_3 include:

(1) the *in situ* reaction of submicrometer particles with NH_3 (gas) (Charlson *et al.*, 1974; this method does not differentiate between NH_4HSO_4 and H_2SO_4),

(2) i.r. spectra of collected samples (Cunningham and Johnson, 1976),

(3) Raman spectra of collected samples (Rosen and Novakov, 1977),

(4) titration of collected samples (Brossett *et al.*, 1974),

(5) the volatility of collected samples (Roberts and Friedlander, 1975),

(6) the *in situ* volatilization at elevated temperatures of light scattering aerosols (Cobourn *et al.*, 1978; the volatility methods do not differentiate between $(NH_4)_2SO_4$ and NH_4HSO_4),

(7) laboratory simulations in which sulfate compounds are produced photochemically (Kuhlman *et al.*, 1977),

(8) chemical model calculations in which H_2SO_4 is a logical product of oxidation of SO_2,

(9) precipitation composition studies in which H_2SO_4 is the dominant free acid.

(2) *Other inorganic sulfate compounds.* (a) $MgSO_4$, magnesium sulfate, is found as part of maritime sea spray aerosol. Sea salt has an $MgSO_4$ content of 5.7% by weight. Hence $MgSO_4$ is a ubiquitous aerosol in maritime areas, especially when the wind is onshore at high speeds.

(b) $CaSO_4$, calcium sulfate, is introduced into the atmosphere by wind erosion of exposed gypsum beds. Hence, it is found mainly in coarse mode particles, i.e. in particles of size greater than about 1–2 μm. Some may be generated by industrial activity (e.g. gypsum processing or scrubbing of SO_2 with $CaCO_3$). $CaSO_4$ is a relatively insoluble mineral and is clearly not a deliquescent salt, although $CaSO_4$ is found naturally in the hydrated form $CaSO_4 \cdot 2H_2O$. Some reports of $CaSO_4$ on filters or in rain may be suspect due to the lack of separation (by size or time) of H_2SO_4 and $CaCO_3$ aerosols which can react with each other during or after sampling. Stoichiometric inference is particularly subject to this error.

(c) Na_2SO_4, sodium sulfate, is produced in the incineration of wastes from some types of paper manufacture. It has been isolated as a dominant compound in particulate matter emitted from combustion of lignin–sulfur compounds (Bosch *et al.*, 1970). It is deliquescent at or above *ca.* 85% r.h. and apparently is very active as a cloud condensation nucleus (Hobbs *et al.*, 1976). It is not acidic and is mainly of concern to problems of visual range in areas where paper is manufactured.

(d) Zinc sulfate, $ZnSO_4$, and ammonium zinc sulfate $(NH_4)_2SO_4 \cdot ZnSO_4$ were identified by Hemeon (1955) in one sample taken in Donora, Pennsylvania. Whereas $ZnSO_4$ is nearly as water soluble as $(NH_4)_2SO_4$, the mixed compound ammonium zinc sulfate is much less water soluble (7 g/100 ml at 0°C). These compounds are expected to be local products of certain metal processing industries.

(e) HSO_4^-, bisulfate ion, is found in acid sulfate aerosols *most* of the time due to the hygroscopic/deliquescent nature of H_2SO_4 and its neutralization products with NH_3. As stated above, it is itself a modestly strong acid, scavenger for NH_3(g) and highly reactive with respect to metal oxides.

(f) SO_4^{2-}, sulfate ion, is also found either in dilute, aqueous acidic aerosol particles at high relative humidity or in any neutralized and dissolved sulfate compound, e.g. $(NH_4)_2SO_4$.

(3) *Organic sulfur compounds.* Organic sulfur compounds, notably the esters of H_2SO_4, should exist in situations where this acid can react with prevailing alcohol functional groups in organic species. Alkyl hydrogen sulfate compounds are known to form from the addition of olefins to H_2SO_4, and, thus, they should also exist since both reactants are abundantly available. To our knowledge, no atmospheric data exist on these or related organic compounds.

(*b*) *Oxidation state* +4

(1) *Chemisorbed* SO_2. Novakov *et al.* (1972) using ESCA has identified S(IV) in fine particles collected on filters in Pasadena, California. He suggests that the S(IV) is chemisorbed and subsequently oxidized on the particle surface. S(IV) concentrations were estimated to be as high as 6 μg^{-3}, although enrichment of S(IV) at the particle surface probably makes this an overestimate of the total mass loading of S(IV). Rosen and Novakov (1977) have recently identified graphitic carbon as the likely adsorption surface.

(2) *Metal ion complexes.* Certain metal cations (e.g. Hg^{2+}, Fe^{3+}, Cu^{2+}) will form metal sulfite or metal hydroxy sulfite complexes in the presence of dissolved SO_2. Depending on the molar ratio of metal to sulfur, the sulfur will be tightly bound to the metal, will be stable to air oxidation and will not out-gas from solution upon droplet evaporation (Hansen *et al.*, 1976).

Eatough *et al.* (1977) using thermometric titration with $K_2Cr_2O_7$ have identified S(IV) in aerosol collected on filters near a copper smelter and in trace amounts in a sample from New York City. They suggest that these are transition metal complexes, based on Mössbauer, and photoelectron spectra of samples taken in the work environment of a copper smelter and on correlation studies of S(IV) concentrations with Fe^{3+} and Cu^{2+} concentrations in ambient samples. They found that more than 80% of the sulfur (IV) is in the coarse mode, accounting for about 1–2% of the total aerosol mass. Assuming a submicron aerosol mass concentration of 50 μg^{-3}, this means that about 0.1 μg^{-3} metal sulfites would be present in ambient air, roughly 1–10% of the mass concentration found by Novakov *et al.* (1972) as discussed earlier. At times the metal ion complexes appear to represent an important fraction of the total S(IV) in the submicron, subsaturated aerosol.

(3) *Dissolved* SO_2 *species.* When gaseous SO_2 contacts aqueous droplet aerosols in the atmosphere, a fraction of the SO_2 dissolves forming aqueous SO_2 ($SO_2 \cdot xH_2O$). The aqueous SO_2 dissociates into bisulfite ions (HSO_3^-) which further dissociate into sulfite (SO_3^{2-}) ions, as described by the following equilibria:

at 25°C

(1) $SO_2 + H_2O \rightleftharpoons SO_2 \cdot xH_2O \quad K_1 = 0.81\ \text{M atm}^{-1}$

(2) $SO_2 \cdot xH_2O \rightleftharpoons HSO_3^-\ H^+ \quad K_2 = 1.72 \times 10^{-2}\ \text{M}$

(3) $HSO_3^- \rightleftharpoons SO_3^{2-}$ H^+ $\quad K_3 = 6.24 \times 10^{-8}$ M

(4) 2 $HSO_3^- \rightleftharpoons S_2O_5^{2-}$ H_2O $\quad K_4 = 7 \times 10^{-2}\ M^{-1}$

As can be seen by Equations (2)–(4), the extent of dissociation increases with decreasing $[H^+]$. These S(IV) species are subsequently oxidized, principally forming sulfate. Since SO_3^{2-} is more readily oxidized by O_2 than is HSO_3^-, both the total amount of SO_2 dissolved and the reactivity toward O_2 are minimal in strongly acid solutions.

With $[H^+] = 10^{-1}$ M in a sulfuric acid aerosol with a liquid water content of 0.1 $\mu l\ m^{-3}$ air at 25°C, about 10^{-9} mole % of the SO_2 is dissolved in the drop. With $[H^+] = 10^{-5}$ M in an ammonium sulfate aerosol with the same liquid water content, about 10^{-4} mole % of the SO_2 is dissolved in the aerosol. Of that latter dissolved fraction (at 100 ppbv SO_2), about 99% is HSO_3^- and the total mass concentration of oxidation state +4 sulfur is about 1 ng m^{-3}. These small amounts of dissolved SO_2 have not been detected in atmospheric aerosols for several reasons: (1) attempts to collect and concentrate the aerosol will disturb the equilibria with gaseous SO_2 and may totally evaporate the aqueous phase; (2) *in situ* methods that would not disturb this equilibria currently do not have enough precision to detect the small fraction of SO_2 dissolved in the aerosol.

Due to the greater amounts of liquid water involved, the amount of SO_2 dissolved in cloud drops is 3–4 orders of magnitude greater than the amounts dissolved in a subsaturated aerosol. These larger amounts are similar in magnitude to the concentrations of S(IV) found by both Novakov *et al.* (1972) and Eatough *et al.* (1977) in subsaturated aerosols.

(4) *Bisulfite addition products.* Bisulfite ion reacts with unhindered carbonyl compounds, such as aldehydes, methyl ketones and cyclic ketones to produce an addition product R-C-OH-SO_3^- (hydroxy sulfonic acid). As with other carbonyl addition reactions, bisulfite addition is reversible. Given a typical bisulfite droplet concentration of 10^{-5} m and an addition equilibrium constant of $10^4\ M^{-1}$ [propionaldehyde (Gubareva, 1947)], as much as 10% of the aldehyde could be addition complex. Since the complex dissociates in strong acids solutions, releasing SO_2 and reforming the aldehyde, it should not be present in acid sulfate aerosol droplets. The rate of dissociation is rapid [$\sim 10^{-1}\ s^{-1}$ for benzaldehyde at 13°C (Sousa and Margerurm, 1960)], implying that these complexes may release SO_2 even in neutral solutions during droplet evaporation. To our knowledge, these complexes have not been identified in atmospheric aerosols.

(c) Other oxidation states

(1) *Metallic sulfide and disulfide compounds.* These compounds constitute a small fraction of the earth's crust and thus might be found in those coarse mode particles generated by wind erosion. Only the alkaline and alkaline earth sulfide compounds are readily soluble in aqueous solution. Thus the other sulfide and disulfide compounds should not affect aqueous aerosol chemistry. To our knowledge, these species have not been specifically identified in coarse mode aerosol.

(2) *Chemisorbed sulfide compounds and sulfur.* Carbon disulfide and elemental sulfur have been shown to form on the surface of carbon exposed to SO_2 at high temperature (Novakov *et al.*, 1974; Barbaray *et al.*, 1977). Therefore these compounds should exist on the surface of carbon containing aerosols produced by combustion.

(3) *Gaseous sulfide compounds and their solution products.* Like SO_2, H_2S also dissolves and dissociates in aqueous solution, forming mainly HS^-. Significant dissociation occurs in neutral or alkaline solutions, but not in strongly acid solutions. Coupled with the relatively low concentrations of H_2S in the atmosphere, this inability to dissociate in acid aerosols implies very low HS^- concentrations. Not surprisingly, these species have not been identified in ambient aerosols.

The small amount of HS^- present in aerosol droplets can react to produce trace amounts of other sulfur species. Oxidation with O_2 can produce a significant fraction of $S_2O_3^{2-}$ (Cline and Richards, 1969). Reaction with dissolved SO_2 species in the presence of O_2 (Wackenroder's solution) can produce elemental sulfur as well as polythionate species ($S_nO_6^{2-}$).

Gaseous organic sulfide compounds such as methyl sulfide, dimethyl sulfide, dimethyl disulfide and methyl mercaptan can also dissolve to some extent in aqueous solution. In general, their expected concentrations should be even smaller than the HS^- concentration.

(4) *Dithionate compounds.* When SO_2 is oxidized by O_2 in aqueous solution, a few per cent of the product is dithionate ($S_2O_6^{2-}$) (Bassett and Parker, 1951). It is also the major reaction product of dissolved SO_2 and MnO_2. Dithionic acid and its salts are stable to air oxidation and readily precipitate with barium ion. Dithionate ion is therefore easily mistaken for sulfate ion and may be one reason why it has not been identified in ambient aerosols.

3. HYGROSCOPIC PROPERTIES OF SULFATE AEROSOLS

Many of the sulfur compounds enumerated above are water soluble and hygroscopic. In aerosol form they exist as solution droplets in water vapor equilibrium with their surroundings over certain ranges of relative humidity and respond to changes in humidity by taking up or releasing water. Concurrent with the identification of SO_4^{2-} as a major constituent of atmospheric aerosol on a regional scale there has been considerable interest and research on the hygroscopic properties of pure sulfate compounds and the effect of sulfate compounds on the hygroscopic

Table 2. Bulk deliquescence points

Compound	r.h. (%)
$MgCl_2$	33
NH_4HSO_4	30–40
$Na\ NO_3$	64
$(NH_4)_3H(SO_4)_2$	68
NaCl	75
$(NH_4)_2SO_4$	81
Na_2SO_4	85–93
$MgSO_4$	91
$Zn\ SO_4 \cdot 7H_2O$	87

properties of atmospheric aerosols (Junge, 1952; 1963; Winkler, 1973; Charlson *et al.*, 1974; Tang and Munkelwitz, in press). Hygroscopic growth of particles containing SO_4^{2-} has a profound effect on their physical and chemical properties (size, shape, refractive index, pH, reactivity) which in turn will influence environmental factors such as human health (toxicity, respiratory deposition efficiency), climate, visibility, and atmospheric sulfur cycles.

A basic understanding of hygroscopic properties can be obtained from observation of aqueous solution properties—solute/solvent mole fraction as a function of equilibrium water vapor pressure or relative humidity, r.h. The dependence of sulfate mole fraction on r.h. for three different sulfate compounds is illustrated in Fig. 2 (Raoult's Law behavior assuming complete dissociation is included for comparison). H_2SO_4 exhibits a monotonic curve characteristic of simple hygroscopic behavior—absorption and desorption in continuous equilibrium. By contrast NH_4HSO_4 and $(NH_4)_2SO_4$ exhibit step changes at 39 and 81% r.h. respectively and monotonic curves at greater r.h. This latter phenomenon, a sudden uptake of water when the r.h. exceeds a certain level is termed deliquescence and is exhibited by these two and a number of other organic and inorganic compounds, primarily salts. The r.h. at which a compound deliquesces—its deliquescence point—is defined as the ratio of water vapor pressure above a plane surface of a saturated solution of the compound to the vapor pressure over a plane surface of pure water. Table 2 lists bulk deliquescence points for a number of compounds commonly found as constituents of atmospheric aerosols. The existence of and differences in deliquescence points between compounds provide a possible basis for molecular identification. The reverse process, a sudden release of water with decreasing r.h. is termed efflorescence.

Hygroscopic/deliquescent compounds in the aerosol phase equilibriate rapidly with their surroundings and exhibit changes in size and other properties in response to r.h. changes. Strictly hygroscopic aerosols exhibit monotonic growth curves while deliquescent aerosols remain dry below the deliquescence point and grow suddenly to hygroscopic solution droplets above that point. Figure 3 shows theoretical growth (increasing r.h.) curves for some sulfate compound aerosols based on vapor pressure data. Although deliquescence and deliquescence points are observed to be well defined for many compounds in aerosol form when subjected to increasing r.h. from a dry crystalline state, efflorescence and efflorescence points are not. In the aerosol phase most compounds seem to remain as solution droplets at concentrations well above saturation, i.e. at an r.h. well below the deliquescence point. This leads to the phenomenon of hysteresis (Orr *et al.*, 1958; Junge, 1963). Once an aerosol is within the hysteresis loop it will not exhibit deliquescent properties. Atmospheric aerosols probably exhibit hysteresis (Winkler and Junge, 1972) but this has not been confirmed *in situ*. If so, they may

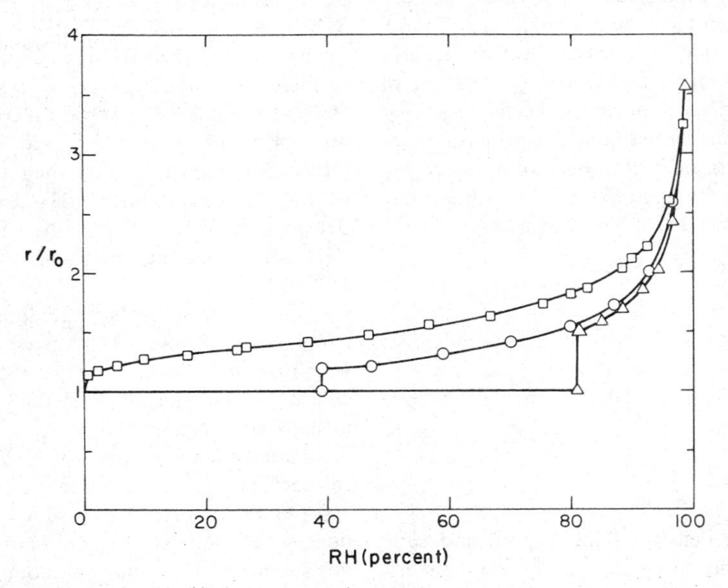

Fig. 3. Theoretical growth curves based on vapor pressure data for three sulfate aerosols exposed to increasing humidity □ H_2SO_4 ○NH_4HSO_4 △ $(NH_4)_2SO_4$

contain considerably more water in the range 30–70% r.h. than would be expected from their deliquescence curves.

Since vapor pressure data are based on bulk measurements of pure compounds, their usefulness in calculating hygroscopic properties of aerosols is limited. Corrections must be made for the increase in vapor pressure over curved surfaces (Kelvin effect) and, in the case of deliquescence, for the small number of crystal units present. Data are available for only a limited number of pure compounds. Simple mixtures of inorganic compounds may have radically different hygroscopic properties. For the complex mixture of compounds generally found in atmospheric aerosols, hygroscopic properties are indeed difficult to model.

A number of techniques have been employed to directly investigate the hygroscopic properties of aerosols (Orr *et al.*, 1958; Tang and Munkelwitz, in press; Kapustin *et al.*, 1973; Sinclair *et al.*, 1974). Only two methods have been applied extensively to atmospheric aerosol studies (Winkler, 1973; Covert, 1974; Charlson *et al.*, 1974).

Our interest in the hygroscopic growth and molecular composition of aerosols stems from their role in particle optics. In turn we have applied optical techniques to the *in situ* measurement of hygroscopic properties and aspects of the molecular form of tropospheric aerosols. The apparatus, termed a humidograph (Charlson *et al.*, 1974), measures the light scattering coefficient of the particulate matter, b_{sp}, as a function of increasing r.h. This technique has the advantage of providing an *in situ* measurement and of measuring a relevant integral aerosol property, b_{sp}. While this technique does not sense all particle sizes equally, its range of sensitivity (for particles 0.1 to 1.0 μm dia) corresponds well with the accumulation mode of atmospheric aerosols where most sulfates and other hygroscopic compounds are found. Although this technique provides a direct measure of the effect of hygroscopic growth on b_{sp}, relation of this parameter to dia., d, or mass, m, as measured by others requires further assumptions primarily concerning the particle size distribution. For a power law number distribution (Junge, 1963, $v = 3$) in the range 0.1–1.0 μm dia, Hänel (1976) has shown that

$$\frac{b_{sp}(\text{r.h.})}{b_{spo}} = \left[\frac{d(\text{r.h.})}{d_0}\right]^2$$

is a close approximation for growth below 90% r.h., where subscript 'o' refers to dry state or low r.h. values. This can be further related to changes in mass with the formula:

$$\frac{d}{d_0} = \left(\frac{m_w}{m_0} \cdot \frac{\rho_0}{\rho_w} + 1\right)^{1/3}$$

where ρ is the bulk density of the aerosol and subscript 'w' refers to pure water values. This relationship requires an assumption of particle density which for submicrometer sized atmospheric particles is roughly between 1–2 g cm^{-3}. This assumption is not critical since density enters the equation only as the cube root.

Data (humidograms) for pure laboratory generated sulfate aerosols which are obtained with this humidograph are presented in Fig. 4.

Figures 4(a–e) show humidograms for submicrometer H_2SO_4 aerosol and for four of its neutralization products with ammonia. Theoretical curves calculated from available vapor pressure data and the simple b_{sp} and d^2 proportionality are included for comparison (radius of curvature effects have been neglected since they are small in the particle size and r.h. ranges of concern here). Two major points of disagreement are in the deliquescence points of NH_4HSO_4 and $(NH_4)_3H(SO_4)_2$. The lack of observed deliquescence for NH_4HSO_4 may be due to hysteresis effects since the aerosol was produced from solution and initially dried to only about 15% r.h. The discrepancy between the expected and observed dliquescence points for $(NH_4)_3H(SO_4)_2$ is as yet unexplained. Tang and Munkelwitz (in press) report consistent experiment results (solution vapor pressure and particle size) and theoretical results for both salts. We believe that for $NH_4^+:SO_4^{2-}$ ratios <1.5 the ionic composition and elimination of hysteresis effects may be critical to the deliquescence of these mixed H_2SO_4, $(NH_4)_2SO_4$ salts.

Humidograms for some other hygroscopic sulfate compounds, Na_2SO_4, $MgSO_4$, $ZnSO_4 \cdot (NH_4)_2SO_4$, are shown in Figs. 4(f–h). The difference in (or lack of) deliquescence points from that observed for bulk samples may be due to strong hysteresis effects or hydrate formation. Na_2SO_4 forms the hydrate $Na_2SO_4 \cdot 10\,H_2O$ at 81% r.h. (Wylie, 1963).

The differences between the humidograms for H_2SO_4 and NH_4HSO_4 are for practical purposes indistinguishable. Those for both $(NH_4)_3H(SO_4)_2$ and $(NH_4)_2SO_4$ are distinguishable from the rest due to the presence of their deliquescence steps. The mixture of these two salts, Fig. 4(e), shows two deliquescence steps at 45 and 80%. These differences coupled with the ability to change the acid sulfate aerosols to $(NH_4)_2SO_4$ (and thus alter their hygroscopic nature) by *in situ* reaction with NH_3 provide a means for detection of sulfate and the distinction between $(NH_4)_2SO_4$ and the more acidic forms (Charlson *et al.*, 1974).

Atmospheric aerosols are seldom comprised of pure compounds, however. The 'mixed aerosol' hypothesis which was forwarded by Junge (1952) and further articulated by Winkler (1973) defines two extreme cases for chemical mixing in aerosols,

(a) internal mixtures in which each individual aerosol particle contains a mixture of compounds,

(b) external mixtures in which each particle is a pure compound, yet the population of aerosol particles consists of various compounds.

All intermediate degrees of mixing are obviously possible. External mixtures will have hygroscopic proper-

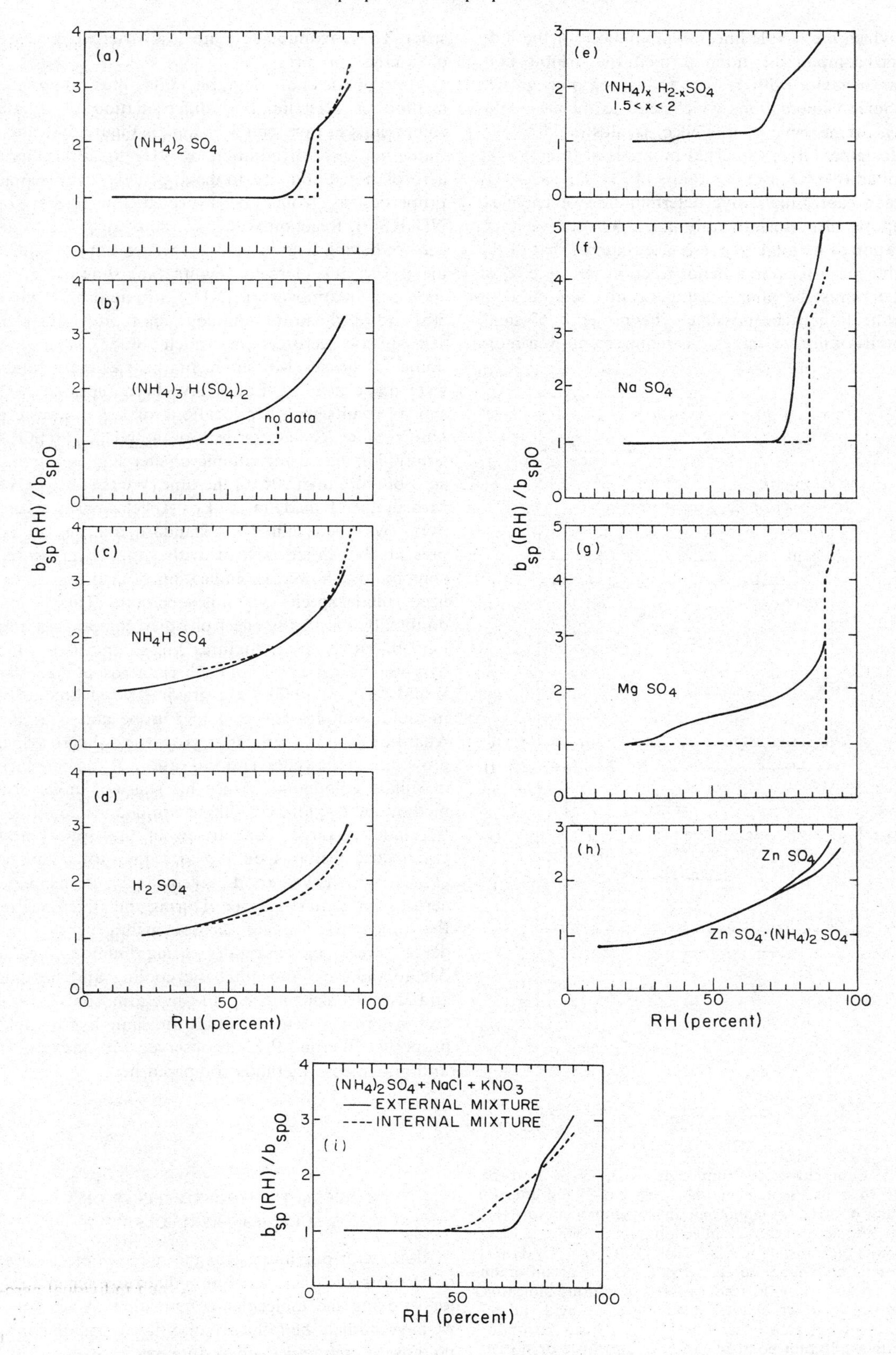

Fig. 4. Humidograph data for pure, laboratory generated sulfate compounds. (a) $(NH_4)_2SO_4$, (b) $(NH_4)_3H(SO_4)_4$, (c) NH_4HSO_4, (d) H_2SO_4, (e) $(NH_4)_xH_{2x}SO_4$ $1.5 < x < 2$, (f) $NaSO_4$, (g) $MgSO_4$, (h) $ZnSO_4$ and $ZnSO_4 \cdot (NH_4)_2SO_4$, (i) $(NH_4)_2SO_4$ + NaCl + KNO_3 internal vs external mixture. Except in (i) dashed curve is calculated.

ties which are simple linear combinations of the individual compounds. Internal mixtures exhibit nonlinear behavior with respect to hygroscopic growth due to ion interactions which tend to increase solubility. An example of this effect is illustrated in Fig. 4(i) for external and internal mixtures of $(NH_4)_2SO_4$, NaCl and KNO_3 in mass ratios of 2:1:1 respectively in each case. Qualitative determination of chemical composition of an internally mixed aerosol by interpretation of its humidity response requires that of the soluble aerosol mass a major fraction (50 mole % or greater) must be of one deliquescent compound for identification to be possible. The presence of small amounts of nitrate makes this requirement even more strict. These requirements are less strict in the case of external mixtures.

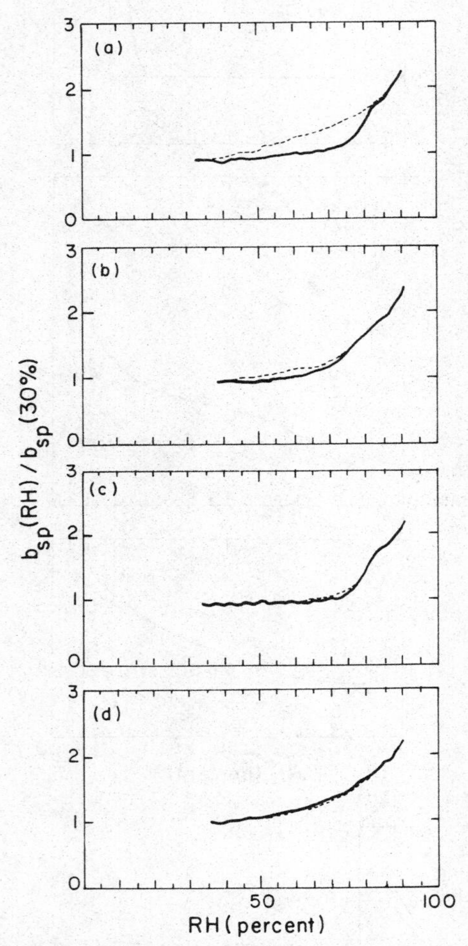

Fig. 5. Four classes of humidograms observed near St. Louis, Missouri, September 1973, before (---) and after (–) addition of NH_3. (a) monotonic (hygroscopic) curve; NH_3 addition to aerosol caused inflection point at 80% r.h. (deliquescence) to appear; indicative of H_2SO_4 or NH_4HSO_4 as dominant compound in aerosol. (b) initial inflection point at 80% r.h. enhanced by NH_3 addition; indicative of aerosol in which sulfates including some acid sulfates are present but as a lesser fraction than cases (a) or (c). (c) initial inflection point at 80% r.h. and little or no enhancement by NH_3 addition; indicative of $(NH_4)_2SO_4$ as dominant compound in aerosol. (d) monotonic curve unaffected by NH_3; indicative of aerosol in which sulfate is not dominant.

Atmospheric humidograph data illustrating the method of detection and differentiation of sulfate compounds are presented in Fig. 5. Figure 5(a) illustrates the effect of addition of NH_3 to atmospheric aerosol noted initially to have strictly hygroscopic properties as would be expected for H_2SO_4 or NH_4HSO_4. Reaction with NH_3 converted this to an aerosol having hygroscopic properties similar to that for $(NH_4)_2SO_4$ aerosol. Figure 5(c) shows the response of an atmospheric $(NH_4)_2SO_4$ aerosol. Figures 5(b) and (d) illustrate humidograms representative of atmospheric aerosols in which SO_4^{2-} is a less dominant species. Measurements such as these which were made near St. Louis have been repeated with similar results in other locations of the midwestern United States. Sulfate aerosols of the NH_4^+, H^+, SO_4^{2-} family dominated the submicrometer light scattering aerosol more than 90% of the time (Weiss *et al.*, 1977). Measurements made in the Los Angeles basin, where SO_4^{2-} was determined by standard analysis to be present though co-existent with other hygroscopic compounds, showed no indication of acid or neutralized sulfate species via this technique. This is undoubtedly due to the effect of other compounds and ions in an internal mixture. Measurements of the hygroscopic nature of atmospheric aerosols made by Winkler (1973; 1976) via a gravimetric technique of impactor samples collected in Europe and over the Atlantic show similar properties—hygroscopic growth, deliquescence and indication of the presence of sulfate compounds. Since his was not an *in situ* method the test for acid sulfate compounds could not be made. A direct comparison of the optical and gravimetric methods on the same aerosol (Winkler *et al.*, 1973) showed good agreement for atmospheric aerosols in central Europe. During their lifetimes in the atmosphere, sulfate aerosols will generally undergo several increasing-decreasing humidity cycles due to radiative and adiabatic cooling and heating and also turbulent mixing. In these atmospheric processes, aerosols seem to exhibit the same hygroscopic properties (Werner, 1972) as observed with the optical and gravimetric methods compared here.

4. REFRACTIVE INDEX EFFECTS OF SULFUR IN AEROSOLS

Along with our interest in the hygroscopic nature of aerosols and the use of humidification techniques for studying the molecular composition of aerosols, we have studied the influence of different sulfate compounds on refractive index and hence on some of their optical properties. Bhardwaja *et al.* (1974) and Bhardwaja (1976) have suggested that the real part of refractive index may be determined *in situ* by measurement of the background hemispheric scattering ratio, R:

$$R = \frac{b_{bsp}}{b_{sp}} = \frac{\int_{\pi/2}^{\pi} \beta(\theta) \sin \theta \, d\theta}{\int_{0}^{\pi} \beta(\theta) \sin \theta \, d\theta}$$

where $\beta(\theta)$ is the angular scattering function and θ is the scattering angle. R may be measured with a modified integrating nephelometer (Charlson *et al.*, 1974a; Heintzenberg and Bhardwaja, 1976). If the atmospheric compounds are sufficiently pure, then chemical information may be inferred from the refractive index.

Figure 6 shows the dependence of R on the real part of refractive index, m_r, with values noted for H_2SO_4 (at 50% r.h.) and $(NH_4)_2SO_4$ (solid). Here, the role of the molecular form of sulfate compounds is expected to act in two ways:
(1) the pure substances have quite different refractive indices and
(2) due to the hygroscopic nature of H_2SO_4 at all values of r.h. and the deliquescent nature of NH_4HSO_4 and $(NH_4)_2SO_4$ at *ca.* 30–40 and 80% r.h. respectively, large amounts of H_2O (liquid) can be included in the particles. Since liquid water has a real refractive index of $m_r = 1.33$, it can substantially lower the apparent value for the aerosol and significantly change R. In doing the Mie calculations for Fig. 6, the imaginary part of the refractive index (light absorption) and the size distribution were allowed to vary throughout the range expected from various atmospheric measurements (Weiss *et al.*, 1976; Willeke and Whitby, 1975). Geometric mean size (by volume) had almost no effect for particle diameters greater than 0.2 μm. Changes in imaginary refractive index decreased the value of R slightly while perserving the fundamental linear relationships of R and m_r. It was, of course, impossible to do these calculations for non-spherical or inhomogeneous particles. These

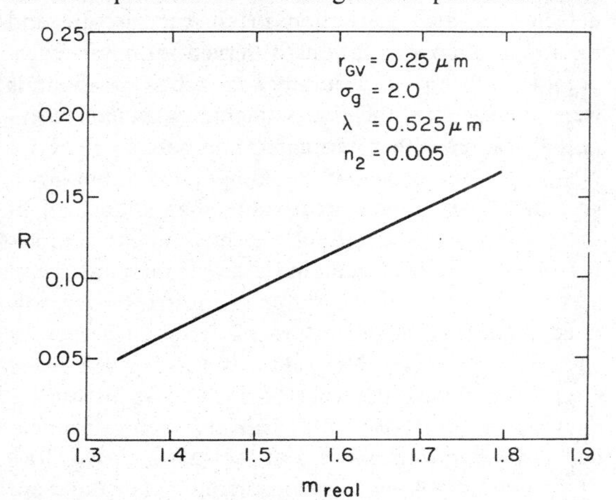

Fig. 6. Variation of the calculated (Mie Theory) hemisphere backscatter, total scatter ratio, R, as a function of the real part of the refractive index, m_r, of the aerosol particle. m_r for H_2SO_4 at 50% r.h. = 1.4, m_r for $(NH_4)_2SO_4$ = 1.52.

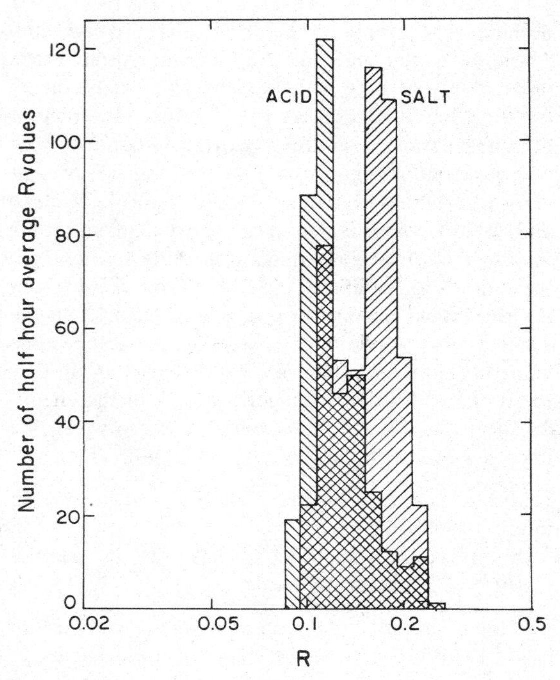

Fig. 7. Histogram of frequency of occurrence of R values for atmospheric sulfate aerosol measured near St. Louis in September 1973, separated according to their salt and acid character as determined by the humidograph.

calculations and those for the accuracy of the backward hemispheric integrating nephelometer (Heintzenberg and Bhardwaja, 1976) suggest that the relatively pure different sulfate compounds found in air should be differentiable on the basis of a measured value of R.

Field measurements of R with the modified integrating nephelometer coincided with our studies near St. Louis in 1973 of the influence of different sulfate compounds on the dependence of light scattering on humidity (Charlson *et al.*, 1974). Figure 7 shows a plot of the frequency of occurrence of values of R, indicating the expected result of a lower mean value of R ($R \cong 0.12$) for acid sulfates than for $(NH_4)_2SO_4$ ($R \cong 0.18$). The acid/salt discrimination was provided by the humidograph. While the relative difference in R values between acid and salt is consistent with Mie calculations (Fig. 6), the observed values of R are about twice those predicted from theory. Although agreement between observation and theory is encouraging for sites such as the eastern third of the U.S. where relatively pure sulfate species exist, the problems of non-sphericity, non-homogeneity and compositional impurities seem to indicate that refractive index is too variable to be used as a general analytical tool (Vanderpol, 1975).

5. SUBSTANCES CO-EXISTING WITH SULFUR IN TROPOSPHERIC AEROSOLS

Tropospheric aerosols are almost always chemically heterogeneous, even though there may be a dominance by one or another sulfate compound. There is a tendency for the accumulation mode (approx 0.1 μm < D_p < 1.0 μm) in which most of the sulfate is found, to exist as a relatively stable fraction of atmospheric aerosols (except in cloud condensation and precipitation processes). According to present aerosol production models (Whitby, 1974; Willeke and Whitby, 1975; Heisler and Friedlander, 1977) aerosol mass in the accumulation mode results from the coagulation of nuclei < 0.1 μm diameter and from gas to particle conversion on existing particles in the accumulation mode. Coagulation processes become relatively slow for particles greater than 1 μm in diameter. As a result, two general features of the chemical heterogeneity of tropospheric aerosols can be hypothesized in the absence of direct sources:

(a) the accumulation mode should *tend* toward an internal mixture, and

(b) the accumulation mode exists as an external mixture with the coarse mode.

While there are some data available (e.g. Flocchini *et al.*, 1976) for judging the chemical homogeneity of the accumulation mode in well aged cases, it is not clear that they are adequate to test how well this hypothesis is met. Most data where there is any size resolution below 1 μm diameter has been obtained with cascade impactors, often with only one size separation within the accumulation mode.

On the other hand, the compositional difference between the accumulation and coarse modes is striking. Dzubay and Stevens (1975) show that virtually always the mechanically produced materials such as soil dust are inertially separable from the sulfur bearing accumulation mode. In particular, most of the sulfur in aerosols is clearly *not* mixed internally with the soil derived substances. It is quite possible, for example, for limestone dust ($CaCO_3$) to co-exist with acid sulfate aerosols. Thus we can assume that sulfur compounds in the 0.1–1.0 μm dia range selectively tend to have impurities other than soil dust and other compounds confined to the coarse mode.

Of inorganic substances, liquid H_2O is probably the most abundant compound found with sulfate. Pb as a halide (possibly PbClBr, Pierrard, 1969) is frequently found with sulfate compounds from car exhaust. Elemental carbon has been identified, both by combustion analysis (Appel *et al.*, 1976) by Raman spectroscopy (Rosen and Novakov, 1977), and inferred optically (Weiss *et al.*, 1976). Nitrate, probably as NH_4NO_3, tends to be found along with sulfate compounds which, due to its hygroscopic properties, adds more liquid water to the aerosol. Trace metals such as Fe, V, Mn from metal processing or from trace elements in fossil fuel possibly start out as oxides, and may be dissolved by free acid in hydrated sulfate particles. Table 3 is a list of elements and ions observed to co-exist with sulfur in the accumulation mode. In situations with low sulfate concentration and/or a local large source of coarse particles (soil or sea salt particles) the sulfate compounds are mixed with the smallest soil and/or sea elements and compounds.

Table 3. Inorganic substances co-existing with sulfates in the 0.1 to 1.0 μm dia range

	Typical concentrations in μg^3
Al	0.008–0.05
As	0.04
Br	0.01–3
C graphitic	0–10
Cd	0.02–0.4
Cl	0.005–5.0
Cr	0–0.002
Cu	0.002–0.14
Fe	0.02–1.0
K	0.01–0.2
Mg	0.006–0.15
Mn	0.002–0.03
NH_3	0.5–5
Ni	0.002–0.1
NO_3^-	3–60
Pb	0.5–6
Si	0.06–0.6
Ti	0.03–0.06
V	0.004–0.007
Zn	0.003–0.06

Organic materials make up the lengthiest and most complex list of compounds found with sulfate compounds in the atmosphere. Some of these organic compounds are themselves secondary particulate matter since they are likely to be formed from gas phase reactants (Heisler and Friedlander, 1977). Others may be formed in or near the source as primary pollutants coated or adsorbed onto substances such as soot (Novakov *et al.*, 1972). Cronn (1975) and Cronn *et al.* (1977) list a large variety of organic compounds (or classes of organics) found to co-exist with sulfate compounds at sites in both the Los Angeles and St. Louis areas. Table 4 is this list of compounds detected by high resolution mass spectroscopy and the range of concentrations observed for a variety of organic substances. A number of other compounds were detected in this work including sulfate compounds, a few other inorganic compounds (NH_4Cl, NH_4NO_3) and some organic compounds of unknown structure. It is important to note that almost all of these compounds could be detected in *all* samples that were analyzed, including a few from rural areas.

A proper view of the sulfur chemistry of aerosols must include a consideration of certain trace gases such as SO_2, H_2S, NH_3, O_3 and NO_2. We have already discussed the role of SO_2, gaseous sulfide compounds and NH_3. NH_3 can also neutralize acid particles after they have been collected on a filter. The possibility of reactions on the filter introduces an ambiguity into simultaneous NH_4^+ and SO_4^{2-} aerosol data taken from filter samples. Neutralization of acidic aerosols by NH_3 can have other chemical consequences. The amount of dissolved SO_2 species, dis-

Table 4. Compounds co-existing with sulfates in the 0.1-1 μm dia range as detected by high resolution mass spectroscopy concentrations in μg^{-3}, Cronn (1975)

Compound	Concentration	Compound	Concentration
alkanes	0.03–4	hexadecanoic acid	0.007–2.9
Alkenes	0.03–5.4	octadecanoic acid	0.005–0.2
alkyl benzenes	0.003–6	dodecanoic acid	0–0.55
pentamethyl benzene	0.0003–0.08	butanedioic acid	0.01–0.15
substituted styrene	0.001–0.8	pentanedioic acid	0.02–0.63
alkyl nitrates	0–0.07	hexanedioic acid	0.003–1.4
alkyl nitriles	0–0.3	heptanedioic acid	0.01–1.3
amides	0.005–1.5	benzaldhyde	0.003–0.16
alkyl piperidenes	0.01–1.12	benzyl alcohol	0.01–0.5
quinoline	0–0.06	benzoic acid	0.001–1.15
2-methyl imidazoline	0–0.05	phenylacetic acid	0.004–0.55
r-substituted piperidines	0–0.04	phenyl propionic acid	0.004–0.2
1-4 benzoquinone	0–0.1	phenyl buteric acid	0.003–0.37
phenol	0–0.6	CHO—CH=CH—CH(CH_3)—CHO	0.009–2.7
mono-r-substituted phenols	0–5.8	CH_2OH—CH=CH—CH=C(CH_3)—CHO	0.01–1.0
2,4,6 trimethyl phenol	0–0.16	5 hydroxy pentanoic acid	0.006–1.1
4-substituted naphthalenes	0–2.5	6 oxo-hexanoic acid	0–0.7
naphthalene	0–0.05	6 hydroxy hexonoic acid	0.002–1.0
methyl naphthalene	0–0.02	oxo-heptanoic acid	0.01–0.5
phenanthrene	0–0.009	pinonic acid	0.15–2.7
anthracene	0–0.01	pinic acid	0.08–0.65
3,4 benzopyrene	0–0.05	norpinonic acid	0.06–2.6
nitronaphthalene	0–0.009		
total organic acid	0.01–5.5		

solved gaseous sulfide species, and bisulfite addition products will increase upon neutralization; the potential for formation of sulfonic acids and sulfate esters will decrease upon neutralization.

Both O_3 and NO_2 are produced along with sulfate compounds in photochemically active air parcels (White *et al.*, 1976). O_3 can further interact with dissolved S(IV) species (Penkett, 1972) to produce sulfate compounds. NO_2 can potentially react with acid sulfate compounds to produce nitrosyl sulfuric acid and other meta-stable sulfur compounds.

From the data mentioned above, it is necessary to conclude that sulfur compounds in the accumulation mode, which are dominated by sulfates compounds, are systematically found to co-exist with other materials. Thus, the physical and chemical properties ascribed to the sulfate compounds must account for the omnipresent impurities. In some cases, the effect of importance might *not* be due to sulfur but rather some of the large number of coincident substances. Surface tension will certainly be influenced by these trace organic substances, and is of importance in controlling particle growth at high r.h. Care must be exercized in any situation where effects are correlated with sulfur or SO_4^{2-} because the coexisting substances may be correlated as well.

6. CONCLUSIONS

We have presented a summary of information regarding sulfur compounds in atmospheric particulate matter. The total picture is still qualitative due to the limitation of the data and to experiments that are not yet completed. It is clear, in spite of such limitations, that sulfur—largely SO_4^{2-}—is a substance which frequently governs the physicochemical nature of submicromater particles. In the eastern third of the United States we can describe the accumulation mode aerosol semi-quantitatively (with respect to its chemistry) as a collection of particles composed of impure sulfate compounds ranging from H_2SO_4 to $(NH_4)_2SO_4$. The impurities include metals, inorganic ions such as nitrate compounds, and a wide variety of trace organic compounds. These substances are on the average mixed internally rather than externally. Sulfate compounds as well as other substances and impurities cause the submicrometer particles to accumulate relatively large amounts of liquid water.

In the future, it will be desirable to establish the molecular composition in a quantitative sense. It will also be useful to study the role of impurities in governing the relevant thermal and physical properties such as hygroscopicity, toxicity, refractive index and acidity. When such information is established it will be possible to relate the chemical properties to the source process and to environmental effects, and hence to an understanding of the need and means for control.

Acknowledgements—This work was supported in part by the Federal Interagency Energy/Environment Research and Development Program through EPA grant No. R800665-12 in part by NSF grant DES-75-13922, and in part by NIEHS grant No. 1P01ES01478-01.

REFERENCES

Appel B. R.. Colodny P. and Wesolowski J. J. (1976) Analysis of carbonaceous materials in southern California atmospheric aerosols. *Envir. Sci. Technol.* **10**, 359–363.

Barbaray B., Contour J. P. and Mouvier G. (1977) Sulfur dioxide oxidation over atmospheric aerosol—X-ray photoelectron spectra of sulfur dioxide adsorbed on V_2O_4 and carbon. *Atmospheric Environment* **11**, 351–356.

Bassett H. and Parker W. (1951) The oxidation of sulfurous acid. *J. chem. Soc.* 1540–1560.

Bhardwaja P. S., Herbert J. and Charlson R. J. (1974) Refractive index of atmospheric particulate matter: an *in-situ* method for determination. *Appl. Opt.* **13**, 731–734.

Bhardwaja P. S. (1976) The backward hemispheric scattering ratio of atmospheric aerosols. Ph.D. Dissertation, University of Washington, Seattle.

Bosch J. C., Pilat M. J. and Hrutfiord B. F. (1970) Size distribution of aerosols emitted from a Kraft mill recovery furnace. *J. Tech. Ass. Pulp and Paper Inst.* **54**, 1871–1875.

Brosset C., Andreasson K. and Ferm M. (1974) The nature and possible origin of acid particles observed at the Swedish west coast. *Atmospheric Environments* **9**, 631–642.

Charlson R. J., Vanderpol A. H. Covert D. S., Waggoner A. P. and Ahlquist N. C. (1974) $H_2SO_4/(NH_4)_2SO_4$ background aerosol: optical detection in the St. Louis region. *Atmospheric Environment* **8**, 1257–1267.

Charlson R. J., Porch W. M., Waggoner A. P. and Ahlquist N. C. (1974a) background aerosol light scattering characteristics: nephelometric observations at Mauna Loa observatory compared with results at other remote locations. *Tellus* **26**, 345–360.

Cline J. E. and Richards F. A. (1969) Oxygenation of Hydrogen sulfide in seawater at constant salinity, temperature and pH. *Envir. Sci. Technol.* **3**, 838–843.

Cobourn W. G., Husar R. B. and Husar J. D. (1978) Monitoring of ambient H_2SO_4 and its ammonium salts by *in-situ* aerosol thermal analysis. *Atmospheric Environment* **12**, 89–98.

Covert D. S. (1974) A study of the relationship of chemical composition and humidity to light scattering by aerosols. Ph.D. Thesis, University of Washington, Seattle.

Cronn D. (1975) Analysis of atmospheric aerosols by high resolution mass spectrometry. Ph.D. Dissertation, University of Washington, Seattle.

Cronn D. R., Charlson R. J., Knights R. L., Crittenden A. L. and Appel B. R. (1977) A survey of the molecular nature of primary and secondary components of particles in urban air by high resolution mass spectrometry. *Atmospheric Environment* **11**, 929–937.

Cunningham P. T. and Johnson S. A. (1976) Spectroscopic observation of acid sulfate in atmospheric particulate samples. *Science*, **191**, 77–79.

David D. J., Willson M. C. and Ruffin D. S. (1977) Measurement of Ammonia in Ambient Air *A.C.S. Abstracts*, Division of Environmental Chemistry, 173rd National Meeting, New Orleans, March 20–27, pp 41–44.

Dzubay T. G. and Stevens R. K. (1975) Ambient air analysis with dochotomous sampler and XRF spectrometer. *Envir. Sci. Technol.* **9**, 663–668.

Eatough D. J., Hansen L. D., Izatt R. M. and Mangelson N. F. (1977) Analysis of Sulfite Species in Aerosols. *A.C.S. Abstracts*, Division of Enviromental Chemistry, 173rd Annual Meeting, New Orleans, March 20–27, pp. 57–60.

Flocchini R. G., Cahill T. A., Shadoan D. J., Lange S. J., Eldred R. A., Feeney P. J., Wolfe G. W., Simmeroth D. P. and Suder J. K. (1976) Monitoring California aerosols by size and elemental composition. *Envir. Sci. Technol.* **10**, 76–82.

Gubareva M. A. (1947) Bisulfite compounds of aldehydes and ketones. I. Equilibrium of the addition reaction in aqueous solution. *J. gen. Chem. U.S.S.R.* **17**, 2259–2264.

Hänel G. (1976) The properties of atmospheric aerosol particles as functions of the relative humidity at thermodynamic equilibrium with the surrounding moist air. *Adv. Geophys.* **19**, 73–188.

Hansen J. D., Whiting L., Eatough D. J., Jensen T. E. and Izatt R. M. (1976) Determination of sulfur (IV) and sulfate in aerosols by thermometric methods. *Analyt. Chem.* **48**, 634–638.

Heintzenberg J. and Bhardwaja P. S. (1976) On the accuracy of the backward hemispheric integrating nephelometer. *J. appl. Met.* **15**, 1092–1096.

Heisler S. L. and Freidlander S. K. (1977) Gas to particle conversion in photochemical smog: aerosol Growth laws and mechanisms for organics. *Atmospheric Environment* **11**, 157–168.

Hemeon W. C. L. (1955) The estimation of health hazards from air pollution. *Arch ind. Hlth* **11**, 397–404.

Hobbs P. V., Radke L. F. and Hindman E. E. (1976) An integrated airborne particle measuring facility and its preliminary use in atmospheric aerosol studies. *J. Aerosol Sci.* **1**, 195–211.

Junge C. E. (1952) Die konsitution der atmospharischen aerosols. *Ann. Met. Hamburg*, **5**, 1–55.

Junge C. E. (1954) The chemical composition of atmospheric aerosols, I: Measurements at Round Hill field station, June–July, 1953. *J. Met.* **11**, 323–333.

Junge C. E. (1963) *Air Chemistry and Radioactivity*. Academic Press, New York, 382.

Kapustin V. N., Lyubovtseva Yu.S. and Rozenberg G. V. (1974) Variability of aerosols under the influence of cloud modulation of the radiation field. *Izvestia, Atmospheric and Oceanic Physics* **10**, 1327–1331.

Kuhlman M. R., Fox D. L. and Jeffries H. E. (1977) The Effect of CO on sulfate Aerosol Formation, *A.C.S. Abstracts*, Division of Environmental Chemistry, 173rd National Meeting, New Orleans, March 20–27, 254–257.

Lau G. and Charlson R. J. (1977) On the discrepancy between background atmospheric ammonia gas measurements and the existence of acid sulfates as a dominant atmospheric aerosol. *Atmospheric Environment* **11**, 475–478.

Low R. D. H. (1969) A theoretical study of nineteen condensation nuclei. *J. Rech. atmos.* **4**, 65–78.

Novakov T., Mueller P. K., Alcocer A. E. and Otvos J. W. (1972) Chemical composition of Pasadena aerosol by particle size and time of day—III. Chemical states of nitrogen and sulfur by photoelectron spectroscopy. *J. Colloid Interface Sci.* **30**, 225–234.

Novakov T., Chang S. G. and Harker A. B. (1974) Sulfates as pollution particulates: catalytic formation on carbon (soot) particles. *Science* **186**, 259–261.

Orr C., Hurd F. K. and Corbett W. J. (1958) Aerosol size and relative humidity. *J. Colloid Interface Sci.* **13**, 472–482.

Penkett S. A. (1972) Oxidation of SO_2 and other atmospheric bases by ozone in aqueous solution. *Nature* **240**, 105–106.

Pierrard J. M. (1969) Photochemical decomposition of lead halides from automobile exhaust. *Envir. Sci. Technol.* **3**, 48–51.

Roberts P. T. and Freidlander S. K. (1975) Conversion of SO_2 to sulfur particulate in the Los Angeles atmosphere. *Envir. Hlth Perspectives* **10**, 103–108.

Rosen H. and Novakov T. (1977) application of Raman scattering to the characterization of atmospheric aerosol particles. *Nature*. **266**, 708.

Sinclair D., Countess R. J. and Hoopes G. S. (1974) Effect of relative humidity on the size of atmospheric aerosol particles. *Atmospheric Environment* **8**, 1111–1118.

Sousa J. A. and Margerurm J. D. (1960) Equilibrium constant of benzaldehyde sodium bisulfite. *J. Am. chem. Soc.* **82,** 3013–3016.

Tang I. N. (1976) Phase transformation and growth of aerosol particles composed of mixed salts. *J. Aerosol Sci.* **7,** 361–371.

Tang I. N. and Munkelwitz H. R. (1977) phase Transformation and Droplet Growth of Sulfate aerosols *A.C.S. Abstracts,* Division of Environmental Chemistry, 173rd Meeting, New Orleans, March 20–27, 168–170; (1977) Aerosol Growth Studies—III "Ammonia bisulfate aerosols in a moist atmosphere". *J. Aerosol Sci.* (in press).

Vanderpol A. H. (1975) A systematic approach for computer analysis of air chemistry data. Ph.D. Dissertation, University of Washington, Seattle.

Vanderpol A. H., Carsey F. D., Covert D. S., Charlson R. J. and Waggoner A. P. (1975) Aerosol chemical parameters and air mass character in the St. Louis region. *Science,* **190,** 570.

Weiss R., Charlson R. J., Waggoner A. P., Baker M. B., Covert D., Thorsell D. and Yuen S. (1976) Application of Directly Measured Aerosol Radiative Properties to Climate Models in Radiation in the Atmosphere Symposium—Garmisch Partenkirchen, Federal Republic of Germany, August.

Weiss R. E., Waggoner A. P., Charlson R. J. and Ahlquist N. C. (1977) Sulfate aerosol: its geographical extent in the midwestern and southern United States. *Science,* **195,** 979–981.

Werner C. (1972) Lidar measurements of atmospheric aerosol as a function of relative humidity. *Opto-electronics,* **4,** 125–132.

Whitby K. T. (1974) Modeling of Multimodal Aerosol Distributions presented at the G.A.F. meeting, Bad Soden, Germany, October 17.

White W. H., Anderson J. A., Blumenthal D. L. Husar R. B., Gillani N. V., Husar J. D. and Wilson W. E., Jr. (1976) Formation and transport of secondary air pollutants: ozone and aerosols in the St. Louis urban plume. *Science* **194,** 187–189.

Willeke K. and Whitby K. T. (1975) Atmospheric aerosols: size distribution interpretation. *J. Air Pollut. Control Ass.* **25,** 529–534.

Winkler P. (1973) The growth of atmospheric aerosol particles as a function of the relative humidity—II. An improved concept of mixed nuclei. *Aerosol Sci.* **4,** 373–387.

Winkler P. and Junge C. (1972) The growth of atmospheric aerosol particles as a function of the relative humidity —I. Methods and measurements of different locations. *J. Rech. atmos.* **6,** 617–638.

Winkler P. (1976) On Production Rates in Marine Aerosols Symposium, in Radiation in the Atmosphere Garmisch-Partenkirchen, Federal Republic of Germany.

Winkler P., Covert D. S. and Heintzenberg J. (1973) unpublished results.

Wylie R. G. (1963) The Properties of water-salt systems in relation to humidity. *Humidity and Moisture* **3,** 507–517 (Edited by Wexler A.). National Bureau of Standards.

Coordinating Research Council (1968) *Handbook of Chemistry and Physics* **49,** D-180, E-37,. Chemical Rubber Co., Cleveland, Ohio.

Atmospheric Environment Vol. 12, pp. 55–68. Pergamon Press 1978. Printed in Great Britain.

SAMPLING AND ANALYSIS OF ATMOSPHERIC SULFATES AND RELATED SPECIES

ROBERT K. STEVENS and THOMAS G. DZUBAY
U.S. Environmental Protection Agency Research, Triangle Park, NC27711, U.S.A.

and

GEORGE RUSSWURM and DWIGHT RICKEL
Northrop Services, Inc., Research Triangle Park, NC27709, U.S.A.

(*First received* 13 *July* 1977 *and in final form* 5 *October* 1977)

Abstract—Sampling and analytical methods to measure atmospheric concentrations of sulfur, sulfates and related species are compared for aerosols collected in New York City; Philadelphia, PA; South Charleston, WV; St. Louis, MO; Glendora, CA; and Portland, OR. Dichotomous samplers equipped with virtual impactors were used to separately collect fine ($<3.5\ \mu m$) and coarse ($>3.5\ \mu m$) particles on membrane filters. Both size fractions were analyzed by energy-dispersive X-ray fluorescence spectroscopy to determine the total amounts of sulfur and other elements, and the samples were analyzed by an Ion Chromatograph and by the thorin spectrophotometeric method to determine sulfate. These analyses reveal that more than 70% of the sulfur occurs in the fine particle fraction for at least 90% of the samples. Sulfate typically accounts for about 40% of the mass of the fine particle fraction, and in some instances it accounts for more than 50%. For the fine particle fraction, the ratio of the sulfate and the sulfur mass concentrations is 2.96 ± 0.15, which is in excellent agreement with the value of 3.00 for sulfate.

To test for the existence of sulfite ions in the samples, a cold extraction procedure was developed, which minimizes the conversion of sulfite to sulfate. Analysis for sulfite using the Ion Chromatograph indicates that less than 2% of the sulfur collected in South Charleston, WV, and Philadelphia, PA, is in the form of sulfite.

A procedure was developed to preserve the acidity of aerosols collected with an automated dichotomous sampler and was recently used to sample aerosols in Research Triangle Park, NC. Analysis of the samples for H^+, NH_4^+, and SO_4^{2-} ions revealed that the sulfate concentrations were typically 10 to 14 μg^{-3} and that H^+ ions accounted for 5 to 60% of the cations associated with sulfate. In addition the sum of NH_4^+ and H^+ equalled the SO_4^{2-} concentration expressed in nanoequivalents.

In the network of stations that was part of the Regional Air Pollution Study (RAPS) the hi-volume sampler gave mass and sulfate concentrations that were consistently higher than values obtained for the automated dichotomous sampler. For samples collected at eight RAPS stations at St. Louis, MO, between September and December 1975, the ratio for sulfate determined for the two types of samplers was 1.32.

INTRODUCTION

Although sulfates have long been known to exist in the atmosphere, reliable sampling and analytical methods to determine their physical and chemical form have not been readily available until recently. For the past 5 years the U.S. Environmental Protection Agency (EPA) has supported projects to develop procedures to collect aerosols in a manner that minimizes artifact formation and to separate the respirable particles (particles $<3.5\ \mu m$ in aerodynamic diameter) from larger particles. The instrumentation that now appears to satisfy these sampling requirements is based on the virtual impactor principle (Conner, 1966) and has been engineered into what has been termed the dichotomous sampler. Both automated and manual versions of the dichotomous sampler have been used in several field studies across the United States. The samples have been analyzed to determine mass, sulfate, nitrate, ammonium ion, hydrogen ion, and elemental composition. The results from these field studies confirm the limited observations of Weiss (1976) that sulfur was the predominant component in the submicron aerosol in the United States.

Concurrent with the development of these samplers was the development of an energy dispersive X-ray fluorescence (EDXRF) system (Jacklevic, 1974, 1976, 1977) and their application to the analysis for as many as 35 elements contained in particles collected on membrane filters with dichotomous samplers (Dzubay, 1975; Loo, 1976; Dzubay *et al.*, 1977).

Because EDXRF only measures elemental composition, additional methods were developed to characterize the chemical form of the samples. Brosset (1976) developed methods to measure the H^+, NH_4^+, and SO_4^{2-} concentrations in aerosols. Recent developments in ion exchange chromatography (Small, 1976) have led to a commercial instrument, the Ion Chro-

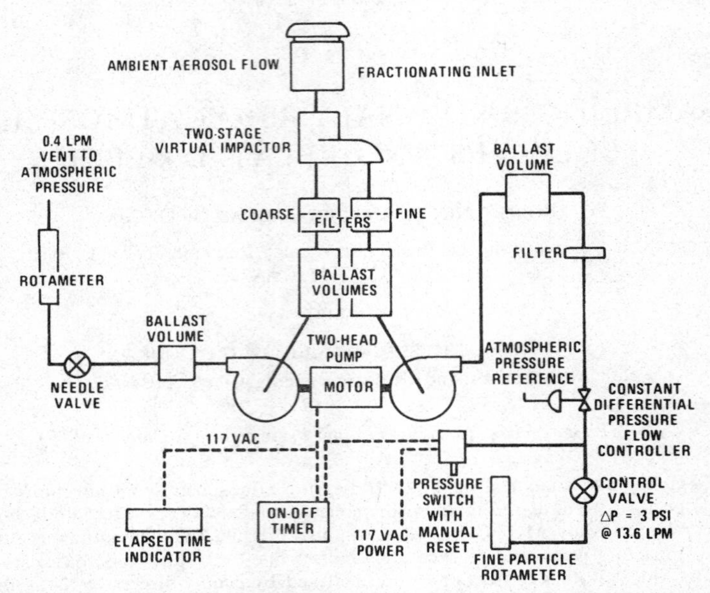

Fig. 1. Schematic view of dichotomous sampler with constant flow rate system having a pressure regulator to maintain a constant pressure differential across a fixed orifice.

matograph*, which simultaneously measures Cl^-, Br^-, SO_3^{2-}, NO_3^-, and SO_4^{2-} (Mulik, 1976). In the present text the analytical capabilities of these methods are compared. The comparison consists of the analysis of aerosols collected in New York City; Philadelphia, PA; South Charleston, WV; St. Louis, MO; Glendora, CA; and Portartland, OR.

These studies not only serve as a basis for selecting sampling and analytical methodology to determine sources of atmospheric sulfates but also demonstrate the importance of automated aerosol sampling for periods of less than 3 hours. Short term sampling and analysis are made possible both by the development of reliable automated aerosol size-fractionation devices and by the improved sensitivity and specificity of methods to measure sulfates and related species.

EXPERIMENTAL

Sampling

Manual dichotomous samplers (MDS) fabricated by Environmental Research Corporation, St. Paul, MN (Stevens, 1975; Dzubay, 1975; Dzubay *et al.*, 1977), were used for field studies in New York City; Philadelphia, PA; South Charleston, WV; and Portland, OR. The samplers were operated at flow rates of 14 l. min^{-1}, and the sampling periods ranged from 24 h in South Charleston to 2 h in New York City. A network of 10 automated dichotomous samplers (ADS) fabricated by Lawrence Berkeley Laboratory (Goulding, 1975; Loo, 1976) was used for the St. Louis studies (Meyers, 1975). The flow rate for the ADS was 50 l. min^{-1}, and the sampling intervals ranged from 2 to 12 h. An ADS was also operated in Glendora, CA, where samples were collected every 6 h. For the MDS the aerosol particles were collected on 37 mm diameter FALP Fluoropore filters (Millipore Corp., Bedford, MA), which consist of 1 μm pore-size Teflon† membranes bonded to a polyethylene support net. In the ADS the filters that were used had 1.2 μm pores and were made of esters of cellulose (Millipore). Fluoropore filters could not be used in the ADS because the polyethylene support net prevented the filters from being adequately sealed in the filter holder.

In the MDS that was operated in Philadelphia, a pneumatic flow rate controller of the type shown in Fig. 1 was used to maintain a constant sampling flow rate. An inexpensive differential pressure regulator (Moore Products Co., Houston, TX, Flow Controller CG-63BD) in the exhaust line of the pump maintained a constant pressure differential across a fixed orifice and thereby maintained a constant flow rate through the system. The differential flow control system has been tested over a −20 to +40°C temperature range, and the flow rate varied less than 5%. In the manual dichotomous samplers operated at the other sites, a Sierra flow controller (Sierra Corp., Carmel Valley, CA), consisting of an anemometer sensor and a variable-speed pump, was used to maintain a constant flow rate. Tests have shown that both types of flow control systems are capable of maintaining the flow rate constant to better than 5% for pressure drops up to 25 cm Hg across the filter.

For other studies designed to measure the H^+ content, an ammonia denuder was coupled to a special ADS sampler fabricated by Cabot Corporation (Billerica, MA). The denuder, shown in Fig. 2, is composed of 16 glass tubes (30 cm length, 0.5 cm i.d.) arranged in parallel and coated with phosphorous acid. The coating was applied by inserting approximately 300 mg of H_3PO_3 into the end of each tube and gently warming and rotating the tube to coat the walls evenly. The efficiency for the removal of NH_3 was measured by flowing 100 ppb NH_3 through the denuder at relative humidities of 20, 50 and 80% and measuring the exit concentration with a chemiluminescent NH_3 monitor (Hodgeson, 1971). The measured denuder efficiency was in agreement with the calculated values of 99.9%. For the acid measurements, the aerosol collection surface consisted of unbacked 1 μm Teflon membrane filters. After the aerosol samples were collected, they were stored in a NH_3 free atmosphere at 20 to 50% relative humidity to prevent neutralization and volatilization of sulfuric acid in the sample.

---

* Mention of product or company name is not intended as an endorsement by the US Environmental Protection Agency.

† Registered trademark of E.I. DuPont De Nemours & Company, Inc., Wilmington, DE.

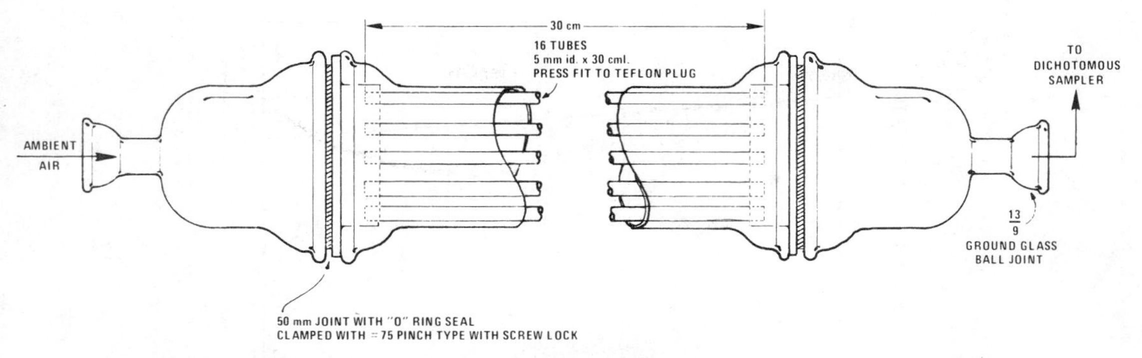

Fig. 2. Cross sectional view of NH_3 gas diffusion denuder used with dichotomous sampler.

X-ray fluorescence analysis

Prior to any chemical extractions, the elemental composition of filter-collected particles was nondestructively measured in our laboratory using an EDXRF spectrometer. In this device, fabricated by the Lawrence Berkeley Laboratory, a pulsed X-ray tube excites a secondary target, which excites the sample with nearly monoenergetic X-rays (Jaklevic, 1974, 1976, 1977; Goulding, 1975). To obtain high sensitivity for a wide range of elements, each sample is excited by means of three different secondary targets. For the K X-rays of elements with atomic numbers in the ranges 13–20, 21–38 and 39–56 the secondary targets consist of titanium, molybdenum and samarium, respectively. The molybdenum target also excites the L X-rays of lead and other heavy elements. The fluorescent X-rays from the sample are detected using a lithium drifted silicon detector, which uses electronic collimation to minimize the background. Because of a compact geometrical arrangement between components of the spectrometer, an X-ray tube power of only 100 W is sufficient to provide the maximum usable count rate for filter samples.

In applications where the amount of collected particulate matter is expected to be small, it is important to estimate the minimum detectable concentration, C'_D. Currie (1977) defines C'_D as 3.29 times the standard deviation for determining the background and expresses it as:

$$C'_D = 3.29\,\eta^{1/2}(St)^{-1}[tR_b + (\epsilon\, t\, R_b)^2]^{1/2} \qquad (1)$$

where: η = measure of extent to which blank is well characterized (see Currie, 1977). S = sensitivity expressed in counts s^{-1} per ng cm^{-2} for the spectral region containing the full width at half maximum of the peak for the element of interest. t = counting time in seconds. R_b = background in counts s^{-1} for the same spectral region used to determine S for the element. ϵ = relative standard error in determining R_b if errors due to counting statistics were negligible. Assuming that $\epsilon = 0.02$, $\eta = 1$ and $t = 100$ s for each secondary target, values of C'_D are given in Table 1 for several elements.

To express C'_D in units of ng m^{-3}, one must divide the values in Table 1 by the volume sampled per unit area of filter. For example, this conversion factor is 3.05 m^3 cm^{-2} for 24 h of sampling at a rate of 14 l. min^{-1} through 6.6 cm^2 of filter area. The corresponding C'_D value for sulfur is 12 ng m^{-3}. If reasonably accurate information on a certain element is needed, then the concentration C of that element must exceed C'_D by at least a factor of 3, to insure that C exceeds the standard deviation by a factor of 10. If one wanted to measure a sulfur concentration of 1 $\mu g\ m^{-3}$, then for the above example, one could shorten the sampling time from 24 to 1 h and still have sufficient sample.

After a sample has been irradiated in our EDXRF spectrometer, the resulting spectrum is analyzed to determine the number of counts that each element contributes. To accomplish this, a least-squares procedure is used to find a linear combination of single-

Table 1. Sensitivities and minimum detectable limits with energy dispersive spectrometer

Element	Secondary target	Sensitivity Counts s^{-1} μg^{-1} cm^2	Background (R_b)	Minimum detectable concentration (C_D), ng cm^{-2}
Al	Ti	7.2	13	203
S	Ti	47	16	36
K	Ti	133	22	16
V	Mo	31	5.8	28
Fe	Mo	67	7.6	15
Se	Mo	199	2.9	3
Sr	Mo	269	6.9	3.6
Sr	Sm	38	1.7	11.7
Cd	Sm	84	1.8	5.4
Sn	Sm	75	3.5	8.8
Ba	Sm	50	45	74
Pb	Mo	76	3.5	8.6

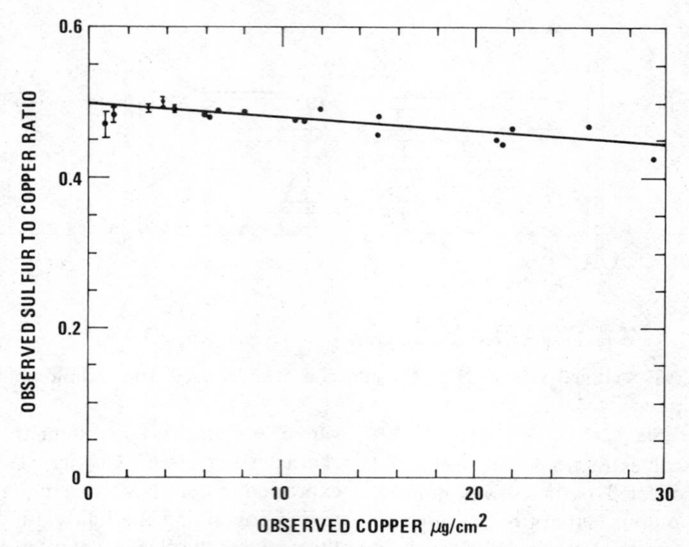

Fig. 3. Ratio of sulfur and copper XRF yields for 0.2 μm copper sulfate deposits of varying mass. The ratios are normalized to 0.505 for zero deposited mass.

element spectra that best describes the unknown spectrum. A library of single-element spectra is obtained for thin standards of each element and is stored in computer memory. The library must contain all elements that could contribute to an unknown sample (Arinc, 1976; Rickel, in press).

The least-squares method is slightly modified for the K X-rays from sulfur to deal with the overlapping M X-rays from lead. Since the concentration of lead can be accurately determined from the emission of L X-rays, it is possible to 'strip' out the interfering lead M peak in the vicinity of the sulfur peak. Such stripping is done prior to performing the least-squares analysis for sulfur. The uncertainty in the result for sulfur because of the presence of lead is estimated to be about 5% of the lead concentration. For example, if the lead concentration were 2 $\mu g\ m^{-3}$, the resulting sulfur uncertainty would be 0.1 $\mu g\ m^{-3}$.

After completion of the spectral analysis, the concentration for each element in the sample is computed from the expression:

$$C = N_i\, a_i/(F_i A_i) \qquad (2)$$

where: N_i = normalization factor to relate counts to concentration for either sulfur, copper, or tin. i = secondary target designation. a_i = number of counts in the spectrum due to each element. F_i = calibration factor for each element relative to $F_i = 1$ for either sulfur, copper or tin. A_i = attenuation correction factor for the element observed on secondary target i.

The EDXRF spectrometer is calibrated to determine the normalization factors, N_i, for copper and tin using standards that consist of vacuum-evaporated deposits of copper and tin on Mylar films. For sulfur, N_i is determined relative to copper using copper sulfate deposits. The latter were prepared by generating an aerosol of 0.2 μm $CuSO_4 \cdot 5\,H_2O$ particles and collecting the particles on Fluoropore filters. The copper sulfate deposits were prepared in a variety of thicknesses, and the results are shown in Fig. 3. All but the thinnest deposits are thicker than a single monolayer, and the layers of particles attenuate the fluorescent X-rays. To obtain a calibration equivalent to a thin standard, the data in Fig. 3 were least squares fitted with a straight line, which was extrapolated to zero deposit thickness. The sulfur calibration was adjusted so that the y axis intercept is equal to 0.505, which is the sulfur-to-copper mass ratio for copper sulfate.

For the remaining elements, the F factors were determined relative to $F = 1$ for sulfur, copper and tin. Figure 4 shows measured F values using dried solution deposits on 1 $mg\ cm^{-2}$ esters of cellulose membranes obtained from Columbia Scientfic, Inc., Austin, TX, and using vacuum-deposited films obtained from Micromatter Co., Seattle, WA. Figure 4 also shows a smooth curve, which is a fit to the data using a method described by Giauque (1973). The curves were normalized to $F = 1$ for $Z = 16$, 29 and 50. The smooth curve represents an improved calibration since random errors in some of the standards would not significantly affect the curve.

For EDXRF analysis of aerosol particles collected on a membrane filter a correction must be made for any X-ray attenuation by the sample matrix. Such attenuation can occur within individual particles, within layers of particles, and within the filter medium into which particles might penetrate.

Attenuation within individual particles is a small effect for sulfur in the fine particle fraction (Dzubay, 1975); however, a typical aerosol sample consists of layers of particles. The attenuation in such layers can be expressed as:

$$A = [1 - exp(-\mu_L M/f)]f/\mu_L M \qquad (3)$$

where: M = average mass per unit area, f = fraction

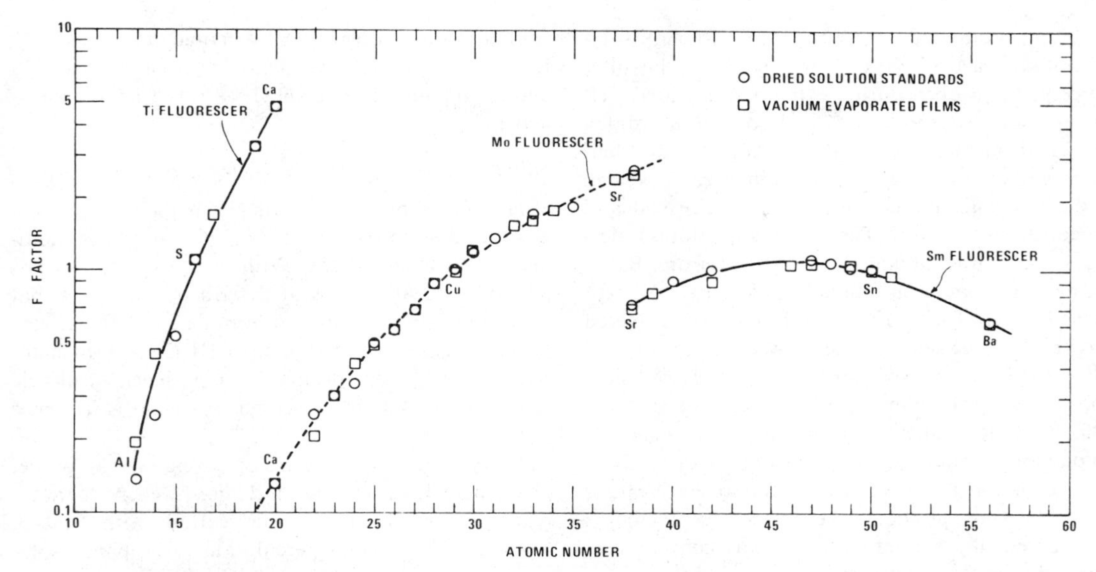

Fig. 4. Calibration data for K X-rays excited using titanium, molybdenum and samarium secondary targets.

of particle deposit area that contains particles, $\mu_L = \mu \sec\theta + \mu' \sec\theta'$, μ,μ' = mass absorption coefficients of incident and fluorescent X-rays in the deposit and θ,θ' = angles between the sample normal and the incident and fluorescent X-ray paths.

The fraction f in Equation 3 is needed to account for any grid pattern or structure in the particle deposit profile. For Fluoropore filters, the polyethylene support net gives a pronounced pattern to the aerosol deposit. By applying Equation 3 to the $CuSO_4$ data in Fig. 3, one deduces the value, $f = 0.45$, for Fluoropore filters. This is consistent with visual observations of aerosol deposits on Fluoropore that indicate that 50 to 60% of the filter is obstructed by the support net.

The mass absorption coefficients are given by the expression:

$$\mu = \Sigma \mu_i W_i$$

where μ_i are the mass absorption coefficient, and W_i are mass fractions of each element in the deoposit. For the fine fraction of a typical urban aerosol, $\mu_L = 0.56\ \mathrm{cm^2\,mg^{-1}}$ for the geometry of our spectrometer (Dzubay, 1975).

Because of the manner in which sulfur is calibrated using copper sulfate aerosol particles, no correction is needed for attenuation of the X-rays by the filter medium. The 0.2 μm mass-median-diameter particles that were used in this calibration are similar in size to atmospheric particles that contain the fine particle sulfur. Thus the X-rays from the atmospheric particles and the calibration have approximately the same attenuation, and the effect cancels. According to measurements by Loo (1977) the attenuation factor for fine particles in 1.2 μm Millipore filters is 0.98 ± 0.02 for particles in the 0.03 to 1 μm size range, which indicates that such an attenuation effect is small and relatively independent of particle size.

In the coarse fraction, the particles are collected on the surface of the membrane filter, and for loadings below 500 $\mu\mathrm{g\,cm^{-2}}$ the particles form a monolayer. The attenuation factor for sulfur in the coarse fraction is $A = 0.64 \pm 0.22$. This value and values used for other elements are given by Dzubay (1975). Since the composition of the coarse particles in which the sulfur occurs is unknown, 0.64 was the median of values that were calculated for a variety of mineral and botanical particle matrices. The uncertainty in A represents the range of possible values for the coarse-particle attenuation factor.

Extraction procedure

After the EDXRF analysis had been completed, certain samples were selected for extraction and subsequent analyses. The extraction process consisted of removing the filter from its holder, loading it into an extraction vessel, filling the vessel with extraction solution, and then extracting by the use of an ultrasonic bath.

The filters arrived in the laboratory either held in holders with snap rings or bound to the holders. The filters were removed from the holders with tweezers in the first case or were cut from the holders with a sharp knife in the second. In either case the implements used were cleaned, rinsed, and allowed to dry after their use on each sample. The removal process was done as quickly as possible in order to prevent any contamination. For acidity measurements care was taken not to breathe on the filters to minimize exposure to exhaled ammonia. The filters were then placed in the extraction vessel with the back of each filter facing the bottom, to prevent any material from being inadvertently removed as the filter was inserted into the vessel.

The extraction vessel was a 30-ml Nalgene polypropylene bottle (Nalge, Inc., Rochester, NY) which had been conditioned by soaking it in extraction solution for at least 8 h before use. To keep the filter submerged and open during extraction, a fluted Teflon pipe was placed in the vessel so that the fluted end rested on the outer, unloaded edge of the filter.

The volume of extraction solution used depended on the type and number of analyses performed and ranged from 8 to 20 ml. The extraction solution, perchloric acid, was obtained as 11.98 N from Baker Chemical Co. and was diluted to 5×10^{-5} N with distilled water. The extraction solution was delivered with 0.05% precision to the vessel by an Oxford Laboratories macro set pipet. The vessel was then capped and placed in an ultrasonic bath (Model 8845-60 Cole-Palmer Instrument Co., Chicago, Illinois) for 20 minutes. The ultrasonic bath tends to produce standing waves in the water so that regions exist where no agitation takes place. To overcome this problem the extraction vessel was continuously moved during the extraction period.

The above extraction procedure was used for analysis of SO_4^{2-}, NO_3^-, NH_4^+ and H^+. To determine the extraction efficiency, a set of 50 randomly selected extracted filters was reanalyzed by EDXRF. For the fine and coarse particle fractions, the extraction efficiencies for sulfur were found to be $98 \pm 1\%$ and $95 \pm 2\%$ respectively.

For those samples on which a sulfite determination was to be made, the extraction procedure was modified to minimize the conversion of sulfite to sulfite. We have demonstrated that carrying out the extraction at room temperature caused complete conversion of sulfite to sulfate. To prevent such conversion, a cold extraction process was developed. In the modified procedure the water bath was made up of a slurry of ice and water that maintained the temperature of the sample at 0°C during the extraction. With this procedure 90% of the sulfite was preserved.

Hydrogen ion analysis (H^+)

Titrimetry and a Gran's function plot were used to determine strong acidity (Brosset, 1976). The instrumentation included an Orion pH meter, a combination electrode, a Gran's function generator, and a Radiometer autoburett ABU 12. The data were recorded on a 1-mV strip-chart recorder. The Gran's function generator is an antilog amplifier (Model 755, Analog Devices, Norwood, MA).

The acid analysis procedure was as follows: (1) standardization of sodium hydroxide titrant, (2) measurement of the concentration of the acid extraction solution; (3) titration of extract using the 0.001 N NaOH, and (4) data analysis. Steps 1 and 2 were done once for each set of filter extracts or at least once each day. Nitrogen was bubbled through the solution being titrated to remove an interference by CO_2. The ionic strength of the volume being titrated was maintained at 0.02 M by the addition of KCl.

The method was characterized using 50 measurements on sulfuric acid standards that had a range of concentrations. The linear dynamic range was 10^{-6} to 10^{-4} M. The relative standard deviation was 2%, and the minimum detectable level was 1 nano-equivalent ml^{-1}.

Sulfate analysis spectrophotometric-thorin

The spectrophotometric-thorin method for sulfates was first described by Persson (1966) and later modified by Brosset (1975). With two exceptions, this method is identical to that described by Brosset; but instead of using dioxanes or acetone as a solvent, isopropanol is used and sulfonazo III (3, 6-*bis*-(*o*-sulfophenylazo)-4, 5-dihydroxy-2, 7-naphthalenedisulfonic acid) (Fisher Scientific) is used to obtain a reference intensity level.

Analysis was performed by using an autopipet (Autochem Instruments, Lindingo, Sweden) to automatically mix 1.8 ml of filter extract with 5 ml of 6.8×10^{-5} M barium perchlorate isopropanol solution and 0.4 ml of 1.37×10^{-3} M thorin solution in a 2 cm optical cuvette. The thorin solution was doped with H_2SO_4 to adjust the calibration to the linear range of the absorbance curve for the barium thorin titration (Brosset, 1975). Absorbance measurements at 520 nm are made using an Hitachi Model 101 spectrophotometer. The linear dynamic range of the method is from 0 to 6×10^{-5} M sulfate with a minimum detectable level of 2.5×10^{-6} M per ml. Analysis of 100 quality control solutions gave a relative standard deviation of 2%.

Ion exchange chromatography

Ion exchange chromatography is a relatively new technique for routine analysis of anions and cations in aqueous extracts of ambient aerosols. Mulik (1976) has described the application of a commercial liquid ion exchange chromatograph (IC) (Dionex Model 14, Sunnyvale, CA) to the analysis of water soluble anions. Before injecting a sample into the IC, 5 ml of the aerosol extract was spiked with a 1:1 mixture of 0.003 M $NaHCO_3$ and 0.0005 M Na_2CO_3 to adjust the molar CO_3^{2-} concentration to the range of 10^{-3}–10^{-5}. This basic mixture was added to neutralize the minute concentrations of perchloric acid used in the extraction procedure so that the injected solutions were approximately the molar concentration of the eluent used in the chromatograph. Because the instrument calibration curve was not linear, multipoint calibrations were performed over the range of concentrations expected in the samples. Since these samples were previously analyzed by EDXRF, the SO_4^{2-} calibration range could be adjusted accordingly. The signal from the IC conductivity detector was processed by a Hewlett Packard 3385A chromatographic control system. Peak areas and retention times were recorded to quantify and identify specific anions.

The relative standard deviations for mixtures of sodium nitrate and ammonium aqueous standards are shown in Table 2. Retention times and relative stan-

Table 2. Relative standard deviations for IC analysis of nitrate and sulfate standards

Molarity	Dionex attenuation setting	Standard deviation* SO_4^{2-}	NO_3^-
1×10^{-4}	10	1%	1%
5×10^{-5}	10	0.3%	1%
1×10^{-5}	3	1.6%	1%

* Based on four consecutive injections of 0.4-ml volumes of standard solutions.

dard deviations were determined daily. The minimum detectable levels for sulfates and nitrates were 10^{-1} neq ml^{-1} and 5×10^{-2} neq ml^{-1} respectively. A typical IC chromatogram of an extract of a fine particle sample collected in Philadelphia is shown in Fig. 5. One notes the low nitrate concentration compared to sulfate. For almost all aerosol samples collected with the dichotomous sampler using Fluoropore filters, nitrate was always a small fraction of the sulfate concentration.

Ammonium analysis

The concentration of ammonium ions was determined with an ion selective NH_3 gas diffusion electrode (Orion Model 95-10). The response was recorded from a Corning-Digital 112 pH meter. Analysis was performed by mixing 7 ml of filter extract and 0.2 ml 5 N sodium hydroxide in a 20-ml glass beaker. The electrode was immersed into the solution with the tip extending approximately 1 cm into the solution. Readings were made between 2 and 3 min after immersing the electrode into the extract. When the sample concentration was less than 1×10^{-5} M, equilibration required as long as 5 min. The electrode was calibrated using dilute NH_4Cl standards (10^{-2} to 10^{-6} M). The minimum detectable level was 3 neq ml^{-1}. At concentrations above 10 neq ml^{-1}, the relative standard deviation was $\pm 5\%$.

RESULTS AND DISCUSSION

Results of elemental and ionic analyses

In accordance with the objectives of this study, the mean concentrations for 18 elements and the mean total suspended particulate (TSP) masses of aerosol particles collected by dichotomous samplers at seven sites in the United States are presented in Table 3. The masses of the elemental concentrations represent unweighted arithmetic averages as determined by EDXRF. The means of the TSP represent unweighted arithmetic averages as determined by gravimetic measurements of the filtered aerosol. In some in-

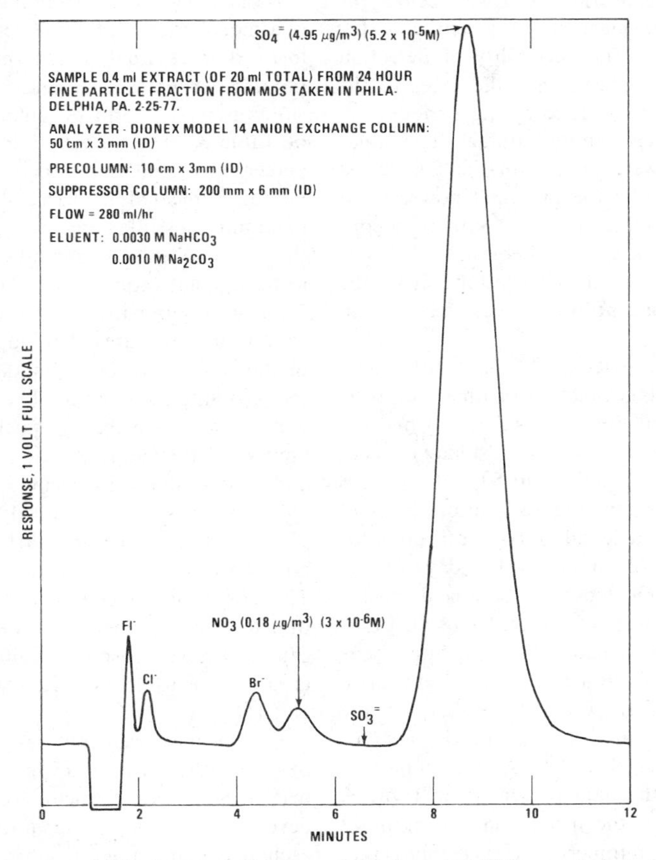

Fig. 5. Chromatogram of 0.4 ml extract of fine particles from a MDS operated on 25 February 1977 in Philadelphia (24-hour sample collection period). Concentrations of the sulfate and nitrate are expressed in molarity and in $\mu g\ m^{-3}$.

Table 3. Mean elemental concentrations (nanograms m^{-3}) and mean total suspended particulate (TSP) (micrograms m^{-3}) for two size fractions at several sites in the United States

Site	New York, NY 2/1977		Philadelphia PA 2–3/1977		Charleston W VA 4–8/1976 1/1977		St. Louis, MO RAMS 105 12/1975		Portland, OR 2/1977		Glendora, CA 3/1977	
Data	F*	C†	F	C	F	C	F	C	F	C	F	C
No‡	39	38	17	17	106	102	60	60	17	18	69	69
Al	129	840	48	642	262	1110	1983	8678	210	1175	<50§	331
Si	358	2018	262	1802	562	2616	412	4303	89	2808	188	995
P	173	120	57	40	42	33	218	583	55	104	24	30
S	5536	400	3096	454	3774	345	2774	752	1395	284	1612	240
Cl	227	607	115	225	144	232	145	180	478	719	72	719
K	273	185	137	165	97	279	200	324	153	190	100	179
Ca	358	1151	168	936	93	831	120	1910	67	765	99	442
Ti	<50§	41	21	61	11	81	36	74	18	146	10	42
V	207	81	60	11	<2§	<2§	23	29	29	19	5	<3§
Cr	<25§	<15§	2	3	<2§	<2§	18	22	14	9	4	3
Mn	55	44	17	14	7	12	40	33	27	21	5	6
Fe	383	957	213	691	163	625	337	1001	187	946	124	360
Hi	57	18	30	7	0.6	0.3	15	10	42	10	14	3
Cu	73	31	25	7	25	29	59	16	62	17	12	6
Zu	373	85	148	38	30	20	147	74	61	30	45	16
Br	218	39	264	51	133	33	215	51	288	48	138	32
Sr	1	5	2	3	1	2	10	14	3	7	1	3
Pb	1057	170	945	170	623	134	833	243	866	174	614	92
TSP ($\mu g\ m^{-3}$)	64.5	42.6	31.4	17.5	33.1	23.4	NA‖	NA	20.9	27.6	NA	NA

* Fine fraction.
† Coarse fraction.
‡ Number of filters analyzed.
§ Average value less than detection limit, individual values exceeded detection limit.
‖ Not available.

stances, the arithmetic average was well below the detection limit for the element, in which case the detection limit is cited. The variability of detection limits from site to site reflects the differences in the flow rates and duration of the sampling periods.

We observe from these limited studies that vanadium concentrations were highest in the East coast areas where oil is used as the principal power plant energy source. For example, the mean vanadium concentration in New York City in February 1977 was 207 ng m^{-3}, while in South Charleston, WV, the mean vanadium concentration was less than 2 ng m^{-3}.

Figure 6 consists of plots of the fine and coarse sulfate concentration as a function of time. For New York City, 2-h concentrations of sulfate as high as 45 $\mu g\ m^{-3}$ were observed. At the peak sulfate concentration, the sulfate represented about 50% of the mass of the fine particle fraction. Figure 7 shows plots of the fine and coarse particle sulfur concentrations for RAPS sites 106 and 122 for nearly 100 days during 1975. Station 106 is located in a botanical garden a few kilometers northwest of the center of St. Louis and about 0.4 km from a major freeway. Site 122 is 45 km north of St. Louis in a rural agricultural area. The ADS sampler data show that for 94% of all the days sampled in St. Louis, the percentage of sulfur in the fine fraction exceeded 70%. A similar pattern appears for sulfur in the data presented in Table 3. Further examination of the mean concentrations of elements with atomic numbers above 12 shown in Table 3 reveals that sulfur is the most abundant element in the fine particle fraction.

Because the environmental effects and health effects of atmospheric sulfur are dependent upon its chemical form, it is essential to determine which sulfur compounds are present. If one assumes that all of the sulfur is in the form of sulfate, then using the data of Table 3, one can estimate the maximum possible percent of the sulfate associated with the various compounds indicated in Table 4. For each metal, the maximum possible percent is computed assuming that the entire observed amount of the metal occurs as a metal-sulfate compound; these results are given in Table 4. A summation of the maximum possible percent values indicates that no more than 10 to 30% of the sulfate can occur in the form of the assumed metal-sulfate compounds. The actual amount of such compounds is probably much less than this upper limit since Ti, Fe and Zn are likely to occur as oxides and lead is likely to occur as either an oxide or a halide. It must be concluded, therefore, that only a small fraction of the fine particle sulfur is chemically bound to elements with atomic numbers above 13. This conclusion is consistent with the measurements that will be described in a later paragraph that indicate that the atmospheric sulfur (assumed to be sulfate) is primarily associated with the cations NH_4^+ and H^+.

Further measurements were made to determine the oxidation state of sulfur in atmospheric particulate matter. Such measurements were made using recently developed IC, which separates anions on an ion exchange column with sufficient resolution to distinguish between SO_4^{2-} and SO_3^{2-}. Samples from South Charleston, Research Triangle Park, Philadel-

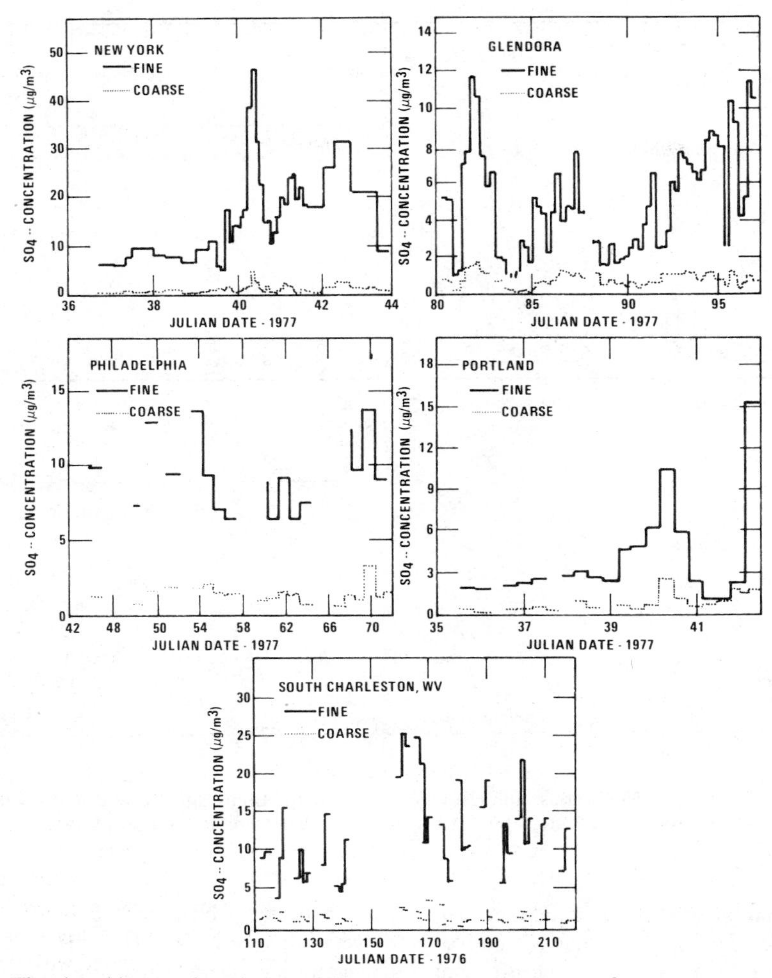

Fig. 6. The time history of equivalent sulfate concentrations in $\mu g\, m^{-3}$ as determined from EDXRF elemental analysis of dichotomous filter samples in two size fractions at five sampling sites in the US. The durations of the sampling periods are represented by the length of the horizontal lines, while the vertical lines connect consecutive sampling periods. The sampling periods were as follows: New York, 2 h; Phil., 24 h; South Charleston, 24 h; Glendora, 6 h; Portland, 4 h.

phia, and New York were analyzed, and in no case did the SO_3^{2-} exceed the minimum detectable level of 8 ng m^{-3}. In comparing the detection limit for sulfite with the mean sulfate concentration, one concludes that less than 0.1% of the sulfur in the filter extracts was in the form of sulfite. From the extraction efficiency studies, we determined that no more than 2% of the fine particle sulfur remained on the filter after extraction. If some of the nonextractable sulfur were a sulfite species, then sulfite could not account for more than 2% of the sulfur in the fine particle fraction.

Figure 8 shows a plot of water soluble sulfate measured by the IC versus total sulfur measured by EDXRF and plotted as sulfate. The data represent 18 randomly selected fine particle samples collected with manual dichotomous samples operated in South Charleston and Philadelphia during 1976 and early 1977. A linear regression analysis of the data reveals that:

$$SO_4^{2-}(IC) = 0.04\ \mu g\, m^{-3} + 0.98\ SO_4^{2-}(EDXRF)$$

and the correlation coefficient is 0.997. The ratio of the mean concentrations measured by IC and XRF is 0.988. This excellent agreement indicates that the measurements of water-soluble sulfate by IC and total sulfur by XRF are equivalent for the fine particle fraction.

A set of 18 randomly selected coarse particle samples was also analyzed for water-soluble sulfate by the IC and for total sulfur by EDXRF. A linear regression analysis reveals that

$$SO_4^{2-}(IC) = -0.30\ \mu g\, m^{-3} + 0.92\ SO_4^{2-}(EDXRF)$$

with correlation coefficient of 0.905. It is tempting to interpret the deviation of slope of the linear regression from unity as characteristic of the chemical species of sulfur found in the course particle mode. However, before such an interpretation can be made, possible sources of systematic error in the two techniques at low concentrations must be investigated. The coarse particle sulfate concentration in the present data are all below 2 $\mu g\, m^{-3}$ whereas the fine

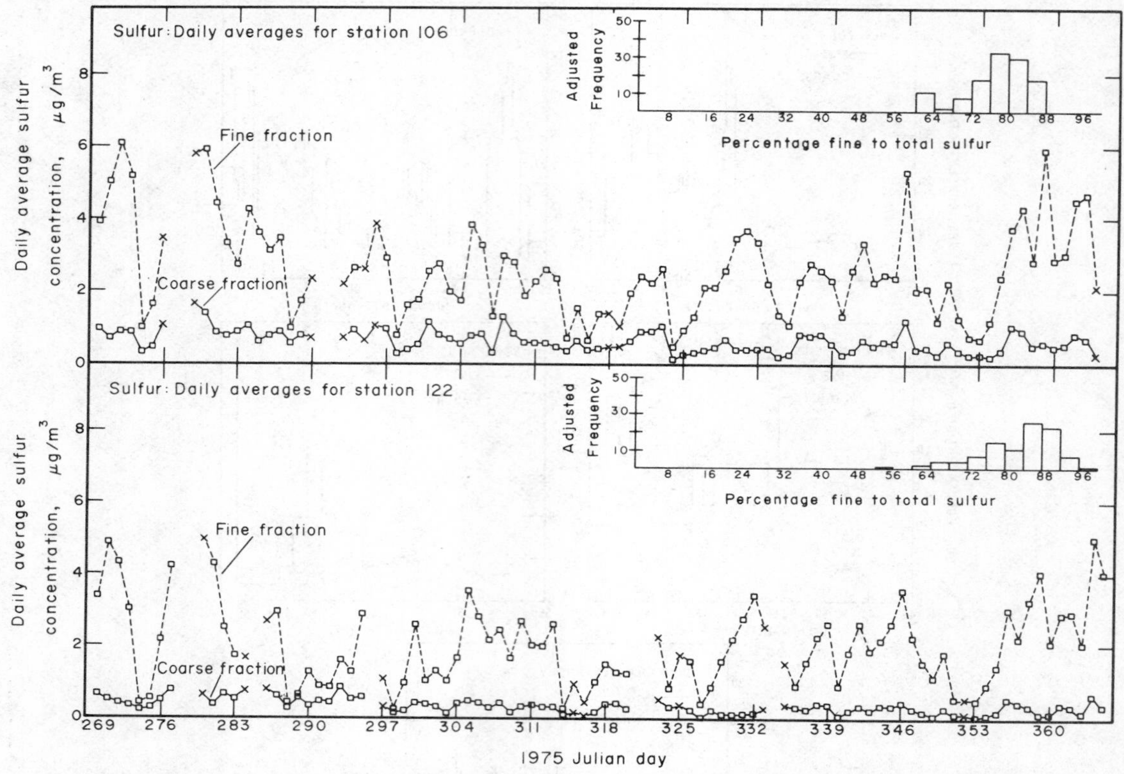

Fig. 7. Fine and coarse dichotomous sulfur fractions at two St. Louis sites. Frequency distribution of the ratio of fine particle sulfur to the fine plus coarse sulfur are also plotted.

particle concentrations were all above that value. The extraction procedures developed for the IC analysis were optimized for the sulfate concentrations expected for the fine particles. The appropriateness of these methods for the concentration ranges found in the coarse particle mode is currently being investigated. The relatively greater variability of the coarse particle data (correlation coefficient = 0.905) as compared to that of the fine particle data (correlation coefficient = 0.997) is indicative of the possibility that the combination of the sulfate with cations other than H^+ or NH_4^+ in the coarse particle mode may make a simple interpretation difficult. Work in progress to investigate these problems will be reported elsewhere.

Measurements of NH_4^+, H^+, SO_4, *and* NO_3^{2-}

Determinations of the H^+, NH_4^+, SO_4^{2-}, and NO_3^- concentrations were made using the Gran titration, ion selective electrode, thorin titration and IC. Such analyses were essential to characterize the sample, since the EDXRF analysis revealed that the metal cations were of minor importance.

Very few samples collected in our field studies over the past 18 months with the dichotomous sampler contained significant amounts of strong acid. In almost all instances the relationship between the NH_4^+ and SO_4^{2-} concentrations indicated that sulfate was in the form of ammonium sulfate. Only one aerosol sample out of over 120 collected in South Charles-

Table 4. Maximum possible percentages that various metal–sulfate compounds could contribute in the fine particle fraction. The percent sulfate values are computed from the data in Table 3 assuming that 100% of each metal is in the form of a sulfate compound

	Maximum possible percent of total sulfate					
Compound	New York	Phila.	Charleston	St. Louis	Portland	Glendora
K_2SO_4	2.0	1.8	1.1	3.0	2.0	2.5
$CaSO_4$	5.2	4.3	2.0	3.5	3.8	4.9
$Ti_2(SO_4)_3$	0.9	0.7	0.3	1.8	1.3	0.6
VSO_4	2.3	1.2	0.1	0.6	1.3	0.2
$Fe_2(SO_4)_3$	5.9	5.9	3.7	10.5	11.5	6.6
$ZnSO_4$	3.3	2.3	0.4	2.6	2.2	1.4
$PbSO_4$	3.0	4.7	2.6	4.6	9.6	5.9
Total	22.6	20.9	10.2	26.6	31.7	22.1

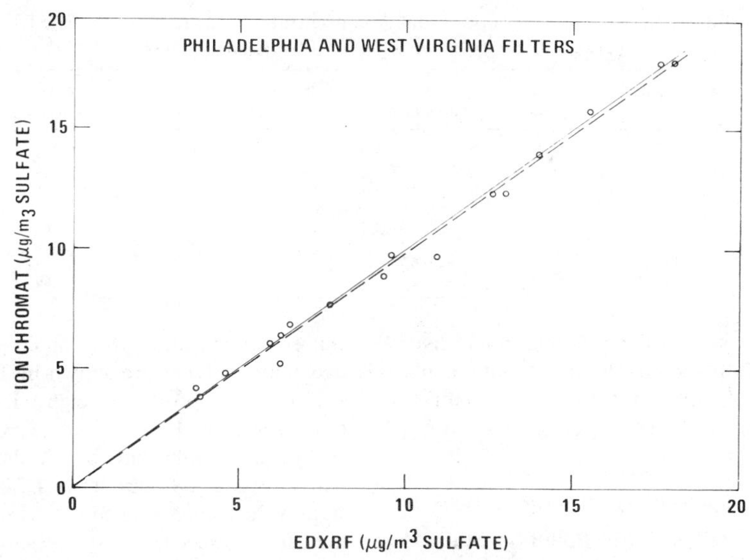

Fig. 8. A comparison of IC and EDXRF sulfate analyses for fine particles in 18 randomly selected samples.

ton contained detectable strong acid, and it only amounted to a few nanoequivalents. Aerosol samples collected during the winter of 1977 in Philadelphia and New York contained no detectable acidity. For these samples, no special effort was made to prevent neutralization of particles either during collection or during storage prior to analysis.

During July of 1977, aerosols were collected at Research Triangle Park with an automated dichotomous sampler designed to preserve the acidity of the samples. This ADS was equipped with a diffusion denuder to remove ammonia at the sampling inlet. After each sample was collected, the filter was immediately transferred within the ADS to a storage chamber that was continuously purged with nitrogen. When the filters were removed from the ADS, they were immediately transferred to a hermetically sealed ammonia-free chamber, where they remained until they were analyzed. The fine particle aerosol fraction was analyzed for H^+, NH_4^+ SO_4^{2-} and NO_3^-. The results, which are plotted in Fig. 9, reveal a significant amount of acidity. We attribute this detection of acidity to the precautions taken to preserve the integrity of the sample. Within the experimental error of the

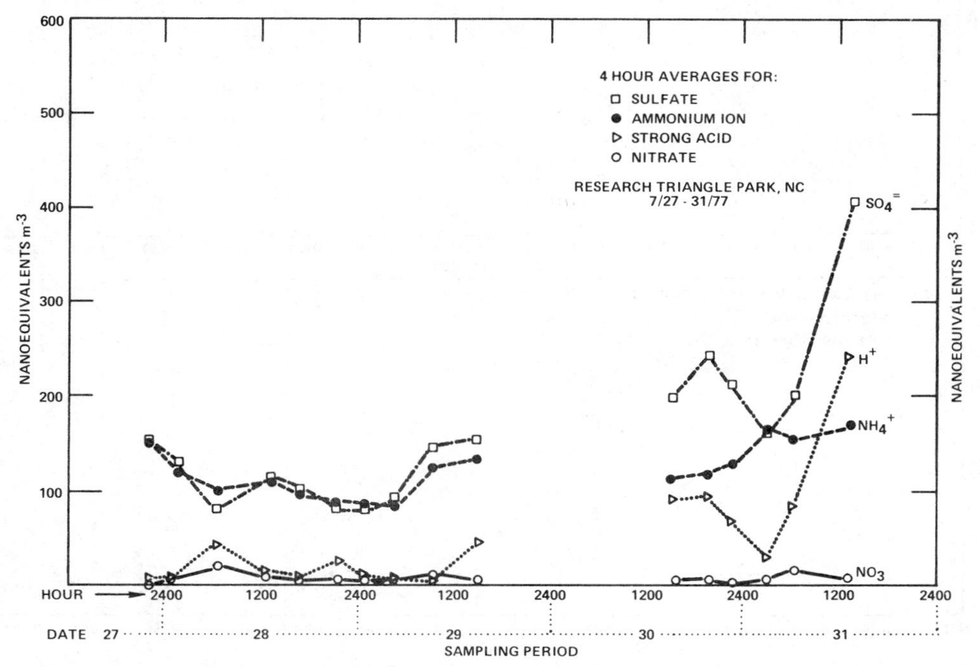

Fig. 9. Time series plot of the H^+, SO_4^{2-}, NO_3^- and NH_4^+ concentrations in fine particles collected with an ADS fitted with an NH_3 gas diffusion denuder and a sample preservation chamber.

Table 5. Summary of St. Louis sulfate concentrations ($\mu g\ m^{-3}$) determined by two methods for period September through December 1975

RAMS station number	Avg Hi-vol	Avg total dichotomous	Ratio Hi-vol total dichotomous	% Days fine S/total S >0.70
106	11.0	8.9	1.23	97
108	12.7	10.0	1.27	79
112	10.6	9.2	1.23	95
115	9.6	7.8	1.23	99
118	9.9	6.6	1.50	83
120	9.9	7.3	1.35	100
122	8.9	6.2	1.44	94
124	9.8	6.4	1.53	99

measurements, the sum of the ammonium and hydrogen ion concentrations equals the sulfate concentration. In these samples, the nitrate concentration is a small fraction of the sulfate concentration.

Hi-volume and dichotomous sampler comparisons

As part of the Regional Air Pollution Study, Hi-volume samplers were operated at the Regional Air Monitoring Stations (RAMS) in St. Louis (Myers, 1975). At 10 RAMS stations, ADS were also operated. In the RAMS network, Hi-vols were operated for a 24-h period once every 3 days and the ADS were operated continuously. The fine and coarse fractions of aerosols collected with the ADS were analyzed for mass and elemental composition by beta gauge and EDXRF (Jaklevic, 1977) while aerosols collected with the Hi-vol were analyzed for mass and sulfate content (methyl-thymol blue method for sulfate). The average concentrations and ratios of concentrations for the two sampler types are shown in Table 5.

For a composite of eight RAMS stations the time-averaged Hi-vol sulfate concentration was 10.3 $\mu g\ m^{-3}$ while that for the dichotomous sampler (fine plus coarse fractions) was 7.8 $\mu g\ m^{-3}$. For every station the time-averaged concentration measured by the Hi-vol exceeded that measured by the dichotomous sampler. The discrepancy between the two methods ranged from 23 to 53%. No firm explanation of this preliminary result is offered. It is suggestive, however, of either aerosol sampling efficiency variations or possible sulfate artifact formation on the glass fiber filters of the Hi-vol. Figure 10 shows corresponding time series plots for total mass collected by ADS and Hi-vol at sites 106 and 122. The time-average value for the ratio of Hi-vol mass to total ADS mass at station 106 was 1.8. The corresponding ratio at station 122 was 1.2.

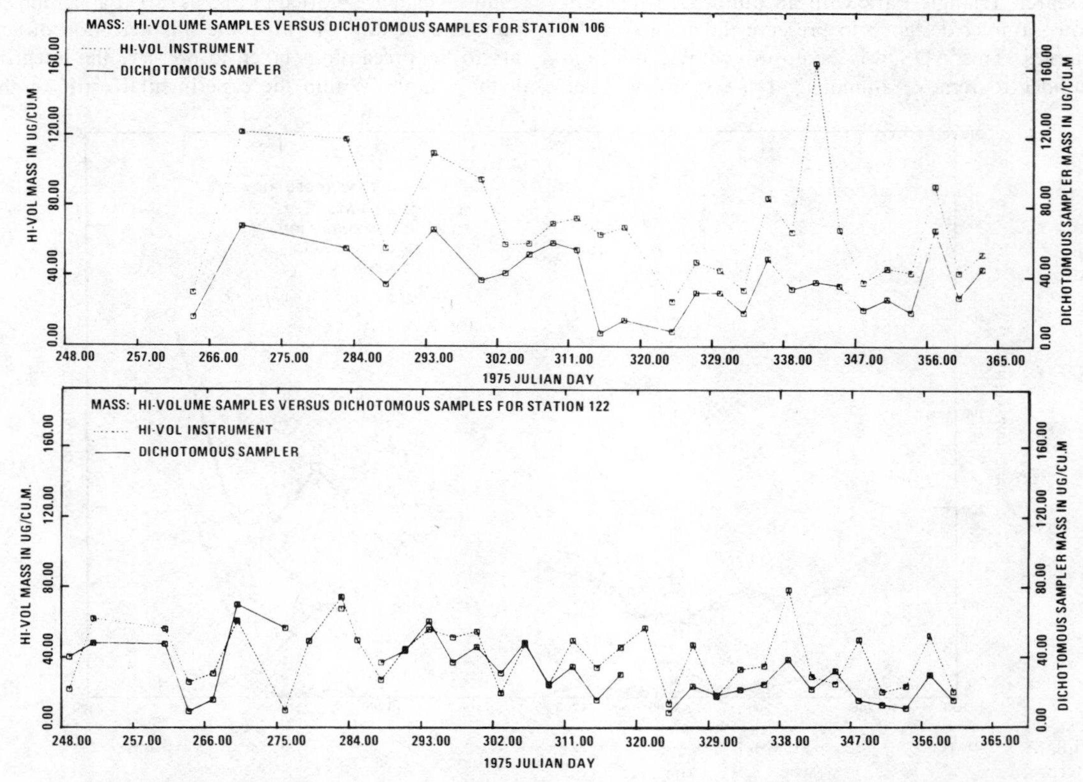

Fig. 10. Comparison of total mass concentrations measured by hi-vol and dichotomous sampler at two St. Louis sites.

In the RAMS stations, Hi-vols were operated from midnight to midnight and then automatically turned off. The filters were generally removed either the next day or in some cases 2 or 3 days later. Chahal (1976) has shown that a Hi-vol operated in this manner tended to pick up mass due to dust being blown into the sampler while it was idle and before the filter was replaced. This may partly explain why the Hi-vols operated in the RAMS network consistently showed higher mass loadings than the ADS.

CONCLUSIONS

The data presented represent the culmination of several years of intensive research to develop procedures and instrumentation to characterize the chemical form of sulfur in the atmosphere. These data definitively determine the chemical form of sulfur and serve to define several areas where further research on aerosols is needed.

Analysis by EDXRF and IC of aerosol samples collected with dichotomous samplers at six locations across the United States revealed that sulfur in particles $<3.5\ \mu m$ in aerodynamic diameter was in the form of sulfate. Analysis of randomly selected fine particle samples shows the ratio of the sulfate measured by IC and sulfur by EDXRF was 2.96 ± 0.15, which is in excellent agreement with the 3.00 value for pure sulfate. Other randomly selected samples analyzed by the IC showed that the sulfite levels were below the minimum detectable level ($8\ ng\ m^{-3}$). Based on these studies, EDXRF methods can now be considered a reliable analytical procedure to measure sulfate (as sulfur) in ambient aerosols.

Strong acid analysis of the fine particle fraction collected at a number of urban and rural areas in the United States consistently resulted in measurements below the minimum detectable level ($0.5\ ng\ m^{-3}$). This is likely due to neutralization of the acid on the filter either during sampling and/or subsequent handling. This conclusion is borne out by the immediately successful analysis for strong acid when the NH_3 denuder was used in conjunction with the modified ADS. It is important therefore when attempting to measure strong acid that precautions be taken to prevent neutralization.

This study shows that classical sampling methods are subject to a number of problems leading to erroneous measurements. For example, artifact formation caused by acid gas reaction with the filter and improper inlet design leading to collection of wind blown material may have caused misleading mass determinations. The dichotomous sampler with its inert filter medium and size fractionation has apparently overcome these difficulties.

Acknowledgements—The authors are indebted to Dr. William J. Courtney for his invaluable contribution in interpreting and formating the data from EDXRF measurements. The authors acknowledge the work of Dennis Reutter who fabricated the NH_3 denuder and was responsible for the collection and analysis of acid aerosols. We are indebted to the Northrop Services staff–particularly Carolyn Owen who assisted in data logging and reduction, and Edward L. Tew and Charles M. Kistler for providing the acidity, ammonium, and IC data cited in this report. In addition, we thank the West Virginia Air Pollution Control Board, William Ballanger, Ronald Patterson, A. P. Waggoner, and Robert Burton, for assisting in the operation of dichotomous samplers from which a majority of the data in this report were derived. Mass and elemental data for samplers collected with ADS in St. Louis were provided by the Lawrence Berkeley Laboratory.

REFERENCES

Arinc F., Gardner R. P., Wielopolski L. and Stiles A. R. (1976) Applications of the least-squares method to the analysis of XRF spectral intensities from atmospheric particulates collected on filters. *Advan. X-ray Analysis* **19,** 367–380.

Brosset C., Andreasson K. and Ferm M. (1975) The nature and possible origin of acid particles observed at the Swedish west coast. *Atmospheric Environment* **9,** 631–642.

Brosset C. and Ferm M. (1976) Man-made airborne acidity and its determination. IVL Publications B-34B. Gothenburg, Sweden.

Chahal H. S. and Romano D. J. (1976) High volume sampling: effect of windborne particulate matter deposited during idle periods. *J. Air Pollut. Control Ass.* **26,** 885–886.

Conner W. D. (1966) An inertial-type particle separator for collecting large samples. *J. Air Pollut. Control Soc.* **16,** 35–38.

Currie L. A. (1977) Detection and quantitation in X-ray fluorescence spectrometry. *X-ray Fluorescence Analysis of Environmental Samples.* pp. 289–306. Ann Arbor, Ann Arbor, Michigan.

Dzubay T. G. and Nelson R. O. (1975) Self absorption corrections for X-ray fluorescence analysis of aerosols. *Advan. X-ray Analysis* **18,** 619–631.

Dzubay T. G. and Stevens R. K. (1975) Ambient air analysis with dichotomous sampler and X-ray fluorescence spectrometer. *Environ. Sci. Technol.* **9,** 663–668.

Dzubay T. G., Stevens R. K. and Petersen C. M. (1977) Application of the dichotomous sampler to the characterization of ambient aerosols. *X-Ray Fluorescence Analysis of Environmental Samples.* pp. 95–105. Ann Arbor, Ann Arbor, Michigan.

Giauque R. D., Goulding F. S., Jaklevic J. M. and Pehl R. H. (1973) Trace element determination with semi-conductor X-ray spectrometers. *Analyt. Chem.* **45,** 671–681.

Goulding F. S., Jaklevic J. M. and Loo B. W. (1975) Fabrication of monitoring system for determining mass and composition of aerosols as a function of time. Environmental Protection Agency, Report No. EPA-650/2-75-048.

Hodgeson J. A., Bell J. P., Rehme K. A., Krost K. J. and Stevens R. K. (1971) Application of a Chemiluminescence Detector for the Measurement of Total Oxides of Nitrogen and Ammonia in the Atmosphere. Joint Conference on Sensing of Environmental Pollutants, Paper Number 71–1067, AIAA, New York, p. 197.

Jaklevic J. M., Goulding F. S., Jarrett B. V. and Meng J. D. (1974) Application of X-ray fluorescence techniques to measure elemental composition of particles in the atmosphere. *Analytical Methods Applied to Air Pollution Measurements* (Edited by R. K. Stevens.). pp. 123–146. Ann Arbor, Ann Arbor, Michigan.

Jaklevic J. M., Landis D. A. and Goulding F. S. (1976) Energy dispersive X-ray fluorescence spectrometry using pulsed X-ray excitation. *Advan. X-ray Analysis* **19,** 253–265.

Jaklevic J. M., Loo B. W. and Goulding F. S. (1977) Photon induced X-ray fluorescence analysis using energy dispersive detector and dichotomous sampler. *X-Ray Fluorescence Analysis of Environmental Samples.* pp. 3–18. Ann Arbor, Ann Arbor, Michigan.

Liu B. Y. H. and Kuhlmey G. A. (1977) Efficiencies of air sampling media. *X-Ray Fluorescence Analysis of Environmental Samples.* pp. 107–119. Ann Arbor, Ann Arbor, Michigan.

Loo B. W., Jaklevic J. M. and Goulding F. S. (1976) Dichotomous virtual impactors for large scale monitoring of airborne particulate matter. Fine Particles (Edited by B. Y. H. Liu), pp. 311–350. Academic Press, New York.

Loo B. W., Gatti R. C., Liu B. Y. H., Kim C. J. and Dzubay T. G. (1977) Absorption corrections for submicron sulfur collected in filters. *X-Ray Fluorescence Analysis of Environmental Samples.* pp. 187–202. Ann Arbor, Ann Arbor, Michigan.

Mulik J., Puckett R., Willians D. and Sawicki E. (1976) Ion chromatographic analysis of sulfate and nitrate in ambient aerosols. *Analyt. Lett.* **9,** 653–663.

Myers R. L. and Reagan J. A. (1975) The regional air monitoring system. St. Louis, MO, Proc. International Conference on Environmental Sensing and Assessment, Las Vegas, NV., 14–19 Sept.

Persson G. A. (1966) Automatic colorimetric determination of low concentrations of sulfate for sulfur dioxide in ambient air. *Air Wat. Pollut. Int. J.* **10,** 845–852.

Rickel D. G. and Dzubay T. G. Fitting of thin sample X-ray fluorescence spectra by a least-squares method. (To be published.)

Small J., Stevens T. S. and Bauman W. C. (1975) Novel ion exchange chromatographic method using conductimetric detection. *Analyt. Chem.* **47,** 1801.

Stevens R. K. and Dzubay T. G. (1975) Recent development in air particulate monitoring. *IEEE Trans. Nucl. Sci.* **NS-22,** 849–855.

Weiss R. E., Waggoner A. P., Charlson R. J. and Ahlquist N. C. (1977) Sulfate aerosol: geographic extent. *Science* **195,** 979–981.

Atmospheric Environment Vol. 12, pp. 69 82. Pergamon Press 1978. Printed in Great Britain.

A QUANTITATIVE METHOD FOR THE DETECTION OF INDIVIDUAL SUBMICROMETER SIZE SULFATE PARTICLES

YAACOV MAMANE* and ROSA G. DE PENA
Department of Meteorology, The Pennsylvania State University, University Park, PA 16802, U.S.A.

(*First received* 13 *June* 1977 *and in final form* 25 *July* 1977)

Abstract—A quantitative method for the analysis of individual submicrometer size sulfate particles has been developed. It is based on the reaction of the sulfate ion with barium chloride and can be applied only to soluble sulfates. The method is specific for sulfates. Carbonates, sulfites and nitrates can be distinguished from sulfates. The size of the halo depends on the size of the particle, the thickness of the barium chloride film and the relative humidity at which the reaction takes place. By fixing the values of the last two parameters, a single correlation can be established between the halo and the particle size. The method requires sampling time in the order of minutes for clean air to seconds for very polluted areas. An example of the application of the method to atmospheric aerosol is shown.

1. INTRODUCTION

Epidemiological studies conducted by the Environmental Protection Agency (EPA) suggested "that adverse human health effects are consistently associated with exposure to suspended sulfates, more so than sulfur dioxide or total suspended particulate matter" (Shy and Finklea, 1973). The former studies indicated that health effects were associated with sulfate concentrations "as low as 6 to 10 $\mu g\,m^{-3}$ (24 h average)" (EPA Report, 1975) although some doubt has subsequently been raised on these conclusions (U.S. House of Representatives Committee Report). Recorded sulfate concentration in a large portion of the Eastern United States and major urban locations in the West are significantly higher than these levels (Altshuller, 1976). However, most of these studies are based on 24 h total sulfate measurement which are imperfect indicators of the presence of respirable toxic sulfate components. In addition, even if a 24 h average of 6–10 $g\,m^{-3}$ is not harmful by itself; an adverse health effect may be caused by a short period peak concentration.

Direct chemical analysis of atmospheric aerosols indicated that sulfates, especially ammonium sulfates, are the most important soluble component of submicrometer size particles in the continental aerosols (Junge, 1963). These results have been obtained all over the U.S. and also in Europe and Japan, over a period of some 25 years (Montefinale *et al.*, 1971; Wilson, 1976).

Atmospheric aerosol has been observed to present a bimodal distribution (Whitby *et al.*, 1972; Wilson, 1976), with one mode roughly within 0.1–1.0 μm and a second one within 1.0–10 μm. Sulfate particles are contained mostly in the first mode.

In addition to the many available techniques for the analysis of a bulk sample, several methods are oriented towards the analysis of individual particles. Such methods are mostly an extension of techniques applied to a bulk sample; however, they are entirely different in the details of their procedure.

There are two general types of single particle techniques: (1) those which investigate the sample without causing any change to the particles, and, (2) those which change the physico-chemical nature of the particle. The former include the use of the optical microscope to determine the shape, color, morphology, refractive index and size distribution of collected particles (McCrone *et al.*, 1967); the use of the transmission electron microscope (TEM) to obtain the size, shape, and diffraction patterns of individual particles; and the application of a scanning electron microscope equipped with an X-ray analyzer to study the elemental composition of the particles.

Among the single particle techniques which change the nature of the particles, we can mention: exposure to water vapor to discriminate between hygroscopic and non-hygroscopic particles (Twomey, 1954), applying specific chemical tests to a single particle (Chamot and Mason, 1940), or treating the entire sample with reagents with the purpose of identifying micrometer-size particles containing a particular ion (Lodge, 1962; Anyz, 1966; Ackerman and Lodge, 1973; Rinehart, 1971). The product of the reaction between the reagent and the ion is a reaction spot, or halo, which size is dependent on the size of the particles. The reaction spot methods have been applied mostly to particles which could be observed in the optical microscope (Seely, 1952; Pidgeon, 1954; Anyz, 1966).

Submicrometer size particles, on the other hand, require observation under the transmission electron

* Present affiliation: Research Institute for Environmental Health Nuisances, Medical School, Tel-Aviv University, Ramat Aviv, Israel.

microscope (TEM). This poses some problems in the TEM: the sample is exposed to a high intensity electron beam, to high vacuum, and to contamination by hydrocarbon vapors.

This is why the use of physical and chemical changes to identify submicrometer-size particles is more complex and more difficult to apply than it is for larger particles. Several methods based on the volatility and hygroscopicity of the particles were reported (see review by Ackerman and Lodge, 1973). However, none of them is specific. For submicrometer size particles Bigg *et al.* (1974) and later Ayers (1976) proposed to modify the method of spot reaction by applying a thin film of reagent on the surface of collection, or on the sample, in a vacuum evaporator. A reaction spot is formed when the sample is exposed to an appropriate relative humidity.

The analysis of individual submicrometer size particles is far from being widely used and certainly it has not been applied in a quantitative way. The aim of this study has been the further development of Bigg's method in order to insure its reproducibility, its specifity, and its quantitative character.

2. EXPERIMENTAL

2.1. *Description of the method*

The procedure developed for identifying submicrometer size sulfate particles is basically a modified spot test. When a sulfate particle is brought into contact with a thin layer of barium chloride and the relative humidity is adequate, a reaction occurs. The characteristic spot or halo formed is composed of insoluble barium sulfate and is recognizable under the electron microscope or, when large, under the optical microscope. Figure 1a is a photomicrograph of the halo obtained when the method was applied to a sulfuric acid droplet. It can be seen that the halo is almost a perfect circle. The halo departs from the circular shape in Fig. 1b obtained with ammonium sulfate particles.

It is desirable to collect the particles directly on carbon covered electron microscope screens. These screens can be placed in most of the jet impactors, cyclones, thermal and electrostatic precipitators. In this study, two or more screens were fixed exactly below the jet nozzle of the third and fourth stage of a Casella cascade impactor (May, 1945). Since the screens are very thin, no significant influence of their presence on the collection efficiency of the particles is expected. By using relocation screens with special marks, it is possible to examine a certain field of view over and over again.

The method requires the particles to be covered with a film of barium chloride. This is done by placing a mass, M, of dry barium chloride, typically a few milligrams, in a tungsten wire boat inside a vacuum evaporator at a distance r from the carbon coated screens. A current of few amperes through the tungsten boat evaporates the barium chloride to form a uniform layer of thickness t. Assuming that the material is spherically distributed around the tungsten boat, and that the boat behaves as a point source, a simple equation relating M, t and r is obtained (Preuss, 1959):

$$t = [(M \sin \theta)/(4\pi r^2 \rho)] \quad (1)$$

where θ is the angle between the screen surface and the tungsten boat, and ρ is the density of the barium chloride.

It should be noted that this is only an 'equivalent thickness' because the barium chloride may not be distributed in a spherical way (distribution would probably be affected by the vacuum evaporator geometry), and the tungsten boat is not strictly a point source. For these reasons the 'equivalent thickness' may be different for each vacuum evaporator. Therefore, the values of M, θ, and r in equation (1) have to be determined to obtain the film of the right thickness in each case.

For the size range considered in the present research, 0.1–1 μm, a film of 'equivalent thickness' of 0.03 ± 0.006 μm proved to be the most convenient. If the film is too thick it masks the reaction spot and it becomes 'non-transparent' to the electron beam. If it is too thin, there is not enough material to react with, and the reaction will not be complete.

The barium chloride coated screen is then exposed in a sealed chamber to a relative humidity (r.h.) of 70 to 75% for a period of one hour and examined under the microscope.

The right r.h. was obtained by placing in the chamber a solution of either $MgCl_2$ or NaCl. The r.h. can be modified by changing the concentration of this solution (Weast, 1974; Low, 1969; Wylie, 1963; Orr *et al.*, 1958). The following concentrations were used to obtain relative humidities of 60, 70, 75 and 80%: 3.9, 3.1, 2.8 and 2.4 M $MgCl_2$. A saturated solution of NaCl was used also for a r.h. of 75%.

The sulfate particles can then be identified by the characteristic halos, or reaction spots. Several photomicrographs, normally of a few hundred particles, are taken and the sulfate halos are sized and counted. By comparing the number density of sulfate halos in the barium coated screen with the total number density of particles in a nontreated screen, the percentage of the sulfate particles in the sample is obtained.

Knowing the magnification factor: ratio between the diameter of the reaction spot and the diameter of the particle, the size distribution of the sulfate particles can be obtained.

2.2 *Determination of the ratio: halo to particle size*

The size of the halo obtained by this method depends on (1) the size of the original particle, (2) the thickness of the barium chloride layer, and (3) the exposure time to the right water vapor pressure. If the thickness and the exposure time are kept constant, then the size of the halo is only dependent on the particle size.

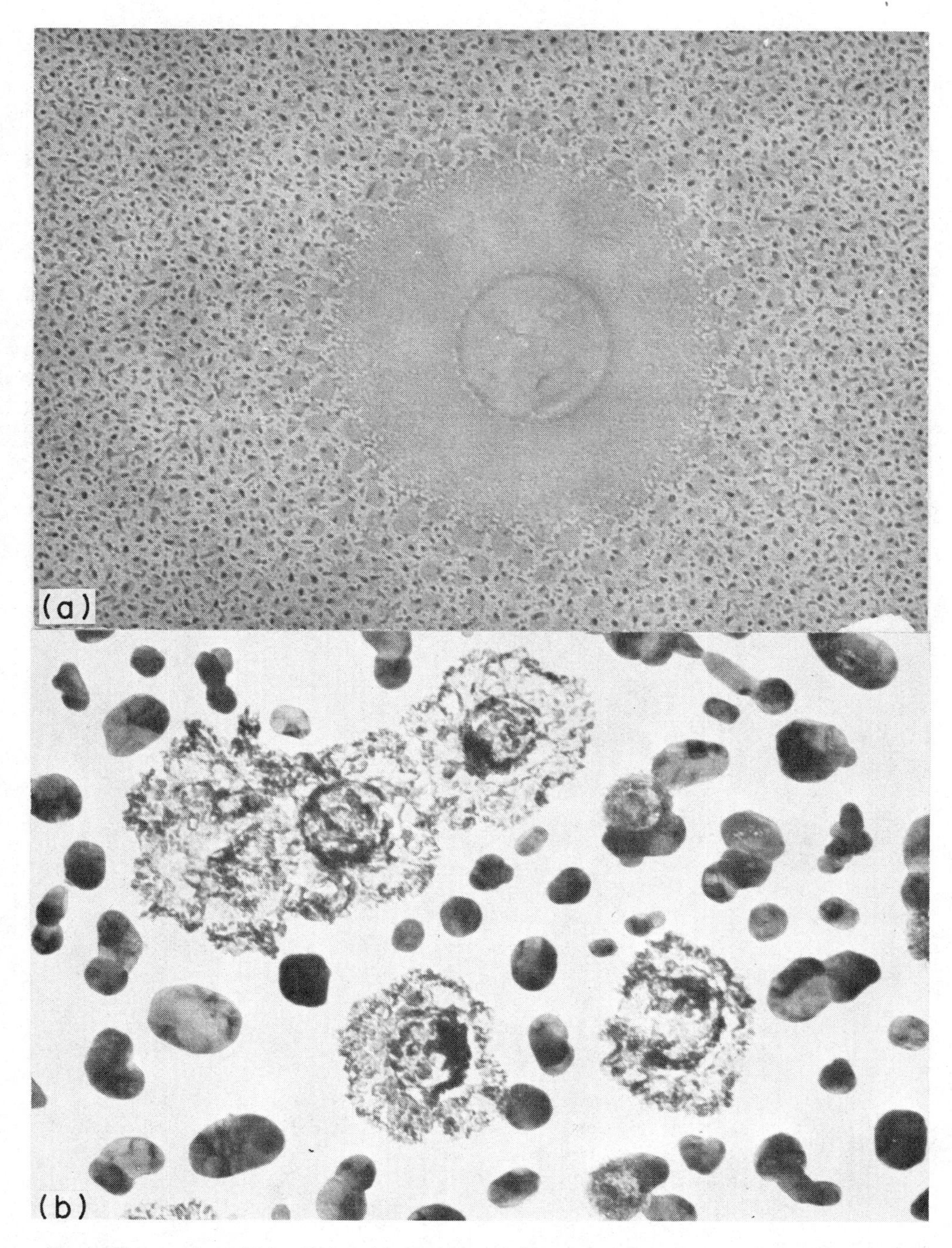

Fig. 1. The reaction of (a) sulfuric acid droplets and (b) ammonium sulfate particles with a barium chloride film.

2.2.1. *Barium chloride film thickness.* The barium chloride film obtained in the vacuum evaporator is in the form of grains whose size is determined by the thickness of the film: the smaller the thickness the smaller the grain. In the process of developing the sample, the barium chloride recrystallizes forming new grains whose size is determined by both the relative humidity used in the process of development and the thickness of the film. As there is a minimum relative humidity to use, below which there is no reaction, the grain size for a given film thickness is controlled by this minimum relative humidity. Figures 2a and 2c are photomicrographs of a film of 0.01 μm and 0.08 μm. The correspondent grain size is about 0.005 and 0.05 μm respectively. The film of 0.01 μm thickness, however, is not sufficient for particles larger than 0.1 μm, as can be seen in Fig. 2b where an ammonium sulfate particle of 0.15 μm diameter did not react completely. Photomicrographs of Figs 2b and 2d were obtained after the sample shown in Figs 2a and 2c

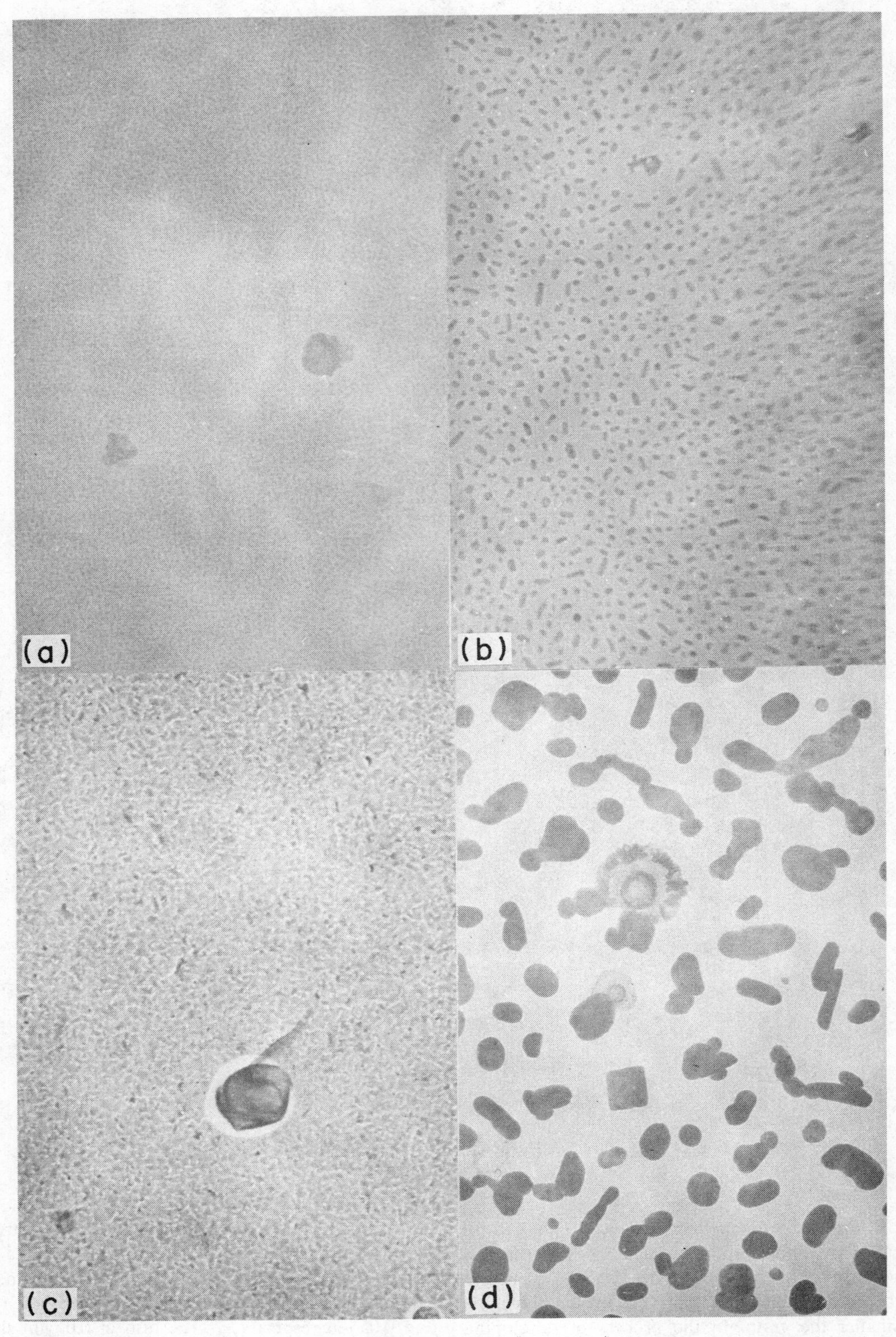

Fig. 2. Photomicrographs of barium chloride films of 100Å and 800Å thickness, before and after 1 h exposure to r.h. of 75%: (a) 100Å before, (b) 100Å after exposure to r.h., (c) 800Å before, and, (d) 800Å after exposure to r.h.

were submitted to the developing process in a relative humidity of 75%. At this relative humidity the final grain size is 0.02, 0.1 and 0.3 μm for film thicknesses of 0.01, 0.03 and 0.08 μm respectively.

For the size range of particles considered in this study, 0.1 to 1.0 μm, a film of 0.03 μm is adequate, since this film with a recrystallized grain size of 0.1 μm allows the detection of particles larger than about $1.0/1.6 \cong 0.06\ \mu$m where 1.6 is the magnification factor as will be shown later.

2.2.2. *The effect of the relative humidity.* In general, the relative humidity necessary for the reaction between a sulfate particle and the barium chloride was lower than that of the respective saturated solution. For example, for ammonium sulfate the equilibrium relative humidity is about 80% (Low, 1969), but for 0.1–1 μm particles the reaction did occur at a relative humidity of 65%. Below this value no halo was observed. Above it the halo size increased proportionally to the relative humidity.

In the process of development, recrystallization of the barium chloride takes place resulting in a film of larger grains. The size of these grains also increases with the relative humidity as can be seen in Figs 3a

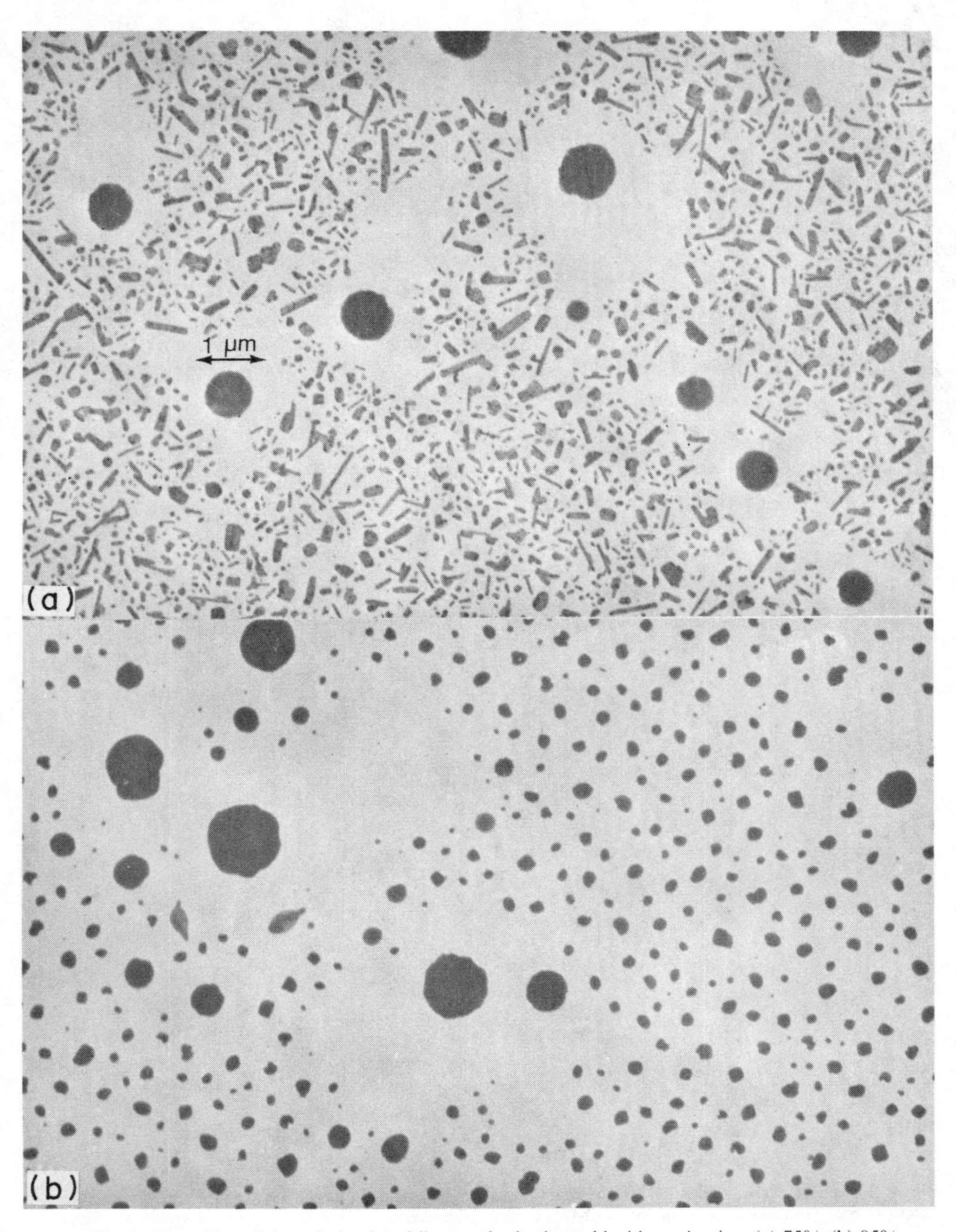

Fig. 3. The effect of the relative humidity on the barium chloride grain size: (a) 75%, (b) 85%.

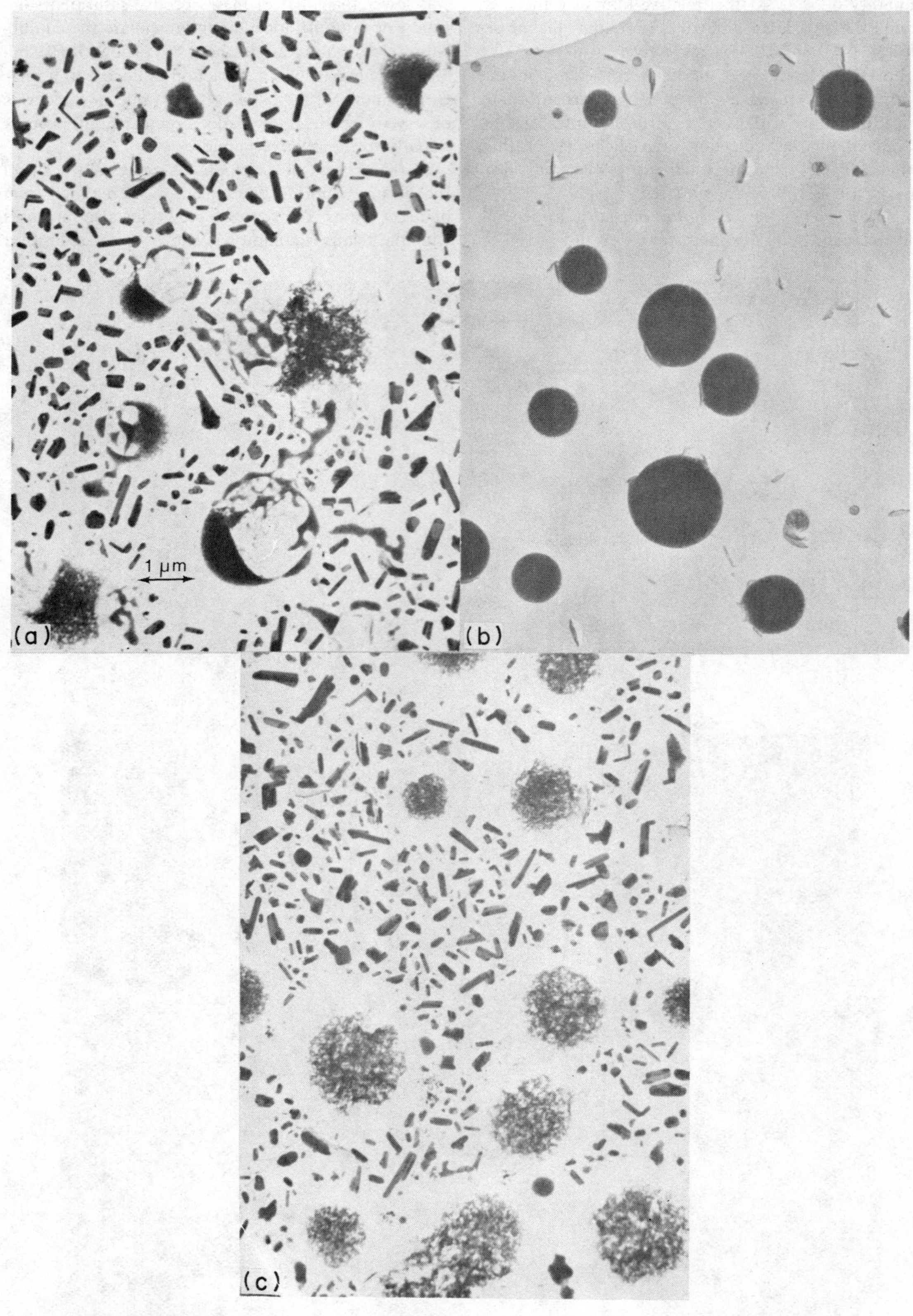

Fig. 4. The effect of 'pre-exposure' on the reaction of sulfates with barium chloride film: (a) incomplete reaction spots obtained with the previously exposed particles seen in (b), (c) reaction spots obtained with particles not previously exposed.

and 3b obtained with a film of 0.03 μm in thickness and submitted to 75 and 80% respectively. High relative humidities and thick films both contribute to the increase of barium chloride grains, which interfere in a quantitative analysis of sulfate particles of the same size range.

To obtain the halo to particle ratio, sulfate particles were generated by spraying dilute solutions of H_2SO_4 and $(NH_4)_2SO_4$ and collected on electron microscope relocation screens. Before the barium chloride treatment, the screens were observed and photographed in the TEM with the lowest intensity beam and the

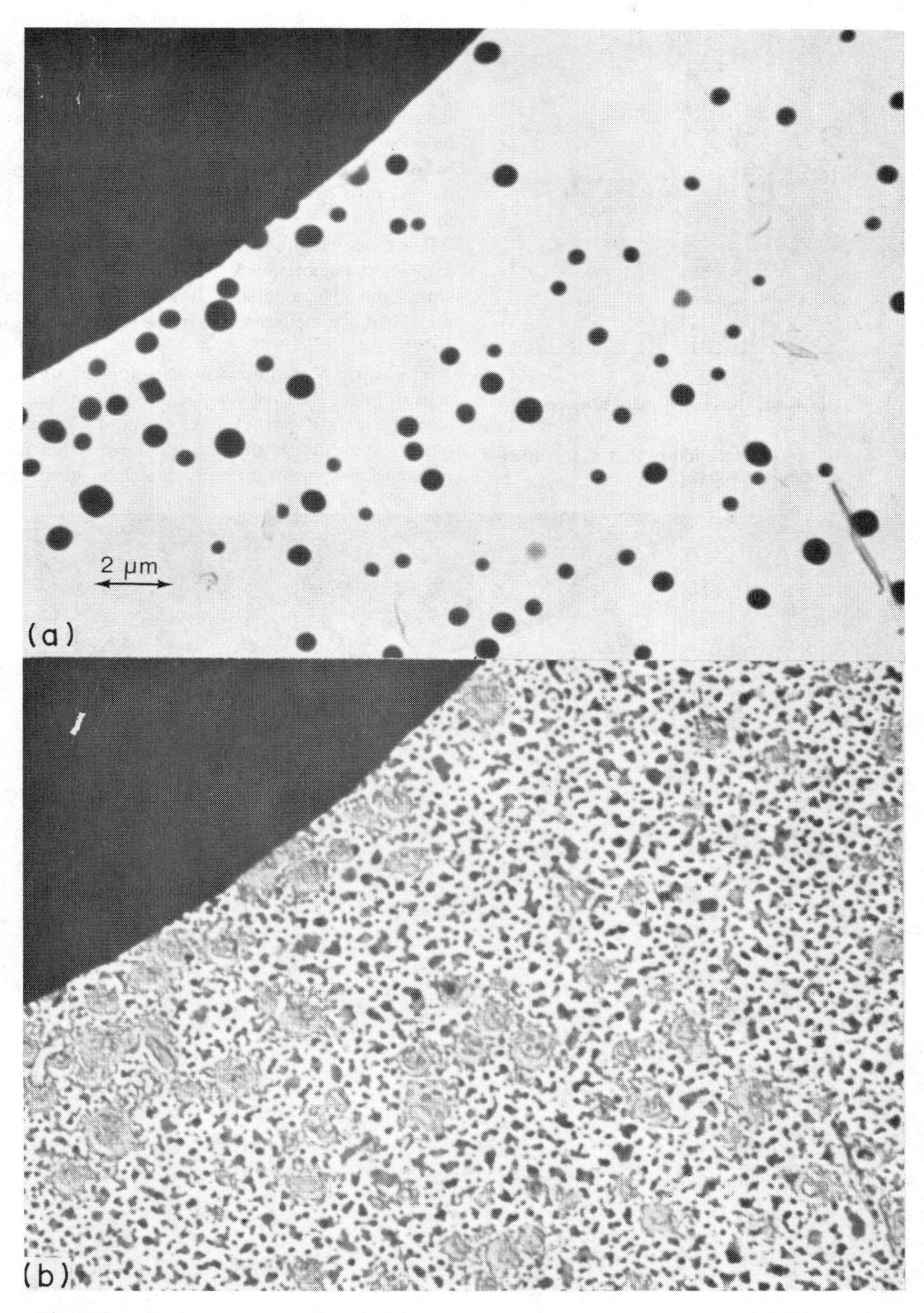

Fig. 5. Sample of ammonium sulfate particles on a 'relocation screen' (a) before and (b) after barium chloride treatment.

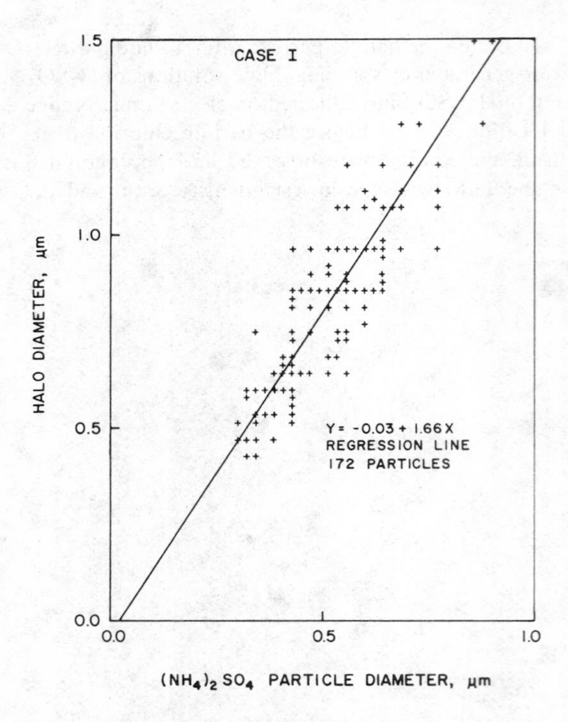

Fig. 6. Halo diameter as a function of that of the original ammonium sulfate particle.

shortest time possible, to prevent any damage to the sample. Usually the focusing was done on a certain field of view, and then moved to another area where the photomicrographs were taken. In this way the field studied was less exposed to the TEM beam. Sulfate particles exposed to a high intensity beam reacted incompletely with the barium chloride film.

Figures 5a and b correspond respectively to the same sample before and after the barium chloride treatment. When Fig. 5b is placed over Fig. 5a in such a way that the grid's marks in the two photomicrographs coincide, it is possible to see that the halos superimpose exactly on the particles, and the ratio of both sizes can be determined one by one. An example of a determination of this type is given in Fig. 6.

The complete procedure was repeated for two more samples. Comparison of the three runs shows that, within the 95% confidence limits, there is practically no difference between the three curves (Mamane, 1977).

The same procedure was also applied to sulfuric acid droplets. The results can be seen in Figs 7 and 8. Sulfuric acid particles were found to evaporate more rapidly under the electron beam, allowing for a much lower beam intensity, which in turn affected

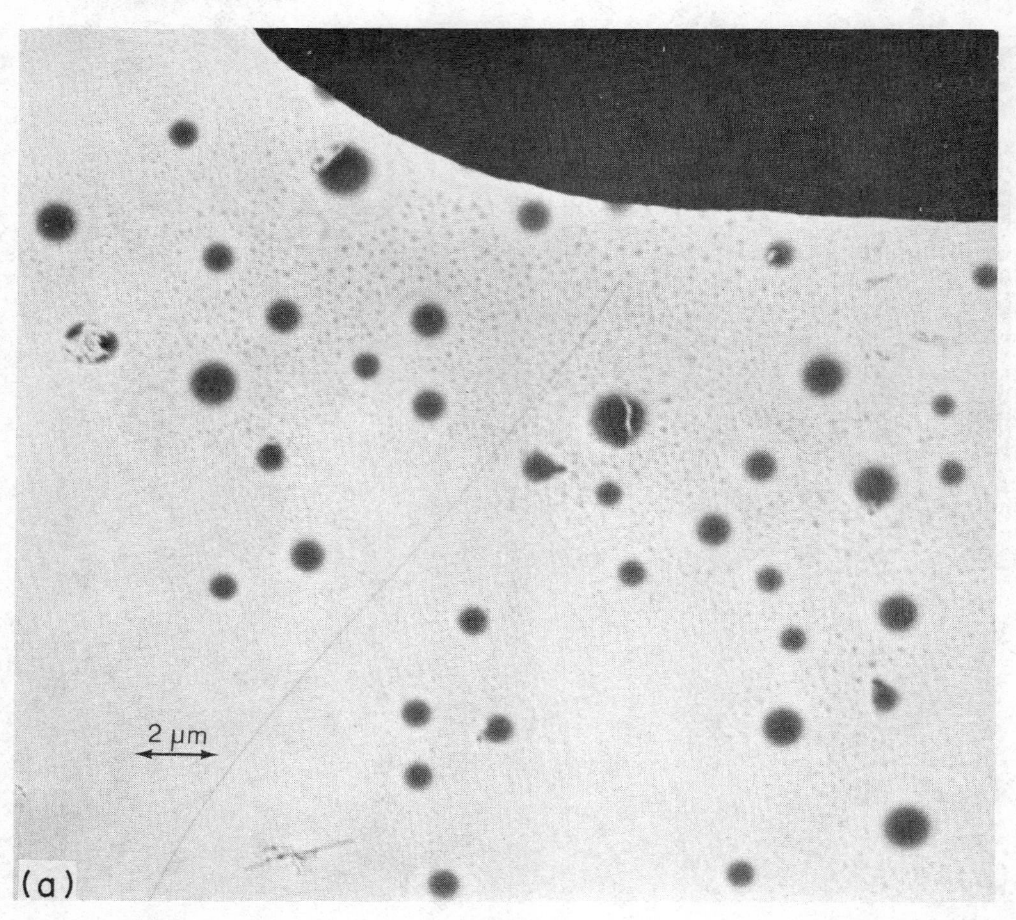

Fig. 7(a).

the quality of the pictures. Figures 7a and b correspond to a sample of sulfuric acid particles before and after treatment, and Fig. 8 is the calibration curve.

The ratio halo to particle diameter for sulfuric acid particles ranges from 2.7 for 1.0 μm, to 3.6 for 0.1 μm particles, which is about twice the ratio for ammonium sulfate. We can see at least two reasons for that difference: (1) Upon impaction, sulfuric acid particles form a central droplet surrounded by many smaller satellites, occupying a larger surface than if the same mass were contained in one single droplet, and (2) in the developing process the sulfuric acid is completely dissolved forming a droplet whose size is determined by the vapor pressure in the chamber, while, as it was already suggested the ammonium sulfate may not be completely dissolved, forming a 'droplet' which is smaller than the equilibrium size droplet.

2.3. *Interference by other anions*

Of all the anions interfering with the sulfates in the reaction with barium chloride (Chamot and Mason, Chapter XI, 1940) we will discuss the carbonates, nitrates, and sulfites because they are the most common in the atmosphere.

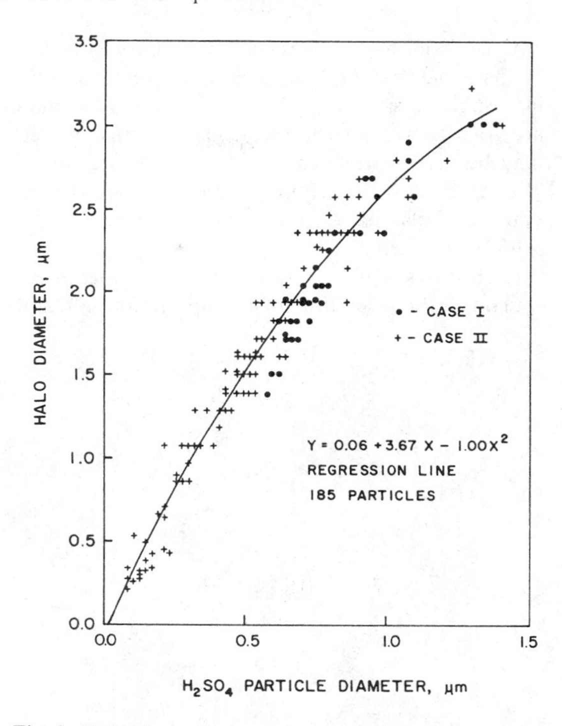

Fig. 8. Halo diameter as a function of that of the original sulfuric acid droplet.

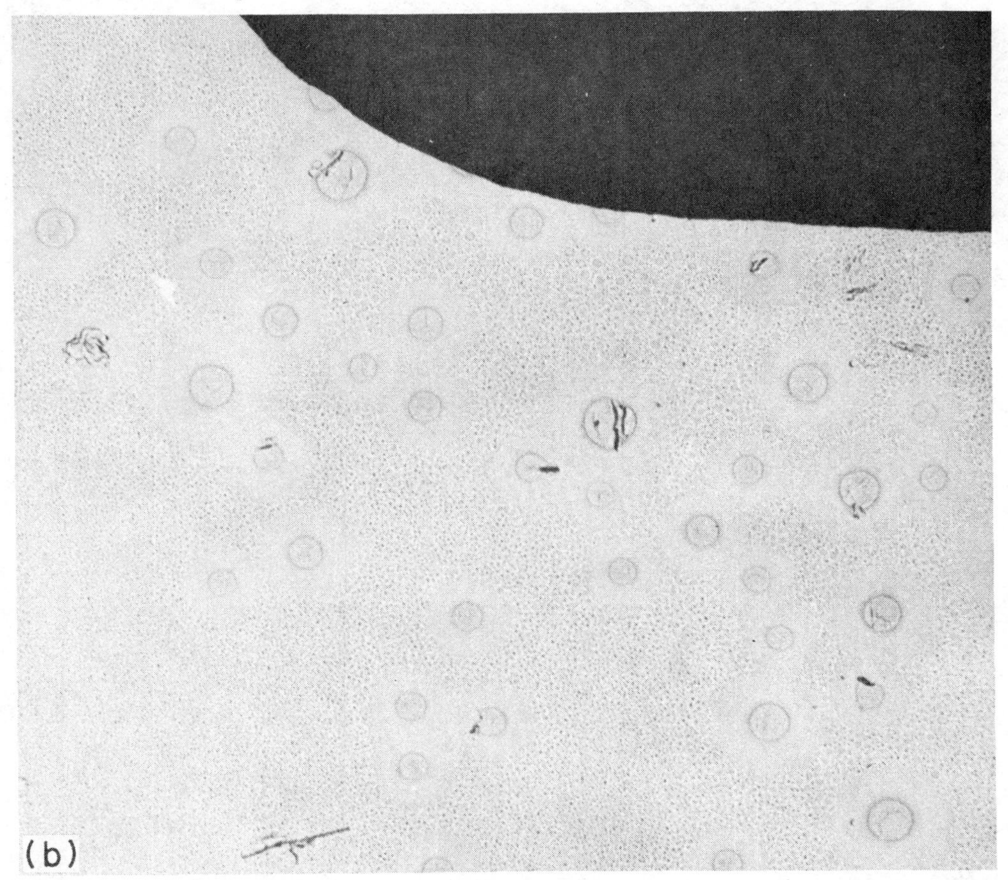

Fig. 7(b).

Fig. 7. Sample of sulfuric acid droplets on a 'relocation screen' (a) before and (b) after barium chloride treatment.

Carbonates and sulfites react with barium to form an insoluble precipitate of barium carbonate and barium sulfite respectively. However, these precipitates are soluble in hydrochloric acid, and this property may be used to differentiate them from the sulfates. Nitrates react with barium to form few crystals rather than a circular halo, since barium nitrate is slightly soluble in water.

Carbonates, sulfites and nitrates were generated by spraying a dilute solution of an appropriate salt. After collection on electron microscope screens, the particles were coated with barium chloride and placed in a chamber of 85, 70 and 75% relative humidity respectively. Carbonates and sulfite halos were then exposed to hydrochloric acid vapor in a chamber which contained a 50% hydrochloric acid solution.

Figures 9a and b and 10a and b show typical halos obtained from the reaction of carbonates and sulfites with barium before and after hydrochloric acid treatment, respectively. By comparing the photomicro-

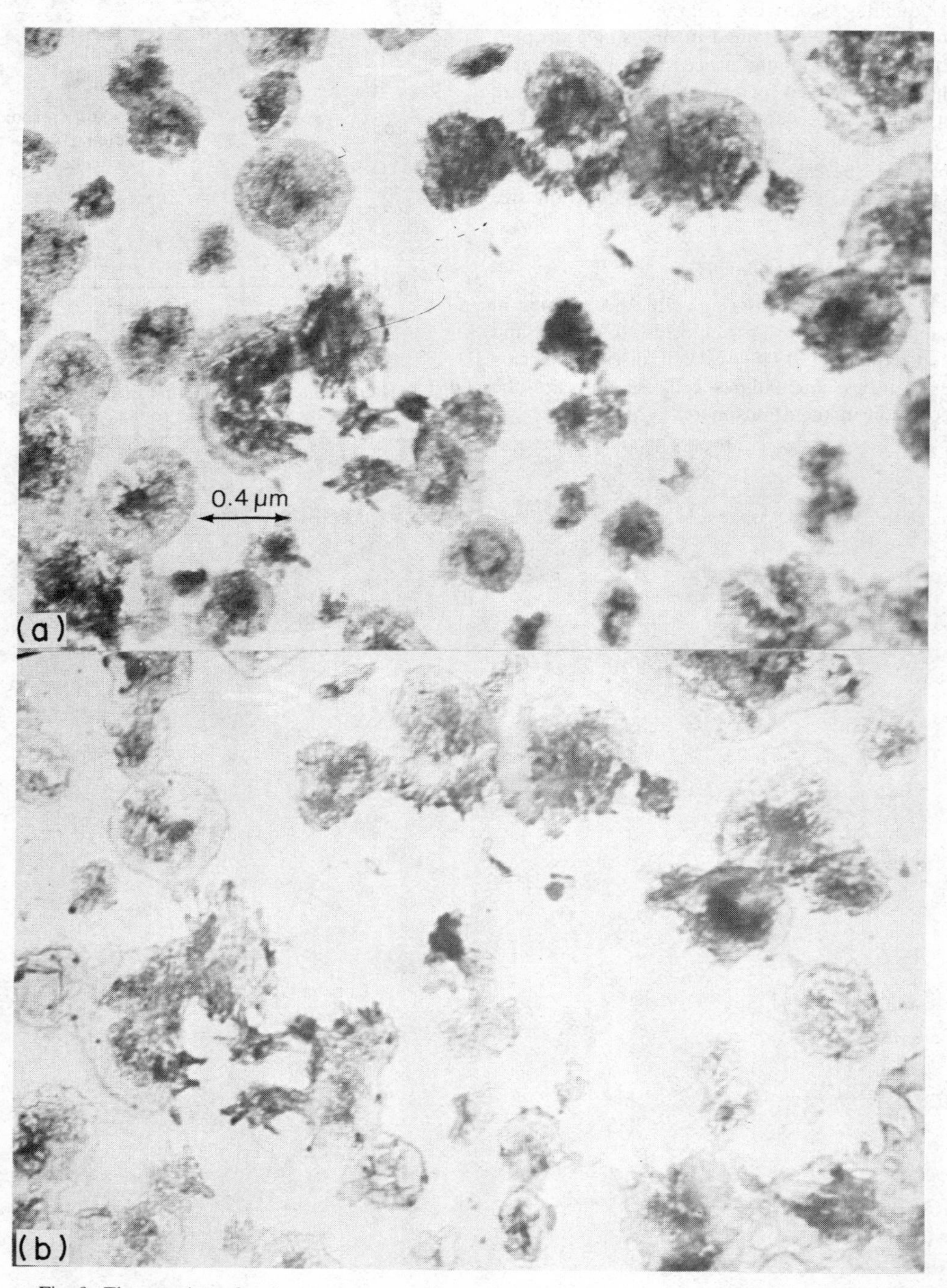

Fig. 9. The reaction of carbonate particles with barium chloride (a) before and (b) after hydrochloric acid treatment.

graphs, b with a, it is evident that the halos formed are partially destroyed, leaving a spot which is different from the original.

Figure 11 is a photomicrograph of a sample of nitrate particles, part of which was coated with barium chloride (right bottom). The comparison of both parts reveals the difference between the nitrate particles and their reaction spots which contain cubical crystals. Because of their appearance the nitrate particles do not cause interference.

3. APPLICATIONS

The method described requires very short sampling times of the order of minutes or seconds. Because

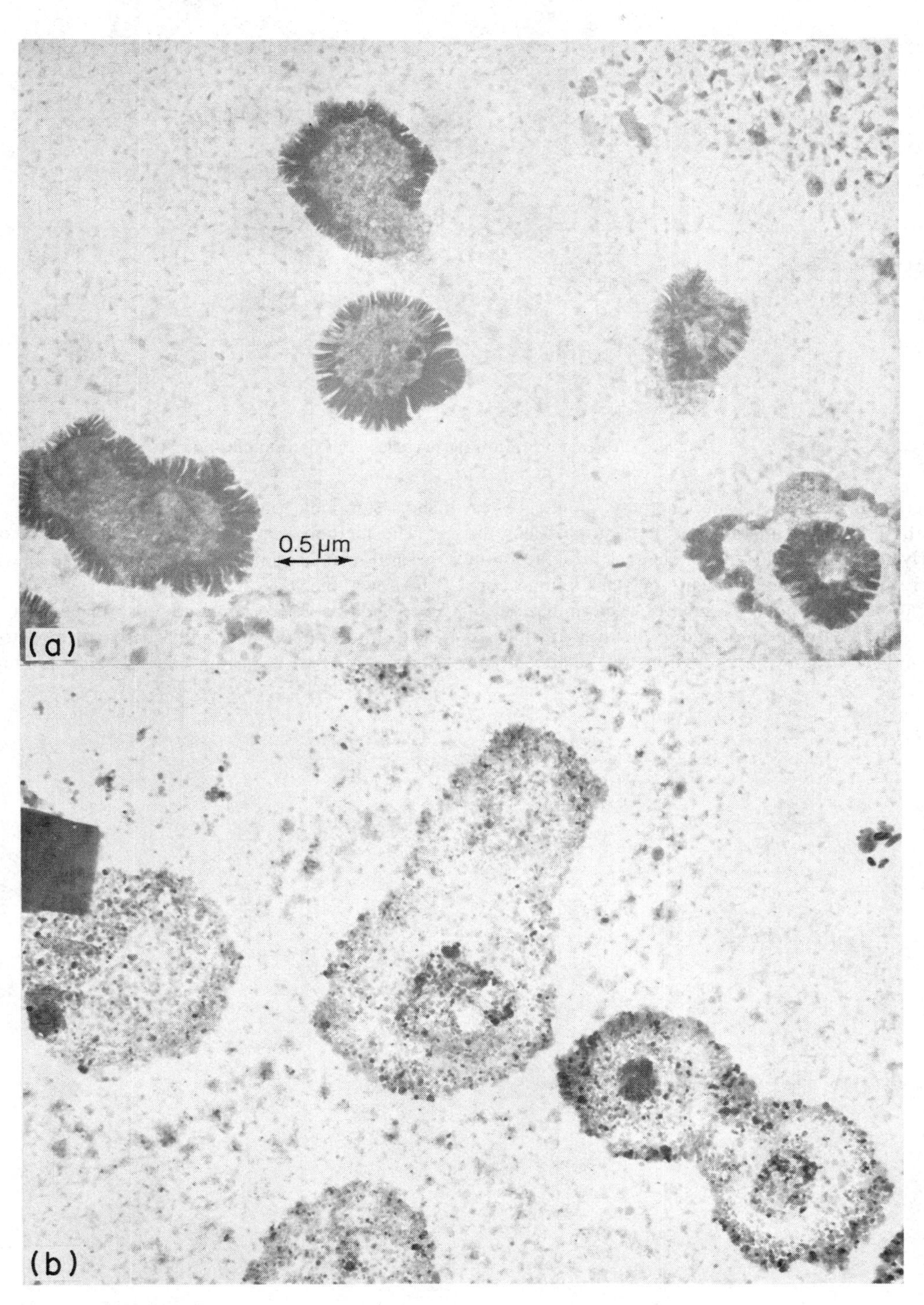

Fig. 10. The reaction of sulfite particles with barium chloride (a) before and (b) after hydrochloric acid treatment.

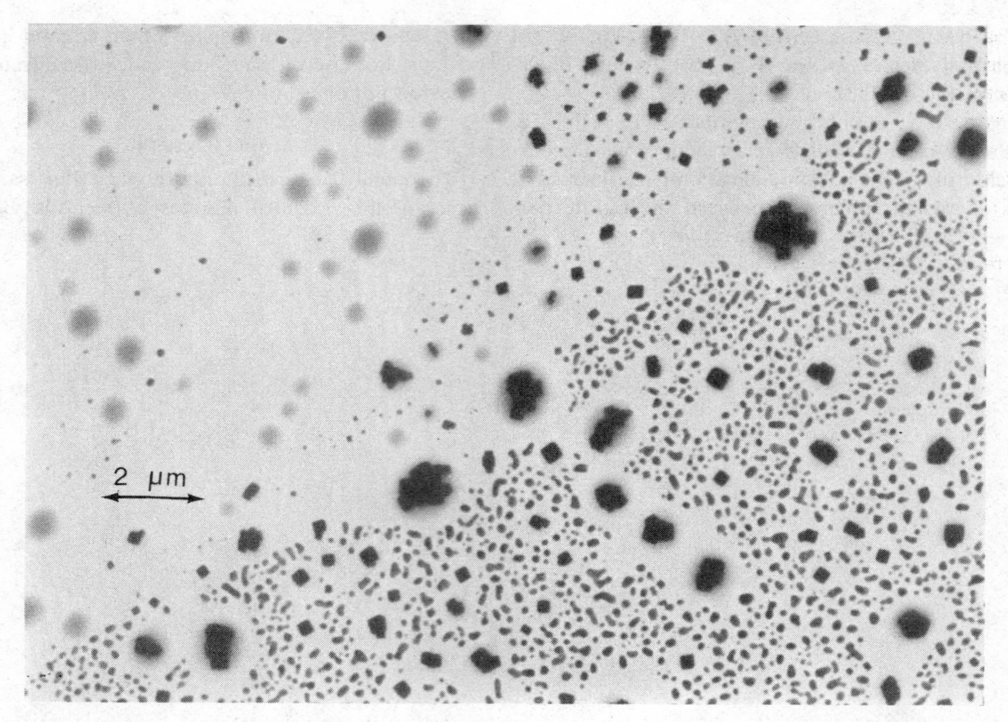

Fig. 11. The reaction of nitrate particles with barium chloride.

of its high resolution, this method should be especially useful in atmospheric sulfate studies such as: hourly and diurnal variations, changes brought about by precipitation, measurements of stack plume concentrations, transport of sulfates, determination of rate of transformation of sulfur dioxide to sulfates, and any laboratory investigations of sulfate where short sampling times are required.

The method was applied to a sample collected in a rural area of central Pennsylvania, four miles from the town of State College. Particles were collected on electron microscope screens (placed on stages three and four of the Casella cascade impactor) and then treated with barium chloride and developed in

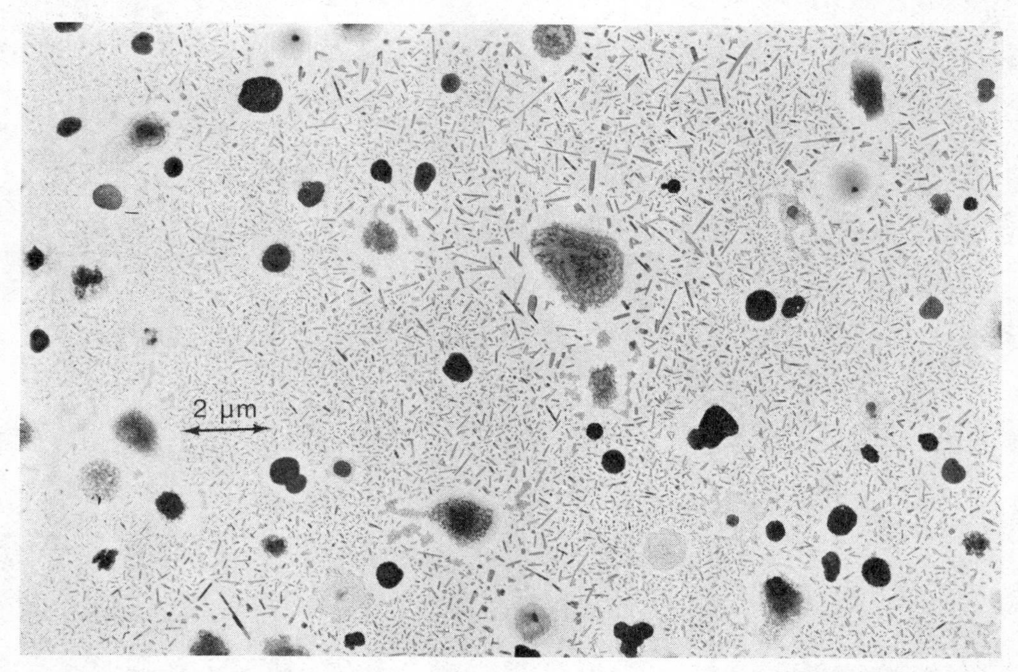

Fig. 12. Typical atmospheric particles collected in a rural area in central Pennsylvania after they have reacted with barium chloride.

the usual way for sulfate analysis. Figure 12 is a photomicrograph of a portion of the sample after treatment. The large fraction of sulfate particles can be appreciated.

The samples were analyzed in order to obtain the percentage of sulfate particles and their size distribution. The results reflect the sample collected on the screen without correcting for collection efficiencies of the impactor. 48% of the particles were found to be sulfate, and only a few particles (of the order of 1–2%) had a mixed nature: an insoluble nucleus surrounded by sulfate, the rest were of a non-identified nature. Examples of the mixed particles are shown in Figs 13a and b where halos are surrounding a spherical particle and a conglomerate of crystals, respectively. The method is currently applied for sulfate studies in power plant plumes. The results will be published elsewhere.

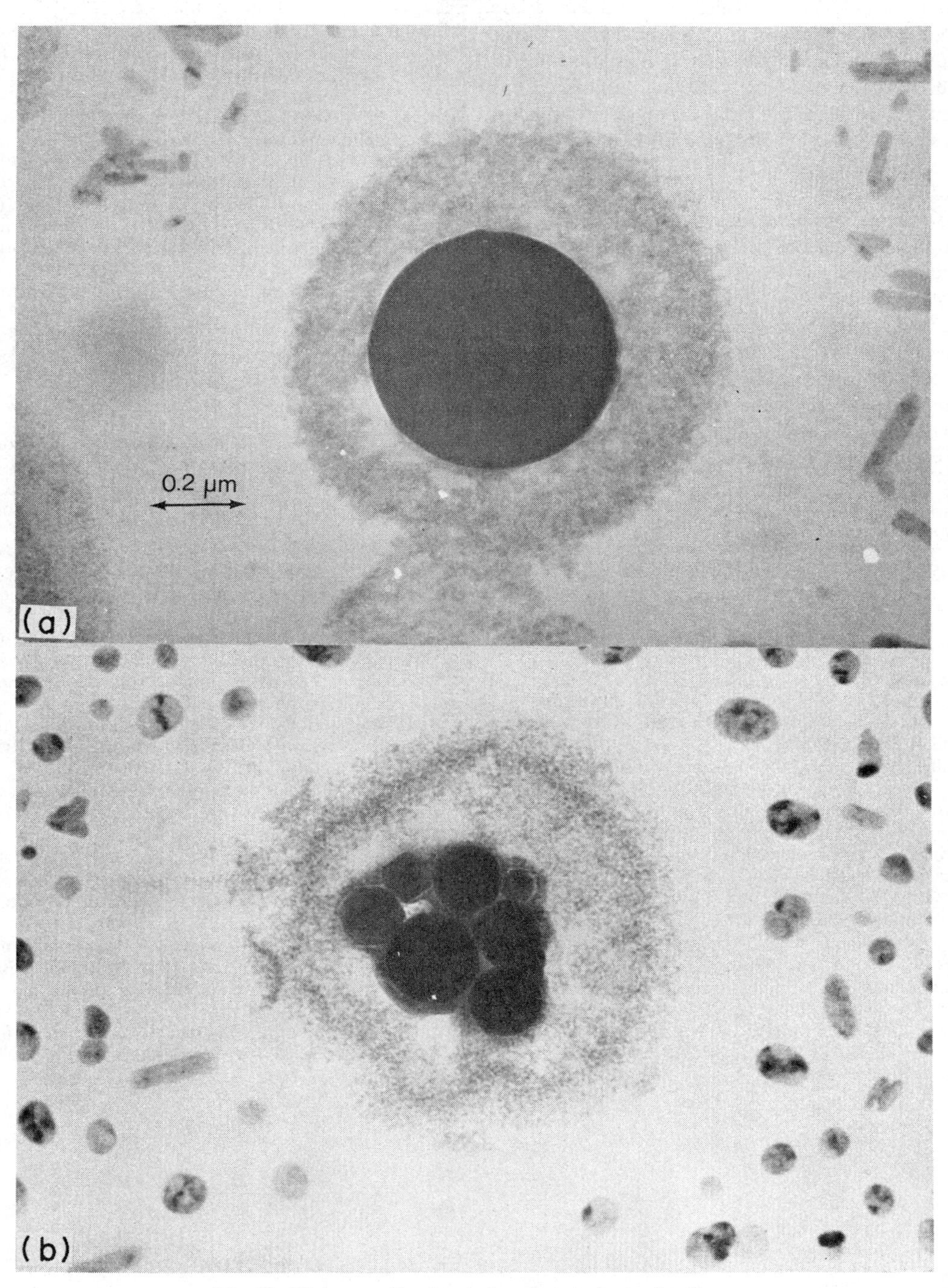

Fig. 13. High magnification halos of typical mixed sulfates.

4. SUMMARY AND CONCLUSIONS

A quantitative method for the identification of individual submicron sulfate particles has been described. In addition, it provides size distribution and in this respect it is a valuable supplement to the bulk analysis. The method is based on the reaction of the sulfate ion with barium chloride and can be applied only to soluble sulfates. The result of the reaction is a characteristic halo easily recognized under the electron microscope.

The principal characteristics of the method are summarized in the following: it is specific for sulfates, reproducible and quantitative, and it is not affected by the background of the collecting surface, in this case a thin film of carbon mounted on electron microscope screen. It also enables us to distinguish between sulfates and sulfuric acid.

The method has been applied to an atmospheric aerosol sample. The analysis provided besides the size distribution, some information about the nature of the sulfate particles. Especially it was possible to observe the existence of mixed particles which could be the result of the heterogeneous nucleation of sulfur dioxide on foreign particles. In this aspect, it is one of the few, if not the only one, to provide such information. As with any other method which employs the microscope for counting and sizing, our method requires time and patience to perform the analysis.

Acknowledgements—This research was supported by funding from the U.S. Environmental Protection Agency under Grant No. R800397 and by E.R.D.A. under grant No. E(11-1)2463.

REFERENCES

Ackerman E. R. and Lodge, Jr. J. P. (1973) Atmospheric microscopy. In *The Encyclopedia of Microscopy and Microtechniques* (edited by P. Grey) pp. 14–28. Van Nostrand Reinhold, New York.

Altshuller A. P. (1976) Regional transport and transformation of sulfur dioxide to sulfates in the U.S.. *J. Air Pollut. Control Assoc.* **26,** 318–324.

Anyz F. (1966) The techniques of membrane filter processing in the chemical detection of individual aerosol particles. *J. Rech. Atmos.* **2,** 441–447.

Ayers G. P. (1977) An improved thin-film sulfate test for submicron particles. *Atmospheric Environment* **11,** 391–395.

Bigg E. K., Ono A. and Williams J. A. (1974) Chemical tests for individual submicron aerosol particles. *Atmospheric Environment* **8,** 1–13.

Chamot E. M. and Mason C. W. (1940) *Handbook of Chemical Microscopy* Vol. 2, 2nd Ed. John Wiley, New York.

EPA Report (1975) *Position Paper on Regulation of Atmospheric Sulfates,* U.S. Environmental Protection Agency, Research Triangle Park, North Carolina, EPA-450/2-75-007.

Glauert A. M. (1972) *Practical Methods in Electron Microscopy.* Elsevier, New York.

Junge C. E. (1963) *Air Chemistry and Radioactivity.* Academic Press, New York.

Lodge J. P., Jr. (1962) Identification of aerosols. In *Advances in Geophysics* (edited by H. E. Landsberg and J. Van Mieghem) Vol. IX, pp. 97–139. Academic Press, New York.

Low R. D. H. (1969) A theoretical study of nineteen condensation nuclei. *J. Rech. Atmos.* **4,** 65–78.

Mamane Y. (1977) A quantitative method for the detection of individual submicron sulfate particles. Ph.D. Thesis, Department of Meteorology, The Pennsylvania State University, University Park, PA 16802.

May K. R. (1945) The cascade impactor: an instrument for sampling aerosol. *J. Sci. Instr.* **22,** 178–195.

McCrone W., Draftz R. and Delly J. (1967) *The Particle Atlas.* Ann Arbor, Michigan.

Montefinale A. C., Montefinale T. and Papée H. M. (1971) Recent advances in the chemistry and properties of atmospheric nucleants: a review. *Pure appl. Geophys.* **91,** 171–210.

Orr C., Jr., Hurd F. K. and Corbett W. J. (1958) Aerosol size and relative humidity. *J. colloid Sci.* **13,** 472–482.

Pidgeon F. D. (1954) Controlling factors in identification of microscopic chloride particles with sensitized gelatin films. *Analyt. Chem.* **26,** 1832–1935.

Preuss L. E. (1959) Shadow casting and contrast. *Sci. Instr. News* **4,** 13–22.

Rinehart G. S. (1971) Sulfates and other water solubles larger than 0.15 μm radius in a continental nonurban atmosphere. *J. Rech. Atmos.* **5,** 57–68.

Seely B. K. (1952) Detection of micron and submicron chloride particles. *Analyt. Chem.* **24,** 576–579.

Shy C. M. and Finklea J. F. (1973) Air pollution affects community health *Envir. Sci. Technol.* **7,** 204–208.

Twomey S. (1954) The composition of hygroscopic particles in the atmosphere. *J. Met.* **11,** 334–338.

U.S. House of Representatives Report prepared for the Subcommittee on Special Studies Investigations and Oversight and the Subcommittee on the Environment and the Atmosphere of the Committee on Science and Technology on the Environmental Protection Agency Research Program, November, 1976. U.S. Govt. Printing Office, Washington, D.C. U.S.A.

Whitby K. T., Husar R. B. and Liu B. Y. (1972) The aerosol size distribution of Los Angeles smog. *J. Colloid Interface Sci.* **39,** 177–204.

Weast R. C. (1974) *Handbook of Chemistry and Physics, 1974–1975* 55th Edn. CRC Press, Cleveland, Ohio.

Wilson W. (1976) Chemical Characterization of Aerosols—Progress and Problems. 8th Materials Research Symposium, Methods and Standards for Environmental Measurement, September 20–24, 1976. National Bureau of Standards, Gaithersburg, Maryland.

Wylie R. G. (1963) The properties of water-salt systems in relation to humidity. In *Humidity and Moisture* (edited by A. Wexler and W. A. Wildhack) Vol. 3, pp. 507–517. Reinhold, New York.

Atmospheric Environment Vol. 12, pp. 83–88. Pergamon Press 1978. Printed in Great Britain.

CONTINUOUS MEASUREMENT AND SPECIATION OF SULFUR-CONTAINING AEROSOLS BY FLAME PHOTOMETRY

JAMES J. HUNTZICKER, ROBERT S. HOFFMAN and CHAUR-SUN LING
Oregon Graduate Center, 19600 N.W. Walker Road, Beaverton, OR 97005, U.S.A.

(*First received* 20 *June* 1977 *and in final form* 26 *August* 1977)

Abstract—A sulfur-specific flame photometer has been used for the real time measurement of sulfur-containing aerosols. Specificity for the aerosols was achieved with a diffusion tube stripper which removed sulfur-containing gases from the air stream by diffusion to an adsorbing wall but transmitted particles to the flame photometer. The sensitivity of the flame photometer to $(NH_4)_2SO_4$ and NH_4HSO_4 aerosols was identical, but a reduced response was found for H_2SO_4 aerosol. This occurred because of the high temperature (145°C) of the flame photometer burner block which caused the evaporation and loss of H_2SO_4 to the wall before reaching the flame. The addition of NH_3 to the sample air just upstream of the burner converted H_2SO_4 to $(NH_4)_2SO_4$ or NH_4HSO_4 and produced the expected increase in the flame photometer response. Heating the aerosol upstream of the diffusion stripper converted the sulfates to sulfur-containing gases over a temperature range characteristic of the aerosol being sampled. These gases were removed in the stripper, thereby decreasing the flame photometer output. The normalized response to aqueous H_2SO_4 aerosol decreased from unity at 50°C to 0.04 at 110°C and for $(NH_4)_2SO_4$ and NH_4HSO_4 aerosols from unity at 115°C to zero at 190°C. When the aerosol contained both H_2SO_4 and $(NH_4)_2SO_4$, the resultant thermogram was a function of both the $(NH_4)_2SO_4/H_2SO_4$ ratio and the manner in which the two components were mixed in the particles comprising the aerosol (i.e. homogeneously with constant $(NH_4)_2SO_4/H_2SO_4$ ratios or heterogeneously with varying ratios).

1. INTRODUCTION

In this paper we report on a laboratory study of the use of a sulfur-specific flame photometer for the continuous measurement and speciation of sulfur-containing aerosols. Preliminary results have been reported elsewhere (Huntzicker *et al.*, 1975, 1977). Flame photometry has been used for a number of years to measure sulfur-containing gases, but despite early measurements of sulfur-containing aerosols (Crider, 1965; Crider *et al.*, 1969; Dagnall *et al.*, 1967), the method has been neither thoroughly investigated nor extensively applied.

The goal of the present work was to provide the information necessary for the design of a practical sulfate monitor capable of the real time measurement of both the sulfuric acid and ammonium sulfate fractions of the aerosol. The principal features of the work, which are discussed below, involved the calibration of the flame photometer response as a function of the concentrations of different sulfur-containing aerosols and the study of a thermal method for distinguishing between sulfuric acid and the ammonium sulfates.

2. EXPERIMENTAL

The aerosol generation and sampling system is shown in Fig. 1. The heart of the system is a Meloy SA-285 sulfur-specific flame photometer which measures the S_2^* chemiluminescence produced when sulfur-containing molecules are introduced into a fuel-rich hydrogen-air flame. Because the formation of S_2^* requires the presence of two sulfur atoms, the intensity of the chemiluminescence is theoretically proportional to the square of the concentrations of such molecules as SO_2, H_2SO_4, or $(NH_4)_2SO_4$. In practice, however, the power law dependence is generally less than two and depends, in particular, on the sulfur compound being analyzed (Dagnall *et al.*, 1967; Crider *et al.*, 1969; Stevens *et al.*, 1971). For all the experiments discussed in this paper, the output of the flame photometer phototube was measured with a Keithley 610C electrometer. The linearized output of the Meloy SA-285 was not used.

The flame photometer was made specific to sulfur-containing aerosols by removing sulfur-containing gases in a diffusion stripper upstream of the flame photometer. The stripper is a tube which removes SO_2 and H_2S by diffusion to an absorbing wall but efficiently transmits particles because of their lower diffusion coefficients. The stripper used in this work was a glass tube (22 cm long, 1 cm i.d.), which contained a cylinder of Whatman 41 filter paper impregnated with $Pb(CH_3CO_2)_2$. Although the long-term effectiveness of the stripper has not been evaluated, short exposures to SO_2 and H_2S concentrations in excess of 100 ppb produced no change in the flame photometer output. If only diffusive losses are considered, the particle transmission through the stripper can be estimated from the Gormley–Kennedy equations (Gormley and Kennedy, 1949) as 88% for 0.01 μm dia particles and 99.3% for 0.1 μm dia particles. The flame photometer baseline was obtained by inserting a low pressure drop glass fiber filter between the stripper and the flame photometer. The diffusion stripper approach has also been used by Durham *et al.* (1976) and Mueller (1976).

The continuous monitor also contains a heater which was used for the sulfate speciation experiments and an ammonia addition port. The heater was a 51 cm long section of 1.3 cm (i.d.) copper tubing. The central 26 cm of the tube were enclosed by a 2.2 cm (i.d.) pyrex tube which was wrapped with heating tape. Temperatures were measured with a glass-enclosed iron–constantan thermocouple located inside the copper tube. The axial temperature profile was such that over a 16 cm length of the copper

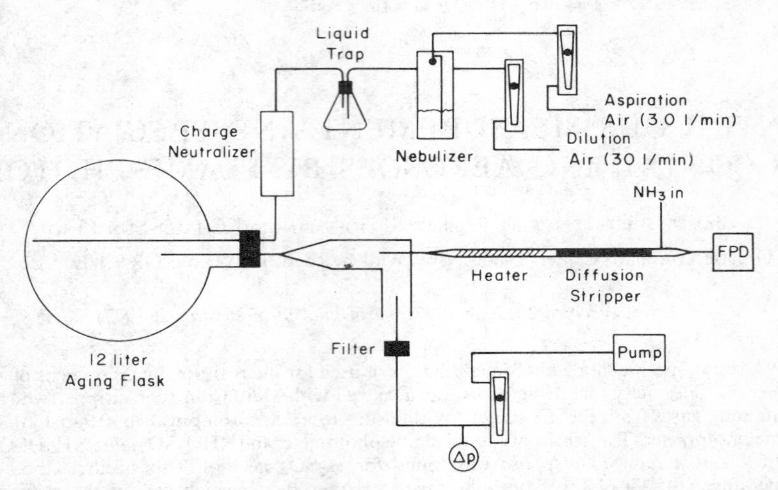

Fig. 1. Aerosol generation and sampling system. The aerosol is sampled by a flame photometric detector (FPD) through a heater and a diffusion stripper, which removes SO_2 and H_2S from the air. Aerosol is also collected for independent chemical analysis by the filter, rotameter, pump system. In this system Δp is a differential pressure gauge.

tube the deviation from the mean temperature was $\pm 5\%$. For some experiments ammonia was added to the system by passing about 3 ml min^{-1} of air over an NH_3 permeation tube and mixing it with the sample air just upstream of the flame photometer. Resultant ammonia concentrations were approx 6 ppm and were sufficient to convert any H_2SO_4 aerosol to $(NH_4)_2SO_4$ or NH_4HSO_4 upstream of the flame.

Polydisperse sulfate aerosols were produced by nebulizing aqueous solutions 10^{-4}–10^{-3} M in sulfate. Excess charge was removed from the aerosol in the ^{85}Kr charge neutralizer, and equilibration with the relative humidity of the dilution air was achieved in the 12 l. aging flask. To calibrate the response of the flame photometer as a function of sulfate concentration, sulfate aerosol was collected by the filter system shown in Fig. 1 simultaneously with the flame photometric measurement. Fluoropore filters (0.5 μm pore size) were used for the collection. The filter-collected aerosol was extracted in 10 ml water and analyzed for sulfate by the flash volatilization-flame photometric method (Roberts and Friedlander, 1976; Husar *et al.*, 1975). In our variation of this method of 4 μl of the solution to be analyzed are placed on a 0.025 cm dia Pt wire connected to the terminals of a 0.3 F capacitor. The capacitor is charged to about 5 V and discharged through the wire which is rapidly heated to greater than 1000°C. Any sulfur on the wire is converted to SO_2 which is measured as a sharp pulse by the flame photometer. Because a difference in the response of this method to H_2SO_4 and $(NH_4)_2SO_4$ was consistently observed, 4 μl of 27 mg ml^{-1} NH_3 solution were added to the filter extracts to neutralize any acid sulfate. All samples were then analyzed as $(NH_4)_2SO_4$ and the response compared to $(NH_4)_2SO_4$ standards.

3. RESULTS

1. *Flame photometer response*

The response of the flame photometer to aerosols of H_2SO_4, $(NH_4)_2SO_4$, NH_4HSO_4, and H_2SO_4 neutralized by NH_3 is shown in Fig. 2. For these experiments the heater was kept at room temperature. Although there was some scatter in the data, particularly at low concentrations, the responses to aerosols of $(NH_4)_2SO_4$, NH_4HSO_4, and H_2SO_4 neutralized by NH_3 were the same, but the response to unneutralized H_2SO_4 aerosol fell systematically lower. This difference was verified by first measuring the flame photometer response to H_2SO_4 without NH_3 and then adding NH_3 with all other conditions remaining the same. This converted any H_2SO_4 to $(NH_4)_2SO_4$ or NH_4HSO_4 and produced an increase in the flame photometer response in each instance.

The cause of this difference was found to be the brief exposure of the aerosol to the hot ($\sim$ 145°C) aluminum burner block before entering the flame. As discussed in Section 3.2, the exposure of H_2SO_4 aero-

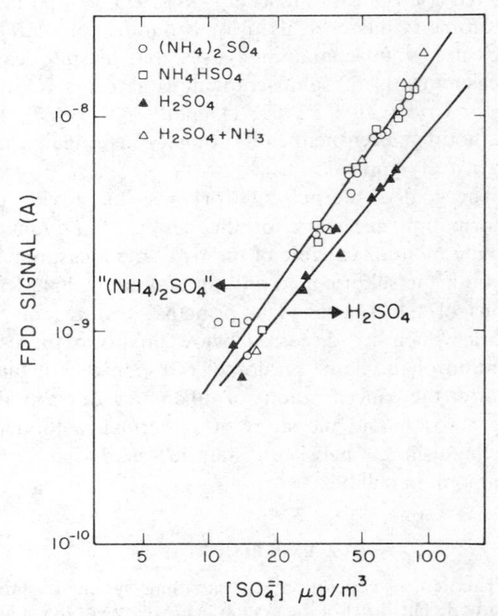

Fig. 2. Flame photometer signal as a function of sulfate aerosol concentration. The '$(NH_4)_2SO_4$' line is the least squares fit to the open symbol points ($(NH_4)_2SO_4$, NH_4HSO_4, and $H_2SO_4 + NH_3$), and the H_2SO_4 line is the least squares fit to H_2SO_4 alone (the solid triangles).

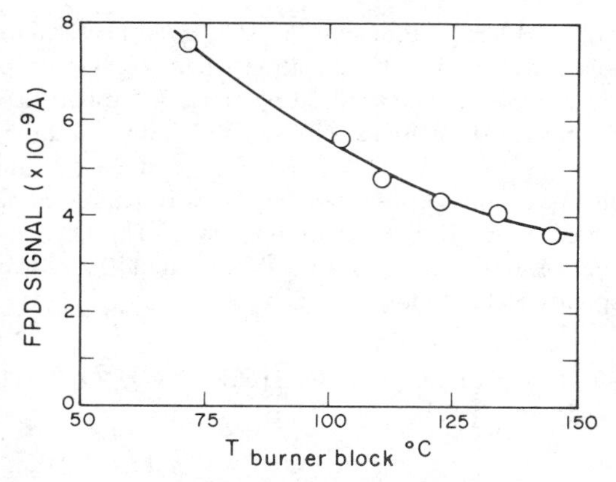

Fig. 3. Flame photometer response to H_2SO_4 aerosol as a function of burner block temperature.

sol to temperatures above 50°C partially vaporizes H_2SO_4 which, because of its reactive nature, is scavenged at any nearby wall. $(NH_4)_2SO_4$, however, is stable with respect to sulfur loss up to about 120°C (cf. Fig. 4). Thus, for equal concentrations of H_2SO_4 and $(NH_4)_2SO_4$ aerosols entering the burner, the H_2SO_4 aerosol would be expected to produce a lower response because of H_2SO_4 vapor losses to the wall. The response of the flame photometer to H_2SO_4 aerosol as a function of burner block temperature is shown in Fig. 3 and indicates a factor of 2 increase between 145°C (the normal burner block temperature) and 72°C (the minimum attainable temperature). Much smaller increases ($\sim 25\%$) were observed in the response to $(NH_4)_2SO_4$ aerosol and SO_2. With the burner block temperature at 72°C the addition of NH_3 to the H_2SO_4 aerosol produced only a small increase in the flame photometer response, and this could be attributed to a small positive interference from the NH_3 itself. Thus at a burner block temperature of 72°C the responses to H_2SO_4 and $(NH_4)_2SO_4$ were essentially identical. At the higher temperatures the response difference was therefore due to the temperature effect rather than a fundamental difference in the flame chemistry of the two compounds. The fact that H_2SO_4 could be measured at all with the burner block at 145°C is due to the short residence time of the aerosol in the burner block and the resultant incomplete vaporization.

Unfortunately, the response time of the flame photometer to a change in the aerosol sulfur concentration also increased at the lower temperatures. To avoid this undesirable effect yet still retain equality in the H_2SO_4 and $(NH_4)_2SO_4$ responses, the burner block temperature was returned to 145°C, and NH_3 was added as shown in Fig. 1 to convert the H_2SO_4 to $(NH_4)_2SO_4$. As indicated in Fig. 2, the response to $H_2SO_4 + NH_3$ was identical to the $(NH_4)_2SO_4$–NH_4HSO_4 response. The addition of NH_3 is thus a convenient method for ensuring consistency in the continuous flame photometric measurement of the H_2SO_4–$(NH_4)_2SO_4$ aerosol system.

The line through the $(NH_4)_2SO_4$, NH_4HSO_4, and $H_2SO_4 + NH_3$ points in Fig. 2 was determined by a least squares fit and has a slope of 1.5, which corresponds to the power law response of the flame photometer. We have not attempted to increase the response toward the theoretical optimum of 2.0, but as Eckhardt *et al.* (1975) have shown, an improved response can be attained by fine tuning of the hydrogen and air flows.

An additional problem with the continuous monitoring of sulfur-containing aerosols is the poor signal to noise ratio at low concentrations. With a 2 s integration time the peak–peak noise was about 7×10^{-11} A, which is equivalent to 2–3 $\mu g\ SO_4^{2-}\ m^{-3}$. Preliminary experiments with digital signal averaging, however, suggest a lower limit of detection (i.e. twice the peak–peak noise) of about 1 $\mu g\ SO_4^{2-}\ m^{-3}$ when the averaging time is 70 s. The continuous monitoring of total particulate sulfur can then be achieved by the diffusion stripper-flame photometer system with digital integration of the flame photometer output. Because of the inevitable drift of the flame photometer baseline, periodic up-dating of the baseline by automatically switching a filter into the flow system is necessary. For the general applicability of such an instrument, the response of the flame photometer to other sulfur compounds (e.g. Na_2SO_4) would have to be the same as to $(NH_4)_2SO_4$. This question is currently being studied.

2. *Thermal speciation*

For the speciation of aerosols in the H_2SO_4–$(NH_4)_2SO_4$ system, the aerosols were heated in the copper tube heater and the response of the flame photometer measured as a function of heater temperature. Ammonia was added, as in Fig. 1, to ensure equivalent responses to H_2SO_4 and $(NH_4)_2SO_4$ aerosols. The concentration of ammonia upstream of the heater was minimized by filtering the air used for the aerosol generation through activated charcoal. The concentration of all the aerosols studied was in the 30–40 $\mu g\ SO_4^{2-}\ m^{-3}$ range.

In Fig. 4(a) thermograms for H_2SO_4, $(NH_4)_2SO_4$, and NH_4HSO_4 are shown. The H_2SO_4 concentration began to decrease above 55°C and by 110°C only 4% of the original sulfate remained. Above 110°C the H_2SO_4 concentration decreased gradually but even at 220°C 1% of the initial concentration remained. The decrease of the H_2SO_4 signal results from the dehydration of the H_2SO_4 aerosol, the accompanying increase of the H_2SO_4 vapor pressure, and the removal of the H_2SO_4 vapor at the wall of the copper tube as the temperature is raised. The persistence of the H_2SO_4 signal to 220°C is not understood but could result from impurity cations (e.g. Na^+) in the droplet capable of stabilizing the sulfate against volatilization.

The thermograms for $(NH_4)_2SO_4$ and NH_4HSO_4 exhibited only slight differences as both began to decrease at about 120°C and fell to zero at about 190°C. The mechanisms for the decomposition of $(NH_4)_2SO_4$

and NH_4HSO_4 are quite complex. Kiyoura and Urano (1970) found that at 100°C $(NH_4)_2SO_4$ lost NH_3 to form NH_4HSO_4. Between 100 and 170°C, however, $(NH_4)_3H(SO_4)_2$ rather than NH_4HSO_4 was formed in the deammoniation reaction. NH_4HSO_4 decomposed to NH_2SO_3H (sulfamic acid) and $(NH_4)_2S_2O_7$ (ammonium pyrosulfate) at about 150°C. Above 200°C complete decomposition to SO_2, SO_3, H_2SO_4, NH_3, N_2, H_2, and H_2O occurred for both $(NH_4)_2SO_4$ and NH_4HSO_4. The observed decrease in the sulfate signal at 120°C in our experiment indicates that for the dynamic situation occurring in the copper heater, the sulfur-containing gases were evolved at lower temperatures than found by Kiyoura and Urano (1970) and were scavenged in the heater or the stripper. Qualitatively similar thermograms to those in Fig. 4a have been found by Kittelson *et al.* (1977), and by Husar (1976), who used temperature-programmed nephelometry.

In Fig. 4(b) thermograms are shown for aerosols produced by nebulizing H_2SO_4–$(NH_4)_2SO_4$ solutions with NH_4^+/H^+ ratios = 0.25, 0.50, 0.75. These aerosols correspond to the situation where H_2SO_4 droplets are partially neutralized by NH_3. Although these droplets can contain molecular H_2SO_4 and the aqueous ions H^+, NH_4^+, HSO_4^-, and SO_4^{2-}, the chemical composition can be formally expressed as a mixture of H_2SO_4 and $(NH_4)_2SO_4$. The chemical change upon heating such a droplet to 110°C corresponds to the following reaction:

$$\underset{\text{liquid}}{a(2H^+ + SO_4^{2-}) + b(2NH_4^+ + SO_4^{2-}) + H_2O} \xrightarrow{110^\circ C} \underset{\text{vapor}}{(a-b)H_2SO_4} + \underset{\text{solid}}{2b(NH_4HSO_4)} + \underset{\text{vapor}}{H_2O} \quad (1)$$

where a and b are the relative molar fractions of H_2SO_4 and $(NH_4)_2SO_4$ and $a \geq b$. Equation (1) predicts that the decrease in the sulfate signal between room temperature and 110°C corresponds to $(a - b)$ and the residual signal at 110°C to $2b$. For the three aerosols studied, the experimental values of b/a derived from Equation (1) were 0.29, 0.43 and 0.72, which compare well with the expected values of 0.25, 0.50 and 0.75. (The experimental values assume that at 110°C only 96% of the H_2SO_4 is evolved as shown in Fig. 4a.) Aerosols with $a \leq b$ will exhibit the $NH_4HSO_4/(NH_4)_2SO_4$ thermogram only. As discussed above, NH_3 rather than H_2SO_4 is evolved, but the exact stoichiometry of the reaction is complex and not well understood.

The above discussion concerned homogeneous sulfate aerosols, i.e. aerosols in which b/a is essentially the same for all the particles. For heterogeneous aerosols with b/a varying from particle to particle, the observed thermogram will be a linear combination of the thermograms of the individual particles. Equation (1) will only apply if $b \leq a$ for all the particles in the aerosol, and in this case the thermogram will yield values of a and b for the total aerosol. In the most general case with no restrictions on a and b, Equation (1) does not apply. An example of such a case is shown in Fig. 4c. Aerosols of H_2SO_4 ($a = 1$, $b = 0$) and $(NH_4)_2SO_4$ ($a = 0$, $b = 1$) were generated in separate nebulizers and mixed in the liquid trap (c.f. Fig. 1) such that 53% of the total sulfate was as H_2SO_4 and 47% as $(NH_4)_2SO_4$. If the aerosols did not coagulate before reaching the heater, the concentration difference between room temperature and 110°C corresponded to the total H_2SO_4 concentration (i.e. a) and the residual signal at 110°C to the total $(NH_4)_2SO_4$ concentration (i.e. b). With such an interpretation the thermogram of Fig. 4c indicates that 42% of the sulfate was as H_2SO_4 and 58% as $(NH_4)_2SO_4$. The difference between this and the prepared concentration could have arisen from partial coagulation of the H_2SO_4 and $(NH_4)_2SO_4$ aerosols upstream of the heater.

In atmospheric aerosols it is not known *a priori* whether the H_2SO_4 and $(NH_4)_2SO_4$ fractions are homogeneously or heterogeneously mixed, and it is therefore not possible to define completely the com-

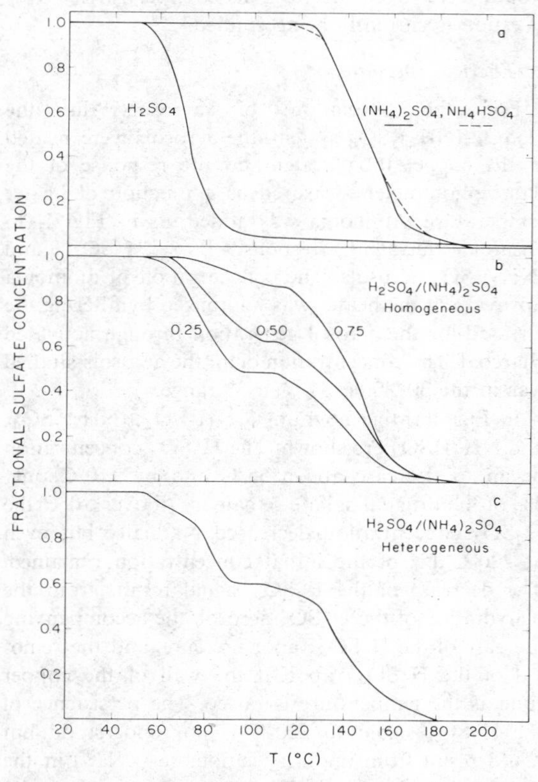

Fig. 4. Thermograms for sulfate aerosols. (a) H_2SO_4, $(NH_4)_2SO_4$ (——), and NH_4HSO_4 (---); (b) homogeneous H_2SO_4–$(NH_4)_2SO_4$ aerosols with b/a = 0.25, 0.50, 0.75 where b and a are the molar fractions of $(NH_4)_2SO_4$ and H_2SO_4 in the particles; (c) heterogeneous mixture of H_2SO_4 droplets and $(NH_4)_2SO_4$ particles with the H_2SO_4 constituting 53% of the total sulfate and $(NH_4)_2SO_4$ 47%.

position of a $H_2SO_4/(NH_4)_2SO_4$ aerosol by the thermal speciation method. For homogeneous mixing the difference in the thermogram between room temperature and 110°C corresponds to $a - b$ (for $a \geq b$) whereas for the heterogeneous case of discrete H_2SO_4 and $(NH_4)_2SO_4$ particles this difference is equal to a. The residual signal at 110°C is equal to $2b$ for homogeneous mixing and b for the discrete heterogeneous case. Thus, when thermograms similar to those in Figs. 4b and 4c are observed, only upper and lower limits to the H_2SO_4 and $(NH_4)_2SO_4$ fractions can be specified. The difference between the limits represents a range of uncertainty which for both species is equal to the value of b determined from the homogeneous mixing assumption (i.e. Equation 1). For example, the thermogram of Fig. 4c, when interpreted by the discrete heterogeneous mixing assumption, gave $a = 0.42$ and $b = 0.58$ (with the total sulfate concentration normalized to unity). If the homogeneous mixing assumption had been used, values of $a = 0.71$ and $b = 0.29$ would have been obtained. The ranges of uncertainty for both a and b are 0.29, and the averages of the two extremes are $a = 0.565 \pm 0.145$ and $b = 0.435 \pm 0.145$. The largest range of uncertainty occurs when only the $NH_4HSO_4/(NH_4)_2SO_4$ thermogram of Fig. 4(a) is observed and is equal to 0.5 (on a normalized scale). In this case the aerosol could be NH_4HSO_4 with composition $a = b = 0.5$, or pure $(NH_4)_2SO_4$ ($b = 1$), or any mixture of NH_4HSO_4 and $(NH_4)_2SO_4$.

A possible design for a thermal speciation instrument would involve the sequential sampling of the aerosol through copper tubes maintained at room temperature, 110°C, and 220°C. Only refractory sulfur-containing aerosols would be transmitted through the 220°C tube and the measurements through the other tubes would indicate the range of H_2SO_4 and $(NH_4)_2SO_4$ concentrations. As in the total particulate sulfur monitor, periodic referencing to the baseline and digital integration would be required. A better definition of the aerosol would be attainable if the evolved NH_3 were also measured—provided, of course, that the stoichiometry of this reaction were known.

4. SUMMARY

The principal findings of this work concern the sensitivity of the flame photometer to different sulfur-containing aerosols and the use of a thermal volatilization method for distinguishing between sulfuric acid and the ammonium sulfates. Aerosols of $(NH_4)_2SO_4$ and NH_4HSO_4 produced an equivalent response in the flame photometer, but a reduced response to H_2SO_4 was found. This occurred because of the high temperature of the flame photometer burner block which caused the evaporation and loss of H_2SO_4 to the burner block wall before reaching the flame. The addition of NH_3 to the sample air just upstream of the burner rectified this problem by converting the H_2SO_4 to $(NH_4)_2SO_4$ or NH_4HSO_4. The thermal speciation method consisted of heating the aerosol upstream of the burner and the diffusion stripper. For aerosols of H_2SO_4, $(NH_4)_2SO_4$, and NH_4HSO_4 thermograms characteristic of the aerosol being sampled were obtained. However, when the aerosol contained both H_2SO_4 and $(NH_4)_2SO_4$, the thermogram was a function of both the $(NH_4)_2SO_4/H_2SO_4$ ratio and whether the two components were mixed homogeneously or heterogeneously. Because the nature of the mixing is not known *a priori*, only upper and lower limits to the H_2SO_4 and $(NH_4)_2SO_4$ concentrations can be obtained for atmospheric aerosols by this method.

Acknowledgements—The support of the U.S. Environmental Protection Agency through Grant No. R804750010 and the Northwest Air Pollution Center, a consortium of Oregon and Washington industries supporting the environmental work of the Oregon Graduate Center, is gratefully acknowledged. The contributions of Lorne Isabelle and John Watson in the early phase of this work are appreciated.

REFERENCES

Crider W. L. (1965) Hydrogen flame emission spectrophotometry in monitoring air for sulfur dioxide and sulfuric acid aerosol. *Analyt. Chem.* **37**, 1770–1773.

Crider W. L., Barkley N. P., Knott M. J. and Slater R. W. (1969) Hydrogen flame chemiluminescence detector for sulfate in aqueous solutions. *Anal. chim. Acta* **47**, 237–241.

Dagnall R. M., Thompson K. C. and West T. S. (1967) Molecular-emission spectroscopy in cool flames—I. The behavior of sulfur species in a hydrogen–nitrogen diffusion flame and in a shielded air–hydrogen flame. *Analyst* **92**, 506–512.

Durham J. L., Wilson W. E. and Bailey E. G. (1976) Continuous measurement of Sulfur in Submicrometric Aerosols. U.S. Environmental Protection Agency Report EPA-600/3-76-088.

Eckhardt J. G., Denton M. B. and Moyers J. L. (1975) Sulfur FPD flow optimization and response normalization with a variable exponential function device. *J. chromatogr. Sci.* **13**, 133–138.

Gormley P. G. and Kennedy M. (1949) Diffusion from a stream flowing through a cylindrical tube. *Proc. R. Irish Acad.* **52A**, 163–169.

Huntzicker J. J., Isabelle L. M. and Watson J. G. (1975) The continuous monitoring of particulate sulfate by flame photometry. In *Proc. Int. Conf. on Envir. Sensing and Assessment*, Vol. 2, paper 23–4, Las Vegas, Nevada.

Huntzicker J. J., Isabelle L. M. and Watson J. G. (1977) The continuous monitoring of particulate sulfur compounds by flame photometry. Presented before the Division of Environmental Chemistry, American Chemical Society, New Orleans, Louisiana, March 20–25.

Husar R. B. (1976) *In situ* thermal analysis, in Summary of Activities in Summer 1975 MISTT Study, unpublished report.

Husar J. D., Husar R. B. and Stubits P. K. (1975) Determination of submicrogram amounts of atmospheric particulate sulfur. *Analyt. Chem.* **47**, 2062–2064.

Kittelson D. B., McKenzie R., Linne M., Dorman F., Pui D., Liu B. and Whitby K. (1977) Total sulfur aerosol detection with an electrostatically-pulsed flame photometric detector system. Presented before the Division of Environmental Chemistry, Am. Chem. Soc., New Orleans, Louisiana, March 20–25.

Kiyoura R. and Urano K. (1970) Mechanism, kinetics and equilibrium of thermal decomposition of ammonium sulfate. *Ind. Engng Chem. Process Des. Develop.* **9**, 489–494.

Mueller P. K. (1976) (private communication).

Roberts P. T. and Friedlander S. K. (1976) Analysis of sulfur in deposited aerosol particles by vaporization and flame photometric detection. *Atmospheric Environment* **10**, 403–408.

Stevens R. K., Mulik J. D., O'Keefe A. E. and Krost K. J. (1971) Gas chromatography of reactive sulfur gases in air at the parts-per-billion level. *Analyt. Chem.* **43**, 827–831.

Atmospheric Environment Vol. 12, pp. 89–98. Pergamon Press 1978. Printed in Great Britain.

CONTINUOUS *IN SITU* MONITORING OF AMBIENT PARTICULATE SULFUR USING FLAME PHOTOMETRY AND THERMAL ANALYSIS

W. G. Cobourn, R. B. Husar and J. D. Husar
Air Pollution Research Laboratory, Washington University, St. Louis, MO 63130, U.S.A.

(*First received* 20 *June* 1977 *and in final form* 5 *October* 1977)

Abstract—The sulfur component of the St. Louis ambient aerosol has been continuously monitored using a flame photometric detector (FPD) to measure particulate sulfur concentration, and *in situ* thermal analysis to chemically analyze the aerosol for H_2SO_4 and its ammonium salts, NH_4HSO_4, $(NH_4)_3H(SO_4)_2$, and $(NH_4)_2SO_4$. During the sixteen day monitoring period, the sulfate aerosol varied with respect to chemical composition, but tended to be in the form of the ammonium salts rather than in the form of sulfuric acid. Comparison of particulate sulfur levels with the light scattering coefficient during the monitoring period indicates that particulate sulfur was a significant constituent of the light scattering aerosol, but its mass accounted for less than half of the mass of the light scattering aerosol.

INTRODUCTION

During a 16-day period, from April 23 to May 8, 1977, the St. Louis ambient aerosol was continuously monitored, and *in situ* thermal analysis was automatically performed every 15 min. Three aerosol detectors were incorporated into the measurement system: a flame photometric detector (FPD), a nephelometer and a charger. The FPD has not been traditionally used to monitor particulate sulfur, but recent improvements in the sensitivity of commercially available instruments has made this possible.

Measurement of particulate sulfur using flame photometry was first reported by Crider (1965), who measured ppm concentrations of H_2SO_4 by using a filter to scrub particles, causing a signal decrease corresponding to the H_2SO_4 concentration. The use of a diffusion tube scrubber to denude sulfur gases from incoming sulfate aerosol was reported by Huntzicker *et al.* (1975) and Durham *et al.* (1976); their FPD instruments were capable of measuring particulate sulfur in ppb levels. Further applications of the FPD are reported in this symposium by Huntzicker and Hoffman (1978) and Kittelson *et al.* (1978). The instrument used in this study is actually capable of measuring particulate sulfur in concentrations down to 0.1 ppb (as equivalent ppb of SO_2).

Thermal analysis is a technique whereby a substance is analyzed by observing its physical or chemical properties as a function of temperature. It is an especially useful technique for analyzing sulfate aerosols, because sulfuric acid exhibits a distinctly different thermal behavior from its ammonium salts.

Thermal analysis has been used for the study of aerosols by a number of workers. Twomey (1968) used an electrically heated silica tube with a thermal diffusion cloud chamber as detector to study the volatility of laboratory generated aerosols and atmospheric aerosols. Twomey found that ammonium sulfate volatilized between 150 and 300°C. Atmospheric aerosols exhibited similar behavior when heated, causing Twomey to speculate that ammonium sulfate could be a major constituent of natural aerosols. In a later work, Twomey (1971) reported that ammonium chloride particles volatilized between 100 and 150°C, which is well below the ammonium chloride sublimation temperature of 340°C. This observation is in accordance with our observations of sulfuric acid aerosols, which in our system volatilize between 100 and 150°C, well below the sulfuric acid boiling point of 338°C.

The evaporation of stratospheric aerosols was studied by Rosen (1971), who used a balloon-borne photoelectric particle counter with a heated inlet tube to study particles in the stratosphere. Rosen observed that at higher altitudes, where the atmospheric pressure is lower, the aerosol volatilized at lower temperatures.

Pueshel *et al.* (1973) used a nephelometer and a condensation nuclei counter to study the volatility of aerosols on the island of Hawaii. Kojima *et al.* (1973) employed an optical particle counter for thermal analysis of urban and maritime aerosols. Husar and Shu (1975) performed thermal analysis of the Los Angeles smog aerosol using a nephelometer and an optical particle counter; they also obtained size spectra of heated (to 200°C) and unheated aerosols.

The purpose of this study was primarily to test the FPD as a continuous monitor of ambient particulate sulfur at concentration levels down to 0.1 $\mu g\,m^{-3}$. In addition, we were interested in observing the short term variability of particulate sulfur concentration throughout the 2-week period, as well as the relative occurrences of sulfuric acid and its ammonium salts.

EXPERIMENTAL

The St. Louis ambient aerosol was continuously monitored for a 16-day period from April 23 to May 8, 1977.

A.E. 12—1/3—G

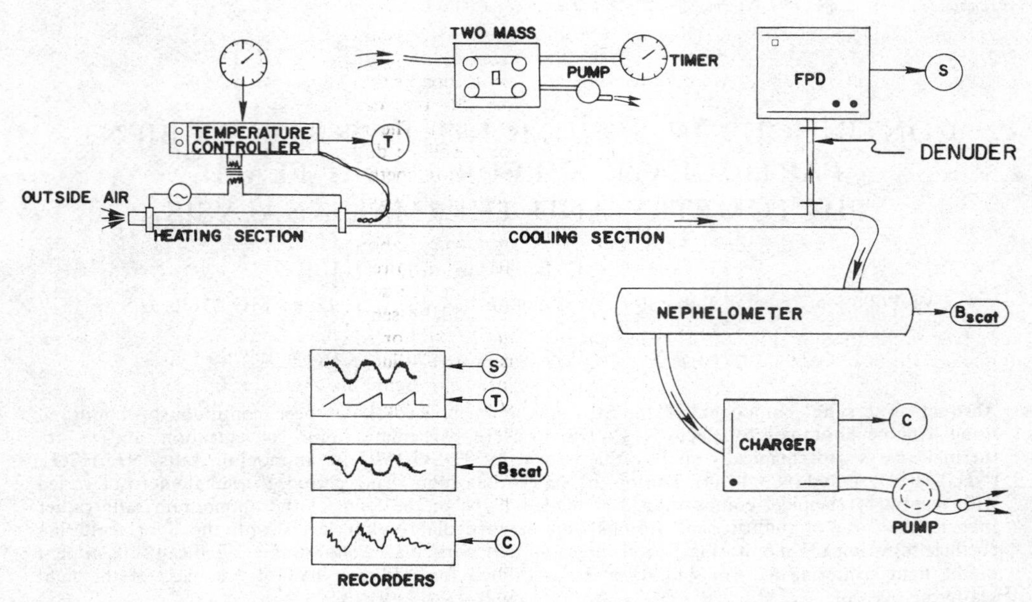

Fig. 1. Diagram of the measurement system.

In situ thermal analysis was performed automatically every 15 min using an FPD sulfur analyzer, a nephelometer and a charger as aerosol detectors. In addition, aerosol was continuously sampled with a TWOMASS filter sampler (Macias and Husar, 1976), which automatically advanced the filter tape every 2 h. The data from the three aerosol detectors, as well as the heater temperature data, were recorded on strip chart recorders. Sampling was done from the fourth floor of Urbauer Hall on the campus of Washington University. The campus is located in a suburban community 10 km west of downtown St. Louis.

Description of the measurement system

The thermal analysis measurement system (see Fig. 1) is a continuous flow sampling system in which the aerosol flows first through a heater tube and then into the various aerosol detectors. The heater tube is a thin walled (0.152 mm) stainless steel tube with an outside diameter of 6.35 mm and a length of 256 cm. Heat is applied to the aerosol by resistance heating, which begins 8 cm from the inlet; the heating takes place over the following 48 cm. The remaining 200 cm of the tube is for cooling the air before entering the instruments. A thermocouple placed 1 cm downstream of the heating section gives a measure of the 'temperature' T_e of the heated aerosol. Since there exists at that point a radial temperature gradient (see Appendix 1), this temperature is merely an indicator of the amount of heating, rather than a unique aerosol property at that point. The amount of heating as indicated by T_e is caused to linearly increase with time by a temperature controller. When the system is used for monitoring, the temperature controller is cycled by a timer so that every 15 min it initiates a 10 min temperature ramp from ambient temperature to 300°C. Thus, in each cycle there is a 5 min period without heating during which the ambient levels are measured, and a 10 min period during which thermal analysis is performed.

The flow rate through the system was maintained at 2.5 l min^{-1}. The FPD analyzer draws 200 ml min^{-1}, and the remaining flow is drawn by an auxiliary pump. At room temperature, this flow corresponds to a Reynolds number of 472, indicating that the flow is laminar throughout. The electrical power for heating is supplied by the temperature controller through a 115V/6V a.c. transformer. At the peak of the temperature ramp ($T_e = 300$°C) the power supplied to the heater is 92.6 W, and the electrical resistance of the heater is 0.152 Ω.

The aerosol is detected by three aerosol detectors, connected as Fig. 1 shows. The charger (Husar *et al.*, 1976) is a particle detector which tends to be most sensitive to particles in the 0.01–0.1 μm size range. The nephelometer (Ahlquist and Charlson, 1967) measures the light scattering coefficient of the aerosol, and is most sensitive to aerosols in the 0.1–1 μm light scattering size class. The flame photometric detector (Lucero and Paljug, 1974) is a Meloy Model 285 continuous monitor, and is highly specific to sulfur compounds. The instrument is made specific to particulates by installing a gas diffusion denuder immediately upstream of the FPD inlet (Fish and Durham, 1971). The denuder is a 1.5 m long stainless steel tube (3.2 mm o.d., 0.152 mm wall) coated on the inside with PbO_2. The denuder scrubs the ambient sulfur gases (SO_2 and H_2S), and in this system also scrubs the various sulfur gases that are created when H_2SO_4 and $(NH_4)_2SO_4$ thermally decompose. For the characteristic tube length of the denuder, the diffusional loss of particles in the size range 0.1–1.0 μm was calculated to be less than 2% (Friedlander, 1977, p. 72). Laboratory experiments have indicated that the sulfur gases evolved by volatilizing H_2SO_4 and $(NH_4)_2SO_4$ aerosols are scrubbed at nearly 100% efficiency. As explained, each aerosol detector is associated with a particular size class or chemical class of particles, so in this system each detector performs a thermal analysis of its associated aerosol component.

Calibration of the system

Each of the aerosol detectors gives a signal which is an indication of the amount of aerosol present. However, exact calibration for mass concentration is very difficult because each of these instruments has a sensitivity which is a specific function of particle size. Other factors such as particle shape and refractive index also enter in, but not as importantly as particle size. The size dependence of the charger (Husar *et al.*, 1976) and the nephelometer (Friedlander, 1976) is known. However, the particle size distribution of the aerosol is still needed in order to calculate the mass concentration.

Urban aerosols in the light scattering size class tend to have size distributions which are roughly similar, even from hour to hour or from day to day. On this basis, the nephelometer manufacturer recommends using a constant value of $c = 38\ (\mu g\ m^{-3})/(10^{-4}\ m^{-1})$ to convert the light scattering coefficient b_{scat} to mass concentration m.

$$m = c \cdot b_{scat}. \tag{1}$$

A 'minimum mass concentration' associated with the light scattering coefficient can be estimated by assuming that all particles are 0.55 μm in diameter (at which point light scattering efficiency is at a maximum) and have an average density of 1.5 $\mu g\ cm^{-3}$. For this case c is about 16.5 $(\mu g\ m^{-3})/(10^{-4}\ m^{-1})$.

Charger size aerosols are smaller in size and have a more transient character than the light scattering aerosol component. The size distribution of these aerosols can change rapidly by formation of nuclei from local sources and by ageing processes such as coagulation, growth and removal. The charger is a good indicator of the smaller particles, but cannot independently be used to estimate the mass concentration of these particles.

Studies of the particle size dependence of the FPD sulfur analyzer have been undertaken in our laboratory, but there is not currently enough data to draw any solid conclusions. Preliminary indications are, however, that there is relatively little size dependence in the size range 0.1–1 μm, where most of the mass of urban sulfate aerosols exists.

The FPD was accurately calibrated for SO_2 using a permeation tube calibration system. A laboratory calibration with pure H_2SO_4 or $(NH_4)_2SO_4$ aerosols is not yet completed. However, filter sampling was done concurrently with the FPD sulfate monitoring during one week of the monitoring period. Subsequent analysis of the filter samples for sulfur using the flash vaporization flame photometric detection (FVFPD) technique (Husar *et al.*, 1976) thus provided a calibration for ambient sulfur aerosols. Figure 2 shows a plot of the calibration points. The filter samples were all 2 h samples, and so in this figure the direct FPD values are the corresponding 2 h average concentrations. Not all of the filter samples were analyzed; some were discarded because it was apparent that collectied aerosol material had come off the spot when the filter tape was unrolled.

As Fig. 2 shows, the direct FPD monitor gives, on the average, about the same measure of particulate sulfur concentration as the filter method. There is quite a bit of scatter to the calibration points (correlation coefficient $r = 0.86$). The reason for this is not quite clear. It may very well be due to errors and uncertainties in the filter collection method, which involves many steps, including handling and extraction of the filter material. A similar calibration of another Meloy 285, done several months ago, using laboratory-generated $(NH_4)_2SO_4$ as the sulfur aerosol, also resulted in scattered calibration points.

There are some indications that the sulfur analyzer is not equally sensitive to all sulfur compounds (Durham *et al.*, 1976). For sulfate aerosols, the difference in sensitivity is probably due largely to the fact that the aerosol is heated just before entering the hydrogen flame. For purposes of stabilizing the flame, the 'burner block' which houses the hydrogen flame is normally maintained at 150°C by a heater and temperature controller system. Since 150°C is well above the volatilization temperature of sulfuric acid aerosol, the H_2SO_4 may be partially volatilized into gases and absorbed on the metal walls of the burner block before reaching the flame. An experiment in which the burner block temperature was varied from 160 to 100°C verified that the sensitivity to H_2SO_4 aerosol does indeed increase with decreasing temperature. A similar experiment with ammonium sulfate aerosol, which has a volatilization temperature above 150°C, indicated very little change in sensitivity when the burner block temperature was varied from 180 to 100°C. Figure 3 illustrates the results of these experiments. Since the FPD is more sensitive to sulfuric acid at lower burner block temperatures, the monitoring was done at a burner block temperature of 110°C, which was the lowest controlled temperature possible without supplemental (fan) cooling.

Thermograms of laboratory generated sulfate aerosols provide a calibration of the thermal analysis system with respect to chemical composition. The aerosols were generated using an ultrasonic nebulizer. The particle size can be varied by using varying strengths of solutions in the nebulizer. Varying the particle size from about 0.3 to 1 μm

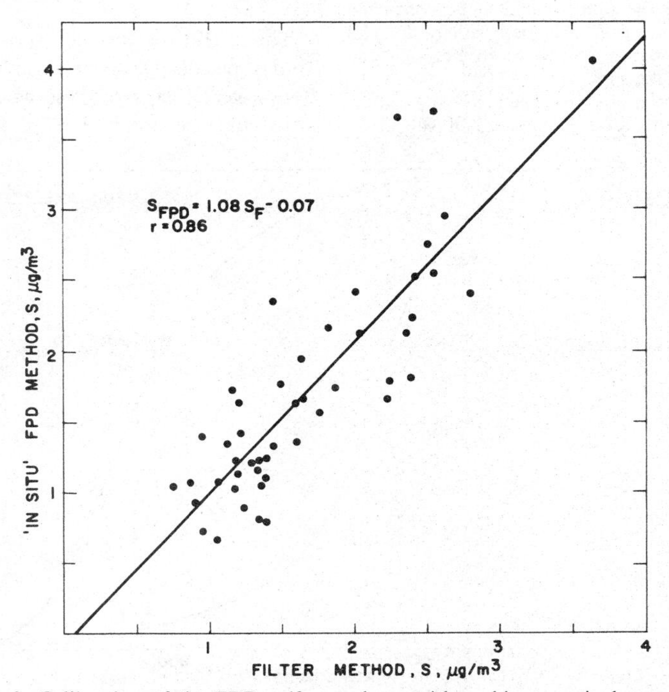

Fig. 2. Calibration of the FPD sulfur analyzer with ambient particulate sulfur.

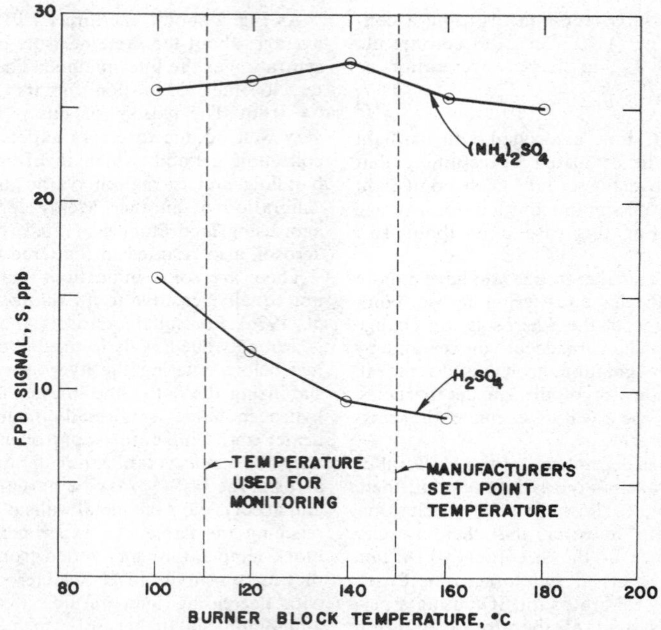

Fig. 3. Effect of the burner block temperature on the FPD sensitivity to H_2SO_4 and $(NH_4)_2SO_4$ aerosols.

had no noticeable effect on the chemical calibrations. Figures 4–6 illustrate the results of the calibrations for each instrument. The thermograms vary from instrument to instrument, but qualitatively they are much alike. Figure 5 shows that the thermograms for ammonium bisulfate NH_4HSO_4 and ammonium sulfate $(NH_4)_2SO_4$ are practically identical. Figure 7 shows the results of a qualitative experiment in which successive syringe injections of gaseous ammonia were added to a large plastic bag of H_2SO_4 aerosol. The experiment demonstrates that sulfuric acid aerosol will react with gaseous ammonia to form NH_4HSO_4 or $(NH_4)_2SO_4$. Apparently, if the NH_3 concentration is insufficient for the sulfuric acid to be converted completely to NH_4HSO_4, an intermediate will be formed which exhibits a thermogram which is in between the H_2SO_4 and NH_4HSO_4 thermograms.

The calibration thermograms presented here should be considered as the response of a given aerosol substance to heating in this particular measurement system only. This is because there are radial and axial temperature gradients in the heater, with the result that the measurement of the heater exit temperature is peculiar to this system. Detailed consideration of the temperature field is given in Appendix 1. The thermogram of a given substance is fundamentally determined by the volatilization of the aerosol in response to the changing temperature field in the heater. The volatilization of particles is discussed in Appendix 2.

RESULTS OF THE MONITORING PROGRAM

The FPD performed very reliably during the 16-day period. Data for two short periods of several hours were lost, but these were due to failures of a strip chart recorder. The particulate sulfur concen-

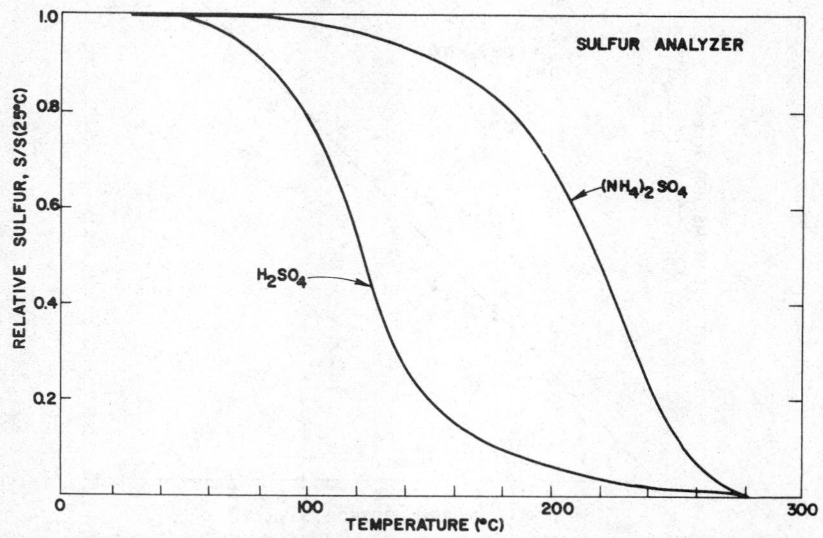

Fig. 4. Chemical calibration of the thermal analysis system using the sulfur analyzer.

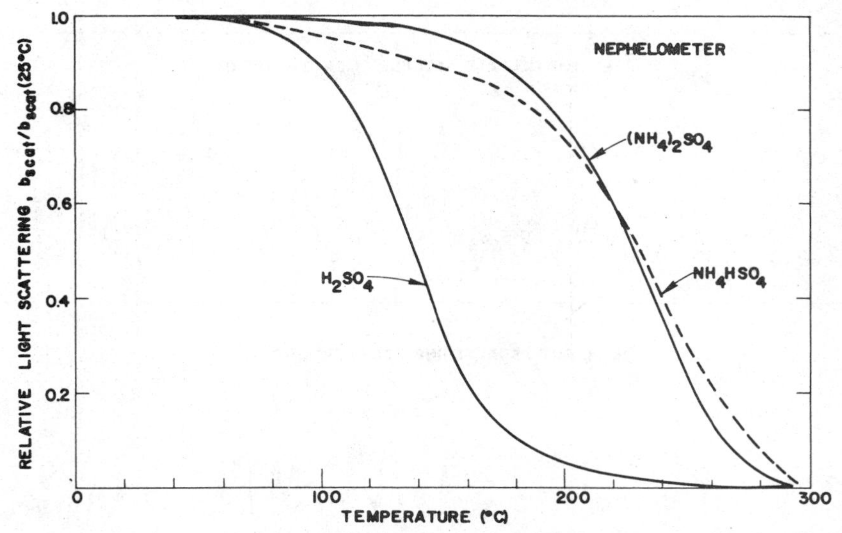

Fig. 5. Chemical calibration of the thermal analysis system using the nephelometer.

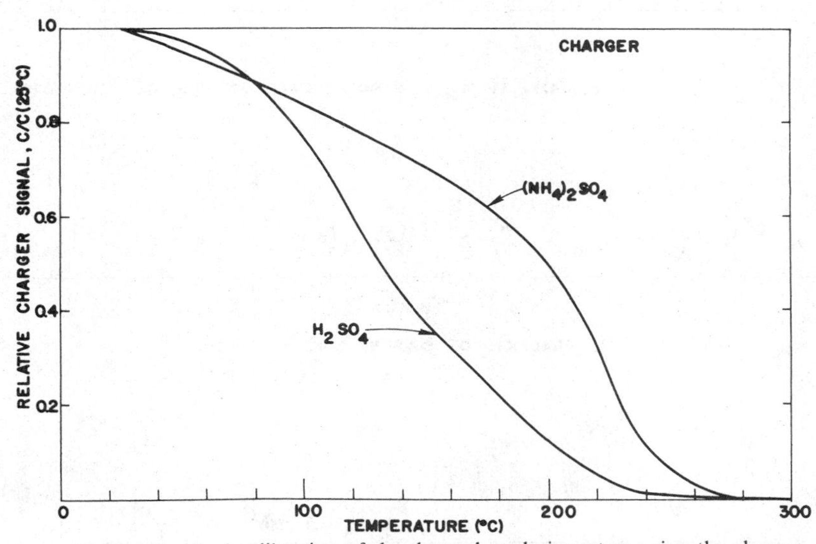

Fig. 6. Chemical calibration of the thermal analysis system using the charger.

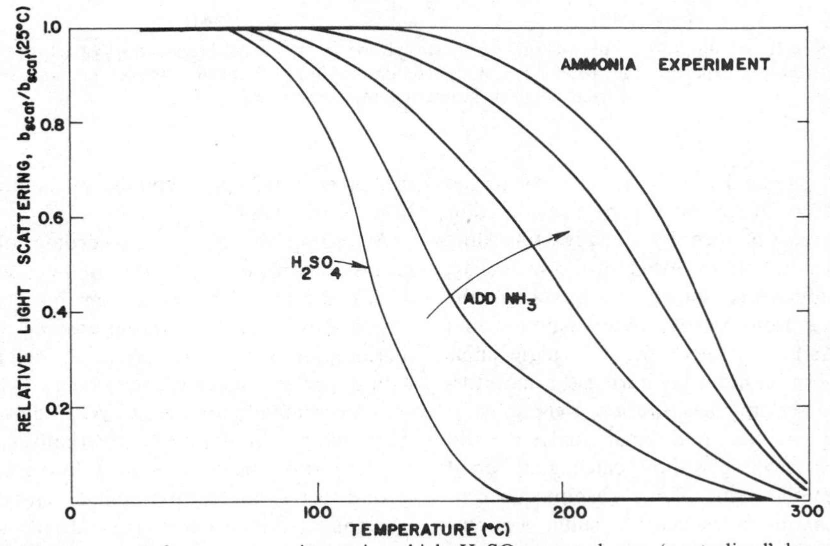

Fig. 7. Thermograms from an experiment in which H_2SO_4 aerosol was 'neutralized' by successive injections of gaseous ammonia into the large plastic bag containing the aerosol.

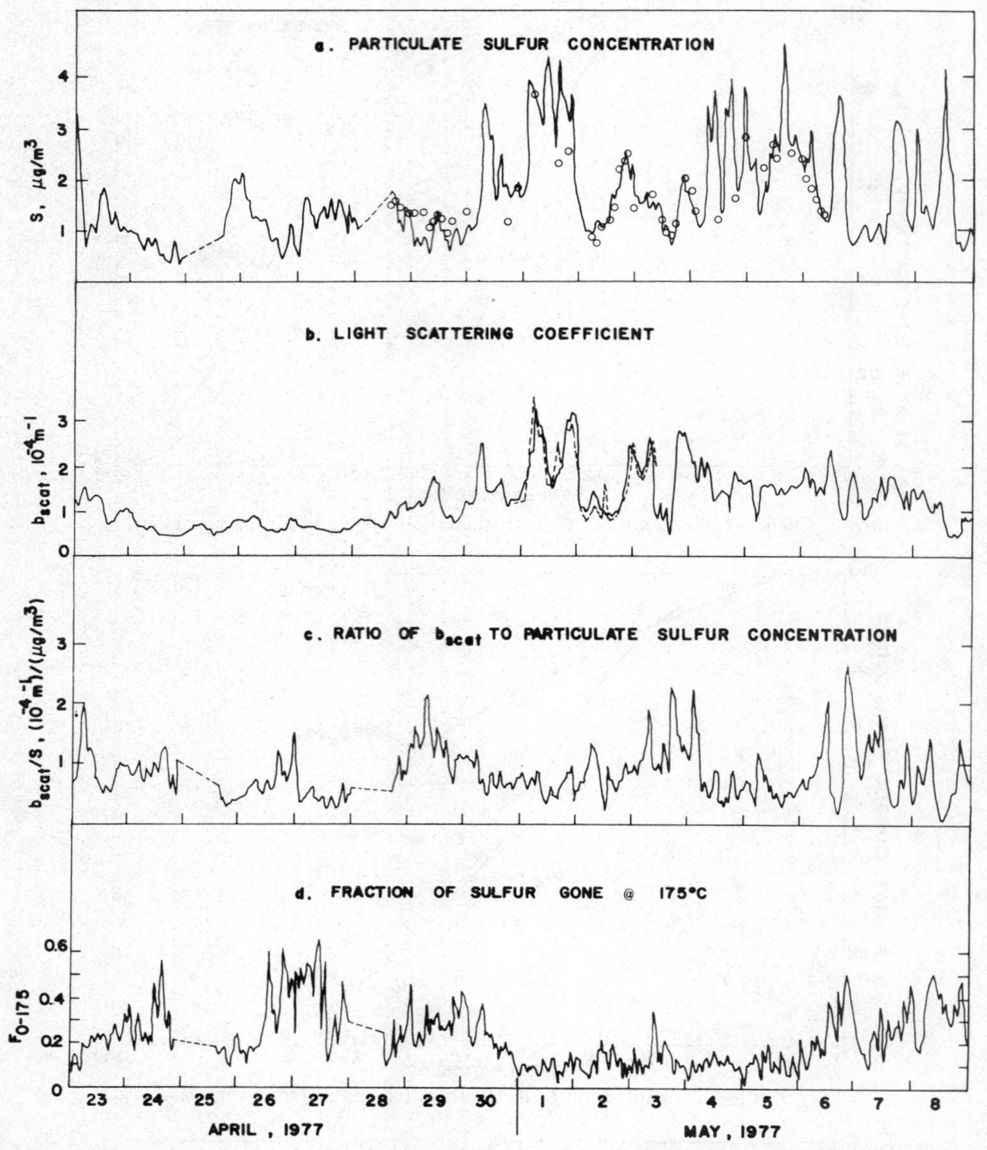

Fig. 8. Results of the 16-day monitoring program, showing (a) particulate sulfur concentration; (b) light scattering coefficient; (c) ratio of b_{scat} to particulate sulfur; (d) the fraction of particulate sulfur volatilized between ambient temperature and 175°C.

tration as given by the FPD for the 16-day period is shown in Fig. 8a. These values of particulate sulfur concentration were obtained by directly converting the FPD readings (in ppb) recorded on the strip chart to $\mu g\ m^{-3}$ of sulfur. Also shown in Fig. 8a are the calibration points from the 2 h filter samples that were analyzed. As Fig. 8a shows, the particulate sulfur concentration varied quite a bit during the monitoring period. Changes in concentration were gradual enough, however, so that a thermal analysis could be performed every 15 min without causing any great uncertainty as to the value of the absolute concentration. The maximum particulate sulfur concentration observed during the monitoring period was 4.4 $\mu g\ m^{-3}$ (3.3 ppb) and the minimum was 0.33 $\mu g\ m^{-3}$ (0.25 ppb), still above the lower detection limit of the FPD.

An example of strip chart recordings of the thermograms is shown in Fig. 9. The figure shows recordings of two different 1-h periods on July 17, 1977. The first group is particularly interesting because each thermogram shows two separate declines in signal with an inflection point in between; this type is only very occasionally observed. We believe that these thermograms are due to an externally mixed aerosol, the first drop being due to sulfuric acid, and the second due to ammonium sulfate salts. Recent calibrations have indicated that externally mixed sulfate aerosols (i.e. mixtures of pure sulfuric acid particles and pure ammonium sulfate particles) exhibit such

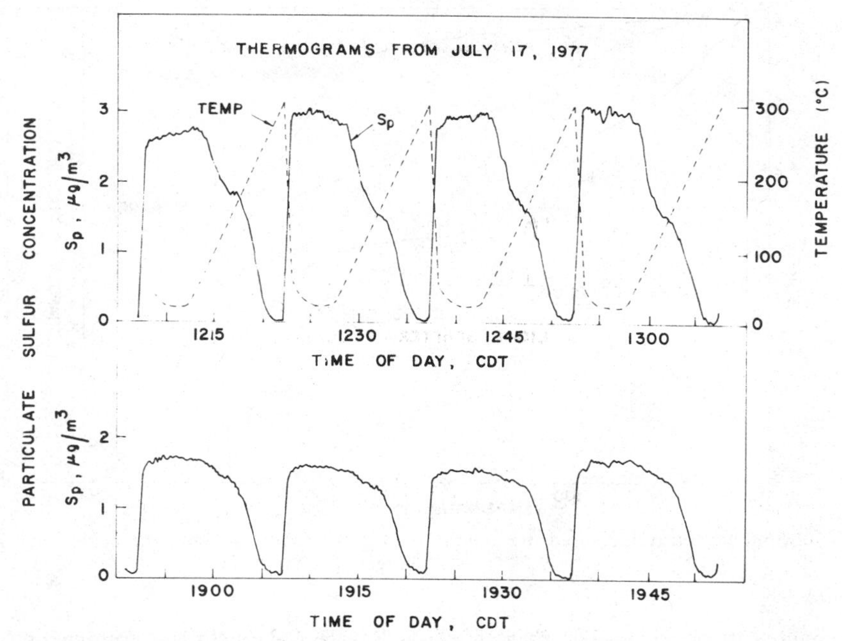

Fig. 9. Examples of strip chart recordings of thermograms from July 17, 1977.

thermograms with an inflection point, whereas internally mixed sulfate aerosols exhibit thermograms varying in the manner of those of Fig. 7.

The particular FPD analyzer that we used unfortunately had a zero drift of about 2–3 ppb over a 24-h period, which is not within the factory specifications. Fortunately, the periodic thermal analyses provided an automatic zero reference every 15 min, as the ambient sulfate aerosol is completely volatilized at 300°C. The zero was also frequently checked with a high efficiency, low pressure drop filter. The filter zero signal and the signal at 300°C were the same, indicating that all of the particulate sulfur was volatilized below 300°C. This indicates that practically all of the ambient particulate sulfur was in the form of volatile sulfates such as H_2SO_4 and $(NH_4)_2SO_4$. The denuder was checked every few days by temporarily replacing it with a new denuder. The signal from the FPD was always found to be the same for both denuders, indicating removal of SO_2 with equal, presumably near 100%, efficiency. The denuders have also been tested for efficiency using a permeation tube system as a known SO_2 source, and have been found to remove SO_2 with an efficiency of practically 100%.

The nephelometer was unfortunately unavailable for most of the 16-day period, because it was being used for another monitoring program. However, it was in the system for about 70 h, during which we obtained nephelometer thermograms. Figure 8b shows the variation in light scattering coefficient during the monitoring period. These data were obtained from a St. Louis County air monitoring station located in an urban area about 2 km west of the Washington University campus. Also shown on Fig. 8b, as a dashed line, is the light scattering coefficient as measured with our nephelometer while it was in the system. As Figs. 8a and 8b show, there is a rough correlation between particulate sulfur and b_{scat} levels, but the correlation is far from perfect. This is to be expected, since the light scattering aerosol contains other substances besides sulfate, and since the light scattering coefficient depends on the particle size. The average b_{scat} during the 16-day period was $1.2\,10^{-4}\,m^{-1}$. This corresponds to an estimated fine particle mass concentration of $20\,\mu g\,m^{-3}$ using the 'minimum' value of $c = 16.5\,(\mu g\,m^{-3})/(10^{-4}\,m^{-1})$ and $44\,\mu g\,m^{-3}$ using the manufacturer's suggested value. The average mass concentration of particulate sulfur during the period, assuming that all of the particulate sulfur was $(NH_4)_2SO_4$, was $7\,\mu g\,m^{-3}$. Thus, particulate sulfur accounted for a substantial fraction of the mass of the light scattering aerosol during this period, but the fraction was clearly less than one-half. This result is in essential agreement with previous studies in which the element sulfur was found to comprise about 10% of the fine particle mass of St. Louis aerosol (Macias and Husar, 1976). Figure 8c shows the ratio of light scattering coefficient to particulate sulfur concentration. The variability can be due to changes in the size distribution of the light scattering aerosol, or to changes in the sulfur fraction of this aerosol.

Thermal analysis provides some scant information about the chemical constituency of the light scattering aerosol, and substantially more information about the chemical constituency of the sulfate aerosol. For example, Fig. 10 shows the light scattering, charge, and sulfur thermograms of aerosol sampled at 2100 h, May 1. As the figure shows, practically all of the decrease in particulate sulfur occurs between 175 and 300°C, which is the type of thermogram displayed

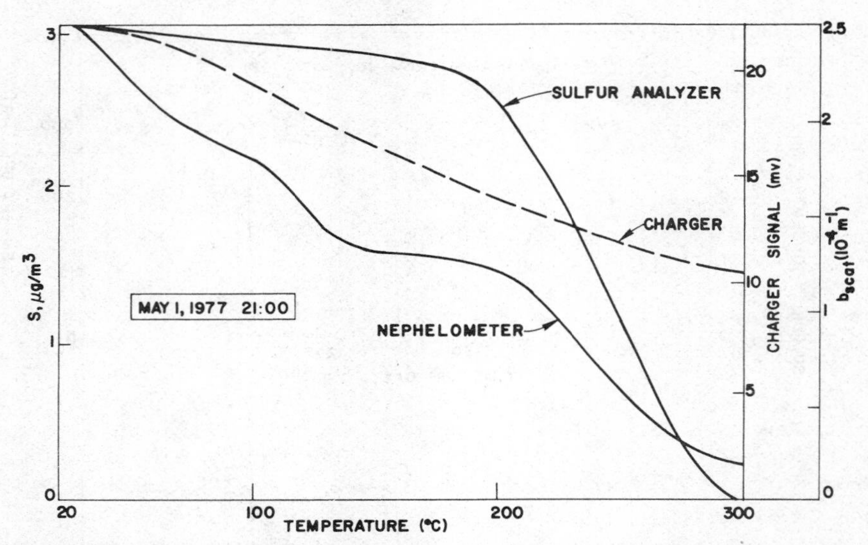

Fig. 10. Simultaneously measured thermograms using the sulfur analyzer, the nephelometer, and the charger.

by the ammonium salts of sulfate. The charger thermogram drops steadily between 25 and 300°C; the thermogram indicates that about half of the charger size aerosol is non-volatile when heated to 300°C. The light scattering thermogram drops in two steps: first dropping between 25 and 140°C, leveling off at 140°C, and then dropping between 180 and 300°C. In the absence of information from the sulfur analyzer, one might have made the conjecture that the first drop is due to sulfuric acid, and the second due to ammonium sulfate salts. The sulfur thermogram, however, indicates that there is practically no sulfuric acid present. The first drop in the light scattering thermogram is therefore due to some other substance. Furthermore, the sulfur analyzer indicates a mass concentration of 12.5 $\mu g\ m^{-3}$ (assuming all of the sulfate particles are $(NH_4)_2SO_4$); this does not account for all of the light scattering aerosol that volatilizes between 175 and 300°C, which must have a mass concentration of at least 20.6 $\mu g\ m^{-3}$. Assuming that our calibration of the FPD for ambient particulate sulfur is essentially correct, this means that the portion of the light scattering aerosol that volatilizes between 175 and 300°C was in this case composed of about 25–50% sulfate.

Figure 11 shows several sulfur thermograms taken at various times during the 16-day period. This figure is somewhat similar to Fig. 7, suggesting that the chemical composition of the ambient sulfate aerosol varies from one time to another, depending on the chemical and physical history of the aerosol. In a manner similar to the plots of Figs. 8a and 8b, we have illustrated the variation in chemical composition in Fig. 8d. The parameter plotted, F_{0-175}, which is

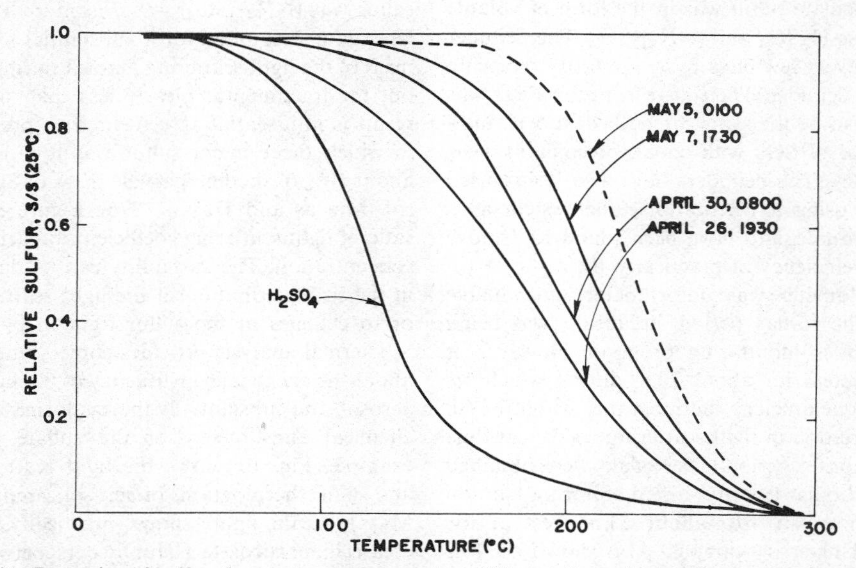

Fig. 11. Examples of sulfur thermograms measured at different times during the monitoring period.

the fraction of the sulfur aerosol volatilized between ambient temperature and 175°C, is an indicator of the 'acidity' of the aerosol. High values of F_{0-175} indicate the presence of sulfuric acid, and low values indicate ammonium sulfate salts (refer to Fig. 4). Although the parameter F_{0-175} is not strictly a quantitative measure of the percentage of sulfate that is sulfuric acid, it could be used as such to make a rough estimate of the amount of sulfuric acid. So for example, referring to Fig. 8, one could estimate that on April 27 at 0800 h roughly 50% of the particulate sulfur was in the form of sulfuric acid, which would correspond to a mass concentration of about 2.3 $\mu g\, m^{-3}$ of sulfuric acid. As Fig. 8d shows, the chemical composition of the sulfate aerosol varied considerably during the monitoring period, often changing significantly from hour to hour. For the most part, the aerosol tended to be in the form of ammonium sulfate salts rather than in the form of sulfuric acid. This fact is illustrated by Fig. 8d, showing that F_{0-175} was 0.2 or less during 54% of the monitoring period. As Fig. 5 shows, a value of 0.2 for F_{0-175} would correspond to pure ammonium sulfate or ammonium bisulfate. Although it appears that the sulfate aerosol tended to be in the form of the ammonium sulfate salts during this period, there are clearly some times when the aerosol is more acidic. Interestingly enough, variation in the acidity of sulfate aerosols has also been observed by Cunningham and Johnson (1976) in the city of Chicago, using i.r. spectroscopy.

As yet we have not attempted to correlate the variation in chemical composition with meteorological factors such as r.h., air mass trajectories, etc., but such work is planned. Several studies have already shown that the concentration and chemical composition of atmospheric aerosols is related to the air mass history. In particular, Brosset (1976) has identified two distinct types of particulate pollution episodes, called 'black episodes' and 'white episodes', each associated with distinctly different air mass trajectories into Sweden, and each having a distinctly different acidity, as determined by analysis for hydrogen ion and ammonium ion. Vanderpol *et al.* (1975), using the 'humidogram method' (Charlson *et al.*, 1973), have examined the chemical character of ambient aerosols in St. Louis during September 1973, and report that the presence of $(NH_4)_2SO_4$ is associated with the arrival of polar air masses. The results from these studies reaffirm the notion that particulate sulfur pollution is due to conversion processes taking place over large regional areas, rather than due to local sources.

CONCLUSIONS

The newer and more sensitive commercial FPD sulfur analyzers can be used as reliable continuous monitors of the ambient sulfur aerosol. The instrument is made specific to sulfur particles by installing a diffusion tube sulfur gas denuder preceding the inlet of the instrument. Our sulfate monitor was apparently capable of measuring particulate sulfur levels as low as 0.1 $\mu g\, m^{-3}$. There is some question as to the accuracy of our sulfate monitor, as there were some problems with calibrating the instrument. Further studies should resolve this issue.

Thermal analysis can be used along with the FPD particulate sulfur monitor to distinguish sulfuric acid from its ammonium salts. *In situ* thermal analysis with the FPD shows promise as a tool for studying the concentration and chemical composition of urban and rural sulfate aerosols, with the eventual aim of understanding the formation and transport processes. Simultaneous thermal analysis of both the light scattering and sulfate components of the ambient aerosol should be useful in assessing the impact of sulfate aerosols on visibility reduction.

Acknowledgements—This research was supported by the Federal Interagency Energy-Environment Research and Development Program, through Environmental Protection Agency Grant No. R803896, and by EPA Grant No. R803115. We wish to thank Prof. Edward Macias and Dr. Richard Delumyea of the Washington University Chemistry Department for their help and suggestions, and Peggy Fuller and Jean Patterson for performing the sulfate analysis of filter samples.

REFERENCES

Ahlquist N. C. and Charlson R. J. (1967) A new instrument for evaluating the visual quality of air. *J. Air Pollut. Control Ass.* **17,** 467.

Brosset C. (1976) Air borne particles: black and white episodes. *Ambio* **5,** 157–163.

Charlson R. J., Vanderpol A. H., Covert D. S., Waggoner A. P. and Ahlquist N. C. (1973) Sulfuric acid–ammonium sulfate aerosol: optical detection in the St. Louis region. *Science* **184,** 156–158.

Cunningham P. T. and Johnson S. A. (1976) Spectroscopic observation of acid sulfate in atmospheric particulate samples. *Science* **191,** 77–79.

Durham J. L. and Wilson W. E. (1976) Continuous Measurement of Sulfur in Submicrometric Aerosols. U.S. Environmental Protection Agency, EPA-600/3-76-088.

Eckert E. R. G. and Drake R. M. (1959) *Heat and Mass Transfer.* McGraw-Hill, New York.

Fish B. R. and Durham J. L. (1971) Diffusion coefficient of SO_2 in air. *Envir. Lett.* **2,** 13–21.

Friedlander S. K. (1976) *Smoke, Dust and Haze: Fundamentals of Aerosol Behavior.* Wiley Interscience, New York, pp. 134–136.

Huntzicker J. J., Isabelle L. M. and Watson J. G. (1975) The continuous monitoring of particulate sulfate by flame photometry. *Int. Conf. on Environmental Sensing and Assessment,* Las Vegas, NE, Sept. 14–19, 1975.

Huntzicker J. J. and Hoffman R. (1978) The continuous monitoring of particulate sulfur compounds by flame photometry. *Atmospheric Environment* **12,** 83–88.

Husar J. D., Husar R. B. and Stubits P. K. (1975) Determination of submicrogram amounts of atmospheric particulate sulfur. *Analyt. Chem.* **47,** 2062–2064.

Husar R. B. and Shu W. R. (1975) Thermal analysis of the Los Angeles smog aerosol. *J. appl. Met.* **14,** 1558–1565.

Husar R. B., Macias E. S. and Dannevik W. P. (1976) Measurement of dispersion with a fast response aerosol detector. *3rd Symp. on Atmospheric Turbulence, Diffusion and Air Quality,* Raleigh, N.C., October 19–22, 1976.

Kittleson D. B., McKenzie R., Vermeersch M., Dorman F., Pui D., Linne M., Liu B. and Whitby K. (1978) Total sulfur aerosol concentration with an electrostatically pulsed flame photometric detector system. *Atmospheric Environment* **12**, 105–111.

Kiyoura R. and Urano K. (1970) Mechanisms, kinetics, and equilibrium of thermal decomposition of ammonium sulfate. *Ind. Engng Chem. Process Des. Develop.* **9**, 489–494.

Kojima H., Sekikawa T. and Tanaka F. (1973) On the volatility of large particles in the urban and oceanic atmosphere. *J. met. Soc., Japan* **53.**

Lucero D. P. and Paljug J. W. (1974) Monitoring sulfur compounds by flame photometry. In *Instrumentation for Monitoring Air Quality.* ASTM Special Publication 555, Philadelphia, PA.

Macias E. S. and Husar R. B. (1976) Atmospheric particulate mass measurement with beta attenuation mass monitor. *Envir. Sci. Tech.* **10**, 904–907.

Pueschel R. F., Bodhaine B. A. and Mendoca B. G. (1973) The proportion of volatile aerosols on the island of Hawaii. *J. appl. Met.* **12**, 308–315.

Twomey S. (1968) The evaporation of cloud nuclei in the northeastern United States. *J. Rech. atmos.* **3**, 281–285.

Twomey S. (1971) The evaporation of submicron aerosol particles. *J. Rech. atmos.* **6**, 1372–1378.

Vanderpol A. H., Carsey F. D., Covert D. S., Charlson R. J. and Waggoner A. P. (1975) Aerosol chemical parameters and air mass character in the St. Louis region. *Science* **190**, 570.

APPENDIX 1

Radial and axial temperature distributions in the heater tube

The heat transfer process in the tube closely resembles the classical problem of laminar, axially symmetric flow with constant heat addition, for which the momentum and energy equations are respectively

$$\frac{\partial u}{\partial t} + u\frac{\partial u}{\partial x} = -\frac{1}{\rho}\frac{\partial P}{\partial x} + \frac{\mu}{\rho}\left(\frac{\partial^2 u}{\partial r^2} + \frac{1}{r}\frac{\partial u}{\partial r}\right) \quad (2)$$

$$\frac{\partial T}{\partial t} + u\frac{\partial T}{\partial x} = -u\frac{\partial T}{\partial x} + \frac{k}{\rho c_p}\left(\frac{\partial^2 T}{\partial r^2} + \frac{1}{r}\frac{\partial T}{\partial r}\right) \quad (3)$$

where u is velocity, x is the axial coordinate, r is the radial coordinate, P is pressure, ρ is density, μ is viscosity, T is temperature, k is thermal conductivity, knd c_p is the specific heat. The terms on the left hand side of each equation can be dropped, since they reflect unsteady and inertial effects. The solution to this set of coupled equations is straightforward (Eckert and Drake, 1959), and is as follows:

$$u(x,r) = -\frac{1}{4\mu}\frac{\partial P}{\partial x}(R^2 - r^2) \quad (4)$$

$$T(x,r) = T_0 + \frac{2q}{Gc_pR}x + \frac{4q}{kR^3}\left(\frac{R^2r^2}{4} - \frac{r^4}{16}\right) \quad (5)$$

where T_0 is the centerline ($r = 0$) temperature, q is the wall heat flux per unit length, G is the mass flow rate, and R is the tube radius. The bulk mean temperature is obtained by flow-averaging the radial temperature profile, and is as follows

$$T_b = T_0 + \frac{7}{24}\frac{qR}{k}. \quad (6)$$

The wall temperature follows directly from Equation (5) with $r = R$

$$T_w = T_0 + \frac{3}{4}\frac{qR}{k}. \quad (7)$$

Corresponding to a flow rate of $2.5\,\mathrm{l\,min^{-1}}$ and a heater exit temperature of 300°C, the temperature differences $T_b - T_0$ and $T_w - T_0$ are calculated as follows;

$$T_b - T_0 = 34°\mathrm{C}$$

$$T_w - T_0 = 88°\mathrm{C}.$$

These calculated temperature values should be considered only as rough estimates, because the actual heater differs in some respects from the idealized problem just considered. For example, since the metal tube becomes very hot, there are appreciable convective and radiative losses on the outside of the tube.

From Equation (5), the magnitude of the radial temperature gradient is proportional to q, which in turn is approx proportional to the flow rate for a given heater exit temperature. Therefore, one way of reducing the radial temperature gradient is to reduce the flow rate.

APPENDIX 2

Discussion of the volatilization of aerosol particles

While we cannot at this point fully explain the complicated problem of the chemistry and physics of volatilizing particles in the heater tube, there are certain facts and observations which are pertinent to aerosol thermal analysis and to the understanding of our measurement system. Sulfuric acid, being a hygroscopic substance, always has water present in its particles, and so those particles are a two-component evaporating system. Ammonium sulfate particles chemically decompose into various gases upon heating, making the analysis of this system a quite difficult problem. An additional difficulty is the fact that there exists a fairly complicated temperature field in the heater (see Appendix 1).

The fact that the thermograms of several aerosol substances have similar 'S' shapes and decline over a rather broad range of temperatures suggests the possibility that the 'shape' and 'width' of the thermograms may have something to do with the evaporation rates and residence times of particles that pass through the heater. There are other effects that also play a role. The radial temperature gradient would itself cause a 'width' effect, since the particles passing near the wall experience relatively higher temperatures. In addition, substances such as ammonium sulfate which undergo chemical decomposition do so over a range of temperatures, even in bulk form (Kiyoura *et al.*, 1970).

We have often observed that the thermograms of a given substance seem to be relatively insensitive to flow rate. In one experiment, the flow rate was increased by a factor of three while the heater exit temperature was held constant. The aerosol signal, which was that of sulfuric acid at 120°C (see Fig. 4), was reduced by about 20%. This trend is the opposite of what would be expected from the reduced residence time, but is the same trend as would be expected by the fact that the radial temperature gradient would increase with increasing flow rate (see Appendix 1). The result indicates that the latter effect is dominant, but does not tell about the relative importance of these two competing effects. In any case, the fact that the thermograms are insensitive to the flow rate is fortunate because it means that precise control of the flow is not critically important in terms of the chemical calibration of the system.

Atmospheric Environment Vol. 12, pp. 99–104. Pergamon Press 1978. Printed in Great Britain.

THE AEROSOL MOBILITY CHROMATOGRAPH: A NEW DETECTOR FOR SULFURIC ACID AEROSOLS

B. Y. H. Liu, D. Y. H. Pui, K. T. Whitby and D. B. Kittelson
Particle Technology Laboratory, Mechanical Engineering Department, University of Minnesota, Minneapolis, MN 55455, U.S.A.

Y. Kousaka
University of Osaka Prefecture, Mozu, Umemachi, Sakai, Osaka, Japan

and

R. L. McKenzie
National Bureau of Standards, Washington, DC 20234, U.S.A.

(*First received* 13 *June* 1977 *and in final form* 26 *August* 1977)

Abstract—A new instrument has been developed for measuring sulfuric acid aerosols. The instrument is called an Aerosol Mobility Chromatograph since it is based on the electrical mobility of aerosol particles and operates in a way similar to that of the conventional liquid or gas chromatograph. The particle diameter range of the instrument is from 0.005 to 0.2 μm and the sensitivity (for detecting monodisperse sulfuric acid aerosols), from 0.01 to 10^{-5} $\mu g\ m^{-3}$, depending upon the specific particle detector used. This paper describes the operating principle of the AMC and the performance characteristics of a prototype device developed at the Particle Technology Laboratory, University of Minnesota.

INTRODUCTION

There has been considerable interest in recent years in the study of sulfur-containing particles in the atmosphere and many techniques have been developed for their measurement. Among the more widely used techniques are those for the measurement of sulfate using wet chemical analysis (Brosset and Fern, 1976; Appel *et al.*, 1977). The sulfur content of particles can also be determined by X-ray fluorescence (Dzubay and Stevens, 1975; Dzubay, 1977). Recently, several techniques based on the use of the flame-photometric detector (FPD) have also been developed. The FPD, used primarily for gas measurements on SO_2 and H_2S, has been adapted for particulate sulfur measurement by several different schemes. In one scheme (Roberts and Friedlander, 1976; Husar *et al.*, 1975; Tanner *et al.*, 1977), the particles are collected, and then flash-volatilized and measured by the FPD. In a second scheme (Huntzicker *et al.*, 1977), the SO_2 and H_2S gases are chemically scrubbed, leaving the sulfur-containing particles to enter the FPD and be detected. In a third scheme (Kittelson *et al.*, 1977), the gaseous sulfur are not scrubbed, but the aerosol particles are electrostatically chopped. The resulting AC signal from the FPD is then measured with a lock-in amplifier. Other techniques that have been used include radioactive tracer (Forrest and Newman, 1977), micro-Raman spectroscopy (Etz *et al.*, 1977), laser-Raman spectroscopy (Stafford *et al.*, 1976), in-beam gamma-ray spectroscopy (Macias, 1976) and ion chromatography (Mulick, 1976). Table 1 summar-

Table 1. Principal techniques for particulate sulfur measurement

Technique	Element or compound measured	Working range* or detection limit**	References	Comment
1. Wet chemical analysis	SO_4^{2-}		Appel *et al.* (1977)	
(a) Methylthymol Blue		6–60 $\mu g\ ml^{-1}$ extract*	Brosset & Fern (1976)	
(b) Barium Chloranilate		13–50 $\mu g\ ml^{-1}$ extract*		
(c) Modified Brosset		3–13 $\mu g\ ml^{-1}$ extract*		
(d) Brosset		0.2 $\mu g\ SO_4^{2-}\ ml^{-1}$**		
2. X-ray fluorescence	S	10^1–10^3 $ng\ cm^{-2}$ or	Dzubay & Stevens (1975)	
		10^1–10^3 $ng\ m^{-3}$**	Jaklevic & Walter (1977)	
3. Flame-photometric detector				
(a) Flash volatilization	S	0.06 $\mu g\ ml^{-1}$ extract**	Husar *et al.* (1975)	
			Roberts & Friedlander (1976)	
	H_2SO_4 & HSO_4^-	1 $\mu g\ m^{-3}$**	Tanner *et al.* (1977)	
(b) Gas scrubbing	S	1 $\mu g\ m^{-3}$ for H_2SO_4**	Huntzicker *et al.* (1977)	
with thermal speciation	SO_4^{2-}, H_2SO_4			
(c) Electrostatic aerosol				
chopping	S	0.3 $\mu g\ m^{-3}$**	Kittelson *et al.* (1977)	Meloy SA285 Detector with 1 ppb min. det. level
with thermal speciation	SO_4^{2-}, H_2SO_4		McKenzie *et al.* (1976)	
4. Radioactive tracer	Soluble sulfate	1 μg S/6 in^2 filter**	Forrest & Newman (1977)	Sulfite is an interference
5. Micro-Raman spectroscopy	HSO_4^-, SO_4^{2-}, SO_3^{2-}		Etz *et al.* (1977)	Lower particle size limit: ~1 μm
6. Laser-Raman spectroscopy	SO_4^{2-}	10 ppb**	Stafford (1976)	
7. Gamma-ray spectroscopy			Macias (1976)	
8. Ion chromatography	SO_4^{2-}	0.3 $\mu g\ ml^{-1}$ extract**	Mulick (1976)	

izes the principal techniques used and their pertinent characteristics.

It is clear that while many techniques have been developed for particulate sulfur measurement, only a few can provide chemical speciation capabilities. Chemical speciation is important because of the different health effects of various sulfur compounds. Among the sulfur compounds of interest in health-effect studies, sulfuric acid is probably one of the most important because of its known irritant potential and effect on the respirable airways. In this paper, a new instrument for measuring sulfuric acid aerosols, known as an Aerosol Mobility Chromatograph, is described.

OPERATING PRINCIPLE AND BASIC FEASIBILITY

The Aerosol Mobility Chromatograph operates by growing sulfuric acid particles in a humid environment and measuring the particle growth with a differential mobility analyzer. The growth of sulfuric acid and other hygroscopic particles in humid atmospheres is well-known and has been studied extensively by the use of the nephlometer (Covert, 1974; Charlson *et al.*, 1974). However, in this paper, a precise particle size measuring device is used to more fully utilize the principle of particle identification by means of its hygroscopic properties.

To demonstrate the basic operating principle and feasibility, the system shown in Fig. 1 has been developed. The purpose of the system is to show that H_2SO_4 particles in an aerosol containing a mixture of particles can indeed be separated from other non-hygroscopic, or less hygroscopic particles and be identified by the AMC principle. In the particular experiments performed, particles of H_2SO_4 and K_2SO_4 were generated by atomization with the use of two syringe-pump atomizers shown. The syringe-pump atomizer (Liu and Lee, 1976) was chosen because of its stability, i.e., its ability to generate a stable aerosol with a constant concentration and particle size distribution. Following atomization, the solution droplets were mixed and allowed to enter a diffusion dryer where the relative humidity was lowered to about 8%. The mixed aerosol was then passed through a Krypton-85 neutralizer (see Liu and Pui, 1974a for a description of the operating principle of the Krypton-85 neutralizer) to obtain a Boltzmann equilibrium charge on the particles, and then introduced into the first of two differential mobility analyzers (DMA). In the first DMA, particles of H_2SO_4 and K_2SO_4 within a narrow size range were extracted according to their electrical mobility. This monodisperse aerosol was then humidified to about 53% relative humidity by means of a packed tower humidifier. The 53% humidity was sufficiently high to cause a significant growth in size for the H_2SO_4 particles, but not sufficiently high to cause the K_2SO_4 to deliquesce and grow. The size distributions of the H_2SO_4 and K_2SO_4 particles in this new humid environment was then measured with the second DMA, which functions as a particle mobility, and size, analyzer. The DMA used in these experiments were developed at the University of Minnesota and had been used for a variety of purposes including particle size classification, and monodisperse aerosol generation (Liu and Pui, 1974b; Liu *et al.*, 1974; Liu and Kim, 1976; Liu *et al.*, 1975). The operating principle of the device has been analyzed in detail by Knutson and Whitby (1975a, b).

A variety of detectors can be used with the above described system to detect particles emerging from the second DMA. In the system shown in Fig. 1, an electrometer current sensor is used which detects the singly charged particles emerging from the device by collecting the particles on an insulated filter and measuring the rate of charge collection by a sensitive electrometer. In Fig. 2, the output of the electrometer current sensor is shown as a function of the applied voltage on the collector rod in the second DMA, the voltage on the collector rod in the first DMA being kept fixed in these experiments. Two curves are

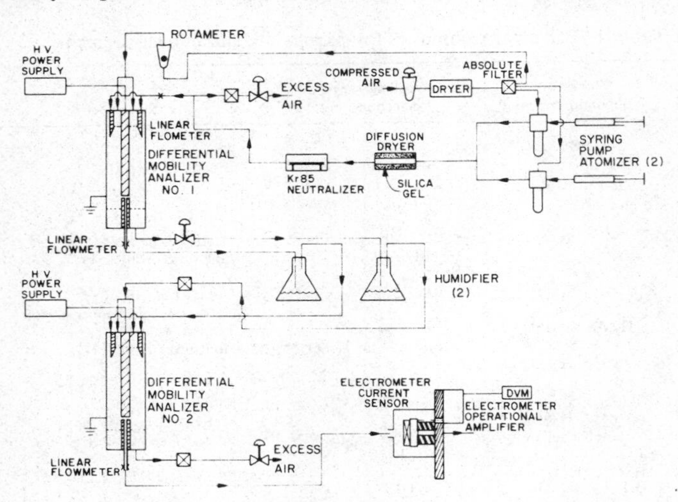

Fig. 1. System demonstrating the principle of the Aerosol Mobility Chromatograph.

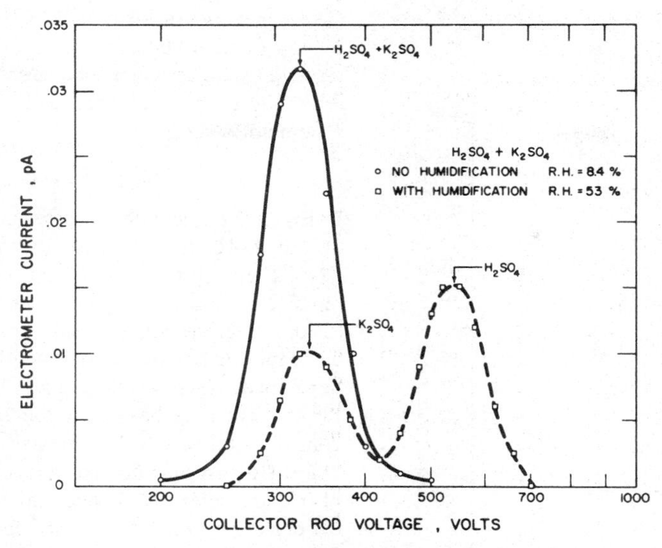

Fig. 2. Output voltage–current curve for the Aerosol Mobility Chromatograph showing the separation of H_2SO_4 and K_2SO_4 particles.

shown. The solid curve corresponds to the case where there is no humidification between the two DMAs. The peak in the curve indicates the monodispersity of the particles being produced by the first DMA and that of the particles being measured by the second DMA. However, if prior to entering the second DMA, the aerosol is first humidified as described above, the dotted curve shown in Fig. 2 is obtained. The two peaks in the curve correspond to the two different sized particles present under the new humidity conditions. The peaks are quite distinct, indicating that the particles are now different in size. It should be further noted that, if the collector rod voltage is set to, say, the value corresponding to the second peak—approximately 550 volts in the case shown—the aerosol emerging from the second DMA would consist of H_2SO_4 particles only. The K_2SO_4 particles, being smaller in size, would be rejected by the second DMA. By this means, the hygroscopic H_2SO_4 particles can be physically separated from other non-hygroscopic or less hygroscopic particles in an aerosol mixture. This particular property of the device is analogous to that of the conventional liquid or gas chromatograph in which physical separation of the molecular species also takes place on account of their different migration velocities through the chromatographic column. In the present case, the DMA functions as a chromatographic column with the different sized particles being separated according to mobility.

PARTICLE SIZE RANGE AND SENSITIVITY

The particle size range of the AMC is estimated to be from 0.005 to 0.2 μm diameter. The upper size limit is determined by the fact that in order for the first DMA to extract a monodisperse aerosol from a polydisperse aerosol according to electrical mobility, the particles must be singly charged. For particles in Boltzmann equilibrium (Liu and Pui, 1974c), multiple charging does not become a significant factor until the particle diameter approaches 0.1 μm. While the instrument will continue to function with appropriate corrections for multiple charging effects for particles up to 0.2 μm diameter, the effect of multiple charging on larger particles will become too large to be satisfactorily corrected. On the other hand, the lower size limit of the instrument is determined by the fact that for small particles, the fraction of particles that are electrically charged is small. Thus, the instrument sensitivity will rapidly diminish as the particle size is reduced. It is estimated that with the particle detectors now available, particles as small as 0.005 μm can be satisfactorily detected by the instrument with reasonable sensitivity.

The sensitivity of the instrument for detecting sulfuric acid particles is determined by the specific particle detector used. For the electrometer current sensor shown in Fig. 1, the current, I(A), indicated by the sensor is related to the particle concentration, N(cm^{-3}) as follows,

$$I = qeNf \tag{1}$$

where q (cm^3 s^{-1}) is the aerosol flow rate entering the electrometer current sensor, f is the fraction of particles carrying either $+1$ or -1 unit of charge and e (1.602×10^{-19} Coulomb) is the elementary unit of charge. For an aerosol flow rate of $q = 50$ cm^3 s^{-1}, which is the standard flow rate used in the above described experiments, and a minimum measurable electrometer current of 0.001 pA, which is typical of the electrometer used in these experiments, the minimum measurable particle concentration is

$$N = 1.25 \times 10^2/f. \tag{2}$$

For particles in Boltzmann charge equilibrium and

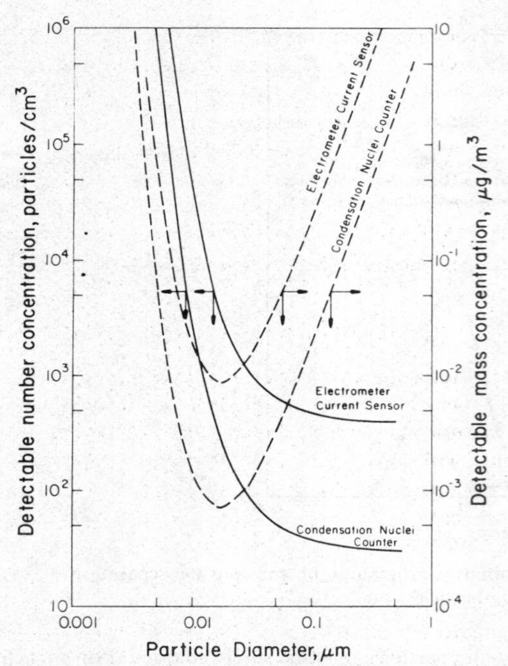

Fig. 3. Minimum number and mass concentrations of monodisperse, H_2SO_4 aerosols detectable by the Aerosol Mobility Chromatograph using an electrometer current sensor or a condensation nuclei counter as the particle detector.

for sufficiently small particle sizes, the particles will be either electrically neutral or carry ± 1 elementary unit of charge, in which case,

$$f = N_1/(2N_1 + N_0) \tag{3}$$

and

$$N_1/N_0 = \exp(-e^2/D_pkT) \tag{4}$$

where N_1 is the concentration of the singly charged particles of one polarity and N_0, the concentration of the neutral particles, the aerosol being assumed to be monodisperse. In equation (4), D_p (cm) is the particle diameter, k is the Boltzmann's constant, and T is the absolute temperature.

Figure 3 shows the theoretical sensitivities calculated by means of the above equations. For the mass sensitivity, the calculations were made for unit density particles. However, the actual density of sulfuric acid will depend on the ambient humidity. For instance, for an ambient humidity of 50%, the density of sulfuric acid is 1.325 g cm^{-3} and the droplet will contain 42.5% of sulfuric acid by weight, in which case, the sensitivity values shown in Fig. 3 should be multiplied by (1.325)(0.425) = 0.563 to obtain the instrument mass sensitivity for sulfuric acid. It is interesting to note that the minimum number sensitivity of the instrument is about 500 particles cm^{-3}, whereas the minimum mass sensitivity is about 0.01 μg m^{-3} for monodisperse aerosols at a particle size of 0.02 μm.

In addition to the electrometer current sensor, other particle detectors can also be used. Using the CNC-2 Condensation Nuclei Counter manufactured by the General Electric Co., and assuming a minimum detectable particle concentration of 10 cm^{-3} for the CNC-2, the detection limit of the AMC for sulfuric acid particles is,

$$N = 10/f, \text{cm}^{-3}. \tag{5}$$

In Fig. 3 the calculated number and mass sensitivities of the AMC for the case of the General Electric CNC-2 are also shown. The minimum detectable mass concentration is seen to be about 0.001 μg m^{-3} for this case. Recently, a continuous flow, single-particle counting condensation nuclei counter has been developed (Agarwal, Sem and Pourprix, 1977). This device, which is based on the earlier work on the continuous flow CNC of Bricard *et al.* (1976), is capable of measuring aerosol concentrations to a level of 0.01 particles cm^{-3}. With this device, the theoretical detection limit of the AMC for sulfuric acid aerosols of 0.02 μm diameter is about 10^{-5} μg m^{-3}. A prototype version of this single-particle counting, continuous flow CNC was used in some preliminary experiments described later to demonstrate the feasibility of this approach.

EXPERIMENTAL VERIFICATION OF THE PARTICLE SIZE GROWTH FACTOR FOR SULFURIC ACID AEROSOLS

The ability of the AMC to identify sulfuric acid particles is based on the known response characteristic of sulfuric acid to ambient humidity. The growth in size of sulfuric acid particles with increasing humidity was studied in a series of experiments using the same apparatus shown in Fig. 1. In these experiments, only sulfuric acid aerosol was generated and the atomizer for K_2SO_4 particles was turned off. By varying the humidity of the aerosol entering the second DMA, the change in particle size of sulfuric acid as a function of humidity could be measured. In Fig. 4 the experimental results are compared with the theoretically calculated values using the equilibrium vapor

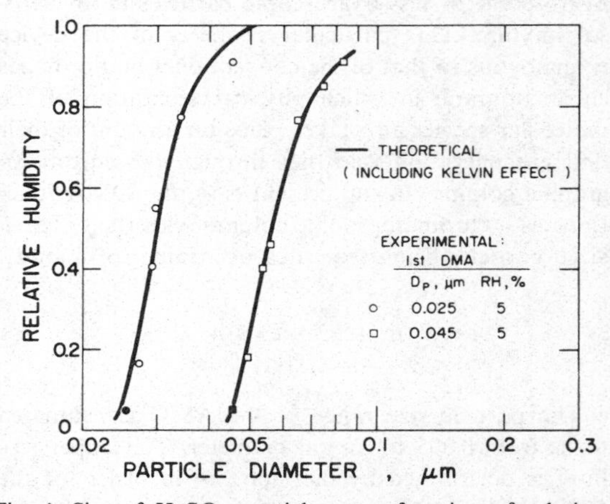

Fig. 4. Size of H_2SO_4 particles as a function of relative humidity—comparison of theory and experiments.

pressure data for sulfuric acid given in the *Chemical Engineers' Handbook* (Perry and Chilton, 1973) and taking into account the Kelvin's effect. It should be noted that for the size range covered in these experiments, the Kelvin's effect was quite small for the curve on the left in Fig. 4, while for the curve on the right, the Kelvin's effect is entirely negligible. It is seen that the agreement between theory and experiment is good. These results show that: (1) the change in particle size of sulfuric acid particles as a function of relative humidity can be accurately predicted from theory, and (2) the experimental apparatus is functioning properly.

DETECTION OF LOW LEVEL SULFURIC ACID AEROSOLS BY THE AEROSOL MOBILITY CHROMATOGRAPH

So far we have not succeeded in detecting the presence of free sulfuric acid aerosols in the ambient atmosphere in Minneapolis. However, in a sequence of experiments performed over a period of one week, ambient aerosols were sampled into a plastic bag and a small quantity of sulfuric acid aerosol was injected into the bag. The presence of this artificially generated sulfuric acid aerosol in the natural, ambient aerosol background was detected by the AMC in a few cases. These experiments are described below.

For these experiments, the single-particle counting, continuous flow condensation nuclei counter was used as the particle detector. The ambient aerosol was sampled into a 0.2 m^3 plastic bag. Then a small amount of sulfuric acid aerosol was generated by the syringe pump atomizer and injected into the bag. The mixed aerosol was then sampled by the AMC and measured. For the experiments that resulted in the detection of this injected H_2SO_4 aerosol, the results shown in Fig. 5 are typical. It is seen that the appearance of the H_2SO_4 peak is quite distinct and that the experimentally determined particle size growth factor of 1.38 agrees well with the theoretically calculated factor of 1.40. It should be mentioned that in a few experiments, the injection of H_2SO_4 aerosol into the bag did not result in the appearance of a new peak. It was postulated that for these experiments, there was sufficient ammonia present in the ambient atmosphere to neutralize the small amount of H_2SO_4 particles injected. However, further experiments are needed in order to confirm this hypothesis.

CONCLUSION AND FUTURE WORK

The Aerosol Mobility Chromatograph described in this paper has been developed as a means for measuring sulfuric acid aerosols on the basis of particle size change with relative humidity. We have demonstrated the basic feasibility of the concept, and verified the theoretical size growth factor for sulfuric acid particles. We have also shown that low level sulfuric acid aerosols can be identified by the AMC in the presence of other particles normally found in the ambient atmosphere. The prototype instrument described in this paper is useful primarily for studying sulfuric acid aerosols in the laboratory. For actual ambient measurements, further improvement is needed. In addition, possible interference effects from other types of particles with varying degree of hygroscopicity must be worked out in detail. Some of these problems will be addressed in forthcoming communications.

Acknowledgements—This research is supported by the U.S. National Bureau of Standards through its grant No. 6-9002 and the U.S. Environmental Protection Agency through its grants No. R804600 and R803851 to the University of Minnesota.

REFERENCES

Agarwal J. K., Sem G. J. and Pourprix M. (1977) A continuous flow condensation nuclei counter capable of counting single particles. Paper prepared for presentation at *Ninth International Conference on Atmospheric*

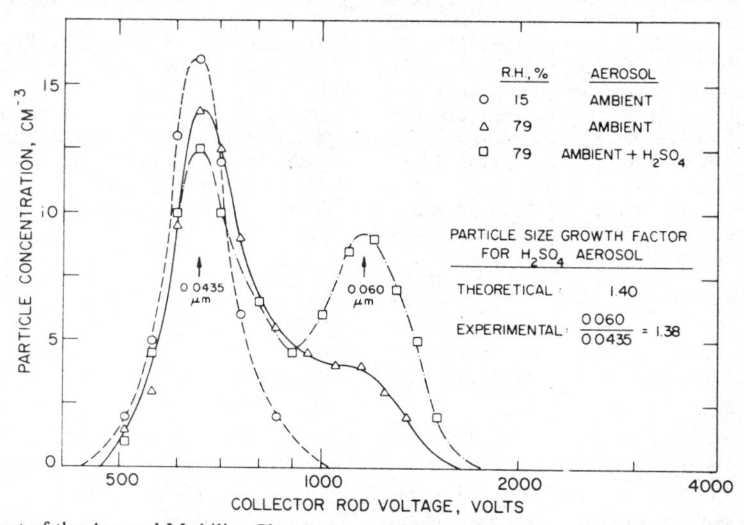

Fig. 5. Output of the Aerosol Mobility Chromatograph showing the detection of H_2SO_4 in the presence of background particles.

Aerosol, Condensation and Ice Nuclei. University College, Galway, Ireland, Sept. 21–27, 1977.

Appel B. R., Kothny E. L., Hoffer E. M., Buell G. C., Wall S. M. and Wesolowski J. J. (1977) A comparative study of wet chemical and instrumental methods for sulfate determination. Presented at the *173rd National Meeting of the American Chemical Society.* New Orleans, Louisiana, March 20–27, 1977.

Bricard J., Delattre P., Madelaine G. and Pourprix M. (1976) Detection of ultra-fine particles by means of a continuous flux condensation nuclei counter. In *Fine Particles: Aerosol Generation, Measurement, Sampling, and Analysis* (edited by B. Y. H. Liu). Academic Press, New York.

Brosset C. and Ferm M. (1976) Man-made airborne acidity and its determination. (IVL Publications B-314A. Gothenburg, Sweden 1976)

Charlson R. J., Vanderpol A. H., Covert D. S., Waggoner A. P. and Ahlquist N. C. (1974) $H_2SO_4/(NH_4)_2SO_4$ background aerosol: optical detection in St. Louis Region. *Atmospheric Environment* **8,** 1257–1267.

Covert D. S. (1974) A study of the relationship of chemical composition and humidity to light scattering by aerosols. Ph.D. Thesis, University of Washington, Washington.

Dzubay T. G. (1977) *X-ray fluorescence analysis of environmental samples.* Ann Arbor Science, Michigan.

Dzubay T. G. and Stevens R. K. (1975) Ambient air analysis with dichotomous sampler and X-ray fluorescence spectrometer. *Environ. Sci. Technol.* **9,** 663–668.

Etz E., Rosasco G. J. and Cunningham W. C. (1977) The chemical identification of airborne particles by Laser–Raman spectroscopy. In *Environmental Analysis* (edited by G. W. Ewing). Academic Press, New York.

Forrest J. and Newman L. (1977) Application of ^{110}Ag to microgram sulfate analysis. Presented at the *173rd National Meeting of the American Chemical Society.* New Orleans, Louisiana, March 20–27, 1977.

Huntzicker J. J., Isabelle L. M. and Watson J. G. (1977) The continuous monitoring of particulate sulfur compounds by flame photometry. Presented at the *173rd National Meeting of the American Chemical Society.* New Orleans, Louisiana, March 20–27, 1977.

Husar J. D., Husar R. B. and Stubits P. K. (1975) Determination of submicrogram amounts of atmospheric particulate sulfur. *Analyt. Chem.* **47,** 2062–2065.

Jaklevic J. M. and Walter R. L. (1977) Comparison of minimum detectable limits among X-ray spectrometers. In *X-ray Fluorescence Analysis of Environmental Samples* (edited by T. G. Dzubay). Ann Arbor Science, Michigan.

Kittelson D. B., Mckenzie R., Vermeersch M., Linne M., Dorman F., Pui D. Y. H., Liu B. Y. H. and Whitby K. T. (1977) Total sulfur aerosol detection with an electrostatically-pulsed flame photometric detector system. Presented at the *173rd National Meeting of the American Chemical Society.* New Orleans, Louisiana, March 20–27, 1977.

Knutson E. O. and Whitby K. T. (1975a) Aerosol classification by electric mobility: apparatus, theory, and applications. *J. Aerosol Sci.* **6,** 443–451.

Knutson E. O. and Whitby K. T. (1975b) Accurate measurement of aerosol electric mobility moments. *J. Aerosol Sci.* **6,** 453–460.

Liu B. Y. H. and Pui D. Y. H. (1974a) Electrical neutralization of aerosols. *J. Aerosol Sci.* **5,** 465–472.

Liu B. Y. H. and Pui D. Y. H. (1974b) A submicron aerosol standard and the primary, absolute calibration of the condensation nuclei counter. *J. Colloid Interface Sci.* **47,** 155–171.

Liu B. Y. H. and Pui D. Y. H. (1974c) Equilibrium bipolar charge distribution of aerosols. *J. Colloid Interface Sci.* **49,** 305–312.

Liu B. Y. H., Marple V. A., Whitby K. T. and Barsic N. J. (1974) Size distribution measurement of airborne coal dust by optical particle counters. *Am. Ind. Hyg. Assoc. J.* **35,** 443–451.

Liu B. Y. H. and Lee K. W. (1975) An aerosol generator of high stability. *Am. Ind. Hyg. Assoc. J.* **36,** 861–865.

Liu B. Y. H., Pui D. Y. H., Hogan A. W. and Rich T. A. (1975) Calibration of the Pollak counter with monodisperse aerosols. *J. Appl. Meteorol.* **14,** 46–51.

Liu B. Y. H. and Kim C. S. (1976) On the counting efficiency of condensation nuclei counters. *Atmospheric Environment,* in press.

Macias E. S. (1976) Gamma-ray analysis of sulfur. In *Proceedings of the 8th Materials Research Symposium: Methods and Standards for Environmental Measurements.* U.S. National Bureau of Standards, Washington, D.C., Sept. 1976 (in press).

McKenzie R. L., Kittelson D. B., Veermersch M., Kwok K., Pui D. Y. H., Liu B. Y. H., Whitby K. T. and Kousaka Y. (1976) Development of sulfur particle analyzer. Progress Report to National Bureau of Standards, Particle Technology Laboratory Publication No. 319, University of Minnesota, Minnesota.

Mulick J., Puckett R., Swicki F. and Williams D. (1976) Ion chromatography—a new analytical technique for the assay of sulfate and nitrate in ambient aerosols. In *Proceedings of the 8th Materials Research Symposium: Methods and Standards for Environmental Measurements.* U.S. National Bureau of Standards, Washington, D.C., Sept. 1976 (in press).

Perry R. H. and Chilton C. H. (1973) *Chemical Engineers' Handbook,* 5th Edn. McGraw-Hill, New York.

Roberts P. T. and Friedlander S. K. (1976) Analysis of sulfur in deposited aerosol by vaporization and flame photometric detection. *Atmospheric Environment* **10,** 403.

Stafford R. G., Chang R. K. and Kindlmann P. J. (1976) Laser-Raman monitoring of ambient sulfate aerosols. In *Proceedings of the 8th Materials Research Symposium: Methods and Standards for Environmental Measurements.* U.S. National Bureau of Standards, Washington, D.C., Sept. 1976 (in press).

Tanner R. L., Garber R. W. and Newman L. (1977) Speciation of sulfate in ambient aerosols by solvent extraction with flame photometric detection. Presented at the *173rd National Meeting of the American Chemical Society.* New Orleans, Louisiana, March 20–27, 1977.

Atmospheric Environment Vol. 12, pp. 105–111. Pergamon Press 1978. Printed in Great Britain.

TOTAL SULFUR AEROSOL CONCENTRATION WITH AN ELECTROSTATICALLY PULSED FLAME PHOTOMETRIC DETECTOR SYSTEM

D. B. KITTELSON, R. MCKENZIE,* M. VERMEERSCH, F. DORMAN, D. PUI, M. LINNE, B. LIU and K. WHITBY

Particle Technology Laboratory, Department of Mechanical Engineering, University of Minnesota, Minneapolis, MN 55455, U.S.A.,
*National Bureau of Standards Analytical Chemistry Division, Washington, DC 20234, U.S.A.

(*First received* 16 *June* 1977 *and in final form* 7 *September* 1977)

Abstract—This paper describes a new system for continuous measurement of ambient aerosol sulfur concentrations. The system consists of a pulsed electrostatic precipitator followed by a total sulfur flame photometric detector (FDP). With this system the output signal of the FPD is composed of a DC signal related to the gas phase sulfur concentration and an AC signal related to aerosol sulfur concentration. The AC signal is separated from the DC signal by a phase and frequency selective lock-in amplifier. Using a standard Meloy SA-285 FPD and operating at a chopping frequency of 0.2 Hz the system has a sensitivity of about 0.2 μg m^{-3} (as sulfur) for ammonium sulfate at a signal to peak-to-peak noise ratio of one and has a system response time of about 1.5 min. Some interference from SO_2 exists but should be minimized by modification of the precipitator and of the linearizer electronics of the FPD.. A thermal speciation system has been developed for use in conjunction with the system to yield information on the chemical composition of the aerosol sulfur measured. Preliminary field evaluations of the instrument were performed at the Los Angeles Roadway Study. The instrument performed well and aerosol total sulfur concentrations measured with it were in good agreement with measurements by other methods.

INTRODUCTION

This paper describes a new measurement technique for aerosol total sulfur concentration. The technique, illustrated schematically in Fig. 1, gives a measure of aerosol or particle sulfur, independent of gas phase sulfur, at ambient level concentrations in real time.

The concentration of the aerosol component of the sample is modulated at some selected frequency (usually 0.2 Hz) by the pulsed electrostatic precipitator. The gas-phase component concentrations are essentially unaffected by the precipitator. Analysis of the sample by the flame photometric detector (FPD) results in a signal having two components: a DC signal resulting from gas phase sulfur and an AC signal resulting from the modulated aerosol sulfur. The AC signal is separated from the DC signal and amplified by the lock-in amplifier. The system is potentially capable of simultaneously giving a measure of both gas phase sulfur concentrations (from the DC signal) and aerosol sulfur concentrations (from the AC signal).

PERFORMANCE OF THE ELECTROSTATIC PRECIPITATOR

One unique feature of our system is the use of a pulsed electrostatic precipitator as a particle chopper. The precipitator is shown in Fig. 2. The precipitator design is conventional except in those features relating to pulsed operations. It consists of a concentrically positioned central corona wire and an outer stainless steel collection surface which is electrically grounded. This surface forms a nozzle to flatten the

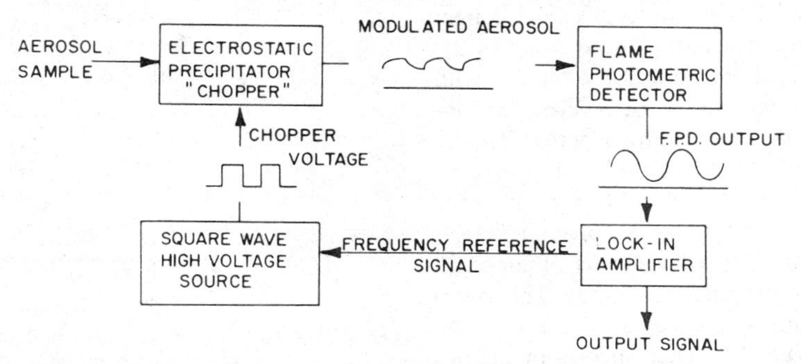

Fig. 1. Schematic diagram of electrostatically pulsed flame photometric detector system.

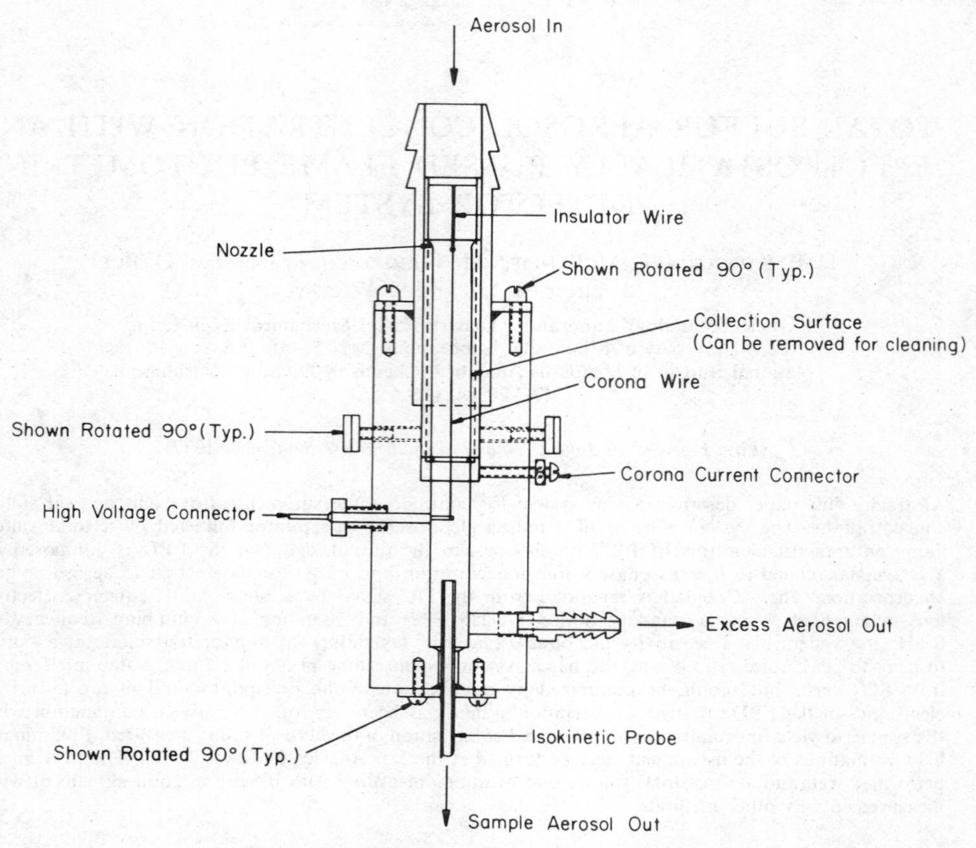

Fig. 2. Pulsed electrostatic precipitator design.

velocity profile within the precipitator. When a high voltage is applied to the corona wire, an electric field is established between the wire and collection surface and a corona discharge occurs at the wire. Particles in this section of the precipitator are charged by the resulting ions and electrons; once charged, they are precipitated by the electric field onto the collector surface. In order to minimize the time constant of the precipitator, only the central section of the flow is drawn into the flame photometric detector via an isokinetic sampling probe. The precipitator has been designed to have a time constant of less than 0.1 s.

The precipitator has been calibrated using essentially monodisperse aerosols. The calibration has been done under steady state operating conditions, however the results should also apply to pulsed operation provided that the characteristic times are much longer than the chopper time constant. Figure 3 shows the variation of the precipitator collector efficiency with corona voltage for 0.3 μm diameter particles. The efficiency is virtually 100% for corona wire voltages greater in magnitude than -2.4 kV or $+4.0$ kV. Figure 4 shows the variation of collection efficiency with particle size for several negative corona voltages. At -2.4 kV virtually 100% efficiency is obtained in the 0.03 μm to 1.0 μm diameter size range. This operating condition was initially selected for use with the complete sulfur instrument. When incorporated into the system the precipitator is driven by a high voltage square wave which is synchronized to a reference signal generated by the lock-in-amplifier.

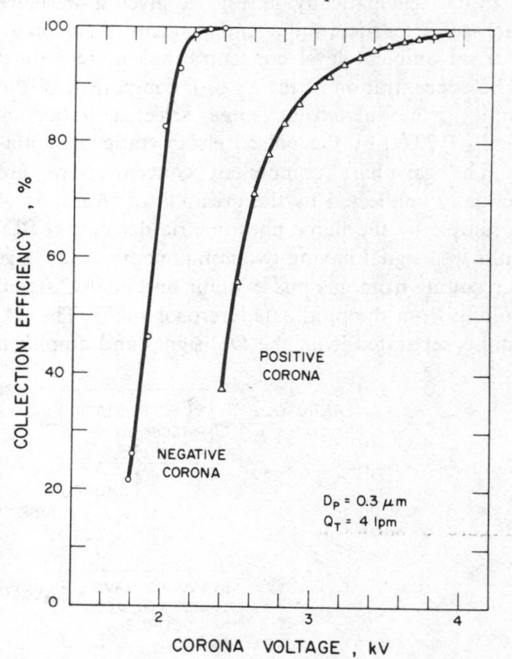

Fig. 3. Electrostatic precipitator collection efficiency as a function of corona wire voltage.

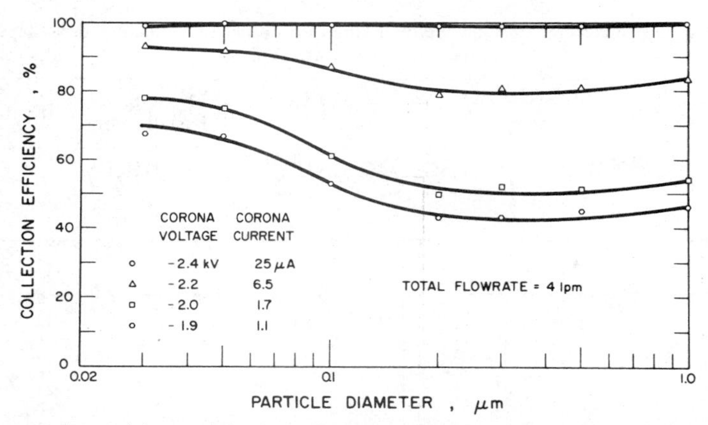

Fig. 4. Electrostatic precipitator collection efficiency as a function of particle size for various negative corona voltages.

PERFORMANCE OF THE FLAME PHOTOMETRIC DETECTOR

The two most important requirements of a flame photometric detector for this application are high sensitivity and rapid response to sudden changes of input aerosol concentration. Most of the work described here has been done with a Meloy SA-285 Total Sulfur Flame Photometric Detector. However, a fast response option has recently become available from Meloy and an instrument with this modification is currently being evaluated. With this option the inlet plumbing is modified to minimize surface interactions and heating of the sample flow before it enters the flame.

In order to determine the response of the detector to sulfur-containing aerosols, ammonium sulfate and sulfuric acid aerosols were generated using a Collison type atomizer. The resulting aerosols were fed directly to the FPD and a Thermo-Systems Model 3030 Electrical Aerosol Analyzer (EAA). For the first set of tests the detector was used in a steady mode (no precipitator). Aerosol volume concentrations were measured using the EAA. These were converted to aerosol sulfur mass concentrations ($\mu g\ m^{-3}$) by multiplying by the product of the particle density and the sulfur mass fraction. Standard handbook values were used for ammonium sulfate and values given by Bray (1970) were used for the sulfuric acid particles which are in equilibrium with ambient water vapor.

Sulfur aerosol mass concentrations were determined with the Meloy SA-285 by assuming that its sensitivity to the sulfur in the aerosol was the same as its sensitivity to sulfur in SO_2. The instrument was calibrated for SO_2 using an SO_2 permeation tube system.

Figure 5 shows a plot of sulfur concentration measured with the EAA against that measured with the Meloy. Both the ammonium sulfate and sulfuric acid measurements fall on the same line which has a slope of 1.04 and essentially zero intercept. The Meloy can thus be used in the steady mode to make quantitative measurements of these sulfates. Since the present application requires that the detector operate in a pulsed mode it was necessary to evaluate the transient response of the detector. This was done by turning the aerosol off or on with the precipitator in a manual switching mode. The results of these experiments are listed in Table 1. The table indicates the times required for the detector signal to reach 90% of its final (steady state) value ('up 90%') when the aerosol is suddenly turned on and the time required for the detector signal to fall to 10% of its initial value when the aerosol is suddenly turned off. Note that the up and down response times are not the same and that the response of the standard SA-285 is much slower for sulfuric acid than for ammonium sulfate. This results in a reduced sensitivity of the pulsed system to sulfuric acid.

PERFORMANCE OF THE COMPLETE SYSTEM

The response of the pulsed precipitator-flame photometric detector system is shown in Fig. 6. The peak to peak signal measured at the Meloy photomultiplier output is plotted against chopping fre-

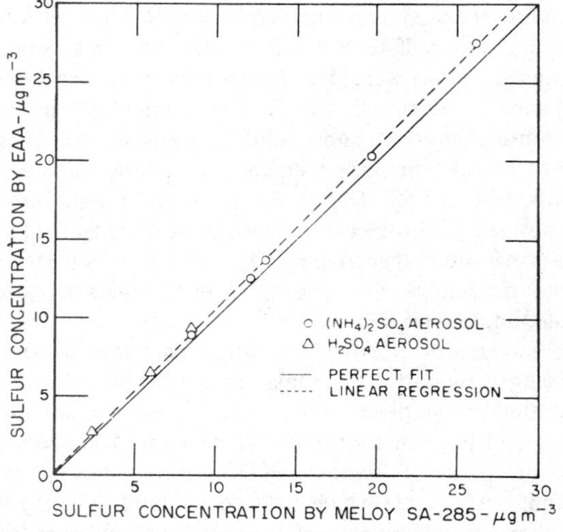

Fig. 5. Comparison of EAA and Meloy SA-285 FPD aerosol sulfur measurements.

Table 1. Time response characteristics of Meloy Flame Photometric Detector to a step change in sulfur aerosol concentration

	Meloy SA-285 'Standard'		Meloy SA-285 'Fast Response'		
Signal change	NH_4HSO_4	H_2SO_4	$(NH_4)_2SO_4$	NH_4HSO_4	H_2SO_4
Up 90%	13.2 s	60 s	3.4	4	4
Down 90%	3.6 s	24 s	5.8	5.7	5.8

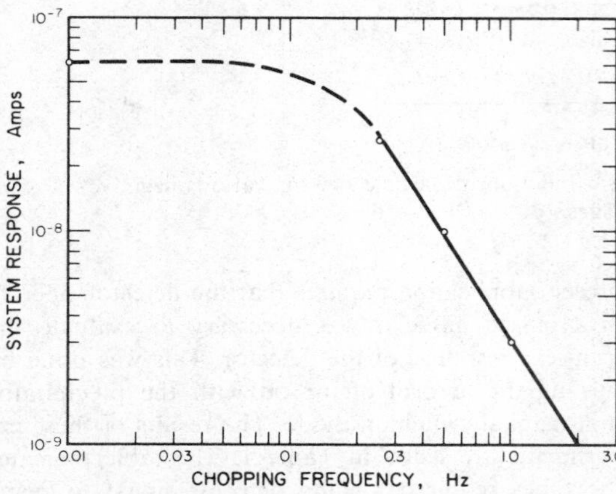

Fig. 6. Meloy SA-285 peak-to-peak photomultiplier current as a function of electrostatic precipitator chopping frequency for an ammonium sulfate aerosol.

quency. These results were obtained with an ammonium sulfate aerosol using the standard Meloy FPD. The influence of the fairly long response times of the FPD are evident. At 0.01 Hz the full peak to peak steady state Meloy response is available. As the frequency is raised the peak to peak response drops. At 0.25 Hz the response has dropped to only about 45% of its value at 0.01 Hz. It drops off rapidly at higher frequencies. An operating frequency of 0.2 Hz was selected as a reasonable compromise between sensitivity and time response. The standard Meloy SA-285 and this frequency were used in all of the experiments described below.

The entire system has been calibrated using both ammonium sulfate and sulfuric acid aerosols. Sulfur concentrations were determined from EAA measurements as described above. The overall system response is plotted against sulfur concentration in Figs. 7 and 8 for ammonium sulfate and sulfuric acid aerosols, respectively. The results fit good straight lines, however two problems are immediately evident: there is some interference from SO_2, and the sensitivity to sulfuric acid is only about 1/3 of the sensitivity to the ammonium sulfate.

Two reasons for the SO_2 interference are currently being explored: (1) modulation of the SO_2 concentration by the precipitator and (2) non-linear interactions in the flame photometric detector. The precipitator apparently removes SO_2 by production of ions and free radicals which cause conversion of SO_2 to sulfuric acid droplets which are subsequently precipitated. SO_2 removal is strongly dependent on corona

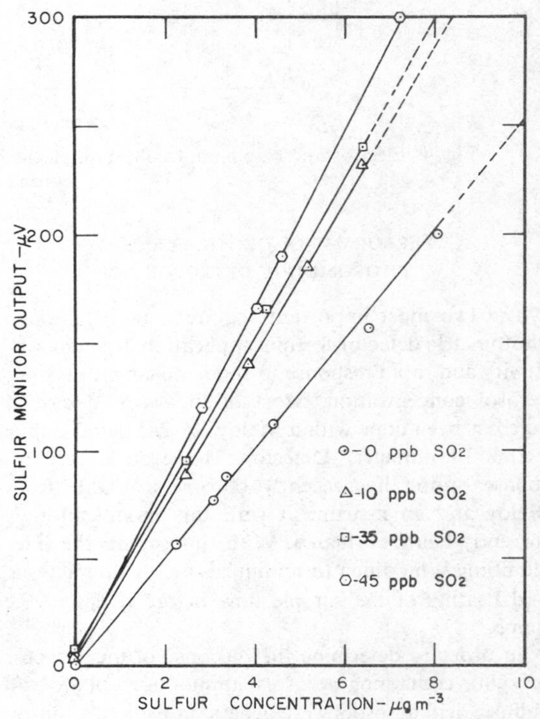

Fig. 7. $(NH_4)_2SO_4$ calibration curves for various levels of SO_2. (1 ppb SO_2 = 1.33 μgS m^{-3} at 20°C, 1 atm.).

voltage. Initial experiments using −2.4 kV demonstrated a strong SO_2 interference. The results shown in Figs. 7 and 8 were obtained using −2.2 kV which is the lowest corona voltage that can be used without

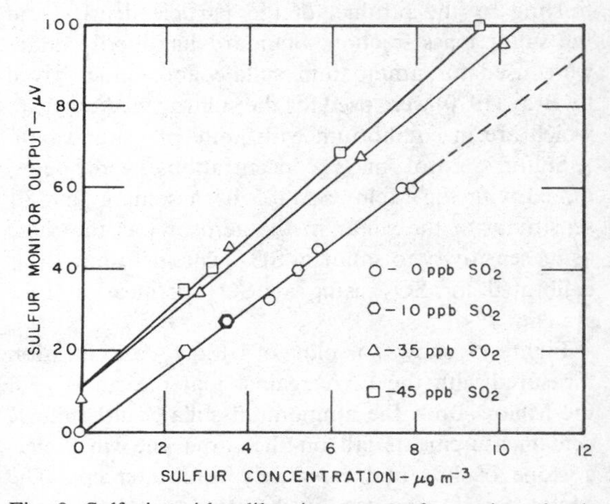

Fig. 8. Sulfuric acid calibration curves for various levels of SO_2. (1 ppb SO_2 = 1.33 μgS m^{-3} at 20°C, 1 atm.).

a significant loss in aerosol collection efficiency. Alternative precipitator designs which would minimize this effect are currently being considered. The change in the slope of the calibration curves that results from SO_2 addition can not be readily explained by SO_2 concentration modulation, however, because this phenomenon should result only in a change in the intercept. The output of the photomultiplier in the flame photometric detector is non-linear in total sulfur (Brody and Chaney, 1966). The Meloy SA-285 has a built-in electronic linearizer so that the instrument output is linear in SO_2. Slight nonlinearities in the overall system which are not evident from the SO_2 calibration curve could appreciably influence the measurement of a small concentration of sulfur aerosol in the presence of a large concentration of sulfur gas. Detailed linearity tests are currently underway to attempt to understand this problem. It would be possible to completely eliminate the SO_2 interference by removing SO_2 from the incoming aerosol with a diffusion denuder. This solution would, unfortunately, make it impossible to simultaneously measure SO_2 and aerosol with the same system.

Despite this interference problem the instrument can be used quite effectively in the field. The use of the pulsed system does not interfere with SO_2 measurements. Thus the calibration curve corresponding to the measured SO_2 concentration may be used to determine the aerosol sulfur concentration.

The difference between the sensitivity to sulfuric acid and the sensitivity to ammonium sulfate can be explained in terms of the relatively slower response time of the FPD to a change in sulfuric acid aerosol concentration discussed earlier in connection with Table 1. It is hoped that this difference in sensitivity will be eliminated by incorporating the fast response detector into the system.

Another problem with the present system is a weak sodium interference. For urban aerosols typical upper limit sodium and lower limit sulfur concentrations are 8 and 2 $\mu g\ m^{-3}$ respectively (Cooper, 1973). Under these conditions sulfur measurements with this system would be in error by about +2% for an ammonium sulfate aerosol. For less extreme urban conditions the interference would be much less. The only common aerosol where this interference might introduce a large error (> 10%) is a background marine aerosol. Similar interferences may result from other metals. This problem is currently being investigated.

The system as described above does not differentiate between sulfuric acid, ammonium sulfate and bisulfate, and other common sulfur-containing compounds. Some compositional information can be obtained, however, by the use of a heated inlet system. This is accomplished through chemical-species-specific thermal decomposition of the incoming aerosol in a heated section of the inlet and the subsequent loss of that sample by diffusion of the resulting gases to the walls. Figure 9 shows the behavior of sulfuric acid and ammonium sulfate and bisulfate in the heated inlet system.

When pure sulfuric acid aerosol is introduced into the system, loss of the aerosol begins to occur at 30°C, with 50% loss of aerosol occurring at 71°C and nearly complete loss at 120°C. Ammonium sulfate and bisulfate losses do not begin until 90°C, with 50% loss of these aerosols not occurring until 142°C, and almost complete loss occurring at 190°C, which is the upper temperature limit in our present system. Although Fig. 9 shows a slight difference in behavior between ammonium sulfate and ammonium bisulfate decomposition curves, in practice it is difficult to distinguish between these two aerosols. The temperatures at which 50% sample loss occurs for sulfuric acid and for ammonium sulfate and bisulfate are specific to those respective chemical species in this system and can therefore be used to identify those species of sulfur. Thus, for these pure aerosols, four temperatures adequately characterize the thermal decomposition: ambient (20–25°C), 71°C (H_2SO_4), 142°C (NH_3 salts of H_2SO_4), and 190°C (only refractory sulfur species remaining).

Experiments are currently underway to examine the behavior of aerosols consisting of mixed salt solutions which are more representative of atmospheric aerosols.

In order to eliminate the long time transients required to reach thermal equilibrium that occur if

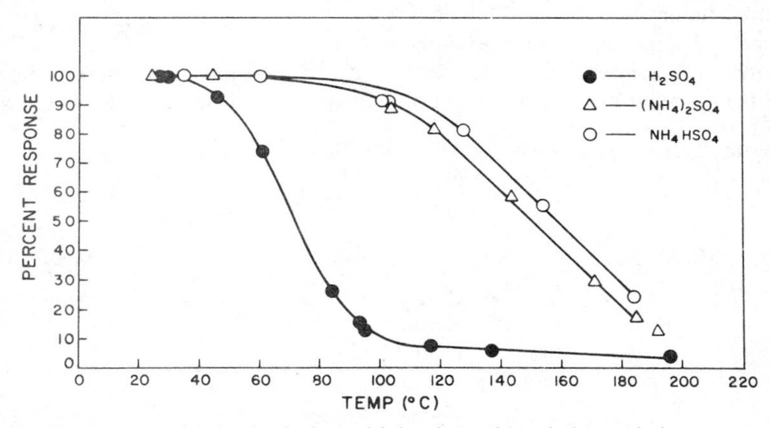

Fig. 9. Aerosol behavior in heated inlet thermal speciation technique.

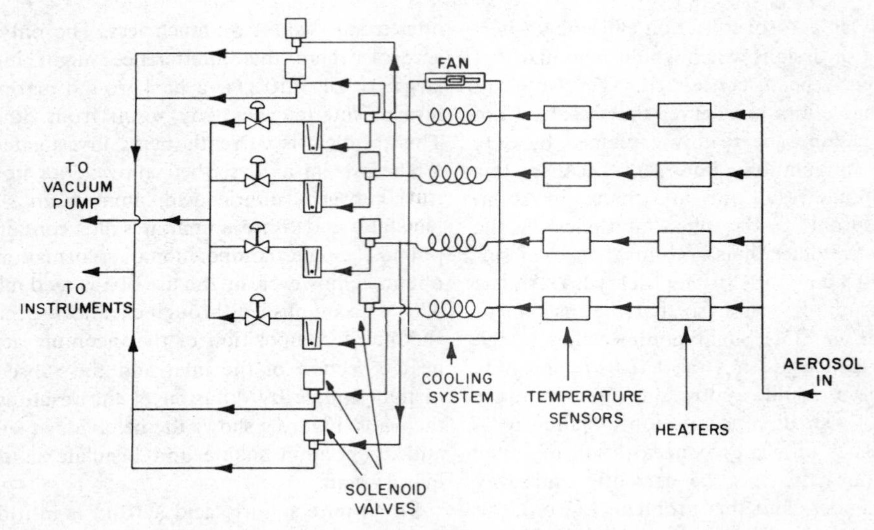

Fig. 10. Schematic diagram of thermal speciation system.

one cycles between the four temperatures indicated above, we developed the thermal speciation system shown in Fig. 10. Each of four identical flow pathways is equilibrated at one of the desired temperatures and each pathway is always at thermal equilibrium. The sample to the aerosol sulfur analyzer is selected from a particular pathway by the solenoid valves. Switching from one pathway to another is essentially instantaneous.

The chopper-FPD-lock-in amplifier system was installed in the University of Minnesota Mobile Atmospheric Measurements Lab. and used during the Los Angeles Roadway Study (LARS) in October, 1976 (Stevens, in preparation). One field test of the heated inlet system was made during that study. Figure 11 shows the strip chart recording of the lock-in amplifier output. One sees very little signal loss between ambient temperature and 71°C, suggesting little sulfuric acid aerosol is present in the sample. The change between 71°C and 190°C is quite large and indicates that about 25% of the signal of the sample behaves in a manner similar to ammonium sulfate and bisulfate. This corresponds to about 0.25 μgS m^{-3} as ammonium sulfate. Approximately 60% of the sample behaves as refractory sulfur compounds. The sampling site was near the coast and the prevailing breezes were offshore which suggests that the refractory sulfur aerosol may contain sodium sulfate. Part of the refractory signal may also be due to sodium interference. Figure 11 also shows the time response of the system and the typical signal noise levels. The response of the system and the typical signal noise levels. The response time to a step change in input aerosol concentration is about 1.5 min and the peak to peak noise level corresponds to about 0.2 μgS m^{-3}.

During the LARS experiments we ran the sulfur analyzer, making real time measurements while two collection systems were also operating—a dichotomus sampler (Loo *et al.*, 1976) and a low-pressure multistage impactor developed by Caltech (Hering *et al.*, 1977). The collected samples were analyzed by sample volatilization techniques and FPD. Figure 12 shows a comparison of our measurements, averaged over the corresponding time intervals during which the LBL or the impactor operated, to the results of analysis of the samples collected by the two integrating techniques. A linear regression line fitted to the data has a slope of 0.97 and an intercept of -0.56 μg m^{-3}. The correlation coefficient is 0.87. In view of the difficulties associated with field measurements of this type, this is considered to be very good agreement.

CONCLUSIONS

In its present configuration the instrument provides a useful new method for the measurement of ambient

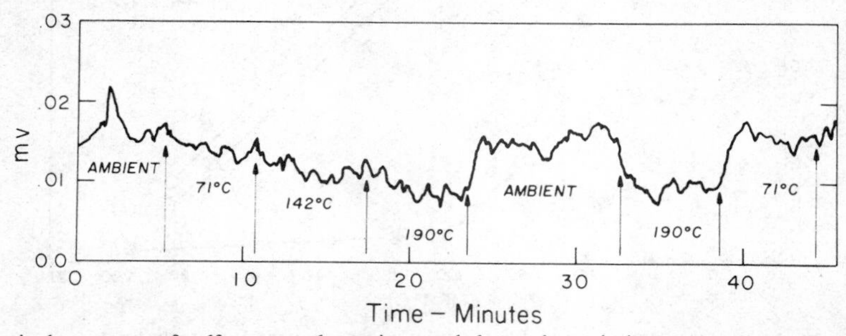

Fig. 11. Typical response of sulfur aerosol monitor and thermal speciation system to ambient aerosols.

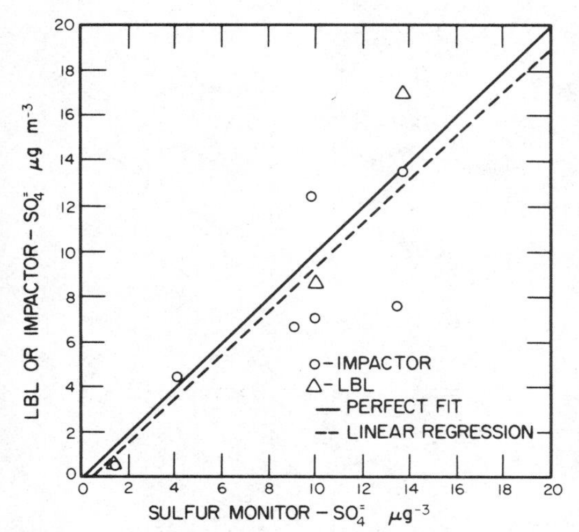

Fig. 12. Comparison of sulfur measurements made with sulfur monitor to other sulfur measurements made during LARS.

sulfates. However its utility will be greatly increased if several problems are solved. They are: (1) SO_2 interference—it is hoped that this problem can be solved by modification of the precipitator design and the electronic signal linearizer in the Meloy. (2) Low sensitivity to sulfuric acid compared to other sulfates—this problem should be solved by using the fast response flame photometric detector. (3) Sodium interference—this should be a minor problem in most applications. However, its elimination would probably require modification of the flame and/or optics of the flame photometric detector. (4) Time response —although the 1.5 min response time of the current system is adequate for stationary monitoring, it is too slow for mobile studies, particularly aircraft plume studies. The use of the fast response flame photometric detector may make it possible to operate at 0.5 Hz with little loss in sensitivity. This should reduce the system response time to 30–40 s. A different type of flame photometric detector is currently under study which should have even shorter time constants. The ultimate goal of the program is to reduce the system response time to under 10 s.

Acknowledgements—The development of this instrument is being supported by NBS grant No. 6-9002. Supplementary equipment support and support for field studies is by EPA grant No. R803851-02.

REFERENCES

Bray W. H. (1970) Water vapor pressure control with aqueous solutions of sulfuric acid. *J. Mater.* **5,** 233–248.

Brody S. S. and Chaney J. E. (1966) Flame photometric detector. The application of a specific detector for phosphorus and for sulfuric compounds-sensitive to subnanogram quantities. *J. Gas Chromatog.* **4,** 42–46.

Cooper J. A. (1973) Battelle Northwest Laboratories Report BNWL-SA-4690.

Hering S. V., Flagen R. C. and Friedlander S. K. (1977) A new low-pressure impactor: determination of the size distribution of aerosol sulfur compounds. Presented at 173 A.C.S. meeting, New Orleans, La., 20–27 March 1977.

Loo B. W., Jacklevic J. M. and Goulding F. S. (1976) Dichotomous virtual impactors for large scale monitoring of airborne particulate matter in fine particles. *Aerosol Generation, Measurement, Sampling, and Analysis* (Edited by B. Y. H. Liu) pp. 311–350. Academic Press, New York.

Stevens R. K. EPA-Los Angeles Roadway Study Report (in preparation).

Atmospheric Environment Vol. 12, pp. 113–125. Pergamon Press 1978. Printed in Great Britain.

TECHNIQUES FOR DETERMINING THE CHEMICAL COMPOSITION OF AEROSOL SULFUR COMPOUNDS*

L. NEWMAN

Brookhaven National Laboratory, Atmospheric Chemistry Laboratory, Upton, NY 11973, U.S.A.

(*First received* 15 *June* 1977 *and in final form* 26 *August* 1977)

Abstract—An extensive critical review is given of the methods presently available for the determination of the chemical composition of aerosol sulfur compounds. Some recent typical results employing the methods are presented. The advantages, limitations and future promises of the techniques are discussed. Potential problems associated with the alteration of chemical composition during sampling are illustrated. The utility of general analytical and wet chemical identification techniques are demonstrated. A sequential selective solvent separation scheme and the use of thermal volatilization for the determination of various sulfates are described and illustrative results given. The application of thermometric titrimetry and electron spectroscopy for the identification of oxidation states is presented and evaluated. A discussion is given for the use and utility of i.r. and laser Raman spectroscopic techniques for the identification of sulfate species. A diffusion battery processor technique for sampling suboptical particles is described and its utility in conjunction with chemical analysis of these size separated particles evaluated.

INTRODUCTION

There is increasing evidence that sulfates in the atmosphere are deleterious to human health (Shy and Finklea, 1973; Lave and Seskin, 1973). Controlled animal studies have shown that the irritant potency of various sulfates can vary by more than an order of magnitude (Amdur, 1969) and that particle size (Amdur and Corn, 1963), and oxidation state (Alarie *et al.*, 1973) can be quite important. Sulfates are generally associated with sizes less than 3 μm (Wagman *et al.*, 1967) and thus have high efficiency for lung retention (Lippman and Albert, 1969). The chemical form of sulfate has been identified as playing an important role in the condensation of aerosols (Junge, 1960). These fine particulates scatter light effectively, can alter the global heat budget, affect cloud formation and precipitation processes.

It is apparent that it is not sufficient to simply measure total sulfate if an understanding and assessment of health and climatological effects are to be forthcoming. It is necessary to measure both the physical and chemical forms of sulfur in atmospheric aerosols. Only through such measurements will an understanding develop of the chemistry of sulfur in the atmosphere and through that understanding the fundamental elements will be established for an appropriate assessment of the importance of aerosol sulfur.

Measurements of the chemical and physical forms of aerosol sulfur compounds are just beginning to be made, not because there has been a lack of interest, but mostly because of a lack of appropriate tools. These tools are now forthcoming and it is the purpose of this review to describe the more promising ones.

Forrest and Newman (1973a) critically reviewed the state of the art that existed at the time for the measurement of sulfur dioxide, particulate sulfate, sulfuric acid, ammonium and metal sulfates, hydrogen sulfide and organic sulfides. Dominated by perceived needs, there ensued a rapid development of the measurement technology for the determination of sulfate compounds. These new developments were critically reviewed by Tanner and Newman (1976a). This review included discussions of techniques for total aerosol and total soluble sulfate determinations. Methods for the determination of the chemical form of sulfur in ambient aerosols were discussed, and these included thermal volatilization, solvent extraction, gas-phase ammonia titration, i.r., ESCA, electron microscopy, acid–base indicators, ring oven, and titratable acidity. In addition, techniques for size discrimination and filter sampling were discussed.

Since the above reviews were completed many new developments have occurred, with emphasis on the utility of these methods. Therefore, this review will only include a discussion of those areas where there are pertinent new developments, insights, and results which are useful for an understanding of the utility and shortcomings of the available methods for determining the chemical composition of aerosol sulfur compounds. A table is presented summarizing the advantages and disadvantages of all practical available methods, including some discussed only in the previous reviews.

SAMPLING

A discussion of sampling for the determination of sulfur in ambient aerosols will be given elsewhere in

* This work was performed under the auspices of the United States Energy Research and Development Administration under Contract No. EY-76-C-02-0016.

this Symposium (Stevens *et al*., 1978). Nevertheless, it is important to discuss some specific observations and concerns relating to sampling for the determination of the chemical composition of sulfur in ambient aerosols.

If sampling is to be performed by filters, it is obvious that the filter should first be evaluated for collection efficiency. This is usually done, however, what is often neglected is a determination of whether the filter matrix itself can spuriously cause sulfur dioxide, either by absorption or oxidation, to appear as sulfate when the sample is analyzed. Membrane filters, e.g. those that are Teflon based, usually do not cause any problem, however, they generally have low flow rates and cannot be used when short time resolution is desired. Furthermore, the techniques employed for molecular composition determinations are selected for their specificity not sensitivity and this is further aggravated by the fact that the various sulfur compounds are determined as parts of the whole. Consequently, recourse often has to be made to high volume filters and these generally consist of glass fibers. That commercial filters are notorious in their ability to spuriously convert sulfur dioxide is well documented (see e.g. Forrest and Newman, 1973b). The problem is generally associated with alkaline sites (Pierson *et al*., 1976).

Acid treatment and high temperature ignition has been found to be useful not only to render fiberglass filters inert to sulfur dioxide, but also to remove appreciable sulfate impurities (Forrest and Newman, 1973b). Recently, Tanner *et al*. (1977a) reported that an ignition at 700°C followed by a phosphoric acid wash can render Pallflex tissue quartz (Pall Products Corp., Putnam, Conn.) useful for sulfate collection and sulfuric acid determination. The filters treated in this manner become quite fragile, but the handling properties can be markedly improved if the procedure is reversed so that an ignition step follows the phosphoric acid wash (Leahy, 1977). Quartz filters treated in this manner are presently the filters of choice when molecular composition determinations are to be made in conjunction with the high volume sampling of atmospheric aerosols.

The spurious appearance of sulfate on a filter matrix has sometimes been suggested to be caused by the collected aerosol (Coffer and Charlson, 1974). In other instances the collected aerosol has not been found to be active in this regard (Forrest and Newman, 1973b; Pierson *et al*., 1976). However, in a unique and convincing study Noll *et al*. (in press) appear to be able to indict co-collected ambient aerosols as causing the spurious formation of sulfate. A diffusion denuder tube was used to permit atmospheric sulfates to be collected without passage of sulfur dioxide over the fresh particulate deposit. Paired samples were compared, in which one filter sampled sulfates and the other sulfates plus ambient sulfur dioxide. The sample exposed to sulfur dioxide yielded an elevated sulfate level. Curiously, samples taken over a five-month period in Chicago, showed a remarkably constant increased sulfate level of 60%, independent of the amount of sulfate, sulfur dioxide or ratio of the two, all of which varied substantially. These important observations, as they might relate not only to sulfate levels but also molecular composition determinations, need to be verified and understood.

Sampling by impaction is often employed to obtain information on chemical composition as a function of particle size, e.g. Cunningham *et al*. (1974). Problems such as particle bounce or sticking efficiency have to be and usually are addressed adequately and meaningful samples can be obtained. Unfortunately, impactors generally have a lower size cutoff at 0.3 μm and a final filter (sometimes incompatible with the method of analysis) has to be employed to obtain information below this size level. However, low pressure impactors can be used effectively to size segregate smaller particles (Herring *et al*., 1977).

Of especial concern in the sampling of atmospheric aerosols is the possibility of obtaining chemical conversions during the sampling procedure. A typical problem to be aware of is, e.g. the reaction of ambient ammonia passing through a filter on which you collected sulfuric acid or for that matter, the reaction between any basic and acidic particles colliding on the collection surface. Cunningham and Johnson (1976) saw evidence for recrystallization of samples on Lundgren impactor surfaces, especially when the relative humidity was high. Harker *et al*. (1977) present evidence that sulfuric acid passing through a filter can release nitrate, presumably as nitric acid. This process not only gives rise to low nitrate values, but in the process obviously must also change the molecular composition of the sulfuric acid, possibly to form ammonium sulfate.

Rational analyses of the molecular composition of ambient sulfur aerosols cannot be made without due concern and further understanding of the problems addressed in this section. Ideally, the collection procedure should involve obtaining as diffuse a sample in as short a time period as possible.

GENERAL ANALYSIS

General analytical methods for the determination of sulfur in collected aerosols will be discussed in another paper in this symposium (Stevens *et al*., 1978). In addition, extensive critical reviews have appeared (Forrest and Newman, 1973; Tanner and Newman, 1976a) and an updated and extended review will soon be published (Tanner *et al*., in press). Some salient features need be discussed on how information obtained from general analytical techniques might impact on the determination of the chemical composition of aerosol sulfur compounds.

All wet chemical analytical determinations are preceded by sample dissolution from a collecting matrix. Sometimes this matrix can cause difficulties for particle removal. This has been observed with a Teflon filter (Dzubay, 1976). More importantly, real differences can arise from solubility considerations and the results might differ according to the solvent selected, whereby you might be measuring water-soluble sulfates, hot-water-soluble compounds, or acid-soluble substances. Of course, once dissolved in water the identity of the molecular species can be lost due to either neutralization or chemical reactions. Furthermore, just hot water dissolution has been observed to permit oxidation of sulfur (IV) compounds collected on filters (Eatough, 1977). Accordingly, Hansen *et al.* (1976) claim that it is necessary to dissolve the sample in an air-free hydrochloric acid solution containing ferric ion to prevent oxidation and loss of sulfite species (see section on thermometric methods).

Consideration should also be given to the characteristics of the method of choice used for the end sulfur determination. Of particular concern would be whether the method determines sulfur in all oxidation states, for otherwise the states must either be preserved if differentiation is desired, or converted to the state which the analytical method measures if only total sulfur is to be determined. Sometimes subtle problems can arise with regard to influences from the chemical composition of the aerosol. As an example, when utilizing the flash volatilization flame photometric determination of particulate sulfur (Husar *et al.*, 1975; Roberts and Friedlander, 1976), the sulfur might not completely volatilize due to the presence of co-collected inorganic salts. This effect has been observed when collecting ambient aerosols near the ocean, whereby the sodium ion from sea spray forms non-volatile sodium sulfate during the evaporation stage in the analysis (Garber, 1977).

The X-ray fluorescence techniques (X-ray, proton, or charged particle induced) purportedly measure total sulfur in the collecting matrix. It is sometimes of interest to ascertain whether water-soluble sulfate is or is not equal to total sulfur. Samples were collected by Dzubay of EPA/ESRL on fluoropore (Teflon) filters. A direct comparison has been made by analyses performed at BNL using a water-soluble turbidimetric determination of sulfate (Sulfate Method VIb—1959) with analyses performed at EPA utilizing an X-ray-induced X-ray fluorescence method (Dzubay and Stevens, 1975). The results yielded a correlation coefficient $R = 0.955$, and averages that differed by 20% with the turbidimetric determination being higher (Tanner *et al.*, 1977a). No significance could be attributed to this difference.

A method employing ^{110}Ag was developed to determine total sulfur (Forrest and Newman, 1977). The filter with its contents are placed in a mixture of hydriodic, hydrochloric and hypophosphorous acids which upon heating will convert all sulfur compounds (including refractory compounds such as barium sulfate) to hydrogen sulfide. The hydrogen sulfide is precipitated as cadmium sulfide which is metathesized to silver sulfide by silver nitrate containing tracer ^{110}Ag. The silver sulfide is then collected on a membrane filter and gamma counted from which the amount of sulfur in the original sample can be calculated. A water extract of the filter can also be analyzed by the above procedure and comparisons made of total vs water-soluble sulfur. Such comparisons were made for samples collected at Brookhaven National Laboratory (rural New York), New York City, and High Point (rural New Jersey) and no differences were found (Forrest and Newman, 1977). Comparisons were also made with the turbidimetric method which at times gave more variable and higher results. The differences are attributed to errors introduced into the turbidimetric procedure by its susceptibility to a variety of interferences.

It can be concluded that no significant differences have been found to date between total and water soluble sulfur for ambient collected aerosols. It would be remarkable if this always proved to be the case and, consequently, further studies in this regard are warranted in order to obtain a better understanding of the chemistry of ambient aerosol sulfur compounds.

WET CHEMICAL IDENTIFICATION

The oldest and still one of the most useful approaches for the determination of the molecular compostion of aerosol sulfur compounds is the utilization of classical wet chemical techniques. A water-soluble sample can be analyzed for specific ions including ammonium and sulfate and for total acidity. The exploitation by Askne *et al.* (1973) of the Gran (1952) equivalence point determination method for the precise measurement of titratable acid has greatly facilitated the utility of the wet chemical approach. Brosset (1975) extends the application of the Gran method for the determination of the dissociation constant and identification of weak acids when they are associated with ambient aerosols.

Utilizing the wet chemical approach Brosset (1975) was able to suggest the existence of ammonium sulfate, letovicite ($(NH_4)_3H(SO_4)_2$), and ammonium bisulfate at a sampling location 40 km south of Gothenburg. He (Brosset, 1976) reports on the occurrence of black and white particles as collected on filter samplers located at the Onsala Peninsula of Sweden. Both black and white episodes are associated with $NH_4^+ + H^+$ equivalent to SO_4^{2-} but the black particles contained NH_4^+/H^+ ratios of 30 whereas the white particles varied between 1 and 5. Brosset suggests that the sulfate in the more neutral black particles was formed by heterogeneous catalytic oxidation of sulfur dioxide, and in the more acidic white particles as having possibly been formed by a photochemical mechanism.

Tanner *et al.* (1977a) also report on the frequent existence of equivalents of $NH_4^+ + H^+$ equal to SO_4^{2-}. Analyzing samples collected with a dichotomous sampler (Dzubay and Stevens, 1975) Tanner *et al.*, found that the fine fraction samples ($< 2\ \mu m$) contained essentially all of the sulfate and more acid than the total particulate sample. This suggests the existence of basic particles in the $>2\ \mu m$ range and that if interpretable data are to be obtained concerning the acidity of sulfate, the large particles should be excluded from the sample. Basic properties have been observed in other instances especially when sampling near a roadway (Tanner and Newman, 1976b) or within cities (Tanner *et al.*, 1977a). In a New York City aerosol study (Tanner *et al.*, 1977c) a high correlation between particulate ammonium and sulfate in all size ranges of respirable particles was observed and in a continuation of this study a molar ratio of NH_4^+/SO_4^{2-} of the order of 1.5 was established (Tanner *et al.*, 1977d). Variations of the strong acid content of the sulfate particles with particle size were only marginally detected.

A serious drawback to the utility of wet chemical identification of sulfur compounds is the problem associated with the fact that obscuring reactions can occur once the sample is dissolved in water. To some extent these can be overcome by adding preserving reagents to maintain oxidation states or size fractionating to eliminate neutralizing components. This simple, straightforward approach should not be overlooked since it can still be one of the more powerful tools in our arsenal of methods for the identification of sulfur-containing aerosols.

SOLVENT SEPARATION

A solvent scheme based on the work of Leahy *et al.* (1975) has been developed which allows for the sequential selective dissolution from a filter of sulfuric acid with benzaldehyde, ammonium bisulfate with iso-propanol (or methanol) and remaining sulfates with water (Tanner *et al.*, 1977b). Extensive laboratory tests have demonstrated the applicability of the method to the determination of $10\ \mu g$ sulfuric acid. The flash volatilization flame photometric detection method (Husar *et al.*, 1975) has been modified so that $4\ \mu g$ of particulate mass per cm^2 filter area may be collected from typical ambient aerosols containing at least $1\ \mu g\ m^{-3}$. The particle density on the filter is sufficiently low that topochemical reactions between acid sulfate and co-resident basic particles are eliminated. Furthermore, an explicit virtue of the benzaldehyde solvent is that sulfuric acid is isolated from the sample without neutralization occurring from the co-collected basic materials including, e.g. ammonium sulfate. Samples collected in the field should be dried prior to analysis, e.g. in a dessicator (Leahy, 1977). Indeed, Barrett *et al.* (1977) found inconsistent selectivity for ammonium bisulfate and ammonium sulfate which they attributed to water present in commercial benzaldehyde. They were able to rectify this problem by vacuum-distilling the benzaldehyde and by storing it under dry nitrogen until use.

Field data are just beginning to become available, through which this technique can be evaluated. Samples near sources have been collected and Lusis *et al.* (1975) have found as much as 86% of the sulfate collected in the plume of a nickel smelter to be present as sulfuric acid. Tanner and Newman (1976b) have also shown the existence of sulfuric acid by collecting samples near a test roadway around which were driven automobiles equipped with catalyst converters.

Samples taken in a suburban St. Louis location have been shown to contain sulfuric acid (Tanner *et al.*, 1977a). In addition, samples, taken on board an aircraft, of the elevated St. Louis urban plume have shown good correspondence between benzaldehyde-extractable sulfuric acid and titratable acid (Tanner and Newman, 1977). The extractable acid at times represented more than 1/3 the amount of sulfate present.

It should be recognized that the presence of titratable acid does not necessarily mean that you have to have sulfuric acid. In fact, a value of the solvent separation method is its ability to detect such situations. Indeed, observations have been made where the acidity is not associated with sulfuric acid and it is even greater than the amount of sulfate present (Tanner *et al.*, 1977a). It is interesting to contemplate how such techniques as the humidograph (Charlson *et al.*, 1974) or the i.r. technique (Cunningham and Johnson, 1976) might respond to this condition. Side by side comparisons would certainly be informative.

Hansen (1977) suggests that it might be desirable to purify the benzaldehyde by vacuum-distillation. He observed that commercial material can extract metal sulfates, in addition to sulfuric acid, from samples collected at a lead smelter operation. He attributes the metal sulfate solubility to the presence of benzoic acid impurity in the benzaldehyde. Pure material did not extract the metal sulfates but would still extract metal bisulfates. This problem might be unique to samples collected near smelter operations, and for routine ambient use the commercial material is probably adequate and selective.

Important results confirming the utility of the benzaldehyde method have just been reported (Barrett *et al.*, 1977). Sulfuric acid aerosol was deposited on Mitex filters (a matted Teflon). Then varying amounts of ammonia were passed through the filter to neutralize the acid. When two moles of ammonia per mole of acid was absorbed and then subjected to benzaldehyde, as expected, no sulfuric acid was found. Even when only one mole of ammonia is absorbed it also might be expected not to find any sulfuric acid by the benzaldehyde method, since all the acid could get converted to ammonium bisulfate. However, the authors found 50% of the acid as sulfuric acid, indicating that on this matted filter the ammonia must first completely neutralize the upper layers of acid, leaving the bottom layers unreacted. However, what is significant is that the experiment clearly demonstrates that the benzaldehyde method is capable of separating sulfuric acid aerosol from coexisting particles of ammonium sulfate with no neutralization occurring during the process—a most important confirmation of the method.

This solvent separation scheme is probably unique in being able to unequivocally identify sulfuric acid especially as distinguishable from ammonium bisulfate. Further applications and extensive use should be strongly encouraged and in conjunction with side-by-side comparisons with other techniques.

THERMAL VOLATILIZATION

Thermal volatilization is an attractive concept for the separation and identification of decomposable sulfate compounds. Tanner and Newman (1976a) critically reviewed the existing literature. The present discussion will be limited to some new developments. Although Leahy *et al.* (1975) were able to establish the parameters for separation and identification of sulfuric acid, ammonium bisulfate and sodium bisulfate, they were unsuccessful in trying to measure sulfuric acid in the presence of ambient aerosols. Mudgett *et al.* (1974) report the use of thermal volatilization of sulfuric acid collected on a filter in conjunction with a flame photometric analyzer. An advanced automated instrumental version employing this principle has recently been deployed (Lamothe *et al.*, 1976). The authors were able to measure sulfuric acid near a test roadway with catalyst-equipped cars. They did not however, observe acid in the ambient background air. Since no details involving this instrument have been reported in the open literature concerning recoveries of sulfuric acid, one might question whether the technique suffers from the same problems associated with the method employed by Leahy *et al.* (1975). Namely, is collected sulfuric acid quantitatively released from a filter or is some lost, possibly due to reaction with ambient aerosols during the volatilization process. Admittedly, the new technique has to its potential advantage a lower volatilization temperature (130 vs 190°C) and the presence of less collected material per unit filter area. However, problems appear to have been demonstrated by the work of Barrett *et al.* (1977). In one experiment they found that predeposited ambient aerosols on a Teflon filter can cause neutralization of sulfuric acid aerosol subsequently collected on the filter. Consequently, as you might anticipate, the flame photometric response on heating the filter was significantly lower than calculated even if the temperature was increased to 200°C. Nevertheless, this new automatic and almost real time instrument deserves further extensive testing and evaluation.

A new concept employing thermal volatilization is described in this symposium (Cobourn *et al.*, 1978). The method consists of linear programmed heating of atmospheric aerosols followed by rapid cooling. The decrease in light scattering is measured as a function of temperature. The response reflects the decomposition of sulfuric acid and ammonium bisulfate or ammonium sulfate. Also employed is a flame photometric detector to directly measure the disappearance of the sulfate compounds. The advantage of the method is that the technique gives real time measurements and the aerosol need not be pre-deposited on any surface. The disadvantages are that there is no differentiation between ammonium bisulfate and ammonium sulfate and furthermore since the method is non-specific, all decomposable aerosols will respond and confound the resulting thermogram. One might question whether you can anticipate obtaining uniquely characteristic ambient thermograms. Coupling the flame photometer is certainly a helpful addition, assuming prior removal of sulfur dioxide possibly with a diffusion denuder (Durham *et al.*, 1972). Adequate sensitivity of commercial flame photometers for this purpose will undoubtedly be a serious limitation. Nevertheless, when the ambient aerosol consists mainly of sulfate compounds the method will be uniquely useful and it will be interesting to compare the measurements obtained herein with the Charlson *et al.* (1974) humidograph.

THERMOMETRIC METHODS

Thermometric titrimetry (Hansen *et al.*, 1974a) can be used as an analytical technique to determine the composition of an unknown by continuous addition or rapid injection of a titrant while monitoring the heat output associated with the resulting reaction. If titrant is added continuously, the analytical end point is detected as a change in the rate of heat production. If an excess of titrant is rapidly injected, the concentration of any unknown is calculated from the total temperature rise. This technique has recently been exploited by members of the Thermochemical Institute at Brigham Young University to quantitatively determine the amount of sulfur in the plus four and plus six oxidation states (Hansen *et al.*, 1976).

Collected ambient aerosols are generally dissolved under air-free conditions in a solution containing 0.1 M HCl and 5 mM $FeCl_3$ (Hansen *et al.*, 1975). The acid serves the function of preventing oxidation of the sulfite species and the iron complexes the sulfite, preventing its loss from the solution. The continuous addition of dichromate in an acid solution allows for the determination of sulfite by the reaction: $3H_2SO_3 + Cr_2O_7^{2-} + 2H^+ = 3SO_4^{2-} + 4H_2O + 2Cr^{3+}$. The equivalence point is determined by observing a change in the rate of heat production. Then, by measuring the overall change in heat production up to this equivalence point, the ΔH value for the reaction can be determined. Finally, by observing breaks in the titration curve (changes in slope) additional equivalence points and thereby other oxidizable substances can be detected, e.g. organic sulfites. The explicit nature of these compounds can be inferred by comparing the calculated ΔH value with reasonable test compounds of known

ΔH. These are best confirmed by analysing the compounds in the identical apparatus.

After the oxidation is complete, an excess of barium (II) is rapidly injected and the heat associated with the precipitation of barium sulfate is measured to determine the total sulfur present (sulfate plus oxidized sulfites). The method has to be calibrated with standard sulfate solutions. The initial sulfate is then determined as the difference between the total sulfate and all oxidizable compounds. In this instance the ferric which was added during the dissolution process serves an additional purpose. Namely, forming as a product of the oxidation, a ferric sulfate complex which is precipitatable by barium instead of allowing a chromium (III) sulfate adduct to form which would be slow to dissociate and consequently not be precipitated by barium (II) (Eatough, 1977). It has not yet been demonstrated that all anticipated oxidized sulfites can be precipitated in this manner.

Samples collected in and around a copper smelter facility (Hansen *et al.*, 1975) clearly show the existence of a sulfite species. The sulfite represented about 2% of the aerosol mass, independent of particle size from <1 to $>7\,\mu$m. However, the sulfate varied between 2 and 30% of the aerosol mass over this particle size range. There appears to be a good correlation between iron and sulfite in the collected aerosols from which the authors suggest the existence of a ferric sulfite complex. Utilizing calorometric, spectrophotometric, and Mössbauer studies the authors have demonstrated in the laboratory that iron (III) and copper (II) sulfite complexes can exist in solution and in solid phases (Hansen *et al.*, 1974b). This hypothesis has recently been explored in more detail (Eatough *et al.*, 1977). The authors conclude that transition metals in aerosols can promote the formation of sulfite complexes from sulfur dioxide in the atmosphere.

Hansen *et al.* (1977) in a New York City aerosol study claim to periodically observe significant amounts of sulfite, iron (II), and organic sulfites. Hansen (1977) also claims to find sulfite and organic sulfites in a coal-fired plume. Their present conjecture, based on ΔH values, is that the organic compounds are sulfite derivatives of aldehydes or quinones. If these results bear up upon critical evaluation, then the findings are most important to our understanding of atmospheric chemistry sulfur and the concomitant implications concerning health effects.

There appears to be an as yet unexplained discrepancy associated with the thermometric method of analysis. Hansen *et al.* (1975; 1976) have shown that their method of sample dissolution (0.1 M HCl), dichromate oxidation, and thermometric titration gives rise to sulfate values which can be 1/2 those determined after a conventional hot water dissolution method (Intersociety Committee, 1972). However, a thermometric analysis of the hot water dissolution sample gives the same results as a conventional spectrophotometric method of analysis. The unlikely suggestion might be made that the hot water can cause the oxidation of sulfite species not oxidized by dichromate. Instead, the authors suggest (Eatough, 1977) that the hot water extract, conducted by refluxing with boiling water, permitted oxidation and consequent dissolution of metal sulfides, but on the contrary, under the argon atmosphere employed during the HCl dissolution procedure the metal sulfides were neither oxidized or dissolved. In addition, Smith *et al.* (1976) report similar results but show that proton induced X-ray fluorescence, which should measure total sulfur in the sample, gave results equivalent to the acid dissolution thermometric method. Here it is argued that Smith analysed the acid extract so that again you would not expect to see the sulfides. However, to further complicate the issue, Hansen (1977) in a New York City aerosol study has found that the X-ray fluorescence method can at times measure significant amounts of sulfur when none is to be observed by the thermometric method.

A resolution of these serious discrepancies will certainly increase our understanding of the chemical composition and chemistry of ambient aerosol sulfur. One point that needs to be resolved in this connection is, do all reduced sulfur species precipitate as $BaSO_4$ when oxidized by dichromate (recalling problem concerning Cr (III) adducts); or conversely, do some substances oxidized by dichromate, deceive us by thermometrically appearing to contain sulfur upon the addition of barium when they really do not. The above problems notwithstanding, the thermometric method is an exciting addition to our new array of tools. Indeed, it is unique in being the only one presently employed with which attempts are made to quantitatively determine the amount of sulfur present in the various oxidation states in ambient aerosols.

ESCA

Electron Spectroscopy for Chemical Analysis (ESCA) also called X-ray photoelectron spectroscopy is a well demonstrated tool for the determination of chemical states in molecules (Hollander and Jolly, 1970). Novakov (1973) gives a good description of the principles of ESCA and its utility for the chemical characterization of atmospheric aerosols.

ESCA consists of the measurement of the kinetic energies of photoelectrons expelled from a sample irradiated with monoenergetic X-rays. The kinetic energy of a photoelectron E_{kin}, expelled from a subshell i, is given by

$$E_{kin} = h\nu - E_i$$

where $h\nu$ is the X-ray photon energy and E_i is the binding energy of an electron in the subshell. If the photon energy is known, the determination of the kinetic energy of the photoelectron peak provides a direct measurement of the electron-binding energy.

The ESCA method is sensitive only to surface composition since the electron escape depth is only a few tens of Ångstroms. The electron binding energies are characteristic for each element, which enables the method to be used for elemental analysis. The binding energies, however, vary when the atom is in different chemical environments. These differences are known as the 'chemical shift'. The chemical shifts are not simply related to the oxidation state but are correlated to a high degree with the effective charge which the atom possesses in the molecule. The ESCA technique can therefore be used for both elemental analysis and for the determination of chemical states.

Lindberg *et al.* (1970) use a simple electrostatic model in which the charges are idealized as point charges on atoms in a molecule. The model predicts a near linear relationship between the binding energy shift and the effective charge. The more oxidized the species, the more positive the shift and the more reduced the more negative. They have obtained good correlations between the estimated charges and binding energy shifts for a large number of sulfur compounds.

Hulett *et al.* (1971) used ESCA to study sulfur compounds in fly-ash and smoke particles. They reported three chemical states of sulfur present on coal smoke particles. These included a single reduced state which was assigned to hydrogen sulfide or a mercaptan, and two species in higher oxidation states corresponding to sulfate and sulfite. The sulfur in the fly-ash was tentatively identified as a sulfate.

Craig *et al.* (1974) analyzed ambient aerosols collected in the San Francisco and Los Angeles areas. Associated with these aerosols they claim to have evidence for the existence of SO_3, SO_4^{2-}, SO_2, SO_3^{2-}, S^0 and two kinds of S^{2-}. However, not all of these species occurred at all times and at all locations. Sulfates were always found to be the dominant species, although reduced forms of sulfur were at times present in comparable amounts.

Novakov *et al.* (1976), performed ESCA analysis on ambient samples collected in West Covina, California and St. Louis, Missouri. By studying the nitrogen and sulfur spectral regions and interpreting the intensities in these regions, they were able to identify three distinctly different cases: ammonium sulfate accounting for the entire ammonium and sulfate content of the sample; ammonium appearing in concentrations above those expected for ammonium sulfate (and nitrate) with the 'excess' ammonium being volatile in vacuum; ammonium appearing mostly in a volatile form independent of sulfate and nitrate. They were, unfortunately, not able to identify the anions corresponding to the volatile ammonium species.

Novakov *et al.* (1974) in a series of laboratory experiments were able to demonstrate that graphite and soot particles could oxidize sulfur dioxide in air. Furthermore, they observed that soot-catalyzed oxidation of sulfur dioxide can occur in the flame and in combustion gases. These observations suggest that primary production of sulfate from fossil fueled power plants could very well be a function of the burning conditions. This might be especially significant in oil-fired operations where excess air is often maintained at a minimum. In addition, the catalytic formation of sulfate on soot particles would be expected to occur near point sources where both sulfur dioxide and soot concentrations are highest (Schwartz and Newman, in press).

Novakov *et al.* (1976) have performed ESCA and i.r. studies to determine the chemical state of the sulfur on the soot particles. Their results indicate that a sulfate salt-like species with a tetrahedral structure is formed on the soot surface and that the surface provides relatively strong basic sites. The product is water-soluble and behaves like sulfuric acid.

Some caution must be exercised in the use of ESCA for the determination of the molecular composition of aerosols. The assignment of specific binding energy shifts to particular chemical species is not straightforward, since the measurements reflect the net core charges and not the specific compounds. Reference compounds with the appropriate cation (usually not known) should be run to obtain a less equivocal assignment. However, it is claimed (Craig *et al.*, 1974) that certain species such as sulfates and sulfides have binding energies which are not very sensitive to the choice of cations.

ESCA by its very nature can only look at the first few layers of the surface. A question of concern would therefore be as to how well representative is a surface analysis of the bulk composition of ambient aerosols. In addition, the surface composition might be over emphasizing the importance of relatively minor constituents. Finally, especial care has to be taken in interpreting the relative concentrations of measured species.

These important questions are being addressed (Novakov, 1977) and it appears that some ambient aerosols are quite uniform in their composition and that with care ESCA can be utilized at least in a semi-quantitative manner. To be sure however, molecular assignments can still be controversial. Nevertheless, ESCA is a most powerful tool, whose exploitation for atmospheric studies is still in its infancy. Presently, the method is possibly unique in its ability to determine the molecular composition of sulfur in aerosols unperturbed by chemical and physical manipulations.

INFRARED

Absorption bands in the i.r. can be assigned to individual sulfate anions, thereby, the use of i.r. spectroscopy to determine the molecular composition of ambient aerosol sulfates has been exploited (Blanco *et al.*, 1968; 1972 and Cunningham *et al.*, 1974; 1976). Both groups collect size fractionated samples with impactors and then scrape the material off, make KBr pellets, and measure the i.r. spectra. Blanco *et al.*, use an Andersen sampler, whereas Cunningham *et al.*, use a Lundgren inertial impactor which recently has been modified to operate unattended for periods of at least one week (Cunningham, 1977). Size and time resolutions of 3 h are possible.

A review of the above works has been given (Tanner and Newman, 1976a). Briefly, Blanco found that ammonium and sulfate could dominate the submicrometer particle region and in addition Cunningham *et al.*, at times found bands attributable to bisulfate and a somewhat less acidic species which could be associated with the presence of letovicite, $(NH_4)_3H(SO_4)_2$. No clearcut evidence has been found for the existence of sulfuric acid by this technique. Since the impaction method does not permit the collection of samples less than 0.2 μm in diameter, Cunningham (1977) is developing a water extraction-KBr technique for the analysis of samples collected on filters.

An aspect of the infrared technique that is often overlooked concerns the fact that the method as presently employed has not been developed to quantitatively determine the amount of any of the identified sulfate species. Development in this regard would certainly be desirable. In addition, a serious limitation is imposed by the utilization of KBr pellets, since neutralization reactions occur during the pelletizing operation. Therefore, if any basic materials are present (including for e.g. ammonium sulfate) they could neutralize sulfuric acid, just as if you were to dissolve the sample in water.

An advance in the measuring aspect has been made using internal reflection infrared spectroscopy (Novakov *et al.*, 1976). This technique to measure the spectra directly on the collection matrix is under development (Cunningham, 1977) and would obviate the problem of neutralization reactions after sample collection. However, as indicated in the sampling section neutralization reactions can still occur during the collection procedure.

The infrared technique has proven most useful in identifying the existence of acidic sulfates, especially those occurring over short time periods (<6 h). Further development along the lines indicated would make this tool one of the most important at our disposal.

LASER RAMAN SPECTROSCOPY

The origin of the Raman effect lies in the ability of chemical species (molecules, ions, etc.) to inelastically scatter incident electromagnetic radiation. This occurs when a dipole moment is induced in the species by the incident light and is a consequence of the difference in the polarizability of the species in their different stationary states, whether they be rotational or vibrational states. In the Raman effect, either some of the energy of the incident light is given-up to the chemical species or the species imparts some of its energy to the light. In either case, the Raman scattered photons are of a different frequency than the frequency of the incident light. The frequency change of the scattered radiation corresponds to the stationary states of the species and is uniquely characteristic of the chemical identity and physical environment (gas, liquid or solid) of the species. Raman spectroscopy is a potentially viable method for not only quantitatively measuring sulfates in the atmosphere but also for differentiating the chemical and physical forms of the sulfates without the inherent uncertainties of techniques that require the prior physical or chemical processing of the samples.

Rosen and Novakov (1976) have measured the Raman spectra of sulfuric acid, ammonium bisulfate and ammonium sulfate. The concentrations are not given. A peak at 976 cm^{-1} is associated with sulfate from the ammonium sulfate spectrum, however, it was not observed in the sulfuric acid or ammonium bisulfate spectra. A sample of ambient aerosol was found to contain the 976 cm^{-1} peak and they therefore ascribed it to the presence of ammonium sulfate (a not unlikely observation). In recent studies at Brookhaven National Laboratory (Adamowicz *et al.*, 1977) the Raman spectra of one molar sulfuric acid and ammonium sulfate solutions have been measured. In this instance a 981 cm^{-1} band was found common to both solutions and is stated to be due to the SO_4^{2-} moiety in solution while a 1054 cm^{-1} band was found only in the sulfuric acid solution and attributed to HSO_4^-. The Raman spectrum of a dry ammonium sulfate powder was measured and found to be strikingly different especially for small Raman shifts.

These observations attest to both the capabilities and difficulties of using Raman spectroscopy for differentiating between different sulfate species in the physical environment in which they reside. However, a major consideration in assessing the viability of Raman spectroscopy for ambient sulfate monitoring is the intensity of the Raman effect, which is notoriously weak. A recent study of the detection limit of Raman spectroscopy for aqueous solutes (Cunningham *et al.*, 1977) indicate that the SO_4^{2-} moiety can be detected down to a concentration of a few mg l^{-1} ($\sim 5 \times 10^{-5}$ M) whic is well below the expected sulfate concentration in wet ambient aerosol.

In another recent work Stafford *et al.* (1976) were able to measure artificially produced ammonium sulfate aerosol down to ~ 8 ppb of SO_4^{2-} (considering the species as a dispersed gas) with a 1000 second integration time. Their laser Raman system included a multiple-pass optical cavity and a 0.34 W argon laser. Together, an effective 31 W of laser power excited the Raman signal.

The ultimate utility of laser Raman spectroscopy for ambient sulfate analysis, as well as for atmospheric pollution monitoring and characterization in general, rests on the application of more powerful lasers, on intra-cavity laser techniques, highly efficient signal collection and processing electronics, multiplex data acquisition, fluorescence rejection techniques, and on the ingenuity of the scientist. Improvements (Hickman and Lang, 1973) in the effective laser power of 100 times and greater can be achieved with present high-gain CW lasers and intra-cavity techniques, providing up to thousands of watts of radiant power for generation of Raman signals. Research in applying these techniques to characterize pollutants by Raman spectroscopy is currently in progress at Brookhaven National Laboratory.

DIFFUSION PROCESSOR

Techniques are available for size classifying particles in the optical region by a variety of impactor devices but no techniques were available for sub-optical size classifying. Marlow and Tanner (1976) have reported a method of diffusion processing of aerosols for filter sampling with utilizes a diffusion battery ('battery' of diffusion cells). This device which selectively removes from smaller to larger aerosol particles by diffusion, has conventionally been used with a condensation nuclei counter (CNC) for the determination of numerical size distributions of suboptically sized particles. The innovation of Marlow and Tanner is that of direct filter sampling of aerosol particles which have been size-classified under ambient atmospheric conditions in a series of highly collimated, uniform, porous diffusion cells of exceptional length (Sinclair, 1972). A diffusion-processing device applicable for field measurement of aerosol particles of $<0.3\ \mu m$ diameter has thus been constructed which is unambiguous in the mechanism of its size separation for these particles has no inherent limitations in its sample-processing rate, and can provide, direct, quantitative, analytical information on chemical composition.

In their most recent configuration (Tanner and Marlow, 1977) a collimated hole structure diffusion battery (Sinclair, 1972) is used through which $29\ l\ min^{-1}$ of air are drawn. After passing through seven segments of the battery, half the air is sampled by drawing through a filter and the rest of the air allowed to pass through the remaining eleven segments and sampled again through a filter. Simultaneously, a sample of untreated air is obtained on a third filter. The sample on the first filter has a 50% fractional penetration diameter of $0.035\ \mu m$ and the second filter has a $0.13\ \mu m$ penetration diameter. A selection of flows and sampling arrangements can be made to alter the fractional penetration diameters. A phosphoric acid treated quartz filter was generally employed and chemical analyses performed usually for, total acidity, sulfate and ammonium. In addition, analyses for iron and nitrate are sometimes included (Tanner *et al.*, 1977a). Of course, other substances could be determined if desired.

In the present configuration, the three samples can be characterized as (a) an untreated sample, (b) one in which most of the Aitken particles have been removed, and (c) one in which most of the suboptical particles have been removed. From the chemical analyses performed, comparisons were made between the results obtained on each of the filters e.g. (a–b) would give the chemical composition of the Aitken particles and (b–c) the composition of the suboptical. Often ambient aerosols are found to contain equivalents of hydrogen ion plus ammonium equal to sulfate although there are some very real exceptions.

In their original study using a slightly different diffusion battery configuration Marlow and Tanner (1976) found instances where at times the airborne sulfate in the suboptical region was in the chemical form of sulfuric acid and at times existed as ammonium bisulfate and even as ammonium sulfate. Little, if any, sulfate was found associated with particles in the Aitken nuclei region. In a second study performed in a rural location outside of St. Louis, Tanner *et al.* (1977a) again report no measurable sulfate in the very small particle size region. However, about 50% of the sulfate mass is found in the suboptical particle region. Utilizing a screen battery design of Sinclair and Hoopes (1975) in conjunction with a condensation nuclei counter these workers were able to construct time averaged size distributions of aerosol particles by number. They were then able to show that essentially none of the aerosol volume exists in the $<0.035\ \mu m$ region. Finally, assuming a density of two for the aerosol they are able to calculate that about 40% of the total aerosol mass was sulfate and that essentially all of the suboptical aerosol was sulfate.

These same investigators found three incidents of high sulfate levels during a nine-day study. Two were associated with high ammonium and low acid levels indicating that the sulfate was predominately present as ammonium sulfate. The third incident had both high ammonium and acidity levels and they were able to conclude that the aerosol was more acidic during the daytime than at night. The suggestion is offered that photochemical production during the daytime leads to sulfuric acid which is neutralized by ammonia during the night.

A criticism of this technique might be made that the size cuts are not as sharp as for other techniques (impactors). These other techniques, with the possible exception of the low pressure impactor (Herring *et al.*, 1977), cannot be used in this size range. To date this is the only tool which has opened up for inspection the suboptical size region and it has very exciting potentials towards helping us to understand the formation and fate of atmospheric sulfur aerosols. Under development (Marlow, 1977) is a sample processor which will have a flow capability of $200\ l\ min^{-1}$. This will further open up this area in allowing for a shorter time scale resolution.

FUTURE CONSIDERATIONS

Reference should be made to Table 1 for an overall grasp of the perceived advantages and problems associated with the available methods. Coupling of various methods should not be overlooked. The immediate future should initially consist of more extensive field utilization of the techniques described in

Table 1. Advantages and shortcomings of methods for determining the chemical composition of aerosol sulfur compounds

Method	Advantages	Shortcomings
Sampling	Isolation from gaseous interferences Concentrating Size fractionating	Reactions on sampling matrix changes composition
Diffusion processor	Suboptical size fractionating. Concentrating	Flat size fractions Need to collect sample
General analysis	Comparison between total and solubilized compounds	Non specific Need to collect sample
Wet chemical identification	Total acidity, sulfate, ammonium Gives compositional information Reduced sulfur possible	Oxidation, hydrolysis, neutralization reactions on dissolution No specificity Need to collect sample
Gas phase reactions	*In situ* real time capability Specificity obtained by introducing gas phase reactions	Premium on inventiveness for reaction selectivities
Solid phase reactions	Specificity obtained by introducing reactions with collected aerosols	Premium on inventiveness for reaction selectivities Need to collect sample
Microscopy	Specific identification of solid phases	Questionable specificity Tedious Need to collect sample
Solvent separation	Selective (especially H_2SO_4) Minimization of induced neutralization and reaction changes	Unchecked for reduced forms Need to collect sample
Thermal volatilization	Selectivity by temperature	Volatilization induced reactions Unchecked for reduced forms Generally need to collect samples
Light scattering	*In situ* real time capability Specificity aided by introducing gas phase chemical reactions and humidity changes	Qualitative Questionable specificity
Thermometric	Reduced and oxidized forms Specificity	Specificity somewhat indirect Need to collect sample
ESCA	Reduced and oxidized forms Specificity	Surface composition only Qualitative Need to collect sample
Infrared	Specificity	Qualitative Neutralization reactions in solvent matrix Need to collect sample
Laser Raman	*In situ* real time capability Molecular specificity	Sensitivity Portability

this review. Additional testing in the laboratory along the lines of Barrett *et al.* (1977) should be employed to ascertain the unique specificity characteristics of the techniques. A more careful evaluation of interferences is also necessary. It would be most desirable to develop cooperative programs whereby as many of the methods as possible are run alongside each other.

A new technique called ion chromatography (Small *et al.*, 1975) is gaining acceptance as a method for the analysis of sulfate in ambient aerosols (Mulik and Wittgenstein, 1977). A novel combination of ion exchange resins are employed to reduce background electrolyte so that a simple sensitive conductivity cell can be used to measure ions of interest as they are eluted off the column. This technique should prove to be especially useful for the determination of reduced sulfur species such as sulfite.

There is a pressing need for a real time sulfate measurement capability and in this regard two techniques are described in this Symposium. Both methods employ the use of a flame photometric detector with Huntzicker and Hoffman (1978) using prior diffusion denuding of sulfur dioxide and Kittelson *et al.* (1978) employing electrostatically pulsed aerosol removal.

Following the need for real time sulfate measurement is the requirement for the real time measurement of the chemical composition of sulfur aerosols. In this regard the utilization of the humidograph with gas phase ammonia titration (Charlson *et al.*, 1974) would be a real advance if it could be made quantitative and its specificity ascertained. Two new techniques are described in this Symposium. An *in situ* thermal analysis method (Cobourn *et al.*, 1978) and an aerosol mobility chromatograph (Liu *et al.*, 1978) both look promising but again specificity will be a question.

The clever use of gas phase chemistry is an untapped resource, and the advent of high sensitive flame photometric detectors and/or nephelometers coupled with diffusion removal of sulfur dioxide should prove to be a rich area for exploration. Finally, laser Raman spectroscopy with its attendant molecular specificity is a most hopeful technique providing that, through higher power, sufficient sensitivity can be obtained and, through more portable laser systems, increased mobility.

As indicated by this review, methods for the determination of the molecular composition of sulfur aerosols are rapidly being established on the basis of sound scientific principles. However, new ideas are still needed for significant future development.

REFERENCES

Adamowicz R., Schwartz S. and Meyers R. E. (1977) Unpublished Brookhaven National Laboratory results.

Alarie Y., Wakisaka I. and Oka S. (1973) Sensory irritation by sulfite aerosols. *Environ. physiol. Biochem.* **3,** 182.

Amdur M. O. and Corn M. (1963) The irritant potency of five ammonium sulfates of different particle sizes. *Ind. Hydg. J.* **24,** 326–333.

Amdur M. O. (1969) Toxocological appraisal of particulate matter oxides of sulfur and sulfuric acid. *J. Air Pollut. Control Ass.* **19,** 638–643.

Askne C., Brosset C. and Ferm M. (1973) Determination of the Proton-Donating Property of Air-Borne Particles. Swedish Water and Air Pollution Research Laboratory (IVL), Report B157, Gothenburg, Sweden.

Barrett W. J., Miller H. C., Smith J. E. and Christina H. G. (1977) Development of a Portable Device to Collect Sulfuric Acid Aerosol. Interim Report. EPA-600/2-77-027, available from National Technical Information Service, Springfield, Virginia 22161.

Blanco A. J. and Hoidale G. B. (1968) Microspectrophotometric technique for obtaining the infrared spectrum of microgram quantities of atmospheric dust. *Atmospheric Environment* **2,** 327–330.

Blanco A. J. and McIntyre R. G. (1972) An infrared spectroscopic view of atmospheric particulates over El Paso, Texas. *Atmospheric Environment* **6,** 557–562.

Brosset C. (1975) Determination of Air-Borne Acidity with Some Examples on the Role of Acid Particles in Acidification. Swedish Water and Air Pollution Laboratory (IVL), Report B226, Gothenburg, Sweden.

Brosset C. (1976) Air-borne particles: black and white episodes. *Ambio* **V,** 157–163.

Charlson R. J., Vanderpol A. H., Covert D. S., Waggoner A. P. and Ahlquist N. C. (1974) H_2SO_4 $(NH_4)_2SO_4$ background aerosol-optical detection in St. Louis region. *Atmospheric Environment* **8,** 1257–1267.

Cobourn W. G., Husar R. B. and Husar J. D. (1978) Monitoring of ambient H_2SO_4 and its ammonium salts by *in situ* aerosol thermal analysis. *Atmospheric Environment* **12,** 89–98.

Coffer J. W. and Charlson R. J. (1974) SO_2 Oxidation to Sulfate Due to Particulate Matter on a High Volume Air Sampler Filter. Extended Abstracts of The Division of Environmental Chemistry, American Chemical Society Meeting, Los Angeles, California, 3 April.

Craig N. L., Harker A. B. and Novakov T. (1974) Determination of the chemical states of sulfur in ambient pollution aerosols by X-ray photoelectron spectroscopy. *Atmospheric Environment* **8,** 15–21.

Cunningham K. M., Goldberg M. C. and Weiner E. R. (1977) Investigation of detection limits for solutes in water measured by laser-Raman spectroscopy. *Analyt. Chem.* **49,** 70–75.

Cunningham P. T., Johnson S. A. and Yang R. T. (1974) Variations in the chemistry of airborne particulate matter with particle size and time. *Environ. Sci. Technol.* **8,** 131–135.

Cunningham P. T. and Johnson S. A. (1976) Spectroscopic observation of acid sulfate in atmospheric particulate samples. *Science* **191,** 77–79.

Cunningham P. T. (1977) Private communication.

Durham J. L., Wagman J., Fish B. R. and Seeley F. G. (1972) Radiochemical analysis of $^{35}SO_2$ adsorbed on PbO_2. *Analyt. Lett.* **5,** 469–478.

Dzubay T. G. and Stevens R. K. (1975) Ambient air analysis with dichotomous sampler and X-ray fluorescence spectrometer. *Environ. Sci. Technol.* **9,** 663–668.

Dzubay T. G. (1976) Private communication.

Eatough D. J. Hansen L. D., Izatt R. M. and Mangelson N. F. (1977) Extended Abstracts of the Division of Environmental Chemistry, American Chemical Society Meeting, New Orleans, Louisiana, 20–25 March.

Eatough D. J. (1977) Private communication.

Forrest J. and Newman L. (1973a) Ambient air monitoring for sulfur compounds: a critical review. *Air Pollut. Control Ass. J.* **23,** 761–768.

Forrest J. and Newman L. (1973b) Sampling and analysis of atmospheric sulfur compounds for isotope ratio studies. *Atmospheric Environment* **7,** 561–573.

Forrest J. and Newman L. (1977) Application of ^{110}Ag Microgram Sulfate Analysis for the Short Time Resolution of Ambient Levels of Sulfur Aerosol. Extended Abstracts, Div. of Environ. Chem. Amer. Chem. Soc. Meet., New Orleans, Louisiana, 121–123. *Analyt. Chem.* **49,** 1579–1584.

Garber R. W. (1977) Unpublished Brookhaven National Laboratory Report.

Gran G. (1952) Determination of the equivalence point in potentiometric titrations, Part II. *Analyst* **77,** 661–671.

Hansen L. D., Izatt R. M. and Christensen J. J. (1974a) Applications of thermometric titrimetry to analytical chemistry. Chapter 1 in *New Developments in Titrimetry, Treatise on Titrimetry* (edited J. Jordon). Vol. 2 Dekker, New York.

Hansen L. D., Eatough D. J., Whiting L., Bartholomew C. H., Cluff C. L., Izatt R. M. and Christensen J. J. (1974b) Transition Metal-SO_3^{2-} Complexes: A postulated Mechanism for the Synergistic Effects of Aerosols and SO_2 on the Respiratory Tract, Trace Substances in Environmental Health, VII. A Symposium. Edited by D. D. Hemphill. Univ. of Missouri, pp. 393–397.

Hansen L. D., Eatough D. J., Mangelson N. F., Jensen T. E., Cannon D., Smith T. J. and Moore D. E. (1975) Sulfur Species and Heavy Metals in Particulates from a Copper Smelter, Proceedings of International Conference on Environmental Sensing and Assessment, Vol. 2, Las Vegas, Nevada.

Hansen L. D., Whiting L., Eatough D. J., Jensen T. E. and Izatt R. M. (1976) Determination of sulfur (IV) and sulfate in aerosols by thermometric methods. *Analyt. Chem.* **48,** 634–638.

Hansen L. D., Major T., Ryder J., Richter B., Mangelson N. F. and Eatough J. (1977) Reduced Species and Acid-Base Components of The New York City Aerosols Presented at The Industrial Hygiene Conference, New Orleans, Louisiana, 24–27 May.

Hansen L. D. (1977) Private communication.

Harker A. B., Richards L. W. and Clark W. E. (1977) The effect of atmospheric SO_2 photochemistry upon observed nitrate concentrations in aerosols. *Atmospheric Environment* **11,** 87–91.

Herring S. V., Flagan R. C. and Friedlander S. K. (1977) A New Low-Pressure Impactor: Determination of the Size Distribution of Aerosol Sulfur Compounds. Extended Abstracts Div. of Environ. Chem., Amer. Chem. Soc. Meet., 20–25 March, pp. 108–111.

Hickman R. S. and Liang L. (1973) Intracavity laser-Raman spectroscopy using a commercial laser. *Appl. Spectrosc.* **27,** 425–427.

Hollander J. M. and Jolly W. L. (1970) X-ray photoelectron spectroscopy *Accounts Chem. Res.* **3,** 193–200.

Hulett L. D., Carlson T. A., Fish B. R. and Durham J. L. (1971) Studies of Sulfur Compounds Adsorbed on Smoke Particles and Other Solids by Photoelectron Spectroscopy, Proceedings of the Symposium on Air Quality, 161st National Meeting, ACS, Los Angeles. Plenum, Washington, D.C.

Huntzicker J. J. and Hoffman R. (1978) The continuous monitoring of particulate sulfur compounds by flame photometry. *Atmospheric Environment* **12,** 83–88.

Husar J. D., Husar R. B. and Stubits P. K. (1975) Determination of submicrogram amounts of atmospheric particulate sulfur. *Analyt. Chem.* **47,** 2062–2065.

Intersociety Committee (1972) Methods of Air Sampling and Analysis. American Public Health Association, p. 296.

Junge C. E. (1960) Sulfur in the atmosphere. *J. geophys. Res.* **65,** 227–237.

Kittelson D. B., Veemersch M., Liu B. Y. H., Pui D. Y. H. and Whitby K. T. (1978) Total sulfur aerosol detection with an electrostatically pulsed flame photometric detector system. *Atmospheric Environment* **12,** 105–111.

Lamothe P. J., Dzubay T. G. and Stevens R. K. (1976) Chemical Characterization of Aerosols Present During the General Motors Sulfate Dispersion Experiment. Appears in EPA-600/3-76-035. The General Motors/Environmental Protection Agency Sulfate Dispersion Experiment, pp. 1–28.

Lave L. and Seskin E. P. (1973) An analysis of the association between U.S. mortality and air pollution. *J. Am. Statistical Ass.* **68,** 284–290.

Leahy D., Siegel R., Klotz P. and Newman L. (1975) The separation and characterization of sulfate aerosol. *Atmospheric Environment* **9,** 219–229.

Leahy D. F. (1977) Unpublished Brookhaven National Laboratory Report.

Lindberg B. J., Hamrin K., Johansson G., Gelius U., Fahlman A., Nordling C. and Siegbahn K. (1970) Molecular spectroscopy by means of ESCA II. Sulfur compounds correlation of electron binding energy with structure. *Physica Scripta* **1,** 286–298.

Lippman M. and Albert R. E. (1969) The effects of particle size on the regional deposition of inhaled aerosols in the human respiratory tract. *Am. Inst. Hydg. Assoc. J.* **30,** 257–275.

Liu B. Y. H., Pui D. Y. H., Whitby K. T. and Kittelson D. B. (1978) The aerosol mobility chromatograph, a new detector for sulfuric acid aerosols. *Atmospheric Environment* **12,** 99–104.

Lusis M., Wiebe H. A., Anlauf K. G. and Sanderson H. P. (1975) The Rate of Oxidation of the Sulfur Dioxide to Sulfuric Acid and Sulfate in the Plume of the INCO Super Stack at Sudbury, Ontario. Report ARQA 26-75, Atoms. Environment Service, Downsview, Ontario.

Marlow W. H. and Tanner R. L. (1976) Diffusion sampling method for ambient aerosol size discrimination with chemical composition determination. *Analyt. Chem.* **48,** 1999–2001.

Marlow W. H. (1977) Private communication.

Mudgett P. S., Richards L. W. and Roehrig J. R. (1974) *A New Technique to Measure Sulfuric Acid in the Atmosphere in Analytical Methods Applied to Air Pollution Measurement* (Edited by R. Stevens and Herget,) Ann Arbor Science, Michigan.

Mulik J. and Wittgenstein E. (1977) Symposium on Ion Chromatographic Analysis of Environmental Pollutants at the Environmental Protection Agency, Research Triangle Park, N.C., 28 April.

Noll K. E., Laskol J., Washeleski M. C. and Allen H. E. Sulfate formation from SO_2 oxidation in ambient aerosol deposits. Submitted to *Atmospheric Environment.*

Novakov T. (1973) Chemical Characterization of Atmospheric Pollution Particulates by Photoelectron Spectroscopy. LBL-5232, Published in the Proceedings of the Conference on Sensing of Environmental Pollutants, Washington, D.C., 10–12 Dec., pp. 197–204.

Novakov T., Chang S. G. and Harker A. B. (1974) Sulfates as pollution particulates: catalytic formation on carbon (soot) particles. *Science* **186,** 259–261.

Novakov T., Chang S. G., Dod R. L. and Rosen H. (1976) Chemical Characterization of Aerosol Species Produced in Heterogeneous Gas-Particle Reactions. LBL-5215, Presented at the Air Pollution Control Association Annual Meeting, Portland, Oregon, 27 June–1 July.

Novakov T. (1977) Private communication.

Pierson W. R., Hammerle R. H. and Brachaczek W. W. (1976) Sulfate formed by interaction of sulfur dioxide with filters and aerosol deposits. *Analyt. Chem.* **48,** 1808–1811.

Roberts P. T. and Friedlander S. K. (1976) Analysis of sulfur in deposited aerosol particles by vaporization and flame photometric detection. *Atmospheric Environment* **10,** 403–408.

Rosen H. and Novakov T. Identification of primary particulate carbon and sulfate species by Raman-spectroscopy. *Atmospheric Environment*, in press.

Schwartz S. E. and Newman L. Processes limiting the oxidation of sulfur dioxide in stack plumes. Submitted to *Environ. Sci. Tech.*

Shy C. M. and Finklea J. F. (1973) Air pollution affects community health. *Environ. Sci. Tech.* **7,** 204–208.

Sinclair D. (1972) A portable diffusion battery. Its application to measuring aerosol size characterisitics. *Am. Ind. Hyg. Ass. J.* **32,** 729–735.

Sinclair D. and Hoopes G. S. (1975) A novel form of diffusion battery. *Am. Ind. Hyg. Ass. J.* **35,** 39–42.

Small A., Stevens T. S. and Bauman W. C. (1975) Novel ion exchange chromatographic method using conductimetric detection. *Analyt. Chem.* **47,** 1801–1809.

Smith T. J., Eatough D. J., Hansen L. D. and Mangelsen N. F. (1976) The chemistry of sulfur and arsenic in airborne copper smelter particulates. *Bull. Environmental Contamination Toxicology* **15,** 651–659.

Stafford R. G., Chang R. K. and Kindlmann P. J. (1976) Laser-Raman Monitoring of Ambient Sulfate Aerosols. Presented at the 8th Materials Research Symposium on Methods and Standards for Environmental Measurement, National Bureau of Standards, Gaithersburg, MD.

Stevens R. K., Dzubay T. G., Russwurm G. and Rickel D. (1978) Sampling and analysis of atmospheric sulfates and related species. *Atmospheric Environment* **12,** 55–68.

Sulfate Method VIb Via Turbidimetry (1959) Technicon Industrial Systems, Tarrytown, NY.

Tanner R. L. and Newman L. (1976a) The analysis of airborne sulfate: a critical review. *J. Air Pollut. Control Ass.* **26,** 737–747.

Tanner R. L. and Newman L. (1976b) Chemical Speciation of Sulfate Emissions from Catalyst Equipped Automobiles Under Ambient Conditions. Appears in EPA-600/3-76-035. The General Motors/Environmental Protection Agency Sulfate Dispersion Experiment, pp. 131–145.

Tanner R. L. and Marlow W. H. (1977) Size discrimination and chemical composition of ambient airborne sulfate particles by diffusion sampling. *Atmospheric Environment*, **11**, 1143–1150.

Tanner R. L. and Newman L. (1977) Chemical-Analytical Techniques for Aerosols. Presented at the Symposium on Aerosol Science and Technology, 82nd Amer. Inst. Chem. Eng. Meeting, Atlantic City, NJ., 30 Aug.–1 Sept. To appear in *Recent Developments of Aerosol Science* (Edited by D. Shaw). Wiley & Sons, New York.

Tanner R. L., Cederwall R., Garber R., Leahy D., Marlow W., Meyers R., Phillips M. and Newman L. (1977a) Separation and analysis of aerosol sulfate species at ambient concentrations. *Atmospheric Environment*, **11**, 955–966.

Tanner R. L., Garber R. W. and Newman L. (1977b) Speciation of Sulfate in Ambient Aerosols by Solvent Extraction with Flame Photometric Detection. Extended Abstracts of the Division of Environmental Chemistry. American Chemical Society Meeting, New Orleans, Louisiana, 20–25 March, pp. 112–113.

Tanner R., Garber R., Marlow W., Leaderer B., Eatough D. and Leyko M. A. (1977c) Chemical Composition and Size Distribution of Sulfate as a Function of Particle Size in New York City Aerosol. Presented at the Industrial Hygiene Conference, New Orleans, Louisiana,

Tanner R. L., Marlow W. M. and Newman L. (1977d) Chemical Composition Correlations of Sulfate as a Function of Particle Size in New York City Aerosol. Submitted for Presentation at the Symposium on Atmospheric Sulfur Compounds: Formation and Removal Processes. Amer. Inst. of Chem. Eng., 7th Annual Meeting, New York, NY., 13–17 Nov.

Tanner R. L., Forrest J. and Newman L. Determination of Atmospheric Gaseous and Particulate Compounds. *Sulfur in the Environment* (Edited by J. O. Niragu) Wiley & Sons, New York (in press).

Wagman J., Lee R. E. and Axt C. J. (1967) Influence of some atmospheric variables on the concentration and particle size distribution of sulfate in urban air. *Atmospheric Environment* **1**, 478–488.

Atmospheric Environment Vol. 12, pp. 127–133. Pergamon Press 1978. Printed in Great Britain.

REMOTE SENSING OF SULPHUR DIOXIDE

P. M. HAMILTON and R. H. VAREY
Central Electricity Research Laboratories, Leatherhead, Surrey, UK
and
M. M. MILLÁN
Atmospheric Environment Service, Dufferin St., Downsview, Ontario, Canada

(*Received* 27 *September* 1977)

Abstract—There are many possible methods for the remote sensing of sulphur dioxide but only two are sufficiently sensitive to measure trace amounts: correlation spectrometry and differential lidar.
The correlation spectrometer measures line integrals of concentration, or burdens (in atm–μm or ppm–m), by analysing incident radiation in the ultraviolet for absorption by sulphur dioxide. It has already been widely used to measure vertical burdens against a skylight background. Measurements of emission rate have been derived from traverses of a plume near its source: they are limited by the accuracy of the associated wind speed rather than by the spectrometer. Comprehensive measurements of horizontal dispersion and its dependence on times of travel and sampling have also been obtained from traverses farther downwind. The instrument has proved particularly valuable as a sensitive plume locator on mobile surveys.
The differential lidar provides range-resolved measurements of concentration by reflecting pulses of laser light at two wavelengths with different absorption coefficients from particles along the line of sight. It offers a sensitivity of a few parts per billion (ppb) to ranges in excess of 1 km with resolution in space and time of 100 m and 10 s. The instrument has already been demonstrated in prototype form and is now being developed for operational use.

1. INTRODUCTION

During the past decade or so interest has grown in obtaining accurate information about the concentrations of sulphur dioxide that occur at both ground level and aloft in the vicinity (0–20 km) of power plants. Several comprehensive ground-based surveys using fixed point samplers have been completed. As power plants increase in size and disperse their emissions at greater heights these surveys have indicated an increasing need for the development of sensitive and mobile remote sensing equipment which can analyse airborne pollution conveniently from the ground. This has led to a spate of international activity in exploiting physical interactions of radiation with gaseous species to make such measurements at ppb levels. The present paper briefly reviews progress in remote sensing of the most important gaseous emission from industrial plant, namely sulphur dioxide.

2. THE OPTIONS

Remote sensing is accomplished by observing changes in a field of radiation produced by interaction with a target. For gases, the most suitable form of radiation is electromagnetic and the interactions are known as 'spectroscopic'. They include absorption, emission, Raman scattering and resonance scattering.

There are two distinct groups of techniques: *passive*—using natural radiation; and *active*—using radiation from a controlled source. There are also two types of measurement: *line-integral*—yielding the integral of concentration along a line of sight; and *range-resolved*—yielding full spatial distributions. Passive techniques are usually simpler and more convenient than active ones but, in the case of sulphur dioxide, they provide only line-integral measurements. Active techniques may be used for both types of measurement. Line-integrals may be obtained either: directly, by viewing a distant source; or, over a folded path, by reflecting radiation from a retroreflector or topographic feature. Range-resolved measurements may be obtained by reflecting radiation from the atmosphere.

The absorption spectrum for sulphur dioxide shows one strong band in the ultraviolet (0.18–0.32 μm) and three weak or moderate bands in the infrared (4.0, 7.4 and 8.7 μm). The various options for remote sensing are set out in Table 1 under each of these wavebands and against each of the basic interactions. The table also lists the strengths of absorption, natural sources and interfering gases in each band. Only two bands have so far been exploited and, of these, the ultraviolet band has received most attention. Both of the techniques listed in this band rely on absorption and both are sensitive to integrated concentrations of 10 atm–μm (= 10 parts per million × 1 m = 28.6 μg m^{-2}) or less. The other band at 8.7 μm offers techniques based on emission and Raman scattering but they are limited to a sensitivity of about 1 atm–mm. More sensitive techniques based on absorption have also been proposed in this band but not yet devel-

Table 1. Options for remote sensing of sulphur dioxide

	Wavebands/μm	0.18–0.32	3.9–4.1	7.1–7.7	8.1–9.4
Characteristics	Strength	Very strong	Weak	Strong	Moderate
	Natural sources	Solar above 0.295 μm	Solar (weak) Thermal (weak)	Thermal	Thermal
	Interference	O_3	None	H_2O	H_2O (slight)
	Absorption	*Correlation Spectrometer Spectroscopic Lidar*	Weak interaction	Interaction subject to interference	(*Spectroscopic Lidar*)
Interactions	Emission	No interaction	Weak interaction	Interaction subject to interference	*Thermal Radiometer Heterodyne Radiometer*
	Raman scattering	No interaction	No interaction	Weak interaction subject to interference	*Lidar*
	Resonance scattering	Interaction quenched	No interaction	No interaction	No interaction

oped. Neither of the other two bands seems promising: that at 4.0 μm is weak; that at 7.4 μm is subject to interference.

The technique in widest current use is the Correlation Spectrometer. This provides line-integral measurements in both passive and active modes. It will be described in detail later but, in essence, it is a device for receiving broadband radiation and analysing it to extract components due to absorption by a particular species. The received spectrum is effectively "correlated" with a reference spectrum represented by the transfer function of the instrument. Most versions use a grating to disperse the incoming spectrum, but others operate non-dispersively by measuring the effects of passing radiation through samples of the relevant species held in the instrument. For sulphur dioxide the technique is used in the ultraviolet band. Instruments are relatively inexpensive ($20,000) and compact (20 kg) and are well suited to mobile use. In active mode, they are mainly used with xenon lamps to measure line-integrals over near-horizontal paths. In passive mode they have been much used to analyse solar radiation scattered from the zenith, thereby determining the vertical line-integral of concentration or 'overhead burden'. Such measurements are, however, liable to interference caused by variations in atmospheric scattering and absorption by other species (especially ozone).

The Spectroscopic or Differential Absorption Lidar is not subject to these interferences. This is an active device giving range-resolved measurements. It is the most powerful technique available but is expensive ($200,000) and bulky (500 kg). In brief, light is emitted from a tunable laser, reflected from atmospheric particles and collected in a fairly simple receiver incorporating a suitable photodetector. Sulphur dioxide concentrations are determined from changes in signals caused by tuning the laser through an absorption line. Although the technique was first demonstrated for water vapour more than ten years ago, its application to sulphur dioxide has only become practicable with the advent of suitable lasers and digital processors. Present systems operate in the ultraviolet but efforts are being made to apply the technique in the 8.7 μm band.

Since they are by far the most sensitive of the available techniques, most of the rest of this paper will be devoted to detailed discussions of the correlation spectrometer and spectroscopic lidar. First, however, a few comments on the other techniques listed in Table 1.

The interaction of radiation with a molecule may lead to scattering at the same frequency (Rayleigh) or a different one (Raman). For Raman scattering, the change in frequency depends on the frequency of a suitable vibration or rotation band. The effect has been demonstrated for sulphur dioxide with the 8.7 μm band in a Raman Lidar (Inaba and Kobayasi, 1972). The interaction is, however, weak and would only be useful for monitoring concentrations near the source.

Scattering may be enhanced by tuning the frequency of incident radiation to that of a suitable absorption line. Although such resonance scattering occurs with sulphur dioxide in the ultraviolet it is suppressed or 'quenched' at tropospheric pressures and is too weak for remote sensing.

Other methods expoit the thermal or black-body emission from a hot gas. Thus, Dupoux *et al.* (1976) used a broad-band Thermal Radiometer operating at 8.9 μm (SO_2 emission) and 4.25 μm (CO_2 emission) to measure the ratio of these gases and infer the concentration of sulphur dioxide at chimney top. Menzies (1972) has made similar measurements with a Heterodyne Radiometer. This offers high spectral resolution and better sensitivity but is also limited to measurements fairly near the source.

3. CORRELATION SPECTROMETRY

3.1 *Principles*

Dispersive correlation spectrometry was used first

to study water vapour in planetary atmospheres (Bottema *et al.*, 1964) and subsequently to measure trace amounts of sulphur dioxide (Kay, 1967). The early development of the technique and its application to air pollution studies were carried out by Barringer Research Limited and led to a series of widely-used mask correlation spectrometers under the trade name COSPEC (Newcomb and Millán, 1970; Moffat *et al.*, 1971; Moffat and Millán, 1971). More recently, similar instruments have been developed elsewhere (Bonafé *et al.*, 1976). The principle of the COSPEC is shown in Fig. 1. First, incident radiation, whose spectrum is the product of a source spectrum [e.g. Skylight (Fig. 1a)] and the sulphur dioxide transmission (Fig. 1b), is received through a telescope. It passes through an entrance mask, a grating spectrometer and an exit mask onto a photomultiplier detector. In recent instruments (COSPEC IVB) the entrance mask has three slits producing three superimposed spectra. There are four exit masks on a rotating disc, each with seven slits, and these sample radiation alternately at wavelengths with high and low sulphur dioxide transmission. This optical processing may be represented by multiplying together the incident spectrum, the transfer efficiency and each of the mask transfer functions (Fig. 1c). The four resulting signals are then processed electronically to produce the signal $S = (S1 - S2)/S1 + k(S3 - S4)/S3$ where k may be adjusted. The sensitivity of the instrument is established from the responses produced by known amounts of gas in reference cells. Detailed expositions of the various considerations for designing and operating mask correlation spectrometers, with particular reference to COSPEC, have appeared in two recent papers (Millán and Hoff, 1977a; Millán and Hoff, 1976).

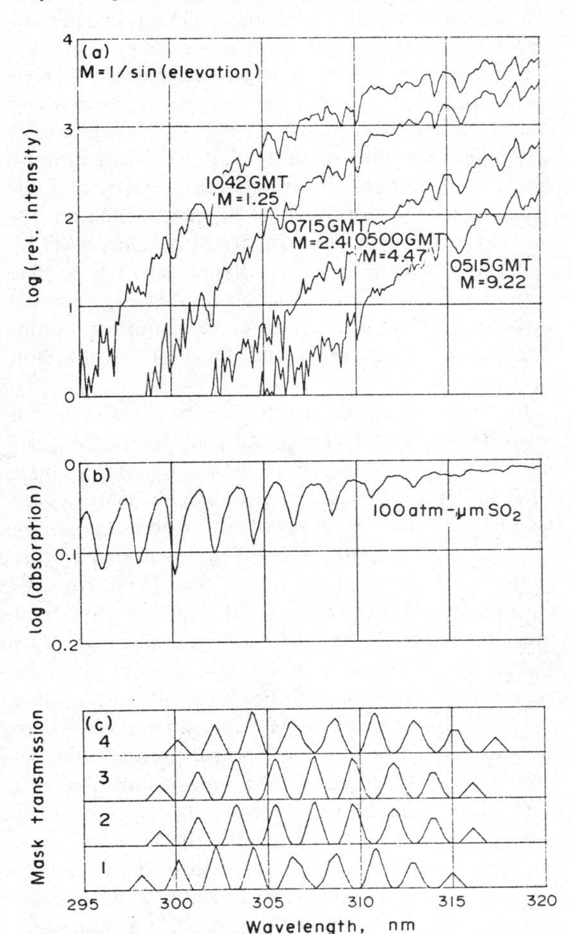

Fig. 1. (a) Radiance of zenith sky, Leatherhead, 31 July 1973. (b) Absorbance of 100 atm–μm SO_2. (c) Mask transfer functions of Barringer COSPEC IVB.

Almost all the work to date has been carried out with dispersive spectrometers, but it is worth mentioning a new sensor based on a somewhat different principle and described as a selective modulation radiometer (Laurent, 1975). This relies on the fact that the overall transmission of a band of radiation through two consecutive quantities of gas is not equal to the product of the individual transmissions. The instrument determines the amount of gas through which incident light has already travelled from the differences produced in transmission through further internal samples. Its sensitivity and performance characteristics are very similar to those of COSPEC instruments.

3.2 *Long path measurements*

Several papers have appeared on the use of correlation spectrometers in conjunction with lamps to measure sulphur dioxide over paths of up to 8 km (Newcomb and Millán, 1970; Moffatt *et al.* 1971; Bonafé *et al.*, 1976; Evangelisti *et al.*, in press; Sandroni and Cerutti, 1977). Such measurements offer several advantages over those made by point monitors: (1) operation over suitable paths yields data better suited to regional studies; (2) the method is very sensitive (better than 1 ppb); and (3) the measurements are made without contact. Despite these advantages, it seems that the method has so far been used only in relatively short term studies.

3.3 *Passive measurements*

Correlation spectrometers have been widely used in the passive mode to measure burdens of sulphur dioxide along the path of scattered radiation. In particular, many surveys have been conducted with instruments looking vertically upwards from vehicles and either upwards or downwards from aircraft. There have been three primary objectives: (1) to measure the emission from specific sources; (2) to study the dispersion of plumes; and (3) to measure depletion due to deposition and conversion. Of course, few surveys have been confined to just one of these goals and, indeed, the majority now combine all three in the rather general aim of collecting data to supplement existing networks of pollution monitors for comparison with predictions in regional models. Nevertheless, it is convenient to look at each objective in turn.

Before doing so, however, a few words about the limitations of the technique. A few examples of skylight spectra measured in the zenith during the course of a clear day are shown in Fig. 1(a). Two features are immediately apparent: (1) the fine structure, principally due to solar absorption, has an amplitude comparable with that produced by transmission through about 500 atm–μm sulphur dioxide; and (2) the overall slope, produced by absorption at short wavelengths in the atmospheric ozone layers, becomes progressively steeper as the sun's elevation decreases. These features, in turn, have two consequences: (1) the background signal observed in the absence of sulphur dioxide is not, in general zero; and (2) it may vary during the course of the day, perhaps by the equivalent of up to 250 atm–μm. Fortunately, in favourable conditions, it is possible to reduce this variation by careful adjustment of the optical and electronic processing to about 25 atm–μm for a period of at least 6 h in the middle of the day (Millán and Hoff, 1977a, 1977b, in press). Even so, the measurements cannot be made absolute, since there is no direct way of establishing the zero.

The first measurements of the emission from a particular source were made from fixed locations by observing the plume just above the chimney top (Moffat and Millán, 1971). However, this procedure depends on the geometry and velocity of the plume in a region where they may be rather difficult to define. Thus, it is now more usual to traverse beneath the plume looking vertically upwards at distances of around 1 km from the chimney (Jepson and Langan, 1974). The flux is then found by measuring burdens in the plume relative to background values outside it, integrating with distance across wind and multiplying by a suitable wind speed. Several traverses will normally be needed, because of random variations introduced by fluctuations in wind speed and direction. In an exhaustive study of this method based on 465 traverses, Sperling (1975) found that an accuracy of $\pm 30\%$ (standard deviation) could be achieved over a 20-min period with an average of 5 traverses. Most of the inaccuracy was probably associated with the wind speed.

A further limitation is that measured burdens relate to some 'effective' path. In an ideal situation, the sulphur dioxide lies in a clear particle-free atmosphere and is illuminated entirely from behind. In practice, the gas will be accompanied by particles which scatter radiation into the line of sight from other angles. If the sulphur dioxide is distributed uniformly in horizontal layers this can only increase the effective path length and inflate measurements of vertical burden. For an isolated plume, however, the measurements could either be increased or reduced (Moffatt and Millán, 1971). If the concentration of particles is high, multiple scattering within the plume or layer could further increase the effective path. During a recent survey, two or more spectrometers simultaneously observed enhanced burdens in a power station plume on several occasions (Moore and Hamilton, 1977). The values were as much as two or three times those calculated from emission and wind data. They were most marked when the plume became mixed with a layer of low cloud or when it contained clearly visible quantities of condensed water vapour. Preliminary results from a theoretical study of the problem indicate that such substantial enhancements can easily occur in plumes, but that measurements in a mixing layer with depth 1 km and visibility 5 km would not be enhanced by more than about 10% (Sutton and Varey, 1977).

Since correlation spectrometers measure vertical integrals they are invaluable for detecting the plumes emitted from tall chimneys and studying their horizontal dispersion especially in conditions where they do not reach the ground. Millán *et al.* (1976) and Millán (1976) have discussed the relevant procedures as they applied them to a study of the plume from a 380-m stack at Sudbury, Ontario. They have shown the value of interpreting crosswind profiles in terms of both Eulerian averages (in a fixed co-ordinate frame) and Lagrangian averages (relative to the plume axis). The latter are particularly useful for studying the effects of wind shear and bifurcation (Fanaki, 1975). Of course, just as with other instruments, the full potential of the correlation spectrometer is only fully realised by using it in conjunction with other sensors. Thus, it is finding a place in mobile survey units together with monitors for measuring ground level concentrations. Furthermore, the study of some aspects of dispersion, such as its dependence on distance of travel, can only be carried out effectively by deploying several units simultaneously. Up to five teams cooperated in this way on campaigns organised recently by the Commission of the European Communities at Lacq (1975) and Drax (1976) (Guillot, in press).

In recent years, much attention has been directed towards the transformation and deposition of sulphur dioxide during its transport over distances of order 1000 km. Since correlation spectrometers can be used to measure flux, it is evidently worth examining whether they might be useful in determining depletion. The variation of sulphur dioxide burden with distance has therefore been calculated for different rates of depletion and for emission from an area source (Fig. 2a) and a point source (Fig. 2b). The area source corresponds to England and Wales, while the point source is a 2000 MWe power station. There are two distinct methods for measuring depletion: (1) from the difference between a known emission and a downwind flux; and (2) from the difference between two downwind fluxes. Probably the most hopeful strategy is to use method (1) with a point source. Even so, because of errors introduced by uncertainties in emission (10%), COSPEC zero drift (5 atm–μm), sampling (10% after 10 traverses) and wind speed (10%), the overall accuracy (20%) would barely be sufficient to distinguish between consecutive curves in

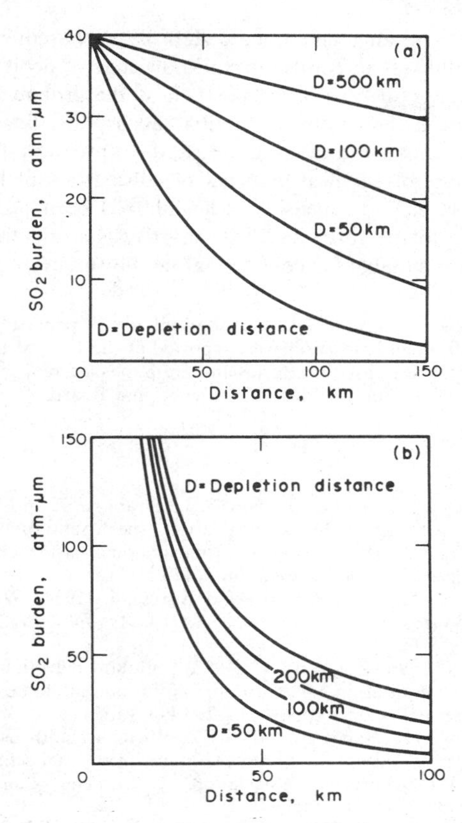

Fig. 2. Dependence of SO_2 burden on distance and depletion. (a) area source: emission = 100 $m^3 s^{-1}$; width = 500 km; wind = 5 $m s^{-1}$. (b) point source: emission = 3 $m^3 s^{-1}$; width = 0.2 × distance; wind = 5 $m s^{-1}$.

Fig. 2b. It is unlikely that this method could be used with an area source because neither the emission nor the instrument zero can be established with the necessary precision. It is also unlikely that method (2) could be used with either source since the differences in slope between consecutive curves are rather small.

These predictions were confirmed in a survey of sulphur transport to 100 km reported by Fisher *et al.* (1977). During this survey, a COSPEC IV was flown across plumes from the English Midlands at several levels in the hope of determining vertical profiles of concentration. This proved impossible both because dispersion is a random process and because the COSPEC zero could not be determined with sufficient accuracy.

Despite its quantitative limitations, the correlation spectrometer has emerged as an almost indispensable tool for the efficient conduct of mobile surveys. It is a very sensitive instrument capable of locating plumes 100 km or more from their source (Millán and Chung, 1977). It is also invaluable in supplementing measurements from fixed networks of ambient monitors (Davies *et al.*, 1975).

4. DIFFERENTIAL ABSORPTION LIDAR

Lidar has long been used for the remote measurement of the trajectories and dispersion of power station plumes (Hamilton, 1969). With the basic technique, a laser pulse is fired from the ground at the plume and light backscattered by particulate matter is detected in a telescope system adjacent to the laser. The distance of the plume is found by measuring the time taken for the light to travel from the laser to the plume and back again. If a wavelength-tunable laser is used, the technique can be adapted to give absolute measurements of the concentration of gases in a plume or in the atmosphere. In the case of SO_2 it is possible to measure concentrations as low as a few ppb at ranges of a kilometre or more. The method works by tuning the wavelength of the laser pulse on and off an absorption line in the spectrum of the gas being measured. For SO_2 the wavelengths used are normally close to 300 nm. The ratio of the intensities of the backscattered radiation at the two wavelengths gives the concentration of the absorbing gas and resolution in range is obtained by measuring the ratio as a function of time.

The differential absorption lidar was first applied to the measurement of water vapour with a ruby laser by Schotland (1966). The method received little attention until tunable lasers became commercially available. The first reported measurements of a pollutant in the atmosphere were performed by Igarashi (1973) who used a dye laser to measure NO_2 over a 300 m path. Using a 1 mJ coumarin dye laser, concentrations of NO_2 down to 0.2 ppm at ranges up to 3.5 km have been measured (Rothe *et al.*, 1974a) and contour maps of NO_2 distributions close to a factory in Cologne have been produced (Rothe *et al.*, 1974b). Calibration cell measurements of NO_2 (Grant *et al.*, 1974) SO_2 and O_3 (Grant and Hake, 1975) have also been carried out. Recently SO_2 has been measured in the atmosphere at concentrations less than 15 ppb out to a range of 1 km (Hoell *et al.*, 1975). The technique has also been extended to the infrared to measure water vapour up to 2 km (Murray *et al.*, 1976). Attempts are being made to use the method on SO_2 at about 9.4 μm but these depend on the development of a sensitive heterodyne receiver. Theoretical studies of differential lidar have been given by Measures and Pilon (1972), Byer and Garbuny (1973), and Schotland (1974).

The detection limits of a system being developed by the Central Electricity Generating Board are shown in Fig. 3. Principally, it is the amount of light which is backscattered and received at the detector which determines the detection limits so that a high energy laser pulse operating under conditions of good visibility provides the best sensitivity. A further factor affecting the detection limits is the amount of background light reaching the detector. With a bright sky background this can be considerable and filters have to be used to cut out radiation at all wavelengths except the narrow region where the SO_2 absorption

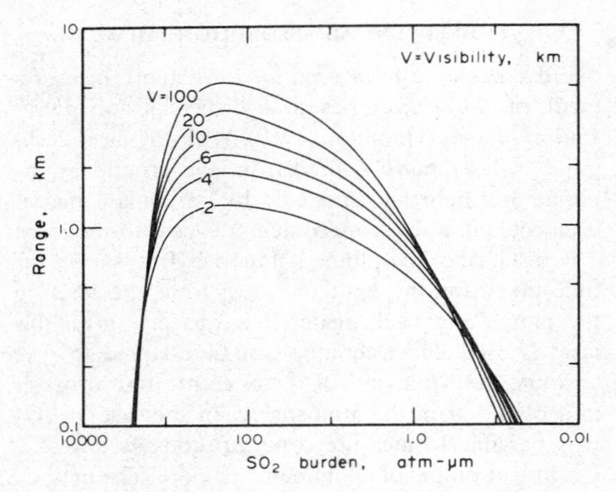

Fig. 3. Calculated detection limits of C.E.G.B. differential lidar system: receiver diameter = 250 mm; pulse length = 1 μs; pulse energy = 0.1 mJ; filter bandpass = 10 nm; background = bright sky; integration = 250 shots; wavelengths = 300.1 nm (high SO_2 absorption); 299.5 nm (low SO_2 absorption).

occurs. It is for this reason that measurements are sometimes carried out at night (Hoell *et al.*, 1975). Because the backscattered light from a large number of pulses has to be integrated to measure concentrations with good sensitivity a mini-computer is usually incorporated in the system. Typically 500 pulses are required to obtain a satisfactory signal-to-noise ratio and these can be produced in 30 s or so. The resolution in range is limited by the duration of the laser pulse, typically, to about 100 m.

Such a system has applications to the study of plumes from large power stations both close to the source and far downwind. It will provide the results obtainable from conventional lidar plus SO_2 concentrations. Operated at different wavelengths it will also give concentrations of NO_2 and O_3. Unlike the correlation spectrometer, it is not limited to day-time use. Indeed, as mentioned above, night-time operation can improve its performance. It has further applications in the study of long-range transport of SO_2 since the detection limits are within the background levels of most industrialised countries.

5. CONCLUSIONS

Of many possible methods for remote sensing of sulphur dioxide, only two have sufficient sensitivity for measuring at the ppb level. The correlation spectrometer measures line integrals of concentration and is relatively small and inexpensive. It has been widely used in surveys to measure source strengths and horizontal dispersion. The emission from a single source can be measured to the accuracy of the associated wind speed. The measurements of dispersion with its dependence on times of travel and sampling are far more comprehensive than any obtained hitherto. However, the correlation spectrometer does have a serious weakness in that it can only measure burdens relative to an indeterminate zero. As a consequence, it is unlikely to contribute to knowledge of depletion processes such as dry deposition at the ground. On the other hand, the differential absorption lidar is both sensitive and absolute, and it provides fully range-resolved measurements of concentration. It is also large and expensive. It has already been proved in prototype form and will shortly be available in a form suitable for conducting mobile surveys.

Acknowledgements—The contributions to this paper by Dr Hamilton and Dr Varey were reported at the Central Electricity Research Laboratories and are published by permission of the Central Electricity Generating Board.

REFERENCES

Bonafe U., Cesari G., Giovanelli G., Tirabassi T. and Vittori O. (1976) Mask correlation spectrophotometry: advanced methodology for atmospheric measurements. *Atmospheric Environment* **10**, 469–474.

Bottema M., Plummer W. and Strong J. (1964) Water vapour in the atmosphere of Venus. *Astrophys. J.* **139**, 1021–1022.

Byer R. J. and Garbuny M. (1973) Pollutant detection by absorption using Mie scattering and topographic targets as retoreflectors. *Appl. Opt.* **12**, 1496–1505.

Davies J. H., van Egmond N. D., Wiens R. and Zwick H. (1975) Recent developments in environmental sensing with the Barringer correlation spectrometer, *Can. J. Remote Sensing* **1**, 85–94.

Dupoux N., Morel P. and Vavasseur C. (1976) Résultats des mesures radiomètriques faites dans l'infrarouge moyen sur les panaches de la centrale de Drax au cours de la campagne organisée par les Communautes Européennes en Septembre 1976, Commisariat a l'Energie Atomique, Saclay, France. Report ARD 8411.

Evangelisti F., Giovanelli G., Orsi G., Tirabassi T. and Vittori O. Application features of mask correlation spectrophotometry to long horizontal paths. *Atmospheric Environment* (in press).

Fanaki F. H. (1975) Experimental observations of a bifurcated buoyant plume, Boundary Layer. *Meteorology* **9**, 479–495.

Fisher B. E. A., Gotaas Y., Maul P., Hamilton P. M., Houlgate R. and Moore D. J. (1977) Observations and calculations of airborne sulphur from multiple sources out to 1000 km. *Atmospheric Environment* **11**, 1163–1170.

Grant W. B. and Hake R. D., Jr. (1975) Calibrated remote measurements of SO_2 and O_3 using atmospheric backscatter. *J. appl. Phys.* **46**, 3019–3023.

Grant W. B., Hake R. D. Jr., Liston E. M., Robbins R. C. and Proctor E. K. Jr. (1974) Calibrated remote measurements of NO_2 using the differential absorption backscatter technique. *Appl. Phys. Lett.* **24**, 550–552.

Guillot P. Report on the 1st CEC Campaign of Remote Sensing of Air Pollution at Lacq, CEC Luxembourg Report (in press).

Hamilton P. M. (1969) The application of a pulsed-light rangefinder (lidar) to the study of chimney plumes. *Phil. Trans. R. Soc. Lond. A*, **265**, 153–172.

Hoell M. J., Jr., Wade W. R. and Thompson R. T. Jr. (1975) Remote sensing of SO_2 using the differential absorption lidar technique, International Conference on Environmental Sensing and Assessment, Las Vegas, Nevada, September, 1975.

Igarashi T. (1973) Laser radar study using resonance absorptions for remote detection of air pollutants. Fifth Conference on Laser Radar Studies of the Atmosphere, Williamsburg, Virginia, June, 1973.

Inaba H. and Kobayasi T. (1972) Laser–Raman radar—laser–Raman scattering methods for the remote detection and analysis of atmospheric pollution. *Opto-electron.* **4,** 101–123.

Jepsen A. F. and Langan L. (1974) Gas flow measurement by plume analysis with a Barringer correlation spectrometer. *Flow.* Vol. 1, Pt 1, pp. 361–367, Instrument Society of America, Pittsburgh, U.S.A.

Kay R. B. (1967) Absorption spectra apparatus using optical correlation for the detection of trace amounts of SO_2. *Appl. Opt.* **6,** 776–778.

Laurent J. (1975) Selective modulation radiometer. ONERA, Chatillon, France, (Unpublished).

Measures R. M. and Pilon G. (1972) A study of tunable laser techniques for remote mapping of specific gaseous constituents of the atmosphere. *Opto-electron.* **4,** 141–153.

Menzies R. T. (1972) Remote sensing with infrared heterodyne radiometers. *Opto-electron.* **4,** 179–168.

Millán M. M. (1976) A note on the geometry of plume diffusion measurements. *Atmospheric Environment* **10,** 655–658.

Millán M. M. and Chung Y. S. (1977) Detection of a plume 400 km from the source. *Atmospheric Environment* **11,** 939–944.

Millán M. M., Gallant A. J. and Turner H. E. (1976) The application of correlation spectroscopy to the study of dispersion from tall stacks. *Atmospheric Environment* **10,** 499–511.

Millán M. M. and Hoff R. M. (1976) The COSPEC remote sensor II, electronic set-up procedures. Atmospheric Environment Services, Downsview, Ontario, Canada, M3H 5T4, Report ARQT-6-76.

Millán M. M. and Hoff R. M. (1977a) How to minimize the baseline drift in a COSPEC remote sensor. *Atmospheric Environment* **11,** 857–860.

Millán M. M. and Hoff R. M. (1977b) Dispersive correlation spectroscopy: a study of mask optimization procedures. *Appl. Optics* **16,** 1609–1618.

Millán M. M. and Hoff R. M. Remote sensing of air pollutants by correlation spectroscopy: instrumental response characteristics. *Atmospheric Environment* (in press).

Moffat A. J. and Millán M. M. (1971) The application of optical correlation techniques to the remote sensing of SO_2 plumes using sky light. *Atmospheric Environment* **5,** 677–690.

Moffat A. J., Robbins J. R. and Barringer A. R. (1971) Electro-optical sensing of environmental pollutants. *Atmospheric Environment* **5,** 511–525.

Moore D. J. and Hamilton P. M. (1977) Report on 2nd CEC Campaign of Remote Sensing of Air Pollution at Drax: estimation of fluxes and plume parameters from surface vehicle traverses C.E.R.L., Leatherhead, England (unpublished).

Murray E. R., Hake R. D. Jr., van der Laan J. E. and Hawley J. G. (1976) Atmospheric water vapour measurements with an infrared (10-μm) differential-absorption lidar system. *Appl. Phys. Lett.* **28,** 542–543.

Newcomb G. S. and Millán M. M. (1970) Theory, applications and results of the long-line correlation spectrometer *IEEE Trans. Geosci. Electron.* **GE-8,** 149–157.

Rothe K. W., Brinkman U. and Walther H. (1974a) Applications of tunable dye lasers to air pollution detection: measurements of atmospheric NO_2 concentrations by differential absorption. *Appl. Phys.* **3,** 115–119.

Rothe K. W., Brinkman U. and Walther H. (1974b) Remote measurement of NO_2 emission from a chemical factory by the differential absorption technique. *Appl. Phys.* **4,** 181–182.

Sandroni S. and Cerutti C. (1977) Long path measurements of atmospheric sulphur dioxide by a Barringer COSPEC III. *Atmospheric Environment* **11,** 1225–1232.

Schotland R. M. (1966) Some observations of the vertical profile of water vapour by means of a laser optical radar. Fourth Symposium on Remote Sensing of the Environment, Ann Arbor, Michigan, Proceedings, 273–283, April, 1966.

Schotland R. M. (1974) Errors in the lidar measurements of atmospheric gases by differential absorption. *J. appl. Met.* **13,** 71–77.

Sperling R. B. (1975) Evaluation of the correlation spectrometer as an area SO_2 monitor. NTIS Springfield, Virginia, U.S.A., Report EPA-600/2-75-077.

Sutton S. and Varey R. H. (1977) The effect of multiple scattering on correlation spectrometer measurements of gas burden. C.E.R.L., Leatherhead, England (unpublished).

Atmospheric Environment Vol. 12, pp. 135–159. Pergamon Press 1978. Printed in Great Britain.

THE PHYSICAL CHARACTERISTICS OF SULFUR AEROSOLS

KENNETH T. WHITBY

Particle Technology Laboratory, Mechanical Engineering Department, University of Minnesota, Minneapolis, MN 55455, U.S.A.

(*First received July* 1977 *and in final form* 20 *September* 1977)

Abstract – A review of the physical characteristics of sulfur-containing aerosols, with respect to size distribution of the physical distributions, sulfur distributions, distribution modal characteristics, nuclei formation rates, aerosol growth characteristics, and *in situ* measurement, has been made.

Physical size distributions can be characterized well by a trimodal model consisting of three additive log-normal distributions.

When atmospheric physical aerosol size distributions are characterized by the trimodal model, the following typical modal parameters are observed:

1. Nuclei mode – geometric mean size by volume, DGV_n, from 0.015 to 0.04 μm, $\sigma_{gn} = 1.6$, nuclei mode volumes from 0.0005 over the remote oceans to 9 $\mu m^3\, cm^{-3}$ on an urban freeway.

2. Accumulation mode – geometric mean size by volume, DGV_a, from 0.15 to 0.5 μm, $\sigma_{ga} = 1.6$–2.2 and mode volume concentrations from 1 for very clean marine or continental backgrounds to as high as 300 $\mu m^3\, cm^{-3}$ under very polluted conditions in urban areas.

3. Coarse particle mode – geometric mean size by volume, DGV_c, from 5 to 30 μm, $\sigma_{gn} = 2$–3, and mode volume concentrations from 2 to 1000 $\mu m^3\, cm^{-3}$.

It has also been concluded that the fine particles ($D_p < 2\ \mu$m) are essentially independent in formation, transformation and removal from the coarse particles ($D_p > 2\ \mu$m).

Modal characterization of impactor-measured sulfate size distributions from the literature shows that the sulfate is nearly all in the accumulation mode and has the same size distribution as the physical accumulation mode distribution.

Average sulfate aerodynamic geometric mean dia. was found to be 0.48 $\pm$ 0.1 μm (0.37 $\pm$ 0.1 μm vol. dia.) and $\sigma_g = 2.00 \pm 0.29$. Concentrations range from a low of about 0.04 $\mu g\, m^{-3}$ over the remote oceans to over 80 $\mu g\, m^{-3}$ under polluted conditions over the continents.

Review of the data on nucleation in smog chambers and in the atmosphere suggests that when SO_2 is present, SO_2-to-aerosol conversion dominates the Aitken nuclei count and, indirectly, through coagulation and condensation, the accumulation mode size and concentration. There are indications that nucleation is ubiquitous in the atmosphere, ranging from values as low as 2 $cm^{-3}\, h^{-1}$ over the clean remote oceans to a high of $6 \times 10^6\ cm^{-3}\, h^{-1}$ in a power plant plume under sunny conditions.

There is considerable theoretical and experimental evidence that even if most of the mass for the condensational growth of the accumulation mode comes from hydrocarbon conversion, sulfur conversion provides most of the nuclei.

NOMENCLATURE

Roman notation

A and k	constants in $dN/d\log D_p = AD_p^k$
ANC	Aitken nuclei count
CCN	cloud condensation nuclei concentration
D_p	particle diameter
DGN	geometric number mean size
DGS	geometric surface mean size
DGV	geometric volume mean size
F	designates N, S, or V in general
FN_n	nuclei number flow rate in plume
FN_a	accumulation mode flow rate in plume
F_{SO_2}	SO_2 flow rate in plume
K_{nn}	homogeneous coagulation coefficient for the nuclei mode
K_{na}	heterodisperse coagulation coefficient, nuclei–accumulation
K_{aa}	homogeneous coagulation coefficient for the accumulation mode
m	mass concentration
N_n	number concentration of particles in the nuclei mode
N_a	number concentration of particles in the accumulation mode
N_c	number concentration of particles in the coarse particle mode
NT	total number concentration of particles in whole distribution
QN_n	rate of nuclei number transfer into the nuclei mode
QN_a	rate of particle number transfer into the accumulation mode
QV_n	rate of aerosol volume transfer into the nuclei mode
QV_a	rate of aerosol volume transfer into the accumulation mode
QVT	$QV_n + QV_a$
r	particle radius or correlation coefficient
Se	equilibrium surface area, Equation (6)
t	time
T	total N, S, or V in Equation (5)
TSP	total suspended particulate
V_n	volume concentration of particles in the nuclei mode
V_a	volume concentration of particles in the accumulation mode
V_c	volume concentration of particles in the coarse particle mode
VT	total volume concentration in whole distribution
V_g	plume gas flow at a given cross-section

Subscripts

n, a, and c	designate the nuclei, accumulation and coarse particle modes, respectively
b	designates background
p	designates plume
$\bar{N}_n$	the bar designates an average

Greek notation

α	N_a/V_a
ρ_p	particle density
σ_g	geometric standard deviation
χ_n^2	normalized chi square, Equation (5)

INTRODUCTION

It is now well established that a significant fraction of sulfur, emitted into the atmosphere either as a primary aerosol or as the product of gas-to-particle conversion, ends up in the atmosphere as a quite stable submicron aerosol. It is also increasingly clear that the fraction of the total fine particle mass aerosol of sulfur-containing, or what are more generally called sulfate, aerosols is increasing over and downwind of industrialized countries.

Other papers at this symposium discuss its spatial concentration, chemistry, optics and aerosol dynamics. Although it is certain that this paper overlaps these reviews in some respects, it is hoped that whatever duplication exists is enlightening and constructive because of the different points of view presented.

The principal objective of this paper will be the review of the physical and sulfur size distributions and concentration of sulfur aerosols, and their relationship to the total size distribution of atmospheric aerosols for the U.S.A. The presentation will be as quantitative as possible, since we are rapidly moving out of the era in which qualitative presentation of aerosol size distributions adds much to our knowledge. Presentation of experimental data will be given precedence over theoretical predictions in most cases, partly because the author is an experimentalist, but also because good experimental data on nucleation and growth of atmospheric aerosols is much scarcer than theoretical results.

To discuss aerosols quantitatively, it is necessary to adopt some interpretive framework. Since the multimodal nature of atmospheric aerosols is now reasonably well established, the basic trimodal distribution described in terms of three additive log-normal distributions will be used to interpret both physical and sulfur size distributions. Where practicable, published sulfur size distribution data has been reanalyzed

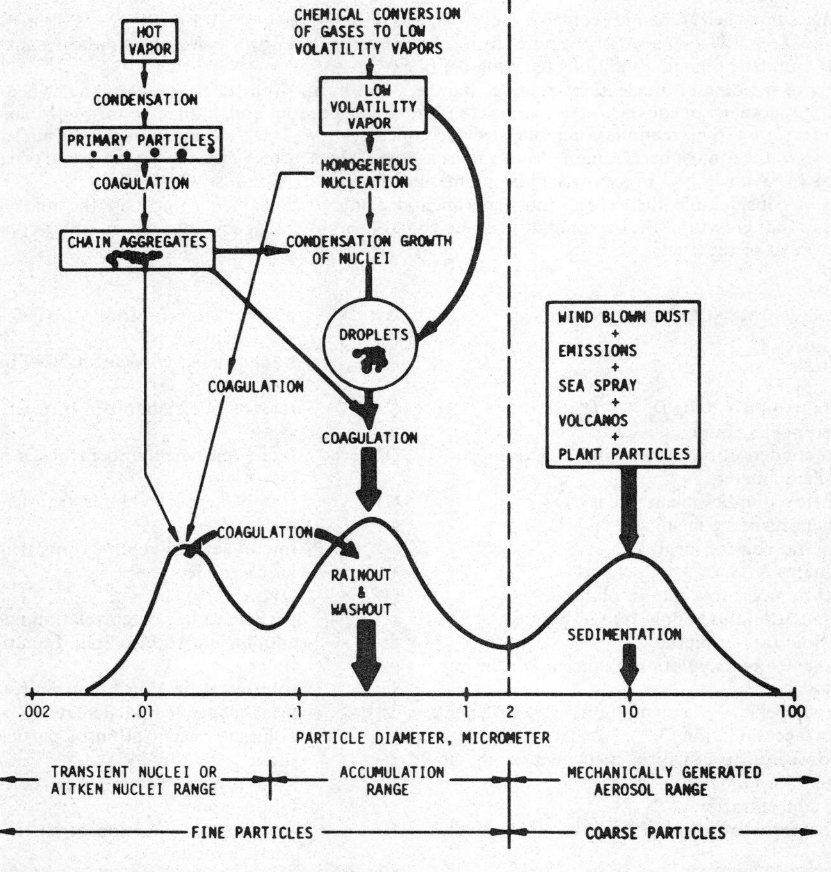

Fig. 1. Schematic of an atmospheric aerosol surface area distribution showing the three modes, main source of mass for each mode, the principal processes involved in inserting mass into each mode, and the principal removal mechanisms.

and reinterpreted in terms of the multimodal model.

The literature on atmospheric aerosols is vast, and even that on sulfur aerosols is now so large that even a cursory mention of all of it would consume too much space. Therefore, results have been selected because of their significance, quality, recent publication, or because they illustrate a significant point.

THE PHYSICAL SIZE DISTRIBUTION OF ATMOSPHERIC AEROSOLS

Nomenclature

For the purposes of discussion and interpretation, the basic trimodal model and the nomenclature shown in Fig. 1 will be used. The rationale behind this model has been previously discussed (Whitby *et al.*, 1972; Whitby, 1974; Willeke and Whitby, 1975; Whitby and Cantrell, 1976).

The distinction between "fine particles" ($D_p < 2\ \mu$m) and "coarse particles" ($D_p > 2\ \mu$m) is a fundamental one. There is now an overwhelming amount of evidence that not only are two modes in the mass or volume distribution usually observed (Whitby *et al.*, 1972a; Lundgren, 1973; Hidy *et al.*, 1976; Graedel and Graney, 1974; Kadowaki, 1976; Mainwaring and Harsha, 1976), but that these fine and coarse modes are usually chemically quite different (Appel *et al.*, 1974; Hidy *et al.*, 1976; Dzubay and Stevens, 1975). The physical separation of the fine and coarse modes originates because condensation produces fine particles while mechanical processes produce mostly coarse particles. As will be explained later, the dynamics of fine particle growth ordinarily operate to prevent the fine particles from growing larger than about 1 μm. Thus, the fine and coarse modes originate separately, are transformed separately, are removed separately, and are usually chemically different. As will be shown later, practically all of the sulfur found in atmospheric aerosol is found in the fine particle fraction. Thus, the distinction between fine and coarse fractions is of fundamental importance to any discussion of aerosol physics, chemistry, measurement, or aerosol air quality standards.

Presentations of size distributions

Figure 2 shows a typical urban-type size distribution presented in five different ways.

Typically, atmospheric aerosol number size distributions have been presented as $\log \mathrm{d}N/\mathrm{d} \log D_p$ vs $\log D_p$, as shown in Fig. 2(a). Because a good portion of such plots could be fitted by a power function of the form $\mathrm{d}N/\mathrm{d} \log D_p = AD_p^k$, this has been used by many investigators to model the number-size distribution (Junge, 1953; Clark and Whitby, 1967). However, when such number distributions are transformed to surface area-size, or volume-size distributions [Figs. 2(c), 2(d) and 2(e)], it is seen that the apparently minor deviations from the power law are actually significant modes.

The form $\log \mathrm{d}N/\mathrm{d} \log D_p$ vs $\log D_p$ is dominated by the nuclei mode for most urban and near-urban situations. For background aerosols, the number in the accumulation mode may occasionally dominate the distribution. This plot usually masks the modal structure of aerosols, and erroneously suggests that the whole distribution can be modeled by a continuous function. Although a power function is often an

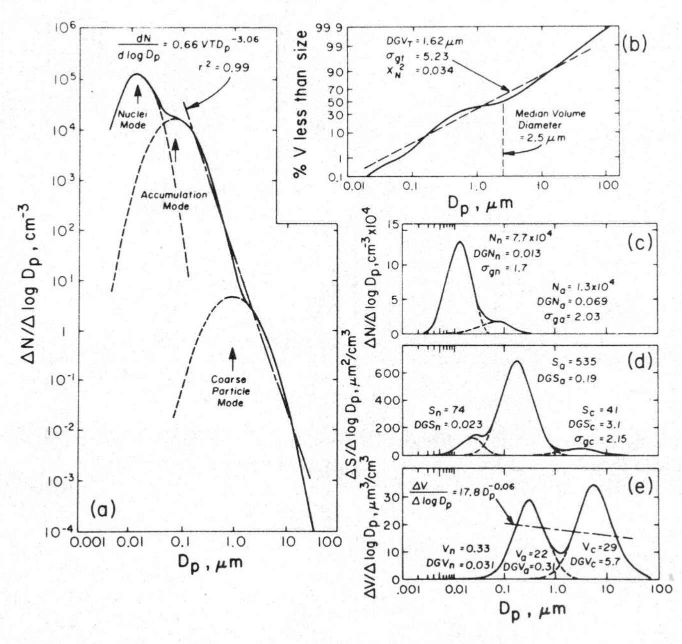

Fig. 2. Average urban aerosol plotted in five different ways. In (a), a power function has been fitted to the number distribution over the size range 0.0–32 μm. In (b), a log-normal distribution has been fitted to the range 0.1–32 μm. It is evident that the modal nature of the aerosol is shown best by the plots of (c), (d) and (e).

adequate model for the number distribution in the 0.1–10 μm range, it is a poor model for the surface and volume distributions. From Fig. 1, it will be noted that for $0.1 < D_p < 32$ μm, a power function correlates very well with the number distribution. However, from Fig. 2(e), it is seen that the correlation of that same function with the volume distribution is poor.

The forms $\log dS/d \log D_p$ and $\log d \log D_p$ vs $\log D_p$ emphasize the accumulation and coarse particle modes more, but the apparent area under the curves is distorted (Friedlander, 1962; Berry, 1967; Whitby *et al.*, 1972a). This form is useful for presenting the full range of data, since log scales are used for concentration.

The linear–log forms shown in Figs. 2(c), 2(d) and 2(e) are useful because the apparent area under the distribution curves is proportional to the integral in the size range. From these plots, it is possible to visually judge the relative *N*, *S* and *V* in the different modes of the distribution.

Cumulative plots of *N*, *S*, *V*, or *m* vs $\log D_p$ [Fig. 2(b)] have been widely used for the presentation of powder size distributions and for the presentation of impactor results (Wagman *et al.*, 1967).

From such a plot of a unimodal distribution on logarithmic probability paper, it is possible to determine the geometric mean by inspection. However, the cumulative plot has several deficiencies. First, the cumulative form masks the modes in the distributions. Second, the median used to determine the geometric mean is a function of the upper and lower cutoffs of the measurement method, since the distribution must be normalized by the total *N*, *S*, or *V*. This is an especially serious problem for the number and mass distributions, since measurement methods seldom sample all of the number or mass that is actually suspended in the air. Sometimes the geometric means determined from cumulative plots are as much a function of the sampling method as of the actual distribution. Means determined from cumulative plots should only be compared with data obtained by identical sampling methods under identical sampling conditions.

Thirdly, the median obtained from a cumulative plot is kind of an average between the fine and coarse modes, and therefore does not have a clear-cut physical significance. For the distribution shown in Fig. 2, the geometric mean diameters by vol. DGV of the accumulation and coarse particle modes are 0.31 and 5.7 μm, respectively. The volume median from the cumulative plot is 2.5 μm.

In retrospect, it is now clear that the fundamentally modal nature of atmospheric size distributions was not seen until about 1970 by the majority of workers in aerosols because most workers, including this one, only plotted data as in Figs. 2(a) and (b). It is now clear that the *N*, *S* and *V* size distributions must be all examined and then plotted in several ways if a proper understanding is to be achieved.

Cumulative plots of "*N*, *S*, or *V* greater than" are useful for determining critical sizes. For example, the condensation critical size has been determined from the size at which the "number greater than" is equal to the cloud condensation nuclei count.

Schaefer (1976) has plotted $\log dN \times A$ vs $\log A$ where $A = (\pi/4)D_p^2$. This form also shows the accumulation and coarse particle modes. However, since the

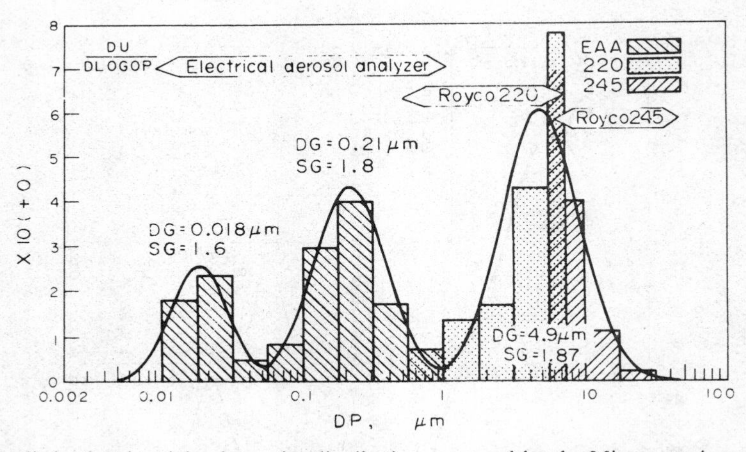

Fig. 3. Example of a distinctly trimodal volume size distribution measured by the Minnesota Aerosol Analyzing System aboard the EPA trailer during the General Motors Sulfate Study, 29 October, 1975. The size range measured by each of the instruments is shown. Such distinctly trimodal distributions are only observed under clean conditions near strong sources of nuclei, such as a roadway.

GM Milford Proving Grounds

Site: GMPG 15 — Date: 10/29/75
Run No: 15 — Time: 8.40 to 9.00 h EDT.

WD:	354.9	CNC:	1.02E 05	V2:	1.35	NO/NOX:	0.30
WSP:	17.0	NTM:	1.03E 06	V3:	2.32	VAN:	1.30
RH:	73.8	ST:	4.53E 02	V3–:	3.67	VAC:	2.76
		VT:	7.16	V4+:	3.49	VCP:	4.09

integrals under the curves are not proportional to N, S, or V, this form is inferior to $dF/d \log D_p$ vs $\log D_p$.

The trimodal nature of atmospheric aerosol size distributions

There is now a considerable body of evidence to suggest that most atmospheric aerosol size distributions are basically trimodal (Whitby, 1973; Whitby, 1974; Whitby and Cantrell, 1975). For most atmospheric aerosols, all three modes are not clearly evident in any one distribution weighting. However, Fig. 3 shows an unusual distribution measured alongside of the test track during the General Motors Sulfate Study in 1975 Wilson *et al.*, 1977; Whitby *et al.*, 1977). Because the background concentration was low when this measurement was made, the H_2SO_4 aerosol emitted by the catalyst-equipped cars in the nuclei mode at 0.018 μm is clearly evident even in the volume distribution.

For most atmospheric aerosol, no more than two distinct modes will be seen in any one weighting. This is illustrated in Figs. 4, 5 and 6, where average

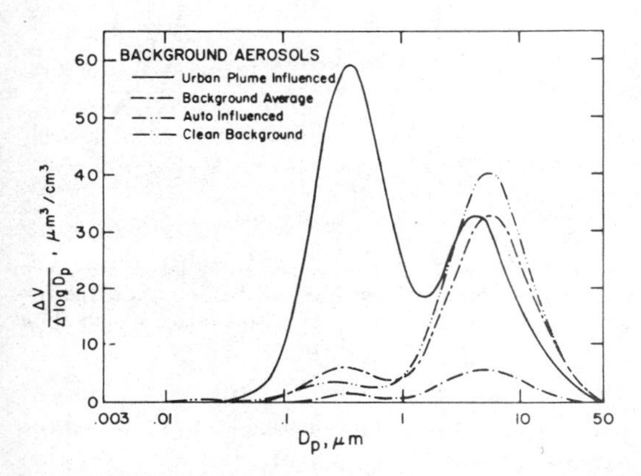

Fig. 4. Linear–log plot of the volume distributions for the four background distributions shown in Fig. 4 and Table 1. Notice how much the urban plume adds to the accumulation mode of the background.

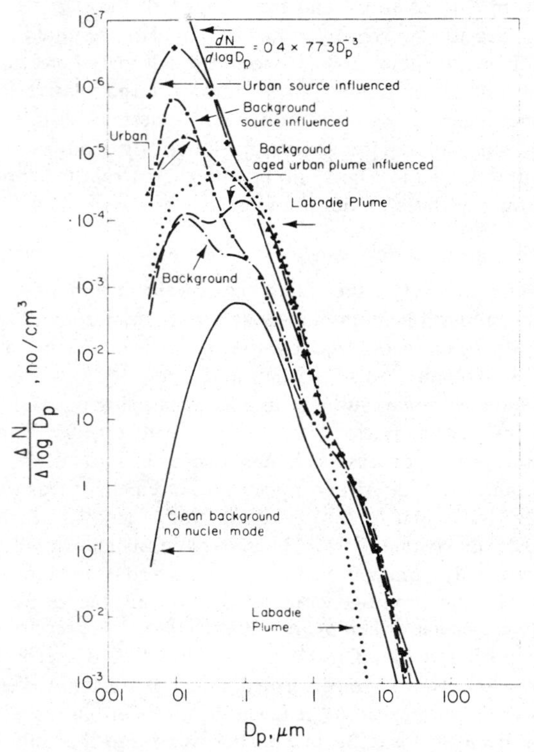

Fig. 5. Number distribution corresponding to the last seven distributions in Table 1. Note that, although the large variations in the nuclei mode are clearly seen, the variations in the accumulation and coarse particle modes are masked by this distribution plot.

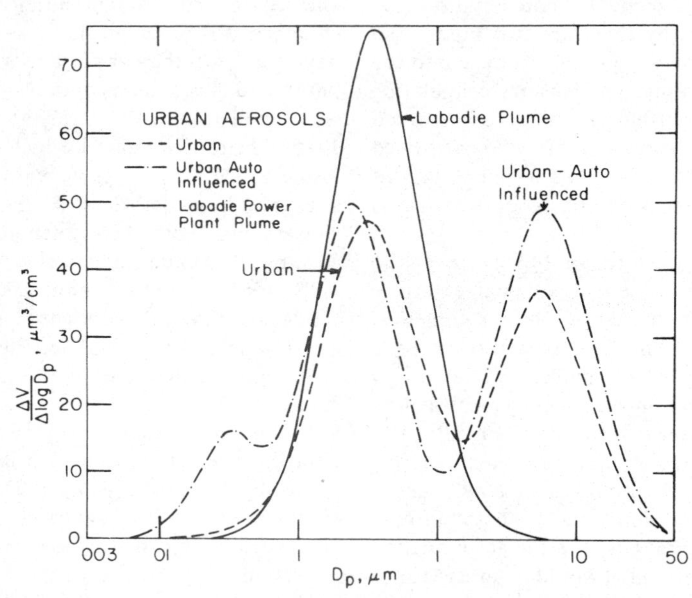

Fig. 6. Linear–log plot of the volume distributions for two urban aerosols and a typical distribution measured in the Labadie coal-fired power plant plume near St. Louis (Whitby *et al.*, 1978a). Size distributions measured above a few hundred meters above the ground generally have a rather small coarse particle mode.

distributions for a variety of location categories have been plotted. From Fig. 5, it is seen that background volume distributions typically only show the accumulation and coarse particle modes. Generally the number distributions only show the nuclei mode, although, from Figs. 2 and 4 and from Fig. 1 of the paper on nucleation by Whitby *et al.* (1978b) also presented at this symposium, at background locations where the accumulation mode concentration is large relative to the nuclei mode, both modes are sometimes visible.

The trimodal model for atmospheric distributions is summarized in Fig. 1, along with the current ideas of how the three modes are formed.

Origin of the three modes

Qualitatively, the differences in the origin of the accumulation mode and coarse particle modes have long been recognized by the use of such terms as "smoke" and "dust" (Green and Lane, 1957), smoke being defined as submicron particles arising primarily from condensation, and dust originating from mechanical processes. It has also been known for a long time that Aitken nuclei measurements measure particles which are predominantly smaller than 0.05 μm (Aitken, 1923). These Aitken nuclei originate primarily from combustion products and from heteromolecular homogeneous nucleation in the atmosphere. Schaefer (1976) and Hogan (1976) have recently reported in detail on the large amount of aerosol research they have performed, much of it using Aitken nuclei counters in a great variety of locations.

Because it is to the fine particle size range that sulfur in the atmosphere contributes its mass, our subsequent discussion will focus almost entirely on this part of the size distribution.

Since both the nuclei and accumulation modes are formed directly or indirectly by condensation, the question arises as to why there are two modes. To answer this question, it is necessary to look into the dynamics of a nucleating, condensing, coagulating aerosol having a size distribution similar to that found in the real atmosphere. Although this will be reviewed by Friedlander in detail in another paper at this symposium, a brief discussion of aerosol dynamics is necessary here.

First, let us address the question of why there is so little interaction between the fine and coarse particle size ranges. Figure 7 shows the result of a numerical calculation of the rate at which volume is transferred to larger size particles from all smaller size ranges by heterodisperse coagulation for three typical atmospheric size distributions. Note that for all three distributions, the rate of transfer decreases sharply above about 0.5 μm and is only about 10^{-9} $\mu m^3\,cm^{-3}\,s^{-1}$ at 1 μm. Since even background aerosols have several $\mu m^3\,cm^{-3}$ in the accumulation mode, calculations show that it would require weeks to transfer this mass to the coarse particle mode by coagulation. Thus, for the usual few days' residence time of fine particles in the atmosphere, for all practical purposes, the fine and coarse ranges form, transform, and are removed independently.

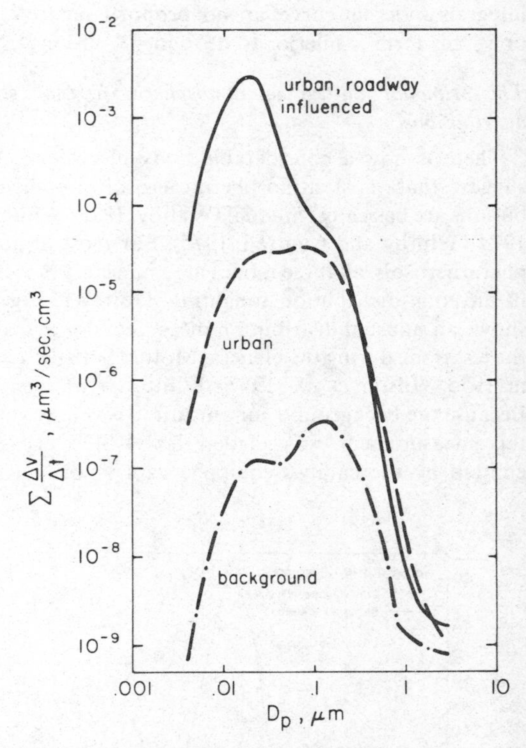

Fig. 7. The rate of volume transfer to all larger sizes from each size for three different size distributions. Note the very low rates of volume transfer for particles larger than about 1 μm.

Now let us look into the reason why the nuclei and accumulation modes are usually separate.

If a nucleating condensing aerosol is formed and allowed to grow in an initially particle-free smog chamber, it is observed that a unimodal distribution having a σ_g less than about 1.5 forms and grows. If the photooxidizing system contains SO_2, the initial size is very small (Smith *et al.*, 1975; Clark and Whitby, 1975). The original nuclei, which are thought to form in the 15–20 Å range, grow within tens of seconds to sizes on the order of 75 Å. Subsequent growth is slower, but both experimental data (Clark and Whitby, 1975) and numerical calculations (Hidy and Lilly, 1965) show that the distribution remains unimodal and that the σ_g does not increase much beyond 1.6. The data of Figs. 8(a) and 8(c) also show that the size remains in the nuclei mode range even after several hours.

If the photooxidation of reactive hydrocarbons contributes most of the condensing mass, then experiments by Kocmond *et al.* (1975) and Heisler and Friedlander (1977) show that DGV of the unimodal size distribution can grow to 0.5 μm under reasonable conditions [Fig. 8(b)].

However, in the real atmosphere, the accumulation mode is the most ubiquitous and persistent part of the size distribution. This means that the coagulation and

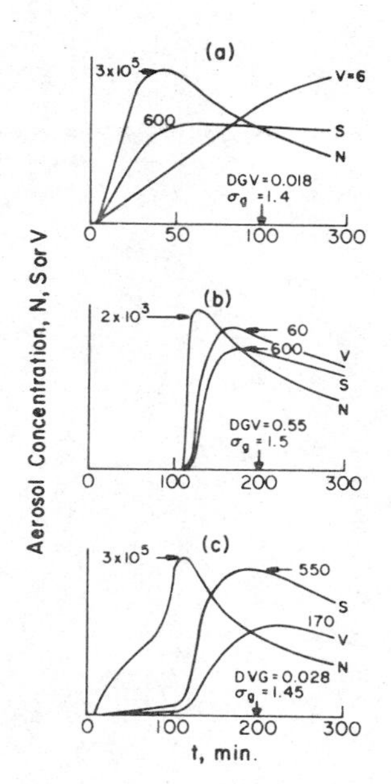

Fig. 8. Time development of number concentration, N, surface area concentration, S, and volume concentration, V, for the photooxidation of mixtures in a chamber; (a) SO_2 = 0.2 ppm in a very slightly hydrocarbon-contaminated chamber, (b) NO = 0.124 ppm and cyclohexene = 0.35 ppm, and (c) SO_2 = 0.045 ppm, NO = 0.13 ppm, and cyclohexene = 0.35 ppm. Note that for (a) and (c), the SO_2 causes the peak number to be two orders of magnitude higher than for the hydrocarbon alone, and that the mean size of the aerosol remains in the subvisible size range.

condensational growth of the nuclei mode always occurs in the presence of the accumulation mode. To understand the limiting effect of the accumulation mode on the growth of the nuclei mode, it is necessary to discuss the dynamical relationship between these two modes.

If the number, surface area and volume are all considered to be lumped at the geometric mean size of each mode, then the following three equations describe the approximate nucleation–coagulation dynamics of the aerosol. Hogan (1975) has also used a grouped size approach for aerosol dynamic calculations, assuming one group at 0.04 and the other at 0.1 μm. Whitby *et al.* (1978b) have shown that if the heterodisperse coagulation coefficients K_{na} and K_{ac} are based on the surface geometric mean diameter of the modes, then the rates of coagulation predicted by Equations (1) and (2) are close to those of the complete numerical calculation.

Nuclei mode:

$$\frac{dN_n}{dt} = -K_{nn}N_n^2 - K_{na}N_nN_a - K_{nc}N_nN_c + QN_n \quad (1)$$

Accumulation mode:

$$\frac{dN_a}{dt} = -K_{aa}N_a^2 - K_{ac}N_aN_c + QN_a \quad (2)$$

Coarse particle mode:

$$\frac{dN_c}{dt} = -K_{cc}N_c + QN_c. \quad (3)$$

The dominant role of the accumulation mode in the coagulation dynamics of an average atmospheric distribution can be illustrated by calculating the ratio of dN/dt to N, the number concentration in the smallest mode of the two coagulating modes. The modal parameters and the results of this calculation for the grand average of the six distributions shown in Table 1 are shown in Table 2 as a coagulation rate matrix. Note that dN_n/dt as a percentage of N_n for coagulation with the accumulation mode is 79% h^{-1} vs 31% h^{-1} for nuclei particles coagulating with themselves. Thus, heterogeneous coagulation of the nuclei mode with the accumulation mode exceeds that of the nuclei mode with itself.

Numerous complete numerical coagulation experiments by various investigators all show that if the nuclei mode is allowed to coagulate with itself until the rate of coagulation is negligible compared to the usual residence time of aerosols in the atmosphere, the maximum increase in nuclei mode mean size will be less than a factor of three.

The rather narrow range of sizes for the nuclei mode in the real atmosphere is further confirmed by the results of Sverdrup (1977), in which he correlated DGN_n with N_n for the 900 size distributions measured during the General Motors Sulfate Study (Whitby *et al.*, 1977). These data included a wide range of background as well as auto-related aerosols. The correlation described by Equation (4) shows that the change in size with observed concentration is only about a factor of three from $N_n = 10^3$–10^7.

$$DGN_n = \left(\frac{4.7 \times 10^{-10}}{N_n}\right)^{0.13}. \quad (4)$$

Modeling of atmospheric size distributions

It is probably a little facetious to suggest that there have been almost as many models for atmospheric size distributions as there have been workers in the field, but a large number have been proposed and used for various purposes. Many of these models have been fitted to only a limited size range for only one weighting, e.g. number. Examples of this are the use of various gamma functions for fitting the portion of atmospheric aerosols that is optically significant in the atmosphere (Tomasi and Tampieri, 1976; McCartney, 1976). Because this author believes that this is not an adequate approach, he has proposed that the following criteria be applied to the development of any generalized size distribution models (Whitby, 1974).

Ideally, a function or combination of functions should have the following characteristics to be suitable as models for particle size distributions.

(1) It should be able to fit the distributions over their entire range of sizes, 0.003–100 μm for atmospheric aerosols.

(2) Number, surface area and volume or mass distributions should fit equally well. For this to be true, the form of the distribution must be independent of the weighting. Some commonly used functions, such as the

Table 1. Modal and integral parameters for eight typical atmospheric size distributions derived from averages of measurements made on sites in each category. For further information see Whitby and Sverdrup (1978

	Nuclei mode			Accumulation mode			Coarse particle mode			Integral parameters		
	DG μm	σ_g	$V, \mu m^3 cm^{-3}$ / $N, cm^{-3} \times 10^3$	DG μm	σ_g	$V, \mu m^3 cm^{-3}$ / $N, cm^{-3} \times 10^3$	DG μm	σ_g	$V, \mu m^3 cm^{-3}$ / N, cm^{-3}	$NT, cm^{-3} \times 10^3$	ST $\mu m^2 cm^{-3}$	VT $\mu m^3 cm^{-3}$
1. Marine, surface **BK**	0.019	1.6*	0.0005	0.3	2*	0.10	12	2.7	12	0.4	30	12.1
	0.01*		0.34‡	0.071†		0.06†	0.62		3.1			
2. Clean continental **BK**	0.03	1.6	0.006	0.35	2.1	1.5	6	2.2	5	1.8	42	6.5
	0.016		1.0	0.067		0.80	0.93		0.72			
3. Average **BK**	0.034	1.7	0.037	0.32	2.0	4.45	6.04	2.16	25.9	8.6	148	30.4
	0.015		6.4	0.076		2.3	1.02		3.2			
4. **BK** and aged urban plume	0.028	1.6	0.029	0.36	1.84	44	4.51	2.12	27.4	16.2	938	71.4
	0.014		6.6	0.12		9.6	0.83		7.2			
5. **BK** and local sources	0.021	1.7	0.62	0.25	2.11	3.02	5.6	2.09	39.1	407	352	42.7
	0.009		402	0.047		4.5	1.10		4.9			
6. Urban average	0.038	1.8	0.63	0.32	2.16	38.4	5.7	2.21	30.8	138	1131	69.8
	0.014		106	0.054		32	0.86		5.4			
7. Urban and freeway	0.032	1.74	9.2	0.25	1.98	37.5	6.0	2.13	42.7	2160	3201	89.4
	0.013		2120	0.062		37	1.08		4.9			
8. Labadie plume (1976)	0.015	1.5	0.1	0.18	1.96	12	5.5	2.5	24	130	576	36
	0.0092		118	0.046		30	0.44		12			
Averages V	0.029 ± 0.007	1.66 ± 0.1	0.26 ± 0.33§	0.29 ± 0.06	2.02 ± 0.1	21.5 ± 20 ¶	6.3 ± 2.3	2.26 ± 0.22	25.9 ± 13	100 ± 148‖	460 ± 440‖	38.4 ± 25‖
N	0.013 ± 0.003		103 ± 172§	0.068 ± 0.02		14.4 ± 16	0.86 ± 0.23		5.2 ± 3.3			

* Assumed.
† Derived from data of Mészáros and Vissy (1974).
‡ Calculated from $N_n = NT - N_a$, $NT = 400$ cm^{-3}.
§ Average of distributions 2 6 only.
¶ Average of distributions 2 8 only.
‖ Without No. 7.

Table 2. Trimodal parameters of the grand average continental size distributions and the number rate of transfer matrix

Parameter	Mode: Nuclei (n)	Accumulation (a)	Coarse particle (c)	Total concentration
σ_g	1.7	2.03	2.15	
DGN, μm	0.013	0.069	0.97	
N, no. cm^{-3}	7.7×10^4	1.3×10^4	4.2	9×10^4
DGS, μm	0.023	0.19	3.1	
S, μm^2 cm^{-3}	74	535	41	650
DGV, μm	0.031	0.31	5.7	
V, μm^3 cm^{-3}	0.33	22	29	51.3

Mode	$dN/N\,dt$, % h^{-1}: n	a	c
n	31		
a	79	4.8	
c	0.5	0.0013	0.0005
Total	110.5	4.8013	0.0005

power law and the various forms of the gamma distribution, do not meet this criterion. The log-normal distribution is one of the simplest functions that does.

(3) The function should have some physical basis. Although there have been attempts to provide a physical basis for the power law, including several by this author, these have not really been successful, and it must be concluded that at best, both the power law and the gamma distributions are only empirical fits to atmospheric distributions.

From the central limit theorem, it may be shown that the addition of distributions, whatever their form, will tend toward some form of the normal distribution. The question is, why should the log-normal distributions fit atmospheric size distributions? From the central limit theorem, it may also be shown that if the distributed variable results from formation mechanisms in which the effect is proportional to the already achieved magnitude of the variable, then the distribution of the variable will be log-normal. Since both coagulation and condensation, the governing mechanisms for the nuclei and accumulation modes, are functions of attained size, it may be that this is the reason that the log-normal distributions fit these modes so well.

For these reasons, we have used three additive log-normal functions to characterize atmospheric physical and chemical distributions. As a measure of quality of fit, we have used the normalized chi square, χ_n^2,

$$\chi_n^2 = \frac{1}{T}\sum \frac{(f_i - f_e)^2}{f_e}, \tag{5}$$

where f_i = the experimental frequency in the size range, f_e = expected frequency calculated from the added log-normal functions, and T = total N, S, or V in the mode being fitted or in the total distribution.

Recently, Kelkar and Joshi (1977) have also found that a similar trimodal model fitted Trombay aerosol well.

Davies (1974) has proposed using seven log-normal components to fit atmospheric distributions. However, after fitting thousands of distributions with three log-normal components, we see little justification for using more than three.

The author plans to write a detailed paper describing the fitting procedures that have been developed in our laboratory over the past four years.

PHYSICAL SIZE DISTRIBUTIONS AND CONCENTRATIONS IN VARIOUS LOCATIONS

Tens of thousands of aerosol measurements have been made by a great variety of experimental methods in almost every possible location in the atmosphere during the last several decades. Much of the background data have been obtained with Aitken nuclei counters, which only sense the nuclei mode near sources of nuclei where nuclei dominate the number distribution, but are relatively more sensitive to the accumulation mode for aged aerosols where the nucleation rate is small. On the other hand, most of the pollution-related concentration measurements have been of mass concentration using filter or impactor samples having variable and often uncertain large particle cutoffs. Only in the last decade have enough complete size distribution measurements been made so that the relative concentrations of the different modes in the distribution under various conditions can be determined.

Since aerosol concentrations may be presented in terms of primary integral properties, such as N, S, V, or m, or as integrals of some property such as light scattering with the size distribution, or in terms of the amounts in different size ranges, e.g. the three modes, meaningful intercomparison of data obtained by different investigators is often difficult. Typically, investigators present their data in one of the forms shown in Fig. 2, with perhaps a median or σ_g, and let it go at that. To overcome this problem, we have begun to reanalyze published size distributions using a digitizer connected to a minicomputer, so that we may reinterpret published data using the trimodal models and fitting procedures described above. The physical size distribution and concentration data of Mészáros and Vissy (1974), Fig. 9, has been reanalyzed in this way to the form shown in Fig. 10. This procedure has also been used to obtain the ammonium and sulfate modal parameters discussed later.

Continental aerosols

To establish a basis for comparison, the results of the trimodal parameterization of a large amount of physical size distribution data that we have collected will be presented first. Most of this data was obtained with various versions of the Minnesota aerosol analyzing system, consisting of an Aitken nuclei counter, electrical aerosol analyzer, and one or two optical particle counters (Whitby *et al.*, 1972b; Cantrell *et al.*, 1978).

Each category, except the marine background aerosols, is the average of ten to several hundred individual size distribution measurements made on several dozen different sites during field studies from 1966 to 1975. A more complete description of the data will be contained in the book describing the Aerosol Characterization Experiment (ACHEX) conducted in California in 1972 and 1973 (Whitby and Sverdrup, 1978).

Average trimodal parameters for eight different locations categories are shown in Table 2, and the corresponding distributions plotted in Figs. 4, 5 and 6. Several observations can be made from these plots and the table.

At the earth's surface over land, there is far less variation in the volume of coarse particle from one location to another than in the accumulation and nuclei modes. This is so because the coarse particle mode is dominated by such ubiquitous natural sources of particles as wind-blown dust, sea spray, or plant-

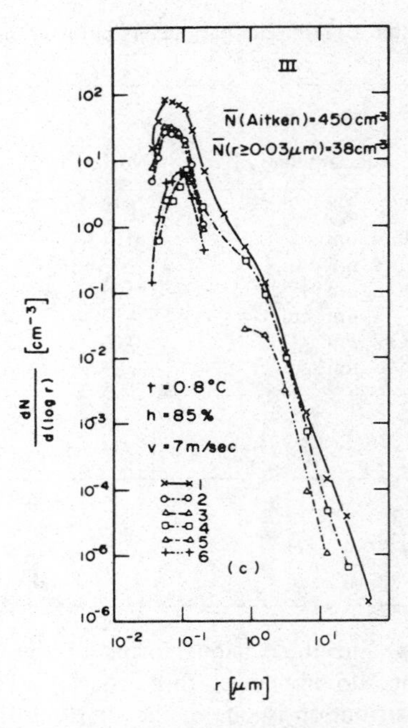

Fig. 9. Number size distributions of (1) all particles, (2) $(NH_4)_2SO_4$, (3) long shaped particles, (4) NaCl, (5) non-cubic crystalline particles, and (6) H_2SO_4 measured by Mészáros and Vissy (1974) in the South Atlantic.

produced particles. Furthermore, a rather unique relationship between coarse particle volume and size exists. From Fig. 11, it is seen that DGV_c is rather independent of V_c up to $V_c \simeq 40\ \mu m^3\ cm^{-3}$, after which there is a linear correlation.

The correlation between DGV_c and V_c for $DGV_c > 40\ \mu m$ (Fig. 11) for continental aerosols is $DGV_c = -15.7 + 5.53 \ln V_c$, $r^2 = 0.82$. The correlation between DGV_c and V_c for the marine aerosols measured

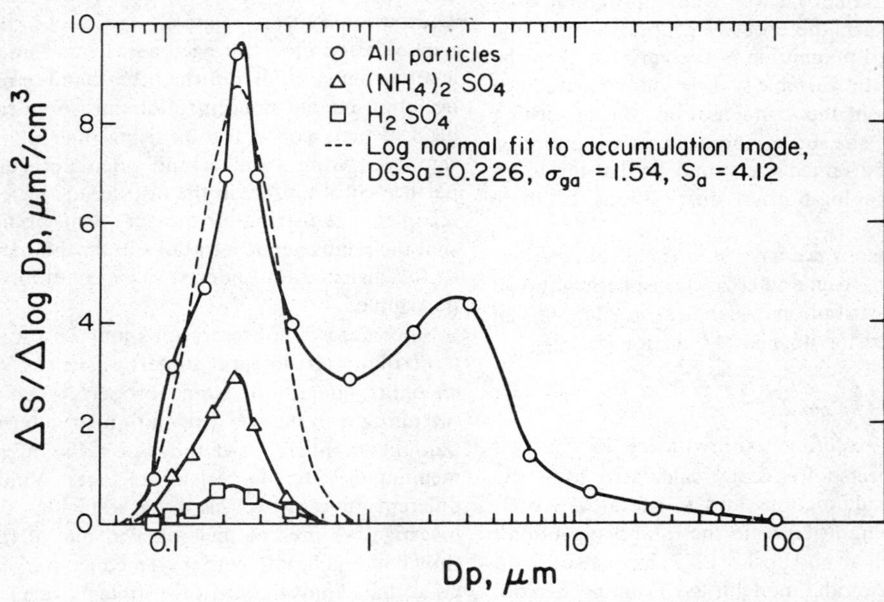

Fig. 10. Data of Fig. 9 shown as a surface area distribution. Note that the mode size and σ_g of the $(NH_4)_2SO_4$ and H_2SO_4 particles are the same as for the physical distribution.

Table 3. Accumulation and coarse particle mode parameters obtained from a log-normal fit to the oceanic data of Mészáros and Vissy (1974)

	Accumulation mode					Coarse particle mode					
Distribution	DGV_a μm	σ_g	N_a no. cm^{-3}	V_a $\mu m^3\ cm^{-3}$	% cf all	DGV_c μm	σ_g	N_c no. cm^{-3}	V_c $\mu m^3\ cm^{-3}$	N_a+N_c no. cm^{-3}	NT no. cm^{-3}
I, all	0.34	1.51	48	0.46		15.9	2.81	3.1	12.1	50.7	600
$(NH_4)_2SO_4$	0.28	1.29		0.28	61						
II, all	0.30	1.45	19.2	0.138		8.3	2.36	1.25	9.4	20.3	450
$(NH_4)_2SO_4$	0.27	1.40		0.046	33						
III, all	0.27	1.54	37.4	0.171		3.8	3.15		1.74	38	300
$(NH_4)_2SO_4$	0.23	1.44		0.0381	22						
H_2SO_4	0.26	1.42		0.0117	6.9						
IV, all	0.28	1.61	34	0.116		19.6	3.42	4.8	24.0	38.8	300
$(NH_4)_2SO_4$	0.25	1.49		0.0214	18						
H_2SO_4	0.31	1.31		0.0166	14						
Average all	0.30±0.03	1.53±0.07	34.7±12	0.22±16		11.9±7.2	2.69±0.56	3.1±1.8	11.81±9.2	37±13	413±144

	dN/dt, % hr^{-1}		
Mode	n	a	c
n	0.30		
a	0.35	0.14	
c	0.50	0.0023	0.0007
Total	1.15	0.1423	0.0007

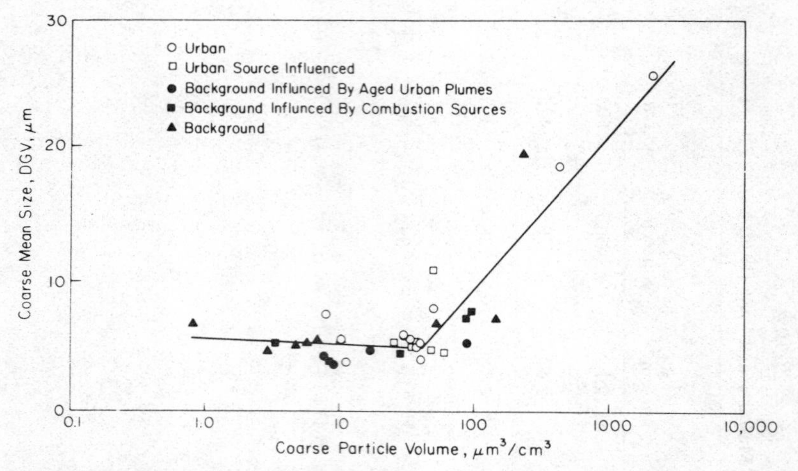

Fig. 11. Relationship between the mean size of the coarse particle mode and the volume in that mode for a variety of continental aerosols (Whitby and Sverdrup, 1978). The change in slope at $V_c = 50\ \mu m^3\ cm^{-3}$ probably indicates a change in the particle suspension mechanisms at about that concentration.

by Mészáros and Vissy (1974), Table 3, is $DGV_c = -0.53 + 5.88 \ln V_c, r^2 = 0.84$. The fact that the slope is nearly the same for these greatly different locations suggests that, on the average, the same classification mechanisms are operative in suspending dusts over the continents as in suspending salt aerosols over the oceans.

In Fig. 12, DGV_a has been plotted vs V_a for the same data. Except for the urban aerosols, there is no clear correlation. Only for the high concentrations of urban aerosol, mostly measured in Los Angeles, is there any significant correlation between DGV_a and V_a. The constant size of the accumulation mode size at about 0.3 μm for so many different locations and for such a wide concentration range is a remarkable and useful fact.

The size DGN_n and σ_g of the nuclei mode are the least accurate of the parameters given in Table 2. Most of the data shown was obtained prior to 1974 using EAA's for which the calibrations in the nuclei mode size range were such that the σ_g's are probably 15–25% too high. Data obtained with improved instruments and calibrations in 1976 (Cantrell and Whitby, 1978) suggest that a $DGN_n = 0.01$ and a $\sigma_g = 1.5$ are closer to the true values in the atmosphere.

Marine aerosols

Because the oceans occupy such a large fraction of the earth's surface, aerosols in the atmosphere over the oceans are important because of the effect on the earth's climate, and also because changes in their concentration are a sensitive indicator of global contamination from anthropogenic sources (Junge, 1972). Because of the low concentration of submicron aerosols over the oceans (Aitken counts in the 300–800 cm^{-3} range), size distribution measurement has

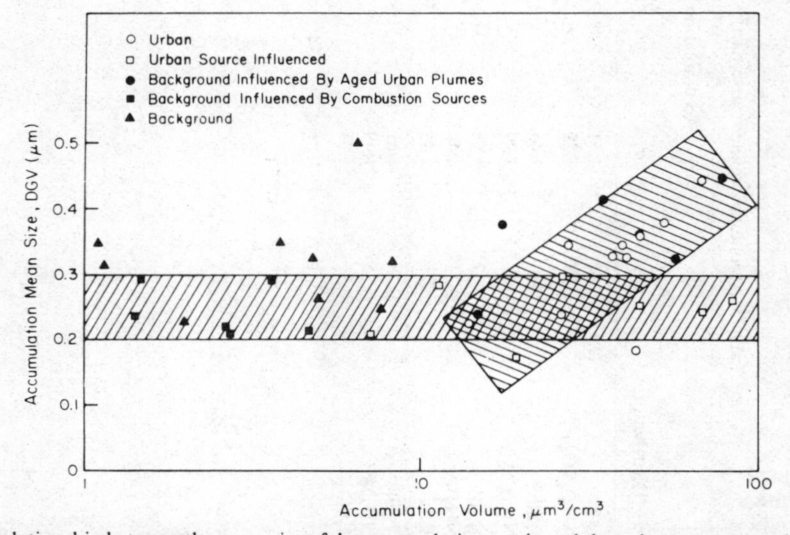

Fig. 12. Relationship between the mean size of the accumulation mode and the volume concentration in the accumulation mode for the same data shown in Fig. 11. Except for a rather weak correlation for the urban aerosol, the lack of correlation indicates that size is independent of concentration.

been difficult. Most data have been obtained with Aitken nuclei counters which cannot distinguish between the particles in the nuclei and accumulation modes. Recently, Hogan (1976) has written a comprehensive report summarizing a large amount of work on background and marine aerosols. From measurements on remote sites such as Pitcairn Island in the South Pacific and at Mauna Loa Observatory, he has concluded that there is clear evidence for gas-to-particle conversion even above the trade inversion, and that between one-third and one-half of the Aitken count consists of particles in the nuclei mode with the rest being in the accumulation mode. From Fig. 13, it is seen that this is in general agreement with other data. Figure 13 will be discussed later during the discussion of nucleation.

Hogan has found a hyperbolic fall-off of ANC from about $10^4\ cm^{-3}$ off of the U.S.A. coast to 300–600 1500 km away.

It now seems probable that nucleation is quite general throughout the atmosphere, and, as will be seen later, gas-to-particle conversion of sulfur probably dominates nucleation. Thus, sulfur plays an important role in determining fine particle size distribution and concentration in the marine environment.

SULFUR AEROSOL SIZE DISTRIBUTION

Whether or not the sulfur converted from the gaseous state to the low vapor pressure state forms new nuclei or condenses directly on existing aerosol particles, most of it ends up in the submicron size range and, in fact, is found in the accumulation mode.

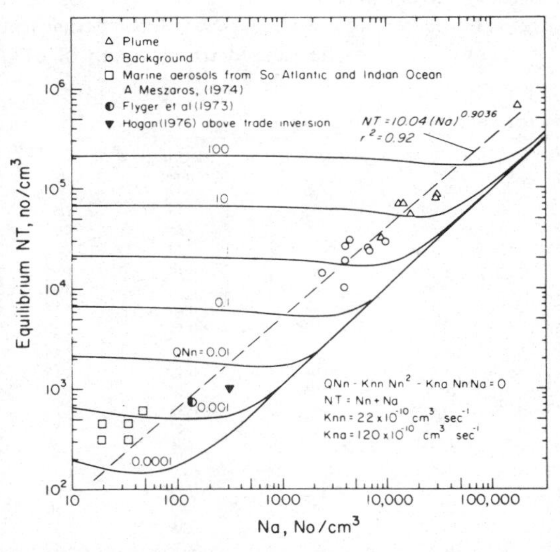

Fig. 13. Comparisons of equilibrium NT values with the number in the accumulation mode, N_a, for aerosols in and adjacent to the Labadie plume in St. Louis in 1976 and for several background aerosols. Note that NT correlates well with N_a over more than three orders of magnitude in concentration. The nucleation rates were calculated from the lumped mode heterodisperse coagulation model (Whitby *et al.*, 1978b).

Qualitatively, this has been known for a long time (Junge, 1960; Ludwig and Robinson, 1965; Roesler *et al.*, 1965), but has only been documented accurately and quantitatively by the large amount of impactor sampling and size-selective chemical analysis of the last few years. This review will concentrate, therefore, on the quantitative description of some of the better recent data and on relating these results to the physical size distributions.

The size distribution of impactor-sampled sulfur aerosols

For this analysis, the cascade impactor data of Wagman *et al.* (1967), Lee and Patterson (1969), Appel *et al.* (1974), Patterson and Wagman (1977), and Kadowaki (1976) were selected. The published data was digitized so that a log-normal distribution function could be fitted to the sulfate and ammonium data.

A typical result from this procedure for a sulfate distribution sampled by Wagman *et al.* (1967) is shown in Fig. 14.

Two fitting procedures were used. In the first, a log-normal distribution function was fitted to all of the mass below about 3 μm. In the second, a partial fit was made to the sulfate mass in the hatched area, but constraining the distribution integral to include all of the mass below about 3μm. Generally, the two procedures produced DGm's and σ_g's which agreed within

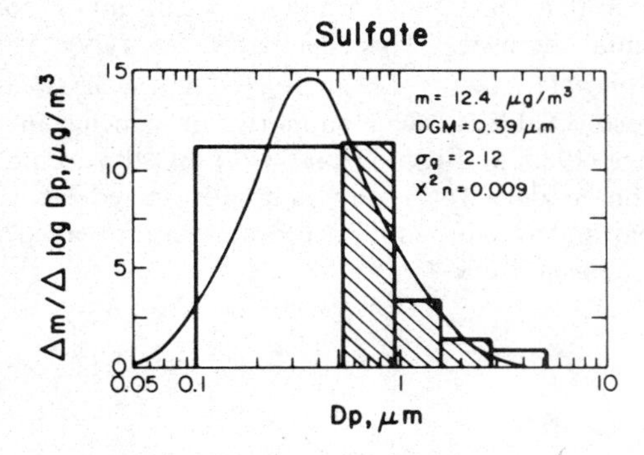

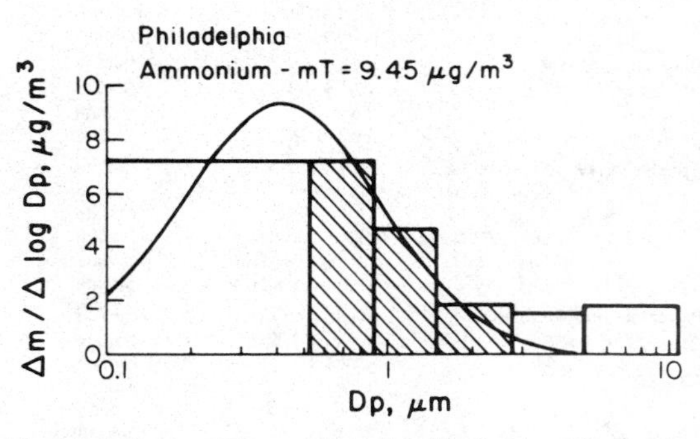

Fig. 14. Sulfate and ammonium size distributions illustrating partial fitting procedure that was used to obtain the geometric mean size and the σ_g. Sulfate was sampled by Wagman *et al.* (1967) in Philadelphia, and ammonium was sampled by Lee and Patterson (1969) in Philadelphia. Fitting was done to hatched area but including mass on after filter. Lower size for filter fraction was arbitrarily assumed to be 0.1 μm

10%. Because it was believed that the second procedure was the best, the results shown in Table 3 were all calculated by the second procedure. The average χ_n^2, as defined by Equation (5), was 0.04 ± 0.07. Considering the variety of sources for these data, this is a good fit.

From Table 3, it is seen that DGma, the geometric mass mean diameter of the accumulation mode, is 0.48 ± 0.1 for the sulfate and 0.38 ± 0.08 for the ammonium distributions. The geometric standard deviations, σ_{ga}, were 2.00 ± 0.29 for the sulfate and 1.95 ± 0.38 for the ammonium. The average percentage that was included in the accumulation mode was 95% for the sulfate and 96.5% for the ammonium.

The results of this analysis, Table 4, support the following conclusions:

(1) On the order of 95% of the ammonium and sulfate is in the accumulation mode.

(2) Within the accuracy of the data and the analysis, there is no significant difference in the size distribution of the sulfate and ammonium aerosol.

(3) The DGma and σ_{ga} for the sulfate and ammonium are essentially equal to the DGma and σ_{ga} of the physical size distribution discussed earlier (Table 2) if the aerodynamic diameters reported for the impactor data are corrected by the square root of some reasonable density. For example, if $\rho_p = 1.7$ gm cm^{-3}, then the accumulation mode DGma should be divided by $\sqrt{(1.7)} = 1.3$ to obtain the true size. From Table 2, it is seen that $\overline{DGV_a} = 0.31$ and $\sigma_{ga} = 2.02$ μm for continental aerosols. This compares favorably with $\overline{DGma/\sqrt{(1.7)}} = 0.37$ μm and $\sigma_{ga} = 2.00$ for the sulfate aerosol (Table 4). This supports the conclusion of Junge (1960) and Kadowaki (1976) that the accumulation mode aerosols are internally mixed for the urban and near-urban surface locations represented by the data of Table 4.

(4) The average sulfate concentrations vary from a low of 1.7 μg m^{-3} for the background data of Patterson and Wagman (1977) to a high of 31.6 for West Covina in the Los Angeles Basin (Appel *et al.*, 1974). Hidy *et al.* (1976) have reported a daily average of 80 μg m^{-3} during an episode on 8 July, 1974 in the central eastern U.S.A.

The relationship of total aerosol to sulfur aerosol size distributions over the ocean

Mészáros and Vissy (1974) have published the results of a rather unique study in which samples collected aboard ship in the south Atlantic and Indian Oceans were analyzed microscopically using morphological identification in such a way that the fraction of $(NH_4)_2SO_4$ and H_2SO_4 particles could be determined. We have reanalyzed their data, characterizing the distributions of total particles ("all" in Figs. 9 and 10), $(NH_4)_2SO_4$ and H_2SO_4 in terms of the trimodal model. The modal parameters are tabulated in Table 3, and the surface area distributions for location III are shown in Fig. 10. The surface area distributions are shown because they show the accumulation and coarse particle modes more clearly. A number of interesting conclusions can be drawn from their data.

(1) DGV_a and σ_{ga} for the accumulation modes are the same for the total distributions, for $(NH_4)_2SO_4$, and for H_2SO_4 particles for all locations (Table 3, columns 2 and 3). The average DGV_a of 0.30 ± 0.03 μm is almost exactly equal to the grand average for the physical accumulation mode, $DGV_a = 0.31$, for the variety of continental locations shown in Table 1, and to the sulfate and ammonium DGm's corrected for density (Table 4).

(2) The percentage of $(NH_4)_2SO_4$-like particles in the accumulation mode ranged from a high of 61%

Table 4. Modal parameters of impactor-collected sulfate aerosol

Reference	Location	Species	$(SO_4^{2-})T$ μg m^{-3}	$(SO_4^{2-})a$ μg m^{-3}	$(SO_4)a/(SO_4)T$ %	DGm μm	σ_g
Lee and Patterson (1969)	Fairfax	Ammonium	5.74	5.34	93	0.43	2.5
	Philadelphia	Ammonium	9.45	9.36	96	0.43	2.4
	Chicago	Ammonium	4.0	3.92	98	0.30	2.2
Wagman *et al.* (1967)	Cincinnati	Sulfate	10.0	9.5	95	0.37	1.97
	Chicago	Sulfate	8.7	8.3	95	0.36	2.40
	Fairfax	Sulfate	7.2	7.1	98	0.40	2.12
	Philadelphia	Sulfate	12.4	11.8	95	0.55	2.28
Appel *et al.* (1975) Sept. 18, 19	Rubidoux	Sulfate	23.0	21.8	95	0.66	1.85
Sept. 5, 6	Rubidoux	Sulfate	18.9	17.9	94	0.48	2.15
Aug. 16, 17	Pomona	Sulfate	27.5	26.3	96	0.58	2.1
July 23, 24	West Covina	Sulfate	32.7	31.6	97	0.52	2.1
Oct. 10, 11	Dominguez Hills	Sulfate	6.10	4.96	81	0.43	2.5
Oct. 4, 5	Dominguez Hills	Sulfate	33.9	32.4	96	0.65	1.95
Patterson and Wagman (1977)	Background	Sulfate	1.7	1.7	100	0.48	1.79
	Level A	Sulfate	12.1	12.1	100	0.37	1.63
	Level B	Sulfate	14.3	13.7	96	0.42	1.50
	Level C	Sulfate	20.9	18.0	86	0.35	1.57
	Background	Ammonium	1.8	1.8	100	0.24	1.8
	Level A	Ammonium	4.7	4.6	98	0.32	1.48
	Level B	Ammonium	9.5	9.1	96	0.35	1.55
	Level C	Ammonium	10.8	10.3	95	0.34	1.75
Kadowaki (1976)	Nagoya	SO_4^{2-}	21.1	21	100	0.50	2.10
	Nagoya	NH_4^+	4.37	4.2	96	0.52	1.95
	Nagoya	Total mass	131	66.3		0.39	2.14
	Average	NH_4^+				0.38 ± 0.08	1.95 ± 0.38
	Average	SO_4^{2-}				0.48 ± 0.10	2.00 ± 0.29

near the equator in the south Atlantic to a low of 18% in the Indian Ocean. Although these percentages cannot be equated exactly with per cent mass in the accumulation mode, they do suggest that the accumulation mode, even in these remote locations, contains a significant fraction of sulfur compounds.

(3) Whereas there is no correlation between DGV_a and V_a, there is a significant correlation between DGV_c and V_c, e.g. $DGV_c = -0.53 + 5.88 \ln V_c$ with $r^2 = 0.84$. Although the constant is different than the one shown in Fig. 9 for continental aerosols, the slope of the log term is about the same.

(4) The average of the sum of the number in the accumulation and coarse particle modes is $37 \pm 13 \text{ cm}^{-3}$, whereas the average NT is estimated to be 413 ± 144. Thus, most of the particle number is in the nuclei mode. As has been pointed out by Whitby *et al.* (1978b) in a separate paper on nucleation included in this symposium, this suggests that there is a significant nucleation rate. From Fig. 13, it can be estimated that the nucleation rate is on the order of 2–5 nuclei $\text{cm}^{-3} \text{h}^{-1}$.

(5) The intermodal coagulation matrix at the bottom of Table 3 shows that the coagulation rates of the nuclei mode with itself, with the accumulation mode, and with the coarse particle mode are about equal, in contrast to that for an average continental aerosol shown in Table 2, where the coagulation with the coarse particle mode is negligible. The total nuclei coagulation rate for the marine aerosol is only 1.15% h^{-1}, compared to 111% h^{-1} for the continental aerosol. This may explain why the accumulation mode of the marine aerosol is externally mixed, whereas the accumulation mode of the continental aerosols is internally mixed. The Mészáros and Vissy (1974) data suggest that in the marine environment at the very low accumulation mode concentrations observed, separate populations of accumulation mode sulfur-containing and non-sulfur-containing aerosol can grow and be removed without significant internal mixing.

Fraction of sulfate in the accumulation mode

Table 3 shows clearly that most of the sulfate in continental aerosols is in the accumulation mode, and Table 4 shows that the same is true for marine aerosols. In Table 5, average sulfate concentrations from a number of recent studies are analyzed in order to estimate the fraction of sulfate in the accumulation mode. Where the accumulation mode mass or volume was not available, it has been estimated by assuming that m_a was equal to one-half of the total suspended particulate mass. Nevertheless, Table 5 shows that, on the average, continental aerosol contains 15–30% sulfate, while marine aerosol contains somewhat more, in the range of 30–60%, even though the mass concentration over the oceans is only about 1 $\mu\text{g m}^{-3}$, compared to an average of about 35 for the continental data. The fact that the accumulation mode has a significant mass fraction of sulfate everywhere provides further evidence of the ubiquitous role of sulfur in the formation of submicron atmospheric aerosols. 15–30% sulfate is enough to make the fine particle aerosols hygroscopic at normal humidities, and to ensure that they behave as optically ideal droplets and are sticky enough not to bounce on dry impactor plates. That is probably why the dry plates used by Wagman *et al.* (1967) and others in impactor studies measured the correct accumulation mode sulfate distribution.

AEROSOL GROWTH

Although Jaenicke and Friedlander are also discussing aerosol growth and nucleation in this symposium, it is desirable to briefly discuss these topics before the data on atmospheric nucleation is reviewed.

Before beginning the discussion of the less understood aspects of aerosol growth, some of the well understood areas should be summarized.

Photooxidation of SO_2 in clean air, in slightly contaminated air and in air with hydrocarbons has been studied extensively in initially aerosol-free chambers. Some of the major features of the aerosol development can be illustrated from the work of Kittelson *et al.* (1977) and Kocmond *et al.* (1973, 1975). The development of the number, surface area and

Table 5. Estimated fraction of sulfate in accumulation mode

Location	Reference	TSP ($\mu\text{g m}^{-3}$)	m_a ($\mu\text{g m}^{-3}$)	SO_4^{2-} ($\mu\text{g m}^{-3}$)	SO_4^{2-}/m_a (%)
New York	Patterson and Wagman (1977)	118	59*	12.4	21
Nagoya	Kadowaki (1976)	131	66*	14.5	22
California	Twiss (1977)				15
Urban East	Altshuller (1973)			13.5	10–20
Urban West				6.4	5–10
Non-urban East				8.1	15–25
Non-urban West				2.6	15–25
East Central U.S.A.	Hidy *et al.* (1976)			12.3	
South Atlantic	Mészáros and Vissy (1974)		0.4	0.14	34
Atlantic background	Winkler (1975)			0.6	60
Greenland	Flyger *et al.* (1976)		1†	0.24	24

* Taken as 1/2 TSP.
† Calculated from Equation (10) for $NT = 600 \text{ cm}^{-3}$.

volume for the aerosol produced by the photooxidation of SO_2 in a slightly contaminated chamber, cyclohexene + NO and cyclohexene + NO + SO_2 are shown in Fig. 8.

For SO_2 alone, aerosol volume is produced, after a short delay of a few minutes, at an approximately constant rate up to the time where chamber losses begin to be significant. Peak number concentration is usually in the 1–10 × 10^{-5} cm^{-3} range. The surface area becomes essentially constant after 40 min or so, and the DGV of the aerosol seldom exceeds 0.02 μm in size. Clark and Whitby (1975) found that the equilibrium surface area, Se, was related to the aerosol volume formation rate, QV, by:

$$Se = 314(QV)^{0.58}. \quad (6)$$

This relation, which has been further confirmed experimentally by the work of Kittelson *et al.* (1977), was recently derived entirely from physical considerations by McMurry (1977).

For a hydrocarbon + NO mixture, without any SO_2, no aerosol is produced until most of the NO has been converted to NO_2. Then a very high volume rate of aerosol formation is observed, but a low nucleation rate. As a result, the aerosol size grows very rapidly into the optical range to DGV's up to 0.6 μm. Heisler and Friedlander (1977) also found that the critical size for condensation for this case was in the 0.2–0.3 μm range, much larger than for SO_2-containing mixtures.

From Fig. 8, it can be seen that when SO_2 is added to the cyclohexene + NO system, the peak number concentration is 3 × 10^5 cm^{-3}, about the same as for SO_2 alone. Therefore, even though the peak volume concentration is greater, 550 vs 66 μm^3 cm^{-3}, than for the cyclohexene + NO, DGV = 0.028 μm is now subvisible. The large effect that sulfur nuclei have on the mean particle size in chamber experiments explains some of the observations of early workers on aerosol formation, who had at their disposal only nuclei counters and light scattering instruments for measuring aerosol concentration (Renzetti and Doyle, 1960; Stevenson *et al.*, 1965; Wilson *et al.*, 1973). These workers interpreted the lack of light scattering aerosol as no aerosol. Because of coagulation, the CNC cannot give a true measure of concentration, and light scattering can only detect the fraction of particles larger than about 0.15 μm. Not until the application of the electrical aerosol analyzer, with its measurement capability in the 0.005–1 μm range, to chamber studies by Clark and Whitby (1975) did the relationship of CNC and light scattering results to the true aerosol growth characteristics become apparent. Therefore, in the chamber experiments, the SO_2 has several important effects on the aerosol formation.

(1) SO_2 participates in the conversion chemistry to increase the aerosol formation rate by a factor of two or so. Second, the large number of nuclei produced by the photooxidation of the SO_2 reduces the size of the growing aerosol by an order of magnitude or more.

Because the atmosphere probably contains both reactive hydrocarbons and SO_2 in most locations, most of the time SO_2 nucleation has a significant effect on the development of the accumulation mode size distribution in the real atmosphere.

(2) Although the initial size of the nuclei produced by SO_2 photooxidation is predicted to be about 0.0012 μm dia., at the usual SO_2 conversion rates of about 1% h^{-1}, growth is very rapid. Calculations indicate that only a few minutes are required for the size to reach a 0.004–0.006 μm number mean size under initially particle-free conditions. Ordinarily investigators (Smith *et al.*, 1975; Clark and Whitby, 1975) have observed an apparent time delay of a few minutes before nuclei are counted by the Aitken nuclei counters. Liu and Kim (1977) have shown that the minimum size detected by the General Electric and Environment/One CNC counters is probably closer to 0.006 μm than to the 0.002 μm usually assumed by investigators doing nucleation experiments. Walter and Jaenicke (1973) also have theoretical and experimental indications that the minimum size detected is near 0.005 μm. Thus, the apparent initial time delay in nuclei formation results from the time required for the nuclei to grow to the 0.004–0.006 μm detection limit of the CNC counters.

Actually, the initial size of the nuclei is of little significance to the size development of nucleating and condensing aerosols because within the usual time frame for chamber experiments, or of interest in the atmosphere, the sulfur-containing nuclei will usually exceed 0.006 μm in size.

(3) If the aerosol has aged to where it has surface areas exceeding about 500 μm^2 cm^{-3}, or if the nuclei are mixed with an aerosol having considerable surface area, then most of the H_2SO_4–H_2O vapor will condense on existing particles (Pich *et al.*, 1970; Husar *et al.*, 1972; Heisler and Friedlander, 1977; Kiang and Middleton, 1977; Suck *et al.*, 1977).

Since the accumulation mode is ubiquitous in the real atmosphere, nucleation in the presence of the accumulation mode will be discussed in detail later in this paper. Later it will also be shown that in the presence of normal atmospheric amounts of accumulation mode aerosol, the volume fraction of aerosol going into the formation of new nuclei is about 5% of the total aerosol volume formed. Thus, while the time development of the Aitken nuclei count in a nucleating–condensing aerosol system is a qualitative indicator of gas-to-particle conversion, it is a useful quantitative index only under special circumstance.

Nuclei formation in the atmosphere

Qualitatively, it has been known for a long time that Aitken nuclei counts in the daytime (Junge, 1953) are higher than at night. Bricard (1974) and Hogan (1976) suggest that the generally observed 50% higher Aitken nuclei concentration during the day, compared to night, is caused by nuclei from SO_2-to-aerosol conversion.

A large number of studies of the time history of Aitken nuclei concentration development have also been made in chambers, e.g. Dunham (1960), Vohra *et al.* (1970), Birenzvig and Mohnen (1975), and Bricard *et al.* (1977). However, little work on the nucleation rates in the presence of realistic atmospheric size distributions and concentrations has been done until recently. Because of its importance, and because this work is just being published, this review will concentrate on this recent work.

From the previous discussion of the dynamics of a trimodal size distribution, it was seen that nuclei mode for a real atmospheric size distribution, with the ever-present accumulation mode, coagulates into the accumulation mode rapidly enough so that, if a pronounced nuclei mode is observed where N_n constitutes one-half or more of NT, a source of new nuclei must exist. Whitby *et al.* (1978b), in a companion paper included in this symposium, have used Equation (1) to calculate nucleation rates, QN_n, in a power plant plume and the adjacent background. N_a was obtained by log-normal fitting to the size distribution of the accumulation mode, and N_n from $N_n = NT - N_a$. NT was assumed equal to the Aitken nuclei count. The total aerosol volume growth rate, QVT, was also calculated. Assuming that $QVT = QV_n + QV_a$, the fraction of aerosol volume going into new nuclei compared to that condensing directly into the accumulation mode could be calculated. Somewhat surprisingly, QV_n/QVT was relatively constant at about 0.05 in the plume. QN_n/QV_n, the number of new nuclei produced per unit volume of aerosol going into the nuclei mode, was somewhat higher for the SO_2–NO–HC smog chamber data of Kocmond *et al.* (1975), $5 \times 10^6\ \mu m^{-3}$, compared to the plume values of 0.6 to $2 \times 10^6\ \mu m^{-3}$. The somewhat lower values in the plume suggest that the presence of the accumulation mode surface area has a depressing effect on the actual nucleation rate. Similar results have been found by McMurry (1977).

In the atmosphere away from sources of nuclei and outside of plumes, $dN/dt = 0$ in Equation (1). If QN_n is then calculated as a function of N_a and NT, and plotted as in Fig. 13, it is seen that NT is a function of QN_n only for a considerable range of N_a. QN_n is therefore related approximately to NT by:

$$NT = 1.725 \times 10^4 \sqrt{(QN_n)}. \tag{7}$$

It is therefore possible to make a first-order estimate of the nucleation rate from the Aitken nuclei counts for air masses which have been aged sufficiently long, and where N_n is greater than N_a. For example, the data of Mészáros and Vissy (1974) shown in Fig. 9 and 10, obtained in a clean oceanic background, suggest nucleation rates of about $0.0004\ cm^{-3}\ s^{-1}$, or only about $2.5\ cm^{-3}\ h^{-1}$. In contrast, Whitby *et al.* (1978a) obtained QN_n's averaging about 2 in the background outside the Labadie power plant plume to an average of about $20\ cm^{-3}\ s^{-1}$ in the plume. Early one morning, a high of $1700\ cm^{-3}\ s^{-1}$ was observed.

Flyger *et al.* (1973) made a variety of aerosol measurements with an aircraft in the vicinity of Greenland. These measurements included Aitken nuclei concentration, cloud condensation nuclei concentration (CCN), and nuclei size. Although it is not possible to match up the different measurements exactly so that N_a, N_n, and QN_n can be calculated from their data for the different flights, if one takes their average $\overline{NT} = 807\ cm^{-3}$ and average $\overline{CCN} = 155\ cm^{-3}$ and assumes that the CCN concentration is equal to the N_a, the number in the accumulation mode,* it is possible to compare their averages with the data of Whitby *et al.* (1978b) (Fig. 13). For the Labadie power plant study, the empirical relation between NT and N_a is given in Equation (8):

$$NT = 10 N_a^{0.904}. \tag{8}$$

For $N_a = \overline{CCN} = 155\ cm^{-3}$, Equation (8) predicts an $NT = 955$, in satisfactory agreement with the observed average Aitken nuclei count of $807\ cm^{-3}$.

On 31 August, 1971, a few days after back trajectory calculations indicated that the air mass being sampled originated from above the Great Lakes region of the North American continent, Flyger *et al.* (1973) observed Aitken nuclei concentrations as high as $9000\ cm^{-3}$ near Greenland. From Equation (8), this suggests a nucleation rate of $0.27\ cm^{-3}\ s^{-1}$, comparable to the lower values we have found in the background near the St. Louis area. They suggest that these high nuclei concentrations are from ships. However, the layering in concentration that they observed, the small size, and the high concentration suggest *in situ* nucleation in polluted air masses transported from the North American continent.

In 1975, Flyger *et al.* (1976) measured SO_2 particle sulfur, as well as Aitken nuclei concentrations, over Greenland and the surrounding seas. They had a grand average $NT = 960\ cm^{-3}$ and an average SO_2 concentration of $0.86\ \mu g\ m^{-3}$. If Equation (8) is used to estimate QN_n from NT, the ratio QN_n/SO_2 may be computed and compared to the coal-fired power plant results obtained by Whitby *et al.* (1978a). This ratio turns out to be 0.00353 for Flyger's data and 0.103 for the plume, where QN_n has the units no. $cm^{-3}\ s^{-1}$ and SO_2 is $\mu g\ m^{-3}$. The much higher ratio in the plume is more than can be accounted for by the assumptions involved in deriving Equation (8), and so it may be concluded that if SO_2 is responsible for the nuclei formation over Greenland, the nucleation rate is not proportional to SO_2 concentration, as it was found to be in the power plant plume.

The fact that the size distributions observed by Mészáros and Vissy (1974) can only be accounted for by the existence of a low but significant nucleation rate,

* At 1% supersaturation, the critical size for cloud condensation is about 0.05 μm. Since, in general, more than 80% of the accumulation mode particles are larger than this size, the *CCN* count is approx. equal to N_a.

and that the Flyger *et al.* (1973) data indicate significant nucleation rates over Greenland, support the suspicions of many investigators that nucleation may be a rather ubiquitous phenomenon which may play a much more important role in governing both the size distribution and nuclei concentration in well-aged air masses remote from land than has been previously thought. Also, the fact that NT and N_a data from such widely different locations as the Indian Ocean, St. Louis and Greenland are correlated by Equation (8) suggests that, in general, not only NT but also N_a is governed by nucleation. If nucleation is high relative to the mass in the accumulation mode, then the size of the accumulation mode will decrease. On the other hand, if QN_n is low relative to V_a, then DGV_a will increase. Over the long term, however, nucleation and coagulation in concert act as a governor to keep the accumulation mode size DGV_a in the range from 0.25 to 0.4 μm most of the time.

If the empirical relationships given by Equations (7) and (8) are combined, V_a can be calculated from QN_n,

$$V_a = (1.725 \times 10^3 \alpha) QN_n^{0.553} \tag{9}$$

where $\alpha = N_a/V_a$, the number of particles in the accumulation mode per unit accumulation mode volume.

Values of α for various locations and kinds of aerosol are given in Table 6. Note that although the range of α shown is one order of magnitude, the data include an extreme range of locations and are the results from a variety of aerosol measurement techniques.

Equation (8) may also be rearranged so that estimates of the volume, V_a, in the accumulation mode can be calculated from the Aitken nuclei count, NT, using appropriate values of α from Table 6:

$$V_a = \frac{1}{\alpha}\left(\frac{NT}{10}\right)^{1.1} \tag{10}$$

Some sources of nuclei mode sulfur aerosols

In addition to the nuclei mode sulfur aerosols resulting from gas-to-particle conversion that have already been discussed in connection with nucleation in the atmosphere, there are a few other sources. Recent research (Pierson *et al.*, 1974; Wilson *et al.*, 1977; Whitby *et al.*, 1977) has shown that the H_2SO_4 aerosol emitted by catalyst-equipped automobiles is of nuclei mode size. During the General Motors Sulfate Study in 1975, nuclei mode aerosol concentrations of up to 23 $\mu m^3\,cm^{-3}$ were observed. Presumably this aerosol was nearly all some form of sulfate, probably H_2SO_4, in equilibrium with water vapor at the ambient humidities. A typical aerosol size distribution showing a pronounced sulfate nuclei mode is shown in Fig. 3.

A more recent study (Richards, 1977) conducted by a team of EPA investigators, grantees and contractors on the freeways of Los Angeles has shown that although about one-third of the cars on the freeway were estimated to be catalyst-equipped, nuclei mode sulfate concentrations were estimated to be less than 2 $\mu g\,cm^{-3}$, even though the nuclei mode aerosol concentrations averaged 12 $\mu m^3\,cm^{-3}$. Since the coagulation of the nuclei mode aerosol is quite rapid on the freeways and in the normal urban atmosphere, within a few minutes the nuclei mode will have coagulated with the accumulation mode. Even if there was free sulfuric acid in the nuclei mode aerosol, after it coagulates it will be internally mixed with all of the other compounds in the accumulation mode, so that it really cannot be considered a free acid anymore.

Thus, the health effects of the nuclei mode sulfate aerosols are only of potential concern to travelers on the freeways, and the low concentrations found in the 1976 Los Angeles Roadway Study may indicate that nuclei mode sulfate exposure may not even be a problem on freeways.

An interesting source of sulfur-related nuclei produced in sulfur-containing propane flame has recently been observed by Barsic (1977). The size distribution of two populations of nuclei at three different relative humidities is shown in Fig. 15. Nuclei mode 1 with a DGV_1 of 0.0085 μm is quite deliquescent. Nuclei mode 2 with a $DGV_2 = 0.02$ μm is insensitive to humidity change, and is thought to consist of carbon or other hydrophobic substances. Whitby *et al.* (1978b) have shown that the number of nuclei produced per unit volume aerosol in the propane flame for nuclei mode 1 is the same as that produced by photooxidation. This suggests that the mechanisms which govern QN_n/QVT and QVT/SO_2 are physical, rather than chemical.

Table 6. Number concentration of accumulation mode particles, N_a, per unit accumulation mode concentration, V_a, for various locations

Location and reference	QN_n no. $cm^{-3}\,s^{-1}$	DGV_a μm	σ_{ga}	DGN_a μm	$\alpha = N_a/V_a$ (μm^{-3})
Plume, Whitby *et al.* (1978a)	1200	0.15	1.6	0.007	1529
Plume, Whitby *et al.* (1978a)	30	0.2	1.8	0.071	1130
Background, Whitby *et al.* (1978a)	3	0.23	1.9	0.067	1002
Grand average (Table 2)	–	0.31	2.03	0.069	612
Impactor average (Table 4)	–	0.36	2.04	0.078	403
Mészáros and Vissy (1974) (Table 3)	10^{-3}	0.30	1.53	0.174	160
				Average	806 ± 506

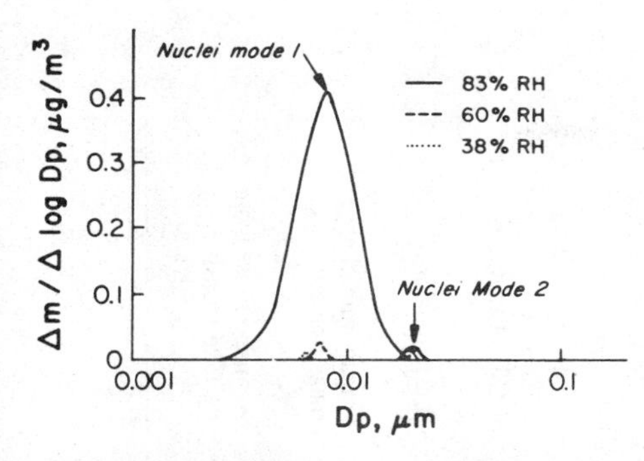

Fig. 15. Nuclei modes measured by Barsic (1977) in the products of combustion from a sulfur-containing propane flame. Nuclei mode 1 is quite hygroscopic, while nuclei mode 2 is not. It is therefore believed that mode 1 is the one containing the sulfur.

IN SITU SIZE DISTRIBUTION MEASUREMENT

Since other reviewers at this symposium discuss collection methods and optical methods for measuring integral aerosol concentration, this section only reviews *in situ* methods for measuring atmospheric aerosol size distribution and concentration. It is assumed that the reader is familiar with the basic principles so that the discussion can be concentrated on the characteristics of the method and its field use.

Because of the breadth of the aerosol measurements field, writings on measurement methods of necessity reflect the interests and limitations of the authors. In general, the best reviews have been the results of conferences.

A recent book which reviews aerosol sampling and measurement techniques in moderate detail and with respectable breadth has been written by Mercer (1973). The symposium proceedings edited by Mercer *et al.* (1972) provide detailed reviews of many of the methods referred to by Mercer (1973).

The proceedings of the conference on fine particle measurement, sampling and analysis, edited by Liu (1976), is an up-to-date review on sampling and fine particle aerosols. The proceedings of another aerosol measurement workshop, edited by Lundgren (1977), contains some excellent discussions of the factors important in applying different methods. The conference on environmental sensing (1976) contains much recent useful information on measurement methods.

In situ *vs collection*

Methods such as filters or cascade impactors collect the aerosol onto a surface. Evaluation must then be done on the collected sample for size and composition. Other methods, such as the optical techniques, sense the aerosol *in situ* without collecting it. As evidence mounts that accumulation mode-sized aerosols are hydroscopic, and therefore contain a substantial fraction of liquid at normal temperatures and humidities, there has been a growing realization that it is desirable to size these fine particles *in situ* without precipitation. Furthermore, evaluation techniques such as electron microscopy, which subject collected particles to a high vacuum, may yield misleading results when used on volatile fine particles.

Integral vs differential measurement of particle size

Many aerosol measuring methods, such as the Aitken nuclei counter, integrating nephelometer, filter collectors, or electric charger, integrate the response function of the instrument with the size distribution which the method senses. For example, although a filter collects all particles, the results depend on the method of evaluation. If particle counting is used, then the response is equal to the integral of the number distribution weighting.

Integral methods are always sensitive to the modification of the size distribution by the sampling inlets and transport lines used. Although size resolving methods are also sensitive to inlet losses, the portion of the distribution affected can at least be isolated.

Characteristics of most important aerosol sampling and measurement methods

Major integral sampling and measurement methods used for measuring atmospheric aerosols are listed and their most important characteristics described in Table 7, along with recent key references. Table 8 shows the corresponding information for size resolving methods.

In the Remarks column, an attempt is made to indicate which part of the size distribution and which modes dominate the sensitivity of the method. The upper and lower size limits are nominal values for the most commonly used forms of the technique.

In assessing the suitability of a method for sizing or sampling the sulfur-containing atmospheric aerosol, it should be kept in mind that most anthropogenic sulfur aerosol is in the accumulation mode, e.g. the 0.05–2 μm size range.

The practical aspects of using optical particle counters (OPC) for the size distribution of atmospheric aerosols has recently been reviewed by Whitby and Willeke (1977). There are several good OPC's on the market which, if operated and calibrated correctly, are capable of providing all of the accuracy that is needed for size distribution determination in the 0.4–7 μm range. The electrical aerosol analyzer (EAA) is capable of resolving the nuclei and accumulation modes under most conditions. Significant advances in calibration and data inversion for the EAA have been made recently by Liu and associates (Liu and Pui, 1975; Liu *et al.*, 1977).

A number of improved condensation nuclei counters (CNC) have been developed in recent years for various research and field applications. Of special importance are the continuous flow counters developed by Sinclair and Hoopes (1975) and by Bricard *et al.* (1976). The latter counter, which will be marketed this year in the U.S.A., will have a lower concentration

Table 7. Major integral sampling and measurement methods used for measuring atmospheric aerosols

Method or instrument	Effective lower size limit (μm)	Effective upper size limit (μm)	Lower concentration limit	Measurement principle
1. Hi Vol filter and other filters	0	25–50	$2\ \mu g\ m^{-3}$	Filter – weighing
Measures all of the non-volatile mass in accumulation mode and most of that in coarse particle mode, depending on sampling efficiency (McKee *et al.*, 1971).				
2. Nephelometer, LIDAR, and other optical methods	0.15	2	Air scattering	Light scattering
Essentially responsive to the *in situ* mass in the accumulation mode only (Stevens *et al.*, 1976).				
3. Condensation or Aitken Nuclei Counter (CNC)	0.0035–0.006	0.05–0.2	$1\ cm^{-3}$	Condensation – high super saturation
For combustion-dominated aerosols, measures the nuclei mode. For aged aerosols, measures no. in accumulation mode (Saxena *et al.*, 1972; Bricard *et al.*, 1976).				
4. Electrostatic charger	0.007–0.02	2–25	$1\ \mu g\ m^{-3}$	Charging and precipitation
With diffusion charging, upper limit is about 2 μm. Senses both nuclei and accumulation mode (Liu and Lee, 1975; Husar *et al.*, 1976).				
5. Cloud Condensation Nuclei Counter (CCN)	0.01	5	$1\ cm^{-3}$	Condensation – low super saturation
Lower size limit depends on super saturation and design. Measures part of nuclei mode and most of accumulation mode (Radke and Turner, 1972).				
6. Ice Nuclei Counter (INC)	?	?	$10^{-5}\ l^{-1}$	Ice nucleation
Measures only those few particles capable of causing supercooled water drops to freeze. Responsive to part of the accumulation mode sizes and most of the coarse particle mode sizes (Hobbs, 1970).				
7. Quartz crystal	0.01 electrostatic 0.3 impactor	10 20	$1\ \mu g\ m^{-3}$	Frequency change of crystal
Electrostatic version measures nuclei and accumulation modes and part of C.P. Impaction version measures 2/3 of accumulation mode and most of C.P. (Lundgren *et al.*, 1976).				
8. β attenuation	0 on filter	20	$2–5\ \mu g\ m^{-3}$	β attenuation
Performance dependent on counting period, collection surface mass (Macias and Husar, 1976).				
9. Contact electrification	10	1000	$100\ \mu g\ m^{-3}$?	Impact and charge measurement
Useful only for high concentrations. Sensitivity varies over 50/1 range for different materials (John, 1976).				
10. Acoustic Counter	5	100	$1\ l^{-1}$	Acoustic pulse from particle transit
Has been used primarily to count ice crystals as a sensor on INC (Karuhn, 1973).				

Table 8. Major size resolving methods used for measuring atmospheric aerosols

Method of instrument	Effective lower size limit (μm)	Effective upper size limit (μm)	Maximum number of size intervals practical	Lower concentration limit	Measurement principle
1. Cascade impactors	0 – filter 0.3 – last stage	15–30	8	1 $\mu g\ m^{-3}$, stage	Inertial impaction
Measures 1/2 to 2/3 of accumulation mode, all of C.P. mode. Particle bounce is a serious problem in some applications (Chuan, 1976; Marple and Willeke, 1976; Rao and Whitby, 1977; Wesolowski *et al.*, 1978).					
2. Dichotomous virtual impactors	0 – filter 3.5 for cut on 2nd filter	30	2	1 $\mu g\ m^{-3}$, stage	Virtual impaction
One filter collects F.P. and one, C.P. No bounce problem (Dzubay and Stevens, 1975; Loo *et al.*, 1976).					
3. Crystal microbalance impactor	0 – filter 0.6 – last stage	30	10	0.02 μg stage^{-1} 1 $\mu g\ m^{-3}$, stage	Impaction, quartz crystal sensing
Measures 1/2 of nuclei mode, all of accumulation mode and C.P. mode. Subject to particle bounce on dry aerosols (Stöber and Mönig, 1975).					
4. Spiral centrifuge	0.06	10	20	10 $\mu g\ m^{-3}$	Centrifuge onto filter
Measures 1/2 of nuclei mode, all of accumulation mode, part of C.P. mode. Sensitivity depends on method of evaluation. Mass determination is difficult (Stöber, 1976).					
5. Optical particle counters	0.3 commercial 0.1 research	3–100	>200 possible 20 useful	1 particle l^{-1}	Light scattering from single particles
Single counter useful for only 1 decade of size. Requires calibration. Sensitive to shape and refractive index (Cooke and Kerker, 1975; Whitby and Willeke, 1977; Willeke and Liu, 1976).					
6. Electric mobility	0.006	1	15	1 $\mu g\ m^{-3}$	Diffusion charging and electric mobility measurement
Can measure nuclei and accumulation modes. Can resolve σ_g of 1.4 without correction and 1.25 σ_g with correction (Liu and Pui, 1976; Whitby, 1976; Liu *et al.*, 1977).					
7. Parametric	0.01	2	Nuclei and accumulation	1 $\mu g\ m^{-3}$	Use of CNC, charger and nephelometer to calculate size distribution
By assuming σ_{gn}, σ_{ga}, and DGN_n, can calculate DGN_a, N_n, and N_a (Sverdrup, 1977).					
8. Diffusion battery	0.002	0.2	1	Sensor sensitivity	Diffusion battery followed by a sensor
Limited to determining DGN and σ_g of the distribution (Sinclair and Hinchliffe, 1972).					
9. Denuder	0.015	1	10	Sensor sensitivity	Electric condenser followed by CNC or S
Depends on Boltzmann charge equilibrium. Can only determine DGN of the 0.01–1.0 μm range (Rich *et al.*, 1959; Hogan, 1976).					

sensitivity of 0.01 cm^{-3} for a two-minute count and an upper limit of about 10^7 cm^{-3}. It will be partially self-calibrating and is especially designed for field use. It should open new opportunities for sizing methods such as the diffusion battery and the denuder, which requires a CNC as a sensor.

As important as the instruments themselves has been the development of mobile instrument systems for physical and chemical aerosol measurement. To make accurate *in situ* size distribution measurements from 40 μm down to the Aitken nuclei range requires at least four instruments. We have used a CNC, an EAA, and one or two OPC's. Figure 3 shows the ranges of these instruments as used aboard a large EPA mobile laboratory in the U.S.A. Recently we have used a CNC, an EAA, and one OPC for mobile measurement aboard an aircraft, in an automobile, and aboard a special University of Minnesota mobile laboratory (Wolf, 1978). To use instruments which have a several-minute measurement cycle, it is necessary to have them sample from an automatic bag sampling system. This bag system has been described by Cantrell *et al.* (to be submitted). A well-instrumented aircraft having a complete aerosol sizing system has also been described recently by Radke *et al.* (1976). This aircraft has been used to obtain SO_2 concentrations and aerosol size distribution data in paper mill plumes, a volcano plume and several power plant plumes (Hindman *et al.*, 1976; Radke *et al.*, 1976).

Sverdrup (1977) has recently developed a measurement which uses a condensation nuclei counter, an electrical charger, an integrating nephelometer and the trimodal size distribution model to deduce V_n, DGV_a, and V_a. This technique has promise for fine particle monitoring, since the integral sensors used can be made much simpler than submicron size resolving instruments.

In situ *chemical aerosol instruments*

Because of the labor and cost of measuring aerosol sulfur from collected samples, there has been considerable effort devoted recently to the development of instruments which use a flame photometric detector (FPD) to continuously measure aerosol sulfur. Since several of these will be discussed in separate papers at this symposium, only the general principles will be mentioned.

Huntzicker *et al.* (1975) have used an SO_2 denuder ahead of the FPD. Kittelson *et al.* (1977) have used a chopped electrostatic precipitator, in conjunction with a lock-in amplifier, to separate the modulated sulfur aerosol signal from the steady gaseous sulfur signal. This latter instrument has achieved better than 1 $\mu g\,m^{-3}$ sulfur sensitivity. In the near future, these instruments should permit the measurement of aerosol sulfur as easily as gaseous sulfur can now be measured.

Acknowledgement – This paper was prepared with the support of EPA Research Grant No. R 803851-03.

REFERENCES

Aitken J. (1923) *Collected Scientific Papers.* Cambridge University Press.

Altshuller A. P. (1973) Atmospheric sulfur dioxide and sulfate distribution of concentration at urban and nonurban sites in United States. *Envir. Sci. Technol.* **7,** 709–713.

Appel B. R., Wesolowski J. J. and Hidy G. M. (1974) *Analysis of the sulfate and nitrate data from the Aerosol Characterization Study.* AIHL Report No. 166, pp. 1–16, Air and Industrial Hygiene Laboratory, Berkeley, CA.

Barsic N. J. (1977) Size distribution and concentrations of fine particles produced by propane–air combustion in a controlled humidity environment. Ph.D. Thesis, Particle Technology Laboratory, Mechanical Engineering Dept., University of Minnesota, Minneapolis, MN 55455.

Berry E. X. (1967) Cloud droplet growth by collection. *J. atmos. Sci.* **24,** 688–701.

Birenzvig A. and Mohnen V. A. (1975) Gas-to-particle conversion of sulfur dioxide reaction products at low humidities. Atmospheric Sciences Research Center, State Univ. of New York, Albany.

Bricard J. (1974) Aerosol production in the atmosphere. *Proc. conf. aerosols in nature, medicine and technology.* GAF Bergweg 7, Vorderhinderlang 8973, W. Germany.

Bricard J., Cabane M. and Madelaine G. (1977) Formation of atmospheric ultrafine particles and ions from trace gases. *J. Colloid Interface Sci.* **58,** 113–124.

Bricard J., Delattre P., Madelaine G. and Pourprix M. (1976) Detection of ultrafine particles by means of a continuous flux condensation nuclei counter. In *Fine Particles: Aerosol Generation, Measurement, Sampling and Analysis* (Edited by Liu B. Y. H.), pp. 565–580. Academic Press, New York.

Cadle R. D. (1975) *The Measurement of Airborne Particles.* Wiley-Interscience, New York.

Cantrell B. K., Brockmann J. E. and Dolan D. F. A mobile aerosol analysis system with applications. *Atmospheric Environment* (To be submitted).

Cantrell B. K. and Whitby K. T. (1978) Aerosol size distributions and aerosol volume formation rates for a coal-fired power plant plume. *Atmospheric Environment* **12,** 323–333.

Chuan R. L. (1976) Rapid measurement of particulate size distribution in the atmosphere. In *Fine Particles: Aerosol Generation, Measurement, Sampling and Analysis* (Edited by Liu B. Y. H.), pp. 763–775. Academic Press, New York.

Clark W. E. and Whitby K. T. (1967) Concentration and size distribution measurement of atmospheric aerosols and a test of the theory of self-preserving size distributions. *J. atmos. Sci.* **24,** 677–687.

Clark W. E. and Whitby K. T. (1975) Measurements of aerosols produced by the photochemical oxidation of SO_2 in air. *J. Colloid Interface Sci.* **51,** 477–490.

Cooke D. D. and Kerker M. (1975) Response calculations for light scattering aerosol particle counters. *Appl. Opt.* **14,** 734–739.

Davies C. N. (1974) Size distribution of atmospheric particles. *J. Aerosol Sci.* **5,** 293–300.

Dunham S. B. (1960) Detection of photochemical oxidation of sulfur dioxide by condensation nuclei techniques. *Nature* **188,** 51–52.

Dzubay T. G. and Stevens R. K. (1975) Ambient air analysis with dichotomous sampler and X-ray fluorescence spectrometer. *Envir. Sci. Technol.* **9,** 663–668.

Dzubay T. G. and Stevens R. K. (1976) The characterization of atmospheric aerosol by physical and chemical methods. *Envir. Sci. Technol.* (in Press).

Flyger H., Hansen K., Megaw W. J. and Cox L. C. (1973) The background level of the summer tropospheric aerosol over Greenland and the North Atlantic. *J. appl. Met.* **12,** 161–174.

Flyger H., Heidem N. Z., Hansen K., Megaw W. J., Walther E.

G. and Hogan A. W. (1976) The background level of the summer tropospheric aerosol, sulfur dioxide and ozone over Greenland and the North Atlantic Ocean. *J. Aerosol Sci.* **7,** 103.

Friedlander S. K. (1962) The similarity theory of the particle size distribution of the atmospheric aerosol. *Proc. first nat. conf. on aerosols.* Liblice.

Graedel T. E. and Graney J. P. (1974) Atmospheric aerosol size spectra: rapid concentration fluctuations and biomodality. *J. geophys. Res.* **79,** 5643–5645.

Green H. L. and Lane W. R. (1957) *Particulate Clouds: Dusts, Smokes and Mists.* Van Nostrand, New York.

Heisler S. L. and Friedlander S. K. (1977) Gas-to-particle conversion in photochemical smog: aerosol growth laws and mechanisms for organics. *Atmospheric Environment* **11,** 157–168.

Hidy G. M. (1975) Summary of the California aerosol characterization experiment. *J. Air Pollut. Control Ass.* **25,** 1106.

Hidy G. M. and Lilly D. K. (1965) Solutions to the Equations fot the Kinetics of Coagulation *J. Colloid Sci.* **20,** 867.

Hidy G. M., Tong E. Y. and Mueller P. K. (1976) Design of the Sulfate Regional Experiment (SURE) – supporting data and analysis. EPRI Report No. EC-125, EPRI, Palo Alto, CA 94304.

Hindman E. E., Hobbs P. V. and Radke L. F. (1975) Airborne Investigations of Aerosol Particles from a Paper Mill. Presented at the 68th Annual Meeting, APCA.

Hobbs P. V. (1970) Comparison of ice nucleus concentrations measured with an acoustical counter and millipore filters. *J. appl. Met.* **9,** 828–829.

Hobbs P. V., Radke L. F. and Hindman E. E. (1976) An integrated airborne particle-measuring facility and its preliminary use in atmospheric aerosol studies. *J. Aerosol Sci.* **7,** 195.

Hogan A. W. (1975) A simplified coagulation model. APCA Paper 75-42.2.

Hogan A. W. (1976) *Physical properties of the atmospheric aerosol.* Atmospheric Sciences Research Center Publ. No. 408, State University of New York, Albany.

Husar R. B., Macias E. S. and Dannevik W. P. (1976) Measurement of dispersion with a fast response aerosol detector. Presented at the Third Symposium on Atmospheric Turbulence, Diffusion and Air Quality, Raleigh, NC.

Husar R. B., Whitby K. T. and Liu B. Y. H. (1972) The physical mechanisms governing the dynamics of Los Angeles smog aerosol. *J. Colloid Interface Sci.* **39,** 211–224.

Huntzicker J. J., Isabelle L. M. and Watson J. G. (1976) The continuous monitoring of particulate sulfate by flame photometry. *Proc. int. conf. on environmental sensing and assessment,* Las Vegas, NV. Paper No. 23–4.

Institute of Electrical and Electronic Engineers (1976) *Proc. int. conf. on environmental sensing and assessment,* Las Vegas, NV.

John W. (1976) Contact electrification applied to particulate matter-monitoring. In *Fine Particles: Aerosol Generation, Measurement, Sampling and Analysis* (Edited by Liu B. Y. H.), p. 649–667. Academic Press, New York.

Junge C. E. (1953) Die rolle der aerosole und der gasformingen beimengungen der luft im spurenstoffhaushalt der troposphere. *Tellus* **5,** 1–26.

Junge C. E. (1954) The chemical composition of atmospheric aerosols – I. Measurements at Round Hill Field Station, June–July 1953. *J. Met.* **11,** 323–333.

Junge C. E. (1960) Sulfur in the atmosphere. *J. geophys. Res.* **65,** 227–237.

Junge C. E. (1972) Our knowledge of the physico-chemistry of aerosols in the undisturbed marine environment. *J. geophys. Res.* **77,** 5183.

Kadowaki S. (1976) Size distribution of atmospheric total aerosols, sulfate, ammonium and nitrate particulates in the Nagoya area. *Atmospheric Environment* **10,** 39.

Karuhn R. F. (1973) The development of a new acoustic particle counter for particle size analysis. *Proc. conf. particle technology,* IIT Research Institute, Chicago, IL.

Kelkar D. N. and Joshi P. V. (1977) A note on the size distribution of aerosols in urban atmospheres. *Atmospheric Environment* **11,** 531–534.

Kiang C. S. and Middleton P. (1977) Formation of secondary acid aerosols in urban atmosphere. *Geophys. Res. Lett.* **3,** 17–20.

Kittelson D. B., Cress W. D. II, Kwok K. C. and Auw P. K. (1977) Photochemical aerosol formation in initially homogeneous systems. Internal Report, Particle Technology Laboratory, University of Minnesota, Minneapolis, MN 55455.

Kocmond W. C., Kittelson D. B., Yang J. Y. and Demerjian K. L. (1973) Determination of the formation mechanisms and composition of photochemical aerosols. Calspan Report No. NA5365-M-1.

Kocmond W. C., Kittelson D. B., Yang J. Y. and Demerjian K. L. (1975) Study of aerosol formation in photochemical air pollution. EPA Report No. EPA-650/3-75-007.

Lee R. E. and Patterson R. K. (1969) Size determination of atmospheric phosphate, nitrate, chloride, and ammonium particulate in several urban areas. *Atmospheric Environment* **3,** 249–255.

Liu B. Y. H. (ed.) (1976) *Fine Particles: Aerosol Generation, Measurement, Sampling and Analysis.* Proc. symp. fine particles, Minneapolis, MN, 1975. Academic Press, New York.

Liu B. Y. H. and Kim C. S. (1977) On the counting efficiency of condensation nuclei counters. *Atmospheric Environment* **11,** 1097–1100.

Liu B. Y. H. and Lee K. W. (1975) An aerosol generator of high stability. *Am. ind. Hyg. Ass. J.* **36,** 861–865.

Liu B. Y. H. and Pui D. Y. H. (1975) On the performance of the electrical aerosol analyzer. *J. Aerosol Sci.* **6,** 249–264.

Lui B. Y. H. and Pui D. Y. H. (1976) Electrical behavior of aerosols. Presented at the 82nd National Meeting of American Institute of Chemical Engineers, Atlantic City, NJ.

Liu B. Y. H., Pui D. Y. H. and Kapadia A. (1977) Electrical aerosol analyzer: history, principle and data reduction. In *Proceedings of the aerosol measurement workshop conference* (Edited by Lundgren D. A.), University of Florida, Gainesville.

Loo B. W., Jaklevic J. M. and Goulding F. S. (1976) Dichotomous virtual impactors for large-scale monitoring of airborne particulate matter. In *Fine Particles: Aerosol Generation, Measurement, Sampling and Analysis* (Edited by Liu B. Y. H.), pp. 311–350. Academic Press, New York.

Ludwig F. L. and Robinson E. (1965) Size distribution of sulfur-containing compounds in urban aerosols. *J. Colloid Sci.* **20,** 571–584.

Lundgren D. A. (1973) Mass distribution of large atmospheric particles. Ph.D. Thesis, University of Minnesota, Minneapolis, MN 55455.

Lundgren D. A. (ed.) (1977) *Proceedings of the Aerosol Measurement Workshop Conference,* University of Florida, Gainesville.

Lundgren D. A., Carter L. D. and Daley P. S. (1976) Aerosol mass measurement using piezoelectric crystal sensors. In *Fine Particles: Aerosol Generation, Measurement, Sampling and Analysis.* (Edited by Liu B. Y. H.), pp. 485–510. Academic Press, New York.

Mainwaring S. J. and Harsha S. (1976) Size distribution of aerosols in Melbourne city air. *Atmospheric Environment* **10,** 57–60.

Macias E. S. and Husar R. B. (1976) A review of atmospheric particulate mass measurement via the beta attenuation technique. In *Fine Particles: Aerosol Generation, Measurement, Sampling and Analysis* (Edited by Liu B. Y. H.), pp. 535–564. Academic Press, New York.

Marple V. A. and Willeke K. (1976) Inertial impactors:

theory, design and use. In *Fine Particles: Aerosol Generation, Measurement, Sampling and Analysis* (Edited by Liu B. Y. H.), pp. 411–446. Academic Press, New York.

McCartney E. J. (1976) *Optics of the Atmosphere: Scattering by Molecules and Particles.* John Wiley, New York.

McKee H. C., Childers R. E. and Saenz O. Jr. (1971) Collaborative study of method for the determination of suspended particulates in the atmosphere (high-volume method). Report from Southwest Research Institute.

McMurry P. H. (1977) On the relationship between aerosol dynamics and the rate of gas-to-particle conversion. Ph.D. Thesis, California Institute of Technology, Pasadena, CA 91125.

Mercer T. T. (1973) *Aerosol Technology in Hazard Evaluation.* Academic Press, New York.

Mercer T. T., Morrow P. E. and Stöber W. (eds.) (1972) *Assessment of Airborne Particles.* Thomas, Springfield, IL.

Mészáros A. and Vissy K. (1974) Concentrations, size distribution and chemical nature of atmospheric aerosol particles in remote oceanic areas. *J. Aerosol Sci.* **5,** 101–109.

Patterson R. K. and Wagman J. (1977) Mass and composition of an urban aerosol as a function of particle size for several visibility levels. *J. Aerosol Sci.* **8,** 269–279.

Pich J., Friedlander S. K. and Lai F. S. (1970) The self-preserving particle size distribution for coagulation by Brownian motion. *J. Aerosol Sci.* **1,** 115–126.

Pierson W. R., Hammerle R. H. and Kummer J. T. (1974) Sulfuric acid aerosol emissions from catalyst-equipped engines. SAE Paper No. 740287.

Radke L. F., Hobbs P. V. and Smith J. L. (1976) Airborne measurements of gases and aerosols from volcanic vents on Mt. Baker. *Geophys. Res. Lett.* **3,** 93–96.

Radke L. F. and Turner F. M. (1972) An improved automatic cloud condensation nucleus counter. *J. appl. Met.* **11,** 407–409.

Rao A. K. and Whitby K. T. (1977) Nonideal collection characteristics of single stage and cascade impactors. *Am. ind. Hyg. Ass. J.* **38,** 174–179.

Renzetti N. A. and Doyle G. J. (1960) Photochemical aerosol formation in sulfur dioxide hydrocarbon systems. *Int. J. Air Pollut.* **2,** 327–345.

Rich T. A., Pollak L. W. and Metnieks A. L. (1959) Estimation of average size of sub-micron particles from the number of all and uncharged particles. *Geofis. pura appl.* **44,** 233.

Richards L. W. (1977) Los Angeles field modeling and measurement study. Interim Report, EPA Contract No. 68-02-2463.

Roesler J. F., Stevenson H. J. R. and Nader J. S. (1965) Size distribution of sulfate aerosols in ambient air. *J. Air Pollut. Control Ass.* **15,** 576–579.

Saxena V. K., Alofs D. J. and Tebelak A. C. (1972) A comparative study of Aitken nuclei counters. *J. Rechs atmos.* **6,** 495–505.

Schaefer V. J. (1976) The air quality patterns of aerosols on the global scale, Parts I and II. Final Report to NSF, Atmospheric Sciences Research Center, State University of New York, Albany.

Sinclair D. and Hinchliffe L. (1972) Production and measurement of submicron aerosols. In *Assessment of Airborne Particles* (Edited by Mercer T. T. *et al.*). Thomas, Springfield, IL.

Sinclair D. and Hoopes G. S. (1975) A continuous flow condensation nucleus counter. *J. Aerosol Sci.* **6,** 1–7.

Smith R., de Pena R. G. and Heicklen J. (1975) Kinetics of particle growth – VI. Sulfuric acid aerosol from the photooxidation of SO_2 in moist O_2–N_2 mixtures. *J. Colloid Sci.* **53,** 202.

Stevens R. K., McFarland A. R. and Wedding J. B. (1976) Comparison of the virtual dichotomous sampler with the hi-volume sampler. Presented at the April Meeting of Am. Chem. Soc., New York.

Stevenson H. J. R., Sanderson D. E. and Altshuller A. P. (1965) Formation of photochemical aerosols. *Int. J. Air Water Pollut.* **9,** 367–375.

Stöber W. (1976) Design, performance and applications of spiral duct aerosol centrifuges. In *Fine Particles: Aerosol Generation, Measurement, Sampling and Analysis* (Edited by Liu B. Y. H.), pp. 351–397. Academic Press, New York.

Stöber W. and Mönig F. J. (1976) Measurements with a prototype mass distribution monitor for particulate air pollution. *Proc. int. conf. environmental sensing and assessment*, Las Vegas, NV.

Suck S. H., Middleton P. B. and Brock J. R. (1977) On the multimodality of density functions of pollutant aerosols. *Atmospheric Environment* **11,** 251.

Sverdrup G. M. (1977) Parametric measurement of submicron atmospheric aerosol size distributions. Ph.D. Thesis, Particle Technology Laboratory, University of Minnesota, Minneapolis, MN 55455.

Tomasi C. and Tampieri F. (1976) Size distribution models of tropospheric particles of different origin in terms of the modified gamma function and relationship between skewness and mode radius. *Tellus* **29,** 66–74.

Twiss S. (1977) Composition of particulate matter in California hi-vol data. Interagency Symp. on Air Monitoring Quality Assurance, AIHL, Berkeley, CA.

Vohra K. G., Vasudean K. N. and Nair P. V. (1970) Mechanisms of nucleus-forming reactions in the atmosphere. *J. geophys. Res.* **75,** 2951–2960.

Wagman J., Lee R. E., Jr. and Axt C. J. (1967) Influence of some atmospheric variables on the concentration and particle size distribution of sulfate in urban air. *Atmospheric Environment* **1,** 479–489.

Walter H. and Jaenicke R. (1973) Remarks about the smallest particle size detectable in condensation nuclei counters. *Atmospheric Environment* **7,** 939–944.

Wesolowski J. J., Alcocoer A. and Appel B. R. (1978) The validation of the Lundgren impactor. Hutchinson Memorial Volume on the ACHEX Studies in California. (To be published).

Whitby K. T. (1973) On the multimodal nature of atmospheric aerosol size distributions. Presented at VIIIth *int. conf. on nucleation*, Leningrad, U.S.S.R.

Whitby K. T. (1974) Modeling of multimodal aerosol distributions. *Proc. conf. aerosols in nature, medicine and technology*, GAF Bergweg 7, Vorderhinderlang 8973, W. Germany.

Whitby K. T. (1976) Electrical measurement of aerosols. In *Fine Particles: Aerosol Generation, Measurement, Sampling and Analysis* (Edited by Liu B. Y. H.), pp. 581–624. Academic Press, New York.

Whitby K. T. and Cantrell B. K. (1975) Atmospheric aerosols: characteristics and measurements. Presented at the *Int. conf. environmental sensing and assessment*, Las Vegas, N.V.

Whitby K. T., Cantrell B. K., Husar R. B., Gillani N. V., Anderson J. A., Blumenthal D. L. and Wilson W. E. (1978a) Aerosol formation in a coal-fired power plant plume. *Atmospheric Environment.* (To be submitted).

Whitby K. T., Cantrell B. K. and Kittelson D. B. (1978b) Nuclei formation rates in a coal-fired power plant plume. *Atmospheric Environment* **12,** 313–321.

Whitby K. T., Husar R. B. and Liu B. Y. H. (1972a) The aerosol size distribution of Los Angeles smog. *J. Colloid Interface Sci.* **39,** 211.

Whitby K. T., Kittelson D. B., Cantrell B. K., Barsic N. J. and Dolan D. F. (1977) Aerosol size distributions and concentrations measured during the General Motors Proving Grounds Sulfate Study. To be published in *Proc. Symp. on Chemical Properties of Automotive Emissions from Catalyst-Equipped Cars and Biological Effects of Auto Emissions-Related Pollutants* (Edited by Stevens R. K.), Ann Arbor Sciences Inc., Ann Arbor, MI.

Whitby K. T., Liu B. Y. H., Husar R. B. and Barsic N. J. (1972b) The Minnesota aerosol analyzing system used in the Los Angeles Smog Project. *J. Colloid Interface Sci.* **39**, 177–204.

Whitby K. T. and Sverdrup G. M. (1978) California aerosols: their physical and chemical characteristics. Hutchinson Memorial Volume on the ACHEX Studies in California. (To be published).

Whitby K. T. and Willeke K. (1977) Light scattering instruments, principles and field use. In *Proc. Aerosol Measurement Workshop* (Edited by Lundgren D. A.), University of Florida, Gainesville.

Willeke K. and Liu B. Y. H. (1976) Single particle optical counter: principle and application. In *Fine Particles: Aerosol Generation, Measurement, Sampling and Analysis* (Edited by Liu B. Y. H.), pp. 697–729. Academic Press, New York.

Willeke K. and Whitby K. T. (1975) Atmospheric aerosols: size distribution interpretation. *J. Air Pollut. Control. Ass.* **25**, 529.

Wilson W. E., Miller D. F., Levy A. and Stone R. K. (1973) *J. Air Pollut. Control Ass.* **23**, 949.

Wilson W. E., Spiller L. L., Ellestad T. G., Lamothe P. J., Dzubay T. G., Stevens R. K., Macias E. S., Fletcher R. A., Husar J. D., Husar R. B., Whitby K. T., Kittelson D. B. and Cantrell B. K. (1977) General Motors Sulfate Dispersion Experiment: summary of EPA measurements. *J. Air Pollut. Control Ass.* **27**, 46.

Winkler P. (1975) Chemical analysis of Aitken particles (< 0.4 diameter) over the Atlantic Ocean. *Geophys. Res. Lett.* **2**, 45–48.

Wolf J. L. (1978) Design, construction and operation of a mobile air pollution laboratory. M.S. Thesis, Particle Technology Laboratory, Mechanical Engineering Dept., University of Minnesota, Minneapolis, MN 55455.

Atmospheric Environment Vol. 12, pp. 161–169. Pergamon Press 1978. Printed in Great Britain.

PHYSICAL PROPERTIES OF ATMOSPHERIC PARTICULATE SULFUR COMPOUNDS

R. JAENICKE
Max-Planck-Institut, Mainz, W. Germany

(*First received* 14 *June* 1977 *and in final form* 22 *September* 1977)

Abstract—Integral properties of the total aerosol—which are usually measured first—tend to be related to certain limited size ranges only and do not represent the aerosol in its entire particle size range. The connection between integral and differential properties will be discussed, as well as the consequences with respect to the presentation of total sulfur mass and the sulfur size distribution. The question of the mixture state will be discussed and data given for the internal and external mixture ratio of ammonium sulfate as the most abundant sulfur compound in aerosols. Finally, a discussion is given of the particle sizes which must be produced to maintain the observed aerosol size distribution by any nucleation processes converting sulfuric gases into particles.

1. INTRODUCTION

The size distribution of aerosols in general is probably the most important physical parameter of the aerosol. It reveals immediately whether the aerosol has optical effects or whether it might be subject to long range transport in the atmosphere. It shows if the aerosol can be adsorbed in the human respiratory system, with all the possible impacts on human health, or if cloud formation might be influenced by it. Even the impact of the atmospheric aerosol on the radiation budget of the earth can only be estimated if the size distribution is known. However, influences of the aerosol can only be estimated effectively if further information is available with regard to chemical composition, bulk density, complex index of refraction, shape of particles and others.

The aerosol in the atmosphere consists of substances of various origin, such as soil, ocean, organic material, and gases. After production the aerosol is mixed and further modified as it ages. This includes mixing of aerosols of different origin, coagulation to form new particles, adsorption of gases, chemical reactions within the particles, mainly in the presence of liquid water and dry and wet removal.

Despite the various sources of particles the ageing and mixing of the aerosol should result in a very uniform aerosol. In such an aerosol the chemical composition as well as some other properties would then be independent of the particle size. The uniformity of the final aerosol, however, depends upon the time available for the ageing and mixing processes. This time must be shorter than the residence time of the aerosol so that sufficient time is available to achieve this uniformity before the particles are removed from the atmosphere. To my knowledge, estimates of the length of this ageing time for atmospheric aerosols do not exist.

Ageing of the aerosol also tends to produce an aerosol with a narrow size distribution. This is caused by two processes: coagulation and removal. Coagulation tends to combine the highly mobile small particles to produce larger particles. Removal of the particles from the atmosphere by dry sedimentation and wet removal tends to influence only the larger particles. Ageing would then remove the lower and the upper end of the atmospheric particle radius range, thus squeezing the distribution. The aerosol size distribution in all parts of the troposphere, however, is a very wide one, covering more than five orders of magnitude in radius. Consequently, the simultaneous presence of Aitken nuclei on one hand and giant particles on the other indicates immediately the rather unaged state of the aerosol. The obvious consequence of the unaged state is the conclusion that the atmospheric aerosol would need much more time to achieve uniform properties than is available within its residence time which, at present, is estimated to be about 5 days.

If information about the internal and external mixture* of an aerosol were available it would be possible to estimate the production processes of the aerosol and conclude how aged the aerosol is.

The rather unmixed state of the aerosol is confirmed by the confinement of substances and elements to certain size ranges of the aerosol (Rahn *et al.*, 1971). That this is also true for sulfur compounds became clear immediately after sulfur compounds were found in the aerosol, and first studies (Junge, 1954) concerned with differential distribution were carried out. Furthermore, it must be concluded that aerosol properties like index of refraction and bulk density are not distributed uniformly within the aerosol. An aerosol with non-uniform properties behaves differently from a uniform aerosol with regard to the

* Explained later in the paper.

physical properties and the chemical reactions to substances in the carrier gas.

At present, we are not in the position to establish the differential distribution of most aerosol parameters. In this case, integral values have to be used. In some cases like organic and mineral compositions the first estimates of differential distributions are now becoming available. Sulfate and sodium chloride certainly belong to the group where distributions can be established. The presentations of those data, however, are still disputable and some necessary information is lacking. This paper will discuss the way integral and differential properties are presented and the consequences on the interpretation of the data. Further thoughts will be given to the concept of internal and external mixture of the aerosol with special emphasis on sulfur compounds. The possible production mechanism of sulfur compounds will be discussed from the standpoint of the observed aerosol size distribution.

2. THE RELATION BETWEEN DIFFERENTIAL AND INTEGRAL PROPERTIES OF THE AEROSOL

Usually the integral value of properties of an aerosol, e.g. total mass, total number concentration, total surface, bulk density, mass of a specific compound (e.g. sulfur compound) within the aerosol, can be measured more easily than the differential distribution of these properties. Normally the availability of collected aerosol mass is the limiting factor. For the same reason most networks measure integral properties, and trends of the atmospheric aerosol were determined with integral parameters only (Cobb *et al.*, 1970; Jaenicke and Schütz, in press).

In this context it must be mentioned, of course, that some questions can be answered using only integral values. For example, the influence of the aerosol on radiation can be estimated by determining the atmospheric turbidity. Atmospheric turbidity is a parameter obtained not only by integrating over-all particle sizes, but also over the vertical extent of the atmosphere. Other questions, however, cannot be answered directly if only integral parameters are known. The removal of sulfur compounds depends on the size of the sulfur-containing particles and their physical and chemical behavior in the carrier gas air, usually in the presence of water. The detailed knowledge of the sulfur compound differential size distribution is, therefore, necessary. The available data, however, are at present inadequate.

The connection between differential and integral or volume properties will now be discussed (Jaenicke, 1973). If we denote the aerosol number size distribution as

$$n^*(r) = \frac{\mathrm{d}N(r)}{\mathrm{d}\lg r} \tag{1}$$

with $n^*(r)$ = differential size distribution, $N(r)$ = cumulative size distribution, any integral effect E of the aerosol is

$$E = E\left(\int_{r=0}^{r=\infty} w(r)\cdot n^*(r)\mathrm{d}\lg r\right) \tag{2}$$

with $w(r)$ = weighing function of the effect E. Integral effect E can mean total mass or volume, density or index of refraction, surface, number, or some other property.

This equation includes any functional dependency of the effect E on the integral. It should be noted that this equation also can be understood as an experimental procedure which usually acts as a filter with a filter function $w(r)$. A well-known example is the collection of aerosols with impactors in which only particles of a certain size are collected.

Equation (2) is only valid if all particles with their total volume contribute to this effect. This is not the case if for instance the total mass of a certain sulfur compound is regarded as the integral effect E. In this case, the internal and external mixture* of the aerosol have to be considered. Its influence is in detail

$$E = E\left(\int_{r=0}^{r=\infty} w(r)\cdot\epsilon(r)\cdot n^*(r)\mathrm{d}\lg r\right) \tag{3}$$

with $\epsilon(r)$ = fraction per number of particles with radius r containing the compound under discussion, as it is the case for an external mixture. If, for example, only 50% of all particles contain some sulfur compounds, $\epsilon(r) = 0.5$.

$$E = E\left(\int_{r=0}^{r=\infty} w(r)\cdot\eta(r)\cdot n^*(r)\mathrm{d}\lg r\right) \tag{4}$$

with $\eta(r)$ = volume fraction of the particles with the compound under discussion, as is the case for internal mixture.† If, for example, the particles consist only of 30% sulfur, $\eta(r) = 0.3$. The combined effect of internal and external mixture as we expect it usually in atmospheric aerosols would be

$$E = E\left(\int_{r=0}^{r=\infty} \epsilon(r)\cdot w(r)\cdot\eta(r)\cdot n^*(r)\mathrm{d}\lg r\right). \tag{5}$$

Equation (5) shows the relationship between the integral property E and the differential property $w(r)\eta(r)\epsilon(r)\cdot n^*(r)$ in its most general form. This differential property of course contains properties which are also regarded as differential, such as the distribution of sulfur-containing particles $\epsilon(r)\cdot n^*(r)$. This will be discussed later. The importance of the number size distribution is clearly demonstrated by these equations. However, we will see that the presentation of aerosol size distribution data in forms other than $n^*(r)$ can be of great benefit.

Examples will now be discussed. These are total particle concentration, electrical conductivity of the

---

* Discussed later in the paper.

† Usually internal mixture would be defined for volume fraction. For simplicity this is not done here in full detail.

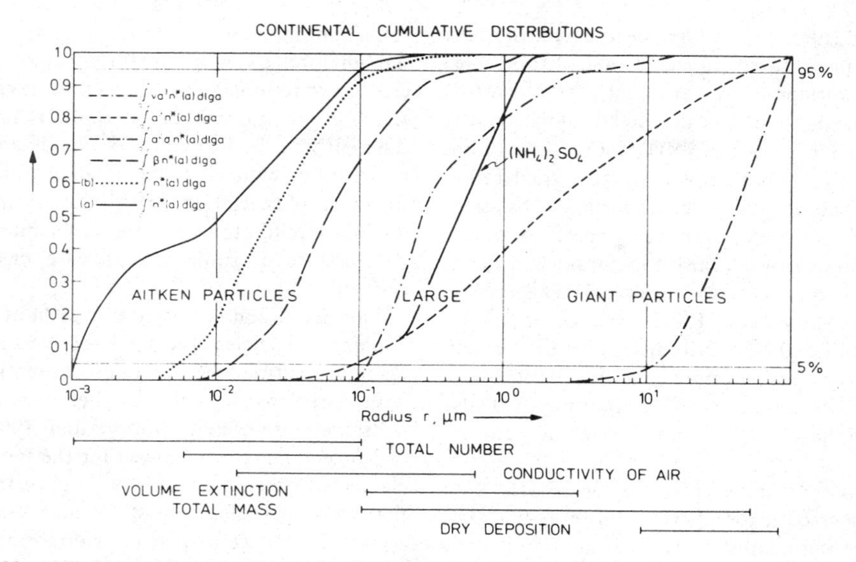

Fig. 1. Normalized cumulative weighed distributions for various properties of an idealized continental aerosol. $n^*(a)$—particle number distribution for the earlier size distribution (b) and the more recent assumptions (a), $\beta \cdot n^*(a)$—electrical conductivity of air, $a^2 \cdot \sigma \cdot n^*(a)$—volume extinction, $a^3 n^*(n)$—volume of particles, $v \cdot a^3 \cdot n^*(n)$—dry deposited particle volume. The mass of $(NH_4)_2SO_4$ is given separately. The radius range of influence of the given properties is indicated.

aerosol, extinction, total mass, sedimented mass, and sulfate mass.

For a generalized size distribution of continental aerosols Fig. 1 shows weighed cumulative distributions typical for some measuring methods. It becomes obvious that in such an aerosol the total particle mass is mainly influenced by particles in the size range 0.1 μm–50 μm if the 5% level of the cumulative distribution is taken as the limit. If results from dry deposition are obtained, only particles greater than 10 μm in radius influence the collected mass. The total particle concentration—as another integral parameter—is influenced only by particles in the range of 0.001–0.1 μm radius if the new data (Jaenicke and Schütz, in press) are used, or 0.01–0.1 μm radius for earlier models of the aerosol size distribution (Junge, 1963). In both cases, the particles larger than 0.1 μm radius do not contribute significantly to the total particle concentration. In the same way the electrical conductivity of air is influenced by particles in the range 0.01–0.4 μm radius and volume extinction by 0.1–2 μm particles in radius. As can be seen from the insert of Fig. 1, the differential properties are all of the form $w(r) \cdot n^*(r)$. In the case of the total particle concentration $w(r) = 1$.

In addition, Fig. 1 contains the cumulative distribution of $(NH_4)_2SO_4$ which was estimated from data discussed later in the text. The $(NH_4)_2SO_4$-mass cumulative distribution departs from the mass distribution only above 0.2 μm radius. This indicates that the Aitken particles over the continent consist probably of $(NH_4)_2SO_4$. The main mass of $(NH_4)_2SO_4$ is in the same range as the large particles. The giant $(NH_4)_2SO_4$-particles practically do not contribute to the total $(NH_4)_2SO_4$-mass. Determination of the total $(NH_4)_2SO_4$-mass gives information about particles in the range 0.1–1.5 μm radius.

In addition, other important information can be concluded from Fig. 1; (1) In observing trends of the aerosol with different methods, they do not necessarily have to exhibit similar effects. If methods cover independent size ranges the results of trend observations can be contradictory if the size distribution changes shape. (2) The presentation of the size distribution in another form than that of number size distribution might be very valuable. Fig. 1 indicates some of them, namely surface $(r^2 \cdot n^*(r))$, volume or mass $(r^3 \cdot n^*(r))$, and even $r^5 \cdot n^*(r)$ in the case of dry deposition. r^5 is obtained because the settling velocity v is a function of r^2. This shows the coexistence of number, surface, volume, and even higher distibutions. The last remark is pointed out in a recent discussion (Jaenicke and Davis, 1976) about the mathematical expressions describing the aerosol size distributions. The presentation of aerosol data as $dM/d lgr$ on a linear scale as a function of radius r on a logarithmic scale, as recently employed by Whitby *et al.* (1972), shows nicely the independent behavior of certain size ranges of the aerosol (called modes in that paper). This has been known since the early days of research in atmospheric aerosols when only total number and total mass were investigated. As Fig. 1 shows, the total particle concentration and the total particle mass practically do not overlap in radius ranges. And indeed, in most atmospheric aerosols it is impossible to observe any correlation between both parameters (Junge *et al.*, 1971; Willeke *et al.*, 1974; Jaenicke and Schütz, in press). On the other hand, optical parameters like visibility and total mass usually show some correlation (Charlson *et al.*, 1968).

As Fig. 1 indicates, this should be expected because the radius range of extinction and that of total mass show some overlapping. Consequently, total particle concentration and visibility are mostly uncorrelated.

Comparing the curves of $(NH_4)_2SO_4$ and volume extinction in Fig. 1, both cover roughly equal radius ranges. In the absence of other important substances, like water, one therefore, should expect to obtain good correlation between both the parameters in the atmosphere. This indeed was observed (Waggoner *et al.*, 1975). To summarise, Fig. 1 and other similar calculations of weighed distributions give the possibility of selecting and designing proper experiments if the aerosol in its totality is to be monitored. It also shows the importance of any weighing on the observed effect.

In this context, the presentation of aerosol data should be discussed if the distribution of a substance with possible internal and/or external mixture is presented. Presently 3 different methods for presenting sulfate data are employed. Georgii *et al.* (1971) present ammonium sulfate size distributions with an equivalent radius. This radius describes the volume of ammonium sulfate in a particle regardless of its physical size. Size distributions of this kind make it possible to calculate the total mass of ammonium sulfate in an aerosol. They may obscure other aerosol properties like inertia effects, sedimentation velocity, long range transport, to name only a few. Gravenhorst (1975b) presents the actual size of a sulfate-containing particle regardless of its sulfate content. This makes it impossible to calculate sulfate mass, but permits estimation of other physical effects. Mészaros *et al.* (1974) give size distributions of morphologically identified particles. To my knowledge it is not yet known what fraction of $(NH_4)_2SO_4$ a particle must contain to exhibit a typical $(NH_4)_2SO_4$ morphology.

All 3 methods will result in different 'sulfate' distributions and it becomes clear that it is insufficient to present data without information about the state of mixture.

3. INTERNAL AND EXTERNAL MIXTURE

The discussion above makes it immediately clear that mixed particles exist in the aerosol. This idea was introduced by Junge (1952).

Winkler (1975) introduced the distinction between internal and external mixture. Internal mixture describes the state of mixing of the various substances within the individual particle. External mixture describes the state where only particles consisting of pure substances are present and the aerosol consists of pure particles of different substances. Any chemical analysis of filter samples of aerosol particles will not differentiate between these mixture states. Winkler (1973) and later Charlson *et al.* (1974) indicated that aerosols with perfect internal mixture show a different behavior in the uptake of water with increasing relative humidities or the release of water with decreasing relative humidities than an aerosol with perfect external mixture. External mixtures should result in increased water uptake as opposed to internal mixtures. Aerosols with internal or external mixture will behave differently in the formation of clouds and the change of visibility with changing relative humidities. It is, therefore, of crucial importance for the aerosol behavior in a humid atmosphere whether internal or external mixture or some intermediate mixture state is present.

This discussion is carried out without regard to the content of particles of water insoluble substances. The insoluble substances are only of minor importance with regard to the deliquescence as long as soluble substances are present in more than 10% per volume, as Junge *et al.* (1971) showed for the forming of cloud elements. Mészaros *et al.* (1974) gave the impression that most particles consist of pure substances. This could be regarded as perfect external mixture, at least with regard to some substances. However, Charlson *et al.* (1974) estimate for $(NH_4)_2SO_4$ that in internal mixture 30–50 mole % would be sufficient to show plainly visible inflection points in a curve of water uptake vs humidity. It cannot be determined if that amount of one substance in one particle is sufficient to modify the morphology of the particle so that it resembles a particle of a pure substance.

Charlson *et al.* (1974) show in a summary of their results that internal and external mixture might be present at times in continental aerosols. It remains open, however, whether their method was aimed at making that distinction, because the inflection point is visible in the presence of $(NH_4)_2SO_4$ in more than 30–50 mole % regardless of whether the aerosol is mixed internally or externally. Winkler (1973) on the other hand concludes that internal mixture predominates. Since he investigated only bulk samples, the fact cannot be excluded that this conclusion is biased to a certain extent by the method of observation.

Winkler (1973) proposed the use of the standard deviation(s) of the relative composition of the individual particles to describe the mixture state. For particles with identical chemical composition $s = 0$; there is no variation in the composition and the internal mixture is perfect. However, in this concept perfect external mixture is not defined by a given value of s. Furthermore, the two following examples result in identical values of s. Assume that a substance like $(NH_4)_2SO_4$ is distributed over 10 particles of equal size. In the first case one particle consists of $(NH_4)_2SO_4$ only, while the other 9 particles do not contain this substance. In this case $s = 0.316$ with perfect external mixture. For the second case it is assumed that one particle consists of $(NH_4)_2SO_4$ only, 5 particles contain a fraction of 0.4 $(NH_4)_2SO_4$ each, and 4 particles contain no trace of $(NH_4)_2SO_4$. This is a case of intermediate mixture state, but it results in $s = 0.316$, the same as above. It is necessary, therefore, to deduce a mixture parameter which has well-defined values for perfect internal and external mixtures. Other mixture states are, therefore, somewhere between these limits. It must be emphasized, however, that this mixture ratio describes the mixture state only with certain assumptions about the aerosol. It is believed, however, that assumptions in this approach do not limit seriously the conclusions drawn later in the paper.

In developing the mixture parameter it is assumed that the aerosol is monodisperse. It consists of two 'substances': the substance S under consideration and the residue. If the substance S is present in only some of the particles it is assumed that all these particles contain the same amount of the substance S. In reality, the substance S is distributed somehow on the particles, probably normally distributed. Our assumption, therefore, is only a first approach to tackling the problem. In the following it will be necessary to define first of all several different mixing parameters with regard to the substance S until finally a master mixing parameter can be developed.

The aerosol is characterized by

n_s = number concentration of particles containing the substance S,
n_T = total particle concentration of the aerosol,
V_s = volume concentration of the substance S in the aerosol,
V_T = total volume concentration of the aerosol.

The use of volume rather than mass is preferred because of

$$V_T = n_T \cdot \tfrac{4}{3} \cdot \pi \cdot r^3 \tag{6}$$

where r is the radius of the aerosol particles. In this case the properties n_T and V_T of the aerosol are related without information about the bulk density that would be required if the aerosol mass is used as parameter.

The volume mixing ratio is defined by

$$M_v = \frac{V_s}{V_T}. \tag{7}$$

This mixture can be determined from collected samples of aerosol particles which are usually used for chemical analysis (e.g. Georgii *et al.*, 1971). M_v describes the volume fraction of the substance S in the aerosol.

The number mixing ratio is defined by

$$M_n = \frac{n_s}{n_T}. \tag{8}$$

M_n describes which fraction by number of the aerosol particles contains the substance S (e.g. Gravenhorst, 1975b). $M_n = 1$ in the case of perfect internal mixture. With perfect external mixture the value $M_n = M_v$ is obtained. In the case of a given value V_s of the substance present M_n might vary in the range $M_v \leqq M_n \leqq 1$.

The internal mixing ratio M_i can now be defined as

$$M_i = \frac{M_n - M_v}{1 - M_v} = \frac{n_s V_T - n_T V_s}{n_T V_T - n_T V_s}. \tag{9}$$

For perfect internal mixture $M_i = 1$ because $M_n = 1$. For perfect external mixture $M_i = 0$ because $M_n = M_v$ as explained above. For all possible states of mixtures M_i might then vary in the limits $0 \leqq M_i \leqq 1$. Consequently, the external mixing ratio M_E is defined as

$$M_E = 1 - M_i \tag{10}$$

The above Equation (9) for internal mixing ratio M_i shows that the two parameters M_n and M_v must be known to describe the state of mixture of an aerosol with regard to one substance S. To my knowledge the simultaneous determination of M_n and M_v has not been done in the past with this object in mind. However, it can be inferred from certain publications.

Figure 2 shows the relationship between the above defined mixture ratios. It can be seen that with a given volume mixture ratio M_v perfect internal mixture $M_i = 1$ is possible as well as perfect external mixture $M_E = 1$, if the number mixture ratio M_n has a suitable value. On the other hand, with a given number mixture ratio M_n only certain ranges of internal M_i and external mixture M_E are possible. This means, e.g. that for $M_n = 0.5$ the internal mixture might vary only in the range $0 \leq M_i \leq 0.5$. In that case volume mixtures $M_v > 0.5$ are impossible.

Figure 2 also shows another interesting feature. For large volume mixtures ($M_v > 0.8$ approx) it is impossible in most practical cases to determine the degree of internal mixture by determining addi-

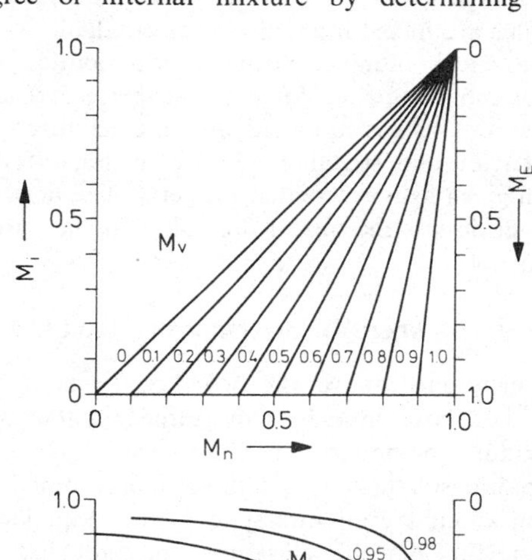

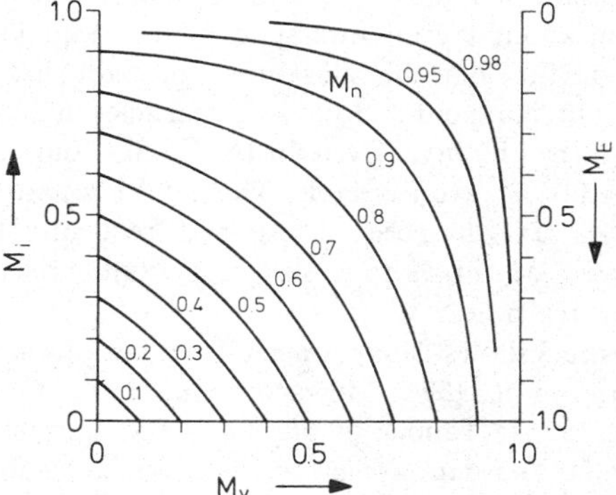

Fig. 2. Relationship of volume mixture ratio M_v, number mixture ratio M_n, internal mixture ratio M_i, and external mixture ratio M_E. For details see text.

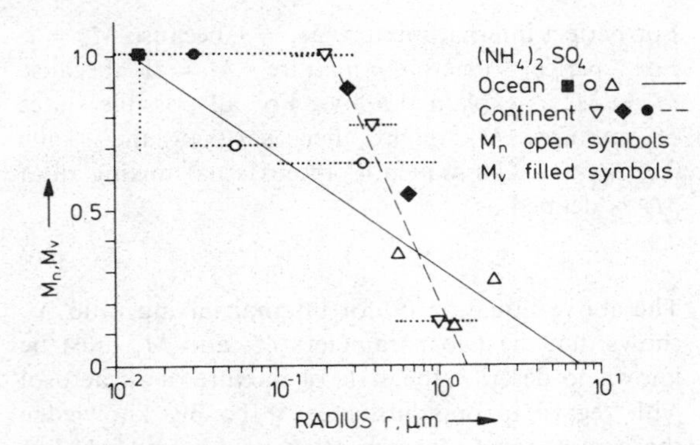

Fig. 3. Internal and external mixture observations for ocean and continental aerosols after Georgii *et al.* (1971); Mészaros *et al.* (1974); Gravenhorst (1975); Winkler (1975); Tanner *et al.* (1977) together with unpublished results (Jaenicke). The best fit for continental and oceanic aerosols is indicated using the assumption $M_n = M_v$. This assumption was made because only few data are available.

tionally the number of mixture M_n. Since M_n must be $M_n > 0.8$ in any case the possible variation is restricted to $0.8 < M_n < 1$ which should also determine the full range of internal mixture $0 < M_i < 1$. With respect to $(NH_4)_2SO_4$ and the present day accuracy of the methods for determination of M_v and M_n, it probably is impossible for values of $M_v > 0.8$ to determine the degree of internal mixture in an aerosol. With respect to continental aerosols—as we will see later in Fig. 3—it probably is not possible to say at present whether the particles smaller than 0.2 µm in radius are mixed internally or externally.

This concept of mixture defined for a monodisperse aerosol can be applied for a polydisperse aerosol if narrow size ranges are used and the mixture state is given for that size interval. The mixing ratio M_i is then given as a differential property. The next section will discuss the mixture of sulfur in the natural aerosol.

4. THE STATE OF MIXTURE OF SULFUR

As mentioned above, the two mixture parameters M_n and M_v have rarely been measured simultaneously for sulfur compounds in the aerosol. Only for ammonium sulfate is it possible to obtain some information about the mixture state. It has been known for some time that this substance is probably the most important compound of the large particles in aerosols undisturbed by man. Even for $(NH_4)_2SO_4$ only a few values for M_n are available. The total concentration of $(NH_4)_2SO_4$ has been determined frequently*), but values of M_v in various size ranges are not abundantly available.

Figure 3 shows data extracted from various sources (Georgii *et al.*, 1971; Mészaros *et al.*, 1974; Gravenhorst, 1975a; Tanner *et al.*, 1977 and unpublished results†). The data are given for ocean aerosols and continental aerosols. In both sets it can be seen that the mixtures M_n and M_v increase as the radius decreases. For the giant particles larger than 1 µm in radius $(NH_4)_2SO_4$ particles are of minor importance. For the Aitken particles ($r < 0.2$ µm), $(NH_4)_2SO_4$ particles probably become the most abundant substance. The large particle range is characterized by a change of mixture state.

The limited data base does not permit the determination of the internal mixture ratio M_i. For continental aerosol particles having radii smaller than 0.5 µm, M_i cannot be determined in any case—as discussed in connection with Fig. 2—because both M_n and M_v are larger than 0.8. For particles larger than 0.5 µm only one data point each for M_n and M_v is available.

As a first approximation, Fig. 3 indicates that M_n and M_v in continental aerosols behave similarly, so that we can state $M_n = M_v$. That means $M_i = 0$ and $M_E = 1$. Thus $(NH_4)_2SO_4$ should be in the state of nearly perfect external mixture. We are aware that this is in disagreement with some earlier results which yielded for soluble substances in general the state of nearly perfect internal mixture. These earlier results were obtained from the water uptake behavior of aerosol samples (Winkler, 1973) and individual particles (Junge, 1952) with relative humidity. However, Charlson *et al.* (1974) admit that external mixture might exist at times in continental aerosols.

The few results of M_n and M_v for oceanic air do not permit any conclusions about the internal mixture of $(NH_4)_2SO_4$. However, the results published by Mészaros *et al.* (1974) based on morphological investigations can only be understood if external mixture exists to a high degree. Figure 3 indicates the important role that $(NH_4)_2SO_4$ plays in the Aitken particle range for oceanic as well as continental aerosols. With $M_n = M_v$ for continental aerosols (Fig. 3) the differential mixture distribution was calculated and applied to 'sulfate' distributions as published by Georgii *et al.* (1971). Figure 4 shows the result. A 'sulfate' distribution takes the radius to be an equivalent radius of a pure $(NH_4)_2SO_4$-sphere. Figure 4 shows clearly that two corrections—one in radius, the other in concentration—have to be applied to the sulfate distribution to obtain a true aerosol distribution. Thus, the curve changes its steepness from $r^{-4.9}$ to $r^{-2.4}$ the latter of which is a rather realistic value for a continental aerosol. This calculation might indicate how realistic the calculated mixture ratios are. It is clear that only the knowledge of both mixing ratios M_n and M_v permits an estimate of the physico-chemical behavior of an aerosol containing a compound like $(NH_4)_2SO_4$ in a certain mixture state.

5. SOURCE SIZE DISTRIBUTION OF SULFUR PARTICLES

The previous discussion including Figs. 3 and 4 showed the important role of $(NH_4)_2SO_4$ particles in

* Usually the ions are determined only.

† Unpublished results by R. J. from observations on the North Atlantic Ocean using data of Winkler (1975).

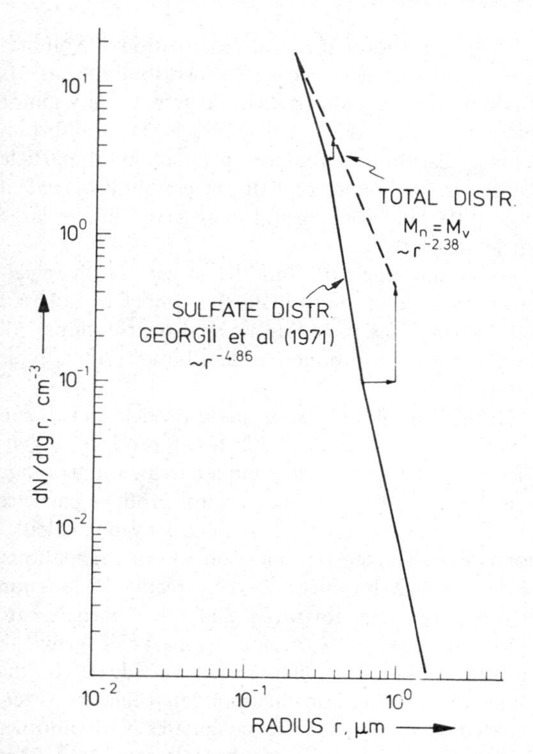

Fig. 4. Using the best fit of Fig. 3 for continental aerosols the 'sulfate'-distribution of Georgii *et al.* (1971) was transformed into the total aerosol size distribution. The arrows indicate the amount of correction for internal mixture (horizontal) and external mixture (vertical).

the aerosol—whether over the oceans or over the continents—for the large aerosol particles with radii larger 0.1 μm. This includes the strong evidence that $(NH_4)_2SO_4$ particles make up the majority of the Aitken particles. Martell (1966) assumed that Aitken particles in the troposphere are sulfate particles. Lodge *et al.* (1966) indicated the possible morphological identification of sulfate particles around 0.1 μm in radius. Mészaros (1968) concluded that many Aitken particles consist of sulfate and ammonium. Winkler (1975) showed that over the ocean practically all Aitken particles contain sulfur, most likely in the form of ammonium sulfate. Tanner *et al.* (1977) explained the mass of Aitken nuclei in rural aerosols on the basis of sulfate ions alone. First size distributions of sulfur compounds in fractions of the Aitken range were published by Mészaros *et al.* (1974).

If we conclude from all this that the Aitken particles consist to a large part of $(NH_4)_2SO_4$, the following discussion can be carried out with the knowledge of the total size distribution of the Aitken particles alone. A number of Aitken particle size distribution measurements were recently carried out in urban, continental, and background aerosols and summarized in Jaenicke (1977).

These size distributions are rather broad covering the whole range of 0.001–0.1 μm in radius. At present it seems impossible to explain how this size distribution is maintained in the atmosphere despite the rather short residence time of the aerosol, which is mainly due to the rapid coagulation of the particles. Nucleation theories of sulfur compounds (Castleman, 1974) usually predict produced particles clustering around a critical radius. Calculations with such a starting distribution and ageing through coagulation (Walter, 1973) tend to produce size distributions of atmospheric Aitken particles with a deep structured shape just opposite to the observations. Nor were Middleton *et al.* (in preparation) able to produce a smoothed Aitken particle distribution by including condensation of vapors on the particles formed by nucleation or coagulation.

Therefore, it seems reasonable at present, to admit that our understanding of the formation of sulfur compound particles fails to explain the observed size distribution. In the following discussion we will present an approach to the problem from a different point of view.

The ageing of an aerosol due to coagulation is usually described by

$$\frac{\partial f(r,t)}{\partial t} = \alpha(r) + \int_0^{r/\sqrt[3]{2}} K(\rho, \sqrt[3]{r^3-\rho^3}) \times \frac{r^2}{(\sqrt[3]{r^3-\rho^3})^2} . f(\rho) . f(\sqrt[3]{r^3-\rho^3}) \mathrm{d}\rho - f(r) \int_0^{\infty} K(r,\rho) f(\rho) \mathrm{d}\rho \quad (10)$$

with $f(r, t)$ = size distribution as function of time t, $\alpha(r)$ = number production rate, K = coagulation constant.

The new approach now assumes a balance between particle production and removal and, thus, no change of the size distribution with time t:

$$\frac{\partial f(r,t)}{\partial t} = 0. \quad (11)$$

This assumption probably is close to reality, if we focus on the background aerosol far away from point sources. The original concept was that the background aerosol was assumed to be produced from a quite constant volume source rather than being exclusively an aged continental aerosol. The assumption in Equation (11) was also made by Whitby (1977). He carried out calculations for the different modes he proposed. With Equations (10) and (11) the source function $\alpha(r)$ can be calculated as a 'source particle size distribution'. In Fig. 5 this source function is calculated for the case of a measured size distribution of the background aerosol over the North Atlantic Ocean. Figure 5a shows the number particle production distribution.

The equilibrium number distribution requires particles to be produced over a wide range of radii from at least 0.001 to 0.02 μm in reasonable numbers. In this logarithmic presentation the number removal can only be indicated and is very small compared to the production rate.

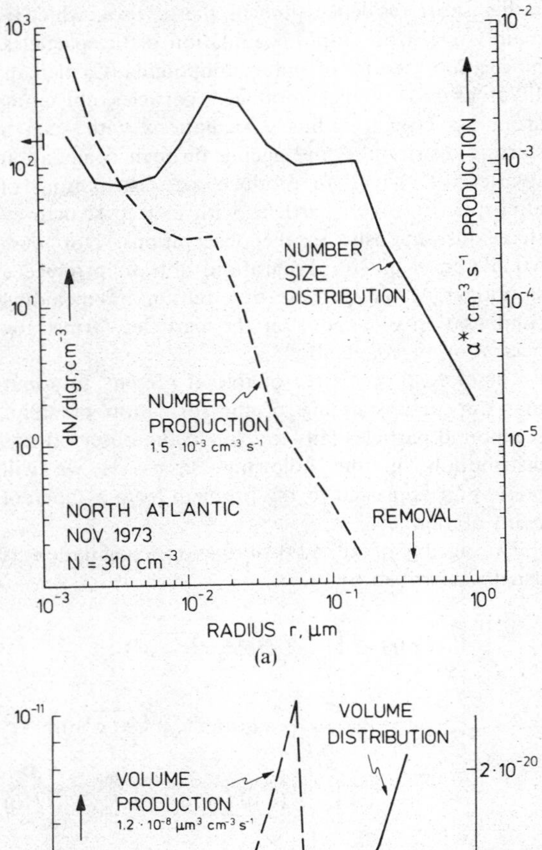

(a)

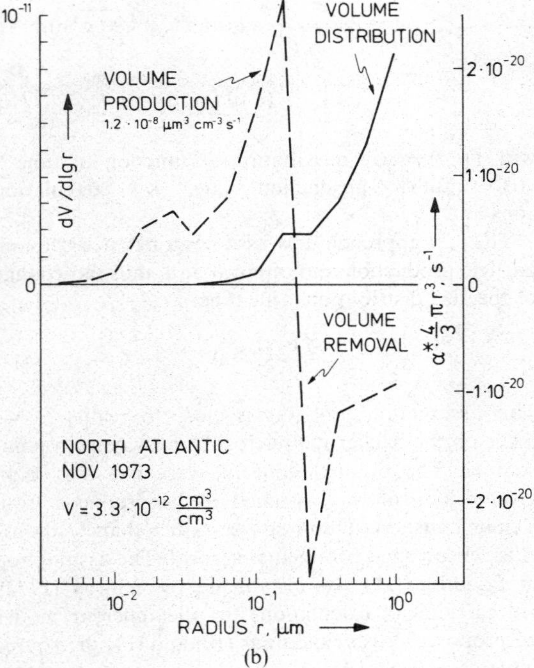

(b)

Fig. 5. A size distribution of the North Atlantic Ocean aerosol of November 1973 was used to calculate the equilibrium production. α^* is the number production rate of Equation (10) modified to the log-axis of the particle radius. (a) Number size distribution and number production. The removal on this log–log graph is indicated only and is minor in number. The left hand arrow indicates the range usually assumed for particle production in theoretical considerations. The total particle concentration is given as N. (b) Volume distribution and volume production on lin–log scale. Removal is expressed as negative production. The total particle volume in the range below 1 μm radius is given as V.

Figure 5b shows the same calculation if volumes are considered. The measured distribution has its maximum in the giant particle range with only minor fractions in the Aitken particle range. The equilibrium volume distribution requires production of particle volume in the range of 0.01–0.1 μm radius, peaked around 0.1 μm. Net removal is observed in the large particle range.

For completeness it should be added that these calculations were carried out for a number of different size distributions including normal distributions. All results indicate a production distribution over a wide radius range.

The nucleation models of gas-to-particle conversion of sulfur compounds available today produce in general sulfur particles clustered in a narrow radius range as indicated in Fig. 5a. They do not produce particles up to 0.02 μm in radius, as required by our calculation results. The term 'production' in our calculations, however, must be understood correctly. Production in this sense does not mean that these particles are produced suddenly. It means the particles 'grow' to this size within their lifetime as individuals. In the coagulation calculation this individual lifetime is terminated after the particle coagulates with another particle. This time span can be calculated and is of the order of one day for the 0.02 μm particles. These particles then have up to one day to achieve the final size and still would be considered as being produced with that size. To my knowledge there is no nucleation theory available which takes this procedure into account to produce production curves as required by the calculations.

In this context, one could think of a statistical approach. The coagulation of particles clustering around a critical radius—as is recommended by the current nucleation theories—is a statistical process: some coagulate immediately, some later and some escape coagulation for a while. This, of course, happens in such a way that the average residence time caused by coagulation is obtained. The few particles escaping the sudden coagulation have then time to 'grow'. Thus, their mobility is reduced and so is the probability of coagulating. It might be possible in this way to obtain the size distribution shown in Fig. 5.

6. CONCLUSION

The discussion shows the necessity of knowing the differential distribution of sulfur compounds for the understanding of physico-chemical processes in the aerosol. The differential distribution is connected with internal and external mixture in the aerosol. It could be shown that number mixture as well as volume mixture must be known to describe the mixture state of the aerosol. Usually for volume mixtures larger than 0.8, as can be observed for sulfur compounds in the large particle range, the characterization of the aerosol as being in internal or external mixture is impossible. Using the few data available, a first crude picture of the differential mixture of $(NH_4)_2SO_4$ could be shown. The data are not accurate enough to calculate the internal mixture. This has to be done in the future. However, all facts known today indicate that the Aitken particles consist mainly of sulfur compounds. For these a number of production mechanisms exist, all of which do not explain the observed size distributions. Assuming an equilibrium in the aerosol, requirements for the production of sulfur particles can be derived resulting in a rather broad production distribution. Since present day nucleation theories do not produce such aerosols, this remains as an aim for future work. Some tentative suggestions as to how this might proceed have been presented.

Acknowledgements—The computer calculations were carried out by G. Keller, E. Kudzus from the Max-Planck-Institute, Mainz. This work was supported by the German Science Foundation through its Sonderforschungsbereich 'Atmospheric Trace Substances'.

REFERENCES

Castleman A. W. (1974) Nucleation processes and aerosol chemistry. *Space Sci. Rev.* **15**, 547–589.

Charlson R. J., Ahlquist N. C. and Horvath H. (1968) On the generality of correlation of atmospheric aerosol mass concentration and light scatter. *Atmospheric Environment* **2**, 455–464.

Charlson R. J., Vanderpol A. H., Covert D. S., Waggoner A. P. and Ahlquist N. C. (1974) $H_2SO_4/(NH_4)_2SO_4$ background* aerosol: optical detection in St. Louis region. *Atmospheric Environment* **8**, 1257–1267.

Cobb W. E. and Wells H. J. (1970) The electrical conductivity of oceanic air and its correlation to global atmospheric pollution. *J. Atmos. Sci.* **27**, 814.

Georgii H. W., Jost D. and Vitze W. (1971) Konzentration und Grössenverteilung des Sulfataerosols in der unteren und mittleren Troposphäre. *Ber. Inst. met. Geoph.* University of Frankfurt **23.**

Gravenhorst G. (1975a) Der Sulfatanteil im atmosphärischen Aerosol über dem Nordatlantik. Ph.D. University of Frankfurt.

Gravenhorst G. (1975b) The sulphate component in aerosol samples over the North Atlantic. *Meteor Forsch. Erg.* **B 10**, 22–31.

Jaenicke R. (1973) Monitoring of Aerosols by measurements of single parameters. Stockholm Tropospheric Aerosol Seminar, University of Stockholm, 29–31.

Jaenicke R. (1977) The size distribution of Aitken nuclei in background aerosols. Paper to be presented at the 9th Int. Conf. Aerosol and Ice Nuclei, Galway, Ireland.

Jaenicke R. and Davies C. N. (1976) The mathematical expression of the size distribution of atmospheric aerosols. *J. Aerosol Sci.* **7**, 255–259.

Jaenicke R. and Schütz L. A comprehensive study of the physical and chemical properties of the surface aerosol in the Cape Verde Islands Region. *J. Geophys. Res.* (in press).

Junge C. (1952) Die Konstitution des atmosphärischen Aerosols. *Ann. Met.* 1952, 1–55.

Junge C. (1954) The chemical composition of atmospheric aerosols—1. Measurements at Round Hill Field Station, June–July 1953. *J. Met.* **11**, 323–333.

Junge C. (1963) *Air Chemistry and Radioactivity*. Academic Press, New York.

Junge C. and Jaenicke R. (1971) New results in background aerosols, studies from the atlantic expedition of the R. V. Meteor, spring 1969, *J. Aerosol Sci.* **2**, 305–314.

Junge C. and McLaren E. (1971) Relationship of cloud nuclei spectra to aerosol size distribution and composition. *J. Atmos. Sci.* **28**, 382–390.

Lodge J. P. and Frank E. R. (1966) Chemical identification of some atmospheric components in the Aitken size range. *J.R.A.* **2**, 139–140.

Martell E. A. (1966) The size distribution and interaction of radioactive and natural aerosols in the stratosphere. *Tellus* **18**, 486–498.

Mészaros E. (1968) On the size distribution of water soluble particles in the Atmosphere. *Tellus* **20**, 443–448.

Mészaros A. and Vissy K. (1974) Concentration, size distribution and chemical nature of atmospheric aerosol particles in remote oceanic areas. *J. Aerosol Sci.* **5**, 101–109.

Middleton P. and Kiang C. S. A kinetic aerosol model for the formation and growth of secondary sulfuric acid particles (in preparation).

Rahn K. and Winchester J. W. (1971) Sources of trace elements in aerosols—an approach to clean air. University of Michigan Techn. Report ORA Project 089030.

Tanner R. L. and Marlow W. H. (1977) Size discrimination and chemical composition of ambient airborne sulfate particles by diffusion sampling. *Atmospheric Environment* **11**, 1143–1150.

Walter H. (1973) Coagulation and size distribution of condensation aerosols. *J. Aerosol Sci.* **4**, 1–15.

Waggoner A. P., Vanderpol A. H., Charlson R. J., Granat L., Trägardh C. and Larsen S. (1975) The sulfate/light scattering ratio: an index of the role of sulfur in tropospheric optics. Dep. Met. Univ. Stockholm Report AC–33.

Whitby K. T., Husar R. B. and Liu B. Y. H. (1972) The aerosol size distribution of Los Angeles smog. *J. Colloid Interface Sci.* **39**, 177.

Whitby K. T. (1977) The physical characteristics of sulfur aerosols. Part. Techn. Lab. Publ. No **335**, 1–51.

Willeke K., Whitby K. T., Clark W. E. and Marple V. A. (1974) Size distributions of Denver aerosols—a comparison of two sites. *Atmospheric Environment* **8**, 609–633.

Winkler P. (1970) Zusammensetzung und Feuchtewachstum von atmosphärischen Aerosolteilchen. Ph.D. University of Mainz.

Winkler P. (1973) The growth of atmospheric aerosol particles as a function of the relative humidity—II. An improved concept of mixed nuclei. *J. Aerosol Sci.* **4**, 373–387.

Winkler P. (1975) Chemical analysis of Aitken particles (<0.2 μm radius) over the Atlantic Ocean. *Geophys. Res. Lett.* **2**, 45–48.

Atmospheric Environment Vol. 12, pp. 171–177. Pergamon Press 1978. Printed in Great Britain.

AEROSOL GROWTH KINETICS DURING SO_2 OXIDATION

D. BOULAUD[1], J. BRICARD[2] and G. MADELAINE[1]

[1]Centre d'Etudes Nucléaires, Fontenay-aux-Roses, France; and [2]Université de Paris, Paris VI, France.

(*First received* 13 *June* 1977 *and in final form* 27 *September* 1977)

Abstract—Consideration is given to the water vapor/sulfuric acid mixture and, after discussing the theoretical results obtained by various authors, they are applied to the case of the atmospheric conditions after having introduced the notion of heteromolecular condensation. With consideration being given to the latter, it is shown that these results lead to orders of magnitude which are comparable to the results of simulation studies currently in progress.

Then the evolution of the aerosol formed in this manner is studied in the case of a continuous addition of sulfuric acid vapor and, at the end of a sufficient period of contact between the particles and the vapor, their subsequent growth through condensation is evaluated.

1. FORMATION OF PARTICLES FROM THE MIXTURE OF TWO VAPORS

The classical theory of nucleation, applies to the case of a pure vapour which can be mixed with a foreign gas which does not react with the vapour and which plays no part in condensation. It has been generalized (Reiss, 1950) in the case of a multicomponent system (see also Zettlemoyer, 1969). The application of such generalizations to the water/sulfuric acid system was carried out by Doyle (1961); Kiang *et al.* (1973); Mirabel and Katz (1974) and Takahashi *et al.* (1975). It is this system with which we will be concerned.

We will not undertake to discuss here the mechanism for the transformation of atmospheric SO_2 into sulfuric acid in moist air; it seems quite well established that it can be attributed to photochemical oxidation of SO_2 by intermediate species originating, in particular, from photolysis of NO_2 under the effect of solar rays, the oxidation rate ranging from 0.1% (only slightly polluted atmosphere) and 10% h^{-1}, depending on the presence of the other impurities (hydrocarbons in particular) (Bricard, 1977).

Let us also point out SO_2 oxidation in moist air by the effect of ionizing radiations (α and β radiolysis) which, it would appear, can be partly attributed to the formation of the intermediate OH radical (Castelman, 1972), but the mechanism of which, it would seem, has not been completely clarified (Bricard *et al.*, 1975).

(a) The variation in free energy ΔG corresponding to the formation of a spherical embryo from a binary mixture of vapors is expressed as follows (Reiss, 1950):

$$\Delta G = n_1(\mu_{11}-\mu_{12}) + n_2(\mu_{21}-\mu_{22}) + A\sigma(n_1, n_2) \quad (1)$$

A, which represents the surface area of the embryo, being expressed as follows:

$$A = (36\,\pi)^{1/3}(v_1 n_1 + v_2 n_2)^{2/3}.$$

In these formulas, v_1 and v_2 represent the volumes occupied by the molecules in the embryo (1) and (2). The n's are the numbers of molecules in the embryo, the μ_1's the chemical potentials in the liquid phase of the same composition and the μ_2's the chemical potentials in the gaseous phase, $\sigma(n_1, n_2)$ which represents the surface tension, is a function of n_2 and n_1. p_1 and p_2 are the real partial pressures of gases 1 and 2. Finally, if $p_{1\infty}$ and $p_{2\infty}$ represent the saturated vapor pressure corresponding to liquids 1 and 2, formula (1) will be written:

$$\Delta G = -n_1 kT \log_e \frac{p_1}{p_{1\infty}} + n_1 \Delta F_1 - n_2 kT \log_e \frac{p_2}{p_{2\infty}} + n_2 \Delta F_2 + A\sigma \quad (2)$$

in which ΔF_1 and ΔF_2 are the partial free molecular energies of the mixture.

In the case under consideration here, let us represent component 1 by the water and 2 by sulfuric acid, or nitric acid, and let S designate the relative humidity, i.e. $S = p_1/p_{1\infty}$ and a the activity, i.e. $a = p_2/p_{2\infty}$.

The formula will finally be written as follows:

$$\Delta G = n_1 kT \log_e S + n_1 \Delta F_1 - n_2 kT \log_e a + n_2 \Delta F_2 + A\sigma. \quad (3)$$

ΔG will be represented in a 3-dimensional space, i.e. $\Delta G(n_1, n_2)$. We find that the corresponding surface has a saddle point directed toward the increasing ΔG values, of which the summit ΔG^* correspond to nucleation (Fig. 1). We will assume that σ is constant when located near the nucleation point.

To determine ΔG^* which is the free enthalpy corresponding to the formation of the critical embryo, we write that $(\delta\Delta G/\delta n_1)_{n_2} = 0$, $(\delta\Delta G/\delta n_2)_{n_1} = 0$ which makes it possible to know the critical numbers of molecules n_1^* and n_2^* as well as radius R^* of the critical embryo.

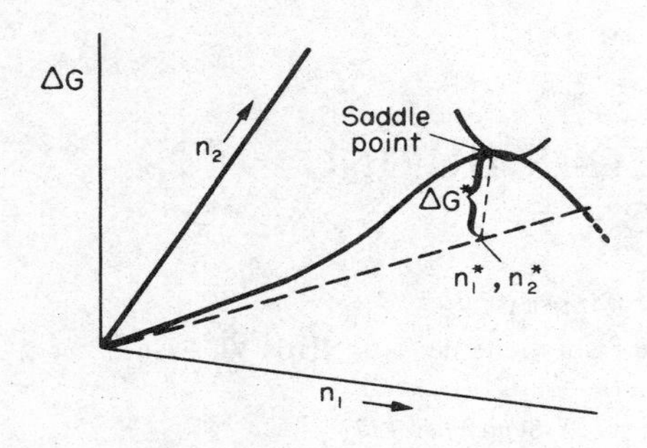

Fig. 1. Free energy of formation for a binary embryo vs n_1 and n_2.

Thus, the critical features are determined for a given activity and a given relative humidity. Now then, even if we know the vapor pressure of the sulfuric acid in the medium, the activity cannot be precisely determined owing to the fact that, at a given temperature, the saturation vapor pressure $p_{2\infty}$ of sulfuric acid is not well known for the time being. Depending on the authors, it would appear to range between 3.5 10^{-4} torr (Mirabel and Katz, 1974) and 10^{-6} torr (Takahashi *et al.*, 1975). On the other hand, the values of ΔF_1 and ΔF_2 are well known (Giauque *et al.*, 1960) as are the density of the water and sulfuric acid mixture (Handbook of Chemistry, 1972) and its surface tension as a function on the composition.

It can be assumed that the atmospheric concentration of H_2SO_4 vapor is of the order of $4 \cdot 10^{-5}$ to $1.3 \cdot 10^{-2}$ ppm. The calculation (Mirabel and Katz, 1974) shows that, by assuming a saturation vapor pressure of $p_{2\infty} = 3.5 \cdot 10^{-4}$ torr, the critical radius depends little on the humidity and is about 7 Å, which would correspond to an acid content of about 10 molecules. A value of the same order of magnitude is found when $p_{2\infty}$ is taken to equal 10^{-6} torr (Takahashi *et al.*, 1975).

Experimental determination of the value of the critical radius of the atmospheric aerosol is a delicate operation because particles of such small dimensions cannot be detected on an individual basis and even escape detection by condensation nucleus counters. Maigné (1977) gave an order of magnitude ranging from 5 to 8 Å by extrapolation toward the small radii of size distribution measured with a diffusion battery of the aerosol produced through radiolysis of the gaseous impurities of the air.

(b) The nucleation rate in the case of a binary mixture can be written as follows:

$$I = C \exp(-(\Delta G)^*/(kT)) \quad (4)$$

if ΔG^* represents the free energy of formation of the critical embryo of radius R^* is expressed by:

$$\Delta G^* = 4\pi\sigma R^{*2}/3 \quad (5)$$

the value of R^* corresponding to critical numbers n_1^* and n_2^*. The factor of proportionality C brings into play the total concentration $N_1 + N_2$ of the molecules of vapor present in the surrounding medium and the rate at which the new molecules are incorporated in the critical embryo. Kiang *et al.* (1973) have simplified these formulas. If the concentration N_2 of one of the components (component 2) is low compared with the concentration N_1 of component (1) over a very long period, Kiang *et al.* consider that the rate of incorporation of the new molecules in the critical embryo is directly proportional to the product of the surface of the latter, i.e. $4\pi R^{*2}$, multiplied by the flux of active gas molecules, i.e. $N_2 kT/(2\pi m_2 kT)^{1/2}$, in which m_2 represents the mass of the corresponding molecules. We will therefore write the following rough formula:

$$I = 4\pi R^{*2} \frac{N_2 kT}{(2\pi m_2 kT)^{1/2}} N_1 \exp\left(-\frac{\Delta G^*}{kT}\right). \quad (6)$$

According to this formula, the nucleation rate depends on both the concentration of sulfuric acid and water molecules in the gaseous mixture and the activity and r.h. through R^* and ΔG^*.

(c) The most recent complete numerical calculations have been made by Mirabel and Katz (1974), directly from formulae (4) and (5), assuming that $p_2 = 3.5 \cdot 10^{-4}$ torr. Figures 2 and 3 show the results of these calculations in the case of H_2SO_4 and HNO_3 in moist air. I (rate of nucleation) has been plotted on the y-axis vs the r.h. for various values of the activity. According to Fig. 4 (Kiang *et al.*, 1973) with a given humidity, the acid concentration required to form the same quantity from HNO_3 vapours is at least 6–8 orders of magnitude greater than that corresponding to H_2SO_4. Such concentrations are not found in nature and it would therefore seem probable that HNO_3 does not intervene directly in the atmosphere (see the last paragraph).

Figure 5 shows the experimental results (marked points) obtained by Boulaud (1977) (we will discuss

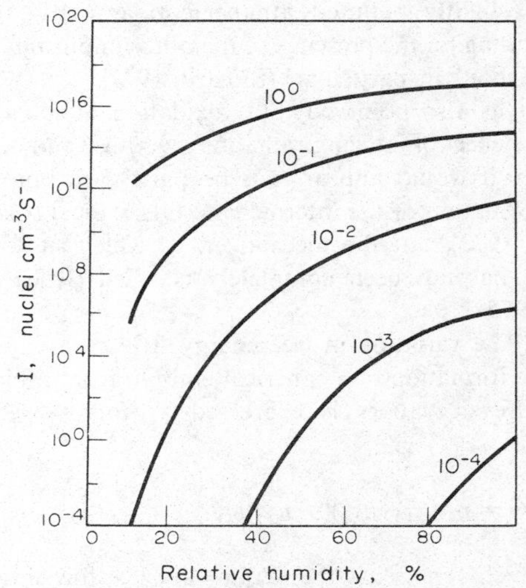

Fig. 2. Nucleation rate I calculated based on the relative humidity for $H_2SO_4 + H_2O$ at 25°C. The parameter plotted was the activity of the acid in the vapor phase.

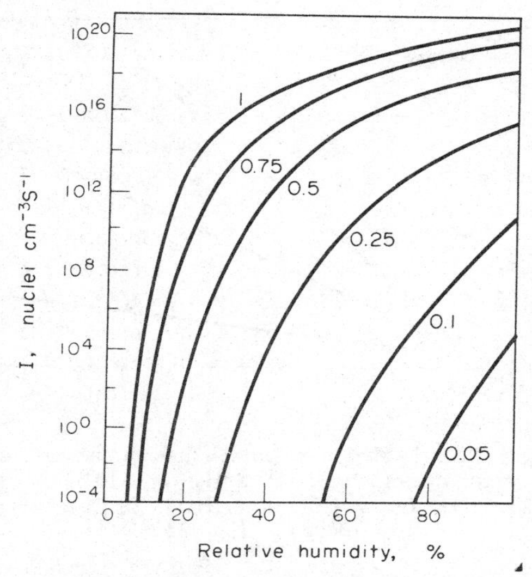

Fig. 3. Nucleation rate I calculated based on the r.h. for $HNO_3 + H_2O$ at 25°C. The parameter plotted was the activity of the acid in the vapor phase.

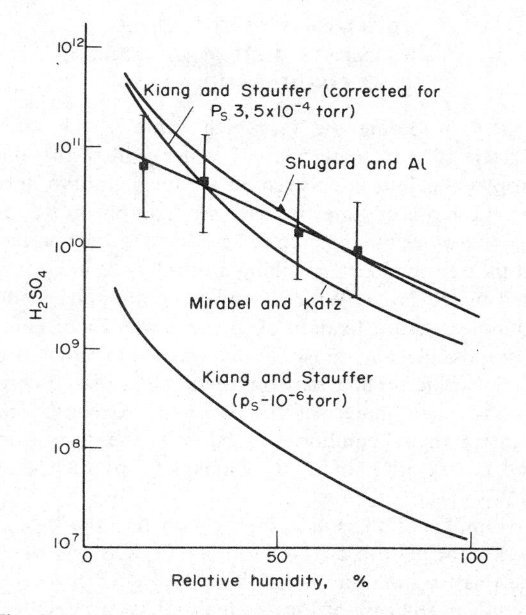

Fig. 5. H_2SO_4 vapor concentration based on the r.h. in % corresponding to $I = 1$ particle $cm^{-3} s^{-1}$ according to the calculation made by Mirabel and Katz and that made by Kiang and Stauffer corrected for $P_{2\infty} = 3.5 \cdot 10^{-4}$ torr. The points plotted represent the experimental points and Shugard and Reiss' calculation.

the experiments used a little later). This involves orders of magnitude obtained from measurements of nucleation rates for the sulfuric acid and water vapour mixture in simulation enclosure. The sulfuric acid concentrations have been plotted on the y-axis and the r.h. on the x-axis for a rate of nucleation $I = 1$ particle $cm^{-3} s^{-1}$. Full lines represents the same results as calculated by Mirabel and Katz on one hand, and on the other hand, with formula (6), $p_{2\infty}$ being equal to $3.5 \cdot 10^{-4}$ torr. Experimental and theoretical results are in good agreement. The results of formula (6) with $p_{2\infty} = 10^{-6}$ torr are also shown in Fig. 5.

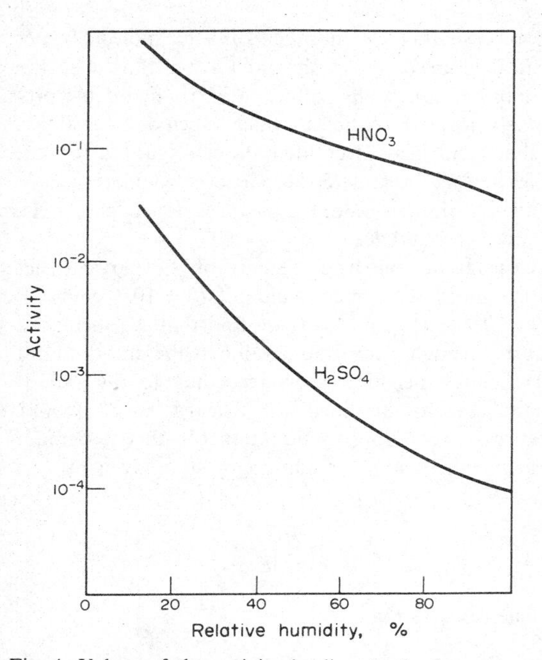

Fig. 4. Values of the activity leading to the formation of 1 particle $cm^{-3} s^{-1}$, based on the r.h. for H_2SO_4 and HNO_3.

(d) Actually, these calculations bring the H_2SO_4 vapor and H_2O mixture into play, but they do not take into account the formation of H_2SO_4 hydrates. This was recently introduced into calculations of the nucleation rate by Shugard *et al.* (1974).

The results indicate two opposing effects which are as follows:

(i) The presence of hydrates has the effect of increasing the nucleation rate since the number of interactions between H_2SO_4 and H_2O required to produce a critical embryo is not as great. When each growing embryo fixes a hydrate, it only fixes one acid molecule but, at the same time, it fixes a certain number of water molecules. (ii) A second effect decreases the nucleation rate since the presence of hydrates has a stabilizing influence on the vapor. If the partial H_2SO_4 pressure is increased, most of the additional acid is converted into hydrates which has the effect of limiting the increase in the vapour pressure of H_2SO_4 molecules. This second, much more important, effect has the result of suppressing the preceding one.

Shugard *et al.* (1974) made the calculation for relative humidities of 50, 200 and 300%. Figure 5 shows the results corresponding to the figure of 50%. It can be seen that the corresponding point is closer to the experimental points than the results of Mirabel and Katz (1974).

3. DISCUSSION OF THE ABOVE MEASUREMENTS—HETEROMOLECULAR CONDENSATION

(a) Considering the very high value of the ratio N_1/N_2, 10^5–10^9 molecules of water will strike the droplet which just appeared on a critical embryo during the lapse of time between the fixation of 2 consecutive molecules of sulfuric acid. Let us assume that all the acid molecules striking a droplet are incorporated in the latter. With an embryo containing some 50 molecules, the fixation of so many water molecules is impossible and most of the water molecules are re-evaporated. Thus, between the fixation of two consecutive acid molecules (i.e. when n_2 remains constant), a virtual equilibrium relating to the dimension and composition of a droplet is set up; such that $(\delta\Delta G/\delta n_1) = 0$.

Hamill (1975) has used this approach in the special case of the growth of H_2O and H_2SO_4 aerosols when the relative humidity is less than 100%, which corresponds to the case of Fig. 5. He obtained curves indicating the radius of the particles vs the growth times for various relative humidities (22, 55, and 88%) and a concentration of 10^{12} molecules of acid cm^{-3}.

The experimental points of Fig. 5 have been established by counting the number of particles formed at the end of 150 s in a closed vessel in the presence of given acid and water vapour concentrations. The accuracy of the experimental results partly depends on the uncertainty existing as to the counts because of the small radii of the corresponding concentration nuclei (radius 100 Å). Particles smaller than this size cannot be counted, and the measured concentration is too low. To estimate this error, it is necessary to know the radius of the particles at the time when this measurement is made.

Since nucleation takes place continuously throughout the duration of the experiments (the results of which are shown in Fig. 5), the fact must be taken into consideration that a certain number of particles which were just formed at the end of the experiment, have not had enough time to grow. Taking this fact into account, and using Fig. 6, Boulaud (1977) demonstrated that the top limits of the counting losses in these experiments range from 6 to 35%. This really is an upper top limit since, in condensation nuclei counters, for particles from which the radius is less than about 100 Å, the loss of counting represents only a fraction of the total concentration of particles, and not all the particles.

4. COAGULATION AND CONDENSATION

Let N_p stand for the concentration of the particles already formed at a given time, including the embryos, N_2 that of the acid molecules in suspension and let us assume that the experiment is conducted in a closed vessel which makes it possible to control the experimental conditions.

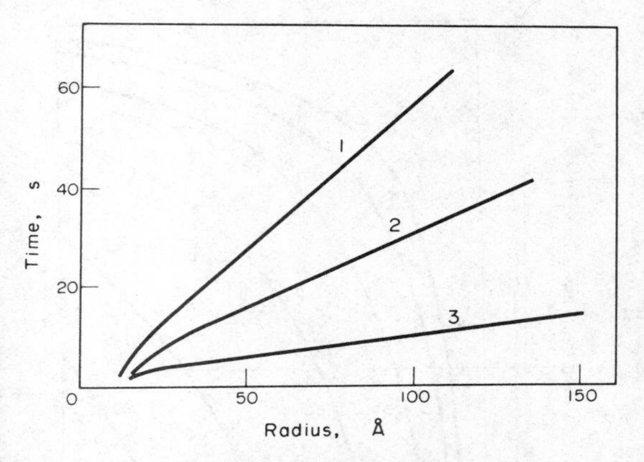

Fig. 6. Particle growth based on time in the case of a nucleation rate of 1 cm^{-3} s^{-1}: 1. H_2SO_4 $8\cdot10^9$ cm^{-3} H_2O 88%; 2. H_2SO_4 $2\cdot10^{10}$ cm^{-3} H_2O 55%; 3. H_2SO_4 $8\cdot10^{11}$ cm^{-3} H_2O 22%.

By assuming that the aerosol is monodisperse, and neglecting the wall effect, we can write the following equations (Takahashi, 1975):

$$\frac{dN_2}{dt} = A - (\alpha + \gamma N_p)N_2$$

$$\frac{dN_p}{dt} = \frac{\alpha}{n_2^*}N_2 - KN_p^2 \quad (7)$$

in which A represents the H_2SO_4 production, α is the rate of conversion of sulfuric acid on an embryo (% sulfuric acid fixed s^{-1}) defined by:

$$\alpha = \frac{In_2^*}{N_2} \quad (8)$$

n_2^* representing the number of acid molecules per critical embryo and I the nucleation rate, γ is the rate of fixation of the sulfuric acid vapor on the previously formed particles, which is easy to evaluate if the latent heat of condensation is ignored (Fuchs, 1959 and 1970), and K the particle coagulation constant. The dimension of the particles depends on time, as well as α and K.

Considering the heterogeneity of the particles, let us assume that the mean value of $K = 10^{-8}$ cm^3 s^{-1}. When $I = 10^{+6}$ cm^{-3} s^{-1}, we find that N_p increases, passes through a maximum value at the end of about 10 min and then decreases, according to the experimental results obtained by Bricard *et al.* (1968), Figure 7. According to Equation (7), this maximum value corresponds to a stationary state given by

$$\frac{dN_p}{dt} = \frac{dN_2}{dt} = 0.$$

The result is that:

$$(N_p)_{max} = \left(\frac{\alpha}{n_2^* K}\right)^{1/2}\left(\frac{A}{\alpha + \gamma N_p}\right),$$

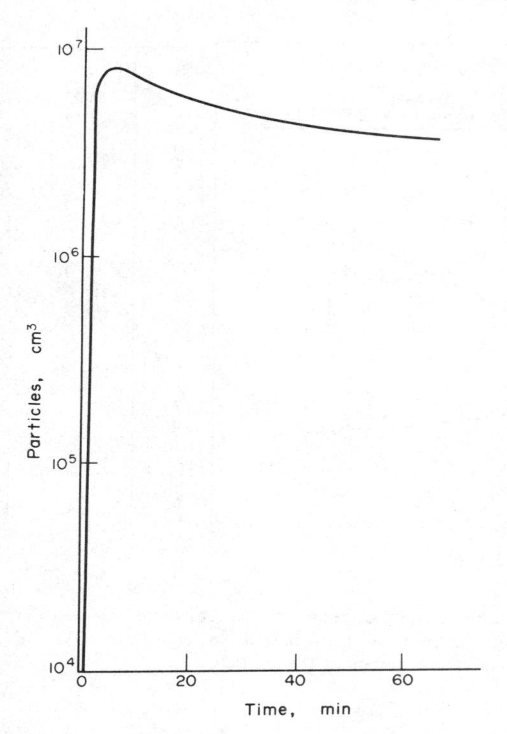

Fig. 7. Variation of the number of particles per cm^3 as a function of time.

which takes the form, if condensation is negligible, $(\alpha \gg \gamma N_p)$:

$$(N_p)_{max} = \left(\frac{A}{n_2^* K}\right)^{1/2} = \left(\frac{I}{K}\right)^{1/2}. \quad (9)$$

The calculations of I (rate of nucleation) made by Takahashi are expressed as a function of the activity and the relative humidity, without bringing the value of the saturated sulfuric acid vapor into play. Figure 8 shows the experimental results of Boulaud *et al.* (1975). In these experiments, the production of H_2SO_4 vapour in a simulation chamber was provided by photolysis of the mixing SO_2, NO_2, H_2O (see discussion in the last paragraph). The maximum number of particles formed per cm^3 is plotted on the y-axis vs the relative humidity. The points marked represent Takahashi's theoretical results for a given value of A in our experimental conditions. We can deduce therefrom, as long as Equation (9) remains valid, i.e. during the first formation phase, that there is negligible condensation of sulfuric acid and that the particles grow through coagulation. This is no longer true when the particles are sufficiently grown. They can then play an important part as centers of condensation.

5. SUBSEQUENT COAGULATION AND CONDENSATION

When a sufficient amount of time has elapsed between the beginning of nucleation in a medium which was initially free of particles, the surface of the

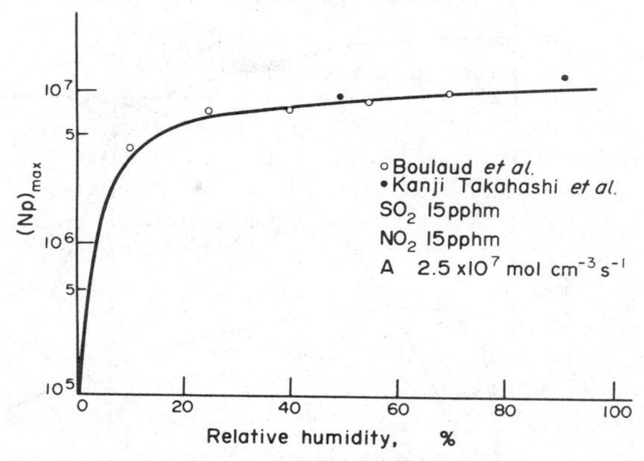

Fig. 8. Variations of the maximum number of particles per cm^3 measured by Boulaud *et al.* The points plotted correspond to the calculation made by Takahashi *et al.*

particles which grow through coagulation and condensation can reach a value such that the nucleus-forming vapour condenses on the previously formed particles, no longer being available for the formation of new embryos.

A series of experiments basically aimed at obtaining qualitative results were carried out in a simulation enclosure using irradiation by photolysis of the SO_2–NO_2–H_2O system as a source of particles. The nucleus-forming vapor concentrations were sufficiently high (about 10 times greater than in the case of Fig. 8) for the particle radius to quickly reach a value of 10^{-6} cm with respect to the duration of the experiments, so that the presence of the particles of lesser dimensions could be reduced to a minimum.

Figure 9 shows the development of the total concentration of the particles as a function of time. This figure also shows the surface area and the total volume of the particles of a diameter in excess of $2R = 5 \cdot 10^{-3}$ and 10^{-2} μm determined with a Whitby analysor. The curves corresponding to the particles of lesser radius are plotted with dash lines. Considering the uncertainties existing as to the measurements which can be taken in this dimension range, these latter indications must be considered as qualitative.

The variations in the total concentration N have been determined with a C.N.C. At the beginning of nucleation, the corresponding N_{max} of Fig. 7 is not seen because the C.N.C. is saturated, due to the fact that the production of particles was voluntarily increased to make them grow rapidly, with the result that they cross the detection threshold in a very short time. On the contrary, part of the curve represents the development of the concentration from the effects of condensation and coagulation.

In the period of time represented by region D, nucleation ends because vapor condensation on the previously formed particles becomes preponderant. Since no more embryos are produced, the development of the concentration is then solely governed by the coagulation of the small embryos still present with the larger particles. The coagulation constant correspond-

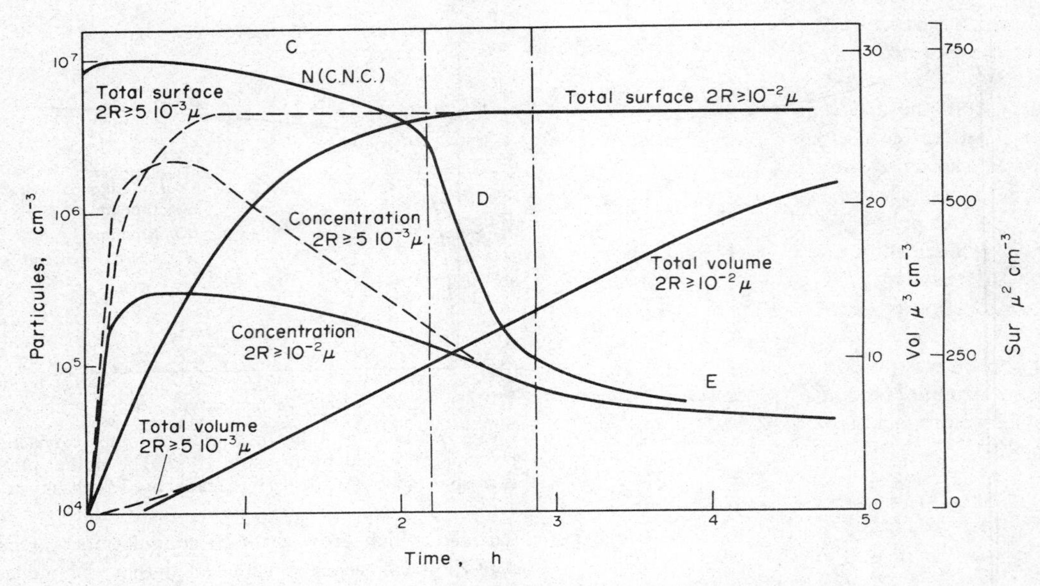

Fig. 9. Variations in time of various parameters characterizing the aerosol: concentration of the condensation nuclei measured with a C.N.C., concentration of the particles of which diameter is greater than $5 \cdot 10^{-3}$ and 10^{-2} μm, total surface areas and volumes of the particles.

ing to region D is found to be on the order of $K = 4.4 \cdot 10^{-9}$ cm^{+3} s^{-1}, which, according to Fuchs (1964) corresponds to the fixation of embryos on the order of $2 \cdot 10^{-3}$ μm on 10^{-2} μm. In their calculations, Walter (1973) and Takahashi (1975) adopted a K value of 10^{-8} cm^3 s^{-1} which is slightly higher than our experimental value.

The period of time corresponding to E represents the development of the aerosol (still under the effect of coagulation), the coagulation constant then being lower ($K = 2.5 \cdot 10^{-9}$ cm^{+3} s^{-1}), the aerosol being less polydispersed after fixation of the embryos on the larger particles.

Comparison of the curves representing the development of the concentration of particles of diameter greater than $5 \cdot 10^{-3}$ and 10^{-2} μm respectively confirms the vapour condensation hypothesis which, by stopping nucleation, indicates the coagulation of the embryos still present with the larger particles. Indeed, the curves, which are quite distinct in part C approach one another in part D and merge into one another in part E.

Study of the curves characterizing the development of the total surface and volume of the aerosol (both being calculated for $2R > 5 \cdot 10^{-3}$ μm) indicates that the surface area is nearly constant in part E when the volume, with respect to the unit of gas volume, regularly increases at a speed of increase of $dv/dt = 1.25 \cdot 10^{-12}$ s^{-1}. Even though no further embryos are produced and, by definition, coagulation conserves the total volume and reduces the total surface area of the aerosol, it is found that this surface area remains constant and that the volume increases, owing to condensation of the vapor on those particles still present.

These results are qualitatively comparable to those obtained by Husar *et al.* (1973) during similar experiments. The greater sensitivity of the electric analyzer they used enabled them to work with a lower particle concentration, permitting the growth period of the particles to be longer.

It might seem paradoxical that heteromolecular nucleation is not stopped by condensation as soon as the total surface area of the aerosol has reached its maximum value. Actually, the term corresponding to condensation does not only depend on the surface area but also on the Knudsen number characterizing each particle and on the difference between the concentration of the acid at an infinite distance and in the immediate vicinity of each droplet, both of these parameters depending on the radius of the droplet.

DISCUSSION

It should be pointed out that these results cannot be directly generalized to the conditions of the atmosphere and only represent a first estimate, for consideration must be given to the presence of the other gaseous impurities in the air (ammonia in particular). It is well known that a large portion of the atmospheric aerosol consists of ammonium sulfate (Megaw, 1966; Meszaros, 1972). It is obvious, therefore, that the above considerations are only valid if the sulfuric acid vapours which appear initially are converted by binary nucleation to give particles which then react with the other impurities in the air (NH_3).

Moreover, let us point out that, in a number of cases, it has been found that the atmospheric particles contain acid nitrosyl sulfate SO_4HNO (Bourbigot *et al.*, 1974) which is absent when the experiment is conducted in the presence of excess ammonia.

The formation of particles containing SO_4HNO may result from the action of NO_2 on SO_2 in the presence of water, followed by SO_4HNO nucleation, which could lead us to show up the weakness of the nucleation process discussed above, would no longer be applicable. At present, the formation of the particles through nucleation of the $H_2SO_4 + H_2O$ mixture is a first estimate and consideration must be given to nucleation of H_2SO_4, H_2O, HNO_3 ternary mixture.

This problem was raised by Kiang *et al.* (1975) and in the case of multicomponent mixture, but, for the time being, we are lacking data on this topic.

Acknowledgement—This work was partially supported by the French Ministry of the Life Quality.

Note added in press—After completing this work, the work by Marvin and Shugard (submitted) has appeared. They made the calculations of the influence of hydrates on the system $H_2SO_4H_2O$ for relative humidities of 20%, 35%, 50%, 65% and 80%. These results, evaluated using Shugard's method which takes account of hydrates in the system, shows the same slope as our experimental results. This result was anticipated in a recent article (Boulaud *et al.*, 1977).

REFERENCES

Boulaud D., Bricard J. and Madelaine G. (1975) Note concernant l'article de Takahashi K., Kasahara M. and Itoh M. I. A kinetic model of sulfuric acid and aerosol formation from oxidation of Sulfur dioxide vapour. *J. Aerosol. Sci.* **6,** 483.

Boulaud D., Madelaine G., Vigla D. and Bricard J. (1975) SO_2 transformation in controlled atmosphere leading to the production of aerosol Particles. *J. Air, Wat. and Soil Pollut.* **4,** 435.

Boulaud D. (1977) Contribution à l'étude des transformations en phase gazeuse de l'anhydride sulfureux. Thesis Paris, 1977.

Boulaud D., Madelaine G., Vigla D. and Bricard J. (1977) Experimental study on the nucleation of water vapor sulfuric acid binary system *J. Chem. Phys.* **66,** 4854–4860.

Bourbigot Y., Bricard J., Madelaine G. and Vigla D. (1973) Identification des Aerosols produits par photolyse en présence d'anhydride sulfureux. *C. R. Acad. Sci., Paris* **276,** 547.

Bricard J. (1977) Aerosol production in the atmosphere (Edited by J. O'M. Bockris). In *Environmental Chemistry*, pp. 313–330. Plenum Press, New York.

Bricard J., Billard F. and Madelaine G. (1968) Formation and evolution of condensation nuclei that appear in air initially free of aerosols. *J. geophys. Res.* **73,** 4487.

Bricard J., Cabane M., Madelaine G. and Vigla D. (1972) Formation and properties of neutral particles and small ions conditioned by gaseous impurities of the air. *J. Colloïd Interface Sci.* **39,** 42.

Bricard J., Cabane M., Madelaine G. et Ollion P. (1975) Rôle des ions dans la production des noyaux de condensation à partir du rayonnement β. *C. R. Acad. Sci. Paris* **B 280,** 761.

Castelman A. W. and Tang I. N. (1972) Rôle of small clusters in nucleation about ions. *J. chem. Phys.* **57,** 629–000.

Doyle G. J. (1961) Self-nucleation in the sulfuric acid water system. *J. chem. Phys.* **35,** 795.

Fuchs N. A. (1959) *Evaporation and Droplet Growth in Gaseous Media*, Pergamon Press, Oxford.

Fuchs N. A. (1970) Basic properties of highly dispersed aerosols. *Topics in Current Aerosol Research* Vol. 2, Pergamon Press, Oxford.

Giauque W. F., Hormung E. W., Kinzler J. E. and Robin T. R. (1960) The thermodynamics properties of aqueous sulfuric acid solutions and hydrates from 150 to 300 K. *J. Am. chem. Soc.* **82,** 62.

Hamill P. (1975) The time dependent growth of H_2O–SO_4H_2 aerosols by heteromolecular condensation. *J. Aerosol. Sci.* **6,** 475.

Handbook of Chemistry and Physics, 1972. C.R.C. Press., Cleveland.

Heist R. H. and Reiss H. (1974) Hydrates in supersaturated binary sulfuric acid water vapour. *J. chem. Phys.* **61,** 573.

Husar R. B. and Whitby K. T. (1973) Growth measurements and size spectra of photochemical aerosols. *Environ. Sci. Technol.* **7,** 241.

Kiang C. S., Stauffer D., Mohnen V. A., Bricard J. and Vigla D. (1973) Heteromolecular nucleation theory applied to gas to particle conversion *Atmospheric Environment* **7,** 1279.

Kiang C. S., Cadle R. A., Mohnen V. A. and Yu G. K. (1975) Ternary nucleation applied to gas particle conversion. *J. Aerosol Sci.* **6,** 475.

Maigné J. P. (1977) Détermination de la granulométrie d'un aérosol au moyen d'une batterie de diffusion. Thesis, Paris.

Marvin D. C. and Shugard W. J. (1977) Nucleation in H_2SO_4–H_2O mixtures. *J. Chem. Phys.* (submitted).

Megaw G. W. (1966) Chemical constitution of atmospheric aerosol. Research Progress Report Health Physics and Medical Division IKEA.

Meszaros E. (1973) Evidence of the role of indirect photochemical process in the formation of atmospheric sulfate particulate. *J. Aerosol. Sci.* **4,** 429.

Mirabel P. and Katz J. L. (1974) Binary homogeneous nucleation as a mechanism for the formation of aerosols. *J. chem. Phys.* **60,** 1138.

Reiss J. J. (1950) The kinetics of phase transition in binary systems. *J. chem. Phys.* **18,** 840.

Shugard W. J., Heist R. H. and Reiss J. J. (1974) Theory of water phase nucleation in binary mixtures of water and sulfuric acid. *J. chem. Phys.* **61,** 5298.

Takahashi K., Kasahara J. and Itoh M. I. (1975) A kinetic model of sulfuric acid aerosol formation from photochemical oxidation of sulfur dioxide vapour. *J. Aerosol Sci.* **6,** 45.

Walter H. (1973) Coagulation and size distribution of condensation aerosols. *J. Aerosol Sci.* **4,** 1.

Zettlemoyer A. C. (1969) *Nucleation.* Decker, New York.

Atmospheric Environment Vol. 12, pp. 179–185. Pergamon Press 1978. Printed in Great Britain.

EXPERIMENTAL AND THEORETICAL EXAMINATION OF THE FORMATION OF SULFURIC ACID PARTICLES

PAULETTE MIDDLETON*

Atmospheric Sciences Research Center, State University of New York at Albany, Albany, New York

and

C. S. KIANG

National Center for Atmospheric Research,† Boulder, Colorado 80307, U.S.A.

(*First received* 14 *June* 1977 *and in final form* 20 *September* 1977)

Abstract—Previous experimental and theoretical studies of sulfuric acid aerosol formation for the SO_2–air system in smog chambers are discussed. Shortcomings in both theory and experiments are outlined. The effects of these uncertainties on calculated and measured nucleation rates are illustrated with numerical kinetic aerosol model calculations. From this examination recommendations for future experimental and theoretical nucleation studies are made.

INTRODUCTION

Field observations of significant amounts of sulfates, nitrates and organics in urban aerosols (Gartrell and Friedlander, 1975; Charlson *et al.*, 1974; Whitby *et al.*, 1978) and rough estimates of global aerosol budgets (Hidy and Brock, 1970) strongly suggest that gas to particle conversion processes are an important source of submicron particles in the atmosphere. The relevance of this mechanism to the production of the potentially harmful sulfuric acid particles has received special attention as the necessity for increased use of sulfur-containing fuels has become apparent. The actual rates of formation or the details of precursor gas phase reactions leading to the formation of this aerosol are, as yet, difficult to assess.

Many experimental and theoretical efforts have been made to understand the details of this formation mechansim. Smog chamber experiments have been used to simulate sulfuric acid formation from the sulfur dioxide–air system by many workers (e.g. Quon *et al.*, 1971; Cox, 1973; Friend *et al.*, 1973; Birenzvige and Mohnen, 1975; Bouland *et al.*, 1975). Heteromolecular nucleation theory has been applied to studies of sulfuric acid aerosol formation from sulfuric acid vapor and water vapor by several groups (e.g. Doyle, 1961; Kiang and Stauffer, 1973; Mirabel and Katz, 1974). Several attempts have been made to compare the measured and calculated formation rate of sulfuric acid droplets. Even when apparently similar 'known' atmospheric conditions are chosen, the disagreement between the experimental and theoretical results can be orders of magnitude (Stauffer *et al.*, 1973; Leifer *et al.*, 1974; Birenzvige and Mohnen, 1975; Bricard *et al.*, 1977). These discrepancies are usually attributed to uncertainties in the nucleation theory itself. However, it should be stressed that uncertainties in the assumed gas phase and aerosol kinetics occurring in the reaction vessel which could also introduce significant variations in the results were not taken into account in these comparisons.

In this paper we do not attempt to resolve these discrepancies. Rather we outline the problems and utilize kinetic aerosol model calculations to illustrate the effect of these uncertainties on nucleation rate estimates. From this examination future recommendations for both experimental and theoretical nucleation studies are made.

PREVIOUS STUDIES

In smog chamber experiments, in general, specified quantities of SO_2 and filtered humidified air (or a mixture of N_2 and O_2) are introduced into a reaction vessel which is being irradiated with light of specified wavelengths and intensities for a known period of time. Total particle number concentrations are then measured. For flow systems the measurement is taken after the species have passed through the flow tube. In this case the total number concentration measured is assumed to be the maximum number of particles formed in the chamber. The nucleation rate is then taken to be the total number concentration divided by the residence time in the flow tube which is a function of flow rate and chamber size. In diffusion systems, total number concentration measurements can be taken as a function of time. In this case the nucleation rate is defined as the positive slope of the

* Currently a visitor at the National Center for Atmospheric Research.

† The National Center for Atmospheric Research is sponsored by the National Science Foundation.

Table 1. Experimental conditions

Investigators	r.h. %	λ Å	SO_2 ppm	Other gases 1 Atm.	Temperature °C	Experimental system	Irradiation time s
Quon *et al.* (1971)	13–77	3000–4000	0.2–0.65	Filtered air	23	Diffusion system	0–240
Cox (1973)	1–80	2900–4000	5–500	N_2:O_2 (4 to 1 ratio)	35	Flow system residence time ≈ 150 s	Constant
Friend *et al.* (1973)	<20	2200–5000	0.1–1.0	Filtered air	−55, ~25	Flow system residence time ≈ 2100 s	Constant
Bouland *et al.* (1975)	0–90	2900–4000	0.2–0.3	Filtered air	~25	Diffusion system	Constant
Birenzvige and Mohnen (1975)	0–18	>2000	0.00348–0.054	Filtered air	~25	Flow system	Constant

total number versus time curve. Gas and aerosol kinetics occurring in the chamber are not precisely known and can only be inferred from the experimental conditions and measurements which are themselves subject to problems of impurities and limitations of instrumentation. A list of experimental conditions used in new particle formation studies is given in Table 1.

The rate of nucleation is usually given by a quasi-equilibrium expression of the form $J = C \exp(-\Delta G^*/kT)$ where C is the kinetic prefactor and ΔG^* is the free energy of formation of the stable cluster. ΔG^* is a function of solution droplet equilibrium vapor pressure and surface tension. Reiss (1950) first derived in detail a kinetic prefactor which takes into account both nucleating species for the case of binary heteromolecular homogeneous nucleation and Doyle (1961) used the Reiss (1950) expression to study sulfuric acid and water nucleation at one r.h. Nucleation rates for sulfuric acid and water mixtures were studied for various sulfuric acid activities and relative humidities by Kiang and Stauffer (1973) using a simplified kinetic prefactor and by Mirabel and Katz (1974) using Reiss' (1950) factor. Recent studies comparing the calculations of Kiang and Stauffer (1973) and Mirabel and Katz (1974) show that the previous discrepancy in the results is due primarily to the different choice of equilibrium vapor pressure for sulfuric acid (Bricard *et al.*, 1977). More recently the effects of hydration—the clustering of water molecules about a sulfuric acid molecule before actual nucleation—have been considered by Shugard *et al.* (1974). At relative humidities below 100% this effect is not significant. A summary of these theoretical studies including the choice of microphysical parameters is given in Table 2.

Although there are uncertainties in both theoretical and experimental approaches, several comparison studies have been reported. Birenzvige and Mohnen (1975) compare their calculated nucleation rates as a function of r.h. with their measured total particle number as a function of time and find that the theoretical prediction is 10^6–10^8 too low.

Bricard *et al.* (1977) compare the amount of sulfuric acid required to give a nucleation rate of 1 particle $cm^{-3}\ s^{-1}$ at various relative humidities for theoretical

Table 2. Summary of theoretical studies

Investigators	r.h. %	H_2SO_4 activity	Equilibrium vapor pressure Torr	Surface tension source	Equation reference for source
Doyle (1961)	50	10^{-5}–10^{2}	10^{-6} (LaMer *et al.*, 1950)	International Critical tables (1928)	Reiss (1950)
Kiang and Stauffer (1973)	0–100	10^{-3}–1	10^{-6} (LaMer *et al.*, 1950)	Sabinia and Terpugov (1935)	Simplified Reiss (1950) expression
Marabel and Katz (1974)	0–100	10^{-3}–1	3.6×10^{-4} (Gmitro and Vermeulen, 1964)	Sabinia and Terpugov (1935)	Reiss (1950)

calculations (Kiang and Stauffer, 1973; Mirabel and Katz, 1974; Shugard *et al.*, 1974) to their experimental results in which known amounts of sulfuric acid and water vapor are directly put into the system. Very close agreement is obtained when the higher equilibrium vapor pressure data are used in the calculations.

Leifer *et al.* (1975) compared the number of sulfuric acid droplets required to form a given number of particles at a given time from measurements with theoretical estimates by Kiang and Stauffer (1973). It was assumed that the sulfuric acid vapor was produced by the reaction of SO_2 with $O(^3P)$ atoms. With this gas kinetic assumption the theory predicted 10^4 more molecules than the experiment required.

Stauffer *et al.* (1974) compared the number of droplets formed in Cox's (1973) experiments with the number of droplets estimated from nucleation theory assuming various relationships between time scales of aerosol kinetics taking place in the chamber. Qualitatively, there is a good agreement between the calculations and measurements. However, quantitatively the difference between the calculation and measurement can reach as much as three orders of magnitude at different relative humidities.

SHORTCOMINGS OF NUCLEATION EXPERIMENTS AND THEORY

In order to have a meaningful comparison between experiment and theory, one must perform the theoretical sensitivity analysis and examine all the possible shortcomings occurring in the experimental set up. In this paper, we single out three possible shortcomings which have not been previously discussed in detail. Experimentally, we discuss the uncertainties of gas and aerosol kinetics occurring in the smog chamber. Theoretically, we investigate the possible error due to the microphysical parameters required for nucleation calculation. The problem of wall loss has been discussed by Cox (1973) and the problem of contamination due to use of certain materials in the seals has been considered by Friend *et al.* (1973). Inaccuracies in the measurement of SO_2, determination of r.h., purity of air and measurement of the range and intensity of radiation as well as procedural differences in smog chamber experiments will not be considered here.

Gas phase kinetics

The rate of production of sulfuric acid vapors which then mix with water vapor to form particles in the reaction chamber is related to the amount of SO_2 present and the wavelength of radiation used in the experiment. Loss of reactive species to the chamber walls and/or reactions with impurities present in the air may also affect the rate. In the smog chamber experiments the amounts of SO_2 and the wavelength of radiation are known reasonably well. The photochemical mechanisms leading to the formation of H_2SO_4 are not known and can only be inferred. Thus the rate of production of H_2SO_4 has not been precisely determined.

For the SO_2–air system this task is made easier if we assume the air is 'pure'—all reactive trace gases have been removed. In that case the oxidation of SO_2 to SO_3 or HSO_3 which then can react rapidly to form H_2SO_4 (Davis and Klauber, 1975) can occur by the possible pathways listed in Table 3.

Smog chamber experiments are usually carried out at specific radiation wavelength ranges. Thus, possible reaction pathways can be categorized according to the wavelengths of radiation used in the experiment. If the wavelength (λ) is > 2454 Å, reactions 3–5 can not occur. In this case the rate of sulfuric acid production depends on the quantum yield of SO_3 (ϕ_{SO_3}), the specific absorption rate of u.v. light by SO_2 (k_a) in the reactor and the concentration of SO_2. It has been estimated by Cox (1972) that the upper bound for the quantum yield is 0.3×10^{-3}—a value which may be high, since the experiments to determine ϕ_{SO_3} were carried out at much higher SO_2 concentrations than are used in the nucleation experiments. For this reason it is difficult to assign a lower bound for the uncertainty in the rate of production of sulfuric acid vapor in the SO_2–air system when $\lambda > 2454$ Å.

For $\lambda < 2454$ Å, oxygen molecules can be dissociated to give $O(^3P)$ which initiates a sequence of reactions leading to the production of OH and HO_2 radicals provided water vapor is present. Assessment of which reaction (3–5) is the dominant oxidation step must be done indirectly since concentrations of intermediate species such as $O(^3P)$, OH and HO_2 are not known from the experimental measurements. It is assumed that reaction (3) is less important than reactions (4) and (5) from the rate constant measurements. If the concentration of HO_2 in pure wet air is less than the OH concentration which is itself estimated as 10^5–10^6 molecules cm^{-3} (Birenzvige and Mohnen, 1975), then reaction (4) is dominant.

If there are trace impurities in the filtered air, the concentrations of OH might possibly resemble estimated global averages for the natural atmosphere which are in the range 10^6–10^7 molecules cm^{-3} (e.g. Davis *et al.*, 1976; Wang *et al.*, 1975; Crutzen, 1975) or more recently, 2–3×10^5 molecules cm^{-3} (Singh, 1977; Crutzen and Fishman, in press).

If, however, we do assume the air is pure, then reaction (4) is the most likely oxidation step and the rate of production of sulfuric acid vapor for this reaction, $R = [OH][SO_2][air]k_4$, will have an uncertainty of about one order of magnitude.

Aerosol kinetics

The formation of new particles in the smog chamber depends on the sulfuric acid vapor concentration, r.h. and the temperature. Once the particles are formed they can grow by condensation of sulfuric acid and water vapor and/or by coagulation with each other. Initially the total number of particles increases rapidly due to nucleation. As the number of

Table 3. Possible gas phase oxidation mechanisms for SO_2 in SO_2–air experiments

Reaction	Rate constant	Reference	Comment
(1) $SO_2 + h\nu \rightarrow SO_2^*$	$5 \times 10^{-4}\ s^{-1}$ *	Cox (1973)	$^1SO_2$: 2900 Å < λ < 3400 Å; $^3SO_2$: 3400 Å < λ < 4000 Å
(2) $SO_2^* + O_2 \rightarrow (SO_4)$†	$0.96 \pm 0.05 \times 10^{-8}$ l. molecules^{-1} s^{-1}	Sidebottom *et al.* (1972)	$SO_2^* = {}^3SO_2$ at 3828 Å 25°C
(3) $SO_2 + O(^3P) + M \rightarrow SO_3 + M$	$(3.4 \pm 0.4) \times 10^{-32}$ $\exp(-2240/RT)$ cm^6 molecules^{-2} s^{-1}	Hampson and Garwin (1974)	$M = N_2$ 220 K < T < 353 K
(4) $SO_2 + OH + M \rightarrow HSO_3 + M$	$0.80 \pm 0.08 \times 10^{-13}$ cm^3 molecules^{-1} s^{-1}	Hampson and Garwin (1974)	$M = N_2$ (5 Torr) 300 K‡
(5) $SO_2 + HO_2 \rightarrow SO_3 + OH$	$(8.7 \pm 1.8) \times 10^{-16}$ cm^3 molecules^{-1} s^{-1}	Payne *et al.* (1973)	300 K

* Depends on wavelength and intensity of light and chamber size. In Cox (1973), 2900 Å < λ = 4000 Å and chamber volume = 5×10^3 cm^3.
† Detailed mechanism not certain. $^3SO_2$ thought to be main excited species in 2400 Å < λ < 3400 Å range.
‡ Rate constant dependence on pressure and M species is listed in reference.

particles begins to build up, coagulation starts to compete with nucleation and the total number of particles begins to decrease. Thus, total particle number counts can be lower than the actual number produced in the chamber, due to the coagulation process. Nucleation rates deduced from these measurements, as a result, will be too low.

The measured total number concentration may also be in error due to the limitations of commercial condensation nucleus counters as absolute aerosol counters. Cooper and Langer (in press) find in their studies that the detection efficiencies of both the Environment/One Rich 100 and the General Electric drop off considerably for particles less than 0.02 μm dia. Whitby *et al.* (1978) report that the accuracy of number/size distributions in the size range 0.0056–0.018 μm using the 1976 version of the Electrical Aerosol Analyzer (EAA) was only good enough to determine mean diameters of 0.009 μm ± 0.001. Since newly formed particles are about 0.001 μm in radius, large numbers of tiny particles may not be detected in the counters, and the total number of particles measured may be much less than the actual number present in the reaction vessel.

Theoretical

The theoretical expression for heteromolecular homogenous nucleation is based on the assumption of quasi-equilibrium. The validity of this assumption is yet to be tested. Given that the assumption is reasonable for the H_2O–H_2SO_2 binary system, then the main uncertainties in the theory lie in the determination of equilibrium vapor pressure and the microscopic surface tension for sulfuric acid solution droplets. Vapor pressures over pure sulfuric acid have been indirectly calculated from other measured thermodynamic quantities. The vapor pressure data set calculated by Gmitro and Vermeulen (1964), and used by Mirabel and Katz (1974), are an order of magnitude higher than that used by Doyle (1961) and Kiang and Stauffer (1973). Surface tension data used by both Mirabel and Katz (1974) and Kiang and Stauffer (1973) were taken from bulk ('macroscopic') surface tension measurements (Sabinia and Terpugov, 1935). The 'microscopic' surface tension could be as much as 15% lower than these values (e.g. Hamill *et al.*, 1974). The nucleation rate is very sensitive to the supersaturation—the ratio of the sulfuric acid vapor pressure to the solution vapor pressure—and to the surface tension. Consequently, for higher supersaturations (corresponds to lower equilibrium vapor pressure) or lower surface tensions, the nucleation rate could be several orders of magnitude higher.

UNCERTAINTIES IN NUCLEATION RATES

Uncertainties in calculated and measured nucleation rates which can be attributed to the above shortcomings are illustrated by numerical kinetic aerosol model calculations (for model details, see Middleton and Kiang, in press; Kiang and Middleton, 1978).

Rate of production of H_2SO_4

In the smog chamber experiment the rate of production of H_2SO_4 must be inferred from our knowledge of the wavelengths of radiation and the SO_2 concentrations used. For λ > 2454 Å the uncertainty in the rate is difficult to assign due to lack of informa-

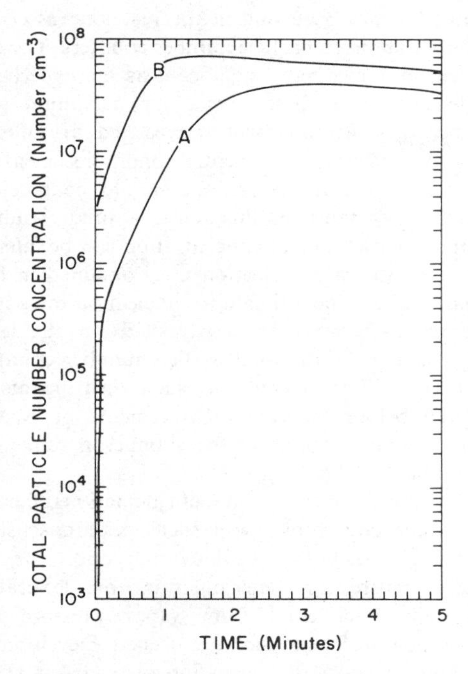

Fig. 1. Total particle number concentration as a function of time and rate of production of H_2SO_4 vapor, R. For A, R = 10^7 molecules cm^{-3} s^{-1}. For B, R = 10^8 molecules cm^{-3} s^{-1}.

tion on quantum yield values for SO_3. At best, we can obtain an upper bound. For $\lambda < 2454$ Å, however, the uncertainty is estimated to be about one order of magnitude. As a result, the total number concentration (see Fig. 1) is uncertain by a factor of five at most and the calculated nucleation rates—given as the slope of the total number concentration versus time curves—will be uncertain to at least an order of magnitude.

Rate of coagulation

In flow systems nucleation rates are obtained from experiments by dividing the measured total number concentration by the residence time of the experiment. If coagulation becomes important before the measurement is taken, the total particle number concentration measured will be less than the maximum number of particles formed and the estimated nucleation rate will be too low. The time at which coagulation becomes the dominant mechanism is a function of experimental conditions, as is illustrated in Fig. 1. For this comparison experimental conditions used in case A and case B are identical except that a factor of ten more SO_2 is introduced into case B system than into case A system. The maximum in total number concentration for case B occurs at 1 min whereas for case A the maximum occurs at 3 min. Thus, the nucleation rate for case B obtained from a total particle number concentration measurement taken at 3 min would be one order of magnitude lower than the rate obtained from a minute measurement.

Total particle number measurements

The possible error due to limitations in particle number counters is illustrated in Fig. 2. If particles < 0.0075 μm dia are not detected, the total particle number measured after 5 min could be almost four orders of magnitude too low.

Theory

Uncertainty in the theoretical calculations due to the uncertainty in the microphysical data for the nucleation calculation is shown in Fig. 3. Comparison of the curves during rapid nucleation (1 min) shows that the greatest uncertainty in total number concentration due to microphysical data uncertainty is two orders of magnitude. After the growth processes become significant (5 min), the uncertainty has decreased to one and a half orders of magnitude.

Examination of the uncertainties in nucleation rates obtained from theoretical calculations and experimental measurements shows that the most serious errors could arise from limitations in total particle number counters. Uncertainty in the gas phase kinetics leading to the production of sulfuric acid vapor results in an order of magnitude uncertainty in the nucleation rates. Uncertainty in the aerosol kinetics also could lead to an order of magnitude uncertainty in the nucleation rates calculated from flow system measurements. The uncertainties in parameter values required for nucleation theory calculations give rise to a maximum uncertainty of two orders of magnitude. Total particle number concentration measure-

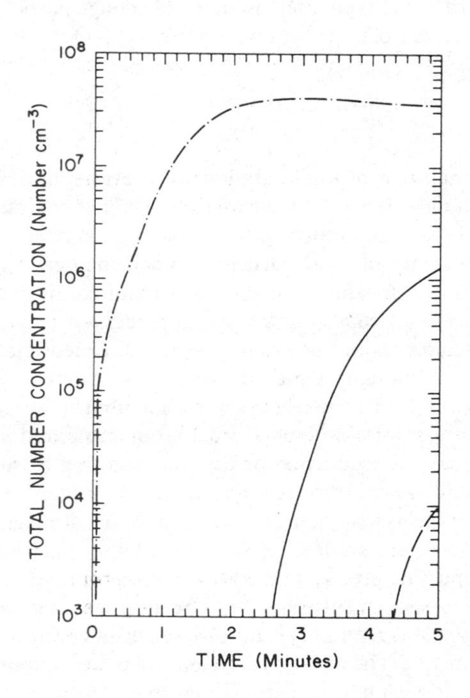

Fig. 2. Total particle number concentration as a function of time and particle counter detection efficiency. All particles greater than D μm in diameter are detected. —·—·— D = 0.001, —— D = 0.005, — — D = 0.0075.

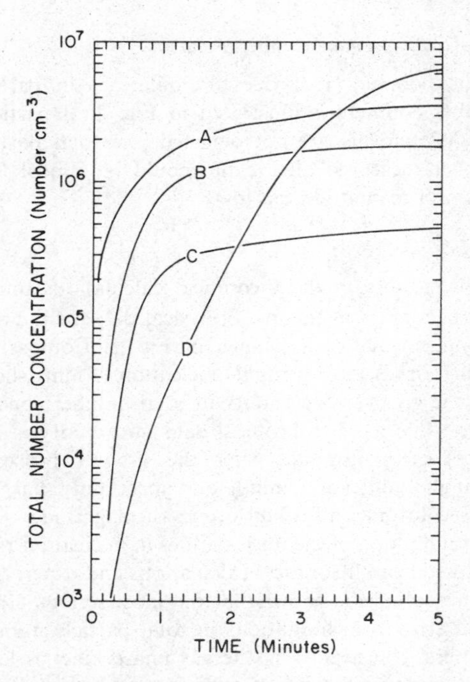

Fig. 3. Total particle number concentration as a function of time and uncertainty in microphysical data. Sabinia and Terpugov (1935) surface tension data (ST), H_2SO_4 equilibrium vapor pressure (VP) = 3.6×10^{-4} torr, curve D. ST lowered by 15%, curve C. VP = 3.6×10^{-5} torr, curve B. ST lowered by 15% and VP = 3.6×10^{-5} torr, curve A.

ments could be, however, four orders of magnitude too low if the counter is not detecting particles < 0.0075 μm dia.

RECOMMENDATIONS FOR FUTURE STUDIES

Examination of the limitations of previous nucleation studies shows that uncertainties in the gas and aerosol kinetics occurring in the smog chamber, in the sensitivity of total particle number counters and in the microphysical parameter data required for nucleation rate calculations can lead to orders of magnitude differences in both experimental and theoretical results. Although these discrepancies cannot be clearly resolved at present, several recommendations for future studies are suggested by our examination.

The rate of production or the concentration of sulfuric acid vapor which is required for nucleation rate calculations is uncertain to at least an order of magnitude. In future studies, the kinetics of the precursor gases must be precisely determined either through accurate measurement of H_2SO_4 or transient species, such as the OH radical, or through well-controlled experiments. The purity of system and the amount of wall loss should be carefully examined.

The uncertainties in nucleation rates deduced from experimental measurements of total particle number concentration arise from lack of knowledge about the aerosol kinetics occurring in the reaction vessel and from limitations in particle number counters. The first problem becomes more significant as the experiment residence time increases. Since the maximum total number concentration will be reached at different times for different experimental conditions, measurements for the same conditions should be taken at different flow rates. In this sense, a time evolution of total particle number concentration can be inferred and more accurate nucleation rates obtained in flow system studies. The estimates of nucleation rates from experiments, however, are most sensitive to the detection sensitivity of the total particle number counters. Continued improvement of such instruments is required before accurate measurements of particle number concentrations in the submicron range can be obtained.

As is illustrated by the model calculations, the formation and growth of new particles is quite sensitive to uncertainities in the calculated nucleation rate. Improved theoretical determination or laboratory measurement of equilibrium vapor pressure and microscopic surface tension are needed. Development of a kinetic theory for nucleation of binary mixtures to examine the validity of the current heteromolecular nucleation theory which is based on the equilibrium assumption is also recommended.

It should be emphasized that the details of gas and aerosol kinetics occurring in the controlled environment of the smog chamber may not bear much resemblance to reactions taking place in the real atmosphere. One important difference is the presence of preexisting particles in the real atmosphere. With particles present to compete with nucleation for the sulfuric acid vapors, nucleation rates may be lower than those predicted in the smog chamber experiment. The main value of the smog chamber experiment, then, is not to predict atmospheric nucleation rates but rather to provide accurate validation for nucleation theory which can then be used to assess the possibility of new particle formation in the atmosphere. Only from concerted efforts of both refined nucleation experiments and theory can we begin to describe with greater accuracy new particle formation under a variety of atmospheric conditions.

REFERENCES

Birenzvige A. and Mohnen V. A. Gas-to-particle conversion of sulfur dioxide reaction products at low relative humidities. Interim Report submitted to NSF, April, 1975, 97 pp. ASRC-SUNY Publ. 363. *J. Aerosol Sci.* (Submitted for publication).

Bricard J., Cabane M. and Madelaine G. (1977) Formation of atmospheric ultrafine particles and ions from trace gases. *J. Colloid Interface Sci.* **58**, 113–124.

Bouland D., Madelaine G., Vigla D. and Bricard J. (1975) SO_2 Transformation in controlled atmosphere leading to the production of aerosol particles. *Water, Air, and Soil Pollut.* **4**, 435–445.

Charlson R. J., Vanderpol A. H., Covert D. S., Waggoner A. P. and Ahlquist N. C. (1974) $H_2SO_4/(NH_4)_2SO_4$ Background aerosol: optical detection in St. Louis region. *Atmospheric Environment* **8**, 1257–1267.

Cooper G. and Langer G. Limitations of commercial condensation nucleus counters as absolute aerosol counters. *J. Aerosol Sci.* (Submitted for publication).

Cox R. A. (1972) Quantum yields for the photooxidation of sulfur dioxide in the first allowed absorption region. *J. phys. Chem.* **76**, 814–820.

Cox R. A. (1973) Some experimental observations of aerosol formation in the photooxidation of sulfur dioxide. *J. Aerosol Sci.* **4**, 473–483.

Crutzen P. J. (1975) A two dimensional photochemical model of the atmosphere below 55 km: estimates of natural and man made caused perturbation due to NO_3. *Proc. of the* IV *CIAP Conf.* Cambridge, Mass.

Crutzen P. and Fishman J. Average concentrations of OH in the Northern Hemisphere troposphere, and the budgets of CH_4, CO and H_2. *Geophys. Res. Lett.* (Submitted for publication).

Davis D. D. and Klauber C. (1975) Atmospheric gas phase oxidation mechanisms for the molecule SO_2. *Int. J. chem. Kinet. Symp.* **1**, 543–556.

Davis D. D., Heaps W. and McGee T. (1976) Direct measurements of natural tropospheric levels of OH via an aircraft-borne turnable dye laser. *Geophys. Res. Lett.* **3**, 331–333.

Doyle G. J. (1961) Self-nucleation in the sulfuric acid–water system. *J. chem. Phys.* **35**, 795–799.

Friend J. P., Leifer R. and Trichon T. (1973) On the formation of stratospheric aerosols. *J. atmos. Sci.* **30**, 465–479.

Gartrell G., Jr and Friedlander S. K. (1975) Relating particulate pollution to sources: the 1972 California aerosol characterization study. *Atmospheric Environment* **9**, 279–299.

Gmitro J. I. and Vermeulen T. (1964) Vapor liquid equilibrium for aqueous sulfuric acid. *Am. Inst. chem. Engng* **10**, 741.

Hamill P., Stauffer D. and Kiang C. S. (1974) Nucleation theory: Fisher's droplet picture and microscopic surface tension. *Chem. phys. Lett.* **28**, 209–212.

Hampson R. F. and Garvin D. (editors) (1974) Chemical kinetics and photochemical data for modelling atmospheric chemistry. *NBS Technical Note* **866**, 17, 55.

Hidy G. M. and Brock J. R. (1970) An assessment of the global sources of tropospheric aerosols. *Proc. 2nd Clean Air Congress*, IUAPPA, Washington, D.C., December, 1088–1097.

International Critical Tables (1928) McGraw-Hill, New York, Vol. 4, p. 464.

Kiang C. S. and Stauffer D. (1973) Chemical nucleation theory for various humidities and pollutants. *Faraday Symp.* **7**, 26–33.

Kiang C. A. and Middleton P. (1977) Formation of secondary sulfuric acid aerosols in urban atmosphere. *Geophys. Res. Lett.* **4**, 17–20.

La Maer V. K., Inn E. C. Y. and Wilson I. B. (1950) The methods of forming, detecting, and measuring the size and concentration of liquid aerosols in the size range of 0.01 to 0.25 microns diameter. *J. Colloid Sci.* **5**, 471–496.

Leifer R., Friend J. P. and Trichon M. (1974) The mechanism of formation of stratospheric aerosols. *Proc. Int. Conf. on Structure, Composition and General Circulation of the Upper and Lower Atmospheres and Possible Anthropogenic Perturbations.* Jan. 14–25, 1974, Melbourne, Australia.

Middleton P. and Kiang C. S. A kinetic aerosol model for the formation and growth of secondary sulfuric acid particles. *J. Aerosol Sci.* (To be published).

Mirabel P. and Katz J. L. (1974) Binary homogeneous nucleation as a mechanism for the formation of aerosols. *J. chem. Phys.* **60**, 1138–1144.

Payne W. A., Stief L. J. and Davis D. A. (1973) A kinetics study of the reaction of HO_2 with SO_2 and NO. *J. Am. chem. Soc.* **95**, 7614–7619.

Quon J. E., Siegel R. P. and Hulburt H. M. (1971) Particle formation from photooxidation of sulfur dioxide in air. *Proc. 2nd Int. Clean Air Congress*, Academic Press, New York, pp. 330–335.

Reiss H. (1950) The kinetics of phase transitions in binary systems. *J. chem. Phys.* **18**, 840–848.

Sabinia L. and Terpugov L. (1935) The surface tension of the system sulfuric acid–water. *Z. phys. Chem.* A **173**, 237–241.

Shugard W. J., Heist R. H. and Reiss H. (1974) Theory of vapor phase nucleation in binary mixtures of water and sulfuric acid. *J. chem. Phys.* **61**, 5298–5305.

Sidebottom H. W., Badcock C. C., Jackson G. E. and Calvert J. G. (1972) Photooxidation of sulfur dioxide. *Envir. Sci. Tech.* **6**, 72–79.

Singh H. (1977) Atmospheric halocarbons: evidence in favor of reduced average hydroxyl radical concentrations in the troposphere. *Geophys. Res. Lett.* **4**, 101–104.

Stauffer D., Mohnen V. A. and Kiang C. S. (1973) Heteromolecular condensation theory applied to gas-to-particle conversion. *J. Aerosol Sci.* **4**, 461–471.

Wang C., Davis L., Wu C. H., Jasper S., Niki H. and Weinstock B. (1975) Hydroxyl radical concentrations measured in ambient air. *Science* **189**, 797–800.

Whitby K. T., Cantrell B. K. and Kittelson D. B. (1978) Nuclei formation rates in a coal-fired power plant plume. *Atmospheric Environment* **12**, 313–321.

Atmospheric Environment Vol. 12, pp. 187–195. Pergamon Press 1978. Printed in Great Britain.

A REVIEW OF THE DYNAMICS OF SULFATE CONTAINING AEROSOLS

S. K. FRIEDLANDER

(*First received* 22 *June* 1977 *and in final form* 5 *October* 1977)

Abstract—Sulfate containing aerosols are generated by irradiating mixtures of SO_2, olefins and NO–NO_2 in air. In smog chamber studies, three time domains are observed. When the system is first irradiated, new particles form at a rate which depends on the concentration of the pre-existing aerosol. If the initial aerosol concentration is low, particles form rapidly and in high concentration. Increasing the initial aerosol concentration (and surface area) suppresses new particle formation because of scavenging in the size range $d_p < 100$ Å. Following new particle formation, the system passes into a transition period in which the onset of coagulation leads to a peaking and then a reduction in the number concentration.

In the third time domain, the aerosol surface area per unit volume of gas, A, approaches a value sufficient to accommodate new condensable molecules and new particle formation becomes negligible. The asymptotic value of A results from the balance between condensation and coagulation, and is a function of the rate of gas to particle conversion, F, the gas temperature and particle density. For $T = 300$ K and a particle density of 1.46 g cm^{-3}, theory predicts that $A = 6.23 \times 10^3 F^{3/5}$, where A has dimensions of cm^{-1} and F s^{-1}. This expression agrees well with smog chamber data for aerosols produced in a variety of chemical systems. Data for the Los Angeles smog aerosol are in fair agreement with this expression, too.

For SO_2, NO–NO_2 and low molecular weight olefins the aerosol which forms by irradiation is primarily sulfuric acid. In smog chamber studies with a pre-existing aerosol, the mass distribution of sulfur with respect to particle size peaks in the size range between 0.1 and 0.2 μm. A theoretical analysis which takes the particle growth law into account is in agreement with these experimental results. In the Los Angeles atmosphere, however, the peak occurs near 0.5 μm. Possible explanations for the discrepancy are discussed.

INTRODUCTION

The oxidation of SO_2 leads to the formation of sulfur containing compounds which accumulate in the aerosol phase. Sulfur in the aerosol phase is present in various oxidation states, depending on the chemical history of the system. Much of it is usually in the form of sulfate and it will be so called in the rest of this paper.

Concern with this problem originated because of irritant effects produced on breathing air polluted by the combustion of coal. Coal contains sulfur compounds which are oxidized to SO_2 during combustion; subsequently, oxidation to particulate phase sulfates takes place. It was early recognized that the sulfates, not the SO_2, were the primary cause of lung irritation (Haldane and Priestley, 1935). Subsequently experiments by Amdur *et al.* (1971) showed that the particle size and type of sulfate compounds strongly influenced irritant effects in animals. Aerosol size distribution and chemical composition also affect the atmospheric behavior of sulfate containing aerosols. Visibility degradation and droplet nucleation both depend on size and chemical nature. Dry deposition and vegetation scavenging of atmospheric sulfates depend on the distribution of sulfate with respect to particle size.

Aerosol dynamics refers to the evolution in space and time of the particle size distribution and of the chemical composition as a function of particle size. The dynamics of an aerosol strongly depends on the mechanism of gas-to-particle conversion. (Friedlander, 1977, Chap. 9.) In the case of sulfur dioxide, reactions may take place in the gas phase, in a solution phase if droplets are present or, possibly, on the surfaces of solid phases. The different mechanisms result in different size distributions and different distributions of sulfate with respect to particle size.

The system which has been most thoroughly studied in the laboratory is the oxidation of SO_2 by gas phase reactions in irradiated mixtures of SO_2, NO–NO_2 and olefins in air. This mixture includes ingredients of the type which go into the formation of photochemical smog. The early work on the chemistry of such systems was reviewed by Leighton (1961); later studies by Cox and Penkett (1972) and others are reviewed by Sander and Seinfeld (1976).

The oxidation of SO_2 in aqueous solution was studied by Junge and Ryan (1958) and van den Heuvel and Mason (1963) who were interested in droplet phase reactions occurring in the atmosphere. Based on the results of such experiments, Mason (1971) concluded that the large sulfate containing particles present in polluted air result from the oxidation of SO_2 in fog and cloud droplets in the presence of NH_3. Trace metal ions catalyze the reaction. The formation of aerosol-sized particles results from evaporation of the droplets. Condensation and evaporation may take place many times in the atmosphere before the sulfate finally precipitates as rain.

The oxidation of SO_2 may also be catalyzed by surface reactions. This is the mechanism by which sulfuric acid is produced industrially over platinum catalysts. The measurements of Novakov *et al.* (1974) have shown that soot particles formed in fossil fuel combustion are also active in this respect.

Of the various mechanisms of gas-to-particle conversion, only the first-gas phase reactions to form a condensable product—have been studied in detail in smog chambers. The results of these experiments are reviewed in this paper and compared with the characteristics of the Los Angeles aerosol.

Photochemical aerosol generation: chamber studies

Many studies of photochemical aerosol dynamics have been carried out in chambers to which small quantities of gaseous aerosol precursors were added. By using sufficiently large chambers, the surface to volume ratio can be kept small thereby minimizing aerosol deposition on the walls. It is also possible to sample enough gas for chemical analysis of the aerosol without significantly changing the volume of the chamber.

In studies of aerosol dynamics, the chamber is usually filled with filtered air to which NO, NO_2 and a reactive organic gas have been added at concentrations in the range between 0.1 and 1 ppm. The system is then irradiated by exposure to an artificial light source or to sunlight. When SO_2 is added to the gas, it is oxidized to form condensable products which accumulate in the aerosol phase. In the chamber experiments, SO_2 concentrations ranged between 0.01 and 2.9 ppm. The chemistry of the reactions of SO_2 in such mixtures has been reviewed by Sander and Seinfeld (1976) and Grosjean (1977), and the subject will not be discussed in depth in this paper. SO_2 is probably oxidized by certain free radicals present in photochemical smog. The most important such radicals are thought to be the hydroxyl, OH:

$$SO_2 + OH + M \rightarrow HSO_3 + M$$

the hydroperoxyl, HO_2:

$$SO_2 + HO_2 \rightarrow SO_3 + OH$$

and peroxyalkyl, RO_2:

$$SO_2 + RO_2 \rightarrow SO_3 + RO.$$

The SO_3 is then rapidly converted to H_2SO_4 by combination with water vapor

$$SO_3 + H_2O \rightarrow H_2SO_4.$$

The fate of HSO_3 radical is not known. Rate constants for reactions involving peroxyalkyl radicals are not well known and more data are needed to assess their importance for the atmospheric oxidation of SO_2. Calculations based on the kinetics of known reactions involving SO_2 have not been able to

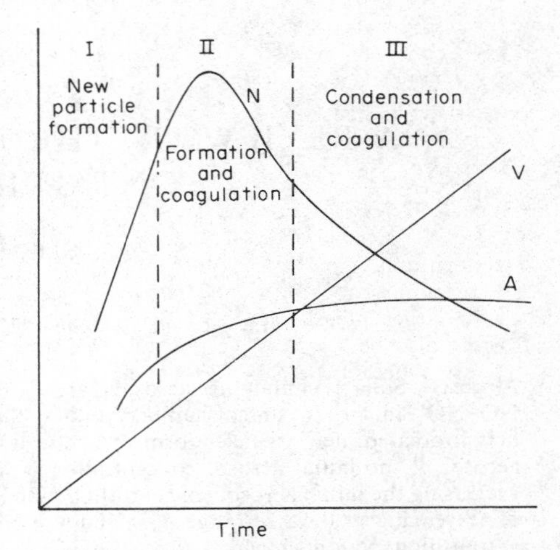

Fig. 1. Aerosol formation in an irradiated chamber (after Husar and Whitby, 1973).

account for SO_2 oxidation rates measured in smog chamber studies (Sander and Seinfeld, 1976). Reactions involving RO_2 type radicals may be sufficiently fast to make up the difference.

Particle size distributions of the aerosols in such chamber studies have frequently been studied with the electrical mobility analyzer (Whitby *et al.*, 1972). By integrating over the size distribution, the aerosol number concentration, N, area concentration, A, and volume concentration, V, can be measured as a function of time. When these parameters are plotted as a function of time, results of the type shown in Fig. 1 are frequently obtained. Three separate time domains can be identified. In Domain I, collisions among the condensable molecules generated by chemical reaction result in new particle formation. In Domain II, coagulation becomes important because the particle concentrations are high; the surface area increases and then approaches an approximately constant value in Domain III in which coagulation and condensation on the existing aerosol are the dominant processes. The asymptotic approach to constant surface area was first predicted by Pich *et al.* (1970) for particles large compared with the mean free path of the gas. For particles small compared with the mean free path, theory indicates that the surface area tends to increase slowly with time as shown in one of the following sections.

New particle formation: Domain I

There is now considerable evidence that new particle formation takes place at certain times in the polluted Los Angeles urban atmosphere (McMurry, 1977), rural atmospheres (Hogan, 1968) and stack plumes (White *et al.*, 1976). Photochemical processes were believed to be responsible in each case. Particles are invariably present in parcels of gas in which new particle formation is initiated. Condensable molecules formed by gas phase reactions may either form small clusters or deposit on pre-existing particles. Clusters

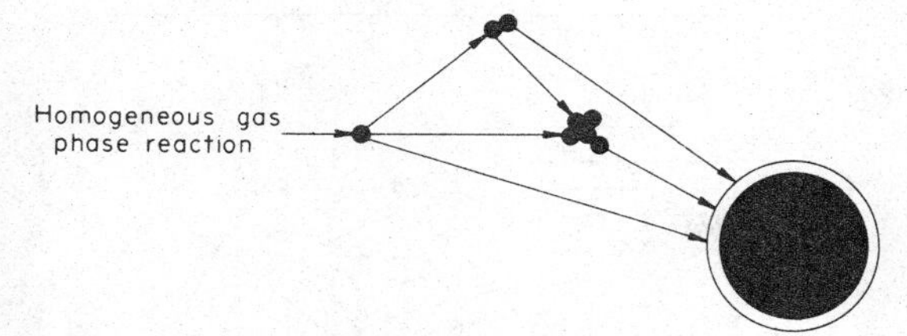

Fig. 2. Condensable molecules (monomer) produced by gas phase reactions can either produce new particles or condense on larger pre-existing particles. New particles (clusters) may also be scavenged by the pre-existing particles (McMurry, 1977).

are scavenged by the pre-existing aerosol. The situation is illustrated in Fig. 2.

If the concentration of the pre-existing aerosol is high, the condensable molecules and small clusters should be scavenged and new particle formation suppressed. *The measured rate of new particle formation will depend on the detection limit of the aerosol counter.*

McMurry (1977) conducted experiments on new particle formation in mixtures of SO_2, propylene and NO–NO_2 in the presence of a pre-existing aerosol. The concentration of the pre-existing aerosol was controlled by filling a Teflon balloon chamber (Roberts and Friedlander, 1976) with ambient air which was passed through a 20 cm (diameter) Gelman Type A glass fiber filter with a small hole in the center; the air entering the chamber consisted of a mixture of filtered and unfiltered air corresponding to a diluted sample of the ambient aerosol. To this system, small amounts of the aerosol precursor gases were added. Natural solar radiation passing through the Teflon promoted the chemical reactions leading to aerosol formation.

The electrical mobility analyzer was used to measure the particle size distribution of the contents of the chamber. The smallest particle accurately detected with this instrument was about 0.01 μm. By integrating over all particle sizes, the increase in the number concentration of particles larger than 0.01 μm (100 Å) was measured.

Partial filtration of ambient air, in itself, resulted in particle formation in the chamber air. To prevent new particle formation until after the reactant gases were added, the chamber was covered with a black plastic sheet during filling. NO was added to consume ambient ozone as the air entered the balloon. The chamber remained covered until after the reactant gases were added and mixed with air.

In Fig. 3, the number concentration of particles larger than 0.01 μm is shown as a function of time for chamber experiments with different initial aerosol concentrations. The rate of oxidation of SO_2, as measured by the decay in SO_2 concentration, was about the same in each case. Increasing the initial aerosol loading for a given rate of SO_2 oxidation reduced the rate of new particle formation. Condensable molecules were scavenged by the larger surface areas associated with the higher initial aerosol concentrations. In experiments with similar initial aerosol concentrations but different rates of SO_2 oxidation, the rate of new particle formation increased as the oxidation rate increased.

New particle formation: discussion

The formation of 100 Å particles in the smog chamber and the atmosphere does not take place directly. Homogeneous nucleation leads to the formation of stable nuclei smaller than 10 Å which then grow by condensation and coagulation to the size range detectable by the instrument, 100 Å in this case. The theory of the formation of stable nuclei has been applied to the sulfuric acid water system by Mirabel and Katz (1974) and Middleton and Kiang (submitted). For the special case in which the critical nuclei are of molecular dimensions, McMurry (1977) has carried out a theoretical analysis for the rate at which molecular clusters much larger than stable nuclei are formed in the presence of a pre-existing aerosol. The results of this analysis compared well with the experimental results of his smog chamber studies. The analysis is based on the following assumptions: (1) All collisions involving the condensable product molecules (monomer) result in sticking. This assumption (zero activation energy) holds best when the saturation ratio is very high as in the case of compounds with low vapor pressure such as $(NH_4)_2SO_4$ or NH_4NO_3. (2) The rate of collision can be calculated from the collision frequency function for hard spheres.

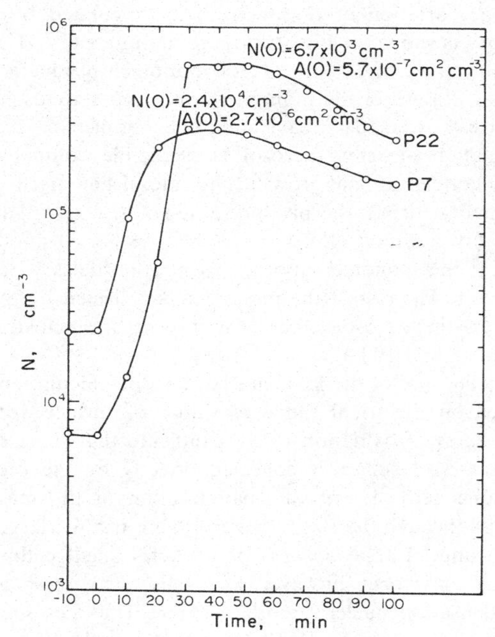

Fig. 3. Aerosol number concentrations as a function of time for experiments with similar rates of SO_2 oxidation (1.8×10^{-3} ppb s^{-1}) but different initial aerosol loadings. The initial aerosol number concentrations, $N(0)$, and surface area concentrations, $A(0)$, are shown. Gas-to-particle conversion started at time zero. Increasing the initial aerosol concentration suppresses new particle formation (McMurry, 1977).

This expression is applied to collisions of monomer molecules with molecular clusters formed by previous collisions, and to collisions of molecules and clusters with the pre-existing aerosol. (3) A steady state exists in the cluster distribution; the rate of formation of clusters of a given size by monomer attachment to the next smallest cluster is equal to the rate of cluster loss by monomer attachment and scavenging by the pre-existing aerosol. (4) Coagulation among the clusters can be neglected.

If the rate of generation of clusters containing k monomer molecules ($cm^{-3} s^{-1}$) is denoted by $G(k)$, the analysis leads to the result:

$$\frac{G(k)}{R} = f(L, K) \tag{1}$$

where the dimensionless group L is given by:

$$L = \frac{\gamma^2 A^2}{\beta_{1,1} R} \tag{2}$$

where $\gamma = (k/T2\pi m)^{1/2}$, A is the surface area of the pre-existing aerosol per unit volume of gas, $\beta_{1,1}$, is the collision frequency between the condensable molecules and R is the rate of gas-to-particle conversion ($cm^{-3} s^{-1}$). The mass of the condensable molecules is m, and k and T are Boltzmann's constant and the absolute temperature, respectively. The parameter L is a measure of the relative rates of heterogeneous and homogeneous processes in removing condensable molecules formed by gas phase reactions.

The variation of $G(k)$ with L as calculated from theory is shown in Fig. 4 for several values of k. For small values of L, the rate of monomer production $R \rightarrow kg(k)$. All of the monomer condenses on clusters of size k or smaller; scavenging of monomer or clusters by pre-existing aerosol is negligible compared with condensational growth. The rate of new particle formation drops sharply for values of L larger than 0.1. For example, for $k = 2{,}500$ ($d_p \approx 100$ Å) about 1% of the monomer appears as new particles when $L = 0.1$. The rest of the monomer is collected by the pre-existing aerosol either as monomer or as clusters smaller than 100 Å.

In the case of the Los Angeles aerosol, calculations based on the usual range of values of surface area and rates of oxidation of SO_2 indicate that most of the *mass* of sulfate is collected directly by the pre-existing aerosol; only a small fraction of the mass passes through the 100 Å size and then into the larger size range. Large *numbers* of particles smaller than 100 Å may form, however, and these may serve as condensation nuclei for other chemical species such as organic compounds. This, in turn, would have a significant effect on the nature of the aerosol.

This behavior differs significantly from that observed in chambers containing a reactive gas mixture from which all of the particles present initially are removed by filtration. In the absence of pre-existing particles, nuclei grow into the larger size range

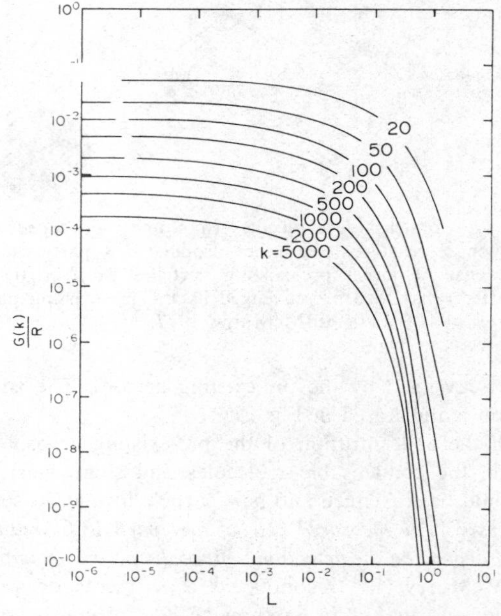

Fig. 4. The rate of formation of particle as a function of particle size for a rate of monomer formation = 1.43 × 10^7 $cm^{-3} s^{-1}$ and $A = 5 \times 10^{-6}$ cm^2 aerosol surface area cm^{-3} air. The monomer volume was taken to be 2.1 × 10^{-22} cm^3 and aerosol density was 1.46 g cm^{-3}, values characteristic of H_2SO_4 at 30% relative humidity. The temperature was 300 K (McMurry, 1977).

by condensation and coagulation. Eventually, of course, enough area is generated and new particle formation becomes very small (Domain III).

Asymptotic behavior: Domain III

The surface area per unit volume of gas of an aerosol composed of spherical particles is

$$A(t) = \int_0^\infty \pi\, d_p^2 n(v, t)\, dv \tag{3}$$

where d_p is particle diameter. The aerosol size distribution function, $n(v, t)$, is defined so that the number of particles with volume between v and $v + dv$ is

$$dN = n(v, t)dv. \tag{4}$$

In the chamber experiments discussed above, the aerosol surface area first increases and then levels off after a period of about one hour. In this final time period, Domain III, coagulation and condensation are of controlling importance. This is the last stage of aerosol evolution, the period in which the size distribution would be expected to reach the self-preserving form. (Friedlander and Wang, 1966). The experimental results of Husar and Whitby (1973) and McMurry and Friedlander (1977) show that the size distributions of such aerosols are indeed self-preserving after sufficiently long periods of time.

In this case, McMurry and Friedlander (1977) have derived the following expression for the dependence of the aerosol surface area concentration on the rate of formation of aerosol volume by gas-to-particle conversion:

$$A = \left(\frac{2}{\alpha_1}\right)^{2/5} \mu\left[\left(\frac{7}{5}\right)^{2/5} (36\pi)^{1/3} \times \left(\frac{4\pi}{3}\right)^{1/15} \left(\frac{\rho}{6kT}\right)^{1/5}\right] F^{3/5} t^{1/5} \quad (5)$$

where α_1 and μ are numerical constants, ρ is the particle density, k is Boltzmann's constant. T is the absolute temperature and $F = dv/dt$ is the rate of formation of aerosol volume. The surface area, A, is a weak function of time, and increases as the 3/5 power of the rate of aerosol formation. The only aerosol property on which A depends is the particle density, and this dependence is weak. In general, α_1 and μ depend on the form of the self-preserving size distribution which, in turn, depends on the rate and mechanism of gas-to-particle conversion. In the absence of condensation, α_1 and μ approximately equal 6.67 and 0.9, respectively (Lai *et al.*, 1972; Graham and Homer, 1973). With these values, and taking time to be one hour, density 1.46 g cm^{-3} and temperature 300 K, Equation (5) becomes

$$A = 6.23 \times 10^3 F^{3/5} \quad (6)$$

where A has dimensions of cm^{-1} and F s^{-1}.

Comparison of experiment with theory: Domain III

The electrical mobility analyzer has been used by a number of investigators to follow the evolution of photochemically generated aerosols in chambers of various types. Figure 5 shows the experimental results of various investigators for aerosol surface areas as a function of rate of aerosol formation. Husar and Whitby (1973) reported values of F for the Los Angeles aerosol with corresponding values of A. These data are also shown in Fig. 5 with the theoretically predicted relationship between A and F, (6).

Agreement between experiment and theory is good for a variety of chemical systems. An exception is the cyclohexene + NO system for which the experimental values are considerably less than predicted by theory. With this system the maximum value of N generally ranges from 10^3 to 10^4 cm^{-3}. Aerosol concentrations generated by other chemical systems typically are 10 to 1000 times larger. As a result, the assumption that $N(t) \ll N(0)$ is poor for the cyclohexene system. Also, since the number of particles generated by the cyclohexene, NO system for a given rate of aerosol formation is small, the particles grow out of the free molecules regime more quickly than particles in systems with higher concentrations, thus invalidating the assumption of free molecule aerosol dynamics.

Data for the Los Angeles smog aerosol fall near the theoretical line indicating that the surface area of some ambient aerosols is regulated by gas-to-particle conversion. However, this does not mean that the particle size distribution of the ambient aerosol is self-preserving; rather, *that portion* of the size distribution near 0.1 μm with which most of the surface area is associated is approximately in the self-preserving form.

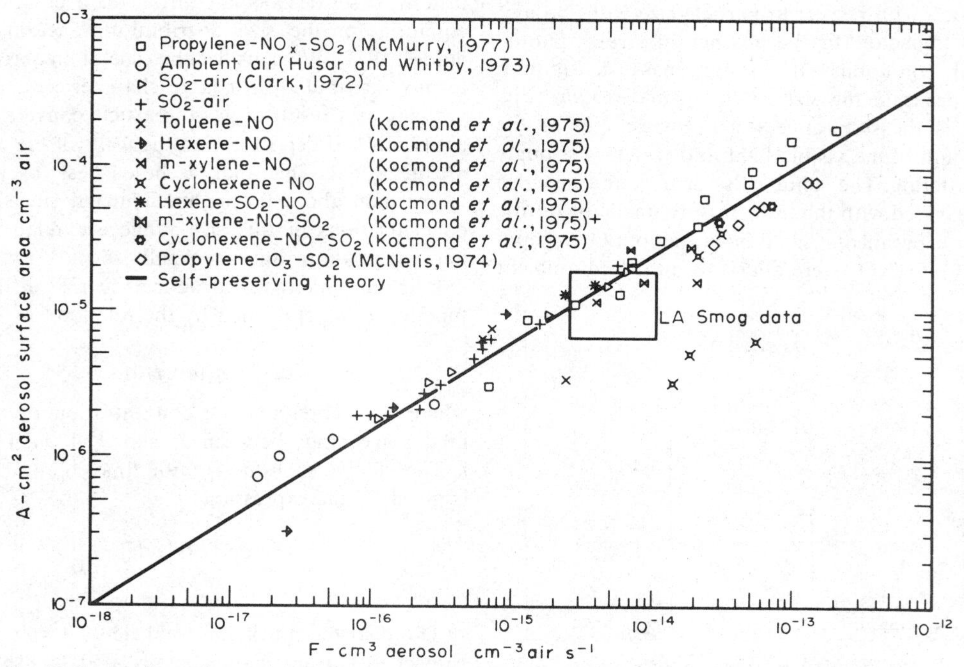

Fig. 5. Comparison of experimentally measured aerosol surface areas for a given rate of aerosol formation, F, with surface areas predicted from theory (McMurry and Friedlander, 1977). A time of one hour was chosen for the theoretical line in accordance with laboratory time scales required to achieve constant surface areas. The range of values for Los Angeles smog was reported by Husar and Whitby (1973).

Distribution of sulfur with respect to particle size

The development of a low pressure impactor capable of fractionating aerosol particles larger than about 200 Å has been described by Hering *et al.* (1977). The impactor has eight single jet stages and a flow rate of 1 l. min^{-1}. The aerodynamic cutoff diameters corresponding to a particle collected with an efficiency of 50% are 4.0, 2.0, 1.0, 0.5, 0.25, 0.11, 0.05, and 0.02 μm. The cutoff diameter for the last two stages are approximate, based in part on calculation and in part on an approximate calibration.

Particles larger than 0.5 μm are sampled at near atmospheric pressure using the first four stages of the impactor. The last four stages operate at pressures from 150 mm down to 8 mm Hg (absolute) to size segregate the smaller particles. Material deposited at each stage can be analyzed for sulfur containing compounds at the nanogram level by flash volatilization followed by flame photometric analysis (Roberts and Friedlander, 1976). Enough material for chemical analysis can be collected by sampling ambient air with sulfate levels of 10 μg m^{-3} for 15 min. The instrument has been checked with ammonium sulfate and sulfuric acid aerosols by comparison with an electrical mobility analyzer. Good agreement between the two instruments was found indicating that the evaporation of these compounds in the impactor was not important. A further check was carried out by sampling simultaneously, in parallel an ammonium sulfate aerosol, with the impactor and a filter. Excellent agreement was obtained.

Measurements have been made of the sulfur mass distribution with respect to particle size with the low pressure impactor. In the absence of strong photochemical conditions, the sulfur mass distribution usually peaks in the size range around 0.5 μm (Fig. 6). This is the size range which corresponds to the wave length of the visible light and which is strongly light scattering. The sulfur mass distribution has also been measured with the low pressure impactor during chamber experiments; small quantities of SO_2, propylene and NO–NO_2 were added to unfiltered ambient

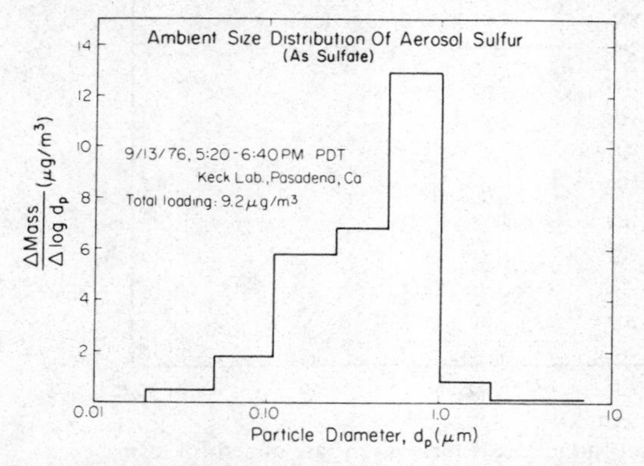

Fig. 6. Distribution of sulfur with respect to particle size for the ambient aerosol in Pasadena, California, measured with a low pressure impactor. The peak occurs in the particle size range between 0.5 and 1.0 μm (Hering, 1977).

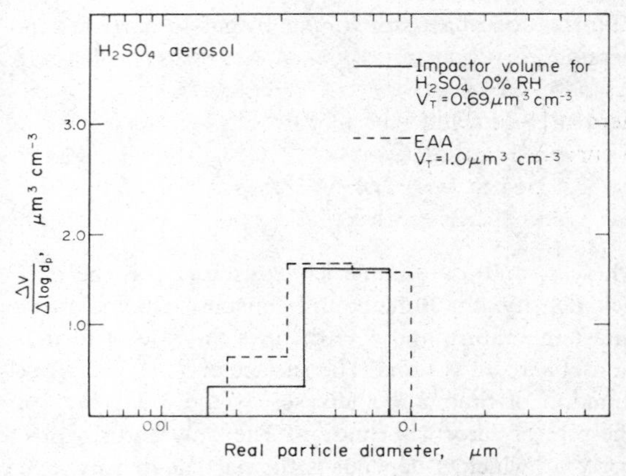

Fig. 7. The distribution of sulfur with respect to particle size for a balloon chamber aerosol, measured with the low pressure impactor. SO_2, propylene and NO–NO_2 were added to unfiltered ambient air and the mixture irradiated. A peak in the sulfur distribution developed in the particle size range between 0.1 and 0.2 μm (Hering *et al.*, 1977).

air exposed to solar radiation. The sulfur mass distribution peaked in the range between 0.1 and 0.2 μm, (Fig. 7), significantly smaller than the peak range for the ambient aerosol. (Significant quantities of sulfur in the 0.1 to 0.2 μm range were also measured in one measurement made on a day of strong photochemical activity.)

Aerosol growth dynamics

The distribution with respect to particle size of a chemical species converted from the gas phase can in some cases be calculated from the general dynamic equation for the size distribution function (Friedlander, 1977). We consider the case of an aerosol uniformly distributed in a large chamber with particles growing as a result of gas-to-particle conversion. Coagulation and deposition on the walls of the chamber are neglected. This analysis holds best for particles larger than about 0.1 μm in diameter, since particle concentrations in this size range are relatively low and coagulation rates are small.

It is convenient to introduce the size distribution function $n_1(d_p, t)$ defined by the relation

$$dN = n_1(d_p, t)d(d_p) \tag{7}$$

where dN represents the concentration of particles in the size range between d_p and $d_p + d(d_p)$ at time t. The change in $n_1(d_p, t)$ with time in the chamber is given by the expression

$$\frac{\partial n_1}{\partial t} + \frac{\partial n_1[d(d_p)/dt]}{\partial d_p} = 0. \tag{8}$$

The distribution of the total aerosol volume with respect to particle diameter is related to the particle size distribution by the expression

$$\frac{\partial V}{\partial \log d_p} = y = \frac{\pi}{6} \ln 10\, d_p^4 n_1. \tag{9}$$

The area under the curve of y plotted as a function of log d_p is proportional to the total particle volume. As a result of gas-to-particle conversion, the total aerosol volume increases with time. Each point on a curve of w vs log d_p shifts to a new value determined by the growth law. The rate at which any point on the volume distribution function changes with time is given by

$$\frac{dy}{dt} = \frac{\partial y}{\partial t} + w\frac{\partial y}{\partial d_p} \tag{10}$$

where the particle growth rate, w, depends on the mechanism of gas-to-particle conversion. The mechanisms are reviewed by Friedlander (1977). By evaluating the right-hand-side of this expression, it is possible to determine the particle size at which the value of w grows most rapidly. Substitution of (9) in (10) gives

$$\frac{dy}{dt} = \frac{\pi}{6}\ln 10\left[d_p^4\left(\frac{\partial n_1}{\partial t} + w\frac{\partial n_1}{\partial d_p}\right) + 4wn_1 d_p^3\right]. \tag{11}$$

If the particles are much smaller than the mean free path of the gas, and they grow by collision with monomer their volumetric growth rate is proportional to their surface area. The rate of increase in particle diameter with time is constant in this case:

$$w = d(d_p)/dt = \text{constant}. \tag{12}$$

When w is constant, substitution in (8) gives:

$$\frac{\partial n_1}{\partial t} + w\frac{\partial n_1}{\partial d_p} = 0. \tag{13}$$

Substituting in (11) the result is

$$\frac{dy}{dt} = \frac{2\pi}{3}\ln 10 w d_p^3 n_1. \tag{14}$$

But the distribution of aerosol surface area with respect to particle size is given by

$$\frac{\partial A}{\partial \log d_p} = \pi \ln 10\, d_p^3 n_1. \tag{15}$$

So

$$\frac{dy}{dt} = \frac{2w\,\partial A}{3\,\partial \log d_p}. \tag{16}$$

Thus the volume distribution tends to grow most rapidly at the particle size corresponding to a peak in the area distribution function $\partial A/\partial \log d_p$. This is observed, approximately, in chamber experiments with SO_2, propylene and NO–NO_2 in air containing ambient aerosol particles (Fig. 7).

Deviation between theory and experiment

Why then does the sulfur mass distribution of the ambient aerosol usually peak at a particle size much larger than the diameter corresponding to the peak in the area distribution function? Although we do not know the answer to this question, there are several possible explanations:

(1) The data of Whitby *et al.* (1972) indicate that coagulation can play a significant role in reducing the number concentration of the Los Angeles aerosol between 2000 and 0400 hours (data of 3 Sept. 1969). Over this time period the concentration dropped by a factor of almost 10 while the volume concentration of the aerosol remained almost constant. Thus the mean particle volume increased by a factor of about 2.15 over that time period. Part of this material remains in the air shed and contributes to the sulfate present in the larger particle sizes. The measurements were made at a fixed point in Pasadena so the effect of advection was not taken into account.

(2) Certain condensable organic compounds generated by gas phase reactions accumulate preferentially in the particle size range around 0.5 μm. This was the case with two cyclic olefins, cyclopentene and cyclohexene, and one diolefin, 1,7 octadiene, which served as aerosol precursors in experiments reported by Heisler and Friedlander (1977). The experiments were carried out in a balloon reactor; the hydrocarbons and oxides of nitrogen were added to unfiltered ambient air in the chamber and exposed to solar radiation. Difunctional organic compounds produced in this way condense on existing aerosol particles. The rates of particle growth could be correlated by a mechanism based on diffusion to the particles of the condensable gases with a critical particle size near 0.2 μm below which growth did not occur (Kelvin effect). This leads to the development of a peak in the volume distribution in the particle size range, $d_p > 0.5$ m (Heisler and Friedlander, 1977). The area distribution would develop in a size range between 0.2 and 0.5 μm.

Simultaneous experiments with SO_2 were not conducted. If, however, sulfuric acid were formed in such a mixture by the oxidation of SO_2, the H_2SO_4 monomer and small clusters might then be scavenged by the developing peak in the area distribution at a larger particle size than observed in chamber experiments without organic aerosols.

(3) If the oxdiation of SO_2 takes place in a droplet phase such as a morning fog, the droplet growth rate $d(d_p)/dt \propto d_p$ (Friedlander, 1977). In this case it can be shown by modifying the analysis of the previous section that the sulfur peak would tend to develop in the size range corresponding to the peak in the *volume* distribution of the fog. When the fog partially evaporates, the sulfate would be found in a larger particle size range.

SUMMARY

Sulfate containing aerosols are dynamic systems; their size distribution and the sulfur mass distribution with respect to size change with time. Effects on human health, visibility, cloud nucleation and dry deposition are linked to the aerosol dynamics.

The dynamics of aerosols formed by photochemically induced reactions in laboratory chambers are now fairly well understood. The behavior of such systems can be divided into three time domains. When the system is first irradiated, Domain I, new particle formation takes place at a rate which depends on the concentration of the pre-existing aerosol. In the transition period, Domain II, the particle concentration reaches a maximum and then decays as coagulation becomes important. New particle formation is small in Domain III; enough surface area is available to accommodate condensable molecules generated by homogeneous gas phase reactions.

In Domain I, increasing the concentration of the pre-existing aerosol suppresses new particle formation. For a given initial aerosol concentration increasing the rate of gas-to-particle conversion increases the rate of new particle formation. The rate of formation of clusters of a given size between 10 and 100 Å can be estimated theoretically. For the Los Angeles aerosol, the calculations indicate that most of the sulfate (mass basis) resulting from the gas phase oxidation of SO_2 collects on particles in the size range around 0.1 μm; little grows through the 100 Å size range into the larger sizes.

In Domain III, new particle formation has ceased and coagulation and condensation take place simultaneously. The surface area of the aerosol per unit volume of gas, A, can be related to the rate of gas-to-particle conversion, F, by a theoretical analysis which leads to the result

$$A = 6.23 \times 10^3 F^{3/5}$$

where A has dimensions of cm^{-1} and F s^{-1} and $T = 300$ K.

This result is in good agreement with the experimental data of many investigators for different chemical systems. It is in approximate agreement with limited data for the Los Angeles aerosol on smoggy days.

Measurements of the distribution of sulfur with respect to particle size have been made with a low pressure impactor. In smog chamber studies, the mass distribution of sulfur peaks in the size range between 0.1 and 0.2 μm. This is in agreement with theoretical predictions for the growth of aerosols by transport from the gas phase. In the Los Angeles atmosphere, the peak occurs near 0.5 μm, a significantly larger particle size. Possible explanations for the discrepancy include (1) coagulation occurring during the evening followed by 'recycle' of the aerosol (2) simultaneous growth with organic compounds which tend to accumulate on larger particles as a result of the Kelvin effect and (3) droplet phase reactions, perhaps involving marine fog.

The dynamics of aerosols in which droplet phase reaction controls SO_2 oxidation has not been studied experimentally. It would be of interest to follow the change with time of the size distribution function and of the distribution of sulfur with respect to particle size for such aerosols. An important goal of such experiments would be to explain the observed atmospheric distribution of sulfur with respect to particle size.

Acknowledgements—This work was supported in part by EPA Grant No. R802160 and NSF-RANN Grant No. ENV76-04179. The contents do not necessarily reflect the views and policies of the Environmental Protection Agency.

REFERENCES

Amdur M. O. (1971) Aerosols formed by oxidation of sulfur dioxide—review of their toxicology. *Arch. Environ. Hlth* **23**, 459.

Clark W. E. (1972) Measurements of aerosols produced by the photochemical oxidation of SO_2 in air. Ph.D. Thesis, Mechanical Engineering, University of Minnesota.

Cox R. A. and Penkett S. A. (1972) Particle formation from homogeneous reactions of sulfur dioxide and nitrogen dioxide. *J. Chem. Soc. Faraday Trans. I* **68**, 1735.

Friedlander S. K. and Wang C. S. (1966) The self-preserving particle size distribution for coagulation by Brownian motion. *J. Colloid Interface Sci.* **22**, 126.

Friedlander S. K. (1977) *Smoke, Dust and Haze: Fundamentals of Aerosol Behavior.* Wiley-Interscience, New York.

Graham S. C. and Homer J. B. (1973) Coagulation of molten lead aerosols. *Faraday Symp.* **7**, 85.

Grosjean D. (1977) *Ozone and Other Photochemical Oxidants.* Chap. 3. Report to National Academy of Science, Washington, DC.

Haldane J. S. and Priestley J. G. (1935) *Respiration* p. 415. Clarendon Press, Oxford.

Heisler S. L. and Friedlander S. K. (1977) Gas-to-particle conversion in photochemical smog: growth laws and mechanisms for organics. *Atmospheric Environment* **11**, 157.

Hering S. V., Flagan R. C. and Friedlander S. K. (1977) A new low pressure impactor: determination of the size distribution of aerosol sulfur compounds. Presented at the American Chemical Soc. Meeting, New Orleans, 20–25 March 1977.

Hering S. V. (1977) personal communication.

Hogan A. W. (1968) An experiment illustrating that gas conversion by solar radiation is a major influence in the diurnal variation of Aitken nucleus concentrations. *Atmospheric Environment* **2**, 599.

Husar R. B. and Whitby K. T. (1973) Growth mechanisms and size spectra of photochemical aerosols. *Environ. Sci. Technol.* **7**, 241.

Junge C. and Ryan T. G. (1958) Study of the SO_2 oxidation in solution and its role in atmospheric chemistry. *Quart J. Roy. Meteor. Soc.* **84**, 46.

Kocmond W. C., Kittelson D. B., Yang J. Y. and Demerjian K. L. (1975) Study of aerosol formation in photochemical air pollution. EPA 650/3-75-007.

Lai F. S., Friedlander S. K., Pich J. and Hidy G. M. (1972) The self-preserving particle size distribution for Brownian coagulation in the free-molecule regime. *J. Colloid Interface Sci.* **39**, 395.

Leighton P. A. (1961) *Photochemistry of Air Pollution.* Academic Press, New York.

Mason B. J. (1971) *The Physics of Clouds* p. 71. Clarendon Press, Oxford.

Middleton P. and Kiang C. S. A kinetic aerosol model for the formation and growth of secondary sulfuric acid particles. *J. Aerosol Sci.* (submitted for publication).

Mirabel P. and Katz J. L. (1974) Binary homogeneous nucleation as a mechanism for the formation of aerosols. *J. Chem. Phys.* **60**, 1138.

McMurry P. H. (1977) On the relationship between aerosol dynamics and the rate of gas-to-particle conversion. Ph.D. Thesis, Environmental Engineering Science, California Institute of Technology.

McMurry P. H. and Friedlander S. K. (1977) Aerosol formation in reacting gases: relation of surface area to rate of gas-to-particle conversion. *J. Colloid Interface Sci.* (in press).

McNelis D. N. (1974) Aerosol formation from gas-phase reactions of ozone and olefin in the presence of sulfur dioxide. EPA 650/4-74-034.

Novakov T., Chang S. C. and Harker A. B. (1974) Sulfates as pollution particulates: catalytic formation on carbon (soot) particles. *Science* **186**, 259.

Pich J., Friedlander S. K. and Lai F. S. (1970) The self-preserving particle size distribution for coagulation by Brownian motion-III. Smoluchowski coagulation and simultaneous Maxwellian condensation. *J. Aerosol Sci.* **1**, 115.

Roberts P. T. and Friedlander S. K. (1976) Photochemical aerosol formation—SO_2, 1-heptene, and NO_x in ambient air. *Environ. Sci. Technol.* **10**, 573.

Sander S. P. and Seinfeld J. H. (1976) Chemical kinetics of homogeneous atmospheric oxidation of sulfur dioxide. *Environ. Sci. Technol.* **10**, 1114.

Van den Heuvel A. P. and Mason B. J. (1963) The formation of ammonium sulphate in water droplets exposed to gaseous sulphur dioxide and ammonia. *Quart. J. Roy. Meteor. Soc.* **89**, 271.

Whitby K. T., Husar R. B. and Lui B. Y. H. (1972) The aerosol size distribution of the Los Angeles smog. *J. Colloid Interface Sci.* **39**, 177.

White W. H., Anderson J. A., Knuth W. R., Blumenthal D. L., Hsuing J. C. and Husar R. B. (1976) Mapping large pollutant plumes by instrumented aircraft: support for project MISTT, 1974. Meteorology Research Inc. Report MRI 76 FR-1414 to EPA, Contract 68-02-1919, Altadena, California.

Atmospheric Environment Vol. 12, pp. 197–226. Pergamon Press 1978. Printed in Great Britain.

MECHANISM OF THE HOMOGENEOUS OXIDATION OF SULFUR DIOXIDE IN THE TROPOSPHERE

JACK G. CALVERT, FU SU
Chemistry Department, The Ohio State University, Columbus, OH 43210, U.S.A.

and

JAN W. BOTTENHEIM and OTTO P. STRAUSZ
Hydrocarbon Research Center, University of Alberta, Edmonton, Alberta, Canada, T6G 2G2

(*Received in final form* 2 *August* 1977)

Abstract—An evaluation has been made of the existing kinetic data related to the elementary, homogeneous reactions of SO_2 within the troposphere. A set of preferred values of the rate constants for these reactions is presented. Simulations using these data provide significant new evidence that the oxidation of SO_2 can occur at substantial rates through these homogenous reaction paths. The direct photo-oxidation of SO_2 by way of the electronically excited states of SO_2 is relatively unimportant for most conditions which occur within the troposphere. The oxidation of SO_2 within the natural troposphere is expected to occur largely by way of reactions 39, 31, and 33, with reaction 39 being the dominant path: $HO + SO_2\ (+M) \rightarrow HOSO_2\ (+M)$ (39); $HO_2 + SO_2 \rightarrow HO + SO_3$ (31); $CH_3O_2 + SO_2 \rightarrow CH_3O + SO_3$ (33). By combining our kinetic estimates with the Crutzen and Fishman calculation of [HO], [HO_2], and [CH_3O_2] for the troposphere, we estimate that the total rates of SO_2 oxidation as high as 1.5% h^{-1} are expected at midday in July in the midlatitudes. Theoretical estimates of the monthly rates averaged over the northern hemisphere vary from a low of 0.1% h^{-1} in January to a maximum of about 0.2% h^{-1} in July. From our computer simulations of the reactions within an SO_2, NO_x, hydrocarbon, CO, aldehyde-polluted lower troposphere, it is predicted that the three reactions, 39, 31, and 33 occur with about equal rates; SO_2 oxidations for this case can proceed homogeneously at rates as high as 4% h^{-1}. Considerations of the reactions in stack gas plumes suggest that a small maximum in the SO_2 photo-oxidation rate may occur during the early stages of the dispersion of a parcel of the stack gases into the air. This should be followed by a short period of slower oxidation. In theory the initial burst is expected to arise from NO_2 and HONO photolysis followed by reaction 39 and the reaction, $O(^3P) + SO_2\ (+M) \rightarrow SO_3\ (+M)$. After the extensive dilution of the stack gases by polluted urban air, the rate of SO_2 homogeneous oxidation is expected to approach that for a typical polluted urban atmosphere ($\sim$4% h^{-1}).

1. INTRODUCTION

The day has arrived when it is possible to include more than an empirically assigned, first order rate constant for SO_2 conversion to 'sulfate' in our atmospheric transport and conversion models for SO_2. Atmospheric scientists have shown a new interest in the quantitative evaluation of the significance of the possible paths which convert SO_2 to its oxidation products, SO_3, H_2SO_4, NH_4HSO_4, $(NH_4)_2SO_4$, etc. It is now generally recognized that the development of scientifically sound and lasting strategy for 'sulfate' aerosol control requires that we understand the nature of the various atmospheric reactants which oxidize SO_2, the chemical nature of the products formed in its reactions, the rates at which these reactions will occur in a given atmosphere, as well as the physical processes which transport these gases and aerosols about the atmosphere and those which remove them from it. Our research groups at the Ohio State University and the University of Alberta are two of the many scientific research groups throughout the world which are actively engaged in the study of the kinetics and mechanisms of the elementary gas phase reactions which can lead to the homogeneous chemical transformation of SO_2 within the troposphere. The understanding of the homogeneous chemistry of SO_2 appears at first sight to be one of the simplest of the several important tasks faced by atmospheric scientists. The evaluation of the mechanisms and rates of the heterogeneous paths of SO_2 oxidation within the troposphere, the significance of surface removal processes, and the transport and diffusion processes are less amenable to laboratory study. However as we shall see in our discussions, there are as many unsolved problems as there are well defined areas of knowledge related to the homogeneous atmospheric chemistry of SO_2.

In any case, the time is right to pause and take note of those aspects of the problem which appear to be well understood, as well as to reconsider some of the apparently conflicting observations, the unexplained results, and the alternative hypotheses which are so common in this area of research. The state of this science has improved greatly in recent years, and some new and definitive conclusions can be formulated from the wealth of existing information. In many cases where confusion remains, at least in our minds, we will offer our speculation on the possible resolutions. We hope that our prejudices will be readily distinguished from the experimental facts we quote and that they will serve to stimulate the reader to an active participation in the future solution to some of the many problems which remain unresolved.

We begin this study with an evaluation of the possible atmospheric reactions of the electronically excited SO_2 molecules. Next we consider the various possible reactions of the ground state SO_2 molecule with the reactive atmospheric components. From this we derive a set of our 'preferred' values for the rate constants of the various elementary reactions. In the final section we attempt to evaluate the relative importance of the various homogeneous SO_2 conversion modes in the several major types of gaseous atmospheres which we encounter within the troposphere: the 'clean' troposphere; the NO_x-hydrocarbon-CO aldehyde-polluted regions of the troposphere, and finally the gaseous mixtures peculiar to the stack gas plumes.

2. TROPOSPHERIC REACTIONS OF ELECTRONICALLY EXCITED SULFUR DIOXIDE

(a) *The nature of the reactive states in* SO_2 *photochemistry*

The photochemistry of sulfur dioxide excited within the lower atmosphere provides, in principle, several reaction pathways which may lead to the oxidation or other transformation of SO_2. There has been general agreement in recent years that several other paths of SO_2 oxidation are probably much more rapid than those involving the excited states of SO_2. However this consensus was reached on the basis of a very limited body of information, and a new quantitative look at the reactions of excited SO_2 is in order. Some new insight into the rates and mechanisms of these reactions is possible from the results of recent studies.

Sulfur dioxide absorbs light within the u.v. region of the solar radiation incident within the troposphere. Its two near-u.v. absorption bands can be seen in Fig. 1. The dashed curve represents the wavelength dependence of the relative quanta $cm^{-2} s^{-1}$ of the solar actinic flux incident near sea level at the solar zenith angle of 40° (Peterson, 1976). The regions of overlap of these curves show that the absorption of solar energy by SO_2 in the troposphere occurs within the long wavelength tail of the relatively strong 'allowed' band of SO_2 (2900–3300 Å) as well as within the much weaker 'forbidden' band in the 3400–4000 Å region. Only when SO_2 is photoexcited at wavelengths less than 2180 Å is photodissociation of SO_2 energetically possible, $SO_2 + h\nu(\lambda < 2180\ \text{Å}) \rightarrow O(^3P) + SO(^3\Sigma^-)$. Thus solar radiation absorbed by SO_2 within the troposphere leads to non-dissociative, excited states of SO_2, and it is the nature of these states and their chemical reactions which concerns us here.

It is clear that excitation within the 'forbidden' band of SO_2 leads to the population of the $SO_2(^3B_1)$ molecule, reaction I (Brand *et al.*, 1970; Merer, 1963; Su *et al.*, 1977a):

$$SO_2(\tilde{X}\,^1A_1) + h\nu(3400 < \lambda < 4000\ \text{Å}) \rightarrow SO_2(^3B_1). \quad \text{(I)}$$

Excitation of SO_2 within its first allowed absorption band leads to the generation of two emitting singlet species (Brus and McDonald, 1974; Su *et al.*, in press), a very short-lived state which may be the $SO_2(^1A_2)$ species, and a long-lived state which is probably the $SO_2(^1B_1)$ molecule:

$$SO_2(\tilde{X}\,^1A_1) + h\nu(2400 < \lambda < 3300\ \text{Å}) \rightarrow SO_2(^1A_2) \quad \text{(II)}$$

$$\rightarrow SO_2(^1B_1). \quad \text{(III)}$$

Collisional perturbation of these singlet states results in an additional efficient pathway for the formation of the $SO_2(^3B_1)$ species:

$$SO_2(^1A_2, {}^1B_1) + M \rightarrow SO_2(^3B_1) + M. \quad (1)$$

In addition to these optically detectable or emitting states of SO_2, indirect chemical and physical evidence has led to the conclusion by some workers that two other non-emitting triplet states, presumably $SO_2(^3A_2)$ and $SO_2(^3B_2)$, are important in the photochemistry of SO_2. For example, see Cehelnik *et al.* (1971) and Kelley *et al.* (1976–77) and the references

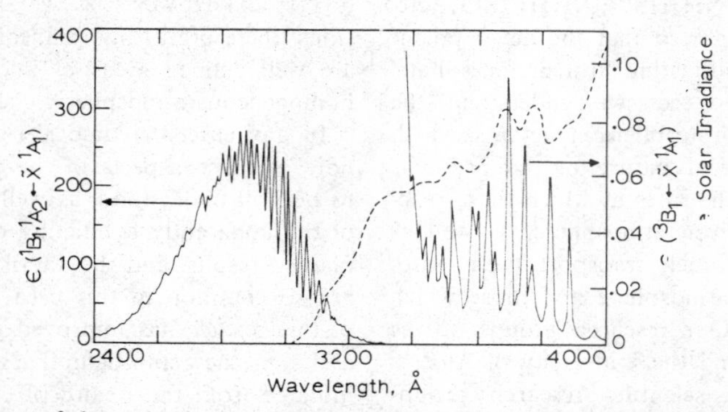

Fig. 1. Comparison of the extinction coefficients of SO_2 within the first allowed band (left), the 'forbidden' band (right), and a typical wavelength distribution of the flux of solar quanta (relative) at ground level (dashed curve).

therein. However recent observations of Rudolph and Strickler (1977), Su *et al.* (1977b), and Su and Calvert (1977) open to serious question this conclusion. It has been observed that the phosphorescent lifetime of the $SO_2(^3B_1)$ molecules at atmospheric pressure deviates markedly from the simple Stern–Volmer behavior characteristic of the low pressure regime. The lifetime of the $SO_2(^3B_1)$ molecule is about 2.4 times longer in one atmosphere of air than extrapolation of the low pressure quenching rate data would suggest. The phenomenon was first observed and rationalized by Rudolph and Strickler (1977) in terms of the saturation of the triplet-quenching at high pressures, an expectation of the recent theory of Freed (1976). A consideration of the details is not in order here, but a qualitative picture will suffice. The rate of triplet molecule quenching to ground state becomes independent of added gas pressure when the nearly degenerate rotational sublevels of the different singlet vibronic levels of the ground electronic state have collision-induced broadening which exceeds their spacings, so that the rovibronic manifold becomes an effective quasicontinuum. In the case of SO_2 this saturation occurs at pressures above about 100 Torr of added nitrogen gas. On the other hand, the chemical quenching of $SO_2(^3B_1)$ molecules follows second order kinetics as expected from theory, and for these cases, the low pressure bimolecular rate constants are applicable even at 1 atmosphere pressure. Su and Calvert (1977) have reevaluated the previous high pressure studies of SO_2–CO and SO_2–C_2H_2 mixture photolyses with SO_2 excitation within the 'forbidden' band. They have found that a simple mechanism involving only a single excited state of SO_2, the $SO_2(^3B_1)$ species, and the newly discovered saturation-quenching effect explain well all of the experimental data. Thus the postulate of Kelly *et al.* (1976/77) that the photolysis of SO_2–C_2H_2-added gas mixtures within the 'forbidden' band involves the reactions of three different excited states of SO_2 appears to be incorrect.

When SO_2 in the atmosphere is excited within the first allowed band, it is unlikely that the newly formed $SO_2(^1B_1)$ and $SO_2(^1A_2)$ species will live to react with impurity molecules, since they are quenched by collisions with the major atmospheric gases so very efficiently (Su *et al.*, in press). The possible chemical interaction of the excited singlet states with O_2 may occur in principle to generate an SO_4 excited triplet species or SO_3 and $O(^3P)$ in the reactions: $SO_2(^1A_2, {}^1B_1) + O_2(^3\Sigma_g^-) \rightarrow SO_4$ (or $SO_3 + O(^3P)$). However the near equality of the rates of quenching of each of the singlet states by the potentially reactive oxygen molecule and the 'unreactive' nitrogen molecule argues against this possibility (Su *et al.*, in press). The hypothetical energy transfer reaction, $SO_2(^1A_2, {}^1B_1) + O_2(^3\Sigma_g^-) \rightarrow SO_2(\tilde{X}\,^1A_1) + O_2(^1\Delta_g, {}^1\Sigma_g^+)$, is spin forbidden and is probably unimportant also (Davidson *et al.*, 1972–73). However the quenching does form the reactive $SO_2(^3B_1)$ molecule. Ground state molecules and conceivably some other unreactive or reactive species are formed in reaction 2.

$$SO_2(^1A_2, {}^1B_1) + M \rightarrow SO_2(\tilde{X}\,^1A_1) + M \text{ (or other products)}. \quad (2)$$

Important for our considerations here is the intersystem crossing ratio, $k_1/(k_1 + k_2)$; this is the fraction of the originally excited singlet species which form the reactive $SO_2(^3B_1)$ molecules. This fraction has been measured for various M species in experiments at relatively low pressures and rather short wavelengths of SO_2 excitation at 2650, 2662, and 2875 Å (Horowitz and Calvert, 1972a, b; Wampler *et al.*, 1973a). In view of the recent saturation quenching effect observed with $SO_2(^3B_1)$, it may be inaccurate to use these low pressure data for our estimates for SO_2 in one atmosphere of air. The density of the vibronic states of the electronically excited singlet and triplet states of SO_2 are significantly less than that of the ground state SO_2 molecule at energies in the vibronic manifold equivalent to the excitation of the $SO_2(^3B_1)$ or the $SO_2(^1B_1, {}^1A_2)$ states. As a result one would expect that the rate constants for the reaction 1 would not pressure saturate at as low a pressure as the rate constant for the reactions 2 and 3:

$$SO_2(^3B_1) + M \rightarrow SO_2(\tilde{X}\,^1A_1) + M. \quad (3)$$

This may lead to an increase in the fraction of the singlets which are transformed into triplets when higher pressures of the quenching gases are employed in the photochemical experiments. Although there is now no direct experimental evidence of the saturation effect in singlet quenching, the possibility exists, and one must exercise caution in the use of the low pressure values for $k_1/(k_1 + k_2)$ to estimate $SO_2(^3B_1)$ formation rates in air at 1 atm. Furthermore recent observations of Su and Calvert (in preparation) suggest that there is an increased efficiency of inter-system crossing for SO_2 singlet excitation at the longer wavelengths. Thus the estimates of $k_1/(k_1 + k_2)$ from the experiments in the 2650 to 2875 Å range are probably inappropriate for SO_2 excited singlets populated at the relatively long wavelengths present in the solar spectrum within the troposphere ($\lambda > 3000$ Å). Both the effects outlined would lead to the more efficient generation of $SO_2(^3B_1)$ on excitation of SO_2 at the long wavelength region of the first excited singlet band than would be anticipated from the low pressure, short wavelength data. In view of this discussion we should accept with great reservations the postulate of the participation of other reactive states of SO_2 to explain the 'excess' triplet observed in singlet excited SO_2-containing systems at high pressures and long wavelength excitation.

The evidence at hand today appears to support the original contention of Okuda *et al.* (1969), that the $SO_2(^3B_1)$ molecule is the major photochemically active species formed in the photoexcitation of SO_2 even within the first allowed singlet absorption band

Table 1. Estimated rates of $SO_3(^3B_1)$ generation by solar radiation in lower troposphere*

Solar zenith angle, deg	Rate by I† / Rate by II, 1			Total rate of $SO_2(^3B_1)$ generation, (% h^{-1})		
	r.h. 0%	50%	100%	0%	50%	100%
A. Intersystem crossing ratios from the low pressure experiments at 2662 Å of Wampler *et al.* (1973); no singlet quenching saturation is assumed‡						
0	0.093	0.085	0.078	4.03	4.36	4.72
20	0.099	0.090	0.083	3.64	3.96	4.28
40	0.125	0.113	0.104	2.59	2.84	3.06
60	0.243	0.220	0.203	1.04	1.15	1.22
80	0.569	0.515	0.475	0.18	0.18	0.18
B. Intersystem crossing ratios for all gases assumed to be 0.11 at 1 atm pressure						
0		0.0273			12.9	
20		0.0289			11.7	
40		0.0365			8.2	
60		0.0710			3.1	
80		0.166			0.43	

* Percentage of singlet SO_2 quenching by the various atmospheric components at 25°C and 1 atm of air were estimated using $k_{N_2}/k_{SO_2} = 0.326$; $k_{O_2}/k_{SO_2} = 0.321$; $k_{Ar}/k_{SO_2} = 0.275$ (Su *et al.*, in press); $k_{H_2O}/k_{SO_2} = 1.2$ (Mettee, 1969); Rates of direct excitation of $SO_2(^3B_1)$ and $SO_2(^1A_2, {}^1B_1)$ are from Sidebottom *et al.* (1972).

† These rates refer to the reaction I and II followed by 1 of the text; the ratio gives the relative population of $SO_2(^3B_1)$ through absorption in the forbidden band to that for the first allowed band.

‡ The values used are: $k_1/(k_1 + k_2) = 0.033$ ($M = N_2$); 0.030 ($M = O_2$); 0.025 ($M = Ar$); 0.09 ($M = H_2O$); 0.095 ($M = SO_2$).

at high pressures. Thus if we assume an average intersystem crossing ratio of 0.10 for $M = SO_2$ in Okuda's experiments, an effective first order rate constant for $SO_2(^3B_1)$ quenching to ground state of $1.5 \times 10^6\ s^{-1}$ for 730 Torr of SO_2 (Su *et al.*, 1977b), the bimolecular chemical rate constant for reaction 4, $k_4 = 7.0 \times 10^{-14}\ cm^3\ mol^{-1}\ s^{-1}$ (Chung *et al.*, 1975).

$$SO_2(^3B_1) + SO_2 \rightarrow SO_3 + SO(^3\Sigma^-) \qquad (4)$$

$\Phi_{SO_3} = 0.026$ from the singlet excited SO_2, we derive the observed value of $\Phi_{SO_3} = 0.08$ reported by Okuda *et al.* (1969).

We may update the earlier estimates of the rate of $SO_2(^3B_1)$ generation in the lower atmosphere (Sidebottom *et al.*, 1972; Penzhorn *et al.*, 1974a) utilizing all of the current data and alternative 'reasonable' estimates of the effective intersystem crossing ratio. These data are shown in Table 1. The calculations have been made for two cases: (1) We have chosen the experimental values for the intersystem crossing ratio $k_1/(k_1 + k_2)$ determined by Wampler *et al.* (1973) from experiments at low pressures of added atmospheric gases; i.e. we assume that there are no pressure saturation or wavelength effects on the intersystem crossing ratio; see Section A of Table 1. (2) In Section B of the Table we have taken the intersystem crossing ratio as 0.11 at 1 atm for all of the atmospheric gases; i.e. we have assumed that pressure saturation effects and the longer solar wavelengths populating the excited singlet SO_2 resulted in an increase in the ratio observed at low pressures. It can be seen from the data of Table 1 that the maximum rate of excitation of $SO_2(^3B_1)$ for a solar zenith angle of 0° (in a clean lower troposphere at 100% r.h.) ranges from 4.7% h^{-1} in the first case to 12.9% h^{-1} in the second case.

(b) *The nature and rate of the* $SO_2(^3B_1)$–O_2 *quenching reaction*

Obviously the rate of excitation of $SO_2(^3B_1)$ does not alone establish the maximum rate of SO_2 photoreaction since quenching by the unreactive gases such as N_2, Ar, CO_2, and H_2O leads to no net chemical change. However there are several types of reactive species in the atmosphere which are of special interest as possible reactants with $SO_2(^3B_1)$ molecules. The first of these reactants is oxygen. Table 2 summarizes the estimated fractions of $SO_2(^3B_1)$ which will be quenched by the various atmospheric components; we have used the newly determined, pressure saturation quenching data in deriving these estimates. It should be noted that the quenching of $SO_2(^3B_1)$ by O_2 does not pressure saturate, as does that by N_2, Ar, CO_2, and other chemically unreactive gases. This points to some type of chemical action which accompanies this quenching action. The rate constant for the quenching of $SO_2(^3B_1)$ by O_2, $k = 1.6 \times 10^{-13}\ cm^3\ mol^{-1}\ s^{-1}$, is very similar to that observed for

Table 2. Percentage of $SO_2(^3B_1)$-quenching by various atmospheric gases in air at 25°C and 1 atm

Component	Percentage of $SO_2(^3B_1)$-quenching r.h. 0%	50%	100%
Nitrogen	55.5	45.7	38.8
Oxygen	44.1	41.7	39.9
Argon	0.4	0.3	0.3
Water	0	12.2	21.1

N_2, $k = 1.4 \times 10^{-13}$ cm^3 mol^{-1} s^{-1} (Sidebottom *et al.*, 1972). We have argued previously that this might indicate a physical, non-chemical, quenching of $SO_2(^3B_1)$ by O_2 as well as N_2. However if this were the case then, contrary to the fact, the saturation of $SO_2(^3B_1)$ quenching by O_2 would be seen at high O_2 pressures. From the data of Tables 1 and 2 we can estimate that the maximum rate of $SO_2(^3B_1)$–O_2 chemical interaction in air at 50% relative humidity (25°C) and solar zenith angle of 40°, will range from 1.8% h^{-1} with mechanism A and 3.4% h^{-1} with mechanism B of Table 1. Obviously these rates are very significant if chemical changes in SO_2 do result from the interaction.

We have considered previously several possible chemical changes which could occur in theory as $SO_2(^3B_1)$ is quenched by O_2 (Sidebottom *et al.*, 1972). Ozone formation from excited $SO_2(^3B_1)$–O_2 interactions is possible through two reaction routes:

$$SO_2(^3B_1) + O_2(^3\Sigma_g^-) \rightarrow SO_4(\text{cyclic}); \quad \Delta H \cong -79 \text{ kcal mole}^{-1} \tag{5}$$

$$SO_4(\text{cyclic}) + O_2 \rightarrow SO_3 + O_3; \quad \Delta H \cong 15 \text{ kcal mole}^{-1} \tag{6}$$

$$SO_2(^3B_1) + O_2(^3\Sigma_g^-) \rightarrow SO_3 + O(^3P); \quad \Delta H = -38 \text{ kcal mole}^{-1} \tag{7}$$

$$O + O_2(+M) \rightarrow O_3(+M). \tag{8}$$

Presumably the singlet SO_4 (cyclic) species could be formed in 5 and stabilized by collisions. However even when vibrationally equilibrated, this species is rather unstable with respect to decomposition into SO_2 and O_2 ($\Delta H = 5$ kcal mole^{-1}). A vibrationally rich SO_4 species initially formed in 5 could in principle react in 6, but the thermally stabilized SO_4 cannot react significantly to generate O_3 and SO_3 in reaction 6 at atmospheric temperatures, since this reaction is about 15 kcal mole^{-1} endothermic. See Section 3a for a more in depth consideration of possible roles of the SO_4 intermediate in atmospheric chemistry. Reaction 7 is energetically possible, but we would expect it to be slow since it involves an electron spin inversion. Indeed there is a great deal of experimental evidence which suggests the relative inefficiency of these reactions in pure air. Thus the photooxidation of dilute mixtures of SO_2 in oxygen leads to very small quantum yields of SO_2 oxidation to SO_3 or H_2SO_4. In those previous SO_2 photooxidation studies where a relatively large fraction of the gaseous mixture was SO_2 (Hall, 1953; Sethi, 1971; Allen *et al.*, 1972a, b), it has been shown by Chung *et al.* (1975) that the quantum yield of SO_3 can be accounted for in large part through reactions 4 and 9:

$$SO_2(^3B_1) + SO_2 \rightarrow SO_3 + SO(^3\Sigma^-) \tag{4}$$

$$SO + SO_3 \rightarrow 2SO_2. \tag{9}$$

Relative large quantum yields of SO_3 formation in SO_2-rich mixtures are observed at short irradiation times (Skotnicki, 1975) or in flow systems in which SO_3 is removed quickly and the reverse reaction 9 is unimportant (Chung *et al.*, 1975). In steady state photolyses in static systems, the rapid build up of the relatively low levels of SO_3 and SO is followed largely by reformation of SO_2 in 9. Much of the SO_3 (or H_2SO_4) which was formed in previous studies of the solar u.v.-irradiated SO_2–O_2 or SO_2–clean air mixtures probably was derived from the reaction 4 and bears no relation to the occurrence of the theoretically possible photooxidation steps 5 and 6 or 7 and 8. Several studies show that there is a small, but not insignificant rate of SO_3 (H_2SO_4) formation in very dilute SO_2–air mixtures; thus for typical solar intensities at ground level in clean air, the following SO_2 oxidation rates (at low [SO_2]) were estimated experimentally (% h^{-1}): 0.65 (Cox and Penkett, 1970: 0.02–0.04 (Cox, 1972, 1973); <0.1 (Junge, 1972); 0.7 (Kasahara and Takahashi, 1976); 0.00 (Friend *et al.*, 1973).

Altwicker (1976) reported recently that the generation of small amounts of O_3 occurs in dilute SO_2-containing purified air samples irradiated in sunlight; e.g. 1–3 pphm more ozone was developed in a pure air mixture with SO_2 (3 ppm) after 40 min irradiation than was formed in the irradiation of a similar SO_2-free sample of purified air (0.6–1.0 pphm). His observations are interesting and in qualitative accord with the occurrence of the reaction sequences 5 and 6 and/or 7 and 8 at the rate of about 0.7% h^{-1} or less. However further more detailed studies of the ozone formation in SO_2-air mixtures must be carried out before meaningful conclusions concerning the mechanism of O_3 generation in Altwicker's experiment can be established. However with the exception of Friend's observation, all of the present studies show a small, but not negligible, rate of the direct photooxidation of SO_2 occurs in pure air. Since it is so difficult to obtain pure air samples which have no small impurities of NO_x, O_3, RH, etc., one should not give great weight to the photooxidation experiments as proof of reactions 5 and 6 and/or 7 and 8 occurring. We conclude that the present data are consistent with an upper limit for the rate constants, $k_5 + k_7 = 3 \times 10^{-15}$ cm^3 mol^{-1} s^{-1}.

At most a few per cent of the $SO_2(^3B_1)$ quenching collisions with O_2 lead to SO_3. Thus some other 'chemical' result must account for most of the quenching. Some further clues about this interaction can be had in consideration of the nature of the oxygen product formed in this process and the seemingly similar organic triplet molecule quenching by oxygen. It appears to us that the major chemical result of $SO_2(^3B_1)$–O_2 quenching encounters is likely energy transfer to generate excited singlet-O_2 species:

$$SO_2(^3B_1) + O_2(^3\Sigma_g^-) \rightarrow SO_2(\tilde{X}\,^1A_1) + O_2(^1\Sigma_g^+) \tag{10a}$$

$$\rightarrow SO_2(\tilde{X}\,^1A_1) + O_2(^1\Delta_g). \tag{10b}$$

The studies of the Abrahamson group (Davidson and Abrahamson, 1972; Davidson *et al.*, 1972–73) have shown convincingly that reaction 10a occurs. The fraction of the reaction which leads to the alternative singlet states of O_2 is not now known. However the experimental data are not inconsistent with $k_{10a} + k_{10b} \cong 1.6 \times 10^{-13}$ cm^3 mol^{-1} s^{-1}, near the total quenching rate constant for $SO_2(^3B_1)$ reacting with O_2. Support for this view is added from a comparison of these data with those obtained from organic triplet molecule quenching by O_2; in this case the analogous energy transfer reactions occur efficiently. In Fig. 2 is shown a plot of the logarithm of the rate constants for the quenching of a series of organic triplet molecules by oxygen versus the energy difference between the first excited triplet and ground state of the various molecules (E_{T_1}–E_{S_0}). Data shown as open squares are from studies in benzene solvent; open circles are from hexane solutions (Patterson *et al.*, 1970; Gijzeman *et al.*, 1973a). Since these reactions at high E_{T_1}–E_{S_0} values are not diffusion controlled, and non-polar solvents were employed, a comparison with gas phase rate constants should be meaningful. It is seen from the positions of the closed circles that our gas phase rate constant data for O_2-quenching of triplets of SO_2 and $(CH_3CO)_2$ follow well the trend observed with the other triplet molecules (Horowitz and Calvert, 1972; Sidebottom *et al.*, 1972). Thus it seems most reasonable that the same triplet energy transfer mechanism invoked for organic triplet $M(T_1)$ quenching by O_2

$$M(T_1) + O_2(^3\Sigma_g^-) \rightarrow M(S_0) + O_2(^1\Sigma_g^+, ^1\Delta_g) \quad (11)$$

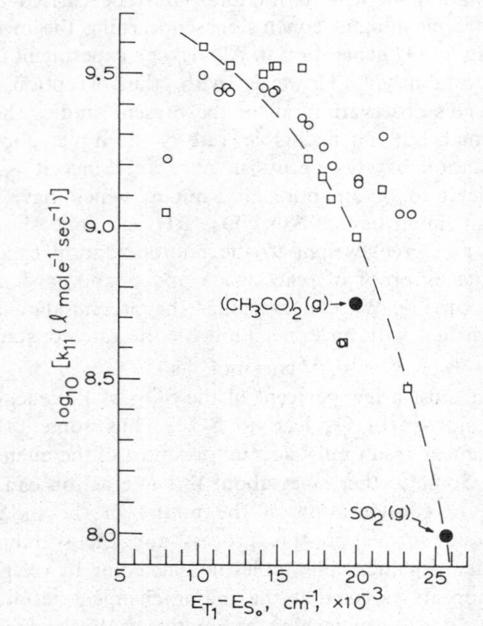

Fig. 2. Relation between the rate constants for the triplet-quenching reaction with oxygen and the energy separation between the first excited triplet and the ground state of a series of molecules: aromatic triplets measured in hexane solution (squares), benzene solution (open circles) from Patterson *et al.* (1970) and Gijzemen *et al.* (1973a); $(CH_3CO)_2$ and SO_2 data from gas phase measurements (Horowitz and Calvert, 1972; Sidebottom *et al.*, 1972).

is descriptive of that for $SO_2(^3B_1)$ as well (Kawaoka *et al.*, 1967; Kearns, 1971; Gijzeman and Kaufman, 1973; Gijzeman, 1974).

In summary the present data are consistent with singlet oxygen formation as the major result of $SO_2(^3B_1)$ quenching by O_2 ($k_{10a} + k_{10b} \cong 1.6 \times 10^{-13}$ cm^3 mol^{-1} s^{-1}), and the oxidation of SO_2 to SO_3 results at most a small fraction of the quenching collisions ($k_5 + k_7 = 3 \times 10^{-15}$ cm^3 mol^{-1} s^{-1}).

(c) *The nature and rates of the $SO_2(^3B_1)$-chemical quenching reactions with the common atmospheric contaminants*

The extensive rate data related to $SO_2(^3B_1)$ quenching reactions points to some potentially significant chemical reactants with $SO_2(^3B_1)$ among the common air pollutants. Thus the very large rate constant for the quenching of the triplets with NO (1.2×10^{-10} cm^3 mol^{-1} s^{-1}; Sidebottom, 1972) suggests some important chemistry at first sight. Although the possible reaction, $SO_2(^3B_1) + NO \rightarrow SO(^3\Sigma^-) + NO_2$ is exothermic by 15 kcal mole^{-1}, there is no evidence that this chemical change or any other of significance occurs. Nitric oxide has been found to quench efficiently the triplet excited states of many molecules. The lowest excited electronic state of NO, a $^4\Pi$ state, lies 35,000 cm^{-1} above the ground state, so energy transfer from $SO_2(^3B_1)$ 25,766 cm^{-1} above $SO_2(\tilde{X}\,^1A_1)$, is not a possible quenching mechanism here. However theoretical studies show that intersystem crossing of an excited triplet molecule ($^3M^*$) to its ground state (^1M) may be catalyzed by its interaction with nitric oxide (Gijzeman *et al.*, 1973):

$$NO(^2\Pi) + {}^3M^* \rightarrow {}^2(M\text{–}NO) \rightarrow NO(^2\Pi) + {}^1M. \quad (12)$$

Gijzeman *et al.* (1973b) have observed a correlation between the magnitude of the rate constant for triplet molecule quenching by NO and the triplet-ground state energy difference of the molecules; see Fig. 3. The open circles are data from hexane solution studies of a variety of organic triplet molecules. They have suggested a theory which can rationalize this correlation in terms of reaction 12. Observe the magnitude of the quenching rate constant increases dramatically with E_{T_1}–E_{S_0} value for energy separations greater than 1.5×10^4 cm^{-1}. This trend is just the opposite to that observed in Fig. 2 and the O_2-quenching of triplets where the mechanism is electronic energy transfer to O_2. Our data from the gas phase studies of the quenching of SO_2 and biacetyl triplet molecules by NO are plotted in Fig. 3 as closed circles. It is seen that these molecules follow the same trend observed for the organic triplet molecules. Thus it is probable that the net effect of $SO_2(^3B_1)$ quenching by NO can be represented simply by reaction 13; no interesting net chemistry is expected here.

$$SO_2(^3B_1) + NO(^2\Pi) \rightarrow {}^2(SO_2\text{–}NO) \rightarrow SO_2(\tilde{X}\,^1A_1) + NO(^2\Pi). \quad (13)$$

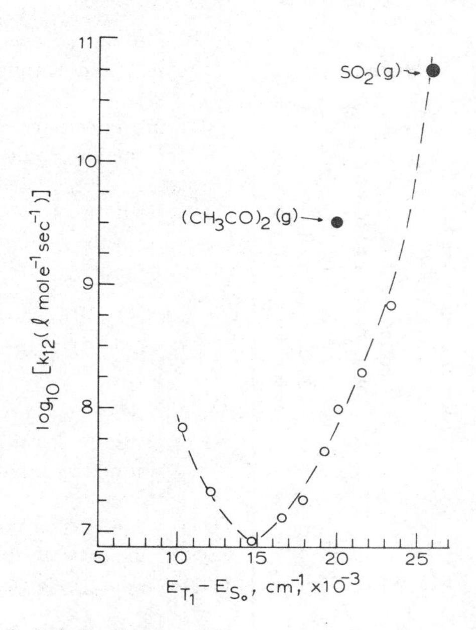

Fig. 3. Relation between the rate constant for the triplet-quenching reaction with nitric oxide and the energy separation between the first excited triplet and the ground state of a series of molecules; aromatic triplets measured in hexane solutions (open circles) from Gijzeman *et al.* (1973b); $(CH_3CO)_2$ and SO_2 data from gas phase measurements (Horowitz and Calvert, 1972; Sidebottom *et al.*, 1972).

Other potential atmospheric reactants with $SO_2(^3B_1)$ are SO_2, CO, C_2H_2, alkenes, and alkanes. The quenching rate constants for the 'chemical' part of the reactions of $SO_2(^3B_1)$ with some representative impurity molecules and atmospheric components are summarized in Table 3. The $SO_2(^3B_1)$ can act as an oxidizing agent to form CO_2 from CO in reaction 14 (Cehelnik *et al.*, 1971, 1973–74; Kelly *et al.*, 1976–77; Jackson and Calvert, 1971; Wampler *et al.*, 1972). Through some undefined intermediate product, CO is formed from C_2H_2 in reaction 15 (Kelly *et al.*, 1976). The alkenes quench $SO_2(^3B_1)$ at rates which approach the collision frequency (Sidebottom *et al.*, 1971; Wampler *et al.*, 1973b). The quenching act appears to be the formation of a transient species resulting from the addition of SO_2 to the double bond of the alkene, reaction 16. This transient allows rotation about the bond, and its rapid dissociation leads to the *cis–trans* isomers of the alkene as the only net chemical effect of the quenching (Demerjian *et al.*, 1974). Although light scattering aerosol is formed in the irradiated SO_2–alkene systems, its formation appears to correlate best with the extent of SO_2 molecule quenching of $SO_2(^3B_1)$; it is thought to be SO_3 (or ultimately H_2SO_4) which causes the scattering observed following reaction 4 in the system. So although the alkenes quench $SO_2(^3B_1)$ with very high efficiency, the net trivial chemistry which results from this process is apparently of no importance to the removal or the chemical transformation of SO_2 in the atmosphere.

The alkanes quench $SO_2(^3B_1)$ reasonably effectively, and the magnitude of the quenching rate constant increases with decreasing strength of the C—H bonds and the increasing number of C—H bonds in the hydrocarbon (Badcock *et al.*, 1971; Wampler *et al.*, 1973b). The quenching act has been related to an H-atom abstraction by the $SO_2(^3B_1)$ molecule (Badcock *et al.*, 1971; Penzhorn *et al.*, 1975). Thus in the case of isobutane–SO_2 mixture photolysis, the measured chemical rate of quenching of $SO_2(^3B_1)$ to form sulfinic acids and other complex products is very nearly equal to the rate of $SO_2(^3B_1)$–isobutane quenching as measured by lifetime studies (Su and Calvert, in preparation).

$$SO_2(^3B_1) + iso\text{-}C_4H_{10} \rightarrow HOSO + C_4H_9. \quad (17)$$

The enthalpy change which accompanies reaction 17 is about -22 kcal mole^{-1}, not very different from that for the analogous HO-radical reaction with isobutane (-26.6 kcal mole^{-1}). If the abstraction of an H-atom occurs by $SO_2(^3B_1)$–alkane interaction in the atmosphere, the ultimate result of the quenching likely will not be the formation of sulfinic acids or other products characteristic of the laboratory studies of SO_2–RH mixtures. Thus the HOSO and R radicals formed in the quenching reaction 18 will probably react with O_2, the domiant reactive species in the air. The HOSO radical may combine with oxygen and ultimately form SO_3 or H_2SO_4 by a reaction sequence such as the following:

$$SO_2(^3B_1) + RH \rightarrow HOSO + R \quad (18)$$

$$HOSO + O_2 \xrightarrow{(a)} HOS(=O)OO \xrightarrow{(b)} HO + SO_3$$

$$HOS(=O)OO \xrightarrow{(c)} HOS(=O)(O)-O \xrightarrow{(d)} HO + SO_3$$

$$HOS(=O)(O)-O \xrightarrow{(+RH)} H_2SO_4.$$

The overall enthalpy change for steps a and b above is about -25 kcal mole^{-1}; $\Delta H_c \cong -65$; $\Delta H_d \cong 40$ kcal mole^{-1}. Thus path d will not occur. It seems less likely to us that the HOSO will react with O_2 in the less exothermic step, $HOSO + O_2 \rightarrow SO_2 + HO_2$; $\Delta H \cong -6$ kcal mole^{-1}. The radical R formed in 18 will give RO_2, RO, aldehydes, etc., in subsequent reactions in the atmosphere. In view of these considerations, net photooxidation of SO_2 could result from the quenching of $SO_2(^3B_1)$ by alkanes.

The predicted rates of the various 'chemical' quenching reactions of $SO_2(^3B_1)$ by impurity molecules in the atmosphere have been estimated for typical solar intensities encountered near ground level for a solar zenith angle of about 40° (see Table 4). The hypothetical atmosphere contains 1 ppm of the different impurity molecules. Various amounts of SO_2 are

Table 3. The $SO_2(^3B_1)$ "chemical" quenching rate constants for various atmospheric components and impurity species in the overall reactions shown

Reactant		Reaction	k, $cm^3\ mol^{-1}\ s^{-1}$	Reference
Oxygen	(5, 6)	$SO_2(^3B_1) + O_2 \rightarrow SO_4 \xrightarrow{O_2} SO_3 + O_3$		
	(7)	$SO_2(^3B_1) + O_2 \rightarrow SO_3 + O(^3P)$	$< 3 \times 10^{-15}$	See text.
	(10)	$SO_2(^3B_1) + O_2 \rightarrow SO_2 + O_2(^1\Delta_g, ^1\Sigma_g^+)$	$\sim 1.6 \times 10^{-13}$	Sidebottom *et al.* (1972)
Sulfur dioxide	(4)	$SO_2(^3B_1) + SO_2 \rightarrow SO_3 + SO(^3\Sigma^-)$	7.0×10^{-14}	Chung *et al.* (1975)
Nitric oxide	(13)	$SO_2(^3B_1) + NO(^2\Pi) \rightarrow {}^2(SO_2\text{–}NO) \rightarrow SO_2 + NO(^2\Pi)$	1.3×10^{-10}	Sidebottom *et al.* (1972)
Carbon monoxide	(14)	$SO_2(^3B_1) + CO \rightarrow SO(^3\Sigma^-) + CO_2$	4.3×10^{-15}	Jackson and Calvert (1971), Su and Calvert (1977)
Acetylene	(15)	$SO_2(^3B_1) + C_2H_2 \rightarrow (?) \rightarrow CO$ + other products	2.7×10^{-12}	Kelley *et al.* (1976), Su and Calvert (1977)
cis-2-Butene	(16)	$SO_2(^3B_1) + cis\text{-}C_4H_8 \rightarrow {}^3(SO_2\text{–}C_4H_8) \rightarrow trans\text{-}C_4H_8$; $\rightarrow cis\text{-}C_4H_8$	2.2×10^{-10}	Sidebottom *et al.* (1971), Demerjian and Calvert (1974)
Isobutane	(17)	$SO_2(^3B_1) + iso\text{-}C_4H_{10} \rightarrow HOSO + C_4H_9$	1.4×10^{-12}	Wampler *et al.* (1972), Badcock *et al.* (1971), Su and Calvert (in preparation)

chosen which range from 500 ppm, common to fresh stack gas emissions, to 0.05 ppm which is more representative of the levels of SO_2 present in ambient polluted air. The rates (ppm h^{-1}) of all of the reactions shown are very slow. Even under the most favorable conditions which one might choose to enhance the importance of these reactions, the rates are far below those which we estimate for the reactions of HO, HO_2, etc., with these impurities and SO_2 in the lower troposphere (see Section 3).

From our considerations we may conclude that the major chemical effect of SO_2 photooxidation by sunlight within the polluted atmosphere is the generation of $O_2(^1\Sigma_g^+)$ and $O_2(^1\Delta_g)$ through reactions 10a and 10b (see Table 4). Although the expected rates are relatively large, the rates of this reaction are about 1/20th of those expected from excited NO_2 reactions with O_2 when NO_2 and SO_2 are at comparable levels (Jones and Bayes, 1971; Frankiewicz and Berry, 1972; Demerjian *et al.*, 1974). The available data do support as well the occurrence of a slow, but not insignificant rate of SO_2 photooxidation, presumably through reactions 5 and 6 or 7 and 8. The maximum rates for these reactions correspond to less than 0.04% h^{-1} (see Table 4). The observed rates of SO_2 oxidation in air are much higher than this. Obviously the oxidation of SO_2 by reactions other than those involving photoexcited SO_2 molecules must be important within the troposphere. We shall consider these in the remaining sections of this study.

3. REACTIONS OF GROUND STATE SO_2 WITH REACTIVE MOLECULES AND TRANSIENT SPECIES IN THE TROPOSPHERE

Many reactive species present in the lower atmosphere are potentially important reactants for SO_2. A number of reviews have appeared in recent years in which the significance of the various reaction pathways for SO_2 conversion in the atmosphere has been estimated (Calvert and McQuigg, 1975; Davis and Klauber, 1975; Sander and Seinfeld, 1976; Bottenheim and Strausz, 1977; Levy *et al.*, 1976). Some significant new rate data related to these systems have become available recently, and it is timely for us to reevaluate the various potentially important SO_2 reaction pathways. The increasing use of these chemical reaction mechanisms in atmospheric models makes a careful, continuing reassessment of the kinetics and rate constant data related to these reactions of special value. We have summarized the thermochemistry of several reactions of interest in Table 5. Most of the reactions shown are exothermic, and from these energy considerations alone, they warrant our attention, since they could occur at significant rates at the temperatures of the troposphere. Also listed in Table 5 are our recommended values for the various rate constants. Each is expressed as an apparent second order rate constant which should be applicable for the pressures of the lower atmosphere near 25°C. In this section we will consider in some detail the data upon which each of these estimates were based.

(a) *The reactions of* $O_2(^1\Delta_g)$ *and* $O_2(^1\Sigma_g^+)$ *with* SO_2

The quenching reactions of $O_2(^1\Delta_g)$ and $O_2(^1\Sigma_g^+)$, reactions 19–24, have been studied quantitatively in recent years. Penzhorn *et al.* (1974) estimated the total quenching rate constant for $O_2(^1\Delta_g)$ by SO_2, $k_{19} + k_{20} + k_{21} = (3.9 \pm 0.9) \times 10^{-20}\ cm^3\ mol^{-1}\ s^{-1}$.

$$O_2(^1\Delta_g) + SO_2 \rightarrow SO_4(\text{biradical}); SO_4(\text{cyclic}) \quad (19)$$

$$O_2(^1\Delta_g) + SO_2 \rightarrow SO_3 + O(^3P) \quad (20)$$

$$O_2(^1\Delta_g) + SO_2 \rightarrow O_2(^3\Sigma_g^-) + SO_2. \quad (21)$$

The endothermicity and spin inversion of the possible reaction 20 excludes it from further consideration. The possible formation of the transient SO_4 species in reaction 19 may account for at least a portion of the total quenching reaction observed. Although many researchers have invoked the intermediate SO_4 in their considerations of SO_2 photooxidation since this suggestion was first made by Blacet (1952), only recently has any rather direct evidence for the existence of the SO_4 intermediate species been observed. The i.r. kinetic studies of the $O(^3P)$–SO_3 reaction in the gas phase by Daubendiek and Calvert (1973, paper in preparation) gave strong indirect evidence that the SO_4 intermediate was involved in the formation of transient molecules S_3O_9 and S_3O_8 observed in their system (see Section 3m). Kugel and Taube (1975) have prepared an SO_4 species in CO_2(s) and Ar(s) matrices at 78 and 15 K, respectively, through the action of O-atoms on SO_3. It is probably significant that Kugel and Taube saw no SO_4 formation when SO_2 was irradiated with 2537 Å in an O_2 matrix at 15 K. Presumably $O_2(^1\Delta_g, {}^1\Sigma_g^+)$ would be formed by interaction of the adjacent partners, $SO_2(^3B_1)$–O_2 under these circumstances, and reactions such as 19 or 22 could then occur. Neither of these reactions appear to be important, and the mechanism of deactivation of $O_2(^1\Delta_g)$ by SO_2 remains unclear. It probably involves no chemical change in the SO_2 molecule, but only electronic relaxation occurs as in 21. In any case for the usual steady state levels of $O_2(^1\Delta_g)$, about 10^8 mol cm^{-3}, which we might anticipate in a typical polluted urban atmosphere (Demerjian *et al.*, 1974), the maximum rate of quenching by SO_2 amounts to an insignificant, 1.4×10^{-6}% h^{-1}.

Kear and Abrahamson (1974–75) determined the rate constant for the quenching of $O_2(^1\Sigma_g^+)$ by SO_2 to be: $k_{22} + k_{23} + k_{24} = 6.6 \times 10^{-16}\ cm^3\ mol^{-1}\ s^{-1}$.

$$O_2(^1\Sigma_g^+) + SO_2 \rightarrow SO_4(\text{biradical}); SO_4(\text{cyclic}) \quad (22)$$

$$O_2(^1\Sigma_g^+) + SO_2 \rightarrow SO_3 + O(^3P) \quad (23)$$

$$O_2(^1\Sigma_g^+) + SO_2 \rightarrow SO_2 + O_2(^1\Delta_g). \quad (24)$$

The rate of $O_2(^1\Sigma_g^+)$ attack on SO_2 in the lower

Table 4. The theoretical rate of reaction (ppm h^{-1}) of $SO_2(^3B_1)$ reactions with various impurity species and O_2 in a hypothetical sunlight-irradiated lower troposphere*

Reactant molecule	Reaction No.		Initial $[SO_2]$, ppm 500	50	5	0.5	0.05
NO	13	A†	3.1×10^{-2}	3.1×10^{-3}	3.1×10^{-4}	3.1×10^{-5}	3.1×10^{-6}
		B†	8.9×10^{-2}	8.9×10^{-3}	8.9×10^{-4}	8.9×10^{-6}	8.9×10^{-6}
CO	14	A	9.9×10^{-7}	9.9×10^{-8}	9.9×10^{-9}	9.9×10^{-10}	9.9×10^{-11}
		B	2.9×10^{-6}	2.9×10^{-7}	2.9×10^{-8}	2.9×10^{-9}	2.9×10^{-10}
C_2H_2	15	A	6.1×10^{-4}	6.1×10^{-5}	6.1×10^{-6}	6.1×10^{-7}	6.1×10^{-8}
		B	1.8×10^{-3}	1.8×10^{-4}	1.8×10^{-5}	1.8×10^{-6}	1.8×10^{-7}
cis-2-C_4H_8	16	A	5.0×10^{-2}	5.0×10^{-3}	5.0×10^{-4}	5.0×10^{-5}	5.0×10^{-6}
		B	1.5×10^{-1}	1.5×10^{-2}	1.5×10^{-3}	1.5×10^{-4}	1.5×10^{-5}
iso-C_4H_8	17	A	3.4×10^{-4}	3.4×10^{-5}	3.4×10^{-6}	3.4×10^{-7}	3.4×10^{-8}
		B	9.7×10^{-4}	9.7×10^{-5}	9.7×10^{-6}	9.7×10^{-7}	9.7×10^{-8}
SO_2	4	A	8.1×10^{-3}	8.1×10^{-5}	8.1×10^{-7}	8.1×10^{-9}	8.1×10^{-11}
		B	2.4×10^{-2}	2.4×10^{-4}	2.4×10^{-6}	2.4×10^{-8}	2.4×10^{-10}
O_2‡	10	A	<9.0	$<9.0 \times 10^{-1}$	$<9.0 \times 10^{-2}$	$<9.0 \times 10^{-3}$	$<9.0 \times 10^{-4}$
		B	<17.0	<1.7	$<1.7 \times 10^{-1}$	$<1.7 \times 10^{-2}$	$<1.7 \times 10^{-3}$
	5,6,7	A	$<1.0 \times 10^{-1}$	$<1.0 \times 10^{-2}$	$<1.0 \times 10^{-3}$	$<1.0 \times 10^{-4}$	$<1.0 \times 10^{-5}$
		B	$<1.9 \times 10^{-1}$	$<1.9 \times 10^{-2}$	$<1.9 \times 10^{-3}$	$<1.9 \times 10^{-4}$	$<1.9 \times 10^{-5}$

* Calculated for a solar zenith angle of 40°, near sea level, 25°C, 1 atm, 50% relative humidity, and 1 ppm of the specific impurity molecule present; in the case of SO_2 impurity, the column heads represent the amount present.
† Values in rows labeled A were calculated for the case A in Table 1 and 50% relative humidity; values in rows labeled B were calculated for case B in Table 1.
‡ Oxygen is assumed to be present in each case at 156.7 Torr (air at 50% relative humidity, 25°C).

atmosphere must be very low also, about $1.4 \times 10^{-7}\%\ h^{-1}$, for typical values of $[O_2(^1\Sigma_g^+)] = 6 \times 10^2$ mol cm^{-3} (Demerjian *et al.*, 1974). Thus we can conclude that the rate of singlet oxygen reactions with SO_2 in the lower atmosphere is insignificant, and we may confidently neglect them in our further considerations.

(b) *The reaction of* $O(^3P)$ *with* SO_2

The earlier kinetic work on the reaction 25 has been reviewed by Schofield (1973), Hampson and Garvin (1975), and Westenberg and deHaas (1975a).

$$O(^3P) + SO_2\,(+M) \rightarrow SO_3\,(+M). \qquad (25)$$

From the data available in 1973, Schofields's choice of constants gave $k_{25} = (1.3 \pm 1.3) \times 10^{-13}$ cm^3 $mol^{-1}\,s^{-1}$ for the apparent second order constant at 1 atm of N_2 (25°C). Hampson and Garvin picked the value of Davis *et al.* (1974a) as their preferred rate constant: $k_{25} = 2.0 \times 10^{-14}$ cm^3 $mol^{-1}\,s^{-1}$ (1 atm N_2, 25°C). Westenberg and deHaas (1975a) have concluded, however, that most of the earlier estimates of k_{25} which were based upon $O(^3P)$ disappearance rate are too high by a factor of 2. Their recent findings show that the reaction, $SO_3 + O(^3P) \rightarrow SO_2 + O_2$, is so very fast that a second O-atom loss is expected to occur almost every time the reaction 25 occurs for the conditions employed in many of the studies. From their estimates, $k_{25} = (4.9 \pm 0.2) \times 10^{-14}$ cm^3 $mol^{-1}\,s^{-1}$ for $M = N_2$ at 1 atm, 25°C. The only studies of k_{25} with O_2 as the third body come from the work of Mulcahy *et al.* (1967); correcting their estimate for the occurrence of reaction 46, in accord with the Westenberg and deHaas suggestion, we derive $k_{25} = (9.2 \pm 1.7) \times 10^{-14}$ for $M = O_2$ (1 atm, 25°C). Combining the best estimates for $M = N_2$ and $M = O_2$, we derive, $k_{25} = (5.7 \pm 0.5) \times 10^{-14}$ cm^3 $mol^{-1}\,s^{-1}$, M = air (1 atm, 25°C). The estimate which we have chosen is in reasonable accord with that observed by Atkinson and Pitts (1974) using N_2O as the third body: $k_{25} = 7.6 \times 10^{-14}$ cm^3 $mol^{-1}\,s^{-1}$ at 1 atm, 25°C; the greater number of internal degrees of freedom of N_2O than for N_2 and O_2 would result in a somewhat higher value of k_{25} for $M = N_2O$. Our present estimate is considerably higher than that of Davis *et al.* (1974) with $M = N_2$; the peculiarly high efficiency of SO_2 as M which was found by Davis *et al.* may have resulted from an overcorrection for the contribution of the reaction. $O(^3P) + SO_2 + SO_2 \rightarrow SO_3 + SO_2$, to the total rate measured in the N_2–SO_2 mixtures.

Our estimate of k_{25} coupled with a fairly normal $[O(^3P)]$ for a sunlight-irradiated, NO_x-polluted, lower troposphere, 2×10^5 mol cm^{-3}, the anticipated oxidation rate for SO_2 by reaction 25 is about $1.2 \times 10^{-2}\%\ h^{-1}$. Thus we expect reaction 25 to be relatively unimportant for the conditions normally encountered in the lower troposphere. We should retain this reaction in our simulations, however, since it is important, at least in theory, during the early stages of stack gas dilution (see Section 4).

(c) *The* O_3–SO_2 *reaction*

In view of the great exothermicity of reaction 26, it is surprisingly slow; the upper limit on its rate constant has been set by Davis *et al.* (1974b) as

$k_{26} < 10^{-22}$ and by Daubendiek and Calvert (1975) as $k_{26} < 8 \times 10^{-24}$ $cm^3\,mol^{-1}\,s^{-1}$.

$$O_3 + SO_2 \rightarrow SO_3 + O_2. \quad (26)$$

Using our upper limit estimate of k_{26} and an $[O_3] = 5 \times 10^{12}$ $mol\,cm^{-3}$, typical of photochemical smog in a highly polluted atmosphere, it can be shown that the SO_2 oxidation rate through reaction 26 will be less than $1.4 \times 10^{-5}\%\,h^{-1}$, truly unimportant for our further consideration.

(d) *The reactions of the oxides of nitrogen with* SO_2

The extrapolation of the high temperature rate data for the NO_2–SO_2 reaction 27 from Boreskov and Illarionov (1940) leads to the estimate, $k_{27} = 8.8 \times 10^{-30}$ $cm^3\,mol^{-1}\,s^{-1}$ for 25°C.

$$NO_2 + SO_2 \rightarrow NO + SO_3. \quad (27)$$

Data of Daubendiek and Calvert (1975) and Davis and Klauber (1975) give $k_{28} < 7 \times 10^{-21}$ and about 10^{-21}, respectively; these workers also report $k_{30} < 4 \times 10^{-23}$ and about 10^{-23} $cm^3\,mol^{-1}\,s^{-1}$, respectively.

$$NO_3 + SO_2 \rightarrow NO_2 + SO_3 \quad (28)$$

$$N_2O_5 + SO_2 \rightarrow N_2O_4 + SO_3. \quad (30)$$

For typical concentrations of NO_2 (5×10^{12} mol cm^{-3}; 0.20 ppm), the concentrations of NO_3 and N_2O_5 which are expected in heavy photochemical smog are about 2.5×10^7 and 2.5×10^9 $mol\,cm^{-3}$, respectively. Thus the rates of attack on SO_2 by NO_2, NO_3, and N_2O_5 should be of the order of: 1.6×10^{-11}, 6.3×10^{-8}, and $3.6 \times 10^{-8}\%\,h^{-1}$. The suggestion of the probable importance of NO_3 in atmospheric reactions of smog regularly reappears (Stephens and Price, 1972; Wilson *et al.*, 1972; Louw *et al.*, 1973). However it seems unlikely that these reactions can be an important source for oxidation of SO_2 during the daylight hours. Hov *et al.* (1977) have made important point for consideration here; during the night time hours the chemistry of NO_3 may become significant since its major formation reaction ($NO_2 + O_3 \rightarrow NO_3 + O_2$) is not strongly reduced in rate at night, while the rate of its major loss reaction during the day ($NO_3 + NO \rightarrow 2NO_2$) is lowered at night due to the decline in the [NO]. However the rates of attack of SO_2 should remain relatively unimportant even under these conditions.

The possible reactant ONOO, the intermediate species formed by the NO–O_2 bimolecular interaction, is presumably in equilibrium with NO and O_2 in air:

$$NO + O_2 \rightleftharpoons ONOO. \quad (48)$$

From the data of Benson (1976) we may estimate that the [ONOO]/[NO] ratio in air at 1 atm and 25°C is about 0.029. If the rate constant for reaction 29 is near equal to that for analogous reaction 28 of the symmetrical NO_3 species which it parallels in enthalpy change, then for the $[NO] = 5 \times 10^{12}$, $[ONOO] = 1.5 \times 10^{11}$ $mol\,cm^{-3}$, and the estimated SO_2 oxidation rate by reaction 29 will be less than $3.8 \times 10^{-4}\%\,h^{-1}$.

$$ONOO + SO_2 \rightarrow NO_2 + SO_3. \quad (29)$$

In summary, current data rule out the significant contribution of all of the oxides of nitrogen (NO_2, NO_3, ONOO, N_2O_5) to the oxidation of SO_2 within the lower troposphere.

(e) *The* HO_2–SO_2 *reactions*

The study of Payne *et al.* (1973) provides the only experimental estimate of reaction 31 of which we are aware. It was based upon a photochemical, competitive, $^{18}O_2$-labeling technique in which rate measurements of reaction 31 versus reaction 49 were made:

$$HO_2 + SO_2 \rightarrow HO + SO_3 \quad (31)$$

$$2HO_2 \rightarrow H_2O_2 + O_2. \quad (49)$$

They derived the estimate $k_{31}/k_{49}^{1/2} = (4.8 \pm 0.7) \times 10^{-10}\,(cm^3\,mol^{-1}\,s^{-1})^{1/2}$; taking the then preferred value of $k_{49} = 3.3 \times 10^{-12}$ $cm^3\,mol^{-1}\,s^{-1}$ (Baulch *et al.*, 1972), they estimated $k_{31} = (8.7 \pm 1.3) \times 10^{-16}$ $cm^3\,mol^{-1}\,s^{-1}$. The error limits shown are those inherent in the measurement of the rate constant ratio alone. The uncertainty in k_{49} is an additional source of error. Thus Hampson and Garvin chose $k_{49} = 5.6 \times 10^{-12}$ at 25°C from a review of the rate data available in 1975, while Lloyd (1974) chose $k_{49} = 3.2 \times 10^{-12}$ $cm^3\,mol^{-1}\,s^{-1}$ at 25° in his evaluation. A brief review of the estimates of k_{49} is important here in that this value determines our k_{31} estimate through the measured ratio, $k_{31}/k_{49}^{1/2}$.

The first estimate of k_{49} was made by Foner and Hudson (1962); they found $k_{49} \cong 3 \times 10^{-12}$ $cm^3\,mol^{-1}\,s^{-1}$ in a fast-flow, mass spectrometric method of HO_2 detection. The later measurements of Paukert and Johnston (1972) and Hochanadel *et al.* (1972), based upon u.v. detection of HO_2, gave $k_{49} = 3.65 \times 10^{-12}$ and 9.58×10^{-12} $cm^3\,mol^{-1}\,s^{-1}$, respectively, at 25°C. Hamilton (1975) noted that the major difference between the reactant systems employed by Paukert and Johnston and Hochanadel *et al.* was the presence of H_2O vapor (21 Torr) in the latter work; no H_2O was present in the former study. He reinvestigated the reaction 49 using varied amounts of H_2O vapor in a 1.5 MeV electron pulsed gaseous mixture of H_2 (2 atm) and O_2 (5 Torr). He followed HO_2 by u.v.-absorption spectroscopy. Indeed he found that the k_{49} was sensitive to the pressure of H_2O vapor present; using Paukert and Johnston's absorption coefficient for HO_2 for experiments at $P_{H_2O} = 0$, he derived a value, $k_{49} = 3.15 \times 10^{-12}$ $cm^3\,mol^{-1}\,s^{-1}$, in excellent agreement with that of Paukert and Johnston. In experiments with added water the k_{49}/ϵ_{HO_2} values derived were near equivalent to those of Hochanadel *et al.* In view of these results Hamilton and Naleway (1976) have proposed that a complex between water and HO_2 radicals forms in the H_2O-containing mixtures:

$$HO_2 + H_2O \rightleftharpoons HO_2 \cdot H_2O. \quad (50)$$

The apparent increase in rate constant for reaction 49 with increasing H_2O seemed to result from the occurence of the more rapid reaction 51 in addition to 49:

$$HO_2 + HO_2 \cdot H_2O \rightarrow H_2O_2 + O_2 + H_2O. \quad (51)$$

They estimated the thermodynamic properties of the $HO_2 \cdot H_2O$ species theoretically, and noted that the ratio $[HO_2 \cdot H_2O]/[HO_2]$ was equal to about 0.037 in air at 100% relative humidity (25°C). About 3.5 and 1.8% of the HO_2 radicals in the atmosphere at 100 and 50% r.h., respectively, are in the complex at 25°C. The observations of Hamilton and Naleway introduce a new and probably significant complication into the already horrendous problems of the atmospheric scientist who desires to simulate the rates of chemical changes in the troposphere. If the interpretation of Hamilton and Naleway is correct as outlined, then modelers should employ a value of k_{49} which varies with the humidity as well as the temperature. We have used the estimates of Hamilton and Naleway to calculate the fraction of the HO_2 which will be in the form of the complex for various humidities and temperatures; see Fig. 4. Although the amount of H_2O in the air at a given relative humidity rises exponentially as the temperature is increased, the stability of the complex decreases somewhat with increasing temperature. The net result seen in Fig. 4 is that there is small increase in the fraction with increasing temperature (constant relative humidity). Although about 1% of the HO_2 is expected to be complexed with H_2O at 0°C and 50% relative humidity, about 4.5% is complexed at 40°C and 100% r.h. The immediate significance of these observations for our purposes is the effect on the choice of value for k_{49} which should be used with the data of Payne *et al.* in deriving k_{31}. Note that 20 Torr of H_2O vapor was employed in all of the experiments of Payne *et al.*, and it may be more appropriate to use the higher value for the rate constant k_{49} (9.6×10^{-12} cm^3 mol^{-1} s^{-1}) in estimating k_{31}. Thus the rate constant k_{31} may be as high as 1.5×10^{-15} cm^3 mol^{-1} s^{-1}. Although we have chosen the lower value in our summary of Table 5, we have shown it as a lower limit which must be increased if further experimentation proves the Hamilton and Naleway hypothesis correct.

A typical $[HO_2]$ expected in a sunlight-irradiated region of the lower atmosphere which is highly polluted is about 6×10^9 mol cm^{-3} (Demerjian *et al.*, 1974). This will lead to a rate of SO_2 oxidation through reaction 31 of about 1.9% h^{-1}. The $[HO_2]$ for a fairly clean atmosphere may be lower by a factor of 10 or so, and the rates of conversion of SO_2 by 31 will be reduced correspondingly. However, it is seen that this reaction can be a major source of SO_2

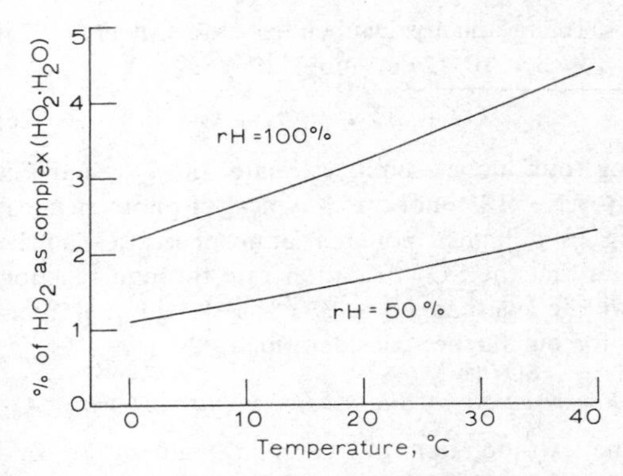

Fig. 4. The theoretical percentage of HO_2 in the gas phase which is complexed with H_2O vapor as a function of the temperature and relative humidity; estimated using the thermodynamic data from Hamilton and Naleway (1976).

conversion in the atmosphere, and it must be included in our further considerations.

Reaction 32 which results in HO_2 addition to SO_2 has not been observed experimentally and only speculation on its possible significance can be made at this time.

$$HO_2 + SO_2 (+M) \rightarrow HO_2SO_2 (+M). \quad (32)$$

Calvert and McQuigg (1975) have estimated by analogy with similar reactions that $k_{32} = 10^{-16}$ cm^3 mol^{-1} s^{-1}. If this magnitude is correct, then reaction 32 will result in only a few tenths of a per cent conversion h^{-1} for the highly polluted atmosphere, conditions which favor it. Benson's enthalpy estimates (in press) suggest that the intermediate adduct HO_2SO_2 formed in 32 will not be stable at 25°, since the decomposition reaction, $HO_2SO_2 \rightarrow HO + SO_3$ will be energetically favorable ($\Delta H = -12$ kcal mole^{-1}). Thus the net effect of the occurrence of reaction 32 in the experiments of Payne *et al.* (1973) may be the formation of HO and SO_3, hence their estimate probably refers to the rate constant sum, $k_{31} + k_{32}$.

(f) *The* CH_3O_2–SO_2 *reaction*

The first preliminary estimates of reactions 33 and 34 have been reported recently by Whitbeck *et al.* (1976).

$$CH_3O_2 + SO_2 \rightarrow CH_3O + SO_3 \quad (33)$$

$$CH_3O_2 + SO_2 \rightarrow CH_3O_2SO_2. \quad (34)$$

They observed the kinetics of the decay of CH_3O_2 spectroscopically following its generation in flash photolyzed mixtures of azomethane and oxygen, and azomethane, oxygen, and sulfur dioxide. The range of initial concentrations of SO_2 which could be employed was very limited ($P_{SO_2} < 0.3$ Torr), since the products of the reaction formed an aerosol which interferred with the optical measurements for experiments at high SO_2 pressures. The rate constant estimate, independent of the absolute extinction coefficient for CH_3O_2, gave $k_{33} + k_{34} \cong (5.3 \pm 2.5) \times$

Table 5. Enthalpy changes and recommended rate constants for potentially important reactions of ground state SO_2 and SO_3 molecules in the lower troposphere

Reaction	$-\Delta H°$*, kcal mole^{-1} (25°C)	k,† cm^3 mol^{-1} s^{-1}
(19) $O_2(^1\Delta_g) + SO_2 \rightarrow SO_4$ (biradical); SO_4 (cyclic)	~25; ~28	
(20) $O_2(^1\Delta_g) + SO_2 \rightarrow SO_3 + O(^3P)$	−13.4	$(3.9 \pm 0.9) \times 10^{-20}$
(21) $O_2(^1\Delta_g) + SO_2 \rightarrow O_2(^3\Sigma_g^-) + SO_2$	22.5	
(22) $O_2(^1\Sigma_g^+) + SO_2 \rightarrow SO_4$ (biradical); SO_4 (cyclic)	~40; ~43	
(23) $O_2(^1\Sigma_g^+) + SO_2 \rightarrow SO_3 + O(^3P)$	1.7	6.6×10^{-16}
(24) $O_2(^1\Sigma_g^+) + SO_2 \rightarrow SO_2 + O_2(^1\Delta_g)$	15.1	
(25) $O(^3P) + SO_2 (+M) \rightarrow SO_3 (+M)$	83.3	$(5.7 \pm 0.5) \times 10^{-14}$
(26) $O_3 + SO_2 \rightarrow O_2 + SO_3$	57.8	$<8 \times 10^{-24}$
(27) $NO_2 + SO_2 \rightarrow NO + SO_3$	10.0	8.8×10^{-30}
(28) $NO_3 + SO_2 \rightarrow NO_2 + SO_3$	32.8	$<7 \times 10^{-21}$
(29) $ONOO + SO_2 \rightarrow NO_2 + SO_3$	~30	$<7 \times 10^{-21}$
(30) $N_2O_5 + SO_2 \rightarrow N_2O_4 + SO_3$	24.6	$<4 \times 10^{-23}$
(31) $HO_2 + SO_2 \rightarrow HO + SO_3$	19.3	$>(8.7 \pm 1.3) \times 10^{-16}$
(32) $HO_2 + SO_2 (+M) \rightarrow HO_2SO_2 (+M)$	~7	
(33) $CH_3O_2 + SO_2 \rightarrow CH_3O + SO_3$	~27	$(5.3 \pm 2.5) \times 10^{-15}$
(34) $CH_3O_2 + SO_2 \rightarrow CH_3O_2SO_2$	~31	
(35) $(CH_3)_3CO_2 + SO_2 \rightarrow (CH_3)_3CO + SO_3$	~26	$<7.3 \times 10^{-19}$
(36) $(CH_3)_3CO_2 + SO_2 \rightarrow (CH_3)_3CO_2SO_2$	~30	
(37) $CH_3COO_2 + SO_2 \rightarrow CH_3CO_2 + SO_3$	~33	$<1.3 \times 10^{-18}$
(38) $CH_3COO_2 + SO_2 \rightarrow CH_3COO_2SO_2$	~37	
(39) $HO + SO_2 (+M) \rightarrow HOSO_2 (+M)$	~37	$<(1.1 \pm 0.3) \times 10^{-12}$
(40) $CH_3O + SO_2 (+M) \rightarrow CH_3OSO_2 (+M)$	~24	$\sim 6 \times 10^{-15}$
(41) $RCH(-O-O-O-)CHR + SO_2 \rightarrow 2RCHO + SO_3$ (O—O—O bridging RCH——CHR)	~89	
(42) $RCH(O\cdot)-CH(O-O\cdot)R + SO_2 \rightarrow 2RCHO + SO_3$	~103	
(43) $R\dot{C}HOO\cdot + SO_2 \rightarrow RCHO + SO_3$	~98	$k_{43a}/k_{43} \cong 6 \times 10^{-5}$
(43a) $R\dot{C}HOO\cdot + H_2O \rightarrow RCOOH + H_2O$	~147	
(44) $RCH(O\cdot)O\cdot + SO_2 \rightarrow RCHO + SO_3$	~52	
(45) $SO_3 + O(^3P) (+M) \rightarrow SO_4$ (biradical); SO_4 (cyclic)	~38; ~41	$\sim 7 \times 10^{-13}$
(46) $SO_3 + O(^3P) (+M) \rightarrow SO_2 + O_2$	35.9	
(47) $SO_3 + H_2O \rightarrow H_2SO_4$	24.6	$(9.1 \pm 2.0) \times 10^{-13}$

* Enthalpy change estimates were derived from the data of Benson (1976, 1977), O'Neal and Blumstein (1973), and Domalski (1971).

† The rate constants are all expressed as second order reactions for 1 atm air at 25°C. See the text for the evaluation of the kinetic data and the references to the original literature upon which they are based.

10^{-15} cm^3 mol^{-1} s^{-1}. If $k_{33} > k_{34}$ then the magnitude of this estimate is in reasonable accord with that for the analogous reaction 31. Taking the difference in entropies of activation for reactions 31 and 34 as ~0.8 eu, then the ratio of the observed rate constants $k_{31}/k_{33} = 0.16$ corresponds to an activation energy difference, $E_{31}-E_{33} \cong 1.3$ kcal mole^{-1}, not unreasonable in view of the differences in enthalpy between these reactions: $\Delta H_{31}-\Delta H_{33} = 7.6$ kcal mole^{-1}.

The CH_3O_2 radical is probably the most abundant of the many organic peroxy radicals in the atmosphere. In a highly polluted atmosphere it is expected to be present at about 10^9 mol cm^{-3} (Demerjian *et al.*, 1974), and the oxidation of SO_2 may occur through 33 and 34 by a rate as large as 2% h^{-1}. In the fairly clean troposphere, rates of 1×10^{-2}% h^{-1} are expected (Crutzen and Fishman, 1977). Obviously we do want to include these reactions in our further considerations.

It is not possible to determine the extent to which each of the reactions 33 and 34 occur from the existing data, but thermochemical arguments favor the reaction 33; thus the $CH_3O_2SO_2$ intermediate formed in 34 may decompose readily into CH_3O and SO_3 after a short delay.

(g) *The tert-butylperoxy radical–SO_2 reaction*

Whitbeck *et al.* (1976) have studied the tert-$C_4H_9O_2$ decay kinetics by flash photolysis of $(CH_3)_3CN{=}NC(CH_3)_3$ mixtures with pure oxygen and in mixtures with O_2 and SO_2. In this case the reactivity of the radicals toward SO_2 was sufficiently low for aerosol formation not to be a serious problem at short times. At long times (1 min) aerosol appeared. This was attributed to the subsequent addition of $(CH_3)_3CO$ radicals to SO_2 following their generation in the major $C_4H_9O_2$ loss reaction in this system: $2C_4H_9O_2 \rightarrow 2C_4H_9O + O_2$. No difference in the decay rate of the tert-$C_4H_9O_2$ radicals could be detected in runs with added SO_2 (0.2 Torr), so in this case only an upper limit can be set for the rate constants: $k_{35} + k_{36} < 7.3 \times 10^{-19}$ cm^3 mol^{-1} s^{-1}.

$$(CH_3)_3CO_2 + SO_2 \rightarrow (CH_3)_3CO + SO_3 \quad (35)$$

$$(CH_3)_3CO_2 + SO_2 \rightarrow (CH_3)_3CO_2SO_2. \quad (36)$$

This value is smaller than the estimate of $k_{33} + k_{34}$ by a factor of 10^{-4}; thus we conclude that reactions 35 and 36, as well as those of the other highly hindered tert-alkylperoxy radicals, will be unimportant in the atmospheric conversion of SO_2. The very marked difference in reactivity of the tert-$C_4H_8O_2$ and the CH_3O_2 radicals noted here with SO_2, is also seen in the RO_2–RO_2 reactions of these species: 2t-$C_4H_9O_2 \rightarrow$ 2t-$C_4H_9O + O_2$; $2CH_3O_2 \rightarrow 2CH_3O$ (or $CH_2O + CH_3OH$) + O_2. The rate constant of the former is 10^{-4} to 10^{-5}-times that of the latter (Parkes *et al.*, 1973; Parkes, 1974; Whitbeck *et al.*, 1976).

(h) *The acetylperoxy radical–SO_2 reactions*

The peroxyacyl radicals are significant participants in photochemical smog formation. Some of them formed in the polluted atmosphere react with NO_2 to form the peroxyacyl nitrates:

$$RCOO_2 + NO_2 \rightarrow RCOO_2NO_2. \quad (52)$$

The possible significance of these radicals in SO_2 oxidation has not been studied directly, but there is some information related to the reactions 37 and 38 which we can extract from existing published kinetic data:

$$CH_3COO_2 + SO_2 \rightarrow CH_3CO_2 + SO_3 \quad (37)$$

$$CH_3COO_2 + SO_2 \rightarrow CH_3COO_2SO_2. \quad (38)$$

Heicklen (1976) has quoted unpublished data of Shortridge which presumably gives evidence that the peroxyacetyl radical reacts readily with SO_2 in reaction 37. He found SO_2 addition to photolyzed mixtures of CH_3CHO and O_2 did not affect the rate of acetaldehyde oxidation, although SO_2 was removed and an aerosol formed. It is not clear how this evidence proves uniquely the occurrence of reaction 37, since many other chain carriers such as HO, HO_2, CH_3O, CH_3O_2, etc., will be present and may also oxidize SO_2 without terminating the chains.

Perhaps somewhat more definitive results related to the reactions 37 and 38 were reported by Pate *et al.* (1976). They observed the rate of reaction of peroxyacetyl nitrate (PAN) with many compounds including SO_2. The apparent bimolecular rate constant at 296° K for the PAN–SO_2 reaction was found to be less than 1.35×10^{-23} cm^3 mol^{-1} s^{-1}. However it has been rather well established that PAN is in equilibrium with CH_3COO_2 and NO_2, and we can estimate from the data of Pate *et al.* rate information related to reactions 37 and 38. The data from the Pitts group and those from Hendry and Kenley (1977) clearly demonstrate the dynamic character of the PAN present in gaseous mixtures:

$$CH_3COO_2NO_2 \rightarrow CH_3COO_2 + NO_2 \quad (53)$$

$$CH_3COO_2 + NO_2 \rightarrow CH_3COO_2NO_2. \quad (54)$$

Hendry and Kenley have estimated the rate constant k_{53} as a function of temperature. For 23°C, the temperature of the experiments of Pate *et al.*, $k_{53} = 2.65 \times 10^{-4}$ s^{-1}. Taking their theoretical estimate of $k_{54} = 1.04 \times 10^{-12}$ cm^3 mol^{-1} s^{-1}, we may estimate the [CH_3COO_2] at equilibrium:

$$[CH_3COO_2]_{eq} = \frac{[CH_3COO_2NO_2](2.6 \times 10^8)}{[NO_2]} \text{ mol cm}^{-3}. \quad (55)$$

It seems reasonable to attribute any reaction of PAN to that of the CH_3COO_2 radicals in equilibrium with it. Thus the observed rate constant (k_{exp}) may be related to the rate constant sum, $k_{37} + k_{38}$, by relation 56:

$$(k_{37} + k_{38})[CH_3COO_2][SO_2] = k_{exp}[PAN][SO_2]. \quad (56)$$

Estimating [CH_3COO_2] from the equilibrium data and relation 55 gives: $k_{37} + k_{38} = k_{exp}[NO_2]/2.55 \times 10^8$ cm^3 mol^{-1} s^{-1}. It is not reported what levels of NO_2 were present in the experiments of Pate *et al.*, but it is likely that no more than 1 ppm (2.5×10^{13} mol cm^{-3}) was present on the average. With this estimate the data of Pate *et al.* lead to a value of $k_{37} + k_{38} < 1.3 \times 10^{-18}$ cm^3 mol^{-1} s^{-1}. For the theoretically estimated levels of the peroxyacetyl radicals expected to be present in heavy photochemical smog ($\sim 2 + 10^8$ mol cm^{-3}), the new rate constant estimate suggests a maximum rate of oxidation of SO_2 through 37 and 38 of about 1×10^{-4} h^{-1}. We conclude that the oxidation of SO_2 by CH_3COO_2 radicals, and peroxacyl radicals in general, is negligible for the usual atmospheric conditions. The results of Fox and Wright (1977) can also be interpreted to support this conclusion. They carried out sunlight-irradiated smog chamber experiments using dilute mixtures of C_3H_6 and NO_x in air and matched mixtures of these compounds but with SO_2 added (0.75 ppm). They observed that SO_2 addition did not inhibit significantly the formation of PAN; the peak concentration was not lowered appreciably. The rate of [PAN] increase with time, if altered at all by SO_2 addition, was somewhat more rapid in the SO_2-containing system. Thus in our opinion all of the definitive evidence at hand point to the unimportance of the reactions 37 and 38 in the tropospheric conversion of SO_2.

(i) *The HO–SO_2 addition reaction*

The kinetics of free radical addition to SO_2 have been observed for a number of free radical species; thus H-atoms (Halstead and Jenkins, 1969), CH_3-radicals (Good and Thynne, 1967a; Calvert *et al.*, 1971; James *et al.*, 1973), C_2H_5-radicals (Good and Thynne, 1967b), fluoroethyl radicals, $CH_2\,^{18}FCH_2$ (Milstein *et al.*, 1974), cyclohexyl and other radicals formed during recoil tritium reactions with cyclohexene (Fee *et al.*, 1972), all react with reasonably large rate constants which approach those of the analogous reaction with oxygen in some cases. Of course the H-atoms and alkyl radicals formed in the troposphere do not live to encounter SO_2 present at low impurity concentrations. They react with oxygen largely, to form alkylperoxy radicals and ultimately alkoxy radicals, HO_2, and other free radicals and molecular products. As we have seen in our earlier discussions, the additions of some alkylperoxy and acylperoxy radicals (reactions 36 and 38) to SO_2 appear to be rather slow reactions, while those of HO_2 and CH_3O_2 additions remain unevaluated. Much of the important homogeneous chemistry of the impurity molecules within the troposphere is dominated by the reactions of the ubiquitous radical pair, HO and HO_2. Indeed it appears that the homogeneous SO_2 removal paths also depend largely on the reactions of these radicals. In particular the HO-addition to SO_2, reaction 39, seems to be the most important of the several homogeneous reaction paths of SO_2 in the troposphere for many different atmospheric conditions.

$$HO + SO_2\,(+M) \rightarrow HOSO_2\,(+M). \qquad (39)$$

There has been a significant and productive effort of several research groups to determine the rate constant for this seemingly important reaction. It was apparently first suggested by McAndrew and Wheeler (1962) as one of the reactions necessary to rationalize the effect of SO_2 on radical chain terminations in propane-air flames. Following the suggestions of Fair and Thrush (1969), the McAndrew and Wheeler rate constant estimations can be used to derive a third order rate constant, $k_{39} = 1.1 \times 10^{-31}$ $cm^6\,mol^{-2}\,s^{-1}$ at 2080°K. Many of the estimates of k_{39} were obtained in competitive rate studies using the HO reaction with CO as a reference.

$$HO + CO \rightarrow H + CO_2. \qquad (57)$$

With this technique Davis *et al.* (1973) reported a third order rate constant $k_{39} \cong 3 \times 10^{-31}\,cm^6$ $mol^{-2}\,s^{-1}$ with $M = H_2O$. In similar experiments Payne *et al.* (1973) found $k_{39} = 1.5 \times 10^{-31}\,cm^6$ $mol^{-2}\,s^{-1}$ using the Stuhl and Niki (1972) estimate of the rate constant for reaction 57. Cox (1974/75, 1975) also carried out competitive SO_2–CO reaction studies by photolyzing HONO in air at 1 atm pressure; they found $k_{39} = (6.0 \pm 0.8) \times 10^{-13}\,cm^3$ $mol^{-1}\,s^{-1}$ in air at 1 atm, using $k_{57} = 1.5 \times 10^{-13}$ $cm^3\,mol^{-1}\,s^{-1}$ as recommended by Baulch and Drysdale (1974). The Castlemen group (Wood *et al.*, 1974) also have made rather extensive competitive rate studies of 39 in experiments which have extended over a number of years. They photolyzed H_2O in a mixture of SO_2, CO, and H_2O in N_2 carrier gas. Their first work gave $k_{39} = 3.8 \times 10^{-13}\,cm^3\,mol^{-1}\,s^{-1}$ for the high pressure limit. They recognized early the probable importance of this reaction in SO_2 conversion in the atmosphere (Castleman *et al.*, 1974; Wood *et al.*, 1975). In a later similar study (Castleman *et al.*, 1975), they found a pseudo-second order rate constant at 1 atm of N_2, $k_{39} = 6.0 \times 10^{-13}$, taking $k_{57} =$ $1.4 \times 10^{-13}\,cm^3\,mol^{-1}\,s^{-1}$ (Hampson and Garvin, 1975). In the most recent report of this group, Castleman and Tang (1976/77) extended their competitive rate study from 20 to 1000 Torr of added N_2 gas. Again accepting k_{57}, they calculated: $k_{35} = 6.0 \times$ $10^{-13}\,cm^3\,mol^{-1}\,s^{-1}$. In experiments at low pressures carried out over a range of temperatures 24 to −20°C, they estimated $E_{35} = -2.8$ kcal $mole^{-1}$.

Some more direct measurements of k_{39} have been made by flash photolysis HO-resonance fluorescence techniques. Using this method Davis and Schiff (1974) reported the reaction 39 to be in the high pressure fall off region above ~10 Torr of He. At 500 Torr of added helium, the effective bimolecular rate constant was reported to be: 2.5×10^{-13} (2.5×10^{-3} shown, must be in error) $cm^3\,mol^{-1}\,s^{-1}$, and the low pressure value of the third order rate constant, $k_{39} = 2 \times 10^{-32}\,cm^6\,mol^{-2}\,s^{-1}$ with He = M (work of Davis and Schiff quoted in Payne *et al.*, 1973). In further experiments Davis (1974b) reported preliminary values for the effective bimolecular rate constants for k_{39}: with He as M, values varied from 0.87×10^{-13} at 50 Torr to 2.7×10^{-13} at 500 Torr; for M = Ar, 1.37×10^{-13} at 50 Torr to 3.7×10^{-13} at 500 Torr; for $M = N_2$, 0.80×10^{-13} at 5 Torr to $2.4 \times 10^{-13}\,cm^3\,mol^{-1}\,s^{-1}$ at 20 Torr. Harris and Wayne (1975) studied reaction 39 using the HO-resonance technique but in a discharge flow system at low pressures. They derived the third order rate constants: $k_{39} = (4.5 \pm 1.5) \times 10^{-31}$ with M = Ar, and $k_{39} = (7.2 \pm 2.6) \times 10^{-31}\,cm^6\,mol^{-2}\,s^{-1}$ with $M = N_2$. Atkinson *et al.* (1976) also utilized H_2O flash photolysis with HO-resonance spectroscopy to determine k_{39} over a wide range of argon pressures. The estimated second order rate constant for 25°C and 1 atm (M = Ar) was $k_{39} = (6.7 \pm 0.7) \times 10^{-13}$, with the high pressure limit (extrapolated) of $8.3 \times 10^{-13}\,cm^3\,mol^{-1}\,s^{-1}$. Gordon and Mulac (1975) employed pulsed radiolysis and absorption spectroscopy to follow the HO radical at 3087 Å; in $H_2O(g)$ at 1 atm and 435 K; they obtained $k_{39} =$ $1.8 \times 10^{-12}\,cm^3\,mol^{-1}\,s^{-1}$.

Recently Sie *et al.* (1976) and Cox *et al.* (1976) have presented strong evidence that the rate constant for the reaction 57, used as a reference value in many competitive studies of reaction 39, was pressure sensitive. Both groups of workers concluded that the rate constant k_{57} was very much larger at 1 atm total gas pressure than had been assumed in view of the earlier low pressure data. A unique test of the pressure dependence of k_{57} was not possible in the work of Cox *et al.* (1976), since their experiments were carried out at a fixed pressure of 1 atm (largely air). However they found the ratio k_{57}/k_{58} to be about twice that observed by other experimentalists who used much lower pressures of reactants and added gas; they suggested the hypothesis of a pressure dependent k_{57} seemed most compatible with their results.

$$HO + H_2 \rightarrow H_2O + H. \quad (58)$$

In the work of Sie *et al.* (1976) the pressure of added gas (H_2, He, SF_6) was varied from 20 to 774 Torr with an observed increase in k_{57}/k_{58} with pressure by over a factor of two in the case of added H_2 or SF_6. Chan *et al.* (1977) confirmed the conclusions of these workers. They reported from competitive HO rate studies with dilute mixtures of CO and iso-C_4H_{10} in air, that the rate constant ratio k_{57}/k_{59} was about a factor of two higher in experiments at 700 Torr of air than in runs with 100 Torr of air.

$$HO + iso\text{-}C_4H_{10} \rightarrow C_4H_9 + H_2O. \quad (59)$$

There is no reason to believe that the pressure dependence observed by these workers resulted from the normal bimolecular H-atom abstraction reactions 58 and 59; it is reasonable to assume that these rate constants remain at their low pressure values: $k_{58} = 7.0 \times 10^{-15}$, (Cox *et al.*, 1976); $k_{59} = 2.34 \times 10^{-12}$ $cm^3\ mol^{-1}\ s^{-1}$ (Greiner, 1967). Accepting this seemingly plausible premise, we have derived estimates from the various competitive experiments for k_{57} and have plotted these together with the data from the direct measurements at low pressures in Fig. 5. Excluding the one very divergent point of Sie *et al.* (1976) at 570 Torr of added SF_6, the general increase of the rate constant k_{57} with increasing pressure of H_2, N_2, O_2, and SF_6 seems clear. Obviously values of the HO–SO_2 reaction rate constants from the data at the higher pressures based upon the low pressure value of k_{57} must be corrected to take into account the pressure sensitive character of k_{57}. Thus the estimates of k_{39} by Davis *et al.* (1973), Payne *et al.* (1973), Cox (1975), Castleman *et al.* (1975), and Castleman and Tang (1976/77) should be corrected. We have made such a correction to the extensive and seemingly accurate data set of Castleman and Tang, using the values corresponding to those of the dashed curve given in Fig. 5 for k_{39}. The apparent second order rate constants derived in this fashion are plotted versus the reciprocal of the pressure of N_2 in Fig. 6 (triangles). Also plotted here are the values of k_{39}^{-1} determined more directly by resonance fluorescence of HO in experiments at various pressures of Ar; the open circles are data from Atkinson *et al.* (1976) and the closed circles are data from Davis (1974b). The two sets of data from the direct HO-measurements do not check very well at low pressures but appear to reach about the same value at high pressures. The

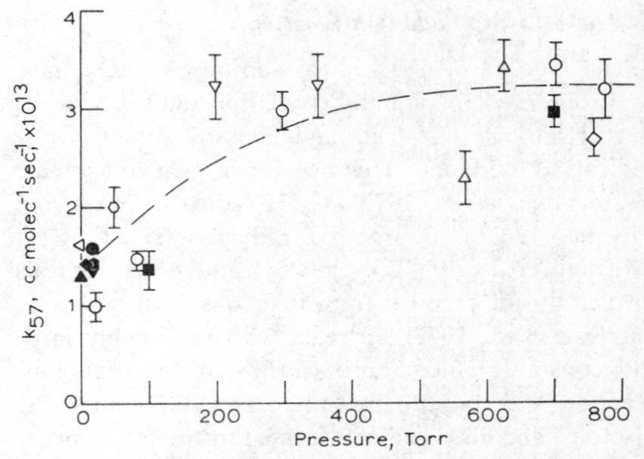

Fig. 5. The variation of the apparent second order rate constant for the reaction 57, $HO + CO\ (+M) \rightarrow H + CO_2\ (+M)$, with pressure of added gases; data for H_2, Sie *et al.*, 1976 (open circles); air, Cox, 1975 (diamond), Chan *et al.* 1977 (closed squares); SF_6, Sie *et al.*, 1976 (open triangles), Overend *et al.*, 1974, and Paraskevopoulos, 1976 (inverted triangles); He, Paraskevopoulos, 1976 (hexagons); low pressure points shown: Stuhl and Niki, 1972 (inverted closed triangle); Greiner, 1967 (closed diamond); Mulcahy and Smith, 1971 (on-end triangle); Davis *et al.*, 1974c (open diamond); Westenberg and deHaas, 1973 (closed triangle); Smith and Zellner, 1973 (closed hexagon).

data of Atkinson *et al.* give k_{39} (M = Ar, 1 atm) = $(6.7 \pm 0.7) \times 10^{-13}$ and a high pressure limit of $8.3 \times 10^{-13}\ cm^3\ mol^{-1}\ s^{-1}$. Since Ar is somewhat less efficient as a third body than N_2 and O_2, the latter number is probably the best estimate from

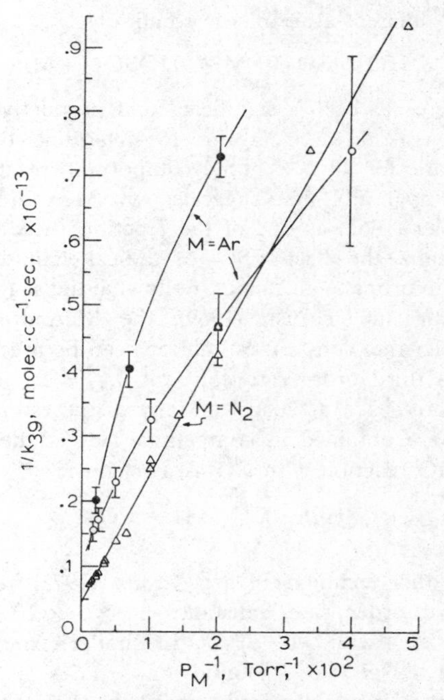

Fig. 6. Plot of the reciprocal of the apparent second order rate constant for the reaction 39, $HO + SO_2\,(+M) \rightarrow HOSO_2\,(+M)$, versus the reciprocal of the pressure of added gas (M); estimates for M = N_2 were calculated from the data of Castleman and Tang (1976/77) using corrected values for k_{57} (triangles); for M = Ar data are from Atkinson *et al.*, 1976 (open circles) and Davis, 1974b (closed circles).

the Atkinson *et al.* data which is applicable to air at 1 atm. The Davis (1974b) data are less abundant, but the extrapolated upper limit to k_{39} for M = Ar is compatible with this estimate as well. The Davis data for k_{39} using $M = N_2$ consisted of only 3 points (not shown in Fig. 5) at low pressures (P < 40 Torr), so the value obtained by extrapolation of these data to 1 atm N_2 ($k_{39} \cong 9 \times 10^{-13}$) cannot be very accurate. We will take as the best estimate of the Davis group the number quoted by Davis and Klauber (1975) for the bimolecular rate constant in the troposphere, $k_{39} \cong 8 \times 10^{-13}\,cm^3\,mol^{-1}\,s^{-1}$. The Castleman and Tang (1976/77) corrected estimates (using our preferred choice for $k_{57} = 3.0 \times 10^{-13}\,cm^3\,mol^{-1}\,s^{-1}$, Chan *et al.*, 1977) gives $k_{39} = 1.4 \times 10^{-12}\,cm^3\,mol^{-1}\,s^{-1}$ for $M = N_2$ (1 atm, 24°C). The Cox (1975) estimate corrected to our preferred choice for k_{57} gives $k_{39} = (1.2 \pm 0.2) \times 10^{-12}\,cm^3\,mol^{-1}\,s^{-1}$. We feel the best current choice for k_{39} in the lower troposphere (1 atm air, 25°C) is derived from an average of the results of the four most extensive studies at the highest pressures, i.e. the data of Davis (1974b), Atkinson *et al.* (1976), and the corrected data of Cox (1975), and Castleman and Tang (1976/77). We suggest the value: $k_{39} = (1.1 \pm 0.3) \times 10^{-12}\,cm^3\,mol^{-1}\,s^{-1}$ which is given in Table 5.

The typical [HO] which is anticipated theoretically in the highly polluted (NO_x, RH, SO_2), sunlight-irradiated, lower atmosphere, is about 7×10^6 mol cm^{-3} (Calvert and McQuigg, 1975). That expected for a relatively clean atmosphere in the midday summer sun in the midlatitudes of the northern hemisphere, is somewhat lower, typically 1×10^6 mol cm^{-3} (Crutzen and Fishman, 1977). Using these estimates we predict that SO_2 removal by reaction with HO in 39 may be as high as 2.7% h^{-1} in the dirty atmosphere and typically 0.4% h^{-1} in the clean troposphere. Our evaluation certainly confirms the current view of atmospheric chemists that reaction 39 is a very important factor in the homogeneous oxidation of SO_2 in the atmosphere.

(j) *The fate of the* $HOSO_2$ *product of reaction* 39

The $HOSO_2$ species formed as reaction 39 occurs is not a stable product, but it is a free radical which will react further to form final products. Calvert and McQuigg (1975), Davis and Klauber (1975), and Benson (in press) have speculated on the subsequent events expected following the formation of $HOSO_2$. It is important to review these possible steps in light of recent information concerning them.

The first step in the sequence of $HOSO_2$ reactions, reaction 60, should be the major fate of the $HOSO_2$ radical in the troposphere. It is exothermic by 16 kcal $mole^{-1}$. The alternative disproportionation reaction, $HOSO_2 + O_2 \rightarrow HO_2 + SO_3$, will be non-competitive with 60 since it is endothermic by about 8 kcal $mole^{-1}$. As with the HO_2 and RO_2 radicals, it is likely that the $HOSO_2OO$ radical will oxidize NO, reaction 61, or react with NO_2 in reaction 62, since NO and NO_2 impurity will be present with SO_2 in most of the impurity-laden air mixtures encountered. Reaction 61 is more exothermic (-25 kcal $mole^{-1}$) than the comparable, analogous, fast reaction, $RO_2 + NO \rightarrow RO + NO_2$ (-17 kcal $mole^{-1}$). The coupling reaction 62 should be somewhat exothermic, and it forms in theory an inorganic analogue to peroxyacetyl nitrate (PAN) as suggested by Calvert and McQuigg (1975). This compound would be expected to be somewhat unstable toward decomposition, and reaction 62 is written as reversible. Benson (in press) has not reported estimates for ΔH_f ($HOSO_2OONO_2$), but he noted that both the adducts of the $HOSO_2OO$ radical with NO and NO_2 will not be stable; the reaction 64 with NO_2 to form $HOSO_2O$ and NO_3 is slightly exothermic. Disproportionation of the $HOSO_2OO$ radical may occur with the HO_2 radical in 65, or, in the absence of other radicals, it may react with another $HOSO_2OO$ radical in 66. It is unlikely that $HOSO_2OO$ will oxidize SO_2 at any significant rate: $HOSO_2OO + SO_2 \rightarrow HOSO_2O + SO_3$ ($\Delta H = -35$ kcal $mole^{-1}$), in view of the slowness of the reaction 37 of near equal ΔH. If this is important,

Reaction	~ΔH, kcal $mole^{-1}$	
$HO + SO_2 (+M) \rightarrow HOSO_2 (+M)$	−37	(39)
$HOSO_2 + O_2 \rightarrow HOSO_2OO$	−16	(60)
$HOSO_2OO + NO \rightarrow HOSO_2O + NO_2$	−25	(61)
$HOSO_2OO + NO_2 \rightleftharpoons HOSO_2OONO_2$	?	(62)
$HOSO_2OONO_2 \rightarrow HOSO_2O + NO_3$	?	(63)
$HOSO_2OO + NO_2 \rightarrow HOSO_2O + NO_3$	−2	(64)
$HOSO_2OO + HO_2 \rightarrow HOSO_2O_2H + O_2$	−43	(65)
$2HOSO_2OO \rightarrow 2HOSO_2O + O_2$	−22	(66)
$HOSO_2O + NO \rightarrow HOSO_2ONO$	−26	(67)
$HOSO_2ONO + h\nu \rightarrow HOSO_2O + NO$		(68)
$HOSO_2O + NO_2 \rightarrow HOSO_2ONO_2$	−22	(69)
$HOSO_2O + HO_2 \rightarrow HOSO_2OH + O_2$	−57	(70)
$HOSO_2O + C_3H_8 \rightarrow HOSO_2OH + iso\text{-}C_3H_7$	−10	(71)
$HOSO_2O + C_3H_6 \rightarrow HOSO_2OCH_2\dot{C}HCH_3$	?	(72)
H_2SO_4 + aerosol (H_2O, NH_3, CH_2O, C_nH_{2n} . . .) → (growing aerosol)		(73)
$HOSO_2ONO_2$ + aerosol (H_2O) → aerosol (H_2SO_4, $HONO_2$ . . .)		(74)
$HOSO_2ONO$ + aerosol (H_2O) → aerosol (H_2SO_4, HONO . .)		(75)

then rates of SO_2 oxidation estimated in this work are minimum rates.

The $HOSO_2O$ radical in this sequence is somewhat analogous to the HO radical. It may form sulfuric acid by abstracting a hydrogen from a hydrocarbon, aldehyde, or other H-containing species such as HO_2 in reactions 70 and 71. However the H-atom abstraction reaction of the HO-radical analogous to the typical reaction 71 is much more exothermic ($\Delta H = -25$ kcal mole^{-1}) than reaction 71, so one may expect the rate constant for H-abstraction by $HOSO_2O$ to be somewhat smaller than the analogous HO reaction. The $HOSO_2O$ radical may add to alkenes and generate organic sulfate containing species in reaction 72. It may react with NO in reaction 67 and form the well known reactive reagent for nitrosation and oxidation of organic compounds and diazotization of amines, nitrosylsulfuric acid, $ONOSO_2OH$. It is interesting to observe that this compound is available now commercially (duPont) in a sulfuric acid solution. Nitrosylsulfuric acid may photolyze in sunlight in 68, react in aerosol solutions to hydrolyze to H_2SO_4 and HONO, or react with organic matter present in the H_2SO_4-rich aerosol in 75. Alternatively the $HOSO_2O$ species may combine with NO_2 in 69 and form the nitrylsulfuric acid. In principle this compound could react on aerosols to hydrolyze to H_2SO_4 and $HONO_2$, or it may act to nitrate certain organic compounds present in the aerosol solution, reaction 74.

It seems probable to us that the reactions of the $HOSO_2O_2$ and $HOSO_2O$ radicals which are shown do occur in the atmosphere, and the reactive species formed in these reactions should be considered as prime candidates for the active forms of the 'sulfate' aerosol of our urban atmospheres. The assignment of rate constants for these reactions of the basis of our present knowledge is very speculative. However recognize that once the HO radical has added to SO_2, the reactions proceed to form sulfuric acid, peroxysulfuric acid, and other compounds which will eventually lead to sulfate and nitrate containing aerosols. If the aerosol is rich in unneutralized H_2SO_4, then much of the nitrate will be lost to the atmosphere as nitric acid.

Davis and Klauber (1975) and Davis *et al.* (1974d) have suggested an alternative reaction route for the $HOSO_2O$ radical which we should consider further here. This is the chain reaction sequence 76, 77, and 61 which can presumably pump NO to NO_2 as in the HO_2, HO cycle in smog.

$$HOSO_2O + O_2(+M) \rightarrow HOSO_2O_3(+M) \quad (76)$$

$$HOSO_2O_3 + NO \rightarrow HOSO_2O_2 + NO_2 \quad (77)$$

$$HOSO_2O_2 + NO \rightarrow HOSO_2O + NO_2. \quad (61)$$

The reaction 76 is analogous to the reaction 78 involving the HO radical:

$$HO + O_2 \rightleftharpoons HOOO. \quad (78)$$

Using Benson's (1976) estimates for $\Delta H_f(HO_3)$, and taking S° $(HO_3) \cong 60$ eu (1 atm, 25°C), we calculate that at equilibrium in the lower troposphere at 25° the ratio $[HOOO]/[HO] = 2.6 \times 10^{-19}$. The greater endothermicity of the reaction 76 compared to 78 will lead to a still lower ratio of $[HOSO_2O_3]/[HOSO_2O]$ in the lower troposphere. Thus it is improbable that the proposed reactions 76 and 77 occur in the atmosphere.

The observed O_3-bulge which Davis and colleagues saw late in the transport of a stack gas plume does not require reaction 76 and 77 for explanation. The conventional reactions involving NO to NO_2 conversion chains of typical smog are the likely O_3 developing mechanism in a plume which is well diluted with the usual contaminants of polluted urban air.

(k) *The methoxy radical addition to* SO_2

There are no kinetic data related to reaction 40 of which we are aware.

$$CH_3O + SO_2 \rightarrow CH_3OSO_2. \quad (40)$$

Calvert and McQuigg (1975) have presented a very rough estimate: $k_{40} \cong 6 \times 10^{-15}$ cm^3 mol^{-1} s^{-1}, which was based upon comparisons with analogous reactions of other radicals. An experimental estimate should be made. In contrast to the CH_3O_2 and other alkyl peroxy radicals, the CH_3O radical and other alkoxy radicals are reactive toward molecular O_2; e.g. $CH_3O + O_2 \rightarrow CH_2O + HO_2$. Thus the steady state concentrations of CH_3O and other alkoxy radicals within a sunlight-irradiated, polluted, lower atmosphere are quite low, about 5×10^6 mol cm^3. Using this estimate the rate of SO_2 reaction in 40 should be about 0.01% h^{-1}. This reaction seems to be an unimportant loss mechanism for SO_2 in the atmosphere. However in view of the reactive nature of the potential alkylating agents which should form eventually following reaction 40, an experimental determination of the rate constants for alkoxy radical reactions with SO_2 should be made to allow an accurate assessment of the significance of this reaction.

(l) *The oxidation of* SO_2 *by products of the* O_3*-alkene reaction*

As we have seen in Section 3(c), the rate of oxidation of SO_2 by O_3, reaction 26, is negligibly slow at ambient air temperatures and at the low levels which are characteristic of O_3 and SO_2 impurities in the troposphere. However when a third reactant, an alkene is present in the O_3–SO_2–air mixture, a fairly rapid oxidation of SO_2 occurs even in the dark. This very significant observation was first reported in detail by Cox and Penkett (1971a, b). For their conditions the rate of removal of SO_2 was 3% h^{-1} with the alkene *cis*-2-pentene and 0.4% h^{-1} with propylene. They considered as a possible reactant in 43 the so-called 'zwitterionic' species, or Criegee intermediate, $R\dot{C}HOO\cdot$, postulated by Criegee (1957) in his classical mechanism of the ozone–alkene reactions in solution:

$$O_3 + RCH{=}CHR \rightarrow \overset{O-O-O}{RCH \text{———————} CHR} \qquad (79)$$

$$\overset{O-O-O}{RCH \text{———————} CHR} \rightarrow RCHO + R\dot{C}HOO\cdot \qquad (80)$$

$$R\dot{C}HOO\cdot + SO_2 \rightarrow RCHO + SO_3. \qquad (43)$$

In a later more extensive study, Cox and Penkett (1972) also considered the original ozonide (molozonide) product of 79 as an alternative reactant in 41:

$$\overset{O-O-O}{RCH \text{———————} CHR} + SO_2 \rightarrow 2RCHO + SO_3. \qquad (41)$$

The rate data from a series of different alkenes (*cis*-2-butene, *cis*-2-pentene, *trans*-2-butene, 2-methyl-1-pentene, and 1-hexene) were rationalized well by a simple mechanism in which an intermediate, presumably the original molozonide or the Criegee intermediate, oxidized SO_2 through the reaction series 79, 80, 43, and/or 41, or underwent decomposition, reaction at the wall, or some other fate unproductive to SO_2 oxidation. Although a strong inhibition of the SO_2 oxidation occurred with increased $[H_2O]$ in the reaction mixture, the effect of H_2O remained puzzling and unexplained by Cox and Penkett. Wilson *et al.* (1974) accepted a similar mechanism in their computer simulation of the Cox and Penkett O_3–alkene–SO_2–air results.

In the reaction schemes considered here for this system, two other reactive entities related to the molozonide and its products should be noted as well as potential reactants: the open form of the original molozonide (reaction 42), and the rearranged Criegee intermediate (reaction 44):

$$\cdot O\!-\!\overset{O\cdot}{RCH}\!-\!\overset{O-O\cdot}{CHR} + SO_2 \rightarrow 2RCHO + SO_3 \qquad (42)$$

$$\overset{O\cdot}{\underset{}{|}}\!\!RCH{-}O\cdot + SO_2 \rightarrow RCHO + SO_3 \qquad (44)$$

Other possible reactants for SO_2 in the alkene–O_3–SO_2–air system have been suggested. Demerjian *et al.* (1974) speculated that the reactant diradical of 44 might be formed from the original Criegee intermediate by reaction with O_2:

$$R\dot{C}HOO\cdot + O_2 \rightarrow \cdot OOC(R)HOO\cdot \qquad (81)$$

$$\cdot OOC(R)HOO\cdot \rightarrow O\langle\overset{O-O}{\underset{RCH}{}}\rangle O \rightarrow O_2 + \cdot OCH(R)O\cdot \qquad (82)$$

Alternatively O'Neal and Blumstein (1973) envisaged a rapid rearrangement of the Criegee intermediate through reaction sequence 83:

$$R\dot{C}HOO\cdot \rightarrow \overset{O}{\underset{O}{|}}\!\!\rangle C(R)H \rightarrow \cdot OCH(R)O\cdot. \qquad (83)$$

Thus at least four different reactants should be considered as potential SO_2 reactants in this system; note in Table 5 that all of the potential reactions 41–44 are considerably exothermic. However there is some uncertainty whether one needs to invoke any of these reactions to explain SO_2 oxidation. Demerjian *et al.* (1974) have argued that the $\cdot OCH(R)O\cdot$ radical will react readily with oxygen to generate other reactive species (RCO_2, HO_2, RO_2, etc.), and Calvert and McQuigg (1975) suggested that these may be responsible for the SO_2 oxidation observed in the alkene–O_3–SO_2 system.

The O_3–alkene reactions are very complex, and the simple Criegee mechanism cannot be the only route of fragmentation and rearrangement to products in the gas phase system. There is abundant evidence today that fragmentation of the ozonide formed with the simple alkenes in gas phase reactions creates highly excited free radicals including HO and carbonyl species (Kummer *et al.*, 1971; Pitts *et al.*, 1972; Finlayson *et al.*, 1974; Atkinson *et al.*, 1973). It is difficult to rationalize the formation of these highly excited species through the Criegee mechanism alone. O'Neal and Blumstein (1973) have suggested several alternative paths for decomposition of the original ozonide which can account better for some of the chemiluminescent products of the gas phase ozone–alkene reactions. Various modified forms of the O'Neal and Blumstein mechanism have been adopted by various groups of modelers of smog chemistry; for example, see Whitten and Hogo (1976); Sander and Seinfeld (1976). However, the extent and the nature of the fragmentation and rearrangement paths which are chosen must necessarily be rather arbitrary at this stage of our knowledge. Use of the Benson (1976) thermochemical–kinetic considerations have been made by O'Neal and Blumstein (1973) and others to derive 'best estimates' of the importance of the alternative paths of reaction of the ozonides. In most detailed modeling schemes involving SO_2 removal which are in use today, reactions such as 41, 42, 43, or 44 are not considered to be important, but the SO_2 oxidation in the alkene–O_3–SO_2 system is made to occur exclusively through partial fragmentation of the ozonide to various free radicals, followed by reactions 31–40 of Table 5 and their various analogues. For examples see Sander and Seinfeld, 1976; Walter *et al.*, 1977.

However it appears to us that the ideas related to the mechanism of the gas phase O_3–alkene reactions must be modified again. Recent FTS-IR spectroscopic observations of the O_3–alkene–air systems have been made by Niki *et al* (1977). They provide striking new evidence for the reasonable stability of many of the gas phase ozonides. Thus the role of reactions 41–44 should not be discarded. Hull *et al.* (1972) had observed through low temperature (-175 to $-80°C$) i.r. studies in the condensed phase, the primary ozonides of C_3H_6, *iso*-C_4H_8, *cis*- and *trans*-2-C_4H_8, cyclopentene, cyclohexene, trimethylethylene, and tetramethylethylene; these decomposed to various products upon warming the frozen mixtures to room temperature. In 1959 Hanst *et al.* concluded from long-path i.r. experiments that they observed ozonide formation from the gas phase reactions of O_3 with 1-hexene and 3-heptene but not from 1-pentene and the smaller alkenes. However the recent work of Niki *et al.* (1975) shows clearly that ozonide formation can be observed in the gas phase from olefins as small as propylene. In their work, Niki *et al.* reacted O_3 (5 ppm), *cis*-2-C_4H_8 (10 ppm), and CH_2O (10 ppm) in 700 Torr of air. The identified products (ppm) included: CO_2 (1.9), CO, CH_4(0.66), CHO_2H(0.25), CH_3OH(0.4), CH_2CO, and, significantly, propylene ozonide (0.88 ppm). Thus they presented unambiguous evidence that the Criegee intermediate $CH_3\dot{C}HOO\cdot$ (or conceivably the rearranged radical, $CH_3CH(\dot{O})O\cdot$) had a much longer lifetime in air than has been suggested in recent years by O'Neal and Blumstein (1973), Demerjian *et al.* (1974), and many others; it lived to react with the added CH_2O and formed propylene ozonide:

$$CH_3\dot{C}HOO\cdot + CH_2O \rightarrow CH_3CH\overset{O-O}{\underset{O}{\langle\rangle}}CH_2. \quad (84)$$

For the purposes of this review another most important observation was made by Niki *et al.* (1977); they found that addition of SO_2 at the 5 ppm level to the O_3, C_3H_6, CH_2O system quenched ozonide formation completely, and the SO_2 was consumed to an extent comparable to the ozonide yield observed in its absence. Furthermore the rearrangement of the $CH_3\dot{C}HOO\cdot$ species to acetic acid was not observed in these experiments, although it is expected in the O'Neal and Blumstein considerations:

$$CH_3\dot{C}HOO\cdot \rightarrow CH_3CH\begin{matrix}O\\ |\\ O\end{matrix} \quad CH_3CH\begin{matrix}O\cdot\\ \\ O\cdot\end{matrix} \rightarrow CH_3COOH \quad (85)$$

Niki (1977) has indicated to us that the ozonide yield is about 20% for all of the alkenes studied except C_2H_4 where no ozonide formation was detected. The 'radical' product yield was less than 50%.

At this writing it is not clear what species we should invoke as the reactants in the O_3–alkene–SO_2 system, but Niki's work demonstrates that the Criegee intermediate is a strong candidate. Some reaction from various free radical fragmentation products may occur as well. With this apparent return to the original mechanism suggested by Cox and Penkett (1971a, b), it is instructive to reconsider the detailed results of Cox and Penkett (1972).

A major problem in the quantitative evaluation of the rates of reactions 41–44, is the lack of information related to the absolute rate constants of the intermediate, SO_2-oxidizing species. In the mixture of gases encountered in the polluted troposphere, NO is commonly present with SO_2. If the Criegee intermediate is formed by ozone–alkene interactions, then SO_2 must compete with NO (reaction 86) and other reactants as well in order to be oxidized by $R\dot{C}HOO\cdot$.

$$CH_3\dot{C}HOO\cdot + NO \rightarrow CH_3CHO + NO_2 \quad (86a)$$

or

$$CH_3CH(\dot{O})\text{—}O\cdot + NO \rightarrow CH_3CHO + NO_2. \quad (86b)$$

The enthalpy changes for these reactions ($\Delta H_{86a} \cong -97$; $\Delta H_{86b} \cong -51$ kcal mole^{-1}) may be compared to that for the oxidation of NO by O_3 ($\Delta H_{87} = -48$ kcal mole^{-1}), a reactive species which is very similar to the Criegee intermediate.

$$O_3 + NO \rightarrow O_2 + NO_2. \quad (87)$$

It seems likely that $k_{86a} > k_{86b} \cong k_{87} = 1.6 \times 10^{-14}$ cm^3 mol^{-1} s^{-1} (Clyne *et al.*, 1964), and a good competition with the possible SO_2 oxidation reactions must be provided by NO in the atmosphere. In the absence of an experimental basis for the evaluation of the inhibition of SO_2 oxidation by NO addition to the O_3–alkene–SO_2 system, it will be impossible to judge accurately the significance of reactions such as 43 and 44 in the atmosphere. It is interesting to note that ΔH_{44} is comparable to ΔH_{26}, and reaction 26 is immeasurably slow at room temperature. This evidence seems to favor the direct involvement of the Criegee intermediate and not the rearranged form, $RCH(\dot{O})\text{—}O\cdot$ in SO_2 oxidation.

Let us adopt for the purposes of our considerations here an expanded version of the Cox and Penkett mechanism for the *cis*-2-butene, O_3, SO_2, H_2O, air system, and focus our attention on the Criegee intermediate as the major source of SO_2 oxidation in these experiments. The competitive reactions of the Criegee intermediate in the NO-free system may occur in principle with SO_2, the alkenes, ozone, oxygen, water, as well as the unimolecular decay to other products:

$$O_3 + cis\text{-}2\text{-}C_4H_8 \rightarrow (\text{molozonide}) \rightarrow CH_3\dot{C}HOO\cdot + CH_3CHO \quad (88)$$

$$\rightarrow (\text{RCHO, RCO}_2\text{H, etc.}) \quad (89)$$

$$CH_3\dot{C}HOO\cdot + SO_2 \rightarrow CH_3CHO + SO_3 \quad (90)$$

$$CH_3\dot{C}HOO\cdot + C_4H_8 \rightarrow CH_3CHO + C_4H_8O \text{ (and other products)} \quad (91)$$

$$CH_3\dot{C}HOO\cdot + O_3 \rightarrow CH_3CHO + 2O_2 \quad (92)$$

$$CH_3\dot{C}HOO\cdot + O_2 \rightarrow CH_3\underset{}{\overset{O\cdot}{\overset{|}{C}}}H\text{—}O\cdot \text{ (other products)} \quad (93)$$

$$\rightarrow O_3 + CH_3CHO \quad (94)$$

$$CH_3\dot{C}HOO\cdot + H_2O \rightarrow CH_3COOH + H_2O \quad (95)$$

$$CH_3\dot{C}HOO\cdot \rightarrow (CH_3COOH) \rightarrow CH_4 + CO_2(CH_3OH, CO, \text{etc.}). \quad (96)$$

Each of these steps is energetically feasible, but the extent of the participation of each is unclear. Since the rate of O_3–alkene reaction and the rate of SO_2 oxidation in alkene–O_3–SO_2 mixtures was relatively unaffected by replacing the air with N_2 (~0.2%) in the experiments of Cox and Penkett (1972), then reactions 93 and 94 are probably unimportant. Deviations of the O_3–alkene stoichiometry from 1 to 1 in reactant mixtures with excess of O_3 or excess of alkene, suggest that 91 and 92 may occur to some extent. The marked inhibition by H_2O which they observed is consistent with some reaction such as 95. One might speculate that H_2O catalyzes the rearrangement of $CH_3\dot{C}HOO\cdot$ to CH_3CO_2H by way of some complex between the Criegee intermediate and water:

$$CH_3\text{—}C(\text{—H})(O\cdots H\text{—}O\text{—}H)(O) \longrightarrow CH_3\text{—}C(=O)\text{—}O\text{—}H \;\; H\text{—}O\text{—}H$$

The Cox and Penkett data for *cis*-2-butene, SO_2, O_3, H_2O, air mixtures can be reconsidered in terms of these reactions. Following Cox and Penkett we may assume that the reactive Criegee intermediate achieves a steady state concentration. Let us assume further that SO_2 oxidation occurs only in reaction 90 and that $R_{92} \ll R_{88} + R_{89}$. Then we expect relation 97 to hold:

$$\frac{-R^0_{O_3}}{R^0_{SO_3}} = \left(\frac{k_{88} + k_{89}}{k_{88}}\right)\left(1 + \frac{1}{[SO_2]k_{90}} \times \{k_{96} + [C_4H_8]k_{91} + [O_3]k_{92} + [H_2O]k_{95}\}\right). \quad (97)$$

The rather limited data cannot provide a test of the functional form for each of the reactants in relation 97, but certain observations can be made. In the Cox and Penkett experiments the $[O_3]^0$ and $[C_4H_8]^0$ were held essentially constant at 0.5 and 1.0 ppm, respectively. For these conditions the form of the relation 97 simplifies to 98, where A and B are constants:

$$\frac{-R^0_{O_3}}{R^0_{SO_3}} = B + \left(\frac{B\{A + [H_2O]k_{95}\}}{k_{90}}\right)\frac{1}{[SO_2]}. \quad (98)$$

From relation 98 we expect a linear relation between $-R^0_{O_3}/R^0_{SO_3}$ and $1/[SO_2]$ for a series of runs made at constant $[H_2O]$. This was observed by Cox and Penkett, and such a plot has been redrawn from the original data; see Fig. 7. Observe that the slope to intercept ratios for the linear plots at each $[H_2O]$ should give the quantity, $(A + [H_2O]k_{95})/k_{90}$, where $A = k_{96} + [C_4H_8]k_{91} + [O_3]k_{92}$. For runs at 76, 40 and 10% relative humidity, the slope/intercept ratios in Fig. 7 are: 1.43 ± 0.26, 0.83 ± 0.21, and 0.30 ± 0.02 ppm. These are plotted vs $[H_2O]$ in Fig. 8 and the expected linear dependence is seen. The intercept and slope in Fig. 8 give $A/k_{90} \cong 0.16 \pm 0.05$ ppm and $k_{95}/k_{90} \cong (6.1 \pm 0.3) \times 10^{-5}$, re-

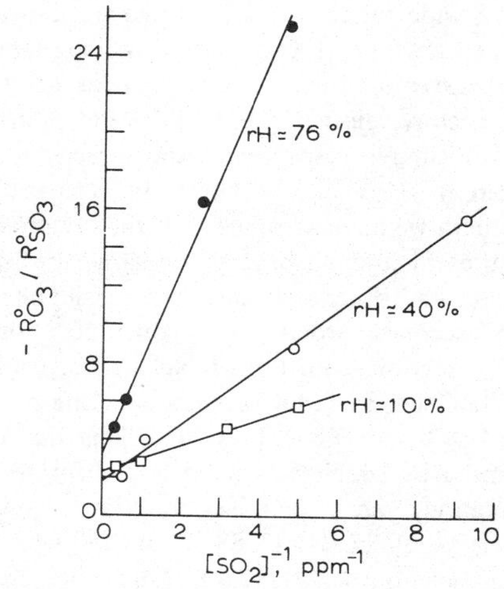

Fig. 7. Plot of the ratio of rate of O_3 loss to rate of SO_3 formation versus $[SO_2]^{-1}$ for the data of Cox and Penkett (1972) from the dark reaction in *cis*-2-butene–SO_2–O_3–H_2O–air mixtures at various relative humidities (22°C).

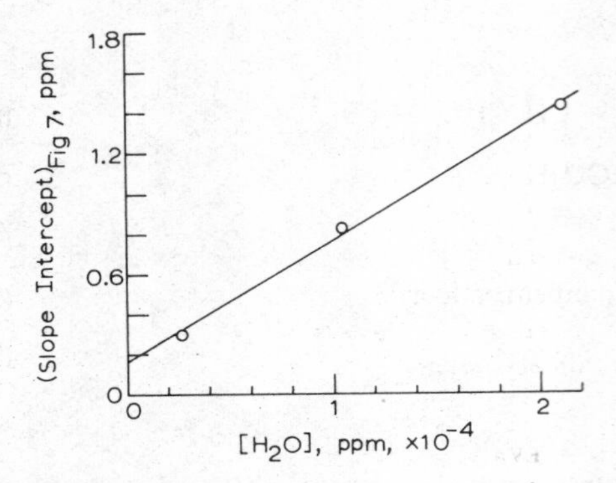

Fig. 8. Plot of the ratio of slope/intercept for the plots of Fig. 7 vs [H_2O] derived from the data of Cox and Penkett (1972); the slope gives the k_{95}/k_{90} estimate used in this study.

spectively. The kinetic treatment given here differs somewhat from that of Cox and Penkett in that we have attempted to include explicitly the reaction of H_2O with the intermediate. The data do suggest that the reaction with water is by far the dominant reaction of the intermediate; the ratio of rate of the reaction with water to that for all other reactions of the intermediate varies from about 4 for experiments at 40% r.h. to about 8 at 76% r.h. This striking effect for water noted by Cox and Penkett suggests indirectly that RO_2, HO_2, HO, and other radical species which may be formed in the alkene–SO_2–O_3–H_2O–air system are not the important oxidizing species present. Their reactivity toward H_2O is thought to be very low, and their concentrations will be altered insignificantly with H_2O increase.

There is one other rather thorough study of the alkene–O_3–SO_2–air system of which we are aware. McNelis (1974) made a kinetic study of the C_3H_6, O_3, SO_2, air system, and it is interesting to compare these results with those of Cox and Penkett. First it must be observed that McNelis did not vary the r.h. over a wide range in his study so a quantitative test of the effect of H_2O is not possible. However he did observe in otherwise similar runs at 20 and 36% r.h., a conversion of 0.071 and 0.066 ppm SO_2/ppm O_3 consumed, respectively, the direction of the trend seen by Cox and Penkett. In retreating the McNelis data we have assumed that the reaction with H_2O is present and have used the same mechanism outlined for the butene studies. We recalculated the apparent second order rate constants for the O_3–C_3H_6 reaction from the McNelis data, using the conventional second order rate law and the observed data for [O_3] and [C_3H_6] vs time. The initial rate method used by McNelis seemed less accurate to us. The seemingly consistent set of data from the McNelis runs 26, 27, 31, 87, 88, 89 and 90 were used. The data gave $k_{95'}/k_{90'} \cong (7.4 \pm 1.5) \times 10^{-5}$ for the propylene data, in reasonable accord with the estimate from the C_4H_8 data. Reactions 95′ and 90′ refer to the reactions of the species $\cdot CH_2OO\cdot$ as well as $CH_3\dot{C}HOO\cdot$ which are both formed in the C_3H_6 system.

We can estimate the rate of SO_2 oxidation by the reactive intermediates formed in the O_3–alkene reactions using the data derived here. Taking a concentration of total alkenes = 0.10 ppm, [O_3] = 0.15 ppm, and [SO_2] = 0.05 ppm, typical of a highly polluted, sunlight-irradiated urban atmosphere, and using Niki's observation that about 20% of the O_3 molecules which react with alkene form the Criegee intermediate, we estimate that the rate of SO_2 oxidation, presumably through 43, will occur at about 0.23 and 0.12% h^{-1} at 50 and 100% r.h. (25°C) when the reactivity of the alkene toward O_3 is typical of that for an alkene with a terminal double bond ($k \cong 1 \times 10^{-17}$ cm^3 mol^{-1} s^{-1}). In the unlikely event that 0.10 ppm of a highly reactive alkene such as *cis*-2-butene were present together with 0.15 ppm of ozone, then much higher rates of SO_2 oxidation would be expected. However we have neglected completely any loss reaction for the Criegee intermediate with NO in these considerations, so the rates estimated represent theoretical upper limits. Although these rates do not seem large in comparison with those expected for HO, HO_2, and CH_3O_2 reactions, they are not insignificant. The O_3–alkene reactions will continue to occur during the night-time hours when both reactants are present, and the SO_2 oxidation from the products of this interaction will carry on as well; however, the rates are not expected to be large.

(m) *The* SO_3 *reactions in the atmosphere*

Reactions 45 and 46 have generally been neglected in kinetic simulations of the tropospheric chemistry of SO_2 and SO_3.

$$SO_3 + O(^3P)(+M) \rightarrow SO_4(+M) \quad (45)$$

$$SO_3 + O(^3P)(+M) \rightarrow SO_2 + O_2(+M). \quad (46)$$

Jacob and Winkler's (1972) estimate of the bimolecular rate constant for the SO_3–$O(^3P)$ reaction, $k_{45} + k_{46} \cong 5 \times 10^{-17}$ cm^3 mol^{-1} s^{-1}, justified this action. However more recently Daubendiek and Calvert (1973, paper in preparation) and Westenberg and deHaas (1975b) have found that these reactions are very much faster than the earlier measurements suggested, and a new look at their potential role in the troposphere should be taken. Daubendiek and Calvert reported i.r. kinetic studies from the O_3 photolysis ($\lambda > 590$ nm) in the presence of SO_3; they noted a rapid reaction (compared to $O + O_3 \rightarrow 2O_2$) which destroyed SO_3, but surprisingly, there was a delay in the SO_2 appearance which is expected if reaction 46 occurs. A metastable product absorbing at 6.78 μ was seen which decayed in the dark ($\tau_{1/2} = 280$ s) to form SO_3 and SO_2. From the stoichiometry of the gases evolved the species appeared to be a mixture of S_3O_8 and S_3O_9. These seemed to form in the reaction between an SO_4 initial product of 45 and SO_2

and SO_3. The O-SO_3 reaction appeared to obey second order kinetics for pressures of SO_3 and O_3 above about 6 Torr. From the competitive rates of SO_3 and O_3 reactions with $O(^3P)$,

$$O(^3P) + O_3 \rightarrow 2O_2 \quad (99)$$

they derived $k_{45} + k_{46} \cong 7 \times 10^{-13}$ cm^3 mol^{-1} s^{-1}, accepting $k_{99} = 8.5 \times 10^{-15}$ cm^3 mol^{-1} s^{-1} at 25°C (Hampson and Garvin, 1975).

Westenberg and deHaas (1975b) reported rate studies of the $O(^3P)$-SO_3 reaction using a discharge flow method with ESR detection of the $O(^3P)$. They found no evidence for an SO_4 intermediate for their conditions, but the fast overall reaction 46 seemed to occur with third order kinetics up to 7 Torr of He. They derived $k_{46} = 1.4 \times 10^{-31} e^{785/T}$ cm^6 mol^{-2} s^{-1} (298–507°K). The third order nature of the reaction which they observed at low pressures suggests that an SO_4 intermediate is formed prior to rearrangement to their observed products, SO_2 and O_2. Indeed the results of both recent studies give strong independent evidence for the rapid occurrence of reaction 45 and 46; where the data do overlap in pressures employed, good agreement is found between the Daubendiek and Calvert and Westenberg and deHaas data. Both sets are consistent with the estimate: $k_{45} + k_{46} \cong 7 \times 10^{-13}$ cm^3 mol^{-1} s^{-1} in air at 1 atm and 25°C.

Consider the possible implications of these findings on the atmospheric chemistry of SO_2 and SO_3. Westenberg and deHaas (1975a) concluded from a consideration of the two reactions 25 and 46 alone, that the main effect of the presence of SO_2 on O in the atmosphere would be to catalyze its recombination with little or no net SO_3 formation.

$$O(^3P) + SO_2\,(+M) \rightarrow SO_3\,(+M) \quad (25)$$

$$O(^3P) + SO_3\,(+M) \rightarrow SO_2 + O_2\,(+M). \quad (46)$$

Thus using our present estimates of the rate constants for air at 1 atm, and assuming that SO_4 always forms $SO_2 + O_2$ ultimately, then we find $[SO_3]/[SO_2] \cong 0.08$ at the steady state. However there is no chance that a steady state of SO_3 and SO_2 will be established involving reactions 25 and 46 in the real atmosphere. The SO_3 molecule encounters with O-atoms will be very much less frequent than those with the abundant water molecule within the lower troposphere. Recent evidence of Castleman *et al.* (1974) confirmed the conclusion of Goodeve *et al.* (1934) that the overall reaction 47 is very fast; they estimated $k_{47} = (9.1 \pm 2.0) \times 10^{-13}$ cm^3 mol^{-1} s^{-1} from flow experiments at low pressures.

$$SO_3 + H_2O \rightarrow H_2SO_4. \quad (47)$$

The rate constant estimate is surprising large for such a complex rearrangement of reactant molecules that must accompany this reaction; the observation of the mass 98 product in the experiments of Castleman *et al.* may reflect the formation of an $SO_3 \cdot H_2O$ adduct which is a precursor to the final rearranged, stable product molecule, H_2SO_4. In any case the ultimate removal of SO_3 by its net reaction 47 with water is so very fast that the reaction of SO_3 with any other species in the moist lower atmosphere should be unimportant. We conclude from the evidence at hand that the usual assumption of modelers of the SO_2 conversion in the troposphere, namely, that H_2SO_4 formation will always follow the generation of SO_3, appears to be sound.

4. EVALUATION OF THE RELATIVE IMPORTANCE OF THE HOMOGENEOUS SO_2 REACTIONS WITH VARIOUS REACTIVE SPECIES IN THE TROPOSPHERE

It is instructive to use the present evaluation of the SO_2 rate data to estimate in greater detail the rates of SO_2 homogeneous oxidation which we expect to occur in various clean and polluted regions of the troposphere. First let us consider the rate of oxidation of SO_2 which is anticipated for the relatively clean troposphere. The concentrations of the important species, HO, HO_2, and CH_3O_2 were estimated for various elevations within the troposphere of the northern hemisphere, and for various latitudes and seasons. If one considers the presence of SO_2 to be at the ppb level or below in these atmospheres, it is a fair approximation to assume very little perturbation will occur in the estimated radical concentrations. Thus we can use the Crutzen and Fishman (1977) estimates to calculate the theoretical SO_2 conversion rates by reactions with HO, HO_2, and CH_3O_2 in the troposphere. Shown in Fig. 9 are the rates averaged over all northern latitudes and the entire depth of the troposphere for each month of the year. Obviously the rate for the HO-radical reaction with SO_2, reaction 39, accounts for most of the

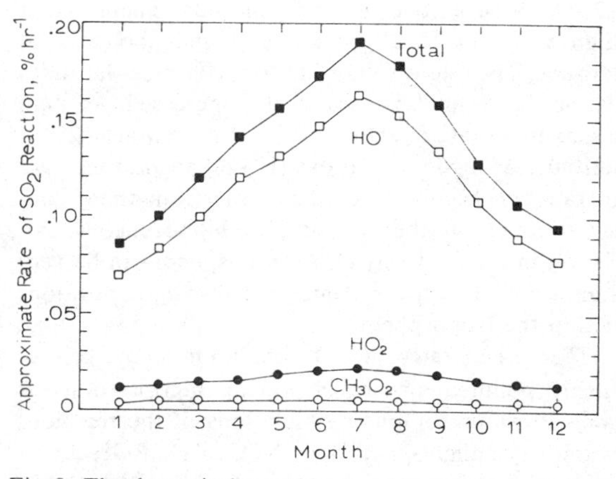

Fig. 9. The theoretical monthly average of the rate (% h^{-1}) of SO_2 oxidation within the northern troposphere as a function of month of the year; rates are shown for the HO reaction 39, the HO_2 reaction 31, the CH_3O_2 reaction 33, and the total of these three rates; calculated using [HO], [HO_2], and [CH_3O_2] estimates of Crutzen and Fishman (1977).

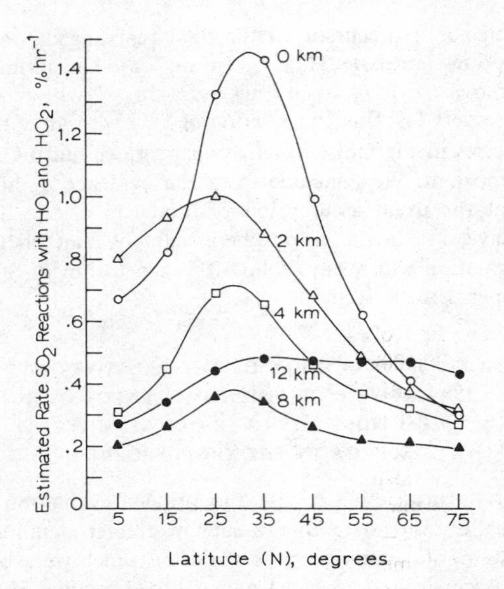

Fig. 10. The theoretical rate ($\% \, h^{-1}$) of SO_2 oxidation by HO (reaction 39) and HO_2 (reaction 31] at various elevations within the troposphere and at various latitudes in the northern hemisphere; data are for the average rates for the 1200 h in the month of July; calculated using the [HO] and [HO_2] estimates of Crutzen and Fishman (1977).

oxidation; a significant fraction is also contributed by the HO_2-radical reaction 31 and a smaller amount from the CH_3O_2 species. The total average rate of SO_2 oxidation from these three most important homogeneous reactions varies from a low of 0.09% in January to a maximum of $0.2\% \, h^{-1}$ in July. Of course these rates are directly related to the solar irradiance, the temperature, and the atmospheric composition at the particular point within the troposphere. One can observe in Fig. 10 the theoretical rates which are representative of the maximum values of the SO_2 oxidation by HO and HO_2 which occur near the 1200 h averaged over each day in July. These are as high as $1.5\% \, h^{-1}$ at ground level and about 32° N latitude. The rate is somewhat lower at most latitudes as the elevation above sea level is increased. At 8 km a maximum rate of about $0.4\% \, h^{-1}$ is seen near 28° N latitude. At higher elevations (12 km) a small increase in rate can be observed. It is clear from these considerations that the rate of SO_2 homogeneous oxidation in the relatively clean troposphere can be very significant at certain time periods and positions within the troposphere.

Theoretical rates of SO_2 oxidation in parcels of highly polluted air are of special interest to us as well. Previous computer simulations of the reactions within a sunlight-irradiated, NO_x, RH, RCHO-polluted air mass have been made by Calvert and McQuigg (1975), Sander and Seinfeld (1976), and Graedel (1976). All workers agreed as to the importance of the HO- and HO_2-radical reaction rates with SO_2. For the simulated polluted atmosphere chosen by Calvert and McQuigg and for an older less accurate set of rate constants, a maximum SO_2 oxidation rate of $1.1\% \, h^{-1}$ was predicted in simulations carried out at a solar zenith angle of 40°. With a somewhat different choice of elementary reactions, rate constants, and initial reactants, Sander and Seinfeld predicted a maximum SO_2 homogeneous oxidation rate in their simulated polluted atmosphere of $4.5\% \, h^{-1}$, with HO and HO_2 radicals again accounting for much of the reaction rate. Graedel's simulations employed the energetically unfavorable and unlikely chain reaction sequence of Davis and Klauber (1975), so comparisons with his results are of questionable value. In Fig. 11 is given an updated version of the Calvert and McQuigg simulations of SO_2 removal rates based upon the new data and the new rate constant choices presented in this study. The simulated atmosphere contained initially (ppm): [NO] = 0.15; [NO_2] = 0.05; [*cis*-2-C_4H_8] = 0.10; [CO] = 10; [CH_4] = 1.5; [CH_2O] = 0; [CH_3CHO] = 0; [SO_2] = 0.05; r.h., 50% (25°C); solar zenith angle, 40°; a stagnant air mass without dilution was considered to simulate the conditions of highest smog-forming potential. The O_3 concentration in this simulation rose to 0.15 ppm at the 120-min point of the irradiation at which time the alkene had decreased to 0.013 ppm. The length of the ordinate within each area of Fig. 11 represents the theoretical $\% \, h^{-1}$ of SO_2 oxidation by the radicals shown. Note that the total oxidation rate from the several species rises to over $4\% \, h^{-1}$ at about 60 min into the irradiation. There is one major difference between these results and those of the previous simulations. The newly estimated rate constant for the CH_2O_2–SO_2 reactions

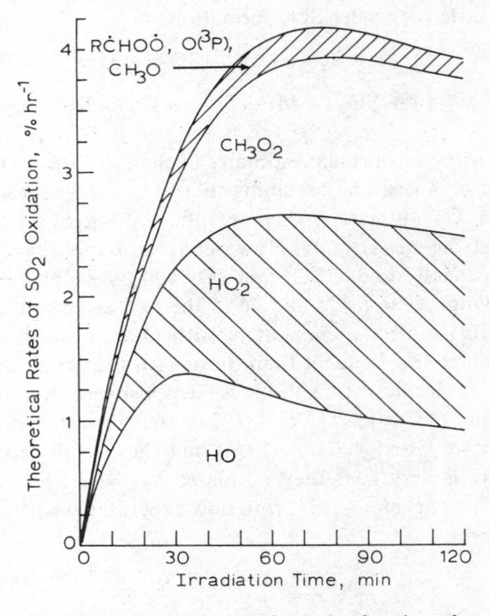

Fig. 11. The theoretical rate of attack of various free radical species on SO_2 [$\% \, h^{-1}$) for a simulated sunlight-irradiated (solar zenith angle = 40°), polluted atmosphere; initial concentration (ppm): [SO_2] = 0.05; [NO] = 0.15; [NO_2] = 0.05; [CO] = 10; [CH_4] = 1.5; [CH_2O] = 0; [CH_3CHO] = 0; r.h., 50% (25°C).

33 and 34, has been included, and this reaction has added an additional increment to the SO_2 homogeneous oxidation rate expected in theory. It appears that about equal rates of oxidation of SO_2 occur in a highly polluted atmosphere through the reactions of the HO_2, CH_3O_2, and the HO species (reactions 31, 33, 39). We have picked conditions, a high concentration of very reactive alkene (*trans*-2-butene) and $k_{86} = 0$, which should accentuate the contribution of the Criegee intermediate. Even so the rates of SO_2 oxidation from reaction 43 are relatively small compared to the rates of the three major reacting species; the contribution from this reaction increases to a maximum of 0.25% h^{-1} at about 60 min into the irradiation period. It is improbable that an actual polluted atmosphere would contain this much reactive alkene and that $k_{86} = 0$. Hence we conclude that the Criegee intermediate is probably never a major reactant for SO_2 oxidation in the highly polluted atmosphere, although during night-time hours this reaction is likely the major homogeneous SO_2 oxidation mechanism which is operative.

There is one other major class of SO_2 polluted atmosphere which is of special interest to us. It is that contained in a stack plume from a power plant or other industrial combustion operation. These gaseous mixtures are of unique composition; the original effluent is oxygen-depleted and NO- and SO_2-rich. Because of the very high NO levels and the very low hydrocarbon and CO levels, the long oxidation chains which can lead to the reasonably rapid conversions of SO_2 to SO_3, H_2SO_4, etc. are suppressed during the first stages of the plume transport. The initial $[NO_2]/[NO]$ ratio is very low so that the $[O_3]$ generated through the reaction sequence 100, 101, and 87,

$$NO_2 + h\nu(\lambda < 4100\,\text{Å}) \rightarrow NO + O \qquad (100)$$

$$O + O_2(+M) \rightarrow O_3(+M) \qquad (101)$$

$$O_3 + NO \rightarrow NO_2 + O_2 \qquad (87)$$

will be very much below the 0.04 ppm values observed in relatively clean ambient air. For the conditions present in the plume after a short mixing period, the O_3 should reach a photostationary state for which the $[O_3]$ will be related to the $[NO_2]/[NO]$ ratio, the apparent first order rate constant for the rate of photolysis of NO_2 in 100 (k_{100}), the rate limiting process for O_3 formation in 101, and the rate constant k_{87} for the dominant O_3 loss reaction 87.

$$[O_3] \cong [NO_2]\,k_{100}/[NO]\,k_{87}. \qquad (102)$$

k_{100} is a function of the solar irradiance in the 4100 to 2900 Å region and a typical value for k_{100}/k_{87} is ~0.02 ppm for the 1000–1400 time period (Calvert, 1976). Thus the ozone level will not reach the 'clean air' background values within the plume until the $[NO_2]/[NO]$ ratio climbs to near 2. Actual plume data show significant depletion of the ambient ozone levels within the plume for very great distances. For examples see Davis *et al.* (1974d) and Wilson *et al.* (1976). Only after the plume gases have mixed with sufficient quantities of reactant hydrocarbons, aldehydes, CO, etc. present in the ambient air can extensive NO to NO_2 conversion be effected and O_3 levels climb. However, there are at least two mechanisms which in theory can contribute to an initial burst of SO_2 oxidation involving the elementary reactions of Table 5.

The first of these involves the photolysis of NO_2 (reaction 100) early in the plume dilution when the $[O_2]$ is relatively low, $[SO_2]$ is high, and the fraction of O-atoms captured by SO_2 can be significant. NO_2 represents a small fraction of the plume gases released to the atmosphere. Additional NO_2 is formed as the ambient air mixes in with the NO-rich gases: $2NO + O_2 \rightarrow 2NO_2$. For these conditions the SO_2 in the plume gases can compete somewhat successfully with NO, NO_2, as well as O_2 for the O-atoms formed from NO_2 photolysis in 100:

$$NO_2 + h\nu \rightarrow O + NO \qquad (100)$$

$$O + O_2(+M) \rightarrow O_3(+M) \qquad (101)$$

$$O + NO_2 \rightarrow O_2 + NO \qquad (102)$$

$$O + NO_2(+M) \rightarrow NO_3(+M) \qquad (103)$$

$$O + NO(+M) \rightarrow NO_2(+M) \qquad (104)$$

$$O + SO_2(+M) \rightarrow SO_3(+M). \qquad (25)$$

The rates of SO_2 oxidation from reaction 25 alone in the sunlight-irradiated stack gases can be significant during the early stages of dilution. Thus if we assume stack gases to be at the concentrations: $[NO] = 500$, $[SO_2] = 500$, $[NO_2] = 20$; $[O_2] = 1 \times 10^5$ ppm, for the midday sunlight we expect an instantaneous rate of SO_2 oxidation by 25 alone to be about 1.4% h^{-1}. This rate will drop fairly quickly as further dilution occurs with the transport of the plume. As the plume is further diluted by factors of 4, 8, 16, and 32, the instantaneous rates of SO_2 oxidation by reaction 25 are expected to fall to 0.43, 0.20, 0.10, and 0.05% h^{-1}.

A similar early peak in the rate of the homogeneous oxidation of SO_2 by reaction 39 is expected with the HO-radicals formed from the photolysis of HONO.

$$HONO + h\nu(\lambda < 4000\,\text{Å}) \rightarrow HO + NO. \qquad (105)$$

The generation of HONO early in the dilution of stack plumes may occur in theory through the homogeneous reaction 106 (Chan *et al.*, 1976) or through heterogeneous paths involving water condensation or particulate surfaces:

$$NO + NO_2 + H_2O \rightarrow 2HONO. \qquad (106)$$

Thus the development of as much as 10 ppm of HONO within a short dilution period of certain stack effluents with very high NO, NO_2, and H_2O concentrations may occur. When HONO is at the 10 ppm level with $[SO_2] = 500$, $[NO] = 500$, $[NO_2] = 20$, $[CO] = 1$ ppm, the instantaneous rate of SO_2 oxidation by 39 is expected to be about $1{\cdot}2\%\,h^{-1}$ in noonday sunlight. In this case O_2 is not a competitor for HO but the NO, NO_2, and CO are:

$$HO + SO_2(+M) \rightarrow HOSO_2(+M) \quad (39)$$

$$HO + NO(+M) \rightarrow HONO(+M) \quad (107)$$

$$HO + NO_2(+M) \rightarrow HONO_2(+M) \quad (108)$$

$$HO + CO(+M) \rightarrow H + CO_2(+M). \quad (109)$$

After dilution of the reaction mixture by a factor of ten with clean air, the rate of SO_2 oxidation by 106 again falls precipitously to $0.1\%\,h^{-1}$.

Recently Miller (1977) has simulated the chemical rates of change within a stack plume; the only SO_2 removal mode which he included was reaction 39. It is interesting that he did observe a maximum rate of SO_2 oxidation (about $0.1\%\,h^{-1}$) early in the dilution of the stack gases with either unpolluted or polluted air.

As we have seen one anticipates in theory that SO_2 may be oxidized at rates up to several per cent h^{-1} for a short period during the early stages of stack gas dilution through the occurrence of the homogeneous reactions 25 and 39 in a sunlight-irradiated plume. It is evident that the apparent order of SO_2 conversion will appear to higher than first during the first stages of plume dilution. Such orders have been observed in stack plumes. They have been rationalized well in terms of heterogeneous reactions alone (Schwartz and Newman, in press).

The homogeneous SO_2 conversion rates at long transport times are expected to increase as dilution of the stack gases with urban air containing hydrocarbons, aldehydes, ozone, etc., occurs, and the rapid chain reaction sequences familiar to smog chemistry take over. The SO_2 to sulfate conversion rates estimated in plumes at long transport times in the study of Wilson *et al.* (1976) were observed to increase from about $1.5\%\,h^{-1}$ during the 0.7–1.5 h transport time period (10–21 km path) to about $5\%\,h^{-1}$ after 2.3–3.2 h of transport (32–45 km path). Gillani *et al.* (1978) observed in recent studies in the St. Louis area, maximum rates of SO_2 conversion to particulate sulfur of the order of $3\%\,h^{-1}$ in plumes with significant transport time and dilution. They also presented some significant evidence that homogeneous photochemical processes may be important in the SO_2 conversions observed.

From the considerations presented in the present study we conclude that the homogeneous oxidation of SO_2 in the troposphere does occur at rates which constitute a major fraction of the rates observed experimentally. It is the hope of the authors that the discussion and the data presented in this study will be useful to the atmospheric scientists in their continued development of more quantitative models of the tropospheric SO_2 oxidation and transport.

Acknowledgements—The authors gratefully acknowledge the support of this work at the Ohio State University through a research grant (R804348010) from the U.S. Environmental Protection Agency, and at the University of Alberta funds were provided by the Research Secretariat of Alberta Environment, Canada. We also thank Sidney W. Benson, Paul Crutzen, Jack Fishman, Art Levy, David Miller, Hiromi Niki, Stephen E. Schwartz, Leonard Newman, Rudolf B. Husar, Kenneth T. Whitby, I. N. Tang, Harvey E. Jeffries, and Donald L. Fox for preprints of papers provided to the authors and valuable discussions related to this study.

REFERENCES

Allen E. R. and Bonelli J. E. (1972b) The photooxidation of sulfur dioxide in urban atmospheres. Abstracts, 6th Informal Photochemistry Conference, Oklahoma State University, p. 6.1–6.2.

Allen E. R., McQuigg R. D. and Cadle R. D. (1972a) The photooxidation of gaseous sulfur dioxide in air. *Chemosphere* **1**, 25–32.

Altwicker E. R. (1976) Ozone formation during sulfur dioxide irradiation. *J. Envir. Sci. Health* All, 439–443.

Atkinson R., Finlayson B. J. and Pitts J. N. Jr. (1973) Photoionization mass spectrometer studies of gas phase ozone–olefin reactions. *J. Am. Chem. Soc.* **95**, 7592–7599.

Atkinson R., Perry R. A. and Pitts J. N. Jr. (1976) Rate constants for the reactions of the OH radical with NO_2 (M = Ar and N_2) and SO_2 (M = Ar). *J. chem. Phys.* **65**, 306–310.

Atkinson R. and Pitts J. N. Jr. (1974) Rate constants for the reaction of $O(^3P)$ atoms with SO_2 ($M = N_2O$) over the temperature range 299–392 K. *Chem. Phys. Lett.* **29**, 28–30.

Badcock C. C., Sidebottom H. W., Calvert J. G., Reinhardt G. W. and Damon E. K. (1971) Mechanism of the photolysis of sulfur dioxide–paraffin hydrocarbon mixtures. *J. Am. chem. Soc.* **93**, 3115–3121.

Baulch D. L. and Drysdale D. D. (1974) An evaluation of the rate data for the reaction $CO + OH \rightarrow CO_2 + H$. *Combust. Flame* **23**, 215–225.

Baulch D. L., Drysdale D. D. and Loyd A. C. (1972) *Evaluated Kinetic Data for High Temperature Reactions*, Vol. 1. Butterworths, London.

Benson S. W. (1976) *Thermochemical Kinetics* (2nd Edn.). John Wiley, New York, pp. 1–320.

Benson S. W. Thermochemistry and kinetics of sulfur containing molecules and radicals. Preprint of a paper to appear in *Chem. Revs.*

Blacet F. E. (1952) Photochemistry in the lower atmosphere. *Ind. Engng. Chem.* **44**, 1339–1346.

Boreskov G. K. and Illarionov V. V. (1940) *J. phys. Chem. Moscow* **14**, 1428; quoted in Kondratiev V. N. (1970). *Rate Constants of Gas-Phase Reactions*. Science Publishing House, Moscow, p. 233.

Bottenheim J. W. and Strausz O. P. (1977) Review of pollutant transformation processes relevant to the Alberta oil sands area and proposals for further studies. Report to Atmospheric Environment Service, Dept. Supply and Services Canada, Contract No. OSU 76–00169, from Hydrocarbon Research Center, University Alberta, Edmonton, Alberta, Canada, T6G 2G2.

Brand J. C. D., deLauro C. and Jones V. T. (1970) Structure of the $^3B_1$ state of sulfur dioxide. *J. Am. chem. Soc.* **92**, 6095–6096.

Brus L. E. and McDonald J. R. (1974) Time-resolved fluorescence kinetics and $^1B_1$ ($^1\Delta_g$) vibronic structure in tunable ultraviolet laser excited SO_2 vapor. *J. chem. Phys.* **61**, 97–105.

Calvert J. G. (1976) Test of the theory of ozone generation in Los Angeles atmosphere. *Envir. Sci. Technol.* **10,** 248–256.

Calvert J. G. and McQuigg R. D. (1975) The computer simulation of the rates and mechanisms of photochemical smog formation. *Int. J. chem. Kinet. Symp.* **1,** 113–154.

Calvert J. G., Slater D. H. and Gall J. W. (1971) The methyl radical-sulfur dioxide reaction. In *Chemical Reactions in Urban Atmospheres* (Edited by Tuesday C. S.). Elsevier, New York, pp. 133–158.

Castleman A. W. Jr., Munkelwitz H. R. and Manowitz B. (1974) Isotopic studies of the sulfur component of the stratospheric aerosol layer. *Tellus* **26,** 222–234.

Castleman A. W. Jr. and Tang I. N. (1976/77) Kinetics of the association reaction of SO_2 with the hydroxyl radical. *J. Photochem.* **6,** 349–354.

Cehelnik E., Heicklen J., Braslavsky S., Stockburger III L. and Mathias E. (1973/74) Photolysis of SO_2 in the presence of foreign gases—IV. Wavelength and temperature effects with CO. *J. Photochem.* **2,** 31–48.

Cehelnik E., Spicer C. W. and Heicklen J. (1971) Photolysis of sulfur dioxide in the presence of foreign gases—I. Carbon monoxide and perfluoroethylene. *J. Am. chem. Soc.* **93,** 5371–5380.

Chan W. H., Nordstrom R. J., Calvert J. G. and Shaw J. H. (1976) Kinetic study of HONO formation and decay reactions in gaseous mixtures of HONO, NO, NO_2, H_2O, and N_2. *Envir. Sci. Technol.* **10,** 674–682.

Chan W. H., Uselman W. M., Calvert J. G. and Shaw J. H. (1977) The pressure dependence of the rate constant for the reaction: $HO + CO \rightarrow H + CO_2$. *Chem. Phys. Lett.* **45,** 240–244.

Chung K., Calvert J. G. and Bottenheim J. W. (1975) The photochemistry of sulfur dioxide excited within its first allowed band at 3130 Å and the 'forbidden' band (3700–4000 Å). *Int. J. chem. Kinet.* **7,** 161–182.

Clyne M. A. A., Thrush B. A. and Wayne R. P. (1964) Kinetics of the chemiluminescent reaction between nitric oxide and ozone. *Trans. Faraday Soc.* **60,** 359–370.

Cox R. A. (1972) Quantum yields for the photooxidation of sulfur dioxide in the first allowed absorption region. *J. phys. Chem.* **76,** 814–820.

Cox R. A. (1973) Some experimental observations of aerosol formation in the photooxidation of sulfur dioxide. *Aerosol Sci.* **4,** 473–483.

Cox R. A. (1974/75) The photolysis of nitrous acid in the presence of carbon monoxide and sulfur dioxide. *J. Photochem.* **3,** 291–304.

Cox R. A. (1975) The photolysis of gaseous nitrous acid—a technique for obtaining kinetic data in atmospheric photooxidation reactions. *Int. J. chem. Kinet. Symp.* **1,** 379–398.

Cox R. A., Derwent R. G. and Holt P. M. (1976) Relative rate constants for the reactions of OH radicals with H_2, CH_4, CO, NO, and HONO at atmospheric pressure and 296 K, *J. Chem. Soc. Faraday Trans. I* **72,** 2031–2043.

Cox R. A. and Penkett S. A. (1970) The photooxidation of sulfur dioxide in sunlight. *Atmospheric Environment* **4,** 425–433.

Cox R. A. and Penkett S. A. (1971a) Oxidation of atmospheric SO_2 by products of the ozone–olefin reaction. *Nature* **230,** 321–322.

Cox R. A. and Penkett S. A. (1971b) Photooxidation of atmospheric SO_2. *Nature* **229,** 486–488.

Cox R. A. and Penkett S. A. (1972) Aerosol formation from sulfur dioxide in the presence of ozone and olefinic hydrocarbons. *J. chem. Soc. Faraday Trans. I* **68,** 1735–1753.

Criegee R. (1957) The course of ozonization of unsaturated compounds. *Record Chem. Progr.* **18,** 111–120.

Crutzen P. J. and Fishman J. (1977) Average concentrations of OH in the northern hemisphere troposphere, and the budgets of CH_4, CO, and H_2. *Geophys. Res. Lett.* submitted for publication May, 1977; in a private communication to one of the authors, computer readouts of [OH], [HO_2], and [CH_3O_2] were provided.

Daubendiek R. L. and Calvert J. G. (1973) A photochemical study involving SO_2, O_2, SO_3 and O_3. Abstracts 165th National Meeting of the Amer. Chem. Soc., Dallas, Texas, Physical Section, paper No. 19.

Daubendiek R. L. and Calvert J. G. (1975) A study of the N_2O_5–SO_2–O_3 reaction system. *Envir. Lett.* **8,** 103–116.

Daubendiek R. L. and Calvert J. G. An infrared study of the $O + SO_3$ reaction. Paper in preparation.

Davidson J. A. and Abrahamson E. W. (1972) The SO_2 photosensitized production of O_2 ($^1\Sigma_g^+$). *Photochem. Photobiol.* **15,** 403–405.

Davidson J. A., Kear K. E. and Abrahamson E. W. (1972/73) The photosensitized production and quenching of O_2 ($^1\Sigma_g^+$). *J. Photochem.* **1,** 307–316.

Davis D. D. (1974a) A kinetics review of atmospheric reactions involving H_xO_y compounds. *Can. J. Chem.* **52,** 1405–1414.

Davis D. D. (1974b) Absolute rate constants for elementary reactions of atmospheric importance: results from the Univeristy of Maryland's gas kinetics laboratory, report No. 1, Chemistry Department, University of Maryland, July, 1974; quoted in Hampson and Garvin (1975), p. 55.

Davis D. D., Fischer S. and Schiff R. (1974c) Flash photolysis resonance fluorescence kinetics study: temperature dependence of the reactions $OH + CO \rightarrow CO_2 + H$ and $OH + CH_4 \rightarrow H_2O + CH_3$. *J. phys. Chem.* **61,** 2213–2219.

Davis D. D. and Klauber G. (1975) Atmospheric gas phase oxidation mechanisms for the molecule SO_2. *Int. J. chem. Kinet. Symp.* **1,** 543–556.

Davis D. D., Prusazcyk J., Dwyer M. and Kim P. (1974b) A stop-flow time-of-flight mass spectrometer kinetics study. Reaction of ozone with nitrogen dioxide and sulfur dioxide. *J. phys. Chem.* **78,** 1775–1779.

Davis D. D. and Schiff R. (1974); unpublished work quoted in Davis D. D. (1974a).

Davis D. D., Schiff R. and Fischer S. (1974a) unpublished work quoted in Davis D. D. (1974a).

Davis D. D., Smith G. and Klauber G. (1974d) Trace gas analysis of power plant plumes via aircraft measurement: O_3, NO_x, SO_2 chemistry. *Science* **186,** 733–736.

Demerjian K. L., Calvert J. G. and Thorsell D. L. (1974) A Kinetic study of the chemistry of the SO_2 ($^3B_1$) reactions with *cis-* and *trans-*2-butene. *Int. J. chem. Kinet.* **6,** 829–848.

Demerjian K. L., Kerr J. A. and Calvert J. G. (1974) The mechanism of photochemical smog formation. *Adv. Envir. Sci. Technol.* **4,** 1–262.

Domalski E. S. (1971) Thermochemical properties of peroxyacetyl (PAN) and peroxybenzoyl nitrate (PBN). *Envir. Sci. Technol.* **5,** 443–444.

Fair R. W. and Thrush B. A. (1969) Reaction between hydrogen atoms and sulfur dioxide. *Trans. Faraday Soc.* **65,** 1550–1570.

Fee D. C., Markowitz S. S. and Garland J. K. (1972) Sulfur dioxide as a radical scavenger in alkene systems. *Radiochim. Acta* **17,** 135–138.

Finlayson B. J., Pitts J. N. Jr. and Atkinson R. (1974) Low pressure gas phase ozone–olefin reactions. Chemiluminescence, kinetics and mechanisms. *J. Am. chem. Soc.* **96,** 5356–5367.

Foner S. N. and Hudson R. L. (1962) Mass spectrometry of inorganic free radicals. *Advances in Chemistry Series* **36,** 34–49.

Fox D. L. and Wright R. S. (1977) Photochemical smog mechanisms—HC–NO_x–SO_2 systems. Paper presented at the 4th Int. Clean Air Congress, May 16–20, Tokyo,

Japan; the authors are grateful for a preprint of this paper.

Frankiewicz T. and Berry R. S. (1972) Singlet molecular oxygen production from photoexcited NO_2. *Envir. Sci. Technol.* **6**, 365–366.

Freed K. F. (1976) Theory of collision induced intersystem crossing. *J. chem. Phys.* **64**, 1604–1611.

Friend J. P., Leifer R. and Trichon M. (1973) On the formation of stratospheric aerosols *J. Atmos. Sci.* **30**, 465–479.

Gijzeman, O. L. J. (1974) Interaction between oxygen and aromatic molecules. *J. chem. Soc. Faraday Trans. II* **70**, 1143–1152.

Gijzeman O. L. J. and Kaufman F. (1973) Oxygen quenching of aromatic triplet states in solution. Part 2. *J. chem. Soc., Faraday Trans. II* **69**, 721–726.

Gijzeman O. L. J., Kaufman F. and Porter G. (1973a) Oxygen quenching of aromatic triplet states in solution. Part 1. *J. chem. Soc., Faraday Trans. II* **69**, 708–720.

Gijzeman, O. L. J., Kaufman F. and Porter G. (1973b) Quenching of aromatic triplet states in solution by nitric oxide and other free radicals. *J. chem. Soc. Faraday Trans. II* **69**, 727–737.

Gillani N. V., Husar R. B., Husar J. D. and Patterson D. E. (1978) Project MISTT: Kinetics of particulate sulfur formation in a power plant plume out to 300 km. *Atmospheric Environment* **12**, 589–598.

Good A. and Thynne J. C. J. (1967a) Reaction of free radicals with sulfur dioxide. Part 1-Methyl radicals. *Trans. Faraday Soc.* **63**, 2708–2719.

Good A. and Thynne J. C. J. (1967b) Reaction of free radicals with sulfur dioxide. Part 2. Ethyl radicals. *Trans. Faraday Soc.* **63**, 2720–2727.

Goodeve C. F., Eastman A. S. and Dooley A. (1934) The reaction between sulfur trioxide and water vapors and a new periodic phenomenon. *Trans. Faraday Soc.* **30**, 1127–1133.

Gordon S. and Mulac W. A. (1975) Reaction of the $OH(X\,^2\Pi)$ radical produced by the pulse radiolysis of water vapor. *Int. J. chem. Kinet. Symp.* **1**, 289–299.

Graedel T. E. (1976) Sulfur dioxide, sulfate aerosol, and urban ozone. *Geophys. Res. Lett.* **3**, 181–184.

Greiner R. (1967) Hydroxyl-radical kinetics by kinetic spectroscopy—I. Reactions with H_2, CO, and CH_4 at 300°K. *J. chem. Phys.* **46**, 2795–2799.

Hall T. C. (1953) Photochemical studies of nitrogen dioxide and sulfur dioxide. Ph.D. Thesis, University of California, Los Angeles.

Halstead C. J. and Jenkins D. R. (1969) Sulfur dioxide-catalyzed recombination of radicals in premixed fuel-rich hydrogen + oxygen + nitrogen flames. *Trans. Faraday Soc.* **65**, 3013–3022.

Hamilton E. J. Jr (1975) Water vapor dependence of the kinetics of the self-reaction of HO_2 in the gas phase. *J. chem. Phys.* **63**, 3682–3683.

Hamilton E. J. Jr. and Naleway C. A. (1976) Theoretical calculation of strong complex formation by the HO_2 radical: $HO_2 \cdot H_2O$ and $HO_2 \cdot NH_3$. *J. phys. Chem.* **80**, 2037–2040.

Hampson R. F. and Garvin D. (1975) Chemical kinetic and photochemical data for modeling atmospheric chemistry. U.S. National Bureau Standards Technical Note 866, U.S. Government Printing Office, 1–113.

Hanst P. L., Stephens E. R., Scott W. E. and Doerr R. C. (1959) Atmospheric ozone–olefin reactions. Paper presented before Division of Petroleum Chemistry, 136th National Meeting American Chemical Society, Atlantic City, Sept. 13, paper No. 28, p. 7R.

Harris G. W. and Wayne R. P. (1975) Reaction of hydroxyl radicals with NO, NO_2 and SO_2. *J. chem. Soc. Faraday I* **71**, 610–617.

Heicklen, J. (1976) *Atmospheric Chemistry*. Academic Press, New York, pp. 343–344.

Hendry D. G. and Kenley R. A. (1977) Generation of peroxy radicals from peroxy nitrates (RO_2NO_2). Decomposition of peroxyacyl nitrates. *J. Am. chem. Soc.* **99**, 3198–3199.

Hochanadel C. J., Ghormley J. A. and Ogren P. J. (1972) Absorption spectrum and reaction kinetics of the HO_2 radical in the gas phase. *J. Chem. Phys.* **56**, 4426–4432.

Horowitz A. and Calvert J. G. (1972a) The SO_2-sensitized phosphorescence of biacetyl in photolyses at 2650 and 2875 Å; the intersystem crossing ratio in sulfur dioxide. *Int. J. chem. Kinet.* **4**, 175–189.

Horowitz A. and Calvert J. G. (1972b) A study of the intersystem crossing reaction induced in gaseous sulfur dioxide molecules by collisions with nitrogen and cyclohexane at 27°. *Int. J. chem. Kinet.* **4**, 191–205.

Horowitz A. and Calvert J. G. (1972c) Emission studies of the mechanism of gaseous biacetyl photolysis at 3450, 3650, 3880, and 4380 Å and 28°C. *Int. J. chem. Kinet.* **4**, 207–227.

Hov Ö., Isaken I. S. A. and Hesstvedt E. (1977) Diurnal variations of ozone and other pollutants in an urban area. Report No. 24, Institutt for Geofysikk, Universitetet i Oslo, Jan., 1977.

Hull L. A., Hisatsune I. C. and Heicklen J. (1972) Low-temperature infrared studies of simple alkene–ozone reactions. *J. Am. chem. Soc.* **94**, 4856–4864.

Jackson G. E. and Calvert J. G. (1971) The triplet sulfur dioxide-carbon monoxide reaction exited within the $SO_2\,(^1A_1) \rightarrow SO_2\,(^3B_1)$ "forbidden" band. *J. Am. chem. Soc.* **93**, 2593–2599.

Jacob A. and Winkler C. A. (1972) Kinetics of the reaction of oxygen atoms and nitrogen atoms with sulfur trioxide. *J. chem. Soc. Faraday Trans. I* **68**, 2077–2082.

James F. C., Kerr J. A. and Simons J. P. (1973) Direct measurement of the rate of reaction of the methyl radical with sulfur dioxide. *J. chem. Soc. Faraday Trans. I* **69**, 2124–2129.

Jones I. T. N. and Bayes K. D. (1971) Energy transfer from electronically excited NO_2. *Chem. Phys. Lett.* **11**, 163–166.

Junge C. E. (1972) The cycle of atmospheric gases—natural and man made. *Q. J. R. met. Soc.* **98**, 711–729.

Kasahara M. and Takahashi K. (1976) Experimental studies on aerosol particle formation by sulfur dioxide. *Atmospheric Environment* **10**, 475–486.

Kawaoka K., Khan A. U. and Kearns D. R. (1967) Role of singlet excited states of molecular oxygen in the quenching of organic triplet states. *J. Chem. Phys.* **46**, 1842–1853.

Kear K. and Abrahamson E. W. (1974/75) Electronic energy transfer in the gas phase: the quenching of $O_2(^1\Sigma_g^+)$. *J. Photochem.* **3**, 409–416.

Kearns D. R. (1971) Physical and chemical properties of singlet molecular oxygen. *Chem. Rev.* **71**, 395–427.

Kelly N., Meagher J. F. and Heicklen J. (1976/77) The photolysis of sulfur dioxide in the presence of foreign gases. VIII. Excitation of SO_2 at 3600–4100 Å in the presence of acetylene. *J. Photochem.* **6**, 157–172.

Kerr J. A. and Parsonage M. J. (1972) *Evaluated Kinetic Data on Gas Phase Reactions*. Butterworths, University of Birmingham, England.

Kugel R. and Taube H. (1975) Infrared spectrum and structure of matrix-isolated sulfur tetroxide. *J. phys. Chem.* **79**, 2130–2135.

Kummer W. A., Pitts J. N. Jr. and Steer R. P. (1971) The chemiluminescent reaction of ozone with olefins and organic sulfides. *Envir. Sci. Technol.* **5**, 1045–1047.

Levy A., Drewes D. R. and Hales J. M. (1976) SO_2 oxidation in plumes: a review and assessment of relevant mechanistic and rate studies. Report EPA-450/3-76-022 prepared for the U.S. Environmental Protection Agency,

Office of Air Quality Planning and Standards, Contract No. 68-02-1982 by Battelle Pacific Northwest Laboratories, Richland, Washington 19352.

Lloyd A. C. (1974) Evaluated and estimated kinetic data for gas phase reactions of the hydroperoxyl radical. *Int. J. chem. Kinet.* **6,** 169–228.

Louw R., van Ham J. and Nieboer H. (1973) Nitrogen trioxide: key intermediate in the chemistry of polluted air? *J. Air Pollut. Control Ass.* **23,** 716.

McAndrew R. and Wheeler R. (1962) The recombination of atomic hydrogen in propane flame gases. *J. phys. Chem.* **66,** 229–232.

McNelis D. N. (1974) Aerosol formation from gas-phase reactions of ozone and olefin in the presence of sulfur dioxide. Final report 650/4-74-034 to the Environmental Protection Agency, Research Triangle Park, North Carolina, for program element 21 AKB, ROAP 38, August, 1974.

Merer A. J. (1963) Rotational analysis of bands of the 3800 Å system of SO_2. *Disc. Faraday Soc.* **35,** 127–136.

Mettee H. D. (1969) Foreign gas quenching of sulfur dioxide vapor emission. *J. Chem. Phys.* **73,** 1071–1076.

Miller D. F. (1977) Simulations of gas-phase reactions in power plant plumes. Preprint of a paper presented at 173rd National Meeting Am. Chem. Soc., New Orleans, March, 1977.

Milstein R., Williams R. L. and Rowland F. S. (1974) Relative reaction rates involving thermal fluorine-18 atoms and thermal fluoroethyl radicals with oxygen, nitric oxide, sulfur dioxide, nitrogen, carbon monoxide and hydrogen iodide. *J. phys. Chem.* **78,** 857–863.

Mulcahy M. F. R., Steven J. R. and Ward J. C. (1967) The kinetics of reaction between oxygen atoms and sulfur dioxide: an investigation by electron spin resonance spectrometry. *J. phys. Chem.* **71,** 2124–2131.

Mulcahy M. F. R. and Smith R. H. (1971) Reactions of OH radicals in the H-NO_2 and H–NO_2–CO systems. *J. chem. Phys.* **54,** 5215–5221.

Niki H. (1977) Private communication to one of the authors.

Niki H., Maker P. D., Savage C. M. and Breitenbach L. P. (1975) Fourier transform spectroscopic studies of organic species participating in photochemical smog formation. Paper at International Conference on Environmental Sensing and Assessment, Las Vegas, Nevada, Sept. 14, 1975, Vol. 2, 24–4.

Niki H., Maker P. D., Savage C. M. and Breitenbach L. P. (1977) Fourier transform IR spectroscopic observations of propylene ozonide in the gas phase reaction of ozone-*cis*-butene–formaldehyde. *Chem. Phys. Lett.* **46,** 327–330.

Okuda S., Rao T. N., Slater D. H. and Calvert J. G. (1969) Identification of the photochemically active species in sulfur dioxide photolysis within the first allowed absorption band. *J. phys. Chem.* **73,** 4412–4415.

O'Neal H. E. and Blumstein C. (1973) A new mechanism for gas phase ozone–olefin reactions. *Int. J. chem. Kinet.* **5,** 397–413.

Overend R., Paraskevopoulos G. and Cvetanovic R. J. (1974) Hydroxyl radical rate measurement for simple species by flash photolysis kinetic spectroscopy. Paper 6-4, Abstracts 11th Informal Conference of Photochemistry, Vanderbilt University, Nashville, Tennessee, 248–252.

Paraskevopoulos G. (1976) Private communication to one of the authors.

Parkes D. A. (1974) The roles of alkylperoxy and alkoxy radicals in alkyl radical oxidation at room temperature. Paper presented at the 15th International Symposium on Combustion, Japan, 1974.

Parkes D. A., Paul D. M., Quinn C. P. and Robson R. C. (1973) The ultraviolet absorption by alkylperoxy radicals and their mutual reactions. *Chem. Phys. Lett.* **23,** 425–429.

Pate C. T., Atkinson R. and Pitts J. N. Jr. (1976) Rate constants for the gas phase reaction of peroxyacetyl nitrate with selected atmospheric constituents. *J. envir. Sci. Health—Envir. Sci. Eng.* All, 19–31.

Patterson L. K., Porter G. and Topp M. R. (1970) Oxygen quenching of singlet and triplet states. *Chem. Phys. Lett.* **7,** 612–614.

Paukert T. T. and Johnston H. S. (1972) Spectra and kinetics of the hydroperoxyl free radical in the gas phase. *J. chem. Phys.* **56,** 2824–2838.

Payne W. A., Stief L. J. and Davis D. D. (1973) A kinetics study of the reaction of HO_2 with SO_2 and NO. *J. Am. chem. Soc.* **95,** 7614–7619.

Penzhorn R.-D., Filby W. G. and Güsten H. (1974a) Die photochemische abbaurate des schwefeldioxids in der unteren atmosphäre mitteleuropas. *Z. Naturforsch.* 29a, 1449–1453.

Penzhorn R.-D., Filby W. G., Günther K. and Stieglitz L. (1975) The photoreaction of sulfur dioxide with hydrocarbons—II. Chemical and physical aspects of the formation of aerosols with butane. *Int. J. chem. Kinet. Symp.* **1,** 611–627.

Penzhorn R.-D., Güsten H., Schurath U. and Becker K. H. (1974b) Quenching of singlet molecular oxygen by some atmospheric pollutants. *Current Res.* **10,** 907–909.

Peterson J. T. (1976) Calculated actinic fluxes (290–700 nm) for air pollution photochemical applications. U.S. Environmental Protection Agency Report 600/4-76-025, p. 21.

Pitts J. N. Jr., Finlayson B. J., Akimoto H., Kummer W. A. and Steer R. P. (1972) The chemiluminescent reactions of ozone with olefins and organic sulfides. *Advances in Chemistry Series* **113,** 246–254.

Rudolph R. N. and Strickler S. J. (1977) Direct measurement of the lifetimes of the $^3B_1$ state of SO_2 in air at atmospheric pressure. *J. Am. chem. Soc.* **99,** 3871–3872.

Sander S. P. and Seinfeld J. H. (1976) Chemical kinetics of homogeneous atmospheric oxidation of sulfur dioxide. *Envir. Sci. Technol.* **10,** 1114–1123.

Schofield K. (1973) Evaluated chemical rate constants for various gas phase reactions. *J. Phys. Chem. Ref. Data* **2,** 25–77.

Schwartz S. E. and Newman L. Processes limiting the oxidation of sulfur dioxide in stack plumes. Manuscript of a paper to be submitted for publication.

Sethi D. S. (1971) Photo-oxidation of sulfur dioxide. *J. Air Pollut. Control Ass.* **21,** 418–420.

Sidebottom H. W., Badcock C. C., Calvert J. G., Rabe B. R. and Damon E. K. (1971) Mechanism of the photolysis of mixtures of sulfur dioxide with olefin and aromatic hydrocarbons. *J. Am. chem. Soc.* **93,** 3121–3128.

Sidebottom H. W., Badcock C. C., Calvert J. G., Rabe B. R. and Damon E. K. (1972a) Lifetime studies of the biacetyl excited singlet and triplet states in the gas phase at 25°. *J. Am. chem. Soc.* **94,** 13–19.

Sidebottom H. W., Badcock C. C., Jackson G. E., Calvert J. G., Reinhardt G. W. and Damon E. K. (1972b) Photooxidation of sulfur dioxide. *Envir. Sci. Technol.* **6,** 72–79.

Sie B. K. T., Simonaitis R. and Heicklen J. (1976) The reaction of OH with CO. *Int. J. chem. Kinet.* **8,** 85–98.

Skotnicki P. A., Hopkins A. G. and Brown C. W. (1975) Time dependence of the quantum yields for the photooxidation of sulfur dioxide. *J. phys. Chem.* **79,** 2450–2452.

Smith I. W. M. and Zellner R. (1973) Rate measurements of reactions of OH by resonance absorption. Part 2. Reactions of OH with CO, C_2H_4, and C_2H_2. *J. chem. Soc. Faraday Trans. II* **69,** 1617–1627.

Stephens E. R. and Price M. A. (1972) Comparison of synthetic and smog aerosols. *J. Colloid Interface Sci.* **39,** 272–286.

Stuhl F. and Niki H. (1972) Pulsed vacuum–u.v. photochemical study of OH with H_2, O_2, and CO using a resonance–fluorescence detection method. *J. chem. Phys.* **57,** 3671–3677.

Su F. and Calvert J. G. (1977a) The mechanism of the photochemical reactions of SO_2 with C_2H_2 and CO excited within the $SO_2(^3B_1) \leftarrow SO_2(\tilde{X}\,^1A_1)$ "forbidden" band. *Chem. Phys. Lett.* **52,** 572–578.

Su F. and Calvert J. G. (1977b) work in preparation for publication.

Su F., Bottenheim J. W., Sidebottom H. W., Calvert J. G. and Damon E. K. Kinetics of fluorescence decay of SO_2 excited in the 2662–3273 Å region. *Int. J. chem. Kinet.* (in press).

Su F., Bottenheim J. W., Thorsell D. L., Calvert J. G. and Damon E. K. (1977a) The efficiency of the phosphorescence decay of the isolated $SO_2(^3B_1)$ molecule. *Chem. Phys. Lett.* **49,** 305–311.

Su F., Wampler F. B., Bottenheim J. W., Thorsell D. L., Calvert J. G. and Damon E. K. (1977b) On the pressure saturation effect of the quenching of SO_2 ($^3B_1$) molecules. *Chem. Phys. Lett.* **51,** 150–154.

Walter T. A., Bufalini J. J. and Gay B. W. Jr. (1977) Mechanism for olefin–ozone reaction. *Envir. Sci. Technol.* **11,** 382–386.

Wampler F. B., Calvert J. G. and Damon E. K. (1973a) A study of the bimolecular intersystem crossing reaction induced in the first excited singlet of SO_2 by collisions with O_2 and other atmospheric gases. *Int. J. chem. Kinet.* **5,** 107–117.

Wampler F. B., Horowitz A. and Calvert J. G. (1972) The mechanism of carbon dioxide formation in 3130 Å-irradiated mixtures of sulfur dioxide and carbon monoxide. *J. Am. chem. Soc.* **94,** 5523–5532.

Wampler F. B., Otsuka K., Calvert J. G. and Damon E. K. (1973b) The temperature dependence and the mechanism of $SO_2(^3B_1)$ quenching reactions. *Int. J. chem. Kinet.* **5,** 669–687.

Westenberg A. A. and deHaas N. (1973) Rates of CO + OH and H_2 + OH over an extended temperature range. *J. chem. Phys.* **58,** 4061–4065.

Westenberg A. A. and deHaas N. (1975a) Rate of the reaction $O + SO_2 + M \rightarrow SO_3 + M$. *J. chem. Phys.* **63,** 5411–5415.

Westenberg A. A. and deHaas N. (1975b) Rate of the $O + SO_3$ reaction. *J. chem. Phys.* **62,** 725–730.

Whitbeck M. R., Bottenheim J. W., Levine S. Z. and Calvert J. G. (1976) A kinetic study of the CH_3O_2 and $(CH_3)_3CO_2$ radical reactions by kinetic flash spectroscopy. Abstracts of 12th Informal Conference on Photochemistry, U.S. National Bureau of Standard, Gaithersberg, Maryland, July, 1976, pp. K1-1–K1-5.

Whitten G. Z. and Hogo H. H. (1976) Mathematical modeling of simulated photochemical smog. Final Report EF 76–126 by Systems Application Inc., to the U.S. Environmental Protection Agency, August 1976.

Wilson W. E., Charlson R. J., Husar R. B., Whitby K. T. and Blumenthal D. (1976) Sulfates in the atmosphere. Paper presented at 69th meeting of the Air Pollut. Control Ass., Portland, Oregon, June, 1976.

Wilson W. E., Dodge M. C., McNelis D. N. and Overton J. (1974) SO_2 oxidation mechanism in olefin–NO_x–SO_2 smog. Paper presented before Division of Environmental Chemistry, American Chemical Society, Los Angeles, April, 1974.

Wilson W. E., Levy A. and Wimmer D. B. (1972) A study of sulfur dioxide in photochemical smog—II. Effect of sulfur dioxide on oxidant formation in photochemical smog. *J. Air Pollut. Control Ass.* **22,** 27–32.

Wood W. P., Castleman A. W. Jr. and Tang I. N. (1974) Mechanisms of aerosol formation from SO_2. Paper presented at 67th annual meeting Air Pollut. Control Ass., Denver, Colorado, June 9–13, 1974.

Wood W. P., Castleman A. W. Jr. and Tang I. N. (1975) Mechanisms of aerosol formation from SO_2. *J. Aerosol Sci.* **6,** 367–375.

Atmospheric Environment Vol. 12, pp. 227–230. Pergamon Press 1978. Printed in Great Britain.

HOMOGENEOUS OXIDATION OF SULPHUR COMPOUNDS IN THE ATMOSPHERE

A. E. J. EGGLETON and R. A. COX
Environmental & Medical Services Division, A.E.R.E., Harwell, Oxon OX11 0RA, England

(*Received* 27 *September* 1977)

Abstract—Work carried out under the European COST 61a Project on the homogeneous oxidation of sulphur compounds in the atmosphere is briefly reviewed. Mechanisms for sulphur dioxide can be divided into three classes; (a) oxidation by free radicals generated photochemically, (b) oxidation by intermediates produced in thermal reactions, and (c) direct photo-oxidation.

Only (a) makes a substantial contribution to SO_2 oxidation with calculated maximum rates of between 2 and 6% h^{-1} in sunlight irradiated urban air during summer months and 1–2% h^{-1} in unpolluted air. Most of the oxidation is brought about by the attack of the OH radical on SO_2 but the contribution of RO_2 radical attack is not well determined due to uncertainties in RO_2 rate constants. H_2S, CH_3SH and $(CH_3)_2S$ react with OH radicals giving atmospheric life-times about 1 day.

1. INTRODUCTION

This paper consists of a review of work carried out on homogeneous oxidation of sulphur compounds in the atmosphere under the recently completed 4-year European COST Project 61A entitled "Research into the physico-chemical behaviour of sulphur dioxide in the atmosphere." As a full version of this paper will be published as part of the scientific report of the Project only a summary version is published here.

Of the 27 national projects which made up COST Project 61A, four were concerned with various aspects of the homogeneous oxidation of sulphur dioxide and other sulphur compounds in the atmosphere. They were

(A) Principal Investigators: A. E. J. Eggleton, R. A. Cox, Environmental & Medical Sciences Division, Harwell, U.K.

(B) Principal Investigators: K. H. Becker, U. Schurath, Institut für Physikalischer Chemie der Universität Bonn, W. Germany.

(C) Principal Investigators: G. Madeleine, J. Bricard, Centre d'Etudes Nucleaires, Fontenay-aux-Roses and Université de Paris, Paris, France.

(D) Principal Invetigators: R. Penzhorn, H. Jordan, Gesellschaft für Kernforschung, Karlsruhe, W. Germany.

The four projects will be referred to hereafter by letter.

Three distinct mechansims have been discussed for homogeneous gas phase oxidation of atmospheric sulphur dioxide. They are (a) oxidation of SO_2 by reactive intermediates, e.g. atoms or free radicals which are generated photochemically; (b) oxidation of SO_2 by reactive intermediates generated in thermal reactions, e.g. ozone–olefin reactions; (c) direct photo-oxidation involving the reactions of excited SO_2 molecules, produced following the absorption of solar u.v. radiation by SO_2.

A full review of these topics must necessarily consider the work of many authors outside Project COST 61A. Here however the emphasis is on results obtained within the Project.

2. OXIDATION OF SO_2 BY PHOTOCHEMICALLY GENERATED SPECIES

Photochemically-generated species which may react with SO_2 in the atmosphere include atoms, free radicals, and excited molecular species. It can now be considered well-established that the atmospheric oxidation of hydrocarbons and related substances, which in polluted air leads to photochemical smog formation, proceeds by a free radical chain process. The free radicals involved in this system are also available for reaction with SO_2. Experimental results relating to the individual reactions of SO_2 with these photochemical intermediates are considered below.

2.1 *Reactions of* SO_2 *with* OH *radicals*

The reaction of OH with SO_2 is a termolecular addition reaction

$$OH + SO_2 + M = HOSO_2 + M. \qquad (1)$$

Evidence obtained in (A) using a ^{35}S technique shows that under simulated atmospheric conditions the $HOSO_2$ free radical reacts further to form a sulphur-bearing aerosol and does not reform SO_2. Using a competitive technique with nitrous acid photolysis as a source of OH radicals, an effective bimolecular rate constant $k_1[M] \simeq 6 \times 10^{-13}$ cm^3 molecule^{-1} s^{-1} for M = air at 1 atmosphere was obtained in reasonable agreement with other workers' values (Cox, 1974).

2.2 *Reaction of organic radicals with* SO_2

Qualitative evidence for the occurrence of the reaction of alkylperoxy radicals with SO_2 was obtained from the study of the photolysis of SO_2–alkene–O_2 mixtures in (B). It was concluded that sulphuric acid is the major product but unfortunately the data do not allow an estimate of the rate constant for any of the alkylperoxy radicals. Work in (A) indicates that acetylperoxy radicals react only slowly with SO_2 in contrast to their rapid reaction with NO and NO_2.

Alkyl radicals react with SO_2 and the products from methyl, ethyl, secondary and tertiary butyl radicals have been characterised in (D).

2.3 *The reaction of* HO_2 *radicals with* SO_2

No direct evidence for this reaction has been obtained in COST 61A studies but results obtained in laboratory simulation studies carried out in (A) and (B) are consistent with its occurrence.

2.4 *Reactions of* SO_2 *with* $O(^3P)$

Indirect evidence of the occurrence of this reaction has been obtained in (C) when $O(^3P)$ was formed by the photolysis of NO_2 in ultra pure air.

2.5 *Reaction of* SO_2 *with* $O_2(\Delta g)$

In a collaborative experiment between (B) and (D) the rate constant for the reaction between SO_2 and singlet oxygen molecules was measured for the first time (Penzhorn *et al.*, 1974). The value obtained ($k = 3.9 \times 10^{-20}$ cm^3 molecule^{-1} s^{-1}) is a factor of ten less than the effective quenching rate constant for singlet oxygen molecules in ambient air, where O_2 is the main quencher.

2.6 *Reaction of* SO_2 *with ozone*

The thermal reaction between SO_2 and O_3 was studied in (B) using a 220 m^3 evacuated sphere as a reaction vessel. No detectable oxidation of SO_2 could be observed, allowing an upper limit estimate of not greater than 2×10^{-20} cm^3 molecule^{-1} s^{-1}.

2.7 *Estimation of* SO_2 *oxidation rates in the atmosphere from reactions with photochemically generated species*

During the atmospheric oxidation of hydrocarbons and related photochemical processes both in clean and polluted atmospheres, a pseudo steady state concentration of atoms and free radicals is achieved. Any trace gas such as SO_2 which can react with free radicals will be removed at a rate proportional to the concentration of the free radical and its rate constant for reaction with the free radical. The concentration of free radicals can be calculated by computer modelling, or in some cases can be measured directly, enabling an estimate to be made of the removal rate or oxidation rate of SO_2. Table 1 gives the results of such calculations with free radicals or reactive species arranged in a descending order of importance for the atmospheric oxidation of SO_2.

It is clear from Table 1 that photochemically-generated reactive species other than OH, HO_2 and RO_2, play a negligible role in the oxidation of SO_2 in the lower atmosphere. The total rate of SO_2 oxidation from OH and RO_2 species in sunlight irradiated

Table 1. Oxidation of SO_2 by photochemically-generated species in an urban atmosphere. Data from COST 61A and other sources

Species	Steady state* concentration (molecule/cm^{-3})	Rate constant† for reaction with SO_2 (cm^3 molecule^{-1} s^{-1})	SO_2 oxidation rate (% h^{-1})
OH	7.1×10^6	$0.5 - 1.0 \times 10^{-12}$	$1.2 - 2.5$
HO_2	2.6×10^9	9×10^{-16}	0.84
CH_3O_2	2.2×10^9	$\leq 10^{-15}$	≤ 0.8
CH_3COO_2	8.1×10^8	$\leq 5 \times 10^{-18}$	$\leq 1.5 \times 10^{-3}$
Total RO_2‡	5.1×10^9	$\leq 10^{-15}$	≤ 1.84
CH_3	3.2×10	3×10^{-13}	3×10^{-6}
$O(^3P)$	5.7×10^4	1.9×10^{-14}	4×10^{-4}
O_3	4.2×10^{12}	$\leq 10^{-23}$	1.5×10^{-5}
NO_3§	1.4×10^8	$\leq 7 \times 10^{-22}$	3.5×10^{-8}
N_2O_5§	7.6×10^{10}	$\leq 4 \times 10^{-23}$	1.0×10^{-6}
$O_2(^1\Delta)$	1.0×10^8	3.9×10^{-20}	1.4×10^{-8}

* Concentrations calculated from a photochemical model of the surface atmosphere (0–1 km) using the following initial trace gas concentrations measured in urban air.

NO =	20 ppb	CH_4 =	1500 ppb
NO_2 =	20 ppb	C_2H_6 =	125 ppb
SO_2 =	100 ppb	C_3 to C_5 paraffins =	105 ppb
			105 ppb
CO =	5000 ppb	C_2H_4 =	20 ppb
H_2 =	750 ppb	C_3H_6 =	5 ppb

Diurnal variation of solar intensity is simulated for 50°N lat. July 1. Concentrations at 12 noon GMT are given here.

† Effective bimolecular rate constant at 1 atm pressure.

‡ Total peroxy radical concentration, excluding HO_2 and CH_3O_2.

§ Reactive species produced in the thermal O_3–NO_2 reaction: see later.

urban air during summer months is calculated to be between approx 2.2 and 6.5% h^{-1}. In order to determine a realistic oxidation lifetime of SO_2 in the atmosphere, oxidation rates in unpolluted air have also been computed with suitable input trace gas concentrations. In unpolluted air the role of OH radicals in SO_2 oxidation becomes more dominant than in the urban situation. The calculated SO_2 oxidation rate in summer lies between 1.4 and 2.2% h^{-1} and in winter between 0.2 and 1.1% h^{-1}.

2.8 *Laboratory simulation studies of photochemical oxidation of* SO_2

A detailed study of the oxidation of SO_2 in a simulated photochemical smog system has been carried out under (B). Mixtures of various olefins and nitric oxide in air were irradiated in a 425 l. glass photochemical chamber. The rate of oxidation of SO_2 was found to follow closely the rate of olefin consumption, reaching peak values as high as 30% h^{-1}. These peak oxidation rates were due mainly to the oxidising intermediate(s) from the reaction of unreacted olefin with ozone produced during the smog reaction (see section 3). However, after correction for this a residual oxidation of SO_2 could be discerned which could be quantitatively explained on the basis of OH as the dominant attacking radical species. These residual SO_2 oxidation rates were of the order of a few per cent per hour and provide a valid basis for extrapolation to the atmosphere indicating a conservative upper limit of about 4% h^{-1} for sunlit polluted urban atmospheres.

Similar results were obtained in (C) where the rate of SO_2 removal and the formation of condensation nuclei were observed using mixtures of ethylene and nitrogen dioxide in air containing SO_2. Observed SO_2 conversion rates were in the range 1–4% h^{-1}. Laboratory studies carried out under (A) and (B) have also provided information about the mechanism of the reaction of OH with SO_2 in the presence of nitrogen oxides.

In the photolysis of nitrous acid–SO_2–air mixtures at 365 nm (Project (A)) it was found that NO is oxidised to NO_2 in the reactions following the addition of OH to SO_2. This provides experimental evidence for the reactions

$$HOSO_2 + O_2 = HOSO_2O_2$$

$$HOSO_2O_2 + NO = HOSO_2O + NO_2$$

which have been proposed as a fate for the addition product from the reaction of OH with SO_2 in the atmosphere. Both in this and in experiments carried out in (B), there was evidence for the participation of $HOSO_x$ radicals in a reaction chain involving the abstraction of hydrogen atoms from hydrocarbons and also for the oxidation of NO to NO_2.

In an attempt to overcome the problems of extrapolation from laboratory systems to the atmosphere, experiments were conducted in project (A) in which SO_2 oxidation was studied in ambient air in the presence of natural sunlight. Ambient air samples containing some pollution from motor traffic etc. were spiked with radioactive SO_2 and then exposed to sunlight in transparent plastic bags. The SO_2 oxidation rates were measured from the rate of production of ^{35}S aerosol, which was collected by filtration. Oxidation rates of between 0.7 and 2.5% were observed during summer months, depending on the degree of pollution of the air sample. Oxidation rates in winter and early spring were a factor of 3–5 lower. Correlation of the observed oxidation rates with a concentration of NO_x initially present in the air sample, gave an empirical relationship which can be used to predict oxidation rates under summer conditions in western Europe. k_r^{max} ($h^{-1} \times 10^2$) = 0.45 + 0.26 [NO_x] (pphm) where k_r^{max} is the maximum rate of SO_2 oxidation.

3. OXIDATION OF SO_2 BY REACTIVE SPECIES PRODUCED IN THERMAL REACTIONS

3.1 *Olefin reactions*

Project (B) has examined in detail the mechanism and kinetics of the ozone–olefin reaction (Becker *et al.*, 1974) and their interaction with SO_2. The nature of the aerosol produced in the reactions has also been investigated in detail (Schulten and Schurath, 1975). This work has revealed that the mechanism of ozone–olefin reactions is extremely complex and it is not possible to identify with certainty the species which attacks SO_2, though considerable progress has been made. The oxidation of SO_2 can be rationalised by a kinetic scheme in which the reactive species produced by the reaction of ozone with olefins, either reacts with SO_2 or is removed by a pseudo-first-order process. This forms a basis for the extrapolation of laboratory measured SO_2 oxidation rates to atmospheric conditions taking propene as a realistic model olefin. It is calculated that the SO_2 oxidation rate only becomes greater than 0.1% h^{-1} if the concentration of propene is greater than 50 ppb, when ozone is at its natural background concentration. Such high olefin concentrations are only encountered in polluted air and oxidation of SO_2 by this mechanism is likely to be significant only in the urban atmosphere or in the vicinity of industrial sources of olefins.

3.2 *Reaction of* SO_2 *with reactive species produced in the thermal* $NO_2 + O_3$ *reaction*

Results from (B) show clearly that SO_2 is not oxidised to a measureable extent by any of the species present in a reacting mixture of NO_2 and O_3, i.e. N_2O_5 and NO_3.

4. DIRECT PHOTO-OXIDATION OF SO_2

The atmospheric photochemistry of SO_2 is determined by the chemical behaviour of the first excited (triplet) state and the second excited (singlet) state. Much of the available information on the excited states is from the work of J. G. Calvert and coworkers. An important contribution to this field is due to work carried out under project (D) (Penzhorn *et al.*, 1974b) with a small contribution from project (A) (Cox, 1973). The singlet SO_2 appears to be chemically unreactive and is rapidly quenched by other gases including the major atmospheric gases forming triplet SO_2. This is more chemically reactive but the available quantum yield evidence suggests that reactions leading to conversion to SO_3 occur with only a very low efficiency under atmospheric conditions. Convincing evidence that the direct photo-oxidation of concentrations of SO_2 is slow, comes from studies made under project (B) and (C). Both showed a negligible rate of oxidation of SO_2 when dilute mixtures of SO_2 in highly purified air were exposed to u.v. light of wavelength above 290 nm. Calculations based on the measured quantum yield at low dilution in air together with the calculated rate of light absorption by SO_2, give oxidation rates of 0.02% h^{-1} in sunlight at a zenith angle of 40°.

5. AEROSOL PRODUCTION FROM THE OXIDATION OF SO_2

5.1 *Nucleation of sulphuric acid and related aerosols*

Theoretical models for the nucleation of H_2SO_4 aerosol have been formulated by several groups and experimental tests of the calculated nucleation rates have been carried out under project (C). Good agreement was obtained between predicted and observed particle production rates from the photolysis of $SO_2 + NO_2 + H_2O$ mixtures. This lends support to the assumptions made in the molecular theory of binary nucleation and condensation processes.

5.2 *Chemical composition of aerosols from* SO_2 *oxidation*

It has usually been assumed that the aerosol initially produced by homogeneous oxidation of SO_2 in the atmosphere is sulphuric acid. However in the real atmospheric system where hydrocarbons and nitrogen oxides are present, the possibility for formation of sulphur containing aerosols other than H_2SO_4 exists. All four projects have found evidence for various aerosol products besides sulphuric acid. These are mainly compounds containing both SO_x and NO_x, such as nitrosyl sulphuric acid $NOHSO_4$ or related compounds, evidence for which was found in project (C) by electron microscopy.

6. HOMOGENEOUS OXIDATION OF HYDROGEN SULPHIDE AND ORGANIC SULPHIDES

6.1 *The reaction of* H_2S *with ozone*

A detailed study of the thermal gas phase reaction between H_2S and ozone was undertaken in project (B) (Becker *et al.*, 1975). A complex reaction mechanism was found proceeding by a chain reaction at the high ozone concentrations required to obtain a measurable rate. The results, however, indicate strongly that the chain process will not occur under atmospheric conditions. The same holds for the methyl derivatives, CH_3SH and $(CH_3)_2S$.

6.2 *Free radical oxidation of* H_2S *and organic sulphides*

The photo-oxidation of H_2S and dimethyl sulphide under laboratory-simulated atmospheric conditions was studied in (A). Evidence was obtained for the attack of OH radicals on both H_2S and $(CH_3)_2S$ and for the attack of $O(^3P)$ on $(CH_3)_2S$ when these compounds were irradiated in air in the presence of nitrogen oxides. Ozone was produced during the oxidation of $(CH_3)_2S$. Estimates of the average tropospheric lifetimes of H_2S, dimethyl sulphide and methyl mercaptan based on their reactivity with OH radicals were determined in (A). The data show that the tropospheric lifetimes in respect to oxidation by OH are very short, considerably less than 1 day in most situations. The concentrations of these sulphur compounds in the atmosphere are therefore likely to be very small and variable.

REFERENCES

Becker K. H., Schurath U. and Seitz H. (1974) Ozone–olefin reactions in the gas phase—rate constants and activation energies. *Int. J. Chem. Kinet.* **4**, 725.

Becker K. H., Inocêncio M. and Schurath U. (1975) The reaction of ozone with hydrogen sulphide and its organic derivatives. *Int. J. chem. Kinet. Symp.* **1**, 205–220.

Cox R. A. (1973) The sulphur dioxide photosensitised *cis–trans* isomerisation of butene-2. *J. Photochem.* **2**, 1–13.

Cox R. A. (1974) The photolysis of nitrous acid in the presence of carbon monoxide and sulphur dioxide. *J. Photochem.* **3**, 291–304.

Penzhorn R. D., Gusten H., Schurath U. and Becker K. H. (1974a) The quenching of singlet molecular oxygen by some air pollutants. *Environ. Sci. Techn.* **8**, 907–909.

Penzhorn R. D., Filby W. G., Gunther K. and Stieglitz L. (1974b) The photoreaction of sulphur dioxide with hydrocarbons—II. Chemical and physical aspects of the formation of aerosols with butane. *Int. J. chem. Kinet. Symp.* **1**, 611–627.

Schulten H. R. and Schurath U. (1975) Analysis of aerosols from the ozonolysis of 1-butene by high resolution field absorption mass spectrometry. *J. phys. Chem.* **79**, 51–57.

Atmospheric Environment Vol. 12, pp. 231–239. Pergamon Press 1978. Printed in Great Britain.

HETEROGENEOUS SO_2-OXIDATION IN THE DROPLET PHASE

S. BEILKE

Umweltbundesamt; Pilotstation Frankfurt, Feldbergstraße 45, 6000 Frankfurt a.M., W. Germany

and

G. GRAVENHORST

Institut für Atmosphärische Chemie, Kernforschungsanlage Jülich, W. Germany

Abstract—In order to determine the rate controlling step for sulfate formation in a heterogeneous droplet system. SO_2-transfer within the gasphase towards and within droplets are calculated. equilibrium between SO_2 in the gasphase and sulfur (IV) in cloud and fog droplets is reached within less than 1 second. Oxidation of sulfur (IV) to sulfate in the droplet phase proceeds slower by orders of magnitutde. Three mechanisms of SO_2-oxidation are discussed: (a) SO_2-oxidation by O_2 in the absence of catalysts; (b) SO_2-oxidation by O_2 in the presence of catalysts, and (c) SO_2-oxidation by strongly oxidizing agents. Mechanism (a) contributes only to a negligible extent to sulfate formation in droplets even in the presence of typical concentrations of ammonia. Indications are strong that the major function of the SO_2–NH_3–H_2O-system is not the oxidation of SO_2 to sulfate. Oxidation mechanism (b) may contribute to a significant extent to sulfate formation in urban fogs in which case the concentrations of catalysts can be sufficiently high. For clouds in remote areas with much lower catalyst concentrations SO_2-oxidation by mechanism (b) seems to be of little importance. The oxidation by strongly oxidizing agents (mechanism (c)) appears to be the dominant mechanism although some experimental discrepancies have to be resolved.

1. INTRODUCTION

There is strong evidence that the predominant part of atmospheric sulfate found in the Federal Republic of Germany is generated by oxidation of atmospheric SO_2 and not directly emitted as sulfate containing particles. The oxidation proceeds by 2 processes: (1) Homogeneous SO_2-gasphase oxidation. (2) Heterogeneous SO_2-oxidation in atmospheric droplets and on aerosol particles. This paper deals with the SO_2-oxidation in the heterogeneous SO_2-air-droplet-system. It is attempted to integrate some of the reported experimental and theoretical investigations into a unified picture and to assess some quantitative data on sulfate formation and SO_2-removal in the atmosphere.

In a real atmosphere we are concerned with a heterogeneous SO_2-oxidation in the aqueous phase whereas in laboratory experiments both homogeneous and heterogeneous aqueous systems were used to study SO_2-oxidation. In order to apply the laboratory results obtained in a homogeneous aqueous phase to the atmospheric heterogeneous system some estimates are made of SO_2-transport rates within both the gaseous and the droplet phase. These calculations are described in the first part of this paper.

In the second part, three mechanisms of SO_2-oxidation will be discussed and their importance for atmospheric sulfate formation assessed: (a) SO_2-oxidation by O_2 in the absence of catalysts, (b) SO_2-oxidation by O_2 in the presence of catalysts and (c) SO_2-oxidation by strongly oxydizing agents. On the basis of this discussion further studies to fill knowledge gaps will be suggested.

2. SO_2-TRANSFER AND ABSORPTION

In a heterogeneous system in which gaseous SO_2 is in equilibrium with sulfur (IV)* in the aqueous phase, the following equations describe the equilibrium,

$$(SO_2)_{gas} + H_2O \overset{H}{\rightleftharpoons} SO_2{\cdot}H_2O \qquad (1)$$

$$SO_2{\cdot}H_2O \overset{K_1}{\rightleftharpoons} H^+ + HSO_3^- \qquad (2)$$

$$HSO_3^- \overset{K_2}{\rightleftharpoons} H^+ + SO_3^{2-} \qquad (3)$$

where $SO_2 \cdot H_2O$ is physically-dissolved SO_2, HSO_3^- is the bisulfite ion, and SO_3^{2-} is the sulfite ion. H = Henry constant; K_1 = first dissociation constant; K_2 = second dissociation constant.

Reaction (2) proceeds very rapidly as can be seen from the corresponding forward reaction rate constant (k_1) and reverse reaction rate constant (k_{-1}) measured by Eigen *et al.* (1961) using a relaxation technique

$$k_1 = 3.4 \times 10^6\ s^{-1}$$
$$k_{-1} = 2.0 \times 10^8\ l.\ mol^{-1}\ s^{-1}\ (20°C).$$

The rate constants of Wang and Himmelblau (1964) seem to be too low by *ca* 8 orders of magnitude

* Sulfur (IV) stands for the sum of $SO_2 \cdot H_2O$, HSO_3^- and SO_3^{2-}.

Table 1. Equilibrium constants for the system SO_2–water as a function of temperature

Temperature (°C)	H (mol l^{-1} atm^{-1})	K_1 (molar)	K_2 (molar)
5	2.72	2.78×10^{-2}	10.0×10^{-8}
15	1.76	2.19×10^{-2}	7.90×10^{-8}
25	1.24	1.74×10^{-2}	6.24×10^{-8}
Author's	Gmelin (1963)	Sillen (1964)	Sillen (1964)

($k_1 = 3.3 \times 10^{-2}\,s^{-1}$; $k_{-1} = 2.7$ l. mol^{-1} s^{-1}—see for example Beilke and Lamb, 1975).

Reaction (3) is a pure ionic reaction and is therefore expected to proceed similarly fast as reaction (2). In Table 1 some values for the thermodynamic constants H, K_1 and K_2 are given as a function of temperature.

Figure 1 shows the mole fractions of the three sulfur (IV) species in equilibrium as a function of solution pH at 25°C. As can be seen, the bisulfite ion HSO_3^- is the most abundant sulfur (IV) species in the pH range of atmospheric droplets (i.e. pH 3–6).

Transport of SO_2 within the gas phase to the droplet proceeds by turbulent and molecular diffusion. The rate determining molecular diffusion can be calculated using the Fickian equation for diffusion along with suitable initial and boundary conditions. The SO_2-concentration profile normal to the droplet surface for maximum diffusion can be calculated assuming zero concentration at the droplet surface (i.e. the droplet is a complete SO_2-absorber) and constant SO_2-concentration in infinity:

$$\frac{\partial(RC)}{\partial t} = D\,\frac{\partial^2(RC)}{\partial R^2} \tag{4}$$

Initial condition:

$$RC = RC_0 \text{ for } t = 0 \text{ and } R > r \tag{5}$$

Boundary condition:

$$RC = 0 \text{ for } t > 0 \text{ and } R = r \tag{6}$$

where R = radius vector; r = droplet radius; C = SO_2-concentration in the gas phase; D = molecular diffusivity of SO_2 in air.*

This problem has been solved by Smoluchowski (1918) for diffusion of aerosol particles into a water droplet and has been applied to SO_2 by Beilke (1965). The SO_2-concentration profiles normal to the droplet are given by

$$C = C_0\left(1 - \frac{r}{R} + \frac{2r}{R\sqrt{\pi}}\int_0^{(R-r)/2\sqrt{Dt}} e^{-x^2}\,dx\right) \tag{7}$$

x = Integration variable

and are shown in Fig. 2 for a droplet with a radius of 10 μm after different times. The quantity of diffusing SO_2, $M_t(g)$, which has entered the droplet in time t under the assumed conditions (4) to (6), is given by:

$$M_t(g) = 4\pi D r C_0\left(t + \frac{2r\sqrt{t}}{\sqrt{\pi D}}\right). \tag{8}$$

The assumption that a water surface is a complete SO_2-absorber (i.e. that the gas concentration is zero at the surface) is approximately valid for the absorption of SO_2 in alkaline solutions like ocean water (Liss, 1971; Liss and Slater, 1974; Brimblecombe and Spedding, 1972; Spedding, 1972; Beilke and Lamb, 1974).

For atmospheric droplets with pH values normally between 3 and 6 with a limited absorption capacity for sulfur (IV), the droplet surface can represent a resistance to SO_2-transfer which is not negligible especially if the droplet pH is low (Liss, 1971; Brimblecombe and Spedding, 1972). If it is taken into account that the concentration of SO_2 increases at the droplet

* D_{gas} was taken as 0.10 cm^2 s^{-1} in our calculations for 10°C.

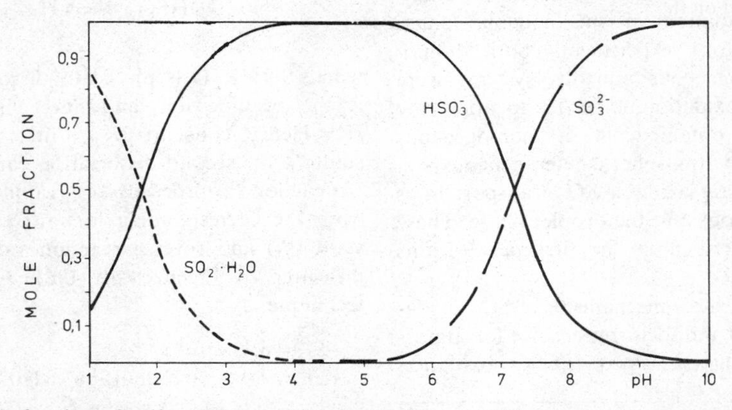

Fig. 1. Mole fraction of sulfur (IV) species in equilibrium at 25°C as a function of aqueous solution pH.

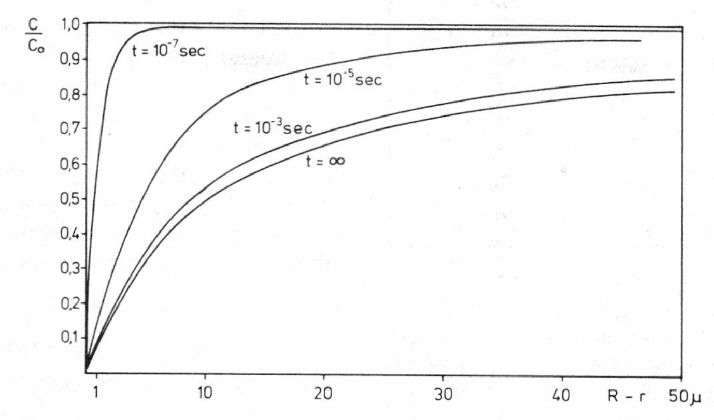

Fig. 2. SO_2-concentration profiles normal to the surface of a droplet (radius 10 μm) caused by molecular diffusion calculated for a complete SO_2-absorber.

surface due to absorbed SO_2 the calculated diffusion fluxes are reduced.

In Fig. 3 the SO_2-flux to a complete absorbing droplet is compared with the SO_2-flux to a droplet with increasing SO_2-concentration at the surface according to the amount of SO_2 absorbed. It was assumed that the absorbed SO_2 was immediately mixed within the droplet. In this figure the ratios between the SO_2-fluxes to the droplet and the equilibrium concentration of S(IV) are plotted as a function of time for various pH values which were kept constant during the absorption process. In a droplet with $r = 10\ \mu$m equilibrium between SO_2 in the gasphase and sulfur (IV) is reached within a few seconds.

Transport of SO_2 within a droplet proceeds by molecular diffusion and turbulent transfer. If transport proceeds within the whole droplet by molecular diffusion only and if the SO_2-concentration at the surface is kept constant, the total amount of diffusing sulfur (IV) from the surface to the interior is given by

$$\frac{M_{t(a)}}{M_\infty} = 1 - \frac{6}{\pi^2}\sum_{n=1}^{\infty}\frac{1}{n^2}\exp(-Dn^2\pi^2 t/r^2) \qquad (9)$$

* D was taken as $1.15 \times 10^{-5}\ \mathrm{cm^2\ s^{-1}}$ in our calculation for 10°C after Broeker and Peng (1974) for all sulfur (IV) species.

† M_∞ can be calculated on the basis of Equations (1)–(3).

(Crank, 1976) where r is droplet radius, D is molecular diffusivity of S(IV) in water,* $M_t(a)$ is the quantity of diffusing S(IV) which has entered the droplet in time t, M_∞† is the concentration of S(IV) in equilibrium with SO_2-gasphase concentration.

As seen in Fig. 4, for cloud and fog droplets smaller than 50 μm radius, the equilibrium between SO_2 in the gasphase and S(IV) within the droplet is reached within approximately 1 s if the SO_2-concentration at the surface is kept constant. For cloud drops both the transport of SO_2 to a droplet and transport of S(IV) within a droplet will be enhanced by turbulent transfer. A comparison of calculated and measured ratios (Barrie, 1975) $M_t(a)/M_\infty$ for a droplet with $r = 1.06$ mm radius shows a good agreement.

On the basis of our estimations of SO_2-transport both in the gaseous- and droplet phase equilibrium between SO_2 in the gas phase and sulfur (IV) in the droplet phase will be achieved within less than *ca* 1 s for atmospheric cloud and fog droplets.

As will be shown in the next section, oxidation of S(IV) to sulfate in a droplet proceeds slower by orders of magnitude. Thus the rate determining step in the overall heterogeneous SO_2-oxidation in cloud and fog droplets smaller than 50 μm is the oxidation of S (IV) to sulfate and not diffusion processes.

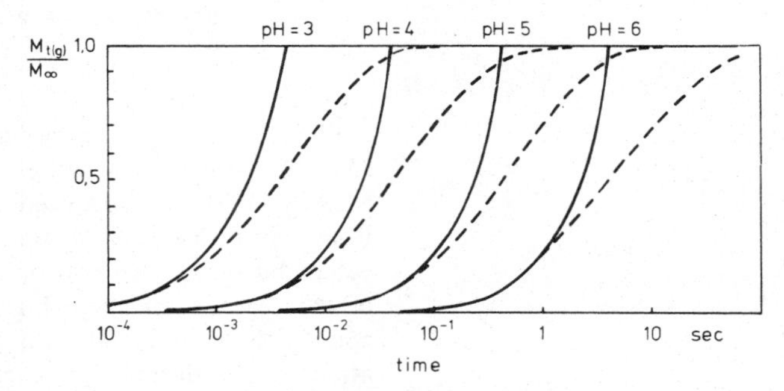

Fig. 3. Ratios between the SO_2-fluxes to the droplet, $M_t(g)$, and the equilibrium concentration of S(IV), (M_∞), in a droplet ($r = 10\ \mu$m) as a function of time for various pH values which were kept constant during the adsorption process. The solid curves represent complete absorption, the dashed ones reduced absorption due to an exponential increase of surface resistance to SO_2-transfer.

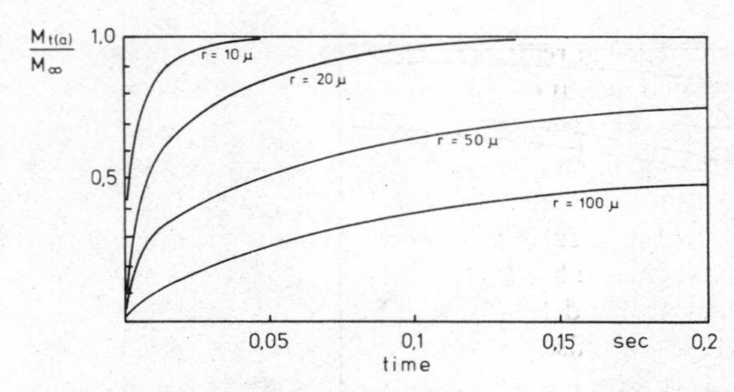

Fig. 4. Ratios between the quantity of diffusing S(IV), $M_t(a)$, which has entered a droplet and quantity of S(IV) in equilibrium, M_∞, as a function of time and droplet radius.

3. OXIDATION OF SO_2 IN THE DROPLET PHASE

One of the most important questions in SO_2-oxidation in the droplet phase is whether the bisulfite ion (HSO_3^-) or the sulfite ion (SO_3^{2-}) is the oxygen carrier or both. A great many discrepancies in the literature concerning the pH dependance of SO_2-oxidation and sulfate formation could be removed if there were a clear answer to this question since in a drop in equilibrium with a fixed SO_2 concentration HSO_3^- varies with $[H^+]^{-1}$ and SO_3^{2-} with $[H^+]^{-2}$. We will interpret the SO_2-oxidation in terms of the more widely accepted view that the sulfite ion SO_3^{2-} is the oxygen carrier and not HSO_3^- or both. This assumption is in accordance for example with Winkelmann (1955) or Rand and Gale (1967) but is in contrast to that of Schroeter (1963) and Penkett *et al.* (1976).

In the following sections we will compare three different mechanisms of SO_2-oxidation and assess their importance for sulfate formation in the droplet phase. In order to compare investigations, we have transformed all rate expressions—whenever possible—into the corresponding expression of our concept with the sulfite ion (SO_3^{2-}) being the oxygen carrier.

3.1 *SO_2-oxidation by O_2 in the absence of catalysts*

According to Winkelmann (1955), the oxidation follows the simple overall reaction:

$$SO_3^{2-} + \tfrac{1}{2}O_2 \xrightarrow{k_{10}} SO_4^{2-}. \qquad (10)$$

Most authors agree that the oxidation is zero order with respect to oxygen over a wide range of O_2-concentrations (for example Titoff, 1903; Riccoboni *et al.*, 1949; Winkelmann, 1955; Penkett *et al.*, 1976). The order with respect to sulfite is unity in most investigations (for example Titoff, 1903; Winkelmann, 1955; Rand and Gale, 1967). The following overall pseudo-first-order rate constants, k_{10}, for reaction (10) are available*:

(a) $k_{10} = 1.7 \times 10^{-3}\ s^{-1}$ at pH 6.8 and *ca* 25°C. Scott and Hobbs (1967), from measurements of Van den Heuvel and Mason (1963) at room temperature.

(b) $k_{10} = 3 \times 10^{-3}\ s^{-1}$ between pH 2–4 at 25°C. Miller and de Pena (1972).

(c) $k_{10} = 3.5 \times 10^{-3}\ s^{-1}$ at pH 7 and 25°C. Winkelmann (1955).

(d) k_{10}† $= 3.7 \times 10^{-3}$–$0.6 \times 10^{-3}\ s^{-1}$ between pH 4–6 at 25°C with 0.6×10^{-3} at pH 6. Brimblecombe and Spedding (1974).

(e) k_{10}‡ $= 0.5 \times 10^{-3}\ s^{-1}$ at pH 6 and 20°C. Penkett *et al.* (1976).

(f) k_{10}§ $=$ *ca* 6×10^{-3}–$0.6 \times 10^{-3}\ s^{-1}$ between pH 7–8 at 25°C. Schroeter (1963).

(g) $k_{10} = 1.2 \times 10^{-4}\ [H^+]^{-0.16}$ between pH 3–6 at 25°C. (k in s^{-1}, $[H^+]$ in mol/l.). Beilke *et al.* (1975).

(h) $k_{10} = 0.013 + 59\ [H^+]^{1/2}$ (pH below 6 at 25°C) (k in s^{-1}, $[H^+]$ in mol/l.). Mc Kay (1971) from measurements of Fuller and Crist (1941).

Figure 5 shows k_{10} values as a function of pH for 25°C. As seen in this figure, the k_{10}-values used by McKay (1971) and deduced from measurements of Fuller and Crist (1941) are higher than all other values by more than two orders of magnitude. One of the reasons for these high values is that Fuller and Crist did not take into consideration the variation of pH during the reaction nor the resulting changes of the ionic composition of the reaction mixture, even though such changes of the mole fraction of HSO_3^- and SO_3^{2-} are clearly indicated in Fig. 1.

* k_{10} is defined by: $d[SO_4^{2-}]/dt = k_{10}\ [SO_3^{2-}]$.

† In order to compare the results of Brimblecombe and Spedding (1974) with the others we transformed their rate constants given in Table 1 of their paper for pH 4 to 6 into our concept.

‡ For Penkett *et al.* (1976) we calculated the corresponding k_{10} value for pH 6.

§ For Schroeter (1963) we calculated k_{10} values for the pH range 7–8.

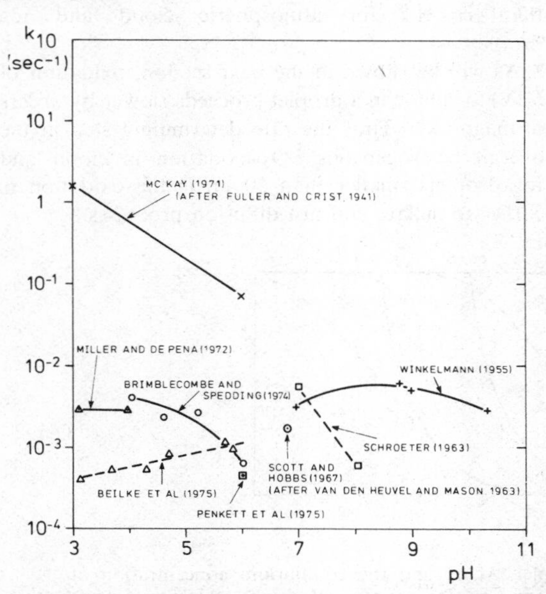

Fig. 5. Pseudo first order rate constant for reaction 10, k_{10}, as a function of pH.

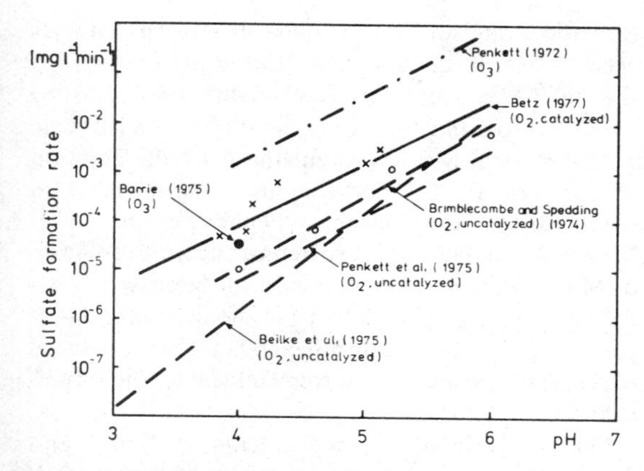

Fig. 6. Sulfate formation rates in the droplet phase as a function of droplet pH for three different SO_2-oxidation mechanisms. The calculated rates apply to the following conditions: gasphase concentrations for SO_2: 1 ppb, O_3: 40 ppb, $T = 10°C$.

For this reason we will confine ourselves to the other k_{10}-values given in the pH range 3 to 7.

Although there is an opposite pH dependance of k_{10} between Beilke *et al.* (1975) and Brimblecombe and Spedding (1974), the values differ by no more than a factor of *ca* 2 in the pH range between 5 and 6. For pH 6 the corresponding value of Penkett *et al.* (1976) is also within this range. Therefore it seems reasonable to accept a k_{10} value of *ca* $10^{-3}\,s^{-1}$ in the pH range 5 to 6 at 25°C.

Although the oxidation can be described by reaction (10), the exact mechanism seems still to be a matter of discussion. It is likely that the oxidation proceeds via two chains with the radicals HO_2 and SO_3^- as chain carriers (Schmittkunz, 1963).

Two results of Beilke *et al.* (1975) who used a heterogeneous system to study SO_2-oxidation should be mentioned here: Sulfate formation in aqueous solution was first order with respect to SO_2 in the gas phase and did not depend on temperature within experimental error. This somewhat surprising result may well have arisen from the compensating effects of decreasing SO_2-solubility (see Table 1) and increasing reaction rate constant with increasing temperature. On the basis of this result an activation energy for reaction (10) has been estimated to be 14 kcal mol^{-1} in the pH range between 3 and 6 using the thermodynamic quantities given in Table 1. Since Beilke *et al.* (1975) used a heterogeneous system, their laboratory results will be directly applied to the atmosphere.

For a cloud with a liquid water content (LWC) of 0.1 g m^{-3}, SO_2-removal rates were calculated to fall between *ca* 10^{-6}% h^{-1} and 1.5% h^{-1} for droplet pH values between 3 and 6, or ten times higher for LWC = 1 g m^{-3}. The lower dashed curves in Fig. 6 show sulfate formation rates due to SO_2-oxidation by O_2 in the absence of catalysts as a function of droplet pH for 10°C and 1 ppb SO_2-gas phase concentration. This SO_2-gas phase concentration was found to be realistic at cloud level (1–5 km) in background air (Georgii and Jost, 1964; Gravenhorst, 1975). As seen in this figure, sulfate formation rates are extremely small except for high droplet pH (above ~6).

In Fig. 6 sulfate formation rates which were calculated on the basis of the experimental data of Brimblecombe and Spedding (1974) and of Penkett *et al.* (1976) are also included. A comparison of the three dashed curves shows that very similar conclusions concerning the role of SO_2-oxidation by O_2 in the absence of catalysts can be drawn.

A special type of the SO_2-oxidation by O_2 in the absence of catalysts is the oxidation in the presence of ammonia (NH_3). In the past, this mechanism has attracted substantial attention. It had previously been accepted by many scientists that the so-called SO_2–NH_3–H_2O-mechanism was the most effective SO_2-oxidation process in atmospheric droplets. The effect of ammonia is to keep the droplet pH high when S(IV) is oxidized to sulfate. This effect was investigated by many scientists, for example by Junge and Ryan (1958), Van den Heuvel and Mason (1963), Scott and Hobbs (1967), Mc Kay (1971), Miller and de Pena (1972), Easter and Hobbs (1974) and Tomasi *et al.* (1975).

The model of Scott and Hobbs was in best agreement with measurements by Beilke *et al.* (1975). It can, however, not be used (as for example Easter and Hobbs (1974) did) in order to calculate sulfate formation rates without further modifications of the physics involved (Beilke and Barrie, 1974). The calculated sulfate formation rates of Easter and Hobbs apply to a droplet pH of 6.6 (Harrison, 1974). It seems to be unlikely that such a high pH value is generally found.

In a real atmosphere NH_3 competes for control over droplet pH with many other soluble substances for example with an acidic background aerosol (Junge and Scheich, 1969; Gravenhorst, 1975) causing in most cases an overall droplet pH between *ca* 4–5. The sulfate formation rates in the droplet phase should therefore be lower than those predicted in the model of Scott and Hobbs by a factor between *ca* 10^2 to 10^4 (Beilke, to be published). Therefore the major function of NH_3 is likely to be no longer an effective enhancement of the SO_2-oxidation but rather the conversion of a pre-existing sulfate containing droplet into a droplet containing ammonium and sulfate ions. However, the absorption of ammonia can shift the pH value from the 3 to 4 range to a range where oxidation of SO_2 by O_2 in the presence of catalysts is more important (4–5). If the source of sulfate in cloud drops is aqueous phase oxidation the oxidation mechanism must be effective between pH 3 and 5.

3.2 SO_2-*oxidation by* O_2 *in the presence of catalysts*

Catalysts are mainly transition metals of the 4th period especially manganese and iron. In the case of catalytic SO_2-oxidation by O_2 in the presence of catalysts we will clearly separate aqueous phase SO_2-oxidation occurring in saline solutions of deliquesced aerosol salts with catalyst concentration of 10^{-1}–10^{-3} M from that occurring in dilute solutions like fog and cloud drops with catalyst concentrations between 10^{-4}–10^{-8} M. We will confine ourselves to the latter.

Since the important work of Bigelow in 1897 more than 200 publications have appeared in the literature dealing with catalytic oxidation of S(IV). Most worked with SO_2- and catalyst concentrations which were higher by orders of magnitude than those occurring in the atmosphere. In addition, most studies were carried out with sodium sulfite solutions which have a pH range between 8–9. The atmospheric pH-range is between 3 and 6. Therefore we will restrict discussion to a few investigations carried out with concentrations of SO_2 and catalysts similar to atmospheric conditions (catalyst concentrations between 10^{-4}–10^{-8} M and (or) SO_2-concentration far below 1 ppm). The effect of catalysts was investigated using both a heterogeneous system (Barrie and Georgii, 1976) and a homogeneous one (Coughanowr and Krause, 1965; Bracewell and Gall, 1967; Tsunogai, 1971; Brimblecombe and Spedding, 1974; Penkett *et al.*, 1975; Lenelle, 1975; Betz, 1977).

The oxidation of S(IV) was found to be proportional to the square of the Mn^{2+}-concentration when the concentration ratios $S(IV)/Mn^{2+} \gg 1$. (Coughanowr and Krause, 1965) which is in agreement with the findings of Bracewell and Gall (1967).

An empirical equation describing the oxidation in the presence of Mn^{2+} was given by Bracewell and Gall:

$$\frac{1}{R} = \frac{A}{[Mn^{2+}]_0^2} + \frac{B}{[S(IV)]_0^{2.2}}$$

where A, B = constants, $[Mn^{2+}]_0$ = initial concentration of Mn SO_4 in aqueous solution, $[S(IV)]_0$ = initial S(IV) concentration in aqueous solution and R = reaction rate.

Brimblecombe and Spedding (1972) investigated the Fe^{3+}-catalyzed SO_2-oxidation in acidic aqueous solutions with concentration for $Fe^{3+} \approx 10^{-6}$ M and $S(IV) \approx 10^{-5}$ M. They found the disappearance rate of S(IV) to be proportioned to both Fe^{3+} and S(IV). At a pH of 4 the oxidation proceeded faster in the presence of 10^{-6} M Fe^{3+} than in the absence of catalysts by *ca* a factor 10^3. Barrie and Georgii (1976) investigated the catalyzed SO_2-oxidation at 25°C and 8°C using single droplets of 2.1 mm diameter consisting of distilled water with heavy metal concentrations in the range between 10^{-6} to 10^{-4} molar for manganese and iron. The droplets were exposed to SO_2-concentrations in air between 0.01–1 ppm. The sulfate formation rate was found to be first order with respect to SO_2 in the gas phase. In the pH range 2–4.5 the most effective catalyst was Mn^{2+} followed by Fe^{2+} and to a much lesser extent Fe^{3+}. The catalytic effectiveness of Mn^{2+} was enhanced by the addition of Fe^{2+} or Fe^{3+}. This synergistic effect was also observed by other authors (Penkett *et al.*, 1975; Lenelle, 1975). The increase of temperature of a solution of Mn^{2+} from 8 to 25°C caused an increase in oxidation rate by a factor 5 to 10 in the pH range 2–4.5. However, the temperature dependance was reduced when Fe^{2+} or Fe^{3+}-ions were added to the Mn^{2+} solution.

On the basis of the experiments of Barrie and Georgii (1976), the conclusion can be drawn that sulfate formation due to catalytic SO_2-oxidation can be significant in urban air clouds or fogs in which case the catalysts concentration can be of the order of 10^{-5} molar or even higher.

A realistic approach with respect to concentrations and composition of catalysts is to use natural rainwater as solvent for studying the catalytic SO_2-oxidation. Betz (1977) has investigated the oxidation in a homogeneous aqueous phase of rainwater collected in the area of Frankfurt/M. (Germany). The heavy metal concentrations were between 10^{-7}–10^{-6} molar for manganese and between 10^{-6}–10^{-5} molar for iron. Of the total manganese 80–90% were dissolved. For iron the corresponding fraction was 60–75%. The rainwater pH was between 3.2 and 5.2. By measuring the concentration decrease of S(IV), sulfate formation rates were determined as a function of pH and temperature.

The activation energy was measured to be 23 kcal mol^{-1} which is in close agreement with the value of 21 kcal mol^{-1} found by Penkett *et al.* (1975). The rate constants are *ca* 1 to 2 orders of magnitude higher than those reported for the SO_2-oxidation by O_2 in the absence of metal catalysts. In order to apply the experimental data of Betz (1977) into the atmospheric heterogeneous system, three assumptions were made: (1) SO_2-transport processes proceed fast compared with oxidation in the droplet phase.* (2) Sulfate formation proceeds first order with respect to SO_2 in the gasphase. (3) The concentrations of heavy metals measured in rainwater are the same as in cloud or fog water. This assumption is to some extent uncertain.

The results of our calculations (solid curve in Fig. 6) for a gas phase concentration of 1 ppb and 10°C show that sulfate formation proceeds faster than in the absence of metal catalysts by *ca* 2 orders of magnitude. It has often been concluded on the basis of experiments with natural rainwater that this type of SO_2-oxidation contributes only little to sulfate formation in the droplet phase at least for country clouds in rural areas. These conclusions are drawn under the assumption that catalyst concentrations and pH in rainwater are the same as in cloudwater which is highly uncertain. Very similar results have been

* For a droplet with 10 μm radius the calculated rate of SO_2-diffusion to the droplet is higher than the measured SO_2-oxidation rate by a factor between *ca* 10^7 to 10^4 for the droplet pH range of interest here (3–5).

obtained by Penkett *et al.* (1975a) and Lenelle (1975).* The oxidation rate measured by Tsunogai (1971) seems to be lower. The mechanism of the catalyzed SO_2-oxidation by O_2 is still a matter of discussion. The oxidation proceeds likely via two chains with the radicals HO_2 and SO_3^- as chain carriers (Schmittkunz, 1963). The radicals are likely produced as follows: (see Barrie and Georgii (1976) who used the proposed mechanism of Schmittkunz (1963)).

$$[Me^{x+}\ S(IV)_3]^{-(6-x)} + O_2 \rightarrow [Me^{x+}\ S(IV)_3]^{-(5-x)} + O_2^- \quad (12)$$

$$O_2^- + H^+ \rightarrow HO_2 \quad (13)$$

$$[Me^{x+}\ S(IV)_3]^{-(5-x)} + SO_3^{2-} \rightarrow [Me^{x+}\ S(IV)_3]^{-(6-x)} + SO_3^- \quad (14)$$

As can be seen, the complex $[Me^{x+}\ S(IV)_3]^{-(6-x)}$ is regenerated. Me^{x+} is transition metal ion of valency x. S(IV) is SO_3^{2-} or HSO_3^-.

In the case of manganese as a catalyst, the results of Barrie and Georgii (1976) seem to indicate that the manganese ion Mn^{2+} formed complexes $[Mn^{2+}(SO_3^{2-})_3]^{-4}$ before taking part in the reaction. Interested readers are referred to the publication of Schmittkunz (1963).

3.3 SO_2-*oxidation by strongly oxidizing agents*

The effect of strongly oxidizing agents was investigated by Penkett (1972); Penkett and Garland (1974); Penkett, Jones and Eggleton (1975); Penkett, Brice and Eggleton (1976) and of Barrie (1975) for ozone (O_3) and by Penkett, Brice and Eggleton (1976) for hydrogen peroxide (H_2O_2).

The reaction of ozone with sodium sulfite in a homogeneous aqueous phase was studied by Penkett (1972) at pH 4.65 and was extended later to a pH range 1 to 5 by Penkett, Jones and Eggleton (1975). The SO_2-oxidation was measured by following the decline in ozone. In these experiments the concentration of total sulfite† was kept in excess of ozone. In this case the reaction was first order with respect to ozone. The order with respect to total sulfite was unity at pH 4 and decreased with increasing pH. The O_3-concentrations used were about 5×10^{-5} M corresponding to a gas phase concentration of ~2700 ppm in equilibrium. According to Penkett and Garland (1974) the results obtained with such high O_3-concentrations can be extrapolated to atmospheric O_3-concentrations as indicated by their measurements on the rate of SO_2-oxidation in a fog chamber using atmospheric concentrations of both ozone and SO_2. The mechanism of the reaction between ozone and total sulfite is likely to proceed via a free radical mechanism (Penkett, Jones and Eggleton, 1975).*

It should be mentioned, however, that Barrie (1975) did not find a much enhanced SO_2-oxidation rate due to ozone at a pH of *ca* 4. He exposed three droplets (1.06 mm radius) of distilled water to a stream of air containing a mixture of 0.4 ppm SO_2 and different O_3-concentrations between 0.01 and 0.5 ppm. In the droplet phase the increase of total sulfur was measured using a new isotope dilution technique (Klockow *et al.*, 1974).

Barrie's measurements with distilled water droplets showed only a very low enhancement of SO_2-oxidation rate due to ozone. He found, however, an enhancement of the oxidation rate due to O_3 when manganese was present. For an O_3-gasphase concentration of 0.14 ppm the rate in a 10^{-5} molar $MnCl_2$-droplet was doubled.

In order to estimate sulfate formation rates for the heterogeneous atmospheric droplet system on the basis of the experiments of Penkett (1972) and Penkett, Jones and Eggleton (1975) using a homogeneous aqueous phase, we used their kinetic data along with the thermodynamic quantities in Table 1 for S(IV) and for ozone (Barrie, 1975). Again, we assumed transport processes for both SO_2-gas and O_3-gas to proceed fast compared with the SO_2-oxidation‡ in the droplet. The uppermost dashed curve in Fig. 6 shows sulfate formation rates due to SO_2-oxidation by ozone after the experimental results of Penkett, Jones and Eggleton (1975) for the following set of conditions: gas phase concentrations are 1 ppb for SO_2 and 40 ppb for ozone and 10°C. The corresponding sulfate formation rate calculated on the basis of Barrie's experiment is lower by a factor of *ca* 10^2 at pH 4.

Another oxidizing agent of importance for SO_2-oxidation in the droplet phase could be H_2O_2 for two reasons: (a) The solubility of H_2O_2 in water is extremely high. (b) The reported steady state concentrations of H_2O_2 in the lower troposphere are about 1 ppb (3 ppb according to model calculations of Levy (1973) and 1 ppb after Crutzen (1977) (personal communication to S. Beilke).

Very recent experiments of Penkett, Brice and Eggleton (1976)§ have shown that the reaction is very fast even in the pH range 3 to 5. However, the importance of this oxidation process for sulfate formation in the droplet phase depends strongly on H_2O_2 gas-phase concentrations which have not yet been measured to our knowledge.

* Since the papers of Penkett *et al.* (1975) and of Lenelle (1975) are not yet published in the open literature, no further details on their work are given here.

† Total sulfite is mainly HSO_3^- in the range investigated by Penkett *et al.* (pH 1–5).

‡ For a droplet with $r = 10\,\mu m$ radius the calculated rate of SO_2-diffusion to the droplet is higher than the measured (Penkett *et al.*, 1975) SO_2-oxidation rate due to ozone by a factor between *ca* 10^5 to 10^2 for the droplet pH range 3–5.

§ Since the results of Penkett, Brice and Eggleton (1976) have not been published yet in the open literature, no details are given here.

4. IMPORTANCE OF HETEROGENEOUS SO_2-OXIDATION IN THE ATMOSPHERE AND FURTHER RESEARCH

Figure 6 shows a comparison of sulfate formation rates due to three SO_2-oxidation mechanisms (a) SO_2-oxidation by O_2 in the absence of catalysts (lower dashed curves), (b) SO_2-oxidation by O_2 in the presence of catalysts (solid curve) and (c) SO_2-oxidation by ozone (upper dashed curve and one single dot).

As seen in Fig. 6, oxidation type (a) is unimportant for sulfate formation unless droplet pH is higher than *ca* 6. Although sulfate formation by type (b) proceeds faster by two orders of magnitude than by type (a), it seems to be at least for background conditions of little importance for atmospheric sulfate formation unless droplet pH is higher than *ca* 5. For urban fog conditions, however, with higher heavy metal concentrations, this mechanism may play an important role. Sulfate formation rates due to SO_2-oxidation by ozone (type (c)) are higher than the corresponding rates for type (b) by *ca* 1 order of magnitude according to Penkett *et al.* (1975). In contrast, however, Barrie (1975) found that at least at pH 4 this rate is lower. The main discrepancies in heterogeneous SO_2-oxidation obvious from Fig. 6 are the pH dependance and the role of ozone.

Therefore we suggest the following further research:

(a) It should be determined whether the SO_3^{2-}-ion or the HSO_3^--ion (or both) carries the oxygen in the pH range 3 to 7. This question is of fundamental importance for the pH dependance of sulfate formation in atmospheric droplets.

(b) Further research is necessary to explain the discrepancy of the experimental results of Penkett *et al.* (1975) and Barrie (1975) concerning the role of ozone in heterogeneous SO_2-oxidation in the absence and presence of catalysts.

(c) In order to calculate sulfate formation rates due to SO_2-oxidation by O_2 in the presence of catalysts, we assumed that the concentrations of heavy metals and the pH value measured in rainwater are the same as in cloudwater. We therefore recommend measurements of the composition and pH of cloud and fog water since the small cloud and fog droplets are more important for sulfate formation due to their relatively long lifetime (minutes to hours) compared with the lifetime of large raindrops (*ca* 1 minute).

In a real atmosphere the three oxidation mechanisms proceed at the same time. We are therefore concerned with a very complex synergistic SO_2-oxidation by O_2, O_3, H_2O_2 in the presence of NH_3, metal catalysts and organic compounds.

Organic substances may either act as inhibitors or catalyst for SO_2-oxidation and can increase or decrease droplet pH. Any ammonia absorbed by a droplet tends to increase the droplet pH which in turn causes higher sulfate formation rates due to oxidation by O_2 in the presence of catalysts since the amount of S(IV) increases for a given SO_2-gas phase concentration. In addition, the catalytic effectivity of any heavy metal of the 4th period depends also on pH. Furthermore, the catalytic effectivity of a special metal ion depends on the presence of other metals. The SO_2-oxidation rate obtained with a mixture of manganese and iron is, for example, higher than the sum of the rates with each ion present alone (Barrie, 1975). In addition to the influence of pH the promoting effect of ozone for SO_2-oxidation, or the inhibiting effect of NO_2 both in the presence of metal catalysts (Barrie, 1975) complicates the situation.

(d) Because the overall effect of SO_2-oxidation in droplets depends on the complex variety of interactions, experiments should be carried out which take all these possibilities into account.

These experiments should complement the studies suggested under (a)–(c) above. Such experiments should be carried out in a heterogeneous system, since in a homogeneous aqueous system compounds reacting in the liquid phase cannot be supplied from the gasphase to replace removed ones. That is for example true for SO_2, O_3, H_2O_2 and NH_3. We suggest that drops consisting of natural rain- or cloudwater be suspended in outside air to simulate actual atmospheric conditions. By measuring the change of droplet pH* and increase of total sulfur† or sulfate, sulfate formation rates can be determined under realistic conditions.

* Droplet pH can easily be measured with a micro-pH cell.

† Total sulfur in a single droplet can be measured with the isotope dilution technique of Klockow *et al.* (1974).

REFERENCES

Barrie L. (1975) An experimental investigation of the absorption of sulfur dioxide by cloud- and rain drops containing heavy metals. Berichte des Institutes für Meteorologie der Universität Frankfurt a.M., Nr. 28.

Barrie L. and Georgii H.-W. (1976) An experimental investigation of the absorption of sulfur dioxide by water drops containing heavy metal ions. *Atmospheric Environment* **10,** 743–749.

Beilke S. and Barrie L. (1974) On the role of NH_3 in heterogeneous SO_2-oxidation in the atmosphere. Paper presented at the European sulfur symposium, Ispra, Italy, October 1974.

Beilke S. (1965) Untersuchungen über 'rainout' und 'washout' von Schwefel-dioxid. Diplomarbeit. Universität Frankfurt a.M.

Beilke S. SO_2-oxidation in the aqueous phase. To appear as a book in late 1978.

Beilke S. and Lamb D. (1974) On the absorption of SO_2 in ocean water. *Tellus* **XXVI,** 268–271.

Beilke S. and Lamb D. (1975) Remarks on the rate of formation of bisulfite ions in aqueous systems. *AIChE Journal* **21,** 402–404.

Beilke S., Lamb D. and Müller J. (1975) On the uncatalyzed oxidation of atmospheric SO_2 by oxygen in aqueous systems. *Atmospheric Environment* **9,** 1083–1090.

Betz M. (1977) Untersuchungen über die Absorption und Oxidation von Schwefeldioxid in natürlichem Regenwasser. Diplomarbeit. Universität Frankfurt. September 1976.

Bracewell J. M. and Gall D. (1967) The catalytic oxidation of sulphur dioxide in solution at concentrations occurring in fog droplets. Proc. Symp. on Physico-chemical Transformation of Sulphur Dioxide Compounds in the Atmosphere and the Formation of Acid Smogs. Mainz, Germany, 1967.

Brimblecombe P. and Spedding D. I. (1972) Rate of solution of gaseous sulphur dioxide of atmospheric concentrations. *Nature* **236,** 225.

Brimblecombe P. and Spedding D. I. (1974) The catalyzed oxidation of micromolar aqueous sulphur dioxide—I. *Atmospheric Environment* **8,** 937–945.

Broecker W. S. and Peng T. H. (1974) Gas exchange rates between air and sea. *Tellus* **XXVI,** 21–35.

Coughanowr D. R. and Krause F. E. (1965) The reaction of SO_2 and O_2 in aqueous solutions of Mn SO_4. *Ind. Eng. Chem. Fund.* **4,** 61–67.

Crank I. (1976) *The Mathematics of Diffusion.* 2nd edn. Clarendon Press, Oxford.

Crutzen P. (1977) Personal communication to S. Beilke.

Easter R. C. and Hobbs P. (1974) The formation of sulphates and the enhancement of cloud condensation nuclei in clouds. *J. atm. Sci.* **31,** 1586–1594.

Eigen M., Kustin K. and Maass G. (1961) Die Geschwindigkeit der Hydratation von SO_2 in wässriger Lösung. *Z. Phys. Chem.* **30,** 130.

Fuller E. C. and Crist R. H. (1941) The rate of oxidation of sulfite ions by oxygen. *J. Am. chem. Soc.* **63,** 1644–1650.

Georgii H. W. and Jost D. (1964) Untersuchungen über die Verteilung von Spurengasen in der freien Atmosphäre. *Pure appl. Geophys.* **59,** 217–224.

Gravenhorst G. (1975) Der Sulfatanteil im atmosphärischen Aerosol über dem Nordatlantik. Berichte des Institutes für Meteorologie der Universität Frankfurt a.M. Nr. 29

Gmelin (1963) Handbuch Schwefel, 8. Auflage, Teil B-Lieferung 3.

Harrison H. (1974) Discussion to paper by S. Beilke, D. Lamb, and J. Müller, The heterogeneous oxidation of SO_2 in relation to atmospheric scavenging. *Precipitation Scavenging.* U.S. Atomic Energy Commission, Oake Ridge, Tenn. ERDA symposium series, CONF-741003.

Junge C. E. and Ryan T. G. (1958) Study of the SO_2-oxidation in solution and its role in atmospheric chemistry. *Q. Jl. R. met. Soc.* **84,** 46–55.

Junge C. E. and Scheich G. (1969) Studien zur Bestimmung des Säuregehaltes von Aerosolteilchen *Atmospheric Environment* **3,** 423–441.

Klockow D., Denzinger H. and Rönicke R. (1974) Anwendung der substöchiometrischen Isotopenverdünnungsanalyse auf die Bestimmung von atmosphärischem Schwefel und Chlor in Background Luft. *Chem. Ing. Techn.* **46,** 18.

Lenelle Y. (1975) Contribution a l'etude de l'oxydation des sulfites dans la pluie. Paper presented at the European sulfur symposium, Ispra, Italy, October 1975.

Levy H. (1973) Photochemistry of minor constituents in the troposphere. *Planet. Space Sci.* **21,** 575–591.

Liss P. S. (1971) Exchange of SO_2 between the atmosphere and natural waters. *Nature* **233,** 327–329.

Liss P. S. and Slater P. G. (1974) Flux of gases across the air–sea interface. *Nature* **247,** 181–184.

McKay H. A. C. (1971) The atmospheric oxidation of sulphur dioxide in water droplets in presence of ammonia. *Atmospheric Environment* **5,** 7–14.

Miller J. M. and de Pena R. G. (1972) Contribution of scavenged sulfur dioxide to the sulfate content of rain water. *J. geophys. Res.* **77,** 5905–5916.

Penkett S. A. (1972) Oxidation of SO_2 and other atmospheric gases by ozone in aqueous solution. *Nature Phys. Sci.* **240,** 105–106.

Penkett S. A. and Garland J. A. (1974) Oxidation of sulphur dioxide in artificial fogs by ozone. *Tellus* **XXVI,** 284–290.

Penkett S. A., Jones B. M. R. and Eggleton A. E. J. (1975a) A study of SO_2-oxidation in stored rainwater samples. Paper presented at the European sulfur symposium, Paris, 1975.

Penkett S. A., Jones B. M. R. and Eggleton A. E. J. (1975b) Rate of oxidation of sodium sulphite solutions by oxygen and by ozone. Paper presented at the European sulfur symposium, Ispra, 1975.

Penkett S. A., Brice K. A. and Eggleton A. E. J. (1976) A study of the rate of oxidation of sodium sulphite solution by hydrogen peroxide and its importance to the formation of sulphate in cloud- and rainwater. Paper presented at the European sulfur symposium, Ispra 1976.

Rand M. C. and Gale S. B. (1967) Kinetics of the oxidation of sulfites by dissolved oxygen. In *Principles and Applications of Water Chemistry* (Edited by Faust S. D. and Hunter J. V.) pp. 380–404. Wiley, New York.

Riccoboni L., Foffani A. and Vecchi E. (1949) Studi di cinetica chimica sui processi di autoossidazione in fase liquida diluita—II. *Gazz. chim. Ital.* **79,** 418–442.

Schmittkunz H. (1963) Chemilumineszenz der Sulfitoxidation. Dissertation of the Naturwissenschaftliche Fakultät der Universität Frankfurt.

Schroeter L. C. (1963) Kinetics of air oxidation of sulfurous acid salts. *J. Pharm. Sci.* **52,** 559–563.

Scott W. D. and Hobbs P. V. (1967) The formation of sulfate in water droplets. *J. atm. Sci.* **24,** 54–57.

Sillen L. G. (1964) *Stability Constants of Metal-Ion Complexes. Section I—Inorganic Ligands,* second edition, Chem. Soc. (London), Special Publication No. 17.

Smoluchowski M. (1918) Versuch einer mathematischen Theorie der Koagulationskinetik. *Z. Phys. Chem.* 151.

Spedding D. J. (1972) Sulfur dioxide absorption by sea water. *Atmospheric Environment* **6,** 583–586.

Titoff A. (1903) Beiträge zur Kenntnis der negativen Katalyse in homogenen System. *Z. Phys. Chem.* **45,** 641–683.

Tomasi C., Guzzi R. and Vittori O. (1975) The 'SO_2–NH_3-solution droplet' system in an urban atmosphere. *J. atm. Sci.* **32,** 1580–1586.

Tsunogai S. (1971) Oxidation of sulfite in water and its bearing on the origin of sulphate in meteoric precipitation. *Geochem. J.* **5,** 175.

van den Heuvel A. P. and Mason B. J. (1963) The formation of ammonium sulfate in water droplets exposed to gaseous sulfur dioxide and ammonia. *Q. Jl. R. met. Soc.* **89,** 271–275.

Wang J. C. and Himmelblau D. M. (1964) A kinetic study of sulfur dioxide in aqueous solution with radioactive tracers. *AIChE Journal* **21,** 402–404.

Winkelmann D. (1955) Die elektrochemische Messung der Oxidationsgeschwindigkeit von Na_2SO_3 durch gelösten Sauerstoff. *Z. Elektrochemie* **59,** 891–895.

Atmospheric Environment Vol. 12, pp. 241–253. Pergamon Press 1978. Printed in Great Britain.

OXIDATION OF SULFUR DIOXIDE IN AQUEOUS SYSTEMS WITH PARTICULAR REFERENCE TO THE ATMOSPHERE

DEAN A. HEGG and PETER V. HOBBS

Atmospheric Sciences Department, University of Washington, Seattle, WA 98195, U.S.A.

Abstract—Laboratory studies of the uncatalyzed liquid phase oxidation of SO_2 by oxygen are reviewed; significant discrepancies exist between the derived pseudo first order rate coefficients. The production rate is independent of the concentration of oxygen. For pH <7 the activation energy for this reaction is 7.1×10^3 J mol^{-1}, and for $7 \lesssim pH \lesssim 9.5$ it is 8.9×10^4 J mol^{-1}. The mechanism of oxidation is probably via a free radical chain involving $SO_3^{\bar{\cdot}}$ and $SO_5^{\bar{\cdot}}$. Sulfates may also be produced in solution by the oxidation of SO_2 by ozone. In this case, the rate of production of sulfates varies linearly with the concentration of ozone and bisulfite and the activation energy is 1.6×10^4 J mol^{-1}. Laboratory studies of the catalyzed liquid phase oxidation of SO_2 suggest that $d[SO_4^{2-}]/dt = K[M^+][H^+][SO_3^{2-}]$, where $[M^+]$ is the concentration of the metal catalyst (e.g. Fe^{3+}, Mn^{2+}, Cu^{2+}). There is some evidence, which needs confirmation, that catalysis by certain 'mixed salts' produces an oxidation rate about ten times greater than with either salt alone. The oxidation of SO_2 in solution is inhibited by a large number of compounds present in the atmosphere. Consequently, the net effect of positive and negative catalysts on the oxidation rate of SO_2 in cloud and rain water could be small. Extrapolation of laboratory results to the atmosphere suggests that in 'clean' (rural) air, the uncatalyzed oxidation of SO_2 by O_3 and possibly O_2 and catalyzed 'mixed salt' oxidation should be competitive. For 'dirty' (urban) air, uncatalyzed oxidation and iron catalyzed oxidation should be competitive, but, if manganese is present in solutions in concentration above 10^{-6} M, the manganese catalyzed oxidation may dominate. However, extrapolation of laboratory results to the atmosphere is fraught with danger. Some field observations of the concentrations of cloud condensation nuclei and sulfates in air entering and leaving clouds suggest that the production of sulfates by the oxidation of SO_2 in clouds is important. Also, preliminary calculations suggest that the liquid phase oxidations of SO_2 in clouds may be the dominant world-wide source of sulfates in the atmosphere. Suggestions are made for further laboratory, field and modeling studies which should improve our understanding of liquid phase SO_2 oxidation in the atmosphere.

1. INTRODUCTION

Although the conversion of sulfur dioxide gas (SO_2) to sulfates (SO_4^{2-}) probably accounts for no more than 10% of the rate of removal of SO_2 from the atmosphere (the dominant processes are precipitation washout and absorption at the earth's surface), it is thought to be the dominant anthropogenic source of atmospheric particles. Moreover, there is considerable interest in sulfates in view of their role in the radiation balance of the atmosphere and cloud microphysical processes, as well as their potential threat to human health.

Estimates of the rate of the conversion of SO_2 to SO_4^{2-} in the atmosphere vary from 0.1 to 10% per hour (Harrison *et al.*, 1975). While some of this variability may be due to lack of observational precision, a large part of it is probably due to the variety of processes which may convert SO_2 to SO_4^{2-} in the atmosphere. In this paper we are concerned with a particular class of atmospheric SO_2-to-SO_4^{2-} conversion processes, namely, those which occur in aqueous solutions by homogeneous–heteromolecular reactions.

We review first the numerous laboratory and theoretical studies which have been carried out over the past fifty years on the oxidation of SO_2 in aqueous solutions. The applicability of these studies to the atmosphere is considered next. Field measurements on the liquid phase oxidation of SO_2 are then discussed. Finally, a number of recommendations are made for further studies which should increase our understanding of the liquid phase oxidation of SO_2 in the atmosphere.

2. LABORATORY AND THEORETICAL STUDIES

The oxidation of SO_2 in aqueous solutions (uncatalyzed and catalyzed) is a classic process which has been studied extensively over the past fifty years. In this section we review laboratory and theoretical studies of the liquid phase oxidation of SO_2 with particular emphasis on the empirical rate expressions which have been derived. Our principal objective is to determine a general form of the rate expression for both the uncatalyzed and catalyzed reactions and to evaluate the conditions under which they should assume importance in the atmosphere.

2.1 *Uncatalyzed liquid phase oxidation of* SO_2 *by oxygen*

(a) *Rate studies.* Table 1 summarizes the rate expressions which emerge from some of the principal

Table 1. Some rate expressions for the uncatalyzed oxidation of SO_2 in aqueous solutions

Reference	Rate expression	Rate constants	pH range	Remarks
(a) *by oxygen*				
Reinders & Vlès (1925)	$\frac{d[SO_4^{2-}]}{dt} \propto [H^+]^{1/2}[SO_3^{2-}]$ Derived from Reinders and Vlès' Fig. 1		<8.4	Buffered rate study in air at 293 K and 298 K.
Fuller & Crist (1941)	$\frac{d[SO_4^{2-}]}{dt} = (K_1 + K_2[H^+]^{1/2})[SO_3^{2-}]$	$K_1 = 1.3 \times 10^{-2}\ s^{-1}$ $K_2 = 6.6\ s^{-1}\ M^{-1/2}$	5.09 to 7.82	Initial rate study in pure O_2. Quartz apparatus. Excellent care taken to ensure purity of all reagents. 298 K. Manometric measurements.
Winkelmann (1955)	$\frac{d[SO_4^{2-}]}{dt} = K_1[SO_3^{2-}]$	$K_1 = (4 \pm 1.5) \times 10^{-3}\ s^{-1}$	7.0 to 10.3	Buffered system in O_2 and N_2 mixtures at 298 K. Polarographic measurements.
Schroeter (1963)	$\frac{d[SO_4^{2-}]}{dt} = K_1[SO_3^{2-}] + K_2[HSO_3^-][H^+]^{-1/2}$ $= K_1[SO_3^{2-}] + K_2'[SO_3^{2-}][H^+]^{1/2}$	$K_1 = 2.9 \times 10^{-3}\ s^{-1}$ $K_2 = 2 \times 10^{-6}\ s^{-1}\ M^{1/2}$ $K_2' \simeq 32\ s^{-1}\ M^{1/2}$	6.95 to 8.20	Initial rate study in air at 298 K. Potentiometric measurements.
Scott & Hobbs (1967)	$\frac{d[SO_4^{2-}]}{dt} = K_1[SO_3^{2-}]$	$K_1 = 1.6 \times 10^{-3}\ s^{-1}$	?	Based on study of Van den Heuval and Mason (1963) which indirectly measured oxidation in drops exposed to SO_2 and NH_3 at 298 K. pH not measured, but with assumptions calculated to be 6.2 to 6.9.
McKay (1971)	Same as Fuller & Crist's	$K_1 = 1.3 \times 10^{-2}\ s^{-1}$ $K_2 = 59\ s^{-1}\ M^{-1/2}$	5.09 to 7.82	McKay simply adjusted the value of K_{2s} used by Fuller and Crist (5×10^{-6} M) to its modern value of 6.24×10^{-8} M at 298 K.
Miller & de Pena (1972)	$\frac{d[SO_4^{2-}]}{dt} = K_1[SO_3^{2-}]$	$K_1 = 3 \times 10^{-3}\ s^{-1}$	?	Similar to those of Van den Heuval and Mason. pH not measured, but with assumptions calculated to be 2 to 4.
Beilke *et al.* (1975)	$\frac{d[SO_4^{2-}]}{dt} = K_1[H^+]^{-0.16}[SO_3^{2-}]$	$K_1 = 1.2 \times 10^{-4}\ s^{-1}\ M^{0.16}$	3.0 to 6.5	Buffered rate study in air at 276 and 298 K. Turbidimetric measurements.
Larson *et al.* (in press)	$\frac{d[SO_4^{2-}]}{dt} = (K_1 + K_2[H^+]^{1/2} + K_3 P_{\lambda_2}[H^+]^{-1})[SO_3^{2-}]$	$K_1 = (4.8 \pm 0.6) \times 10^{-3}\ s^{-1}$ $K_2 = (8.9 \pm 1)\ s^{-1}\ M^{-1/2}$ $K_3 = (3.9 \pm 0.6) \times 10^{-12}\ s^{-1}\ M\ atm^{-1}$	4 to 12	Buffered rate study in air at 278 and 298 K. Iodimetric titration measurements. Rates given are for 298 K.
(b) *by ozone*				
Penkett (1972)	$\frac{d[SO_4^{2-}]}{dt} = K_1[O_3]_{aq.}[HSO_3^-]$	$K_1 = 3.3 \times 10^5\ s^{-1}\ M^{-1}$	4.65	Ozone absorption measurements in stopped flow apparatus at 282.6 K.
Larson *et al.* (in press)	$\frac{d[SO_4^{2-}]}{dt} = K_1 K_{HO_3} P_{O_3}[HSO_3^-][H^+]^{-0.1 \pm 0.02}$	$K_1 = (4.4 \pm 2) \times 10^4\ s^{-1}\ M^{-0.9}$	4 to 6	Same as for Larson *et al.*'s study in oxygen. Rate constant is for 298 K.

laboratory experiments which have been carried out on the uncatalyzed oxidation of SO_2 by oxygen. It can be seen that most investigators have expressed the rate of production of SO_4^{2-} as a function of the sulfite (SO_3^-) concentration in the aqueous solution* and that this rate varies linearly with the concentration of SO_3^-. The rate of production of SO_4^{2-} is independent of oxygen concentration for acidic solutions of atmospheric interest ($3 < pH < 7$) but it varies linearly with the concentration of oxygen at high (>9) pH's.

The variation of sulfate production with the pH of the solution is best demonstrated by considering the rate of production to be proportional to the concentration of SO_3^{2-}, and plotting the pseudo first-order rate coefficient (K_0) for this reaction as a function of pH. This is done in Fig. 1 where it can be seen that the results fall into the following three groups: (i) Those of Fuller and Crist (1941), Schroeter (1963) and Larson *et al.* (in press) for which

$$\frac{d[SO_4^{2-}]}{dt} = \{K_1 + K_2[H^+]^{1/2}\}\,[SO_3^{2-}] \qquad (1)$$

(ii) Those of Winkelmann (1955), Miller and dePena (1972) and Scott and Hobbs (1967). (iii) Those of Beilke *et al.* (1975) which indicate that the rate coefficient increases with increasing pH in acidic solutions. The relatively low values reported by Beilke *et al.* may be due to the plastic vessel employed which could have been a source of organic inhibitors at low pH. However, another possible explanation for the discrepancies between the various results is that in the experiments which lead to Equation (1), Na_2SO_3 was employed as the reduced sulfur species, while Beilke *et al.* used H_2SO_3. The Na_2SO_3 may have contained trace metal catalysts which enhanced the oxidation of low pH's. However, Brimblecombe and Spedding (1974) used gaseous SO_2 and measured a pH dependence for the uncatalyzed oxidation rate similar to Larsen *et al*'s. Clearly, more work is needed to clarify these discrepancies.

The temperature dependence of the uncatalyzed liquid phase oxidation rate of SO_2 by oxygen in the acidic region if fairly weak; Larson *et al.* report an activation energy of only $7.1 \times 10^3\ J\ mol^{-1}$ (1.7 kcal mol^{-1}) in this region. However, for $7 \lesssim pH \lesssim 9.5$,

* This is due to the widely held view that the rate limiting process in the formation of SO_4^{2-} is the oxidation of SO_3^{2-} ions. While rate expressions such as those listed in Table 1 do not establish this *per se*, there is independent evidence that this is indeed the case. It should be noted that it is actually reduced sulfur (sulfur IV) which is commonly measured. The SO_3^{2-} concentration is then generally calculated on the basis of the sulfurous acid equilibria and the measured pH.

Larson *et al.* report an activation energy of 8.9×10^4 J mol^{-1} (21.3 kcal mol^{-1}).

(b) *Mechanisms of oxidation.* The first question which arises in considering the mechanisms by which SO_2 may be oxidized in aqueous solutions in the presence of oxygen is the origin of the oxygen actually involved in the reaction. Work by Winter and Briscoe (1951) and Halperin and Taube (1952a), in which an ^{18}O tracer was used, indicated that the oxidation involved gaseous O_2 dissolved in solution rather than O_2 from the solvent water.

Recent oxygen isotope studies by Cunningham *et al.* (1975) indicate that the oxidation of SO_2 involves oxygen from the water rather than from dissolved oxygen. However, these results do not clearly differentiate oxygen exchange between the sulfite ion and water, and oxidation of the sulfite ion to sulfate. In the atmosphere these processes will be competitive (Halperin and Taube, 1952b; Lloyd, 1968) and the isotopic composition of the sulfate will be a function of their relative rates. If the sulfite is oxidized by the dissolved O_2 (which is suggested by Winter and Briscoe, and Halperin and Taube), the results of Cunningham *et al.* can be explained by the above two competing processes, rather than by O_2 uptake from solvent water.

Early work by Backstrom (1927), Alyea and Backstrom (1929) and Haber and Willstatter (1931) indicated that the actual oxidation mechanism is via a free radical chain involving $SO_3^{\cdot-}$ and $SO_5^{\cdot-}$. The most widely accepted chain is the following, which was proposed by Backstrom (1934):

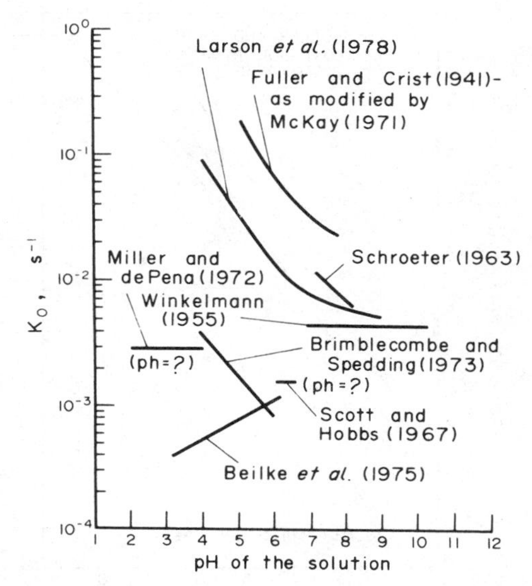

Fig. 1. Pseudo first-order rate coefficient (K_0) as a function of pH for the uncatalyzed liquid phase oxidation of SO_2 in the presence of oxygen.

these studies, could explain the scatter in the experimental observations. It should be noted however that the above equations also predict an oxygen dependence in acid solutions which has not been observed experimentally. Moreover, a chain which is initiated by a metal ion is quite similar to a chain which is catalyzed by a metal ion, particularly if the metal radical produced in the initiation of the chain can be reoxidized in

(Chain initiation)	$SO_3^{2-} + M^+$	$\rightleftharpoons SO_3^{\cdot-} + M\cdot$		(2)
(Chain propagation)	$SO_3^{\cdot-} + O_2$	$\rightleftharpoons SO_5^{\cdot-}$		(3)
	$SO_5^{\cdot-} + HSO_3^-$	$\rightleftharpoons HSO_5^- + SO_3^{\cdot-}$	(pH $\lesssim$ 7)	(3a)
	$SO_5^{\cdot-} + SO_3^{2-}$	$\rightleftharpoons SO_5^{2-} + SO_3^{\cdot-}$	(pH $\gtrsim$ 7)	(3b)
(Oxidation)	$SO_3^{2-} + HSO_5^-$	$\rightleftharpoons HSO_4 + SO_4^{2-}$	(pH $\lesssim$ 7)	(4a)
	$SO_3^{2-} + SO_5^{2-}$	$\rightleftharpoons 2SO_4^{2-}$	(pH $\gtrsim$ 7)	(4b)
(Termination)	$SO_3^{\cdot-} + SO_5^{\cdot-}$	$\rightleftharpoons S_2O_6^{2-} + O_2$		(5a)
	$SO_5^{\cdot-}$ + Inhibitor	$\rightleftharpoons$ non-reactive products		(5b)

The inhibitor in Equation (5b) can be any one of a number of inorganic or organic compounds (Schroeter, 1966).

Work by Hayon *et al.* (1972), in which $SO_3^{\cdot-}$ and $SO_5^{\cdot-}$ radicals have been detected in oxidizing Na_2SO_3 solutions by flash photolysis, provides additional support for the above chain mechanism. It is interesting to note that the above mechanism predicts that the rate of oxidation is proportional to the square-root of the concentration of the metal ion (M). Since metal ions are not considered explicitly in Table 1, variations in the trace concentrations of metals, which may have been present in the solutions used in the chain itself. Larson *et al.* (in press) have proposed a modified (but more complex) Backstrom mechanism which does not require initiation by a metal ion, is not dependent on oxygen in the acid region, and which varies with the oxygen concentration in the alkaline region in the manner which they observed (see Table 1). However, the Backstrom mechanism can also be modified to eliminate the oxygen dependence in the acid region while still retaining the metal ion initiated chain (Larson, 1976), therefore, the strictly uncatalyzed reaction proposed by Larson *et al.* should be very sensitive to traces of certain contaminants.

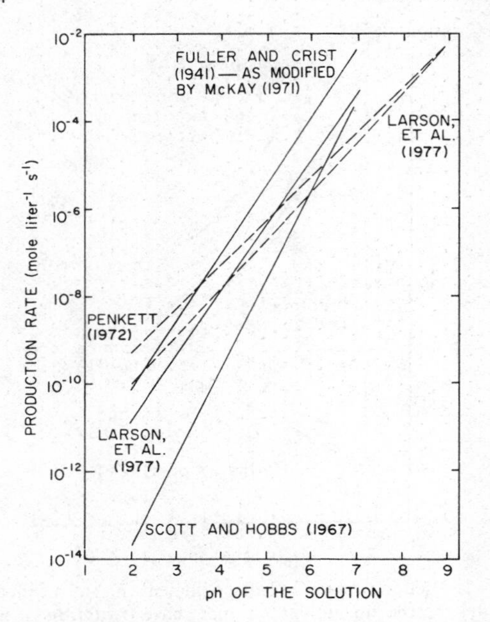

Fig. 2. Rate of production of sulfate versus pH of the solutions for the uncatalyzed liquid phase oxidation of SO_2 by oxygen (solid lines) and by ozone (dashed lines). For comparison purposes the experimental results have been extrapolated over a range of pH values using the derived rate laws shown in Table 1.

2.2 *Uncatalyzed liquid phase oxidation of* SO_2 *by ozone*

SO_2 may also be oxidized in solution by ozone. Rate laws for this reaction have been studied in the laboratory by Penkett (1972) and Larson *et al.* (in press); the results are summarized in Table 1. The rate of production of SO_4^{2-}, derived by both of these workers, varies linearly with the concentration of ozone and bisulfite. Larson *et al.* observed a weak dependence on the explicit hydrogen ion concentration, and Penkett observed no dependence on explicit hydrogen ion concentration. These results suggest that the liquid phase oxidation of SO_2 by ozone will be less prone to retardation in acid solutions than oxidation by oxygen. It is noteworthy that the rates derived from these two studies agree within a factor of two.

Shown in Fig. 2 are the rates of production of SO_4^{2-} as a function of pH for both oxygen and ozone oxidation. Using the two sets of results from the studies of Larson *et al.* (which should be comparable), we see that for an ozone concentration of 50 ppb oxidation by ozone should exceed that by oxygen for pH's below about 4.2. Ozone oxidation of SO_2 in aqueous solutions should therefore be of importance in the atmosphere.

The temperature dependence of the ozone oxidation reaction is relatively weak. Larson *et al.* measured an activation energy of 1.6×10^4 J mol^{-1} (3.8 kcal mol^{-1}) over a pH range of 4–6. This is not significantly different from the activation energy for oxidation by oxygen, therefore, the relative importance of these two oxidizing agents should not vary greatly with temperature.

2.3 *Catalyzed liquid phase oxidation of* SO_2 *by oxygen*

(a) *Rate studies.* Most of the numerous laboratory studies of the catalyzed oxidation of SO_2 by oxygen in aqueous solutions have been carried out either at high concentrations of dissolved SO_2 (e.g. Hoather and Goodeve, 1934a; Junge and Ryan, 1958), at high catalyst concentrations (e.g. Cheng *et al.*, 1971), or, most commonly, at both high SO_2 and high catalyst concentration (e.g., Bassett and Parker, 1951; Matteson *et al.*, 1969). Also, many of these studies have failed to produce actual rate expressions. Shown in Table 2 are results from a representative selection of the studies which have produced rate expressions, including those studies which were carried out with concentrations of reactants approaching those in the atmosphere (e.g., Brimblecombe and Spedding, 1974).

It can be seen from Table 2 that most workers have found that the rate of production of SO_4^{2-} varies linearly with the sulfur species which they have chosen to use. Also, most studies suggest that the production rate is dependent on the concentration of SO_3^{2-} or HSO_3^- (e.g. Fuller and Crist, 1941; Barrie and Georgii, 1976). The following comments can be made concerning those studies listed in Table 2 which appear to disagree with these conclusions. The rate expression derived by Bracewell and Gall (1967) is for the *initial* production rate of SO_4^{2-} expressed in terms of the *initial* SO_2 concentration. The $[SO_3^{2-}]^{3/2}$ dependence given by Barron and O'Hern is based on data with considerable scatter which could plausibly support a linear dependence on sulfite. Freiberg's (1974) squared dependence on SO_3^{2-} is based on the same data as those used by Karraker (1963), but the latter results give a linear dependence on reduced sulfur (Table 2) if data points for reaction times greater than 30 mins are disregarded. (Karraker feels that other oxidation mechanisms or SO_2 degassing may have affected the measurements for $t > 30$ min).

With the exceptions of work done at very high sulfite concentrations (e.g. Miyamoto and Kaya, 1930), the catalyzed liquid phase oxidation of SO_2 has been found to be independent of the concentration of oxygen (at least for pH $\lesssim 8$).

Few studies have investigated the explicit dependence of the reaction rate on hydrogen ion concentration. However, some tentative conclusions can be reached from the results shown in Table 2. For purposes of comparison we shall consider that the rate of production of SO_4^{2-} is proportional to the concentration of sulfite. Studies with sufficient data to permit inference of the H^+ dependence (Karraker, 1963; Brimblecombe and Spedding, 1974; Freiberg, 1974; Barrie and Georgii, 1976; Sutherland, 1977), yield rates which generally show a first-order dependence on H^+. The exception is the study by Barrie and Georgii; from the data shown in their Fig. 2 we deduce that the rate of production of sulfate was pro-

Table 2. Some rate expressions for the catalyzed oxidation of SO_2 in aqueous solutions

Reference	Rate expression	Rate constants	pH range	Remarks
Fuller & Crist (1941)	$\frac{d[SO_4^{2-}]}{dt} = \{K_1 + K_3[Cu^{2+}]\}[SO_3^{2-}]$	$K_1 = 0.013\ s^{-1}$ $K_3 = 2.68 \times 10^8\ s^{-1}\ M^{-1}$	$\leqslant 10.6$	Same as the uncatalyzed experiment but with 10^{-9} to 10^{-4} M cupric ion added as copper sulfate.
Karraker (1963)	$\frac{d[H_2SO_3]}{dt} = K_1 \frac{[Fe^{3+}]}{[H^+]}[H_2SO_3] + K_2[Fe^{2+}][HSO_3]$	—	$\leqslant 0.6$	O_2 removed from ferric solutions (Fe^{3+} in perchloric acid) to which H_2SO_3 was added at 298 K. Note that this is oxidation by Fe^{3+}, *not* O_2 catalyzed by Fe^{3+}.
Karraker (1963)	$\frac{-d[H_2SO_3]}{dt} = K_1 \frac{[Fe^{3+}]}{[H^+]}\left(1 - K_2 \frac{[Fe^{2+}]}{[O_2]}\right)[H_2SO_3]$	—	1.39 to 2.6	Derived from the data of Neytzell–deWilde and Tavener (1958) which was taken at 10 C. 10^{-4} M $<[Fe^{3+}]<10^{-1}$ M and $[SO_2]_{aq} \simeq 10^{-3}$ M.
Barron & O'Hern (1966)	$\frac{-d[O_2]}{dt} = K_1[Cu^{2+}]^{1/2}[SO_3^{2-}]^{3/2}$	$K_1 = 1.2 \times 10^3\ s^{-1}\ M^{-1}$ at 298 K	$\simeq 7$ to $\simeq 9$	Rapid Mixing Method of Hartridge and Roughton (1923). Data taken at 293 and 303 K. Activation energy of $7.6 \times 10^4\ J\ mol^{-1}$ (18.3 kcal mol^{-1}) estimated. 10^{-7} M $<[Cu^{2+}]<10^{-4}$ M.
Bracewell & Gall (1967)	$R_0 = \frac{[SO_2]^{2.2}_{initial}[MnSO_4]^2}{K_1[MnSO_4]^2 + K_2[SO_2]^{2.2}_{initial}}$	$K_1 = 1.32 \times 10^{-3}\ s\ M^{-1}$ $K_2 = 3.2 \times 10^{-4}\ s\ M^{-1}$	—	Conductometric measurements at 298 K. Note that the rate is in terms of the initial aqueous SO_2 concentration. 1×10^{-5} M $<[SO_2]_{initial}< 3.4 \times 10^{-5}$ M 1×10^{-6} M $<[Mn^{2+}]< 7.5 \times 10^{-5}$ M
Matteson *et al.* (1969)	$\frac{d[SO_2]_{aq}}{dt} = K_1[Mn^{2+}][SO_2]_{aq} + (K_2[Mn^{2+}]_{initial} - [Mn^{2+}]\}$	$K_1 \simeq 4.2 \times 10^3\ s^{-1}\ M^{-1}$ $K_2 = 0.17 \times s^{-1}$	—	Complex acidic aerosol–SO_2 gas mixture. SO_4^{2-} determined by filter probe analysis. Experiments at 298 K, relative humidity 95%. Deliquesced aerosol study.
Cheng *et al.* (1971)	$\frac{d[SO_4^{2-}]}{dt} = K_1[\text{Mass of } Mn^{2+}][SO_2]$	—	—	Packed bed reactor at 295–298 K over which an air–SO_2 mixture was passed at various relative humidities. Deliquesced aerosol study.
Brimblecombe & Spedding (1974)	$\frac{-d[S(IV)]}{dt} = K_1[Fe(III)][S(IV)]$	$K_1 = 100\ s^{-1}\ M^{-1}$	$\leqslant 4.9$	Bulk solution study. Ph held constant for various initial pH's by automatic pH titrator. Temperature 25°C. 10^{-7} M $<[SO_3]<10^{-2}$ M, 5×10^{-8} M $<[Fe^{3+}]< 5 \times 10^{-6}$ M.
Freiberg (1974)	$\frac{d[SO_4^{2-}]}{dt} = \{K_1 + K_2[Fe^{3+}]\}[Fe^{3+}][H^+][SO_3^-]^2$	$K_1 = 3.8 \times 10^{16}\ s^{-1}\ M^{-3}$ $K_2 = 5.7 \times 10^{18}\ s^{-1}\ M^{-4}$	1.39 to 2.6	Based on Neytzell–deWilde, and Tavener (1958). This rate law is what Freiberg's general expression *reduces* to if $[Fe^{3+}]<10^{-2}$ M.
Barrie & Georgii (1976)	$\frac{d[SO_4^{2-}]}{dt} = K_1[SO_3^{2-}]$	$K_1 = 9.4 \times 10^2\ s^{-1}$ with $[MnCl_2] = 10^{-4}$ M	$\simeq 1.8$ to 3.0	SO_2 absorption by droplets of dilute metal solution (10^{-6}–10^{-4} M) exposed to air with traces of SO_2 (10–1000 ppb). Measurements at 281 and 298 K.
Sutherland (1977)	$\frac{d[SO_4^{2-}]}{dt} = \frac{[HSO_3^{2-}][H^+]^n[Fe^{3+}]}{K_1 + K_2[HSO_3^-]}$	$n \ll 1.0$ $K_1 \gg K_2[HSO_3^-]$ for small $[HSO_3^-]$	<4.5	—

portional to $[H^+]^{0.46}$ to $[H^+]^{0.65}$. This dependence is closer to that for the uncatalyzed reaction than to a first-order dependence on H^+.

The dependence of the rate of production of sulfate on the concentration of the catalyst is most difficult to resolve from the available data. The difficulty arises because of the variety of catalysts employed, the differences in the concentrations of both catalyst and sulfite employed in various studies, and the different anions of the catalyst salts. We restrict our discussion to the positive catalysts which have been found to be most efficient, namely, Mn^{2+}, Fe^{3+} (or Fe^{2+}) and Cu^{2+}. Some information on these catalysts is given in Table 2 and further (more specific) information is contained in Table 3.

Before discussing these catalysts we would like to clarify the effect of NH_3 on the sulfite oxidation process, since it has commonly been referred to as a positive catalyst. NH_3 enhances the oxidation of sulfite solutions but does so by acting as a buffer to maintain the solutions at relatively high pH. Its effect can be evaluated for both the uncatalyzed and catalyzed reactions simply by determining the effect on the solution pH of the ammonia present and then evaluating the rate at this modified pH (Scott and Hobbs, 1967).

The general results indicated by the catalyzed studies summarized in Tables 2 and 3 are that the rate of production of sulfate appears to be linearly dependent on the concentration of the catalyst. However, there is considerable disagreement on the relative efficiencies of the catalysts and the minimum amounts to catalyze the reaction (see Table 3). For example, Junge and Ryan (1958) found that Mn^{2+} was a more efficient catalyst than Cu^{2+}, and Cu^{2+} was more efficient than Fe^{3+}, but Bracewell and Gall (1967) rank these three metals in the order Fe^{3+}, Mn^{2+} and Cu^{2+}. Also, Bassett and Parker (1951) found a definite anion effect on the catalytic efficiency. In view of these uncertainties, great care must be taken in extrapolating these laboratory results to the atmosphere.

An interesting catalytic effect reported by Barrie

Table 3. Summary of the behavior of catalysts in the oxidation of SO_2 in aqueous solutions

References	Catalyst(s)	Molar concentration of catalyst	Molar concentration of sulfur species	Minimum molar concentration of catalyst required for catalysis	Reaction order with respect to catalyst
Reinders & Vlès (1925)	$CuSO_4$-? (and various negative catalysts)	10^{-8}–1.4×10^{-3}	$\sim 1 \times 10^{-2}$	10^{-12} (estimated)	~ 1
Hoather & Goodeve (1934a)	$MnSO_4$	10^{-5}–10^{-4}	$\geq 10^{-3}$	10^{-5}	~ 1.7
Hoather & Goodeve (1934b)	Mn^{2+} and Fe^{3+} (from ground glass in solution)	$\sim 5 \times 10^{-7}$ (Mn^{2+}) 5×10^{-7} (Fe^{3+}) to 2×10^{-6}	$\geq 10^{-3}$	10^{-7} (?)	~ 1
Fuller & Crist (1941)	$CuSO_4$	10^{-9}–10^{-4}	$\sim 10^{-2}$	10^{-9}	1
Bassett & Parker (1951)	$CuCl_2$, $CuSO_4$, $FeCl_2$, $FeCl_3$, $Fe_2(SO_4)_3$, $FeSO_4$, $MnCl_2$, $MnSO_4$	10^{-2}–0.3 (Mn^{2+}) 10^{-2}–3 (Fe^{3+}) 10^{-3}–3 (Fe^{2+}) $5 \times 10^{-5} - 2$(Cu^{2+})	$\sim 10^{-2}$	10^{-4}	?
Johnstone & Coughanour (1958)	$MnSo_4$ $MnSO_4 + Fe_2(SO_4)_3$ $MnSO_4 + CuSO_4$ $Fe_2(SO_4)_3$	5×10^{-3} 5×10^{-3} 5×10^{-3} ~ 0.1	10^{-2}–10^{-3}		2 (assuming zero order in SO_2)
Junge & Ryan (1958)	$FeCl_2$, $CuCl_2$, $MnCl_2$	10^{-5}	$\sim 10^{-2}$		<1
Brimblecombe & Spedding (1974)	Fe^{3+} (anion not stated)	10^{-8}–10^{-6}	10^{-5}	10^{-8}	1
Barrie & Georgii (1976)	$FeCl_2$, $FeCl_3$, $MnCl_2$, $MnCl_2 + FeCl_2$	10^{-6}–10^{-4}	$\sim 10^{-4}$	10^{-6}	~ 1

and Georgii (1976) should be mentioned here. These investigators found that equimolar concentrations (on the order of 10^{-5} M) of $MnCl_2$ and $FeCl_3$ (or $FeCl_2$) produced an oxidation rate about one order of magnitude greater than that for the same molar concentration of $MnCl_2$ or $FeCl_3$ alone. If this 'mixed salt' effect were to occur under atmospheric conditions it could be of great importance in the oxidation of SO_2 in the atmosphere. A somewhat similar effect was reported by Hoather and Goodeve (1934b), but there is some ambiguity in this case because of their heterogeneous experimental conditions. However, Johnstone and Coughanour (1958) did not observe this effect in salt mixtures. Therefore, at the present time, this effect should be considered as peculiar to certain conditions and extrapolation to the atmosphere carried out with extreme caution.

Turning now to the temperature dependence of the catalyzed reaction, an activation energy of 7.7×10^4 J mol^{-1} (18.3 kcal mol^{-1}) is most often quoted (Barron and O'Hern, 1966). This measurement, however, was made on mildly alkaline solutions ($7 \lesssim pH \lesssim 9$). Hoather and Goodeve (1934a) reported a 'critical increment' for the overall process of 11.4×10^4 J mol^{-1} (27.3 kcal mol^{-1}) $\pm 4\%$ in acidic solutions (with $MnCl_2$ as catalyst), and Sutherland has recently reported an activation energy of $(14.2 \pm 0.8) \times 10^4$ J mol^{-1} (34 ± 2 kcal mol^{-1}) for pH ≤ 4.5 (with Fe^{3+} as catlyst). The higher values in the acid region cast some doubt on the importance of the catalyzed reaction in the atmosphere. However, Barrie and Georgii found that their 'mixed salt' catalyst displayed a negligible temperature dependence, although the activation energies for the single salts were similar to that found by Hoather and Goodeve. Hence, a determination of the range of conditions over which the 'mixed salt' effect occurs is doubly important.

To compare various catalyzed rates with each other and with uncatalyzed rates, we will again plot the normalized rate of production of sulfate against the pH of the solution. To bracket the range of atmospheric conditions, we will consider two cases: air typical of urban ('dirty') conditions, and air typical of rural ('clean') conditions. The data we have chosen to plot are those of Larson *et al.* for the uncatalyzed rate, and Brimblecombe and Spedding and Barrie and Georgii for the catalyzed rate. The latter two studies were selected because the concentrations of both catalyst and sulfur species which were used are closest to atmospheric conditions. Also, these studies were carried out in bulk solutions and should therefore be comparable with the uncatalyzed studies of Larson *et al.* We have taken an activation energy of 14.3×10^4 J mol^{-1} (34 kcal mol^{-1}) for the Brimblecombe and Spedding data, one of 11.5×10^4 J mol^{-1} (27.3 kcal mol^{-1}) for Barrie and Georgii's $MnCl_2$ catalyst data, and an activation energy of zero for Barrie and Georgii's 'mixed salt' data. For the urban atmosphere we have assumed the following values: $[SO_2] = 1$ ppm, $[M^+] = 10^{-5}$ M, $[O_3] = 50$ ppb, and for the rural atmosphere: $[SO_2] = 1$ ppb, $[M^+] = 10^{-8}$ M, $[O_3] = 50$ ppb. The temperature has been taken as 8°C in both cases.

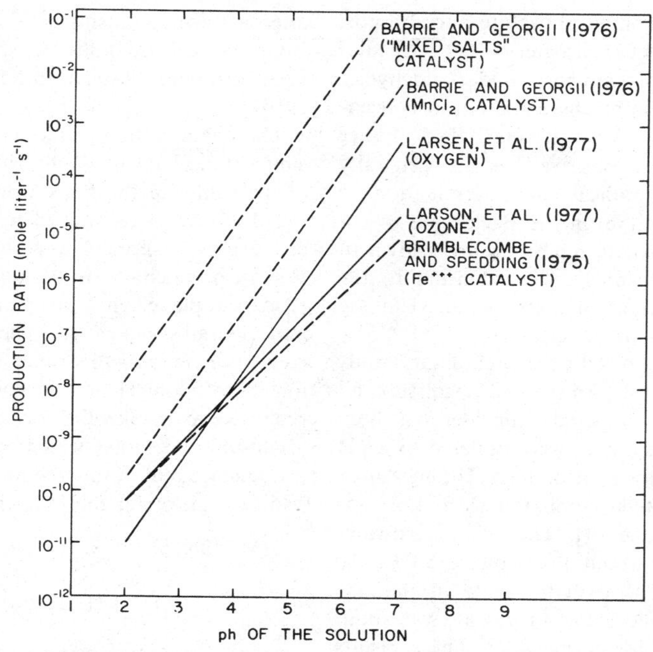

Fig. 3. Comparison of laboratory measurements of uncatalyzed (solid lines) and catalyzed (dashed lines) SO_2 oxidation rates in aqueous solutions applied to a 'dirty' (urban) atmosphere. The experimental results have been extrapolated to cover the pH range of atmospheric interest.

For urban conditions (Fig. 3), Barrie and Georgii's $MnCl_2$ and 'mixed salt' catalytic rates dominate the rate of production of sulfate. For the rural case (Fig. 4), the 'mixed salt' reaction appears to dominate. However, it is quite doubtful whether Barrie and Georgii's rates are applicable to the rural atmosphere (with 10^{-8} to 10^{-6} M metal ion concentration), since they observed negligible oxidation with 10^{-6} M of $MnCl_2$ and there is no clear evidence that the 'mixed salt' effect will act at lower metal concentrations. Therefore, the uncatalyzed oxidation of SO_2 by ozone (or oxygen, if Fuller and Crist and Larson *et al's* measurements are correct) may dominate in rural atmospheres.

A large number of both inorganic and organic compounds have been found which, even at quite low concentrations, inhibit the oxidation of SO_2 in aqueous solutions. For example, inorganic anions,

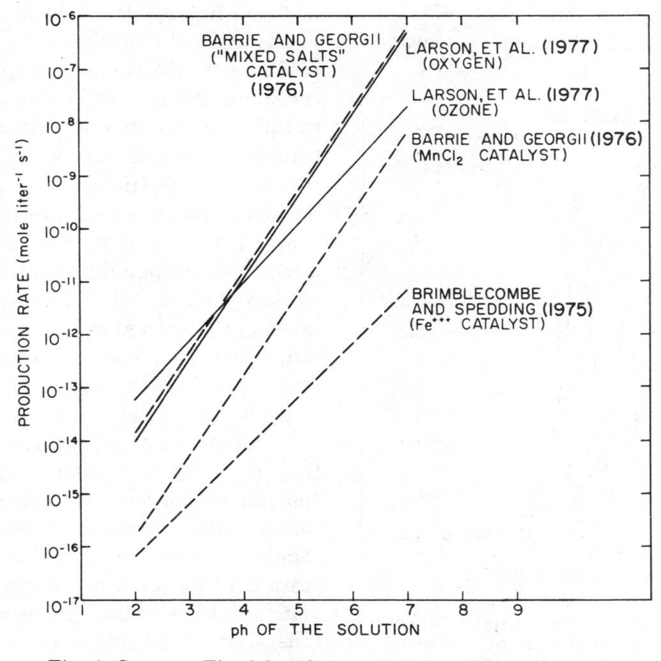

Fig. 4. Same as Fig. 3 but for a 'clean' (rural) atmosphere.

such as arsenite, antimonite, phosphite and cyanide, have an inhibitive effect (Schroeter, 1966). Organics such as alcohols, glycols, amines, amides, aldehydes, ketones and phenols, inhibit the oxidation at concentrations as low as 10^{-6} M (Schroeter, 1966; Fuller and Crist, 1941). These observations are generally explained in terms of radical chain termination as illustrated by Equation (5b) in the Backstrom mechanism. Since many of these inhibitors are known to be present in the atmosphere, the extension to the atmosphere of SO_2 oxidation rates measured in the laboratory is a precarious undertaking.

Larson (1976) has reported on a preliminary study in which the net effect of positive and negative catalysts found in urban and rural rainwater has been investigated. The rain samples were buffered to a pH of 5.5 and sulfite was then introduced. The measured oxidation rates were then compared with those for clean water of the same pH. The urban rainwater oxidized the sulfite at about the same rate ($\pm 25\%$) as the clean water. The oxidation rate in the one sample of rainwater collected at a rural site was about 30% slower than that for clean water. These results suggest that in some cases (at least for oxidation at pH 5.5) the net effect on the oxidation of SO_2 of the positive and negative catalysts in rainwater may be small. This work should be extended to lower oxidation pH's.

(b) *Kinetic mechanisms.* The kinetic mechanisms involved in the catalytic oxidation of SO_2 in solutions is so complex that it is only possible here to give the atmospheric scientist a brief overview and an indication of some of the problems involved.

Most of the recent studies of the catalyzed oxidation of SO_2 in solution have proposed a free radical chain mechanism similar to that for the uncatalyzed reaction (Brimblecombe and Spedding, 1974; Barrie and Georgii, 1976). It should be noted, however, that

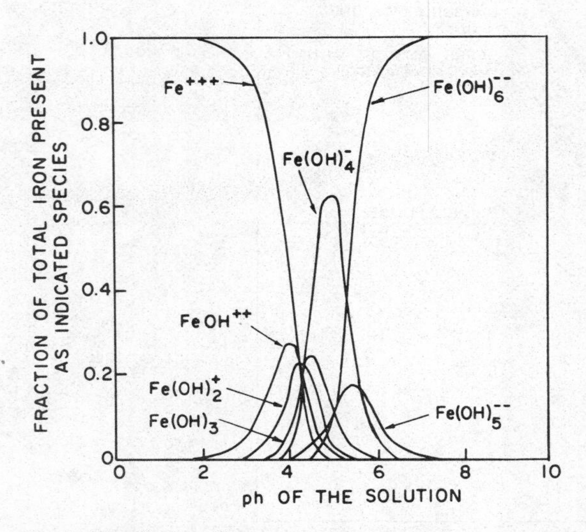

Fig. 5. Distribution diagram for iron III species in solution as a function of pH. Total iron present = 1×10^{-5} M. (From Singley and Sullivan, 1969).

alternative mechanisms exist, for example, the autocatalysis of metal sulfite (or bisulfite) complexes has been suggested (Bassett and Parker, 1951; Pritchett, 1960).

The chain initiating mechanism is uncertain. The simplest mechanism would be the metal ion oxidation of the sulfite in the Backstrom mechanism, with the metal ion being reoxidized by one of the chain radicals (e.g. $SO_5^{\cdot-}$) or by some other metal ion in solution with which the initiating ion forms a favorable redox couple. Such a reaction could explain Barrie and Georgii's 'mixed salt' effect, since either the ferrous or ferric ion will oxidize the reduced manganous ion in solution. However, several investigators have preferred to invoke the formation of a metal-sulfite (or bisulfite) complex which acts as the initial radical producing agent in the free radical chain. Thus Barrie and Georgii prefer the following mechanism:

$$(M^{+x}SA^{-(6-x)} + O_2 \rightleftharpoons (M^{+x}SA)^{-(5-x)} + O_2^{\cdot-} \quad (6)$$

$$O_2^{\cdot-} + H^+ \rightleftharpoons O_2H \quad (7)$$

$$(M^{+x}SA)^{-(5-x)} + SO_3^{2-} \rightleftharpoons (M^{+x}SA)^{-(6-x)} + SO_3^{\cdot-} \quad (8)$$

Where,

$$SA = (SO_3^{2-})_3 \text{ or } (HSO_3^-)_3$$

followed by propagation in two parallel chains, one with $SO_3^{\cdot-}$ and the other with O_2H. Matteson *et al.* (1969) and Pritchett (1960), on the other hand, have proposed an even more complicated mechanism involving autocatalysis of up to four intermediate metal–sulfite complexes. Finally, Bassett and Parker (1951) proposed the formation of complex sulfite ions carrying coordinated oxygen molecules, such as $[O_2 \rightarrow Mn(SO_3)_2]^{2-}$, which rapidly undergoes self-oxidation and reduction.

There are difficulties in applying either the simple metal ion initiated chain, or the more elaborate metal sulfite complex initiated chains and complex autocatalysis processes, to the atmosphere. With respect to metal ion initiation, one of the most widely studied catalysts, which is also most prevalent in the atmosphere, is ferric iron. In solution at low pH's ferric iron forms the hexaquo ion characteristic of the transition metals. This ion then rapidly undergoes hydrolysis as the pH rises and at any given pH several hydroxylated ferric complexes will coexist. Fig. 5, though not correct in detail, illustrates the complexity of this system. It can be seen that the ferric-hydroxyl species in solution are a strong function of pH, particularly in the pH range of atmospheric interest. The redox potential of the ferric ion changes as hydroxyl ligands are added, therefore, if metal ion initiated kinetics is applied to SO_2 oxidation in the atmosphere great care must be taken to employ only those laboratory rate constants which have been determined for pH's characteristic of those present in the atmosphere. Generally this has not been done.

The evidence for the formation of metal complexes in some oxidizing sulfite solutions, although indirect, is convincing. However, most of the studies which have produced this evidence have been concerned with rather concentrated sulfite solutions. Thus, the studies of Bassett and Parker (1951) and Hoather and Goodeve (1934b) were carried out with sulfite concentrations $\geq 10^{-3}$ M. Also, as noted previously, Hoather and Goodeve's study was of a heterogeneous system. That the formation of metal complex may not be the same in concentrated and dilute solutions is suggested by the work of Hansen *et al.* (1976). These investigators found evidence of Fe^{3+}-SO_3^{2-} complex formation in solutions of 0.1 M HCl if the concentration ratio of Fe^{3+} to SO_3^{2-} was 5:1 ($[SO_3^{2-}] = 10^{-3}$ M), but when the ratio was 1:1 there was no evidence of complex formation. This implies a 'solubility product' of $\sim 10^{-6}$. Hence, there may have been complex formation in the solutions studied by Bassett and Parker since the sulfite concentration was $\sim 10^{-2}$ M and the metal salt concentration varied from 10^{-3} to 3 M for the manganese and iron ions. However, since the concentrations of metal ions in the atmosphere are $\leq 10^{-4}$ M and sulfite concentrations are $\leq 10^{-4}$ M, metal complexes are unlikely to form. Sulfite complex formations would be even less likely at atmospheric pH's where considerable metal ion hydrolysis will have taken place (at least in the case of Fe^{3+}).

An additional complication is an anion effect. Bassett and Parker found that $MnSO_4$ solutions oxidized sulfite twelve times faster than $MnCl_2$ solutions, but for iron the chloride salts produced faster oxidation rates than the sulfates. This certainly suggests complex formation, at least in concentrated solution. It also indicates that before laboratory measured oxidation rates can be extrapolated to the atmosphere, various anion, as well as cation, concentrations must be known.

To summarize: we conclude that the kinetic mechanisms involved in the catalyzed oxidation of SO_2 in aqueous solutions are complex and depend on the concentrations and natures of both the reactants and the catalysts. Empirical studies suggest that the rate law is of a form:

$$\frac{d[SO_4^{2-}]}{dt} = K[M^+][H^+][SO_3^{2-}]. \qquad (10)$$

However, it is doubtful whether numerical values of K determined from laboratory experiments can be extrapolated with any reliance to the atmosphere.

3. EXTRAPOLATION OF LABORATORY RESULTS TO THE ATMOSPHERE

In the course of our review of laboratory studies of the oxidation of SO_2 in aqueous solutions, we have pointed out some of the difficulties in extrapolating the results to the atmosphere. A further complication is that more than one oxidation mechanism may be important in the atmosphere in any given situation*. For example, it can be seen from Fig. 4 that, for the rural case, uncatalyzed O_2 and O_3 oxidation and 'mixed salt' catalyzed oxidation are all competitive for $3 \leq pH \leq 4.5$. For the urban case (if the concentration of manganese in solution is negligible), uncatalyzed O_2 and O_3 oxidation and iron catalyzed oxidation are competitive for $3 \leq pH \leq 5$ (Fig. 3).

Most workers who have extrapolated laboratory results on the liquid phase oxidation of SO_2 to the atmosphere have made extremely simplifying assumptions. This is particularly true of the catalyzed rate studies. For example, Barrie and Georgii assumed an aerosol concentration of manganese in the urban atmosphere of 0.2 $\mu g\ m^{-3}$, and they considered a certain percentage of this to be soluble in cloud droplets present in concentrations of 0.1 g of liquid water per cubic meter of air. Even granting the low liquid water content assumed, there are problems associated with the distribution of manganese in cloud droplets. Lee and von Lehmden (1973) have pointed out that 60–87% of the manganese-containing particles in the air are greater than 1 μm dia. But the total concentration of particles with diameters greater than 1 μm is only about 10 cm^{-3} even in heavily polluted air (Whitby *et al.*, 1972). Concentrations of cloud droplets, on the other hand, are on the order of 100 cm^{-3}. Therefore, only a small fraction of cloud droplets will contain manganese.

Extrapolation of laboratory measurements on the uncatalyzed liquid phase oxidation of SO_2 to the atmosphere are on somewhat firmer ground because fewer variables are involved. However, to our knowledge, there has been only one attempt (Easter and Hobbs, 1974) to couple a definite oxidation mechanism with a cloud model. In this study several different uncatalyzed rate constants were used (Scott and Hobbs, 1967; McKay, 1971; Miller and de Pena, 1972) which encompass the range of values measured in the laboratory. As a result of their theoretical studies, Easter and Hobbs concluded that the production of sulfates in clouds could be the major sources of atmospheric sulfates. However, the rate laws which they employed did not all contain an $[H^+]^{1/2}$ dependent term. Furthermore, the activation energy which was chosen (7.7×10^4 J mol or 18.3 kcal mol^{-1}) and the ambient gas concentrations which were used (1 ppb $\leq [SO_2] \leq$ 3 ppb; 3 ppb $\leq [NH_3] \leq$ 6 ppb), can be criticized in the light of experimental data which have since become available. For example, SO_2 concentrations at cloud level down to 0.3 ppb have been observed and NH_3 concentrations of 1–2 ppb above altitudes of 2 km have been reported (Georgii

* In addition to the liquid phase oxidation which is the subject of this paper, SO_2 in the atmosphere is oxidized by heterogeneous processes and by homogeneous gas-phase reactions. For example, in power plant plumes, homogeneous gas-phase oxidation (Cox and Penkett, 1972), heterogeneous gas-solid oxidation (Novakov *et al.*, 1974) and heterogeneous liquid phase oxidation (Bassett and Parker, 1951) may all play a role.

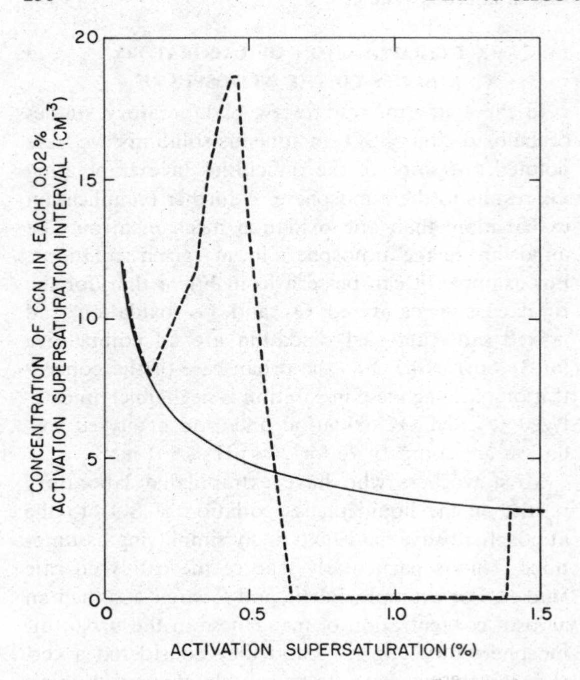

Fig. 6. Model calculations of cloud condensation nuclei (CCN) activated at a given supersaturation in air entering (solid lines) and leaving (dashed line) a wave cloud. (From Easter and Hobbs, 1974).

and Muller, 1974; Jost, 1974). Nevertheless, the general approach taken by Easter and Hobbs is sound and provides a foundation upon which future interactive chemical-cloud models can be based.

4. FIELD MEASUREMENTS

Direct evidence from field measurements of sulfate production due to the oxidation of SO_2 in droplets is very scanty. However, there have been a number of indirect observations which may be explained in terms of this mechanism.

Radke and Hobbs (1969), Dinger *et al.* (1970), Saxena *et al.* (1970) and Radke (1970) have all observed that the concentrations of cloud condensation nuclei (CCN) active at a given supersaturation are often higher in air from evaporating clouds than in ambient air which has not been involved in cloud formations. Radke and Hobbs suggested that the enhancement in CCN activity is due to sulfates, produced in the cloud droplets by the oxidation of SO_2, being deposited on the CCN within the cloud droplets, so that when the droplets evaporate the original CCN are left coated with an additional layer of sulfate. Measurements of sulfate particle concentrations in the vicinity of clouds by Jost (1974) showed peak concentrations below cloud base which could have been due to the sulfate particles from evaporating drops.

The field observations of Radke and Hobbs (1969) and Radke (1970) provided the stimulus for Easter and Hobbs' cloud model. In this model the changes in CCN activity due to SO_2 oxidation in small wave clouds are calculated. The results of one set of such calculations are shown in Fig. 6. In this case the lifetime of a droplet as it moved through the cloud was only about 4 mins, yet the change in the CCN activity is dramatic and readily measureable. Clearly, measurements of the CCN activity of aerosol before and after they are involved in the formation of clouds provides a very sensitive test of whether or not sulfates are being produced within the cloud droplets. However, in addition to such observations, measurements of sulfate concentrations in air before and after it has been involved in cloud formation are required to provide a direct test of the hypothesis that sulfate production in clouds is important in the atmosphere.

Field observations of SO_2 to sulfate conversion in the absence of clouds also provide indirect evidence for the importance of the liquid phase. It can be seen from Figs 7 (a) and (b) that the rate of conversion increases sharply as the relative humidity increases above about 70%. Since this is approximately the

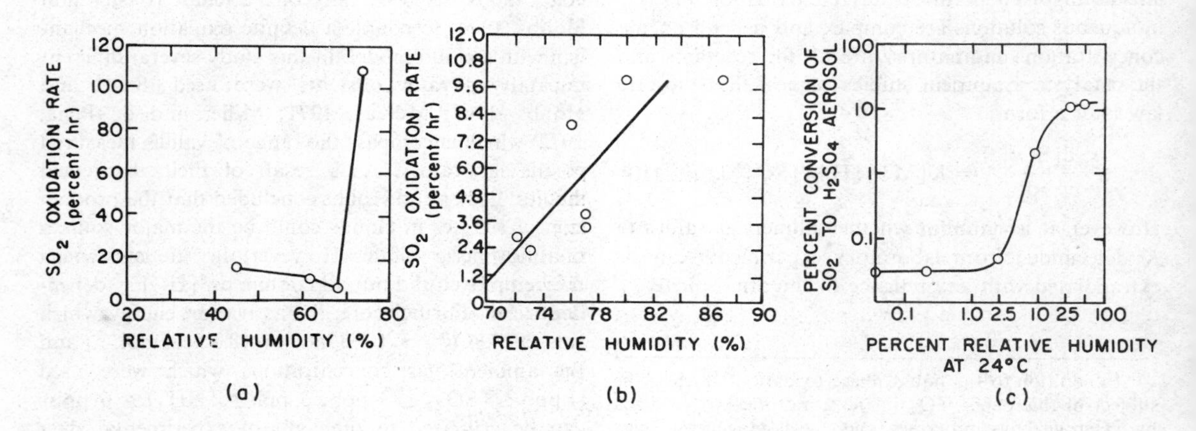

Fig. 7. SO_2 oxidation rates as a function of the relative humidities of the air (a) measurements in a coal power plant plume (from Gartrell *et al.*, 1963), (b) measurements in the atmosphere over the North Sea (from Smith and Jeffery, 1975), and (c) laboratory measurements of homogeneous gas phase reactions (from Cox, 1974).

relative humidity at which sea-salt and ammonium sulfate-rich continental aerosol deliquesce, it is plausible to associate the enhanced conversion rate at relative humidities above 70% with the liquid water of deliquescence. We recognize, of course, that the production of sulfates by homogeneous gas phase reactions also increases with increasing relative humidity (Wood *et al.*, 1975), but this mechanism shows an inflection point at much lower humidities and not much variation with relative humidity above about 50% (Fig. 7 (c)). We note, however, that since liquid-phase oxidation decreases sharply in acidic solutions, some haze droplets may not be very effective in oxidizing SO_2.

5. RECOMMENDATIONS FOR FURTHER RESEARCH

Our review of the oxidation of sulfur dioxide in aqueous systems suggests a number of avenues for future research which need to be pursued if we are to increase our understanding of the role which this mechanism plays in oxidizing sulfur dioxide in the atmosphere. These areas for future research are summarized below under laboratory, field and modeling studies.

5.1. *Laboratory studies*

We suggest the following laboratory studies (not listed in order of importance).

(a) Experiments to resolve the discrepancies between the measured values of the rate coefficients for the uncatalyzed oxidation of SO_2 in the presence of oxygen.

(b) Further catalyzed rate studies to determine the range of SO_2 and catalyst concentrations, pH's and temperatures over which the 'mixed salt' effect of Barrie and Georgii is effective.

(c) Barrie and Georgii (1976) mention, in passing, that the manganese catalyzed oxidation of SO_2 in water droplets is enhanced by ozone. However, we know of no published work on this reaction except at extremely high catalyst and ozone concentrations. Studies of this reaction are required at catalyst and ozone concentrations similar to those in the atmosphere.

(d) Further experimental studies are required to determine the actual kinetic mechanism(s) responsible for the catalyzed reaction over a wide range of SO_2 and catalyst concentrations and temperatures. As we have pointed out, the efficiences of the various proposed mechanisms (e.g., oxidation by metal-ion complexes) will be strongly dependent on the pH and the concentrations of various chemical species in the solution, as well as on temperature. Hence, under various conditions, different mechanisms may dominate. Until the actual chemical species in solution have been identified (e.g., by u.v. absorption), and the kinetic mechanisms elucidated, the extension of laboratory results to the atmosphere will remain uncertain.

(e) Many of the laboratory studies of SO_2 oxidation in solution have been carried out on bulk solutions rather than on clouds of small droplets which are of interest in the atmosphere. The principal reason for this is the difficulty of suspending droplets for long periods of time. The Atmospheric Cloud Physics Laboratory aboard the SPACELAB will contain an expansion chamber for forming a cloud of droplets, as well as aerosol generating and measuring equipment. The near zero-g conditions aboard the SPACELAB should be ideal for carrying out studies of chemical reactions in clouds of suspended droplets under controlled conditions for long time periods. (Experiments of this type have been proposed to NASA by the writers).

5.2. *Field studies*

We suggest the following field studies (not listed in order of importance).

(a) Many more measurements are required in the atmosphere of sulfate production which can be associated unambiguously with liquid phase oxidation. An ideal situation for such measurements are wave clouds, where aerosol measurements can be made in the air entering and leaving the wave clouds. Through controlled releases of SO_2 and NH_3 into a wave cloud, 'laboratory-type' experiments can be carried out. (The authors are now undertaking experiments of this type).

(b) Most studies to date have assumed that the chemistry of atmospheric liquid water (e.g., its pH) is controlled by the concentrations of sulfite, sulfate and ammonia. While this is true for laboratory studies on bulk solutions, the chemistry of small cloud droplets may be strongly affected by hygroscopic salts such as NaCl. Measurements of the pH of cloud water, and of the concentrations of relevant chemical species (e.g., manganese, arsenite) at cloud levels, are urgently required.

(c) It can be seen from Fig. 4 that in the 'clean' atmosphere the uncatalyzed oxidation of SO_2 should dominate. Larson and Harrison (1977) calculate, on the basis of laboratory experiments, that the ozone oxidation could produce sulfate with an NH_4^+/SO_4^{2-} molar ratio down to 1.5, whereas, oxidation by oxygen produced sulfate with an NH_4^+/SO_4^{2-} molar ratio of about 2. Hence, in principle, it should be possible to determine which reaction is dominant in the atmosphere at the time of sampling by determining the NH_4^+/SO_4^{2-} molar ratio in atmospheric sulfate samples. (This technique could also be extended to differentiate between homogeneous gas-phase oxidation, which presumably may produce sulfates with an NH_4^+/SO_4^{2-} molar ratio of less than 1.5, and liquid-phase oxidation).

5.3. *Modeling studies*

(a) The various proposed mechanisms for liquid phase oxidation of SO_2 need to be incorporated into realistic cloud models in order to predict sulfate production and the related effects on various properties of the atmospheric aerosol. Comparisons between model predictions and field measurements may be used to deduce the dominant oxidation mechanism(s).

(b) As confidence is developed in such interactive chemical-cloud models (through field testing) they should be used to predict sulfate production on regional and global scales.

Acknowledgements—Preparation of this review paper was supported under Contract RP330-1 from the Electric Power Research Institute. We wish to thank Drs. T. Larson and H. Harrison for useful discussions. Contribution No. 445 Atmospheric Sciences Dept.

REFERENCES

Alyea H. and Backstrom H. (1929) Inhibitive action of alchohols on the oxidation of sodium sulfite. *J. Am. chem. Soc.* **51,** 90–109.

Backstrom H. (1927) Chain theory of negative catalysis *J. Am. chem. Soc.* **49,** 1460–1472.

Backstrom H. (1934) The chain mechanism in the autoxidation of sodium sulfite solutions. *Z. physik. Chem.* **B25,** 99–121.

Barrie L. A. and Georgii H. W. (1976) An experimental investigation of the absorption of sulphur dioxide by water drops containing heavy metal ions. *Atmospheric Environment* **10,** 743–749.

Barron C. H. and O'Hern H. A. (1966) Reaction kinetics of sodium sulfite oxidation by the rapid mixing method. *Chem. Engng. Sci.* **21,** 397–404.

Bassett H. and Parker W. (1951) The oxidation of sulfurous acid. *J. chem. Soc.* **47,** 1540–1560.

Beilke S., Lamb D. and Müller J. (1975) On the uncatalyzed oxidation of atmospheric SO_2 by oxygen in aqueous systems. *Atmospheric Environment* **9,** 1083–1090.

Bracewell J. M. and Gall D. (1967) The catalytic oxidation of sulphur dioxide in solution at concentrations occurring in fog droplets. *Proceedings of the Symposium on the Physio–Chemical Transformation of Sulphur Compounds in the Atmosphere and the Formation of Acid Smogs*, Mainz, Germany.

Brimblecombe P. and Spedding D. J. (1974) The catalytic oxidation of micromolar aqueous sulphur dioxide—I (oxidation of dilute solutions of iron (III)). *Atmospheric Environment* **8,** 937–945.

Cheng R. T., Corn M. and Frohliger J. O. (1971) Contribution to the reaction kinetics of water soluble aerosols and SO_2 in air at ppm concentrations. *Atmospheric Environment* **5,** 987–1008.

Cox R. A. (1974) Particles formation from homogeneous reactions of sulphur dioxide and nitrogen dioxide. *Tellus* **26,** 235–240.

Cox R. A. and Penkett S. A. (1972) Aerosol formation from sulfur dioxide in the presence of ozone and olefinic hydrocarbons. *J. chem. Soc.* **68,** 1735–1753.

Cunningham P. T., Halt B. D., Hubble B. R., Johnson S. A., Siegel S., Wilson W. I., Cafasso F. A. and Burris L. (1975) Chemical Engineering Division Environmental Chemistry Annual Report, July 1974–June 1975. Argonne National Laboratory Report ANL-75-51.

Dinger J. E., Howell H. B. and Wojciechowski T. A. (1970) On the source and composition of cloud nuclei in a subsident airmass over the North Atlantic. *J. atmos. Sci.* **27,** 791–797.

Easter R. C. and Hobbs P. V. (1974) The formation of sulfates and the enhancement of cloud condensation nuclei in clouds. *J. atmos. Sci.* **31,** 1586–1594.

Freiberg J. (1974) The mechanism of iron catalyzed oxidation of SO_2 in oxygenated solutions. *Atmospheric Environment* **9,** 661–672.

Fuller E. C. and Christ R. H. (1941) The rate of oxidation of sulfite ions by oxygen. *J. Am. Chem. Soc.* **63,** 1644–1650.

Gartrell J. E., Thomas J. W. and Carpenter S. B. (1963) Atmospheric oxidation of SO_2 in coal-burning power plant plumes. *Amer. ind. Hyg. Ass. Quart.* **24,** 113–120.

Georgii H. W. and Müller W. J. (1974) On the distribution of ammonia in the middle and lower troposphere. *Tellus* **26,** 180–184.

Haber F. and Willstatter R. (1931) Autoxidation (VI): action of light on sulphite solutions in absence and presence of oxygen. *Z. phys. Chem.* **B18,** 203–210.

Halperin J. and Taube H. (1952a) The transfer of oxygen atoms in oxidation–reduction reactions, IV. *J. Am. Chem. Soc.* **74,** 380–382.

Halperin J. and Taube H. (1952b) The transfer of oxygen atoms in oxidation–reduction reactions, III. *J. Am. Chem. Soc.* **74,** 375–379.

Hansen L. D., Whiting L., Eatough D. J., Jensen T. E. and Izatt R. M. (1976) Determination of sulfur (IV) and sulfate in aerosols by thermometric methods. *Analyt. Chem.* **48,** 634–638.

Harrison H., Larson T. and Hobbs P. V. (1975) Oxidation of sulfur dioxide in the atmosphere: a review. *Proceedings, International Conference on Environmental Sensing and Assessment.* Las Vegas, Nevada, Sept 1975.

Hartdridge H. and Roughton F. J. W. (1923) Measurement of the rates of oxidation and reduction of hemoglobin. *Nature* **111,** 325–326.

Hayon E., Treinin A. and Wilf J. (1972) Electronic spectra, photochemistry, and autoxidation mechanism of the sulfite–bisulfite–pyrosulfite systems. The SO_2^-, SO_3^-, SO_4^- and SO_5^- radicals. *J. Am. chem. Soc.* **94,** 47–57.

Hoather R. C. and Goodeve C. F. (1934a) The oxidation of sulphurous acid. III Catalysis by manganous sulphate. *Trans. Faraday Soc.* **30,** 1149–1156.

Hoather R. C. and Goodeve C. R. (1934b) The oxidation of sulphurous acid. IV Catalysis by a glass powder containing manganese and iron. *Trans. Faraday Soc.* **30,** 1156–1161.

Johnstone H. F. and Coughanour D. R. (1958) Absorption of sulfur dioxide from air. *Ind. Eng. Chem.* **50,** 1169–1172.

Jost D. (1974) Aerological studies on the atmospheric sulfur budget. *Tellus* **26,** 206–212.

Junge C. and Ryan T. G. (1958) Study of SO_2 in Oxidation in solution and its role in atmospheric chemistry. *Q. J. R. met. Soc.* **84,** 46–55.

Karraker D. G. (1963) The kinetics of the reaction between sulfurous acid and ferric ion. *J. phys. Chem.* **67,** 871–874.

Larson T. (1976) The kinetics of sulfur dioxide oxidation by oxygen and ozone in atmospheric hydrometeors. *Ph.D. Thesis, University of Washington*, Seattle, W.A.

Larson T. and Harrison H. (1977) Acidic sulfate aerosols: formation from heterogeneous oxidation by O_3 in clouds. *Atmospheric Environment.* **11,** 1133–1141.

Larson T., Horike N. and Harrison H. Oxidation of sulfur dioxide by oxygen and ozone in aqueous solution: a kinetic study with significance to atmospheric rate processes. *Atmospheric Environment* (in press).

Lee R. E. and von Lehmden D. J. (1973) Trace metal pollution in the environment. *J. Air. Pollut. Control Assoc.* **23,** 853–857.

Lloyd R. M. (1968) Oxygen isotope behavior in the sulphate-water system. *J. geophys. Res.* **73,** 6099–6110.

Matteson M. J., Stöber W. and Luther H. (1969) Kinetics of the oxidation of sulfur dioxide by aerosols of manganese sulfate. *I. & E.C. Fund.* **8,** 677–687.

McKay H. A. C. (1971) The atmospheric oxidation of sulfur dioxide in water droplets in the presence of ammonia. *Atmospheric Environment* **5,** 7–14.

Miller J. M. and dePena R. (1972) Contribution of scavenged sulfur dioxide to the sulfate of rainwater. *J. geophys. Res.* **30,** 5905–5916.

Miyamoto S. and Kaya T. (1930) The velocity of solution of oxygen in water, I. *Bull. Chem. Soc. Japan* **5,** 123–136.

Neytzell-deWilde F. G. and Taverner L. (1958) Experiments relating to the possible production of air oxidizing acid leach liquor by autooxidation for the extraction of uranium. *2nd U.N. International Conference on the Peaceful Uses of Atomic Energy Proceedings* **3,** 303–317.

Novakov T., Chang S. G. and Harker A. B. (1974) Sulfates as pollution particulates: catalytic formation on carbon (soot) particles. *Science* **186,** 259–261.

Penkett S. A. (1972) Oxidation of SO_2 and other atmospheric gases by ozone in aqueous solutions. *Nature (Phys. Sci.)* **240,** 105–106.

Pritchett P. W. (1960) The catalyzed reaction of oxygen with sulfurous acid and its effect. Ph.D. thesis, University of Delaware, DE.

Radke L. F. (1970) Field and laboratory measurements with an improved automatic cloud condensation nucleus counter. *Preprints Amer. Meteor. Soc. Conference on Cloud Physics,* Fort Collins, CO.

Radke L. F. and Hobbs P. V. (1969) Measurement of cloud condensation nuclei, light scattering coefficient, sodium-containing particles, and Aitken nuclei in the Olympic Mountains of Washington. *J. atmos. Sci.* **26,** 281–288.

Reinders W. and Vlès S. I. (1925) The catalytic oxidation of sulfites. *Rec. Trav. Chem.* **44,** 249–268.

Saxena V. K., Burford J. W. and Kassner J. L. (1970) Operation of a thermal diffusion chamber for measurements on cloud condensation nuclei. *J. atmos. Sci.* **27,** 73–80.

Schroeter L. C. (1963) Kinetics of air oxidation of sulfurous acid salts. *J. pharm. Sci.* **52,** 559–563.

Schroeter L. C. (1966) *Sulfur dioxide; application in foods, beverages, and pharmaceuticals,* p. 56. Pergamon Press, Oxford.

Scott W. D. and Hobbs P. V. (1967) The formation of sulfate in water droplets. *J. atmos. Sci.* **24,** 54–57.

Singley J. E. and Sullivan J. H., Jr. (1969) Reactions of metal ions in dilute solution: recalculation of hydrolysis of iron (II) data. *J. AWWA* **61,** 190–192.

Smith F. B. and Jeffery G. H. (1975) Airborne transport of sulphur dioxide from the U.K. *Atmospheric Environment* **9,** 643–659.

Sutherland J. (1977) Private communication.

Van der Heuval P. P. and Mason B. J. (1963) The formation of ammonium sulphate in water droplets exposed to gaseous sulphur dioxide and ammonia. *Q. J. R. meteor. Soc.* **89,** 271–275.

Whitby K. T., Husar R. B. and Lin B. Y. H. (1972) The aerosol size distribution of Los Angeles smog. *J. Colloid Interface Sci.* **39,** 177–204.

Winkelmann D. (1955) Die electrochemische messung von exodationsgeschwindigkeit von Na_2SO_3 durch gelasten sauerstaff. *Z. Electrochemie* **59,** 891–895.

Winter E. R. S. and Briscoe H. V. A. (1951) Oxygen atom transfer during the oxidation of aqueous sodium sulfite. *J. Am. chem. Soc.* **73,** 496–497.

Wood W. P., Castleman A. W., Jr. and Tang I. N. (1975) Mechanisms of aerosol formation from SO_2. *J. Aerosol. Sci.* **6,** 367–374.

Atmospheric Environment Vol. 12, pp. 255–261. Pergamon Press 1978. Printed in Great Britain.

ADSORPTION AND OXIDATION OF SULFUR DIOXIDE ON PARTICLES

A. LIBERTI, D. BROCCO and M. POSSANZINI

Laboratorio Inquinamento Atmosferico C.N.R., Via Montorio Romano, 36 Rome, Italy

(*Received* 10 *August* 1977)

Abstract—The role of atmospheric particulated matter in affecting atmospheric SO_2 and its reactions has been investigated. A variety of dusts of various sources (urban particulated matters and stack emissions of industrial plants) have been characterized in terms of their physical and chemical properties and submitted to a SO_2 adsorption process at room temperature and desorption at 175°C.

The interaction between particles and SO_2 can occur through two processes: adsorption and conversion to sulphate. The extent of these processes depends upon the particles, chemical composition and their nature, which can be defined in terms of pH, titratable acidity, surface area, humidity and degree of surface coverage by adsorbed components.

SO_2 adsorption by particles is the primary process which occurs in two steps, only the first one being apparently of environmental significance. Humidity has an important role in the adsorption, the higher its value the higher results the amount of adsorbed SO_2. The behaviour of atmospheric dusts collected in different areas and seasons is very similar, the reaction constant of the first order process being 4–5 $\times$ 10^{-2} min^{-1}.

Both 'fresh' particles coming from stacks, which do not carry SO_2, as well as 'aged' atmospheric particles adsorb SO_2 the relative extent of the process being mainly determined by the dust reaction. The former in most cases do not release SO_2 by heating, this behaviour being taken as an example of chemisorption, whereas the latter lose SO_2 by heating, SO_2 being retained by only a physical bond.

Conversion to sulphate occurs with a very high rate on particles coming from industrial emissions, the alkaline reaction being the determining factor, whereas it does not take place on urban atmospheric dusts to an appreciable extent.

This mechanism is supported by measurements by differential thermal analysis and by X-ray photoelectron spectroscopy.

Though, in the atmosphere, it is impossible to discriminate various effects due to homogeneous and heterogeneous reactions, the main interaction between SO_2 and particulated matter is adsorption, most catalytic reactions occurring at high temperature and most probably at the chimney outlet.

Particulate matter is the most abundant pollutant and its composition and effect on the environment has been the object of many investigations. The behavior of particulate matter is usually related to its concentration and to date little weight has been put on reactions which either might occur on it or might be affected by its presence. The role of atmospheric particulate matter vs SO_2 has been long debated and it has been observed that in the presence of several finely ground particles the sulfur dioxide concentration decreases; a catalytic effect is usually attributed to this material. Reactions of sulfur dioxide with particles have been assumed to yield sulfate and sulfuric acid (Cheng *et al.*, 1971; Corn and Cheng, 1972). A theoretical treatment of particle-catalyzed oxidation of atmospheric pollutants including SO_2 has been given to assess the possible importance to the overall pollution problem (Judeikis and Siegel, 1973).

However, in an open atmosphere it is impossible to discriminate between homogeneous and heterogeneous reactions as well as gas-to-particle conversion, including heteromolecular nucleation, condensation and thermal coagulation, as all processes occur simultaneously and are greatly influenced by humidity and by meteorological conditions.

It must be stressed that it is incorrect to discuss atmospheric particulate matter as a unique system with identical properties in all areas, as its origin can be quite different as a result of dispersion and condensation. The former process occurs by grinding or atomization of solids and liquids and by the transfer of powders in suspension through the air stream, whereas the latter takes place when supersaturated vapour condenses to give formation of non-volatile products. Both processes occur simultaneously and it is obvious that specific industrial emissions can affect the nature and composition of atmospheric particulate matter.

On account of its complex composition and nature, it seemed essential to define the characteristics of particulate matter and to submit dusts of various sources to an SO_2-adsorption and desorption process to obtain information of general validity on the interaction of SO_2.

The aim of this investigation was the study of the behavior of particulate matter collected from the urban environment as well from stacks, in terms of

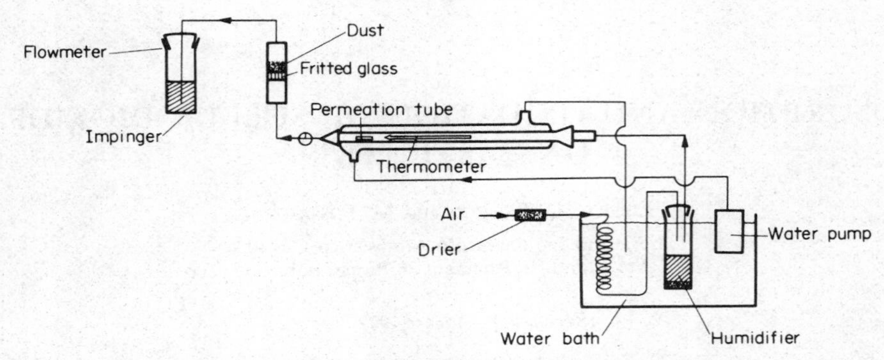

Fig. 1. Experimental set-up for SO_2 adsorption by dusts.

sulfur dioxide adsorption and oxidation at ambient temperature.

EXPERIMENTAL

The experimental set-up is shown in Fig. 1. A dust sample (100 mg) was placed into a reactor consisting of a glass tube 15 cm length, 1 cm i.d., with a sintered glass soldered in the middle part. The reactor was set in a vertical position and air flowing through the fritted glass kept the dust floating giving conditions similar to a fluid bed.

Purified air at a constant flow (50 ml min^{-1}) at atmospheric pressure was passed into a hygrostat to acquire a constant humidity. Sulfur dioxide was added as the gas stream flowed through a thermostated condenser at 20°C containing a permeation tube. In most experiments the rate of permeation was 1.15 $\mu g\ min^{-1}$ and the SO_2 concentration in the gas stream became 8.6 ppm.

Measurements were carried out either with a continuous monitoring of SO_2 with a flame photometric analyzer (Monitor Labs. mod. 8450) or discontinuously by adsorption of SO_2 into a sodium tetrachloromercurate solution using the West and Gaeke method. With the former procedure the intensity of the current versus time was recorded and the amount of SO_2 obtained by integration. With the second method the amount of SO_2 adsorbed was calculated as the difference between the amount measured and that obtained by running a blank with the reactor empty. A typical graph obtained in the adsorption process with a continuous monitoring of SO_2 is shown in Fig. 2. Plots of this type have been used to obtain kinetic information on the adsorption process.

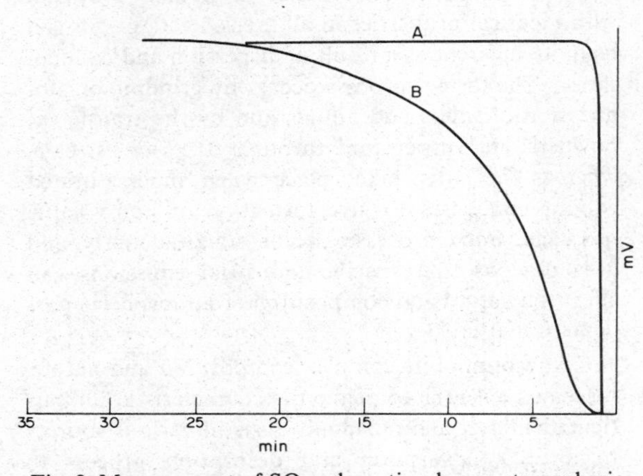

Fig. 2. Measurement of SO_2 adsorption by an atmospheric dust (sample 1) vs time, SO_2 being analyzed by flame photometric detector. Operating conditions: influent SO_2 conc. 8.6 ppm; gas flow rate—50 ml min^{-1}; temperature 20°C; r.h.—35%. Curve A: empty reactor, curve B: reactor containing 100 mg of atmospheric dust.

In all experiments, the SO_2 concentration was kept constant whereas gas streams with various humidities were obtained by placing solutions of various composition in the hygrostat at 20°C (35% Hum: sat. solution $CaCl_2 \cdot 6H_2O$—55% Hum: sat. solution $Ca(NO_3)_2 \cdot 4H_2O$—76% Hum: sat. solution NaCl).

To obtain information on the release of SO_2 by atmospheric particles, desorption experiments were carried out by using the set-up shown in Fig. 3. Nitrogen was made to flow (50 ml min^{-1}) through a heater kept at 175°C and then into the absorbing solution where SO_2 was determined.

Airborne samples collected from various sources were investigated by evaluating their behaviour in terms of the above described adsorption–desorption process. They can be classified as:

(a) urban particulate matter collected directly from the atmosphere by means of an electrostatic precipitator or from ventilation plants of buildings in various seasons;

(b) black emissions from various sources: fly ash from oil and coal electric power furnaces, from a cement factory, and from iron and steel plants. Some of them were sampled in the chimney and some from electrostatic filters before the chimney.

The above materials were characterized in terms of the following parameters:

pH; titratable acidity or alkalinity; surface area and relative porosity; humidity; degree of coverage of the surface by other adsorbed components or reaction products of former reactions.

The pH of a dust can be measured with a glass electrode by stirring a sample of dust (150 mg in 100 ml de-ionized water). The titratable acidity or alkalinity was measured by electrometric titration of the water suspension using Gran's plots according to Brocco *et al.* (1976).

The surface area could be determined by nitrogen adsorption (Devitofranceso and Liberti, 1966). No general

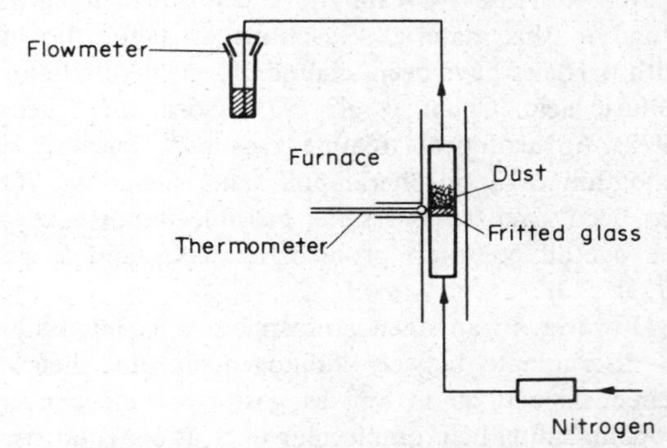

Fig. 3. Apparatus used for desorption measurements.

Table 1. pH, titratable acidity, surface area, S%, C% and free SO_2

Particulate matter	pH	H^+ ($meq\ g^{-1}$)	OH^- ($meq\ g^{-1}$)	Surface area ($m^2\ g^{-1}$)	S% (W/W)	C% (W/W)	SO_2 ($mg\ g^{-1}$)
1 Atmospheric dust (Rome Summer 74)	7.27	—	0.357	8.1	3.3	22.6	0.7
2 Atmospheric dust (Rome Winter 74–75)	6.68	—	0.350	1.2	2.8	22.4	0.9
3 Atmospheric dust (Milan STMI 139)	6.00	—	0.142	1.1	4.7	42.0	1.2
4 Atmospheric dust (Turin Winter 75)	5.50	—	0.066	1.2	2.8	—	1.4
5 Dust from cement factory stack (Bescia)	10..77	—	0.470	2.7	1.3	5.7	0
6 Dust from manganese industry stack (Yugoslavia)	8.66	—	0.088	1.8	1.3	—	0
7 Dust from blast furnace PHF_2 (France)	9.93	—	0.210	1.0	0.5	15.0	0
8 Dust from iron agglomeration plant (France)	8.10	—	0.152	3.8	0.6	3.5	0
9 Dust from LD furnace, PAC (France)	10.24	—	0.238	2.0	0.4	0.3	0.2
10 Soot from experimental oil furnace (Pisa)	3.41	0.446	—	19.0	7.4	69.3	2.4
11 Ash from oil burner in electric power station (Monfalcone)	2.07	1.880	—	0.6	13.7	—	1.4
12 Ash from naphtha combustion (Rome)	1.50	7.100	—	1.4	18.5	—	0.7
13 Fly ash from a naphtha burner (Rome)	6.80	—	0.015	15.8	0.2	—	0.1
14 Fly ash from coal fired power plant (Mannheim D)	6.40	—	—	—	—	—	—

procedure was followed to evaluate the degree of coverage of the surface: solvent extraction was used to determine organics of a dust and gas elution in a closed system to strip off volatile compounds.

RESULTS AND DISCUSSION

Analytical results are collected in Table 1. With the exception of samples 7, 8 and 9, which have a high iron content, other metals such as manganese and aluminum, which are usually catalytically active, are present in the dusts only as minor components. Atmospheric dusts collected in summer time are almost neutral whereas those sampled in winter have an acid reaction. The pH and the surface activity of particulates from stack emissions differ noticeably from each other. They can be definitely basic as cement and LD furnace dusts or acid as most fly ashes.

Adsorption–desorption process

Results for the adsorption process described are collected in Table 2. They refer to standard, controlled experimental conditions where air with a relative humidity of 55% at 20°C containing 8.6 ppm SO_2 flowed for 30 min through a 100 mg sample.

Table 2. Adsorption and desorption of SO_2 by particulate matter

Dust No.	$\mu g\ SO_2\ g^{-1}$* adsorbed	SO_2 adsorbed (%)	SO_2 desorbed (%)†
1	75	23	100
2	152	44	100
3	127	37	100
4	105	30	100
5	345	100	0
6	96	28	—
7	345	100	0
8	345	100	0
9	345	100	0
10	18	6	25
11	7	2	0
12	13	4	—
13	331	96	—
14	60	18	0

* After 30 min flowing air containing 8.6 ppm SO_2 and r.h. 55%.

† Desorption carried out at 125°C.

The pH and the particulates' acidity are the most important factors which affect SO_2 adsorption, dust with an alkaline reaction not being saturated in the experimental conditions described. However, all materials adsorb SO_2, the amount adsorbed increasing with increase in the humidity.

A plot of SO_2 adsorption on winter atmospheric dust measured at various humidities, is shown in Fig. 4.

Measurements carried out with the same dust after benzene extraction at different humidities are also shown in the same graph. In both cases the higher the humidity, the higher is the amount of SO_2 adsorbed.

Desorption experiments indicate that urban atmospheric dust desorbed SO_2 with an almost quantitative yield, whereas in most dusts collected from stacks no desorption occurs on heating at 175°C under a nitrogen flow.

A comparison between the results of Table 2 and the content of free SO_2 allows particulate matter to be classified in terms of its 'aged' or 'fresh' nature. A stack emission particulate can be defined as a 'fresh' system which is going to equilibrate in the atmosphere, whereas atmospheric particulate matter, unless

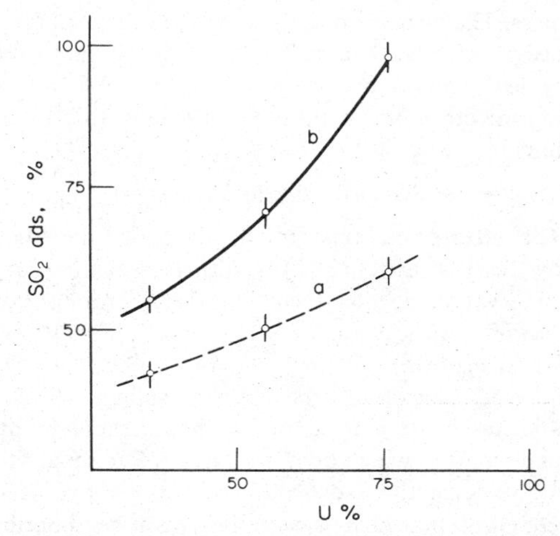

Fig. 4. SO_2 adsorption by atmospheric dust (sample 2) at various humidity levels: (a) untreated sample; (b) the same sample after benzene extraction.

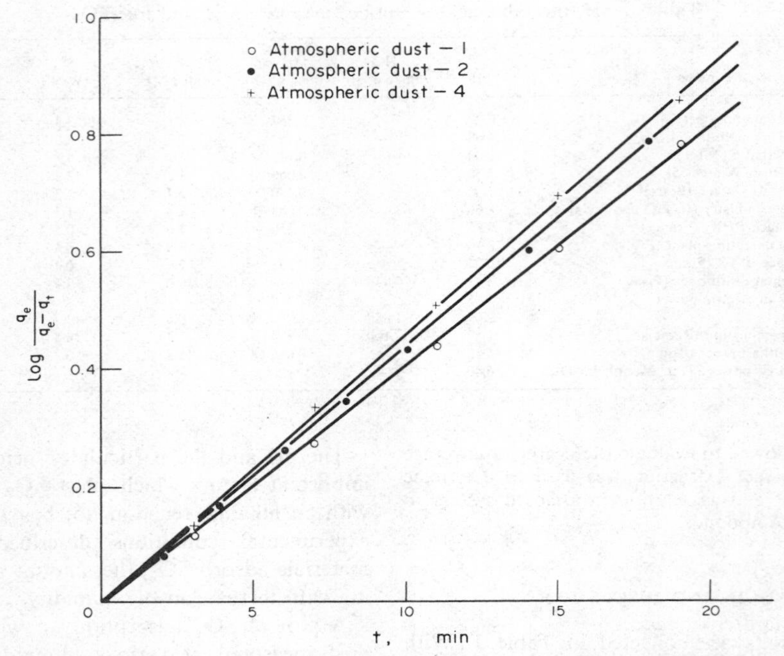

Fig. 5. SO_2 adsorption rate by various atmospheric dusts; conditions as in Fig. 2.

directly affected by a specific emission, is an 'aged' system where equilibrium with air components has been reached. Particulate matter sampled from a stack does not show the presence of SO_2, which is always found in 'aged' systems. Both dusts might be reactive but reactions in a 'fresh' system proceed faster and are much more extensive.

The extent of SO_2 adsorption depends mainly upon the titratable alkalinity and, accordingly, samples 5–9, which are 'fresh' dust, consisting mainly of various metal oxides, have a high capacity for SO_2 uptake which is taken directly from the atmosphere. However, adsorption also occurs on industrial dust having a weak alkalinity or an acid surface. Fly ashes should in no sense be considered homogeneous material and they can be assumed to consist of acid and alkaline centers. The former do not take part in the adsorption process, whereas the latter, though comparatively very few, should be considered responsible for the reaction with SO_2. Samples 10–14 exhibit this specific behavior.

Interaction of SO_2 *with particulate matter*

The adsorption experiments described in Fig. 2 allow the interaction of SO_2 with particulate matter to be evaluated; this seems to occur through two processes:

(1) adsorption;

(2) oxidation of adsorbed SO_2 to sulfate.

Measurements carried out with atmospheric particulate matter, which does not adsorb SO_2 quantitatively, indicate that adsorption occurs in two steps. The first step, which is very fast, can be described by the equation:

$$lg\,[q_e/(q_e - q_t)] = K_I t$$

where q_e is the amount of SO_2 adsorbed at the equilibrium per g of particulate matter and q_t the amount of SO_2 adsorbed at time t. Constant K_I is almost the same for atmospheric dusts, being in the range $4–5 \times 10^{-2}\,\text{min}^{-1}$ as is shown in Fig. 5. The reaction seems to be of the first order and it appears that adsorption is a quite fast event.

To this process, which is believed to be of environmental significance, a second rather slow process occurs which is rendered evident by the additional uptake of SO_2 by a sample dust previously treated with this gas. It has been observed that after equilibration with SO_2 a dust sample adsorbs a further amount of SO_2 after a certain time (1–4 days). This process is described by the relationship:

$$lg\, q_{ads} = K_2 t$$

where q_{ads} is the amount of SO_2 adsorbed at time t, $K_2 = 6 \times 10^{-5}\,\text{min}^{-1}$.

Given the dynamics of the atmosphere, this process may be of no relevance for the environment.

In all dusts where the adsorbed SO_2 is quantitatively desorbed by heating at 175°C, no chemisorption occurs and no sulfate formation was observed. On the other hand, in dust samples where SO_2 adsorbed is not released at 175°C, such as the alkaline dust sampled in a cement factory stack, no free SO_2 was detected and the sulfate weight increase corresponds to the amount of sulfur dioxide adsorbed.

The information obtained from the adsorption–desorption process are supported by the application of t.g.a. and d.t.a. and by X-ray photo-electron spectroscopy. Thermal analysis of atmospheric particulate matter shows an endothermic process due to desorption of SO_2, which occurs at about 125°C. The energy

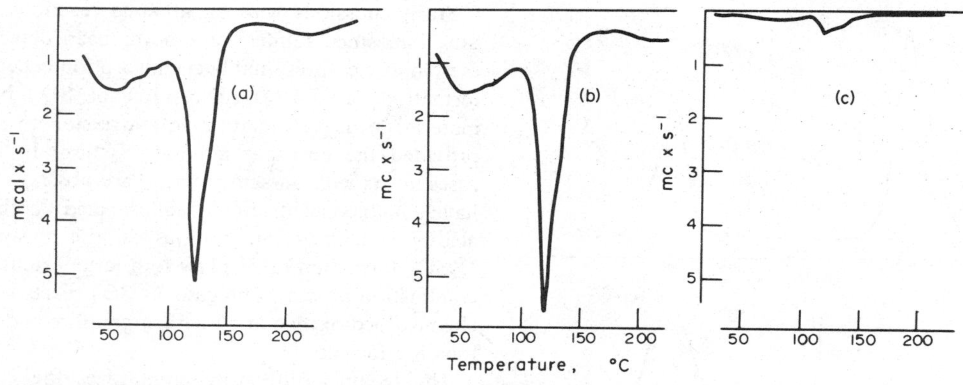

Fig. 6. (a) thermal analysis of an atmospheric dust (sample 1 of Table 1). (b) thermal analysis of the same sample after SO_2 adsorption (120 $\mu g\, g^{-1}$). (c) differential thermal analysis (DTA) of (b) vs (a).

involved in this process corresponds to a physical adsorption. By exposing the sample to sulfur dioxide the desorption peak increases, as is shown in Fig. 6.

The desorption peak is not observed in samples taken from stacks of various emissions, whereas it is found in soot and combustion ashes.

By keeping urban atmospheric dusts or fly ashes either in a desiccator or in the open atmosphere for long periods of time (three months) thermal analysis shows exactly the same desorption peak, indicating that the SO_2 content does not change appreciably. This indicates that in most dusts oxidation of sulfur dioxide does not occur to an appreciable extent. If, however, a dust is alkaline, the conversion of adsorbed SO_2 to sulfate is very fast.

Additional information was obtained by using X-ray photoelectron spectroscopy (XPS). The study of the kinetic energy of photoelectrons expelled from a dust sample irradiated with mono-energetic X-ray provides a direct measurement of the electron binding energy of one element. Since the binding energies are modified by the valence electron distribution, it is possible to obtain a general picture of the various sulfur containing species (Allegrini and Mattogno, 1977).

The XPS spectrum of particulate material from an atmospheric dust (sample 3) shows a peak corresponding to a binding energy of 154 eV, due to Si_{2s}, a small band at 162 eV due to S(−2), a barely visible break assigned to *S*(+4) and a large peak at 168 eV characteristic of S(+6) (Fig. 7).

The XPS spectrum of another urban dust has a similar shape (sample 1) which shows only variations of the peak size due to differences in surface concentration. On the other hand, the spectrum of a 'fresh' dust collected from a cement factory (sample 5) does not show the shoulder due to S(+4), in agreement with our findings (Fig. 8(a)). In agreement with this, other stack emissions have a similar behavior, as appears from sample 6 (dust from a manganese plant) where the S(+4) peak is not detected (Fig. 8(b)).

ROLE OF ORGANICS IN THE ADSORPTION PROCESS

Adsorption experiments carried out with dusts, which were first extracted with benzene to bring most

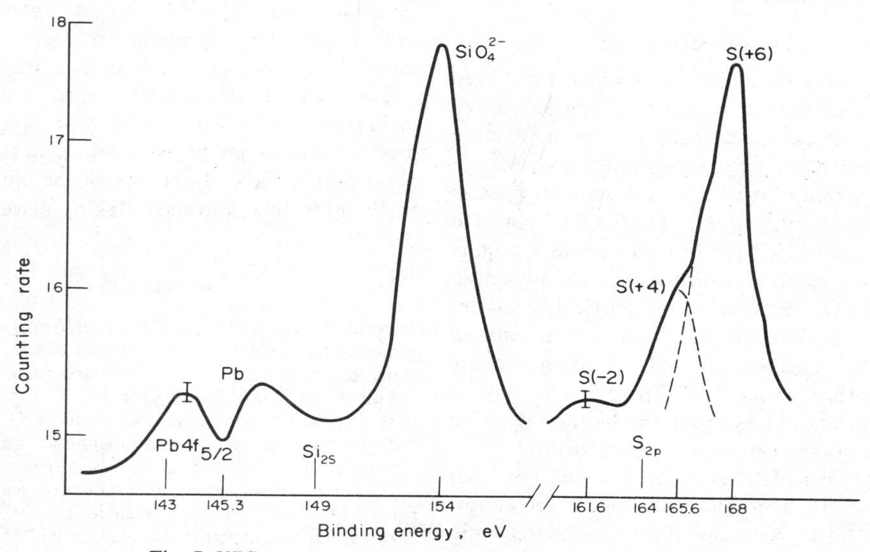

Fig. 7. XPS spectrum on an atmosphere dust (sample 3).

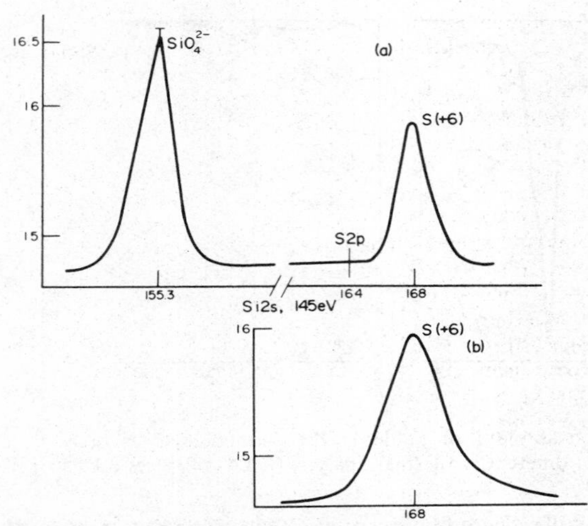

Fig. 8. XPS spectra: (a) dust from a cement factory stack (sample 5); (b) dust from a manganese plant stack (sample 6).

organics into solution, show that in most cases a larger amount of sulfur dioxide is adsorbed. Experiments on dusts 1–4 show that more than twice the amount of SO_2 is adsorbed by the same dust after extraction. The larger amount of SO_2 adsorbed means that a larger number of active sites occupied by organic compounds becomes available for SO_2 adsorption. Since several organic compounds have been detected in atmospheric particulate matter (Brocco *et al*., 1975), the problem arises whether, besides oxidation to sulfate, SO_2 might also react with other species. However, the concentration of various sulfur-containing compounds is small and it appears that they might come from exhaust gases and stack emissions, rather than through combination of SO_2 with organic compounds from various sources. Though it is also reported (Conte *et al*., 1976) that by equilibrating particulate matter with SO_2, sulfonic acid derivatives are formed, the formation of these compounds seems to be of minor importance in the conversion phenomena of SO_2.

CONCLUSIONS

The variety of atmospheric dusts which have been examined and of the experimental conditions tested allows some general conclusions on the interaction between atmospheric particulate matter and SO_2 at ambient temperature to be drawn. The main process SO_2 undergoes is adsorption, which takes place to the extent of several $\mu g\, g^{-1}$ of dust, through a quite rapid first order reaction, which is strongly influenced by humidity. SO_2 remains on the particulate matter as such, or in a hydrated form, so that it is introduced into the living organism not only in the gas phase but also adsorbed on particles. These results may be of some importance in assessing the biological action of dust, as it always carries a layer of free SO_2.

Conversion to sulfate at room temperature does not take place to a considerable extent on atmospheric dusts unless aerosols due to industrial emissions of a specific nature affect their composition.

Stack emissions with an alkaline reaction adsorb SO_2 but, since sulfites have never been detected in these particles, presumably chemisorption occurs with formation of sulfate. Adsorption of SO_2 by this material leads to quantitative conversion to sulfate, provided the dust has a basic reaction. In certain cases, dusts with an acid reaction are observed which have a high concentration of sulfuric acid. As the high acidity is usually found in ashes, this means that H_2SO_4 formation takes place to a large extent in the combustion process. The gaseous SO_2 reacts with the chemisorbed oxygen at a high temperature and sulfate ions are formed.

The technical literature emphasizes the catalytic action of atmospheric particulate matter, in connection with carbon particles (Novakov and Chang, 1974), vanadium pentoxide (Barbaray *et al*., 1977) and ferric oxide (Chung and Quon, 1973). In most cases, however, experiments have been carried out at fairly high temperature and it is likely that these conditions occur only at the outlet of a chimney where the high reactants concentration and the high temperature may favor the oxidation rate of SO_2. At ambient temperature, however, the heterogeneous non-photochemical sulfate formation reaction strongly depends upon the reaction of the aerosol surface.

An acid–base reaction is in most cases the determining factor, whereas on carbon it seems that sulfur dioxide adsorbs at active sites of oxygen complex on carbon, the main surface species being $\overset{|}{\text{—C}}\text{=O}$, —C—OH and —COOH (Smith, 1959). An oxidation mechanism, which involves surface radicals (especially the hydroxyl group) might be therefore introduced (Yue *et al*., 1976).

It can be concluded that whereas in an open atmosphere SO_2 adsorption is by far the most important process, a competitive reaction occurs at the chimney outlet. Sulfate-bearing particles are emitted into the atmosphere, together with SO_2, as a primary pollutant. The ratio SO_4^{2-}/SO_2 depends upon the combustion regime, the type of the emissions, the size and especially the nature of the surface area of particles.

These conclusions are in accord with the results of Forrest and Newman (1977) who pointed out that most oxidation might occur in a coal-fired power plant during the early history of the plume with virtually no further conversion taking place down wind.

REFERENCES

Allegrini I. and Mattogno G. (1977) Analysis of environmental particulate matter by means of ESCA. Paper prepared for the Seminar on Fine Particulates (ECE-ONU) Villach (Austria) 17–21 October.

Barbaray B., Contour J. P. and Mouvier G. (1977) Sulfur dioxide oxidation over atmospheric aerosols; X-ray photoelectron spectra of sulfur dioxide adsorbed on V_2O_5 and carbon. *Atmospheric Environment* **11**, 351–356.

Brocco D., Liberti A. and Possanzini M. (1976) Determinazione dell'acidità del materiale particolato in atmosfere urbane e industriali. *Ann. Ist. Super. Sanità* **12**, 49–55.

Brocco D., Liberti A. and Possanzini M. (1975) Adsorption desorption of sulfur dioxide by air-borne particulate matter. Presented at the 3th Technical Symposium COST Project 61a, Ispra, 18–20 November.

Cheng R. T., Corn M. and Frohliger J. O. (1971) Contribution to the reaction kinetics of water-soluble aerosols and SO_2 in the air at ppm concentrations. *Atmospheric Environment* **5**, 987–1008.

Chun K. C. and Quon J. E. (1973) Capacity of ferric oxide particles to oxidize sulfur dioxide in air. *Envir. Sci. Technol.* **7**, 532–538.

Conte C., Devitofrancesco G. and Starace G. (1976) Behaviour of SO_2 in the atmosphere-interactions with the particulate matter. *Atmospheric Pollution*, pp. 243–253. Elsevier, Amsterdam.

Corn M. and Cheng R. T. (1972) Interaction of sulfur dioxide with insoluble suspended particulate matter. *J. Air Pollut. Control Ass.* **11**, 870–875.

Devitofrancesco G. and Liberti A. (1966) Determining the dust concentration by surface measurement. *Staub* **26**, 13–15.

Forrest J. and Newman L. (1977) Further studies on the oxidation of sulfur dioxide in coal-fired power plant plumes. *Atmospheric Environment* **11**, 465–474.

Judeikis H. S. and Siegel S. (1973) Particle-catalyzed oxidation of atmospheric pollutants. *Atmospheric Environment* **7**, 619–631.

Novakov T. and Chang S. G. (1974) Catalytic oxidation of SO_2 on carbon particles. Presented at the 76th National AICHE Meeting, Tulsa, Oklahoma, 10–13 March.

Smith R. M. (1959) The chemistry of carbon–oxygen surface compounds. *Q. Rev.* **13**, 287–395.

Yue G. K., Mohnen A. V. and Kiang C. S. (1976) Sulfur dioxide to sulfate conversion in the atmosphere. 12th Int. Colloquium Poll. Atm, Paris.

Atmospheric Environment Vol. 12, pp. 263–271. Pergamon Press 1978. Printed in Great Britain.

THE FORMATION AND STABILITY OF SULFITE SPECIES IN AEROSOLS

D. J. EATOUGH, T. MAJOR, J. RYDER, M. HILL, N. F. MANGELSON, N. L. EATOUGH and L. D. HANSEN

Thermochemical Institute, Contribution No. 119, and Departments of Chemistry and Physics, Brigham Young University, Provo, UT 84602, U.S.A.

and

R. G. MEISENHEIMER and J. W. FISCHER

Lawrence Livermore Laboratory, Livermore, CA 94550, U.S.A.

(*First received* 15 *June* 1977 *and in final form* 5 *October* 1977)

Abstract—Recent epidemiological and animal toxicological studies indicate that reactions between SO_2 and metal containing aerosols result in the formation of respiratory irritants. These studies point out the importance of understanding in detail the chemical species formed by such interactions. Using a combination of thermometric, ESCA and PIXE analysis techniques, it has been demonstrated that both inorganic and organic S(IV) species are stable constituents of aerosols associated with pollution sources containing SO_2 and transition metals or with pollution sources resulting from the combustion of fossil fuels. The data indicate the inorganic sulfite species are present as complexes with Fe(III), Cu(II), Zn(II), and possibly Pb(II). The concentration of these inorganic sulfite species is 10 to 30% of the sulfate concentration in primary aerosols produced by smelters. These inorganic sulfite species tend to be evenly distributed over the various particle sizes. In contrast, the inorganic sulfite in primary aerosols produced by fossil fuel burning sources tends to exist in the $< 3\,\mu m$ size range and can vary from a negligible to a major fraction of the sulfur species produced. The factors which control this variability are presently unknown. The principal mode of formation of such species in the ambient atmosphere appears to be via SO_2 absorption. Oxidation of S° or S(-II)species to form inorganic sulfite complexes or oxidation of the sulfite species to sulfate are both extremely slow, with time constants on the order of months. Aerosol samples collected from the plume, stack, or flue lines of coal burning facilities or collected in New York City or rural Utah produce sulfite when hydrolyzed with dilute aqueous acid. It is postulated this sulfite is produced from organic-SO_2 adducts in the sample. These organic S(IV) containing species are predominantly found in the respirable size range and are present at from 5 to 50% of the sulfate concentration. It is probable that some of these S(IV) species play an important role in the removal of SO_2(g) from the atmosphere to form sulfur containing aerosol species.

INTRODUCTION

The existence of S(IV) species in ambient aerosols has been suggested by photo electron spectroscopic analysis of samples collected in the Los Angeles basin (Novakov, 1972) and generally throughout California urban areas (Craig, 1974), and by wet chemical analysis of samples collected in a copper smelter (Hansen, 1975; Smith, 1976). Based on the results obtained at the copper smelter, it has been postulated by Hansen *et al.* (1975) that the observed species in aerosols produced by primary copper smelters are present due to the formation of Fe(III)–SO_3^{2-} complexes which should be stable to air oxidation. From existing data it is not possible to determine if the S(IV) species formed by smelter and fossil fuel facilities are chemically similar, or if chemical agents other than Fe(III) may stabilize the S(IV) oxidation state. The relative rate of adsorption of SO_2 to form S(IV) compounds and of oxidation of these S(IV) species in the ambient atmosphere are also unknown.

The identification of stable S(IV) compounds in aerosols will possibly alter our interpretations of both epidemiological and plume SO_2 removal studies. Stable sulfite species could be expected to have etiological significance in studies relating respiratory health problems to measured pollutant levels (Colucci, 1976) and may provide one plausible mechanism for purported SO_2-aerosol synergisms (Hansen, 1974). In addition, most studies on SO_2 removal by aerosols in plumes measure SO_2 and/or sulfate aerosol concentrations only. Even though the S(IV) species can be expected to constitute a minor fraction of the aerosol sulfur component, failure to identify changes in concentrations of these species or interferences by these species in analysis for sulfate may lead to erroneous conclusions regarding the fate of the SO_2 which is converted to aerosol sulfur species during transport of a plume.

Using a combination of thermometric, proton induced X-ray emission (PIXE), and photoelectron spectroscopy (ESCA) analysis techniques, we have studied the formation of S(IV) species in aerosols collected from the flue lines or stacks of copper and lead smelters, and coal and oil fired power plants, from the plume of coal and oil fired power plants, from the in-house and nearby ambient environment of a copper smelter, and from New York City. The

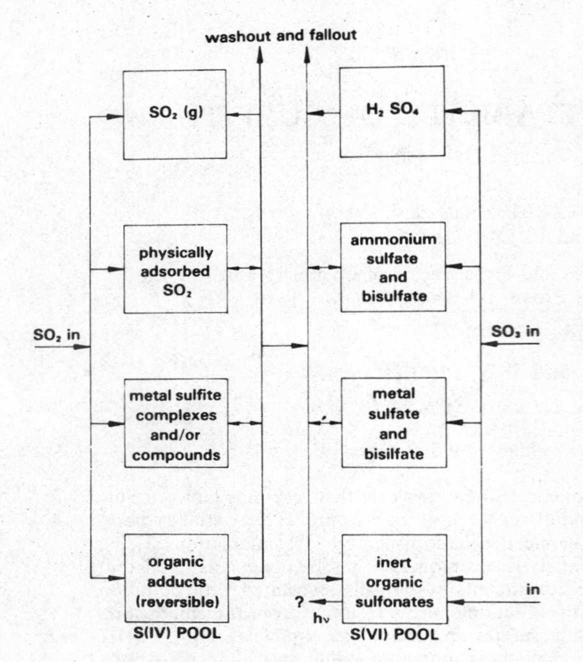

Fig. 1. Flow diagram of possible SO_x reactions in the atmosphere.

data indicate that stable S(IV) species may form in aerosols from each of these sources. Aerosols from the smelters, coal fired stations, and urban area, but not the oil fired stations, appear to contain stable inorganic sulfite species. In addition, the data suggest organic-SO_2 compounds are present in aerosols from coal fired stations and the urban area.

A flow diagram of sulfur chemistry resulting from the interaction of SO_x with particulate matter suggested by these results is given in Fig. 1. The S(IV) POOL is composed of SO_2(g), physically adsorbed SO_2 and aerosol S(IV) species resulting from the reaction of SO_2 with metal ions or organic functional groups. These complexes are stable and are only slowly (on the time scale of months) converted to S(VI) compounds. The rate of formation of the stable S(IV) species from the interaction of SO_2 with aerosols is presently not known. These species often represent a significant fraction of the total S(IV) + S(VI) content of samples originating from the smelters (0–25%), coal fired stations (0–70%), or urban area (10–50%) studied, and should not be ignored in the study of SO_2 reactions or sulfur balance in these types of aerosols.

METHODOLOGY

Analytical techniques

Calorimetry. The analytical techniques used for the analysis of sulfite and other reducing agents and of sulfate in collected aerosol samples have been described by Hansen *et al.* (1976). The portion of an aerosol sample extractable with 0.1 M HCl, 2.5 mM $FeCl_3$ is analyzed for sulfite by a thermometric redox titration with dichromate. Other dichromate oxidizable species are each independently detected by this method. Sulfate is determined calorimetrically by precipitation with $BaCl_2$. The samples were extracted at room temperature in an ultrasonic bath for twenty minutes. All solutions used in the analysis were prepared, stored and used under an argon atmosphere.

PIXE. Elements present in the extractant solution with atomic number 16 (sulfur) or greater were analyzed by proton induced X-ray emission, PIXE (Johansson, 1975; Nielsen, 1976). 5 μl of the extractant solution was placed on a mylar target, dried in air, and analyzed. The methods used for data reduction have been described by Nielsen *et al.* (1976).

ESCA. Samples for analysis were prepared by sprinkling the collected particulate matter onto a double face scotch tape on a sample platen and obtaining the spectrum with a Hewlett-Packard X-ray photoelectron spectrometer. High resolution data of the energy region represented by the sulfur 2p photoelectron lines were obtained to determine S(-II), S°, S(IV), and S(VI) species present *on the surface* of the sample.

Other techniques. Sulfite concentrations in some of the smelter flue dust samples were determined by West–Gaeke and ion chromatographic techniques through collaborative arrangements with other laboratories. Sulfite was determined by extraction of the sample with a tetrachloromercurate solution, at room temperature in an ultrasonic bath for 2–5 minutes, followed by conventional analysis for S(IV) using the procedure described by West and Gaeke (1956). The ion-chromatographic analyses were performed on a Dionex Ion Chromatograph. The samples were extracted with buffered solution using a cold ultrasonic bath and analyzed as rapidly as possible. Analyses of Na_2SO_3 spiked solutions were used to assist in identifying the SO_3^{2-} peak. Details of the analytical procedure have been given by Stevens *et al.* (in press).

Sample sources and collection

Smelter samples. Size classified particulate samples were collected from within the workroom environment (Smith, 1976) and vicinity (Hansen, 1975) of a copper smelter on glass fiber filters (without organic binder) mounted in high volume samplers equipped with Anderson cascade impactor sampling heads. In addition, flue dust samples were obtained in bulk from the flue lines of five different primary copper and lead smelters.

Fossil fuel samples. A Sierra high volume sampler with a Model 235 Sierra Cascade impactor sampling head operating on the rooftop of a building adjacent to a small coal fired heating plant with no controls was used to collect size classified particulate samples on quartz fiber filters when meteorological conditions favored collection of the plume. Flue dust samples were also collected in bulk from the flue line of the same plant. In addition, particulate matter was sampled onto quartz fiber filters isokinetically from the

Table 1. Results of analyses of particulate or flue dust samples from smelters, given as mmol g^{-1} in the total sample or wt% of the total sample in parentheses*

Sample type	Smelter type	S(VI), SO_4^{2-}	S(IV), SO_3^{2-}	Fe	Cu	Zn	Pb
Particulate	Cu1	0.61 (5.9)	0.24 (1.9)	0.54	1.09	0.18	0.04
Flue dust	Cu2	1.61 (15.5)	0.09 (0.7)	3.04	3.43	0.54	0.11
Flue dust	Cu3	0.83 (8.0)	0.19 (1.5)	2.06	1.01	0.31	0.23
Flue dust	Pb1	1.06 (10.2)	0.25 (2.0)	0.77	0.05	0.15	0.94
Flue dust	Pb2	0.18 (1.7)	0.00 (0.0)	0.36	0.03	1.44	1.80
Flue dust	Pb3	0.72 (6.9)	0.11 (0.9)	0.97	0.46	1.77	1.00

* Uncertainties in the values, based on replicate analysis of samples, are approximately $\pm 0.1 \pm 0.2\,x$, $0.02 \pm 0.15\,x$, and $0.05 \pm 0.1\,x$ mmol g^{-1} for x = S(VI), S(IV), and metal concentrations, respectively.

stack of a large coal fired generating station equipped with a wet scrubber and in bulk from the flue line of the same facility. Both facilities burn low sulfur (0.5%) western coals. Bulk samples were also obtained from the flue lines of two oil fired power plants which burn high sulfur, high vanadium oil. Samples from the plume of an oil fired power plant were collected by aircraft equipped with a high volume sampler with quartz fiber filters.

Ambient samples. Ambient samples collected in midtown New York City (Hansen, 1977) with Anderson cascade impactor sampling heads or with high volume total particulate matter samplers were obtained for analysis.

Sample analysis

All collected samples were extracted with a 0.1 M HCl, 2.5 mM $FeCl_3$ solution and analyzed calorimetrically for $K_2Cr_2O_7$ oxidizable species and for sulfate. Extracts from all smelter samples, all oil fired station samples, and selected urban and coal fired station samples were analyzed by PIXE to determine elemental content. ESCA analyses were performed on the smelter flue dust samples and on one of the plume samples from a coal fired heating plant. Comparative analyses by the West–Gaeke method and by ion chromatography were performed on the smelter flue dust samples. The effect of shelf storage in ambient air on the reduced sulfur species was studied for the flue dust samples from the smelter and coal fired stations, and on plume samples collected from the coal fired heating plant. The possible presence of acid hydrolyzable sulfite adduct compounds in samples associated with the combustion of fossil fuels was also studied by allowing extracted species to age for several days in the extractant solution after undissolved solids were removed by filtering or centrifugation.

RESULTS

Sulfite species were commonly seen in samples collected from primary lead and copper smelters, as summarized in Table 1. Extensive cross checks of the sulfite species in these samples have been made by several techniques, as summarized in Table 2. It may be noted that the concentrations of SO_3^{2-} given in Tables 1 and 2 for these samples are not identical. This is because the data in Table 1 were obtained when the sample was first collected, and the results in Table 2 were obtained when the samples were several months old. Changes in SO_3^{2-} concentration with time for these samples are summarized in Fig. 2. Similar changes in sulfur species have also been shown by ESCA analysis of the >7 μm cut from one of the filters in sample Cu1. One month after collection ESCA analysis indicated the ratios of S(VI): S(IV):S(-II) in the sample were 1.0:1.0:0.8. After storage of the target for six months, the spectrum was again determined and the measured ratio was 1.0:1.7:0.0.

Information on the particle size distribution of sulfite, sulfate and trace elements in aerosols produced by a copper smelter were obtained from a series of 8 samples collected within the copper smelter located near the southern tip of the Great Salt Lake and three

Table 2. Comparative analysis of smelter particulate and flue dust samples*

Sample	SO_4^{2-}, wt% by calorimetry	SO_3^{2-}, wt% by calorimetry	SO_3^{2-}, wt% by West–Gaeke	SO_3^{2-}, wt% by ion chromatography	Sulfur species seen by ESCA	Mole ratio of species in ESCA
Cu1	5.9 ± 0.4	1.9 ± 0.2	NA†	NA	S(VI), S(IV), S(-II)	1:1:1
Cu2	15.5 ± 0.6	1.10 ± 0.15	0.00 ± 0.00	0.00 ± 0.00	S(VI)	1
Cu3	11.0 ± 0.4	1.3 ± 0.3	0.00 ± 0.00	NA	S(VI), S(IV), S(-II)	5:1:1
Pb2	5.0 ± 0.6	0.00 ± 0.00	0.02 ± 0.01	0.00 ± 0.00	S(VI), S(-II)	5:1
Pb3	4.6 ± 0.1	1.00 ± 0.08	0.85 ± 0.07	0.5 ± 0.2	S(VI), S(IV), S(-II)	9:2:1

† NA = not analyzed.

* The uncertainties are given as the standard deviation of the mean for replicate analysis. Uncertainties in the ESCA ratios are estimated to be ±0.2.

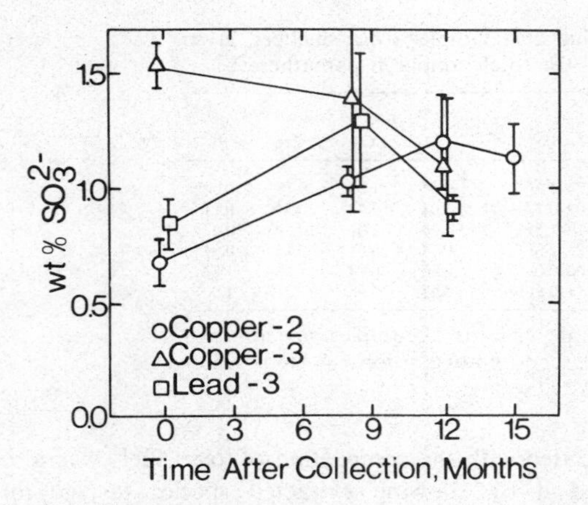

Fig. 2. Sulfite concentrations in smelter flue dust samples as a function of time.

samples collected in Magna, a small community 7 km from the smelter, as summarized in Figs. 3 and 4. The error bars in the figure are the standard deviations from the mean for the various samples. In addition to the samples obtained in Magna, samples were also obtained at sites located 18, 25 and 45 km from the smelter on different days when the wind direction was from the smelter towards the individual sampling site. The levels of SO_3^{2-} in the smelter were correlated with Fe(III) (correlation coefficient 0.87, $p < 0.001$) as shown in Fig. 5. In contrast, the levels of SO_3^{2-} in the samples obtained in the surrounding area were not correlated with Fe(III) but were correlated with the sum of Fe(III) and Cu(II) (correlation coefficient 0.50, $p < 0.01$) as shown in Fig. 6. Significantly, in none of these samples did the concentrations of SO_3^{2-} exceed the concentrations of Fe plus

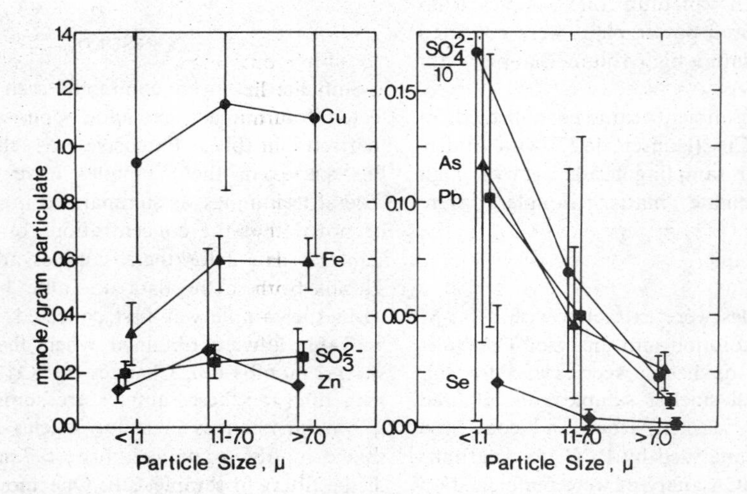

Fig. 3. Concentrations of species in particulate matter collected from within a copper smelter as a function of particle size. The variation in the collected samples is given as the standard deviations and result mainly from inhomogeneity of the sampled air. Analysis uncertainty for any given sample is ±10–20%.

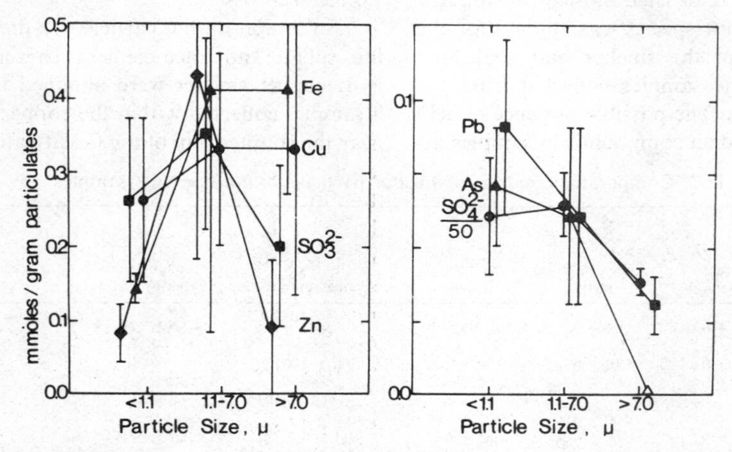

Fig. 4. Concentrations of species in particulate matter collected from the plume of a copper smelter as a function of particle size. The variation in the collected samples is given as the standard deviations and result mainly from inhomogeneity of the sampled air. Analysis uncertainty for any given sample is ±10–20%.

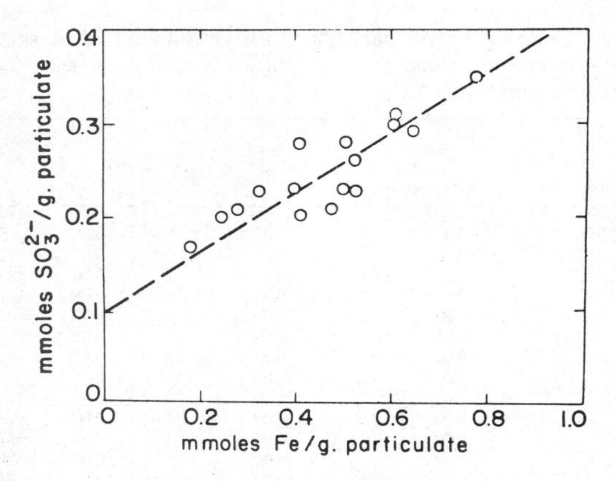

Fig. 5. Plot of SO_3^{2-} vs Fe(III) in particulates collected from the workroom of a smelter.

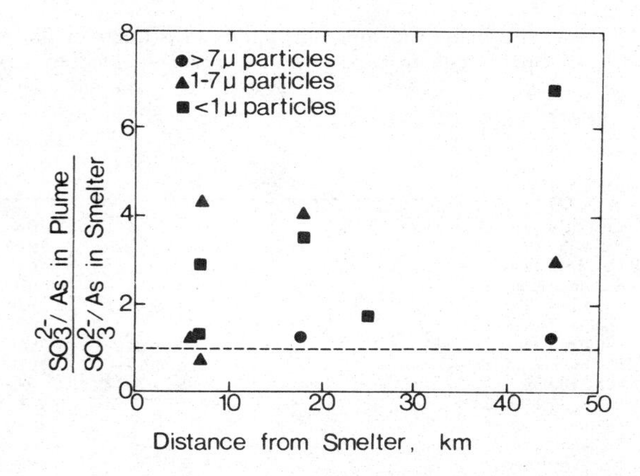

Fig. 7. Ratio of SO_3^{2-}/As in plume collected samples to SO_3^{2-}/As in smelter workroom collected samples as a function of distance from the smelter.

Cu. The set of samples obtained in the vicinity (and presumably representative of the plume) of the copper smelter may give an indication that SO_2 adsorption occurs outside the source. If it is assumed that the arsenic content of aerosols produced by the smelter is constant and much higher than background arsenic levels, a comparison of the S(IV):As ratio found in the plume to that found in the smelter should show whether or not the particulate sulfite concentrations change as a function of distance from the source and hence time in the plume. This comparison is plotted vs distance of the sampling site from the smelter in Fig. 7. The results indicate increased levels of sulfite in ambient compared to in-smelter collected samples. Furthermore, there may be an indication that the relative concentrations of sulfite are increasing in the $<1\ \mu m$ particle size range as residence time in the plume increases. It should be stressed that these ambient samples were all collected on different days and do not represent monitoring of a single stable plume as a function of residence time of plume aerosol. Nonetheless, the results do suggest that increases in aerosol sulfite levels may be occuring with time. More detailed plume transformation studies should be conducted to determine if this is the case.

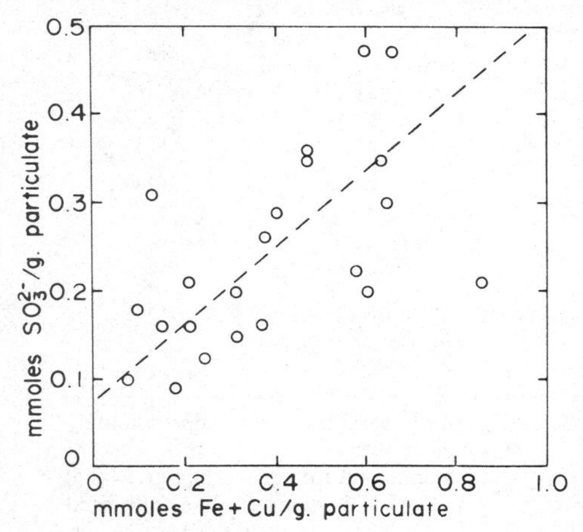

Fig. 6. Plot of SO_3^{2-} vs Fe(III) + Cu(II) in particulates collected from the plume of a smelter.

The presence of sulfite in particulate samples from fossil fuel burning sources was investigated by analysis of flue line, stack and plume samples from coal and oil burning stations, and of atmospheric samples collected in New York City and in a rural area in Utah effected by coal burning sources but not by the smelter. In addition to the thermometric redox titration (by $Cr_2O_7^{2-}$) of SO_3^{2-} ($\Delta H_{ox} = -23$ kcal/eq) and Fe(II) ($\Delta H_{ox} = -30$ kcal/eq) seen in some of the samples, other species were commonly seen with ΔH_{ox} values of -18, -14, and -11 kcal/eq. These other species could not be explained by the inorganic components of the aerosol nor by such organic compounds as olefins, aldehydes, or alcohols. It is possible, however, that some of the species are polyhydroxy aromatic compounds. For example, we found that hydroquinone and catechol (1,4- and 1,2-dihydroxybenzene) are stoichiometrically oxidized by dichromate with ΔH_{ox} values of -17.6 and -18.3 kcal/eq, respectively. Resorcinol (1,3-dihydroxybenzene) is oxidized in dilute solution with a ΔH_{ox} value of -10 kcal/eq. Polynuclear aromatic analogs of these compounds would be expected to have similar ΔH_{ox} values. To determine if organic SO_2 adduct compounds might also be present, aerosol samples were extracted in 0.1 M HCl, 2.5 mM $FeCl_3$ solution, separated from the undissolved sample and the extracted compounds stored in this hydrolyzing solution for several days at room temperature and reanalyzed by thermometric titration with dichromate. The results, summarized in Table 3, indicate that the species present with a ΔH_{ox} value of -14 kcal/eq is removed by this process and sulfite is formed. In most cases this balance is stoichiometric, in other cases sources of sulfite appear to be present.

Atmospheric concentrations of both inorganic sulfite and the apparent organic SO_2 adduct are highest in the respirable size range of particulate samples collected from the plume of a coal fired heating plant and in New York City as summarized in Table 4. It should be noted that the size data for aerosols from

Table 3. Formation of inorganic sulfite in extracts of flue dust or atmospheric particulate matter under hydrolysing conditions. As sulfate concentrations were observed to be constant with time, the concentrations are expressed as the mole ratio to sulfate*

Sample	Time after extraction, days	Inorganic sulfite $\Delta H = -23$ kcal/eq	Other dichromate oxidizable species, ΔH −18 kcal/eq	−14 kcal/eq	−11 kcal/eq
Coal fired heating	0	0.00	0.00	0.08	0.05
Plant, flue line	3	0.05	0.00	0.01	0.03
Coal fired Power	0	0.00	0.00	0.00	0.00
Station, flue line-1	7	0.01	0.07	0.00	0.08
Coal fired power	0	0.00	0.00	0.00	0.00
Station, flue line-2	7	0.04	0.08	0.00	0.00
Oil fired power	0	0.00	0.00	0.00	0.01
Station, flue line-1	7	0.00	0.00	0.00	0.01
Oil fired power	0	0.00	0.00	0.00	0.00
Station, flue line-2	7	0.00	0.00	0.00	0.00
Oil fired power	0	0.00	0.00	0.00	0.00
Station, plume	9	0.00	0.00	0.00	0.00
New York ambient	0	0.00	0.00	0.12	0.00
Aerosol-1	7	0.12	0.12	0.00	0.10
New York ambient	0	0.07	0.13	0.00	0.09
Aerosol-2	6	0.31	0.11	0.00	0.10
New York ambient	0	0.00	0.00	0.26	0.00
Aerosol-3	7	0.20	0.00	0.00	0.00
New York ambient	0	0.00	0.00	0.22	0.00
Aerosol-4	9	0.49	0.00	0.00	0.20
New York ambient	0	0.00	0.19	0.00	0.00
Aerosol-5	8	0.24	0.17	0.00	0.19
New York ambient	0	0.00	0.00	0.20	0.00
Aerosol-6	8	0.28	0.00	0.00	0.24
Utah ambient	0	0.00	0.00	0.16	0.00
Aerosol	8	0.12	0.13	0.07	0.00

* The uncertainties in the values are approximately 0.01 ± 10% of the given value.

New York City do not include the <1.1 μm particles which were not collected in a manner suitable for analysis. In a set of TSP collections from New York City, inorganic sulfite was seen in 6 of 24 samples with a range of 0.1–1.5 and a mean of 0.4 $\mu g\ m^{-3}$ and the organic adduct was seen in 19 of 24 samples with a range of 2–18 and a mean of 6 $nmol\ m^{-3}$. The sulfur composition of the respirable range of aerosols from the plume of the coal fired heating plant proved to be highly variable as summarized in Table 5.

DISCUSSION

As illustrated by the data in Table 1, Table 4 and Figs 3, 4, 5 and 6, inorganic sulfite species in particulate matter are consistently associated with the metal ion fraction of the sample. It has been previously postulated by Hansen *et al.* (1975) and Smith *et al.* (1976) that this association is due to the formation of $Fe(III)–SO_3^{2-}$ complexes in the aerosol. The comparative data in Table 2 suggest that other metal ion complexes are also formed. This conclusion may be

Table 4. Ambient levels of sulfate, inorganic sulfite, SO_2-Organic adduct, and Fe + Cu in the vicinity of a coal fired-heating plant and in New York city*

Sample	Sampler† stage	Sulfate $\mu g\ m^{-3}$, SO_4^{2-}	Inorganic sulfite $\mu g\ m^{-3}$, SO_3^{2-}	Inorganic sulfite $nmol\ m^{-3}$	SO_2-Organic adduct $nmol\ m^{-3}$	Fe + Cu $nmol\ m^{-3}$‡
Coal fired	1	3.3	0.7	8.5	8.0	65.0
Heating plant	2	3.1	0.5	6.4	7.2	36.0
Ambient	3	2.3	1.6	20.5	6.8	30.0
	4	6.1	3.4	42.2	3.2	—
New York city	1	3.0	0.0	0.0	3.6	3.2
Ambient-A	2	0.2	0.0	0.0	3.6	3.4
	3	3.2	0.0	0.0	2.4	3.9
	4	4.4	0.0	0.0	9.2	3.2
New York city	1	0.0	0.0	0.0	0.4	10.0
Ambient-B	2	0.0	0.0	0.0	0.5	3.2
	3	0.3	0.1	1.0	0.5	0.4
	4	2.2	0.0	0.0	4.7	0.3
New York city	1	0.1	0.0	0.0	0.0	3.7
Ambient-C	2	2.3	0.0	0.0	2.7	5.6
	3	0.3	0.0	0.0	7.1	9.2
	4	0.5	0.5	6.1	0.0	8.8

* Uncertainties in the values, based on replicate analysis of samples, are approximately $0.2 \pm 0.2 \times \mu g\ m^{-3}$ for sulfate and $0.1 \pm 0.1 \times \mu g\ m^{-3}$ for the other analysis.

† For the coal fired heating plant sample. The calibrated particle size cut off ranges are (sp. gr. = 2.5) >3.1, 3.1–1.7, 1.7–0.4, and <0.4 μm for the four stages. For the New York City samples. The calibrated particle size cut off ranges are (sp. gr. = 2.5) >7.7, 7.7–3.3, 3.3–2.2, and 2.2–1.1 μm for the four stages.

‡ The majority of the sum was Fe (usually >90%) in the coal fired heating plant sample and Cu (usually >90%) in the New York City samples.

Table 5. Chemical composition of 0.4–1.7 μm ambient particles collected in the vicinity of a coal fired heating plant, expressed as the mole ratio to sulfate*

Sample	Inorganic sulfite	Organic-SO_2 adduct
1	0.12	0.05
2	0.00	0.23
3	1.71	0.57
4	0.06	0.05
5	0.00	0.11

* Uncertainties in the values are 0.02 ± 10% of the stated value.

reached from a consideration of the thermodynamics and equilibria involved in the various solution analytical techniques used. Available data (Hansen, 1974; Sillen, 1964) indicate that the log K values for the reaction $M^{n+} + SO_3^{2-} = MSO_3^{(n-2)}$ in aqueous solution are 19, 12, 8 and 3 for $M^{n+} = Fe^{3+}$, Hg^{2+}, Cu^{+} or Cu^{2+}, and Cd^{2+} (or presumably Zn^{2+}, Pb^{2+}, etc.), respectively. The thermometric method of analysis is based on the addition of Fe(III) to the extraction solution, conversion of any inorganic sulfite species to the Fe(III) complex and titration of this complex with dichromate. As Fe(III) forms the most stable metal ion–SO_3^{2-} complex known, this technique would be expected to determine all inorganic sulfite species. The West–Gaeke method is dependent on formation of the $HgSO_3$ complex. As the Fe(III) complex is more stable than the Hg(II) complex, any Fe(III) species present would be missed by this technique but all other inorganic sulfite species would be detected. Ion chromatography is dependent on identification of the SO_3^{2-} or HSO_3^- anions. Thus, metal ion complexes which would be undissociated in solution (those with Fe^{3+}, or Cu^{2+}) would be missed but weaker complexes, i.e. with Zn^{2+}, Pb^{2+}, etc., would be detected. Sulfite is detected by all three techniques in the lead smelter samples, suggesting that the sulfite is present as a weak complex of Pb^{2+} or Zn^{2+}. The formation of Zn^{2+}–SO_3^{2-} complexes has also been suggested from results of SO_2 adsorption studies (Dyson, 1976). In contrast, only the thermometric technique identifies the sulfite in the copper smelter samples, indicating the presence of Fe(III) complexes. This is also consistent with the excellent correlation between Fe(III) and sulfite observed in samples collected from inside the smelter, Fig. 5. In contrast, the levels of sulfite in samples collected in the vicinity of the same smelter are not correlated with iron and often exceed the levels of Fe(III). The sum of Fe(III) and Cu(II) are correlated with sulfite, Fig. 6, however, suggesting that reactions take place in the flue system or the ambient in addition to those reactions occurring in the workroom environment. Additional adsorption of SO_2 to form sulfite in the ambient is also suggested by the data in Fig. 7. Insufficient data are available from the samples associated with power plant plumes or New York City to make similar correlations between metal ion content and the presence of sulfite. It is, however, noted that in all samples from coal fired power plants or New York City where inorganic sulfite is seen, the concentrations of iron and copper are always higher than the concentrations of sulfite. Also, in the oil fired station flue dusts where no inorganic sulfite species were seen, the levels of Fe, Cu, Zn and Pb extracted were very low (<0.1% by weight).

At least four reactions can affect the concentration of inorganic sulfite in ambient particulates with time. These are:

$$SO_3^{2-}\text{ (aerosol)} \xrightarrow{O_2} SO_4^{2-}\text{ (aerosol)} \quad (1)$$

$$S^\circ \text{ or } S^{2-}\text{ (aerosol)} \xrightarrow{O_2} SO_3^{2-}\text{ (aerosol)} \quad (2)$$

$$SO_3^{2-}\text{ (aerosol)} \longleftrightarrow \text{organic S compound (aerosol)} \quad (3)$$

$$SO_2\text{(g)} \longrightarrow SO_3^{2-}\text{ (aerosol)}. \quad (4)$$

Reaction (1) would result in a loss of sulfite in the particulate phase. Reactions (2) and (4) would result in a gain of sulfite. Depending on the direction of equilibrium, reaction (3) could result in an increase or a decrease in inorganic sulfite with time. Analysis of the smelter flue dust samples by calorimetry, Fig. 2, and of an air borne particulate sample by ESCA during long term storage indicates reactions (1) and (2) are slow, with $t_{1/2}$ values on the order of months. Analysis of stored samples of ambient aerosols collected near the coal fired heating plant over a time period of 40 days indicates that no changes occur between the inorganic sulfite and organic S(IV) compounds, reaction (3), during that time period. Additional work needs to be done to establish the kinetics of reaction (4) in the atmosphere.

The analysis of stored extractant solutions from coal fired power plants, from rural Utah, and from New York City, Table 3, indicates that several strong reducing agents in addition to inorganic sulfite are found in these samples. The appearance of inorganic sulfite in these solutions during storage, coupled with the disappearance of the species with $\Delta H_{ox} = -14$ kcal/eq, suggests dichromate oxidizable SO_2 adducts exist in the sample. In samples NYAA-2, 4 and 5, and CFPS-1 and 2 more sulfite was produced than can be attributed to the decrease in the other reducing agents, suggesting these samples also contain SO_2 organic adducts which are not oxidizable by dichromate but are hydrolyzable in HCl solution. While the calorimetric data do not provide positive identification of these compounds, reasonable assumptions about the structures of these postulated compounds can be made based on known reactions of SO_2 with organic compounds (Schroeter, 1966; Roberts, 1965).

Table 6. Frequency of identification of S(IV) compounds in non-smelter associated aerosol

Type of sample	Source	No. samples	Inorganic sulfite no. detected	Organic-SO_2 adduct no. detected
Coal fired heating plant	ambient	7	6	7
	flue line	4	2	4
Coal fired power plant	stack, elec. precip.	2	0	0
	stack, wet scrubber	2	1	0
	flue line	2	0	0
Oil fired power plants	flue line	5	0	0
	plume	2	0	0
New York City	ambient	31	11	23

Likely adducts which are oxidizable by dichromate with a ΔH value of approximately -14 kcal/eq are addition compounds of aldehydes.

$$R{-}\underset{}{\overset{H}{|}}{C}{=}O + H_2SO_3 \rightarrow R{-}\overset{OH}{\underset{H}{C}}{-}SO_3H$$

Likely adducts which would not be oxidizable by dichromate but which would hydrolyze to produce sulfite in cold dilute aqueous acid are addition compounds of polyketo or hydroxy aromatics, e.g.

$$\text{p-benzoquinone} + H_2SO_3 \longrightarrow \text{HO, }SO_3H\text{ adduct}$$

$$\text{2-naphthol (OH)} + H_2SO_3 \longrightarrow \text{(H, H; OH, }SO_3H\text{ adduct)}$$

Compounds which would fall into the three classes of these precursers have been identified in environmental samples (Pellizzari, 1976; Cautreels, 1976; Pierce, 1976) but usually at low levels. The GC–MS techniques usually used would not identify these adduct compounds since both extraction solvent polarity and temperature profiles would not allow dissolution or volatilization of the organic precurser or the SO_2 adduct (Pellizzari, 1976; Gold, 1975).

It should be emphasized that the calorimetric data only indicate that SO_2 adduct compounds exist in the samples studied. The suggested identity of these compounds is tentative and needs to be verified by appropriate analytical techniques. The hydrolyzable sulfite containing species in ambient samples are commonly present at from 5 to 49% of the sulfate levels as indicated in Tables 3–6. In contrast, samples collected directly from the flue lines of fossil fuel burning facilities, Table 3, contain little or none of these species. This is not unexpected since most organic compounds would be volatile in the flue line and would not be associated with particulate matter at temperatures normally seen in flue systems.

The various reactions of sulfur oxides suggested by this work to be important in atmospheric particulates are summarized in Fig. 1. Emitted SO_3 will be converted rapidly to H_2SO_4 aerosol as hot gases emitted by a source are cooled. The H_2SO_4 can be neutralized by metal oxides, NH_3, etc. to form various sulfates or bisulfates, or possibly react with organics to form inert organic sulfonates. No mechanism has been reported for the production of H_2SO_4 from these latter compounds, nor for the reduction of sulfate species to produce S(IV) species in the atmosphere. Emitted SO_2 may be oxidized to form a sulfate aerosol or may react with aerosols to form weakly adsorbed SO_2, stable inorganic sulfite species or organic adducts. The identity and stability of the inorganic sulfite species are fairly well understood. In contrast, the identity of the organic pool and the relative importance of SO_2 reactions which lead to stable S(IV) inorganic or organic compounds vs the reactions which lead to sulfate compounds are presently ill defined.

Acknowledgements—Appreciation is expressed to L. Newman, T. Kneip, W. Henery, and J. Fischer for furnishing the oil fired plume, New York City, oil fired power plant flue, and coal fired power plant flue particulate samples. The West–Gaeke and ion chromatography analysis were obtained through B. Appel, California Department of Health, and R. Stevens, Environmental Protection Agency. Appreciation is expressed to Dr. Jones and B. Richter for technical assistance. This work was supported by USERDA Contract EY765–02-2988.

REFERENCES

Cautreels W. and Cauwenberghe K. V. (1976) Determination of organic compounds in airborne particulate matter by gas chromatography–mass spectrometry. *Atmospheric Environment* **10**, 447–457.

Colucci A. V. and Eatough D. J. (1976) Determination and possible public health impact of transition metal sulfite aerosol species. EPRI EC-184.

Craig N. L., Harker A. B. and Novakov T. (1974) Determination of the chemical states of sulfur in ambient pollution aerosols by X-ray photoelectron spectroscopy. *Atmospheric Environment* **8**, 15–21.

Dyson W. L. and Quon J. E. (1976) Reactivity of zinc oxide fume with sulfur dioxide in air. *Environ. Sci. Technol.* **10**, 476–481.

Gold A. (1975) Carbon black adsorbates: Separation and identification of a carcinogen and some oxygenated polyaromatics. *Analyt. Chem.* **47**, 1469–1471.

Hansen L. D., Eatough D. J., Whiting L., Bartholomew C. W., Cluff C. L., Izatt R. M. and Christensen J. J. (1974) Transition metal–SO_3^{2-} complexes. A postulated mechanism for the synergistic effects of aerosols and SO_2 on the respiratory tract. *Trace Substances in Environ. Health VII* pp. 393–397. University of Missouri Press, Columbus, Mo. U.S.A.

Hansen L. D., Eatough D. J., Mangelson N. F., Jensen T. E., Cannon D., Smith T. J. and Moore D. E. (1975) Sulfur species and heavy metals in particulates from a copper smelter. Proceedings, International Conference on Environmental Sensing and Assessment, Las Vegas.

Hansen L. D., Whiting L., Eatough D. J., Jensen T. E. and Izatt R. M. (1976) Determination of sulfur (IV) and sulfate in aerosols by thermometric methods. *Analyt. Chem.* **48**, 634–638.

Hansen L. D., Eatough D. J., Mangelson N. F., Izatt R. M., Hill M., Lee M. L., Major T., Ryder J. and Richter B. (1977) Reduced species and acid-base components of New York City aerosols. Amer. Ind. Hyg. Conf. New Orleans, 23–27 May.

Johansson T. B., Van Grieken R. E., Nelson J. W. and Winchester J. W. (1975) Elemental trace analysis of small samples by proton induced X-ray emission. *Analyt. Chem.* **47**, 855–860.

Nielson K. K., Hill M. W. and Mangelson N. F. (1976) Calibration and correction methods for quantitative proton induced X-ray emission analysis of autopsy tissues. *Advan. X-ray Analysis* **19**, 511–520.

Novakov T., Mueller P. K., Alcocer A. E. and Otvos J. W. (1972) Chemical composition of Pasadena aerosol. Chemical states of nitrogen and sulfur by photoelectron spectroscopy. *J. Collid Interface Sci.* **39**, 225–234.

Pellizzari E. D., Bunch J. E., Berkley R. E. and McRae J. (1976) Determination of trace hazardous organic vapor pollutants in ambient atmosphere by gas chromatography/mass spectrometry/computer. *Analyt. Chem.* **48**, 803–807.

Pierce R. C. and Katz M. (1976) Chromatographic isolation and spectral analysis of polycyclic quinones. Application to air pollution analysis. *Envon. Sci. Technol.* **10**, 45–51.

Roberts J. D. and Caserio M. C. (1965) *Basic Principles of Organic Chemistry* pp. 908–909. Benjamin, New York.

Schroeter L. C. (1966) *Sulfur Dioxide*. Pergamon Press, London.

Sillen L. G. and Martell A. E. (1964) *Stability Constants of Metal-Ion Complexes*, Spedial Publication No. 17, pp. 230–232. The Chemical Society, London.

Smith T. J., Eatough D. J., Hansen L. D. and Mangelson N. F. (1976). The chemistry of sulfur and arsenic in airborne copper smelter particulates. *Bull. Env. Cont. Toxicol.* **6**, 651–659.

Stevens R. K., Russwurm G. M. and Kistler C. M. (1977) Ion exchange chromatographic analysis for sulfates and sulfites in atmospheric aerosols. Proceedings, Symposium on Ion Chromatographic Analysis of Environmental Pollutants, EPA, 28 April, in press.

West P. W. and Gaeke G. C. (1956) Fixation of sulfur dioxide as disulfitomercurate(II) and subsequent colorimetric estimation. *Analyt. Chem.* **28**, 1816–1819.

Atmospheric Environment Vol. 12, pp. 273–280. Pergamon Press 1978. Printed in Great Britain.

PRECURSOR EFFECTS ON SO_2 OXIDATION

D. F. MILLER
Battelle-Columbus Laboratories, 505 King Avenue, Columbus, OH 43201, U.S.A.

(*First received* 20 *June* 1977 *and in final form* 16 *September* 1977)

Abstract—Smog chamber experiments were conducted to determine the relationships between SO_2 oxidation and the gaseous precursors common in polluted air: NO_x, NMHC (a mixture of 17 hydrocarbons) and SO_2. SO_2 oxidation to sulfate aerosols was first-order in SO_2. The maximum rate of oxidation was strongly related to initial NMHC/NO_x ratios, but over 6-h irradiation intervals, the conversion of SO_2 to sulfate aerosol was only weakly related to initial NMHC/NO_x ratios. For constant NMHC/NO_x ratios, SO_2 conversion was independent of NMHC and NO_x concentrations. Typical SO_2 oxidation rates for polluted air ranged from 2–8%/h, but the higher rates were sustained for only a few hours. SO_2 lifetimes > 100 h are predicted from these experiments, in accord with kinetic simulations of photochemical smog and lifetimes derived from tropospheric data.

1. INTRODUCTION

Sulfur dioxide is a major environmental problem in many parts of the world, particularly where the occurrence of SO_2 is elevated by the combustion of fossil fuels. Photochemical oxidation of SO_2 results in the formation of submicron-size aerosols (sulfuric acid and sulfate salts) which scatter solar radiation, increase the acidity of rainwater and which may influence precipitation patterns and adversely affect human health. In view of the present and foreseeable problems associated with sulfate aerosols and the pressing demands for coal utilization, it is necessary that we fully understand the factors involved in SO_2 oxidation for a wide variety of atmospheric conditions.

The formation and transport of ozone, as well as that for sulfate aerosols, is a concern in the United States. Generally speaking, ozone and sulfate aerosols have much in common, although it is well recognized that the processes governing their concentrations are very complex and linked indirectly. For example, in relatively unpolluted air, the photolysis of ozone leads to the production of OH which is the free radical most likely involved in oxidizing SO_2. Ozone and sulfate aerosols formed in polluted air are believed to be related to common gaseous precursors; namely nonmethane hydrocarbons (NMHC), nitrogen oxides (NO_x), and of course SO_2 in the case of sulfate aerosols. The objective of the studies (Battelle Reports, 1977) summarized in this paper is to determine experimentally the quantitative relationships between sulfate aerosols and their precursors. These relationships will hopefully provide guidance for the design of precursor-control strategies to help reduce the elevated sulfate concentrations and the mutual accompaniment of ozone that is so widespread in the United States.

2. EXPERIMENTAL METHODS

2.1 *Experiments*

Laboratory experiments were conducted in Battelle-Columbus' 17.3 m^3 smog chamber having a surface/volume ratio of 0.8 m^{-1}; the surface is polished aluminum and FEP Teflon. Direct irradiation through 5-mil Teflon windows is provided by a bank of 95 fluorescent blacklamps and 15 fluorescent sunlamps. The photon flux of the blacklamps is distributed unimodally in the u.v. region, with peak intensity at 370 nm; the sunlamp peak intensity occurs at 310 nm. The lamps yield a NO_2 photo-dissociation rate (often referred to by k_1) of 0.27 min^{-1}. Constant light intensity at that level for 10 h is approximately equal to a diurnal radiation dosage with a peak intensity of 1.14 cal cm^{-2} min^{-1}. Background air for the chamber is passed through a purification system which includes permanganate pellets, a charcoal filter system and an absolute filter. The air is humidified with double distilled water to provide relative humidities in the 40–50% range at 30°C. When nitrous acid is added to the chamber to provide a source of OH, the combined reactivity of the background impurities in the air correspond theoretically to the reactivity for 0.046 ppm butane. The gas phase chemistry of the atmosphere was monitored with conventional instrumentation. NO_2 was determined by the automated Saltzman method to avoid interferences to NO_2 analyses that are encountered using reducing catalysts in conjunction with chemiluminescence analyzers. Detailed hydrocarbon analyses were obtained hourly with two FID gas chromatographs; one equipped with a capillary column and temperature programming from −100 to 136°C.

Results are also presented for smog chamber experiments conducted under natural irradiation. The experiments were performed in July, 1976 in St. Louis, MO, at RAM stations 106 and 103. A 13.5 m^3, 5-mil Teflon chamber with a surface/volume ratio of 1.7 m^{-1} was used to confine early morning air. The RAM stations provided information about the air parcels prior to confinement and study. During the course of the experiments, direct measurements of the chamber atmosphere were made for SO_2, NO, NO_x, O_3, aerosol size distributions, and sulfate aerosol by filtering. General conditions in St. Louis during the study period were high temperatures (90–105°F), high humidity, variable winds and clear skies or light clouds.

2.2 *Methods for calculating* SO_2 *oxidation rates*

For the experiments conducted under both laboratory and ambient conditions, SO_2 oxidation rates were derived from analyses of aerosol formation. Two subordinate methods were used. A nearly continuous record of total aerosol growth was obtained with a Thermo Systems Electrical Aerosol analyzer (Liu *et al.*, 1974). Aerosol volume concentrations were calculated from the particle size distributions and the number concentrations given by the analyzer. To obtain more specific information on sulfate concentrations, intermittent samples were obtained as filter collections and subsequently analyzed by ion chromatography (Small *et al.*, 1975; Mulik *et al.*, 1976). Pallflex QAST filters and 1 μm Fluoropore FA Teflon filters were used depending on the amount of sample available. It has been demonstrated (Coutant, in press) that these filter types, unlike glass filters, do not yield artifact sulfate due to SO_2 sorption.

Without going into detail, special experiments were conducted in the laboratory and in St Louis for which it was certain that sulfate was the exclusive constituent of the aerosol being monitored by the electrical aerosol analyzer (EAA). Comparison of the EAA concentration curves with the chemical analyses for sulfate established a scaling factor between the volumetric concentrations given by the EAA and the sulfate mass concentrations. The average scaling factor used to convert EAA data to sulfate mass concentrations is 1.76; a value that is nearly identical to the density of $(NH_4)_2SO_4$. The nearness of the factor to the density of $(NH_2)_2SO_4$ is most likely fortuitous since the calibration of the EAA changes from time to time. Also, in many instances the NH_4^+/SO_4^{2-} ratio was not 2 but ranged anywhere between 1/1 and 2/1, and the apparent ratio between sulfate mass and EAA volume did not vary accordingly. Thus the true density and stoichiometry of the sulfate aerosols being monitored is quite uncertain, and we merely refer to it as sulfate.

For the laboratory experiments, additional information was available to help make interpretations between total particulate, as indicated by the EAA, and sulfate particulate, as indicated by chemical analyses. In a previous study (Miller and Joseph, 1976) a similar matrix of experiments was conducted with identical hydrocarbon and NO_x mixtures, but *without* SO_2. Data from those experiments were used to construct nonsulfur aerosol growth curves for the corresponding experiments with SO_2. Thus in the exemplary profiles shown in Fig. 1, the sulfate concentration (shown as the hatched area above the nonsulfate aerosol curve) corresponds to the sulfate concentration determined by chemical analysis and the total aerosol volume concentration inferred from the EAA.

SO_2 conversion rates were computed by comparing sulfate mass concentrations, converted to ppb (vol./vol.)S with the ppb concentrations of S in the gas phase. SO_2 conversions derived from the laboratory experiments have been corrected for the dilution losses and wall losses in the chamber. Dilution was not a factor for the outdoor experiments, but wall losses were substantial due to static charging of the Teflon surfaces, and the losses were not taken into consideration in computing the oxidation rates for the experiments in St. Louis. By convention, the SO_2 oxidation rates appearing in this paper are expressed as fractional rates (units of $\% \, h^{-1}$) rather than absolute rates.

3. RESULTS AND DISCUSSION

3.1 *Laboratory experiments*

Initial experimental conditions and a summary of experimental results are presented in Tables 1 and 2, respectively. As indicated in Table 1, each experiment was conducted with a constant mixture of 17 reactive hydrocarbons chosen to be representative of the hydrocarbons present in polluted urban air (Stephens, 1973). Figure 1 illustrates 3 typical photochemical-smog profiles for experiments with similar initial concentrations of NO_x, but with 3 different concentrations of NMHC. The profiles show the oxidation of NO to NO_2, the formation of ozone, and the formation of sulfate and nonsulfur aerosol as a function of the irradiation time. Note that while sulfate aerosol formation precedes ozone formation, the two concentration curves nearly parallel each other. In each case the maximum rate of SO_2 oxidation (sulfate aerosol formation) occurs during the first 3 h of irradiation—a result we will later show to be consistent with kinetic theory and results obtained under natural radiation conditions and ambient air. In the top profile in Fig. 1 where the $NMHC/NO_x$ ratio is 20/1, SO_2 oxidation occurs quite rapidly and then nearly ceases as the reactive hydrocarbons and NO_x

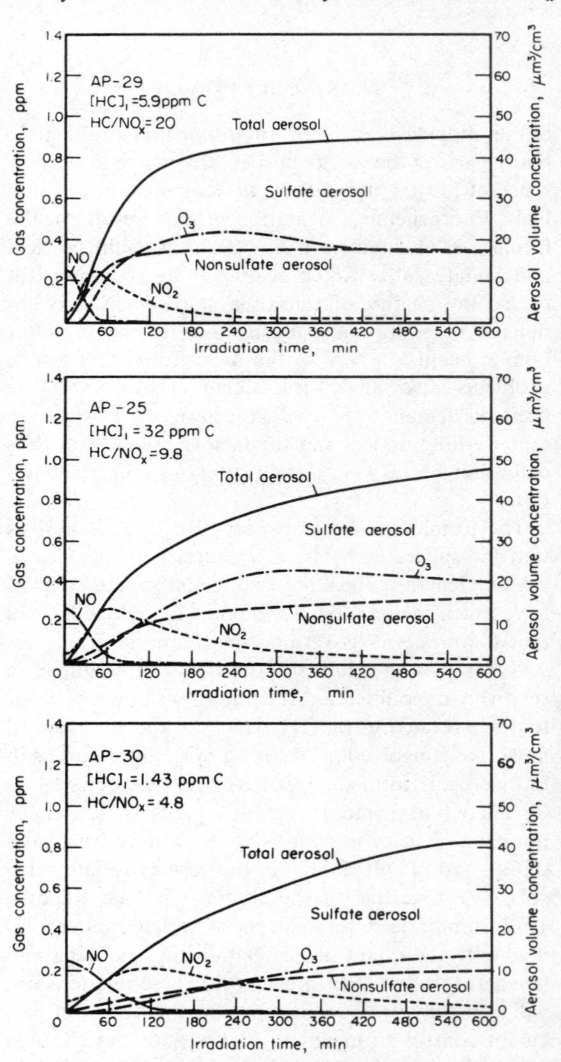

Fig. 1. Profiles of photochemical smog formation and SO_2 conversion to sulfate aerosols for variable hydrocarbon concentrations and constant NO_x concentrations.

Table 1. Initial pollutant concentrations

	AP-23	AP-25	AP-26	AP-27	AP-28	AP-29	AP-30	AP-31	AP-34	AP-35	AP-36	AP-37	AP-38	S-324	S-325	S-326	S-327
Carbon monoxide	15	15	15	15	15	15	15	15	15	15	15	15	15	15	15	15	15
Sulphur dioxide	0.100	0.100	0.100	0.100	0.100	0.100	0.100	0.100	0.100	0.100	0.100	0.100	0.100	0.100	0.100	0.100	0.100
Nitrogen oxides	0.600	0.324	0.150	0.590	1.130	0.297	0.299	1.248	0.885	0.304	0.150	1.200	0.750	0.150	0.060	0.075	0.075
Nitric oxide	0.499	0.270	0.125	0.490	0.930	0.250	0.249	1.044	0.736	0.252	0.125	1.000	0.620	0.125	0.500	0.062	0.062
Nitrogen dioxide	0.101	0.054	0.025	0.100	0.194	0.047	0.050	0.204	0.149	0.052	0.025	0.200	0.130	0.025	0.100	0.013	0.013
Nonmethane hydrocarbons (as CH_4)	12.368	3.176	1.568	1.493	2.970	5.895	1.434	2.981	5.839	12.176	5.922	11.824	3.686	2.929	6.365	9.570	3.184
acetylene	1.013	0.238	0.124	0.127	0.249	0.454	0.104	0.242	0.466	0.935	0.460	0.922	0.269	0.180	0.457	0.715	0.280
ethylene	0.992	0.254	0.107	0.109	0.235	0.438	0.110	0.218	0.480	0.905	0.458	0.855	0.241	0.230	0.483	0.693	0.288
propylene	0.316	0.076	0.032	0.028	0.099	0.144	0.018	0.086	0.171	0.354	0.159	0.322	0.084	0.019	0.124	0.208	0.049
trans-2-butene	0.487	0.087	0.021	0.025	0.077	0.197	0.019	0.087	0.188	0.435	0.194	0.440	0.114	0.045	0.183	0.323	0.047
2-methyl-2-butene	0.169	0.033	0.173	0.016	0.030	0.056	0.012	0.031	0.051	0.151	0.489	0.148	0.059	0.016	0.045	0.078	0.015
ethane	0.487	0.141	0.082	0.067	0.104	0.226	0.052	0.099	0.196	0.467	0.254	0.473	0.117	0.090	0.283	0.310	0.127
propane	0.325	0.094	0.054	0.045	0.069	0.150	0.035	0.066	0.131	0.311	0.169	0.315	0.078	0.060	0.188	0.207	0.085
n-butane	1.156	0.378	0.156	0.154	0.293	0.540	0.144	0.300	0.524	1.058	0.529	1.041	0.325	0.287	0.572	0.855	0.329
2-methylpropane	0.306	0.132	0.037	0.041	0.093	0.144	0.042	0.075	0.157	0.292	0.159	0.297	0.097	0.050	0.164	0.192	0.054
n-pentane	0.573	0.147	0.071	0.074	0.143	0.286	0.072	0.146	0.281	0.579	0.280	0.565	0.181	0.141	0.305	0.462	0.142
2-methylbutane	1.118	0.301	0.144	0.144	0.284	0.572	0.146	0.291	0.556	1.150	0.562	0.117	0.374	0.324	0.626	0.930	0.288
2-methylpentane	0.823	0.201	0.104	0.104	0.203	0.414	0.100	0.208	0.403	0.917	0.403	0.812	0.243	0.192	0.412	0.675	0.206
benzene	0.563	0.141	0.072	0.075	0.140	0.290	0.072	0.146	0.280	0.577	0.291	0.564	0.197	0.142	0.301	0.490	0.148
toluene	1.370	0.349	0.184	0.178	0.333	0.607	0.175	0.347	0.677	1.390	0.675	1.348	0.417	0.351	0.744	1.155	0.368
m-xylene	1.631	0.384	0.216	0.193	0.390	0.819	0.205	0.398	0.800	1.650	0.801	1.615	0.528	0.486	0.893	0.397	0.461
p-ethyltoluene	0.620	0.130	0.088	0.069	0.141	0.290	0.078	0.146	0.287	0.606	0.288	0.597	0.222	0.184	0.353	0.526	0.185
1,2,4-trimethylbenzene	0.419	0.091	0.058	0.044	0.088	0.188	0.049	0.094	0.190	0.400	0.189	0.393	0.140	0.131	0.232	0.353	0.112

(1) All concentration units are ppm (vol./vol.); hydrocarbon units are expressed as ppm CH_4 or ppm C.

Table 2. Summary of experimental results

Run No.	Initial conditions HC, ppmC	NO$_x$, ppm	HC/NO$_x$	SO$_2$, ppm	NO$_2$-t_{max},* min	NO$_2$ rate,† ppb min^{-1}	HC rate,‡ % h^{-1}	HC conversion,§ %	Peak O$_3$,‖ ppm	SO$_2$ Rate,¶ % h^{-1}	SO$_2$ Conversion,** %	Sulfur aerosol,†† $\mu g/m^{-3}$	Non-sulfur aerosol,‡‡ $\mu g/m^{-3}$	Total aerosol,§§ $\mu g/m^{-3}$
AP 23	12.4	0.60	21	0.10	42	9.9	12.0	37	0.61	6.5	12	65	47	112
AP 25	3.2	0.32	9.8	0.10	63	3.4	11.4	29	0.45	3.4	8.6	46	25	71
AP 26	1.56	0.15	10.4	0.10	52	1.9	11.6	39	0.33	3.6	10.0	53	16	69
AP 27	1.49	0.59	2.5	0.10	210	1.1	8.3	27	0.24(a)	1.8	7.7	41	10	51
AP 28	3.0	1.13	2.6	0.10	300	1.3	9.5	30	0.30(a)	1.2	5.3	28	9	37
AP 29	5.9	0.30	20	0.10	43	4.7	11.4	33	0.44	6.8	8.6	46	30	76
AP 30	1.43	0.30	4.8	0.10	110	1.5	12.0	39	0.28(a)	2.0	8.3	44	16	60
AP 31	3.0	1.25	2.4	0.10	240	1.0	10.6	32	0.04(b)	1.1	4.3	23	7	30
AP 34	5.8	0.89	6.6	0.10	165	2.6	11.0	37	0.43(a)	2.4	6.0	32	26	58
AP 35	12.2	0.30	40	0.10	25	8.5	5.8	22	0.49	7.5	6.7	36	37	73
AP 36	5.9	0.15	39	0.10	28	3.8	8.8	33	0.31	7.2	6.6	35	22	57
AP 37	11.8	1.20	9.9	0.10	105	6.5	12.1	40	0.68	4.5	8.6	46	37	83
AP 38	3.7	0.75	4.9	0.10	135	2.4	13.8	41	0.34(a)	1.9	6.4	34	17	51
S-324	2.9	0.15	19	0.10	52	2.1	11.8	38	0.25	4.9	9.4	50	21	71
S-325	6.4	0.60	11	0.10	77	4.2	12.0	10	0.41(b)	3.3	6.6	35	34	69
S-326	9.6	0.075	130	0.10	30	1.3	6.8	29	0.17	11.0	6.6	35	25	60
S-327	3.2	0.075	42	0.10	27	2.1			0.18	7.5	7.7	41	7	48

* Time to reach the maximum [NO_2].

† {$[NO_2]_{max} - [NO_2]_{initial}$}/time to $[NO_2]_{max}$.

‡ Overall nonmethane hydrocarbon oxidation rate corrected for smog-chamber dilution.

§ Percent of initial nonmethane hydrocarbon converted to products during the first 6 h of irradiation.

‖ Maximum [O_3]: (a) [O_3] still increasing after 10 h. (b) [O_3] still increasing after 6 h.

¶ Maximum fractional oxidation rate for SO_2.

** Percent of the initial [SO_2]converted to sulfate aerosol during the first 6 h of irradiation. Conversions are corrected for smog-chamber dilution.

†† Mass concentration of sulfate aerosol at 6 h assuming that all of the sulfate is $(NH_4)_2SO_4$ with density of 1.77 g cm^{-3}. Aerosol concentrations are corrected for chamber dilution.

‡‡ Mass concentration of non-sulfur aerosol at 6 h assuming that the aerosol density is 1.77 g cm^{-3}.

§§ Mass concentration of total aerosol at 6 hr assuming that the aerosol density is 1.77 g cm^{-3}.

(i.e. NO and NO_2) become converted to less reactive products. In the middle profile at a lower $NMHC/NO_x$ ratio of 9.8/1, the maximum rate of SO_2 oxidation is less than for the previous case, but it appears that SO_2 continues to be oxidized for the entire irradiation period. Trends are similar for the bottom profile in Fig. 1. There the initial $NMHC/NO_x$ ratio is even lower (4.8/1), the maximum rate of SO_2 oxidation is appreciably slower, but the maximum sulfate concentration (or the total conversion of SO_2 to sulfate) at the end of the irradiation period is approximately the same as for the other 2 experiments having higher maximum rates of SO_2 oxidation. The results of these 3 experiments illustrate 2 important points that are supported further in the paragraphs to follow: (1) the maximum rate of SO_2 oxidation is strongly related to the initial $NMHC/NO_x$ ratio, and (2) the maximum sulfate concentration and thus the total conversion of SO_2 to sulfate in a diurnal irradiation period is relatively independent of the initial $NMHC/NO_x$ ratio.

The precursor effects of NMHC and NO_x on SO_2 oxidation for all laboratory experiments are shown in Figs. 2, 3 and 4. The solid circles in Figs. 2 and 4 indicate the NMHC and NO_x coordinates, while the values above the circles correspond to the SO_2 oxidation parameters; maximum SO_2 rates are plotted in Fig. 2, and maximum sulfate aerosol concentrations are plotted in Fig. 4. In both figures, we have attempted to draw isopleths or contours for certain ranges of the oxidation parameters, but the boundaries indicated may not be statistically significant.

It is clear from the data in Fig. 2 that the maximum rate of SO_2 oxidation varies directly with the initial $NMHC/NO_x$ ratio, whether NMHC is increased with respect to NO_x or NO_x is decreased with respect to NMHC. The relationship is illustrated further in Fig. 3 where all of the SO_2 rate data are plotted against $NMHC/NO_x$ ratios, regardless of absolute precursor

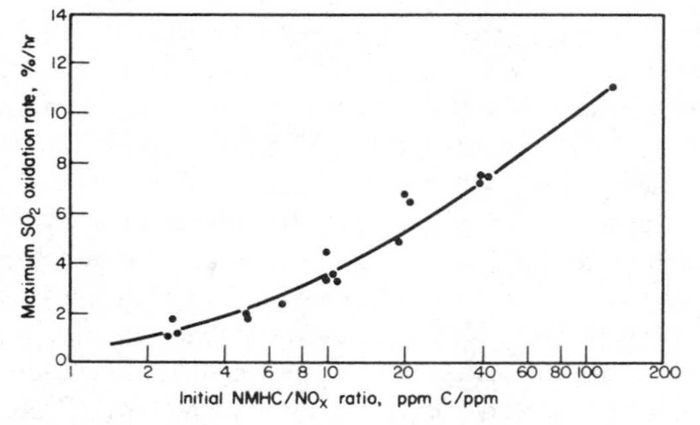

Fig. 3. Effect of initial $NMHC/NO_x$ ratio on the maximum rate of SO_2 oxidation.

concentrations. It must also be emphasized, however, that for constant $NMHC/NO_x$ ratios the maximum rate of SO_2 oxidation is relatively constant over a large range of initial precursor concentrations, as shown in Fig. 2.

Figure 4 is a plot of sulfate aerosol mass concentrations after 6 h irradiations, for the same initial concentrations of NMHC and NO_x as in Fig. 2. In this case, a region of maximum sulfate aerosol concentrations appears for $NMHC/NO_x$ ratios in the range from 12/1 to 25/1. In comparison to the precursor relationships with respect to maximum SO_2 oxidation rates, total SO_2 conversion to sulfate is much less sensitive to initial concentrations of NMHC and NO_x, as mentioned earlier in reference to the profiles in Fig. 1. In fact, SO_2 conversion after 6 h had no correlation with the maximum fractional rate of SO_2 oxidation, even though the experiments were all conducted with constant SO_2 concentrations.

Among the reactivity parameters included in Table 2, the maximum rate of SO_2 oxidation correlated best ($R = 0.78$) with the normalized parameter for NO oxidation (NO_2-t_{max}). The second best correlation was with the overall hydrocarbon depletion rate ($R = 0.50$). The fractional rates of SO_2 oxidation did not

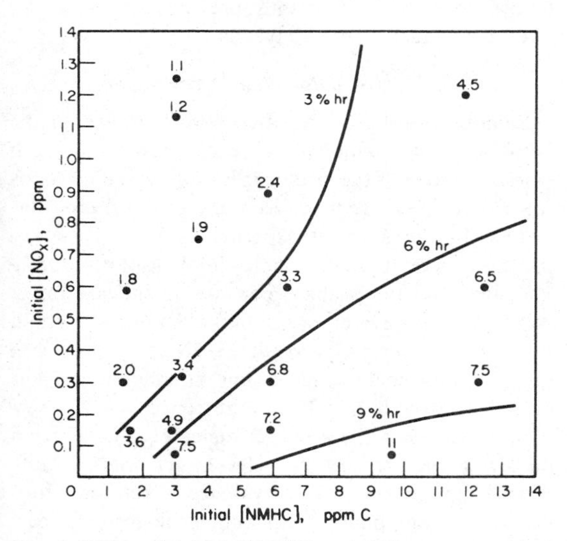

Fig. 2. Effect of initial NMHC and NO_x concentrations on the maximum rates of SO_2 oxidation.

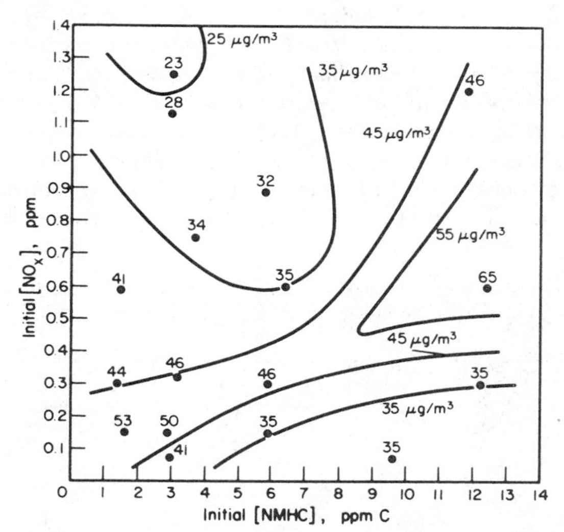

Fig. 4. Effect of initial NMHC and NO_x concentrations on the total conversion of SO_2 sulfate aerosol in a 6 h irradiation interval.

correlate with peak ozone concentrations. SO_2 conversion at 6 h was moderately correlated with the NO_2 rate ($R = 0.47$), peak ozone concentrations ($R = 0.50$), and the peak nonsulfur aerosol concentrations ($R = 0.47$).

Another series of experiments with identical hydrocarbon conditions was conducted to determine the effect of SO_2 concentrations on SO_2 oxidation (Battelle Reports, 1977). Those experiments are not discussed in this paper, but the pertinent finding should be noted. For experiments conducted at $NMHC/NO_x$ ratios of 5/1 and 20/1, the rate of sulfate aerosol formation was directly proportional to the SO_2 concentration over a SO_2 concentration range from 33 to 900 ppb. Thus all of the data from these studies have indicated that photochemical oxidation of SO_2 is indeed first-order with respect to SO_2.

3.2 *Ambient air experiments*

As described in Section 2, outdoor irradiation experiments were conducted in the Summer of 1976 in St. Louis. In addition to determining SO_2 oxidation rates representative of that area, the objectives were to continue to explore NMHC and NO_x precursor relationships to SO_2 oxidation, and to study SO_2 oxidation under conditions more authentic than those for the laboratory. A Teflon smog chamber was mounted on an open bed truck parked adjacent to the RAM stations. A tubular aluminum skeleton external to the chamber permitted air to be rapidly exhausted from the chamber without collapsing it. A 20 cm diameter port in the top of the chamber allowed ambient air to fill and purge the chamber while a blower exhausted air from the opposite end of the chamber. This procedure had the advantage of flowing ambient air into the chamber without the risk of contamination from a pump, and it preserved the integrity of the primary particulate in the ambient sample. Experiments usually began between 0700 and 0800 Eastern Standard Time. Sulfur dioxide was added to the air mixtures to improve the accuracy for determinating sulfate concentrations.

Experimental conditions and results for 6 days at site 103 are presented in Table 3. There was not much variation in the early morning pollutant concentrations during this period; NMHC, for example, ranged from 0.10 ppmC to 0.41 ppmC, while the

Table 3. Initial pollutant concentrations and the maximum rates of SO_2 oxidation for ambient air experiments in St. Louis

Mo/day	Initial concentrations, ppm			HC/NO$_x$ ratio	SO_2 Oxidation, % h^{-1}
	NMHC	NO_x	SO_2*		
7/22	0.31	0.039	0.125	7.9	5.5
7/23	0.26	0.033	0.140	7.8	4.2
7/26	0.10	0.019	0.048	5.2	3.4
7/27	0.30	0.030	0.051	10	4.0
7/28	0.18	0.036	0.095	5	3.5
7/30	0.41	0.109	0.085	3.7	1.6

* SO_2 was added to the ambient air.

$NMHC/NO_x$ ratios varied from 3.7/1 to 10/1. Time–concentration profiles of sulfate aerosol formation in these experiments were very similar to the profiles obtained in the laboratory experiments, in spite of considerable differences in light intensity, temperature and humidity. For the ambient air experiments, maximum SO_2 oxidation rates occurred during the first 2 hours of irradiation and aerosol formation diminished markedly after 3 hours. In accord with the laboratory results, the maximum rate of SO_2 oxidation appeared to depend on the $NMHC/NO_x$ ratio, although other unmonitored factors may have also been effectual. The trends observed are quite consistent with the results of a similar study (Clark *et al.*, 1976) conducted downwind of a Los Angeles freeway. The found the aerosol volume concentrations to be directly proportional to the SO_2 concentration in the air; a finding with which we definitely concur. However, they also concluded that the aerosol formation rate was linearly related to the initial NMHC concentrations, after normalizing for variations in initial SO_2 concentrations and light intensity. Clark *et al.* did not consider the significance of NO_x in arriving at that conclusion. Using their data, we plotted the aerosol formation rates vs the initial $NMHC/NO_x$ ratios, normalizing as they did for $[SO_2]$ and light intensity. In so doing, the linear correlation coefficient between the $NMHC/NO_x$ ratios and the aerosol formation rates turned out to be $R = 0.999$.

The main point to be made regarding the study in St Louis is that the rate of SO_2 oxidation at relatively low pollutant concentrations (i.e. NMHC in the range from 0.1 to 0.3 ppmC) was essentially the same as the rates observed in the laboratory for much higher pollutant concentrations (NMHC in the range from 1–12 ppmC). These data seem to confirm the precursor relationships developed from the laboratory experiments, and they serve to extend the relationships into precursor concentration regions where the laboratory results could only previously be extrapolated with much uncertainty.

3.3 *Aerosol size distributions and growth features*

Sulfate aerosol growth patterns were similar for the laboratory and ambient-air experiments. In both cases the mode of the size distribution of sulfate aerosol volume occurred between 0.1 and 0.2 μm diameters. In the ambient experiments where primary particles were present, sulfate condensation usually took place on the primary particles in the size range <0.1 μm. On several occasions when only a small amount of primary particulate existed in this size range, homogeneous nucleation of sulfate aerosols occurred in the 0.01–0.02 μm size range.

In no case was there any evidence of condensation occurring on primary particles larger than 0.2 μm. Furthermore, there was no evidence of heterogeneous growth in conjunction with either the primary particulates or the sulfate aerosols. The observations are consistent with earlier laboratory results with auto

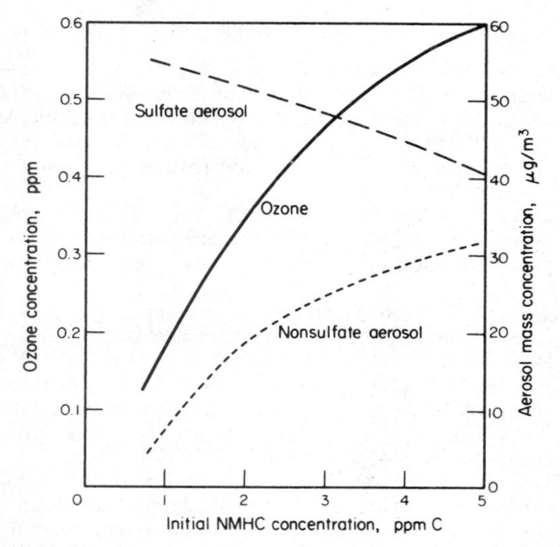

Fig. 5. Smog chamber predictions of the relative effects of NMHC and NO_x control strategies on secondary pollutant concentrations.

exhausts (Miller and Levy, 1973; Miller *et al.*, 1976) where it was shown that primary exhaust particulates act as nuclei for the subsequent condensation of photochemically derived aerosols, but they do not significantly affect the gas-to-aerosol conversion rates.

3.4 *Precursor control strategies*

Data developed from smog chamber experiments have been used to predict the relative influence of hydrocarbons and nitrogen oxides on ozone formation (Dimitriades, 1977) and nonsulfate aerosol formation (Miller and Joseph, 1976). Presumably the data developed in this study can likewise be applied to suggest the effectiveness or ineffectiveness of precursor controls for limiting sulfate formation. An example of the precursor relationships with secondary pollutants is presented in Fig. 5. In this case the relationships between peak O_3 concentrations, peak sulfate concentrations, and peak nonsulfur aerosol concentrations are plotted against the initial precursor concentrations (represented by NMHC) for a constant $NMHC/NO_x$ ratio of 10/1. Peak concentration data are taken from 6 h irradiation intervals for the experiments with the mixture of 17 hydrocarbons and a constant initial SO_2 concentration of 0.1 ppm. The relationships show that both ozone and nonsulfur aerosol concentrations can be reduced by reducing the precursor concentrations. The same strategy will not, however, be beneficial in reducing sulfate concentrations.

3.5 *Interpretations of* SO_2 *oxidation*

It is of interest to examine some of the theoretical expectations and atmospheric data on SO_2 oxidation in comparison to the data presented here. Because of incomplete knowledge of the chemistry and kinetics of photochemical smog (particularly that for the involvement of hydrocarbons), it is only possible to attempt to model smog-chamber experiments using generalized mechanisms. Some satisfactory progress has been made, but the results to date must be considered preliminary. With that precaution it can be stated that the kinetic models we've applied indicate that hydroxyl radicals (OH) account for at least one-half of the total SO_2 conversion, while RO_2 and HO_2 radicals together account for the remaining fraction of the conversion. It can also be shown that diradicals or 'zwitterions' resulting from olefin–ozone reactions make a minimal contribution to the overall SO_2 oxidation process.

If OH is needed a principal, if not predominant, radical precursor of SO_2 oxidation, then it is worthwhile to review information on the measurements and predictions of OH concentrations in polluted and unpolluted air. Kinetic simulations of hypothetically polluted air (Calvert and McQuigg, 1975) show that the maximum OH concentration attained during an irradiation is a function of the initial $NMHC/NO_x$ ratio, and that for a very reactive hydrocarbon such as 2-butene, the function goes through a maximum with respect to $NMHC/NO_x$ ratios. A striking feature of the kinetic simulations, however, is the relatively narrow range of attainable OH concentrations over a wide range of $NMHC/NO_x$ ratios. For example, Calvert and McQuigg's simulations with 2-butene predicted peak OH concentrations ranging between 0.7 and 1.1×10^{-7} ppm for all HC/NO_x ratios between 0.5 and 20.

The kinetic models that we have applied to the smog chamber system with 17 hydrocarbons generally predicts peak OH concentrations between 2 and 5×10^{-7} ppm. For most of the experimental conditions, the OH concentration goes through a fairly broad maximum during the first 2 h of the simulations, in parallel with the formation of sulfate aerosol. We have also found that the peak OH concentrations and the total OH integrals (i.e. $\int_0^{600} [OH]\, dt$) are nearly constant with respect to the initial precursor concentrations of NMHC and NO_x, at constant $NMHC/NO_x$ ratios. This prediction is consistent with the laboratory results which showed that for constant $NMHC/NO_x$ ratios, the maximum rates of SO_2 oxidation and total SO_2 conversion were quite insensitive to the initial precursor concentrations.

Estimates of OH concentrations in smog have been derived through the analysis of Lagrangian data for hydrocarbons and NO_2 (Calvert, 1976). His estimates of pre-noon OH concentrations in Los Angeles averaged near 1×10^{-7} ppm. Applying a rate constant of 1700 ppm^{-1} min^{-1} for the $SO_2 + OH$ reaction (Chan *et al.*, 1977) yields an SO_2 oxidation rate of about 1% h for the corresponding conditions. Direct measurements of OH in the atmosphere suggest substantially higher values. Wang *et al.* (1975) reported peak OH concentrations in Detroit of about 20×10^{-7} ppm. More recent measurements by Davis (1977) in New Mexico place the daytime average OH concentration near 4×10^{-7} ppm. In their account for the CO balance in the atmosphere, Weinstock and Chang (1974) estimated the average global daytime OH concentration at 2.28×10^{-7} ppm. Based on that estimate, the average residence time for SO_2 in the troposphere would be approximately 85 h. The results of Prahm *et al.* (1976) are essentially in agreement. From transport studies over 1000 kilometers, they determined the transformation rate for SO_2 to sulfate to be close to the eddy deposition rate for sulfate, half-lives of 60 and residence times of 90 h for each.

A variety of other reports have suggested both shorter and longer residence times for SO_2. The largest discrepancy pertains to urban pollution conditions. For example, Mészáros *et al.* (1977) estimated SO_2 half lives in the range of 2–10 h for the daytime period in an area south of Budapest. In comparison, our smog chamber results (normalized for diurnal radiation periods) suggest upperlimit daytime conversion for SO_2 of 20% which corresponds to a daytime half-life of 37 h and a residence time of over 100 h, assuming no oxidation at night. These upperlimits for gas phase SO_2 oxidation are consistent with the recent findings from the MISTT Program (Gillani, 1978).

4. CONCLUSIONS

Nitrogen oxides suppress or delay photooxidation of SO_2 in the atmosphere while reactive hydrocarbons promote oxidation. In polluted air, the ratio of these precursors ($NMHC/NO_x$), rather than their absolute concentrations, dominate the rate of oxidation. It is also clear that the rate of SO_2 oxidation to sulfate aerosol is directly proportional to the SO_2 concentrations. While NO_x is seen to delay SO_2 oxidation in much the same way as it delays ozone formation, the lifetime of NO_2 is so short compared to that of SO_2 that controls resulting in lower $NMHC/NO_x$ ratios are not expected to effectively reduce the SO_2-to-sulfate transformation process. From the standpoint of precursor control strategies, the most reliable approach for reducing ambient sulfate levels appears to be one that reduces SO_2 emissions.

A variety of atmospheric data exist which would indicate relatively short conversion times for SO_2 in polluted air. However, theoretical considerations of smog kinetics, together with the smog chamber results in the laboratory and outdoors, indicate that daytime conversions of SO_2 in urban areas may be limited to 10–20%. This suggests that other mechanisms, such as dry deposition, should be competitive with gas-phase oxidation mechanisms for removing SO_2 from the atmosphere.

Acknowledgements—The author gratefully acknowledges the assistance of Battelle associates; D. W. Joseph, G. W. Keighley, J. Kouyoumjian and G. F. Ward. The research was supported by the U.S. Environmental Protection Agency, Chemistry and Physics Laboratory, under the supervision of Dr. Basil Dimitriades (Contract Nos. 68-02-1720 and 68-02-2240).

REFERENCES

Battelle-Columbus Reports to the U.S. EPA (in preparation, 1977). A study of aerosol formation as a function of reactant concentrations (Contract 68-02-2240). A smog-chamber study of the rate of conversion of SO_2 as a function of reactant concentrations (Contract 68-02-1720).

Calvert J. G. and McQuigg R. D. (1975) The computer simulation of the rates and mechanisms of photochemical smog formation. J. Chem. Kinet., Symp. **1**, 113–154.

Calvert J. G. (1976) Hydrocarbon involvement in photochemical smog formation in Los Angeles atmosphere. *Environ. Sci. Technol.* **10**, 256–262.

Chan W. H., Uselman W. M., Calvert J. G. and Shaw J. H. (1977) The pressure dependence of the rate constant for the reaction: $HO + CO \rightarrow H + CO_2$. *Chem. Phys. Lett.* **45**, 240–244.

Clark W. E., Landis D. A. and Harker A. B. (1976) Measurements of the photochemical production of aerosols in ambient air near a freeway for a range of SO_2 concentrations. *Atmospheric Environment* **10**, 637–644.

Coutant R. W. Effect of environmental variables on the collection of atmospheric sulfate. *Environ. Sci. Technol.* (in press).

Davis D. D. (1977) Power plant plume study: direct measurements of the OH radical. Final Report to Electric Power Research Institute, Palo Alto, Ca.

Dimitriades B. (1977) Oxidant control strategies. Urban oxidant control strategy derived from existing smog chamber data. *Environ. Sci. Technol.* **1**, 80–88.

Gillani N. V. (1978) Project MISTT: Mesoscale plume modeling off the dispersion, transformation and ground removal of SO_2. *Atmospheric Environment* **12**, 569–588.

Liu B. Y. H., Whitby K. T. and Pui D. Y. H. (1974) A portable electrical analyzer for size distribution measurements of submicron aerosols. *J. Air Pollut. Control Ass.* **24**, 1067–1072.

Mészáros E., Moore D. J. and Lodge J. P. (1977) Sulfur dioxide–sulfate relationships in Budapest. *Atmospheric Environment* **11**, 345–349.

Miller D. F. and Levy A. (1973) Aerosol formation in photochemical smog. The effect of humidity and small particles. *Proceedings of the Third International Clean Air Congress*, VDI-Verlag GmbH, Dusseldorf.

Miller D. F., Levy A., Pui D. Y. H., Whitby K. T. and Wilson W. E. (1976) Combustion and photochemical aerosols attributable to automobiles. *J. Air Pollut. Control Ass.* **26**, 576–581.

Miller D. F. and Joseph D. W. (1976) Smog chamber studies on photochemical aerosol precursor relationships. EPA-600/3-76-080, NERC, Research Triangle Park, N.C.

Mulik J., Puckett R., Williams D. and Sawicki E. (1976) Ion chromatographic analysis of sulfate and nitrate in ambient aerosols. Analyt. Lett. **9**, 653–663.

Prahm L. P., Torp U. and Stern R. M. (1976) Deposition and transformation rates of sulphur oxides during atmospheric transport over the Atlantic. *Tellus* **28**, 355–372.

Small H., Stevens T. S. and Bauman W. C. (1975) Novel ion exchange chromatographic method using conductimetric detection. Analyt. Chem. **47**, 1801–1809.

Stephens E. R. (1973) Hydrocarbons in polluted air, Summary Report to the Coordinating Research Council (Project CAPA-5-68), Statewide Air Pollution Research Center at Riverside, Ca.

Wang C. C., Davis L. I., Wu C. H., Japar S., Niki H. and Weinstock B. (1975) Hydroxyl radical concentrations measured in ambient air. *Science* **189**, 797–800.

Weinstock B. and Chang T. Y. (1974) The global balance of carbon monoxide. *Tellus* **26**, 108–115.

Atmospheric Environment Vol. 12, pp. 281–287. Pergamon Press 1978. Printed in Great Britain.

EXPERIMENTAL INVESTIGATION OF THE AEROSOL-CATALYZED OXIDATION OF SO_2 UNDER ATMOSPHERIC CONDITIONS

G. HAURY*, S. JORDAN and C. HOFMANN

Kernforschungszentrum Karlsruhe, Germany, Laboratorium für Aerosolphysik und Filtertechnik, Postfach 3640, 75 Karlsruhe, W. Germany

(*First received* 17 *June* 1977 *and in final form* 26 *August* 1977)

Abstract—One of the most important processes affecting the chemical and physical form of atmospheric sulfur dioxide is the interaction of gaseous SO_2 with solid or liquid aerosol particles. The investigation of the catalytic activity of some synthetic aerosols and dust particles of different industrial origin under real atmospheric conditions (i.e. a wide range of relative humidity and temperature) have been investigated in the course of an experimental research program at Karlsruhe Nuclear Research Center, Germany. The experiments were carried out in a 4.5 m^3 climate-controlled reaction chamber especially designed for SO_2-experiments in the ppm-concentration range. Starting with SO_2-concentrations of 3–5 mg m^{-3} the removal of SO_2 from gas phase in the presence of moist aerosol-free air and moist aerosol-polluted air was investigated as was the change of chemical composition of aerosols due to SO_2-adsorption and catalytic oxidation. The results of these experiments with synthetic $MnSO_4$ aerosols and fly-ash particles of a hard coal fired power plant are presented. They indicate a strong dependence of the catalytic oxidation rate on the relative humidity but no disappearance of the catalytic activity at low relative humidities.

INTRODUCTION

Sulfur dioxide is an important air pollutant mainly released into the atmosphere by fossil fuel fired heat and power plants. To estimate the impact of SO_2-emission it is necessary to know both atmospheric dispersion and deposition on the ground and chemical reactions of SO_2 in the atmosphere during transport.

According to the present state of knowledge three processes are mainly responsible for the oxidation of SO_2 in the atmosphere: (a) Photoreactions caused by sunlight and the presence of many other pollutants. (b) Heterogeneous reactions of SO_2 adsorbed on catalyzing particles. (c) Combined reactions with particles, sunlight and other pollutants. The literature on the catalytic oxidation of SO_2 using real aerosol systems and taking into account physics of aerosols is very limited. Up to now the influence of aerosol systems has been either neglected or only theorized about. The first experiments on catalytic oxidation of SO_2 were made with solutions of $MnSO_4$. Junge and Ryan (1958) found the oxidation of SO_2 to be much dependent on the pH-value of the solution. Johnstone and Caughanowr (1958) investigated SO_2 reaction with O_2 in big droplets of $MnSO_4$ solution. They found the oxidation rate proportional to the square of $MnSO_4$ concentration. Johnstone and Moll (1969) used airborne particles. Experiments were performed at relative humidities above 86% and SO_2 concentrations of 250 ppm. Cheng *et al.* (1971) deposited aerosols effective as catalysts on filters and exposed these filters to SO_2-polluted gas. They found that $MnSO_4$, $MnCl_2$ and $CuSO_4$ are 12.2, 3.5 and 2.4 times respectively more effective for the oxidation of SO_2 than NaCl. It was established that the influence of relative humidity is (qualitatively) very important. Most experiments were carried out at r.h.'s above 80%. Chun and Quon (1973) investigated the oxidation of SO_2 through iron oxide particles using a technique similar to Cheng's.

From these studies it can be concluded that many heavy metal compounds which are known to be present in atmospheric aerosols have an important catalytic effect on SO_2 oxidation, especially at high relative humidities. Furthermore, the efficiency of catalysts was found to decrease with decreasing pH. Important questions still remain to be solved, especially with respect to problems of reaction kinetics of SO_2 with dry aerosols. The following experiments were carried out to investigate the catalytic oxidation of SO_2 by genuine airborne natural aerosols and conditions which allow a transfer of results to atmospheric simulation calculations.

EXPERIMENTAL FACILITIES

The experiments were carried out in a controlled environment reaction chamber especially designed for SO_2 experiments. The vacuum tight vessel had a capacity of 4500 l. and was coated inside with an inert plastic material (Penton: Chlorinated Polyether) (Jordan, 1973) to prevent SO_2 wall reactions. The

* Any communications to: Babcock-Brown Boveri Reactor GmbH, 68 Mannheim 41, POB 410323.

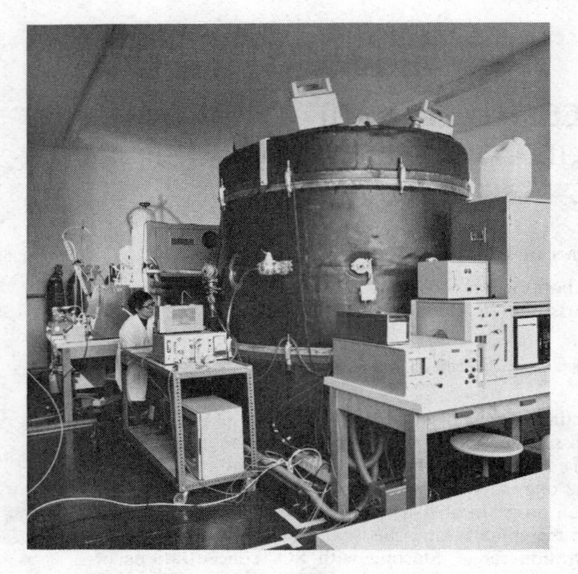

Fig. 1. Photograph of the reaction chamber.

chamber was temperature controlled between −15 and +50°C. The relative humidity of gas flow into the chamber was controlled between 10 and 95%. Figure 1 is a photograph, Fig. 2 shows the flow diagram of the reaction chamber.

For measurement of SO_2 concentration the SO_2 monitors Picoflux I and Picoflux 2T which measure changes in conductivity through oxidation of SO_2 in H_2O_2 solution and Tracor Analyzer System 1200 (a flame ionisation detector selectively detecting different sulfur compounds) were used. The dew point in the reactor was measured continuously by a LiCl-detector. Aerosol mass concentration was measured before entering the vessel and in the chamber itself by a microbalance of Thermo-Systems Corp. (range between 1.10^{-6} g m^{-3} to 0.1 g m^{-3}). Particle number concentration was measured by Condensation Nuclei Counter (CNC) from Environmental One Corp. (range up to 1.10^{13} particles m^{-3}).

Some of the experiments were performed with artificial aerosols. These aerosols were produced by spraying salt solution (0.01 and 0.1 M) with an ultrasonic aerosol generator and subsequent dehumidification of the droplets by mixing with dry air. The size of primary droplets was measured to be 1.3 ± 0.2 μm. The diameter of the dried particles depended on the concentration of the solution and aerosol deposition and coagulation processes in pipes before entering the reaction chamber. The physical state of $MnSO_4$ and other salt particles as a function of relative humidity was discussed in detail by Haury and Jordan (1973). The particle number concentration at the inlet of the reactor was measured to be 5×10^{12} particles m^{-3}. By sampling aerosols and SO_2 at different locations in the reactor vessel at short time intervals it was found that both aerosols and SO_2 were homogeneously mixed in the bulk volume after a few minutes.

Several experiments were performed with genuine dust from a coal fired power plant (Großkraftwerk Mannheim) obtained from the stack gas behind the electrofilters, i.e. with the same particles normally released to the atmosphere. The sampled dust was reheated to stack temperature (140°C), fluidized by a magnetic stirrer and dispersed by a jet of dry gas. Before entering the reaction chamber the aerosol was stored in a buffer vessel to allow deposition of clustered particles. By this means, the investigated aerosol in the reaction vessel was very similar to the original

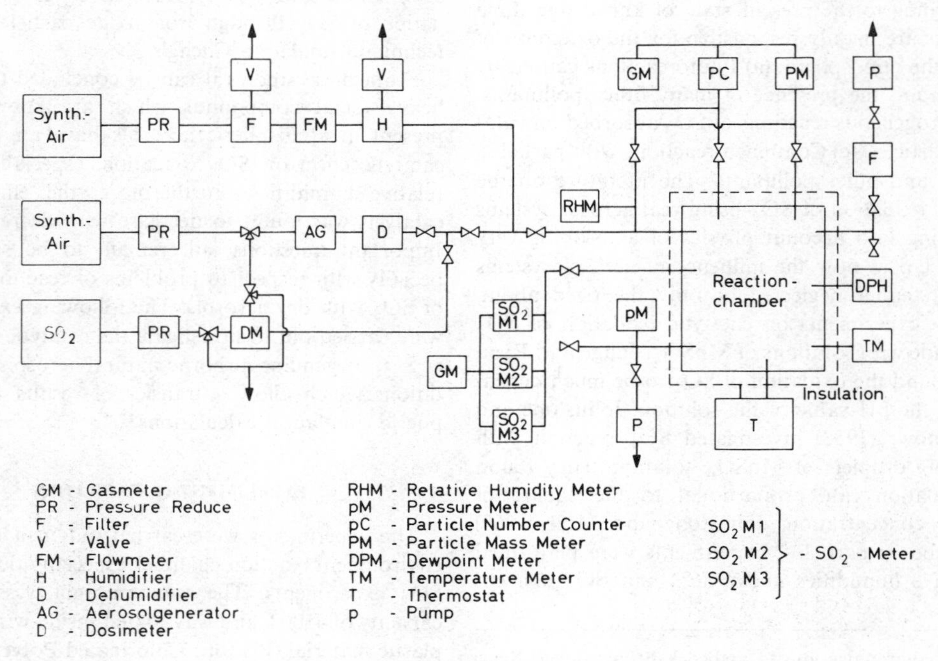

Fig. 2. Flow diagram of test facility.

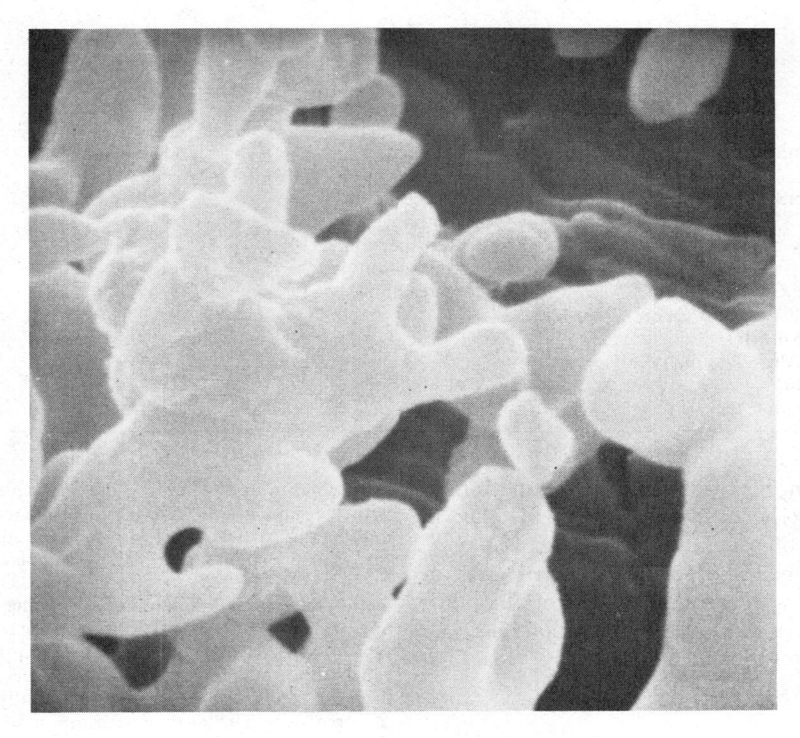

Fig. 3. Photograph of $MnSO_4$ aerosols.

flue gas aerosol. Form and structure of the particle surfaces were determined with a scanning electron microscope. Figure 3 is a photograph of $MnSO_4$ aerosols (crystals) and Fig. 4 is a picture of fly-ash particles both used for experiments.

The surface area of particles was determined by a nitrogen adsorption method (BET) described by Devitofrancesco and Liberti (1966) with a Sorptometer from Perkin–Elmer Corp. Results of these measurements are compiled in Table 1 and compared

Fig. 4. Photograph of fly-ash particles.

Table 1. Specific surface areas of different aerosols

Aerosol	State	Specific Surface Area ($m^2 g^{-1}$)
$MnSO_4$	before reaction low r.h.	4.3 ± 1
$MnSO_4$	after reaction low r.h.	4.15 ± 1
Fly-ash	before dispersion	2.8 ± 1
Fly-ash	after dispersion	3.1 ± 1
Fe_2O_3 (Chun and Quon, 1973)	before reaction	$247\ m^2\ g^{-1}$
Coal ash (Bickelhaupt, 1975)		$0.3\ m^2\ cm^{-3}$
Dust (Liberti, 1970)	urban area	$2\text{–}2.5\ m^2\ g^{-1}$

with results from other authors. The specific surface of $MnSO_4$ particles and fly-ash was found to be in the same order of magnitude of about $3\text{–}4\ m^2\ g^{-1}$. The chemical composition of the fly-ash is given in Table 2 together with chemical analysis of other dust samples. The samples under investigation therefore seem to be representative of flue ash of coal-fired plants.

EXPERIMENTS

Experiments were carried out as follows: after cleaning and evacuating the reaction vessel it was filled with humidity-controlled artificial air filtered by charcoal and fiberglass filters, SO_2 was dosed into the gas stream to a concentration of $3\text{–}5\ mg\ m^{-3}$ (initial SO_2 concentration) by a special dosimeter. Each experiment consisted of two runs. Depending on experimental conditions each run took 20–24 h. During this time SO_2 concentration decreased to about $0.1\ mg\ SO_2\ m^{-3}$. The first run the chamber was filled with SO_2 polluted air only without aerosols. Decreasing SO_2 concentration in the chamber caused only by withdrawing gas through instruments and through wall effects was measured. The second run followed under the same conditions after evacuating and refilling the chamber with the same gas mixture (and SO_2 concentrations). During this run particles were blown continuously with a fixed rate into the chamber with the gas replacing the air used by SO_2 and particle monitors. Decreasing SO_2 concentration with time was compared with the first run. Figure 5 represents the measured SO_2 concentration during a typical experiment. The upper curve represents the first run without catalytic aerosol; the lower curve shows the decreasing SO_2 concentration additionally caused by particle reaction. The dust concentration in the vessel as a function of time for experiments with different relative humidities is shown in Fig. 6.

Table 2. Chemical composition of fly-ash collected behind electrofilters. 1, 2, 3: samples from Großkraftwerk Mannheim (1975), 4: mean values of samples from U.S. coal-fired plants (Bickelhaupt, 1975)

Element	Proportion in % 1	2	3	4
SiO_2	44.0	47.33	—	48.2
Al_2O_3	28.4	31.21	1.39	23.8
Fe_2O_3	8.44	8.12	19.9	8.93
TiO_2	3.05	2.31	0.31	1.55
CaO	1.94	4.23	0.56	7.4
MgO	1.81	1.91	1.57	2.4
Na_2O	1.24	0.15	6.87	0.65
K_2O	4.52	0.19	0.29	1.88
ZnO	0.89	0.76	0.85	—
MnO	0.08	0.07	0.15	—
NiO	0.15	0.10	27.4	—
CuO	0.22	0.15	0.18	—
Cr_2O_3	0.11	0.15	0.34	—
V_2O_5	0.09	0.24	34.6	—
SO_3	1.05	2.65	4.92	—
P_2O_5	1.86	0.31	—	0.36
Residue	2.15	0.12	0.77	4.83

By evaluating the kinetics of these two runs the disappearance of SO_2 by aerosol reaction was detected. To obtain information on the parameters of heterogeneous chemical reactions, experimental data, time functions of SO_2 concentrations, dust-input rate and concentration in the vessel had to be freed from influences which were not determined by the chemical reaction itself. In a series of preliminary experiments the following observations were made which were confirmed by all later experiments and which therefore were used to derive a concept for mathematical evaluation of results.

The wall reaction increased only very slowly from experiment to experiment although a great part of aerosols inserted into the chamber during the foregoing experiment stuck to the walls due to sedimentation and diffusion. Consequently the SO_2 adsorbed by aerosols must have changed the aerosol surface chemically because after each

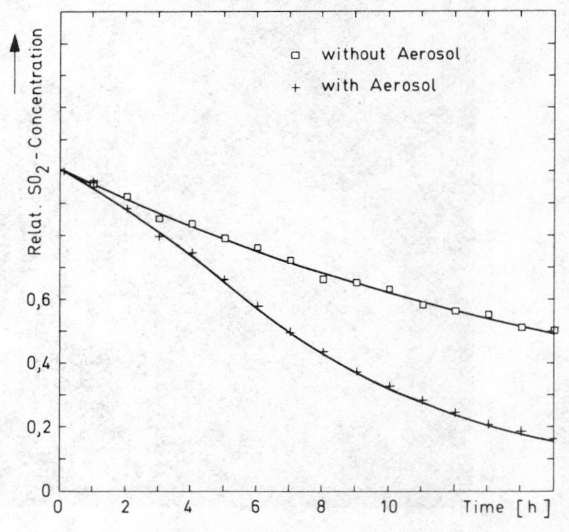

Fig. 5. SO_2-concentration during experiment.

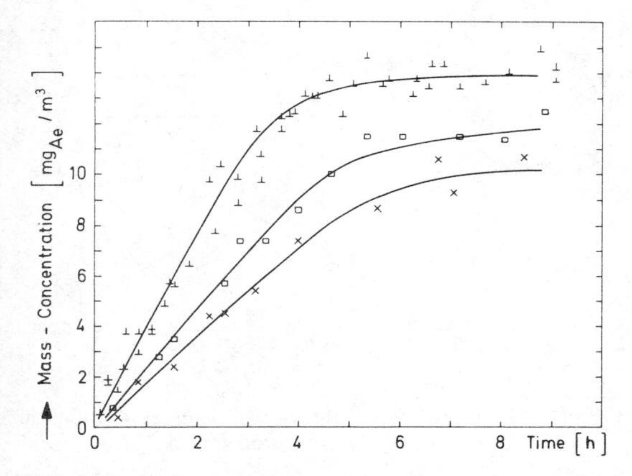

Fig. 6. Aerosol mass concentration in reaction chamber at r.h. = 60% (—); r.h. = 87% (□); r.h. = 91% (×).

experiment the chamber was evacuated and therefore SO_2 which had been only adsorbed would have been desorbed during evacuation. (The aerosols sticking to the walls would have increased the wall reaction at the next experiment.)

The rate of SO_2 conversion during experiments with aerosols was constant after a short initial phase if the rate of aerosol addition, temperature, humidity and flow rate of gas samples were kept constant. Because the amount of aerosol material, which is airborne as well as adsorbed at the walls, increased during the experiment due to a constant addition rate and only a small sample rate this effect can only be explained by the assumption that the catalytic activity of each particle decreased as a consequence of the chemical reaction with SO_2. This is understandable if the possible products of SO_2 oxidations (sulfurous and sulfuric acid or some of their salts) decrease the activity of aerosol particles so that further reaction which is known to be limited by pH values cannot take place.

The reaction seemed to follow first or zero order in SO_2 concentration depending on the type of aerosol material and on the relative humidity.

RESULTS AND DISCUSSIONS

On the basis of these observations and their logical consequences and by means of the well-known kinetics of sample gas flow and aerosol input the following differential equations were derived to closely describe the kinetics of experiments: zero order reaction:

$$\frac{d}{dt}c(t) = -[v(t) + w(t)]\cdot c(t) - \frac{a_0 \cdot z}{(v + \alpha)}(1 - \exp\{-(\alpha + v)\cdot t\}). \quad (1)$$

First order reaction:

$$\frac{d}{dt}c(t) = -\left[v(t) + w(t) + \frac{a_1 \cdot z}{(v + \alpha)} \times (1 - \exp\{-(\alpha + v)\cdot t\})\right]\cdot c(t). \quad (2)$$

The exponential terms in Equations (1) and (2) represent the decrease of catalytic activity of each individual particle during the reaction (Haury, 1976).

Integration of Equations (1) and (2) leads to the time functions of SO_2 concentrations for experiments with reaction of zero order (Equation 3) and of first order (Equation 4). The rate of wall reaction $w(t)$ was assumed not to be time dependent (further represented therefore by the symbol w only) (Haury, 1976).

Zero order reaction:

$$c(t) = \left[c_0 - \frac{a_0 z}{(v + w)(w - \alpha)}\right]\exp\{-(v + w)\cdot t\} + \frac{a_0 z}{(\alpha + v)(w - \alpha)}\cdot\exp\{-(\alpha + v)\cdot t\} - \frac{a_0 z}{(v + w)(\alpha + v)}. \quad (3)$$

First order reaction:

$$c(t) = c_0 \exp\left\{-(v + w) + \frac{a_1 z}{(\alpha + v)}\cdot t\right\} \times \exp\left\{-\frac{a_1 z}{(\alpha + v)^2}(\tau - 1)\right\} \quad (4)$$

with

$$\tau = \exp\{-(\alpha + v)\cdot t\}.$$

The symbols have the following meanings:

$c(t)$ [$g\,m^{-3}$]	Time dependent SO_2 concentration.
$t[h]$	Reaction time.
c_0[$g\,m^{-3}$]	Initial SO_2 concentration.
$v[h^{-1}]$	Rate of dilution by gas and aerosol sampling kept constant during one experiment.
$w(t)$ [h^{-1}]	Rate of wall reaction, generally time dependent by increasing aerosol deposition during the experiment. As $w(t)$ is small compared with v and reaction rate, $w(t)$ can be assumed to be nearly constant.
a_0[$g_{SO_2}\, g_{aerosol}^{-1}\, h^{-1}$] a_1 [$g_{aerosol}^{-1}\, h^{-1}$]	Initial reaction rates of zero or first order reaction respectively on freshly added aerosols.
z[$g_{aerosol}^{-1}\, m^{-3}\, h^{-1}$]	Rate of aerosol addition.
$\alpha[h^{-1}]$	Rate of decrease of catalytic activity.

Especially parameters a_0, a_1 and α in Equations (1)–(4) can be used to characterize the catalytic reaction of different types of aerosols. One of the most interesting findings was the fact that the capacity of each particle to oxidize SO_2 molecules decreased with reaction time and was totally depleted after some time. Therefore, in addition to Equation (5) parameters k_0 and k_1 are defined to be a measure of the conversion capacity of a given aerosol material:

$$k_1 = \int_{t=0}^{t=\infty} a_i \exp(-\alpha t)\,dt = \frac{a_i}{\alpha}$$

$$k_0 = \frac{a_0}{\alpha}\,[g_{SO_2}\, g_{aerosol}^{-1}] \quad (5)$$

$$k_1 = \frac{a_1}{\alpha}\,[g_{aerosol}^{-1}]$$

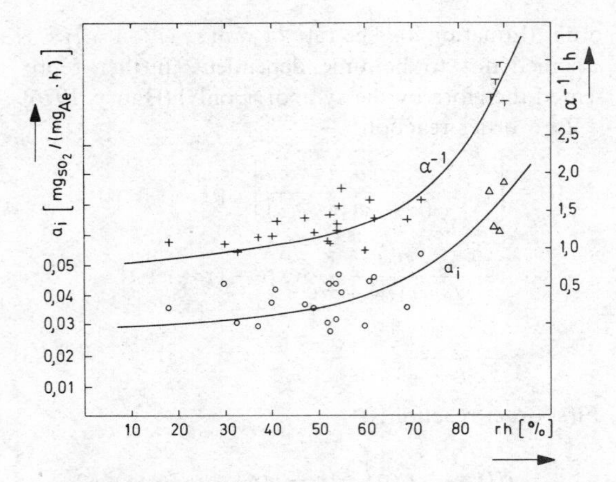

Fig. 7. Dependence of reaction rate constant a_i and half-life time α^{-1} for $MnSO_4$ aerosols from r.h.

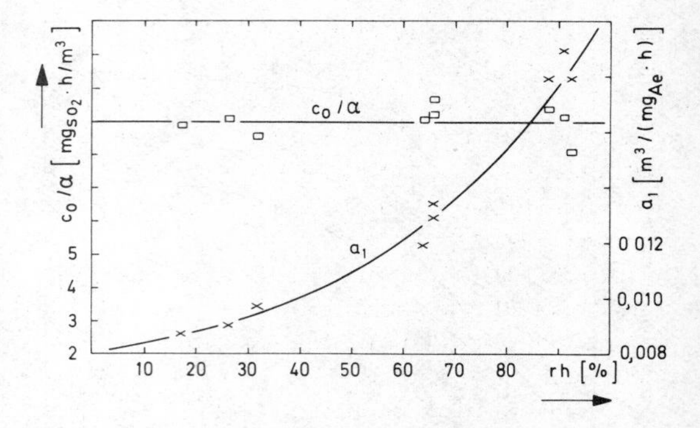

Fig. 9. Relative reaction rate a_1 and c_0/α as a function of r.h. for fly-ash reaction with SO_2.

In the case of first order reaction k_1 is of relative importance only. By multiplication with c_0 another k'_1 is obtained which has the same dimension as k_0 [$g_{SO_2}\, g^{-1}_{aerosol}$]. Also in the case of the direct comparison of reactions with zero order and first order the relative parameter a_1 for the first order was multiplied by c_0 to obtain the same dimension (a_1: $g_{SO_2}\, g^{-1}_{aerosol}\, h^{-1}$). The dependence of α and k_1 on c_0 in this case is described in detail by Haury (1976). These parameters were evaluated from experimental data of each experiment and they will be used for discussion of results in the following.

The compiled values for parameters a_0, a_1, α, k_i are presented as a function of relative humidity in Figs. 7 to 10 for the two investigated aerosol systems ($MnSO_4$ aerosols and fly-ash aerosols). Dependence of these parameters on temperature was not found in the temperature range 18–55°C if relative humidity was kept constant.

The main results which can be obtained from this compilation of data are the following: (1) *Manganese salt aerosols.* Reaction on manganese sulfate aerosols (Figs. 7 and 8) changed from zero order at low relative humidity to first order at high relative humidities. Essentially it was found that (by introducing parameters of the same dimension): the initial reaction rates increased slowly for relative humidities under 70% but increase faster for higher humidities; the half life times for catalytic activities (that means α^{-1}) show the same behavior; the reactive capacities of droplets of $MnSO_4$ solution are nearly four times as great as those of the equivalent $MnSO_4$ crystals at lower humidities. (2) *Fly-ash aerosols.* The results of experiments with fly-ash particles are presented in Figs. 9 and 10. The essentials are: the reaction was found in all cases to be of first order in SO_2 concentration; the initial reaction rates a_1 increased by about the square of relative humidity (Fig. 9); the half life times (α^{-1}) of aerosol activities depend very much on initial SO_2 concentration and not on relative humidity. The products of α^{-1} and c_0 are nearly constant for the whole range of relative humidity; the total reactive capacity of fly-ash particle (Fig. 10) for SO_2 conversion increases from 0.08 at 20% r.h. to about 0.2 $mgSO_2\, mg^{-1}$ dust at about 90% r.h. proportional to about the square of r.h.

A comparison of the experimental results for $MnSO_4$ aerosols with other authors' data shows in some cases good accordance. A direct comparison of reaction rates was not possible because of great differences in the definition of these rates due to different experimental kinetics. A comparison of reactive capacity, i.e. the parameter of greatest interest in view of environmental problems was possible and showed good agreement. Cheng *et al.* (1971) found a reactive capacity of 0.231 mg mg^{-1} MnCl for r.h. of 95% and a temperature of 23°C whereas the authors' own

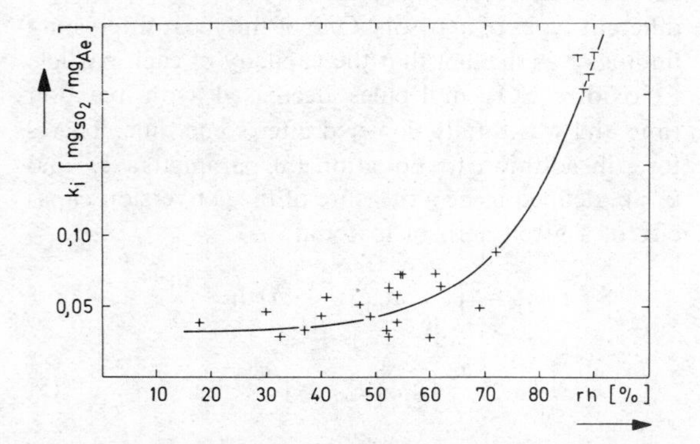

Fig. 8. Reaction capacity of $MnSO_4$ aerosols as a function of r.h.; +, zero order, T: first order reaction.

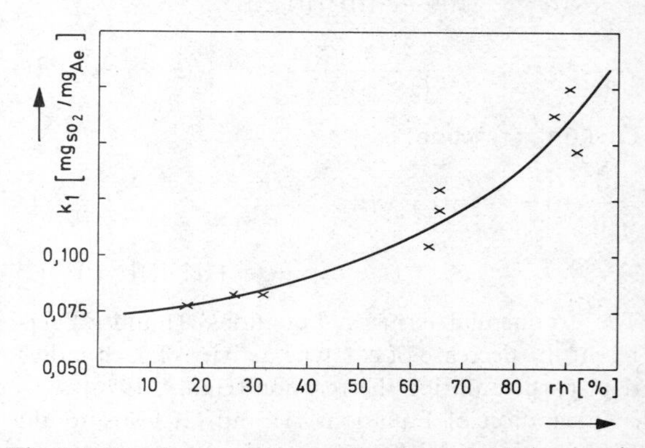

Fig. 10. Reaction capacity as a function of relative humidity for fly-ash reaction with SO_2.

results show a reactive capacity of about 0.2 mg mg^{-1} aerosol. In the case of lower humidities Cheng *et al.* (1971) give data for NaCl aerosols which indicate a strong decrease of all reaction parameters for decreasing r.h. from 81 to 60% and a weaker decrease for transition to 45 or 36% respectively. In each case, however, Cheng *et al.* (1971) confirm the result of this work, i.e. that oxidation of SO_2 does not stop at lower humidities as was discussed by Junge and Ryan (1958), Johnstone and Coughanowr (1958) and Anderson and Johnstone (1955).

A comparison of results of $MnSO_4$ experiments with those of dust particle experiments on the other hand shows some interesting facts: though the manganese content of the dust particles (Table 2) is very low compared with manganese sulfate particles, the parameters of reaction are in good agreement for both cases. It seems, therefore, that not only manganese but other metals or metal compounds in dust particle as well have good catalytic activities. As can be seen in Figs. 3 and 4 the surface structures of dry $MnSO_4$ particles and dust particles are quite different whereas the specific adsorption surfaces are comparable in both cases (Table 1). On the other hand the reaction parameters in both cases are similar so that it seems that from all physical properties of the aerosols the adsorption surface seems to have the greatest influence on catalytic SO_2 reaction.

Acknowledgement—The authors wish to thank the Großkraftwerk Mannheim for providing samples of fly-ash and chemical analysis.

REFERENCES

Anderson L. B. and Johnstone H. F. (1955) Gas adsorption and oxidation in dispersed media. *Am. Inst. Chem. J.* **1**, 135–141.

Bickelhaupt R. E. (1975) Surface resistivity and the chemical composition of fly-ash. *J. Air Pollut. Control Ass.* **25**, 148–152.

Cheng R. T. *et al.* (1971) Contribution to the reaction kinetics of water soluble aerosols and SO_2 in the air at ppm concentration. *Atmospheric Environment* **5**, 987–1008.

Chun K. C. and Quon J. E. (1973) Capacity of ferric oxide particles to oxidize sulfur dioxide in air. *Environ. Sci. Technol.* **7**, 532–538.

Devitofrancesco G. and Liberti A. (1966) Bestimmung der Staubkonzentration durch Oberflächenmessung. *Staub* **26**, 194–196.

Großkraftwerk Mannheim AG (1975) Private communication.

Haury G. and Jordan S. (1973) Katalytische Oxidation von SO_3 an Aerosolen. Proceedings of the Annual Meeting of German Aerosol Society, Bad Soden, Germany.

Haury G. (1976) Untersuchungen zur katalytischen Oxidation von Schwefeldioxyd an Aerosolen unter atmosphärischen Bedingungen, KFK 2318 UF.

Johnstone H. F. and Coughanowr D. R. (1958) Adsorption of sulfur dioxide from air. *Ind. Engng Chem.* **50**, 1169–1172.

Johnstone H. F. and Moll A. J. (1960) Formation of sulfuric acid in fogs. *Ind. Engng Chem.* **52**, 861–863.

Jordan S. (1973) Messungen der Permeabilität einiger Kunststoffe gegenüber Schwefeldioxyd. *Staub* **33**, 36–38.

Junge C. E. and Ryan T. G. (1958) Study of the SO_2-oxidation in solution and its role in atmospheric chemistry. *Q.J.R. Met. Soc.* **84**, 46–55.

Liberti A. (1970) The nature of particulate matter. *Pure appl. Chem.* **24**, 631–642.

Atmospheric Environment Vol. 12, pp. 289–293. Pergamon Press 1978. Printed in Great Britain.

INTERACTIONS BETWEEN SO_2 AND CARBONACEOUS PARTICULATES*

R. TARTARELLI, P. DAVINI, F. MORELLI and P. CORSI
Institute of Industrial and Applied Chemistry of the University of Pisa, Italy

(*First received* 6 *June* 1977 *and in final form* 27 *September* 1977)

Abstract – Interactions between SO_2 at low concentrations and carbonaceous particulates have been investigated. The amount of SO_2 adsorbed by the solid has been related to its surface area and valued at different compositions of the gaseous mixture in the temperature range 20–150°C. A distinction between two types of adsorption has been made: a part of the adsorbed SO_2 is weakly adsorbed, SO_2(a) and the residual part is strongly bound to the solid, SO_2(b). The fraction of SO_2(b) is enhanced by the presence of O_2 and H_2O. Evolution of SO_2(a) to SO_2(b) on the solid occurs and its rate appears to be increased by the presence of adsorbed O_2 and H_2O. By contacting the particulate containing SO_2 with a flux of air at room temperature, the SO_2(a) desorbs into the gaseous stream, but a part of it may be also transformed to SO_2(b).

INTRODUCTION

The interactions between SO_2, present in the flue gases from the combustion of oils containing sulfur compounds, with the carbonaceous particulates (soot and coke particles) generated by incomplete combustion may be expected to contribute to the evolution of SO_2 in the atmosphere around power plant installations. In this connection, Novakov *et al.* (1974) showed the important role played by finely divided carbon particles in the catalytic oxidation of sulfur dioxide to sulfate in polluted atmospheres.

The SO_2-particulate contact conditions change from the chimney exit through the following dispersion of SO_2 and solid particles into the environment. Thus, the concentration of SO_2 in contact with solid will become lower and lower and, according to the different conditions, SO_2 may be adsorbed, retained or released by particulate.

In the present laboratory investigation the interactions between SO_2 and carbonaceous particulates are analysed in the SO_2 adsorption from gaseous mixtures and desorption by air.

* Work supported by the CNR Program "Improvement of the Environment Quality".

EXPERIMENTAL SECTION

The different samples of carbonaceous particulate have been collected from flue ducts of oil-burning power stations. Their contents of carbon, hydrogen, sulfur, and ashes (determined as residual product from combustion of carbonaceous part) and specific surface areas have been reported in Table 1.

Sulfur may be present under different compounds as metallic salts, sulfuric acid or sulfur oxides, formed or deposited on the particles in the combustion chamber or along the flue duct. The acidity of the particulate, indicated by the pH of the suspension of solid (1 g) in water (50 cm^3) (Garten and Weiss, 1955), may be attributed mostly to the acid sulfur compounds, as it appears in Fig. 1, where the acidity increases generally with the sulfur content.

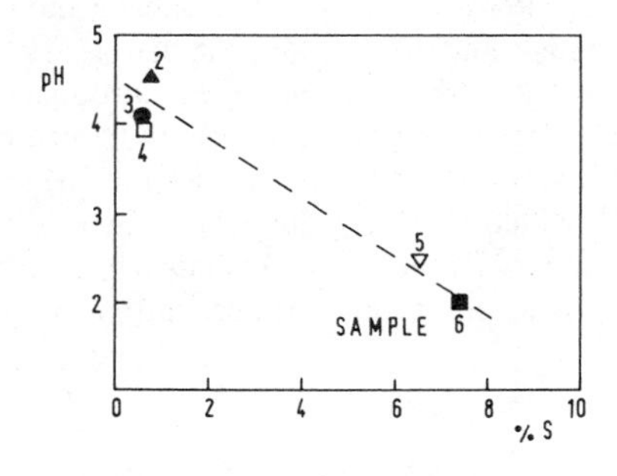

Fig. 1. Particulate acidity against sulfur content.

Table 1. Carbon, hydrogen, sulphur, and ashes contents and specific surface area of particulates

Sample	C (%)	H (%)	S (%)	Ashes (%)	Specific surface ($m^2\ g^{-1}$)
1	94.86	1.41	0	1.53	100
2	94.24	1.66	0.90	0.23	52
3	93.01	1.12	0.58	1.08	827
4	88.39	1.16	0.55	0.54	925
5	86.68	0.40	6.52	3.58	110
6	76.77	0.62	7.42	7.45	29

The sulfur dioxide has been adsorbed from a mixture of nitrogen ($9\,l.\,h^{-1}$), with different contents of O_2 and H_2O, flowing through a fixed bed of carbonaceous particulate. The removal of SO_x from the solid has been carried out by a stream of nitrogen or air ($\simeq 1.5\,l.\,h^{-1}$) at a controlled temperature. The exit stream bubbles in a trap containing quartz wool impregnated with a 5% H_2O_2 water solution, which absorbs SO_x completely. The collected H_2SO_4 is titrated by 0.02N NaOH and bromophenol as indicator.

It has been reported (Davini *et al.*, 1977a) that active carbon, on which SO_2 has been previously adsorbed, releases only a part of the SO_2 in a nitrogen stream at temperatures below 120°C, the desorption of the residual part starting above 200°C with a behavior quite like that of a carbon containing sulfuric acid. On these bases, a distinction between two types of SO_2 adsorption has been made: weakly adsorbed sulfur dioxide, which will be indicated as SO_2(a), and sulfur dioxide strongly bound to the carbonaceous matrix, SO_2(b). For active carbons, SO_2 was determined by treating the samples with a nitrogen stream for 3 h at 160°C; the complete desorption of SO_2 as SO_2(b) was accomplished by a treatment for 15 h at 350°C (Davini *et al.*, 1977b). For the particulates, containing originally sulfur compounds, SO_2(a) has been determined by the treatment at 160°C but SO_2(b) has been evaluated as the difference between the total adsorbed SO_2 and SO_2(a). The last method follows from the fact that fresh particulate containing sulfur compounds releases negligible amounts of SO_2 at 160°C, but desorption of SO_x occurs more markedly at 350°C.

RESULTS AND DISCUSSION

Adsorption of SO_2 *at* 20°C

The different samples of particulate produce a quite wide range of surface area. It appears from Fig. 2, where SO_2 has been adsorbed for 1 h from a mixture N_2–SO_2 (0.24% SO_2) on 0.5 g of solid, that the total amount of adsorbed SO_2 may be related directly to the specific surface. This behavior, previously pointed out for carbons (Siedlewski, 1964; Davini *et al.*, 1977b) is in agreement with the prevailingly physical type of SO_2 adsorption at low temperatures (Ficai, 1928; Lederer, 1932; Remy and Hene, 1932; Beebe and Dell, 1955; Kachanak and Jona, 1961; Siedlewski, 1965a).

The SO_2 adsorbed from the carbonaceous particulates may be compared with that adsorbed from an active carbon, having a specific surface of 871 $m^2\,g^{-1}$, which was 5.5 g SO_2/100 g carbon.

In Fig. 3 the total amount of SO_2 adsorbed by 0.3 g of Sample 3 is given at different adsorption times for

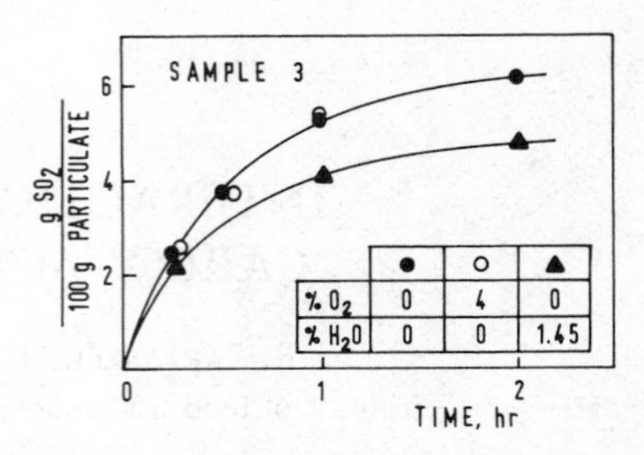

Fig. 3. Total amount of adsorbed SO_2 against adsorption time.

gaseous mixtures containing 0.24% SO_2 and different concentrations of O_2 and H_2O. All the following runs have been carried out under the same conditions, i.e. adsorption on 0.3 g of particulate for 1 h.

In Fig. 4 the total amount of adsorbed SO_2 is plotted against the SO_2 concentration.

From Figs. 3 and 4 it appears that the presence of moisture hinders the SO_2 adsorption, but the effect becomes weaker as the O_2 concentration rises.

The presence of O_2 and H_2O increases the fraction of strongly adsorbed SO_2, determined immediately after adsorption (Fig. 5). The fraction of SO_2(b) may be related to the acidity of the particulates, as it appears in Fig. 6 where the ratio SO_2(b)/SO_2 increases generally with the acidity; this behavior is in agreement with

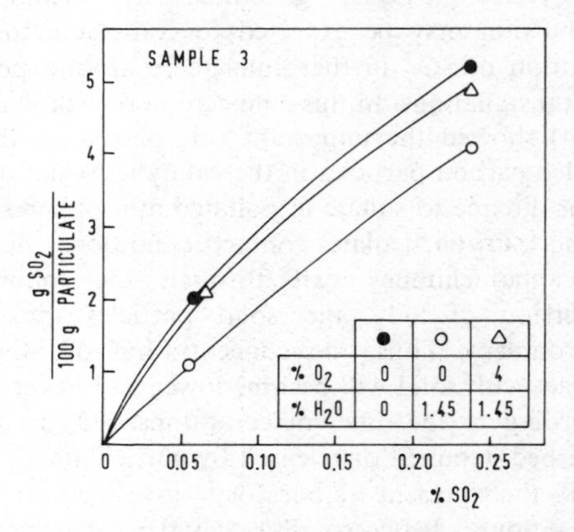

Fig. 4. Total amount of adsorbed SO_2 against SO_2 concentration.

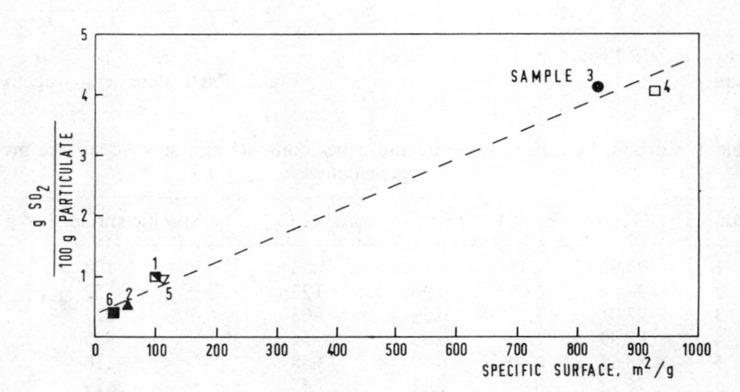

Fig. 2. Total amount of adsorbed SO_2 against specific surface of particulates.

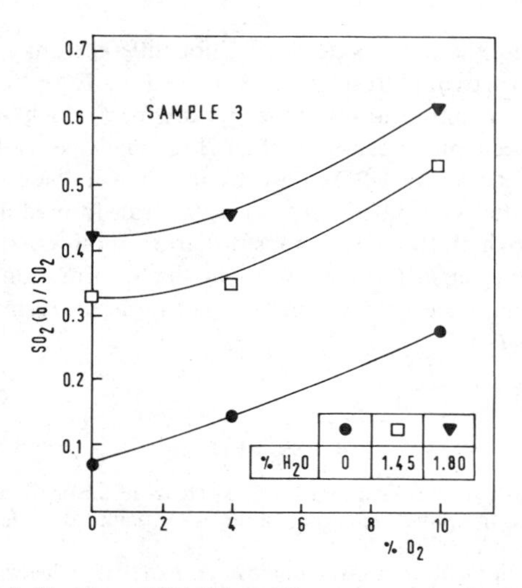

Fig. 5. Fraction of $SO_2(b)$ at different concentrations of O_2 and H_2O in a mixture containing 0.24% SO_2.

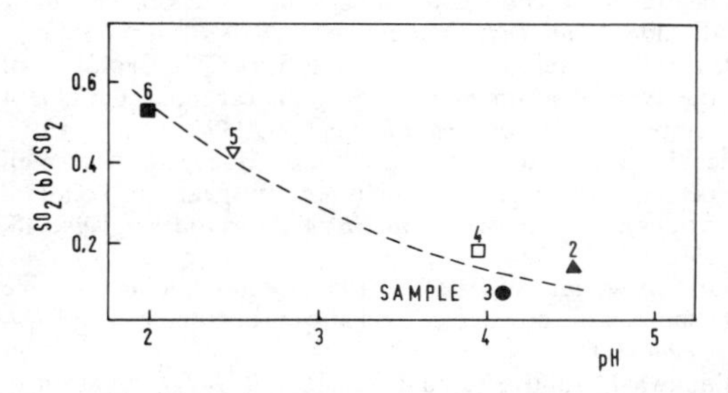

Fig. 6. Fraction of $SO_2(b)$ against particulate acidity. SO_2 adsorbed from a mixture 0.24% SO_2–N_2.

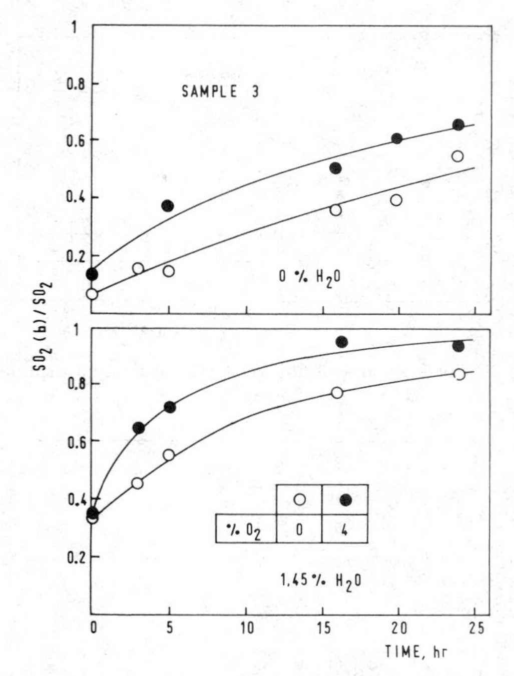

Fig. 7. Fraction of $SO_2(b)$ at different times after adsorption. SO_2 adsorbed from a mixture containing 0.24% SO_2.

that of a carbon containing sulfuric acid, on which SO_2 was adsorbed prevailingly as $SO_2(b)$ (Davini *et al.*, 1977a).

If the amount of $SO_2(a)$ is determined on samples kept isolated at 20°C for different times after adsorption, the weakly adsorbed SO_2 appears to be transformed into $SO_2(b)$ with a rate which increases with the contents of O_2 and H_2O in the mixture from which SO_2 has been previously adsorbed (Fig. 7).

These results may be interpreted by assuming that the surface oxygen, both adsorbed from the gaseous mixture and originally present on the solid, causes oxidation of adsorbed SO_2 with formation of complexes characterized by different stabilities up to SO_3 (Davtyan and Ovchinnikova, 1955; Ovchinnikova and Davtyan, 1956, 1961a, 1961b; Johswich, 1962; Siedlewski, 1965a, 1965b; Ishikawa *et al.*, 1969). Such products, which do not desorb spontaneously at room temperature but are fixed on the carbonaceous matrix, prevent further adsorption of gaseous SO_2 (Siedlewski, 1965a). While the total amount of adsorbed SO_2, in the absence of moisture, is practically independent of O_2 content in the gaseous mixture, a higher amount of adsorbed oxygen causes a higher fraction of strongly bound SO_2 and higher rate of $SO_2(a)$ oxidation (Figs. 5 and 7).

The unfavourable effect of moisture on SO_2 adsorption, especially at low content of O_2, may be attributed to a competition with SO_2 for the same adsorption sites and/or to the presence of a liquid film, formed on the solid surface, which hinders the penetration of SO_2 to the sites.

On the other hand, the presence of adsorbed H_2O improves the transformation $SO_2(a) \rightarrow SO_2(b)$; this behavior may be attributed to the removal of adsorbed product from the active sites, which will be restored for further $SO_2(a)$ oxidation.

The favourable effect of both O_2 and H_2O in promoting the formation of strongly adsorbed SO_2 may be considered consistent with the results by Novakov *et al.* (1974), where the presence of O_2 and H_2O in the gaseous mixture containing SO_2 increased the intensity of the sulfate peaks in the ESCA spectra of soot.

Adsorption of SO_2 *at higher temperatures*

Figures 8 and 9 show the effect of temperature in the range 20–150°C on the adsorption of SO_2 from mixtures containing 0.24% SO_2, 4% O_2 and different contents of H_2O.

The total amount of adsorbed SO_2 falls as the temperature rises (Fig. 8). The fraction of strongly adsorbed SO_2, determined immediately after adsorption in Fig. 9, increases with temperature; this behavior may be attributed to different effects, as preferential adsorption of sulfur dioxide as $SO_2(b)$, increased rate of the transformation $SO_2(a) \rightarrow SO_2(b)$, and surface modifications involving number and type of active sites. In addition, the fraction of $SO_2(b)$ appears to be practically independent of moisture

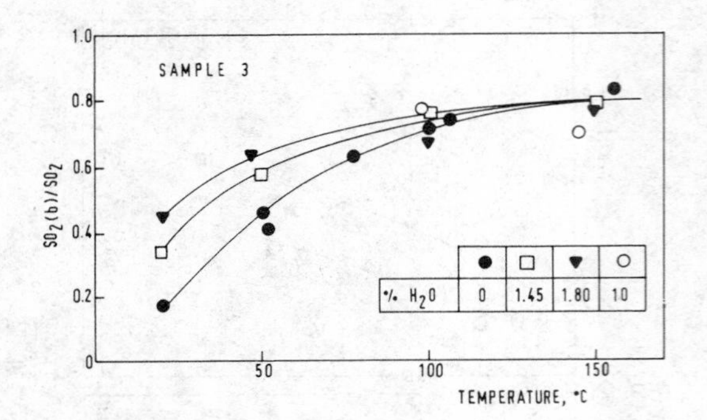

Fig. 8. Total amount of adsorbed SO_2 against adsorption temperature.

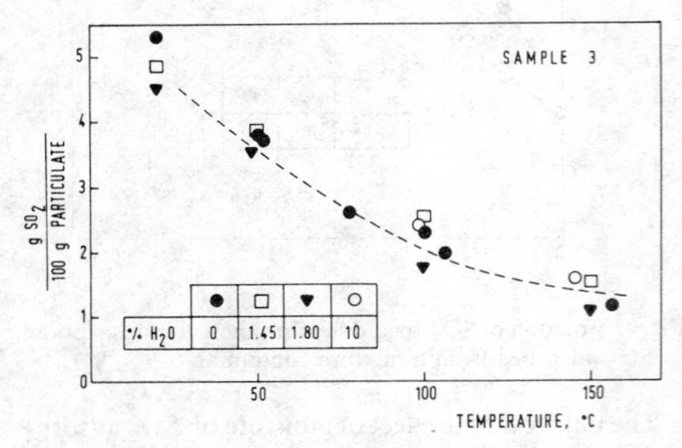

Fig. 9. Fraction of SO_2(b) against adsorption temperature.

content above 100°C. The weaker effect of moisture with increasing temperature may be attributed to a strong drop in H_2O adsorption.

Desorption of SO_2

Desorption of adsorbed SO_2 has been carried out by treating the particulate at 20°C for different times by an air stream. Results are reported in Fig. 10 at different compositions of the mixture from which SO_2 has been previously adsorbed. Practically, only the weakly adsorbed SO_2 desorbs into the air stream; nevertheless, a part of SO_2(a) is also transformed into SO_2(b) in the time. This transformation is improved by the oxygen in the air, owing to likely formation of oxygenated active sites on the solid surface (Davini *et al.*, 1977b).

REFERENCES

Beebe R. A. and Dell R. M. (1955) Heats of adsorption of polar molecules on carbon surfaces – I. Sulfur dioxide. *J. phys. Chem.* **59,** 746.

Davtyan O. K. and Ovchinnikova E. N. (1955) Chemisorption and the oxidation of sulfur dioxide on solid catalysts at room temperatures. *Dokl. Akad. Nauk S.S.S.R* **104,** 857.

Davini P., Morelli F. and Tartarelli R. (1977a) Behavior on heating of carbon containing sulfur acid and/or sulfur dioxide. *Chim. Ind. (Milan)* **59,** 11.

Davini P., Morelli F. and Tartarelli R. (1977b) Evolution of the type of adsorption of SO_2 on carbons at ambient temperature. *Chim. Ind. (Milan)* **59,** 235.

Ficai D. (1928) Adsorption of sulfur dioxide, present in small percentages in gaseous mixtures, by means of colloidal oxides and of active carbon. *Giorn. Chim. Ind. Applicata* **10,** 199.

Garten V. A. and Weiss D. E. (1955) Quinone-hydroquinone character of activated carbon and carbon black. *Austral. J. Chem.* **8,** 68.

Ishikawa I., Inouye K. and Suzuki S. (1969) Oxidation of sulfur dioxide with surface oxygen on carbon black. *Kogyo Kagaku Zasshi* **72,** 2488.

Johswich F. (1962) Sulfur removal from flue gases – importance and possibilities. *Brennst.–Wärme-Kraft* **14,** 105.

Kachanak S. and Jona Z. (1961) Kinetics of sulfur dioxide adsorption on active carbon. *Chem. Prum.* **11,** 127.

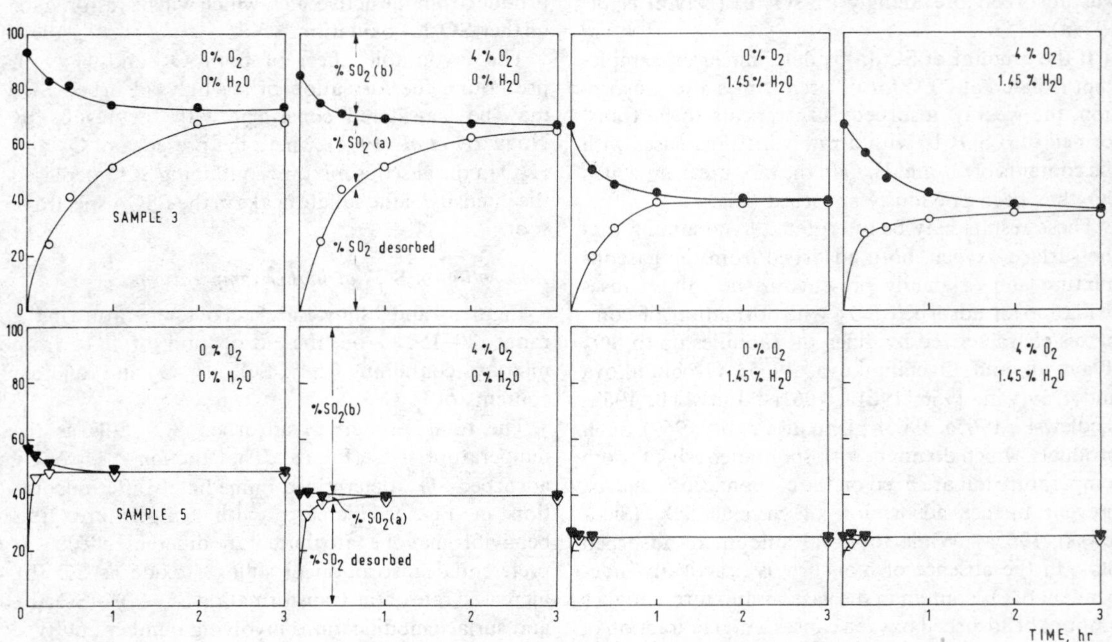

Fig. 10. Per cent of SO_2(a), SO(b), and SO_2 desorbed into the air stream at different times. SO_2 adsorbed at 20°C from a mixture containing 0.24% SO_2.

Lederer E. (1932) A test of adsorption equations on the basis of adsorption measurements on active charcoal. *Kolloid Z.* **61,** 323.

Novakov T., Chang S. G. and Harker A. B. (1974) Sulfates as pollution particulates: catalytic formation on carbon (soot) particles. *Science* **186,** 259.

Ovchinnikova E. N. and Davytyan O. K. (1956) Oxidation of SO_2 on activated charcoal by liquid-contact. *Zh. Fiz. Khim.* **30,** 1735.

Ovchinnikova E. N. and Davtyan O. K. (1961a) Investigation of the mechanism of oxidation, hydrogenation, and electrochemical combustion on solid catalysts – I. Oxidation of sulfur dioxide on activated carbon at 20°C in the presence of water vapor. *Zh. Fiz. Khim.* **35,** 713.

Ovchinnikova E. N. and Davtyan O. K. (1956) Oxidation of of oxidation, hydrogenation, and electrochemical combustion on solid catalysts – II. Catalytic peculiarities of surface "oxides" on carbon. *Zh. Fiz. Khim.* **35,** 992.

Remy H. and Hesse W. (1932) The adsorption of gases by active charcoal. *Kolloid Z.* **61,** 313.

Siedlewski J. (1964) Mechanism of catalytic oxidation on activated carbon – IV. Influence of free radicals of carbon on SO_2 adsorption. *Roczniki Chem.* **38,** 1539.

Siedlewski J. (1965a) Mechanism of catalytic oxidation on activated carbon – V. The contribution of carbon free radicals in the SO_2 to SO_3 oxidation process. *Roczniki Chem.* **39,** 263.

Siedlewski J. (1965b) Mechanism of catalytic oxidation on activated carbon – VI. Determination of the number of active centers on the carbon surface by the electrochemical method in oxygen chemisorption. *Roczniki Chem.* **39,** 425.

Atmospheric Environment Vol. 12, pp. 295–296. Pergamon Press 1978. Printed in Great Britain.

AIRBORNE INVESTIGATIONS OF SO_2 OXIDATION IN THE PLUMES FROM POWER STATIONS

H. FLYGER, E. LEWIN, E. LUND THOMSEN, J. FENGER, E. LYCK and S. E. GRYNING

Danish Air Quality Laboratory, Risø National Laboratory, DK-4000 Roskilde, Denmark

(*First received* 7 *June* 1977 *and in final form* 12 *July* 1977)

Abstract—The Danish contribution to the EUROCOP COST 61a project is described. Work concerned the physical and chemical reactions of sulphur dioxide released from a power station. The investigation was based on the application of two tracers. Inactive, inert SF_6 is used to monitor the dispersion of and deposition from the plume; it was intended to use radioactive $^{35}SO_2$ to determine the degree of oxidation of sulphur released from the stack; so far, however, public reaction has prevented the use of a release of activity in field experiments. Some attempts have been made to base the investigations of analyses of inactive samples only. Previously published results obtained at an oil-fired power station suggest a half-life of about 30 min for SO_2 in the first part of the plume. New results obtained at a mainly coal-fired power station give a half-life of about 90 min.

INTRODUCTION

One of the important stages in the fate of pollutant sulphur in the atmosphere is the oxidation of sulphur dioxide to sulphate. This process is influenced by a series of factors: light intensity, humidity, concentrations of various compounds, catalysts, etc. Therefore the oxidation rate is greatly dependent on the general circumstances. In the EEC project COST 61a 11 European countries collaborated in clarifying the physico-chemical behaviour of SO_2 in the atmosphere. Denmark participated with an airborne study of the conversion of SO_2 in the plume from an oil-fired power station stack.

The changes in sulphate concentration are generally small compared with the background—especially on the borders of the plume; therefore it is desirable to label the effluent sulphur compounds with a radioactive sulphur isotope. The obvious choice is sulphur-35, which has been used as a tracer in many studies of SO_2. Preparations for such an experiment have been described in detail in a technical report (Flyger *et al.*, 1977). However, when the experiments were mentioned in the course of the nuclear power debate, objections were raised by the public; consequently, the Minister of Education intervened and the use of radioactive tracers was postponed.

Some early attempts to base the investigation only on the analysis of inactive sulphur have been described in a recent paper (Flyger and Fenger, 1976). The results suggest that in the first part of the plume from an oil-fired power station the SO_2 may disappear with a rate corresponding to a half-life of only 30 min.

EXPERIMENTAL AND RESULTS

A new series of experiments is now being carried out at the Asnæs power station on the west coast of Zealand. The power station is divided into four blocks, each with a 122 m stack, and it is fired with a mixture of coal and oil. Sulphur hexafluoride can be added to the effluent of one of the stacks at a controlled rate.

In the experiments the plume is crossed by an airplane at various distances from the station; all traverses are made at distances so large that a complete mixture of individual plumes can be assumed. During the flight the concentration of sulphur hexafluoride is recorded continuously by means of a special electron capture detector (Thomsen and Lovelock, 1976; Flyger *et al.*, 1977). The sulphur dioxide concentration is recorded with a flame spectrophotometer (Bendix 8100). In analogy with the earlier experiments (Flyger and Fenger, 1976), the recorded plume profiles are used to determine the ratio between the SO_2 and the SF_6 concentrations; a decrease with distance in this ratio represents a loss of SO_2 from the plume.

In practice the experiments can only be carried out under weather conditions allowing a well defined plume. During an experiment performed on 15 June 1977 a north-east wind was blowing and therefore the measurements were mainly carried out over the sea; the wind velocity was about 8 m s^{-1}. The atmosphere was misty with occasional rain showers at the end of the experiment.

The station was fired with 13 t oil and 104 t coal h^{-1}, containing a weighed mean of 1.8% sulphur.

A total of 36 kg SF_6 was released in 90 min, and the plume was crossed a couple of times at various distances. The ratio between the SO_2- and the SF_6-concentrations was determined, in arbitrary units, to be:

12 km:	0.54;	0.45;	0.67
23 km:	0.47;	0.43;	0.32
47 km:	0.30;	0.31.	

A one-way analysis of variance yields a level of significance of 0.06. The regression line is:

$$y([SO_2]/[SF_6]) = 0.61 - 6.84 \times 10^{-3} \times x$$

(distance in km).

Thus, half of the SO_2 is lost from the plume within 45 km; this corresponds to a travel time of about 90 min.

CONCLUSION

Both our old and our new experiments indicate that it is difficult to obtain accurate results for sulphur conversion without the use of an unambiguous tracer. So far, we have only been able to demonstrate that the disappearance of SO_2 whether by oxidation or deposition, can be quite rapid in a plume. The reactions seemed to be slower for a coal-fired plant than for an oil-fired plant. This is in agreement with the results of Newman *et al.* (1975a, b).

REFERENCES

Flyger H. and Fenger J. (1976) Conversion of sulphur dioxide in the atmosphere. *Z. anal. Chem.* **282,** 297–300.

Flyger H. *et al.* (1977) Physical and chemical processes of sulphur dioxide in the plume from an oil-fired power station. Risø Report No. 328, 56 pp.

Newman L., Forrest J. and Manowitz B. (1975a) The application of an isotopic ratio technique to a study of the atmospheric oxidation of sulphur dioxide in the plume from an oil-fired power plant. *Atmospheric Environment* **9,** 959–968.

Newman L., Forrest J. and Manowitz B. (1975b) The application of an isotopic ratio technique of sulphur dioxide in the plume from a coal fired power plant. *Atmospheric Environment* **9,** 969–977.

Thomsen E. L. and Lovelock J. E. (1976) A continuous and immediate method for the detection of SF_6 and other tracer gases by electron capture in atmospheric diffusion experiments. *Atmospheric Environment* **10,** 917–920.

Atmospheric Environment Vol. 12, pp. 297–306. Pergamon Press 1978. Printed in Great Britain.

A STUDY OF PRODUCTION AND GROWTH OF SULFATE PARTICLES IN PLUMES FROM A COAL-FIRED POWER PLANT

ALLEN C. DITTENHOEFER and ROSA G. DE PENA
Department of Meteorology, The Pennsylvania State University, University Park, PA 16802, U.S.A.

(*First received* 13 *June* 1977 *and in final form* 5 *August* 1977)

Abstract—A number of airborne plume sampling experiments designed to examine the importance of sulfate particle-generating chemical reactions within coal-burning power station plumes are described. The flights were conducted downwind of the Keystone Generating Station in western Pennsylvania, with The Penn State University research aircraft, an Aerocommander 680E. On-board aerosol sampling instrumentation included a condensation nucleus counter, an optical particle counter, and an electrical aerosol analyzer. A casella cascade impactor containing electron microscope copper grids coated with carbon film was used to collect particles at varying distances from the stacks. These samples were analyzed for sulfate content and particle size distribution. Measurements of SO_2 were made with a rapid-response pulsed fluorescent analyzer. Atmospheric pressure, temperature, dewpoint, winds and aircraft position were also monitored. For each flight, a vertical spiral aircraft sounding was made upwind of the power station to determine atmospheric stability and background aerosol particle and SO_2 concentrations. Downwind, the flight pattern consisted of a series of cross wind and longitudinal plume penetrations out to distances at which SO_2 reached background levels. During the case in which cooling tower plume and stack plume merger occurred, sampling continued out to regions where the liquid plume had dissipated. It was found that when relative humidity was low, stability near-neutral, and solar radiation intense, the production of new Aitken particles was the primary mechanism of SO_2 oxidation. In the case of merger between the stack plume and the cooling tower plume, the formation of sulfate on pre-existing particles predominated over the formation of new particles. During cases with intermediate meteorological conditions both processes were of equal importance.

1. INTRODUCTION

In recent years there have been several attempts to measure the extent of the atmospheric oxidation of sulfur dioxide to sulfate within the plumes of large power plants utilizing instrumented aircraft (Gartrell *et al.*, 1963; Dennis *et al.*, 1969; Stephens and McCaldin, 1971; Newman *et al.*, 1975a, 1975b; and Hegg *et al.*, 1976). The wide assortment of techniques used, however, have in general suffered from inherent limitations as to their accuracy. This, coupled with a general deficiency of measurements compiled during highly varying meteorological conditions of temperature, relative humidity, and stability, have created large uncertainties concerning the chemical mechanism(s) involved. Some authors, notably Gartrell *et al.* (1963), have reported a strong dependence of the oxidation rate on relative humidity, while in the investigation made by Newman *et al.* (1975a) this dependency was at best obscure. Freiberg (1974) postulated that both temperature and relative humidity individually exert sizeable influences on the iron-catalyzed hetereogeneous oxidation of atmospheric SO_2. Furthermore, it has been pointed out (Newman *et al.*, 1975b) that the oxidation rate is probably highly dependent upon fuel type and atmospheric particulate loading. This paper describes the results of a number of airborne plume sampling experiments designed to examine the importance of particulate-generating chemical reactions within power station plumes under widely varying meteorological conditions. The effect of merger between the stack plume and the cooling tower plume is also assessed. The flights were conducted in the vicinity of the coal-burning Keystone Generating Station, near Indiana, Pennsylvania, during March 1976 and December 1976 through March 1977.

2. EXPERIMENTAL

The Keystone Power Plant, which has been the site of previous plume dispersion observational programs (e.g. Schiermeier and Niemeyer, 1968 and Johnson and Uthe, 1971), is located 305 m above sea level in a shallow rural valley in Western Pennsylvania roughly 50 km ENE of Pittsburgh. The station has a generating capacity of 1800 MW produced by two boiler units, each of which is equipped with an electrostatic precipitator of 99.5% efficiency rating. Boiler effluent is emitted through two 244-m stacks, while four 99-m tall natural draft cooling towers supply recirculation water for steam condensation. Both mine-mouth and truck coal are used, with an average sulfur content of about 2%. The ash, which comprises approximately 20% of the coal, generally contains 50% of SiO_2, 26% of Al_2O_3, 15% of Fe_2O_3, 2% of Cao, 3% of K_2O, 2% of Na_2O, 2% of TiO_2, 1% MgO and other components in concentrations lower than 1%.

Flights were made using The Pennsylvania State University research aircraft, an Aerocommander 680E. On-board aerosol sampling instrumentation included an Environment One model Rich 100 Condensation Nuclei Counter (CNC), a Royco model 225 Optical Particle Counter

(OPC) and a TSI model 3030 Electrical Aerosol Analyzer (EAA). In the CNC the particles are exposed to supersaturations such that particles of diameter $\geq 0.0025\ \mu m$ are activated and can be counted. The CNC was calibrated against a Pollack counter. The OPC is sensitive only to particles larger than about 0.3 μm dia. The OPC was calibrated with latex spherical particles. For these particles the instrument separates the concentrations into five size ranges: 0.34–0.72; 0.72–0.95; 0.95–2.55; 2.55–4.25 and $>4.25\ \mu m$, respectively. The EAA, which was not available for the March 1976 flights, supplies aerosol number concentrations in eight distinct size ranges between 0.01 and 1.00 μm dia. An isokinetic intake probe (Pena *et al.*, 1977) installed at the top of the airplane decelerates the air flow from that of the aircraft speed ($\simeq 70\ m\ s^{-1}$) to $4\ m\ s^{-1}$ and thus insures accurate particle sampling.

Measurements of SO_2 were made with a Thermo Electron model 43 SO_2 pulsed fluorescent analyzer. This instrument measures the intensity of the fluorescent emission of SO_2 excited by pulsating u.v. light. Measurements of SO_2 obtained during the flights conducted in March 1976 were not used, as the instrument had a response time considerably larger than that of the particle instruments. Prior to the series of flights commencing in December 1976 the instrument was modified such that its response time (5 s to obtain 95% of the actual value) agreed well with the other instruments on-board the aircraft.

A Casella cascade impactor containing electron microscope copper grids coated with carbon film was used to collect particles upwind and at varying distances downwind of the stacks. The grids were processed for sulfate particle analysis by coating them with a layer of barium chloride and exposing them to a relative humidity of 70% for a period of one hour (Mamane, 1977). The sulfate particles react with the barium chloride film forming a characteristic halo easily recognizable under the electron microscope. By carefully counting and measuring a representative sample of sulfate particles on each grid, comparisons of sulfate particle size and concentration at various distances from the stacks were made.

Aircraft meteorological observations included pressure, temperature, dewpoint and winds, Integrated Doppler radar and recorded navigational DME and VORTAC data were used to determine aircraft position. Flight data were recorded at 0.5 s intervals on magnetic tape, thus establishing horizontal and vertical spatial resolutions of 35 m and 2.5 m respectively (aircraft flight speed $\sim 70\ m\ s^{-1}$ in horizontal and $\sim 5\ m\ s^{-1}$ in climb or descent).

For each flight a vertical spiral aircraft sounding was made upwind of the power station to determine atmospheric stability and background particle and SO_2 concentrations. Downwind, the flight pattern consisted of a series of crosswind and longitudinal plume penetrations out to distances at which SO_2 reached background levels. During the case when significant cooling tower plume and stack plume merger occurred, sampling continued out to regions where the liquid plume had dissipated.

3. RESULTS

The flights chosen for this study are listed in Table 1, along with the meteorological conditions which prevailed. Plumes were classified as merging and non-merging. Temperature, relative humidity, and wind speed and direction were averaged at plume height downwind of the stacks. Data obtained from the vertical spiral flight patterns were used to compute the average lapse rate within the plume layer.

In only one case, flight No. 4, did significant interaction between the stack plume and cooling tower plume occur. During this flight, the visible merged plume was detected out to 50 km downwind of the plant (over 110 min plume travel time) and could be attributed to high atmospheric relative humidity, a stable lapse rate and low wind speed. Although similar conditions of stability and wind speed prevailed during flight No. 7, the slightly lower relative humidity and higher temperature caused the liquid plume to evaporate very near the towers and hence no *visible* merging was observed. The remaining flights represented cases of low relative humidity ($\leq 70\%$). Flight 1 was conducted under a sunny March afternoon sky during conditions of low humidity, high wind speed and neutral stability. Similar atmospheric conditions existed during flight No. 2, which took place in the late morning hours. Flights 3, 5 and 6 were made during early morning immediately after dawn and encompass a wide range of atmospheric temperature, wind speed and stability.

The data from a number of flights had to be excluded from this study due to a variety of reasons. During several flights, the wind direction was such that contamination from nearby power plants and town plumes occurred. There were also cases in which high wind speeds and lack of atmospheric stability resulted in well-mixed and ill-defined plumes. During such instances the plume was extremely difficult to locate or too short to allow adequate sampling. Finally, a case of low cloud cover during another flight prohibited plume sampling beyond 4 km from the stacks.

Table 1. Summary of flights

Flight	Date	Time of day	Plume type	Sky condition	Temperature (°C)	Relative humidity	Winds Speed ($m\ s^{-1}$)	Winds Direction	Ave. lapse rate (°C $100\ m^{-1}$)	Mean plume elevation (m a.s.l.)
1	3/6/76	14:02–15:39	non-merging	mostly sunny	−0.7	42	12.4	272	0.95	750
2	3/7/76	09:24–10:30	non-merging	clear–sunny	−3.0	70	12.3	240	0.83	880
3	12/14/76	07:04–09:59	non-merging	clear–sunny	−5.6	55	14.6	222	−0.20	720
4	2/8/77	06:51–08:57	merging	clear–some ground fog	−16.8	>95	7.5	250	0.17	860
5	3/30/77	06:33–08:01	non-merging	clear	16.7	70	12.7	239	0.59	770
6	4/18/77	05:39–07:10	non-merging	some high clouds, haze	9.9	59	4.9	308	0.87	1210
7	3/10/76	06:55–08:47	non-merging	overcast	−4.5	89	6.2	232	0.51	770

In this study concentrations of Aitken particles, as measured by the CNC, large particles, measured by the OPC, and sulfur dioxide concentrations were categorized with respect to position relative to the stacks and represent in-plume time averages. The EAA was used to obtain particle size frequency distributions upwind and at several distances downwind of the stacks. Background concentrations were obtained at plume level upwind of the plant. Stack plume boundaries were assumed to be located at the regions where SO_2 concentrations exceeded background levels. All downwind SO_2 and particle concentrations presented in this report, including the size distribution supplied by the EAA, have been corrected for background contributions.

Aerosol particle concentrations measured downwind of the plant are not only dependent upon plume dynamics and chemistry, but also upon the position in the plume where the sampling takes place. Laser studies (e.g. Martin and Barber, 1973) have revealed that plumes originating from large power stations are highly fragmentary in nature. It is, therefore, common to find irregular variations of gas and particle concentrations within such plumes. By computing ratios of simultaneous concentrations of various pairs of plume components (e.g. large particles, Aitken particles, SO_2) it is possible to eliminate the effects of plume inhomogeneity and thus isolate the effects attributable solely to plume chemistry.

To permit a realistic comparison of these ratios between flights, the results are expressed as a function of plume travel time, defined as the distance downwind of the power plant divided by the average wind speed at plume height.

It was found that the rate of SO_2 consumption due to chemical reaction was low enough to enable use of SO_2 as a conservative plume tracer. This is consistent with recent findings of Hegg *et al.* (1976), in which the SO_2 oxidation was found to be only of the order of 0.3% h^{-1}. Results obtained by Newman *et al.* (1975) downwind of the Keystone Power Station also support this finding. The amount of SO_2 converted to sulfate over plume transport times exceeding two hours was generally less than 4%. Aerosol particle concentrations measured during this study, therefore, have been normalized with respect to simultaneous readings of SO_2.

3.1 *Particle concentration*

Aitken particles and large particle concentrations normalized with respect to simultaneous readings of sulfur dioxide concentrations are presented in Figs 1 and 2. An increase of the ratio of Aitken particles to SO_2 with time would indicate a net production of particles within the plume. In Fig. 1 there appears no sizeable difference between flights 3, 4 and 5 despite the wide range of ambient temperature, relative humidity and stability. In the case of stack and cooling tower plumes merger (flight No. 4) the ratio underwent a significant increase only between 89 and 106 min travel time. The particle generating chemical reactions within each of these plumes seems to be minimal.

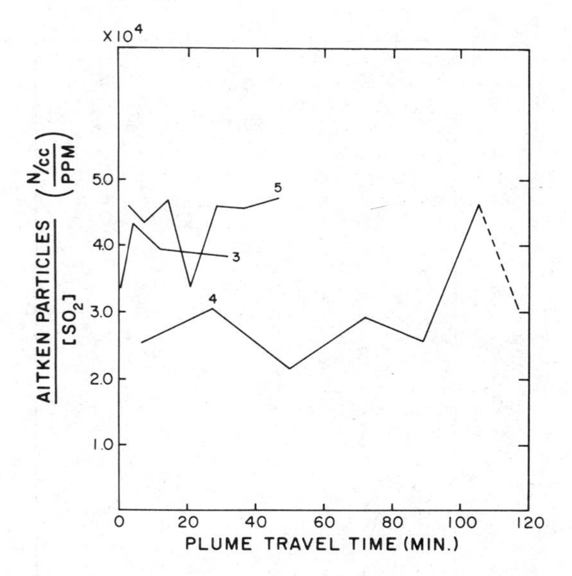

Fig. 1. Ratios of Aitken particles to SO_2 concentration as a function of plume travel time for flights 3, 4 and 5. The dashed line for flight 4 represents the region where the liquid had evaporated.

An increase of the ratio of large particles to SO_2 would indicate growth within the plume of pre-existing particles to sizes larger than 0.34 μm. Again no major difference between flights 3, 4 and 5 over comparable plume travel times seems to exist (Fig. 2). However, between 89 and 106 min plume transport time for flight 4 a very rapid increase in large particle concentrations occurred. Within the region where the

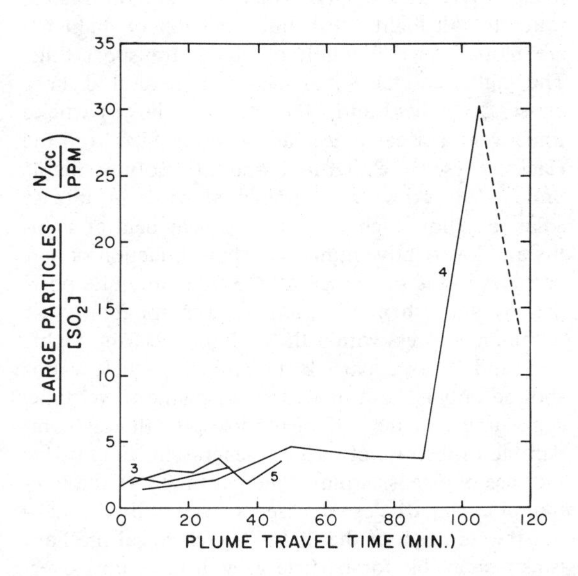

Fig. 2. Ratios of large particles to SO_2 concentration as a function of plume travel time for flights 3, 4 and 5. The dashed line for flight 4 represents the region where the liquid had evaporated.

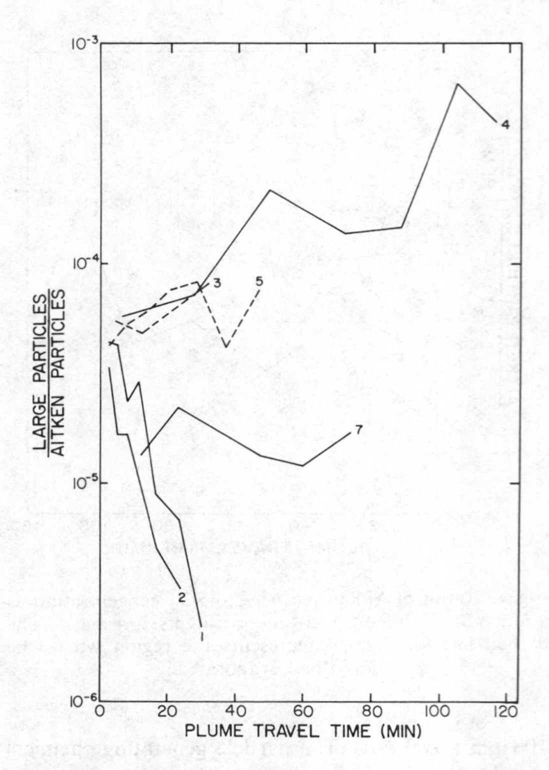

Fig. 3. Ratios of large to Aitken particle concentrations as a function of plume travel time.

liquid plume had dissipated (106–117 min) large particle concentrations decreased; however, these concentrations were still almost one order of magnitude larger than the original values.

Ratios of large/Aitken particles can be interpreted as the fraction of the total number of particles within the plume that are larger than 0.34 μm. In Fig. 3, ratios for all flights, with the exception of flight 6*, are plotted as a function of plume transport time. The flights can be subdivided into several distinct classes. In flights 1 and 2 the fraction of large particles underwent a sizeable decrease with distance from the stacks. These were afternoon and late morning flights under clear skies and represented cases of intense solar radiation, high wind speeds, near neutral stability and low relative humidity. The production of new particles in the size range 0.0025–0.34 μm, quite possibly by photochemical reactions, was apparently the dominant process within these plumes. During flights 3, 5 and 7 large particle to Aitken particle ratios showed only a slight increase with plume travel time, suggesting that the rate of particle growth was comparable to the rate of particle generation. Finally, for the case of extensive plume merger (flight 4), the fraction of large particles increased substantially throughout the duration of the flight. The chemical mechanisms responsible for particle growth have thus overwhelmed the processes which generate new particles,

* For flight 6 data from the OPC were not available.

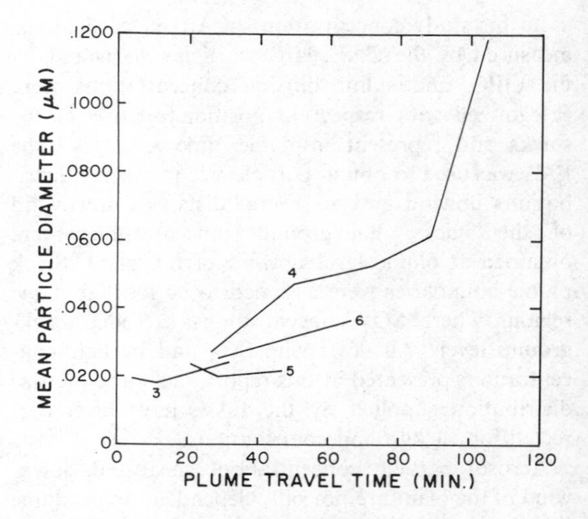

Fig. 4. Mean particle diameter as obtained from the Electrical Aerosol Analyzer as a function of plume travel time.

and appear related to the presence of liquid water in the plume.

3.2 *Mean particle diameter*

Mean particle diameters have been computed as a function of plume transport time using the data from the EAA for flights 3, 4, 5 and 6. Mean diameter is defined as

$$\bar{d} = \Sigma_i[(ni\ di)/n]$$

where ni/n is the fraction of particles within size range i and di is the class mark dia. Fig. 4 shows that for

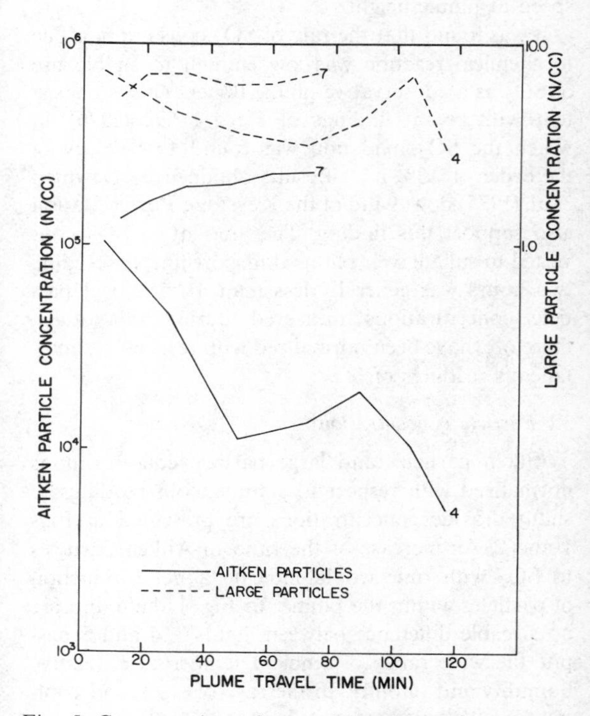

Fig. 5. Concentration of —— Aitken particles; ----- large particles for flights 4 and 7.

flights 3, 5 and 6, cases of low relative humidity, the mean particle dia undergoes a slight increase with travel time. During flight 4, however particle sizes grow continuously at a much greater rate especially within 89 and 106 min. These results agree well with the trends observed for normalized large particle concentrations (Figs 2 and 3). For the merging plume situation, particle growth is continuous with time; this growth initially occurs mainly within droplets too small to be detected by the OPC. As particle growth continues, however, the droplets eventually attain sizes greater than 0.34 μm; hence, the rapid sudden increase in large particle concentrations between 89 and 106 min travel time.

In Fig. 5, flight 4 is compared with flight 7. Both flights were made in comparable atmospheric conditions of high relative humidity, light winds and stable lapse rate. However, in flight 4 a long visible merged plume was observed, while in flight 7 the liquid plume was very short and visible merging was not present.

Both large and Aitken particles were higher during flight 7 and actually increased with distance from the plant. We have reasons to believe that the chemical mechanisms responsible for the particle production and growth were very active during this flight and have overwhelmed the effects of plume dilution, although without adequate SO_2 data we cannot be sure. In flight 4, concentrations of both large and Aitken particles generally decreased with plume travel time, but at a considerably different rate, such that the ratio of large to Aitken particles increased with time, as discussed earlier.

3.3 *Sulfate analysis*

Samples collected on the electron microscope grids were treated with barium chloride for sulfate analysis according to Mamane (1977). We were primarily concerned with the relative number of sulfate, mixed and non-sulfate particles for each size range and how these quantities changed with plume travel time. Therefore, we did not attempt to make corrections for collection efficiencies nor for the distortion of the flow caused by the grids.

Figures 6–10 are photomicrographs of barium chloride treated samples collected on the fourth stage of the impactor, upwind (Fig. 6) and at 3 different locations downwind of the power plant during flight 4 (Figs 7 to 10). The very numerous small oblong shaped particles are barium chloride crystals. The dark circular spots are particles other than sulfates. Some of them appear surrounded by the halo resulting from the reaction of sulfate with the barium ion; these we classified as mixed particles. Finally, there are numerous halos which are the result of the reaction of particles formed solely by sulfate. The ratio of the size of the halo to the size of the sulfate particles is approx two (Mamane, 1977).

In the sample collected upwind, the sulfate particles were small and uniform in size (Fig. 6). They compose roughly 95% of the total number of particles found in the grid. Mixed and non-sulfate particles made up only 3 and 2%, respectively. Within the plume, 12.2 km from the stack (Figs 7 and 8), the sulfate particles collected were much larger and comprised only 69% of the total. Mixed particles (24%) and non-sul-

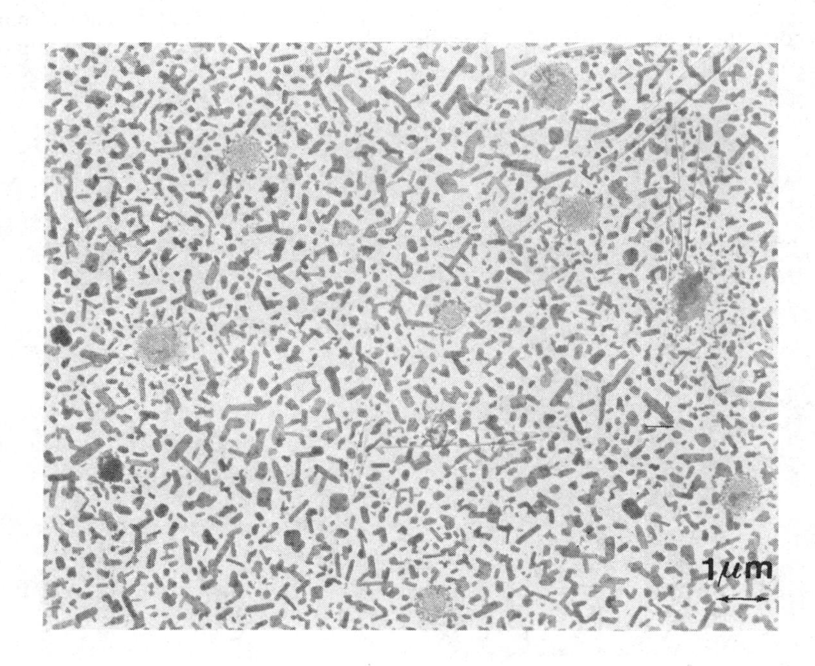

Fig. 6. Electron microscope micrographs of barium chloride-treated samples collected on the 4th stage of the impactor upwind.

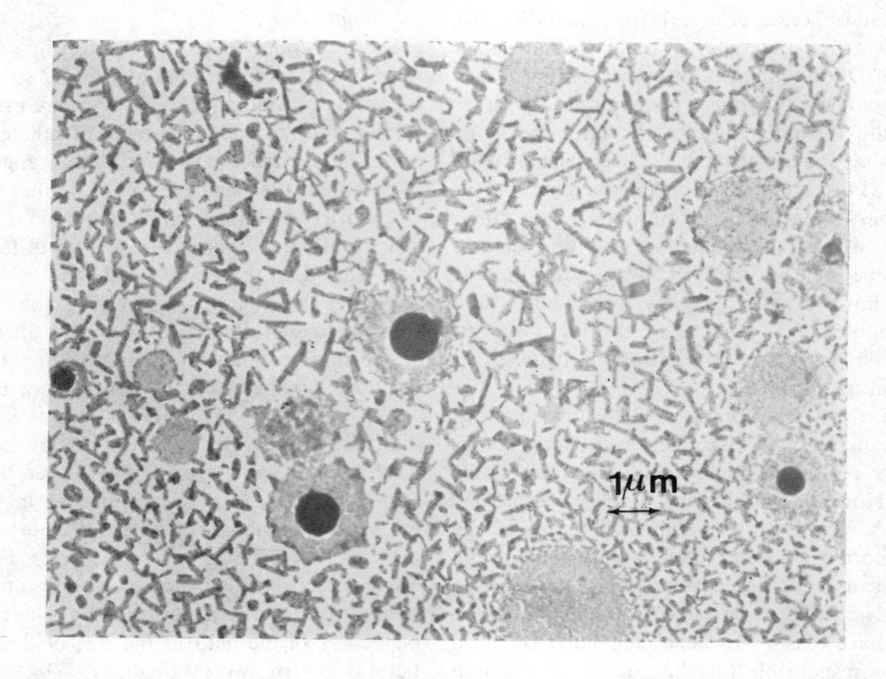

Fig. 7. Electron microscope micrographs of barium chloride-treated samples collected on the 4th stage of the impactor 12.2 km downwind.

fate particles (7%) accounted for the remaining 31%. Further downwind at 32.7 km and 49 km the fraction of sulfate particles rose to 89 and 95%, respectively (Figs 9 and 10). These results are summarized in Fig. 11 in which sulfates, mixed and non-sulfate particles frequencies have been plotted as a function of plume travel time.

Figure 12 shows the frequency of sulfate particles as a function of their size, for flight 4. It can be clearly seen that at all locations, including upwind, the very small particles are primarily sulfate. For particles of dia $\leq 0.5\ \mu m$, sulfate particles comprised at least 94% of the total number of particles. As particle diameter increased, the fraction of sulfate particles generally decreased, while the frequency of mixed and non-sulfate rose. Upwind of the plant, almost all of the sul-

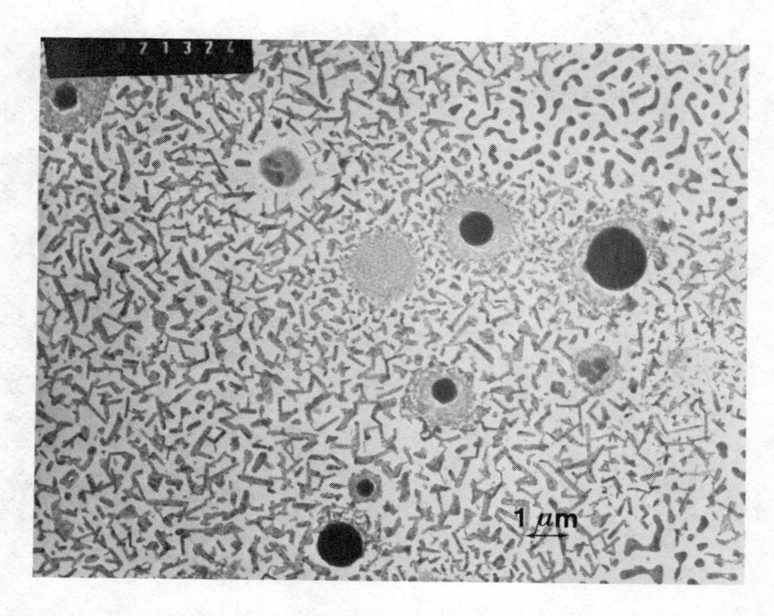

Fig. 8. Electron microscope micrographs of barium chloride-treated samples collected on the 4th stage of the impactor 12.2 km downwind.

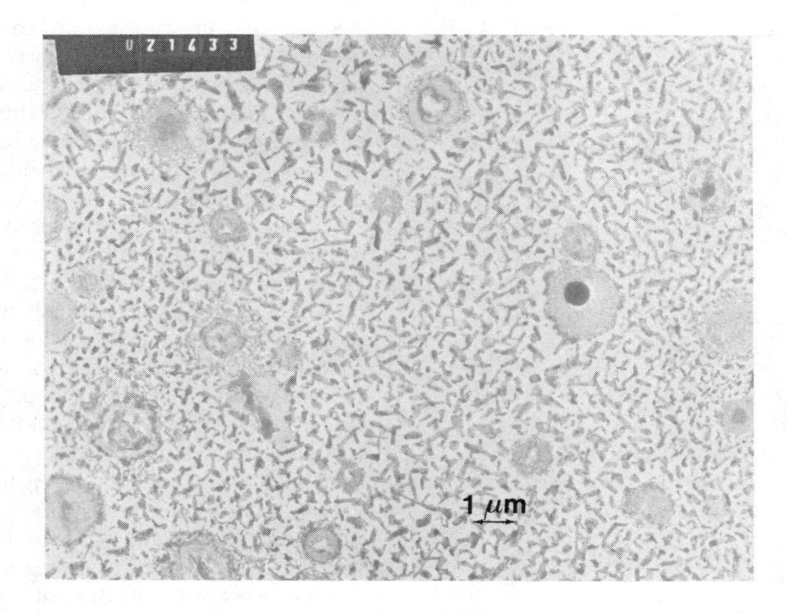

Fig. 9. Electron microscope micrographs of barium chloride-treated samples collected on the 4th stage of the impactor 32.7 km downwind.

fate particles were concentrated in the smaller sizes and very few sulfates larger than 0.8 μm were found in the background air. Downwind of the plant, however, sulfate particles made up a much larger fraction of the total number of particles in the larger sizes. Furthermore, sulfate particle frequency appears to be related to distance downwind of the plant. For particles of dia greater than 1.33 μm, the fraction of sulfate particles increased monotonically with distance, suggesting that particle growth has occurred. Inspection of the grids in stage 3 of the impactor, which collects larger particles, verifies this (Table 2).

Table 2. Particle frequency for flight 4 on stage 3 of the impactor

	Upwind	12.2 km	32.7 km	49.0 km
Non-sulfate	0.88	0.91	0.88	0.30
Mixed	0.04	0.07	0.06	0.16
Sulfate	0.08	0.02	0.06	0.54

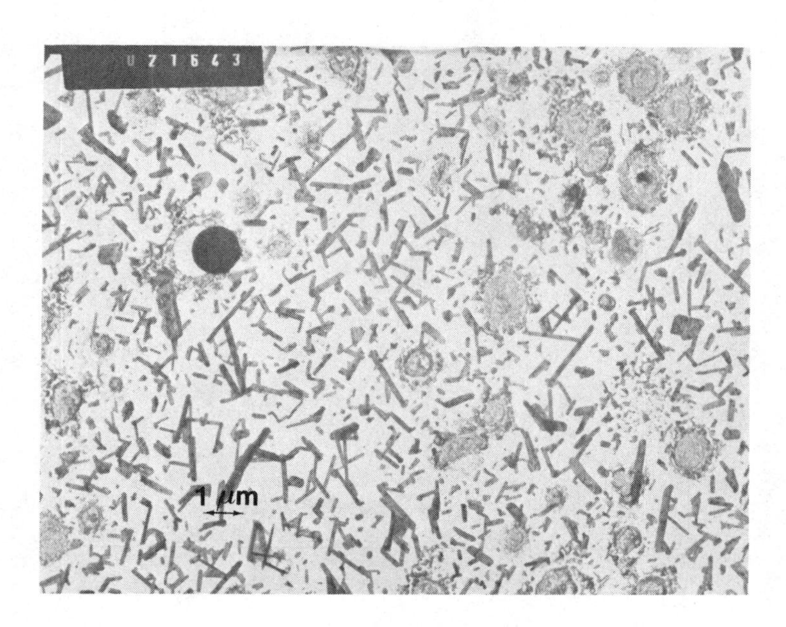

Fig. 10. Electron microscope micrographs of barium chloride-treated samples collected on the 4th stage of the impactor 49.0 km downwind.

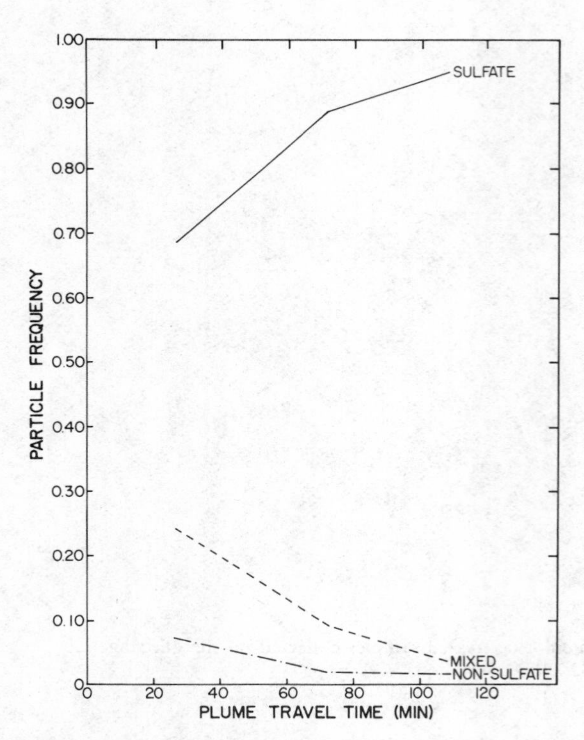

Fig. 11. Particle frequency expressed as the ratio of the number of particles to the total number of particles as a function of plume travel time for flight 4.

The sudden increase in the frequency of sulfate particles between 32.7 and 49.0 km (73 and 109 min plume travel time) appears responsible for the previously observed rapid increase of normalized large particle concentrations measured by the OPC (Figs 2 and 3) and the rise in mean diameter computed from the EAA (Fig. 4) between these travel times. This significant growth of particle size, which was not observed in any of the other flights, was apparently linked to the presence of liquid water supplied by the cooling tower plume and involved the absorption of SO_2 contained within the stack plume.

Further evidence for the production of sulfate within the droplets of the plume for flight 4 also exists. The volume fraction of sulfate contained within each of the mixed particles found on stage 4 was computed and averaged as a function of distance downwind from the stack. At 12.2 km downwind, 75% of the volume was sulfate, while at 32.7 km and 49.0 km this fraction increased to 80 and 87%, respectively.

An attempt was made to calculate the SO_2 conversion rate for this flight, in which the rate of particle growth was highest. Particle size frequency distributions given by the EAA and the OPC were combined with the results of the sulfate analysis, in which a relation between the fraction of particles containing sulfate and particle size was found (Fig. 12). The curves of Fig. 12 were extended to include particles too small to be collected by the impactor. Finally, we assumed the particles to be droplets of sulfuric acid at 95% r.h., and the actual mass of solute contained within each size interval was found by use of the Köhler curves for sulfuric acid. We were thus able to compute the fraction of H_2SO_4 to SO_2 at each location and the conversion rate. The value obtained was of the order of $0.5\%\,h^{-1}$. We realize that this constitutes a very crude estimate and a more complete model will be developed in the future. However, the order of magnitude is correct and it justifies the use of SO_2 as a plume tracer.

Finally, the sulfate particle spectra for flights 3 and

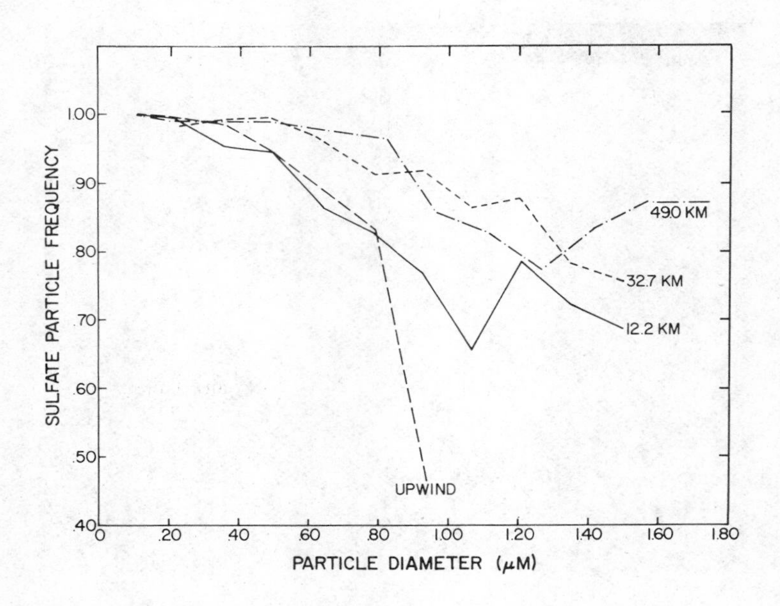

Fig. 12. Sulfate particle frequency expressed as the ratio of the number of sulfate particles to the total number of particles as a function of particle dia for flight 4.

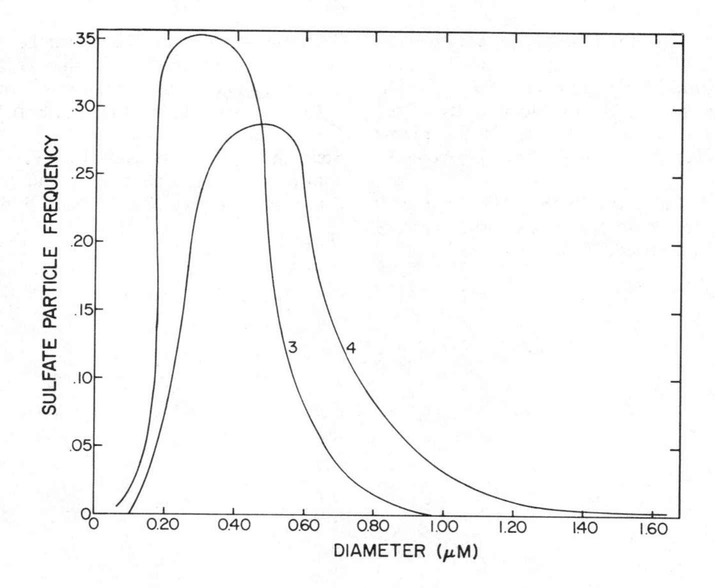

Fig. 13. Sulfate particle frequency expressed as the ratio of the number of sulfate particles of dia d to the total number of sulfate particles.

4 at comparable plume travel times (32 and 27 min, respectively) are compared in Fig. 13. The sulfate particles collected during flight 3, a case of low relative humidity and a non-merging plume, were of significantly smaller size than those found during the merging plume case. There was an increase in modal sulfate particle dia between flights 3 and 4 of approx 0.3–0.5 μm.

4. CONCLUSION

Our results suggest that there exists more than one mechanism acting in the conversion of SO_2 to sulfate within power plant stack plumes. Under conditions of low r.h., near-neutral stability, and intense solar radiation, it was found that production of new particles within the stack plume was the dominant chemical process. It is likely that photochemical reactions involving SO_2 are a major source of these particles. Within a merged plume (r.h. near 100%) net production of new particles was of relatively minor importance compared to the growth of pre-existing droplets by absorption and oxidation of SO_2. Many of these droplets grew to sizes of 0.3 μm or greater after 90 min of plume travel. Analysis of particles collected on stages 3 and 4 of a cascade impactor revealed that sulfate production accounted for much of this particle growth.

No mention has been made of the possible role temperature plays in the conversion of SO_2 into sulfate within power plant plumes. Freiberg (1974) has postulated that the iron-catalyzed oxidation of SO_2 in atmospheric water droplets is inversely related to temperature. We found the highest rate of particle growth (flight 4) when average plume temperature was at its lowest (-16.8°C) and least growth (flight 5) during the case of highest temperature (16.7°C, Fig. 4). Flights 3 and 6, in which we observed similar rates of particle growth under similar conditions of relative humidity, had vastly different temperatures (-5.6°C and 9.9°C, respectively). More flights will be needed to justly evaluate the importance of temperature. It is likely that several meteorological parameters, including relative humidity, stability, and temperature, all exert vital controls on plume chemistry.

Acknowledgements—This work was supported by ERDA grant E(11-1)2463. We would like also to show our appreciation to Yaacov Mamane for his help in sample preparation for electron microscope analysis.

REFERENCES

Dennis R., Billings C. E., Record F. A., Warneck P. and Arin M. L. (1969) Measurements of sulfur dioxide losses from stack plumes. APCA Paper No. 69–156, 62nd Meeting of the Air Pollution Control Association, New York.

Freidberg J. (1974) Effects of relative humidity and temperature on iron-catalyzed oxidation of SO_2 in atmospheric aerosols. *Environ. Sci. Technol.* **8,** 731–733.

Gartrell F. E., Thomas F. W. and Carpenter S. B. (1963) Atmospheric oxidation of SO_2 in coal-burning power plant plumes. *Am. Ind. Hyg. J.* **24,** 113–120.

Hegg D. A., Hobbs P. V. and Radke L. F. (1976) Reactions of nitrogen oxides, ozone and sulfur in power plant plumes. Electric Power Research Institute EA-270, Palo Alto, California.

Johnson W. B. and Uthe E. E. (1971) Lidar study of the Keystone stack plume. *Atmospheric Environment* **5,** 703–724.

Mamane Y. (1977) A quantitative method for the detection of individual submicron sulfate particles. Ph.D. Thesis, The Pennsylvania State University, University Park, PA.

Martin A. and Barber F. R. (1973) Further measurement around modern power stations—I–III. *Atmospheric Environment* **7,** 17–37.

Newman L., Forrest J. and Manowitz B. (1975a) The application of an isotopic ratio technique to a study of the atmospheric oxidation of sulfur dioxide on the plume from an oil-fired power plant. *Atmospheric Environment* **9**, 959–968.

Newman L., Forrest J. and Manowitz B. (1975b) The application of an isotopic ratio technique to a study of the atmospheric oxidation of sulfur dioxide in the plume from a coal fired power plant. *Atmospheric Environment* **9**, 969–977.

Pena J. A., Norman J. M. and Thomson D. W. (1977) Isokinetic sampler for continuous airborne aerosol measurements. *APCA Journal* **27**, 337–341.

Schiermeier F. A. and Niemeyer L. E. (1968) Large power plant effluent study—Volume 1—Instrumentation, procedures and data tabulations. National Air Pollution Control Administration Publication No. APTD 70-2, Raleigh, N.C.

Stephens N. T. and McCaldin R. O. (1971) Attenuation of power station plumes as determined by instrumented aircraft. *Environ. Sci. Technol.* **5**, 615–621.

Atmospheric Environment Vol. 12, pp. 307–312. Pergamon Press 1978. Printed in Great Britain.

CLOUD NUCLEUS FORMATION IN A POWER PLANT PLUME

R. F. PUESCHEL and C. C. VAN VALIN
Atmospheric Physics and Chemistry Laboratory, Environmental Research Laboratories, NOAA, Boulder, Colorado, U.S.A.

Abstract—A rate of production of *ca* 10^{16} particles s^{-1} of H_2SO_4 aerosol that acts as cloud condensation nuclei at 1% supersaturation has been estimated to take place in the emission plume from the Four Corners power plant near Farmington, N.M. This generating plant produces about 2175 MW at full capacity, and emits about 3.7 kg s^{-1} of SO_2, 2.2 kg s^{-1} of NO_x, and 0.9 kg s^{-1} of particulates. The site area is desert; annual precipitation averages around 15 cm, daytime relative humidity is typically 10–20%, and annual sunshine is 75% of possible.

An instrumented aircraft was utilized during five investigative periods over a 2-year span to measure plume parameters and to collect plume aerosol samples.

The estimated production rate of 10^{16} particles s^{-1} in the plume is equivalent to a gas phase oxidation of SO_2 and homogeneous nucleation-condensation of about 5 $\mu g\ m^{-3}\ h^{-1}$ of H_2SO_4, and is consistent with this mechanism being the primary route for removal of SO_2.

1. INTRODUCTION

Sulfur dioxide (SO_2), as a man-made pollutant, may have effects hundreds of kilometers away from its source, either by acidifying rainfall or, when oxidized to particulate sulfate, by influencing cloud formation and short-wave radiation. According to projections made by the U.S. Environmental Protection Agency in its recent Northern Great Plains Resource Program (1975), coal production in the northern Great Plains will rise from 5×10^{10} kg in 1975 to over 35×10^{10} kg by the year 2000. Part of the fuel development in this area includes utilization of the coal for power generation at mine mouth or at sites providing easy access to the coal supplies. Accordingly, a proportionately larger amount of aerosols and gases will be emitted into the atmosphere. Because of their hygroscopic nature, sulfur aerosols can act as centers around which cloud drops form at relative humidities less than 100%. The measurements reported in this paper reflect the number of cloud droplets that form at 101% relative humidity. This is about the upper limit of humidity encountered in clouds. An increase in the numbers of cloud condensation nuclei (CCN) i.e. particles that function as nuclei for the formation of water droplets could lead to an increase in the frequency with which clouds form, an increased colloidal stability of the clouds, and a difference in the frequency and/or intensity of rainfall.

The economic impact of changes in rainfall can be assessed from the results of previous studies which consider the benefits to be derived from enhancement of precipitation through deliberate weather modification. The change in agricultural production from one inch of incremental rainfall has been estimated at several hectoliters of grain or a few hundred kilograms of forage per hectare (Johnson, 1974; Bauer, 1972). These figures demonstrate the importance of small changes in amounts of rainfall for agriculture in a semi-arid region. Any such changes in precipitation due to inadvertent effects of energy development would lead to modification of the Great Plains economy.

The oxidation of SO_2 in plumes has been studied extensively, e.g. Newman *et al.* (1975), Lusis and Wiebe (1976). In the following discussion we show that a climatologically significant aerosol is found among the particulate effluents of a coal-fired power plant; CCN efficient at 1% supersaturation are formed in the plume at a rate of 10^{15} to 10^{16} per second at distances up to 150 km from the source. By relating the aerosol formation rate to a SO_2 depletion rate due to chemical reaction, a gas (SO_2) to particle (SO_4^{2-}) conversion process has been isolated as one mechanism by which these nuclei are formed.

2. EXPERIMENT LOCATION AND EQUIPMENT

The data on CCN formation that we present here were obtained from investigations that we have conducted since the summer of 1975 in the plume of the Four Corners power plant near Farmington, New Mexico. The plant consists of five pulverized-coal-burning generating units with a total full-load output of 2175 MW at a coal consumption rate of 25.08×10^6 kg day^{-1}. Emissions include 3.2×10^5 kg day^{-1} of SO_2, 1.9×10^5 kg day^{-1} of NO_x, and 8.0×10^4 kg day^{-1} of particulates. The average power output is about 73% of maximum; emissions are proportionately reduced. The power plant is situated in the broad valley of the San Juan River at an elevation of 1675 m. The elevation of uplands to the south is about 2450 m. Mountains to the east and north rise to 3350 m and 4250 m, respectively. Nocturnal inversions occur on about three-fourths of the nights during the year, and the associated drainage winds flow to the west-northwest along the river valley. Wind velocity is usually about 1–2 m s^{-1}, and the drainage winds persist for 10 to 14 hour periods, typically.

Table 1. Parameters measured by aircraft during the four-corners power plant study

Parameter	Units	Method of recording	Frequency of recording	Instrument	Precision (%)	Sensitivity
Cloud condensation nuclei	m^{-3}	photography	preselected times	thermal diffusion chamber (Allee, 1970)	± 5	2×10^{-6}
Aitken nuclei	m^{-3}	strip chart	continuous	expansion chamber	± 10	1×10^{-4}
Ice nuclei (filter sample)	m^{-3}	visual crystal counts	preselected times	thermal diffusion chamber (laboratory)	± 50	1×10^{-4}
Light scattering coefficient	m^{-1}	strip chart	continuous	integrating nephelometer	± 5	0.15×10^{-4}
Aerosol size distribution (filter sample)	m^{-3}	visual count log	preselected times	transmission EM (laboratory)	± 20	
Aerosol elemental composition (filter sample)		visual count log mag tape	preselected times	scanning EM X-ray energy spectrometry (laboratory)		
Temperature	°C	strip chart	continuous	thermistor	± 2	0.25
Relative humidity	%	strip chart	continuous	hygrometer	± 3	0.5
Ozone	ppb	strip chart	continuous	chemiluminescent	± 5	5.0
Nitrogen oxides	ppb	strip chart	continuous	chemiluminescent	± 5	5.0
Sulfur dioxide	ppb	strip chart	continuous	polarographic	± 10	10.0

A Cessna single-engine, 6-place airplane serves as our sampling platform. Table 1 shows the parameters measured with the aircraft instrument package, including the methods, frequency of measurement, precision and sensitivity.

3. RESULTS AND DISCUSSION

Figure 1 presented here as a typical situation, shows the variation in the plume axis of the SO_2 concentration, the light scattering coefficient (b_{SCAT}), O_3 concentration and CCN concentration as functions of the distance downwind from the power-plant stacks. These parameters are plotted as ratios of the measured values at the plume axis to the values measured outside the plume. Typical background values are as follows: SO_2, 5–10 ppb; O_3, 25–40 ppb; CCN, 500–1000 cm^{-3}; b_{SCAT}, 1.7×10^{-5} m^{-1}. The de-

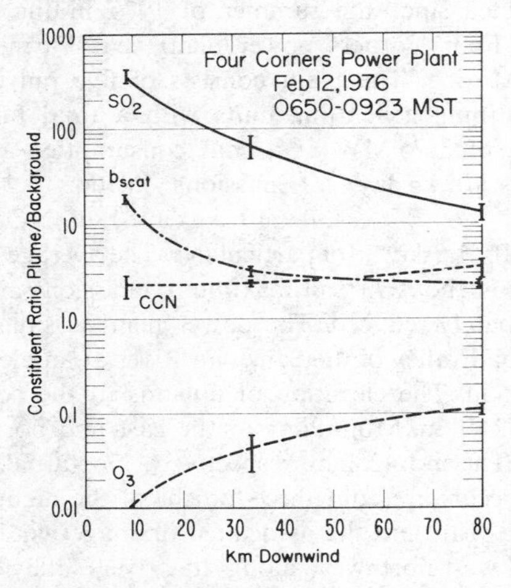

Fig. 1. Variations with distance L from the stacks of SO_2, b_{SCAT}, CCN, and O_3.

crease of the plume axis concentration of SO_2 is contrasted to a slight increase in the concentration of CCN. If we fit the data of Fig. 1 to a power curve, we find relationships between constituent concentration C and distance L of the kind $C_L = C_0 \times L^K$. C_0 is the constituent concentration at distance zero, obtained by extrapolating the measured data back to the point where the plume could be said to have stabilized following dissipation of the thermal and kinetic energy contained at exit from the stacks. The case illustrated in Fig. 1 has values of C_0 for SO_2 of 7.58×10^3, and for CCN of 1.34. The value of K that determines the change of constituent ratio with distance is -1.42 for SO_2 and 0.18 for CCN. It can be seen that a relatively strong decrease of SO_2 with distance is compared to a virtual independence of CCN with distance. Such a condition was found during each of the five measurement periods from July 1975 through October 1976. The values for each experiment of the parameters C_0 and K for SO_2 and CCN are shown in Table 2. The averages for SO_2 of 17 experiments yield $C_0 = 27.4 \pm 18.9$, $K = -1.35 \pm 0.25$, and for CCN, $C_0 = 5161 \pm 4852$, $K = 0.08 \pm 0.29$. The strong variability in both C_0 and K reflect variations in source strength, diffusion and deposition of plume constituents, as well as experimental error whose magnitudes are indicated by the vertical bars in Fig. 1. However, the relatively large error margin notwithstanding, the SO_2 is strongly depleted, while CCN are slightly increased.

Figure 2 illustrates the model that we assume for calculation of losses of constituents due to diffusion, deposition, and reaction. The plume is assumed to be emitted at height Z above the ground and to travel distance L with velocity u. In time $t_1 = L_1/u$ it will pass cross section 1, and at later time $t_2 = L_2/u$ it will pass cross section 2. The contribution of diffu-

Table 2. Power curve fit $C_L = C_0 \times L^K$ between constituent C measured in the plume axis at distance L from the stacks and L. r^2 is the coefficient of determination

Date	SO$_2$ C_0 (ppm)	SO$_2$ K	SO$_2$ r^2	CCN C_0 (cm^{-3})	CCN K	CCN r^2
7-29-75			1.0	1461*	−0.07	0.04
7-30-75				3745*	−0.35	0.92
8- 1-75				4475*	−0.16	1.00
10-14-75				3358*	−0.01	1.00
10-16-75	36.61	−1.43	0.92	5228*	−0.02	0.78
10-17-75	40.95	−1.53	0.96	8077*	−0.18	0.89
2-11-76	26.05	−1.13	1.00	8078†	−0.09	0.50
2-11-76	50.57	−1.49	0.99	6288†	−0.02	
2-12-76	23.35	−0.98	0.88	1740	+0.23	0.31
2-13-76	8.11	−1.13	0.96	20631†	−0.38	0.81
6-25-76	63.5	−1.75	0.94	237	+0.45	0.93
6-26-76	6.72	−1.14	1.00	824	+0.03	0.01
6-27-76	19.0	−1.46	1.00	638	+0.28	1.00
10- 8-76	−2.26*	−1.50	1.0	2187*	−0.07	0.28
10- 9-76	24.99*	−1.39	0.96	8546*	−0.48	0.54
10-10-76	9.31*	−1.14	1.0	5009*	−0.28	1.00
10-11-76	45.10*	−1.59	1.0	7212*	−0.28	1.00

* r.h. < 20%.
† r.h. > 50%.

sion, deposition and reaction to the measured change of SO_2 concentration per unit time, i.e.

$$-\frac{\Delta SO_2}{\Delta t}\bigg|_{MEAS} = \frac{-\Delta SO_2}{\Delta t}\bigg|_{DIF} - \frac{\Delta SO_2}{\Delta t}\bigg|_{DEP} - \frac{\Delta SO_2}{\Delta t}\bigg|_{REAC}, \quad (1)$$

can be calculated if the plume dimensions at cross sections 1 and 2, the wind velocities at times t_1 and t_2, and the vertical and horizontal SO_2 concentration gradients are known. Then the change of SO_2 with time, due to chemical reaction $\Delta SO_2/\Delta t|_{REAC}$ can be estimated quite accurately. In a first approximation we assume that the plume travels in a thin layer downwind to about 8 km from the stacks. At this point it becomes mixed homogeneously in the space between ground and the temperature inversion layer. This condition is frequently in agreement with observations; the 'Hogback', a sharp ridge at right angles to the direction of plume travel about 5 km downwind, induces turbulence in the flow that results in fumigation within a short additional distance.

Table 3 gives the pertinent plume data for some measurements that were performed in October 1976.

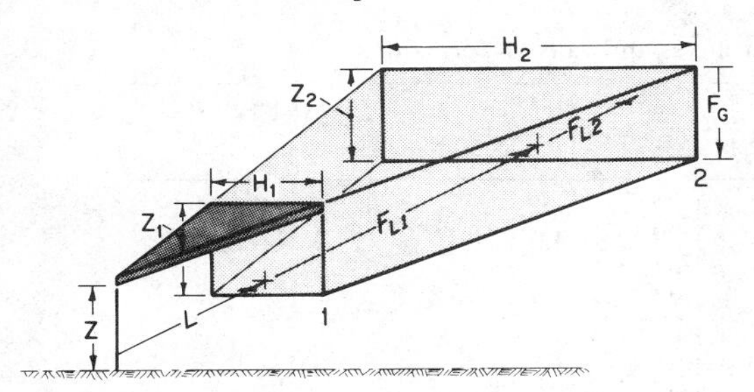

Fig. 2. Schematic of the plume model employed in the SO_2 and CCN flux calculations.

The first column shows the date. The second column shows the distances, L, in km at which the measurements were performed. The third and fourth columns, respectively, show the horizontal, H (km), and vertical, Z (m), extends of the plume at distances L. Columns five and six give the vertical, A_V (m^2), and horizontal, A_H (m^2) cross-sectional extents of the plume. Wind speed, temperature and relative humidity are given in columns seven to nine. Table 4 gives the measured mean SO_2 concentration in column three, and the SO_2 transport through the plume cross sections at the various distances,

$$F_L = \bar{u} \cdot [SO_2] \cdot A_V \quad (2)$$

in column four, where $\bar{u}$ is the mean horizontal windspeed and $[SO_2]$ is the mean SO_2 concentration. The vertical mass transport, F_G, in column five, follows as

$$F_G = V_G \cdot [SO_2] \cdot A_H \quad (3)$$

where V_G is settling velocity for SO_2 over bare soil (Garland, 1976), $[SO_2]$ is SO_2 concentration and A_H is horizontal plume area. The net change of SO_2 concentration due to chemical reaction is

$$\frac{\Delta SO_2}{\Delta t}\bigg|_{REAC} = F_{L_1} - F_{L_2} - F_G, \quad (4)$$

i.e. the differences of SO_2 transports through the plume cross sections at L_1, L_2 and the ground. The relative SO_2 uptake by chemical reaction per hour is shown in column six.

The rate of change of CCN with time is

$$\frac{\Delta CCN}{\Delta t}\bigg|_{MEAS} = \frac{\Delta CCN}{\Delta t}\bigg|_{COAG} - \frac{\Delta CCN}{\Delta t}\bigg|_{DIF} - \frac{\Delta CCN}{\Delta t}\bigg|_{DEP} + \frac{\Delta CCN}{\Delta t}\bigg|_{REAC}. \quad (5)$$

Table 3. Relevant plume parameters

Date	L (km)	H (km)	Z (m)	A_V (m^2)	A_H (m^2)	$\bar{u}$ (m s^{-1})	T (°C)	r.h. (%)
10- 8-76	8	12	163	1.96×10^6	4.80×10^7	1.0	7.8	20
	32	17.8	163	2.90×10^6	2.84×10^8	1.0	5.6	17
10- 9-76	8	3.1	240	7.44×10^5	1.24×10^7	1.1	12.2	12
	80	28.4	152	4.32×10^6	1.14×10^9	0.9	12.2	10
10-10-76	8	11.6	229	2.65×10^6	4.64×10^7	1.6		
	32	20.0	193	3.86×10^6	3.20×10^8	1.8	14.4	12
	80	26.7	200	5.34×10^6	1.07×10^9	1.6	14.4	12
10-11-76	8	2.7	224	6.04×10^5	1.08×10^7	2.0	12.2	10
	32	9.8	183	1.79×10^6	1.56×10^8	2.2	16.7	8

Table 4. Sulfur dioxide fluxes and rate of conversion

Date	L (km)	SO_2 (μg m^{-3})	F_L (μg s^{-1})	F_G (μg s^{-1})	$\Delta SO_2/\Delta t\|_{REAC}$ (% h^{-1})
10- 8-76	8	2.1×10^3	4.2×10^9		
	32	5.24×10^2	1.5×10^9	1.2×10^9	5.4
10- 9-76	8	4.19×10^3	3.4×10^9		
	80	1.38×10^2	5.4×10^8	1.5×10^9	2.0
10-10-76	8	2.20×10^3	9.5×10^9		
	32	7.58×10^2	5.3×10^9	2.1×10^9	6.0
	80	1.05×10^2	9.3×10^8	8.6×10^8	8.0
10-11-76	8	3.67×10^3	4.4×10^9		
	32	7.86×10^2	3.1×10^9	1.2×10^9	3.3

The diffusion and deposition terms are negative, that is, they represent losses of fine particles. The coagulation and reaction terms are positive in that they indicate the formation of CCN from coagulation of molecular agglomerates and from gas phase and gas-solid reactions. The diffusion and deposition losses for CCN can be calculated similarly to those for SO_2 with the proper settling velocity in Equation (3) (Fuchs, 1964).

Table 5 summarizes the results of the CCN measurements. The dates in column one correspond to the dates in Tables 4 and 5. Column 2 lists the distances from the stacks at which the experiments were performed. Column 3 shows the CCN concentration, column 4 gives the transport through vertical plume cross sections at distances L, and column 5 gives the transport of cloud nuclei to the ground as calculated by Equation (3) with a settling velocity $v_G = 2.2 \times 10^{-6}$ m s^{-1}, corresponding to sulfuric acid (H_2SO_4) droplets of size $r = 10^{-6}$ cm. The net increase of CCN per unit time in column 6 is the transport out of the box of Fig. 2, minus the transport into the box, i.e.

$$\frac{\Delta CCN}{\Delta t} = F_{L_2} + F_G - F_{L_1}. \tag{6}$$

The size range of the CCN that are of interest in this discussion is 10^{-6} cm $< r < 10^{-5}$ cm, where r is the particle radius. The upper size limit follows from simultaneous nephelometer measurements that verified the inability of the freshly formed nuclei to scatter visible light. The lower size limit is based on observations by Junge and McLaren (1971), who have shown that the critical radius of a water soluble cloud nucleus must be at least $r = 10^{-6}$ cm in order to grow unstably at 1% supersaturation.

If all the nuclei that are formed are made up to H_2SO_4, this would correspond to a formation mechanism.

$$2SO_2 + O_2 \rightarrow 2SO_3 \qquad SO_3 + H_2O \rightarrow H_2SO_4. \tag{7}$$

Table 5. Cloud condensation nuclei fluxes and rates of formation

Date	L (km)	CCN (m^{-3})	F_L (CCN s^{-1})	F_G (CCN s^{-1})	$\Delta CCN/\Delta t$ (s^{-1})
10- 8-76	8	2.0×10^9	6.7×10^{15}		
	32	2.0×10^9	9.9×10^{15}	1.0×10^{12}	3.2×10^{15}
10- 9-76	8	4.0×10^9	4.5×10^{15}		
	80	1.5×10^9	9.7×10^{15}	3.7×10^{12}	5.2×10^{15}
10-10-76	8	2.0×10^9	1.5×10^{16}		
	32	1.9×10^9	1.9×10^{16}	1.1×10^{12}	4.0×10^{15}
10-11-76	8	4.0×10^9	1.7×10^{16}		
	32	2.7×10^9	2.9×10^{16}	8.8×10^{11}	1.2×10^{16}

If, on the average (see Table 4), 2% per hour of 700 $\mu g\,m^{-3}$ of SO_2 are converted to H_2SO_4 at $L_2 = 32$ km, then 14 $\mu g\,m^{-3}\,h^{-1}$ of SO_2 are consumed to form 21.4 $\mu g\,m^{-3}\,h^{-1}$ of H_2SO_4. With an average plume volume of $6 \times 10^{10}\,m^3$ at $L_2 = 32$ km, the formation rate of H_2SO_4 is $3.6 \times 10^2\,g\,s^{-1}$. This corresponds to a formation rate of $10^{16}\,s^{-1}$ of CCN of size $r = 10^{-5}$ cm, which is in good agreement with the experiment.

In this discussion we have ignored the direct or subsequent formation of NH_4HSO_4 or $(NH)_2SO_4$. The end result in any event would be the formation of soluble sulfate particles.

The simple model (Fig. 2) on which the rate calculations are based probably overemphasizes the deposition fluxes, since in reality a smaller portion of the plume has contact with the ground than has been assumed in the model. This is compensated for in part by the assumption that the lower SO_2 concentration at distance L_2 prevails over the overall distance $L_1 < L < L_2$. In the case of CCN, partial compensation is achieved by basing the flux calculation to the ground on the settling velocity for $r = 10^{-6}$ cm sized particles.

The low rate of about 1% per hour of SO_2 that is converted to sulfates is compatible with the low ozone concentration that has been measured (see Fig. 1), and with the probable absence of other oxidizing species. Furthermore, it supports the notion that a mechanism as shown in Equation (7) is responsible for H_2SO_4 formation. Note that in Table 4 the higher conversion rate was observed when the humidity exceeded 10%. In this connection one should also note that the highest formation rate of CCN was found during the 13 February 1976 experiment (Table 2), when the relative humidity exceeded 50%. This lends support to an accelerated liquid phase oxidation mechanism of SO_2 (Scott and Hobbs, 1967).

The source strength listed in Table 5 for CCN from a coal-fired power plant is significantly less than the global anthropogenic source strength of 10^{20} to $10^{21}\,s^{-1}$. It is comparable, however, with the strengths of other industrial activities and medium-sized industrial–urban complexes (Radke and Hobbs, 1976). Comparing the source strength for CCN of a coal-fired power plant with the production rate of natural CCN from land of $5 \times 10^6\,m^{-2}\,s^{-1}$ (Squires, 1966) and realizing that the SO_2 conversion time is on the order of hours to days, and the residence times of fine particles in the atmosphere are on the order of days to weeks, it becomes apparent that the atmospheric CCN budget in the regional or even short mesoscale surrounding power generating stations is dominated by the nuclei from this single anthropogenic source. In view of the effects that CCN can have on cloud structure and precipitation processes, immediate attention should be given to possible changes in the microstructure of clouds in the northern Great Plains downwind from power plants.

4. SUMMARY AND CONCLUSIONS

Measurements of SO_2 gradients in the plume of the Four Corners coal-fired power plant led to an estimate of SO_2 conversion rates on the order of a few percent per hour. This value is compatible with an observed formation rate of 10^{15} to $10^{16}\,s^{-1}$ of CCN under the assumption that they are initially composed of H_2SO_4.

The source strength for CCN of coal-fired power plants is of similar magnitude to the strengths of large Kraft pulp mills, aluminum smelters, and intermediate urban–industrial areas (Radke and Hobbs, 1976).

Most important for climatic implications is the fact that the number of CCN formed in the plume of one isolated power plant is equivalent to the number formed from land surfaces of areas equal to plume area, at least to distances of 100 km. This doubling of the atmospheric CCN concentration indicates that the inadvertent modification of clouds and precipitation downwind from the plants is possible in the regional scale, perhaps even in the mesoscale. The amount and frequency of rainfall, as well as its acidity, can be affected.

The gas-to-particle conversion process as a mechanism by which cloud nuclei are formed acts in addition to the one discussed elsewhere (Pueschel, 1976), where it was established that the condensation on siliceous fly ash particles of water-soluble sulfates renders these particles efficient in the condensation of water at supersaturations less than 1%. The gas-to-particle formation process is probably the more important one because of the slow conversion rate of SO_2 in the absence of excessive moisture and without high oxidant concentrations. This process also takes place in those cases where the particulate removal is more efficient, resulting in a larger amount of gaseous sulfur to be emitted to the atmosphere.

Acknowledgements—It is a pleasure to acknowledge the valuable contributions of S. Fisher of Western Flight Training, Boulder, and of R. Proulx and D. Wellman of NOAA.

REFERENCES

Bauer A. (1972) Effect of water supply and seasonal distribution on spring wheat yields. North Dakota State University, Agricultural Experiment Station Bulletin 490, Fargo, N.D.

EPA-NGPRP (1975) Effects of coal development in the Northern Great Plains. A review of major issues and consequences at different rates of development. Northern Great Plains Resources Program, U.S. Environmental Protection Agency, Region VIII, Denver, Colo.

Fuchs N. A. (1964) *The Mechanics of Aerosols.* Pergamon Press, New York.

Garland J. A. (1976) Dry deposition of SO_2 and other gases. Atmosphere–surface exchange of particulate and gaseous pollutants, ERDA Technical Information Center, Oak Ridge, Tenn.

Johnson J. E. (1974) The effects of added rainfall during the growing season in North Dakota, Final Report. Agricultural Experiment Station Research Report No. 52.

Junge C. E. and McLaren E. (1971) Relationship of cloud nuclei spectra to aerosol size distribution and composition. *J. atmos. Sci.* **28,** 382–390.

Lusis M. A. and Weibe H. A. (1976) The rate of oxidation of sulfur dioxide in the plume of a nickel smelter stack. *Atmospheric Environment* **10,** 793–798.

Newman L., Forrest J. and Manowitz B. (1975) The application of an isotopic ratio technique to a study of the atmospheric oxidation of sulfur dioxide in the plume from a coal fired power plant. *Atmospheric Environment* **9,** 969–977.

Pueschel R. F. (1976) Aerosol formation during coal combustion: condensation of sulfates and chlorides on fly ash. *Geophys. Res. Lett.* **3,** 651–653.

Radke L. F. and Hobbs P. V. (1976) Cloud condensation nuclei on the Atlantic seaboard of the U.S. *Science* **193,** 999–1002.

Scott W. D. and Hobbs P. V. (1967) The formation of sulfates in water droplets. *J. atmos. Sci.* **24,** 54–57.

Squires P. (1966) An estimate of the anthropogenic production of cloud nuclei. *J. Rech. Atmos.* **3,** 297–308.

Atmospheric Environment Vol. 12, pp. 313–321. Pergamon Press 1978. Printed in Great Britain.

NUCLEI FORMATION RATES IN A COAL-FIRED POWER PLANT PLUME

K. T. WHITBY, B. K. CANTRELL and D. B. KITTELSON

Particle Technology Laboratory, Mechanical Engineering Department, University of Minnesota, Minneapolis, MI 55455, U.S.A.

(*First received* 13 *June* 1977 *and in final form* 15 *August* 1977)

Abstract—From total aerosol concentration data obtained with a condensation nuclei counter (CNC) and size distribution information obtained with an electrical aerosol analyzer (EAA), in and adjacent to the Labadie power plant plume near St. Louis, Missouri, nucleation rates, QN, and the fraction of aerosol mass from gas-to-particle conversion going into the nuclei mode and accumulation mode have been calculated. Nucleation rates in the plume ranged from 1700 $cm^{-3} s^{-1}$ at 0630 in the morning close to the stack to 1 at night and long distances. Background rates ranged from 2 to 4 in the middle of the day to essentially zero at night. Approximately 5% of the new mass from gas-to-particle conversion in the plume formed new nuclei while the remainder condensed directly on accumulation mode particles. Nucleation rates calculated from the photooxidation of SO_2 in smog chamber experiments were found to compare favorably with those in the plume. Furthermore, the number of new nuclei formed per unit volume of aerosol formed by the gas-to-particle conversion were found to be comparable in the plume and in the smog chamber experiments, being about $5 \times 10^6\ \mu m^{-3}$ in the smog chamber and 0.6 to $2.6 \times 10^6\ \mu m^{-3}$ in the plume. The ratio of the volume formation rate of aerosol to the SO_2 concentration in the plume and in the smog chamber were found to be comparable, 38 $\mu m^3 cm^{-3} h^{-1} ppm^{-1}$ in the smog chamber compared to 57 in the plume. These ratios were somewhat smaller than those calculated for aerosol formed in a sulfur-containing propane flame by Barsic (1977) and alongside of a Los Angeles freeway by Clark *et al.* (1976). A simple correlation between the total number concentration of the aerosol and the number of particles in the accumulation mode (Na) has been found to be $NT = 10\ (N_a)^{0.904}$ (NT, Na cm^{-3}). Using the lumped mode model, a further simple relationship between NT and the nucleation rate, QN_n, has been found to be $NT = 1.725 \times 10^4\ (QN_n)^{1/2}$. These later correlations would only apply to well-aged aerosol several hours away from fresh injections of combustion nuclei (QN_n no. $cm^{-3} s^{-1}$).

NOMENCLATURE

DGN	Geometric number mean size.
DGS	Geometric surface mean size.
DGV	Geometric volume mean size.
F	Designates N, S, or V in general.
FN_n	Nuclei number flow rate in plume.
FN_a	Accumulation mode flow rate in plume.
F_{SO_2}	SO_2 flow rate in plume.
K_{nn}	Homogeneous coagulation coefficient for the nuclei mode.
K_{na}	Heterodisperse coagulation coefficient, nuclei-accumulation.
K_{aa}	Homogeneous coagulation coefficient for the accumulation mode.
N_n	Number concentration of particles in the nuclei mode.
N_a	Number concentration of particles in the accumulation mode.
N_c	Number concentration of particles in the coarse particle mode.
NT	Total number concentration of particles in whole distribution.
QN_n	Rate of nuclei number transfer into the nuclei mode.
QN_a	Rate of particle number transfer into the accumulation mode.
QV_n	Rate of aerosol volume transfer into the nuclei mode.
QV_a	Rate of aerosol volume transfer into the accumulation mode.
QVT	$QV_n + QV_a$.
V_n	Volume concentration of particles in the nuclei mode.
V_a	Volume concentration of particles in the accumulation mode.
V_c	Volume concentration of particles in the coarse particle mode.
VT	Total volume concentration in whole distribution.
Vg	Plume gas flow at a given cross-section.
Subscripts	
n, a, and c	Designate the nuclei, accumulation, and coarse particle modes, respectively.
b	Designates background.
p	Designates plume.
$\bar{N}_n$	The bar designates an average.
σ_g	Geometric standard deviation.

Particle Technology Laboratory Publication No. 327.

INTRODUCTION

It is well-known that when air containing SO_2 is photolized, either in a smog chamber or in the atmosphere, large numbers of nuclei smaller than 0.01 μm diameter may be produced. These nuclei result from the heteromolecular nucleation of H_2SO_4 and H_2O under conditions where competing condensation on accumulation mode-sized particles (0.05 to 0.6 μm) is small enough to allow a significant fraction of the vapor to form new nuclei.

Although some studies of nucleation rates have been made in smog chambers starting with particle-free atmospheres so that there is initially no competing condensation on existing accumulation mode-sized particles, experimental difficulties have so far precluded similar determination in the real atmosphere. However, advances in aerosol size distribution measurement instrumentation (Cantrell *et al.*, submitted) and aerosol size distribution modeling (Whitby, 1974) during the past few years have now made it possible to calculate nucleation rates from aerosol size distribution data obtained aboard the MRI Cessna 206 used in the 1976 MISTT measurements in a coal-fired power plant plume near St. Louis.

Data for these studies were obtained using a Thermo-Systems Model 3030 electrical aerosol analyzer (EAA) having improved response and calibration in the 0.0056 to 0.018 μm size range. Special care was taken to calibrate the Environment/One condensation nuclei counter (CNC) so that its absolute accuracy for particles larger than about 0.006 μm would be on the order of $\pm 20\%$.

The aircraft (Blumenthal *et al.*, 1978), experiment design (Whitby *et al.*, 1978), University of Minnesota instrument system (Cantrell *et al.*, submitted), and methods for calculating plume flows and size distribution and aerosol formation rates (Whitby *et al.*, 1978) are described in other papers at this symposium.

This paper reports on nucleation rate calculations from the 1976 MISTT experiment and compares the results with SO_2 smog chamber experiments performed by the University of Minnesota and Calspan Corporation (Kocmond *et al.*, 1973; 1975).

No comparisons with classical nucleation theory are included in this paper because classical theory can only predict the rate of appearance of embryos on the order of 0.0014 μm in size, whereas the instrumentation used in these studies could only sense particles that had grown to sizes larger than about 0.006 μm. Since it would take about 75 embryos to make one 0.006 μm particle, if coagulation of embryos was the only nuclei growth mechanism, it is to be expected that classical theory would substantially overpredict the observed nucleation rates. Some preliminary comparisons of results using McMurry's (1977) heterodisperse nucleation model with our data indicate order of magnitude agreement. This work will be reported in a future paper.

DATA ANALYSIS

This paper contains only the data analysis specifically used to calculate nuclei formation rates. The basic size distribution data, plume flow calculations, and meteorological descriptions of the two days analyzed are described in the companion paper included in this symposium (Cantrell and Whitby, 1978).

Aerosol dynamics

For a typical plume trimodal size distribution undergoing dilution (Fig. 1), where all particles are considered to be lumped at one mean size for each mode, the rate of change of the number concentration in the three modes may be described by the following set of three equations.

Nuclei mode:

$$\frac{d\bar{N}_n}{dt} = Q\bar{N}_n - K_{nn}\bar{N}_n^2 - K_{na}\bar{N}_n\bar{N}_a - K_{nc}N_nN_c - (\bar{N}_{np} - \bar{N}_{nb})\frac{dVg}{Vg\,dt}. \quad (1)$$

Accumulation mode:

$$\frac{d\bar{N}_a}{dt} = Q\bar{N}_a - K_{aa}\bar{N}_a^2 - K_{nc}\bar{N}_a\bar{N}_c - (\bar{N}_{ap} - \bar{N}_{ab})\frac{dVg}{Vg\,dt}. \quad (2)$$

Coarse particle mode:

$$\frac{d\bar{N}_c}{dt} = Q\bar{N}_c - K_{cc}\bar{N}_c^2 - (\bar{N}_{cp} - \bar{N}_{cb})\frac{dVg}{Vg\,dt} + S(t) \quad (3)$$

where $\bar{N}$ = the number concentration, $Q\bar{N}$ = the average rate insertion of new particles, Vg = the volume flow of gas in the diluted plume vs plume age at each cross-section, $\bar{N}_p$ = the average number concentration in the plume, $\bar{N}_b$ = the average number concentration in the background air diluting the plume, $S(t)$ = a settling function for the coarse particles, and K = a coagulation coefficient ($cm^3\,s^{-1}$).

The subscripts n, a, and c designate the nuclei, accumulation, and coarse particle modes, respectively. The subscripts p and b denote the plume and background, respectively.

Calculations for a typical atmospheric size distribution showed that the rates of coagulation of the nuclei and accumulation modes with the coarse particle mode were less than 1% of the nuclei–nuclei mode and nuclei–accumulation mode rates. Therefore, co-

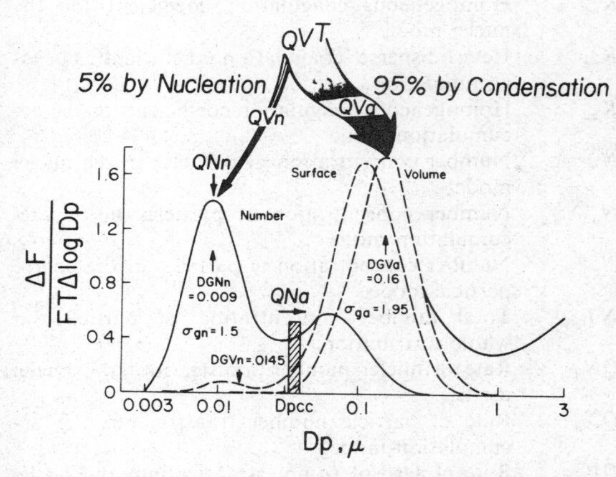

Fig. 1. Typical plume number, surface, and volume distributions showing how the new volume formed from gas-to-particle conversion is inserted into the distribution. Modal parameters are typical of those observed in the Labadie plume in 1976. From modeling studies, the critical size for condensation appears to be about 0.035 μm.

agulation with the coarse particle mode has been neglected.

Since the modes are actually polydisperse, the nuclei mode typically having $\sigma_{gn} = 1.5$ and the accumulation mode, $\sigma_{ga} = 1.9$, the coagulation coefficients are somewhat greater than the monodisperse values based on the number geometric mean size of the modes, DGN_n and DGN_a. Investigation showed that if the monodisperse values for K_{nn} and K_{aa} were about doubled, and that if K_{na} was based on DGS_n, the geometric surface mean diameter of the nuclei mode, and DGS_a, the geometric surface mean diameter of the accumulation mode, the value of the total number concentration NT where $NT = N_n + N_a$ calculated from numerical integration of

$$\frac{dN_n}{dt} = K_n N_n^2 + K_{na} N_a^2 \tag{4}$$

and

$$\frac{dN_a}{dt} = K_{aa} N_a^2 \tag{5}$$

agreed quite well with a complete numerical calculation developed by Dolan (1977) using a coagulation coefficient matrix calculated from Fuchs' equations (Fuchs, 1964). Comparison of the two calculations in Fig. 2 shows that they agree within a few percent for the kinds of size distributions encountered in the plume and for coagulation rates typical of those for which equations (1) and (2) have been used to calculate QN_n.

In order to use equation (1) to calculate QN_n, the gas volume as a function of age $Vg(t)$ must be specified. Although more complex dispersion models could be used, a simple power function, fitted to the plume gas flows at each cross-section (equation 6), was found to be adequate for this purpose. The procedures used for calculating Vg are described by Whitby *et al.* (1978).

The accumulation mode number concentration, N_a, was calculated from a log-normal fit to the accumulation mode of the volume size distribution (Cantrell and Whitby, 1978). The nuclei mode number concentration, N_n, was calculated by subtracting N_a from the NT measured by the Environment/One condensation nuclei counter.

Typically, in each pass, the aircraft obtained several background size distributions adjacent to, but outside of, the plume. These three to six measurements per cross-section were averaged together to get $\overline{N}_{nb}$ and $\overline{N}_{ab}$. The plume average values of $\overline{N}_{np}$ and $\overline{N}_{ap}$ were calculated two different ways. In the first method, the total number flow of nuclei, FN_n, or, total number flow of accumulation mode particles, FN_a, was divided by the gas flow Vg (Cantrell and Whitby, 1978), e.g.

$$\overline{N}_{np} = \frac{FN_n}{Vg} \quad \text{and} \quad \overline{N}_{ap} = \frac{FN_a}{Vg}. \tag{6}$$

Because this method averages over the whole plume cross-section, including the relatively dilute plume edges, it results in lower values of $\overline{N}_{np}$, $\overline{N}_{ap}$, and hence, lower QN_n and QN_a than the second method.

In the second method, the several values of N_{np} and N_{ap} for the individual size distributions measured in the plume at each cross-section were averaged together. Because these individual distributions were usually obtained close to the plume center, $\overline{N}_{np}$ and $\overline{N}_{ap}$ obtained this way tend to be higher than by the first method, sometimes 50% higher.

Because the first method is more presentative and less influenced by the exact location of the sampling points, it has been used in all of the results reported here.

The average rate of change of N_n with plume age $d\overline{N}_n/dt$ was obtained by fitting a power function or an exponential to $\overline{N}_{np}$ at each cross-section and then calculating the derivative at each cross-section (see Cantrell and Whitby, 1978) for tabulation of $\overline{N}_{np}$ vs t). For 5 July, there were three cross-sections available, and it was found that $d\overline{N}_{np}/dt = -5.83$ at each.

For 14 July, each of the two runs consisted of only two cross-sections. After examination of the data, it was decided to fit an exponential to the two points and calculate $d\overline{N}_{np}/dt$ from it. This introduces somewhat more uncertainty into the calculations of QN_n for the 14th than for the 5th.

The rate of change of the volume of aerosol in the accumulation mode, dV_{ap}/dt, needed to calculate QV_{ap}, the aerosol volume formation rate per unit gas volume from equation (7), was calculated in a similar way. V_{ap} was plotted as a function of time of travel and the derivative obtained graphically or by fitting an appropriate function to the data:

$$QV_{ap} = \frac{dV_{ap}}{dt} + (V_{ap} - V_{ab})\frac{dVg/dt}{Vg}. \tag{7}$$

Mean size of the nuclei mode, DGN_n

Although the EAA used in 1976 was improved with respect to its sizing capabilities below 0.01 μm, and a special set of data reduction constants (Whitby *et*

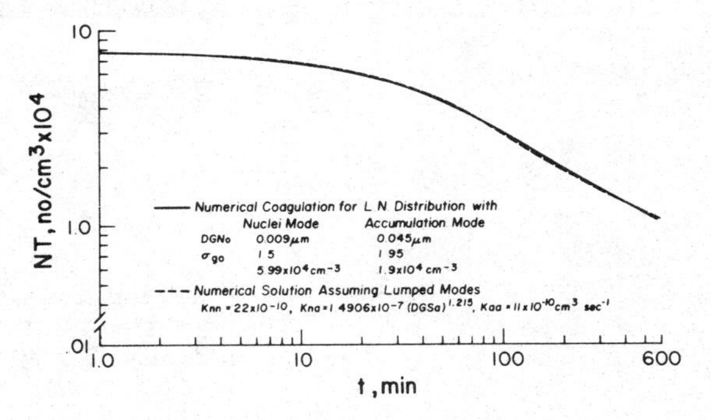

Fig. 2. Comparison of the total number of particles in the nuclei and accumulation modes of a coagulating aerosol as calculated using a complete numerical coagulation and the lumped mode model. The excellent agreement shows that the lumped mode model using the surface mean diameters of the modes to determine K_{na} is adequate for calculations of the nucleation rates.

Table 1. Summary of aerosol parameters used in the calculation of plume and background nucleation rates for 5 and 14 July 1976

Date	Section	Time CDT		Travel t (h)	Vg ($m^3 s^{-1} \times 10^6$)	RH (%)	$\bar{V}_{ap}$ ($\mu m^3 cm^{-3}$)	$\bar{N}_n$ (no. $cm^{-3} \times 10^4$)	$\bar{N}_a$ (no. $cm^{-3} \times 10^4$)	$\bar{D}\bar{G}\bar{S}_a$ (μm)	$\bar{Q}\bar{N}_n$ (no. cm^{-3} s^{-1})	$\bar{Q}\bar{V}_{np}$ ($\mu m^3 cm^{-3}$ h^{-1})	$\bar{Q}\bar{V}_{ap}$ ($\mu m^3 cm^{-3}$ h^{-1})	$\bar{Q}\bar{V}\bar{T}_p$ ($\mu m^3 cm^{-3}$ h^{-1})	$\bar{Q}\bar{V}_{np}/\bar{Q}\bar{V}\bar{T}_p$
5	1	1230	p	0.73	30.8	47	5.86	5.80	1.34	0.105	26.8	0.153	2.99	3.14	0.0512
			b			50		2.59	0.44	0.133	2.9				
5	2	1340	p	1.75	123	52	5.65	3.90	1.69	0.099	7.3	0.042	1.58	1.62	0.0264
			b			52		2.24	0.40	0.138	2.3				
5	3	1730	p	2.61	189	53	6.57	2.28	0.88	0.122	−1.04	—	2.01	2.01	—
			b			48		1.42	0.40	0.142	1.2				
14	1	0630	p	0.72	2.44	54	86.6	45.0	17.6	0.123	1729	9.90	182	191.9	0.058
			b			55		1.69	0.68	0.121	1.95				
14	2	0800	p	1.24	12.2	54	31.7	17.8	4.25	0.122	119	0.68	7.25	7.93	0.094
			b			58		1.90	0.65	0.115	2.13				
14	1′	0920	p	1.19	8.52	57	25.3	5.95	3.04	0.125	65	0.37	32.8	33.17	0.011
			b			46		0.63	0.39	0.126	0.38				
14	2′	1200	p	2.20	69.1	53	8.8	5.18	2.95	0.098	30	0.17	2.3	2.47	0.074
			b			50		1.93	0.97	0.115	0.82				
14	1″	1415	p	0.21	16.8	61	5.6	5.56	1.43	0.116	108	0.62	19.3	19.9	0.031
			b			56		1.20	0.24	0.141	0.71				
														Average =	0.0524
														σ =	0.03

al., 1978) were used, the accuracy of the number size distribution in the size range 0.0056 to 0.018 μm was only good enough to determine $DGN_n = 0.009\ \mu$m $\pm$ 0.001 and $\sigma_{gn} = 1.5 \pm 0.1$. Therefore, these values were used for all calculations as well as the determination of K_{nn}. For $DGN_n = 0.009\ \mu$m, it was determined from the coagulation matrix that $K_{na} = 1.4906 \times 10^{-7}\ (DGS_a)^{1.2148}$.

The surface mean size of the accumulation mode DGS_a was calculated from the log-normal fit to the accumulation mode.

Effect of stack nuclei concentration

The concentration of nuclei mode and accumulation mode aerosols emitted by the stack will be on the order of 10^7 and 10^5 cm^{-3}, respectively. To see if enough of these high initial concentrations could be left at the first cross-section measured to affect the nucleation calculations, numerical calculations including heterodisperse coagulation and the observed plume dilution rates were made. These calculations showed that at the first measurement time of about 0.73 h, dilution and coagulation had so reduced the initial concentrations that nucleation dominated both the nuclei and accumulation modes.

RESULTS AND DISCUSSION

Nucleation rates

The various parameters entering into the calculations of the nucleation rate, QN_n, and the calculated average nucleation rates at each plume cross-section are presented in Table 1. $Q\bar{N}_{np}$ and $Q\bar{N}_{nb}$ were calculated from equation (1). QV_{np} was calculated from $Q\bar{V}_{np} = Q\bar{N}_{np}\ (\pi/6)\ DGV_n^3$ with $DGV_n = 0.0145\ \mu$m. $Q\bar{V}_{ap}$ was calculated from equation (7) and $Q\bar{V}\bar{T}_p = Q\bar{V}_{np} + Q\bar{V}_{ap}$.

The last column shows the ratio $Q\bar{V}_{np}/Q\bar{V}\bar{T}_p$. Although it varies from cross-section to cross-section, the variation is probably within the accuracy of the calculations so that only the average 0.054 $\pm$ 0.03 is significant. This ratio shows that about 5% of the new aerosol mass being formed in the plume goes to form new nuclei at about 0.01 μm, and the remainder condenses directly on the accumulation particles.

A standard error analysis for the nucleation rate, QN_n, was made, based on equation (1). The results are shown in Fig. 5 as error bars. The calculated error averaged $\pm$70%. However, because the calculations of QN_n are quite complex and equation (1) does not account for all the error, the real uncertainty is probably double the calculated value.

It will be noted that $Q\bar{N}_{np}$ is highest nearest the stack and decreases rapidly at the greater distances. It will also be noted that $Q\bar{N}_{np}$ for the cross-section at 1730 on 5 July is slightly negative. Because of the uncertainties in the calculations, it is estimated that the smallest value of Q_{nn} in the plume which is significant is ± 2 cm^{-3} s^{-1}. In the background, it is estimated that the smallest value which is significant is $\pm$0.5 cm^{-3} s^{-1}.

The highest nucleation and volume aerosol formation rates observed so far were observed on 14 July at 0630. This early in the morning, the atmosphere was still quite stable, with the plume trapped between very stable layers. However, for the well-mixed conditions on 5 July, corresponding to a stability class of B or C, nucleation rates were much lower. It is estimated from ground measurements made on a similar type of clear day in July, 1975 that the ultraviolet light intensity was 6% of noonday sun. Could this be the morning aerosol formation effect that has often been speculated to exist?

In Table 2, several other ratios have been calculated. The ratio $Q\bar{N}_{np}/Q\bar{V}\bar{T}_p$ is the number of new nuclei that are produced per cubic μm of new aerosol volume formed. The average value is 3.1×10^5 particles per μm^{-3}. Considering the accuracy of the calculations involved in obtaining this number, it may be considered as constant within the accuracy of the calculations. It is striking that although the nucleation rate varies by three orders of magnitude, the fraction of mass going into the nuclei mode is essentially constant.

Since the conversion of SO_2 to particle mass is approximately first order for SO_2, although undoubtedly influenced by relative humidity and other factors,

Table 2. Ratio of nucleation rate, $Q\bar{N}_{np}$, to the total volume formation rate, $Q\bar{V}\bar{T}_p$, and of $Q\bar{V}\bar{T}_p$ to $\bar{SO}_2$ for the measurements of 5 and 14 July 1976 in the Labadie plume

Date	Time CDT	$Q\bar{N}_{np}$ (no. cm^{-3} s^{-1})	$Q\bar{V}\bar{T}_p$ (μm^3 cm^{-3} s^{-1})	$Q\bar{N}_{np}/Q\bar{V}\bar{T}_p$ (no. μm^{-3})	$\bar{SO}_2$ (ppm)	$Q\bar{V}\bar{T}_p/\bar{SO}_2$ (μm^3 cm^{-3} h^{-1} ppm^{-1})
5	1230	26.8	8.94×10^{-4}	3.0×10^4	0.128	9.0
5	1340	7.3	3.83×10^{-4}	1.91×10^4	0.0253	19.6
5	1730	~0	5.60×10^{-4}	~0	0.0191	106
14	0630	1729	0.0533	3.24×10^4	1.42	135
14	0800	119	0.00220	5.41×10^4	0.35	22.7
14	0920	65	0.00921	0.71×10^4	0.45	73.7
14	1200	30	0.00069	4.35×10^4	0.072	34.3
14	1415				0.24	
			Averages =	3.1×10^4		57.2
			σ =	1.7×10^4		49

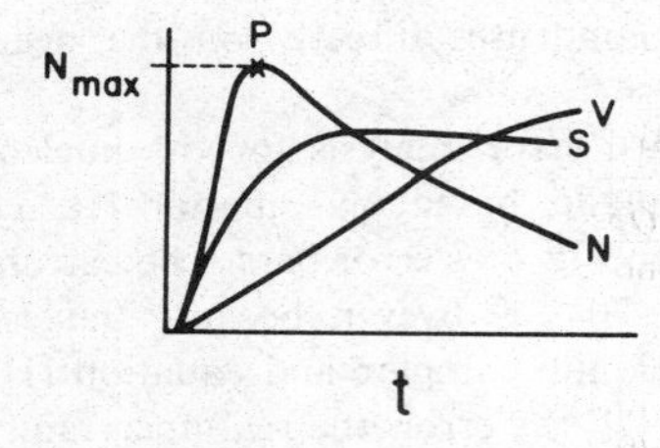

Fig. 3. Typical aerosol number, surface, and volume development for the aerosol produced by photooxidation of SO_2 in a smog chamber. At point P, $dN/dt = 0$ and the nucleation rate is equal to the coagulation rate.

the ratio of the formation rate of new particle volume, $\bar{Q}\bar{V}\bar{T}_p$, to the average $\overline{SO}_2$ concentration in the plume should be more or less constant. The average $\overline{SO}_2$ concentration has been calculated from the SO_2 flow, F_{SO_2}, e.g., $\overline{SO}_2 = F_{SO_2}/Vg$. From Table 2 it is seen that this ratio varies from a low of 9 at 1230 on the 5th to a high of 135 $\mu m^3\,cm^{-3}\,s^{-1}\,ppm^{-1}$ at 0630 on the 14th. The increase from 9 at 1230 on the 5th to 106 at 1730 on the 5th suggests that the aerosol formation rate is staying more or less constant even though the SO_2 concentration in the plume is being reduced by dilution. Although SO_2 conversion may account for most of the aerosol volume formed near the stack, at greater distances other precursors may be more significant. As will be shown below, the average value of $\bar{Q}\bar{V}\bar{T}_p/\overline{SO}_2 = 57 \pm 49$ agrees with that calculated from SO_2 photooxidation experiments conducted in smog chambers by the University of Minnesota and Calspan Corporation.

Smog chamber nucleation rates

Clark (1972), Clark and Whitby (1975), Kocmond *et al.* (1973), and Kocmond *et al.* (1975) used an instrument system consisting of a CNC and an EAA to measure the aerosols produced from the photooxidation of SO_2–air and SO_2–NO–hydrocarbon mixtures in a smog chamber. From the time history of the aerosol size distribution, it is possible to calculate the aerosol volume, surface area, and number concentrations. A typical curve for N, S, and V is shown in Fig. 3. The above investigators then calculated dV/dt from these curves. Their dV/dt is equivalent to the QV_n calculated earlier for the plumes. At point P (Fig. 3), $dN_n/dt = 0$ and $QN_n = K_{nn}N^2_{nmax}$.

In a smog chamber which is aerosol free at $t = 0$, there is no accumulation mode, and all of the aerosol mass is in the nuclei mode.

Average values for $\bar{Q}\bar{N}_n$, $\bar{Q}\bar{N}_n/\bar{Q}\bar{V}_n$, and $\bar{Q}\bar{V}_n/\overline{SO}_2$ for 13 series of runs with SO_2 in air only and with several different hydrocarbons are tabulated in Table 3. Although most of these experiments were done at lower relative humidities than those in the plume, they are close enough so that relative humidity is probably not a major factor except for those of Clark at 15%. Average values of $\bar{Q}\bar{V}_n/\bar{Q}\bar{V}\bar{T}$, $\bar{Q}\bar{N}_n/\bar{Q}\bar{V}_n$, and $\bar{Q}\bar{V}\bar{T}/\overline{SO}_2$ for the smog chamber data are compared with the plume averages in Table 4. $\bar{Q}\bar{N}_n/\bar{Q}\bar{V}\bar{T}$ and $\bar{Q}\bar{V}\bar{T}/\overline{SO}_2$ for the smog chamber data have been corrected to noon sunlight for comparison with the plume data. The average value of $\bar{Q}\bar{N}_n/\bar{Q}\bar{V}_n$ for the smog chamber data is about 7 times larger than in

Table 3. Smog chamber nucleation and $dV/dt/SO_3$ ratios

Reference	Experiments conducted at	Remarks	$\overline{RH}$ (%)	$\overline{SO}_2$ (ppm)	$(\bar{Q}\bar{N}_n/\bar{Q}\bar{V}_n)$ (no. μm^{-3})	$\bar{Q}\bar{V}_n/\overline{SO}_2)$ (no. $cm^{-3}\,s^{-1}\,ppm^{-1}$)
Clark and Whitby (1975)	U. of Minnesota	SO_2 only	15	*	28×10^4	8*
Kocmond *et al.* (1973)	U. of Minnesota	SO_2 only	37	0.26	209×10^4	16.7
Kocmond *et al.* (1973)	Calspan	SO_2 only	32	0.47	212×10^4	1.33
Kocmond (1975) Table X	U. of Minnesota	SO_2 + NO + toluene	36	0.20	19.4×10^4	23.9
Kocmond (1975) Table X	U. of Minnesota	SO_2 + NO + hexene	38	0.15	52×10^4	9.4
Kocmond (1975) Table X	U. of Minnesota	SO_2 + *m*-xylene + NO	40	0.134	88×10^4	10.2
Kocmond (1975) Table X	U. of Minnesota	SO_2 + NO + cyclohexene	28	0.133	144×10^4	9.7
Kocmond (1975) Table IX	Calspan	SO_2 + NO + toluene	25	0.05	26.7×10^4	15.6
Kocmond (1975) Table IX	Calspan	SO_2 + NO + hexene	37	0.07	196×10^4	8.7
Kocmond (1975) Table IX	Calspan	SO_2 + NO + *m*-xylene	29	0.055	508×10^4	16.7
Kocmond (1975) Table IX	Calspan	SO_2 + NO + cyclohexene	30	0.05	396×10^4	14.8
Kocmond (1975) Table IX	Calspan	SO_2 only	27	0.53	60.1×10^4	9.5
Kocmond (1975) Table IX	Calspan	SO_2 only	39	0.05	109×10^4	17.0

* SO_2 varies from 0.088 to 2.88 ppm: $QVT = 7.85\,(SO_2)^{0.85}$, $r^2 = 0.91$.

Table 4. Comparison of aerosol nucleation rates and formation rates for U of Minnesota/Calspan experiments, propane combustion, and photochemical formation near a freeway with those observed in the Labadie plume on 5 and 14 July 1976

	$\bar{Q}\bar{V}_{np}/\bar{Q}\bar{V}\bar{T}_p$	$\bar{Q}\bar{N}_n/\bar{Q}\bar{V}_{np}$ (no. μm^{-3})	$\bar{Q}\bar{V}\bar{T}_p/\bar{S}\bar{O}_2$ ($\mu m^3\, cm^{-3}\, h^{-1}\, ppm^{-1}$)
Smog chamber	1	4.4 (4.6) $\times 10^6$*	38 (13.7)*
Plume—$DGV_n = 0.0145$	0.0524 (0.03)	0.63 (0.06) $\times 10^6$†	57 (49)
—$DVG_n = 0.009$	0.0524 (0.03)	2.63 (0.06) $\times 10^6$	
Flame—quasi-heterogeneous—40% r.h.‡	1	5.5 $\times 10^6$	30
—heterogeneous—90% r.h.	1	2.2 $\times 10^6$	100
Near freeway			136§

* Averages from Table 3 corrected to noonday sun by dividing by 0.35. Figures in parentheses are standard deviations.

† $QN_{np}/QV_{np} = (6/\pi)\,(DGV_n)^{-3}$, $DGN_n = 0.009$, $DGV_n = 0.0145\ \mu m$.

‡ Sulfur-containing propane flame, Barsic (1977).

§ Clark *et al.* (1976) photochemical aerosol growth in a bag of an air sample taken near a freeway. $QVT = [1.51 + 136\,(SO_2)_0]\,(NMHC/1.5)_0$. NMHC = non-methane hydrocarbons.

the plume, e.g. 4.37×10^6 vs $0.63 \times 10^6\ \mu m^{-3}$. However, since $\bar{Q}\bar{V}_n$ has been calculated for a size $DGN_n = 0.009\ \mu m$, and the actual size may be as small as 0.006, the ratio $\bar{Q}\bar{N}_n/\bar{Q}\bar{V}_n$ may be as large as $2 \times 10^6\ \mu m^{-3}$.

Average values of $\bar{Q}\bar{V}\bar{T}/\bar{S}\bar{O}_2$ for the smog chamber and the plumes are almost the same, 38 vs 57 $\mu m^{-3}\, cm^{-3}\, h^{-1}\, ppm^{-1}$. This suggests that the aerosol volume formation rates measured in the smog chamber are a reasonable indication of those that exist in the plume, but that nucleation rates observed in the smog chamber in the absence of an accumulation mode are higher.

Two values of $\bar{Q}\bar{N}_n/\bar{Q}\bar{V}_{np}$, calculated from nuclei formation rates observed by Barsic (1977) in a sulfur-containing flame, are also shown in Table 4. They agree well with the smog chamber values, but are higher than observed in the plume. McMurry (1977) has postulated that the presence of the accumulation mode may actually suppress the heteromolecular nucleation process.

$\bar{Q}\bar{V}\bar{T}_p/\bar{S}\bar{O}_2$ values for the propane flame and for a value obtained by Clark *et al.* (1976) near a freeway are also shown. Clark's results were obtained by irradiating an air sample taken near a freeway in Los Angeles with sunlight and observing the rate of aerosol formation as a function of the initial concentration of SO_2, NO_x, and non-methane hydrocarbons. They found that the correlation coefficient of QVT on SO_2 concentration was 0.99.

Nucleation rates in the background outside the plumes

In the background air outside the plumes, $dN_n/dt \simeq 0$ and there is no dilution. Thus, $QN_n = K_{nn}N_n^2 + K_{na}N_aN_n$.

The values of NT for various values of N_a and QN_n have been calculated for these equilibrium conditions and are plotted in Fig. 4 along with the actual values of NT vs N_a observed in the background and in the plume. It will be noted that NT correlates with N_a quite well for these two days.

From the figure, it is seen that the line of regression crosses the line of constant QN_n where the latter lines are relatively horizontal. Where QN_n is nearly horizontal, NT is relatively independent of N_a. Therefore, a simple relationship between QN_n and NT exists. By fitting, this was determined to be $NT = 1.725 \times 10^4\,(QN_n)^{1/2}$. Since NT is equivalent to the number measured by an Aitken nuclei counter, it is possible to use the above relationship to estimate the nucleation rate from CNC measurements under conditions where significant aerosol is being produced by photochemistry and the concentration of accumulation mode aerosol is not too high. This procedure would probably only be useful in relatively clean background conditions.

Several data points calculated from the data obtained by Meszaros and Vissy (1974) in the Atlantic and Indian Oceans are also shown. These are discussed in the review paper by Whitby (1978).

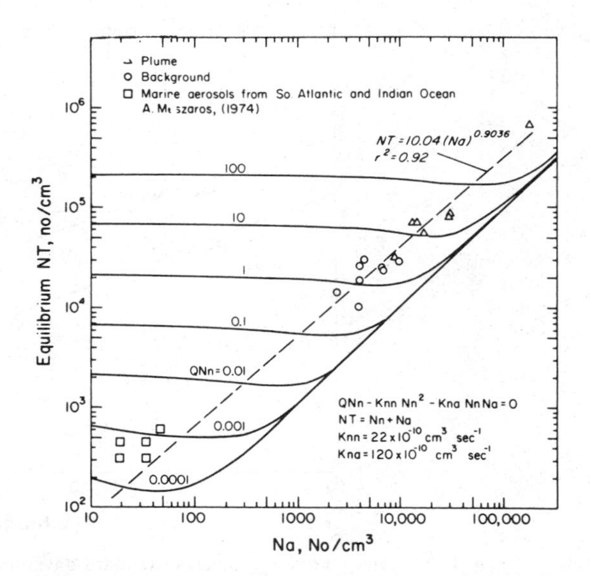

Fig. 4. Calculated nucleation rates, QN_n, as a function of NT and N_a for a two-mode aerosol. The data points are actual values of NT and N_a. The dashed line was fitted only to the 1976 plume and background data.

Although no general averages have been calculated yet, values of QN_n for each size distribution measured in the background have been scrutinized for interesting features. It appears that when the background is not freshly contaminated by the urban plume or other plumes within 50–100 km upwind, QN_n is in the range 2–4 $cm^{-3}\,s^{-1}$ during the middle of the day. In the urban plume, values may be as high as 10.

On 5 and 6 July, an attempt was made to characterize the plume at night. Although only three aerosol size distribution measurements were obtained in the plume, these measurements are of interest. $\bar{Q}\bar{N}_n$ for two measurements at about 2200 on the 5th averaged 2 ± 0.2, and the one at 0320 on the 6th was 1.05. Average background for 2200 was 1.1 ± 0.2 and at 0300 was 0.7 ± 0.1. $\overline{SO}_2$ concentration for the three hits was 0.15 ppm ± 0.03. These results suggest that there might be a little nuclei formation in the plume at night, although the levels are certainly so close to the accuracy limits of this method of calculation that it is not certain whether they are meaningful or not.

Effect of coagulation on accumulation mode size

The relationship between σ_{ga} and QN_n for the plume and background is plotted in Fig. 5. This figure shows that as QN_n increases from 0, σ_{ga} decreases from an initial value of about 2 to a value of about 1.6. When $QN_n = 0$, σ_{ga} approaches 2 in agreement with observations at night and under conditions where photochemical conversion would be expected to be small.

It will be noted from Table 1 that at the closest distances to the stack, and near noon after the aerosol has grown awhile, DGS_a, the surface geometric mean size of the accumulation mode, is about 1/3 smaller in the plume than in the background. Early in the morning on the 14th, the plume and background size are about the same.

This reduced DGS_{ap} in the plume is undoubtedly due to the coagulation of nuclei into the smaller size end of the accumulation mode. Preliminary investigations have been made using the Dolan program for aerosol growth, including nuclei reinforcement, condensation, and coagulation. Size distributions similar to those observed in the plume are calculated by the model if a condensation critical size of about 0.035 μm is used. These investigations are continuing and will be reported in a later paper.

Future work

We have analyzed only a fraction of the data available. We are planning on reanalyzing the data of 1974, which were taken on a day when the background aerosol concentration was much higher, $V_{ab} = 20\ \mu m^3\,cm^{-3}$ vs 2 to 4 in 1976. Also, several other plume runs from 1975 and 1976 have not been analyzed because the plume data are less complete and will therefore require much more care in analysis to produce useful results.

SUMMARY

(1) Nucleation rates in the coal-fired power plant plume near St. Louis ranged from 1700 $cm^{-1}\,s^{-1}$ at 0630 in the morning to values near 1 when the plume was well-mixed at some distance from the stack and at night.

(2) Background nucleation rates near but outside the plume ranged from 2 to 4 during the day to 0.7 at night.

(3) The aerosol volume fraction from gas-to-particle

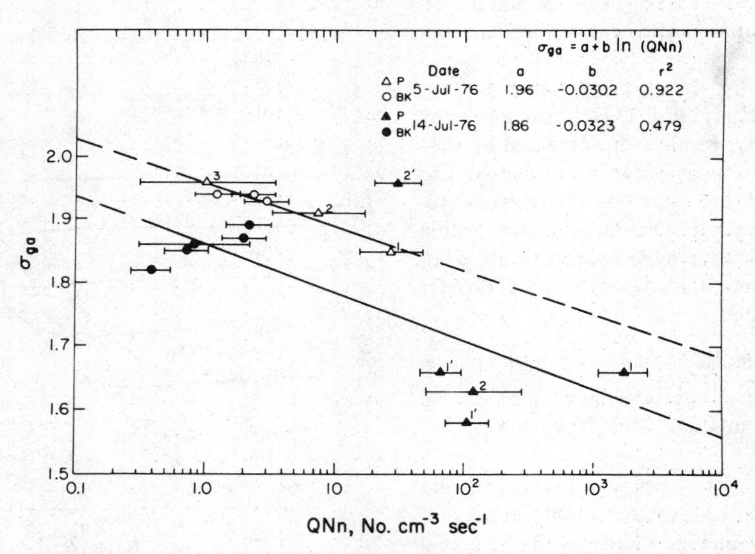

Fig. 5. Correlation of the geometric standard deviation of the accumulation mode, σ_{ga}, with the nucleation rate, QN_n, for the 1976 Labadie data. The decrease in σ_{ga} with QN_n results from nuclei coagulating into the lower size range of the accumulation mode where they can grow by condensation. Error bars for QN_n are shown based on a standard error analysis using equation (1). The error in σ_{ga} is estimated to be about ± 0.05.

conversion going into the nuclei mode was about 5%, and was independent of the nucleation rate.

(4) Comparison of the ratio of aerosol volume being formed to the SO_2 concentration in the plume and for comparable smog chamber data gave a ratio of $\bar{Q}\bar{V}\bar{T}/\bar{SO}_2 = 38\ \mu m^3\, cm^{-3}\, h^{-1}\, ppm^{-1}$ for the smog chamber data and 57 for the plume, a remarkably good agreement. There are indications from the data on 5 July that this ratio increases with increasing distance from the stack, going from 9 for a travel time of 0.73 h to 106 at 2.61 h. This suggests a change in the dominant precursors resulting in aerosol formation as the plume becomes more dilute.

(5) The fact that the $\bar{Q}\bar{V}\bar{T}/\bar{SO}_2$ ratios for the smog chamber experiments were equal to those observed in the plume around noon on these sunny days, under conditions where photochemical conversion was presumably the dominant mechanism of aerosol formation, suggests that most of the aerosol was photochemically formed on 5 and 14 July.

(6) The relationship between the total number concentration of the aerosol, NT, and the number in the accumulation mode, N_a, could be given by $NT = 10.04\ (Na)^{0.904}$ for both plume and background.

Acknowledgements—This research was performed under EPA Research Grant No. R 803851-02, and presents partial results from Project MISTT, an integrated multidisciplinary program directed by Willian E. Wilson, ESRL, EPA. Special thanks are due to the many students and staff at the University who cooperated to carry out this project.

REFERENCES

Barsic N. J. (1977) Size distribution and concentrations of fine particles produced by propane-air combustion in a controlled humidity environment. Ph.D. Thesis, Particle Technology Laboratory, Mechanical Engineering Dept., University of Minnesota, Minneapolis, MN 55455, U.S.A.

Blumenthal D. L., Ogren J. A. and Anderson J. A. (1978) Airborne sampling system for Project MISTT. *Atmospheric Environment* **12**, 613–620.

Cantrell B. K., Brockman J. E. and Dolan D. F. A mobile aerosol analysis system with applications. (Submitted to *Atmospheric Environment*).

Cantrell B. K. and Whitby K. T. (1978) Aerosol size distributions and aerosol volume formation rates for a coal-fired power plant plume. *Atmospheric Environment* **12**, 323–333.

Clark W. E. (1972) Measurements of aerosols produced by the photochemical oxidation of SO_2 in air. Ph.D. Thesis, Particle Technology Laboratory, Mechanical Engineering Dept., University of Minnesota, Minneapolis, MN 55455, U.S.A.

Clark W. E., Landis D. A. and Harker A. B. (1976) Measurements of the photochemical production of aerosols in ambient air near a freeway for a range of SO_2 concentrations. *Atmospheric Environment* **10**, 637–644.

Clark W. E. and Whitby K. T. (1975) Measurement of aerosols by the photochemical oxidation of SO_2 in air. *J. Colloid Interface Sci.* **51**, 477–490.

Dolan D. F. (1977) Investigation of Diesel Exhaust Aerosol. Particle Technology Laboratory, Mechanical Engineering Dept., University of Minnesota, Minneapolis MN 55455, U.S.A.

Fuchs N. A. (1964) *The Mechanics of Aerosols*. Pergamon Press, New York.

Kocmond W. C., Kittelson D. B., Yang J. Y. and Demerjian K. L. (1973) Determination of the formation mechanisms and composition of photochemical aerosols. Calspan Report No. NA5365-M-1.

Kocmond W. C., Kittelson D. B., Yang J. Y. and Demerjian K. L. (1975) Study of aerosol formation in photochemical air pollution. EPA Report No. EPA-650/3-75-007.

McMurry P. H. (1977) On the relationship between aerosol dynamics and the rate of gas-to-particle conversion. Ph.D. Thesis, California Institute of Technology, Pasadena, CA 91125, U.S.A.

Meszaros A. and Vissy (1974) Concentration, size distribution, and chemical nature of atmospheric aerosol particles in remote oceanic areas. *J. Aerosol Sci.* **5**, 101–109.

Smith T. B., Blumenthal D. L., Anderson J. A., Vanderpol A. H. and Husar R. B. (1978) Transport of SO_2 in power plant plumes: day and night. *Atmospheric Environment* **12**, 605–611.

Whitby K. T. (1974) Modeling of multimodal aerosol distributions. Proceedings of Conference on Aerosols in Nature, Medicine and Technology, GAF Berweg 7, Vorderhinderlang 8973, West Germany.

Whitby K. T. (1978) The physical characteristics of sulfur aerosols. *Atmospheric Environment* **12**, 135–159.

Whitby K. T., Cantrell B. K., Husar R. B., Gillani N. V., Anderson J. A., Blumenthal D. L. and Wilson W. E. Aerosol formation in a coal-fired power plant plume. (Submitted to *Atmospheric Environment*).

Atmospheric Environment Vol. 12, pp. 323–333. Pergamon Press 1978. Printed in Great Britain.

AEROSOL SIZE DISTRIBUTIONS AND AEROSOL VOLUME FORMATION FOR A COAL-FIRED POWER PLANT PLUME

B. K. CANTRELL* and K. T. WHITBY

Particle Technology Laboratory, Mechanical Engineering Department, University of Minnesota, Minneapolis, Minnesota 55455, U.S.A.

(*First received* 30 *June* 1977 *and in final form* 27 *September* 1977)

Abstract—During the summer of 1976, the plume from the Labadie power plant near St. Louis, Missouri was mapped on two days using an instrumented aircraft. The mapping consisted of measuring horizontal and vertical concentration profiles of both aerosol and gas contaminates for cross-sections of the plume. Measurements were made for plume travel times of up to 2 h and distances downwind of up to 50 km.

Aerosol size distribution determinations were made in both the plume and surrounding background air mass. Characterization of these distributions was done in terms of two additive log-normal functions: one describing nuclei less than 0.03 μm in size, and the second describing accumulation mode aerosol between 0.03 and 1.0 μm. On the average, results show that distribution of the nuclei volume for aerosol size less than 0.03 μm in both plume and background can be parameterized with a geometric mean size of 0.021 ± 0.005 μm and a geometric standard deviation of 1.5 ± 0.1. Total nuclei volume is found to be about 1% of total submicron aerosol volume. For aerosol larger than 0.03 μm, the geometric mean size of aerosol volume distributions in a daytime plume under uniform meteorological conditions was found to be 0.18 ± 0.02 μm. This is significantly different from the 0.23 ± 0.01 μm measured for background air. The dispersion of the distribution in this size range, measured by a geometric standard deviation of 1.92 ± 0.04, does not change significantly from in-plume to background. For a plume, measured in a morning transition period where average aerosol volume concentrations reached 87 $\mu m^3\,cm^{-3}$, average volume mean geometric size for in-plume aerosol was 0.15 ± 0.01 μm vs 0.18 ± 0.01 μm in the background. Here, dispersion of the size distribution was narrower in the plume, with $\sigma_g = 1.63 \pm 0.04$ vs 1.86 ± 0.03 for background aerosol.

Together with the integral Aitken nuclei concentration and b_{scat} values, the aerosol size distribution measurements have been used to reconstruct the aerosol time history, up to two hours, for the various plumes investigated. Total aerosol volume flow through each cross-section of the plume has been calculated. When corrected for background levels, this flow is generally found to increase with time from levels as low as 60 $cm^3\,s^{-1}$ to about 400 $cm^3\,s^{-1}$. Assuming the excess volume formed is sulfuric acid in equilibrium with water vapor at the ambient r.h., we obtain an average SO_2 conversion rate of $0.7 \pm 0.4\%\,h^{-1}$ for the first 2 h of plume development.

NOMENCLATURE

ANC	Aitken nuclei count, no. cm^{-3}
b_{scat}	optical light scattering coefficient, m^{-1}
DGN	geometric number mean size, μm
DGS	geometric surface mean size, μm
DGV	geometric volume mean size, μm
$f_{H_2SO_4}$	mass fraction of aerosol H_2SO_4
$F_{B_{scat}}$	plume-integrated optical light scattering coefficient, $m^2\,s^{-1}$
FN	number flow rate in plume for specified mode, no. $cm^{-3}\,s^{-1}$
F_{SO_2}	SO_2 flow rate in plume, $m^3\,s^{-1}$
FNP	flow of total plume-associated submicron aerosol number, no. s^{-1}
FST	flow of total submicron aerosol surface in plume, $cm^2\,s^{-1}$
FVP	flow of plume-associated submicron aerosol volume in plume, $cm^3\,s^{-1}$
h	altitude in plume, m (m.s.l.)
Δh	increment of altitude used in plume mapping, m
k	fraction of noontime u.v. radiation at the latitude of St. Louis, Missouri
N	number concentration of particles for specified mode, cm^{-3}
NT	total number concentration for submicron aerosol derived from the plume weighted averages of the modal fit parameters and background measurements, no. cm^{-3}
S	surface concentration of particles for specified mode, $\mu m^2\,cm^{-3}$
Se	equilibrium surface concentration for plume aerosol, $\mu m^2\,cm^{-3}$
SO_2	concentration of atmospheric sulfur dioxide, ppm
ST	total surface concentration in submicron aerosol distribution derived from the plume weighted averages of the modal fit parameters and background measurements, $\mu m^2\,cm^{-3}$
t	plume age, i.e. time from release, s.
$v_\perp(h)$	component of wind velocity at altitude h normal to plume cross-section, $m\,s^{-1}$
V	volume concentration of particles in specified mode, $\mu m^3\,cm^{-3}$
VT	total volume concentration in submicron aerosol distribution derived from the plume weighted averages of the modal fit parameters and background measurements, $\mu m^3\,cm^{-3}$
Vg	plume gas flow at a given cross-section, $m^3\,s^{-1}$
V_s	power plant stack gas flow, $m^3\,s^{-1}$
y	distance along aircraft plume pass, m

* Present address: Stanford Research Inst., 333 Ravenswood Ave, Menlo Park, CA 94025, U.S.A.

Subscripts

$n, a,$ and c	designate the nuclei, accumulation, and coarse particle modes, respectively
b	designates background
p	designates plume
$\bar{N}_n$	the bar designates an average
$\rho_{H_2SO_4}$	density of H_2SO_4 aerosol, g cm^{-3}
σ_g	designates geometric standard deviation of a log-normal fit

INTRODUCTION

During the summer of 1976, a United States Environmental Protection Agency (EPA) sponsored team of investigators used an instrumented Cessna 206 aircraft, furnished by Meteorology Research, Inc. (MRI), to map concentrations of SO_2, NO_x, O_3, and total aerosol in coal-fired power plant plumes. In addition to these, filter-integrated aerosol sulfate and aerosol size distribution measurements were also taken during the plume mappings. This study was conducted as part of the EPA Midwest Interstate Sulfur Transformation and Transport (MISTT) program. The power plant used in the study is a pulverized coal combustion type, located at Labadie, Missouri, approx 40 km west of St. Louis. Intent of the study was to provide more extensive information on coal-fired power plant plumes to augment findings of previous experiments conducted in 1974 on the same power plant (Whitby *et al.*, in press).

This paper reports the results of aerosol size distribution measurements for two experiments during the study conducted on July 5 and 14, 1976. Use is made of the data to reconstruct the time history of the plumes for aerosol flow and to provide a base for subsequent estimation of aerosol volume formation and nucleation rates. The latter are presented in a separate paper by Whitby *et al.* (1978). Some summary features of SO_2 concentrations and flows in the plume are also treated, and associated estimates for SO_2 conversion are calculated. Details of all the gas and meteorological measurements, as well as a description of the aircraft used during the experiments, are reported by Blumenthal *et al.* (1978) and Smith *et al.* (1978). Results of the sulfate aerosol measurements are reported in a companion paper by Husar *et al.* (1978).

PLUME MEASUREMENTS

Plume concentration profiles were obtained from horizontal cross-wind measurement passes which were made with the aircraft at several altitudes to give a cross-sectional mapping of the plume for a given plume age and distance downwind of the plant. An attempt was made to make these measurements at times and distances that would yield approximate Lagrangian treatment of the air parcel under investigation. In addition to in-plume measurements during an experiment, background characterization was also done, both upwind of the power plant and on each side of the plume at the location of each of the cross-sectional mappings. Such extensive measurements permitted us to obtain a dispersion model-independent parameterization of the plume. A more complete description of the experimental design is presented by Whitby *et al.* (in press).

For each pass through the plume, aerosol size distribution measurements were made at points both in the plume and outside in background air. These were of discrete air samples, limited to one or two measurements per pass by the length of time required for the instrumentation to analyze each sample. These single measurements were complementary to the continuously recorded integral aerosol parameters such as optical light scattering coefficient (b_{scat}) and Aitken nuclei count (*ANC*), gas concentrations such as SO_2, and the pass-integrated filter sulfate measurements. Figure 1 illustrates, for a typical pass through a plume, both measured integral aerosol parameters and SO_2 concentrations. It also shows the location and time of the in-plume size distribution measurement.

The aerosol size distribution measurements were made over a range of 0.0056 to 10.0 μm, using a mobile aerosol analysis system designed for use with the Cessna 206. The system is described in detail by Cantrell *et al.* (submitted). It consists of two size-selective aerosol sensors integrated with an automated 'grab-bag' sampling system designed to collect and hold a discrete static air sample for analysis by the instrumentation. The sensors are an electrical aerosol analyzer (EAA) (Thermo-Systems, Inc. Model 3030), used for aerosol in the 0.0056–1.0 μm range, and a modified optical particle counter (OPC) (Royco 218/MCA, UM-1) to provide size information for aerosol in the 0.5–10.0 μm range. However, because of sampling system transport efficiency limitations, the range of usable aerosol data is 0.01–2.5 μm. Aerosol data was recorded, together with gas concentrations and meteorological data, by the aircraft data acquisition system (DAS) for subsequent off-line processing. b_{scat} and *ANC* measurements made at the time the 'bag' sample is taken are used in the reduction and analysis of the size distribution data.

DATA REDUCTION AND ANALYSIS

Continuous parameters

Reduction of the continuously measured parameters during the 1976 plume study was performed by MRI using the aircraft DAS record. From this, time vs concentration profiles in engineering units were constructed for each of the measurement passes made by the aircraft. A first step in the analysis of this data for any single experiment is the determination of the flow of the various measured pollutants through each cross-section in the plume. This is done using a standard technique for determining the flow of material in the atmosphere, described by Gillani and Husar (1976). In this, the product of the com-

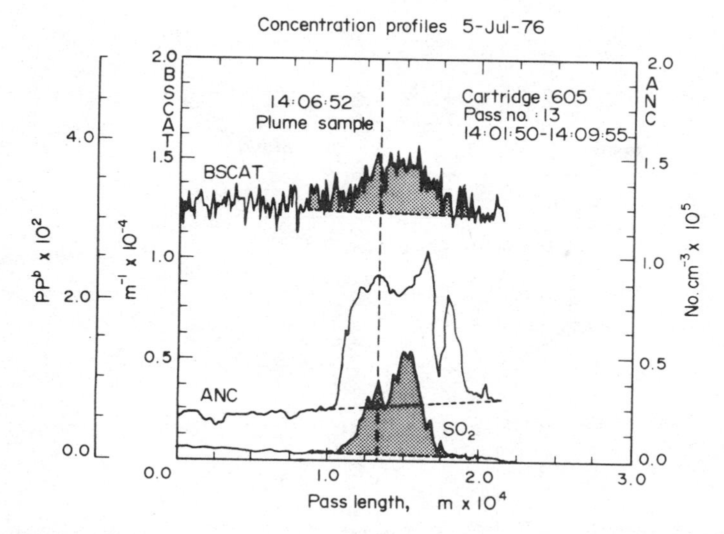

Fig. 1. Concentration profiles for SO_2, b_{scat}, and *ANC* are plotted here vs distance along a pass through the plume on July 5, 1976. Marked on each plot are the positions where aerosol 'bag' samples were taken.

ponent of wind velocity perpendicular to the plume cross-section and a measured concentration profile is integrated over the cross-section to yield the flow associated with the concentration. The numerical form of the calculation used is given in Equation (1):

$$F_c = \Sigma_i \bar{v}_{\perp_i}(h)\Delta h_i \int_{\text{pass}_i} C(y, h)\,\mathrm{d}y. \qquad (1)$$

Here, F_c is the concentration-associated flow of material, y is the horizontal distance in meters along a pass through the plume, $C(y, h)$ is the measured concentration as a function of pass distance and altitude h, $\bar{v}_{\perp_i}$ is the average component of wind velocity perpendicular to y, and Δh_i is the increment of altitude determined by the number (i) and altitude of passes used to map a plume cross-section. The shaded areas of Fig. 1 illustrate the horizontal portion of this integration for one pass. For SO_2, the calculated flow, F_{SO_2}, is expressed in terms of $m^3\,s^{-1}$ at standard temperature and pressure of the gas flowing through the measured plume cross-section. From the *ANC*, calculated aerosol number flow in the plume, *FNP*, is expressed as total number s^{-1}. Note that the integrations are usually of the excess of a concentration in the plume over measured background values. In this way, the flow of material associated only with the plume can be determined. An important parameter for the plume is the total gas flow (*Vg*). This is calculated like the other material flows in the plume using $C(y, h) = 1$ in-plume and $C(y, h) = 0$ outside in Equation (1). Using *Vg*, plume weighted averaged values can be determined from the calculated plume flow for the various concentrations mapped for each cross-section; i.e. $\bar{C} = F_{c/Vg}$. These average concentrations are just the increment added to the background concentration due to the presence of the plume. To recover net average concentrations in the plume, background must be added.

Table 1 summarizes the results of the flow calculations for the two days of interest. Values obtained for these plume flows using the pass integrations are very sensitive to the exact geometry of the plume cross-section. This is particularly true of *Vg*. Plume geometry for each cross-section is defined by the spatial dimensions over which both vertical and horizontal concentration profiles deviate consistently by more than 10% from background during each pass. For example, a plume width of approximately 1 km is indicated in Fig. 1 for Pass 13 on the 5th. If an arbitrary width is chosen for the plume, i.e. the entire pass length of 2.0 km, resulting errors in the pass integrations can be as large as 50% for parameters such as b_{scat}, and a factor of two for *Vg*.

Aerosol size distributions

Reduction of the aerosol size distribution data from the aircraft DAS record was accomplished using reduction procedures for the EAA measurements described by Whitby *et al.* (in press), and reduction procedures for the OPC/MCA measurements described by Cantrell *et al.* (submitted). The information contained in the resulting submicron aerosol size distributions measured both in the plume and background was characterized from these distributions using a log-normal parameterization technique described by Whitby *et al.* (in press). This method is based on the assumption that the submicron aerosol size distribution can be described by two additive log-normal functions, one fitted to the number distribution smaller than 0.03 μm (nuclei mode) and the other fitted to the surface distribution in the 0.03 to 1 μm size range (accumulation mode). The resulting log-normal parameters, geometric mean size (*DG*), geo-

Table 1. This table summarizes the meteorology and integral plume flow parameters obtained for both plume experiments reported in this paper. Flow of plume aerosol number *FNP*, total surface *FST*, and plume aerosol volume *FVP* are listed, together with SO_2 flow F_{SO_2}, integral b_{scat} $F_{B_{scat}}$, and total gas volume flow *Vg* associated with the plume. Plume age at each cross-section, together with average wind speed v, relative humidity RH, and fraction of maximum ultraviolet radiation (k) at the latitude of St. Louis, are also listed

Experiment date	Cross-section	Time of day CDT	Plume age h	Distance downwind km	Avg. wind velocity $m\ s^{-1}$	Relative humidity %	k	F_{SO_2} $m^3\ s^{-1}$	$F_{B_{scat}}$ $m^2\ s^{-1}$ $\times 10^3$	*FNP* no. s^{-1} $\times 10^{16}$	*FST* $cm^2\ s^{-1}$ $\times 10^7$	*FVP* $cm^3\ s^{-1}$	*Vg* $m^3\ s^{-1}$ $\times 10^6$
5-Jul-76	1	1220	0.73	8.8	4.5	47	0.95	3.94	3.37	140	7.4	55	30.8
	2	1340	1.75	26.7	5.3	52	0.95	3.12	12.0	413	31.4	167	123
	3	1740	2.61	45.0	5.9	53	0.30	3.74	19.6	328	41.4	444	189
14-Jul-76	1	0630	0.72	29.0	11.2	54	0.06	3.47	1.07	147	10.1	204	2.4
Morning	2	0800	1.25	49.0	10.6	54	0.30	4.14	3.25	238	17.2	344	12.2
14-Jul-76	1′	0920	1.19	29.0	6.8	57	0.40	3.80	1.82	68	8.8	199	8.5
Daytime	2′	1200	2.20	49.0	6.2	53	0.90	4.99	7.46	359	36.4	346	69.1
	0	1415	0.21	4.0	5.2	61	0.90	3.52	2.50	94	6.6	65	16.8

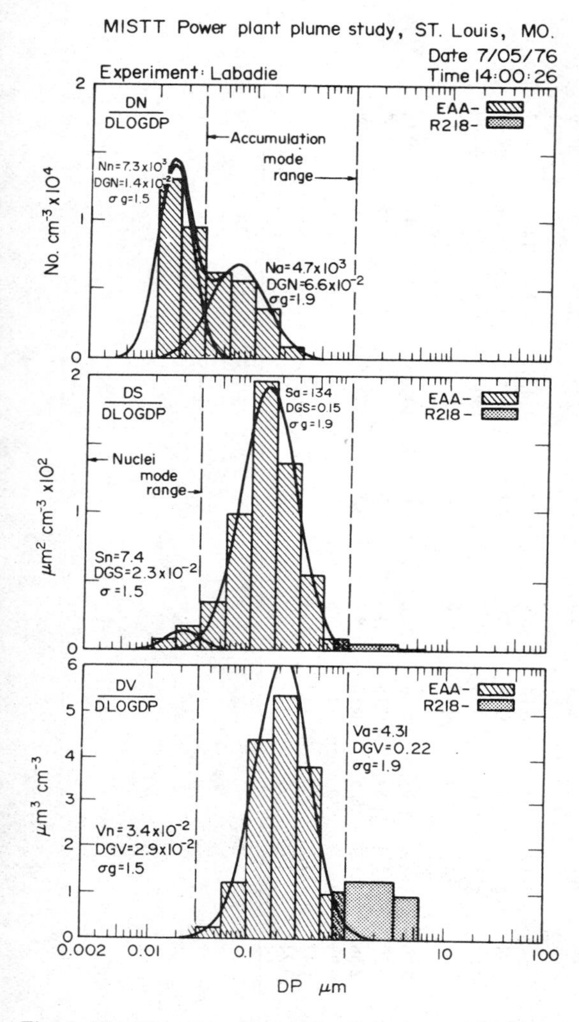

Fig. 2. This illustrates the result of the log-normal fit to a typical submicron aerosol size distribution obtained during the plume experiment of July 5, 1976. The graphs are computer-generated number, surface, and volume plots of data with the fit function superimposed.

metric standard deviation (σ_g), and integral concentration for the various weightings of the data, e.g. number (N), surface (S), and volume (V), are used in the following analysis of plume aerosol.

Figure 2 illustrates the results of the fitting process for a typical measured plume size distribution and denotes the aerosol size range associated with each mode. Note that the fitting process can supply information by extrapolation about portions of size distributions where data is missing, provided constraints can be placed on the log-normal parameters of the mode involved. For most of the plume distributions, a nuclei mode σ_g in the range of 1.5 ± 0.1 is assumed.

Table 2 summarizes averages of the modal parameters determined from the fit procedure for each measured plume cross-section during the two experimental days. With the exception of N_n, which is plume averaged $A\bar{N}C - \bar{N}_a$, values quoted for the modal aerosol concentrations are actually plume weighted averages obtained from the calculated aerosol flow in the plume plus the measured background concentrations; for example, $\bar{V}_a = {}^{FVP}a/Vg + \bar{V}_a$ (background). To obtain values for the aerosol flow in the plume, the fit information obtained from the size distribution measurements made with the aerosol sampling system at discrete points in the plume must be extrapolated over each pass across the plume. This is done for the accumulation mode using the theoretical relationship between the measured b_{scat} and the aerosol size distribution greater than 0.15 μm to convert the integral of b_{scat} over a horizontal pass to the integral of accumulation mode number, surface, or volume over the pass. In turn, these integrals can be used with Equation (1) to calculate flow of accumulation mode aerosol in the plume. Details of the calculation are given by Whitby *et al.* (in press). A similar calculation using the *ANC* measurements yields pass-integrated values for nuclei mode aerosol number, surface, and volume, and hence, aerosol flow for this mode. Summed, the nuclei and accumulation mode flows are used to characterize aerosol development in the plume. Values for plume aerosol flow for the days of interest are included in summary Table 1.

RESULTS AND DISCUSSION

Meteorology

The experiment on the 5th, providing information on daytime plume characteristics, started near midday at 1100 CDT and extended throughout the afternoon until 1600 CDT. In this period, three plume cross-sections were mapped for plume ages of 0.75–2.6 h. During this time, relatively uniform meteorological conditions prevailed and the plume was well-mixed, exhibiting fairly uniform vertical concentration profiles from the surface to the mixing height.

On the 14th, the experiment was designed to obtain information on the night-to-day transition period when night-time stratification starts to break up with the onset of convective mixing. Two cross-sections, at 29 km and 49 km downwind of the plant, were mapped alternately, starting at 0630 CDT and extending to 1400 CDT. During the morning mapping, the plume was 'trapped' in a stratified layer between 350 and 700 m m.s.l. In this layer, a low-level wind jet of 11 m s^{-1} persisted until convective mixing started, about 0900 CDT. The second plume mapping, made after the onset of convective mixing, is considered separately, since the measurements were done in the transition period between night and day as the mixing height rose. Both cross-sections were measured again during this period. A third cross-section, 4 km downwind, measured after the mixing height had stabilized at 1200 CDT, is grouped with this later pair because it was measured at 1415 CDT, close to the end of the period. Both days were clear during the entire sampling period, with the exception of the early

Table 2. This table summarizes averages of the mode parameters obtained using log-normal mode functions for the nuclei (n) and accumulation (a) mode aerosol. Values quoted for the mode integral aerosol concentrations are actually cross-plume weighted averages obtained from the aerosol flow results and background measurements. Also listed are average values for SO_2 concentration and b_{scat}

		Nuclei mode*					Accumulation mode					Totals				
Experiment date/time	Cross-section	DGN_n μm $\times 10^{-3}$	$\bar{N}_n$ $\times 10^4$ cm^{-3}	$\bar{V}_n$ $\times 10^{-2}$ μm^3 cm^{-3}	DGN_a $\times 10^{-2}$ μm	σ_{ga}	$\bar{N}_a$ $\times 10^4$ cm^{-3}	DGS_a μm	$\bar{S}_a$ μm^2 cm^{-3}	DGV_a μm	$\bar{V}_a$ μm^3 cm^{-3}	$\overline{NT}$ $\times 10^4$ cm^{-3}	$\overline{ST}$ μm^2 cm^{-3}	$\overline{VT}$ μm^3 cm^{-3}	$\overline{SO_2}$ ppb	$\overline{b_{scat}}$ $\times 10^{-4}$ m^{-1}
5-Jul-76																
1220	1	8.3	6.52	4.09	5.6	1.85	1.37	0.11	220	0.17	5.86	7.89	240	5.90	128	1.09
	Bk	8.4	2.82	1.83	5.9	1.93	0.52	0.14	128	0.22	4.08	3.34	136	4.10	5	0.90
1340	2	9.7	4.72	4.73	4.8	1.91	1.03	0.11	236	0.17	5.65	5.75	255	5.70	25	0.97
	Bk	10.0	1.85	2.03	6.0	1.94	0.55	0.15	126	0.23	4.32	2.40	134	4.34	5	0.91
1740	3	8.6	2.19	1.93	5.5	1.96	0.91	0.13	212	0.20	6.57	3.10	219	6.59	20	1.04
	Bk	8.9	0.85	0.66	6.2	1.94	0.52	0.15	123	0.23	4.23	1.37	125	4.24	6	0.95
14-Jul-76																
0630	1	5.6	50.2	9.67	7.3	1.66	13.5	0.12	4082	0.16	86.6	63.7	4143	86.7	1422	4.37
	Bk	6.7	1.7	0.53	5.5	1.87	0.68	0.12	132	0.18	3.2	2.4	135	3.2	24	0.79
0800	2	7.0	17.4	6.55	7.5	1.63	4.69	0.12	1368	0.15	30.7	22.1	1406	30.8	350	2.66
	Bk	6.1	1.9	0.53	5.2	1.89	0.65	0.12	145	0.18	2.6	2.6	148	2.6	37	0.71
14-Jul-76																
0920	1′	7.5	5.80	2.66	7.4	1.66	3.20	0.13	1025	0.16	26.6	9.0	1040	26.6	447	2.14
	Bk	7.9	0.6	0.37	6.2	1.82	0.39	0.13	84	0.18	3.3	1.0	86	3.3	35	0.62
1200	2′	—	5.52	2.56	4.8	1.96	2.58	0.10	515	0.16	8.8	8.1	527	8.8	72	1.08
	Bk	8.7	1.9	1.96	5.3	1.86	0.97	0.12	165	0.17	3.8	2.9	172	3.8	34	0.79
1415	0	—	5.39†	2.50	7.6	1.58	1.43	0.12	380	0.14	5.7	7.0	393	5.7	239	1.49
	Bk	7.0	1.2	0.51	6.6	1.85	0.24	0.14	67	0.21	2.0	1.4	70	2.0	2	0.56

* A $\sigma_{gn} = 1.5 \pm 0.1$ was assumed in obtaining these values.

† Derived using $N_{np} - N_{nb} = 6.25 \times 10^5 (k\, V_s/Vg)^{0.282}$.

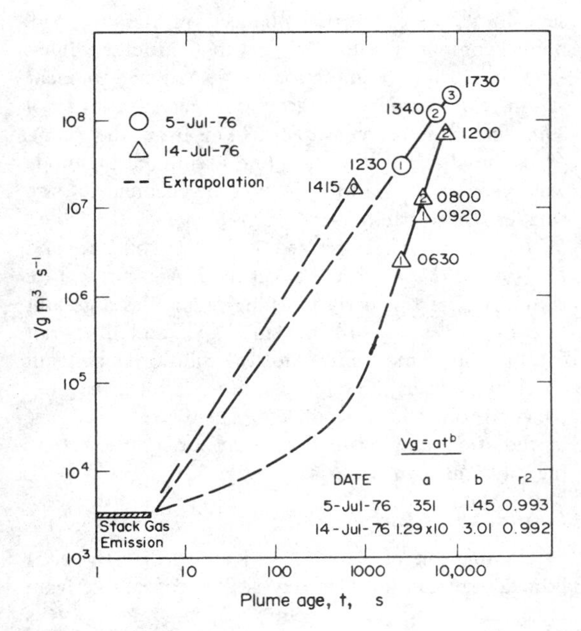

Fig. 3. Total gas volume flow in the plumes of July 5 and 14 are plotted here, together with the power law fit obtained for each day. The extrapolation lines are presented only as an indication of the possible continuity with the stack gas emission rate of $3100\ m^3\ s^{-1}$ for each day. The fits were used to calculate dVg/dt, used in the companion paper by Whitby *et al.* (1978b).

morning flight on the 14th, when haze and patchy ground fog was prevalent.

Plume dispersion

Using Vg as a measure of dispersion in the plumes, the experiments of the 5th and 14th can be compared. The result, as summarized in Fig. 3, shows two different dispersion characteristics for the plumes, evidently controlled by the meteorology during the experiments. No attempt has yet been made to associate these with any dispersion model. However, Vg for each experiment can be fitted using a power function of plume age (t). On the 5th, extrapolation of this function back to zero plume age yields a result comparable with calculated volume flow from the Labadie stacks (V_s), suggesting that relatively uniform dispersion conditions apply during the entire time. The fit for the 14th, however, cannot be extrapolated simply, and indicates relatively little dilution near the stacks followed by dilution that proceeds at a more rapid rate than that of the 5th. Such an increase implies a greater rate of lateral dispersion for the transition plume, since wind velocity decreased significantly during this period. This is in keeping with the fact that the mixing height was rising throughout most of this period. After 1200 CDT, indications are that plume dispersion characteristics are similar to those measured for the 5th.

Aerosol size distribution

Despite the obvious difference in dispersion between the 5th and 14th, plume averages of daytime nuclei mode concentration on both days exhibit similar dependence on the fraction of noontime u.v. radiation available at this latitude (k), and the amount of plume dispersion as measured by Vg/V_s. This is expressible in terms of an empirical fit to nuclei concentrations in the plume by a power function given by:

$$\bar{N}_{np} - \bar{N}_{nb} = a\left(\frac{k}{Vg/V_s}\right)^b. \qquad (2)$$

Here, $\bar{N}_{np}$ and $\bar{N}_{nb}$ are the average nuclei mode number concentrations in both the plume (p) and background (b), respectively. Results of the fit for both days are summarized in Table 3. For the values of b listed, i.e. $b > 0$, an increase in k is reflected by an increase in plume nuclei concentration. For example, an increase in k from 0.5 to 1 on the 5th for given Vg results in an increase in $\bar{N}_{np}-\bar{N}_{nb}$ of 46%. This implies a direct connection between radiation and nuclei formation in the plume. Such an increase in number is directly offset by dilution processes increasing Vg. The combined effect of these factors on N_{np} for the measured daytime plumes is a slight linear decrease in nuclei concentration with plume age of -5.8 no. $cm^{-3}\ s^{-1}$ on the 5th and -2.1 no. $cm^{-3}\ s^{-1}$ on the 14th.

Daytime accumulation mode aerosol in the plume and background accounts for 20–30% of total submicron number. The shape of distributions in this size range, as measured by a σ_{ga} of 1.92 ± 0.04, does not change significantly from in-plume to background. There is, however, a difference in the geometric mean size. In the plume on the 5th, the average geometric mean size of the accumulation volume distribution is $DGV_{ap} = 0.18 \pm 0.02$. For background air, the average volume mean size was significantly larger, $DGV_{ab} = 0.23 \pm 0.01$. Furthermore, this difference between DGV_{ap} and DGV_{ab} decreases with a decrease in radiation from about 0.05 μm for cross-sections 1 and 2, flown at midday, when u.v. radiation is near a peak, to 0.03 μm at 1800 CDT when u.v. radiation is about 30% of midday values. Evaluation of measurements taken on the evening of the 5th shows that the difference in geometric mean size between background and plume disappears after sunset. On the 14th, accumulation mode volume mean size in the plume is also less than that of the background, $DGV_{ap} = 0.15 \pm 0.01$ vs $DGV_{ab} = 0.18 \pm 0.01$.

Taken together, the difference in size of the in-plume accumulation mode aerosol during the daytime

Table 3. Plume-associated nuclei, $N_{np} - N_{nb}$, for the daytime plumes on both July 5 and 14, 1976, were fit using a power law of the form $a(k\ V_s/Vg)^b$

Date	a $\times 10^5$ no. cm^{-3}	b	r^{2}*
5-Jul-76	9.76	0.346	0.942
14-Jul-76	6.25	0.282	—

* Correlation coefficient.

periods and the existence of a relationship between nuclei concentrations in the plume and u.v. radiation over a considerable portion of plume age indicate that a significant fraction of new aerosol volume and number in the plume can be associated with photochemical processes. Indeed, the two effects are directly related, since nuclei thus formed would tend to decrease the mean size of the accumulation mode via polydisperse coagulation and subsequent condensation on the smaller sizes in the accumulation mode.

Whitby *et al.* (1978), in a companion paper, have shown that about 5% of the aerosol volume being formed in the 1976 plumes go to form new nuclei of nuclei mode size, while the remainder condenses directly into the accumulation mode. The fraction going into the nuclei mode was found to be relatively independent of the overall volume formation rate. Thus, the tendency of condensation to cause the accumulation mode to increase in size as the aerosol volume concentration increases is offset by the flow of nuclei coagulating into the small size range of the accumulation mode. Therefore, the accumulation mode size usually remains more or less constant. Under some conditions, such as those that existed on the 5th and 14th in 1976, DGV_{ap} actually decreases significantly relative to the size in the background DGV_{ab} as the plume aerosol volume concentration increases.

Using a program developed by Dolan (1977) of our laboratory, we have studied aerosol growth on the computer with a model that includes a nuclei formation rate of 14 no. $cm^{-3} s^{-1}$, a critical size for condensation in the accumulation mode of 0.035 μm, and aerosol volume formation rates in the range of those actually found in 1976. These model calculations yielded DGV_a size shifts similar to those observed in the plume. This work is continuing and will be reported later.

Contrasting behavior was noted for the early morning plume measurements on the 14th, compared to those later in the day. Then, in-plume aerosol concentrations were greatly enhanced in the 'trapped' plume compared with very low background values. Peak in-plume accumulation mode volumes of near 90 $\mu m^3 cm^{-3}$ and background concentrations of 3 $\mu m^3 cm^{-3}$ were measured 29 km from the plant. Also, the shape of the in-plume accumulation mode was very narrow, with a volume geometric size smaller than background, e.g. a σ_g of 1.65 and $DGV_a = 0.155\ \mu m$, compared with a σ_g of 1.88 and DGV_a of 0.178 μm in the background. Another feature is that $\bar{N}_{np}$ for this early morning plume has a greater rate of decrease, -144 no. $cm^{-3} s^{-1}$, and does not exhibit the same radiation- or dilution-dependent characteristics as the daytime plumes. Nuclei number concentrations here seem less sensitive to a change in radiation and more dependent on dilution than the daytime plume.

Integral plume parameters

Examination of Table 1 will show that the SO_2 flow determined for each day is very repeatable from cross-section to cross-section. Such uniformity is possible only if the Labadie power plant was consistent in its operation, and if information on plume geometry used in the flow integrations is accurate. The resulting constancy of calculated SO_2 flow thus serves as a check on both the concentration profiles and the meteorological information used to determine plume geometry. On the two days of interest, average SO_2 flow in the plume is 10.3 ± 1.2 kg s^{-1} for the 5th and 11.4 ± 1.7 kg s^{-1} for the 14th. Calculation of the source emission of SO_2, based on operation data for the experimental days and a monthly average for coal sulfur content of 3.17%, yields a value of 12.9 ± 0.2 kg s^{-1}. A slightly lower value for the 5th reflects the fact that the plume has fumigated to the ground, and dry removal of the SO_2 may be occurring. The amount of loss is in line with measurements reported by Husar *et al.* (1978). Taking this into account, all of the plume sulfur can be accounted for on the days studied. Overall, the precision of the SO_2

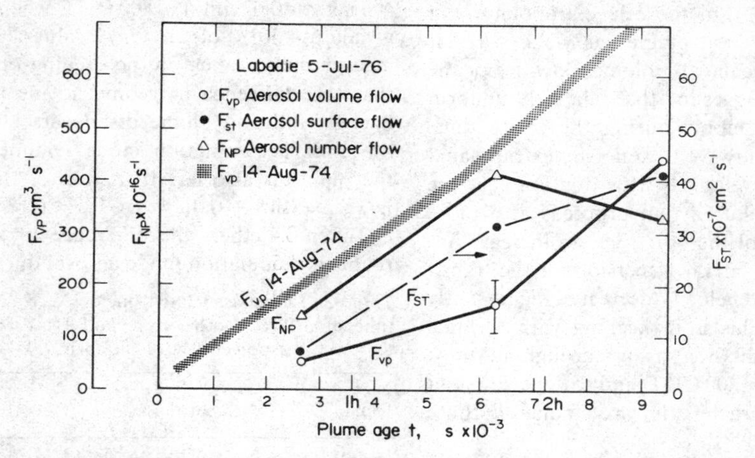

Fig. 4. Plotted here vs plume age are plume aerosol number flow, total aerosol surface flow, and plume aerosol volume flow for the experiment of July 5, 1976. For comparison, plume aerosol volume flow from the August 14, 1974 measurements are also presented.

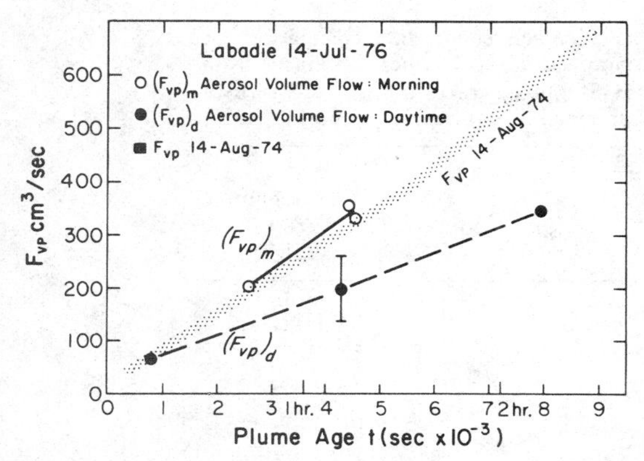

Fig. 5. Plotted here vs plume age are plume aerosol volume flow obtained on July 14, 1976 for morning $(F_{vp})_m$ and daytime $(F_{vp})_p$ periods. For comparison, plume aerosol volume flow from the August 14, 1974 measurements are also presented.

flow averages imply that error in flow calculations similar to that for SO_2 due to uncertainty in plume geometry and meteorology should be no more than 20%.

Figures 4 and 5 illustrate the time history of the aerosol flow in the 1976 plumes. Included in each for comparison are the results for plume aerosol volume flow measured in 1974. For the daytime plume of the 5th, illustrated by Fig. 4, aerosol volume flow, *FVP*, is seen to increase by a factor of 9 over the 2-h period of plume age measured. Total number flow, *FNP*, peaks at about 1.5 h of plume age, and then declines due to a decrease in nuclei production coupled with number concentration loss due to coagulation and continued dilution in the plume. Total aerosol surface flow, *FST*, in the plume approaches a maximum value after about 1.5 h. If this is interpreted as a net equilibrium surface *Se* for the plume aerosol, as defined by Clark and Whitby (1975), we find good agreement between the resulting average equilibrium surface concentration value, $Se = 310$ $\mu m^3\ cm^{-3}$, obtained for the plume, and that predicted by Clark's empirical relation between equilibrium surface and the aerosol volume formation rate, $QV_a = 1.4\ \mu m^3\ cm^{-3}\ h^{-1}$, reported by Whitby *et al.* (1978b). This relation, $Se = 314\ (QV_a)^{0.575}$, has been found to hold in a wide variety of gas-to-particle conversion experiments (Kocmond *et al.*, 1973).

Comparison of plume aerosol volume flow of $80\ cm^3\ s^{-1}$ after one hour of plume age on the 5th with a flow of $250\ cm^3\ s^{-1}$ observed for the 1974 plume experiment shows a difference of about a factor of three in aerosol growth. Two factors are different between the 1976 and 1974 plumes. First, relative humidity is lower in 1976, 51% as opposed to 75% for 1974. The other factor, probably the more important, is the amount of aerosol in the background on the two days. Whereas in 1976, the concentration of aged aerosol in the background was low, the 1974 plume has an order of magnitude higher concentration. This provides a large amount of surface in the accumulation mode for aerosol volume growth by condensation as background air mixes with the plume. In turn, this should deplete condensable species in the plume to the point where very little homogeneous nucleation could take place. Evidence for this is found in the amount of nuclei mode aerosol encountered in the plume. In 1974, the accumulation mode accounted, on the average, for all the aerosol number concentrations measured. In the 1976 experiments, the accumulation mode accounted for only 30% of the number.

As noted above, the plume measurements on the 14th are separable into two distinctly different parts by the meteorology as the experiment progressed. The results of the two sets of measurements for aerosol volume flow are illustrated in Fig. 5. The early morning plume aerosol volume flow, $F(VP)_m$, is quite different from that of the later transition daytime plume, $F(VP)_d$, and compares with the aerosol volume flow observed during the 1974 daytime plume. Unlike the daytime plume, the plume is trapped in a stratified layer. This maintains both aerosol and aerosol precursor concentrations at a high level, apparently resulting in a higher rate of aerosol volume formation than later that same day.

Regardless of the plume dynamics, the general increase in volume flow over the 2-h plume age period for the 1976 plumes is a consistent feature of all the plumes so far investigated. Compared together, the 1974 plume results yield the highest aerosol volume flow, and the result for the 5th (1976), the lowest yet measured.

SO_2 *conversion rate estimate*

Aerosol volume flows for the 1976 plumes have been used to estimate conversion rates of SO_2 using a calculation described by Whitby *et al.* (in press). This estimation is based on the assumption that all of the aerosol volume increase in the plume is sulfuric acid in equilibrium with water vapor at the ambient relative humidity. Table 4 summarizes the results of this calculation for both the 1974 and 1976 experiments.

From the conversion rates obtained for July 5, 1976 and August 14, 1974, there is evidence of significant increase in the conversion rate with increasing plume age for plumes under relatively uniform meteorological conditions. For the two hours mapped in 1976, the conversion rate increases from $0.41 \pm 0.2\%\ h^{-1}$ to $1.12 \pm 0.4\%\ h^{-1}$. In 1974, the increase over a 3-h period is $1.5 \pm 0.5\%\ h^{-1}$ to $4.9 \pm 1.7\%\ h^{-1}$. This is not so for the plume measurement of July 14, 1976 made during the changing meteorology of the morning transition period. Here, comparing $0.43 \pm 0.2\%\ h^{-1}$ in the first hour with $0.48 \pm 0.2\%\ h^{-1}$ in the second, no significant increase in conversion rate is seen. As has been noted, the primary difference between these plumes is the rate at which dilution is occurring. It would seem that the more rapid dilution brought about by changing meteorology, such

Table 4. This table summarizes results of the SO_2 conversion rate calculation for July 5 and 14, 1976. Also included for comparison are values obtained from measurements of the Labadie plume in 1974. H_2SO_4 mass fraction $f_{H_2SO_4}$ and aerosol density $\rho_{H_2SO_4}$ used in the calculation were obtained from Nair and Vohra (1975)

Experiment date	Distance downwind km	Plume age h	$f_{H_2SO_4}$ %	$\rho_{H_2SO_4}$ g cm^{-3}	SO_2^* converted %	SO_2 conversion rate % h^{-1}
5-Jul-76	8.8	0.73				
			42.5	1.32	0.41	0.41 ± 0.2†
	26.7	1.75				
			42.5	1.32	0.96	1.12 ± 0.4
	45.0	2.61				
14-Jul-76						
Morning	29.0	0.72				
			41.0	1.31	0.54	1.06 ± 0.4
	49.0	1.25				
Daytime	4.0	0.21				
			41.0	1.31	0.42	0.43 ± 0.2
	29.0	1.19				
			41.0	1.31	0.48	0.48 ± 0.2
	49.0	2.20				
					1976 Avg.	0.70 ± 0.36
14-Aug-74	10.0	0.70				
			30	1.21	1.1	1.5 ± 0.5
	21.0	1.49				
			30	1.21	2.3	1.8 ± 0.6
	32.0	2.29				
			30	1.21	3.7	4.9 ± 1.7
	45.0	3.17				
					1974 Avg.	2.7 ± 1.9

* Based on measured SO_2 flow in the plume.
† Errors are quoted on the basis of a worst-case error for flow calculations of 20%.

as the rising mixing height of the 14th, acts to suppress volume formation in the plume and, hence, conversion rate.

The number of plumes measured so far does not permit generalization about more than two hours of plume age. The measurements for July 14, 1974, however, indicate that, under relatively uniform meteorological conditions, the increasing conversion rate seems to hold through at least the third hour of plume lifetime.

Considering the differences in conversion rate behavior noted for the two days reported for 1976, it would seem desirable to report such data in the future on a plume-by-plume basis or, at most, averaged over periods with similar meteorology. Averages over several such different days would tend to obscure changes in conversion rate with plume age. Also, any attempt at measurements, such as on the plume center line, that are not adjusted for the difference in dispersion on such days would yield very different results.

For example, if plume dispersion characteristics measured on the 5th are used on the average values for SO_2 obtained on the 14th, the resultant integral sulfur flow calculated for the plume shows an apparent decrease in plume sulfur of 47% h^{-1}. In light of this, care should be taken in using dispersion models to obtain integral flow in a plume from single measurements.

Results of plume measurements reported here and summarized by the calculated SO_2 conversion rates compare very well in magnitude with results obtained by other investigators for both sulfur-emitting sources and gas-to-particle conversion experiments. Hegg *et al.* (1976) have reported conversion rates as low as 0.34% for daytime power plant plumes emitted in relatively clean background air. Lusis and Wiebe (1971) obtain a more or less constant conversion rate of about 1% h^{-1} for the INCO smelter plume at Sudbury, Ontario. Newman *et al.* (1976) report rates of slightly less than 2% h^{-1} for coal-fired power plant plumes. Although in reasonable agreement with the results reported here, as are the others, he finds the conversion rate for SO_2 is fairly independent of distance up to 50 km from the power plant. Indeed, it would seem that this rate reaches a maximum 5–10 km from the plant, and then decreases slightly accompanied with a depletion of plume sulfur. This is clearly at odds with our findings and remains a point that should be cleared up.

Smog chamber studies of SO_2 photooxidation, although not an accurate simulation of plume conditions, can shed some light on the expected gas-to-particle conversion process as a function of time. Kocmond *et al.* (1975) report values of 0.4% h^{-1} in the first hour of development, increasing to 0.8% h^{-1} in the second hour of development. These measurements were done at relative humidities of 50% in a contaminated chamber and, as such, come as close as any to the conditions observed during the 1976 study.

SUMMARY

(1) Nuclei concentrations in the plume are directly correlatable with ultraviolet radiation for both plumes investigated. Together with a smaller

measured mean size for plume accumulation mode aerosol compared to background, this provides evidence suggesting homogeneous nucleation of a photochemical-generated precursor species in the plume. This and the subsequent growth and transformation processes for this aerosol are explored further in a companion paper by Whitby *et al.* (1978).

(2) Virtually all sulfur emitted by the power plant can be accounted for in the plume within 15% for the lifetimes measured. The exception to this was the plume measured on July 5, 1976, where a decrease of $\sim 25\%$ was in line with estimates for loss due to dry deposition of SO_2.

(3) Estimated SO_2 conversion rates for a daytime plume under relatively uniforn meteorological conditions show a significant increase with plume age up to two hours: $0.41 \pm 0.2\% \, h^{-1}$ in the first hour and $1.12 \pm 0.4\% \, h^{-1}$ in the second. In a night-to-day transition period during which the mixing height was rising with a resulting enhanced rate of dilution in the plume, the conversion rate was constant at $0.5 \pm 0.2\% \, h^{-1}$ over the same period of plume age.

Acknowledgement—This research was performed under EPA Research Grant No. R 803851-02, and presents partial results from Project MISTT, an integrated multidisciplinary program directed by William E. Wilson, ESRL, EPA. Special thanks are due to the many students and staff at the University who cooperated to carry out this project.

REFERENCES

Blumenthal D. L., Ogren J. A. and Anderson J. A. (1978) Airborne sampling system for Project MISTT. *Atmospheric Environment* **12,** 613–620.

Cantrell B. K., Brockmann J. E. and Dolan D. F. A mobile aerosol analysis system with applications (submitted to *Atmospheric Environment*).

Clark W. E. and Whitby K. T. (1975) Measurement of aerosols by the photochemical oxidation of SO_2 in air. *J. Colloid Interface Sci.* **51,** 477–490.

Dolan D. F. (1977) Experimental and theoretical investigations of diesel exhaust particulate matter. Ph.D. Thesis, University of Minnesota, Minneapolis, MN, 55455.

Gillani N. V. and Husar R. B. (1976) Analytical–numerical model for mesoscale transport, transformation and removal of air pollutants. Presented at the 7th International Technical Meeting on Air Pollution Modeling and its Application.

Hegg D. A., Hobbs P. V. and Radke L. F. (1976) Reactions of Nitrogen Oxides, Ozone, and Sulfur in Power Plant Plumes. EPRI Report No. EPRI EA-270.

Husar R. B., Gillani N. V., Patterson D. E. and Husar J. D. (1978) Sulfur budget in large plumes. *Atmospheric Environment* **12,** 549–568.

Kocmond W. C., Kittelson D. B., Yang J. Y. and Demerjian K. L. (1973) Determination of the Formation Mechanisms and Composition of Photochemical Aerosols. Calspan Report No. NA5365-M-1.

Kocmond W. C., Kittelson D. B., Yang J. Y. and Demerjian K. L. (1975) Study of Aerosol Formation in Photochemical Air Pollution. EPA Report No. EPA-650/3-75-007.

Nair P. V. N. and Vohra K. G. (1975) *J. Aerosol Sci.* **6,** 265.

Newman L., Forrest J. and Manowity B. (1975) The application of an isoloptic ratio technique to the study of the atmospheric oxidation of sulfur dioxide in the plume from a coal fired power plant. *Atmospheric Environment* **9,** 969.

Smith T. B., Blumenthal D. L., Anderson J. A., Vanderpol A. H. and Husar R. B. (1978) Transport of SO_2 in power plant plumes: day and night. *Atmospheric Environment* **12,** 605–611.

Whitby K. T., Cantrell B. K., Husar R. B., Gillani N. V., Anderson J. A., Blumenthal D. L. and Wilson W. E. Aerosol formation in a coal-fired power plant plume (submitted to *Atmospheric Environment*).

Whitby K. T., Cantrell B. K. and Kittelson D. K. (1978) Nuclei formation rates in a coal-fired power plant plume. *Atmospheric Environment* **12,** 313–321.

Atmospheric Environment Vol. 12, pp. 335–337. Pergamon Press 1978. Printed in Great Britain.

THE RELATIONSHIP BETWEEN SULPHATE AND SULPHUR DIOXIDE IN THE AIR

M. FUGAŠ and M. GENTILIZZA
Institute for Medical Research and Occupational Health,
Zagreb, Yugoslavia

(*Received* 21 *July* 1977)

Abstract—The relationship between the sulphate in suspended particulates and sulphur dioxide in the air was studied in various urban and industrial areas. The relationship is best described by the equation $y = ax^b$, where y is the percentage of the sulphate S in the total S (sulphate and sulphur dioxide) and x is the concentration of the total S in the air. The regression coefficients a and b seem to be characteristics of the area. In urban areas studied so far a was between 316 and 378 and b between -0.74 and -0.83. In industrial areas polluted by dust which contains elevated concentrations of metals a was between 91 and 107 and b between -0.35 and -0.49. In the area polluted by cement dust there was practically no correlation between the sulphate S (%) and the total S, but a relatively high correlation between absolute amounts of the sulphate S and the total S. The relations indicate that the limitation of SO_2 conversion is influenced by aerosol composition. Aerosols containing certain metals may promote the conversion by a catalytic effect while alkaline substances by increasing the pH. Whether this can only happen in the plume or in the air as well remains to be clarified.

Recently there have been several attempts to establish a relationship between the concentration of sulphur dioxide and sulphates in the air (Altschuller, 1973, 1976; Sandberg *et al.*, 1976).

Within an investigation into the possible catalytic effect of metal aerosols on SO_2 conversion we also tried to establish a relationship between sulphate and SO_2 concentration in the air. Simultaneous samples of sulphur dioxide, ammonia and suspended particulate matter were collected over 24 h. In the samples of particulate matter sulphate, ammonium ion and metals (Mn, Fe, Pb, Cu at all sites and Ca in the cement dust polluted area) were determined. Size–weight distribution of particulate components was determined in a limited number of samples with modified Andersen cascade impactors (Lee and Flesch, 1969). Only data on sulphur dioxide and sulphate will be discussed in this paper.

Sulphur dioxide was collected in 1% hydrogen peroxide solution and sulphate on glass fiber filters washed before use according to the procedure described by Scaringelli *et al.* (1969). Both SO_2 and water soluble sulphates were determined by titration with barium perchloride in the presence of thorin indicator.

In the first run simultaneous samples were collected for 20 consecutive days at two industrial and two urban sites. One of the industrial sites was a manganese polluted coastal area with a comparable coastal urban area as a control. The other was a lead polluted continental area and a continental area as a control. Since the possible content of sea salts in the air, the intensity of solar radiation and heating habits were practically the same in the two matched areas any differences in the relationship between sulphur dioxide and sulphate could be connected with the presence of the critical pollutant in the industrial area.

The analysis of data could not prove a correlation between sulphate and SO_2 concentration in both coastal and continental urban area ($SO_2 < 100\ \mu g\ m^{-3}$). A good correlation was found in the lead polluted area, ($SO_2 < 200\ \mu g\ m^{-3}$), but since in this area all other pollutants were also in good mutual correlation it can be assumed that this was due to the common origin of all pollutants (a lead smelter as a major source in the area).

When sulphate S (as SO_2) was correlated with the 'total S' (sulphate + SO_2) a better correlation was obtained in the continental urban and coastal industrial area. However, since the change in sulphate concentration was much smaller than in the corresponding SO_2 concentration, a percentage of sulphate in the total S decreased with the increasing total S concentration. This last relationship seemed to be characteristic of the type of area. For the sake of simplicity the terms 'sulphate S' and 'total S' will be used in all further considerations although equivalent SO_2 concentrations were used in all calculations.

In order to verify the findings from the short preliminary investigation, data collected through 50 days over a 6-month period in the manganese polluted area as well as in a coastal area polluted with cement dust were also analysed. For rough comparison data reported by Altschuller (1976), were analysed in the same way in all instances where both sulphate and SO_2 data were available although these were not fully comparable being 3-year moving averages. The range of total S concentrations, correlation coefficients and regression equations (1) for the relationship between sulphate S ($\mu g\ m^{-3}$) and the total S, with standard

Table 1. Range of total S concentrations and the relationship between sulphate and total S ($\mu g\ m^{-3}$)

Site	N	Total S as SO_2 Range	r_1	Relation between sulphate (y_1) and total S (x) Regression equation (1) $y_1 = a_1\,(s_{a_1}) + b_1\,(s_{b_1})x$
1. Continental urban site	20	2–91	0.40	$y_1 = 6.122\,(\pm 0.993) + 0.053\,(\pm 0.028)x$
2. Coastal urban site	20	14–67	0.12	$y_1 = 6.106\,(\pm 1.167) + 0.018\,(\pm 0.034)x$
3. 15 urban sites in the U.S.A.*	67†	16–429	0.80	$y_1 = 7.616\,(\pm 0.315) + 0.024\,(\pm 0.002)x$
4. Continental lead polluted site	20	26–198	0.74	$y_1 = 4.644\,(\pm 1.076) + 0.049\,(\pm 0.011)x$
5. Coastal manganese polluted site	20	9–58	0.37	$y_1 = 4.905\,(\pm 2.772) + 0.136\,(\pm 0.081)x$
6. Coastal manganese polluted site II	50	6–60	0.43	$y_1 = 4.212\,(\pm 1.667) + 0.187\,(\pm 0.054)x$
7. Coastal cement dust polluted site	50	7–76	0.65	$y_1 = -2.541\,(\pm 2.662) + 0.349\,(\pm 0.059)x$

* Calculated from data reported by Altschuller (1976).
† 3-year moving averages.

Table 2. The relation between relative content (%) of sulphate S (y_2) and concentration ($\mu g\ m^{-3}$) of total S (x) expressed as SO_2

Site	r_2	$\log y_2 = \log a_2(+s_{a_2}) - b_2(+s_{b_2}) \log x$	$y_2 = ax^{-b_2}$
1. Continental urban site	−0.87	$\log y_2 = 2.501\,(\pm 0.132) - 0.737\,(\pm 0.097) \log x$	$y_2 = 316\,x^{-0.74}$
2. Coastal urban site	−0.73	$\log y_2 = 2.549\,(\pm 0.275) - 0.826\,(\pm 0.187) \log x$	$y_2 = 355\,x^{-0.83}$
3. 15 urban sites in the U.S.A.*	−0.96	$\log y_2 = 2.577\,(\pm 0.053) - 0.781\,(\pm 0.027) \log x$	$y_2 = 378\,x^{-0.78}$
4. Continental lead polluted site	−0.73	$\log y_2 = 1.965\,(\pm 0.221) - 0.491\,(\pm 0.116) \log x$	$y_2 = 91\,x^{-0.49}$
5. Coastal manganese polluted site	−0.42	$\log y_2 = 2.029\,(\pm 0.293) - 0.391\,(\pm 0.197) \log x$	$y_2 = 107\,x^{-0.39}$
6. Coastal manganese polluted site II	−0.31	$\log y_2 = 1.996\,(\pm 0.214) - 0.351\,(\pm 0.154) \log x$	$y_2 = 98\,x^{-0.35}$
7. Coastal cement dust polluted site	0.014	$\log y_2 = 1.414\,(\pm 0.279) - 0.018\,(\pm 0.175) \log x$	$y_2 = 26\,x^{0.02}$

errors of regression coefficients are shown in Table 1. Correlation coefficients and regression equation (2) for the relationship between the relative content of sulphate S (%) and total S, with standard errors of regression coefficients are shown in Table 2. To make the comparison easier the exponential form of equation (2) is also given.

The relation between the absolute concentrations of sulphate S and total S (Table 1) shows a very slow rise of sulphate S with the total S in the three groups of urban data (slope I_{1-3}: ≤ 0.05). The two runs of data in the manganese polluted area show that a higher concentration of sulphate S than in urban areas for the same total S concentration is obtained in the presence of manganese aerosol (slope $I_{5,6}$: 0.14–0.19). In the presence of cement dust sulphate formation is further increased (slope I_7: 0.35).

As the slope of equation (1) increases the slope of

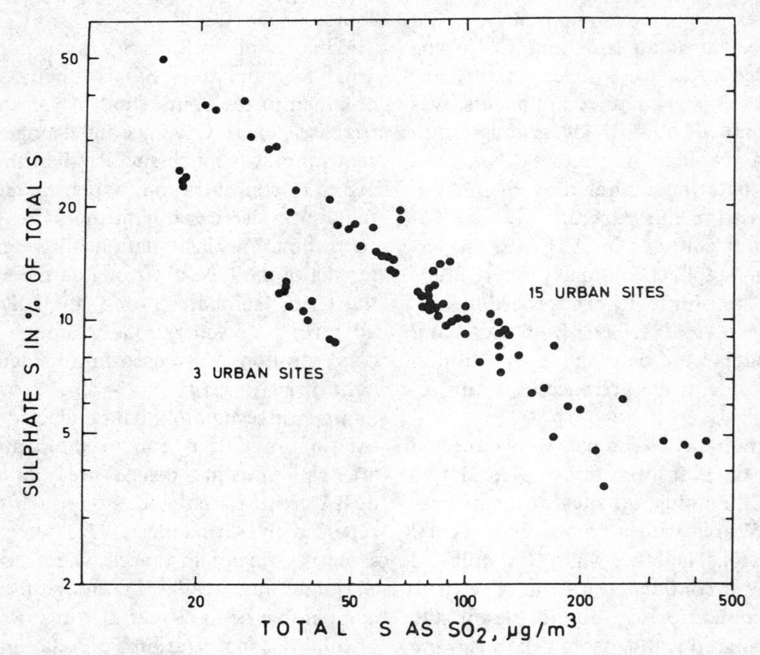

Fig. 1. The relation between the relative content of sulphate S and total S for 18 urban sites in the United States based on data reported by Altschuller (1976).

equation (2) (Table 2) decreases and so does the correlation between the relative portion of sulphate S and the total S, ending with no correlation at all in the cement dust polluted area.

As shown in Table 2, a high negative correlation was obtained between the relative content of sulphate S and the absolute amount of the total S in all urban areas. A very similar regression equation (2) was obtained in both continental and coastal urban area in spite of differences in population density, geographic and climatic characteristics.

A strikingly similar regression equation (2) was obtained for the group of 15 out of 18 urban sites in the United States. The division of the sites into two groups was made on the ground of graphical analysis. The relative content of sulphate S (%) was plotted against the total S on the log/log scale (Fig. 1). Two distinct relationships emerged indicating a different association of data from the two groups.

With the exception of the data from the U.S.A., which represent averages from a large number of data and give highly significant relationships, at all other sites a relatively small number of rather dispersed data collected so far do not permit far-reaching conclusions for a single set of data. However, the fact that the same tendencies were observed repeatedly for the same or similar type of area increases the significance of the findings.

The tendencies indicate that the limitation of SO_2 conversion is significantly influenced by aerosol composition. The mechanism of these processes could be associated with those occurring in the water according to experimental data reviewed or obtained by Barrie (1975). These show that both SO_2 absorption and oxidation are pH dependent and slow down below pH 6 unless catalysts like manganese salts are present. Alkaline substances keep both processes going simply by increasing the pH.

To what extent our findings can be attributed to the SO_2 conversion at the relatively low concentrations in the ambient air and to what extent they reflect the results of the processes within the plume is still to be answered.

Acknowledgements—This study was supported partly by U.S. Environmental Protection Agency Foreign Research Agreement No. 02-513-6 and partly by the Council for Scientific Research of the Republic of Croatia. It also makes a part of the COST 61 a Project.

REFERENCES

Altschuller A. P. (1973) Atmospheric sulfur dioxide and sulfate. Distribution of concentration at urban and non-urban sites in United States. *Environ. Sci. Technol.* **7**, 709–712.

Altschuller A. P. (1976) Regional transport and transformation of sulfur dioxide to sulfates in the United States. *J. Air Pollut. Control Ass.* **26**, 318–324.

Barrie L. A. (1975) An experimental investigation of the absorption of sulphur dioxide by cloud and rain drops containing heavy metals. Berichte des Institutes für Meteorologie und Geophysik der Universität Frankfurt/Main No. 28.

Lee R. E. and Flesch J. P. (1969) A gravimetric method for determining the size distribution of particulates suspended in air. Presented at the 62nd Annual Meeting of the Air Pollution Control Association, New York, N.Y.

Sandberg J. S., Levaggi D. A., DeMandel R. E. and Siu W. (1975) Sulfate and nitrate particulates as related to SO_2 and NO_x gases and emission. *J. Air Pollut. Control Ass.* **26**, 559–564.

Scaringelli F. P. and Rehme K. A. (1969) Determination of atmospheric concentrations of sulfuric acid aerosol by spectrometry, coulometry and flame photometry. *Analyt. Chem.* **41**, 707–713.

Atmospheric Environment Vol. 12. pp. 339–347. Pergamon Press 1978. Printed in Great Britain.

CONVERSION LIMIT AND CHARACTERISTIC TIME OF SO_2 OXIDATION IN PLUMES

JOHNNY FREIBERG

Department of Environmental Science, Cook College, Rutgers University, P.O. Box 231, New Brunswick, NJ 08903, U.S.A.

(*First received* 13 *June* 1977 *and in final form* 28 *July* 1977)

Abstract—The oxidation of sulfur dioxide to sulfate in expanding plumes is modeled for three oxidation reactions: a first order direct homogeneous oxidation, a heterogeneous catalytic oxidation and a second order homogeneous oxidation. The model is general enough to include the commonly used pseudo-first order oxidation as a particular case, but simple enough to yield analytical solutions for the sulfate yield and the characteristic reaction time. The model predicts that, for all oxidation reactions except the first order direct homogeneous, the conversion proceeds to a fractional asymptotic limit, a fact observed in many field investigations. The values of the fractional asymptotic limit as well as the values of the half lives of reaction depend on the ratios of the 'chemical' parameters to the 'dispersion' parameters. The model shows that the values of the half life of reaction are bounded above and below by limits which are constant for any dispersion pattern rather than dependent on rate constants and other chemical parameters. It also shows that most of the oxidation will occur early in the plume's life, a conclusion which is consistent with the findings of a number of field investigations. In addition the model suggests that the heterogeneous SO_2 oxidation in plumes is better characterized as a quasi-second order process than as a pseudo-first order process. The applicability of the model as it is affected by the concentration of ammonia gradient in the plume is discussed.

NOMENCLATURE

K_1	As defined by Equation (11); $m^3\,min^{-1}$.
T_{K_1}, T_{k_2}	Half lives of the first order heterogeneous oxidation and second order homogeneous oxidation, respectively, in an expanding plume; min.
V, W	Volumes of air and water, respectively, in the expanding elliptical ring; m^3.
r.h.	Relative humidity; %
b	Constant used in defining V; $m^3\,min^{-2}$.
k_a, k_w	Dissociation constants for ammonium hydroxide and water, respectively; $mol\,m^{-3}$, $mol\,m^{-6}$, respectively.
k, k_1', k_1'', k_2	Rate constants of SO_2 oxidation (Equations (2), (3), (4) and (5), respectively); min^{-1}, min^{-1}, min^{-1}, $mol^{-1}\,m^3\,min^{-1}$, respectively.
t_0	The time at which $V = V_0$, where V_0 is the initial volume of the elliptical ring; min.
β_s, β_a	Ostwald constants (proportional to Henry constants) for sulfur dioxide and ammonia, respectively.
λ_z	Coefficient relating concentration of sulfate in solution to the relative humidity (Raoult's law); $m^3\,mol^{-1}$.
σ_y, σ_z	Variances (related to dispersion coefficients in the cross-plume horizontal and vertical directions, respectively); m.
$[\,]$; $[\,]_0$, $[\,]_t$	Brackets which indicate concentration; same at time t_0 and t respectively; $mol\,m^{-3}$.
$(\,)$; $(\,)_0$, $(\,)_t$, $(\,)_\infty$	Brackets which indicate amount; same at time t_0, t and ∞, respectively; moles.
$(\,)_{K_1t}$, $(\,)_{k_2t}$	Yield as defined by Equations (12) and (13), respectively; moles.

1. INTRODUCTION

In the past few years, evidence has been accumulating (Amdur *et al.*, 1972; Shy and Finklea, 1973; CHESS, 1974) which indicates that acid sulfates are more harmful to human health than is sulfur dioxide. Moreover, the sulfates produced in the air increase the acidity of rain, thereby lowering the pH of soils and lakes and other water bodies and injuring animals and plants (Bolin, 1971). They also harm buildings and various other objects by accelerating corrosion (Yocum and McCaldin, 1968).

Accordingly, the emphasis in atmospheric sulfur polution research has shifted from the problem of predicting and understanding SO_2 concentrations downwind in plumes (Gifford, 1958; Symposium, 1966) to the problem of predicting and understanding SO_4^{2-} concentrations downwind in plumes (Levy *et al.*, 1976; Alkezweeny and Powell, 1977). The problem of sulfur dioxide in plumes is customarily treated as a dispersion problem. Atmospheric dispersions may be dispersions from instantaneous point sources or dispersions from continuous point sources. The expressions which have been developed to describe concentrations of conservative (i.e. non-reacting) substances in plumes from continuous sources relate to concentrations at each fixed point in an average plume emanating from a continuous source. The expressions estimate concentrations for a plume which is assumed to spread around a non-meandering center-line, with concentrations at each fixed point equivalent to a time-averaged concentration as the meandering loops of the plume pass the point. For example, the statistical theory of turbulent diffusion gives the Gaussian formula for concentration as a function of position (Slade, 1968):

$$\frac{X(x, y, z)}{Q'} = \frac{1}{2\pi\sigma_y\sigma_z\bar{u}}\exp\left[-\left(\frac{y^2}{2\sigma_y^2} + \frac{z^2}{2\sigma_z^2}\right)\right] \quad (1)$$

where $\sigma_y = (\overline{y^2})^{1/2}$ and $\sigma_z = (\overline{z^2})^{1/2}$. Q' is the continuous source strength in $g\,s^{-1}$ and the quantities σ_y, σ_z, regarded as functions of x, are the coefficients of dispersion.

The problem of sulfate in plumes is much more difficult than the problem of sulfur dioxide, because in modeling sulfate formation in plumes we must take into account not only the process of plume dispersion but also the process of SO_2 oxidation. SO_2 oxidation may occur in several ways. These include: homogeneous, photochemically-initiated reactions in the gas phase (Hall, 1953; Davis, 1974); heterogeneous gas–solid interactions (Novakov, 1974; Newman, 1975); and heterogeneous gas–liquid interactions (Gartrell, 1964; Barrie and Georgii, 1976).

From an examination of the literature (Levy *et al.*, 1976) it is seen that in the typical plume field investigation of SO_2-to-sulfate conversion rates, the data obtained, $[SO_2]$ and $[SO_4^{2-}]$, are usually fit to expressions of the type

$$X_r(x, y, z, t) = X(x, y, z)e^{-kt} \tag{2}$$

where $X(x, y, z)$ is given by Equation (1), $X_r(x, y, z, t)$ is $[SO_2]$ downwind, and a pseudo-first order oxidation with a corresponding rate constant k is postulated (in some cases after dry deposition has been accounted for (Alkezweeny and Powell, 1977)).

The use of Equation (2) (which yields k) carries an implicit assumption that the processes of dispersion and SO_2 oxidation are independent of each other. Some investigators (Berger *et al.*, 1968; Mészáros *et al.*, 1977) justify the use of Equation (2) by pointing to the literature of SO_2 oxidation, which deals overwhelmingly with laboratory observed oxidations of SO_2 that are first order processes (Fuller and Crist, 1941; Hall, 1953; Junge and Ryan, 1958; McKay, 1971; Chun and Quon, 1973).

Alkezweeny and Powell (1977) and Levy (1976) have noted that the values for k obtained in different investigations span a range of two orders of magnitude (0.37–55% h^{-1}). A wide range of variation is to be expected, considering the difficulties inherent in making measurements of $[SO_4^{2-}]$ in plumes (Levy *et al.*, 1976) and the degree of variation from study to study in conditions whose influence k is intended to subsume (e.g. r.h., temperature, and concentrations of metal oxides, hydrocarbons, and other pollutants) (Gartrell, 1964; Berger *et al.*, 1969; Smith and Jeffrey, 1975; Alkezweeny and Powell, 1977). Recent advances in the field measurement of SO_4^{2-} (Newman *et al.*, 1975; Wilson *et al.*, 1976) indicate that sulfate data obtained in the future will be more reliable than the data hitherto obtained.

But neither improved accuracy of sulfate data nor a knowledge of the values of the parameters which are thought to produce variations in k will be of much help in harmonizing the observed variations in k unless the functional dependencies between k and these parameters can be quantified. To date, no progress in ascertaining the exact relationships has been reported (except for a few empirical observations about the effect of relative humidity on k—and even these do not seem conclusive (Levy *et al.*, 1976). Consequently, little insight can be derived from k, except to say that the average rate of oxidation is faster under certain sets of conditions than under certain other sets of conditions.

2. THE MODEL

A stack plume from a combustion process typically comes from a continuous point source that emits a mass of hot gases and particulate matter (ash). The gases are primarily nitrogen, carbon dioxide and water vapor, with small amounts of carbon monoxide, nitrogen oxides, and sulfur dioxide. The mass rises and cools as it expands adiabatically. The water vapor condenses into small droplets around particulate matter in the plume.

The plume is transported with the mean wind and subjected to the forces of the irregular eddies associated with turbulent motions in the wind. Consequently, the plume effluent mixes with the surrounding air. This results in continuous dilution of the SO_2. At the same time the SO_2 is being oxidized to sulfate via one or more of the processes mentioned in the introduction. Thus the problem of sulfate formation in plumes can be characterized as a problem of chemical reaction coupled with turbulent diffusion. The process of a chemical reaction occurring in a dispersing system has been described by Friedlander and Seinfeld (1969), who modeled the formation of photochemical smog pollutants in an air shed, by Freiberg (1972, 1976), who modeled the sulfate formation by a second order heterogeneous oxidation in a power plant plume, and by Hilst *et al.* (1973), who modeled the ozone depletion in an SST plume. All the above models advanced the state of knowledge about their particular systems but were too complex to permit analytical solution.

The model presented in this paper is specifically geared to SO_2 oxidation in plumes. It is general enough for considering three characteristic oxidation processes (a reaction which is homogeneous and first order in $[SO_2]$; a reaction which is heterogeneous and first order in $[SO_2]$; and a reaction which is homogeneous and second order in $[SO_2]$) and for including as a particular case the pseudo-first order decay described by Equation (2). It is simple enough to permit analytical solution for the sulfate yields and the characteristic times of reaction. These expressions will facilitate the analysis of the similarities and differences between the overall oxidation rates, with a view towards obtaining insights into SO_2 oxidation in plumes.

2.1 *Chemical submodel*

Consider the following expressions: (a) an expression for a first-order catalytic oxidation of SO_2 in a buffered, oxygenated aqueous solution,

$$\frac{d[SO_4^{2-}]}{dt} = \frac{k_1'[H_2SO_3][Fe^{3+}]}{[H^+]_{buff}} \quad (3)$$

(which is like the ones suggested by Junge (1958), Karraker (1963) and Brimblecombe and Spedding (1974)); (b) an expression for a first-order homogeneous oxidation of SO_2,

$$\frac{d[SO_4^{2-}]}{dt} = k''[SO_2] \quad (4)$$

(which is like the ones suggested by Gerhard and Johnstone (1955) and Hall (1953)); and (c) an expression for a second order oxidation of SO_2,

$$\frac{d[SO_4^{2-}]}{dt} = k_2[SO_2]^2 \quad (5)$$

(which is like the one suggested by Newman *et al.*, (1975)).

For the first-order reactions expressed by Equations (3) and (4), the expressions for the sulfate yield at time t and $t = \infty$ and for the half life are given by the classical expressions

$$x = a[1 - e^{-k_1 t}] \quad (6)$$

$$(x/a)_{t=\infty} = 1 \quad (6a)$$

and

$$\tau_1 = \ln 2/k_1 \quad (6b)$$

respectively, (Moore, 1963) where x, a and k_1 stand for $[SO_4^{2-}]$, $[H_2SO_3]_0$ and $k_1'[Fe^{3+}]/[H^+]_{buff}$, respectively, of Equation (3) and for $[SO_4^{2-}]_t$, $[SO_2]_0$ and k_1'', respectively, of Equation (4).

For the second-order reaction of Equation (5), the sulfate yield and half life expressions are, respectively,

$$[SO_4^{2-}]_t = [SO_2]_0/(1 + 1/(k_2 t[SO_2]_0)) \quad (7)$$

$$[SO_4^{2-}]_\infty/[SO_2]_0 = 1 \quad (7a)$$

and

$$\tau_2 = 1/k_2[SO_2]_0. \quad (7b)$$

2.2. *Coupling the oxidation to the plume dispersion model*

First consider an oxidation of the Equation (3) type which takes place in a droplet solution in an expanding plume. The SO_2 in the gas phase diffuses to the droplet and dissolves. The ash or other nucleation particle within each droplet contains small amounts of Fe^{3+}. Upon dissolving in the droplet these ions catalyze the oxidation of the SO_2 to sulfuric acid.

Because of the formation and hygroscopicity of H_2SO_4 within the droplets, the vapor pressure at the droplet surface is lowered below that of the surrounding atmosphere. This induces water vapor to diffuse towards and condense into the droplets. Thus, while the plume is expanding, an interrelationship exists between the rate of diffusion of water vapor to the droplet, the rate of H_2SO_4 production and the plume dispersion rate.

This mechanism for SO_2 oxidation in a flowing system is analogous to a chemical reaction occurring in an expanding volume reactor, whose mechanism was first explained by Benton (1931). Like Freiberg (1976), who has utilized Benton's methodology for another chemical system of SO_2 oxidation, we shall equate the expanding volume reactor with an elliptical plume ring which is concentric about the plume's meandering centerline (see Fig. 1).

The mechanics of the oxidation process described in the previous paragraphs will be the same in each concentric elliptical plume ring although the reaction rate will decrease from the plume's center to its periphery because of the decrease in $[SO_2]$. (For the details of the ring concepts see Freiberg (1976)).

Applying Benton's method to Equation (3) we obtain the following rate expression for SO_2 oxidation in an elliptical plume ring:

$$\frac{1}{W}\frac{d(SO_4^{2-})}{dt} = \frac{k_1'[H_2SO_3][Fe^{3+}]}{[H^+]} \quad (8)$$

where W is the volume of water solution in the elliptical plume ring.

Since $[Fe^{3+}] = (Fe^{3+})/W$, and $[SO_2] = (SO_2)/V$ where V is the volume of the expanding reactor (or the elliptical plume ring), and $[H_2SO_3] = \beta_s[SO_2]$ where β_s is the Ostwald constant for SO_2, we can rewrite Equation (8) as:

$$\frac{1}{W}\frac{d(SO_4^{2-})}{dt} = \frac{k_1'\beta_s}{[H^+]}\frac{(SO_2)(Fe^{3+})}{WV}. \quad (9)$$

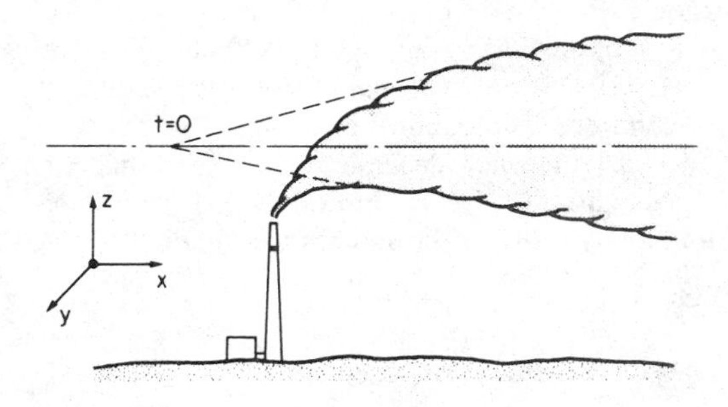

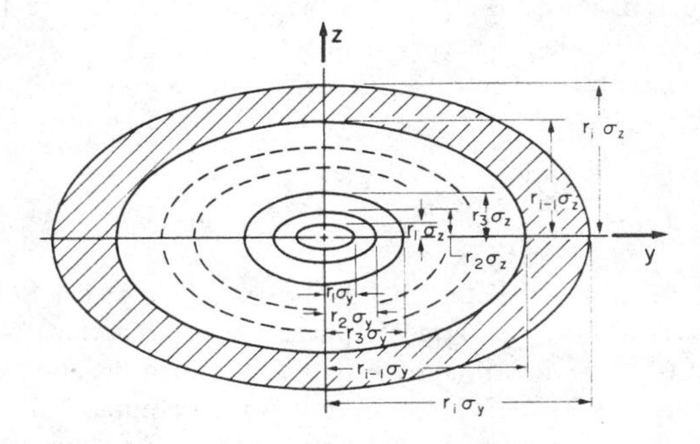

Fig. 1. The plume dispersion model; side view and a transversal section depicting the elliptic rings.

There is sufficient ammonia in the ambient atmosphere to maintain the pH at buffered values at which the reaction can occur at reasonable rates (Junge and Ryan, 1958; Junge and Scheich, 1969). We shall assume that the pH is so buffered and utilize Freiberg's (1974) expression

$$[H^+] = \frac{2[1 - RH.]k_w}{\beta_a k_a \lambda_z [NH_3]} \tag{10}$$

where $[NH_3]$ is the ambient concentration of ammonia. (The implications of introducing this assumption will be discussed in Section 3.2).

By substituting Equation (10) in Equation (9) and separating the variables we obtain:

$$\int \frac{d(SO_4^{2-})}{(SO_2)_0 + (SO_4^{2-})_0 - (SO_4^{2-})_t} =$$

$$\int \frac{k_1' \beta_s \beta_a k_a [NH_3] \lambda_z (Fe^{3+})}{2(1 - RH.)k_w} \frac{dt}{V} \equiv \int K_1 \frac{dt}{V} \tag{11}$$

where $(SO_2)_0$ and $(SO_4^{2-})_0$ are the amounts of SO_2 and SO_5^{2-} present in the elliptical plume ring at time t_0. In the integral on the right-hand side, only V is time dependent. The rest of the expression is independent of time and may thus be denoted by K_1.

The functional form of V depends on the variances σ_y, σ_z (see Equation (1)), which are functions of t. Slade (1968) found that for plume dispersions σ^2 varies with t^{2-f} where $0.0 \leq f \leq 0.25$ and f depends on meteorology. In their plume studies at Northport, New York, Newman *et al.* (1975) use in place of a Gaussian expression a dispersion equation of the form $(SO_2) = at^{-m}$ where a and m are dependent on source rate and on meteorology. They found that the expression fit their data best when they set $m = 2$.

In view of the findings of Slade (1968) and Newman *et al.* (1975), it seems reasonable to postulate that V is a second order function of the form $V = bt^2$ where b is a constant which depends on the local meteorology. Substituting $V = bt^2$ in Equation (11) and integrating analytically on both sides we obtain:

$$(SO_4^{2-})_{K_1 t} \equiv (SO_4^{2-})_t - (SO_4^{2-})_0$$

$$= (SO_2)_0 \left\{1 - \exp\left[\frac{K_1}{b}\left(\frac{1}{t} - \frac{1}{t_0}\right)\right]\right\} \tag{12}$$

where $(SO_4)_t$ is the total amount of sulfate in moles at time t, and $(SO_4^{2-})_{K_1 t}$ is the amount of sulfate in moles produced between time t_0 and t.

Taking the limit of $(SO_4^{2-})_{K_1 t}$ as $t \to \infty$ we obtain from Equation (12)

$$\frac{(SO_4^{2-})_{K_1 \infty}}{(SO_2)_0} = [1 - \exp(-K_1/bt_0)]. \tag{12a}$$

This says that if the chemical process by which SO_2 is oxidized in a plume has a first order reaction rate of the type described by Equation (3), then the conversion of SO_2 to SO_4^{2-} proceeds not to completion (as in Equations (6a) and (7a)) but to a fractional asymptotic limit which depends on the overall rate constant K_1 (a constant which incorporates the chemical parameters—see Equation (11)) and on the dispersion parameters b and t_0, where T_0 is the time at which $V = V_0$ where V_0 is the initial volume of the elliptical ring.

Substituting $(SO_4^{2-})_{K_1 t} = (SO_2)_0/2$ in Equation (12) and solving for t we obtain the half-life expression:

$$\tau_{K_1} = \frac{K_1 t_0}{K_1 - bt_0 \ln 2}. \tag{12b}$$

We shall now proceed to couple the oxidation reactions described by Equations (4) and (5) to the plume dispersion model. We can utilize again the elliptical plume ring concept. After performing the same steps as we used on Equation (3) we obtain the sulfate yield at time t and $t = \infty$ and the half life expressions:

$$(SO_4^{2-})_{k_2 t} = \frac{k_2 (SO_2)_0{}^2 (t - t_0)}{btt_0 + k_2 (SO_2)_0 (t - t_0)} \tag{13}$$

$$\frac{(SO_4^{2-})_{k_2 \infty}}{(SO_2)_0} = \frac{k_2 (SO_2)_0}{bt_0 + k_2 (SO_2)_0} \tag{13a}$$

$$\tau_{k_2} = \frac{k_2 (SO_2)_0 t_0}{k_2 (SO_2)_0 - bt_0} \tag{13b}$$

for the second order homogeneous oxidation (Equation (5)), whereas for the first order homogeneous oxidation (Equation (4)) we obtain

$$(SO_4^{2-})_{k_1'' t} = (SO_2)^0 [1 - e^{-k_1'' t}] \tag{14}$$

$$(SO_4^{2-})_{k_1'' \infty}/(SO_2)_0 = 1 \tag{14a}$$

$$\tau_{k_1''} = \ln 2/k_1''. \tag{14b}$$

3. DISCUSSION

Equations (14), (14a) and (14b) are functionally identical to Equations (6), (6a) and (6b) respectively. Therefore, the oxidation reactions described by Equation (4) (which is a linear process) are, as expected, independent of the dispersion process and proceed the same way whether they take place in a laboratory vessel of fixed volume or in an expanding plume. Moreover since $[SO_2] = (SO_2)/V$ (where V is related to Equation (1) through the variances σ_y, σ_z), Equation (2) is obviously a particular case of the model described above.

3.1 *Fractional asymptotic limit of conversion*

The first order homogeneous oxidation proceeds to completion in an expanding plume (Equation (14a)), as it does in a laboratory vessel of fixed volume (Equation (6a)).

The first order heterogeneous and the second order homogeneous oxidations proceed to completion in a vessel of fixed volume (Equations (6a) and (7a) respectively), whereas in an expanding plume they proceed to a fractional asymptotic limit of conversion (Equations (12a) and (13a) respectively). The limit in Equation (12a) is functionally different from Equation (13a). But for both cases the magnitude of the limit depends on the ratio of chemical parameters (K_1, k_2, $(SO_2)_0$) to dispersion parameters (b and t_0) (viz. K_1/bt_0 and $k_2(SO_2)_0/bt_0$ for Equations (12a) and (13a) respectively). In both cases, an increase in the ratios will cause increase in the limit, so that when $K_1 \gg bt_0$ and $k_2(SO_2)_0 \gg bt_0$,

$$\frac{(SO_4^{2-})_{K_1\infty}}{(SO_2)_0} \to 1 \quad \text{and} \quad \frac{(SO_4^{2-})_{k_2\infty}}{(SO_2)_0} \to 1 \text{ respectively;}$$

and *vice versa*, i.e., when $bt_0 \gg K_1$ and $bt_0 \gg k_2(SO_2)_0$,

$$\frac{(SO_4^{2-})_{K_1\infty}}{(SO_2)_0} \to 0 \quad \text{and} \quad \frac{(SO_4^{2-})_{k_2\infty}}{(SO_2)_0} \to 0.$$

Thus the ratios K_1/bt_0 and $k_2(SO_2)_0/bt_0$ may be considered indicators of the relative importance of the chemistry compared with the dispersion. Magnitudes of the limit for different values of the ratios are illustrated in Figs. 2 and 3.

The existence of a fractional asymptotic limit of SO_2 conversion in plumes has already been found by Freiberg (1976) for a second order heterogeneous oxidation and has been utilized by Forrest and Newman (1977) for explaining their field data. Equations (12a) and (13a) show that this result is characteristic of many more processes and is more general than was previously thought.

3.2 *The characteristic reaction time*

One of the customary expressions for the characteristic reaction time is the half life expression (see Equations (6a) and (7a)). Some investigators (Berger *et al.*, 1969; Weber, 1970; Stephens and McCaldin, 1971) use values of the half life as indications of how high the SO_2 oxidation rates in plumes are.

In the previous section we saw that the oxidation reaction does not always proceed to completion but may instead proceed to a limit of conversion (Equations (12a) and (13a)). Moreover, as can easily be seen, for $K_1 \leq bt_0 \ln 2$ and $k_2(SO_2)_0 \leq bt_0$, the denominators of the half life expressions (12b) and (13b) become negative and Equations (12b) and (13b) lose their physical meaning altogether, since the yield of the reaction is less than 50%.

Therefore, for these reactions, it is appropriate to focus not on a half-life of the *reactants* (as do Equations (12b) and (13b)), but rather on the half-life of the *reactions*. To obtain the half-life of the reactions, we replace $(SO_4^{2-})_{K_1t}$ of Equation (12) with $(SO_4^{3-})_{K_1\infty}/2$ and $(SO_4^{2-})_{k_2t}$ of Equation (13) with $(SO_4^{2-})_{k_2\infty}/2$. ($(SO_4^{3-})_{K_1\infty}/2$ and $(SO_4^{2-})_{k_2\infty}/2$ represent half of the limits of conversion given by Equations (12a) and (13a) respectively). Solving for the reaction half-lives $t = T_{K_1}$ and $t = T_{k_2}$ we get

$$\frac{1}{T_{K_1}} = \frac{1}{t_0} + \frac{b}{K_1} \cdot \ln\left(\frac{e^{-K/bt_0} + 1}{2}\right) \tag{15}$$

and

$$T_{k_2} = t_0\left(\frac{2bt_0 + k_2(SO_2)_0}{bt_0 + k_2(SO_2)_0}\right). \tag{16}$$

We see that T_{K_1} and T_{k_2} vary inversely with

$$\frac{K_1}{bt_0} \quad \text{and} \quad \frac{k_2(SO_2)_0}{bt_0},$$

that is, they decrease when the chemical parameters K_1, k_2 and $(SO_2)_0$ increase with respect to the dispersion parameters b and t_0.

Now let us ascertain what the limits of the reaction half-lives will be when the 'chemistry' operates much faster than the 'dispersion' and what they will be when the 'dispersion' occurs much faster than the 'chemistry'.

We can easily obtain from Equations (15) and (16) that $T_{K_1} = T_{k_2} = t_0$ when $bt_0 \ll K_1$ and $bt_0 \ll k_2(SO_2)_0$. It is also evident that $T_{k_2} = 2t_0$ when $bt_0 \gg k_2(SO_2)_0$. As for the value of T_{K_1} when $bt_0 \gg K_1$, we must substitute NK_1 for bt_0 in Equation (15) and take the limit as $N \to \infty$. We obtain:

$$\lim_{bt_0 \gg K_1} \frac{1}{T_{K_1}} = \lim_{N\to\infty}\left[\frac{1}{t_0} + \frac{N}{t_0}\ln\left(\frac{1 + e^{-1/N}}{2}\right)\right]. \tag{17}$$

Substituting

$$e^{-1/N} = 1 - \frac{1}{N} \cdot \frac{1}{2!\,N^2} \cdots + \frac{(-1)^n \cdot 1}{n!\,N^n} \quad \text{and}$$

$$\ln\left(1 - \frac{1}{2N}\right) = -\frac{1}{2N} - \frac{1}{8N} \cdots - \frac{1}{n(2N)^n}$$

in Equations (17) and (17a) respectively, and taking the limit, we obtain:

$$\lim_{bt_0 \gg K_1} \frac{1}{T_{K_1}} = \frac{1}{t_0} + \lim_{N\to\infty}\left[\frac{N}{t_0}\ln\frac{1}{2}\left(2 - \frac{1}{N} + \frac{1}{2!N^2} \cdots + \frac{(-1)^n}{n!N^n}\right)\right] = \frac{1}{t_0} + \lim_{N\to\infty}\left[\frac{N}{t_0}\ln\left(1 - \frac{1}{2N}\right)\right] = \frac{1}{2t_0}. \tag{17a}$$

Thus T_{K_1} and T_{k_2} have the same upper limit t_0 and the same lower limit, $2t_0$, that is:

$$t_0 < T_{K_1} < 2t_0 \tag{18}$$

and

$$t_0 < T_{k_2} < 2t_0 \tag{18a}$$

This says that half the final sulfate yield of SO_2 oxidation will be obtained close to the source and in an amount of time whose limits are constant for any given dispersion pattern and do not depend on the magnitude of the rate constants and other chemical parameters or on whether the reaction is first order heterogeneous or second order homogeneous.

Figures 2 and 3 depict the sensitivity of the sulfate yield to changes in time for different ratios of 'chemistry' to 'dispersion'. They show that the sulfate yield is especially sensitive to changes in time within the interval $t_0 \leq t \leq 2t_0$ and that the sensitivity decreases

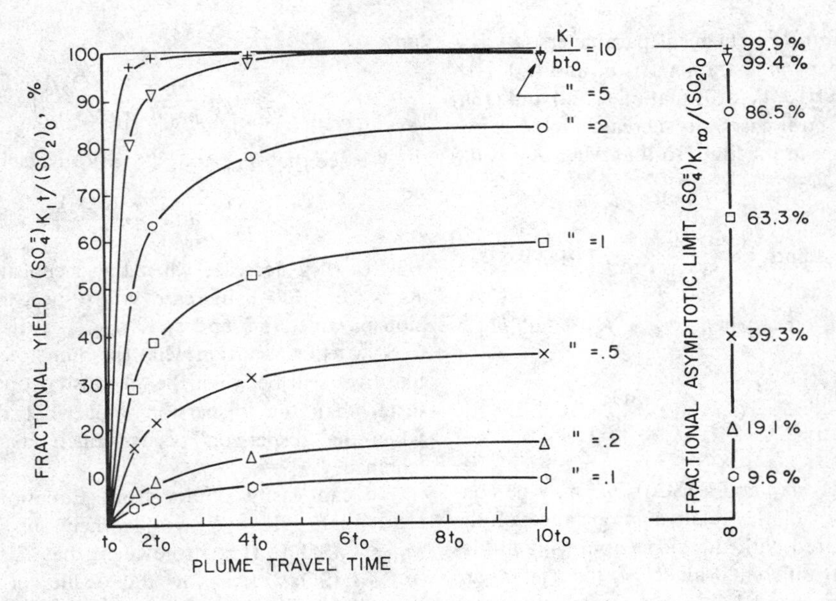

Fig. 2. Fractional yield vs plume travel time.

with increasing time, so that a limit of conversion is reached at about $t = 4t_0$.

The existence of a fractional asymptotic limit and a corresponding characteristic time is consistent with the results of a number of recent field investigations of SO_2 oxidation in plumes. For example, Smith and Jeffrey (1975) state that their results "seem to imply that possibly most of the conversion takes place very early on, near the source region where the concentrations of both SO_2 and other industrially produced pollutants that may be involved in the transformation process will be high".

Dana *et al.* (1975) state that their observed sulfate concentrations "indicate a significant rapid reaction to sulfate near the stack" and that "the in-plume SO_2 oxidation process proceeds very rapidly as the plume leaves the source and alternates as distance downwind increases, *leaving higher amounts of* SO_2 *at larger distances than would be predicted on the basis of linear kinetics in conjunction with the given rate constants*" (emphasis added). They continue by elaborating on the limited usefulness of their pseudo-first-order expression (which is similar to Equation (2)) for interpretations of their field results.

Summarizing the SO_2 oxidation data on the plumes at Labadie, Charleston and Muscle Shoals, Forrest and Newman (1977) state that "the oxidation occurs during the early history of the plume with virtually no further conversion taking place downwind" and that the oxidation is "eventually reaching an

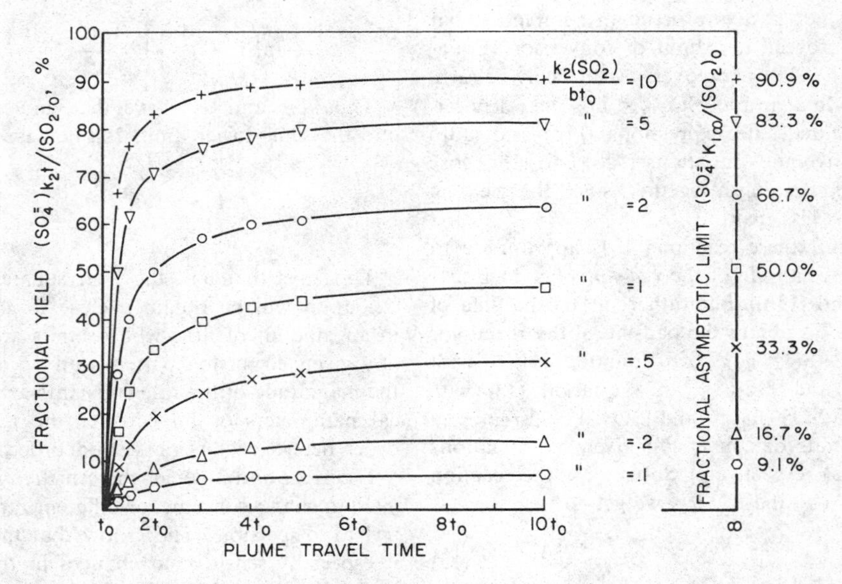

Fig. 3. Fractional yield vs plume travel time.

asymptotic limit of SO_2 conversion". Similar results have been previously observed by Berger *et al.* (1969), Gartrell (1964), Weber (1970) and Stephens and McCaldin (1971).

The model presented above brings together considerations of the simultaneous processes of chemistry and dispersion to derive analytic values for the asymptotic limit. It shows that such limits are characteristic of the second order oxidations in an expanding plume and of the first order heterogeneous oxidations in such a plume.

The model, and thus its results, are grounded on the assumptions that $[NH_3]$ is constant in each elliptical plume ring during plume expansion and that the plume volume varies quadratically with time.

But actually $[NH_3]$ varies from point to point, since the progressive diffusion of the plume into the ambient air and vice versa creates an $[NH_3]$ gradient within the plume, (a gradient whose magnitude depends on the rate of SO_2 oxidation in the plume, on the concentration of NH_3 in the ambient atmosphere, and on the meteorological conditions present). To the extent that the oxidation process requires the buffering influence of ammonia and the $[NH_3]$ gradient is large (little or no ammonia having penetrated the plume), the oxidation will occur only on the periphery of the plume (Wilson *et al.*, 1976) and the model developed above will have little applicability. But to the extent that the role of ammonia in facilitating the oxidation is small, the applicability of the model will be large, regardless of the magnitude of the gradient. And to the extent that the gradient is small (i.e. insofar as NH_3 has penetrated the plume), the applicability of the model will be great, whether or not the oxidation process requires ammonia as a buffer.

As for the assumption about rate of expansion of the plume, Freiberg (1972; 1976) has explained that for evaluating chemical rates in plumes, one should consider concentrations in the instantaneous meandering plumes rather than concentrations in averaged plumes. He has also explained why the data of Gartrell *et al.* (1964) are among those results which best represent instantaneous plumes. The volume of an instantaneous plume can be expressed as $V = bt^n$; the exponent n is strongly dependent on meteorology. An inspection of the data of Gartrell *et al.* (1964) reveals that n ranges from 0.5 in highly stable conditions to 1.35 in neutral conditions. Thus the value $n = 2$ which was utilized in the calculations in this paper must correspond to unstable conditions in which the plume effluent mixes quickly with the ambient atmosphere.

However, calculations similar to those in Equations (11), (12) and (15) show that not only when $n = 2$ but also indeed *whenever* n > 1 (i.e. in unstable conditions and in neutral conditions), a fractional asymptotic limit of conversion will be obtained, and the bounds of the characteristic time of reaction will vary only with the type of dispersion regime.

For $n \leq 1$, the oxidation reactions represented by Equations (3), (4) and (5) will all proceed to completion (although heterogeneous oxidations which are second order in SO_2 will proceed to an asymptotic limit even for $n \leq 1$, as has been shown by Freiberg (1976)). However, even in these cases, there persists the distinction that the first order homogeneous (or pseudo-first order) oxidation (Equation (4)) is independent of the type of dispersion involved, whereas the heterogeneous oxidation (Equation (3)) and the second-order homogeneous oxidation (Equation (5)) are both dependent on the dispersion regime involved.

For example, for $n = 1$ (i.e. $V = bt$) the yield expression for Equation (4) will be Equation (14) (as for $n = 2$), which is entirely independent of the dispersion parameters b and t_0: but the yield expressions for Equations (3) and (5) will be, respectively,

$$[SO_4^{2-}]_{K_1t} = (SO_2)_0[1 - (t_0/t)^{K_1b}] \qquad (19)$$

and

$$(SO_4)_{k_2t} = (SO_2)_0\left[1 - \frac{b}{b - k_2(SO_2)_0 \ln(t/t_0)}\right] \qquad (20)$$

which are dependent on the dispersion parameters b and t_0.

In sum, it will be true for *all* values of n (i.e. for stable, unstable, and neutral conditions) that the heterogeneous and second-order homogeneous oxidations are coupled to the dispersion regime whereas the first order homogeneous (pseudo-first order) oxidation is not.

3.3 *Heterogeneous oxidation in plumes: a quasi-second order process*

In an expanding plume, though the SO_2 chemistry within a droplet or on the surface of a carbon particle is linear, the process by which the SO_2 reaches the droplet or particle where it reacts is similar to a second order process. On the intuitive level, this concept is straightforward and has been implied previously (Newman *et al.*, 1975; Levy *et al.*, 1976). But no published work to date has formalized the concept mathematically, nor have the implications of the concept with respect to the custom of describing the oxidation by a pseudo-first order equation been examined.

Equations (12), (12a), (12b), (15) and (17a) express mathematically the mechanics of the dispersion-coupled chemistry of sulfate formation. The half-life expressions (Equations (12b) and (15)) are independent of $(SO_2)_0$, which is typical of a first order reaction. But as we have shown above, Equations (12), (12a), (15) and (17a), which express the sulfate yield and the characteristic reaction time, have functional analogies to Equations (13), (13a), (16) and (18), which relate to the second order oxidation. Therefore, the heterogeneous oxidation in a plume, being hybrid in nature, is more appropriately characterized as a quasi-second order process than as a pseudo-first order process. And since most SO_2 oxidation processes in plumes are either heterogeneous or second order or both (Smith, 1974), the character of SO_2 oxidation in plumes can be viewed on the whole as quasi-second order.

4. CONCLUSIONS

This paper describes a model which couples the reaction rates of three characteristic SO_2 oxidations to dispersion in a plume. The model is general enough to include the much-utilized pseudo-first order decay as a particular case. But it is simple enough to permit analytical solution for the sulfate yield and the characteristic time of reaction. The solutions obtained show: (1) that for almost all types of SO_2 oxidation the conversion does not proceed to completion but to a fractional asymptotic limit whose value depends on the values of the meteorological and chemical parameters involved; (2) that half of the final sulfate yield be obtained close to the plume source and in an amount of time which varies with the ratio of chemistry to dispersion and which is bounded above and below by limits which are constant for any given dispersion pattern, rather than dependent on the rate constant and other chemical parameters or on whether the reaction is first-order heterogeneous or second-order homogeneous; (3) that the heterogeneous oxidation of SO_2 in a plume is a hybrid of first-order and second-order processes and is better characterized as a quasi-second order process than as a pseudo-first order process.

There is qualitative agreement between these results and field observations.

REFERENCES

Amdur M. O., Lewis T. R. *et al.* (1972) *Toxicology of Atmospheric Sulfur Dioxide Decay Products*, U.S.E.P.A. AP-111.

Alkezweeny A. J. and Powell D. J. (1977) Estimation of transformation of SO_2 to SO_4 from atmospheric concentration data. *Atmospheric Environment* **11**, 179–182.

Brimblecombe P. and Spedding D. J. (1974) The catalytic oxidation of micromolar aqueous SO_2. *Atmospheric Environment* **8**, 937–945.

Barrie L. A. and Georgii H. W. (1976) Experimental investigation of the absorption of SO_2 by water drops containing heavy metals. *Atmospheric Environment* **10**, 743–749.

Bolin B. (1971) *Air Pollution Across National Boundaries. The Impact on the Environment of Sulfur in Air and Precipitation.* Kungl. Boktryckeriert P. A. Norstedt and Soner 710396 Stockholm.

Benton A. J. (1931) The kinetics of gas reaction at constant pressure. *J. Am. chem. Soc.* **53**, 2984–2997.

Berger A. W. and Billings C. E. *et al.* (1969) *A Study of Reactions of Sulfur in Stack Plumes.* First Annual Report, Contract No. PH-86-67-125, GCA Corporation, Report GCA-TR-68-19G.

Chun K. C. and Quon J. K. (1973) Capacity of Fe_2O_3 particles to oxidize SO_2 in air. *Environ. Sci. Technol.* **7**, 532–538.

CHESS, A Report of Health Consequences of Sulfur Dioxide 1970–1971 (1974) U.S.E.P.A. EPA-650/1-74-004.

Dana M. T., Hales J. M. and Wolf M. A. (1975) Rain scavenging of SO_2 and sulfate from power plant plumes. *J. geophys Res.* **80**, 4119–4126.

Davis D. D., Smith G. and Klauber G. (1974) Trace gas analysis of power plant plumes via aircraft measurement; O_3, NO_x and SO_2 chemistry. *Science* **186**, 733–736.

Freiberg J. (1972) Ph.D. Thesis, The Johns Hopkins University.

Freiberg J. (1974) Effects of relative humidity and temperature on iron-catalyzed oxidation of SO_2 in atmospheric aerosols. *Environ. Sci. Technol.* **8**, 731–734.

Freiberg J. (1976) The iron catalyzed oxidation of SO_2 to acid sulphate in dispersing plumes. *Atmospheric Environment* **10**, 121–130.

Friedlander S. K. and Seinfeld J. H. (1969) A dynamic model of photochemical smog. *Environ. Sci. Technol.* **3**, 1175–1181.

Fuller E. C. and Crist R. H. (1941) The rate of oxidation of sulfite ions by oxygen. *J. Am. chem. Soc.* **63**, 1644–1650.

Forrest J. and Newman L. (1977) Further studies on the oxidation of sulfur dioxide in coal-fired power plant plumes. *Atmospheric Environment* **11**, 517–520.

Gifford F. A. (1958) Smoke plumes as quantitative air pollution indicators. *Int. J. Air Wat. Pollut.* **2**, 43.

Gerhard E. R. and Johnstone H. F. (1955) Photochemical oxidation of SO_2 in air. *Ind. Eng. Chem.* **47**, 972–976.

Gartrell F. E., Thomas F. W. and Carpenter S. B. (1964) *Full Scale Study of Dispersion of Stack Gases.* Tennessee Valley Authority Report.

Hall T. C. Jr. (1953) Ph.D. Thesis. University of California at Los Angeles.

Hilst G. R. and Donaldson C. du P. (1973) *The development and preliminary application of an invariant coupled diffusion and chemistry model.* NASA CR-2295 Report.

Junge C. E. and Ryan T. G. (1958) Study of the SO_2 oxidation in solution and its role in atmosphere chemistry. *Q.J.R. met. Soc.* **84**, 46–55.

Junge C. and Scheich G. (1971) Determination of the acid content of aerosol particles. *Atmospheric Environment* **5**, 165–175.

Karraker D. G. (1963) The kinetics between sulfurous acid and ferric ion. *J. phys. Chem.* **67**, 871–874.

Levy A., Drewer D. R. and Hales J. M. (1976) SO_2 oxidation in plumes: a review and assessment of relevant mechanistic and rate studies. U.S.E.P.A. 450/3-76-022.

McKay H. A. C. (1971) The atmospheric oxidation of SO_2 in water droplets in presence of ammonia. Atmospheric Environment **5**, 7–14.

Moore W. J. (1963) *Physical Chemistry.* Longmans. London.

Meszaros E., Moore D. J. and Lodge J. P. (1977) Sulfur dioxide—sulfate relationship in Budapest. *Atmospheric Environment* **11**, 345–349.

Newman L., Forrest J. and Manowitz B. (1975) The application of an isotopic ratio technique to a study of the atmospheric oxidation of sulfur dioxide in the oil fired power plant. *Atmospheric Environment* **9**, 959–968.

Novakov T. (1974) Sulfates as pollution particulates: catalytic formation on carbon particles. *Science* **186**, 259–261.

Shy C. M. and Finklea J. F. (1973) *Environ. Sci. Technol.* **7**, 204.

Smith F. B. and Jeffrey G. H. (1975) Airborne transport of SO_2 from the U.K. *Atmospheric Environment* **11**, 179–182.

Smith J. R. (1974) *Summary Report on Suspended Sulfates and Sulfuric Acid Aerosols*, U.S.E.P.A., Washington, D.C.

Slade H. D. (1968) *Meteorology and Atomic Energy.* U.S. Atomic Energy Commission.

Stephens N. T. and McCaldin R. (1971) Attentuation of power station plumes as determined by instrumented aircraft. *Environ. Sci. Technol.* **5**, 615–621.

Symposium of Plume Behaviour (1966) *Int. J. Air Wat. Pollut.* **10**, 393.

Wilson W. E., Charlson R. J., Husar R. B., Whitby K. T. and Blumenthal D. (1976) *Sulfates in the Atmosphere*, 69th Annual Meeting of A.P.C.A. Portland, Oregon.

Weber E. J. (1970) Contribution to the residence time of sulfur dioxide in a polluted atmosphere. *J. geophys. Res.* **75**, 2909–2914.

Yocum J. E. and McCaldin R. O. (1968) Effects of air pollution on materials and economy. In *Air Pollution* (edited by Stern A. C.). Academic Press, New York (1968).

Atmospheric Environment Vol. 12. pp. 349–362. Pergamon Press 1978. Printed in Great Britain.

DRY AND WET REMOVAL OF SULPHUR FROM THE ATMOSPHERE

J. A. GARLAND

AERE, Harwell, Didcot, Oxon., England

(*First received* 20 *June* 1977 *and in final form* 2 *September* 1977)

Abstract—Dry deposition and removal in precipitation of SO_2 and of particulate sulphate are considered in turn. Many assessments of the dry deposition of SO_2 to various surfaces give deposition velocities of about 0.8 cm s^{-1}, although variations with season and weather conditions are important. The deposition velocity of the sulphate aerosol is probably about 0.1 cm s^{-1}. Removal of SO_2 in rain is also a rather inefficient process, and theoretical and experimental results suggest that the sulphur in precipitation results chiefly from the rainout of cloud condensation nuclei. The removal time constants for SO_2 and sulphate by moderate rain are probably of order 10^{-5} and 10^{-4} s^{-1} respectively. A much simplified model suggests that about a half of the SO_2 emitted to the atmosphere is removed by dry deposition, the remainder is oxidised to sulphate and removed in precipitation, and the atmospheric residence time is about 5 days for sulphur. The method and climatological statistics for a more realistic treatment do not yet appear to be available.

NOMENCLATURE

(numbers in parentheses indicate equations introducing or defining the symbols).

a_p, a_d	particle and drop radii
$c(t)$	the concentration of dissolved SO_2 in a droplet after exposure time t
d	displacement height (8)
D	diffusion coefficient of SO_2 in air
f_g	the fraction of SO_2 removed by dry deposition
F	downward flux of SO_2 per unit area of ground surface
h	mixing height for SO_2 (15)
λ and h'	the time constant and distance of fall for adjustment of c following a change in concentration of SO_2 in air (21, 22)
H	the ratio of the concentration (mass per unit volume) of dissolved undissociated SO_2 to the concentration in air
H_{eff}	the total solubility of SO_2 including the dissociated forms (19)
k	von Kármán's constant ($\simeq 0.41$)
$k_1\, k_2$	equilibrium constant for the first and second ionic dissociations of H_2SO_3 (16, 17)
$K(z)$	gas-phase diffusion coefficient (including molecular and eddy diffusion) (2)
$K_M(z)$	diffusivity for momentum transport (11)
M	the mol. wt of SO_2
$r(z)$	total resistance to SO_2 deposition from height z (6)
$r_g(z)$	gas-phase resistance to SO_2 transport (3, 4)
r_s	surface resistance to SO_2 deposition (1)
$r_M(z)$	resistance to momentum deposition (10)
Sh	the Sherwood number = $[2w\, a_d/(4\pi a_d^2)\chi D]$, where w is the mass diffusing to the droplet surface per unit time
t	time
t_M	time for mixing in a falling raindrop
t_d	mean residence time for SO_2, considering dry deposition only (14)
t_{dc}	residence time for SO_2 considering dry deposition and chemical conversion
$u(z)$	wind speed at height z
u_*	friction velocity (9)
$v_g(z)$	deposition velocity (7)
V_t	terminal velocity of a raindrop
w	vertical component of the instantaneous velocity of the air
W_{SO_2}	the washout ratio for SO_2
z	height above the ground
z_0	roughness length of the surface (8)
z_r	a reference height (15)
Λ_{SO_2}	the fraction of airborne SO_2 removed by falling rain per second
ρ_a	the density of air
$\chi(z)$	mean concentration of SO_2 at height z above the ground
χ_s	concentration of SO_2 immediately adjacent to the surface
τ	shear stress at the surface due to the wind (9)
μ_a, μ_w	viscosity of air and water respectively.

1. INTRODUCTION

Deposition processes limit the lifetime of sulphur compounds in the atmosphere, control the distance travelled before deposition, and limit atmospheric concentrations. Thus, their understanding is essential for a proper assessment of the environmental significance of natural and man-made emissions of sulphur compounds, and they have received increasing attention in recent years.

The processes for removal from the atmosphere to the surface are usually considered under two main headings: wet deposition comprises the incorporation of sulphur compounds in cloud droplets (rainout) and removal by falling precipitation (washout); dry deposition denotes the direct collection of gaseous and particulate species on land or water surfaces.

Collection and analysis of rainwater is an obvious way of measuring wet deposition and has been pursued systematically since the early 1950s when the European Air Chemistry Network was established. Some early studies suggested that dry deposition might be as important as wet deposition (Thomas *et al.*, 1943; Meetham, 1950; Olsen, 1957) but the significance of the former has become generally appreciated only during the 1970s, when a number of field studies have confirmed its importance. During this period the results of extensive field measurements of the concentrations of sulphur compounds in air and rain have become available as a result of the Long Range Transport of Air Pollutants programme (LRTAP) of the OECD. The E.E.C. COST 61A programme, 'Physico-chemical behaviour of SO_2 in the atmosphere', has stimulated investigations of processes involved in the transformation and deposition of sulphur compounds. Consequently it is now possible to discuss the mechanisms and rates of the deposition of sulphur compounds in Europe with considerable confidence.

Sulphur dioxide and particulate sulphate are the predominant forms of atmospheric sulphur in the developed countries. Reduced sulphur compounds, including H_2S, dimethyl sulphide and methyl mercaptan are present in the atmosphere at much smaller concentrations, at least in Europe, and they are most likely deposited only after oxidation to SO_2 or sulphate. Therefore, only SO_2 and sulphate are discussed below. In Sections 2 and 3 dry deposition of these species will be considered. Sections 4 and 5 are devoted to descriptions of processes leading to wet deposition of SO_2 and sulphate respectively, and section 6 attempts to compare the significance of these processes in the atmospheric travel of sulphur compounds in Europe.

2. DRY DEPOSITION OF SO_2

2.1. *Theory of gaseous dry deposition*

The exchange of SO_2 between the atmosphere and soil, vegetation or water is governed both by transfer in the gas-phase and by sorption at the surface, which is assumed here to be irreversible. If the concentration of SO_2 is not too high the rate of sorption at the surface is expected to be proportional to the concentration, χ_s, immediately adjacent to the surface, so that the flux can be written

$$F = \chi_s/r_s \tag{1}$$

where r_s, the surface resistance, depends only on the affinity of the surface for SO_2.

Conditions in the air depend on distance from the surface. Close to the surface viscous drag impedes mixing and transport is determined by molecular diffusion. At greater heights advection or mixing by turbulent eddies produced by the movement of the wind over the surface will dominate. Most analyses of transport to the surface assume that the surface and the air flow are horizontally homogeneous, at least in a statistical sense, so that advection can be ignored in comparison with vertical transport due to turbulence. Then the downward flux can be related to a diffusion coefficient K (representing molecular or eddy diffusion)

$$F = K(z)\,\frac{d\chi}{dz}, \tag{2}$$

where z is the height above the ground. Integrating,

$$F = \frac{\chi(z) - \chi_s}{r_g(z)}, \tag{3}$$

where the gas-phase resistance

$$r_g(z) = \int_0^z \frac{d\xi}{K(\xi)} \tag{4}$$

and from Equations (1) and (3)

$$F = \frac{\chi(z)}{r_s + r_g(z)}. \tag{5}$$

Thus the total resistance to deposition is expressed as the sum of two terms

$$r(z) = r_s + r_g(z). \tag{6}$$

The rate of deposition will be discussed in terms of these resistances and the deposition velocity

$$v_g(z) = \frac{F}{\chi(z)} = \frac{1}{r(z)}. \tag{7}$$

The surface resistance of a chemically and geometrically complex system, such as a soil surface or a vegetation canopy cannot readily be predicted and will be discussed later on the bases of experimental results. However, it is possible to predict r_g from the well-established framework of micrometeorology, provided the condition of horizontal homogeneity is satisfied.

In turbulent flow, where eddy diffusion dominates, the transport is expected to be the same for all gases and vapours and similar for heat and momentum. Thus, experience of the transport of other entities can reasonably be applied to SO_2. When there is no potential temperature gradient the wind speed u varies logarithmically with height:

$$u(z) = \frac{u_*}{k}\ln\left(\frac{z-d}{z_0}\right). \tag{8}$$

Here the roughness length, z_o, and displacement height, d, are characteristic of the surface, approx 0.1 and 0.6 of the canopy height for most vegetated surfaces. The friction velocity u_* is defined in terms of the surface drag τ due to the wind

$$\tau = \rho_a u_*^2 \tag{9}$$

(ρ_a being the density of air) and k, von Kármán's constant, is found to be universally constant. Consideration of the drag as a flux density of momentum, and $\rho_a u(z)$ as the density of momentum per unit volume allows the resistance to momentum transport

$$r_M(z) = \frac{\rho_a u(z)}{\rho_a u_*^2} = \frac{1}{k u_*}\ln\left(\frac{z-d}{z_0}\right) \tag{10}$$

and the diffusivity for momentum,

$$K_M = \frac{\tau}{\rho_a du/dz} = k u_*(z-d) \tag{11}$$

to be obtained. Experiment suggests that the transport of trace gases is also represented by these expres-

Table 1. Examples of calculated values of the gas-phase resistance

Surface and z_0 (m)	Wind speed at 100 m (m s^{-1})	Monin–Obukhov length, L* (m)	u_* (m s^{-1})	$r_g(1)$ (s cm^{-1})	r_g (100) (s cm^{-1})
Grass	3	∞	0.13	1.25	2.1
(0.01)		−11	0.18	0.89	1.21
	10	∞	0.45	0.41	0.67
		−215	0.49	0.38	0.57
		+630	0.41	0.44	0.77
Cereal crop	3	∞	0.18	0.87	1.5
(0.1)		−27	0.25	0.66	0.93
	10	∞	0.59	0.32	0.50
		−480	0.64	0.30	0.45
		+1700	0.57	0.33	0.54
				$r_g(5)$	
Forest	3	∞	0.27	0.26	0.54
(1.0)		−85	0.36	0.20	0.43
	10	∞	0.89	0.11	0.24
		−1500	0.93	0.10	0.22
		+6100	0.88	0.11	0.24

* Unstable conditions (L negative) assume a heat flux of 50 W m^{-2} and stable conditions (L positive) assume a downward heat flux of 10 W m^{-2}. $L = \infty$ when conditions are neutral. In stable conditions at the lower wind speed, the surface is isolated from the 100 m level, preventing diffusion from this height and making resistances from 1 m unpredictable.

sions in the absence of a significant temperature gradient.

More generally, turbulence is significantly modified by buoyancy effects, and the resistance and diffusivity for gases and vapours may then differ significantly from these expressions. A further difference arises close to surface asperities where momentum is conveyed readily by pressure forces but gases must diffuse through laminar layers of air. Garland (1977) indicates the magnitude of these effects and shows how r_g can be estimated from micrometeorological data.

Table 1 shows predicted values for a range of surfaces, meteorological conditions, and for two heights, 1 m being a typical measurement height, and 100 m as representative of the heights from which gaseous pollutants must diffuse to the surface during deposition. As the total resistance can be found from measurements of the rate of deposition and the concentration, the surface resistance can be obtained as the difference

$$r_s = r(z) - r_g(z), \qquad (12)$$

provided sufficient micrometeorological observations are made to estimate $r_g(z)$.

2.2. *Measurement of* SO_2 *dry deposition rate*

Methods used for the measurement of SO_2 dry deposition appear to fall into three categories: the concentration gradient method, tracer methods and mass balance methods. A further possibility, the eddy correlation method, holds some promise for future application.

The concentration gradient method has been widely applied over grass, soil, water and wheat, and is described in detail by several authors (e.g. Garland, 1977; Shepherd, 1974). It is based on Equation (2) and requires simultaneous measurement of the gradient of SO_2 concentration with height and sufficient micrometeorological variables to determine the eddy diffusivity. The method can give accurate results only when measurements are made in the constant flux layer over an extensive, horizontally homogeneous surface, and cannot be used to assess uptake by hedges, trees and other irregular landscape freatures. In practice measurements of concentration, wind speed and temperature have usually been made at several heights within a few metres of the surface. SO_2 concentrations have usually been obtained by absorbing the gas in suitable solutions in bubblers over periods of an hour or so and then analysing for sulphate.

Tracer methods enable the sulphur deposited during a brief exposure to be distinguished from sulphur naturally present in soil or vegetation. Several measurements have been made using radioactive ^{35}S as tracer (Garland, 1977; Garland and Branson, 1977b; Owers and Powell, 1974) and sulphur enriched in the inactive isotope ^{34}S has also been used (Belot *et al.*, 1974). The exposure and sampling techniques in these experiments differed considerably, but the principle in each was to expose the surface to a known concentration of labelled SO_2 and afterwards to assess the uptake by analysis of a known area for the tracer isotope used. The tracer method is well-suited to laboratory experiments and small-scale field experiments, but has the disadvantage that contamination of the surface prevents a succession of measurements over the same area.

Mass balance experiments have been made in the laboratory, measuring the rate of removal of SO_2 from the air in an enclosure (Payrissat and Beilke, 1975). The method has also been used in the field, observing the accumulation of sulphur in pots of soil with or without vegetation (Johansson, 1959) or in the leaves of trees exposed to enhanced SO_2 concentrations (de Cormis *et al.*, 1975). Each of these field experiments gives rise to difficulties of interpretation when the results are used to estimate the rate of removal of SO_2 from the atmosphere, since the pot plants are more openly exposed than plants in a crop canopy and may receive some sulphur from rain, while the tree leaves may have lost sulphur by translocation to other parts of the plant, or by the washing action of rain.

The eddy correlation method has been applied to the measurement of fluxes of heat, momentum and water vapour (Hicks, 1970; Yap and Oke, 1974). Its potential for the measurement of fluxes of trace gases has been demonstrated by Desjardins and Lemon (1974). As the gas is transported by eddies, the vertical flux can be calculated from simultaneous observations of the instantaneous concentration, χ and vertical component of the wind velocity w. The mean flux is simply

$$F = \overline{\chi w} = \overline{\chi' w'} + \overline{\chi}\overline{w} \qquad (13)$$

where the bar denotes the average and the prime the deviation from the average. The total flux is thus considered as the sum of two terms, the first being the eddy flux and the second the flux due to the mean motion. $\overline{w}$ averaged spatially over a uniform surface should be zero, but there may be a spatial correlation of concentration and vertical wind. However, the flux due to the mean motion is usually taken to be zero and subject to the uncertainty in this assumption, the flux can be measured if instruments fast enough to respond to the eddies carrying the eddy flux are available. Desjardins and Lemon consider that sensors with a response time of 0.5 s are adequate as long as they are mounted at least 1 m above the surface. An experiment to demonstrate the feasibility of this technique was performed over a forest in England early this year by Galbally, Wilson and Garland using a high speed flame photometric detector for SO_2 developed by Wilson (unpublished).

2.3. *Results of measurements of the dry deposition of* SO_2 *and the surface resistance*

Table 2 summarizes results from a number of investigations from the last seven years. To remove the dependence of flux on concentration the results are expressed in terms of the deposition velocity v_g (1) relative to 1 m [see Equation (7)]. A wide range of surfaces is listed, ranging from water and soil surfaces through short grass to forest, but the mean deposition velocities are surprisingly similar. It is interesting to look further into the mechanism of deposition to enquire why this should be, and in this section the influence of the surface on deposition rate, as expressed in the surface resistance, r_s, will be discussed.

The surfaces to be considered fall into three groups, comprising respectively water surfaces, soils, and vegetation, and these will be discussed in turn. The exchange of gases at water surfaces, particularly the oceans, has been discussed in terms of a simple model in which the surface resistance is described in terms of diffusion through a thin, laminar film which covers the water body (Broeker and Peng, 1974; Hoover and Berkshire, 1969). Below this film the water is assumed to be well mixed to a great depth. For reactive gases such as SO_2 the diffusion of reaction products (sulphite and bisulphite ions) must be considered in addition to the dissolved gas. Liss (1971) showed that the rate of uptake is enhanced to such a degree that the water surface resistance should be negligible compared with typical values of the gas-phase resistance, provided the pH of the liquid is above 4. The sea, and most inland water bodies have far higher values of pH. It is implied in this treatment that there is some reaction occurring in the bulk of the water which renders the uptake irreversible.

Laboratory experiments (e.g. Brimblecombe and Spedding, 1972) have produced results in close agreement with these predictions.

Two field experiments, using the gradient method over fresh water bodies are reported in the literature. Whelpdale and Shaw (1974) measured deposition to Lake Ontario as well as over grass and snow. The steepest gradient was found over the lake and it was concluded that water was the most efficient sink. The other series of measurements (Garland, 1977) were obtained at a reservoir where the pH is about 8.0. Wind speed and temperature gradients were measured so that the gas-phase resistance could be calculated, and the surface resistance was obtained by difference. Of 14 measurements, 8 showed very low values of r_s and are consistent with the prediction of Liss. The remaining 6 exhibited $r_s \gg 1\ \mathrm{s\ cm^{-1}}$ and are clearly not consistent with this prediction. These results may reflect the periodic formation of an organic film over the surface.

There has been no attempt to predict the surface resistance of soil, but for wet soils of high pH it might be expected that there would be no appreciable surface resistance. Moderate or low pH soils might have appreciable surface resistances when wet, while the surface resistance of any dry soil is problematical. Laboratory investigations (Payrissat and Beilke, 1975; Garland, 1977) have indicated low surface resistances for soils with pH > 7, and rather larger resistances for soils of low pH. The surface resistance of all soils investigated varied in an inverse sense with water content. A field investigation over chalky soil (pH ~ 8.0), reported by Garland (1977), found no appreciable surface resistance even when the soil surface appeared quite dry.

Experiments regarding deposition to vegetation are

Table 2. The results of several experimental investigations of SO_2 deposition

Surface	Method etc.	V_g (cm s^{-1})	Mean r_g (s cm^{-1})	r_s (s cm^{-1})	Author
Water	Gradient, Lake Ontario	2.2			Whelpdale and Shaw (1974)
	Gradient, Reservoir	0.41	1.86	0.56	Garland (1977)
Calcareous soil	Laboratory, mass balance			0.24–0.39	Payrissat and Beilke (1975)
Acid soil	Laboratory, mass balance			1.2–3.8	Payrissat and Beilke (1975)
Calcareous soil	Gradient, field	1.2	0.83	0.01	Garland (1977)
Grass	Gradient, field Summer	0.8		0.8	Shepherd (1974)
	Autumn	0.3		3.0	
Grass	$^{35}SO_2$ tracer, field	0.8			Owers and Powell (1974)
Short grass	Gradient, field	0.85	1.2	0.34	Garland (1977)
Medium grass	Gradient, field	0.89	0.46	0.66	Garland (1977)
Medium grass	Tracer, field	1.19	0.38	0.45	Garland (1977)
Moorland with cotton grass	Gradient, field	0.7	—		Holland *et al.* (1974)
Wheat	Gradient, field	0.74	0.5	1–2.5	Fowler (1976)
Wheat		0.44	0.28	2.0	Dannevik *et al.* (1976)
Soybean		1.25	0.11	0.69	Dannevik *et al.* (1976)
Pineforest (dry)	Tracer	0.1–0.6	0.1	1.5–5.0	Garland and Branson (1978)
Pineforest (dry)	Tracer	1.0			Belot *et al.* (1976)

more numerous than those regarding water or soil. Spedding (1969a) found that relative humidity affected uptake of $^{35}SO_2$ by barley shoots and interpreted this effect in terms of uptake via the stomatal pores in the leaf cuticle. Using the gradient method, Shepherd (1974) and Fowler (1976) found diurnal and seasonal variations in uptake by grass and wheat. Garland and Branson (1977b), with $^{35}SO_2$ and Belot *et al.* (1974) with $^{34}SO_2$ found that the resistances for uptake and for transpiration from pine needles were strongly correlated in the ratio of the diffusivities of SO_2 and water vapour. These results provide strong evidence of irreversible uptake through stomata. According to Fowler (1976) there may also be some uptake on or through the cuticle. Uptake by wheat is enhanced by dew, and uptake on forest canopies carrying intercepted water may be important and may account for the enhancement in sulphate of precipitation penetrating forest canopies (Garland and Branson, 1977b).

Several authors have noted movement of stomata after exposure to SO_2 but the direction of movement seems variable (Unsworth *et al.*, 1972; Bonte *et al.*, 1977). This response may influence the sensitivity of the plant to damage by SO_2, but is probably of minor consequence to SO_2 deposition since large changes in stomatal opening occur only at the high concentrations observed near major sources.

Returning now to the question of the small variation of deposition rate between widely different surfaces, it is apparent from Table 2 that for most of the smoother surfaces the surface resistance is rather small and uptake is largely determined by r_g. On the other hand, surfaces with taller vegetation, such as wheat and forest, exhibit a higher r_s, which counteracts the reduced r_g. This trend may perhaps reflect an increase in stomatal resistance in tall vegetation, developed to control water loss.

By contrast, the height of vegetation may have a marked effect on the deposition rate when dew or intercepted rainwater are present. Fowler (1976) found that the surface resistance of a wheat crop vanished when the crop was wet. Such behaviour would increase the v_g for forest by an order of magnitude, but would produce only a minor change in the value for grass. Greatly increased deposition when the canopy is wetted is expected to compensate for the rather low mean deposition velocity to forest during dry periods, so that the annual deposits of SO_2 on forest and agricultural land in north-west Europe are expected to be closely similar (Garland and Branson, 1977b).

In addition to the natural surfaces described above, a few studies of uptake by building materials have been reported. Spedding (1969b), Braun and Wilson (1970) and Judeikis and Stewart (1976) report uptake of SO_2 by limestones, cement, asphalt and other building materials. Most of these surfaces seem able to maintain a deposition velocity of several millimetres per second. Chamberlain (1960) derived a value of 0.7 cm s^{-1} for London from concentration measurements during the fog of December 1952. While this value might be influenced by the high humidity conditions obtaining at the time, it suggests that urban and rural areas are comparable as sinks for SO_2.

2.4. *Consequences of dry deposition of* SO_2

The results discussed above and summarised in Table 2 suggest that a mean deposition velocity of about 0.8 cm s^{-1} might be applicable to large areas of Europe. This estimate must clearly be treated with caution because of the diurnal and seasonal variations and the paucity of data for forest and some other surfaces. However this figure is probably correct to better than a factor of two, and will be used in discussing the behaviour of atmospheric SO_2.

Consideration of the transport of the gas, confined to a mixed layer of height h above an absorbing surface and in the absence of other removal processes, suggests removal will proceed exponentially with a mean residence time

$$t_d = \frac{h}{v_g(z_r)}. \tag{14}$$

Measurements of the concentration as a function of height (Rodhe, 1972; Jost, 1974; Garland and Branson, 1976, OECD, in press) have shown that the concept of a well-defined mixed layer is frequently misleading. Equation (14) can be applied, however, if h is defined as

$$h = \int_0^\infty \chi(z)\,dz/\chi(z_r) \tag{15}$$

where z_r is the height at which v_g is specified.

Observations yield estimates of h of about 1 km in source areas and rather larger values remote from source areas. Thus, with $v_g \sim 0.8$ cm s^{-1} the expected residence time is about 40 h, and if the wind speed is 25 km h^{-1} the mean transport distance before deposition is roughly 1000 km. These are probably underestimates because of the continued upward diffusion of SO_2 as it leaves source areas. The problem of residence time with both dry and wet deposition will be considered in Section 6.

A further effect of deposition is the reduction of concentration close to the surface. This was observed by Garland and Branson, who found that the highest concentration was frequently observed several hundred metres above ground. This effect is expected to be most marked at night, when an inversion prevents replenishment from aloft of the SO_2 removed at the surface. Numerical modelling (Derwent and Garland, submitted) predicts marked depletion of the surface layer after the inversion forms and this effect probably accounts for the diurnal cycle in the concentration of SO_2 and of some other trace gases observed near the ground. The inversion also limits the amount of SO_2 deposited during the night and this effect may significantly increase the mean residence time of SO_2, depending on the frequency of strong nocturnal inversions.

3. DRY DEPOSITION OF THE SULPHATE AEROSOL

There have been several studies of the deposition of particulate material to natural surfaces (Sehmel, 1974; Chamberlain, 1967; Möller and Schuman, 1970; Little and Wiffen, 1977). Very large particles deposit chiefly by sedimentation, but particles in the range 1–100 μm are also borne towards the surface by turbulence where sedimentation is supplemented by impaction on roughness elements. Submicron particles diffuse by Brownian motion through the thin laminar layers close to the surface elements. As a result, the deposition velocity shows a minimum between 0.1 and 1 μm where none of these processes is efficient. As the deposition velocity depends on particle size, estimation of the rate of deposition of the sulphate aerosol requires a knowledge of the size distribution.

Several measurements of the size distribution of the sulphate aerosol near the ground have been reported (Heard and Wiffen 1969; Cawse 1974; Georgii *et al.*, 1971; Mészáros, 1973). Most of the sulphate mass falls between 0.1 and 1 μm dia although a minor fraction is sometimes found between 1 and 20 μm. The latter fraction may contain sea-salt sulphate, and sulphate in windblown soil. Chemical conversion of SO_2 results in an aerosol formed by condensation of involatile products in the submicron region. Growth of the submicron aerosol by coagulation is too slow to account for a significant fraction reaching the larger size interval, but a small fraction of the oxidised sulphur species may condense on the pre-existing atmospheric aerosol (Cox, 1974) and contribute the large particle sulphate. Thus, it is not clear whether the sulphate aerosol above 1 μm is of anthropogenic origin. Table 3 shows the derivation of the mean deposition velocity to grass using the size distribution of Cawse (1974). If the submicron and larger sizes are treated separately they are found to have deposition rates of 0.025 and 0.56 cm s^{-1} respectively.

The direct measurement of dry deposit on artificial surfaces such as filter paper has been reported by

Table 3. The size distribution of sulphate in air at Chilton, England, measured with an Anderson impactor (Cawse, 1974) and the derivation of the mean deposition velocity of the sulphate aerosol

Effective aerodynamic cut-off dia (lower size limit of each fraction)	% on stage	v_g (cm s^{-1})	Mean v_g of each range (cm s^{-1})	Mean v_g (cm s^{-1})
Backing filter (μm)	29.0	3×10^{-2}		
0.43	22.0	2×10^{-2}	0.025	
0.66	21.0	2×10^{-2}		
1.0	14.0	3×10^{-2}		0.10
2.1	5.0	6×10^{-2}		
3.3	2.8	0.1		
4.7	2.2	0.4	0.56	
7.1	1.8	1.0		
10	1.7	2.5		

Cawse (1974). Such measurements for sulphate are of doubtful interpretation because of the substantial concentrations of SO_2, but other elements with similar size distributions and no significant gaseous species, such as Pb, As, Sb, V, Cs show small deposition velocities, consistent with the above deductions. Further evidence that the deposition velocity of sulphate does not exceed 0.1 cm s^{-1} was obtained in field experiments over grass, in which it was shown that the vertical concentration gradient was undetectably small (Garland and Atkins, unpublished data).

Dry deposition is clearly capable of removing the larger particles from the atmosphere in 2 or 3 days but would require several weeks to remove the more important submicron fraction. There are mechanisms for the removal of such particles in rain (see Section 5) and these determine the lifetime of sulphate in the atmosphere. Dry deposition of these particles is of little consequence.

4. WET DEPOSITION OF SO_2

4.1. *Washout of* SO_2 *by falling rain: theoretical aspects*

The exchange of SO_2 between the atmosphere and falling rain drops has been studied by a number of authors. Chamberlain (1960, 1953) used Frösling's Equation to calculate the rate of transfer of SO_2 to raindrops, assuming irreversible absorption. The calculations show that the concentration in rain decreases with drop size. This factor compensates to some degree for increases in rainfall rate, since raindrops are generally larger in heavy rain. Consequently, the rate of removal of SO_2 from a volume of air, expressed as the proportion removed per second increases from 10^{-4} to 3.10^{-4} as the rainfall rate increases from 1 to 10 mm h^{-1}.

Postma (1970) considers the effect of hydrolysis and ionisation when SO_2 dissolves in water, to deduce the equilibrium concentration in rain falling through air of constant SO_2 concentration in the absence of irreversible chemical change. On dissolution, bisulphite and sulphite ions form:

$$H_2SO_4 \overset{k_1}{\rightleftharpoons} HSO_3^- + H^+ \tag{16}$$

and

$$HSO_3^- \overset{k_2}{\rightleftharpoons} SO_3^= + H^+ \tag{17}$$

$$k_2 = \frac{[SO_3^=][H^+]}{[HSO_3^-]} \text{ is of order } 10^{-7} \text{ mole l}^{-1}$$

and for pH < 6 it follows that $[SO_3^=] \ll [HSO_3^-]$.

$$k_1 = \frac{[HSO_3^-][H^+]}{[SO_{2\,aq}]} \tag{18}$$

is about 0.018 mole l^{-1}, however, and most of the SO_2 present is bisulphite unless pH < 2. In the absence of other ions the acidity is determined by the dissociation of H_2SO_3 and the effective solubility can be predicted.

Hales (1972) has pointed out that diffusion in the liquid phase may be limiting if the drops are not well mixed, and has treated uptake by unmixed raindrops rigorously. However the following suggests that mixing should be rapid. The skin friction component of drag on a raindrop radius a_d, falling at terminal velocity V_t, through air of viscosity μ_a, must be of order $6\pi \mu_a V_t a_d$ even at large Reynolds numbers. Then the shear on the surface, of order $\mu_a V_t / a_d$, must create velocities of order $\mu_a V_t / \mu_w$ within the drop giving a circulation time t_M of order $a_d \mu_w / V_t \mu_a$ (where μ_w is the viscosity of water). For a 1 mm diameter drop $t_M \sim 10^{-2}$ s, very short compared with the time taken by the drop falling to the ground.

Assuming uniformity of concentration within the drop, the approaches of Chamberlain and Postma can be combined to obtain an expression for the concentration $c(t)$ g $l.^{-1}$ of dissolved SO_2 in a drop of initially pure water after falling for time t through air containing SO_2 at concentration χ, g $l.^{-1}$. According to Postma the effective solubility

$$H_{\text{eff}} = \frac{c(\infty)}{\chi} = H\left\{1 + \left(\frac{k_1 M}{H\chi}\right)^{\frac{1}{2}}\right\} \simeq \frac{Hk_1 M}{c(\infty)}, \tag{19}$$

where H is the ratio of dissolved undissociated SO_2

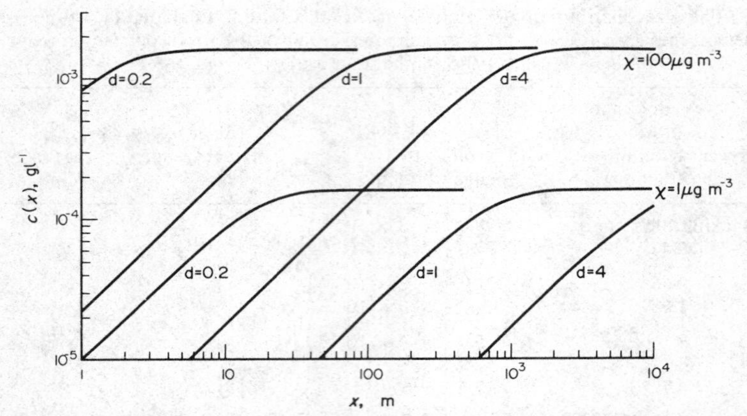

Fig. 1. Concentration $c(x)$ of dissolved SO_2 in rain drops of diameter, d, (mm) after falling a distance x through polluted air of SO_2 concentration χ assuming reversible absorption.

to the concentration in air at equilibrium, and M the mol. wt of SO_2.

Diffusion of SO_2 to the drop surface is proportional to the difference between the ambient concentration χ and the air concentration at equilibrium with the SO_2 already dissolved in the drop. It follows that

$$\frac{dc}{dt} = \frac{3\,\mathrm{Sh}\,D}{2a_d^2}\left(\chi_\infty - \frac{c^2}{Hk_1M}\right). \tag{20}$$

The solution is

$$c = (Hk_1M\chi)^{\frac{1}{2}}\left\{\frac{1-\exp(-2\lambda t)}{1+\exp(-2\lambda t)}\right\} \tag{21}$$

where

$$\lambda = \frac{3\,\mathrm{Sh}\,D}{2a_d^2}\left(\frac{\chi}{Hk_1M}\right)^{\frac{1}{2}}. \tag{22}$$

c approaches the equilibrium value $(Hk_1M\chi)^{\frac{1}{2}}$ in a fashion which is rather similar to a simple exponential of time constant λ (see Fig. 1). The washout ratio for SO_2,

$$W_{SO_2} = \frac{\text{Concentration of dissolved } SO_2 \text{ per unit mass of rain}}{\text{Concentration of } SO_2 \text{ per unit mass of air at ground level}} \tag{23}$$

depends on the vertical distribution of SO_2. Measurements quoted in Section 2.4 indicate that most of the SO_2 is within one or two km of the surface and a uniform concentration to a height of 1 km with clean air was assumed in calculating the values in Table 4. Also included in the table are values of λ and $h' = V_t/\lambda$. Equilibrium is attained when the drop has fallen a distance of 2 or $3\,h'$. The washout ratio assuming irreversible uptake is tabulated for comparison. Very large drops do not reach equilibrium concentration in falling through the assumed km layer, and values of W_{SO_2} are much the same whether reversible or irreversible uptake is assumed. Moderate and small drops achieve equilibrium quickly and much smaller values of W_{SO_2} are predicted when reversible uptake is assumed. W_{SO_2} does not much exceed 200 for any situation unless irreversible uptake is assumed to occur in drizzle.

Chamberlain (1960) integrated irreversible uptake rates over typical raindrop spectra to obtain estimates of the fraction Λ_{SO_2} removed per second. This quantity increases from $10^{-4}\,s^{-1}$ at a rainfall rate of 1 mm h^{-1} to about $3.5 \times 10^{-4}\,s^{-1}$ at 10 mm h^{-1}. Corresponding integrations have not been performed for the assumption of reversible absorption, but as $\Lambda_{SO_2} \propto W_{SO_2}$, Λ_{SO_2} must be expected to be several

Table 4. Washout of SO_2 by falling rain from a polluted layer 1 km deep assuming (a) irreversible absorption (b) reversible absorption with well mixed drops [Equation (21)]

Drop diameter (mm)	V_t ($cm\,s^{-1}$)	(a) W_{SO_2}	Concentration in air ($\mu g\,m^{-3}$)	λ [Eq. (22)] (s^{-1})	(b) h' ($=V_t/\lambda$) (m)	W_{SO_2}
			1	0.043	16.5	210
0.2	71	12600	10	0.14	5.2	66
			100	0.43	1.7	21
			1	5.5×10^{-3}	730.0	180
1.0	403	290	10	0.017	230.0	66
			100	0.055	73.0	21
			1	9.2×10^{-4}	9600.0	21
4.0	883	22	10	2.9×10^{-3}	3030.0	21
			100	9.2×10^{-3}	960.0	16

times smaller in all but the heaviest rain when large drops are predominant. Removal times for SO_2 due to rain are thus expected to be several hours in moderate rain.

Chemical considerations may be of overwhelming importance in controlling uptake. Penkett *et al.* (in press) have shown that in favourable conditions oxidation by ozone or H_2O_2 may be fast enough that SO_2 sorption will be limited by gaseous diffusion. If the concentrations of these reactants are sufficient, the irreversible uptake rates of Chamberlain should apply. On the other hand, acid material present in rain leaving the cloud base may depress the solubility of SO_2 so that, in the absence of rapid oxidising mechanisms, rates below the reversible rates are expected.

4.2. *Uptake of* SO_2 *within clouds*

In the absence of chemical conversion and ionic effects due to condensation nuclei, cloud droplets would reach the equilibrium value defined by Equation (19) with a rate constant λ given by Equation (22). The time required for cloud droplets to equilibrate would be measured in seconds and there is no doubt that cloud water would normally reach the equilibrium concentration before leaving the cloud. As the washout ratio at equilibrium would be of the order of 100 a few per cent of the SO_2 would be incorporated in cloud water in typical clouds with a few tenths of a gramme of water per m^3.

However, with the long residence time of cloud droplets, chemical processes are likely to be very significant. There have been many papers over the years concerned with the oxidation of SO_2 by O_2 in the presence of catalysts and ammonia, but recent work using rainwater (Penkett *et al.*, in press) shows that catalysis has a limited effect. More recently, oxidation by ozone and hydrogen peroxide has received attention (Penkett *et al.*, in press) and it has been shown that these mechanisms dominate the O_2 mechanism, at least in rain, over Britain. About 1 ppb of H_2O_2 is expected to be present in tropospheric air and will rapidly oxidise an equivalent amount of the SO_2. The ozone process may also convert a substantial fraction of the SO_2, depending on the pH of cloud droplets and ozone concentration.

Thus, a substantial fraction of the SO_2 in the cloud volume will be taken up reversibly or irreversibly into the cloud water. The consequences to rain chemistry will be considered in Sections 5 and 6.

5. MECHANISMS FOR THE DEPOSITION OF SULPHATE IN RAIN

The behaviour of hygroscopic aerosols in cloud and rain is complicated by the variation in size that occurs with changing relative humidity. Most atmospheric sampling is performed at moderate or low relative humidity. Georgii *et al.* (1971) showed that most of the sulphate aerosol at cloud forming altitudes is present in the size range 0.02–0.5 μm radius. At saturation the methods of Garland (1969) show that such particles are expected to form droplets of 0.2–10 μm radius. Thus, the larger particles achieve sizes in the range of cloud droplets even before experiencing supersaturation.

Within cloud, water vapour continues to condense on some of the particles resulting in cloud droplets in the range 5–20 μm radius. Subsequently, precipitation elements grow by coagulation and, below freezing point, by the Wegener–Bergeron distillation process onto ice crystals (Mason, 1971). As the water content of raining clouds is typically 0.3 gm^{-3} the ratio

$$\frac{\text{Concentration of sulphate per unit mass of cloud water}}{\text{Concentration of sulphate per unit mass of air}}$$

may approach 3000 if condensation occurs on all the sulphate particles. However, the efficiency of collection of cloud droplets by growing raindrops and the concentration of solute both vary with droplet size, and additionally the effectiveness of the Wegener–Bergeron process varies with ambient temperature and the supply of freezing nuclei. Thus the concentrations in rain and cloud water may differ substantially, and in a manner difficult to predict.

In addition to the rainout of condensation nuclei, described above, several other physical processes may contribute to sulphate in rain. Diffusiophoresis and Brownian diffusion may result in the collection of small particles to the cloud droplets and raindrops may collect further particles by impaction, interception or diffusion. Several authors have examined the effectiveness of these mechanisms theoretically (e.g.

Table 5. Washout ratios for elements with similar size distribution to sulphate at seven locations in the British Isles (Cawse, 1974)

	Chilton	Leiston	Lerwick	Plynlimon	Styrrup	Trebanos	Wraymires
Pb	315	265	1000	100	100	165	205
As	410	325	—	410	330	365	420
Sb	205	230	1730	110	210	130	165
V	450	240	—	200	370	300	455
Cs	415	290	1250	140	370	—	335
Se	335	200	2800	370	300	235	320
NO_3	1800	1500	9200	860	1820	690	1150
SO_4	1050	930	24,000	570	1500	1050	670

Table 6. Mechanisms contribution to sulphur in rainwater

Mechanism	Pollutant lifetime during rain (h) (*a*)	(*b*)	Sulphate concentration in rainwater (mg l^{-1}) (*a*)	(*b*)	Basis of calculation
Particulate sulphate:					
Diffusophoresis			10^{-2}	10^{-3}	Goldsmith *et al.* (1963)
Brownian diffusion to cloud droplets	100	4000	0.2	3×10^{-3}	Mason (1971)
Brownian diffusion to raindrops	10^4	2×10^5	10^{-3}	2×10^{-5}	Frössling's Equation
Impaction and interception by raindrops	400	1	0.02	1.2	Berg (1970)
Rainout of condensation nuclei			3–10		
SO_2:					
Solution and oxidation in cloud droplets			3		
Uptake by falling raindrops	10		1		

Assuming (i) 10 $\mu g\ m^{-3}$ of sulphate composed of (a) 80% submicron particles of typical diameter 0.2 μm and (b) 20% larger particles of typically 4 μm dia.
(ii) 10 $\mu g\ m^{-3}$ of SO_2.
(iii) rain falling at 1 mm h^{-1} as 1 mm drops from a cloud containing 100 drops cm^{-3} each of 10 μm radius.

Goldsmith *et al.*, 1963). Table 6 shows estimates of the possible contribution of each mechanism taking as an example a cloud with 100 droplets of 10 μm radius per cm^3 and precipitation at 1 mm h^{-1} falling as 1 mm dia raindrops in a moderately polluted atmosphere. Only the rainout of condensation nuclei appears capable of explaining the concentrations of several mg $l.^{-1}$ observed in practice.

The washout of large particles by raindrops may make a significant contribution, but this minor fraction of the aerosol will be exhausted by the first few millimetres of rain and may therefore account for the enhancement in sulphate concentration observed at the beginning of some periods of rain (Meurrens, 1974).

Diffusion and interception may be of greater significance in snow because of the larger surface area of the precipitation elements. In addition, the concentration of condensation nuclei collected in precipitation may be much reduced if distillation from the liquid to solid-phase dominates the aggregation of cloud droplets in the growth of snowflakes. However, most rain clouds extend above the freezing level in temperate latitudes and it is not clear that there should be a marked difference between the chemistry of rain and snow as observed at the ground.

Because of the difficulties mentioned previously in describing the rainout of condensation nuclei, use will be made of washout ratios for particulate tracers with size distributions similar to sulphate but without the complication of gaseous atmospheric components. Several suitable tracers were mentioned in Section 3 and Table 5 quotes values from Cawse (1974) of the washout ratios (i.e. the ratios of concentration per unit mass of rain to the concentration per unit mass of air at ground level). Stable trace elements originating from fossil fuel burning or from the soil are preferred to nuclear fallout since the height distribution of the former is more likely to resemble that of sulphate. The washout ratios are reasonably consistent. The highest ratios occur at the island site of Lerwick, where sea spray has probably influenced the results, and the lowest are found on the Welsh mountain, Plynlimon, where heavy rainfall dilutes the collections. A value of 300 seems typical of most elements. Nitrate and sulphate yield values of about 1000, but both may be subject to interference by soluble gases. However Martens and Harris (1973) point out that soluble particles such as sulphate are more efficient condensation nuclei than insoluble particles of the same size, and may have larger washout ratios. We conclude, therefore, that rainout of sulphate particles (with minor contributions from other mechanisms) may result in a washout ratio exceeding 300, but not greater than 1000.

Table 6 shows that the probable contribution of dissolved SO_2 is smaller than the contribution due to rainout of sulphate. However, oxidation of SO_2 in clouds may make a substantial contribution to the sulphate in rain.

6. DEPOSITION OF SULPHUR FROM THE ATMOSPHERE

6.1. *Observations of sulphur compounds in rain*

The last two sections suggest that both SO_2 and sulphate may contribute significantly to dissolved sul-

Table 7. Summary of six months' data (January–June 1973) for sulphate in rain at the LRTAP stations U.K. 1 (Cottered) and U.K. 2 (Eskdalemuir)

Station	Number of days with rain	C_1	C_2	C_3	Correlation coefficients: $[SO_4]_{rain}$ on $[SO_4]_{air}$	$[SO_4]_{rain}$ on $[SO_2]$
U.K. 1	48	0.016 ± 0.05	1.1 ± 0.2	0.62	0.62	0.19
U.K. 2	70	0.014 ± 0.01	0.77 ± 0.10	1.4	0.68	0.03

$$\underset{(\text{mg l}^{-1})}{[SO_4]_{rain}} = C_1 \underset{(\mu g\, m^{-3})}{[SO_2]} + C_2 \underset{(\mu g\, m^{-3})}{[SO_4]_{air}} + C_3$$

phur in rain. The contribution of sulphate appears inevitable as sulphate particles serve as cloud condensation nuclei, but the incorporation of SO_2 may be suppressed if the condensation nuclei are acid. Thus it is necessary to use field data to determine whether SO_2 is removed by rain.

Several measurements of deposition from discrete plumes have been carried out for this purpose, on both sides of the Atlantic. Högström (1973) found that sulphur emitted, largely as SO_2, from Uppsala during rain was deposited with a removal time of one or two hours. However Granat and Rodhe (1973) found that less than 6% of sulphur emitted during rain from the tall chimneys of a power station was removed by rain in 15 km travel. Granat and Söderlund (1975) found that deposition in rain totalled about 5% within 35 km in a later series of measurements. Larson *et al.* (1975) deduced that only 8% of the sulphur emitted from a smelter while rain was falling deposited within 60 km and the deposition rates of Dana *et al.* (1975) imply no more than 3% deposition in rain from a power plant plume in the first 10 km.

Davies (1976) measured dissolved SO_2 in rainwater in Sheffield and found about 3 mg l.$^{-1}$ when the air concentration was about 100 μg m^{-3}. Meurrens (1974) and Meurrens and Lenelle (1976) studied field data from the environs of Brussels and concluded that sulphur compounds in rain originate chiefly from the sulphate aerosol.

With the exception of Högström's results, these studies imply time constants of the order of 10^{-5} s^{-1} for SO_2 removal in rain. Högström's data may indicate rapid oxidation of SO_2 in the Uppsala plume. Confirmation for the slow scavenging of SO_2 by rain may be found in the extensive body of data obtained in the LRTAP project. Table 7 shows the results of an analysis of the daily mean concentrations for two British stations. The data show a significant correlation of sulphate in rain with sulphate in air, but no correlation with SO_2. The coefficients imply washout ratios of 1320 and 920 for sulphate at the two stations.

6.2. *Removal of sulphur from the atmosphere*

The conclusions of the previous sections can be summarized briefly as follows:

(a) SO_2 is removed by dry deposition with a deposition velocity of about 0.8 cm s^{-1} for most surfaces of practical significance;

(b) the deposition velocity for the sulphate aerosol is no larger than 0.1 cm s^{-1};

(c) sulphate is removed by rain with a time constant of order 10^{-4} s^{-1};

(d) SO_2 removal by rain is about an order of magnitude less efficient.

In addition, SO_2 is oxidised to sulphate by photochemical processes in the absence of cloud, and in the aqueous phase when cloud is present, at rates of the order of 0.01 h^{-1}. More rapid oxidation appears to occur in concentrated plumes close to sources, but this will be ignored for the present. We now enquire as to the residence time for SO_2 in the atmosphere with these conversion and removal rates. Ideally, it is desirable to derive the probability distribution of residence time in the manner adopted by Rodhe and Grandell (1972) for particulate matter, but sufficiently flexible models have not yet been developed. It is possible to make some deductions from simple models and from field measurement. Rodhe (1976) and Garland (1977) used field data from the LRTAP project to derive a residence time of about 2 days (Table 8). Dry deposition of SO_2 was found to be of equal or greater significance to precipitation, in depositing sulphur in N.W. Europe. Similar conclusions result from the modelling study carried out during the LRTAP project (OECD, in press).

These studies, however, are limited to the proximity of emission areas. Continued upward diffusion, though frequently hindered by temperature inversions, will increase the mixing depth of SO_2 with distance from the source, reducing the effectiveness of dry deposition. The much simplified model of Garland and Branson (1977a) gives estimates of the residence time of SO_2 free from the bias entailed in using results for source regions. In this model removal of SO_2 in rain is ignored, diffusion throughout the troposphere is permitted and the fraction removed by dry deposition in competition with oxidation to sulphate is calculated. As dry deposition of sulphate is slow, the sulphate fraction is essentially the fraction removed in precipitation. The model shows (Table 9) that a considerable range of parameter values results in a residence time of about 2 days for SO_2, before dry deposition or conversion to sulphate. About a half of the SO_2 is removed by dry deposition. As

Table 8. The atmospheric sulphur budget over N.W. Europe according to Garland (1977), Rodhe (1976) and OECD (in press)

Area included (km^2)	Source strength ($Mr\ S\ a^{-1}$)	Dry deposit ($Mt\ S\ a^{-1}$)	Wet deposit ($Mt\ S\ a^{-1}$)	Wind loss ($Mt\ S\ a^{-1}$)	Author
4×10^6	16–17	6.3 (45%)	2.7 (19%)	5.0 (37%)	Garland
4.2×10^6	14.5	2.9–5.9 (20–41%)	4.1 (28%)	3.1–6.2 (21–43%)	Rodhe
9×10^6	20	11 (55%)	5.6 (27%)	3.4 (17%)	OECD

the residence time of atmospheric particulate is about 5 days, the residence time for sulphur in the atmosphere will exceed 4 days.

These methods and the current state of knowledge are probably adequate to describe deposition on the continental and national scale. However, if the variation of deposition on a smaller scale is at issue, more detailed knowledge is required, particularly concerning:

(1) the variation of dry deposition rate with vegetation type, time of day, and the influence of crop wetness;

(2) the statistical nature of temperature stratification and its effect on vertical diffusion and on deposition, and the nature of the stratification over land and sea near coasts;

(3) the statistics of cloud volume, the quantity of air passing through clouds, its chemical composition and the changes occurring in cloud, the fractions of cloud water removed by precipitation and the fraction evaporating.

These appear to the author to be the chief areas of uncertainty in describing the fate of sulphur compounds in the atmosphere.

Acknowledgement—Work on this topic at Harwell formed part of a programme supported by the Department of the Environment.

Table 9. Residence time, t_{dc} of SO_2 and fraction removed by dry deposition, f_g, in competition with oxidation, rate constant k, in a model with constant diffusion coefficient K up to the tropopause at 10 km. l is the source hieght

l (m)	v_g ($mm\ s^{-1}$)	k (h^{-1})	K ($m^2\ s^{-1}$)	t_{dc} (h)	f_g
30	7.0	0	10	405	1
		0.005		71	0.65
		0.01		43.6	0.56
		0.02		26.3	0.47
		0.01	20	52.7	0.49
	3.5	0.01	10	60.7	0.39
300	7.0	0	10	479	1
		0.01		51.3	0.49
		0.005		60.0	0.70
		0.02		29.1	0.42
	3.5	0.01		59.3	0.41
	7.0	0.01	20	44.1	0.56

REFERENCES

Belot Y., Baille A. and Delmas J.-L. (1976) Modèle numérique de dispersion des polluants atmosphériques en presence de converts végétaux. *Atmospheric Environment* **10**, 89–98.

Belot Y., Bourreau J. C., Dubois M. L. and Pauly C. S. (1974) Mesure de la vitesse de captation du dioxide de souffre sur les feuilles des plantes au moyen du souffre-34. FAO/IAEA Isotope ratios as pollutant source and behaviour indicators. 18–22 Nov. Vienna (IAEA-SM-191-18).

Berg T. G. O. (1970) Collection efficiency in washout by rain. Precipitation Scavenging (1970) AEC Symposium series 22. CONF. 700601. NTIS, Springfield Virginia, U.S.A.

Bonte J., Bonte C., de Cormis L. and Louguet P. (1977) Contribution à l'étude des caractères de résistance de *Pelargonium* à un polluant atmosphérique, le dioxide de soufre *Physiol. Vég.* **15**, 15–27.

Braun R. C. and Wilson M. J. C. (1970) The removal of atmospheric sulphur by buidling stones. *Atmospheric Environment* **4**, 371.

Brimblecombe P. and Spedding D. J. (1972) Rate of solution of gaseous sulphur dioxide at atmospheric concentrations. *Nature* **236**, 225.

Broeker W. S. and Peng T. H. (1974) Gas exchange rates between air and sea *Tellus* **26**, 21–35.

Cawse P. A. (1974) *A Survey of Atmospheric Trace Elements in the U.K.* (1972-73) AERE-R 7669. H.M.S.O. London, U.K.

Chamberlain A. C. (1953) *Aspects of Travel and Deposition of Aerosol and Vapour Clouds.* AERE HP/R 1261. H.M.S.O. London.

Chamberlain A. C. (1960) Aspects of the deposition of radioactive and other gases and particles. *Int. J. Air Pollut.* **3**, 63–88.

Chamberlain A. C. (1967) Deposition of *Lycopodium* spores and other small particles to rough surfaces. *Proc. R. Soc., Lond. A*, **296**, 45–70.

de Cormis L., Bonte J. and Tisne A. (1975) Technique expérimentale permettant l'étude de l'incident sur la vegétation d'une pollution par le dioxide de soufre appliquée en permanence et à dose subnéctrotique. *Pollution Atmosphérique* **66**, 103–107.

Cox R. A. (1974) Particle formation from homogeneous reactions of sulphur dioxide and nitrogen dioxide. *Tellus* **26**, 235–240.

Dana M. T., Hales J. M. and Wolf M. A. (1975) Rain scavenging of SO_2 and sulphate from power plant plumes. *J. geophys. Res.* **80**, 4119–4129.

Dannevik W. P., Frisella S., Granat L. and Husar R. B. (1976) SO_2 Measurements in the St. Louis Region. 3rd Symposium on atmospheric turbulent diffusion and air quality. Raleigh, N.C. U.S.A.

Davies T. D. (1976) Precipitation scavenging of sulphur

dioxide in an industrial area. *Atmospheric Environment* **10**, 879–890.

Derwent R. G. and Garland J. A. Absorption at the ground and the diurnal cycle of concentration of gaseous pollutants. (To be published).

Desjardins P. L. and Lemon E. R. (1974) Limitations of eddy-correlation techniques for the determination of the carbon dioxide and sensible heat fluxes. *Boundary-Layer Met.* **5**, 475–488.

Fowler D. (1976) Uptake of sulphur dioxide by crops and soil Ph.D. Theses at Univerisity of Nottingham.

Garland J. A. (1969) Condensation on ammonium sulphate particles and its effect on visibility. *Atmospheric Environment.* **3**, 347–354.

Garland J. A. (1977) The dry deposition of sulphur dioxide to land and water surfaces. *Proc. R. Soc. Lond., A.* **354**, 245–268.

Garland J. A. and Branson J. R. (1976) The mixing height and mass balance of SO_2 in the atmosphere above Great Britain. *Atmospheric Environment* **10**, 353–362.

Garland J. A. and Branson J. R. (1977a) Reply to discussion by H. Rodhe of "the mixing height and mass balance of SO_2 in the atmosphere over Great Britain" *Atmospheric Environment* **11**, 659–661.

Garland J. A. and Branson J. R. (1977b) The deposition of sulphur dioxide to pine forest assessed by a radioactive method. *Tellus* **29**, 445–454.

Georgii H-W., Jost D. and Vitze W. (1971) Konzentration und Grössenverteilung des Sulphataerosols in der unteren und mittleren Troposphäre. Berichte des Institutes für Meteorologie und Geophysik der Universität Frankfurt/Main. Nr. 23.

Goldsmith P., Delafield H. J. and Cox L. C. (1963) The role of diffusiophoresis in the scavenging of radioactive particles from the atmosphere. *Q. J. R. met. Soc.* **89**, 43–61.

Granat L. and Rodhe H. (1973) Study of fallout by precipitation around an oil-fired power plant. *Atmospheric Environment* **7**, 781–792.

Granat L. and Söderlund R. (1975) Atmospheric Deposition due to Long and Short Distance Sources. Report AC-32, University of Stockholm.

Hales J. M. (1972) Fundamentals of the theory of gas scavenging by rain. *Atmospheric Environment* **6**, 635–660.

Heard M. J. and Wiffen R. D. (1969) Electron microscopy of natural aerosols and the identification of particulate ammonium sulphate. *Atmospheric Environment* **3**, 337–340.

Hicks B. B. (1970) The measurement of atmospheric fluxes near the surface: a generalised approach. *J. appl. Met.* **9**, 386–388.

Högström U. (1973) Residence time of sulphurous air pollutants from a local source during precipitation. *Ambio* **2**, 37–41.

Holland P. K., Sugden J. and Thornton K. (1974) The direct deposition of SO_2 from atmosphere Pt. 2. Measurements at Ringinglow Bog compared with other results. Report No. NW/SSD/RN/PL/1/74 C.E.G.B. North Western Region.

Hoover T. E. and Berkshire D. C. (1969) Effects of hydration of carbon dioxide exchange across an air-water interface *J. geophys. Res.* **74**, 456–464.

Johansson O. (1959) On sulphur problems in Swedish agriculture *Ann. R. Agr. Coll. Sweden* **25**, 57–169.

Jost D. (1974) Aerological studies on the atmospheric sulphur budget. *Tellus* **26**, 206–212.

Judeikis H. S. and Stewart T. B. (1976) Laboratory measurement of SO_2 deposition velocities on selected building materials and soils. *Atmospheric Environment* **10**, 769.

Larson T. V., Charlson R. J., Knudson E. J., Christian G. D. and Harrison H. (1975) The influence of a single sulphur dioxide point source on the rain chemistry of a single storm in the Puget Sound Region. *Wat. Air Soil Pollut.* **4**, 319–328.

Liss P. S. (1971) Exchange of SO_2 between the atmosphere and natural waters *Nature* **233**, 327–329.

Little P. and Wiffen R. D. (1977) Emission and deposition of petrol engine exhaust Pb—1. Deposition of exhaust Pb to plant and soil surfaces. *Atmospheric Environment* **11**, 437–447.

Martens C. S. and Harriss R. C. (1973) Chemistry of aerosols, cloud droplets and rain in the Puerto Rican marine atmosphere. *J. geophys. Res.* **78**, 949–957.

Mason B. J. (1971) *The Physics of Clouds.* Clarendon Press, Oxford.

Meurrens A. (1974) Sulfites et sulfates dans la pluie. Paper presented at the COST Technical Symposium, Ispra.

Meurrens A. and Lenelle Y. (1976) Sulfite et sulfates dans la pluie. Paper presented at the COST Technical symposium, Ispra.

Meetham A. R. (1950) Natural removal of pollution from the atmosphere. *Q. J. R. met. Soc.* **76**, 359–371.

Mezaros E. (1970) Seasonal and diurnal variations of the size distribution of atmospheric sulphate particles. *Tellus* **22**, 235–238.

Möller U. and Schumann G. (1970) Mechanisms of transport from the atmosphere to the earth's surface. *J. geophys. Res.* **75**, 3013–3019.

OECD (1977) The OECD programme on long range transport of air pollutants. Measurements and findings. OECD Paris.

Olsen R. A. (1957) Absorption of sulphur dioxide from the atmosphere by cotton plants. *Soil Sci.* **84**, 107–111.

Owers M. J. and Powell A. W. (1974) Deposition velocity of sulphur dioxide on land and water surfaces using a ^{35}S tracer method. *Atmospheric Environment* **8**, 63–68.

Payrissat M. and Beilke S. (1975) Laboratory measurements of the uptake of sulphur dioxide by different European soils. *Atmospheric Environment* **9**, 211–217.

Penkett S. A., Jones B. M. R. and Brice K. A. Rate of oxidation of sodium sulphite solutions by oxygen, ozone and hydrogen peroxide and its relevance to the formation of sulphate in cloud and rainwater. *Atmospheric Environment* (in press).

Penkett S. A., Jones B. M. R. and Eggleton A. E. J. A study of SO_2 oxidation in stored rainwater samples. *Atmospheric Environment* (in press).

Postma A. K. (1970) Effect of solubility of gases on their scavenging by raindrops. *Precipitation Scavenging* (1970). AEC Symp. Ser. 22, CONF-700601, 247–259. NTIS Springfield, Virginia.

Rodhe H. (1972) Measurements of sulphur in the free atmosphere over Sweden 1969–1970. *J. geophys. Res.* **77**, 4494–4499.

Rodhe H. (1976) An atmospheric sulphur budget for N.W. Europe in nitrogen, phosphorus and sulphur—global cycles. SCOPE Report No. 7, NFR, Stockholm. *Ecol. Bull.* **22**, 123–134.

Rodhe H. and Grandell J. (1972) On the removal time of aerosol particles from the atmosphere by precipitation scavenging. *Tellus* **24**, 442–454.

Sehmel G. A. and Hodgson W. H. (1974) Predicted dry deposition velocities, Atmosphere–Surface Exchange of Particulate and Gaseous Pollutants Symposium, CONF-74091. NTIS, Springfield, Virginia.

Shepherd J. G. (1974) Measurements of the direct deposition of sulphur dioxide onto grass and water by the profile method. *Atmospheric Environment* **8**, 69–74.

Spedding D. J. (1969a) Uptake of sulphur dioxide by barley plants at low sulphur dioxide concentrations. *Nature* **224**, 1229–1231.

Spedding D. J. (1969b) Sulphur dioxide uptake by limestone. *Atmospheric Environment* **3**, 683.

Thomas M. D., Hendricks R. H., Collier T. R. and Hills G. R. (1943) The utilisation of sulphate and sulphur dioxide for the sulphur nutrition of alfalfa. *Pl. Physiol.* **18**, 345–371.

Unsworth M., Biscoe B. V. and Pinckney H. R. (1972) Stomatal response to sulphur dioxide. *Nature* **239**, 458–459.

Whelpdale D. M. and Shaw P. W. (1974) Sulphur dioxide removal by turbulent transfer over grass, snow and water surfaces. *Tellus* **26**, 196–205.

Yap D. and Oke T. R. (1974) Eddy-correlation measurements of sensible heat fluxes over a grass surface. *Boundary-Layer Met.* **7**, 151–163.

Atmospheric Environment Vol. 12, pp. 363–367. Pergamon Press 1978. Printed in Great Britain.

DRY DEPOSITION OF SO_2

U. Platt

Kernforschungsanlage Jülich, Institut für Chemie III, D-517 Jülich Postfach 1913, W. Germany

(*First received* 15 *June* 1977 *and in final form* 27 *September* 1977)

Abstract—SO_2 concentration gradient measurements are reported, which have been carried out between May 1975 and September 1976 at the Kernforschungsanlage Jülich. Most of the measurements were made with differential optical absorption spectroscopy.

Mirrors at various heights (30, 80, 120 m) on the meterological tower were used to reflect the light beam back to the laboratory placed at the ground. In this way absorption paths of about 600 m were obtained. In some cases, the optical data were compared with chemical measured concentration profiles (after West & Gaeke, 1976) showing an agreement within $\pm 20\%$. From simultaneously measured temperature and wind profiles the transfer resistance of the turbulent boundary layer was calculated.

Average values of the vertical SO_2-flux obtained from SO_2 concentration gradient and atmospheric transfer resistance varied from about 0.6 $\mu g\ m^{-2}\ s^{-1}$ in winter and 0.1 to 0.4 $\mu g\ m^{-2}\ s^{-1}$ in summer, corresponding to a lifetime for SO_2, if dry deposition is taken as the only sink mechanism, of half a day up to several days, respectively.

INTRODUCTION

There are essentially three processes removing SO_2 from the atmosphere: Oxidation (gas phase homogeneous photochemical oxidation or reaction with other trace constituents in cloud droplets), wet deposition of SO_2 (rainout or washout), and direct deposition of SO_2 at the ground.

It has been shown recently, that in polluted areas direct deposition of SO_2 may well be the dominant removal process (Garland, 1976). In those sites the vertical flux of SO_2 or the deposition velocity are of particular interest. However measurements of the deposition velocity so far have been mainly limited to measuring the vertical flux of SO_2 over flat surfaces like water or grass (Whelpdale, 1974; Shepherd, 1974; Garland, 1974).

The roughness length under those conditions was a few cm or less so that constant flux layers from 10 to 20 m above the surface resulted which could be used for gradient measurements. However such conditions are not representative of the surface structure in polluted areas.

The present study was performed in an agricultural area with interspersed forests and buildings being rather characteristic for the countryside in middle Europe. From windspeed measurements at 6 different heights between 20 and 120 m (showing a correlation between windspeed and the logarithm of height in neutral conditions usually better than 99%) a roughness length of several meters was calculated. There is no larger SO_2 source within 5 km of the site of measurement. On rare occasions very high SO_2-concentrations (several hundred $\mu g\ m^{-3}$) indicated the plume of a power plant located 8 km southwest. Those measurements were omitted. From SO_2 concentration gradients measured up to heights of 100 m the vertical fluxes of SO_2 were calculated assuming an equal turbulent transport for SO_2 and momentum.

DETERMINATION OF SO_2 FLUX

Dry deposition of SO_2 at the ground is determined by the transport to the surface and the rate of the uptake at the surface itself. These processes can be described by a transfer resistance R, which consists of two terms:

$$R = R_T + R_B.$$

(R_T: transfer resistance in the atmosphere, R_B: "surface resistance"). From the simultaneous measurements of temperature and wind profiles, the turbulent transfer resistance R_T (z_2, z_1) between two heights, can be calculated. With the assumption of a constant vertical SO_2 flux, which should hold fairly well for altitudes ≤ 100 m, the corresponding vertical flux of SO_2 can be simply calculated by:

$$J = \frac{C(z_2) - C(z_1)}{R_T(z_2, z_1)} \tag{1}$$

[with $C(z)$: concentration at height z]. Thus the vertical flux J can be determined from measurements of concentration, wind speed, and temperature in different heights. Knowledge of the mechanism of SO_2 uptake at the surface is not required.

Calculation of the turbulent transfer resistance

Inside the turbulent boundary layer, (under neutral conditions) the turbulent diffusion constant K_M of momentum is given by:

$$K_M = u_* \cdot l = u_*^2(\partial u/\partial t). \tag{2}$$

Near the ground $l = kz$, where

k = Karman constant and $u_* = \sqrt{\tau/\rho}$ = friction velocity.

Integrating over K_M, the 'transfer resistance' of

momentum between two heights z_1 and z_2 can be obtained:

$$R_T(z_1, z_2) = \frac{1}{\int_{z_1}^{z_2} \frac{dz}{K_M}}. \quad (3)$$

Depending on the values of the surface roughness and the horizontal pressure gradient in the atmosphere, the shearing stress may decrease up to 20–30% between the surface and 100 m height. Taking as constant with height, therefore will result in a slight underestimation of the turbulent diffusion constant at higher levels above the ground. Due to the fact that R_T is calculated by integrating K_M over a height interval, the error in R_T from the assumption of a constant shearing stress with height should be less than 15%.

Assuming that the turbulent transfer resistance describes not only the turbulent transport of momentum, but also the transfer of conservative quantities of the atmosphere, R_T calculated from Equation 3 can also be used to calculate the vertical flux of SO_2 according to Equation 1.

Assuming the vertical flux as constant with height, the transfer velocity $v_{15,0}$ between the lowest level (15 m) of data collection and the actual surface with a SO_2 concentration $C_0 = 0$ can be calculated:

$$v_{15,0} = \frac{J}{C_{15}}. \quad (4)$$

The velocity is considered as a deposition velocity, which describes the atmospheric transfer of SO_2 below 15 m and the uptake at the surface itself.

Under non-neutral conditions the flux of sensible heat from or to the ground (under lapse or stable conditions) has to be taken into account in addition to the flux of momentum to the ground. Under stable conditions for example additional energy is needed to mix the stratified atmosphere, thus the windspeed profile becomes steeper. After Dyer-Businger (Roth 1975), the influence of the temperature gradient is expressed by a 'stability-function' $\phi(z/L_*)$, which depends on the normalized height z/L_* (L_*: Monin–Obukhow-length):

$$K_M = u_* \cdot k \cdot z\, \phi(z/L_*). \quad (5)$$

As under neutral conditions, the transfer resistance $R_T(z_2, z_1, u_*, L_*)$ can be calculated from Equation 3, with u_* and L_* derived from the measured temperature- and wind speed profiles.

SO_2 CONCENTRATION MEASUREMENTS

The SO_2 concentration is measured by optical absorption spectroscopy (in the range of 300 nm), using the meteorological tower of the Kernforschungsanlage Jülich. The light beam of a Xe-high-pressure-lamp is directed from the laboratory to mirrors mounted on platforms at 30, 80, and 120 m height and reflected back to the laboratory (Fig. 1). In this way, absorption paths of about 600 m are obtained. In the receiving system (Fig. 2) a concave mirror focusses the light on the entrance slit of a spectrograph. In the focal plane of the spectrograph a portion of 4.5 nm of the dispersed spectrum, which contains an absorption band of the SO_2 molecule, is selected for analysis. This portion of the spectrum is scanned by a slotted disk, which serves as exit slit, and rotates at 180 rev/min. A photomultiplier converts the received light to an electric signal, which is fed to a signal averager. After integration over about 40,000 scans, the light intensities in the center of the absorption band $I(\lambda_1)$ and besides the absorption band $I(\lambda_2)$ are compared and the SO_2 concentration is calculated from Lambert–Beer's-Law:

$$C = \frac{A(\lambda_1, \lambda_2)}{L} \log \frac{I(\lambda_2)}{I(\lambda_1)} \quad (6)$$

$A(\lambda_1, \lambda_2)$ denotes the differential absorption coefficient, which was determined from calibration measurements, while L is the lengths of the absorption path. With this differential absorption technique, the concentration is calculated only from quotients of

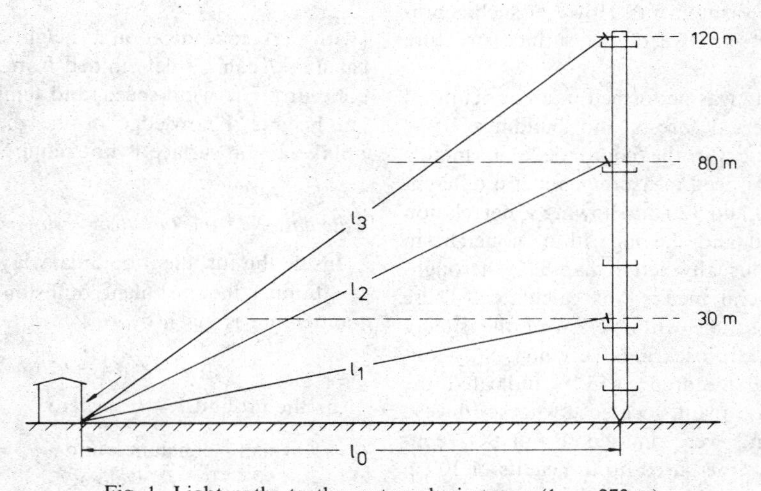

Fig. 1. Light paths to the meteorologic tower (l_0 = 273 m).

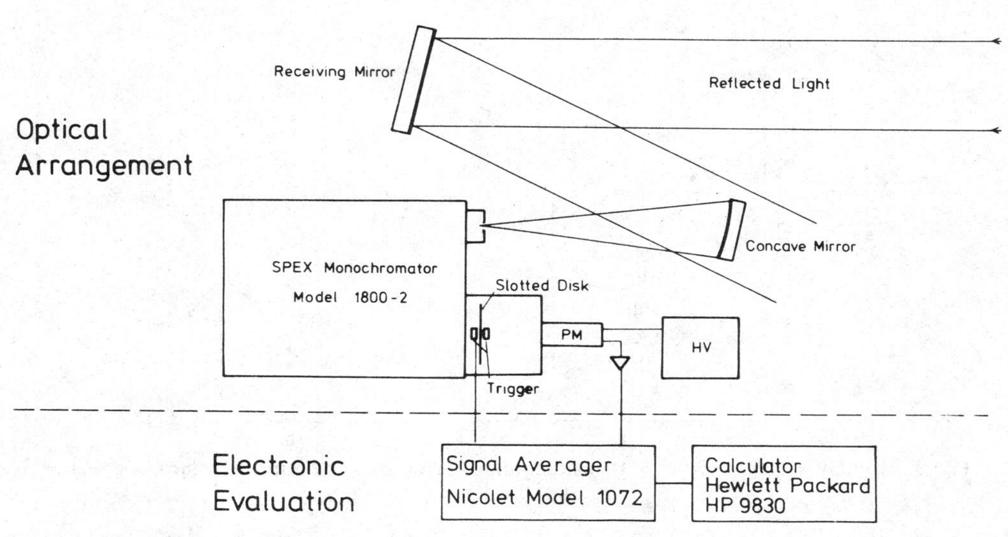

Fig. 2. Setup for SO_2 concentration measurement by optical absorption spectroscopy.

measured light intensities, consequently a continuous light absorption of other atmospheric constitutents does not affect the measurements.

The precision of the optical SO_2 measurements depends on the measuring-time. Averaging over 100 s of measuring-time, the error will be less than 1 $\mu g\ m^{-3}$.

OPTICALLY MEASURED CONCENTRATION PROFILES

In order to measure an SO_2 concentration profile, the light beam is sequentially directed to the mirrors at 30, 80, and 120 m height. This results in averaged concentrations for the three height intervals from 0 to 30 m, 0 to 80 m, and 0 to 120 m, respectively. By differencing the average concentration between two platforms can be calculated. Assuming a linear variation of the concentration between two platforms, the average value between two platforms can be taken as the concentration in the middle of that interval.

COMPARISON WITH CHEMICAL MEASUREMENTS

In addition to the optical profile measurements, on the platforms at 30, 50, 80 m height, chemical SO_2 concentration measurements (after West and Gaeke 1956) are carried out. To obtain a sufficient accuracy in less than one hour of sampling time, a special absorption-device is used, which allows an air-flow of 3 m^3 air h^{-1}. By remote-control the sampling devices are started and stopped from the ground, thus achieving simultaneous measurements at the three levels.

Comparison with the results of the optical method shows differences of simultaneously measured concentrations of generally less than $\pm 20\%$.

The optical method offers several advantages over the customary sampling techniques. The most important advantage is that wall adsorption problems are avoided by the optical measurements, others are short analysis time (about 100 s) and the possibility of real time measurement. The method has a slight disadvantage for the present application because it averages the concentration over height intervals rather than measuring it at distinct levels.

THE SO_2 MEASUREMENTS

Between May 1975 and September 1976 about 430 SO_2 concentration profiles were measured, among them 33 profiles with simultaneous measurements using the chemical method. All measurements were made in daylight. The optically measured profiles are usually averaged over several individual profile measurements covering a time interval of 1–2 h.

Figure 3 shows the monthly averages of the SO_2 concentration. During May to July generally low concentrations (10–40 $\mu g\ m^{-3}$) were observed, higher values during winter (60–80 $\mu g\ m^{-3}$). The error bars in Fig. 3 indicate the S.D. of the data to show the spread of the measured values. The error of the mean would be smaller by about a factor 2–5. From the concentration measurements at the three heights, the concentration gradient was calculated by plotting the measured concentrations vs the logarithm of the height, and fitting a straight line through the three points. The histograms in Fig. 4 show the frequency distributions of the concentration gradients normalized to the lowest measuring level. The distributions are given for unstable, neutral and stable conditions, and inversions respectively. Under unstable conditions ($\Delta\theta < 0$) the average gradient is close to zero ($+8\%$) showing a variance ($\pm 30\%$), which cannot be explained by the error of the concentration measurements, which is about 10%. This fluctuation could be caused by lateral or temporal inhomogeneities of the SO_2 concentration, which can simulate vertical concentration gradients, because the optical measure-

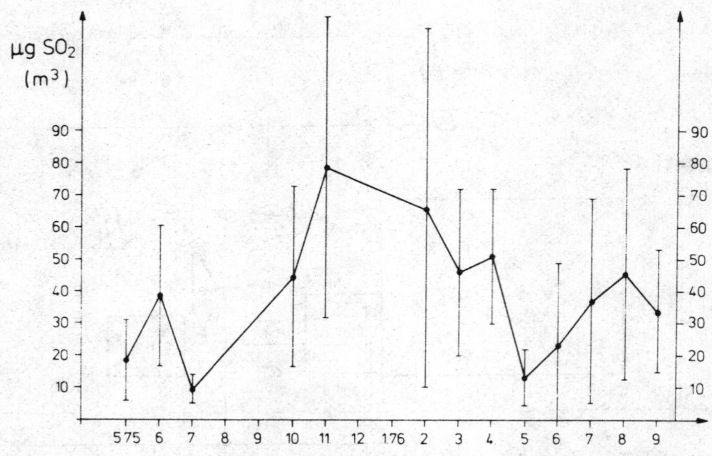

Fig. 3. Monthly averages of the SO_2 concentration. Error bars indicate the standard deviation of the data.

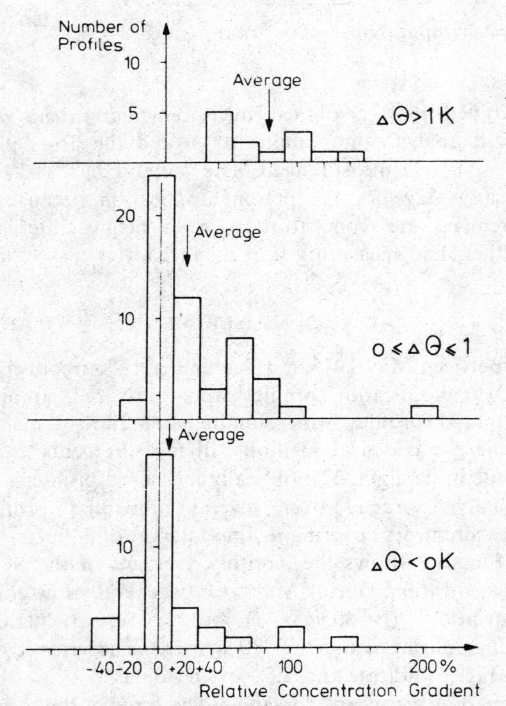

Fig. 4. Frequency distribution of the SO_2 concentration gradients.

ments at the different heights are carried out sequentially. Another source of error could be due to the turbulent transport mechanism itself, because the measuring time is comparable with the time a parcel of air would need to move through the boundary layer. Under neutral and stable conditions ($0 \leq \Delta\theta \leq 1$) the average gradient is about 22% which is caused by the increased turbulent transfer resistance R_T under such conditions. During inversions ($\Delta\theta > 1K$) only positive gradients were observed. At temperature gradients $> 1K/100$ m turbulent transfer becomes very small and to maintain an SO_2 flux at all, high concentration gradients are required.

AVERAGE VERTICAL FLUXES

From the concentration gradient the vertical flux of SO_2 was calculated according to Equation 1. Since under unstable conditions the turbulent transfer resistance and therefore the concentration gradient as well are very small compared to the total resistance, only the profiles measured under neutral and stable conditions are selected for the calculation of the monthly averages of the vertical flux. The vertical flux (Fig. 5) shows a maximum during winter, probably due to the increased SO_2 emission and concentration during that season.

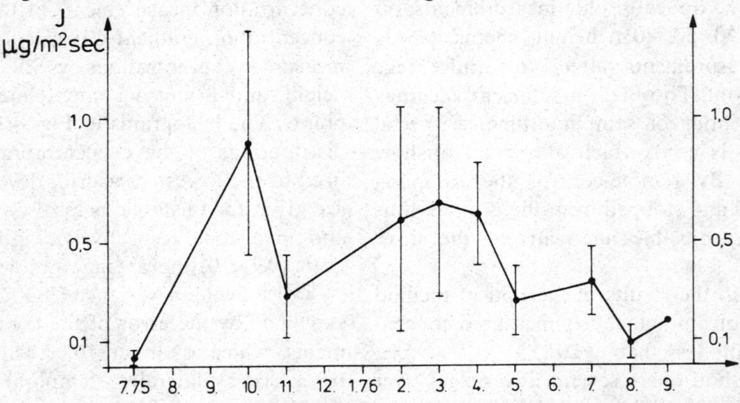

Fig. 5. Monthly averages of the vertical flux of SO_2.

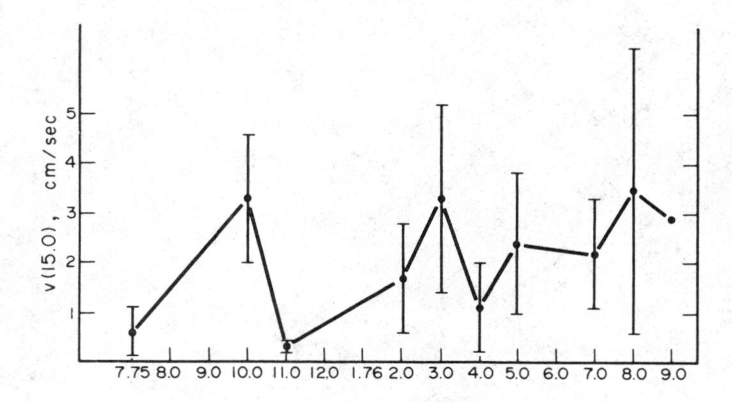

Fig. 6. Monthly averages of the transfer velocity between 15 m height and the surface.

Table 1. Average values of the turbulent transfer resistance $R(100, 15)$, grounding resistance ($R_{15,0} = 1/v_{15,0}$), fractions of the turbulent transfer resistance from the total resistance, and the vertical flux of SO_2 during summer 1975, winter 1975/1976, and summer 1976

Time	$R_{100,15}$	$R_{15,0}$	$\frac{R_{100,15}}{R_{100,15} + R_{15,0}}$ (%)	J $\mu g\,m^{-2}\,s^{-1}$
May–October 1975	4.8	63	7.1	0.52
November 1975–March 1976	9.1	48	16.0	0.41
April–August 1976	8.3	143	5.5	0.18

Figure 6 shows the monthly averages of the deposition velocity $v_{15,0}$ calculated from the vertical flux and the concentration at the 15 m level according to Equation 5. The results show no significant seasonal variations. Expected variations with the vegetation period or dependence on wet or dry surface are not observed. That might be caused by the fact that the decrease of surface resistance during the winter season due to moist surface is compensated by a smaller effective surface area during that season, since the vegetation has lost its leaves.

In Table 1 the seasonal average values of the turbulent transfer resistance $R_{100,15}$ for momentum (which is assumed to apply for SO_2 as well) is given together with the deposition resistance $R_{15,0} = 1/v_{15,0}$ for SO_2. The transfer resistance for momentum $R_{m15,0}$ is about an order of magnitude smaller.

REFERENCES

Garland J. A., Atkins, Readings Caughey (1974) Deposition of gaseous sulphur dioxide to the ground. *Atmospheric Environment* **8**, 75.

Garland J. A. and Branson J. R. (1976) The mixing height and mass balance of SO_2 in the atmosphere above Great Britain. *Atmospheric Environment* **10**, 353.

Lumley J. L. and Panofsky H. A. (1964) *The Structure of Atmospheric Turbulence*. John Wiley & Sons, New York. pp. 99.

Roth R. (1975) Der vertikale transport von luftbeimengungen in der Prandtl-Schicht und die deposition-velocity. *Met. Rdsch.* **28**, 65–71.

Shepherd J. G. (1974) Measurements of the direct deposition of sulphur dioxide onto grass and water by the profile method. *Atmospheric Environment* **8**, 69.

Sutton O. G. (1953) *Micrometeorology*, McGraw-Hill, New York.

West P. W. and Gaeke G. C. (1956) Fixation of sulfur dioxide as disulfitomercurate and subsequent colorimetric estimation. *Analyt. Chem.* **28**, 1816.

Whelpdale D. M. and Shaw R. W. (1974) Sulphur dioxide removal by turbulent transfer over grass, snow, and water surfaces. *Tellus* **26**, 196.

Atmospheric Environment Vol. 12, pp. 369–373. Pergamon Press 1978. Printed in Great Britain.

DRY DEPOSITION OF SO_2 ON AGRICULTURAL CROPS

D. FOWLER

Institute of Terrestial Ecology, Bush Estate, Edinburgh EH26 0QB, U.K.

(*First received 9 June* 1977 *and in final form* 26 *August* 1977)

Abstract—A micrometeorological method has been used to estimate dry deposition rates of sulphur dioxide on agricultural crops from vertical gradients of SO_2 concentration, windspeed and air temperature above the crop surface. Field measurements in a wide range of atmospheric and surface conditions enabled analysis of the results to separate the various atmospheric and surface processes controlling the flux. For wheat, deposition velocity (*Vg*) 1 m above the surface varied between 0.1 and 1.5 cm s^{-1} and was controlled primarily by surface processes, surface resistance generally contributing 70% of total resistance (*rt*). Values of surface resistance are determined essentially by deposition at two sinks, the sub-stomatal cavity, and leaf cuticle acting in parallel. With stomata open 1/3 of the total SO_2 flux was to the leaf surface and 2/3 to the sub-stomatal cavity. When foliage was wet with rain or dew, provided the pH of the liquid was >3.5 surface resistance is negligible and *Vg* then controlled by atmospheric resistance may exceed 1 cm s^{-1}. For agricultural areas of Britain (14×10^6 ha) dry deposition has been estimated at 72 kg SO_2 ha^{-1} annually, 60% of which is deposited during the winter (October–March, inclusive), equivalent to a deposition velocity of 0.6 cm s^{-1} for the area considered.

INTRODUCTION

During the last five years an increasing interest in the removal of sulphur from the atmosphere has resulted in a number of direct measurements of rates of dry deposition of SO_2 on vegetation (Garland *et al.*, 1973; Shepherd, 1974; Owers and Powell, 1974). The measurements indicated rates of deposition, expressed as velocities of deposition (*Vg*), generally between 0.5 to 1.0 cm s^{-1} for grass. In most instances measurements were few and restricted to a small range of surface and atmospheric conditions. Smith and Jeffrey (1975), suggested a mean deposition rate of 0.8 cm s^{-1} for Britain as a whole and this value has been widely accepted.

Assuming the general applicability of a constant velocity of deposition of 0.8 cm s^{-1} some workers have applied it to large areas of countryside not always dominated by grass, and used it to calculate regional sulphur budgets over protracted periods (Dovland *et al.*, 1976; Fisher, 1975). While mean deposition rates are valuable for sulphur budget analysis they provide little insight to the mechanism of deposition and the acceptance of a single value may lead to large errors in estimating sulphur deposition in particular circumstances.

To check the possible errors incurred by others, dry deposition on cereal crops was assessed and the component processes identified. Results of an extensive series of field measurements are then used to show how better estimates of dry deposition on short vegetation may be made from a few easily obtained surface and atmospheric characteristics.

THEORY

Above suitable horizontal and uniformly rough vegetation the one dimensional flux of SO_2, CO_2, H_2O or any other gas with a source or sink at the surface is proportional to the vertical gradient in volumetric concentration in turbulent boundary layer above such surfaces, the constant of proportionality being called the eddy diffusivity.

i.e. for SO_2

$$F_s = -K_s(Z)\,\delta\chi/\delta Z \qquad (1)$$

F_s flux density for SO_2 $\mu g\ m^{-2}\ s^{-1}$

K_s eddy diffusivity for SO_2 $m^2\ s^{-1}$

χ_s concentration of SO_2 $\mu g\ m^{-3}$

Z height above surface, m.

Using micrometereological techniques described elsewhere (Fowler and Unsworth, in preparation) eddy diffusivity for SO_2, assumed identical to that for heat may be estimated from measurements of vertical gradients of air temperature and wind speed above the surface by the aerodynamic method (Monteith, 1973). Measurements of vertical gradients in SO_2 concentration over the same surface enable a measurement of the flux of SO_2 to the surface to be made (Equation 1).

To study individual steps in the deposition process, wind, temperature and SO_2 gradient data may be further analysed in terms of a resistance analogy.

Integrating (1) with respect to height between heights Z_1 and Z_2 yields

$$F_S = \frac{\chi s(Z_2) - \chi s(Z_1)}{\int_{Z_1}^{Z_2} \frac{dZ}{K}} \qquad (2)$$

a form analogous to Ohm's law (flux = potential difference/resistance). If the lower limit of the integral is the surface and assuming that the surface concentration is zero, Equation 2 shows that the total resistance r_t (defined as ${}_{z_1}\int^{z_2} 2\,dz/K$) to SO_2 transfer between a height Z and the surface is

$$r_t = \chi_s(Z)/F_s \qquad (3)$$

The reciprocal of total resistance has dimensions of velocity and is identical to the velocity of deposition (Vg). SO_2 molecules absorbed by the surface must first overcome a resistance to transfer through the turbulent boundary layer, generally estimated from windspeed and temperature gradients. Making assumptions similar to those for Equations 2 and 3

$$r_a(Z) = U(z)/U_*^2 \quad (4)$$

Where U is horizontal windspeed, m s^{-1}, and U_* is friction velocity, m s^{-1}. Because normal pressure or bluff-body forces acting on the roughness of elements of the surface augment momentum transfer by molecular diffusion, whereas simultaneous transfer of SO_2 molecules is restricted to molecular diffusion, the aerodynamic resistance to momentum transfer to a rough surface (r_a) is smaller than the aerodynamic resistance to SO_2 deposition on the same surface. In practice a correction for this is made by adding a further boundary layer resistance (r_b) in series with r_a

$$r_a = (B\,U_*)^{-1} \quad (5)$$

where B is the dimensionless sub-layer Stanton number, (Owen and Thompson, 1963). Experimental evidence suggests that the factor B^{-1} varies little with windspeed or plant geometry and Chamberlain (1974) concluded that an appropriate value was 7 for SO_2 and 5 for water vapour.

If the total resistance r_t is found to be equal to $r_{(a+b)}$ then the surface is behaving as a perfect sink and all molecules arriving at the surface are being absorbed. The oceans are an example of this situation, where deposition rates are controlled by atmospheric rather than surface processes. If a residual resistance is found, it is termed the surface or canopy resistance r_c and its value is determined by the affinity of the different sinks for the gas acting together. For a simple vegetative canopy comprising one layer of uniform vegetation above bare soil there are three sites at the surface for absorption of SO_2 molecules, the sub-stomatal cavity, the leaf cuticle and the soil surface. These can be considered as resistances in parallel, rc_1, rc_2 and rc_3 respectively, the total of these resistances at the surface rc acting in series with $r_{(a+b)}$ (Fig. 1).

EXPERIMENTAL

By positioning instrument masts in a predominantly farmland area about 10 km SW of the city of Nottingham, U.K., they were surrounded on most sides with a uniform wheat crop extending for more than 200 m, i.e. there was adequate fetch for measurements up to a height of 2 m for most wind directions (Monteith, 1973). Within the 2 m depth of the turbulent boundary layer above the crop the following measurements were made.

(1) SO_2 concentrations at 5 levels, using the bubbler method described by Garland (1974).

(2) Air temperature at 7 levels using a differential thermocouple system (Biscoe *et al.*, 1974).

(3) Windspeed at 6 levels using sensitive cup-anemometers, three mounted in each of two masts to avoid mutual interference.

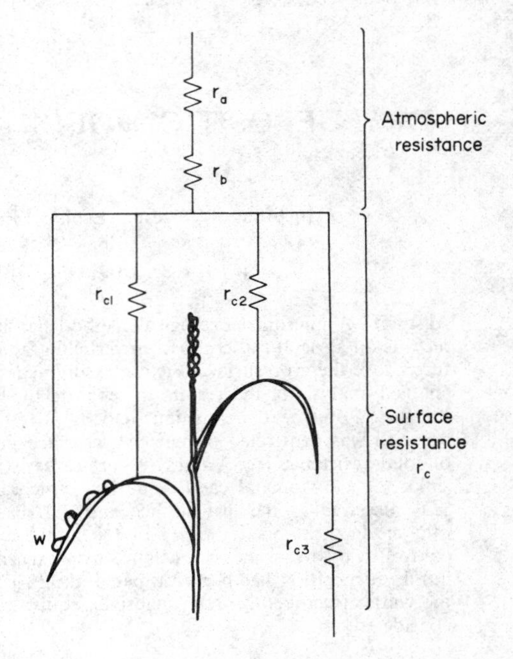

		Average value
r_a	aerodynamic resistance	0.25 s cm^{-1}
r_b	extra boundary layer "	0.25 s cm^{-1}
r_{c1}	stomatal component	1.0 s cm^{-1}
r_{c2}	leaf surface (cuticle)	2.5 s cm^{-1}
r_{c3}	soil component	10.0 s cm^{-1}
w,	for wet surface r_c = 0	

Fig. 1. Resistances to SO_2 deposition on a cereal crop indicating the relative importance of the three main sites for SO_2 uptake.

Gradients of SO_2 concentration, windspeed and air temperature averaged over periods of 60 to 120 minutes were used to estimate the flux of SO_2 to the wheat crop and the total and aerodynamic resistances to deposition. Details of the methods used are given by Fowler (1976).

RESULTS AND DISCUSSION

If the mechanism of SO_2 uptake at the surface were independent of SO_2 concentration, then the velocity of deposition should also be independent of concentration; in this instance the latter might be used to represent rates of SO_2 deposition rather than the measured vertical flux. Using Vg, the influence of changing surface and atmospheric conditions may be observed without the confounding effects of changes in SO_2 concentrations. From a total of 84 estimates, made during 1973 and 1974, Vg was not found to be related to ambient SO_2 concentrations in the range of 10 to 150 μg m^{-3}.

The values of Vg were found to vary diurnally (Figs. 2 and 3). Daytime values of Vg during the 8th, 9th May 1974 ranged from 0.8 to 1.5 cm s^{-1} and exceeded those at night-time by factors of 3 to 5. From measurements of soil water deficit and net radiation,

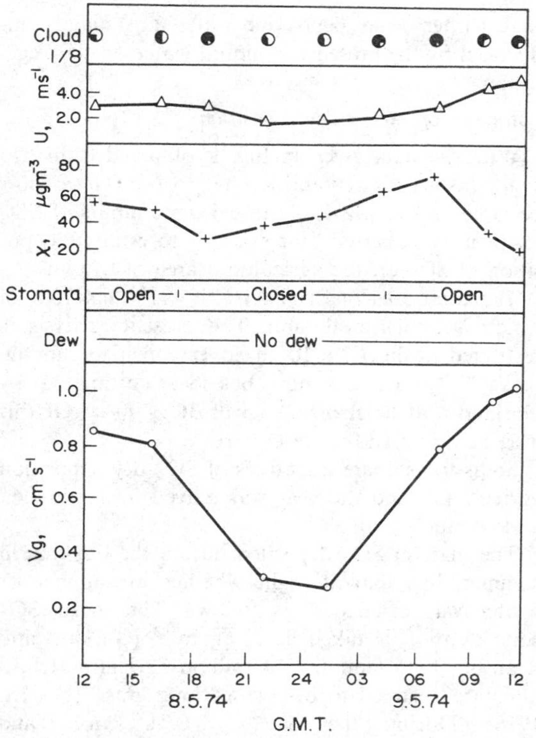

Fig. 2. Dry deposition rates of SO_2 on wheat and other surface and atmospheric variables, during the period of rapid vegetative growth.

made concurrently with the SO_2 flux measurements, it was concluded that the stomata of the crop were (i) open during the period 1200–1600 (all times are G.M.T.) on the 8th and 0800–1200 on the 9th, and (ii) closed during the hours of darkness 2200–0300. Later in the growing season (24–25 July) when the plants were beginning to senesce, deposition velocities were unexpectedly larger at night than during the day (Fig. 3). By this stage in crop development, with stomata closed permanently and leaves dying, daytime deposition velocities of 0.3–0.4 cm s^{-1} were similar to night-time values earlier in the season (e.g. on the 8–9 May 1974). The larger values during the night occurred in the presence of a layer of dew on the foliage.

Early in the growing season, the large daytime deposition velocities are probably due to SO_2 absorption in sub-stomatal cavities which are not open to the atmosphere during the night when smaller deposition rates were measured, a set of observations consistent with the conclusions of Spedding (1969) who measured rates of uptake of SO_2 by barley leaves in laboratory experiments.

Only small amounts of SO_2 are deposited on soil beneath the crop canopy (Fowler and Unsworth, in preparation). The flux during the night to a dry crop is therefore to cuticular material on leaf (and stem) surfaces. Later, as crops mature and stomata fail to open, SO_2 'uptake' is primarily by the leaf cuticle both day and night. As plant cuticles are less efficient sinks than the sub-stomatal cavities, rates of deposition decrease (0.3 to 0.4 cm s^{-1}). If however the surface is wetted with dew then canopy resistance decreases to zero. If the quantity of dew is small the ionization and oxidation of dissolved SO_2 (to yield H^+ ions) may lower the pH of the surface moisture to an acidity at which the solubility of SO_2 is significantly decreased (Liss, 1971).

Contrary to the general view it seems that changing surface and atmospheric conditions cause Vg to range from 0.3 to 1.5 cm s^{-1}. Thus a small number of measurements in a restricted range of surface and atmospheric conditions would only fortuitously yield a mean deposition velocity applicable for annual estimates of SO_2 deposition over the whole country. Further, it seems that dry deposition is usually controlled by the affinity of the surface for SO_2 rather than the conductivity of the atmosphere.

When data for two field seasons were pooled r_c accounted for 65% of r_t on average. Additionally there appear to be two main sites for SO_2 uptake in plant canopies, the sub-stomatal cavities and plant cuticles. With stomata open the sub-stomatal cavity is the major sink for SO_2 and for this reason rates of dry deposition of SO_2 are closely linked to stomatal behaviour except in the presence of a liquid water on the leaves which effectively short-circuits the deposition process (Fig. 1).

For the greater part of the growing season of an annual crop, rates of uptake by leaf cuticles are fairly

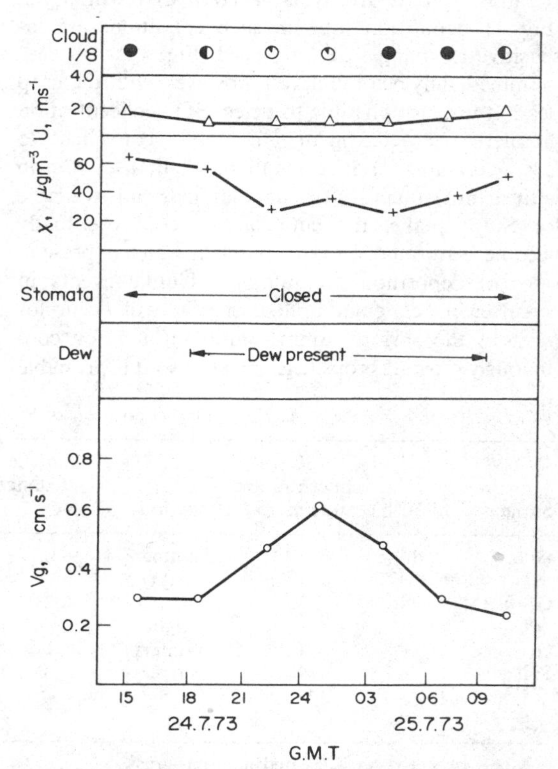

Fig. 3. Dry deposition rates of SO_2 on wheat and other surface and atmospheric variables, during senescence of the crop.

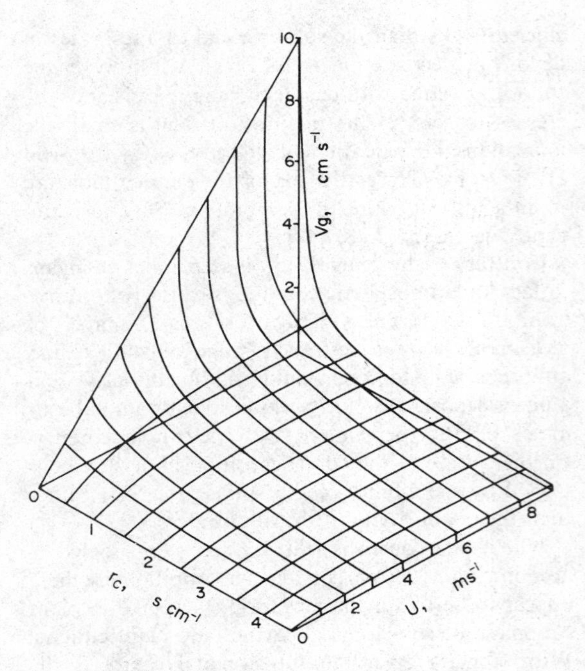

Fig. 4. Predicted variation of deposition velocity with windspeed and canopy resistance for a crop height of 1.0 m.

constant. When leaf surfaces are dry and stomata are closed, the deposition velocity is 0.3 to 0.4 cm s^{-1}. Because r_{c2} is relatively large compared with $r_{(a+b)}$ (Fig. 1), deposition rates in such conditions are insensitive to changes in windspeed (Fig. 4).

Immediately before harvest, however, with the crop dead, it was not possible to detect SO_2 concentration gradients (i.e. $<0.5\ \mu g\ m^{-3}\ m^{-1}$ and equivalent to a $Vg < 0.1$ cm s^{-1}). It seems therefore that for a crop with a finite quantity of cuticular material available for SO_2 uptake, the cuticular material eventually becomes saturated, and except when water is present, rates of deposition are minimal. Unlike events in cereal crops, cuticular uptake of SO_2 will occur for most of the spring, summer and autumn for continuously growing crops (e.g. grass). Also it is probable that winter time deposition rates are mainly influenced by the presence of liquid water on the vegetation.

Estimates of annual dry deposition

With the data given in Fig. 1, obtained from two field seasons, measurements over wheat, it is now possible more realistically to calculate inputs of SO_2 for protracted periods, for example to estimate deposition of SO_2 on the agricultural area of Britain.

The total area of Britain is 2.7×10^7 ha, 73% of which is agricultural land. The present analysis is restricted to the 1.7×10^7 ha over which the annual mean SO_2 concentration has been estimated by Garland and Branson (1976) at 36 $\mu g\ m^{-3}$. Of this area 82% is used for agriculture.

So as to estimate quantities of SO_2 dry deposition agricultural land the year was halved into 'summer' and 'winter'.

The mass of SO_2 deposited during the daytime in summer, like that at night whether in summer or winter was calculated as follows. The mean SO_2 concentration is taken as 30 $\mu g\ m^{-3}$ (Garland and Branson, 1976) and the 14 million agricultural hectares are assumed to behave as 'long grass' (Fowler, 1976). During the day $r_c = 0.7$ s cm^{-1} and $r_{(a+b)} = 0.5$ s cm^{-1}, thus $Vg = 0.8$ cm s^{-1}. Over a 6 month period taking a mean day length of 12 hours this yields a daytime deposition of 19 kg SO_2 ha^{-1}, or 2.7×10^5 tonnes of SO_2 over the entire area.

Table 1 summarizes this example along with estimates of deposition during other periods and to other surfaces. The table shows that the estimated annual dry deposition on agricultural districts is about 1 million tonnes of SO_2, which compares with 0.6 million tonnes of SO_2 (equivalent) deposited by rain on the same area, assuming homogeneous distribution of rain over agricultural and non-agricultural areas. The 1 million tonnes of SO_2 deposited annually is equivalent to a mean Vg of 0.6 cm s^{-1}, smaller but of a similar order to the widely accepted value of 0.8 cm s^{-1}.

There are many uncertainties in the analysis due partly to the difficulty in generalising a very variable

Table 1. Dry deposition of SO_2 on agricultural areas of Britain

Surface	$\times 10^6$ha	SO_2 concentration ($\mu g\ m^{-3}$)	Period	Surface resistance (s cm^{-1})	$r_{(a+b)}$ (s cm^{-1})	Vg* (cm s^{-1})	Flux (kg ha^{-1})	Mass of SO_2 deposited $\times 10^6$ tonnes of SO_2
Grass*	14	30	Summer-day†	0.7	0.5	0.8	19	0.27
Grass	14	30	Summer-night	2.0	0.5	0.4	9.5	0.14
							Summer total	0.41
Grass	10	42	Winter‡	1.5	0.5	0.5	34	0.32
Soil	4	42	Winter	0§	1.0	1.0	66	0.28
							Winter total	0.60
							Annual total	1.01

* The 14×10^6 ha is treated as 'long grass'.
† 12 h per day, 183 days.
‡ 24 h, 183 days.
§ Soil assumed wet, data from Fowler (1976).

quantity like the countryside. In particular no account in the present analysis is made of the effect of hedgerows nor has the influence of dew and rain been satisfactorily incorporated in the estimates of deposition. Thus it seems that the widely accepted value of $0.8\ \mathrm{cm\ s^{-1}}$ provides a reasonable estimate of annual deposition for countryside dominated by grass. The same may not be true of forested areas for which the components of resistance have yet to be evaluated. For the shorter term, however, much better dry deposition estimates may be made from a knowledge of stomatal activity, SO_2 concentration and appropriate use of the components of resistance given in Fig. 1.

Also the use of a resistance analogy to separate the processes governing rates of SO_2 uptake by vegetation provides a method for studying the immediate fate of deposited SO_2, of value to parallel studies of the effects of atmospheric sulphur compounds on plants.

Acknowledgements—This work forming part of a study of SO_2 deposition on agricultural crops in the environmental physics unit at the University of Nottingham, was supported financially by the Central Electricity Generating Board.

REFERENCES

Biscoe P. V., Clark J. A., Gregson K., McGowan M., Monteith J. L. and Scott R. K. (1975) Barley and its environment. *J. appl. Ecol.* **12**, 227–259.

Chamberlain A. C. (1974) In *Heat and Mass Transfer in the Biosphere.* Part 2 (edited by De Vries D. and Afgan N. pp. 561–583. J. Wiley, New York.

Dovland H., Joranger E. and Semb A. (1976) In *Impact of Acid Precipitation on Forest and Freshwater Ecosystems in Norway.* Research Report 6/76 of S.N.S.F. Project, N.I.S.K., 1432 Aas-NLH, Norway, pp. 15–35.

Fowler D. (1976) Ph.D. Thesis, University of Nottingham.

Fowler D. and Unsworth M. H. (1974) Dry deposition of sulphur dioxide on wheat. *Nature* **249**, 389–390.

Fowler D. and Unsworth M. H. (1977) In preparation.

Fisher B. E. A. (1975) The long range transport of sulphur dioxide. *Atmospheric Environment* **9**, 1063–1070.

Garland J. A., Clough W. S. and Fowler D. (1973). Deposition of sulphur dioxide on grass. *Nature* **242**, 256–257.

Garland J. A. (1974) Dry deposition of SO_2 and other gases. Proceedings of the Conference on Deposition and Resuspension of Atmospheric Pollutants, Richland, September 1974.

Garland J. A. and Branson J. R. (1976) The mixing height and mass balance of SO_2 in the atmosphere above Great Britain. *Atmospheric Environment* **10**, 353–362.

Liss P. S. (1971) Exchange of SO_2 between the atmosphere and natural waters. *Nature* **237**, 327–329.

Monteith J. L. (1973) *Principles of Environmental Physics.* Edward Arnold, London.

Owen P. R. and Thompson W. R. (1963) Heat transfer across rough surfaces. *J. Fluid Mech.* **15**, 321–324.

Owers M. J. and Powell A. W. (1974) Deposition velocity of SO_2 on land and water surfaces using a ^{35}S tracer method. *Atmospheric Environment* **8**, 63–67.

Spedding D. J. (1969) Uptake of sulphur dioxide by barley leaves at low sulphur dioxide concentrations. *Nature* **224**, 1229–1230.

Smith F. B. and Jeffrey G. H. (1975) Airborne transport of SO_2 from U.K. *Atmospheric Environment* **9**, 643–659.

Shepherd J. G. (1974) Measurements of the direct deposition of sulphur dioxide onto grass and water by the profile method. *Atmospheric Environment* **8**, 69–74.

Atmospheric Environment Vol. 12, pp. 375–377. Pergamon Press 1978. Printed in Great Britain.

INPUT OF ATMOSPHERIC SULFUR BY DRY AND WET DEPOSITION TO TWO CENTRAL EUROPEAN FOREST ECOSYSTEMS

R. MAYER and B. ULRICH

Institut für Bodenkunde und Waldernährung der Universität Göttingen,
Büsgenweg 2, D-3400 Göttingen, F. R. Germany

(*First received* 13 *June* 1977)

Abstract—The total input of atmospheric sulfur to a beech and a spruce forest in the Federal Republic of Germany has been measured over a period of 6 years. The contribution of dry deposition to the total input was determined indirectly by comparing seasonal changes in the sulfur flux coupled with precipitation beneath the canopy of the deciduous beech forest. As a result of these investigations seasonal and annual sulfur fluxes are reported corresponding to removal rates of atmospheric sulfur. The experimental data show clearly that the removal rates depend upon the quality of the atmosphere/land interface, in forested areas from the tree species forming the canopy. The 6-years average of total deposition on bare soil is 23 kg S ha^{-1} y^{-1}, on a beech forest 47–51 kg S ha^{-1} y^{-1}, on a spruce forest 80–86 kg S ha^{-1} y^{-1}. Based upon the experimental results the role of the forest vegetation in the removal of sulfur from the atmosphere in the area of the Federal Republic of Germany is considered. The figures indicate, that at least 50% of the total sulfur deposition takes place on forested areas which cover only 28% of the total land surface.

INTRODUCTION

Deposition of most ion species, including SO_4^{2-}, measured in the precipitation below the canopy of a forest stand exceeds that measured above the canopy. The increase in the element flux during canopy passage of precipitation is due to dissolution of substances when plant surfaces (leaves, needles, bark) are wetted, or to dry fallout reaching the forest floor subsequently being dissolved in the rain water within the collector. The origin of these substances has been attributed to biological and meteorological sources:

(1) Metabolites from the internal turnover of the trees as, e.g. solutes carried by the transpiration stream to the leaf surfaces, substances resulting from assimilation and respiration processes, or litter particles being decomposed.

(2) Dry deposition of airborne particles, aerosols and gases. This flux depends upon the deposition velocity of substances present in the atmosphere which is controlled by the surface resistance of the canopy, i.e. its physical structure and surface conditions. It has therefore been called 'filtering' by ecologists, since the vegetation cover acts as a filter for atmospheric substances, and these may contribute significantly to the element balance of an ecosystem.

DESCRIPTION OF SITES AND METHODS

The data presented here are resulting from investigations on the element balance of forest ecosystems in central Europe. Research sites are a beech forest (*Fagus silvatica*, age: 125 years) and a spruce forest (*Picea abies*, age: 85 years) located at an altitude of about 500 m a.s.l. in a densely forested mountainous area without large industries or cities about 40 km northwest of Göttingen. Mean annual precipitation is about 1000 mm. Average SO_2-concentrations in the atmosphere range from 5 to 10 μg SO_2 m^{-3} during summer and from 10–20 μg SO_2 m^{-3} during winter. A detailed site description is given by Ellenberg (1971), the methods for measuring the amount and chemical composition of precipitation were: (a) 15 raingauges below the canopy of both forest stands and on a non-forested area close to the forest sites, (b) 3 rain-gutters of 1.5 m^2 surface area below the canopy of both forest stands, (c) stemflow collectors fixed around the beech stems on a total area of 800 m^2.

No measurements were taken to separate wet and dry deposition, but raingauges under the canopy collecting precipitation for chemical analysis were covered by a screen to keep out coarse litter. During winter larger collectors were used to get snow samples.

Samples were taken from the collectors after each precipitation period and stored in a deep-freezer. Chemical analysis was done on monthly samples. Sulfur was determined after separation of phosphate, fluoride and Fe^{3+} as SO_4^{2-} by colorimetry on a Technicon Auto Analyzer using SPADNS as complexing agent. The investigations have been made without interruption since November 1968.

RESULTS

In Table 1 the deposition (wetfall plus dryfall) of S measured on the open plot and under canopy is

Table 1. Deposition of sulfur (dryfall plus wetfall) and precipitation

		1969 W	1969 S	1970 W	1970 S	1971 W	1971 S	1972 W	1972 S	1973 W	1973 S	1974 W	1974 S
BS on a non-forested area (bare soil)†	Kg Sha^{-1}	12.3	12.5	14.0	12.9	9.2	10.5	9.5	14.1	13.6	8.0	10.4	12.5
CP below beech canopy*	kg Sha^{-1}	39.2	33.8	27.6	25.0	18.1	16.9	24.7	22.5	21.9	21.0	21.8	32.2
CP below spruce canopy	kg Sha^{-1}	66.5	41.2	46.7	41.3	34.0	29.6	42.4	34.4	55.6	29.5	44.1	53.1
Precipitation	l. m^{-2}	436	530	730	681	399	370	346	580	435	450	386	602

* Stemflow included. W = Winter (November–April), S = Summer (May–October).
† Soil adsorption neglected.

given for different seasons and years. Sulfur was found to be nearly exclusively in the form of SO_4^{2-}.

The sulfur flux below the canopy considerably exceeds the input from the atmosphere to the non-forested area (bare soil). In the case of the beech forest the difference between both of these fluxes may be taken, during the winter months, as an additional input from the atmosphere. Leaching of metabolites should be negligible when the canopy is defoliated and the physiological activity of the beech as well as litter fall is largely reduced. The only source remaining for the observed flux increase is dry deposition from the atmosphere.

Mayer and Ulrich (1974) have estimated dry deposition ('filtering') during the vegetation period (May–October) by using the same ratio between total deposition on bare soil and additional dry deposition on the canopy for the winter as well as for the summer months. The same flux ratio was later applied to an evergreen spruce forest within the same area (Ulrich *et al.*, 1977). With this approach, assuming the same 'filter efficiency' of the canopy whether leaves are present or not, the additional dry deposition to the beech forest during the summer is most likely underestimated.

In the 6-year period under investigation the estimated input of sulfur by dry deposition accounted, during the vegetation period, for more than 75% of the sulfur flux increase during canopy passage of precipitation (Mayer and Ulrich, 1977). It follows, that, within the period May through October, leaching of metabolites from the canopy contributes less than 25% of the flux increase.

Systematic errors may lead to an overestimate of the input from the atmosphere when, during the winter months, decomposition of plant parts or litter particles occurs. In the case of sulfur this source of error is most probably negligible due to the relatively small S content of the biomass. The total annual litter fall of the beech forest (3.2 kg S ha^{-1} y^{-1}; Mayer and Ulrich, 1974) yields only about 6% of the S-flux coupled with precipitation below the canopy.

For the same reason it may be concluded, that uptake of S (or SO_2) through the stomata does not play an important role in the sulfur balance of the forest stands provided there is no transfer of sulfur from the leaves to the roots and to the soil against the transpiration stream—which again is very unlikely. The annual increment in the biomass of the beech forest is less than 1 kg S ha^{-1} y^{-1} (Mayer and Ulrich, 1974). Based upon this reasoning, the mean annual sulfur input from the atmosphere by wet and dry deposition, $IN_{t(a)}$, to the beech and the spruce stand within the period November 1968 through October 1974 was calculated in the following way:

(1) Maximum values:

$$IN_{t(a)} = CP_{(a)} = CP_{(w)} + CP_{(s)}$$

where $CP_{(a)}$ is the annual sulfur flux below the canopy. Contribution of internal turnover as well as net uptake of S by the leaves and translocation to the tree are assumed to be negligible.

(2) Minimum values:

$$IN_{t(a)} = IN_{t(w)} + IN_{t(s)}$$

with

$$IN_{t(w)} = CP_{(w)}$$

and

$$IN_{t(s)} = BS_{(s)} + 0.75\,(CP_{(s)} - BS_{(s)})$$

where the subscripts *s* and *w* indicate summer (May–October) and winter (November–April) periods, *CP* is the sulfur flux below canopy, *BS* is the sulfur flux on bare soil. The difference ($CP - BS$) gives the flux increase during canopy passage, 75% of which is taken as (minimum) dry deposition.

Table 2. Mean deposition of sulfur (wetfall plus dryfall), $IN_{t(a)}$, within the period November 1969 through October 1974

		kg S ha^{-1} y^{-1}	mg S m^{-2} d^{-1}
On a non-forested area (bare soil)		23.3	6.4
On a beech forest	maximum	50.8	13.9
	minimum	47.4	13.0
On a spruce forest	maximum	86.4	23.7
	minimum	79.8	21.9

Precipitation (Nov. '69–Oct. '74): 991 l. m^{-2}. 30-year average: 1030 l. m^{-2}

Results are given in Table 2. The data may be used for a rough estimate of the annual sulfur deposition on the area of the Federal Republic of Germany:

Total surface F.R.G.	24.7×10^6 ha
including	
non-forested area	17.7×10^6 ha
forested area	
—coniferous forests	4.2×10^6 ha
—deciduous forests	2.8×10^6 ha

If we take the mean deposition rates from Table 2 the annual sulfur deposition is calculated:

Deposition on non-forested areas	410×10^6 kg S
on forested areas	490×10^6 kg S
including	
—on coniferous forests	350×10^6 kg S
—on deciduous forests	140×10^6 kg S

It must be stated, that this estimate can only be approximative for a larger deposition is found close to the emittents (industries, cities), and the distinction of the total land surface into three surface types is an oversimplification. Nevertheless it may give a good idea of the relative importance of the forest in the removal of sulfur from the atmosphere. The total anthropogenic emission of sulfur within the Federal Republic of Germany in 1970 is estimated to 1800×10^6 kg (VDI-Berichte, 1972). Estimates of the annual emission of gaseous sulfur into the atmosphere by natural processes (mainly biological decay) are very uncertain. Taking the figures reported by Granat *et al.* (1976) we calculate 6×10^6 kg S from natural sources for the land surface of the Federal Republic of Germany, i.e. an amount negligible compared to the industrial emission.

REFERENCES

Ellenberg H. (1971) Integrated experimental ecology *Ecological Studies* **2**, 214.

Granat L., Rodhe H. and Hallberg R. O. (1976) The global sulphur cycle. *Ecological Bulletins/NFR (Stockholm)* **22**, 89–134.

Mayer R. (1971) Bioelement-Transport im Niederschlagswasser und in der Bodenlösung eines Wald-Ökosystems. *Gött. Bodenkundl. Ber.* **19**, 1–119.

Mayer R. (1972) Untersuchungen über die Freisetzung der Bioelemente aus der organischen Substanz der Humusauflage in einem Buchembestand. *Z. Pflanzenern. Bodenkunde* **131**, 261–273.

Mayer R. and Ulrich B. (1974) Conclusions on the filtering action of forests from ecosystem analysis. *Oecol. Plantarum* **9**, 157–168.

Mayer R. and Ulrich B. (1977) Acidity of precipitation as influenced by the filtering of atmospheric sulphur and nitrogen compounds—its role in the element balance and effect on soil. *Wat., Air, Soil Pollut.* **7**, 409–416.

Ulrich B., Mayer R., Khanna P. K., Seekamp G. and Fassbender H. W. (1977) Input, Output und interner Umsatz von chemischen Elementen in einem Buchen- und einem Fichtenbestand. *Ber. Tag. Ges. f. Ökologie, Göttingen, Sept. 1976*, 17–28.

VDI-Berichte (1972) Die Emission von Schwefelverbindungen. *Berichte des Vereins Deutscher Ingenieure* 186, 60 p.

Atmospheric Environment Vol. 12. pp. 379–387. Pergamon Press 1978. Printed in Great Britain.

PSEUDOSPECTRAL SIMULATION OF DRY DEPOSITION FROM A POINT SOURCE

RUWIM BERKOWICZ and LARS P. PRAHM

Danish Air Quality Laboratory, Ministry of Environment, Meteorological Institute, Lyngbyvej 100, DK-2100 Copenhagen, Denmark

(*First received* 9 *June* 1977 *and in final form* 27 *September* 1977)

Abstract—A pseudospectral, two-dimensional model for dispersion and dry deposition of atmospheric pollutants is developed on the basis of gradient-transfer theory (K-theory). A symmetrical transform is developed for the vertical direction satisfying the condition of mass conservation. The deposition to the ground is represented by a sink term at the surface, and according to this the model is called the surface depletion model. Comparison with analytical solutions is performed in the case of constant wind and diffusivity profiles. Agreement between numerical and analytical results is within 2–5% even with only 17 grid points in the vertical direction. The error can be further reduced by application of more grid points. The pseudospectral method is more accurate than finite difference methods, especially with respect to advection. Multiple sources and time-dependent physically realistic, e.g. measured wind and diffusivity profiles can easily be treated. Sources between gridpoints can be accurately represented.

The pseudospectral model is used for calculation of deposition rates with diffusivity and wind profiles for different atmospheric stability conditions. Comparison is made with the conventional, Gaussian source depletion method for estimates of dry deposition from a point source. The discrepancy between the two models increases with increasing atmospheric stability. In the stable case with a deposition velocity of $1\ \text{cm}\ \text{s}^{-1}$ and a point source at a height of 25 m, the Gaussian source depletion model underestimates the suspension ratio by a factor of 1.5 at a downwind distance of 22 km from the source. The surface concentration is overestimated by nearly a factor of 2 at the same distance. Contrary to other, more simple surface depletion models which do not take wind and diffusivity profiles into account, it is found that the suspension ratio is smaller for the surface depletion than the source depletion model at short distances, while the opposite relation occurs only at larger distances. This effect is ascribed to the low wind velocity at the surface which results in stronger deposition close to the source, while at larger distances the vertical diffusive transport becomes more important for the rate of dry deposition. The present results are especially relevant for dispersion and deposition in cases of low diffusivities, where the difference between the two-dimensional pseudospectral surface depletion model and other less sophisticated deposition models is most pronounced.

1. INTRODUCTION

Deposition of pollutants from local and distant sources is of increasing interest in relation to the effects of e.g. sulphur pollutants from urban areas, industrial areas and single power plants. But also deposition of other pollutants such as radioactive material from nuclear power plants in case of leaks has drawn attention.

Dispersion and dry deposition of material from a point source can be treated by the gradient transfer theory on the basis of the diffusion equation. Gravitational forces are neglected when only gases and small particles are considered. Dry deposition is represented by a boundary condition at the ground. Neglecting the dry deposition, simplified solutions of the diffusion equation leads to the Gaussian dispersion model, which is easy to handle and to compare with tracer studies. This model is usually applied in engineering approaches to dispersion from point sources. The loss by dry deposition was adapted to the Gaussian dispersion model (Chamberlain, 1953), as discussed by Van der Hoven (1968), referred to as the source depletion model. This method is easy to handle when the source reduction factors are available as function of source height, distance and atmospheric stability. The source depletion model does not treat the vertical diffusive transport in accordance with the gradient transfer theory and neglects the effect of variation of diffusivity and wind with height.

Monin (1958) gave an analytical solution of the diffusion equation including dry deposition, but he assumed constant diffusivity and wind. Inspired by Calder (1961), Smith (1962) developed analytical solutions with different simplified diffusivity profiles. Since then various analytical solutions were presented e.g. by Scriven and Fisher (1975). However, solutions based on empirical, time-dependent diffusivity and wind profiles cannot be given by analytical methods.

Horst (1977) recently reported on a surface depletion method based on the Gaussian dispersion model. He found that results from the less realistic source depletion model deviate significantly from the Gaussian surface depletion model. The source depletion model overestimates the total deposition between source and receptor at all distances from the source and consequently underpredicts the amount of remaining airborne material. The source depletion model consistently overpredicts the surface air concentration and the deposition at downwind locations close to the source and is, as a consequence, biased in opposite directions for locations far from the source. The largest differences appear for low sources in stable stratification over surfaces with relatively large deposition velocities. Some parameters can be in error by factors of 3–4 at downwind distances of 10 km. The analysis by Horst (1977) is, however, still limited by the assumption of a Gaussian dispersion. Draxler and Elliott (1977) solved the one-dimensional diffusion equation numerically by a finite difference method and found substantial differences between a non-Gaussian source and a surface depletion model. Considering the amount of airborne material after one day's travel during neutral stratification and with a deposition velocity of 3 cm s^{-1}, the source depletion model underestimates the total amount of airborne material with a factor of 10 compared with the surface depletion model. It is shown that the results are sensitive towards the diurnal variation of the stability, although Draxler and Elliott (1977) might overestimate this effect, using a velocity of deposition which is not a function of atmospheric stability.

Experimental investigation of the vertical SO_2 profile by Gotaas (1975), also, showed the importance of the use of gradient transfer theory. The inadequacy of the source depletion method and the serious effects of deposition of pollutants causes the need for future studies of deposition based on physical realistic meteorological parameters.

The present research introduces a two-dimensional gradient-transfer model for dry deposition including wind and diffusivity as a function of height. The model is based on the recently developed pseudospectral method, shown to be more accurate than finite difference methods, especially with respect to advection (Christensen and Prahm, 1976). This method was appled for regional dispersion by Prahm and Christensen (1977). The present study describes a pseudospectral method specially applicable for the boundary conditions considered here. Numerical accuracy of the model is tested by comparisons with analytical solutions of the diffusion equation in simplified cases. Comparison is performed in tests with constant diffusivity incuding deposition. Numerical tests including variable diffusivity but no deposition were compared with analytical solutions (Smith, 1957), giving satisfactory results. These tests will be reported in details elsewhere. The pseudospectral numerical method is here compared with the Gaussian source depletion model for different atmospheric stabilities. The effect of a realistic wind profile is discussed. The results of the models are evaluated on the basis of ground level air concentrations and the suspension ratio i.e. the fraction of material which remains airborne. Discussion of discrepancies between the results from the pseudospectral deposition model and other deposition models is presented. All formulas and results are normalized as usual, representing an infinite homogeneous cross-wind line source.

2. NUMERICAL METHOD

The mathematical basis of our approach to the dry deposition problem is the time-dependent, two dimensional dispersion equation

$$\frac{\partial c}{\partial t} = -u(z)\frac{\partial c}{\partial x} + \frac{\partial}{\partial z}\left(K(z)\frac{\partial c}{\partial z}\right) + Q(x, z) + S(x, z) \quad (1)$$

where c is the concentration of the airborne material, u is the horizontal wind velocity and $K(z)$ is the eddy diffusion coefficient. $Q(x, z)$ and $S(x, z)$ represent sources and sinks, respectively. Both u and K are usually dependent on the vertical coordinate. The first term on the right hand side of (1) describes advection in the downwind direction (the x-coordinate) and the second term describes the vertical eddy diffusion.

When the transported material is subject to deposition on the ground, the boundary conditions assumed is a vertical flux F_z at the ground proportional to the concentration at the ground level.

$$F_z = \left\{K(z)\frac{\partial c}{\partial z}\right\}_{z=0} = v_d c(z = 0). \quad (1a)$$

The factor of proportionality, v_d has dimensions of a velocity and is called the deposition velocity. Analytical solution of Equation (1) can be found only for some certain form of wind velocity and diffusion coefficient profiles (Smith, 1962). In general, only a numerical solution is available.

In order to obtain the numerical solution to (1) we shall employ the pseudospectral approximation. This method is described in detail by Christensen and Prahm (1976), and we shall here only briefly review the results and discuss the modifications necessary to account for the deposition effect. In the pseudospectral approximation the space derivatives on the right-hand side of (1) are computed by means of finite Fourier Transforms, that is, in spectral space, whereas the local products and the time integration are evaluated in physical space. In order to evaluate the z-component of the spatial derivatives we expand the concentration in the finite Fourier series

$$c(x, z, t) = \sum_k A(k, x, t)\, e^{ikz}. \quad (2)$$

The values of the 'wave vectors' k are given by

$$k_i = \frac{2\pi}{\Delta z N} n_i \quad (3)$$

where n_i assumes integer values within limits $-N/2 < n_i \leqq N/2$ and N is the number of the grid points. Δz is the spacing of the grid points.

The space derivatives can now be evaluated on grid points using (2)

$$\frac{\partial c}{\partial z} = \sum_k ik\, A(k, x, t)\, e^{ikz}. \quad (4)$$

Similar procedure can be used to evaluate the x-component of the derivatives.

The product $K(z)\partial c/\partial z$ is computed in the physical space and the derivative $\partial/\partial z(K(z)\partial c/\partial z)$ is again evaluated by applying the finite Fourier Transform. Fourier transformation requires periodical boundary conditions. In the physical problem treated here, the periodical boundary conditions are not exactly applicable. Let us first discuss the situation when the deposition velocity v_d is zero, i.e. the ground is unpenetrable for the diffusing material. From (1a) it follows that $K\partial c/\partial z = 0$ at the boundary. In order to satisfy this condition the term $K(z)\partial c/\partial z$ must be expanded in sine-series and this again implies that $c(z)$ should be represented by cosine-series. In fact it means that we can still use (2) but with symmetrical form for $c(z)$, i.e.

$$c(z) = c(-z). \tag{5}$$

The time integration need to be made only for the region $0 \leqq z \leqq H$, where H is the height of the vertical boundary (required by the finite Fourier transformation), and (5) is applied after each time step. The consequence of an upper "mathematical" boundary will be discussed at the end of this section. At this moment it is assumed that the flux at the upper boundary is vanishing too. Unpenetrable boundaries are equivalent to the condition of the conservation of the material flux. Assuming a steady-state ($\partial c/\partial t = 0$), a point source and no sink terms, integration of (1) with respect to z yields

$$\frac{\partial F_x}{\partial x} = 0 \tag{6}$$

where

$$F_x = \int_0^H u(z)\, c(z)\, \mathrm{d}z \tag{7}$$

is the total horizontal flux.

When the deposition on the ground cannot be neglected the condition (6) is not satisfied and must be replaced by

$$\frac{\partial F_x}{\partial x} = -\left\{K(z)\frac{\partial c}{\partial z}\right\}_{z=0} = -v_d c(x,0). \tag{6a}$$

In order to account for the removal of the material by deposition we introduce in Equation (1) a sink term which is acting only on the lower boundary ($z = 0$) and which is defined as

$$S(x,0) = -2v_d c(x,0)\cdot 1/\Delta z \qquad \text{for } z = 0 \tag{8}$$

$$S(x,z) = 0 \qquad \text{for } z \neq 0.$$

The factor 2 in (8) appears because of the symmetrical form of all source and sink terms required by the condition (5). The term $1/\Delta z$ is representation of the Dirac-delta, $\delta(z)$ in the finite grid point system.

Substituting (8) in (1) and providing integration with respect to z yields (a steady-state is again assumed)

$$\frac{\partial F_x}{\partial x} = -v_d c(x,0) \tag{9}$$

equivalent to the requirement stated in (6a). A sink term at the ground boundary yields thus a correct result for mass removal in spite of $\partial c/\partial z = 0$ at this boundary. It is obvious that the smaller the grid point spacing the better will the numerical solution match the analytical one obtained with the boundary condition (1a).

The idea of representing the deposition by a sink term located at the ground surface was recently used by Horst (1977), and he developed a model called a surface depletion model in contradiction to the traditional method based on reduction of the source strength as a function of downwind distance (Van der Hoven, 1968) and referred to here as the source depletion model. We shall use the name surface depletion model to our method, however, there is a considerable difference between the model of Horst and the present one. Horst assumes a Gaussian distribution of the plume in the vertical in the absence of the deposition and uses the sink term given as a negative Gaussian area source to account for the loss of the airborne material due to the deposition. We solve the dispersion equation (1) which in general case yields results quite different from the Gaussian form also in the absence of deposition.

As mentioned before, the finite Fourier Transform requires an upper boundary surface. It is not always physically relevant. When the source is located far from this 'mathematical' surface and the diffusion is small, the effect of the boundary can be neglected. However, in the case of a strong diffusion and at downwind distances at which the plume 'reaches' the upper boundary, a considerable error can be introduced due to mass accumulation at this boundary. Because of the cosine-transform used in our numerical method the vertical flux is zero at boundaries and thus

$$\left\{K(z)\frac{\partial c}{\partial z}\right\}_{z=H} = 0. \tag{10}$$

In order to remove the mass accumulation we introduce a sink term located at the upper boundary. This term is of the form

$$S_H = 2\cdot K(H)\frac{\Delta c}{\Delta z}\cdot 1/\Delta z \tag{11}$$

where $\Delta c = c(H) - c(H - \Delta z)$ and $\Delta c/\Delta z$ approximates thus the concentration gradient at the boundary.

3. NUMERICAL TEST

Only few analytical solutions of the dispersion equation (1) with the deposition effect accounted for are available in literature. Usually a simple power law variation of u and K with z is assumed and

Heaviside operational methods are used to solve (1). Solutions for some certain K-profiles but constant u are given by Smith (1962). Beside deposition on the ground, also the gravitational sedimentation of particulate matter is treated in this paper. In the case of u and K constant with height and a continuous source located at the ground level, the solution given by Smith (1962) is

$$c(x,z) = \frac{Q_0}{\sqrt{\pi u K x}} \exp\left\{-\frac{uz^2}{4Kx}\right\} - \frac{v_d Q_0}{uK} \times \exp\left\{\frac{v_d z}{K} + \frac{v_d^2 z}{Ku}\right\} erfc(t) \tag{12}$$

where

$$t = z\sqrt{\frac{u}{4Kz}} + v_d\sqrt{\frac{x}{Ku}}$$

and $erfc$ is defined by

$$erfc(t) = \frac{2}{\sqrt{\pi}}\int_t^x e - s^2 ds.$$

Q_0 is strength of the continuous source. The first term represents the solution when the deposition velocity v_d is zero, the second term being the modification due to deposition. The most important feature of this solution is that near the ground the gradient of concentration with height becomes positive, as confirmed by various experiments. The traditional Gaussian source depletion model does not reveal this feature.

We choose the analytical solution (12) to test the accuracy of our numerical model. The Fourier transformations were made on a 32 × 32 grid point system. However, due to the symmetry condition (5) only 17 points in the vertical direction represent the physical realistic solution. In the horizontal direction the first four and last five points were used to avoid the periodical boundary conditions as proposed by Christensen and Prahm (1976). The method consists in locating a sink term at the boundary points and the sink is of the form

$$S = -c \cdot u/\Delta x. \tag{13}$$

Due to that sink term, the concentration at the boundaries decreases exponentially and the mass transported across one of the boundaries is not appearing at the opposite boundary, as would be the case without the sink term. The real computational region consists thus only of 17 × 23 grid points.

Dimensionless quantities are used in the numerical computations. The relation between the dimensionless and real quantities is as follows:

$$u' = u\frac{T}{\Delta x};\quad K' = K\frac{T}{(\Delta z)^2};\quad v_d' = v_d \cdot \frac{T}{\Delta z};$$
$$t' = t\frac{1}{T};\quad c' = cu_0 H/Q \tag{14}$$

where T is the time unit, Δx and Δz the spacing of the grid points in the respective directions, and u_0 is a representative wind velocity, here the value at the source height. The dimensionless variables are denoted by primes.

We made our calculations with the following values of the dimensionless variables.

$$u' = 1; K' = 1; v_d' = 0.1; t' = 1. \tag{15}$$

If e.g. $\Delta x = 1000$ m, $\Delta z = 100$ m and $T = 1000$ s, it follows from (14) that $u = 1\ \mathrm{m\,s^{-1}}$, $K = 10\ \mathrm{m^2\,s^{-1}}$ and $v_d = 0.01\ \mathrm{m\,s^{-1}}$ which represent realistic values.

Steady state solution is assumed to be reached when there are no substantial changes in the values of concentration with time.

Some vertical profiles of the concentration are shown in Fig. 1. The solid lines represent the analytical solution (12), while the numerical results are given by crosses. The downwind distance from the source and height are given in grid point units. The agreement between the numerical and analytical result seems to be very good. The relative error is less than 5%, being somewhat larger only at the surface. No attempt was made to remove the mass accumulation at the upper boundary. This results in higher concentration values far from the source (the curve labeled $x = 22 \cdot \Delta x$). In Fig. 2 the surface concentration is shown as function of the downwind distance. The case of $v_d = 0$ is also shown in this figure. The solid lines represent the analytical solutions while crosses the numerical. Additionally, results are shown of the surface concentration as predicted by the source depletion model (the dashed line). According to Chamberlain (1953), the surface concentration corrected for deposition can be written

$$c(x,0) = \sqrt{\frac{2}{\pi}} \cdot \frac{Q(x)}{\bar{u}\sigma_z} \exp\left(-\frac{h^2}{2\sigma_z^2}\right) \tag{16}$$

where the depleted source strength $Q(x)$ is related to

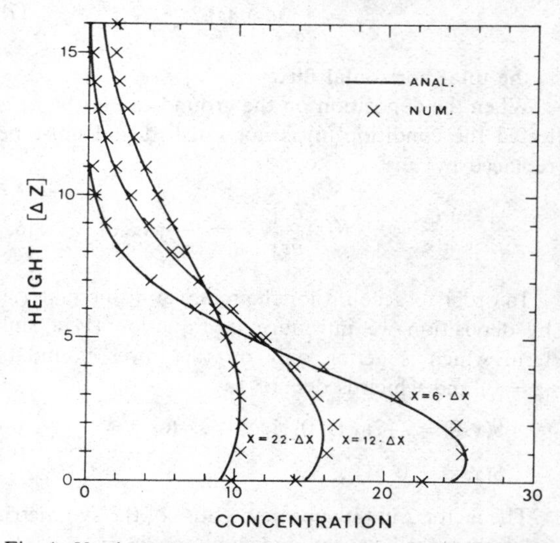

Fig. 1. Vertical profiles of the concentration computed by the pseudospectral numerical method compared with the analytical solution (Smith, 1962), in the case of constant u and K and a source located at ground level. The height and the downwind distance from the source is in grid point units. The concentration unit is $Q(u_0 H)^{-1}$.

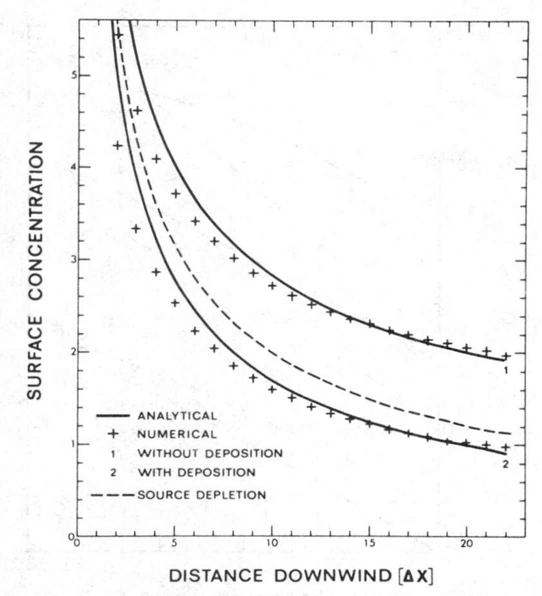

Fig. 2. Surface concentration with and without deposition computed by the numerical method compared with the analytical solution (Smith, 1962). Surface concentration predicted by the source depletion model is also shown. The concentration unit is $Q(u_0H)^{-1}$.

the original source strength Q_0 by the following expression

$$Q(x) = Q_0 \exp\left\{-\sqrt{\frac{2}{\pi}}\cdot\frac{v_d}{\bar{u}}\int_0^x \frac{dx}{\sigma_z \exp(h^2/2\sigma_z^2)}\right\}. \quad (17)$$

$Q(x)$ is equivalent to F_x defined in Equation (7), $\bar{u}$ is a characteristic wind velocity, usually represented by the value at the source height. h is the source height and σ_z is the standard deviation parameter of the Gaussian vertical distribution. When u and K are constant with height, σ_z is given by

$$\sigma_z^2 = 2Kx/u. \quad (18)$$

Substituting (18) in (17) and with $h = 0$ (ground source) we obtain

$$Q(x) = Q_0 \exp\left\{-\frac{2}{\sqrt{\pi}}\frac{v_d}{uK}\sqrt{x}\right\}. \quad (19)$$

Expression (19) together with (16) was used to calculate the dashed curve in Fig. 2. In Fig. 3 the suspension ratios are shown calculated on the basis of the analytical solution (solid curve) given by Smith (1962) together with the numerical results (crosses). Suspension ratio is the fraction of the material which is still airborne and is given by

$$R(x) = \left\{\int_0^\infty uc(x,z)\,dz\right\}/Q_0. \quad (20)$$

As seen from Fig. 3 the agreement between the analytical and numerical results is very good. The relative error is less than 2%. Also the source depletion model values are shown in Fig. 3. Expressions (19) and (17) can directly be used to evaluate the suspension ratio

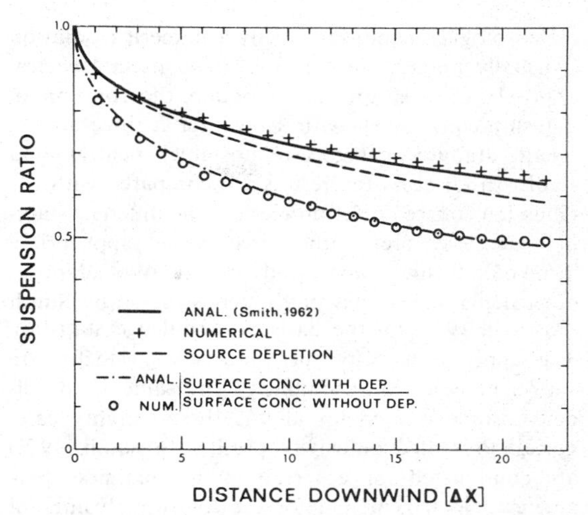

Fig. 3. Suspension ratio computed by the numerical method compared with the analytical solution (Smith, 1962) and source depletion model. The ratio of surface concentration with deposition to surface concentration without deposition is also shown.

in this case, as follows from (17) by integration with respect to z.

$$R(x) = Q(x)/Q_0. \quad (21)$$

Integration of (12) with respect to z yields the suspension ratio predicted from Smith's analytical solution.

$$R(x) = \exp\left(\frac{v_d^2 x}{Ku}\right) erfc\left(v_d\sqrt{\frac{x}{Ku}}\right) \quad (22)$$

or

$$R(x) = \{c_0(x,0) - c_d(x,0)\}\cdot\frac{uK}{v_d Q_0} \quad (23)$$

where $c_0(x, 0)$ is the surface concentration when $v_d = 0$ and $c_d(x, 0)$ is the surface concentration with deposition.

According to the source depletion model, the concentration at all heights is reduced by the same factor, the suspension ratio R. In Fig. 3 the ratio c_d/c_0 is shown as calculated from the analytical solution (the dotted line) and from the pseudospectral numerical model (open circles). A considerable difference is found between values of suspension ratio and surface concentration ratio c_d/c_0, this is in contradiction to the result predicted on the basis of the source depletion model.

4. DEPOSITION AT DIFFERENT ATMOSPHERIC STABILITY CATEGORIES

Comparison between the numerical and analytical solutions presented in the last section show that the pseudospectral approximation can be used for accurate simulation of dry deposition from a point source. The quite obvious discrepancy between the source and surface depletion models, stated here for the simplified case of constant wind and diffusivity profiles, inspires to study the deposition at more realistic meteorological conditions. Only a numerical solution is usually possible in this case. The pseudospectral approximation is applied to obtain the solution of diffusion equation (1) with deposition at three atmospheric stability categories: unstable, neutral and stable. In all cases the results are compared with the Gaussian source depletion model. The different stability categories are characterized by an appropriate form of diffusivity and wind profiles. We adapt to our calculations values of $K(z)$ constructed by Smith (Pasquill, 1975) on the basis of turbulence statistics. The appropriate K-profiles and wind velocity are shown in Fig. 4. For simplicity, the same wind velocity profile is used for all the three stability categories. Presented diffusivity profiles (Pasquill, 1975) are constructed for a terrain with roughness parameter $z_0 = 0.03$ m and for the following values of surface heat flux (in mW cm^{-2}): 26 (unstable), 0 (neutral, -2 (stable). The wind profile is constructed somewhat arbitrarily. It has the geostrophic value of 4 m s^{-1} at 1000 m and asymptotically approaches the neutral logarithmic form at low level. A value of 1.5 m s^{-1} is associated with the ground level (here taken as 1 m above the earth surface) and 3 m s^{-1} corresponds to the height of 25 m, which we chose as the height of the source (the same for all the three stability categories). Deposition calculations are made with deposition velocity $v_d = 0.01$ m s^{-1}, as a value commonly reported in experiments on gaseous pollutants. Diurnal, seasonal and spatial variations in the meteorological parameters such as deposition velocity, diffusivity and wind profiles can be included in the model without any significant changes.

Expressions (16) and (17) are used in connection with the source depletion model. In order to relate the parameters of the source depletion model to parameters of gradient-transfer theory, on which the present model is based, we compute the values of σ_z from the vertical distribution of the concentration obtained from the numerical solution of Equation (1) with $v_d = 0$. The standard deviation parameter σ_z is computed from the following expression

$$\sigma_z^2 = \frac{\int_0^\infty c(x, z)z^2\,dz}{\int_0^\infty c(x, z)\,dz} - h^2. \quad (24)$$

Here h is the height of the source. Integration in (24) is in fact limited by the height of the 'numerical' upper boundary. However, the concentration at this boundary at the downwind distances examined here is small, and the error introduced by this limitation can be neglected. In the case of the unstable category a sink term is applied at the upper boundary. The values of σ_z obtained from (24) are shown in Fig. 5 (solid lines). For comparison also the experimental standard deviation curves, reported by Pasquill (Turner, 1969) and corresponding to the appropriate Pasquill stability categories are shown in Fig. 5. Note the agreement between the theoretical and the experimental slopes of the α_z-curves in the case of the stable (category F) and neutral (category D) conditions. However, no correlation can be observed for the unstable category.

In Fig. 6 surface concentrations calculated from surface and source depletion models are shown for the three stability categories. Both the case with and

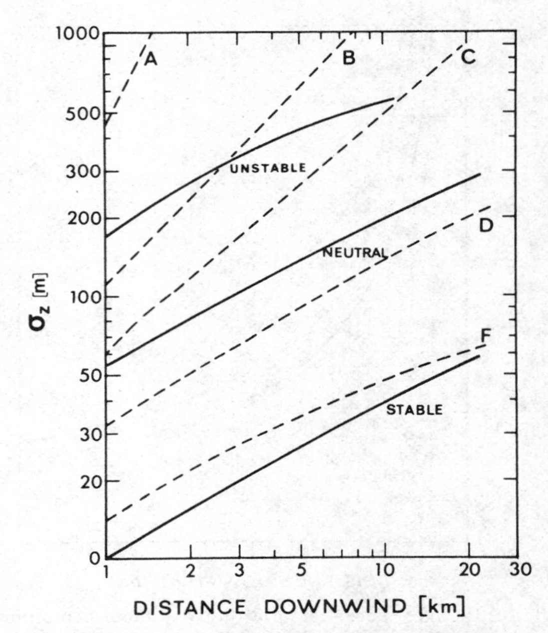

Fig. 5. Standard deviation parameters computed on the basis of the vertical concentration profiles obtained from the numerical solution of the dispersion equation with wind velocity and diffusivity profiles as shown in Fig. 4. The source height, $h = 25$ m for all the three stability categories. For comparison, standard deviation parameters reported by Pasquill (Turner, 1969) are also shown. The letters A to F label the Pasquill stability categories, A being the most unstable and F the most stable category.

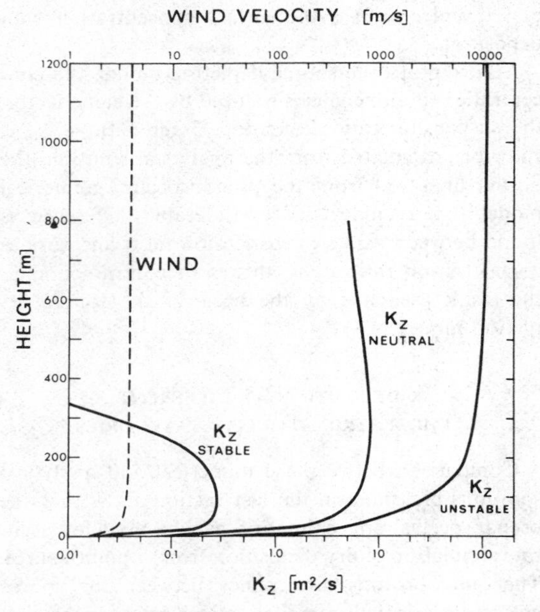

Fig. 4. Wind velocity and diffusivity profiles used in the numerical model for the three considered atmospheric stability categories.

without deposition is presented. Suspension ratio and the ratio c_d/c_0, i.e. the ratio of the surface concentration with deposition to the surface concentration without deposition are shown in Fig. 7. Again the results are presented for both models. Note that the source depletion model predicts the same value for c_d/c_0 as for the suspension ratio.

Different grid point spacing was used in the numerical calculations for the three stability categories. In the case of the neutral and stable category the horizontal grid point spacing $\Delta x = 1000$ m, while for the unstable case $\Delta x = 500$ m was used. Spacing in the vertical direction was: 25 m, 50 m and 75 m for the stable, neutral and unstable category, respectively. The distance scale in Fig. 6 and Fig. 7 can easily be changed by use of the relations in Equation (14). As one can see comparing the source height (25 m) with the grid point spacing in the vertical direction, it is required in the neutral and unstable cases that a point source should be located "between" the grid points. The pseudospectral approximation gives the possibility of such a subgridscale representation. The spectral components of a δ-like source function are calculated analytically by the continuous Fourier transformation. Next they are used to convert the function to the physical space by the finite Fourier transformation. The source field thus acts as a point source at the appropriate height. This way of representing subgridscale information is suggested by Berkowicz and Prahm (submitted).

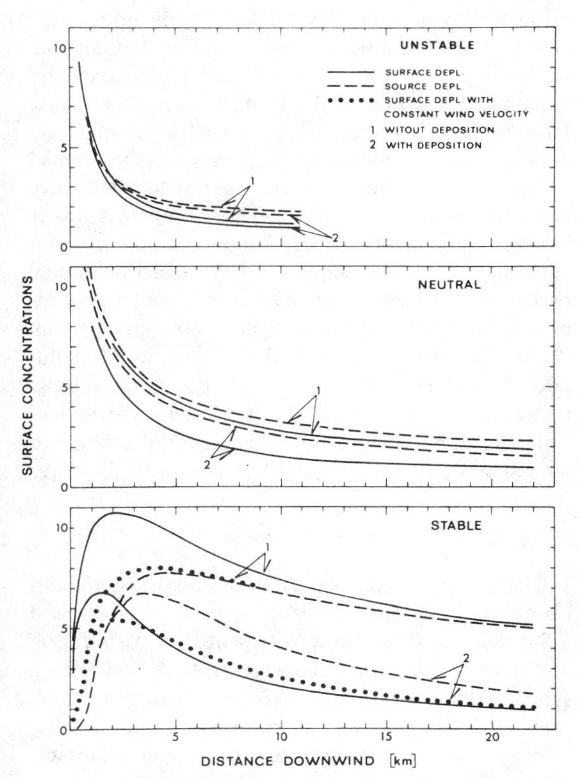

Fig. 6. Surface concentration with and without deposition for the three stability categories computed by the numerical pseudospectral surface depletion model and the Gaussian source depletion model. Source height, $h = 25$ m and deposition velocity, $v_d = 0.01$ m s^{-1} for all the three stability categories. In the surface depletion model, the wind and diffusivity profiles of Fig. 4 are used. For the stable category the case of constant wind velocity ($u = 3$ m s^{-1}) is also shown. The concentration unit is $Q(u_0H)^{-1}$.

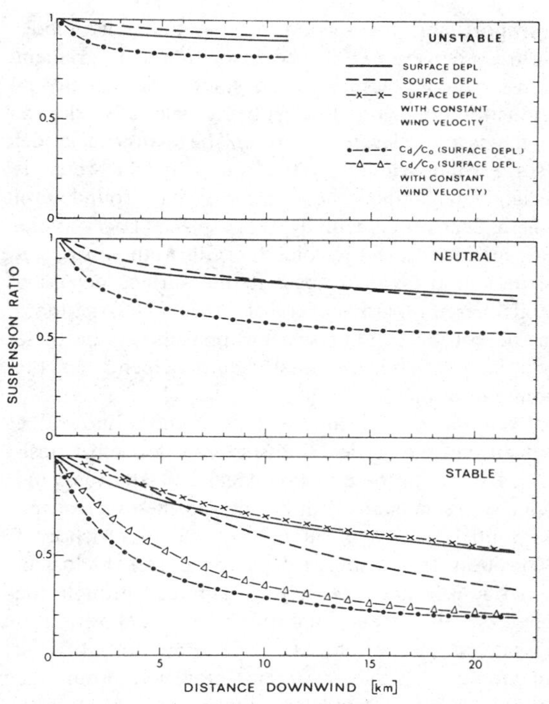

Fig. 7. Suspension ratio and the ratio of surface concentration with deposition to surface concentration without deposition (c_d/c_0). The case of constant wind velocity is shown for the stable category. All parameters as in Fig. 6.

5. DISCUSSION

We have presented a new numerical method to simulate the effect of dry deposition from a point source. Finite Fourier transformation is used to evaluate the space derivatives in the two-dimensional diffusion equation. In order to ensure the condition of mass-conservation, a cosine-transform is adapted for concentrations and a sine-transform for the vertical flux. The analytical boundary conditions are replaced by a sink term at the surface. Comparison with existing analytical solutions (Smith, 1962) showed that this method yields satisfactory results and can be used for realistic deposition calculations.

Computations with variable wind and diffusivity profiles corresponding to different atmospheric stability categories yield the possibility to compare the present method with the commonly used, Gaussian source depletion model, under physically realistic conditions. Two important factors determine the difference between the Gaussian source depletion model and the present one, based on the gradient-transfer theory. The first is, that the source depletion model does not properly account for the vertical diffusive transport and the second one is, that this model (source depletion), in contradiction to gradient transfer theory, assumes a constant wind velocity as a function of height—the last being especially relevant in the case of elevated sources. The Gaussian model uses a wind velocity at the source height which is usually higher than the velocity at the ground level where deposition actually takes place. Each of the two mentioned factors bias the Gaussian source depletion model with respect to the surface depletion model based on gradient transfer theory. Furthermore the deposition is influenced in opposite way by these two factors, and the final results depend on the dominating one.

Examination of results in Figs. 2 and 3 shows the influence of the vertical diffusive transport on deposition. Here both the diffusion coefficient and the wind velocity are constant with height and the two respective models are identical in the case of absence of deposition. In the source depletion model the loss of material is homogeneously distributed through the height of the plume, and the rate of deposition is consequently overestimated. The amount of the remaining airborne material, predicted from the source depletion model is lower, compared with results of the surface depletion model (Fig. 3).

The situation is somewhat more complicated when both wind velocity and diffusivity are functions of height. Examination of the stable case in Fig. 7 shows that the surface depletion model (the solid line) yields lower suspension ratio than the source depletion model (dashed line) at distances close to the source (up to 6 km), but the relation is reversed at longer distances. At a distance of 22 km the source depletion model underestimates the suspension ratio by nearly a factor 1.5. The different behaviour in the two regions can be explained in the following way. Close to the source the major part of the dispersed material is remaining in the lowest layer where the wind velocity is low. The results presented here for the source depletion model are computed with the wind velocity at the source height (3 m s^{-1}). The wind velocity at the surface is, on the other hand, only 1.5 m s^{-1}. This low wind velocity at the surface results in a fast rate of deposition as predicted by the surface depletion model. At larger distances from the source, where the material is more dispersed, the effect of the variable wind velocity becomes less dominating and the vertical diffusive transport determines the difference between the Gaussian source depletion model and the present surface depletion model. In order to examine the effect of variable wind velocity on the relation between the source and surface depletion models, the last model was computed with a constant wind velocity ($u = 3 \text{ m s}^{-1}$). The results for the stable case are shown in Fig. 6 and Fig. 7. The suspension ratio becomes higher and the 'cross-over' point is shifted closer to the source. For small distances, the suspension ratio predicted by the surface depletion model is still lower than that of the source depletion model, but the difference is not so pronounced and can be attributed to the numerical accuracy. At longer distances the effect of the variable wind velocity becomes less important due to the increased vertical mixing of the airborne material. When the atmospheric stability decreases (increasing diffusivity), the difference between the source and surface depletion models decreases. As seen in the case of the neutral stability (Fig. 7) the surface depletion model yields lower suspension ratio than the source depletion model in the initial stage of plume growth (up to 12 km), and the situation is reversed at larger distances. For the unstable case, the 'crossover' point is not reached in the region covered by the present calculations. The effect of variable wind velocity is predominant in this case.

The most pronounced difference between the source and surface depletion models is for all stability categories in the ratio c_d/c_0. The surface depletion model predicts values of c_d/c_0, which are considerably lower than the value of the suspension ratio, while the source depletion model does not differentiate between the two ratios.

Comparison between source and surface depletion models was recently reported by Horst (1977) and by Draxler and Elliot (1977). As already mentioned, Horst (1977) compared the traditional Gaussian source depletion model with a model still based on a Gaussian plume but with a sink term at the surface. Draxler and Elliot (1977) presented a comparison between a one-dimensional surface depletion model based on the gradient transfer theory and a source depletion model based also on the gradient transfer theory. The present study shows a comparison between a two-dimensional surface depletion model based on the gradient transfer theory and the commonly used Gaussian source depletion model (Van der Hoven, 1968). Not only the deposition effect but also the whole problem of material dispersion is treated in a different way by the latter two models. This has to be kept in mind when results of the two models are compared.

6. CONCLUSION

Dispersion and dry deposition is treated by the diffusion equation based on gradient-transfer theory and a flux boundary condition represented by a sink term at the surface. A new pseudospectral method with a symmetrical Fourier Transform in the vertical direction satisfying the mass conservation condition is developed. Numerical tests compared with analytical solutions in special cases reveals the high accuracy of the pseudospectral method.

Comparison with the Gaussian source depletion model shows that this model either underestimates or overestimates the suspension ratio depending on the distance from the source and the meteorological conditions. The wind velocity profile is shown to be important for this effect.

At larger distances from the source the results are qualitatively in agreement with previous less sophisticated surface depletion model computations showing that the Gaussian source depletion model is considerably biased, especially in stable atmospheric conditions, while there is a qualitative discrepancy between our and previous surface depletion models at shorter distances. The present numerical accurate 2-D pseudospectral surface depletion model with realistic wind and diffusivity profiles might be recommended especially in detailed studies of dispersion and deposition up to distances of about 10 km from the source.

Acknowledgements—This study was activated as a part of the project: 'Mesoscale dispersion modelling', coordinated by the Scandinavian council for applied research. The project was supported by the Danish National Technical Scientific Research Council. Computations were performed at NEUCC, Technical University of Denmark.

We would like to thank the Danish Meteorological Institute for making this research possible. The study took place at this institution, the manuscript was typed by Lajla Mittet, and drawings prepared by A. Skjoldbo. Special thanks to O. Christensen, the University of Copenhagen, and G. Thomsen, the Technical University of Denmark, for useful discussion on the numerical methods. Thanks is also given to F. B. Smith, Bracknell, England, for informative correspondence.

REFERENCES

Berkowicz R. and Prahm L. P. Sub-grid scale information in the pseudospectral model (submitted to *J. Atmos. Sci.*).

Calder K. L. (1961) Atmospheric diffusion of particulate matter, considered as a boundary value problem. *J. Met.* **18,** 413–416.

Chamberlain A. C. (1953) Aspects of Travel and Deposition of Aerosol and Vapour Clouds. British Report. AERE-HP/R-1261, HMSO.

Christensen O. and Prahm L. P. (1976) A pseudospectral model for dispersion of atmospheric pollutants. *J. appl. Met.* **15,** 1284–1294.

Draxler R. R. and Elliott W. P. (1977) Long-range travel of airborne material subjected to dry deposition. *Atmospheric Environment* **11,** 35–40.

Gotaas Y. (1975) Dry deposition and its influence on the vertical concentration profile. Lecture No. 39/76. Norwegian Institute for Air Research, Norway.

Horst T. W. (1977) A surface depletion model for deposition from a Gaussian plume. *Atmospheric Environment* **11,** 41–46.

Monin A. S. (1958) On the boundary condition on the earth surface for diffusing pollution. *Adv. Geophys.* **6,** 435–436.

Pasquill F. (1975) The dispersion of material in the atmospheric boundary layer—the basis for generalization. Lectures on Air Pollution and Environmental Impact Analyses, 1975 (edited by Haugen D. A.), pp. 1–34, American Meteorological Society.

Prahm L. P. and Christensen O. (1977) Long range transmission of pollutants simulated by the 2-D pseudospectral dispersion model. *J. appl. Met.* **16,** 896–910.

Scriven R. A. and Fisher B. E. A. (1975) The long range transport of airborne material and its removal by deposition and washout. *Atmospheric Environment* **9,** 49–68.

Smith F. B. (1957) Convection-diffusion process below a stable layer. A paper of the Meteorological Research Committee (London), MRP No. 1048.

Smith F. B. (1962) The problem of deposition in atmospheric diffusion of particulate matter. *J. atmos. Sci.* **19,** 429–434.

Turner D. B. (1969) *Workbook of Atmospheric Dispersion Estimates*. USDHEW, PHS Pub. No. 995-AP-26.

Van der Hoven I. (1968) Deposition of particles and gases. *Meteorology and Atomic Energy*, 1968 (Edited by Slade D.), pp. 202–208, USAEC, TID-24190.

Atmospheric Environment Vol. 12, pp. 389–399. Pergamon Press 1978. Printed in Great Britain.

WET REMOVAL OF SULFUR COMPOUNDS FROM THE ATMOSPHERE

J. M. HALES

Battelle, Pacific Northwest Laboratories, Richland, WA 99352, U.S.A.

(*First received* 28 *June* 1977 *and in final form* 28 *September* 1977)

Abstract—This paper presents a brief overview of our current capability to calculate sulfur scavenging rates. The general wet removal process can be decomposed into several individual pathways. These include direct sulfur dioxide scavenging, direct sulfate scavenging, and combined scavenging and chemical reaction. Modeling approaches for these pathways are discussed, and pertinent research areas for improvement of our present modeling capability are recommended. At the present time the calculation of direct sulfur dioxide scavenging appears to be well in hand, although more careful network measurements of dissolved sulfur dioxide concentrations in rainwater are needed to establish the relative importance of this phenomenon on a regional scale. Direct sulfate scavenging presents a more difficult calculational problem, and much more data regarding particle-size relationships of sulfate-containing aerosols is required before an adequate understanding of this pathway can be achieved. Sulfur scavenging via the pathway of sulfur dioxide absorption followed by aqueous-phase conversion is expected to be rate limited by the chemical conversion step under most circumstances. Much additional information regarding aqueous-phase transformation chemistry is necessary before a reliable calculational basis for describing this pathway can be obtained.

INTRODUCTION

The importance of sulfur compounds in precipitation

Sulfur occurs in the atmosphere in a number of different compounds and oxidation states, the most important of which are gaseous sulfur dioxide (S IV) and particulate sulfate (S VI). Because of this abundance and also because of their generally high solubility in water, these compounds have a profound influence on precipitation chemistry, their wet removal constituting an important component of the global sulfur balance.

Although it is likely that other oxidation states of sulfur play significant atmospheric roles, their direct wet removal is expected to be comparatively unimportant. Budgetary considerations (Junge, 1960) and indirect measurements (Hitchcock, 1976), for example, suggest that naturally emitted sulfur in the −II oxidation state (hydrogen sulfide, mercaptans, organic sulfides) may be an important component of the atmospheric sulfur balance. The observed low concentrations of these substances, however, imply that they are rapidly oxidized to other compounds in the atmosphere. Moreover, their limited solubilities render them of only marginal interest insofar as precipitation chemistry is concerned. Similarly, elemental sulfur (oxidation state 0) occurs in the atmosphere at concentrations that are much lower than those observed to be typical of the compounds in higher oxidation states (Haagen *et al.*, 1968). From these considerations it may be concluded that scavenging of sulfur dioxide and sulfate aerosol constitute the predominant pathways for wet removal of sulfur from the atmosphere, and the emphasis of this paper will focus on these two classes of compounds accordingly.

Some idea of the significance of sulfur removal by wet processes can be obtained by considering the several material-balance studies that have been published over the past three decades (Conway, 1943; Eriksson, 1959; Junge, 1960; Robinson and Robbins, 1968; Kellogg *et al.*, 1972; Friend, 1973; Bolin and Charlson, 1976 and Rasmussen *et al.*, 1974). While these have by no means been in total agreement, there is rather general accord that the global input of sulfur to the atmosphere is of the order of $1\text{–}2 \times 10^{11}$ kg sulfur y^{-1}. Combining this with an estimated tropospheric volume of $3 \times 10^{18}\,m^3$ suggests that, if wet and dry removal processes did not exist, the sulfur concentration would increase by about $70\,\mu g\,m^{-3}\,y^{-1}$. This may be compared with the current United States annual standard for SO_2, which is $40\,\mu g\,m^{-3}$ (expressed as S).

Because of difficulties in obtaining precise measurements of dry deposition rates the relative importance of dry and wet deposition of sulfur compounds is rather uncertain. The budget studies cited above estimate that 50–90% of the atmospheric sulfur burden is removed by wet processes. The higher of these estimates were based on radioactive fallout measurements and can be expected to overestimate the wet component for near-surface sources. In addition, some of the more specific examinations (Whelpdale, 1975; Chamberlain, 1973) have indicated dry deposition to be of relatively greater importance. Regardless of these features, however, there seems little doubt that precipitation scavenging constitutes an important, and in some areas the dominant mechanism for sulfur removal.

As a consequence of its effective nature as an atmospheric cleansing agent, precipitation scavenging is also of concern with regard to the delivery of sulfur compounds to surface ecosystems. Adverse effects to soils, forests and fisheries in several parts of Europe and North America have been extensively documented, and while there is continuing conjecture over specific effects and mechanisms, there appears to be little doubt at present that such negative impact can be severe under specific circumstances. Many of these detrimental effects can be associated directly with the tendency of dissolved sulfur compounds to displace the acid–base chemistry of precipitation, resulting in significant increases in hydrogen-ion concentration. Indeed, dissolved sulfate often has been cited as the primary contributor to the acidity of rain (Granat, 1972; Likens, 1976). Depending upon the buffering capacity of specific soils and surface waters, corresponding increases in acidity can be induced in these media, often with severe detrimental results.

The distribution of sulfur compounds in precipitation

Sulfur compounds have been measured in samples collected on a variety of precipitation chemistry networks over the past three decades. These networks have focused primarily on the measurement of sulfate, which is undoubtedly the most abundant sulfur-containing species present in precipitation under most circumstances. Typical concentrations of sulfate in precipitation range from 10 to 100 μmoles l.$^{-1}$, with much higher values occurring in specific samples on occasion.

Sulfate concentrations in precipitation in general tend to increase in regions downwind of areas of large sulfur emission; there is also some evidence for their increase with time during the last two decades. Nisbet (1975), for example, has utilized United States network data to indicate that a 60–65% increase in sulfate content of precipitation has occurred during the 10-year period between 1955 and 1965. Although his results indicate a partial leveling off subsequent to 1966, the concentrations appear to be resuming their upward trend at the present time. Sulfur concentrations in rain are estimated to increase by another 40% by 1980, if current trends continue. Somewhat in contrast to Nisbet's findings, however, Granat (1978) has completed a careful analysis of European network data to conclude that no discernable trend in sulfate concentrations is apparent within the yearly variability of data, which is rather large. This short-term variability, Granat believes, is caused mainly by variations in the meteorological systems leading to precipitation events.

Comparatively little is known with regard to the abundance of sulfite (IV oxidation state) in precipitation. This species tends to oxidize in solution and generally occurs at concentration levels lower than those characteristic of sulfate. Because of sample preservation problems, reliable measurements of sulfite levels in rain are more difficult to obtain, and most of the usable data in this regard has come from short-term measurement programs where special sample-preservation practices were implemented. Sulfite levels in rain tend to show an inverse relationship with rain acidity and thus with sulfate content. Typical sulfite levels range between 0 and 20 μmoles l^{-1}, with higher values occurring under conditions of high ambient sulfur dioxide concentration. A recent analysis for the northeastern United States (Hales *et al.*, in press) has suggested that from 10 to 30% of the sulfur in rain approaching the surface exists in the form of sulfite ion.

Objectives

In view of the above aspects of sulfur deposition by wet processes, it appears important to possess the capability to calculate the wet deposition rates of sulfur compounds on the basis of source and rainfall information. This capability is valuable for estimating the spatial distribution of sulfate concentrations in both air and precipitation; even more importantly, however, such calculations are essential for predicting the long-term effect of increased industrialization and energy production. Such a capability can be visualized as one component of a comprehensive model to predict the combined effects of transport, deposition, and conversion on the fate of atmospheric sulfur compounds (Bolin and Persson, 1975).

Unfortunately, a totally satisfactory method for calculating scavenging rates does not exist at the present time. In reflection of the multiple pathways for sulfur removal, scavenging calculations tend to be multifaceted, with some aspects well understood and others remaining at relatively unsatisfactory stages of development. Accordingly, the objectives of this paper are to examine the most important of these pathways, to indicate in rather uncomplex terms the (currently) best ways for describing them mathematically, and to describe specific areas where research is needed to upgrade the calculational techniques to a more satisfactory level. A general description of scavenging pathways is given in the next section, followed by detailed discussions of individual mechanisms.

GENERAL SCAVENGING PATHWAYS

Because of the variety of mechanisms contributing to wet removal of sulfur compounds, it is appropriate at this point to examine the pathways for their delivery from the atmosphere to the ground. These are shown in a general fashion in Fig. 1, which depicts the atmospheric interactions between SO_2 gas, sulfate aerosol and precipitation.

As indicated by the figure, gas-phase SO_2 (as well as other sulfur-containing gases) can become absorbed by precipitation and delivered directly to the surface (Steps 1–3–5). Alternatively, absorbed SO_2 may react to form sulfate (Step 3–4) and subsequently approach the surface via Step 4–6. SO_2 also may be converted to sulfate in the gas phase (Step 1–2) and then may be assimilated, along with existing sulfate

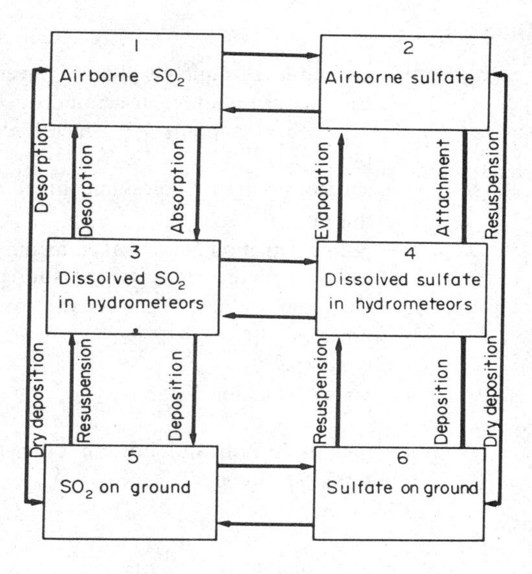

Fig. 1.

aerosol, by precipitation and delivered to the surface (Steps 2–4–6).

The above sequences describe the major pathways for wet removal of sulfur compounds. Reverse processes are of course possible, and can contribute significantly to scavenging behavior under appropriate atmospheric conditions. SO_2 may, for example, desorb from precipitation back into the gas phase; and sulfate aerosol may be regenerated via droplet evaporation. Reverse reactions to form reduced sulfur compounds from sulfate, while possible, are unfavored thermodynamically, and can be considered inconsequential in the atmosphere.* Also depending upon conditions, SO_2 and other sulfur-containing gases may desorb from the surface back into the gas phase. In addition, sulfate aerosol may be resuspended from the surface to the atmosphere. A notable example of this latter effect is the input of sulfate to the atmosphere by the evaporation of sea spray, a mechanism which is thought to account for a significant fraction of the atmospheric sulfur burden.

Two major aspects of this brief discussion concerning scavenging pathways should be emphasized. The first of these is the suggestion that scavenging processes generally occur in a rather dynamic fashion, where a given sulfur atom may undergo several cycles of pathways before deposition. Obviously, this complicates calculations of the scavenging process. The second important point is the indication that processes 1–3–5, 2–4–6, 1–2–4–6, and 1–3–4–6 constitute the major pathways for sulfur compound removal. These will be examined individually in the following sections of this paper.

* This is not always the case in surface waters and soils, however, where reducing conditions often prevail.

PATHWAY 1–3–5: DIRECT SO_2 SCAVENGING

Because of sulfur dioxide's properties as a gas, its direct scavenging can be calculated in a manner that tends to be much more simple and reliable than corresponding calculations involving sulfate. This relative simplicity occurs for several reasons. First, gas-scavenging calculations do not involve size-distributed aerosol systems. This advantage is doubly beneficial because, in addition to being much less cumbersome mathematically, gas-scavenging calculations do not require dealing with the uncertainties usually encountered in specifying size spectra of aerosol systems.

A second simplifying feature of SO_2 scavenging calculations arises from the relatively high diffusivity of this gaseous material. This property permits one to neglect a number of rather complicated transport mechanisms, and allows reliable estimation of interphase transport by physical analogy.

Finally, direct SO_2 scavenging calculations tend to be simpler than corresponding calculations for sulfate because of reversible absorption. Since SO_2 readily absorbs and desorbs from precipitation, its aqueous phase concentration tends to reflect only the gas-phase concentrations experienced during its recent history. This usually makes definition of the total concentration history of the hydrometeor unnecessary, and often is an overwhelming simplifying feature in associated calculations.

Basic gas scavenging theory

A general theory of gas scavenging has been presented elsewhere (Hales, 1972). The basic precept of this development is that gas molecules may absorb or desorb from an individual hydrometeor depending upon the direction of the concentration 'driving-force' between the hydrometeor and its surroundings. For spherical raindrops this behavior may be expressed by a material balance for pollutant within a hydrometeor;

$$\frac{dc_x}{dz} = \frac{3K_y}{v_z ac}(c_y - H'c_x) \tag{1}$$

where

dc_x/dz = change of dissolved gas concentration in hydrometeor with height above ground,

K_y = mass-transfer coefficient for interchange of gas between droplet and its surroundings,

a = hydrometeor radius,

$v_z = dz/dt$ = vertical velocity of hydrometeor,

c = total gas-phase concentration (nominally 1/22,400 moles ml^{-1}),

y, x = subscripts denoting gas and aqueous phase, respectively, and

H' = solubility parameter.

When the distribution of SO_2 between the gaseous and aqueous phases is in equilibrium, the

$$c_y = H' c_x. \tag{2}$$

Inspection of (1) reveals that there will be no net interphase exchange under such circumstances.

Scavenging of gases can be calculated in a relatively straightforward manner using Equation (1), provided physical—or 'pseudophysical'—absorption is the dominant uptake mechanism,* and the spatial distribution of the gas-phase concentration is known. The usual procedure is to discretize the raindrop size spectrum and integrate (1) over the fall path of the raindrop to the ground. Combining repeated calculations of this type for separate drop-size ranges leads to the delivery rate of scavenged material to the surface.

Computer codes currently exist to perform this integration for Gaussian (Dana *et al.*, 1973) and step-function plumes (Adamowicz and Hill, 1977). Under simplified conditions of vertical rainfall and linearized solubility behavior, these situations may be integrated analytically. The results are (cf. Hales *et al.*, 1973).

(1) Gaussian plume:

$$c_x(x, y, o, a) = -\frac{QF\xi}{2\sqrt{(2\pi)}\sigma_y \bar{u}} \exp\left(-\frac{y^2}{2\sigma_y^2} + \frac{\sigma_z^2\zeta^2}{2}\right) \times \left\{\exp(\zeta h)\left[1 - \mathrm{erf}\left(\frac{-\sigma_z^2\zeta - h}{\sigma_z\sqrt{(2)}}\right)\right] + \exp(-\zeta h)\left[1 - \mathrm{erf}\left(\frac{-\sigma_z^2\zeta + h}{\sigma_z\sqrt{(2)}}\right)\right]\right\}. \tag{3}$$

(2) Step-function plume:

$$c_x(x, y, o, a) = S_n \frac{\xi}{\zeta} F(e^{\xi b} - e^{\xi d}). \tag{4}$$

$$\zeta = \frac{3K_y H'}{cv_z a},$$

$$\xi = \frac{3K_y}{v_z a}$$

Here

$c_x(x, y, o, a)$ = ground-level aqueous-phase concentrations of pollutant in a raindrop of radius a at a point x, y beneath the plume,

Q = source strength (moles/unit time) of the plume,

F = plume depletion factor (taken as unity for SO_2 plumes close to the source),

σ_y, σ_z = Gaussian plume dispersion parameters,

$\bar{u}$ = wind speed,

h = virtual emission height,

b, d = heights at top and bottom of step-function plume

and

Sn = step function.
$= o$ for $z > b$ or $z < d$
$= c_y/c$ for $b > z > d$.

Mass-transfer coefficients for use in the above equations depend upon mixing in both the gas and liquid phases. For moderately soluble gases such as SO_2, however, it is acceptable to approximate mixing behaviour by assuming that the majority of transport resistance is in the gas phase. This permits calculation of K_y using the Froessling equation (Bird *et al.*, 1960),

$$\mathrm{Sh} = 2 + 0.6\,\mathrm{Re}^{1/2}\mathrm{Sc}^{1/3}, \tag{5}$$

where

$$\mathrm{Sh} = \frac{2K_y a}{D_{Ay} c_y} = \text{Sherwood number},$$

$$\mathrm{Re} = \frac{-2av_z}{\nu} = \text{Reynolds number},$$

and

$$\mathrm{Sc} = \frac{\nu}{D_{Ay}} = \text{Schmidt number}.$$

Here D_{Ay} and ν denote the diffusivities of mass and momentum in the gas phase.

Solubility of sulfur dioxide in water varies as a function of concentration, temperature and acidity of the hydrometeors. This can be expressed in the combined form† (Hales and Sutter, 1973)

$$c_x = \frac{1}{H'} c_y = \frac{c_y}{H} + \frac{-[H_3O^+]_{ex} + \sqrt{[H_3O^+]_{ex}^2 + 4K_1 c_y/H}}{2} \tag{6}$$

* In this and the following discussion, we will use the convenient chemical engineering concept of 'pseudophysical absorption'. By this we mean that a reactive gas such as sulfur dioxide has absorption behavior somewhat similar to that of a non-reactive gas, and can be treated in roughly the same manner mathematically. This is acceptable whenever the significant chemical reactions involved in the absorption process are reversible and occur rapidly compared to the purely physical steps. The dissociation reactions

$$2H_2O + SO_{2_{aq}} \rightleftarrows HSO_3^- + H_3O^+$$

$$H_2O + HSO_3^- \rightleftarrows SO_3^= + H_3O^+$$

are very rapid and satisfy these conditions.

† This equation is based on the assumption that sequestering of dissolved sulfur dioxide by foreign contaminants such as trace metals and organics is negligible. Our experience indicates this assumption to be generally a valid approximation for both clean and polluted atmospheres, although recently some highly interesting arguments to the contrary have been advanced by Eatough (1977).

where $[H_3O^+]_{ex}$ denotes the concentration, in moles l^{-1}, of hydrogen ion donated by sources other than dissolved sulfur dioxide. Under most conditions it is reasonable to approximate $[H_3O^+]_{ex}$ by equating it to the total free hydrogen ion concentration, i.e.

$$[H_3O^+]_{ex} \cong 10^{-pH}. \qquad (7)$$

The temperature-dependent parameters K_1 and H have been given by Johnstone and Leppla (1934), and are presented in Table 1.*

The above discussion presents a reasonably reliable and simple approach for calculating direct SO_2 scavenging in the locality of sources (Hales, Wolf and Dana, 1973; Hutcheson and Hall, 1974), and similar techniques can be utilized for additional gaseous pollutants if desired. For small hydrometeors or for larger distances from the source, however, the approach is even simpler. This is made possible by the fact that, if the gas-phase concentration varies sufficiently slowly in the vicinity of a hydrometeor, the aqueous-phase concentration should approach equilibrium with its gas-phase surroundings; in short, it should be given by Equation (2).

The above circumstances are known as *equilibrium-scavenging* conditions. Equilibrium scavenging prevails whenever the relaxation time for uptake by the hydrometeors is small in comparison with the rate of change in gas-phase concentration. Equilibrium is favored by small drop sizes, small changes in gas-phase concentration with height and low solubilities. Utilizing elementary mass-transfer considerations these conditions can be combined conveniently into a single dimensionless criterion. For general source conditions this is given by (Hales, 1972).

$$\frac{-3K_yH'c_y|_{\text{ground level}}}{cv_za\left|\frac{dc_y}{dz}\right|_{\max}} \gg 1, \qquad (8)$$

which reduces to

$$\frac{-3K_yH'\sigma_z e^{1/2}}{cv_za\exp(h^2/2\sigma_z^2)} \gg 1 \qquad (8a)$$

for the special case of a Gaussian, point-source plume. Here $|dc_y/dz)|_{\max}$ is the absolute value of the vertical concentration gradient experienced by the falling drop.

Experience has indicated (Dana *et al.*, 1973) that equilibrium scavenging conditions are satisfied whenever the values of these groups are greater than about 15; for SO_2 plumes from tall stacks this suggests that equilibrium scavenging should prevail beyond about 20 stack heights downwind. This extremely fortunate circumstance allows the calculation of SO_2 delivery to regional areas simply on the basis of gas-phase concentrations, rain rates and solubility. As mentioned previously, such an exercise has been recently completed by Hales, Dana and Glover (in press) for the northeastern United States.

Table 1. Values of the parameters K_1 and H

Temperature (°C)	K_1 (moles l^{-1})	H Dimensionless
0	0.0232	0.0136
10	0.0184	0.0196
18	0.0154	0.0270
25	0.0130	0.0332
35	0.0105	0.0445
50	0.0076	0.0673

Insofar as comprehensive regional models of sulfur pollution are concerned, it appears totally acceptable to incorporate the equilibrium-scavenging expression (Equation 2) as a means of calculating the direct SO_2 scavenging component. For local plumes expressions such as (3) or the available computer codes (Dana *et al.*, 1973) should be applied. If repeated calculations are to be made, the computer codes (which are more elaborate but require only small execution times) are recommended to avoid the tedium of hand calculation.

In summary, it is apparent that the calculation of direct SO_2 scavenging is reasonably well in hand, and little further research is needed in this general area. As mentioned earlier, the question of sulfite sequestering in solution was disregarded in this development; if this does occur to an appreciable extent then this paper's results should be reviewed accordingly. Further research in this specific area should be conducted to evaluate this effect; if it is found to be significant, the developments described in this paper should be restructured to reflect such behavior. Because of the generally favorable comparison between field calculations and measurements, however, the chances of any real changes in the calculational procedure in reflection of this effect appear to be small.

A second specific area where more information is needed is that of SO_2 concentrations in regional precipitation. As mentioned previously, such data are practically nonexistent at present and are badly needed to evaluate the real role of direct SO_2 scavenging in regional and global sulfur balances.

Finally, a cautionary note with regard to SO_2 scavenging calculations should be emphasized: one should never attempt to apply 'irreversible' or washout-coefficient theory to describe the scavenging of SO_2. Use of this approach in the past has resulted in the calculation of scavenging rates several orders of magnitude higher than the appropriate values; for most modeling applications it is far better simply to ignore SO_2 washout altogether than to make this conceptual error. Fortunately, the rather simple tech-

* It should be noted here that a variety of measurements of K_1 have been published (e.g. Sillen, 1964), resulting in evaluations of K_1 that are both higher and lower than those given by Johnstone and Leppla. In view of this uncertainty the numbers in Table 1 may be considered more appropriately as parameters in a curve fit rather than as absolute physical properties.

niques described in this section make this choice unnecessary, and provide a comparatively reliable method for assessing SO_2 concentrations in rain.

PATHWAY 2-4-6: DIRECT SULFATE SCAVENGING

Material balances: generalized computational approach

As stated in the previous section, the direct scavenging of sulfate aerosol is usually difficult to describe mathematically, both because it involves distributed particle-size systems and because it typically requires integrations over longer time periods of the hydrometeor's history.

In spite of these differences, however, these calculations share a fair degree of commonality with those for gas scavenging. The aqueous-phase material-balance Equation (1), describing transport of pollutant to a hydrometeor, can be expressed for the aerosol case as

$$\frac{dc_x}{dz} = \frac{3K_y}{v_z ac}(c_y), \tag{1a}$$

where the absence of the c_x term on the right reflects the irreversible nature of particulate scavenging. In principle, (1a) may be integrated over the raindrop's trajectory in a manner similar to that applied for gas scavenging; and indeed this may be completed in a straightforward fashion under conditions when c_y is well defined (e.g. Engelmann, 1968). The fact that c_y (and also the mass-transfer coefficient K_y) are usually unknown, however, leads to the typically low confidence levels that often characterize aerosol scavenging calculations.

The spatial and temporal distributions of sulfate aerosol can be described in terms of a microscopic material balance for the gas phase, given by the form (Hidy, 1973; Hales, 1972, 1976)

$$\frac{\partial c_y}{\partial t} = -\nabla \cdot c_y \mathbf{v} - w + r_y. \tag{9}$$

Here the time rate of change of pollutant concentration in a differential volume element of space is equated to the sum of contributions from divergence, scavenging, and reaction terms: $\mathbf{v}$ represents the velocity vector of airborne pollutant and $-w$ accounts for its removal by scavenging. In physical terms w is the amount of pollutant transferred from the gas phase to the aqueous phase per unit volume per unit time, at a point x, y, z in space. It is related to the local scavenging coefficient by the form

$$w = \Lambda c_y. \tag{10}$$

Equation (10) may be considered a *definition* of the scavenging coefficient.

Under specific conditions Equation (9) reduces to simple forms commonly encountered in published scavenging calculations. For a nonreactive aerosol in a well-mixed volume, for example,

$$\frac{\partial c_y}{\partial t} = \frac{dc_y}{dt} = -w = -\Lambda c_y. \tag{9a}$$

It is important to note that, in contrast to Equation (10), special forms such as (9a) should not be considered as general definitions of Λ. Failure to recognize this fact has led to severe errors in several previous scavenging analyses.

A general procedure for direct sulfate scavenging analysis using Equation (9) can be stated as follows: (1) Obtain transport information to define velocity vector $\mathbf{v}(x, y, z, t)$ in Equation (9). This information may be in the form of measured wind and diffusion data, or it may be generated as output of previous modeling calculations. (2) Obtain quantitative information regarding Λ and (if necessary) r_y. (3) Solve (9), subject to appropriate initial and boundary conditions, to obtain $c_y(x, y, z, t)$. Such solutions may be numerical, or (if possible) analytical in nature. (4) If concentration data in the aqueous phase are required, obtain solutions to (1a) corresponding to the computed values of c_y. This involves decomposition of the overall scavenging coefficient for application to specific drop sizes. This algorithm is rather unsatisfying in a pragmatic sense, because it gives very little specific guidance in the performance of actual scavenging calculations. In view of the diversity of situations occurring in the atmosphere, however, it is impossible to provide a more definitive treatment without loss of generality. Obviously, a great deal of meteorological judgement must be exercised in applying this procedure to obtain calculations of scavenging rates. Observance of this general procedure is extremely important, however, because it provides a systematic approach that helps to avoid many of the conceptual pitfalls that have plagued some previous attempts at scavenging calculation.

Examples where the above procedure has been applied to obtain scavenging rates for relatively simple situations are abundant (e.g. Engelmann, 1968). This algorithm has been applied as well for more complex systems, involving detailed numerical analyses. Key examples of this type of application are the convective-storm scavenging model by Hane (1974) and the frontal-storm scavenging model of Kreitzberg (in press).

Microphysical attachment mechanisms

The variety of mechanisms leading to the transport of aerosol to the aqueous phase (and thus contributing to the magnitude of Λ) is a subject of continuing conjecture (Hidy, 1973). The individual mechanisms most commonly thought to contribute to the interphase transport process are: impaction; interception; diffusion (Brownian motion); turbulent transport; nucleation; thermo- and diffusiophoresis and electrical effects. The first four of these mechanisms depend primarily upon the size and density of the aerosol particles, and thus can be expected to be similar for aerosols of similar size and mass distributions, regardless of their chemical composition. Thermophoresis and electrical effects may be expected to depend upon chemical composition to some extent, owing to their respective dependence on thermal conductivity and dielectric constant. Nucleation is extremely composition-dependent, and can be expected to be a highly effective transport mechanism for hydroscopic materials, such as sulfuric acid and its common salts.

The relative significance of the various attachment mechanisms can be expected to vary markedly as a function of meteorological circumstances, and accordingly, it is convenient to classify the present discussion according to whether transport is occurring within a condensing cloud or in areas populated by large falling raindrops with relatively few cloud particles.

The latter case has been reviewed by Slinn (1974), who has concluded that—given a fixed aerosol size distribution—the primary interphase transport mechanisms are impaction, interception, and diffusion. Slinn estimates the impaction contribution to be

$$E_m = [(S - S_*)/(S + C)]^{3/2}$$

where $C = 2/3 - S_*$. $S = \tau v_z/R$ is the Stokes number, in which τ is the particle stopping time and v_z is the drop's settling speed, and where $S_* = [(12/10) + (1/12)\ln(1 + \mathrm{Re})]$ is the critical Stokes number. There is considerable uncertainty about the interception contribution: Slinn suggests that Fuchs' value,

$$E_n = 3a/R,$$

where a is the particle radius and R is the drop radius may be approximately correct but for incorrect reasons. Finally, Slinn gives for the Brownian diffusion contribution to the collision efficiency

$$E_b = 4\mathrm{Sh}/\mathrm{Pe}$$

where Sh is given by the Froesling correlation (Equation 5) and $\mathrm{Pe} = Rv_z/D$, in which D is the particle diffusivity is the Peclet number.

Total capture efficiencies $E(a, R)$ can be estimated by summing the above contributions. These then can be integrated over the raindrop and aerosol size distributions (provided these are known) to obtain the appropriate scavenging coefficient (Dana and Hales, 1975). For example, the mass-average scavenging rate is given by

$$\Lambda_3 = \pi \int_0^\infty \int_0^\infty v_z(R) R^2 E(a, R) N(R) f(a)\, \mathrm{d}R\, \mathrm{d}a.$$

Here f is the probability-density function for the mass distribution of the aerosol particles, and $v_z(R)$ and $N(R)$ are the terminal velocity and number density function respectively for the raindrops.

Example integrations of (12) for log-normally distributed spectra have been completed by Dana and Hales (1975), leading to plots of Λ_3 vs geometric mean particle size as shown in Fig. 2. This figure pertains to particular meteorological circumstances and care should be exercised in its application to individual situations. Its primary importance in the present context is its indication of how seriously in error one can be by applying inappropriate particle statistics to scavenging calculations. Consider, for example, the calculation of the washout coefficient of an aerosol with a 0.1 geometric mean particle radius and geometric standard deviation of $\sigma_g = 2.5$. As a first approximation one might be tempted to assume the aerosol can be characterized by a homogeneous ($\sigma_g = 1$) ensemble of particles having the indicated geometric mean radius. As can be seen from Fig. 2, such an assumption would lead to an estimate that is almost three orders of magnitude low. Figure 2 also indicates the potential for an aerosol size distribution to be modified via the scavenging process—an effect which can be expected to change the magnitude of the effective washout coefficient as scavenging progresses. This process must be considered competitive with other mechanisms of aerosol growth and removal however, and aside from some rather interesting preliminary studies of these effects (e.g. Radke *et al.*, 1974; Graedel, 1974; Peters 1977) there is relatively little direct experimental information currently available.

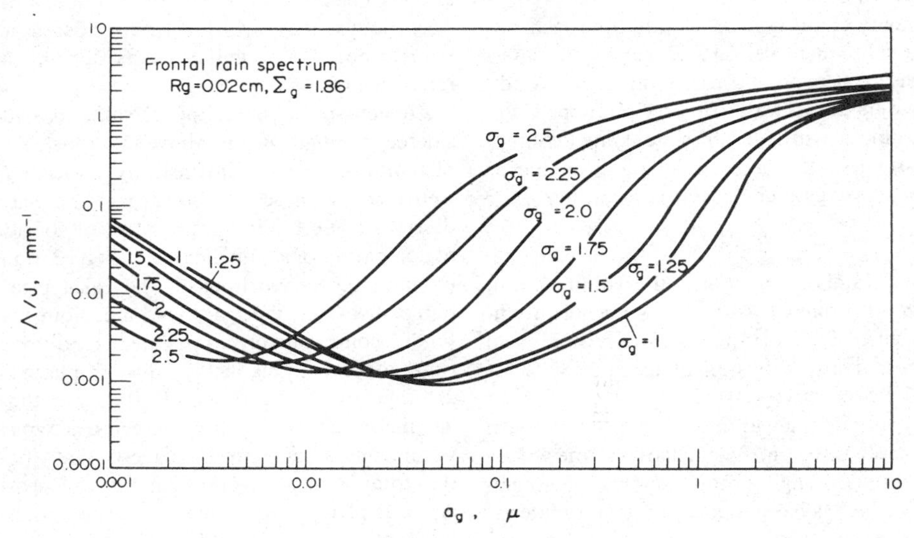

Fig. 2.

The propensity for error arising from the above effects is compounded with our present lack of knowledge with regard to the size-distribution spectra of atmospheric sulfate. Aside from knowing relatively little concerning the size spectra of sulfate aerosols under low r.h. conditions, their state under humidities approaching saturation is almost totally uncertain. Certainly, much valuable insight in this regard has been provided by studies such as those of Orr *et al.* (1958); Covert *et al.* (1972); Ludwig and Robinson (1967); Jaenicke (1978); Charlson *et al.* (1978) and Whitby (1978) but much more detailed and definitive information is needed, particularly with regard to aerosol particles of mixed composition. Given the significant nucleating capability of sulfate aerosol it seems reasonable to assume that the dominant pathway for below-cloud scavenging is first for the particles to grow by condensation of water, and subsequently be removed (with a correspondingly modified efficiency) by the mechanisms discussed. Some interesting evidence in support of this contention is given by Radke *et al.* (1974).

Adding the nucleation step to the scavenging pathway introduces the rather distasteful prospect of dealing with joint distributions of condensation droplet size and sulfate concentration. Much additional research remains to be completed in this area before reliable scavenging calculations can be achieved.

Insofar as scavenging within condensing clouds is concerned, similar mechanisms prevail. Here, however, one should expect nucleation processes to become more dominant (Hidy, 1973), especially for sulfate particles, which can be expected to compete effectively for cloud water with other types of aerosols. Attempts have been made to account rigorously for all in-cloud attachment mechanisms (e.g. Dingle and Lee, 1973); these have tended to be so complex, however, as to discourage widespread application. In reflection of this complexity a number of gross parameterizations have been suggested. Slinn (1977), for example, has reasoned that interphase transport efficiencies in many cloud systems should approach unity, and that subsequent delivery of scavenged material to the surface depends primarily upon the cloud's ability to lose its liquid water by precipitation. Combining these ideas with further simplifying assumptions, Slinn suggests that a reasonable approximation to the mean scavenging coefficient is given by

$$\bar{\Lambda} = J\bar{\epsilon}/2Rm,$$

where J is the rainfall rate, Rm is the volume–mean raindrop size at ground level and $\bar{\epsilon}$ is the mean efficiency with which the storm removes its water. Field data have been reasonably well fit using $\bar{\epsilon} = 1/3$ for scavenging by convective storms.

Similar types of rough approximations have been developed elsewhere; a particularly noteworthy example is the scavenging-ratio concept of Engelmann (1971), which is based upon an integral material balance over a storm system, and provides a convenient rule-of-thumb for estimating pollutant concentrations in precipitation on the basis of their corresponding values in air.

On the basis of this discussion one can conclude that direct sulfate-scavenging calculations are not at a satisfactory state of development at the present time. Although there are several factors contributing to this situation, it seems apparent that the most severe deficiency is our lack of knowledge regarding the character of sulfate-containing aerosols. Specifically, additional research emphasis should be focused on the following areas: (1) size distributions of sulfate-containing aerosols, (2) composition vs size relationships for sulfate-containing aerosols, and (3) size relationships of sulfate-containing aerosols as related to humidity.

Once adequate knowledge in these particular areas has been obtained, our capability to predict direct sulfate scavenging can be expected to advance significantly.

PATHWAYS 1–2–4–6 AND 1–3–4–6: SULFUR DIOXIDE REMOVAL BY THE COMBINED EFFECTS OF SCAVENGING AND CHEMICAL REACTION

Since the preponderance of anthropogenic sulfur emitted to the atmosphere is released in the form of sulfur dioxide, and since the majority of sulfur in precipitation occurs as sulfate, it is logical to presume that pathways 1–2–4–6 and 1–3–4–6 in Fig. 1 constitute the most important wet-removal routes for atmospheric sulfur. Conversion to sulfate by dry processes (or wet processes followed by droplet evaporation) with subsequent wet removal is simply the previously-discussed pathway 2–4–6, preceded by chemical reaction. Gas-phase reactions of sulfur dioxide have been considered in several recent reviews (Granat, 1975; Harrison *et al.*, 1976; Levy *et al.*, 1976) and are discussed at length elsewhere in this volume. Accordingly, these will not be considered here, other than to note their significance as steps in the overall removal process.

Aqueous-phase oxidation of sulfur dioxide is considered at length in the above citations as well. This phenomenon is more intrinsically associated with the wet-removal process, however, and accordingly deserves some brief mention here. In considering the net oxidation of sulfur dioxide, it is important to bear in mind the proportions of the gas typically distributed between the gaseous and aqueous phases. Under comparatively unpolluted conditions equilibrium concentrations in the aqueous phase (Equation 2) often are of the order of 10^4 times those in the gas phase; and combining this with a typical cloud-water content of $1\ g\,m^{-3}$ suggests that about 1% of the total sulfur dioxide loading can exist in the aqueous phase. This implies that even under relatively favorable conditions, bulk reactions in the aqueous phase must be moderately rapid if they are to account for a significant portion of the total conversion to sulfate.

Oxidation of sulfur dioxide in pure aqueous solution in the presence of oxygen atmospheres was investigated by Fuller and Crist (1941), who demonstrated that although oxidation definitely does occur under these circumstances, its rate is normally too slow to be significant as an atmospheric process. These and numerous subsequent authors (cf. Levy *et al.*, 1976) have investigated the effects of added trace materials as promoters for this reaction both in bulk and dispersed aqueous phases. Ammonia has received special attention in this regard because of its rather ubiquitous presence in the atmosphere, and because of its noted effectiveness in increasing the solubility of sulfur dioxide. Other trace materials investigated as promoters include copper, manganese, lead, vanadium, sodium, cobalt, and iron. The role of dissolved ozone and hydrogen peroxide as an oxidant also have been examined (Penkett *et al.*, 1977).

While these studies generally have supported the idea that aqueous-phase conversion can account for a significant fraction of total sulfate formation, they have not come to any real consensus with regard to reaction mechanism(s); nor have they provided rate expressions that can be employed with confidence in scavenging calculations. Considerable divergence of opinion exists, for example, with regard to the true influences of temperature, hydrogen-ion concentration, and sulfur-dioxide concentration on the conversion process. These laboratory studies have been complimented by some limited investigations of sulfur dioxide oxidation in actual rain samples (Hales *et al.*, in press). Sulfur dioxide in these samples was observed to decay slowly at temperature in the vicinity of the freezing point, and to become rapid at temperatures approaching 20°C. Aside from providing a semiquantitative indicator, however, these measurements cannot be considered to be of any large benefit to the general problem of modeling wet-removal processes.

If reliable parameterizations of aqueous-phase conversion rates were available, then incorporating them into a description of the scavenging process could be conducted in a relatively straightforward manner. One fortunate circumstance in this respect is the fact that, in cloud environments, at least, the chemical reaction step is rate limiting, and one can use (2) as a steady-state approximation to describe the aqueous-phase sulfur dioxide concentration. A criterion for chemical-reaction limited scavenging has been derived previously (Hales, 1972) as

$$\left|\frac{ak_1 c}{3K_y H'}\right| \ll 1 \tag{14}$$

and (8) satisfied,

where k_1 is a psuedofirst-order rate constant for the reaction process.* Applying (14) for the case of sulfur dioxide suggests reaction-limited conditions prevail in clouds for reactions having half-times of the order of a few seconds or more.

Reaction-limited conditions may not prevail in the case of rain scavenging, especially in situations involving local plumes where (8) is unsatisfied. Even under these conditions, however, scavenging can be calculated in a rather straightforward manner using extensions of gas-scavenging theory. The computer codes described by Adamowicz and Hill (1977) and by Dana *et al.* (1976) are key examples of such applications.

In conclusion it can be stated that the primary research need in this area is that of obtaining reliable parameterizations of aqueous-phase conversion rates. Because of the relative simplicity involved in composing serviceable scavenging models with simultaneous mass transfer and chemical reaction, future laboratory measurements in this area should be conducted with bulk, rather than dispersed aqueous systems. The latter type of experiment (water-aerosol systems, supported droplets, etc.) contributes nothing additional to the elucidation of the major problem, and usually involves considerable unnecessary experimental difficulty.

It seems apparent that much research remains to be completed in the reaction-rate measurement and parameterization area. Once this is brought to a satisfactory stage, however, a substantially improved basis for sulfate scavenging calculations can be expected to evolve.

Acknowledgements—This work was completed under support of the Division of Biomedical and Environmental Research of the U.S. Energy Research and Development Administration. The author expresses his sincere appreciation to this agency for their support. Special thanks are due also to Dr. W. G. N. Slinn and Dr. Lennart Granat for their constructive and helpful comments, which were utilized in preparing the final draft of this paper.

* K_1 is defined here in terms of the rate equation

$$r_x = -k_1 C_x$$

where r_x is the rate of production (moles/volume of aqueous phase, time) the pollutant species in the aqueous phase.

REFERENCES

Adamowicz R. F. and Hill F. B. (1977) A model for the reversible washout of sulfur dioxide, ammonia and carbon dioxide from a polluted atmosphere, and the production of sulfates in raindrops. *Atmospheric Environment* **11**, 917–927.

Bird R. B., Stewart W. E. and Lightfoot E. N. (1960) *Transport Phenomena.* Wiley, New York.

Bolin B. and Charlson R. J. (1976) On the role of the tropospheric sulfur cycle and the shortwave radiation climate of the earth. *Ambio* **5**, 1–9.

Bolin B. and Persson C. (1975) Regional dispersion and deposition of atmospheric pollutants with particular application to sulfur pollution over western Europe. *Tellus* **27**, 281–308.

Chamberlain A. C. (1973) Deposition of SO_2 by gaseous diffusion. Paper presented at Symposium on Turbulent Diffusion, Charlottesville, North Carolina, April 1973.

Charlson R. J., Covert D. S., Larson T. V. and Waggoner A. P. (1978) Chemical properties of tropospheric sulfur aerosols. *Atmospheric Environment* **12**, 39–53.

Conway E. J. (1943) Mean geochemical data in relation to oceanic evolution. *Proc. R. Irish Acad., A* **48**, 119–159.

Culvert D. S., Charlson R. J. and Ahlquist N. C. (1972) A study of the relationship of chemical composition and humidity to light scattering by aerosols. *J. appl. Met.* **11**, 968–976.

Dana M. T. and Hales J. M. (1975) Statistical aspects of the washout of polydisperse aerosols. *Atmospheric Environment* **10**, 45–50.

Dana M. T., Hales J. M. and Wolf M. A. (1972) Natural Precipitation Washout of Sulfur Dioxide. Final Report to EPA, Division of Meteorology, Contract BNW 389.

Dana M. T., Hales J. M. and Wolf M. A. (1975) Rain scavenging of SO_2 and sulfate from power plant plumes. *J. geophys. Res.* **80**, 4119–4129.

Dana M. T., Hales J. M., Slinn W. G. N. and Wolf M. A. (1973) Natural Precipitation Washout of Sulfur Compounds from Plumes. Final Report to EPA Meteorology Laboratory, EPA-R3-73-047.

Davenport H. M. and Peters L. K. (1977) Field studies of atmospheric particulate concentration changes during precipitation. *Atmospheric Environment* (in press).

Dingle A. N. and Lee Y. (1973) An analysis of in-cloud scavenging. *J. appl. Met.* 1294–1302.

Eatough D. (1977) Brigham Young University, Provo, Utah. Personal Communication.

Engelmann R. J. (1971) Scavenging prediction using ratios of concentrations in air and precipitation. *J. appl. Met.* **10**, 493–497.

Engelmann R. J. (1968) The Calculation of Precipitation Scavenging. In *Meteorology and Atomic Energy* (Edited by Slade D. H.). USAEC, Oak Ridge.

Eriksson E. (1963) The yearly circulation of sulfur in nature. *J. geophys. Res.* **68**, 4001–4008.

Friend J. P. (1973) The Global Sulfur Cycle. In *Chemistry of the Lower Atmosphere* (Edited by Rasool S. I.). Plenum Press, New York.

Fuller E. C. and Crist R. H. (1941) The rate of oxidation of sulfite ions by oxygen. *J. Am. chem. Soc.* **63**, 1644.

Graedel T. F. and Franey J. P. (1974) Field measurements of submicron aerosol washout by rain. *Precipitation Scavenging—*1974. R. W. Beadle and R. G. Semonin, Coords. ERDA Symposium Series CONF—741014. In press.

Granat L. (1978) Sulfate in precipitation as observed by the European Atmospheric Chemistry Network. *Atmospheric Environment* **12**, 413–424.

Granat L. (1972) On the relationship between pH and the chemical composition of atmospheric precipitation. *Tellus* **24**, 550–560.

Granat L. (1975) Conversion Rate of Sulfur Dioxide Under Various Atmospheric Conditions. Final Report on Nordforsk Research Project, University of Stockholm, Stockholm, Sweden.

Haagen-Smit A. J. and Wayne L. G. (1968) Atmospheric Reactions and Scavenging Processes. In *Air Pollution* (Edited by Stern A.) Vol. 1, 2nd edn. Academic Press, New York.

Hales J. M. (1972) Fundamentals of the theory of gas scavenging by rain. *Atmospheric Environment* **6**, 635–659.

Hales J. M. (1975) Atmospheric Transformations of Pollutants. In *Lectures on Air Pollution and Environmental Impact Analyses.* (Edited by Haugen D. A.). American Met. Soc., Boston.

Hales J. M., Dana M. T. and Glover D. W. Sampling for volatile trace constituents on natural precipitation. *Atmospheric Environment* (in press).

Hales J. M. and Sutter S. L. (1973) Solubility of sulfur dioxide in water at low concentrations. *Atmospheric Environment* **7**, 997–1001.

Hales J. M., Wolf M. A. and Dana M. T. (1973) A linear model for predicting the washout of pollutant gases from industrial plumes. *Am. Inst. chem. Engng J.* **19**, 292–297.

Hane C. E. (1974) Precipitation Scavenging in a Squall Line. Pacific Northwest Laboratory Annual Report to USAEC/DBER, BNWL-1850.

Harrison H. T., Larson T. V. and Hobbs P. V. (1976) Oxidation of sulfur dioxide in the atmosphere: a review. IEEE Annals No. 75 (H4004-1 23-1) p. 1–7.

Hidy G. (1973) Removal of gaseous and particulate pollutants. In *Chemistry of the Lower Atmosphere* (Edited by Rasool S. I.). Plenum Press, New York.

Hitchcock D. R. (1976) Atmospheric sulfates from biological sources. *J. Air Pollut. Control Ass.* **26**, 210–215.

Hutcheson M. R. and Hall F. P. (1975) Sulfate washout from a coal-fired power plant plume. *Atmospheric Environment* **8**, 23–28.

Jaenicke R. (1978) Physical properties of atmospheric particulate sulfur compounds. *Atmospheric Environment* **12**, 161–169.

Johnstone H. F. and Leppla P. W. (1934) Solubility of SO_2 at low partial pressures—ionization constant and heat of ionization of H_2SO_3. *J. Am. chem. Soc.* **56**, 2233–2247.

Junge C. E. (1960) Sulfur in the atmosphere. *J. geophys. Res.* **65**, 227–237.

Junge C. E. (1963) *Air Chemistry and Radioactivity.* Academic Press, New York.

Kellogg W. W., Cadle R. D., Allen E. R., Lazrus A. L. and Martell E. A. (1972) The sulfur cycle. *Science* **175**, 587–596.

Kreitzberg C. W. Precipitation cleansing computation in a numerical weather prediction model. *Fate of Pollutants.* Wiley, New York (in press).

Levy A., Drewes R. R. and Hales J. M. (1976) SO_2 Oxidation in Plumes: A Review and Assessment of Relevant Mechanistic and Rate Studies. Final Report to EPA. EPA-450/3-76-022.

Likens G. E. (1976) Acid precipitation. *Chem. Engng News* (Nov. 22, 1976) 29–44.

Ludwig F. L. and Robinson E. (1968) Variations in the size distributions of sulfur containing compounds in urban aerosols. *Atmospheric Environment* **2**, 13–23.

Nisbet I. (1975) Sulfates and Acidity in Precipitation Air Quality and Stationary Source Control. NAS/NAE Report to U.S. Senate Committee on Public Works Ser. 94–4.

Orr C., Hurd F. K. and Corbett W. J. (1958) Aerosol size and relative humidity. *J. Colloid Interface Sci.* **13**, 472–482.

Penkett S. A., Jones B. M. R. and Brice. K. A. (1977) Rate of oxidation of sodium sulphite solutions by oxygen, ozone, and hydrogen peroxide and its relevance to the formation of sulphate in cloud and rainwater. AERE-R 8584.

Radke L. F., Hindman E. E. and Hobbs P. V. (1974) A case study of rain scavenging from a kraft process paper mill plume. *Precipitation Scavenging*—1974. R. W. Beadle and R. G. Semonin, Coovds. ERDA Symposium Series CONF-741014. In press.

Rasmussen K. H., Taheri M. and Kabel R. L. (1974) Sources and Natural Removal Processes for Some Atmospheric Pollutants. Report to EPA, No. EPA—650-/4-74-032.

Robinson E. and Robbins R. L. (1968) Sources, Abundance and Fate of Gaseous Atmospheric Pollutants. Stanford Research Institute Report to API Project PR-6755.

Sillen L. G. (1964) *Stability Constants of Metal-Ion Complexes,* 2nd edn. Chem. Soc. (London) Special Publication No. 17.

Slinn W. G. N. (1977) Some approximations for the wet and dry removal of particles and gases from the atmosphere. *J. Wat. Air Soil Pollut.* **7**, 513–543.
Whelpdale D. M. (1975) Dry Deposition over the Great Lakes. Proc. of Atom.-Surface Exchange Symp., Richland, WA.
Whitby K. T. (1978) The physical characteristics of sulfur aerosols. *Atmospheric Environment* **12**, 135–159.

Atmospheric Environment Vol. 12, pp. 401–406. Pergamon Press 1978. Printed in Great Britain.

SULPHUR AND NITROGEN CONTRIBUTIONS TO THE ACIDITY OF RAIN

A. R. W. Marsh

Central Electricity Research Laboratories, Leatherhead, Surrey, England

(*First received* 16 *June* 1977 *and in final form* 27 *September* 1977)

Abstract—The dominant ions contributing to the acidity of precipitation are usually SO_4^{2-}, NO_3^- and NH_4^+ after allowance for sea salts.

In remote areas with low concentrations of SO_2 theoretical estimates suggest that the washout of the gas is unlikely to produce marked changes in the pH of rain although the presence of NH_3 enhances the final sulphate concentration.

The contribution of washout of SO_2 gas and sulphate aerosol is unlikely to contribute more than 50% to the final concentration of SO_4^{2-} in precipitation in remote areas.

The chemical composition of precipitation in remote areas shows a correlation between the ions SO_4^{2-}, NO_3^- and NH_4^+. Some physical mechanisms such as the variation in composition with rain density clearly contribute to the correlation but it appears that there are chemical mechanisms correlating these ions. A similar correlation is also found in aerosols nearer source areas. The role of NH_3 is straightforward acid neutralisation but the relationship between NO_3^- and SO_4^{2-} is more difficult to define. It appears that the nitrate is formed after sulphate production.

INTRODUCTION

The acidity of rain is governed by the overall charge balance equation of the ions in solution. All ions contribute to determine the final pH but normally the pH is dominated by ions at greater or equal concentration to the hydrogen ions. Acid rain observations by the recent O.E.C.D. project, 'Long Range Transport of Air Pollutants, 1972–1975', have shown that acid rain is widespread in Europe and that the range of pH found at O.E.C.D. sites in a relatively remote area, e.g. in S. Norway, is comparable with that found at relatively urban sites in the U.K. Figure 1 shows a histogram comparison of daily pH readings in the two countries over the period October 1972 to March 1975. There is relatively little difference between the distributions. To account for the surprising acidity in the case of S. Norway, the total ionic composition of the precipitation has to be examined. The average composition of precipitation collected at O.E.C.D. sites in S. Norway is shown in Fig. 2. The dominant ions after correction for sea salts are SO_4^{2-}, NO_3^- and NH_4^+. The mechanisms by which these species are scavenged by precipitation determine the acidity of rain.

Precipitation scavenging is usually divided into two types of process, those associated with cloud phenomena called rainout and those concerned with removal by falling precipitation called washout.

It is at present possible to describe theoretically washout processes but no adequate theory of rainout exists. Easter and Hobbs (1974) have, however, estimated the increase in cloud condensation nuclei after an air mass containing SO_2 and NH_3 passes through a lee wave cloud. This represents the first stages of quantifying rainout. Despite the lack of quantitative estimates of rainout the relative contribution of washout and rainout to the final ionic concentrations can be estimated by quantifying the washout processes.

This paper describes some theoretical aspects of washout of SO_2, NH_3 and sulphate aerosols to compare calculated concentrations with observations. A feature of the observations is a close relationship between the ions SO_4^+, NH_4^+ and NO_3^- and some possible reasons for this relationship are discussed.

THEORETICAL

The washout of a soluble gas involves gas and liquid phase mass transfer rates and the solution equilibria of the gas. For SO_2, these solution equilibria involving sulphurous acid are pH dependent, and low pH suppresses the solubility of SO_2 and the ionisation of the acid. The washout of SO_2 by rain falling through a uniform concentration of SO_2 is a function of; (i) the size spectrum of droplets and hence the rate of raining; (ii) the initial pH of the rain; (iii) the height of the SO_2 concentration and (iv) the absolute magnitude of the SO_2 concentration. In general small rain drops, <1 mm dia, are saturated with SO_2 while larger droplets do not achieve saturation.

Hales (1972) described the fundamental theory of gas scavenging by rain and all the theoretical calculations described below used his equations. The well-mixed drop model equations have been used because both Hales (1972) and these calculations have confirmed that at rates of raining $\lesssim 2\,\text{mm h}^{-1}$, the liquid phase resistance to mass transport is insignificant. The terminal velocities and size spectrum of rain drops used in the calculations are those given by Best (1950).

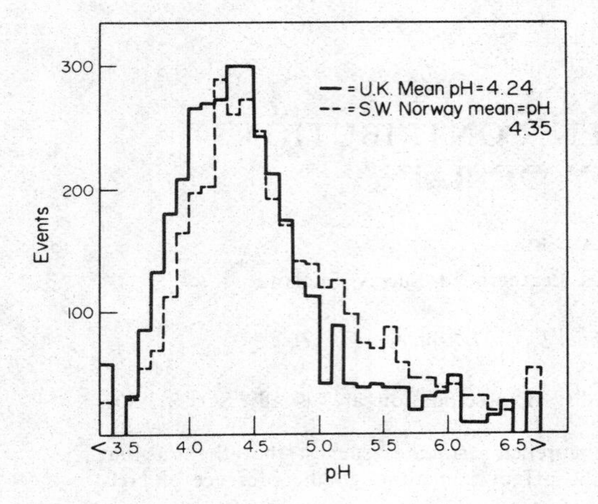

Fig. 1. Comparison of pH of precipitation at O.E.C.D. sites in U.K. and S.W. Norway, (weighted for equal area).

Figure 3 shows the theoretical changes in the pH of rain falling 200 metres through various uniform SO_2 concentrations. If the rain has an initial pH of 4.0 at cloud level this will not be changed by the washout of concentrations of SO_2 less than $100\ \mu g\ m^{-3}$. In remote areas where SO_2 concentrations are say $\sim 10\ \mu g\ m^{-3}$ rain initially of pH 5.0 would be decreased to ~ 4.9. The relatively few measurements of the pH of cloud water show a dependence on cloud type but a value of ~ 5.0 in remote areas has been observed (Petrenchock and Selezneva, 1970).

The SO_2 collected by the precipitation will be in the form of bisulphite and oxidation to sulphate will occur in minutes (Beilke, Lamb and Müller, 1975 and references therein), increasing the acidity as sulphuric is a stronger acid than sulphurous.

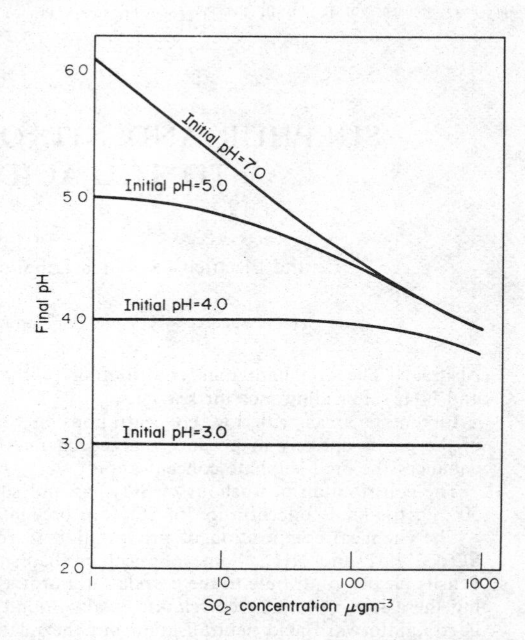

Fig. 3. The variation of the final pH of rain with the concentration of SO_2 as predicted by the well-mixed drop model for washout. (Height 200 m, rain 1 mm h^{-1}.

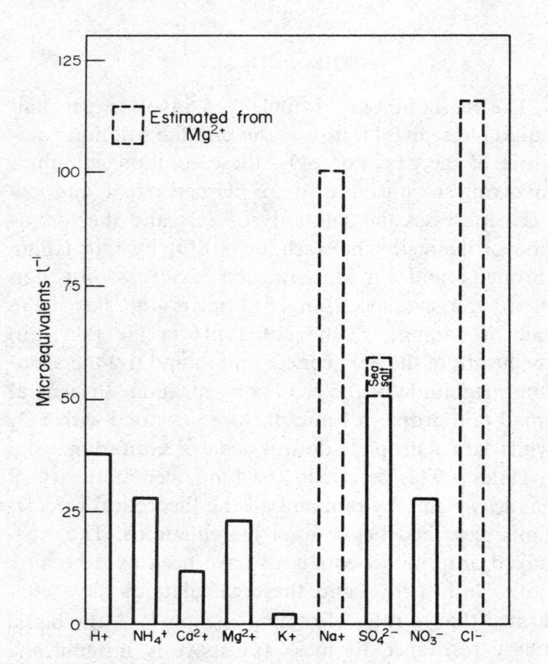

Fig. 2. Average composition of precipitation over S.W. Norway. Period September 1974 to March 1975, Sites N01, N08, N09, N10.

The washout process can be described by a first order equation:

$$-dm/dt = \lambda m$$

where m is the mass of pollutant in the atmosphere at time t, $-dm/dt$ is its rate of removal and λ is the washout coefficient. For the washout of soluble gases such as SO_2 and NH_3, λ itself is not a constant but is dependent in a complex fashion on gas phase concentration. However, λ can be treated as a constant over narrow ranges of gas phase concentration and can be used to illustrate the synergistic effect of the joint washout of NH_3 and SO_2.

Figure 4 shows the calculated changes in λ for washout of $50\ \mu g\ m^{-3}$ of SO_2 in the presence of different amounts of NH_3. The presence of NH_3 enhances the rate of removal of SO_2 and in the presence of an equimolar or greater concentration of ammonia, the rate of removal of SO_2 approaches the limit determined by the gas phase resistance to mass transfer.

The chemistry of the absorption of NO_x in aqueous solutions is complex (Koval and Peters, 1960; England and Corcoran, 1974 and references therein), and estimates of final concentrations due to washout have not been calculated but any resultant nitrite or nitrate from the washout of NO_x will increase ammonia removal.

The joint washout of low concentrations of NH_3 and SO_2 differs from that of SO_2 alone in that the small droplets do not reach saturation although SO_2 uptake is enhanced being dependent on the rate of uptake of NH_3.

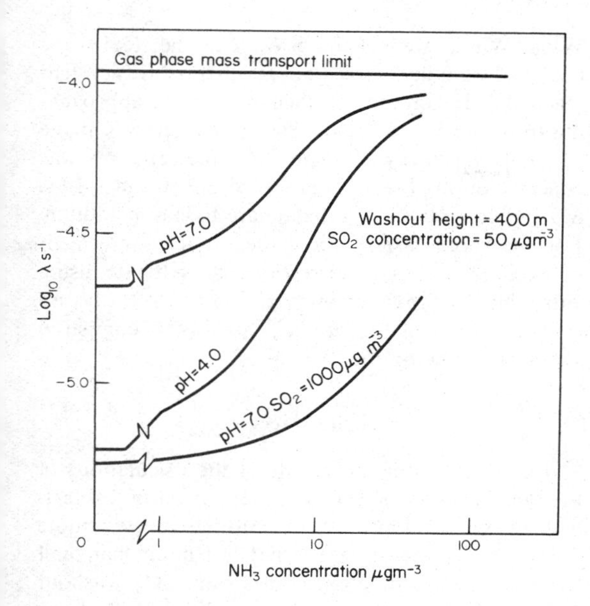

Fig. 4. The variation of the washout coefficient, λ, with the concentration of NH_3 at various initial rain pH's. (Rain $1\ mm\ h^{-1}$).

Table 1 shows the resultant concentration of sulphate and ammonium ions calculated from the joint washout of SO_2 and NH_3 at rainfall rates of 1 and $5\ mm\ h^{-1}$. The final sulphate concentration is calculated assuming all bisulphite is converted to sulphate and the pH is adjusted accordingly. It can be seen that although the concentration of sulphate rises in the presence of ammonia the pH remains almost the same. At higher rates of raining the concentration of both species is reduced.

Actual observations in remote areas such as S.W. Norway as observed in the O.E.C.D. project suggest that while SO_2 concentrations average about $5\ \mu g\ m^{-3}$, the sulphate in precipitation averages about 70 μEquiv/l and can be as high as 200 μEquiv/l. Ammonium concentrations average about 30 μEquiv/l and again values in excess of 100 μEquiv/l have been observed. Clearly washout of SO_2 and NH_3 gases alone cannot account for these concentrations.

Table 1. Washout of gases by rain for various SO_2 and NH_3 gas phase concentrations and an initial rain pH of 5.0 (Rain $1\ mm\ h^{-1}$ and $5\ mm\ h^{-1}$, height 1000 m)

Input $\mu g\ m^{-3}$			Output μEquiv/l		
SO_2	NH_3	pH	SO_4^{2-}	NH_4^+	pH
7.5	0	5.0	12	0	4.66
50	0	5.0	44	0	4.27
0	1.0	5.0	0	9	5.85
7.5	1.0	5.0	34	21	4.64
		$5\ mm\ h^{-1}$			
7.5	1.0	5.0	18	11	4.77

Table 2. Washout of sulphate aerosols by rain assuming uniform size distributions of 1 μm and 2 μm

SO_4^{2-} Aerosol $7.5\ \mu g\ m^{-3}$	SO_4^{2-} μEquiv/l $1\ mm\ h^{-1}$	$5\ mm\ h^{-1}$
1 μm	7	6
2 μm	22	11

The washout of aerosols is a strong function of size. Table 2 gives the calculated concentration of sulphate resulting from the washout of sulphate aerosol assuming all the mass to be of two different uniform sizes, 1 μm and 2 μm. The simple theory of Chamberlain (1953) was used and no hydroscopic or charge effects were considered. The same mass concentration of sulphate aerosol was used as the mass concentration of SO_2 used in Table 1 so that direct comparison of resultant concentrations can be made. In the atmosphere sulphate and ammonium aerosols normally exist with a size distribution such that the mass function has a peak $<1\ \mu m$ (Hidy, 1975) and it appears that the washout of a concentration of $7.5\ \mu g\ m^{-3}$ of such particles is unlikely to contribute more than $\sim 10\ \mu$Equiv/l of SO_4^{2-}.

OBSERVATIONS

Table 3 shows the very high correlations observed in the remote area of S. Norway between the ions H^+, NH_4^+, SO_2^{4-} and NO_3^- in the O.E.C.D. measuring programme over the period September 1974 to March 1975. Most of the O.E.C.D. data are confined to H^+ and SO_4^{2-} ion concentrations but for 400 site days at 4 sites N01, N08, N09 and N10 analysis was made for NH_4^+ and NO_3^- as well. Table 3 is based on these 400 results. The regression coefficients were

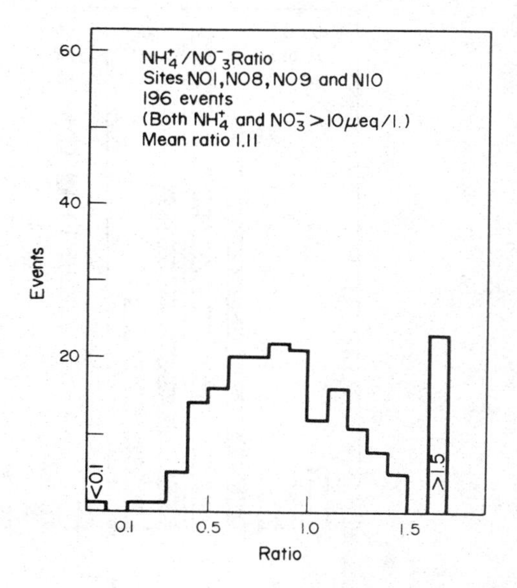

Fig. 5. The range of ratios of NH_4^+ to NO_3^- observed in precipitation over S.W. Norway.

Table 3. Correlation and regression coefficients for SO_4^{2-}, NO_3^- and NH_4^+ ions observed in precipitation over S.W. Norway

Dep. var.	Indep. var.	Corr. coefft.	b_{yx}	C
H^+	SO_4^{2-}	0.80	0.62	5.8
H^+	NO_3^-	0.74	0.93	12.2
NH_4^+	SO_4^{2-}	0.78	0.64	5.0
NH_4^+	NO_3^-	0.66	0.87	4.1
NO_3^-	SO_4^{2-}	0.78	0.49	2.2

determined using micro-equivalent units of concentration. The relationships between SO_4^{2-}, NO_3^- and NH_4^+ are not adequately represented by the average concentrations and Fig. 5 shows the range of ratios of NH_4^+/NO_3^- excluding very low concentrations < 10 μEquiv/l. This particular ratio is chosen because the average concentrations are approximately equal at about 28 μEquiv/l and the mean ratio is ~ 1.1. However, this apparent equality between NO_3^- and NH_4^+ only represents the mean of a very flat distribution of ratios with relatively few events actually having NH_4^+ equal to NO_3^-.

Figure 6 shows some results for aerosol and rain composition obtained at Leatherhead, just south of London. The concentrations of SO_4^{2-}, NO_3^- and NH_4^+ in different aerosol size ranges are compared with the composition of precipitation for three days in February 1977. The aerosol was collected by an Andersen high volume cascade impactor using Whatman 41 filter paper and rain samples were collected in a gauge with a lid operated by a conductive switch. 24 h samples were taken from noon to noon. On the 15/16 a depression was centred to the south of Ireland and an occluded front was approaching from the West. Winds were light S.W. On the 17/18 the occluded front had passed but another depression had reached S. Ireland and further fronts were approaching from the West. Winds were moderate S. On the 24/25 it was foggy all night and precipitation was collected during the foggy period at 6–7.30 a.m. GMT. Sea salt corrections to sulphate are based on sodium. For the three cases there is reasonable agreement between the sea salt corrections to sulphate using either Na^+ or Mg^{2+}, however, there was always an excess of Cl^- relative to Na^+ or Mg^{2+} compared with sea salt composition.

DISCUSSION

A comparison of the results of the calculations of the joint washout of SO_2 and NH_3 given in Table 1, with the O.E.C.D. project observations in the remote area of S. Norway, suggest that not more than half the SO_4^{2-} in precipitation comes from SO_2 washout and that this contribution is dependent on local gas phase NH_3 concentrations. While the calculations of the washout of sulphate aerosols is not so rigorous the results shown in Table 2 show that the washout of aerosols appear to contribute little to the observed concentrations especially when it is remembered that the mass distribution of sulphate is below 1 μm and the average concentration in remote areas is much less than 7.5 $\mu g\ m^{-3}$. These calculations are in agreement with the observations of Petrenchok and Selezneva (1970) who concluded that in remote areas washout accounted for $\sim 50\%$ of the final concentrations.

As washout in remote areas does not dominate the final concentrations in precipitation the factors affecting rainout processes need to be defined. As emissions disperse, conversion of gases such as SO_2 and NO_x

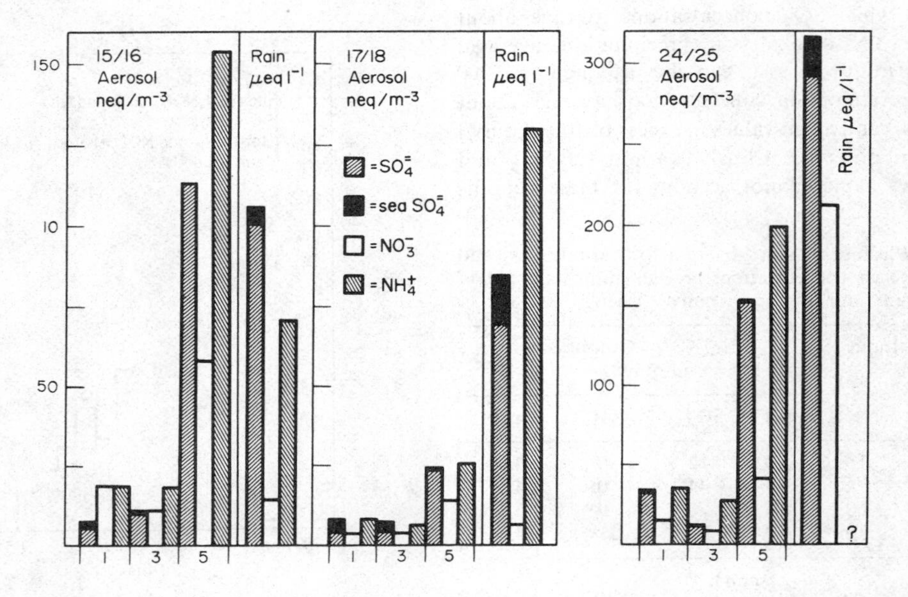

Fig. 6. The composition of aerosol and precipitation at Leatherhead for the days 15/16, 17/18 and 24/25 February 1977. Aerosol size ranges are 1 = >7 μm, 3 = 2.0–3.3 μm and 5 = <1.1 μm.

to aerosols occurs and at long distance from the source the aerosol phase should dominate. Aerosols should be more readily scavenged by rainout processes, such as condensation, than they are by the washout process of capture by falling droplets. The role of aerosols in rainout processes in remote areas should be more important than in the case of washout and the more remote the observation the better should be the correlation between the aerosol and precipitation composition. In less remote areas with higher gas phase concentrations the composition of precipitation should be regarded as a mixture of washout and rainout processes involving both local and distant sources and the composition of precipitation need not correlate strongly with local gas and solid composition.

Although insufficient data have been collected at Leatherhead to merit statistical analysis several features are apparent. There are generally good correlations between SO_4^{2-}, NO_3^- and NH_4^+ in both aerosol and precipitation. There is, however, a poor correlation between the composition, both ratios of ions and absolute magnitude, of aerosols and precipitation except for sea-salt components. The aerosol has large amounts of NH_4^+ and SO_4^+ in agreement with earlier observations (Lee *et al.*, 1974) and most is found in the size range $<1\ \mu m$. There is relatively less nitrate although again the greatest mass is found in the $<1\ \mu m$ size range. Harker *et al.* (1977) have reported that sulphuric acid aerosol will volatilise nitrate aerosol and lead to underestimates of nitrates. As the sulphuric acid aerosol is expected to be $<1\ \mu m$ one would expect the greatest underestimate in this size range. Similarly any adsorbtion of NH_3 from the gas phase is more likely in this size range.

There is an interesting increase in nitrate in precipitation falling through fog when washout of fog droplets should be fairly efficient. The aerosol collected shows no great increase in nitrate compared with previous observations and an easterly wind from the Continent may be more important, the nitrate being scavenged by the rainout of aerosols formed from the emissions of distant sources. Any effects of the fog on the cascade impactor would be expected to manifest themselves on the early stages and the slight increase in concentrations relative to the middle stages may be significant. However, even under these conditions the greatest amount of SO_4^{2-}, NO_3^- and NH_4^+ appears in the smallest size range of aerosol.

The precipitation results of the O.E.C.D. project in the area of S. Norway show a high correlation between the ions sulphate, nitrate and ammonium. A correlation between ions in precipitation may arise from two main physical processes. The first of these is dispersion of a common source. One can imagine an urban plume being dispersed and all components will be diluted by 'clean' air in transit from source to removal. Consequently concentrations of these species in precipitation will rise and fall together depending on the degree of dispersion before removal. The second process is that as the rate of precipitation increases the concentration of all species scavenged decreases.

These two processes are probably not the only factors; a chemical mechanistic relationship involving aerosols could be contributing. The relation between NH_4^+ and SO_4^{2-} and NO_3^- can be seen simply as acid neutralisation but the role of nitrate is intriguing.

The strong correlation between NO_3^- and NH_4^+ might indicate a common source e.g. NH_4NO_3, possibly in the large aerosol size range for easy washout by precipitation. There is some evidence for and some evidence against this suggestion. Thus within the O.E.C.D. project results for S. Norway one finds; (i) the ratio NH_4^+/NO_3 is rarely equal to one, (Fig. 5), although the average ratio is close to unity; (ii) ammonium nitrate aerosol would not account directly for the high correlation of NO_3^- with SO_4^{2-} although one could postulate involvement in the oxidation of SO_2 to SO_4^{2-}; (iii) ammonium nitrate should not be stable in the atmosphere; (iv) if the ions NH_4^+ and NO_3^- resulted from the washout of large particles one might expect to find large variations in concentrations at different sites and generally this does not appear to be the case. Leatherhead results show the greatest mass of nitrate in the smallest size range and indicate that nitrate is not as large a component as NH_4^+ and SO_4^{2-}. There are, however, two questions concerning aerosol measurements, namely the volatility of nitrate compounds (Harker *et al.*, 1977) and the capture efficiency of the samplers for giant particles (Wedding *et al.*, 1977).

A further study of aerosols in remote areas including more detailed size distribution with chemical analysis is required to resolve the possibility of ammonium nitrate aerosol being a significant contribution to the composition of precipitation.

There are other possible mechanisms for correlation the ions NO_3^-, NH_4^+ and SO_4^{2-} which involve the mechanisms of conversion of SO_2 to SO_4^{2-}. This can occur by two main routes: photochemical and aqueous. Both routes produce sulphuric acid which can be partially neutralized by ammonia. Brossett (1975) claims to have distinguished ammonium sulphate of varying degrees of neutralization as the products of both of these routes. A correlation between NH_4^+ and SO_4^{2-} would follow from these mechanisms but this does not explain the role of NO_3^-.

The predominant photochemical route for the oxidation of SO_2 is the reaction with OH radicals. The concentration of these radicals is controlled by the ambient illumination, O_3, NO, NO_2, CO and hydrocarbon concentrations. The OH radical is also effective in producing HNO_2 and HNO_3 in the vapour phase. Thus high conversion of SO_2 to SO_4^{2-} by this route would parallel NO_3^- formation. Calvert and McQuigg (1975) have suggested intermediates containing nitrogen and sulphur as the products of the OH and SO_2 reaction. These decompose with water to form a growing aerosol of H_2SO_4 and HNO_2 or HNO_3 which could then be stabilized by NH_3.

The aqueous oxidation of SO_2 is critically dependent on pH because the solubility of SO_2 depends on the ionisation of sulphurous acid (Beilke *et al.*, 1975). Thus NH_3 would enhance the rate of oxidation by maintaining a relatively high pH, but NO_2 although relatively insoluble would retard the oxidation. The aqueous oxidation of SO_2 can be catalysed by traces of transition metals such as iron but whether NO_2 would have any affect on these rates, except by a pH effect, is not known and merits a further study.

Both photochemical and aqueous routes can produce sulphuric acid aerosol which would be in equilibrium with the water vapour in the air; the photochemical route directly by nucleation and the aqueous route by subsequent evaporation. These acid aerosols can be up to 40–60% acid by weight, at relative humidities of $\sim 50\%$ and NO_2 is known to be soluble in such concentrated solutions. If the resulting solution absorbed NH_3 a stable aerosol of SO_4^{2-}, NO_3^- and NH_4^+ could result.

Although observations of nitrate aerosol concentrations are questionable it appears that in aerosols sulphate and ammonium ions predominate and there is an increase in relative nitrate concentrations in precipitation suggesting that nitrate is formed after sulphate production.

CONCLUSIONS

The dominant ions contributing to the acidity of precipitation are usually SO_4^{2-}, NO_3^- and NH_4^+ after allowance for sea-salts.

In remote areas with low concentrations of SO_2 theoretical estimates suggest that the washout of the gas is unlikely to produce marked changes in the pH of rain although the presence of NH_3 enhances the final sulphate concentration.

The contribution of washout of SO_2 gas and sulphate aerosol is unlikely to contribute more than 50% to the final concentration of SO_4^{2-} in precipitation in remote areas.

The chemical composition of precipitation in remote areas shows a correlation between the ions SO_4^{2-}, NO_3^- and NH_4^+. Some physical mechanisms such as the variation in composition with rain density clearly contribute to the correlation but it appears that there are chemical mechanisms correlating these ions. A similar correlation is also found in aerosols nearer source areas. The role of NH_3 is straight forward acid neutralisation but the relationship between NO_3^- and SO_4^{2-} is more difficult to define. It appears that the nitrate is formed after sulphate production.

Acknowledgements—The author would like to thank Dr G. M. Glover, Mr A. H. Webb and Mr R. J. Hards for their assistance with measurements made at Leatherhead.

REFERENCES

Beilke S., Lamb D. and Müller J. (1975) On the uncatalysed oxidation of SO_2 by oxygen in aqueous systems. *Atmospheric Environment* **9**, 1083–1090.

Best A. C. (1950) The size distribution of rain drops. *Q. J. R. met. Soc.* **76**, 16–30.

Brosset C., Andreasson K. and Ferm M. (1975) The nature and origin of acid particles observed at the Swedish West Coast. *Atmospheric Environment* **9**, 631–642.

Calvert J. G. and McQuigg R. D. (1975) The computer simulation of the rates and mechanisms of photochemical smog formation. Symp. on Chemical Kinetic data for the Upper and Lower Atmosphere, Warrenton, Virginia, 1974. *Int. J. chem. Kinet.* 113–154.

Chamberlain A. C. (1953) Aspects of travel and deposition of aerosol and vapour clouds, A.E.R.E. Report No. HP/R 1261.

Easter R. C. and Hobbs P. V. (1974) The formation of sulphates and the enhancement of cloud condensation nuclei in clouds. *J. atmos. Sci.* **31**, 1586–1594.

England C. and Corcoran W. H. (1974) Kinetics and mechanisms of the gas phase reaction of water vapour and nitrogen dioxide. *Ind. Eng. Chem. Fund.* **13**, 373–384.

Hales J. M. (1972) Fundamentals of the theory of gas scavenging by rain. *Atmospheric Environment* **6**, 635–650.

Harker A. B., Richards L. W. and Clarke W. E. (1977) The effect of atmospheric SO_2 photochemistry upon observed nitrate concentrations in aerosols. *Atmospheric Environment* **11**, 87–91.

Hidy G. M., 1975, Summary of the California aerosol characterisation experiment *J. Air. Pollut. Control Ass.* **25**, 1106–1114.

Koval E. J. and Peters M. S. (1960) Reactions of aqueous nitrogen dioxide. *Ind. Eng. Chem.* **52**, 1011–1014.

Lee R. E., Caldwell J., Akland G. G. and Fankhauser R. (1974) The distribution and transport of airborne particulate matter and inorganic components in Great Britain. *Atmospheric Environment* **8**, 1095–1109.

O.E.C.D. Long Range Transport of Air Pollutants, 1972–1975, data is issued by the Norwegian Institute for Air Research and given in a series of reports LRTAP 4/74, 4/75, 18/75, 19/75, 20/75 and 2/76.

Petrenchuk O. P. and Selezneva E. S. (1970) Chemical composition of precipitation in regions of the Soviet Union. *J. geophys. Res.* **75**, 3629–3634.

Wedding J. B., McFarland A. R. and Cermak J. E. (1977) Large particle collection characteristics of ambient aerosol samplers. *Enivor. Sci. Technol.* **11**, 387–390.

Atmospheric Environment Vol. 12, pp. 407–412. Pergamon Press 1978. Printed in Great Britain.

AN IMPROVED MODEL OF REVERSIBLE SO_2-WASHOUT BY RAIN

L. A. BARRIE
Atmospheric Environment Service, 4905 Dufferin Street, Downsview, Ontario, Canada

(*Received for publication* 22 *September* 1977)

Abstract—Removal of sulphur dioxide from the atmosphere by rain is an important feature of atmospheric sulphur budgets. The exchange of sulphur dioxide between a falling rain drop and air is complicated not only by the complex circulation inside the drop but also by the chemical reactivity of the gas with water. The latter characteristic results in a non-linear relationship between the liquid and gas phase SO_2 concentrations. This relationship is derived.

An improved method of modelling reversible SO_2 exchange between a falling rain drop and air is presented, which takes into account phenomena associated with the solution chemistry of SO_2 (e.g. ion-enhanced diffusion) as well as the microphysical effects of internal circulation.

The model is used to calculate the redistribution and washout of an SO_2 plume by raindrop spectra characteristic of drizzle and heavy rain. The fractional plume washout rate (% mm^{-1} rain) is inversely related to plume concentration and thickness. In a heavy rain (25 mm h^{-1}) washout from a 1000 ppb(v) SO_2 plume of 20 m thickness occurs at a rate of 56% h^{-1}. The relative importance of gas phase and liquid phase mass transfer is dependent on SO_2 atmospheric concentrations. At plume concentrations lower than 300 ppb(v) gas phase control dominates. Beneath a plume, harmful ground level SO_2 concentrations are unlikely to occur as a result of reversible washout by rain.

INTRODUCTION

Absorption and desorption of sulphur dioxide by rain drops plays an important role in its removal and redistribution in the atmosphere. In a general review of the theory of gas scavenging by rain, Hales (1972) pointed out the difficulties involved in calculating the rate of gaseous uptake or loss by a raindrop falling at terminal velocity with a complex internal circulation.

For sulphur dioxide, calculating gaseous exchange is complicated because its solubility in water is non-linearly related to the gas phase concentration. Attempts to use a linearized solubility-relationship in a washout model have proven useful in producing order-of-magnitude estimates (Hales *et al.*, 1973). However, it is generally agreed that for accurate predictions a more exact treatment is required. A washout model (EPAEC model) was developed by Dana *et al.* (1975), which successfully handled this problem but did not rigorously treat the transport of dissolved SO_2 into the droplet. Instead, two extreme situations were considered: (a) absorption by a stagnant drop; (b) absorption by a well-mixed drop. From one extreme case to the other the predicted concentration of SO_2 in rain varied by as much as an order of magnitude. Thus, there is clearly a need for a model that contains a more accurate description of SO_2 absorption and desorption by falling raindrops.

In this paper such a model is presented. It takes into account not only phenomena associated with the solution chemistry of SO_2 but also the micro-physical effects of internal circulation on mass transfer. The improved description of SO_2 gas exchange is employed in a treatment of plume redistribution by reversible washout.

SOLUTION CHEMISTRY OF SO_2

Sulphur dioxide exists in solution as physically dissolved SO_2 ($SO_2 \cdot H_2O$), bisulphite (HSO_3^-) and sulphite (SO_3^{2-}) (Scott and Hobbs, 1967). The equilibrium which is established between gaseous SO_2 and total-dissolved SO_2 is described by the following three equations.

Across the gas–liquid interface

$$(SO_2)_g + (H_2O)_L \rightleftarrows (SO_2 \cdot H_2O)_L$$

$$K_H = \frac{[SO_2 \cdot H_2O]_L}{[SO_2]_g}. \quad (1)$$

In the liquid

$$SO_2 \cdot H_2O \rightleftarrows H^+ + HSO_3^- \quad K_2 = \frac{[H^+][SO_3^{2-}]_L}{[HSO_3^-]_L}. \quad (2)$$

$$HSO_3^- \rightleftarrows H^+ + SO_3^{2-} \quad K_2 = \frac{[H^+][SO_3^{2-}]_L}{[HSO_3^-]_L}. \quad (3)$$

K_H, K_1, K_2 are equilibrium constants and the square brackets represent concentrations (mole cm^{-3}). The dissociation of $SO_2 \cdot H_2O$ and HSO_3^- are very rapid and equilibrium is for all intents and purposes instantaneously established (Beilke and Lamb, 1975).

A linear Henry's Law relationship does exist between physically dissolved SO_2 and gas-phase SO_2 (Equation 1). However, since $SO_2 \cdot H_2O$ dissociates, the concentration of total-dissolved SO_2 in solution

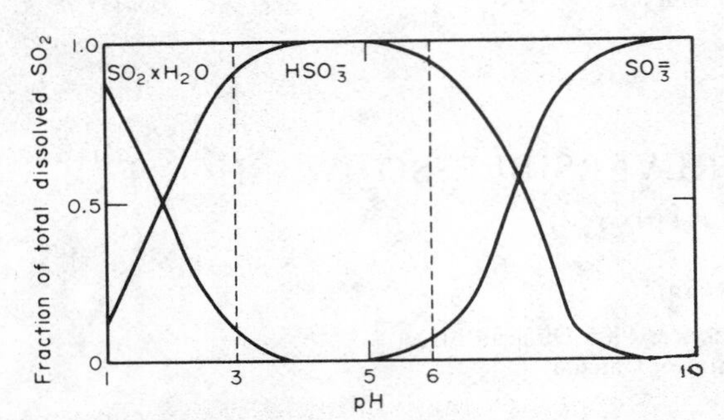

Fig. 1. Abundance of dissolved SO_2 species as a function of pH (25°C).

is related to the gas-phase concentration in a non-linear fashion. The exact form of the relationship is derived below.

SO_2 in solution is distributed amongst the three forms $SO_2 \cdot H_2O$, HSO_3^- and SO_3^{2-} in a manner that depends strongly on pH (Fig. 1) and weakly on temperature (at lower temperatures, curves shift slightly to the left). Between pH 3 and 6, the precipitation-pH range, over 90% of dissolved SO_2 exists as bisulphite. Below pH 6 less than 10% is present as sulphite. Equation 3, describing sulphite formation, can therefore be neglected in precipitation washout calculations since Equations (1) and (2) adequately describe the gas–liquid phase equilibrium.

Oxidation of SO_2 in raindrops during scavenging can also be neglected owing to short reaction times (0–5 min) and low pH's encountered (Beilke *et al.*, 1976). One exception might be in the presence of high concentrations (10^{-4} M) of manganous and ferric ions which rapidly form complexes with dissolved SO_2 (Barrie and Georgii, 1976). This paper does not attempt to deal with such situations.

In cases where SO_2 is the predominant acidic substance (viz. an SO_2 plume containing little or no particulate matter scavenged by neutral precipitation), electroneutrality in solution requires that

$$[H^+] = [HSO_3^-]_L. \tag{4}$$

By combining Equation (4) with Equations (1) and (2) and eliminating $[H^+]$ and $[SO_2 \cdot H_2O]$ one obtains an expression relating the gas-phase SO_2 concentration to the bisulphite ion concentration in equilibrium with the gas.

$$[HSO_3^-]_L^{eq} = \{K_H K_1 [SO_2]_g^{eq}\}^{1/2} \tag{5}$$

(superscript 'eq' denotes equilibrium). Equilibria in Equations (1) and (2) are established so rapidly that the above relationship holds true at the airdrop interface at any time.

MODELLING SO_2 EXCHANGE BETWEEN AIR AND A FALLING RAINDROP

The system modelled is a raindrop (radius 0.01–0.20 cm) falling at terminal velocity in air containing traces of sulphur dioxide. Raindrops contain a well developed internal circulation of 20–30 rev. s^{-1} (Pruppacher and Beard, 1970) which helps to distribute pollutants diffusing into them.

The two-layer model of Whitman (1923) for gas exchange between a well-stirred liquid and well-mixed air was used. Transfer of a gas is imagined to be limited by molecular diffusion through a laminar layer on the gaseous side and a laminar layer on the liquid side of the interface (Fig. 2). The thickness of the respective layers, δ_g and δ_L, depends on the nature of the flow field in the liquid and gas. In general, this approach, although over-simplified, predicts gaseous exchange as well as more sophisticated treatments (Danckwerts, 1970; Liss and Slater, 1974).

The flux of SO_2 through the air layer, F_g, is defined by:

$$F_g = \frac{D_g}{\delta_g}\{[SO_2]_g - [SO_2]_g^I\}. \tag{6}$$

D_g is the diffusivity of SO_2 in air and the gas-phase diffusion-layer thickness (δ_g) is given by

$$\delta_g = \frac{2r}{N_{Sh}} \tag{7}$$

where r is the droplet radius and N_{Sh} is the Sherwood number, the ratio of convective to diffusive mass transfer. This has been accurately determined experimentally for a rain drop falling at terminal velocity in a wind tunnel (Pruppacher and Beard, 1971):

$$N_{Sh} = 1.56 + 0.616\, N_{Re}^{1/2}\, N_{Sc}^{1/3}. \tag{8}$$

N_{Re} is the Reynolds number based on drop diameter and terminal velocity and N_{Sc} is the Schmidt number (ν/D_g) based on the viscosity of air ν. Equation 8 is applicable to droplets of radius 0.002–0.06 cm. For larger drops the Froessling equation

$$N_{Sh} = 2 + 0.6\, N_{Re}^{1/2}\, N_{Sc}^{1/3} \tag{9}$$

is applicable (Hales, 1972).

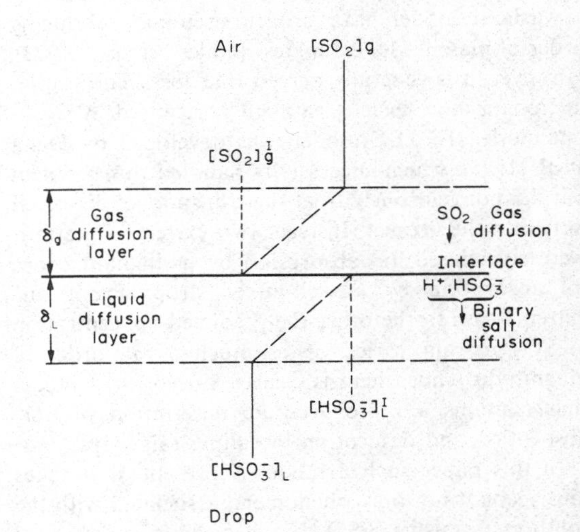

Fig. 2. Double-layer model applied to the air–drop interface.

In solution, SO_2 diffuses as bisulphite across the liquid-diffusion layer. The flux, F_L, is given by

$$F_L = \frac{D_L}{\delta_L}\{[HSO_3^-]_L^I - [HSO_3^-]_L\}. \quad (10)$$

D_L is the binary-salt diffusion coefficient calculated from the mobility of the hydrogen and bisulphite ions (Robinson and Stokes, 1970).

δ_L values were chosen for the model by considering extreme values it could have. Internal flow patterns (Le Clair *et al.*, 1972) suggest an upper limit of 20% of the drop radius. At the other extreme, δ_L cannot be less than the electrical-double-layer thickness (0.0003 cm) determined experimentally by Matteson and Giardina (1974). Thus, a value equal to 10% of the drop radius and falling between the above limits was used in this model. The sensitivity of absorption-desorption to this parameter is discussed below. The values of other model input parameters at 10°C are listed in Table 1.

Assuming that transport of gas across the interface is a steady-state process, it follows that

$$F_L = F_g \quad (11)$$

$$\frac{D_L}{\delta_L}\{[HSO_3^-]_L^I - [HSO_3^-]_L\} = \frac{D_g}{\delta_g}\{[SO_2]_g - [SO_2]_g^I\}. \quad (12)$$

At the interface $[SO_2]_g^I$ and $[HSO_3^-]_L^I$ are related by Equation 5 which can be used to eliminate $[SO_2]_g^I$ from Equation (12) and to obgain an expression for $[HSO_3^-]_L^I$ in terms of the ambient air concentration $[SO_2]_g$ and the droplet's bisulphite concentration $[HSO_3^-]_L$:

$$[HSO_3^-]_L^I = \frac{-1 + \sqrt{(1 + 4ZB\{[HSO_3^-]_L + Z[SO_2]_g\})}}{2ZB} \quad (13)$$

where

$$Z = \frac{D_g \delta_L}{\delta_g D_L}, \quad B = \frac{1}{K_1 K_H}.$$

Given initial gas-phase $[SO_2]_g$ and liquid-phase concentrations $[HSO_3^-]_L$, $[HSO_3^-]_L^I$ can be calculated from 13. Once $[HSO_3^-]_L^I$ is known, the flux of SO_2 into the drop can be calculated using Equation 10 and then from the flux the incremental increase (or decrease) in bisulphite ion concentration determined. This procedure was used to calculate the concentration of bisulphite ion in a rain drop as a function of time during absorption and desorption of SO_2.

Table 1. Constants in the model at 10°C

D_g	$0{\cdot}141\ cm^2\ s^{-1}$
D_L	$1.83 \times 10^{-5}\ cm^2\ s^{-1}$
ν	$0.141\ cm^2\ s^{-1}$
K_H	51.1
K_1	2.42×10^{-5} mole cm^{-3}

In general, SO_2 transfer to and from a raindrop is controlled by both gas and liquid phase diffusion processes. This is demonstrated in Fig. 3. Absorption and desorption curves are shown for a drop entering and leaving an atmosphere containing 1000 ppb(v) SO_2 at 10°C. Gaseous exchange predicted by the model is significantly different from that calculated assuming no liquid-phase resistance (dashed curve, Fig. 3). This conclusion is not altered when δ_L is varied within reasonable limits (factor of 2). Thus predictions of washout models in which liquid-phase resistance is neglected (e.g. Peters, 1976) must be considered to be first approximations whose accuracy will be discussed later.

Various aspects of drop absorption and desorption were investigated using the double-layer model. The following general characteristics were revealed:

(a) SO_2 absorption rates are strongly dependent on drop size (Fig. 4). The time to reach equilibrium decreases as drop size decreases.

(b) SO_2 desorption occurs more slowly than absorption (Fig. 3(a), (b)).

(c) For a given drop size, absorption (desorption) curves are strongly dependent on SO_2 concentration (Fig. 5). Such a dependence is caused by the non-linear solubility relationship of SO_2 (Equation 5). Such behaviour has been previously found (e.g. Hales, 1972).

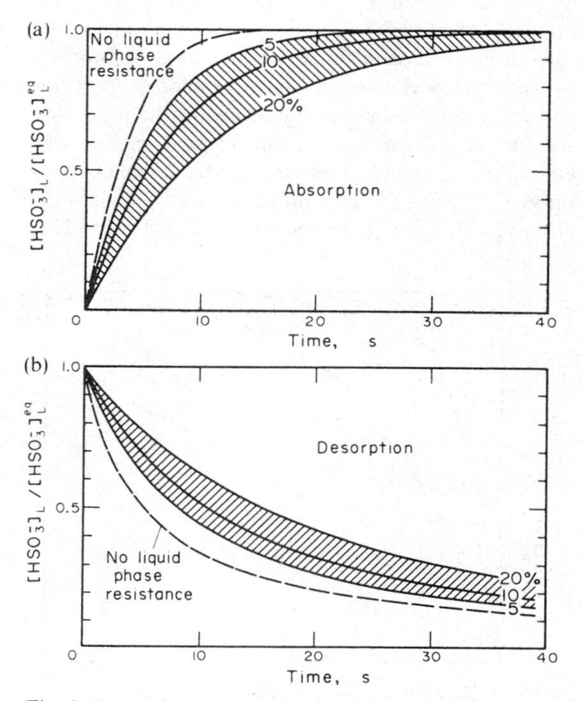

Fig. 3. Dependence of SO_2 uptake and loss by a 0.05 cm radius drop on liquid-phase resistance; shaded area curves are for δ_L = 5, 10 and 20% of the radius (1000 ppb(v) SO_2, 10°C).

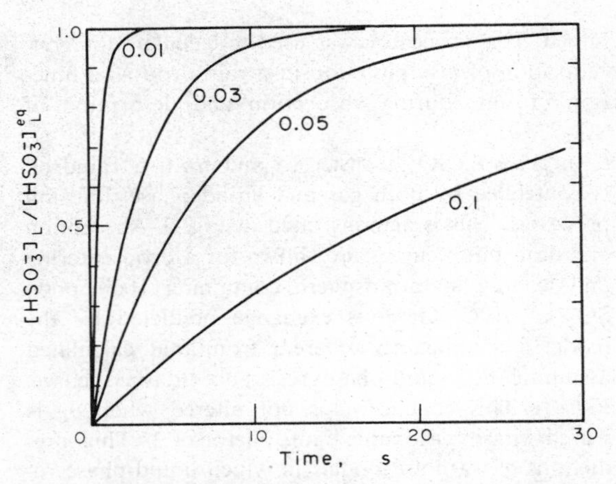

Fig. 4. Dependence of SO_2-absorption by raindrops on drop radius (cm) (1000 ppb(v) SO_2, 10°C).

PLUME WASHOUT

The need for more information on gaseous plume redistribution by rain was recognized by Slinn (1974). In an attempt to obtain order-of-magnitude estimates of plume washdown rates, he solved the convective diffusion equation analytically by making several simplifications. These included: (a) characterization of an entire raindrop size distribution by a single raindrop size, (b) linearization of the solubility relationship of the gas in question. The latter simplification is not strictly applicable to SO_2 since it obeys a non-linear solubility relationship (Equation 5). This in turn makes it very difficult to find a weighting function to calculate the drop size representative of an entire raindrop spectrum.

In order to circumvent these difficulties, a numerical approach was taken in which a raindrop spectrum and the double-layer model outlined in the preceding section were used. A one-dimensional model was employed in which rain fell vertically and continuously through a column of air 200 m thick with an initial SO_2 distribution shown in Fig. 6. For computational purposes, the column was divided into 0.1 m thick layers with which SO_2 concentrations were associated. Reversible SO_2 exchange was permitted between the drops and each layer.

A number density was assigned to 25 drop-size ranges in accordance with the following formulae (Best, 1952). These have been found to describe empirically a great number of measured raindrop spectra:

$$1 - S = \exp\{-(2r/a)^{2.5}\} \tag{14a}$$

$$a = 1.30\, I^{0.232} \tag{14b}$$

$$W = 67\, I^{0.846}. \tag{14c}$$

S is the fraction of rainwater in drops of radius less than r. I (mm h^{-1}) is precipitation intensity and W (mm^3 m^{-3}) the concentration of rainwater in air.

Plume washout was calculated for precipitation intensities of 0.5 (drizzle) and 25 (heavy rain) mm h^{-1} at a temperature of 10°C. The vertical distribution of SO_2 after 30 min of washout (Fig. 6a) confirms Slinn's (1974) conclusion that washdown of an SO_2 plume is very slow. Even in heavy rain the bulk of the plume was scarcely shifted. Resultant ground-level SO_2 concentrations were 2.8 and 66 ppb(v) in drizzle and heavy rain, respectively. Harmful ground-level concentrations of SO_2 are more likely to be caused by turbulent diffusion and vertical air currents induced by rainfall than by reversible washout.

Distribution of washed-out SO_2 amongst various raindrop sizes at the ground depends on the vertical distribution of the gas in the atmosphere. For the

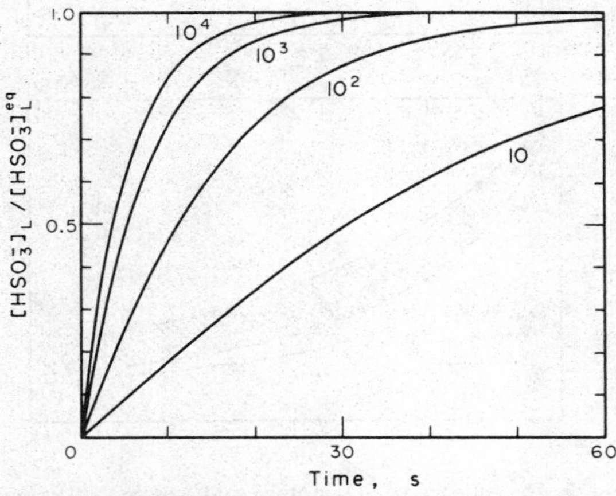

Fig. 5. Dependence of SO_2 absorption by 0.05 cm radius raindrops on SO_2 atmospheric concentration in ppb(v) (10°C).

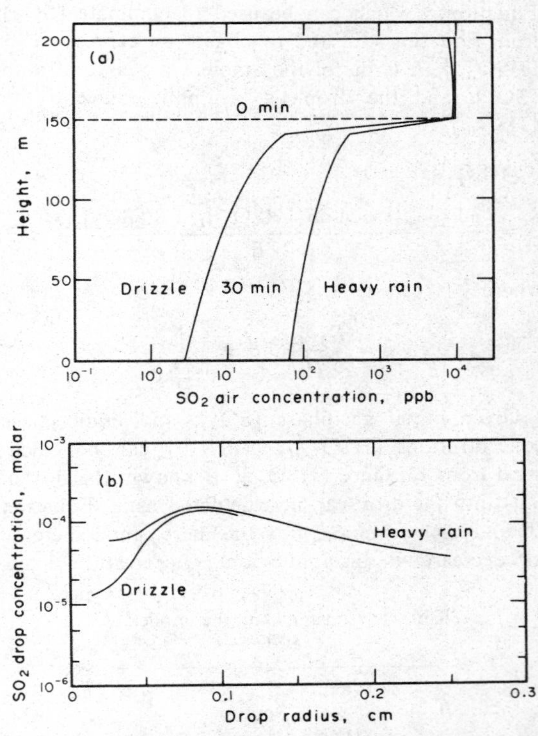

Fig. 6. SO_2 plume redistribution by drizzle and heavy rain (a) and the accompanying SO_2 concentration in rain at the ground after 30 min (b).

plume configuration used here, SO_2 is most concentrated in 0.085 cm drops (Fig. 6b). Smaller drizzle drops that equilibrate rapidly with their surroundings reflect ground-level concentrations. In this case, they have lower SO_2-concentrations because SO_2 ground-level concentrations are low. Bulk precipitation collected at ground level, 150 m below the 10^4 ppb(v) SO_2 plume contains 10^{-4} M SO_2 (3.2 ppm-S) and has a pH of 4. Even though it has passed through a plume this rain water sulphur concentration is not very high (sulphate levels of 1–8 ppm-S are commonly found in precipitation at rural sites in Eastern Canada).

The dependence of initial plume washout rate L (% h^{-1}) on precipitation intensity I (mm h^{-1}), plume thickness H (m, topped at 200 m) and initial plume concentration C (ppb SO_2) was investigated. In all cases, L is constant for the first 15 min of rain. In addition, it is almost linearly dependent on I for fixed C and H. The error involved in using an average normalized washout rate $\bar{Q} = \{(L/I)_{I=0.5} + (L/I)_{I=25}\}/2$ is less than 30% for ranges of I from 0.5 to 25 mm h^{-1}, H from 20 to 80 m and C from 300 to 10^4 ppb(v). The normalized washout rate $\bar{Q}$ (% mm h^{-1}), increases with decreasing C and H (Fig. 7). Under optimum conditions SO_2 removal by washout can be substantial. For instance, in heavy rain (25 mm h^{-1}), a plume of 1000 ppb(v) SO_2 and 20 m thickness is stripped at 14% of its SO_2 in 15 min.

A comparison of the washout rate of a 50 m thick plume (topped at 200 m) from this model and one assuming no liquid-phase resistance was done at a precipitation intensity of 0.5 mm h^{-1} and various plume concentrations (Table 2). At low SO_2 concentrations (300 ppb(v)) both calculations yield the same washout rate (i.e. complete gas-phase control). At higher concentrations plume washout is underestimated by as much as a factor of two when liquid phase resistance is neglected.

A switch from gas-phase control to partial liquid-phase control of transfer as SO_2 concentrations increase is due to the non-linear solubility relationship of SO_2. Using Equation 5, it can be shown that to a first approximation the overall mass transfer coefficient K (Hales, 1972) from air to a drop is related to transfer coefficients in the liquid and gas phases (k_L and k_g, respectively) as follows:

$$\frac{1}{K} \simeq \frac{1}{k_g} + \left\{\frac{[SO_2]_g^I}{K_1 K_H}\right\}^{1/2} \frac{1}{k_L}. \quad (15)$$

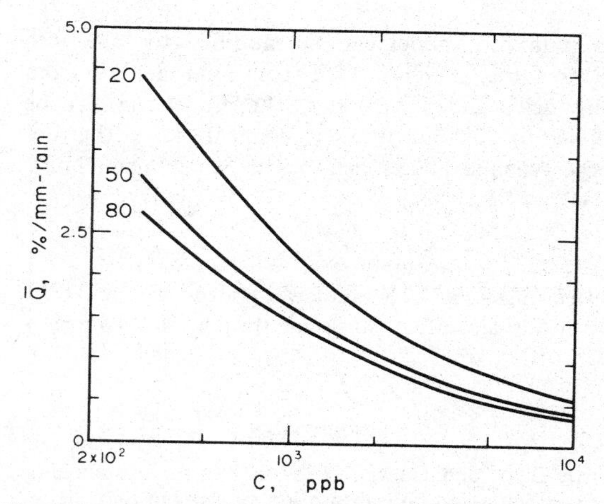

Fig. 7. Dependence of average-normalized washout $\bar{Q}$ on plume concentration C and thickness H (m).

At low SO_2 concentrations ($[SO_2]_g^I \rightarrow 0$) the second term on the right vanishes and transfer becomes gas phase controlled ($K = k_g$). This limit is approached in the plume washout model at concentrations below 300 ppb(v) SO_2 (Table 2).

CONCLUSION

The relationship between gas- and liquid-phase sulphur dioxide at the interface of a water drop has been derived (Equation 5) and utilized with a double-layer Whitman model to calculate SO_2 transfer to and from a falling raindrop. In applying this approach to calculate washout of an elevated SO_2 plume it was found that:

(a) at plume concentrations above 300 ppb(v) liquid-phase transfer resistance cannot be neglected with respect to that in the gas phase. Consequently, actual plume washout rates are generally higher than those predicted by models assuming complete gas-phase control of transfer;

(b) the fractional SO_2-removal rate by washout is inversely related to plume concentration and thickness.

It should be emphasized that the model presented above applies to situations in which sulphur dioxide is the major contributor to raindrop acidity. This special case of a more general situation in which other atmospheric substances (e.g. NH_3, sulphuric acid particulate) influence pH can be handled theoretically in a straightforward fashion. In future, the approach

Table 2. Comparison of washout rates (% h^{-1}) predicted by this model with those predicted assuming no liquid-phase resistance (10°C, I = 0.5 mm h^{-1})

Plume concentration (ppb)	Washout rate L (% h^{-1}) This model	No liquid-phase resistance
300	1.99	1.99
1000	0.93	0.83
3000	0.44	0.33
10000	0.20	0.11

taken in this model can be modified to take into account the presence of background acidity from other acids and particulates. The model can also be applied to washout of other chemical species that undergo chemical dissociation in water (e.g. NO_2, HNO_3, NH_3).

Acknowledgements—The author is grateful to Drs. D. M. Whelpdale and M. A. Lusis for their helpful comments.

REFERENCES

Barrie L. A. and Georgii H. W. (1976) An experimental investigation of the absorption of sulphur dioxide by water drops containing heavy metal ions. *Atmospheric Environment* **10,** 743–749.

Beilke S. and Lamb D. (1975) Remarks on the rate of formation of bisulfite ions in aqueous solution. *Am. Inst. chem. Engng.* **21,** 402–404.

Beilke S., Lamb D. and Müller J. (1975) On the uncatalyzed oxidation of atmospheric SO_2 by oxygen in aqueous system. *Atmospheric Environment* **9,** 1083–1090.

Best A. C. (1950) The size distribution of raindrops in rain. *Q. J. R. met. Soc.* **76.**

Dana M. T., Hales J. M. and Wolf M. A. (1975) Rain scavenging of SO_2 and sulphate from power plant plumes, *J. geophys. Res.* **80,** 4119–4125.

Danckwerts P. V. (1970) *Gas-Liquid Reactions*, p. 99. McGraw-Hill, London.

Hales J. M. (1972) Fundamentals of the theory of gas scavenging by rain. *Atmospheric Environment* **6,** 635–659.

Hales J. M., Wolf M. A. and Dana M. T. (1973) A linear model for predicting the washout of pollutant gases from industrial plumes. *Am. Inst. chem. Engng.* **19,** 292–297.

Le Clair B. P., Hamielec A. E., Pruppacher H. R. and Hall W. D. (1972) A theoretical and experimental study of the internal circulation in water drops falling at terminal velocity in air. *J. atmos. Sci.* **29,** 728–740.

Liss P. S. and Slater P. G. (1974) Flux of gases across the air–sea interface. *Nature* **247,** 181–184.

Matteson M. J. and Giardina P. J. (1974) Mass transfer of SO_2 to growing droplets: role of surface electrical properties. *Envir. Sci. Techn.* **8,** 50–55.

Peters L. K. (1976) Some considerations on the washout of the sulphate from stack plumes. *Wat., Air Soil Pollut.* **6,** 303–319.

Pruppacher H. R. and Beard K. V. (1970) A wind tunnel investigation of the internal circulation and shape of water drops falling at terminal velocity in air. *Q. J. R. met. Soc.* **96,** 247–256.

Pruppacher H. R. and Beard K. V. (1971) A wind tunnel investigation of the rate of evaporation of small water drops falling at terminal velocity in air. *J. atmos. Sci.* **28,** 1455–1464.

Robinson R. A. and Stokes R. H. (1970) *Electrolyte Solutions*, pp. 571 Butterworths, London.

Scott W. D. and Hobbs P. V. (1967) The formation of sulfate in water drops. *J. atmos. Sci.* **24,** 24–27.

Slinn W. G. N. (1974) The redistribution of a gas plume caused by reversible washout. *Atmospheric Environment* **8,** 233–240.

Whitman W. G. (1923) *Chem. metallurg. Engng* **29,** 146.

Atmospheric Environment Vol. 12, pp. 413–424. Pergamon Press 1978. Printed in Great Britain

SULFATE IN PRECIPITATION AS OBSERVED BY THE EUROPEAN ATMOSPHERIC CHEMISTRY NETWORK

L. GRANAT

Department of Meteorology, University of Stockholm, Arrhenius Laboratory, S-106 91 Stockholm, Sweden

(*First received* 17 *June* 1977 *and in final form* 20 *September* 1977)

Abstract—The temporal variation in sulfur concentration/deposition involves long term fluctuations which show striking similarities within certain areas but are different between areas in different positions relative to major sulfur emission areas in Europe. As an example, the deposition has been constant or decreasing during the last ten years in most of the area covered by the network, indicating that an increasing amount of the sulfur emitted in Europe is transported and deposited elsewhere, possibly in an eastward direction.

The seasonal variation in deposition resembles that for amount of precipitation and shows a maximum during the summer or autumn for most areas. The concentration shows a maximum in the spring and a minimum in the autumn at most stations.

The concentration field has a maximum approximately over Belgium and Holland with decreasing concentrations towards SW to NE (over W) but with comparatively high levels extending up over Finland. (Areas to the east and south are not covered by the network.)

The deposition field was, based on a brief discussion of the relation between concentration and amount of precipitation, obtained as the product of these latter fields. High deposition rates are found in the south of Norway and in the middle of Britain in addition to the areas with high concentration.

Sulfur and hydrogen ion are the dominating ions in precipitation followed by ammonium and nitrate except in coastal areas and in places where the soil is bare.

Sources of errors in the data base are discussed and the results from a large number of additional sampling sites around and between the regular network sampling sites are most helpful in this regard. These latter measurements also permit an estimate of the uncertainty in areal concentration averages which are due to local and mesoscale variability.

Finally, past and future importance of continuous measurements (as in the EACN) is discussed briefly.

INTRODUCTION

Rainwater has routinely been collected and analysed for 11 parameters for more than 20 years in several European countries. A substantial and rather unique data base has thus been accumulated which we shall here use to show some characteristic features in the spatial and temporal variability of sulfate in precipitation (wet deposition). When using this data base, which has been produced by many individuals in several groups over a long time period, it is important to have a solid foundation from which the quality of the data can be judged. Such a basis is being established by a series of additional measurements now undertaken in Sweden.

Sulfate is one of the major ionic compounds in precipitation and is of considerable concern because of the hydrogen ion that is liberated during the oxidation of sulfur dioxide to sulfate. From an ecological point of view it is important to assess the quantities of sulfate, acid and other components which are brought from the atmosphere to the ground but it can also be useful for the assessment of transfer and transport processes in the atmosphere as well as other kinds of impact, for instance, corrosion.

Data from the network has been of great help for an early understanding of long distance transport of air pollutants. With some modifications, as discussed below, such long term continuous supervision is motivated also for the future.

THE NETWORK

A systematic collection of air and rainwater samples for chemical analysis was started around 1946 by Professor Egnér at the College of Agriculture, Uppsala. The main purpose was to study the transport from the atmosphere to the ground of some nutrients, above all nitrogen compounds. Under the leadership of Egnér, Rossby and Eriksson this activity was gradually given a more general atmospheric chemical direction and from about 1955 stations were put into operation also in Austria, Belgium, Finland, France, Holland, Iceland, Ireland, Italy, Norway, the U.K. and West Germany. Equipment was supplied from Sweden and to begin with the chemical analyses were also made here. The responsibility for the maintenance of the stations and the performance of analytical analyses has then gradually been taken over by a number of analytical centres in the different countries.

The number of stations grew to a maximum of about 120 in 1959 but then decreased and is at present about 50. The stations that have been in operation for shorter or longer periods are given in Fig. 1. A more detailed description of the laboratories involved in this effort, periods of operation for the individual stations and their location has been given by Granat (1972).

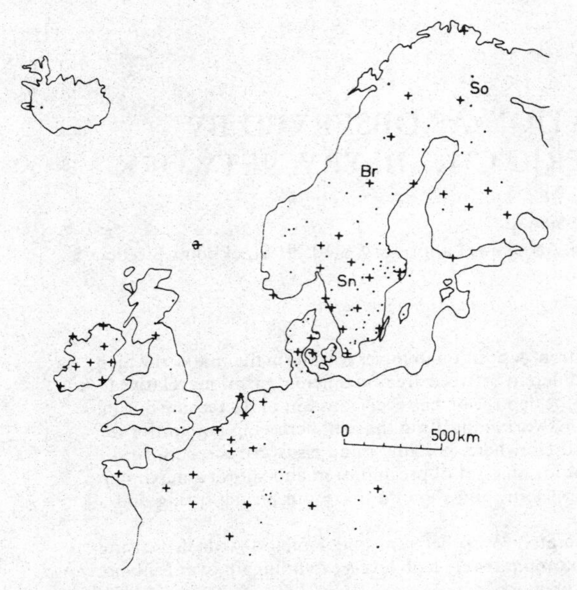

Fig. 1. The so-called European Atmospheric Chemistry Network (EACN). Sampling sites that have been in operation for 13 years and more are denoted by crosses, others by big dots. Symbols refer to stations mentioned in the text.

The equipment is, in principle, a funnel connected to a bottle and thermostated to collect either rain or snow. The air samples have been collected in a fritted glass wet absorber, in recent years preceded by a filter for collection of aerosols while the gases are collected in the absorber. A description of the main type of equipment which has been in operation, from the original so-called Egnér cabinet up to the present day equipment, including specially built instrument houses for the air sampling and rainwater collectors with a lid that opens automatically during precipitation, is given by Granat *et al.* (1977).

Rainwater samples have usually been analysed with regard to the amount of precipitation, sulfate, nitrate, chloride, ammonium, sodium, potassium, magnesium, calcium, pH, hydrogen ion/bicarbonate and electrolytical conductivity. When an aerosol filter is used it is usually leached and determined for the same components as precipitation and the absorbing solution is analysed for some of these constituents including sulfur (to give SO_2).

SUPPLEMENTARY MEASUREMENTS

A series of measurements is being made in Sweden with the purpose of obtaining a direct control of the data from the ordinary network of stations and also to get a better understanding of certain properties in the concentration/deposition field which are of importance when one wishes to interpolate between network stations and to assign confidence limits to the results obtained. Three sets of measurements have so far been compiled, referred to below as REP 1, REP 2, and REP 3 and another two sets are being planned. Each set involves about 50 collectors placed according to a special scheme, and the aim is to collect monthly samples during approximately one year. The areas covered are indicated in Fig. 2. The samples are analysed in the same manner as those from the ordinary network. The collectors are very simple but appropriate, and have been located with extreme care to avoid local contamination.

QUALITY OF THE DATA BASE

One should be aware of the fact that there are errors in the main data base, and it is therefore important to account for them in order to make the information as useful as possible. We can, in principle, divide the errors into three different classes depending on their origin.

(1) Errors in chemical analysis, improper storage and handling of samples, computational errors and similar.

(2) Local conditions at the sampling sites which will either contaminate the samples (for instance dust falling into the collectors during dry periods) or special characteristics of the sites which make them unrepresentative of surrounding areas (for instance a mountain top, an exposed location near the coast etc.).

(3) The structure and density of the network is such that important features in the deposition/concentration field are overlooked when a simple interpolation is performed between the stations. Such errors will probably arise when interpolation is performed over large water bodies, over mountainous areas and,

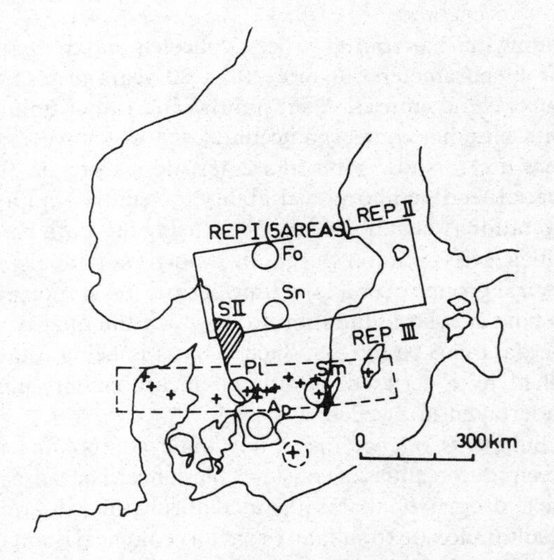

Fig. 2. Location of the sampling sites for the supplementary measurements for the Swedish IMI network. The three sets of measurements so far completed are: REP 1: about 10 collectors placed within each of the areas indicated and with an ordinary network station in the middle here marked Fo, Sn, Pl, Ap, Sm. REP 2: where about 50 collectors were evenly spaced within the indicated area including four sites on islands well outside the mainland. REP 3: at each of the 16 locations indicated in the figure usually three collectors were placed within a few km from each other. Data given in Fig. 3 are obtained from the area indicated by S II.

perhaps most important, over large industrialized areas particularly when they are situated in valleys.

It might be inferred that the temporal variability (*cf.* Figs 4, 5 and 7) is an artefact, due to systematic analytical errors from time to time at different analytical laboratories. From a first glance at the data in the figures this appears possible, but a closer look shows satisfactorily that this is not the case. Since a systematic evaluation of the quality of the whole data body from the EACN network is presently under way at this department and will be reported elsewhere, it may here be enough to recall that none of the characteristic "jumps" in the time series occurs when analyses are changed from one laboratory to another (as given in Granat, 1972) and also that Munn and Rodhe (1971) were able to account for the minimum in the late fifties and the maximum in the middle sixties from a meteorological interpretation of data from two stations belonging to groups denoted "△" and "□" in Fig. 4. Although one should not underestimate the possibility that systematic errors in analytical methods can prevail during long periods of time, it is my impression that errors of the other two kinds listed above are much more important, in particular, unsuitable conditions at the sampling site.

The sulfate concentration seems to be comparatively little affected by local contamination, and the large scale concentration field also seems to be rather smooth. This is apparently due to a low sulfate concentration in soil dust and a comparatively slow removal rate of sulfur dioxide and sulfate containing aerosols which reduces the effect of anthropogenic sources near sampling sites. In contrast, the hydrogen ion concentration, to mention one component, can be severely affected by contamination from locally produced soil dust. Interpretation of the distribution of acidity of precipitation is therefore much more uncertain than that of the sulfate concentration. It is likely that too low values (too high pH) are obtained in areas where the ground is not completely covered by vegetation.

A growing awareness of this problem has led to a partial reconstruction of the network, particularly in Scandinavia, and the introduction of collectors with lids that are automatically kept open only during rain.

DATA SCRUTINIZATION

There are some indications of obviously contaminated samples, and methods for filtering the data have accordingly been developed. For sulfate data dealt with in this report it was considered sufficient to remove the monthly deposition values which showed exceptional deviations from the average for each station. The procedure was as follows: Data from each single station were searched for deviations exceeding $\pm 3.6 \times s$ where s is the logarithmic standard deviation for the particular station. If found, such data were excluded and the procedure repeated. To avoid the problem where deposition values of zero were reported, 1 mg sulfur m^{-2} and month – minimum resolution in the data base – was added to the monthly deposition values before they were converted to logarithmic values and tested for large deviations. Monthly deposition values show approximately "log normal" distribution and the filter applied would thus remove about 1 per mille of correct data or two values per station which has been in operation for 20 years.

CALCULATION OF THE "EXCESS" AND SEA SALT FRACTION OF SULFATE IN PRECIPITATION

In the area covered by the station network there are essentially three sources of sulfur emissions to the atmosphere: man-made sulfur dioxide, sulfur-containing gases of natural origin and sulfate from sea spray.

As the residence time in the atmosphere for man-made sulfur compounds is very different from the residence time for compounds originating from sea spray and as the two source areas are well separated, it is useful to estimate the amount of sulfate originating from sea spray and deduct that from the total sulfate concentration in rainwater to give so-called "excess" sulfur. As can be seen in more detail below, the excess sulfate shows a rather uniform deposition pattern over Europe, while sulfate from sea spray plays a dominating role only within a narrow band (never more than 50 km wide) along the sea coast.

Sea salt sulfur in precipitation is estimated from the sodium concentration in rain and the ratio between sodium and sulfate in ocean water, assuming that all sodium in precipitation is coming from sea salt and that no fractionation is occurring between sodium and sulfur when sea spray is created. Probably neither of these assumptions is strictly correct but one must consider that the correction for sea salt is of importance only for a few coastal stations while it is insignificant for most other stations. Some measurements on the west coast of Sweden, where it was possible to compare excess sulfur with amount of acid – which is not originating from or affected by components coming from sea spray – indicated that this procedure is well justified. The ratio S/Na in sea water is taken to 0.084 g/g or 0.061 mole/mole. The following discussion will deal mainly with excess sulfur.

CHEMICAL COMPOSITION OF PRECIPITATION

Table 1 gives an example of the chemical composition of precipitation. The data are obtained from station Sjöängen in the south of Sweden ("Sn" in Fig. 2). This is one of the two stations in Sweden from where data is reported to WMO and the relative composition is typical for the south and middle of Scandinavia except for coastal areas.

There is a good balance between the sum of positive and negative ions and between calculated and measured conductivity, especially for data from recent

Table 1. An example of chemical composition of precipitation from station Sjöängen in Sweden ("Sn" in Fig. 2) during the years 1973 through 1975. The relative composition is typical for the south and middle of Scandinavia except in areas near the coast

Concentration in μ equiv l^{-1}			
SO_4	69	H^+	52
Cl	18	NH_4	31
NO_3	31	Na	15
		K	3
		Mg	7
		Ca	13
Sum	118		121

pH (estimated from hydrogen ion concentration):	4.3
Measured conductivity:	$28.0 \cdot 10^{-6} \Omega^{-1} cm^{-1}$
Calculated conductivity: (both at 22°C)	$28.9 \cdot 10^{-6} \Omega^{-1} cm^{-1}$

Concentration relative to sulfur (eq/eq)
H^+/SO_4 : 0.75
NO_3/SO_4 : 0.45
NH_3/SO_4 : 0.45
K + Mg + Ca/SO_4 : 0.33

years where analytical methods are well tuned, as seen in the given example. When the whole data base is examined one obtains a frequency distribution for the difference between the two sets of parameters which show a considerable scatter but with the majority of the data well centered around and near zero difference.

We can therefore conclude that all major ionic species are analysed for. In addition there may also be undissolved inorganic as well as dissolved and undissolved organic compounds. Little is known about their concentrations and no information is available from EACN data.

There are four major ions, sulfate, nitrate, ammonium and hydrogen ions. Near the coast, sea salt components may have equally high or higher concentrations, in particular NaCl.

It is of special interest to see how the hydrogen ion relates to other substances. According to current knowledge, one could consider the excess sulfur as originating from sulfur dioxide and nitrate from oxides of nitrogen (NO, NO_2). When these gases are converted to sulfate and nitrate respectively, an equivalent amount of hydrogen ion is also created in a stochiometric sense (in reality it may partly react with a base). Ammonium has entered into the system as ammonia neutralizing an equivalent amount of hydrogen ion.

If the emission of NO_x and SO_2 was completely independent of emission of NH_3 to the atmosphere one could attribute the acidity found in precipitation to sulfur and NO_x sources in proportion to the concentration of sulfate and nitrate in precipitation. If NO_x and NH_3 were introduced into the atmosphere by the same process (say agriculture, for the sake of discussion, which might perhaps partly be the case) one must attribute all the acid to the sulfur compounds. In areas with large soil erosion, some calcareous compounds may also contribute as neutralizing compounds.

With regard to the above point of view and with a typical chemical composition as shown in Table 1, it is evident that the incorporation of the excess sulfur in precipitation gives rise to the major part of the free hydrogen ion found in rainwater. It should, however, be observed that this holds for acidity of precipitation – with regard to acidifying effects after deposition on different eco-systems the effect of hydrogen ion concentration can be modified by other compounds in the rain, for instance NH_4 and NO_3.

RELATION BETWEEN SULFATE CONCENTRATION AND AMOUNT OF PRECIPITATION

With some knowledge of the relation between

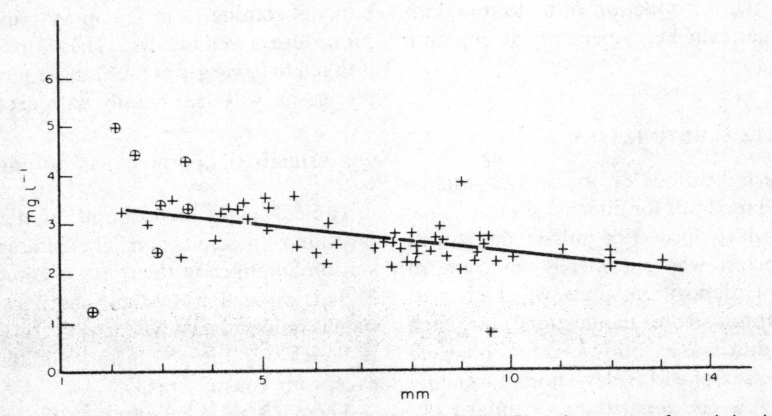

Fig. 3. Relation between concentration of excess sulfate (as SO_4–S) and amount of precipitation during a single event. The location is marked "S II" in Fig. 2. Symbol ⊗ denotes samples from collectors just at the shore line

sulfate concentration and amount of precipitation we would be able to make a better estimate of the deposition at points between network stations where precipitation is regularly measured. The following examples give some information in this regard.

During a single rain event rainwater was collected in an area of about 40 km dia (for location, see S II in Fig. 2) where the amount of rain was very unevenly distributed. Fig. 3 shows how the excess sulfate concentration decreases relatively slowly as the amount of precipitation increases from 2 to 13 mm.

From five sampling places near the west coast of Sweden (REP 3, see Fig. 2) located within a distance of 50 km a relation was obtained for monthly samples between relative deviation in sulfate concentration and in amount of rain. The relation is approximately linear and shows tentatively about 7% decrease in sulfate concentration when the amount of precipitation increases by 50%.

We can therefore conclude, at least tentatively, that concentration shows a much smaller areal variability than deposition and this information will be used below when a deposition field is obtained as the product of the concentration field and the precipitation amount field.

TEMPORAL VARIATIONS OF SULFATE CONCENTRATION/DEPOSITION

It is well known that the concentration of many elements in precipitation, among them sulfate, varies considerably from one rain event to another, sometimes by more than a factor of 10. For monthly averages some of this variability is damped out but may still be considerable and with abrupt changes from one month to the next. This is obviously an effect of occasional transport of pollutants from source areas in combination with precipitation which is of synoptic scale in time and space.

Here we shall consider in some detail the long term temporal variation and, in the following paragraph,

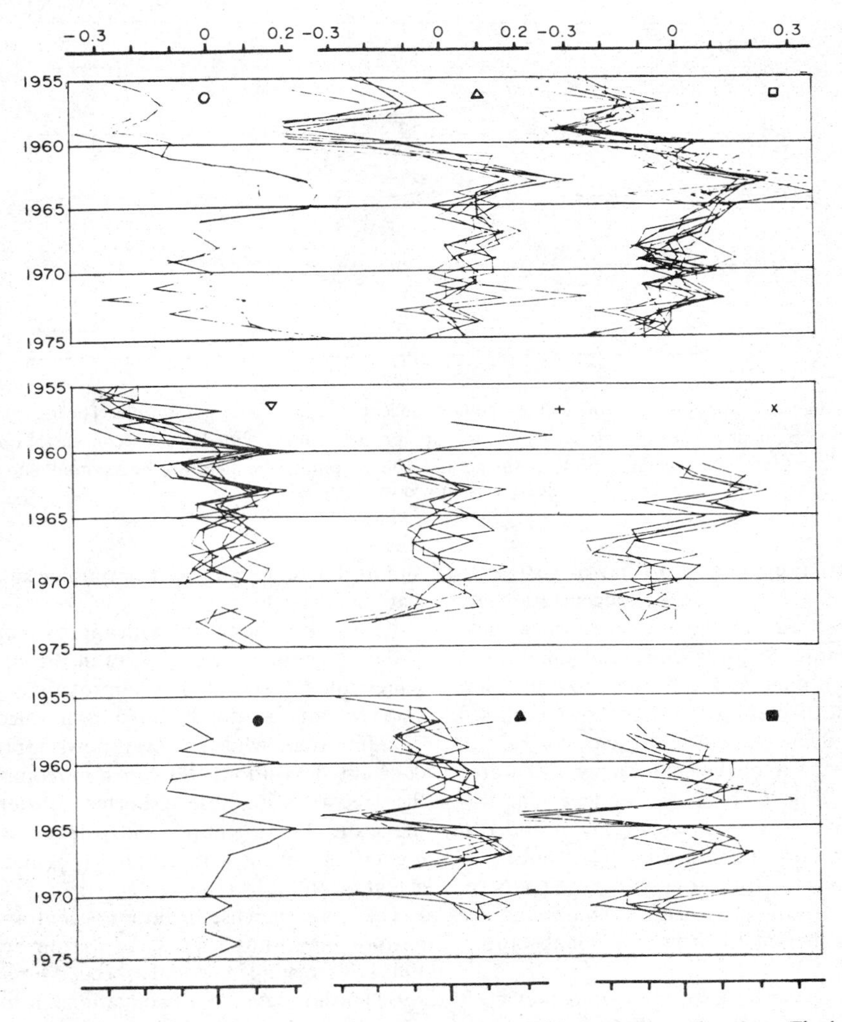

Fig. 4. Long term temporal variation in deposition of excess sulfate for groups of stations. The logarithm values given are in principle $\log d - \overline{\log d}$ where d is yearly deposition and $\overline{\log d}$ the average for the 21-year period. Stations are identified by a symbol and their location is shown in Fig. 6.

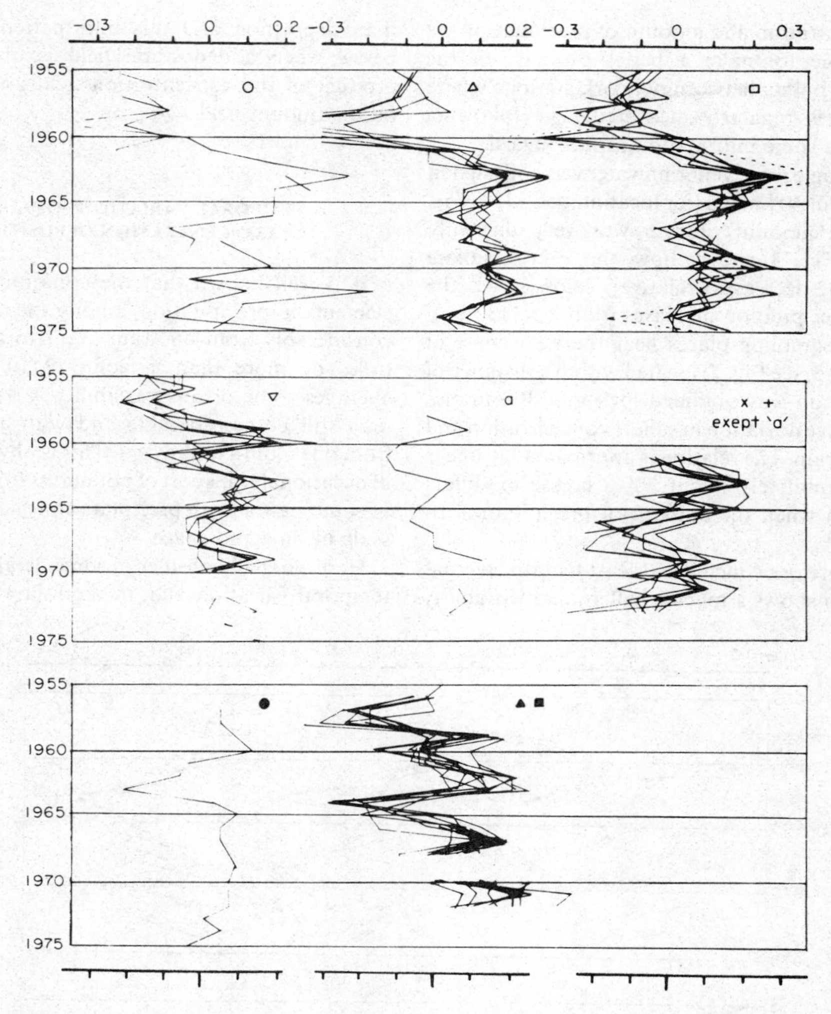

Fig. 5. Long term temporal variation in excess sulfate concentration for groups of stations. The logarithmic values given are in principle $\log c - \overline{\log c}$ where c is yearly precipitation weighted average concentration and $\overline{\log c}$ the average for the 21-year-period. See further in the text. Stations are identified by a symbol and their location is shown in Fig. 6.

the seasonal variation. The logarithm of yearly deposition and of yearly average concentration respectively was plotted vs time for all stations with 13 years' data or more. Records from different stations were then combined and it was possible to form a few groups with quite different patterns as shown in Figs. 4 and 5 with only a few stations being rejected. Each of these patterns seems to be typical for a particular area as indicated by Fig. 6. It must be clear that this stratification involves subjective judgement and is a compromise between minimizing the number of groups and the permissible scatter within each pattern. The apparent "trend" will also be somewhat dependent on how the records from individual stations are combined.

Despite these uncertainties it is obvious that both concentration and deposition of sulfate in rain shows considerable long term fluctuations which are markedly different between areas in different directions and at different distances from major sulfur-emitting areas in Europe.

If we look at the groups with approximately 20 years of data (denoted ○, △, □, ∇, ● in Fig. 4) we can fit a regression line which has approximately the same slope as that for sulphur emissions in Europe (see Fig. 7) in four cases while one (●) shows approximately a constant deposition. We can now compare part of these records with those of shorter duration and notice that – after 1963 – groups 5 and 6 (denoted +, ×) show a good agreement with group 3 (□) and, to a lesser extent, group 2 (△).

The rapid increase in both concentration and deposition for groups 2 and 3 (△, □) from about 1959 to 1963 is interesting in several aspects. Munn and Rodhe (1971), using data from two stations in these groups, were able to show that the increase could be explained almost completely as a meteorological effect expressed as a change in frequency of winds from major sulfur

Fig. 6. An indication of areas with different patterns in long term temporal variation of sulfate concentration/deposition. Symbols refer to groups of stations according to Figs. 4 and 5.

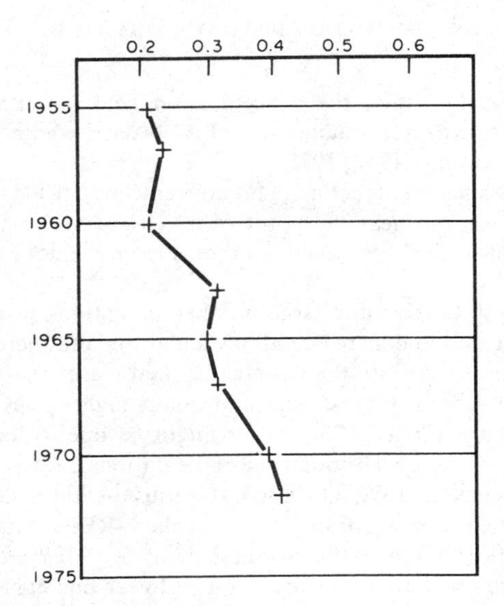

Fig. 7. Sulfur emissions in Europe. Values are given as log $E-7$ where E is annual sulfur emission in ton. Data obtained from Fjeld (1976).

emission areas associated with precipitation at these stations. Without such a meteorological interpretation a similar steep increase could be misjudged as due to changing emission rates. Similarly one can also misjudge a decreasing concentration/deposition as due to decreasing emissions.

From an ecological point of view we can notice that even in large areas (such as Sweden and Norway) the amount of sulfur deposited will not be closely related to changes in sulfur emissions rates in Europe, even for periods as long as ten years.

If we look at the deposition pattern during the ten-year period from 1965 to 1975, we can find a relative increase corresponding to that for sulfur emissions in only a small fraction of the area covered by the network and with otherwise constant or decreasing deposition. A very obvious question is then to ask: where has the increase in sulfur emission rate during these years been deposited? As there is no reason to question the emission rate numbers, and as the dry deposition of sulfur dioxide (which is not measured by the network) would rather decrease than increase due to gradual increase in the average height of injection into the atmosphere, a very probable answer is, therefore, that increasing amounts of sulfur have been deposited in areas not covered by the network, probably eastward of major sulfur emission areas due to the general circulation over the area.

SEASONAL VARIATIONS

The seasonal variation in deposition and amount of precipitation is shown in Fig. 8 and for concentration

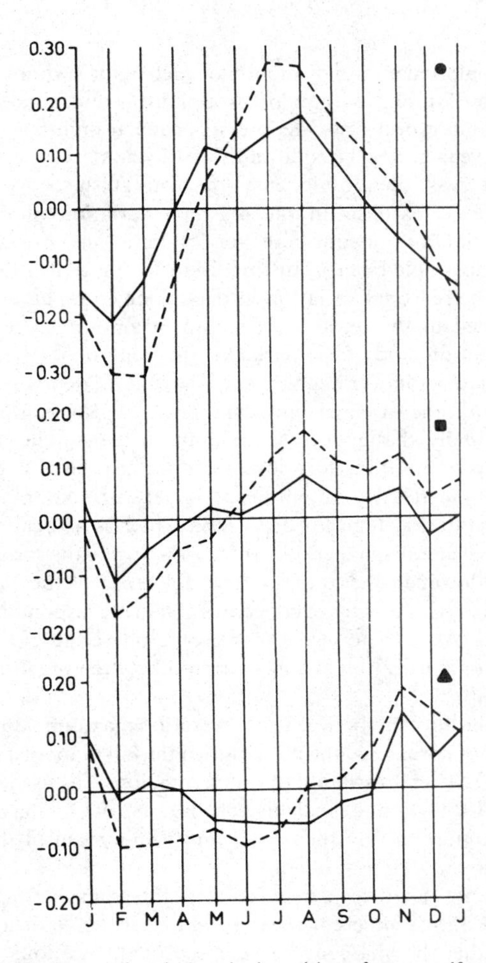

Fig. 8. Seasonal variations in deposition of excess sulfate (—) and of amount of precipitation (- - -) for three groups of stations. Values are in logarithmic units. Stations are identified by a symbol and their location is shown in Fig. 10.

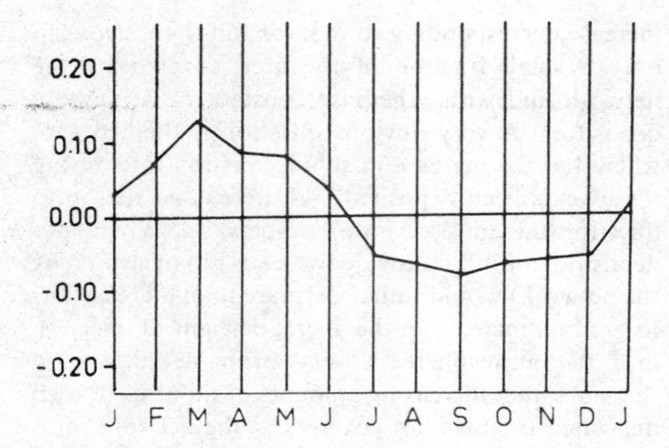

Fig. 9. Seasonal variation in excess sulfate concentration. Values are in logarithmic units and are means for all stations shown in Fig. 10.

in Fig. 9, based on data from all stations with a ten-year record or more. The quantity

$$\left(\sum^{n} \log a_{ij}\right)/n - \left(\sum^{12,n} \log a_{ij}\right)/(n \cdot 12)$$

was plotted month by month for each of the stations (a is deposition, amount of precipitation and concentration respectively, i denotes a particular month and j the year and n the total number of years).

It was then quite easy to distinguish between different patterns in the seasonal variation of deposition and of course also in precipitation. No such groups could be easily distinguished for concentration.

The seasonal variation in deposition is apparently related to the seasonal variation in amount of precipitation and thus markedly different in different climatic zones ranging from maximum deposition during the summer in Scandinavia which is then gradually changing to a minimum in the summer at extreme maritime locations. It could also be noticed that the relative amplitude is larger for amount of precipitation than for deposition which is associated with generally higher concentrations during the spring and lower in the autumn. The difference of the two curves is a measure of concentration in corresponding units. Also, there seems to be some phase lag in the seasonal variation in concentration between the three groups.

The quotient between concentration maximum during the spring and the minimum in the fall is about 1.6. The seasonal variation of sulfur emissions – as used in the LRTAP model calculations (Ottar, 1978) – has a maximum during the winter and a minimum in the summer, with a quotient of about 2.

More detailed interpretation of the transport pattern during different seasons will probably indicate whether the phase lag between seasonal variations in the emissions and in concentration of precipitation can be explained as a meteorological phenomenon, or if different removal processes and rates also play a part.

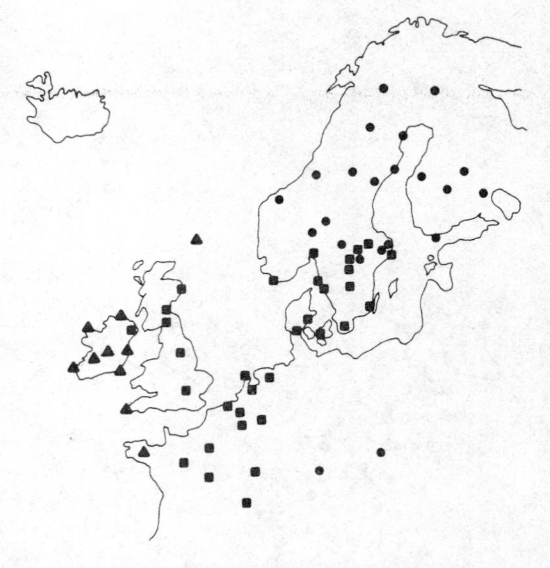

Fig. 10. Areas with different patterns in seasonal variation in deposition of sulfate. Symbols refer to Fig. 8.

CONCENTRATION FIELD FOR SULFATE IN PRECIPITATION

Fig. 11 shows the concentration field for excess sulfate in rain as obtained from EACN data, nominally for the years 1968–1972.

The general structure of the concentration field with a maximum near the point of gravity of sulfur emissions is easily obtained but there are uncertainties in details.

Despite the quite large number of stations in the north and middle of Scandinavia and Finland there is some doubt if the gradients here are mainly north–south or west–east and about high values in northern Finland. This latter might be due to local contamination, although one of the stations (see Fig. 1) is selected as a WMO station. It is thus useful to make detailed investigations (the so-called REP studies mentioned above) also in these areas. In other areas where the station density is much lower one can, of course, only obtain a rough picture.

DEPOSITION FIELD FOR SULFATE IN PRECIPITATION

Deposition can often be more directly used than concentration. This is for instance the case for the sulfur budget over Europe and in some ecological applications where the accumulated effect is of concern (for instance from hydrogen ion). The easiest way to obtain deposition field is of course to start from deposition values obtained at the network stations. A useful alternative is however, to first construct a concentration field and then convert this to deposition from information on amount of precipitation – ideally from the much more dense network of ordinary rain gauges. The latter procedure is to be preferred es-

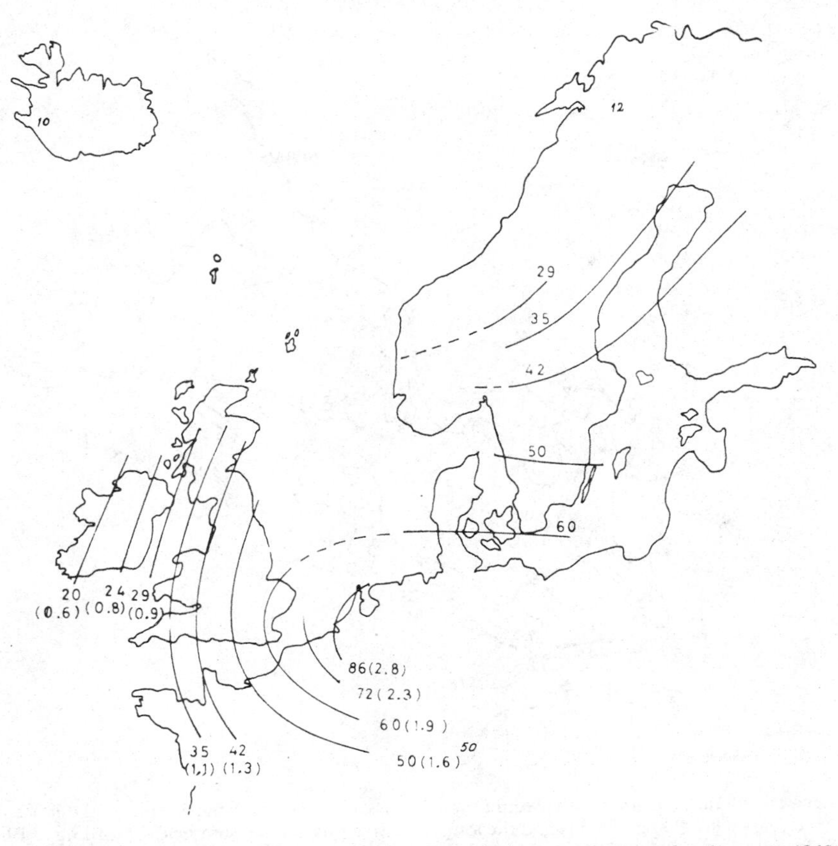

Fig. 11. Excess sulfate concentration in precipitation. Values given are averages for the years 1968–1972 based on yearly precipitation weighted concentration averages. Values are given in μmol l^{-1} with mg SO_4-S l^{-1} within parentheses.

pecially where the amount of precipitation has a large areal variability and was used in the preparation of Fig. 12.

Since data on precipitation field over Europe for the 1968–1972 period were not available at the preparation of this paper, a long term annual precipitation mean (after Blütgen, 1958) was used instead merely to illustrate the shape of the deposition field.

No reliable data on precipitation over large water bodies are available and therefore 400 mm y^{-1} is assumed for the North Sea and 300 mm y^{-1} for the Baltic, with appropriate adjustment to inland values near the coast.

The most striking feature in the deposition field (Fig. 12) compared to the concentration field (Fig. 11) is the large amount of sulfur deposited in the south of Norway, the middle of Britain and, to a lesser extent, the west coast of Sweden. This is almost completely overlooked if the interpretation is made directly from deposition figures from the network stations, simply because mountainous areas with high amounts of precipitation are not properly covered by the network. It must, however, be remembered that Fig. 12 is based on the assumption that concentration is relatively little affected by amount of precipitation, something which has still to be verified in the areas with very large amounts of precipitation.

The deposition field shown here is in reasonable agreement with the results from the LRTAP study (given for 1974, Ottar, 1978) both with regard to absolute levels and the shape of the field, considering the different methods used to obtain the results. The only major deviation is over the middle and northern parts of Britain.

ERRORS IN ESTIMATED CONCENTRATION DUE TO SMALL SCALE AREAL VARIABILITY

The data from the study REP 2 (see Fig. 2) allows a rough estimate of the uncertainty due to the areal variability in estimated concentration values for locations between network stations. The uncertainty is predicted from calculated root mean square deviation (RMSD). Although far from mathematically stringent, it is assumed that this will give a reasonably good estimate.

If we thus remove the large scale areal gradient and calculate an RMSD for yearly average concentration values at the approx 50 stations in the 400 × 200 km area we obtain a value of 14% for sulfate. As a comparison RMSD is about 35% for elements like calcium, potassium and magnesium. The variability of the latter is thus considerably larger than for sulfate for reasons discussed in a previous chapter.

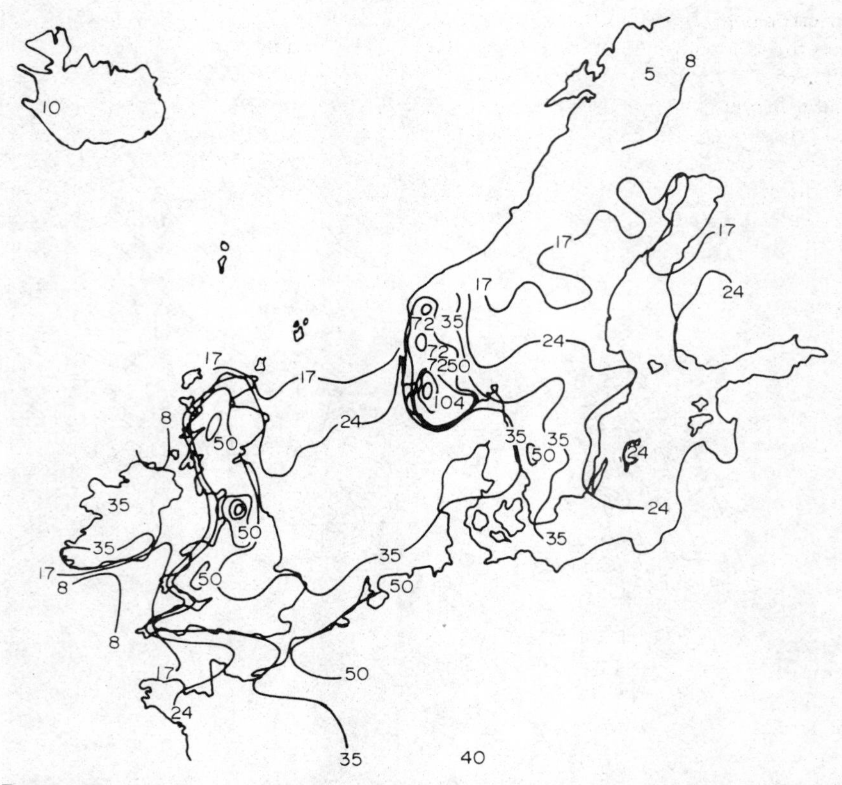

Fig. 12. Excess sulfate deposition with precipitation. Values are obtained as the product of the concentration field for the years 1968–1972 (Fig. 11) and the long term precipitation amount field (according to Blüthgen, 1958). Values are given in m mol SO_4^{2-} m^{-2} y^{-1} (for conversion to mg S m^{-2} y^{-1} *cf.* Fig. 11). Note that this figure is only intended to give a general picture of the deposition field, as explained in the text, despite its detailed structure.

We are usually not interested in point estimates but rather averages for larger areas, and the RMSD for sub-areas of different sizes was consequently calculated (with large scale gradients first removed as for the point estimate). In a real case the concentration field is obtained from a few ordinary network stations instead of the 50 used in REP 2. We account for the uncertainty in these point data – which is due to the areal variability – with RMSD of 14% as obtained from the REP 2 study, but as the concentration field is often obtained through a weighing procedure with support from several neighbouring stations, we can reduce this number by say a factor of $1/\sqrt{2}$ and we then obtain the measure of the uncertainties in areal deposition averages (Table 2).

Table 2. An estimate of uncertainties in areal average for yearly concentration when this is estimated from interpolation between network stations. The estimate is valid for the uncertainty that is due to small and mesoscale areal variability for conditions described in the text and is approximately valid with 90% confidence

Size of the area (dia.)	Confidence limit (%)
less than 300 m	27
50 km	21
100 km	18
200 km	16

The uncertainty here decreases as the area increases. If, on the other hand, one station is in the middle of an area, the uncertainty decreases with decreasing area – see Table 3.

One should remember that these estimates are valid for an area where errors are expected to be smaller than in many other parts of Europe because the terrain is reasonably flat, major sources are at long distances and there are no exceptional differences in amount of precipitation.

When converting from concentration to deposition we introduce errors both from the precipitation amount measurements and due to the fact that the concentration is slightly dependent on amount of precipitation.

Table 3. An estimate of uncertainties in areal average for yearly concentration when this can be estimated from a station in the middle of the area. 90% confidence (*cf.* Table 2)

Size of the area	Confidence limit
less than 300 m	less than a few per cent
50 km	6%

The former error is probably on the 10% level for point estimates for rain and may be much higher for snow. The other kind of error is rather small, about 5% for a 50% change in amount of precipitation provided the amount of precipitation is known. If interpolation was instead to be performed directly in the deposition field, we might instead introduce an error of about 40% for the same change in amount of precipitation. The errors discussed above depend upon small and meso-scale areal variability. In addition, there might also be errors due to unsuitable location etc. as discussed in a previous chapter.

WASHOUT RATIOS

For the last few years a determination of air concentration of sulfur dioxide and sulfate in aerosols is performed parallel to collection and analysis of precipitation.

The quotient (washout ratio) between sulfur in rainwater and sulfur in air (SO_2–S + SO_4–S) was calculated for eight stations. Data for single months showed a substantial scatter which is not surprising as the air concentration is obtained for the whole period but rain is falling only about 10% of the time.

A two-year average (1975, 1976) gave a value of 1900 for Br (see Fig. 1) while seven other stations south of this all showed about the same value with a mean of 770.

The quotient (SO_2–S)/(SO_2–S + SO_4–S) was respectively 0.42 and 0.60.

It is not possible to interpret this difference as due to different scavenging rates between SO_2 and SO_4 as other factors such as co-variance between rain event and high pollution level, scale height and temperature (rain/snow) may change from south to north.

PAST AND FUTURE IMPORTANCE OF THE NETWORK

From Figs. 4 and 5 one can learn that measurements of air and rainwater composition maintained over a few years of time and covering only a part of that area which is under direct influence of industrial emissions in Europe can give comparatively little information on the general development of the regional air pollution situation. Only when the measurements extend over decades and cover an area where most of the industrial emissions (from Europe) are deposited will give such a perspective.

From the same figures it is further evident that a simple extrapolation might give serious errors when estimating future burdens. Even if changing meteorological conditions are taken into account, it is doubtful if a good prediction can be obtained as those changes which are related to changing technology, changing structure in the supply of energy etc., cannot be predicted in this way. Starting from a knowledge of the relation between emissions and deposition in various parts of Europe, although far from complete, combined with estimates of future industrial and agricultural activities would inherently give a much more reliable estimate.

Data from a network of this kind (EACN) are thus best suited to show past and present input of a number of substances from the atmosphere to the ground and thereby the changing burden on different eco-systems and will thus help in explaining observed changes in them.

A good example of this kind is the assessment of the acid rain problem. Already Barret and Brodin (1955) noticed the rather high sulfur levels in rainwater and attributed them to sulfur from fossil fuel combustion. This matter came under more intense debate when Odén (1968) combined observations of ecological changes (increasing acidification of some lakes, and decreasing fish populations) and related that to the acid in precipitation. These arguments were dramatically underlined by a very steep increase of sulfate and hydrogen concentrations in rainwater during a few years around 1960.

In the light of the results from the OECD–LRTAP study (Ottar, 1978) one can now see how well an early rough estimate of transport and deposition of sulfur and acid over Europe could be obtained based on EACN data (see for instance Bolin *et al.*, 1971; Bolin and Persson, 1975).

The EACN data have been and are being used in a number of other interpretations which it is outside the scope of this paper to discuss.

There is, not least in Sweden, a support for maintaining a long term continuous supervision of the input from the atmosphere similar to the EACN network but with analysis for many more components. This will also maintain and extend the accumulated reference material against which investigations of shorter duration can be judged (as for instance the OECD–LRTAP study). Such a reference network is especially valuable if it covers most of Europe – for reasons indicated previously – say with about 100 well located stations.

With regard to sampling equipment and location it is important not merely to continue the sampling with the present network but also to carefully consider the representativeness of the stations and change the location accordingly. The location of the stations should then be based on the measured criteria that good areal deposition averages (the deposition field) could be obtained from them. A collector with a lid that is (automatically) opened only during rain is now gradually being taken into operation and will eliminate much of the possible contamination and increase the representativeness of the samples.

The major importance of such measurements will be with regard to pollutants other than sulfur. With regard to sulfur one would however, with such improvements, expect a more detailed and accurate picture of the deposition field. This should provide a more reliable budget involving a comparison between emitted and deposited sulfur, especially if present research on dry deposition will provide means to estimate dry deposition regularly at the ordinary network stations.

Acknowledgements—This paper is prepared as partial fulfilment of contracts from the Swedish Natural Science Research Council contract No. G 0223-065 and the National Environment Protection Board contract No. 7/117-76.

The assistance of Rolf Söderlund and Henry Larsson for running the necessary computer programs is gratefully acknowledged, as are the suggestions of Bert Bolin and Henning Rodhe on the manuscript.

REFERENCES

Barret B. and Brodin G. (1955) The acidity of Scandinavian precipitation. *Tellus* **7**, 251–257.

Blüthgen J. (1958) Annual precipitation map of Europe. In *Atlas Troll*, Grosser Herder, Freiburg; and as given in: *World Survey of Climatology* **5**, 17–18 (1970) Elsevier, Amsterdam.

Bolin B. *et al.* (1971) *Air pollution across national boundaries. The impact on the environment of sulfur in air and precipitation.* Sweden's case study to the U.N. conference on the human environment. Royal Ministry of Foreign Affairs and Royal Ministry of Agriculture, Stockholm 1971.

Bolin B. and Persson C. (1975) Regional dispersion and deposition of atmospheric pollutants with particular application to sulfur pollution over western Europe. *Tellus* **27**, 281–310.

Fjeld B. (1976) Consumption of fossil fuels in Europe and emission of SO_2 during the period 1900–1972 (in Norwegian). Technical Note No. 1/76, Norwegian Institute for Air Research.

Granat L. (1972) *Deposition of sulfate and acid with precipitation over northern Europe.* Report AC 20. University of Stockholm, Department of Meteorology/International Meteorological Institute, Stockholm.

Granat L., Söderlund R. and Bäcklin L. (1977) *The IMI network in Sweden. Present equipment and plans for improvement.* Report AC 40. University of Stockholm, Department of Meteorology/International Meteorological Institute, Stockholm.

Munn R. E. and Rodhe H. (1971) On the meteorological interpretation of chemical composition of monthly precipitation samples. *Tellus* **23**, 1–13.

Odén S. (1968) *The acidification of air and precipitation and its consequences on the natural environment* (in Swedish). Ecology Committee, Bull. 1, National Science Research Council of Sweden, 1968. Translated by Translation Consultants, Ltd., Arlington, Virginia, No. TR-1172.

Ottar B. (1978) The OECD study of long range transport of air pollutants. *Atmospheric Environment* **12**, 445–454.

Atmospheric Environment Vol. 12, pp. 425–444. Pergamon Press 1978. Printed in Great Britain.

METEOROLOGY OF LONG-RANGE TRANSPORT

D. H. PACK
Meteorology Program, University of Maryland, College Park, Maryland, U.S.A.
and
G. J. FERBER, J. L. HEFFTER, K. TELEGADAS, J. K. ANGELL, W. H. HOECKER, and L. MACHTA
Air Resources Laboratories, NOAA, Silver Spring, Maryland, U.S.A.

(*First received* 15 *May* 1977 *and in final form* 26 *August* 1977)

Abstract—The increasing attention to regional-scale transport of pollution has resulted in numerous air quality networks, models and large scale field studies to relate the sources of pollution to subsequent air quality measurements and thence to the effects of pollution. We compare observed air quality and trajectory data to calculations to quantitatively evaluate the differences. Methods of preparing air trajectories are described with emphasis on the Air Resources Laboratories' trajectory model and an accompanying diffusion model. The comparison of observed atmospheric transport with calculated trajectories shows that large computational errors can occur, and more importantly that these may be systematic, but different, depending on the type of advection (cold vs warm) and whether the transport is over land or over the sea. We describe the use of tetroon recovery locations in trajectory analysis and show that such data can be obtained over distances up to 4000 km. Computed trajectories based on the geostrophic assumption and on the observed wind field are compared to observations to determine the adjustments required to obtain the best comparisons. Directional adjustments of up to 40° and changes in speed by a factor of two are sometimes necessary. We make suggestions for studies to improve the capability of calculating trajectories, including experiments using balloons and controlled tracer releases. Both of these techniques are applicable over regional scales. Finally we show the global distribution of sulfur from the industrial areas of the Northern Hemisphere, as calculated by an efficient computer model, as a step in the determination of the global sulfur budget.

INTRODUCTION

Over the past 20 years there has been a simultaneous growth of interest in the quality of the environment and the ability to measure the chemical constituents with ever greater accuracy and precision. Accompanying these developments has been the widespread use of computers in meteorology. Air quality measurements have shown that industrial effluents have increased to the point where they affect not only the immediate areas near the emissions but are also carried, in significant concentrations, over large regions and, for some long-lived materials, throughout the global atmosphere.

Studies of the long-range transport of pollution are attempting to better define the movement and removal processes affecting atmospheric concentrations and the amount of material deposited on the ground. The OECD Long-Range Transport of Air Pollutants (LRTAP) study (Ottar, 1976), the Canadian regional study (Whelpdale, 1976), and the Sulfate Regional Experiment (SURE) in the U.S.A. (Hidy *et al.*, 1976) typify the extension of environmental assessments for single sources to the problems posed by the aggregate of emissions over large areas.

A vital part of these and future studies is the inclusion of estimates of the transport and diffusion of pollutants, whether natural or man-made. The possible permutations and combinations of atmospheric variables which control the movement and dilution of material is far too great to enable the collection of statistics on air quality alone to provide the understanding needed for effective environmental protection measures. In addition, we are dealing with a 'moving target' as the character and amount of pollutant emissions change from year to year.

As a consequence of the perception of the problem and the development of advanced technologies, attempts are being made to become more precise in determining the location and time resolution of pollution emission and to install air quality monitoring networks with greater time and space resolution. Accompanying these improvements has been the construction of mathematical models of the pollution system of ever increasing detail and complexity. For example, the grid size used in the LRTAP is 127 × 127 km; that proposed for SURE is 80 × 80 km; while Denmark has reported on a model with a 50 × 50 km grid size (Prahm and Christensen, 1976). The evident variability of pollutant removal by precipitation has led to chemical analysis of individual precipitation events and, in a few instances, of continuous chemical analysis during the precipitation. These are in addition to analyses of collections for longer periods, weeks to a month.

Each reduction in the space and time scale within which we attempt to relate emissions–dilution–removal effects requires commensurate precision in the determination of atmospheric influences. Basically this requirment is for ever greater precision and detail in atmospheric transport and dispersion, in three dimensions.

ATMOSPHERIC TRANSPORT ESTIMATES FOR REGIONAL SCALES

As the precision of regional models of pollution increases, it becomes necessary to include specific emissions from large 'point' sources such as cities and industrial complexes (Veltishcheva, 1976). The scale of interest is indicated from a listing of a few distances between cities in Europe and the U.S.A. For example: Paris–London, 350 km; New York–Boston, 300 km; Dusseldorf–Brussels, 180 km; Pittsburgh–Washington, D.C., 300 km; Warsaw–Moscow, 1050 km; Chicago–Washington, D.C., 1000 km. Thus the space scale ranges from less than 200 to more than 1000 km. It is sobering to examine the accuracies needed in the estimates of atmospheric transport to insure that the source-receptor relations be correctly determined. Figure 1 shows the lateral position error that would result from a given initial trajectory direction error, with the assumption of a straight line trajectory and that the initial error persists. If we recall that the wind directions are reported to the nearest 10°, there is the very likely chance that the report can differ from the true wind by five degrees just due to coding procedures. A five degree initial error gives rise to a 65 km error after a travel distance of 750 km. It is instructive to examine the problem in terms of the grid sizes previously mentioned. For an initial error of 10°, an error which exceeds a 127 km grid dimension occurs after 750 km of travel. For a 50 km grid, the error exceeds the grid dimension after only 285 km of travel.

It could be hypothesized that such initial errors may be random and the use of a sufficiently large ensemble of trajectories would, in the mean, correctly relate emission sources and receptors. However, we may not be permitted this luxury, even if the errors could be shown to be random. Increasingly we find that much of the insult from pollution is concentrated in 'episodes' lasting for only a few days; or that there are significant short-term differences in removal processes within the atmosphere. Summers and Hitchon (1973) show that there are major seasonal differences in the sulfur (as sulfate) in precipitation, with the summer rainfall containing 2–3 ppm of sulfate while the winter value is only 0.5 ppm. They calculate that in summer 32–46% of the emitted sulfur is deposited locally, while in winter 98% of the emissions are carried out of the region for later deposition elsewhere. When such an analysis is combined with the fact that precipitation occurs during only about 10% of the total hours, the need for accuracy in calculating individual trajectories is readily apparent.

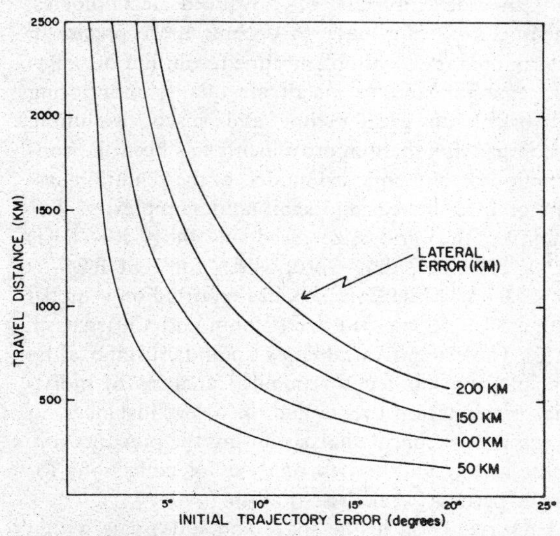

Fig. 1. Trajectory errors created by an initial direction error.

The major purpose of this paper is to examine actual data on measured atmospheric transport over ranges of 50 to 200 km and compare these data with calculated trajectories. However we will first review methods for computing trajectories.

COMPUTATION OF ATMOSPHERIC TRANSPORT

The construction of atmospheric trajectories has been a useful tool in meteorology for decades (e.g. Petterssen, 1940). Estimates of the advection of warm or cold air and of moisture are essential in weather prediction. Changes in air mass characteristics due to passage over differing surfaces are best evaluated along a trajectory. Application of trajectory estimation to the transport of pollutants, whether from a forest fire or an industrial complex, is a logical and necessary extension of the use of meteorological data.

The computational methods, until recently, involved manual determination of the transport wind from analyzed weather charts. The calculations were of two kinds; dynamic (from pressure, or pressure-height charts) or kinematic (from streamline and isotach analyses of the observed winds). In the former case, the instantaneous geostrophic wind was estimated from the pressure patterns. With the streamline-isotach charts, a best estimate of the existing wind was made for the desired location from the analyses.

With the advent of computers in the meteorological services and the storage of pressure and wind data in accessible form in these machines, programs have been devised to compute trajectories. The computer techniques provide for much more flexibility in the type and amount of computation and permit the routine computation of dozens or hundreds of trajectories to relate pollution movement forward in time from sources of emissions, backwards in time from sampling arrays, or both in combination. Such calculations would be a prohibitive manual task.

Computer methods are still based, however, on the two types of information, observed pressures or observed winds. Some techniques analyze these fields and interpolate from the non-symmetric observing network into a regularly spaced grid system from which, by further interpolation, the transport vectors at any arbitrary point can be evaluated. Other methods interpolate the observed data directly to the desired trajectory location.

The early work of Durst *et al.* (1959) is an example of geostrophic computations in the upper atmosphere. The recent article by Sykes and Hatton (1976) provides a comprehensive discussion of the two alternative approaches and describes an advanced geostrophic computational technique (using recorded surface pressure observations) applied to the surface boundary layer.

The kinematic approach has been used in the OECD LRTAP (Ottar, 1976; Eliassen, 1976) employing winds at the surface and at the 850 mb level (approx 1.5 km).

Air Resources Laboratories (ARL) trajectory model

Here we will describe a generalized transport–diffusion–deposition model which, although it primarily uses the kinematic techniques, can also accommodate derived winds (from the Northern Hemisphere analyses made at the NOAA National Meteorological Center). This hybrid, multi-purpose model was described by Heffter *et al.* (1975) but is presented here in a shortened and revised version for greater availability to the interested community. This model is primarily for use in calculating the transport, diffusion, and deposition of effluents on regional and continental scales.

The model was designed to efficiently calculate the large number of trajectories often needed for a variety of pollution problems and also to be used as a research tool to investigate long-range transport and dispersion.

The operational model determines transport: (1) From any origin in the Northern Hemisphere; (2) For four trajectories per day for a month or season; (3) Forward or backward in time; (4) For one to 10 days duration for each trajectory; (5) From observed and/or derived winds averaged in a layer above the average terrain.

The model has an option which permits input of observed winds, derived winds, or both. Observed winds are used whenever possible. However, in areas of sparse wind observations, such as over the oceans, derived winds are used. The use of both types of data is often desirable when the trajectories are near the edge of land areas (i.e. near the edge of dense observing networks) as well as for trans-ocean calculations.

The observed winds used in the model are stored on magnetic tapes. These tapes constitute a unique set of meteorological data ordered by time sequence-data for all reporting stations are grouped together for each observation time (00, 06, 12 and 18Z). One month of upper air wind, temperature and humidity data is packed onto two magnetic tapes. Analyzed meteorological information, also on magnetic tape, in the form of two data analyses per day (00 and 12Z) provide data at 4225 grid points for the Northern Hemisphere at the standard pressure levels of 1000, 850, 700, 500 etc. mb. One month of such data is packed onto two magnetic tapes. For use in the model the winds are extracted from these archive tapes and written on operational tapes so that each tape contains a year of observed winds or three months of derived winds.

Trajectories calculated from observed winds are composed of a series of 3 h segments. Each segment is computed assuming persistence of the winds reported closest to the segment time. For example, the segment from 00 to 03Z is computed from the 00Z winds; the segment from 03 to 06Z and 06 to 09Z are both computed from the 06Z observations, etc.

The observed winds used in the computations are averaged over a specified layer above the average terrain height. For most pollution applications we have assumed that, over regional to continental distances, pollution originating near the surface is distributed through the 'mixed layer', which is determined from vertical temperature profiles along each trajectory.

Trajectories are computed from each point of origin four times daily with the starting times coinciding with the meteorological observing times of 00, 06, 12 and 18Z.

The amount and type of computer output is extremely flexible and can be designated to suit the problem under study. Some of the options are: individual calculated trajectory positions in tabular form; and computer-plotted trajectories on maps. Mercator and polar stereographic projections for the Northern Hemisphere are available. Map scales are variable and can be specified as required. The plotting subroutine for the individual trajectories plots the four trajectories for each day on a computer page. An example is shown in Fig. 2. The individual trajectories are coded, A = 00Z, B = 06Z, etc. The trajectory durations (i.e. travel times) replace the code letters at 6 h intervals along each trajectory with 1 = 6 h of travel, 2 = 12 h, 3 = 18 h, etc. Experience has shown that, for the longer trajectories and for complex flow patterns, the accompanying tabular data is very useful. These data provide the latitude and longitude of the trajectory segment endpoints at specified intervals (≥ 3 h).

The model has been used in a wide variety of applications, some of which we will discuss in subsequent sections.

Diffusion–deposition model

In addition to calculating transport, as described by the trajectories, a diffusion–deposition model has been developed as the basis for long-term surface air concentration calculations from the source

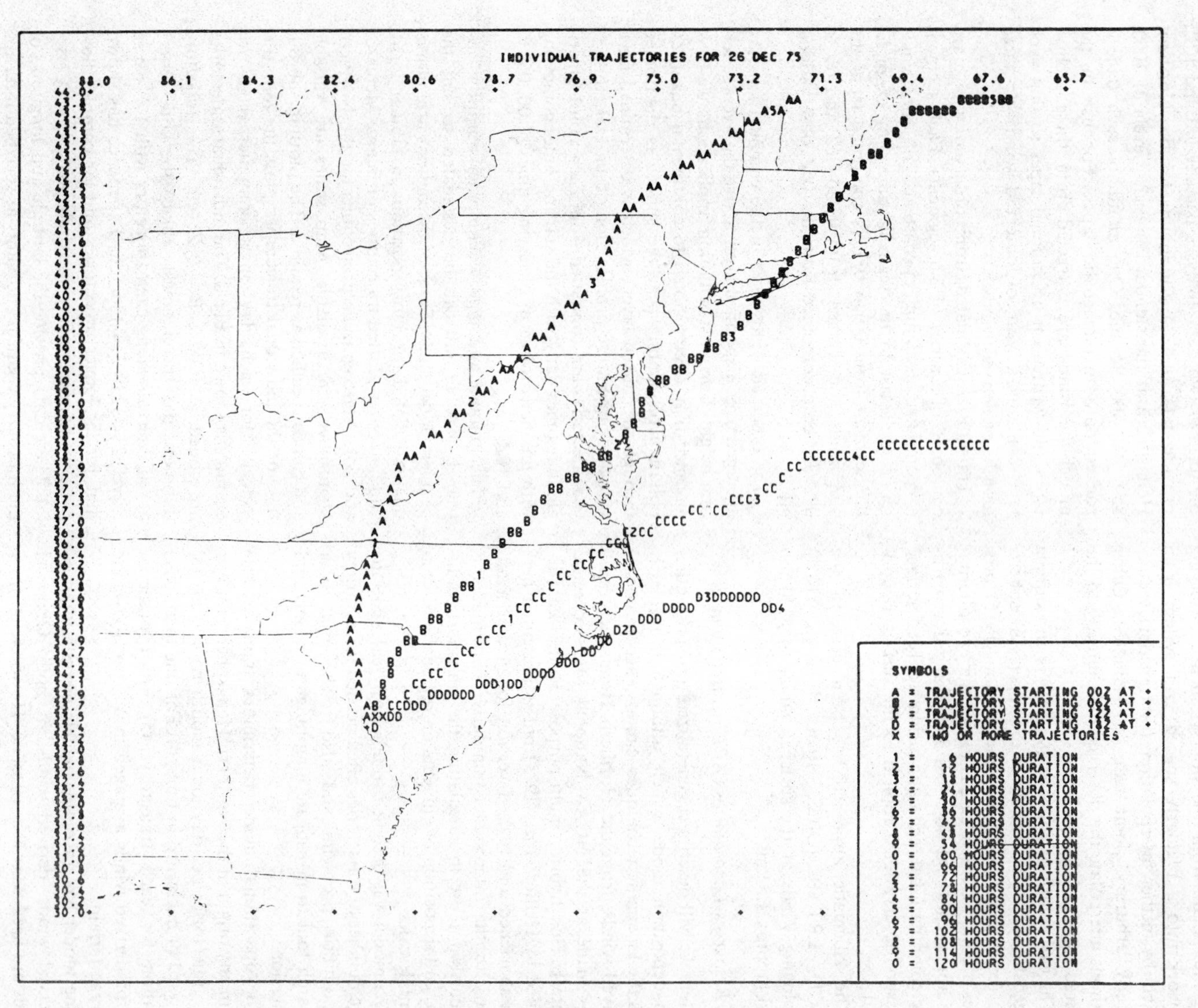

Fig. 2. Computer mapped trajectories. (Lat. and Long. indicated along margins).

to several thousand km. The model calculates mean monthly or seasonal surface air concentrations and deposition amounts. The output is presented as maps of these quantities. This model is intended for long-range use and currently does not provide detailed calculations in the immediate vicinity of the pollutant source. The deposition calculations include removal by both precipitation and by dry deposition.

Long-term (monthly or seasonal) concentration levels are determined in two steps. First, trajectories are calculated and then dispersion–deposition calculations are made along each trajectory. Mean surface air concentrations and deposition amounts are accumulated at grid points for any desired time interval. A 'plume' is represented by a series of puffs (instantaneous point sources) travelling along the computed trajectories.

Each puff is diffused according to:

$$C = \frac{2Q}{(2\pi)^{3/2}(\sigma_h)^2(2K_z t)^{1/2}} \exp\left(-\frac{R^2}{2\sigma_h^2}\right) \quad (1)$$

where: C = surface air concentration ($g\,m^{-3}$, $Ci\,m^{-3}$, etc.)
Q = puff emission (g, Ci, etc.)
σ_h = horizontal standard deviation (m)
K_z = vertical diffusion coefficient ($m^2\,s^{-1}$)
t = puff travel time(s)
R = distance from the puff center (m).

The value of K_z may be specified as required. At present ARL is using $5\,m^2\,s^{-1}$ as an average for the lower troposphere. The model also uses the approximation that $\sigma_h(m) = 0.5t$, where t is the travel time in seconds. This appears to be a reasonable approximation for several days travel time based on a variety of diffusion data as summarized by Heffter (1965).

In some applications it may be preferable to assume that the effluent is uniformly dispersed through some specified mixing layer rather than assuming a Gaussian distribution in the vertical. If this option is chosen the surface air concentrations in the mixed layer are calculated by:

$$C_m = \frac{Q}{2\pi\,\sigma_h^2\,Z_m} \exp\left(-\frac{R^2}{2\sigma_h^2}\right) \quad (2)$$

where Z_m, the height of the mixing layer, in meters, is an input parameter to be specified.

This model is designed primarily for calculations from about 100 km to a few thousand km from the source and might be undesirable for calculations in the immediate vicinity of a pollution source. For example, the height of emission is an important parameter in determining air concentrations near a source but this is not incorporated. Another version of the model has been developed for calculations within about 100 km of a source (Draxler, in press).

Removal of material from the diffusing plume is treated in the following manner. The mass in a mixed layer of unit cross-section area is given by $M_m = C_m Z_m$ where C_m is the mixed layer air concentrations and Z_m is the height of the mixed layer.

The concept of a deposition velocity is used to calculate dry deposition amounts along a trajectory assuming $M_d = CV_d\Delta t$, where M_d is the mass deposited from the mixed layer, C is the surface concentration (equal to C_m in the mixed layer), V_d is the dry deposition velocity and Δt is the time interval over which puff concentrations are calculated.

Deposition velocity is dependent on particle size, wind speed, surface roughness, and other parameters. Most experimentally determined values are in the range from about 0.1 to a few $cm\,s^{-1}$. The user must specify an appropriate value for his particlar problem.

The fraction of mass removed from the mixed layer by dry deposition is $M_d/M_m = V_d\Delta t/Z_m$. The air concentration in the mixed layer depleted by dry deposition, C'_m, is then given by:

$$C'_m = C_m(1 - (V_d\Delta t/Z_m)). \quad (3)$$

Precipitation scavenging is a second mechanism for deposition along a trajectory. Model calculation of wet deposition is based on an empirically derived average scavenging ratio (Engelmann, 1970), $E = C_r/C_p$, where C_r is the concentration in rainwater at the ground, and C_p is the average air concentration in the precipitation layer.

The choice of an appropriate value for E is difficult since the scavenging ratio varies by more than a factor of ten, depending on rainfall rate, atmospheric stability, particle characteristics and other variables, some of which have not yet been thoroughly explored. A value of 500 (mass ratio) was chosen as a reasonable average since it is about in the middle of the range of measured values. This is equivalent to a ratio of 4.2×10^5 by volume, the value currently used in the model. The user may substitute any other value he thinks appropriate.

Wet deposition along the trajectory is then:

$$M_w = C_r P\Delta t = C_p EP\Delta t \quad (4)$$

where M_w is the mass deposited from the precipitation layer and P is the precipitation rate. The average concentration in the precipitation layer is assumed to be $C_p = C_m Z_m/Z_p$, where Z_p is the height of the precipitation layer. Thus, $M_w = C_m EP\Delta t Z_m/Z_p$. The fraction of mass removed from the mixed layer by wet deposition is $M_w/M_m = EP\Delta t/Z_p$. The air concentration in the mixed layer depleted by wet deposition is then given by:

$$C_m^* = C_m(1 - EP\Delta t/Z_p) \quad (5)$$

The expression used to calculate air concentrations depleted by both wet and dry deposition, $C_m'^*$, is:

$$C_m'^* = C_m(1 - V_d\Delta t/Z_m)(1 - EP\Delta t/Z_p). \quad (6)$$

Typical deposition parameter values now being used in the model are:

$V_d = 3 \times 10^{-3}\,m\,s^{-1}$
$E = 4.2 \times 10^5$ (by volume)
$P = 3.2 \times 10^{-8}\,m\,s^{-1}$
$Z_p = 4000$ m.

```
AVERAGE SURFACE AIR CONCENTRATION (UNITS/CUBIC METER) WITH DEPOSITION FOR TRAJECTORIES STARTING          1 DEC 75 TO 31 DEC 75
     88.0       86.1       84.3       82.4       80.6       78.7       76.9       75.0       73.2       71.3       69.4       67.6       65.7
44.0+            +          +          +          +          +          +          +          +          +          +          +          +
43.7 0000000000000000000000000000000000000001A2A4A5A8A1B1B2B2B2B2B2B2B2B2B2B2B2B2B2B2B2B2B2B2B2B2B2B2B2B1B1B1B9A7A6A5A4A
43.3 000000000000000000000000000000001A2A2A4A6A8A1B1B2B2B2B3B3B3B3B2B2B2B2B2B2B2B2B2B2B2B2B1B1B2B2B2B1B1B9A8A7A6A5A4A3A3A
43.0 000000000000000002A3A4A4A5A5A4A4A5A6A8A1B1B2B2B2B3B3B3B3B3B3B3B3B3B3B2B2B2B2B2B2B1B1B1B1B1B1B1B8A6A6A5A5A5A4A4A4A4A
42.7 00000000000001A3A6A1B1B2B2B2B2B2B2B2B2B3B4B5B5B4B3B3B3B4B4B4B3B3B3B3B3B2B2B2B2B2B2B1B1B1B1B1B1B1B7A6A6A6A6A6A7A7A7A7A
42.3 00000000001A3A7A1B2B3B4B5B5B5B4B4B5B5B5B7B8B7B6B4B4B4B4B4B3B3B4B3B3B2B2B2B2B2B2B1B1B1B1B2B2B1B1B8A8A8A9A1B1B1B1B1B1B
42.0 0000000002A3A7A1B2B4B6B8B9B9B8B7B7B7B7B8B9B9B8B6B5B5B5B4B4B4B4B4B3B2B2B2B2B2B2B1B1B1B2B2B2B1B1B1B1B1B2B2B2B2B2B2B2B
41.6 000000002A4A7A1B2B4B6B9B1C1C1C9B8B9B9B9B9B8B7B6B6B6B6B5B4B4B5B5B4B2B2B2B3B3B2B2B2B2B2B3B3B2B2B2B2B2B2B3B3B3B3B3B2B2B
41.3 0000002A4A7A1B2B3B5B7B1C1C1C9B8B8B9B9B9B7B6B5B5B6B6B5B4B4B5B5B5B3B2B2B3B4B3B2B2B3B3B4B3B3B3B3B3B3B3B3B4B3B3B3B3B3B2B2B
40.9 001A2A4A7A1B2B2B3B5B9B1C1C1C8B8B8B9B8B7B5B4B5B5B6B5B5B5B5B6B6B4B3B3B4B5B4B3B3B4B4B5B4B4B4B4B4B4B4B4B4B4B4B3B3B3B2B2B
40.6 2A3A6A9A1B2B2B3B4B6B1C1C1C1C9B9B9B8B6B4B3B4B5B6B5B4B5B6B8B7B5B3B4B5B6B6B5B5B5B6B6B5B5B4B4B5B5B5B5B4B4B4B4B3B3B3B3B2B2B2B
40.2 4A7A1B2B2B3B3B3B4B6B C2C2C1C1C1C8B5B4B3B4B6B5B4B4B4B6B7B6B5B5B6B7B8B7B6B6B7B7B7B6B5B5B5B6B6B5B5B4B4B4B3B3B3B3B3B2B2B2B
39.9 9A1B2B3B3B4B4B3B4B6B C2C2C2C1C1C5B3B3B4B6B5B4B3B4B6B7B6B5B6B8B9B9B8B8B8B9B8B8B7B6B6B6B6B6B6B5B4B4B4B4B4B3B3B3B2B2B2B2B1B
39.5 2B2B3B4B4B5B4B4B4B6B C2C2C2C2C8B4B3B4B6B6B4B3B4B7B7B6B6B7B1C1C1C9B1C1C1C9B7B6B6B6B7B7B7B6B6B5B5B5B4B4B4B4B3B3B2B2B2B1B1B
39.1 2B3B4B4B5B5B5B5B5B7B C2C2C2C2C6B3B4B7B7B4B3B5B7B7B7B9B1C1C1C1C1C1C1C1C1C8B7B7B7B8B8B7B6B6B6B6B6B6B5B5B5B4B3B3B2B2B1B1B9A
38.8 2B3B4B5B5B6B6B6B7B8B1C2C2C2C1C8B4B8B8B4B4B6B8B7B7B1C1C2C2C1C1C1C1C2C1C1C1C1C1C1C IC 3B7B7B7B7B7B7B6B5B4B3B2B2B1B1B8A7A
38.4 2B3B4B5B6B6B7B7B8B8B1C2C3C2C1C7B8B8B4B5B8B8B6B7B1C2C2C2C1C1C1C2C2C2C2C2C1C1C1C1C1C1C1C9B9B9B9B9B8B7B6B4B3B2B1B1B7A6A5A
38.1 2B3B4B5B6B7B7B8B8B9B1B1C3C2C1C1C1C5B7B9B7B7B9B1C2C3C2C2C1C1C2C2C2C2C2C2C2C2C2C1C1C1C1C1C1C9B9B9B8B7B5B4B3B2B1B7A5A4A4A
37.7 2B3B4B5B6B7B8B8B9B1C1C1C4C3C1C1C5B6B1C6B7B1C2C3C3C2C2C2C2C2C2C3C4C4C3C3C2C2C2C2C2C1C1C1C1C9B8B7B7B6B5B4B3B2B1B7A4A3A3A2A
37.3 2B3B4B5B6B7B8B9B1C1C1C1C4C3C1C1C5B4C1B1C1C2C3C3C3C3C3C2C2C3C4C5C5C5C4C3C3C2C2C2C2C1C1C9B8B7B6B5B5B4B3B2B2B1B6A4A2A2A1A1A
37.0 2B3B4B5B7B8B9B1C1C1C1C1C4C4C1C1C7B2C1C2C3C4C3C2C3C4C4C3C4C6C6C6C5C4C4C3C3C2C2C2C1C1C1C8B7B6B5B4B4B3B2B1B8A5A3A2A1A000000
36.6 2B3B4B5B7B8B9B1C1C1C1C2C5C6C1C2C1C2C2C3C5C3C3C3C4C5C5C6C7C7C6C6C5C4C3C2C2C2C2C1C1C1C8B8B7B6B4B3B2B2B1B8A4A2A1A0000000000
36.2 2B3B4B6B8B9B1C1C1C1C1C2C3C7C1C2C3C3C5C5C4C3C4C7C7C7C8C8C8C8C6C5C4C3C3C2C2C2C1C1C8B6B5B5B6B6B4B3B1B1B7A4A2A00000000000000
35.8 1B2B4B6B9B1C1C1C1C1C2C2C3C6C2C4C6C6C6C5C4C7C1D1D9C1D9C8C7C6C5C4C4C4C3C2C2C2C2C1C8B5B4B3B4B4B4B2B8A4A2A1A0000000000000000
35.5 1B2B4B7B9B1C1C1C1C1C2C2C3C6C5C6C7C7C6C5C1D2D1D1D1D1D1D8C7C6C5C4C4C4C3C2C2C2C1C1C9B7B4B3B3B3B2B9A3A1A0000000000000000000
35.1 1B2B4B7B1C1C1C1C2C1C2C2C3C5C1D1D1D8C6C1D2D2D1D1D2D1D1D1D8C6C5C5C5C4C3C2C2C1C1C1C8B6B5B3B2B2B9A4A1A000000000000000000000
34.7 9A2B3B6B C1C2C2C2C2C2C2C2C3C4C2D2D1D1D2D4D2D2D2D2D2D2D1D1D9C8C7C5C5C4C3C2C2C1C1C1C7B5B4B2B1B8A4A1A0000000000000000000000
34.3 7A1B3B5B1C2C2C2C2C2C2C2C3C3C4C2D2D2D3D6D3D3D4D3D2D2D2D2D1D1D1D8C6C5C4C3C2C1C1C1C1C6B3B2B1B5A3A1A000000000000000000000000
33.9 6A1B2B4B8B1C2C3C2C3C3C2C3C6C1D4D3D9D4 IE 5D4D3D2D2D2D1D1D1D9C7C5C3C3C2C2C1C9B8B7B4B2B1B5A2A1A000000000000000000000000000
33.5 4A8A2B3B6B1C2C3C4C4C5C5C7C5C1D0 0 0 1E6D4D3D2D2D2D1D1D1D9C7C6C5C3C2C2C2C1C8B6B4B2B1B6A3A2A1A000000000000000000000000000
33.2 3A6A1B2B4B7B1C2C3C5C7C1D1D2D4D0 0+0 2D3D2D2D1D1D1D ID 8C7C6C5C4C3C2C2C1C1C7B5B3B2B9A5A3A2A1A000000000000000000000000000
32.8 2A4A8A1B2B4B7B1C2C3C4C5C6C5C5D0 0 0 1E5D3D2D2D1D1D1C8C6C6C5C4C3C3C2C2C1C9B7B5B3B2B9A5A3A2A1A0000000000000000000000000000
32.4 2A3A5A9A2B3B4B7B1C1C2C2C2C3C3C5C2D4D5D5D4D2D1D1D9C7C6C6C5C5C4C3C3C2C2C1C8B5B4B2B1B9A5A3A2A1A000000000000000000000000000
32.0 1A2A3A6A1B2B3B4B6B8B1C1C1C2C2C3C9C2D3D3D3D3D2D1D3C6C5C5C4C4C3C3C2C2C2C1C7B5B3B2B1B8A5A3A2A1A00000000000000000000000000000
31.6 1A2A3A4A6A1B2B2B3B5B6B8B9B1C1C2C5C1D2D2D2D2D2D1D1C6C5C4C4C3C3C3C2C2C1C1C7B4B2B2B1B7A5A3A2A1A00000000000000000000000000000
31.2 1A2A2A3A4A6A1B1B2B3B4B4B6B7B9B1C3C7C9C1D2D2D2D1D1D7C5C4C3C3C3C2C2C2C1C9B6B3B2B1B8A5A4A2A1A00000000000000000000000000000000
30.8 1A2A2A2A3A4A6A9A1B2B2B3B3B4B6B1C2C5C7C9C1D1D1D1D9C7C5C4C3C2C2C2C2C1C1C7B4B2B1B8A5A4A2A2A1A0000000000000000000000000000000
30.4 1A2A2A2A3A3A4A6A7A1B1B2B2B3B4B7B2C3C5C6C9C9C8C9C8C7C5C4C3C2C2C1C1C1C7B5B3B2B8A5A3A2A1A000000000000000000000000000000000000
30.0+1A2A2A2A3A3A4A5A6A7A8A1B1B2B3B5B1C2C3C4C5C7C7C6C7C6C5C3C3C2C1C1C9B6B4B3B2B9A4A2A1A0000000000000000000000000000000000000000
                         IC
```

Fig. 3. Computer-mapped monthly average air concentrations. Plotted code: $2E = 2 \times 10^{-11}$, $2D = 2 \times 10^{-12}$, etc.

A puff, representing the effluent plume for a specified time interval (e.g. 6 h for four trajectories per day), diffuses over points on a gridded map as it progresses, at 3-h segment intervals, along a computed trajectory. The grid spacing and area coverage are selected to fit the problem. The concentration from all puffs which pass over a grid point are accumulated at each point and averaged over some specified long-term period (month or season). The results are mapped as average surface air concentrations in a convenient code (e.g. $2E = 2 \times 10^{-11}$ Ci m^{-3}) at each grid point as illustrated in Fig. 3 for a hypothetical source at Savannah River, South Carolina. In this example a uniform emission rate of 0.01 Ci s^{-1} was used.

When deposition is appropriate to the effluent under study a second map is plotted to show the calculated surface deposition pattern with wet and dry deposition combined.

Another option in the model provides for linear interpolation of additional trajectories between those initiated at 6-h intervals. The use of trajectories at three-hourly intervals (or less) is recommended for calculating concentrations over an area less than one day's travel from the source.

VERIFICATION OF CALCULATED TRAJECTORIES

Data from which a comparison of calculated trajectories with 'real' trajectories can be made are not numerous over the regional scale. The ideal would be a continuous point source of a unique material whose emission was accurately known over short time intervals, say hourly, and which could be quantitatively detected at low cost over distances up to 5000 km. Instead we have numerous short-range tracer experiments, a rather large mass of data linking estimated emissions (generally based on annual data) from large area sources to air quality measurements, and a few experiments designed to test long-range tracers. In addition there is a limited amount of data from constant volume balloon flights, some of which were specifically designed to test trajectory calculational procedures.

Short-range trajectories

Although transport over distances much less than 200 km is not the focal point of this Symposium, it is still useful to examine a few experiments. Since experience indicates that trajectory errors increase with increasing distance, a measure of accuracy at short distances is relevant to the larger scale. However, a note of caution is indicated. Most tracer experiments are designed to maximize the probability of a fixed sampling network detecting the tracer material. For this reason such experiments are usually carried out in wind conditions of maximum steadiness, often with quite high wind speeds. Such conditions reduce the probability of error regardless of calculating technique.

Tetroon data over distances up to about 100 km have been compared to trajectories calculated from observed winds and, in two experiments, from the geostrophic assumption.

Druyan (1968), using tetroon data obtained by Hass *et al.* (1967), compared trajectories derived from a 30–60 station network of surface winds and from a much less dense network of pilot balloon observations. He found that surface trajectories were inferior to the use of winds aloft observations. He also found that adjusting the surface data by means of an Ekman wind profile *increased* the error in about one-third of the comparisons. He also determined that, while the total vector error increased with distance, the percentage error decreased from 54% of the length of the trajectory for two hours of travel to 35% after 4 h and to 31% after 6 h. (A similar result will be shown later for much longer distances).

Tetroon flights in the Los Angeles Basin at a height of a few hundred meters (Angell *et al.*, 1972) were compared to trajectories derived from the continuously operating surface wind network within the Basin. These data show that during daylight hours the surface wind trajectory computations were quite good with the differences between calculated and observed transport only 10–15% of the tetroon trajectory length. At night the correspondence was not as good during the stable conditions of this complicated flow regime. During the hours from 2100 to 0900 local time the differences between calculations and observations ranged from about 35 to more than 50% of the travel distance. These data also show a decrease in the *percentage* difference with increased travel distance.

The preceding data were in coastal regimes and were often influenced by sea breeze circulations. However, data were obtained in the mid-western U.S.A. for night-time flow across a medium-sized city (Columbus, Ohio) (Angell *et al.*, 1971). The hourly tetroon travel was compared to surface geostrophic winds calculated from two sets of hourly surface pressure data (Pack, 1973). Each set of pressure data was obtained from four weather observing stations surrounding the tetroon flight area. One set of stations was, on the average, separated by 266 km, the other represented a larger 'grid' size of 364 km. The hourly pressure values (both 'station' and 'sea-level' data were tested) from each station set were used to calculate an hourly surface geostrophic wind comparable in time to the tetroon trajectory segment.

The tetroons were usually set to fly at about 100 m above the surface. Comparisons were made between the calculated geostrophic winds and those tetroon data where the average flight altitude was 125 m. The results are shown in Table 1. The differences are in the expected direction with the calculated wind directed to the right of the observed transport. The size of the 'grid' makes little difference in the comparisons.

Table 1. Comparison of tetroon observations, V_T, and calculated geostrophic wind, V_G

	Pressure 'grid' size	
	364 km	266 km
Direction difference ($V_G - V_T$)		
Mean	+29°	+30°
Standard deviation	21°	32°
Speed difference (m s^{-1})		
Mean	+0.8	+0.3
Standard deviation	3.9	3.6

The most notable feature is the large standard deviations which reflect the uncertainty of any individual trajectory calculation. A similar analysis was made using all of the tetroon data regardless of flight altitude. The mean speed differences remain about the same; the direction differences are reduced to 15°, *but* the standard deviation of the direction differences is doubled, to 63°.

Intermediate range trajectories

At intermediate distances two quite different experiments should be mentioned. In the first, a plume of Argon-41 was tracked to beyond 250 km on a single day (Peterson, 1968) and compared with trajectories calculated from surface wind, applying a speed correction (increase). In this instance the comparisons were excellent, the differences being less than 20 km in more than 250 km of travel.

A study of atmospheric transport and diffusion is being conducted by the Air Resources Laboratories in collaboration with the Savannah River Laboratory (SRL) of E.I. duPont de Nemours and Co. This experiment takes advantage of the Krypton-85 plume produced at the Savannah River Plant (SRP) near Augusta, Georgia. A significant advantage is obtained from the nearly continuous release, allowing transport and diffusion of the effluent to be studied under a wide variety of meteorological conditions. Since this tracer is an inert gas, verification of transport and diffusion calculations can be separated from the action of wet and dry deposition and chemical transformations. Weekly average air samples were taken at 13 locations at distances from 30 to 150 km from the plant.

One objective of the study is to provide a test of the ARL transport model and to compare it to the use of a single-point stability-wind rose model (Turner, 1970). A technique for categorizing standard hourly surface weather observations into Pasquill stability classes, as presented by Turner (1964), was programmed at the National Climatic Center NOAA (Doty *et al.*, 1976). This program (STAR) was run with data from Augusta, Georgia to obtain monthly, seasonal and annual wind distributions for 6 stability classes. These data were then used in the stability-wind rose model to calculate average annual concentrations at all the sampling locations. A simplified version of the ARL transport and diffusional model, referred to as Model A, was also used to determine average concentrations at the same location.

The observed mean annual concentration pattern and those calculated with the stability-wind rose model and Model A are shown in Fig. 3. Weekly samples from March 1975 to February 1976 were averaged (and background removed) in the determination of the observed mean annual concentrations. The lower right diagram of Fig. 3 shows a plot of both the stability-wind rose and Model A calculations versus the observed concentrations. Model A calculates the concentrations within a factor of two while the stability-wind rose model overpredicts by a factor of four on the average. The over-prediction appears to be due mainly to the assumed persistence of a stability category during plume travel and the use of the surface wind speed.

Long-range trajectories

At longer distances of travel, 500 km and beyond, which are now commonly included in the 'regional' scale, we have several examples of air quality data and related atmospheric trajectories. In these examples the source area could be fairly well defined. The work of Prahm *et al.* (1975) relating SO_2, SO_4^2, and other compounds to industrial or non-industrial sources was accomplished by careful selection of sampling periods and the construction of geostrophic trajectories. These data, combined with estimates of mixing heights and rates of diffusion, were successful in permitting estimates of deposition rates and atmospheric residence times, and in determining representative 'clean' air background concentration levels of selected particulates and gases. It is of interest to note, however, that the defined sector from which atmospheric transport was expected to bring anthropogenic material was 50° wide. This reduced the effect of even large trajectory errors and was possible only because of the carefully selected and isolated siting of the sampling station.

That such analyses are not always as straightforward is reported by Smith and Jeffrey (1975). They state "... one of the hardest problems, and one of the most important, is to obtain reasonably accurate back trajectories ... ". Through an ingenious sampling design they were able to minimize the problem in a manner that, effectively, widened the source sector, much in the manner of Prahm. Their experiment also included the much more difficult problem of predicting the trajectories in order to take successive samples in the same volume of air, a Lagrangian approach. The results were inconclusive and did not permit separation of the roles of chemical transformations, diffusion, or incorrect trajectory estimates.

A short series of air quality-trajectory comparisons are available from the analysis of ozone transport from the European continent to England and Ireland (Cox *et al.*, 1975). In this study the ARL model was used to construct trajectories backwards from Sibton,

Table 2. Comparison of observed CCl_3F at Adrigole, Ireland with trajectories from 'source' and 'non-source' areas

	High value	Low value	Total
Source area	36 (42%)	45 (4%)	81
Non-source area	50 (58%)	1105 (96%)	1155
Total	86	1150	1236

England and Adrigole, Ireland. The source, as near as could be judged, was the north-western European land mass. The trajectories and the air quality data were well related. The length of the trajectory, as well as the indicated origin and the paths, correlated with the observed levels of ozone and chlorofluorocarbon concentrations.

The ARL transport model was again used in the analysis of three years of chlorofluorocarbon data obtained at Adrigole, Ireland (Pack *et al.*, 1977). Over the period from May 1973 through December 1975 it appears that a rectangle approx 700 × 1600 km extending from the French coast northeastward to Denmark and the northern Federal Republic of Germany constituted the source region.

The 1236 CCl_3F concentration observations (one every 6 h) during this period were divided into two categories; high concentrations, defined as values 25% or more above the mean for each month; and low concentrations, those less than 25% above the mean. This was done for each month in which sufficient observations were obtained for an adequate sample. Individual monthly mean values were used due to the steady growth of the CCl_3F concentrations during the period of record. The trajectories calculated for each of the 6-h observations were then ascribed to the 'source' area or to non-source regions depending on the observed concentrations. The results are shown in Table 2.

We find that for an unequivocal anthropogenic emission, sampled at a remote location 500 to 1400 km from the most probable source area, 96% of the time the trajectories correctly stipulate transport from a non-source region. However, 'high' concentration levels were correctly ascribed to the source area only 42% of the time.

While the ARL model proved useful in this study, the data do not provide a very good test of the model. First, we know far too little about the sources and source areas. Second, the analysis takes no account of the time of travel, hence the dilution which, over long travel times, can reduce concentrations to near background levels even though the air originally could have contained elevated concentrations. Third, and a serious deficiency, the trajectories were not mapped south of 35°N, although there is evidence that many of these trajectories represented earlier transport southward from the source area. It appears that about one-third of the 'high' concentration values not assigned to the source were of this character.

This data analysis was not designed to evaluate the accuracy of individual trajectories but rather to examine spatial variations and time trends of average monthly data. Our experience is that off-continent trajectories occurred in sets or 'episodes', very often beginning and terminating abruptly. An appropriate verification study should collect more information on sources and consider the concentration-trajectory combinations in a series of case studies. Even so there will remain uncertainties associated with variations in emissions and in rates of atmospheric diffusion.

Another type of tracer experiment which also takes advantage of a 'source of opportunity', but where a suitable substance is injected into the atmosphere in known amounts from a well defined source, is much more easily applied to verification studies. The sampling program at Savannah River is an example of this type. However, attempts to study plumes at much greater distances have had only limited success. For example, an attempt was made to sample the Kr-85 plumes emitted from a source in Idaho, at weather stations in the mid-western U.S.A., 1500–2500 km from the source (Ferber *et al.*, 1976). Samples were collected twice-daily for four months at the locations shown in Fig. 4. Most sample concentrations were within the range of background variability and it proved very difficult to unambiguously identify the plume at these distances. However, the plume from Savannah River was seen at Indianapolis and Detroit, about 1000 km distant from the source, on several occasions. These data are still being analyzed but one example will be shown here.

On 21–22 April 1974 the Kr-85 concentrations in samples at Indianapolis (IND) and at Detroit (DET) were sufficiently above background to initiate an analysis of the transport and diffusion. Mean winds in the layer 0-1500 m above the terrain were used to calculate plume transport with the ARL model.

On April 18 the southeastern portion of the U.S.A. was dominated by a high pressure system centered over Tennessee. Another anticyclone was centered over Hudson Bay in Canada. The flow over the Savannah River area was light and variable. For the following two days (19–20 April) the winds remained light and variable until the two high pressure systems merged into one large anticyclone centered off the Virginia coast. The flow of air over most of the southeastern U.S.A. became predominantly southerly.

Figure 6 shows three trajectories from the Savannah River plant beginning at: 2000LST, 17 April; 0200LST, 18 April; and 0800LST, 18 April. The model calculations show most of the excess Kr-85 in the 10-hour sample taken between 2100, 21 April and 0700, 22 April at both IND and DET. The observed and calculated concentrations close to the time of trajectory arrival are shown in Table 3. All three trajectories pass near the stations during the sampling period when the highest concentrations occurred and there is good agreement between calculated and observed concentrations.

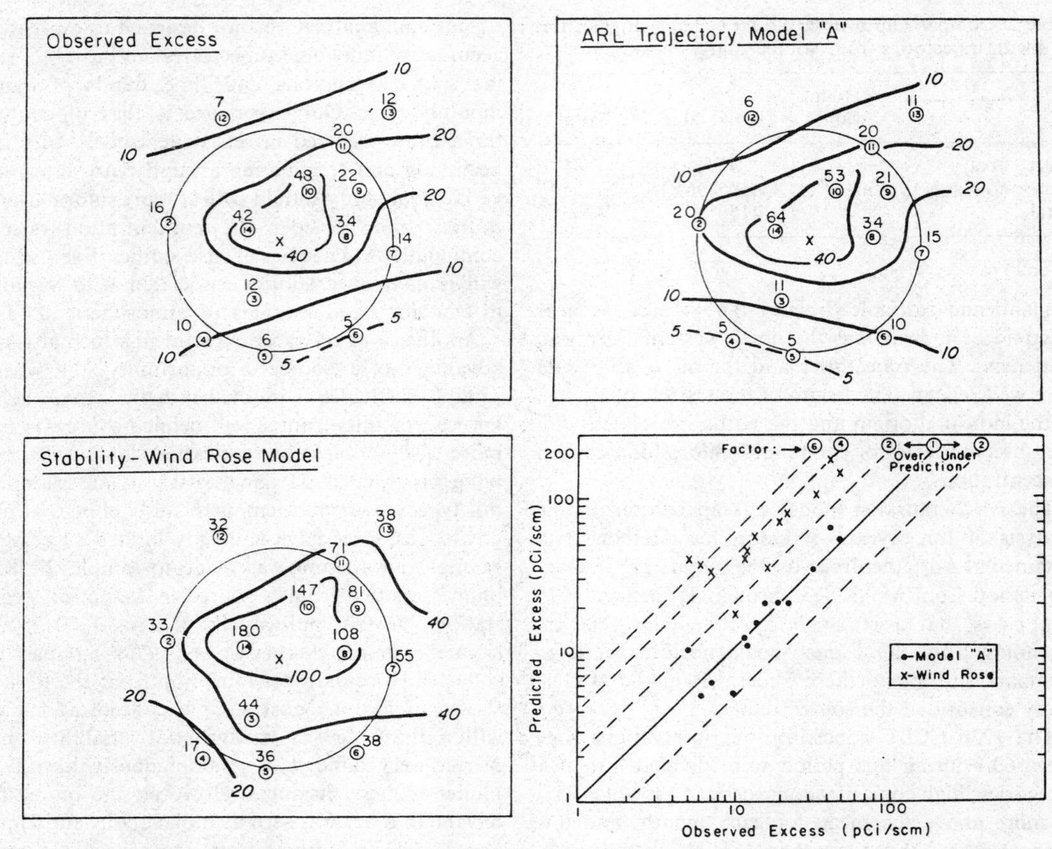

Fig. 4. Comparison of observed and calculated mean annual excess Kr-85 concentrations. × is source location; circled numerals are sampler locations. Large circle denotes radius of 100 km from source.

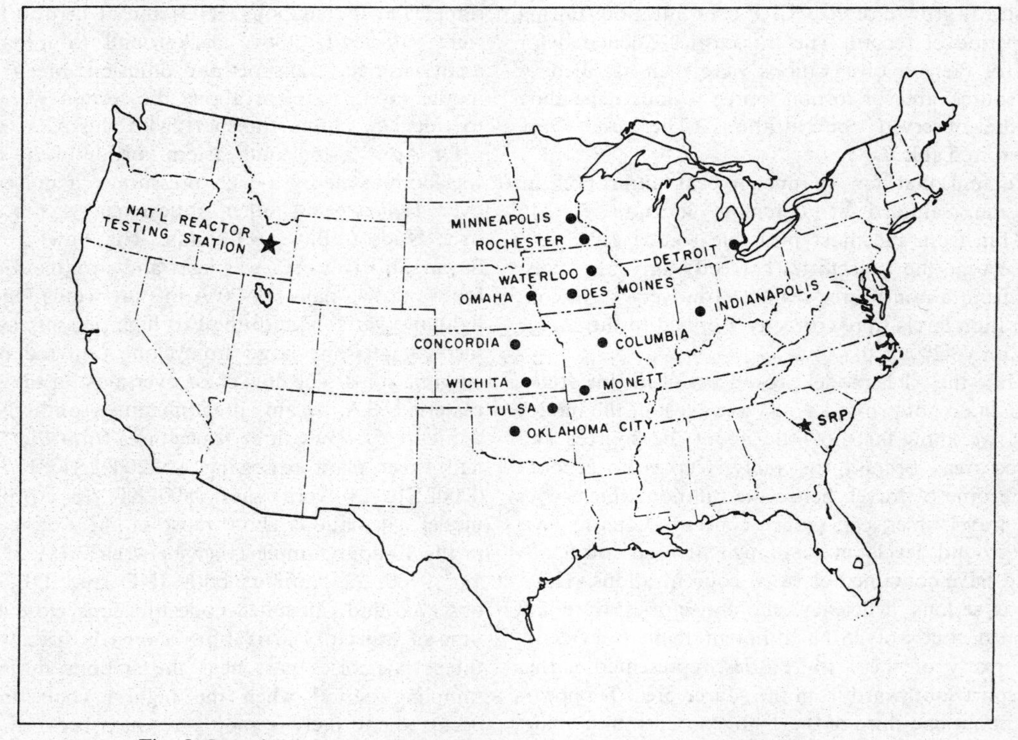

Fig. 5. Location of two sources (stars) and mid-west Kr-85 sampling network.

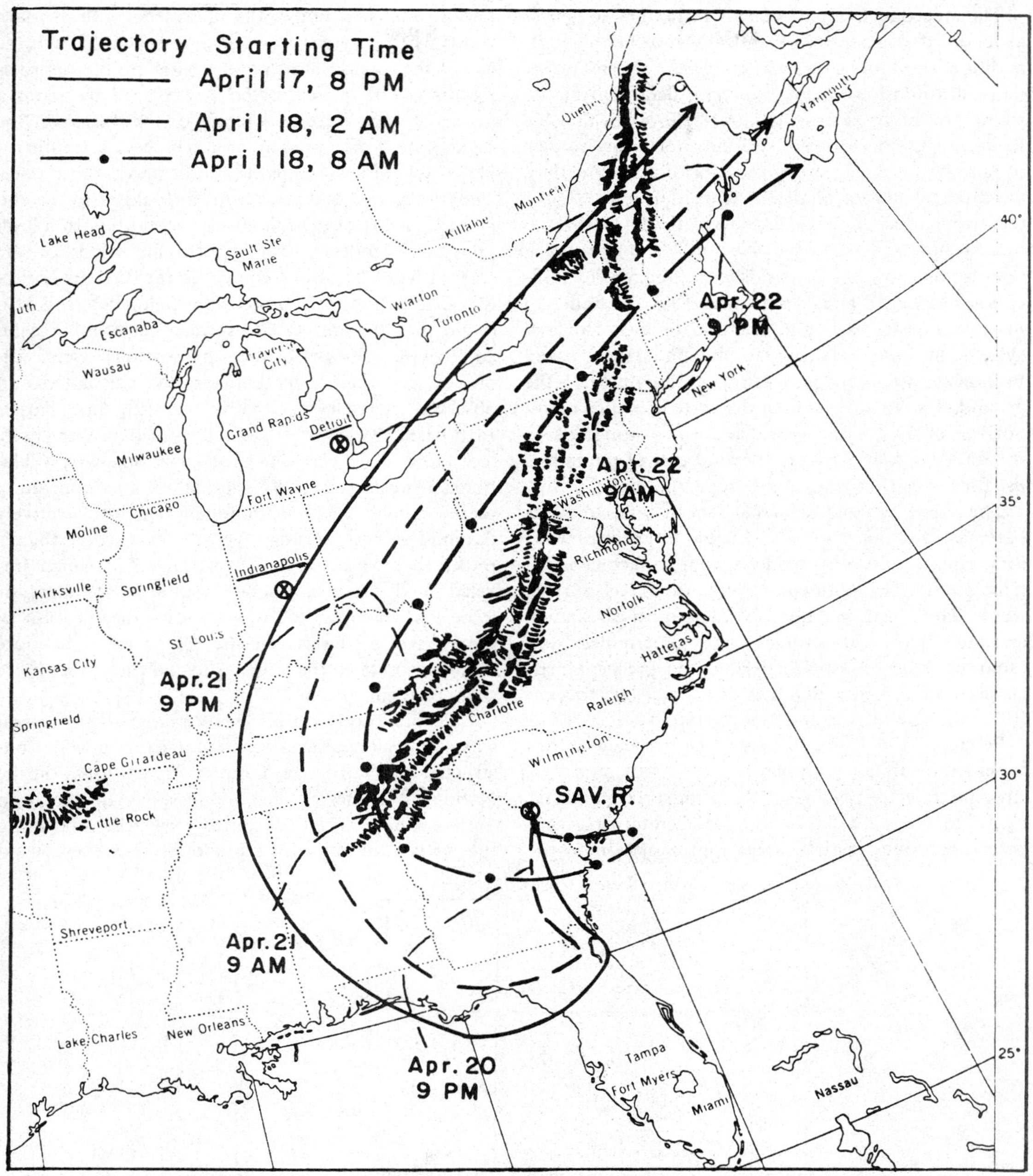

Fig. 6. Calculated trajectories of plumes emitted from Savannah River, using mean-layer winds from 0–1500 m above average terrain. Thin dashed lines and associated date/time indicate calculated time of arrival of plumes.

Table 3. Observed and calculated excess Kr-85 concentrations (pCi/SCM), for 10 h samples, 20–22 April 1974

Sample Began		Indianapolis (IND)		Detroit (DET)	
Date	Time	Observed	Calculated	Observed	Calculated
20	2100 LST	0.7	0	0	0
21	0900 LST	1.8	1.1	1.0	0
21	2100 LST	3.5	4.9	2.6	3.7
22	0900 LST	0	0.4	1.2	1.2

The preceding comparisons are rather few to generalize on the accuracy of trajectory techniques. In addition they are complicated by the spreading of material (diffusion) or the imprecise definition of the source, or both. During the past 15 years, the Air Resources Laboratories have conducted many studies of air motions at numerous locations launching hundreds of tetroons. Almost from the onset 'return tags' were attached to the tetroon or the tetroonborne radar transponder or radar reflector. The tags are self-addressed, stamped, post cards with the flight identification number. Blanks are provided for the finder to insert the time, date, and location of the recovered system. In some experiments, the finder was asked to indicate if the tetroon was observed coming to the ground. Use of this simple idea extended the useful horizon of these atmospheric tracers to hundreds or thousands of kilometers, far beyond that of the radars used for precision tracking early in the flight.

The recovery-point data have been used to compare the end point of the tetroon trajectories with atmospheric transport deduced from surface geostrophic wind calculations or from observed surface winds. The time of tetroon release is precisely known and the initial flight altitudes can be determined from radar tracking or estimated from the precision ballasting and launch techniques developed by Hoecker (1975) and by Delver and Booth (1965).

Peterson (1966) used manually constructed trajectories from surface and upper air wind data, and geostrophic trajectories constructed from surface weather maps, to compare with 19 tetroon return-tag end points. He found that the best approximation to the tetroon recovery point was provided by the surface winds when they were 'adjusted' in speed through the use of the 'one-seventh power' wind profile approximation (which increases the 300 m wind by about a factor of two), and by veering (clockwise shift) the surface wind direction by one degree of azimuth for every 30 m of tetroon flight altitude above the surface. The trajectories so constructed came closer to the end point than any of the following: second reported level upper air winds; surface geostrophic wind; or the 1500 m wind. Figure 7 compares the various trajectory calculations for one tetroon flight. Peterson also found that by shifting the surface geostrophic wind 20° counter-clockwise the comparisons were improved, but the scatter between the calculated and observed trajectory end points was still much larger than when the adjusted surface wind data were used.

A more recent experiment at Oklahoma City, Oklahoma (Angell *et al.*, 1973) provided an opportunity to re-examine Peterson's findings over the relatively flat mid-western terrain. Of the 55 tetroon flights, return tags were received from 27, 13 of which traveled a distance considered suitable for analysis (Hoecker, in press). Figure 8 shows the location of the recoveries. Twelve of the recoveries were more than 500 km from the launch site and three were more than 1000 km distant.

Calculated trajectories were prepared in several ways: (1) Surface geostrophic trajectories were constructed manually from the pressure contours on the National Weather Service 3-Hourly North American Surface Charts; (2) Trajectories were generated using the ARL trajectory model and layer-average wind

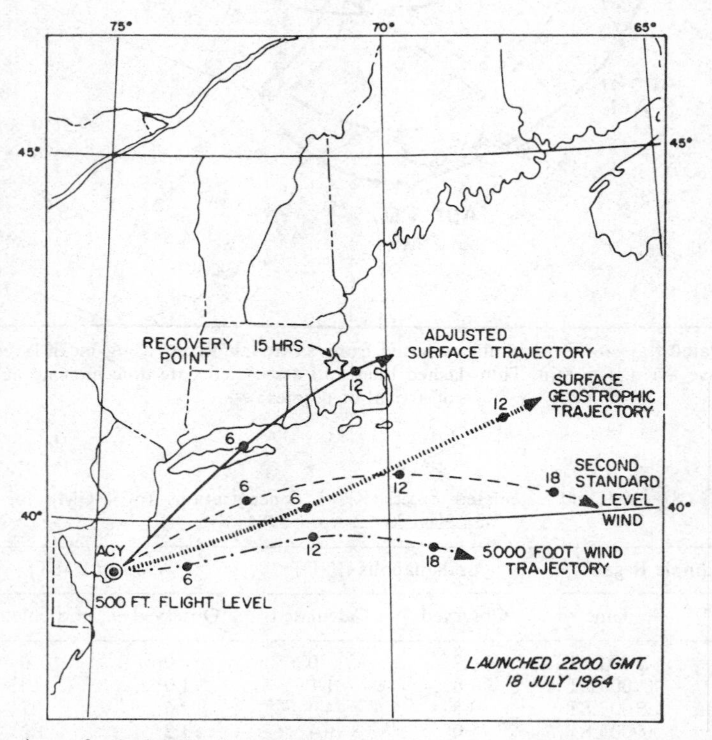

Fig. 7. Comparison of calculated trajectories to tetroon flight end point (star). (After Peterson, 1966).

through the 1000 m layer above the surface; (3) Using the ARL model with surface wind data only with the speeds increased by a factor of two; the directions remaining unchanged; (4) as in (3) but with the directions veered by 10°.

The four types of trajectories were compared to the tetroon recovery point by determining the minimum distance between that point and the calculated trajectory. These data were then plotted as a function of downwind distance for the four different calculations—'geostrophic', 'layer-average', 'adjusted surface' no veering, and 'adjusted surface' veered, both of the latter with the speeds doubled.

Figure 9 shows the results for flights in southerly (S), westerly (W), and northerly (N) flow with the angular deviations also indicated. It is seen that: (1) For the surface geostrophic trajectories the results depend strongly on wind direction. In southerly flow the tetroon trajectories had apparently moved towards low pressure at an angle of about 20° to the isobars, while in the northerly flow this angle doubled, to near 40°. These large and quite different deviations suggest that the direct use of the surface geostrophic calculation to estimate boundary layer trajectories, even though a *fixed* correction may be applied, can lead to large discrepancies. (2) The layer-average (0–1000 m) trajectories were about 10° to the right (looking downwind) of the tetroon end points in southerly and westerly flow but in the four flights

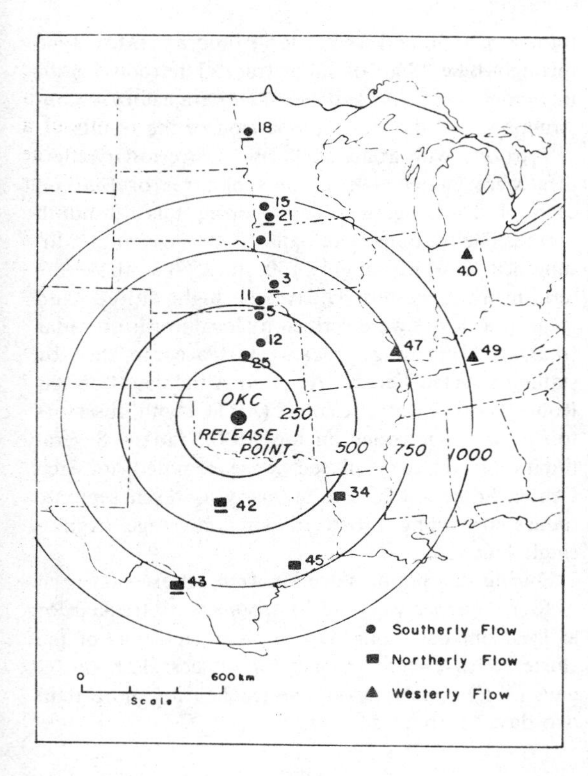

Fig. 8. Location of tetroon recovery sites. All flights originated at Oklahoma City, Oklahoma (OKC). Underlined symbols designate flights recovered within 24 h of launch.

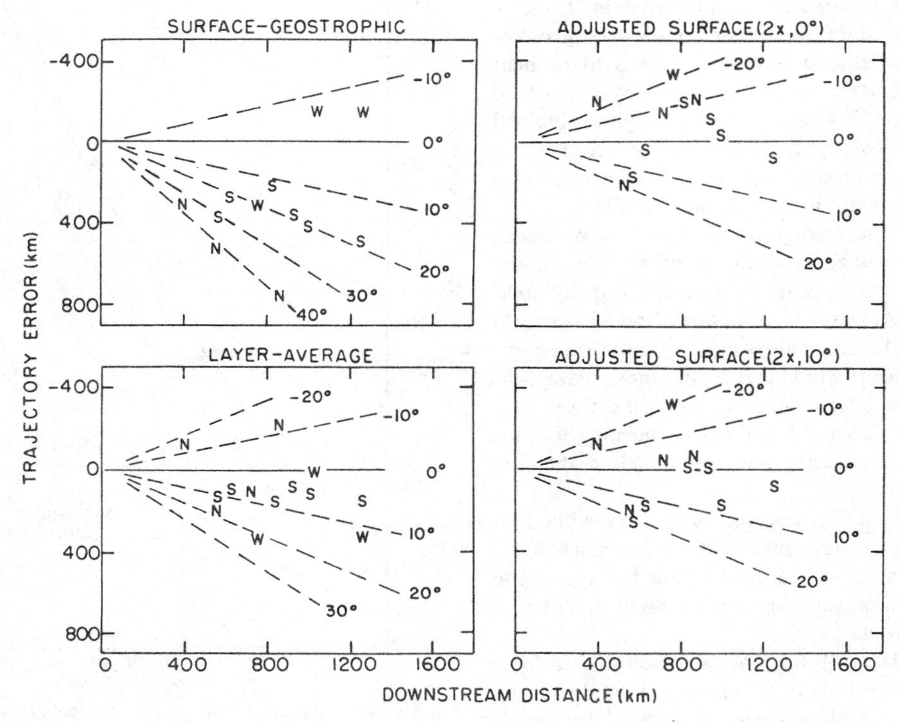

Fig. 9. Comparison of trajectories calculated by four different methods with location of tetroon return tag recovery sites. Flow directions: N = north, S = south, W = west. Angular deviations of calculated trajectories shown by dashed lines.

Table 4. Long-term average veering of wind direction over land (surface to 2.5 km) from 19 rawinsonde stations (after Hoxit, 1974)

	Advection		
Time	Warm	Cold	Neutral
00Z	32°	65°	46°
12Z	34°	42°	38°

in northerly flow the calculated path was to the right twice, and to the left twice. (3) For the adjusted surface wind data with the speed doubled but the direction left unchanged, in two-thirds of the cases the estimated trajectories were to the left of the recovery site as would usually be expected. However the remaining one-third of the calculated trajectories were to the right of the site. (4) The trajectories calculated using surface wind data with the speed doubled and the direction veered by 10° gives the best estimate, on the average, in the sense that about one-half of the trajectories went to the left of the recovery point and about one-half to the right. However, in seven of the 11 cases the angular deviations were more than 10° and in three of these cases the deviation was as large as 20°. Thus even in this case more than one-half of the trajectories had angular deviations greater than 10°.

These results can be compared to the numerous data of Hoxit (1974) who compiled statistics on the wind direction change in the planetary boundary layer from hundreds of upper wind observations. The data for land stations are summarized in Table 4.

These studies did not estimate the sea-level geostrophic vector directly. It is often assumed to be near or slightly above the top of the mixed layer. Garland and Branson (1976) report mixing heights obtained by numerous investigators. These range from about 700 m above the surface over Great Britain to 2000 m over Hungary, with an average near 1000 m.

Using the same average profile data that produced Table 4, the change in wind direction between the surface and 1000 m was determined. For the 00Z observations these are 17° for warm, and 45° for cold advection. At 12Z the values are 34° and 42° respectively. The mean wind speed for these data was 12 m s^{-1} in the lowest one km; hence, these were not just light wind cases. At 12Z (early morning for this area) all of the veering was completed in the first 1000 m.

The Oklahoma City tetroon-trajectories, while few, are in qualitative agreement with these much more numerous observations and illustrate the systematic differences in trajectory directions created by different synoptic situations.

The final data set that we will examine is from an elegant but little known experiment conducted by the ARL office at Las Vegas, Nevada (Allen *et al.*, 1967; Jessup, 1967; Allen and White, 1969). This office, in an attempt to better understand atmospheric transport, mounted a year-long program (May 1965 through June 1966) of radar-tracked tetroon flights, trajectory forecasts, and *post-facto* trajectory reconstruction, and the inter-comparison of the results.

Tetroons, with attached 95 cm passive radar reflectors, were launched in clusters of three or four on each of 224 scheduled days during this 13-month period. The tetroons were ballasted to float at an altitude near 3700 m, roughly 1500 to 2500 m above the terrain in the western U.S.A. The flight altitude and local trajectories were verified by local precision radar tracking. Long range tracking was accomplished by staffing selected Federal Aviation Administration Air Route Traffic Control Centers (ARTC) with observers to follow the tetroons on the ARTC radars. Several radars, strategically placed, were remoted to each Center. Some ARTC had as many as seven separate radars, providing broad tracking coverage from a single location.

During this period tetroons were released on each of four days per week. 552 tetroons were tracked for at least one-half hour. One or more tetroons of the cluster release were tracked for at least 12 h on 72 days (32%). One tetroon was tracked for more than two days to about 2400 km.

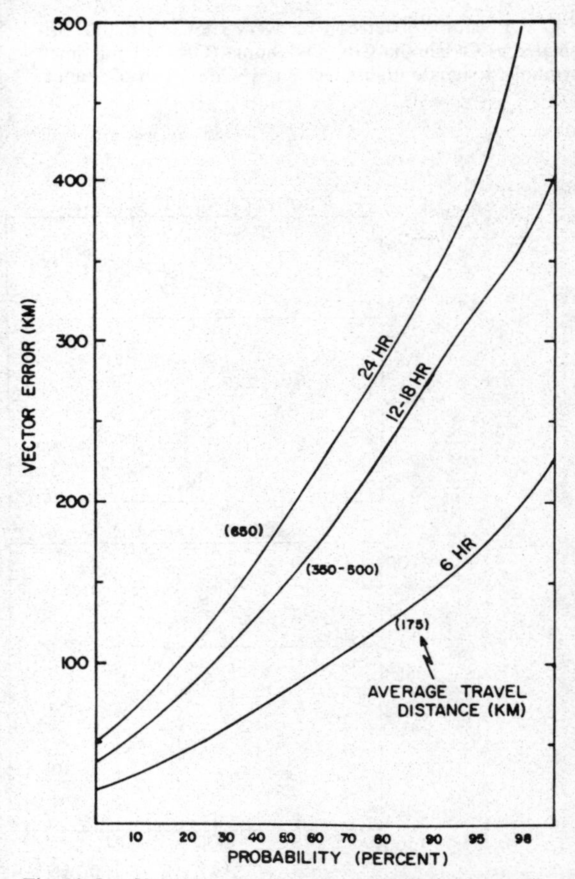

Fig. 10. Las Vegas tetroon-trajectory comparisons—probability of reconstructed trajectory error $\lesssim$ indicated distance after 6, 12 and 24 h of travel. (Derived from Allen *et al.*, 1969).

The tetroon trajectories were compared to forecast trajectories prepared by the National Meteorological Center, NOAA, which was at that time using a 3-level barotropic prediction model. Trajectory forecasts were also prepared by manual, subjective methods from streamline and isotach charts independent of the NMC predictions. Finally, the streamline charts were carefully re-analyzed and *post-facto* trajectories prepared. The comparison of the forecast trajectories with the tetroon paths was not good. There was a systematic bias in the sense that, in the mean, based on 444 trajectory comparisons the predictions were too short and were to the right of the actual paths. The standard vector deviations were large, between 55 and 60% of the distance traveled by the tetroons.

Of even more concern was the comparison of the tetroon transport with the *post-facto* trajectories. These differences, while smaller, were still large, about 21% of the trajectory length. More disturbing was the continued systematic bias. Even using the observed winds the reconstructed trajectories were, in the mean, about as far to the left of the tetroon path as the forecast trajectories were to the right. The trajectory error probability as a function of travel time (and average travel distance) is shown in Fig. 10. These data, although the most extensive known to the authors, provided only a few comparisons for the longer times. Comparisons for 24-h travel times were possible in only 20 cases. Furthermore, the terrain in this area is mountainous and the wind data not numerous. Even with these caveats, these unique data suggest considerable restraint in attributing high accuracy to trajectory reconstruction.

FUTURE TRANSPORT AND DIFFUSION STUDIES—SOME SUGGESTIONS

Our previous discussion has centered on the uncertainties revealed when calculated trajectories are compared to the available data tracing out actual air motions. However, it is evident that in many cases trajectory estimates perform well. They are at their best in relating air quality data at isolated locations to pollution transported from large area sources. Also, during steady conditions, with a well-defined wind field and slowly moving meteorological systems, more reliance can be placed on the calculations.

However, there is much more that needs to be known about the variability in atmospheric transport over regional distances. This issue will become increasingly acute as the pollution models become more detailed and emission inventories more precise and both are applied to smaller space-time scales.

It is not enough just to recognize the problem since there are means at hand to improve our knowledge. First, experiments can be mounted such as those at Oklahoma City and Las Vegas. Figure 11 shows the return tag recovery pattern from this latter experiment. Even though the initial paths were over some of the least populated areas of the U.S.A., tags from 117 of the 552 flights were returned (21%). Even if radar tracking is not feasible much atmospheric transport data can be obtained in this fashion.

Before leaving the possibilities for using balloons in tracing air motions, we should mention an early experiment which does not seem to be widely known (Sakagami, 1961). He released 48,900 rubber balloons to evaluate transport and diffusion. Although he was

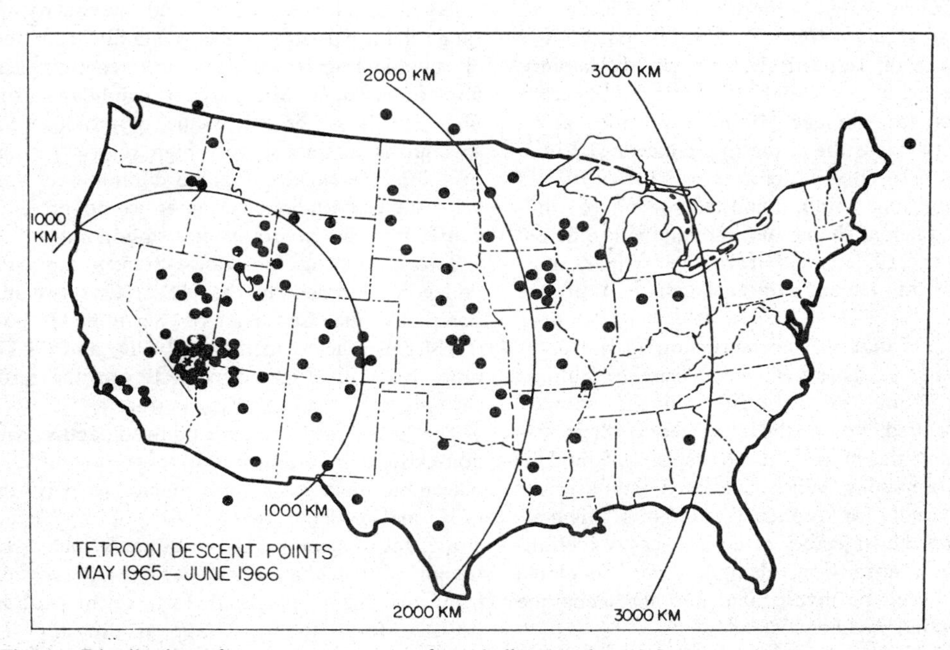

Fig. 11. Distribution of tetroon return tags from balloons released from Las Vegas, Nevada (star). 117 (21%) recovered.

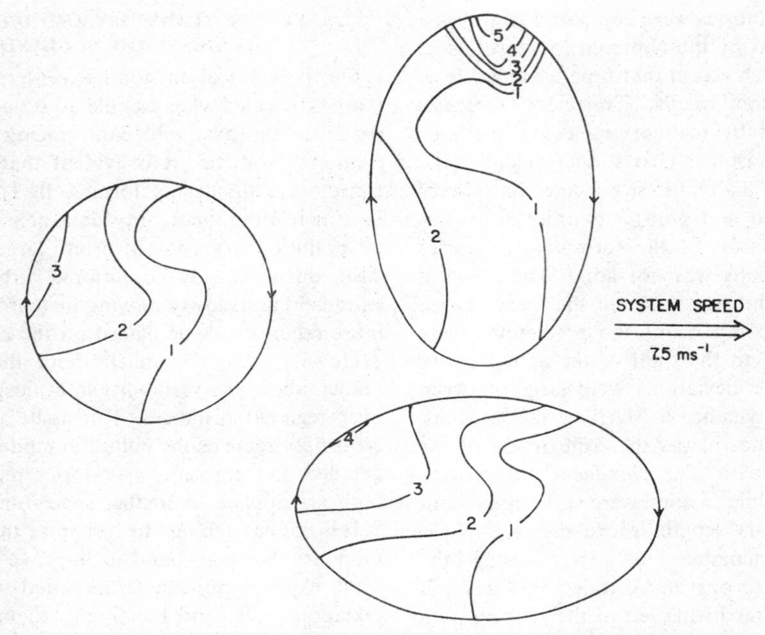

Fig. 12. 'Residence' time within hypothetical anticyclones. Numbered lines represent length of time, in days, air remains in the system. (After Vukovich *et al.*, 1977).

unable to fly the balloons at a particular level as can now be done with tetroons, encasing inexpensive rubber balloons within equally inexpensive plastic 'envelopes' would enable massive cluster releases with a degree of altitude control. Use of return tags and large numbers of balloons should guarantee a considerable data return at minimum expense.

A second general approach, especially for anticyclone-induced stagnation episodes, can be inferred from a series of recent reports (e.g. Ferber *et al.*, 1976; Husar *et al.*, 1976a; b). These reports show the coherence of, respectively, dew point, visibility, and ozone within well-defined regions, generally when under the influence of high pressure areas. Coupled with this is the interesting concept of Vukovich *et al.* (1977). These latter authors have calculated the 'residence' time within (admittedly idealized) anticyclones of different shapes and moving with different speeds. Figure 12, adapted from their paper, illustrates one set of systems, and the calculated residence times.

It is obvious that the reconstruction of the paths of anticyclones is easier than reconstructing multiple trajectories from the various pollutant sources. Further, the prediction of the movement of these systems is one of the more accurate aspects of modern numerical forecasting. When large-area estimates are needed, especially for secondary pollutants, this concept of using the trajectory of a large, easily identifiable entity as a moving, changing, but calculable 'reservoir' should be investigated and the technique tested on actual weather systems.

Finally, the best approach would be the quantitative study of *known* amounts of material placed into the atmosphere specifically to study transport and diffusion. Tracer substances deliberately released to the air to simulate long-range dispersion of pollutant plumes should meet the following criteria: (1) non-toxic, (2) conservative, i.e. non-reactive and non-depositing, (3) low cost of tracer, (4) low cost of collection and analysis procedures, (5) detectable at very low concentrations, and (6) background in the atmosphere should be below detection limits. No tracer in use today meets all these criteria and the paucity of long-range field experiments stems mainly from the lack of suitable tracers. Particles and reactive gases are non-conservative. Many other candidates, such as SF_6, already are present in the air in concentrations so high as to impose very high release amounts to overcome the background and dilution over long travel distances. Radioactive gases would present difficulties from an environmental standpoint.

Two promising long-range tracers are currently under development in the U.S.A. One system, proposed by the Los Alamos Scientific Laboratory (LASL), uses heavy methanes, $^{13}CD_4$ and $^{12}CD_4$; the other, being investigated by ARL, uses the perfluorocarbons, C_6F_{12} and C_8F_{16}, as tracers.

The heavy methanes are non-radioactive, virtually non-existent in the atmosphere at present, and are detectable on a mass spectrometer to parts in 10^{11} of normal methane (parts in 10^{17} of air). To demonstrate the feasibility of long-range meteorological tracing with methane-21 ($^{13}CD_4$), 84 g were released from the Idaho National Engineering Laboratory (INEL), formerly the National Reactor Testing Station, in southeastern Idaho over a three-hour period on 14 May 1974. In this first experiment

(Cowan *et al.*, 1976) the cryogenic samplers set up for the Kr-85 experiment (see Fig. 5) were used to concentrate the methane in the ambient air.

Samples were collected twice-daily at each station, from 9 p.m. to 7 a.m. (P) and 9 a.m. to 7 p.m. (A). On the basis of ARL model calculations, about 25 samples from four locations (Minneapolis, Minn.; Rochester, Minn.; Waterloo, Iowa and Detroit, Mich.) were chosen as likely to contain fractions of the released tracer. Eight additional samples were chosen for background measurements, with the sampling done either before the release or at a location (Tulsa, Oklahoma) far from the predicted trajectories.

Methane was separated from the air samples and purified by chromatographic desorption, and the isotopic composition of the methane was measured by mass spectrometry. The ratios of the intensities at the mass-21 region characteristic of methane-21 to those

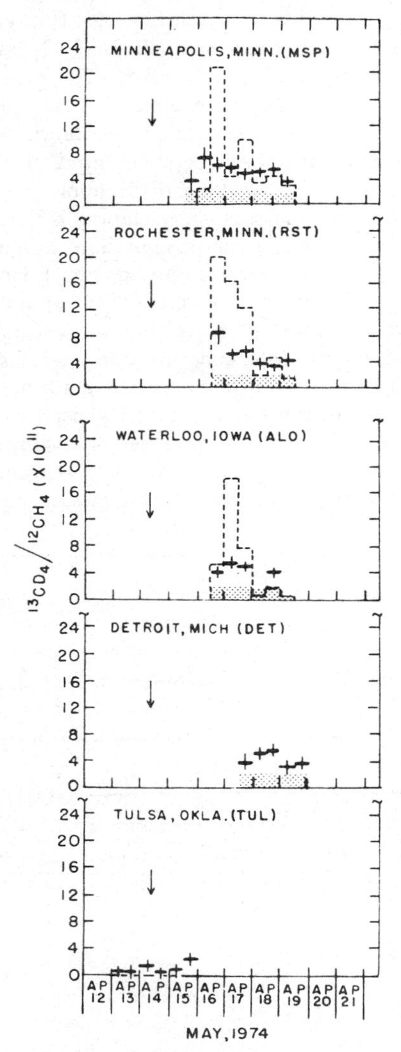

Fig. 13. Calculated and observed ratios of methane-21 to methane-16. Crosses are observed values; dashed bar graphs are calculated values (ARL model). Calculated values for DET were $<1 \times 10^{-11}$. Stippled area indicates background levels in samples unaffected by tracers. Time of tracer release in Idaho indicated by arrows.

at mass-16 (normal methane) for the different samples are presented in Fig. 13. These data show that the eight samples representative of background levels had methane-21/methane-16 ratios of less than 3×10^{-11} (average $= 0.7 \pm 1.2 \times 10^{-11}$). All the other samples, selected as likely to contain the tracer, show higher values of this ratio, with more than half having ratios in excess of 4×10^{-11}. The average value observed in these samples, 4.5×10^{-11}, corresponds to the detection of about 1 part in 2×10^{16} parts (by volume) of methane-21 in the atmosphere. Figure 13 shows clearly that methane-21 has been detected at distances of 1500 to 2500 km from the point of release. The experiment demonstrates that isotopically labelled methane can be used on a continental scale to study atmospheric transport and diffusion.

Certain perfluorocarbons may be ideal as long-range atmospheric tracers (Lovelock, 1975). These electron-absorbing gases may be measurable, with preconcentration of the sample, at concentrations of parts in 10^{15} of air by electron-capture gas chromatography. At present, ARL is investigating two of these compounds, perfluorodimethylcyclohexane (C_8F_{16}) and perfluorodimethylcyclobutane (C_6F_{12}). They appear to have two chief advantages over other electron-absorbing tracers in current use, such as SF_6. The existing background concentrations are very low, less than 2 parts in 10^{14} for C_8F_{16} and much less than that for C_6F_{12}. Much smaller releases would be required than of SF_6 where a background of about 5 parts in 10^{13} of air must be overcome. The second advantage of the perfluorocarbons lies in their great stability at high temperatures which permits the destruction of other interfering compounds before the sample is introduced to the detector. Their principal advantage over the heavy methanes lies in the potential for inexpensive automated sampling and analysis.

Lovelock has developed three prototype sampling systems for ARL. Two of these prototypes are automated sequential samplers and the third is designed to provide continuous real-time measurement of air concentration.

The first field test of the perfluorocarbon tracer system took place in April 1977. The two perfluorocarbons were released simultaneously with the two heavy methane tracers and SF_6. Preliminary analysis indicates that the sequential samplers performed well but the continuous instrument requires some modification. Results will be reported when analysis is completed.

Plans are being developed for full scale experiments in which one or more of these new tracers would be released during each day and night over a period of several weeks. Sequential samples would be taken continuously at many locations to distances of several hundred kilometers from the release point. If resources permit, 25–50 samplers will be constructed during the coming year and the tracer releases would be integrated with the Energy Research and Development Administration's SO_2/SO_4^{2-} studies in the northeastern U.S.A. Concurrent studies of power plant plumes and of the conservative tracers should permit separation of transport and diffusion effects from chemical transformation and deposition.

GLOBAL SCALE TRANSPORT

Although global scale movement of man-made pollutants is not the central theme of the Symposium, experience shows that early recognition and evaluation of the very large-scale distribution of material is prudent since it provides a background of knowledge available should this become a 'problem'. Witness the current interest in carbon dioxide growth and the studies being made of the behavior of the halocarbons in the atmosphere. Here we present one model calculation of global sulfur distribution for information and to provoke further thought.

Global transport models exist in various stages of sophistication and complexity. The simplest is a 'box' model in which either the entire atmosphere or one hemisphere is characterized by a single average concentration. The second degree of complexity uses a number of boxes in the north–south and vertical directions; a two-dimensional (2-D) model. In this case, the concentration is uniform around a circle of latitude, but north–south and vertical gradients of concentration can occur. Finally, the most complete models allow for variation of concentrations and transport in all three dimensions on a spherical earth. We will illustrate the sulfur distribution and deposition using a 2-D model. At this time, the sulfur removal characteristics and validating observations, on a global scale, are too incomplete to justify the high cost of running the more realistic three-dimensional model.

In 2-D models sulfur can be distributed in the north–south and vertical directions by two physical processes; first, by advection using a climatologically estimated circulation which can be found in the meteorological literature and; second, by turbulent diffusion employing a set of eddy diffusion coefficients derived by trying to match them to real tracers. Removal from the bottom box simulates dry deposition (a deposition velocity of 0.3 cm s^{-1} is used here), while precipitation scavenging acts through the lower atmosphere over a depth (estimated) where rain-bearing clouds are present. A fraction of the pollutant is removed by precipitation during each model time-step.

The height of each box in these calculations is two km except for the bottom and top boxes which are one km deep. The north–south box dimension is 10° of latitude. A quantity of sulfur (in this case at the rate of 16 g s^{-1}) is inserted each time-step ($\frac{1}{2}$-day here) in the ground level box between 40 and 60°N latitude and then transported and removed. The organized advective transport has been omitted in the present calculation. After about 10 years, an equilibrium state is approximately achieved between input, transport and deposition. The resulting distribution is shown in Fig. 14. There is a seasonal variation in the diffusion coefficients and in the precipitation scavenging; Figure 14 reflects the model conditions on 1 January.

The highest concentration is, of course, at and near the source latitudes, 40–60°N, but away from the 30–70° latitude band the concentration increases with altitude. This reversal of the expected gradient is especially noticeable in the Southern Hemisphere. This increase of the sulfur concentration or mixing ratio is due to the low-level removal by both wet and dry deposition. The structure of the isolines of concen-

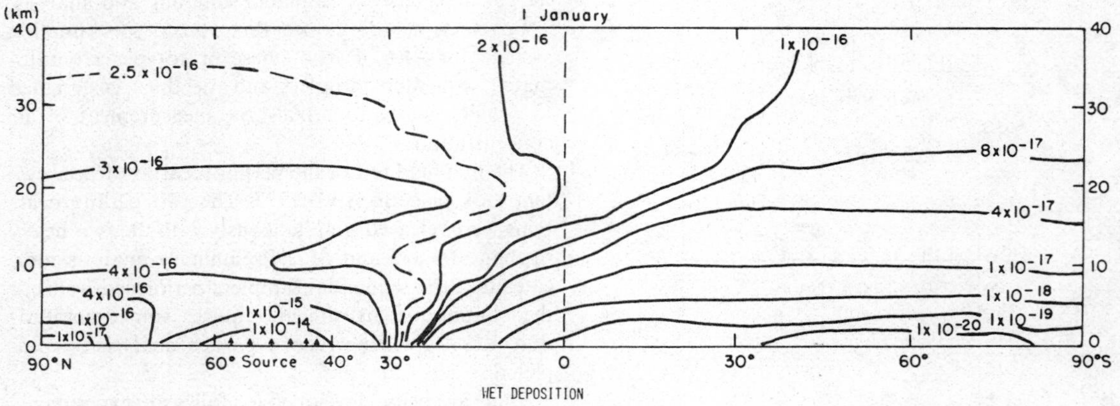

WET DEPOSITION																	
1.1E-22	6.5E-22	5.5E-21	4.7E-20	4.2E-20	3.0E-21	2.5E-22	1.5E-22	5.2E-23	1.5E-23	4.0E-24	9.9E-25	9.0E-25	1.1E-24	1.5E-24	1.4E-24	8.3E-25	6.0E-25
DRY DEPOSITION																	
1.1E-23	3.0E-22	5.6E-21	5.2E-20	4.1E-20	3.1E-21	1.3E-22	8.9E-24	5.4E-25	5.6E-26	3.0E-26	4.8E-26	3.5E-26	2.4E-26	2.0E-26	2.0E-26	2.7E-26	6.2E-26
WET PLUS DRY DEPOSITION																	
1.2E-22	9.5E-22	1.1E-20	9.9E-20	8.3E-20	6.1E-21	3.8E-22	1.6E-22	5.3E-23	1.6E-23	4.0E-24	1.0E-24	9.3E-25	1.1E-24	1.5E-24	1.4E-24	8.6E-25	6.6E-25

UNITS OF CONCENTRATION GRAMS OF SULFUR/GRAMS OF AIR, UNITS OF DEPOSITION GRAMS OF SULFUR M^{-2} S^{-1}, E = POWER OF 10, E.G., E-20 = 10^{-20}, SOURCE: 16 GRAMS OF SULFUR S^{-1} IN 40-60°N AT GROUND LEVEL.

Fig. 14. Global sulfur distribution from a two-dimensional model incorporating wet and dry deposition.

tration at about the 10 km altitude in the vicinity of 30°N is due to the model distribution of precipitation scavenging which varies with latitude and altitude.

Deposition is greatest in the source latitudes and falls off sharply in the Southern Hemisphere. Also of interest is the rough equality between wet and dry deposition near the source latitudes but the much greater wet deposition elsewhere.

Global transport of sulfur can, thus, be calculated. The main source of uncertainty lies in the model simulation of removal; for example, the use of a single deposition velocity for all times and places. Nevertheless such calculations assist in assessing the widespread distribution and deposition of anthropogenic sulfur and can guide measurement programs to better refine our understanding.

SUMMARY

We have indicated the need for increasing accuracy in our knowledge of, and ability to calculate, atmospheric transport. We have examined the character and accuracy of the present methods of estimating transport using measurements from point and area sources, and from tetroons, to describe the actual transport. We find that, by and large, the use of observed wind data, particularly surface wind data that has been 'adjusted' for shear effects, or use of winds averaged through the transport layer, usually provide the most favorable comparisons with measured transport within the planetary boundary layer.

Use of the geostrophic assumption to calculate transport vectors, particularly from surface data, is very attractive, since it is readily available at hourly intervals and can now be processed in quite precise ways at minimum expense. However, we have shown that use of such data to obtain accurate atmospheric transport estimates will entail more study to relate these calculations to meteorological behavior, especially the role of cold and warm air advection.

We suggest that the required additional skill in computing trajectories can be developed by carefully designed experiments using balloon trajectories or gaseous tracers. At present, a proven technology is available for the use of balloons to study long-range transport. For the near future, development and use of long-range atmospheric tracers, such as heavy methanes and perfluorocarbons, appears to be a very promising approach to the improvement and verification of transport and diffusion models.

Although the global distribution of sulfur in the atmosphere has not been raised as a particular problem, its behavior has been modelled as a first step in the compilation of a total sulfur budget.

Acknowledgements—The ARL staff wish to acknowledge the support of the Energy Research and Development Administration, Division of Biomedical and Environmental Research. The illustrations were drafted by Mrs. Marguerite Hodges, ARL and Ms. Clare Villanti, University of Maryland.

REFERENCES

Allen P. W., Jessup E. A. and White R. E. (1967) Long-range trajectories. Proc. of U.S.A.E.C. Symposium Meteorological Information Meeting, Spet. 11–14, 1967, Chalk River, Ontario, Canada, AEC L-2787, 176–190.

Allen P. W. and White R. E. (1969) A method of obtaining long-range air trajectories at low altitudes using constant volume balloons. Presented at the WMO Technical Conf. on Upper-Air Instruments, Neuilly, France, 8–12 Sept., 1969.

Angell J. K., Pack D. H., Dickson C. R. and Hoecker W. H. (1971) Urban influence on night-time airflow estimated from tetroon flights. *J. appl. Met.* **10**, 194–204.

Angell J. K., Pack D. H., Machta L., Dickson C. R. and Hoecker W. H. (1972) Three-dimensional air trajectories determined from tetroon flights in the planetary boundary layer of the Los Angeles Basin. *J. appl. Met.* **11**, 451–471.

Angell J. K., Hoecker W. H., Dickson C. R. and Pack D. H. (1973) Urban influence on a strong daytime air flow as determined from tetroon flights. *J. appl. Met.* **12**, 924–936.

Cowan G. A., Ott D. G., Turkevich A., Machta L., Ferber G. J. and Daly M. R. (1976) Heavy methanes as atmospheric tracers. *Science* **191**, 1048–1050.

Cox R. A., Eggleton A. E. J., Derwent R. G., Lovelock J. E. and Pack D. H. (1975) Long-range transport of photochemical ozone in northwestern Europe. *Nature* **255**, 118–121.

Doty S. R., Wallace B. L. and Holzworth G. C. (1976) A climatological analysis of pasquill stability categories based on "STAR" summaries. NOAA, Environmental Data Service, National Climatic Center, Federal Bldg., Asheville, NC 28801, 51 p.

Delver N. F. and Booth H. G. (1965) The deployment of super-pressured balloons. U.S.W.B. Research Station, Las Vegas, Nev. (NTIS Acc. No. COM-75-1045/AS), 59 p.

Draxler R. R. A mesoscale transport and diffusion model. NOAA Tech. Memo (in press), 31 p.

Druyan L. M. (1968) A Comparison of low-level trajectories in an urban atmosphere. *J. appl. Met.* **7**, 583–590.

Durst C. S., Crossley A. F. and Davis N. E. (1959) Horizontal diffusion in the atmosphere as determined by geostrophic trajectories. *J. fluid Mech.* **6**, 401–422.

Eliassen A. (1976) The Trajectory Model: A Technical Description. Ms. Norwegian Inst. for Air Research, Kjeller, Norway, 14 p.

Engelmann R. J. (1970) Scavenging Prediction Using Ratios of Concentration in Air and Precipitation. Proc. Symposium on Precipitation Scavenging, A.E.C. Symposium Series 22, 475–485.

Ferber G. J., Telegada K., Heffter J. L. and Smith M. E. (1976) Air concentrations of krypton-85 in the mid-west United States during January–May, 1974. *Atmospheric Environment* **10**, 379–385.

Garland J. A. and Branson J. R. (1976) The mixing height and mass blalance of SO_2 in the atmosphere above Great Britain. *Atmospheric Environment* **10**, 353–362.

Hass W. A., Hoecker W. H., Pack D. H. and Angell J. K. (1967) Analysis of low-level constant volume balloon (tetroon) flights over New York City. *Q. J. R. Meteor. Soc.* **93**, 483–493.

Heffter J. L. (1965) The variation of horizontal diffusion parameters with time for travel periods of one hour or longer. *J. appl. Met.* **4**, 153–156.

Heffter J. L., Taylor A. D. and Ferbert G. J. (1975) A Regional-Continental Scale Transport, Diffusion, and Deposition Model. NOAA Tech. Memo. ERL ARL-50, 28 p.

Hidy G. M., Tong E. Y. and Mueller P. K. (1976) Design of the Sulfate Regional Experiment (SURE). Electric Power Research Institute, EC-125.

Hoecker W. H. (1975) A universal procedure for deploying constant volume balloons and for deriving vertical air speeds from them. *J. appl. Met.* **14,** 1118–1124.

Hoecker W. H. (1977) Accuracy of various techniques for estimating boundary layer trajectories. *J. appl. Met.* (In press).

Holzworth G. C. (1972) Mixing Heights, Wind Speeds, and Potential for Urban Air Pollution Throughout the Contiguous United States. Environmental Protection Agency, Office of Air Programs, Pub. No. AP-101, Research Triangle Park, NC, 35 p.

Hoxit L. R. (1974) Planetary boundary layer winds in baroclinic conditions. *J. Atmos. Sci.* **31,** 1003–1020.

Husar R. B., Patterson D. E., Paley C. C. and Gillani N. V. (1976a) Ozone in Hazy Air Masses. Proc. Intnl. Conf. on Photochemical Oxidant and its Control, Sept. 12–20, 1976, Environmental Protection Agency, Research Triangle Park, NC.

Husar N. V., Gillani J. D., Paley C. C. and Turcu P. N. (1976b) Long Range Transport of Pollutants Observed Through Visibility Contour Maps, Weather Maps, and Trajectory Analyses. Third Symp. on Atmospheric Turbulence, Diffusion, and Air Quality. Oct. 19–22, 1976, Raleigh, NC.

Jessup E. A. (1967) Tetroon Trajectories Originating from Southern Nevada. ESSA Tech. Memo. RLTM-ARL 2, Dept. of Commerce, Air Resources Field Research Office, Las Vegas, Nevada.

Lovelock J. E. (1975) Personal communication.

Pack D. H. (1973) Unpublished data.

Pack D. H., Lovelock J. E., Cotton G. and Curthoys C. (1977) Halocarbon behavior from a long time series. *Atmospheric Environment* **11,** 329–344.

Ottar B. (1976) Monitoring long-range transport of air pollutants: the OECD study. *Ambio* **6,** 203–206.

Peterson K. R. (1966) Estimating low-level tetroon trajectories. *J. appl. Met.* **5,** 553–564.

Peterson K. R. (1968) Continuous point source plume behavior out to 160 miles. *J. appl. Met.* **7,** 217–226.

Petterssen S. (1940) *Weather Analysis and Forecasting*, pp. 221–227. McGraw-Hill, New York.

Prahm L. P., Torp U. and Stern R. M. (1975) Deposition and Transformation Rates of Sulfur Oxides During Atmospheric Transport over the Atlantic. Contribution for OECD Technical Meeting 28–30 May, 1975, Danish Meteorological Institute, Air Pollution Section, Lyngbyvej 100, DK-2100, Copenhagen, Denmark.

Prahm L. P. and Christensen O. (1976) Long-Range Transmission of Sulfur Pollutants Computed by the Psuedospectral Model. Ms. Danish Meteorological Inst., Air Pollution Section, Copenhagen, Denmark, 38 p.

Sakagami J. (1961) Diffusion Experiments Using Balloons. Natural Science Report, Ochanomizu Univ., 12, No. 2.

Smith F. B. and Jeffrey G. H. (1975) Airborne transport of sulfur dioxide from the U.K. *Atmospheric Environment* **9,** 643–659.

Summers P. W. and Hitchon B. (1973) Source and budget of sulfate in precipitation from Central Alberta, Canada. *J. Air Pollut. Control Ass.* **23,** 194–199.

Sykes R. I. and Hatton L. (1976) Computation of horizontal trajectory based on the surface geostrophic wind. *Atmospheric Environment* **10,** 925–934.

Turner D. B. (1964) A diffusion model for an urban area. *J. appl. Met.* **3,** 83–91.

Turner D. B. (1970) Workbook of Atmospheric Dispersion Estimates (Rev.). Environmental Protection Agency, Office of Air Programs, Research Triangle Park, NC

Veltishcheva N. S. (1976) Numerical Method for Quantitative Estimates of Long-Range Transport of Sulfur Dioxide. Ms. Hydrometeorological Center, U.S.S.R., 26 p.

Vukovich F. M., Bach Jr. W. D., Cressman B. W. and King W. J. (1977) On the relationships between high ozone in the rural boundary layer and high pressure systems. *Atmospheric Environment* **11,** 967–983.

Whelpdale D. (1976) Private communication.

Atmospheric Environment Vol. 12, pp. 445–454. Pergamon Press 1978. Printed in Great Britain.

AN ASSESSMENT OF THE OECD STUDY ON LONG RANGE TRANSPORT OF AIR POLLUTANTS (LRTAP)*

B. OTTAR

Norwegian Institute for Air Research, P.O. Box 130,
N-2001 Lillestrøm, Norway

(*First received* 17 *June* 1977 *and in final form* 2 *September* 1977)

Abstract—Based on a survey of the sulphur emission in Europe, and measurements at about 70 ground stations and by aircraft, atmospheric dispersion models have been used to evaluate the long range transfer of sulphur pollutants in Europe. The annual mean sulphur dioxide concentrations range from $\sim 20\,\mu g\ m^{-3}$ in rural areas close to the major source regions to $\sim 2\,\mu g\ m^{-3}$ in remote areas of the northern and western Europe. The annual mean aerosol sulphate concentrations are lower, $\sim 10\,\mu g\ m^{-3}$, and fall off more gently to $\sim 0.5\,\mu g\,m^{-3}$ in remote areas. The pattern of wet deposition shows enhanced values in areas exposed to polluted air masses and locally increased precipitation. Qualitatively the main features of the concentration fields are reproduced by the model calculations, but full quantitative agreement cannot be expected because of the many approximations made. However, on an annual basis, a correlation coefficient of 0.9 was obtained between observed and calculated values.

Estimates of the annual transfer of pollutants from one country to other countries are given for 1974. These amounts are, however, dependent on variations in the annual weather pattern. Future plans for monitoring and evaluation of the long range transport of air pollutants in Europe are outlined. Experience indicates that considerable improvements could probably be obtained by introducing more detailed formulations of the chemical reactions and deposition processes in the atmospheric dispersion models.

BACKGROUND AND GENERAL DEVELOPMENT

The acidification of the precipitation in Europe was first noticed in samples collected from the European Air Chemistry Network, established in the middle of the 1950s. A survey by Odén (1968) showed that a central area with highly acid precipitation (pH 3–4) was expanding from year to year, and related this to the acidification of rivers and lakes observed in Scandinavia. The main acid component was sulphuric acid, and the obvious source seemed to be the increasing use of sulphur-containing fossil fuels in Europe. The acidity was accompanied by soot, fly-ash and tar-like substances which on occasions could give the snow a greyish tint, and the health authorities warned people against the consumption of water from melted snow and cisterns.

These observations caused much alarm, and in 1969 the OECD, after having examined the available evidence on a request from the Nordic countries, recommended a full investigation. In May 1970 a project plan was presented to the OECD by NORDFORSK (The Scandinavian Council for Applied Research). After further elaboration and preparatory studies, a co-operative programme to investigate the Long Range Transport of Air Pollutants (LRTAP) with special emphasis on the acidification of the precipitation, was approved by OECD in 1972.

The LRTAP Programme was joined by 10 countries (Austria, Denmark, the Federal Republic of Germany, Finland, France, the Netherlands, Norway, Sweden, Switzerland, the United Kingdom). Canada participated as an observer and special agreements on exchange of data were made with Iceland and Italy. In 1974 Belgium also joined the study. The Programme was supervised by a Steering Committee with representatives of the participating countries, and a Central Co-ordinating Unit (CCU) was established at the Norwegian Institute for Air Research, which had co-ordinated the preparatory work. Preparatory studies had shown that the acidification of the precipitation in Scandinavia was a large scale phenomenon related to southerly winds and enhanced by orographic precipitation. This was confirmed during the first measurement phase, and similar situations were reported from other parts of Europe. In September 1973, the Steering Committee agreed that the long range transport of air pollutants caused a significant contribution to the local pollution level in remote areas and was of considerable importance for the acidification of the precipitation (LRTAP, 1973).

In the second measurement phase, the Programme was expanded to include co-ordinated aircraft observations and more extensive measurements at the ground stations, in order to provide data for a quantitative assessment of the transport of sulphur pollutants and some information on the importance of other chemical components for the acidification of the

* This paper is the author's personal assessment of the results of the OECD/LRTAP study. Although it draws heavily on the conclusions and summary of the final report of that study and much of it is reproduced verbatim in the paper, the paper is not intended to represent the opinion of the Study Steering Committee, for which the reader is referred to the OECD/LRTAP final report.

precipitation. The measurement programme was completed by March 1975.

Concurrent analysis of the data during the measurement period had shown that depending on the weather conditions large amounts of pollutants could be transferred to remote parts of the region. The summing up of the observations showed that, although the countries with the largest emissions also received the largest depositions, parts of the region with low emissions received more pollution from other countries than they received from their own sources. The general picture developed is approximate, as the examination of confidence limits, methods of evaluation, etc. in the following sections will show. However, after a thorough examination of the results, the final report was agreed by the Steering Committee and approved by the OECD Council in July 1977 (OECD, in press). In the following, the main results of the Programme are summarized.

TECHNICAL APPROACH

The co-operative programme was based on national measurement programmes in the participating countries. The CCU was responsible for intercalibration and data quality control, evaluation of the data, and development of the necessary atmospheric dispersion models. The Programme required careful preparations, and assistance from other research institutes was necessary. A CCU modelling group was established in co-operation with the Norwegian Meteorological Institute, which placed their experience and computer facilities at disposal. Methods for sampling and chemical analysis, meteorological evaluations and aircraft measurements were prepared in a co-operative NORDFORSK Programme supported by the major Nordic research institutes concerned (Ottar, 1975). Also laboratories in other European countries participated in the preparatory work on a voluntary basis. The general plan for the Programme is illustrated in Fig. 1 (Ottar, 1976).

On the basis of emission data for sulphur dioxide and air trajectories, one-layer atmospheric dispersion models were used to calculate concentration and deposition fields in a grid system of side length 127 km, covering the northern and western parts of Europe. Deposition velocities and parameters describing the chemical transformation of sulphur dioxide to sulphate were adjusted by calibrating the models. The results were compared with chemical analysis of air and precipitation samples from the network of ground stations and aircraft.

Emission survey

The emission survey was mainly based on information from the participating countries and is discussed in detail by Semb (1978). A main basis was the fuel consumption data regularly collected by OECD. For countries not participating in the Programme other available information and population density were used to estimate the emissions in the grid system. Figure 2 shows that the major emission areas are situated in a belt from England across the European continent, the same areas where the coal deposits originally formed the basis for the industrialization. The annual emission of sulphur dioxide amounts to about 50 million tonnes for Europe as a whole, and the emissions are 60–70% higher in winter than in summer. The total figures for the participating countries are accurate to within 10–25%, while the emissions of the individual grid elements are naturally less certain.

Ground level measurements

The network of ground stations comprised about 70 stations for daily sampling of air and precipitation (Fig. 3). The sites were selected by the countries in co-operation with the CCU to be representative of larger areas within the countries and relatively uninfluenced by local sources. To check the last point, a special inventory of the sulphur dioxide emissions within a 100 km radius of each station was made. This survey showed that the emissions per unit area in the surroundings of the stations are generally smaller than the rates of total sulphur deposition, except for some stations in the United Kingdom and on the Continent. For these stations, however, emissions in the surroundings are less than the typical emissions for the adjacent emission grid elements. Resources were not available for any detailed examination of the representativity of the stations with respect to local topography and prevailing weather conditions. It is, however, felt that much uncertainty in the evaluation of the results could be removed if a

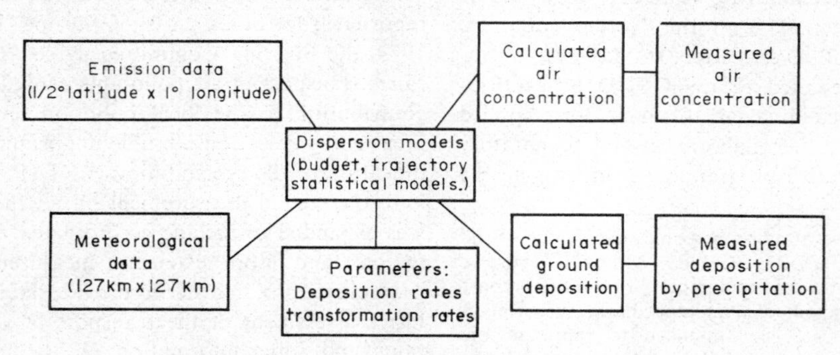

Fig. 1. Block diagram representing operations in the LRTAP Programme.

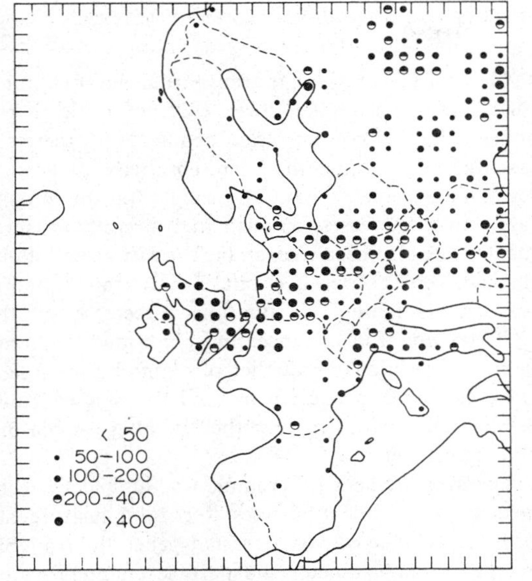

Fig. 2. Estimated emission of sulphur dioxide in 1973 (10^3 tonnes SO_2 y^{-1}).

detailed assessment of these factors was made in similar programmes (Granat, 1975).

The methods for sampling and chemical analysis were selected in co-operation with the participating laboratories. A pilot station programme had been established by OECD in 1964. Under this programme, pilot stations for air measurements in background areas had been established by the Deutsche Forschungsgemeinschaft (DFG) in the Federal Republic of Germany, the Institut de Recherche Chimique Appliquée (IRCHA) in France, the Institute for Water and Air Pollution Research (IVL) in Sweden, and later by the Swiss Federal Laboratory for Testing of Materials (EMPA) in Switzerland. Thus, a group of experienced laboratories was available for discussion and testing of the sampling and analysis methods to be used at the ground stations.

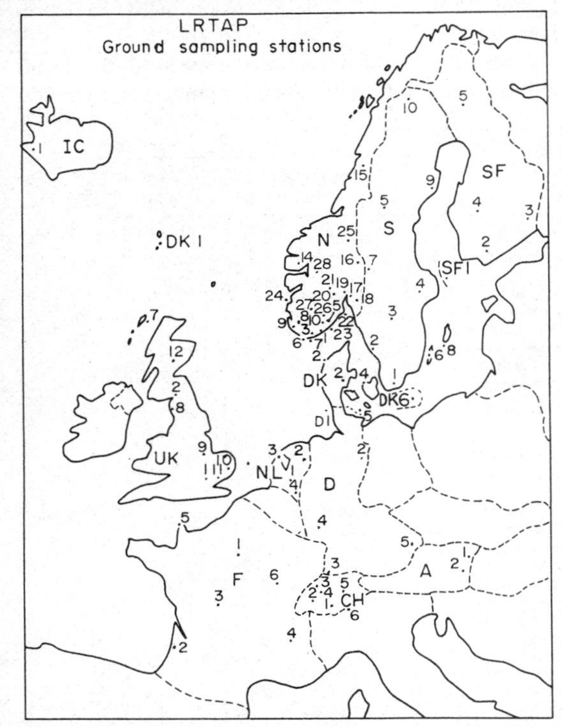

Fig. 3. LRTAP ground sampling network (OECD, in press).

Sulphur dioxide was determined by absorption in acid hydrogen peroxide, and chemical analysis by the barium perchlorate–thorin method (Persson, 1966). The sulphate aerosol was collected on Whatman filters and analysed by X-ray fluorescence (Brosset *et al.*, 1971) or the thorin method. Precipitation acidity was determined by a titration procedure (Liberti *et al.*, 1972) or by pH measurements. Sulphate in precipitation was determined by the thorin method after removal of interfering cations by ion-exchange. The precision of the analysis method is typically ± 0.1–$0.4\ mg\,l^{-1}$ for SO_4, which should be compared with typical values of sulphate in precipitation which range from 0.5 to $10\ mg\,l^{-1}$. For strong acid, the typical precision is 5–$10\ \mu eq\,l^{-1}$.

The sensitivity of the method for sulphur dioxide is limited by the volume of the air sample and the sensitivity of the thorin method for sulphate resulting in a detection limit for SO_2 of 3–$5\ \mu g\ SO_2\ m^{-3}$ for an air sample of 3–$4\ m^3$. This is a serious limitation when a large fraction of the daily mean concentrations at some stations are below $5\ \mu g\ SO_2\ m^{-3}$. However, the yearly mean concentrations are determined by relatively few high values and are therefore more reliable. The X-ray fluorescence determination of sulphate collected on aerosol filters gives excellent reproducibility combined with a very low detection limit ($\sim 0.1\ \mu g\ SO_4\ m^{-3}$).

For technical and economic reasons, the first measurement phase of the Programme was limited to these measurements. However, a NORDFORSK Programme of 100-days with extended measurements showed that other ions, particularly the ammonium- and nitrate-ions have considerable influence on the acidity of particles and precipitation. Therefore, an analysis of these ions was included in two 45-day periods of the second measurement phase.

Standardized procedures for sampling and chemical analysis were used (LRTAP, 1971–1973); and laboratory inter-calibrations carried out to ensure comparability of the results and adequate data quality (Schaug *et al.*, 1975). At coastal stations, corrections for sea salt were based on magnesium and sodium analysis.

In some countries, different methods which had been established on a routine basis, were accepted when comparison over a period of time had shown that equivalent results were obtained. Experience has shown that careful inter-calibration between laboratories and meticulous quality control of the data are important. Also confidence limits for the data are necessary in order to be able to evaluate the usefulness of different modelling approaches. The determination of low background values is another impor-

Table 1. Institutions and laboratories responsible for the ground level measurements

Abteilung für Lufthygiene, Bundesstaatliche bakteriologisch-serologische Untersuchungsanstalt, Wien, Austria
Eidgenössische Materialprüfungs- und Versuchunganstalt für Industrie, Bauwesen und Gewerbe (EMPA Duebendorf, Switzerland
Umweltbundesamt, Berlin, Federal Republic of Germany
Danish Meteorological Institute, Copenhagen, Denmark
Institut National de Recherche Chimique Appliquée (IRCHA), Département Pollution Atmosphérique, Vertle-Petit, France
Vedurstofa Islands, Rannsoknastofnun rikisins, Reykjavik, Iceland
Norwegian Institute for Air Research (NILU), Lilletrom, Norway
National Institute for Public Health (RIV), Bilthoven, The Netherlands
Swedish Water and Air Pollution Research Laboratory (IVL), Gothenburg, Sweden
Finnish Meteorological Institute, Helsinki, Finland
Department of Trade and Industry, Warren Spring Laboratory, United Kingdom

tant point. It sets a limit for the accuracy of budget calculations. It is also necessary if the importance of natural sources and possible sources outside the region are to be evaluated. The limited sensitivity of the method used for sulphur dioxide (3–5 $\mu g\, m^{-3}$) caused problems in this connection.

To calculate the oxidation of sulphur dioxide to sulphate, one average rate constant was used for the whole area at all seasons. Introduction of a quadratic term in the chemical reaction rate did not improve the results significantly. In view of later experiences this is not surprising. The catalytic oxidation rate, among others, depends on the presence of manganese and ammonia (Brosset, 1976). The photochemical oxidation depends on the radiation conditions and the concentrations of nitrogen oxides (Cost Project 61a, 1976). The temporal and spatial variations of all these factors extend beyond 24 hours and 127 km, the resolution of the grid system used. This is a point where the Programme could possibly have been substantially improved, if the necessary measurements and emission data had been available.

Aircraft measurements

Most of the long range transport of air pollutants takes place within the lower 1–2 km of the troposphere. Vertical concentration profiles were therefore essential to examine the representativity of ground level measurements, and the height of the mixing layer. In a few cases, co-ordinated aircraft measurements were also used to examine the flux of pollutants through vertical cross-sections. The aircraft sampling was mainly limited to situations forecasted by the CCU 48 hours in advance, and confirmed 24 hours later. A survey of the number of flights by the different countries and equipment used is given in Table 2. Inter-calibration between the aircraft gave reasonable agreement.

The measurement programme was limited by technical and economic difficulties. Recording instruments with sufficient sensitivity were not generally available except for nephelometer and particle counters which were helpful as indicators of polluted air masses. The chemical analysis of sulphur dioxide and particles had to be made on samples collected over periods of 20–30 min. In this case, sulphur dioxide was collected on impregnated filters and analysed by the thorin or equivalent methods (Johnsen and Atkins, 1975).

To obtain more detailed and instant information about variations in the pollution level during sampling flights, for instance the occurrence of distinct plumes and vertical stratification of polluted air masses, continuous recording instruments are needed. A coulometric cell operated on a limited number of flights, formed a valuable extension of the measurement capacity.

Operational difficulties with respect to the availability of aircraft reduced the number of measure-

Table 2. Participating laboratories and sampling techniques (OECD, in press)

Country	Laboratory	Method	Type of aircraft	Number of flights
Denmark	Aerosol Research Laboratory, Atomic Energy Commission Research Establishment, Riso	Filter in series isotope dilution analysis	DC-3 (Dakota)	3
Federal Republic of Germany	Institut für Meteorologie und Geophysik der Johann Wolfgang Goethe-Universität, Frankfurt am Main	Wet absorption—tetrachloromercurate solution (SO_2)—West and Gaeke	Beechcraft Queen Air	8
France	Etablissement d'Etudes et de Recherches Meteorologiques, Meteorologie Nationale, Observatoire de Magny-les-Hameaux	Filters in series—barium perchlorate–thorin	Cessna 206	5
Norway	Norwegian Institute for Air Research, Lillestrom	Filters in series—barium perchlorate–thorin	Piper Aztec	37
Sweden	Department of Meteorology, University of Stockholm	Before 14 March 1974: wet absorption, peroxide solution (SO_2) glass fiber filter (SO_4) After 14 March 1974: filters in series—barium perchlorate–thorin	Beechcraft Travel Air	35
United Kingdom	Meteorological Office, Bracknell, Atomic Energy Research Establishment, Harwell	Filters in series—barium perchlorate–thorin	Varsity Hercules	14(+6 in 1971) 1
	Warren Spring Laboratory, Stevenage	SO_2: tetrachloromercurate—West and Gaeke SO_4: Filter—barium perchlorate–thorin	Hastings	8

ments. Most of the laboratories had to rely on rented aircraft which were primarily used for other purposes. When interesting situations were forecasted by the CCU, the aircraft were often not available. Towards the end of the Programme, flights initiated by the participating countries were therefore added in order to obtain more data.

Models

The models used in the Programme are described in detail by Eliassen (1978). A number of different models were tried, based on trajectories and Eulerian as well as Lagrangian advection schemes. A Lagrangian model was used for running daily calculations of the concentration fields for sulphur dioxide and sulphate all through the study. A slightly different model, the 'Trajectory model' was used to quantify contributions from a given country to another.

With these models, correlations from insignificant to 0.8 were obtained for daily values at single stations. Comparison of calculated values with 6 hourly measurements actually reduced the correlation, showing that the observed diurnal variations could not be reproduced with the resolution selected for this system. An attempt to use the observed ground level wind near the emission sources and the 850 mb wind further away, improved the correlation in places where there was a marked difference between these two winds (Nordø, 1976). The long term correlation between the observed data and calculated concentration fields (including all stations) was high ($\simeq 0.9$).

In the course of the LRTAP Programme and later, a number of different atmospheric dispersion models have been tested. The general experience is that there is relatively little to gain by refining the advection schemes unless a more extensive measurement programme is undertaken which makes it possible to describe the chemical reactions taking place during the transport and the different deposition processes in greater detail.

RESULTS

The measurements of the LRTAP ground sampling network and the model estimates show that the highest values of sulphur dioxide (Fig. 4) are found near to, but slightly displaced from, the main emission source areas (Fig. 2). The displacement is in the direction of the prevailing wind. The concentration field for particulate sulphate (Fig. 5) shows a similar distribution, but with relatively more gentle gradients and lower values, because of the time required for the chemical transformation of sulphur dioxide.

Measured annual mean sulphur dioxide concentrations range from about 20 $\mu g\ m^{-3}$ in rural areas in the Federal Republic of Germany, the Netherlands, and the United Kingdom, to 2 $\mu g\ m^{-3}$ or lower in the remote areas of northern and western Europe.

Annual mean aerosol sulphate concentrations show a similar distribution with values between 10 and

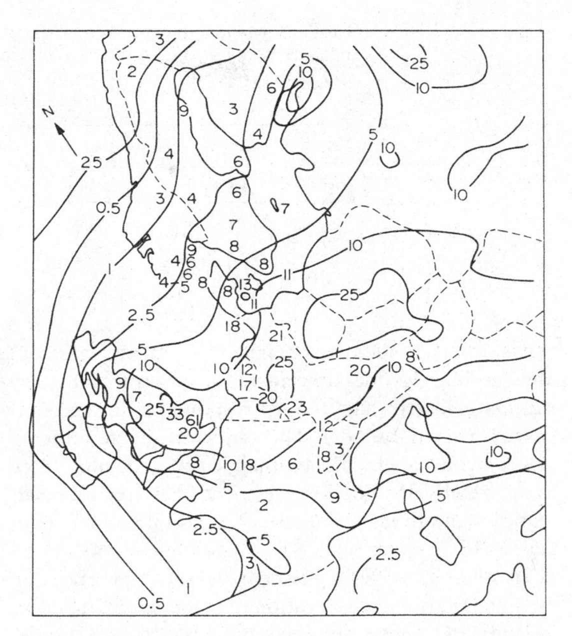

Fig. 4. Estimated mean concentration field for sulphur dioxide for 1974 and observed mean concentrations. Unit: $\mu g\ SO_2\ m^{-3}$ (OECD, in press).

0.5 $\mu g\ m^{-3}$. The accuracy of these annual means is about $\pm 15\%$, but the sulphur dioxide measurements are less accurate at sites with low concentrations ($< 5\ \mu g\ m^{-3}$). A comparison between values observed by aircraft, model estimates and measured values at ground level is given in Table 3.

The aircraft measurements have been separated into 4 height categories. The main part of the sulphur dioxide is seen to be contained in the layer below 1000 m. The distribution of sulphate aerosols is some-

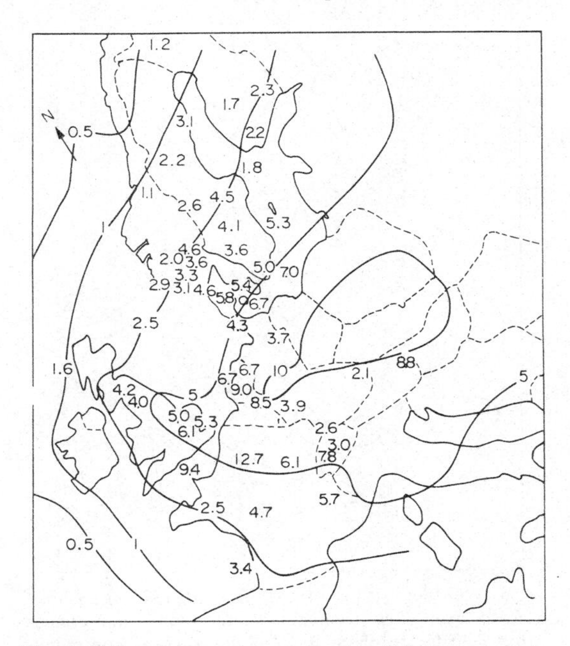

Fig. 5. Estimated mean concentration field for particulate sulphate for 1974 and observed mean concentrations. Unit: $\mu g\ SO_2\ m^{-3}$ (OECD, in press).

Table 3. Mean concentrations ($\mu g\ SO_2\ m^{-3}$ and $\mu g\ SO_4\ m^{-3}$) from aircraft measurements at different altitudes, compared with dispersion model estimates and 24-hour ground observations. Number of data points are given in parentheses (OECD, in press)

Height (m)	SO_2			SO_4		
	Observed	Estimated	Ground	Observed	Estimated	Ground
< 500	20 (245)	22 (245)	16 (228)	11 (243)	6 (243)	9 (230)
501–1000	13 (233)	18 (233)	—	9 (227)	6 (227)	—
1001–1500	7 (132)	—	—	8 (127)	—	—
> 1500	6 (77)	—	—	5 (75)	—	—

what less dependent on height. The mean values of the sulphur dioxide concentrations measured by aircraft are lower than the corresponding means estimated from the model. The observed mean sulphate concentrations are higher than estimated, and there is a relatively good agreement between aircraft samples and model estimates of the total sulphur content.

In theory, little agreement should be expected between surface observations, made over a 24-hr period, and those obtained by aircraft at a higher altitude and in a much shorter time period. However, Table 3 shows that the 24-hour surface concentrations of both sulphur dioxide and sulphate are generally similar to, although slightly lower than, the average of the aircraft measurements made below 500 m. The discrepancy could be explained by dry deposition at the surface. However, there is a wide scatter of the individual data.

The precipitation chemistry data from the LRTAP Programme has verified that sulphuric acid is the main acid component in precipitation. However, a relative contribution of 20 to 50% from nitric acid (on an equivalent basis) is generally found. The acidity is determined mainly by the balance between the sulphate, nitrate and ammonium ions. Annual mean concentrations of sulphate in precipitation range from about 5 mg SO_4 l.$^{-1}$ at sites close to the major source areas to less than 1 mg SO_4 l.$^{-1}$ in the far north of Scandinavia.

With respect to the wet deposition pattern (Fig. 6) it may be noted that the amounts deposited in remote orographic precipitation areas are comparable with the wet deposition in the central part of the region. Particularly exposed are the slopes of mountain ranges or higher terrain frequently faced with wind from highly polluted areas. Examples are the southern part of Scandinavia, the Scottish Highlands, and the northern and southern slopes of the Alps.

The mean pH in precipitation, and the corresponding deposition of acid are illustrated in Figs. 7 and 8. The pH values were selected rather than concen-

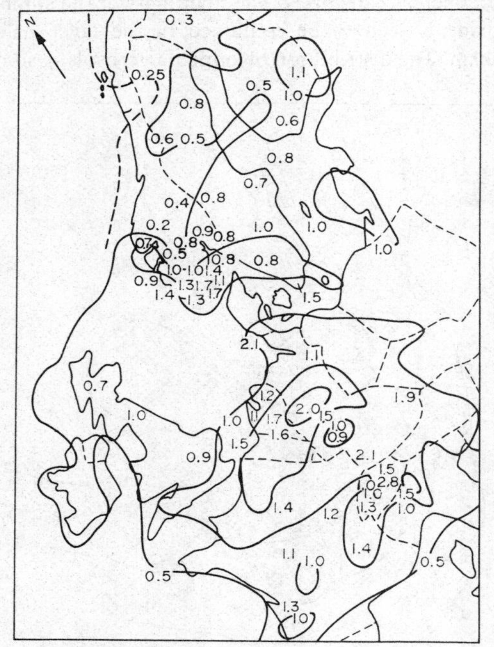

Fig. 6. Wet deposition of sulphate in 1974 (g S m^{-2}) obtained from measurements. An annual mean concentration field for sulphate in precipitation was constructed on basis of the measured values, and thereafter multiplied by the annual precipitation field (OECD, in press)

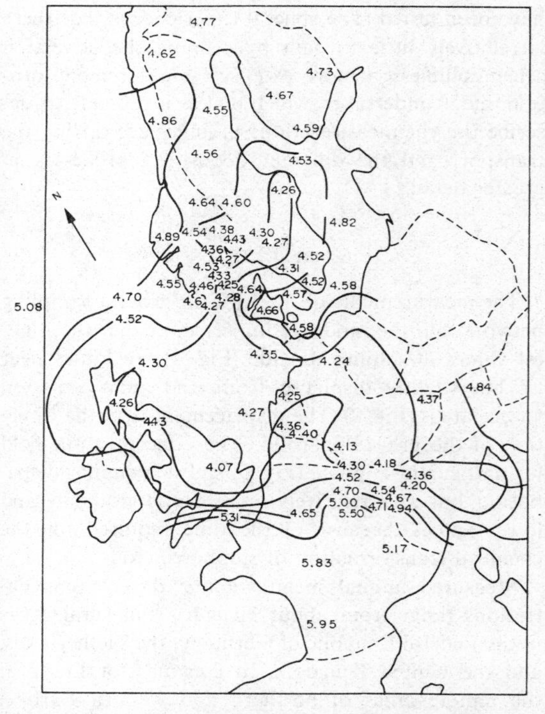

Fig. 7. Objective analysis of measured mean pH in precipitation for 1974 (OECD, in press)

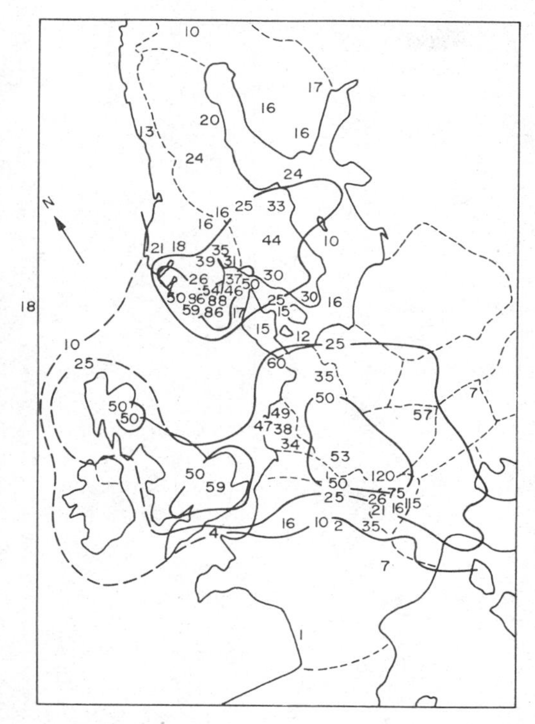

Fig. 8. Wet deposition of acid in 1974 (neq m^{-2}) obtained by multiplying the concentration field of Fig. 7 with the precipitation field. Values measured at the stations included (OECD, in press)

trations of strong acid, because of more complete data series and because there is generally good agreement between pH and the concentration of strong acid when the pH is less than 5. The pH isopleths were derived from a standard computer programme for objective analysis, and the deposition of acid by multiplying by the amount of precipitation.

The spatial distribution of acidity in precipitation is reasonably similar to the concentration field for excess sulphate in precipitation, but the acidity is less than what would correspond to the sulphate concentrations in some areas, notably in France, Denmark and the southernmost part of Sweden, and to some extent in Finland and the Netherlands. This confirms that the pH of precipitation is modified by other substances, mainly ammonia.

As the atmospheric dispersion models include several assumptions concerning methods and parameters used, the data were also examined by sector analysis. In Fig. 9, the arrows from each observation station represent the average concentrations of sulphate aerosol for air trajectories within the corresponding sectors. Evidently, the sectors with high concentrations are not randomly oriented but directed towards the areas of major sulphur emission. This is consistent with the model calculations for areas where the results can be compared.

Similar plots for the acidity in precipitation (Fig. 10) illustrate that the acidity to a large extent is governed by the availability of ammonia. As a result,

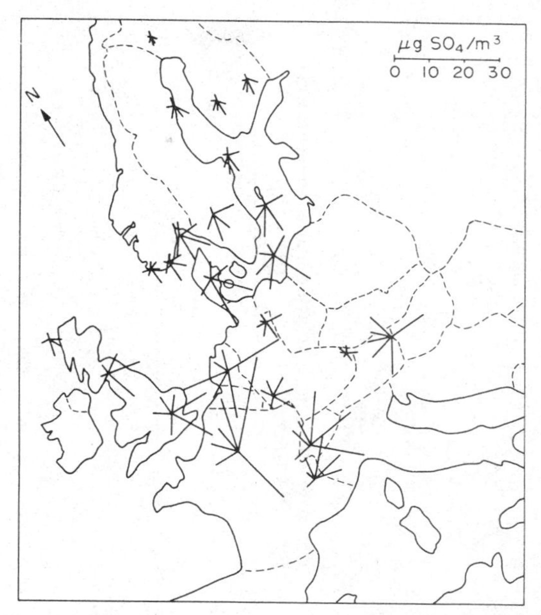

Fig. 9. Mean concentrations of aerosol sulphate for the transport sectors (OECD, in press)

air masses which have passed over the sea show the least degree of neutralization (Brosset, 1976).

A detailed analysis of trajectories and the concentrations observed in precipitation revealed that with trajectories from the Atlantic, a background of 0.8 mg SO_4 $l.^{-1}$ is found on the western coast of Norway. As the trajectory calculations go only 48 hours back in time, this background value may be due to pollutants picked up earlier, or to natural emissions of sulphur. An examination by Nyberg (1976) of precipitation measurements at the weather ships in the

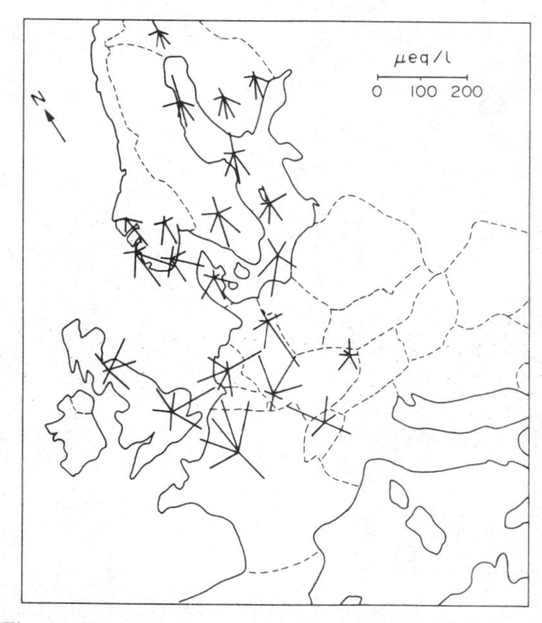

Fig. 10. Mean concentrations of strong acid in precipitation for 6 transport sectors (OECD, in press).

Table 4. Estimated budget for dry plus wet deposition of sulphur for 1974. Unit 10^3 tonnes S. Numbers and sums rounded to one significant figure and accurate to within $\pm 50\%$. The sums are calculated from unrounded figures and thereafter rounded separately (OECD, in press)

Emitters / Receivers	Austria	Belgium	Denmark	Federal Republic of Germany	Finland	France	The Netherlands	Norway	Sweden	Switzerland	United Kingdom and Ireland	Czechoslovakia	German Democratic Republic	Italy	Poland	Other areas	Undecided*	Sum	Annual emission (Table 2.3)
Austria	60	6	0	40	0	20	2	0	0	5	20	20	20	30	7	20	30	300	221
Belgium	0	100	0	20	0	30	5	0	0	0	30	1	4	0	0	1	10	200	499
Denmark	0	1	60	6	0	3	1	0	2	0	10	1	6	0	2	2	10	100	312
Federal Republic of Germany	8	60	7	700	0	100	40	0	2	7	100	20	80	7	10	10	90	1300	1964
Finland	0	2	8	10	100	4	2	2	30	0	10	7	30	0	20	80	70	400	274
France	2	40	1	50	0	600	10	0	0	6	100	5	20	30	2	30	150	1000	1616
The Netherlands	0	10	1	10	0	10	60	0	0	0	30	1	4	0	1	0	10	150	391
Norway	0	4	8	10	1	9	4	30	9	0	60	3	10	0	5	4	100	250	91
Sweden	0	7	30	30	10	10	6	6	100	0	40	8	50	0	20	30	100	500	415
Switzerland	1	2	0	7	0	20	1	0	0	30	10	2	1	6	1	2	20	100	76
United Kingdom and Ireland	0	8	2	10	0	20	4	0	0	0	800	2	9	0	2	1	100	1000	2883†
Czechoslovakia, German Democratic Republic, Italy, Poland and other areas	60	60	80	400	40	200	40	9	50	10	600	900	1300	900	1000	4500	1000	11000	—
Sum	100	300	200	1300	150	1000	200	40	200	60	1800	1000	1500	1000	1100	4600	1900	17000	

* Represents the part of the sulphate in precipitation which cannot be related directly to emission by the model.

† Including $80 \cdot 10^3$ tonnes S from Ireland.

North Atlantic indicate that some 2/3 of this background may possibly originate from the east coast of North America. Analysis of yearly snow layers from the inland ice of Greenland (Koide and Goldberg, 1971) show that the sulphate content is increased from 0.060 mg $l.^{-1}$ before the 19th century to a value of 0.2 mg $l.^{-1}$ today. This is also supported by analyses of aerosol samples from Spitsbergen and Northern Norway which indicate that with southerly winds, polluted air masses from the European and American continent are transported into this part of the Arctic.

The episodic character of the acid precipitation was much stressed at the beginning of the study, and a number of 'episodes' have been described (Nordø, 1974, 1976). In some remote areas, as much as 30% of the annual sulphate deposition could come down in one rain storm. As more data were collected, it became evident that these episodes in remote areas are mainly related to the wind directions. Stagnant weather conditions with low wind speeds and low mixing heights can lead to increased pollution levels over regions of continental scale. Subsequently, a considerable burden of pollutants may be transported to remote areas.

Studies of wet deposition of sulphate have shown that the precipitation scavenging is very effective and large amounts of pollutants can be deposited within limited periods. Observations from southern Norway show that deposition rates of the same order of magnitude as the emission rates in major source areas of Europe can occur without any preceding accumulation of pollutants over continental Europe.

The LRTAP Programme has shown that an extensive exchange of air pollution takes place between all the European countries, and that about 20% leaves the area. Table 4 gives an estimate of the amounts received by some countries from others in 1974. The basis for the calculations is discussed by Eliassen (1978). According to these figures, the deposition due to foreign emissions in many countries is comparable both to the deposition from indigenous sources or even to the sulphur emissions themselves. For five of the countries, the total deposition exceeds the indigenous emission. The estimated total depositions are more certain than the deposition contributions from individual countries, since the calculated concentrations of sulphur in air and precipitation have been compared with measured values. Data of this type are, however, highly dependent on year to year variations in the weather pattern. It should also be noted that in some countries most of the pollutants are deposited in certain particularly exposed regions.

FUTURE DEVELOPMENT

In order to obtain a common basis for future air pollution abatement, it has recently been agreed to establish a permanent monitoring and evaluation programme for the long range transport of air pollutants in Europe. This programme includes both eastern and western European countries. It is organised under the auspices of the United Nations' Economic Commission for Europe in cooperation with the World Meteorological Organization and the United Nations' Environment Programme (ECE, 1976). In its first phase, this programme will be limited to sulphur dioxide and related compounds. As soon as possible, this European Monitoring and Evaluation Programme (EMEP) will hopefully be expanded to include other pollutants, particularly the nitrogen oxides and fine particulates, as recent experience indicates that important interactions take place between these compounds in the atmosphere.

Also special measurement programmes at some stations and for limited periods of time will be important in order to provide a basis for refining the description of the chemical reactions and deposition processes in the atmospheric dispersion models. Recent findings also indicate that the air pollutants are transported in considerable quantities into the very remote Arctic region. More importance should therefore be attached to the precise analysis of low concentration in background areas. Finally, it may be stated that the importance of careful inter-calibration and meticulous data control can hardly be overestimated in programmes of this type.

Acknowledgements—During the LRTAP Programme considerable help and advice was received by the Central Coordinating Unit from the participating laboratories. In the elaboration of the final report, constructive criticism and valuable contributions were received from the countries, particularly from scientists in the United Kingdom and Sweden. I am indebted to the staff of the Norwegian Institute for Air Research for their assistance in preparing the present assessment.

REFERENCES

Brosset C. (1976) Black and white episodes. *Ambio* **5**, 157–163.

Brosset C., Grennfelt P. and Åkerström A. (1971) Determination of filter-collected airborne matter by X-ray fluorescence. *Atmospheric Environment* **5**, 1–6.

Cost Project 61a (1976) Research into the physico-chemical behaviour of SO_2 in the atmosphere. CEC, EUCO/SO_2/59/76.

ECE (1976) Co-operative programme for monitoring and evaluation of the long range transmission of air pollutants in Europe, ENV/WP. 1/5.26, United Nations Economic Commission for Europe.

Eliassen A. (1978) The OECD study on long range transport of air pollutants, long range transport modelling. *Atmospheric Environment* **12**, 479–487.

Granat L. (1975) On the variability of rain water composition and errors in estimates of areal wet deposition. University of Stockholm. Dept. of Meteorology. Report AC-30.

Johnsen D. A. and Atkins D. H. F. (1975) An airborne system for the sampling and analysis of sulphur dioxide and atmospheric aerosols. *Atmospheric Environment* **9**, 825–829.

Koide M. and Goldberg E. D. (1971) Atmospheric sulfur and fossil fuel combustion. *J. geophys. Res.* **76,** 6589–6595.

Liberti A., Pozzanzini M. and Vicedomini M. (1972) The determination of nonvolatile acidity of rain water by a coulometric procedure. *Analyst* **97,** 352–356.

LRTAP (1973) Report from the specialist meeting on the atmospheric dispersion of air pollutants in Gausdal. The OECD Project LRTAP. Norwegian Institute for Air Research, Sept. 1973.

LRTAP (1971–73) Procedures for sampling and chemical analysis. Norwegian Institute for Air Research.

LRTAP (1971–73) Spectrophotometric determination of sulphate by the barium perchlorate–thorin method, LRTAP 4/71. Norwegian Institute for Air Research, Sept. 1971, rev. January 1974.

LRTAP (1971–3) Coulometric titration of strong acid in precipitation (with Appendix: Plotting of Gran's function by means of an electronic analog element). LRTAP 5/71. Norwegian Institute for Air Research, Sept. 1971.

LRTAP (1971–3) Determination of sulphur dioxide in air and airborne sulphate in the particulate phase. LRTAP 2/72. Norwegian Institute for Air Research, July 1972.

LRTAP (1971–3) Determination of strong acid and sulphate in precipitation. LRTAP 3/72. Norwegian Institute for Air Research, July 1972.

LRTAP (1971–3) Determination of particulate sulphate collected on Whatman 40 air filters by X-ray fluorescence. LRTAP 4/72. Norwegian Institute for Air Research, July 1972.

LRTAP (1971–3) Reporting and distribution of results. LRTAP 6/72. Norwegian Institute for Air Research, July 1972, rev. June 1974.

LRTAP (1971–3) Procedures for aircraft sampling and chemical analysis. LRTAP 2/73, Norwegian Institute for Air Research, Dec. 1973, rev. Jan. 1974.

LRTAP (1971–3) Sampling of water-soluble aerosol constituents. Determination of nitrate, ammonium, potassium, calcium, and magnesium in aerosol and precipitation samples. LRTAP 3/73, Norwegian Institute for Air Research, Nov. 1973.

Nordø J. (1974) Quantitative estimates of long range transport of sulphur pollutants in Europe. *Annaln Met. Hamberg* **9,** 71–77.

Nordø J. (1976) Long range transport of sulphur pollutangs in Europe and acid precipitation in Norway. *Water Soil Pollut.* **6,** 199–217.

Nyberg A. (1976) On transport of sulphur over the North-Atlantic. Report of the Swedish Meteorological and Hydrological Institute RMK 6.

Odén S. (1968) Nederbördens och luftens försurning, dess orsaker, förlopp och verkan i olika miljöer. Statens Naturvetenskapliga Forskningsråd, Ekologikommitéen. Bull. no. 1, Stockholm.

OECD (1972) Decision of the OECD Council C(72)13 (Final).

OECD (1977) The OECD Programme on long range transport of air pollutants. Measurements and findings (in press).

Ottar B. (1975) Årsakene til nedbörens forsurning. Rapport fra et samnordisk forskingsprosjekt. NORDFORSK, Miljøvårdssekretariatet, Publikation 1975:10.

Ottar B. (1976) Organisation of long range transport of air pollution monitoring in Europe. *Water, Air Soil Pollut.* **6,** 219–229.

Persson G. A. (1966) Automatic colorimetric determination of low concentrations of sulphate for measuring sulphur dioxide in ambient air. *Int. J. Air Wat. Pollut.* **10,** 845–852.

Schaug J., Semb A. and Gram F. (1975) Remarks on the quality of the LRTAP ground sampling data. LRTAP 16/75. Norwegian Inst. for Air.

Semb A. (1978) The OECD study on long range transport of air pollutants. Source inventory. *Atmospheric Environment* **12,** 455–460.

Atmospheric Environment Vol. 12, pp. 455–460. Pergamon Press 1978. Printed in Great Britain.

SULPHUR EMISSIONS IN EUROPE

ARNE SEMB

Norwegian Institute for Air Research, P.O. Box 130, 2001 Lillestrøm, Norway

(*First received* 17 *June* 1977 *and in final form* 27 *September* 1977)

Abstract—Natural and man-made emissions of sulphur to the atmosphere are discussed. Within Europe the man-made emissions, which are closely associated with the consumption of fossil fuels, are overwhelmingly dominant. Emissions from the 11 countries participating in the OECD study were 8.8 M tonnes S in 1973, and have been estimated at about 25 M tonnes for the whole of Europe. Based on information from the participating countries, and from various other sources, an emission survey has been worked out. This gives the emissions in ½° latitude × 1° longitude geographical reference system which has been transferred to a 127 × 127 km grid used in dispersion model calculations.

1. INTRODUCTION

Within the last two decades there has been a considerable increase in the consumption of energy in the form of liquid fuel, while the consumption of coal has been more stable. Fjeld (1976) has combined available data on fuel production and consumption with estimated sulphur emission factors to provide an estimate of the European emissions of sulphur dioxide for the period 1900–1972 (Fig. 1). It should be added that after 1973 the rate of increase in the consumption of liquid fuel has been smaller than for the previous years, and that the increases in the sulphur dioxide emissions have been significantly smaller for some of the countries.

When the OECD Co-operative Technical Programme to Measure the Long Range Transport of Air Pollutants (LRTAP) was started in 1972, information on the emissions of sulphur dioxide from the various countries was already available from the Economic Commission for Europe (1971) and from on OECD study (1973). For the purposes of the LRTAP study, however, a better space resolution comparable with the grid system of the dispersion model calculations was needed, and the emission data were to be as correct as possible for the years 1973 and 1974.

The present paper is a review of the sulphur dioxide emissions to the atmosphere in Europe, based on a discussion of the literature on natural emissions of sulphur compounds and information from the countries participating in the LRTAP study with regard to the anthropogenic sulphur dioxide emissions.

2. SOURCES OF SULPHUR COMPOUNDS IN THE ATMOSPHERE

Natural sources

Many authors have discussed natural sources in connection with global sulphur budgets. The estimates of Eriksson (1963) and Robinson and Robbins (1968) were based on measured 'background' concentrations and on budget calculations. They concluded that, on a global basis, natural emissions of sulphur compounds, probably mostly gaseous sulphides, are of the same order of magnitude as man-made emissions.

Robinson and Robbins (1968) derived the natural emissions from the difference between estimated dry and wet deposition and the man-made emissions. However, the scanty measurements of air concentrations and precipitated amounts, on which the estimates were based, were not very reliable. Moreover, applying to sulphur compounds the argument of Galbally (1975), who discussed the natural emissions of nitrogen compounds, it can be said that the natural emissions calculated by Robinson and Robbins cannot be explained merely in terms of upward diffusion

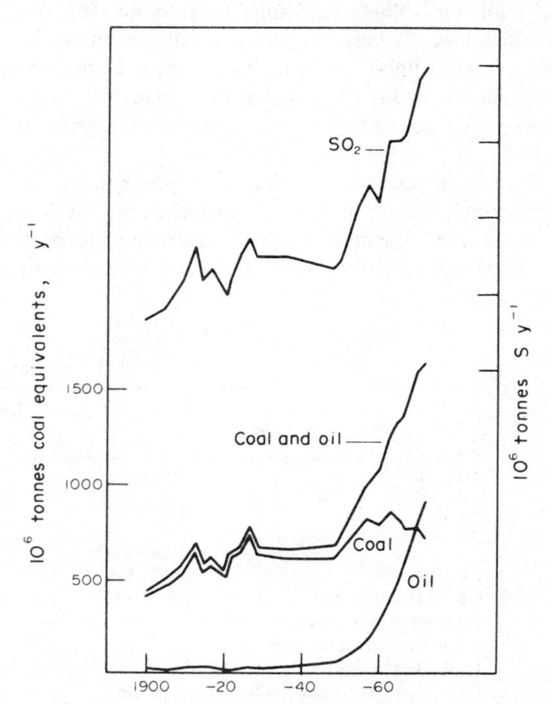

Fig. 1. Fossil fuel consumption and estimated anthropogenic SO_2 emissions in Europe 1910–1972 (after Fjeld, 1976).

through the atmosphere using reasonable estimates of hydrogen sulphide or dimethyl sulphide concentrations.

Granat *et al.* (1976) have presented a global atmospheric budget which concludes that the natural emissions of sulphur compounds are about 40 M tonnes S, or less than half of Robinson and Robbins' figure. This report contains a comprehensive review of worldwide precipitation chemistry data from remote areas.

Information about the actual sources of emissions of gaseous sulphide emissions is very sparse. Very few measurements of actual ground-level concentrations of hydrogen sulphide have been made. In certain areas, such as polluted tidal flats, canals, and estuaries, concentrations may reach the odour limit, i.e. about 200 $\mu g\ m^{-3}$ (Junge, 1963). Breeding *et al.* (1973) measured concentrations as low as 0.1–0.5 $\mu g\ m^{-3}$ at rural sites in the U.S.A., while Junge (1960) found concentrations of 2–20 $\mu g\ m^{-3}$ in Bedford, Mass. and in New York. However, these latter data are at least 17 years old and the analytical techniques used are not as reliable as those now available. It is not possible from these figures to calculate the corresponding emission rates without knowing the concentration and wind profile above the surface. Georgii *et al.* (1976) measured concentrations of 0.1–0.2 μg $H_2S\ m^{-3}$ at a height of 1 m above the shallow water near Sylt in Schleswig-Holstein, increasing to 0.6 $\mu g\ m^{-3}$ just above the surface. These preliminary measurements indicate an hourly emission rate of only 3–4 μg $H_2S\ m^{-2}$, which is negligible compared with man-made emissions. This appears to be true also for the diffuse emissions from cultivated and grassland which may only be a small percentage of the organic sulphur in agricultural crops. If they are assumed to occupy one third of the surface, 10% correspond to an emission of about 0.2 M tonnes of $S\ y^{-1}$.

While considerable reduction of sulphate to hydrogen sulphide takes place on the sea floor, there is no likely mechanism by which the dissolved hydrogen sulphide can reach the surface without oxidation by the sea water.

Lovelock *et al.* (1972) and Rasmussen (1974) have demonstrated that dimethyl sulphide is emitted by plants, soils, and by marine algae, but their method is not sufficiently sensitive to allow measurement of concentrations in ambient air. Hitchcock (1975) has calculated the global emission of dimethyl sulphide to be less than 5 M tonnes of $S\ y^{-1}$.

It appears that if the estimated global emissions of sulphur compounds from natural sources are distributed equally on an areal basis, natural emissions within Europe would be only 1–2 M tonnes $S\ y^{-1}$. This conclusion is not altered by the measured and estimated emissions of gaseous sulphur compounds from biological processes. Table 1 gives a summary of estimated global natural and man-made emissions of sulphur to the atmosphere, compared with the estimated emissions in Europe.

Thus, although more quantitative information on the extent, magnitude and distribution of the natural emissions is highly desirable, natural emissions within Europe are not likely to exceed 10% of the man-made emissions.

Man-made sources

The main man-made source of airborne sulphur compounds is the sulphur content of fossil fuels released to the atmosphere by combustion. Most of the emission is as sulphur dioxide but some 2–3% of the emission is in the form of sulphur trioxide or sulphates. For petroleum products virtually all the sulphur content of the fuel is emitted. For hard coal some 10% of the sulphur is retained in the ash in the case of industrial usage and about 25% in the case of domestic. Corresponding figures for lignite are about 30 and 50% respectively. In certain processes, particularly steel making, gas production and for some cement kilns, much more of the sulphur is retained in the slag or end product. Provided corrections are made for such factors, the total emission of sulphur compounds can be estimated with reasonable accuracy from fossil fuel consumption statistics and typical sulphur contents. In addition, small contributions arise from the smelting of sulphidic ores.

The yearly emissions of sulphur dioxide in Europe

Table 1. Annual sulphur emissions

Area		Man-made M tonnes S	Natural M tonnes S	Reference
Global	510 M km^2	40	280	Eriksson (1960)
	510 M km^2	70	100	Robinson and Robbins (1968)
	510 M km^2	65	35	Granat (1976)
Europe	10 M km^2	25	—	
LRTAP countries	2.4 M km^2	9	—	
Biogenic emissions in Europe				
Hydrogen sulphide from tidal flats			0.02–0.03 t km^{-2}	Georgii *et al.* (1976)
Diffuse emissions of H_2S, $(CH_3)_2S$ etc from land areas: 10% of sulphur content in agricultural crops			0.1–0.2 t km^{-2}	—

(including the European part of the U.S.S.R.) were estimated as almost 25 M tonnes as S in 1970 (ECE 1971, 1976) whilst those emissions from the 11 countries participating in the LRTAP Programme were estimated as 8.8 M tonnes (Rystad *et al.*, 1974).

Table 2 gives the main sources of crude oil to Western Europe and some typical values for the sulphur contents. Most of the sulphur is associated with the residual or 'heavy' fuel oil which makes up 30–50% of the crude. Some of this is used for asphalt making, wax and lubricants, which must be taken into account when estimating the emissions from consumption of crude. Part of the sulphur-rich residual oil is also re-exported in the form of bunker oil for ships.

If emissions are to be estimated for individual countries, the import and export of distillates must also be taken into account. Detailed statistics by sectors are available for the OECD countries. The corresponding sulphur contents are generally not given, but can be deduced from existing legislation and regulations (CONCAWE, 1976). The sulphur contents of the fuel oils may, however, be somewhat lower than specified in the regulations. Another problem is that regulations often apply only to specific highly polluted areas within the countries.

The sulphur content of coal is also rather variable and a major uncertainty in an emission inventory. The average sulphur content of British coals are 1.6% (Wandless, 1959), Ruhr coals contain about 1.2% and Polish coals also generally between 1 and 2% sulphur, while Yugoslavian hard coals lie between 5 and 6% in sulphur content (ECE, 1971). Within OECD Europe 42% of the hard coal is used in thermoelectric power plants. Another large fraction (34%) is used in the production of coke, which is chiefly used in the iron and steel industry.

Brown coals and lignite are significant energy sources in the Federal Republic of Germany, the German Democratic Republic, Czechoslovakia and several other countries. Within OECD Europe more than 80% of the brown coal is used in thermoelectric power plants. The brown coal sulphur content is very variable. Herrmann (1952) gives analysis figures for 19 Saxonian and 6 Rhineland brown coals. Sulphur contents range between 0.48% and 4.75% for the former, and between 0.26% and 2.68% for the latter group. The mean sulphur contents are 2.9 and 1.2, respectively. However, the mean concentration of sulphur in the brown coals that are actually burned may be lower than this.

Within the Velokom brown coal deposit in Northern Bohemia, up to 16% S has been found in part of the seams, and an estimated 1/7 of the deposit consists of coal with more than 3% S. The rest has a sulphur content of between 1 and 3% S (Macak and Vcelak, 1975).

The sulphur contents above are all given in percent by weight. Emission factors giving the amount of sulphur per unit energy include not only an assessment of the heat value of the coals, but also the retention of sulphur in slag, ash, etc.

Sulphur dioxide is also emitted to the atmosphere by the roasting of sulphidic ores, in the production of copper, lead, zinc, and nickel. These contributions are relatively small because of regulations which require smelters to reduce the sulphur dioxide emissions. In fact, most of the sulphur dioxide is converted to commercial sulphuric acid. In addition come pro-

Table 2. Sources of liquid fuel and sulphur content of some relevant crude oils

Origin	1974 Exports of crude to OECD Europe*, million metric tonnes		Sulphur contents in percent†	
Middle East	Saudi Arabia	146	Aramco light	1.8
			Aramco medium	2.7
			Aramco heavy	2.9
	Abu Dhabi	12		
	United Emirates	10	Murban	0.8
	Kuweit	36		2.5
	Qatar	10	Land	1.2
			Marine	1.5
	Iraq	34	Basrah	1.9–2.1
	Iran	96	Iranian light	1.4
			Iranian heavy	1.60
Africa	Algeria	16	Hassi Massoud	0.14
			Arzew	0.14
	Libya	55	Brega	0.2
			Es Sidr	0.5
	Nigeria	41	Medium	0.2–0.3
	Gabon	2		
Central America	Venezuela	8		(3.0)
Others	U.S.S.R.	8	Ural blend	1.7
	North Sea	~1	Ekofisk	0.16
			Forties	0.40

* OECD Quarterly Oil Statistics: 4th Quarter 1976.
† Information obtained from Norwegian oil companies.

cess emissions from sulphuric acid production (from pyrites), cellulose manufacture by the sulphite process, and incineration of solid waste.

For Norway, Sweden and U.S.S.R. emissions from sources not associated with fossil fuels are about 10–20% of the national total. The fraction is lower for the other European countries.

The LRTAP Emission Survey

It was originally intended to use data from an OECD study of emissions from stationary sources (OECD, 1973), as input for the atmospheric dispersion model calculations. When the space resolution of these data were to be improved and updated to 1973–74, it was natural to do this in co-operation with the participating countries.

The dispersion models required a space resolution of 127 × 127 km. In order to make the data more universally applicable, it was decided to use geographical co-ordinates as a basis for the emission grid. The grid elements chosen were $\frac{1}{2}°$ latitude × 1° longitude, or approx 55 × 60 km. Since the elements of the emission grid were smaller, the data can readily be transferred to the 127 × 127 km grid system (of different orientation) used in the dispersion models.

In order to be able to take into account the seasonal variation, the emissions were divided in two categories, a variable and a continuous part. Diurnal variations were not considered.

For the countries participating in the LRTAP Programme, except Belgium, the emissions in the different categories were given by the participating countries, using available national fuel consumption statistics by districts and sectors as the main source of information. In some cases, data for administrative regions were transformed to the grid elements by NILU. The emission data for Belgium and Luxembourg were estimated by NILU using available statistical and economical information. Some of the results are summarised in Table 3 (Rystad *et al.*, 1974). It should be noted that the emission figures were collected for the specific purpose of being used in dispersion calculations, and that no particular effort was made to ensure comparability of the national figures, other than obtaining the best possible estimate of the emissions in each case. The national figures are at best assumed to be accurate to within 10–15% for countries with good statistical data. Data for the individual grid elements are, naturally, less certain.

For surrounding areas, emissions by countries given by ECE was taken as input, and population statistics and various other available information were used to distribute the emissions in the grid system (Fig. 2). These data are not supposed to be as accurate as those given by the participating countries.

The variable component of the emissions is mainly due to room heating in the domestic and commercial sectors, including electricity. From Table 2 it is seen that this component is between 25 and 55% of the total emission of sulphur dioxide for the countries. It is generally between these limits also for the individual grid elements. In order to simplify the model calculations, the seasonal variations have been represented by a sinusoidal variation of ±33%.

A separation was also made between low-level emissions and emissions occurring from tall stacks (>100 m). In most countries the latter account for about half of the emissions.

3. CONCLUSION

The emissions of sulphur dioxide in Europe can, in principle, be obtained from the detailed statistics available with respect to the consumption of fossil

Table 3. Emission of sulphur dioxide by countries for the year 1973

Country	Population (million people)	Annual emission of SO_2 per inhabitant (kg S)			Total emission of SO_2 per country (10^3 tonnes S)		
		C	V	C + V	C	V	C + V
Austria	7.456	25.1	4.5	29.6	187	34	221
Belgium	9.650	36.1	15.6	51.7	348	151	499
Denmark	5.010	28.4	33.8	62.2	142	169	312
Federal Republic of Germany	61.166	23.1	9.0	32.1	1413	551	1964
Finland	4.711	43.4	14.9	58.3	204	70	274
France	49.509	22.7	10.0	32.7	1121	495	1616
Netherlands	13.266	20.4	9.1	29.5	271	120	391
Norway	3.866	17.0	6.5	23.5	66	25	91
Sweden	7.976	37.0	15.0	52.0	295	120	415
Switzerland	6.272	7.9	4.2	12.1	49	26	76
United Kingdom	54.237	28.7	23.0	51.7	1556	1247	2803

C: Continuous component
V: Variable component
The emission data from the different countries are not strictly comparable, because of different basic statistical information and assumptions made in the compilation of the data.

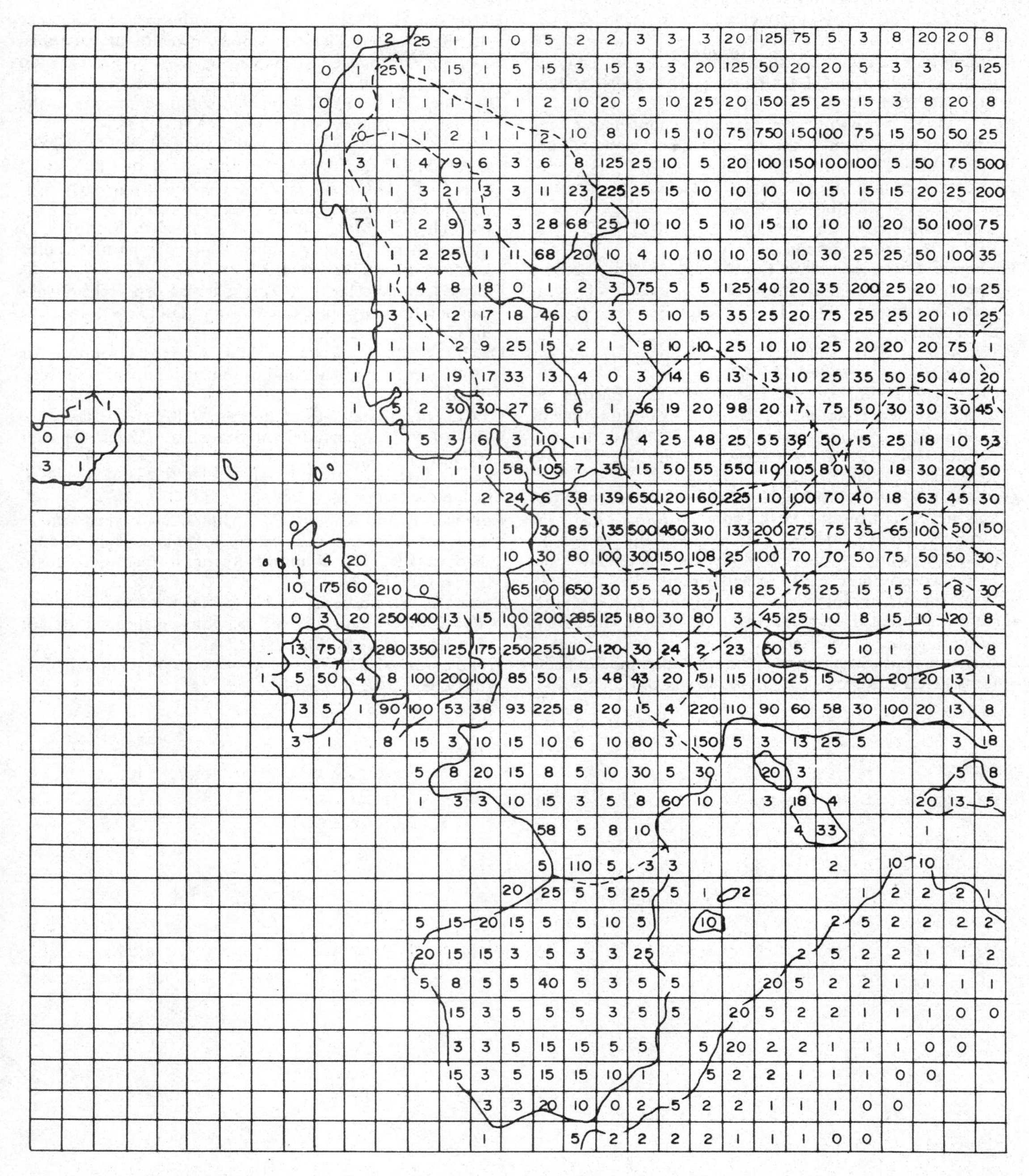

Fig. 2. Estimated annual emissions of SO_2 (10^3 tonnes S) in grid elements with length 127 km at 60° N.

fuels and a knowledge of the sulphur contents, making allowances for the retention of sulphur in different combustion and industrial processes. The sulphur content, particularly that of solid fuels, is a major uncertainty factor.

The LRTAP emission survey gives the emission of sulphur dioxide in 1973 in a $\frac{1}{2}°$ latitude × 1° longitude grid system. The survey is based on information obtained from the countries participating in the LRTAP Programme. Emissions from the other European countries are given in considerably less detail. It is therefore desirable to extend and update the present knowledge of the emissions in connection with future studies of the large-scale dispersion and deposition of sulphur compounds and related substances.

Acknowledgement—The present work was carried out in connection with the OECD Co-operative Technical Programme to Measure the Long Range Transport of Air Pollutants, and the meticulous work of compiling and distributing the emissions in grid squares was conducted by the participating countries, and by S. Strømsøe, B. Rystad and J. Saltbones at the Norwegian Institute for Air Research. The author is indebted to Dr. R. A. Barnes and Dr. L. Granat for helpful discussions.

REFERENCES

Breeding R. J. *et al.* (1973). Background trace gas concentrations in the Central United States. *J. geophys. Res.* **78,** 7057–7064.

CONCAWE (1976). Published regulatory guidelines of environmental concern to the oil industry in western Europe. Stichting Concawe, The Hague.

Economic Commision for Europe (1971). Desulphurization of Fuels and Combustion Gases. Proceedings from a seminar in Geneva, 16–20 November 1970. ST/ECE/AIR POLL/1, United Nations, New York.

Economic Commission for Europe (1976). The Second Seminar on Desulphurization of fuels and combustion gases. Washington D.C. 11–17 November 1975 (ENV/SEM. 4/3).

Eriksson E. (1963). The yearly circulation of sulfur in nature. *J. Geophys. Res.* **76,** 4001–4019.

Fjeld B. (1976). Forbruk av fossilt brensel i Europa og utslipp av SO_2 i perioden 1900–1972. (Consumption of fossil fuels in Europe and emissions of SO_2 during the period 1900–1972.) Norwegian Institute for Air Research, Teknisk Notat No. 1/76.

Galbally I. E. (1975). Emission of oxides of nitrogen (NO_x) and ammonia from the earth's surface. *Tellus* **27,** 67–70.

Georgii H. W., Claude H., Jaeschke W. and Malewski (1976). Messungen von H_2S in Bodennähe und der unteren Troposphäre. In: Arbeitsbericht des SFB 73 'Atmosphärische Spurenstoffe', Frankfurt a.M.—Mainz.

Granat L., Rodhe H. and Hallberg R. O. (1976). The global sulphur cycle. In *Nitrogen, Phosphorus and Sulphur—Global Cycles.* Svensson B. H. and Söderlund R. (Editors). SCOPE Report 7, *Ecol. Bull.* (*Stockh.*) **22,** 89–134.

Herrmann H. (1952) Brown coal with Special Reference to the State of Victoria, Commonwealth of Australia, 612 pp (Melbourne: The State Electricity Commission of Victoria, 1952).

Hitchcock D. R. (1975). Dimethyl sulfide emissions to the global atmosphere. *Chemosphere* **4,** 137–138.

Junge C. E. (1960). Sulphur in the atmosphere. *J. geophys. Res.* **65,** 227–237.

Junge C. E. (1973). *Air Chemistry and Radioactivity.* Academic Press, New York.

Lovelock J. E., Maggs R. J. and Rasmussen R. A. (1972). Atmospheric dimethyl sulfide and the natural sulfur cycle. *Nature* **237,** 452–453.

Macak J. and Vcelak V. (1975). The use of Czechoslovakian sulphuruous coal. *Czech. Unli.* **23,** 13–18 (orig. text in Czech.).

OECD (1973). Report of the Joint Ad Hoc Group of Air Pollution from Fuel Combustion in Stationary Sources, Paris.

OECD (1974). Statistics of Energy 1958–1973, Paris.

OECD (1977). Quarterly Oil Statistics, Fourth Quarter 1976, Paris.

Rasmussen R. A. (1974). Emission of biogenic hydrogen sulphide. *Tellus* **26,** 254–260.

Robinson E. and Robbins R. C. (1968). *Sources, abundance and fate of gaseous atmospheric pollutants.* Final Report, Project PR 6755. Stanford Research Institute, Menlo Park. California.

Rystad, B., Strømsøe S., Amble E. and Knudsen T. (1974). The LRTAP emission survey. Norwegian Institute for Air Research, LRTAP 2/74 (1974).

Wandless A. M. (1959). The occurrence of sulphur in British coals. *J. Inst. Fuel* **32,** 258–266.

Atmospheric Environment Vol. 12, pp. 461–477. Pergamon Press 1978. Printed in Great Britain.

METEOROLOGICAL ASPECTS OF THE TRANSPORT OF POLLUTION OVER LONG DISTANCES

F. B. SMITH and R. D. HUNT

Boundary Layer Research Branch, Meteorological Office, Bracknell, Berkshire, England

(*First received* 13 *June* 1977 *and in final form* 21 *July* 1977)

Abstract—Since sulphur pollution is carried by the atmosphere over considerable distances it is important to have an adequate understanding of those aspects of the atmosphere's structure which are important in this process and to know not only the mean conditions but also have a good feel for the statistical variations and their significance to both wet and dry depositions to the ground. This paper discusses several aspects of the role of the atmosphere and starts by considering the nature of the mixing layer and gives statistics of its depth from measurements made at Cardington, England.

It goes on to summarise conclusions reached from aircraft sampling-flights over the North Sea (1971–75) in terms of the velocity of deposition of SO_2 (0.8 cm s^{-1} over land, 0.5 cm s^{-1} over sea) and the conversion rate of SO_2 to sulphate (roughly $1\% \text{ h}^{-1}$). The next section considers particular states of the atmosphere which can lead to episodes of very high deposition of sulphate in precipitation. Such episodes are given a formal definition and their cause and geographical distribution in western Europe are investigated. Two case-studies of notable episodes are presented and a tentative conclusion reached that average statistics of the state of the atmosphere should be used only with caution when modelling long-term pollution transport in view of the relative dominance of episodes and their own rather distinctive meteorology.

Finally the paper reviews some of the rather more important aspects of meteorology which have a direct bearing on transport and which require further theoretical and experimental study.

1. THE NATURE OF THE BOUNDARY LAYER AND ITS EVOLUTION

The transport of any pollutant such as sulphur dioxide over long distances in the atmosphere is strongly influenced by the time-dependent depth of the layer over which it is dispersed. The deeper the layer, the lower the conversion rate to sulphate (other things being equal) and the more likely wind-direction variations-with-height will be important in spreading out the plume in the horizontal lateral direction.

This layer is often called the mixing layer and depends on the existence of turbulent eddies capable of transporting the pollutant through the vertical. This dependence on the vertical structure of turbulence equates the mixing layer to the meteorologically-defined boundary layer. The latter is the lowest layer of the atmosphere under direct influence of the underlying surface (Smith and Carson, submitted). The sources and sinks of turbulent energy which characterise this layer of variable depth all result directly or indirectly from the interactions between the atmosphere and the surface. These sources and sinks are:

(i) The removal of momentum by viscous drag, the so-called shearing stress, which provides a source of turbulent kinetic energy not only very near the ground but at all levels wherever the momentum loss is effective in creating velocity shear.

(ii) The input to, or extraction from, the atmosphere of sensible heat resulting from a net imbalance of radiational fluxes to and from the surface giving a temperature difference between air and surface. The flow of heat results in density differences; when heat flows into the air, buoyancy forces provide a source of turbulent energy and the boundary layer is 'unstable' whereas when heat is extracted, as on a calm night, the direction of the buoyancy forces is reversed and any turbulent motions are damped and the layer is then 'stable'. A stable boundary layer is clearly not a mixing layer in the limit when turbulence is completely quenched.

(iii) The input of moisture by evaporation from the surface also creates buoyancy forces which are a source of turbulent energy. Condensation may occur at some level and when it does so not only may radiational cooling effects be important (Deardorff, 1976) in the local energy balance but latent heat of vaporisation will be released.

(iv) Radiation may cause warming and cooling at any level and although this effect is not restricted to the boundary layer it has to be allowed for on occasions in the overall balance.

The boundary layer then can be either unstable if the sources of energy outweigh the sinks or stable when the reverse is true. In the unstable situation the boundary layer extends up to a height characteristically between 400 and 2000 m (see later for more details). The turbulent energy does not continue to increase indefinitely of course but is cascaded from the larger eddies to the smallest eddies where viscous dissipation returns the energy to heat. This separation of wavelengths between input and output of energy is equally true in the stable situation and means that even in the latter case when the sink of energy is very strong some vestige of turbulence (and hence mixing) can exist. The stable layer, slightly turbulent or quiescent, may be only a few tens of metres to about 400 m deep.

The most important question we face is how to define the upper limit of the boundary layer. This is a surprisingly difficult question to answer. Sometimes it is obvious; a very well marked inversion at several hundred metres above ground capping a well-mixed layer provides an indisputable upper limit to the boundary layer, everything above is unaffected in a direct sense by the underlying surface whereas everything below is under its influence. On other occasions however no clearly marked top exists and the effect of the surface decreases only very gradually with increasing height. This is often true when buoyancy forces are very nearly zero when the boundary layer is called 'neutral', a transitional state between an unstable and a stable regime.

Perhaps the best we can do is to define the turbulent boundary layer as that part of the atmosphere extending from the underlying surface up to a height at which all turbulent and radiative flux-divergences resulting from surface action have fallen to some prespecified small percentage (say 1%) of their surface values. However sometimes even this simple notion is confused by processes such as penetrative convection where a lower layer, in which the influence of the underlying surface is felt in a widespread direct and continuous way, is overlaid by an upper layer influenced locally and spasmodically in, say, deep cumulus clouds but otherwise uninfluenced and non-turbulent. Rather arbitrarily we restrict the boundary layer to cover the lower layer only.

To sum up then, the true boundary layer is a region dominated by a balance of turbulent-energy sources and sinks all derived from the interactions of atmosphere and underlying surface. It has a variable depth ranging typically from a few tens of metres in stable conditions to about 2000 m in very unstable situations. The layer is characterised by significant vertical flux divergences of momentum and frequently of heat and moisture. The wind speed tends to fall below its free-stream value found above the layer, eventually to zero at the surface although the fall-off may not always be monotonic, especially in stable conditions. Wind direction also changes with height with the biggest deviation from the geostrophic direction occurring at the surface, and whilst the deviation tends to be quite variable, influenced as it is by changes with time in stability, by local topography and meso- and synoptic-scale accelerations, on the average the surface wind is backed some 5° over the sea and 10° over land in unstable conditions and 30° or more in stable conditions. The resulting flow down the pressure gradient provides a significant source of momentum to balance the shearing stress of the ground and on a synoptic scale is important in generating horizontal convergence and upward motions. The turning is also potentially important in enhancing lateral dispersion of plumes coming from major source regions of sulphur pollution.

The boundary layer frequently plays another role, that of a buffer zone into which the heat, moisture and pollution coming from sources at the ground are stored until they are released at a later date into the main body of the atmosphere. Typically in temperature latitudes the boundary layer takes several days (2–4 days) to reach this 'break-down' situation. Breakdown may occur as a result of large scale convective instability, at fronts when upsliding occurs, in mountainous regions where rotors and 'hydraulic jumps' may cause considerable vertical stirring, and by continuous synoptic-scale vertical motion aided by the diurnal cycle in the depth of the boundary layer which 'pumps' the contents of the boundary layer upwards in the middle of the day, leaving it behind thereafter to be acted upon by the steady synoptic upward motions.

Boundary layers are almost always in a state of change; they are either responding to a change in the synoptic pressure field or to some change in the underlying surface. For gradual changes the structure of the layer maintains a quasi-equilibrium but coast lines, sharp mountains and other sudden discontinuities pose very special problems of adjustment. Fortunately these changes do not normally have a very pronounced effect on the main factor affecting the long range transport of pollution; namely the mixing depth, or boundary layer height z_i.

The mixing depth may be estimated in the following ways:

(i) When z_i is relatively small, as at night, turbulence probes, or sensitive anemometers, situated on tall masts can indicate at what level the vertical mixing process ceases.

(ii) Temperature and humidity profiles obtained from instruments supported on the cable of a tethered-balloon which is systematically raised and lowered can clearly indicate the change-over from boundary layer conditions to 'free' atmosphere conditions. The so-called BALTHUM operated at Cardington in England is a good example of this system.

(iii) Radio-sondes perform a very similar function although with slightly less accuracy, and are invaluable for monitoring higher inversions capping the boundary layer, beyond the normal reach of tethered balloon systems.

(iv) In the absence of synoptically-induced capping-inversions the evolutionary-induced inversions produced at the top of the layer are weaker and rather harder to detect in only moderately unstable conditions. Boundary layer theory can then be useful to complement any measurements that are available and help to decide which of perhaps several slight changes in lapse rate really marks the true top of mixing. Allowing for all the assumptions and approximations inherent in any scheme that is based on very limited input data, Fig. 1 shows a simple nomogram for esti-

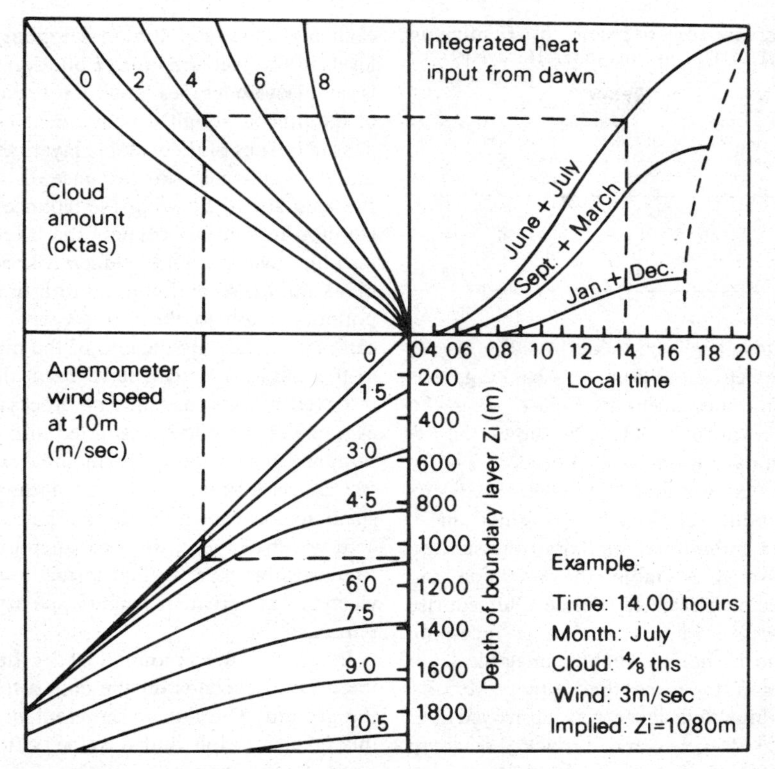

Fig. 1. A nomogram for estimating the depth of the boundary layer in the absence of marked advective effects or basic changes in weather conditions. The marked example shows how the diagram is to be used.

mating z_i in day-time conditions based on time of day, month, cloud cover and wind speed at 10 m. Essentially it says that the first three parameters determine the integrated heat input into the boundary layer from dawn. This heat will have modified the vertical temperature profile most likely to have been associated with that cloud cover (assumed constant in time), and time of year, based on a very detailed analysis of BALTHUM ascents made by Hardy (internal Met Office Note). The nature of the modification is according to the theory of Carson (1973). The nomogram also recognises the two main sources of energy for boundary layer development: the generation of turbulence by the wind and the influence of sensible heating.

(v) In statistical modelling of the long range transport of pollution it may be sufficient to know the statistical behaviour of z_i as a function of time of day. Hardy (1973) has analysed over 4000 ascents made by the BALTHUM system. Table 1 gives the essential statistics for near neutral or unstable boundary layers.

Equally important are stable boundary layers where the inversion is at ground level. Although incorporated in Table 1, Hardy makes a special study of them and his results are summarised in Table 2. They are particularly relevant to the long-range transport problem because they 'insulate' the main body of the pollution cloud from the effect of ground absorption (dry deposition).

From what has already been said it is clear that the evolution in time of the boundary layer and its

Table 1. Percentage of ascents showing an inversion (intensity $\geqslant 0.5^\circ C$) within the indicated height range. Z_{im} is the median inversion height, i.e. on 50% of all occasions an inversion lies at or below this height. Considering the mean inversion height when a capped layer exists, at midday $\bar{z}_i = 800$ m. However a strong seasonal trend is very clear; during the three summer months $\bar{z}_i = 1300$ m, whereas during winter $\bar{z}_i = 500$ m

Time	Range in Z_i (m)						% with inversions at top of mixing layer	Mean height (m) of inversion	
	0–1	1–75	75–300	300–600	600–900	900–1200		$\bar{z}_i$	Z_{im}
00Z	59.5	9.5	5.8	3.5	2.8	2.2	85	100	0
06Z	32.9	12.2	16.2	6.1	4.3	2.9	78	200	130
12Z	2.4	5.6	4.5	8.1	9.7	5.2	50	800	1700
18Z	34.2	4.6	3.0	2.8	3.5	3.7	58	200	1000
All hours	33.2	8.0	7.6	5.0	4.4	4.0	—	—	

Table 2. Details of surface inversions at Cardington obtained from BALTHUM ascents (after Hardy, 1973)

Time	No. of ascents	No inversions at ground (%)	Inversion at ground level with top Below 900 m (%)	Above 900 m (%)	Mean height of inversion top (m)
00Z	1972	30.5	67	2.5	170
06Z	2074	55.4	43	1.6	160
12Z	1695	89.5	10.4	0.1	30
18Z	1847	51.4	48	0.7	70

variation from unstable well-mixed conditions to stable almost quiescent conditions and back again is a very important phenomenon. Figure 2 is an example of such a diurnal cycle and shows the response of the boundary layer to variations in surface sensible heat flux. Before sunrise the heat flux was downward at about $20\,\mathrm{W\,m^{-2}}$ ($= 2\,\mathrm{mW\,cm^{-2}}$) required to balance radiational cooling at the surface. The boundary layer was stable, the inversion top being at about 400 m. Some short time after sunrise the heat flux was reversed in response to incoming solar radiation and a shallow slightly unstable layer formed which began to erode the stable layer. By about 9.30 MST the stable layer was destroyed and the boundary layer deepened more quickly reaching a maximum in early afternoon some 2–3 h after the midday maximum in heat flux. Just before sunset radiational cooling of the ground exceeded solar warming and a stable layer reformed near the ground. Any pollution emitted into the atmosphere during the day now finds itself in a layer which is 'insulated' from the surface and in which the turbulence gradually decays, unable to be replaced since the sources of energy are no longer available. Wind directional shear can now play a much more important role than hitherto since the lack of vertical mixing prevents each molecule of pollution sampling the wind at all levels in the polluted layer. Pollution within the stable layer also undergoes minimal vertical mixing most of the time although it is one of the more interesting characteristics of these stable layers that internal wind gradients increase in response to the removal of shearing stress throwing the balance of forces out of equilibrium, until ultimately the layer becomes dynamically unstable and sudden bursts of turbulence break out for several minutes at a time, bringing fresh pollution close to the ground where dry deposition can play its role. By the end of the night the pollution in the stable layer can have been almost completely removed by this intermittent process. One very clear example of this effect was noted in one of our aircraft sampling flights described in the next section, where the concentration of SO_2 at about 100 m was too small to measure after the air had crossed over the cold water of the North Sea (thereby forming a very similar stable layer to that formed over land at night) whereas at greater heights plenty of SO_2 still remained.

In general the amount lost by dry deposition to the ground depends on the concentration just above the ground. The flux is represented as a product of this concentration and a velocity, the so-called velocity of deposition v_d. Since concentration varies with height, v_d strictly must also be height dependent, although in the first ten metres or so, the variations are relatively small provided v_d is less than a few centimetres per second and the airflow is turbulent. Since the deposition depends on concentration, its magnitude is greatest within the first few kilometres of a low-level source. In the long-range transport problem it is convenient to treat the pollutant as though it were uniformly mixed through the boundary layer (the so-called box-model) and to add an

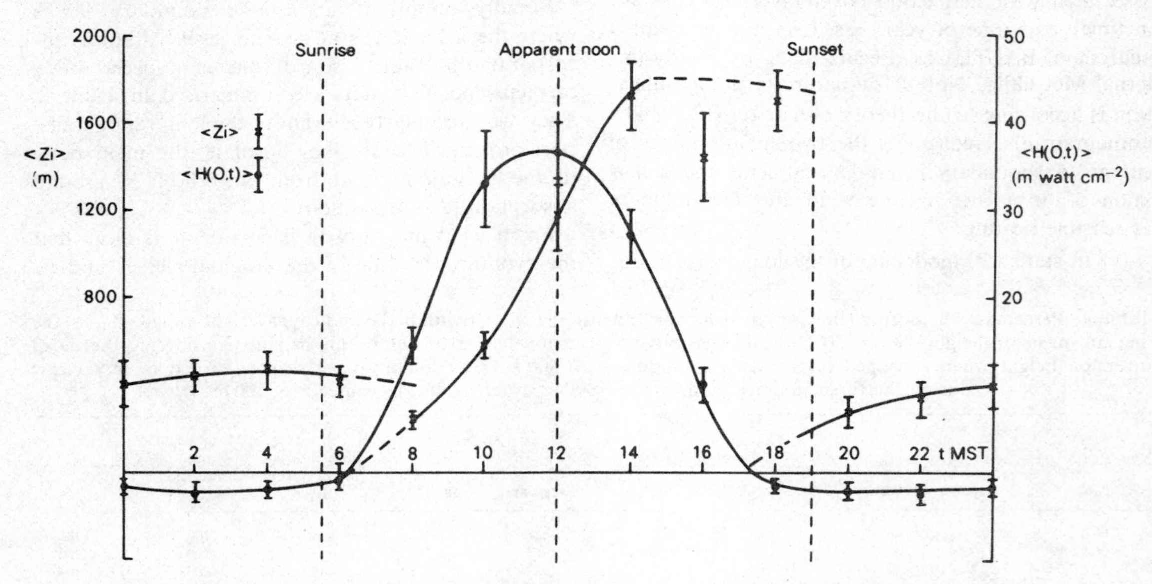

Fig. 2. The mean boundary layer thickness $\langle z_i(t)\rangle$ and sensible heat flux at the surface, $\langle H(o, t)\rangle$, deduced the O'Neill data and plotted with standard errors as functions of time of day, t, in Mean Solar Time.

Table 3. The table shows typical frequencies and mean winds appropriate to the various stability categories. It has been assumed that P for an urban area is one class more unstable than for a rural area (on account of the increased roughness of an urban area and the contribution of anthropogenic heat sources). The various values of λ_i follow from Hanna and Gifford (1973). Λ_i ($i = 3$ or 4) represents the sum of the enhanced ground-level concentrations over and above the concentration appropriate to the box model:

$$\Lambda_i = (\lambda_0 + \lambda_1 + \ldots.. + \lambda_i) - (i + 1)\lambda_{i+1}$$

where by assumption λ_{i+1} corresponds to the box-model concentration.

F_i is the total percentage lost by dry deposition over and above that given by the box model, on the assumption that $v_d = 0.01 \text{ m s}^{-1}$ for SO_2.

$$F_i(P) = 0.01 \times \Lambda_i Q/u \times l^2 \times 1/Q \times 100 = 25 \times \Lambda_i(P)/u(P) \text{ if } l = 5 \text{ (km)}$$

Finally the weighted average loss $W_i \equiv \Sigma_P(f_p/100)F_i(P)$. Other values of W_i are: $W_5 = 12.89$, $W_6 = 13.63$, $W_7 = 14.23$, $W_8 = 14.95$, $W_9 = 15.51$

(P) Stability category	Mean wind u at 10 m (m s^{-1})	% frequency f_p	Concentration Coefficients λ_0	λ_1	λ_2	λ_3	λ_4	15 km Λ_3	15 km F_3%	20 km Λ_4	20 km F_4%
A	1.4	7	0.28	0.10	0.05	0.04	0.03	0.35	6.2	0.38	6.8
B	3.5	17	0.53	0.23	0.13	0.10	0.08	0.68	4.9	0.74	5.3
C	5.6	50	0.88	0.47	0.27	0.20	0.17	1.16	5.2	1.28	5.7
D	2.8	17	1.17	0.69	0.41	0.31	0.26	1.56	13.9	1.74	15.5
E	1.4	7	1.64	1.11	0.69	0.53	0.44	2.21	39.5	2.50	44.6
F	0.7	2	2.46	1.98	1.29	1.01	0.85	3.32	100	3.85	100
				Weighted average loss: $W_3 = 11\%$						$W_4 = 12\%$	

extra initial deposition to allow for the error arising from the fact that the plume from a source takes a finite time to diffuse upwards. During this diffusive process ground-level concentrations and depositions are higher than given by the box-model. Various estimates have been made of this initial deposition and the current OECD model uses for sulphur dioxide a mean value of 15%. This value is supported by very approximate calculations based on the well-known Hanna–Gifford urban pollution model (1973) which says that the concentration C in $\mu g\, m^{-3}$ can be expressed as

$$C = \lambda_i(P)\, Q/ul^2$$

where Q is the area source strength within a grid-square (side l km) in units of tons per year, u is the 10 m wind speed (m s^{-1}), λ is a coefficient dependent on i, the number of gridlengths (each l km) downwind from the source, and P, the Pasquill stability category, which classifies the range of boundary layer stabilities from A (very unstable) to F (very stable). If we can assume that, normally, uniform concentration is achieved in say 10–20 km downwind and if we take a gridlength l of 5 km, then Table 3 sets out the basic analysis, appropriate to an urban area in the U.K.

The model is therefore not sensitive to the assumption as to where uniform concentration is achieved, and the trend in values of W_i gives surprisingly good support for the 15% value selected in the OECD study.

The aircraft sampling flights data summarised in the next section confirm a value of this general magnitude.

2. METEOROLOGICAL OFFICE AIRCRAFT SAMPLING FLIGHTS

Since 1971 the Meteorological Office in collaboration with the Environmental and Medical Sciences Division of AERE, Harwell, have been making occasional flights to measure sulphur dioxide and sulphate aerosol in the mixing layer, and above, off the East Coast of the U.K. in relatively simple meteorological conditions when the air had previously crossed the country. The main purpose of these flights has been to establish the average rate of dry deposition to the ground over typical mixed countryside. Information has also been deduced on the initial deposition referred to at the end of the previous section and the 'initial' conversion of SO_2 to sulphate in the diffusing plume. Flight information, techniques and results have been described in detail in Smith and Jeffrey (1975) and Johnson and Atkins (1975).

Starting in 1974 a few flights have also been made in rather similar conditions off the Danish and Norwegian coasts when the air had earlier crossed the U.K. with the intention of obtaining sulphur dioxide-to-sulphate conversion-rate data once the plume had substantially diffused through the mixing layer.

(i) *Brief summary of measurements to the end of* 1973

In these experiments the concentrations of SO_2 (and sulphate in later flights) were measured at several levels over a path about 150 km long. The measurements were combined with a measured mean wind to give the flux of pollutant crossing the track. Back trajectories were estimated using hourly synoptic charts assuming the relations between measured winds on track and the geostrophic wind held at all times along the trajectory. These trajectories were used to define the area of the U.K. from which the SO_2 emissions originated, which were determined by a small scale map of all emitting source strengths in the U.K. and allowing for seasonal and diurnal variations of emission. Relating the estimated emission to the measured fluxes using a simple model enabled

Table 4

The sulphur budget

Symbol	Meaning	1 October 1971	22 October 1971	2 November 1971	9 August 1973	4 September 1973	7 September 1973	8 October 1973
h	depth of mixing layer (m)	1050	450	600	1200	1700	1200	1600
V	mean speed of sampled air (m s^{-1})	10.0	20.5	17.3	14.7	8.1	12.5	6.9
x	distance from major upwind source (km)	100	220	140	190	125	87	90
t	time of travel from major source (h)	4.25	3.5	2.25	3.5	4.25	2	3.5
σ_z	r.m.s. vertical extent of SO_2 plume on sampling track (m)	600	260	346	580	760	280	540
σ_z (neut.)	theoretical value for neutral atmosphere (m)	680	1100	800	900	750	640	645
r	indicator of significant rain							r
Emissions								
E_2	estimated emission flux of SO_2 from sources (t h^{-1})	238	158	370	149	320	67	78
Sulphur dioxide								
F_2	measured SO_2 flux on sampling track (t h^{-1})	122	112	193	85	187	32	27
F_2/E_2	fraction of emitted SO_2 remaining off E. Coast	0.51	0.71	0.52	0.57	0.58	0.48	0.34
V_d	deduced velocity of dry deposition (cm s^{-1})				0.8	0.8	0.6	
	assumed V_d implied by comparison (cm s^{-1})	0.8	0.8	0.8				0.7
D_2/E_2	fraction of emitted SO_2 lost by dry deposition	0.33	0.20	0.31	0.31	0.33	0.32	0.32
C_2	average SO_2 concentration on sampling track (μg m^{-3})	30	15	25	10	22	6	7
Sulphate (converted to equivalent SO_2)								
F_4	measured sulphate flux on sampling track (t h^{-1})				18	30	13	6
	deduced sulphate flux on sampling track (t h^{-1})	38	14	63				
F_4/E_2	fraction of emitted SO_2 appearing as sulphate on track	0.16	0.09	0.17	0.12	0.09	0.20	0.08
R_4/E_2	R_4 = amount of sulphate removed by rain	0	0	0	0	0	0	0.26
R_4/P_4	P_4 = total production of sulphate from SO_2	0	0	0	0	0	0	0.75
P_4/E_2	$(P_4 = F_4 + R_4)$	0.16	0.09	0.17	0.12	0.09	0.20	0.34
Relative humidity								
	mean relative humidity (%) within plume	78	73	78	78	74	81	87
	range about mean	±6	±7	±4	±5	±3	±3	±3
P_4^*/E_2	P_4^* = production of sulphate implied by empirical curve (Fig. 7)	0.15	0.08	0.15	0.15	0.09	0.21	0.33
Total budget								
$100[F_2 + D_2 + P_4]/E_2 = 100$		100	100	100	100	100	100	100
$100[F_2 + D_2 + P_4^*]/E_2$		99	99	98	103	100	101	99

(a) all fluxes are measured in metric tons h^{-1}.

(b) the basic known emission and flux data are E_2, F_2 and (in 1973 only) F_4. All other items are deduced from budget requirements and (in 1971 and on 8 October 1973) by taking the assumed value of V_d.

(c) on all occasions except 22 October 1971 the SO_2 plume had not completely filled the mixing layer and the loss by dry deposition to the ground was deduced using a developing plume model (see text). On 22 October 1971 the shallow mixing layer was completely filled and a simple 'box-model' was used instead.

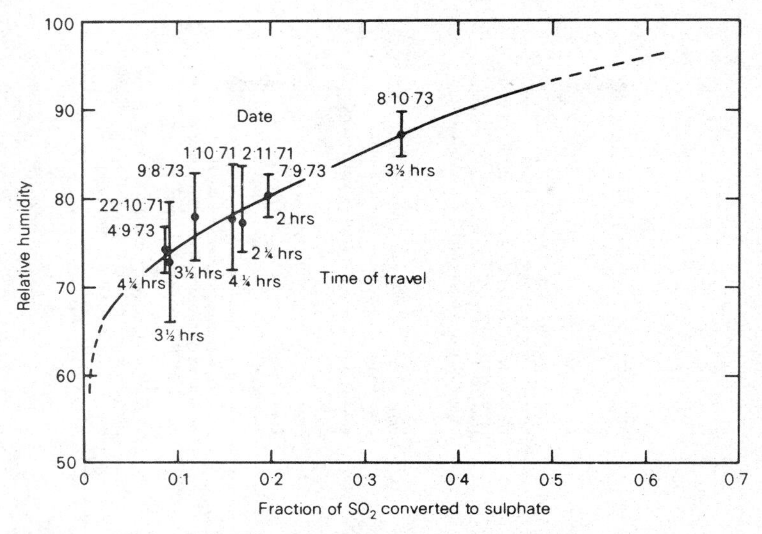

Fig. 3. The implied fraction of the emitted SO_2 converted to sulphate, as measured along the flight track off the East Coast of the U.K., as a function of relative humidity.

estimates of the deposition rate, and conversion rate near the source to be made.

All the analysed flights off the East Coast are consistent in estimating the velocity of deposition of SO_2 to the ground at about $0.8 \pm 0.2\ \mathrm{cm\ s^{-1}}$. However the measured sulphate seemed to be poorly correlated with the distance of travel, implying that most of the conversion takes place near the source where the concentrations of SO_2 and other industrially produced pollutants are high. These data are set out in Table 4 which is taken from Smith and Jeffrey (1975). The only other component involved in the conversion process likely to vary in an important way in these flights is the relative humidity. Figure 3 suggests that there is a significant relationship between the implied initial conversion rate and the relative humidity.

That this initial conversion dominated the total conversion at downwind distances only of the order 100–200 km from the major source region, led us to plan flights at greater range, and the width of the North Sea seemed ideal for this especially since it clearly contained no significant anthropogenic sources of SO_2.

(ii) *A summary of measurements made in* 1974–75

Table 5 sets out the main results for flights during these two years, using the same notation as in Table 4. Five of these were made over the eastern North Sea either just off S.W. Norway or off W. Denmark. Originally it was intended to sample the same air late in the afternoon off the U.K. East Coast and early the next morning on the other side of the North Sea but this posed too many difficulties and constraints on the meteorology, the aircraft logistics and on trajectory forecasting. Implications concerning conversion rates have therefore to be taken with caution. In fact without knowing the 'initial' conversions, which could have been implied by complementary measurements along the East Coast, it is virtually impossible to deduce this rate. Only in one flight on 22 August 1974 can an estimate be made. On this occasion there were two layers; a lower layer with an inversion extending up to about 260 m in which nearly all the SO_2 had been deposited or converted to sulphate, and an upper layer in which the loss of SO_2 was assumed to be due to conversion to sulphate only. Assuming that both layers had identical SO_2 and sulphate concentrations on leaving the U.K. East Coast 18 h earlier, time constants for the exponential decay of SO_2 by deposition (7.3 h) and conversion (99 h) could be deduced. The latter is equivalent to a conversion rate of $1\%\ \mathrm{h^{-1}}$.

Such a slow rate reflects back on the whole concept of double sampling the same air on both sides of the North Sea. The limits of precision of the aircraft system for measuring SO_2 and sulphate are such that a conversion rate as low as $1\%\ \mathrm{h^{-1}}$ would be detected with difficulty and would be subject to much uncertainity. There seems little hope therefore of establishing a relationship between this rate and other atmospheric variables such as relative humidity with our present sampling system.

It is interesting that $1\%\ \mathrm{h^{-1}}$ agrees very well with other estimates. Eliassen and Saltbones (1975) have analysed ground-level measurements in terms of a simple model in which the SO_2 concentrations are subject to a first order decay due to deposition and conversion. They found decay rates varying between 1.7 and $0.3\%\ \mathrm{h^{-1}}$. Recently Lusis and Wiebe (1976) in a study of the INCO nickel smelter-stack plume at Sudbury in Canada, which emits an average of 3500 tons SO_2 per day, deduced an average $1\%\ \mathrm{h^{-1}}$ conversion rate. They failed to find any dependence on temperature or relative humidity.

Returning to Table 5, one flight is clearly anomalous, namely that of 20 February 1975 when very

Table 5. Results of 7 flights in 1974–5. The notation is identical to that explained in Table 4.

Date	Estimated emission E_2 (t h^{-1})	Flux of SO_2 F_2 (t h^{-1})	% SO_2 remaining F_2/E_2	Flux of SO_4 F_4 (t h^{-1})	% SO_2 as SO_4 remaining F_4/E_2	% of E_2 deposited D_2/E_2	Mixing depth h (m)	Mean speed V (m s^{-1})	Distance from source x (Km)	Time from source t (h)	Remarks
21 May 1974	35.4	8.3	23.4	14.7	41.5	35.1	2000	9.3	650	23	Track off W. Denmark. Air accelerated over North Sea
22 August 1974 above inversion		conc. SO_2 = 16.3 μg m^{-3}		conc. SO_4 = 7.3 μg m^{-3}			h = 2700 height of inversion	7.2	700	18	Track off S. W. Norway. On assumption that concentrations of SO_2 were identical above and below
within surface		= 1.4 μg m^{-3}	—	= 5.3 μg m^{-3}	—	—	top ≈ 260 m	7.2	700	18	260 m off the U.K. East Coast. then the measured values imply an $SO_2 \rightarrow SO_4$ conversion of 1% h^{-1}
28 August 1974	323	131	40.5	36	11.1	48.4	1700	9.3	150	4.5	Track off U.K. East Coast. A rather high apparent loss D_2/E_2
20 December 1974	173	53	30.4	25	14.5	55	1600	23	460	5.5	Track over eastern part of North Sea.
5 February 1975	515	244	47.4	60	11.7	40.9	900	13	over Europe		Track west of Cornwall. Trajectory back-tracked to N. Italy where precipitation was falling.
20 February 1975	534	~3		~5			500	12	700	> 14	Track off S.W. Norway. Hand and computer back-tracks crossed major sources in U.K. However h was low and track strongly affected by surface inversion over U.K. the previous night.
						low values confirmed by Norwegian surface data					
26 February 1975	—	—	—	—	—	—	—	—	—	—	An example of the very occasional run which yielded inconsistent and poor quality analytical data. largely due to a very poor batch of filter paper.

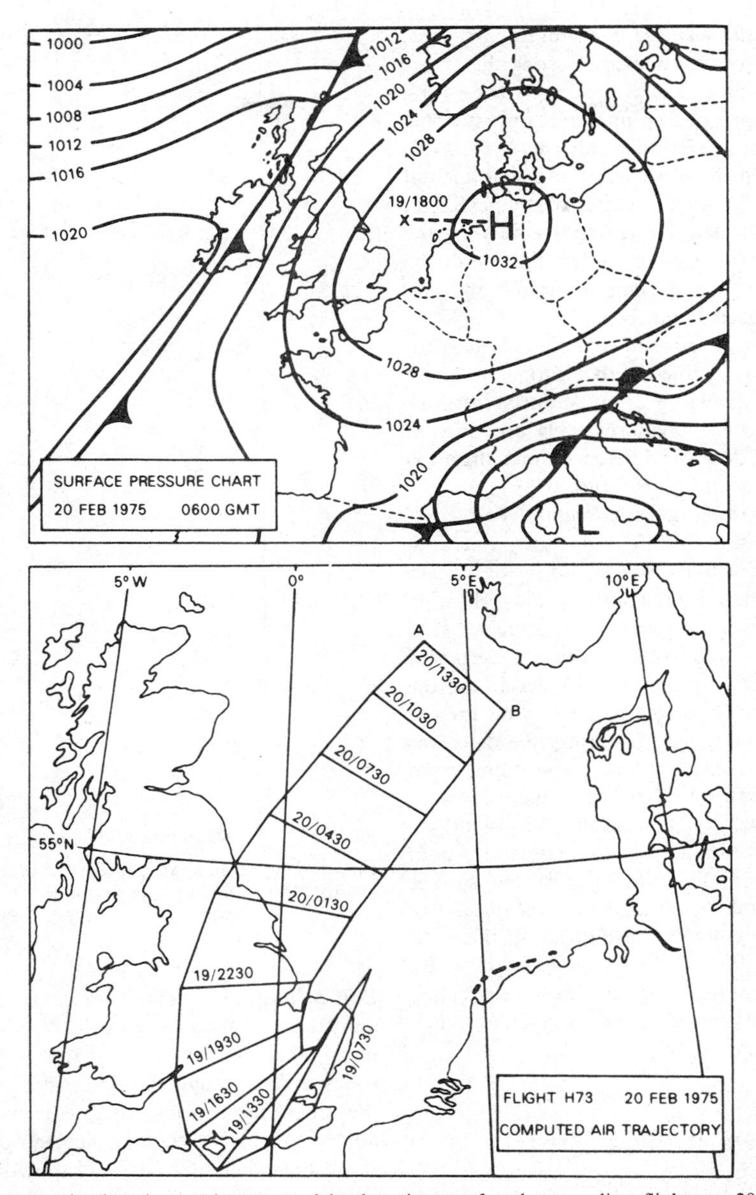

Fig. 4. The synoptic situation and computed back-trajectory for the sampling flight on 20 February 1975.

small concentrations were recorded. Nearly all the apparent emission had been lost, and whether this can be explained by a widespread and fairly deep nocturnal surface inversion over England during the passage of the air that was later sampled, is a matter of conjecture (see Fig. 4). At least the small measured concentrations appear to be genuine since they are consistent with very low measurements recorded at ground level in S.W. Norway.

Flights on 21 May, 28 August, 20 December and 5 February provided complete and useful data. Being in general at longer range from the major sources upwind than those in Table 4, it is consistent that the estimated percentage losses by deposition are that much greater. Subtracting the average estimated 'initial' loss of 15% from both sets of losses and taking average fluxes and distances of travel, the figures imply a velocity of deposition over land of 0.8 cm s^{-1} and very approximately 0.5 cm s^{-1} over the sea.

(iii) *Advecting winds*

One of the more important problems in attempting the determination of back trajectories is to relate the 'advecting wind', that is the mean wind affecting the pollution in the mixing layer at every stage along the track, to some known wind such as the surface geostrophic wind as deduced from surface pressure patterns. Any such relationship is clearly going to be subject to error, especially if the track runs into an area of strong isobaric curvature near the centre of a depression or if the airmass is strongly baroclinic. In maritime areas such as western Europe the errors are normally acceptable as is evidenced by the strong similarity of trajectories based on surface geostrophic winds and those using 850 mb winds (see the OECD LRTAP Report). On the other hand, over continental regions in winter, say over Canada or the U.S.S.R., it is very apparent that the strong stability of the lower layers of the atmosphere renders the use of the surface geostrophic wind inappropriate and an 850 mb wind approach is much better.

The great advantage of using the surface geostrophic wind where it is justified is that surface pressure charts are usually available at one-hourly intervals whereas winds at other pressure levels are only at six-hourly intervals, and this difference must affect the relative accuracies of the trajectories.

In the aircraft sampling experiments, real winds were measured by the aircraft and were related to the local surface geostrophic winds and to the surface 10-m winds when these were available. The following relatively simple procedure was used; a mean wind normal to the track was formed from the measured winds weighting the value at each level by the measured concentration to get a mean wind for the sulphur dioxide plume. The resulting speed V was compared with the local surface geostrophic wind speed G, and the ratio V/G was assumed constant along the back trajectory. The mean plume-wind direction θ_P on the other hand was compared with both the geostrophic wind direction θ_G and the surface 10-m wind direction θ_S and the ratio $(\theta_G - \theta_P)/(\theta_G - \theta_S)$ was also assumed constant along the back trajectory. For the 1971–73 flights the ratio V/G and the angular difference $(\theta_G - \theta_P)$ are both well related to the overall stability of the boundary layer, for when this is near neutral or slightly unstable:

$$V/G \approx 1 \text{ and } (\theta_G - \theta_P) \approx 10°$$

whereas when the layer is slightly stable:

$$V/G \approx 0.9 \text{ and } (\theta_G - \theta_P) \approx 20°.$$

(iv) *Rain-out*

Problems associated with precipitation are discussed later in the paper. In this section we simply wish to note that on the one sampling flight significantly affected by rain (Table 4: 8 October 1973), the implications are that the scattered showers were very effective in washing out the sulphur. Whilst it is virtually impossible to assess what fraction of the airmass had been affected by showers prior to sampling, the apparent loss of 26% of the original emission of SO_2 is not inconsistent with the idea that those areas had been very effectively cleaned.

3. EPISODES

It becomes apparent when studying data for the wet deposition of sulphate that over many parts of Europe a quite large and important part of the total annual amount is deposited on relatively few days of the year. These occasions are referred to as episodes and, although the exact mechanisms by which they affect the various ecosystems are uncertain, they appear to be significant, particularly during the spring snow melt when the most acid snow layers apparently determine the pH of the first melt-water.

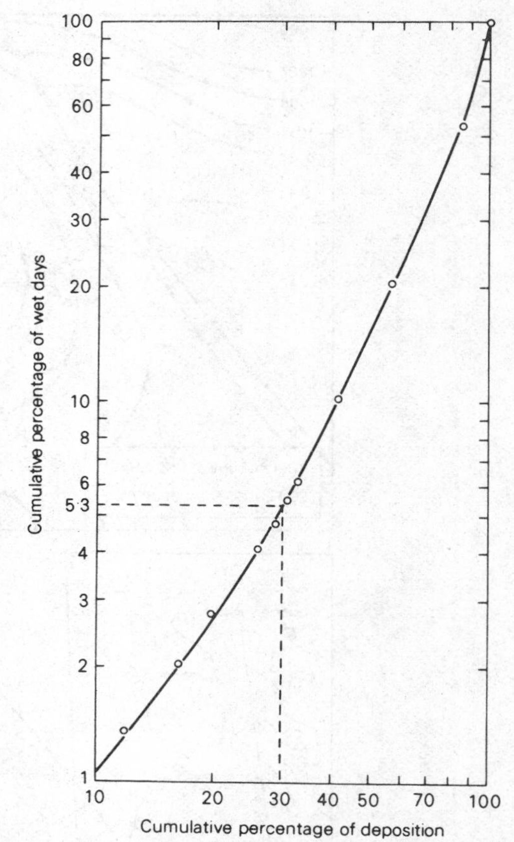

Fig. 5. The cumulative percentage of deposition plotted against the cumulative percentage of days, plotted on logarithmic scales, at Cottered.

Episode definitions are to some extent arbitrary. We have chosen to define 'episode-days' at a particular place as those days with the highest wet depositions which, when summed, make up 30% of the annual wet deposition total. Figure 5 illustrates the definition. At a particular site in England, 30% of the wet sulphate deposition occurs on 5.3% of wet days while 50% occurs on about 15% of wet days (taking data for 1974). Every place has, according to this definition, a certain number of episode-days regardless of the total deposition and of its distribution through the year. However, we can further define the term 'episodicity' as the ratio, expressed as a percentage, of the number of episode-days to the annual number of wet days and then describe an area as 'highly episodic' if the episodicity is less than 5%. As the number of wet days in an area in a year is normally between about 80 and 200, this means the

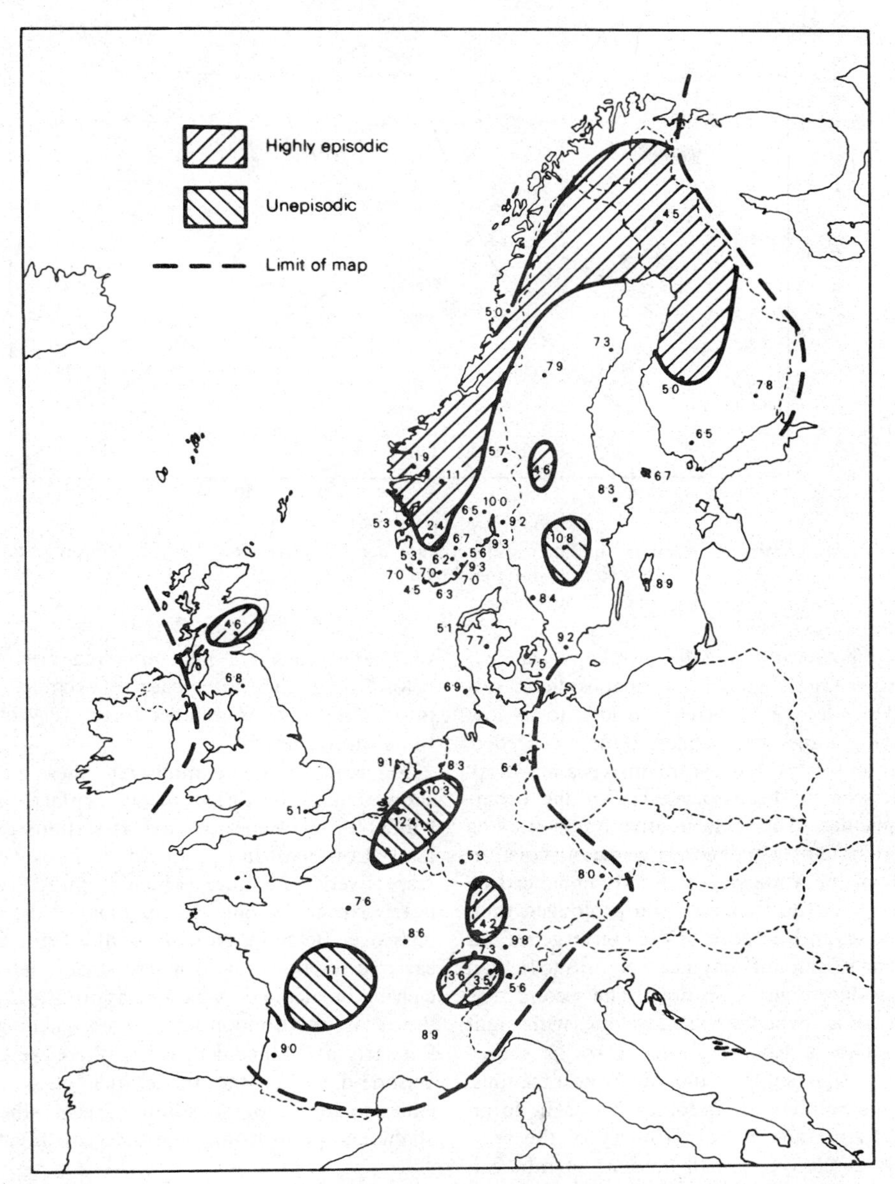

Fig. 6. The map of western Europe showing the distribution of episodicity for 1974.

number of episode-days in a highly episodic area is typically between 4 and 10 even though it may vary considerably from year to year. For the site used in the example above 5.3% of wet days represents nine episode-days in the year and, of course, it just fails to be classified as highly episodic. In a similar fashion, an area can be classified 'unepisodic' if its episodicity is greater than 10%. Figure 6 shows that the highly episodic areas in those parts of Europe connected with the OECD study, again based on 1974 data, are western Norway, and south-west Switzerland, whereas the main unepisodic area covers parts of Belgium and Holland.

One explanation of the episodic nature of wet deposition could lie in the fact that rainfall itself is episodic in character, occurring as it does between about 5–20% of the time. Episode-days and episodicity for rainfall can be defined in an exactly analogous way to that of wet deposition; for example rainfall episode-days are those days with the highest rainfall which contribute 30% of the annual rainfall, but analysis of the 1974 rainfall records reveals that rainfall episodicity has less variation geographically than wet deposition episodicity. Figure 7 shows the smoothed frequency distributions of both quantities, the mean values being 8.3% for rainfall and 6.9% for wet deposition while the standard deviations are 1.7 and 2.4 respectively. Also, while the correlation between rainfall and wet deposition (calculated at some of the LRTAP monitoring stations) is positive, it is not particularly high. Inevitably the highest values occur at stations with wet deposition episodici-

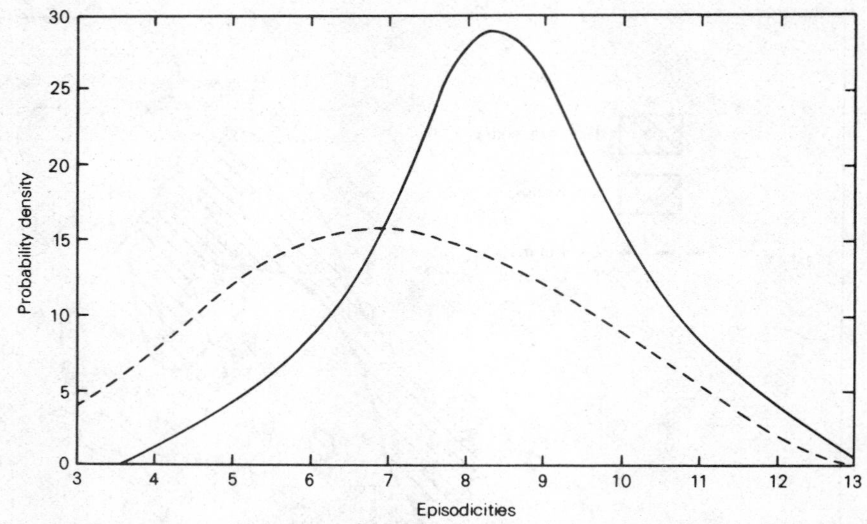

Fig. 7. Episodicities for rainfall (full line) and sulphate deposition (pecked line), based on data for the stations shown in Fig. 6.

ties close to the average rainfall episodicity. At sites which are either highly episodic or unepisodic in sulphate wet deposition, the correlation falls to as low as 0.2, implying some other explanations are needed.

It has been observed that two main types of synoptic situation seem to be associated with the occurrence of episodes. The first involves slow-moving active frontal systems where convergence of moist air at low-levels of the atmosphere, usually enhanced by mountain ranges, produce heavy and prolonged rainfall. In this case, and because of the convergence, air concentrations of sulphur may need be no more than average to produce high depositions. The second type of situation arises when an anticyclone with light winds persists for a few days over a major source area amassing high sulphur dioxide concentrations. If some of this polluted air becomes drawn in to an active frontal zone along the periphery of the anticyclone, only a relatively small amount of rain can lead to a high sulphate deposition.

Referring back to Fig. 6, some comments can be made about the large-scale geographical distribution of episodicity, whilst accepting that they cannot explain the local and yearly variations which occur. It would seem that the main unepisodic areas in Europe are close to major emission regions but situated to the west of them. Here, most of the heavier rainfall will fall with westerly winds when the air will be relatively unpolluted. Easterly winds however, while less common and bringing for the most part lighter rainfalls, will be associated with air having high sulphur concentrations. Depositions, being a product of rainfall amount and concentration, will therefore be quite uniform in magnitude and episodicity will be low. Highly episodic areas on the other hand would seem to be mainly remote from the chief industrial areas. As it passes over these parts, air will be usually relatively unpolluted, but occasionally it will have crossed a high emission area some time before. If, in this latter case, it becomes introduced into frontal rain or thunderstorms very high depositions can occur.

Figures 8 and 9 demonstrate these points. They show back-trajectories (using surface geostrophic winds) for 1974 episode-days at stations in S.E. England (Cottered) and southern Norway (Vatnedalen) respectively. The latter is highly episodic, the former nearly so. At the English site, most of the trajectories pass over Belgium, Holland or the Ruhr area to the east; some, however, originate from the south-west curving in such a way as they approach England that they cross the London area before reaching Cottered. Similarly at Vatnedalen, where there were only four episode-days, all the trajectories pass over major emission areas to the south or east albeit arriving at the end-point from quite different directions.

CASE STUDIES OF TWO EPISODES

On the 26th August 1974, Vatnedalen reported a 24-h^{-1} wet deposition of 151 mg m^{-2} SO_2, very much greater than any other deposition in that year. As we now show, this is a very special case where both types of synoptic situation mentioned above acted together. The back-trajectory calculated on this occasion (one of those shown in Fig. 9) is produced in detail in Fig. 10 together with the relevant part of an approximately 127 × 127 km sulphur emission grid (used in the OECD study) and surface geostrophic wind speeds at various points along the route deduced from surface pressure charts. Winds from any nearby radio-sonde ascents were used as supplementary information and as a check. The trajectory finishing at 0000Z on 27 August originates from the German Democratic Republic and western Poland some 48 h before where a light east-south-easterly

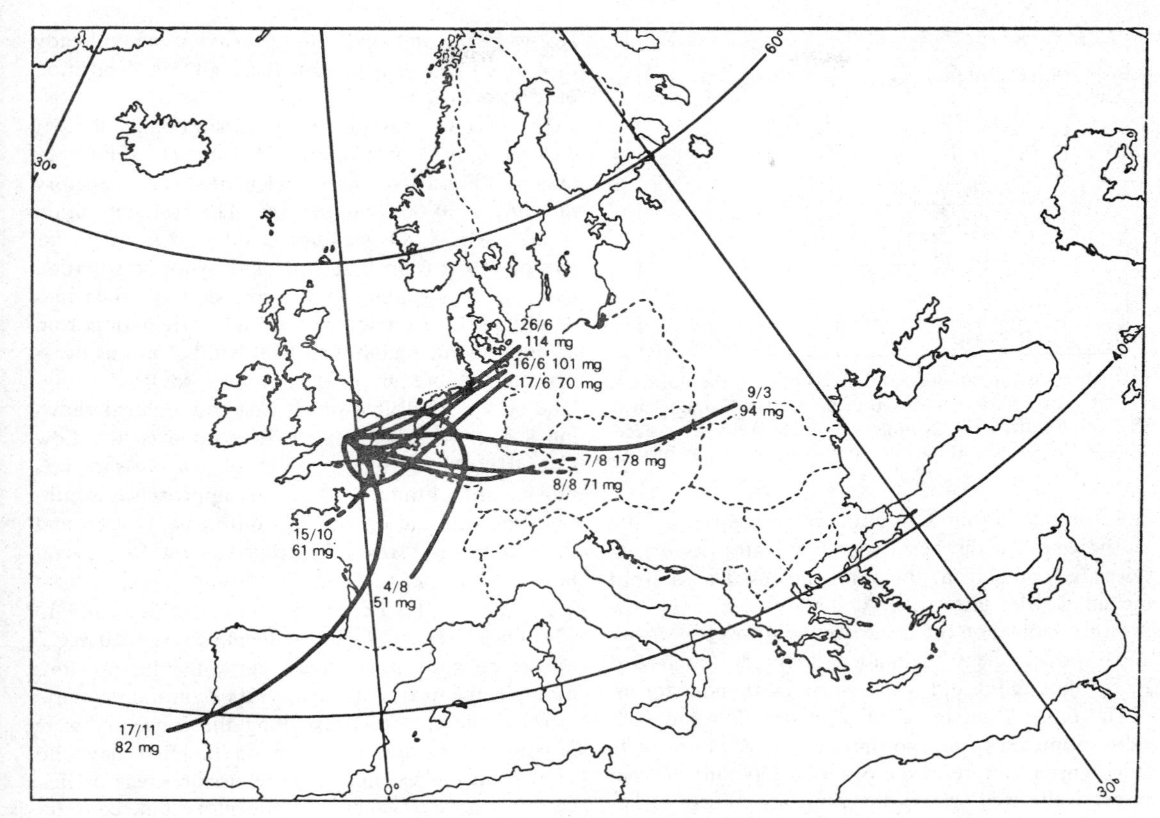

Fig. 8. Back-trajectories for the episode-days at Cottered, 1974.

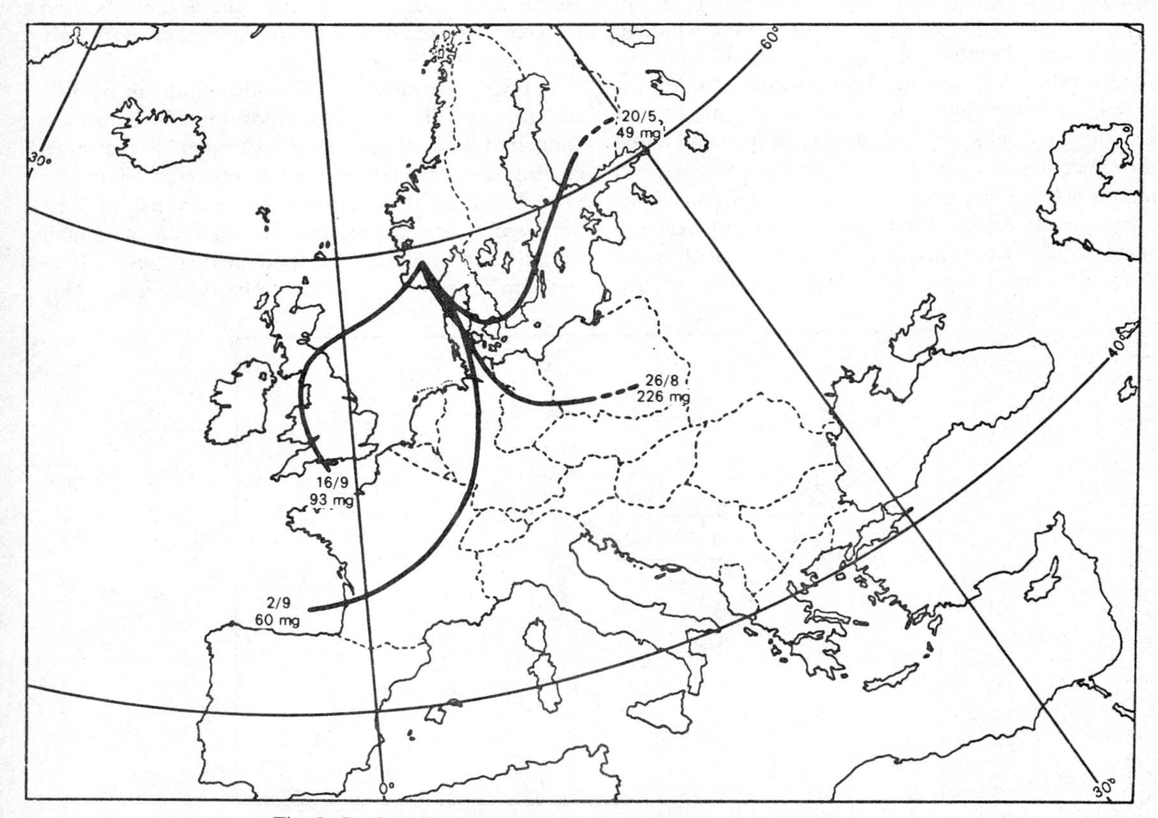

Fig. 9. Back-trajectories for the episode-days at Vatnedalen, 1974.

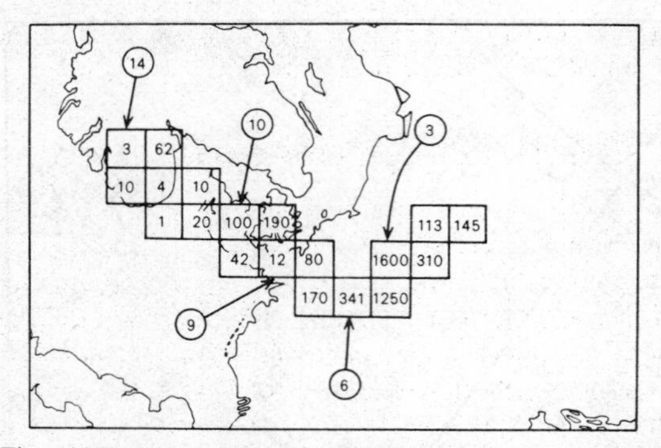

Fig. 10. Emission squares, and emissions (in thousands of t y^{-1}), crossed by the back-trajectory from Vatnedalen on 00Z, 27 August 1974. Numbers in circles refer to surface geostrophic winds, at various points on route, in m s^{-1}.

flow prevailed around an anticyclone centred a little to the east. As the heavily polluted air was drawn towards Scandinavia, the wind increased as a front moved slowly eastwards. A shallow centre of low pressure formed on the front over south-west Norway and produced heavy rainfall undoubtedly intensified by the mountains and with associated thunderstorms.

The daytime mixing depths on the 25th and 26th were estimated from appropriate 1200Z radio-sonde soundings and were of the order of 1500 and 1000 m respectively, well within the ranges outlined earlier (exact magnitudes are not critical). Wind speed was more important however in two ways. Firstly, it can be seen from Fig. 10 that the wind was light (3–4 m s^{-1}) over the high emission areas in the German Democratic Republic giving an initial high air concentration of sulphur dioxide: concentration is inversely proportional to wind speed at source. Secondly, the subsequent increasing of the wind along the trajectory to reach 14 m s^{-1} at the receptor area minimised dry deposition to the ground, since this loss depends on the total travel time and hence is small for high wind speeds. Quite high concentrations were therefore still present as the trajectory reached Norway and combined with the heavy rainfall already mentioned can then explain the high wet deposition on that occasion.

The second example for consideration is that of 7 August 1974 at Cottered. Figure 11, similar in layout to Fig. 10, shows the calculated back-trajectory finishing at 0000Z on the 8th. The wet deposition of 119 mg m^{-2} SO_2 was significantly greater than for any other occasion that year. The synoptic situation over the days leading to the episode has similarities with the first example, the trajectory originating from an anticyclonic region with light winds. On this occasion, an area of high pressure close to south-east England on the 6th drifted slowly east and declined allowing a south-easterly flow to develop over the Low Countries. Meanwhile a trough of low pressure was moving into France and as this approached south-east England late on the 7th, outbreaks of rain and thunderstorms started to occur lasting for several hours. Wind speeds along the trajectory ranged from 3 m s^{-1} over the Federal Republic of Germany to 13–14 m s^{-1} over Belgium reducing to near 10 m s^{-1} at the receptor area. Again as with the previous example, the mixing depths at 1200Z on the two days prior to the episode back along the trajectory were unremarkable—about 1500 m on the first day and 1100 m on the second. Major emission areas in Belgium were traversed by the trajectory but, contrary to the first example, the wind speed here was quite high. The lighter winds earlier were over only moderate emission areas. Clearly then the arguments used to account for the deposition in the first case are only partly effective in the second.

Applying the statistical trajectory model in its general form predicted a wet deposition of 85 mg m^{-2} compared with an observed 119 mg m^{-2}. However, correcting the model for the real observed winds, for probable diurnal variations in emissions, for the observed mixing depths and the duration of rainfall at Cottered, reduces the predicted deposition to 59 mg m^{-2}, roughly half the observed value. One

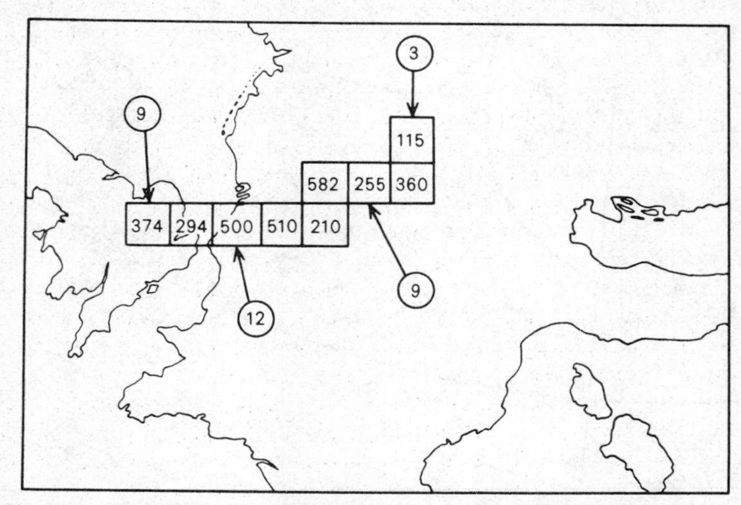

Fig. 11. As Fig. 10 with back-trajectory from Cottered on 00Z 8 August 1974.

probable explanation for this discrepancy is that the individual storm cells represent active centres of convergence into which polluted boundary layer air can be continually drawn and the sulphate removed by washout. This convergence, necessary for the maintenance of the storm, can produce up-draught regions 1–2 km across in the lower parts of the cloud with vertical velocities frequently exceeding 1 m s^{-1} and often 10 m s^{-1}, implying normal local horizontal convergence rates of 10^{-3}–10^{-2} s^{-1}.

On this occasion the cells were moving slowly in a NW direction. Without convergence the sulphate within and below the cloud would rapidly be removed, and by the time Cottered was reached very little deposition would be expected. In order to give twice the model predicted deposition it is necessary that during the passage of an individual cell an amount of air equalling twice the mixing-depth volume below the cloud would have to be drawn into the cell from surrounding areas. If the radius of the active cell is R (m) then the convergence C is directly related to the inflow of air:

$C\pi R^2 h$ = inflow around the boundary within the mixing layer depth h, s^{-1}.
= 2 × (volume of this sub cloud air)/(time of passage of cloud)
$= 2 \times \pi R^2 h \times (V/2R) = \pi R h V$

i.e. $C = V/R$.

On this occasion $V \approx 3 \text{ m s}^{-1}$ and taking R as being very roughly 10^3 m (1 km)
then: $C \approx 3 \times 10^{-3} \text{ s}^{-1}$, falling well within the expected range.

When considering episodes in this manner, it should be noted that errors in calculating trajectories assume a much greater importance than when considering, say, annual statistics of pollution budgeting. Only small errors in the direction of the wind near to the receptor area can and often do make a large difference in determining which regions have been crossed a day or two beforehand. In fact on the occasion of the first example trajectories calculated from the 850 mb wind, at a level strictly above the mixing layer (OECD, 1977) indicate the air having come from as far west as the English Channel area. It should also be recalled that vertical mixing in the polluted layer is assumed sufficiently good that the plume moves with a uniform vector velocity and wind shear effects are fairly minimal. The importance of this latter effect is considered below.

Although these two episode situations were chosen for further study because they had the highest depositions at their sites in 1974, and are therefore in that sense extreme, they nevertheless show how the various assumptions incorporated in the current statistical models can be inappropriate on certain occasions. In fact this conclusion, if valid to at least some degree for all episode situations, must open to question the application of these models even in a statistical sense since episodes by definition constitute a sizeable fraction of the total annual deposition.

4. IMPORTANT PROBLEMS FOR FURTHER STUDY

There are many areas where further study in the meteorological field could lead to an improved understanding of the processes involved in the long-range transport of pollution. Among the most complex and hard to handle is that of rainfall. This is known to be a highly effective scavenger of SO_2 in the air and during precipitation washout is easily the most dominant process for its removal. Chamberlain (1960) discusses rates of removal during precipitation which can be approximated (Scriven and Fisher, 1975) by

$$\lambda = 10^{-4} \times R^{1/2}$$

where λ is the fraction of pollutant washed out s^{-1}
R is rainfall in mm h^{-1}.

Over an area the determination of R, and hence the efficiency of the washout process, becomes difficult. Thunderstorms for example can lead to heavy rainfall, even of an episodic nature, at one point yet leave another point a short distance away completely dry. Studies of rainfall over quite small areas using a fine network of raingauges and assisted by radar have been carried out (e.g. Harrold *et al.*, 1972), while the mesoscale features of the precipitation patterns of a particular frontal system have been studied with a view to improved forecasts, again with radar playing an important role (Hill *et al.*, 1977). Further investigations will be needed over any region, particularly a mountainous one, if the deposition distribution over that region is to be assessed with any accuracy. Objective analyses of rainfall over large areas will be required if deposition is to be represented on a daily or otherwise operational basis. In particular, analysis of rainfall over sea areas has created a problem in earlier work (OECD, 1977) and improved data here are essential.

Mention has been made earlier of the effect of convergence in rain areas. Models have been and are being developed to study convection in cumulus and cumulonimbus clouds (e.g. Miller and Pearce, 1974; Hill, 1974) and, from these, the distance from which air may be drawn into an individual cloud can be assessed. Salter and Richards (1974) are among those who have discussed the effects of low-level convergence in producing heavy frontal rainfall (on a particular episodic occasion in their case). The amount of wet deposition at any point will be modified considerably by the coming together of air from different emission regions at such a front but fine scale three-dimensional models are required to describe this quantitatively.

The relationship between the SO_2 to SO_4 transformation rate and relative humidity at the source region has already been mentioned. More data are needed not only to confirm the precise nature of this dependence but also to determine the effects of fog and

Table 6. Showing division of 16898 observations of surface (V_0 m s^{-1}) and 900 mb (V_{900} m s^{-1}) winds over sea and over land by day and night into classes of lapse rate and V_{900}. P is the percentage of the total number of observations falling within each class, $\bar{\alpha}$ is the mean angle between V_0 and V_{900} taken as positive when V_0 is backed with respect to V_{900}

Lapse rate C/100 m \ V_{900} m s^{-1}		5–10 P	5–10 $\bar{\alpha}$	10–15 P	10–15 $\bar{\alpha}$	15–20 P	15–20 $\bar{\alpha}$	20–25 P	20–25 $\bar{\alpha}$	≥ 25 P	≥ 25 $\bar{\alpha}$
	sea	2	0	3	0	3	0	1.5	0	0.5	0
≥ 1	land (day)	3	−4	1	5	0.2	4	—	—	—	—
	land (night)	0.2	10	0.6	9	0.2	9	0.1	9	0.1	9
		15	5	16	6	12	6	6	5	2	5
0.9–1.0		22	5	13	10	5	11	1	13	0.2	14
		0.5	20	0.2	19	0.2	18	0.3	16	0.1	14
		7	10	8	12	5	14	3	12	2	12
0.55–0.9		13	20	12	19	7	18	3	17	1	16
		6	27	9	26	8	24	3	22	1	18
		2	15	3	17	3	18	1	18	1	19
0·2–0.55		5	30	5	29	3	26	1	24	0.3	21
		18	34	18	33	9	32	2	29	0.4	24
		1	20	1	20	1	20	0.4	21	0.1	23
−0.12–0.2		2	35	1	33	0.4	32	0.2	30	0.1	27
		9	40	5	39	2	35	1	32	—	—
		0.2	25	0.2	24	0.1	21	—	—	—	—
≤ −0.12		1	40	0.3	38	0.2	36	—	—	—	—
		5	45	3	42	1	38	0.1	35	—	—

light drizzle on both the deposition and transformation rates.

Most of the work which has been carried out so far on long-range transport neglects the effects of vertical wind shear through the mixing layer (although there have been exceptions, e.g. Veltischeva, 1976). Acknowledging the existence of such shears has implications for the deposition rates of SO_2 and SO_4 and the calculation of back-trajectories for which many objective techniques (e.g. Sykes and Hatton, 1976; OECD, 1977) use just one wind velocity to describe motion in the layer. Data concerning the relationship between the difference in wind direction of the surface and 900 mb, the lapse rate per 100 m deduced from the temperatures at the surface and 900 mb and the 900 mb wind speed are tabulated below (Table 6).

The figures are essentially taken from Findlater *et al.* (1966) although mean values have been taken for the various land and sea stations used in their study and some smoothing has been carried out.

For unstable conditions (i.e. lapse rate ≥ 1°C per 100 m) and for neutral conditions, applications of Rossby similarity theory and experimental results from the Minnesota experiment can produce likely values of eddy diffusivity which can be employed in the relationship obtained by Saffman [see e.g. Pasquill (1974)] for horizontal spread, σ_y, in terms of the distance travelled by the pollution from the source. Approximate results are listed below in Table 7.

Table 7. The magnitude of horizontal spread, σ_y, due to wind shear alone, related to distance from source, x. A mixed layer of 1 km is assumed. The average wind speed is 7.5 m s^{-1} and the stability is assumed constant (in fact, an implausible assumption beyond 100 km)

x (km)	σ_y (km) unstable and neutral	σ_y (km) Stable
1	0.1	0.2
5	0.3	0.8
10	0.5	1.7
50	1.0	8.3
100	1.5	16.7
1000	4.7	167

For very stable cases (lapse rate < −0.12°C 100 m^{-1}), a turning in wind direction of 40° over the layer from the surface to 900 mb can be taken as representative. The calculations assume that pollution uniformly fills the relatively deep layer even though the layer is now non-turbulent having been diffused during the previous day-time period when vertical turbulence was relatively vigorous. Since the 'particles' of pollution now remain at virtually a constant level, wind direction shear has a maximum effect. On the Gaussian assumption that σ_y is approx 1/6th of the total width, the figures in the final column of Table 7 are readily deduced.

Clearly, the effect of wind shear on horizontal spread is greater, using these figures and assumptions, in very stable conditions than otherwise. Nevertheless, even in this case, the width of the resulting plume after 100 to 1000 km is similar to that which would be achieved with the amount of synoptic swinging of the back-trajectories normally encountered over several hours. In unstable or neutral conditions, the effect of wind shear is found to be very small. In conditions of penetrative convection, however, pollutants could be taken to heights much greater than 900 mb provided precipitation is not falling and washing out the sulphate. Once any convective cloud evaporates, the remaining sulphur could be strongly affected by the different wind velocities found at these greater heights and would lead to a much greater horizontal spread than the values given above. The fate of pollutants once above the mixing layer has been largely ignored in long-range transport studies, but this may need to be remedied in the future.

Both the path and the spread of pollution released into the atmosphere at or near ground level are also modified drastically by the topography of the underlying surface especially in stable conditions. Numerical techniques such as those described by Mason and Sykes (submitted) are currently being developed to study the effect of mountain ranges on boundary layer flow and they could be also valuable in the long-range transport field.

To sum up then, studies of the general precipitation problem, of the effects of wind shear, combined with the diurnal cycle of stability, on lateral dispersion and of the general effects of topography are the most pressing problems which require further study. Meteorological aspects also enter more indirectly into other problems, such as the calculation of depositon and transformation rates both of sulphur compounds and other chemicals, but these are outside the scope of this paper.

REFERENCES

Carson D. J. (1973) The development of a dry inversion-capped convectively unstable boundary layer. *Q. J. R. met. Soc.* **99**, 450–467.

Chamberlain A. C. (1960) Aspects of the deposition of radioactive and other gases and particles. *J. Air. Pollut.* **3**, 63–88.

Deardorff J. (1976) Clear and cloud-capped mixed layers, their numerical simulation, structure, growth and parameterisation. Proc. of Seminars on the treatment of the boundary layer in numerical weather prediction. European Centre for Medium Range Weather Prediction, 234–284.

Eliassen A. and Saltbones J. (1975) Decay and transformation rates of sulphur dioxide as estimated from emission data, trajectories, and measured air concentrations. *Atmospheric Environment* **9**, 425–429.

Findlater J., Harrower, T. N. S., Hawkins G. A. and Wright H. L. (1966) Surface and 900 mb wind relationships. Meteorological Office Scientific Paper No. 23.

Hanna S. R. and Gifford F. (1973) Modelling urban air pollution. *Atmospheric Environment* **7**, 131–136.

Hardy R. N. (1973) Statistics of inversions at Cardington in Bedfordshire. Investigations Division Memorandum No. 108. UK Meteorological Office.

Harrold T. W., Bussel R. and Grinsted W. A. (1972) The Dee Weather Radar Project. Symposium on the distribution of precipitation in mountainous areas. Geilo, Norway. WMO No. 326, 47–61.

Hill F. F., Whyte K. W. and Browning K. A. (1977) The contribution of a weather radar network to forecasting frontal precipitation; a case study. *Met. Mag.* **106**, 70–89.

Hill G. E. (1974) Factors controlling the size and spacing of cumulus clouds as revealed by numerical experiments. *J. atmos. Sci.* **31**, 646–673.

Johnson D. A. and Atkins D. H. F. (1975) An airborne system for the sampling and analysis of sulphur dioxide and atmospheric aerosols. *Atmospheric Environment* **9**, 825–829.

Lusis M. A. and Wiebe H. A. (1976) The rate of oxidation of sulphur dioxide in the plume of a nickel smelter stack. *Atmospheric Environment* **10**, 793–798.

Mason P. J. and Sykes R. I. (1977) On the interaction of topography and Ekman boundary layer pumping in a stratified atmosphere. (Submitted for publication to *Q. J. R. met. Soc.*)

Miller M. J. and Pearce R. P. (1974) A three-dimensional primitive equation model of cumulonimbus convection. *Q. J. R. met. Soc.* **100**, 133–154.

OECD (1977) The deposition and distribution in European countries of pollutants from distant and local sources. Final Report of the OECD 'Long Range Transport of Air Pollutants' Project.

Pasquill F. (1974) *Atmospheric Diffusion.* Ellis Horwood, Chichester.

Salter P. M. and Richards C. J. (1974) A memorable rainfall event over southern England. *Met. Mag.* **103**, 288–300.

Scriven R. A. and Fisher B. E. A. (1975) The long-range transport of airborne material and its removal by deposition and washout. *Atmospheric Environment* **9**, 49–58.

Smith F. B. and Carson D. J. (1977) Some thoughts on the specification of the boundary layer relevant to numerical modelling. (Accepted for publication in *Boundary-Layer Met.*)

Smith F. B. and Jeffrey G. H. (1975) Airborne transport of sulphur dioxide from the United Kingdom. *Atmospheric Environment* **9**, 643–659.

Sykes R. I. and Hatton L. (1976) Computation of horizontal trajectories based on the surface geostrophic wind. *Atmospheric Environment* **10**, 925–934.

Veltischeva N. S. (1976) Numerical Method for quantitative estimates of longrange transport of sulphur dioxide. Unpublished Report to ECE Working Group, Lillestrøm, Norway.

Atmospheric Environment Vol. 12. pp. 479–487. Pergamon Press 1978. Printed in Great Britain.

THE OECD STUDY OF LONG RANGE TRANSPORT OF AIR POLLUTANTS: LONG RANGE TRANSPORT MODELLING

ANTON ELIASSEN

Norwegian Institute for Air Research, P.O. Box 130, 2001 Lillestrøm, Norway

(*First received* 14 *June and in final form* 26 *August* 1977)

Abstract—A Lagrangian-type model has been used to quantify long range transport of sulphur over Europe on an annual basis (1974). Basic model assumptions are discussed. Calculated concentrations of sulphur dioxide and particulate sulphate are compared with observations. The agreement is variable for daily values, but better for annual averages. A sulphur dry deposition pattern for Europe for 1974 is estimated from the calculated annual mean concentration field. A corresponding wet deposition pattern is estimated by an indirect method, in which the annual mean concentration of sulphate in precipitation is approximated by calculated air concentrations of particulate sulphate. The estimated total amounts of dry and wet deposition within the area considered (9 M km^2) are about 11 Mt and 6 Mt S respectively, as compared to a total emission of 20 Mt.

1. INTRODUCTION

The recent years have seen an increasing interest concerning the transport of air pollutants over long distances and the modelling of this phenomenon. A variety of approaches to the modelling have been suggested, and include the statistical models formulated by Bolin and Persson (1975), Fisher (1975). Results obtained with a simple Eulerian model were reported by Nordø *et al.* (1974) and by Nordø (1974). The improved Eulerian advection scheme suggested by Egan and Mahoney (1972) was tested by Pedersen and Prahm (1974). Recently, Christensen and Prahm (1976) have presented an Eulerian model where the time integration is performed using spectral methods. A Lagrangian model has been formulated by Nordlund (1975). Another Lagrangian-type model and some results have been presented by Eliassen and Saltbones (1975) and by Nordø (1976).

The latter model has been used to quantify the long-range transport of sulphur over Europe, in connection with the OECD study Long Range Transport of Air Pollutants (LRTAP). In the following, the model is presented, and results from a model run covering the year 1974 are given.

2. BASIC ASSUMPTIONS AND EQUATIONS

The model describes the horizontal dispersion of sulphur within an atmospheric boundary layer which is assumed to have a constant thickness of 1000 m. This layer is assumed to be completely mixed, i.e. sulphur concentrations and wind are assumed to be independent of the vertical co-ordinate. The model is of the Lagrangian type, in which the horizontal dispersion is described by a large number of trajectories. In the calculations, trajectories calculated from the observed wind in the 850 mb surface were used.

To calculate these trajectories, the observed wind data were analysed numerically to obtain gridpoint wind values in a 127 km grid. Wind observation hours are 00, 06, 12 and 18 GMT, the number of stations reporting varying from about 50 to 110 within the area considered. In regions where wind observations are scarce, such as the Atlantic ocean, the observations were completed with values of a quasi-geostrophic balanced wind produced on a 300 km grid by the Norwegian Meteorological Institute on a routine basis. Wind fields between observation hours were obtained by separate linear time interpolation of the two wind components. Wind values in between gridpoints were interpolated from the four nearest gridpoints. The trajectories were then calculated by a method due to Petterssen (1956).

The calculations are based on a sulphur emission field for Europe, prepared for the LRTAP-programme (Rystad *et al.*, 1974; Semb, 1978). The animal emissions are given in the 127 km grid in which the wind fields were analysed. The emissions have been seasonally adjusted by a sinusoidal function with an amplitude of 30%, and the maximum occurring in January. The same adjustment has been applied to all grid elements. No diurnal or week-end variation factor has been applied.

The model predicts the concentrations q and s of sulphur dioxide and particulate sulphate respectively. The mass-balance equations for sulphur dioxide and particulate sulphate for the assumed 1000 m thick mixing layer are:

$$\frac{Dq}{dt} = E_q + F_q \quad (1)$$

$$\frac{Ds}{dt} = E_s + F_s \quad (2)$$

where E_q, E_s, F_q, F_s are source and sink terms for sulphur dioxide and particulate sulphate. The operator D/dt denotes the total time derivative along the trajectories. Since the sulphur initially is assumed to be completely dispersed within the large grid volumes, no further diffusion mechanism is included.

In reality, the dispersion of sulphur within the mixing layer is not instantaneous, as assumed in the model, but proceeds gradually after emission. During the first phase of dispersion, therefore, ground level concentrations are underestimated and so dry deposition will be too small close to the sources. To account for this, a fraction α of the emitted sulphur is assumed to be deposited locally, i.e. in the same grid element as it is emitted. The rest of the sulphur takes part in long range transport, and is subject to a gradual deposition en route. Of the emitted amount, one part β of the total is assumed to be emitted directly as sulphate, and the remainder as sulphur dioxide. The subsequent removal of sulphur dioxide and of particulate sulphate and the transformation of sulphur dioxide to sulphate are assumed to be proportional to the concentrations.

If Q is the seasonally adjusted emission of sulphur per unit area and time (given as sulphur dioxide), then from the above assumptions we have:

$$\begin{aligned} E_q &= (1 - \alpha - \beta)\frac{Q}{h} \\ F_q &= -kq \\ E_s &= \frac{3}{2}\left(\beta\frac{Q}{h} + k_t q\right) \\ F_s &= -\kappa s \end{aligned} \tag{3}$$

where h is the assumed mixing height, k and κ the decay rates for sulphur dioxide and particulate sulphate, and k_t the rate of transformation from sulphur dioxide to sulphate. The factor 3/2 is the ratio of the mol. wts of sulphate and sulphur dioxide. The parameters α, β, h, k_t and κ are assumed constant, whereas k is given a higher value during precipitation than under dry conditions, according to the prepared precipitation fields (see Section 5). The parameter values used are listed in Table 1.

The mass-balance equations (1) and (2), combined with the expressions (3) for the source and sink terms, may be integrated numerically along the calculated trajectories. In addition, for a section of a trajectory where precipitation either does not occur or occurs all the time, k is constant, and analytical solutions for the concentrations can be obtained instead (Eliassen and Saltbones, 1975). Since the equations are linear in the concentrations q and s, the contributions from the emissions at individual positions along a trajectory may be considered separately, and added to give complete concentrations. For a trajectory section consisting of N positions or timesteps Δt one obtains:

$$q(N\Delta t) = q(0)\,e^{-kN\Delta t} + \sum_{i=0}^{N-1}(1 - \alpha - \beta) \times \frac{Q_i\Delta t}{h}e^{-ki\Delta t} \tag{4}$$

$$s(N\Delta t) = s(0)\,e^{-\kappa N\Delta t} + \frac{3k_t}{2(k-\kappa)}\sum_{i=0}^{N-1}\frac{Q_i\Delta t}{h} \times (1 - \alpha - \beta)(e^{-\kappa i\Delta t} - e^{-ki\Delta t}) + \frac{3}{2}\sum_{i=1}^{N-1}\beta\frac{Q_i\Delta t}{h}e^{-\kappa i\Delta t}. \tag{5}$$

Here $q(N\Delta t)$ and $s(N\Delta t)$ are the concentrations at the end of the trajectory section, and $q(0)$ and $s(0)$ are the concentrations at the start of the section. Integration along a trajectory consisting of many sections, each with a constant value of k, may be performed by making the initial concentration of one section equal to the final concentration of the preceding section.

Two slightly different versions of the model have been employed, the difference arising mainly from a different organisation of the trajectories in space and time. In the first type, referred to as the 'Lagrangian' model, trajectories are initiated at the centre of each emission grid element, and followed for 12 h. The concentrations of sulphur dioxide and particulate sulphate associated with each trajectory are computed by simple forward time integration of equations (1), (2) and (3), with a timestep of 1 h. After 12 h, new

Table 1. Parameter values applied in the dispersion model calculations

Parameters for concentration estimates		Value
α	part of sulphur emission deposited locally	0.15
β	part of sulphur emission transformed directly to sulphate	0.05
k	decay rate SO_2 — rain	$4\cdot10^{-5}\,s^{-1}$
	— dry	$1\cdot10^{-5}\,s^{-1}$
k_t	transformation rate $SO_2 \rightarrow SO_4$	$3.5\cdot10^{-6}\,s^{-1}$
κ	overall loss rate SO_4	$4\cdot10^{-6}\,s^{-1}$
h	mixing height	1000 m
Additional parameters for dry deposition estimates		
v_d	deposition velocity for sulphur dioxide	$0.8\,cm\,s^{-1}$
	deposition velocity for particulate sulphate	$0.2\,cm\,s^{-1}$

trajectories are initiated from each emission grid element, with initial concentrations obtained by spatial interpolation from the concentrations associated with the previous trajectories. In this model, the air is followed for its complete residence time within the area considered.

The second type, referred to as the 'Trajectory' model, is based on trajectories followed for 48 h arriving at the centre of all emission grid elements every 6 h. The equations (4) and (5) are applied along the trajectories with a timestep of 2 h, to obtain concentration estimates at all gridpoints every 6 h. In this model, emissions encountered more than 48 h ago are neglected.

3. PARAMETER VALUES

3.1. *Calculation of concentrations*

The parameter values used for calculating concentration fields of sulphur dioxide and particulate sulphate are given in the first part of Table 1.

The assumed values of the decay rate k for sulphur dioxide compare with those estimated by Eliassen and Saltbones (1975) for long range transport of sulphur dioxide. The value of $4 \cdot 10^{-5}\,s^{-1}$ assumed to apply during precipitation is about half the value estimated by Högström (1973). As it is, the time resolution of the precipitation data available (6 and 12 h) results in precipitation duration being overestimated. To compensate for this, the smaller loss rate for sulphur dioxide is applied during times of precipitation.

Atmospheric sulphate particles are mostly submicron. Particles of this size have considerably smaller deposition velocities than the corresponding values found for sulphur dioxide, see, for example, Chamberlain (1966). Therefore, a small overall removal rate κ of $4 \cdot 10^{-6}\,s^{-1}$ has been assumed for particulate sulphate. The same value for κ has been assumed to apply in both wet and dry periods. This at first may seem surprising, but the increased removal rate of airborne sulphate during precipitation is assumed balanced by an increased production rate of sulphate from sulphur dioxide. In the model this situation is described by its net effect: unaltered production and removal rates of sulphate during precipitation, but an increased decay rate k for sulphur dioxide. The disadvantage of this apporoach is that in the event of an air mass travelling for some time, so that most of the sulphur dioxide has been either dry deposited or converted to sulphate, a precipitation event would not remove an appropriate amount of sulphur from the air mass. However, this drawback was considered acceptable in view of the problem which would be encountered due to the poor resolution of the precipitation data, if wet deposition of sulphate were considered separately.

The atmospheric chemistry of sulphur involves a number of different chemical reactions. A detailed, quantitative description of the transformation $SO_2 \rightarrow SO_4$ is hardly feasible at present. Average transformation rates for long range transport (transport time typically one day) estimated by Eliassen and Saltbones (1975) ranged from $7.8 \cdot 10^{-7}\,s^{-1}$ to $4.8 \cdot 10^{-6}\,s^{-1}$, with a mean value of $2 \cdot 10^{-6}\,s^{-1}$. Prahm *et al.* (1976) have estimated the transformation rate for a case with transport over a distance of 1000 km from the U.K. to the Faroe Islands to about $4 \cdot 10^{-6}\,s^{-1}$. Smith and Jeffrey (1976) give results from sampling of sulphur dioxide and sulphate from aircraft off the east coast of England. Their results indicate that about 10% of the sulphur emissions are present as sulphate after a transport time of some 3 h, corresponding to a transformation rate of about $2 \cdot 10^{-5}\,s^{-1}$, which is considerably faster than the values found for long range transport. The transformation rate thus appears to be faster close to the source than at greater distances. In the model this is accounted for by a relatively low constant transformation rate ($k_t = 3.5 \cdot 10^{-6}\,s^{-1}$) and an initial instantaneous conversion to sulphate (β) equal to 0.05 of the emitted sulphur.

The fraction α of the emitted sulphur assumed to be deposited in the same grid element as it is emitted (see Section 2), has been taken as 0.15 for all grid elements. In reality α would be different for different elements, depending on the structure of the emission field within the element. The value of 0.15 represents a first approximation, giving calculated sulphur budgets for individual countries comparable to independent estimates. In some cases, especially in highly industrialized areas, it might represent an underestimate. On the other hand, it might represent an overestimate in areas dominated by emissions from very tall stacks.

3.2. *Calculation of dry deposition*

Annual dry deposition was estimated by applying deposition velocities to the previously calculated annual mean concentration fields of sulphur dioxide and particulate sulphate (see Section 5). Deposition velocities of $0.8\,cm\,s^{-1}$ and $0.2\,cm\,s^{-1}$ were employed for sulphur dioxide and sulphate respectively (Table 1). The value of $0.8\,cm\,s^{-1}$ is identical to the mean value estimated by Owers and Powell (1974) for the British Isles at 0.2 m above the surface. Other investigations (Garland *et al.*, 1974; Shepherd, 1974; Whelpdale and Shaw, 1974; Dovland and Eliassen, 1976) show that the deposition velocity for sulphur dioxide usually is of the order $1\,cm\,s^{-1}$, but can be much smaller under conditions of weak turbulence.

From equations (1)–(3) it is seen, however, that the decay rate k in principle includes all removal processes for sulphur dioxide, and in particular the transformation to sulphate. In the absence of precipitation, the assumed values of k and k_t given in Table 1 therefore imply a removal rate of $6.5 \cdot 10^{-6}\,s^{-1}$ which would be due to dry deposition alone. Under the assumed conditions of complete and instantaneous mixing up to a mixing height (h) of 1000 m this would be equivalent to a deposition velocity $v_d = k\,h$ of

0.65 cm s^{-1} instead of the employed value of 0.8 cm s^{-1}. However, with the removal and transformation rates which have been assumed, the concentration of particulate sulphate is overestimated by the Lagrangian model (see Section 4). Better agreement between observed and estimated sulphate concentrations, would be obtained, for example, by using a smaller transformation rate, thereby reducing the calculated sulphate concentrations. The calculated concentrations of sulphur dioxide can be kept unchanged by keeping the decay rate k constant. A smaller part of k would then be due to transformation, and a larger part to dry deposition. Thus, when the dry deposition patterns of sulphur were estimated from the calculated annual mean concentration fields, the larger value of 0.8 cm s^{-1} was used for the depositions velocity of sulphur dioxide, consistent with a transformation rate k_t of $2 \cdot 10^{-6}$ s^{-1} (see Section 5). The alternative use of a parameter set with $k_t = 2 \cdot 10^{-6}$ s^{-1}, but otherwise identical to the set given in Table 1, would produce exactly the same calculated sulphur dioxide concentrations, but lower concentrations of particulate sulphate, in better agreement with the observed data.

The contribution of particulate sulphate to the dry deposition pattern for sulphur was estimated from the calculated annual mean concentration field using a deposition velocity of 0.2 cm s^{-1}. To be consistent with the deposition velocity of 0.8 cm s^{-1} applied to the sulphur dioxide concentrations, this contribution should have been calculated from particulate sulphate concentrations obtained with a transformation rate k_t of $2 \cdot 10^{-6}$ s^{-1} rather than the $3.5 \cdot 10^{-6}$ s^{-1} employed in the calculations. However, the deposition of particulate sulphate amounts to only about 10% of the total estimated dry deposition, and the overestimation of particulate sulphate concentrations arising from the use of the higher transformation rate is therefore unimportant for the total dry deposition estimates.

4. ESTIMATED CONCENTRATIONS COMPARED WITH OBSERVATIONS

76 ground LRTAP sampling stations have been in operation for various periods between August 1972 and March 1975. They were located, as far as possible, to be representative of rural conditions within a geographical region, and not unduly affected by nearby emissions. The annual mean sulphur dioxide and sulphate aerosol concentrations are accurate to about ±15%, but sulphur dioxide measurements are less accurate at sites with low concentrations (<5 μg m^{-3}).

Unfortunately, the chemical data for the whole 33 months of ground level sampling could not be utilized because detailed precipitation data (see Section 5) were not readily available for more than 15 months. Using the parameter values given in Table 1, model calculations with both the Lagrangian and the Trajectory model have been performed for 1974. For the Lagrangian model, the estimated daily mean concentrations are averages of hourly concentrations, whereas in the Trajectory model, daily mean concentrations are taken as averages of instantaneous concentrations obtained every 6 h.

The correlation coefficients between observed and estimated daily mean concentrations, while generally poor (as might be expected with a simple large scale advection model), were higher for particulate sulphate than for sulphur dioxide. For the Lagrangian model, the coefficients varied between 0.4 and 0.8 for sulphate, and between insignificant and 0.7 for sulphur dioxide (sample size 180). Similar figures were

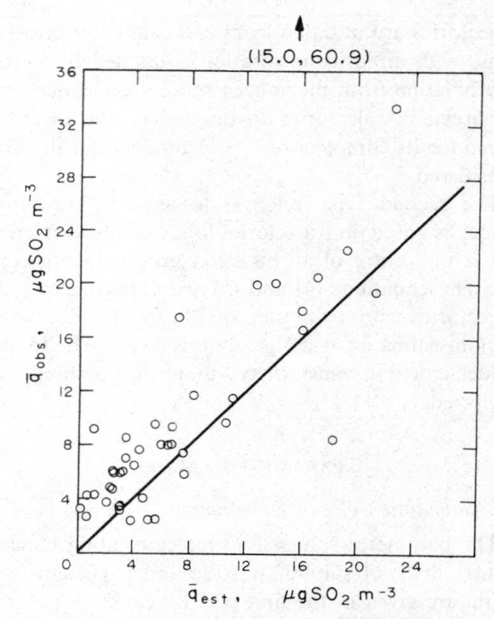

Fig. 1. Observed versus estimated annual 1974 mean concentrations of sulphur dioxide (Trajectory model). The straight line represents the ideal relationship (OECD, in press).

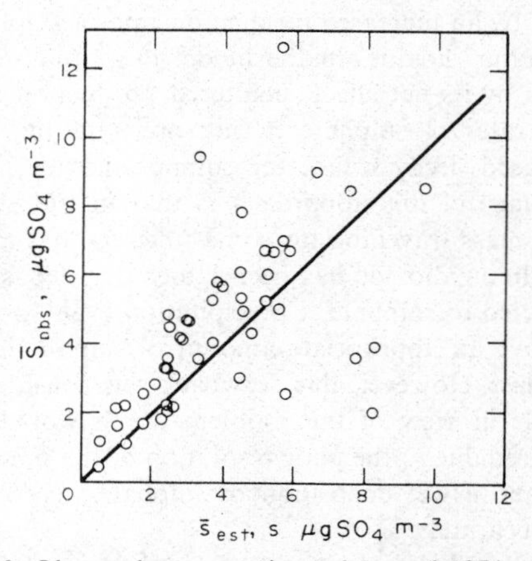

Fig. 2. Observed versus estimated annual 1974 mean concentrations of particulate sulphate (Trajectory model). The straight line represents the ideal relationship (OECD, in press).

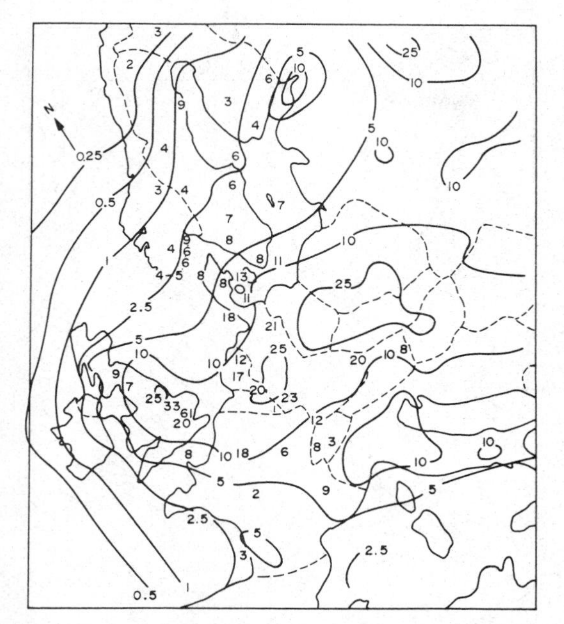

Fig. 3. Estimated and observed mean concentrations of sulphur dioxide for 1974 (Trajectory model). Unit: $SO_2\ m^{-3}$ (OECD, in press).

obtained with the Trajectory model. The better correlations obtained for particulate sulphate could be related to the sulphate having had a longer residence time in the atmosphere than sulphur dioxide. The sulphate concentration field would therefore not contain as much small-scale variation as the concentration field for sulphur dioxide.

Figures 1 and 2 show that the observed and calculated (Trajectory model) annual mean concentrations of sulphur dioxide and particulate sulphate are related, although, in both cases, there is a bias to low calculated values. The estimated sulphur dioxide concentrations are 2–3 $\mu g\ SO_2\ m^{-3}$ lower and the aerosol sulphate concentrations 1–1.5 $\mu g\ SO_4\ m^{-3}$ lower than the observed values. At 6 stations, the observed sulphur dioxide concentrations are considerably higher than the estimated ones. Most of these high observed values are presumably due to local sulphur emissions.

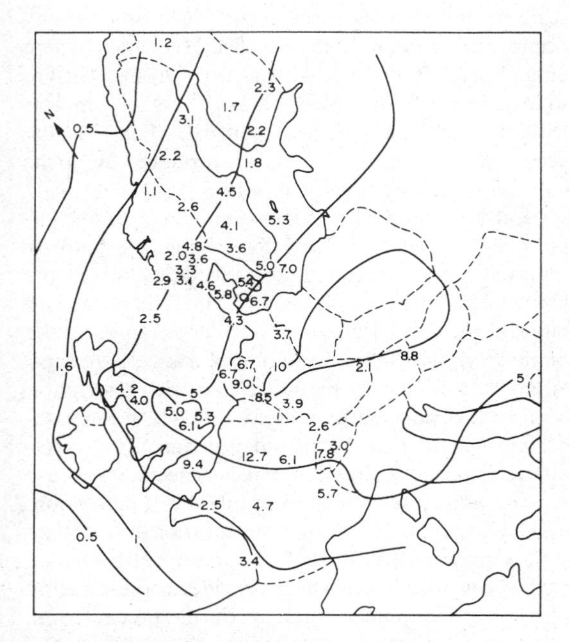

Fig. 4. Estimated and observed mean concentrations of particulate sulphate for 1974 (Trajectory model). Unit: $\mu g\ SO_4\ m^{-3}$ (OECD, in press).

The estimated annual mean concentration fields of sulphur dioxide and particulate sulphate for 1974 are shown in Fig. 3 and 4, respectively. The concentration patterns both reflect the structure of the emission field. However, since most of the particulate sulphate is produced gradually from the emitted sulphur dioxide, the concentration field for the sulphate is smoother.

The sulphate dioxide estimates of the Trajectory model are typically 10% lower than the estimates of the Lagrangian dispersion model. For particulate sulphate concentrations, the Trajectory model estimates are about 60% of the estimates of the Lagrangian dispersion model. Except for these differences in amplitude the estimates of two model versions are very similar. Since the parameter values (Table 1) are identical for the two models, this shows the effect of using only 48-h trajectories in the Trajectory model. Naturally, the discrepancy is smallest for the sulphur dioxide concentrations, since the assumed residence time for sulphur dioxide is considerably shorter than 48 h, while the residence for sulphate is longer. However, the dry deposition patterns obtained with the two models are very similar, because most of the dry deposition is accounted for by sulphur dioxide.

5. ESTIMATED DRY AND WET ANNUAL SULPHUR DEPOSITION PATTERNS

5.1. *Dry deposition*

The dry deposition pattern was estimated by applying a deposition velocity of 0.8 $cm\ s^{-1}$ to the sulphur dioxide concentration field of Fig. 3, and a value of 0.2 $cm\ s^{-1}$ to the particulate sulphate concentration field of Fig. 4. The additional deposition due to local emissions is also included in the dry deposition pattern. The resulting total dry deposition pattern is shown in Fig. 5.

Since the equations (4) and (5) are linear with respect to the emission field Q, and the dry deposition also is estimated by a linear process, the deposition field can be split up in contributions arising from the emissions in different countries or emission areas. As an example, Fig. 6 shows the dry deposition pattern due to emissions in Norway.

5.2. *Wet deposition*

Using precipitation observations from daily weather reports received at the Norwegian Meteoro-

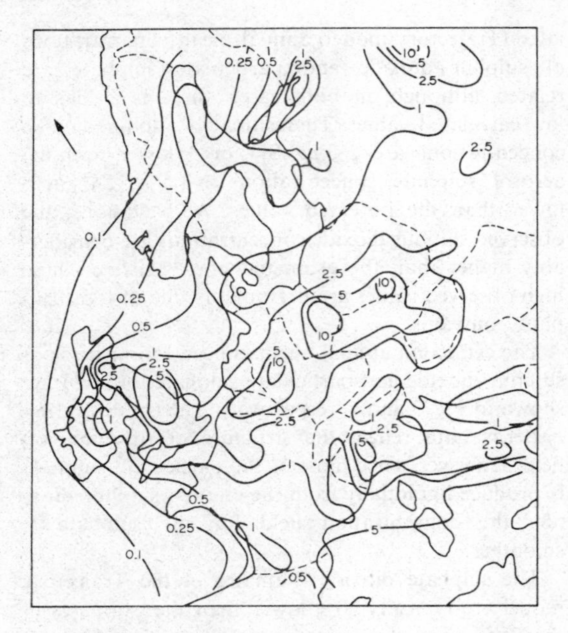

Fig. 5. Estimated pattern for total dry deposition of sulphur (Trajectory model). Unit: g S m^2. Deposition velocity for sulphur dioxide: 0.8 cm s^{-1}. Deposition velocity for particulate sulphate: 0.2 cm s^{-1}. Additional local deposition of 15% also included (see text) (OECD, in press).

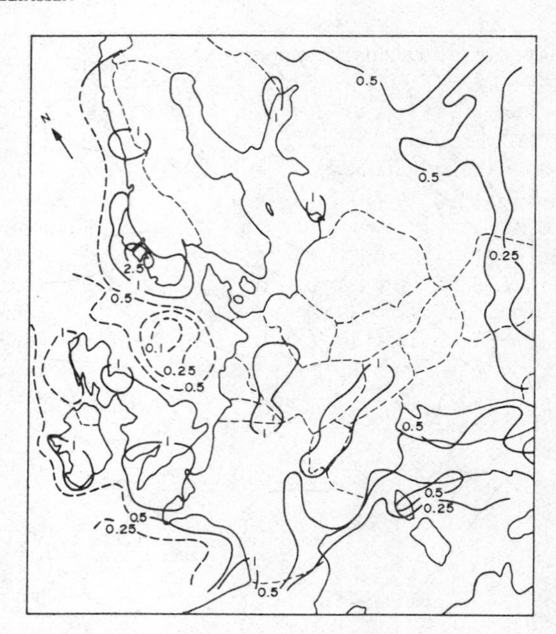

Fig. 7. The sum of the individual 12- or 6-hourly precipitation fields for 1974, representing the estimated annual precipitation amount. Unit: m. The too low precipitation amount in the middle of the North Sea is caused by incomplete data coverage (see text) (OECD, in press).

logical Institute, precipitation fields were constructed in the 127 km grid. Up to 25 September 1974, this was done subjectively, using additional information from surface weather maps, on a 12-hourly basis. For the last part of 1974, 6-hourly precipitation fields were constructed, applying the method of objective analysis given by Cressman (1960) with a 'distance of influence' of 100 km over land and 300 km over sea. The number of stations reporting data varied from about 500 up to 1000. The sum of the individual precipitation fields gives the precipitation field for 1974, shown in Fig. 7.

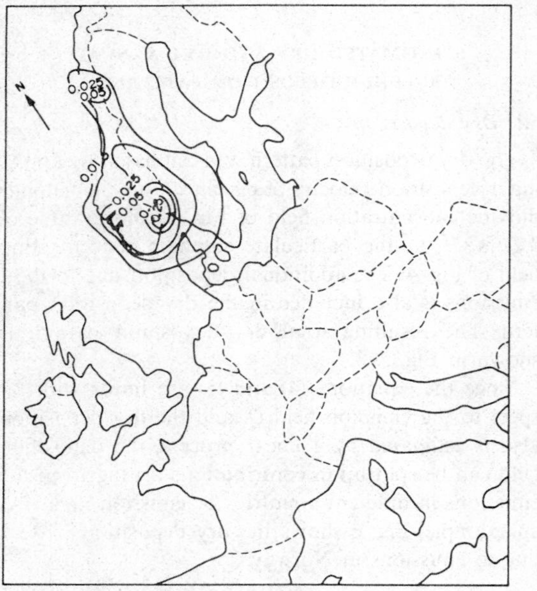

Fig. 6. Dry deposition pattern for 1974 due to emissions in Norway (Trajectory model). Unit: g S m^{-2} (OECD, in press).

The use of a 127 km grid, necessary for model input, inevitably produces a smoothing effect of small-scale precipitation features. Major inaccuracies may occur in areas such as the Scottish Highlands, where annual precipitation amounts changes from over 4 m to around 1 m in less than 100 km. Another problem with the techniques used to construct Fig. 7 is the very low precipitation amounts produced in areas with no data, e.g. the North Sea.

Ideally, the annual wet deposition pattern for sulphur should be calculated by storing the sulphur removed for each precipitation episode. However, the limited knowledge of the complicated chemical and physical processes leading to wet deposition, and the limited time and space resolution of available precipitation data have only permitted a model description of the wet deposition by a decay rate which increases during precipitation, but is independent of the precipitation rate. This situation makes it necessary to use a more indirect technique to estimate wet deposition, in which the annual mean concentration of sulphate in precipitation $\hat{c}$ is related to some quantity which can be calculated and which provides some measure of the sulphur content of the air during precipitation. The best near linear relationship was found when $\hat{c}$ is approximated by the annual weighted mean $\hat{s}$ of particulate sulphate concentrations from the Trajectory model calculations, given as:

$$\hat{s} = \frac{1}{P} \sum_i p_i s_i \qquad (6)$$

where p_i is the amount of precipitation observed on day i; s_i the corresponding calculated daily mean concentration of particulate sulphate, and P the total amount of precipitation for 1974 at the station. Days without precipitation obviously do not contribute to $\hat{s}$. The annual mean concentration $\hat{c}$ of sulphate in precipitation can be calculated for each sampling station and is given by:

$$\hat{c} = \frac{1}{P} \sum_i p_i c_i \qquad (7)$$

where c_i is the observed concentration at day i. Fig. 8(a) shows $\hat{c}$ plotted against $\hat{s}$ for 50 of the LRTAP stations. It should be emphasized that the use of a linear relationship between $\hat{c}$ and $\hat{s}$ makes it possible to split up the resulting wet deposition in contributions arising from emissions in different areas or countries.

From the data of Fig. 8(a), the regression line of $\hat{c}$ on $\hat{s}$ is:

$$\hat{c} = a\hat{s} + b$$

$$a = 0.69 \text{ mg l}^{-1} \text{ m}^3 \mu\text{g}^{-1}, b = 1.19 \text{ mg l}^{-1}. \qquad (8)$$

As can be seen, the scatter of points is quite large ($r = 0.73$) and is thought to reflect the effect of local

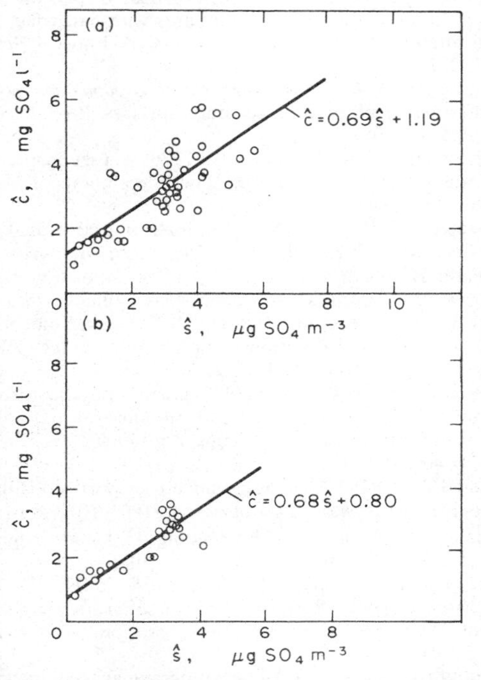

Fig. 8. Annual 1974 mean concentrations $\hat{c}$ of sulphate in precipitation plotted against weighted mean values $\hat{s}$ of calculated concentrations of aerosol sulphate (Trajectory model). Also shown is the regression line of $\hat{c}$ and $\hat{s}$. (a) Values for 50 sites. (b) Finnish and Norwegian sites alone. See text for further explanation (OECD, in press).

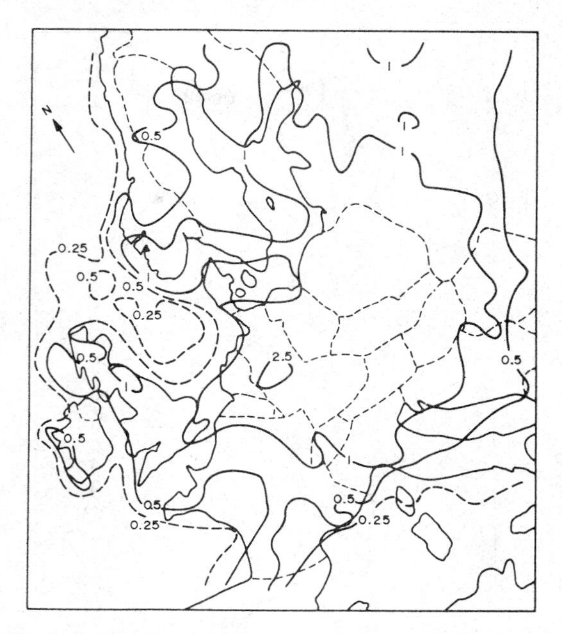

Fig. 9. Estimated sulphur wet deposition pattern for 1974. Unit: g S m^{-2} (OECD, in press).

(within each grid element) deposition. Support for this theory is provided by Fig. 8(b), constructed for the 26 Finnish and Norwegian sites, where the influence of local emissions is low. The regression line has the same slope as that of Fig. 8(a) and equation (8), but a correlation coefficient of 0.90 and an intercept on the $\hat{c}$-axis of 0.80 mg l^{-1}. It was therefore thought appropriate to use the regression line shown in Fig. 8(b) to calculate the wet sulphate deposition in Finland and Norway. Unfortunately, there are insufficient data points to construct regression lines for other countries with larger but varying local emissions. If a third graph were to be prepared, using the points shown on Fig. 8(a) but excluded from 8(b), the considerable variation in local emissions within these countries could be reflected in a correlation coefficient of only 0.56. As would be expected, the local sources are also reflected in a $\hat{c}$-intercept of 2.0 mg l^{-1}. However, the scatter is too great to use this 'relationship' for calculation of wet deposition, and, in the absence of any satisfactory alternative, the regression line of Fig. 8(a) has been used instead.

When calculating the wet deposition pattern, $\hat{c}$ is estimated from calculated $\hat{s}$-values for each grid element. The sulphate wet deposition pattern for 1974 is then obtained by multiplication with the annual precipitation amount. The resulting wet deposition pattern is shown in Fig. 9. Figure 10, giving the wet deposition pattern obtained from objective analysis of the measurements of sulphate in precipitation, is represented for direct comparison. Although there are some minor differences between the two patterns, the agreement is generally very good.

In general, both the dry and wet deposition are at maximum in the major emission areas and decline with increasing distance from them. However, certain

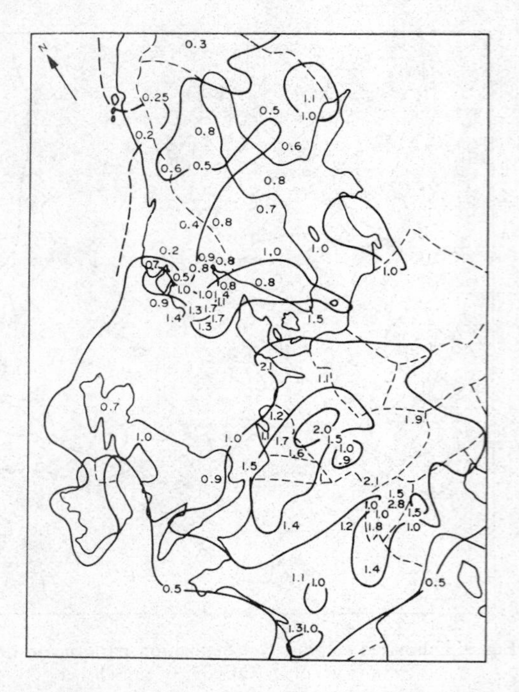

Fig. 10. Wet deposition of sulphate in 1974 ($g\,S\,m^{-2}$) obtained from measurements. An annual mean concentration field for sulphate in precipitation was constructed on basis of the measured values, and thereafter multiplied by the annual precipitation field (Fig. 6) (OECD, in press).

localised areas, like southern Scandinavia, have higher amounts of wet deposition than in neighbouring areas. This is due to a greater incidence of precipitation, caused mainly by orographic effects.

From the emission survey, the total amount of sulphur emitted annually within the area considered (9 Tm^2) is 20 Mt S. The total amount of dry deposition is 11 Mt S or around 50% of that emitted. The total wet deposition is 6 Mt S or around 30% of that emitted. These results agree well with those obtained with the Lagrangian model by similar methods.

6. CONCLUSIONS

As might be expected, this simple one-layer large-scale advection model cannot predict in detail the daily mean concentrations of sulphur dioxide or particulate sulphate observed on specific days. However, the calculated annual mean concentrations agree fairly well with observed values. Uncertainties in the estimated amounts of annual dry deposition, therefore, arise mainly from the calculation of dry deposition during the first phase of dispersion and from uncertainties in the deposition velocity of sulphur dioxide. The uncertainty of the dry deposition estimates, is assumed to be within ±40%.

The uncertainty in the estimated wet deposition is partly due to the unavoidable smoothing of small-scale precipitation and partly to the method used for estimating mean sulphate concentrations in precipitation. There is, however, a reasonable agreement between calculated and measured wet deposition. The general picture represented by the total dry and wet deposition patterns should be accurate to within ±50%.

The deposition patterns relate to 1974 only. Analysis performed for other years would produce different patterns, the difference depending on emissions and meteorological variables. The wet deposition will probably vary more from year to year than the dry deposition pattern.

To improve the spatial resolution of the calculated concentration and deposition fields, not only a finer emission grid is needed, but also a model describing the atmospheric dispersion and deposition processes with compatible accuracy. This requires much more information on the state of the boundary layer, and also a more detailed and physically correct description of the processes relating to wet deposition.

Acknowledgements—The reported work on atmospheric modelling took place within a group consisting of Oddvar Jensen, Norwegian Institute for Air Research (NILU), Jack Nordø (Norwegian Meteorological Institute), Jørgen Saltbones (NILU) and the present author, who is indebted to his collaborators. The author further wishes to thank Harald Dovland, Brynjulf Ottar and Arne Semb (NILU), Dr. R. A. Barnes, Department of Environment, U.K., and Dr. B. Fischer, Central Electricity Research Laboratories, for valuable suggestions and comments.

REFERENCES

Bolin B. and Persson C. (1975) Regional dispersion and deposition of atmospheric pollutants with particular application to sulfur pollution over western Europe. *Tellus* **27**, 281–310.

Chamberlain A. C. (1966) Transport of *Lycopodium* spores and other small particles to rough surfaces. *Proc. R. Soc. A* **296**, 45–70.

Christensen O. and Prahm L. P. (1976) A pseudospectral model for dispersion of atmospheric pollutants. *J. appl. Met.* **15**, 1284–1294.

Cressmann G. P. (1960) An operational objective analysis system. *Mon. Weat. Rev. U.S. Dep. Agric.* **87**, 367–374.

Dovland H. and Eliassen A. (1976) Dry deposition on a snow surface. *Atmospheric Environment* **10**, 783–785.

Egan B. A. and Mahoney J. R. (1972) Numerical modeling of advection and diffusion of urban area source pollutants. *J. appl. Met.* **11**, 312–322.

Eliassen A. and Saltbones J. (1975) Decay and transformation rates of SO_2, as estimated from emission data, trajectories and measured air concentrations. *Atmospheric Environment* **9**, 425–429.

Fisher B. E. A. (1975) The long range transport of sulphur dioxide. *Atmospheric Environment* **9**, 1063–1070.

Garland J. A., Atkins D. H. F., Readings C. J. and Caughey S. J. (1974) Deposition of gaseous sulphur dioxide to the ground. *Atmospheric Environment* **8**, 75–79.

Högström U. (1973) Residence time of sulfurous air pollutants from a local source during precipitation. *Ambio* **2**, 37–41.

Nordlund G. G. (1975) A quasi-Lagrangian cell method for calculating long-distance transport of atmospheric pollutants. *J. appl. Met.* **14**, 1095–1104.

Nordø J., Eliassen A. and Saltbones J. (1974) Large-scale transport of air pollutants. *Adv. Geophys.* **18B**, 137–150.

Nordø J. (1974) Quantitative estimates of long range transport of sulphur pollutants in Europe. *Ann Met.* **9**, 71–77.

Nordø J. (1976) Long range transport of air pollutants in Europe and acid precipitation in Norway. *Wat., Air Soil Poll.* **6**, 199–217.

OECD The OECD Programme on Long Range Transport of Air Pollutants. Measurements and Findings (in press).

Pedersen L. B. and Prahm L. P. (1974) A method for numerical solution of the advection equation. *Tellus* **26**, 594–602.

Petterssen S. (1956) *Weather Analysis and Forecasting.* McGraw-Hill, New York.

Prahm L. P., Torp U. and Stern R. M. (1976) Deposition and transformation rates of sulphur oxides during atmospheric transport over the Atlantic. *Tellus* **28**, 355–372.

Rystad B., Strømsøe S., Amble E. and Knudsen T. (1974) The LRTAP emission survey. Norwegian Institute for Air Research report LRTAP 2/74.

Semb A. (1978) The OECD study of long range transport of air pollutants. Source Inventory. *Atmospheric Environment* **12**, 455–460.

Shepherd J. G. (1974) Measurement of the direct deposition of sulphur dioxide onto grass and water by the profile method. *Atmospheric Environment* **8**, 69–74.

Smith F. B. and Jeffrey G. H. (1976) Airborne transport of sulphur dioxide from the U.K. *Atmospheric Environment* **9**, 643–660.

Whelpdale D. M. and Shaw R. W. (1974) Sulphur dioxide removal by turbulent transfer over grass, snow and water surfaces. *Tellus* **26**, 196–205.

Atmospheric Environment Vol. 12, pp. 489–501. Pergamon Press 1978. Printed in Great Britain.

THE CALCULATION OF LONG TERM SULPHUR DEPOSITION IN EUROPE

B. E. A. FISHER

Central Electricity Research Laboratories, Kelvin Avenue, Leatherhead, Surrey, KT22 7SE, U.K.

(*First received 9 June* 1977 *and in final form* 2 *September* 1977)

Abstract—A model based on statistical distributions of windspeed, wind direction and dispersion categories and of rainfall is used to calculate the annual wet and dry deposition of sulphur over Europe, using a detailed emission inventory. The model explicitly allows for vertical diffusion in the mixing layer and thereby the effect of source height. Wet and dry removal are taken into account as well as the conversion of sulphur dioxide to sulphate. In regions of Europe with heavy annual precipitation, such as southwest Norway, corrections to the deposition pattern are made.

Other factors, such as the diurnal variation of mixing depth and large scale synoptic vertical motions, are investigated and are not thought to change the overall pattern of annual deposition. The calculated pattern compares favourably with measurements and calculations based on trajectory models. The results show that the highest total deposition rates occur over the major emission areas ($10\,g\,m^{-2}\,y^{-1}\,SO_2$) and decrease to low values ($1\,g\,m^{-2}\,y^{-1}\,SO_2$) in outlying areas of the U.K., France, Italy and in Scandinavia. The wet deposition does not show such pronounced maxima close to emissions. The decrease in wet deposition is also modified in regions of heavy annual precipitation, such as southern Norway.

A sulphur budget is constructed for a region of southern Norway. The calculated total deposition, which agrees fairly well with the measured total deposition, is analysed according to the sectors containing the main source areas. The contribution from each source sector is only estimated approximately (to within a factor of two) but shows the expected dependence on source strength and frequency of wind direction. The main uncertainty is a large contribution from the background concentration of sulphate in precipitation not directly attributable to man-made sources in Europe.

The model calculation excludes the contribution from sources within about 100 km of a receptor site. The deposition at any sampling site is dependent on details of the source emissions within the neighbourhood of the site. However the average regional deposition from sources within the receptor region can be assessed. It is small for southern Norway because of the low emissions in the region. Source height strongly influences the local deposition, but on a larger scale is less important than other factors, such as the frequency of wind direction, rainfall and uncertainties in the emission inventory.

1. INTRODUCTION

The purpose of this paper is to describe further results from one type of model which determines the long-term average dispersion of airborne material over long distances. The important factors influencing dispersion are outlined and the ways in which different models treat them are discussed.

There are two approaches to the problem of modelling the long range transport of air pollutants, both equally important. One emphasises processes that influence the transport of sulphur such as dry deposition, rainout/washout, dispersion etc. The second involves meteorological factors which determine where material goes to on specific occasions, its speed, whether it is raining etc. In the former category, one is interested in whether long range transport occurs in typical meteorological conditions and then quantifying it, rather than following individual air masses on any particular day. A statistical approach can then be used to obtain the long term average of the deposition pattern. Some models use this approach (Bolin and Persson, 1975; Fisher, 1975). In the latter category are trajectory models, which follow the transport of individual air masses on a daily basis (Eliassen and Saltbones, 1975; 1976). The events of each day have to be summed in order to give the annual distribution. Both models rely on the same sort of emission grid.

An earlier paper (Fisher, 1975) describes a statistical model for obtaining the SO_2 budget for any area of north west Europe. The United Kingdom and Sweden are compared in the paper. The model had certain shortcomings and these are remedied in the latest programmed version of the model.

The main assumption behind the model is that although details of all fluctuations of meteorological and physical conditions are not taken into account, the influence of these fluctuations is small on the overall average and therefore the model is an efficient way of obtaining average distribution patterns. Our knowledge of many of the important parameters affecting the long range transport of air pollutants is not good anyway. A sensitivity analysis using the model has confirmed that better knowledge of a number of parameters would not greatly alter the long term average distribution patterns.

A.E. 12—1/3—FF

The model is statistical in the following sense. The average distribution of SO_2 is calculated directly from the frequency distributions of values of parameters in the model. The model is close to that of Bolin and Persson (1975). For example, it also assumes that periods of rain and dry weather occur randomly, so that expressions for the average removal of SO_2 by scavenging, as a function of travel time can be used. However, unlike the approach of Bolin and Persson (1975) the model relies entirely on analytic expressions for the vertical distribution of SO_2 above the ground.

The advantages of using analytical expressions are twofold. Firstly with analytic expressions one can investigate directly the influence of the various parameters in the model to see, for example, where during travel deposition velocity or vertical eddy diffusivity are important. Secondly the use of analytic expressions makes the model cheap to use and easy to program.

The main features of the model are described in Section 2. One improvement over the earlier version is the facility to take account of regional variations of rainfall. Originally the probability of rain was assumed to be uniform throughout Europe and so higher rainfall, in say, south west Norway, was not taken into account. Norway is now divided into three regions according to rainfall, and different rainfall patterns are used in each region (see Section 5).

The model also has the advantage of treating the vertical dispersion of material within the mixing layer explicitly. The profile of concentration within the mixing layer can therefore be calculated and the effect of source height on the long range transport of SO_2 can be investigated.

2. THE MODEL USED IN THE CALCULATIONS

The most useful quantities required from any model of the longterm average distribution of sulphur are the annual average distributions of sulphur dioxide and sulphate concentration at the ground, and the rates of wet and dry deposition. The basis for the model used in these calculations has been described earlier (Fisher, 1975). To extend it to allow for the conversion of sulphur dioxide to sulphate, it is assumed that a fraction F_4 of the SO_2 is converted to SO_4 during the early stages after emission; and that thereafter the conversion rate is W_4 (Smith and Jeffrey, 1975). Then the annual average concentration at the ground at a distance x from a point source of height, h, is given by

$$C(x) = \sum_{u,a,K} (1 - F_4) \frac{Q}{2\pi x} G(0, h, x/u) \times F(\theta) F(u, a, K) e^{-W_4 x/u} \quad (1)$$

if the effects of precipitation scavenging are neglected for the time being. In equation (1) Q is the emission of sulphur dioxide from the source. G is a factor depending on the vertical diffusion of sulphur dioxide in the atmosphere given in the earlier paper. G is a function of the meteorological variables: windspeed, u, mixing depth, a, and vertical eddy diffusivity, K, in the mixing layer. $F(\theta)$ is the ratio of the frequency with which the wind blows in the direction θ to the frequency assuming all wind directions are equally likely (see Section 3). $F(u, a, K)$ is a weighting factor to allow for the different frequencies with which different windspeed, mixing depth and diffusivity categories occur during any year. The summation in equation (1) is over these categories. Since we always deal with quasi-steady state situations in the model, time since emission does not appear explicitly as a parameter in equation (1). However within the approximations of the model it is satisfactory to interchange x/u with time of travel, whenever this is convenient. Detailed knowledge of the crosswind dispersion and the turning of the wind with height is not required in order to use equation (1).

The dry deposition, D, of sulphur dioxide is given by $v_g C$. The exact calculation of the sulphate concentration in air would have to take account of the vertical diffusion. The treatment of the vertical diffusion of sulphate is much more difficult as it is being created all the time, at all heights in the mixing layer. However, as it is deposited at the ground very slowly and precipitation scavenging is the only important removal mechanism, it is a reasonable approximation to assume that the sulphate is uniformly mixed in the atmosphere. In this case the concentration of sulphate is given by an equation of the form of equation (1) but with $Ge\, W_4 x/u$ replaced by G_1, where

$$G_1 = \frac{1}{ua}\left[F_4 + \sum_n \frac{2\cos\phi_n(1 - h/a)\sin\phi_n}{\phi_n(1 + \sin 2\phi_n/2\phi_n)} \times \left[\frac{W_4(1 - F_4)}{W_4 + K\phi_n^2/a^2}\right] \times [1 - e^{-}(W_4 + K\phi n^3/a^2)t] \right] \quad (2)$$

where ϕ_n are the roots in ascending order of $\phi \tan \phi = av_g/K$. v_g is the deposition velocity of sulphur dioxide and the final term in equation (2) allows for the gradual build-up of sulphate from sulphur dioxide as it travels downwind. When it rains both SO_2 and sulphate in the mixing layer are removed. For both species the scavenging coefficient has been set equal to $10^{-4}\,s^{-1}$. As the scavenging coefficient is the same, it does not matter whether the SO_2 is converted to sulphate before possibly being removed by rain. Hence the order of rain periods is not important and the simple stochastic model to take account of scavenging described in Section 5 can be applied. The effect of rain is just to introduce an additional multiplicative factor into equations (1) and (2). Severe complications would also arise if the rate of conversion were varied according to wet and dry periods. In practice one might expect a faster conversion rate during wet periods, but the parameter values used in the model suggest that during precipitation most of the sulphur is removed anyway and therefore any error is small. There may also be a small error when there is large scale ascent of material which is not removed directly in the precipitation. The importance of transport outside the mixing layer is that the process of dry deposition, which normally depletes concentrations to one third over distances of order 1000 km, is no longer operational. Material in this situation therefore has the potential for very long range transport, particularly in the case of large scale vertical motion when it might rise to a height of 10 km and is involved in the general atmospheric circulation. It will only return to the boundary layer in the next area of large scale descending motion.

3. DIRECTIONAL VARIATION OF TRAJECTORIES

Daily average measurements of sulphur at a sampling site can be grouped into sectors according to the trajectory of the air mass bringing the sulphur. It is then possible to obtain the average sulphur burden for winds blowing in different general directions. Førland (1973) used this technique to interpret the origin of acidity in precipitation readings collected at a station in south west Norway. He found that most trajectories were only slightly curved, so an allocation of trajectories to broad general sectors avoids problems over the accuracy of drawing trajectories.

One must put into the statistical model a directional weighting factor $F(\theta)$ which represents the frequency with which trajectories lie in different directions from a source or receptor. The directional weighting factor should be different for each source and receptor region of Europe and should be based on the best information available. In the absence of better information, the directional weighting factor in the earlier paper was based on measured surface wind directions in the United Kingdom. A later analysis of wind directions 390 m above ground, measured from a tower in the east Midlands, gave approximately the same weighting factor. A directional weighting factor for southern Sweden can be derived from the factor used by Munn and Rodhe (1971). Munn and Rodhe analysed wind directions causing precipitation in the Flahult–Plönninge area of Sweden.

One of the benefits of sector analyses is that they give further information on these direction weighting factors. For example the sector analyses undertaken by NILU and the British Meteorological Office, as part of the OECD project on the long range transport of air pollutants, provide valuable information on the frequency and directions of winds bringing rain into south west Norway and other parts of Europe. Examples of the directional weighting factor to be put on winds into southwest Norway are shown in Fig. 1. The statistical models are unable to deal with stagnant conditions as a definite 'trajectory' rose has to be defined within the model.

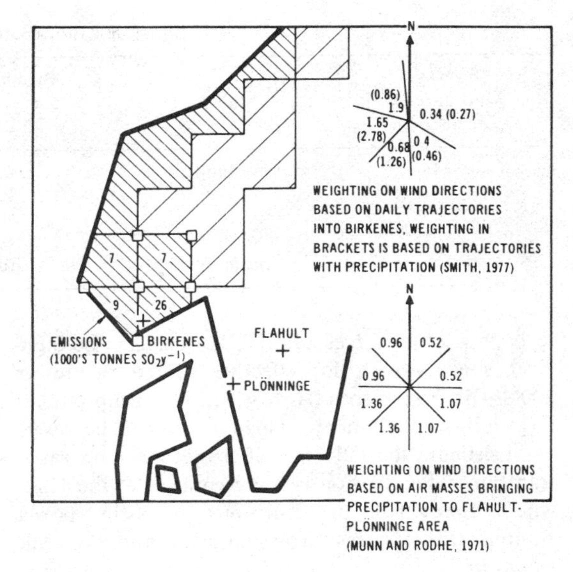

Fig. 1. Rainfall pattern and weighting on wind direction over Norway.

The limitation of the sector analysis is that the contribution from one country to another country can only be assessed in certain well specified cases. In order to undertake this task modelling techniques were developed in the OECD programme. Models should include as many of the factors influencing long range transport as is practical. The essential data base for any model is an accurate source inventory.

3.1 *Source inventory*

The model makes use of a grid of 127 × 127 km squares over north-western Europe. The centre point of each grid square is taken to be a source of SO_2, whose strength is equal to the total emission of SO_2 within the square. The emission of SO_2 p.a. within each square has been estimated by Eliassen and Saltbones (1975) as part of the OECD project. Although all sources in Eliassen and Saltbones' distribution map of Europe are used, it is often only necessary to calculate deposition at grid points over Scandinavia. Figure 1 shows this part of the grid.

Since all sources are assumed to be at the centre of grid squares, while deposition is calculated at the corners of grid squares, the contribution to deposition is neglected from sources within about 100 km of a grid point. Estimates of the contribution of local sources can be made when a more detailed inventory is available.

The uncertainty for countries, for which good information on source emission is available, is probably ±10%, and for other countries is considerably worse. In the present calculation half the sources are taken

Table 1. Simple classification of boundary layer conditions

Average wind speed, u ($m s^{-1}$)	Stability class, S	Probability of occurrence (all wind directions)	Depth of mixing layer, a (m)	Vertical eddy diffusivity ($m^2 s^{-1}$), K
4	mechanical	0.18	450	2
	convective	0.10	1500	30
10	mechanical	0.30	600	6
	convective	0.10	1500	30
16	mechanical	0.32	960	15

to be low-level sources and half are taken to be high-level sources with an effective source height of 200 + (1000/wind speed) (m), which is appropriate for large industrial chimneys. This fraction can be varied to investigate the influence of height on long range transport. This division is representative of the U.K., where about half the emissions are from power stations with the rest from industry and domestic sources.

3.2 *Specification of dispersion categories*

The annual average concentrations and distributions derived from equations (1) and (2) depend on the frequency distribution $F(u, a, K)$ of the parameters windspeed, u, mixing depth, a, and diffusivity, K, determining the dispersion within the mixing layer and thus indirectly the form of G in equation (1). The mixing is derived from two types of processes. For a mixing layer dominated by mechanical turbulence, the depth of the mixing layer is proportional to windspeed. These layers occur in overcast conditions with moderate or strong winds. The diurnal variation is negligible.

By contrast the convective mixing layer is unsteady and depends on the history of the heat input into the boundary layer. Since in convective conditions the concentration is almost uniform, details of the internal structure of the boundary layer are not important to long range transport.

The climatology of dispersion categories in any region can be based solely on windspeed and a stability class S. The stability class, S, depends on whatever meteorological information is readily available i.e. it could be based on temperature stratification, if representative temperature profiles are measured in the region, or angular wind fluctuations or on insolation, cloud cover and windspeed. The parameters a and K are specified functions of windspeed and stability.

A climatology of stability classes based on temperature profiles up to a height of 380 m is available in the U.K. from the Belmont Tower in Lincolnshire. In Fisher (1975) a broad classification of stability classes was based on measurements over a shorter 180 m tower. The frequency distributions of windspeed and stability classes are clearly functions of wind direction (determining air mass characteristics) and fetch over land (Moore, 1977). However, for the purposes of this paper a broad classification is made which does not vary with position in Europe or wind direction. This is unsatisfactory but is adopted for several reasons. Firstly, long range transport is fairly insensitive to detailed meteorological conditions. Secondly, information on dispersion categories over Europe is limited. Thirdly, the present model has been kept as simple as possible.

Five categories have been chosen in this paper. They are summarised in Table 1.

The depth of the mixing layer is taken to be $60\,u$ in the mixing layer dominated by mechanical turbulence (Moore, 1975) except when the average wind speed is $4\ m\,s^{-1}$. In that case $a = 450$ m to ensure all plumes remain within the mixing layer. The mixing layer has a depth of 1500 m in convective conditions. The average vertical eddy diffusivity is taken to be a fraction ($\frac{1}{2}$ or 1) of $0.4\,u_* l$ for the mechanical mixing layer, where the friction velocity, $u_* = u/20$ and the length scale, $l = 0.1\,a$. A large value of the diffusivity is adopted for convective mixing layers.

The insensitivity to choice of dispersion categories will become apparent by comparison of results in this paper with those with a different selection (Fisher, 1975).

4. SPECIFICATION OF REMOVAL RATES

Values of the parameters v_g, deposition velocity of SO_2, W_4, conversion rate of SO_2 to sulphate and F_4, initial fraction of SO_2 converted to sulphate, must be chosen. Experimental evidence on the deposition velocity of sulphur dioxide indicates a deposition velocity of order $10\ mm\,s^{-1}$ with an average value of $8\ mm\,s^{-1}$ often adopted in practice.

An average transformation rate can be estimated from the parameter in long range transport models representing the oxidation rate (Eliassen and Saltbones, 1975) and is found to be of order $2.10^{-6}\ s^{-1}$. For aircraft flights, Smith and Jeffrey (1975) found the oxidation rate over shorter time scales to be of the same magnitude, but had to take into account an initial 10% oxidation in the vicinity of the source. Prahm *et al.* (1976) estimated the transformation rate for transport from U.K. to Faroes to be about $4.10^{-6}\ s^{-1}$. The oxidation rate of sulphur dioxide in the absence of detailed information is therefore assumed to be several times slower than the removal rate by dry deposition. It is assumed that 10% of the SO_2 is converted to SO_4 during the early stages after emission; thereafter the conversion rate is $10^{-6}\,s^{-1}$.

The sulphate particles themselves are mainly submicron in size. Experimental work (Chamberlain, 1966; Sehmel and Sutter, 1974) indicates that submicron sulphate particles have a much smaller deposition velocity over land and sea than sulphur dioxide. The deposition velocity of SO_2 is taken to be $5\,mm\,s^{-1}$, while the deposition velocity of SO_4 has been set equal to zero. The reason for the slightly low value of the deposition velocity of SO_2 is discussed in the next section.

4.1 *Decrease in turbulent length scale close to the ground*

The assumption of a uniform eddy diffusivity K to describe mass transfer throughout the mixing layer is not strictly valid. Close to the ground the presence of a rigid boundary suppresses the maximum length scale of eddies. To be consistent with the profile method for obtaining the deposition, $K(z)$ in neutral conditions should be taken as a linear function of height, of the form $K(z) = 0.4\,u_* z$, where u_* is the friction velocity. The vertical diffusivity $K(z)$ increases linearly with height in neutral conditions in the surface layer of the atmosphere up to a height of around 30–100 m. Scriven and Fisher (1975) have considered analytic solutions of the diffusion approximation with this diffusion profile. It is shown that a useful approximation can be made by assuming that the vertical flux of material is constant throughout the surface layer, provided the depth of the surface layer is much less than the depth of the mixing layer, one is sufficiently far downwind and the source height is above the surface layer. This result is just an extension of the constant flux argument from a level about 1 m above the ground, where most of the measurements are made, to further into the mixing layer. The effect of the surface layer is incorporated in the same way as the atmospheric resistance below 1 m. The effective deposition velocity v_{g1} defined at a height z_1, at the top of the surface layer, is given in neutral conditions by

$$\frac{1}{v_{g1}} = \frac{1}{v_{g0}} + \frac{1}{0.4u_*}\ln z_1/z_0 \qquad (3)$$

where v_{g0} is the deposition velocity at the surface. Because of the extra resistance of the surface layer, the effective deposition velocity used in long range transport models is slightly less than the measured values at 1 m.

As the effective deposition velocity depends to some extent on the diffusion in the lowest layers of the atmosphere, it should be a function of windspeed and stability. However, for simplicity this effect has been neglected in this paper and the deposition velocity is taken as constant.

4.2 *Time and space variations along a trajectory*

The transport of material over long distances depends on the meteorological conditions during travel. The decay rate depends on wind speed, mixing layer depth, deposition velocity and eddy diffusivity as can be seen from equations (1) and (2). For estimating the daily average concentration one would require the best estimate of these parameters, while for an annual average one requires the appropriate statistical distribution of meteorological conditions. However most models do not take into account variations in space and time along the trajectory of the air mass carrying the material. For transport times of order 24 h or more (which are the order of transport times involved in long range transport in the absence of precipitation), the diurnal cycle of the boundary layer over land will interfere significantly (Högström, 1975).

A simple method of allowing for these fluctuations is to imagine material to be frozen during very stable night-time periods and then to scale the deposition velocity accordingly, by taking a lower value than an average $8\,mm\,s^{-1}$ for daytime periods. This is adequate when the annual average behaviour is required, rather than a particular case study of long range transport.

Changes in stratification of the atmospheric boundary layer can be taken into account by dividing the travel path into segments; each segment representing, say, 12 h of transport under steady uniform conditions. From Scriven and Fisher (1975) the fractional loss of sulphur dioxide due to deposition alone in a mixing layer of height, a, is approximately

$$P_2(\beta) = \left[1 - \frac{2(1+0.4\beta)^2 \exp - v_g T/a(1+0.4\beta)}{\{(1+\beta)(1+0.4\beta)+1\}}\right] \qquad (4)$$

during an interval $T(>a^2/K)$ assuming that sulphur dioxide is uniformly mixed at the start of the interval. The parameter β in equation (4) is equal to av_g/K.

For convective periods the vertical eddy diffusivity is very high and β is approx zero. In this case P_2 reduces to $P_2(0) = 1 - \exp(-v_g T/a_c)$, where a_c is the height of the convective mixing layer. During a following interval of stable conditions the fraction removed is

$$\left(\frac{a}{a_c}\right)P_2(\beta) \qquad (5)$$

where β is defined by the values of a and K during stable conditions and the factor (a/a_c) arises because only part of the day time mixing layer is in contact with the ground during stable conditions. The amount of material surviving a number of periods of convective and stable conditions is therefore

$$(1-P_2(0))\left(1-\frac{a}{a_c}P_2(\beta)\right)(1-P_2(0))\left(1-\frac{a}{a_c}P_2(\beta)\right)\text{etc.} \qquad (6)$$

Thus it is possible to treat diurnal variations fairly simply within a steady state model. During a 24 h period divided into two equal segments with $a_c = 1500$ m in the unstable period and $a = 300$ m, $K = 3\ \mathrm{m^2\ s^{-1}}$ during the stable period, and using $v_g = 10\ \mathrm{mm\ s^{-1}}$, the fraction of material surviving the unstable period is 0.75 and the fraction surviving the stable period is 0.87. Over the complete 24 h cycle, 0.65 of the original material remains. If steady conditions throughout the 24 h are assumed with $a = 1000$ m, $K = 10\ \mathrm{m^2\ s^{-1}}$ and $v_g = 10\ \mathrm{mm\ s^{-1}}$ then 0.56 of the original material survives. If a value of $5\ \mathrm{mm\ s^{-1}}$ for the deposition velocity is adopted, then 0.79 of the material survives. These examples, with the parameter values chosen fairly arbitrarily, indicate that using a value of deposition velocity lower than the usual value based on daytime measurements can compensate for the effects of diurnal variations in the mixing layer. The shear of wind speed and direction with height can cause considerable horizontal dispersion of material although it is possible to define a mean advective wind for determining the trajectory of the centre of mass of part of a plume. The additional spread is not usually important for wide continuous sources of SO_2 (especially if annual average concentrations are required). However, the back trajectory of material from high sources, which is mixed to the surface only during the daytime convective period of a diurnal cycle, is different from that of material which has remained in contact with the surface at all times.

4.3 *Effect of large scale ascending and descending motion*

The effect of a non-zero divergence in the mean horizontal wind field can be assessed simply in terms of the concentration in a box with sides of length L_1 and L_2 and height a. If the divergence $(\partial u/\partial x) + (\partial v/\partial y)$ equals γ then the area of the box becomes $L_1 L_2\ e^{\gamma t}$ after a time t. γ can be written $-w/a$ where w is the vertical velocity at the top of the mixing layer. The height of the box or top of the mixing layer will be assumed independent of w, although this is not strictly correct (see e.g. Carson, 1973). With this assumption, pollutant concentration decreases as $e^{wt/a}$, if w is negative. Since $|w|$ is of the same order as the deposition velocity, its effect on the concentration in an air mass could be the same as removal by deposition. However, in this case the material is not being removed from the atmosphere, but diluted with air from above the mixing layer and transferred into areas of convergence (γ negative).

Where γ is negative, there is an overall upward velocity (w positive) and material leaves the top of the mixing layer at a rate w/a over the whole box, but the concentration within the (shrinking) box remains unaffected.

For relatively high vertical windspeeds ($\sim 0.05\ \mathrm{m\ s^{-1}}$) this removal mechanism is much more important than dry deposition. However, this situation is usually associated with frontal precipitation, which is already assumed to remove sulphur oxides fairly efficiently anyway. The net effect of vertical motions is therefore likely to be an increase of wet deposition at the expense of dry deposition in the fair weather areas.

5. INCLUSION OF PRECIPITATION SCAVENGING

The method proposed by Rodhe and Grandell (1972) is used to include precipitation scavenging in the annual deposition of sulphur dioxide. To represent random transitions from periods of rainfall to dry periods and back again, it is assumed that the onset of a wet (or dry) period occurs with a constant transition probability. Thus it is assumed that during an arbitrary interval of length Δt in a dry period, the probability of a transition to the next wet period is $\Delta t/t_0$ and conversely the probability of transition during an arbitrary interval Δt in a wet period is $\Delta t/t_1$. The times t_0, t_1 represent the mean lengths of wet and dry periods, and to start with were taken to be independent of position over Europe.

This does not accurately represent orographic rain for example, but attempts have been made to describe daily rainfall occurrences in terms of simple probability models. To take account of the variation in rainfall in different regions of Europe, t_0 and t_1 can be allowed to take different values in different regions. In practice, adjustments to the rainfall pattern were made only over Norway.

The average influence of precipitation scavenging, in any region in which t_0 and t_1 are constant, is given in terms of the 2×2 matrix $G_{ij}(t)$, where the subscript, $i = 0$, refers to a dry period and the subscript, $i = 1$, refers to a wet period. The general element $G_{ij}(t)$ is the probability of a particle surviving until a period of type, i, after a time, t, having started at $t = 0$ in a period of type j. $G_{ij}(t)$ is equal to

$$\frac{\exp E_0 t}{(E_0 - E_1)}\begin{pmatrix} \Lambda + t_1^{-1} + E_0 & t_1^{-1} \\ t_0^{-1} & t_0^{-1} + E_0 \end{pmatrix} + \frac{\exp E_1 t}{(E_1 - E_0)}\begin{pmatrix} \Lambda + t_1^{-1} + E_1 & t_1^{-1} \\ t_0^{-1} & t_0^{-1} + E_1 \end{pmatrix} \quad (7)$$

where Λ is the scavenging coefficient during wet periods and

$$2E_{0,1} = -(\Lambda + t_1^{-1} + t_0^{-1}) \pm \sqrt{(\Lambda + t_1^{-1} + t_0^{-1})^2 - 4\Lambda t_0^{-1}}. \quad (8)$$

If a particle has passed through several regions with different rainfall properties, the overall probability function $G_{ij}(t)$ is given by the matrix product of $G(t - t')\ G'(t' - t'')\ G''(t'')$ etc., where the particle has had to spend a time, t'', in the first region time, $t' - t''$, in the second and $t - t'$ in the third and G, G' and G'' are the probability functions for each region.

In many cases, one is only concerned with the probability $P(t)$ of surviving to a time, t, assuming that the time of emission is arbitrary. $P(t)$ is given by

$$P(t) = (G_{00}(t) + G_{10}(t))p_0 + (G_{11}(t) + G_{01}(t))p_1 \quad (9)$$

where $p_0 = t_0/(t_0 + t_1)$ is the probability that the particle is emitted during a dry period and $p_1 = t_1/(t_0 + t_1)$ is the probability that a particle is emitted during a wet period. When precipitation scavenging is included the expressions for the dry deposition of SO_2 and annual average concentrations of SO_2 and sulphate in air given by equations (1) and (2) must be multiplied by $P(t)$, given by equation (9). Since $P(t)$ is less than one, the dry deposition is less when scavenging is included. The total deposition of sulphur dioxide is the sum of the dry deposition, calculated in this way, plus the wet deposition. In the case of constant rainfall probabilities throughout the lifetime of the particle $P(t)$ reduces to

$$P(t) = (E_1 - E_0)^{-1}[(E_1 + \Lambda t_1/(t_1 + t_0))e^{E_0 t} - (E_0 + \Lambda t_1/(t_1 + t_0))e^{E_1 t}]. \quad (10)$$

The wet deposition of sulphur dioxide is given by

$$\Lambda P_1(t)\left(\int_0^a dz \sum_{u,a,K} (1 - F_4)\frac{Q}{2\pi x} G(z, h, t) \times F(\theta)\,F(u, a, K)\,e^{-W_4 t}\right). \quad (11)$$

The term in brackets in this equation is concentration integrated with height. $P_1(t)$ is the probability that SO_2 survives to a time t and that it is raining after this time assuming the time of emission is arbitrary. $P_1(t)$ is given by

$$G_{10}(t)p_0 + G_{11}(t)p_1. \quad (12)$$

In the case of constant rainfall probabilities throughout the lifetime of the particle $P_1(t)$ reduces to

$$P_1(t) = \frac{t_1/(t_0 + t_1)}{E_1 - E_0}[(E_1 + \Lambda)e^{E_0 t} - (E_0 + \Lambda)e^{E_1 t}]. \quad (13)$$

It is thought that the decay time of sulphur dioxide during scavenging is probably much shorter than the average duration of a wet period which itself is much shorter than the average duration of a dry period. The physical interpretation of equations (10) and (13) becomes clear under these conditions. The conditions imply that once a period of rain starts sulphur dioxide is almost certainly removed during that period. Under the conditions $\Lambda^{-1} \ll t_1 \ll t_0$, E_0 reduces to $-t_0^{-1}$ approx and E_1 reduces to $-\Lambda$ approx. Thus

$$P(t) \sim t_0/(t_1 + t_0)e^{-t/t_0} + t_1/(t_1 + t_0)e^{-\Lambda t}. \quad (14)$$

The first term represents the probability that it first starts raining after time t, after which the sulphur dioxide is rapidly removed. The second term, which is only significant at short distances from the source, allows for the possibility of rain during emission.

Similarly $P_1(t)$, the probability that SO_2 survives until a time t and that it is raining after this time, becomes (using $E_1 \sim -\Lambda - t_1^{-1}$)

$$P_1(t) \sim \frac{t_1}{t_0 + t_1}\left(\frac{1}{\Lambda t_1}e^{-t/t_0} + e^{-\Lambda t}\right) \quad (15)$$

$t_1/(t_0 + t_1)$ is the probability that it is raining at time t. Then e^{-t/t_0} is the probability that there have been no other wet periods between emission and time t and the factor allows for some of the material to have already been removed during the present wet period. The second term is only significant a short time after emission and gives the correct form for P_1 if it has been raining continuously since emission.

Figure 1 shows how Norway has been divided into three regions according to rainfall. The mean length of a dry period in Region 1, where the rainfall is lightest, is taken to be 69 h and the corresponding mean length of a wet period is 6.9 h (Rodhe and Grandell, 1972). This rainfall distribution is taken to represent the average distribution of rain over the rest of Europe apart from Norway. The same rainfall pattern was used over the whole of Europe in the earlier paper.

In Region 2 the ratio of wet periods (mean length = 2.8 h) to dry periods (mean length = 28 h) is the same, but allowance is made for more frequent changes from wet to dry and back again. Region 3 represents the region where orographic rain occurs. The duration of orographic rain is taken to be 1 h which ensures some rain in the time it takes for an air parcel to cross the region, as the mean period of dry weather is taken to be 5 h, and gives twice the rainfall of other areas. Changes in the rainfall pattern along the line joining a source and a grid point are put into the model by considering separately transport along each segment of the line over which rainfall conditions are uniform.

In orographic rain one would expect the intensity of rain as well as the duration of rain to be higher than in other areas. However, quantitative knowledge of the effect this has on the removal of sulphur in rain is so limited that it will not be introduced at this stage.

When it rains both SO_2 and SO_4 are removed. For both species the scavenging coefficient, Λ, has been set equal to $10^{-4}\,s^{-1}$. The washout rate of SO_2 in pure rain, falling at a rate of 1 mm h^{-1} has been predicted to be about $10^{-5}\,s^{-1}$ (Billingsley *et al.*, 1976), though the rate varies with the SO_2 and ammonia concentration in the air. This calculation only applies to washout of SO_2. Using $\Lambda = 10^{-5}\,s^{-1}$ as a combined rainout and washout removal rate gives values of the wet deposition over Norway which are too low to be consistent with actual measurements. Taking the scavenging coefficient to be higher, i.e. $10^{-3}\,s^{-1}$, does not affect the wet deposition in most areas, because the mean length of a wet period is long enough for all the sulphur to be removed anyway. However, in the region where orographic rain occurs, wet periods occur so frequently that with $\Lambda(SO_2) = 10^{-3}\,s^{-1}$ *all* sulphur is removed within 200 km of the coast, leading to an overestimate of wet deposition.

6. MODEL CALCULATIONS OF THE PATTERN OF WET AND DRY DEPOSITION OVER EUROPE

The method described in the earlier sections has

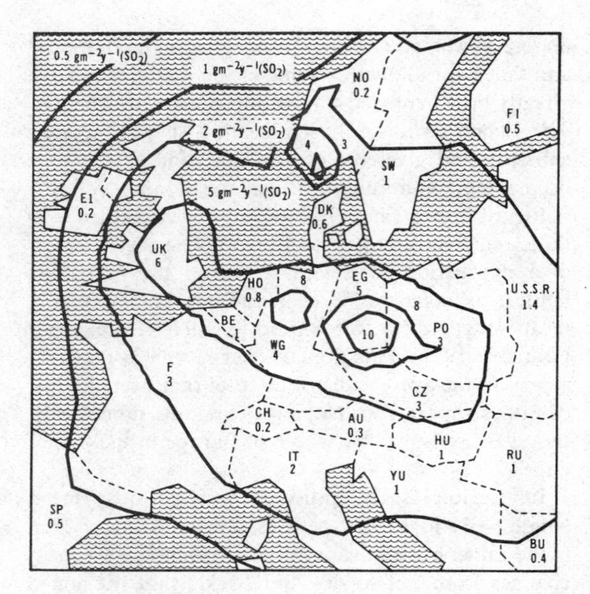

Fig. 2. Total annual deposition of sulphur over Europe ($gm^{-2}y^{-1}SO_2$) from man-made sources in Europe. Emissions (by country) are given in millions tonnes $SO_2\,y^{-1}$.

been applied to the calculation of the pattern of total and wet deposition over Europe. In Fig. 7 of an earlier paper (Fisher, 1975) the pattern of total deposition (wet plus dry) over Europe was plotted. The calculation leading to this pattern was subject to a number of limitations which have been outlined in earlier sections. The main correction, the inclusion of orographic rain over Scandinavia, has now been made. The corrected forms of the earlier deposition pattern are now presented as Figs. 2 and 3 of this paper. The categories of windspeed and mixing depth are as in Table 1.

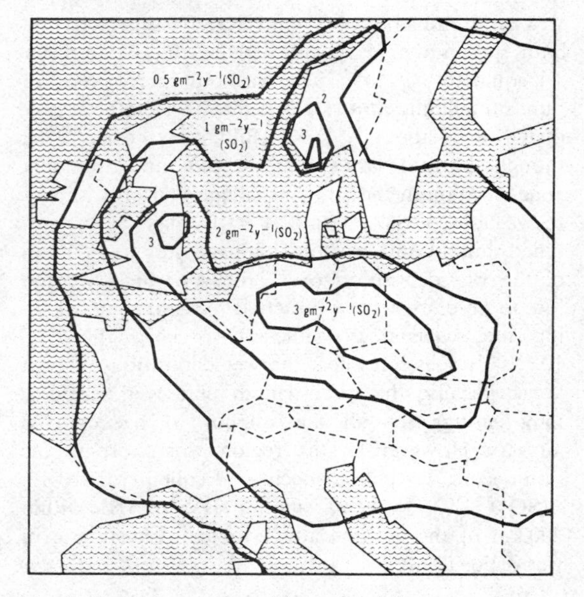

Fig. 3. Annual wet deposition of sulphur over Europe ($gm^{-2}y^{-1}SO_2$) from man-made sources in Europe. There may be an extra background contribution to the wet deposition.

A single directional weighing factor has been used to represent the frequency with which trajectories lie in different directions, based on the analysis by Munn and Rodhe (see Fig. 1). The use of a single weighting factor for all of Europe is rather limiting. However the overall distribution is very similar to that obtained in other calculations. Eliassen and Saltbones have obtained rather similar distributions based on a Lagrangian model (reproduced by Ottar (1976) and Dovland *et al.* (1976)). Bolin and Persson (1975) derive a similar distribution pattern over Europe, apart from areas of Norway where Bolin and Persson take no account of orographic rain.

The main features of the patterns are that usually the largest rates of deposition ($\sim 10\,g\,m^{-2}\,y^{-1}\,SO_2$) are associated with the main emission areas with a gradual decline to lower levels ($\sim 1\,g\,m^{-2}\,y^{-1}\,SO_2$) in remote areas. Dry deposition is generally greater than the wet deposition over most of the areas. The wet deposition does not show such pronounced maxima in the regions closest to the sources. The reason for this is that the ground level concentration of SO_2, which determines the dry deposition, decreases rapidly in the neighbourhood of sources because of vertical dispersion. In the far field sulphur is more uniformly mixed in the mixing layer and the decline in both the dry and wet deposition patterns is determined by removal processes, which are rather slow on average. This effect is included in the model.

There are other physical effects, not included specifically in the model, which will emphasise this behaviour (Rodhe, 1972). Sulphur deposited in rain has probably been involved in synoptic scale vertical motion and is therefore much less influenced by local emissions than sulphur which has remained near the ground. Secondly much of the sulphur dioxide is probably not deposited, until it has become oxidised to sulphate, which gives a time delay, in which further dilution occurs.

Polluted areas, such as cities, if included in these patterns would appear as small circles enclosing areas of considerably higher deposition ($>20\,g\,m^{-2}\,y^{-1}\,SO_2$). The calculation is limited in detail, because it is based on a square grid of side 127 km and therefore peaks in the deposition pattern due to local source variations (within a grid square) are automatically smoothed. This is true of all the large scale dispersion calculations. The exact calculation of the concentration or deposition at a sampling site would need to combine the influence of near field sources and medium range sources with the long range calculation presented here. Over the medium and short range, dispersion and source distribution are the main factors determining concentrations, not removal and transformation rates and this is confirmed by aircraft measurements (Fisher *et al.*, 1977).

The exception to the broad pattern outlined above is that in areas of pronounced orographic precipitation the decline in deposition out from the main source areas is modified. Over southern Norway, for example, the regional annual average ground level concentrations ($\sim 5\ \mu g\ m^{-3}\ SO_2$) is probably lower by a factor of ten than the regional average concentration in the main emission areas. However the wet deposition of sulphate in southern Norway is comparable with that over the main emission areas.

6.1 *Accuracy and comparison with other calculations and measurements*

The calculations are subject to a large number of uncertainties. The accuracy is hard to determine objectively. The emission inventory (Eliassen and Saltbones, 1975) is probably at best accurate to $\pm 10\%$, as far as individual countries are concerned. There will be year to year variability in emissions and also year to year variability in weather conditions. This should be reflected in the frequency distributions of the meteorological parameters in the model.

In the earlier paper it was shown that at long distances from its source the deposition was insensitive to parameters which determine the removal rate of sulphur dioxide. These parameters are physical parameters, such as deposition velocity, and meteorological parameters, such as windspeed and mixing layer depth. Closer to the source the deposition is more sensitive to the choice of parameters. This is why the calculated distribution patterns may not agree so well regarding maximum deposition rates in the main source areas but do give the same sort of radial decrease with distance.

The dependence of the spatial distribution on angle depends directly on the directional weighting factor, $F(\theta)$, in equation (1). This weighting factor varies with position in Europe, with type of trajectory used to define the path of air masses, and with long term climatological trends. However, the weighting in any one sector is unlikely to vary by more than a factor of two from the weighting assuming all wind directions are equally likely.

As a general guide therefore the annual average distributions give values of the wet and dry deposition which are probably accurate to within a factor of 2. The aim of this type of calculation is not to obtain accurate estimates of the deposition at any site. It is rather to be able to interpret measurements made at sites remote from sources and interpolate between the sites.

Measurements from the European Air Chemistry Network (see Blokker (1970) and Granat (1972)) and the OECD study on the long range transport of air pollutants (Dovland *et al.* 1976) can be used to check the model calculations. However, the density of sampling sites is usually not accurate enough to be able to draw an unambiguous distribution pattern based on measurements. Moreover the measured deposition at a site will contain short-range contributions, as well as long-range contributions, and may therefore not give representative values for the regional distribution.

Bolin and Persson (1975) compared the distribution pattern they obtained with values from the European Air Chemistry Network. Eliassen and Saltbones (1976) compared values from their Lagrangian dispersion model with measurements from the OECD survey. This was reasonable for deposition over Norway because of the good network of sampling sites in southern Norway. They found that the maximum precipitation zone of southern Norway was along the west coast, whereas the maximum sulphate wet deposition is found along the southern coast. Precipitation on the west coast is thought to occur during westerly winds with fairly clean air, whereas on the south coast precipitation is thought to occur during southerly winds and is associated with air masses which are more likely to have crossed high SO_2 emission areas. Agreement with the measurements was fairly good. Figure 4 shows the wet deposition over Scandinavia using the wind direction weighting for Birkenes (see Fig. 1).

The positions and measurements at sampling stations in the OECD network are shown. The data for the deposition at Norwegian sites is based on the period July 1972 to July 1975. The data for the Danish stations refer to 1973 only and for the German station to 1974 only. The deposition at other sites is based on 1973 and 1974.

Sites marked by a circle are ones for which the sea salt correction has been performed while at sites marked by a square no sea salt correction has been made. Over Scandinavia the Lagrangian model and the present calculations agree fairly well. The main differences occur over the North Sea, where in the Lagrangian model the frequency of rainfall has been underestimated, and to the northeast of Scandinavia, where the Lagrangian model gives about twice the

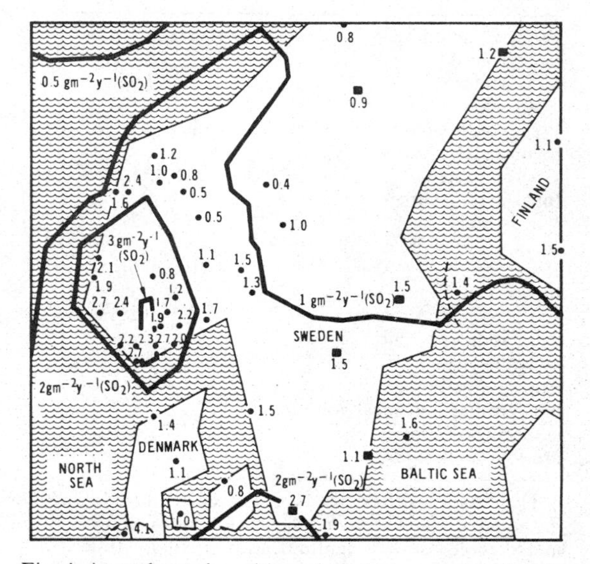

Fig. 4. Annual wet deposition of sulphur over Scandinavia from man-made sources in Europe. There is an additional background contribution of approximately $0.5\ gm^{-2}\ y^{-1}$ (SO_2).

wet deposition of the present model. The reason may be because the Lagrangian model is unable to ensure that once sulphur is removed along the path of a trajectory by rain, it does not appear in wet deposition further along the trajectory. This is the way Eliassen and Saltbones (1976) explain discrepancies at some of the sampling sites.

There is one important difference in the methods of calculating the wet deposition, particularly in remoter areas, such as Norway. The Lagrangian model includes a contribution which is not directly related to the emissions in Europe. This 'background' contribution is important in areas of high rainfall well away from the main sources. The calculation in this paper only contains deposition directly related to sources. Hence the calculated wet deposition attributable directly to manmade sources is somewhat higher than that obtained by Eliassen and Saltbones (1976). This may be because of the deliberately chosen extreme rainfall pattern in the higher rainfall region in Norway (see Fig. 1). The background contribution is discussed further in the next section, where an example is given of how the model can be used to obtain the sulphur budget for a specific region.

7. SULPHUR BUDGET FOR SOUTHERN NORWAY

A sulphur budget for southern Norway has been constructed using the model and conditions described in earlier sections. The weighting of the deposition according to wind direction is based on an analysis by the British Meteorological Office (Smith, 1977) of trajectories arriving at Birkenes over a three year period. This analysis was undertaken in connection with the OECD project on the long range transport of air pollutants. Two back trajectories per day were plotted and allocated to one of five broad sectors. The sectors can be classed according to the main source areas lying in each, namely a United Kingdom sector, a west continental sector, an east continental sector, a northeast sector and an Atlantic sector. The occurrence of stagnant periods is small ($<10\%$) and has been neglected. There is a distinct difference between the angular distribution of trajectories bringing precipitation to the Birkenes region and the angular distribution of all trajectories, regardless of whether there is precipitation. Since the model does not take account of the dependence of rainfall frequency on direction (t_0, t_1 should be functions of θ) neither weighting factor is really appropriate. Air masses in the U.K., east continental and west continental sectors are wetter, on average, when they reach Birkenes, while the air masses in the Atlantic and northeast sectors are drier, on average, when they reach Birkenes. Because the influence of topography is dependent on site position in southern Norway it is preferable to use the weighting based on all air trajectories into the region. The weighting using only air trajectories bringing precipitation is helpful as a guide to the accuracy of these budget estimates. The budgets are shown in Table 2.

The U.K. appears on its own because of its geographical position. It is possible to break down the contribution from each country but this has not been

Table 2. Contributions to the total annual deposition of sulphur over southwestern Norway

Source region	Total deposition g m^{-2} y^{-1} (SO_2)	Wet deposition g m^2 y^{-1} (SO_2)	Ground level SO_2 $\mu g\ m^{-3}$	Concentration SO_4 $\mu g\ m^{-3}$
Europe (including background + local)	3.0 (4.2)	2.4 (3.3)	3.8 (5.8)	2.2 (3.4)
U.K. sector (man-made sources only)	1.0 (1.6)	0.7 (1.2)	5.6	3.4
West continental (man-made sources only)	0.6 (1.0)	0.4 (0.8)	11.2	6.9
East continental (man-made sources only)	0.5 (0.55)	0.3 (0.4)	13.1	7.4
North east (man-made sources only)	0.15 (0.2)	0.1 (0.15)	2.8	1.5
Background (present in all sectors)	0.6 (0.6)	0.6 (0.6)	—	—
Extra local contribution from sources within receptor region	0.2 (0.2)	0.2 (0.2)	—	—
Europe (including background and local sources assuming all emissions are low level)	(4.0)	(3.1)	(5.4)	(3.2)

The Atlantic sector has been chosen to include virtually no man-made sources so that this sector does not appear in the table. The calculated deposition for all sources in Europe includes a background contribution. The values shown in the table use a weighting based on daily trajectories into Birkenes and the values in brackets use a weighting based on trajectories causing precipitation (see Fig. 1). The spread in values gives an indication of the uncertainties. The concentrations of SO_2 and sulphate are the average concentrations in the receptor area when the trajectory lies in the appropriate sector. The annual average is obtained from the weighted average of the concentration in each sector.

done because the accuracy of the calculation does warrant such refined interpretation. The values of the deposition represent an average over an area of Norway bounded by latitude 60° north, longitude 10° east and the sea.

The calculated wet deposition (in Table 2) due to all man-made emissions in Europe ranges from 1.6 to 2.9 $g\,m^{-2}\,y^{-1}\,SO_2$. A value 1.6 is somewhat less than the measured wet deposition ($\sim 2.0\,g\,m\,y^{-1}\,SO_2$ over the region) while 2.9 $g\,m^{-2}\,y^{-1}\,SO_2$ is somewhat higher. The present model, which relates man-made sulphur emissions via transport in the mixing layer to measured deposition, works fairly well. However the error from neglecting the transport of material out of the top of the mixing layer will tend to show itself as a global background on top of the direct long range contribution and will most likely be detectable in precipitation. Natural emissions of sulphur oxides will supplement the longer lived man-made sulphur contribution. An estimate of the natural background can be made from the various sulphur budgets in the literature, which are summarised by Rasmussen *et al.* (1975). Over the earth as a whole the natural sulphur cycle dominates over man-made sulphur emissions. Taking 1.0×10^{14} g as the annual natural emission and a global annual precipitation of 5.10^{17} l of water, the natural background in rain is 0.2 $mg\,l^{-1}$ S (0.6 $mg\,l^{-1}\,SO_4$). This is equivalent to an average background wet deposition of 0.4 $g\,m^{-2}\,y^{-1}\,SO_2$. Rodhe (1972) on the basis of these arguments took the background in Northern Europe to be 0.5 $g\,m^{-2}\,y^{-1}\,SO_2$ in his study of the sulphur budget over Europe.

A background concentration of about 0.6 mg $l^{-1}\,SO_4$ from either natural or longlived man-made sulphur, is consistent with measurements of sulphur in remote areas made during the OECD project on the long range transport of air pollutants. It is also consistent with the empirical relationship used by Eliassen and Saltbones (1976) to derive the wet deposition over Norway from the calculated concentration of SO_2 in air. The intercept of 0.6 $mg\,l^{-1}$ on the empirical curve is the concentration of sulphur in rain not directly attributable to man-made sources within the area of the model. This then corresponds to a background wet deposition over southwest Norway of 0.6 $g\,m^{-2}\,y^{-1}\,SO_2$ taking the average rainfall over the area to be 1500 $mm\,y^{-1}$. This contribution to the sulphur budget has been added to Table 2, and is important.

7.1 *Local sources*

The other contribution which has been omitted from the calculation is the contribution of local sources. Since all sources are at the centre of grid squares while deposition is calculated at the corners of the grid squares the contribution to deposition is neglected from sources within about 100 km of a grid point. The solution of the diffusion equation far downwind a point source of height, h, is approximately

$$G(z,h,t) \simeq \frac{1}{a} S_0(z,h) \exp\left(-t\Big/\left(\frac{a}{v_g} + 0.4\frac{a^2}{K}\right)\right] \quad (16)$$

and this gives the vertical concentration profile at long distances.

It should be noted that the decay time is increased by a factor $0.4\,a^2/K$ compared with the decay time a/v_g used in box models and the Lagrangian model. This extra term represents the time it takes for SO_2 to diffuse down through the mixing layer and can increase decay distances considerably.

The concentration profile has a shape $S_0(z, h)$ which depends on the rate of removal at the ground and the vertical diffusivity.

Equation (16) only applies to the behaviour of sulphur dioxide far downwind of the source. The exact solution is a sum of terms like equation (16) which for short distances tends to the usual Gaussian solution for vertical spread (allowing for partial reflection at the ground and perfect reflection at the top of the mixing layer). Because the profile in the near field does not take the simple form $S_0(z, h)$ there may be greater or lesser deposition in the near field. For a ground level source, the fraction deposited within 100 km of the source is

$$1 - \frac{1}{a}\int_0^a S_0(z,0)\,dz \exp - X_1/u(a/v_g + 0.4a^2/K)$$

$$\simeq 1 - \frac{2(1+0.4\beta)\exp - v_g X_1/ua(1+0.4\beta)}{1+(1+\beta)(1+0.4\beta)} \quad (17)$$

where $\beta = a\,v_g/K$ and $X_1 = 100$ km. With conditions close to the average conditions in Table 1 ($a =$ 1000 m, $K = 10\,m^2\,s^{-1}$, $v_g = 0.005\,m\,s^{-1}$, $u = 10$ m s^{-1}) $\beta = 0.5$ and fraction deposited within 100 km of the source is 0.19. Equation (17) is a simple formula which can be used to estimate the fraction of material which is deposited in dry conditions within a grid square.

The extra amount deposited in the near field is also dependent on the height at which the sulphur is released. For a high level release within the mixing layer, material will be distributed more uniformly before deposition starts to remove material. The amount deposited in the near field is therefore smaller than the amount calculated by equation (17). The correction for source height is obtained by multiplying the second term in equation (17) by the factor

$$\frac{\cos\phi_0(1 - h/a)}{\cos\phi_0}$$

(see equation (2) for definition of ϕ_0). As a first approximation the effect of source height will be neglected in calculating the near field deposition over southern Norway in dry conditions (see next section). The fraction 0.19 is therefore an overestimate. The fraction deposited when it is raining at the time of emission is 1-exp-$\Lambda X_1/u$ ($\simeq 0.63$); and it rains about

10% of the time. Hence a rough estimate of the average deposition over southern Norway can be made. The emissions in the grid squares in southern Norway (Eliassen and Saltbones, 1975) are indicated in Fig. 1. The total deposition over the region is therefore (0.1.0.63 + 0.9.0.19) (7 + 7 + 9 + 26).10^6 kg y^{-1}, leading to an average deposition of 0.2 g m^{-2} y^{-1} SO_2. This term is included in Table 2.

There is an additional contribution from Norwegian sources outside the region. This amounts to a total deposition of 0.1 g m^{-2} y^{-1} SO_2 and a wet deposition of 0.05 g m^{-2} y^{-1} SO_2 which are already included in Table 2 in the other sectors (i.e. north east sector). The contribution from sources within a grid element to the measured deposition at a sampling site is not given by this calculation. The measurements at a sampling site depend on the details of the source distributions within a grid element and is therefore not part of the present calculation.

7.2 *Effect of source height*

The importance of source height has already been discussed in terms of its effect on the average deposition within a grid square from sources within a grid square. The factor by which far field concentrations are increased when the source height is h, compared with the concentration for a ground level source, is $\cos(\phi_0(1 - h/a))/\cos\phi_0$. The correction factor for elevated sources ranges from 1, for low level emissions with an effective source height at the top of the surface layer (see Section 4.1), to $1/\cos\phi_0$ for sources at the top of the mixing layer. Using the approximations leading to equations (4) and (17), the upper bound is $(1 + \beta(+ 0.4\beta))^{1/2} \simeq 1.26$ under typical transport conditions ($\beta \sim 0.5$). This defines an upper limit to the effect of source height on increasing concentrations or depositions at long distances. This result is conditional on material remaining within the mixing layer.

The calculation leading to Fig. 4 was based on the assumption that half the sources in Europe are high level sources ($\sim$ 300 m) and half are low level sources. The same calculation was repeated assuming all sources were low level sources ($\sim$ 30 m). The effect of changes in source height on the smoothed wet deposition (Table 2) is seen to be less than about 10%. Of course source height is much more important on the concentration within a grid square from sources within the same grid square.

8. CONCLUSIONS

The purpose of the model presented in this paper is to devise a fairly rapid and efficient method of determining the pattern of total and wet annual deposition of sulphur over Europe based on the estimated man-made emissions in the area. The resulting patterns are useful, not because they are thought to be highly accurate (no better than a factor of two), but because they can be used to interpret measurements and to interpolate between widely separated measuring sites. Errors in the calculation arise from the simple statistical distributions of windspeed, wind direction, stability and rainfall occurrences assumed in the calculation. Errors from an incorrect choice of values for the removal and transformation parameters are probably not very important, because of the insensitivity of the model to these parameters.

Other factors which are not included in the model, such as large scale vertical motions which might change the relative effectiveness of wet and dry deposition and time and space variations along the trajectory, are not thought to be an important influence in the overall pattern of annual deposition because the calculated pattern compares favourably with measurements. The calculations also agree with those based on other models. The agreement with models based on trajectory computations is especially pleasing, as these models do not rely on a simple weighting factor for wind direction.

The model is based on a system of 127 km by 127 km grid squares which limits the detail in the deposition pattern. Emissions are defined with the same resolution and anyway are not known with very great accuracy (at best to $\pm 10\%$ and probably much worse). Regions of very high concentration, such as urban areas, do not appear in the calculated deposition patterns because they usually occur over a length scale shorter than the grid length (127 km) in the model calculation.

The calculated patterns of annual deposition show that the highest total deposition rates (10 g m^{-2} y^{-1} SO_2) occur within distances of order 100 km from the main emissions. The wet deposition does not show such pronounced maxima in regions closest to the sources. The decrease in wet deposition is also modified in regions of high annual precipitation.

One such area is southern Norway. The model takes into account wet removal by a statistical treatment of periods of precipitation. For most areas of Europe the probability of precipitation is kept the same, but a method for including pronounced precipitation is presented. Because of the special features of the wet deposition over southern Norway and the large number of sampling sites in the area, it is used in an example of applying the model to a specific region.

The calculated wet deposition agrees fairly well with the measured deposition over southern Norway. The contributions from sources within broad direction sectors to the annual deposition show the expected dependence on source strength and wind direction. The contribution from individual countries could be shown, but the accuracy of the calculation is insufficient to justify such refinement. The main deposition arises from sources outside the receptor region, because the region contains very few sources. There are two additional factors which are difficult to assess within the model and apply to the budget for southern Norway.

Firstly the model excludes the contribution to deposition within a grid square from sources within the same grid square. A separate calculation is necessary to make an estimate of this contribution. Even this is insufficient to assess the contributions at a sampling site since this depends on details of the source distribution and dispersion in the neighbourhood of the site. Apart from this local contribution, the model does maintain a sulphur balance between emissions and deposition over Europe. Secondly the contribution from the background concentration of sulphate in rain, not directly attributable to man-made sources, is important in remote areas of high precipitation. This background contribution is an important part of the budget and warrants further investigation.

One advantage of the model is that it includes vertical dispersion within the mixing layer. The behaviour of material in the short to medium range is automatically taken into account. Simple approximate formulae are derived for the fraction of emitted material which is deposited in the short to medium range. This fraction depends on source height. However, as the fraction is usually small most material has the opportunity of long range transport and the effect of source height is not as important as other factors. Source height thus strongly influences concentrations and deposition in the grid square surrounding the emission, but is less important than the wind direction distribution and the uncertainty in source strength.

Acknowledgement—This work was carried out at the Central Electricity Research Laboratories and is published by permission of the Central Electricity Generating Board.

REFERENCES

Blokker P. C. (1970) The atmospheric chemistry and long range-drift of sulphur dioxide. *J. Inst. Petroleum* **56**, 71–79.

Billingsley J., Kallend A. S. and Marsh A. R. W. (1976) Washout of sulphur dioxide by rain, CEGB Report RD/L/N 85/76.

Bolin B. and Persson C. (1975) Regional dispersion and deposition of atmospheric pollutants with particular application to sulphur pollution over Western Europe. *Tellus* **27**, 281–309.

Carson D. J. (1973) The development of a dry inversion-capped convectively unstable boundary layer. *Quart. J. Roy. Met. Soc.* **99**, 450–467.

Chamberlain A. C. (1966) Transport of Lycopodium spores and other small particles to rough surfaces. *Proc. Roy. Soc.* **A296**, 48–70.

Dovland H., Joranger E. and Semb A. (1976) Deposition of air pollutants in Norway, in summary report on the research fresults from phase I of the SNSF project (1972–1975), pp. 15–35.

Eliassen A. and Saltbones J. (1975) Decay and transformation rates of SO_2 as estimated from emission data, trajectories and measured air concentrations. *Atmospheric Environment* **9**, 425–430.

Eliassen A. and Saltbones J. (1976) Concentration of sulphate in precipitation and computed concentrations of sulphur dioxide. In *Atmospheric Pollution*, pp. 123–133. Elsevier, Amsterdam.

Fisher B. E. A. (1975) The long range transport of sulphur dioxide. *Atmospheric Environment* **9**, 1063–1070.

Fisher B. E. A., Gotaas Y., Hamilton P. M., Houlgate R., Maul P. and Moore D. J. (1977) Observations and calculations of airborne sulphur from multiple sources out to 100 km. *Atmospheric Environment* **11**, 1163–1170.

Førland E. J. (1973) A study of the acidity in the precipitation in southwestern Norway. *Tellus* **25**, 291–299.

Granat L. (1972) Deposition of sulfate and acid with precipitation over northern Europe. University of Stockholm, Institute of Meteorology Report AC-20.

Högström U. (1975) Further comments on the long range transport of airborne material. *Atmospheric Environment* **9**, 946–947.

Moore D. J. (1975) A simple boundary layer model for predicting time mean ground-level concentration of material emitted from tall chimneys. *Proc. Inst. Mech. Engineers* **189**, 33–43.

Moore D. J. (1977) Possible substitutes for tall tower meteorological data in plume dispersion predictions. ASTM Conference on Air Quality Meteorology, Boulder, Colorado, August, 1977.

Munn R. E. and Rodhe H. (1971) On the meteorological interpretation of the chemical composition of monthly precipitation samples. *Tellus* **23**, 1–13.

Ottar B. (1976) Monitoring long-range transport of air pollutants: the OECD study. *Ambio* **5**, 202–206.

Prahm L. P., Torp U. and Stern R. M. (1976) Deposition and transformation rates of sulphur oxides during atmospheric transport over the Atlantic. *Tellus* **18**, 355–372.

Rasmussen K. H., Taheri M. and Kabel R. L. (1975) Global emissions and natural processes for removal of gaseous pollutants. *Water, Air Soil Pollut.* **4**, 33–64.

Rodhe H. (1972) A study of the sulphur budget for the atmosphere over northern Europe. *Tellus* **24**, 128–138.

Rodhe H. and Grandell J. (1972) On the removal time of aerosol particles from the atmosphere by precipitation scavenging. *Tellus* **24**, 442–454.

Scriven R. A. and Fisher B. E. A. (1975) The long range transport of airborne material and its removal by deposition and washout. *Atmospheric Environment* **9**, 49–68.

Sehmel G. A. and Sutter S. L. (1974) Particle deposition rates on a water surface as a function of particle diameter and air velocity. *J. Rech. Atmos.* **8**, 911–920.

Smith F. B. and Jeffrey G. H. (1975) Airborne transport of sulphur dioxide from the United Kingdom. *Atmospheric Environment* **9**, 643–659.

Smith F. B. (1977) private communication.

Atmospheric Environment Vol. 12, pp. 503–509. Pergamon Press 1978. Printed in Great Britain.

TRANSFORMATION AND REMOVAL PROCESSES FOR SULFUR COMPOUNDS IN THE ATMOSPHERE AS DESCRIBED BY A ONE-DIMENSIONAL TIME-DEPENDENT DIFFUSION MODEL*

GUNNAR OMSTEDT and HENNING RODHE

Department of Meteorology, University of Stockholm, Arrhenius Laboratory, S-106 91 Stockholm, Sweden

(*First received* 14 *June* 1977 *and in final form* 10 *August* 1977)

Abstract—The model is used to simulate vertical profiles of H_2S, SO_2 and SO_4^{2-} in the atmosphere up to about 8 km. Transformation and wet removal processes are treated as first order reactions with constant rate coefficients and the dry deposition is estimated using deposition velocity parameters. From a systematic study of the sensitivity of the model to variations in some of the key parameters and from a comparison with aircraft measurements of SO_2 and SO_4^{2-} over Scandinavia the following results are derived. An increase in the value of the deposition velocity in long range transport models beyond about 1 cm s^{-1} has little effect on the estimated dry deposition. As the rate coefficient for the transformation of H_2S to SO_2 varies from 10 to 0.01 h^{-1} the scale height of H_2S varies from 40 to 1400 m and that of SO_2 (applicable to background air) from 1300 to 3400 m. An average value for the first 30 hours of the rate of transformation of man-made SO_2 to SO_4^{2-} is probably in the range 0.007 to 0.04 h^{-1} for European conditions. It is difficult to simulate the observed pattern of wet deposition of sulfur over northern Europe if the wet removal of SO_2 is neglected.

1. INTRODUCTION

In this paper we address ourselves to the question of how the relative abundance in the atmosphere and the observed deposition of the various sulfur compounds is interpretable in terms of transformation and removal processes. We apply a one-dimensional time-dependent diffusion model similar to the one used by for example Draxler and Elliott (1977). Natural sulfur emissions are assumed to be in the form of H_2S (or other volatile reduced sulfur compounds) and take place at the surface. Further assumptions include the transformation of H_2S to SO_2 through a first order process, a similar transformation of SO_2 to SO_4^{2-} and wet and dry removal of the latter two compounds. Man-made emissions are assumed to be in the form of SO_2 only and to take place at levels between 50 and 300 m.

Admittedly, such a model can only give a very crude description of the actual conditions and the model results must be used with caution particularly in situations when the sources and sinks have a non-uniform distribution horizontally. We believe, however, that some useful information can still be derived from such a model and that the rather limited observational data presently available may not fully justify the use of a more sophisticated model for simple considerations of this kind. We shall, at least partially, avoid some of the uncertainties introduced by the model by concentrating on the *relative abundance* of the different sulfur compounds. The inadequacies in the description of the transport and diffusion processes are then less critical.

Bolin *et al.* (1974) used a similar approach to study the influence on residence times of source characteristics, diffusion processes and sink mechanisms. Their study was however limited to a single pollutant and to steady state conditions. The former limitation also applies to the work by Draxler and Elliott (1977).

2. THE MODEL

The basic equation for the model is given by

$$\frac{\partial q}{\partial t} = \frac{1}{\rho}\frac{\partial}{\partial z}\left(\rho D \frac{\partial q}{\partial z}\right) - k_i q + S \qquad (1)$$

where q is the ratio of the mass of the sulfur (H_2S-S, SO_2-S or SO_4^{2-}-S) to the mass of the air, ρ = density of air, D = vertical diffusion coefficient, k_i = rate coefficients for transformation or removal processes and S = source term.

We assume that the diffusion coefficient D increases linearly from the ground up to a height H_0 above which it remains constant

$$D = \begin{cases} \kappa u_* z & z_0 \le z \le H_0 \\ \kappa u_* H_0 & z > H_0 \end{cases}$$

z_0 is here the roughness height, u_* the friction velocity and κ von Karman's constant. At the upper boundary no flux is permitted ($\partial q/\partial z = 0$) whereas at the height z_0 the boundary conditions for SO_2 and SO_4^{2-} are

* Contribution No. 355.

given by

$$\rho \kappa u_* z_0 \frac{\partial q}{\partial z} = v_d q \rho. \tag{2}$$

The parameter v_d—the deposition velocity—is thus defined as the ratio of the flux and the concentration at the height z_0. For H_2S the upward flux is specified at the height z_0.

Equation (1) is applied simultaneously to the three sulfur compounds. The transformation terms thus appear as a sink in one equation and a source in another. We use the following implicit Crank–Nicolson finite difference scheme which has good stability characteristics and a second order accuracy.

$$\frac{q_j^{n+1} - q_j^n}{t_{n+1} - t_n} = \frac{(\rho D)_{j+1/2}\left(\dfrac{q_{j+1}^{n+1} + q_{j+1}^n - q_j^{n+1} - q_j^n}{z_{j+1} - z_j}\right) - (\rho D)_{j-1/2}\left(\dfrac{q_j^{n+1} + q_j^n - q_{j-1}^{n+1} - q_{j-1}^n}{z_j - z_{j-1}}\right)}{\rho_j(z_{j+1} - z_{j-1})} - k\frac{q_j^{n+1} + q_j^n}{2} + S \tag{3}$$

where n is the time index and j the height index. The lower boundary condition is expressed as

$$D_{1\frac{1}{2}} \frac{q_2^{n+1} - q_1^{n+1}}{z_2 - z_1} - \frac{v_d}{2}(q_2^{n+1} + q_1^{n+1}) = 0. \tag{4}$$

The grid system is logarithmically spaced with six points per decade and covers the range 0.001 to 10.000 m. In order to minimize numerical disturbances the time step is initially kept at only 1 s. It is gradually increased to 10 min. The total mass has been shown to be well conserved: to within 2% after 300 h.

3. THE CHOICE OF PARAMETER VALUES

Even a simple model like the present one includes several parameters whose numerical values are not easily assigned. The philosophy adopted here is to postulate a set of 'test values' and a range of variability and/or uncertainty around those values. These values are given in Table 1. When nothing else is mentioned the parameters have been given the test values. No attempt is made to present a comprehensive discussion about the rationale behind the choice of these values. We make only a few comments.

u_* *and* H_0. H_0 corresponds to the height of the surface layer in which the profiles of wind and temperature are close to logarithmic under neutral stability conditions. With $\kappa = 0.4$ the value of the diffusion coefficient D above the surface layer will be in the range 4–40 $m^2\ s^{-1}$. In a general sense these limits correspond to a stable and an unstable situation, respectively.

z_0. The lower limit of the roughness height applies to, for example, short grass whereas the upper limit corresponds to a rough surface like a forest or a city.

v_d. For a discussion about different estimates of the deposition velocity over various surfaces reference is made to the review by Garland (1978).

k_i. We use the following rate coefficients: k_{c1}—transformation H_2S to SO_2, k_{c2}—transformation SO_2 to SO_4^{2-}, k_{w2}—precipitation scavenging of SO_2, k_{w3}—precipitation scavenging of SO_4^{2-}.

All rate coefficients are assumed to be independent of height and time. In reality, particularly k_{w2} and k_{w3} are likely to have a pronounced variation up through the troposphere, but the complication of allowing for such variations in the model was not considered desirable at this stage. Variations due to meteorological conditions also had to be neglected. Because of the idealized model the rate coefficients may not have exact physical interpretations. This will contribute to making the appropriate numerical values quite uncertain. Concerning previous attempts to estimate such coefficients reference is made to works by, for example, Cox (1975) (k_{c1}), Rodhe and Grandell (1972) (k_{w2} and k_{w3}) and Alkezweeny and Powell (1977) (k_{c2}). For a further discussion see Section 4.4. It is worth pointing out that the time scales derived by Rodhe and Grandell (1972) for wet removal do not strictly correspond to first order reaction coefficients. Their coefficients represent averages over residence time distributions that are not quite exponential as the assumption of first order reactions requires. However, for our present purposes the difference is not significant.

Table 1. Model parameters used in the study. For further explanation see text

Parameter	Test value	Range	Unit
u_*	0.25	0.1–1	$m\ s^{-1}$
H_0	100	—	m
z_0	0.1	0.01–1	m
v_d-SO_2	1	0–4	$cm\ s^{-1}$
v_d-SO_4	0.3	0–4	$cm\ s^{-1}$
k_{c1}^{-1}	10	0.1–100	h
k_{c2}^{-1}		25–150	h
k_{w2}^{-1}	variable	50 and above	h
k_{w3}^{-1}	cf. Table 4	50 and above	h

4. RESULTS

In this section we first present some results obtained by varying one of the parameters at a time within the given range. Subsequently we study the effect of various combinations of the key parameters (k_{c2}, k_{w2} and k_{w3}) while keeping the others at some typical values. By comparison with observations we try to draw some general conclusions about the actual transformation and removal rates.

4.1 *Dependence on* z_0 *and* u_*

From Equation (2) it would seem as if the dry deposition was proportional to z_0. However, the concentration gradient decreases as z_0 increases so that the

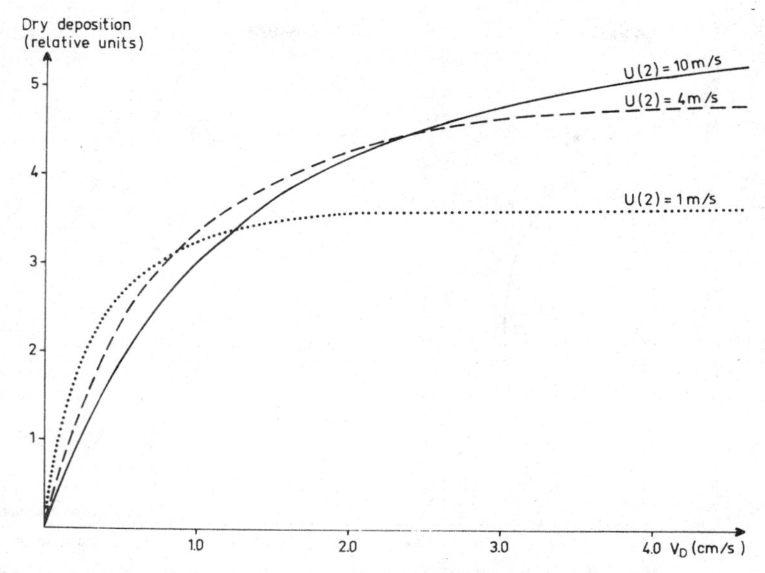

Fig. 1. Dry deposition of SO_2 as a function of the deposition velocity for three cases with u_* values adjusted to give windspeeds at 2 m equal to 1, 4 and 10 m s^{-1} respectively. The model has been run to a steady state with continuous emission taking place at heights between 50 and 300 m.

total effect of changes in z_0 on the model calculations is small.

u_* is an important parameter which determines the values of the diffusion coefficient. For example, a change of u_* from 0.1 to 1 m s^{-1} increases the scale height (cf. Section 4.3) of H_2S by a factor 3.6 and of SO_2 by a factor 2.5 in a steady state situation.

4.2 *Dependence on* v_d

The value of the deposition velocity parameter determines the uptake rate at the surface. However, this dependence is not proportional because an increased v_d will decrease the concentration close to the surface thereby limiting the increase in the deposition. For deposition velocities greater than about 1 cm s^{-1} this compensating effect is very pronounced. For such large values of the deposition velocity the aerodynamic resistance in the surface layer will limit the uptake rate.

To illustrate this effect Fig. 1 shows the dry deposition of SO_2 as a function of the deposition velocity for three cases with u_*-values adjusted to give windspeeds at 2 m equal to 1, 4 and 10 m s^{-1}. A continuous emission of SO_2 is taking place at heights between 50 and 300 m and the model has been run to a steady state. It is seen that, particularly at low wind speeds, deposition velocities above 1 cm s^{-1} will produce only a very slight increase in the dry deposition.

It may be worth pointing out that the above conclusion applies to model calculations. When estimating the actual dry deposition from observed values of the surface air concentration, the deposition will of course be proportional to the chosen value of the deposition velocity. Results similar to those mentioned in Sections 4.1 and 4.2 are discussed in somewhat more detail in the study by Bolin *et al.* (1974).

4.3 *Dependence on the oxidation rate of* H_2S

Granat (Granat *et al.*, 1976, p. 115) presents some arguments about how the relative importance of wet and dry removal of natural SO_2 depends on the oxidation rate of H_2S. The point in question is that a short turn-over time (i.e. a rapid oxidation) will keep H_2S confined to the lowest layers of the atmosphere and the effective source height for the SO_2 produced by this process will be low. In such a case the dry deposition of SO_2 will be relatively more important than if H_2S has had time to spread to higher elevations before the oxidation.

In order to evaluate this effect quantitatively we have run the model with the inverse value of the transformation rate H_2S to SO_2 (k_{c1}^{-1}) equal to 0.1, 1, 10 and 100 h. The result is shown in Fig. 2 and Table 2. In these examples the dry deposition velocity of SO_2 was equal to 0.6 cm s^{-1} and the inverse value of the combined rate of the other removal processes 55 h. The scale heights H have been defined as

$$H = \frac{\int_0^\infty \rho q \, dz}{(\rho q)_{8\mathrm{m}}}.$$

It is seen that the effect of variations in k_{c1} on the scale height of H_2S is very pronounced. The scale height of SO_2 as well as the ratio of dry deposition of SO_2 to other removal processes is much less dependent on this rate coefficient. By comparison with Granat's arguments it is seen that even a very wide range of k_{c1} values does not produce such large differences in the scale height of SO_2 and in the relative importance of dry deposition and other removal processes as implied in his study.

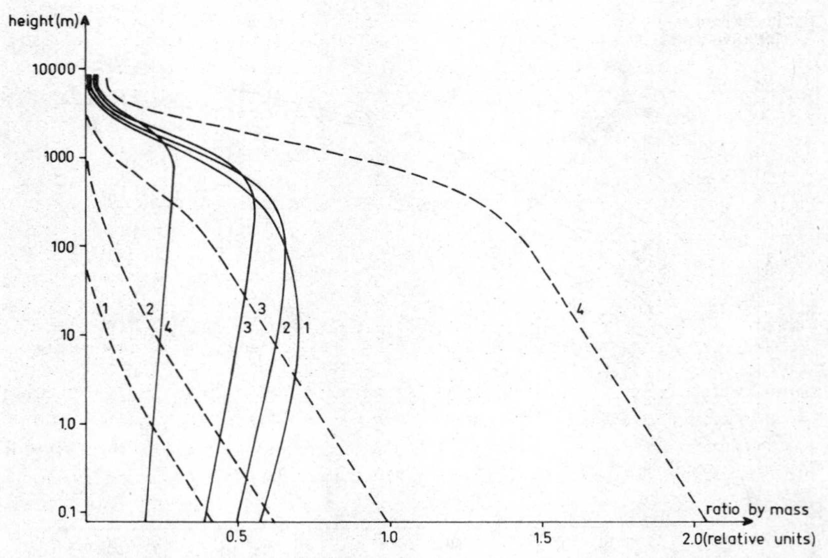

Fig. 2. Vertical distributions of H_2S-S (dashed lines) and SO_2-S (solid lines) under steady state conditions for the following values of the rate coefficient (k_{c1}) for the transformation of H_2S to SO_2: 10 h^{-1} (1), 1 h^{-1} (2), 0.1 h^{-1} (3) and 0.01 h^{-1} (4). The deposition velocity for SO_2 in these cases has been taken as 0.6 cm s^{-1} and the removal rate of SO_2 equal to 0.018 h^{-1}.

4.4 *Possible combinations of* k_{c2}, k_{w2} *and* k_{w3}

In this section we shall compare some observational evidence with model calculations and attempt to indicate likely ranges for some of the rate coefficients (k_{c2}, k_{w2} and k_{w3}) for man-made SO_2 emissions over Europe.

As demonstrated by Rodhe and Grandell (1972) wet removal rates are determined not only by the incloud or subcloud scavenging rates but also by the frequency of occurence of dry and wet spells. They showed that even for a very rapid incorporation of sulfur into the raindrops the average inverse value of the removal rate coefficient will be about 90 hours in summer and 40 hours in winter. Although these estimates were based on a rather limited amount of data (rainfall records from Stockholm) it seems fair to assume that as an annual average k_{w2}^{-1} and k_{w3}^{-1} both must be greater than or about equal to 50 h for the European climate.

On the other hand, if the rate of wet removal of both SO_2 (k_{w2}) and SO_4^{2-} (k_{w3}) were very low it would be hard to explain the observed maximum of wet deposition of sulfur over Europe. The estimate of the sulfur budget over NW Europe made by Rodhe (Granat *et al.*, 1976) indicates that within the area studied roughly 30% of the emitted sulfur had been removed by precipitation scavenging. By comparison, the dry deposition within the same area was estimated to be 20–45% of the emissions.

Other observational data include measurements of SO_2 and SO_4^{2-} concentrations in air. Aircraft measurements obtained within the LRTAP project (Gotaas, 1975) provide a useful set of data for our purposes.

Lagrangian box models with linear transformation and removal processes have been used for example in connection with the LRTAP project in Europe (Eliassen and Saltbones, 1975) and the SURE project in the U.S. (EPRI, 1976) to study the long range transport of sulfur compounds. Attempts were made to determine model parameters including transformation and removal rates by a systematic comparison with observed values. Unfortunately, wet removal processes were not included in these comparisons and therefore the results are not directly applicable to our model.

Let us envisage a reasonably large 'parcel of air' which moves across the major source regions of man-made SO_2 emissions over the European continent or England. After having been loaded with SO_2 it moves out over less polluted areas, the sulfur being subject to transformation and removal processes. In the model we simulate this situation by applying a constant emission rate for a period of 10 h and thereafter running the model without emissions. Because of the

Table 2. Calculated values of the scale heights for H_2S and SO_2 and of the ratio between dry deposition and other removal processes for SO_2 for different values of the rate of transformation between H_2S and SO_2

k_{c1}^{-1} (h)	H_{H_2S} (m)	H_{SO_2} (m)	Ratio of dry deposition to other removal processes for SO_2
0.1	40	1330	0.7
1	130	1620	0.6
10	490	2280	0.4
100	1440	3360	0.3

Table 3. Measurements of the ratio SO_2-S to SO_4^{2-}-S made over southern Sweden at heights ranging from 350 to 2000 m. The travel time refers to the time since the corresponding 850 mb trajectory left the indicated source region

Date	Travel time (h)	$\alpha = (SO_2\text{-S}/SO_4^{2-}\text{-S})$ Average value	Individual values	Source region
731105	18	3.1	0.85, 0.96, 3.0, 3.4, 3.8, 6.5	North Western European continent
740924	20	2.3	1.7, 1.95, 1.95, 2.2, 2.4, 2.9, 3.3	North Western European continent
741204	20	1.5	1.2, 1.7	England
730613	27	1.2	0.73, 0.88, 2.1	England

crudeness of our model and of the difficulty in prescribing an appropriate source strength, we pay little attention to the absolute values of the SO_2 and SO_4^{2-} concentrations but focus on the ratio α of the amounts of sulfur occuring in the form of SO_2 and SO_4^{2-}

$$\alpha = (SO_2\text{-S}/SO_4^{2-}\text{-S}).$$

This ratio is large initially but decreases as more and more SO_4^{2-} is formed in the parcel of air as it moves away from the source regions. In order to make a fair comparison with observed ratios we have avoided surface air measurements where an influence from local SO_2 sources may have affected the values. Situations when the observed concentrations of either SO_2 or SO_4^{2-} was below 1 μg S m^{-3} were considered to be uncertain and were also excluded from the analysis.

From the aircraft measurements made by Trägårdh (1977) over southern Sweden during the LRTAP project we have selected only those situations when air trajectories (calculated at the 850 mb level) leave no doubt that the air in which the measurements were made had passed over England or the European continent a well defined period of time earlier. Table 3 summarizes the observed α-values and the corresponding travel times from the source region. Because of the constraints mentioned above the amount of data is unfortunately quite small. The heights at which the measurements were made varies from 350 to 2000 m.

In Table 4 we summarize the results of the model calculations for various combinations of the parameters k_{c2}, k_{w2} and k_{w3}. The numbers refer to the inverse of the rate coefficients and the unit is hours. The numerical values of the other model parameters are the 'test values' given in Table 1. The α values are evaluated at a height of 1 km and the time of evaluation is taken as the time since the emissions were turned off in the model. The β values represent the ratio between dry deposition (both SO_2-S and SO_4^{2-}-S) and wet deposition (also both SO_2-S and SO_4^{2-}-S) accumulated up to the indicated time. The β values may be used for a comparison with the regional scale budget estimates mentioned above.

There are certainly other possible combinations of the parameters but we believe that the fifteen cases include some of the most important possibilities. By comparing the model calculations shown in Table 4 with the observational evidence discussed above the following tentative conclusions may be drawn.

(i) For values of k_{c2}^{-1} equal to 25 h (case 1) the α values for travel times 20–30 h are low compared to the averages of the observed values (cf. Table 3). Despite a rapid removal of SO_4^{2-} (k_{w3}^{-1} = 50 h) the SO_4^{2-} levels are high compared to the SO_2 levels. Thus, a value of k_{c2}^{-1} around or below 25 h (i.e. $k_{c2} \gtrsim 0.04\ h^{-1}$) does not seem to be likely.

(ii) k_{c2}^{-1} equal to 150 h (cases 14 and 15) implies a slower conversion of SO_2 to SO_4^{2-}. Even with no wet removal of SO_4^{2-} (case 15) the α values are high compared to the observations. Consequently, k_{c2}^{-1} should probably be less than 150 h.

(iii) If the wet removal of SO_2 is negligible, as in cases 7 and 13, even a reasonably rapid transformation of SO_2 to SO_4^{2-} and a subsequent rapid wet

Table 4. Values of the ratios α and β (for definitions see text) as functions of the travel time for various combinations of the model parameters describing conversion between SO_2 and SO_4^{2-} (k_{c2}) and wet removal of SO_2 (k_{w2}) and SO_4^{2-} (k_{w3})

Case No	1	2	3	4	5	6	7	8	9	10	11	12	13	14	15
k_{c2}^{-1} (h)	25	50	50	50	50	50	50	100	100	100	100	100	100	150	150
k_{w2}^{-1} (h)	∞	50	50	50	100	100	∞	50	50	50	100	100	∞	50	50
k_{w3}^{-1} (h)	50	50	100	∞	50	100	50	50	100	∞	50	100	50	100	∞
α(15 h)	0.96	1.9	1.7	1.5	2.1	1.9	2.4	4.3	3.8	3.4	4.7	4.2	5.2	5.9	5.3
α(20 h)	0.71	1.4	1.2	1.1	1.6	1.4	1.8	3.2	2.8	2.4	3.7	3.2	4.2	4.4	3.8
α(25 h)	0.55	1.1	0.92	0.76	1.3	1.1	1.5	2.6	2.2	1.8	3.0	2.6	3.5	3.5	2.9
α(30 h)	0.44	0.87	0.70	0.57	1.1	0.86	1.3	2.1	1.7	1.4	2.5	2.1	3.0	2.8	2.3
β(20 h)	2.7	1.0	1.1	1.3	1.6	1.9	5.1	1.0	1.1	1.2	1.8	2.0	9.8	1.1	1.1
β(50 h)	1.3	0.75	0.88	1.2	1.1	1.4	2.2	0.79	0.86	1.0	1.3	1.4	4.0	0.86	0.95

removal of SO_4^{2-} (case 7) would imply a large ratio β of dry deposition to wet deposition. The estimate of the sulfur budget over Europe (Granat *et al.*, 1976) gave values of this ratio of 0.7 to 1.5 for travel times of roughly 30–50 h. Comparing these figures we conclude that it seems unlikely that the wet removal of SO_2 can be neglected.

(iv) Judging only from the model calculations of α and β the neglect of wet removal of SO_4^{2-} (cases 4 and 10) would not seem to lead to any contradictory results. However, this possibility may be excluded for other reasons such as, for example, the estimates of life times of radioactive particles in the troposphere (Martell and Moore, 1974).

(v) In order to have a rapid enough wet removal of sulfur (i.e. to have a β value comparable to that derived from the budget estimates) at least one of the wet removal rates has to be larger than or about equal to $0.01\ h^{-1}$. In cases 6 and 12 where this is not the situation it is seen that the β values are quite high. Similar high values of β occur in case 11 where the more rapid wet removal of SO_4^{2-} is compensated by a slow rate of transformation from SO_2.

We believe that it is not possible with the present model to draw any firmer conclusions regarding these parameters and to separate between the more reasonable cases (2, 3, 5, 8 and 9).

In order to see how sensitive the results are to the assumed vertical profile of the diffusion coefficient D, we have also run the model with D decreasing to $2\ m^2\ s^{-1}$ above 1000 m. This results in a slight increase in the relative importance of dry deposition—β increases less than or equal to 10%—but it doesn't affect the above conclusions.

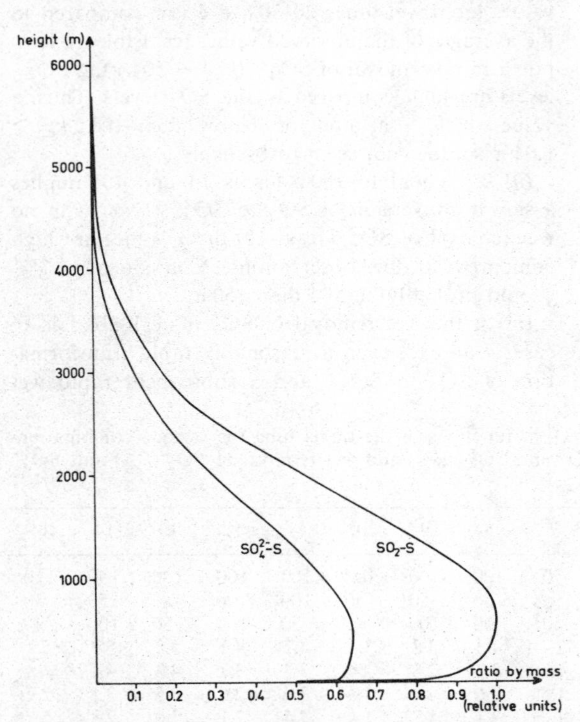

Fig. 3. Vertical distribution of SO_2-S and SO_4^{2-}-S 20 hours after the emissions have been turned off. The parameter values are those of case 5 in Table 4.

An additional possible way to check the model results would be to look at the vertical profiles of SO_2, SO_4^{2-} and α. In Fig. 3 we show the model profiles of SO_2-S and SO_4^{2-}-S after a travel time of 20 hours with parameter values according to case 5 in Table 4. Unfortunately, the observations referred to in Table 3 are too few and too variable to make a comparison with the model profiles meaningful. For example, only 3 out of the 18 measurements were taken at heights above 1 km.

5. CONCLUDING REMARKS

We wish to emphasize that, in addition to the uncertainties introduced by the simplified treatment of the transport and dispersion, the assumption about first order transformation and removal processes may be a serious limitation. In particular, the comparison made in Section 4.4 between model calculations and observations were made only for intermediate travel times of 20–30 h. The inferences about the magnitudes of the time scales of the transformation and removal processes may therefore be representative neither for conditions close to the source nor for a full life cycle of the sulfur in the atmosphere (cf. Rodhe, 1978).

It is not unlikely that one can learn more about dispersion, transformation and removal processes by further experiments with a one-dimensional model such as the one described in this paper. For example it would be interesting to study the effect of a time varying boundary layer. However, for many situations more detailed quantitative results can probably only be arrived at with the aid of dispersion models with at least one additional space dimension. Much better observations of concentrations of the various sulfur compounds are also required particularly from the troposphere above the boundary layer—in polluted air as well as in background air—before reliable estimates of the overall transformation and removal rates can be made.

Acknowledgements—This work has been sponsored by the Swedish Natural Science Research Council under contract No G3922-001.

REFERENCES

Alkezweeny A. J. and Powell D. C. (1977) Estimation of transformation rate of SO_2 to SO_4 from atmospheric concentration data. *Atmospheric Environment* **11**, 179–182.

Bolin B., Aspling G. and Persson C. (1974) Residence time of atmospheric pollutants as dependent on source characteristics, atmospheric diffusion processes and sink mechanisms. *Tellus* **26**, 185–195.

Cox R. A. (1975) Atmospheric photo-oxidation reactions. A.E.R.E. R-8132, Harwell, England.

Draxler R. R. and Elliott W. P. (1977) Long-range travel of airborne material subjected to dry deposition. *Atmospheric Environment* **11**, 35–40.

Eliassen A. and Saltbones J. (1975) Decay and transformation rates of SO_2, as estimated from emission data, trajectories and measured concentrations. *Atmospheric Environment* **9**, 425–429.

EPRI (1976) Design of the sulfate regional experiment (SURE). Final report of research project 485. Vol. 1: Supporting data and analysis. Electric Power Research Institute, California, USA.

Garland J. A. (1978) Dry and wet removal of sulphur from the atmosphere. *Atmospheric Environment* **12**, 349–362.

Gotaas Y. (1975) Aircraft sampling of sulphur dioxide and sulphates—discussion of results obtained within the OECD programme (Preliminary report). LRTAP 7/75, NILU, Oslo.

Granat L., Hallberg R. O. and Rodhe H. (1976) The global sulphur cycle. In *Nitrogen, Phosphorus and Sulphur—Global Cycles* (edited by B. H. Svensson and R. Söderlung). SCOPE Report 7. Ecol. Bull. (Stockholm) **22**, 89–134.

Martell E. A. and Moore H. E. (1974) Tropospheric aerosol residence times: a critical review. *J. Rech. Atmos.* **8**, 903–910.

Rodhe H. (1978) Budgets and turn-over times of atmospheric sulfur compounds. *Atmospheric Environment* **12**, 671–680.

Rodhe H. and Grandell J. (1972) On the removal time of aerosol particles from the atmosphere by precipitation scavenging. *Tellus* **24**, 442–454.

Trägårdh C. (1977) Department of Meteorology, University of Stockholm. Personal communications.

Atmospheric Environment Vol. 12, pp. 511–527. Pergamon Press 1978. Printed in Great Britain.

LONG TERM REGIONAL PATTERNS AND TRANSFRONTIER EXCHANGES OF AIRBORNE SULFUR POLLUTION IN EUROPE

W. B. JOHNSON, D. E. WOLF and R. L. MANCUSO
Atmospheric Sciences Laboratory, SRI International, Menlo Park, California 94025, U.S.A.

(*First received* 28 *July* 1977)

Abstract—This paper reports on progress to date of an ongoing effort to develop, evaluate, and apply a European Regional Model of Air Pollution (EURMAP). This model is capable of calculating long-term (monthly, seasonal, and/or annual) averages of the contributions from SO_2 in individual emittor countries to SO_2 and SO_4^{2-} concentrations, dry deposition, and wet deposition in receptor countries. The model covers all of western and central Europe, a geographical area 2100 km × 2250 km in size. A trajectory-type approach is used, which involves the tracking of pollutant 'puffs' released from each emissions cell in an extensive 32 × 36 grid. Meteorological data in the form of wind and precipitation values from some 45 upper-air and 535 surface stations are input at 6-hourly intervals for use in the calculations of puff transport and wet deposition. A wet deposition coefficient is used that depends upon precipitation rate.

The preliminary model has been used to calculate annualized as well as monthly mean maps for January, April, July, and October 1973 of SO_2 and SO_4^{2-} concentration, dry deposition, and wet deposition patterns resulting from SO_2 emissions in 13 countries in western and central Europe. The dry and wet deposition patterns are presented, along with values of calculated international exchanges of SO_2 and SO_4^{2-} wet and dry deposition among these various countries. The EURMAP results are compared with those from Fisher's (1975) model and the LRTAP model (Ottar, 1978; OECD, 1977). In many (but not all) respects the results from the three models are similar. The possible reasons for the differences revealed by this comparison are examined.

1. INTRODUCTION

There has been in recent years, a growing awareness that air pollution can have significant effects on areas located long distances (thousands of km) away from source regions. The quality of the background air coming into a given country may be critically important in terms of the ability of that country to meet its air quality goals.

Thus there has developed a need to be able to more accurately identify source areas affecting air quality in a given region, and to quantify those effects. This need has prompted considerable efforts, particularly in Europe, for the development and evaluation of regional-scale air pollution simulation models. The most ambitious study has been the cooperative OECD-sponsored LRTAP* project, led by the Norwegian Institute for Air Research (Ottar, 1978; OECD, 1977). Other major efforts include the work of Prahm and Christensen (1977), Wendell *et al.* (1976), Fisher (1975), Bolin and Persson (1975), Rao *et al.* (1976), Smith and Jeffrey (1975), Renne and Elliott (1976), McMahon *et al.* (1976), and Sheih (1977), among others.

The work discussed in this report involves the development of an innovative trajectory-type regional air pollution model, called 'EURMAP', which stands for 'European Regional Model of Air Pollution'. The overall objectives of this study are to develop, evaluate, and apply an accurate and practical quantitative technique for assessing the current and future effects of air pollution emissions in each country in central and western Europe on long-term air quality in each of the other countries. This report is intended to be an updating of our earlier progress report (Johnson *et al.*, 1976) on this continuing project.†

The model developed for this purpose (EURMAP) has a number of distinctive features, among which are these:

(1) It uses as input long sequences of historical meteorological data at 6-h intervals, and preserves all of the temporal and spatial detail inherent in these data.

(2) It calculates long-term (monthly, seasonal, and annual) SO_2 and SO_4^{2-} concentration and dry and wet deposition patterns and international exchanges resulting from SO_2 emissions in each of 13 countries in central and western Europe, taken both individually and jointly. These long-term values are obtained by totalling and/or averaging short-term values calculated at 3-h time steps.

This approach is in contrast with a number of other models (e.g. Rao *et al.*, 1976) which are not designed to obtain long-term values or interregional exchanges.

* Long Range Transport of Air Pollutants.

† Sponsored by the Federal Environmental Office of the Federal Republic of Germany.

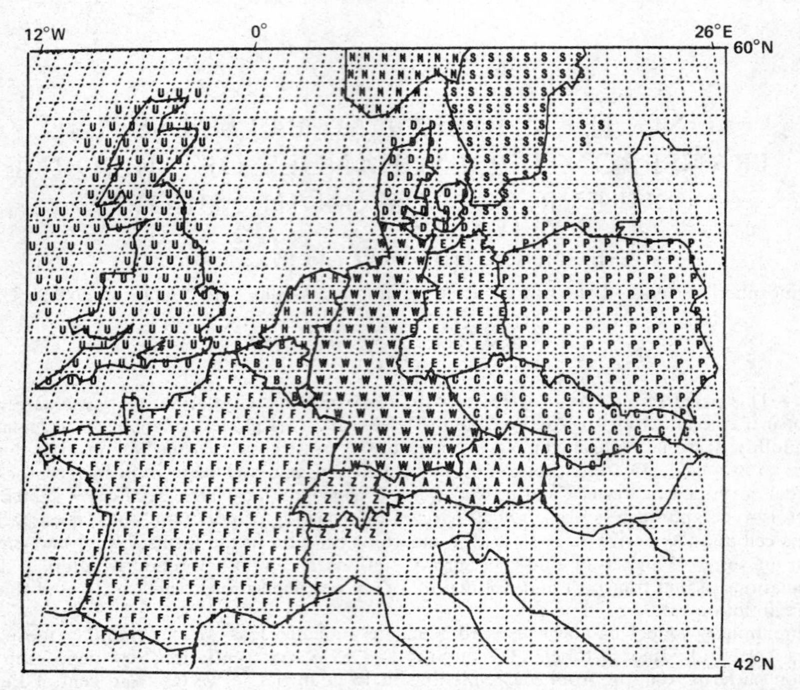

Fig. 1. Emissions data grid used in the model runs; letters represent countries (see below) for which 1973 SO_2 data were available from Rystad *et al.* (1974)

A—Austria
B—Belgium/Luxembourg
C—Czechoslovakia
D—Denmark
E—German Democratic Republic
F—France
H—The Netherlands (Holland)
N—Norway (South of 60°N)
P—Poland
S—Sweden (South of 60°N)
Z—Switzerland
U—United Kingdom and Ireland
W—Federal Republic of Germany

2. DESCRIPTION OF EURMAP DESIGN AND APPLICATION

A. *Model rationale and design goals*

In the design and development of any air quality simulation model, there are usually two conflicting goals: maximum realism and accuracy on the one hand, and minimum computing requirements on the other. The achievement of better realism and accuracy usually requires more detailed and sophisticated formulations, which in turn requires more computer time and memory capacity.

In designing EURMAP, it was clear from the outset that the computing requirements could become severe, for two reasons:

(1) The model must treat a very large geographical area (2100 km N–S by 2250 km E–W), and yet preserve acceptable spatial resolution (50 × 50 km).

(2) The model must compute monthly and annual mean concentration and deposition fields while preserving the original temporal resolution (6 h) of standard meteorological data, and thus must make repetitive calculations for long sequences of input data.

Accordingly, as a first step, we set out to design a very simple model—one that would have minimum computing requirements (and thus be practical and economical to use), while at the same time offering acceptable realism in simulating the most important processes involved in the transfrontier air pollution problem. Our approach is to add sophistication only when it is proven to be needed. The result of this effort is the current basic version of EURMAP, which is described in the following sections.

B. *Pollutant emissions*

The SO_2 emissions inventory used in this study was taken from that developed by Rystad *et al.* (1974) as part of the OECD-sponsored LRTAP project. As shown in Fig. 1, annual total SO_2 emissions data for 1973 were available for an array of cells 0.5° in latitude by 1.0° in longitude (approx 55 km N–S by 70 km E–W), which includes most countries in central and western Europe. Letter identifiers are used in Fig. 1 to designate those countries for which emissions data are available. Unfortunately, emissions data for Spain, Italy, Yugoslavia, Rumania, Bulgaria, and the USSR are not yet available for this detailed grid.

C. *Pollutant transport*

Figure 2 illustrates the basic principles behind the operation of the model.* Normally the model is run

* The basic concept is similar in principle to that developed by Start and Wendell (1974) in the MESODIF model, except that MESODIF was designed to be used with only one or a few point sources.

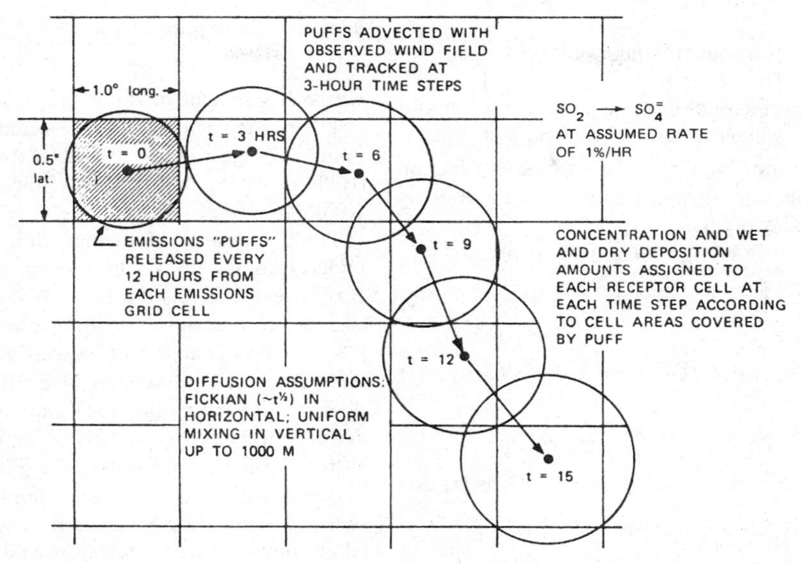

Fig. 2. Emissions puff advection and diffusion scheme used in EURMAP.

separately for each country of interest. Thus only the array of emissions cells associated with the particular country of interest is used at any one time.

For each of these cells, the annual-average emissions are divided into discrete 12-h-average emission increments, or 'puffs', which are released at 12-h intervals and tracked at 3-h time steps until either they move outside the overall study region (see Fig. 1) or their concentration drops to an insignificant level. The reason for selecting the 12-h intervals is that these intervals correspond to the frequency at which most synoptic upper-air wind observations are available. Thus, 730 puffs/year, or 61 puffs/month, are released and tracked for each of the cells. The largest country, France, contains 146 emissions cells, while West Germany, an average-size country, contains 67 cells. Thus, during the course of a month 4,087 puffs from West Germany are tracked, and at an average of 28 tracking steps/puff, a total of 114,436 puff positions/month are calculated for this country alone. This indicates the magnitude of the computational requirements.

The individual puffs are transported according to the wind field applicable at that time, as interpolated objectively from the available upper-air wind observations at 850 mb (approx 1500 m a.s.l.) from about 45 sounding stations for the year 1973. Although we recognize that significant vertical wind shears occur between the surface and the 1500-m level, especially in winter and at night, we believe that the use of a single wind to represent the transport within this layer is justified as a first approximation.

To account in an approximate fashion for surface-friction effects that cause wind shear in the boundary layer, the 850-mb winds have been adjusted as follows to give a more representative layer average:

$$\bar{U} = 0.75\, U_{(850\text{ mb})}$$
$$\bar{\theta} = \theta_{(850\text{ mb})} - 15^\circ$$

where $\bar{U}$ and $\bar{\theta}$ are the assumed layer-average wind speed and direction, respectively.

D. *Model formulation*

The model assumes that the pollutant concentration is always uniform throughout each puff at any given time. Accordingly, the SO_2 concentration (C_1) in a puff at any time (t) is determined as follows:

$$C_1 = M_1/V = (M_0 - M_T - M_{d1} - M_{w1})/V \quad (1)$$

where

M_1 = mass of SO_2 in puff at time t
M_0 = initial mass of SO_2 in puff
M_T = mass of SO_2 lost by transformation to SO_4^{2-} from $t = 0$ to $t = t$ $(0 \rightarrow t)$
M_{d1} = mass of SO_2 lost by dry deposition $(0 \rightarrow t)$
M_{w1} = mass of SO_2 lost by wet deposition $(0 \rightarrow t)$
V = volume of puff at time t

Similarly, the SO_4^{2-} concentration (C_2) becomes

$$C_2 = M_2/V = (3M_T/2 - M_{d2} - M_{w2})/V \quad (2)$$

where the factor of 3/2 is the ratio of the molecular weight for SO_4^{2-} to that for SO_2.

Assuming that the rates of change of pollutant mass in a puff due to transformation and deposition processes are proportional to mass (puff concentration times volume), we have

$$(dM/dt)_{\text{transformation }(T)} = k_T C\, V \quad (3)$$

$$(dM/dt)_{\text{wet deposition }(d)} = k_d C\, V \quad (4)$$

$$(dM/dt)_{\text{wet deposition }(w)} = k_w C\, V \quad (5)$$

where

k_T = $SO_2 \rightarrow SO_4^{2-}$ transformation coefficient (h^{-1})
k_d = dry deposition coefficient (h^{-1})
k_w = wet deposition coefficient (h^{-1})
$= \lambda \cdot R(t)$

and

λ = washout/rainout coefficient [h^{-1} (mm $h^{-1})^{-1}$, or mm^{-1}]

$R(t)$ = precipitation rate (mm h^{-1} of liquid water) at location of puff at time t.

The wet deposition is thus taken to be a function of precipitation rate, in accordance with findings by Bielke (1970) and others.

Equations (3)–(5) can be integrated to obtain

$$M_T = \int_0^t k_T CV \, dt = k_T \int_0^t CV \, dt \quad (6)$$

$$M_d = \int_0^t k_d CV \, dt = k_d \int_0^t CV \, dt \quad (7)$$

$$M_w = \int_0^t k_w CV \, dt. \quad (8)$$

Substituting these expressions into Equations (1) and (2), we have

$$C_1 = \frac{M_0}{V} - \frac{(k_T + k_d)}{V} \int_0^t C_1 V \, dt - \frac{1}{V} \int_0^t k_w C_1 V \, dt \quad (9)$$

$$C_2 = \frac{3k_T/2}{V} \int_0^t C_1 V \, dt - \frac{k_d}{V} \int_0^t C_2 V \, dt - \frac{1}{V} \int_0^t k_w C_2 V \, dt \quad (10)$$

E. *Pollutant diffusion*

Since diffusion on the regional scale is less important than the transport and removal processes, a very simple treatment of diffusion is used in the current version of EURMAP. Upon release, each puff is assumed to immediately diffuse vertically to give a uniform concentration in the layer between the surface and the 1000-m level. Aircraft measurements reported by Jost (1973, 1974) indicate that an average mixing depth of 1000 m is reasonable for Central Europe. However, to be more realistic, the mixing depth should vary with latitude and with season. Incorporation of this refinement into EURMAP is planned, but a constant mixing depth has been used for the computer runs so far. Thus the seasonal variations in the results to be presented in the next section reflect only the seasonal changes in wind and precipitation patterns, rather than changes in mixing depth.

Horizontal diffusion is also treated very simply. The initial lateral extent of the puff is approximated by a disk whose diameter (about 60 km) is adjusted so that the cross-sectional area of the disk is the same as that of the emissions cell. Fickian diffusion is then assumed during the transport of the puff, so that the puff radius (r) at any time (t) is given by

$$r = (r_0^2 + Kt)^{1/2} \quad (11)$$

where

r_0 = initial puff radius

K = horizontal eddy diffusivity $\equiv 10^8 \, cm^2 \, s^{-1} = 36 \, km^2 \, h^{-1}$

and the puff volume (V) is

$$V = \pi H (r_0^2 + Kt) \quad (12)$$

where H = height of the mixed volume $\equiv$ 1000 m.

The particular magnitude selected for K was based upon average values available from the literature (Heffter, 1965; Hilst, 1968) for travel times of 10–100 h.

The assumption of Fickian diffusion may lead to underestimates of the horizontal puff spread, since current evidence (Gifford, 1977) indicates that $r \sim t$ is a better descriptor of atmospheric diffusion than $r \sim t^{1/2}$. This change will be considered for incorporation into the next version of EURMAP. When this is done, however, the resulting concentration and deposition fields are unlikely to differ much from the current results. This is because the use of a faster horizontal diffusion rate will simply cause adjacent puffs to merge together faster, resulting only in a smoothing of the concentration and deposition fields.

F. *Pollutant transformation and deposition*

Recognizing that the processes of SO_2 conversion to sulfate aerosol and SO_2 removal by dry and wet deposition are particularly significant over the long transport distances involved in this study, we have included these effects in the model formulation (Equations (9) and (10)), but in a relatively simplified manner. Clearly these processes are very complex, depending upon such variables as surface and vegetation types (Whelpdale and Shaw, 1974); meteorological conditions (humidity, clouds, rain, fog, atmospheric stability, sunlight intensity, etc.); and concentrations of other chemical species in the air, such as ammonia and metal catalysts (Weber, 1971). However, until these processes are better understood, the simple treatment used here seems reasonable for this type of model.

A review of recent field, laboratory, and theoretical studies was conducted to find the best values available for the transformation coefficient (k_T) and the wet and dry deposition coefficients for SO_2 (k_{w1}, k_{d1}) and SO_4^{2-} (k_{w2}, k_{d2}). The values selected are compared in Table 1 with those used by other investigators. The value of 0.029 h^{-1} selected for k_{d1} corresponds to a dry deposition velocity for SO_2 of 0.8 cm s^{-1}, which is within the generally accepted range of values (cf. Eliassen and Saltbones, 1975).

Two important points should be noted here:

(1) Coefficient values used by various investigators show a wide scatter, reflecting the relative lack of reliable experimentally derived values.

(2) The outputs of all models (including EURMAP) which simulate SO_2/SO_4^{2-} transformation and deposition processes are highly dependent on the values of these coefficients.

Thus all model results, including those from EURMAP, should be treated with caution. Values of the transformation and deposition coefficients used in EURMAP are based upon best available information,

Table 1. Comparison of selected transformation and deposition coefficients with those used by other investigators

Investigator	$SO_2 \rightarrow SO_4^{2-}$ Transformation rate (h^{-1})	Deposition coefficients (h^{-1}) Dry		Wet*	
		SO_2	SO_4^{2-}	SO_2	SO_4^{2-}
Ottar (1978); OECD (1977)	0.007	0.036	0.0144	0.144	—
Renne and Elliott (1976)	0.050	0.036	0.0036	0.500	0.050
Rao *et al.* (1976)	0.015	—	—	—	—
McMahon *et al.* (1976)	0.010–0.020	0.054	0.0070	0.216	0.108
Johnson, Wolf, and Mancuso (1977)†	0.010	0.029	0.0070	0.216	0.070

* Values shown are of washout coefficient λ, where $k_w = \lambda R$, and rainfall rate R is in mm h^{-1}.
† Current paper.

but nevertheless these are uncertain, and accordingly the model results should be considered tentative at this time.

For the purpose of calculating the SO_2/sulfate removal by wet deposition, daily total precipitation observations during 1973 at a total of about 535 surface stations are input to the model. Interpolated precipitation values are used to calculate the mass of SO_2 and SO_4^{2-} removed from each puff at each location on the grid during each 3-h time step.

The computer program keeps track of each puff's location and duration of exposure to the variable precipitation intensity, calculates the pollutant mass removed during each time step based upon the assumed values of k_{w1} and k_{w2}, and assigns these amounts to the appropriate cell (geographical location) in the receptor grid. The precipitation rate is calculated by dividing the observed values of accumulated precipitation by the number of hours in the accumulation period. This infers an assumption that the precipitation was uniform during this period.

Dry deposition is handled in a similar, but simpler, manner. Based on the assumed values of k_{d1} and k_{d2}, the model removes appropriate fractions of the SO_2 and SO_4^{2-} mass in each puff at each time step, and assigns these amounts as dry deposition to the appropriate cell (geographical location) in the receptor grid.

G. *Calculation of monthly-mean concentration and deposition distributions*

EURMAP is designed to calculate long-term (monthly, seasonal, and/or annual) mean distributions of SO_2 and SO_4^{2-} concentrations, dry deposition, and wet deposition over all of western and central Europe, resulting from SO_2 emissions from each of the 13 countries in this area taken individually. Accordingly, each computer run of EURMAP involves reading in long sequences of surface and upper-air meteorological data at 6-hourly intervals. Each of the SO_2 puffs released at 12-hourly intervals from each of the emissions cells for the country of interest is moved every 3 h in accordance with the interpolated field of observed winds applicable for that time.

Values of concentration, dry deposition, and wet deposition are summed up at each 3-hourly time step over a 41 × 44 grid of receptor cells, each of which is 50 km on a side. This is carried out by adding the contributions of each of the puffs at each of the receptor cells.

Concentrations are assigned to cells around the location of the puff at the beginning of the time step. Deposition amounts are evenly divided between the cells at the starting location and those at the ending location of the puff for each time step. Both for concentration and deposition, pollutant amounts are assigned to the cell containing the center of the puff and to the 8 surrounding cells in amounts proportional to the area of the puff relative to the area of the central cell.

Long-term mean concentration distributions are obtained simply by averaging the concentrations assigned to each cell at each time step over the total period (number of time steps) of interest. Corresponding deposition distributions are obtained by totalling all amounts deposited in each cell during the period of interest.

3. RESULTS FROM EURMAP COMPUTER RUNS

A. *Description of results*

For the purpose of developing a set of model results for evaluation against available observations, the basic model has been run for the months of January, April, July and October, 1973 in the manner described in the last section, using detailed meteorological data at 6-hourly intervals. These particular months were selected in order to examine seasonal variations in the results. Separate runs have been made for each of the 13 countries for which emissions were available (Fig. 1).* For each of the four months, maps of SO_2 and SO_4^{2-} concentrations, dry deposition, and wet deposition resulting from the SO_2 emissions in each of the 13 individual countries have been prepared. Space limitations preclude the presentation

* For convenience, Luxembourg has been combined with Belgium and denoted simply as 'Belgium'.

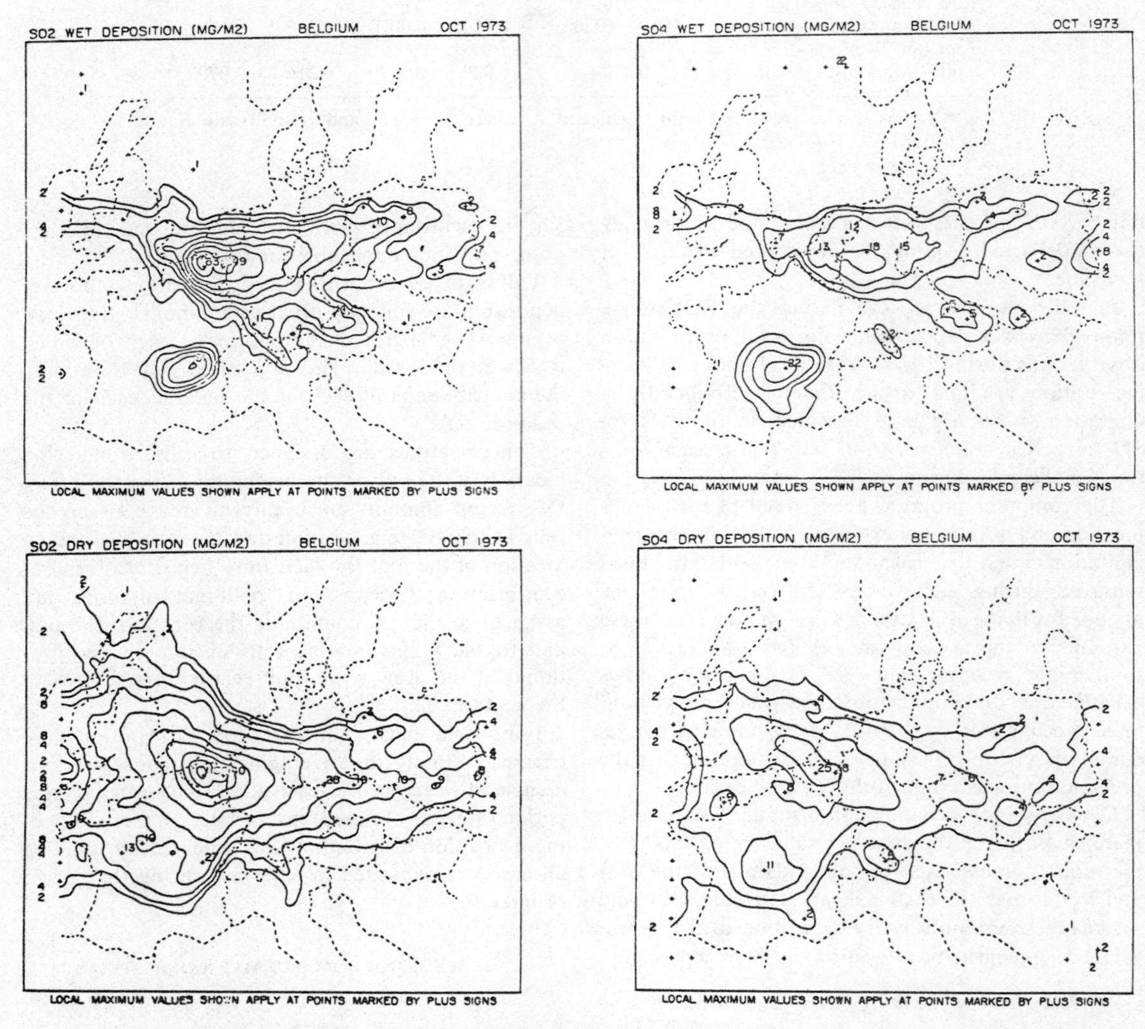

Fig. 3. Estimated SO_2 (left) and SO_4^{2-} (right) wet and dry deposition (top to bottom) from emissions in Belgium (monthly means and totals for October, 1973).

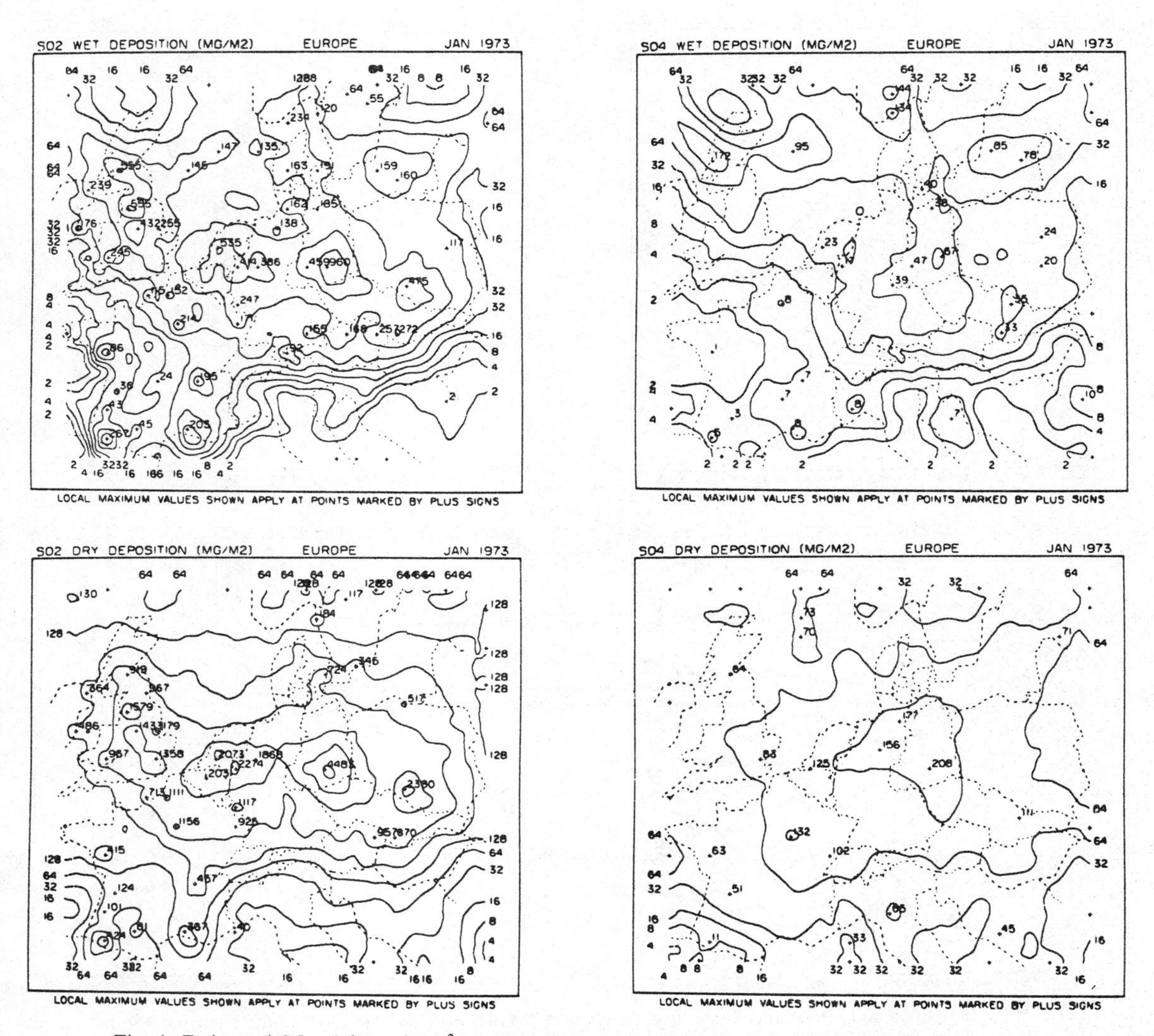

Fig. 4. Estimated SO_2 (left) and SO_4^{2-} (right) wet and dry deposition (top to bottom) from emissions in central and western Europe (monthly means and totals for January, 1973).

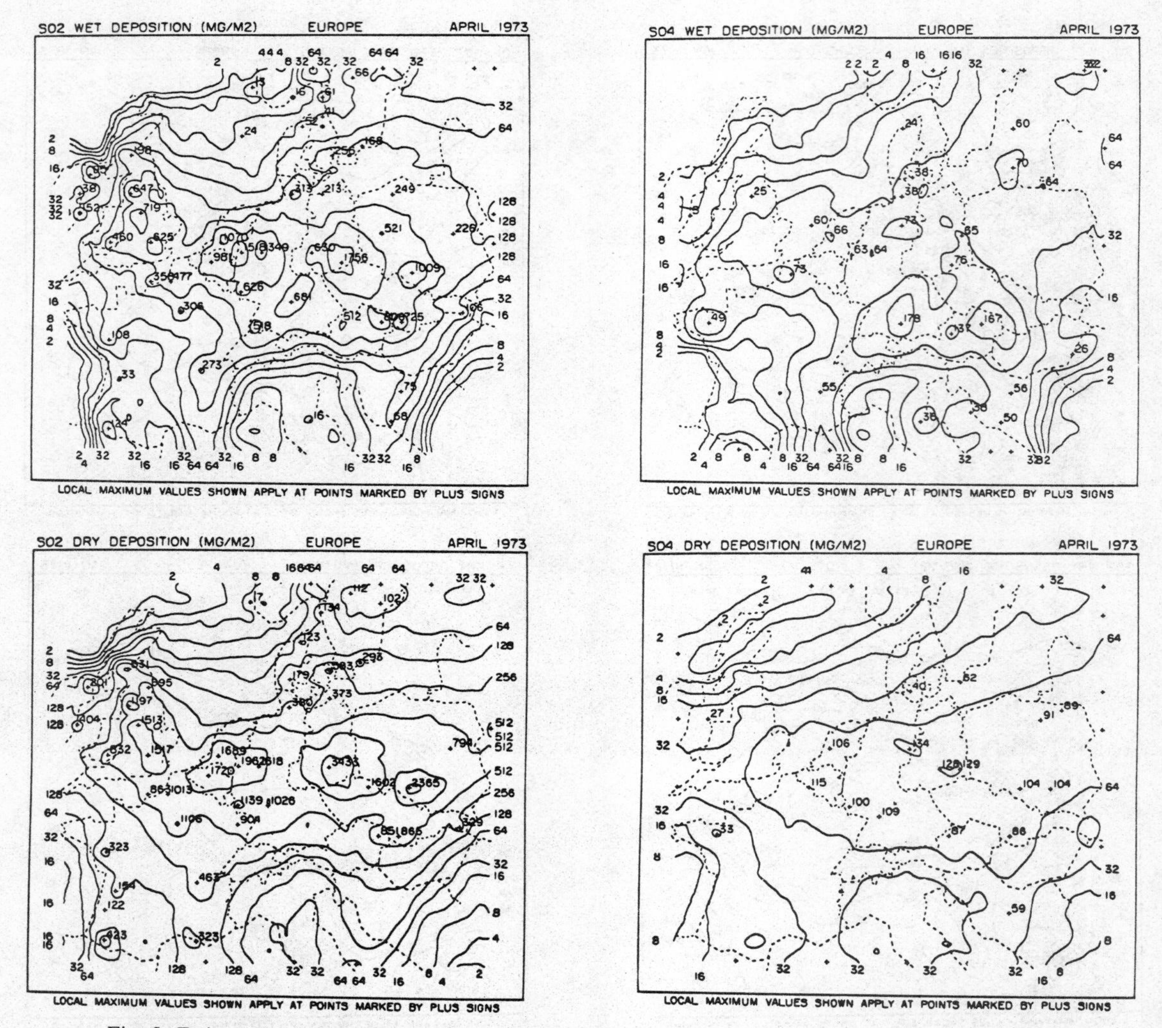

Fig. 5. Estimated SO_2 (left) and SO_4^{2-} (right) wet and dry deposition (top to bottom) from emissions in central and western Europe (monthly means and totals for April, 1973).

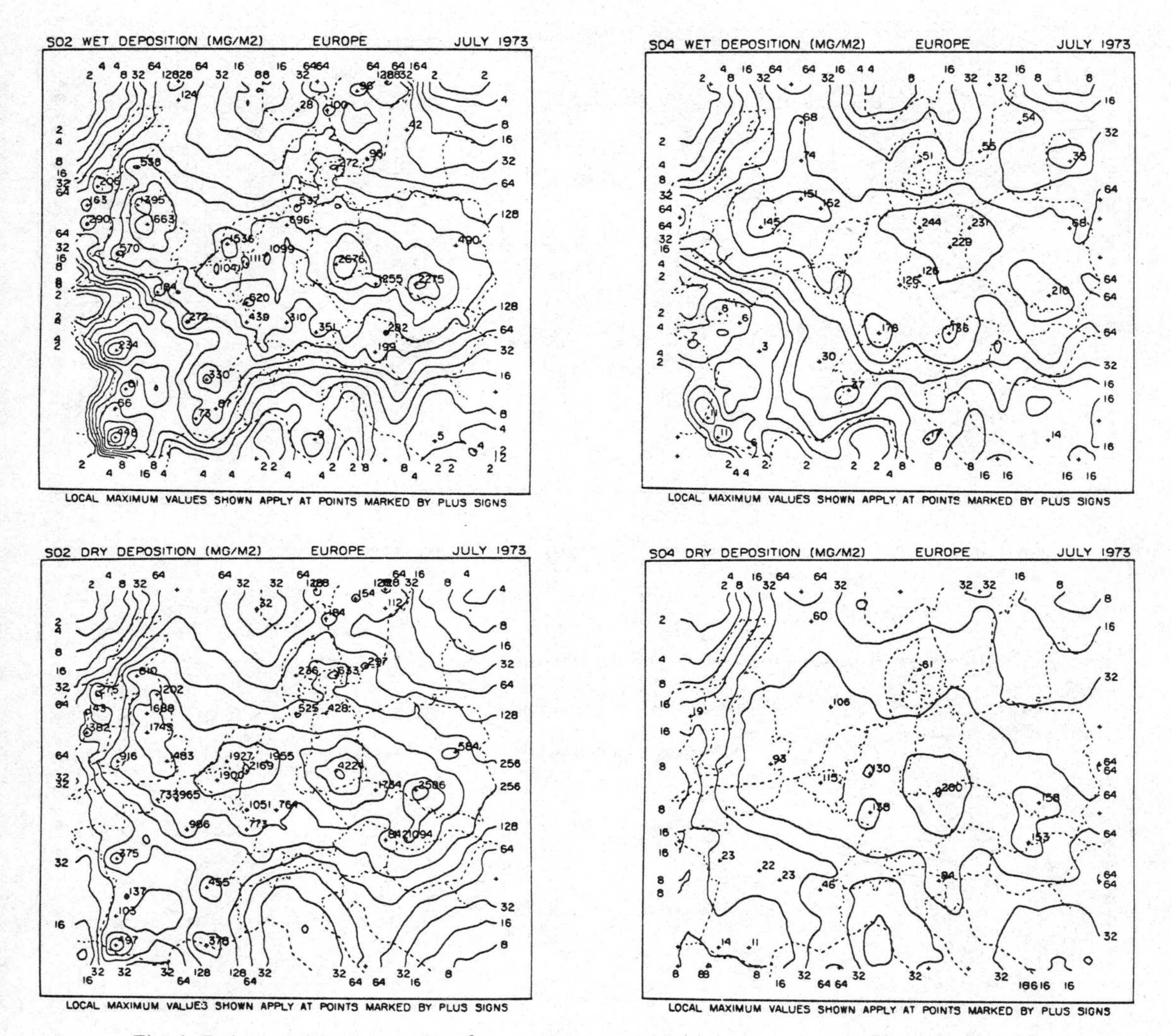

Fig. 6. Estimated SO_2 (left) and SO_4^{2-} (right) wet and dry deposition (top to bottom) from emissions in central and western Europe (monthly means and totals for July, 1973).

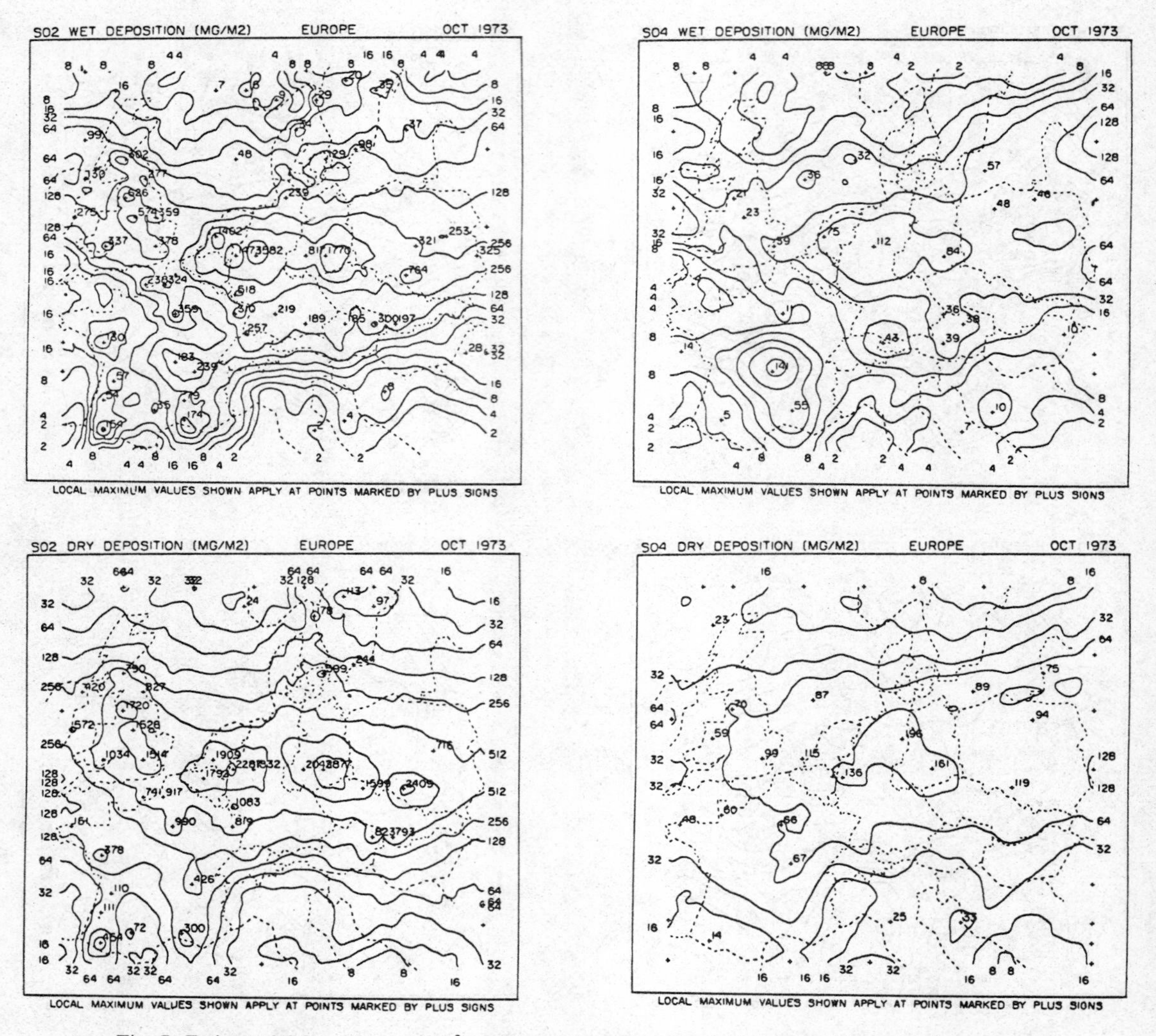

Fig. 7. Estimated SO_2 (left) and SO_4^{2-} (right) wet and dry deposition (top to bottom) from emissions in central and western Europe (monthly means and totals for October, 1973).

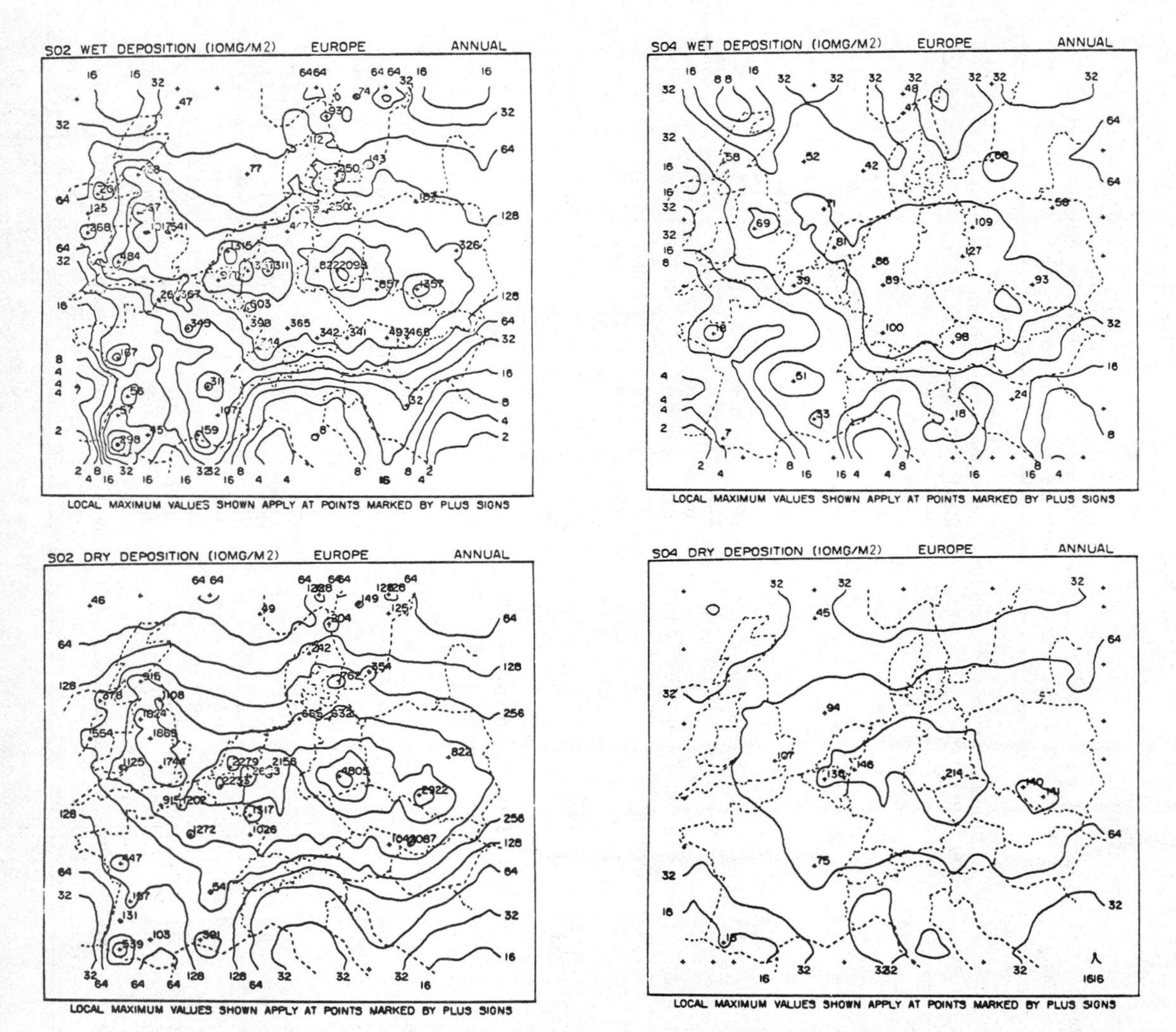

Fig. 8. Estimated SO_2 (left) and SO_4^{2-} (right) wet and dry deposition (top to bottom) from emissions in central and western Europe (annual means and totals for 1973).

Table 2. Percent contributions to SO_2 and SO_4^{2-} wet and dry deposition within receptor countries from SO_2 emissions in emittor countries (Annual totals for 1973; totals for each receptor country are also given in metric kilotons)

(a) SO_2 Annual (1973)

Emittor country	Percent contributions to SO_2 wet depositions within receptor countries												
	FRG	GDR	France	Poland	Czech.	Denmark	Holland	Belgium	U.K.	S. Norway	S. Sweden	Austria	Switz.
W. Germany (FRG)	47.5	12.7	3.8	3.7	8.0	12.0	9.8	4.6	1.3	6.6	4.9	17.4	11.3
E. Germany (GDR)	8.9	73.3	0.3	30.1	28.0	15.6	1.3	0.1	0.3	23.1	10.7	13.2	4.9
France	10.6	1.6	73.9	0.7	1.3	1.0	3.5	20.7	4.3	0.7	0.4	7.0	34.5
Poland	0.3	1.5	0.0	46.1	9.1	2.3	0.0	0.0	0.0	7.1	6.1	2.7	0.0
Czechoslovakia	1.8	3.6	0.1	15.8	46.5	2.2	0.2	0.0	0.0	4.4	1.6	14.3	1.1
Denmark	0.2	0.5	0.0	0.6	0.1	45.3	0.0	0.0	0.0	12.8	18.5	0.0	0.0
Holland	12.1	2.5	1.4	0.7	1.5	4.9	53.8	6.3	1.5	1.0	3.3	2.4	1.3
Belgium	8.9	1.9	4.4	0.6	0.6	1.0	16.4	50.4	1.7	0.5	0.5	3.9	3.7
U.K.	8.5	2.1	15.1	0.7	1.1	14.4	14.9	17.8	90.7	11.9	6.4	5.0	10.3
S. Norway	0.0	0.0	0.0	0.0	0.0	0.2	0.0	0.0	0.0	24.8	3.7	0.0	0.0
S. Sweden	0.0	0.0	0.0	0.1	0.0	1.1	0.0	0.0	0.0	6.9	43.7	0.0	0.0
Austria	0.5	0.2	0.0	0.8	3.7	0.1	0.0	0.0	0.0	0.3	0.1	31.9	0.8
Switzerland	0.7	0.1	0.9	0.0	0.1	0.0	0.0	0.0	0.0	0.0	0.0	2.3	32.0
Total (%)	100.0	100.0	100.0	100.0	100.0	100.0	100.0	100.0	100.0	100.0	100.0	100.0	100.0
Total (KT)	1166	732	499	1080	559	78	254	179	742	31	85	181	54

Emittor country	Percent contributions to SO_2 dry depositions within receptor countries												
	FRG	GDR	France	Poland	Czech.	Denmark	Holland	Belgium	U.K.	S. Norway	S. Sweden	Austria	Switz.
W. Germany (FRG)	48.9	9.8	7.6	3.1	5.1	9.8	17.9	12..3	1.5	5.6	3.6	11.1	14.7
E. Germany (GDR)	17.6	76.6	1.7	28.2	27.4	17.0	2.6	2.0	0.3	8.4	8.9	13.6	4.2
France	7.4	0.8	60.0	0.5	0.9	0.6	2.8	15.3	3.4	0.9	0.2	2.6	29.8
Poland	0.6	2.1	0.1	48.2	11.3	3.7	0.1	0.0	0.0	4.0	6.4	4.7	1.0
Czechoslovakia	4.0	5.6	0.6	16.7	48.6	2.2	0.3	0.2	0.0	1.2	1.2	21.8	2.8
Denmark	0.4	0.9	0.0	0.6	0.2	54.5	0.1	0.0	0.1	16.8	21.5	0.1	0.0
Holland	9.3	1.6	3.0	0.6	0.7	2.5	51.2	9.7	2.1	2.0	1.8	1.0	1.6
Belgium	6.0	1.0	8.0	0.4	0.5	0.6	15.0	46.8	2.0	1.1	0.4	1.4	3.5
U.K.	4.4	1.1	18.0	0.5	0.6	6.7	10.0	13.6	90.6	8.4	3.8	1.6	9.7
S. Norway	0.0	0.0	0.0	0.0	0.0	0.4	0.0	0.0	0.0	41.1	5.4	0.0	0.0
S. Sweden	0.0	0.1	0.0	0.2	0.0	1.9	0.0	0.0	0.0	10.4	46.8	0.0	0.0
Austria	0.8	0.4	0.1	0.9	4.6	0.1	0.0	0.0	0.0	0.1	0.1	40.6	2.1
Switzerland	0.5	0.0	0.9	0.0	0.1	0.0	0.0	0.0	0.0	0.0	0.0	1.3	30.5
Total (%)	100.0	100.0	100.0	100.0	100.0	100.0	100.0	100.0	100.0	100.0	100.0	100.0	100.0
Total (KT)	2325	1787	1730	2589	1388	196	523	449	1910	35	181	320	93

(b) SO_4^{2-} Annual (1973)

Emittor country	FRG	GDR	France	Poland	Czech.	Denmark	Holland	Belgium	U.K.	S. Norway	S. Sweden	Austria	Switz.
	Percent contributions to SO_4^{2-} wet depositions with receptor countries												
W. Germany (FRG)	19.0	18.8	11.8	9.9	12.8	9.3	15.4	7.5	11.8	7.3	10.6	13.9	12.9
E. Germany (GDR)	17.2	31.1	2.4	27.6	24.8	19.8	9.4	3.3	9.5	26.7	17.9	25.9	22.7
France	14.9	7.6	45.7	5.2	7.2	3.5	13.0	27.3	17.6	1.6	2.6	12.6	27.7
Poland	1.8	5.9	0.1	25.3	11.9	7.5	0.2	0.0	0.9	21.8	12.0	3.6	0.5
Czechoslovakia	4.7	7.5	0.6	17.1	20.7	5.9	2.4	0.6	2.3	12.6	5.1	10.9	4.5
Denmark	0.4	0.8	0.1	1.4	0.4	7.1	0.2	0.1	0.4	5.2	6.3	0.4	0.2
Holland	8.6	5.6	5.3	2.6	4.3	6.6	19.2	5.0	5.6	1.6	7.0	3.6	1.9
Belgium	8.6	8.2	6.3	2.6	3.0	2.7	11.9	18.7	6.8	0.7	1.7	5.4	3.7
U.K.	22.7	12.6	25.8	5.3	10.1	35.8	27.8	36.9	44.2	16.7	28.9	17.5	17.0
S. Norway	0.0	0.0	0.0	0.1	0.0	0.2	0.0	0.0	0.0	1.5	1.2	0.0	0.0
S. Sweden	0.1	0.1	0.0	0.3	0.1	0.7	0.0	0.0	0.1	2.7	5.6	0.1	0.0
Austria	1.2	1.4	0.3	2.3	4.1	0.6	0.1	0.1	0.6	1.4	0.8	4.9	3.1
Switzerland	0.7	0.4	1.7	0.3	0.4	0.2	0.2	0.6	0.2	0.2	0.1	1.2	5.7
Total (%)	100.0	100.0	100.0	100.0	100.0	100.0	100.0	100.0	100.0	100.0	100.0	100.0	100.0
Total (KT)	204	103	102	203	86	24	24	14	110	19	44	61	17

Emittor country	FRG	GDR	France	Poland	Czech.	Denmark	Holland	Belgium	U.K.	S. Norway	S. Sweden	Austria	Switz.
	Percent contributions to SO_4^{2-} dry depositions within receptor countries												
W. Germany (FRG)	23.9	16.6	13.3	9.5	10.5	10.8	20.6	15.5	10.1	14.0	10.8	12.6	17.2
E. Germany (GDR)	26.9	43.0	7.2	30.8	30.3	24.4	14.0	13.8	6.1	15.6	18.0	30.2	14.2
France	8.7	3.7	33.5	3.3	4.3	2.4	8.7	14.6	10.6	3.9	2.1	6.4	19.3
Poland	2.9	6.8	0.5	26.4	14.5	11.3	1.1	0.5	0.8	13.8	14.0	8.9	4.3
Czechoslovakia	8.1	9.8	2.4	16.4	23.5	7.8	3.2	2.5	1.2	5.6	5.1	17.5	7.7
Denmark	0.9	1.9	0.3	1.5	1.0	11.8	0.4	0.3	0.9	11.9	9.7	0.7	0.5
Holland	7.8	4.8	5.7	2.7	3.1	5.1	20.2	7.5	6.4	4.6	6.2	2.9	4.6
Belgium	6.2	3.5	8.1	2.1	2.2	1.9	10.3	19.4	6.0	3.1	1.9	3.0	4.8
U.K.	12.5	7.5	27.2	4.3	5.8	22.0	20.8	25.0	57.4	18.2	21.3	8.5	17.7
S. Norway	0.0	0.1	0.0	0.1	0.0	0.4	0.0	0.0	0.0	3.6	2.0	0.0	0.0
S. Sweden	0.2	0.4	0.1	0.5	0.4	1.5	0.1	0.0	0.1	4.7	8.1	0.3	0.1
Austria	1.4	1.7	0.5	2.2	4.2	0.5	0.4	0.5	0.2	0.7	0.6	8.2	4.1
Switzerland	0.5	0.3	1.1	0.2	0.3	0.1	0.2	0.2	0.1	0.1	0.1	0.8	5.5
Total (%)	100.0	100.0	100.0	100.0	100.0	100.0	100.0	100.0	100.0	100.0	100.0	100.0	100.0
Total (KT)	324	161	295	315	154	38	44	36	168	12	50	67	25

of these results in this paper. However, Fig. 3 gives an example of the dry and wet deposition results for Belgium for October, 1973. The results from the individual runs have also been combined into maps for each month showing the total concentration, dry deposition, and wet deposition distributions resulting from SO_2 emissions from all 13 countries taken together. The resulting dry and wet deposition patterns are presented in Figs 4–7. (Concentration patterns are not presented since they are similar to those for dry deposition.)

Finally, assuming that the results for each of the 4 months are representative of seasonal values, annualized depositions have been estimated by totalling the 4-monthly deposition values and multiplying by 3. Similarly, estimates of annual average concentrations have been obtained by averaging values for the 4 months. Annualized deposition patterns calculated in this way are presented in Fig. 8. Note that the units of deposition in Fig. 8 are $10\,mg\,m^{-2}$ ($10^{-2}\,g\,m^{-2}$), while those in Figs 3–7 are $mg\,m^{-2}$ ($10^{-3}\,g\,m^{-2}$).

Logarithmic contour intervals (2, 4, 8, 16, 32, ...) are used in the maps shown in Figs 4–8, and only the open contours are labelled. The other numbers on the contour maps and the adjacent crosses (+) are the magnitudes and locations of maximum values in the concentration and deposition fields.

The results shown in Figs. 3–8 are fairly self-explanatory, but a few general comments are in order:

(a) The wet deposition patterns reflect the normal inhomogeneities in the spatial distribution of precipitation, and thus are not as smooth as the dry deposition patterns.

(b) Particularly for SO_2, the patterns are strongly influenced by the emissions distribution, with maxima generally corresponding to maxima in the emissions field. The patterns thus show a broad band of high values corresponding to the heavily industrialized region extending from central England east-southeastward to southern Poland.

(c) The occurrence of large horizontal gradients in the deposition fields in the southern and eastern portions of the maps is an artifact caused by the fact that emissions from Spain, Italy, Yugoslavia, Hungary, Rumania, and the U.S.S.R. have not yet been included.

(d) The results for the individual countries (see Fig. 3) are much more revealing of transport directions, etc. than are the combined results from all countries.

(e) The SO_4^{2-} deposition maxima are generally displaced to the eastward of SO_2 source areas (i.e. in the mean downwind direction), as would be expected considering the finite time required for transformation of SO_2 to SO_4^{2-}.

(f) SO_4^{2-} wet deposition maxima frequently occur in regions far from source areas, such as the maximum in southern France from Belgium emissions (Fig. 3), and the maxima from European emissions in January over the Baltic Sea and western Scotland (Fig. 4), in April over eastern Austria (Fig. 5), and in October over the northwestern U.S.S.R. and southern France (Fig. 7).

(g) Particularly for SO_2, the shapes and magnitudes of the deposition patterns for the four months are rather similar. This may be caused by the lack of seasonally-dependent mixing depths and emission rates in the model. (Specification of the seasonal dependencies in these two variables involves a fair degree of arbitrariness, although this is an obvious refinement that will be included in our future model runs. For these initial calculations, however, the variations in mixing depth and emissions with time of year were deliberately excluded to permit the effects of seasonal changes in the wind fields to be isolated and examined.)

Table 2 lists the calculated 1973 international exchanges of SO_2 and SO_4^{2-} wet and dry deposition among the 13 countries included in the model runs. Similar tables are also available for the months of January, April, July and October, but these do not differ greatly and are not presented here. The annualized SO_2 wet and dry deposition exchanges generally are similar, although some significant differences do occur (e.g. the results indicate that southern Norway receives 23% of its SO_2 wet deposition from East Germany, but only 8% of its SO_2 dry deposition from this source). In comparing the SO_2 to the SO_4^{2-} depositions, however, a different picture emerges. For example, the results indicate that southern Sweden receives 29% of its SO_4^{2-} wet deposition from the U.K., but only 6% of its SO_2 wet deposition from the same source.

B. *Comparison of EURMAP results with those from other models*

Obviously the results just described must be evaluated and verified against available observations before they can be considered credible. This effort is underway but not yet completed. In the meantime, it is useful to compare the EURMAP results with those from other models to get an idea of their reasonableness.

Figure 9 shows Fisher's (1975) results for annual total deposition in Europe. The shape of the deposition pattern calculated by Fisher is very similar to the EURMAP SO_2 wet and dry deposition patterns (Fig. 8), and the magnitudes agree well over most of the map (when the EURMAP wet and dry deposition values are added together). However, the $10\text{-}g\,m^{-1}$ contour in central Europe on Fisher's map corresponds to approx $20\,g\,m^{-2}$ as calculated by EURMAP. Possible reasons for this disagreement between maximum values include the following:

(a) EURMAP has a basic temporal resolution, in terms of grid size, of 50 × 70 km (emissions grid) and 50 × 50 km (receptor grid), while Fisher's model uses a 127 × 127-km grid size for both emissions and receptor areas. Thus, the EURMAP calculations should contain more spatial detail (less areal smooth-

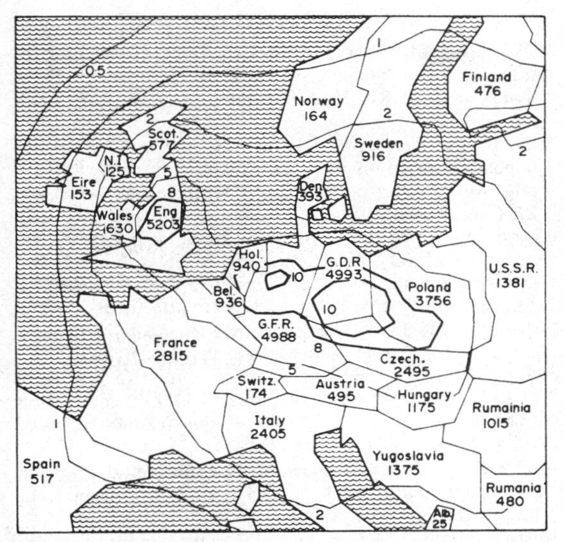

Fig. 9. Fisher's (1975) results for dry plus wet deposition in $g\,m^{-2}\,SO_2\,y^{-1}$ from all the European sources shown. Diffusivity = $50\,m^2\,s^{-1}$; deposition velocity = $5\,mm\,s^{-1}$; maximum mixing depths = 1000 and 2000 m. Scavenging coefficient = $10^{-4}\,s^{-1}$. Emission strengths for each country are shown in units of kilotons $SO_2\,y^{-1}$.

ing), which accordingly should result in higher maximum values.

(b) Fisher used a dry deposition velocity of $0.5\,cm\,s^{-1}$, compared with our value of $0.8\,cm\,s^{-1}$.

(c) A value of 1000 m is used in EURMAP for the mixing depth, while Fisher used 1000 m for half the year, and 2000 m for the other half of the year.

A detailed comparison between the EURMAP results and the model results from the LRTAP Project (Ottar, 1978; OECD, 1977) has also been made. In general, the shapes of all concentration and deposition patterns are quite similar. In terms of magnitudes, the EURMAP dry depositions (SO_2 and SO_4^{2-}) agree closely with the LRTAP values. The wet deposition results for EURMAP are lower than those from the LRTAP by about a factor of two near the boundaries of the modeled area, probably because the LRTAP model includes emissions from Italy, the U.S.S.R., and other countries near the boundaries. However, the EURMAP values are higher than those of the LRTAP model by about a factor of two in the regions of maximum values. For SO_2 and SO_4^{2-} concentrations, the EURMAP results are again approximately twice as high as the LRTAP values in the regions near maxima.

The reasons for this lack of agreement are not fully understood at this time, but again the differences in spatial resolution between the two models may be involved. The LRTAP model has the same 127 × 127 km resolution as Fisher's, and thus the point previously raised may also hold here. Another,

Table 3. Comparison of EURMAP estimates of sulfur exchanges between various European countries with those from Fisher (1975) and the LRTAP project (Ottar, 1978; OECD, 1977) (Values are annual deposition in $g\,m^{-1}$ as SO_2)*

		Dry deposition			Wet deposition			Total deposition		
Emittor	Receptor	Fisher	LRTAP	EURMAP	Fisher	LRTAP	EURMAP	Fisher	LRTAP	EURMAP
UK	UK	4.0	4.31	5.28	—	0.24	2.08	5.0	4.55	7.36
Sweden	Sweden	0.3	1.67	0.61	—	0.24	0.27	0.4	1.91	0.88
Sweden	UK	—	0.00	0.00	—	0.00	0.00	—	0.00	0.00
UK	Sweden	0.2	0.15	0.10	—	0.58	0.10	0.2	0.73	0.20
FRG	Sweden	—	0.11	0.07	—	0.35	0.05	—	0.46	0.12
Denmark	Sweden	—	0.24	0.30	—	0.21	0.12	—	0.45	0.42
GDR	Sweden	—	0.27	0.16	—	0.44	0.10	—	0.71	0.26
Poland	Sweden	—	0.15	0.11	—	0.13	0.06	—	0.28	0.17
FRG	FRG	—	4.70	4.53	—	0.82	2.21	—	5.52	6.74
FRG	Holland	—	0.64	2.66	—	0.11	0.73	—	0.75	3.39
FRG	UK	—	0.05	0.12	—	0.02	0.05	—	0.07	0.17
FRG	GDR	—	—	1.79†	—	—	0.99†	—	—	2.78†
FRG	France	—	0.16	0.29	—	0.04	0.05	—	0.20	0.34
Holland	FRG	—	0.17	0.89	—	0.12	0.58	—	0.29	1.47
UK	FRG	—	0.27	0.50	—	0.55	0.50	—	0.82	1.00
GDR	FRG	—	0.50	1.78	—	0.14	0.48	—	0.64	2.26
France	FRG	—	0.34	0.72	—	0.38	0.55	—	0.72	1.27
Holland	Holland	—	3.20	7.31	—	0.22	3.73	—	3.42	11.04
Holland	UK	—	0.02	0.14	—	0.01	0.05	—	0.03	0.19
UK	Holland	—	0.69	1.57	—	0.64	1.13	—	1.33	2.70
France	France	—	1.90	2.05	—	0.24	0.74	—	2.14	2.79
France	UK	—	0.07	0.23	—	0.05	0.13	—	0.12	0.36
UK	France	—	0.26	0.68	—	0.20	0.17	—	0.46	0.85
Averages			0.90	1.34		0.26	0.63		1.13	2.00

* For convenience in obtaining sums, values are carried to three significant figures rather than rounded, but this does not imply a similar level of accuracy.

† Average does not include FRG to GDR amounts since LRTAP values are not available.

but probably minor, factor may be the difference in years modeled: the EURMAP results apply for 1973, while those for the LRTAP are for 1974.

Finally, Table 3 presents a comparison of the international exchanges of sulfur deposition as calculated by the three models for various pairs of countries selected as examples. EURMAP appears to agree a bit better with Fisher's model than does LRTAP (except for the U.K.-to-U.K. value). LRTAP gives much lower values than does EURMAP for exchanges involving Holland. Again, the poorer spatial resolution of the LRTAP model may be the cause of this. In the LRTAP model, Holland is covered by only about 4 emissions grid cells, with several of these overlapping into high-emission regions of West Germany, while EURMAP uses 11 emissions cells and 15 receptor cells for Holland. If the Holland values are excluded, the LRTAP and EURMAP dry deposition exchanges are similar.

With regard to wet deposition exchanges, the LRTAP values on the average are considerably lower than those of EURMAP. Although the EURMAP results may be overestimates, some of the LRTAP values for wet deposition exchanges are surprisingly low relative to those for dry deposition: 0.24 g m^{-2} (wet) vs. 4.31 (dry) for U.K.-to-U.K.; 0.22 (wet) vs 3.20 (dry) for Holland-to-Holland; 0.82 (wet) vs 4.70 (dry) for FRG-to-FRG.

C. *Conclusions*

Results from the current version of EURMAP are generally similar to those from other comparable models, but significant differences do exist. In those cases, it is not clear at this time which model is more correct. An ongoing effort to evaluate EURMAP with available observations should yield more definitive information. Based upon the results of this evaluation, appropriate refinements will be incorporated into the model.

Ultimately, information in the form shown in Table 2 from a suitably validated version of EURMAP should be useful in the management of air quality on an international basis in western and central Europe. Using information on planned or potential future changes in the emissions distribution over the region, EURMAP could also be used at some later stage to prepare air quality projections for future years.

Acknowledgements—This study is being conducted under a contract from the Federal Environmental Office (Umweltbundesamt) of the Federal Republic of Germany (FRG), under the direction of Dr. J. Pankrath and Dr. D. Jost of that office and Dr. E. Weber of the Federal Ministry of the Interior (FRG). We wish to acknowledge with thanks the assistance of Dr. Pankrath, Dr. Jost, Dr. Weber, and Dr. B. Ottar of the Norwegian Institute of Air Research in offering useful suggestions and furnishing valuable technical information necessary for this project. We are also grateful for the assistance of Miss Joyce Kealoha, Mrs. Linda Jones, and Mrs. Evelyn Freisheim in the preparation of this report.

REFERENCES

Beilke S. (1970) Laboratory investigations on washout of trace gases. Precipitation Scavenging (1970)—Proc. Symp. held at Richland, Washington, June 2–4, 1970. AEC Symposium Series 22, 1970.

Bolin B. and Persson C. (1975) Regional dispersion and deposition of atmospheric pollutants with particular application to sulfur pollution over Western Europe. *Tellus* **27**, (No. 3) 281–310.

Eliassen A. and Saltbones J. (1975) Decay and transformation rates of SO_2, as estimated from emission data, trajectories and measured air concentrations. *Atmospheric Environment* **9**, 425–429.

Fisher B. E. A. (1975) The long range transport of sulphur dioxide. *Atmospheric Environment* **9**, 1063–1070.

Gifford F. A. (1976) Tropospheric relative diffusion observations. Unpublished paper, ATDL Contribution File No. 76/10, NOAA, Oak Ridge, Tennessee, May 1976.

Heffter J. L. (1965) The variation of horizontal diffusion parameters with time for travel periods of one hour or longer. *J. appl. Met.* **4**, 153–156.

Hilst G. R. (1968) Meteorological management of air pollution. *Air Pollution* (Edited by Arthur C. Stern) Vol. 1, Chap. 10 (2nd Edn) pp. 321–347. Academic Press, New York.

Johnson W. B., Wolf D. E. and Mancuso R. L. (1976) The European Regional Model of Air Pollution (*EURMAP*) and its application to transfrontier air pollution. Proceedings of the 7th Int. Tech. Meeting on Air Pollution Modeling and Its Application, 7–10 Sept. 1976, Airlie House, VA, a report of the Air Pollution Pilot Study, NATO Committee on the Challenges to Modern Society.

Jost D. (1973) Zue Husbreitung von Luftbeimengungen über Industriegebieten. *Ann. Meterol.* **6**, 151–156.

Jost D. (1974) Aerological studies on the atmospheric sulfur budget. *Tellus* **26**, (No. 1–2) 206–212.

McMahon T. A., Denison P. J. and Fleming R. (1976) A long-distance air pollution transportation model incorporating washout and dry deposition components. *Atmospheric Environment* **10**, 751–761.

OECD (1977) The OECD programme on Long Range Transport of Air Pollutants (Measurements and Findings). Organisation for Economic Co-operation and Development, 2 rue André-Pascal, 75775 Paris Cedex 16, France.

Ottar B. (1978) The OECD study on Long Range Transport of Air Pollutants (LRTAP). Proceedings of the Int. Symp. on Sulfur in the Atmosphere, 7–14 Sept. 1977, Dubrovnik, Yugoslavia. *Atmospheric Environment* **12**, 445–454.

Prahm L. P. and Christensen O. (1977) Long range transmission of pollutants simulated by the 2-D pseudospectral dispersion model. Danish Meterorological Institute Report 77/1, Denmark, Feb. 1977.

Rao K. S., Thomson T. and Egan B. A. (1976) Regional transport model of atmospheric sulfates. Proc. Annual Meeting of APCA, June 1976, Portland, OR, Paper No. 76–34.3.

Renne D. S. and Elliott D. L. (1976) Regional air quality assessment for northwest energy scenarios. Presented at the 69th Annual Meeting of the Air Pollution Control Association, June 27–July 1, 1976, Portland, OR, Paper No. 76–23.5.

Rystad B., Stromsoe S., Amble E. and Knudsen T. (1974) The LTRAP emission survey. Norwegian Inst. for Air Research, Norway, 19 p.

Sheih C. M. (1977) Application of a statistical trajectory model to the simulation of sulfur pollution over northeastern United States. *Atmospheric Environment* **11**, 173–178.

Smith F. B. and Jeffrey G. H. (1975) Airborne transport

of sulphur dioxide from the U.K. *Atmospheric Environment* **9,** 643–659.

Start G. E. and Wendell L. L. (1974) Regional effluent dispersion calculations considering spatial and temporal meteorological variations. NOAA Tech. Memo ERL ARL-44, Air Resources Laboratories, Idaho Falls ID, 63p.

Weber E. (1971) Removal of sulfur dioxide from the atmosphere. Proc. of the Second Meeting, NATO/CCMS Panel on Modeling, July 1971, Paris, France, Report No. 5, Chapter XII.

Wendell L. L., Powell C. D. and Drake R. L. (1976) A regional scale model for computing deposition and ground level air concentration of SO_2 and sulfates from elevated and ground sources. Proc. of the Third Symposium on Atmospheric Turbulence, Diffusion and Air Quality, Oct. 19–22, 1976, Raleigh, NC, published by the AMS, Boston, MA. Also in Proceedings of the 7th Int. Tech. Meeting on Air Pollution Modeling and Its Application, 7–10 Sept. 1976, Airlie House, VA, a report of the Air Pollution Pilot Study, NATO Committee on the Challenges to Modern Society.

Whelpdale D. H. and Shaw R. W. (1974) Sulphur dioxide removal by turbulent transfer over grass, snow, and water surface. *Tellus* **26,** 196–205.

Atmospheric Environment Vol. 12, pp. 529–535. Pergamon Press 1978. Printed in Great Britain.

TRANSMISSION OF SULFUR DIOXIDE ON LOCAL, REGIONAL AND CONTINENTAL SCALE

D. J. SZEPESI
Institute of Atmospheric Physics, P.O. Box 39, H-1675 Budapest, Hungary

(*First received* 13 *June* 1977 *and in final form* 27 *September* 1977)

Abstract—The transport, dispersion, transformation and removal, i.e. the transmission of sulfur dioxide, depends on the scale of the pollution process and on the type of the source, too. To take these considerations into account, working formulas were developed to simulate the transmission on different scales. The overall decay and transformation rates were separated into terms of dry deposition, transformation during low (r.h. less than 90%) and high (r.h. more than 90%) humidity conditions and wet deposition, together with the respective time intervals during which the separate mechanisms were effective. The intervals were evaluated along the trajectories on surface and upper air charts.

To simulate the transmission for continental scale pollution processes, 72 h, 850 mb, isobaric, backward trajectories were constructed. For regional and local scale pollution processes trajectories were constructed on the basis of the surface wind pattern. The dispersion of the sulfur dioxide for the larger scale processes was taken into account by air trajectory box models, for the local process by Gaussian model.

The removal processes were taken into account by using scale-dependent rate constants from the available current literature.

By applying the transmission model presented here, four case studies showed good agreement between the measured (63 and 1 $\mu g\,m^{-3}$) and calculated (63 and 7 $\mu g\,m^{-3}$) sulfur dioxide concentrations, respectively, and between measured turbidity parameters ($B = 0.308$ and 0.072) and calculated particulate sulfate concentrations (22 and 4 $\mu g\,m^{-3}$) respectively.

1. INTRODUCTION

For areas where the contribution from larger scale pollution processes is considerable, estimation of regional and continental scale transmission is an indispensable part of any local or urban scale modeling effort. For the validation of long-range transmission models the measurements of such stations are recommended where the WMO Regional Background Pollution criteria are met.

Long range transmission models were reported by Bolin (1975); Eliassen (1975); Ferber (1975); Smith (1975), Husar (1976); Veltischeva (1976) and Prahm (1976). The models operate with a time resolution better than 24 h. For the validation 24 h observational data are required. The models use the following routinely available meteorological (surface and/or 850 mb winds, mixing heights, precipitation) and emission data. 925 mb winds may be preferable in all models.

Present models of long range transmission can be further improved by taking into account the main meteorological factors (precipitation, humidity, temperature, solar radiation, etc.) of removal and transformation correctly.

According to the findings of a recent workshop on long-range transmission (Ottar, 1976) some improvement could be achieved by incorporating known dry deposition rates as a function of location and season. Wet deposition of sulfur dioxide and sulfate could also be incorporated in the general decay term, or as a discontinuous process. The latter requires knowledge of the extent and duration of precipitation within the grid size and 6-h time resolution. This approach may be capable of simulating the actual deposition which in turn can be compared with measurements on an event basis.

The approach so far used to incorporate the chemical transformation of sulfur dioxide to sulfate by applying decay time, is only adequate in the present models for long-range averages. Further refinements will require the inclusion of reactions with other relevant substances, the concentration measurements of these other substances, and compiling of source inventories for these materials. Advanced short-term modeling is needed in order to improve long-term statistics.

Following this concept, a modified version of Eliassen's method will be presented, taking into account the variation of removal mechanisms along the trajectories and estimating the separate contributions of continental, regional and local sources to the regional background pollution.

2. THE AIR TRAJECTORY BOX MODEL

The transport, dispersion, transformation and removal of pollutants emitted by anthropogeneous sulfur dioxide sources will be simulated by a transmission model. The basic idea of the approach is as follows;

A representative air volume Z m^3 ($1 \times 1 \times Z$ m, where Z is the mixing height) is transported along the air trajectory, which is supposed to be characteristic of the air flow. The emission and transmission

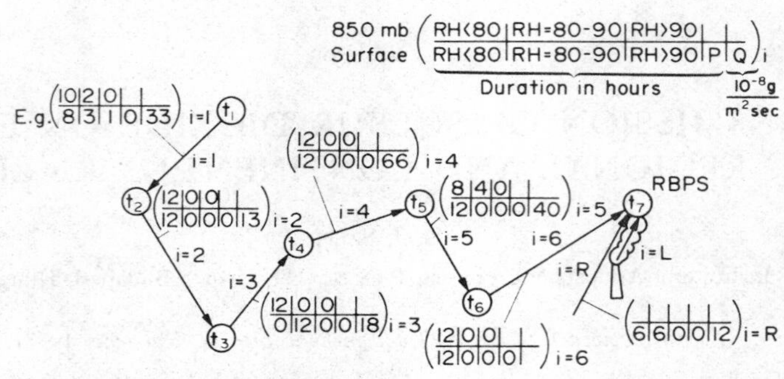

Fig. 1. Lower part: trajectories of continental ($i = 1, \ldots, 6$) regional ($i = R$) and local ($i = L$) scale pollution processes. RBPS = Regional Background Pollution Station. Upper part: form to evaluate time intervals with different humidity, precipitation (P) conditions and emissions data (Q) along the ith trajectory interval from surface and 850 mb charts.

of the representative air volume depend on the average emission, mixing height, humidity and precipitation conditions along each trajectory interval passed during 12-h periods according to Fig. 1. The parameters of transmission for each trajectory interval during the passage of the representative air volume can be evaluated by using surface and upper air charts.

The next assumption of the investigation is that the polluting effects of continental, regional and local scale pollution processes are superimposed, as shown as by Fig. 1.

A short definition of the different scales of pollution processes follows (Szepesi, 1977).

A pollution process termed continental in the upwind range of transmission between 200 and 3000 km of the receptor point, regional or local between 0 and 200 km. The emission inventory for continental scale process includes pollutants emitted from all types of sources, for regional scale process only pollutants emitted from low level ($h < 100$ m) sources are taken into account. Continental and regional processes can be simulated by box models. Local scale elevated sources, higher than 100 m are taken into account separately, one by one, by using the Gaussian model. This is to avoid constructing unnecessarily detailed emission inventories, which are not supposed to give results with higher resolution than the method presented here. Another reason is that near field dry deposition is considerably smaller for pollutants emitted from tall stacks, compared to that from low level sources.

2.1 *Transmission of sulfur dioxide*

Assuming horizontal and incompressible air flow the sulfur dioxide mass balance equation of a representative air volume can be written:

$$\Delta\chi_{SO_2}(t) \approx -\frac{0.693}{\tau_i}\chi_{SO_2}(t)\,\Delta t \qquad (1)$$

where $\chi_{SO_2}(t)$ is the time dependent concentration of SO_2 in the air volume, τ_i the overal half life of SO_2, depending on the scale of pollution process, $0.693 = -\log_e(0.5)$.

If $\Delta t \to 0$ and at t_i

$$\chi_{SO_2}(t_i) = \frac{Q_i\,\delta t}{Z_{max}},$$

the sulfur dioxide concentration at t_7 originating from one trajectory interval will be given by

$$\chi_{SO_2}(t_7) = \frac{Q_i\,\delta t}{Z_{max}}\exp - 0.693\int_{t_i}^{t_7}\frac{dt}{\tau_i}$$
$$\approx \frac{Q_i\,\delta t}{Z_{max}}\exp - \left[0.693\sum_i^6 \times\left(\frac{\delta t_1}{\tau_1}+\frac{\delta t_2}{\tau_2}+\frac{\delta t_3}{\tau_3}+\frac{\delta t_4}{\tau_4}\right)_i\right] \qquad (2)$$

where Q_i g m^{-2} s^{-1} the average SO_2 emission along trajectory interval i, $\delta t = 12$ h resolution time of trajectories, Z_{max} m the maximum mixing height at the receptor point, $(\delta t_1, \delta t_2, \delta t_3, \delta t_4)_i$ h partial time intervals during which in a representative air volume the pollutant was affected by dry deposition, transformation at low and high humidity conditions and wet deposition respectively, and evaluated along a certain part of the trajectory, $(\tau_1, \tau_2, \tau_3, \tau_4)_i$ are partial half lifes of sulfur dioxide, taking into account the effect of dry deposition, transformation during low and high humidity conditions and wet deposition respectively, for certain scale of pollution processes (further details are given by Equation (18)).

Because the differential equation for $\chi_{SO_2}(t)$ is linear, the sulfur dioxide concentration at a receptor point at t_7 will be given by summing up the contributions originating from different trajectory intervals, transported by the same representative air volume.

According to these assumptions at the receptor point the sulfur dioxide concentration originating from continental, regional and local scale pollution processes are given by the following algorithms:

$$\chi_{SO_2}(t_7) = \chi_{SO_2}^{Cont}(t_7) + \chi_{SO_2}^{Reg}(t_7) + \chi_{SO_2}^{Loc}(t_7) \qquad (3)$$

where

$$\chi_{SO_2}^{Cont}(t_7) = \frac{\delta t}{Z_{max}}\sum_{i=1}^{5} Q_i \exp - \left[0.693\sum_i^6 \times\left(\frac{\delta t_1}{\tau_1}+\frac{\delta t_2}{\tau_2}+\frac{\delta t_3}{\tau_3}+\frac{\delta t_4}{\tau_4}\right)_i\right] \qquad (4)$$

$$\chi_{SO_2}^{Reg}(t_7) = \frac{\delta t}{Z_{max}} Q_{i=R} \exp - \left[0.693 \times \left(\frac{\delta t_1}{\tau_1} + \frac{\delta t_2}{\tau_2} + \frac{\delta t_3}{\tau_3} + \frac{\delta t_4}{\tau_4} \right)_{i=R} \right] \quad (5)$$

$$\chi_{SO_2}^{Loc}(t_7) = \sum_j \frac{2.03 Q_j f_j}{\sigma_{zj} u_j x_j} \exp - \left[\frac{(h_j + \Delta h_j)^2}{2\sigma_{zj}^2} + 0.693 \times \left(\frac{\delta t_1}{\tau_1} + \frac{\delta t_2}{\tau_2} + \frac{\delta t_3}{\tau_3} + \frac{\delta t_4}{\tau_4} \right)_{i=L} \right] \quad (6)$$

where j denotes the elevated sources, Q_j g s^{-1} the average SO_2 emission of the elevated source, f_j % fraction of 12-h period during which the jth stack affected the receptor point, σ_{zj} m vertical dispersion coefficient in a layer representative for the plume of the elevated source, x_j m distance between the elevated source and the receptor point, $h_j + \Delta h_j$ m effective height of the plume.

This work in some part is based on the original concept of Eliassen (1975), who by taking into account $T = 48$-h trajectory analysis, constant mixing height h and constant overall decay rate k for the estimation of SO_2 at the sampling site on the arrival of the trajectory reported the following working formula

$$q(T) = \sum_{i=1}^{N} \frac{Q_i \Delta t}{h} e^{-k(i-1)\Delta t}. \quad (7)$$

2.2. *Transmission of particulate sulfate*

Assuming first order transformation from sulfur dioxide to particulate sulfate and first order removal of sulfate particles, the sulfate mass balance equation of a representative air volume can be given as follows:

$$\Delta\chi_{SO_4}(t) \approx - \frac{0.693}{\tau_{5i}} \chi_{SO_4}(t) \Delta t + 1.53 \times 0.693 \left(\frac{1}{\tau_{2i}} + \frac{1}{\tau_{3i}} \right) \chi_{SO_2}(t) \Delta t \quad (8)$$

where $\chi_{SO_4}(t)$ g m^{-3} the particulate sulfate concentration of the representative air volume, τ_{5i} half life of wet deposition of particulate sulfate for certain scale of pollution processes, 1.53 is the ratio of mol. wt of sulfate and sulfur dioxide.

The first member of the right side of Equation (8) is the sink term, the second one is the source term of particulate sulfate.

If $\Delta t \to 0$ and after some rearrangements

$$d\chi_{SO_4}(t) + 0.693 \frac{dt}{\tau_{5i}} \chi_{SO_4}(t) = 1.53 \times 0.693 \left(\frac{1}{\tau_{2i}} + \frac{1}{\tau_{3i}} \right) \chi_{SO_2}(t) dt. \quad (9)$$

This is a linear, first order, non-homogeneous differential equation. Assuming, that at $t = t_i$, $\chi_{SO_4}(t_i) = 0$, and that

$$\frac{\dfrac{1}{\tau_{2i}} + \dfrac{1}{\tau_{3i}}}{\dfrac{1}{\tau_i} - \dfrac{1}{\tau_{5i}}} = \frac{\sum_i^6 \left(\dfrac{\delta t_2}{\tau_2} + \dfrac{\delta t_3}{\tau_3} \right)_i}{\sum_i^6 \left(\dfrac{\delta t_1}{\tau_1} + \dfrac{\delta t_2}{\tau_2} + \dfrac{\delta t_3}{\tau_3} + \dfrac{\delta t_4}{\tau_4} - \dfrac{\delta t_5}{\tau_5} \right)_i}$$

which means that along trajectory intervals $i \ldots i = 6$ half lives τ_{2i}, τ_{3i}, τ_i and τ_{5i} are taken into account according to the respective partial time intervals δt_{2i}, δt_{3i}, δt and δt_{5i} during which the separate removal processes were active.

For one trajectory interval the solution of Equation (9) is:

$$\chi_{SO_4}(t_7) = \frac{1.53\, \delta t}{Z_{max}} C_i Q_i \left\{ \exp - \left[0.693 \sum_i^6 \left(\frac{\delta t_5}{\tau_5} \right)_i \right] - \exp - \left[0.693 \sum_i^6 \left(\frac{\delta t_1}{\tau_1} + \frac{\delta t_2}{\tau_2} + \frac{\delta t_3}{\tau_3} + \frac{\delta t_4}{\tau_4} \right)_i \right] \right\} \quad (10)$$

where for $i = 1, \ldots, 5$

$$C_i = \frac{\sum_i^6 \left(\dfrac{\delta t_2}{\tau_2} + \dfrac{\delta t_3}{\tau_3} \right)_i}{\sum_i^6 \left(\dfrac{\delta t_1}{\tau_1} + \dfrac{\delta t_2}{\tau_2} + \dfrac{\delta t_3}{\tau_3} + \dfrac{\delta t_4}{\tau_4} - \dfrac{\delta t_5}{\tau_5} \right)_i}. \quad (11)$$

According to the previous assumption, at any receptor point the particulate sulfate concentration originating from continental, regional and local scale pollution processes by using Equation (10) for the whole trajectory can be given by the following algorithms:

$$\chi_{SO_4}(t_7) = \chi_{SO_4}^{Cont}(t_7) + \chi_{SO_4}^{Reg}(t_7) + \chi_{SO_4}^{Loc}(t_4) \quad (12)$$

$$\chi_{SO_4}^{Cont}(t_7) = \frac{1.53\, \delta t}{Z_{max}} \sum_{i=1}^{5} C_i Q_i \left\{ \exp - \left[0.693 \sum_i^6 \left(\frac{\delta t_5}{\tau_5} \right)_i \right] - \exp - \left[0.693 \sum_i^6 \left(\frac{\delta t_1}{\tau_1} + \frac{\delta t_2}{\tau_2} + \frac{\delta t_3}{\tau_3} + \frac{\delta t_4}{\tau_4} \right)_i \right] \right\} \quad (13)$$

$$\chi_{SO_4}^{Reg}(t_7) = \frac{1.53\, \delta t}{Z_{max}} C_{i=R} Q_{i=R} \left\{ \exp - \left[0.693 \left(\frac{\delta t_5}{\tau_5} \right)_{i=R} \right] - \exp - \left[0.693 \left(\frac{\delta t_1}{\tau_1} + \frac{\delta t_2}{\tau_2} + \frac{\delta t_3}{\tau_3} + \frac{\delta t_4}{\tau_4} \right)_{i=R} \right] \right\} \quad (14)$$

$$\chi_{SO_4}^{Loc}(t_7) = \sum_j \frac{2.03 \times 1.53}{\sigma_{zj} u_j x_j} C_{i=L} Q_j f_j \exp - \left[\frac{(h_j + \Delta h_j)^2}{2\sigma_{zj}^2} \right] \times \left\{ \exp - \left[0.693 \left(\frac{\delta t_5}{\tau_5} \right)_{i=L} \right] - \exp - \left[0.693 \left(\frac{\delta t_1}{\tau_1} + \frac{\delta t_2}{\tau_2} + \frac{\delta t_3}{\tau_3} + \frac{\delta t_4}{\tau_4} \right)_{i=L} \right] \right\} \quad (15)$$

where $C_{i=R}$ and $C_{i=L}$

$$C_{i=R/L} = \frac{\left(\dfrac{\delta t_2}{\tau_2} + \dfrac{\delta t_3}{\tau_3} \right)_{i=R/L}}{\left(\dfrac{\delta t_1}{\tau_1} + \dfrac{\delta t_2}{\tau_2} + \dfrac{\delta t_3}{\tau_3} + \dfrac{\delta t_4}{\tau_4} - \dfrac{\delta t_5}{\tau_5} \right)_{i=R/L}}. \quad (16)$$

By taking into account the constant transformation rate from SO_2 to sulfate k_t, and constant decay rate of sulfate κ, originally the mass balance equation for sulfate was set up and solved by Eliassen (1974). For the estimation of particulate sulfate concentration at the sampling site on arrival of the trajectory he gave the following algorithm:

$$p(T) = \frac{3}{2} \frac{k_t}{k - \kappa} \left[\sum_{i=1}^{N} \frac{Q_i \Delta t}{h} \left(e^{-\kappa(i-1)\Delta t} - e^{-k(i-1)\Delta t} \right) \right]. \quad (17)$$

If the particulate sulfate concentration originated from natural sources and anthropogeneous sources at earlier stages of the trajectory could be taken into account, especially in synoptic situations without precipitation, this might further improve the approximation.

3. TRAJECTORIES

For one year (1 April 1974–31 March 1975) 72-h backward trajectories were hand-constructed by the Petterssen (1956) method on 850 mb level charts at 00.00 and 12.00 h GMT by using RAWIN sonde data. Because in most cases observed wind data were used, speed and curvature accelerations were not eliminated, as in the case when the geostrophic wind approximation is applied. However, because 12-h resolution time was used in certain cases, it can result in other types of error, which numerical methods with shorter time resolution do not have.

The regional and local scale transport (Bell, 1964) was taken into account by using the surface wind pattern at 00.00, 06.00, 12.00 and 18.00 h GMT.

4. EMISSION DATA

For the investigation of the continental scale transmission sulfur dioxide inventories prepared by Stromsoe (1973) and recent emission data compiled by Brodsky (1975) were used. On the basis of these data seasonal maps were constructed, assuming that the variable part of emission is in summer 0%, in spring and fall 25% and in winter 50% of the yearly total for space heating.

A regional scale emission inventory was prepared by using the yearly sulfur dioxide emission total (Brodsky, 1975). The total was tentatively divided into 25×25 km squares according to the rate of industrialization and number of flats. The seasonal breakdown of the yearly data was calculated by assuming a similar 0–25–50–25% distribution of the variable part of emission from space heating.

5. FACTORS OF TRANSFORMATION AND REMOVAL

To take into account the transformation and removal processes by applying proper rate constants or half lives of the pollutant is one of the main tasks simulating transmission.

According to Gillani and Husar (1976), humidity and the presence of ammonia and ozone, for example, are believed to be factors in the oxidation of SO_2 to sulfates. Since ozone itself is a secondary pollutant whose concentration generally changes with x within the plume, the half life τ should be free to vary with plume transport time.

To apply time-dependent half lives has to be the main goal of efforts simulating transmission, but presently it encounters both practical and theoretical difficulties.

In the present method the overall factor of decay is separated into partial terms which allow the time intervals during which the individual transformation and removal mechanisms are effective to be evaluated along the trajectories, and properly taken into account in the transmission model.

The selection of half-life values for the transformation and removal processes on different scales was based on a thorough review of the current literature of this topic. The findings of this review are listed in the lower part of Table 1. For the sake of simplicity and comparability all the rate constants selected were converted to half-life values in h.

Because of the diversity of the half-life values found it was judged better to fit them into categories comprised of ranges instead of presenting them as discrete values. These tentatively selected half-life values are shown by the upper part of Table 1.

The basic idea of the selection is that by following the recent views it was assumed that beside other factors, the transformation of SO_2 to sulfates depends mostly on the concentration of pollutants and on the r.h. Based on this assumption the half-lives are chosen, depending on the values of these routinely measured factors.

Half-lives of dry deposition τ_1 are calculated as follows:

$$\tau_1^C = 30\%/18\text{ h},\ 30\%/66\text{ h}, \qquad \tau_1^R = 25\%/6\text{ h},$$
$$\tau_1^L = 10\%/3\text{ h},\ 10\%/6\text{ h}.$$

Selection of these parameters is based on the following principles; (a) 30, 25 and 10% of the sulfur dioxide is removed from the emission of continental, regional and local elevated sources respectively, by dry deposition; (b) to simulate time-dependent dry deposition, the travel time between the point of emission and reception was taken as 18–66, 6 and 3–6 h for continental, regional and local scale pollution processes, respectively; (c) the depth of the plume and its variation with time were not taken into account in the selection of dry deposition half-life values.

As a most probable selection, half-life values underlined in Table 1 were used for the calculations, but computer evaluation for 18 combinations is under way to optimize the selection.

6. EVALUATION OF PARTIAL TIME INTERVALS

Partial time intervals can be defined as time scales during which, in a representative air volume, the pollutant was affected by dry deposition (δt_1), transformation at low humidity (δt_2) and high humidity conditions (δt_3) and wet deposition (for sulfur dioxide δt_4 and for particulate sulfate δt_5) and were evaluated along certain (ith) parts of the trajectory.

In other words along the 850 mb and surface trajectories, on the basis of dew point depression and precipitation data, zones with different r.h. were evaluated and expressed in time scale of h. Precipitation zones

Table 1. Half-lives of transformation and removal mechanisms in continental (*C*), regional (*R*), urban (*U*) and local (*L*) scale pollution processes. (*) personal communication

Scale of pollution processes \ Half life in hours	SULFUR DIOXIDE: Dry deposition τ_1	SULFUR DIOXIDE: Transformation r.h. < 90% τ_2	SULFUR DIOXIDE: Transformation r.h. ≥ 90% τ_3	SULFUR DIOXIDE: Wet deposition τ_4	Sulfate τ_5
Continental $i = 1,\ldots,6$ (200–3000 km)	42–153	92	8.6	0.2–0.4–1.2	5.0–46
Regional $i = R$ ($X < 200$ km)	17	14	6.9	0.4–0.8–1.6	5.0–46
Local (i = L) X < 200 km h ≥ 100 m	20–42	8.0	5.8	0.6–1.2–1.9	5.0–46
Benarie 1973		$\tau_{1-3}^{R,U} = 2.7$–12			
Bolin* 1975			$\tau_3^{C,R} = 9.0$	$\tau_4^{C,R} = 0.4$	
Chamberlain 1953				$\tau_4^{C,R} = 1.9$	
Eliassen 1975			$\tau_{2-3}^{C,R} = 90$	$\tau_{1-4}^{C,R} = 4.6$	$\tau_5^{C,R} = 46$
Eliassen 1976		$\tau_{1-3}^{C,R} = 13$–18			
Granat* 1975		$\tau_{2-3}^{C,R} = 4.3$–32			
Högström 1974				$\tau_4^R = 1.9$	
Husar 1976a		$\tau_{1-3}^L = 13$–43			
Husar 1976b		$\tau_{1-3}^{R,U} = 3.0$			
Makhonyko 1967				$\tau_4 = 0.2$–1.9	
Mészáros 1977		$\tau_{2-3}^{R,U} = 1.9$–9.6			
Newman 1975		$\tau_{1-3}^L = 5.0$–16			
Olsson* 1975			$\tau_3^{C,R} = 6.4$		
Prahm 1975	$\tau_{1-3}^{C,R} = 18$	$\tau_{2-3}^{C,R} = 140$			$\tau_S^{C,R} = 70$
Smith 1975		$\tau_2^{C,R} = 6.0$–60	$\tau_3^{C,R} = 3.2$		
Smith 1976	$\tau_{1-3}^{C,R} = 28$	$\tau_{2-3}^{C,R} = 100$			

were evaluated the similar way only for areas where the 850 mb dew point depression was ≤ 1°C and on the surface maps persistent rain was reported.

Partial time intervals in Equations (2), (4)–(6), (10), (11), (13)–(16) can be interpreted by the following symbol:

$$\int_{t_i}^{t_7} \frac{dt}{\tau_i} = \sum_i^6 \left(\frac{\delta t_1}{\tau_1} + \frac{\delta t_2}{\tau_2} + \frac{\delta t_3}{\tau_3} + \frac{\delta t_4}{\tau_4} \right)_i \quad (18)$$

overall decay	dry dep.	transformation r.h. < 90%	transformation r.h. ≥ 90%	wet dep.

and were evaluated according to the following principle:

T–T_d °C	> 3	2–3	0–1	Prec. dur.	$Q \dfrac{10^{-8}\,\text{g}}{\text{m}^2\,\text{s}^{-1}}$
r.h. %	< 80	80–90	> 90		

For example at the trajectory interval i the following data were evaluated:

$$\begin{matrix}\text{at 850 mb} \\ \text{at surface}\end{matrix} \left(\frac{6}{6} \Big/ \frac{2}{6} \Big/ \frac{4}{2} \Big/ \frac{}{1} \Big/ \frac{}{64} \right)_i \quad \text{duration in h.}$$

This evaluation form makes it possible to improve the simulation technique of transmission by applying the upper air and the surface humidity conditions in combination with weighting factors to optimize the correlation between calculated and measured values. Presently, as a most probable combination, the following selection was applied:

$$\begin{pmatrix} 9 & 3 & 1 & 64 \end{pmatrix}$$

which means, that (a) as a threshold value between low and high r.h. 90% was chosen, and (b) the humidity conditions were characterized by the average values of the 850 mb and the surface conditions.

7. RESULTS

Working formulae developed to calculate the sulfur dioxide concentration (Equations (3)–(6)) and the particulate sulfate concentration (Equations (12)–(15)) originating from continental, regional and local sources were validated by four case studies selected so that the measured concentration at the regional Background Pollution Station (RBPS) in Hungary were extremely high or low.

The RBPS in Hungary has been in full operation since 1973 in the central part of the country. 24-h sulfur dioxide measurements are carried out from 7 a.m. to 7 a.m. by using TCM method. Because the calculated values are available twice daily, measured values were compared with calculated ones averaged from 12.00 h GMT and 00.00 h GMT values. The yearly mean sulfur dioxide concentration was 11.0 μg m^{-3}, the 24 h absolute maximum concentration was 84.3 μg m^{-3}.

Turbidity measurements are made at the RBPS by pyreheliometer with RG-2 filter according to the recommendations of the WMO Manual. Particulate sulfate measurements were not carried out.

Case 1

63 $\mu g\ m^{-3}$ 24-h sulfur dioxide concentration was measured at the RBPS from 7 a.m. December 14 until 7 a.m. December 15 1974. The synoptic pattern was anticyclonic along the trajectory, the high pressure system centered over Western Europe and Poland. Relative humidity most of the time was higher than 80%. Precipitation occurred on some intervals over Czechoslovakia. $Z_{max} = 720$ m, stratification was neutral, high continental scale emission occurred. Regional wind pattern in the Carpathian Basin was NNW–NW, $u = 3$–$10\ m\ s^{-1}$, local elevated sources affected the RBPS.

Calculated mean concentrations are: $\chi_{SO_2}^{Cont} = 25\ \mu g\ m^{-3}$, $\chi_{SO_2}^{Reg} = 32\ \mu g\ m^{-3}$, $\chi_{SO_2}^{Loc} = 6\ \mu g\ m^{-3}$. The total calculated value $\chi_{SO_2}(t_7) = 63\ \mu g\ m^{-3}$ is in good agreement with the measured one.

Case 2

1 $\mu g\ m^{-3}$ 24-h sulfur dioxide concentration was measured at the RBPS from 7 a.m. September 25 until 7 a.m. September 26 1974. The weather pattern along the trajectory was cyclonic, the low pressure system centered N of Hungary. A cold front passed the measuring station during the sampling period. Relative humidity along the last part of the trajectory was over 80–90%. Precipitation along the first trajectory occurred in traces, along the second one during 6 h. $Z_{max} = 1500$ m, stratification was isothermic, low to medium emission occurred. Regional wind pattern was SSW, $u = 2\ m\ s^{-1}$ later NW, $u = 5\ m\ s^{-1}$. Local elevated sources affected the RBPS.

Calculated mean contibutions are: $\chi_{SO_2}^{Cont} = 2\ \mu g\ m^{-3}$, $\chi_{SO_2}^{Reg} = 3\ \mu g\ m^{-3}$, $\chi_{SO_2}^{Loc} = 2\ \mu g\ m^{-3}$. The total calculated value is $\chi_{SO_2}(t_7) = 7\ \mu g\ m^{-3}$ somewhat higher than the measured one.

Case 3

High turbidity (Schüepp turbidity parameter $B = 0.380$) was measured at the RBPS at 12.00 h GMT September 14 1974. The weather pattern was anticyclonic over Hungary, low r.h. prevailed, no precipitation occurred. $Z_{max} = 2320$ m, stratification was superadiabatic, continental scale emission was medium. Regional flow pattern was SW, $u = 2\ m\ s^{-1}$. Local elevated source affected the RBPS.

Calculated contributions are: $\chi_{SO_4}^{Cont} = 18\ \mu g\ m^{-3}$, $\chi_{SO_4}^{Reg} = 1\ \mu g\ m^3$, $\chi_{SO_4}^{Loc} = 3\ \mu g\ m^3$. The total calculated value was $\chi_{SO_4}(t_7) = 22\ \mu g\ m^{-3}$.

Case 4

Low turbidity (Schüepp turbidity parameter $B = 0.072$) was measured at the RBPS at 12.00 h GMT November 15 1974. Strong prefrontal SW winds prevailed $Z_{max} = 1730$ m. The sky was clear, r.h. was medium. Along one trajectory interval short term precipitation occurred. Continental scale emission was low. Regional wind pattern was S, $u = 5\ m\ s^{-1}$. No local elevated source affected the RBPS.

Calculated contributions are: $\chi_{SO_4}^{Cont} = 3\ \mu g\ m^{-3}$, $\chi_{SO_4}^{Reg} = 1\ \mu g\ m^{-3}$, $\chi_{SO_4}^{Loc} = 0\ \mu g\ m^{-3}$. The total calculated value was $\chi_{SO_4}(t_7) = 4\ \mu g\ m^{-3}$.

The ratio of two calculated particulate sulfate concentrations of Cases 3 and 4 is 1/7, and the ratio of the two turbidity parameters is 1/5.

8. MAJOR CONCLUSIONS

(a) The overall factor of decay has to be separated into partial terms which allow the time intervals during which the individual transformation and removal mechanisms were effective, to be evaluated on synoptic maps (or from weather data) and taken into account in the transmission model properly.

(b) According to recent views, beside other factors, the transformation of SO_2 depends mostly on the concentration of pollutants and on the r.h. The present method attempts to choose the rate constants or half-lives as parameters depending only on these routinely measured factors.

(c) Case studies showed, that the Regional Background Stations are capable of detecting the long-range transmission of SO_2 if the continental, regional and local scale pollution processes are taken into account separately.

(d) Such an approach makes it possible to prepare country-wide transmission models to check the environmental impact of alternative national abatement policies, taking into account the effect from long-range transmission.

(e) Case studies support the current thinking on the relation between particulate sulfate concentration and the rate of turbidity by showing that a 7-fold increase in calculated particulate sulfate concentration was accompanied by a 5-fold increase of the Schüepp turbidity parameter of *B*.

(f) The model in its present stage does not attempt to interpret the physics of dry and wet deposition, but provides a method to take into account the separate effects of removal mechanism by properly selected half-life values. More physical interpretation can be given after the optimalization of the surface and upper air conditions has been completed.

REFERENCES

Bolin B. and Persson C. (1975) Regional dispersion and deposition of atmospheric pollutants with particular application to sulfur pollution over western Europe. *Tellus* **27**, 281–310.

Béll B. (1954) *Climatology of the Troposphere over Hungary*. Academic Press, Budapest.

Benarie M. (1973) Étude de la transformation de l'anhydride sulfureux en acide sulfurique en relation avec les données climatologiques, dans un ensemble urbain à caractère industriel, Rouen. *Atmospheric Environment* **7**, 403–421.

Brodsky J. N. (1975) Preliminary report on the emission of sulfur dioxide from sources in the U.S.S.R. and other member countries of COMECON. 5th Meeting of the Working Group on Air Pollution. ECE. Jan. 13–17, 1975 Geneva.

Chamberlain A. C. (1953) Aspects of travel and deposition of aerosol and vapour clouds. British Report AERE-HP/R-1261.

Eliassen A. and Saltbones J. (1975) Decay and transformation rates of SO_2, as estimated from emission data, trajectories and measured air concentrations. *Atmospheric Environment* **9**, 425–429.

Eliassen A. (1976) *The Trajectory Model: A Technical Description.* NILU, Kjeller, Norway.

Ferber G. J., Elliott W. P., Machta L. and Heffter J. L. (1975) Deposition parameters in a continental-scale dispersion model (In preparation).

Högström U. (1974) Wet fallout of sulphurous pollutants emitted from a city during rain or snow. *Atmospheric Environment* **8**, 1291–1303.

Husar R. B. and Gillani N. V. (1976a) Mesoscale model for pollutant transport, transformation and ground removal. Presented at 3rd Symp. on Atm. Turb. Diff. and Air Qual. Oct. 19–22, 1976 Raleigh, N.C.

Husar R. B., Wilson W. E., Charlson R. J. and Whitby K. T. (1976b) Sulfates in the atmosphere. Presented at the 69th Annual Meeting of APCA, June 1976, Portland, Oregon.

Husar R. B. and Gillani N. V. (1976c) Mathematical modeling of air pollution—a parametric study. *Proc. of the 2nd Federal Conf. on the Great Lakes.* Great Lakes Basin Commission.

Makhonyko K. P. (1967) Simplified theoretical notion of contaminant removal by precipitation from the atmosphere. *Tellus* **19**, 467–476.

Mészáros E., Moore D. J. and Lodge J. P. (1977) Sulfur dioxide–sulfate relationship in Budapest. *Atmospheric Environment* **11**, 345–349.

Newman L., Forrest J. and Manowith B. (1975) The application of an isotopic ratio technique to a study of the atmospheric oxidation of sulfur dioxide in the oil fired power plant. *Atmospheric Environment* **9**, 959–968.

Ottar B. (1976) Report of the meeting of modeling experts, 18–19 Oct. 1976, Lillestrøm, Norway.

Petterssen S. (1956) *Weather Analysis and Forecasting.* McGraw-Hill, New York.

Prahm L. P. and Christensen O. (1976) Long-range transmission of sulphur pollutants computed by pseudospectral model. ECE Task Force, Lillestrøm, Norway.

Smith F. B. (1975) Airborne transport of sulphur dioxide from U.K. *Atmospheric Environment* **9**, 643–660.

Smith F. B. (1976) Inter-model comparison using selected test data. Meeting on Intercomparison of Dispersion Models, Oct. 18–19, 1976. Lillestrøm, Norway.

Strosoe S., Amble E. and Knudsen T. (1973) The LRTAP Emission Survey. Norwegian Institute for Air Research, Kjeller, Norway.

Szepesi D. (1977) *Application of Meteorology to Atmospheric Pollution Problems on a Local and Regional Scale.* Final Report of the Rapporteur of CoSAMC. WMO.

Veltischeva N. S. (1976) Numerical method for quantitative estimates of long-range transport of sulphur dioxide. ECE Task Force, Lillestrøm, Norway.

Atmospheric Environment Vol. 12, pp. 537–547. Pergamon Press 1978. Printed in Great Britain.

SULFATES IN THE ATMOSPHERE: A PROGRESS REPORT ON PROJECT MISTT*

WILLIAM E. WILSON

U.S. Environmental Protection Agency, Aerosol Research Branch, MD-57, Research Triangle Park, NC 27711, U.S.A.

(*First received* 17 *June* 1977 *and in final form* 10 *October* 1977)

Abstract—The size and sulfate content of atmospheric aerosols and the rate and mechanisms for sulfate formation from sulfur dioxide in power plant plumes are reviewed. Emphasis is given to results from the recent USEPA study, Project MISTT (Midwest Interstate Sulfur Transformation and Transport). The rate of conversion of sulfur dioxide to sulfate aerosol in power plant plumes is low near the point of emission, but increases to several per cent h^{-1} as ambient air mixes with the plume. Tall stacks reduce ground level concentrations of sulfur dioxide, resulting in a reduction of the amount removed by dry deposition. In urban plumes, which are well-mixed to the ground near the source, sulfur dioxide is removed more rapidly by dry deposition. Thus, tall stacks increase the atmospheric residence time of sulfur dioxide, which leads to an increase in atmospheric sulfate formation. These sulfate aerosols may be transported over distances of several hundred kilometers and produce air pollution episodes far from the pollution source.

INTRODUCTION

Sulfate is a pollutant of note, having been linked in epidemiological and laboratory studies with adverse effects on human health (Amdur *et al.*, 1972; EPA, 1974). Epidemiological studies have, in fact, indicated that sulfate may be more toxic than sulfur dioxide or total suspended particulates. Sulfates are also known to be major contributors to reductions in visual range caused by atmospheric aerosols (Waggoner *et al.*, 1975). Studies of acid precipitation in Scandinavia (Bolin, 1971) have implicated sulfuric acid in a variety of adverse ecological effects.

The known adverse health effects of sulfur dioxide (SO_2) led to the control of this pollutant (PHS, 1969). Reduction in urban SO_2 emissions and concentrations effected by the mandatory use of low-sulfur fuels, however, were not accompanied by a proportional decrease in urban sulfate (EPA, 1975).

The observed lack of a proportional decrease has four possible explanations: (1) sulfates can be biogenic in origin, resulting from transformations of hydrogen sulfide, methyl disulfide and methyl mercaptans, which are natural products; (2) measured sulfate values are not real but anomalous, resulting from conversion of SO_2 to sulfates on filters used in sampling; (3) observed sulfates are primary pollutants produced from the combustion of high-vanadium oil or from combustion in small, inefficient furnaces; and (4) observed sulfates may be explained by the transformation–transport theory. Reductions in urban SO_2 emissions have been accompanied by increases in rural SO_2 emissions from new power plants located outside cities. SO_2 from these power plants may be transformed to sulfate in the atmosphere and transported over long distances to urban areas.

The fourth possibility, the transformation–transport theory, is supported by Scandinavian (Bolin, 1971) and U.S. studies (Wilson, submitted) and could account for the increased sulfate levels observed in rural areas and the static sulfate levels observed in urban areas in the U.S. (EPA, 1975). Early information on the rate of conversion of SO_2 to sulfate in power plant plumes suggested that the rate of conversion was too low to result in significant formation, and subsequent transport, of sulfate (Wilson, submitted). However, studies of the flux of sulfur out of various areas, e.g. studies by Smith and Jeffrey (1975) and of Waggoner *et al.* (1976), successfully showed that some of the sulfur transported from source areas like the United Kingdom arrived in Scandinavia as sulfates.

Interest by the Environmental Protection Agency in the transformation–transport theory, and the relevance of this theory to energy usage, led to a major expansion of existing studies with the establishment of Project MISTT. Project MISTT is funded with Federal Interagency Energy/Environmental Research and Development funds. The technical approach of Project MISTT is to study the transformations of SO_2 to sulfate in polluted air masses undergoing transport. The intent is to measure pertinent chemical and meteorological parameters with sufficient accuracy that they may be used with physical and mathematical models to derive rate parameters which characterize the transformation processes. This research should also give insight into transformation mechanisms and serve to guide laboratory studies of rates and mechanisms. The initial plans called for the

* Midwest Interstate Sulfur Transformation and Transport.

study of power plant and urban plumes. However, during the program another type of pollutant transport was recognized and the program was expanded to include the 'blob'. This term refers to large-scale polluted air masses, apparently associated with anticyclones, in which pollution can build up during stagnant periods and be carried for long distances as the weather system moves.

This paper will discuss the technical and management approach to Project MISTT, describe the type of information obtained, review the important experimental findings, and discuss their implications for air pollution control policies.

TRANSFORMATION MECHANISMS

Though they are not well understood, the mechanisms by which SO_2 is oxidized to sulfates are important because they determine the rate of formation of sulfate, the influence of the concentration of SO_2 on the reaction rate and, to some extent, the final form of sulfate. Atmospheric SO_2 may be oxidized to sulfur trioxide (SO_3) and converted to sulfuric acid aerosol, or it may form sulfite ions that are then oxidized to sulfate. Subsequent to the oxidation, sulfuric acid or sulfate may interact with other materials to form other sulfate compounds. The most important sulfate formation mechanisms identified to date are summarized in Table 1.

A key concern for air pollution control strategy is to determine whether SO_2 conversion is first order or zero order in SO_2. If the conversion is first order, then an 80% decrease in SO_2 emissions could lead to an 80% decrease in sulfate. If, on the other hand, we assume the conversion is zero order in SO_2 and that at the present time 10% of emitted SO_2 is converted to sulfate, then it could require a 98% reduction in SO_2 emissions to produce an 80% reduction in sulfate. In the latter case it becomes very important to determine if there are other pollutants that influence SO_2 conversion and that could be more easily controlled. For example, if heavy metal catalysis is important (Mechanism 5), better fly ash controls to limit primary aerosol emissions could be used to reduce sulfate.

Mechanisms 1 and 2 are first order in SO_2 and in sunlight. Mechanism 2 also depends on the composition of the background air because this will influence reactive species such as OH, HO_2, and CH_3O_2. Mechanisms 3, 4, 5 and 6 may be near zero order in SO_2, but depend strongly on the concentration of ammonia, oxidants, soluble catalytic metal ions, or particulate surface area. Mechanisms 3, 4, and 5 will be important only when liquid droplets can exist, i.e. under conditions of high water vapor content and high relative humidity.

It has been the intention of Project MISTT to study conditions conducive to the various types of mechanisms. However, experimental problems and weather conditions during field studies were such that we have more results applicable to the homogeneous mechanisms (1 and 2) than to the heterogeneous mechanisms (3–6).

SIZE DISTRIBUTION

Studies (Whitby, 1975; 1978) over the past 5 years of the size distribution of particles in both sulfate aerosols and general atmospheric aerosols have led to important changes in our understanding of the behavior of aerosols such as sulfates in the ambient atmosphere. A schematic diagram of a typical atmospheric aerosol size distribution is shown in Fig. 1. The three principal modes (particle-size ranges), the

Table 1. Mechanisms by which sulfur dioxide is converted to sulfates (10)

Mechanism	Overall reaction	Factors on which sulfate formation primarily depends
1. Direct photooxidation.	$SO_2 \xrightarrow[\text{water}]{\text{light, oxygen}} H_2SO_4$	Sulfur dioxide concentration, sunlight intensity
2. Indirect photooxidation.	$SO_2 \xrightarrow[\text{organic oxidants, hydroxyl radical (OH), } HO_2, RO_2 \text{ radicals}]{\text{smog, water, } NO_x} H_2SO_4$	Sulfur dioxide concentration, organic oxidant concentration, OH, NO_x, HO_2, RO_2.
3. Air oxidation in liquid droplets.	$SO_2 \xrightarrow{\text{liquid water}} H_2SO_3$ $NH_3 + H_2SO_3 \xrightarrow{\text{oxygen}} NH_4^+ + SO_4^{2-}$	Ammonia concentration.
4. Catalyzed oxidation in liquid droplets.	$SO_2 \xrightarrow[\text{heavy metal ions}]{\text{oxygen, liquid water}} SO_4^{2-}$	Concentration of heavy metal (Fe, V, Mn) ions.
5. Catalyzed oxidation on dry surfaces.	$SO_2 \xrightarrow[\text{carbon particle}]{\text{oxygen, water}} H_2SO_4$	Carbon-particle concentration (surface area).

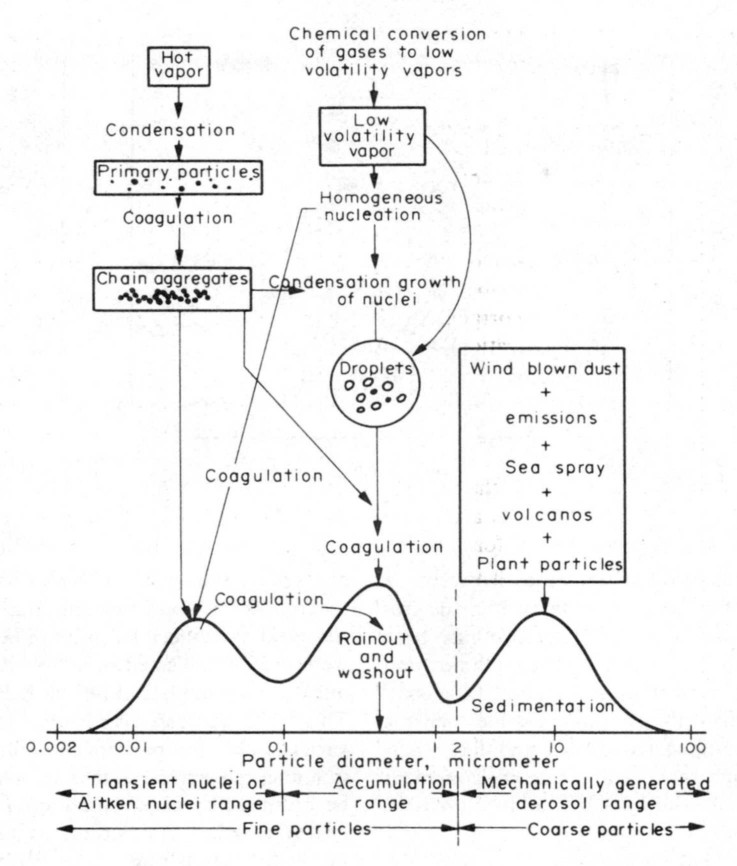

Fig. 1.

main sources of mass for each mode and the principal processes involved in inserting and removing mass from each mode are shown. Particles in the Aitken nuclei mode, 0.005–0.05 μm dia, are formed by condensation of vapors produced either by high temperatures or chemical processes. The accumulation mode, which includes particles from 0.05 to 2 μm dia, is formed by coagulation of particles in the nuclei mode and by growth of particles in the nuclei mode through condensation of vapors onto the particles. Coarse particles are formed by mechanical processes such as grinding or rubbing—e.g. soil, street dust and rubber tire wear—and by evaporation of liquid droplets. Accumulation mode particles do not continue to grow into coarse particles, however, because the more abundant small particles have a higher gas-aerosol collision rate and dominate the condensational growth process. Sulfates formed by the conversion of SO_2 are found in the accumulation mode; $MgSO_4$ from sea salt, Na_2SO_4 from paper pulping and $CaSO_4$ from gypsum are found in the 'coarse particle mode'.

REVIEW OF PLUME STUDIES

Information on the rate of conversion of SO_2 to sulfate in power plant plumes was needed to quantify the contributions of power plants to atmospheric sulfates. A critical review of plume studies (Wilson, submitted) revealed no reliable information on conversion rates, and only two studies that provided information on the amount of SO_2 converted to sulfate. The early work of Gartrell *et al.* (1963) indicated little conversion of SO_2 to sulfate except at relative humidities greater than 75%. However, at high r.h., conversion of SO_2 to sulfate may occur either on the filter surface or within the aerosol particles collected on the filter. Therefore, Gartrell's results cannot be accepted unless verified by other techniques.

Investigators at the Brookhaven National Laboratory used two techniques to study sulfate formation in power plant plumes (Newman *et al.*, 1975a, b): the sulfur isotope technique and direct measurements of SO_2 and sulfate. The isotope technique is considered erroneous because of the presence of several competing reactions having different isotope effects (Wilson, submitted). The direct measurement of SO_2 and sulfate is considered to yield valid concentration ratios but this may not be adequate for the determination of conversion rates. The Brookhaven workers measured conversion from 0 to 8.5% h^{-1} in coal-fired power plant plumes (Newman *et al.*, 1975a) and 0–26% h^{-1} in oil-fired power plant plumes (Newman *et al.*, 1975b). The Brookhaven work has previously been interpreted to indicate an average conversion rate in coal-fired plumes of 1% h^{-1} or less and a sub-

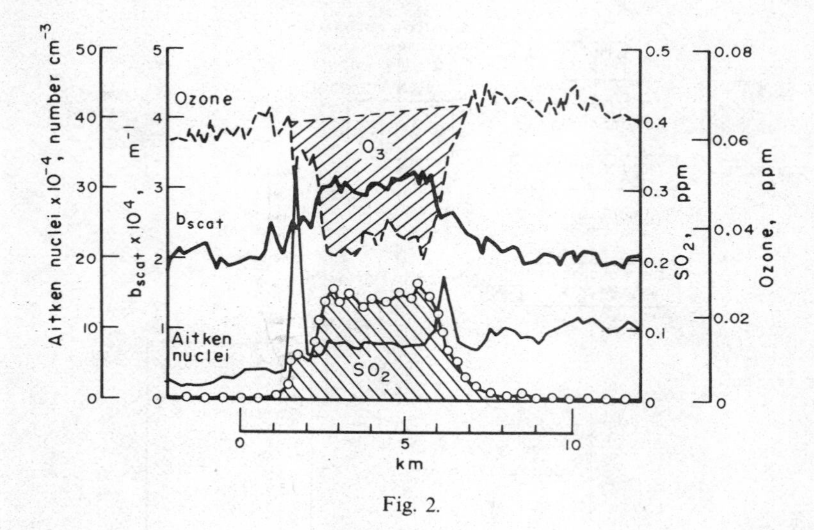

Fig. 2.

stantially higher rate in oil-fired plumes. However, a careful analysis of the data, after discarding the isotope results, indicates that no difference has been established between oil- and coal-fired plumes and that, depending on conditions, conversion rates substantially greater than $1\% \, h^{-1}$ are possible (Wilson, submitted). The sampling techniques and flight patterns used were such that an accurate measurement of the conversion rate could not be obtained (Wilson, submitted).

EPA PLUME STUDIES

Program plan

In order to obtain a better understanding of the physical and chemical processes occurring in power plant plumes, a new series of field studies were undertaken. During July and August of 1976 and July, August and February of 1975, extensive studies involving three-dimensional mapping of large plumes were carried out in the St. Louis area as part of the Midwest Interstate Sulfur Transport and Transformation Study (MISTT). Preliminary studies were conducted during July and August of 1973 and 1974.

These plume studies differed from earlier ones in that: (1) more gas and aerosol parameters were measured; (2) horizontal and vertical profiles were measured; (3) data were interpreted in terms of mass flows instead of concentration ratios; (4) the background air mixing with the plume was characterized; (5) the chemical composition and size distribution of the aerosols in the plume were determined; (6) measurements were made in approximately the same air mass as it moved downwind, and (7) measurements were made at the same distance downwind as the plume shape and structure changed with changes in meteorological conditions.

In previous studies, the fractional conversion of SO_2 to sulfate was calculated from the ratio of SO_2 to sulfate collected as integrated filter samples while circling in random portions of the plume at various downwind distances. This technique can overestimate conversion if any SO_2 is removed by dry deposition to ground surfaces. For this reason and because of the need to collect large amounts of SO_2, samples were usually collected in cohesive plumes which had undergone minimal dilution with background air. This technique can also lead to errors if conversion varies in different parts of the plume or if conversion is influenced by the extent to which the plume has been diluted by background air. That this can indeed be the case is demonstrated in Fig. 2, in which the nuclei formation rate is seen to be much greater at the edges of the plume than in the center. Although such a dramatic difference was not often observed, current interpretation of sulfate formation mechanisms suggests that the conversion rate will be higher in those portions of the plume, such as the edges, in which dilution has led to conversion of plume NO to NO_2 by background O_3 (Wilson *et al.*, 1975). To overcome these problems the EPA plume mapping studies were designed to measure SO_2 and sulfate mass flow rates at increasing distances from the source in both cohesive and well-mixed plumes.

Another problem which occurs when studies are limited to cohesive plumes has to do with the effect of sunlight on the SO_2 conversion process. The key reactant for conversion of SO_2 to sulfate is thought to be the hydroxical radical, with HO_2 and RO_2 becoming important in polluted air (Mechanism 2). The concentration of these species will be proportional to the sunlight intensity. Cohesive plumes occur in the morning and evening when, due to low sunlight intensity, there is insufficient thermal energy to cause rapid mixing of the plume with background air. Thus the lack of sunlight means no direct or indirect photochemical conversion (Mechanism 1–2). The lack of dilution results in low input of ammonia into the plume, causing the aerosols to become acidic, if liquid, or coated with SO_2 if solid, until no further SO_2 can be absorbed. Therefore heterogeneous reactions (Mechanisms 3, 4 and 5) can occur only to a limited extent. Studies of conversion in cohesive

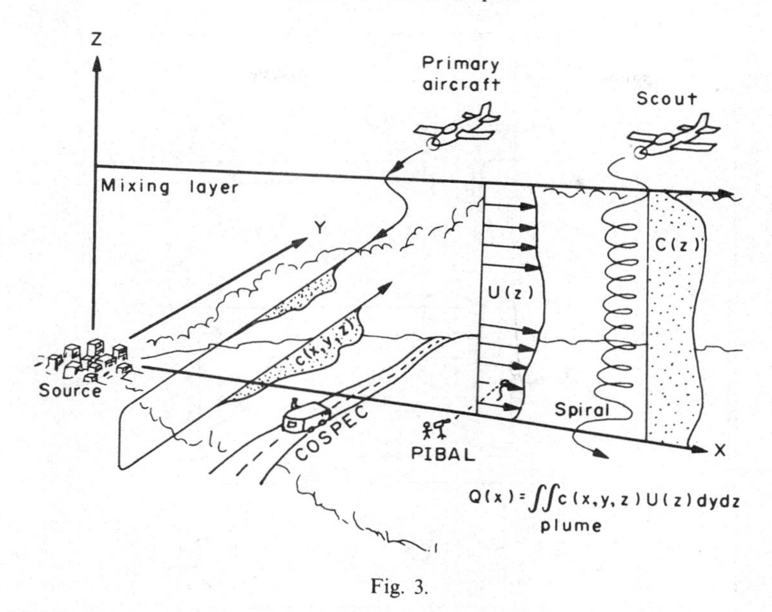

Fig. 3.

plumes therefore would be expected to show low reaction rates.

Plume mapping program

Two instrumented aircraft, an instrumented van, and three mobile single-theodolite pilot-balloon units were used in a coordinated measurement program (Fig. 3). The primary sampling platform was a single-engine aircraft equipped for continuous monitoring of the gaseous pollutants (O_3, NO, NO_x, SO_2) (Blumenthal *et al.*, 1978); three aerosol parameters (condensation-nuclei count, light-scattering coefficient and aerosol charge); several meteorological variables (temperature, r.h., dew point and turbulent dissipation); and navigational parameters. Particulate sulfur samples were gathered by a sequential filter-tape sampler equipped with a respirable-particle size separator. An array of four cascade impactors collected size-differentiated aerosol samples for later chemical and microscopic analyses. An optical counter and an electrical-mobility analyzer provided details of the *in situ* particle-size distribution of grab samples (White *et al.*, 1976a). During the summer 1976 study a twin engine aircraft was equipped for continuous monitoring of O_3, SO_2, light scattering coefficient, aerosol charge and navigational parameters, and for collection of aerosol samples with a sequential filter tape sampler.

An instrumented 'scout' aircraft located the plume, laid out the sampling path of the primary aircraft, and directed the three mobile pilot-balloon (pibal) units to their respective positions in the center and at the edges of the plume. Coordinated with the aircraft sampling, a van equipped with a correlation spectrometer (COSPEC) made lateral traverses under the plume and measured the integrated overhead burden of SO_2 and NO_2 (Vaughan, 1975). During summer 1975 a ground sampling station was located 100 km north of St. Louis. In cooperation with Project MISTT measurements were made by Argonne National Laboratory to characterize the structure and dynamics of the boundary layer in the area north of St. Louis. During summer 1976 a ground mobile sampling unit made measurements under the plume and COSPEC measurements were conducted from an airplane platform. The operations headquarters coordinated the mobile units by radio communication.

Data collected aboard the two aircraft and in the vans were recorded automatically at 1- to 10-s intervals on magnetic tapes for subsequent computer processing. The aircraft's instruments were calibrated before and after each flight. The aircraft data, which were originally recorded on magnetic tape cartridges, were transferred to nine-track tapes immediately after the flight. Calibrated constants were inserted, engineering units calculated, and strip charts generated within 24 h so that the data would be available for inspection and discussion prior to subsequent flights. This also ensured that any instrument malfunctions or errors in sampling design would be discovered immediately and could be corrected with minimal loss of needed data.

The flight pattern of the primary aircraft was designed to enable characterization of the plume at discrete distances downwind from the source. At each distance, horizontal traverses were made in the plume perpendicular to the plume axis at three or more elevations. These were supplemented by vertical spirals within and outside the plume. The continuous instruments monitored the distribution of pollutants along each pass. From the three-dimensional pollutant concentration field obtained in this manner, together with the vertical profiles of wind velocity measured every half-hour by the three pibal units, the horizontal flow rates of pollutants at each downwind distance were directly calculated. Since no continuous monitor for sulfate was available, the continuous light-scattering (b_{scat}) measurement was used as an indication of the

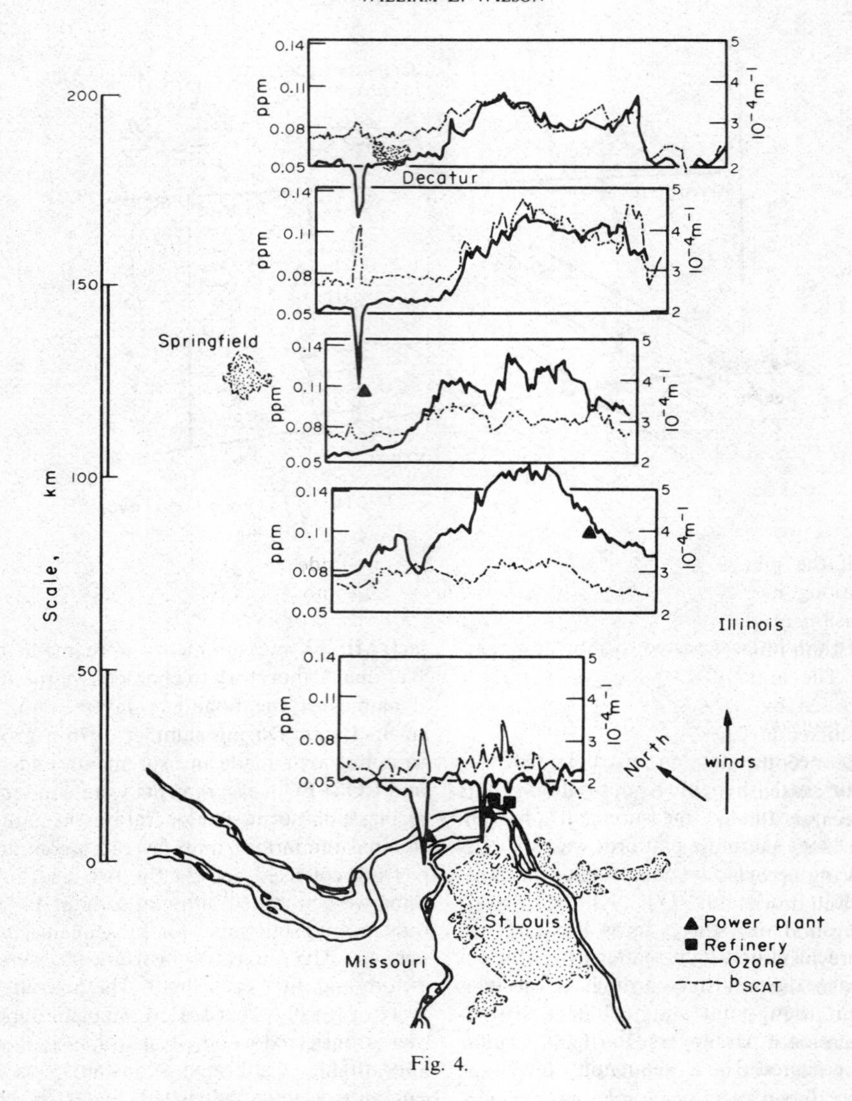

Fig. 4.

sulfate concentration distribution. Sulfate, collected in filter samples during each pass through the plume, was analyzed by the flash vaporization–flame photometric detection method—a new, highly sensitive technique modified for aircraft use (Husar *et al.*, 1975, 1977). The sulfate flow was calculated from the b_{scat}/sulfate concentration ratio, integrated over a pass, and the b_{scat} flow. Similarly, b_{scat}-to-aerosol volume ratios were used to calculate an aerosol volume flow. From the change in flow rate with distance, it is possible to calculate transformation and removal rates for individual pollutants (Husar *et al.*, 1974, 1976). Under well-defined wind conditions (unidirectional, steady wind field), the reproducibility of the pollutant flow measurement was $\pm 20\%$.

Urban plumes

The techniques just described were used to map the three-dimensional flow of aerosols and trace gases in power plant plumes and in the air leaving the St. Louis area. It was found that under certain summer, daytime meteorological conditions the aggregate pollutant emissions from metropolitan St. Louis often formed a cohesive, well-defined 'urban plume' downwind of the city (Husar *et al.*, 1974, 1976; White *et al.*, 1976b). Such a plume can be identified in Fig. 4, which shows ozone concentrations and light-scattering coefficients measured within the mixing layer northeast of St. Louis on 18 July 1975 (White *et al.*, 1976b). The application of the plume mapping technique to this plume will be described as an example of the type of data and results obtained from Project MISTT.

On that day, the St. Louis plume incorporated the emissions of major coal-fired power plants at Labadie (Mo.) and Portage des Sioux (Mo.) and of a refinery complex near Wood River (Ill.), in addition to emissions from industries and automotive traffic of the St. Louis–East St. Louis urban area itself. The concentrated individual plumes immediately downwind of large combustion sources can be identified in Fig. 4 by their ozone 'deficits'; the depressed concentrations of O_3 in these plumes reflect the high concentration of NO in effluents.

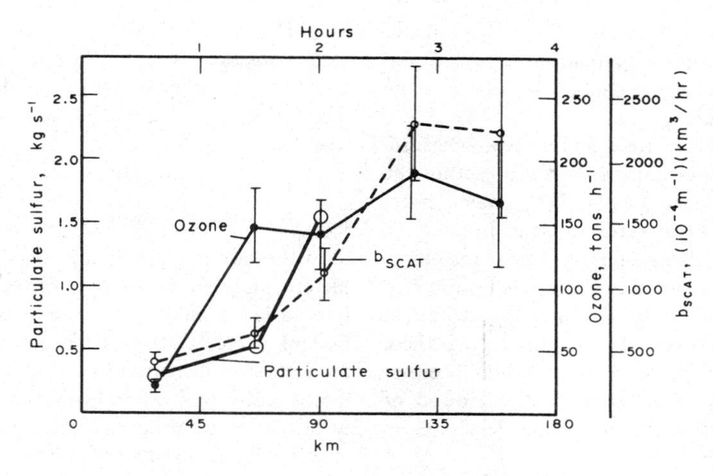

Fig. 5.

The width of the plume, approx 40 km, did not change much along its 150-km length. Apparently, horizontal diffusion at the boundaries of the urban plume is slower than might be predicted by Gaussian plume models. The amount of plume spreading is probably controlled by the amount of wind shear within the well-mixed layer. It appears likely that the elevated ozone concentrations in this plume and the reduced visibility caused by the plume were exported well beyond Decatur, Ill.

Unlike the primary pollutants NO and SO_2, ozone and light-scattering aerosols attained their maximum concentrations well downwind of St. Louis on 18 July 1975. The ozone and most of the aerosol were not emitted but were products of chemical and physical transformations in the atmosphere; their secondary origin is evident in Fig. 5, which shows flow rates of ozone and aerosol in the St. Louis plume increasing with distance from the city. The flow rates in Fig. 5 represent only the contribution of the St. Louis plume, and are calculated from the measured winds within the plume and from the difference between pollutant concentrations inside and outside the plume.

Based on a number of experiments in 1974 and 1975, the following features of the urban plume were observed to be typical: (1) The St. Louis plume significantly degraded the air quality of communities as far as 150 km from the city. (2) The most conspicuous components of the St. Louis plume 50 km or more downwind of the city were the reaction products formed along the way. (3) Most of the aerosol responsible for the high b_{scat} and resulting decrease in visibility within the St. Louis urban plume was formed in the atmosphere.

The sulfate concentrations in the 18 July 1975 urban plume attained a maximum of 20 $\mu g\, m^{-3}$. This is not an extremely high value. On 29–30 July 1975 the urban plume was followed for 18 h and measurements were made both in the evening and the following morning. During this experiment sulfate concentrations of 60 $\mu g\, m^{-3}$ were measured, as shown in Fig. 6. This set of measurements demonstrates that at least under some conditions sulfate formation can exceed removal and dilution and can lead to the accumulation of very high sulfate concentrations.

Sulfate budget

By examining the total flow of sulfur—both SO_2 and sulfate—we can draw some interesting conclusions regarding the transformation and removal of SO_2 in the two classes of plumes, that is, power plant plumes and urban plumes (Husar *et al.*, 1976). Data were examined for those days in which the wind direction was such that the plumes of the rural power plants (tall stacks, > 200 m) were well separated from the urban plume which contained emissions from low stacks ($\leqslant$ 100 m) and near ground level sources. In such plumes, the total sulfur flow rate (gaseous plus particulate) has been found to decrease with distance. The pollutant flow rate, as determined from both the aircraft-measured concentrations and correlation spectrometer data, shows a decay in excess of the estimated experimental uncertainty of the methods ($\pm$20%).

Initial sulfur depletion in the urban plumes measured appeared to be equivalent to an exponential rate of decay with a characteristic (1/e) decay time of 3–4 h. Formation of new sulfate aerosol is undetec-

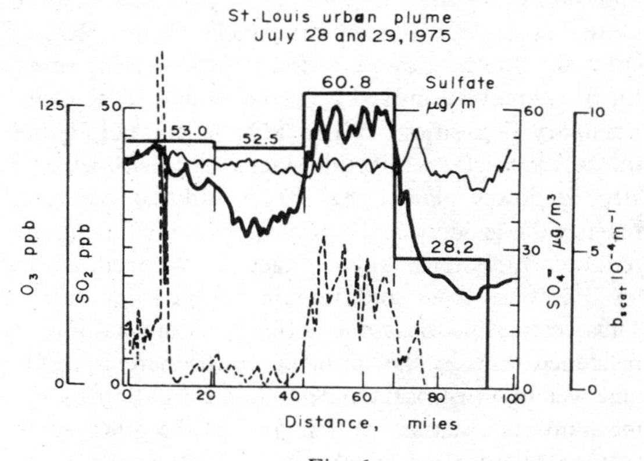

Fig. 6.

table up to a distance of 50 km. There is evidence that during the daytime the formation of aerosol, including sulfate in an urban plume, begins after an ageing time of 1 or 2 h.

In power plant plumes total sulfur flow measurements frequently showed no loss of sulfur within the limits of experimental uncertainty. There was conversion of SO_2 to sulfates at rates varying from $\frac{1}{2}\%$ to $5\% \, h^{-1}$ (Husar *et al.*, 1978). These power plant plume measurements demonstrate that there was little removal of the sulfur emitted from tall stacks within a region of 50–100 km from the plant. During these measurements, dry deposition was minimal either because the plumes had not touched the ground or because they were very dilute at ground level.

Removal processes

It is important to understand that most sulfate aerosol is secondary, i.e. formed by gas-to-particle conversion in the atmosphere after emission, and occurs in the accumulation mode. Particles in this size range have very long lifetimes. Some confusion has existed regarding the relative removal rates for SO_2 and sulfate aerosol by deposition. The deposition velocity of aerosols depends on the particle size and reaches a minimum in the accumulation size range of 0.1–1.0 μm. The deposition velocity of gaseous SO_2 depends on the chemical reactivity of the surface. Both deposition velocities depend on the friction velocity and the surface roughness. However, since the diffusion of accumulation mode aerosols through the surface boundary layer (0.001–1 cm) is much slower than the diffusion and surface reaction of an SO_2 molecule, the deposition velocity of sulfate aerosol is usually smaller than for gaseous SO_2 (Sehmel, 1971, 1973; Sehmel and Hodgson, 1974). Consequently, though SO_2 is removed fairly rapidly by dry deposition on soil, vegetation and other surfaces, the removal processes become slower once the SO_2 has been converted to sulfate. Therefore, sulfates can travel long distances before their removal by fallout, rainout, or deposition.

Effects of tall stacks

These aspects of reaction mechanisms, aerosol size and deposition rates are critical to an understanding of the possible effects of tall stacks. When SO_2 is emitted near the ground, as from home heating units, the SO_2 can be removed by surface removal mechanisms (dry deposition). When SO_2 is emitted higher in the air, as from the tall stacks of fossil-fuel-fired electric power plants, the SO_2 is diluted before it reaches the ground, and the surface removal rates are reduced. Emissions may, in fact, be trapped above an inversion layer and remain trapped for hours. Thus, elevated stacks theoretically permit a longer residence time in the ambient atmosphere for SO_2 and promote fine particulate sulfate formation by the mechanisms discussed previously. On the other hand, they provide for increased dilution of the sulfate and SO_2 and thus reduce the impact of emissions in the vicinity of the source. Continued studies are planned to study this phenomenon as a function of time of day and extent of mixing.

SCIENTIFIC RESULTS

Reaction rates and mechanisms

In the power plant plumes which have been completely analyzed the rate of conversion of SO_2 to sulfate varied from $\frac{1}{2}\%$ to $5\% \, h^{-1}$ (Husar *et al.*, 1978; Gillani *et al.*, 1978). Condensation nuclei counts and aerosol size distribution profiles indicate that the major pathway is a homogeneous reaction, first order in SO_2, probably involving the OH radical (Whitby *et al.*, 1978; Cantrell and Whitby, 1978). The pollutant profiles, in power plant and urban plumes, resemble those observed in chamber studies and suggest that the current chemical kinetic models can be used to calculate sulfate formation (Miller *et al.*, 1978). The reaction rate certainly depends on sunlight intensity and appears to also depend on water vapor concentration, background ozone levels and the extent to which the plume has mixed with background air (Husar *et al.*, 1978).

Heterogeneous reactions may be important at night, in clouds, or other conditions during which high water vapor content and high r.h. may lead to the existence of liquid droplets. Attempts to make night-time measurements during summer 1975 were frustrated by difficulties in locating the plume. The use of lidar during summer 1976 allowed us to locate the plume but unusually dry conditions led to night-time relative humidities substantially lower than normal. Therefore, the data analyzed to date do not allow any conclusions regarding heterogeneous reactions.

Plume study techniques

One of the most important advances has been the realization that a plume measurement must be treated as a multi-dimensional problem. In addition to its extent in the horizontal and vertical direction and the downwind distance we must consider time as a fourth dimension. We must be concerned not only with the time at which the plume is measured but also the time at which the plume was emitted and the history of the plume since it was emitted. For example, it may have been a cohesive plume prior to measurement or it may have been highly diluted with background air; it may have been isolated above the mixing layer or it may have been well mixed to the ground; it may have traveled at night, under cloud cover, or in bright sunlight. Much of the early work on plumes has yielded misleading values because the measurements were made only in cohesive plumes early in the morning or late in the evening.

The use of mass flow measurement, in addition to SO_2/sulfate ratio measurements, permits a determination of the loss of SO_2 by ground deposition. This technique makes it possible to determine rates during periods when the plume is well mixed to the ground and the SO_2/sulfate ratio measurements would give erroneously high rates. The development of a sulfate analytical technique with sufficient sensitivity for a measurement integrated over one pass through a plume made possible the determination of sulfate mass flows. The measurement of size distribution profiles permitted measurements of aerosol volume flows for comparison with sulfate mass flows and gave insight into the type of reaction mechanism.

Transport distances

The time and distance over which an air mass maintains its integrity depends on its initial size and the meteorological conditions. Power plant and urban plumes have been tracked for 300 km (Gillani *et al.*, 1978). These plumes maintain their integrity and high pollutant concentrations for much longer times and farther distances than was originally expected. During stable night-time conditions, the cohesive plume is frequently caught in a nocturnal jet which carries it along at as much as twice the normal wind speed (Smith *et al.*, 1978). Gaussian plume models are satisfactory for the first few tens of km but beyond that wind shear seems to play the dominant role in determining dilution. In urban plumes wind shear is clearly the determining factor. Using visibility isopleths and long range trajectories it has been possible to track 'blobs' (hazy air masses associated with stagnating anticyclones) for over a 1000 km (Lyons *et al.*, 1978; Husar *et al.*, 1976).

Models

Models, of several levels of complexity, have been developed for the calculation of secondary pollutant concentrations in power plant and urban plumes. These include a reacting plume model with relatively simple mixing parameters but with provisions for aerosol formation, coagulation and growth and a multi-step chemical kinetic model (Brock *et al.*, 1978). In addition, a model has been developed using a sulfate formation rate which is a function of sunlight intensity and which has more sophisticated meteorological terms including multi-layers for vertical diffusion and dry deposition (Gillani, 1978).

Future work

The present study has concentrated mainly on sulfate, ozone and light scattering. It is planned to extend the EPA plume studies to include measurements of organic aerosols and vapors and nitrate aerosols and vapors, such as nitric acid. In order to determine the importance of heterogeneous reactions, more work is needed under conditions of high r.h. and high water vapor content which are conducive to heterogeneous reactions. In general, more information is needed to provide better statistics on the parameters which influence reaction rates in power plant and urban plumes.

The 'blob' is a special challenge. New techniques are required to develop models and to study transformation and transport in such systems.

SUMMARY

Earlier work had shown that the rate of conversion of SO_2 to sulfate in power plant plumes was slow except at very high r.h. In the recent EPA program, consistent conversion of from $\frac{1}{2}$ to 5% h^{-1} at normal levels of r.h. has been measured. The conversion of SO_2 to sulfate aerosol in power plant plumes is slow in the early part of the plume; that is, close to the point of emission. As ambient air mixes with the plume, the rate of conversion increases. Thus tall stacks reduce ground-level concentrations of SO_2 but increase sulfate aerosol formation by reducing surface losses of SO_2 and by increasing the atmospheric residence time, which results in increased SO_2-to-sulfate conversion. In urban plumes, which are well-mixed to the ground, SO_2 may be removed by reaction with plants and by deposition. The SO_2 dry deposition rates vary with vegetation, with the nature of the surface and with time of year.

Power plant and urban plumes have been sampled out to 300 km from their sources. Sampling at these distances revealed that sulfate, generated from SO_2 in power plant plumes, and ozone, generated from hydrocarbons and nitrogen oxides in urban plumes, may be transported at least hundreds of km and may cause air pollution episodes far from the source of pollution. Air pollution, due to secondary pollutants such as sulfates and ozone, cannot be controlled by the government entity where the air pollution impact actually occurs. Therefore, current concepts of air quality control regions must be revised to take into account the long-range transport of secondary pollutants.

Acknowledgements—The author would like to thank Stephen J. Gage, Deputy Assistant Administrator, Office of Energy, Minerals, and Industry, for his continuing interest in this program and for providing financial support through the Interagency Energy/Environment R & D Program, and Gregory D'Alessio, OEMI, for his administrative assistance.

REFERENCES

Amdur M. O., Lewis T. R., Fitzhand M. P. and Campbell K. I. (1972) Toxicology of Atmospheric Sulfur Dioxide Decay Products. Publication No. AP-111, U.S. Environmental Protection Agency, Research Triangle Park, NC.

Air pollution across national boundaries. The impact on the environment of sulfur in air and precipitation. (1971) Sweden's Case Study for the United Nations Conference on the Human Environment. (edited by Bolin B.) Kungl. Koktryckeriert P. A. Nortsedt & Soner 710396, Stockholm, Sweden. 1971.

Blumenthal D. L., Ogren J. A. and Anderson J. A. (1978) Airborne sampling system for Project MISTT. *Atmospheric Environment* **12**, 613–620.

Cantrell B. K. and Whitby K. T. (1978) Aerosol size distribution and aerosol volume formation for a coal-fired power plant plume. *Atmospheric Environment* **12,** 323–333.

Health Consequences of Sulfur Oxides: A Report from CHESS 1970–1971 (1974) EPA-650/1-74-004, U.S. Environmental Protection Agency, Research Triangle Park, NC.

Position Paper on Regulation of Atmospheric Sulfates. (1975) EPA-450/2-75-007, U.S. Environmental Protection Agency, Research Triangle Park, NC.

Gartrell F. W., Thomas F. W. and Carpenter S. B. (1963) Atmospheric oxidation of SO_2 in coal-burning power plant plumes. *Am. ind. hyg. J.* **24,** 113–120.

Gillani N. V. (1978) Project MISTT: mesoscale plume modeling of the dispersion transformation and ground removal of SO_2. *Atmospheric Environment* **12,** 569–588.

Gillani N. V., Husar R. B., Husar J. D., Patterson D. E. and Wilson W. E. (1978) Project MISTT: kinetics of particulate sulfur formation in a power plant plume out to 300 km. *Atmospheric Environment* **12,** 589–598.

Husar R. B., Blumenthal B. L., Anderson J. and Wilson W. E. (1974) The urban plume of St. Louis. In *Proc. 167th Natl ACS Meeting, Div. envir. Chem.*, Los Angeles, CA.

Husar J. D., Husar R. B. and Stubits P. K. (1975) Determination of submicrogram amounts of atmospheric particulate sulfur. *Analyt. Chem.* **47,** 2062.

Husar R. B., Husar J. D., Gillani N. V., Fuller S. B., White W. H., Anderson J. A., Vaughan W. M. and Wilson W. E. (1976) Pollutant flow rate measurement in large plumes: sulfur budget in power plant and area source plumes in the St. Louis region. In *Proc. 171st Natl ACS Meeting, Div. envir. Chem.*, New York, N.Y.

Husar J. D., Husar R. B., Macias E. S., Wilson W. E., Durham J. L., Shepherd W. K. and Anderson J. A. Particulate sulfur analysis: application to high time resolution aircraft sampling in plumes. *Atmospheric Environment* **10,** 591–595.

Husar R. B., Gillani N. V., Husar J. D., Paley C. C. and Turcu P. N. Long range transport of pollutants observed through visibility contour maps, weather maps and trajectory analysis. In *Preprints third symp. atmospheric turbulence, diffusion and air-quality. Am. Met. Soc.*, Boston, MA. U.S.A.

Husar R. B., Patterson D. E., Husar J. D., Gillani N. V. and Wilson W. E. (1978) Sulfur budget of a power plant plume. *Atmospheric Environment* **12,** 549–568.

Lyons W. A., Dooley J. C. and Whitby K. T. (1978) Satellite detection of long range pollution transport and sulfate aerosol hazes. *Atmospheric Environment* **12,** 000–000.

Miksad R. W., Bower K. K. and Brock J. R. (1978) Chemically reactive power plant plume models. *Atmospheric Environment* (submitted).

Miller D. F. (1978) Precursor effects on SO_2 oxidation. *Atmospheric Environment* **12,** 273–280.

Newman L., Forrest J. and Manowitz B. (1975) The application of an isotopic ratio technique to the study of the atmospheric oxidation of sulfur dioxide in the plume from an oil fired power plant. *Atmospheric Environment.* **9,** 954.

Newman L., Forrest J. and Manowitz B. (1975) The application of an isotopic ratio technique to the study of the atmospheric oxidation of sulfur dioxide in the plume from a coal fired power plant. *Atmospheric Environment.* **9,** 969.

Air quality criteria for sulfur oxides. (1969) Publication No. AP-50, U.S. Department of Health, Education, and Welfare, Public Health Service, Washington, D.C.

Smith F. B. and Jeffrey G. H. (1975) Airborne transport of sulfur dioxide from the U.K. *Atmospheric Environment* **9,** 643–659.

Smith T. B., Blumenthal D. L., Anderson J. A. and Vanderpol A. H. (1978) Transport of SO_2 in power plant plumes: day and night. *Atmospheric Environment* **12,** 605–611.

Sehmel G. A. (1971) Particle diffusivities and deposition velocities over a horizontal smooth surface. *J. Colloid Interface Sci.* **37,** 891–906.

Sehmel G. A. (1973) Particle eddy diffusivities and deposition velocities for isothermal flow and smooth surfaces. *J. Aerosol Sci.* **4,** 125–138.

Sehmel G. A. and Hodgson W. H. (1974) Particle and Gaseous Removal in the Atmosphere by Dry Deposition. Battelle Pacific Northwest Laboratories, Richland, WA 99352, BNWL-SA-4941.

Vaughan W. M., Sperling R. B., Gillani N. V. and Husar R. B. (1975) Horizontal SO_2 Mass Flow Measurements in Plumes: A Comparison of Correlation Spectrometer Data with a Dispersion and Removal Model. Paper No. 75-17.2, *68th Ann. Mtng of the Air Pollut. Control Ass.*, Boston, Mass.

Waggoner A. P., Vanderpol A. H., Charlson R. J., Granat L., Trägard C. and Laisen S. (1975) Sulfates as a Cause of Tropospheric Haze. Report AC-33, Dept. of Meteorology, University of Stockholm, Sweden. UDC 551.510. 4:535.33.

Waggoner A. P., Vanderpol A. H., Charlson R. J., Granat L., Trägard C. and Laisen S. (1976) The sulfate light scattering ratio. An index of the role of sulfur in tropospheric optics. *Nature* **261,** 120–122.

Whitby K. T. (1975) Modeling of Atmospheric Aerosol Size Distribution. Progress Report, EPA Research Grant No. R800971.

Whitby K. T. (1978) The physical characteristics of sulfur aerosols. *Atmospheric Environment* **12,** 135–159.

Whitby K. T., Cantrell B. K. and Kittelson D. B. (1978) Nuclei formation rates in a coal-fired power plant plume. *Atmospheric Environment* **12,** 313–321.

White W. H., Anderson J. A., Knuth W. R., Blumenthal D. L., Hsiung J. C. and Husar R. B. (1976) Midwest Interstate Sulfur Transformation and Transport Project: Aerial Measurements of Urban and Power Plant Plumes, Summer 1974. EPA-600/3-76-110, U.S. Environmental Protection Agency, Research Triangle Park, NC.

White W.H., Anderson J. A., Blumenthal D. L., Husar R. B., Gillani N. V., Husar J. D. and Wilson W. E., Jr. (1976) Formation and transport of secondary air pollutants: ozone and aerosols in the St. Louis urban plume. *Science* **194,** 187–189.

Wilson W. E., Husar R. B., Whitby K. T., Kittleson D. B. and White W. H. (1976) Chemical reactions in power plant plumes. In *Proc. 171st Natl ACS Mtng, Div. envir. Chem.*, New York, N.Y.

Wilson W. E. Sulfate formation in power plant plumes: a critical review. (Submitted).

APPENDIX

Program management

Project MISTT is an integrated scientific research program based on the formation of a team of the most competent scientists in the field. From the nucleus of a few aerosol scientists who first worked together in Los Angeles in 1969, a team comprising many disciplines and interests has slowly been developed. Over the years expertise in atmospheric chemistry, meteorology, and chemical and diffusion modelling has been added to the group. Although maximum freedom in program planning and execution is encouraged, strong central direction is maintained to ensure that the Environment Protection Agency's needs are satisfied.

The senior investigators are listed below.

William E. Wilson, Project Director; Environmental Protection Agency.

R. B. Husar, Field Director; Washington University.

Don Blumenthal, Aircraft Operations; Meteorology Research, Inc.

K. T. Whitby, Aerosol Dynamics; University of Minnesota.

Seventeen institutions have been involved in the program at various levels of participation. The institutions, the principal investigators, and a brief description of their responsibilities are given below:

Argonne National Laboratory: Paul Frenzen, boundary layer structure and dynamics; Bruce Hicks, dry deposition rates (ANL work was cosponsored by ERDA.)

Battelle Columbus Laboratories: David Miller, outdoor smog chamber measurements of sulfate formation rates in St. Louis; Peter Jones, Gas chromatographic–mass spectrometric measurements of organic vapors.

California Institute of Technology: Sheldon Friedlander; Development of a supersensitive sulfate measurement technique, sulfate measurements—1974.

Environmental Measurements, Inc.: William Vaughan; COSPEC measurements.

Environmental Quality Research: William Dannevik; Forecasting and other meteorological support, Dry deposition studies.

EPA-Las Vegas: Roy Evans; Helicopter measurements, winter 1976, aircraft lidar observations, summer 1976.

EPA-Research Triangle Park: W. E. Wilson and Jack Durham; Program management; Tom Ellestad; Instrument calibration, data transfer, measurements in EPA mobile lab.

Florida State University: John Winchester and William Nelson; Aerosol measurements using the FSU 'Streaker' sampler with PIXIE analysis.

IIT Research Institute: Ron Draftz; Optical and electron microscopy.

Meteorology Research, Inc.: Don Blumenthal, Jerry Anderson, Warren White and Ted Smith; Aircraft measurements, data analysis, meteorological interpretation.

University of Minnesota: K. T. Whitby and Bruce Cantrell; Aerosol size distribution measurements, aerosol dynamics, ground measurements, data analysis and interpretation.

Rockwell International Science Center: Al Jones; Pilot balloon operations.

Stanford Research Institute: Ed Uthe; Ground-mobile lidar operations.

University of Texas: Jim Brock; Effects of charge on aerosol deposition, reactive plume models.

Washington State University: R. Rasmussen and R. Chatfield; Detailed hydrocarbon analysis, interpretation of ozone and hydrocarbon data.

Washington University: R. B. Husar; Field Director, data analysis and interpretation; Noor Gillani; Data manager and model development; Janja Husar; Sulfate determinations.

University of Washington: R. Charlson and A. Waggoner; Ground-based measurements and data interpretations, sulfate species measurement, air mass trajectories.

Atmospheric Environment Vol. 12, pp. 549–568. Pergamon Press 1978. Printed in Great Britain.

SULFUR BUDGET OF A POWER PLANT PLUME

R. B. HUSAR, D. E. PATTERSON, J. D. HUSAR, N. V. GILLANI

Air Pollution Research Laboratory, Department of Mechanical Engineering, Washington University, St. Louis, MO 63130, U.S.A.

and

W. E. WILSON, JR.

Aerosol Research Branch, U.S. Environmental Protection Agency, Research Triangle Park, NC 27711, U.S.A.

(*First received* 28 *June* 1977 *and in final form* 3 *October* 1977)

Abstract—As part of the Midwest Interstate Sulfur Transformation and Transport (MISTT) study, the summer sulfur budget of the plume of the 2400 MW coal-fired Labadie power plant near St. Louis, Missouri is assessed via aircraft data, ground monitoring network data, and a two-box model. The particulate sulfur (S_p) formation rate is obtained from three-dimensional plume mapping combined with a high time-resolution S_p sampling technique. During noon hours the SO_2 conversion rate is found to be 1–4% per hour, compared to night rates below 0.5% per hour. Plume excess light scattering coefficient (b_{scat}) and excess S_p correlated well ($r = 0.87$), indicating most S_p is formed in the light-scattering size range.

During daytime the well-mixed plume is transported at $5\ m\ s^{-1}$ on the average; at night the July average wind speed at plume height is $12\ m\ s^{-1}$ due to the low-level jet. The nocturnal plume is less than 100 m thick at 400 m above ground, and is decoupled from the surface until morning. Ground monitoring data from the Regional Air Pollution Study (RAPS) show that plume entrainment into the rising mixing layer is completed by 1000 Central Daylight Time (CDT). Due to daytime vertical mixing and nocturnal decoupling, the dry removal rate for the elevated plume is highest near noon. In a daily cycle, the plume sequentially passes through a reservoir regime, dissociated from delivery to the ground, and then enters the mixing-removal regime.

A two-box model representing the two regimes, with diurnally periodic rate constants for transformation and removal, is employed to estimate plume sulfur budgets. Ignoring wet removal, 30–45% of the SO_2 is estimated to be converted to S_p, half within the first day. Particulate sulfur is formed unevenly: the afternoon plume contributes more than its share because it rises so high that it has more time to react before removal begins. In short: transformation and removal occur mainly during the daytime, while transport is fastest at night. After a hard day of convection, reaction and deposition, the lower atmosphere relaxes at dusk while the midwestern plume takes off overnight on a jetstream and begins the next day's work 300–400 km from the stack.

INTRODUCTION

Laboratory studies of transformations and of dry and wet removal are designed for the systematic parameter by parameter examination of these processes. Regional (or 1000 km) scale studies, on the other hand, have the objective to determine the overall transport, transformation and removal (transmission) processes. The gap between laboratory simulation experiments and regional scale studies may be bridged by mesoscale studies of sulfur transmission. Such experiments are inherently restricted to a few hundred kilometers and less than a day of airmass aging. The current work is a part of Project MISTT (Midwest Interstate Sulfur Transformation and Transport), initiated in 1974 by the United States Environmental Protection Agency with the aim of understanding sulfur transformation and removal processes in well-defined plumes. The specific objectives, approach, and philosophy of this experiment are described by the Project MISTT director (Wilson, 1978).

This paper is an attempt to systematically analyze the transmission processes that determine the sulfur budget of a power plant plume. In this effort we are utilizing results of experiments, monitoring data from the Regional Air Monitoring System (RAMS) in St. Louis, Missouri, and a simple box model. Most of the supporting field data come from the MISTT field mapping results from the years 1974 and 1976. The emphasis is on the extraction of the key parameters and their synthesis into a coherent picture while maintaining the conceptual simplicity of box models. In companion papers the dispersion and removal processes are considered in the framework of a time-dependent transport, transformation and diffusion model (Gillani, 1978) and two long range plume sampling missions to over 300 km are described by Gillani *et al.* (1978).

Results of this analysis are generated from (and therefore strictly applicable to) the coal fired (3.2% S, and about $6\ kg\ s^{-1}$ sulfur output) Labadie power plant (2400 MW, base load), which has a tall stack (218 m), and is located 60 km west of downtown St. Louis in gently rolling terrain. All mapping of the Labadie plume was performed during summer (June–July–August) non-stagnant periods, in the absence of precipitation. The general features of the MISTT field program are described by Wilson (1978); specific description of the Washington University operations is given in Appendix 1.

TRANSFORMATION

In the MISTT plume mapping program, SO_2 conversion is assessed by continuously monitoring the SO_2 concentration and three aerosol parameters: light scattering coefficient (b_{scat}), aerosol charge, and the total condensation nuclei concentration. The aerosol size spectrum is measured in grab samples (Cantrell *et al*., submitted). The particulate sulfur concentration, averaged over individual plume traverses, as well as in the background air, is obtained from analysis of filter samples (Husar *et al*., 1976). In the MISTT data analysis emphasis has been placed on the consistency of these aerosol detectors as indicators of aerosol formation.

In the present paper we shall utilize primarily the particulate sulfur and the light-scattering coefficient data in estimating conversion rates. Valid conversion rate estimates have two major problems. First, the assessment of slow reactions requires substantial aging time of the plume before appreciable conversion will be seen. The second problem is a competing interference from the other major budget process, removal. Therefore the focus in this section is on conversion estimates in power plant plumes over time periods sufficiently large to allow substantial conversion, yet sufficiently short to permit isolation of a 'pure' transformation process.

Examination of a plume sulfur budget in practice requires subtraction of background levels from the in-plume concentrations of SO_2, sulfate, and b_{scat} levels. By assuming this superposition principle one is limited to a search for linear, quasi-first order processes. The MISTT plume sampling is Eulerian, but evaluation of a kinetic process must be done in a Lagrangian framework. Therefore downwind distance must be translated into plume age.

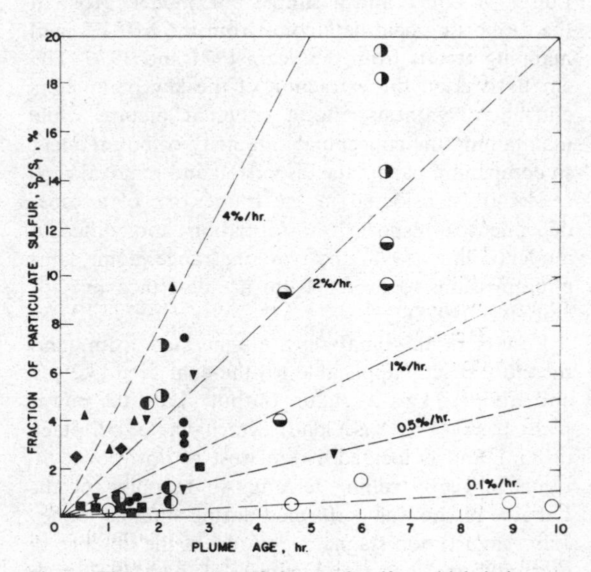

Fig. 1. The percentage of the total sulfur which is in particulate form, S_p/S_t, for nine sampling missions. Both S_p and S_t are excesses over the background. The symbols in Figs. 1–4 are defined in Table 1.

The results of power plant plume conversion obtained as part of MISTT in 1974 and 1976 are shown in Fig. 1 and tabulated in Table 1. The data points represent the average excess particulate sulfur S_p as obtained from integrated filter samples, normalized to the excess average gaseous and particulate sulfur concentration ($S_t = S_g + S_p$) across the same path. In the nine sampling experiments, in which the plume age ranged up to ten hours, the fraction of the particulate sulfur, compared to the total was between 0.2 and 20%. Generally S_p/S_t increased with increasing age. It is apparent, however, that this rate of increase was quite variable from one sampling event to another, ranging between 0.1 and 4% h^{-1}. Another qualitative measure of conversion is the excess b_{scat}, since most of the aerosol is formed in the light scattering size range (Whitby, 1978). It is apparent that the b_{scat} per unit of total sulfur also increases with age, but it too is quite variable. The wide scatter in the S_p/S_t rate of increase (i.e. conversion rate) immediately raised the question of experimental errors. A consistency check was therefore sought between excess S_p/S_t and excess b_{scat}/S_t, as shown in Fig. 2. The comparison, considering the error bars for S_p measurement (Husar *et al*., 1976), shows a comforting correlation ($r = 0.87$) of the two aerosol parameters; this, supplemented with data from the aerosol charger, demonstrates to our satisfaction that the apparent variability in the conversion rate is not an artifact of experimental uncertainty but rather a manifestation of actual changes in the rate. The linear regression gives $\sim 20\ \mu g\ m^{-3}$ sulfate per unit b_{scat}; however, the high correlation coefficient is primarily due to high and consistent readings on a single day. This relationship suggests that the excess b_{scat} is a fair surrogate for the sulfur aerosol formed. Conversely, the sulfate formed in the Labadie power plant causes visibility degradation roughly in proportion to the converted S_p.

Initial inspection was directed toward the role of background aerosol, ozone, temperature, etc. as possible causes for the conversion rate variability. This line of analysis led us to a common underlying parameter, the time of day. High conversion rates (greater than 0.5% h^{-1}) were observed only between 0800 and 2000 Central Daylight Time (CDT). Conversion rates below 0.5% are limited to the nighttime hours, 2000–0800 CDT. The diurnal pattern of the conversion rate data is shown in Fig. 3, where each horizontal bar extends from plume release time to plume sampling time. Conversion rate is obtained as $\%S_p$ divided by plume age; therefore, the area under each bar is proportional to the total observed ratio S_p/S_t. During daytime, the observed conversion rates range between 1 and 4% h^{-1}. The low conversion rates and small S_p/S_t ratios in the early morning and dusk hours are consistent with the plume conversion measurements of Forrest and Newman (1977). Plume samples with less than one hour of aging were not plotted in Fig. 3 because of the large error bars which

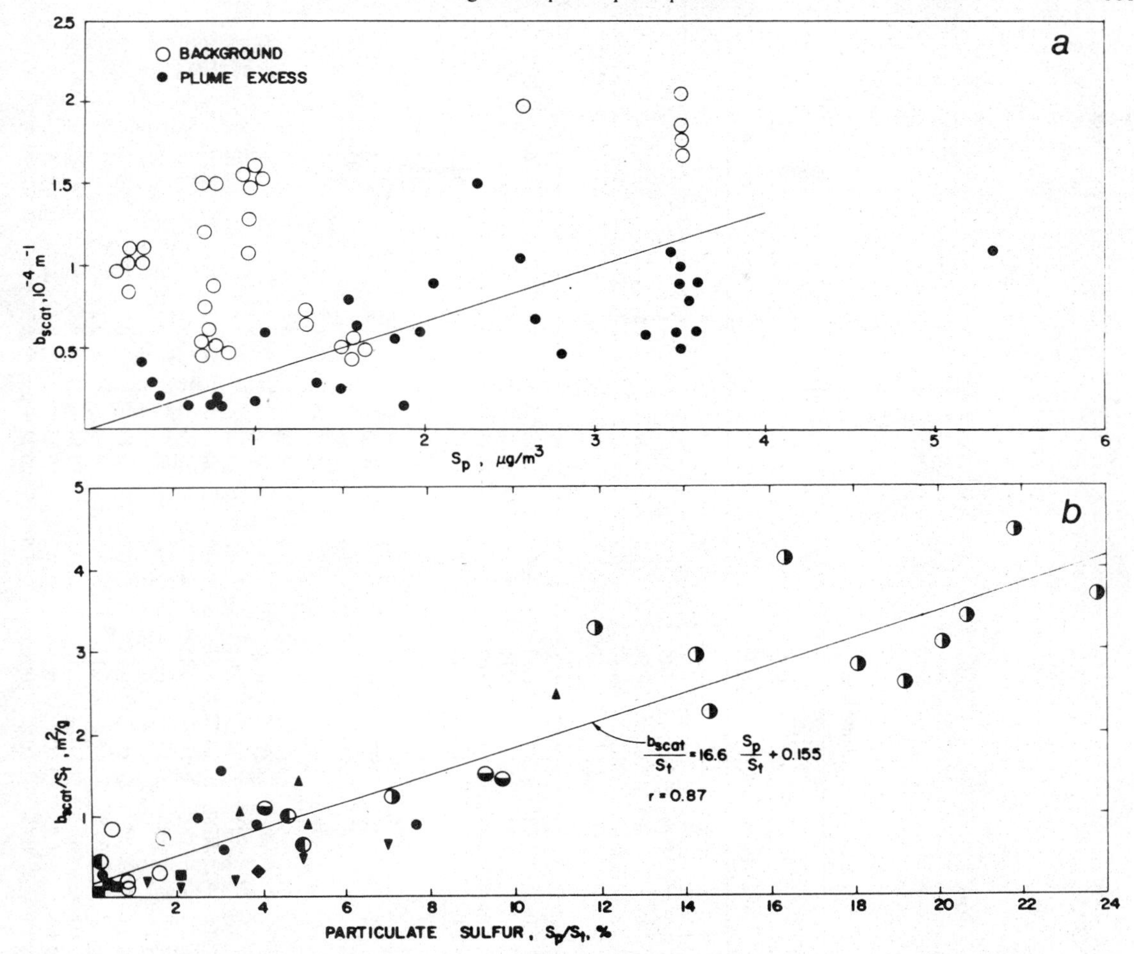

Fig. 2. Comparison of light scattering coefficient and particulate sulfur values: (a) b_{scat} versus particulate sulfur averaged over plume traverses. The plume excess b_{scat}/S_p is approximately 0.3. The higher b_{scat}/S_p in the background can be attributed to non-sulfur aerosol compounds or to a characteristic particle size which is more efficient for light scattering; (b) Plume excess light scattering coefficient per total excess sulfur versus percentage of particulate sulfur.

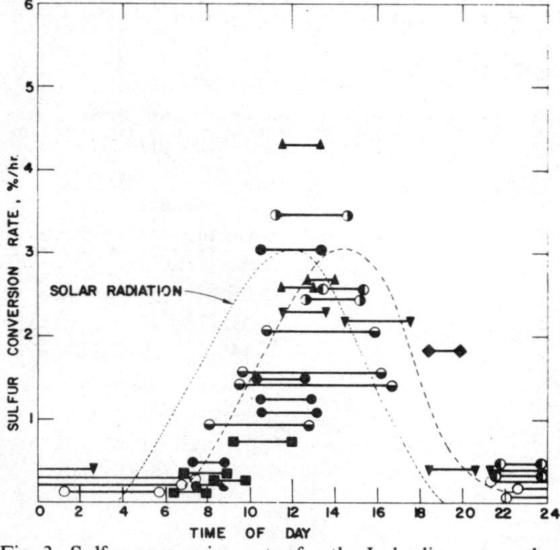

Fig. 3. Sulfur conversion rates for the Labadie power plant plume for nine sampling missions. The points to the left of each bar are the release times and to the right the sampling times. The dashed line is the diurnal conversion rate used in the two box model.

result from the division of two small numbers. Conversion rates for samples with plume age such that ground removal became important were also omitted; two such days are discussed in detail by Gillani *et al.* (1978). The sampling day of the WU aircraft on 14 July 1976 was included even though the plume age was over ten hours (360 km). The use of a special sampling pattern indicated that the nocturnal plume had not touched ground; hence, the absence of dry removal was assured. The conversion rate data shown in Fig. 3 constitute one of the main results of our particulate sulfur formation studies as part of MISTT.

The corresponding rates of light scattering aerosol formation, obtained by dividing excess b_{scat}/S_t by plume age, are shown in Fig. 4. The diurnal pattern of the light scattering aerosol formation is characterized by high daytime rates compared to low values overnight. It is worth noting that the variability of noontime light-scattering aerosol formation rate is about a factor of five, compared to the spread of a factor of three in sulfate formation rate. It is therefore apparent that the formation of light scattering aerosol

Table 1. Summary of MISTT power plant data

Date	Symbol	Craft	Dist. $\bar{x}$ (km)	Age Δt (h)	Decimal Time Sample (h)	Decimal Time Release (h)	Background S_p $\mu g\ m^{-3}$	Background b_{scat} $10^{-4}\ m^{-1}$	Excess S_p $\mu g\ m^{-3}$	Excess S_g $\mu g\ m^{-3}$	Excess b_{scat} $10^{-4}\ m^{-1}$	S_p/S_t %	b_{scat}/S_t $m^2\ g^{-1}$	$S_p/S_t \cdot \Delta t$ $\%h^{-1}$	$b_{scat}/S_t \cdot \Delta t$ $m^2\ g^{-1} \cdot h^{-1}$
8/5/74	◆	MRI	20	1.2	19.5	18.3	2.52	1.9	1.81	44.2	0.12	3.9	0.27	3.3	0.22
8/14/74	▲	MRI	8	0.5	11.8	11.3	3.50	1.8	2.70	50.1	0.46	5.1	0.87	—	—
			16	1.0	12.4	11.4	3.50	1.6	1.80	50.1	0.54	3.5	1.04	3.5	1.06
			24	1.5	13.2	11.7	3.50	1.8	1.98	38.2	0.57	4.9	1.42	3.3	0.96
			36	2.2	13.3	11.1	3.50	2.1	2.60	22.1	0.61	10.5	2.47	4.8	1.11
7/5/76	▼	MRI	14	0.8	12.5	11.7	0.57	1.0	1.20	91.1	0.16	1.3	0.18	1.6	0.24
			33	1.8	13.8	12.0	0.57	1.0	1.46	27.7	0.13	5.0	0.45	2.8	0.26
			55	2.7	17.3	14.6	0.57	1.0	1.22	16.2	0.11	7.0	0.61	2.6	0.23
			72	2.5	20.8	18.3	2.18	1.0	0.63	29.4	0.03	2.1	0.11	0.8	0.04
			145	5.5	2.5	21.0	2.18	1.0	1.87	58.3	0.11	3.4	0.18	0.6	0.03
7/9/76	◒	WU	70	3.0	11.5	8.5	2.25	2.1	0.70	24.3	0.17	2.8	0.70	0.9	0.23
			105	4.5	12.7	8.2	2.25	2.0	1.74	40.7	0.45	4.1	1.07	0.9	0.24
			135	6.5	15.5	9.0	2.00	1.7	3.66	35.7	0.39	9.3	1.50	1.4	0.23
			135	6.5	15.8	9.3	2.00	1.7	3.47	27.0	0.49	11.4	1.57	1.7	0.24
			135	6.5	16.5	10.0	2.00	1.7	2.16	20.1	0.32	9.7	1.43	1.5	0.22
7/14/76	■	MRI	27	0.7	6.5	5.8	0.25	1.0	2.57	684.0	1.08	0.4	0.16	0.5	0.22
			47	1.3	7.5	6.2	0.25	1.0	1.08	684.0	0.68	0.2	0.10	0.1	0.08
			50	1.4	8.3	6.7	0.25	1.0	3.44	700.0	1.01	0.5	0.14	0.3	0.10
			27	1.3	9.5	8.2	0.25	1.0	1.58	475.0	0.63	0.3	0.13	0.3	0.10
			50	2.8	12.0	9.2	0.25	1.1	3.48	169.0	0.45	2.0	0.26	0.7	0.09
			4	0.4	14.3	13.9	0.25	0.9	0.60	229.0	0.42	0.3	0.18	—	—
7/14/76	○	WU	165	4.6	5.9	1.3	0.25	1.0	0.58	83.0	0.15	0.7	0.18	0.2	0.04
			275	7.6	6.4	22.8	0.25	1.0	0.78	49.5	0.15	1.5	0.30	0.2	0.04
			325	9.0	7.0	22.0	0.25	0.8	0.75	105.0	0.15	0.7	0.14	0.1	0.02
			360	10.0	7.3	21.3	0.75	0.8	0.75	17.0	0.15	4.2	0.85	0.4	0.09
			215	6.0	9.7	3.7	0.25	1.0	0.45	27.0	0.20	1.6	0.73	0.3	0.12

7/18/76	◐	MRI	30	2.4	15.4	13.0	0.71	0.6	1.47	28.0	0.13	5.0	0.45	2.1	0.19
			30	2.3	15.6	13.3	0.71	0.6	1.47	25.3	0.12	5.3	0.45	2.0	0.19
			30	2.2	16.0	13.8	0.71	0.6	1.47	34.4	0.16	4.1	0.45	1.9	0.20
			50	1.2	23.0	21.8	0.68	1.5	3.52	700.0	1.00	0.5	0.14	0.4	0.12
			75	2.2	0.0	21.8	0.68	1.5	2.27	280.0	0.50	0.8	0.18	0.4	0.08
7/18/76	◑	WU	30	2.5	14.5	12.0	1.30	0.6	1.14	13.5	0.09	7.8	0.62	3.1	0.25
			30	2.1	14.7	12.6	1.30	0.7	1.00	13.0	0.17	7.1	1.21	3.4	0.58
			110	6.5	15.8	9.3	0.71	0.5	2.25	10.8	0.34	17.3	2.86	2.7	0.44
			110	6.5	16.1	9.6	0.71	0.5	3.28	15.0	0.48	17.9	2.62	2.7	0.40
			110	6.5	16.2	9.7	0.71	0.5	2.37	13.4	0.47	15.0	2.98	2.3	0.46
Spiral			110	6.5	16.5	10.0	0.71	0.5	3.57	18.2	0.90	16.4	4.13	2.5	0.64
			160	8.3	18.3	10.0	1.53	0.5	3.37	16.7	0.62	16.8	3.10	2.0	0.37
Spiral			160	8.9	18.7	9.8	1.53	0.6	2.81	14.0	0.60	16.7	3.57	1.9	0.40
			160	8.9	19.0	10.1	1.53	0.5	2.39	11.7	0.63	17.0	4.45	1.9	0.50
			160	8.9	19.3	10.4	1.53	0.4	2.00	9.6	0.43	17.2	3.68	1.9	0.41
			220	10.0	20.0	10.0	1.53	0.5	2.46	12.5	0.34	16.4	2.25	1.6	0.22
			320	12.5	0.2	11.7	1.53	0.5	2.04	11.6	0.44	15.0	3.27	1.2	0.26
7/30/76	●	MRI	31	1.4	8.5	7.1	0.96	1.1	0.31	154.0	0.42	0.2	0.26	0.1	0.18
			29	1.3	6.8	5.5	0.96	1.7	0.40	154.0	0.30	0.3	0.19	0.2	0.15
			31	1.4	8.7	7.3	0.96	1.7	1.37	210.0	0.30	0.6	0.14	0.5	0.10
			27	2.5	12.6	10.1	0.95	1.6	3.30	100.0	0.60	3.2	0.58	1.3	0.23
			27	2.5	12.8	10.3	0.95	1.6	3.55	89.0	0.80	3.8	0.86	1.5	0.35
			27	2.5	13.0	10.5	0.95	1.6	2.05	79.0	0.80	2.5	0.99	1.0	0.40
			27	2.5	13.1	10.6	0.95	1.6	1.55	49.0	0.80	3.1	1.58	1.2	0.61
			27	2.5	13.3	10.8	0.95	1.5	3.65	44.0	0.90	7.7	1.89	3.1	0.76
			45	4.2	14.2	10.0	0.95	1.4	5.36	35.0	1.10	13.3	2.73	3.2	0.65

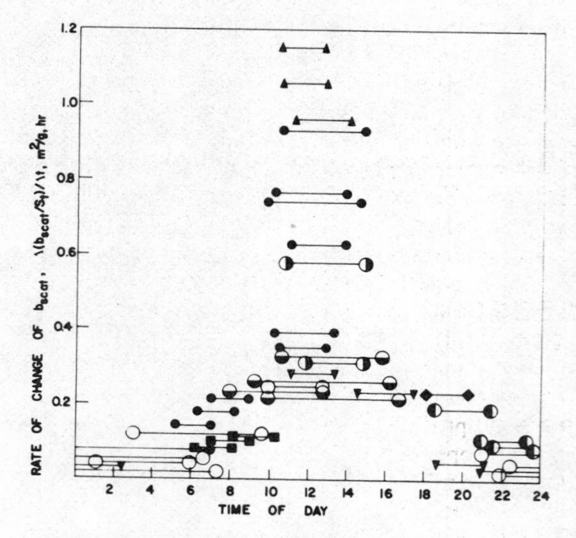

Fig. 4. Diurnal pattern of the rate of formation of light scattering aerosol.

depends on parameters in addition to those governing the sulfate formation, e.g. particle size spectra. Finally a comparison is made between the observed S_p formation rate and the rate of change of light scattering aerosol, Fig. 5. The correlation coefficient is $r = 0.64$ and the least squares regression line is $b_{scat}/(S_t\,\Delta t) = 19.4\ S_p/(S_t\,\Delta t) + 0.029$.

The above conversion data show that the rate controlling parameters of SO_2 oxidation are associated with time of day. The informed reader could be tempted to interpret the daytime conversion as a clue that photochemistry is the dominant mechanism. However, the extent of our current data analysis is insufficient to exclude in-cloud conversion, mixing with background air or other mechanisms as being rate controlling.

PLUME TRANSPORT

The sulfur budget in plumes is influenced by meteorology, most obviously by providing the horizontal and vertical transport mechanisms. Horizontal transport, characterized by the wind speed field over the plume height, provides dilution and 'long range transport'; the turbulent eddy exchange in the vertical direction delivers the matter to surfaces for dry deposition. The mixing of plume material with the potentially reactive background air is facilitated by both horizontal and vertical dispersion.

Horizontal transport

The speed at which plume material is transported over a given location is a function of both height and time. For the St. Louis region, the Regional Air Pollution Study (RAPS) provides an extensive data base for the time–height dependence of wind speed and direction as well as temperature and dewpoint (Myers and Reagan, 1975). The average vertical profiles of wind speed and their standard deviation from July 1976 are shown in Fig. 6a. Vertical velocity soundings at 1300 show on the average a uniform speed of 4–5 m s^{-1} below about 1000 m, i.e. less than the geostrophic wind speed (6 m s^{-1}) above, due to the drag of buoyant thermals. At 1900 there is evidence of increased speed between 300 and 600 m which becomes more pronounced at 0100, when an average speed of 12 m s^{-1} is recorded at 450 m in a low-level jet. This phenomenon is illustrated in more detail by Blackadar (1957), Bonner *et al.* (1968) and Smith *et al.* (1978). By 0700 the vertical profile has become more homogeneous at reduced speeds, with the exception of the lowest 200 m, which is already subject to drag from the nascent mixing layer. The diurnal pattern of the wind speed obtained from hourly averages at seven heights is shown in the Fig. 6c. The ground level (10 m) wind speed is obtained

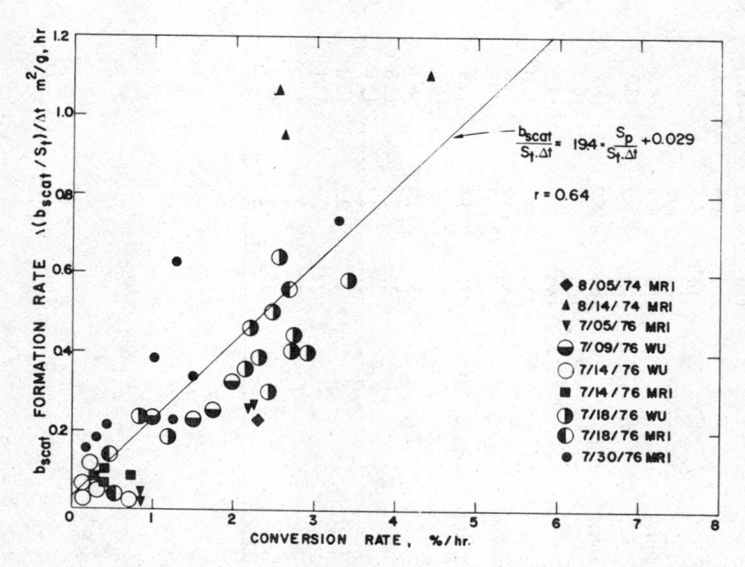

Fig. 5. Comparison of the formation rate of the light scattering aerosol and sulfur conversion rate.

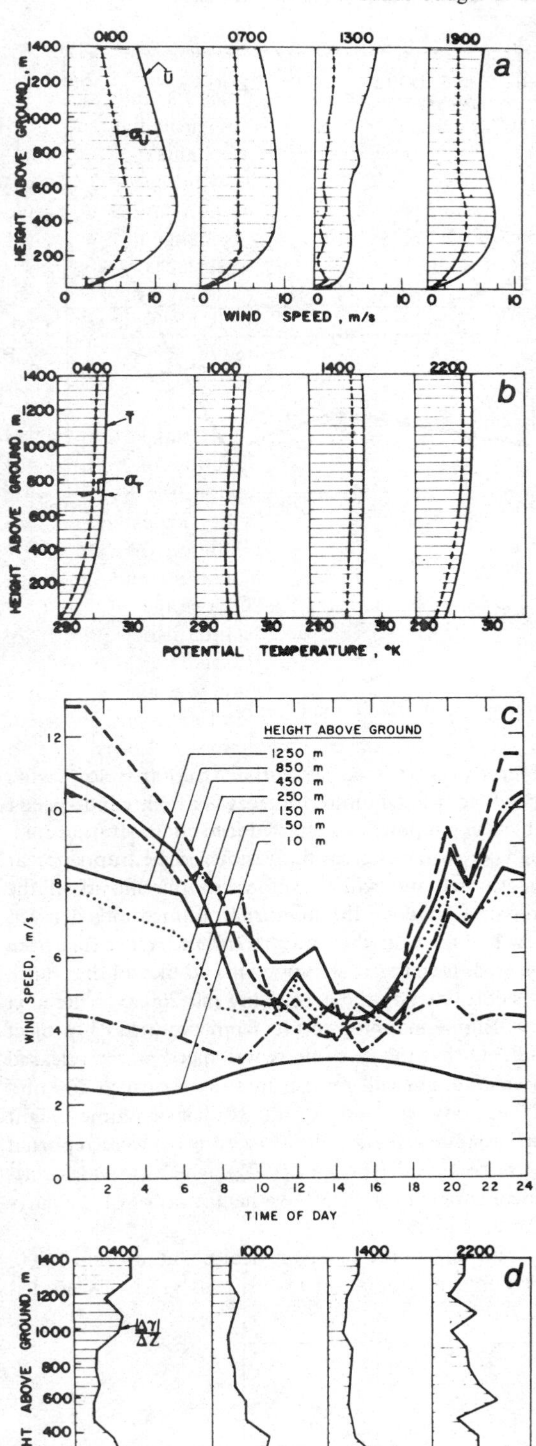

Fig. 6. Wind data obtained from hourly releases by the RAPS balloon sounding network, averaged for July 1976: (a) Vertical profiles of velocity for 0100, 0700, 1300 and 1900 CDT releases; (b) Vertical profiles of potential temperature (referenced to 1000 mb) and its standard deviation; (c) Diurnal pattern of the wind speed at seven different heights; (d) The absolute change of wind direction with height, °/100 m.

from RAPS surface measurements. The figure illustrates that within the diurnal cycle in the St. Louis region, nighttime plume transport is greater than at midday by a factor of nearly two. Nocturnal plumes emitted from tall stacks are affected by 10–20 m s^{-1} low level jets in two ways: their effective stack height is reduced due to the crosswind and the distance travelled overnight is increased to as much as 500 km. Since such summertime plumes in the midwestern U.S. are subject to minimal vertical mixing, aircraft sampling is feasible up to several hundred kilometers.

On 14 July 1976 the Washington University aircraft followed a thin nocturnal power plant plume in a low level jet to 360 km from the stack. In this mission the concept of detailed plume mapping at fixed distances (Appendix 1) has been abandoned; instead the flight path was along the easily detectable plume edge in a zig-zag manner, occasionally interrupted for a conventional traverse perpendicular to the plume axis. The key plume dimensions, vertical profiles of wind speed and temperature, the maximum gaseous sulfur concentration and the fraction of particulate sulfur are plotted in Fig. 7. The lack of any appreciable dispersion is evidenced by the fact that even at 300 km distance and 10 hours plume age, the maximum S_g was about 200 μg m^{-3} (400 μg m^{-3} as SO_2). During the night the particulate sulfur fraction remained at about 1% of total sulfur, but had increased to 4% when sampling terminated at 0800.

In addition to the longitudinal and vertical transport, plume dispersion also occurs by 'veer', i.e. the absolute wind directional change with height. The vertical profiles of veer for a given hour were obtained by taking absolute value of the directional change per 100 m. Of the four profiles shown in Fig. 6d, the 1400 readings showed the least average veer. Evidently the vigorous daytime mixing inhibits the development of such lateral dispersion. Soundings at 0400 exhibited the strongest veer effects near the surface, supporting the notion that the low level nocturnal jet is decoupled from the surface stable layer.

Therefore, while transformation in the midwestern United States in summer appears to be a daytime mechanism, the long-range transport is primarily a nocturnal phenomenon.

Vertical transport

Pollutants within the mixed layer are delivered by turbulent mass transfer from the plume height to the roughness height of the canopy level, where dry deposition takes place. The height of the mixed layer has a characteristic diurnal pattern. For the St. Louis region, a typical pattern was constructed based on MISTT vertical aircraft soundings and RAPS radiosonde data. This pattern is illustrated in Fig. 8, which shows that the mixed layer height rapidly increases from near ground level in the morning to 1000 meters at noon and then levels off at about 1200 meters by mid-afternoon. After 1800, the convective mixing rapidly dies out due to the development of a cold

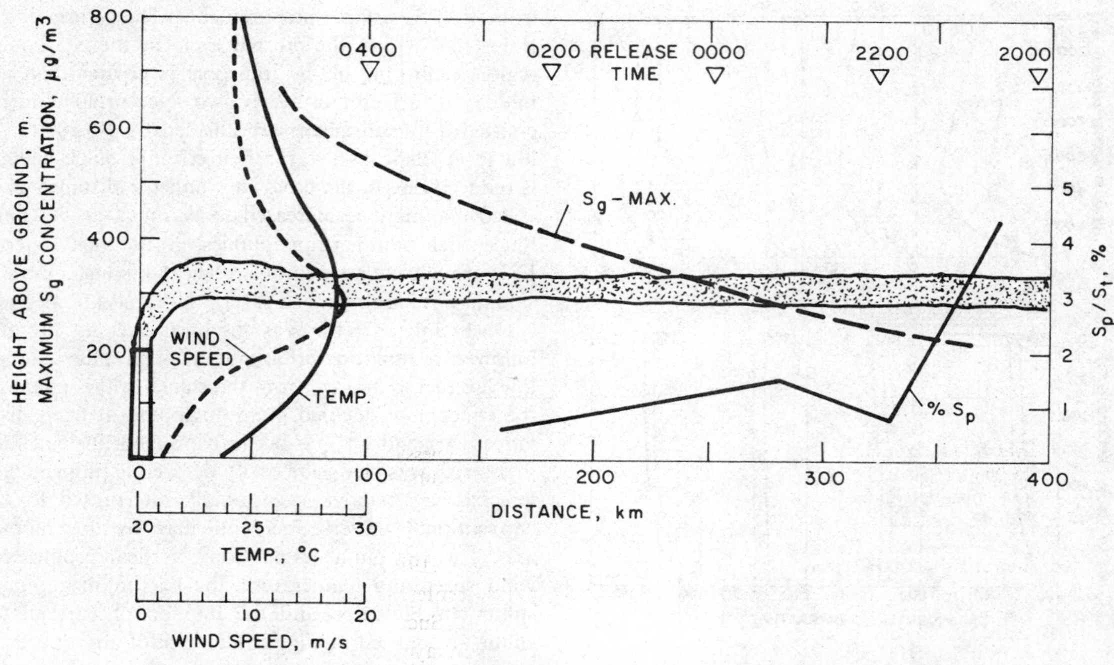

Fig. 7. Plume schematics of the morning sampling flight of WU on 14 July 1976.

stable air layer near the ground. The former mixed layer becomes layered and pollutants tend to remain in the layers into which they were introduced.

Due to radiation cooling, the surface temperature decreases at night, on the average about 9°C below the maximum daytime temperature in St. Louis. Consequently, a radiation inversion develops in the first 100 to 200 meters above the surface (Fig. 6b), stabilizing the nocturnal surface layer. This surface based inversion causes a decoupling of the surface layer from the bulk of the mixed layer, inhibiting exchange of matter (and momentum and heat) across the inversion interface. In spite of the surface based inversion, a shallow mixing layer develops at night caused entirely by the mechanical turbulence introduced by the surface roughness elements. In the St. Louis region in the summer, terrain is essentially open country with low vegetation. Under clear or moderately cloudy night skies, the height of the nocturnal mixed layer can vary from essentially zero for low wind speeds to a few hundred meters for high wind speeds. Therefore, a pollutant emitted into the nocturnal mixing layer without great buoyancy will be trapped near the surface and will be subjected to removal. If the emission is above the nocturnal surface mixed layer, it will remain in the stratum into which it has been emitted, forming a shallow, ribbon-like plume, until it is entrained by the developing mixing layer the next day. Plume mapping results from the MISTT project indicate that the Labadie power plant plume released after midnight will remain in strata between 200 and 500 m above ground. A valuable set of plume height data for the Labadie plume (see Fig. 8) were reported by Forrest and Newman (1977), which provides statistical support for the above height range of the early morning plume.

With the onset of solar heating at around 0500, the nocturnal radiation inversion is gradually eroded

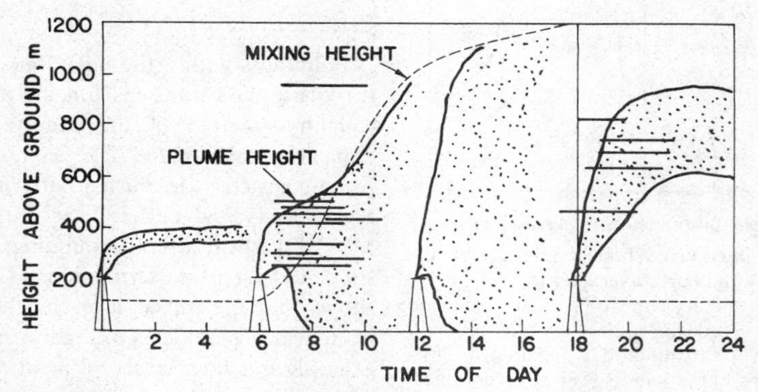

Fig. 8. Schematics of plume geometry at 4 different parts of the diurnal cycle. The horizontal bars are plume height data reported by Forrest and Newman (1977) for the Labadie plume in the summer.

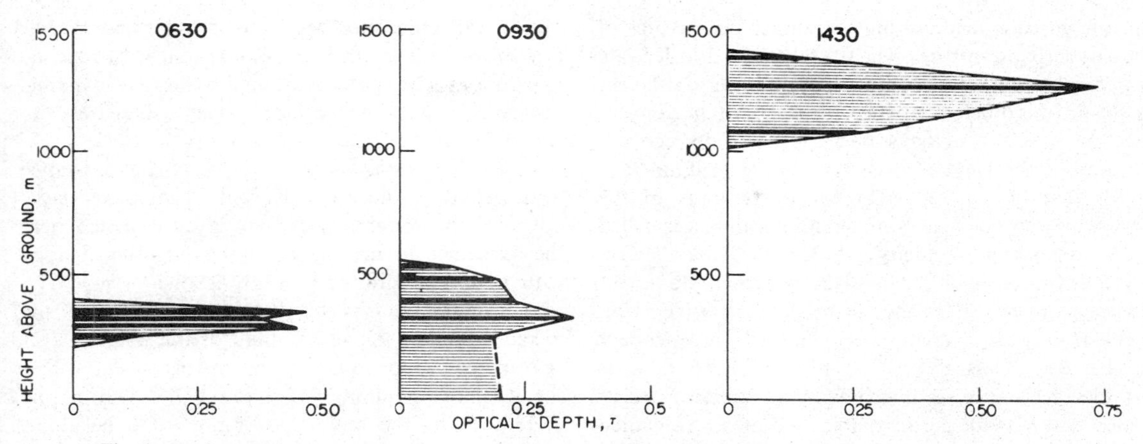

Fig. 9. Horizontal integrals of plume excess b_{scat} (optical thickness τ) at 20 km obtained 14 July 1976, at 0630, 0930 and 1430 CDT by MRI aircraft. The heavy horizontal lines show the heights of traverses; the contours of the vertical profile of τ are obtained by interpolation and extrapolation of the data.

by the buoyant thermals rising from the surface. Typically, by 1000 the rising thermals penetrate the residual radiation inversion and rise up to the second inversion layer at about 1000–1500 meters. With the reduction of solar radiation in late afternoon the intensity of thermal convection reduces, and the mixed layer becomes quiescent. This time also corresponds to the formation of the nocturnal surface-based inversion layer. Plumes emitted during the daytime convective hours, between 0800 and 1600, mix more or less uniformly with respect to height (with characteristic eddy diffusion coefficient K_z) throughout the mixed layer of height H and acquire a uniform vertical concentration within a time scale of $t = H^2/K_z$ (Gillani, 1978) which ranges from 30 min at 0900 to about one hour at noon. The vertical concentration profile that develops in the Labadie plume during midday is shown in Fig. 9.

In the afternoon the low wind speed and slight instability of the mixed layer permit thermally buoyant plumes to rise almost to the inversion layer, which usually persists at 1000 to 1500 meters. The large effective stack height of afternoon buoyant plumes has important consequences regarding sulfur transformations and removal, i.e. sulfur budget.

The rise of thermals above cities, industrial areas, and power plants is well documented in the 'heat island' literature. In St. Louis, development of thermals has been frequently observed over the Wood River refinery complex, causing industrial cumulus formation sometimes penetrating the inversion layer (e.g. Auer, 1976). Similar cumulus formation has also been noted over the 2400 MW Labadie power plant. It is recognized that such cloud formation is only possible under light wind conditions. The average wind speed in the mixed layer during the convective hours is between 4 and 5 m s^{-1} (Fig. 6c). When there is an appreciable wind, the rising buoyant plume is bent in the direction of the wind flow and reaches the top of the mixed layer without penetrating it. Such a plume persists overnight as 'convective debris' (Edinger, 1973) at 1000–1500 meters. Hence, the plume centerline is just below the upper inversion lid (Fig. 9). Such a plume will remain in an elevated stratified layer until the next day when the rising mixing layer reaches the plume.

There are two consequential phenomena of the diurnal vertical transport pattern of the midwestern plume worth emphasizing: (1) the decoupling of the daytime mixed layer from the underlying surface via the nocturnal inversion and (2) the lifting of late afternoon plumes which remain in high strata at least until the following noon when they may or may not be entrained in the mixing layer.

DRY REMOVAL

Unlike the field measurement of sulfur transformation which has a measurable product in the form of particulate sulfur, the estimation of removal rates is inherently more difficult. Removal rate is obtained from the difference of S_t at two heights (gradient method) or at two plume ages (plume budget method). The difficulties which appear in practice in such estimates arise from taking the difference of two values of comparable magnitudes, each with uncertainties comparable to the difference. This is particularly true for elevated plumes emitted from tall stacks, where removal is often not appreciable for the first several hours of plume age. This necessitates the combined use of deposition velocity estimates, ground level concentration data in the vicinity of power plants, and sulfur budget data from plume mapping experiments.

The mass transfer coefficient

Dry removal is a mass transfer process whereby gaseous or particulate sulfur is first transported to surfaces by turbulent and molecular diffusion, and then removed by absorption or adsorption of gases and by particle adhesion due to van der Waals forces. The removal rate is determined by the diffusion rate in the gaseous medium and by the characteristics of the underlying surfaces. The overall mass transfer rate (Q) can be characterized by a mass transfer coefficient (v_d) and the difference between the pollutant concentration just above the surface (C_0) and its concentration at the surface of the absorbing medium (C_s), such that $Q = v_d(C_0 - C_s)$. Since the units of the mass transfer coefficient are length per time it is called 'the deposition velocity', following Chamberlain (1966). Conceptually, it is also convenient to introduce the overall resistance to mass transfer, $r = 1/v_d$, which is the sum of the several largely independent resistances. Thus $r = r_s + r_b + r_a + r_m$, where r_s is surface resistance, r_b is the laminar sublayer resistance, and r_a is the aerodynamic resistance. A detailed discussion of the concept and range of the first three resistances is given by Garland (1978). The delivery of material from the plume height to the top of the surface layer is governed by the intensity of mixing and will be referred to as the resistance to mixing, r_m. It is included for use of v_d at plume height.

For gases, r_b is believed to be small due to the high molecular diffusivity of the common pollutants. Therefore, r_b for SO_2 is lumped into a single resistance, $(r_s + r_b)$. Due to the uncertainties of actual diurnal r_s pattern, a constant value $r_s + r_b = 0.3\ \mathrm{s\,cm^{-1}}$ is taken (Fig. 10).

The transport in the lowest few meters to a surface is controlled by the aerodynamic resistance, r_a. It is dependent on the wind speed $\bar{u}$ at the reference height, and the friction velocity u^*, modified by the Richardson number, reflecting the influence of temperature gradients on the transfer of vapor and momentum. The average diurnal pattern of the aerodynamic resistance was calculated using data from RAPS, following the procedure described by Garland (1974); it is shown in Fig. 10. We estimate that with increasing atmospheric stability at night, the aerodynamic resistance in the surface layer increases to $r_a = 1.5\ \mathrm{s\,cm^{-1}}$.

Resistance in the mixed layer, r_m. The resistance to turbulent transfer from the bulk of the mixed layer down to the roughness canopy layer is called here the resistance to mixing, r_m. Following the analogy with molecular diffusion this resistance is given by $r_m = h/K_z$, where h is the characteristic height of the mixed pollutant. K_z in the bulk of the mixing layer is roughly proportional to the mixing height times the Richardson number (Gillani, 1978). During the midday hours the bulk K_z is reported to be about $50\ \mathrm{m^2\,s^{-1}}$, while during the night it is on the order of $0.6\ \mathrm{m^2\,s^{-1}}$ or less. Within the daytime and nocturnal mixed layer the characteristic plume height, h, is taken as one half of the mixing height, H. Hence the diffusional resistance in the mixed layer is $r_m = H/2K_z$; the crude estimate of diurnal variation of r_m is displayed in Fig. 10. It was obtained such that at night $H = 120$, $K_z = 0.6$, which gives $1\ \mathrm{s\,cm^{-1}}$. The minimum value for noon with $H = 1000$ m, $K_z = 50$, is $0.1\ \mathrm{s\,cm^{-1}}$. Assumption of constant K_z/H leads to a level value for r_m over the daytime, with a subjective interpolation across the morning and evening transition hours.

The diurnal pattern of the overall resistance to mass transfer, r (the sum of individual resistances) is also shown in Fig. 10 along with its inverse, the deposition velocity, v_d. In the above synthesis the nocturnal resistance to transfer is dominated by the aerodynamic resistance in the surface layer and by the resistance to mixing in the nocturnal mixed layer. During

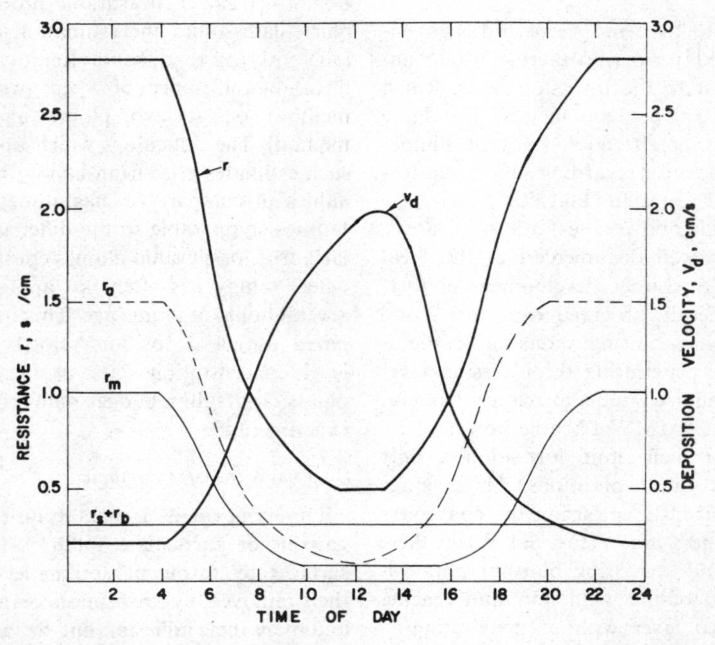

Fig. 10. Diurnal pattern of resistances to mass transfer and of the deposition velocity for SO_2.

the convective daytime hours r_a and r_m diminish so that the surface resistance becomes the rate-controlling parameter for removal.

The above exercise with the mass transfer coefficient for dry deposition of SO_2 suggests the importance of daytime deposition and points to surface resistance, controlled by vegetation, as the key rate-limiting parameter. We wish to stress however that (1) the above conclusions are based on a weak data base and intuitive arguments and therefore require empirical validation, (2) the deposition velocity as defined above includes the resistance to mixing, and (3) the diurnal pattern of the overall removal rates depend on the mass transfer coefficient v_d as well as the concentration of SO_2. Lacking reliable data on the deposition velocity of particulate sulfur, for the purposes of modeling we have assigned to the aerosols a deposition velocity one-tenth of the v_d for SO_2.

Surface concentrations and removal estimates from RAPS data

The dry deposition rate may vary during the diurnal cycle either due to changes in the mass transfer coefficient or due to the diurnal variations of the concentration near the surface, which in turn is controlled by vertical dispersion. Concentration monitoring near the ground can therefore provide valuable clues regarding the vertical transport as well as the daily pattern of dry depostion rate. The utility of a monitoring network in the vicinity of major sources is that sufficient data can be gathered for a statistical evaluation. For this analysis, an aggregate of five RAPS stations (108, 113, 114, 115, 121) was utilized from June, July and August 1975. These stations are within a triangle with corners demarked by three major sources: (1) Portage des Sioux power plant (1800 MW, 200 m stack); (2) Alton-Wood River refinery complex, including two small power plants; and (3) the urban center of St. Louis. The impact of these three sources on the five station network was examined first by wind sector analysis. The SO_2 concentration rose by wind direction octant is shown in Fig. 11. The monitoring data were further stratified by hour of day. Since each major source could be uniquely associated with a wind direction sector, the diurnal pattern of each plume impact was determined in this manner. The daily impact of the Portage de Sioux plant on the five stations located 10 to 30 km downwind is shown in Fig. 12a; the main feature is the relatively narrow peak of about 30 $\mu g\ m^{-3}$ around 1000 and the low average nighttime concentrations of 3 $\mu g\ m^{-3}$. The peak concentration at 1000 is consistent with the previous analysis, namely that the plume is contained in a shallow mixed layer, and thus the concentrations are highest. Evidently before 0900 the plume has not touched down at all. The concentration decay after 1000 can be interpreted as dilution due to the rising mixed layer. At night the concentrations are low, confirming that the power plant plume is passing over the network without touchdown.

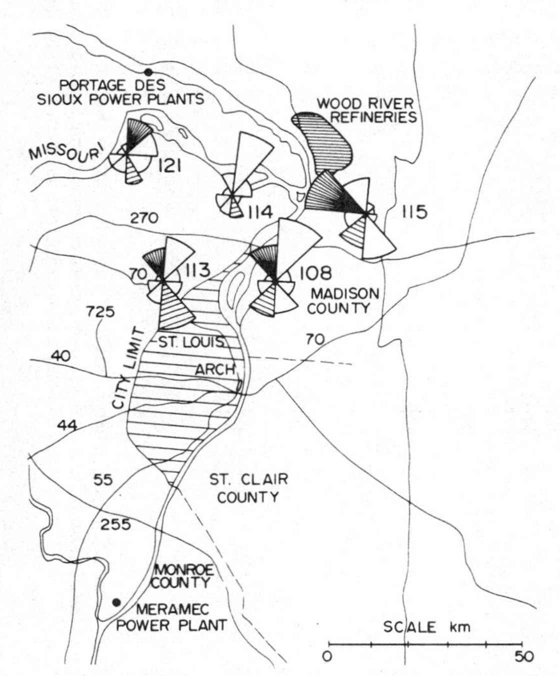

Fig. 11. Summer average SO_2 roses at five RAPS stations (108, 113, 114, 115, 121) located between Portage des Sioux power plant, Alton-Wood River refinery/small power plant complex, and the St. Louis central city. Shaded sectors point to the assumed source: radial lines for Portage, dotted area for Wood River, cross lines for the city.

The impact of the Alton-Wood River refinery complex on the five station network also ranging from 10 to 30 km from the source is shown in Fig. 12b. Qualitatively the diurnal pattern of this source impact is similar to that of the tall stack power plant. The peak concentration here is also attained at 1000; however, the concentrations tend to rise earlier in the morning (by 0600) and the decay occurs later at night (1900). Evidently the lower emission heights of the refinery–power plant complex causes earlier entrainment and plume touchdown. It is interesting to note that the concentrations remain low at night indicating clearly that these emissions are also above the nocturnal mixing layer.

The effect of the downtown St. Louis area on the diurnal pattern of the sampling network is shown in Fig. 12c. Unlike the other two major sources, the average concentration in wind direction segments pointing toward the city has a peak at midnight. The daytime concentrations between 1200 and 1800 are the lowest. This reversed diurnal pattern is interpreted as evidence that the central city emissions are trapped within the shallow nocturnal mixed layer leading to high concentrations at night. Vigorous daytime convection and dilution explain the low afternoon concentrations. In this context, it is instructive to compare this diurnal pattern of SO_2 with the NO_x patterns (Fig. 12d) which are known to be dominated by automotive sources emitting within the nocturnal mixing layer. The high NO_x concentrations are at

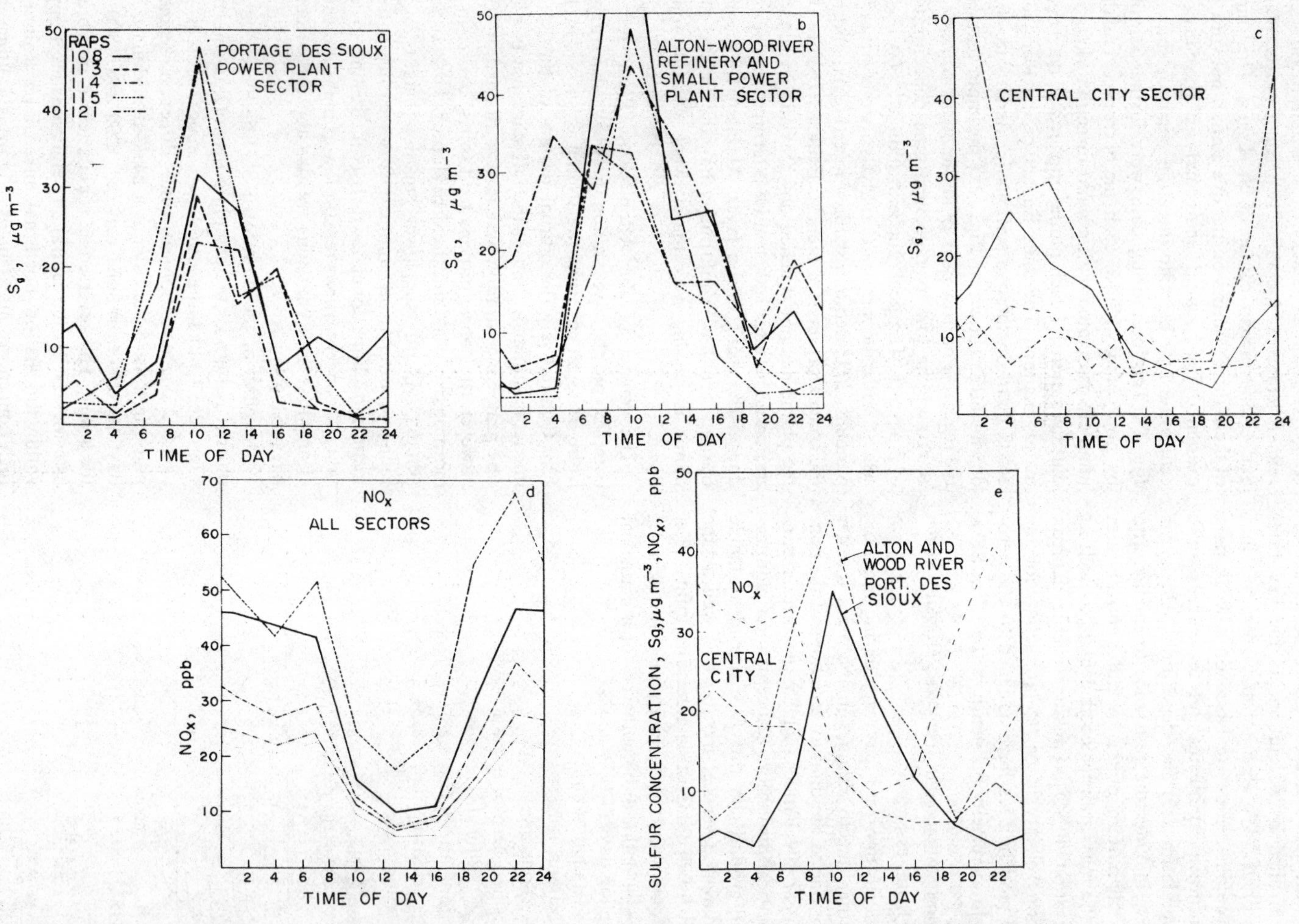

Fig. 12. Diurnal pattern of SO_2 concentration at five RAPS stations: (a) wind direction sector of the Portage des Sioux power plants; (b) wind direction sector of Alton-Wood River refinery/small power plant complex; (c) central city sector; (d) NO_x concentrations from all sectors; (e) comparison of the diurnal pattern of the four source contributions, averaged for the 5 stations.

night and the lowest at midday, confirming that the night time NO_x emissions from automobiles are trapped within the low mixing layer, and they are rapidly diluted during the convective daytime hours.

The preceding sector analysis of RAPS data and the arising diurnal pattern gives support to the notion of nocturnal decoupling of elevated emissions from the surface along with the trapping of surface based emissions within a shallow mixing layer. Table 2 contains the 2-hourly averages of the S_g concentration contributed by the Portage des Sioux power plant over June–July–August 1975. The product of v_d with the surface concentration at the 'average' station (Fig. 12a) and the circular area of influence with radius of 30 km (corresponding to the most distant station) yields an estimate of the sulfur removal. The average power plant emission rate during the period was about 3 kg s^{-1} of sulfur, and is used to find the fraction of S_g removed within 30 km, or equivalently about 2 hours plume age. The resulting rates of removal range from essentially zero at night to about 3.5% h^{-1} at 1000 CDT.

Removal estimates from MISTT plume data

Extensive plume mapping of the Labadie power plant emissions could not be carried out at distances over 200 km due to the width and vertical extent which exist after passage through a day of mixing and several hours of plume age; unfortunately it is precisely those factors which enhance removal. Approximate values of removal were obtained, however, for six sampling days in 1974 and 1976 (Fig. 13). With sufficiently detailed sampling in the crosswind and vertical directions, it is possible to integrate over the plume dimensions and estimate the total amount of sulfur as gas and aerosol remaining in any air parcel. The 'missing' sulfur (as compared to the original mass) is then attributed to the only remaining mechanism—removal. The 'mass flow' method of plume budgeting is described in a forthcoming paper

Table 2. Average concentration at five RAPS stations due to the Portage des Sioux power plant

Time (h)	S_g^* $\mu g\,m^{-3}$	v_d (cm s^{-1})	Flux F $S_g v_d A^{*\dagger}$ (kg s^{-1})	Fraction of S_g lost (F/3 kg s^{-1})	Loss rate (% h^{-1})
0100	0.634	0.2	0.004	0.12	0.06
0400	0.371	0.2	0.002	0.07	0.04
0700	1.495	0.85	0.036	1.20	0.60
1000	4.456	1.65	0.208	6.93	3.46
1300	2.737	2.00	0.155	5.16	2.58
1600	1.414	0.95	0.038	1.27	0.63
1900	0.651	0.43	0.008	0.26	0.13
2200	0.336	0.20	0.002	0.08	0.04

* The average contribution from the power plant regardless of wind direction is obtained by multiplying the sector concentration by the fraction of the time the wind came from the sector.

† An area with radius of 30 km is used to calculate the deposition rate.

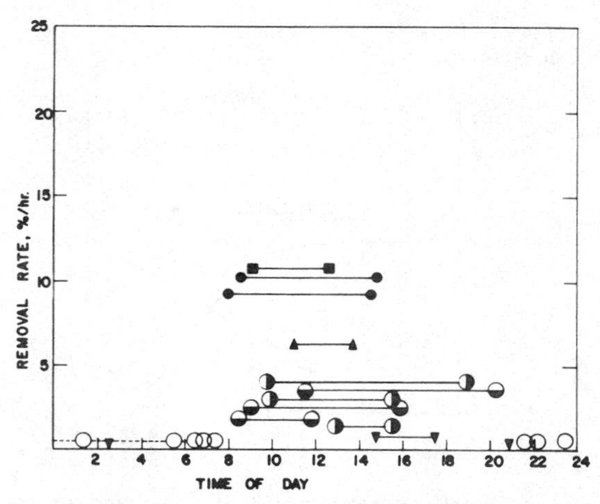

Fig. 13. Removal rates estimated from MISTT plume mapping data.

on the sulfur budget of low level sources, where dry removal in the first few hours of plume age is appreciable.

Budgets for the Labadie power plant were constructed on 9 July and 18 July 1976 for plumes at downwind distances up to 190 km (Gillani *et al*., 1978). In that paper the details of the procedures involved in estimation of the budgets are illustrated for these two important sampling missions. The removal rates in Fig. 13 were obtained by dividing the fraction of the missing sulfur estimated for a given plume cross-section by the plume age. The area under each bar (representing the plume age) represents the fraction of sulfur removed. In Fig. 13 we have also entered the symbols for 5 July and 14 July 1976 missions of the WU plane without estimating a mass balance; on these two night flights vertical mapping of the plume gave assurance that the plume had not touched ground, and hence no dry deposition could occur. The error bar for these removal rate estimates is large, so we do not consider it appropriate to extract characteristic removal rates without the use of a diagnostic diffusion model (Gillani, 1978). However, the above described data do depict detectable downwind daytime dry deposition, as perceived from the power plant plume passing planes projected perpendicular to the parcel's path. Further evidence of ground removal is seen in the vertical SO_2 profile obtained within the Labadie plume on 18 July 1976 (Fig. 14). The SO_2 gradient near the surface indicates 'missing' SO_2 as compared to the light scattering aerosol parameter, b_{scat}, and the suboptical aerosol concentration measured by the charger. Lacking reliable eddy diffusion coefficient estimates, the observed gradient cannot be connected to an SO_2 flux value.

RATE CONSTANTS FOR THE LABADIE PLUME BUDGET

In the present analysis, the SO_2 oxidation rate will be taken as a time-dependent quasi-first order process

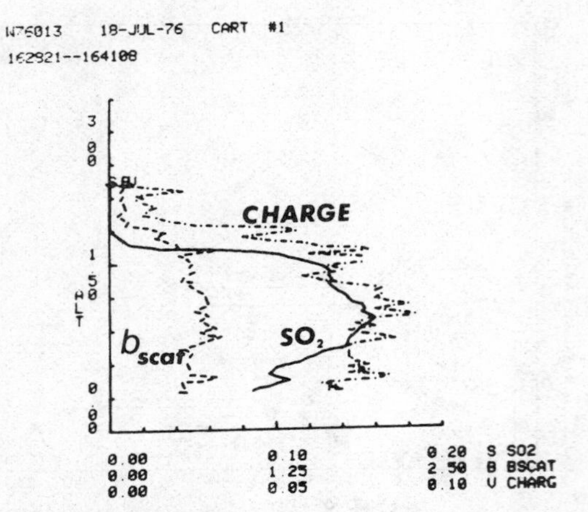

Fig. 14. Sample vertical profile of SO_2, b_{scat} and aerosol charge taken at 1635 CDT, 18 July 1976. The decrease of SO_2 concentration near ground vs lack of gradient in the aerosol parameters is evidence of SO_2 removal at the surface.

with rate constant k_t. Thus, empirical account can be made for the diurnal changes of external parameters which may influence the conversion rate. The empirical values of k_t are extracted from the MISTT program data, Fig. 3. The dashed line in Fig. 3 results from an approximate deconvolution of the longer averages. The same dashed curve for the diurnal pattern of k_t is shown in Fig. 15 along with the dry removal rate constants.

The removal rate constant for gaseous sulfur, d_g, is calculated from the deposition velocity v_d and the mixing height H, each having its own diurnal pattern. In a well-mixed box the removal rate $d_g = v_d/H$; its pattern is shown in Fig. 15. Note that the high values early in the morning and late in the evening result from the assumed shallow mixing height of 120 m overnight. Also, d_g explicitly operates only upon matter within the nocturnal and daytime mixed layer; the discontinuity arises from the assumption that a decoupling of the mixed layer from the surface occurs at 1800 and a nocturnal layer of about 120 m appears instantaneously. As noted earlier, summertime dry deposition in the midwestern U.S. is a daytime phenomenon for elevated plumes and therefore the important portions of the d_g function are the day hours, when the bulk of the plume is contained within the mixed layer. After the first day of plume age, an equivalent curve acting on a constant 1200 m mixed layer may be used instead; this is the k_r curve of the figure. The d_g values at night only operate on the content of the shallow nocturnal mixing layer while the plume is conserved above the inversion. Hence the plume sulfur budget is insensitive to the value of the nocturnal removal rate constant. The above statements do not hold for the sulfur budget of surface emissions, over rough terrain, or in a cloudy, windy climate.

TWO BOX MODEL OF THE SULFUR BUDGET IN LARGE PLUMES

The essence of the previous considerations regarding the diurnal behaviour of the mixing layer is that the plume has two distinctly different regimes during the diurnal cycle. The first regime develops when the plume is decoupled from the underlying surfaces by radiation inversion or by other stable strata. In this regime material is lost due to entrainment into a second, mixing-deposition regime. In the second regime, the plume is dispersed rapidly within the mixed layer, loses material at the surface, and gains material by entrainment from above. Conversion of gaseous sulfur to particulate sulfur occurs in each

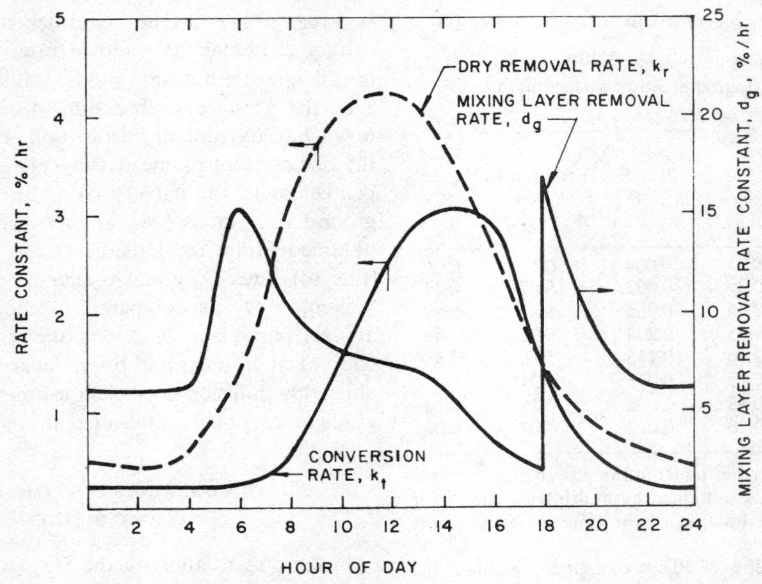

Fig. 15. Overall rate constants used in the two-box model.

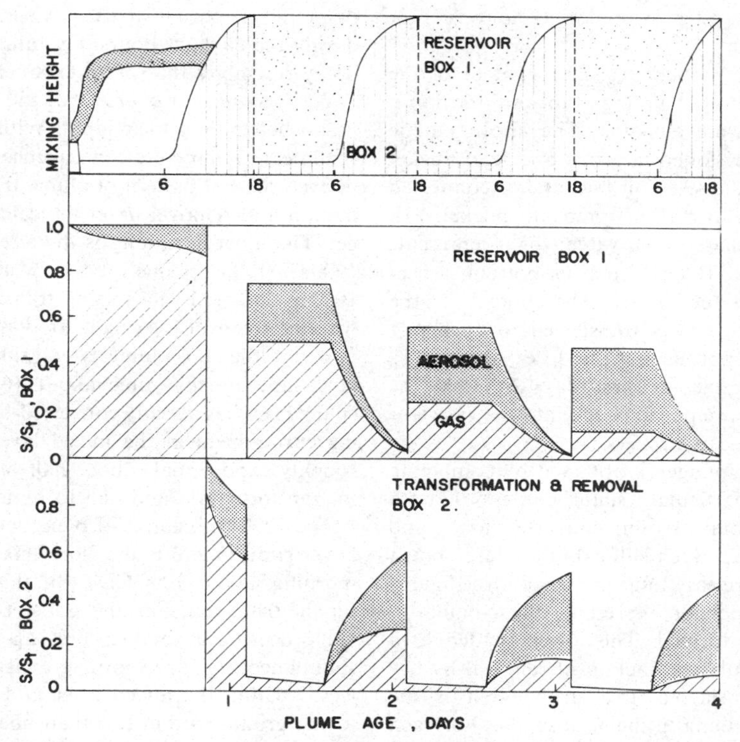

Fig. 16. Schematics of plume transition from the quiescent reservoir to the mixing, transformation and removal regime (top), gaseous and particulate sulfur in the reservoir box (center) and the removal regime (bottom).

regime. During its lifetime, the plume sequentially passes through these two regimes every day, as illustrated schematically in Fig. 16a.

For estimating the sulfur budget in a plume passing through such a diurnal cycle, it is instructive to describe the system in terms of a 'two box model'. The sulfur budget at any given time can then be obtained by carrying out the mass balance in each of the boxes. Box I shall be the quiescent 'aging reservoir', decoupled from the surface. The mass balance equation for the gaseous sulfur S_g and for particulate sulfur S_p in box I can be written as follows:

$$\frac{\mathrm{d}}{\mathrm{d}t} S_g^{\mathrm{I}} = -\frac{1}{H'}\frac{\mathrm{d}H}{\mathrm{d}t} S_g^{\mathrm{I}} - k_t S_g^{\mathrm{I}}$$

$$\frac{\mathrm{d}}{\mathrm{d}t} S_p^{\mathrm{I}} = -\frac{1}{H'}\frac{\mathrm{d}H}{\mathrm{d}t} S_p^{\mathrm{I}} + k_t S_g^{\mathrm{I}}.$$

The first terms on the right hand sides of the equations signify the transfer of material from box I to box II due to the rising mixed layer height, H. The second terms are the conversion terms.

The material balance for the mixing-transformation–removal box II can be written as follows:

$$\frac{\mathrm{d}}{\mathrm{d}t} S_g^{\mathrm{II}} = \frac{1}{H'}\frac{\mathrm{d}H}{\mathrm{d}t} S_g^{\mathrm{I}} - k_t S_g^{\mathrm{II}} - d_g S_g^{\mathrm{II}}$$

$$\frac{\mathrm{d}}{\mathrm{d}t} S_p^{\mathrm{II}} = \frac{1}{H'}\frac{\mathrm{d}H}{\mathrm{d}t} S_p^{\mathrm{I}} + k_t S_g^{\mathrm{II}} - d_p S_p^{\mathrm{II}}.$$

The terms involving d_g and d_p are the dry removal terms for gaseous and particulate sulfur, respectively (see Appendix 2 for a mathematically complete formulation).

During the nighttime, the elevated plume considered here is emitted into the reservoir box and resides there until entrainment into the next day's mixed layer. Daytime emissions are also held in the reservoir box for a characteristic 'mixing time', following which perfect mixing is assumed. All plume material is then transferred into the mixing-transformation-removal box (Fig. 16b). The transformation and removal within box II will proceed until 1800 in accordance with the rate constants given in Fig. 15. At 1800 the collapse of the mixing height, followed by the decoupling of the surface layer from the former mixed layer, is imposed as a 'catastrophe'. The consequence of this abrupt transition is to transfer the bulk (90%) of the plume material back into the reservoir box. A token amount (10%) of the material is retained in the mixing, transformation and removal box of the nocturnal mixing layer, and is subject to dry deposition. By morning the stable layer near the ground is nearly depleted of SO_2, while the mass in the reservoir is conserved. At 0600 the rising mixed layer starts to entrain material from box I to box II and subjects an increasing amount of matter to dry removal. By late afternoon, practically all material has been transferred to the mixing box II, followed by a repeat of the catastrophic transfer from box II to box I at 1800.

The gaseous and particulate sulfur content of the two boxes are displayed in Fig. 16b and 16c for a 4-day period following emission. The above kinetic process may be envisioned in terms of a 'two bucket analogy', in which the reservoir bucket, I, is connected to a mixing-transformation-removal bucket, II, through a pipe regulated by a valve which represents entrainment. Bucket II has a porous bottom, reflecting dry removal at the surface. The contents of the reservoir bucket are slowly transferred to bucket II which through its porous bottom loses about 40% of its contents during the first day. At 1800 the remaining contents are poured back into the reservoir and the process repeats itself.

The plume sulfur budget is obtained by combining the gaseous and particulate sulfur content of both boxes to get the total amount converted to S_p and that remaining as S_g. Such sulfur budgets are shown in Fig. 17 for four release times. In the budget figures the dotted upper portion represents the cumulative percent of aerosol formed. The lower shaded area stands for the cumulative fraction of removal by dry deposition. The unshaded area in between is the gaseous sulfur remaining in the atmosphere. For each of the four emission types both aerosol formation and

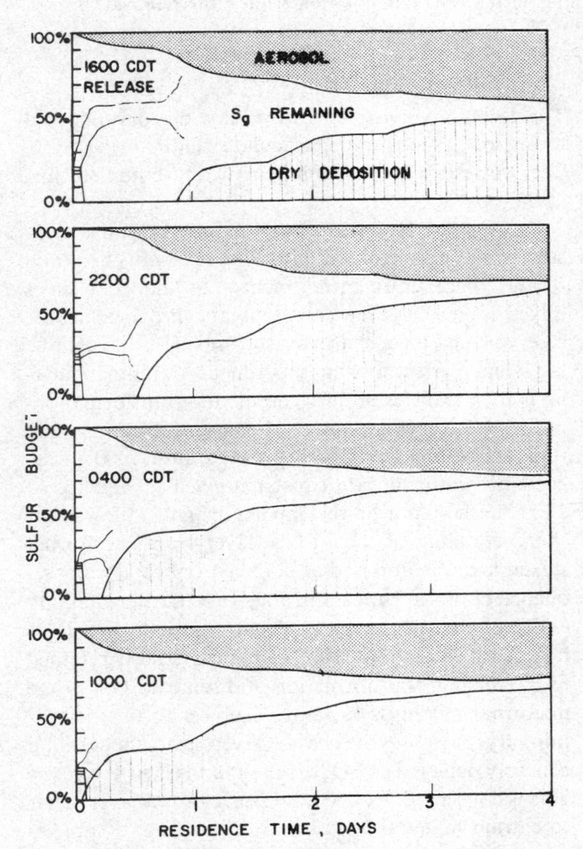

Fig. 17. Sulfur budget over 4 days for the power plant emission obtained from two-box model, for four emission times.

dry removal occurs in daily waves with peak effects during the daylight hours. Overall, however, the SO_2 decay resembles that of an exponential function.

We recognize of course that the model results can be no better than their input. With this in mind, we shall next examine the consequences of the previously derived diurnal pattern of plume dispersion, transformation and removal upon the calculated sulfur budget. The most intriguing is the afternoon emission at 1600 CDT. The plume leaving the stack rises to nearly 1000 m and undergoes some transformation ($\sim 10\%$) without removal overnight. It is not until noon the next day that the plume is entrained into the mixed layer, and by that time about 16% of the original sulfur mass has already converted to S_p form. Following entrainment in the mixed layer, the S_g decay is roughly exponential: about half of the decay is due to transformation and half to removal.

The 2200 emission is also emitted into a stable but lower stratum and is therefore mixed earlier the next morning, about 1000 CDT. In this case, as well as for the 0400 emission, the transformation and removal processes have their first major impact almost simultaneously. The morning emission at 1000 is injected within the mixed layer and therefore exposed to the ground within less than one hour. Here again SO_2 depletion by transformation and removal starts simultaneously.

Figure 18a shows the fraction of total gaseous sulfur which is converted to sulfate aerosol at plume ages of 6 hours to 8 days. The total fraction converted ranges between 30 and 45%, which is attained within 8 days. For the average daily emission about half of the ultimate particulate sulfur is formed within 24 hours after emission. The highest fraction converted (45%) belongs to the afternoon emissions, due to their head start in conversion. This exercise points to the late afternoon releases of the mid-western plume as most conducive to sulfate formation, and provokes the thought that any reduction in the afternoon emissions would yield a most efficient sulfate reduction.

The formation time for particulate sulfur, expressed in terms of the time required to form 50 and 66% of the ultimate S_p (Fig. 18b), has a characteristic value between 25 and 40 hours. There is a significant variation of the 1/e decay time of S_g ranging between 26 hours for 1000 CDT emission to 46 hours for the 1600 CDT emission (Fig. 18c). The SO_2 half-life exhibits even more variability, varying from 10 hours for the 0800 CDT emission to 38 hours for the 1600 CDT emission. The corresponding residence time distribution functions, as defined by Rodhe (1978) are shown for two emission times (Fig. 19). The damped periodicity results from the assumed diurnal cycle of both removal and transformation rates.

The characteristic transport distances associated with different plume ages are estimated as follows: (1) a plume height or range of heights was assigned to each hourly emission throughout its mixing history; (2) the horizontal wind speed was obtained from

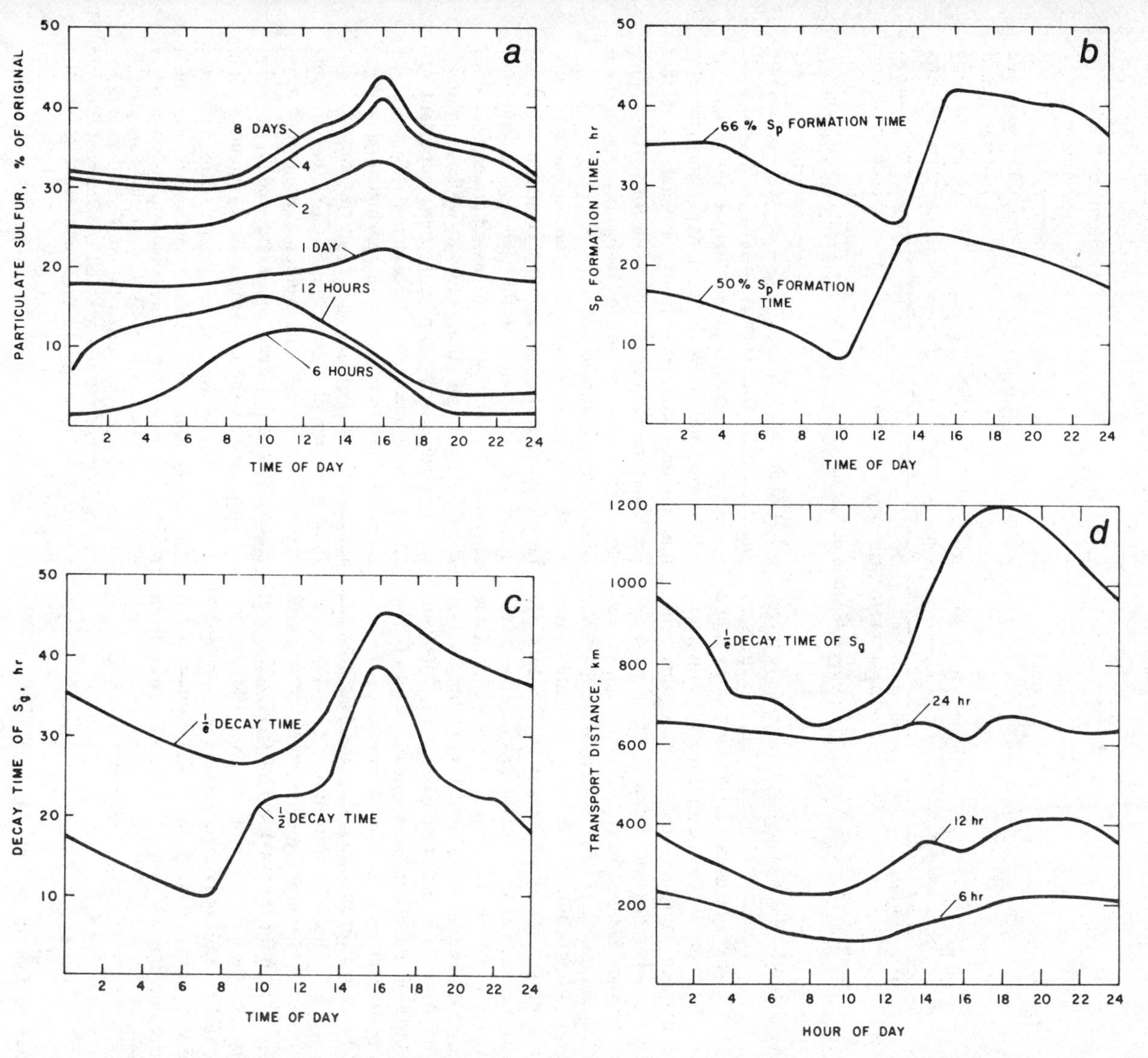

Fig. 18. Results of model calculations to explore the role of emission time upon: (a) particulate sulfur fraction for different plume ages; (b) particulate sulfur formation time; (c) S_g decay time; (d) transport distance.

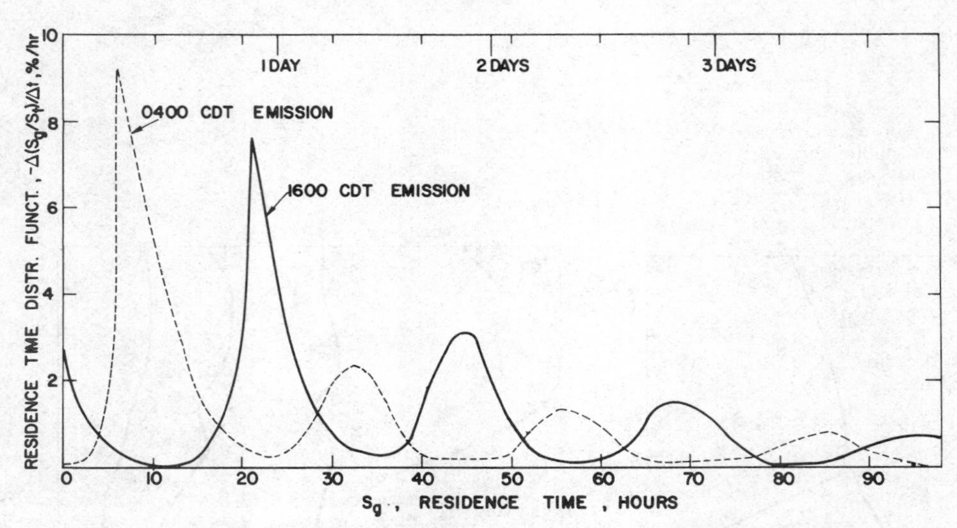

Fig. 19. Residence time distribution function of S_g emitted at 0400 and 1600 CDT.

the RAPS data of Fig. 6. The travel distance is therefore a numerical integral of $\bar{u}(t)\,\Delta t$ from the emission time. The transport distance ($\bar{x}$) obtained is the range to which the plume would travel if the wind direction and wind speed were invariant in space; this being unlikely, our characteristic distance may at best approximate the distance travelled along a curved synoptic-scale trajectory. The calculated distances for 6, 12 and 24 hours and for 1/e decay time are shown in Fig. 18d. The low $\bar{x}$ of 200 km at 12 hours age for 0600 emissions arises from the low daytime transport velocity within the mixing layer. The plume emitted at 2000 travels to $\bar{x} = 410$ km within the first 12 hours. After 24 hours of plume age $\bar{x}$ is ~ 600 km irrespective of emission time. From the point of view of plume impact it is also worth estimating the distance the plume travels when the modelled SO_2 has decayed to a fraction 1/e of its initial value (Fig. 18d). Due to the low 1/e decay time the 'impact' distance is about 700 km for the 0600 to 1200 CDT emissions; the distance then rises rapidly to over 1000 km for the 1400–2400 emissions. Hence, the afternoon plume not only appears to have higher sulfate formation but it seems that its SO_2 impact is further away from the plant.

SUMMARY AND CONCLUSION

So where does all the sulfur go?

In the coal-fired power plant in St. Louis in the summer, about 30–45% of the SO_2 is evidently converted to sulfate aerosol, about half of that within the first day. The other 55–70% is removed as SO_2. But where does the sulfate go? We have not even speculated on it. It appears that the sulfate is formed unevenly: the afternoon plume contributes more than its share because it rises so high that it cooks for hours before it is mixed to the ground. We also believe that both transformation and removal occur mainly during the daytime, while the transport is fastest at night. After a hard day of convection, reaction and deposition, the lower atmosphere relaxes at dusk, while the mid-western plume takes off overnight on a jetstream and starts the next day's work three or four hundred kilometers from the stack. Note that the above is not believed to hold for surface emissions, in rough terrain, in colder seasons or in a windy, cloudy climate.

The synthesis of clues about transformation, transport and removal lured us into a great deal of mental acrobatics. The value of this exercise depends not only on the validity of our assumptions, but even more on the implications of what we have ignored: wet removal, which could be the most important sulfate removal mechanism, occasional lifting of the plume beyond the reach of the normal mixing layer, variations of individual days from the averages, etc. In order to complete bridging the gap between laboratory and regional scale studies, the missing or incomplete modules need to be replaced or modified as warranted by the evolving state of the art.

Acknowledgements—We thank participants of project MISTT. The data from the Regional Air Pollution Study were provided to us by U.S. EPA for which we are grateful. This research was supported by the Federal Interagency Energy/Environment Research and Development Program through U.S. Environmental Protection Agency Grant No. R803896.

REFERENCES

Auer A. H., Jr. (1976) Observations of an industrial cumulus. *J. appl. Meteorol.* **15**, 406–413.

Blackadar A. K. (1957) Boundary layer wind maxima and their significance for the growth of nocturnal inversions. *Bull. Amer. Meteor. Soc.* **38**, 283–290.

Bonner W. D., Esbensen S. and Greenberg R. (1968) Kinematics of the low-level jet. *J. appl. Meteorol.* **7**, 339–347.

Cantrell B. K., Brockmann J. E. and Dolan D. F. A mobile aerosol analysis system with applications. (Submitted for publication.)

Chamberlain A. C. (1966) Transport of gases to and from grass and grass-like surfaces. *Proc. Roy. Soc. A* **296,** 45–70.

Cobourn W. G., Husar R. B. and Husar J. D. (1978) Continuous *in situ* monitoring of ambient particulate sulfur using flame photometry and thermal analysis. *Atmospheric Environment* **12,** 89–98.

Edinger J. G. (1973) Vertical distribution of photochemical smog in Los Angeles basin. *Environ. Sci. Tech.* **7,** 247–252.

Forrest J. and Newman L. (1977) Further studies on the oxidation of sulfur dioxide in coal-fired power plant plumes. *Atmospheric Environment* **11,** 465–474.

Garland J. A. (1974) Dry deposition of SO_2 and other gases. Proceedings of the Symposium on Atmospheric Surface Exchange of Particulate and Gaseous Pollutants, Richland, WA.

Garland J. A. (1978) Dry and wet removal. *Atmospheric Environment* **12,** 349–362.

Gillani N. V. (1978) Project MISTT: mesoscale plume modeling of the dispersion, transformation and ground removal of SO_2. *Atmospheric Environment* **12,** 569–588.

Gillani N. V., Husar R. B., Husar J. D. and Patterson D. E. (1978) Project MISTT: kinetics of particulate sulfur formation in a power plant plume out to 300 km. *Atmospheric Environment* **12,** 589–598.

Husar J. D., Husar R. B., Macias E. S., Wilson W. E., Durham J. L., Shepherd W. K. and Anderson J. A. (1976) Particulate sulfur analysis: application to high time resolution aircraft sampling in plumes. *Atmospheric Environment* **10,** 591–595.

Husar R. B., Macias E. S. and Dannevik W. P. (1976) Measurement of dispersion with a fast response aerosol detector. Proceedings of the Third Symposium on Atmospheric Turbulence, Diffusion and Air Quality, Raleigh, N.C.

Myers R. L. and Reagan J. A. (1975) The regional air monitoring system, International Conference on Environmental Sensing and Assessment, Las Vegas, NV. IEEE Annals, No. 75 CH1004-1 5–2.

Rodhe H. (1978) Budgets and turn-over times of atmospheric sulfur compounds. *Atmospheric Environment* **12,** 671–680.

Smith T. B., Blumenthal D. L., Anderson J. A., Vanderpol A. H. and Husar R. B. (1978) Transport of SO_2 in power plant plumes: day and night. *Atmospheric Environment* **12,** 605–611.

Whitby K. T. (1978) Physical properties of sulfur aerosols. *Atmospheric Environment* **12,** 135–159.

Wilson W. E. (1978) Sulfates in the atmosphere: A progress report on project MISTT (Midwest Interstate Sulfur Transformation and Transport). *Atmospheric Environment* **12,** 537–547.

APPENDIX 1

MISTT field program by Washington University

The MISTT field program (Wilson, 1978) consists of two airborne sampling platforms and associated ground support, as illustrated in Fig. A1. Washington University maintained one aircraft and supplied field direction as well as data evaluation for the project.

The 680 FL Aero-Grand Commander built by Rockwell International is a turbo-charged, twin engine aircraft normally equipped with 10 seats. All of the seats were removed for instrumentation except for the three needed for the pilot and instrumentation personnel. The aircraft has a minimum sampling speed of 160 km h^{-1} and a maximum speed of 320 km h^{-1}. For the weight and fuel capacity of 233 gallons, the sampling flight time is 3.0 and 3.7 hours.

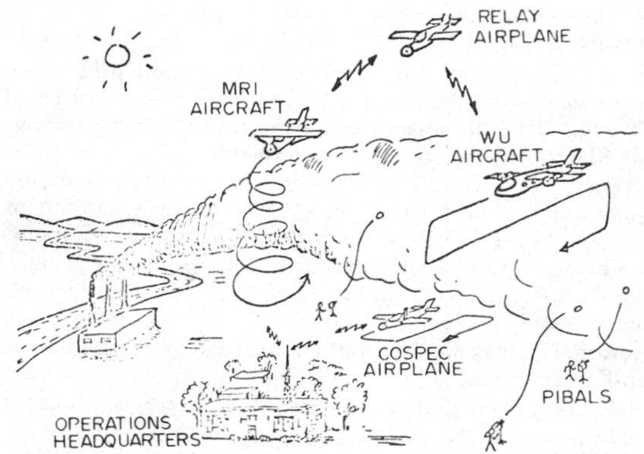

Fig. A1. Schematic drawing of the 1976 MISTT plume program. The coordinated efforts of two instrumented aircraft, three pilot balloon units and an airborne correlation spectrometer are utilized. Communication between the mobile units and the operations headquarters (OHQ) is maintained by a radio relay airplane.

The normal rate of climb is 100 ft min^{-1} at 125 mph, while the maximum full-power climb rate is about 1500 ft min^{-1}. The on-board instrumentation package may be divided into three groups: aerosol sensors, gas monitors and position sensors.

The concentration and approximate size distribution of fine particles are monitored with three fast-response detectors: an integrating nephelometer (manufactured by Meteorology Research, Inc.), a condensation nuclei counter (Environment One), and an aerosol charge detector (Husar *et al.*, 1976). Each of the three detectors responds to a different window of the atmospheric fine particulate size spectrum: the nephelometer is most sensitive in the 0.2–1.0 μm size range; the aerosol charger responds to the size window 0.05–0.2 μm; the condensation nuclei count indicates the total number of particles, which is dominated by 0.01–0.1 μm size range. The three detectors provide three integral moments of the aerosol size spectrum.

Filter samples for chemical analysis are collected by a two-stage filter-tape sampler that collects separate deposits for coarse (>3 μm) and fine (<3 μm) particles (Husar *et al.*, 1976). The gaseous sulfur concentration is monitored with the high sensitivity Meloy SA-285 Sulfur Analyzer, which has high sensitivity (noise-equivalent signal 0.05 ppb) and high temperature stability (Cobourn *et al.*, 1978). The ozone monitor is a Dasibi UV-absorption instrument, which requires no reaction or combustion gas. The standard aircraft position parameters, including pressure-altitude (above sea level), VOR and DME are recorded. Analog and digital data from the above instruments are sampled by a METRODATA Model DL640 digital data logger in 2 s increments and are simultaneously recorded on a four channel stripchart recorder. When necessary, data are transmitted via the UHF radio to the ground operations headquarters.

The operations and communications headquarters (OHQ) is equipped for radio voice and telemetry communications with the mobile field units, and also houses a complete computerized data processing system. Data received in real time from the WU aircraft are displayed on aeronautical maps, such that the plume concentration of a selected pollutant may be viewed while the sampling mission is on. Guided by this along with meteorological data information, the acting field director at OHQ makes operational decisions, such as setting of the end points and flight altitudes for horizontal traverses, location and transfer of pilot balloon units, termination of the plume mapping, etc.

Single-theodolite pibal releases before each sampling mission are used to establish the current local wind field to indicate strong wind shear and veer, and other flow features. During the mission, three pibal units are directed through OHQ commands to take positions at the horizontal edges of the plume and one underneath the plume centerline at the vertical plane being traversed. Each of the three pibal units maintains a stepwise forward motion in the downwind direction, in coordination with the developing flight pattern. Pibals are released by each unit at half-hourly intervals except while moving between the traversed plume sections. Upon request, the observed azimuthal and elevation angles are transmitted verbally to the OHQ for on-line processing and decision-making. The on-site meteorologist prepares daily weather forecasts to assist flight planning; he also prepares daily weather summaries for all sampling days for use in the data analysis phase of the program.

Communication between the mobile units and OHQ is maintained by two radio systems. Normal voice communication is conducted through very high frequency (VHF) aircraft radios installed in all units. In addition, the WU airplane and the OHQ are equipped with UHF radios for long distance data or voice transmission. A relay airplane equipped with a repeater serves as a communication link between the mobile field units themselves and with the OHQ when the sampling operations are being conducted at distances beyond 50–100 km from the source, when air–ground communication with low-flying aircraft through regular aircraft radio is not possible. With the relay airplane positioned at one-third the distance between OHQ and the sampling aircraft, and at an altitude of 3500 m, the maximum communication range was 500 km. The pilot and the observer in the relay airplane also serve as voice-communication link between the pilot balloon units and the OHQ.

The flight patterns of the aircraft are designed for detailed characterization of the plume at discrete distances downwind of the source. At each distance, horizontal traverses are made in the plume perpendicular to the plume axis at three or more elevations. The traverses are augmented by vertical spirals outside the plume and near the plume center-line, extending up to at least 100 meters above the mixing layer height or the maximum plume height. The plume mapping begins near the source and proceeds downwind, with the two sampling aircraft making measurements, whenever feasible following a single air parcel. Based on the weather forecast, several plume mapping scenarios are elaborated at least 12 hours in advance of each sampling mission at a planning meeting. A typical sampling mission consists of two or three sampling flights, each lasting 2–4 hours. The pre-mission pibal data are used as the final input in making the decision to implement a specific flight plan for the mission.

Before each mission, the calibrations of SO_2, NO/NO_x and O_3 instruments are performed by EPA using calibration gases. During the flights, these instruments, as well as the aerosol instruments, are zero-checked at frequent intervals. At the end of each mission, the instrumentation in each aircraft undergoes detailed, multi-point calibration. Between missions, the instruments operate by external power provided in the home base hanger.

Data processing is performed by a minicomputer-based system. The main features of the system are (1) interactive graphics capability for data display, editing and analyses, and (2) output of graphical data displays within a few hours after collection, thereby permitting data review prior to the next mission. An alphanumeric/graphics terminal provides capability for interactive on-line data display and data editing. A hard copy unit reproduces on paper the graph displayed on the terminal. A Metrodata cassette reader transfers digital data logged on Metrodata cartridges during in-flight data acquisition to the computer system.

The data processing software permits rapid data retrieval and graphical display of selected parameters on the screen. On-line editing is possible for offsets, glitches (due to radio interference), scaling, flagging of instrument malfunctions, etc. Graphical plots of horizontal traverses and vertical spirals, the pibal data, etc. with scale annotations in engineering units are produced routinely. Figure 14 shows a sample plot for a spiral. A position–concentration plot produced automatically from recorded WU aircraft position (DME, VOR) data is shown in (Gillani *et al.*, 1978) Fig. 1.

Post-flight processing of the aircraft data commences immediately after the landing at the home base. The raw data are read and stored on magnetic tapes in the airport EPA trailer. The next step is to reduce the data by averaging over about 8 s intervals, to introduce calibration corrections, and to write out the results on a new data file. Graphical outputs are inspected to check for any indications of instrument malfunctions or any other abnormalities. The results of data processing and preliminary analyses are presented at a review meeting on the day following the sampling mission, attended by the pilots, instrument operator/observers, data processing personnel, the field director and other participants of the program. The flight pattern, the meteorology and the data are reviewed in order to identify the weak points and possible improvements in the program.

APPENDIX 2: TWO-BOX MODEL

The two-box model differs from a single-box approach only for the crucial first day of plume age. In the initial stage all matter is kept in box I; at mixing, all matter is instantaneously transferred to box II. Subsequently, the governing equations are:

$$\frac{d}{dt}\begin{bmatrix} S_g^{I} \\ S_g^{II} \\ S_p^{I} \\ S_p^{II} \end{bmatrix} = \begin{bmatrix} -k_t(t) - E(t) & 0 & 0 & 0 \\ E(t) & -k_t(t) - d_g(t) & 0 & 0 \\ k_t(t) & 0 & -E(t) & 0 \\ 0 & k_t(t) & E(t) & -d_p(t) \end{bmatrix} \begin{bmatrix} S_g^{I} \\ S_g^{II} \\ S_p^{I} \\ S_p^{II} \end{bmatrix}$$

where the entrainment function of box I into box II is

$$E(t) = \frac{1}{H_{max} - H(t)} \frac{d}{dt} H(t).$$

All rates are assumed to be diurnally periodic. The upper box I is emptied by 1800 h and is refilled each day by a new initial condition:

$$\begin{bmatrix} \bar{S}_g^{I} \\ \bar{S}_g^{II} \\ S_p^{I} \\ S_p^{II} \end{bmatrix} = \begin{bmatrix} 0 & 1-p & 0 & 0 \\ 0 & p & 0 & 0 \\ 0 & 0 & 0 & 1-p \\ 0 & 0 & 0 & p \end{bmatrix} \begin{bmatrix} S_g^{I} \\ S_g^{II} \\ S_p^{I} \\ S_p^{II} \end{bmatrix}$$

where p is the fraction trapped in the newly formed stable layer.

Atmospheric Environment Vol. 12, pp. 569–588. Pergamon Press 1978. Printed in Great Britain.

PROJECT MISTT: MESOSCALE PLUME MODELING OF THE DISPERSION, TRANSFORMATION AND GROUND REMOVAL OF SO_2

N. V. GILLANI

Air Pollution Research Laboratory, Mechanical Engineering Department, Washington University, St. Louis, MO 63130, U.S.A.

(*First received* 4 *July* 1977)

Abstract—A diagnostic β-mesoscale (25–250 km) plume model is developed using an existing steady-state model as a building block. This quasi-steady, Lagrangian model incorporates the diurnal variability of the planetary boundary layer (PBL) structure and of the parameters governing the chemical conversion and ground removal of SO_2. The vertical inhomogeneity of atmospheric dispersion is simulated by the use of an assumed height- and stability-dependent profile of the eddy diffusion coefficient. Two important dimensionless system parameters are identified which govern pollutant dilution and ground removal. Model inputs are derived from Project MISTT aircraft data and the ground monitoring data of the St. Louis Regional Air Pollution Study (RAPS). On 9 and 18 July 1976, the plume of the 2400 MW, coal-fired Labadie power plant near St. Louis was sampled from aircraft out to 300 km. Model application is considered specifically for the data of these two days, and corresponding quantitative information about the dispersion, transformation and ground removal of SO_2 is extracted. The results show that peak daytime SO_2 conversion rates reached 1.8 and 3.0% h^{-1} on 9 and 18 July, respectively; the corresponding peak dry deposition velocities were between 1.5 and 2.0 $cm\,s^{-1}$. The model is used to investigate the effects of source height, time of SO_2 release and eddy diffusion on the overall sulfur budget of the plume. The mid- and late-afternoon plumes appear to have the highest potential for long range transport and sulfate formation. Ground removal is strongly influenced by the profile of vertical eddy diffusion in the surface layer, and much less by the profile shape and magnitude higher up.

1. INTRODUCTION

The prevalence of high and increasing levels of atmospheric sulfates over wide regions of the eastern U.S. has now received official recognition (EPA, 1975) and its causes, nature and effects are the subject of several large and well-known research programs in the U.S. Of particular concern is the gas-to-particle conversion of sulfur compounds released to the atmosphere (primarily as SO_2) from large sources such as tallstack power plants and large industrial complexes. The amount of secondary sulfate formation depends on two primary factors—the rate of the conversion process and the atmospheric residence time of SO_2. Under average atmospheric conditions, the conversion rates are believed to be of the order of only 1% h^{-1}. Consequently, observations of substantial amounts of sulfates over large regions are explained as the result of long range transport of SO_2 lasting a day or longer. Such a possibility immediately raises two questions of paramount practical significance. First of all, is the proliferation of modern tall-stack power plants which burn fossil fuels largely responsible for the observed distribution of sulfates in the U.S.? Secondly, what is the role of diurnal variations in emissions and ambient atmospheric and ground conditions in the overall formation of secondary sulfates?

Numerous efforts are under way in the U.S. and in Europe to develop mathematical models to simulate aspects of long range (synoptic scale) transport, transformation and removal of SO_2. Thus, the existing steady-state models developed for application over a range of a few tens of kilometers are now being joined by a new class of 'regional' models whose range of applicability is of the order of 1000 km. Conspicuously rare in the existing family of models are those designed specifically for application in the intermediate β-mesoscale range, roughly between 25 and 250 km.* This dearth is particularly surprising in the case of models for isolated large plumes, because much of the current activity of airborne sampling of large plumes is within that range. It is also in the mesoscale range that the plume becomes dispersed throughout the mixing layer and first comes into contact with the ground. The effect of source height is also most pronounced in this range. The amount and distribution of the pollutant which become available to synoptic scale transport are determined by the mesoscale effects on emissions.

For the transport of large plumes over a short range (<50 km), the variability of the PBL and of SO_2 conversion and removal parameters is usually small, and steady-state models can provide an adequate simulation. During long range transport, source

* In meteorological literature, the mesoscale is often subdivided into γ-mesoscale (0–25 km), β-mesoscale (25–250 km) and α-mesoscale (250–500 km).

effects are no longer important and simple box-model simulations are valid. In the intermediate β-mesoscale range (25–250 km), source effects, spatial inhomogeneities of the PBL structure, and the diurnal variability of the PBL as well as of SO_2 conversion and ground removal parameters all exert important influences on the sulfur budget of the plume. This complexity, coupled with a general lack of suitable air quality and meteorological data, has discouraged past efforts at plume modeling in the β-mesoscale.

In this paper, the formulation and application of a diagnostic β-mesoscale plume model are described. This Lagrangian model incorporates the diurnal variability of the planetary boundary layer (PBL) structure and of the parameters governing the chemical conversion and ground removal of SO_2. It is a quasi-steady model which uses an existing steady-state model to perform the solution over one hour at a time, with inputs which are variable from one hour to the next. The vertical inhomogeneity of atmospheric dispersion is simulated by the use of an assumed height-dependent profile of the eddy diffusion coefficient. Application of the model using the field data of the transport and kinetics of SO_2 in the plume of a large coal-fired power plant yields quantitative information about the dispersion, transformation and ground removal of SO_2. This information is then used in the model to investigate the parts played by source height, eddy diffusion and time of emission in determining the overall mesoscale sulfur budget of the plume.

The model is designed specifically for implementation on a minicomputer with 56K bytes of core storage. The cost of running this model is thus insignificant in comparison with many existing grid models which require the use of a large computer.

The work reported here was performed as part of Project MISTT (Midwest Interstate Sulfur Transformation and Transport), an integrated, interdisciplinary research program of field, laboratory and theoretical studies sponsored by the U.S. Environmental Protection Agency, and carried out under the technical direction of William E. Wilson, Jr. The primary objective of the program is to obtain a physical understanding and quantitative assessment of the nature and fate of atmospheric sulfur released in large power plant and urban-industrial plumes. An intensive field program of increasing scope and duration, centered around aircraft sampling of the urban plume of St. Louis and of the plume of the Labadie power plant near St. Louis, has been conducted each summer since 1973. Details concerning the overall Project are given elsewhere in this symposium (Wilson, 1978). Detailed descriptions of the field program and a study of the sulfur pudget in a power plant plume are given by Husar *et al.* (1978). The spatial range covered in these plume mapping experiments increased steadily from 50 km in 1973 to more than 300 km in 1976. On 9 and 18 July 1976, the Labadie plume was sampled out to 300 km. The data for these two days are used in the model application. Specific details of these data of plume transport and sulfur kinetics are discussed in a companion paper by Gillani *et al.* (1978).

1.1 *Objectives of diagnostic plume models*

The primary function of contemporary air pollution models is to identify and quantitatively characterize pollutant emission at its source, and its subsequent transmission through the atmosphere subject to meteorological transport, physical and chemical transformations, and a variety of wet and dry removal processes. All present air pollution models are diagnostic in that they all take certain observed fields of meteorological variables as inputs and attempt to diagnose a consistent distribution of the pollution variables. The emphasis in this paper is on the use of plume models as a diagnostic tool in an integrated program of field and laboratory work. The specific objectives of such a model are: (i) to provide a quantitative framework for the physical interpretation of the findings of field experiments; (ii) to identify the most important system parameters influencing the atmospheric fate of pollutants, and to estimate the sensitivity of primary pollutant variables to changes in these parameters; (iii) to aid in the design of future field experiments by identifying types of data needed for more realistic model inputs, and finally (iv) to validate and upgrade the simulation capabilities of the model by incorporating new physical insights and better data acquired from the field program.

2. MODEL FORMULATION

2.1 *Basic equation*

The mathematical formulation of a model requires the choice of a frame of reference (Eulerian or Lagrangian), the degree of resolution (integral or differential approach), and finally, the specific mathematical simulation of the physics, chemistry and initial and boundary conditions. The Lagrangian frame moves with the system of interest along the trajectory described by the mean flow field. It is necessary to compute the mean trajectory of the air volume of interest. Spatial resolution is partly lost in Lagrangian models because it is necessary to consider a homogeneous volume of definition. At least in the unstable, capped mixing layer of the daytime, the necessary assumption of uniform wind speed with height represents a reasonably good approximation (Townsend, 1967; Deardorff, 1970; Husar *et al.*, 1978). In the U.S. Midwest and Great Plains, the approximation may be poor during the night hours when a low-level jet blows at a height of about 500 m (Blackadar, 1957; Husar *et al.*, 1978; Smith *et al.*, 1978).

The differential formulation of the principle of conservation of mass employed most commonly in mesoscale models pertaining to the transport over flat terrain of a plume from a single, isolated source is (for a single component such as SO_2)

$$\frac{\partial \bar{c}}{\partial t} + \bar{u}\frac{\partial \bar{c}}{\partial x} = K_H \frac{\partial^2 \bar{c}}{\partial y^2} + \frac{\partial}{\partial z}\left(K_z \frac{\partial \bar{c}}{\partial z}\right) + r \qquad (1)$$

where x, y, z denote the downwind, crosswind and the vertical co-ordinates, respectively. This formulation is based on the following assumptions: (a) molecular diffusion is negligible compared with eddy diffusion; (b) the vertical component of the velocity is negligible; (c) the x-coordinate is always aligned with the direction of the horizontal component ($\bar{u}$) of the wind; (d) the horizontal and vertical turbulent mass fluxes are assumed to be describable in terms of eddy diffusion, with K_H and K_z representing the corresponding eddy exchange coefficients; (e) eddy diffusion in the longitudinal direction is negligible compared to downwind advection. In the above formulation, r represents the rate of internal generation or loss of the pollutant. In our model, it describes the conversion loss of SO_2.

The validity of assumption (b) diminishes with increasing transport time, and becomes questionable during long range transport (Pasquill, 1974). Also, in regions of local vertical transport such as a heat island, this assumption will fail. Assumption (c) neglects directional wind shear with height and can result in significant errors in the estimation of crosswind plume spread, particularly during long range transport. In a recent detailed analysis of the validity of the eddy diffusion approximation (d), Corrsin (1974) concluded that "the partial success of gradient transport models in turbulence is largely fortuitous, and certainly surprising". The principal limitations of the gradient transfer approach are that the length and time scales of the effective turbulent eddies must be sufficiently small in comparison with the corresponding scales of the plume. The eddy diffusion concept is thus least valid near elevated point sources where the scale of atmospheric turbulence is generally large and the plume dimensions are small. In practice, therefore, the exchange coefficients are sometimes expressed in power law functions of the distance x from the source, in order to account for progressively larger eddies as they take part in the diffusion process, until the action of the entire spectrum of eddies has been felt (Peters and Klinzing, 1971). During the initial stage of plume spread, the gradient transfer theory cannot simulate the translation of the whole plume in the vertical or crosswind directions due to eddy motion larger than the plume cross-section. The assumption of negligible longitudinal eddy diffusion is generally valid for $\bar{u} > 1\ \mathrm{m\,s^{-1}}$ (Herrmann, 1976). In spite of their obvious shortcomings, eddy diffusion models have been relatively successful in simulating mesoscale plume transport whenever appropriate time and height-dependent models of K_z have been employed.

The validity of the above formulation is perhaps most restricted by the combined implication of assumptions (c) and (d) that crosswind plume spread may be adequately described by a lateral diffusivity. This approximation is frequently made in plume dispersion models because it greatly simplifies model solution. Its validity is least questionable during the daytime convective conditions when the directional shear of the wind with height is smallest (Husar *et al.*, 1978). The application of the β-mesoscale model is considered here only for the daytime hours. Furthermore, as is shown later (Section 3.3), model calculations and comparisons with the field data are considered only for crosswind integrals of the concentration of SO_2 rather than for crosswind distributions. Thus, the model is not actually applied to simulate the details of crosswind plume spread.

2.2 *Non-dimensionalization*

An important step in the mathematical formulation of the model is the proper non-dimensionalization of all variables, and it permits the identification of the system parameters which govern the physical processes. If τ, l, u_c, K_c and C_c are chosen to denote characteristic values of time, length, speed, eddy diffusivity and pollutant concentration with respect to which the real variables are normalized, then equation 1 may be rewritten in the following non-dimensional form:

$$\frac{\partial \chi}{\partial T} + U\left(\frac{u_c l}{K_c}\right)\frac{\partial \chi}{\partial \xi} = g_H \frac{\partial^2 \chi}{\partial Y^2} + \frac{\partial}{\partial Z}\left(g_z \frac{\partial \chi}{\partial Z}\right) + R, \quad (2)$$

where $T = t/\tau$, $\xi = x/l$, $Y = y/l$, $Z = z/l$, $U = \bar{u}/u_c$, $g_H = K_H/K_c$, $g_z = K_z/K_c$, $R = r(\tau/C_c)$ and $\chi = \bar{c}/C_c$.

The above formulation yields one important system parameter, $(u_c l/K_c)$, which has been identified as the Peclet number (Pe) of turbulent mass transfer (Gillani and Husar, 1975). Pe is an important meteorological parameter related to plume dilution. It may be used in the further reduction of the downwind distance parameter as $X = \xi/\mathrm{Pe}$.

As will be demonstrated in Section 3.1, the use of X permits the direct comparison of the St. Louis urban-industrial plume data gathered on different days under quite different meteorological conditions. The application of similar non-dimensionalization procedures to the surface boundary condition applicable to equation 2 yields another important system parameter, as will be shown later.

The proper choice of the characteristic quantities τ, l, u_c, K_c and C_c is also important. For the problem of mesoscale dispersion within a confined mixing layer, $l = h$ (the mixing height), and $\tau = h^2/K_c$ (characteristic time of vertical turbulent dispersion of a particle through the mixing height) are the natural length and time scales relevant to the physical situation. The particular choice of the characteristic transport velocity u_c is less obvious; the geostrophic wind speed, the average transport speed over the mixing height, or the wind speed at effective stack height at the time of emission (u_0) are all appropriate choices. $u_c = u_0$ is chosen in the present model application. K_c may be chosen as the maximum value of K_z in the mixing layer. $C_c = (Q_0/u_c h^2)$, where Q_0 is the pollutant emission rate, is a meaningful choice.

In the Lagrangian formulation of the model, application of Taylor's hypothesis, $t = x/\bar{u}$, leads to the equality $X/U = T$, and the left hand side of equation 2 reduces to just the first term.

2.3 *The source and near-source effects*

Anthropogenic sources of air pollution are generally classified as point sources (e.g. tall stack), line sources (roadway), or area sources (urban area). Area sources are usually made up of multiple stationary and moving point sources whose emissions rapidly become mixed. A short distance downwind of the horizontal area source, the emissions are distributed over a vertical plane. Consequently, it is sometimes convenient to consider an urban plume source to be a vertical area source of finite lateral width and a given vertical profile of emission distribution (see illustration in Fig. 5). If Q_0 is the overall rate of pollutant emission, $\bar{C}_0(z)$ is the corresponding vertical distribution of the pollutant source (assumed uniform in the lateral direction), and the source area is assumed to be bounded by $|y| \leq b$ and $z_1 \leq z \leq z_2$, then the pollutant mass balance may be expressed by the following 'boundary' condition

$$Q_0 = \int_{z_1}^{z_2} \int_{-b}^{b} \bar{C}_0(z) \bar{u}_0(z) \, dy \, dz, \tag{3}$$

where $\bar{u}_0$ denotes the mean wind speed at the source. For sufficiently small values of b and z_2–z_1, such an area source model may be applied to point sources.

The emission of the pollutant is accompanied by an efflux of heat and momentum at the stack exit. Consequently, the plume is a buoyant jet. The interactions between plume buoyancy, the wind, and atmospheric turbulence result in plume rise, plume bending and plume spread by the entrainment of ambient air. Final plume rise is often much greater than the physical stack height. Thus, the knowledge of the effective stack height is of considerable importance in the plume model. There are basically three methods of determining plume rise: using empirical formulas, solving numerical models which simulate the buoyant plume behavior, and by direct observation of the plume, as from an aircraft. Of these, the last method is perhaps the most reliable, but generally the least available. Fortunately, appropriate aircraft data indicating the approximate location of the plume centerline at a distance of several kilometers from the source are quite often available in Project MISTT, and are used in this model study. The method of calculating plume rise from empirical formulas is often frustrating because there are many such formulas, and the results of using different formulas are quite different—often by a factor of 10 or more (Briggs, 1975).

2.4 *Eddy exchange coefficients*

The realistic simulation of the vertical dispersion of the plume is a principal requirement of a mesoscale dispersion model. It governs the rate of delivery of the pollutants to or away from the ground sink, depending on whether the source is elevated or near the ground. Thus, it influences the residence time of SO_2 in the air, and thereby indirectly the amount of sulfate formation. More directly, however, atmospheric mixing is responsible for the entrainment of background air into the plume. For the power plant plumes, such entrainment could make available air mass constituents such as ozone, hydrocarbons, and free radicals which may promote sulfate formation.

Turbulent flux terms appear in the equations of conservation of mass, momentum and energy as cross-correlations of stochastic quantities. Thus $\overline{w'c'}$ represents the vertical turbulent mass transfer of the pollutant. In the plume model, such quantities must be expressed in analytical closed-form approximations which provide an adequate parametrization of turbulent mixing in the planetary boundary layer. Perhaps the most promising of the existing approaches is that using second and higher order closure schemes (Donaldson, 1973) in which dynamic equations modeling the subgrid fluxes are solved. Deardorff (1973) and Orlanski *et al.* (1974) have developed eddy coefficients which depend respectively on local turbulent energy and on local vertical gradients of potential temperature. For the exchange coefficients to be 'local', this method of sub-grid scale parametrization requires the use of a fine grid. Both of the above methods of parametrization of turbulence require extensive computations on large computers.

The use of semi-empirical assumed vertical profiles of eddy diffusivities represents only a first-order closure approximation of turbulent transfer, but it gives reasonable simulations with substantially less effort and cost. The success of this approach depends on the choice of proper functional forms for K_H and K_z. The functional forms of K_H and K_z may be chosen as:

$$K_H = K_H(t) \cdot (x)^{\gamma_0} \tag{4a}$$

$$K_z = K_{zm} \cdot (x/x_r)^{\gamma} \cdot g(t, z) \tag{4b}$$

According to Islitzer and Slade (1968), $\gamma = 0.42$ for the neutral case. In our model, we choose the following stability-dependent representation

$$\gamma = \begin{cases} \gamma_0 & x \leq x_r \\ 0 & x > x_r \end{cases}, \tag{5}$$

where $\gamma_0 = 0.4\,(1 - 3/L)$, $|L| > 3$. x_r is a reference distance within which the action of the entire spectrum of turbulence has been felt by the spreading plume; $x_r = 3h$ is used in our model. L is the Monin–Obukhov similarity length.

K_{zm} is the highest value attained by K_z in the mixing height during the entire duration of application of the model. Based on the application of the Lagrangian similarity theory in the constant-flux surface layer,

$$K_z = \frac{\kappa u_* z}{\phi(z/L)}, \qquad z_0 \leq z \leq z_{SL} \tag{6}$$

where ϕ is a stability correction factor and z_0 and z_{SL} are the roughness length and surface layer height respectively. For the case of heat and mass transfer, Businger *et al.* (1971) have suggested the following empirical expressions for ϕ:

$$\phi_n = 0.74, \text{ neutral case, } L = \infty \tag{7a}$$

$$\phi_s = 0.74 + 4.7\, z/L, \text{ stable case, } L > 0 \tag{7b}$$

$$\phi_u = 0.74(1 - 9\, z/L)^{1/2} \text{ unstable case, } L < 0. \tag{7c}$$

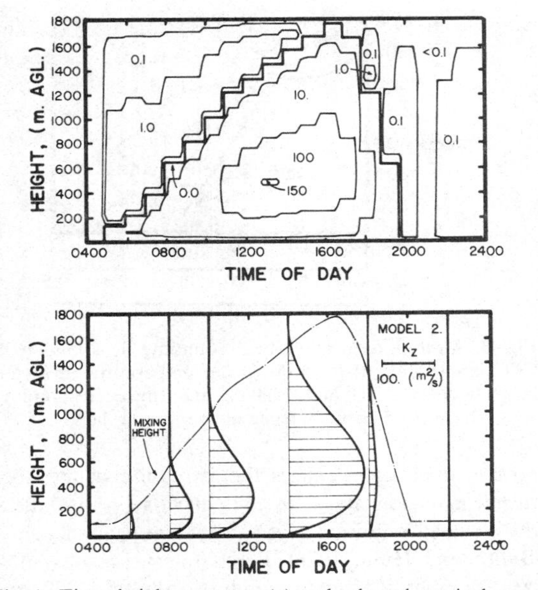

Fig. 1. Time–height contours (a) and selected vertical profiles (b) of K_z for 18 July 1976 according to the O'Brien model (K-Model 2). The contour for $K_z = 0$ in (a) represents the simulated mixing height variation during unstable conditions.

Above the surface layer, the similarity theory is not well developed and there is much greater uncertainty about the magnitude and height-dependence of K_z. Several models for K_z have been used for this region. The models of O'Brien (1970) and Shir (1973) are well-known. They are derived to satisfy specified realistic boundary conditions. O'Brien's model is a cubic polynomial above the surface layer, and matches the surface layer value and slope of K_z at the interface. Shir's profile is based on a single continuous function which fits the specified boundary conditions at the ground and at the inversion base, and is also forced to attain a maximum value (K_m) at $z/h = 0.25$. The expressions for these two models are as follows:

$$\text{O'Brien:}\quad K_z = K_h + \left(\frac{h - z}{h - z_{SL}}\right)^2 \times \left\{K_{SL} - K_h + (z - z_{SL}) \times \left[K'_{SL} + 2\frac{K_{SL} - K_h}{h - z_{SL}}\right]\right\}, \quad z \geq z_{SL} \tag{8}$$

$$\text{Shir:}\quad \frac{K_z}{K_m} = 0.73(4Z)\left[e^{1-4Z} + \frac{1}{1 + 16Z^{1.6}}\right], \quad 0 \leq Z = z/h \leq 1. \tag{9}$$

In the O'Brien model, K_h and K_{SL} are values of K_z at $z = h$ and z_{SL} (height of surface layer), and K'_{SL} is the derivative of K_z at z_{SL}. K_{SL}, K'_{SL} are determined from surface layer theory. The application of this model requires *a priori* knowledge of the thickness of the constant-flux layer and is very sensitive to this choice. O'Brien's model is tested in our β-mesoscale model, and is henceforth referred to as K-Model 2. Time–height contours and vertical profiles of K_z, based on the O'Brien model, were constructed with $z_{SL} = |L|$ for each hour of the day for 18 July 1976 in St. Louis, and are shown in Fig. 1. Shir's model was used in our steady-state, short range plume model.

For application in the present β-mesoscale plume model, the author has developed the following simple model for K_z which varies with time and height and is responsive to changes in atmospheric stability. The model also permits adequate flexibility in varying the maximum value of K_z at each time in such a way as to provide a reasonable fit for the aircraft data. In the surface layer, the model employs the basic expressions of equations 6 and 7, with the choice $z_{SL} = |L|$. The evaluation of the Monin–Obukhov stability length L is based on hourly ground data, and is performed as described in Appendix A. The changing stability from hour to hour is thus incorporated into the model explicitly. For unstable stratification, the model uses the following expression for 'free convection' immediately above the surface layer:

$$K_z = A\kappa u_* z \left|\frac{z}{L}\right|^{1/3}, \quad -L \leq z \leq z_{ml} \tag{10}$$

where A is chosen such that equations 6 and 10 give matching values of K_z at $z = -L$. This value is obtained as 4.2733. The above expression is based on the scheme proposed by Priestly (1954), and subsequently used by Yordanov (1972) in a plume model for emission from a high source. Above z_{ml}, $K_z = K_m = K_z(z_{ml}) =$ constant is used for $z_{ml} \leq z \leq 0.9\,h$. Such a constant K above 50 m has been suggested by Deardorff (1967). The value of K_m is directly related to the stability parameter $-h/L$ of unstable conditions (Deardorff, 1970) by

$$K_z = -f_m h/L, \; z_{ml} \leq z \leq 0.9\,h \tag{11}$$

where the constant f_m is chosen to fit the observed data. The choice of f_m determines the height z_{ml} (by equating equations 10 and 11 at $z = z_{ml}$) above which the constant K value ($=K_m$) is used. Values of f_m are estimated to vary from 0.15 on a strongly unstable day (18 July) to 0.25 on a weakly unstable day (9 July). Above $z = 0.9\,h$, the value of K is allowed to decrease linearly to 0.2 K_m at $z = h$.

In stable conditions, surface layer values of K_z are obtained according to equations 6 and 7. The height of the surface layer is chosen as L. Above the surface layer, K_z is approximated by a slight variation of the O'Brien model for the case of $K_h = K_{SL}$. The resulting expression is

$$K_z = K_{SL}\left[1 + \beta\left(\frac{h - z}{h - L}\right)^2\left(\frac{z}{L} - 1\right)\right], \quad z \geq L > 0 \tag{12}$$

where $\beta = 0.136$ ensures that the slope of K at $z = L$ matches its value based on surface layer formula. Pro-

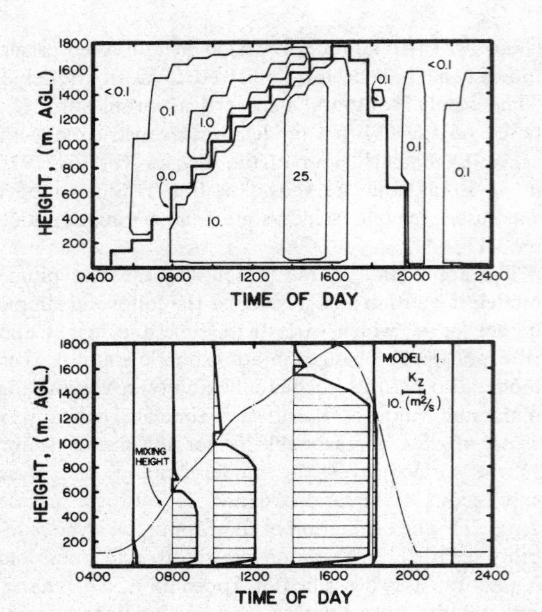

Fig. 2. Time–height contours (a) and selected vertical profiles (b) of K_z for 18 July 1976 according to the author's model (K-Model 1). Notice that the scales for K_z profiles in (b) are different in Figs 1 and 2 by a factor of about 4.

files of K_z based on the above model (henceforth referred to as K-Model 1) have been computed for each hour of 9 and 18 July. Selected vertical profiles and the time–height contours of K_z for 18 July are shown in Fig. 2.

2.5 *Dynamics of the mixing layer*

The fate of any given pollutant release within the lowest kilometer or so of the atmosphere is influenced by the evolution of the daily mixing layer. The existence of a well-defined daytime mixing layer in the continental U.S. is well documented (Holzworth, 1972). Such a mixing layer is commonly characterized by an adiabatic or superadiabatic thermal gradient capped by a stable inversion layer, and good turbulent mixing within the layer which drops off sharply at the top of the layer. Aircraft vertical soundings reveal a rather uniform vertical profile of pollutant variables in the daytime mixing layer with a drop near the mixing height (see Fig. 3).

During the night, the air close to the ground is cooled by conduction and radiation, forming a shallow nocturnal inversion. Some degree of turbulent transfer will continue to prevail in the lowest layer near ground. Above the inversion, the PBL is stably stratified. After the sun rises, convection develops due to surface heating, and gradually the night inversion is eroded. As insolation increases, so does the magnitude of the ground-to-air sensible heat flux, and an unstable well-mixed convective layer grows rapidly upwards, typically reaching heights of 1 to 2 km in the afternoon. Our understanding of the structure and behavior of the PBL is perhaps the weakest for the evening transition period when the ground-to-air heat flux rapidly ceases and radiational cooling of the

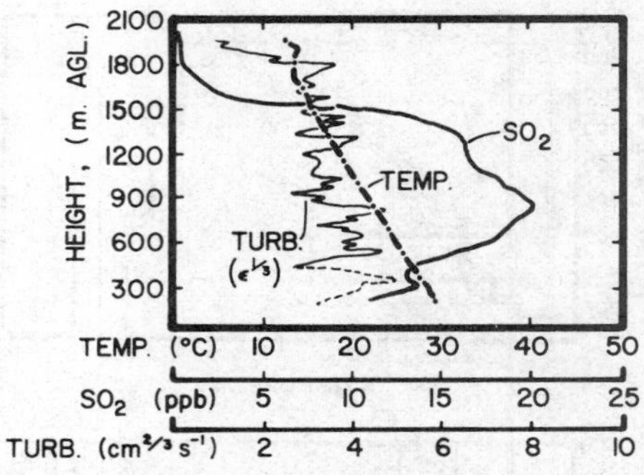

Fig. 3. Aircraft data of vertical sounding at about 1630 CDT on 18 July 1976. Notice the temperature inversion layer between 1700 and 1900 m, and the corresponding sharp drop in turbulence at the mixing height.

ground once again causes the formation of the nocturnal inversion above it, thus effectively 'insulating' the elevated pollution layers from the ground below (Smith and Hunt, 1978). Based on the results of a large number of aircraft soundings, Sholtes (1972) has concluded that during the late afternoon, atmospheric turbulence and the mixing height undergo a very rapid decrease. The boundary layer measurements of the Sangamon Experiments (Argonne, 1975) revealed the existence of a transient multilayered, stably-stratified structure of the PBL during morning and evening transition periods. Presumably, the dispersive ability of the atmosphere diminishes drastically and the corresponding transient mixing layer structure is complex.

The simulation of the daytime PBL structure in the present β-mesoscale model is as a three-layer model: the surface layer of height $|L|$ (or 25 m, whichever is smaller), the bulk mixing layer capped by a rising elevated inversion of finite thickness (one grid spacing), and a stable layer above. The variable mixing height is denoted by $h(t)$ and the fixed height of the three-layer PBL (including the top stable layer) is H, chosen to be equal to the maximum value attained by h during mid-afternoon. The mixing layer and the stable layer above it are effectively decoupled, without the introduction of an explicit boundary condition at the interface, by the insertion of an inversion layer of finite thickness Δ within which $K_z = 0$. In the stable layer, K_z decreases linearly from $K_h/2$ ($K_h = K_z$ at $z = h$) at $z = h + \Delta$ to zero at $z = H$.

For the sampling missions of 9 and 18 July 1976, direct estimates of h (Fig. 4a) were made from the vertical aircraft sounding data, such as shown in Fig. 3, collected in the afternoon and evening hours. These data were augmented by the temperature radiosonde data of the Regional Air Monitoring Systems (RAMS) of the St. Louis Regional Air Pollution Study (RAPS) of the U.S. E.P.A. (Myers and Reagan, 1975). Figure 4b shows the corresponding discrete simulation of the PBL. The mixing height shows a stepwise increase after each hour. The evening transition is modelled by discrete elevated inversion layers, which no longer exist after 1900. Subsequently, the entire PBL is very

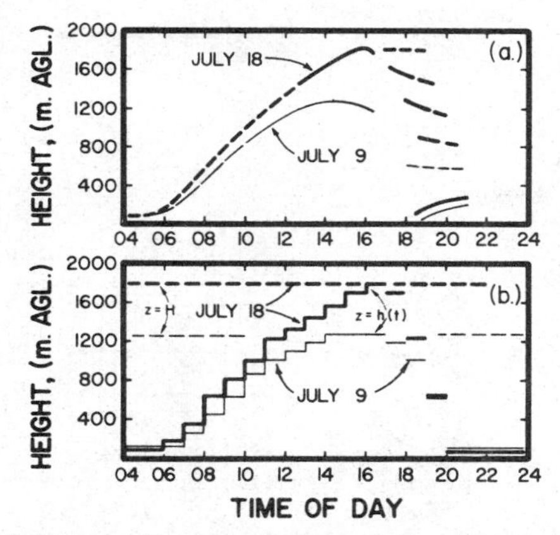

Fig. 4. Temporal variation of the observed (a) and simulated (b) mixing height, h, on 9 and 18 July 1976. Observe the multi-layered PBL structure (a) during the evening transition hours (1800–2000 CDT).

stable. Values of K_z drop by at least two orders of magnitude between 1600 and 2000 (see Fig. 2).

The present model application is considered only for plume releases within the daytime mixing layer. Calculations of plume concentrations of SO_2 are made during the hours of mixing layer development as well as during the evening transition. Only during the latter case does the plume vertical spread extend into the stable region between $h(t)$ and H. The structure of K_z in this layer thus enters into the model only during the evening transition hours.

2.6 *Modeling of reactive plumes*

Past field and laboratory research on SO_2 oxidation rates have implicated temperature, humidity, turbulent mixing, intensity of solar radiation, in-cloud residence time of the plume, catalytic heavy metals, soot, ozone, free radicals, hydrocarbons, ammonia, background haziness, air mass history, and various other factors as rate-controlling in the conversion kinetics. Reliable *in situ* measurements of these factors and of the concentrations of some of the reacting species are sparse or even non-existent. A wide range of uncertainty exists concerning the values of many of the reaction rate constants (Davis and Klauber, 1975). The complexity of the situation raises serious questions regarding the relevance of detailed numerical modeling of plume chemistry with the quality and quantity of available input data. One class of models is devoted to a detailed simulation of the formation of photochemical smog, with a focus on multi-component systems of hydrocarbons (HC) and oxides of nitrogen (NO_x). Recently some of these models have been extended to include reactions involving SO_2 (Liu *et al.*, 1976; Isaksen *et al.*, 1978). Another class uses the concept of 'overall' conversion rate, α, in a formulation of the following form for a single component

$$r = -\alpha c^a. \tag{14}$$

In general, α is a function of time and space, and through its time dependence, in particular, includes the bulk effect of all possible factors influencing the rate of conversion under given conditions. A proper application of this simple approach in a diagnostic model can probably provide results which compare favourably with those from more complex reactive plume models. Many long range model applications use a constant average value of α (Sheih, 1976; Wendell *et al.*, 1976; Rao *et al.*, 1976; Eliassen and Saltbones, 1975) and a linear conversion process ($a = 1$) as a representation of conditions during long range transport. Our β-mesoscale model is based on the concept of 'overall' conversion rate.

The eddy diffusion equation is an attractive vehicle for modeling simple as well as complex chemical situations while retaining significant rigor in dispersion simulation. If r can be modeled as a linear term, i.e.

$$r = -\alpha(t)\bar{c}, \tag{15}$$

then the solution may be simplified by the following separation of variables:

$$\bar{c}(t, x, y, z) = \exp\left(-\int \alpha(t)\, dt\right)\chi(t, x, y, z) \tag{16}$$

where χ represents the solution of the pure-diffusion problem. This decoupling of the dispersion and conversion problems means that the pure-dispersion concentration profiles remain self-similar even in the presence of conversion. It was shown by Friedlander and Seinfeld (1969) that such an assumption is valid as long as the rate of change of the concentration due to chemical reactions is much smaller than that due to turbulent diffusion. This condition is satisfied in the early hours of plume transport for SO_2 for which the rate of conversion is slow. Our β-mesoscale model uses the formulation of equation 15 and makes the assumption of equation 16.

2.7 *Modeling ground removal*

Surface dry deposition is most easily treated in the eddy diffusion model as the lower boundary condition, thus

$$K_z \frac{\partial \bar{c}}{\partial z} = v_d \bar{c}, \qquad z = z_r \tag{17}$$

where each side represents the vertical downward flux of the pollutant crossing a reference plane at z_r, commonly about 1 m from the ground. v_d is the so-called deposition velocity introduced by Chamberlain and later discussed by him in terms of a resistance analogy (Chamberlain, 1966). It is a turbulent mass transfer coefficient for the vertical transfer of the pollutant from z_r to its final absorption within the leaf, if the ground cover is vegetation. The mass transfer below z_r is subject to three resistances: the aerodynamic resistance to turbulent transfer above the surface, r_a; a laminar sublayer resistance, r_b; and the resistance at the surface itself, r_s. Then, $v_d = (r_a + r_b + r_s)^{-1}$. During the well-mixed daytime conditions, the surface resistance predominates as r_a is negligible. At night, the atmosphere is stable, and r_a may be expected to be high. The surface resistance is believed to be higher at night, except possibly in the presence of dew

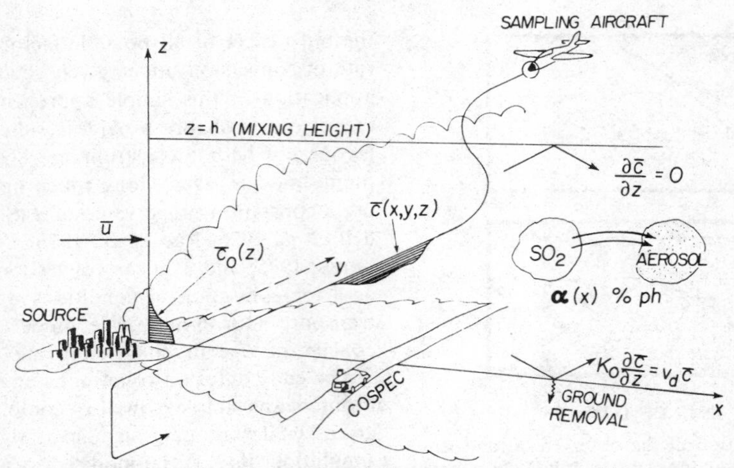

Fig. 5. Illustration of aircraft sampling and modeling of the transport, diffusion, transformation and ground removal of an urban emission within a finite mixing layer.

(Fowler and Unsworth, 1974). Consequently, v_d is probably smaller at night.

The above formulation of the surface boundary condition was introduced by Monin (1959) and discussed by Calder (1961). The left side of the equation represents the turbulent delivery of the material from above. Consequently, it is really the ratio (v_d/K_0), where $K_0 = K_z$ at z_r, which is the effective ground removal parameter. The non-dimensional formulation may be represented as

$$\frac{\partial \chi}{\partial Z} = \text{Sh}.\,\chi \quad \text{at } Z = 0, \tag{18}$$

where $\text{Sh} = v_d l/K_0$ = turbulent Sherwood number of vertical surface layer mass transfer. l denotes any convenient length scale which may be chosen as h, the mixing height. This dimensionless system parameter governs the ground removal process. Since v_d, h and K_0 are all time-dependent quantities, so is Sh, in general. Typically, $v_d = 0.01\ \text{m s}^{-1}$, $h = 1000$ m, $K_0 = 0.1\ \text{m}^2\,\text{s}^{-1}$, and Sh = 100.

3. PROJECT MISTT PLUME MODELS

The principal objective of Project MISTT field studies and related activities is to determine the fate of sulfur emissions in the plumes of the large power plants and the urban–industrial complex of St. Louis. Specifically, the rates of SO_2-to-sulfate conversion and of the ground removal of SO_2 must be extracted from field data. Project MISTT plume modeling efforts started with an analytical, steady-state model with constant K_z (Gillani and Husar, 1975). The model was made numerical with the introduction of a height-dependent K_z (Gillani and Husar, 1976a, b). The β-mesoscale model is still in the developmental stage. It is a quasi-steady numerical model. The two numerical models are described here.

3.1 *Steady-state model*

Figure 5 shows a schematic drawing of an urban plume of near-ground origin, dispersing within a finite mixing layer of constant height, h. A constant steady wind speed is assumed over the mixing layer. Gas-to-particle conversion is modeled as a first order process with the conversion rate being a function of plume age. Vertical eddy diffusion is height-dependent. The applicability of such a model is believed to be valid for periods of up to four or five hours of plume transport under relatively steady, near-neutral conditions. The model formulation is as follows:

$$\bar{u}\frac{\partial \bar{c}}{\partial x} = K_H\left(\frac{x}{x_r}\right)^{\gamma}\frac{\partial^2 \bar{c}}{\partial y^2} + K_m\left(\frac{x}{x_r}\right)^{\gamma}\frac{\partial}{\partial z}\left(g(z)\frac{\partial \bar{c}}{\partial z}\right) - \alpha(x)\bar{c}. \tag{19}$$

Applying the Taylor hypothesis $t = x/\bar{u}$, and the transformations

$$\chi = \frac{\bar{c}\exp\left(-\int \alpha(t)\,\mathrm{d}t\right)}{(Q_0/\bar{u}h^2)} \quad \text{(equation 16)},$$

and

$$T' = \int \left(\frac{T}{T_r}\right)^{\gamma} \mathrm{d}T,$$

and carrying out the non-dimensionalization procedure outlined previously (Section 2), we get the following non-dimensional formulation

$$\frac{\partial \chi}{\partial T'} = g_H \frac{\partial^2 \chi}{\partial Y^2} + \frac{\partial}{\partial Z}\left(g(Z)\frac{\partial \chi}{\partial Z}\right). \tag{20}$$

The source is treated as a vertical area source, symmetric laterally about the plane $Y = 0$ and of half width B and vertical distribution $\chi_0(Z)$. The distribution of the emission is assumed to be uniform laterally. Hence, the boundary conditions are

$$\int_0^1 \chi_0(Z)\,\mathrm{d}Z = \frac{1}{2B}, \qquad |Y| \le B$$

$$\chi \to 0 \text{ as } Y \to \pm\infty, \quad \frac{\partial \chi}{\partial Z} = 0 \text{ at } Z = 1,$$

and

$$\frac{\partial \chi}{\partial Z} = \text{Sh}\cdot\chi \text{ at } Z = 0.$$

The following separation of variables solution is assumed

$$\chi(T', Y, Z) = \eta(T', Y) \cdot \zeta(T', Z) \quad (21)$$

for which η may be determined analytically as

$$\eta(T', Y) = \frac{1}{4B}\left[\operatorname{erf}\left(\frac{Y+B}{2\sigma}\right) - \operatorname{erf}\left(\frac{Y-B}{2\sigma}\right)\right], \quad (22)$$

where $\sigma = \sqrt{g_H T'}$.

The factor $\zeta(T', Z)$ is a solution of the following two-dimensional problem:

$$\frac{\partial \zeta}{\partial T'} = \frac{\partial}{\partial Z}\left(g(Z)\frac{\partial \zeta}{\partial Z}\right) \quad (23)$$

$$\zeta(0, Z) = \zeta_0(Z)$$

$$\frac{\partial \zeta}{\partial Z} = 0 \text{ at } Z = 1 \text{ and } \frac{\partial \zeta}{\partial Z} = \mathrm{Sh} \cdot \zeta \text{ at } Z = 0.$$

Shir's model (equation 9) for $K_z(Z)$ is used for $g(Z)$ in the present model application. The above plume model is solved by the so-called method of lines (Madsen and Sincovec, 1974), a semi-discrete numerical procedure in which the spatial term on the right hand side is discretized by a centered finite difference approximation over a fixed, non-uniform grid in the vertical direction. The semi-discretization yields the following system of simultaneous first order ordinary differential equations:

$$\frac{\partial \zeta_1}{\partial T} = \frac{2}{Z_2 - Z_1}\left[g_{1+\frac{1}{2}}\frac{\zeta_2 - \zeta_1}{Z_2 - Z_1} - g_1 \cdot \mathrm{Sh} \cdot \zeta_1\right]$$

$$\text{at } Z = Z_1 = 0,$$

$$\frac{\partial \zeta_i}{\partial T} = \frac{2}{Z_{i+1} - Z_{i-1}}\left[g_{i+\frac{1}{2}}\frac{\zeta_{i+1} - \zeta_i}{Z_{i+1} - Z_i} - g_{i-\frac{1}{2}}\frac{\zeta_i - \zeta_{i-1}}{Z_i - Z_{i-1}}\right], \quad i = 2, 3, \ldots, (N-1)$$

$$\frac{\partial \zeta_N}{\partial T} = \frac{2}{Z_N - Z_{N-1}}\left[-g_{N-\frac{1}{2}}\frac{\zeta_N - \zeta_{N-1}}{Z_N - Z_{N-1}}\right]$$

at $Z = Z_N = 1$.

In the above equations, the subscript i refers to the value of the variable at the node of the ith grid line above the ground. $i = 1$ corresponds to $Z = 0$, and $i = N$ to the top of the mixing layer at $Z = 1$. The subscript $i + 1/2$ refers to the value of the variable at the mid point between nodes i and $i + 1$. The equations for $i = 1$ and $i = N$ include the boundary conditions at the ground and the mixing layer top, respectively. The above set of N simultaneous first-order differential equations at each T are integrated numerically by means of Hammings modified predictor-corrector method (Ralston and Wilf, 1969).

Figure 6 illustrates the role of the ground removal parameter, Sh, for the case of a broad source near the ground, such as an urban–industrial complex. For such an emission, the conversion process is slow in the initial stages of plume transport (White *et al.*, 1976), and dry deposition may be expected to predominate. Consequently, SO_2 conversions are neglected. The plot shows the fractional mass of the gaseous sulfur emission which continues to remain airborne (M_g/M_0) as a function of dimensionless time of transport. For such a ground source, ground removal is seen to be quite rapid initially, and increases with increasing Sh. The data points shown are for the transport of the urban–industrial plume of St. Louis on three summer days in 1973 and 1974. The use of the dimensionless time coordinate includes the effect of meteorology and permits the direct comparison of data of different days for such a ground source. The Sherwood number values are seen to range between 40 and 260, and correspond to a midday range of values of v_d between 0.6 and 2.6 cm s^{-1}. Application of this model with height-dependent K_z to the urban plume data of 9/6/73 gave deposition velocities between 2.1 and 2.6 cm s^{-1}. The application of the analytical model with constant K to the same data gave corresponding values of 3.9 and 6.9 cm s^{-1}. The more realistic simulation of the surface layer resistance to vertical mass transfer in the height-dependent model thus has the effect of lowering the value of v_d significantly.

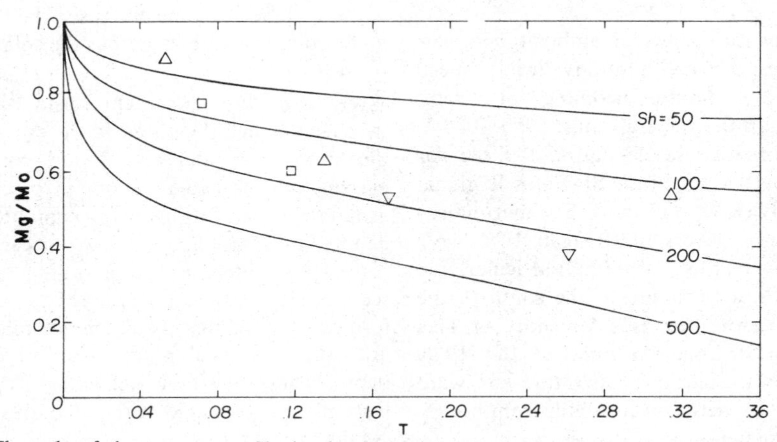

Fig. 6. The role of the parameter, Sh, in the ground removal of a surface level emission. The data points are for three different summer days. M_g/M_0 denotes the airborne fraction of the sulfur emission from the urban–industrial complex of St. Louis.

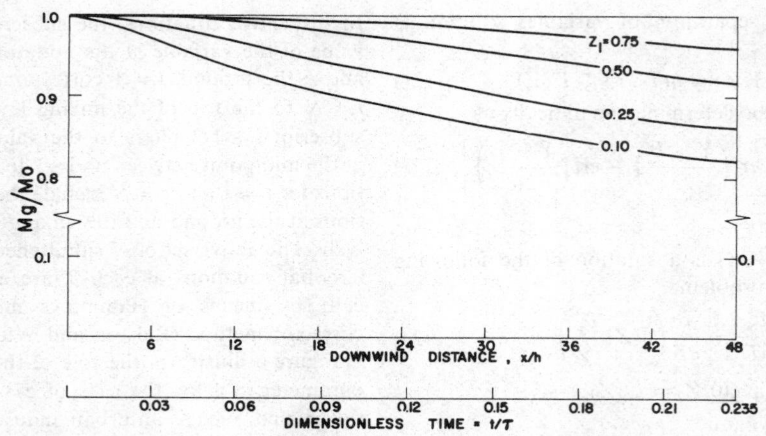

Fig. 7. The role of source height ($Z_1 = z_1/h$) in the depletion of SO_2 by ground removal during short range plume transport (<50 km).

Figure 7 illustrates the role of source height in the ground removal of SO_2. $Z_1 = z_1/h$ denotes the dimensionless source height. Four cases are considered, viz. $Z_1 = 0.1$, 0.25, 0.50, 0.75. The results plotted are for plume transport with $Pe = \bar{u}h/K_m = 200$ and $Sh = v_d h/K_0 = 100$. The ground loss of sulfur is seen to be about 18% at $x/h = 48$ for the lowest source and less than 4% for the highest source. These results clearly point out the need to make this mesoscale correction in applications of the box model in transport over a longer range.

3.2 *β-Mesoscale model: the data base*

The data of particular relevance in the present application of the β-mesoscale model are those pertaining to the Labadie power plant plume of 9 and 18 July 1976, and have been discussed in detail by Gillani *et al.* (1978). There are two main sources of data: aircraft sampling and the St. Louis RAPS network. Data related to power plant characteristics and operations, fuel usage and sulfur emissions were supplied by Union Electric Company. Details related to the aircraft sampling procedures are included in our companion paper (Husar *et al.*, 1978).

Ground level measurements of ambient and dew point temperature, relative humidity, wind speed (10 m), total (direct + diffuse) incoming solar radiation intensity, R_T, in the spectral range 3950–7000 Å, and ozone monitored at the 25-station Remote Air Monitoring System (RAMS) of the St. Louis Regional Air Pollution Study (RAPS) of the U.S. Environmental Protection Agency (Myers and Reagan, 1975) were obtained in hourly-average form. In particular, the data of Station 118, a rural site to the south of the urban area, have been utilized (see Appendix A). This site was upwind of St. Louis on 9 and 18 July 1976. In addition, vertical profiles of temperature and wind at four RAPS stations were also available from hourly radiosonde measurements.

The above data constitute the level 1 data set. Using this data set, a level 2 data set is generated which consists of quantities directly required in the main plume model or in one of the sub-modules. The components of the level 2 data set are described below.

(a) *Meteorology.* Transport, $\bar{u}$—backward plume trajectories were constructed from the point of sampling to the source. In this manner, plume identification is verified and plume age is determined. These trajectories were computed by Dr. T. B. Smith of MRI utilizing the MISTT mobile pibal data and the RAPS upper air wind data. The upper air wind data of RAPS Station 142 located near the Labadie were used to estimate u_0, the wind speed at emission time and height. Mixing height, h—mixing heights were determined from inspection of MISTT aircraft data of vertical spirals, and the RAPS vertical temperature profiles. Aircraft data parameters used for this purpose were turbulence, temperature, light scattering coefficient, aerosol charge, SO_2 and ozone. Eddy diffusivity, K_z—the quantities needed to determine $K_z(z)$ at each hour are u_*, L and h. The determination of u_* and L are described in Appendix A.

(b) *Kinetics.* Conversion and removal parameters, α, Sh—these parameters use the above level 2 meteorological data as well as the solar radiation data from level 1.

(c) *Air quality.* For each constant-altitude aircraft traverse through the plume in the crosswind direction, the crosswind integrals of the concentrations of gaseous and particulate sulfur in excess of the background concentrations are computed. These are denoted as S_g and S_p ($\mu g\ m^{-2}$). For each set of aircraft traverses at different altitudes at a given plume cross-section, the vertical integrals of S_g and S_p are performed to yield the instantaneous mass contents M_g and M_p in vertical plume slabs of unit downwind depth. These line and area integrals of the concentrations are compared with the corresponding model results.

(d) *Source.* The Labadie power plant of the Union Electric Company is located about 60 km west of the

St. Louis Arch. It is a coal-fired plant with four tangentially fired units feeding into one double and two single stacks of 214 m height. This base-load plant has a rated capacity of 2400 MW. The actual power outputs during the hours of interest were about 2250 MW on 9 July and about 1700 MW on 18 July. Based on the monthly average of 3.17% sulfur content of the fuel for July 1976, the corresponding total sulfur emissions averaged about 13.6 and 10.2 kg s^{-1} (as SO_2) respectively on 9 and 18 July.

3.3 *β-Mesoscale model: description*

For a proper simulation of plume transport beyond 50–100 km and 4–5 hours, the effect of unsteady winds and the dynamics of the mixing layer cannot be ignored. Project MISTT data analyses for the Labadie plume indicate that the overall rate of conversion of SO_2 varies diurnally in close relationship with the intensity of solar radiation (Gillani *et al.*, 1978). The ground removal rate is also believed to be higher during the daytime (see Section 2.7). The model presented here is a quasi-steady model in which all time-dependent variables are allowed to change in a stepwise manner after each hour. The steady state model described in Section 3.1 is used as a building block. Hourly-averaged values of the input variables are used at any given hour to obtain a steady-state solution for plume dispersion, transformation and ground removal during that hour. At the end of the hour, the output solution (SO_2 concentration distribution) is treated as the initial condition for the next hour, the values of all input variables are updated, and the steady state solution is now computed for the new hour. In this manner, the solution is advanced as the mixing layer evolves, atmospheric stability changes, the transport wind changes, and the pollutant kinetic parameters (α and Sh) change.

The solution is performed in a Lagrangian manner by following a given pollutant release from the source in the quasi-steady manner described above. At any given time, all vertical variations are considered on a fixed Eulerian grid of 21 points extending from the ground ($z = 0$) to the top of the PBL ($z = H = h_{max}$). The use of a fixed coarse grid in the vertical plane is not suitable in the early stages of plume spread. Therefore, the analytical solution of Gillani and Husar (1975) is used in the first hour of transport following plume release. This solution uses K_z which is a constant with respect to height, and is a power law function of time in this early stage of plume spread. In the numerical solution of subsequent hours, the chosen grid is coarse and approximately uniform above $Z = 0.05$, and relatively fine in the surface layer below $Z = 0.05$.

The evolution of the mixing layer is considered within the context of the three layer structure of the PBL described in Section 2.5. Because the mixing height is variable, the constant PBL height, H, is used in the non-dimensionalization of the spatial coordinates, instead of h. The 21-point fixed grid extends to H, and the upper reflecting boundary condition is applied there. During the rise of the mixing layer in the daytime, the pollutant released within the mixing layer is effectively decoupled from the stable layer above because $K_z = 0$ in the inversion layer. Thus, the dispersion is confined within the mixing layer. In the evening, following the break-down of the well-defined capped mixing layer, the pollutant in the upper part of the former mixed layer continues to be transported in the stable boundary layer. The model for K_z developed by the author and described in Section 2.4 (K Model 1) is used in the model solutions.

Based on observations, the conversion rate of SO_2 is chosen of the form

$$\alpha(t) = A_1 + A_2 R_T(t), \tag{24}$$

where A_1 and A_2 are empirical constants and R_T is the intensity of solar radiation expressed in kW m^{-2}. Based on the analysis of Project MISTT data for the Labadie plume, nighttime conversion of SO_2 during the summertime is slow compared to the daytime rate (Gillani *et al.*, 1978; Husar *et al.*, 1978). $A_1 = 0$ is used in the present model computations. Solar radiation data are presented in Appendix A.

The ground removal parameter, Sh, is defined as

$$\mathrm{Sh} = \frac{v_d H}{K_0} = \frac{v_d H}{K_{zm}}\frac{K_{zm}}{K_0} = \mathrm{Sh}_m \cdot \frac{1}{g_1}.$$

In the model formulation, the removal rate is chosen of the following form

$$\mathrm{Sh}_m = S_1 + S_2 \cdot R_T(t) \tag{25}$$

on the assumption that ground removal is more efficient during the daytime than at night. Here, S_1 denotes the nighttime conversion rate. S_1 and S_2 are empirical constants whose values are extracted from the comparison of the model results and field data.

The general solution of the model has the form

$$\frac{\bar{c}(t, y, z)}{Q_0/u_0 H^2} = \exp\left(-\int \alpha(t)\,\mathrm{d}t\right) \cdot \eta(T, Y) \cdot \zeta(T, Z). \tag{26}$$

The crosswind factor $\eta(T, Y)$ is such that

$$\int_{-\infty}^{\infty} \eta(T, Y)\,\mathrm{d}Y = 1.$$

Consequently, as long as the desired outputs are the crosswind integral (S_g), and the cross-sectional area integral (mass M_g) of the excess plume concentration of gaseous sulfur, the factor $\eta(T, Y)$ need not be computed at all. For this reason, the model solution is carried out only for the two-dimensional time-height plane.

The numerical solution remains stable and well-behaved for all hours under neutral and unstable conditions for which the solutions are carried out. The largest truncation error results from the finite differencing of derivatives with respect to height, and is limited to about 6%. The round-off error is limited by the double-precision accuracy of the computations.

3.4 *β-Mesoscale model: comparison with field data*

The model application is specifically considered for the field data of 9 and 18 July 1976. Sampling of the Labadie plume was performed beyond 300 km on each day. However, only single traverses were made at each of the sampling distances beyond 200 km. In this study, only those sampled plume cross-sections are considered for which the plume age was greater than 2 hours, and at which multiple crosswind traverses at different altitudes were made. There were three such sampled cross-sections on each of the two days. The corresponding sampling summary is as follows:

	9 July	18 July
Downwind distance (km):	70, 135, 190	30, 110, 160
Approximate plume release time (CDT):	0800, 0900, 1200	1400, 1000, 1000
Plume age at sampling (h):	3, 6.5, 8.5	2, 6, 9

On 18 July, the 30 and 110 km measurements were made at about the same time by two sampling aircraft. It is pointed out that, in general, the sampling of a given day was not Lagrangian, i.e. different air parcels were sampled at different downwind distances. The trajectory and time history of each plume sample is thus independent and model simulation requires an independent run corresponding to each sampled distance. The last two samplings of 18 July, at 110 and 160 km, however, approximated a Lagrangian air parcel.

The model is used as a diagnostic tool, in conjunction with the field data, to extract information about three sets of empirical model inputs, viz. $K_z(t, z)$, $\alpha(t)$ and Sh(t). An attempt is made to choose these inputs such that a good match of the model results with the appropriate measured sulfur data is obtained. The corresponding sulfur data to be matched are S_g (the crosswind integral of gaseous sulfur for each traverse), M_p/M_0 (the fraction of the emission which has undergone conversion to particulate sulfur at each sampled plume section), and M_R/M_0 (the fraction of the emission which has been removed at the ground). More details concerning these mass ratios and the manner in which they are determined are given in the companion paper by Gillani *et al.* (1978). Figure 8 shows the comparison of S_g for both days corresponding to the following choice of model inputs:

	9 July	18 July
K-model	Model 1	Model 1
K_{zm}($m^2 s^{-1}$)	33.6	13.4
(A_1, A_2)	(0, 0.020)	(0, 0.032)
(S_1, S_2)	(0.5, 1.5)	(0.25, 0.75)

The above choices of A_1, A_2 and S_1, S_2, when combined with the solar radiation data in accordance with equations 24 and 25, give the following values for the kinetic parameter values:

	9 July	18 July
α, nighttime (%h^{-1})	0	0
α, noon (%h^{-1})	1.82	3.02
Sh, nighttime (−)	88	110
Sh, noon (−)	112	146
v_d, nighttime (cm s^{-1})	0.53	0.47
v_d, noon (cm s^{-1})	1.99	1.79

The comparison between the observed and computed values of S_g (Fig. 8) at each plume section is considered to be good. The comparisons between the observed and computed values of the conversion fraction, (M_p/M_0), and the ground removal fraction, (M_R/M_0), corresponding to the above choice of the dispersion, conversion and removal parameters are given in Table 1. The comparisons, particularly for the conversion fraction, are good. However, the reader is cautioned that the error margin on the observed data of mass ratios, particularly in the case of ground removal, is quite large.

The rationale for the particular choices of the kinetic rates used in the above comparisons is illustrated in Figs 9 and 10. Figure 9 shows the simulated sulfur mass balance for 9 hours of plume transport corresponding to the 1000 CDT plume release on 18 July. This release was sampled in a quasi-Lagrangian mode at ages 6 and 9 hours. The sampled data points are also shown in the figure. Three model runs for this plume transport situation were carried out, each with $(A_1, A_2) = (0, 0.032)$, and each with a different set of values of (S_1, S_2), as shown in the figure. The particular choice of $(S_1, S_2) = (0.25, 0.75)$ is made as an approximate representation of the prevailing removal conditions for the day. Near the top of Fig. 9, the time variation of the chosen α is also shown. In Fig. 10, the effect of different values of A_2, and hence of α, is illustrated. The range of $A_2 = 0$ to 10 corresponds to peak daytime values of α in the range 0 to 9.4% h^{-1}. For the choice of $(S_1, S_2) = (0.25, 0.75)$, the temporal behavior of v_d and the crosswind-integrated ground level concentration of gaseous sulfur are shown in the lower portion of Fig. 10. Comparison of the model results and the data show that the conversion rates on 9 and 18 July attained peak midday values of about 1.8% h^{-1} and 3.0% h^{-1}, respectively. The corresponding peak inferred values of v_d are between 1.5 to 2.0 cm s^{-1}.

3.5 *The effect of diurnal variability on sulfur budget*

The pronounced diurnal variability of the PBL has important consequences with respect to the ground removal and the gas-to-particle conversion of atmospheric sulfur. In particular, the existence of a nocturnal inversion near the ground is believed to cause an effective decoupling of the surface sink from the stable layers aloft within which the pollutant contents

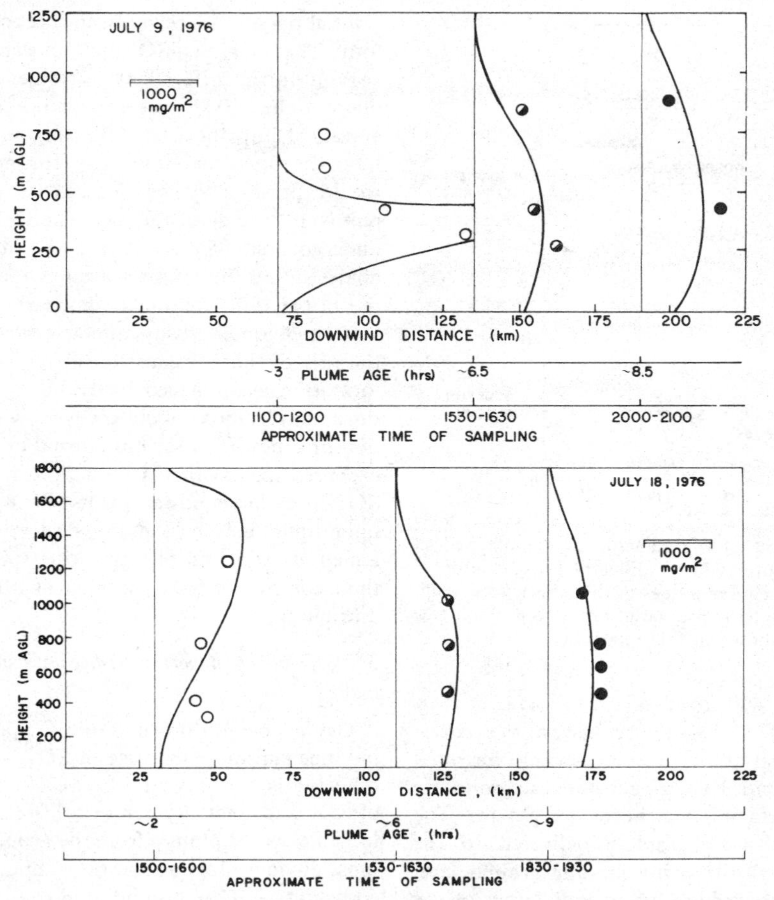

Fig. 8. Comparison of the observed (data points) and calculated values of the crosswind-integrated concentration of gaseous sulfur (S_g) for 9 and 18 July 1976 for the Labadie power plant plume.

of the previous evening remain trapped and become available for overnight long range transport. The relatively longer atmospheric residence of nighttime emissions of SO_2, however, is accompanied by a substantially reduced nighttime sulfur conversion rate. It is thus difficult to determine intuitively the time in the diurnal cycle corresponding to which the SO_2 emission from a large source will normally experience the maximum amount of conversion.

To investigate the role of time of emission on the overall fate of sulfur in the mesoscale, three plume releases of the Labadie plume on 18 July are considered at 0600, 1000 and 1400 CDT. For the latter two cases, plume measurements were made and the necessary inputs are available. The 0600 release is simulated on the assumption that the plume became trapped under the overlying rising inversion ($h = 280$ m at 0600, stack height $= 214$ m). The same kinetic parameters (A_1, A_2, S_1, S_2) are used for all three releases. The effective source heights are observed to increase during the day as the vertical mixing domain grows. The results of model computations for the

Table 1. Sulfur kinetics in the Labadie plume

		Per cent of emission			
	Downwind distance	Conversion (M_p/M_0)		Ground removal (M_R/M_0)	
Date	(km)	Observed*	Calculated	Observed*	Calculated
7/9/76	70	2.8	3.9	5	3.5
	135	9.7	7.6	16	22.2
	190	6.3	5.9	23	22.9
7/18/76	30	4.8	4.1	3	3.0
	110	13.8	12.7	17	11.1
	160	13.0	13.0	27	24.2

* The error range for these observed ratios is estimated to be at least 25%. The error range is estimated to be significantly higher for the removal mass ratios.

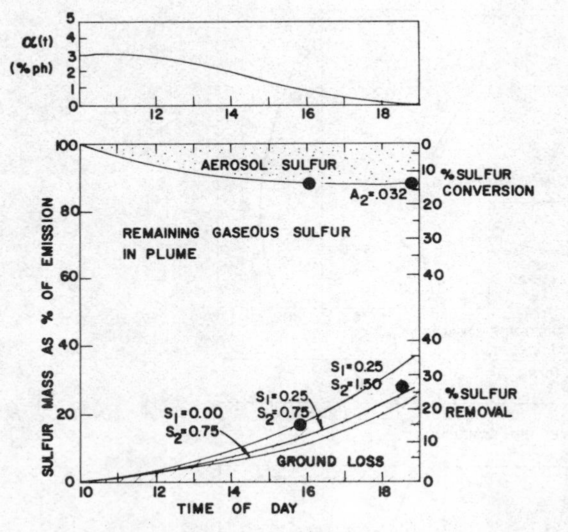

Fig. 9. Sulfur budget of the Labadie plume released at 1000 CDT on 18 July 1976 for different ground removal parameters. For the chosen values of the conversion coefficient A_1 and A_2, the temporal variation of conversion rate α is shown in the upper box.

0600, 1000 and 1400 releases are shown in Fig. 11a, b, c, respectively. Figure 12 shows the corresponding temporal plots of crosswind-integrated ground level concentration (a), and the fraction of the sulfur emission remaining airborne as gas (b). The early morning release is rapidly delivered to the ground (fumigation), the corresponding ground level concentration remains high for several hours of the morning, and the ground sink effectively removes about 50% of the sulfur mass by 1300. The 1000 release is subjected to the intense midday mixing, becomes effectively diluted over the entire deep midday mixing layer, with a smaller delivery of the pollutant to the ground compared to the 0600 case, and correspondingly smaller ground removal. Finally, the 1400 release rises to an effective stack height of over 1100 m, experiences a short period of intense mixing, which contributes more towards SO_2 accumulation near the top of the mixing layer than towards downward delivery to the ground. The result is that at 1900 when the residual atmospheric distributions of SO_2 for each of the three releases are compared, the early morning release is mostly depleted (to about $1/e$), the 1000 release has also undergone much depletion (to 60%) and is well-mixed. In contrast, the 1400 release shows a preferential accumulation aloft, having experienced little depletion (about 10%). As these remaining pollutant burdens are made available for nighttime long range transport, the implications for sulfate formation within each of these air masses become a little more clear. The bulk of the 1400 release may be expected to remain aloft and not become entrained into the mixing layer of the next day until perhaps noon or later. Consequently, a significant portion of the sulfur in this air parcel may still remain airborne on the second evening, a substantial part of it already in the secondary particulate form. The remaining SO_2 may undergo yet more conversion during a third day of residence in the atmosphere. The 1000 release would start becoming entrained into the next day's mixing layer from the time the nocturnal inversion becomes eroded, and would presumably be subject to efficient ground removal throughout the day. The 0600 release would undergo a similar fate. It may be inferred from this analysis, that the 1400 release has the greatest potential of the three for maximum particulate sulfur formation. A qualitatively similar conclusion regarding the role of time of emission in overall plume sulfate formation was reached by Husar *et al.* (1978) based on a simpler box model analysis which was carried out for a period of several diurnal cycles. Such model exercises suggest that a reduction in the emission of SO_2 from large power plants during the late afternoon hours may be a more effective control measure aimed at reduction of large scale sulfate distribution than controlling SO_2 emission at other times of the diurnal cycle.

3.6 *The effect of vertical eddy diffusion on the sulfur budget*

Dry deposition on the ground is a major scavenging mechanism in the case of SO_2, and vertical eddy diffusion is the only link between elevated emissions of SO_2 and their dry removal. The effective decoupling of elevated plumes from the ground at night is the most obvious illustration of the importance of vertical eddy diffusion to ground removal. The box model assumes instantaneous delivery of matter to the ground from above. Scriven and Fisher (1975) showed

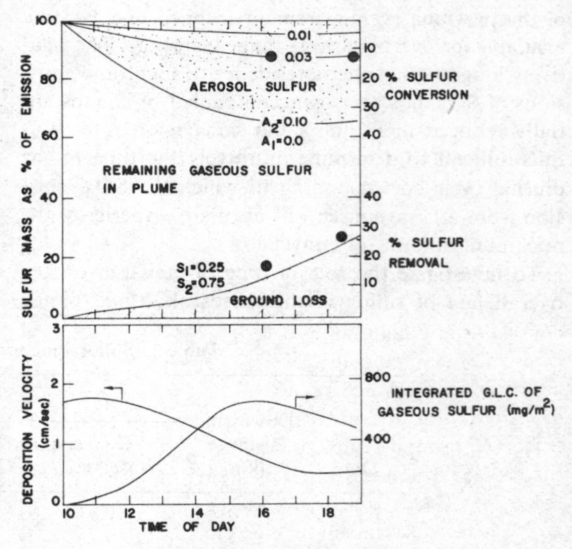

Fig. 10. Sulfur budget of the Labadie plume released at 1000 CDT on 18 July 1976 for different values of the conversion coefficient A_2. For the chosen values of the ground removal coefficients S_1 and S_2, the temporal variations of deposition velocity and crosswind-integrated ground level concentration of gaseous sulfur are shown in the lower box.

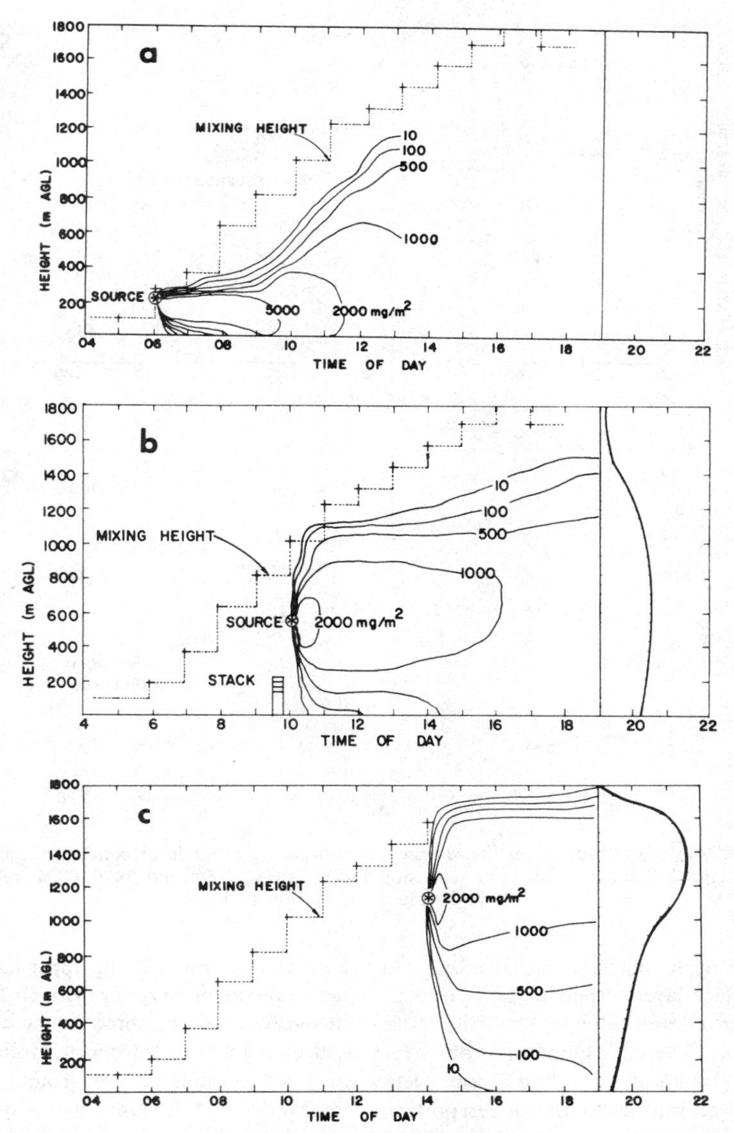

Fig. 11. Time–height contours of the crosswind-integrated concentration of gaseous sulfur, S_g, for (a) 0600 CDT release, (b) 1000 CDT release, and (c) 1400 CDT release of the Labadie plume of 18 July 1976.

that the effect of introduction of finite eddy diffusion (as compared to box model) was to reduce the effective value of v_d by the factor $(1 + 4/\pi^2 \, v_d h/K_z)$, for constant K_z. Gillani and Husar (1975) present a comparison of dry removal of surface emissions for finite and infinite K_z, which points out a dual role of eddy diffusion: near the ground source, a finite K_z gives increased ground removal; far from the source, when the pollutant concentration is higher aloft than near the ground, eddy diffusion has the opposite effect, in agreement with the finding of Scriven and Fisher. Scriven and Fisher also showed that the introduction of a more realistic K-profile in the surface layer resulted in a further reduction in the effective value of v_d. Gillani and Husar (1976a, b) later confirmed this finding in their numerical solution using Shir's vertical profile of K_z. They pointed out that the effect of the surface layer resistance to vertical mass transfer was to slow down the delivery of elevated emissions to the ground as well as to slow down the diffusion away from the ground of surface-level emissions. In the present study, we investigate the effect on ground removal and conversion of SO_2 of different magnitudes and profile-shapes of K_z in the region above the surface layer.

The transport of the 1000 release of 18 July is considered in two runs of the plume model with different models of $K_z(t, z)$. Model 1 (Fig. 2) is that presented in Section 2.4 as the author's formulation. The maximum value of eddy diffusion in the afternoon is $K_{zm} = 33.6 \, m^2 \, s^{-1}$. Model 2 (Fig. 1) is O'Brien's model also presented in Section 2.4. This cubic polynomial profile reaches a maximum value of K_z of $150 \, m^2 \, s^{-1}$ in the mid-afternoon. The two models are

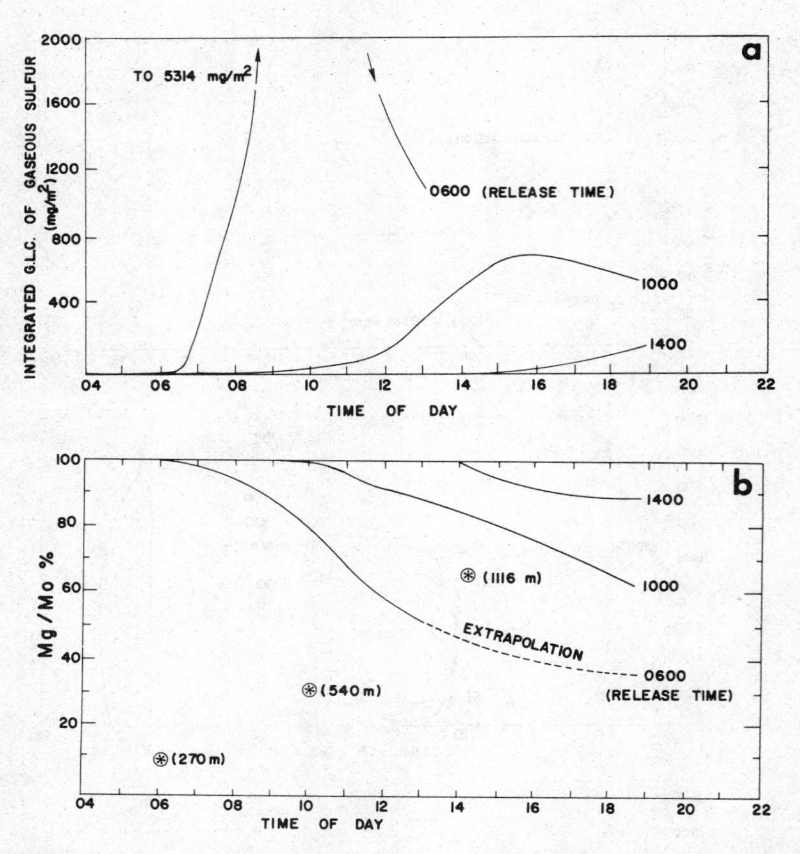

Fig. 12. Temporal variations of (a) the crosswind-integrated ground level concentration, and (b) the airborne fraction of the gaseous sulfur emission for the 0600, 1000 and 1400 CDT releases of the Labadie plume of 18 July 1976.

thus vastly different in the bulk mixing layer, but are identical in the surface layer of height $|L|$. Figure 13 shows the vertical profiles of S_g at two-hour intervals using Model 1 for K_z. The corresponding profiles for Model 2 (not shown) are only slightly different. Delivery to the ground as well as to the upper portion of the mixing layer is a little more rapid for Model 2 during early transport. The effective difference in the sulfur budget results obtained using Models 1 and 2 is shown in Fig. 14. The upper figure shows a higher net removal of SO_2 for Model 2, mostly occurring in the first two or three hours of transport. At the end of 9 hours of transport, Model 2 gives 31% loss of gaseous sulfur to the ground compared to 27% for Model 1. A fraction of this difference is a result of a small difference in values of v_d for which the models were run (see lower figure). There is about 1% difference in the amounts of net sulfate formation

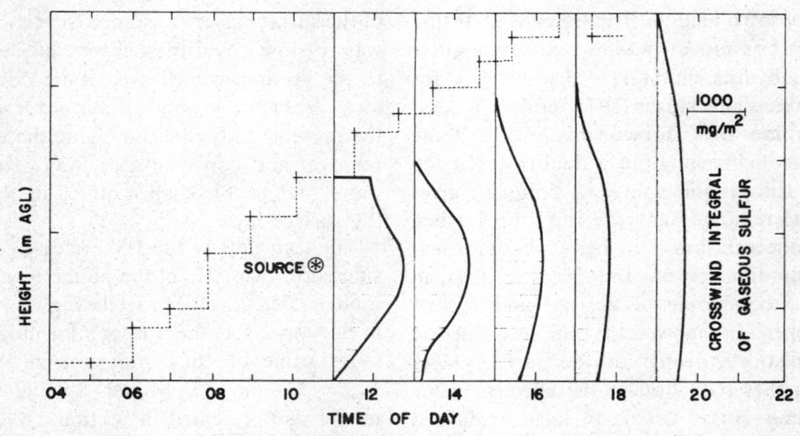

Fig. 13. 2-hourly vertical profiles of the crosswind-integrated concentrations of gaseous sulfur (S_g) calculated using K-Model 1.

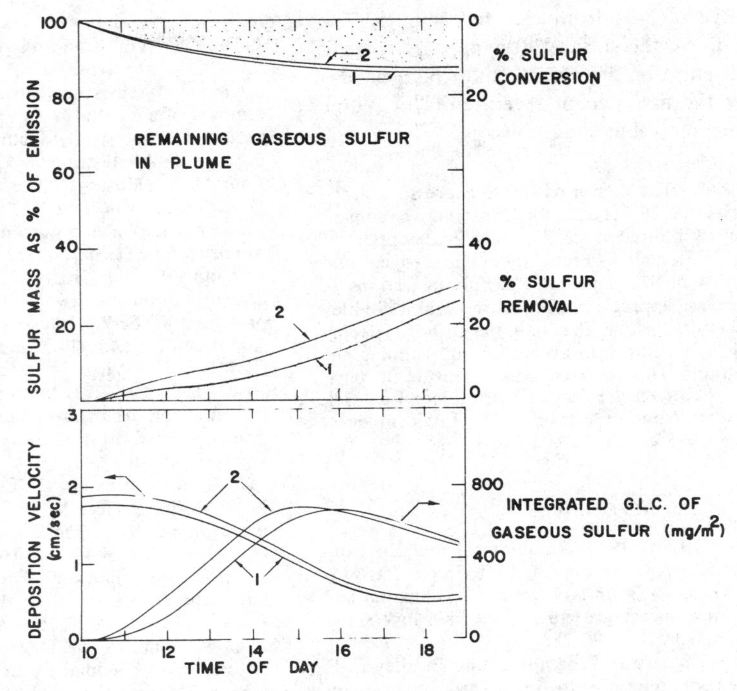

Fig. 14. Comparison of the calculated sulfur budget of the power plant plume using K-Model 1 (author) and K-Model 2 (O'Brien).

for the two cases after 9 hours of daytime transport. This surprisingly small difference in dispersion, removal and conversion of SO_2 for the two different K-Models indicates that large differences in K-profiles in the bulk mixing layer are of small consequence in the long range dispersion of pollutants. The profile of K_z in the surface layer is of dominant significance in the role of vertical eddy diffusion on ground removal. This finding lends further support for the use of a constant K_z in the bulk mixing layer, at least for unstable conditions.

4. CONCLUSIONS

As part of the summer 1976 field program of Project MISTT, the plume of the 2400 MW, coal-fired Labadie power plant was sampled from instrumented aircraft to downwind distances exceeding 300 KM on 9 and 18 July. Both SO_2 and particulate sulfur were sampled. Detailed meteorological data were also available. These data provided the motivation for the development of the diagnostic β-mesoscale plume model described in this work. This quasi-steady, Lagrangian model was developed using an existing steady-state model as a building block. Its present application covers a transport range of 200 km over 9 hours. The model incorporates the diurnal variability of the PBL structure, and of the parameters governing the chemical conversion and ground removal of SO_2. A semiempirical, dynamic, height-dependent submodel is developed for the vertical eddy diffusion coefficient. Two important dimensionless system parameters are identified: the Peclet number, $Pe = \bar{u}H/K_z$, governs plume dilution by transport and eddy diffusion; the Sherwood number, $Sh = v_d H/K_z$, governs ground removal.

The plume model is applied to the field data of 9 and 18 July 1976, and quantitative information about the dispersion, transformation and ground removal of SO_2 is extracted. The variable SO_2 conversion rates are shown to reach peak values of 1.8 and 3.0% h^{-1} at noon on 9 and 18 July, respectively. Peak values of deposition velocity over open country are estimated at between 1.5 and 2.0 cm s^{-1}, also at noon.

The model is used to investigate the roles of Sh, source height, time of emission and eddy diffusion on the overall sulfur budget of the plume. Of the daytime emissions of SO_2, the early morning release becomes depleted to $1/e$ by 1900, while the mid-afternoon release tends to accumulate aloft and undergoes little depletion by 1900. The mid- and late-afternoon plume releases appear to have the highest potential for long range transport and sulfate formation. This suggestion based on model calculations has important implications regarding the formulation of sulfate control strategy. However, no direct evidence based on Labadie plume measurements has been presented here in support of this suggestion. The need for further measurements of the plume over the entire diurnal cycle is clearly indicated in order to assess the significance of time of plume release on overall sulfate formation.

The effect of finite eddy diffusion, in general, is to reduce the effective deposition velocity. A surface layer (~ 50 m) of high resistance to vertical mass transfer is of dominant importance in governing the ground delivery of SO_2 from elevated plumes. The profile of K_z above the surface layer appears to have relatively little effect on the overall sulfur budget, particularly after the first two or three hours of plume transport under unstable conditions.

Acknowledgements—The author wishes to express his gratitude to Professor R. B. Husar for constructive consultations ranging over three years. The technical support of Dr. William E. Wilson, Jr. is gratefully acknowledged, and the patient review of this paper and useful suggestions of David Patterson are appreciated. To Morris Kahn, Stanley Nowak and Chandrashekhar Pradhan, the author is deeply indebted for their assistance in preparing and running the computer programs. This research was supported (in part) by the Federal Interagency Energy/Environment Research and Development Program through U.S. Environmental Protection Agency Grant No. R803896.

REFERENCES

Argonne National Lab. (1975) Radiological and Environmental Res. Div. Annual Report ANL-75-60, Part IV.

Blackadar A. K. (1957) Boundary layer wind maxima and their significance for the growth of nocturnal inversions. *Bull. Amer. Meteor. Soc.* **38,** 283–290.

Businger J. A., Wyngaard J. C., Izumi Y. and Bradley E. F. (1971) Flux-profile relationships in the atmospheric surface layer. *J. atmos. Sci.* **28,** 181–189.

Calder K. L. (1961) Atmospheric diffusion of particulate material considered as a boundary value problem. *J. Meteor.* **18,** 413–415.

Chamberlain A. C. (1966) Transport of gases to and from grass and grass-like surfaces. *Proc. Roy. Soc., Series A,* **290,** 236–265.

Corrsin S. (1974) Limitations of gradient transport models in random walks and in turbulence. *Advan. Geophys.* **18A,** 25–60.

Davis D. D. and Klauber G. (1975) Atmospheric gas phase oxidation mechanisms for the molecule SO_2. *Int. J. Chem. Kinetics, Symp.* No. 1, 543–556.

Deardorff J. W. (1967) Empirical dependence of the eddy coefficient for heat upon stability above the lowest 50 m. *J. appl. Meteor.* **6,** 631–643.

Deardorff J. W. (1970) Preliminary results from numerical integrations of the unstable planetary boundary layer. *J. Atmos. Sci.* **27,** 1209–1211.

Deardorff J. W. (1973) Three-dimensional numerical modeling of the planetary boundary layer. In *Workshop on Micrometeorology,* Amer. Met. Soc. pp. 271–311.

Donaldson C. duP. (1973) Construction of a dynamic model of the production of atmospheric turbulence and the dispersal of atmospheric pollutants. In *Workshop on Micrometeorology,* Amer. Met. Soc., pp. 313–392.

Eliassen A. and Saltbones J. (1975) Decay and transformation rates of SO_2 as estimated from emission data, trajectories and measured air concentrations. *Atmospheric Environment* **9,** 425–429.

EPA (1975) *Position Paper on Regulation of Atmospheric Sulfates,* U.S. Environmental Protection Agency, EPA 450/2-75-007, Research Triangle Park, N.C.

Fowler D. and Unsworth M. H. (1974) Dry deposition of sulfur dioxide on wheat. *Nature* **249,** 389–390.

Friedlander S. K. and Seinfeld J. H. (1969) A dynamic model of photochemical smog. *Environ. Sci. Technol.* **3,** 1175–1181.

Gillani N. V. and Husar R. B. (1975) Mathematical modeling of air pollution—a parametric study. Proc. 2nd. Federal Conf. on the Great Lakes, Argonne, Ill.

Gillani N. V. and Husar R. B. (1976a) Mesoscale model for pollutant transport, transformation and ground removal. Preprints, 3rd Symposium on Atmospheric Turbulence, Diffusion and Air Quality, Raleigh, NC.

Gillani N. V. and Husar R. B. (1976b) Analytical–numerical model for mesoscale transport, transformation and removal of air pollution. Paper presented at the 7th International Technical Meeting on Air Pollution Modeling and Its Application, NATO/CCMS, Airlie, Virginia.

Gillani N. V., Husar R. B., Husar J. D. and Patterson D. E. (1978) Project MISTT: Kinetics of particulate sulfur formation in a power plant plume out to 300 km. *Atmospheric Environment* **12,** 589–598.

Herrmann K. (1976) The role of turbulent diffusion in the direction of the mean wind in a numerical AQSM. Paper presented at the 7th International Technical Meeting on Air Pollution Modeling and Its Applications, NATO/CCMS, Airlie, Virginia.

Holzworth G. C. (1972) Mixing heights, wind speeds, and potential for urban air pollution throughout the contiguous United States. U.S. Environmental Protection Agency, AP-101.

Husar R. B., Patterson D. E., Husar J. D. and Gillani N. V. (1978) Project MISTT: Sulfur budget of a power plant plume. *Atmospheric Environment* **12,** 549–568.

Isaksen I. S. A., Hov O. and Hesstvedt E. (1978) A chemical model for urban plumes: Test for ozone and particulate sulfur formation in St. Louis urban plume. *Atmospheric Environment* **12,** 599–604.

Islitzer N. F. and Slade D. H. (1968) in *Meteorology and Atomic Energy* (edited by D. H. Slade) chapter 4.

Liu M. K., Yocke M. A. and Mundkur P. (1976) Numerical simulation of reactive plumes. 1976 Air Series, Am. Inst. Chem. Engrg.

Madsen N. K. and Sincovec J. F. (1974) The numerical method of lines for the solution of non-linear partial differential equations. In *Computation Methods in Nonlinear Mechanics* (edited by J. T. Oden *et al.*) Texas Inst. for Computational Mechanics, Austin, Texas.

Monin A. S. (1959) On the boundary condition on the earth surface for diffusing pollution. *Advan. Geophys.* **6,** 435–436.

Myers R. L. and Reagan J. A. (1975) The Regional Air Monitoring Systems, St. Louis, Missouri, U.S.A. Paper presented at the International Conference on Environmental Sensing and Assessment, Las Vegas, 15–19 September.

O'Brien J. J. (1970) A note on the vertical structure of the eddy exchange coefficient in the planetary boundary layer. *J. atmos. Sci.* **27,** 1213–1215.

Orlanski I., Ross B. B. and Polinsky L. J. (1974) Diurnal variation of the planetary boundary layer in a mesoscale model. *J. atmos. Sci.* **31,** 965–989.

Pasquill F. (1974) Limitations and prospects in the estimation of dispersion of pollution on a regional scale. *Advan. Geophys.* **18B,** 1–13.

Peters L. K. and Klinzing G. E. (1971) The effect of variable diffusion coefficients and velocity on the dispersion of pollutants. *Atmospheric Environment* **5,** 497–504.

Priestly C. H. B. (1954) Convection from a large horizontal surface. *Austral. J. Phys.* **6,** 279–290.

Ralston A. and Wilf H. S. (1960) *Mathematical Methods for Digital Computers,* pp. 95–109. Wiley, New York.

Rao K. S., Lague J. S. and Egan B. A. (1976) An air trajectory model for regional transport of atmospheric sulfates. Preprints, 3rd. Symp. on Atmos. Turbulence, Diffusion, and Air Quality. Amer. Met. Soc., Raleigh, NC., pp. 325–331.

Scriven R. A. and Fisher B. E. A. (1975) The long range transport of airborne material and its removal by deposition and washout—II. The effect of turbulent diffusion. *Atmospheric Environment* **9,** 59–68.

Sheih C. M. (1976) Application of a Lagrangian statistical trajectory model to the simulation of sulfur pollution over northeastern United States. Preprints, 3rd. Symp.

Atmos. Turbulence, Diffusion, and Air Quality. Am. Meteor. Soc., Raleigh, NC., pp. 311–317.

Shir C. C. (1973) A preliminary numerical study of atmospheric turbulent flows in the idealized planetary boundary layer. *J. atmos. Sci.* **30,** 1327–1339.

Sholtes R. S. (1972) The growth and decay of turbulent mixing in the planetary boundary layer. Environmental Monitoring Series EPA-R4-72-001.

Smith T. B., Blumenthal D. L., Anderson J. A. and Vanderpol A. H. (1978) Transport of SO_2 in power plant plumes; day and night. *Atmospheric Environment* **12,** 605–611.

Smith F. B. and Hunt R. D. (1978) Meteorological aspects of the transport of pollution over long distances. *Atmospheric Environment* **12,** 461–477.

Townsend A. A. (1967) Wind and the formulation of inversions. *Atmospheric Environment* **1,** 173–175.

Wendell L. L., Powell D. C. and Drake R. L. (1976) A regional scale model for computing deposition and ground level air concentration of SO_2 and sulfates from elevated and ground sources. Preprints, 3rd Symp. Atmos. Turbulence, Diffusion, and Air Quality, Amer. Met. Soc., Raleigh, NC., pp. 318–324.

White W. H., Anderson J. A., Blumenthal D. L., Husar R. B., Gillani N. V., Husar J. D. and Wilson W. E., Jr. (1976) Formation and transport of secondary air pollutants: ozone and aerosols in the St. Louis urban plume. *Science* **194,** 187–189.

Wilson W. E. (1978) Sulfates in the atmosphere: A progress report on Project MISTT. *Atmospheric Environment* **12,** 537–547.

Yordanov D. (1972) Simple approximation formulae for determining the concentration distribution of high sources. *Atmospheric Environment* **6,** 389–398.

APPENDIX A

Determination of Monin–Obukhov stability length, L

By definition,

$$L = \frac{\rho c_p T_0 u_*^3}{\kappa g H_0} \quad \text{(A1)}$$

where

ρ = density of air = 1.20 kg m^{-3},
c_p = specific heat of air at constant pressure = 1.00 kJ/kg °K,
T_o = ground level temperature, °K
u_* = friction velocity, m s^{-1},
κ = von Karman's constant = 0.4,
g = acceleration due to gravity = 9.81 m s^{-2}
H_0 = sensible heat flux from the ground, KW m^{-2}.

In the surface layer, the shape of the wind profile is given by

$$(\bar{u}/u_*) = (1/\kappa)[\ln(z/z_0) - \psi_m(z/L)], \quad \text{(A2)}$$

where z_0 is the ground roughness length, and the stability correction function for momentum, ψ_m is given by (Argonne, 1975).

$$\psi_m(z/L) = 0, \quad \text{Neutral}$$

$$\psi_m(z/L) = -5(z/L), \quad \text{Stable}$$

$$\psi_m(z/L) = 2\ln[(1+\chi)/2] + \ln[(1+\chi^2)/2] - 2\tan^{-1}\chi + \pi/2, \quad \text{Unstable}$$

where $\chi = 1 - 16(z/L)^{0.25}$.

Substitution of ψ_m into equation (A2) gives an implicit relationship between u_* and L which is solved iteratively.

Equations A1 and A2 require five empirical inputs, viz. z_0, z, $\bar{u}(z)$, T_0 and H_0. For the open country terrain over which the Labadie plume was transported on 9 and 18 July 1976, z_0 = 5 cm is taken as a representative value. z = 10 m is chosen since the hourly data for $\bar{u}$ at 10 m are available for RAPS station 118, a rural ground monitoring station of the Regional Air Pollution Study in St. Louis (Myer and Reagan, 1975). The measured value of ambient temperature at 5 m height at the RAPS station is utilized for T_0. No direct measurements are available for H_0. Its hourly value is derived from values of R_T, the intensity of total (direct + diffuse) solar radiation in the spectral range 3950–7000 Å, as measured by a hemispherical pyranometer apparatus at RAPS station 118. Thus, $H_0 = (H_0/R_T)\,R_T$. The ratio H_0/R_T is determined based on hourly measured values of H_0 and R_T on 10 August 1975 at the Sangamon Experiment site (a rural location approximately 100 km NE of St. Louis) of Argonne National Laboratory (Argonne 1975). The values of the ratio are adapted to the conditions of 9 and 18 July 1976 by assuming that H_0/R_T is directly proportional to the mixing height under peak midday convective conditions. These values of h_{max} are available from direct measurements, using the WHAT (wind, height and temperature) system at Sangamon on 10 August, and from MISTT aircraft soundings on 9 and 18 July.

The hourly values of T_0, $\bar{u}_{10}$, R_T, H_0, h, u_* and L for the hours 0500 to 2400 for 9 and 18 July are listed in Tables A1 and A2, respectively.

Table A1. Meteorological data for 9 July 1976

Time (CDT)	T_0 (C)	$\bar{u}_{10}$ (m s^{-1})	R_T (KW m^{-2})	H_0 (KW m^{-2})	h (m)	u_* (m s^{-1})	L (m)
0500	21.6	1.9	0.022	−0.030	100	0.127	6.2
0600	22.5	2.0	0.225	0.001	150	0.155	−258.2
0700	24.6	2.2	0.533	0.041	275	0.198	−17.3
0800	26.3	2.2	0.501	0.048	450	0.200	−15.5
0900	28.2	2.4	0.652	0.090	650	0.226	−11.7
1000	29.6	3.4	0.802	0.123	850	0.301	−20.6
1100	30.4	4.1	0.888	0.175	1000	0.358	−24.3
1200	31.0	4.8	0.909	0.215	1100	0.411	−30.1
1300	31.6	5.2	0.869	0.211	1200	0.438	−37.2
1400	32.1	4.8	0.760	0.177	1250	0.406	−35.3
1500	32.0	4.7	0.600	0.131	1250	0.392	−42.9
1600	31.8	4.9	0.426	0.088	1220	0.398	−67.2
1700	31.1	4.5	0.233	0.032	1120	0.355	−130.7
1800	29.8	3.0	0.070	0.000	920	0.226	∞
1900	27.9	2.6	0	−0.018	600	0.191	35.8
2000	26.1	2.6	0	−0.036	100	0.186	16.4
2100	25.1	3.1	0	−0.050	100	0.224	20.5
2200	24.3	3.0	0	−0.050	100	0.216	18.3
2300	23.9	3.0	0	−0.050	100	0.216	18.2
2400	23.9	3.0	0	−0.050	100	0.216	18.2

Table A2. Meteorological data for 18 July 1976

Time (CDT)	T_0 (C)	$\bar{u}_{10}$ (m s^{-1})	R_T (KW m^{-2})	H_0 (KW m^{-2})	h (m)	u_* (m s^{-1})	L (m)
0500	14.1	1.6	0.046	0.003	100	0.130	−64.8
0600	15.8	1.5	0.210	0.019	150	0.140	−12.7
0700	19.7	1.7	0.406	0.047	350	0.165	−8.6
0800	22.9	2.1	0.591	0.089	600	0.204	−8.7
0900	25.3	2.2	0.744	0.152	800	0.222	−6.6
1000	26.8	2.6	0.842	0.190	1000	0.255	−8.0
1100	27.7	3.1	0.931	0.272	1200	0.300	−9.1
1200	28.2	3.6	0.944	0.347	1350	0.342	−10.7
1300	28.7	3.3	0.886	0.330	1500	0.319	−9.1
1400	29.2	2.8	0.776	0.287	1600	0.280	−7.0
1500	29.3	2.6	0.623	0.192	1700	0.255	−8.0
1600	29.1	2.7	0.443	0.123	1800	0.253	−12.1
1700	28.5	2.7	0.247	0.046	1700	0.234	−25.8
1800	27.2	2.8	0.075	0.000	1200	0.211	∞
1900	24.9	2.2	0	−0.018	650	0.159	20.3
2000	23.3	2.8	0	−0.036	100	0.202	20.9
2100	22.5	2.3	0	−0.050	100	0.155	6.8
2200	21.6	2.5	0	−0.050	100	0.173	9.4
2300	21.0	3.0	0	−0.050	100	0.216	18.0
2400	20.0	3.0	0	−0.050	100	0.216	18.0

Atmospheric Environment Vol. 12, pp. 589–598. Pergamon Press 1978. Printed in Great Britain.

PROJECT MISTT: KINETICS OF PARTICULATE SULFUR FORMATION IN A POWER PLANT PLUME OUT TO 300 kM

N. V. GILLANI, R. B. HUSAR, J. D. HUSAR and D. E. PATTERSON
Air Pollution Research Laboratory, Mechanical Engineering Department, Washington University, St. Louis, MO 63130

and

W. E. WILSON, JR.
U.S. Environmental Protection Agency, Research Triangle Park, NC 27711, U.S.A.

(*First received* 20 *June* 1977 *and in final form* 24 *August* 1977)

Abstract—As part of the 1976 field program of Project MISTT, the plume of the coal-fired Labadie power plant near St. Louis was positively identified and sampled from aircraft over a range exceeding 300 km and 10 h of transport during day and night on July 9 and July 18. Measurements were made of SO_2, NO_x, ozone, particulate sulfur and various other pollutant and meteorological parameters. For both days, it is found that the gas-to-particle conversion of sulfur occurred mostly during daylight hours. The ratio of particulate to total sulfur was related linearly with the total solar radiation dose experienced by the plume. The maximum rate of particulate sulfur formation was less than $3\%\,h^{-1}$ on both days. Production of ozone was also observed within the plume on both occasions. Ground removal of total sulfur was found to be about 25% in the first 200 km, and its magnitude is compared to that of the gas-to-particle conversion.

1. INTRODUCTION

Gas-to-particle conversions of atmospheric sulfur compounds are important because the secondary products have a longer lifetime in the air, and because they are believed to be more harmful than the precursor gases. The formation of particulate sulfur by the oxidation of SO_2 in fossil-fuel burning power plant plumes is believed to be a slow process, and only during long-range transport of the plume may the particulate component be expected to reach a high proportion of the total sulfur content of the plume air. As a result, there is much current interest in understanding plume transport and pollutant transformation over a range of several hundred km during day and night. We are not aware of any quantitative plume measurements of rates of SO_2 oxidation for power plant plumes beyond 100 km. The most comprehensive published results of a previous field study of airborne measurements of SO_2 conversions in coal-fired power plant plumes are for a range of 70 km and transport times of up to 3 h (Forrest and Newman, 1977). These results indicate that SO_2 conversion seldom exceeds 5% in up to 3 h of transport during early morning and evening hours at relative humidities ranging between 30 and 85%. The observations have been interpreted as being indicative of a heterogeneous mechanism of conversion in the presence of catalysts. Conclusions based on measurements over such a short range, however, cannot be generalized *per se* to the more significant plume conversions and ground losses at long distances from the source.

In this paper, we report a detailed analysis of the kinetics of particulate sulfur formation in the plume of the Labadie power plant near St. Louis, Missouri, for two days in the summer of 1976, viz. July 9 and July 18. On both occasions the plume was tracked and sampled from aircraft to distances exceeding 300 km from the source. Application of a plume model to the data presented here is reported in a separate companion paper (Gillani, 1978). The measurements were made as part of the 1976 summer field program of Project MISTT (Midwest Interstate Sulfur Transformation and Transport), an integrated, multidisciplinary program sponsored by the U.S. Environmental Protection Agency.

The principal objective of Project MISTT is the quantitative estimation of the kinetics of SO_2 oxidation and ground removal in large plumes, and hence to determine the sulfur mass balance during plume transport. The scope, design and major findings of the Project are summarized elsewhere (Wilson, 1978). In a companion paper by Husar *et al.* (1978), the data of SO_2 conversion rates from nine sampling days are synthesized with those of plume transport and SO_2 ground removal in order to estimate the summertime sulfur budget of the Labadie plume. Major aspects of the field program are also described in that paper.

1.1. *Plant description*

The Labadie power plant of the Union Electric Company is located about 60 km west of the St. Louis

Arch. It is a coal-fired plant with four tangentially fired units feeding into one double and two single stacks of 214 m height. This base-load plant has a rated capacity of 2400 MW. The actual power outputs during the hours of interest were about 2250 MW on July 9 and about 1700 MW on July 18. Based on the monthly average of 3.17% sulfur content of the fuel for July 1976, the corresponding total sulfur emissions averaged about 13.6 and 10.2 kg s^{-1} (as SO_2) respectively on July 9 and 18.

1.2. *Sampling experiment*

Two primary sampling aircraft were used in the plume measurement missions of July 9 and 18—a twin-engine Rockwell AeroCommander operated by Washington University (WU) and a single-engine Cessna 206 operated by Meteorology Research Inc. (MRI). The WU aircraft was equipped for continuous measurements of SO_2, O_3, b_{scat}, aerosol charge, condensation nuclei count, time and aircraft position (altitude, VOR, DME). The MRI aircraft measured, in addition, NO, NO_x, turbulence and ambient as well as dew-point temperatures. Both primary aircraft also collected short-term (5–20 min) sequential filter samples for subsequent analysis for particulate sulfur. Detailed descriptions of the aircraft systems are given by Husar *et al.* (1978) and Blumenthal *et al.* (1978). On July 9, only the WU aircraft was sampling. The July 9 and July 18 data reported here are based on the WU aircraft measurements, except when otherwise stated. Aircraft sampling consisted of horizontal crosswind traverses and vertical soundings at discrete downwind plume sections at intervals ranging from 30 to 100 km. Background measurements were made well outside the plume during transit between successive plume sections sampled. In addition to the aircraft data, co-ordinated horizontal wind field measurements were made at half-hour intervals by three mobile single-theodolite pilot balloon units. These data were supplemented by extensive ground-level measurements and temperature and wind radiosonde data collected by the Regional Air Monitoring Systems (RAMS) of the St. Louis Regional Air Pollution Study (RAPS) of the U.S. E.P.A. (Myers and Reagan, 1975).

During both flight missions reported here, extensive use was made of a long-range communications system including a separate information relay aircraft which facilitated constant contact between the sampling aircraft, the pibal units, and the ground Operation Headquarters (OHQ). Selected aircraft and pibal data were received and plotted in real time at the OHQ. These plots as well as synoptic weather information served as the basis for flight direction from the OHQ (Husar *et al.*, 1978). The role of the ground mission control was of crucial importance in the positive identification of the plume at far distances. The use of a high sensitivity sulfur analyzer (Meloy SA-285, noise-equivalent signal 0.05 ppb) proved to be of similar significance. Finally, post-flight backward trajectory analyses were performed to confirm (or repudiate) the preliminary plume identifications.

2. PLUME TRANSPORT

Summaries of the plume samplings of July 9 and 18 are given in Tables 1 and 2. It is seen that, on both occasions, the first half of the plume transport occurred during daylight and the second half at night. The age of the sampled plumes ranged up to 12 h, during which the extremes of the diurnal variability of the meteorology were experienced. For such a

Table 1. Summary of plume sampling—Labadie: July 9, 1976

Downwind distance (km)	Traverse altitude (m AGL)	Median sampling time (CDT)	Plume release time* (CDT)	Wind speed ($m\ s^{-1}$)	$(c_b)_p$† ($\mu g/m^3$)	Background b_{scat} ($10^{-4}\ m^{-1}$)	Ozone (ppm)	$\bar{S}_p/\bar{S}_{tot}$‡ (%)
70	307	1051	0800		2.25	2.20	0.06	2.4 ± 0.4
	455	1107	0810		2.25	2.16	0.06	N/A
	614	1130	0825	6.2	2.25	2.04	0.06	2.6 ± 0.4
	753	1145	0850		2.25	2.04	0.06	4.4 ± 0.7
	909	1156	0850		2.25	1.93	0.06	N/A
105	305	1239	0810	6.1	2.25	1.96	0.07	4.1 ± 0.6
135	295	1533	0845		2.00	1.74	0.07	9.3 ± 0.4
	441	1551	0915	7.5	2.00	1.74	0.07	11.4 ± 1.7
	894	1631	1015		2.00	1.74	0.07	9.7 ± 1.5
190	452	2002	1110	8.0	2.00	1.49	0.06	7.7 ± 1.1
	914	2040	1225	11.6	2.00	1.49	0.06	N/A
250	463	2145	1200	15.5	2.00	1.49		7.1 ± 1.0
310	464	0005	1445	19.0	2.00	1.03		2.8 ± 0.4

* Estimated from backward trajectory analysis utilizing mobile pibal data (MISTT) and RAPS radiosonde wind data from several sites in the St. Louis area.

† $(c_b)_p$ = particulate sulfur concentration of the background air mass.

‡ $\bar{S}_p$ and $\bar{S}_{tot}$ are respectively the excess plume concentrations of particulate and total sulfur averaged over the traverse.

Table 2. Summary of plume sampling—Labadie: July 18, 1976

Downwind distance (km)	Traverse altitude (m AGL)	Median sampling time (CDT)	Plume release time* (CDT)	Wind speed (m s^{-1})	$(c_b)_p$† (μg/m^3)	Background b_{scat} (10^{-4} m^{-1})	Ozone (ppm)	$\bar{S}_p/\bar{S}_{tot}$‡ (%)
30	300§	1525	1300		0.71	0.63	0.10	5.0 ± 0.7
	406	1435	1200	4.0	1.30	0.55	0.10	7.8 ± 1.2
	747§	1540	1320		0.71	0.63	0.10	5.5 ± 0.8
	1219§	1600	1400		0.71	0.65	0.10	4.1 ± 0.6
110	441	1545	1000		0.71	0.50	0.10	17.3 ± 2.6
	740	1600	1000	4.7	0.71	0.50	0.10	17.9 ± 2.7
	1018	1615	1100		0.71	0.50	0.10	15.0 ± 2.2
160	432	1805	1000		1.53	0.49	0.09	N/A
	587	1820	1000	6.1	1.53	0.49	0.09	16.8 ± 2.5
	738	1900	1000		1.53	0.49	0.08	17.0 ± 2.5
	1047	1920	1000		1.53	0.44	0.08	17.2 ± 2.6
220	447	2000	1000	7.5	1.53	0.49	0.08	16.4 ± 2.4
320	581	2400	1200	12.0	1.53	0.49	0.07	15.0 ± 2.2

* See Table 1.
† See Table 1.
‡ See Table 1.
§ Data collected aboard MRI aircraft. All other data reported in this paper, except otherwise stated, were collected aboard the WU aircraft.

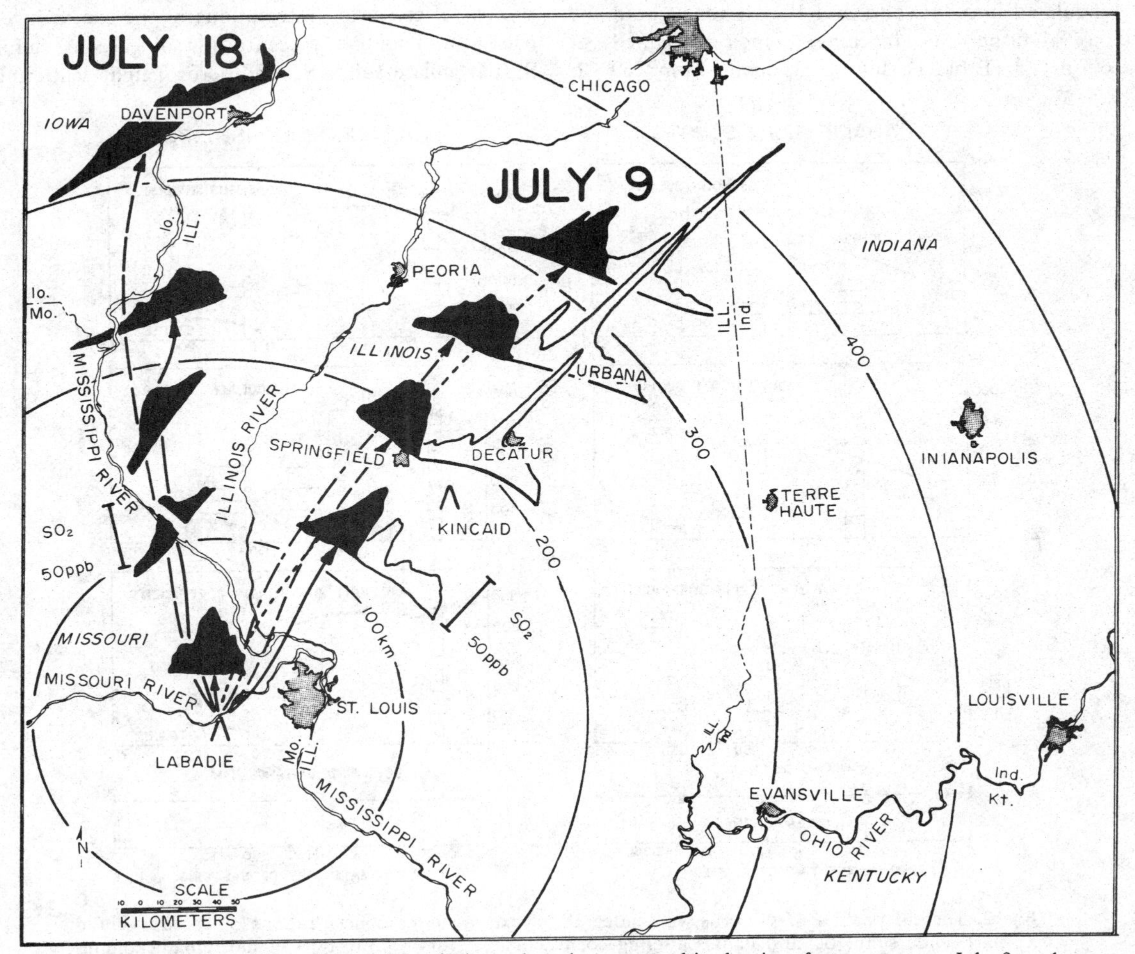

Fig. 1. Horizontal profiles of SO_2 during selected constant-altitude aircraft traverses on July 9 and July 18, 1976. July 9 traverses are at about 450 m AGL, and July 18 traverses are at about 750 m AGL. The Labadie plume sections are shaded. Also shown are backward trajectories for the Labadie plume.

range of transport, no meaningful analysis of the plume data can be performed without the support and use of adequate meteorological measurements (spatial and temporal variations of the wind field and atmospheric stability).

The transport of the Labadie plumes of July 9 and 18 are illustrated graphically in Fig. 1, which shows crosswind profiles of SO_2 concentration during selected plume traverses by the aircraft flying at a constant altitude. On July 9, the Labadie plume as well as other clearly identifiable St. Louis area plumes were mapped while being transported by south-westerly winds almost to the doorstep of Chicago. On July 18, the Labadie plume was followed up the Mississippi River valley into Iowa. Figure 1 also shows some of the backward plume trajectories computed by MRI (Smith *et al.*, 1978) at the appropriate approximate altitudes of the traverses shown.

On July 9, low altitude air flow patterns in the Mississippi River valley followed a trajectory from the central Gulf Coast region to a warm front moving through northern Illinois and southern Michigan. Therefore, warm and moist, and somewhat stable air circulated into Missouri and Illinois throughout the day. During the morning, winds remained at 6–8 m s^{-1} (Table 1) shifting gradually from 260° to 235°. During the afternoon and evening, the wind blew steadily from about 220° at speed increasing to about 20 m s^{-1} by midnight. A large cool continental high pressure area had spread over much of the eastern U.S. during the night preceding the July 18 mission. A stationary front, extending from Texas to Georgia, prevented the northerly flow of the warm, moist Gulf air. Thus, the background air mass of July 18 was significantly different from the warm, moist air of July 9. Prevailing winds were southerly in the morning and evening, shifting southeast during the afternoon as far as 145°. During the daytime, a well-mixed layer capped by an inversion extended as high as 2 km, and wind speeds were about 3–7 m s^{-1} (Table 2). In the late evening hours, winds of about 15 m s^{-1} blew near the sampling location in a very stable, stratified atmosphere. Directional wind shear of about 10°/100 m at plume level caused substantial lateral spread of the plume during the evening hours.

The vertical structure of the mixing layers on the two days is viewed most meaningfully by reconstructing the vertical spread of the plumes based on aircraft measurements (Fig. 2). The figure shows the vertical profiles of crosswind integrations of the excess (over background) plume concentration of gaseous sulfur S_g, particulate sulfur S_p, and aerosol light scattering

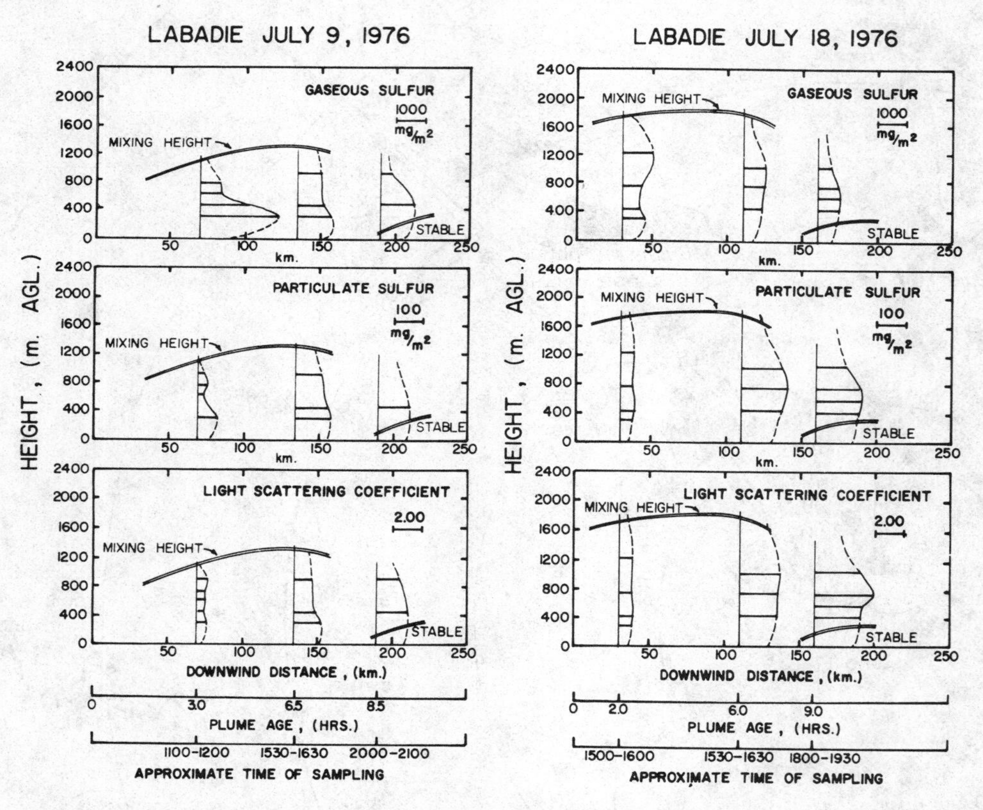

Fig. 2. Vertical profiles of the crosswind integrals of excess plume concentrations of gaseous sulfur S_g, particulate sulfur S_p, and light scattering coefficient b_{scat}, for the transport of the Labadie plume of July 9 and 18, 1976. Broken portions of the vertical profiles represent extrapolations. Vertical integrals of the profiles of S_g and S_p denote the masses M_g and M_p, respectively, of gaseous and particulate sulfur in vertical plume stabs of unit downwind depth. The region marked 'stable' represents the inversion layer forming near the ground during the evening.

coefficient b_{scat}. The mixing heights attained their maximum values of 1250 and 1800 m AGL* in the mid-afternoons of July 9 and 18 respectively. The plume of July 18, as indicated by the sulfur gas or the aerosol population, is seen to become well-mixed throughout the mixing layer much more rapidly than on July 9, The vertical stability of the atmosphere and its diurnal variation are shown plotted for both days in Fig. 3. The parameters used for the stability of the bulk mixing layer are $h/|L|$ for the unstable case and $\sqrt{u_*/fL}$ for the stable case, where h is the height of the mixing layer and L is the Monin–Obukhov length, u_* is the friction velocity and f is the Coriolis parameter. Details of the determination of these parameters are given by Gillani (1978). The figure shows that during the daytime hours, plume transport was under strong to moderately unstable conditions on July 18, and under weakly unstable conditions on July 9. The night-time atmosphere was stratified stable on both occasions.

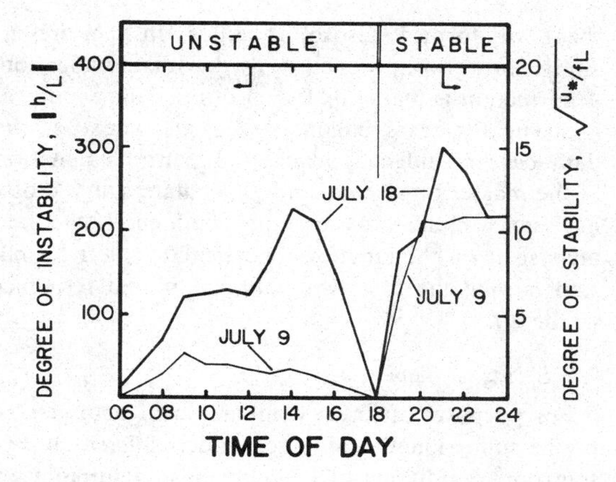

Fig. 3. Comparison of atmospheric stability on July 9 and July 18 based on data monitored at RAMS site 118 and the aircraft vertical sounding data. Prior to 1800 CDT, the stability parameter value is a measure of mixing layer instability; thereafter, the value is a measure of stability.

3. KINETICS OF PARTICULATE SULFUR FORMATION

Figure 2 shows clearly that there was a substantial formation of particulate sulfur as well as of total light scattering aerosol during plume transport on July 9 and 18. Furthermore, the amounts of these aerosols formed were considerably greater on July 18 than on July 9. In contrast, the background levels of both were significantly higher on July 9 (Tables 1 and 2).

The problem of particulate pollutant formation is essentially one of kinetics, i.e. the rate at which the primary matter is converted to secondary matter and the rate at which it is removed from the atmosphere. The assumption made in some power-plant plume studies that total sulfur is conserved in the plume can lead to significant errors in the estimation of SO_2 conversion rates following plume touchdown, because of neglecting the ground removal. The effect of ground removal on the gas-to-particle conversion of atmospheric sulfur is of particular concern in long-range plume transport over vegetation during the daytime.

3.1. *The method of data analysis*

Four basic approaches to field program design and data analysis have generally been followed in the study of atmospheric transformations of SO_2: the tracer method (Newman *et al.*, 1971), the particulate-to-total sulfur concentration ratio (S_p/S_{tot}) method (Newman *et al.*, 1975), the sulfur mass balance method (Husar *et al.*, 1978), and the aerosol size spectrum kinetics method (Whitby, 1978). In the tracer method, the rates of changes of the primary and secondary matter relative to that of a conservative tracer substance such as sulfur hexafluoride are used to estimate the conversion and removal rates. The method is difficult to implement in practice, particularly during long-range transport. The sulfur ratio method requires the measurement of the excess plume concentrations S_g and S_p, averaged over any part of the plume (e.g. a crosswind traverse), and the subsequent evaluation of the ratio S_p/S_{tot}, where $S_{tot} = S_p + S_g$. The ratio is assumed to be invariant with respect to position in the plume. In general, changes in the ratio during plume transport do not provide a direct measure of plume sulfur conversion. The ratio may increase either due to an increase in S_p or due to a decrease in S_g, or due to a combination of the two. For July 9 and 18, the increase in S_p is substantial enough to ensure its significant contribution to any corresponding increase in the value of the ratio. The use of the ratio method in estimating sulfur conversion rates is most appropriate whenever ground removal of sulfur is negligible compared to its chemical conversion. Otherwise, the effect of ground removal must be specifically considered in some appropriate manner. In the third method, the sulfur mass balance at successive downwind plume cross-sections is performed explicitly in terms of airborne gaseous and particulate sulfur, and the total sulfur removed from the plume. This approach is the most fundamental one, and is particularly suitable for estimation of plume ground removal rates; however, it requires detailed characterization of the two-dimensional concentration distribution for both gaseous and particulate sulfur at each of several sampled plume sections. Such detailed measurements are often impractical at far downwind distances where the plume is very wide, and the accuracy of the mass determinations also generally diminishes with downwind distance. In the aerosol-size spectrum kinetic method, the growth of the aerosol volume concentration is interpreted as evidence of sulfur conversion. The assumption is made that a known fraction of

* The mixing height is obtained by inspection of the vertical soundings of temperature as well as turbulence and aerosol concentration.

the newly-formed aerosol is sulfur. In the present study, the ratio approach is used with modification. A correction is made for the effect of ground removal by using the mass balance technique whenever the data base is judged adequate to permit estimation of the masses of gaseous and particulate sulfur crossing a given plume cross-section. The focus of the data analysis is on the kinetics of particulate sulfur formation *beyond* the first two hours of plume residence in the air.

3.2. S_p/S_{tot} *vs plume age*

For purposes of direct comparison of data sets of plume sulfur kinetics obtained under different plume transport conditions, it is more meaningful to view the change in S_p/S_{tot} as a function of plume age rather than downwind distance of transport. Figure 4 shows plots of $\overline{S_p}/\overline{S_{tot}}$ vs plume age for July 9 and 18. The bar lines in the ratio indicate concentrations averaged over a crosswind plume traverse at a constant altitude. The percentage of $\overline{S}_p$ at age 2 h is about 4% of total on July 18, and 2% or less on July 9. Its value increases to a maximum of nearly 18% on July 18 after about 6 h of transport, and to about 12% on July 9, also after 6 or 7 h of transport. Curiously, the percentage of S_p then declines on both days. This decline does not necessarily indicate a net depletion of particulate sulfur after its formation, because the sampling generally does not follow a single air parcel in a Lagrangian manner. The data points refer to sampling of air parcels with independent time histories. On both days, the first 6 h of transport were during the daylight hours prior to 1700 CDT, when solar irradiance is significant, and the remaining hours were in the evening and at night. The daylight hours appear to be most conducive to aerosol formation and ground removal, while evening hours appear to be less conducive.

3.3. *The relation of conversion to ambient conditions*

Past field and laboratory research on SO_2 oxidation rates have implicated temperature, humidity, solar radiation, in-cloud residence time, catalytic heavy metals, soot, ozone, ammonia, hydrocarbons, free radicals and various other factors as rate-controlling in the conversion kinetics. No single factor is dominant under all atmospheric conditions. The identification of a rate of conversion is of particular value when it is related to the particular prevailing conditions.

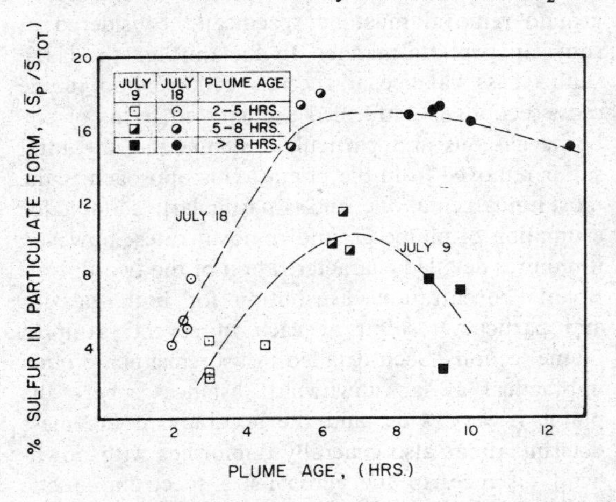

Fig. 4. The changing ratio of particulate-to-total sulfur concentrations (averaged for time of plume traverse) as a function of plume residence time in the atmosphere.

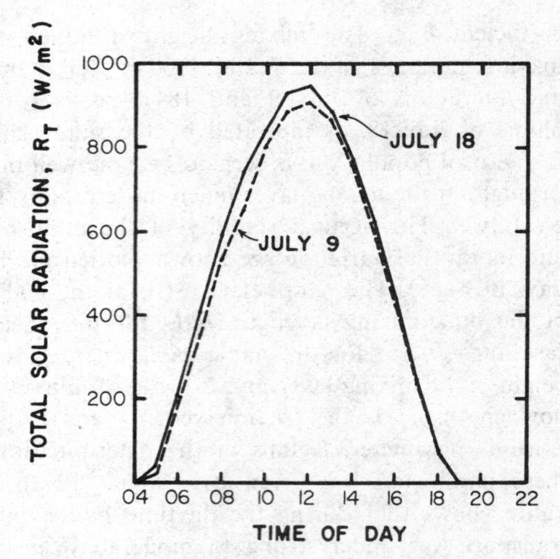

Fig. 5. Temporal variations of the incoming total (direct and diffuse) solar radiation flux, R_T, in the spectral range 3950–7000 Å as measured at RAMS site 118.

Observations of ground-level relative humidity, monitored at RAPS station 118, ranged between 36 and 53% on July 9, and between 19 and 33% on July 18 during the maximum SO_2 conversion hours of 1100–1700 CDT. During the hours of interest, the absolute temperatures remained consistently about 3°C higher on July 9, reaching a maximum of 33°C during the afternoon. The dew-point temperature on July 9 remained between 20 and 23°C throughout the day, and was between 5 and 10°C above the corresponding values for July 18. Background b_{scat} as measured in the aircraft, was about three to four times higher on July 9 than on July 18 (Tables 1 and 2). Thus, the day with the higher conversion rate was also the cooler, less humid and less hazy day.

The poor correlation between % S_p and plume age over the full range of transport, and the higher rate of increase of the particulate sulfur fraction during daytime transport motivates an investigation of the role of the cumulative solar exposure of each plume parcel in sulfur conversion kinetics. Figure 6 shows the result of plotting % S_p vs the cumulative dose, D_R, of solar radiation to which each air parcel has been exposed during its transport up to the point of sampling. The actual time variation of the solar radiation for the two days is shown in Fig. 5. The solar radiation data represent the hourly average of the total (direct + diffuse) solar radiation intensity, R_T, in the spectral range of 3950 Å to 7000 Å, measured with a pyranometer at one of the RAMS

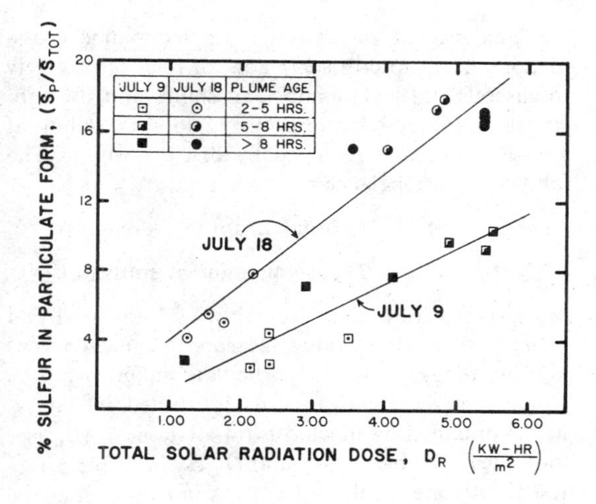

Fig. 6. Ratio of particulate-to-total sulfur concentrations (average for traverse) as a function of total plume solar radiation dose, D_R (time integral of R_T between plume release time and plume sampling time). The least square regression lines shown have the following equations:

July 9: $y = 2.05x - 0.89, r = 0.92$
July 18: $y = 3.22x + 0.75, r = 0.96$.

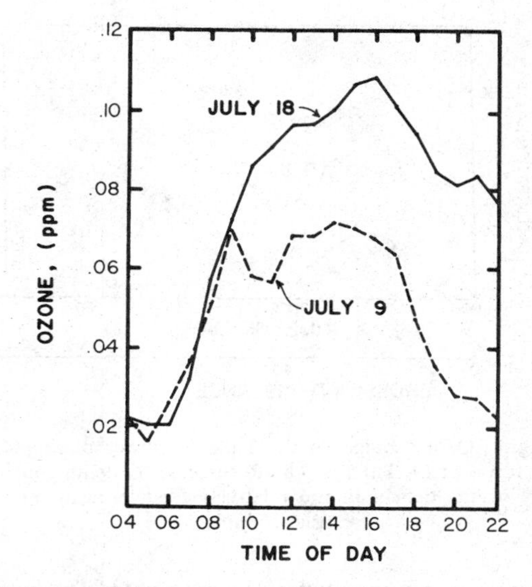

Fig. 7. Comparison of ozone concentrations for July 9 and July 18 as recorded at RAMS site 118.

sites in St. Louis (Station 118, a non-urban site to the south of St. Louis).

Figure 6 shows that a linear relationship exists between % S_p content and the radiation dose of the sampled plume air parcels on both days. There was little significant difference in the insolation (R_T) for the two days. For a given radiation dose D_R, the particulate sulfur fraction in the plume was higher on July 18 than on July 9. The highest hourly rates of increase of the particulate fraction of total sulfur implied by the regression lines of Fig. 6 are about 1.8% on July 9 and 3.2% on July 18, and occur around the noon hour. Since total sulfur is not necessarily conserved in the plume, these percentages represent the upper limits of sulfate formation rates.

The production of aerosols has been observed to be enhanced near the edges of a power-plant plume (Wilson *et al.*, 1976), presumably as a result of reactions with the entrained background air. This suggests a role of background ozone in the oxidation of SO_2, since there is commonly more ozone near plume edges than at the centerline. Figure 7 shows the time variation of ozone levels monitored at RAMS station 118 on July 9 and 18. On both occasions, this station was well upwind of the St. Louis urban area, and at least during the midday hours when the boundary layer is well-mixed, the ozone data may be considered representative of the background air mass within which the Labadie plume mixed. The July 9 and 18 ozone data of station 118 are similar to those recorded at other non-urban stations of the RAMS network. The air mass ozone concentration remains substantially higher at all times on July 18 as compared to July 9. The ozone comparison is presented for ground level rather than aircraft data because the ozone monitor in the WU aircraft was inoperative on July 18.

In addition to the entrainment of ozone from the air mass, a net production of ozone within the Labadie plume is also observed on both days of interest. As shown in Fig. 8, there is unmistakable ozone bulge in the 9-h old Labadie plume of July 9 at 190 km. By comparison, there is still a strong ozone deficit in the fresh plume of the Kincaid power-plant near Springfield, Illinois (Fig. 1). On July 9, ozone deficits persisted in the Labadie plume up to about 100 km, but substantial ozone production within the plume was observed thereafter. On July 18, relevant MRI aircraft data show that ozone bulges appeared at every plume traverse as close as 30 km from the source, in the 2- to 3-h old plume.* Ozone production in power-plant plumes beyond 30 km has previously been reported by Davies *et al.* (1974). Excess plume ozone has been shown to occur almost simultaneously with sulfate formation in the St. Louis urban plume (White *et al.*, 1976). Evidently, ozone and the formation of sulfate belong to the same chemical system.

Another quantity of interest which has been analyzed is atmospheric mixing. As evidenced both by the rapid vertical spread of the pollutants (Fig. 2) and by the magnitude and temporal behaviour of the stability parameter (Fig. 3), atmospheric mixing was much more intense on July 18. Atmospheric mixing is responsible for the entrainment of air mass constituents into the plume air. For power-plant plumes, the entrainment of background air may contribute constituents such as ozone and hydrocarbons, or soot particles, which can enhance the sulfur conversion

* A more detailed description of observations of ozone formation in the Labadie plume during the 1976 MISTT program will be presented in a separate paper presently under preparation.

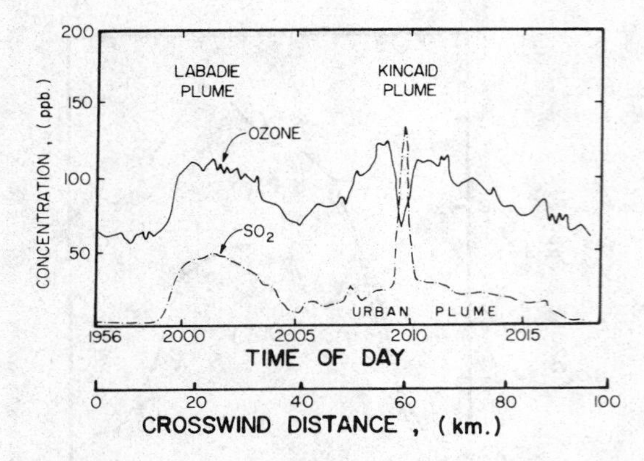

Fig. 8. Ozone 'bulge' in the Labadie power plant plume of July 9 at 190 km and 9 h after release. Note, in contrast, the ozone 'deficit' in the relatively fresh Kincaid power-plant plume.

process within the plume. It is also noted that under the better mixing conditions of July 18, the transition from ozone deficit to ozone bulge in the Labadie plume occurred in less than 2 h, while such a transition occurred only after about 5 h of plume transport under the relatively stable conditions of July 9. On both days, the first indication of ozone production within the plume was observed in the afternoon hour around 2 p.m.

3.4. *Relative magnitudes of* SO_2 *oxidation and dry deposition*

The flat terrain over which the plume was transported is predominantly covered with low vegatation, which is a known sink for SO_2 (Hill, 1971). Thus, the increase in the ratio $(\overline{S}_p/\overline{S}_{tot})$ with increasing exposure to sunlight is, at least partly, due to sulfur uptake by the vegetation. In order to determine the rates of SO_2 conversion, it is therefore necessary to separate the effect of the ground removal of total sulfur by performing the sulfur mass balance at each sampled plume section. Table 3 summarizes the sulfur mass balance at successively sampled plume cross-sections. The subscripts g, p and tot refer respectively to gaseous, particulate and total sulfur, and the subscripts 0 and R refer respectively to total sulfur at emission and that removed by dry deposition. The following mass balance relations hold:

$$M_{tot} = M_g + M_p \text{ (total airborne sulfur mass)}$$

$$M_0 = M_{tot} + M_R \text{ (total sulfur at emission).}$$

M_g and M_p are obtained directly by horizontal and vertical integration of the measured plume concentrations of gaseous and particulate sulfur over the plume cross-sectional area, as illustrated in Fig. 2. M_0 is obtained by upwind extrapolation of M_{tot} (see footnotes of Table 3). M_0 and Q_0/U_0 in Table 3 represent estimates of the sulfur emission mass obtained in two different ways: M_0 is based on aircraft measurements and Q_0/U_0 is based on emission data supplied by the power plant utility company. Table 3 also shows the fractions of the sulfur emission mass which have undergone gas-to-particle conversion (M_p/M_0) and ground removal (M_R/M_0). M_R in our analysis refers to the ground removal of total sulfur. Of this, only a small fraction may be expected to be for particulate sulfur. Whitby (1978) has observed that the sulfate aerosol formed in the Labadie plume is predominantly in the submicron size range. The mean deposition velocity of particulate sulfur in this size range is estimated to be about 0.025 cm s^{-1}, or about 3% of the corresponding value for SO_2 estimated to be about 0.8 cm s^{-1} (Garland, 1978). Furthermore, the observed concentration of particulate sulfur is much less than that of gaseous sulfur. Hence, the mesoscale ground loss of sulfur from the Labadie plume is believed to be mostly due to the uptake of SO_2 at the ground.

On July 9, maximum sulfur conversion is about 10% at the last daylight sampling (1600 CDT, 135 km), and ground removal attains a value of about 23% at about 200 km. On July 18, maximum conver-

Table 3. Sulfur budget in the Labadie plume

		Sulfur mass* (kg m^{-1})					Mass ratios	
		At sampling section			At emission		Conversion	Removal
Date	Downwind distance (km)	Gas M_g	Aerosol M_p	Total M_{tot}	M_0	Q_0/U_0†	M_p/M_0 (%)	M_R/M_0 (%)
7/9/76	70	1.062	0.032	1.094	1.15‡	0.96	2.8	5.0
	135	0.880	0.115	0.995	1.18§	0.98	9.7	15.7
	190	0.829	0.075	0.904	1.18§	0.98	6.3	23.4
7/18/76	30	1.335	0.070	1.405	1.45¶	1.28	4.8	3.0
	110	1.000	0.200	1.200	1.45¶	1.28	13.8	17.2
	160	0.873	0.188	1.061	1.45¶	1.28	13.0	26.8

* The error margin on sulfur mass values is estimated at about ±25%.

† Q_0 and U_0 are, respectively, emission rate of sulfur (kg s^{-1}) and wind speed (m s^{-1}) at effective stack height at plume release time. Q_0 ranged between 6.7 and 6.875 kg s^{-1} for July 9 and was nearly constant at 5.1 kg s^{-1} on July 18.

‡ Based on assumption of 5% ground removal of total sulfur between source and 70 km.

§ Obtained by scaling previous value (1.15) according to increase in emission rate from 6.7 to 6.875 kg s^{-1} at corresponding release times.

¶ Based on assumed ground removal of 3% of the total sulfur between source and 30 km.

sion is about 14% after 6 h (1600 CDT, 110 km), and ground removal exceeds 26% in 160 km. Both conversion and ground removal of sulfur are thus seen to be greater on July 18. The higher ground removal rate of July 18 may be attributed to the corresponding better mixing which brings about more efficient delivery of the pollutant to the ground.

In recent years, results of several aircraft measurement programs aimed at investigation of plume sulfur conversion and removal kinetics have been reported. However, field studies of power plant plumes have generally neglected ground losses of SO_2. Total SO_2 removal rate estimates are only available for urban plumes (Breeding *et al.*, 1976; Alkezweeny and Powell, 1977) and for regional plumes (Smith and Jeffrey, 1975).

In aircraft plume studies aimed at estimating the sulfur kinetic parameters, no direct measurements of SO_2 uptake at the ground are usually made. The use of indirect methods must therefore be invoked in order to assess the effect of ground loss. If both gaseous and particulate concentrations are measured at discrete downwind plume sections, then the ground loss may be estimated by determining the decrease in total sulfur mass between successive plume sections. If the observed data are insufficient to permit estimation of the sulfur masses at the sampled plume sections, then assumptions must be made regarding the vertical distribution of the mass.

In this paper, we use the sulfur mass results given in Table 3 in a simple box-model analysis to separate the relative contributions of sulfur conversion and ground removal in the ratio $\bar{S}_p/\bar{S}_{tot}$, which includes both effects. We define $\bar{S}_0$ as the average concentration of total sulfur for any given traverse if ground removal were zero. A plot of $\bar{S}_p/\bar{S}_0$ would then provide a direct representation of the formation of particulate sulfur by the oxidation of SO_2 in the plume (i.e. 'pure' conversion). Plume transport between 70 and 190 km on July 9, and between 30 and 160 km on July 18, is largely during the daytime hours of good mixing, and the approximation of uniform vertical plume profiles of S_g and S_p is reasonably valid. Based on this approximation,

$$\frac{M_p}{M_0} \approx \frac{\bar{S}_p \cdot A}{\bar{S} \cdot A} = \frac{\bar{S}_p}{\bar{S}_0} = r_0, \text{ say}$$

and

$$\frac{M_p}{M_{tot}} \approx \frac{\bar{S}_0 \cdot A}{\bar{S}_{tot} \cdot A} = \frac{\bar{S}_0}{\bar{S}_{tot}} = r_T, \text{ say}$$

where A denotes the cross-sectional area of the well-mixed plume. Hence,

$$r_0 \approx \frac{M_{tot}}{M_0} \cdot r_T.$$

Thus, the factor (M_{tot}/M_0) is the correction factor to be applied to r_T in order to obtain the ratio r_0, which characterizes sulfur conversion alone. No specific assumptions regarding mixing height or deposition velocities are necessary in this approach. The ratios

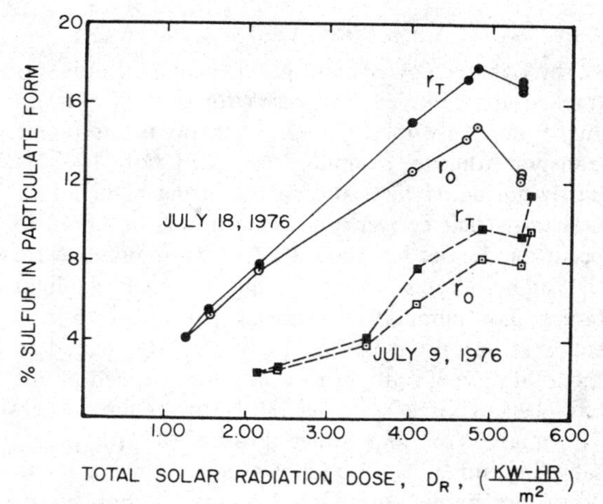

Fig. 9. Relative contributions of sulfur gas-to-particle conversion (r_0) and ground removal ($r_T - r_0$) to the ratio of plume particulate-to-total sulfur (r_T).

r_0 and r_T are both plotted simultaneously as functions of solar radiation dose, D_R, in Fig. 9. The difference between r_T and r_0 corresponds to the ground removal contribution to r_T. It is observed that this difference increases with D_R on both days, indicating the dependence of ground uptake on insolation as well as transport time. More importantly, it is seen that even when the magnitude of ground uptake of sulfur is greater than that of gas-to-particle conversion (Table 3), its contribution to r_T represented by $(r_T - r_0)$ is smaller than that due to conversion (represented by r_0), at least for mesoscale transport under conditions similar to those considered. The correction due to ground removal, however, is quite significant, and increasingly so for longer ranges of plume transport. The principal source of error in the sulfur mass method arises in the vertical integration of the crosswind integrals of plume sulfur concentrations, since these values are usually available at only three or four discrete height levels. On July 9 and 18, these integrals were evaluated only for those plume sections where multiple traverses were made. The procedure is first to complete the vertical profile of S over the entire mixing layer by extrapolation near the upper and lower boundaries. As a guide in this step, the observed vertical profiles obtained during vertical aircraft soundings through the plume section are utilized. For the data shown in Fig. 2, the average error in the integrations is estimated to be at least 25%, and it is the largest source of error in the sulfur budget estimate.

After making the above corrections for ground removal, and assuming all of the primary emission of sulfur to be as gas, it is found that the average estimated rate of particulate sulfur formation during the first 6 h of transport is $1.6 \pm 0.5\%\ h^{-1}$ on July 9 and $2.4 \pm 0.8\%\ h^{-1}$ on July 18. For the 10–12 h of transport ending at midnight, the average rate of SO_2 conversion is estimated at under $1\%\ h^{-1}$ on July 18 and under $0.3\%\ h^{-1}$ on July 9.

4. CONCLUSIONS

The Labadie power plant plume near St. Louis was tracked and sampled by aircraft for over 300 km on July 9 and July 18, 1976. On both occasions plume transport during sampling extended for 10–12 h, partly during daylight and partly during night hours. Gas-to-particle conversion of sulfur was observed to occur at rates of less than $3\%\ h^{-1}$ at all times, with the highest rates corresponding to peak sunlight hours. The cumulative ratio of particulate-to-total sulfur at any point is seen to be linearly related to the total dose of sunlight to which the sampled plume has been exposed since its release from the source. The analysis of nine other Project MISTT plume sampling missions also shows SO_2 conversion to occur preferably during the daytime (Husar *et al*., 1978). Informed readers of this paper will be inclined to interpret the above data of SO_2 conversion as evidence favoring photo-oxidation as the rate-controlling mechanism. While our own inclination is also in that direction, the amount of data analysis performed so far is insufficient to rule out the other possible mechanisms.

For the two days of data reported, temperature, dew-point and relative humidity were lower on the day with the higher observed rate of conversion. The ozone level and the intensity of atmospheric mixing were higher on the day of more rapid conversion. A net production of ozone within the plume was observed on both days. Ground removal of SO_2 was greater than the conversion on both days, particularly for plume transport beyond 100 km, and caused a significant reduction in particulate sulfur formation. The relative magnitudes of sulfur conversion and removal are compared. It is suggested that the transport of power-plant plumes over long distances during summer in the U.S. midwest may often be accompanied by substantially more sulfate formation during the daylight hours as compared to night hours.

Acknowledgements—The authors wish to express their appreciation to Dr. T. B. Smith of MRI for his helpful meteorological consultation, to Mr. John Wooten of the Union Electric Co. for supplying data related to the plant operating conditions, and to the U.S. EPA for making the RAPS data available to us. This research was supported (in part) by the Federal Interagency Energy/Environment Research and Development Program through Environmental Protection Agency Grant No. R803896.

REFERENCES

Alkezweeny A. J. and Powell D. C. (1977) Estimation of transformation rate of SO_2 to SO_4 from atmospheric concentration data. *Atmospheric Environment* **11,** 179–182.

Blumenthal D. L., Ogren J. A. and Anderson J. A. (1978) Airborne sampling system for Project MISTT. *Atmospheric Environment* **12,** 613–620.

Breeding R. J., Klonis H. B., Lodge J. P. Jr., Pate J. B., Sheesley D. C., Englert T. R. and Sears D. R. (1976) Measurements of atmospheric pollutants in the St. Louis area. *Atmospheric Environment* **10,** 181–194.

Davis D. D., Smith G. and Klauber J. (1974) Trace gas analysis of power plant plumes via aircraft measurements: O_3, NO_x, SO_2 chemistry. *Science* **186,** 733–736.

Forrest J. and Newman L. (1977) Further studies on the oxidation of sulfur dioxide in coal-fired power plant plumes. *Atmospheric Environment* **11,** 465–474.

Garland J. A. (1978) Dry and wet removal of sulfur from the atmosphere. *Atmospheric Environment* **12,** 349–362.

Gillani N. V. (1978) Project MISTT: Mesoscale plume modeling of the dispersion, transformation and ground removal of SO_2. *Atmospheric Environment* **12,** 569–588.

Hill A. C. (1971) Vegetation: A sink for atmospheric pollutants. *J. Air Pollut. Control Ass.* **21(6),** 341.

Husar R. B., Patterson D. E., Husar J. D., Gillani N. V. and Wilson W. E. (1978) Project MISTT: sulfur budget in large plumes. *Atmospheric Environment* **12,** 549–568.

Myers R. L. and Reagan J. A. (1975) The Regional Air Monitoring Systems, St. Louis, Missouri, U.S.A. Paper presented at the International Conference on Environmental Sensing and Assessment, Las Vegas, September 15–19, 1975.

Newman L., Manowitz B. and Tucker W. D. (1972) Two nuclear techniques for stack plume studies: Isotope ratio and sulfur hexafluoride. *Trans. Am. nucl. Soc. Meeting*, Las Vegas, Nevada.

Smith F. B. and Jeffrey G. H. (1975) Airborne transport of sulfur dioxide from the U.K. *Atmospheric Environment* **9,** 643–659.

Smith T. B., Blumenthal D. L., Anderson J. A. and Vanderpol A. H. (1978) Transport of SO_2 in power plant plumes: day and night. *Atmospheric Environment* **12,** 605–611.

White W. H., Anderson J. A., Blumenthal D. L., Husar R. B., Gillani N. V., Husar J. D. and Wilson W. E. Jr. (1976) Formation and transport of secondary air pollutants: ozone and aerosols in the St. Louis urban plume. *Science* **194,** 187–189.

Wilson W. E., Charlson R. J., Husar R. B., Whitby K. T. and Blumenthall D. L. (1976) Sulfates in the atmosphere. Paper presented at the 69th Annual Meeting of A.P.C.A., Portland, ORE, June 27–July 1, 1976.

Wilson W. E. (1978) Sulfates in the atmosphere: A progress report for Project MISTT. *Atmospheric Environment* **12,** 537–547.

Whitby K. T. (1978) Physical properties of sulfur aerosols. *Atmospheric Environment* **12,** 135–159.

Atmospheric Environment Vol. 12. pp. 599–604. Pergamon Press 1978. Printed in Great Britain.

A CHEMICAL MODEL FOR URBAN PLUMES: TEST FOR OZONE AND PARTICULATE SULFUR FORMATION IN ST. LOUIS URBAN PLUME

I. S. A. ISAKSEN, E. HESSTVEDT and Ö. HOV

Institute of Geophysics. University of Oslo, Norway

(*First received* 14 *June* 1977 *and in final form* 24 *August* 1977)

Abstract—A chemical model with anthropogenic sources of nitrogen oxides and hydrocarbons is applied to simulate the chemical behaviour of pollutants in the St. Louis urban plume. It is suggested that a substantial increase in peroxy radical concentrations (HO_2, $R\dot{O}_2$) in the polluted air mass outside its source region leads to an effective formation of secondary pollutants like ozone and sulfate particles. The model indicates characteristic time for ozone generation in the plume of a few hours. Maximum ozone mixing ratio of 115 ppb is predicted after 4 h transport time outside the source region. Conversion rates of SO_2 to H_2SO_4 through gas phase reactions with hydroxyl and peroxy radicals are estimated to be 1–5% h^{-1}. This leads to an approx 25% conversion of SO_2 to particulate sulfur in the plume during the day. Agreement with measured ozone concentrations and flow rates of ozone and particulate sulfur in the St. Louis plume on 18 July 1975 can be taken as strong indications that ozone and sulfate particle formation in the plume proceeds through the suggested mechanisms.

INTRODUCTION

Several recent measurements of ozone and sulfate concentrations in polluted air masses, outside the source regions, show a strong correlation between elevated ozone and sulfate particle concentrations. White *et al.* (1976) performed measurements from an aircraft in the St. Louis urban plume downwind of the urban area. Pronounced increase in both ozone and particulate sulfur concentrations was observed in the plume during the first few hours after passage over the city, whereafter the concentrations of these pollutants remained at high levels for several hours. Atkins *et al.* (1972) observed enhanced levels of sulfate over England (Harwell) during periods with meteorological conditions which favoured photochemical formation of oxidants. Grennfelt (1976) found strong correlation between high concentrations of ozone and sulfate in the south-western part of Sweden during episodes when polluted air masses arrived from remote industrial areas.

All these observations can be taken as evidence that analogue chemical reaction mechanisms are involved in the conversion of SO_2 to sulfate and in the generation of ozone in polluted air masses. Further support is obtained from a smog chamber experiment performed by Cox and Penkett (1971) where SO_2 was mixed with nitrogen oxide and hydrocarbons (*cis*-2-pentene) and irradiated with u.v. radiation. The conversion rate of SO_2 to sulfate was shown to depend strongly on the presence of NO_x and hydrocarbons, which are also the main precursors for ozone in smog situations. Conversion rates of approximately 10% h^{-1} were obtained in their experiment; this is considerably faster than usually observed in the free atmosphere (Eliassen and Saltbones, 1975).

The observations from St. Louis, on 18 July 1975, already mentioned, involve data for the extent of the plume and for ozone and particulate sulfate at different positions downwind of the source. In this paper we present a theoretical simulation of this case. On the basis of our calculation of the generation of secondary pollutants in the plume, we are able to calculate rates of conversion of SO_2 to sulfate due to photochemical processes.

PLUME MODEL

We assume that the gases are mixed instantly within the plume whose mixing height is limited by the inversion layer and the Earth's surface. The time variation of the average concentration C of a gas in an air parcel transported along in the plume is obtained from the following continuity equation

$$\frac{dC}{dt} + \alpha(C - C_0) = P_c + P_s - L_c \cdot C - L_a \cdot C. \quad (1)$$

α is a factor expressing the dilution of the plume due to mixing with outside air where the concentration of the compound in question is C_0. Observations of the St. Louis urban plume show that the width of the plume remained nearly constant throughout the day. Thus, horizontal mixing is negligible. On the other hand, a marked deepening of the plume occurred. The mixing height (h) increased from approx 300 m in the morning (9.00 a.m.) to more than 1 km in the afternoon. Accordingly, we will assume that the dilution of a gas is determined solely by the increase in mixing height, which leads to the following expression

$$\alpha = \frac{1}{h}\frac{dh}{dt}. \quad (2)$$

Based on the observations in this actual case dh/dt can be approximated by a constant value 3.6 cm s^{-1} between 9.00 a.m. and 5.00 p.m. This leads to values of α in the range 0.3–1.1 $\times$ 10^{-4} s^{-1}, low enough to consider α as constant during each time step (Δt = 30 s). Before 9.00 a.m. and after 5.00 p.m., dh/dt is taken as zero. P_c and $L_c \cdot C$ represent photochemical production and loss terms. P_s expresses the emission rate of the primary pollutants: hydrocarbons (HC), nitrogen oxides (NO_x), carbon monoxide, and sulfur dioxide. A HC mixture thought to be representative for urban emissions is adopted, given by 10% as CH_4, 20% as C_2H_2, 20% as C_2H_4, 10% as C_3H_6, 10% as C_4H_{10}, 10% as C_6H_{14}, and 20% as C_8H_{10}. The release of NO_x is assumed to consist of 99% NO and 1% NO_2. The total release of non-methane hydrocarbons in the St. Louis area is estimated to be 36 ton h^{-1} during morning hours when the air parcel we consider passes over the urban area.

The chemical activity in the plume depends on the ratio between the abundance of HC and NO_x in the emissions (Hesstvedt *et al.*, 1977b). In these calculations a ratio equal to 2 (by vol.) for the HC/NO_x release is adopted. This is found to be representative values for urban areas (Roth *et al.*, 1974; OECD, 1976). The rates of release of CO and SO_2 are assumed to be ten times, and twice, the rate of release of NO_x (by vol.) in the urban area. These assumptions are discussed more thoroughly elsewhere (Hov *et al.*, 1977).

We assume that the sources of pollutants are homogeneously distributed over the urban area. The production term P_s of a compound due to emission is obtained from the expression

$$P_s = \frac{\Psi_{\text{poll}}}{A \cdot h_0} \tag{3}$$

where Ψ_{poll} denotes the total rate of release of the compound. The area A is equal to 2000 km^2 (40 $\times$ 50 km). The mixing height h_0 is constant and equal to 300 m during the period when the air parcel we consider passes over the urban area.

If we adopt a wind speed of 5 m s^{-1} and a 50 km traverse over the urban area, the air parcel is exposed to pollutant emissions for a period of 3 h, between 6.00 a.m. and 9.00 a.m.

The term $L_a \cdot C$ represents the loss due to absorption on the ground. L_a is given as

$$L_a = \frac{v_a}{h} \tag{4}$$

where v_a is the absorption velocity at the Earth's surface.

In these calculations we assume that SO_2, NO_2 and O_3 are removed through ground absorption. Measurements of the absorption velocity for SO_2 over the British Isles for different vegetation types give average values somewhat less than 1 cm s^{-1} (Owers and Powell, 1974; Smith and Jeffrey, 1975). We therefore regard the use of an absorption velocity equal to 0.7 cm s^{-1} for SO_2 to be justified. The absorption of NO_2 by vegetation is assumed to be less effective and a value of v_a = 0.3 cm s^{-1} is adopted for NO_2.

It is important to note that a proper determination of the absorption velocity of NO_2 is not critical for the determination of NO_2-concentrations, because in the present model the loss of NO_x is dominated by the gas-phase conversion of NO_2 to HNO_3. This is in contrast to the loss mechanisms of SO_2, where deposition on the ground may dominate and thus indirectly determine the sulfate formation.

The absorption velocity of ozone is set equal to 0.4 cm s^{-1} in agreement with experimental values obtained by Aldaz (1969). This is somewhat on the lower side of recent values obtained by Garland and Penkett (1976), but again ozone loss in the plume is dominated by photochemical reactions, and the resulting ozone concentrations will therefore not depend critically on the absorption velocity adopted. Furthermore, the assumption of complete vertical mixing in the plume, where the averaged concentration is thought to represent the value at the ground level, leads to an overestimation of ground absorption.

NUMERICAL PROCEDURE

The continuity equation (1) may be rewritten as

$$\frac{dC}{dt} = P - L \cdot C \tag{5}$$

where P and $L \cdot C$ are the production and loss terms.

The numerical integration procedure applied to solve (5) is a version of a QSSA (quasi-steady state approximation) method. It is described in detail elsewhere (Hesstvedt *et al.*, 1977a). The main line of reasoning goes as follows:

If P and L are assumed to be constant over the timestep Δt, (5) can be integrated to give

$$C_{t+\Delta t} = \frac{P}{L} + \left(C_t - \frac{P}{L}\right) e^{-L\Delta t}. \tag{6}$$

The quantity $\tau = 1/L$ denotes the timescale of variation or lifetime of the component. According to the length of the lifetime, all components in the model are divided into three groups.

(1) If $\tau \ll \Delta t$, $C_{t+\Delta t}$ is calculated according to

$$C_{t+\Delta t} = \left(\frac{P}{L}\right)_{t+\Delta t} \tag{7}$$

and the compound is assumed to be in instant equilibrium with all other species. In this case the photochemical terms are the dominating ones in P and L.

(2) If $\tau \gg \Delta t$, the variation of the concentration of the component is very slow and close to linear from

timestep to timestep, and $C_{t+\Delta t}$ is calculated according to the simple Euler expansion formula

$$C_{t+\Delta t} = C_t + (P - L \cdot C_t)\Delta t. \quad (8)$$

(3) If $\tau \approx \Delta t$, the complete expression (6) is utilized to compute $C_{t+\Delta t}$. The lifetime of any compound is in principle checked at each timestep to see whether the classification according to 1–3 changes or not.

There are strong couplings between key compounds, in particular O_3, NO, and NO_2. Such couplings cause severe numerical instability problems when a timestep as long as e.g. 30 s is applied. To get around such problems, we introduce several substitutions whereby the system of equations undergoes linear transformations. These substitutions are described in detail elsewhere (Hesstvedt *et al.*, 1977a).

To increase the computational accuracy, an iteration procedure is deviced to secure that all short-lived compounds are in mutual equilibrium at any timestep. In order to test the accuracy of our numerical scheme, a series of computational runs have been made (Hesstvedt *et al.*, 1977a), where the results from this method are compared with the generally accepted Gear's method (Gear, 1971; Whitten and Meyer, 1976), which is an automatic procedure designed to solve sets of stiff differential equations. Comparisons were made with a propylene–ethylene–NO_x–air mixture, with a reaction scheme, including about 100 steps and more than 50 compounds. Conditions with respect to initial values, solar radiation and emission fluxes are varied over a range of cases thought to cover most situations encountered in atmospheric pollution problems. The results differ only 1–2% or less in models relevant to atmospheric studies of the kind discussed here: plume models where an air parcel is exposed to emission of pollutants whereafter the chemical development is studied for a period of e.g. 12 h.

OZONE AND SULFATE GENERATING MECHANISMS

Generation of ozone in polluted air is determined by the rate at which nitric oxide is converted to NO_2 without consumption of odd oxygen. This conversion is known to take place through reactions with HO_2 and organic peroxy radicals

$$NO + HO_2 \xrightarrow{k_1} NO_2 + OH \quad (R1)$$

$$NO + R\dot{O}_2 \xrightarrow{k_2} NO_2 + R\dot{O}. \quad (R2)$$

This leads to formation of ozone through the reactions

$$NO_2 + h\nu \xrightarrow{k_3} NO + O \quad (R3)$$

$$O + O_2 + M \xrightarrow{k_4} O_3 + M. \quad (R4)$$

Recent measurements by Cox *et al.* (1976) indicate that reactions (R1) and (R2) (with methyl peroxide) proceed with rate constants considerably faster ($k = 1.3 \times 10^{-12}\,cm^3\,molecule^{-1}\,s^{-1}$) than previously used (Hampson and Garvin, 1975) leading to a more effective generation of ozone in polluted air. In this model we adopt the rate constants given by Cox *et al.* (1976) for NO reacting with HO_2 and $CH_3\dot{O}_2$, but apply the rate constant $k_2 = 2.9 \times 10^{-13}\,cm^3\,molecule^{-1}\,s^{-1}$ (Demerjian *et al.*, 1974) for the reaction of NO with higher peroxy radicals. HO_2 and $R\dot{O}_2$ are formed at various stages in the decomposition chains of the hydrocarbons, initiated mainly through reactions of the type

$$HC + OH \xrightarrow{k_5} Products. \quad (R5)$$

A detailed discussion on how the peroxy radical concentrations are determined in our model is given in two papers (Hesstvedt *et al.*, 1977b; Hov et al., 1977). The scheme adopted for the decomposition of paraffins and olefins is largely based on the work by Demerjian *et al.* (1974).

Rather high concentrations of aromatic compounds are found in urban air (Calvert, 1976), and in the paper by Hov *et al.* (1977) we propose a scheme for the decomposition of aromatic compounds where approx 6 peroxy radicals are formed through the decomposition of one HC-molecule. Thus, aromatic compounds contribute heavily to the ozone formation.

All chemical reactions and rate constants adopted in the model are given in the paper by Hov *et al.* (1977).

Gas-phase conversion of SO_2 to sulfate is thought to take place through reactions with free radicals

$$SO_2 + HO_2 \xrightarrow{k_6} SO_3 + OH \quad (R6)$$

$$SO_2 + R\dot{O}_2 \xrightarrow{k_7} SO_3 + R\dot{O} \quad (R7)$$

$$SO_2 + OH(+M) \xrightarrow{k_8} HSO_3(+M). \quad (R8)$$

A rate constant of $k_6 = 1 \times 10^{-15}\,cm^3\,molecule^{-1}\,s^{-1}$ has been measured by Payne *et al.* (1973), while Castleman *et al.* (1975) give the value of $8 \times 10^{-13}\,cm^3\,molecule^{-1}\,s^{-1}$ for k_8. Preliminary results of laboratory measurements for the reaction with $CH_3\dot{O}_2$ obtained by Calvert (private communication, 1977) indicate a value approx three times faster than the one reported for reaction (R6) ($k_7 = 3.3 \times 10^{-15}\,cm^3\,molecule^{-1}\,s^{-1}$). Sander and Seinfeld (1976) suggest rate constants for other types of free radical reactions similar to k_6.

The formation of SO_3 through reactions (R6) and (R7) will rapidly be followed by the formation of H_2SO_4 molecules through the reaction

$$SO_3 + H_2O(+M) \xrightarrow{k_9} H_2SO_4(+M) \quad (R9)$$

where a rate constant $k_9 = 9 \times 10^{-13}\,cm^3\,molecule^{-1}\,s^{-1}$ is measured by Castleman *et al.* (1975). H_2SO_4 molecules are converted to aerosols.

The fate of the HSO_3 molecules is less clear. Several mechanisms have been suggested for its further oxidation (Cox, 1976), with possible feedback to the ozone generation mechanisms (through NO to

NO_2 conversion and free radical formation). At present there is insufficient evidence to draw conclusions of this kind. It is, however, likely that HSO_3 is converted to H_2SO_4. In the calculations presented here we will therefore assume that the rate of sulfate aerosol formation is equal to the initial production rates of SO_3 and HSO_3 from reactions (R6)–(R8). (It should be noted that the free radical concentrations in the plume are only insignificantly reduced by these reactions.)

It should follow that agreement between observed and calculated generation of ozone and conversion of SO_2 to particulate sulfate offers solid evidence for the involvement of free radicals (OH, HO_2, $R\dot{O}_2$) in both these chemical processes.

MODEL CALCULATIONS

The concentrations of peroxy radicals in the plume are calculated to increase drastically when the polluted air has passed over the urban area (Fig. 1). There is more than five-fold increase in their concentrations from the time the polluted air leaves the urban area (9.00 a.m.) until maximum concentrations are reached approx 5 h later (2.00 p.m.). Concentrations of a few tenths of a ppb in the plume agree well with observations obtained by Ehhalt and coworkers (Ehhalt, 1977).

This development of the radical concentrations in the plume outside the urban area is a direct result of the decrease in NO concentrations shown in the figure, since the peroxy radical concentrations are proportional to the inverse values of the NO concentrations (R1 and R2). Several factors cause the drastic decrease in NO concentrations. The release of pollutants due to diffuse rural sources is small compared to the emissions in the urban area. The generation of high ozone concentrations leads to an efficient conversion of NO to NO_2; at the same time nitric acid is formed at an increasing rate through the reaction

$$NO_2 + HO(+M) \xrightarrow{k_{10}} HNO_3(+M) \quad (R10)$$

due to enhanced OH concentrations (see Fig. 1). With a rate constant $k_{10} = 10^{-11}\,cm^3\,molecule^{-1}\,s^{-1}$, conversion rates for $NO_x(NO_2 + NO)$ to HNO_3 of nearly $20\%\,h^{-1}$ are obtained around noon. The formation of PAN is also favoured as the ratio NO_2/NO increases. In addition, the deepening of the mixing layer reduces the NO_x concentrations considerably.

The calculations made clearly show the potential for peroxy radical formation inherent in the urban plume outside the source region of pollutants. As a consequence, the gas-phase conversion of SO_2 to particulate sulfur, which is proportional to the free radical concentrations, is very efficient (Fig. 2). Reaction with OH(R8) dominates the loss of SO_2 inside the urban area and during the first hour when the emissions are abolished, whereafter the peroxy radical reactions take over as the main loss mechanism. Maximum conversion rate of $5\%\,h^{-1}$ is obtained around 2.00 p.m.

It is important to note that the SO_2 to H_2SO_4 conversion through reactions (R6)–(R8) is rather sensitive to the level of the NO concentrations, due to the strong coupling between NO and peroxy radicals.

Based on the estimated gas-phase conversion rates of SO_2 to H_2SO_4, the relative amounts of gaseous (SO_2) and particulate sulfur in the plume are given in Fig. 3. During the early stage of the plume SO_2 is lost mainly by absorption at the ground. The dilution of the plume due to the increase in mixing height during the day implies that ground absorption loses

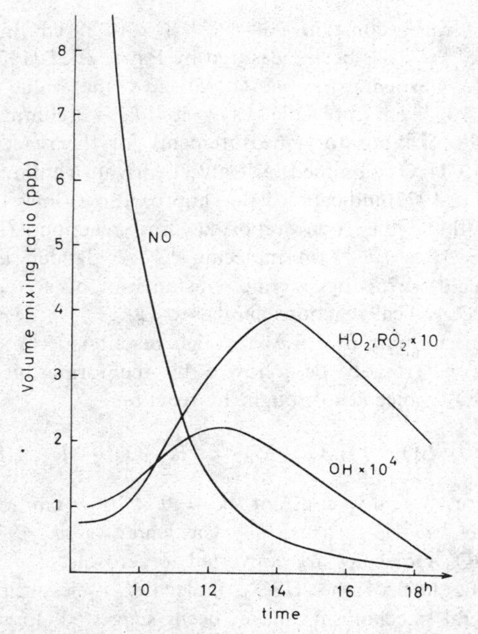

Fig. 1. Calculated volume mixing ratios of nitric oxide, hydroxyl and peroxy radicals vs time in the urban plume.

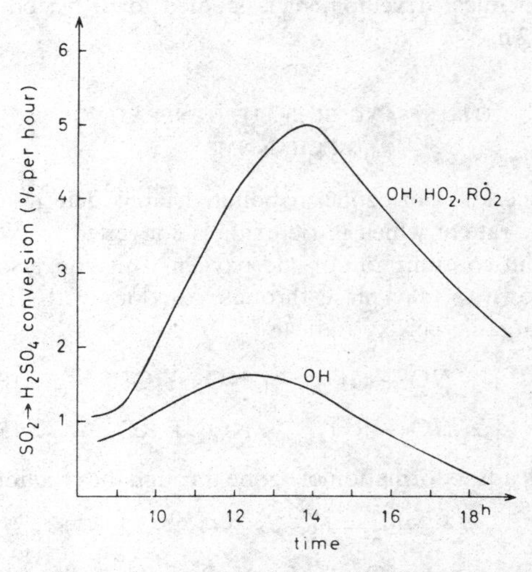

Fig. 2. Conversion rates of SO_2 to H_2SO_4 vs time in the plume. The curve denoted OH shows the gas-phase conversion rate through the reaction of sulfur dioxide with hydroxyl (R8), and the curve denoted OH, HO_2 and $R\dot{O}_2$ the similar rate through reactions of SO_2 with both hydroxyl and peroxy radicals (R6–R8).

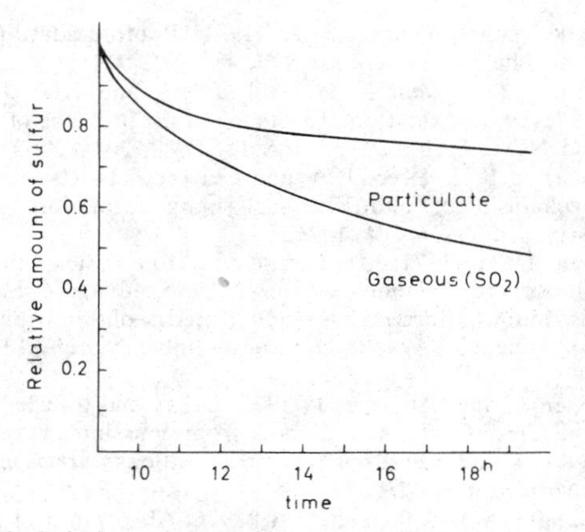

Fig. 3. Relative amounts of gaseous (SO_2) and particulate sulfur vs time in the plume.

importance (see Equation 4), while gas-phase conversion to sulfate occurs with increased efficiency. After 6.00 p.m. less than 50% of the total burden of SO_2 present in the plume originally is still there in the form of gaseous SO_2. Approximately 25% is converted to particulate sulfur, and 25% is absorbed at the ground. The peroxy and hydroxyl radicals formed through photochemical processes involving HC and NO_x from anthropogenic sources speed up the rate of sulfate formation. In this way a larger part of the total SO_2 emitted is transformed to sulfate before the bulk of SO_2 is lost through deposition. The effective conversion of SO_2 to particulate sulfur demonstrated in these calculations therefore is of importance to the total budget of sulfate particles in the troposphere.

Figure 4 demonstrates the strong effect on the ozone concentrations in the plume (through reactions R1 and R2) from urban release of pollutants. The emission rates of HC and NO_x from the St. Louis area adopted here, cause an increase in concentrations above the recommended upper limit of 80 ppb after a transport time of approx 1 h over rural area, and a maximum ozone concentration of 115 ppb is reached at 1.00 p.m. This contrasts strikingly with the calculated rural background concentrations of 60 ppb (curve denoted $\Psi_{poll} = 0$) resulting from diffuse rural sources, where the source strength is assumed to be approx 5% of that in the urban area. This figure shows that the calculated ozone variations agree well with the observed concentrations, although the latter seem to indicate a higher ozone maximum in the plume. It is interesting to note that even though the production of ozone decreases rapidly after 2.00 p.m. due to decrease in the abundance of NO and $R\dot{O}_2$ (cf. Fig. 1), the ozone concentrations remain at elevated levels throughout the afternoon, indicating a rather long chemical lifetime for ozone, once it is formed in the plume.

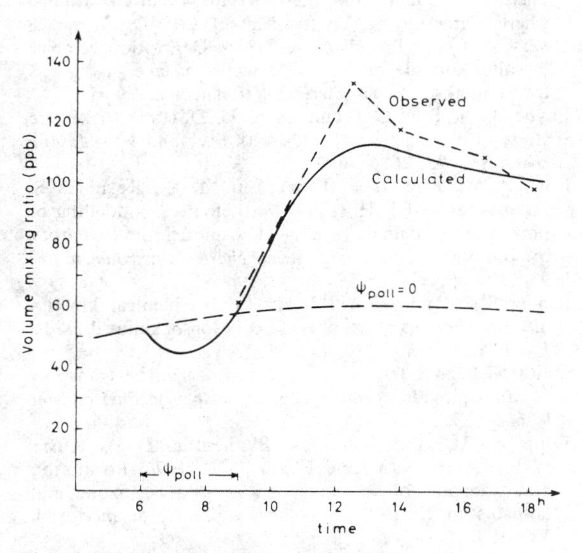

Fig. 4. Volume mixing ratios of ozone vs time. Primary pollutants (denoted by Ψ_{poll}) are emitted into the plume between 6.00 and 9.00 a.m. The curve denoted by $\Psi_{poll} = 0$ shows ozone variations outside the plume. The observed ozone values are taken from White *et al.* (1976).

FLOW RATES OF OZONE AND PARTICULATE SULFUR

The horizontal mass flow rate of ozone and particulate sulfur in the plume is obtained from the equation

$$F_h = u \cdot (C - C_0) \cdot h \cdot L \tag{9}$$

u is the wind velocity, which in the model is assumed to have a constant value of $5\,m\,s^{-1}$ throughout the day. This may lead to considerable underestimations of the flow rates because the observations show a marked increase in u throughout the day. $L = 40$ km is the width of the plume. $C - C_0$ gives the difference between concentrations in the plume and in the background rural air. Equation (9) therefore expresses the net flow in the plume.

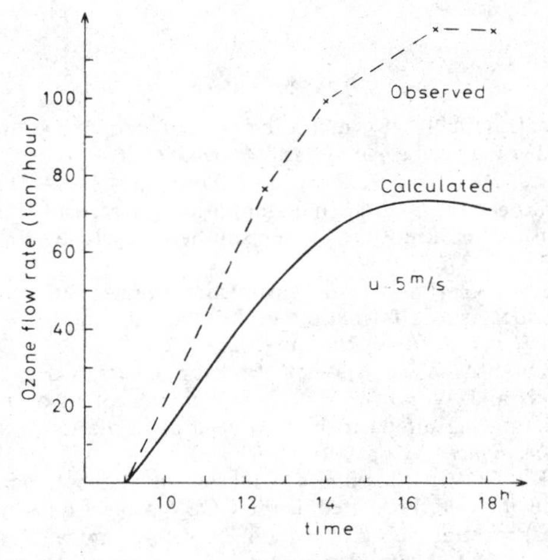

Fig. 5. Ozone flow rates vs time in the plume, for a constant wind velocity of $5\,m\,s^{-1}$. The observed flow rates are taken from White *et al.* (1976).

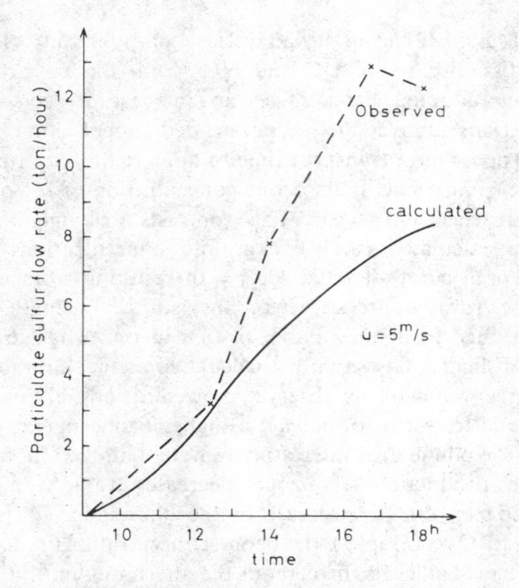

Fig. 6. Particulate sulfur flow rates (S_p) in the plume, for a constant wind velocity of $5\ m\,s^{-1}$. The observed flow rates are taken from White *et al.* (1976).

Figures 5 and 6 show the estimated flow rates of ozone and particulate sulfur respectively. The corresponding flow rates obtained from observations by White *et al.* (1976) are also shown on the figures.

In both cases the calculated mass flow rates in the plume increase rapidly as the polluted air is transported over areas with small emissions, in agreement with observations. The calculated values are, however, approx 30% lower than the observed ones. This difference is not surprising when we consider the uncertainties in the release rates of the primary pollutants included in the chemical scheme, and the simplifications made in the model adopted for the plume.

Acknowledgements—We wish to thank Dr. Rudolf Husar for valuable discussions. Our work has partly been sponsored by The Norwegian Research Council for Science and the Humanities.

REFERENCES

Aldaz L. (1969) Flux measurements of ozone over land and water. *J. geophys. Res.* **74,** 6943–6946.

Atkins D. H. F., Cox R. A. and Eggleton A. E. J. (1972) Photochemical ozone and sulphuric acid aerosol formation in the atmosphere over southern England. *Nature* **235,** 372–376.

Calvert J. G. (1976) Hydrocarbon involvement in photochemical smog formation in Los Angeles atmosphere. *Envir. Sci. Technol.* **10,** 256–262.

Castleman A. W. Jr., Davis R. E., Munkelwitz H. R., Tang I. N. and Wood W. P. (1975) Kinetics of association reaction pertaining to H_2SO_4 aerosol formation. *Int. J. chem. Kinet.* (Symp. No. 1), 629–635.

Cox R. A. (1976) Gas-phase oxidation of atmospheric sulphur dioxide. Presented at the COSY project 61 A technical meeting sub-group I on Physico-chemical Behaviour of SO_2 in the Atmosphere, Rome, April, 1976.

Cox R. A. and Penkett S. A. (1971) Photo-oxidation of atmospheric SO_2. *Nature* **229,** 486–488.

Cox R. A., Derwent R. G., Holt P. M. and Kerr J. A. (1976) Photooxidation of methane in the presence of NO and NO_2. *J. chem. Soc.* (Faraday I) **72,** 2044–2060.

Demerjian K. L., Kerr J. A. and Calvert J. G. (1974) The mechanism of photochemical smog formation. *Adv. envir. Sci. Technol.* **4,** 1–262.

Ehhalt, D. H. (1977) Measurements of free radicals in the atmosphere. Presented at the NATO Advanced Study Institute on 'Spectroscopy and Kinetics of Atmospheric Constituents'. Arabba/Dolomiti, Italy, March 13–26, 1977.

Eliassen A. and Saltbones J. (1975) Decay and transformation rates of SO_2, as estimated from emission data, trajectories and measured air concentrations. *Atmospheric Environment* **9,** 425–429.

Garland J. A. and Penkett S. A. (1976) Absorption of peroxy acetyl nitrate and ozone by natural surfaces. *Atmospheric Environment* **10,** 1127–1131.

Gear C. W. (1971) The automatic integration of ordinary differential equations. *Commun. Ass. Comput. Mach.* **14,** 176–179.

Grennfelt P. (1976) Ozone episodes on the Swedish west coast. Presented at the Int. Conf. on Photochemical Air Pollution and its Control, Raleigh, North Carolina, 12–17 Sept., 1976.

Hampson R. F., Jr. and Garvin D. (1975) Chemical kinetic and photochemical data for modelling atmospheric chemistry. U.S. Dept. of Commerce, National Bureau of Standards, Tech. Note 866.

Hesstvedt E., Hov Ö. and Isaksen I. S. A. (1977a) Quasi-steady state approximation in air pollution modelling. Comparison of two numerical schemes for oxidant prediction. Report No. 25 Institutt for Geofysikk, Universitetet i Oslo.

Hesstvedt E., Hov Ö. and Isaksen I. S. A. (1977b) On the chemistry of mixture of hydrocarbons and nitrogen oxides in air. *Geophys. Norv.* **31,** No. 6.

Hov Ö., Isaksen I. S. A. and Hesstvedt E. (1977) Diurnal variations of ozone and other pollutants in an urban area. Report No. 24 Institutt for Geofysikk, Universitetet i Oslo.

OECD (1976) Second report on the problem of photochemical oxidants and their precursors in the atmosphere. Direction de l'environnement, OECD, Paris.

Owers M. J. and Powell A. W. (1974) Deposition velocity of sulfur dioxide on land and water surfaces using ^{35}S tracer method. *Atmospheric Environment* **8,** 63–67

Payne W. A., Stief L. J. and Davis D. D. (1973) A kinetic study of the reaction of HO_2 with SO_2 and NO. *J. Am. chem. Soc.* **95,** 7614–7619.

Roth P. M., Roberts P. J. W., Liu M. K., Reynolds S. D. and Seinfeld J. H. (1974) Mathematical modelling of photochemical air pollution—II. A model and inventory of pollutant emissions. *Atmospheric Environment* **8,** 97–130.

Sander S. P. and Seinfeld J. H. (1976) Chemical kinetics of homogeneous atmospheric oxidation of sulfur dioxide. *Envir. Sci. Technol.* **10,** 1114–1123.

Smith F. B. and Jeffrey G. H. (1975) Airborne transport of sulfur dioxide from the U.K. *Atmospheric Environment* **9,** 643–659.

White W. H., Anderson J. A., Blumenthal D. L., Husar R. B., Gillani N. V. and Husar J. D. (1976) Formation and transport of secondary air pollutants: ozone and aerosols in the St. Louis urban plume. *Science* **194,** 187–189.

Whitten G. Z. and Meyer J. P. (1976) CHEMK: a computer modelling scheme for chemical kinetics. Systems Applications Inc., San Rafael, California.

Atmospheric Environment Vol. 12, pp. 605–611. Pergamon Press 1978. Printed in Great Britain.

TRANSPORT OF SO_2 IN POWER PLANT PLUMES: DAY AND NIGHT

T. B. SMITH, D. L. BLUMENTHAL, J. A. ANDERSON, and A. H. VANDERPOL
Meteorology Research, Inc. Altadena, CA 91001, U.S.A.

(*First received* 23 *June* 1977 *and in final form* 10 *October* 1977)

Abstract – As part of the July 1976 Midwest Interstate Sulfur Transformation and Transport experiment (MISTT) in St. Louis, Missouri, U.S.A., the long-range transport of SO_2 and other pollutants emitted from the 216 m stack of a 2000 Mw coal-fired power plant have been documented during both day and night summer conditions.

Aircraft measurements and trajectory analyses showed that SO_2 emitted during the late morning hours of 18 July, 1976, underwent rapid mixing near the plant. Once mixed through the mixing layer, however, the plume maintained its identity. In the evening, a surface radiation inversion formed, thermally driven mixing ceased, and the lapse rate aloft became more stable. Although winds less than 4 m s^{-1} existed at the surface, from 300 to 800 m msl, the winds were 8–15 m s^{-1} for most of the night. SO_2 emitted during the day became decoupled from the ground and was shown to be transported more than 300 km by midnight in the high wind speed region aloft. The SO_2 emitted at night remained decoupled from the ground and experienced much less dilution. At 75 km in the night-emitted plume, the peak SO_2 concentration was greater than 0.85 ppm.

The high wind speeds seen aloft in stable air on the night of 18 July are a common occurrence in the U.S. and presumably in other parts of the world during the summer months. In July 1976, conditions were favorable for the long-range transport of SO_2 in layers aloft on at least 18 out of 31 nights. The stable night regime is important for the long-range transport of sulfur compounds in the air. Without the ground as a sink, or solar heating to drive mixing, the SO_2 can be transported hundreds of kilometers. Accumulation and transport of SO_2 and its reaction products is a synoptic scale, multiday problem.

1. INTRODUCTION

As part of the July 1976, Midwest Interstate Sulfur Transformation and Transport (MISTT) experiment in St. Louis, Missouri, U.S.A., the distribution and transport of SO_2 and other pollutants emitted from the 216 m stack of the 2000 Mw coal-fired Labadie power plant were documented during both day and night summer conditions. The plume was sampled using two aircraft.

The primary sampling aircraft was a single-engine Cessna 206, instrumented to measure gaseous pollutants, including SO_2, aerosol size distribution and chemical composition, and several meteorological and position parameters (Blumenthal *et al.*, 1978). The 206 was generally used to provide detailed measurements of the plume out to about 100 km from the stack. The flight pattern of the 206 was designed to characterize the plume at discrete distances downwind. At each distance, horizontal traverses were made through the plume perpendicular to the plume axis at several elevations (Fig. 1). The second aircraft, a twin-engine Aerocommander, was used to locate the plume and sample at far downwind distances (up to 300 km). The Aerocommander was equipped to measure SO_2, ozone and some aerosol and position parameters (Husar *et al.*, 1978).

The aircraft were supported by three mobile pilot balloon (pibal) crews who obtained winds aloft data hourly at various positions along the path of the

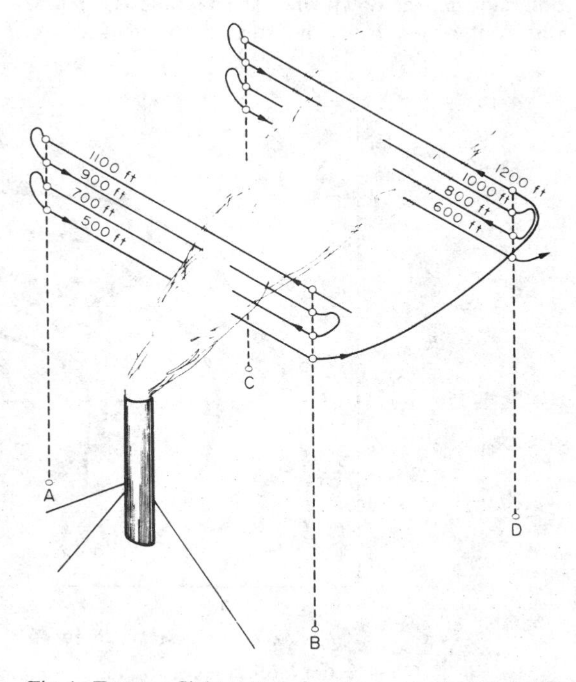

Fig. 1. Traverse flight pattern for plume sampling. Points A, B, C, and D are ground reference points.

plume. The pibal wind data were radioed to the aircraft to aid in tracking the plume. In addition, hourly radiosonde measurements of wind and temperature were made available from several locations in the St. Louis area by the Environmental Protection Agency's (EPA) Regional Air Monitoring System (RAMS). These data were used for extensive meteorological analysis to identify the source of the measured plumes (see Appendix). Unless otherwise noted, all data discussed in this paper were taken from air parcels which were traced to the vicinity of the Labadie Power Plant.

Descriptions of the MISTT program are presented by Wilson *et al.* (1978) and Husar *et al.* (1978). Analyses of the chemical transformation data and of the aerosol size distribution data are presented by Gillani *et al.* (1978) and Whitby *et al.* (1978). In this paper we present data from 18 July, 1976, that document the long-range transport of SO_2.

2. RESULTS

On 18 July, 1976, the St. Louis area, along with much of the eastern United States, was under the influence of a large, dry, high pressure system. During the afternoon, the temperature lapse rate was adiabatic, and strong thermal mixing existed. However, since the r.h. at the surface was only 35%, no clouds developed. Winds aloft were southerly at about 4–6 m s^{-1} from the surface to 900 m above mean sea level (msl), shifting slightly southwesterly and decreasing to about 3 m s^{-1} from 900–2000 m msl. Although transport was slow, the strong mixing rapidly dispersed the plume, both vertically and horizontally, as the pollutants moved downwind. The daytime SO_2 plume was well mixed from the surface to greater than 1400 m msl and had a maximum measured concentration of 60 ppb at 25 km. (All concentrations are reported by volume.)

Figure 2 is a cross section of the daytime plume SO_2 concentration between 1520 and 1608 Central Daylight Time (CDT, Greenwich Mean Time minus 5 h), at 25 km downwind of the plant, about $2\frac{1}{2}$ hours' travel. This figure, as well as Figs. 5 and 6, was constructed from horizontal traverses perpendicular to the plume axis. It is not a "snapshot", but a composite constructed from data collected by the Cessna 206 aircraft.

To draw the contours, instantaneous data values (one each two seconds or ~60 m) were superimposed on the traverse path (dashed lines). Using these values to fix the location of contour lines along each traverse path, smooth contours were extrapolated, based on meteorological constraints and an approximate match of the flux from the power plant. Contours outside the traverse range and those which intersect only one level are presented as dashed lines to clearly indicate the extrapolation.

During the same afternoon and night, the Aerocommander sampled at distances as far as 300 km and measured SO_2 emitted during the day. The left-hand side of Fig. 3 shows the SO_2 concentrations measured about 600 m msl. The right-hand side shows the approximate traverse path, the time of the traverse, and the calculated trajectory of the air parcels sampled (see Appendix). High values of SO_2 measured on each traverse were shown to originate near the Labadie Plant, well within the errors of the wind field calculations.

The calculated times that the plume was emitted from the plant are also indicated for each trajectory. For the 2000 CDT traverse about 220 km from the

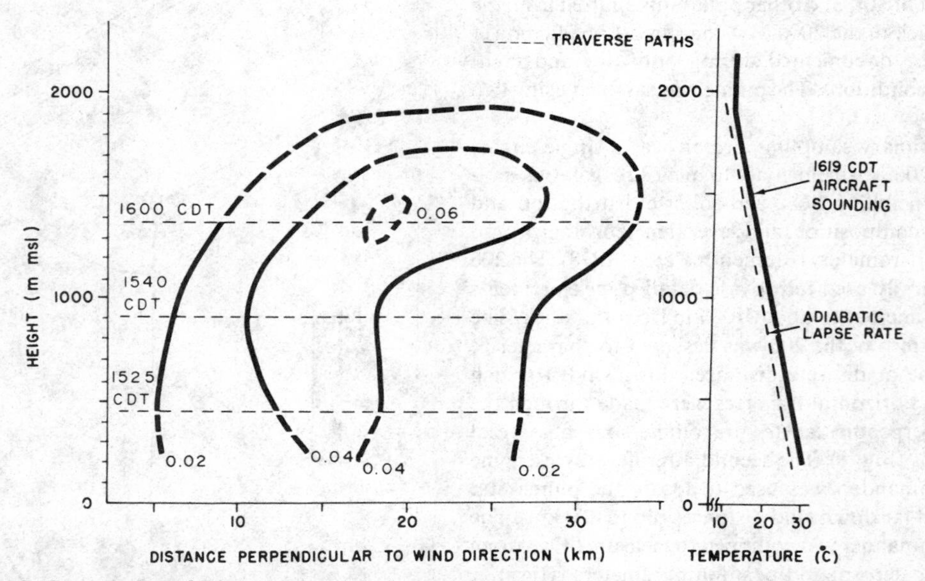

Fig. 2. Cross section of SO_2 concentrations (ppm) in the Labadie plume made from sampling traverses between 1525 and 1600 CDT at 25 km downwind (ground level 150 m). Contours outside the traverse range and those which intersect only one level are represented as dashed lines to indicate extrapolations.

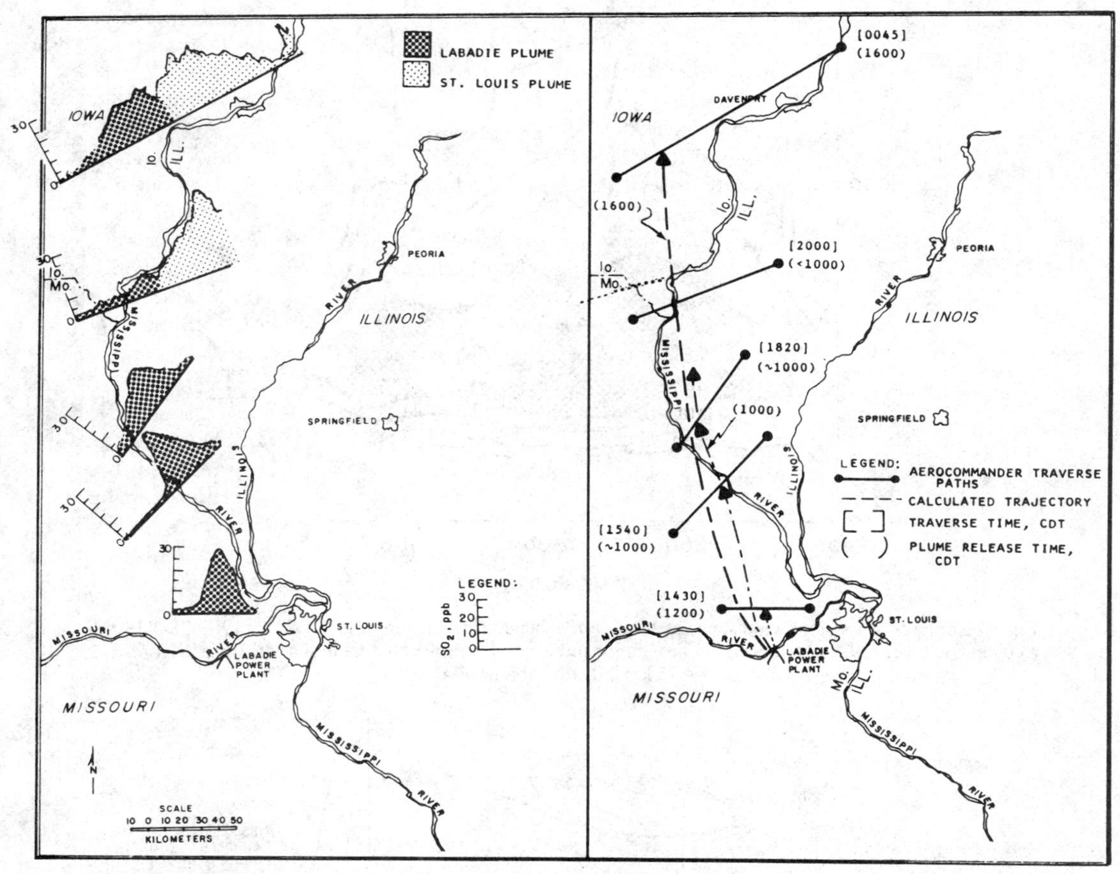

Fig. 3. Horizontal profiles of SO_2 and plume trajectories for selected passes, July 18, 1976. The left-hand side shows the SO_2 concentrations measured at about 600 m msl. The right-hand side shows the approximate traverse path, the time of the traverse, the calculated trajectory from the plant, and approximate plume release time (see Appendix). An exact determination of the trajectory for the 2000 CDT traverse is not available because of insufficient data before 10 a.m.

plant, the remnants of the SO_2 emitted about 1000 CDT in the morning are still distinguishable (An exact determination of the trajectory for the 2000 CDT traverse is not available because of insufficient data before 10 a.m.) For the 0045 traverse at about 300 km from the plant, the SO_2 peak is more visible than on the 2000 CDT traverse because the plume was emitted later in the day and was dispersed less.

These data document an SO_2 plume emitted during the day that was relatively dilute, but maintained its identity through the day and well into the night. In the process, SO_2 released by the plant and its reaction products (see Gillani *et al.*, 1978) travelled at least 300 km.

In the evening, a surface radiation inversion formed, thermally driven mixing ceased, and the lapse rate aloft became more stable. Although the wind speed at the surface was less than 4 m s^{-1} during the night, a low level jet gradually formed aloft. Figure 4 shows a time–altitude cross section of wind speed for the night of 18 July. It was constructed from radiosonde data taken from Station 142 of the RAMS network. The maximum velocity in the jet was about 15 m s^{-1} at about 520 m msl (350 m agl). The jet maintained velocities greater than 12 m s^{-1} from 2300 CDT on 18 July to 0800 CDT on the morning of 19 July.

Figure 5 is a cross section of the SO_2 plume measured at night at 25 km downwind from the plant, the same distance as the daytime measurement of Fig. 2. Two distinct SO_2 plumes are shown, separated by a traverse at 900 m msl with considerably lower concentrations.

Trajectory analysis using pibal and radiosonde data show that the upper one had been emitted from the plant 2–3 h prior to sampling. It is in a region of light winds and was released before the atmosphere had become well stabilized and the jet had become established. The lower SO_2 plume was emitted from the same plant, only about an hour prior to sampling (1–2 h after the upper plume). It became confined in a more stable atmosphere and in a region of much higher wind velocity.

The height of each plume and difference between the two plumes is attributable to interaction of buoyancy effects of the plume with the wind speed and stability regime at release time (see Moore, 1973; and Briggs,

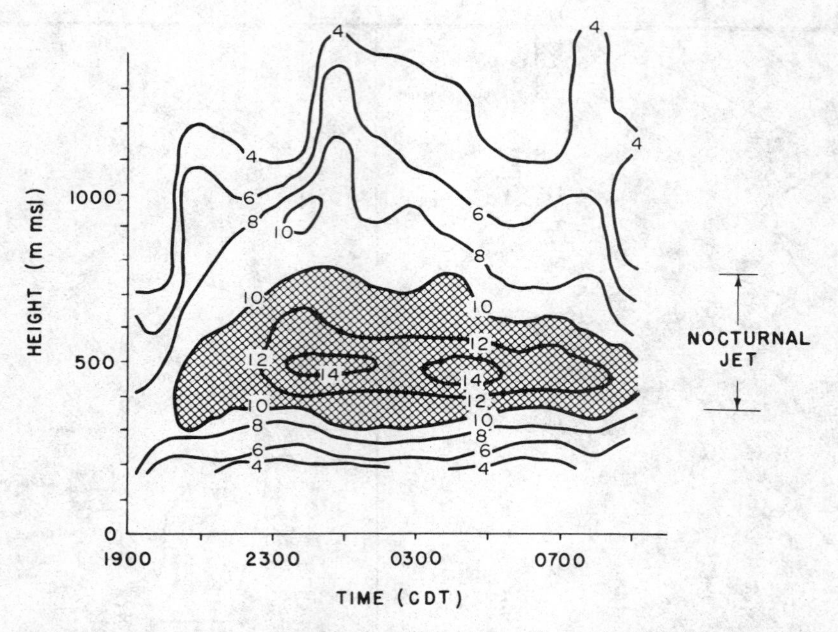

Fig. 4. Isopleths of constant wind speed as a function of height (m msl) and time (CDT) for 1900 CDT 18 July to 0900 CDT July 19, 1976. (Speeds in m s^{-1}, ground level 150 m.) Plotted from RAMS Data, Station 142 ~ 15 km NE of Labadie Plant.

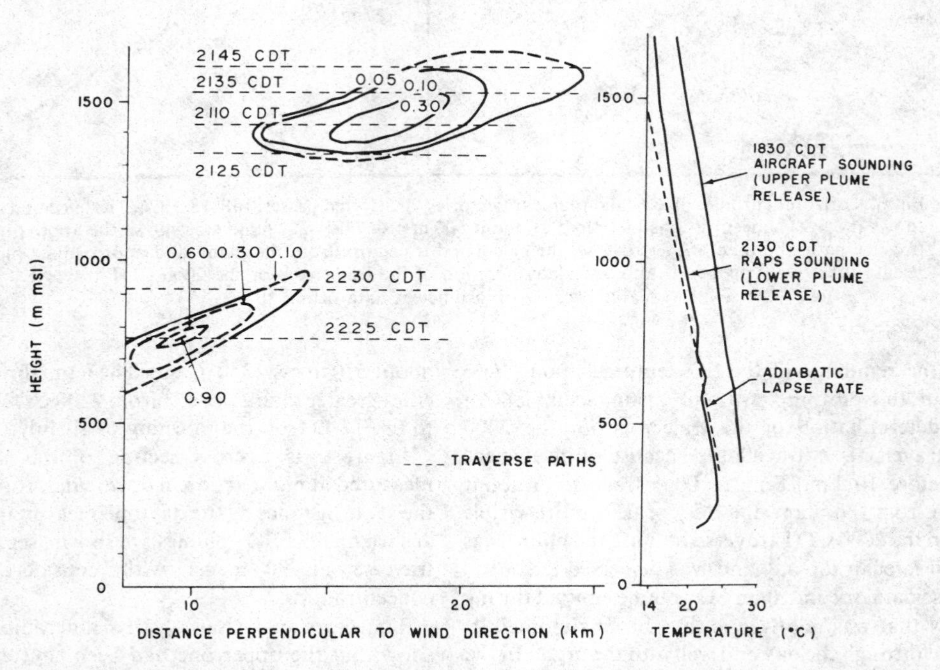

Fig. 5. Cross section of SO_2 concentrations (ppm) in the Labadie plume made from sampling traverses between 2107 and 2233 CDT at 25 km downwind. (Upper plume emitted about 1900 CDT; lower plume emitted about 2130 CDT.) Contours outside the traverse range and those which intersect only one level are represented as dashed lines to indicate extrapolations.

1969). The SO_2 plumes are offset from one another due to directional shear. At a single distance from the plant, we were able to detect both the remnants of a late daytime SO_2 release and a night release. At the same distance, the late daytime SO_2 release had dispersed to one-third the concentration of the lower night-time SO_2 release.

Figure 6 is an SO_2 cross section obtained between 2246 and 2318 CDT that is made up of three passes, varying slightly in distance from the plant, but all within 5 km of 50 km. The SO_2 plume is seen to be elongated due to slight directional shear. The peak instantaneous value measured at this distance exceeded 1 ppm.

Figure 7 shows the measured SO_2 concentrations for three traverses in the stable night plume, ranging

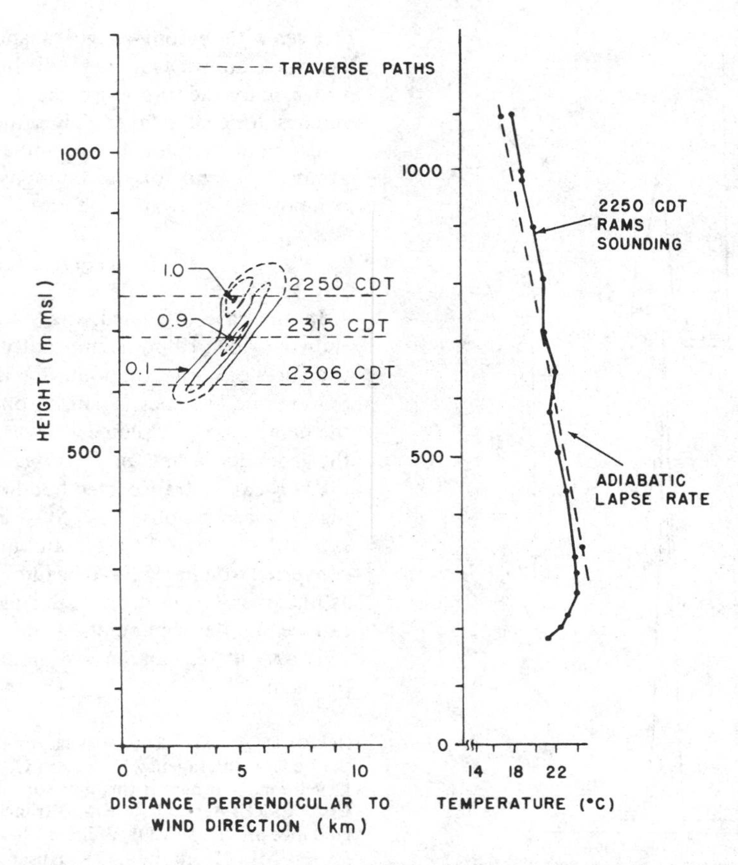

Fig. 6. Cross section of SO_2 concentrations (ppm) in the Labadie plume made from sampling traverses between 2246 and 2318 CDT at 50 km downwind. (The plume was emitted around 2150 CDT.) Contours outside the traverse range and those which intersect only one level are represented as dashed lines to indicate extrapolations.

from 25 to 75 km downwind. All three traverses were made at 760 m msl. (Note that the 25 and 50 km SO_2 traces are the data used for the contour levels at 760 m in Figs. 5 and 6, respectively.) At 75 km, the SO_2 concentration in Fig. 7 still exceeds 0.5 ppm, and an orbit at 75 km indicated a maximum concentration of over 0.85 ppm. The SO_2 is seen to remain very concentrated with little dilution other than that due to initial buoyancy effects.

Even with the high wind speeds, mixing was minimal. A good indicator of mixing is the turbulence measured in the aircraft. The aircraft turbulence probe indicated between 0.5 and 1 $cm^{2/3}\ s^{-1}$ in the plume. For comparison, "light" turbulence registers 2–3 $cm^{2/3}\ s^{-1}$ (MacCready, 1966).

Although the stable, concentrated SO_2 plume shown in Fig. 7 was not actually followed further than 75 km, the wind data in Fig. 4 suggest that much greater transport was likely. From Fig. 4, at 760 m msl, the wind speed remained about 10 $m\ s^{-1}$ (36 $km\ h^{-1}$) until 0400 CDT, and then averaged at least 8 $m\ s^{-1}$ (29 $km\ h^{-1}$) until 0600 CDT. Four hours of additional transport at 10 $m\ s^{-1}$ and two hours at 8 $m\ s^{-1}$ would put the plume at a downwind distance of about 275 km by 0600 CDT. Although the wind data in Fig. 4 were obtained near the plant, the synoptic meteorology of the 18th and pilot balloon data taken downwind during the night suggest that the vertical wind structure was not appreciably different over the potential 275 km distance.

3. DISCUSSION

The data presented here show a dramatic contrast between day and night dispersion and transport of SO_2. During the day the SO_2 became vertically well mixed through the mixing layer allowing removal of SO_2 by deposition. With relatively constant wind direction and minimal shear, the SO_2 plume maintained its identity, but concentrations were relatively low. The light daytime wind speeds contributed little to long-distance transport. As the stable night conditions set in and wind speeds above the ground level increased, the residual daytime SO_2 plume became decoupled from the ground and was transported with little dilution over long distances.

The SO_2 emitted at night remained decoupled from the ground and experienced much less dilution. High concentrations of SO_2 and other pollutants were transported up to 300 km before dawn. The SO_2 in these night plumes is isolated from deposition and is

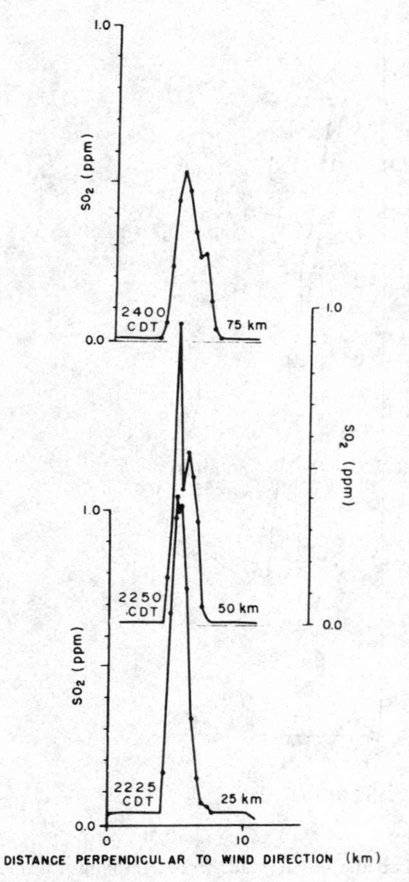

Fig. 7. SO_2 concentrations at 760 m msl at downwind distances of 25, 50 and 75 km for traverses made north of the plant in Labadie plume between 2225 and 2400 CDT, 18 July, 1976

thus available for conversion to sulfate at least during the night and into the next morning.

Sampling on other days showed that, as solar heating eroded the radiation inversion, these stable plumes were mixed to the ground and added to surface SO_2 concentrations over 200 km from their source. Stable plumes of this type have also been reported to impact on high terrain at night, causing extensive vegetation damage (Long and Williams, 1977).

Blackadar (1957; 1976) has indicated that nocturnal low level jets, such as the one which occurred on 18 July, 1976, are a common occurrence over much of the world. He has shown that the jets are a result of interaction between the geostrophic wind field and the radiationally induced daily cycle of the turbulence intensity in the boundary layer.

In July 1976, conditions in the St. Louis area were favorable for the long-range transport of SO_2 in stable plumes aloft on at least 18 out of 31 nights. On at least three different occasions, power plant plumes were actually tracked to distances greater than 200 km during the night and early morning hours. The night-time regime thus provides a mechanism for the transport of SO_2 from a single source over a synoptic scale region.

Even without long-range transport, conditions were favorable almost every night during the July sampling program for the trapping of stable high concentration plumes in elevated layers. Thus, the night-time regime is also important for the accumulation of sulfur in the atmosphere and for the multiday buildup of sulfur compounds in an air mass.

4. CONCLUSIONS

In the case discussed, the stable night regime is shown to contribute significantly to the long-range transport of sulfur compounds in the air. The daytime SO_2 plume becomes isolated from the ground while the night-time SO_2 plume is always isolated. Without the ground as a sink, or solar heating to drive mixing, the SO_2 can be transported hundreds of kilometers at high concentrations. The SO_2 in these plumes is available through the night and morning hours to be converted to sulfate and to be fumigated to the ground as the mixing layer deepens during the day. Thus, we can see that the accumulation and transport of SO_2, as well as sulfate, can be a synoptic scale, multiday problem.

Acknowledgements – This work has been supported in part by the Federal Interagency Energy/Environment Research and Development program through various contracts with the EPA, ESRL, Aerosol Research Branch. We appreciate the guidance provided by Dr. William Wilson (EPA), Director of Project MISTT and by Dr. R. Husar, Project coordinator. We also appreciate the technical support of Dr. Noor Gillani for his assistance in data processing, and that of Messrs Stephen Howard and William Knuth, who performed most of the meteorological analysis.

REFERENCES

Blackadar A. K. (1976) Modeling the nocturnal boundary layer. Presented at the Third Symp. on Atmospheric Turbulence Diffusion and Air Quality, Raleigh, North Carolina, October 19–22.

Blackadar A. K. (1957) Boundary layer wind maxima and their significance for the growth of nocturnal inversions. *Bull. Am. met. Soc.* **38,** 283–290.

Blumenthal D. L., Ogren J. A. and Anderson J. A. (1978) Airborne sampling system for Project MISTT. *Atmospheric Environment* **12,** 613–620.

Briggs G. A. (1969) *Plume Rise.* Prepared for the Nuclear Safety Information Center, Oak Ridge National Laboratory, U.S. Atomic Energy Commission, Division of Technical Information.

Gillani N. V., Husar R. B., Husar J. D. and Patterson D. E. (1978) Project MISTT: Kinetics of particulate sulfur formation in a power plant plume out to 300 km. *Atmospheric Environment* **12,** 589–598.

Husar R. B., Gillani N. V., Patterson D. E. and Husar J. D. (1978) Sulfur budget in large plumes. *Atmospheric Environment* **12,** 549–568.

Long J. H. and Williams D. R. (1977) Aerial photographic survey of vegetation damage caused by an air pollution incident. Presented at the Aerial Techniques for Environmental Monitoring, Topical Symp. of the American Nuclear Society, Las Vegas, Nevada, March 7–11.

Moore D. J. (1973) Predicting the concentration of effluent material within a plume emitted from a tall chimney. From the *Faraday Symp. Chem. Soc.* **7,** 222–228.

MacCready P. B., Jr. (1966) Operational application of a universal turbulence measuring system. AMS/AIAA Paper No. 66–364.

Whitby K. T., Cantrell B. K. and Kittleson D. B. (1978) Nuclei formation rates in a coal-fired power plant plume. *Atmospheric Environment* **12,** 313–321.

Wilson W. E., Jr. (1978) Sulfates in the atmosphere: a progress report on Project MISTT. *Atmospheric Environment* **12,** 537–547.

APPENDIX

The trajectories from the plant were calculated using the MISTT and RAMS pibal and RAMS radiosonde data. The initial step was to use the data from the hourly pibal and radiosonde releases to plot the three-dimensional wind fields for most hours on July 18, 1976. Each three-dimensional field was constructed by incorporating several two-dimensional planes – one at each 300 to 400 m increment in height up to about 1400 m msl.

Constant altitude trajectories for the plume measurements were then calculated from these flow fields. Each hourly trajectory segment was developed from the hourly flow fields. Trajectories were calculated starting from the plant and carried until they intercepted a given traverse path. For most of the July 18 traverses shown in this paper, the trajectories starting from the plant were traced to the SO_2 maxima in the plume traverses within the errors inherent in the data. For the 300 km traverse, trajectories starting from the plant ended near the secondary SO_2 peak. An exact determination of the trajectory for the 2000 CDT traverse is not available because of insufficient data before 10 a.m. For this traverse, assuming a steady-state flow field before 1000 CDT would project the secondary peak to the Labadie plant.

Atmospheric Environment Vol. 12, pp. 613–620. Pergamon Press 1978. Printed in Great Britain.

AIRBORNE SAMPLING SYSTEM FOR PLUME MONITORING

D. L. BLUMENTHAL, J. A. OGREN and J. A. ANDERSON
Meteorology Research, Inc., Altadena, CA 91001, U.S.A.

(*Received for publication* 13 *October* 1977)

Abstract—A single-engine Cessna 206 was modified and instrumented for plume sampling. This aircraft was used as one of the primary sampling platforms for Project MISTT (Midwest Interstate Sulfur Transformation and Transport) as well as other studies. On board aerosol instrumentation included a condensation nuclei monitor, aerosol charge acceptance monitor, integrating nephelometer, electrical aerosol analyser, optical particle counter, size-segregated filter sampler, and a wing-mounted impactor system. The efficiency of the size distribution sample inlet system was greater than 90% for aerosols between 0.009 and 2.5 μm dia. Ozone, sulfur dioxide, nitric oxide and total oxides of nitrogen were monitored continuously. Hydrocarbon concentrations and composition were obtained by gas chromatographic analysis of air samples collected in stainless steel canisters. Temperature, dewpoint, turbulence, pressure (altitude), and bearing and distance from an aircraft navigation station were also measured. Data from the aircraft system were processed in a preliminary fashion in the field within a few hours of a flight. Final data processing was performed at the various laboratories involved and included application of instrument response time corrections when necessary. The large number of simultaneous measurements possible with the system and the operational procedures which allow rapid feedback of sampling results to the investigators have greatly facilitated the study of pollutant transformation and transport in plumes.

INTRODUCTION

The *in situ* measurement of the physical and chemical characteristics of urban and point source plumes requires an airborne sampling system. Ideally, such a system should be capable of monitoring simultaneously a wide variety of gas, aerosol and meteorological parameters with a spatial resolution adequate to resolve the gradients existing in the plumes. The sampling system and operational procedures used should allow rapid processing of the data and timely feedback of the results to the field crew.

As part of Project MISTT (Midwest Interstate Sulfur Transformation and Transport (see Wilson, 1978), extensive airborne urban and power plant plume measurements were made to determine plume transport and dispersion properties and secondary pollutant formation rates. Several airborne platforms were used during the program. A highly modified Cessna 206 and its predecessor, a Cessna 205, were instrumented especially for plume monitoring and were used during almost all the MISTT field programs from 1973 to 1977. Details of the Cessna 206 system as it was configured and used in 1976 are presented in this paper. An instrumented Rockwell Aero-Commander was also used during the 1976 program and is described by Husar *et al.* (1978).

The purpose of this paper is to document the sampling system used to obtain the data on which several other papers are based. Analyses of data obtained by the Cessna 206 (or 205) during Project MISTT and other programs can be found in: Blumenthal *et al.* (1974); Blumenthal *et al.* (in press); Cantrell and Whitby (1978); Gillani *et al.* (1978); Husar *et al.* (1978); Husar *et al.* (1977); Ogren *et al.* (1976); Smith *et al.* (1978); Whitby *et al.* (1976); Whitby *et al.* (1978); White *et al.* (1976a); White *et al.* (1976b); and Wilson (1978).

AIRCRAFT DESCRIPTION

The Cessna 206 used in Project MISTT is a six-passenger, single-engine, turbocharged, short takeoff and landing (STOL) aircraft. Modifications for airborne sampling included: (1) installation of the sample inlets in a dummy window; (2) installation of instrument racks and mounting pads; (3) installation of electrical and vacuum supply systems for the instruments; and (4) installation of a shackle underneath the wing for mounting an impactor system. Aircraft specifications relevant to airborne sampling are presented in Table 1.

Table 1. Cessna 206 specifications

Crew:	2
Instrument payload:	400 kg
Normal flight duration:	4 h
Minimum sampling speed:	30 m s^{-1}
Maximum sampling speed:	60 m s^{-1}
Available instrument power:	1000 VA at 115 VAC 560 VA at 28 VDC
Vacuum sources:	3–50 lpm at 12 cm Hg venturi exhausts 1–300 lpm at 30 cm Hg engine-driven pump

MEASUREMENT SYSTEM DESCRIPTION

The Cessna 206 sampling system was designed for easy operation and maintenance. All equipment is easily removable from the aircraft for maintenance purposes, and the instrument package can be operated on ground power for calibration, warmup, or

Table 2. Cessna 206 measurement systems.

Parameter	Sampler Manufacturer and Model	Analysis Technique	Measurement Ranges	Time Response (to 90%)	Approximate Resolution
SO_2	Theta Sensors LS400	Electrochemical Cell	1.0*, 10.0 ppm	5-20 s	0.01 ppm
NO/NO_x	Monitor Labs 8440	Chemiluminescence	0.2, 0.5*, 1.0, 2.0, 5.0 ppm	5-10 s	0.01 ppm
O_3	REM 612	Chemiluminescence	0.5*, 2.0 ppm	5 s	0.005 ppm
Hydrocarbons, Halocarbons**	Washington State University	Stainless Steel Canisters/Gas Chromatography		20 s	~1 ppb ~1 ppt
Sulfate	MRI TWO-MASS (Prototype)	Flash vaporization/ flame photometric	< 3 μm dia.	~10 min	~3 $\mu g m^{-3}$
Aerosol Composition***	MRI Airborne Impactor System	Microscopy, IEXE	3 stages & filter D_{50}'s ≈15, 3, 0.4 μm	10-30 min	
Light Scattering Coefficient	MRI 1550	Integrating Nephelometer	10*, 40 × $10^{-4} m^{-1}$	1 s	0.1 × $10^{-4} m^{-1}$
Aerosol Charge Acceptance***	Washington University	Ionizer/Electrometer	0.01-1.0 μm	3 s	
Condensation Nuclei	Environment One Rich 100	Light Attenuation	(1, 3, 10, 30, 100*, 300*, 10K)× 10^3 CN/CC	3 s	1000 CN/CC*
Aerosol Size Distribution	TSI 3030 Royco 218	Electrical Mobility Optical Particle Counter	0.006-1.0 μm dia. 1.0-4.5 μm dia.	3 min analysis time per distribution	
Turbulence	MRI 1120	Pressure Fluctuations	0-10 $cm^{2/3} s^{-1}$	3 s (to 60%)	0.1 $cm^{2/3} s^{-1}$
Temperature	YSI/MRI	Bead Thermistor/ Vortex Housing	-5° to +45°C	5 s	0.5°C
Dew Point***	Cambridge Systems 137	Cooled Mirror	-50° to +50°C	0.5 s/°C	0.5°C
Altitude	Validyne	Absolute Pressure Transducer	0-3000 m msl	1 s	6 m
Indicated Airspeed	Validyne	Differential Pressure Transducer	20-65 ms^{-1}	1 s	0.1 ms^{-1}
Position	King KX170B/ Metrodata M8	Aircraft VOR/DME	0-150 km from station	1 s	1° (bearing), .2 km (distance)
Data Logger (includes time)	Metrodata 620	Tape Cartridge - 4 hour capacity in continuous operation	0-9.99 VDC	0.8 s/40 channel scan	0.01VDC
Stripchart Recorder**	Linear Instruments	Dual channel	.01, 0.1, 1, 10VDC	<1 s	

* - Usual Range(s)
** - 1976 only
*** - 1975, 76 only

MRI - Meteorology Research, Inc.
TSI - Thermo Systems, Inc.
YSI - Yellow Springs Instruments

checkout. The system is highly automated and can be operated by a single observer.

Instrumentation on the aircraft can be grouped into five categories: gas, aerosol, meteorological, position and data acquisition systems. Relevant specifications of the aircraft measurement system are listed in Table 2, locations of external sensors and inlet tubes are shown in Fig. 1, and the interior layout of the aircraft is depicted in Fig. 2. Most of the instrumentation used are standard commercial items and are not discussed further. Details of nonstandard parts of the system are presented below.

Sample inlet system

Air samples were brought into the aircraft cabin through inlet tubes mounted on the dummy window (Fig. 1). A schematic of the sample flow plumbing is shown in Fig. 3. Primary design considerations for the inlet system included flow streamlines, placement with respect to engine exhaust and sample residence time.

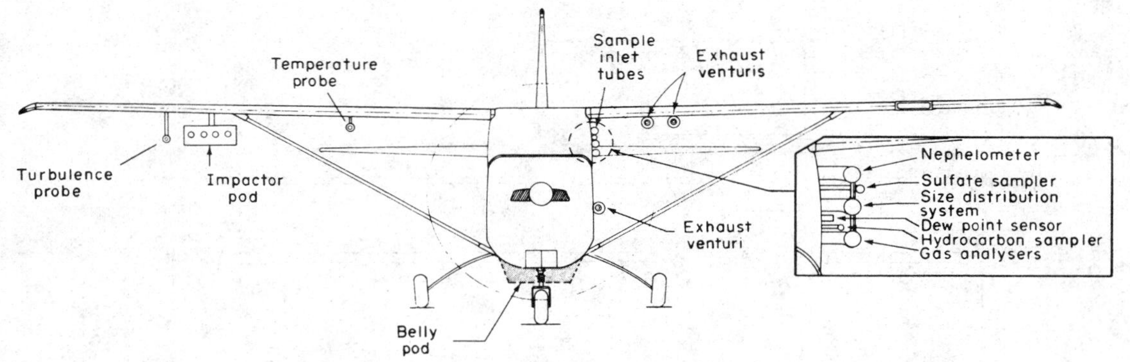

Fig. 1. Locations of external sensors and inlets for Cessna 206 plume sampling aircraft.

Adams and Koppe (1969) studied the effects of engine exhaust on a similar inlet system using carbon monoxide as an indicator of engine exhaust, and concluded that no effects could be observed. The instrument package on the Cessna 206 allowed a more detailed evaluation. In clean air, concentrations of condensation nuclei, acetylene, and nitrogen oxides were regularly measured to be at concentrations below the sensitivity of the instruments or at accepted clean air levels, indicating that contamination by engine exhaust was not a problem.

Aerosols of primary interest to the MISTT program were those in the nuclei and accumulation modes (0.005–2 μm dia). Sample losses in this size range were minimized by keeping sample lines as short and straight as possible. The size distribution measurement system is described in more detail by Cantrell *et al.* (in press). Their calculations indicate that the size distribution measurement system was greater than 90% efficient for aerosols between 0.009 and 2.5 μm dia. Some concern has been expressed to the authors about the possibility of the propeller affecting the aerosol size distribution or concentrations which are perceived by the sampling system. We believe this effect is minimal. For the size range of interest, the aerosol particles tend to follow the air streamlines and will enter the sampling inlets with little change from their ambient concentrations. In

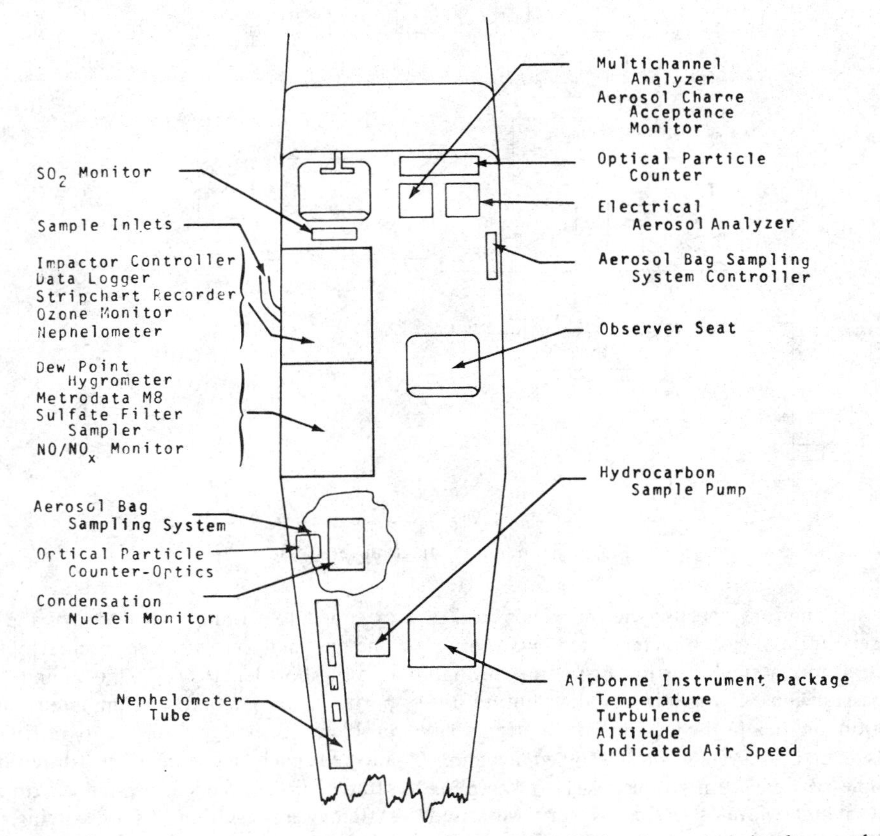

Fig. 2. Instrument layout in Cessna 206 (all instruments shown are carried and operated simultaneously).

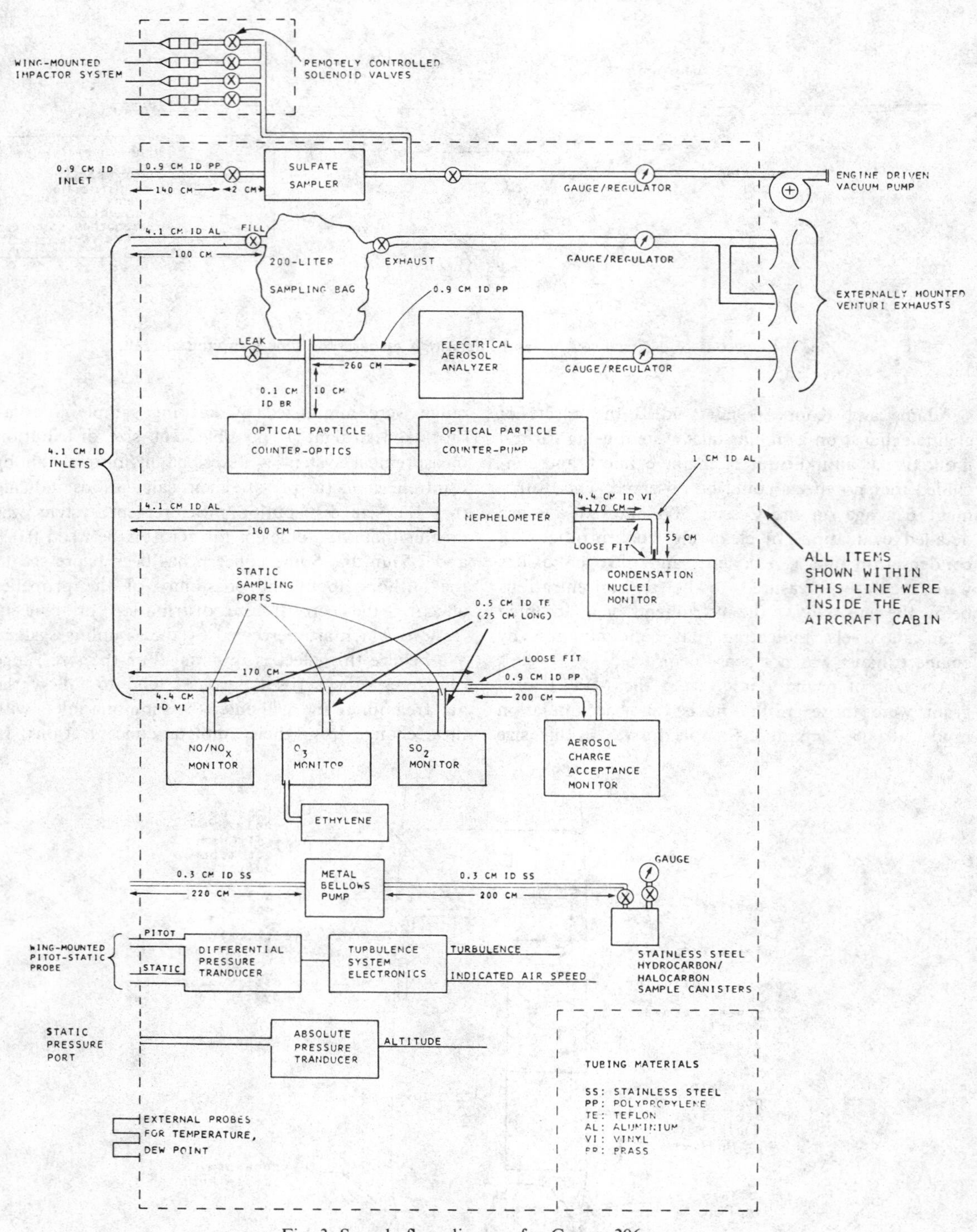

Fig. 3. Sample flow diagram for Cessna 206.

addition, at sampling speeds, the propeller makes only one revolution for each meter of forward motion and contacts only a small portion of the air through which it passes. Thus, changes in size distribution due to impaction on the propeller should be negligible.

The gas inlet flow system was designed for minimum sample residence time by using short sampling tubes and high flow rates. Ram pressure was used to move air through the sample manifold, which extended to within several cm of the gas analyzers. The gas analyzers were connected to the inlet manifold by short lengths of Teflon tubing. Static sampling ports in the manifold eliminated sample pressure problems caused by the ram air inlet system and allowed each instrument to draw only as much sample air as needed. Residence times in the inlet system were less than 1 s for all instruments except for the aerosol charge acceptance monitor, which was about 5 s (residence time is the time needed for outside air to reach the instrument). These times were low enough so that measurement response times were limited primarily by the instruments themselves.

Gas monitors

The continuous gas analyzers were commercially available instruments. The ozone and NO_x monitors required only trivial modifications for aircraft use. The SO_2 monitor utilized an electrochemical cell with a semipermeable membrane responsive to the partial pressure of SO_2. Various modifications to the cells were made during the program which resulted in cell response times of 5–20 s with varying degrees of reliability.

During the program, instrument response corrections were applied to the SO_2 data when the instrument response time was slow compared to the time required to traverse large concentration gradients, e.g. during sampling close to the source (see Markowski, 1975). Response time corrections are important to determine peak concentrations in narrow plumes for which the traverse time might be similar to the instrument response time. For determination of flow rates, however, the integral of the concentration across the plume is used. For linear instruments, this integral is not affected by slow response times, and the response correction is unnecessary. Response corrections were not applied to the 1975 and 1976 data.

Air samples for gas chromatographic analysis of hydrocarbons and halocarbons were collected in stainless steel canisters using a system designed at Washington State University. These 6 l. cans were specially treated to minimize the effects of wall reactions, and were pressurized to about 1 atm by a Metal Bellows MB-158 pump. Stainless steel tubing was used throughout this system to minimize contamination problems since it had been found that valid samples could not be obtained with Teflon tubing (Rasmussen, 1976).

Aerosol measurements

Continuous aerosol measurements included condensation (or Aitken) nuclei (CN), light scattering coefficient (b_{SCAT}), and aerosol charge acceptance. The CN monitor was described by Rich (1970) and responded primarily to particles between 0.005 and 0.8 μm dia (Sverdrup, 1977). An integrating nephelometer (Charlson *et al.*, 1969) was used to measure b_{SCAT}; this instrument was sensitive primarily to particles in the 0.1–1.0 μm dia size range.

The aerosol charge acceptance monitor was a prototype instrument developed at Washington University. Air samples were passed through an ionizer, and the charged particles were detected by an electrometer. This instrument responded primarily to particles in the 0.01–1.0 μm dia size range (Sverdrup, 1977).

Aerosol size distribution measurements were obtained using an electrical aerosol analyser (EAA) and an optical particle counter multichannel analyzer system (OPC). Together, the EAA and OPC resolved the aerosol size distribution between 0.006 and 4.5 μm dia into 11 discrete intervals. The largest and smallest size ranges were limited by the efficiency of the sample inlet system, but could be used by applying a correction factor to the data. Both instruments were operated in conjunction with a grab bag sampling system. Ram air from one of the sample manifolds was used to fill the 200 l. plastic bag from which the instruments drew their samples; about 2 s were required to fill the bag. After the size distribution analyses were complete, the bag was evaculated in preparation for the next sample. Approximately 150 s were required for a complete cycle.

Particulate sulfur concentrations were measured by chemical analysis of aerosol samples collected on glass-fiber filters. The sulfur sampler segregated particles into two size ranges: greater than and less than 3 μm dia (Macias and Husar, 1976). A flash vaporization–flame photometric analysis system was used to determine particulate sulfur concentrations in the smaller size range (Husar *et al.*, 1976). For the typical operating conditions in the aircraft (10 min sample collection time), this system was capable of measuring particulate sulfur concentrations of 3 μg m^{-3} with an accuracy of about 15%.

A wing-mounted impactor system was used to collect aerosol samples for microscopic and trace element analyses. Up to four remotely controlled impactors were mounted in a pod attached to the underside of the wing. Isokinetic flow rates and nozzles extending into undisturbed air in front of the wing allowed collection of representative aerosol samples.

The inertial cascade impactors provided three size cuts plus a backup filter. Approximate 50% cutoff diameters for the three stages were 15, 3 and 0.4 μm. Samples were collected on substrates mounted on stainless steel collection dishes. The substrate materials and coatings depended on the type of analysis to be performed on the samples. For 1975, when the system was optimized for analysis by optical microscopy, gold-coated Nuclepore substrates were used. In 1976, the system was optimized for determination of elemental composition using ion-excited X-ray emission (IEXE) techniques described by Flocchini *et al.* (1976). Therefore, mylar substrates coated with Apiezon-L grease were used.

Meteorological, position measurement and data acquisition systems

All meteorological and position measurements were made with commercially available equipment listed in Table 2. The primary data acquisition system was a Metrodata 620, which recorded 38 analog data channels plus time of day (2 channels) onto magnetic tape cartridges. Several of the analog data channels were used to record instrument status (e.g. ranges, valve status) and flight status (i.e. type of sampling run) information. Each of the 40 channels were scanned each 0.8 s, corresponding to a distance of about 25 m at minimum sampling speeds. The tape cartridges had a capacity of about 4 h of data. A dual-channel stripchart recorder served as a supplementary data acquisition system for the 1976 program, although its primary purpose was to aid the aircraft crew in plume tracking.

OPERATIONAL PROCEDURES

To obtain valid data from a system as complex as that described here, we have found that a formalized operational procedure and an extensive series of checklists are necessary. The procedures specific to the operation of the Cessna 206 for a single sampling mission are outlined below. A description of operational procedures and flight patterns relevant to the overall MISTT program can be found in Husar *et al.* (1978).

Outline of operational procedures

Time relative to takeoff	Procedure
−150 min	Visually inspect sample lines, electrical connections, external sensors, instruments. Preflight aircraft. Turn on instruments for warmup.
−60 min	Using preflight checklist, perform checks of switch settings, flow rates, span and zero settings, operation of data system and grab sampling systems. Load grab samplers (filters, impactor substrates, hydrocarbon canisters). Fill out preflight documentation forms.
−10 min	Start engine and switch from ground to aircraft power.
+5 min	Perform checks on instruments dependent on venturi operation and 28V power.
Twice during flight	Fill out instrument checksheet documenting ranges and any available instrument cross checks—e.g. altimeter reading vs pressure sensor, data logger reading vs instrument meter reading.
During each sampling pass	Record sampling times, routes, altitudes, instrument ranges, grab sample identification numbers, general comments, etc. on flight record forms.
After landing	Program Manager reviews flight documentation for completeness. Sampling maps are prepared. Data tapes, forms, and samples are copied, packaged and distributed as appropriate

CALIBRATION

A separate crew provided frequent calibrations of the continuous gas analyzers. The calibration system provided NO calibration gas using gas phase dilution of a 100 ppm reference gas bottle, and NO_2 and O_3 using gas phase titration techniques. Sulphur dioxide calibrations were performed using a permeation tube operated in a temperature-controlled water bath. Critical parts of the calibration system (e.g. flow controllers, NO span gas) were checked before and after the field program to verify their stability. In general, multipoint calibrations were performed on the continuous gas analysers after each sampling mission. Calibration gas concentrations were chosen to reflect the range of concentrations encountered on sampling missions (SO_2: 0.15–2.4 ppm, NO: 0.04–0.5 ppm, O_3: 0.02–0.3 ppm). The efficiency of the NO_2 converter in the NO/NO_x analyzer was checked during each calibration.

The aerosol size distribution measurement system, condensation nuclei monitor, and aerosol charge acceptance monitor were calibrated by the University of Minnesota before the field program using the monodisperse aerosol generation techniques described by Marple (1974). Calibrations of the integrating nephelometer were performed by MRI before and after the field program, using Freon-12 as the reference gas. Span and zero checks of the nephelometer using its internal calibrator were performed as part of the routine pre-flight procedures. The technique used to calibrate the sulfate analysis system was described by Husar *et al.* (1975).

Although a complete systems check involving a fly by comparison with a comparable ground station was not performed with this system, there have been opportunities to compare measurements of individual parameters with ground measurements of the same parameter. In general, when spatial and temporal variations are small, the aircraft and ground gas measurements agree within about 20% for upscale readings. Other investigators with comparable sampling systems have made formal cross comparisons with similar results (Research Triangle Institute, 1975).

The aerosol monitors have not been cross-checked directly against ground monitors. However, in studies where ground data were available, but not simultaneously, the concentrations and size distributions measured by the ground stations were similar to those measured by the aircraft under similar ambient conditions (see Russel, 1976).

The temperature and dewpoint measurements have been cross-checked with ground measurements on various occasions and usually agreed to within 1°C.

Many of the instruments in the 206 are sensitive to changes in pressure. For the altitude range over which sampling has been performed in Project MISTT (1500 m msl), these effects are generally less than 15% of full-scale. The altitude response of most of the instruments has been determined, however, and corrections were applied to the data during processing when necessary.

In general, we believe that the accuracy of the data

obtained by the aircraft system is comparable to that obtainable by most ground stations. However, the analyses performed on the data and the conclusions drawn from them are mostly dependent upon the relative concentrations at various locations and not on the absolute concentration. Systematic errors in the gas and aerosol concentration data of about 25% or less would have a minimal impact on the conclusions drawn from the data.

DATA PROCESSING AND ANALYSIS

Initial processing of the aircraft data was performed in the field and the preliminary results were used to guide further sampling missions (Husar *et al.*, 1978). This processing involved copying the Metrodata cartridges to a 1/2-in. magnetic tape and producing plots of important parameters as a function of time. Final processing was performed at MRI and other laboratories involved in the project, and included application of zero drift and calibration corrections to the data. The outputs of the final data processing step were 1/2-in. magnetic data tapes, as well as plots of parameters measured by the aircraft for each sampling run.

An example of data obtained by the Cessna 206 during a sampling pass perpendicular to a power plant plume is shown in Fig. 4. This figure is indicative of the complex interactions which can occur in plumes and points out the necessity of measuring several parameters simultaneously in order to be able to interpret the data. Plots such as those shown in Fig. 4 were used for making day-to-day sampling decisions in the field and for subsequent detailed analysis of the data. Other examples of data obtained by the 206 and analyses of these data can be found in the references mentioned in the introduction.

CONCLUSIONS

The Cessna 206 measurement system described in this paper has proven to be reliable, highly mobile, and easy to operate. The large number of simultaneous measurements possible with the system and the operational procedures which allow rapid feedback of the sampling results to the investigators have greatly facilitated the study of pollutant transformation and transport in plumes.

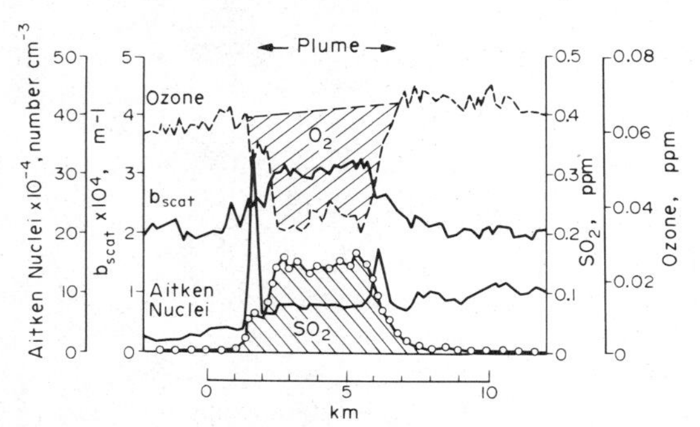

Fig. 4. Profiles of Aitken nuclei concentration (AN), light scattering aerosol (b_{SCAT}), sulfur dioxide (SO_2), and ozone (O_3) obtained on a pass through a power plant plume perpendicular to the wind direction (also presented in Wilson, 1978).

Acknowledgements—This work has been supported in part by the United States Federal Interagency Energy/Environmental Research and Development program through various contracts with the EPA, ESRL, Aerosol Research Branch. We appreciate the guidance provided by Dr. William Wilson, Director of Project MISTT. We also appreciate the technical support of Dr. R. Husar, Washington University, and Dr. B. Cantrell, University of Minnesota, in designing and fabricating the aircraft system and data processing techniques described here.

REFERENCES

Adams D. F. and Koppe R. K. (1969) Instrumenting light aircraft for air pollution research. *J. Air Pollut. Control Ass.* **19,** 410–415.

Blumenthal D. L., Anderson J. A. and Sem G. J. (1974) Characterization of Denver's urban plume using an instrumented aircraft. *67th Ann. Mtg of the Air Pollution Control Association*, Denver, Colorado, June 9–13.

Blumenthal D. L., White W. H. and Smith T. B. Anatomy of a Los Angeles smog episode: pollutant transport in the daytime sea breeze regime. *Atmospheric Environment* (in press).

Cantrell B. K., Brockman J. E. and Dolan D. F. A mobile aerosol analysis system: with applications. *Atmospheric Environment* (to be submitted).

Cantrell B. K. and Whitby K. T. (1978) Aerosol size distributions and aerosol volume formation rates for a coal-fired power plant plume. *Atmospheric Environment* **12,** 323–333.

Charlson R. J., Ahlquist N. C., Selvidge H. and MacCready P. B., Jr. (1969) Monitoring of atmospheric aerosol parameters with the integrating nephelometer. *J. Air Pollut. Control Ass.* **19,** 937–942.

Flocchini R. G., Cahill T. A., Shodoan D. J., Lange S. J., Eldred R. A., Feeney P. J., Wolfe G. W., Simmeroth D. C. and Suder J. K. (1976) Monitoring California's aerosols by size and elemental composition. *Envir. Sci. Technol.* **10,** 76–82.

Gillani N. V., Husar R. B., Husar J. D. and Patterson D. E. (1978) Project MISTT: kinetics of particulate sulfur formation in a power plant plume out to 300 km. *Atmospheric Environment* **12,** 589–598.

Husar R. B., Gillani N. V., Patterson D. E., Husar J. D. and Wilson W. E. (1978) Sulfur budget in large plumes. *Atmospheric Environment* **12,** 549–568.

Husar J. D., Husar R. B., Macias E. S., Wilson W. E., Durham J. L., Shepherd W. K. and Anderson J. A. (1976) Particulate sulfur analysis: application to high time resolution aircraft sampling in plumes. *Atmospheric Environment* **10,** 591–595.

Husar J. D., Husar R. B. and Stubits P. K. (1975) Determination of submicrogram amounts of atmospheric particulate sulfur. *Analyt. Chem.* **47** 2062–2065.

Husar R. B., Patterson D. E., Blumenthal D. L., White W. H. and Smith T. B. (1977) Three-dimensional distribution of air pollutants in the Los Angeles Basin. *J. appl. Met.* **16,** 1089–1096.

Macias E. S. and Husar R. B. (1976) A review of atmospheric particulate mass measurement via the beta attenuation technique. *Fine Particles.* Academic Press, New York, p. 535.

Markowski G. R. (1975) A useful method for transient response analysis and data correction applied to aircraft plume sampling. Presented at the 68th Ann. Mtg of the Air Pollution Control Association, Boston, Massachusetts, June.

Marple V. A. (1974) Development, calibration and application of size distribution instruments at the University of Minnesota. National Bureau of Standards Special Publication 412. Aerosol Measurements, *Proc. of Seminar on Aerosol Measurements*, Gaithersburg, Maryland, May.

Ogren J. A., Blumenthal D. L. and White W. H. (1976) Study of ozone formation in power plant plumes. Presented at the Air Pollution Control Association Speciality Conference on Ozone/Oxidants: Interactions with the Total Environment, Dallas, Texas, March.

Rasmussen R. A. (1976) personal communication.

Research Triangle Institute (1975) Investigation of rural oxidant levels as related to urban hydrocarbon strategies. EPA-4503-75-036.

Rich T. A. (1970) Particles in air pollution. Colloquium on the Physics and Chemistry of Aerosols, Fontenay-Aux-Roses, France, September.

Russell P. A. (ed) (1976) Denver air pollution study-1973. Proceedings of a symposium, Environmental Sciences Research Laboratory, Office of Research and Development, Vol. I, EPA-600/9-76-007a; Vol. II, EPA-600/9-77-001.

Smith T. B., Blumenthal D. L., Anderson J. A. and Vanderpol A. H. (1978) Transport of SO_2 in power plant plumes: day and night. *Atmospheric Environment* **12,** 605–611.

Sverdrup G. M. (1977) Parametric measurement of submicron atmospheric aerosol size distributions, PhD Thesis, University of Minnesota.

Whitby K. T., Cantrell B. K. and Kittleson D. B. (1978) Nuclei formation rates in a coal-fired power plant plume. *Atmospheric Environment* **12,** 313–321.

Whitby K. T., Cantrell B. K., Husar R. B., Gillani N. V., Anderson J. A., Blumenthal D. L. and Wilson W. E., Jr. (1976) Aerosol formation in a coal-fired power plant plume. Presented before the Division of Environmental Chemistry, American Chemical Society, New York, New York, April.

Whitby K. T. (1976) Electrical measurement of aerosols. In *Fine Particles.* Academic Press, New York, p. 581.

White W. H., Anderson J. A., Blumenthal D. L., Husar R. B., Gillani N. V., Husar J. D. and Wilson W. E., Jr. (1976a) Formation and transport of secondary air pollutants: ozone and aerosols in the St. Louis urban plume. *Science* **194,** 187–189.

White W. H., Anderson J. A., Knuth W. R., Blumenthal D. L., Hsiung J. C. and Husar R. B. (1976b) Midwest interstate sulfur transformation and transport project: aerial measurements of urban and power plant plumes, summer 1974. EPA-600/3-76-110.

Wilson W. E., Jr (1978) Sulfates in the atmosphere: a progress report on Project MISTT. *Atmospheric Environment* **12,** 537–547.

Atmospheric Environment Vol. 12, pp. 621–631. Pergamon Press 1978. Printed in Great Britain.

SATELLITE DETECTION OF LONG-RANGE POLLUTION TRANSPORT AND SULFATE AEROSOL HAZES

W. A. LYONS and J. C. DOOLEY, JR.
MESOMET, Inc., 3415 University Avenue S.E., Minneapolis, MN 55414
and
K. T. WHITBY
Department of Mechanical Engineering, University of Minnesota, Minneapolis, MN 55455, U.S.A.

(*First received* 23 *June* 1977 *and in final form* 20 *September* 1977)

Abstract—While not designed for such a task, meteorological satellites now play a growing role in our understanding of long range transport of secondary pollutants. This paper reports on a demonstration project showing that currently available synchronous satellite data can detect the aerial extent and motion of large-scale "hazy" air masses associated with sulfate and ozone episodes. An interactive computer graphics system is utilized showing that digital satellite data can obtain precise measurements of upward scattered solar radiation which is correlated to aerosol optical thickness and therefore to sulfate concentrations. Measurements over Lake Michigan for instance, reveal over-water image brightness enhanced fully 60–70% as visibility estimates of b_{scat} increased from $1.8 \times 10^{4}\,m^{-1}$ to $5.7 \times 10^{-4}\,m^{-1}$. Digital satellite data is shown to have great promise in mapping sulfate haze areas, especially over water.

INTRODUCTION

As has been pointed out by Altshuller (1976) and the U.S. Environmental Protection Agency (1975), the emphasis on pollution control and abatement has shifted from the local control strategies espoused by the Clean Air Act. Mesoscale and long-range transport and chemical transformation, which a decade ago were the concern of relatively few scientists and even fewer regulatory officials, now have assumed almost paramount priority with regard to secondary pollutants such as photochemical oxidants and sulfate aerosols.

Significant inter-regional and international transport of primary and secondary pollutants has been repeatedly illustrated. One of the first studies of a synoptic scale air pollution episode was discussed by Hall *et al.* (1973) suggesting a massive pall of hazy, smoky air drifted from the Ohio Valley into the Great Plains. Lyons and Cole (1976) demonstrated that ozone levels at any given midwestern U.S. location are the combined effects of local emissions, mesoscale transport (time periods less than one day), and synoptic advection, all intermingled in a manner making it most difficult to separate the fraction resulting from each mechanism. Coffey and Stasiuk (1975) show ozone being transported into urban locales from rural areas, the apotheosis of earlier conventional thinking. Lyons and Rubin (1976) detected the Chicago plume advecting SO_2 across the Wisconsin–Illinois border at more than 22,000 kg h^{-1}. White *et al.* (1976) summarize some MISTT program measurements of the St. Louis plume up to 240 km downwind with increasing levels of ozone and visibility-reducing aerosols forming *in situ*, especially beyond 50 km. Sulfate transport for over 1000 km in Europe and the North Atlantic has been documented by Prahm *et al.* (1976) among many others.

While urban SO_2 levels have decreased in the U.S. by 50% in the decade 1960–1970, sulfate aerosol levels have not fallen in urban areas and indeed have risen in rural locations. The reduction of near-source ground level concentrations of SO_2 by increasing effective stack heights has played a major role in the "sulfate problem". Over 400 million tons of coal are burned yearly in the U.S. Power plants account for approx 55% of the total U.S. SO_2 emissions of 33 million tons. U.S. population and SO_2 emissions, while concentrated in the northeast, are not colocated (Fig. 1). Also, while the highest sulfate concentrations appear to occur in West Virginia and Pennsylvania, SO_2 emissions and resultant effects such as acid rain and reduced visibility at great distances downwind are interrelated through very complex meteorological processes. Sulfate aerosol concentrations an order of magnitude above natural background have been monitored in many locales, and Husar *et al.* (1976) estimated that perhaps 40% of the mass occurred as sulfuric acid.

Since the sulfate aerosols form through complex gas-phase reactions, with particle sizes being mostly in the accumulation mode (0.05–2.0 μm), one would expect such hazes to be visible as a result of light scatter to an observer in space.

This paper reviews (1) how meteorological satellite data have furthered our knowledge of long-range transport, and (2) the feasibility of providing semi-quantitative evaluations of aerosol concentrations from spacecraft.

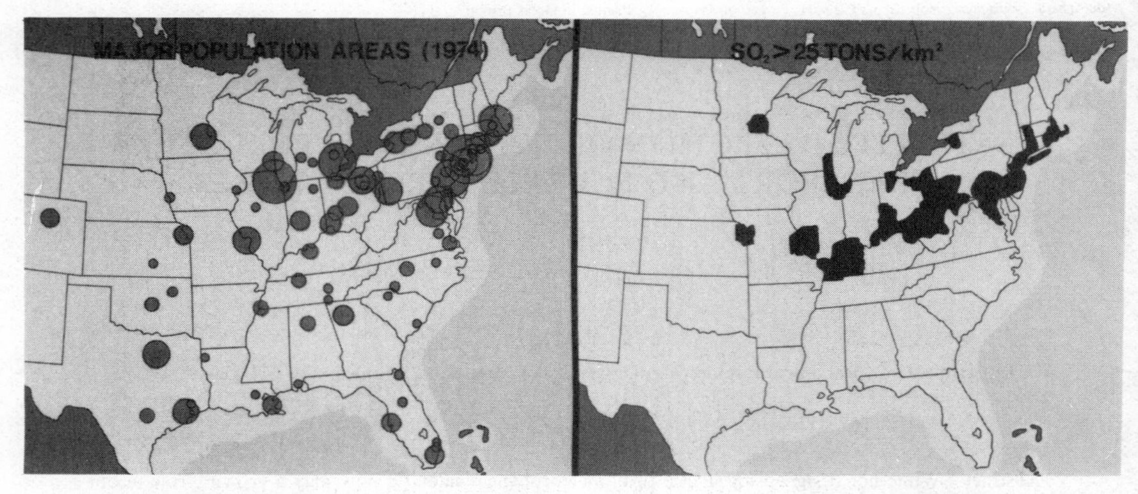

Fig. 1. Location of major U.S. population areas east of the Rocky Mountains and the source inventory for sulfur dioxide showing areas with emissions greater than 25 tons km^{-2} y^{-1} (22,675 kg km^{-2} y^{-1}).

SATELLITE MONITORING

Since the launching of TIROS I in 1960, the resolution of satellite sensors has improved from more than 15 km to less than 100 m (the multi-spectral Landsat scanner). The launching of the synchronous ATS-1 satellite in 1967 produced visible images of Saharan dust clouds traversing the Atlantic into the southeastern U.S. More recently NOAA-5 revealed dust clouds blowing from New Mexico eastward to Bermuda on 23–24 February 1977 (Vermillion, 1977). Figure 2 shows a 4500 m deep dust cloud in western Texas, indicated by a milky white haze being blown northeastward into a low pressure system over Kansas. It is notable that considerable very fine brown particulate matter settled with the snowfall in southern Minnesota 12 h later. Figure 3 shows a massive plume of smoke several hundred kilometers long emanating from forest and brush fires in the Los Angeles area during the autumn of 1976. Snyder *et al.* (1976) used DMSP and NOAA-3 high resolution visible data to monitor smoke plumes extending more than 200 km away from a brush fire in the Florida Everglades.

Landsat-1 detected numerous smoke plumes from industries at the southern end of Lake Michigan drifting more than 150 km downwind (Lyons and Pease, 1973). Lyons (1975) summarized numerous satellite observations of both manmade and natural aerosols. Landsat imagery taken over the eastern end of Lake Ontario during a cloud-free fresh polar air mass allowed easy delineation of small surface features such as roadways. The exact same frame on a late summer cloud-free day when the area was influenced by a stagnant maritime tropical high pressure system, showed ground features obscured by an extremely thick haze. This day corresponded to a major sulfate aerosol episode in the northeastern U.S.

McLellan (1971) was among the first to attempt to use satellite data to measure anthropogenic pollution by examining the digital brightness values in ATS visible data over the Los Angeles basin. Griggs (1973) and Mekler *et al.* (1977) have suggested using Landsat radiance measurements in the visible bands to infer aerosol optical thicknesses and therefore the columnar aerosol density over low albedo (water) surfaces. While the technique shows great promise, the Landsat images are made over any given region on the earth only every 9 or 18 days in a narrow strip 180 km wide, and thus do not make it a likely candidate for an operational monitoring system.

Figure 4 is a print from a color 35 mm slide taken from a commercial jet aircraft flying at 11,000 m over the Ohio Valley during a sulfate haze episode. There are no water clouds in the scene. An inspection of archival images by one of the authors (Lyons) revealed that such extensive "smog blobs" could be frequently detected in synchronous satellite pictures. Figure 5 shows such an episode over the Mississippi and Ohio Valleys on 30 June 1975 (Lyons and Husar, 1976). Ernst (1975) also found a similar hazy area drifting from the eastern U.S. over the western Atlantic Ocean.

THE SMA/GOES SATELLITE SYSTEM

The first truly operational synchronous meteorological satellite (SMS) was launched on 17 May 1974. This was the beginning of a global monitoring program which will eventually total five synchronous satellites to be maintained by the U.S. (GOES-West, and GOES-East), ESRO, Japan and the U.S.S.R. The current U.S. satellites have equitorial subpoints located at roughly 135° West and 75° West longitude. Images are acquired by a visible-i.r. spin-scan radiometer (VISSR) which scans the earth at 100 rev min^{-1}, providing simultaneous visible and i.r. images of the earth. Full hemispheric views are acquired every 30 min. The visible sensor operats in the 0.55–0.70 μm

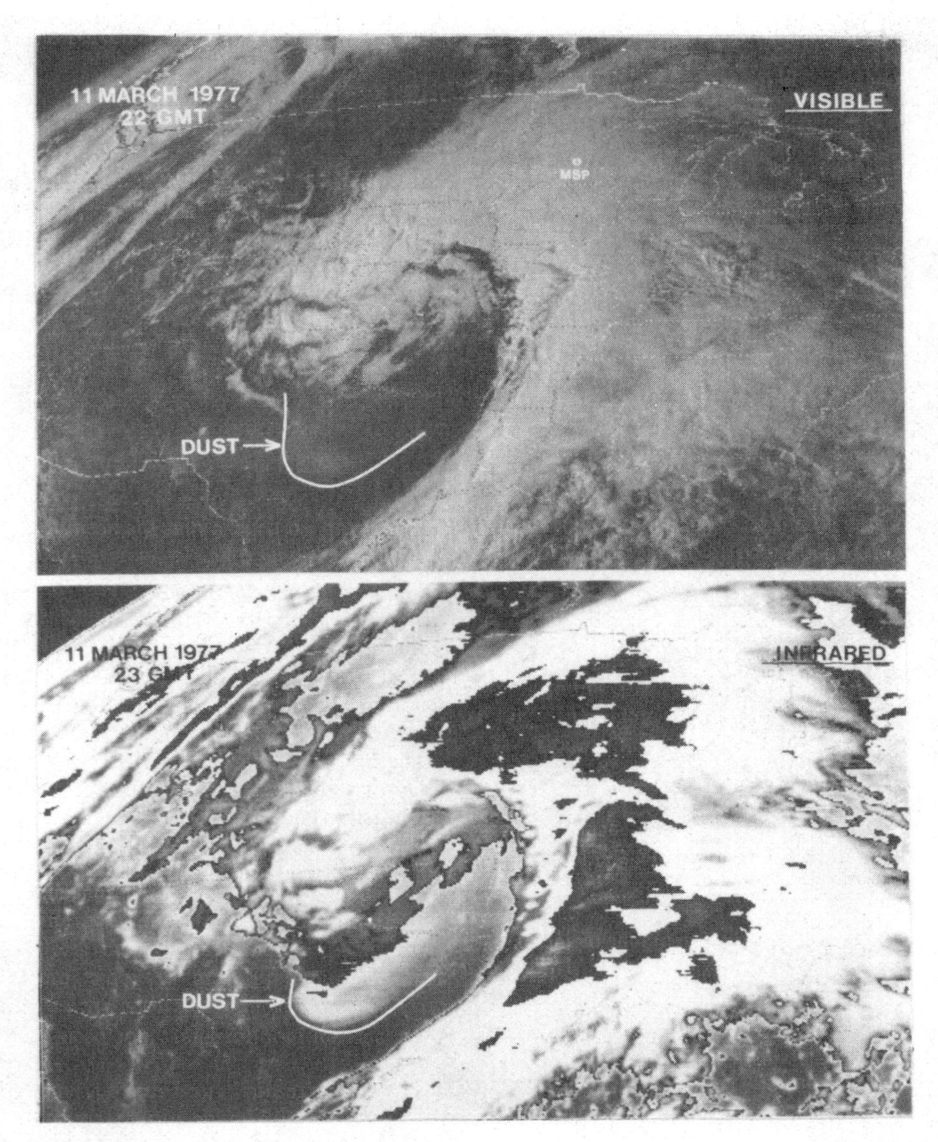

Fig. 2. GOES-East Satellite images showing dust cloud over west Texas on 11 March 1977. Top image is visible 2 km resolution picture at 2200 GMT, and lower shows computer enhanced 8 km i.r. image at 2300 GMT.

wavelength with a 1 km subpoint resolution. The SMS satellite produces an enormous data stream, totaling almost 75 billion bits daily.

In order to more efficiently manipulate and process SMS data, the University of Winsconsin–Madison Space Science and Engineering Center (SSEC) has developed over the past several years the Man-Computer Interactive Data Acquisition System, known as MCIDAS. Digital satellite data are archived locally on tape and can be displayed in real-time on a color television monitor, with interactive manipulation of the digital data. MCIDAS can provide image enhancement, digital densitometry, and overlay observed meteorological data directly from teletype observations. Successive images can be linked on an analog video disk and played in animation mode to perform motion analyses of image features. MCIDAS was utilized to determine the feasibility of using real-time digitally processed SMS data to detect the aerial extent and movement of aerosol laden air masses over the United States and adjoining oceanic areas.

DATA

SMS data were recorded on 9-track tape during much of the summer of 1976, partially in support of the MISTT program. Equipment malfunctions caused major data gaps on 20, 21 and 22 August 1976. I.r. and 2-km resolution visible images were archived from 1230 to 2230 GMT each day, in an area encompassed by 30° North and 50° North latitude and 100° West

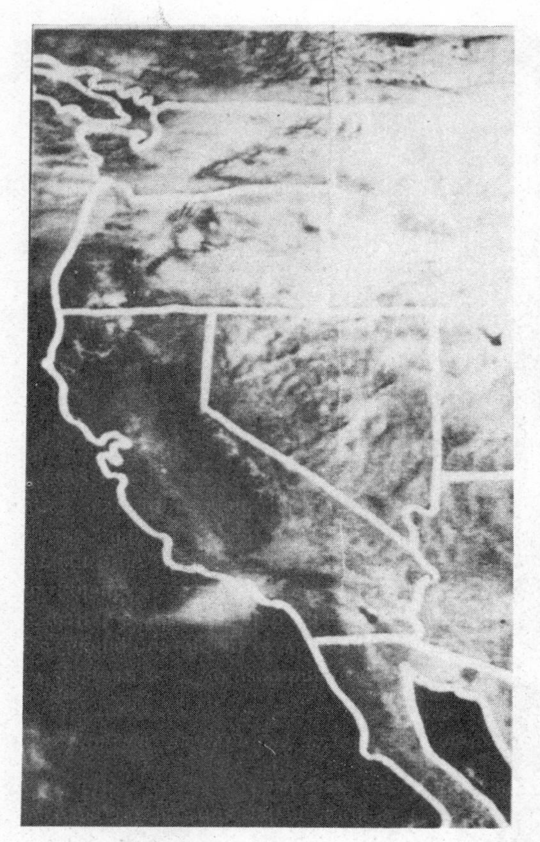

Fig. 3. Plume from 1976 forest and brush fires in hills surrounding Los Angeles, extending several hundred kilometers into the Pacific, GOES-West Satellite.

and 65° West longitude. Photographic hard copy for the August 1976 sulfate episode was also acquired from the National Environmental Satellite Service, which provided 10″ × 10″ duplicate negatives produced from archival masters. All conventional teletype and facsimile charts were retained. Sulfate measurements were obtained from the NASN Network for 16, 22 and 28 August 1976. Those data, as well as all available ozone data in the eastern two-thirds of the U.S., were requested through the U.S. Environmental Protection Agency. Unfortunately, even a year after the episode all the data had still not been processed, and we had to rely on sketchy and incomplete data for the research reported herein.

DATA ANALYSES, 16–29 AUGUST 1976

On 16 August 1976 it appeared that a stagnating anticyclone over the eastern U.S. would produce a major sulfate episode. Figure 6 shows successive midday synoptic charts of the eastern U.S. Areas of reduced visibility determined from about 300 hourly airport observations are plotted with a light stipple denoting 3–5 nm (5.5–9.2 km) and darker stipple, less than 3 nm (5.5 km).

The first 4 days (17–20 August) saw patchy local areas of reduced visibility developing through the southeastern and Mississippi Valley states. Conditions rapidly worsened on the 21st and 22nd of August, with turbid air extending in a broad band from St. Louis through New England. On the 23rd and 24th, a weak

Fig. 4. Photograph taken from commercial jet aircraft at 11,000 m over Ohio Valley during a major sulfate haze episode. There were no water clouds present in this view, only turbid air.

Fig. 5. GOES-East Satellite visible image, 2 km resolution, 1430 GMT, 30 June 1975, showing major haze area over central U.S., discussed by Lyons and Husar (1976). Image photographically enhanced from Datalog original.

cold front pushed south from Canada. In the center of the high pressure cell, visibilities showed marked improvement over the Great Lakes. However, the surface cold front traveled *through* the reduced visibility zone. In effect, it appears that the undercutting front, which was quite shallow, forced the aerosol-laden air mass aloft to a height of perhaps 1000–3000 m, and during the following days convective mixing brought the material back through the weak frontal inversion surface into the layer close to the ground. The episode reached full strength on 25, 26, 27 August. By the 27th a strong cold front began moving east with dry, clean polar air behind it. By the 29th the entire turbid air mass had been pushed into the Atlantic. Both the photographic prints and the digitally-produced video display suggested the hazy air mass was best defined about 2–3 h before sunset. Relatively low solar elevation angles seem to produce the maximum upward scatter of visible radiation at this time. As expected the hazy areas were most visible over the low albedo water surfaces such as oceans, Great Lakes, or large lakes and rivers. However a satellite meteorologist familiar with the appearance of the landscape during clear air periods could easily recognize marked dimunition of surface feature contrasts as the smog blob passed by. Figure 7 is a reproduction of a portion of one of the prints used. The hazy areas seen on the photographs and in the video display were found to coincide extremely well with the areas of reduced visibility obtained from the aviation weather network plotted in Fig. 6.

The McIDAS system was employed to track the motion of the leading edge of the smog blob. Successive daily images at 1430, 1800 and 2130 GMT were

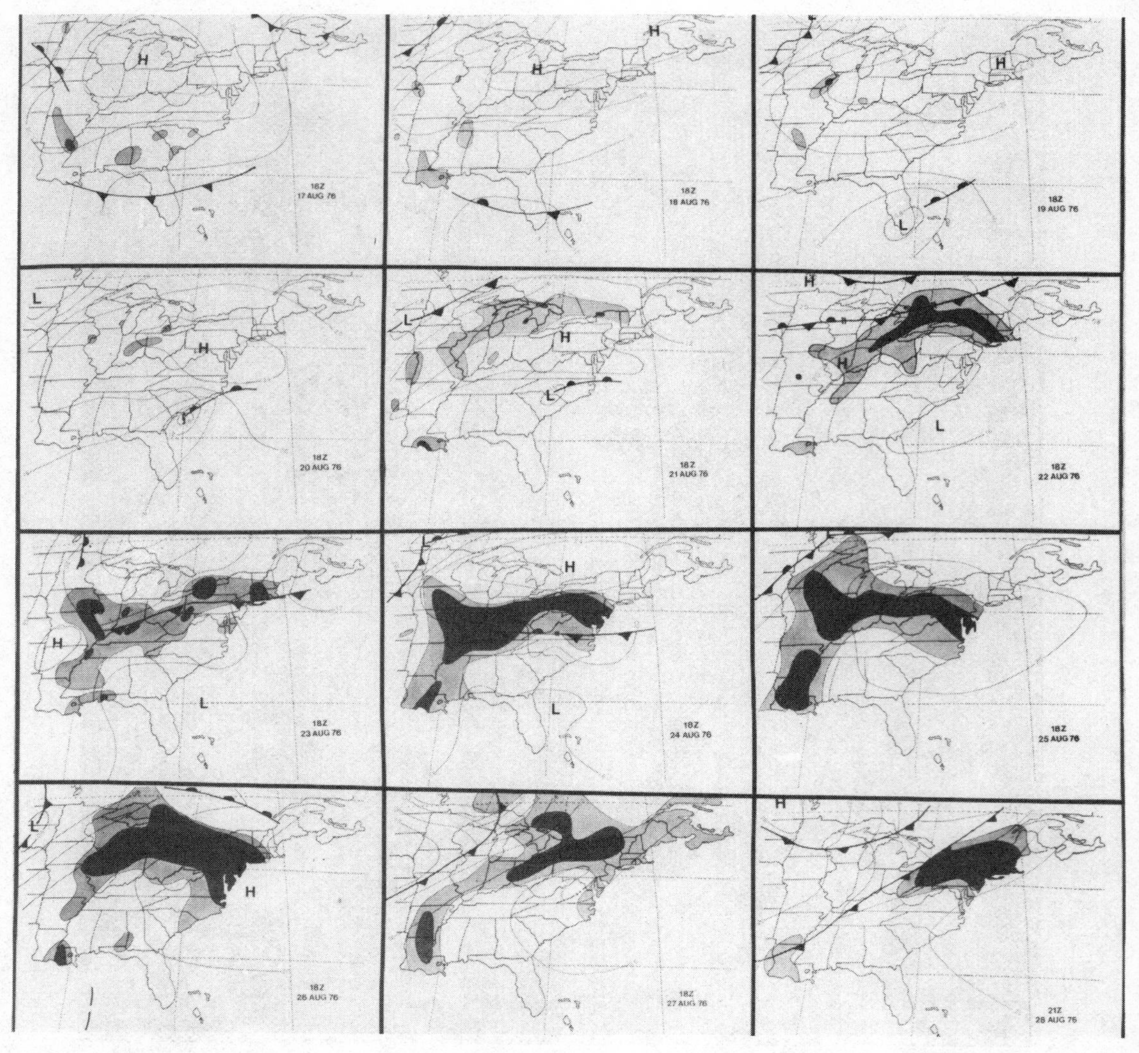

Fig. 6. Surface synoptic maps, 1800 GMT, 17–28 August 1976, with low visibility areas indicated. Isobars at 4 mb intervals.

transferred from the digital tapes to the analog video disk. This allowed for an instant time-lapse image sequence on a color TV screen by which the movement of the smog blob from day to day became clearly evident, and these are summarized in Fig. 8. The southwest motion across the Atlantic of the smog blob from the 22nd to the 26th of August revealed highly turbid air arriving over the Florida peninsula, but disappearing on the 27th as the air mass became entangled in intense thunderstorm convection causing significant wash-out. Thus polluted air, originating from sources in the Ohio Valley and the Great Lakes states traveled several thousand kilometers over the ocean to approach Miami, reminiscent of polluted air measured in that Florida city in May 1972.* The movement of the northern edge of the smog blob from 25 to 29 August corresponded very well with the advance of the cold front as well as an increased southwesterly flow of dry air from the desert that preceded the front on 26 and 27 August.

Once a particular satellite image is destroyed on the MCIDAS system, a digital densitometric analysis of a given area or along a scan line can be made via cursor control (Fig. 9). An example of such an analysis was made over central Lake Michigan (Fig. 10). On 17 August the smog blob had not yet invaded the area. The visibility during mid-day reported at airports surrounding the southern basin of Lake Michigan averaged 22 km. The maximum ozone reported for any hour at any station was only 0.10 ppm. This then is characteristic of "clean" summertime air masses in the upper midwest. Fig. 10 shows brightness values of approx 65–70 units† overland, dropping to between 44 and 48 over water, quite typical of deep water areas.

* Hall F. (1973) The Miami air pollution of May 1972: the case of long range transport unpublished manuscript). Environmental Protection Agency.

† Full scale range is 1 to 256 units.

Fig. 7. Visible GOES-East Satellite image, 1430 GMT, 26 August 1976, 2 km resolution with hazy area extending from Great Lakes to mid-Atlantic states and offshore.

On 26 August Lake Michigan was once again cloud free, but embedded within the smog blob. The mean afternoon regional visibility had decreased to only 6.8 km and the peak hourly ozone reading from surrounding stations had risen to 0.16 ppm. The image brightness trace shows significantly higher values, particularly over water, where brightness values approach 80, or a 60–70% increase.

As a first approximation, the atmospheric extinction coefficient, b_{ext}, can be related to visual range by

$$b_{ext} = 3.92/\text{visual range}.$$

Thus the 5 nm (9.2 km) and the 3 nm (5.5 km' visibility isopleths shown in Fig. 16 respond to extinction coefficient values of $4.2 \times 10^{-4}\ m^{-1}$ and $7.1 \times 10^{-4}\ m^{-1}$ respectively. Based upon preliminary results, Husar *et al.* (1976) obtained a relationship between the extinction coefficient and sulfate concentration as follows:

$$b_{ext}\ (10^{-4}\ m^{-1}) = 3.24 + 0.11\ SO_4^{-2}\ (\mu g\ m^{-3}).$$

The 16 samples used were uncorrected for r.h., but still showed a correlation coefficient as 0.7. If we adopt the above relationship, then the visibility contours shown in Fig. 6 correspond to approximate sulfate aerosol concentrations of 9 $\mu g\ m^{-3}$ and 35 $\mu g\ m^{-3}$ respectively. Classes I, II and III are defined according to visibility/sulfate classes (see caption, Table 1).

At the time of writing of this paper only preliminary data from the nationwide ozone and sulfate network were available through EPA, preventing significant statistical analyses. However the data are summarized in Table 1 and do in fact suggest some meaningful interrelationships. Fig. 8 shows the various locations discussed. For instance, Location 1, South Central Ohio, was found to be cloud-free at 2230 GMT for all days on which digital data were analysed on the MCIDAS system. During high visibility periods upward directed radiance expressed as digital brightness units were between 53 and 61 units. Sulfate readings measured in northwestern Ohio at Location 2 were

Table 1. Relationships among data at various locations, 16–29 August 1977

ID	Place	Parameter	Unit	16	17	18	19	20	21	22	23	24	25	26	27	28	29
1	SC Ohio	SMS bright.	Arb.	–	54	61	58	–	–	–	76	79	72	80	70	68	53
1	SC Ohio	Vis. class	Arb.	I	I	I	I	I	I	II	III	III	III	III	III	II	I
2	NW Ohio	Sulfates	$\mu g\ m^{-3}$	8.6	–	–	–	–	–	24.5	–	–	–	–	–	63.9	–
3	Wash.	SMS bright	Arb.	–	54	54	54	–	–	–	67	77	82	77	C	67	44
3	Wash.	SMS bright.	Arb.	–	54	54	54	–	–	–	67	77	82	77	C	67	44
3	Wash.	Max. ozone	ppm	0.10	0.08	0.07	0.09	0.11	0.19	0.20	0.11	0.11	0.13	0.13	0.11	0.08	0.08
3	Wash.	Vis. class	Arb.	I	I	I	I	I	I	I	II	III	III	III	I	II	I
4	Chspk. Bay	SMS bright.	Arb.	–	44	44	44	–	–	–	68	72	72	72	C	67	44
4	Chspk. Bay	Vis. class	Arb.	I	I	I	I	I	I	I	II	III	III	III	II	II	I
5	E. St. Louis	Max. ozone	ppm	0.05	0.09	0.09	0.08	0.12	0.14	0.08	0.08	0.13	0.18	0.15	0.15	0.05	0.06
5	E. St. Louis	Vis. class	Arb.	I	I	I	I	I	II	I	III	III	III	II	II	I	I
6	SW Conn.	Max. ozone	ppm	0.06	0.05	0.08	0.06	0.07	0.11	0.09	0.11	0.06	0.10	0.14	0.07	0.17	0.09
6	SW Conn.	Vis. class	Arb.	I	I	I	I	I	I	II	III	I	I	III	II	I	I
7	NE Penn.	Sulfates	$\mu g\ m^{-3}$	6.3	–	–	–	–	–	32.0	–	–	–	–	–	40.8	–
7	NE Penn.	Vis. class	Arb.	I	I	I	I	I	I	II	II	I	II	III	III	III	I

(–) means missing data (C) means cloud obscuration

Visibility Class I refers to greater than 5 NM (9.2 km) and estimated less than 9 $\mu g\ m^{-3}$ sulfate. Class II is for visibilities between 3 and 5 NM (5.5–9.2 km) and sulfate concentrations between 9 and 35 $\mu g\ m^{-3}$, and Class III refers to visibilities less than 3 NM (5.5 km) and sulfate concentrations higher than 35 $\mu g\ m^{-3}$.

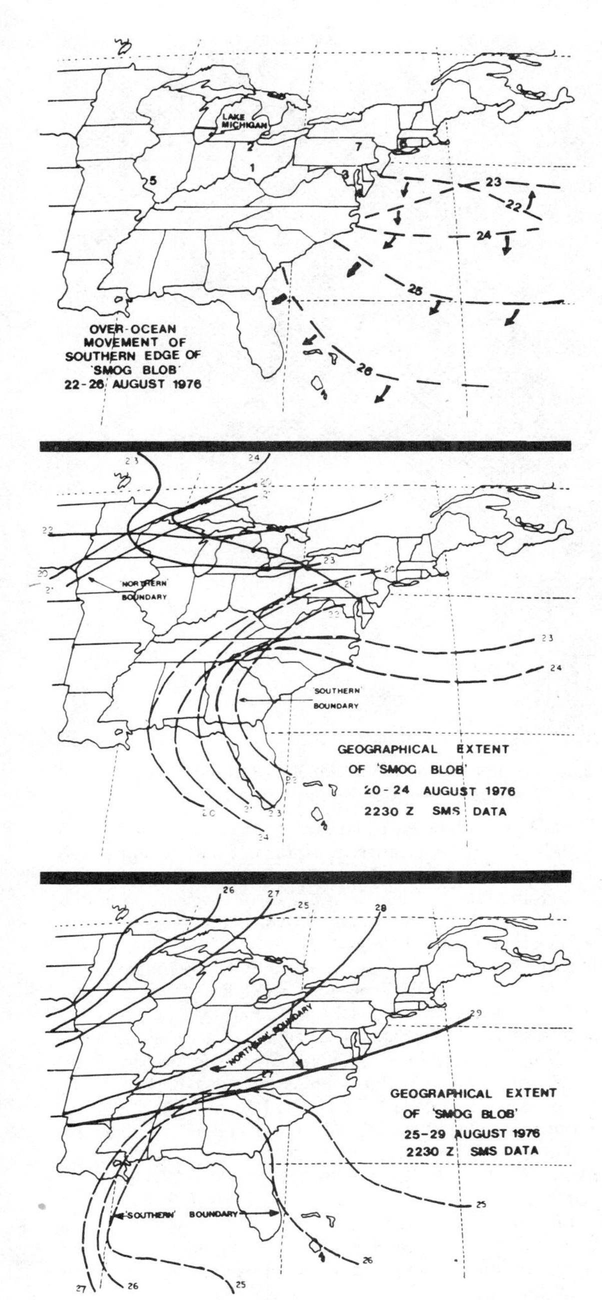

Fig. 8. (Top) Southward movement of the edge of the hazy "smog blob" from 22 to 26 August 1976 at 2230 GMT, derived from image motion analysis on the McIDAS system. Also shown are location identifiers for Table 1 and the scan line used in Fig. 10. (Middle) Boundaries of "smog blob" at 2230 GMT, 20–24 August 1976 and (Bottom) for 25–29 August 1976.

apparently below 10 μg m^{-3}. As the smog blob advected over the area the visibility class degraded to II and then III, while brightness values rose to near 80 and a high sulfate reading of 63.9 μg m^{-3} was attained on 28 August (representing 24-h average conditions prior to the cold front passage). Image brightness over land near Washington D.C. (Location 3) and over water at Chesapeake Bay (Location 4) likewise show clear correlation to the visibility class. The highest hourly ozone readings in urban East St. Louis (Location 5) and semi-rural southwest Connecticut (Location 6) show a general increase to above federal standards as the smog blob passed. At Location 7 in northeastern Pennsylvania, measured sulfate concentrations likewise were related to visibility class.

CONCLUSIONS AND SUMMARY

Sulfate aerosol concentrations, observed visibilities, and the upward scatter of solar radiation as determined by satellite, are significantly interrelated. It has been demonstrated that SMS data both in digital and photographic form are very useful as ancillary aids in the study of widespread sulfate aerosol episodes. SMS data have a distinct advantage inasmuch as they can be obtained in real time. An interactive digital graphics system based upon the MCIDAS concept could monitor the development, movement, and semi-quantitatively, the intensity of a sulfate episode. Atmospheric radiance above selected pre-calibrated land and water targets could be monitored daily. Cirrus cloud interference could be detected by using the i.r. channel. Since hourly aviation weather data can be superimposed over the satellite images and r.h. easily calculated, the computer operator could discriminate areas above 70% r.h. where hygroscopic effects could yield spurious values. Real-time detection and short-range forecasting of "smog fronts" could be realized through video display motion analyses. It would seem that as plans for the SURE project develop (Hidy *et al.*, 1976) consideration should be made for utilization of satellite data. Furthermore, satellite designers should realize that their systems, especially if multi-spectral, could play an important role in environmental monitoring in the 1980s. While there will never be a substitute for *in situ* measurements of sulfate aerosols, the spatial coverage and immediacy of data available from satellites designates it to be a most useful tool in solving this national and international problem.

Acknowledgements—Thanks are due to J. T. Young and G. Chatters of the Space Science and Engineering Center, University of Wisconsin–Madison, for their assistance in digital SMS data tape analysis. Additional meteorological analyses were performed by E. M. Rubin and R. W. Dixon. Manuscript preparation handled by K. R. Walker. This work was supported in part by the U.S. Environmental Protection Agency under Grand R-80385103 to the University of Minnesota.

REFERENCES

Altshuller A. P. (1976) Regional transport and transformation of sulfur dioxide to sulfates in the U.S. *J. Air Pollut. Control Ass.* **26,** 318–324.

Coffey P. E. and Stasiuk W. N. (1975) Evidence of atmospheric transport of ozone into urban areas. *Envir. Sci. Techn.* **9,** 59–62.

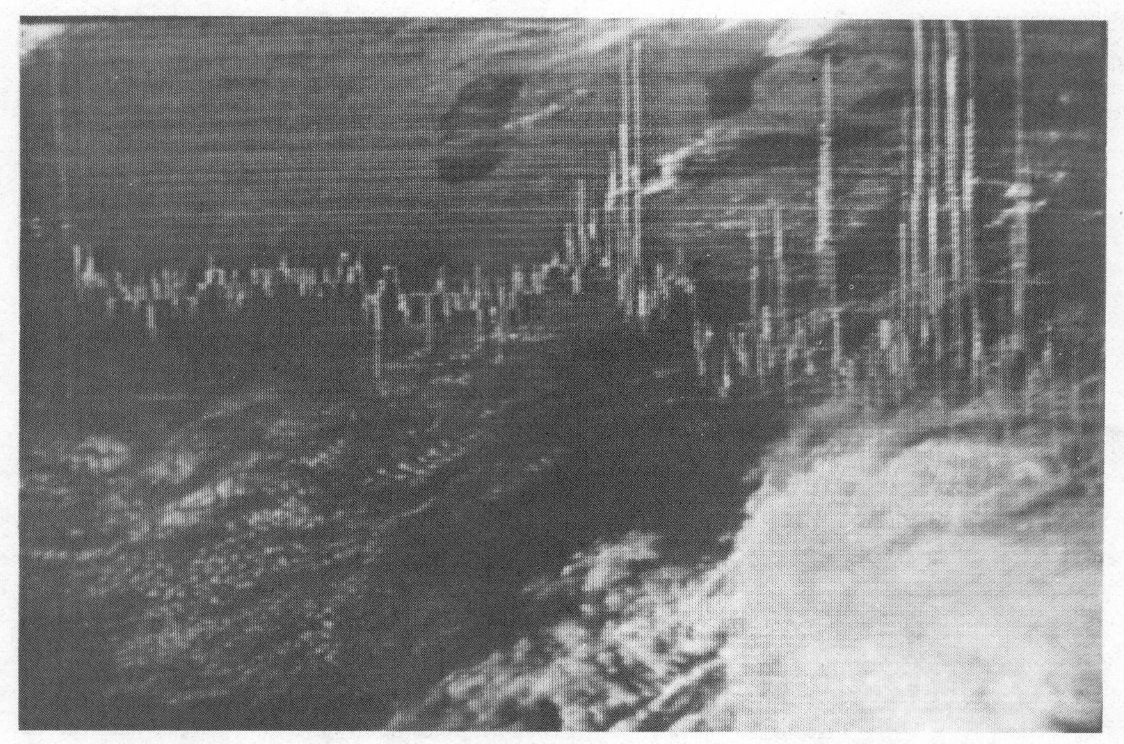

Fig. 9. Photograph of television screen showing GOES image with superimposed MCIDAS plot of digital image brightness along as a selected scan line. Image quality is severely degraded by re-photography.

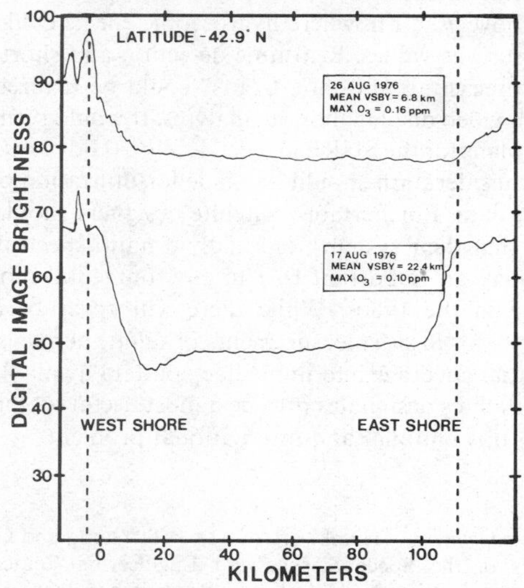

Fig. 10. Plot of digital image brightness values using MCIDAS system on identical west-to-east scan lines across central Lake Michigan, at 2130 GMT, 17 August and 26 August 1976. Skies were free of clouds in both cases, and differences are attributed to increased aerosol loading.

Ernst J. A. (1975) A different perspective reveals air pollution. *Weatherwise* **28,** 215–216.

Griggs M. (1973) Determination of aerosol content of the atmosphere, in *Symp. on significant results obtained from the earth resources technology satellite*, NASA, Goddard Space Flight Center, Greenbelt, MD.

Hall F. P., Jr., Duchon C. E., Lee L. G. and Hagon R. R. (1973) Long range transport of air pollution : A case study, August 1970. *Monthly Weath. Rev.* **101,** 404–411.

Husar R. B., Gillani N. V., Husar J. D., Paley C. C. and Turcu P. N. (1976) Long range transport of pollutants observed through visibility contour maps, weather maps and trajectory analysis. In *Preprints, third symp. atmospheric turbulence, diffusion and air quality*, American Meteorological Society, Boston, MA, 344–347.

Hidy G. M., Tong E. Y. and Mueller P. K. (1976) Design of the Sulfate Regional Experimental (SURE), Electric Power Research Institute, EPRI EC-125, Palo Alto, Ca.

Lyons W. A. (1975) Satellite detection of air pollutants. In *Remote Sensing Energy-Related Studies* (edited by T. N. Vaziroglu) pp. 263–290, John Wiley, New York.

Lyons W. A. and Cole H. S. (1976) Photochemical oxidant transport: Mesoscale lake breeze and synoptic-scale aspects. *J. appl. met.* **15,** 733–743.

Lyons W. A. and Husar R. B. (1976) SMS/GOES visible images detect a synoptic-scale air pollution episode. *Monthly Weath. Rev.* **103,** 1623–1626.

Lyons W. A. and Pease S. R. (1973) Detection of particulate air pollution plumes from major point sources using ERTS-1 imagery. *Bull. Am. Met. Soc.* **54,** 1163–1170.

Lyons W. A. and Rubin E. M. (1976) Aircraft measurements of the Chicago urban plume at 100 km downwind. In *Preprints, third symp. atmospheric turbulence, diffusion and air quality*, 358–365. American Meteorological Society, Boston, MA.

McLellan A. (1971) Satellite remote sensing of large scale local atmospheric pollution. In *Proc. 2nd Int Clean Air Congr.* 1380–1388, Air Pollution Control Association, Pittsburgh, PA.

Mekler Y., Quenzel H., Ohring G. and Marcels I. (1977) Relative atmospheric aerosol content from ERTS observations. *J. geophys. Res.* **82,** 967–970.

Prahm L. P., Torp U. and Stern R. M. (1976) Deposition and transformation rates of sulphur oxides during atmospheric transport over the Atlantic. *Tellus* **28,** 355–372.

Snyder J. F., Ashman J. P. and Brandli H. W. (1976) Meteorological coverage of Florida Everglades fires. *Monthly Weath. Rev.* **104,** 1330–1332.

U.S. Environmental Protection Agency (1975) Position paper on regulation of atmospheric sulfates. EPA-450/2-75-85 U.S.EPA, Research Triangle Park, NC.

Vermillion C. H. (1977) NOAA-5 views dust storm. *Bull. Am. met. Soc.* **58,** 330.

White W. H., Anderson J. A., Blumenthal D. L., Husar R. B., Gillani N. V., Husar J. D. and Wilson W. E., Jr. (1976) Formation and transport of secondary air pollutants: ozone and aerosols in the St. Louis urban plume. *Science* **194,** 187–189.

Atmospheric Environment Vol. 12. pp. 633–639. Pergamon Press 1978. Printed in Great Britain.

ASSESSMENT OF THE FATE OF SULFUR DIOXIDE FROM A POINT SOURCE

J. NEGUS DE WYS
Biology Department, Environment Studies Lab, University of Utah, Salt Lake City, Utah, 84108, U.S.A.

A. CLYDE HILL
Biology Department, University of Utah, Salt Lake City, Utah, 84108, U.S.A.

and

ELMER ROBINSON
Chemical Engineering Department, Washington State University, Pullman, Washington, 99163, U.S.A.

(*First received* 15 *June* 1977)

Abstract—A simplified plume transport model is adopted to assess reaction and scavenging processes. Experimental data on the conversion rate of SO_2 to SO_4^{2-} and the uptake rates by vegetation and soil are used to calculate the relative importance of the various fates of sulfur dioxide. Meteorological parameters include wind speed, day and night mixing heights, night diffusion above the mixing height and plume dispersion angle. Other parameters considered are length of day and night, region, season, type of canopy, and emission rate of SO_2 from a point source.

1. INTRODUCTION

The purpose of this paper is to relate depositional velocity measurements of sulfur dioxide into vegetation canopies and soils to the other fates of this pollutant and thus assess the importance of surface cover absorption. The experimental data may better be interpreted within a modeling framework that makes certain assumptions based on laboratory and field data. A model was developed and several calculations have been made in this report assuming a single point source of sulfur dioxide and certain typical sets of meteorological and diurnal conditions.

This is not a predictive transport model, but rather a simplified expanding box model that permits consideration of the relative loss percentages due to various fates of sulfur dioxide, and conversion to submicron sulfate particles. Through calculations of percent loss of sulfur dioxide, and accumulation of submicron sulfate particles, the residence time of sulfur dioxide is then calculated from the percentage of submicron sulfate particles remaining in the atmosphere.

2. MODEL

Many meteorological transport models exist in the literature for analysis of long range transport of gaseous pollutants. The simplified plume transport model used here is designed to provide a framework for the study of the effects of in-plume reactions, in-plume scavenging, loss from the boundary layer by upward diffusion, and surface deposition losses. These processes are typically relatively slow processes. Thus, the model design emphasizes mechanisms occurring over considerable distances from the source and the scale of the considered process is large. The emphasis is on what is happening within the plume and the effect on sulfur dioxide and submicron sulfate particles remaining in the atmosphere. Residence times and relative loss percentages are the aim rather than emphasizing the simulated transport model.

The formula used for calculating the concentration of sulfur dioxide at a distance from the source is:

$$C_x(SO_2) = \frac{QP}{2Y(X_n - X_{n-1})h} \tag{1}$$

where

$C_x(SO_2)$ = SO_2 concentration at distance X from the source ($\mu g\ m^{-3}$)
Q = total SO_2 emission from the source covered by travel period $X_n - X_{n-1}$ (μg)
P = the % SO_2 remaining in the atmosphere
$2Y$ = the plume width at distance X (meters)
$(X_n - X_{n-1})$ = the length of model box segment at distance X (meters)
h = mixing height at distance X (meters)

For the remaining pollutant at the end of box 1:

$$P_{x_1} = \left(1 - \exp\left[-\frac{V_d(SO_2) + V_u}{h}\right]\frac{x}{\bar{u}}\right) - Ct - \frac{V_f}{t_1}t. \tag{2}$$

A.E. 12—1 3—00

where

V_d = depositional velocity ($m\ s^{-1}$) of SO_2
V_f = precipitation frequency (episodes per t_1 times % washout per episode)
V_u = upward diffusion ($m\ s^{-1}$)
C = conversion rate of SO_2 to SO_4^{2-} ($\%\ s^{-1}$)
X = distance (meters)
$\bar{u}$ = mean wind speed (meters s^{-1})
t_1 = time over which precipitation is averaged (s)
t = time for model segment being calculated (s)

and

$$P_{x_2} = P_{x_1} \exp\left[-\frac{V_d(SO_2) + V_u}{h} \right] \frac{x}{\bar{u}} - Ct - \frac{V_f}{t_1} t \quad (3)$$

where

P_{x_2} = remaining % SO_2 at the end of box 2.

This mathematical expression incorporates the various transformation and scavenging mechanisms proposed for SO_2. The surface loss time and the transfer of material out of the boundary layer, V_d and V_u, includes an exponential approach that would permit the incorporation of changes in these parameters, specifically V_d, on a more detailed basis within a given box sector. The atmospheric reaction and precipitation scavenging parameters were included in a linear fashion to simplify calculations and because only longer term average scavenging factors were expected to be used. In the several preliminary forms of this model various exponential and linear terms were used. The form presented above was the one selected to carry out the calculations presented in this paper.

The angular expansion of the box units was based upon the angular wind shear between the surface and the wind and upper part of the mixing layer.

The term expressing a loss or transfer of SO_2 out of the mixing layer and into the remaining portion of the troposphere has not been usually included in transport models. It has been included here to permit estimates to be made of this possible aspect of multi-day transport and to recognize that there is some exchange between the lower layers of the atmosphere and layers aloft, by considering this factor as a sort of negative or upward deposition velocity. The magnitude of this term was estimated by Wooldridge (1977) from a consideration of the nature of vertical diffusion factors for various stability conditions. This loss out of the mixing layer was considered to be active only at night and thus resulted in a reduction of the flux into the next day's diffusion processes. In practice this loss might be achieved by large scale atmospheric motions but is unlikely to exceed the values in this model.

The time period in which the Gaussian plume model applies is a very short period to the 1–3 day transport period here considered. Thus, it is disregarded in the calculations and model presented in this report. In the sample calculations a plume angle of 20° is assumed. Here Y is the distance from the plume axis to the plume edge and thus the plume width is $2Y$. Wind speed determines the length of a box segment on the X axis. Appropriate mixing heights were selected for day and night (Holzworth, 1972). For these calculations it is assumed that during the daytime the pollutant is uniformly mixed from the surface to the mixing height.

It is assumed that total pollutant emitted from the source in a unit of time, e.g. 1 h (100%), moves into a box volume defined by the plume dimension $2Y$ (dependent upon the 20° plume angle assumption), and the vertical depth of the mixing layer, h. The total emission of the source per unit time, Q, moves through this plane at Y_1, and in the specified unit time mixes uniformly into a box of downwind length $\bar{u}t$, where t is the unit time interval being used and $\bar{u}$ is the mean wind speed. One hour time intervals are used in these calculations, as well as full day boxes, or full night boxes of varying duration, depending on the season.

The box model phase of the model proceeds in step-wise fashion to distances X_2, $X_3, \ldots X_n$ where each segment is defined by $\bar{u}t$. At each point a value for the plume width $2Y_{x\delta}$ may be measured from a projection with the assumed plume angle geometry or computed from the assumed plume angle. In calculations with a changing night time mixing height, the daytime mixing height is projected through the night time box segment, but depositional velocities operate only on the lower portion below the night mixing height. Similarly the upward diffusion process at night only operates on that portion of the night box segment above the night mixing level to a level above the daytime mixing height (see Fig. 1). The loss of SO_2 to upper diffusion is not re-entrained in the following day. It is effectively a vertical diffusion loss to the deeper atmosphere and due to non-re-entrainment.

The most important factor in determining the amount of SO_2 that is converted to SO_4^{2-} is deposition into vegetation and soil since conversion to SO_4^{2-} is concentration dependent and the most important factor affecting concentration downwind is deposition into vegetation. Deposition can vary considerably depending on such factors as density, type, and conditions of vegetation; moisture content of the soil surface; degree of snow cover; and degree of atmospheric turbulence which brings the pollutant in contact with the surface. This laboratory has measured the deposition rate of SO_2 into several typical plant canopies and hydrogen fluoride has been used as a tracer to measure deposition into other canopies in field studies. From the fluoride data and its relationship to SO_2 uptake, estimates can be made of deposition of SO_2 (Hill, 1971).

The daytime deposition velocity measured for SO_2 in alfalfa in chamber studies is 2.81 $cm\ s^{-1}$ and that obtained in field studies is 2.3 $cm\ s^{-1}$ (Hill, 1971). Based on fluoride field studies the estimated deposi-

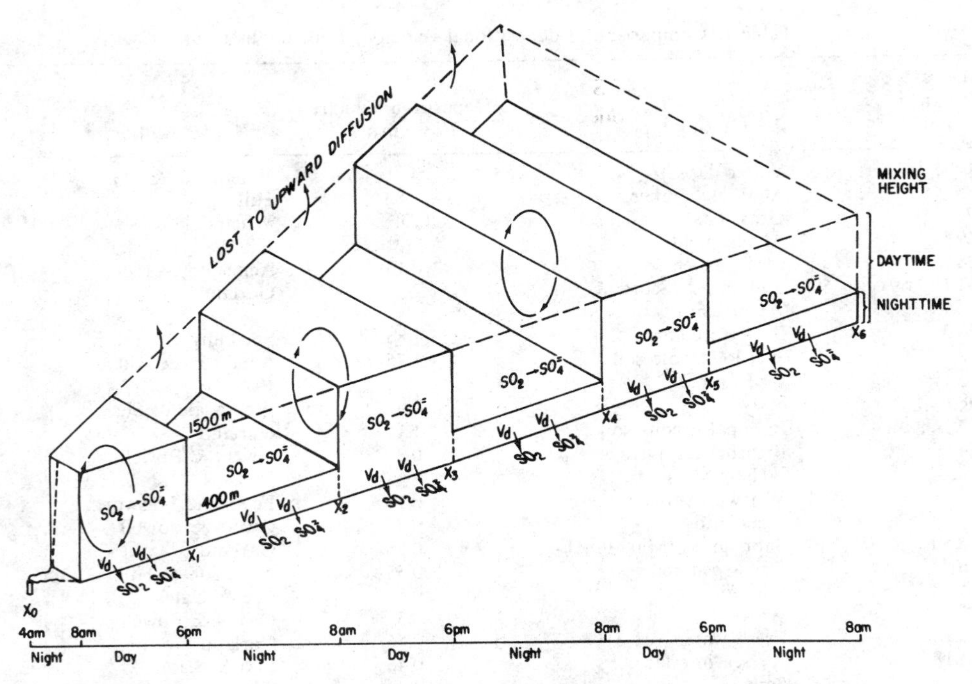

Fig. 1. Isometric schematic of the summer transport model. The initial part of the first box section represents emission at night (4 a.m.–8 a.m.) with no mixing. Complete mixing is assumed during the day periods, but very little mixing at night, with a reduced mixing height.

tion of SO_2 into alfalfa is 2.6 cm s^{-1}. The deposition velocity estimated for SO_2 into other canopies is 5 cm s^{-1} for corn, 1.1 cm s^{-1} for dry scrub oak, 1.1 cm s^{-1} for lawn grass, and 2.1 cm s^{-1} for aspen. Soil values from a recent chamber study in this laboratory range from 0.53 cm s^{-1} for dry aspen soil to 2.13 cm s^{-1} for moist aspen soil (Anderson and Hill, 1977). Chamber studies were all performed in specially designed chambers (Hill, 1971).

Israel measured an average (day and night) deposition velocity for fluoride into alfalfa of 1.94 cm s^{-1}. This would be equivalent to 1.5 cm s^{-1} for SO_2 for an average 24 h exposure.

Data reported in the literature for deposition velocities for SO_2 by vegetation canopies or by soil, snow and water, are given in Table 1. Uptake rates at 5 pphm SO_2 concentration are shown for comparison.

For the calculations presented here the day values used for SO_2 deposition range from 0.5 cm s^{-1} to 2.5 cm s^{-1}. The night value used is 0.2 cm s^{-1}. The values for SO_4^{2-} deposition are 0.2 cm s^{-1} for day, and 0.1 cm s^{-1} for night. These values were not chosen arbitrarily, but were chosen on the basis of the experimental values for surface cover most similar to conditions assumed and with consideration given to turbulence, stability class, percentage of ground cover, soil absorption, and season.

3. CALCULATIONS

3.1 *Typical summer agricultural area conditions*

Calculations of sulfur dioxide loss processes and of sulfate concentration were made for a 3-day transport from a single point source of SO_2 in a typical agricultural area during summer. The data are presented in graphic form in Fig. 2. For these calculations it is assumed that during the first 4 h, 4 a.m. to 8 a.m. no mixing to the ground occurs. The day mixing height is assumed to be 1500 meters, with complete mixing to that height. The night mixing height is 400 meters with more stable conditions assumed. The wind speed is 16 km/h. This parameter plus length of day or night controls the distance parameter of the day or night box segment. The length of day is 10 h, and the length of night is 14 h. No precipitation is assumed.

The day deposition velocity of 2.5 cm s^{-1} chosen for this 3-day transport model would be comparable to a vegetation canopy of alfalfa. Deposition into corn would be higher than in alfalfa and deposition into certain other crops such as pasture grasses might be lower but an average value of 2.5 cm s^{-1} appears to be realistic for an agricultural area such as the midwest during summer when soil moisture is adequate. A night depositional value used was 0.2 cm s^{-1}. Heavy night dews could result in a higher potential for uptake at night compared to day as was found in a study over a hedge in England (Martin *et al.*, 1971), but the rate of vertical mixing in the atmosphere is the major limiting factor in deposition at night in this model.

A relatively high SO_2-to-ammonium sulfate conversion rate of 0.5% h^{-1} was used for these calculations.

The data shown in Fig. 2 show that under these conditions only about 6% of the SO_2 is left in the atmosphere at the end of 3 days. Deposition into vegetation and soil was responsible for 72% of the

Table 1. Comparison of depositional velocities from the literature

Uptake rate $\mu l\ m^{-2}\ min^{-1}$ at 5 pphm	Canopy type	Deposition velocity $V_d = cm\ s^{-1}$	Author	Date
120	Water (lapse)	4	Whelpdale & Shaw	1974
85	Alfalfa (chamber)	2.83	Hill	1974
72	Grass (stability, lapse)	2.4	Whelpdale & Shaw	1974
69	Alfalfa (field)	2.3	Hill	1974
64	Moist aspen soil	2.13	Anderson & Hill	1977
57 (V_d 1.9)	Pine forest (Scotts & Corsican)	<2	Garland	1974
56.3	Barley seedling	1.85	Spedding	1969
52.52	Wet lodgepole soil pH ~6.2	1.75	Anderson & Hill	1977
48	Snow	1.6	Whelpdale & Shaw	1974
33	Bare calcareous soil	1.1	Garland	1974
24	British Isles average (Feb–Mar)	0.8	Owers & Powell	1974
23	Wet wheat (maximum)	0.76	Fowler & Unsworth	1974
	Grassland	0.7	Owers & Powell	1974
16.50	Short grass (Mar–June)	0.55	Garland	1974
16	Dry aspen soil	0.53	Anderson & Hill	1977
15.18	Pine forest	0.5	Lee & Gates	1964
15	Water	0.5	Owers & Powell	1974
15.18	Fresh water (pH 8)	0.46	Garland	
15.18	Air sea interface	0.45	Liss & Slater	1974
6	Soils	0.2	Slinn	1970
6	Dry sandy loam	0.2	Seim	1970
4.8	Water (stable)	0.16	Whelpdale & Shaw	1974
3.34	Dry wheat (minimum)	0.11	Fowler & Unsworth	1974
2	Snow (stable low wind)	0.1	Dovland & Eliassen	1976

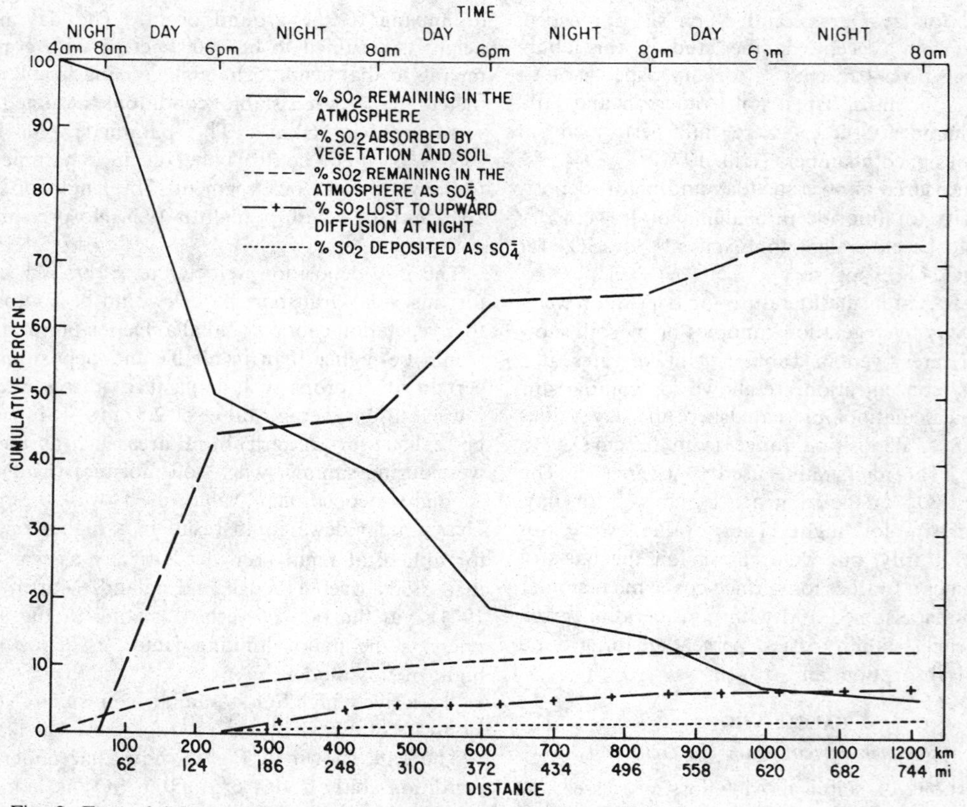

Fig. 2. Fate of SO_2 as a function of time and distance for summer conditions in an agricultural area. All percents are in terms of initial SO_2 emitted from a point source.

SO_2 loss and 45% was removed by this process during the first day. Dry deposition of SO_4^{2-} into soil and vegetation accounted for 2% of the SO_2 loss. The SO_4^{2-} deposition velocity value used for these calculations is probably high, perhaps by as much as a factor of 10. The loss of SO_4^{2-} to deposition even with such a value is small. It can be concluded that dry deposition of SO_4^{2-} is negligible. In this calculation loss to the atmosphere at night above the daytime mixing height accounts for 8% of the SO_2. This SO_2 would not have any effect on sulfate concentrations in the mixing height and it would eventually be removed by precipitation. If the same rate of loss to upward diffusion were assumed for SO_4^{2-} as is assumed for SO_2 about 1% less SO_4^{2-} would remain in the atmosphere at the end of the 3 days as compared to that shown in Fig. 2. About 12% of the initial SO_2 is in the form of SO_4^{2-} at the end of the 3-day period.

3.2 *Arid desert conditions*

Calculations of sulfate concentration were made for a 3-day transport from an SO_2 point source in an arid desert area during the summer. The data are presented in graphic form in Fig. 3. The changes from the previously discussed model are the mixing height and depositional velocities. Again, no mixing to the ground is assumed for the first 4 h, 4 a.m.–8 a.m. The day mixing height is assumed to be 3600 m, with complete mixing to that height. The night mixing height is 300 m with more stable conditions assumed. The wind speed is again assumed to be 16 km h^{-1}; the length of day is 10 h and the length of night is 14 h. No precipitation is assumed.

The day depositional velocity of 0.5 cm s^{-1} chosen for this model is comparable to a sparse vegetation canopy and dry soil. The choice of a night value of 0.2 cm s^{-1} compares to the value for dry sandy loam and also reflects the effect of more stable night conditions. A conversion rate of SO_2 to ammonium sulfate of 0.5% h^{-1} is used and no SO_4^{2-} deposition is assumed.

Loss to the upper troposphere at night is assumed as a depositional velocity of 0.3 cm s^{-1}.

The data graphed in Fig. 3 shows that under these desert conditions about 47% of the initial SO_2 remains in the atmosphere at the end of 3 days.

Loss of SO_2 by upward diffusion at night is about 8%. As in the other summer model this SO_2 would not have any effect on sulfate concentrations in the mixing height and it would eventually be removed by precipitation. About 28% of the SO_2 is in the form of SO_4^{2-} at the end of the 3-day period. Of all models presented here this model shows the largest percent of SO_4^{2-} in the atmosphere.

3.3 *Typical winter conditions*

Calculations were also made for typical winter conditions (see Fig. 4). Winter mixing heights were assumed to be 700 m during the day and 350 m at night. During the initial period of 4 h from 4 a.m. to 8 a.m. no mixing was assumed and during the day complete mixing was assumed. During the night more stable conditions were assumed with a loss to upward diffusion at a rate of 0.3 cm s^{-1} from only the upper portion of the night model box segment. The day period was assumed to be 8 h and the night period

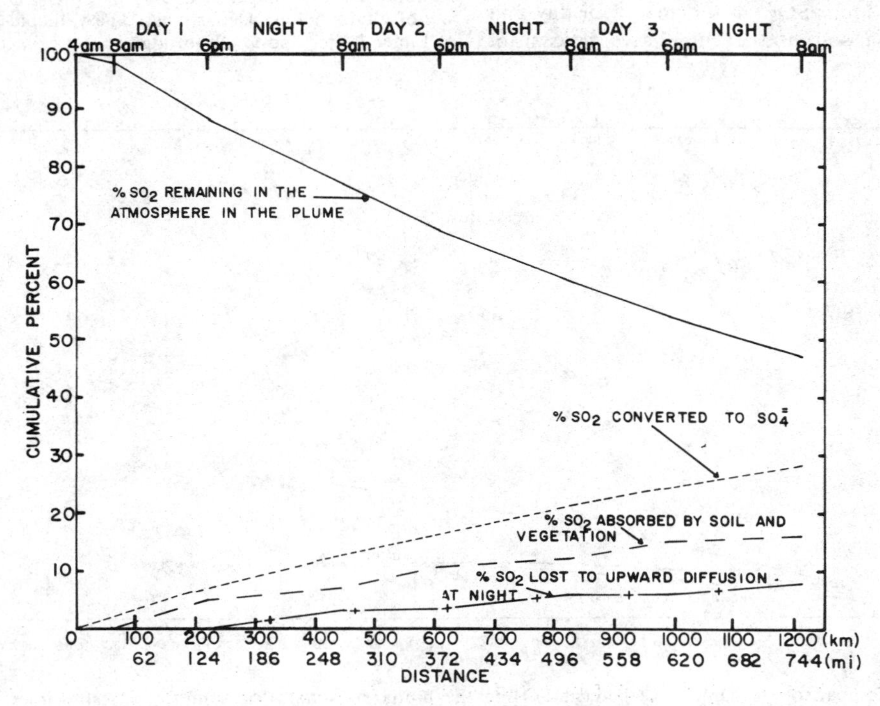

Fig. 3. Fate of SO_2 as a function of time and distance for arid desert conditions.

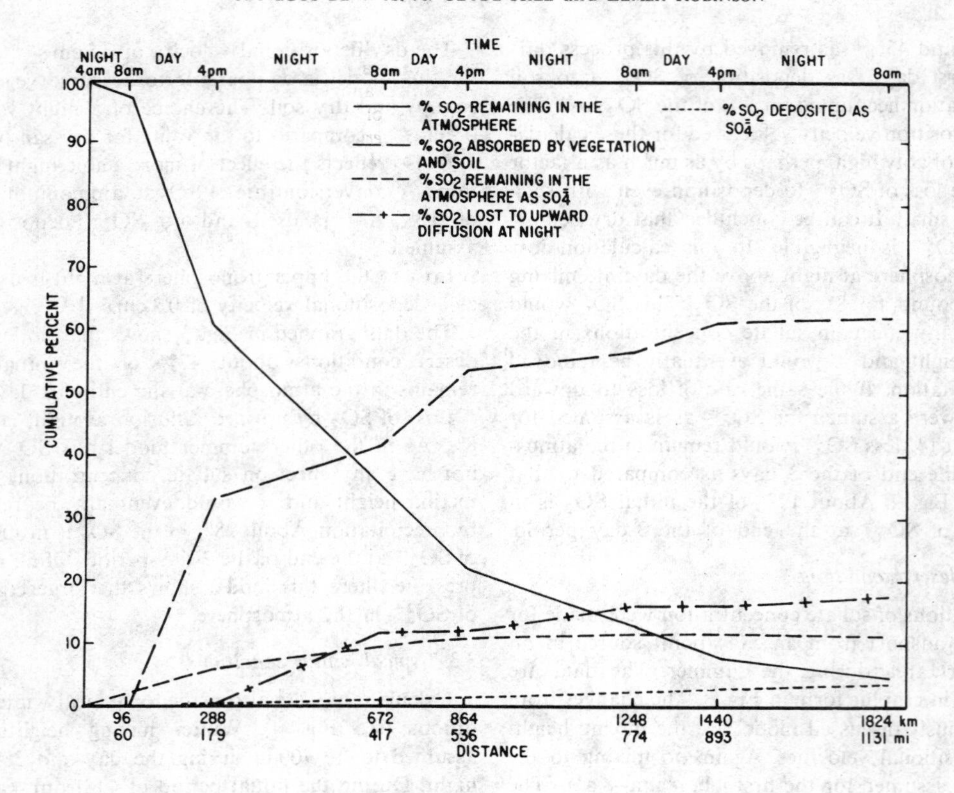

Fig. 4. Fate of SO_2 as a function of time and distance for winter conditions.

was assumed to be 16 h. A typical wind speed of 24 km h^{-1} was assumed. Dry deposition velocities for SO_2 were assumed to be 1 cm s^{-1} for day and 0.2 cm s^{-1} for night. Dry deposition during the winter would be primarily into snow and soil. The deposition velocity used for SO_4^{2-} is 0.1 cm s^{-1} for day and night. A comparison of the cumulative percent of SO_4^{2-} in the plume plotted against distance for these three models is shown in Fig. 5.

Acknowledgements—This study was performed under EPA Research Grant Number R 802967-03-1 with The University of Utah Biology Department. The valuable assistance of Elmer Reiter and G. Woolridge is appreciated.

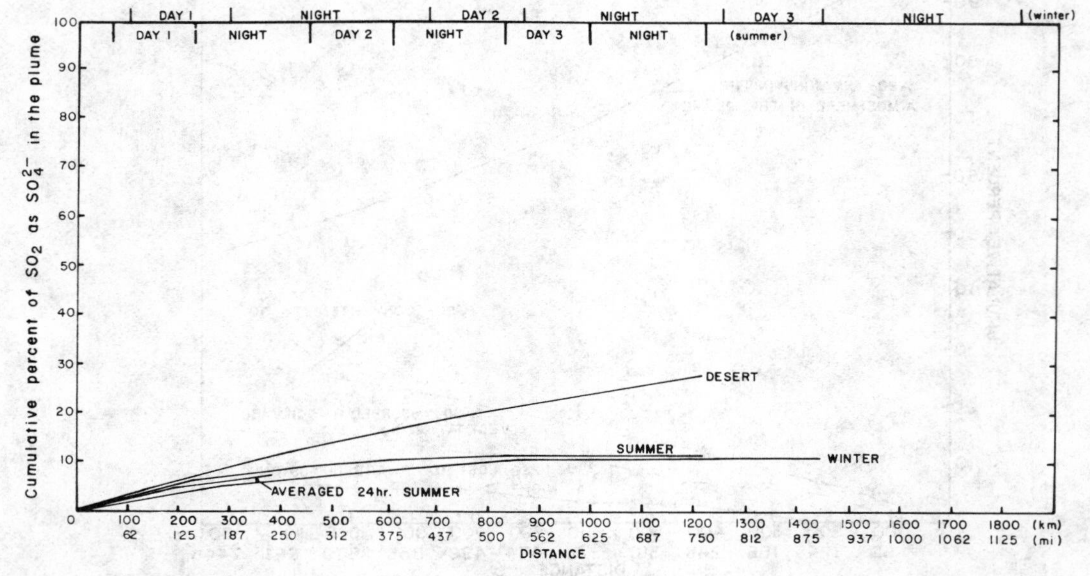

Fig. 5. Comparison of cumulative percent SO_2 in the plume as sulfate for summer, desert, winter, and summer averaged 24 hourly calculations.

REFERENCES

Anderson F. A. and Hill A. C. (1977) Unpublished data.

Dovland H. and Eliassen A. (1976) Dry deposition on a snow surface. *Atmospheric Environment* **10,** 792–795.

Fowler, D. and Unsworth M. H. (1974) Dry deposition of SO_2 on wheat. *Nature* **249,** 389–390.

Garland F. A., Atkins D. J. F., Readings C. J and Caughey J. S. (1974) Deposition of gaseous sulphur dioxide to the ground. *Atmospheric Environment* **10,** 75–79.

Hill A. C. (1967) A special purpose plant environmental chamber for air pollution studies. *J. Air. Pollut. Control Ass.* **17,** 743–748.

Hill A. C. (1971) Vegetation: a sink for atmospheric pollutants. *J. Air Pollut. Control Ass.* **21,** 341–346.

Holzworth, G. C. (1972) Mixing Heights, Wind Speeds, and Potential for Urban Air Pollution throughout the Contiguous United States. EPA Office of Air Programs Public No. AP-101, Research Triangle Park, North Carolina, pp. 27, 29, 32, 34.

Liss, P. S. and P. G. Slater (1974) Flux of gases across the air–sea interface, *Nature* **247,** 181–184.

Martin, A. and F. R. Barber (1971) Some measurements of loss of atmospheric sulphur dioxide near foliage. *Atmospheric Environment* **5,** 345–352.

Owers, M. J. and A. W. Powell (1974) Deposition velocity of sulphur dioxide on land and water surfaces using a ^{35}S Tracer model. *Atmospheric Environment* **8,** 63.

Seim, E. C. (1970) Sulfur Dioxide Absorption by Soil; University of Minnesota, Ph.D. Thesis.

Slinn, W. G. N. (1975) Some approximations for the wet and dry removal of particles and gases from the atmosphere. Article submitted for publication in the *J. Air, Wat. Soil Pollut.*

Spedding, D. J. (1969) Uptake of sulfur dioxide by barley leaves at low sulfur dioxide concentrations. *Nature* **224,** 1229–1231.

Whelpdale, D. M. and R. W. Shaw (1974) Sulphur dioxide removal by turbulent transfer over grass, snow, and water surfaces. *Tellus* **XXVI,** 196–204.

Atmospheric Environment Vol. 12, pp. 641–647. Pergamon Press 1978. Printed in Great Britain.

SULFATE REGIONAL EXPERIMENT IN NORTHEASTERN UNITED STATES: THE 'SURE' PROGRAM

RALPH M. PERHAC

Electric Power Research Institute, Box 10412, Palo Alto, CA 94303, U.S.A.

(*First received* 17 *June* 1977 *and in final form* 26 *August* 1977)

Abstract—The Electric Power Research Institute (EPRI) is supporting a research program to define the relation between emitted primary pollutants (e.g. SO_2) and regional, ambient concentrations of secondary products (e.g. sulfates). Emphasis will be on identifying the contribution of the electric power industry to ambient sulfate levels in the northeastern United States. This project, the Sulfate Regional Experiment (SURE) will be conducted over three years at a cost of about $6 m. Part of the stimulus for SURE is the concern that utilities may be required to reduce SO_2 emissions in order to meet an ambient sulfate standard. The relationship, however, between SO_2 emissions and ambient sulfate concentrations is obscure, at best. Present studies seem to indicate that the formation of sulfates is a regional problem, tied not just to SO_2, hence no simple relation exists between SO_2 emissions and regional sulfate levels. If so, a control on SO_2 emissions may not be a realistic means of effecting sulfate reduction. The SURE program comprises four main elements: (1) a ground monitoring network of 54 randomly distributed stations throughout northeastern United States; (2) a program of measurements of air quality using airplanes; (3) a detailed emissions inventory; and (4) a modeling program. Nine ground stations will operate continuously over a 19-month period. The remaining 45 will operate continuously for one month out of each of the four seasons of the year. At all ground stations, a wide range of chemical and meteorological parameters will be measured. Emphasis is on sulfur compounds; however, measurements will also include nitrogen species, ozone, hydrocarbons and some trace metals. The airborne work will serve as a limited supplement to the ground network in an attempt to give a 3-dimensional validity to any conclusions drawn from the measurements made on the ground. The emissions inventory will include SO_x, NO_x, suspended particles and hydrocarbons from power plants, other industry, homes and surface transportation. Emissions will be reported as daily averages for each season. The modeling effort will develop a simulation system which predicts regional concentrations of primary and secondary pollutants in terms of: (1) multiple sources; (2) pollutant transport and chemical reaction; and (3) meteorological parameters, viz., temperature and humidity. Data for the model will be drawn from other studies as well as from SURE. The program is closely coordinated with activities of U.S. government agencies. Field measurements began in July 1977.

INTRODUCTION

The Sulfate Regional Experiment (SURE) comprises an extensive program of measuring atmospheric pollutants throughout northeastern United States. This $6 m project, funded by the Electric Power Research Institute (EPRI), is aimed primarily at defining the regional, ambient concentrations of secondary pollutants (e.g. sulfates) in terms of local emissions of primary precursors (e.g. SO_2). A number of pollutants will be measured; however, a particular concern is the contribution of the electric power industry to regional sulfate levels. The basic elements of the project are: (1) a ground-based measurement program; (2) a series of measurements from airplanes; (3) an emissions inventory; and (4) development of a model to predict regional concentrations as a function of multiple sources.

EPRI's concern with sulfur pollution arises from the anticipated tremendous increase in coal burning during the next few decades. Almost certainly, an increased use of coal will be a major means of satisfying the nation's energy needs. And even with extensive use of newly mined low-sulfur coals, a marked increase in sulfur loading of the atmosphere can be expected. A number of inhalation toxicology studies have already shown that sulfur aerosols at sufficiently high levels can produce respiratory difficulties (McJilton *et al.*, 1973). And EPA (U.S. Environmental Protection Agency) has implicated sulfates as potential health hazards (EPA, 1974). The levels at which sulfur compounds in the atmosphere can affect man and the exact nature (chemical speciation) of such compounds is far from known; nevertheless, concern is sufficiently widespread so that EPRI has a major program, the physical chemistry of sulfur in the atmosphere, as well as the health effects of sulfur compounds.

A major concern of the electric power industry in the United States centers on the possibility that an ambient sulfate standard may be promulgated. The impact of such a standard on the power industry would be severe. The Sierra Club (U.S.A.) has already brought suit against EPA to impose such a standard—perhaps at a level of $10\,\mu g\,m^{-3}$. Most sulfate is apparently formed in the atmosphere by chemical conversion from a precursor (SO_2). The amount of sulfate produced by direct combustion is generally small—probably not more than a few per cent of the total emitted sulfur. For this reason, the only way to meet an ambient sulfate standard seems to be by control of SO_2 emissions. Unfortunately, we have no evidence to suggest that any sort of simple relations exists between the amount of emitted SO_2 and concentrations of ambient sulfate, hence reducing SO_2 emissions by some factor may really have little bearing on sulfate levels. Some evidence suggests quite the contrary. In a number of cities, for example, SO_2 emissions have been reduced considerably over the past few years, yet sulfate levels have remained nearly constant (Altshuller, 1976). And throughout much of the northeast, sulfate levels reach maxima in summer whereas SO_2 emissions are highest in winter. Of course, eliminating all SO_2 from power plants would almost certainly reduce sulfate levels but total elimination of anthropogenic SO_2 emissions is drastic and, for all practical purposes, unattainable. With our present knowledge, we have no way of knowing how much to reduce SO_2 emissions in order to effect a specific reduction in ambient sulfate concentrations. And we will not know until a firm relation between emissions and ambient concentrations is established —hence SURE.

BACKGROUND

A number of studies, e.g. Lioy *et al.* (1976), suggest that the sulfate problem (i.e. formation of sulfates in the atmosphere) is regional in character and may be intimately related to atmospheric transport phenomena and chemical reactions occurring over distances of hundreds of kilometers. Evidence for the regional nature of the problem comes from a number of sources. What few chemical kinetic data are available (EPA, 1976; Halstead *et al.*, 1977) suggest transformation rates (SO_2 to sulfate) of 1 or 2% h^{-1}, hence considerable distances would be involved in generating significant quantities of sulfates. Analysis of the abundant data from the U.S. National Air Surveillance Network also supports the idea that great distances are involved in sulfate formation. Evidence is also drawn from the regional extent of visibility reduction and the occurrence of acid rain, both of which phenomena may be related, in part, to atmospheric sulfate particles (Galloway *et al.*, 1976; Cass, 1976).

Within the United States, the highest sulfate levels occur in the northeast (Hakkarinen and Hidy, 1976). Concentrations ranging from 5–25 $\mu g\ m^{-3}$ are typical. (During 'episodes' values up to 80 $\mu g\ m^{-3}$ have been recorded). This is in marked contrast to the 3–4 $\mu g\ m^{-3}$ commonly measured in western U.S.A. Because of the high northeast values, EPRI decided to initiate a regional sulfate study—focusing the effort on the populous northeast.

Before embarking on a major project, EPRI first supported a 1 year planning study whose goal was to evaluate if, in fact, the sulfate problem is regional and what sort of measurement program might resolve the issue. The study involved mainly use of existing data: meteorological, emissions, and SO_2 concentrations. In addition, over 3000 existing filters (which had been stored) from northeastern monitoring stations were analyzed for sulfate content.

The planning study yielded four principal findings relative to the northeast: (1) Local, ambient sulfate levels correlate poorly with local and sub-regional SO_2 levels. (2) Long-term averages of sub-regional sulfate concentrations show irregular, geographic distribution patterns. The regional distribution of sulfates, therefore, cannot be explained by any simple theory of slow and steady conversion from SO_2 and by assuming long residence time in the atmosphere. (3) The sulfate maxima in summer and minima in winter correlate negatively with SO_2 maxima and minima. (4) Sulfate concentrations seem to correlate with both temperature and dew point and with maritime tropical air masses.

Findings 1 and 2 suggest that the dynamic and kinetic processes of sulfate formation cover hundreds of kilometers and involve time scales on the order of hours to days. Findings 3 and 4 imply a significant role for temperature and absolute humidity in the conversion of SO_2 to sulfates. All four observations do suggest that an extensive regional study aimed at measuring both chemical and meteorological parameters over a period of many months is a means of resolving the northeast sulfate question.

THE SURE PROGRAM

SURE comprises four distinct, yet interwoven, elements: an extensive ground-level measurement program, a more limited aircraft program, a detailed emissions inventory, and a modeling effort. The program began late in 1976 and will be completed early in 1980.

Ground network

The heart of SURE is a network of 54 ground stations distributed somewhat randomly (geographically) throughout the northeastern United States (Fig. 1). Nine of the stations are designated as Class I stations, the remaining 45 as Class II. These two classes of stations differ both in the types of measurements made and in the frequency of sampling. The Class I stations will operate daily and continuously for 19 months and will gather the greater variety of data. In contrast, Class II stations will take measurements during only four months of each year—the four being the middle months (January, April, July, October) of each season. During these four months, the so-called 'intensive' periods, Class II stations will operate daily.

Class I stations will contain both hi-vol and sequential samplers. Samples from the hi-vols will be used for measurement or particle mass, sulfate, sulfite, nitrate, chloride, ammonium and water-soluble

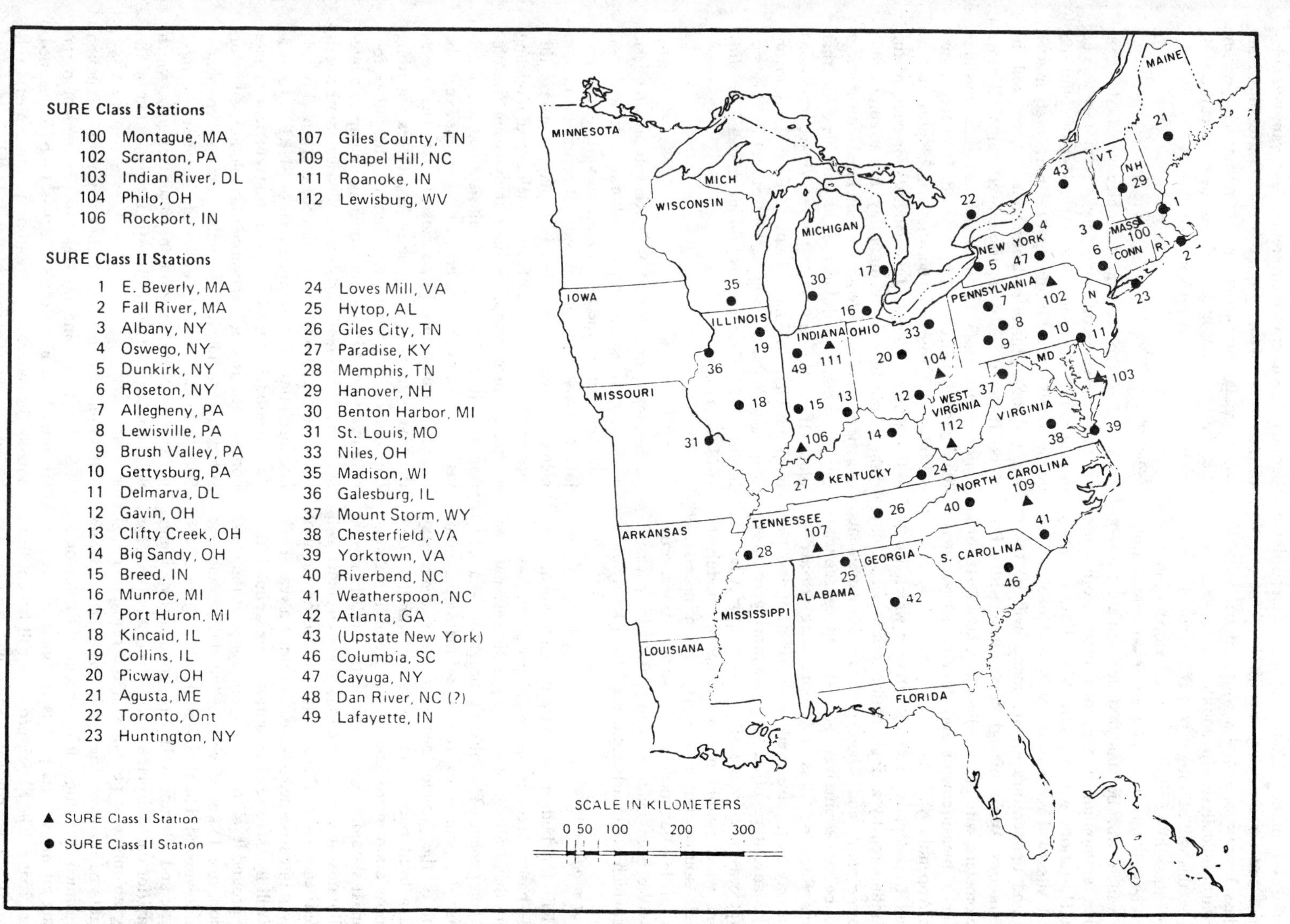
SURE Class I Stations
100 Montague, MA
102 Scranton, PA
103 Indian River, DL
104 Philo, OH
106 Rockport, IN
107 Giles County, TN
109 Chapel Hill, NC
111 Roanoke, IN
112 Lewisburg, WV
SURE Class II Stations
1 E. Beverly, MA
2 Fall River, MA
3 Albany, NY
4 Oswego, NY
5 Dunkirk, NY
6 Roseton, NY
7 Allegheny, PA
8 Lewisville, PA
9 Brush Valley, PA
10 Gettysburg, PA
11 Delmarva, DL
12 Gavin, OH
13 Clifty Creek, OH
14 Big Sandy, OH
15 Breed, IN
16 Munroe, MI
17 Port Huron, MI
18 Kincaid, IL
19 Collins, IL
20 Picway, OH
21 Agusta, ME
22 Toronto, Ont
23 Huntington, NY
24 Loves Mill, VA
25 Hytop, AL
26 Giles City, TN
27 Paradise, KY
28 Memphis, TN
29 Hanover, NH
30 Benton Harbor, MI
31 St. Louis, MO
33 Niles, OH
35 Madison, WI
36 Galesburg, IL
37 Mount Storm, WY
38 Chesterfield, VA
39 Yorktown, VA
40 Riverbend, NC
41 Weatherspoon, NC
42 Atlanta, GA
43 (Upstate New York)
46 Columbia, SC
47 Cayuga, NY
48 Dan River, NC (?)
49 Lafayette, IN
▲ SURE Class I Station
● SURE Class II Station
MINNESOTA
WISCONSIN
MICH
MICHIGAN
IOWA
ILLINOIS
INDIANA
OHIO
MISSOURI
KENTUCKY
WEST VIRGINIA
VIRGINIA
PENNSYLVANIA
NEW YORK
MAINE
VT
NH
MASS
CONN
R
N
MD
NORTH CAROLINA
TENNESSEE
ARKANSAS
S. CAROLINA
GEORGIA
ALABAMA
MISSISSIPPI
LOUISIANA
FLORIDA
SCALE IN KILOMETERS
0 50 100 200 300

Fig. 1.

organics. The sequential sampler will provide material for analyzing for total particle mass, fine particle ($<2\,\mu m$ dia) mass, sulfate, nitrate, water-soluble organics and certain heavy metals (Fe, Mn, Pb, V). At Class I stations, samples will also be taken for hydrocarbon analysis. In addition, a large suite of heavy metals will be sought by PIXE (proton induced X-ray emission) analysis through cooperation with EPA. Class II stations will have only hi-vols, samples from which will be analyzed for sulfate, nitrate, chloride, ammonium and particle mass. Both Class I and II stations will measure SO_2; however, only Class I sites will record NO/NO_2 and ozone. Dew point and temperature measurements will be taken at all stations. Again through cooperation with EPA, nephelometers will be installed in Class I stations. (Details of measurements and data recovery criteria are in Appendix A).

Class I stations were placed at sites which would be uninfluenced by a single major source (viz., a city or power plant) and where pollutant measurements would be representative of rural, ambient concentrations. Sites for Class II stations were selected in such a way as to give random distribution and also to take advantage of existing monitoring sites. An attempt was also made to use sites for Class II stations which would not be influenced unduly by a single source, but not every station necessarily meets this last qualification. For such stations, provisions have been made to compensate for measurements which are clearly influenced by a local source.

Aircraft program

SURE, for budget reasons, depends mainly on ground-based stations. Such a program has the obvious disadvantage of lacking a 3-dimensional (vertical) component. The aircraft program is a compromise. Its goal is to give a 3-dimensional component to the ground measurements and yet stay within a fixed budget. The program will not solve the upper atmosphere question of chemical transformation. It should, however, define the vertical profile of pollutants at selected stations. Two airplanes will be used, each at a Class I station, during each of the seven intensive periods, i.e. four times a year. Each plane will fly three vertical spirals (from about 170 m above ground level to 3300 m), one spiral at the Class I station, one 15 km upwind of the station, and one 15 km downwind. The three spirals will be flown both morning and afternoon. This program of two flights per day (for each airplane) will be repeated six days during an intensive period with the restriction that no flights be made on successive days. The program is not aimed at flying during specific episodes or at tracking air masses. Its goal is routine measurement at Class I sites. The SURE program will benefit from additional aircraft coverage which will be provided from the sulfate program of the U.S. Department of Energy (DOE). Two DOE airplanes will fly at two additional Class I stations during the intensive periods, thus a total of four of SURE's nine Class I stations will have aircraft coverage during intensive periods. The aircraft measurement program will complement that of the ground stations. Measurements will be made for particle mass, fine particle ($<2\,\mu m$ dia) mass, sulfate, hydrocarbons, heavy metals, SO_2, NO/NO_2 and ozone. (Appendix A).

Emissions inventory

Critical to the fundamental objective of SURE is an adequate emissions inventory. Inventories will be prepared for four pollutants: (1) oxides of sulfur, (2) oxides of nitrogen, (3) total emitted particulate matter and (4) non-methane hydrocarbons. Sources will include: (1) fossil-fueled electric power plants, (2) other major industries, (3) non-manufacturing and business (commercial) sources, (4) homes and (5) surface transportation. All source inventories will be presented as seasonal averages (e.g. June–August) for each of the eight 3-hour periods during a day. For example, SO_x emissions calculated from an individual power plant during summer will yield eight numbers for that plant, the eight representing the average emissions from midnight to 3 a.m., 3 a.m. to 6 a.m., and so on. The basic inventory will be for sources in existence in July 1977. Calculations will be updated for July 1978 sources.

Modeling

The measurements part of SURE will provide basic data on the distribution of pollutants. It will not, however, explain the formation and variation of pollutant levels in terms of source contribution and meteorological variabilities. In order to achieve this interpretative step, SURE also includes a modeling program which comprises data analysis and simulation modeling. Data analysis and simulation modeling are inextricably intertwined but they can be broadly distinguished by considering data analyses as the development and testing of hypotheses regarding the component parts of, for example, the sulfate formation system. Simulation modeling attempts to integrate the component parts into a physical–mathematical representation which mimics the real world of sulfate formation. The ultimate goal of SURE is to produce a model capable of predicting ambient sulfate concentrations as a function of multiple SO_2 emission sources, particularly emissions from power plants. Such a model can be developed; however, we have no way yet of knowing the magnitude of error which will be associated with the estimate generated by the model.

It should be apparent, from the general discussion of SURE, that some key elements (component parts) are missing if a realistic model is to be developed, namely the biogenic contribution to atmospheric sulfur levels, the chemistry of the SO_2 sulfate transformation, and the removal of sulfur compounds through dry deposition. Other programs supported by EPRI are aimed at elucidating some aspects of these key elements. Specifically, one project to measure biogenic emissions is already underway, as is a program to study sulfur chemistry in plumes. And plans are being formulated for a major study of dry deposition to begin in 1978. The modeling effort will also take advantage of findings from non-EPRI projects, viz. the DOE sulfate program. Close coordination is being maintained between EPRI and a number of governmental agencies.

Expected results

Firstly, and most importantly, SURE will provide a vast amount of routine data on pollutant levels over a large part of the United States. The data will be defined in terms of accuracy and overall quality control (Appendix A). They should be useful, therefore, for a number of independent studies for a number of years. If nothing else were to develop from SURE, the data bank will be valuable as a contribution to the scientific community. It will contain numbers whose reliability can be specified. Secondly, SURE will yield a high quality emissions inventory for northeastern United States. Finally, a predictive model will be generated—a model which predicts the impact of emissions of a precursor on the ambient concentrations of a secondary pollutant. Admittedly, we cannot yet predict the error in the estimate from the model, but a model will be developed which examines the important relation between emissions and regional concentrations of pollutants. We envision the SURE study as being important for: (1) answering, in part, the question of SO_2 control to meet ambient sulfate standards, (2) possibly re-evaluating use of supplemental control systems, and (3) serving as a guide to improved health-effects studies.

Acknowledgements—The Electric Power Research Institute is appreciative of the cooperation of agencies of the U.S. Government. Drs. Paul Altshuller and William Wilson (Environmental Protection Agency) and Dr. Michael Mac Cracken (Lawrence Livermore Laboratory-DOE), have devoted considerable time to working with EPRI staff on the SURE program.

REFERENCES

Altshuller A. P. (1976) Regional transport and transformation of sulfur dioxide to sulfates in the U.S. *J. Air Pollut. Control Ass.* **26**, 318–324.

Cass G. R. (1976) The relationship between sulfate air quality and visibility at Los Angeles: Calif. Inst. Tech., EQL Memorandum 18, 39 p.

EPA (1974) Health consequences of sulfur oxides: a report from CHESS, 1970–71: U.S. Envr. Protection Agency, Rpt. EPA-650/1-74-004, 368 p.

EPA (1976) SO_2 oxidation in plumes: A review and assessment of relevant mechanistic and rate studies: U.S. Envrn. Protection Agency Rpt. EPA-450/3-76-022, 96 p.

EPRI (1976) Design of the sulfate regional experiment: Elec. Power Res. Inst. Rpt. EPRI EC-125, v. 1. 274 p.

Galloway J. N., Likens E. G. and Edgerton E. S. (1976) Acid precipitation in the northeastern United States: pH and acidity. *Science* **194**, 722–723.

Hakkarinen C. and Hidy G. M. (1976) Atmospheric sulfates in the Ohio Valley: Results of the sulfate regional experiment planning study: Am. Power Conf. Proc., Vol. 38, pp. 825–829.

Halstead H., Larson T. V. and Hobbs P. V. (1977) Oxidation of sulfur dioxide in the atmosphere: a review: Proc. Symposium on Aerial Techniques for Environmental Monitoring (U.S. Energy Res. & Develop. Admin.), Las Vegas, March 1977, 7 p.

Lioy P. J., Wolff G. T., Szachor J. S., Coffey P. E., Stasiuk W. N. and Romano D. (1976) Evidence of high concentrations of sulfates detected at rural sites in the Northeast: paper presented at Am. Chem. Soc. Albany, NY meeting, Aug. 1976.

McJilton E. E., Frank N. T. and Charlson R. (1973) Role of relative humidity in the synergistic effect of a sulfur dioxide–aerosol mixture in the lung. *Science*, **182**, 503–504.

APPENDIX A

Data recovery criteria

Priority A: essential for the objectives of SURE. 90% of the data must be recovered with an accuracy of ±10%.
Priority B: necessary for the objectives of SURE. 90% of the data must be recovered with an accuracy of ±10%.
Priority C: desirable for the objectives of SURE. 80% of the data must be recovered with an accuracy of ±10%.
Priority D: exploratory within the objectives of SURE. Recovery rate and accuracy to be specified.

Data recovery rate is defined as the percentage of the total number of possible observations which are actually recovered, and these must be within the accuracy limits. Example: SO_2, a Priority A parameter, is to be measured for each consecutive 3 h period for the duration of the SURE measurements program at Class I stations. 1 July, 1977 through 31 January, 1979 = 580 days = 4640 3 h periods. $0.90 \times 4640 = 4176$ required values/Class I station.

Accuracy is defined as the percentage departure of a measurement from either its 'true' value or from that of a primary standard when the measured value is greater than 10 times the threshold of detectability. When the measured value of the parameter is less than 10 times the measurement threshold, acceptable accuracy is defined by a linear decrease from 100% at the threshold value to the prescribed precision at ten times the threshold value.

Example: the threshold of detectability of SO_2 is 4 ppb. If 4 ppb are measured, the true value should lie between zero and 8 ppb. When 10 ppb are observed, the accuracy should be 75%, or the true value should lie between 2.5 and 17.5 ppb. When the observed value is 40 ppb the true value should lie between 36 and 44 ppb (±10%).

Any or all of the following techniques will be used to assure the appropriate measures of data recovery and accuracy. (1) Detailed logs of instrument or analyses systems malfunctions and downtime for repair, calibration, or replacement. (2) Calibration of instruments or analyses systems against primary standards at suitable intervals of time. (3) Replicate *in-situ* sampling and comparative analyses in sufficient numbers and ranges of conditions (parameter values) to assure statistical significance.

MEASUREMENTS AT SURE GROUND STATIONS

Parameter	Class I	Class II
SO_2 (gas analyzer)	Priority A Continuous measurement Averaging time 3 h	Priority A Continuous during intensives Averaging time 24 h
NO/NO_2 (gas analyzer)	Priority A Continuous measurement Averaging time 3 h	Not measured
O_3 (gas analyzer)	Priority B Continuous measurement Averaging time 3 h	Not measured
SO_4^{2-} (seq. sampler)	Priority A Continuous during intensives Averaging time 3 h	Not measured
SO_4^{2-} (hi-vol)	Priority A Continuous measurement Averaging time 24 h	Priority A Continuous during intensives Averaging time 24 h
NO_3^- (hi-vol)	Priority A Continuous measurement Averaging time 24 h	Priority A 8 days during intensives Averaging time 24 h
Particle mass (TSP) (hi-vol)	Priority A Continuous measurement Averaging time 24 h	Priority A Continuous during intensives Averaging time 24 h
Fine particle mass	(RSP) Priority B Continuous during intensives Averaging time 3 h	Not measured
Hydrocarbons (C_2–C_5, $>C_6$)	Priority B 8 days during intensives Averaging time 3 h	Not measured
Water-soluble organics (seq. sampler)	Priority C 3 days/month Averaging time 3 h	Not measured
Water-solbule organics (hi-vol)	Priority C 8 days/month Averaging time 24 h	Not measured
Dew-point	Priority B Continuous measurement Averaging time 1 h	Priority B Continuous measurement Averaging time 1 h
Air temperature	Priority B Continuous measurement Averaging time 1 h	Priority B Continuous measurement Averaging time 1 h
Fe, Mn, Pb, V (seq. sampler)	Priority C 3 days during intensives	Not measured
NH_4^+, Cl^-, H^+	Priority C 8 days/month Averaging time 24 h	Priority C 8 days during each intensive Averaging time 24 h

SURE AIRCRAFT MEASUREMENTS

Parameter	Requirements
SO_2 (gas analyzer)	Priority A Continuous measurement Average height 17 m @ 170–1700 m; 33 m @ 1700–3300 m
SO_4^{2-}	Priority A Continuous measurement Averaging heights: 170–1700 m; 1700–3300 m
NO/NO_2 (gas analyzer)	Priority A Continuous measurement Averaging height: 17 m (@) 170–1700 m; 33 m @ 1700–3300 m
Particle Mass (TSP) (dichotomous)	Priority A Continuous measurement Averaging heights: 170–1700 m; 1700–3300 m
O_3 (gas analyzer)	Priority B Continuous measurement Averaging height: 17 m @ 170–1700 m; 33 m @ 1700–3300 m
Hydrocarbons (C_2–C_5, $>C_6$) (cannister)	Priority B Grab sample Averaging heights: 170–1700 m; 1700–3300 m
βscat, βext (Nephelometer)	Priority B Continuous measurement Averaging height: 33 m @ 170–1700 m; 66 m @ 1700–3300 m
Condensation nuclei	Priority C Continuous measurement Averaging heights: 170–1700 m; 1700–3300 m
Temperature, relative humidity	Priority A Continuous measurement Averaging height: 17 m @ 170–1700 m; 33 m @ 1700–3300 m
Altitude, airspeed location, time/date	Priority A Continuous measurement Time checks corresponding to height intervals of averaging for other parameters

Atmospheric Environment Vol. 12. pp. 649–659. Pergamon Press 1978. Printed in Great Britain.

MAP3S: AN INVESTIGATION OF ATMOSPHERIC, ENERGY RELATED POLLUTANTS IN THE NORTHEASTERN UNITED STATES

MICHAEL C. MACCRACKEN

Lawrence Livermore Laboratory, University of California, Livermore, CA 94550, U.S.A.

(*First received* 13 *June* 1977)

Abstract—The Multi-State Atmospheric Power Production Pollution Study (MAP3S) is a major new atmospheric research program of the U.S. Energy Research and Development Administration. The goal of the MAP3S program is to develop and demonstrate an improved, verified capability to simulate the present and potential future changes in pollutant concentration, atmospheric behavior and precipitation chemistry as a result of pollutant releases to the atmosphere from large-scale power production processes, primarily coal combustion. A major motivation of this program is to be able to provide those agencies charged with the task of meeting the nation's energy needs with the knowledge required to assess alternative strategies for generating power while ensuring ample protection of human health and adequate preservation of the natural environment. Since coal is the most abundant domestic fossil energy resource and since electric power production is a major and growing sector of our energy economy, this study focuses on the effects of emissions from coal fired electric power plants, particularly sulfur oxide emissions. The study domain is the high population, energy intensive northeastern quadrant of the United States. Research projects are underway to measure present sulfur oxide concentrations and composition, to assess the potential for long range transport, to investigate transformation processes in plumes from point and urban sources, to sample precipitation chemistry and improve understanding of scavenging mechanisms, and to develop numerical models that can simulate future air quality on sub-continental scales given patterns of anticipated combustion emissions.

1. INTRODUCTION

Will the acidity of precipitation and atmospheric turbidity increase in the United States with increased coal combustion? Can atmospheric concentrations of particulate sulfur be reduced by reducing sulfur oxide emissions? To provide the basis for answering such questions in energy and environmental planning, ERDA's Division of Biomedical and Environmental Research (DBER) is undertaking the MAP3S program to provide the knowledge required to assess alternative strategies for generating power while ensuring ample protection of human health and adequate preservation of the natural environment (MacCracken, 1977). Over the next several years, in cooperation with programs of other organizations within and outside the United States, the MAP3S program will seek to improve scientific understanding of the transport, transformation and fate of atmospheric energy-related pollutants as a basis for an improved, verified capability to simulate present and potential future changes in pollutant concentration, atmospheric behavior and precipitation chemistry. Since coal is the most abundant domestic fossil-fuel resource and since electric power production is a major and growing sector of our energy economy, this study focuses on the effects of emissions from coal-fired electric power plants, particularly in the high population, energy intensive northeastern quadrant of the United States where both average and episode condition sulfate concentrations have been observed to be significant.

The extended term interests of the MAP3S program encompass the entire spectrum of pollutants that may be ascribed to fossil-fuel electric power production, including sulfur and nitrogen oxides, oxidants, hydrocarbons, trace inorganic elements and particles. However, a number of reasons have led to assigning priority to study of sulfur oxides and their associated cations during the first three-year phase of the MAP3S program. These reasons include: (1) Sulfur oxides are a major pollutant from the combustion of coal and research into atmospheric effects of these pollutants is of increasing international interest with the prospect for combined efforts of many groups. (2) The potential health and environmental consequences of sulfur oxides have been of interest for a number of years. Although uncertainty exists and more research is needed (e.g. Comar and Nelson, 1975; OTA, 1976), there are certainly preliminary indications that high concentrations of particulate-sulfur can be an important health parameter (e.g. EPA, 1974; Coffin and Knelson, 1976). In addition, precipitation chemistry effects extensively documented in Europe (e.g. Braekke, 1976) are of increasing concern in the United States (Likens, 1976). (3) The setting of an air quality standard for particulate-sulfur appears likely within several years (EPA, 1975), whereas the health research will probably not be adequate to justify standards for the other pollutants until the 1980s. Implementation of an adequate and effective control strategy requires understanding, rather than simple parameterization, of the atmospheric sulfur cycle. (4) ERDA's contribution to atmospheric research is centralized in the capabilities of the major ERDA national laboratories and several contract research organizations and universities. Research in atmospheric sulfur compounds has become of increasing import to them as the ERDA focus on nuclear energy has broadened to include other energy technologies. The MAP3S program includes researchers at Brookhaven (BNL), Argonne (ANL), Battelle Pacific Northwest (PNL) and the ERDA Health and Safety (HASL) Laboratories, other government agencies, the Illinois State Water Survey (ISWS) and several universities. Total support for MAP3S exceeds $3 M per year. (5) The MAP3S program can interface with health, ecology and assessment studies within ERDA. These studies include animal toxicology studies at the Inhalation Toxicology Research Institute, Oak Ridge, Battelle Northwest, among others and ecological effects studies of acid rain on crops, ferns and tree seedlings at Argonne, Brookhaven and Oak Ridge National Laboratories.

In Europe, the OECD study (Ottar, 1977) has clearly focused international interest on the problems of long range transport of air pollutants. In the United States, similar problems are becoming more apparent and the need for atmospheric research on scales beyond 100 km is now recognized by EPA, EPRI and ERDA. Although these organizations and ERDA may have slightly different long-range objectives, the scientific research which is necessary to develop the needed understanding of problems in North America has much in common. Thus, in addition to building on the work of the European scientific community, MAP3S will coordinate its research program and especially the field experimentation and observation programs, with related activities of EPRI's SURE program (Perhac, 1978), EPA's MISTT study (Wilson, 1978) and STATE (EPA, 1977) program and Canada's sulfate pollution study (Whelpdale, 1978).

2. MAP3S SCIENTIFIC PROGRAM—RESPONDING TO UNCERTAINTY

Although desirable to develop atmospheric research priorities from the requirements of studies of health and ecological effects, available data on some effects are too sparse (e.g. on the response of typical U.S. forest ecosystems to particular characteristics of modified precipitation chemistry) and uncertainties in such effects are too large (e.g. on what pollutant species cause what health effects) either to define particular questions or to establish research priorities. Specific examples where data are lacking include the relative importance of such factors as long-term average and short-term episodic concentrations, pollutant particle size and composition, and instantaneous or storm average precipitation acidity. Indeed, there seems to be no way to rule out any one process or pollutant species as being significantly more or less important than any other. Therefore, in planning MAP3S the approach has been to identify those aspects of the problem where uncertainties remain, and to direct theoretical and experimental studies to the potentially significant pollutants and atmospheric mechanisms controlling the transport, transformation, and removal of energy-related emissions.

Conceptually, the tasks that must be carried out in order to alleviate uncertainty and improve understanding of relevant atmospheric behavior are divided into three sub-programs:*

(a) Characterization Sub-program: measurement of the chemical and meteorological variables that determine the distribution of pollutant species and that will be needed as input for and as a means of verifying the predictions of numerical models.

(b) Field Experiments Sub-program: design and execution of those atmospheric research experiments necessary to understand the mechanisms and related processes that must be included in simulation models in order to improve their accuracy.

(c) Simulation Sub-program: development, verification and demonstration of the capability to simulate the atmospheric behavior, pollutant concentrations and precipitation chemistry effects of emissions from fossil-fuel electric power production relevant to human health and welfare.

It is anticipated that these several sub-programs will be carried out simultaneously, so that, for example, a repertoire of characterization measurements will be being developed even as numerical models are being constructed and tested against measurements being gathered in the field program.

The goal of MAP3S includes the development of a verified simulation capability in order to understand the physical processes taking place during the transport, transformation and eventual removal of pollution from the atmosphere by both wet and dry processes. As the models are improved and verified, they will be exercised for various scenarios of fossil-fuel electric power production, including different geographical distributions of fuel use, fuel characteristics, abatement strategies, etc. The results of these studies will be made available for assessment studies which consider effects upon human health and ecology.

Ten main elements or tasks in the work program have been identified, covering the three sub-programs mentioned above, and these are set out below. A more complete description of the status of recent research in these areas and the extent of uncertainty is contained in the MAP3S Program Plan (MacCracken, 1977).

* A laboratory sub-program is not included in MAP3S, except to the extent that substantial efforts will be made to improve instrument techniques within the characterization and field experiments sub-programs.

Task 1. Specification and quantification of the emissions of atmospheric, energy-related (AER) pollutants resulting from present power production processes and consideration of the spectrum of pollutants that may be emitted as a result of introduction of new processes

The Brookhaven National Laboratory (BNL) is using existing data from the Federal Power Commission (FPC, 1972), Environmental Protection Agency (NEDS, 1976) and state emission inventories to develop an emissions data base. The inventory has been divided into the largest 500 point sources (not all of which are power plants) and several thousand area sources representing more than 55,000 smaller point sources in the eastern United States. The inventory includes emissions of sulfur and nitrogen oxides, hydrocarbons and particles.

Task 2. Identification and quantification of sources of AER pollutants that do not result directly from power production and of other substances that may affect the concentration, distribution, transformation and fate of AER pollutants

The effort discussed in Task 1 is also developing an area source inventory in which more than 55,000 small point sources are aggregated into elements, each roughly 30 by 30 km. Emissions from the biosphere are not included, although efforts are underway to identify potential source regions (e.g. marshlands) by identifying land types. Work in this area by other researchers (e.g. Semb, 1978; Mészáros, 1978) and organizations (e.g. EPA, EPRI) will be utilized where possible.

Task 3. Characterization of the physical and chemical properties of AER pollutants, including particle size, oxidation state, derivative compounds, molecular form, etc. that are commonly found in the atmosphere

Directly emitted gaseous pollutants (e.g. SO_2, NO_x) have been sampled for many years, have been followed out to tens of kilometers downwind from power plants, are reasonably well characterized, and in most cases have been controlled to such an extent that air quality standards are being met. For secondary, or derivative, compounds (such as ammonium sulfate and bisulfate, sulfuric acid mist, sulfite, etc.) most observations have been reflected in measurements of the total suspended burden, with relatively few analyses looking at such properties as molecular form, particle size distribution, acid sulfate speciation, etc.

Table 1. Particulate sulfur analysis techniques to be applied as part of MAP3S program

Ion chromatography
X-ray fluorescence
Methyl-thymol blue
Silver-110 (^{110}Ag) precipitate
Gran titration
Infrared spectroscopy
Benzaldehyde extraction
Thermochemical titration
Photo-electron spectroscopy

A variety of approaches (see Table 1) are being used in MAP3S to characterize atmospheric sulfur compounds collected throughout the region. The techniques, described more fully by Newman (1978), include: i.r. spectroscopy (Cunningham *et al.*, 1975), X-ray photoelectron spectroscopy (Craig *et al.*, 1974) and Gran titrimetric analysis (Tanner and Newman, 1976) to look at such characteristics as particle acidity and molecular form; thermochemical titration (Eatough, 1978) to look for S(IV) compounds (e.g. sulfites); and a diffusion battery (Marlow and Tanner, 1976) to provide samples for analyses of composition in various particle size categories. Figure 1 shows the present network, for example, of Lundgren impactors, filters from which will be analyzed routinely by i.r. spectroscopy and occasionally by other methods.

Task 4. Determination of the spatial and temporal distribution of AER pollutants under both average and extreme conditions

For data on surface concentrations of pollutants, MAP3S will rely largely on the SURE network (Perhac, 1978; Hidy and Mueller, 1978) and available measurements from state agencies. MAP3S will utilize the BNL and PNL aircraft to provide data on the horizontal and vertical patterns of pollutant concentration in the planetary boundary layer. When possible, these flights will be in conjunction with aircraft sampling (primarily in the vertical) by the SURE aircraft. Flight patterns for the first SURE intensive period (early August 1977) and areas for possible future flights are shown on Fig. 2. We anticipate data from these flights to provide information on the time and space scales of sulfate variability and the presence or absence of plumes of sulfate from major industrialized areas (e.g. the Ohio River Valley).

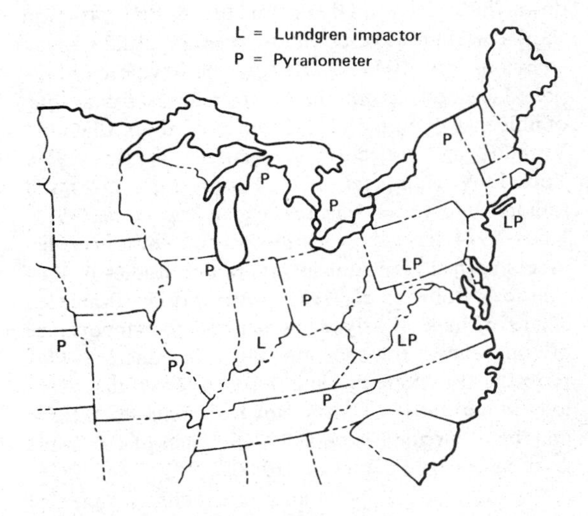

Fig. 1. Location of special turbidity and sulfate acidity instruments sponsored by MAP3S.

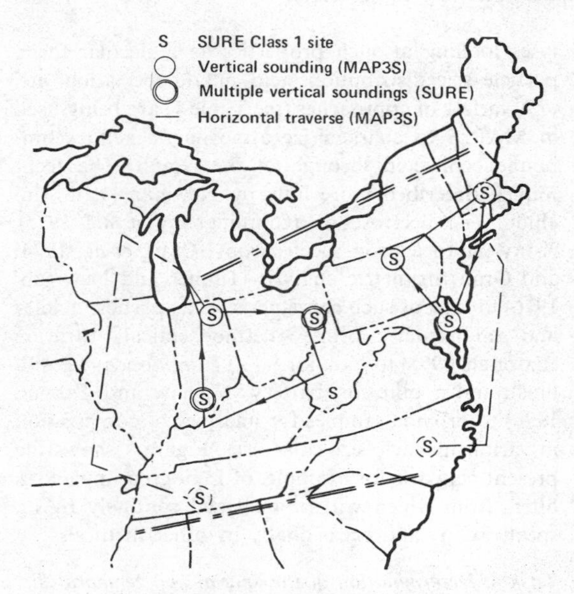

Fig. 2. Aircraft flight patterns for intensive measurement periods. Solid lines indicate tentative flight plans for first intensive period (August 1977). Dashed lines indicate possible future flight patterns to measure fluxes in and out of region. Double barred lines indicate additional regional boundaries where measurements will be needed.

Task 5. Determination of the processes and parameters governing the vertical and horizontal transport of AER pollutants

In using the atmosphere as a disposal medium for combustion products, dependence is placed on dispersal in both the horizontal and vertical dimensions. In the U.S., studies downwind of St. Louis (Breeding *et al.*, 1973; Lowry *et al.*, 1974; Changnon and Semonin, 1975; White *et al.*, 1976; Zak, 1976; Alkezweeny and Powell, 1977; Wilson, 1978), of Milwaukee (Alkazweeny, 1977) and of the East Coast (Brown and Garber, 1976) have indicated that pollutants emitted by cities form plumes extending at least 100 km downwind. Early analyses for the SURE program show correlations of pollutant level and sulfur oxide emissions from 100 to 300 km, which have been interpreted by Hidy *et al.* (1976) as evidence for a zone of influence of an individual source of about that size. Visibility and trajectory studies (e.g. Wexler, 1950; Volz, 1969; Hall *et al.*, 1973; Husar *et al.*, 1976) and radionuclide tracer releases (e.g. Knox *et al.*, 1971; Knox, 1974) have shown transport to even longer distances in the lower atmosphere. Major studies in Europe (e.g. Bolin *et al.*, 1971; Ottar, 1978; Prahm *et al.*, 1976) have clearly demonstrated the importance of long range transport in the atmospheric sulfur problem. Preliminary data also indicate that very long range transport from North America to Europe may be occurring (Nyberg, 1976). Much of this work is discussed by Pack *et al.* (1978).

The dispersal of pollutants is equally complex in the vertical dimension. Diurnal changes in atmospheric stability lead to the vertical dispersal of pollutants up to several kilometers during daytime mixing periods and then isolation of pollutants aloft as low level nocturnal inversions are formed (e.g. see Hess and Hicks, 1975). Depending on wind speed and direction, these isolated layers can be transported long distances during night-time hours, as was documented by a daVinci balloon experimental flight in 1976 (Zak, 1976), before daytime mixing again brings the pollutants into contact with the surface.

For investigation of horizontal dispersal mechanisms, ERDA is supporting a cooperative effort among ARL, BNL, HASL and LASL to develop a practical system for sampling and analysis of special tracers at very low concentrations. Field tests of prototype instruments in April at Idaho Falls intercompared five tracers (SF_6, two deuterated methanes and two perfluorocarbons) using the new perfluorocarbon detection instruments developed by J. Lovelock. Preliminary results show good agreement between SF_6 and the perfluorocarbons at 50 km.

A recent field program at ANL has looked separately at early morning inversion break-up and early evening reformation and has been expanded to study a several day continuous period in the late summer of 1977. A number of earlier field experiments have indicated a high degree of stratification associated with inversions, including layers of local maxima in pollutant concentration. It is likely that in stable flow, such as is characteristic of nights in the midwestern United States, such layers of pollutants may be free to travel for considerable distances without being subject to significant vertical dispersion.

To complement these empirical approaches to understanding atmospheric transport phenomena, numerical modeling studies are underway. Meyers *et al.* (1976) have developed a diagnostic mesoscale model employing mass and total energy conservation constraints which provides self consistent wind field and mixing height fields in a layered structure through the troposphere. This work should provide improved wind field and vertical mixing information for use in trajectory and grid models. Field experiments with increased meteorological data and including tracer experiments are being considered for use in validating the performance of this approach.

Task 6. Identification of the chemical and physical transformation processes affecting AER pollutants and determination of the rates and mechanisms controlling these processes

Both reductions in emissions and a variety of indirect control measures (e.g. tall stacks, rural location of power plants, etc.) have been used to attain acceptably low atmospheric concentrations of those pollutants which are directly emitted by power plants. Unfortunately, this approach to improving air quality has not succeeded in reducing some of the secondary species (e.g., sulfate) formed as a result of chemical and physical transformations taking place in the atmosphere after emission. This non-linearity, for example, between emissions of SO_2 and concentrations of SO_4^{2-} (as shown in data discussed by Altshuller, 1976) requires that an understanding of detailed mechanisms and rates be established. The MAP3S program is carrying out point source and urban plume studies in order to broaden the data base upon which the varied hypotheses of transformation mechanisms may be tested.

Sampling of point source plumes by BNL will continue (some in conjunction with EPRI) under an expanded range of atmospheric conditions. Forrest and Newman (1977) have reported results from sampling the Labadie power plant on 21 different days that provide little evidence to distinguish between conversion mechanisms (Fig. 3). Sampling under conditions involving higher humidities, winter sunlight, more particulate matter (as recently done at the AnClote plant in Florida), increased mixing, and a larger range of temperatures will apparently be needed to allow correlation of conversion rates with atmospheric conditions and to deduce information on conversion mechanisms.

Studies of conversion rates of SO_2 to SO_4^{2-} deduced in urban plumes indicate that such conditions may be more conducive to conversion than in point source plumes. To expand the available data base, a study of the Milwaukee urban plume is underway by PNL. Under west wind conditions this plume is carried across Lake Michigan and, when the air is warmer than the lake surface, can be isolated from both interaction with the surface and introduction of fresh pollutants. At about 100 km, the plume again comes over land. This can initiate increased mixing and contact of the polluted air with the surface, thus providing altered conditions from which to draw conclusions.

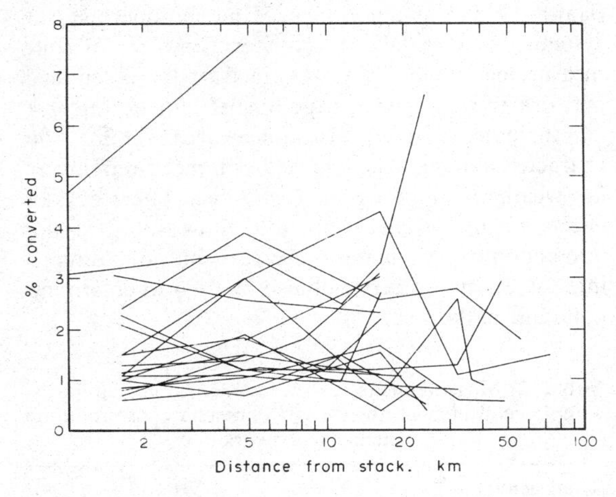

Fig. 3. The observed conversion rate of SO_2 to sulfate with distance on twenty-one different days for the Union Electric Company power plant in Labadie, Missouri (Forrest and Newman, 1977). Possibly because of a limited range of atmospheric conditions, no distinct correlation could be found between the extent of SO_2 conversion and distance, travel time, temperature, relative humidity, time of day, or atmospheric stability.

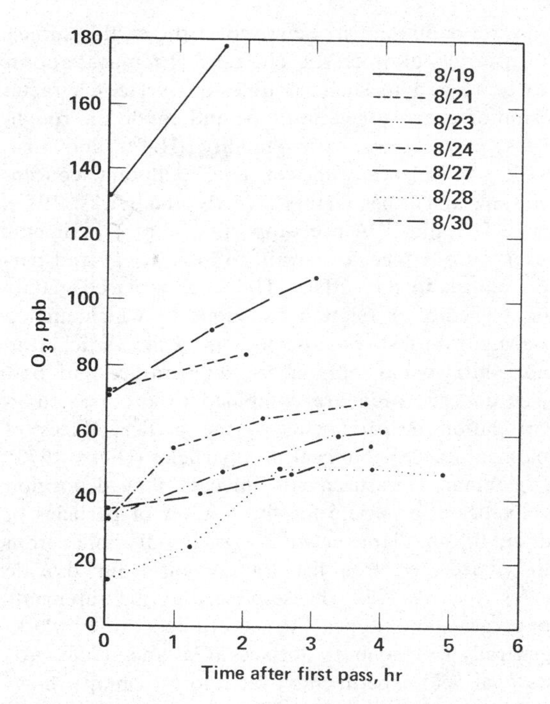

Fig. 4. Vertically averaged ozone concentrations in the Milwaukee urban plume versus time (proportional to distance) in August 1976.

Results from the August 1976 sampling of the Milwaukee plume indicate that ozone and sulfate are forming as the air mass ages (Fig. 4), while SO_2, NO, and NO_2 are decreasing. The SO_2 to SO_4^{2-} transformation rate on 27 August reached $6.8\%\,h^{-1}$ with a peak O_3 concentration of 108 ppb, whereas on 28 August the conversion rate was approximately zero with 49 ppb peak O_3 (Alkezweeny, 1977). Lead concentrations were almost four times as great on 27 August, indicating greater contributions of automobile exhaust products to the pollutant mix and suggesting the importance of chemical radicals in the reaction scheme (Walter *et al.*, 1977; Isaksen *et al.*, 1978). Layers of increased sulfate concentration were also found at about 3000 m on days when the wind was from the northeast, perhaps indicative of long range transport. Results from flights during the summer of 1977 in the same region are now being analyzed.

In addition to the plume studies, direct mechanistic studies have begun at ANL in which determination of the oxygen isotope ratio in SO_2 and sulfates is used to determine the chemical paths that occurred during transformation processes (Cunningham and Holt, 1976). Laboratory testing on the technique is being augmented by analysis of field samples.

Task 7. *Determination of the rates of physical and biochemical mechanisms governing removal of AER pollutants from the atmosphere at the earth's surface (dry deposition)*

As discussed by Garland (1978), gaseous and particulate pollutants are removed at the earth's surface by a variety of processes. The rates of removal appear to be related to such variables as surface character (land or water, vegetation type and condition, roughness, etc.), atmospheric stability (Hicks and Liss, 1976), turbulence, time of day, pollutant concentration, and other factors (Wesely and Hicks, 1977).

ERDA and EPA are supporting joint experiments over various types of terrain to look at gas and particle fluxes to the surface. The work is based mainly on the eddy correlation technique in which analog signals from fast-response sensors of the vertical and horizontal wind components, temperature and particle concentration are combined to produce eddy-correlation measurements of the vertical fluxes of momentum, sensible heat and particles (Hicks, 1970). Preliminary measurements indicate that deposition velocities appropriate for the transfer of particles of about 0.1 μm diameter are not greatly dissimilar from those accepted fror the transfer of sulfur dioxide (Wesely *et al.*, 1977). These direct flux measurements over grassland appear to contradict earlier work, generally over smooth surfaces (Garland, 1978). Additional field experiments over a forest canopy, however, seen to confirm these higher estimates for sulfate deposition (Wesely, 1977).

Because, preliminary model experiments (e.g. Sheih, 1977) indicate that variation of the rates of these dry deposition processes can have substantial effects on downwind concentrations of pollutants and the concentrations of their derivative products, further experiments are planned to evaluate deposition velocities for a variety of atmospheric contaminants and size distributions over a range of natural surfaces. Measurement techniques will include conventional eddy-correlation and profile methods, as well as less familiar variance and spectral density approaches.

Task 8. *Identification of the mechanisms and rates governing the removal of AER pollutants by precipitation scavenging and determination of the effects of AER pollutants on trace material balances and precipitation chemistry, specifically including the acid-base relationships*

Although precipitation is episodic and highly spatially variable, the great effectiveness of the wet scavenging process causes precipitation to be an important sink process for atmospheric pollutants. In addition, the relatively high levels of pollution in the northeastern United States can have an important effect on the chemical balance of the precipitation, bodies of water, and the earth's surface. Papers by Garland (1978), Hales (1978) and Granat (1978) provide recent reviews of the present state of knowledge and the extent of seriously acidified precipitation (see also FISAPFE, 1976).

As part of MAP3S, a complementary program of observations and field experiments has been initiated to address both what is occurring and what mechanisms are leading to the observed results. The observation program will involve scales ranging from regional to interstate. MAP3S is currently supporting ISWS in the completion of data analysis for the multi-year METROMEX project 100 km around Chicago. As part of the new Chicago Area Program (CAP), MAP3S will be supporting analysis of contaminants in precipitation in order to better understand pollutant mass balances from the large Chicago metropolitan area.

On the inter-state scale, a high-quality prototype precipitation chemistry network is being established in order to evaluate the chemical composition and properties of precipitation in the northeastern states where there are indications that 'acid-rain' is becoming an increasingly serious problem. Unlike Scandinavia, there has not been a long-term record of precipitation quality developed over a wide region. The MAP3S network is envisioned as the first step in establishing a more complete, more extensive, and longer term U.S. network (Cowling, 1975). As a prototype network, special care is being taken to evaluate needed procedures (Hales *et al.*, to be submitted), compare collection techniques, and investigate the need for collecting on an event basis as opposed to a monthly basis. This last factor will allow correlation of pH with the trajectories of the storms, thus permitting an analysis of the source of precipitated pollutants. A list of species for which the precipitation samples are being analyzed is given in Table 2. In late 1976, samples were located along a north–south line of four university-operated sites from upper New York State to Virginia (see Fig. 5). Four additional sites will be established in 1977.

In order to improve understanding of scavenging processes, PNL and cooperating universities are investigating the mechanisms by which pollutants are scavenged from the atmosphere by precipitation (Hales, 1978). The initial focus is on stationary storm systems, such as lake effects storms, where aircraft measurements can be combined with a surface network in well-defined experiments. For example, it is anticipated that the Milwaukee urban plume, the characteristics of which have been measured under fair weather conditions for Task 6, will intersect lake effects storm systems. This will allow study of the interaction of scavenging processes with pollutants of interest. Special tracers will also be used to determine pollutant pathways.

Table 2. Measurements to be made on precipitation samples collected in the MAP3S interstate precipitation chemistry network

Total acidity
Conductivity and pH(H^+)
SO_4^{2-}, SO_3^{2-}
NO_3^-, NO_2^-
NH_4^+, Na^+, K^+, Mg^{2+}, Ca^{2+}
PO_4^{3-}
F^-, Cl^-
Dissolved Al

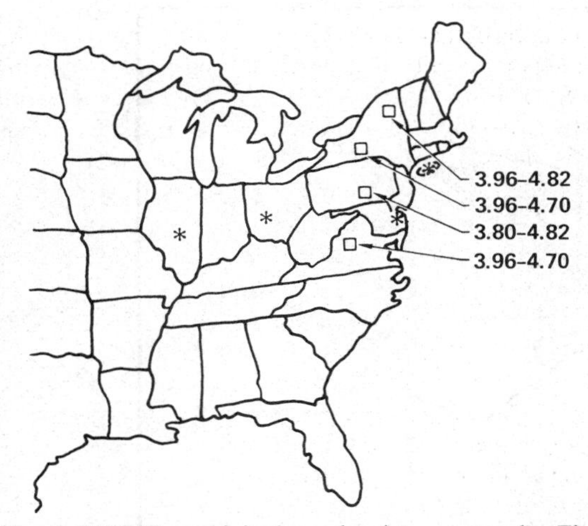

Fig. 5. MAP3S precipitation chemistry network. The boxes □ indicate sites established in 1976 and the asterisks * sites being established in 1977. The range of observed pH at the four existing sites is shown for the period November 1976 to April 1977 based on laboratory measured concentrations of hydrogen ion.

In association with these observation and field studies, closely related numerical modeling studies are underway at PNL, ANL, BNL and ISWS, in order to develop insight into parameterization of the precipitation scavenging mechanism. Chemical transformation within raindrops is also being numerically modeled at BNL.

Task 9. *Determination of the effects of AER pollutants upon weather and climate, including effects on visibility, radiation transport, and the amount and extent of precipitation*

Although effects of urban areas on local scale precipitation and weather are beginning to become apparent, the impact of pollutants on the day to day weather, and therefore ultimately possibly on the climate, over the regional and continental scale is very poorly understood and the proposed climatic impacts remain largely speculative. Reduction of visibility is probably the most noticeable effect, and there is growing evidence relating energy-related pollutants to such effects (e.g. Waggoner *et al.*, 1976). It has been suggested that injection of pollutants higher into the boundary layer by use of tall stacks has led to deeper layers of polluted, low visibility air. Then, in turn, the atmospheric residence time of pollutants is extended because elevated layers can be isolated by low-level nocturnal inversions. This lengthened opportunity for pollutants to be transported and transformed may contribute to visibility obstruction over large areas.

While visibility reduction can be an aesthetic and air safety problem, it is probably the impact of the aerosol on the atmospheric heating and cooling patterns that may have the most significant effects. Changnon *et al.* (1975) and Bolin and Charlson (1976) have suggested that the reduction in solar radiation reaching the surface caused by increased levels of tropospheric aerosols may reduce the length of the growing season by from several days to several weeks. In addition to changes in total solar radiation, the ratio of diffuse to direct radiation is changed (Wesely and Lipschutz, 1976), which may also lead to responses in plants. Whether the redistribution in solar energy absorption induced by aerosols leads to further climatic effects remains uncertain. In addition, the extent to which these effects compare with the normal variability of such factors on a year to year basis has not yet been determined.

The potential effects of increased atmospheric aerosols on the precipitation mechanism has been considered at the Chemist/Meteorologist Workshop in 1975 (Slade *et al.*, 1975). Because mechanisms are poorly understood, details are not clear, but the potential effects appear significant. Not only is precipitation chemistry affected, as discussed in Task 8, but cloud processes (including coagulation, nucleation, cloud condensation nucleii formation, and other aerosol surface phenomena) can be altered, leading to suspected changes in precipitation patterns and amount. Possible changes in dew frequency and fog formation, which in turn catalyze various plant diseases, also may occur.

Although these weather and climate problems are potentially as significant as somewhat better defined health and ecological effects, they are not yet the focus of adequate research attention. The only MAP3S project in this area involves location of network of silicon cell pyranometers in the northeast to evaluate the impairment of direct solar radiation by tropospheric aerosols (Wesely, 1975). The sites presently instrumented are shown in Fig. 1. If persistent and large effects are seen, further investigation of the changes in the radiation pattern will be encouraged using numerical radiation transport models.

Task 10. *Development, verification and demonstration of methods (numerical models) that will make possible accurate assessment of the atmospheric transport and transformation of AER pollutants and of various strategies for generation of power while minimizing atmospheric pollution*

Assimilation of the knowledge about the various processes and mechanisms controlling air quality is essential if predictive methods (e.g. numerical models) are to be developed for use in simulating changes in future air quality due to changes in emissions. Review papers by Pack *et al.* (1978), Eliassen (1978) and Fisher (1978) provide a thorough discussion of the major approaches to development of numerical models in the U.S. and Europe.

The MAP3S modeling sub-program will pursue a two-pronged approach to improving numerical models. The first focuses on upgrading present trajectory models. Such Lagrangian models provide the opportunity to undertake a wide variety of sensitivity studies. For example, Wendell *et al.* (1976) have

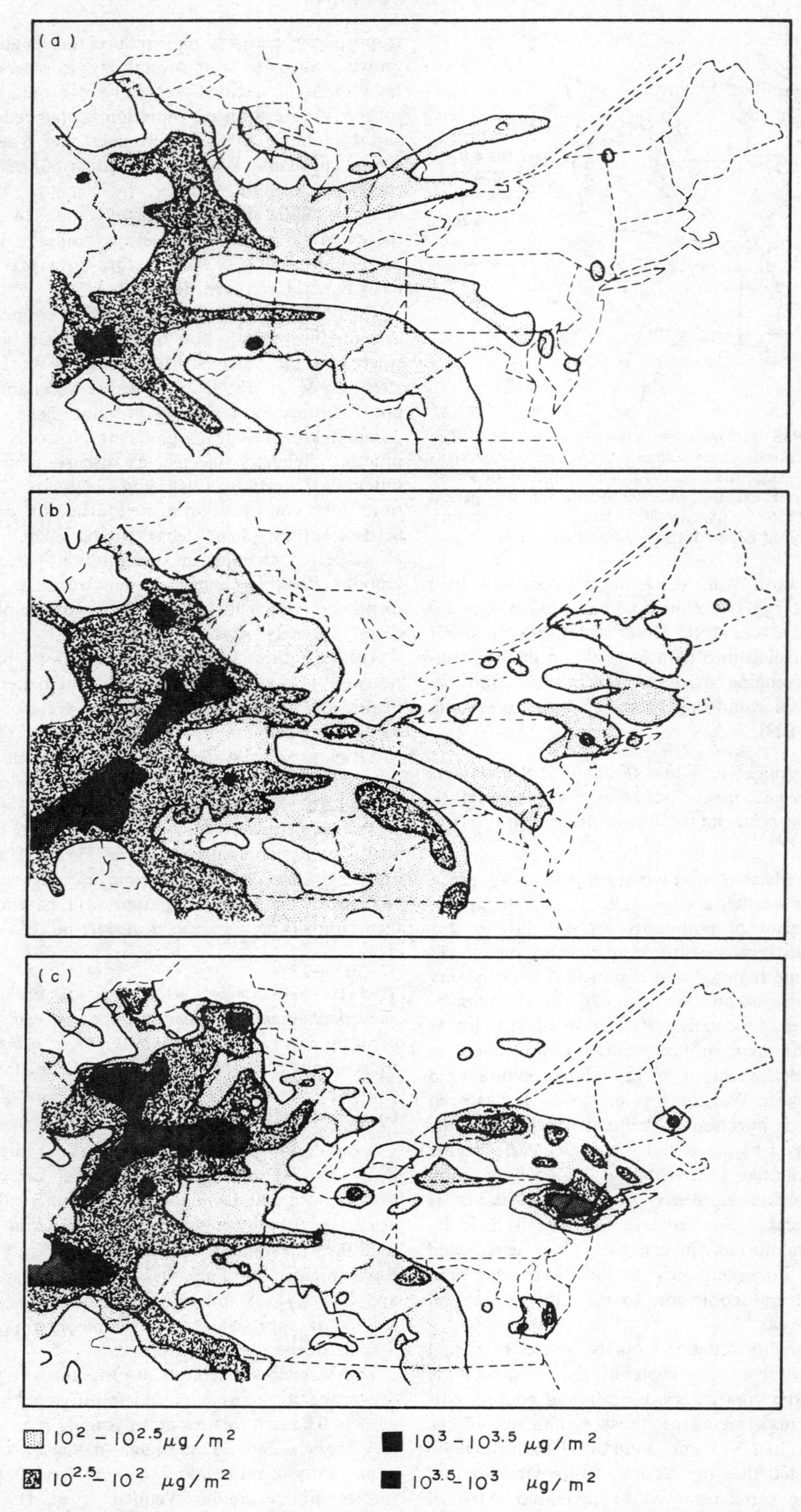

Fig. 6. Sulfate deposition in $\mu g\,m^{-2}$ for an 18 day period in April 1974 for (a) dry removal only, (b) wet and dry removal using average precipitation, and (c) wet and dry removal using real-time precipitation.

shown that different patterns of deposition on the surface result if rain is assumed to be continuous (as might be done in models attempting to calculate long-term average concentrations) rather than episodic (Fig. 6).

The second thrust in model development will be an Eulerian grid model which treats the region as a number of fixed cells approximately 40 km on a side. Vertically, the model will be designed to simulate the diurnal variation in depth of the surface mixed layer, surface and elevated emissions, topography, wind shear, detailed chemistry and other factors important in multi-day simulations. These models will not be predictive, boundary-layer models, although they may be interfaced with such models. Rather they will be diagnostic, relying on observed meteorological data to drive the transport, transformation and deposition mechanisms.

An important aspect of model development is verification. In addition to using specific field experiments to assist in developing detailed parameterizations of specific processes, integrated model results for both the Eulerian and Lagrangian models will be compared with the observations provided by the SURE network and MAP3S characterization flights. Comparisons will be made for both average and episode conditions. Further, MAP3S envisions a few very detailed 'box budget' experiments which will attempt to document the flux and fate of pollutants in a well-defined region. These experiments will be compared to model simulations to assess model accuracy.

As a final aspect of the modeling program, application of the models to various scenarios of future energy use will be continued. The objective is to have a capability useful in formulating energy policy; thus model application will focus on such questions as the impact of dramatically increasing coal use in the Midwest, the effect of widespread application of flue-gas desulfurization, the impact of Ohio River basin emissions on the New York megalopolis, etc. Although point source plume models will be developed to assist in interpreting field experiment data, the focus will not be on the impact of a single plant, but rather on the effect on a national or regional energy policy.

3. SUMMARY AND CONCLUSIONS

Because the research needed to resolve the scientific uncertainties is substantial, the resources required to meet these needs are also large. It will require the combined resources and talents of researchers around the world acting with cooperation and coordination if an efficient and effective approach is to be successful in addressing the scientific uncertainties. In the United States this involves particularly the cooperation of EPA, EPRI and ERDA. Because North America also has international frontiers, coordination between U.S. and Canadian programs is anticipated. The emphasis in MAP3S supported research will be on field experiments aimed at improving understanding of atmospheric processes, numerical modeling aimed at simulating what is happening, and, with substantial reliance on the SURE surface network, in assisting to characterize present air quality.

Acknowledgements—As project director for MAP3S, I have been assisted by a steering committee selected from Project participants including Drs. Leonard Newman, Paul Cunningham, Jake Hales, and Larry Wendell and Bruce Hicks and Ron Meyers. ERDA's program in sulfur studies is due in large part to the support of Mr. David Slade, Deputy Program Manager for Environmental Programs, ERDA. This work was performed under the auspices of the U.S. Energy Research and Development Administration under contract No. W-7405-Eng-48.

REFERENCES

Alkezweeny A. J. (1977) private communication.

Alkezweeny A. J. and Powell D. C. (1977) Estimation of transformation rate of SO_2 to SO_4 from atmospheric concentration data. *Atmospheric Environment* **11**, 179–182.

Altshuller A. P. (1976) Regional transport and transformation of sulfur dioxide to sulfates in the U.S. *J. Air Pollut. Control Assoc.* **26**, 318–324.

Bolin B. *et al.* (1971) *Air Pollution Across National Boundaries, The Impact on the Environment of Sulfur in Air and Precipitation, Sweden's Case Study for the U. N. Conference on the Human Environment.* Royal Ministry of Foreign Affairs and Royal Ministry of Agriculture, Stockholm.

Bolin B. and Charlson R. J. (1976) On the role of the tropospheric sulfur cycle in the shortwave radiative climate of the earth. *Ambio* **5**, 47–54.

Braekke F. H. (1976) *Impact of Acid Precipitation on Forest and Fresh-water Ecosystems in Norway.* Research Report June 1976, Agricultural Research Council of Norway.

Breeding R. J. *et al.* (1973) Background trace gas concentrations in the central United States. *J. geophys. Res.* **78**, 7057–7064.

Brown R. M. and Garber R. W. (1976) Airborne measurements of aerosol and sulfate concentration discontinuities in vertical and horizontal profiles, *Proceedings of the Third Symposium on Atmospheric Turbulence Diffusion and Air Quality.* pp. 340–343. American Meteorological Society, Boston.

Changnon S. A., Jr. *et al.* (1975) The role of aerosols in producing inadvertent weather and climate modification, in *Chemist/Meteorologist Workshop, 1975.* U. S. ERDA Report WASH 1217–55, 37–50.

Changnon S. A. and Semonin R. G. (1975) Studies of selected precipitation cases from METROMEX, Illinois State Water Survey, Urbana.

Coffin D. L. and Knelson J. H. (1976) Acid precipitation: effects of sulfur dioxide and sulfate aerosol particles on human health. *Ambio* **5**, 239–242.

Comar C. L. and Nelson N. (1975) Health effects of fossil fuel combustion products: report of a workshop. *Environ. Health Perspectives* **12**, 149–169.

Cowling E. B. (1975) Statement to the House subcommittee on the environment and the atmosphere. *Hearings on Research and Development Related to Sulphates in the Atmosphere, July 1975*, 408–440.

Craig N. L., Harker A. B. and Novakov T. (1974) Determination of the chemical states of sulfur in ambient pollution aerosols by X-ray photoelectron spectroscopy. *Atmospheric Environment* **8**, 15–21.

Cunningham P. T. *et al.* (1975) *Chemical Engineering Division Environmental Chemistry Annual Report.* Argonne National Laboratory Report ANL 75-51.

Cunningham P. T. and Holt B. D. (1976) Stable isotope ratio measurement in atmospheric sulfate studies. In *Measurement, Detection, and Control of Environmental Pollutants, Proceedings of a Symposium, Vienna, 15–19 March 1976*. IAEA-SM-206 (in press).

Eatough D. J., Major T., Ryder J., Mangelson N. G., Eatough N. L., Hill M. and Hansen L. D. (1978) The formation and stability of sulfite species in aerosols. *Atmospheric Environment* **12**, 263–271.

Eliassen A. (1978) The OECD study of long range transport of air pollutants: long range transport modeling. *Atmospheric Environment* **12**, 479–487.

EPA (1974) *Health Consequences of Sulfur Oxides: A Report from CHESS, 1970–1971*. EPA-650/1-74-004, U.S. Environmental Protection Agency, Research Triangle Park.

EPA (1975) *Position Paper on Regulation of Atmospheric Sulfates*. EPA-405/2-75-007, U. S. Environmental Protection Agency, Research Triangle Park.

EPA (1977) *Statement of Sulfates Research Approach*. U.S. Envrionmental Protection Agency Report EPA-600/8-77-004.

FISAPFE (1976) First international symposium on acid precipitation and the forest ecosystem. *Water, Air Soil Pollut.* **6**, 135–514. Also (1977) **7**, 279–550.

Fisher B. E. A. (1978) OECD–LRTAP: long range transport modeling. *Atmospheric Environment* **12**, 489–501.

Forrest J. and Newman L. (1977) Further studies on the oxidation of sulfur dioxide in coal-fired power plant plumes. *Atmospheric Environment* **11**, 465–474.

FPC, Federal Power Commission (1972) *Steam Electric Plant Air and Water Quality Control, Data for the Year Ended December 1, 1972, Based on FPC Form No. 67, Summary Report*. FPC-S-246. (Also reports for years 1969 1970, and 1971 numbered S-229, S-233, and S-239, respectively).

Garland J. A. (1978) Dry and wet removal. *Atmospheric Environment* **12**, 349–362.

Granat L. (1978) European rain chemistry network. *Atmospheric Environment* **12**, 389–399.

Hales J. M. (1978) Wet and dry removal of sulfur compounds. *Atmospheric Environment* **12**, 413–424.

Hales J. M., Dana M. T. and Glover D. W. (1977) Sampling for volatile trace constituents in natural precipitation, Battelle Northwest Laboratory Report BNWL-SA-6316, to be submitted for publication.

Hall F. P., Jr., Duchon C. E., Lee L. G. and Hagan R. R. (1973) Long range transport of air pollution: a case study, August 1970. *Mon. Wea. Rev.* **101**, 404–411.

Hess G. D. and Hicks B. B. (1975) A study of PBL structure: the Sangamon experiment of 1975. Argonne National Laboratory Report ANL 75–60, Part IV, 1–4.

Hicks B. B. (1970) The measurement of atmospheric fluxes near the surface: a generalized approach. *J. appl. Meterol.* **9**, 386–388.

Hicks B. B. and Liss P. S. (1976) Transfer of SO_2 and other reactive gases across the air-sea interface. *Tellus*, **28**, 348–354.

Hidy G. M. and Mueller P. K. (1978) Regional sulfur oxides air pollution in the United States.

Hidy G. M. *et al.* (1976) *Design of the Sulfate Regional Experiment (SURE)*. Electric Power Research Institute Report EC-125, Palo Alto, CA, Vol. 1–4.

Husar R. B., Gillani N. V., Husar J. D., Paley C. C. and Turco P. N. (1976) Long-range transport of pollutants observed through visibility contour maps, weather maps, and trajectory analysis. *Proceedings of the Third Symposium on Atmospheric Turbulence, Diffusion and Air Quality*. American Meteorological Society, 344–347.

Husar R. B., Patterson D. E., Husar J. D., Gillani N. V. and Wilson W. E. (1978) Sulfur budget of a power plant plume. *Atmospheric Environment* **12**, 549–568.

Isaksen I. S. A., Hesstvedt E. and Hov O. (1978) A chemical model for urban plumes: test for ozone and particulate sulfur formation in St. Louis urban plume. *Atmospheric Environment* **12**, 599–604.

Knox J. B. (1974) Numerical modeling of the transport, diffusion, and deposition of pollutants for regions and extended scales. *J. Air Pollut. Control Assoc.* **24**, 660–664.

Knox J. B. Crawford T. V., Peterson K. R. and Crandall W. K. (1971) *Comparison of U.S. and USSR Methods of Calculating the Transport, Diffusion, and Deposition of Radioactivity*. Lawrence Livermore Laboratory Report UCRL-51054.

Likens G. E. (1976) Acid precipitation. *Chem. and Eng'g. News* November 22, 29–44.

Lowry W. P., *et al.* (1974) Project METROMEX. *Bull. Amer. meteoro. Soc.* **55**, 86–121.

MacCracken M. C. (1977) *MAP3S: An Investigation of the Transport, Transformation and Fate of Atmospheric Energy-Related Pollutants*. Lawrence Livermore Laboratory, Livermore, CA.

Marlow W. H. and Tanner R. L. (1976) Diffusion sampling method for ambient aerosol size determination with chemical composition determination. *Analyt. Chem.* **48**, 1999–2002.

Mészáros E. (1978) Concentration of sulfur compounds in remote continental and oceanic areas. *Atmospheric Environment* **12**, 699–705.

Meyers R. E., Cederwall R. T., Ohmstede W. D. and aufm Kampe W. (1976) Transport and diffusion using a diagnostic mesoscale model employing mass and total energy conservation constraints. *Proceedings of the Third Symposium on Atmospheric Turbulence, Diffusion and Air Quality*. American Meteorological Society, Boston, 90–97.

NEDS (1976) *National Emission Data System*. Monitoring and Data Analysis Division, Environmental Protection Agency, Research Triangle Park, N.C.

Newman L. (1978) Techniques for determining the chemical composition of aerosol sulfate. *Atmospheric Environment* **12**, 113–125.

Nyberg A. (1976) on transport of sulphur over the North Atlantic. submitted to *Tellus*.

OTA Office of Technology Assessment (1976) *A Review of the U.S. Environmental Protection Agency Environmental Research Outlook, FY-1976 through 1980*. OTA Report OTA-E-32.

Ottar B. (1977) *Long range transport of air pollutants: Final report*. Norwegian Institute for Air Research.

Ottar B. (1978) The OECD study of long range transport of air pollutants, a general survey. *Atmospheric Environment* **12**, 445–454.

Pack D. H. *et al.* (1978) Meteorology of long range transport. *Atmospheric Environment* **12**, 425–444.

Perhac R. M. (1978) Sulfate regional experiment (SURE). *Atmospheric Environment* **12**, 641–647.

Prahm L. P., Torp U. and Stern R. M. (1976) Deposition and transformation rates of sulphur oxides during atmospheric transport over the Atlantic. *Tellus* **28**, 355–372.

Semb A. (1978) The OECD study of long range transport of air pollutants: sulfur dioxide emissions in Europe. *Atmospheric Environment* **12**, 455–460.

Sheih C. M. (1977) Application of a statistical trajectory model to the simulation of sulfur pollution over northeastern United States. *Atmospheric Environment* **11**, 173–178.

Slade D. H., Beadle R. W. and Newell R. (1975) *Chemist-Meteorologist Workshop—1975*. US-ERDA Report WASH-1217-75, Washington, D.C.

Tanner R. L. and Newman L. (1976) The analysis of airborne sulfate, a critical review. *J. Air Pollut. Control Assoc.* **26**, 737–747.

Volz F. E. (1969) Some results of turbidity networks. *Tellus* **21,** 625–630.

Waggoner A. P., Vanderpol A. J., Charlson R. J., Larsen S., Granat L. and Träjårdh C. (1976) Sulphate-light scattering ratio as an index of the role of sulphur on tropospheric optics. *Nature* **261,** 120–122.

Walter T. A., Bufalini J. J. and Gay R. W., Jr. (1977) Mechamism for olefin-ozone reactions. *Environ. Sci. Tech.* **11,** 382–386.

Wendell L. L., Powell D. C. and Drake R. L. (1976) A regional scale model for computing deposition and ground level air concentration of SO_2 and sulfates from elevated and ground sources. *Proceedings of the Third Symposium on Turbulence, Diffusion and Air Quality.* American Meteorological Society, Boston, 318–324.

Wesely M. L. (1975) Measurements of atmospheric turbidity in an arc downwind of St. Louis. ANL Report 75–60, Part IV, 22–30.

Wesely M. L. (1977) personal communication.

Wesely M. L. and Hicks B. B. (1977) A review of some factors that affect the deposition rates of sulfur dioxide and similar gases on vegetation. *J. Air Pollut. Control Assoc.* **27,** 1110–1116.

Wesely M. L., Hicks B. B., Dannevik W. P., Frisella S. and Husar R. B. (1977) An eddy correlation measurement of particulate deposition from the atmosphere. *Atmospheric Environment* **11,** 561–563.

Wesely M. L. and Lipschutz R. C. (1976) An experimental study of the effects of aerosols on diffuse and direct solar radiation received during the summer near Chicago. *Atmospheric Environment* **10,** 981–987.

Wexler H. (1950) The great smoke pall, September 24–30, 1950. *Weatherwise* **3,** 129–133.

Whelpdale D. M. (1977) Canadian sulfate study. *ISSA.*

White W. H., Anderson J. A., Blumenthal D. L., Husar R. B., Gillani N. V., Husar J. D. and Wilson W. E., Jr. (1976) Formation and transport of secondary air pollutants: ozone and aerosols in the St. Louis urban plume. *Science* **194,** 187–189.

Wilson W. E. (1978) MISTT: Midwest interstate sulfur transformation and transport. *Atmospheric Environment* **12,** 537–547.

Zak B. (1976) Long distance transport and transformation experiments using a Lagrangian measurement platform. *EOS* **57,** 924.

Atmospheric Environment Vol. 12, pp. 661–670. Pergamon Press 1978. Printed in Great Britain.

LARGE-SCALE ATMOSPHERIC SULFUR STUDIES IN CANADA

D. M. WHELPDALE

Atmospheric Environment Service, Department of Fisheries and Environment, 4905 Dufferin Street, Downsview, Ontario, Canada

(*First received* 21 *June* 1977 *and in final form* 20 *September* 1977)

Abstract – This review describes efforts underway in Canada to investigate large-scale atmospheric sulfur pollution. An assessment of the susceptibility of various parts of the country, in terms of receptor sensitivity, meteorological regime, and source–receptor configuration, as well as evidence of ecosystem damage in North America and Europe, has led the federal Department of Fisheries and Environment to undertake a comprehensive program to investigate the long-range transport of air pollutants in eastern Canada. The program includes studies of emissions, atmospheric phenomena, and aquatic and terrestrial ecosystem effects. Atmospheric aspects of particular interest include long-range transport, regional particulate-sulfate levels, and the acidity and sulfate content of precipitation.

Results from an intensive, month-long study of particulate and precipitation sulfate in eastern Canada indicate that episodes of elevated particulate sulfate concentration occur over a large part of the region, usually in association with the southerly flows behind large areas of high pressure situated over the eastern portion of the continent. Most severely affected is the lower Great Lakes region, with 24 h average sulfate concentrations in the 40–50 $\mu g\, m^{-3}$ range and averages for the period near 10 $\mu g\, m^{-3}$. Precipitation chemistry measurements from this study and other ongoing monitoring programs indicate that the southeastern portion of the country is also affected by high precipitation-sulfate concentrations and low pH values, typically 5 $mg\, l^{-1}$ and 4.3 respectively.

The geographical distributions of both particulate-sulfate and precipitation acidity are consistent with those found in the eastern United States. Air pollution problems associated with sulfur compounds are of a large-regional nature in eastern North America, affecting both Canada and the United States. International approaches to their further study and eventual solution are a necessity.

INTRODUCTION

As control strategies become more and more effective in reducing air pollution in cities and near large sources, interest in the study of pollution problems is shifting from the local to the regional and larger scales in many parts of the world. Pollution problems on the larger scale result from the atmospheric dispersion of emissions over distances of hundreds to thousands of kilometers downwind of large industrialized regions. Frequently, control strategies, such as tall stacks, which are intended to alleviate local pollution, contribute to a worsening of the large-scale situation.

Compounds of sulfur have been most studied as regional-scale pollutants, both because they comprise the bulk of emissions, and because they can have adverse effects on human health (EPA, 1975), the environment, and possibly climate (Bolin and Charlson, 1976). For example, sulfur dioxide and acid precipitation, whose main acid-contributing component is sulfate, are known to cause acidification of some freshwater lakes (Wright and Gjessing, 1976) with resulting damage to fish species (Schofield, 1976) and aquatic organisms (Hendrey *et al.*, 1976). Acid precipitation is also known to cause changes in the properties of some soils (Malmer, 1976), but effects on forest vegetation have not yet been convincingly demonstrated. In Canada, investigations of the past few years have shown evidence of both regional-scale pollution in eastern Canada (e.g. Munn, 1973; Summers and Whelpdale, 1976) and adverse effects on some lakes and fish species in the Canadian Shield region by acid precipitation (e.g. Beamish *et al.*, 1975). This has led to a closer examination of sulfur-related regional pollution and long-range transport in the eastern portion of the country.

Three factors determine the susceptibility of an area to large-scale pollution problems: emission source strength and proximity, meteorological regime, and receptor sensitivity. The part of eastern North America shown in Fig. 1, an area of approx 1500 × 1500 km^2, accounts for industrial SO_2 emissions of about 24 × 10^6 t annually. Stationary fossil fuel combustion for electric power generation is the main emission source in the northeastern states, and primary smelting is the main source in eastern Canada. The SO_2 produced in this area accounts for approx 18% of the man-made emissions globally (Table 1).

During the summer months southeastern Canada frequently experiences southerly flows of warm, moist, tropical air that have traversed major source regions to the south. In addition to the relatively high pollution levels encountered in these situations, pollutant deposition by convective precipitation can be substantial. However, orographic precipitation and intense, localized acid deposition do not occur to the same extent as in parts of southern Scandinavia. During other parts of the year the more frequent

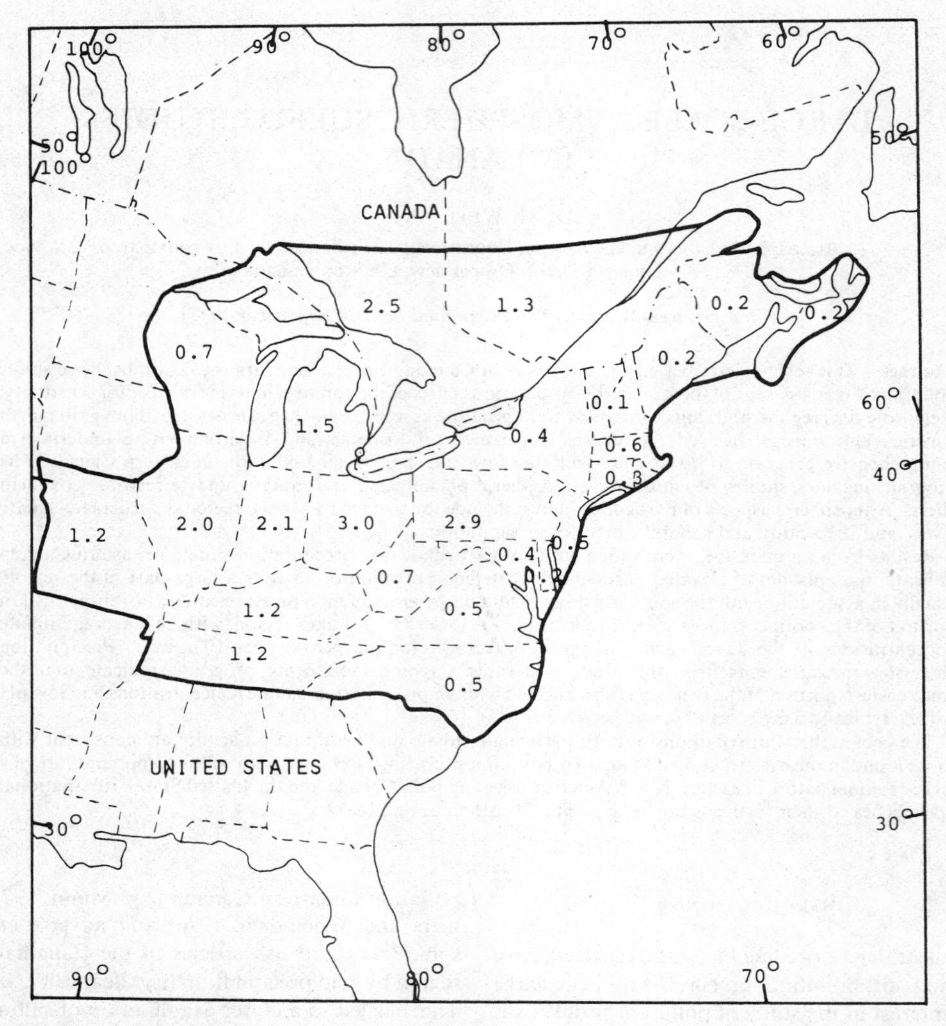

Fig. 1. Annual man-made sulfur dioxide emissions for high emission area of eastern North America. Units are 10^6 metric ton SO_2 y^{-1}. Total for area is 24×10^6 t y^{-1}. Data sources: for Canada, EPS (1973, 1976); for U.S.A., Fox (1976).

Table 1. Man-made sulfur dioxide emissions. Units are 10^6 metric ton SO_2 y^{-1}

Area	Year*	Emission	Percentage of global
Global†	1970	130	100
Europe‡	1973	50	38
United States§	1972	30	23
Canada•	1972	6	5
Eastern North America‖	1972	24	18

* Base year for emission figures.
† Friend (1973).
‡ Ottar (1976).
§ EPA (1975).
• EPS (1976).
‖ Area defined in Fig. 1.

westerly and northerly flows tend to result in a lower influx of pollution.

In terms of receptor sensitivity, the lower Great Lakes–St. Lawrence Valley region is the area of greatest population density: approx 50% of the country's 22.5 million inhabitants live in the $10^3 \times 10^2$ km^2 Windsor to Quebec City corridor. Poorly buffered freshwater lakes and non-calcareous soils, which are susceptible to acid input, are found in some areas of the Canadian Shield in Ontario and Quebec, and in parts of the Atlantic Provinces. Fish species, e.g. trout and salmon, and coniferous forest species, e.g. white pine, which are sensitive to changes in their chemical environment caused by acidification, are common in this region. (See Fig. 2 for locations.)

Thus, southeastern Canada – which includes the area, approximately, east of Lake Superior and south of 49°N latitude – has all three factors which can lead to a regional-scale pollution problem: major sources of emissions, a meteorological regime conducive to source–receptor transport, and sensitive receptors – both ecological and human.

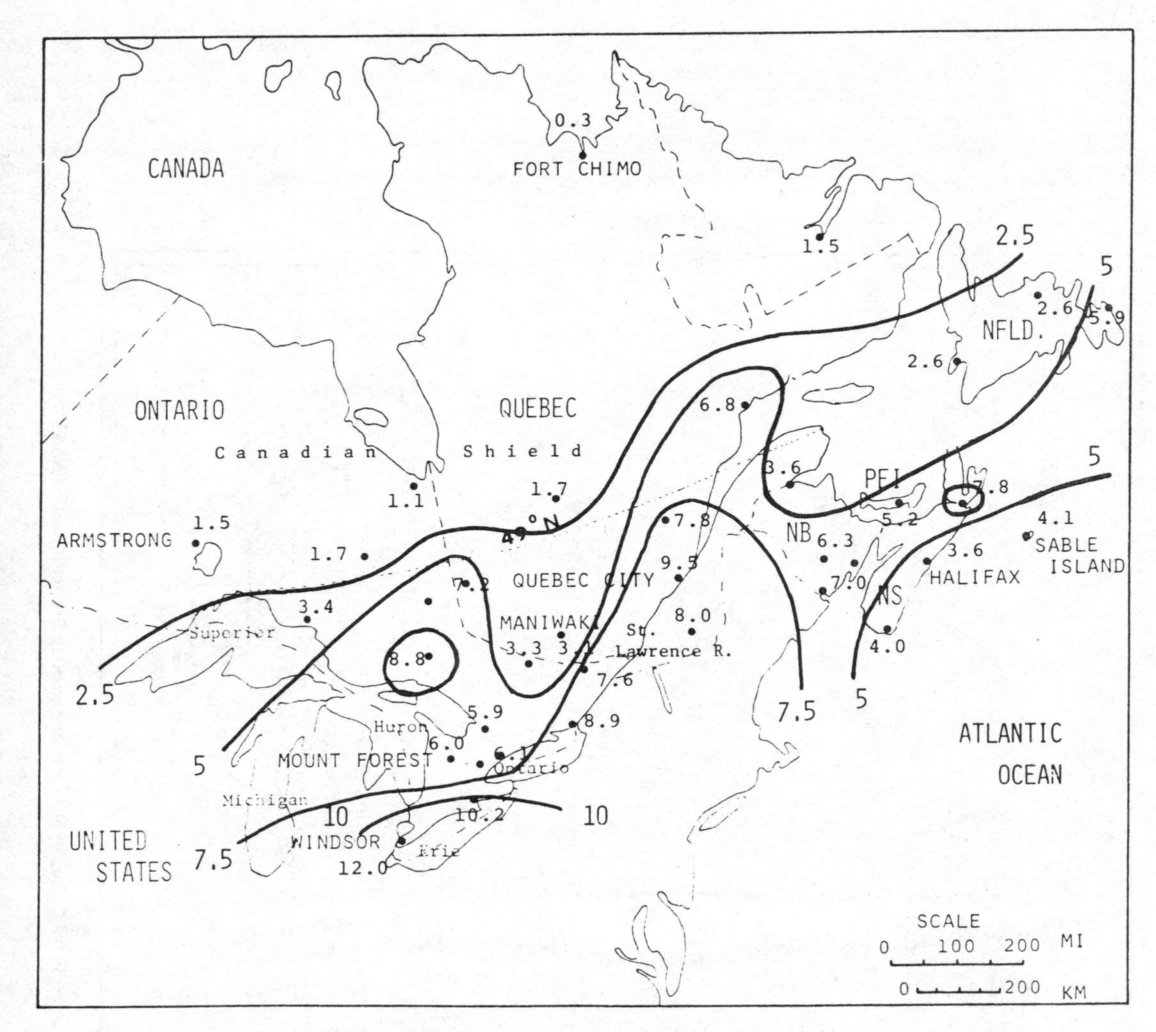

Fig. 2. Map of Intensive Sulfate Study area in eastern Canada showing geometric mean soluble particulate-sulfate concentrations for study period. Units are $\mu g\ SO_4\ m^{-3}$.

CANADIAN LONG-RANGE TRANSPORT OF AIR POLLUTANTS PROGRAM

In response to the above concerns, a comprehensive program was established by the Canadian Department of Fisheries and Environment in mid-1976. Its purpose was to assess the occurrence and effects of regional pollution and long-range transport of air pollutants in eastern Canada. Initially, emphasis is on sulfur compounds and related substances, although some work on effects is being done for mercury and synthetic organics (e.g. PCBs). The program is multidisciplinary and includes source emission inventories, atmospheric dispersion modeling, pollutant concentration and deposition fields, and pathways and effects studies in aquatic and terrestrial ecosystems. The eventual goal is to evaluate the socio-economic costs of the pollution situation in order that any necessary regulatory action may be undertaken.

The atmospheric portion of the program has drawn heavily on the experience of the OECD Long-Range Transport of Air Pollutants Program. The initial objectives of the atmospheric program are (i) to determine current, large-scale concentration and deposition fields for sulfur dioxide, and sulfate in particulates and precipitation; (ii) to determine the frequency and geographical extent of episodes due to long-range transport in the region; and (iii) to identify and determine the relative importance of local and distant sources of pollution.

An inventory of SO_2 emissions for Canada and the United States east of the Rockies is being compiled. Major point sources are identified separately and area sources are assigned to the centre of a representative area in a latitude/longitude grid. Initially, spatial resolution is 1/2 degree latitude and 1 degree longitude (approx $50 \times 50\ km^2$ at 60°N), and time resolution is annual. As the inventory is refined, space and time resolution will be improved, stack parameters added, and natural emissions included. A three-dimensional trajectory analysis model has been developed which uses objectively-analyzed wind fields at the 1000, 850,

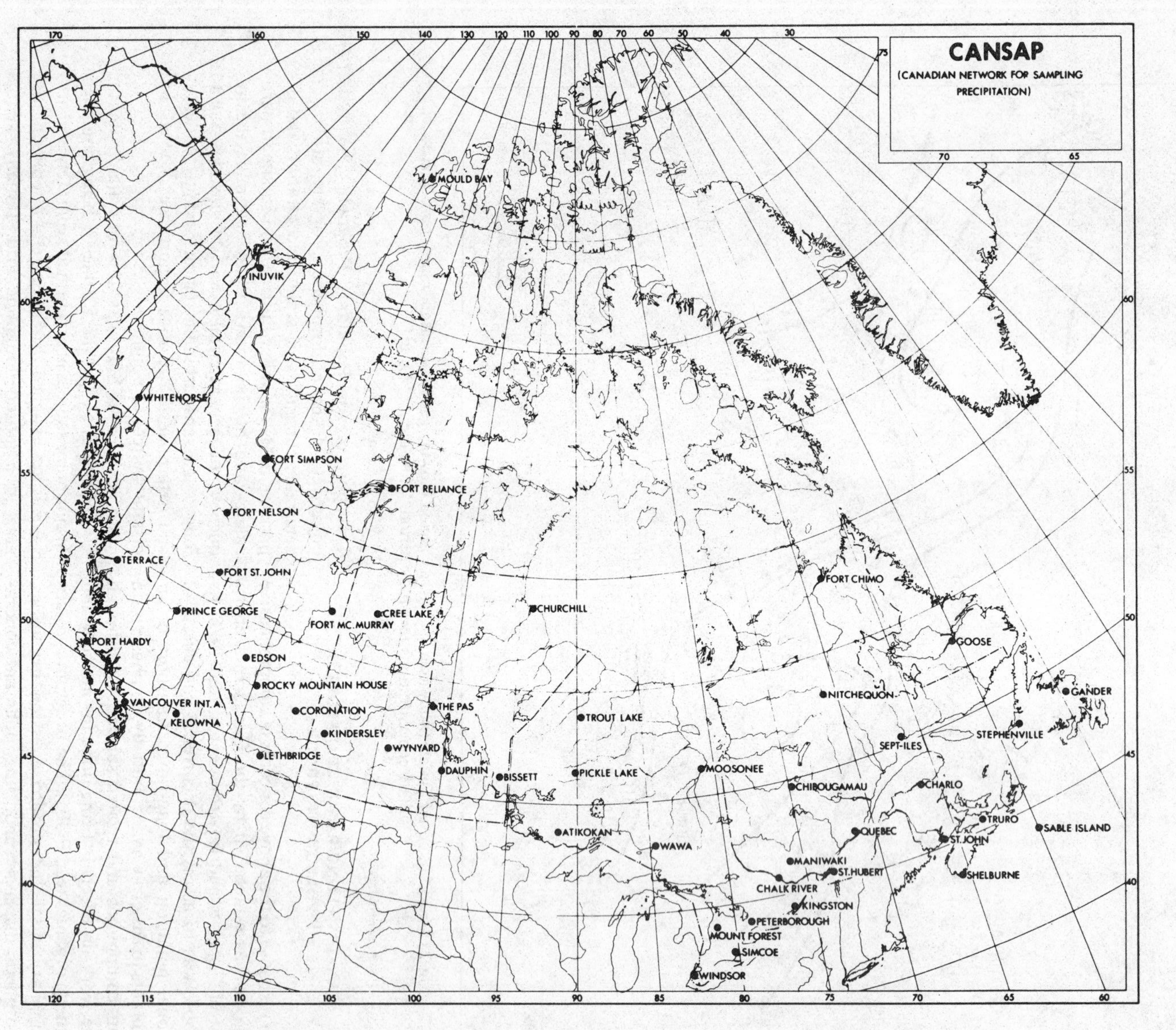

Fig. 3. Map of Canada showing stations in the Canadian Network for Sampling Precipitation (CANSAP).

700 and 500 mb levels. Analysis is based on the standard meteorological grid of 381 km, with the capability of operating on sub-grid scales down to 95 km. A concentration/deposition box model, based on the trajectory model is being developed to provide supporting information for episode measurements and to compile statistics of events and fields. Efforts are also underway to develop a three-dimensional, time-dependent, pseudo-Lagrangian model that can more realistically represent the many physical and chemical processes, and that can eventually provide a reliable predictive capability.

Precipitation chemistry measurements are being made in a fifty-station, country-wide network, the Canadian Network for Sampling Precipitation (CANSAP), which includes ten stations from the WMO Background Air Pollution Network (Fig. 3). The full network went into operation 1 April, 1977 using automatic precipitation collectors; the WMO stations have been in operation for 2–4 years. Samples are collected for month-long periods and shipped to a central laboratory for analysis of major ions, pH, and conductivity. Five "background" air and particulate sampling stations are being established across the eastern portion of the country to provide measurements of regional levels of SO_2, particulate-sulfate, and other substances of interest. They will serve as points for model verification, and as research sites where multi-disciplinary investigations can be undertaken. Research projects underway which are particularly relevant to the long-range transport program include studies of tall-stack dispersion, regional ozone pollution, SO_2 plume oxidation, and pollutant deposition.

PRELIMINARY RESULTS

Previous investigations have indicated that southeastern Canada is subject to higher levels of pollution than other parts of the country. For example, Munn (1973) found a significant increase in the number of hours of reported haze, smoke and/or dust associated with south to southwest winds during summer at synoptic observing stations in the Atlantic Provinces during the past two decades. Because particulate emissions were being reduced at the time, he suggested that the increase in haziness was the result of photochemical activity from greater emissions of gases such as SO_2 and NO_x along the United States eastern seaboard. These results are consistent with the findings of Miller *et al.* (1972), who found significant increases in the frequencies of reduced summer visibility during the period 1962 – 1969 at Akron, Ohio, Lexington, Kentucky and Memphis, Tennessee.

(a) *Pilot study*

The Environmental Protection Service of the Canadian Department of Fisheries and Environment conducted a special study in spring, 1975 (EPS, 1975) to examine sulfate levels across the country. (High-volume filter samples from the National Air Pollution Surveillance (NAPS) network, are not routinely analyzed for sulfate content.) Measurements made on 8 days over a six-week period, at approx 70 predominantly urban sites, showed that the southeastern portion of the country experienced relatively higher sulfate levels than did other areas. Concentrations were typically below 5 $\mu g\ m^{-3}$ in all of Canada except for a region south of approximately the 49th parallel and east of Lake Superior. Here, levels were typically in the 10–15 $\mu g\ m^{-3}$ range, and occasionally they exceeded 20 $\mu g\ m^{-3}$. These results showed qualitative agreement with data from the United States National Air Surveillance Network (NASN) presented by Frank (1974). Three-year, running-mean sulfate concentrations were lowest in the northwestern states (similar to western Canada) and highest in the northeastern quadrant of the United States, where geometric means were in the 10–25 $\mu g\ m^{-3}$ range. This is directly south of the area of highest concentrations found in Canada.

(b) *Intensive sulfate study*

One of the first projects in the Long-Range Transport of Air Pollutants program was undertaken during August, 1976, specifically to examine the spatial and temporal variations in particulate-sulfate and precipitation composition in eastern Canada. Thirty-three stations were established throughout the region (Fig. 2), equipped with high-volume samplers, funnel-and-bottle precipitation collectors, and standard rain-gauges. Most were located at existing NAPS sites or at Atmospheric Environment Service weather stations.

The most striking results of the particulate measurements were the frequent episodes of elevated total suspended particulate and particulate-sulfate levels, and the large geographical extent of each episode (Lafleur and Whelpdale, 1977). Five well-defined episodes of elevated sulfate levels occurred during the 31-day study. Soluble particulate-sulfate concentrations ranged from high values of 40–50 $\mu g\ m^{-3}$ immediately adjacent to the north shores of Lakes Ontario and Erie, to low values of a few $\mu g\ m^{-3}$ at Fort Chimo in northern Quebec. Between episodes, values throughout the region were usually below 10 $\mu g\ m^{-3}$, and near 1 $\mu g\ m^{-3}$ at northern stations. Geometric mean values for the entire period are shown in Fig. 2.

Sulfate concentrations for the period 8–16 August, a typical episode, are shown in Fig. 4a for four widely spaced, rural stations: Armstrong in northwestern Ontario, Mount Forest in southern Ontario, Maniwaki in western Quebec, and Halifax on the east coast. (The Armstrong–Maniwaki and Maniwaki–Halifax distances are each approx 900 km.) Episodes usually occurred in the southerly and southwesterly flows of warm moist air behind centres of high pressure which often dominate the summer weather over the eastern portion of the continent. Maximum concentrations in this episode occurred on consecutive days at stations from west to

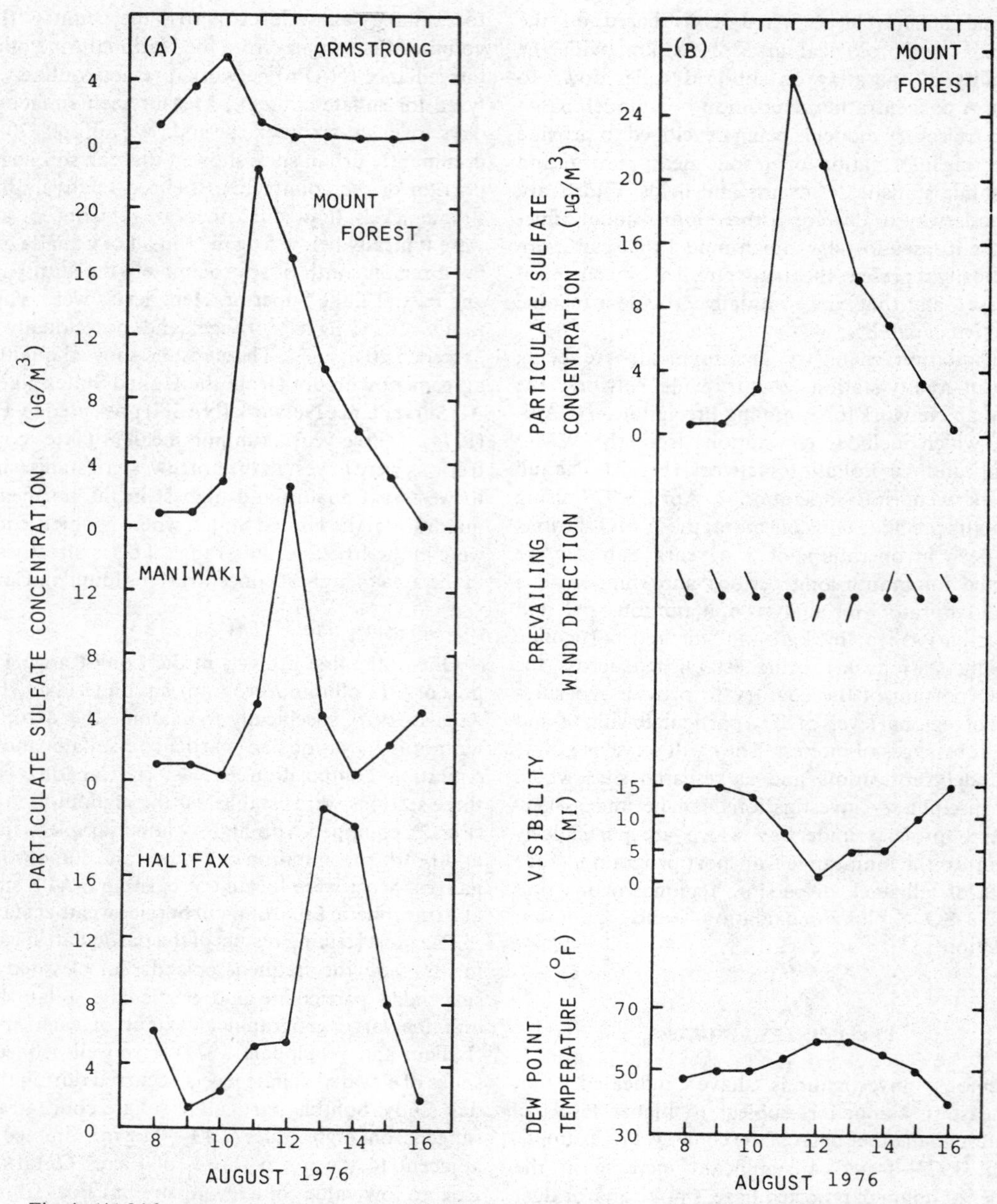

Fig. 4. (A) 24-h average particulate-sulfate concentration records for the episode of 8–16 August, 1976 at 4 stations: Armstrong, Mount Forest, Maniwaki and Halifax. Samples are collected from midnight to midnight local time; dates are marked at local noon. (B) Particulate-sulfate concentration, surface wind, visibility, and dew point temperature data at Mount Forest for the same period.

east as synoptic systems moved across the northern edge of the high pressure area. Elevated sulfate values were well correlated with southerly surface winds, decreased visibility, and elevated dew point temperatures, as might be expected in maritime tropical air masses. These data are shown in Fig. 4b for Mount Forest.

Air parcel trajectories were calculated backward in time for the four stations (Fig. 5) at synoptic hours coinciding as nearly as possible with sulfate maxima, and with the preceding minima. For each station, trajectories corresponding to concentration maxima (solid lines) crossed regions of relatively higher emissions, than did those corresponding to the concentration minima (broken lines). This supports the idea that episodes of high sulfate concentration detected at the various measurement sites were a result of pollution transport from major source regions, often considerable distances away, and were not locally generated. This approach does not, however, provide information on the importance of local or en route photochemical production of sulfate aerosol. (850 mb, 3-day, back-trajectories are assumed to be representative of regional-scale transport in this situation. Surface and 925 mb trajectories were also calculated, and their use would not alter the conclusions above.)

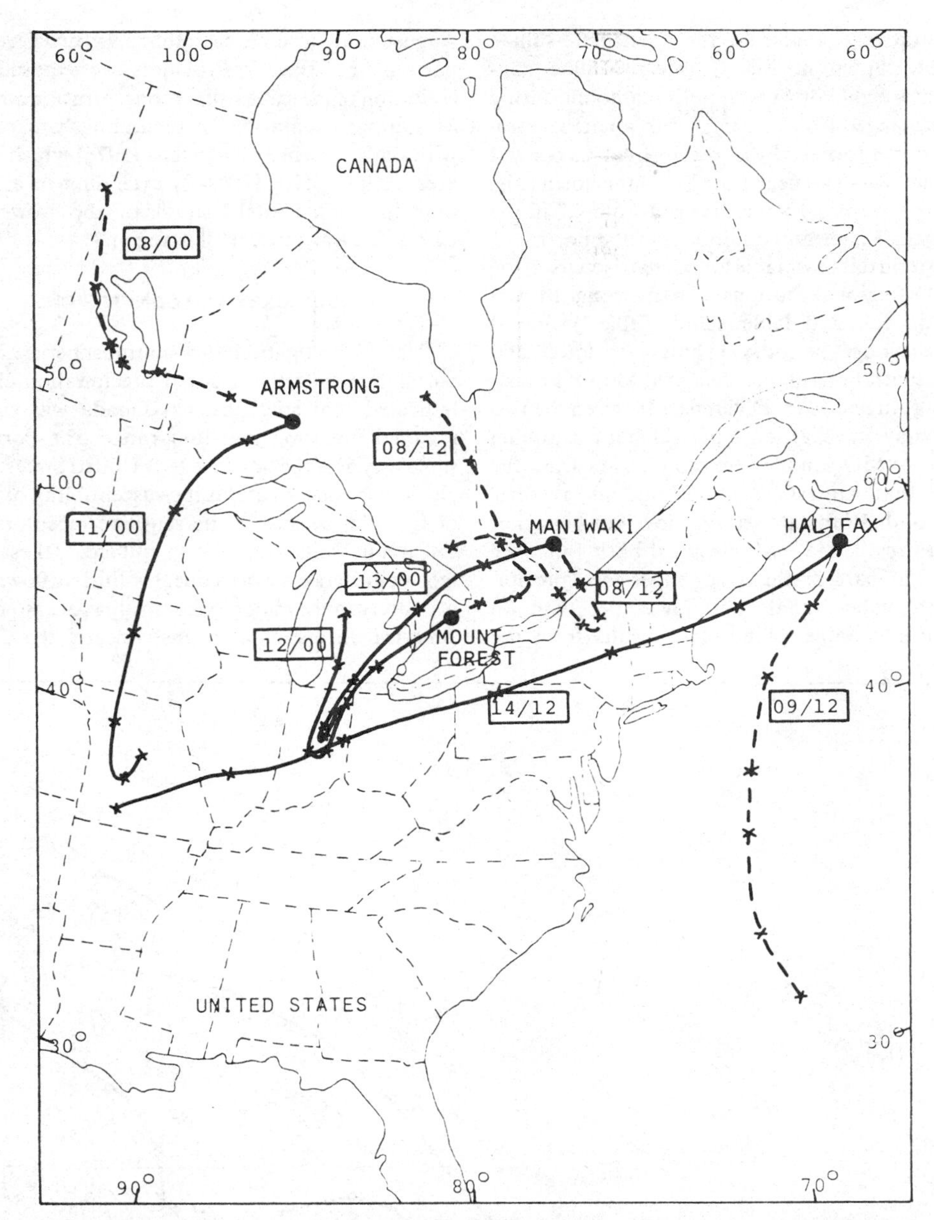

Fig. 5. 850 mb, 3-day, back-trajectories for Armstrong, Mount Forest, Maniwaki, and Halifax for synoptic hours corresponding as nearly as possible to concentration maxima (solid lines) and minima (dashed lines) for the episode shown in Fig. 4. Date and GMT hour of trajectory arrival at stations are in boxes beside trajectories. Crosses on trajectories mark off 12-h intervals.

These results, and studies in the United States (e.g. Altshuller, 1976; EPRI, 1976; Husar *et al*., 1976), have shown that a large portion of southeastern Canada and the northeastern states is frequently subjected to episodes of high particulate-sulfate concentration, and that these episodes often occur when the eastern part of the continent is under the influence of a large area of high pressure.

(c) *Precipitation chemistry measurements*

A recent survey of the composition and acidity of precipitation in Canada (Summers and Whelpdale, 1976) showed that southern Ontario experienced precipitation with relatively high sulfate content and low pH. Typically, sulfate concentrations were in the 2–7 $mg\,l^{-3}$ range, and pH values were usually below 4.5. They suggested that the affected area extended through southern Quebec and into the Atlantic Provinces, although few data were available to confirm this.

In order to investigate the temporal and spatial variability of precipitation composition throughout eastern Canada, event precipitation samples were collected in the network (Fig. 2) established for the intensive sulfate study. Average values for the period gave patterns for precipitation sulfate and pH which were qualitatively similar to the particulate-sulfate distribution shown in Fig. 2. Mean sulfate concentrations, weighted by precipitation amount during each event, ranged from >5 mg l^{-1} in a narrow strip along the north shores of the lower Great Lakes and St. Lawrence Valley to near 1 mg l^{-1} in the north (Fig. 6). Weighted mean pH values ranged from 4.2 in the south to near 5.6 (clean conditions) in the north.

Longer-term data available for comparison from the four WMO Regional Stations at Armstrong, Mount Forest, Maniwaki and Sable Island (Table 2) showed that, over the longer period, pH values were lower, and sulfate concentrations higher (except at Mount Forest) than during August 1976. Differences between the two data sets may have arisen from different sampling techniques (WMO samples are monthly) or from the possibility that August 1976 was not an 'average' summer month. Both data sets do show, however, that Mount Forest and Maniwaki, both in the south–central part of the area, experience precipitation with higher sulfate and lower pH than do Armstrong and Sable Island. The results from the August study also confirm that southern Quebec and parts of the Atlantic Provinces are exposed to precipitation of the same sulfate concentration and acidity as southern Ontario. These findings are consistent with data presented by Likens (1976) which show an area of low pH ($pH < 4.2$) extending in a SW–NE direction immediately south of the Lower Great Lakes–St. Lawrence Valley region.

SUMMARY AND CONCLUSIONS

Past investigations of atmospheric haziness, particulate-sulfate levels and precipitation chemistry indicated that southeastern Canada was subject to regional pollution and long-range transport, in the same way as is the northeastern United States. There is particular concern about the susceptibility of this part of Canada, because of its sensitive receptors – non-calcareous soils and poorly buffered lakes; its meteorological regime of frequent southerly flows; and its proximity to the major emission areas of eastern North America. In response to this concern, the Canadian

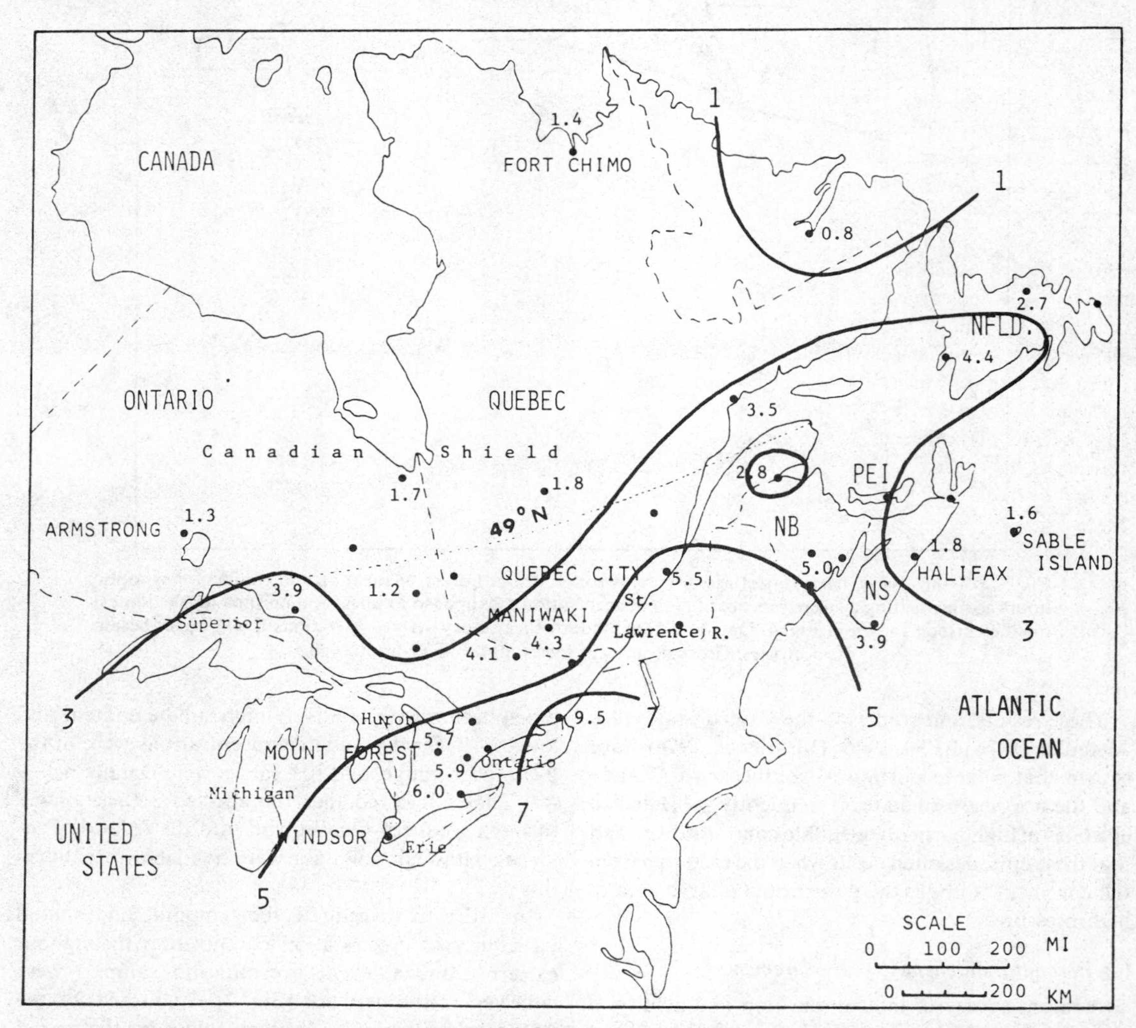

Fig. 6. Mean sulfate concentration in precipitation events, weighted by precipitation amount, during the August 1976 intensive sulfate study. Units are mg l^{-1}.

Table 2. Comparison of precipitation SO_4^{2-} concentration and pH for August 1976 study and previous summer (June–September) months. August 1976 figures are mean values of several events weighted by precipitation amount, and other figures are means of monthly sample values weighted by monthly precipitation amount

Station	Number of months	Precipitation (mm)	SO_4^{2-} concentration ($mg\,l^{-1}$)	pH
Armstrong	August 1976	76	1.3	5.7
	11	86	2.2	4.8
Mount Forest	August 1976	41	5.7	5.0
	12	75	4.9	4.2
Maniwaki	August 1976	75	4.3	4.4
	6	63	5.8	4.2
Sable Island	August 1976	117	1.6	5.6
	5	60	2.2	4.6

Department of Fisheries and Environment has undertaken a multi-component program to determine the present extent, severity and effects of regional-scale pollution and long-range transport, and to assess the importance of atmospheric pathways, particularly from distant sources, relative to other pollutant inputs. The program includes atmospheric, and aquatic and terrestrial ecosystem components.

First results of the program, from a month-long, intensive study of particulate-sulfate and precipitation composition conducted during August 1976 in eastern Canada, have shown that episodes of elevated sulfate levels occur over a large area in southeastern Canada. These episodes usually occur in association with the southerly flows behind large high pressure centres situated over eastern North America. In addition, precipitation sulfate and acidity were found to be elevated in the same region over the study period; these data are supported by results from the previous few years at four WMO Regional Stations in the area. In the case of both particulate-sulfate and precipitation composition, the results tend to complement more numerous data from investigations in the United States.

Over the next few years as the program develops, we expect to be able to better quantify the occurrence and effects of regional pollution and long-range transport, as well as to gain a better understanding of the physical, chemical and biological processes involved. This should form a sound scientific basis for any regulatory action that may be required. This pollution situation is common to both Canada and the United States, and is known to occur in other areas, notably western Europe. Thus, considerable benefit will accrue if a free and energetic cooperation is achieved between the various programs and the scientists involved.

Acknowledgements – The assistance of R. L. Berry, R. J. Lafleur and M. P. Olson at various stages of the work is gratefully acknowledged, as are the suggestions of Drs. L. A. Barrie, R. E. Munn and G. A. McBean on the manuscript.

REFERENCES

Altshuller A. P. (1976) Regional transport and transformation of sulfur dioxide to sulfates in the U.S. *J. Air Pollut. Control Ass.* **26,** 318–324.

Beamish R. J., Lockhart W. L., VanLoon J. C. and Harvey H. H. (1975) Long-term acidification of a lake and resulting effects on fishes. *Ambio* **4,** 98–102.

Bolin B. and Charlson R. J. (1976) On the role of the tropospheric sulfur cycle in the shortwave radiative climate of the earth. *Ambio* **5,** 47–54.

EPA (1975) *Position paper on regulation of atmospheric sulfates.* EPA-450/2-75-007, US Environmental Protection Agency, Research Triangle Park, N.C.

EPRI (1976) Design of the Sulfate Regional Experiment (SURE) Volume I. Prepared by Environmental Research and Technology, Inc. for the Electric Power Research Institute, Palo Alto, California, February 1976. NTIS No. PB 251–701.

EPS (1973) *A nationwide inventory of air pollutant emissions* 1970. Report EPS 3-AP-73-2, Environmental Protection Service, Department of Fisheries and Environment, Ottawa K1A OH3, Ontario.

EPS (1975) NAPS special sulfate study data, April–May, 1975. Air Pollution Control Directorate, Environmental Protection Service. Unpublished data.

EPS (1976) *A nationwide inventory of air pollutant emissions: summary of emissions for* 1972. Report EPS 3-AP-75-5, Environmental Protection Service, Department of Fisheries and Environment, Ottawa K1A OH3, Ontario.

Fox D. G. (1976) Modeling atmospheric effects – an assessment of the problems. *Wat., Air, Soil Pollut.* **6,** 173–198

Frank N. H. (1974) Temporal and spatial relationships of sulfates, total suspended particulates and sulfur dioxide. Paper No. 74-245 presented at the 67th Annual Meeting of the Air Pollution Control Association in Denver Colorado, June 1974.

Friend J. P. (1973) The global sulfur cycle. In *Chemistry of the Lower Atmosphere* (Edited by Rasool S.I.), pp. 177–201. Plenum Press, New York.

Hendrey G. R., Baalsrud K., Traaen T. S., Laake M. and Raddum G. (1976) Acid precipitation: some hydrobiological changes. *Ambio* **5,** 224–227.

Husar R. B., Gillani N. V., Husar J. D. and Patterson D. E. (1976) A study of long range transport from visibility observations, trajectory analysis and local air pollution monitoring data. Paper presented at 7th Int. Techn. Mtng on Air Pollution Modeling and its Application, NATO/CCMS in Airlie Virginia, September 1976.

Lafleur R. J. and Whelpdale D. M. (1977) Spatial distribution of sulfates over eastern Canada during August 1976. Paper No. 77–48.3 presented at the 70th Ann. Mtng of the Air Pollut. Control Ass. in Toronto Canada, June 1977.

Likens G. E. (1976) Acid Precipitation. *Chem. Engng News* **54,** 29–37.

Malmer N. (1976) Acid precipitation : chemical changes in the soil. *Ambio* **5,** 231–234.

Miller M. E., Canfield N. L., Ritter T. A. and Weaver C. R. (1972) Visibility changes in Ohio, Kentucky and Tennessee from 1962 to 1969. *Monthly Weath. Rev.* **100,** 67–71.

Munn R. E. (1973) Secular increases in summer haziness in the Atlantic Provinces. *Atmosphere* **11,** 156–161.

Ottar B. (1976) Monitoring long-range transport of air pollutants: the OECD study. *Ambio* **5,** 203–206.

Schofield C. L. (1976) Acid precipitation: effects on fish. *Ambio* **5,** 228–230.

Summers P. W. and Whelpdale D. M. (1976) Acid precipitation in Canada. *Wat., Air., Soil Pollut.* **6,** 447–455.

Wright R. F. and Gjessing E. T. (1976) Acid precipitation: changes in the chemical composition of lakes. *Ambio* **5,** 219–223.

Atmospheric Environment Vol. 12, pp. 671–680. Pergamon Press 1978. Printed in Great Britain.

BUDGETS AND TURN-OVER TIMES OF ATMOSPHERIC SULFUR COMPOUNDS

HENNING RODHE
Department of Meteorology, University of Stockholm, Arrhenius Laboratory, S-106 91 Stockholm, Sweden*

(*First received* 13 *June* 1977 *and in final form* 20 *September* 1977)

Abstract—The paper starts by defining the time concepts 'residence time' 'age' and 'turn-over time' and by discussing their inter-relations briefly. Some basic comments about box models as applied to sulfur in the atmosphere are also presented. It is suggested that when using budgets to estimate overall fluxes the optimum size for such boxes should be about four times the product of the turn-over time and an average transport velocity.

A brief review is made of previous estimates of atmospheric sulfur budgets of scales ranging from global down to about 100 km. Particular emphasis is put on the implied turn-over times (i.e. average residence times) for the different sulfur compounds. These estimates are compared with estimates of time scales made from observations and experiments and from the use of long range transport models. Such comparisons are complicated by possible deviations from exponential residence time distributions which make, for example, rate coefficients estimated for short transport times unrepresentative for the full life cycle of the molecules in the atmosphere.

Based on the presently available data an estimate is presented of the overall turn-over times in the midlatitude atmosphere of man-made SO_2 and SO_4^{2-} and of the turn-over times with respect to the different transformation and removal processes. The estimated overall turn-over times of SO_2, SO_4^{2-} and sulfur ($SO_2 + SO_4^{2-}$) in this climate are roughly 25, 80 and 50 h respectively. The turn-over time with respect to the different removal and transformation processes are generally somewhat more uncertain. For SO_2 the dry deposition, wet deposition and transformation to SO_4^{2-} correspond to turn-over times of about 60, 100 and 80 h respectively. These figures would imply that about 30% of the SO_2 is transformed to SO_4^{2-} before being deposited. The time scales mentioned here represent average conditions for a particular type of climate and may not be applied directly to specific situations (e.g. seasons) in similar climates or to other climatic zones.

Most of the sulfur budgets estimates utilized in this study apply to conditions in Europe. A plea is made for similar studies to be undertaken in other parts of the world. In North America there may already be enough data available for such studies.

1. INTRODUCTION

When describing the circulation of chemical compounds in the atmosphere it is essential to have an idea about time scales that characterize the transformation and removal processes. If such time scales are known and if measurements of the concentration in the air are available, the amounts that are transformed or removed can be estimated. In connection with dispersion calculations a knowledge of these time scales will also in principle enable us to calculate the patterns of air concentration and fallout provided the sources and the dispersive characteristics of the atmosphere are known. On the other hand, if, for example, the deposition has been measured, a knowledge of the time scale of deposition will make it possible to infer something about the amounts of this compound in the atmosphere and about a possible build up of the concentrations.

There are basically two approaches to the problem of estimating the time scales of transformation and removal processes. One is to study the processes themselves by measurements, either in the atmosphere itself or in a laboratory, or possibly by theoretical calculations. The other approach, and the one I shall be mainly concerned with in this paper, is to use budget calculations to estimate the space and time averages of such time scales. The wish to assess man's impact on the cycle of various compounds is another important reason to attempt to make budget estimates i.e. systematic comparisons of sources, sinks and mean concentrations over reasonably large portions of the atmosphere. Most previous atmospheric sulfur budgets have dealt with global totals of sources and sinks. Since at least the man-made sulfur sources are very unevenly distributed over the globe and because of the limited residence time of sulfur in the atmosphere (see below) it is obvious that possible ecological impacts of man's interference with the sulfur cycle most likely are of a local or regional rather than global character. Consequently, it should be more logical to consider budgets of limited regions of the globe (Granat *et al.*, 1976).

The purpose of this paper is to discuss how budget estimates (i.e. systematic comparisons of sources, sinks and concentrations over reasonably large time and space scales) can be used to estimate the time scales that characterize transformation and removal processes of sulfur compounds in the atmosphere. The

* Contribution No. 350 from International Meteorological Institute in Stockholm.

first sections contain a general discussion about different ways of defining time scales and about some fundamental concepts of box models. In the following sections various sulfur budget estimates are described and the time-scales obtained are compared with other independent estimates. Finally, an attempt is made to combine the information available in an estimate of likely values of the various time scales associated with the circulation of sulfur in polluted regions in midlatitudes.

2. DEFINITIONS OF TIME SCALES

The terminology used here follows largely the one used by Bolin and Rodhe (1973). *Steady state* condition is assumed to prevail.

1. *Residence time* is the time spent by a molecule from the moment of injection into or formation in the atmosphere to the moment of transformation or removal. *Transit time* could be used as an alternative expression for the same quantity. Particularly in cases when the sources and sinks are situated at the boundary of the reservoir (in this case at the ground) this latter expression may be preferable.

Since the residence time (transit time) for individual molecules of a given compound may be quite different it is useful to introduce a frequency function $\phi(\tau)$ describing the frequency distribution of residence times.

If we are interested in the fate of a SO_2 molecule emitted into the atmosphere we must separate between the residence time of the SO_2 molecule itself, τ_{SO_2}, and the residence time of the sulfur atom, τ_S. A transformation of SO_2 to SO_4^{2-} implies a sink for SO_2 (and a source for SO_4^{2-}), but not for the S atom. A sulfur atom that is introduced as SO_2 and removed as SO_4^{2-} will thus have a residence time $\tau_S = \tau_{SO_2} + \tau_{SO_4}$.

2. *Average residence time*, τ_r, is the average value of the residence times (transit times) of all molecules:

$$\tau_r = \int_0^\infty \tau\phi(\tau)\,d\tau.$$

This is the quantity, often refered to simply as the 'residence time', commonly used to characterize the circulation of a compound in the atmosphere. If the frequency distribution of residence times is approximately exponential such a simple characterization may suffice. In other cases, it may be necessary to add more information about the frequency function such as higher order moments.

3. The *age* of a molecule in a reservoir is the time since it was introduced. The corresponding frequency function we denote by $\psi(\tau)$. The age of a molecule *when it leaves the reservoir* is equal to the residence time of that molecule. In general, the age and residence time are of course quite different from each other.

4. *The average age*, τ_a, is the average value of the ages of all molecules found in the reservoir at any one time:

$$\tau_a = \int_0^\infty \tau\Psi(\tau)\,d\tau.$$

τ_a may be larger or smaller than τ_r depending upon the shape of the frequency function (*cf.* Bolin and Rodhe, 1973). They are equal when the frequency functions have an exponential shape.

5. *Turn-over time*, τ_0, is the ratio of the total mass of the compound in the reservoir to the total flux out of (or into) it: $\tau_0 = M/F$. This is a bulk quantity which contains no explicit information about the shape of $\phi(\tau)$ or $\psi(\tau)$. It can be shown (Bolin and Rodhe, 1973) that τ_0 is identical with τ_r so that the turn-over time is actually a measure of the average residence time spent by molecules of the compound in the reservoir.

The age concept has a natural application in problems involving radioactive material. In the following discussion I will confine myself to the residence time concept and the turn-over time.

In cases where there are two or more competing sink mechanisms one may also apply the turn-over time concept to an individual sink mechanism. As an example we can think of a pollutant that is removed either by dry or wet deposition (precipitation scavenging). Let Q be the emission rate and D and W the rates of dry and wet deposition respectively. If M is the steady state mass of the pollutant in the atmosphere (or in a well defined fraction of the atmosphere) we have $\tau_0 = M/(D + W) = M/Q$. The turn-over times with respect to dry and wet deposition separately may now be defined as $\tau_{0_D} = M/D$ and $\tau_{0_W} = M/W$ respectively. It follows that $1/\tau_0 = (1/\tau_{0_D}) + (1/\tau_{0_W})$. In the special case when the removal process may be described by a first order reaction, the turn-over time with respect to this particular process as defined above is equal to the inverse of the rate coefficient.

It is worth noting that whereas dry deposition is a process which is generally continuous in time, the wet removal is only effective during certain periods of time. Therefore the two time scales τ_{0_D} and τ_{0_W} have a somewhat different character. When discussing the physical interpretation of τ_W it may be advantageous to think of it as determined by two distinct processes;

(a) the frequency of occurrence of clouds or rain,
(b) the scavenging process inside the cloud or the rain.

Rodhe and Grandell (1972) showed that under quite general conditions one may write

$$\tau_{0_W} = A + B\lambda^{-1}, \qquad (1)$$

where A and B are functions that describe the frequency of occurrence of rain and λ is the scavenging coefficient in the rain. Thus even with very effective scavenging in rain ($\lambda^{-1} \approx 0$), τ_{0_W} has a finite value (A) corresponding basically to the time it takes to experience the first rain. It may be mentioned that in the model used by Rodhe and Grandell (1972) the frequency function of residence times ($\phi(\tau)$) is not a simple exponential function. It is therefore not quite appropriate to use the inverse of their turn-over time as a decay constant in a first order reaction.

3. BOX MODELS

Before the discussion of the various sulfur budget estimates (see section 4) it may be useful to clarify some basic concepts regarding box models.

In a box model the spatial distribution of the compound inside the box (reservoir) is not explicitly considered. In most cases it is rather assumed that the conditions inside the box are reasonably homogeneous. This is in contrast to analytical or numerical dispersion models where the spatial (and possibly temporal) distribution is dealt with explicitly. The advantage with a box model is that it may enable an overall assessment to be made of the time scales involved and of the relative importance of various transformation and removal processes without laborious computations. As is normally done, we shall limit the discussion to steady state situations where sources and sinks balance each other.

It is often useful to consider a system of reservoirs rather than a single one. This may be the case when the conditions are spatially non-uniform or when one is interested in the circulation of a compound, between different media. As an example one may take the model of the global sulfur cycle as formulated by, for example, Eriksson (1960). Another situation when a series of reservoirs may be considered is when one is considering the circulation of an element in a medium and the element occurs in the form of different chemical compounds. In this case each compound may be taken as a reservoir for the element. As a simple example let us consider anthropogenic sulfur emissions into the atmosphere and let the atmospheric content of SO_2 and SO_4^{2-} be two sulfur reservoirs (Fig. 1).

In this figure Q is the emission of SO_2, C the conversion of SO_2 to SO_4^{2-} and D and W the dry and wet deposition processes. M_1 and M_2 are the amounts of sulfur in the forms of SO_2 and SO_4^{2-} respectively. To simplify the notation we shall drop the subscript 'o' for the turn-over time τ_0.

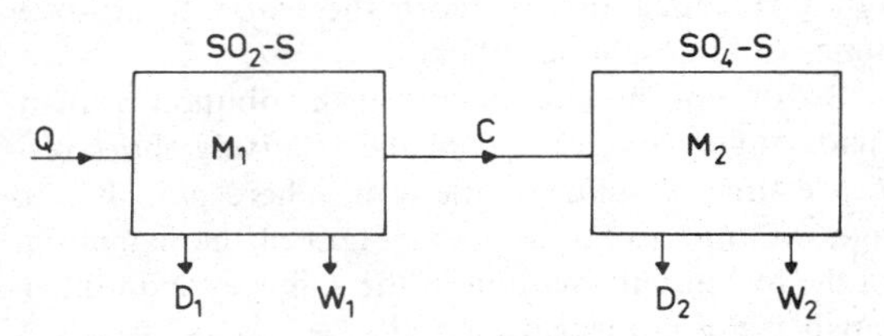

Fig. 1. Simplified diagram of atmospheric cycle of man-made sulfur. Q represents the emission of SO_2, C the conversion of SO_2 to SO_4^{2-} and D and W the dry and wet deposition respectively. M_1 and M_2 are the amounts of SO_2–S and SO_4^{2-}–S present in the atmosphere.

The turn-over time of sulfur in this model is given by

$$\tau_S = \frac{M_1 + M_2}{D_1 + D_2 + W_1 + W_2} = \frac{M_1 + M_2}{Q}. \quad (2)$$

The turn-over times for SO_2 and SO_4^{2-} are

$$\tau_{SO_2} = \frac{M_1}{Q} = \frac{M_1}{D_1 + W_1 + C} \quad (3)$$

$$\tau_{SO_4} = \frac{M_2}{C} = \frac{M_2}{D_2 + W_2}. \quad (4)$$

It follows that

$$\tau_S = \tau_{SO_2} + \alpha\tau_{SO_4}, \quad (5)$$

where $\alpha = C/Q$ is the fraction of the sulfur that is converted to sulfate before being deposited.

When applying a box model to study the sulfur budget of a portion of the atmosphere the choice of an appropriate box size, L, is not a trivial matter. On the one hand, a sulfur budget for the global or hemispheric atmosphere entails an averaging over very different conditions: highly polluted industrial areas and more or less clean remote areas. Evidently such an average will be representative neither for the polluted regions nor for the clean. On the other hand, if too small a box size is chosen the fluxes across the boundaries will be so large that it may be impossible to make any meaningful estimates of the importance of emissions, transformations and removal mechanisms. In principle, it is of course possible to keep track of the fluxes across the lateral boundaries of the box, but in practice the estimates of such fluxes may be so uncertain that subsequent deductions are of limited value if the box size is small.

For studies of the fate of man-made emissions of a compound like sulfur the optimum box size should be such that the distance from the major source region to the edge of the box is somewhat larger, by a factor 2, say, than the scale of decay of the sulfur in the atmosphere ($R = \tau_s \times \bar{V}$ where $\bar{V}$ is an average transport velocity). If we let L_{opt} denote the diameter of an appropriately chosen circular box with a point source near the center we then have

$$L_{opt} \sim 4\,\tau_s\,\bar{V}. \quad (6)$$

Since τ_S is a quantity which may not be known *a priori* but which one wants to determine from the budget estimate it follows that the choice of L may have to be done through trial and error.

It goes without saying that other factors of a practical nature, such as the availability of measurements, also influence the choice of box size.

4. PREVIOUSLY PUBLISHED ATMOSPHERIC SULFUR BUDGETS

No attempt is made to make a complete review of all previous estimates. I will rather limit myself to making a few comments about some of these estimates, partly in the light of what was said about time scales in the foregoing sections. I will also try to discuss the reasons for some of the differences between various estimates.

4.1 *Global budgets*

The basic purpose for making estimates of the global sulfur budget can be formulated as seeking answers to the following questions (*cf.* Kellogg *et al.*, 1972):

(a) What are the sources (natural and man-made), sinks and residence times of sulfur compounds in the atmosphere–ocean–soil system?

(b) How do man's activities compare with nature's in determining the overall budget of atmospheric sulfur compounds?

The determination—directly or indirectly—of the annual fluxes is normally the starting point in these budgets (Eriksson, 1960, 1963; Junge, 1960, 1963; Robinson and Robbins, 1970; Kellogg *et al.*, 1972; Friend, 1973; Granat *et al.*, 1976).

The turn-over time is either determined by a comparison with estimated burdens or it is taken from independent estimates and used to calculate the burdens. Of all global fluxes, only the man-made emissions into the atmosphere can be considered known with a sufficient accuracy: maybe $\pm 25\%$. The wet deposition, extrapolated from measurements primarily in Europe and the U.S., has generally been considered also to be reasonably well known. However, a critical review of all available data made by Granat (Granat *et al.*, 1976) indicated that the global wet deposition might have been overestimated in all earlier estimates. The dry deposition cannot be measured directly but is normally estimated with the aid of (a few) measurements of the background surface air concentrations and a deposition velocity. The most uncertain post in the atmospheric part of these budgets is the flux of volatile reduced sulfur compounds (H_2S, dimethylsulfide, etc.) from land or sea surfaces. This flux is generally derived as a balance for the atmospheric reservoir. There is little independent quantitative evidence to support estimates of this flux.

A comparison between the different authors' estimate of this flux is given in Table 1. The difference between the first two estimates and the following five is mainly due to a likely overestimate of the dry deposition on soil, vegetation and water surfaces by the former authors.

Table 1. Estimates of the annual global emission of reduced volatile sulfur compounds into the atmosphere by natural processes.

Author	Flux in Tg sulfur y^{-1}
Eriksson, 1960	267
Junge, 1963	230
Robinson and Robbins, 1970	98
Kellogg *et al.*, 1972	89
Friend, 1973	106
Granat *et al.*, 1976	33 and 37

The low sulfur budget presented by Granat *et al.* (1976) is actually based on two independent estimates both of which show similar low values.

1. The first one (due to Hallberg) assumes that there was a pre-industrial balance for the sulfur in the soil and that the emission of H_2S (or any other volatile reduced sulfur compound) from land areas is small (this assumption was based on the work by Hitchcock, 1976). From estimates of weathering, volcanism and river run-off the pre-industrial (natural) deposition of sulfur on land was derived and subsequently the required emission from coastal and ocean areas. This budget does not rely explicitly on any atmospheric deposition data. It is, however, critically dependent on the assumption about small H_2S emission from land areas. If such emissions can be shown to be larger, this budget must be revised to account for a larger turn-over.

2. The second estimate (due to Granat) is based entirely on atmospheric data. The low flux of H_2S is mainly a result of a downward revision of the global wet deposition. Earlier estimates relied almost entirely on data from Europe and the U.S. which were probably not unaffected by man-made emissions. Therefore, the extrapolation of those deposition values to a global figure might have given too large values. Although this second budget estimate of Granat *et al.* took into account more recent deposition measurements from remote areas than previous authors, the data base is still very inadequate and the global deposition estimate accordingly quite uncertain. The interpretation of the data in this case was deliberately slightly biased towards the low side.

The difference between the budgets by Granat *et al.* (1976) and the three previous ones may possibly be taken as a measure of the uncertainty involved in such estimates.

The following comments may serve as concluding remarks to this subsection.

Estimates of the global sulfur budget are very uncertain and will remain so until much more data is available on deposition and air concentrations of different sulfur compounds. The most uncertain part is the emission of reduced volatile sulfur compounds into the atmosphere.

Because of these uncertainties, global budgets will not give much useful information about turn-over times of atmospheric sulfur.

Because of the uneven spatial distribution of man-made sulfur sources and of the relatively short residence time of sulfur in the atmosphere global averages will not give a fair description of the magnitude of the human intervention in the sulfur cycle in different parts of the world.

4.2 *Regional sulfur budgets*

In this section I will consider sulfur budgets made for more limited parts of the atmosphere, from continental scales down to one or a few hundred kilometers. Areas chosen for such budgets have generally been those significantly affected by man-made sulfur emissions. This is, of course, due to the interest in the fate of the pollution sulfur and to the availability of concentration and deposition measurements in such areas.

Many attempts have been made to establish the balance of sulfur compounds on even smaller space scales such as in cities or in chimney plumes. Interesting as they may be these studies are normally representative for very special conditions and must not be extrapolated to represent a full life cycle of the sulfur in the atmosphere.

An early attempt to establish a sulfur balance on a continental scale was made by Junge (1960). Based on measurements of wet deposition in the U.S. and estimates of the man-made emission in the same area he concluded that the turn-over time of the pollution sulfur with respect to wet deposition ($(\tau_S)_w$) was about 6 days. With an estimated dry deposition a total turn-over time τ_S of 3–4 days was derived (Junge, 1963).

By applying similar arguments to an area in northern Europe Rodhe (1972) arrrived at a figure of 2–4 days for the same quantity (τ_S). It seems likely that, primarily because of a rather high estimate of the background sulfur deposition in the area and an underestimate of the dry deposition, this value of the turn-over time is slightly high.

Another budget estimate for northern Europe, based on the more detailed information then available, was presented by Rodhe (Granat *et al.*, 1976). The area studied is bounded by latitudes 45°N and 65°N and by longitudes 10°W and 20°E and has a characteristic horizontal dimension of about 2000 km (Fig. 2). The input data included the source inventory and measurements of surface air concentrations performed within the LRTAP-project (Eliassen and Saltbones, 1975) and wet deposition data from the European Air Chemistry Network (Granat, 1978). By comparing the mass of sulfur in the air with the total man-made emissions, a turn-over time of 2–3 days was derived. The net export of pollution sulfur from the area was estimated to 23–46% of the emission (see Table 2). In view of the comments about a suitable box size made in section 3 ($L_{opt} \approx 4\tau_0\bar{V}$), the results show that it would have been advantageous to study a somewhat larger area. With $\tau_0 \approx 50$ h and $\bar{V} \approx 25$ km h^{-1} L_{opt} would be about 5000 km.

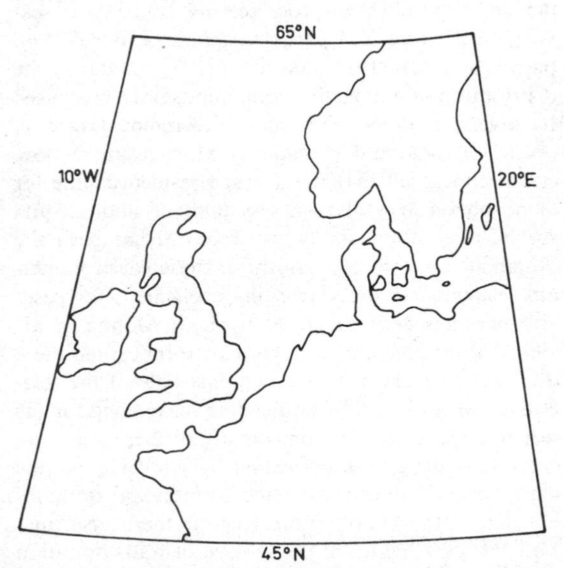

Fig. 2. The area used by Rodhe (in Granat *et al.*, 1976) to study the atmospheric sulfur budget.

Table 2. Balance sheet for atmospheric sulfur over NW Europe around the year 1973 (Granat *et al.*, 1976) (unit, Tg y^{-1})

	Pollution sulfur	Non-pollution sulfur	Total
Emission	13	?	13+?
Deposition			
by precipitation	4.1	0.5	4.6
by direct uptake	2.9–5.9	0.2–0.3	3.1–6.2
Total	7.0–10.0	0.7–0.8	7.7–10.8
Net export of pollution sulfur	3.0–6.0		
Net export as percentage of emission	23–46		

A similar area was used by Garland (1977) in a study of the sulfur balance over western Europe. This estimate was based on essentially the same data and the conclusions are not very different: a net export out of the area equal to about 30% of the man-made emissions implying a turn-over time of this sulfur of roughly 2 days. The estimate of the total dry deposition was slightly higher and that of the total wet deposition slightly lower than the corresponding estimates made by Rodhe (1976).

It should be stressed here that this type of budget estimate depends critically on average values of a few key parameters notably the deposition velocity, v_d, and the scale height, H, of atmospheric sulfur, none of which can be considered very accurately determined. The value of the deposition of sulfur from natural sources and from man-made sources outside the area is also an uncertain quantity.

Several budgets on a smaller scale such as for individual countries have been presented. The first attempt in this direction seems to have been made by Meetham (1950) who estimated the balance of soot and SO_2 over Great Britain. His estimate of the turn-over time of SO_2 was very low (about 10 h) essentially because of an incorrect assumption about its vertical distribution. Garland and Branson (1976) used much more reliable data including measurements of the vertical distribution of SO_2 to set up a SO_2 balance for Great Britain. Their study shows that the deposition within the country is well below 50% of the emitted amounts (see also Garland, 1977). No attempt was made to determine a turn-over time of sulfur over the country but it was suggested that the dry deposition alone would limit the mean life time of SO_2 to about 2 days. With our notation (see Table 3) this corresponds to $(\tau_{SO_2})_D \approx 2$ days.

This estimate was based on a value of the SO_2 scale height of about 1200 m measured over rural Great Britain. Since the scale height would probably increase by diffusion as the sulfur is transported away from the country it is probable that $(\tau_{SO_2})_D$ is somewhat larger than 2 days when a full life cycle is considered (*cf.* discussion of the paper by Garland and Branson, 1977).

Sulfur balance estimates have recently been presented for Hungary (Várhelyi, in press) and for an area covering parts of Austria, Czechoslovakia and Hungary (Mészáros and Várhelyi, 1977). No attempt was made to derive any turn-over time. In the latter study the authors used the balance equation to estimate the emissions inside that area. Because of the small size of these areas the fluxes across the boundaries become the dominant terms in the balance and the inferences about the remaining flux terms become more uncertain.

Smith and Jeffrey (1975) used aircraft measurements of SO_2 and SO_4^{2-} off the east coast of England to make a sulfur balance for an area roughly 200 km wide covering the major source region upwind over the country. They found that within 2–4 h about 30% of the SO_2 had been deposited and 10–20% converted to SO_4^{2-}. The remaining 50–60% was observed as an export of SO_2 from the area. The implied deposition velocity for SO_2 was around 0.8 cm s^{-1}. No number for the transformation rate was given but it was suggested that most of the transformation might occur very rapidly after the emission and that the subsequent transformation was small.

5. OTHER WAYS OF ESTIMATING TIME SCALES ASSOCIATED WITH THE TRANSFORMATION AND DEPOSITION OF SULFUR COMPOUNDS

We have seen that budgets can be used to estimate the importance of different transformation and deposition processes. On the other hand, independent estimates of the importance of these physical and chemical processes are often used as important input parameters in the budgets. It is therefore of interest to discuss briefly some other evidence about such processes.

5.1 *Observations and experiments*

It is beyond the scope of this paper to make a review of the large amount of work on the transformation and removal of sulfur compounds that has been performed. The reader is referred to the other review papers in this volume. Only some brief comments will be made.

(a) *Dry deposition.* Measurements of dry deposition of SO_2 by the profile method and by some other methods have been made mainly for grass and water surfaces. Deposition velocities—defined as the ratio of the flux and the surface air concentration—normally lie in the range 0.5 to 2 cm s^{-1} with an average value of slightly less than 1 cm s^{-1} (Garland, 1978). With a scale height of 1.5 km, a deposition velocity of 0.8 cm s^{-1} corresponds to a turn-over time with respect to dry deposition of roughly 50 h.

It is not yet clear how effective forests are in removing SO_2 under different conditions and the extrapolation of the measured deposition values to large areas with different surface characteristics is far from straight forward. Furthermore, most deposition measurements have been made in temperate climate and may not be representative for other climatic regimes. Some measurements indicate that the deposition of SO_2 on snow might be as low as 0.1 cm s^{-1} (Dovland and Eliassen, 1976).

Regarding the dry deposition of sulfate particles it is generally believed (e.g. Garland and Branson, 1976) that it is small compared to the wet deposition. Deposition velocities for submicron particles of less than 0.1 cm s^{-1} have been mentioned (Chamberlain, 1967). Somewhat larger values (0.2–0.5 cm s^{-1}) were reported by Van der Hoven (1975).

(b) *Wet deposition.* As was pointed out in section 3 the average rate of removal by precipitation is determined both by the frequency of occurrence of precipitation events and by the scavenging coefficient during the precipitation. The former quantity is evidently highly dependent on the prevailing climatic condition. Based on a very limited amount of data Rodhe and Grandell (1972) found that the *lower limit* of the turn-over time with respect to precipitation scavenging—i.e. the term A in Equation 1—was about 35 h in winter and 90 h in summer for Scandinavian climate. As an annual average this lower limit may be set at about 50 h for the climate of northern Europe.

The incorporation of sulfur into the rain-drop can take place either inside the cloud during the formation of raindrops or when the raindrop is falling through the cloud-free air below the cloud. For details of the physical processes involved in the scavenging of gases and particles, reference is made to papers by Hales (1972) and Slinn (1977). Data on the distribution of radioactive compounds have been used by several authors to estimate the removal rate. A review of such studies made by Martell and Moore (1974) concluded that the average residence time for tropospheric aerosols was less than or about equal to one week. Presumably, wet removal has been the dominant removal mechanism in these cases. Generally, estimates of the scavenging coefficients for gases and particles vary widely (see e.g. McMahon *et al.*, 1976). As far as sulfur compounds are concerned there does not even seem to be a consensus about the relative importance of SO_2 and sulfate scavenging for the wet removal of sulfur from the atmosphere.

(c) *The rate of transformation* SO_2 *to* SO_4^{2-} If the experimental evidence for removal rates was quite inconclusive, this is even more true for the transformation rates. Estimates of the inverse of transformation rates vary from a few hours to at least hundred hours (EPA, 1975).

There are reasons to believe that the transformation rates vary not only between different climatic conditions but also depending upon the concentrations of various other man-made compounds. It may therefore not be possible to compare rates deduced from studies made in chimney plumes with rates applicable to large scale budgets (see section 6.1).

Husar *et al.* (1978) reported detailed studies of the sulfur budget of a power plant plume in the St Louis area in the U.S. performed as part of the MISTT Project. The plume was sometimes identified up to a distance of about 300 km (Gillani *et al.*, 1978). Reaction rates for the conversion of sulfur dioxide to sulfate were found to be typically 0.01–0.04 h^{-1} during noon hours and less than 0.005 h^{-1} at night. Similar values have been reported from measurements in the plume from the Sudbury Smelter Plant in Canada (Lusis and Wiebe, 1976; Forrest and Newman, 1977). Alkezweeny and Powell (1977) report somewhat higher values (around 0.10 h^{-1}) in the urban plume of St Louis for travel times up to a few hours.

5.2 *Long-range transport models*

Provided that the description of the transport and dispersion processes in such models is trustworthy, a comparison between calculated and observed concentrations can give useful information about transformation and/or removal processes. Since such a condition is not always fulfilled, a certain amount of scepticism about interpreting model parameters as physical parameters is well motivated.

A further complication arises in those cases when attempts are made to determine more than one parameter at the same time. It is then not enough to find a set of parameters which gives a reasonable agreement between calculated and observed values. One has to show that such a set is unique i.e. that other sets may not also give a fair agreement. This problem was encountered and commented upon by Eliassen and Saltbones (1975). In their study, these authors compare the measurements of surface air concentrations (daily means) obtained in the LRTAP network with values calculated with the aid of a dispersion model based on trajectories. The data were taken from 6 stations in northern Europe and covered two periods of each two months' duration. Three parameters were estimated by regression analysis: the decay rate of SO_2 (including dry removal, wet removal and conversion to sulfate), the conversion rate SO_2 to sulfate and the mixing height. Eliassen and Saltbones found average values of these parameters around 0.07 h^{-1}, 0.007 h^{-1} and 1200 m respectively. Because of the neglect in the model of any removal of sulfate by precipitation the derived conversion rate is likely to be smaller than the real value. The decay rate of SO_2 is more likely to have some physical significance, but the comments made above must be kept in mind.

In the calculations referred to in the LRTAP report (Eliassen, 1978) the values of the parameters are partly based on the above estimates by Eliassen and Saltbones (1975). If, in the model, precipitation scavenging were allowed to act on sulfate particles instead of on SO_2 higher values of the transformation rate and of the decay rate for SO_4^{2-} would have been implied.

A similar type of model was used in the SURE experiment in the U.S.A. (EPRI, 1976) to calculate SO_2 and SO_4^{2-} concentrations for comparison with observed 24-h values. Transformation rates (SO_2 to SO_4^{2-}) in the range 0.02 h^{-1}–0.01 h^{-1} seemed to give a fair agreement but the limited amount of data prevented any firm conclusions from being drawn.

Prahm *et al.* (1976) applied a simple trajectory box model to study the transport of SO_2, SO_4^{2-} and trace metals from the British Isles to the Faroes where measurements had been made. Values of the deposition velocities for SO_2 and SO_4^{2-} (2 and 0.4 cm s^{-1}, respectively) and of the transformation rate (0.011 h^{-1}) were derived. These values are, however, dependent on several assumptions the most critical probably being a 'dilution factor' and they are therefore uncertain to at least 50%. For example, a larger dilution would decrease the deposition velocity and at the same time increase the transformation rate.

Omstedt and Rodhe (1978) used a simple one-dimensional (height) dispersion model to study the ratio of SO_2 to SO_4^{2-} in a polluted region. By comparison with observations of this ratio they concluded that the average conversion rate is likely to be in the range 0.04–0.007 h^{-1} for European conditions. They further found that it was difficult to explain the observed wet deposition pattern if all sulfur had to be transformed to SO_4^{2-} in the air before being scavenged by precipitation. In order to have a rapid enough wet removal of sulfur the scavenging of SO_2 is likely to be significant.

6. DISCUSSION

In this section I shall first make some comments about the comparison between the various time scales referred to in the previous two sections. When making such comparisons it is essential to be very clear about the definitions and significances of the time scales used. I finally summarise the information about turnover times available from the budget estimates and attempt an estimate of likely values as applied to a European type climate.

6.1 *The relation between reaction rates and turn-over times*

One of the first things to consider is to which extent it may be justified to equate the inverse of estimated rate coefficients with turn-over times (i.e. average residence times). Almost all estimates of reaction rates from measured data are based on the assumption that the reaction is of first order with a constant coefficient and that therefore the decay corrected for the disper-

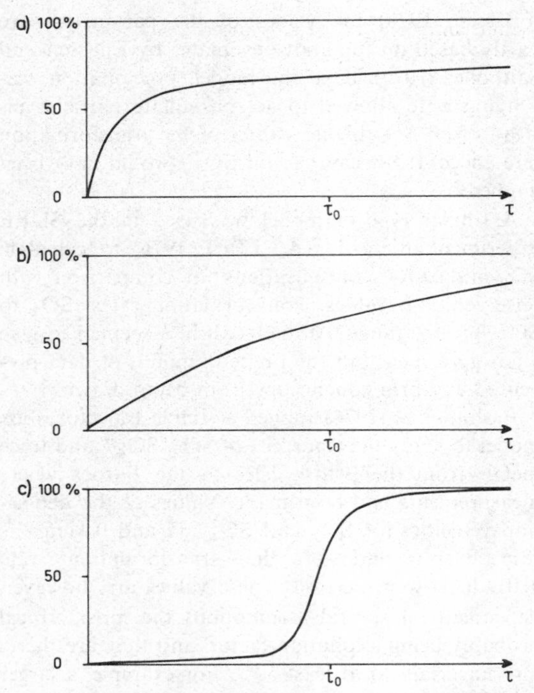

Fig. 3. Three examples of distribution functions for residence times. The turn-over time τ_0 (the average residence time) is about equal in the three cases. If rate coefficients are determined from measurements close to the source (i.e. short times) the inverse of such rate coefficients may differ very markedly from the turn-over time (see text).

sive dilution is exponential. If this is not valid the inverse of the estimated rate coefficient may be a poor measure of the turn-over time. In order to exemplify such situations let us consider the distribution function, $F(\tau)$, for the residence times of individual molecules (Bolin and Rodhe, 1973). $F(\tau)$ is thus the fraction of all molecules that have a residence time less than τ and $1 - F(\tau)$ is the fraction which is still in the air the time τ after they have been emitted. Figure 3 shows three different examples of such distribution functions.

(a) The rate of removal is rapid in the beginning but decreases later so that a substantial fraction of the molecules have a long residence time. This situation may occur for example when high concentration of other pollutants near the source contribute to a rapid reaction. The values of SO_2 oxidation rates quoted in Table 8 in the review by McMahon *et al.* (1976) indicate that this is a normal situation.

Removal by dry deposition at the surface may also give rise to a similar shape. This is because the proportion of the molecules that quickly get in contact with the surface will decrease as the pollutant is spread to higher elevations.

If situations such as these apply, a rate coefficient estimated from measurements made at short distances from the source may seriously underestimate the real turn-over time ($k_{est}^{-1} < \tau_0$). Even if measurements are extended up to a point when 50% of the molecules have been transformed or removed, the inverse of the estimated rate coefficients may be much smaller than the real turn-over time. When modelling a situation like this a decay proportional to the square of the concentration might be considered.

(b) This is the exponential case when $k_{est}^{-1} = \tau_0$.

(c) The rate of removal is slow in the beginning and increases later on. Here the inverse of the rate coefficient estimated close to the source may be larger than the turn-over time ($k_{est}^{-1} > \tau_0$). A physical situation when this may apply is decay due to wet removal during a precipitation situation if the pollutant has to disperse up to cloud levels before it is being significantly removed.

Measurement of SO_2 oxidation rates in a power plant plume reported by Husar *et al.* (1978) indicate a variation of the rate coefficient during the first few hours which would give rise to a similar distribution of residence times.

6.2 *Summary of turn-over time estimates*

Table 4 summarizes the information about turn-over times obtained from the budget estimates discussed in section 4. All estimates discussed here refer to pollution sulfur, emitted mainly as SO_2, in a temperature climate (Europe or U.S.A.). The definition of the various time scales used in Table 4 are given in Table 3.

Since the time scales in the table are interrelated there is actually more information about each time scale than indicated by the numbers in the particular column. These interrelations follow from the definitions of the different time scales as given in Table 3. For example, we have

$$\frac{1}{\tau_{SO_2}} = \frac{1}{(\tau_{SO_2})_D} + \frac{1}{(\tau_{SO_2})_W} + \frac{1}{(\tau_{SO_2})_C},$$

$$\frac{1}{\tau_{SO_4}} = \frac{1}{(\tau_{SO_4})_D} + \frac{1}{(\tau_{SO_4})_W},$$

$$\frac{1}{\tau_S} = \frac{1}{(\tau_S)_D} + \frac{1}{(\tau_S)_W}$$

and $\tau_S = \tau_{SO_2} + \alpha\, \tau_{SO_4}$ where α is the fraction of the emitted sulfur that is transformed to sulfate before

Table 3. Definitions of the turn-over times used in Table 5. The symbols of the right hand sides are defined in Fig. 1.

$(\tau_{SO_2})_D = \dfrac{M_1}{D_1}$	$(\tau_{SO_4})_W = \dfrac{M_2}{W_2}$
$(\tau_{SO_2})_W = \dfrac{M_1}{W_1}$	$\tau_{SO_4} = \dfrac{M_2}{D_2 + W_2} = \dfrac{M_2}{C}$
$(\tau_{SO_2})_C = \dfrac{M_1}{C}$	$(\tau_S)_D = \dfrac{M_1 + M_2}{D_1 + D_2}$
$\tau_{SO_2} = \dfrac{M_1}{D_1 + W_1 + C} = \dfrac{M_1}{Q}$	$(\tau_S)_W = \dfrac{M_1 + M_2}{W_1 + W_2}$
$(\tau_{SO_4})_D = \dfrac{M_2}{D_2}$	$\tau_S = \dfrac{M_1 + M_2}{D_1 + W_1 + D_2 + W_2} = \dfrac{M_1 + M_2}{Q}$

Table 4. Summary of estimates of turn-over times of sulfur compound in the atmosphere (unit, hours)

Authors	$(\tau_{SO_2})_D$	$(\tau_{SO_2})_W$	$(\tau_{SO_2})_C$	τ_{SO_2}	$(\tau_{SO_4})_D$	$(\tau_{SO_4})_W$	τ_{SO_4}	$(\tau_S)_D$	$(\tau_S)_W$	τ_S	Comments
Junge, 1960, 1963									150	75–100	Based on regional budgets
Rodhe, 1972										50–90	
Granat *et al.*, 1976										50–70	
Garland, 1977										50	
Eliassen and Saltbones, 1975			140(a)	15							(a) Precipitation scavenging not considered
EPRI, 1976			50–100								limited data material
Omstedt and Rodhe, 1978		< [illegible]	25–150								based essentially on rate coefficient estimates for the first 30 h
Prahm *et al.*, 1976			90	12	70						sensitive to assumptions about dilution
Smith Jeffrey, 1975	(b)		(c)			(d)					(b) about 30% deposited within 3 h (c) about 15% converted within 3 h (d) effective scavenging
Garland and Branson, 1976	40										based on 'local' scale height
Husar *et al.*, 1978				35							based on plume budgets no precipitation scavenging
Martell and Moore, 1974							100–150				based on radioactive tracers
Rodhe and Grandell, 1972		> ~50				> ~50			> ~50		lower limits only
Present estimate	60	100	80	25	large	80	80	120	90	50	

deposition. α is related to the time scales involved by the relation

$$\alpha = \frac{\tau_{SO_2}}{(\tau_{SO_2})_C}.$$

Taking into account such relations and the comments made in Section 4 about the various estimates we may make a rough guess of likely values of the different time scales for the particular conditions prevailing in northern Europe. Such an estimate is given at the bottom of Table 4. This estimate is of course uncertain (no attempt is made to estimate any ranges) but at least it gives an internally consistent picture. The figures imply that about 30% of the SO_2 is transformed to SO_4^{2-} before being deposited.

None of the estimated values seems to disagree too much with the data given in the table. Minor differences are by and large within the uncertainty ranges of the different estimates (*cf.* section 4).

The major uncertainties in the set of values given at the bottom of Table 4 seem to be associated with the three turn-over times $(\tau_{SO_2})_W$, $(\tau_{SO_2})_C$ and $(\tau_{SO_4})_W$. For example, from the arguments presented here there is no serious contradiction in assuming a $(\tau_{SO_2})_W$ value well below 100 h. Similarly $(\tau_{SO_4})_W$ may be taken to be somewhat less or larger than 80 h.

It should again be emphasized that the time scales estimated here represent average conditions for a particular type of climate. For specific situations (e.g. seasons) in the same region or for other climatic zones the values may be quite different.

7. CONCLUDING REMARKS

Budget estimates may be useful for tying together various pieces of information into a consistent picture. They may also help in identifying the most significant uncertainties in our knowledge about the circulations of the various compounds. Global sulfur budgets have the common weakness that they form an average over extremely different conditions in different parts of the globe. Furthermore, concentration and deposition data are only available from very limited areas.

These factors point to the advantages of making sulfur budgets over regions where data are available and where the conditions (climatewise and emissionwise) may be reasonably homogeneous. It has been shown that such regional budgets—most of them applied to European conditions—may produce some useful estimates of the importance of those processes that transform and remove sulfur compounds from the atmosphere. It is recommended that similar studies be undertaken in other parts of the world. In particular, North American may offer an interesting possibility in view of the large man-made emissions and of the measurements already available.

One may also caution against the use of annual averages in budget estimates. Pronounced seasonal variations in the rates of emission, transformation and/or removal may make such averages quite misleading.

In the discussion above it has become apparent

that there are definite problems associated with a comparison between time scales derived from budgets and those derived in other ways. In particular, the difference between turnover times (i.e. average residence times) and the inverse of rate coefficients should be kept in mind. A need for careful definitions of the time scales used is obvious. This paper is an attempt to formulate a basis for the discussion of associated problems.

Acknowledgement—This work was done under contract No G3922-003 of the Swedish Natural Science Research Council. I wish to thank L. Granat for useful discussions.

REFERENCES

Alkezweeny A. J. and Powell D. C. (1977) Estimation of transformation rate of SO_2 to SO_4 from atmospheric concentration data. *Atmospheric Environment* **11**, 179–182.

Bolin B. and Rodhe H. (1973) A note on the concepts of age distribution and transit time in natural reservoirs. *Tellus* **25**, 58–62.

Chamberlain A. C. (1967) Transport of *Lycopodium* spores and other small particles to rough surfaces. *Proc. R. Soc.* **296**, 45–70.

Dovland H. and Eliassen A. (1976) Dry deposition on a snow surface. *Atmospheric Environment* **10**, 783–785.

Eliassen A. (1978) The OECD study of long range transport of air pollutants. Long range transport modelling. *Atmospheric Environment* **12**, 479–487.

Eliassen A. and Saltbones J. (1975) Decay and transformation rates of SO_2 as estimated from emission data, trajectories and measured concentrations. *Atmospheric Environment* **9**, 425–429.

EPA (1975) Position paper on regulation of atmospheric sulfates. U.S. Environmental Protection Agency, Research Triangle Park, U.S.A.

EPRI (1976) *Design of the sulfate regional experiment (SURE). Final report of research project 485, Vol. 1: Supporting data and analysis.* Electric Power Research Institute, California, U.S.A.

Eriksson E. (1960) The yearly circulation of chloride and sulfur in nature; meteorological, geochemical and pedological implications—Part II. *Tellus* **12**, 63–109.

Eriksson E. (1963) The yearly circulation of sulfur in nature. *J. geophys. Res.* **68**, 4001–4008.

Forrest J. and Newman L. (1977) Oxidation of sulfur dioxide in the Sudbury smelter plume. *Atmospheric Environment* **11**, 517–520.

Friend J. P. (1973) The global sulfur cycle. In *Chemistry of the Lower Atmosphere.* (Edited by Rasool S. I.) pp. 177–201. Plenum Press, New York.

Garland J. A. (1977) The dry deposition of sulphur dioxide to land and water surfaces. *Proc. R. met. Soc. Lond.* A **354**, 245–268.

Garland J. A. (1978) Dry and wet removal of sulphur from the atmosphere. *Atmospheric Environment* **12**, 349–362.

Garland J. A. and Branson J. R. (1976) The mixing height and mass balance of SO_2 in the atmosphere above Great Britain. *Atmospheric Environment* **10**, 353–362. Discussion of this paper by H. Rodhe. *Atmospheric Environment* **11**, 659–661.

Gillani N. V., Husar R. B., Husar J. D. and Patterson D. E. (1978) Project MISTT: Kinetics of particulate sulfur formation in a power plant plume out to 300 km. *Atmospheric Environment* **12**, 589–598.

Granat L. (1978) Sulfate in precipitation as observed by the European Atmospheric Chemistry Network. *Atmospheric Environment* **12**, 413–424.

Granat L., Hallberg R. O. and Rodhe H. (1976) The global sulphur cycle. In *Nitrogen, Phosphorus and Sulphur—Global Cycles.* (Edited by Svensson B. H. and Söderlund R.) SCOPE report 7, Ecol. Bull. (Stockholm) **22**, 89–134.

Hales J. M. (1972) Fundamentals of the theory of gas scavenging by rain. *Atmospheric Environment* **6**, 635–659.

Hitchcock D. R. (1976) Microbiological contributions to the atmospheric load of particulate sulfate. In *Environmental Biochemistry* (edited by Nriagu J. O.) pp. 351–367. Ann Arbor Science, Ann Arbor.

Husar R. B., Patterson D. E., Husar J. D., Gillani N. V. and Wilson W. E. (1978) Sulfur budget of a power plant plume. *Atmospheric Environment* **12**, 549–568.

Junge C. E. (1960) Sulfur in the atmosphere. *J. geophys. Res.* **65**, 227–237.

Junge C. E. (1963) *Air Chemistry and Radioactivity.* pp. 70–74. Academic Press, New York.

Kellogg W. W., Cadle R. D., Allen E. R., Lazrus A. L. and Martell E. A. (1972) The sulfur cycle. *Science* **175**, 587–596.

Lusis M. A. and Wiebe H. A. (1976) The rate of oxidation of sulfur dioxide in the plume of a nickel smelter plant. *Atmospheric Environment* **10**, 793–798.

Martell E. A. and Moore H. E. (1974) Tropospheric aerosol residence times: a critical review. *J. Rechs. atmos.* **8**, 903–910.

McMahon T. A., Denison P. J. and Fleming R. (1976) A long-distance air pollution transportation model incorporating washout and dry deposition components. *Atmospheric Environment* **10**, 751–761.

Meetham A. R. (1950) Natural removal of pollution from the atmosphere. *Q. Jl. R. met. Soc.* **76**, 359–371.

Mészáros E. and Várhelyi G. (1977) An attempt to estimate the continental sulphur emissions on the basis of atmospheric measurements. *Atmospheric Environment* **11**, 169–172.

Omstedt G. and Rodhe H. (1978) Transformation and removal process for sulfur compounds in the atmosphere as described by a one-dimensional time-dependent diffusion model. *Atmospheric Environment* **12**, 503–509.

Prahm L. P., Torp U. and Stern R. M. (1976) Deposition and transformation rates of sulphur oxides during atmospheric transport over the Atlantic. *Tellus* **28**, 355–372.

Robinson E. and Robbins C. (1970) Gaseous sulfur pollutants from urban and natural sources. *J. Air Pollut. Control. Ass.* **20**, 233–235.

Rodhe H. (1972) A study of the sulfur budget for the atmosphere over northern Europe. *Tellus* **24**, 128–138.

Rodhe H. and Grandell J. (1972) On the removal time of aerosol particles from the atmosphere by precipitation scavenging. *Tellus* **24**, 442–454.

Slinn W. G. N. (1977) Some approximations for the wet and dry removal of particles and gases from the atmosphere. *Wat. Air. Soil Pollut.* **7**, 513–543.

Smith F. B. and Jeffrey G. H. (1975) Airborne transport of sulphur dioxide from the U.K. *Atmospheric Environment* **9**, 643–659.

Van der Hoven I. (1975) Deposition of particles and gases. USAEC TID-24190, Clearinghouse FSTI, Springfield, Va.

Várhelyi G. An attempt to estimate the sulfur budget over Hungary. *Atmospheric Environment* **12**, (in press).

Atmospheric Environment Vol. 12, pp. 681–690. Pergamon Press 1978. Printed in Great Britain.

LARGE SCALE SPATIAL AND TEMPORAL DISTRIBUTION OF SULFUR COMPOUNDS

H.-W. GEORGII

Department of Meteorology and Geophysics, University of Frankfurt, W. Germany

(*First received* 28 *June* 1977 *and in final form* 27 *September* 1977)

Abstract—The paper contains new results on the large-scale distribution of SO_2, H_2S and sulfate-aerosols. Recently developed analytical methods permit the measurement of SO_2 in the upper troposphere and lower stratosphere as well as first measurements of the background-concentration of H_2S. Measurements of H_2S close to the ground and in relation to the air-temperature above the ground shows a temperature-dependence of H_2S-release from the soil by biogenic activity. The measurement of sulfate-aerosols over the ocean clarified the picture of the formation of maritime sulfate-aerosols and explained the existence of excess sulfate in the maritime atmosphere. Decay of SO_2 and transformation of SO_2 to sulfate was also studied in a meso-scale study over the continent which supplied further information on the influence of anthropogeneous SO_2 sources on the large-scale sulfur budget. Vertical profiles of SO_2 resulting from aircraft measurements over the continent and the Atlantic ocean are compared. They are of importance to evaluate the transport and decay of SO_2 over long distances and also indicate the sink function of the ocean surface for the uptake of SO_2.

1. INTRODUCTION

Sulfur plays an important role in the tropospheric and stratospheric budget of atmospheric trace-substances. Much attention has been paid in the past to the growing amount of sulfur-components emitted by different anthropogeneous sources into the atmosphere and leading to regionally high accumulations of SO_2 and sulfate-aerosols. In spite of increasing activity to understand the global sulfur budget, our knowledge is still rudimentary. There exist still insufficient data on the distribution of the different sulfur-components in unpolluted air with a nearly complete lack of information from the Southern Hemisphere.

We shall reduce the discussion here to the distribution of SO_2, H_2S and sulfate-aerosols and only touch the different conversion mechanisms leading from one to the other. We shall emphasise the discussion on measurements in the free atmosphere as resulting from aircraft measurements. We also want to refer to the paper by Mészáros (1978) in this issue dealing with similar aspects.

In Fig. 1 we have summarized the different sources and sinks of atmospheric sulfur-components as well as the main homogeneous and heterogeneous reactions transforming the gaseous sulfur components into sulfate-containing aerosols. The scheme follows a diagram presented by Junge (1972) some years ago but has been brought up to date mainly by including the release of organic sulfur-components which, according to a paper by Crutzen (1976), may be a source of stratospheric SO_2. Following the discussion by Davis and Klauber (1975) and considering the OH measurements by both Perner and Ehalt *et al.* (1976)

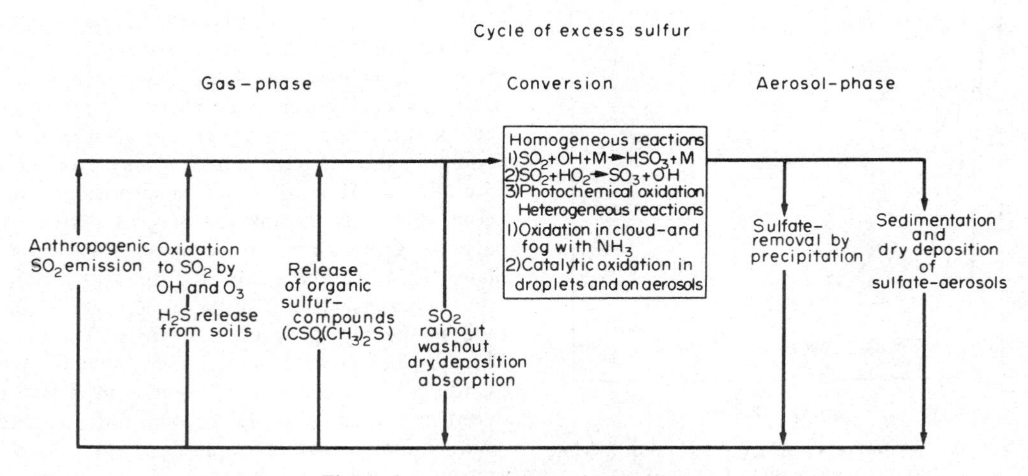

Fig. 1. Sources and sinks of excess sulfur.

A.E. 12–1 3—RR

the homogeneous reactions $H_2S + OH$ and $SO_2 + OH + M$ lead to a rapid oxidation of these compounds and are of greater importance for the sulfur-budget than considered earlier. The SO_2-removal rate by catalytic oxidation of SO_2 in droplets was found to be 0.8–3.0% h^{-1} by Barrie and Georgii (1976) and is of the same order as the oxidation-rate of SO_2 with OH-radicals.

Quantitative information of the different figures of the sulfur budget is still unsatisfactory. Later in this paper we shall present first results of H_2S measurements and it is hoped that some progress toward the complete understanding of the sulfur budget can be made when the importance of H_2S and organic sulfur-compounds can be estimated with more confidence.

2. DISTRIBUTION OF SO_2

Measurements of the horizontal distribution of SO_2 show a decrease of the concentration from the continent to the oceans. Details will not be given here. We only want to summarize the present knowledge and give further details on the vertical distribution of SO_2. Over western Europe the average SO_2-concentration in unpolluted air decreases with altitude from about 20 g m^{-3} near the ground during winter and 10 g m^{-3} during summer, reaching 5 g m^{-3} in 2–3 km altitude and 1 g m^{-3} in 5 km altitude, the half concentration of the ground value is reached in 1.5 km above ground (Jost, 1974). It is of interest to observe that the difference between winter and summer values is only found in the lower troposphere up to 3 km altitude. The higher winter values are caused by additional anthropogeneous sources and

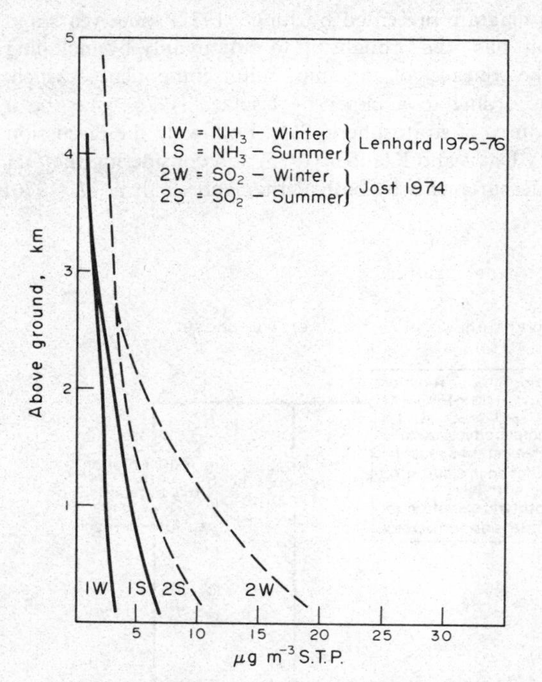

Fig. 2. Vertical distribution of SO_2 and NH_3 during winter and summer over western Germany.

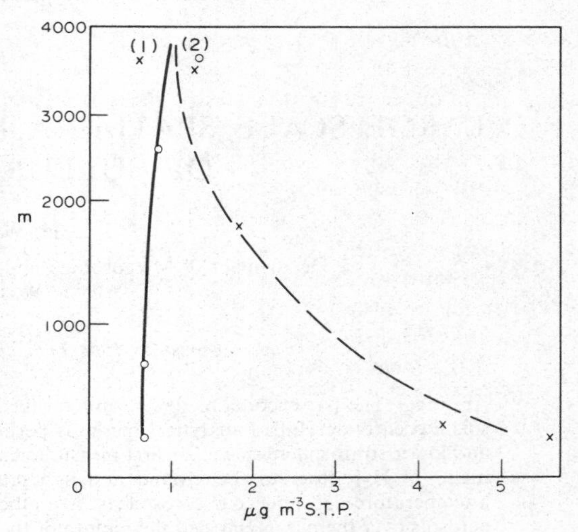

Fig. 3. Vertical distribution of SO_2 over the Atlantic Ocean (1) and the island of Teneriffa (2).

secondly by lower mixing-heights during winter. In Fig. 2 we have also entered the vertical NH_3-distribution as measured by Lenhard (1976) in the same region.

In the discussion of sulfate-formation in cloud-droplets in the presence of NH_3 it is important to know the NH_3-concentration present in cloud-level. Also in the case of NH_3 we find a seasonal difference of the concentration near ground-level. But in contrast to SO_2, the NH_3 values are higher during summer, reaching approx 5–7 $\mu g\, m^{-3}$ while the winter values were always below 5 $\mu g\, m^{-3}$. The decrease of NH_3 with altitude is greater in summer than in winter. This behaviour of the NH_3-concentration strongly reflects the source-function of the soil where high microbiotic activity during the warm season leads to an increased NH_3 production rate when the temperature of the ground is higher. Fig. 2 indicates clearly that only during summer months SO_2- and NH_3-concentrations are comparable in the lower troposphere while during the cold season there is a high excess concentration of SO_2 in comparison to NH_3.

Above the ocean nearly constant SO_2-concentration with altitude was found during aircraft measurements above the Atlantic as is shown in Fig. 3. A background-concentration below 1 $\mu g\, m^{-3}$ is indicated and there are indications that in pure maritime conditions the SO_2-concentration may even increase with altitude. This would not be surprising since the ocean-water acts as sink for SO_2 as outlined in a paper by Georgii and Gravenhorst (1977). Liss and Slater (1974) assume a flux of 1.5×10^{14} g $SO_2\, y^{-1}$ into the oceans which is of the same order of magnitude as the SO_2-production by fossil-fuel burning. This may not exclude the possibility that some SO_2 found above the ocean may originate from H_2S produced in coastal areas. H_2S would, after oxidation, act as an indirect source for maritime SO_2. The development of a new chemiluminescence method to analyze SO_2 in the sub-ppb range by Stauff and Jaeschke (1975) permitted the performance of SO_2 measurements in the upper troposphere and the lower stratosphere. This may open the door to a better understanding of the formation and persistence of the stratospheric aerosol layer during periods with little volcanic activity. We have no direct evidence yet of the diffusion of gaseous sulfur-components from the troposphere into the stratosphere on a larger scale which would be necessary to prove the hypothesis that the conversion of tropospheric gas constituents leading to the formation of sulfate-aerosols with an average particle radius of 0.5 μm takes place within the stratosphere. The stratospheric aerosol would then be returned to the troposphere by sedimentation and vertical mixing (Junge, 1974). First measurements in the upper troposphere and lower stratosphere were carried out during March 1976 as described by Jaeschke *et al.* (1976). A cross-section of the flight path with the analytical results is shown in Fig. 4. In Fig. 5 we have plotted the vertical distribution of SO_2 relative to the tropopause-level. The SO_2-concentration on that particular day in the lower stratosphere was found to be 0.14 $\mu g\ m^{-3}$ S.T.P. In the middle and upper troposphere SO_2-values between 0.23 and 0.88 $\mu g\ m^{-3}$ S.T.P. were found except one singular value of 0.14 $\mu g\ m^{-3}$ S.T.P. close to the tropopause.

The lack of a sudden drop of the SO_2-concentration when the tropopause was crossed may seem surprising. This can be explained by the specific weather situation of the particular day. The aircraft passed an occlusion with a tropopause break when climbing up to stratospheric level. In this region the tropopause did not prevent the vertical mixing of tropospheric and stratospheric air. This may have resulted in the smoothing out of differences in the SO_2-concentration between these two reservoirs. It is quite obvious that these results must be confirmed by further investigations. However, they could indicate a small-scale transfer of SO_2 from the troposphere into the stratosphere which would supply gaseous sulfur compounds necessary for the formation of stratospheric sulfate particles. The concentrations found during this flight were close to those assumed by Junge (1974) in the calculations for the stratospheric SO_2-oxidation model, but one must be careful not to take our preliminary results as representative. It cannot be excluded that a certain but quantitatively unknown fraction of the SO_2 found within the stratosphere originates from the photodissociation of carbonyl sulfide (CSO) as suggested by Crutzen (1976). CSO has been discovered in tropospheric air where it is probably produced from biological and industrial sources and behaves as a relatively inert gas. We can therefore expect that its mixing ratio does not decrease significantly within the troposphere. Several photochemical reactions in the stratosphere may eventually lead to the formation of SO_2. Very recent measurements by Sandalls and Penkett (1977) indicate an average concentration of 0.51 ppb which is higher than the value 0.2 ppb given by Hanst *et al.* (1975).

At present we have the following understanding of the large-scale SO_2 distribution which is schematically drawn in Fig. 6. The continental SO_2-concentration is higher than the SO_2-concentration above the ocean, resulting in a horizontal gradient of SO_2 from the land to the sea. Maximum values of the SO_2-concentrations are found in densely inhabited areas and in industrialized regions which are indicated by the maximum near the ground showing

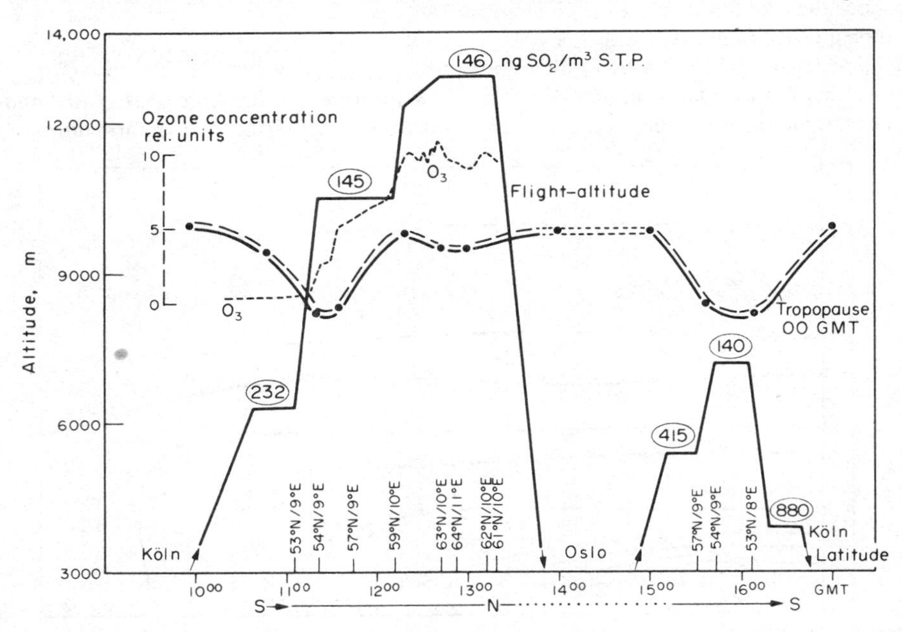

Fig. 4. Profile of stratospheric SO_2-measurements during flight from Cologne to Oslo on 27 March 1976.

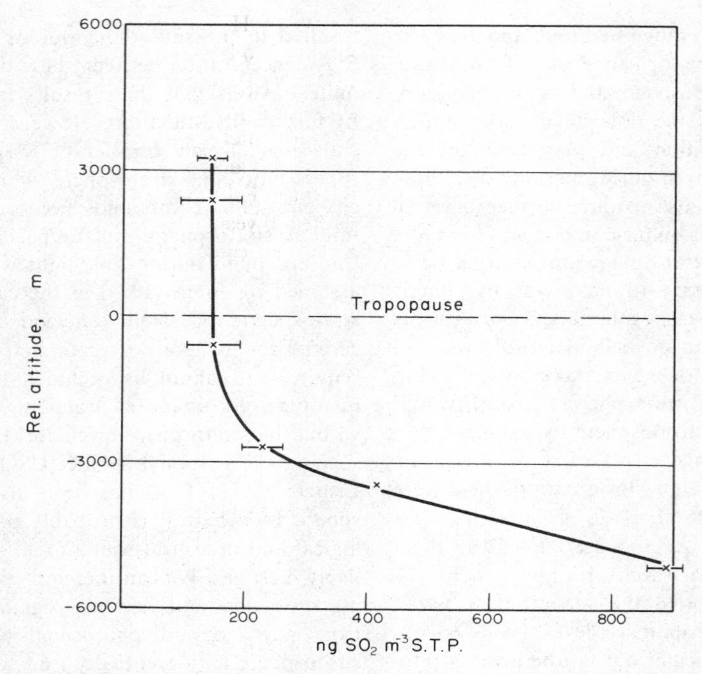

Fig. 5. Vertical distribution of SO_2 above and below the tropopause on 27 March 1976.

200 μg SO_2 m^{-3}. However, aircraft ascents above industrial areas show that the dispersion of pollutants restricts the accumulation of SO_2 to the lower troposphere. In a level from 1500 m upwards, these regional concentration maxima cannot be observed any more. The production of SO_2 near the ground with sinks and dispersion in the troposphere leads to a strong decrease of SO_2 with altitude above the continents. At about 3–5 km (changing a little with the season) altitude a background-concentration of approx 0.5 μg m^{-3} S.T.P. is reached. It appears from preliminary investigations that the SO_2-concentration at and above the tropopause is around 0.1 μg m^{-3} S.T.P.

Measurements carried out on research ships and by aircraft above the Atlantic Ocean show a decrease of the SO_2-concentration near the ocean surface reaching the background of 0.5–1 μg m^{-3} S.T.P. in the middle of the ocean and nearly constant concentration with altitude. The SO_2-concentration decreases also from temperate to tropical latitudes, the background-concentration in the tropical region of the Atlantic being less than 0.3 μg m^{-3} (Büchen and Georgii, 1971). In some instances we find even a slight increase of the concentration with altitude which may be an indication of the sink function of ocean-water. We can therefore conclude that SO_2 found over the oceans is of continental origin.

3. DISTRIBUTION OF H_2S

With respect to the large-scale spatial and temporal distribution of H_2S, little information is available.

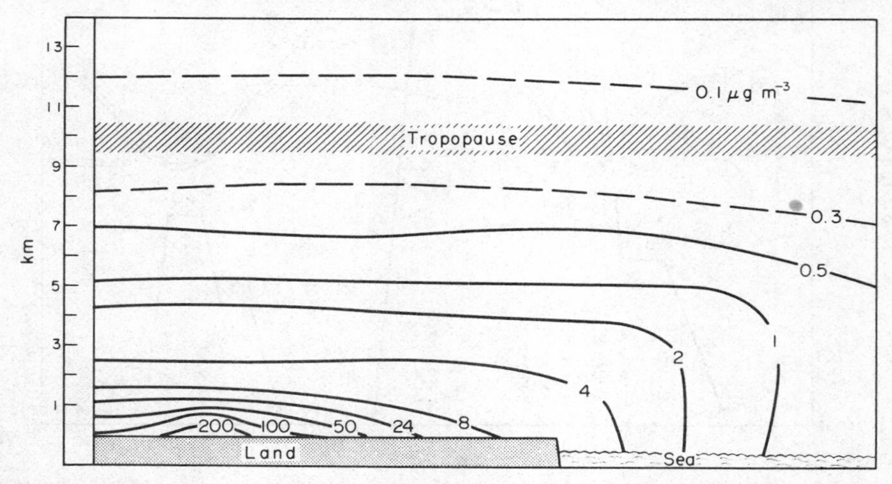

Fig. 6. Schematic distribution of SO_2 in temperate latitudes; concentrations reduced to S.T.P.; distance across section of order 1000 km.

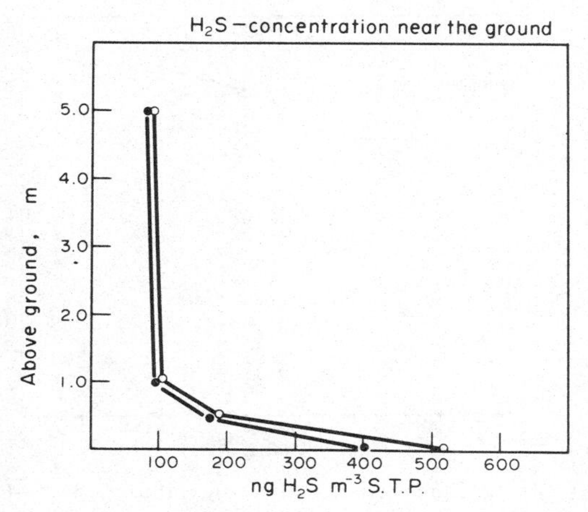

Fig. 7. H_2S-concentration above tidal flats over the island of Sylt (North Sea).

Microbiological activity in the soil and in coastal waters are the main natural sources of H_2S, but the quantitative production rate is not well known. The figures given in the literature by several authors differ greatly. Kellogg *et al.* (1972) assume 268×10^6 t y^{-1} as production rate from soils and coastal waters. More recently, smaller figures are given for instance 116×10^6 t y^{-1} by Friend (1973) and 103×10^6 t y^{-1} by Robinson and Robbins (1975). The latter figure includes organic sulfur compounds. All authors agree that little H_2S comes from industrial sources. Robinson and Robbins suggest as an average value of the tropospheric H_2S-concentration 0.28 g m^{-3} S.T.P. Latest figures by Granat *et al.* (1976) indicate a reduced production of volatile sulfur from biological decay of 34×10^6 t y^{-1}.

An improved analytical method by Grunert, Ballschmitter and Tölg (1968) which was adapted for atmospheric measurements by Jaeschke and Haunold (1977) permits for the first time the detection of H_2S in the concentration range below 1 μg m^{-3} S.T.P. With the special sampling and analytical procedure applied during these measurements and with the flow rate of 1 m^3 h air it was possible to detect concentrations as low as 10 ng m^{-3}. Preliminary measurements of H_2S were performed in areas which are assumed to be source-regions of H_2S, namely swamps in upper Bavaria and tidal flats around the island of Sylt in the North Sea. These measurements were supplemented by aircraft ascents and by measurements in unpolluted air on a mountain station (Kleiner Feldberg/Taunus 800 m a.s.l.). The results gained revealed that the H_2S-concentration was in nearly all cases below 1 μg m^{-3}. The average H_2S-concentration in the tidal flats was 380 ng m^{-3} with a strong decrease with altitude. Figure 7 shows two profiles of the H_2S-concentration near the ground at the island of Sylt and it can be seen that already at 1 m above the ground the concentration is reduced to only 100 ng m^{-3}. Advection of H_2S-free air and decay of H_2S are possible explanations for this rapid decrease which seems too large to be due to a surface source effect. Another example in Fig. 8 shows the result of a simultaneous H_2S- and SO_2-measurement at the same location, by an aircraft ascent. The profile shows over the whole range of altitude a great difference between the H_2S- and SO_2-concentration. At 800 m above ground the SO_2-concentration was still 4 μg m^{-3}, while the H_2S-concentration, in contrast to that, amounted to only 100 ng m^{-3}. This result emphasises that even in unpolluted air in the vicinity of natural H_2S-sources, the SO_2-concentration dominates by far as major gaseous sulfur component. Measurements at the ground supported by aircraft-ascents showed that in the Rhine–Main area the influence of anthropogeneous sources determines the regional H_2S distribution. The summary of aircraft measurements carried out above the island of Sylt and above the Rhine–Main area is presented in Fig. 9. Above the Rhine–Main area a strong decrease of the H_2S-concentration from about 1 μg m^{-3} near the ground to 0.2 μg m^{-3} in 1 km altitude, slowly further decreasing to 0.1 μg m^{-3} in about 3 km altitude was observed. In contrast to that, above the tidal flats the H_2S-concentration decreased from only 0.25 μg m^{-3} near the ground to approx 0.1 μg m^{-3} in 500 m above the ground, remaining at that constant background within the lower troposphere. This result was

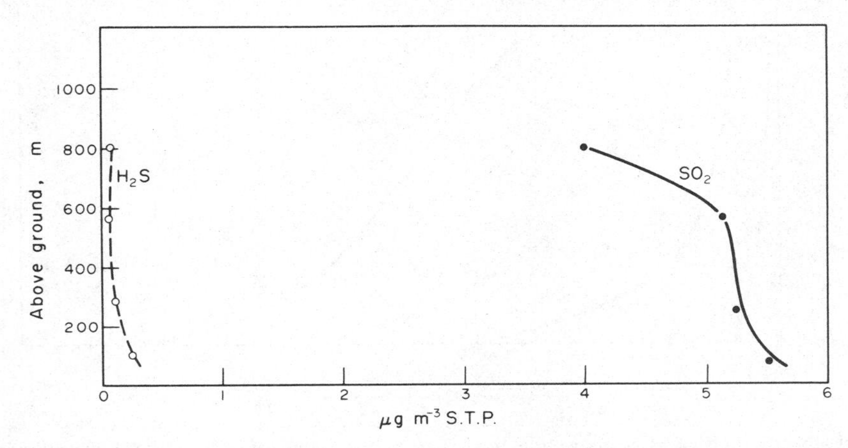

Fig. 8. Distribution of H_2S and SO_2 above the island of Sylt (North Sea), 1 October 1975.

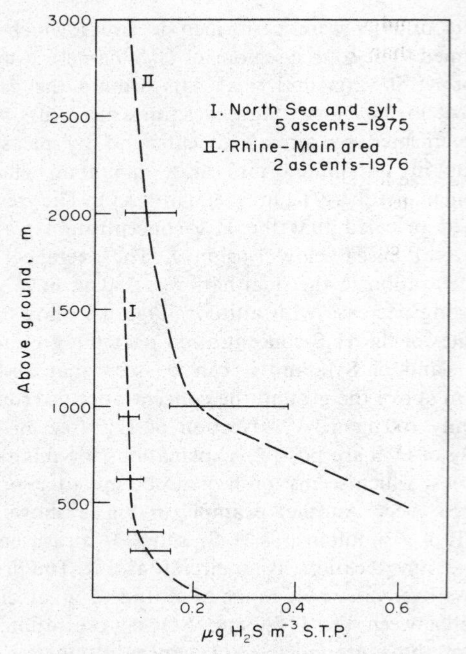

Fig. 9. Vertical distribution of H_2S over the North Sea and the Rhine–Main area.

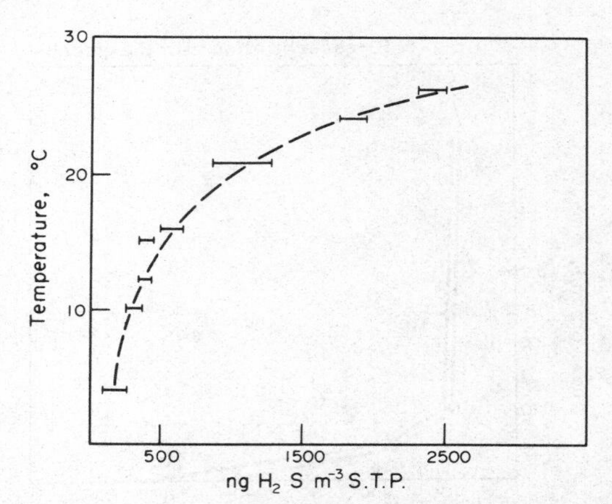

Fig. 10. Relation between H_2S-concentration near the ground and air temperature.

confirmed during practically all aircraft measurements. These measurements show that anthropogeneous and industrial sources cannot be neglected in the budget of H_2S and that their contribution compared to that of natural sources may be greater than expected. At present we investigate whether sulfides other than H_2S or organic sulfur compounds might contribute to this background-concentration of 0.1 μg m^{-3}. The SCEP Report (1970) states that hydrogen sulfide is so rapidly converted to sulfur dioxide over land that the detection of any H_2S over the oceans and over the polar regions is unlikely. This seems to be supported by our preliminary measurements. A study made by Malewski (1977) shows that the biogenic production-rate of H_2S increases during the summer with rising average temperatures as is to be expected from microbiological activities.

Figure 10 shows the dependance of the concentration of H_2S near the ground on air temperature. The air samples for analysis of H_2S were collected a few cm above the ground. It was interesting to note that the difference of the H_2S-concentration close to the ground was very small compared with the two natural source-areas in upper Bavaria and the North Sea. Within the temperature-range between 5° and 15°C the increase of the H_2S-release from the soil is not very large. At temperatures above 15°C the H_2S-emission from the soil increases rapidly, resulting in concentrations between 1 and 2.5 g m^{-3}. This is probably in indication of the temperature-sensitivity of H_2S-producing bacteria rather than an effect of changing stability or wind speed. The influence of the temperature on the H_2S concentration in the ground-layer of the atmosphere is also reflected in the vertical distributions of H_2S which were measured simultaneously at four altitudes on the island of Sylt, namely in 5 cm, 25 cm, 50 cm and 1 m. The left side of Fig. 11 summarizes seven vertical profiles measured

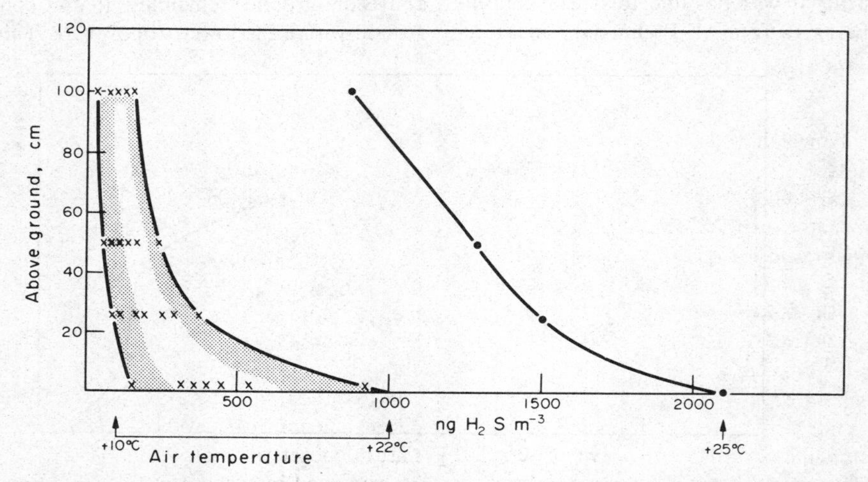

Fig. 11. Distribution of H_2S above the tidal flats of the island of Sylt. Vertical distribution near the ground with respect to air temperature.

at temperatures between 10° and 22°C; the profile on the right side of the same figure was measured at 25°C and very light wind (<3 m s^{-1}). The wind velocity during the measurements entered on the left side was ≥ 5 m s^{-1}. As mentioned above the H_2S-concentration at 1 m above the ground and higher does generally not show great differences except for the vertical distribution on the right side of Fig. 11. These results lead into the problem of the decay of H_2S in the atmosphere. The first step of the conversion mechanism H_2S—SO_2—SO_3—H_2SO_4 is of great importance for the atmospheric sulfur budget. It was previously assumed that reactions of H_2S occur mainly with O and O_3. Stuhl calculated a chemical lifetime of 30 h for H_2S when the average tropospheric concentration of OH-radicals is 2.3×10^6 cm^{-3}. We do not want to go into details of the H_2S-decay mechanism in this context. If the oxidation of H_2S with OH is the main sink in the lower troposphere, the diurnal trend of the concentration of OH-radicals with a minimum during the night should also find a corresponding variation of H_2S with a high production as well as decay-rate during the day and a reduced production- and decay-rate during the night.

In order to understand more fully the large-scale distribution of H_2S it appears necessary to perform H_2S-measurement in tropical latitudes which we must assume to be the main source-areas for H_2S. This is already indicated by measurements of the hydrogen-sulfide content of surface-waters of the Amazonas rivers carried out by Brinkmann and Santos (1974).

However, it appears from our measurements that sinks besides the oxidation-mechanism with OH-radicals are necessary to explain the rapid decrease of H_2S above the ground and the low background-concentration in the unpolluted atmosphere.

It is assumed that the surface acts as a secondary sink by action of H_2S-oxidizing bacteria and that the source area of H_2S is located in some depth in the soil and that H_2S is only able to escape into the atmosphere when the surface is damaged.

4. DISTRIBUTION OF SULFATE-AEROSOLS

The ultimate result of the oxidation-mechanism of gaseous sulfur-compounds in the atmosphere is the formation of sulfate-containing aerosols. In this paper we shall summarize some recent results of aircraft measurement on the distribution of sulfate-aerosols over the continent and the ocean. The large scale distribution near the ground and near the ocean surface will be dealt with elsewhere (Mézáros, 1978). Recent investigations by Gravenhorst (1976) into the composition of sulfate-aerosols in the unpolluted air over the ocean have shown that the sulfate-concentration of the aerosols in marine air is always in excess compared with the sulfate content of ocean water. The contribution of non-maritime-sulfate to total sulfate in maritime aerosols is about 60–80%.

The simultaneous measurement of SO_2 and SO_4^{2-} shows a change of the ratio SO_2/SO_4^{2-} from unpolluted maritime areas to polluted continental sites with the following average values:

	SO_2/SO_4^{2-}
Maritime locations	1:5
Unpolluted continental air	1:1
Polluted continental air	40:1 (ranging from 5:1 to about 100:1).

In polluted atmospheres close to anthropogeneous and industrial sources of SO_2, this pollutant dominates, while over the ocean a fraction of sulfate originates from the ocean surface. The contribution from the maritime source decreases, however, with altitude. This is in line with the assumption that in higher levels above the ocean the continental influence dominates. In this paper we can only touch the conclusions drawn from detailed studies of the sources of maritime aerosols and must refer to the literature for a more extensive discussion (see e.g. Georgii and Gravenhorst, 1972; Gravenhorst, 1975, 1976; Mészáros A. and Vissy, 1974; Mészáros E. and Varhelyi, 1975). The studies made over the Atlantic ocean show that the ocean surface can only be the source for particles in the size-range above 1 μm radius. The dominant number of smaller particles must be advected over long distances from continental sources or produced in the atmosphere over the ocean by gas-to-particle reactions. An indication that such a reaction occurs is the discovery by Mészáros and Vissy that the Aitken particle concentration measured over the South Atlantic had a diurnal trend with a maximum at noon. The mass-distribution of the maritime sulfate in relation to the size-spectrum of aerosol-particles shows that particles below 1 μm radius contain hardly any sulfate originating from the sea surface while the contribution of sea-salt particles increases in the particle-fractions with radii above 2 μm. This investigation showed also that the total concentration of excess sulfate found in aerosol-particles over the ocean is 0.96 μg m^{-3}, while the total sea-salt sulfate amounts to only 0.14 μg m^{-3}. This means that not only the predominant number of particles but also the majority of the sulfate-mass over the Atlantic is of continental origin. The high amount of sulfate-containing particles in the size-range below 0.5 μm radius suggests a continuous production of small sulfate-particles. This statement is supported by X-ray fluorescence analysis of Aitken nuclei collected over the Atlantic by Winkler (1974). Aircraft measurements made by Gravenhorst (1976) over the Bay of Biscay and the Canary Islands show a decrease of chloride-containing particles with altitude while the concentration of sulfate-containing particles remains nearly constant with altitude. It should be mentioned that the observed excess-sulfate concentration cannot be explained by present H_2S measurements and their preliminary results. Two mechanisms seem more

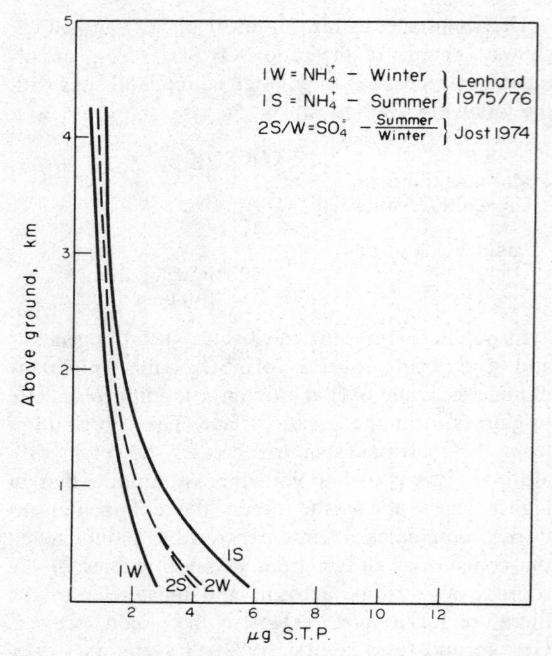

Fig. 12. Vertical profiles of NH_4^+ and SO_4^{2-} in aerosol particles.

plausible: (1) Sulfate particles are formed by homogeneous gas-reaction as sulfuric-acid which subsequently absorbs ammonia. This mechanism seems mainly responsible for the particle-formation in the size-range below 0.3 μm radius. (2) Part of the excess sulfate is formed by direct absorption of acid sulfur-containing components on the surface of existing sea-salt particles. Mechanism (2) would explain the existence of excess sulfate in the particle size-range above 0.5 μm radius. In both particle fractions the NH_4-content is too small to account for ammonium sulfate particles. These results support the hypothesis that the majority of small particles found over the ocean has been formed in maritime air by homogeneous gas-reaction. The production-rate of $(NH_4)_2SO_4$-particles over the ocean is small.

Over the continent the ammonium-concentration of aerosol particles shows a seasonal difference between winter and summer up to several km above ground with higher values during summer. In Fig. 12 we have entered ammonium- and sulfate-concentrations of continental aerosols from measurements by Jost (1974) and Lenhard (1976). The decrease of the sulfate-mass with altitude does not vary with season. This and also the absolute mass-concentration are in agreement with the results of Mészáros E. and Varhelyi (1975). The ammonium-concentration is lower than the sulfate-concentration during winter but higher during summer. This discrepancy again would suggest that the NH_3–SO_2–liquid water process leading to the formation of $(NH_4)_2SO_4$-particles is not important during the winter months. A comparison of the mass-ratio SO_4^{2-}/NH_4^+ in aerosol-particles gained by different authors shows that in most cases an excess of sulfate compared with the ratio SO_4^{2-}/NH_4^+ in $(NH_4)_2SO_4$-particles was found. This excess sulfate has the highest values under maritime conditions and in the middle troposphere (Fig. 13) as revealed by aircraft measurements. This does not yet explain the dominant mechanism of sulphate-particle formation but emphasizes that sulfate-containing aerosols have certainly a complex composition. Besides $(NH_4)_2SO_4$-particles, the presence of H_2SO_4-particles must be attributed to be of great importance.

5. INFLUENCE OF REGIONAL TRANSPORT AND TRANSFORMATION ON THE SULFUR BUDGET

During an intensive regional investigation of the transport and transformation of sulfur-components between the Ruhr area in western Germany and the

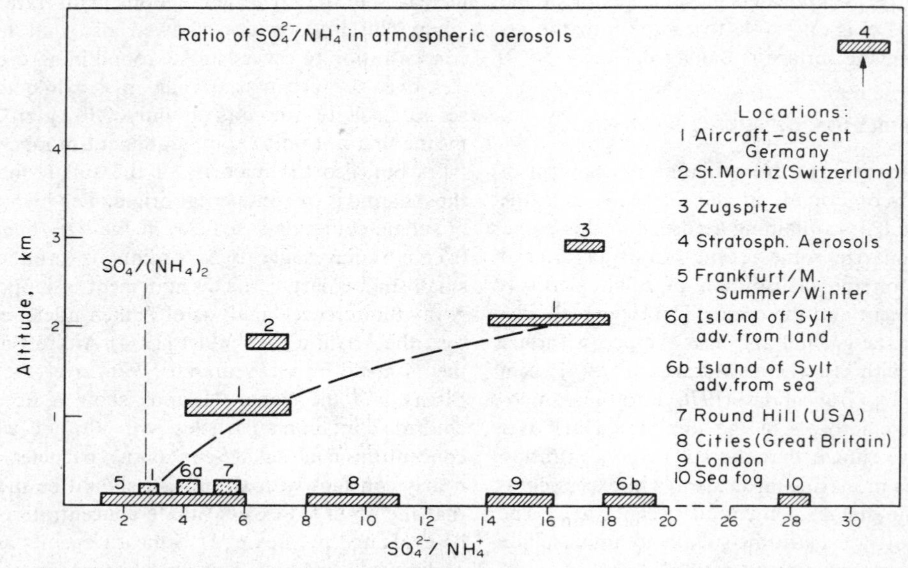

Fig. 13. The ratio SO_4/NH_4^+ in atmospheric aerosols according to different investigators.

industrial areas in the Netherlands and Belgium the problem of SO_2-decay and sulfate-aerosol formation was also dealt with. In this paper we can only briefly discuss those results of the study which are of general importance for the sulfur budget on a larger scale. Aircraft measurements of SO_2 and SO_4 showed an effect from a large industrial area on the SO_2-distribution at a distance of about 200 km (Bingemer, 1977). At distances of more than 300 km downwind of the source area, the vertical distribution of SO_2 did not reveal any significant deviations caused by the industrial SO_2-emissions.

In contrast to that, the distribution of the SO_4^{2-}-concentration was only very weakly influenced by the SO_2-sources. The SO_4^{2-}-distribution is much more conservative, with little spatial variation. With increasing distance from the source area of SO_2, there existed no significant transformation of SO_2 to sulfate aerosol. Sulfate-formation could only be observed close to the SO_2-sources. These results show a very weak relation between the SO_2- and sulfate-concentration. The decay of the SO_2-concentration with growing distances from the source-area can therefore only to a smaller degree be explained by oxidation-mechanisms. In the absence of precipitation it is mainly caused by dry deposition and in the case of advection with westerly winds by convective transport moving the sulfur-components from the boundary layer into higher layers of the atmosphere. This assumption is also supported by measurements of the vertical distribution of SO_2 and sulfate-aerosols with growing distances from the SO_2-sources. SO_2-sinks and convective transport leads to a pseudo-decay rate of SO_2 to 14–18% h^{-1} ($4\text{–}6 \times 10^{-5}\ s^{-1}$) in the presence of convective systems and of 10–14% h^{-1} ($3\text{–}4 \times 10^{-5}\ s^{-1}$) under stable temperature conditions in the lower troposphere. This decay-rate would be equivalent to a residence time of SO_2 of 7–10 h. For the reaction of SO_2 with OH-radicals, decay-rates of 2–3% h^{-1} are reported in the literature. However, in this case, the decay of SO_2 is accompanied by a corresponding increase of SO_4^{2-} which was not observed. Measurements and calculations of the meso-scale apparent-decay rate of SO_2 in the literature vary between $2 \times 10^{-5}\ s^{-1}$ (Prahm *et al.*, 1976) and $4.8 \times 10^{-4}\ s^{-1}$ (Lewin *et al.*, 1975). We want to restrict our discussion of the results of meso-scale transformation and decay of SO_2 to the above brief statements. They show that in polluted atmospheres the decay of SO_2 in the lower troposphere is much faster than can be explained by oxidation-mechanisms alone. It is evident that this has consequences for the atmospheric sulfur-budget on a larger scale. A detailed publication of these findings is in preparation.

Our knowledge of the different components of the atmospheric sulfur budget, particularly on their distribution in unpolluted air and on the different transformation-mechanisms active in the atmosphere has been greatly increased during recent years. However, there is still a lack of knowledge which prevents the exact formulation of a quantitative sulfur budget. We require better data on the H_2S-distribution and on organic sulfur-components but also more research on the relevance and dominance of certain reactions under atmospheric conditions which had until now only been studied in the laboratory.

Acknowledgement—The research reported in this article has been supported by 'Deutsche Forschungsgemeinschaft' through 'Sonderforschungsbereich 73, Atmospheric Trace-Substances.'

REFERENCES

Barrie L. A. and Georgii H.-W. (1976) An experimental investigation of the absorption of SO_2 by water drops containing heavy metals. *Atmospheric Environment* **10,** 743–749.

Bingemer H. (1977) Transport und Abbau schwefelhaltiger Luftverunreinigungen im Lee großer Flächenquellen. Diplomarbeit, Universität Frankfurt/Main.

Brinkmann W. and Santos U. de M. (1974) The emission of biogenic hydrogen sulfide from Amazonian flood-plains lakes. *Tellus* **26,** 261–268.

Brinkmann W. and Santos U. de M. (1970) Mans Impact on the global Environment. SCEP Report, MIT Press, Cambridge, Mass.

Büchen M. and Georgii H.-W. (1971) Ein Beitrag zum atmosphärischen Schwefelhaushalt über dem Nordatlantik. *Meteor-Forschungs-Erg.* **B7,** 71–77.

Crutzen P. (1976) The possible importance of CSO for the sulfate layer of the stratosphere. *Geophys. Res. Lett.* **3,** 73–76.

Davis D. D. and Klauber G. (1975) Atmospheric gas-phase oxidation mechanism for the molecule SO_2. *Int. J. chem. Kinet. Symp.* **1,** 543–556.

Friend J. P. (1973) The global sulfur cycle. In *Chemistry of the Lower Atmosphere* (Edited by Rasool S. I.). Plenum Press, New York.

Georgii H.-W. and Gravenhorst G. (1972) Untersuchungen zur Konstitution des Aerosols über dem Atlantischen Ozean. *Met. Rdsch.* **25,** 180–181.

Georgii H.-W. and Gravenhorst G. (1977) The ocean as source or sink of reactive trace-gases. PAGEOPH **115,** 503–511.

Granat L., Rohde H., Hallberg R. O. (1976) The global sulfur cycle. Scope Report No. 7, 89–134.

Gravenhorst G. (1975) The sulfate component in aerosol samples over the North Atlantic. *Meteor—Forschungs-Erg.* **B10,** 22–31.

Gravenhorst G. (1976) Der Sulfatanteil im atmosphärischen Aerosol über dem Nordatlantik. *Ber. Inst. Met. Geophys.* Univ. Frankfurt/Main Nr. 30.

Grunert A., Ballschmitter K. H. and Tölg G. (1968) Fluoreszenzanalytische Bestimmungen von Sulfidionen im Nanogrammbereich. *Talanta* **15,** 451.

Jaeschke W. and Haunold W. (1977) New methods and first results of measuring Atmospheric H_2S and SO_2 in the ppb-range. WMO-Special Environmental Rep. No. 10 193–199.

Jost D. (1974) Aerological studies on the atmospheric sulfur-budget. *Tellus* **26,** 206–213.

Junge C. (1972) The cycle of atmospheric gases—natural and man made. *Q. J. R. met. Soc.* **98,** 711–720.

Junge C. (1974) Sulfur budget of the stratospheric aerosol layer. Proc. IAMAP-Conf. Melbourne. Vol. 1, 85–97.

Kellogg W. W., Cadle R. D., Allen E. R., Lazrus A. L. and Martell E. (1972) The sulfur cycle. *Science* **175,** 587–596.

Lenhard U. (1977) Messung von Ammoniak in der unteren Troposphäre und Untersuchung der NH_3-Quellstärke von Böden. Diplomarbeit, Univ. Frankfurt/Main.

Lewin E., Gryning S. E. and Lyck E. (1975) Airborne measurements in the plume of a power station stack. Danish contribution of COST project 61 a report.

Liss P. S. and Slater P. G. (1974) Flux of gases across the air/sea interface. *Nature* **247**, 181–184.

Malewski H. (1977) Untersuchung der Konzentrationsverteilung von Schwefelwasserstoff in der bodennahen Luftschicht. Diplomarbeit, Univ. Frankfurt/Main.

Mészáros A. and Vissy K. (1974) Concentration, size distribution and chemical nature of atmospheric aerosol particles in remote oceanic areas. *J. Aerosol Sci.* **5**, 101–109.

Mészáros E. and Varhelyi G. (1975) On the concentration, size distribution and residence time of sulfate particles in the lower troposphere. *Idöjaras* **79**, 267–273.

Mészáros E. (1978) Concentration of sulfur compounds in remote continental and maritime areas. *Atmospheric Environment* **12**, 699–705.

Perner D., Ehalt D., Pätz H. W., Platt U., Pöth E. P. and Volz A. (1976) OH-Radicals in the lower troposphere. *Geophys. Res. Lett.* **3**, 466–468.

Prahm L. P., Torp U. and Stern R. M. (1976) Deposition and transformation of sulfuroxides during atmospheric transport over the Atlantic. *Tellus* **28**, 355–372.

Robinson E. and Robbins R. L. (1975) Gaseous atmospheric pollutants from urban and natural sources. In *The Changing Global Environment* (Edited by Singer S. F.). D. Reidel Dordrecht, Holland.

Sandalls F. S. and Penkett S. A. (1977) Measurements of carbonyl-sulphide and carbon disulphide in the atmosphere. *Atmospheric Environment* **11**, 197–199.

Stauff J. and Jaeschke W. (1975) A chemluminescence technique for measuring atmospheric trace concentrations of sulfur dioxide. *Atmospheric Environment* **9**, 1038–1039.

Winkler P. (1975) Chemical analysis of Aitken–particles over the Atlantic Ocean. *Geophys. Res. Lett.* **2**, 45–48.

Atmospheric Environment Vol. 12, pp. 691–698. Pergamon Press 1978. Printed in Great Britain.

THE INFLUENCE OF CLOUDS AND RAIN ON THE VERTICAL DISTRIBUTION OF SULFUR DIOXIDE IN A ONE-DIMENSIONAL STEADY-STATE MODEL

G. GRAVENHORST, TH. JANSSEN-SCHMIDT and D. H. EHHALT
Institut für Atmosphärische Chemie der Kernforschungsanlage
Jülich, D-5170 Jülich, West Germany

and

E. P. RÖTH
Universität Essen, GHS

(*First received* 26 *August* 1977 *and in final form* 27 *September* 1977)

Abstract—It was attempted to include wet chemical removal rates for sulfur dioxide in a one-dimensional steady-state model. The interactions with liquid water were separated into the removal of absorbed sulfur dioxide by rain and into the formation of sulfate in cloud- and rain-water. The chemical reaction rates and the gas-phase diffusion are fast compared to cloud- and rain-formation so that equilibrium conditions between gas-phase and liquid-phase were assumed. The most sensitive parameters affecting the wet chemical removal of SO_2 seem to be the pH value of rain-water and the formation rate of sulfate in atmospheric water. The gas-phase destruction proceeds predominantly through the oxidation by OH radicals. The calculated SO_2 volume mixing ratio decreases from 1 ppb at the ground level to *ca.* 0.01 ppb in 15 km altitude. Integrated over a vertical column the gas-phase destruction is about 2 times larger than wet chemical removal. The relative proportions, however, depend strongly on the chosen parameters for rain- and cloud-water. The direct SO_2 deposition onto the ground seems to be larger than the sum of the removal rates within the atmosphere.

INTRODUCTION

In order to describe the vertical distribution of soluble components of the atmosphere, wet removal mechanisms have to be considered. This paper represents an attempt to include wet chemical interactions of SO_2 in an existing one-dimensional steady-state model.

The basic reason for using a full one-dimensional chemical model was to generate vertical profiles of the gaseous species with which SO_2 reacts in a consistent manner. The 1-D model also includes a water cycle to generate a water-vapour profile and precipitation rates in a self-consistent way. Since very few data exist on the vertical distribution of SO_2 in the troposphere, such an attempt may appear premature. On the other hand, as we shall see below, such a calculation can identify the parameters which dominate the vertical SO_2-distribution and thus guide the design of the experiments, so that all the data required for an explanation are indeed collected simultaneously. In the following, the aspects of the model pertinent to the vertical SO_2-distribution are presented and discussed.

The SO_2 balance equation

In the model, the SO_2-concentration profile is derived by solving the steady-state balance equation of all SO_2 formation and destruction processes. In a vertical one-dimensional steady-state model the balance of SO_2 at any altitude is given by:

$$\frac{\partial}{\partial z}\rho K \frac{\partial[SO_2]}{\partial z} - \sum_i k_i X_i [SO_2] - \sum_j r_j [SO_2] = 0 \quad (1)$$

ρ: air density, K: vertical eddy diffusion coefficient, $[SO_2]$: SO_2 gas-phase concentration, z: height, k_i: effective rate coefficient of component i having a volume mixing ratio of X_i for the gas-phase reaction with SO_2, r_j: effective first-order rate coefficient for wet chemical removal mechanism j.

In Equation (1) the divergence of the vertical eddy flux of SO_2 (first-term) is balanced by the net loss due to gas-phase reactions (second term) and by the net loss due to wet chemical reactions (third term).

Production of SO_2 is assumed to occur only in the layer close to the ground. It is, therefore, included in the boundary condition at 1 km and a production term of SO_2 does not appear in Equation (1). This implicitly assumes that oxidation of H_2S, if important at all, takes place close to the ground.

It is characteristic of a 1-D steady-state model, and therefore of Equation (1) accordingly, that it describes average conditions and all the processes: transport, gas-phase and wet removal always take place simultaneously. In the model the sun is shining and it rains at the same time. This is, of course, also true in the real atmosphere but there these conditions are separated spatially, whereas in the model, due to its average nature, all processes go on simultaneously at any given point. In the following we discuss the three terms of Equation (1) separately.

Divergence of the eddy SO_2 *flux*

To calculate the eddy flux, the vertical profile of the eddy diffusion coefficient has to be known. We assumed an eddy diffusion coefficient constant at $2 \times 10^5\ cm^2\ s^{-1}$ throughout the troposphere decreasing to a stratospheric value of $2 \times 10^3\ cm^2\ s^{-1}$ at the tropopause (Ehhalt *et al.*, 1975). A larger eddy diffusion coefficient in the troposphere would increase the SO_2 delivery from the ground and thus maintain a higher SO_2 concentration. To start the iterative solution of Equation (1) an initial profile of the SO_2 concentration is prescribed (naturally this estimated profile has no influence on the final solution).

Gas-phase reactions

The major homogeneous gas-phase reactions are:

R1: $SO_2 + OH + M \rightarrow\rightarrow HSO_3 + M$

$$k_1 = \frac{1.8 \times 10^{-14} \exp(1104/T)M}{1.9 \times 10^{18} + M}\ cm^3\ s^{-1}$$

R2: $SO_2 + HO_2 \rightarrow\rightarrow SO_3 + HO$

$$k_2 = 9 \times 10^{-16}\ cm^3\ s^{-1}$$

followed by

R3: $SO_3 + H_2O \rightarrow\rightarrow H_2SO_4$

$$k_3 = 9 \times 10^{-13}\ cm^3\ s^{-1}.$$

The effective bimolecular reaction rate constant k_1 was estimated from data given by Castleman *et al.* (1975) and Cox (1976) for 220 K and 295 K. k_2 and k_3 were taken from Hampson and Garvin (1975). As mentioned above the required concentrations of the reactions, OH, HO_2 and H_2O are calculated in the model for annual average conditions. The destruction rate of SO_2 due to R2 amounts to less than 1% of the rate of R1. Oxidation by CH_3O_2 was estimated to be less effective than R2. Since the rate constant is not well known this SO_2 removal was not considered. The rate constant estimate of $5.3 \pm 2.5 \times 10^{-5}$ cited by Calvert *et al.* (1978), however, makes this oxidation rate nearly as important as the OH oxidation. The fate of HSO_3 was not followed because exact reaction mechanisms are unknown. As the model is not intended to simulate polluted atmospheres, most of the reactions discussed by Sander and Seinfeld (1976) are not included. Finally, the absorption of SO_2 on aerosol surfaces was neglected. One part of the aerosol surfaces in the free atmosphere consists of H_2SO_4 which hardly absorbs SO_2, and the absorption efficiency for SO_2 molecules of the other part is not known.

Wet removal of SO_2

The moment water vapour begins to condense in ascending and cooling air mass, the SO_2 present will be distributed between the gas- and the condensed phase. The uptake of SO_2 by the liquid-phase proceeds in several steps. The first part of the SO_2 simply dissolves in the cloud droplets. This dissolved SO_2 then further reacts with water to form bisulfite ions HSO_3^- which in turn partly dissociate to form sulfite ions, SO_3^{2-}. These fast steps are eventually followed by a slow oxidation to sulfate ions (Beilke and Gravenhorst, 1978). All these steps incorporate SO_2 into cloud-droplets which partly coagulate to precipitation and deposit their sulfur content on the ground.

The calculation of wet removal of SO_2 from the atmosphere is greatly aided by the observation that the first steps in the dissolution of SO_2 are in fact so fast that we can assume thermodynamic equilibrium between the gas-phase and the droplet phase in a cloud. The reaction of physically dissolved SO_2 to bisulfite (Eigen *et al.*, 1961) and the pure ionic reaction of HSO_3^- to SO_3^{2-} proceed very fast, so that gas-phase diffusion of SO_2 to a cloud-droplet becomes the rate-limiting step in the initial absorption. This step is still reasonably fast, equilibrium between gas-phase SO_2 and dissolved SO_2, HSO_3^- and SO_3^{2-} is reached within seconds. This time is quite short compared to the dynamical processes and mixing within the cloud. Thus the concentration of the dissolved SO_2, HSO_3^- and SO_3^{2-} are governed by the following thermodynamic equilibria:

$$[SO_2 \times H_2O] = S[SO_2],$$

$$S = 7.1 \times 10^{-4} \exp(3145/T) \frac{cm^3\ \text{gas-phase}}{cm^3\ \text{aqueous-phase}} \quad (2)$$

$$[HSO_3^-][H^+] = K_I[SO_2 \times H_2O],$$

$$K_I = 1.9 \times 10^{-5} \exp(2022/T)\ mol/l \quad (3)$$

$$[SO_3^{2-}][H^+] = K_{II}[HSO_3^-],$$

$$K_{II} = 2.4 \times 10^{-10} \exp(1671/T)\ mol/l. \quad (4)$$

The Henry constant S was taken from Gmelin (1963) and the first and second dissociation constants K_I and K_{II} from Sillen (1964). They were extrapolated to negative temperatures for super-cooled cloud-droplets.

The above absorbed sulfur compounds all possess the oxidation state IV. Moreover they quickly interconvert into each other, so that they can be treated together. The sum of their concentrations can be expressed with the help of Equations (1), (2) and (3) as a function of the SO_2 concentration in the gas-phase, SO_2:

$$S(IV) = S(1 + K_I[H^+]^{-1} + K_I K_{II}[H^+]^{-2})SO_2. \quad (5)$$

During the condensation, the gas-phase concentration of SO_2 is somewhat reduced, which introduces a small correction into Equation (5). Allowing for the conservation of the total sulfur mass during condensation we obtain

$$S(IV) = \frac{S(1 + K_I[H^+]^{-1} + K_I K_{II}[H^+]^{-2})}{1 + S(1 + K_I[H^+]^{-1} + K_I K_{II}[H^+]^{-2})F} SO_{2_0} \quad (6)$$

for the amount of absorbed S (IV) per unit volume of water, where SO_{2_0} is the gas-phase concentration of SO_2 before condensation and F is the volume of liquid water per gas volume. F is not the actual liquid water content measured in a cloud but the annual average liquid water content present at one point in the atmosphere. For average conditions F is of the order of 10^{-7}. Thus in the case of SO_2 and at pH values below 6 the second term in the denominator can be neglected compared to 1 and Equation (6) reduces essentially to Equation (5).

To obtain the wet removal of SO_2 from the atmosphere in the form of absorbed S (IV), we simply have to multiply Equation (5) or Equation (6) by the rate, U, with which precipitation is formed in a given volume element. Unfortunately, at higher altitudes the situation is not so simple. There, condensed water may exist in the liquid (droplets) or solid-phase (ice crystals). Even if we assume that the absorption of SO_2 in ice crystals is small, a fact also indicated by the low deposition velocity of SO_2 on snow-covered ground and thus neglect the uptake of SO_2 by the solid-phase, we still have to convert the precipitation formation for the presence of ice crystals. We do this by multiplying, U, by the fraction of condensed water present as liquid, g. g is a dimensionless function of altitude $g = g(z)$, with $0 < g < 1$. At 6 km 90% of the condensed water, at 8 km 50% and at 10 km 0.3% are present as liquid. Accordingly, the effective first-order rate coefficient for the wet removal of SO_2 in the form of S (IV) absorbed in rain has the form:

$$r_1 = S(1 + K_I[H^+]^{-1} + K_I K_{II}[H^+]^{-2})\,g \cdot U$$

where U, the rate of precipitation formation (molecules $cm^{-3} \cdot s^{-1}$) is derived in the model, r_1 represents only one path of wet removal of SO_2. The other is the irreversible wet oxidation of SO_2 to SO_4^{2-}. It is treated here in the sense that the sulfite ion SO_3^{2-} is the oxygen-carrier and not the bisulfite ion HSO_3^- or both:

$$SO_3^{2-} + 1/2\,O_2 \xrightarrow{r_2} SO_4^{2-} \quad (8)$$

r_2, the rate coefficient for the conversion of SO_3^{2-} is not well known for atmospheric water containing a complex set of metal compounds, ions and dissolved oxidizing agents and inhibitors. An empirical relation

$$r_2 = 1.95\;10^{16}\exp(-11676/T) \quad (9)$$

can be deduced for sulfate formation per second from sulfite ions in natural rain water from data of Betz (1977) for the pH range 3–5. Betz (1977) measured the concentration decrease of S(IV) with time at different temperatures of rain water samples. For calculations outside this range, the same value was used due to lack of other data. This value is about one hundred times larger than in distilled water. Since the measurements were made in a closed system, the continuous supply of oxidizing agents like H_2O_2 was not possible in contrast to the situation in the atmosphere. The true formation of sulfate from S (IV) in cloud- and rain-water may well be even faster.

Since the oxidation to SO_4^{2-} via reaction (8) is irreversible the wet removal rate of SO_2 in form of SO_4^{2-} could be obtained simply by multiplying r_2 by the SO_3^{2-} concentration in liquid water and by the amount of liquid water present at any altitude. We would suggest that the oxidation to SO_4^{2-} separates into two parts, one taking place in rain or droplets coagulating to rain and the other taking place in the non-precipitating liquid water. The reason for this is that rain- and cloud-droplets may have different pH values and different concentrations of catalysts which means that the overall oxidation, which depends on the pH, may proceed at different rates in these two cases. We therefore propose to use:

$$R_1 = r_2\,S\,K_I K_{II}[H^+]^{-2}(1 - f)\,g\,F$$

for the oxidation rate coefficient in cloud-water, and

$$R_2 = r_2\,S\,K_I K_{II}[H^+]^{-2} f g\,F$$

for the oxidation rate coefficient in rain water.

f is the fraction of condensed water present as precipitation. As mentioned above, these rate coefficients are essentially given by the product of r_2 with the SO_3^{2-} concentration and the respective fractions of liquid water per volume of air.

The three removal rates given by r_1, R_1 and R_2 represent our wet removal. They are annual averages covering fair weather and overcast conditions. It is obvious from the respective expressions that they all depend strongly on the pH value. In the model calculations the pH value of the liquid water was kept constant (although different for clouds and rain water). It is therefore not affected by the uptake of SO_2, which greatly simplifies the calculations. Such an assumption can be justified, however, by noting that the pH value of cloud water is mainly determined by the chemical composition of incorporated aerosol particles and not by dissolution of gases other than CO_2. The uptake of NH_3 seems to counterbalance the effect of SO_2 absorption and not to increase the pH value remarkably. Even if the total amount of ammonium found in rain water would be the result of gaseous NH_3 uptake the average concentration of 0.5×10^{-5} mole/l for the northern hemisphere (Böttger *et al.*, 1977) could have changed the pH only slightly from 5.0 to 5.3 or from 4.5 to 4.6 when all other interactions are neglected. In fact considerations which do not include the chemical composition of cloud-active nuclei may lead to conclusions which are unrealistic for atmospheric situations (Scott and Hobbs, 1966; Easters and Hobbs, 1974). In order to calculate the pH value of atmospheric water the acidity and/or alkalinity of cloud-active nuclei and their buffer capacity have to be known. Since only a little information is available (Junge and Scheich, 1969; Gravenhorst, 1978) we hope to circumvent that problem by fixing the pH value at an arbitrary value. Keeping the pH value constant has another consequence for our considerations: the concentrations of the *S* (IV) compounds depend only on the SO_2 concentration in the air and on the temperature. Thus any SO_3^{2-} removed by oxidation to SO_4^{2-} via reaction (8) is immediately replaced by absorbing more SO_2 from the gas-phase. After a few seconds r_2 controls the further uptake of SO_2 by the droplets, and a constant flow of SO_2 molecules passes through the reactions (2), (3), (4) and (8).

Choice of parameters

The pH value of rain water depends on the geographical location. In continental regions at middle latitudes, pH values around 4 are quite common (Likens *et al.*, 1972). Under maritime influence, pH values are shifted to higher values because of the alkaline sea salt (Gravenhorst, 1975). The pH values in precipitating clouds show the same trend (Oddie, 1962; Petrenchuk and Drozdowa, 1966; Fricke *et al.*, in press) and a similar range of pH values as in rain. A pH value of 4.5 was chosen as the most likely average for rain water. The acidity of cloud- and rain-droplets depends to some extent on how much soluble components of cloud-active nuclei are diluted by condensation. Non-precipitating clouds may therefore have a lower pH than rain water because they are expected to experience less condensation and therefore produce a smaller amount of liquid water than precipitating clouds. Hence their dilution effect would be smaller. Even over the ocean, a part of the cloud water could have lower pH values as predicted from H_2O–CO_2 interactions, since maritime aerosols smaller than 0.45 μm radius seem to be acidic (Gravenhorst, 1975). A pH value of 3 was chosen for the cloud-droplets. This value is highly speculative.

The parameters related to condensed water, F and U, are estimated in the following way: from data given by de Bary and Möller (1960), an average vertical frequency distribution can be calculated for the occurrence of condensed water as shown in Fig. 1. The distribution reflects the situation in Central Europe since the observations were made here. The cloud cover will be less in subtropical regions and larger in areas where low pressure systems occur more frequently. Liquid water contents for various clouds are given by Fletcher (1964). They range from 0.11 to 0.64 g m^{-3} with most frequent, but not representative, data between 0.1 and 0.2 g m^{-3}. Assuming an average liquid water content of 0.1 g m^{-3} at all altitudes, for all clouds precipitating or not, a similar vertical distribution of the condensed water is obtained for annual average. The actual value for precipitating clouds is higher than 0.1 g m^{-3}. This difference in liquid water content is taken into account in attributing a high fraction of 10% of the liquid water in the atmosphere to rain ($f = 0.1$).

Formation rate of precipitation, U, was calculated in the model as the divergence of water vapour fluxes resulting from a standard H_2O vapour profile and a constant eddy diffusion coefficient profile.

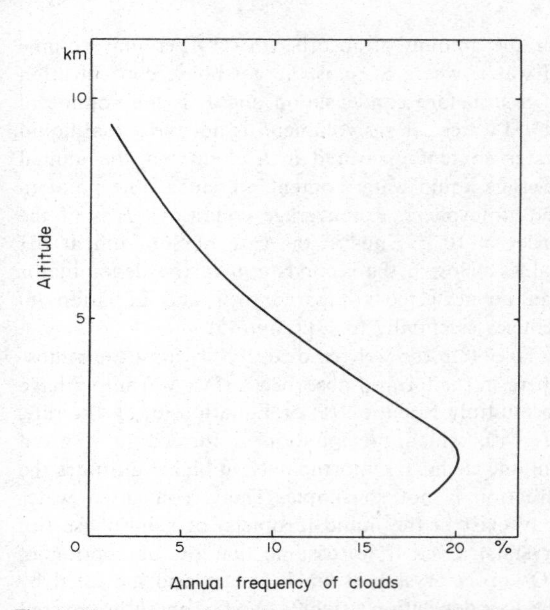

Fig. 1. Frequency of clouds (annual average) at different altitudes over Central Europe (calculated from data given by de Bary and Möller, 1960).

For the SO_2 boundary value at 1 km altitude 1 ppb was chosen. In industrial areas annual averages of the order of 50 ppb are measured whereas in rural areas concentrations of a few ppb are found. 0.08 ppb were reported by Büchen and Georgii (1971) and Prahm *et al.* (1976) for maritime background concentrations. Over the North Atlantic preliminary results show a value of 0.1 ppb (Gravenhorst, 1975). The SO_2 boundary value allows implicitly for SO_2 deposition on the ground and destruction in the layer close to the ground.

RESULTS

The wet removal rate coefficients r_1, R_1, R_2 as function of altitude calculated with the above choice of parameters is shown in Fig. 2. It shows that the incorporation of SO_2 as *S* (IV) into rain water is by far the major wet chemical removal mechanism whereas the oxidation rate to sulfate in cloud-water can be neglected. Oxidation to sulfate in rain water, R_2, becomes significant at lower altitudes. The rate coefficient for SO_2 removal as *S* (IV) has a pronounced maximum between 5 and 7 km altitude. Up to this altitude the increase in the rain formation rate, U, combines with an increase in uptake-capacity for *S* (IV) caused by the decrease in temperature. Above 7 km the rapid decrease in U determines the overall effect. The rate coefficients for the sulfite oxidation, R_1, and R_2 decrease rapidly with altitude. The activation energy of r_2 is much larger than the negative activation energies in K_I and K_{II} and causes a decrease with temperature and altitude [of Equations (3), (4), (9)].

The sum of all wet removal rates (curve 1) is compared with the gas-phase destruction of SO_2, (curve

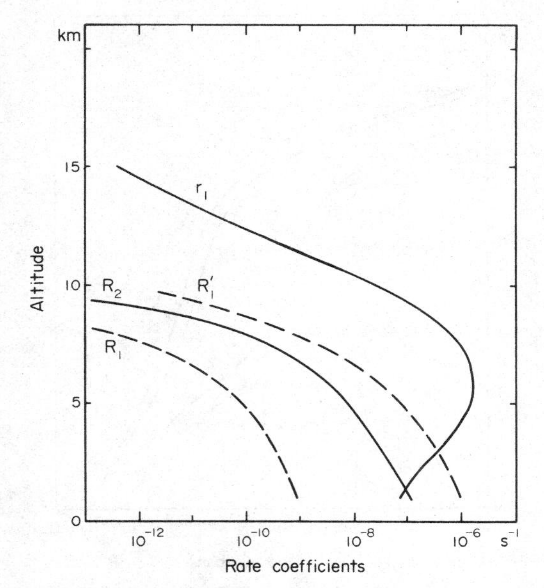

Fig. 2. Effective first-order rate coefficients for reactions between SO_2 and the liquid phase in the atmosphere averaged over space and time. (r_1:S (IV)-removal in rain water for pH 4.5; R_1: sulfate formation in cloud-water for pH 3 and R'_1 for pH 4.5; R_2: sulfate formation in rain water for pH 4.5).

2) in Fig. 3. The OH concentration profile which was generated within the model is also given in Fig. 3. Obviously the altitude dependence of the gas-phase destruction is dominated by the shape of the OH profile. The comparison between curves 1 and 2 suggests that gas-phase destruction dominates below 3 km and above 9 km altitude, in the region in between, wet removal is greater by about a factor of 3. Integrated over all altitudes roughly one third of the SO_2 is wet-removed, the bulk is destroyed by gas-phase reaction. The vertically-integrated conversion rates are 0.14% h^{-1} for liquid water reactions and 0.25% h^{-1} for gas-phase reactions. For wet chemical conversion rates within clouds or fogs, values of 0.002–0.2% h^{-1} were estimated by Beilke and Gravenhorst (1977) and between 0.1 and 3% h^{-1} depending on a variety of cloud conditions (Barrie, 1975). In both these calculations, however, it was neither attempted to calculate average global conditions nor to consider the eddy diffusion transport of SO_2 into the clouds.

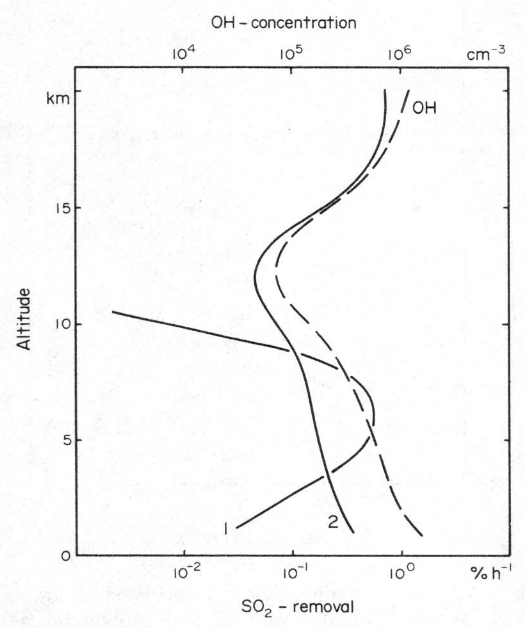

Fig. 3. Removal rates of SO_2 in percent per hour for interactions with cloud- and rain water (1) and gas-phase reactions (2). [pH value for rain water 4.5, for cloud-water 3, $r_2 = 1.95 \times 10^{16} \exp(-11676/T)$]. The OH concentration (upper scale) determines the gas-phase destruction rate.

Other derivations of parameters

The parameters determining the wet chemical removal were chosen to the best of our knowledge. Since only few experimental data are available, parameters which seem to control wet removal were varied in certain ranges. Most uncertain is the pH value of non-precipitating clouds which represent *ca.* 90% of liquid water in the atmosphere. The pH value influences the sulfite- and bisulfite-concentrations which in their turn determine the formation rate of sulfate and thus the SO_2 uptake rate. The effective oxidation rate coefficient, R_1 is compared in Fig. 2 for cloud water pH values of 3 and 4.5. At the higher pH value R_1 is faster than the rate coefficient R_2 for rain water of the same pH value since more cloud water than rain water is present in the atmosphere. Whether in reality the SO_2 uptake due to sulfate formation is higher in cloud water than in rain water depends on the pH values in the atmosphere. At similar pH values for rain and cloud water, the oxidation mechanism of absorbed SO_2 to sulfate seems to be more important than *S* (IV) removal in precipitation in the lower few kilometers. The SO_2 absorption in cloud and rain water can furthermore be altered by a change of the rate coefficient r_2 for the sulfite oxidation. In Fig. 4, SO_2-concentration profiles calculated for different r_2 values are shown. A 100-fold increase of r_2 does not change the SO_2 concentration drastically when pH values of 3 for cloud water and 4.5 for rain water are chosen. However if the pH value of cloud water is raised to 4.5 simultaneously, the SO_2 concentration decreases rather steeply. Hydrogen peroxide and ozone oxidation of absorbed SO_2 as suggested by Penkett *et al.* (1977) seems to justify an increase of r_2. The synergistic effect of oxidizing agents, catalysts and inhibitors is, however, not yet known.

The calculations indicate that the SO_2 concentration, already remarkably below the tropopause level, decreases due to the sharp increase of OH concentrations (see Fig. 3) above 12 km altitude. The tropopause should therefore not generally represent a discontinuity for SO_2. The SO_2 half-life at this altitude is *ca.* 20 days so that the SO_2 distribution will be more determined by gas-phase reactions than by transport mechanisms.

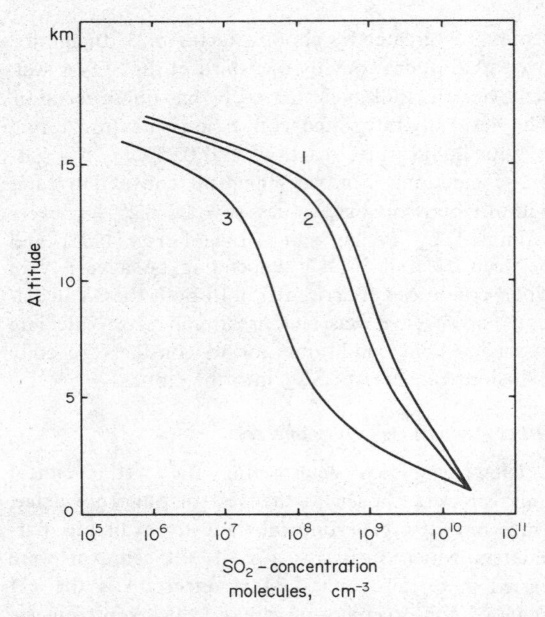

Fig. 4. Vertical SO_2-concentration profiles for different sulfit oxidation rate coefficients and pH values of cloud-water: 1: $r_2 = 1.1 \times 10^{-16}$ exp $(-1400/T)$ $cm^2\,s^{-1}$ (deduced from Betz's (1977) data) and pH = 3.0; 2: $r_2' = 100\,r_2$, pH = 3.0; 3: $r_2' = 100\,r_2$, pH = 4.5. The pH value of rain is always 4.5.

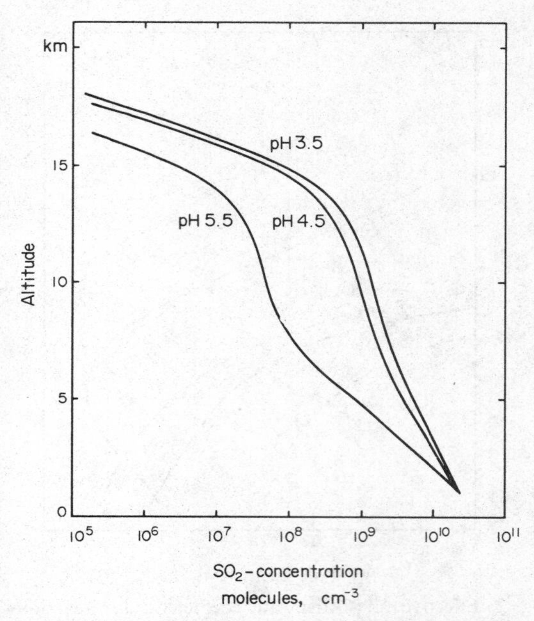

Fig. 5. Vertical distributions of SO_2, calculated for different pH values for rain water (pH value for cloud-water: 3.0).

The most effective wet chemical sink mechanism for SO_2 above 2 km is for this model the removal of S (IV) in rain water (Fig. 2). A striking change in the SO_2 profile is caused by a variation of the pH value of rain water (Fig. 5). It is therefore important to determine representative values for rain water. As appropriate values a pH of 4.5 for rain water and of 3 for cloud water were chosen to calculate vertical SO_2 distribution for the case when wet and gas-phase removal were taken into account (Fig. 6). For comparison pure gas-phase removal was considered. In both situations the same boundary value (1 ppb) was used. The SO_2 concentration has decreased at the tropopause level (15 km) due to gas-phase reactions by a factor of 250 (mixing ratio by a factor of 55 –). The influence of liquid water reduces the SO_2 concentration in addition, by a factor of 1.6 to 5.3×10^7 molecules cm^{-3} (*ca.* 10^{-2} ppb). This value is one order of magnitude lower than the one suggested by Junge (1974) discussing the formation of the stratospheric sulfate layer.

The calculated vertical SO_2-distribution can be related to a few measured data (Fig. 6). Until today, only preliminary results can be reported because of difficulties in measuring SO_2 in low concentrations in remote areas and in the middle and upper troposphere. In polluted continental air masses, the SO_2 concentration decreases more steeply with altitude (curve a: Georgii and Jost, 1964; curve b: Jost, 1974; curve c: Gravenhorst, 1975) because local sources increase the concentration near the ground. Over Colorado a distribution similar to the modelled one was found (Curve d: Georgii, 1970). At 3.5 km above sea-level in Switzerland, about 1 ppb was measured as an average over three years (single dot: Eidgenössisches Amt für Umweltschutz, 1976). A rather homogeneous distribution was indicated in the lower 4 km in pure maritime air with a trend to increasing mixing ratios with altitude (curve e: Gravenhorst, 1975). But here also a comparison is restricted since the ocean does not act as a SO_2 source. First preliminary

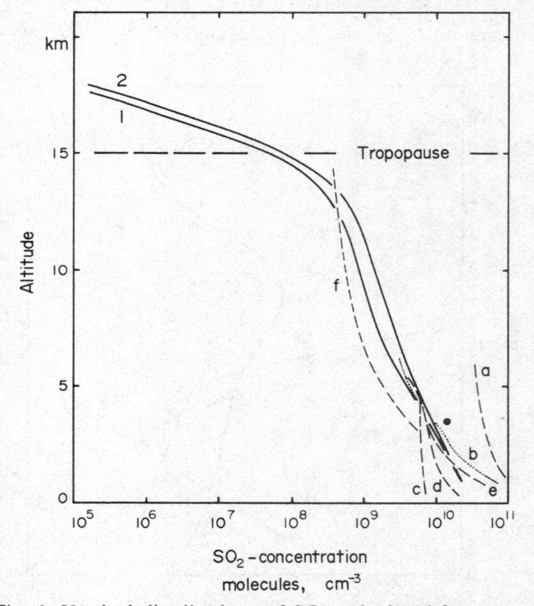

Fig. 6. Vertical distributions of SO_2 calculated for appropriate average conditions with (1) and without (2) wet chemical reactions. Experimental values are added: (a) Georgii and Jost, 1964; (b) Jost, 1974; (c) Gravenhorst, 1975; (d) Georgii, 1970 (e) Gravenhorst, 1975; (f) Jaeschke *et al.*, 1976; single dot ●: Eidgenössisches Amt für Umweltschutz, 1976.

measurements in tropopause levels gave concentrations of similar magnitude as in the model (curve f: Jaeschke *et al.* 1976). Mass spectrometric observations of SO_2 in the stratosphere (Sagawa and Itoh, 1977) show unrealistically high values in the lower stratosphere (*ca.* 20 ppm).

To determine whether gas-phase or wet chemical mechanisms are more important in this model for the SO_2 removal from the atmosphere, weighted rate coefficients for the troposphere were calculated. The SO_2-removal of 0.14% h^{-1} for wet chemical interaction (pH cloud-water 3, pH rain water 4.5) was smaller than the value of 0.25% h^{-1} for gas-phase destruction. This means that during the same time *ca.* half as many SO_2 molecules are absorbed in cloud- and rain water than destroyed by gas-phase reactions. However, if the pH value of cloud water is raised to 4.5, the wet chemical removal increases to 0.33% h^{-1} whereas the gas-phase destruction remains essentially the same, namely 0.26% h^{-1}. For this new condition interactions with the liquid-phase are more effective in the removal of SO_2 from the atmosphere than homogeneous gas-phase destruction. The relative importance of these removal mechanisms depends, therefore, quite sensitively on the chosen pH values. Since most SO_2 is emitted into the atmosphere within the boundary layer, SO_2 can be deposited directly at the ground. Assuming a SO_2 deposition velocity of 0.8 cm s^{-1} the SO_2 removal from the troposphere by dry deposition amounts to *ca.* 1.0% h^{-1}. Compared with wet chemical and gas-phase reactions in this model, dry deposition on the ground seems, therefore, to be the most important SO_2 sink mechanism.

This model approach does not reflect conditions in small scale phenomena such as urban plumes or precipitation systems. Due to its average nature the model should, however, represent the situation above the atmospheric mixing layer since there the influence of SO_2 sources and SO_2 sinks do not vary so much with time and space as close to the ground. More confidence could be placed in the model predictions when better data, especially on chemical properties of cloud- and rain water and their climatological variations, become available.

REFERENCES

Barrie L. A. (1975) An experimental investigation of the absorption of sulfur dioxide by cloud and rain drops containing heavy metals, Promotionsarbeit, Institut für Meteorologie und Geophysik der Universität Frankfurt.

Beilke S. and Gravenhorst G. (1977) A contribution to the formation of atmospheric sulfate and its removal from the atmosphere. Seminar 'Fine Particulate Air Pollution' Villach, Oct. 77, United Nations Economic and Social Council, Economic Commission for Europe, Genf.

Beilke S. and Gravenhorst G. (1978) Heterogenous SO_2 oxidation in the droplet phase. *Atmospheric Environment* **12,** 231–239.

Betz M. (1977) Untersuchungen über die Absorption und Oxidation von Schwefeldioxid in natürlichem Regenwasser, Diplomarbeit, Institut für Meteorologie und Geophysik der Universität Frankfurt.

Böttger A., Gravenhorst G. and Ehhalt D. (1977) Deposition rates of ammonium and nitrate in the northern hemisphere, paper presented at *9th Int. Conf. on Atmospheric Aerosols, Condensation and Ice Nuclei, Galway, Ireland.*

Büchen M. and Georgii H. W. (1971) Ein Beitrag zum atmosphärischen Schwefelhaushalt über dem Atlantik. *"Meteor" Forsch. Ergebnisse B.* **7,** 71–77.

Calvert J. G., Su Fu, Bottenheim Jan W. and Strausz Otto P. (1978) Mechanism of the homogeneous oxidation of sulfur dioxide in the troposphere. *Atmospheric Environment* **12,** 197–226.

Castleman A. W. Jr., Davis R. E., Tang J. N. and Bell J. A. (1975) Heterogenous processes and the chemistry of aerosol formation in the upper atmosphere, *4th Conf. on the climatic impact assessment program.* Massachusetts, Feb. 1975.

Cox R. A. (1976) Gas-phase oxidation of atmospheric sulfur dioxide, EMS Division, AERE, Harwell, OXII, ORA, U.K., (Unpublished manuscript).

de Bary E. and Möller F. (1960) Die mittlere vertikale Verteilung von Wolken in Abhängigkeit von der Wetterlage, Berichte des Deutschen Wetterdienstes, Nr. 67, Offenbach a.M.

Easter R. C. and Hobbs P. (1974) The formation of sulfates and the enhancement of cloud condensation nuclei in cloud, *Atmos. Sci.* **31,** 1586–1594.

Ehhalt D. H., Heidt L. E., Lueb R. H. and Pollock W. (1975) The vertical distribution of trace gases in the stratosphere, *Pure appl. Geophys.* **113,** 389–402.

Eidgenössisches Amt für Umweltschutz (1976) Immissionsangebote an Staub und Schwefelverbindungen bei den OECD Messorten "Payerne" und "Jungfraujoch" 1973 bis 1975, Bern, EMPA, Nr. 17000.

Eigen M., Kustin K. and Maas G. (1961) Die Geschwindigkeit der Hydration von SO_2 in wässriger Lösung, *Z. Phys. Chemie (Neue Folge),* **30,** 130.

Fletcher N. H. (1966) *The Physics of Rain Clouds.* University Press, Cambridge.

Fricke W., Georgii H. W. and Gravenhorst G. (1976) Application of a new sampling device for cloud water analysis, A. M. Borovikor Memorial Volume, Moscow. (In press).

Georgii H. W. (1970) Contribution to the atmospheric sulfur budget. *J. geophys. Res.* **75,** 2365–2371.

Georgii H. W., and Jost D. (1964) Untersuchung über die Verteilung von Spurengasen in der freien Atmosphäre, *Pure appl. Geophys.* **59,** 217–224.

Gmelin (1963) Handbuch Schwefel, 8. Auflage, Teil B-Lieferung 3.

Gravenhorst G. (1975) Der Sulfatanteil im atmosphärischen Aerosol über dem Nordatlantik, Berichte des Instituts für Meteorologie und Geophysik der Universität Frankfurt, Nr. 30.

Gravenhorst G. (1978) Maritime sulfate over the North-Atlantic. *Atmospheric Environment* **12,** 707–713.

Hampson R. F. Jr. and Garvin D. (1975) NBS Techn. Note 866 June, 1975.

Jaeschke W., Schmitt R. and Georgii H. W. (1976) Preliminary results of stratospheric SO_2 measurements. *Geophys. Res. Lett.* **3,** 417–419.

Jost D. (1974) Aerological studies on the atmospheric sulfur budget. *Tellus* **26,** 1–2, 206–212.

Junge C. (1974) Sulfur budget of the stratospheric aerosol layer, *Proc. Int. Conf. Structure, composition and general circulation of the upper and lower atmospheres and possible anthropogenic perturbations,* Melbourne, **1,** 85–97.

Junge C. and Scheich G. (1969) Studien zur Bestimmung des Säuregehaltes von Aerosolteilchen. *Atmospheric Environment* **3,** 423–441.

Likens G. E., Bormann F. H. and Johnson N. M. (1972) Acid rain. *Environment* **14,** 33–40.

Oddie B. C. V. (1962) The Chemical composition of precipitation at cloud levels, *Q. Jl R. met. Soc.* **88,** 378, 535–538.

Penkett S. A., Jones B. M. R. and Brice K. A. (1977) Estimation of the rate at which sulfur dioxide is oxidized in cloud and fog droplets, AERE, PREMS 4, Harwell, pp. 82–83.

Petrenchuk O. P. and Drozdowa V. M. (1966) On the chemical composition of cloud water, *Tellus* **18,** 280–286.

Prahm L. P., Torp U. and Stern R. M. (1976) Deposition and transformation rates of sulfur oxides during atmospheric transport over the Atlantic. *Tellus* **28,** (4) 355–372.

Sagawa E., and Itoh T. (1977) Mass spectrometric observation of SO_2 in the stratosphere, *Geophys. Res. Lett.* **4,** (no. 1) 29–32.

Sander S. P., and Seinfeld J. H. (1976) Chemical kinetics of homogeneous atmospheric oxidation of sulfur dioxide, *Envir. Sci. Tech.* **10,** (12) 1114–1123.

Scott W. D. and Hobbs P. (1966) The formation of sulfate in water droplets, *J. atmos. Sci.* **24,** 54–57.

Sillen L. G. (1964) *Stability Constants of Metal–Ion Complexes.* 2nd Edn. Chem. Soc. Spec. Publ. 17, London.

Atmospheric Environment Vol. 12, pp. 699–705. Pergamon Press 1978. Printed in Great Britain.

CONCENTRATION OF SULFUR COMPOUNDS IN REMOTE CONTINENTAL AND OCEANIC AREAS

E. MÉSZÁROS

Institute for Atmospheric Physics H-1675 Budapest, P.O.B. 39, Hungary

(*Received for publication* 20 *September* 1977)

Abstract—The results of concentration measurements of atmospheric sulfur dioxide and sulfate particles carried out near the Earth's surface under background conditions are compiled. The data for continental (mainly Europe and North America) and oceanic (mainly Atlantic Ocean) areas are discussed. On the basis of the available information the sulfur quantities in an air column and in the whole atmosphere over continents and oceans are calculated. Some data on the H_2S concentration in the air are also presented.

1. INTRODUCTION

It is obvious that for the calculation of the global residence time of the sulfur compounds in the atmosphere as well as for the determination of some components (e.g. deposition velocity, transformation rate) of their budget, the measurement of the concentration of gaseous and particulate sulfur in the air is indispensable (see e.g. the recent publications of Kellog *et al.*, 1972; Friend, 1973; Granat *et al.*, 1976). The aim of the present paper is to compile the information available at present. The results presented make it possible to re-calculate, for further budget calculations, the value of sulfur quantity in the atmospheric reservoir.

Unfortunately, for the time being, in spite of some new measurements, our knowledge about the spatial and temporal distribution of sulfur compounds in the atmosphere is rather scanty. This is particularly true in the case of sulfur in reduced state. However, even for sulfur dioxide and sulfate particles, the amount of information seems to be sufficient only for Europe, North America and for the Atlantic Ocean.

In the paper, first the continental level of SO_2 and SO_4^{2-} will be discussed. After this, the oceanic concentration of these sulfur compounds will be presented, mostly on the basis of measurements carried out over the Atlantic Ocean. For the calculation of the atmospheric sulfur burden, the results of some aircraft measurements will also be used. The concentration of H_2S will only be mentioned at the end of the paper, in spite of the fact that this compound is a precursor of SO_2. This is because our knowledge of its atmospheric level is very uncertain.

2. CONCENTRATION OF SULFUR DIOXIDE AND SULFATE IN CONTINENTAL ENVIRONMENT

The aim of this paper is to present sulfur concentrations for large scale and budget considerations. Thus, the urban and local level of SO_2 and sulfate particles is not discussed. It goes without saying that this does not mean that the measurement of urban concentrations would not be important for purposes connected with the problems of the local pollution.

Table 1 contains the concentrations of SO_2 and SO_4^{2-} measured by various workers under remote and regional background conditions (the latter is defined according to recommendations of WMO, 1974) over the continents. All concentrations are expressed in $\mu g\ m^{-3}$. The ratios of SO_4^{2-}/SO_2 are also given. The concentrations of SO_2 and SO_4^{2-} over the Greenland ice cap were recently determined by Flyger *et al.* (1976), who made some aircraft measurements up to 1400 ft. Thus, these values can probably be considered as lower limits for surface level concentrations. The data given in the table for eastern and western parts of U.S.A. were published by Altschuller (1973) on the basis of nonurban results gained by a network of the Environmental Protection Agency of this country. A map concerning the spatial distribution of sulfate obtained by this network can be found in the paper of Bolin and Charlson (1976). It is an interesting finding that there is a difference in SO_4^{2-} concentrations between eastern and western nonurban sites. Since the SO_2 concentrations in the original paper are averaged for these two territories, the calculation of the SO_4^{2-}/SO_2 ratio was made by using average sulfate value calculated from the two figures given. Altschuller stated that the higher sulfate concentrations at eastern sites of the U.S.A. are due to the transformation of greater anthropogenic emissions of SO_2 in these areas. On the other hand, Hitchcock (1976) recently speculated that nonurban sulfate levels were caused by biogenic sources. Altshuller also mentioned that "sulfate concentrations at sites of farmlands and forests west of the Mississippi ranged from 1.5–5.0 $\mu g\ m^{-3}$". The SO_2 data for Central U.S.A. were measured by Breeding *et al.* (1973) in clean air in rural Illinois and Missouri. They publish four series of SO_2 measurements made on four different days. The value given is an average of all their measurements.

Table 1. Concentration of sulfur dioxide and sulfate particles in continental background areas according to different authors (see text). All values are expressed in $\mu g\ m^{-3}$

	SO_2	SO_4^{2-}	SO_4^{2-}/SO_2
North America			
Greenland ice cap	0.72	0.28	0.39
E. nonurban sites, U.S.A.	10	8.1	0.53
W. nonurban sites, U.S.A.		2.6	
Central U.S.A.	5.6	—	—
Central America			
Panama	2.1	—	—
South America			
Brazil	0.87	—	—
General tropical value	2.6	—	—
Europe			
North Europe	0.75	0.29	0.39
Sweden (south)	7.6	3.0	0.40
Western Europe	20; 15; 18	7.5; 9.0	0.37; 0.50
Central Europe	12.6	5.4	0.43
Eastern Europe	$\lesssim 10$	4.7	≥ 0.47
Asia			
Central Asia, U.S.S.R.	—	16.3	—
Africa			
Ivory Coast (tropics)	6.1	1.5	0.25
Haute-Volta (steppe)	—	1.8	—
Antarctica	<0.87–4.1	—	—

Very little information is available for the remote areas of Central and South America. The author of this paper is aware only of the data published by Lodge *et al.* (1974) for the American humid tropics. The general tropical value in Table 1 is also given by these workers. It is to be noted that this level is higher than those for continental areas since it also contains the results of measurements carried out over the Caribbean Sea. It is questionable whether the higher maritime values are attributable to maritime SO_2 sources (e.g. oxidation of H_2S). The results of Lodge *et al.* (1974) also showed that sulfur dioxide concentrations were higher in Panama than in Amazonia (Brazil). It would be very difficult to explain this difference, mainly if one takes into consideration that in tropical Africa (see later) much higher SO_2 concentrations were found than in Brazil.

The best concentration information about continental SO_2 and SO_4^{2-} is that for Europe. The SO_2 value for North Europe in Table 1 is the average of total gaseous sulfur (considered as SO_2) measured in the clean air of six northern stations in the Swedish network (de Bary and Junge, 1963). The sulfate value, on the other hand, was gained by Beliashova *et al.* (1970) over the European part of the Soviet Union with a latitude higher than 62–63°. These Soviet authors made aircraft measurements. The value given refers to an air layer between 250 and 1000 m (thus it is a lower limit of surface air concentration). The concentrations for southern Sweden are also taken from aircraft measurements carried out by Rodhe (1972). From his profiles the average concentrations up to 500 m were estimated.

The number of measurements made in the regional background areas of western Europe is relatively high. The first SO_2 and sulfate values in Table 1 were averaged from the data published by Eliassen and Saltbones (1975). These data were measured at nine stations operated in the OECD program (here all concentrations published are used except those measured in Norway). The second sulfur dioxide concentration is estimated from the map of Rodhe (see Granat *et al.*, 1976) giving the spatial distribution of SO_2–S for the north-western part of Europe for surface air, outside of cities and industrial areas. Finally, the third value for SO_2 and the second one for sulfate are estimated from the profiles of Jost (1974) for an air layer between 0 and 500 m above the ground. The ratio of sulfate to sulfur dioxide for western Europe is calculated from the data of Eliassen and Saltbones and those of Jost.

The results for central Europe refer partly to a WMO regional station (SO_2, unpublished) and partly to a rural area (SO_4^{2-}, Bónis, 1968), both in Hungary. The eastern European values, on the other hand, were measured by Soviet workers. Petrenchuk (1971) measured the vertical profile of SO_2 concentrations over a rural area near Leningrad. She mentioned that the sulfur dioxide concentration at the ground was usually around 10 $\mu g\ m^{-3}$ or sometimes even lower. Furthermore, the sulfate concentration given is an average value calculated from the data of Beliashova *et al.* (1970) measured in the layer of 250–1000 m over the European part of the U.S.S.R. (except northern territories).

Concerning the other continents we have very little information. Thus in Central Asia (U.S.S.R.), Andreiev and Lavrinenko (1968) measured, among other aerosol parameters, the concentration of sulfate ions. They found a mean value of 16.3 $\mu g\ m^{-3}$ for surface air. Unfortunately no SO_2 samplings were made. Andreiev and Lavrinenko attributed this high sulfate concentration to the mineral dust due to local natural sources. It is interesting to note that under similar conditions in Africa (Haute-Volta, see later) much lower concentrations were found. This high value for central Asia, however, was confirmed by the work of Khusanov *et al.* (1974) carried out over the same area. In Africa the concentration of SO_2 and SO_4^{2-} was recently measured by Delmas *et al.* (in press). They made samplings at two tropical places on the Ivory Coast (the figures given in Table 1 are averages for these two places) and in Ouagadougou (Haute-Volta) under dusty conditions. Finally the SO_2 level for Antarctica given in Table 1 is taken from the publication of Fisher *et al.* (1968). No SO_4^{2-} mass concentration results are published by these American authors.

It is well demonstrated that, except for sea salt, the natural and anthropogenic sulfur emission is in gaseous form (e.g. Friend, 1973). This means that a great majority of sulfate particles in continental atmosphere is formed by chemical conversion from sul-

Table 2. Model concentrations for various territories of Europe. The concentrations for southern Europe are estimated values (see text). All concentrations are expressed in $\mu g\ m^{-3}$ sulfur level

	SO_2–S	SO_4–S	% of territory
Clean N. territories	0.35	0.1	18
Territories with low concn. (e.g. S. Sweden)	3.5	1.0	22
Moderately polluted territories (e.g. central and E. Europe)	4.0	1.7	34
Territories with high concn. (e.g. W. Europe)	7.5	3.0	12
Southern Europe	1.0	3.5	14

fur dioxide (except near the ground level over special areas e.g. central Asia). Thus, apart from SO_2 source strength, the SO_4^{2-}/SO_2 ratio can be considered as a measure of this transformation and of the difference of the removal rate of sulfur dioxide and sulfate particles. One can see from Table 1 that, except for African values, this ratio varies between 0.37 and 0.53 that is, the variations are surprisingly small. In this way we can conclude that under continental background conditions near the ground level an average value of about 0.5 can be considered as representative.

On the basis of the above data characteristic surface air concentrations were constructed for areas of Europe with no local pollution. The results are tabulated in Table 2. The only problem was that no information was found for southern Europe. For this reason it was arbitrarily assumed that this part of the continent has the same sulfur level as southern Sweden ('low concentration'). It was supposed, however, that the majority of this sulfur is sulfate. This idea was based on the work of Nguyen Ba Cuong *et al.* (1974b) who measured much more sulfate than sulfur dioxide over the Mediterranean Sea. By using information in Table 1 and the results of de Bary and Junge (1963) the territory of Europe was divided according to these model concentrations. The relative territory, expressed in per cent of the total territory, for these sulfur levels is given in the last column of Table 2. With the figures in Table 2 a $3.2\ \mu g\ m^{-3}$ SO_2–S and a $1.7\ \mu g\ m^{-3}$ SO_4–S concentration can be calculated (both weighed according to the relative territory).

For the calculation of the sulfur quantity in an air column with unit surface the determination of the so-called scale height (H) is necessary, which is defined as

$$H = \int_0^\infty c(z)\,dz/c(0),$$

where $c(z)$ is the concentration at a level z, while $c(0)$ is the same parameter for surface level air. The vertical profiles of SO_2–S and SO_4–S over Europe were measured by Rodhe (1972) in Sweden and by Jost (1974) in Germany. These authors did not give, however, the values of $c(0)$ for their profiles. For this reason the results of measurements for Hungary were accepted for further calculations. According to Várhelyi (unpublished) the value of H for SO_2–S is about 0.6 km, while the same figure for SO_4–S is about 2 km (Mészáros E. and Várhelyi, 1975). Thus in an air column $1.9 \times 10^3\ \mu g\ m^{-2}$ SO_2–S and $3.3 \times 10^3\ \mu g\ m^{-2}$ SO_4–S can be determined. Multiplying these values with the territory of Europe ($\sim 10 \times 10^6\ km^2$) the result gives the total sulfur quantity in the atmosphere over Europe: 1.9×10^4 t SO_2–S and 3.3×10^4 t SO_4–S.

Obviously it is very speculative to use these European sulfur values for other continents, except the industrialized North America where the surface air concentrations are very similar to those found in Europe. However, on the basis of the above sulfur quantities in an air column, it is proposed with caution that the total sulfur (except reduced S) quantity in the atmosphere over the continents ($\sim 150 \times 10^6\ km^2$) is 0.78×10^6 t. Of this quantity, 0.28×10^6 t is composed of SO_2–S. It goes without saying that these values can be considered as upper limits since they are generalized from the sulfur quantities of an industrialized continent. However, sufficient information is not available at present for less polluted continents like South America, Asia, Australia and Africa.

3. SULFUR CONCENTRATIONS OVER THE OCEANS

Concerning remote oceanic areas, we have a great deal of information for the Atlantic Ocean. Figure 1 summarizes the results on SO_2 and sulfate concentrations obtained by different authors. Thus Prahm *et al.* (1976) measured the sulfur concentrations over the Faroe Islands. Taking into account that their values refer to island conditions, they are probably upper limits for surface air concentrations over oceans. Gravenhorst's data (Gravenhorst, 1975) were obtained during two travels of the German research vessel '*Meteor*'. His sulfate values (including sea salt sulfate) were averaged to the latitudes by the present author. Nguyen Ba Cuong *et al.* (1974a) observed the SO_4^{2-} concentrations between Europe and central America ($10° < \phi < 30°$ N) over the Atlantic. They also measured and averaged SO_4^{2-} and SO_2 levels according to geographical latitudes around Africa. In

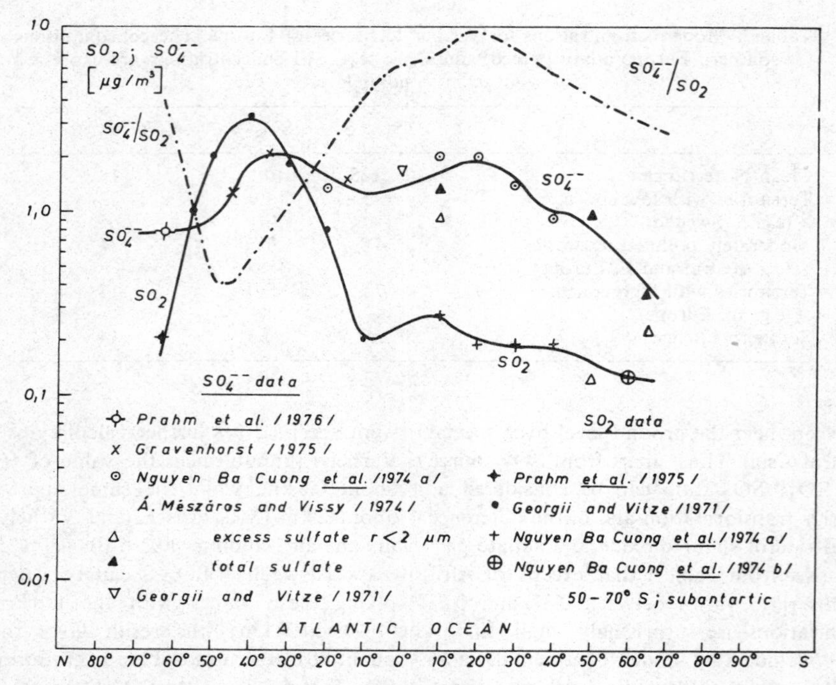

Fig. 1. Concentration of sulfur dioxide and sulfates as a function of the latitude over the Atlantic Ocean according to different authors.

Fig. 1 only the values referring to the Atlantic Ocean are used. These samplings were made in the proximity of the continent. However, considering the very low sulfur dioxide concentrations these results were considered to be representative for more remote oceanic territories. The same French authors (Nguyen Ba Cuong *et al.*, 1974b) took SO_2 samples in sub-Antarctic areas, too. In spite of the fact that these samples were taken between Australia and Antarctica, their results are used in the figure since it is assumed that at these latitudes the longitudinal variations can be neglected.

Furthermore in Fig. 1 the SO_2 results of German workers (Georgii and Vitze, 1971) are also included. The points are taken from their curve and not from the individual measurements. The latter authors also give SO_4^{2-} levels for the Atlantic Ocean near the Equator. Their values, measured between $\phi = 10°$ N and $\phi = 10°$ S, were averaged and plotted in Fig. 1 as equatorial concentration. Finally in Fig. 1 the sulfate mass concentration of ammonium sulfate particles (excess sulfate) as well as the total sulfate concentration (excess sulfate + sea salt sulfate) calculated from the number size distribution of Mészáros Á. and Vissy (1974), obtained on board a Soviet research ship, are also plotted. These Hungarian authors measured the size distribution of ammonium sulfate particles up to 0.25 µm and that of sea salt nuclei up to 64 µm. To calculate the mass concentration of $(NH_4)_2SO_4$–sulfate up to 2 µm, the results of Gravenhorst (1975) for excess sulfate in the large size range received near the Equator were taken into account in the case of the value for $\phi = 10°$ S. For other points it was assumed that the difference between size distributions of all particles and sea salt gives the spectrum of excess sulfate particles. This latter assumption seems a little bit arbitrary. However, the values calculated in this way are in good agreement with the results of direct chemical observations. They also show (see the values for 10° S and 50° S) that the relative quantity of excess sulfate may be very variable from sample to sample.

The most interesting fact emerging from Fig. 1 is the maximum in the SO_2 distribution around $\phi =$ 40° N. Georgii and Vitze (1971) attributed these high concentrations to the SO_2 advection from continental sources. Considering that the German and French workers used practically the same chemical method (West–Gaeke) we accept for further speculation that the distribution in Fig. 1 is real. This means that without the continental (probably anthropogenic) sources in the Northern Hemisphere the oceanic SO_2 level would be rather constant (about 0.1–0.2 µg m^{-3}).

The sulfate distribution has two maxima. The first of them is due to the high SO_2 level around 40° N. The second one is in the proximity of the Equator (a little bit shifted to the south). It follows from this situation that without the anthropogenic disturbance in the sulfur dioxide distribution in the Northern Hemisphere there would be probably only one maximum in the sulfate level at low latitudes. In spite of the fact that the same picture is somewhat reflected in the sulfur dioxide distribution (see the secondary maximum at $\phi = 10°$ S), it is postulated that the SO_4^{2-} variations are mainly due to changes in chemical conversion rate. This view is supported by the

Table 3. Sulfur concentrations in remote oceanic areas expressed in $\mu g\,m^{-3}$. The values for Atlantic Ocean are estimated from Fig. 1, while the other figures are given by various authors

	SO_2–S	SO_4–S	source
Atlantic Ocean	0.1	0.45	Fig. 1
Pacific Ocean	0.15	1.0	(Lodge *et al.*, 1960)
	—	0.43	(N. B. Cuong, 1974)
Indian Ocean	0.1	0.50	(N. B. Cuong, 1974)

fact that at the territories of tropical Atlantic, according to the data of Mészáros Á. and Vissy (1974), the quantity of sea salt sulfate is relatively small. If we accept this explanation for the form of the sulfate distribution in Fig. 1 it can be assumed that the majority of SO_2 transforms over the oceans by photochemically induced thermal reactions, the efficiency of which is greatest under equatorial conditions.

Finally it can be seen from Fig. 1 that, except for the North Atlantic disturbance, the SO_4^{2-}/SO_2 ratio over the oceans (cleaner air) is much higher than over the continents. Under oceanic conditions its value is between 1 and 10.

The first line of Table 3 gives idealized SO_2–S and SO_4^{2-}–S concentrations for the Atlantic Ocean, excluding territories with $\phi > 70°$ and with latitudes between 20 and 60° N. One can estimate from Fig. 1 that for $\phi > 70°$N the SO_2–S is 0.05 $\mu g\,m^{-3}$, while the SO_4–S is about 0.2 $\mu g\,m^{-3}$. In the southern hemisphere ($\phi > 70°$) the corresponding values are about the half of these figures. Moreover, for 20°N $< \phi <$ 60°N the SO_2–S level is around 0.5 $\mu g\,m^{-3}$ (the sulfate–sulfur is about the same as in the table). In Table 3 for comparison some results for other oceans are also included. These figures were measured by Lodge *et al.* (1960) over the Pacific between San Francisco and Honolulu. These latter authors give a 1.0 $\mu g\,m^{-3}$ median value for sulfate–sulfur concentration. On the other hand, the 0.15 $\mu g\,m^{-3}$ sulfur dioxide–sulfur level means that in 42% of the samples the concentration was smaller than this value. The other concentrations in Table 3 are averages taken from the works of Nguyen Ba Cuong *et al.* (1974a, b) made over the South Pacific Ocean and Indian Ocean. It follows from the figures listed that the sulfur level in the air near the oceanic surface is low and practically constant.

Unfortunately, very little information is available concerning the vertical distribution of sulfur compounds over the oceans. According to Gravenhorst (1975) the SO_2 level over the Bay of Biscay practically does not change with altitude. Georgii (1975) also mentioned that SO_2 concentration is constant at least in the lower half of the troposphere over the Atlantic. On the basis of these aircraft results in first approximation, a scale height of 5 km was chosen, which permits some concentration decrease up to the tropopause. Furthermore, Gilette and Blifford (1971) made aerosol samplings for elemental analysis over the Pacific offshore. According to the profile reported, a scale height of 2 km can be calculated for sulfur particulate. The difference with SO_2 scale height is due to the presence of sea salt sulfate, the concentration of which decreases rapidly with increasing altitude (see e.g. the pioneering work of Woodcock, 1953).

To calculate the mean sulfur level over the Atlantic Ocean its territory was divided as follows:

territory with $\phi > 70°$N: 10% of the total,
territory with $\phi > 70°$S: 1% of the total,
territory with 20°N $< \phi <$ 60°N: 25% of the total,

which means that the remaining territory is 64%. It should be mentioned that in the northern hemisphere the Polar Sea was considered as 'ocean'. By averaging the above listed concentrations according to this division one yields 0.19 $\mu g\,m^{-3}$ for SO_2–S and 0.41 $\mu g\,m^{-3}$ for SO_4–S. Thus the SO_2–S and SO_4–S quantities in an air column over the Atlantic, using the above scale heights, are $0.95 \times 10^3\ \mu g\,m^{-2}$ and $0.82 \times 10^3\ \mu g\,m^{-2}$, respectively. Multiplying these figures with the territory of the Atlantic Ocean ($\sim 1.1 \times 10^8$ km²) we receive 0.10×10^6 t SO_2–S and 0.09×10^6 t SO_4–S burden for the atmosphere over the Atlantic Ocean. For the calculation of the total sulfur burden in the atmosphere over the oceans we did not use directly the sulfur quantity from SO_2 in an air column over the Atlantic, in spite of the fact that sulfur concentrations are rather constant over the oceans (see Table 3). For this calculation the high SO_2 concentrations between 20 and 60°N were omitted, since there is no indication for this high sulfur dioxide level over the other oceans. If the SO_2–S concentration for these territories of the Atlantic Ocean were also 0.1 $\mu g\,m^{-3}$ then the average Atlantic SO_2–S level would be 0.094 $\mu g\,m^{-3}$. This latter value gives a $0.48 \times 10^3\ \mu g\,m^{-2}$ SO_2–S quantity in an oceanic air column. Using this latter figure as well as the above value for SO_4–S ($0.82 \times 10^3\ \mu g\,m^{-2}$) the SO_2–S and SO_4–S global burden in the air above all oceans ($\sim 3.6 \times 10^8$ km²) were estimated to be 0.17×10^6 and 0.30×10^6 t, respectively.

Table 4 contains all the values calculated. For comparison the sulfur quantities in the atmospheric reservoir published by Friend (1973) are also given. It can be seen that SO_2–S burdens resulting from these two estimations are in good agreement. In the case of SO_4–S, however, there is a difference of more than three times, which can be explained as follows. Firstly, Friend did not take into consideration the excess sulfate (not sea salt) over the oceans. Secondly, he used a rather small sulfate concentration (1.5 $\mu g\,m^{-3}$) for continental surface air. Thirdly, for the calculation of the scale height he assumed that there is no difference between a model aerosol profile of all particles and that for sulfate particles.

Thus, the present estimation suggests that the global SO_4–S burden in the atmosphere (without the stratospheric sulfate layer) is greater than the SO_2–S loading, since its value is larger than earlier estimates

(Friend, 1973; Granat *et al.*, 1976). It also follows from data in Table 4 even over the oceans there is about two times more excess sulfate–sulfur than sea salt sulfur. These facts must be taken into consideration in further budget calculations.

4. H_2S CONCENTRATION MEASUREMENTS

Up to date, very few H_2S measurements were carried out in remote areas. Thus Lodge and Pate (1966) measured the atmospheric H_2S level in Panama. Except for one isolated sample on the Caribbean coast the concentration was smaller than $1.5\,\mu g\,m^{-3}$. Lodge and his associates (see Breeding *et al.*, 1973) also took air samples for hydrogen sulfide analysis in the central United States. The majority of the measurements gave concentrations between 0.15–0.23 $\mu g\,m^{-3}$. Breeding *et al.* (1973) believe, on the basis of these results, that the background continental concentration of this compound in the air near the ground surface is less than $0.3\,\mu g\,m^{-3}$, probably between 0.075 and $0.2\,\mu g\,m^{-3}$. They also think that the H_2S concentrations over the oceans are not likely to be greater.

Accepting this latter idea, on the basis of the above results a $0.15\,\mu g\,m^{-3}$ global H_2S–S concentration can be estimated for air near the Earth's surface (oceans + continents). Without any information concerning H_2S scale height, all calculations on the H_2S–S burden can be considered as mere speculation. Assuing that the scale height is between 1 and 10 km, it follows from these extreme figures that the global atmospheric H_2S–S burden is probably between 0.075 and 0.75×10^6 t. The upper limit is in an acceptable agreement with the value (0.99×10^6 t) of Friend (1973) based on an average concentration of 0.2 ppbm given by Robinson and Robbins (see Friend, 1973).

It follows from these speculations that much more research is needed in this field. This is particularly true if one takes into consideration that some other reduced sulfur species (e.g. dimethyl sulfide) were proposed (Rasmussen, 1974; Lovelock *et al.*, 1972; Lovelock, 1974) to explain the biological part of the atmospheric sulfur cycle. The concentration of these compounds in the atmosphere is an open question.

5. CONCLUSIONS

The above calculations based on the concentration measurements of sulfur compounds in remote continental and oceanic areas suggest that the global atmospheric sulfur burden is between 1.35×10^6 and 2.00×10^6 t. The uncertainty comes first of all from our insufficient information on the concentration and vertical profile of reduced sulfur species. The global loading of oxidized sulfur is estimated to be 1.25×10^6 t. The larger part of this, about 2/3, is sulfate–sulfur.

Table 4. Global atmospheric sulfur burden over the oceans and continents according to Friend (1973) and to present calculations. All values are expressed in 10^6 t

	SO_2–S		SO_4–S	
	Friend	This paper	Friend	This paper
Oceans	—	0.17	0.09	0.30
Continents	—	0.28	0.16	0.50
Total	0.52	0.45	0.25	0.80

REFERENCES

Altschuller A. P. (1973) Atmospheric sulfur dioxide and sulfate distribution of concentration at urban and non-urban sites in United States. *Envir. Sci. Technol.* **7,** 709–712.

Andreiev B. G. and Lavrinenko R. F. (1968) Some data on the chemical composition of atmospheric aerosols over Central Asia (in Russian). *Meteorologia i Gidrologia* **4,** 63–69.

Beliashova M. A., Petrenchuk O. P. and Selezneva E. S. (1970) Investigation on the physico-chemical and electrical properties of atmospheric aerosols (in Russian). *Proc. Conf. on Cloud Physics and Weather Modification.* Gidromet. Izdat. Leningrad, 84–92.

Bolin B. and Charlson R. J. (1976) On the role of tropospheric sulfur cycle in the shortwave radiative climate of the Earth. *Ambio* **5,** 47–54.

Bónis, K. (1968) A légköri aeroszolban lévő, vizben oldódó anyagokról. *Időjárás* **72,** 104–110.

Breeding R. J., Lodge J. P., Pate J. B., Sheesley D. C., Klonis H. B., Fogle B., Anderson J. A., Englert T. R., Haagenson P. L., McBeth R. B., Morris A. L., Pogue R. and Wartburg A. F. (1973) Background trace gas concentrations in the central United States. *J. geophys. Res.* **78,** 7057–7064.

de Bary E. and Junge C. (1963) Distribution of sulfur and chlorine over Europe. *Tellus* **15,** 370–381.

Delmas R., Baudet J. and Servant J. Etude des sources naturelles de sulfate en milieu tropical humide. To be published in *Tellus.*

Eliassen A. and Saltbones J. (1975) Decay and transformation rates of SO_2, as estimated from emission data, trajectories and measured air concentrations. *Atmospheric Envrionment* **9,** 425–429.

Fisher W. H., Lodge J. P., Pate J. B. and Cadle R. D. (1969) Antarctic atmospheric chemistry: preliminary exploration. *Science* **164,** 66–67.

Flyger H., Heidam N. Z., Hansen K., Megaw W. J., Walther E. G. and Hogan A. W. (1976) The background level of the summer tropospheric aerosol, sulfur dioxide and ozone over Greenland and the North Atlantic Ocean. *J. Aerosol Sci.* **7,** 103–140.

Friend J. P. (1973) The global sulfur cycle. In *Chemistry of the Lower Atmosphere* (Edited by Rasool S. I.) pp. 177–201. Plenum Press, New York.

Georgii H. W. (1975) The ocean as source or sink for SO_2, NH_3 and NO_x. Presented at XVI General Assembly of IUGG, Grenoble.

Georgii H. W. and Vitze W. (1971) Global and regional distribution of sulfur components in the atmosphere. *Időjárás* **75,** 294–299.

Gilette D. A. and Blifford I. H. (1971) Composition of tropospheric aerosols as a function of altitude. *J. atmos. Sci.* **28,** 1199–1210.

Granat L., Rodhe H. and Hallberg R. O. (1976) The global sulphur cycle. *Ecol. Bull.* **22,** 89–134.

Gravenhorst G. (1975) Der Sulfatanteil im atmosphärischen Aerosol über dem Nordatlantik. *Berichte des Instituts für Meteorologie un Geophysik der Universität Frankfurt/Main. No. 30.*

Hitchcock D. R. (1976) Atmospheric sulfates from biological sources. *J. Air Pollut. Control Ass.* **26,** 210–215.

Jost D. (1974) Aerological studies on the atmospheric sulfur budget. *Tellus* **26**, 206–212.

Kellogg W. W., Cadle R. D., Allen E. R., Lazrus A. L. and Martell E. A. (1972) The sulfur cycle. *Science* **175**, 587–596.

Khusanov G. K., Petrenchuk O. P. and Drozdova V. M. (1974) Chemical composition of atmospheric aerosols over some territories of Central Asia (in Russian). *Proc. Main Geophysical Observatory* (Leningrad). *No. 314*, 192–200.

Lodge J. P., MacDonald A. J. and Vihman E. (1960) A study of the composition of marine atmosphere. *Tellus* **12**, 184–187.

Lodge J. P. and Pate J. B. (1966) Atmospheric gases and particulates in Panama. *Science* **153**, 408–410.

Lodge J. P., Machado P. A., Pate J. B., Sheesley D. C. and Wartburg A. F. (1974) Atmospheric trace chemistry in the American humid tropics. *Tellus* **26**, 250–253.

Lovelock J. E. (1974) CS_2 and the natural sulfur cycle. *Nature* **248**, 625–626.

Lovelock J. E., Maggs R. J. and Rasmussen R. A. (1972) Atmospheric dimethyl sulphide and the natural sulphur cycle. *Nature* **237**,

Mészáros Á. and Vissy K. (1974) Concentration, size distribution and chemical nature of atmospheric aerosol particles in remote oceanic areas. *J. Aerosol Sci.* **5**, 101–110.

Mészáros E. and Várhelyi G. (1975) On the concentration, size distribution and residence time of sulfate particles in the lower troposphere. *Időjárás* **79**, 267–273.

Nguyen Ba Cuong, Bonsang B., Pasquier J. L. and Lambert G. (1974a) Composantes marine et africaine des aérosols de sulfates dans l'hémisphère sud. *J. Rechs atmos.* **8**, 831–844.

Nguyen Ba Cuong, Bonsang B. and Lambert G. (1974b) The atmospheric concentration of sulfur dioxide and sulfate aerosols over antarctic, subantarctic areas and oceans. *Tellus* **26**, 241–249.

Rodhe H. (1972) Measurements of sulfur in the free atmosphere over Sweden 1969–1970. *J. geophys. Res.* **77**, 4494–4499.

Petrenchuk O. P. (1971) Some data on sulphur dioxide content in the boundary layer of the atmosphere. *Időjárás* **75**, 300–302.

Prahm L. P., Torp U. and Stern R. M. (1976) Deposition and transformation rates of sulfur oxides during atmospheric transport over the Atlantic. *Tellus* **28**, 355–372.

Rasmussen R. A. (1974) Emission of biogenic hydrogen sulfide. *Tellus* **26**, 254–260.

WMO (1974) Operations manual for sampling and analysis techniques for chemical constituents in air and precipitation. *World met. Org.*, Geneva, *No. 299*.

Woodcock A. H. (1953) Salt nuclei in marine air as a function of altitude and wind force. *J. Met.* **10**, 362–371.

Atmospheric Environment Vol. 12, pp. 707–713. Pergamon Press 1978. Printed in Great Britain.

MARITIME SULFATE OVER THE NORTH ATLANTIC

GODE GRAVENHORST*

Institut für Meteorologie und Geophysik der Universität Frankfurt, D 6000 Frankfurt, W. Germany

(*First received* 8 *August* 1977 *and in final form* 12 *October* 1977)

Abstract—Laboratory and field experiments were undertaken to determine sulfate sources in the aerosol over the North Atlantic. In artificially produced sea salt aerosol no fractionation between sodium and sulfate was found. This result is supported by a calculation of Gibbs' surface adsorption which showed a negligible enrichment of sodium compared to sulfate in the surface film of seawater. In the maritime aerosol, however, *ca.* three times as much sulfate was found as would be calculated from the seawater ratio and concentrations. Sulfur isotope ratios in the aerosol were smaller than the seawater value. Half of the excess sulfate and almost all ammonium belonged to particles smaller than 0.45 μm. In this size range the aerosol was acidic. Based on these findings it is suggested that most of the maritime excess sulfate near the sea surface is not produced by ion fractionation in the surface film of seawater nor by interactions with cloud droplets but by gas phase reactions.

INTRODUCTION

Sulfate is one of the main components of atmospheric aerosols in remote areas. Over the ocean its concentration was attributed to the sea salt component (Junge, 1960). The sea salt fraction, however, appeared to be smaller, so that the additional amount of sulfate was assumed to result from ion fractionation at the sea surface (Bruyevich and Kulik, 1967; Bruyevich and Korzh, 1970) or was interpreted as anthropogenic sulfate (Koide and Goldberg, 1971; Weiss *et al.*, 1975). Investigations on metal concentrations in maritime aerosol samples revealed that these components are not enriched in the sea salt aerosol with respect to seawater but that higher concentrations could be caused by advection from the continent (Gravenhorst and Jendricke, 1974; Hoffman *et al.*, 1974). The excess sulfate can therefore be transported from other sources or be formed over the ocean within the atmosphere. Field and laboratory measurements were therefore undertaken to characterize the sulfate component over the North Atlantic and to explain possible deviations of the aerosol composition from sea water by comparison with other aerosol components. Field measurements were performed on board of R. V. '*Meteor*' during cruise 32 starting from Hamburg and returning via Santo Domingo, Fort de France and Las Palmas (Fig. 1). On the Canary Islands and in the Bay of Biscay concentration profiles of sulfate

* Present affiliation: Institut für Atmosphärische Chemie, Kernforschungsanlage Jülich, D 5170, Jülich.

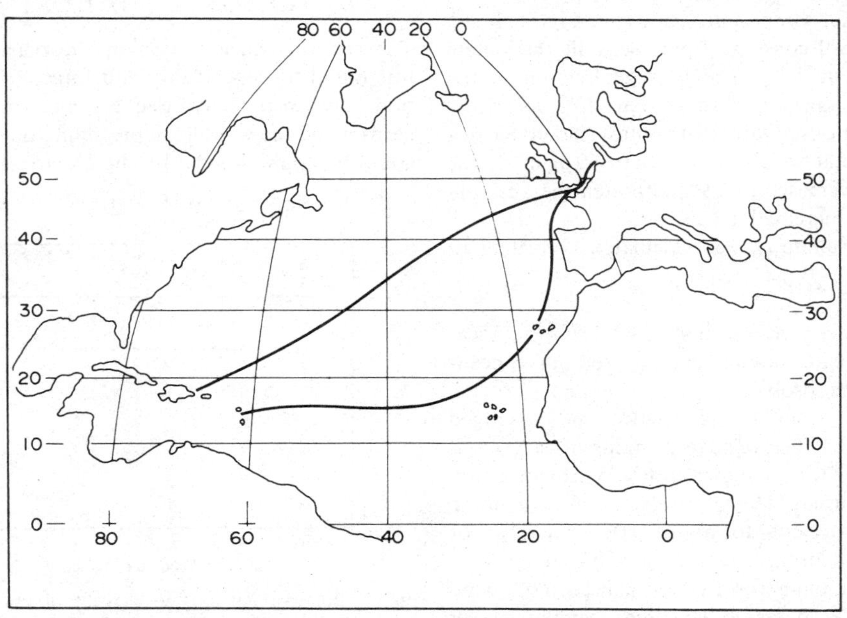

Fig. 1. Route of R. V. '*Meteor*' on cruise 32. Airborne Sahara dust was sampled between 40°W16°N and 20°W24°N.

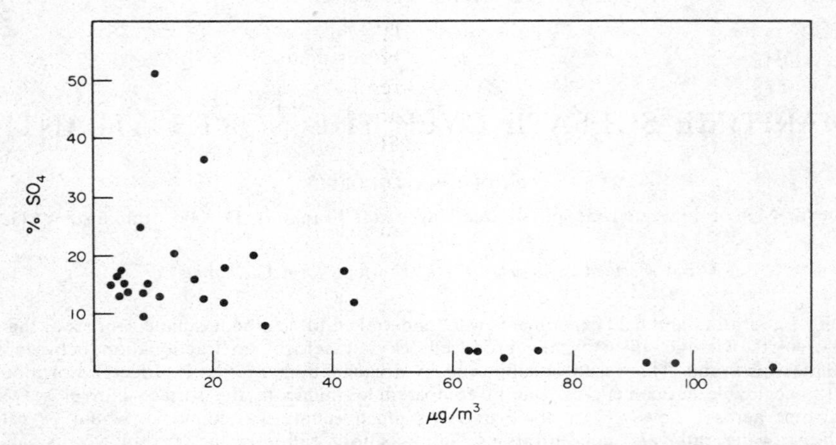

Fig. 2. Relative amount of sulfate mass in the total aerosol. Samples with a total mass larger than 60 $\mu g\ m^{-3}$ were collected in trade winds from the Sahara.

containing particles were measured by aircraft ascends.

TOTAL SULFATE MASS IN THE AEROSOL

The total maritime aerosol was collected nearly isokinetically at *ca.* 15 m a.s.l. by high volume samplers (28 $m^3\ h^{-1}$) on microsorban filter (Delbag, 98/99 S) for chemical analysis, and on glasfibre filter (Schleicher-Schüll) for weight and sulfur isotope analysis. Sulfate was determined by isotopic dilution (Klockow *et al.*, 1974), ammonium by an indophenol-blue method (Georgii *et al.*, 1973) and sodium by means of atomic emission (Perkin Elmer 305).

The average weight at about 45% r.h. of the maritime aerosol was 9.4 $\mu g\ m^{-3}$ for background conditions and 74.7 $\mu g\ m^{-3}$ under the influence of Sahara dust. The sulfate concentration averaged 1.2 and 2.8 $\mu g\ m^{-3}$, respectively. The sulfate mass fractions of the total aerosol samples are shown in Fig. 2. If the maritime aerosol consists of pure sea salt the sulfate should represent 7.7%. The average fraction in the total aerosol, however, amounted to 13% for background conditions. Only in the dust laden Sahara trade wind (total weight higher than 60 $\mu g\ m^{-3}$) the sulfate mass decreased to 2.9% although the absolute concentration increased 1.8 times.

The excess sulfate c_0 was calculated according to

$$c_0 = C - (SO_4^{2-}/Na^+)_{sea} \times Na^+_{aerosol}. \quad (1)$$

(C total sulfate concentration, $(SO_4^{2-}/Na^+)_{sea}$ seawater ratio sulfate/sodium, $Na^+_{aerosol}$ sodium concentration in the aerosol).

The excess sulfate concentration averaged $0.9 \pm 0.5\ \mu g\ m^{-3}$. The relative portions of excess sulfate in total sulfate are shown in Fig. 3. For measurements made during '*Meteor*' cruise 23 (Gravenhorst, 1975a) they amount to about 80% whereas for samples taken during cruise 32 a value of *ca.* 65% was found. For the aerosol in the Sahara trade wind the total sulfate increased 1.8 times. Similar higher values in Sahara air were found by Büchen and Georgii (1971). However, no larger amount of excess sulfate was calculated. It can be explained by the fact that the concentration of water soluble sodium increased from 2.0 to 4.3 $\mu g\ m^{-3}$.

The wind velocity was not significantly higher in the Sahara trade wind than during the other sampling times (7.0 $m\ s^{-1}$ compared to 6.6 $m\ s^{-1}$). The sodium concentration from sea water should therefore not be higher. Under this assumption the total water soluble sodium in the Sahara air consists of two parts: the sea salt sodium (*ca.* 2.0 $\mu g\ m^{-3}$) and the non-sea salt sodium (*ca.* 2.3 $\mu g\ m^{-3}$). The true excess sulfate in the Sahara air (the non-sea salt sulfate) of *ca.* 2.3 $\mu g\ m^{-3}$ is 3.3 times higher than in air not directly influenced by continental advection.

Concentrations of maritime aerosol components in excess of the sea salt fraction are often characterized with an enrichment factor E defined for sulfate as

$$E = (SO_4^{2-}/Na^+)_{aerosol}/(SO_4^{2-}/Na^+)_{sea} - 1. \quad (2)$$

The sodium concentration in the maritime aerosol was found to depend on wind force V. The correlation between the measured Na-concentration in the aerosol and the wind force prevailing during sampling could be approximated by the empirical relation:

$$Na^+_{aerosol} = 0.31 \times 10^{-0.15 + 0.25 V} \quad (3)$$

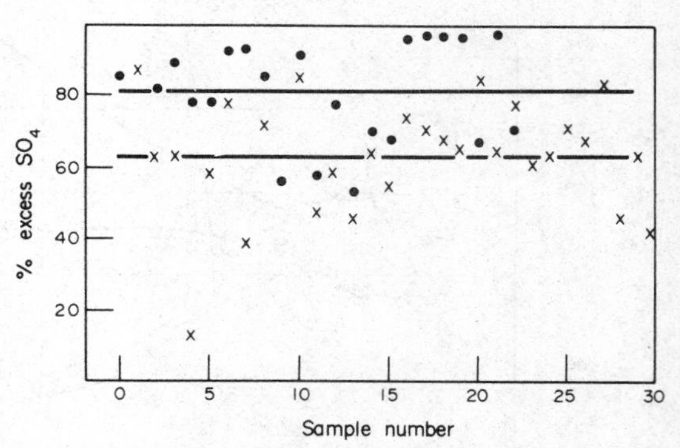

Fig. 3. Relative portions in percent of excess sulfate in total aerosol sulfate. Circles '*Meteor*' cruise 32; crosses '*Meteor*' cruise 23 (Gravenhorst, 1975a).

(Na^+ in g m^{-3}, V is the wind speed on the Beaufort scale). Assuming in Equation (2) the SO_4^{2-}–Na^+ seawater ratio to be 0.25 and inserting for the total sulfate concentration the expression for C in Equation (1) and for $Na^+_{aerosol}$ Equation (3), E can be calculated for constant excess sulfate c_0 as a function of V:

$$E = c_0 \times 13 \times 10^{0.15 - 0.25V}. \quad (4)$$

For constant c_0, E can vary by three orders of magnitude depending on wind force. It is therefore not useful to describe non-sea salt components in maritime aerosols with E. Parallel trends of E for different components may be caused by changing V and do not inevitably reflect a causal relationship between aerosol sources.

SULFATE–SODIUM FRACTIONATION

The sulfate–sodium concentration ratio in the surface film of seawater was calculated according to Gibbs' surface adsorption (Gravenhorst, 1975b).

The enrichment factor E for sulfate in the surface film was calculated as a function of the film thickness from which the sea salt aerosol may be generated. No increase of sulfate but rather an enrichment of sodium was found. The E values decrease from -5×10^{-5} for a film thickness of 0.1 μm to -5×10^{-3} for a thickness of 10^{-3} μm. The film thickness from which sea salt particles are formed is estimated to be *ca.* 1 μm (MacIntyre, 1972). This calculated small decrease of the sulfate–sodium ratio in the seawater surface film compared to bulk seawater can hardly be detected in atmospheric aerosol measurements. The structure of the ocean surface film, however, may be influenced by a variety of organic substances concentrated in the microlayer (Duce and Hoffman, 1976).

In laboratory experiments (Gravenhorst, 1975b) it was therefore attempted to find a possible separation of sulfate and sodium ions in sea salt aerosols. From artificial seawater an aerosol was produced having a similar size distribution as marine sea salt aerosol. Concentration ratios between sulfate and sodium in the aerosol and in the bulk seawater showed no significant difference (Fig. 4) that could explain high sulfate enrichment. An analytical precision of 10% for both sulfate and sodium concentration measurements results already in sulfate–sodium ratio between 0.22 and 0.28. Thus no indication could be found that SO_4^{2-} and Na^+ ions are fractionated at the seawater–atmosphere interface to such extent that this difference could be detected in maritime aerosol measurements.

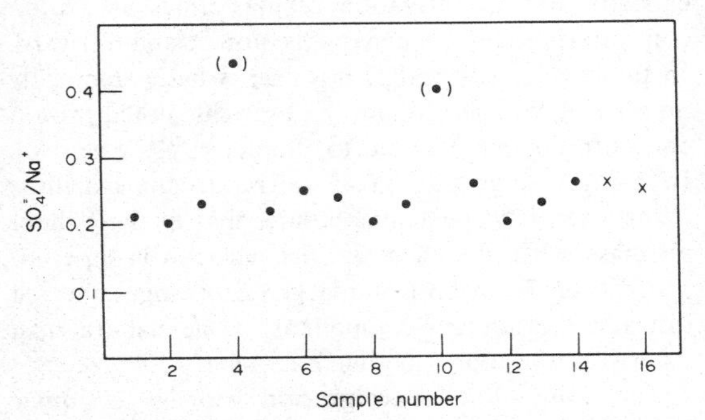

Fig. 4. Sulfate–sodium ratios of aerosols (radius > 0.8 μm) generated with artificial seawater. Measurements of the seawater ratio are marked with crosses.

SULFUR ISOTOPE RATIOS

It is probable that in model calculation and laboratory measurements real situations are not simulated. Therefore, a property of the sampled maritime aerosol was measured which allows us to differentiate between sea salt sulfate and other sources. The isotopic ratio of $^{34}S/^{32}S$ in sea water has a rather uniform and specific value (Nielsen, 1974). Any deviation from it in atmospheric maritime aerosols is a strong indication for sulfur sources other than sea salt since during particle formation no major isotopic fractionation occurs (Lücke and Nielsen, 1972). The isotopic sulfur ratios (calculated in the commonly used $\delta^{34}S$ notation) were found to be considerably lower in the North Atlantic aerosol than in seawater. This is in agreement with analysis of rainwater samples (Östlund, 1959; Mizutani and Rafter, 1971). The $\delta^{34}S$ values for the excess sulfur were determined by considering the contribution of the sea salt sulfur. Using the measured sodium concentration to substract the sea salt sulfur, the following $\delta^{34}S$ values were calculated (the $\delta^{34}S$ values for total sulfur are in parentheses): +7‰ and +9‰ (+10‰, 12‰) for maritime aerosol influenced directly by continents, −12‰ and +10‰ (+2‰, +14‰) for aerosol over the southwest part of the North Atlantic; +11‰ and +13‰ (+14‰, +15‰) for the maritime Sahara aerosol and +11‰ (+13‰) for aerosol samples taken on a frequently passed sea way. Besides the one negative value they are higher than the $\delta^{34}S$ values in New Zealand rain water for non-sea salt sulfur which was depleted by *ca.* 20‰ with respect to seawater (Mizutani and Rafter, 1971). The average $\delta^{34}S$ value of 2.6‰ for sulfate in the stratosphere of both hemispheres (Castleman *et al.*, 1974) is lower than for the excess sulfur in maritime aerosol samples over the North Atlantic (neglecting the −12‰ value). A transport of the maritime sulfate from the tropical North Atlantic into the stratosphere via upwelling motions within the innertropical convergence zone is not supported by the isotope measurements. In addition the ammonium concentrations found in stratospheric aerosol samples are too low (Lazrus *et al.*, 1971) to explain aerosol transport from the marine boundary layer.

MASS DISTRIBUTION OF MARITIME SULFATE

The formation of excess sulfate can be more easily understood when its mass distribution over the aerosol size range is known. The size distribution of sulfate containing particles showed an increase in concentration down to radii of *ca.* 0.2 μm radius (Georgii and Gravenhorst, 1972). A transformation into a mass distribution revealed that a large fraction of the excess sulfate could be attributed to particle sizes smaller than the sea salt aerosol (Gravenhorst, 1975a). Chemical and morphological analysis, too, indicate that sulphur is a main component in the lower size range of large particles, 1.0 μm $> r >$ 0.1 μm (Mészáros and Vissy, 1974; Winkler, 1974). The mass distributions for sea salt sulfate and excess sulfate were therefore measured by means of a three stage impactor. The mass distribution deduced from number concentration measurements of sulfate containing particles (Gravenhorst, 1975a) and the impactor measurements were used to determine an average cumulative mass distribution for the excess sulfate over the North Atlantic (Fig. 5). The 50% radius of 0.4 μm for the excess maritime sulfate resembles much more closely the continental sulfate than the sea salt sulfate having a 50% radius of *ca.* 2.5 μm. 50% of the excess sulfate can be attributed to the size range below the sea salt aerosol. This excess sulfate should therefore not originate in seawater.

Its mass distribution per logarithmic radius interval (Fig. 6) resembles in its shape the distribution for the continental sulfate. The peak value is only slightly shifted to a larger radius and the slope in the giant particle range is not so steep. The first effect could be caused by lower concentrations of gas phase precursors so that less primary particles are produced and the second effect may reflect the influence of the sea salt particles. The principal similarity between both distributions could indicate that the essential formation mechanisms for continental sulfate and maritime excess sulfate may be the same.

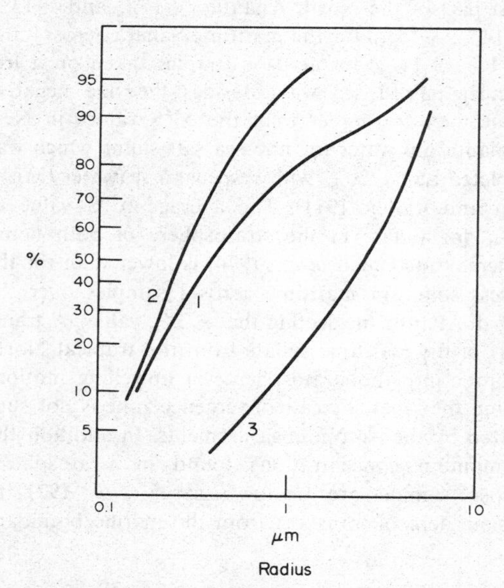

Fig. 5. Cumulative mass distribution as a function of particle size for different sulfate aerosols: (1) maritime excess sulfate over the North Atlantic, (2) sulfate of continental aerosols in the lower troposphere over Germany (after Georgii *et al.*, 1971), (3) sea salt sulfate.

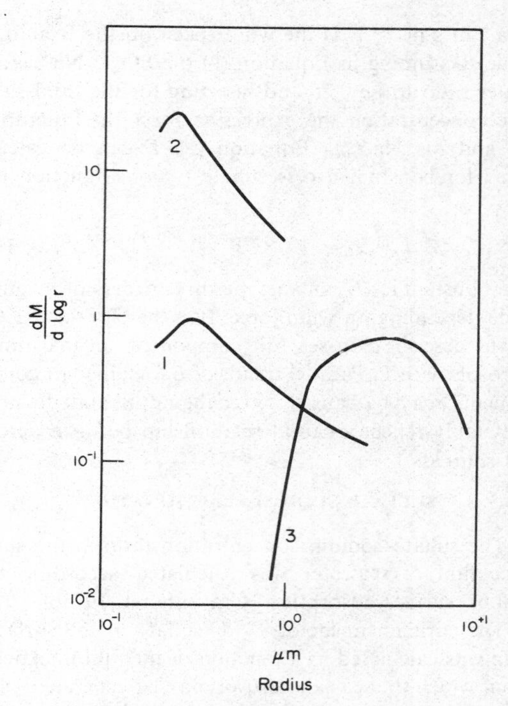

Fig. 6. The differential distribution of sulfate mass M (μg m^{-3}) as a function of particle radius for different sulfate aerosols: (1) maritime excess sulfate over the North Atlantic, (2) sulfate of continental aerosols in the lower troposphere over Germany (deduced from data given by Georgii *et al.*, 1971), (3) sea salt sulfate.

AMMONIUM AND PROTON CONCENTRATIONS

Possible counter ions of excess sulfate in maritime aerosol samples are ammonium ions and protons. Their concentrations in the impactor samples were therefore determined. pH values of aerosol–water solutions were measured and under consideration of blank samples the proton concentration calculated. Since the pH value of the aerosol solution depends on the amount of aerosol dissolved and its buffer capacity, the interpretation of pH values as proton concentration in the aerosol is not straightforward. In the smaller size range, however, where practically no sea salt could be found, the thus calculated proton concentration can be used to characterize the aerosol. In the size range below 0.45 μm radius the maritime aerosol seems to be acidic, besides that in the Sahara air mass, when the effect of dust particles is superimposed (Fig. 7). In both the larger size ranges the sea salt and the mineral component of the Sahara dust determine the more alkaline character.

The ammonium concentration showed a similar dependence on particle size (Fig. 8). Nearly all of the total ammonium was found in the smallest size range. For this size fraction the simultaneously measured

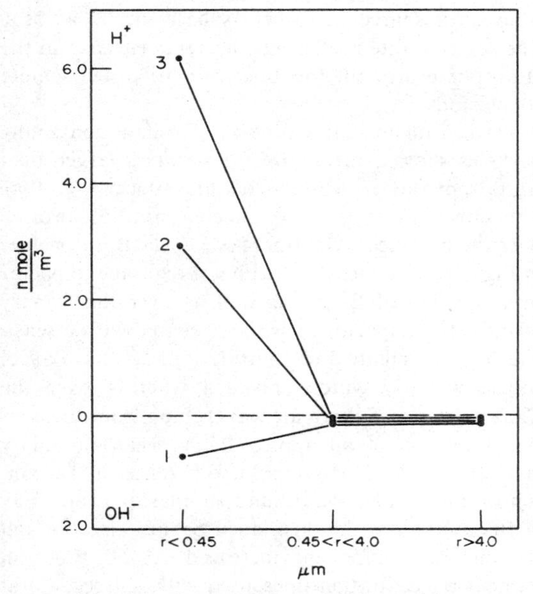

Fig. 7. The H^+ or OH^- concentration in maritime aerosol samples reduced to 1 m^3 of air for three aerosol size ranges and different air masses: (1) Sahara trade wind, (2) 'background' maritime air, (3) maritime air with direct influence from the continent.

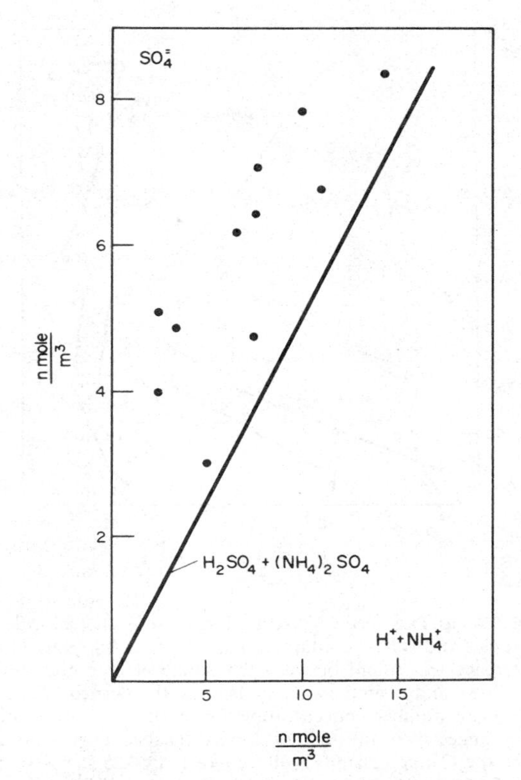

Fig. 9. The excess sulfate concentration in the aerosol size range $r < 0.45$ μm as a function of the sum of ammonium ion and proton concentration (n-mole $\triangleq 10^{-9}$ mole).

concentrations for both NH_4^+ and H^+ are plotted in relation to excess sulfate (Fig. 9). In this size range the NH_4^+ as well as the H^+ concentrations increase with increasing concentration of excess sulfate. The sum of the NH_4^+ and H^+ equivalents nearly compensates sulfate. The calculated remaining difference may be caused by a small fraction of alkaline components which can still be present in this size range. Nitrate does not seem to contribute as HNO_3 to the acidity in a significant amount. In the aerosol range below 0.45 μm radius its concentration was smaller than 0.005 $\mu g\ m^{-3}$. For particle sizes between 0.45 and 4.0 μm radius the nitrate concentration was *ca.* 0.05 $\mu g\ m^{-3}$. Those findings confirm the data of Junge (1963) that nitrate is mainly associated with the sea salt fraction of the maritime aerosol. It seems therefore possible to interpret the excess sulfate in the size range below 0.45 μm radius as a mixture of ammonium sulfate and sulfuric acid both contributing about the same portion. For particles with larger radii than 0.45 μm less than 10^{-3} $\mu g\ m^{-3}$ ammonium could be detected. It seems therefore possible that this fraction of the excess sulfate has a different composition than the one below 0.45 μm radius. Assuming that the excess sulfate in the large size range is not associated with protons, the weight fractions of sulfate each in the form of sulfuric acid and ammonium sulfate amount to *ca.* 10–15% with respect to total sulfate and *ca.* 25% with respect to excess sulfate.

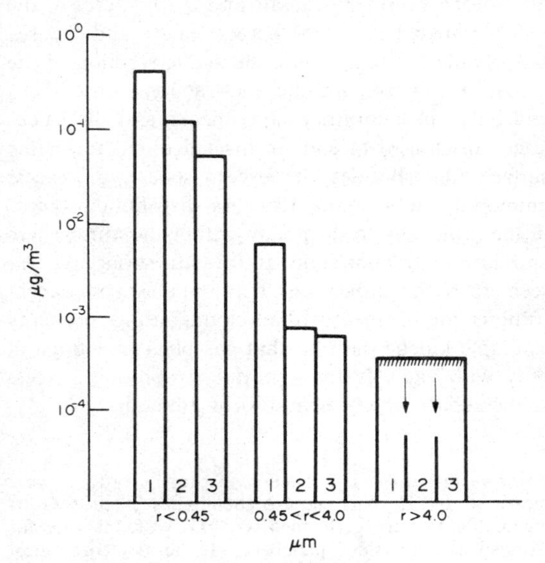

Fig. 8. Ammonium concentration in maritime aerosol samples for three particle size ranges and different air masses: (1) maritime air with direct influence from the continent, (2) Sahara air, (3) 'background' maritime air. In the size range $r > 4.0$ μm the NH_4^+ concentrations were close to the detection limit or lower.

From the bulk analyses of particles larger than 0.45 μm radius it could not be concluded whether the excess sulfate is combined with the sea salt or forms separate particles. Since there seems to be practically no ammonium associated with the excess sulfate in this range, two formation mechanisms may be possible: either sulfur dioxide is directly absorbed by sea salt particles or H_2SO_4 molecules or clusters are attached to the sea salt before they can absorb

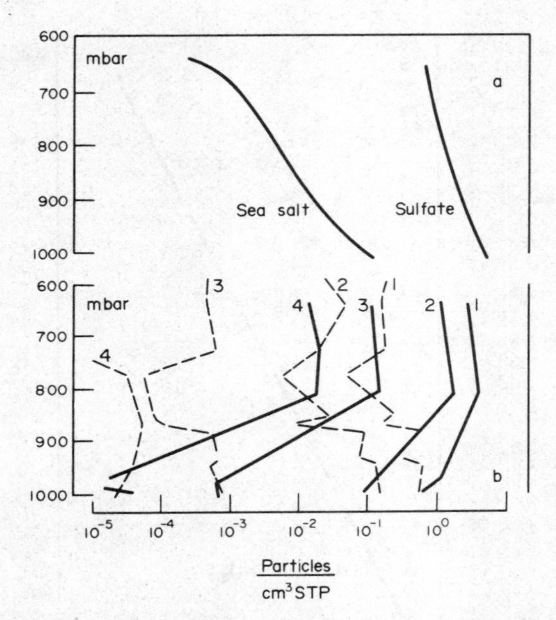

Fig. 10. (a) The vertical distribution of sea salt (chloride) particles and sulfate containing particles ($r > 0.15$ μm). No difference was found between the situation over the Bay of Biscay and over the Canary Islands. (b) Vertical distribution of number concentration for particles in different size ranges over the Bay of Biscay (dashed curves) and over the Canary Islands (solid curves): 1: $0.25 < r < 0.7$; 2: $0.7 < r < 1.5$; 3: $1.5 < r < 2.5$; 4: $r < 2.5$ (r = particle radius in μm).

ammonia. In both cases there should be no separate sulfate particle population besides sea salt to which excess sulfate is attached. That would, however, mean that the coagulation process of ammonium and sulfate-containing particles with the sea salt fraction is not very effective or, because of the high pH value of the sea salt particles, NH_4^+ is liberated from the liquid phase into the gas phase as ammonia when sea salt particles and ammonium sulfate–sulfuric acid particles coagulate.

The acidity of the aerosol below 0.45 μm radius suggests that the particles are not in equilibrium with gas phase ammonia. The ammonia concentration should be in the range of 0.1–1 ppb according to measurements and calculations of the equilibrium pressure over the ocean (Georgii and Gravenhorst, in press). The chemical composition of the aerosol therefore indicates that particles do not represent an aged aerosol but rather a dynamical system in which the acid formed cannot be neutralized by ammonia absorption.

Sulfate containing particles could, however, be formed by reactions of SO_2 with liquid water in clouds. The smaller size range of the maritime aerosol (< 0.45 μm) seems to be too acidic to absorb, according to a model of Beilke and Barrie (1974), under maritime conditions enough SO_2 to produce the measured sulfate. The sulfate–ammonium ratio in the larger size ranges on the other hand appears to be too large to support SO_2/NH_3 droplet phase reactions as a source for excess sulfate in the aerosol. The reaction rate coefficients, however effective in the atmosphere are still too uncertain to draw definite conclusions.

Vertical mixing ratio profiles of sulfate containing particles show a maximum for particles larger than *ca.* 0.1 μm radius near the ocean surface (Fig. 10a). The same feature is shown in concentration profiles over the continent (Georgii *et al.*, 1971). Both profiles in Fig. 10a were measured with a three stage impactor over the Bay of Biscay as well as over the Canary Islands. The impactor plates were coated with a sensitive film for sulfate and chloride ions. In the Bay of Biscay westerly winds prevailed whereas over the Canary Islands, Sahara air was drifting from the continent on top of an approx 2 km ocean boundary layer (Fig. 10b). No systematic difference in the concentration profiles was found in maritime air (Bay of Biscay) and in the dust-laden Sahara air although the mineral component increased *ca.* 50 times in number concentration (measured with Royco 245) at 800 mb. Because of the maximum concentration of sulfate containing particles near the ground gas phase reactions of SO_2 seem to be a more likely source of maritime excess sulfate.

CONCLUSION

In the maritime aerosol over the North Atlantic a higher sulfate mass fraction is found than is expected from sea salt concentration. Its concentration is rather constant with values averaging between 0.4 and 1.4 $\mu g\, m^{-3}$. The excess sulfate does not enter into the atmosphere as sea salt. Some aspects suggest this view: the difference between sulfur isotope ratios in sea salt and in the aerosol, the separation of half of the excess sulfate and the sea salt in different size ranges, the acidic reaction of the aerosol below 0.45 μm, the non-fractionation of Na^+ and SO_4^{2-} in laboratory experiments, and the calculated enrichment of sodium instead of sulfate in the surface film. Besides the excess sulfate, an excess ammonium was found. Its mass distribution points in the same way to formation within the atmosphere and not to fractionation at the sea surface. There seem to be no indications that both excess concentrations are formed within clouds using presently available kinetic data, so that gas phase reactions of SO_2 with radicals may prevail to produce excess acidic sulfate which then absorbs ammonia.

Acknowledgements—The sulfur isotope ratios were measured by H. Nielsen, Geochemisches Institut Göttingen; flights were performed with DFVLR-Institute for Atmospheric Physics; members of the Meteorological Institute in Frankfurt gave support during laboratory and field measurements. I wish to thank them for their helpful assistance.

This work received partial funding through Deutsche Forschungsgemeinschaft, Sonderforschungsbereich 73 'Atmospheric Trace Gases'.

REFERENCES

Beilke S. and Barrie L. (1974) On the role of NH_3 in heterogeneous SO_2 oxidation. Presented at Sulfur Symposium, Ispra/Italy, by S. Beilke, Umweltbundesamt, Pilotstation D 6000 Frankfurt, Feldbergstr. 47.

Bruyewich S. V. and Kulik Y. Z. (1967) Changes of principal salt constituents as it passes into the atmosphere. *Dokl. Akad. Nauk SSSR* **175**, 190–192.

Bruyevich S. V. and Korzh V. C. (1970) Main patterns of salt exchange between the ocean and the air. *Dokl. Akad. Nauk SSSR* **190**, 208–212.

Büchen M. and Georgii H. W. (1971) Ein Beitrag zum atmosphärischen Schwefelhaushalt über dem Nordatlantik '*Meteor*' Forsch. *Ergebnisse* B. **7**, 71–77.

Castleman A. W., Munkelwitz H. R. and Manowitz B. (1974) Isotopic studies of the sulfur component of the stratosphere aerosol layer. *Tellus* **26**, 222–234.

Duce R. and Hoffman E. J. (1976) Chemical fractionation at the air/sea interface. *Ann. Rev. Earth planet. Sci.* **4**, 187–228.

Georgii H. W. and Gravenhorst G. (1972) Untersuchungen zur Konstitution des Aerosols über dem Atlantischen Ozean. *Met. Rdsch.* **25**, 180–181.

Georgii H. W. and Gravenhorst G. The ocean as source or sink of reactive trace gases. *Pure appl. Geophys.* (in press).

Georgii H. W., Jost D. and Müller W. J. (1973) Erprobung eines Verfahrens zur Messung von NH_3 und NH_4^+ im ppb-Bereich, Ber. d. Inst. für Met. u. Geophys. der Universität Frankfurt, Nr. 25.

Georgii H. W., Jost D. and Vitze W. (1971) Konzentration und Größenverteilung des Sulfataerosols in der unteren und mittleren Troposhäre, Berichte des Inst. f. Met. u. Geophys. der Universität Frankfurt, Nr. 23.

Gravenhorst G. (1975a) The sulfate component in aerosol samples over the North Atlantic. '*Meteor*' Forsch. *Ergebnisse* **B. 10**, 22–31.

Gravenhorst G. (1975b) Der Sulfatanteil im atmosphärischen Aerosol über dem Nordatlantik, Berichte des Inst. f. Met. und Geophys. der Universität Frankfurt, Nr. 30.

Gravenhorst G. and Jendricke U. (1973) Konzentrationsverhältnisse von Aerosolkomponenten über dem Nordatlantik, '*Meteor*' Forsch. *Ergebnisse* B. **9**, 67–77.

Hoffman E. J., Hoffman G. L. and Duce R. (1974) Chemical fractionation of alkali and alkaline earth metals in atmospheric particulate matter over the North Atlantic. *J. Rech. atmos.* **8**, 675–688.

Junge C. E. (1960) Sulfur in the atmosphere. *J. geophys. Res.* **65**, 227–237.

Junge C. E. (1963) *Air Chemistry and Radioactivity*. Academic Press, New York, pp. 171–172 and 177.

Klockow D., Denzinger H. and Rönicke G. (1974) Anwendung der substöchiometrischen Isotopenverdünnungsanalyse auf die Bestimmung von atmosphärischem Sulfat und Chlorid in "Background" Luft. *Chemie-Ing. Techn.* **46**(19), 831.

Koide M. and Goldberg E. D. (1971) Atmospheric sulfur and fossil fuel combustion. *J. geophys. Res.* **76**, 6589–6596.

Lazrus A. L., Gandrub B. and Cadle R. D. (1971) Chemical composition of air filtration samples of the stratospheric sulfate layer. *J. geophys. Res.* **76**, 8083–8088.

Lücke W. and Nielsen H. (1972) Isotopenfraktionierung des Schwefels im Blasensprüh. *Fortsch. Mineral.* (Beiheft 3), 36–37.

MacIntyre F. (1972) Flow patterns in breaking bubbles. *J. geophys. Res.* **77**, 5211–5228.

Mészáros A. and Vissy K. (1974) Concentration, size distribution and chemical nature of atmospheric aerosol particles in remote oceanic areas. *J. Aerosol Sci.* **5**, 101–109.

Nielson H. (1974) Isotopic composition of the major contributors to atmospheric sulfur. *Tellus* **26**, 213–221.

Östlund G. (1959) Isotopic composition of sulfur in precipitation and sea water. *Tellus* **11**, 478–480.

Weiss H. V., Bertine K., Koide M. and Goldberg E. D. (1975) The chemical composition of a Greenland glacier. *Geochim. Cosmochim. Acta* **39**, 1–10.

Winkler P. (1974) Chemical analyses of Aitken particles ($< 0.2\ \mu m$ radius) over the Atlantic Ocean. *Geophys. Res. Lett.* **2**, 45–48.

Atmospheric Environment Vol. 12, pp. 715–721. Pergamon Press 1978. Printed in Great Britain.

NEW METHODS FOR THE ANALYSIS OF SO_2 AND H_2S IN REMOTE AREAS AND THEIR APPLICATION TO THE ATMOSPHERE

WOLFGANG JAESCHKE
Department of Meteorology and Geophysics, University of Frankfurt/Main, W. Germany

(*Received in final form* 20 *September* 1977)

Abstract—New techniques have been developed and examined for measuring H_2S and SO_2 in the atmosphere. The calibration curves of the methods are reported and the detection limits are discussed by considering the deviation of the blank values. The atmospheric detection limits are demonstrated to be 0.01 $\mu g\ H_2S\ m^{-3}$ and 0.03 $\mu g\ SO_2\ m^{-3}$. Results of test measurements dealing with the reproducibility of the measured values and the sampling efficiency are reported. A short summary of some results which have been obtained by applying the methods to the atmosphere is given.

INTRODUCTION

During the last decade several attempts have been made to describe the global sulfur cycle (Georgii, 1975; Junge, 1972). One thing all these considerations, discussed by different authors, have in common is the uncertainty in estimating the amounts of natural gaseous sulfur compounds released from the surface of the earth. In order to achieve a balance in the sulfur budgets discussed, large parts of the surface of the earth suggest themselves as natural sources of H_2S and perhaps SO_2. However, the atmospheric H_2S and SO_2 concentrations which were assumed to be produced by natural sources are too small to be established by techniques previously available. For instance, the oceans have been proposed to be sources of H_2S, but little is known about the source strength of open seas, continental surface water or tidal flats.

Therefore, new analytical methods and fresh ways of handling the sampling technique have been developed to detect the trace gases in the sub-ppb range (Jaeschke, 1977). Preliminary results obtained with these new methods have been published (Jaeschke *et al.*, 1976; Georgii, 1978), without, however, describing the sampling technique, the effects on which the analysis is based, and the calibration procedure, in detail. This is now assayed in the present paper. Furthermore, some results are reported in order to show the efficiency of these methods when applied in the atmosphere.

DESCRIPTION OF METHODS

Sampling method

In view of the low concentration levels which are to be detected, it has been necessary to concentrate the analysis to a high degree. For this purpose an air sampling device has been developed, which may be generally applied for the sampling of all atmospheric trace gases of interest. Instead of wet scrubbing, chemically impregnated filters are used. Uncontrolled contamination during the preparation and transport of the impregnated filter is avoided by a new way of handling. The apparatus shown in Fig. 1 makes it possible to carry out the critical steps of impregnation and washing out under field conditions in ambient air.

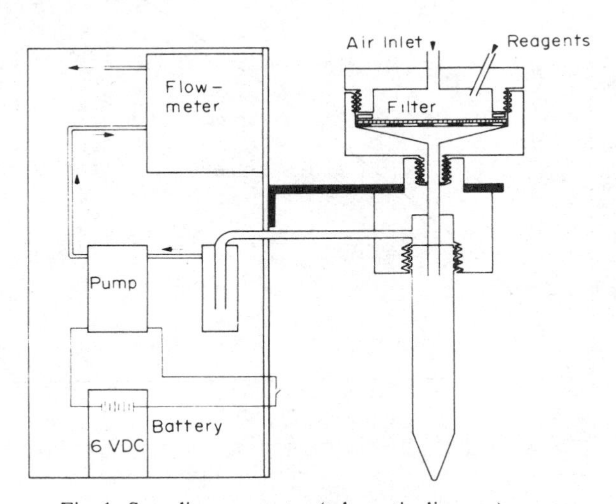

Fig. 1. Sampling apparatus (schematic diagram).

A pump powered by a battery with a flowmeter at its air outlet is placed in a 19 in frame. The air inlet of the pump is connected with the so-called "sampling head" on the front panel. The sampling head consists of a Teflon filter holder which is placed on the top of a Teflon connection. On the bottom of this connection, a sampling tube is fixed with a screw thread. The filter material (Delbag Microsorban 98) is placed on a Teflon mesh in the filter nolder. The filter is impregnated immediately before taking the air sample.

In the case of an H_2S sampling, 5 ml of 0.01 N nitric acid (HNO_3) containing 2% $AgNO_3$, 20% ethanol and 2% glycerol are sprayed on the filter surface according to the method of Axelrod *et al.* (1972). When the entire filter is covered with liquid, the solution is sucked through the filter and captured in the sample tube by a short operation of the pump. In order to make sure that the whole filter is soaked with $AgNO_3$ solution, the procedure is repeated twice. After replacing the sample tube with a clean one, the air sampling starts. During this sampling between 50 and 1000 l of air may be drawn through the filter at a rate of 1000 l h^{-1}. After the sampling has been finished, the filter is rinsed in the filter holder. The Ag_2S built up during the sampling is dissolved by injecting 5 ml of 0.1 N sodium cyanide in sodium hydroxide on the filter. This washing solution is then sucked through the filter and collected in the clean sample tube. Again, this procedure is repeated twice to ensure that the 15 ml of washing solution in the clean sample tube do contain all the sulfide generated on the filter. Consequently this filter, which is now free of H_2S, is used for the next air sampling. The filter is first treated with nitric acid and then again impregnated with $AgNO_3$ solution. To sample atmospheric SO_2 in the ppb range, basically the same filter technique is used. The filter is impregnated with a solution of 0.1 M tetrachloromercurate (TCM) in order to form dichlorosulfitomercurate by reaction with the atmospheric SO_2 (West and Gaeke, 1956). Because this sulfitocomplex is soluble, the filter is rinsed with a fresh solution of TCM after the sampling of air has been finished. Thus, in this case the preparation fluid is also the washing solution. The sampling tubes with the previously captured washing solution are then transferred from the sampling device in the field to the analysis apparatus in the laboratory. Since the filter material is being repeatedly used, it is possible to get a blank value of the filter between each pair of measurements. If these values correspond to the blank of the analyzing method, one can be sure that the filters are clean, and the detection limits of the very sensitive analyzing methods used can be applied for the whole detection method consisting of sampling and analysis.

The sampling device is also used for the collection of sulfate. Here a dry filter is (only once) used. Therefore the deviation of the filter blanks determines the limit of detection, which in this case is much higher than the detection limit of the analyzing method used. The analyzing method is an isotopic dilution technique, developed by Klockow (1974). It has a detection limit of 0.5 $\mu g\ SO_4\ m^{-3}$ STP.

Analysis of H_2S

In order to analyze the H_2S content of the washing solution a fluorescence method is applied which is based on the quenching effect of traces of sulfide on the fluorescence of fluorescein mercuric acetate (FMA) (Grünert *et al.*, 1968). 3 ml of the washing solution are mixed with 10 μl or FMA-solution and pumped through a flow cell of a Farand Fluorometer. The fluorescence signal of the pure washing solution without sulfide is set at 100% fluorescence, or 0% quenching. By measuring standard samples of sulfide the relation between the % quenching and the sulfide concentration is determined and the calibration curve shown in Fig. 2 is obtained.

The detection limit can be discussed by considering the deviation of the blank. The blank fluorescence is uncertain within $\pm$ 2%. This means that a quenching above $3\sigma = 6\%$ is caused by traces of sulfide with a certainty of 99%. This quenching effect corresponds to a standard concentration of 0.7 ng H_2S/ml or, because 15 ml washing solution are used, to 10 ng H_2S/filter. This means that when 1 m^3 of air is drawn through the filter, an atmospheric detection limit of 0.01 μg $H_2S\ m^{-3}$ is obtained.

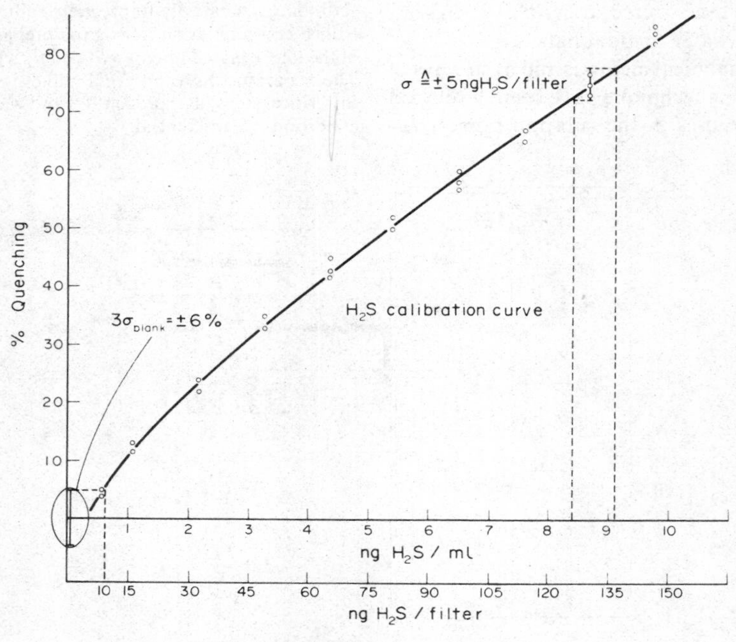

Fig. 2. H_2S calibration curve.

Table 1. Reproducibility of the measured H_2S values

No.	Sample volume (1.) STP	H_2S (μg/Filter)	H_2S ($\mu g\ m^{-3}$)	Deviation (%)
1	353	0.084	0.240	+ 17.1
2	272	0.055	0.204	− 0.5
3	367	0.067	0.184	− 10.2
4	483	0.100	0.208	+ 1.5
5	154	0.029	0.190	− 7.3
6	545	0.698	0.202	− 1.5
			$H_2S = 0.205$	± 6.3

The detection efficiency of this method was examined in several test-measurements under field conditions. The reproducibility of a measured value can be observed by considering the results of six parallel measurements which were taken on top of a hill named Kleiner Feldberg, Taunus, Germany. The data are shown in Table 1. At the mean concentration of 0.205 $\mu g\ H_2S\ m^{-3}$ STP a relative deviation of ± 6.3% occurred (Claude, 1977). The collection efficiency was examined by drawing a volume of 500–1000 l. air through two consecutive filters. In all tests taken, the washing solution of the second filter was identical with blank.

The results of another experiment are shown in Table 2. A volume of about 100 l. air was drawn through eight filters in eight consecutive time intervals. Simultaneously 1100 l. air were drawn through one filter within 80 min. The sum of the H_2S amounts collected on the eight filters was 0.213 μg/837 l. corresponding to 0.254 $\mu g\ m^{-3}$ STP. This value corresponds well with the analysis of the reference filter. The deviation of 15% may have been caused by the variations of the H_2S concentration in the ambient air during the experiment. Such diurnal variations of the H_2S concentration are shown in Fig. 3. They have been measured by taking 24 samples per day (Malewski, 1977).

Analysis of SO_2

Because in applying the sampling procedure for SO_2 the washing solution is also used as impregnation liquid, one can unite the sampling device with the analyzing equipment in one apparatus, as shown in Fig. 4. The analyzing method, which is a new development for detecting SO_2 in the ppb range, is based on a chemiluminescence effect (Stauff and Jaeschke, 1975). The complexed sulfite is oxidized with potassium permanganate. This oxidation process is accompanied by chemiluminescence. 5 ml of the washing solution are injected into a lightproof, opaque chamber with an automatic syringe. In a following step, 1 ml of a solution of 2×10^{-4} M $KMnO_4$ is injected into the chamber. The movement of this automatic syringe starts a photoncounter which counts the light yield from the oxidation of the sulfite present in the chamber during 100 s. The radiation yield of the chemiluminescence is a function of the oxidizable sulfite contained in the washing solution. By measuring standard samples, the relation between the light yield and the SO_2 concentration can be determined. To discuss the detection limit the deviation of the blank values has again to be considered. Here it is the chemiluminescence signal which occurs during the oxidation of 5 ml of TCM solution in the absence of any sulfite. A mean blank value of 14,600 impulses in 100 s is obtained with a standard deviation of $1\sigma = 570$ impulses. Signals above 15,200 impulses are caused by traces of sulfite with a certainty of 33%. This signal corresponds to a standard concentration of 2 ng SO_2/ml or – because 15 ml of washing solution are used – to 30 ng SO_2/filter. This means that when 1 m^3 of air is drawn through the filter, an atmospheric detection limit of 0.03 $\mu g\ m^{-3}$ is obtained.

Beside the sulfite calibration in the liquid phase, we

Table 2. Sampling efficiency of H_2S

No.	Sample volume (1.) STP	H_2S (μg/Filter)	H_2S ($\mu g\ m^{-3}$)
1	110	0.036	
2	140	0.027	
3	99	0.024	
4	114	0.034	
5	100	0.021	
6	117	0.030	
7	101	0.022	
8	92	0.018	
	$\sum = 837$	$\sum = 0.213$	0.254
9	1100	0.328	0.298

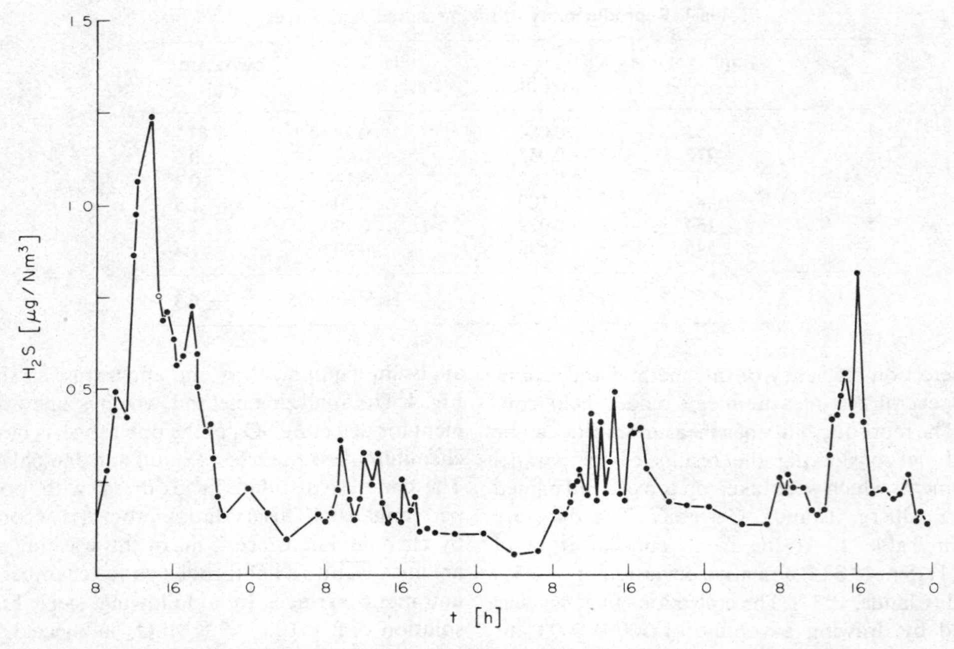

Fig. 3. Daily variation of the H_2S concentration at Kleiner Feldberg, Germany.

were successful in calibrating the sampling and analyzing method together by a SO_2 calibration source in the gas phase. A surprising harmony between the chemiluminescence effect caused by standard solutions of sulfite and by washing solutions of impregnated filters which were treated with known amounts of SO_2 in the gas phase was found (Fig. 5). Nevertheless, some further experiments were conducted in order to test the sampling efficiency. The calibration gas was drawn through two consecutive filters. The content of SO_2 was always 100 μg SO_2 m^{-3}. Using several volumes, six different amounts of SO_2 were accumulated on the first filter. When the washing solution was analyzed, the analyzed and the theoretically expected values were found to be in good agreement, as indicated in Table 3. The washing solution of the second filter showed the blank when the SO_2 concentration on the first filter was 2 μg or lower. Only in the cases of 3, 4 and 5 μg on the first filter we found traces of SO_2 on the second filter. But even there, deviation was minor.

Results of a further experiment are shown in Table 4. 500 l. of a calibration gas with a very low concentration of 0.4 μg m^{-3} were drawn through the filter. It took us three runs until the calibration system was stable enough to allow us to obtain the theoretically expected value. In three more runs the SO_2 traces in the dry calibration gas were mixed with a very high excess of ozone. In each case, values close to the theoretically expected level were discovered on the filter. This means that ozone does not affect the fixation of SO_2 on the

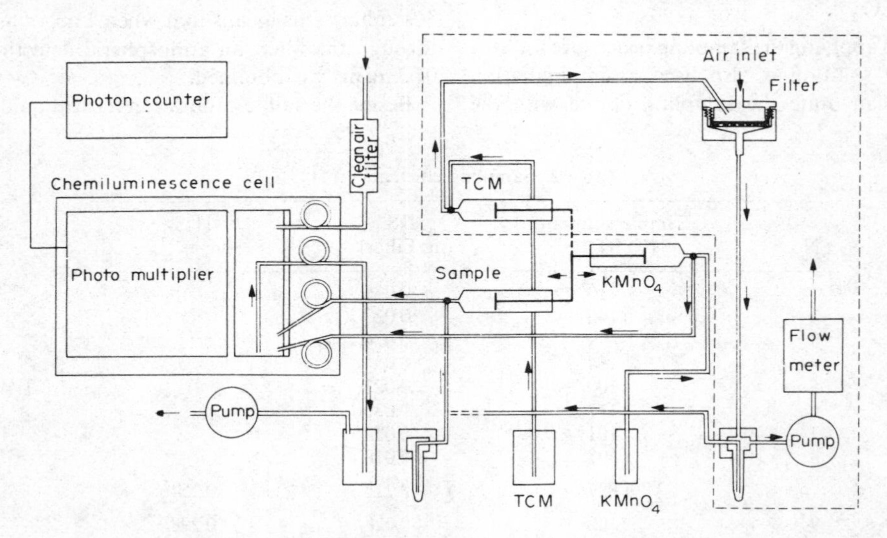

Fig. 4. SO_2 sampling and analyzing apparatus (schematic diagram).

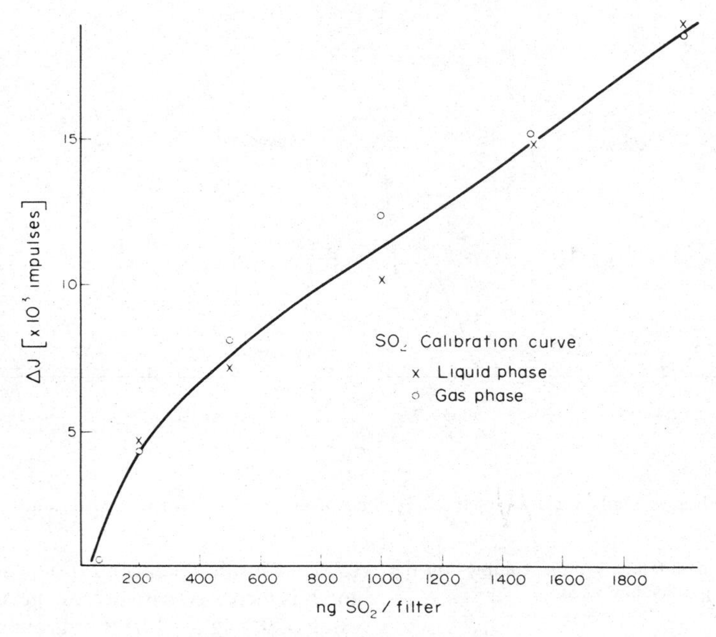

Fig. 5. SO_2 calibration curve.

filter. This was most important since the method was planned to be used in the stratosphere.

To examine the collection efficiency of the SO_2 method under field conditions, some measurements had to be performed in unpolluted areas. Therefore, experiments were conducted in the clean air of St. Moritz, Switzerland. The results may be seen on Table 5. In a first attempt, 603 and 300 l. (STP) of air were drawn through the filters, simultaneously, in two sampling devices. In a second run, on a different day, 1064 and 300 l. (STP) were used. The good agreement of the two concentration values indicates that the collection efficiency is not affected by increasing the air volume which is drawn through the filter. The deviation of the values obtained may have been caused by the variations of the SO_2 concentration in the ambient air of St. Moritz. These variations were measured over several days during which 12 samples per day were taken. The data arrived at are shown in Fig. 6.

Table 3. Sampling efficiency of SO_2 examined with two consecutive filters

Concentration of calibration gas (μg SO_2 m^{-3})	Volume of calibration gas (l.)	SO_2 on filter I (μg)	SO_2 on filter II (μg)	Deviation (%)
100	5	0.5	blank	—
100	10	1.0	blank	—
100	20	2.0	blank	—
100	30	3.0*	0.05	1.6
100	40	4.0*	0.07	1.7
100	50	5.0*	0.17	3.4

* values only calculated

Table 4. The influence of O_3 on the sampling efficiency of SO_2

Mixture of calibration gas (μg SO_2 m^{-3})+(μg O_3 m^{-3})			Volume of calibration gas (l)	SO_2 found (μg/filter)
0.4		—	500	(0.11)
0.4		—	500	(0.31)
0.4		—	500	0.22
0.4	+	880	500	0.25
0.4	+	880	500	0.25
0.4	+	880	500	0.22

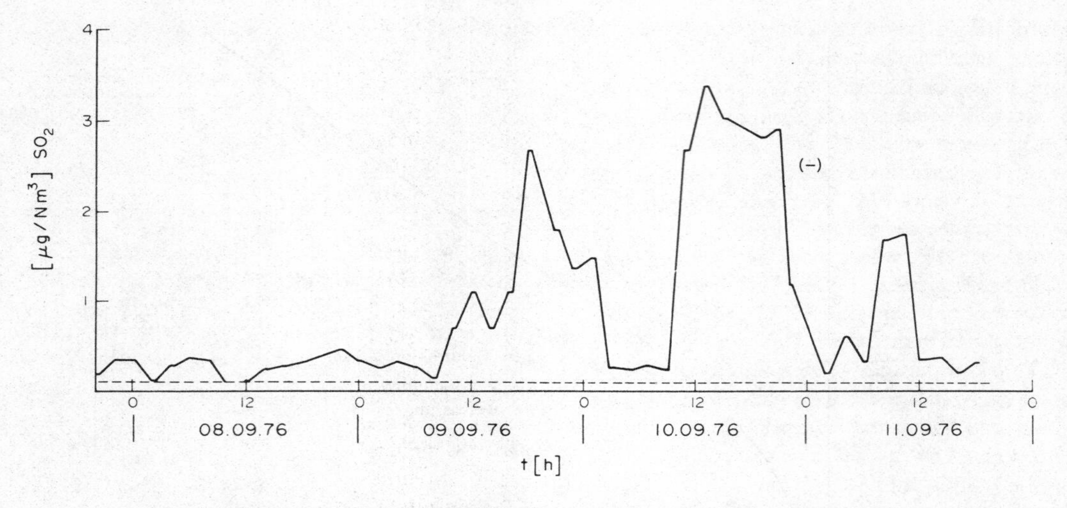

Fig. 6. Daily variation of the SO_2 concentration at St. Moritz, Switzerland.

Table 5. The influence of different sample volumes on the sampling efficiency of SO_2

Sample volume (l.) (STP)	SO_2 (μg/Filter)	SO_2 (μg m^{-3})
300	0.046	0.154
603	0.112	0.186
306	0.112	0.366
1064	0.340	0.319

APPLICATION OF THE METHODS

The methods have been applied in tidal flat regions at the isle of Sylt in the North Sea, and during aircraft ascents into the troposphere and lower stratosphere. In the stratosphere and in the upper troposphere no H_2S could be detected, but demonstrable amounts of SO_2 could be found. The SO_2 concentrations of 0.14 μg m^{-3} STP which were found in the lower stratosphere and upper troposphere were quite close to the estimates of Junge (1974) who suggested 0.1 ppb = 0.280 μg m^{-3} STP as a reasonable SO_2 background concentration for these altitudes (Jaeschke *et al.*, 1976).

With decreasing altitude, between the stratosphere and an altitude of 4300 m, concentration increases. This vertical distribution of the SO_2 concentration, which was discovered during a flight on 27 March 1976 with the aircraft HS 125 over the North Sea, is shown in Fig. 7. In this same figure the results obtained during several previous ascents over the North Sea in the lower troposphere are indicated. These ascents were made with a Piper 28 which is able to reach altitudes up to 4000 m. In this altitude, concentrations of 0.8 μg m^{-3} STP and 1.2 μg m^{-3} STP were found during two different ascents. These concentrations correspond to the concentration of 0.88 μg m^{-3} STP which was measured at nearly the same altitude during the flight to the stratosphere.

It is surprising that no SO_2 could be traced in altitudes of 3000 m and 2000 m. In an altitude of 1500 m a low SO_2 concentration was again detectable, and this increased with decreasing altitude, reaching a value of 9.3 μg m^{-3} STP at ground level.

Although no SO_2 could be discovered during our experiments between 2000 and 3000 m of altitude, minor SO_2 concentrations were again found between 4000 m and the tropopause. We therefore believe an atmospheric source of SO_2 to exist in this region. Following the estimations by Crutzen (1976), the

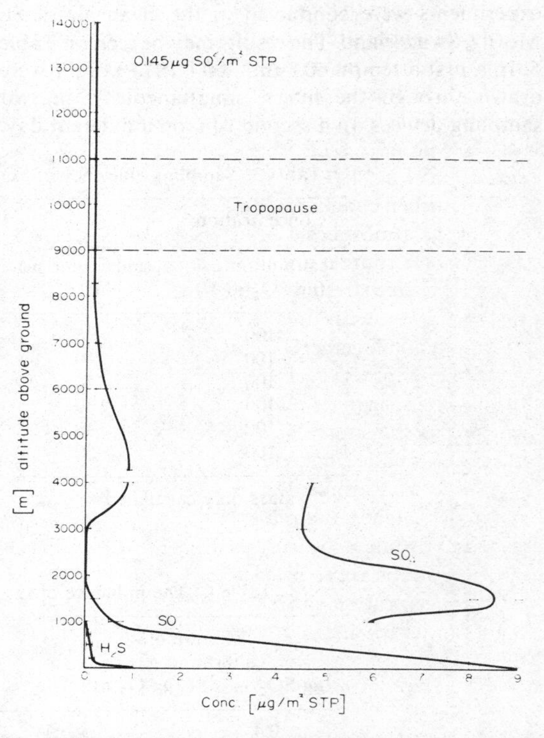

Fig. 7. Vertical distribution of the H_2S, SO_2 and sulfate concentration over the North Sea. The concentration profiles up to an altitude of 4000 m have been measured during aircraft ascents with a PA 28 in autumn, 1975. The SO_2 concentration profile between an altitude of 4000 and 14 000 m was obtained during a flight with a HS 125 in spring, 1976.

photolysis of CS_2 and COS could be the cause of such a source, since the dissociation products of these compounds may be oxidized to SO_2 in this altitude.

During the ascents into the troposphere the sulfate and hydrogen sulfide concentrations were also measured. The vertical distribution of these compounds may be seen from Fig. 7. Between altitudes of 1000 m and 1500 m the decrease of SO_2 is accompanied by an increase of SO_4^{2-}, which may be explained as a result of the atmospheric SO_2 oxidation. Referring to the sulfur contents of both compounds, an excess of sulfate is to be recognized in this layer. This excess could be caused by the sulfate particles released from the ocean surface as sea spray. In the level where no SO_2 could be traced, a sharp decline of the sulfate concentration was also observed. The parallel measured concentration profiles of SO_2 and SO_4^{2-} during these ascents show that in this situation the SO_2 concentration has an influence on the vertical distribution of the sulfate.

On the other hand it can easily be seen from Fig. 7 that the H_2S- concentration, which was also measured during these ascents, has no influence on the vertical SO_2 distribution. It is to be assumed that the SO_2 has reached the air above the North Sea by long range transport from Middle Europe (Buch *et al.*, 1976). Therefore, the SO_2 is no oxidation product of the H_2S stemming from local sources. As such sources, tidal flats have been surmised, the soils of which contain anaerobic sulfate-reducing bacteria.

We have seen that the tidal flats near the isle of Sylt yielded a source strength with regard to H_2S amounting to about 3–4 $\mu g\ m^{-2}\ h^{-1}$. The vertical distribution of the H_2S concentration was characterized by a sharp decrease from 1–2 $\mu g\ m^{-3}$ STP at ground level to 0.1 and less $\mu g\ m^{-3}$ STP at 100 m and higher altitudes (Jaeschke *et al.*, 1977).

For further conclusions about the contribution of H_2S to the atmospheric sulfur cycle, measurements of tropical H_2S sources are necessary. In tropical regions a higher source strength is to be expected, and in these remote areas with unpolluted air the interaction between the biogenic H_2S and the atmospheric SO_2 can be better observed. Furthermore, measurements of the biogenic organic sulfur compounds must be performed, since the source strength of anaerobic soils for dimethyl sulfide and dimethyl disulfide has been measured by Rasmussen (1974) and was found to be ten times higher than that for H_2S. In the upper troposphere some measurements of the CS_2 and COS concentrations are to be desired, because the reported SO_2 concentration profile seems to indicate that in higher altitudes these compounds may be a source for SO_2.

Acknowledgement—The research reported in this document has been sponsored by the Deutsche Forschungsgemeinschaft, Sonderforschungsbereich 73 "Atmospheric Trace Gases".

REFERENCES

Axelrod H. D., Natusch D., Klonis H. B., Teck R. J. and Lodge J. P. (1972) Sensitive method for measurement of atmospheric hydrogen sulfide. *Analyt. Chem.* **44,** 2067.

Buch H., Bjerregaard H., Pedersen H. and Hamer J. S. (1976) The variations in the concentrations of airborne particulate matter with wind direction and wind speed in Denmark. *Atmospheric Environment* **10,** 159–162.

Claude H. (1977) Untersuchungen zur atmosphärischen H_2S Konzentrationverteilung. Diplomarbeit, Inst. of Meteorologie und Geophysik, University of Frankfurt/Main, W. Germany.

Crutzen P. J. (1976) The possible importance of CSO for the sulfate layer of the stratosphere. *Geophys. Res. Lett.* **3,** 73–76.

Georgii (1978) Spatial and temporal distribution of sulfur compounds. *Atmospheric Environment* **12,** 681–690.

Grünert A., Ballschmitter K. H. and Tölg G. (1968) Fluoreszenzanalytische Bestimmungen von Sulfidionen im Nanogrammbereich. *Talanta* **15,** 451–457.

Jaeschke W. and Haunold W. (1977) New Methods and First Results of Measuring Atmospheric H_2S and SO_2 in the ppb Range. WMO Special Env. Rept. N0. 10, pp. 193–198.

Jaeschke W., Schmitt R. and Georgii H.-W. (1976) Preliminary results of stratospheric SO_2 measurements. *Geophys. Res. Lett.* **3,** 517–519.

Jaeschke W., Georgii H.-W., Claude H. and Malewski H. (1977) Contributions of H_2S to the atmospheric sulfur cycle. Paper presented at ISIBA (*Int. Symp. on the Influence of the Biosphere upon the Atmosphere*) July 1977 Mainz, Germany; To be published in *Pure appl. Geophys.* 116.

Junge C. (1972) The cycle of atmospheric gases – natural and man-made. *Q. Jl. R. met. Soc.* **98,** 711–729.

Junge C. (1974) Sulfur budget of the stratospheric aerosol layer. *Proc. IAMAP Conf.*, Melbourne Vol. 1, pp. 85–100.

Klockow D., Denzinger H. and Rönike G. (1974) Anwendung der substöchiometrischen Isotopen-Verdünnungsanalyse auf die Bestimmung von atmosphärischem Sulfat und Chlorid in "Background" Luft. *Chem. Ing. Techn.* **46,** 831.

Malewski H. (1977) Untersuchung der Konzentrationsverteilung von Schwefelwasserstoff in der bodennahen Luftschicht. Diplomarbeit, Institut für Meteorologie und Geophysik, University of Frankfurt/Main, W. Germany.

Rasmussen R. A. (1974) Emission of biogenic hydrogen sulfide. Tellus **26,** 254–260.

Stauff J. and Jaeschke W. (1975) A chemiluminescence technique for measuring atmospheric trace concentrations of SO_2. *Atmospheric Environment* **9,** 1038–1039.

West P. W. and Gaeke G. C. (1956) Fixation of sulfur dioxide and disulfitomercurate and subsequent colorimetric estimation. *Analyt. Chem.* **28,** 1816–1819.

Atmospheric Environment Vol. 12, pp. 723–728. Pergamon Press 1978. Printed in Great Britain.

SULFATE IN ANTARCTIC SNOW: SPATIO-TEMPORAL DISTRIBUTION

R. DELMAS and C. BOUTRON

Laboratoire de Glaciologie, 2, rue Très-Cloîtres, 38031 Grenoble-Cedex, France

(*First received* 24 *June* 1977 *and in final form* 26 *August* 1977)

Abstract—In this study, we report sulfate content measurements of eighty snow samples collected in East Antarctica. Stringent contamination free techniques were used both for sampling and analysis (ionometric titration with a lead selective electrode after preconcentration). Sulfate concentrations (range: 50–100 . 10^{-9} g g^{-1}, accuracy $\pm 10\%$) do not show large variations along a 1100 km coast interior axis, if we except a narrow coastal area. The analysis of well dated snow samples (time range: 1950–1975) collected at South Pole Station and Dome C (124°E, 75°S) suggest a main marine contribution for the sulfate content, to which are added sporadic stratospheric injections linked with major volcanic events, such as Mt. Agung (1963), whose eruption is clearly recorded. There is no evident anthropogenic contribution to the measured values.

INTRODUCTION

Sulfate is a dominant contaminant of precipitation. In remote areas, far away from the direct influence of the sea, it can even be regarded as the most abundant ion.

SO_4 is generally considered as the ultimate step in the oxidation of different natural or artificial gaseous sulfur compounds such as H_2S, DMS, SO_2.... Their origins are multiple: sea-spray, volcanic gases, biological decay or human activity. The acidity of rain or snow seems to be mainly linked with the atmospheric sulfuric acid (Likens *et al*., 1974). During the last decades the problem of acid rains over Europe has enhanced the interest of measuring sulfate concentrations in different regions of the world, of determining the various sources and sinks of sulfur. Despite this international effort, the global sulfur cycle is still not fully understood (Kellogg *et al*., 1972; Granat *et al*., 1976). There is a lack of reliable data on the background level of the atmospheric contamination in sulfur compounds. Polar regions seem to be particularly adequate to study this problem because they are remote from major industrial areas and generally free of natural or anthropogenic sources. Furthermore the polar ice sheets are especially attractive for such studies because the successively deposited snow layers have recorded the changes in the chemistry of the atmosphere over a long period of time.

Numerous investigations relative to the air chemistry have been carried out in polar snows. As early as 1960 Junge (1960) analysed snow and ice samples from the Greenland ice cap in order to obtain information on the general circulation of sulfur compounds. There are two practical difficulties in obtaining reliable data on the sulfate concentrations of polar snows. Because of the low levels of sulfate concentrations generally found in these regions, utmost precautions must be taken for sampling and ultra sensitive methods of determination have to be developed. For these reasons the number of data concerning the concentration of sulfate in Antarctic snow is very limited and no study on its origin in this region has been made.

In our laboratory, we have developed adequate methods for the sampling of polar snow and the determination of sulfate at low levels. We present here the results of the analysis of eighty snow samples (0.8 kg each) collected in the surface layers of East Antarctica. The reported values give the total sulfate deposition and represent likely the sum of the dry deposition and of the sulfate falling with snow owing to the snowout and eventually washout processes.

SAMPLE COLLECTION

All samples were collected using plexiglas or Teflon tubes (diameter 8 cm, length 40 cm) with screwed caps, which had been cleaned in laminar flow clean work stations inside a class 100 clean room, using ultrapure hydrofluoric and nitric acids and water, and then placed in double sealed polyethylene bags.

To study the possible influence of sea derived air masses, surface snow samples were collected along a coast interior axis Dumont d'Urville (139°E, 66°S, near the sea) towards Dome C (124°E, 75°S, 1070 km from the sea, altitude 3240 m) (Table 1, Fig. 1). Sampling was carried out about 500 m upwind from the vehicles by directly hammering the tubes down a horizontal surface rasped with a clean Teflon sheet. The operators wore clean room clothes and particle masks. The time intervals integrated by this surface sampling depend on the accumulation rates, and range from one year near the coast to about 4 years at Dome C.

Samples were also collected in two 5 m deep hand dug pits at South Pole Station (5 km from the station, in the scientific sector) and Dome C (time range: 1950

Table 1. Sulfate concentrations along the Dumont d'Urville–Dome C axis, East Antarctica

Sampling location*	Distance from the coast (km)	Altitude (m)	Sampling date	Sulfate concentration (10^{-9} g g^{-1})	Sampling location	Distance from the coast (km)	Altitude (m)	Sampling date	Sulfate concentration (10^{-9} g g^{-1})
D 5	2	190	02/72	87	D 66	293	2320	01/72	64
			02/72	51				01/72	64
D 10	5	270	02/73	101	D 72	353	2360	12/72	43
			02/73	86				12/72	68
			02/73	90				12/71	42
D 23	16	580	11/72	94				12/71	72
			11/72	87				12/72	48
			11/72	98				12/72	49
			11/72	120	D 80	433	2430	12/72	41
D 33	26	730	02/73	97				12/72	64
			02/73	133				12/72	32
			02/73	84				12/71	41
								12/71	58
D 40	33	850	02/72	56	D 89	520	2631	12/71	87
			02/72	51				12/71	42
D 42	53	1095	11/72	44	D 100	633	2810	01/72	40
D 45	83	1410	01/73	75				01/72	35
			01/73	78				01/72	64
D 46	93	1450	11/72	44	D 110	733	2960	01/72	75
			11/72	46				01/72	55
D 50	133	1730	11/71	89				01/72	135
D 53	163	1910	01/73	64	D 120	833	3010	01/72	55
			01/73	45				01/72	144
D 54	173	1935	11/71	67	Dôme C	1070	3240	01/75	79
			11/71	69				01/75	80
D 59	223	2220	11/72	74					
			11/72	39					
			11/72	119					
			11/72	59					

* Accumulation range = 50 to 3.5 g.cm^{-2} y^{-1}.

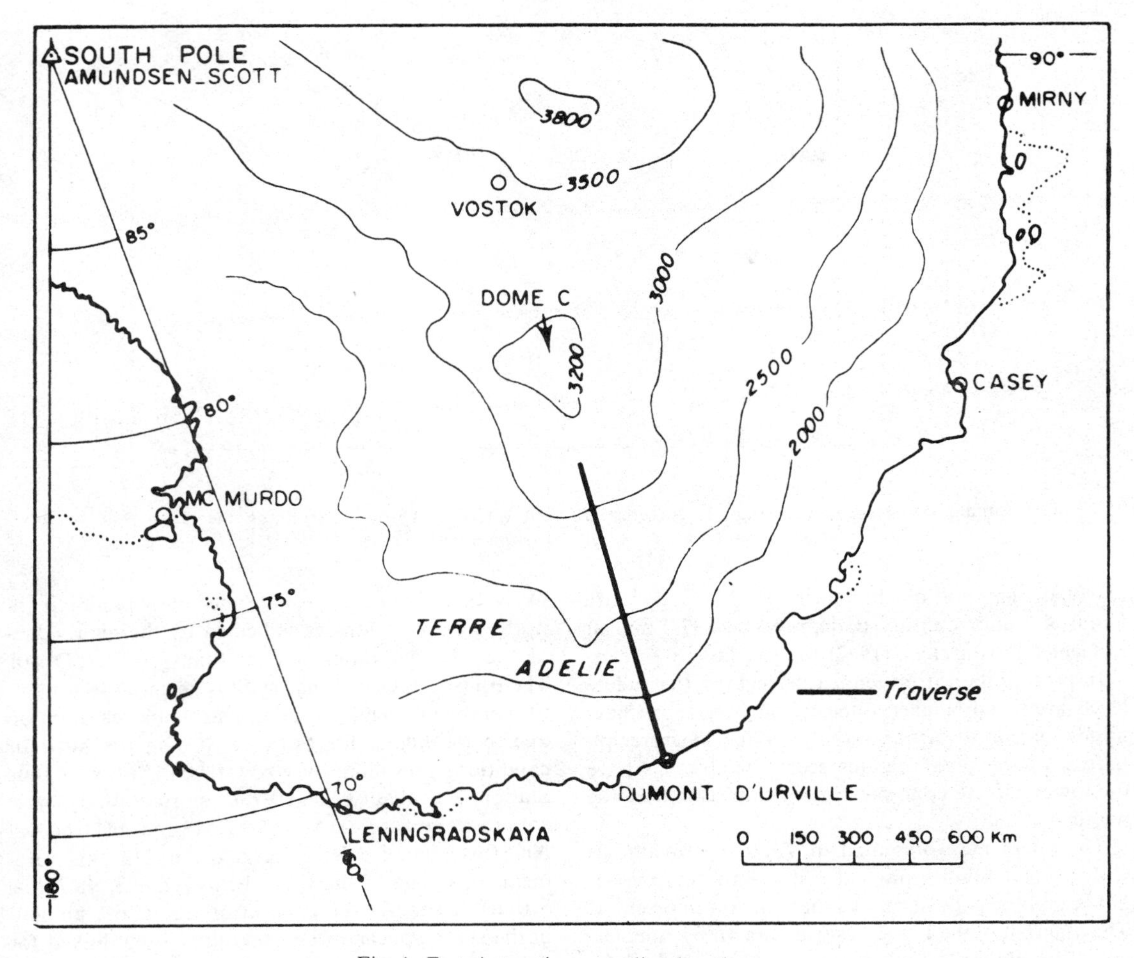

Fig. 1. East Antarctica: sampling locations.

to 1975). The age of the snow samples has been determined by various independent techniques (Lorius, 1977) such as stratigraphy, stable isotopes composition, radioactive fallout. The results which will be published elsewhere indicate an accuracy of ± 0.5 (South Pole) and ± 1 (Dome C) year.

All the samples were shipped back to the laboratory in the frozen state, and were never allowed to melt before the preconcentration step of the analysis.

ANALYTICAL PROCEDURES

Before titration, samples (0.6 to 1 kg) are preconcentrated (factor: 50 to 100) by non boiling evaporation in Teflon bulbs (Boutron, 1977) inside a laminar flow clean work station placed in a class 100 clean room.

The preconcentrated aliquot necessary for the titration of sulfate is 0.25 ml (concentration range $1\text{–}10.10^{-6}\ g\,g^{-1}$). The potentiometric titration of sulfate by lead perchlorate has been improved (Delmas, in press). Lead sulfate is precipitated in a 50% water dioxan mixture. The excess of lead is determined with an ion selective electrode using the known addition and Gran's plot methods. Orion and Mettler ionometric equipment is used. The titrant is added μl per μl with the aid of an automatic Mettler burette.

Preconcentration and titration have been globally calibrated on 20 standard solutions from $10\ \mu l$ to $180.10^{-9}\ g\,g^{-1}$ of sulfate. Standard deviation is $\pm 7.10^{-9}\ g\,g^{-1}$ or about 10% of typical Antarctic values. The contamination from the sampling tubes and analytical processes can be considered as negligible as shown by blank measurements.

RESULTS AND DISCUSSION

The analysis of the 50 samples collected along the traverse Dumont d'Urville towards Dome C has given the values reported in Table 1. The overall mean concentration calculated from these data is $71.10^{-9}\ g\,g^{-1}$ with a standard deviation of 38% from the mean value. This mean value is in essential agreement with the one $(60.10^{-9}\ g\,g^{-1})$ found at Byrd Station by Cragin *et al.* (1974) for the last 10,000 years. At Vostok Station, Doronin (1975) measured about $120.10^{-9}\ g\,g^{-1}$ of sulfate during the year 1971, but no indication is given by the author as to the sampling and analytical precautions taken for these measurements. In Greenland also, the concentrations

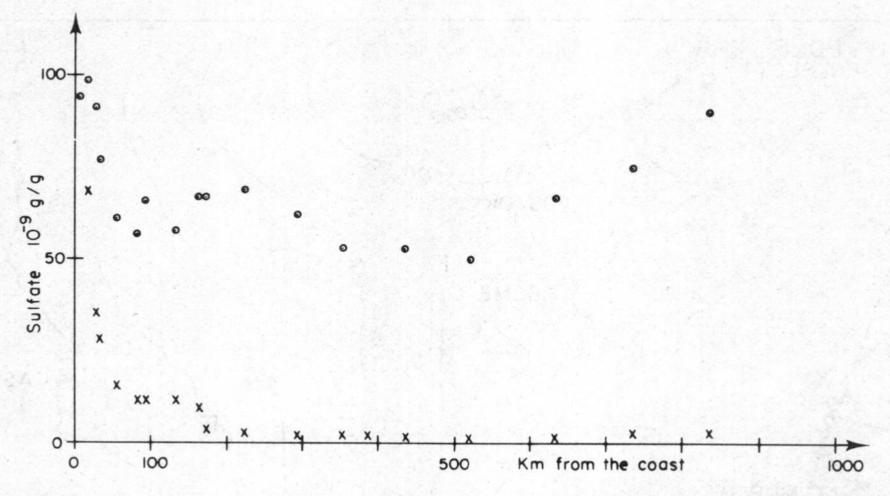

Fig. 2. Sulfate concentrations along the traverse Dumont d'Urville–Dome C (running mean on 3 stations). ○ measured; × sea salt sulfate (calculated).

of sulfate are not much higher: 80.10^{-9} g g^{-1} are found at Camp Century (Cragin *et al.*, 1974) and at Milcent (Herron *et al.*, 1977) for the last 800 years.

In precipitation from temperate regions, the sulfate levels are mostly higher, often by one degree of magnitude or more (Granat *et al.*, 1976). The precipitations in the polar regions seem therefore to have the lowest sulfate concentrations yet measured in the world.

The set of data presented in Table 1 provides us with a good opportunity to evaluate the marine influence on the deposition of sulfate in Antarctic snow. The diagram of the Fig. 2 has been drawn by calculating a running mean on 3 stations using all these data. Generally speaking, the sulfate concentrations in snow appear to be rather constant along the axis except over the first 50 km where a significant decrease of about 40% is observed. There is also a slight increase near the end of the axis. The geographical effect is then much less than for Na, for which previous studies (Briat *et al.*, 1973; Boutron *et al.*, 1975) indicated a very marked decrease from the coast towards the inland (sodium is 50 times lower at D 59 than at D 5).

According to Nguyen *et al.* (1974) the sulfur compounds SO_2 and SO_4 over the Antarctic ocean are nearly of pure marine origin. As indicated by the low level of aluminium in this region (Boutron *et al.*, 1975), the contribution from the continents can be neglected. Moreover there seems to be two distinct sources for the sulfate found over the sea (Nguyen *et al.*, 1977): the sea-salt sulfate and the sulfate formed by a partial oxidation of marine SO_2 ('excess sulfate'). Using the SO_4/Na ratio in bulk sea water (0.25) and our sodium data, we have calculated an estimation of the sea-salt sulfate (Fig. 2). Near the coast its contribution is important, indicating probably its deposition in the more coastal areas. After 200 km, this sulfate represents only 10% or less of the total sulfate deposition by assuming no fractionation. It seems likely that the 'excess sulfate' becomes predominant in the central regions as indicated by the high values of the SO_4/Na ratios (higher than 10 after D 80). The transportation of marine SO_2 far into the Antarctic continent seems to be a reasonable assumption due to the higher life times of SO_2 in the pure air conditions prevailing over Antarctica. At South Pole Station, Maenhaut *et al.* (1976) reported atmospheric concentrations of 0.05 μg S/SCM and of 0.003 μg Na/SCM, giving a S/Na ratio of 16. Our measurements of sulfate in snow (see below) give S/Na ratios 5 times lower. As the concentrations in the air and in the snow appear to be generally closely linked for the elements analysed at this station (Boutron *et al.*, 1977) it seems likely, as detected by Fischer *et al.* (1969) that a sulfur compound other than SO_4 does exist in the Antarctic atmosphere. This compound could be gaseous SO_2 coming from the Antarctic ocean.

Another possible source of sulfate in Antarctica could be the stratospheric reservoir close to an altitude of 20 km known as the 'Junge layer' (Hofmann *et al.*, 1973). The artificial radioactivity detected in polar snows (gross β activity, tritium) has proved that there is a yearly input of stratospheric products into the Antarctic atmosphere. We could therefore expect to observe in snow the variations of the stratospheric sulfate reservoir, if however the fallout of this element is not negligible in comparison with its other sources. The sulfate concentration of the Junge layer has been strongly enhanced during the years 1963–64 by the volcanic eruption of Mt. Agung (8°S, 115°E) on Bali. The arrival of the volcanic debris of this eruption was observed in November 1963 by Viebrock *et al.* (1968) over South Pole Station. An extensive review of the effects of Agung on the Antarctic troposphere and stratosphere has been made by Reiter (1971).

The two continuous profiles that we have obtained at South Pole Station and Dome C cover this period of time and are presented in Fig. 3. As tropospheric

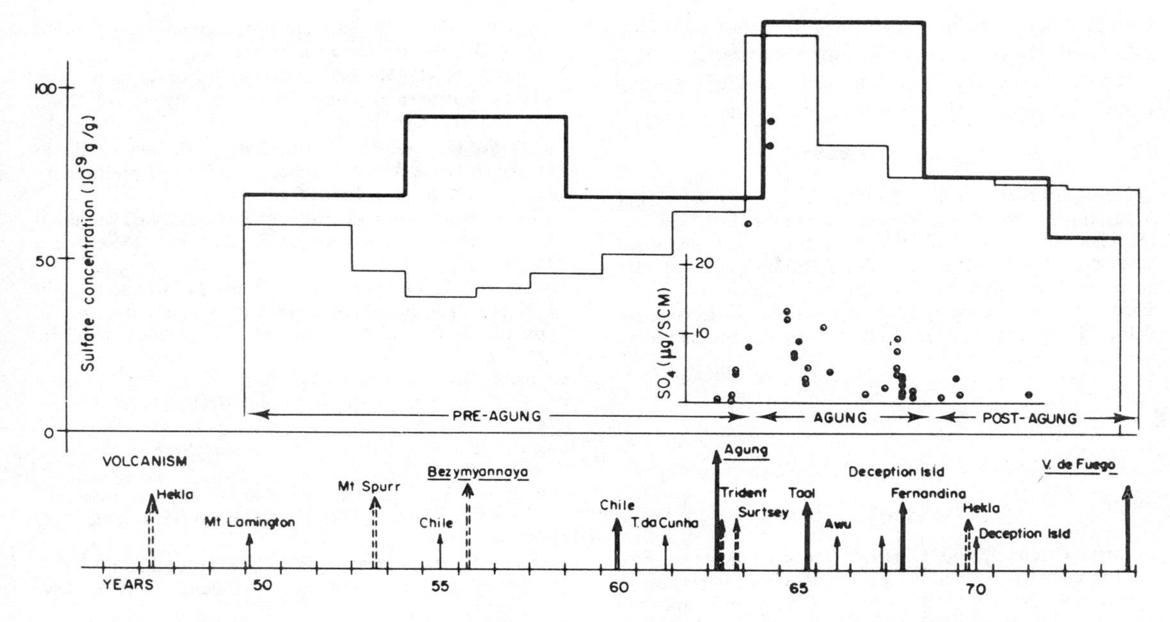

Fig. 3. Temporal variations of sulfate concentration is snow at South Pole (——) and Dome C (——). Stratospheric concentrations of sulfate, in μg SCM^{-1}, are also plotted for the years 62–71 (after Castleman *et al.*, 1974). Volcanism is indicated as follows: (a) volcanic eruptions with injections into the stratosphere: ⇑ Southern hemisphere, ⇑ Northern hemisphere. (b) ↑ other important volcanic eruptions in the southern hemisphere.

transportation of the volcanic sulfur from Bali to Antarctica is most unlikely, our results suggest a stratospheric origin for the high values corresponding to Agung. On the same figure we have added the measurements of Castleman *et al.* (1974) relative to the sulfate concentration of stratospheric aerosols from 1962 to 1971 (in Southern Hemisphere). Castleman's points have been shifted by six months to take into account the time of transportation of the aerosols to the South Polar regions. The impact of Agung on the sulfate concentration is however much stronger in the stratosphere than in the snow. This seems to indicate that the source of antarctic sulfate is not solely stratospheric nor volcanic. During the decade preceding 1963, the explosive volcanic activity (Fig. 3) can be considered as low in the Southern Hemisphere and the stratospheric concentrations at a minimum (Cronin, 1971). As volcanic emissions represent the major source disturbing the sulfate concentration of the stratosphere (Castleman *et al.*, 1974), the volcanic contribution to the 'pre Agung' sulfate level in Antarctic snows is most likely negligible.

As suggested by the analysis of the sulfate samples along the Dumont d'Urville–Dome C traverse, the marine source seems to be dominant but an anthropogenic contribution could also be considered as has been the case for Pb, Zn, Cu and Ag deposition in Antarctic snow (Boutron *et al.*, 1977).

During the 'pre' and 'post Agung' periods (see Fig. 3) some minor accidents are observed in the sulfate profile (for example at Dome C in 1955–58). Such a variability exists for certain other elements too like Na, Al, etc... (Boutron *et al.*, 1977) and could be linked with natural phenomena which have until now not been well understood. At Dome C the mean values before ($77.10^{-9}\,g\,g^{-1}$) and after ($66.10^{-9}\,g\,g^{-1}$) Agung are close. Further measurements concerning the period 1920–1950 and not reported here, give a mean value of $73.10^{-9}\,g\,g^{-1}$ at Dome C. At South Pole Station the post Agung level ($73.10^{-9}\,g\,g^{-1}$) is higher than the pre Agung level ($50.10^{-9}\,g\,g^{-1}$). The natural variability or a local contamination by the activity of the South Pole Scientific Station, seems to be more likely than the global anthropogenic pollution as an explanation for this increase.

A possible contribution of some stratospheric volcanic emissions during the post Agung period (see 'volcanism index' of Fig. 3) must also not be discarded. The absence of a significant anthropogenic contribution is not contradictory with one possible interpretation of the measurements of metals by Boutron *et al.* (1977) and of sulfate in Greenland snows by Herron *et al.* (1977). More than 90% of the anthropogenic sulfur compounds are emitted in the northern Hemisphere and their relatively short residence times in the atmosphere (Kellogg *et al.*, 1972; Granat *et al.*, 1976) limit their influence to the troposphere of this hemisphere.

Our results suggest that the main source of the sulfate deposited in Antarctic snow would be marine. The possibility of sporadic and important stratospheric injections of volcanic sulfate is shown. Stoiber *et al.* (1973) reevaluated the sulfur contribution to the atmosphere by volcanoes by excluding the major eruptions. Our work shows that the stratospheric sulfate, produced by these eruptions, must not be underestimated. On the other hand, our results suggest that a record of the great volcanic eruptions of the past could be found in the Antarctic ice layers.

Acknowledgements—We are grateful to Terres Australes et Antarctiques Françaises, Ministère de la Qualité de la Vie, INAG, Expéditons Polaires Françaises and National Science Foundation, Office of Polar Programs, which supported part of this work. We thank C. Lorius for helpful discussions. The collaboration of F. Gillet, C. Lorius, M. Paillet, F. Pinglot and M. Pourchet during the sampling was greatly appreciated.

REFERENCES

Boutron C. and Lorius C. (1975) Trace elements in East Antarctica snow samples. In *Isotopes and Impurities in Snow and Ice.* Proceedings of the Grenoble Symposium 1975. IAHS Publ. **118,** 164–171.

Boutron C., Martin S. and Lorius C. (1977) Composition of aerosols deposited in snow at the South Pole. Time dependency and sources. *Proc. 9th Int. Conf. on Atmospheric Aerosols, Condensation and Ice Nuclei, Galway, Ireland,* 21–27 September. 1977. Pergamon Press, Oxford.

Boutron C. (1977) Preconcentration at the ng/kg level by non boiling evaporation in Teflon: simultaneous determination of 12 elements in Antarctic snows. International Symposium on Microchemical Techniques, May 1977, Davos, Switzerland, Organized by IUPAC.

Briat M., Boutron C. and Lorius C. (1974) Chlorine and sodium content of East Antarctica firn samples. *J. Rech. Atmos.* **8,** 895–900.

Castleman A. W. Jr., Munkelwitz H. R. and Manowitz B. (1974) Isotopic studies of the sulfur component of the stratospheric aerosol layer. *Tellus* **26,** 222–234.

Cragin J. H., Herron M. M., Langway C. C. Jr. and Klouda G. (1974) Interhemispheric comparison of changes in the composition of atmospheric precipitation during the late cenizoic era. Conference on Polar oceans. Montreal, May 1974. Organized by ISCU/SCOR/SCAR.

Cronin J. F. (1971) Recent volcanism and the stratosphere. *Science* **172,** 847–849.

Delmas R. (1978) Microdosage potentiométrique de l'ion sulfate. *Mikrochim. Acta* (in press).

Doronin A. N. (1975) Chemical composition of snow near Vostok Station and along the axis Mirny-Vostok. *Bull. Sov. Antarct. Exp.* **91,** 62–68.

Fischer W. H., Lodge J. P. Jr., Pate J. B. and Cadle R. D. (1969) Antarctic atmospheric chemistry; preliminary exploration. *Science* **164,** 66–67.

Granat L., Rodhe H. and Hallberg R. O. (1976) The global sulphur cycle. SCOPE Report 7. *Ecol. Bull.* (*Stockholm*) **22,** 89–134.

Herron M. M., Langway C. C., Weiss H. V. and Cragin J. H. (1977) Atmospheric trace metals and sulfate in the Greenland Ice Sheet. *Geochim. Cosmochim. Acta* **41,** 915–920.

Hofmann D. J., Rosen J. M., Pepin J. J. and Pinninck R. G. (1973) Particles in the Polar stratospheres. *Nature* **245,** 369–371.

Junge C. E. (1960) Sulfur in the atmosphere. *J. geophys. Res.* **65,** 227–237.

Kellogg W. W., Cadle R. D., Allen E. R., Lazrus A. L. and Martell E. A. (1972) The sulfur cycle. *Science* **175,** 587–596.

Likens G. E. and Bormann F. H. (1974) Acid rain: a serious regional environmental problem. *Science* **184,** 1175–1179.

Lorius C. (1977) Accumulation rates measurements on cold polar glaciers. In *Isotopic and Temperature Profiles in Ice Sheets* (edited by G. de Q. Robin). Cambridge University Press, London.

Maenhaut W. and Zoller W. H. (1976) Determination of the chemical composition of the South Pole aerosol by instrumental neutron activation analysis. Modern trends in activation analysis, Munich, 1976.

Nguyen B. C., Bonsang B. and Lambert G. (1974) The atmospheric concentration of sulfur dioxide and sulfate aerosols over antarctic, subantarctic areas and oceans. *Tellus* **26,** 241–249.

Nguyen B. C., Bonsang B., Gaudry A. and Lambert G. (1977) Oxidation processes of sulfur components in the marine atmosphere. Presented IAGA/IAMAP. Joint Assembly, Seattle, Sept. 1977.

Reiter E. R. (1971) Atmospheric transport processes Part 2. Chemical tracers 195–218. AEC Critical Review series, Division of technical information.

Stoiber R. E. and Jepsen A. (1973) Sulfur dioxide contributions to the atmosphere by volcanoes. *Science* **182,** 577–578.

Viebrock H. J. and E. C. Flowers (1968) Comments on the recent decrease in solar radiation at the South Pole. *Tellus* **20,** 400–411.

Atmospheric Environment Vol. 12, pp. 729–733. Pergamon Press 1978. Printed in Great Britain.

AEROSOL CHARACTERIZATION FOR SULFUR OXIDE HEALTH EFFECTS ASSESSMENT

A. C. D. LESLIE, M. S. AHLBERG, J. W. WINCHESTER and J. W. NELSON

Department of Oceanography and Department of Physics, Florida State University, Tallahassee, FL 32306, U.S.A.

Abstract—The mean concentrations of sulfur in six particle size fractions, from $<0.25\ \mu m$ to $>4\ \mu m$ aerodynamic dia, have been determined in Florida during July–August and December, 1976 sampling seasons. A trend of decreasing concentration along the length of Florida, at ten sites from Pensacola to Miami, is found for sulfur in the fine particle mode, $<2\ \mu m$ dia, but not for $>2\ \mu m$ coarse mode sulfur. The trend is most regular for 0.5–1 μm particles, and urban concentrations show no consistent departure from the nonurban trend in this size range. Smaller and larger particle concentrations are more locally variable. The trends are interpreted in terms of a flow of continental air, containing a pollution-derived fine particle sulfur component, into Florida during northerly wind regimes and of less polluted maritime air with southerly flow. The magnitude of the apparent decrease in average concentration from north to south is dependent on the mix of air flow regimes at various locations in the state during the sample collection periods. The qualitative finding of the decreasing trend, however, may be typical in Florida over much of the year and may be useful for designing epidemiological studies of the relation between fine particle sulfur and public health.

INTRODUCTION

Florida is situated between the Atlantic–Caribbean–Gulf of Mexico maritime region and the eastern U.S. continental region in which pollution sources of sulfur oxides are prominent. Therefore, geographic trends of sulfur in the atmosphere of Florida may be indicative in part of mixing of air from these two regions. In a companion paper (paper 1, Ahlberg *et al*., 1978) we have examined some aspects of the particle size distribution of sulfur in Florida. In this paper (paper 2) we shall examine some geographic trends of sulfur in the fine particle mode ($<2\ \mu m$ diameter) for sampling periods 28 July–7 August and 5–18 December 1976, at 10 sites from Pensacola to Miami.

RESULTS

Sulfur concentrations in fine particle size fractions sampled by cascade impactor, $<2\ \mu m$ aerodynamic dia (stages 3–6), generally exceeded by several-fold the concentrations in coarse particle fractions, $>2\ \mu m$ dia (stages 1 and 2), throughout Florida during the summer and December 1976 sampling periods. Fig. 1 presents the arithmetic mean concentrations for fine and coarse particle modes found at the different locations, arranged as a sequence from Pensacola to Miami. The coarse particle concentrations were variable among the samples collected and without apparent geographic or meteorological relationships. The fine particle concentrations, on the other hand, exhibited significant trends with sampling location, and these are examined in detail here.

Tables 1 and 2 display the fine aerosol particle sulfur concentrations, in $ng\ m^{-3}$ for stages 3–6 combined, according to sampling date and location. The relative significance of concentrations is 30% or better, although the absolute values may be uncertain by up to a factor of 2 owing to possible variations in efficiency for particle retention by the impactors. Shorter sampling times and cleaner air tend to decrease the variations.

Also shown on the tables are the mean values for air temperature (in °F) and wind direction (as arrows, from north at the top, from east at the right, etc.). For the summer period wind directions varied from west to north, especially for the north Florida stations, to southerly, especially for the south Florida stations. Wind direction shifts, which occurred on 2 or 3 August at Gainesville and Jacksonville, are associated with a frontal passage which also passed Tampa a day later. The west to north air flow samples appear to have higher fine particle sulfur concentrations than the southerly flow samples; for example, the 10 samples collected on 1 August show higher concentrations at the first 3 stations, which experienced west to north winds, than at the remaining 7 stations which were under the influence of southerly air flow. The mean concentrations of the six or so observations made at each station are affected by the proportion of southerly and west to north wind cases at the given station. In general, the south Florida stations had a greater proportion of southerly flow observations than did the north Florida stations. This accounts qualitatively for their lower mean fine particle sulfur concentrations.

In the December 1976 period, Table 2, a similar association of wind direction and fine particle sulfur concentration is observed. Unlike the summer, however, there were several fronts, with shifts in wind direction and air temperature, which passed during the December period. As a rule, with some exceptions, northerly flow sulfur concentrations were greater than

A.E. 12—1 3—UU

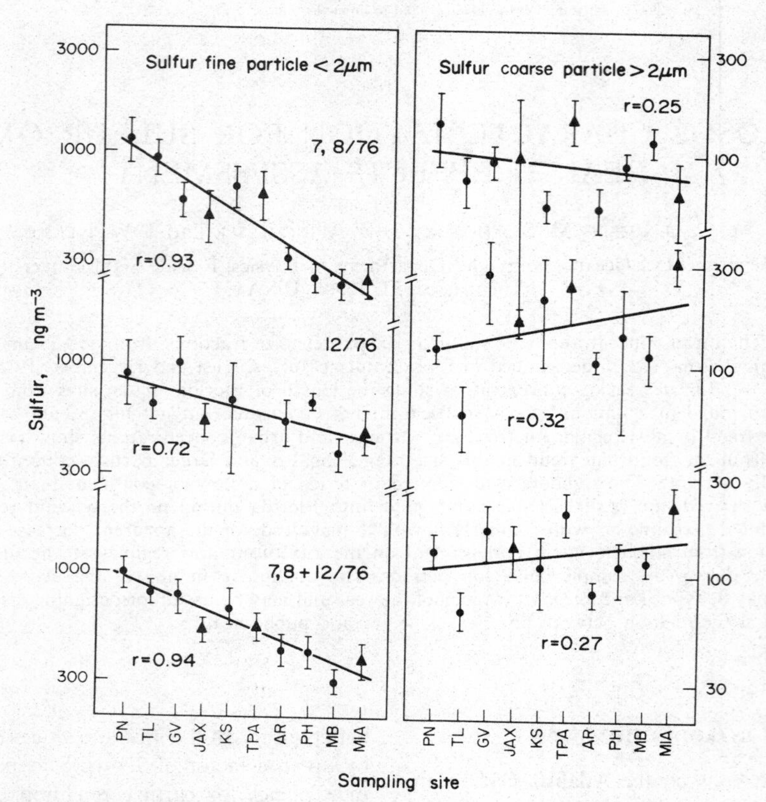

Fig. 1. Sulfur in fine particle (impactor stages 3–6) and coarse particle (stages 1–2) modes during the summer and December 1976 periods, expressed as the arithmetic mean, with its standard deviation, for each group of approximately six samples taken in each period. Points are plotted against sampling site index from Pensacola to Miami. Values of r represent correlation coefficients of log concentration with this arbitrary index. Absolute calibration of concentrations in air is significant to a factor of 2.

southerly flow concentrations, and the northerly air penetrated further south in Florida than it apparently did during summer. When averages are made of the six or so observations at each station, the proportions of southerly and northerly flow cases are not as different among the different stations as was the case for the summer measurements. Consequently, the mean sulfur concentrations for the different stations were more similar than during the summer period, showed less variation in the winter than in the summer, thus accounting qualitatively for the trends of Fig. 1.

In Figs 2 and 3 we present in detail the sulfur con-

Table 1. Sulfur concentrations*, temperatures and wind directions† for fine particle mode $<2\ \mu m$ dia

Sampling site	Date‡, July–August 1976: 28	29	30	31	1	2	3	4	5	6	7	8
PN			707 80° →	1309 79° →	867 81° ↘		3125 80° ↙	1757 79° ↙				
TL		548 81° ↗	526 79° ↗		1154 82° →	743 76° ↗		1514 78° ↓	1259 81° ↙			
GV			298 81° →	496 81° ↗	887 83° ↗	458 80° →	526 77° ←	944 78° ↖				
JAX	399 81° ↗	286 82° ↗			254 85° ↗	762 81° ↗	889 77° ↘					
KS	1123 78° ↗	1105 81° ↑		724 82° ↗	460 84° ↗	399 81° ↖	412 78° ↖					
TPA				399 82° ↗	465 82° ↗	335 81° ↑	471 77° ↖	1082 81° ↙	1155 82° ↘			
AR		196 79° ↘	225 81° ↖		345 80° ↗	380 80° ↑	361 79° ←	419 78° ↖				
PH	83 79° ↓	373 81° ↖		284 79° ↖	243 76° ↗	276 81° ↖	294 78° ↗					
MB		194 79° ↘	328 81° ↖	346 81° ↗	284 80° ↗	169 80° ↗	136 79° ←					
MIA							151 83° ↙	248 82° ↙	348 81° ↘	272 82° ↓	969 84° ↗	660 85° ↖

* Concentrations, $ng\ m^{-3}$, significant to 30% (relative), factor of 2 (absolute).
† 24 hour means of hourly temperature (°F) and wind direction (arrow).
‡ Sampling times 24 hours starting on date shown.

Table 2. Sulfur concentrations*, temperatures and wind directions† for fine particle mode $<2\ \mu m$ dia

Sampling site	5	6	7	8	9	10	11	12	13	14	15	16	17	18
	Date‡. December 1976													
PN	1745 49° ↓		467 53° ↘					607 66° ↗		587 50° ↙		1033 53° ↘		
TL	870 51° ↙		684 60° ↖		773 42° ↙			368 69° ↗		331 51° ↙		852 54° ↘		
GV	1190 55° ↓						638 71° ↖						1645	584
JAX	460 53° ↓				669 43° ↙			453 70° ↗	318 60° ↙	225 56° ↙	818 61° ↘	869 53° ↘		
KS						342 71° ↖						386 56° ↘	1296 54° ↘	
TPA				365 48° ↓	387 53° ↙			408 73° ↖	556 69° ↙	508 67° ↙	539 69° ↗	518 64° ↘		
AR	813 63° ↙		313 70° ↗		1008 57° ↙			279 73° ↖		301 72° ↙		520 68° ↗		
PH	731 68° ↘		1569 74° ↗			566 74° ←						775 66° ↘	590 58° ↘	
MB	567 63° ↙		206 70° ↗					266 73° ↖		413 72° ↙		442 68° ↗		
MIA	689 69° ↘		303 77° ↗		797 63° ↙			201 77° ↖		345 75° ←		555 75° ↗		

* Concentrations, ng m^{-3}, significant to 30% (relative), factor of 2 (absolute).
† 24 hour means of hourly temperature (°F) and wind direction (arrow).
‡ Sampling times 24 hours starting on date shown.

centration trends for the separate particle size fractions in each sampling period. In both figures the trends of impactor stage 4 (0.5–1 μm dia particles) are smoother than the trends of larger or smaller particle sizes. During summer all impactor stages, except stage 1, show a trend of decreasing sulfur concentration with sampling site from Pensacola to Miami. However, additional variability is present in sulfur

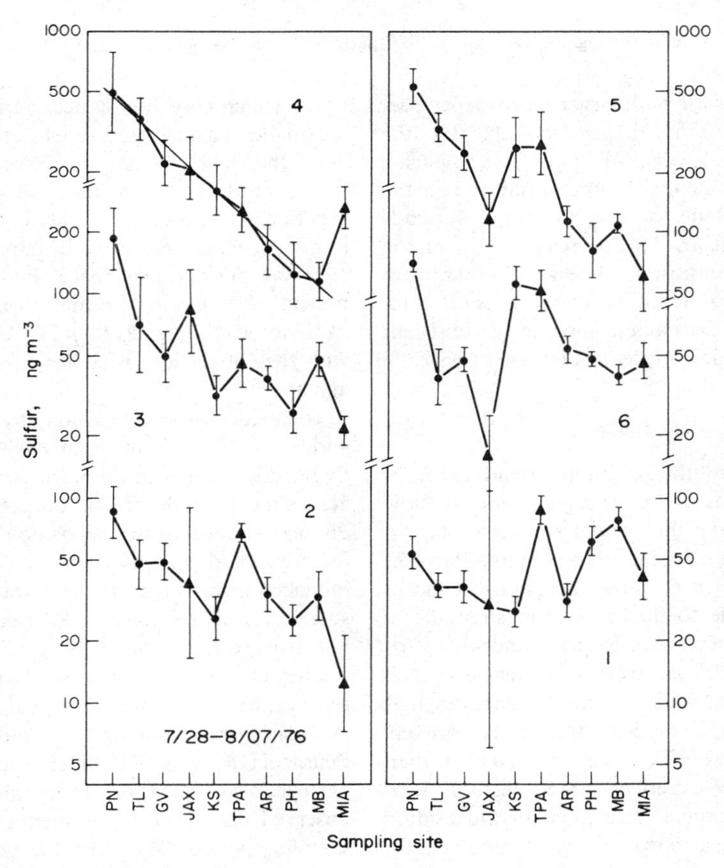

Fig. 2. Sulfur during the summer, July–August 1976, period in each particle size range: 1, $>4\ \mu m$; 2, 2–4 μm; 3, 1–2 μm; 4, 0.5–1 μm; 5, 0.25–0.5 μm; 6, $<0.25\ \mu m$ aerodynamic dia. Points represent geometric means, with the standard deviations of the means, for groups of approx six samples taken. Points are plotted against sampling site index. Absolute calibration of concentrations is significant to a factor of 2.

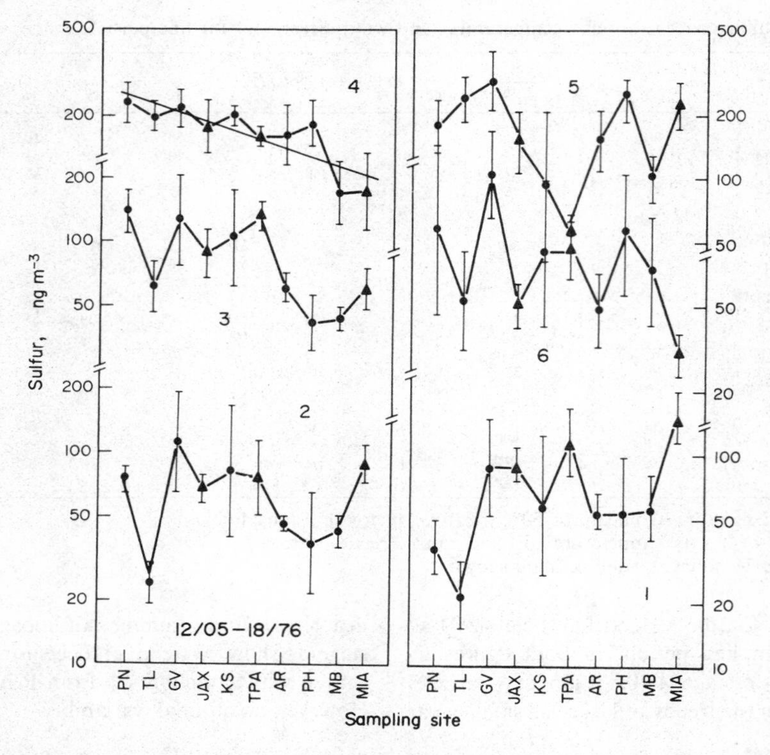

Fig. 3. Sulfur during the December 1976 period in each particle size range, as in Fig. 2.

concentration trends for both larger and smaller particles than the 0.5–1 μm range. Consequently, the trends shown in Fig. 1, where the data from impactor stages 3–6 are combined, are more variable than the trends of stage 4 alone. In fact, the stage 4 trends are remarkably regular. The decrease in geometric mean summer concentration of stage 4 sulfur from Pensacola to Marco Beach is nearly 10-fold, compared to 2-fold for December, apparently consistent with the meteorological evidence discussed above.

DISCUSSION

In order to interpret the geographic trends in sulfur over Florida in terms of source regions and air flow patterns, we may take the regular impactor stage 4 (0.5–1 μm) trends as a point of departure. Particles larger than this size range may experience concentration variation due to fluctuations in strengths of dispersion sources, such as sea spray, and also variations in wet and dry deposition efficiencies which depend on impaction and nucleation processes that favor larger particles. The 0.5–1 μm range is relatively less sensitive to these effects. Particles smaller than this size range may experience concentration variation due to fluctuations in strengths of condensation sources, such as the atmospheric conversion of SO_2 gas to sulfates, and also variations in particle coagulation efficiencies. The 0.5–1 μm range is relatively less sensitive to these effects. Consequently, the stage 4 size range for sampling is a range of relative stability in concentration when compared to locally perturbing effects which may alter concentrations in either larger or smaller particle size ranges. The 0.5–1 μm range may, therefore, contain particles with long atmospheric residence times, and the particles may have experienced more atmospheric aging than either larger or smaller particles. In paper 1 (Ahlberg *et al.*, 1978) we presented evidence that the principal component of sulfur in the fine particle mode is more likely to be $(NH_4)_2SO_4$ than H_2SO_4; this is consistent with the conversion of H_2SO_4 to $(NH_4)_2SO_4$ during aging.

The evidence points strongly to fine particle sulfur, and 0.5–1 μm sulfur in particular, originating primarily outside Florida to the north and west and moving across the state, depending on prevailing winds, with dilution and wet or dry deposition occurring en route. Superimposed on this trend are local effects which influence instantaneous concentrations at any point, such as freshly generated sulfate particles in the smallest sizes, coarse sea spray particles, transfer from smaller to larger sizes by coagulation, and wet or dry removal to the ground. The trend of decreasing average concentration along the length of Florida from Pensacola to Miami is most evident in the 'stable' particle size range 0.5–1 μm, and the trend was observed during both summer and December 1976 sampling periods. Whether it is typical on an annual basis may be predictable on the basis of detailed meteorological analysis of past air flow records over the state.

The potential public health interest in the discovery of a fine particle sulfur aerosol concentration gradient in Florida may be considerable. Because of national interest in the possible health effects of sulfates (EPA, 1975), as contrasted with gaseous SO_2, it is desirable to conduct epidemiological studies of human populations residing in a pollution gradient, such as been done in the CHESS studies in Utah (EPA, 1974). Sulfate levels in the eastern U.S. have been considered to exceed those in the west, but regular sulfate gradients have not previously been demonstrated over geographical areas comparable to the size of Florida. Therefore, Florida may be an ideal location to conduct epidemiological studies of the health effects of long term exposure to fine particle sulfur aerosol.

Acknowledgements—We are indebted to Stephen L. Cohn and Scott Rheingrover for assistance in meteorological analysis and to William E. Wilson, ESRL, EPA, Director of Project MISTT, for technical support. The study was financed in part by the Florida Sulfur Oxides Study, Inc., the National Institutes of Health, the National Science Foundation, and the Federal Interagency Energy/Environment Research and Development Program through EPA grant R803887. The conclusions reached do not necessarily represent endorsement by FSOS, Inc.

REFERENCES

Ahlberg M. S., Leslie A. C. D. and Winchester J. W. (1978) Characteristics of sulfur aerosol in Florida as determined by PIXE analysis. *Atmospheric Environment* **12,** 773–777.

U.S. Environmental Protection Agency (1974) Health Consequences of Sulfur Oxides: A Report from CHESS, 1970–71. EPA-650/1-74-004.

U.S. Environmental Protection Agency (1975) Position Paper on Regulation of Atmospheric Sulfates. EPA-450/2-75-007.

Atmospheric Environment Vol. 12, pp. 735–752. Pergamon Press 1978. Printed in Great Britain.

SPATIAL AND TEMPORAL DISTRIBUTIONS OF AIRBORNE SULFATE IN PARTS OF THE UNITED STATES

G. M. HIDY, P. K. MUELLER and E. Y. TONG

Environmental Research and Technology, Inc., 2625 Townsgate Road, Westlake Village, CA 91361, U.S.A.

(*First received* 23 *June* 1977 *and in final form* 12 *October* 1977)

Abstract—The current knowledge of the spatial and temporal distribution of airborne, water soluble sulfate is surveyed for two "scales" of atmospheric activity. The "urban" scale with episodes extending over a day or two over distances of about 100 km is illustrated for two comparable American cities, Los Angeles and New York. The regional scale with episodes extending up to several days over distances of 1000 km is exemplified by case studies in the greater eastern United States. Examination of available data reveals several features of the spatial and temporal variation in sulfate occurrence, including seasonal changes, and correlations with aerometric parameters. The importance of water vapor and air mass character on sulfate concentrations is assessed in both the urban and regional conditions. The results of initial attempts to simulate the impact of sulfur oxide emissions on ambient sulfate distributions are compared with an episode case extending over several days in July 1974.

1. INTRODUCTION

Since the early 1960s, workers have recognized that the concentrations of airborne particulate sulfate present in the atmosphere cannot be accounted for from natural sources such as sea spray, or from primary man-made emissions. The deductions that most of the sulfate comes from oxidation of sulfurous gases in the atmosphere is now widely accepted. Sulfate is of concern because of the epidemiological allusion to possible adverse effects on human health and welfare, the occurrence of acid precipitation, and visibility impairment in the eastern United States. The potential for increased sulfur oxide emissions has motivated considerable effort to elucidate the behavior of both sulfur dioxide and sulfate in the air.

Prior to recognizing the importance of atmospheric chemical transformations, the analysis of impact from sulfur emissions was concentrated within a few tens of km of their origin. The pioneering work of investigators in northern Europe provided important evidence for regional effects wherein cumulative emissions from concentrations of large sources can influence ambient air conditions hundreds of km away. A recent review of observations in the United States has suggested that large parts of the eastern United States have high (i.e. $> 10\,\mu g\,m^{-3}$ annual average) sulfate concentrations in both the cities and the rural areas (EPA, 1975). The growth of sulfur oxide emissions outside cities over the eastern United States has been ascribed by some to produce regional effects analogous to those identified in Europe. In contrast, the sulfate concentrations observed in the western United States are much lower than in the East except near isolated regions of high emission density. In the West, evidence of long range transport of pollution has not been found despite the existence of very large, but isolated rural sulfur oxide sources. Observations of this kind have led us to the hypothesis that the "regional" problem in the eastern United States is more related to the cumulative impact of major sources closely spaced along a direction parallel to the prevailing wind as distinct from chemical processes and atmospheric accumulation during long range transport (Hidy *et al.*, 1976).

The spatial and temporal behavior of materials in the atmosphere can be classified according to spatial and temporal "scales" as summarized in Table 1. Historically, the work leading to control regulations in the United States has focused on the local and urban scale problems, where end products of emissions from man's activities have their most intense impact. However, the potential for global pollution has been demonstrated with nuclear bomb debris, and with transport of lead-containing particles over very great distances from their origin. Elucidation of the urban and regional scales of pollution phenomena requires the deployment of a widespread network of monitoring stations. This year, such a network has been initiated in the eastern United States. Recent surveys of historical information which served as a basis for the new measurement effort are summarized in this paper.

The geographical areas surveyed included:

(a) The greater metropolitan New York City,
(b) The California South Coast Air Basin (metropolitan Los Angeles),
(c) The greater northeastern United States.

The observed temporal and spatial distributions are discussed in relation to:

(a) Meteorological parameters,

Table 1. Spatial and temporal scales for trace contaminant episode behavior in the atmosphere

Pollution class	Meteorological class	Horizontal spatial range	Temporal range	Comments
Local	Microscale	0–10 km	$\lesssim$1 day	"Classical" pollution impact – basis for non-reactive pollutant reactions.
Urban	Mesoscale	10–100 km	1–2 days	Identified in the 1950s largely with reactive pollutant products such as photochemical oxidant.
Regional	Synoptic scale	1000 km	1 – 5 days	Recognition of problems related to haze and visibility degradation, acid rain, sulfate.
Global	Planetary scale	> 1000 km	3–5 days	Extensively surveyed in relation to dispersal of radio nuclides from bomb tests.

(b) Emission distribution of sulfur oxides, assumed to be SO_2,

(c) The occurrence of other air pollutants,

(d) The current knowledge of atmospheric transport and oxidation chemistry.

The data used for this study have been derived from monitoring observations published by:

(a) The California Air Resources Board,

(b) The Los Angeles County Air Pollution Control District,

(c) The National Aerometric Data Bank (NADB),

(d) The U.S. Environmental Protection Agency's (EPA) Community Health Air Monitoring Program (CHAMP).

The sulfates were analyzed from water extracts of airborne particles collected for 24 hours on a glass fiber filter substrate. The potential limitations of these measurements were not recognized until recently, but have been assessed and discussed widely (e.g. EPA, 1975; USHR, 1976; Mueller, 1976). Some laboratory simulations indicate the potential for large errors, especially at low sulfate and high sulfur dioxide concentrations. However, comparisons of several methods and various kinds of field samples indicate the uncertainties are ±25% or less in reported sulfate data. Thus, we have assumed that the data are a useful representation of the ground level sulfate distributions in the United States.

2. THE URBAN PROBLEM

To illustrate the character of urban sulfate distributions, two metropolitan areas of a contrasting nature have been selected, New York and Los Angeles. Some statistics for these areas are listed in Table 2. The aerometric data are based on daily data obtained between January 1974 and May 1975 from CHAMP and related monitoring stations (Hidy *et al.*, 1977b).

The surface areas considered for the metropolitan regions are similar, but the population densities of these urban areas differ. The population density in the New York area is approximately twice that of Los Angeles. The cities represent two of the largest metropolitan areas in the United States, with a diversity in industry using fossil fuels for heating, power generation, transportation, refineries, and for chemical processing. Their total sulfur oxide emissions are similar, but New York's emission density is 40% larger than that of Los Angeles. Both cities are situated on a sea coast, and both are subject to a mesoscale oscillatory land–sea breeze flow superimposed on synoptic scale processes. Los Angeles has a mild, sunny, semi-arid, Mediterranean coastal climate, while New York is subject to strong seasonal influences modulated by alternate intrusions of continental polar and maritime tropical air masses.

The air flow into the Los Angeles area is normally dominated by a westerly component which brings in relatively fresh air from the Pacific Ocean. The air flows into a large basin which is surrounded by mountains that inhibit transport to the east.

Stable air conditions with low inversion height combined with the blocking of the mountains create conditions of stagnation within the basin. The oscillating, diurnal sea–land breeze regime leads to conditions of pollution accumulation of 1–4 day extent, especially during the late summer and early fall.

The air flow into the New York area is variable, but a westerly synoptic component is prevalent. New York lies on the 'downwind' side of the large eastern region of high sulfate identified by EPA (1975). It may be exposed to sulfate from long distance transport of pollutants as far away as western Pennsylvania and the upper Ohio Valley. Frequent southwesterly air flow along the coast can transport pollution from distant source concentrations in the Washington–New York corridor. Terrain features and the intensity of mesoscale weather disturbances cause stronger winds and greater mixing heights in New York compared with Los Angeles. An annual ventilation factor based on

Table 2. Some characteristics of pollution in the New York and Los Angeles areas. Numbers in parentheses are standard deviations of data

Parameter	Los Angeles	New York
Surface area considered* (km^2)	21,000	17,000
Population estimate (1970)	9,000,000	12,000,000
Population density (No. km^{-2})	430	710
SO_2 emissions (tons y^{-1})†	238,000	266,000
SO_2 emission density (kg km^{-2} y^{-1})	10,300	14,200
Maximum temperature, T_{max} (°C)‡	22.8 (5.5)	15.0 (7.4)
Minimum temperature, T_{min} (°C)‡	10.8 (4.6)	9.3 (8.4)
Relative humidity, r.h. (%)§	50.2 (17.0)	59.6 (16.5)
Normal precipitation (cm)	36	106
Mean wind speed, U (m s^{-1}) §§	3.3 (1.4)	5.8 (2.3)
Mixing height, H (m)‖	849 (472)	1290 (906)
Ventilation, U.H (m^2 s^{-1})	2690 (2160)	7460 (6200)
Sulfur dioxide, SO^2 (μg m^{-3})**	12.5 (19.9)	42.9 (45.0)
Water soluble sulfate, SO_4^{2-} (μg m^{-3})**	10.1 (7.9)	8.9 (5.7)
Nitrogen dioxide, NO_2 (μg m^{-3})**	83.9 (44.3)	67.6 (36.0)
Water soluble nitrate, NO_3^- (μg m^{-3})	9.1 (7.7)	2.6 (2.1)
Ozone, O_3 (μg m^{-3})**	52 (34)	20 (22)
Total particulate mass concentration (TSP) less sulfate and nitrate, TSPM (μg m^{-3})	64.5 (27.4)	40.4 (19.9)

* Greater metropolitan areas; Los Angeles–South Coast Air Basin; New York–tristate metropolitan area.
† Based on EPA Air Quality Control Regions.
‡ Annual mean of daily maximum or minimum hourly temperature.
§ Annual mean of daily minimum humidity.
§§ Annual mean of noon wind speed at surface.
‖ Defined by annual mean of daily midday radiosonde sounding.
** Annual mean of 24-h averaged values 1974–1975; Los Angeles 7 stations, New York 4 stations (see Hidy *et al.*, 1977b for details).

midday conditions is more substantial in New York than in Los Angeles.

Los Angeles has higher annual maximum temperatures and minimum temperatures than New York, on the average. The annual average of daily minimum r.h. is greater, and much more precipitation normally is recorded in the New York area than Los Angeles.

The air quality in the two cities exhibits interesting parallels, and striking differences, as indicated in Table 2. The mean particulate sulfate levels are similar, but New York has significantly higher sulfur dioxide concentrations than Los Angeles. Sulfur dioxide concentrations in New York are more than three times higher than in Los Angeles despite a much smaller difference in emission density and much higher ventilation in New York. Nitrogen dioxide concentrations are somewhat higher and the nitrate levels are much higher in Los Angeles than in New York. There is no complete explanation for such differences as yet. The relatively large sulfate and nitrate concentrations in Los Angeles have been ascribed to the highly oxidative nature of photochemical smog, and its influence on the production of aerosols in the atmosphere (e.g. Hidy and Burton, 1975). The average concentration of ozone in Los Angeles is more than twice that in New York. The total mass concentration of suspended particles, less the sulfate and nitrate contribution, is about 50% greater in Los Angeles than in the New York area. This may be related to higher airborne concentrations of organic material in the former city, and the intensity of photochemical smog in southern California.

2.1 *Geographical distribution*

The geographical distribution of sulfate concentrations in the two cities can be seen from the annual means of 24-h average observations from available air monitoring stations. The distribution in the New York metropolitan area is shown in Fig. 1 for 1972 data (Lynn *et al.*, 1975).

The 1972 annual average data are higher than the average in 1974–1975 reported for the four CHAMP stations in the same area. However, the CHAMP stations are located in areas peripheral to the zone of high sulfate concentrations. From Fig. 1, it is seen that there are significant differences in sulfate concentrations over the urban area, with the largest values observed in a strip extending from Staten Island northeastward to Brooklyn. Heavy local concentrations of sulfur oxide emissions are in the east-southeast of New Jersey, in Staten Island, Brooklyn and the high population density areas of Manhattan. Within a distance of a 10—50 km from the area of heavily concentrated sources, sulfate concentrations have decreased by 30–40% of the maximum values.

The distribution of the means of the annual average concentrations derived from 24-h values in the Los Angeles area are shown in Fig. 2. The distribution is relatively uniform over a large part of the South Coast

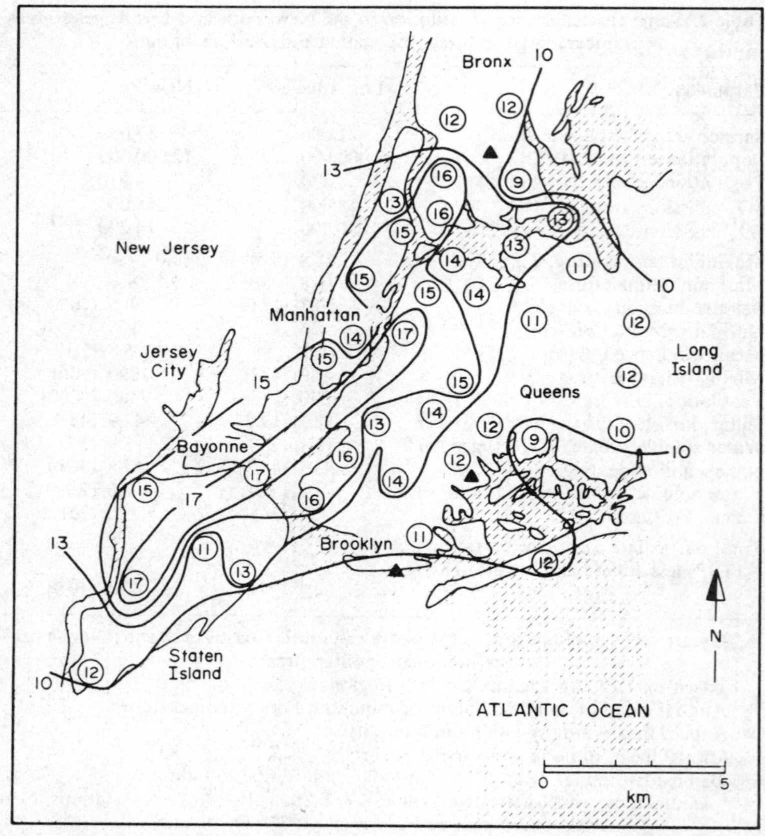

Fig. 1. Annual mean 24-h averaged sulfate levels ($\mu g\ m^{-3}$) in the New York area. Based on 1972 data (from Lynn *et al.*, 1975). Triangles are locations of three CHAMP stations. The fourth station is at the tip of Long Island about 160 km from Manhattan.

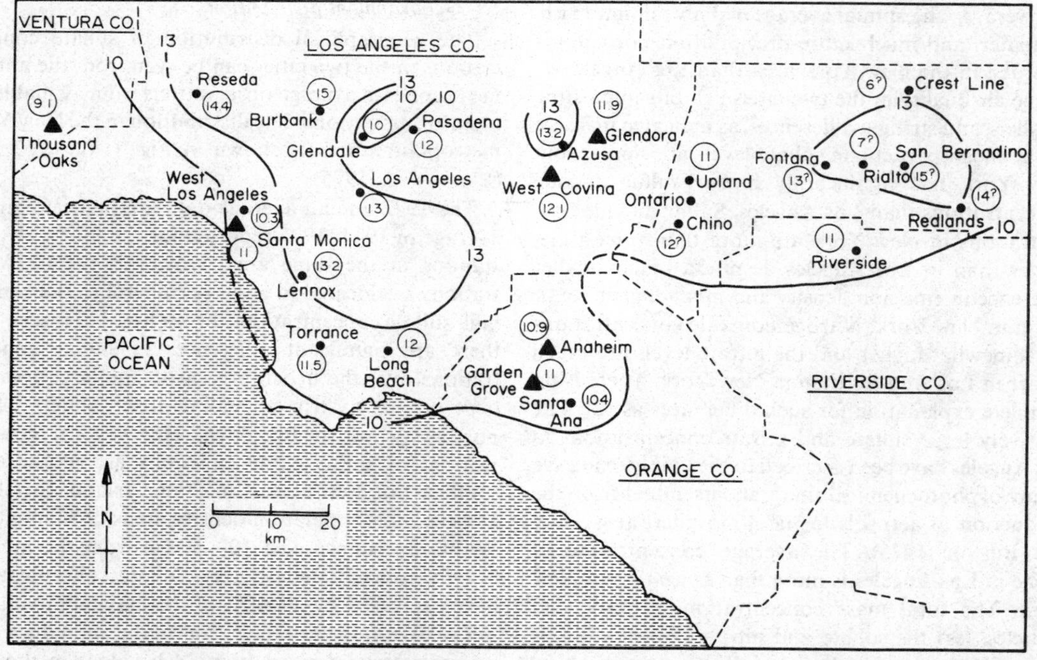

Fig. 2. Distribution of annual average sulfate observations in the greater Los Angeles area, based on 1972–1974 data. Circled numbers are station data; the triangles are CHAMP stations. Question marked data are uncertain because of discrepancies identified between agency analytical methods (from Kurosaka, 1976).

Air Basin, with a weak maximum in the central commercial district and the east San Fernando Valley around Burbank. Other maxima appear near Azusa, and in the San Bernardino area. The regions of highest localized concentration of sulfur oxide emissions are located in the El Segundo and Long Beach–Torrance areas, and in Fontana. Like the New York example, the sulfate concentrations appear to decrease significantly at distances exceeding 50 km or so from the areas of high emission density. This evidently reflects a characteristic scale for the intensity of impact of major sulfur oxide sources.

2.2 *Seasonal variation*

An interesting feature of the available monitoring data was reported for some non-urban locations in the New York–Connecticut area. Hitchcock (1976) found that a summer maximum in 24-h average sulfate concentrations existed in such data. Hidy *et al.* (1976) reported similar findings for several rural locations in the northeastern United States, and in California. But these workers found a much less well defined seasonal effect in 1973–1974 NADB observations in the southeast and in the midwest.

The seasonal variation in sulfur oxides has also been investigated in New York and Los Angeles. The seasonal patterns for 1972 were reported for New York by Lynn *et al.* (1975), and are shown in Fig. 3 along with SO_2 emissions in the area. The ambient sulfur dioxide concentrations correspond reasonably well with the sulfur dioxide emissions, showing a minimum in summer. However, the average sulfate concentrations displayed the summer maximum.

Seasonal changes in ambient sulfate at the downtown Los Angeles monitoring station in 1973 are shown in Fig. 4. Again the summer seasonal maximum is observed, which is distinct from the variations in monthly Los Angeles County power plant emissions. Analysis of the sulfur oxide emissions in Los Angeles

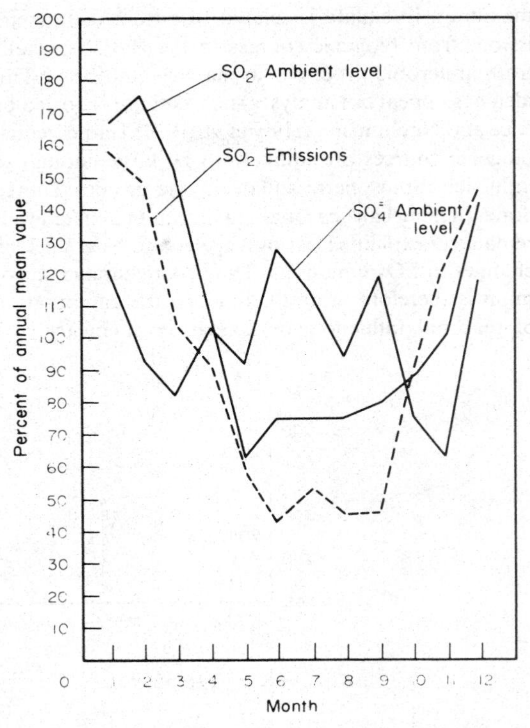

Fig. 3. 1972 Seasonal patterns of SO_2 emissions, and 24-h averaged SO_2 and SO_4 ambient levels in the New York area normalized to the annual average values (from Lynn *et al.*, 1975).

County indicates that daily changes in sulfur oxides typically are largely ascribable to power plant emissions (e.g. Joint Project, 1977); other sources such as the chemical plants, refineries and transport are reasonably steady from day to day. If this were the case in 1973, then the seasonal sulfate maximum is not correlated with sulfur oxide emissions as in the New York example.

Hitchcock (1976) has speculated that the seasonal

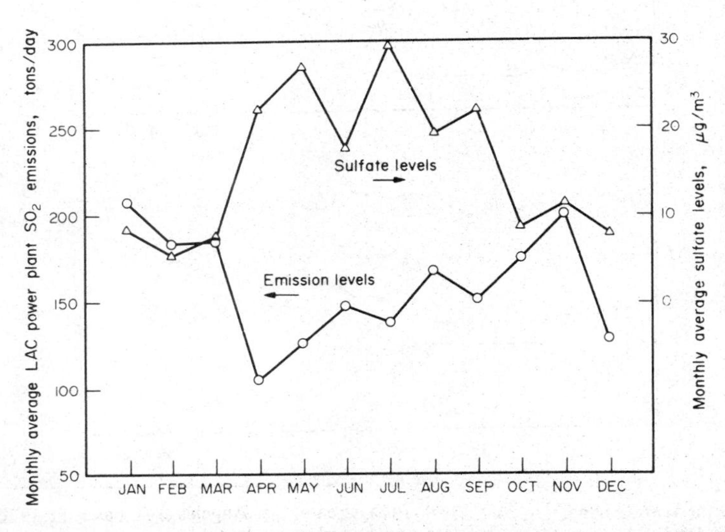

Fig. 4. Monthly variation in monthly mean of 24-h averaged sulfate concentration at downtown Los Angeles, compared with monthly averaged 1973 Los Angeles County power plant emissions (Mirabella, 1977).

maximum in sulfate is related to natural sulfur emissions from biogenic sources in the East, especially from anaerobic processes in marshes and bogs. This idea has appeal but analyses such as those reported by Rice and Nochumson (Hidy *et al.*, 1977a) indicate that biogenic sources are unlikely to be large enough to influence atmospheric sulfate in the eastern United States. The urban seasonal maximum in sulfate levels remains unexplained in Los Angeles and New York by changes in SO_2 emissions. The urban summer maximum is therefore more likely to be the combination of seasonal influences on oxidation chemistry and meteorology. In particular, the summer maximum in photochemical smog intensity may be related to high sulfate levels in summer (e.g. CPUC, 1976).

2.3 *Sulfate and other aeromatic variables*

Detailed study of the changes in atmospheric sulfate and other particle chemistry over intervals of less than 24 h has been reported only in the Los Angeles area. A series of measurements were taken in 1969, 1972 and 1973 during the sampling programs in Los Angeles. The two latter years were undertaken as part of major air chemistry study called the California Aerosol

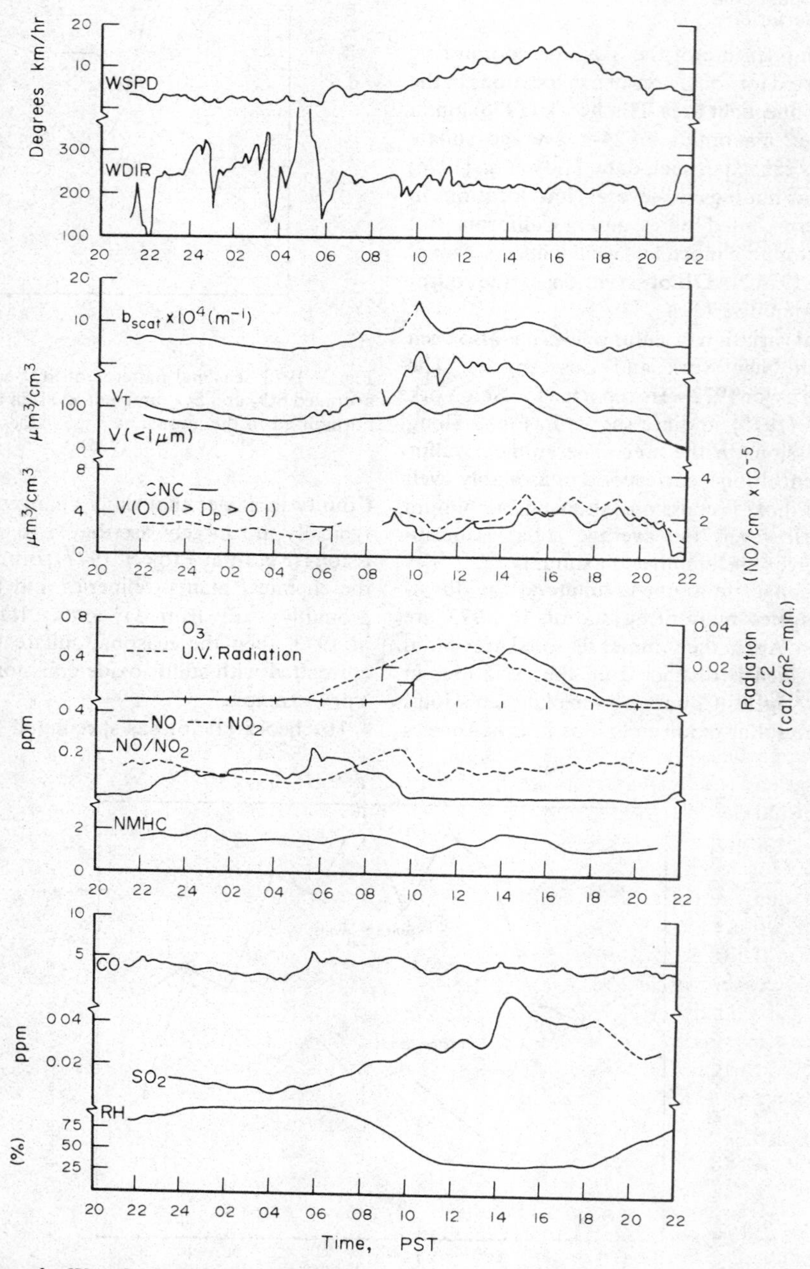

Fig. 5. Diurnal patterns for West Covina, CA, 23–24 July, 1973, V_T = total particle volume $V(< 1\ \mu m)$ = particle volume less than 1 μm dia., $V(0.01 > Dp > 0.1)$ = particle volume in the range 0.01 = 0.1 μm dia. (D_p), and CNC = condensation nuclei count.

Characterization Experiment (ACHEX) (Hidy *et al.*, 1975). These investigations were aimed at elucidating the relationships between aerosol particles and gases in photochemically associated hazes (Hidy *et al.*, 1975). Figure 5 exemplifies diurnal changes in gases and particle size distribution or light scattering parameters. Here b_{scat} is the light scattering coefficient for suspended particles, and CNC is the condensation nuclei concentration measured with an Env 1-Rich counter. The distribution ratios are defined as:

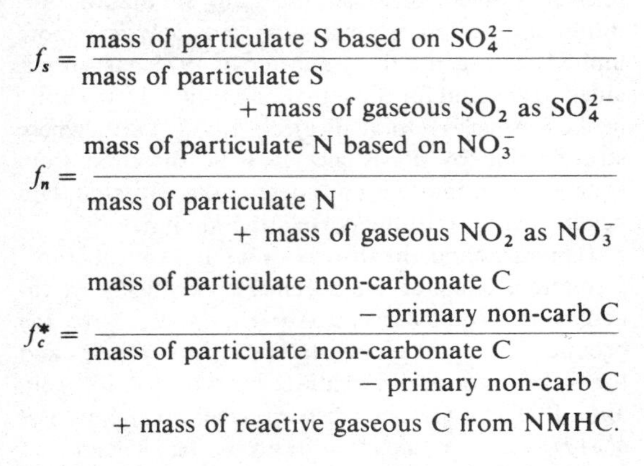

$$f_s = \frac{\text{mass of particulate S based on } SO_4^{2-}}{\text{mass of particulate S} + \text{mass of gaseous } SO_2 \text{ as } SO_4^{2-}}$$

$$f_n = \frac{\text{mass of particulate N based on } NO_3^-}{\text{mass of particulate N} + \text{mass of gaseous } NO_2 \text{ as } NO_3^-}$$

$$f_c^* = \frac{\text{mass of particulate non-carbonate C} - \text{primary non-carb C}}{\text{mass of particulate non-carbonate C} - \text{primary non-carb C} + \text{mass of reactive gaseous C from NMHC.}}$$

The variation in reactive gases at West Covina (Fig. 5) is the "classical" picture for photochemical smog. The NO builds up at night but depletes during the day with an early morning maximum from automobile sources. Later, in mid-morning, an NO_2 peak is observed followed by an ozone peak in early afternoon. The non-methane hydrocarbons are depleted in the morning as smog reactions proceed. The CO concentration remains roughly constant while the SO_2 peaks with the O_3 in mid-afternoon during transport of air eastward across the air basin. The humidity decreases to a minimum by midday.

The diurnal variations in particulate parameters depend on particle size. The CNC count (mostly less than 0.1 μm dia particles) follows the particle volume in this size range and reaches maximum late in the day. However, the total particle volume and the less than 1 μm dia particle volume follow the b_{scat} pattern, and peak at midday just before the ozone maximum.

Particulate sulfur, nitrate and non-carbonate carbon patterns in two size factions are shown in Fig. 6 for samples taken in the same West Covina episode. The particulate sulfur, carbon and the light scattering peak, at midday to early afternoon. However, particulate nitrate reaches a maximum in mid-morning. In contrast to sulfur and carbon, much less nitrate was found in the submicron particle fraction.

* Primary non-carbonate C is estimated to be mainly from automobile emissions in Los Angeles, and is derived from a proportionality to airborne lead concentration (Hidy and Burton, 1975). NMHC corresponds to non-methane hydrocarbon vapors, and was estimated as a propane equivalent for mass concentration estimation. The $\geqslant C_6$ vapor concentration was estimated from chromatographic analysis and was calculated on the basis of a mass density of twice that of propane.

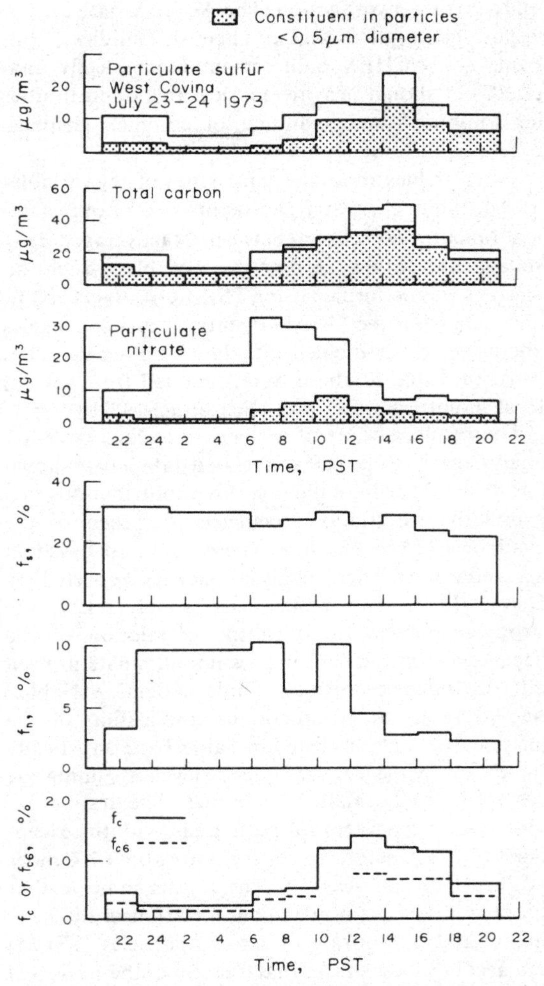

Fig. 6. Aerosol diurnal pattern for West Covina, CA, 23–24 July, 1974 (from Hidy and Burton, 1975).

The distribution ratios provide a measure of the conversion of gases to particulate material. The variations in distribution ratios during the day are indicated in Fig. 6. The value of f_s remained very uniform at approx 30%. The f_n maximum was at night and morning at about 10% with the period of high humidity. The f_c was maximum in mid-afternoon at about 1.5%.

If there are no losses or gains of contaminants, and particles disperse analogously to gases, the changes in distribution ratios can be interpreted in terms of the extent of completion of chemical transformation processes. However, the injection of precursors at varying times in a moving air parcel, combined with differences in rates of SO_2 and particulate removal at the ground, preclude any simple interpretation of the distribution ratios.

Examined with concentrations of the precursors and products, as well as other aerometric data, the marked differences in diurnal behavior of the distribution ratios offer clues about the interactions of source distributions, air transport and chemistry which require further investigation. The ACHEX data set is a useful one for additional air chemistry studies of this kind. The ACHEX data remain incompletely analyzed, but should provide considerable opportunities for improved understanding of complex chemical processes in smog.

Further clues about the consistency of key variables in the sulfate chemistry can be obtained from examination of the correlation between 24 h averaged daily sulfate and other aerometric variables. Such an analysis was performed using TSP, SO_4^{2-}, NO_3^-, NO_2, and SO_2 from the CHAMP stations in Los Angeles and New York combined with data for other variables listed in Table 3. These were obtained from several local monitoring sites (see Hidy *et al*., 1977b).

The results of part of a stepwise linear regression analysis of the CHAMP and related data set are shown in Table 3. Listed are the first three contributions to a cumulative correlation coefficient for each of the available CHAMP stations. These results are based on a standard statistical analysis package reported by Dixon (1973). The method successively selects the strongest contributor(s) to the correlation of the dependent variable, but it does not eliminate in itself interdependencies between "independent" variables. According to the protocol for application of the method, the rankings listed in Table 3 are statistically significant to the 99% confidence level, assuming the variables are normally distributed. The results are surprisingly consistent for both areas, with the exception of Vista, California, a community about 100 miles southeast of Los Angeles. The results indicate that ozone, r.h. and total mass concentration of the non-sulfate and non-nitrate of aerosol particles (TSPM) can account for a substantial fraction of the statistical variability in sulfate at the listed receptor sites. The 24-hour averaged ozone (O_3) or daily maximum of ozone concentration (OX) are indicators of photochemical oxidation. The r.h. at noon (r.h.) is an indicator of water content in the air, and TSPM is an indicator of "reactions" involving suspended particles. The local SO_2 concentrations do *not* enter into the correlation sequence as one of the three "principal" variables.

It is not known what makes up the remainder of the variability in sulfate after consideration of all of the selected ambient variables, including ventilation. Addition of variations in emissions in an analysis of more limited data in the third quarter of 1975 has, so far, failed to account for the remaining residual variability in Los Angeles (Joint Project, 1977). Only where stratification by persistence in wind direction from sources to receptors is considered does emission data begin to enter into the correlation analysis.

To understand the dispersion and chemical transformations of sulfur, measurements need to be extended from ground level to heights well above the ground. In several recent studies in Los Angeles and the St. Louis area, observations have been made from aircraft. These are described elsewhere (e.g. Smith *et al*., 1978; Blumenthal *et al*., in press). The behavior of aerosol particles aloft is as varied and complex as at the ground. Observations suggest that an urban plume containing high concentrations of sulfur oxides, can be tracked up to 200 km from the source. The plume may be trapped in stable air at heights from 300–1000 m, without appreciable mixing to the ground, and can travel for considerable distances over a period of a day. These layers may be mixed to the ground with midday inversion breakup, but their impact at ground level has not been documented well, as yet (see also Martin and

Table 3. Primary ranking of variables for correlating of airborne sulfate in two cities based on a stepwise linear regression analysis of 15 variables from CHAMP and related monitoring stations (from Hidy *et al*., 1977b)

A. Los Angeles area

Variable	Anaheim	Garden Grove	West Covina	Glendora	Santa Monica	Thousand Oaks	Vista*
1	O_3	O_3	O_3	TSPM	O_3	TSPM	T_{min}
2	TSPM	TSPM	TSPM	r.h.	r.h.	r.h.	OX†
3	r.h.	r.h.	r.h.	O_3	TSPM	OX	r.h.
Correlation coefficient (R)	0.71	0.77	0.79	0.79	0.79	0.72	0.56

B. New York

Variable	Brooklyn	Queens	Bronx	Riverhead, L.I.
1	TSPM	TSPM	TSPM	TSPM
2	RH	RH	O_3	RH
3	O_3	O_3	RH	O_3
Correlation coefficient (R)	0.60	0.63	0.54	0.62

* Located 50 km north of San Diego and 16 km inland from the coast.
† Ox = 1 h daily maximum ozone value.

Barber, 1973). The transport of contaminants in stable air layers aloft, without the influence of surface removal processes, may provide a mechanism for high non-urban concentrations at the ground. However, it is difficult to visualize from our knowledge of dispersion processes how high concentrations of pollutants could remain intact for more than 12–24 h in the troposphere (e.g. Rodhe, 1972; Smith and Jeffrey, 1975). At wind speeds of 5–10 m s^{-1}, the significant exposure to polluted air should occur within a maximum of 200–800 km of the source.

3. REGIONAL SCALE PHENOMENA

Until about a year ago, little information of any kind was available which characterized pollution on a regional scale in the United States. The review of the U.S. EPA in 1975 suggested long range transport could be a major factor in the high sulfate concentrations observed in the greater northeastern United States. However, there was little objective basis for such a simple interpretation of the early monitoring data. The spatial and the temporal scales for transport and oxidation chemistry were (and still are) poorly understood; the distinction between urban scale influences and a regional scale influence could not be made. In the monitoring data available by the early 1970s it was evident, however, that the observed high *non-urban* sulfate levels were concentrated in geographical zones of high emission density, such as the Ohio Valley region and eastward. The U.S. national distribution of annual averaged airborne sulfate corresponded to the distribution of intense emissions (see, e.g., EPA, 1975).

During the 1960s, SO_2 emissions and ambient concentrations decreased in cities, while urban and non-urban sulfate levels remained nearly constant over several years, although rural SO_2 emissions increased in the eastern United States. Analogous trends in sulfur oxides are not well documented in the west, except possibly in the cities such as Los Angeles. The interpretation of the trends in monitoring data simply by variations in emissions is confused by the recent evidence of inaccuracies in both the sampling and the analytical methods for sulfur oxides. It is now known that the SO_2 bubbler method is subject to serious uncertainties resulting from temperature effects, and reproducibility of data is of concern (e.g. Barnes, 1973). The sampling of particulate sulfate can be confounded by SO_2 absorption on the filter medium (e.g. Mueller, 1976). The analytical method for sulfate was modified at least once during the past five years. Interpretation of the historical sulfur oxide data can only be made in terms of emission trends if the sampling and analytical methods were reproducible. This reproducibility is difficult to verify from historical records of the data sources.

As a result of the use of available daily monitoring data taken from a dozen rural stations in a strip roughly 1000 km in extent, an improved picture of the nature and extent of the regional sulfur oxide behavior in the eastern United States was obtained by Hidy *et al.* (1976). These results provided part of the basis for the design and implementation of the Electric Power

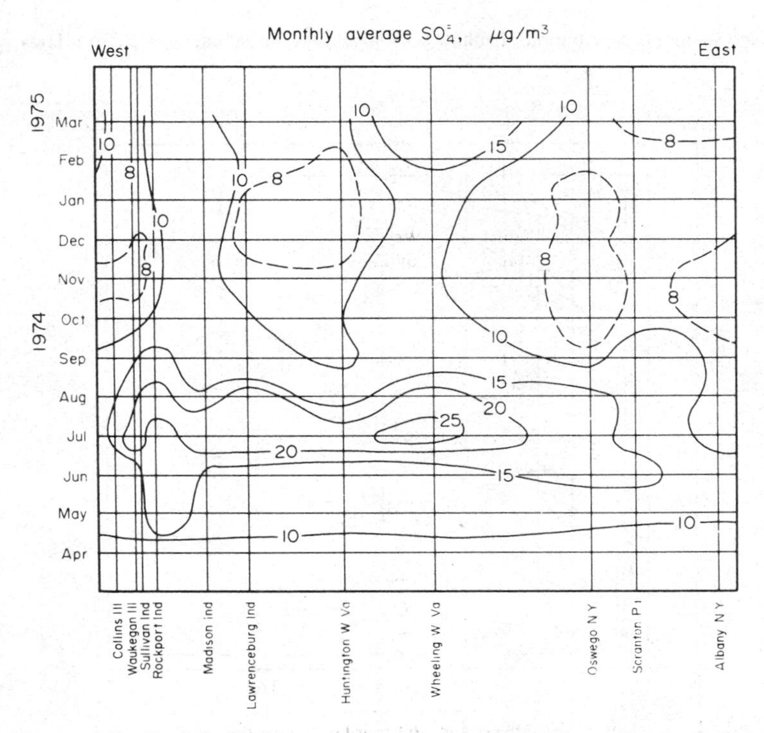

Fig. 7. Regional and seasonal distribution of sulfate concentrations over the Northeast Quadrant of the United States (from Hidy *et al.*, 1976).

Research Institute's (EPRI) Sulfate Regional Experiment (SURE). This experiment is aimed at a detailed characterization of the regional scale pollution phenomenon in the greater northeastern United States using combined ground and aircraft observations over an extended observational period of approximately one and a half years.

3.1 *Seasonal variations*

The seasonal variations similar to those observed in the two cities appear in rural data over a wide region. Daily observations of sulfur oxides were obtained in 1974 and 1975 from stations in Indiana and Illinois eastward to Albany, New York. The monthly averaged sulfate time–distance transect from west to east over a year period is shown in Fig. 7. The data show a widespread summer maximum in 24-h average sulfate concentrations which is prevalent over a distance of 1600 km.

The seasonal differences in sulfate concentration

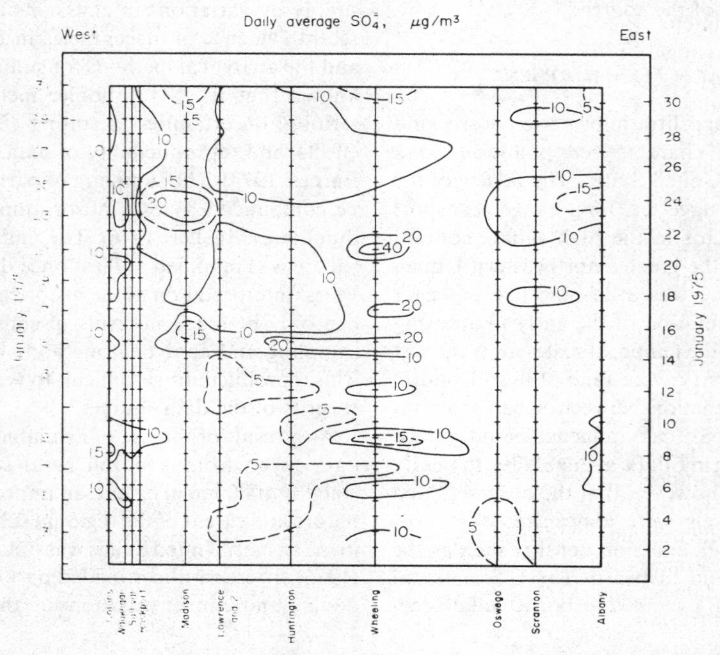

Fig. 8. Spatial and temporal distribution of sulfate concentrations for January, 1975 (from Hidy *et al.*, 1976).

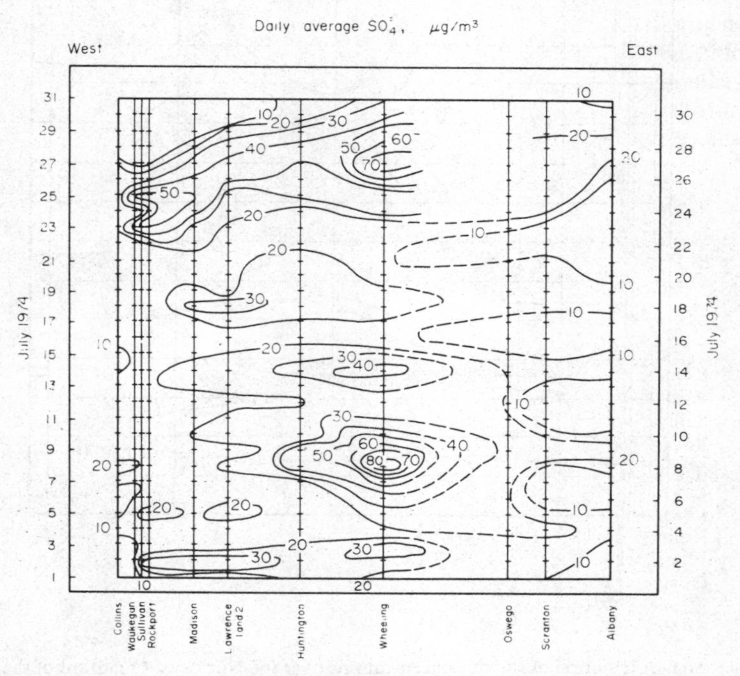

Fig. 9. Spatial and temporal distribution of sulfate concentrations for July, 1974 (Hidy *et al.*, 1976).

and distribution are elucidated in more detail by examining the daily variation by station in the summer and the winter. The time–distance transects for daily rural observations are shown in Fig. 8 for January 1975, and for July 1974 in Fig. 9. In the winter, the high sulfate concentrations are localized and generally the maxima are lower in winter than those in the summer. Winter episodes appear to last only a day or two without a persistent interval between occurrence. The summer case shows intense episodes that appear to evolve over several hundred km and have durations of 1–5 days. A cycle of intense episodes may occur every 5–10 days on a cycle roughly similar to the passage of synoptic scale weather disturbances.

The differences between summer and winter levels are too large to be explained by seasonal variations in industrial sulfur oxide emissions as reported by the estimates of the EPA National Emissions Data System (NEDS). Again, the biogenic sulfur processes may be a factor, but the combination of atmospheric chemistry and meteorology are important to this phenomenon.

An analysis of the rural sulfate data showed a systematic variation in sulfate concentrations with temperature and dew point, and to a lesser extent, total mass concentration of particles. SO_2 concentration correlated weakly with sulfate measured at the same station (Tong *et al.*, 1976a). Temperature and humidity are not only classical indicators of air mass category, but also appear to be important in the oxidation chemistry of SO_2. Sulfate concentrations have been classified by synoptic air mass category. The results for the 1974–1975 rural data set are listed in Table 4. The highest sulfate concentrations over the region covered by the twelve stations correlated with maritime tropical air masses (mT), bringing moist warm air inland from the south over areas of high emission density. The lowest sulfate concentrations were correlated with continental polar (cPk) air masses with penetration of relatively dry, air entering the United States over regions of high localized sulfur oxide emissions in the Great Lakes area and eastward.

3.2 *Case studies of regional behavior*

As extensions of earlier analyses, case studies of the regional behavior of sulfates have been developed by Tong *et al.* (1976b). Some 27 cases in 1974 and 1975 were identified where the non-urban sampling programs produced sufficient concurrent data for a wide area analysis. Sulfate data from more than a hundred stations were used. In their paper, these investigators described the synoptic conditions for four different situations in some detail.

The results of this investigation provide an invaluable insight into the complicated nature of regional scale sulfate behavior. Two extreme situations are outlined here for illustration. One in July 1974 has the appearance of a significant regional scale phenomenon associated with widespread stagnation of air over the eastern United States. A second in October 1974, shows a more localized "urban" scale sulfate buildup.

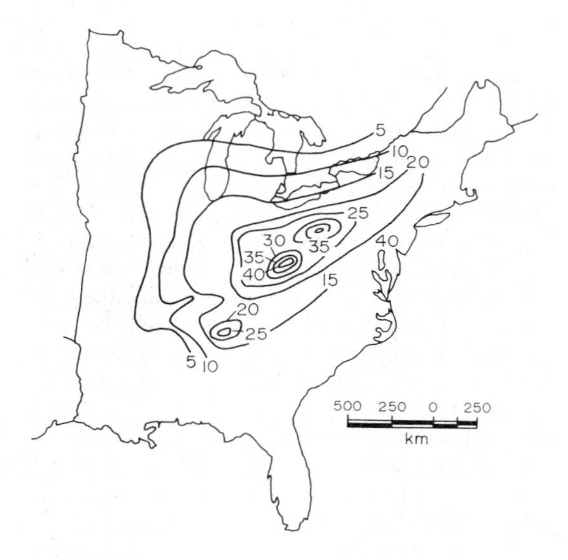

Fig. 10. Atmospheric sulfate distribution ($\mu g\ m^{-3}$) for 8 July, 1974.

Table 5. Average sulfate concentration categorized by air mass (sample size in parentheses) in $\mu g\ m^{-3}$. Transitional conditions are situations not identifiable by distinct cPk or mT air mass character (from Tong *et al.*, 1977a)

Station	cPk	Transitional	mT
Collins, Illinois	8.2 (54)	9.8 (203)	16.5 (29)
Waukegan, Illinois	6.4 (44)	10.0 (157)	22.8 (14)
Sullivan, Indiana*	6.4 (24)	9.3 (118)	7.1 (1)
Rockport, Indiana	9.6 (44)	11.9 (191)	21.9 (33)
Madison, Indiana	4.2 (32)	12.4 (199)	17.6 (60)
Lawrenceburg, I, Indiana	7.0 (45)	10.6 (223)	19.9 (67)
Lawrenceburg, II, Indiana	7.3 (41)	11.1 (211)	19.7 (65)
Huntington, West Virginia	8.8 (41)	8.6 (217)	16.0 (62)
Wheeling, West Virginia	12.9 (32)	15.2 (221)	18.8 (28)
Oswego, New York*	8.4 (18)	7.9 (86)	7.3 (20)
Scranton, Pennsylvania	6.7 (55)	10.5 (227)	14.6 (26)
Albany, New York	4.7 (49)	8.8 (233)	12.6 (23)

* Summer data missing.

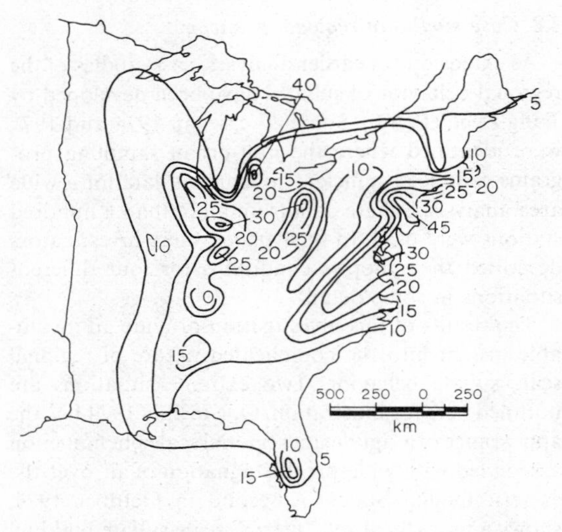

Fig. 11. Spatial distribution of atmospheric sulfate concentration ($\mu g\ m^{-3}$) for 10 July, 1974 (from Tong *et al.*, 1976b).

July 8–11, 1974. As suggested in Fig. 9, this episode actually began on 4 July engulfing much of the Ohio Valley. The episode reached its peak west of the Appalachian Mountains on 8 July, with levels exceeding 40 $\mu g\ m^{-3}$ over a large area. The extent of the area covered is shown by the sulfate distribution in Fig. 10. By 10 July, this intense, widespread zone of high sulfate had subsided and reduced into localized clusters of high sulfate west of the Appalachian mountains and to the south. However, an extensive zone of high sulfate concentration appeared east of the mountains with a maximum in the New York–Philadelphia area and southward, to North Carolina (Fig. 11). The highest sulfate of 45 $\mu g\ m^{-3}$ also appeared in southern Michigan. Lower peak concentrations of 30 $\mu g\ m^{-3}$ were observed in Ohio and Indiana. The widespread regions of high sulfate concentration roughly correspond to

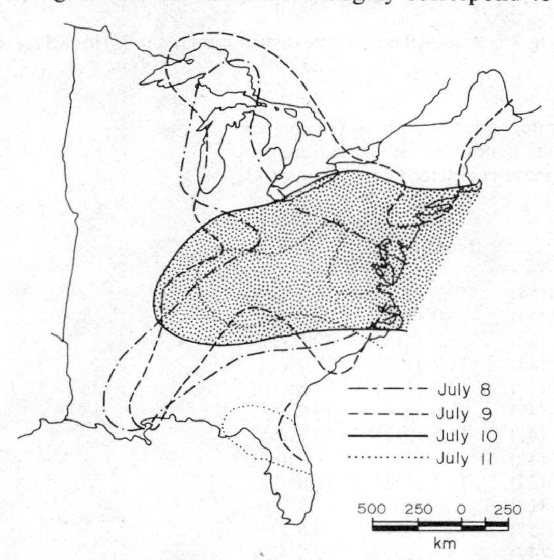

Fig. 12. Isopleths of four-mile midday visibility for 8–11 July, 1974 (from Tong *et al.*, 1976b).

areas having low visibility as reported by the Federal Airways network. Zones of less than four mile visibility at midday are shown in Fig. 12.

These zones cover the same general areas as the zones of elevated sulfate concentration. However, the correspondence between the two variables cannot be considered quantitative in either the 8 July or 10 July case.

The synoptic meteorology during this four-day period was dominated by warm humid maritime tropical air over much of the entire region. On 8 July, a slow moving cold front extended from Minnesota to Quebec and advanced southwards. The position of the front on successive days is shown in Fig. 13. By and large, the areas of high sulfate were formed south of the front in warm moist air. On 8 and 9 July, the synoptic scale air flow was light ($\lesssim 18.5\ km\ h^{-1}$ at the surface) and southerly to southwesterly in direction. At 850 mb the winds were about 28 $km\ h^{-1}$ south to southwesterly direction inland and between northwest and northeast along the coast. As the weak anticyclone moved southward on 10 and 11 July the winds remained light and southerly at the surface but westerly aloft. The winds, though weak, might have been persistent enough to move the polluted air northeastward in one to two days, which could explain the shift in regional high concentrations of sulfate. However, this "simple" explanation for the observed change in sulfate distribution is complicated by the distribution of large amounts of moist air which extended up to 850 mb. Fig. 14 gives the 21°C (70°F) isodrosotherms (dew point). The high sulfate concentrations were found within an area of dew point temperature $\gtrsim 17°C$ (62°F). Moist air had entered far up the Ohio Valley during the episode as well as along the east coast to New Jersey. The penetration of moist air appears to have extended further north and west by 10 July than on the 8 July. However, the influence of moisture content on the oxidation of sulphur dioxide would accelerate locally the sulfate formation in zones of high water content.

It appears that this episode was characterized by a

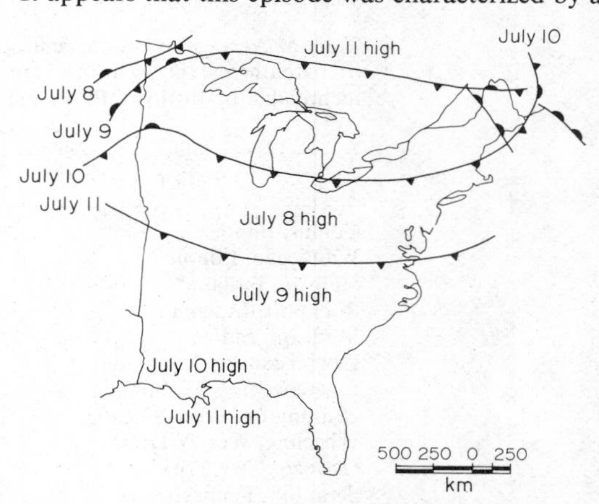

Fig. 13. Synoptic map showing positions of fronts and significant pressure centers, 8–11 July, 1974 (from Tong *et al.*, 1976b).

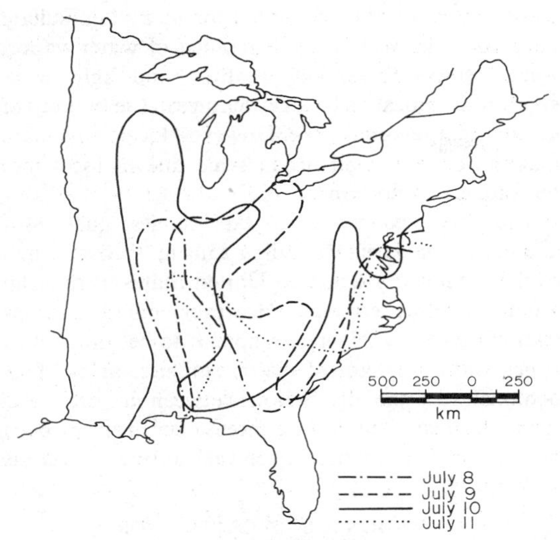

Fig. 14. 17°C (62°F) isodrosotherms for 8–11 July, 1974 (from Tong *et al.*, 1976b).

local, urban scale build up of sulfate around zones of high emission density with nearly stagnant air on which was superimposed a broad scale movement of polluted air eastward.

30 *Septenber–2 October*, 1974. The sulfate distribution during this period is in strong contrast to the intense episode build up in July. The distribution for 2 October is shown in Fig. 15. Sulfate concentrations are uniformly below 10 $\mu g\,m^{-3}$ over the entire eastern United States except for a localized high zone around Chicago. Other small areas of high concentration may have existed without identification between the available monitoring sites.

The weather patterns at the end of September and early October, are shown in Fig. 16, wherein the entire eastern half of the United States was under the influence of a continental polar air mass. An anti-

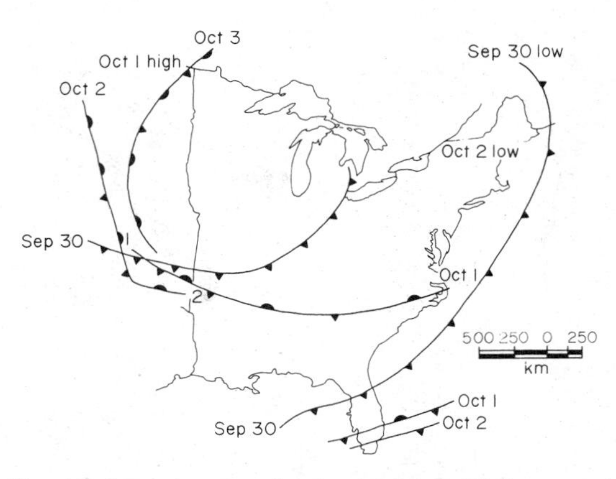

Fig. 16. Synoptic map showing positions of fronts and significant pressure centers for 30 September–3 October, 1974 (from Tong *et al.*, 1976b).

cyclone that was over southern Manitoba on 1 October moved southeast into Wisconsin the next day, and strongly influenced the air circulation over the entire region. By 3 October, the centre of the high pressure region moved further southeast into Ohio.

Visibility (Fig. 17) was less than 4 miles only in highly localized areas; e.g. an east-west strip from Massachusetts to New York state, and in the mid-Ohio Valley. The correspondence with elevated sulfate levels in the Chicago area was not particularly good, nor did the high sulfate zones in the east or the Ohio Valley show reduced visibility.

During this period, the region of dew point encompassed by the 17°C (62°F) isodrosotherm remained far south in Florida (Fig. 18). At 850 mb, large dew point depressions were present over the southern part of the region.

On 30 September, surface winds were blowing from the northwest at speeds between 9 and 18.5 $km\,h^{-1}$. Light southerly or southwesterly flow was observed between the Appalachians and the cold front, while

Fig. 15. Spatial distribution of atmospheric sulfate concentrations ($\mu g\,m^{-3}$) for 2 October, 1974 (from Tong *et al.*, 1976b).

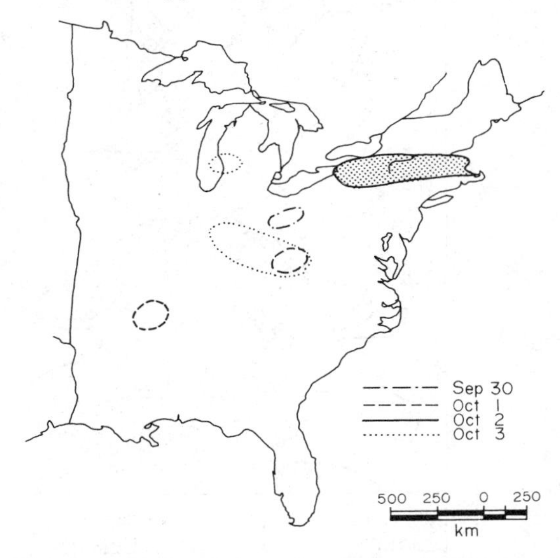

Fig. 17. Isopleths of four-mile midday visibility for 30 September–3 October, 1974 (from Tong *et al.*, 1976b).

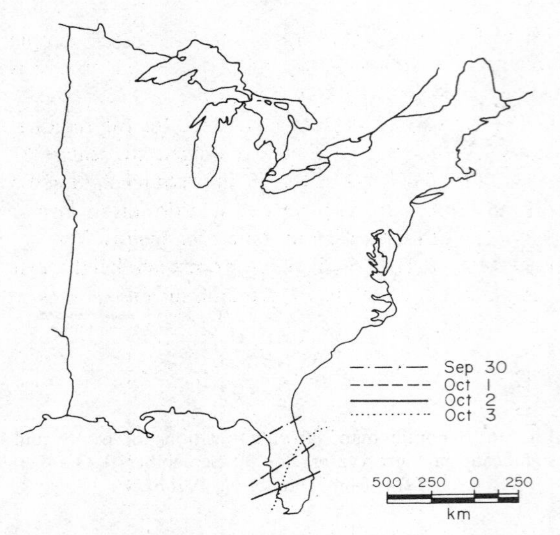

Fig. 18. 17°C (62°F) Isodrosotherms for the period 30 September–3 October, 1974.

stronger northwest winds of 28 km h^{-1} were present behind the front. On 2 October, the winds increased in places to 28 km h^{-1} with a shift from northeast to northwest. The analysis of the winds suggests a broad southward advection of pollutants from the Chicago area, the Ohio Valley and the northeast states. However, aerometric conditions must have precluded any widespread accumulation or buildup of sulfate to the south despite an (assumed) emission level and distribution equivalent to the July episode.

The results of the case studies reported by Tong *et al.* suggest that the interplay between the transport processes in the atmosphere, and the atmospheric chemistry has to be accounted for to explain sulfate behavior. The widespread presence of water vapor, warm temperatures, fair weather, and light wind conditions appear to be a key feature in the buildup of sulfate on a regional scale. However, localized sulfate maxima occur on an urban scale due to local meteorological conditions.

The cases examined to date suggest that 24-h average sulfate levels exceeding 25μg m^{-3} cover much of the greater northeastern United States during the intense summer episodes. However, sulfate concentration exceeding 25 μg m^{-3} appear to be confined to zones within 300 km of major emission areas. This demonstrates the distinction between a local and regional phenomenon. The intense summer episodes occurred only a few times each year during the period 1974–1975.

4. REGIONAL SOURCE-RECEPTOR RELATIONSHIPS

The quantitative analysis of the relationships between the distribution of sulfur oxide emissions and the ambient sulfur oxide conditions is facilitated by the use of an air quality model. Methods using both Lagrangian air mass trajectory calculations, and Eulerian grid techniques have been employed in Europe for several years. Similar approaches are now being adopted in the United States.

One promising approach to the regional analysis was reported as part of the preliminary study leading to the SURE (Hidy *et al.*, 1976). A quasi-Lagrangian calculation scheme developed by Egan and Mahoney (1972) was adapted by S. Rao to cover a broad zone of the greater northeastern United States, 1700 × 2300 km in extent. In its present form, this simulation

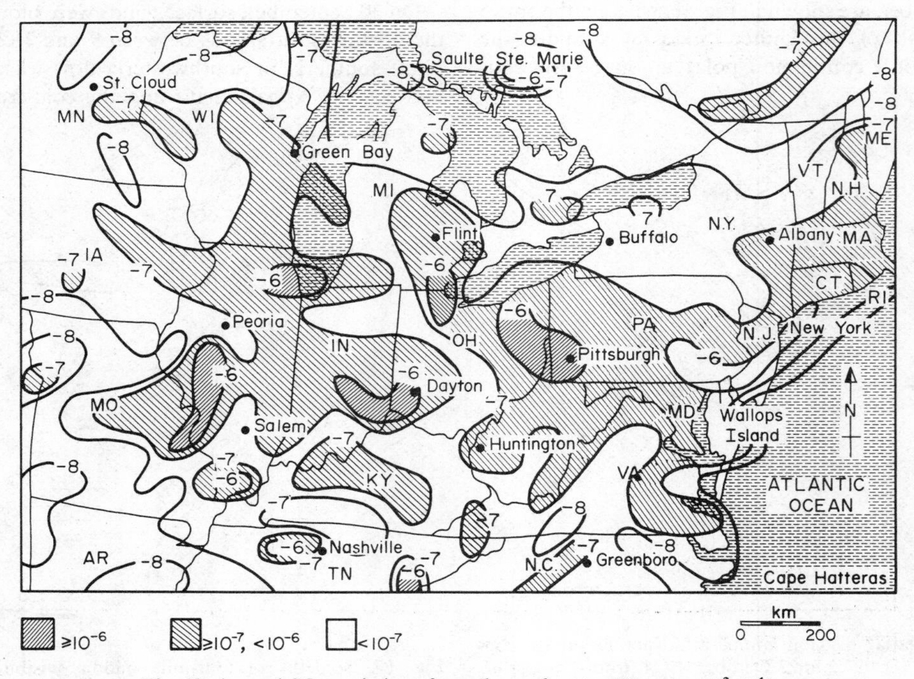

Fig. 19. Annual SO_2 emissions from the surface to 300 m (g m^{-2} s^{-1}).

method, EGAMA/SULFA3D, calculates both sulfur dioxide distributions and sulfate distribution in three vertical layers, 0–300, 300–700 and 700–1500 m, with a resolution of 80 × 80 km, given an equivalent sulfur oxide emission distribution. The model employs a wind analysis based on interpolation of upper air soundings every 12 hours from the U.S. National Weather Service. It uses an eddy diffusion for turbulent dispersion, in which the eddy diffusion coefficient can be set from atmospheric stability analysis in each of the vertical layers. The simulator includes only simple oxidation chemistry, and surface removal processes which are first order in sulfur oxides.

The emission inventory developed for the regional analysis was prepared and averaged for each 80 × 80 km square from point and area source data in the NEDS, as well as supplementary information from a variety of government and industrial origins. Vertical injection height was calculated from stack height, exit gas velocity and temperature, assuming neutral atmos-

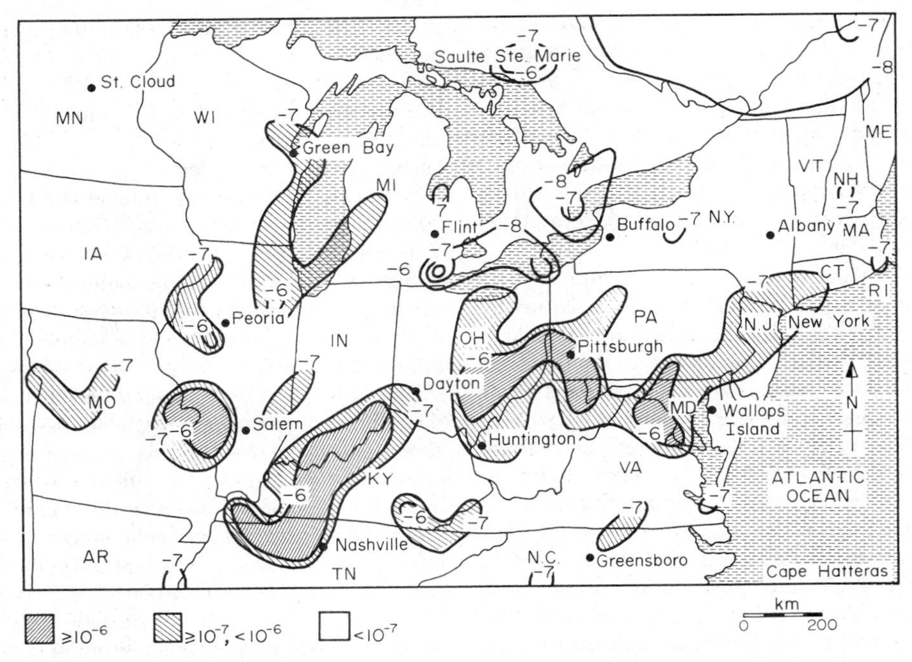

Fig. 20. Annual SO_2 emissions in the height range 300 m–700 m ($g\,m^{-2}\,s^{-1}$).

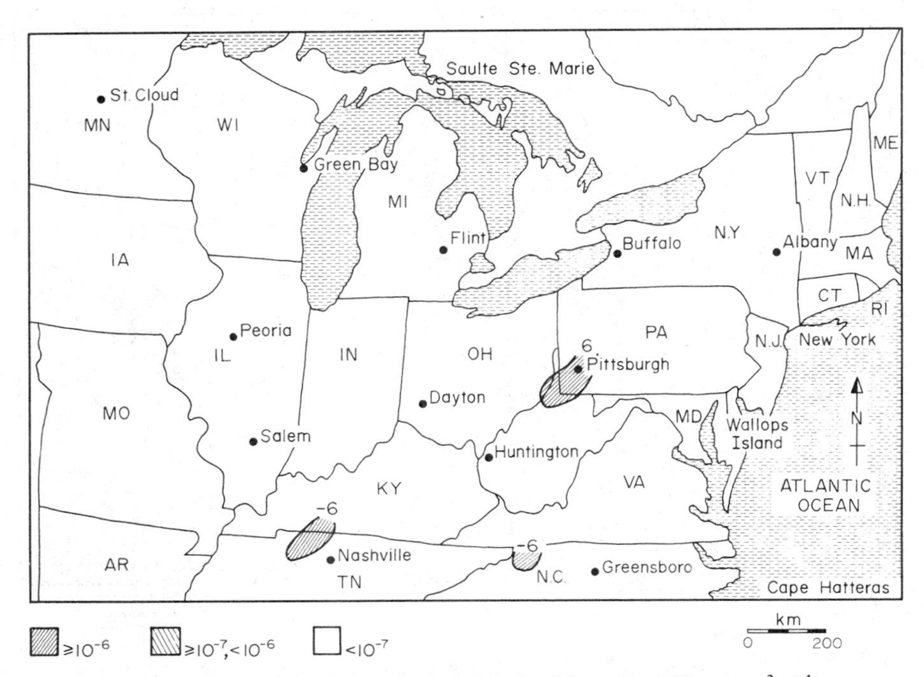

Fig. 21. Annual SO_2 emissions in the height level exceeding 700 m ($g\,m^2\,s^{-1}$).

pheric conditions and 5 m s^{-1} wind speed. The emission distributions for the three layers are shown in Fig. 19–21. As expected, the bulk of the emission is confined to the ground layer. However, there are significant contributions to the 300–700 m height layer, and a few isolated sources input into the highest layer (≥700 m).

As a performance test of the EGAMA regional sulfate model, Hidy *et al.* (1977b) have applied the scheme to simulate sulfur oxide behavior on 10 July, 1974. The details of the simulation are reported elsewhere; only the results for grid boundary conditions of zero, and initial conditions within the grid of $SO_2 = 40\ \mu g\ m^{-3}$ and $SO_4^{2-} = 5\ \mu g\ m^{-3}$ are presented here for illustration. The calculations were made assuming a spatially and temporally uniform oxidation rate of 1.5% h^{-1}. Based on information in the literature, a deposition velocity was selected for SO_2 and SO_4^{2-} of 1.0, and 0.1 cm s^{-1}, respectively with a mixing depth of 1000 m.

The results of the 24-h averaged concentration from the simulation are shown in Fig. 22 and 23 for the lowest layer. The analysis indicates a localized build up of SO_2 near major source zones in the Tennessee–Kentucky area, eastern Ohio, western Pennsylvania, Chicago and the Philadelphia–New York corridor. The isolated intense source at Sudbury, Ontario also is identifiable. The extent of intense exposure associated, or the "zone of influence", with major source areas is confined generally to less than 300 km downwind for SO_2, as indicated in Fig. 22. The corresponding calculated distribution of sulfate at ground level is shown in Fig. 23. The zone of elevated sulfate is a broad band extending northeast from the Ohio Valley–Tennessee source area to the east coast over New England. The calculated concentrations of sulfate exceed 10 $\mu g\ m^{-3}$ over most of the greater northeast. However, the zones of highest sulfate concentration are localized around source rich areas in the lower Ohio Valley – Nashville area, Pittsburgh, and Chicago. Effects from Washington–Boston corridor appear to be primarily along the coast, extending offshore south of Connecticut. The Sudbury area does not show strong sulfate impact at the ground. Nevertheless, it appears from the simulation that the intensity of impact of this large source again remains confined to 100 km from the source. The simulation results suggest that the regional buildup of sulfate for the specified meteorological conditions is a diffuse widespread increase of 10–15 $\mu g\ m^{-3}$ with localized, "urban" or mesoscale effects influence increases which may be two to four times as intense within a radius of 200 km of major sources. Comparison with the observed isopleths (Fig. 11), shows a broad qualitative correspondence in the Midwest and North, but failure to estimate the broad band of high sulfate in the mid-Atlantic coastal zone. Using the uniform initial conditions $SO_2 = 40\ \mu g\ m^{-3}$, and $SO_4^{2-} = 5\ \mu g\ m^{-3}$, the sulfate concentration estimates are generally lower than those observed, especially in the high sulfate zones. The model results shown here correspond to an estimate of the buildup in sulfate over a one day interval from uniform baseline initial conditions. Although the results did not account adequately for the intense sulfate concentration zone along the east coast, they did account for the isolated, localized buildup around areas of high emission density. The model may suffer from over-simplification in several areas includ-

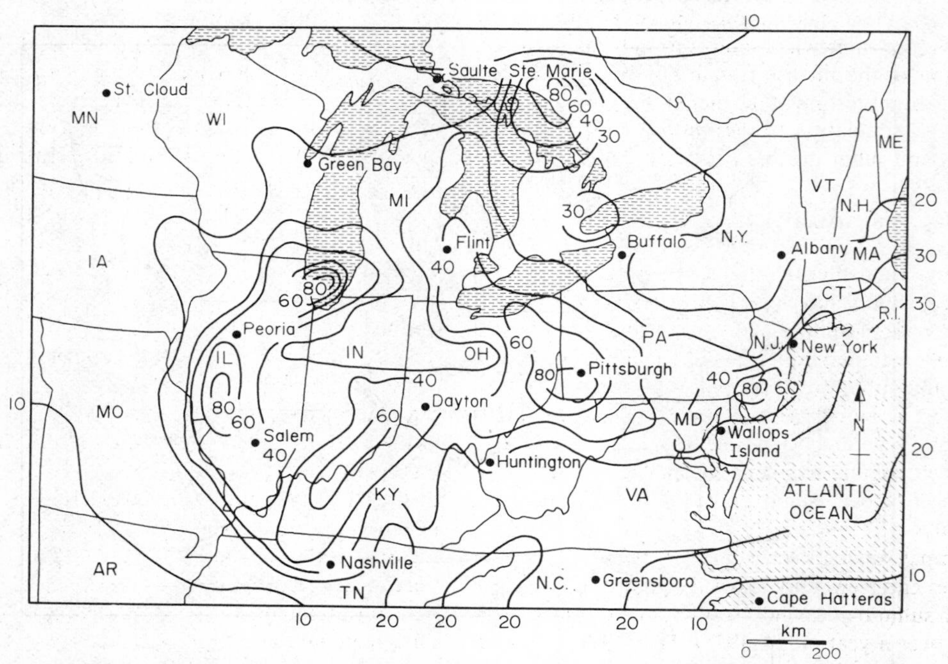

Fig. 22. Level 1 distribution of 24-h averaged SO_2 concentrations ($\mu g\ m^{-3}$) estimated for greater Northeast on 10 July, 1974. Simulation for 0 boundary condition and uniform initial conditions, $SO_2 = 40\ \mu g\ m^{-1}$, $SO_4 = 5\ \mu g\ m^{-3}$.

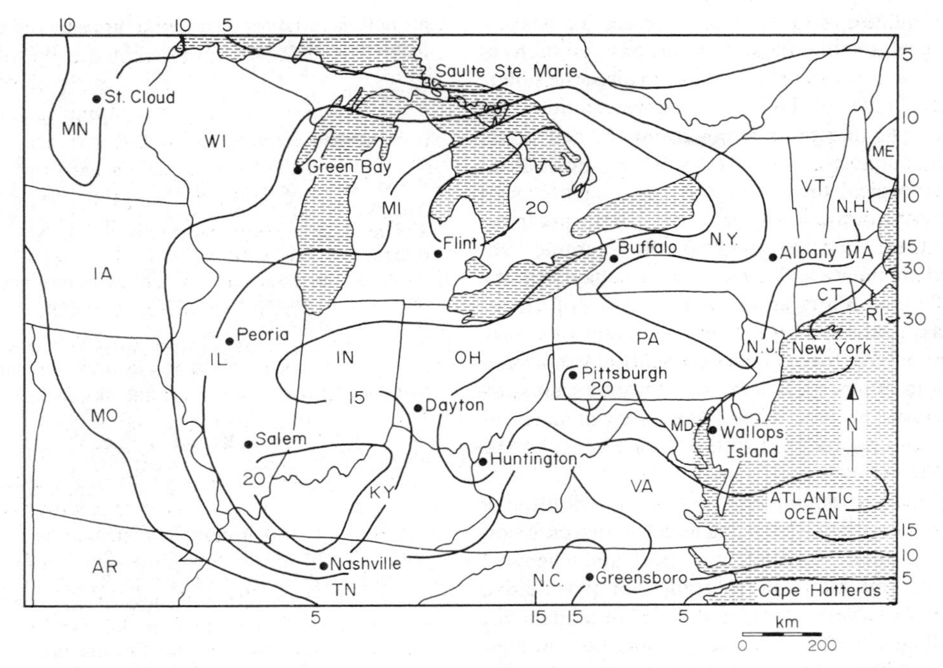

Fig. 23. Level 1 distribution of 24-h averaged SO_4 concentrations ($\mu g\,m^{-2}$) in the greater Northeast for 10 July, 1974. Simulation is for 0 boundary conditions and uniform initial conditions, $SO_2 = 40\,\mu g\,m^{-1}$ and $SO_4 = 5\,\mu g\,m^{-3}$.

ing multi-day buildups, reaction rates, or inadequate transport simulation.

To investigate the influence of regional transport eastward of the polluted air mass observed on 8 July, additional calculations were made for 9 July, whose results were used as initial conditions for 10 July. This would crudely simulate the multi-day buildup of sulfate during the episode, and take better account of the regional transport of pollution eastward. The results of these calculations improved the model performance somewhat, and showed better representation of the intense sulfate buildup on the east coast. Further testing of the model is required, including better simulation of the multi-day pollution behavior, and sulfur dioxide oxidation chemistry.

5. SUMMARY AND CONCLUSIONS

Analysis and interpretation of available air monitoring observations have revealed several features about the behavior of sulfur oxides on an urban and regional scale in the United States. There remains considerable uncertainty about regional pollution effects involving transport and air chemistry over ranges of 1000 km or more and several days in time. However, information is now becoming available which better defines the nature of "urban" sulfur oxide pollution on scales of 100 km.

Comparison between sulfate distributions in Los Angeles and New York showed significant differences in 24-h sulfur oxide concentrations and nitrogen oxide levels over a year (1974–1975). These differences are believed to be related to differences in the intensity in photochemical smog processes between the two cities. Examination of the geographical distribution of annual average sulfate distributions in the two cities suggests that the intensity of elevated pollutant levels from major local sulfur oxide sources is confined to areas only 50–100 km from the sources for both sulfur dioxide and particulate sulfate. A seasonal maximum in sulfate concentration is documented for both cities which cannot be explained in terms of emission changes. Limited data on the diurnal changes in sulfate with other pollutants in Los Angeles suggests that an interaction exists with nitrate and particulate carbon, as well as aerosol forming precursor gases in the evolution of photochemical haze. Correlation between 24-h averaged sulfate data and other air quality variables by stepwise linear regression suggests that 24-h averaged ozone levels, r.h. and particle mass concentration (less the sulfate and nitrate contributions) are principal factors in explaining sulfate variability. This is consistent with expectations from present knowledge of sulfur dioxide oxidation chemistry.

Some rural observations characterizing regional sulfate pollution in the eastern United States have been examined. The widespread extent of sulfate pollution in the eastern United States reported by EPA, based on pre-1970 data, was borne out, at least qualitatively, in later observations taken from 1973 to 1975. The summer maximum was documented in 1974 as a regional scale phenomenon extending from Indiana northeastward to New York state. Analysis of daily 24-h averaged sulfate data revealed significant differences on a regional scale between sulfate episode development in summer and winter. The summer months appear to have appreciable occurrences of intense regional sulfate episodes lasting more than one day, in which rural maximum concentrations as high as 80 $\mu g\ m^{-3}$ are observed. Variations in airborne sulfate in the rural areas of the United States were found to correspond to changes in temperature and dew point. High sulfate has been observed in warm, moist air of maritime tropical air masses; the lowest sulfate levels have been found in cool, dry, continental polar air masses. Case studies reporting the synoptic meteorology linked with observed sulfate behavior in the eastern United States show further details of the air mass classification. The regional character of sulfate pollution is most prevalent in multi-day episodes observed in summer which involve synoptic scale stagnation conditions with widespread inflow of maritime tropical air into the midwest and east coast of the United States.

Initial efforts to simulate the regional distribution of sulfur oxides are reported using an estimated emission grid and a model of meteorological and chemical effects. The simulations show promise for quantitative assessment of the regional pollution potential, but require further testing to elucidate the complicated interactions between local and long range transport and chemical processes.

The key requirements for improved understanding of regional scale pollution processes in the eastern United States is the acquisition of reproducible, systematic, daily observations over the entire region at the ground and aloft. Substantial improvement in simulation for regional sulfate distribution awaits the accumulation of observations from new field studies initiated in the eastern United States.

Acknowledgements – Much of the work discussed here is based on our colleagues' intense effort over the past year. We have selected portions of research undertaken by R. Batchelder, S. L. Heisler, and R. C. Henry to illustrate new detail of sulfate behavior. We are indebted to the Electric Power Research Institute, the Southern California Edison Co., and the American Petroleum Institute for their sponsorship of portions of this work.

REFERENCES

Barnes R. A. (1973) Duplicate measurements of low concentrations of smoke and sulfur dioxide using two national survey samplers with a common inlet. *Atmospheric Environment* **1,** 901–904.

Blumenthal, D. *et al.* (1977) The evaluation of the three-dimensional distribution of air pollutants during a Los Angeles smog episode. In *The Character and Origins of Smog Aerosol*, J. Wiley, New York (in press).

Californian Public Utilities Commission (1977) Petition of the California Air Resources Bd. Seeking Natural Gas Reallocation to Southern California. California Air Resources Board and Southern California Edison Co. Testimony, CPUC Case No. 9642, Sacramento, CA.

Dixon G. J. (ed.) (1973) Statistics Program, BMD-02R, Health Science Computer Facility (UCLA) (3rd Edn). University of California Press, Berkeley, CA.

Egan B. and Mahoney, J. R. (1972) Application of a numerical air pollution transport model to dispersion of the atmospheric boundary layer. *J. appl. Met.* **11,** 1023–1039.

Hidy G. M., Rice H. and Nochumson D. H. (1977a) Perspective on the Current Airborne Sulfate Problem. Sulfate Task Force Report 406-76T-1, American Petroleum Institute, Washington, D.C.

Hidy G. M., Martinez J. R., Tong, E. Y., Henry R., Papineau R. M. and Tran K. (1977b) Preliminary Air Quality Analysis of Urban and Regional Sulfate Distributions. Sulfate Task Force Report 406-76T-2, American Petroleum Institute, Washington, D.C.

Hidy G. M. and Burton C. S. (1975) Atmospheric aerosol formation by chemical reactions. *Int. J. chem. Kinet.* Symp. No. 1, pp. 509–541.

Hidy G. M. *et al.* (1975) Characterization of Aerosols in California. Report SC 524.25 FR, Vol. 4. California Air Resources Board, Contract 358, Rockwell Science Center, Thousand Oaks, CA.

Hidy G. M., Mueller P. K. and Tong E. Y. (1976) Design of the Sulfate Regional Experiment (SURE), Vols. 1 and 2. Electric Power Research Institute, Report EC-125, Palo Alto, CA.

Hitchcock D. (1976) Atmospheric sulfates from biological sources. *J. Air Pollut. Control Ass.* **26,** 210–215.

Joint Project (1977) Joint project for Relating Emissions of SO_2 to Air Quality: Interim Report. Rept. P-5088, prepared by Environ. Res. & Technology, Inc., for the South Coast Air Quality Management District, Westlake Village, CA.

Kurosaka D. (1976) Sulfate Concentrations in the South Coast Air Basin. Report DTS-76-1, Calif. Air Resources Board, Sacramento, CA.

Lynn D. A., Epstein B. S. and Wilcox C. K. (1975) Analysis of New York City Sulfate Data. Final report EPA Contr. 68-02-1337, Task 10. GCA Corp., Bedford, Mass., 76p.

Martin A. and Barber F. R. (1973) Further measurements around modern power stations. *Atmospheric Environment* **7,** 17–37.

Mirabella V. (1977) Testimony presented in hearings for natural fuel reallocation. California Public Utilities Commission Case No. 9642, Exhibit No. 126.

Mueller P. K. (1976) Sulfate Measurement Technology. In Design of the Sulfate Regional Experiment (SURE), Vol. 3, Chap. D. Electric Power Research Institute, Report EC-125, Palo Alto, C.A.

Rhode H. (1972) A study of the sulfur budget for the atmosphere over northern Europe. *Tellus* **24,** 128–138.

Smith F. B. and Jeffrey (1975) Airborne transport of sulfur dioxide from the United Kingdom. *Atmospheric Environment* **9,** 643–659.

Smith T. B., Blumenthal D. L., Anderson J. A. and Vanderpool A. H. (1978) Transport of SO_2 in power plant plumes: day and night. *Atmospheric Environment* **12,** 605–611.

Tong, E. Y. *et al.* (1976a) Regional and Local Aspects of Atmospheric Sulfates in the Northeast Quadrant of the United States. Presented at 3rd Symposium on Atmospheric Turbulence, Diffusion and Air Quality, American Meteorological Society, Raleigh, N.C.

Tong E. Y., Battel G. and Batchelder R. B. (1976b) Case Studies of Atmospheric Sulfate Distribution over the Eastern United States. Presented at the 15th Air Quality Conference, Indianapolis, Indiana.

U.S. House of Representatives (1976) The Environmental Protection Agency's Research Program – Community Health and Environmental Surveillance Systems (CHESS): An Investigative Report. Subcommittee on the Environment and the Atmosphere, Report SS, 2nd session, 94th Congress, U.S. Gov't Printing Office, Washington, D.C.

U.S. Environmental Protection Agency (1975) Position paper on regulation of atmospheric sulfate. Report No. EPA-450/2-75-007, Office of Air Quality Planning and standards, Research Triangle Park, N.C.

Atmospheric Environment Vol. 12, pp. 753–757. Pergamon Press 1978. Printed in Great Britain.

SULFURIC ACID PARTICLES IN SUBSIDING AIR OVER JAPAN

AKIRA ONO

Water Research Institute, Nagoya University, Chikusa-ku,
Nagoya, 464, Japan

(*First received* 3 *June and in final form* 8 *July* 1977)

Abstract—Direct investigations were made of physical and chemical properties of aerosol particles, collected at the summit of Mt. Fuji (3776 m) in subsiding air, based on the appearance and chemical tests of individual particles under the electron microscope.

From electron microscopy of aerosol particles collected on a carbon film and that on a thin film of reacting materials by vacuum deposition, it is found that most of the aerosol particles in subsiding air are present in the form of free sulfuric acid droplets. They are predominantly small, the common radius of the equivalent sphere being smaller than 0.1 μm and their concentrations are in the order of 300 cm^{-3}.

Some of the implications of the present findings in relation to sources of sulfuric acid particles are discussed.

INTRODUCTION

In the study of atmospheric aerosol particles for potential and expected effects on weather and climate, it is essential to attempt to determine the chemical nature of individual aerosol particles, as well as sizes and concentrations. The difficulties in the determination of the chemical nature of individual aerosol particles can be appreciated from considerations of their size and mass. In the atmosphere, the most numerous and most important aerosol particles have radii in the range of 0.01–0.1 μm. With typical radius less than 0.1 μm typical mass will be less than 10^{-15} g, far below the size accessible to optical microscope and to ordinary wet chemical analysis. So far the chemical nature of aerosol particles have usually been obtained from chemical analysis of a bulk sample. Information that is completely lacking at present and is essential in future studies is the size distribution and concentration of aerosol particles containing given chemicals.

Over the last few years, we have developed a vapor method of chemical analysis for individual aerosol particles used in conjunction with a vacuum deposited thin film of reagents in the electron microscope (Bigg *et al.*, 1974). This method has been applied to stratospheric aerosol particles and has shown considerable promise for making the sort of measurements required (Bigg and Ono, 1974; Bigg, 1975).

In the present paper, results are presented of direct investigations of the physical and chemical nature of aerosol particles in subsiding air over Japan based on the appearance and chemical tests of individual aerosol particles in the electron microscope.

SAMPLING PROCEDURES

Atmospheric aerosol particles were collected at the summit of Mt. Fuji (3776 m) in August 1975 by impaction and electrostatic precipitation on to standard 3-mm electron microscope screens. The screens were covered with a nitrocellulose film strengthened with a carbon film and some were also precoated with a thin film of reacting materials by vacuum deposition. The sampling of aerosol particles was divided into a daytime and a night-time period for about 8 h in each period with the intention of seeing the photochemical oxidation effect on aerosol particle chemistry. This sampling period coincides with a daytime

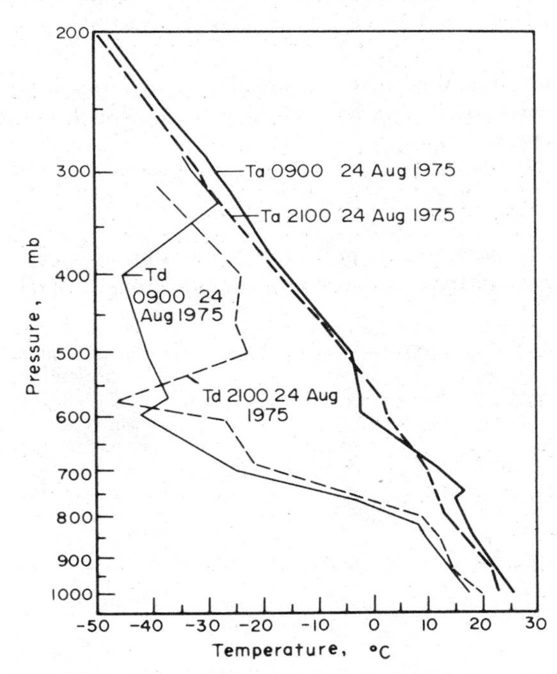

Fig. 1. Vertical profiles of temperature and dew-point observed at Tateno on 24 August 1975. Temperature (*Ta*) and dew-point (*Td*) profiles show a level of a subsidence inversion layer at about 800 mb.

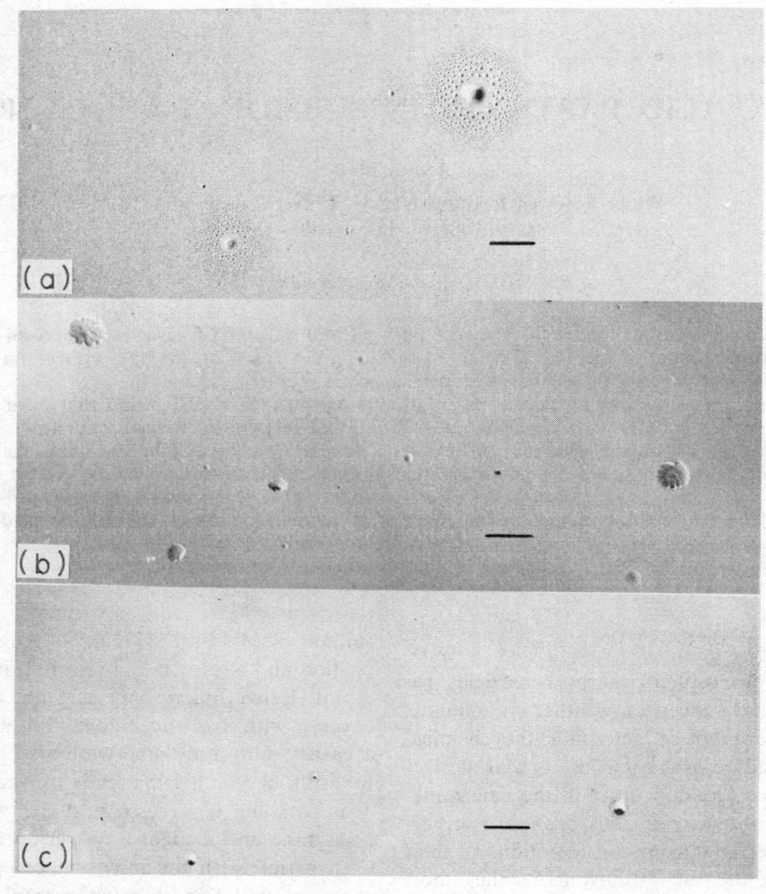

Fig. 2. Classification of aerosol particles collected at the summit of Mt. Fuji based on the appearance in the electron microscope. (a) A central particle surrounded by discrete widespread concentric rings of satellites of graded size. (b) A hemispherical particle surrounded by a circular outline without satellites. (c) A transition type surrounded by small satellites.

up–slope wind period and a night-time down slope wind period of the thermally induced mountain circulation.

RESULTS

Typical vertical profiles of temperature and dewpoint observed at Tateno (60 km away from Mt. Fuji) are shown in Fig. 1. During this observational period a subsidence inversion was formed at an altitude of 2 or 3 km which was readily recognized by the pronounced dryness above the inversion. This drying of air can be interpreted as being indicative of subsidence of extremely dry air from the upper troposphere. A portion of this dry air may have subsided from the lower stratosphere.

Table 1. Summary of sampling data, mean concentration of sulfate particles and dominant types of aerosol particles collected at the summit of Mt. Fuji in August 1975

Sample No.	Sampling date	Mean concentration of sulfate particle (cm^{-3})	Dominant types of aerosol particle	Remarks (subsidence inversion yes or no)
F-2-75	23–24 August 1975 22:55–04:00	67	$(NH_4)_2SO_4$	No
F-3-75	24 August 1975 07:30–18:00	300	$(NH_4)_2SO_4$	Yes
F-4-75	24–25 August 1975 19:00–06:00	380	H_2SO_4	Yes
F-5-75	25–26 August 1975 18:00–18:00	360	H_2SO_4. $(NH_4)_2SO_4$	Yes
F-6-75	29–30 August 1975 22:00–06:55	250	H_2SO_4	Yes

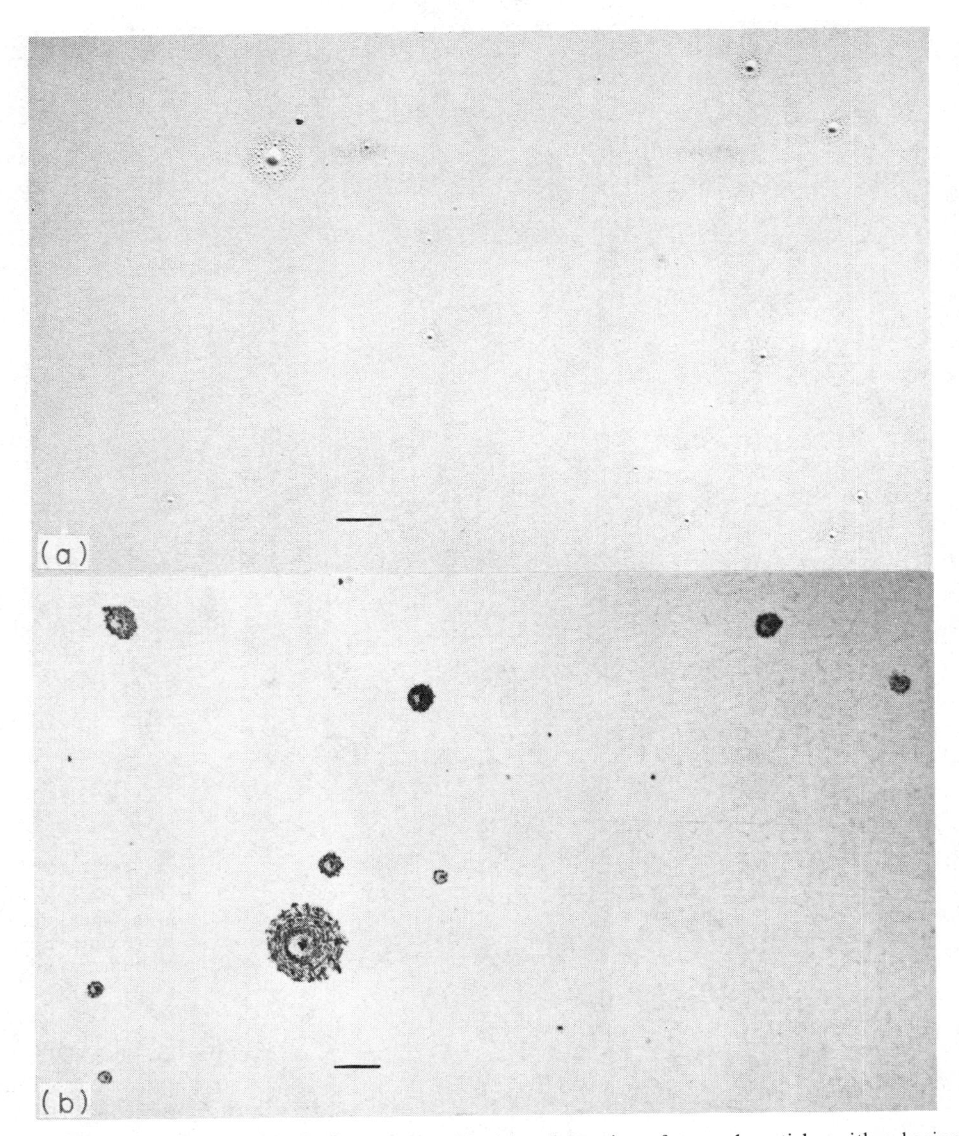

Fig. 3. Aerosol particles collected on a carbon film (a) and reaction of aerosol particles with a barium chloride film (b) without humidification showing the presence of sulfuric acid in particles. Both from a night-time collection in subsiding air on 24–25 August 1975 (comparison scale: 1 μm).

Direct morphological examinations of aerosol particles collected on carbon films revealed that most of the aerosol particles could be classified into three types as shown in Fig. 2. Aerosol particles like those in Fig. 2(a) have discrete widespread concentric rings of satellites of graded size surrounding a central particle. This type dominates the particles collected during night-time sampling in subsiding air as summarized in Table 1. Hemispherical aerosol particles with a circular outline but no satellite rings like those in Fig. 2(b) are unstable in the electron beam and rapidly evaporate to leave a thin cracked shell. They dominate in the aerosol particles collected during a daytime upslope wind period. Transition types surrounded by small satellites rings like those in Fig. 2(c) were occasionally found.

From comparisons in the electron microscope of aerosol particles collected on a carbon film with that on a barium chloride film, it was confirmed that aerosol particles showing a central particle and discrete widespread concentric satellites rings produced a reaction of barium sulfate on a barium chloride film without humidification and indicate that these aerosol particles are present in the form of free sulfuric acid droplets (see Fig. 3). After the electron microscopy examination, we floated specimen screens on pure water. The application of water dialysis of the specimen screen is to minimize the confusing background due to the recrystallization of excess unreacted barium chloride which tends to obscure small aerosol particle reaction and to confirm that the reaction product around the particle is a water insoluble barium sul-

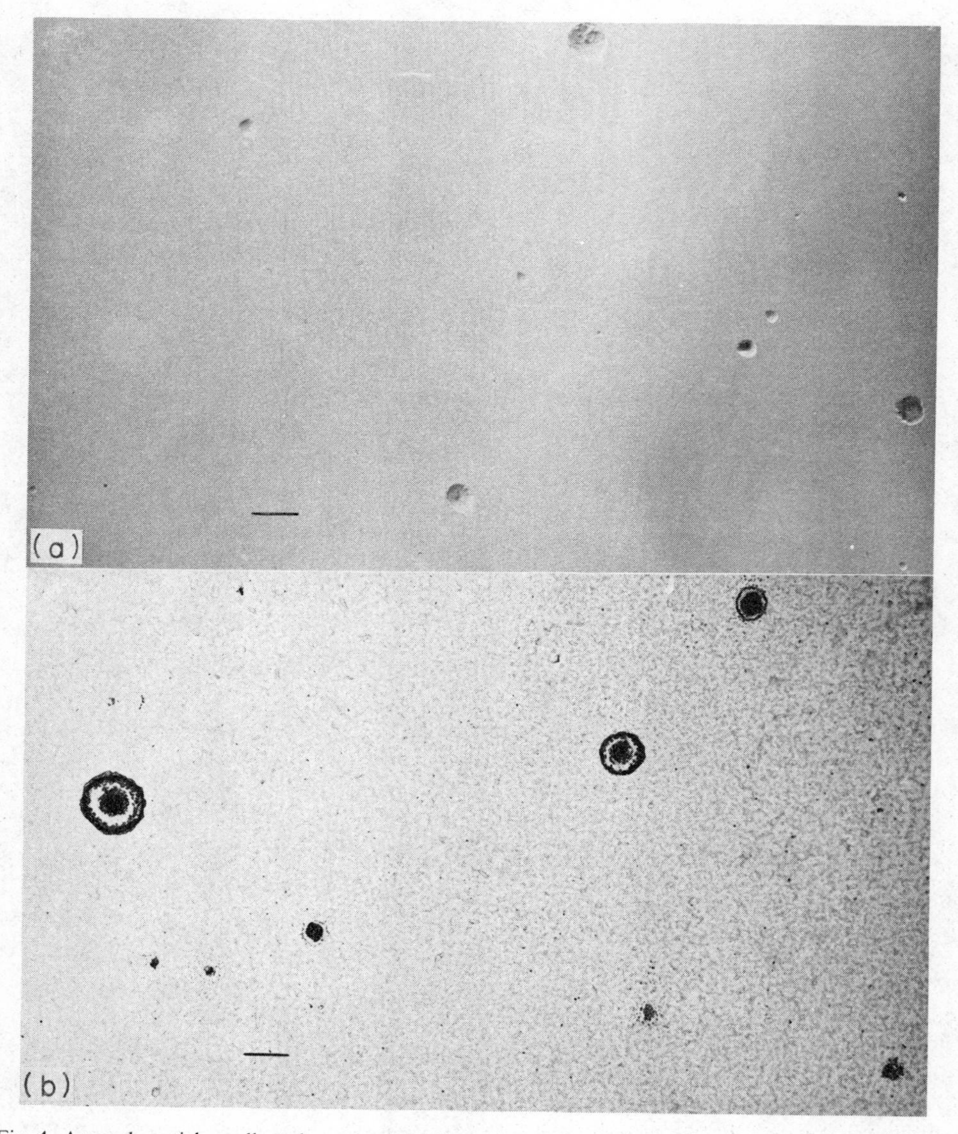

Fig. 4. Aerosol particles collected on a carbon film (a) and reaction of aerosol particles with a barium chloride film in octanol saturated (b) atmosphere showing the presence of sulfate in particles. Both from a daytime collection in subsiding air under an upslope air flow condition on 24 August 1975 (comparison scale: 1 μm).

fate. These particles also produced reactions on copper thin films without humidification indicating the presence of oxy-acid in the particles.

In these ways, we established that aerosol particles in subsiding air were most commonly present in the form of free sulfuric acid droplets.

Most aerosol particles bearing free sulfuric acid are predominantly small, the common radius of the equivalent sphere being less than 0.1 μm (see Fig. 5) and their concentrations are in the range of 300 cm^{-3} as shown in Table 1.

It is interesting to note that although sulfuric acid was the most common particle type collected during night-time in subsiding air, hemispherical aerosol particles without satellites rings were the dominant particles collected during daytime periods when clouds were lifted up with the thermally induced upslope wind and cloud tops were evaporated at the level of sampling site. These particles produced no reactions without humidification and produced the reaction of barium sulfate on a barium chloride film exposed in octanol saturated atmosphere (see Fig. 4) and on a copper film produced reaction if it is exposed in humidity exceeds about 80%. They are almost certainly ammonium sulfate particles since a copper thin film reacts only to oxy-acids and ammonium compounds in solution.

This clearly indicates that sulfuric acid particles and ammonium sulfate particles originated from different sources. Ammonium sulfate particles originate

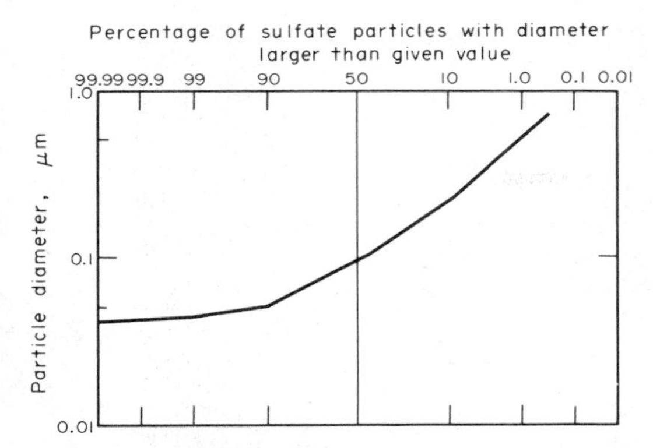

Fig. 5. Size distribution of sulfuric acid aerosol particles in subsiding air collected by electrostatic precipitation at the summit of Mt. Fuji plotted on log-normal probability paper.

in sub-inversion air probably through reactions occurring within cloud droplets or transported as pre-existing aerosol particles from sub-inversion air by an upslope wind.

ORIGIN OF SULFURIC ACID PARTICLES

The present finding that the dominant aerosol particles, especially under clean night-time downslope air flow conditions, are sulfuric acid particles is a rather surprising result. This finding clearly indicates that these sulfuric acid particles are not produced *in situ* by the photochemical oxidation process. Although the mechanism by which sulfuric acid particles are produced in the atmosphere is not well understood, it is generally agreed that photo-chemical oxidation and a subsequent hydration of sulfur containing gases is the most probable gas- to-particle conversion process for atmospheric sulfuric acid particles.

From comparisons of the wet bulb potential temperature profiles prepared from radio-sonde sounding at Tateno (60 km away from Mt. Fuji) and that of subsiding air at the sampling site it is suggested that the origin of the subsiding air mass in which aerosol particles were collected is at an altitude of about 10–11 km.

Based on the considerations mentioned above, we come to the conclusion that the sulfuric acid particles arrived at the sampling site in air subsiding from the upper troposphere or from the lower stratosphere. It is therefore quite resonable to conclude that these sulfuric acid particles observed in a subsiding air mass are representative of the upper tropospheric or stratospheric aerosol particles.

DISCUSSION

From indirect evidence based on the aerosol particle's sensitivity to heating, Dinger *et al.* (1970) suggested that in a subsiding air mass over the North Atlantic Ocean, aerosol particles acting as cloud condensation nuclei have chemical compositions similar to ammonium sulfate or sulfuric acid. Our present direct investigation based on the appearance and the chemical tests of individual aerosol particles in a subsiding air mass over Japan indicate clearly that they are sulfuric acid particles. While on the other hand in the lower troposphere the common aerosol particles containing sulfur are ammonium sulfate.

The present results suggest the possibility that in certain circumstances the sulfuric acid particles generated in the stratosphere and suspended there because of the lack of efficient removal mechanisms could act as appreciable sources of effective cloud condensation nuclei after sedimentation or mixing into the troposphere.

It requires, therefore, a more intensive study of the nature of aerosol particles present in the middle troposphere to see what proportion are of recent stratospheric origin and to see whether human activity will produce marked changes in the proportion.

The present results indicate clearly the importance of simultaneous studies of aerosol chemistry of individual particles as well as size and concentration in order to determine their origin. Air exchanges between the stratosphere and the troposphere imply that we should study aerosol particles in the atmosphere from the ground level up to a height of about 30 km as a whole.

CONCLUSION

The results presented in this paper have led to the conclusion that the dominant aerosol particles in a subsiding air mass over Japan are mainly composed of sulfuric acid, which are not formed *in situ* near the sampling site from sulfur bearing gases by photochemical oxidation processes but could be transported from the aerosol layer in the lower stratosphere.

REFERENCES

Bigg E. K. (1975) Stratospheric particles. *J. atmos. Sci.* **32,** 910–917.

Bigg E. K. and Ono A. (1974) Size distribution and nature of stratospheric aerosols. *Proc. Int. Conf. on Struct. Comps. and Gen. Circ. of the Upper and Lower Atmos. and Possible Anthropogenic Perturbations,* Vol. 1, 144–157.

Bigg E. K., Ono A. and Williams J. (1974) Chemical tests for individual submicron aerosol particles. *Atmospheric Environment* **8,** 1–13.

Dinger J. E., Howell H. B. and Wojciechowski T. A. (1970) On the source and composition of cloud nuclei in a subsident air mass over the North Atlantic. *J. atmos. Sci.* **27,** 791–797.

Atmospheric Environment Vol. 12, pp. 759–771. Pergamon Press 1978. Printed in Great Britain.

LARGE-SCALE MEASUREMENT OF AIRBORNE PARTICULATE SULFUR*

B. W. Loo, W. R. French, R. C. Gatti, F. S. Goulding, J. M. Jaklevic, J. Llacer and A. C. Thompson

Lawrence Berkeley Laboratory, University of California, Berkeley, California 94720, U.S.A.

(*First received* 21 *June* 1977 *and in final form* 26 *August* 1977)

Abstract—We describe an aerosol sampling and analysis system which represents an integral approach to large-scale monitoring of airborne particulate matter. During our two-year participation in the St. Louis, Missouri, Regional Air Pollution Study (RAPS), 34,000 size-fractionated samples were collected by automated dichotomous samplers characterized by a particle size cutpoint of 2.4 μm. The total mass of the particulate matter was measured by beta-particle attenuation and the elemental composition, including sulfur, was determined by photon-excited X-ray fluorescence. The long-term performance of the system will be reported. Potential systematic effects related to the sampling and analysis of sulfur particles are treated here in detail. Both the accuracy and precision of sulfur measurement are estimated to be 2%. While the X-ray attenuation correction required is typically only a few per cent, a larger correction is required for a small fraction of the samples due to the migration of the sulfur into the filter. This correction is derived from the ratio of sulfur determinations made on the front and back surfaces of the membrane filter. Laboratory and field experiments have shown insignificant gaseous SO_2 conversion on the type of filters employed in the study. Preliminary data on the composition and the temporal and spatial distribution of the St. Louis aerosol are presented. About 90% of the sulfur was found in the fine-particle fraction. Sulfur variations were significantly slower than those of the trace elements. Sulfates usually constitute about 35% of the total fine-particle mass, but may rise to 41% during an 'episode'. The long-term (4 month average) sulfur data indicate that the background air masses arriving at St. Louis from the west and north were about 30% lower in particulate sulfur than those from the east and south. Also, an urban station may experience local increases in sulfur level up to a factor of two greater than the general background. Short-term (6 h average) data indicate that the effects of stationary SO_2 sources extend for long distances, (at least 40 km) and are highly directional in character.

INTRODUCTION

In the course of the St. Louis, Missouri, Regional Air Pollution Study (RAPS), we have collected 34,000 size-fractionated aerosol samples at ten of the regional monitoring sites. Total particulate mass and elemental compositions have been measured separately for the coarse- and fine-particle fractions, which have aerodynamic diameters above and below 2.4 μm, respectively. The results provide an unprecedented opportunity to study spatial and temporal variations of aerosol particles in a large metropolitan area, particularly those of sulfur-containing particles which are a major and growing component of the aerosol mass.

Fine airborne particulate sulfur is present in the air almost entirely in the respirable particle-size range and is therefore a threat to human health. Furthermore, it is a secondary aerosol from its gaseous SO_2 precursor; thus, its space and time behavior may hold much of the secret to understanding the process of atmospheric transformation, transport and depletion of aerosols. In view of these factors, which make sulfur so important, emphasis in this paper will be on the problems associated with sulfur analysis. Preliminary data, illustrating the temporal and spatial distribution of sulfur and aerosol mass, are presented. The regional variation of particulate sulfur as related to wind direction and major stationary SO_2 emission sources is also examined.

* This work was supported by the Environmental Protection Agency under Interagency Agreement with the United States Energy Research and Development Administration.

METHOD

When laboratory techniques are used in the field, reliability becomes a crucial problem. The large scope of the RAPS program requires a systematic approach, with rigorous quality assurance measures applied during the complete cycle of sample collection, sample measurement and data processing. We have chosen methods that are nondestructive, and which require no sample preparation and are easily automated for a minimum amount of operator handling.

The urban aerosol size distribution is bimodal, with the fine (respirable) combustion-generated particles being distinct from the coarse aerosol generated by mechanical and natural processes (Whitby, 1972, 1973). In order to sample these fractions separately, we developed the automated dichotomous sampler, which uses the virtual impactor principle (Loo, 1976). This sampler inertially separates particles according to their aerodynamic size—either above

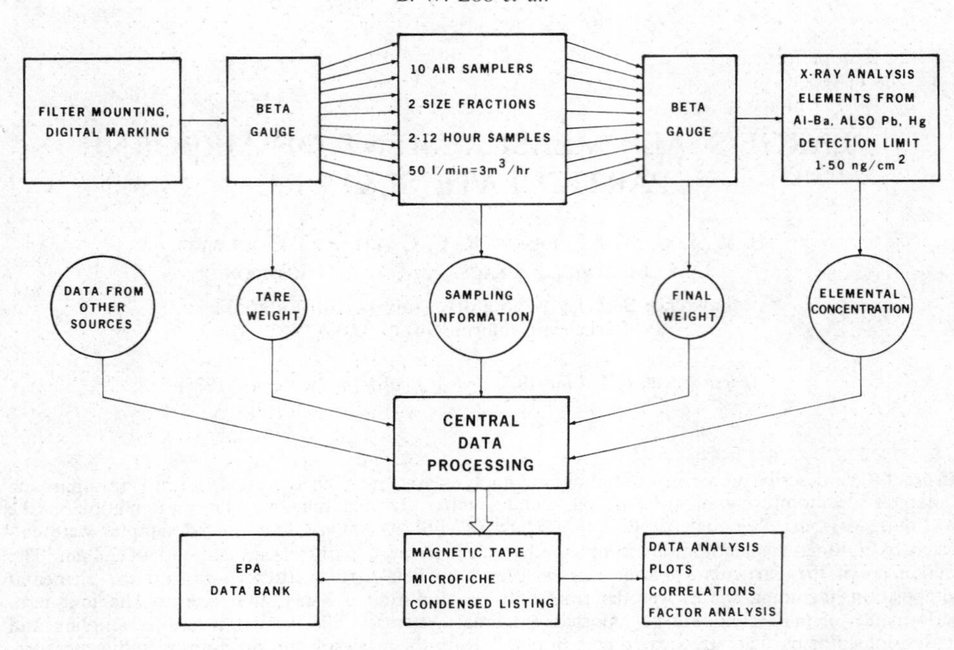

Fig. 1. Block diagram of sampling, measurement and data analysis system.

or below 2.4 μm—and collects each fraction uniformly on a separate filter, with minimum particle loss and re-entrainment. An automated flow controller including fault sensors were built into the sampler so that precise flow conditions and the timing sequence were thoroughly checked and maintained.

The filters used (1.2 μm pore size cellulose ester such as Millipore Type RAWP) have a fairly low impurity content and high filtration efficiency (Liu, 1977). The 37 mm diameter filter discs were mounted in 5 cm square plastic frames and loaded in standard photographic slide projector cartridges. The cartridges, which hold 36 slides in a linear array, were then automatically sequenced through both the sampling and the analysis equipment. Thus, the exact chronological order of the samples was easily maintained.

The mass deposited on the filter was measured with a beta-gauge utilizing a ^{147}Pm source and a large-area semiconductor detector (Goulding, 1975). The mass difference on a typical 4.5 mg cm^{-2} filter before and after particle collection could be determined to a precision of 10 μg cm^{-2} by using a 30 s counting period and making corrections for relative humidity and other systematic effects. This corresponds to an error of 2 μg m^{-3} in the measured atmospheric aerosol concentration for a 12 h sample.

Elemental compositions were measured using energy-dispersive X-ray fluorescence (XRF) analysis. Concentrations of 27 elements, ranging from A1 to Ba and including Hg and Pb, are routinely measured in a time of less than 5 min per sample. Extensive investigations and validation studies of XRF measurements have established the value of this method for the analysis of aerosol specimens (Appel, 1977; Camp, 1975; Dzubay, 1977). Since ambient aerosol sulfur is present overwhelmingly in the form of sulfates, the result of elemental sulfur measurement by XRF can be closely related to atmospheric sulfates (Stevens, 1978). The XRF method is sensitive and specific and, when fully automated, is well-suited for handling a large number of samples. The nondestructive nature of the method is important from the standpoint of data validation and assessment of long-term systematic effects (Jaklevic, 1977a).

Figure 1 is a block diagram of the entire sampling and analysis system. Mounted filters were labeled and beta-gauged for tare weights before they were dispatched to the RAPS sampling sites. The exposed filters were then returned to us for analytical measurements and subsequent data analysis.

SYSTEM PERFORMANCE

The success of the sampling program is illustrated by examining the reliability of the field operation. Of 34,000 samples collected in the RAPS network, 97.2% were validated, uninterrupted, synchronous samples. About 0.7% of the samples were rejected due to sampler malfunctions; 1.7% were rejected because of computer interference and ac power failures during the early months of operations; and 0.4% were excluded because of excessive loading of the filter—a problem which is largely avoided in the later phases of the study by reducing the sampling periods at the susceptible stations.

To illustrate the precision of the XRF technique, a typical St. Louis fine-aerosol sample was measured repeatedly. Table 1 lists the mean values and standard deviations of six periodic measurements over a two-month period. Results are given for all elements routinely measured. The minimum detectable limits (MDL) (Jacklevic, 1977b) (using a typical analysis time of 1.2 min for each of the three excitation energies employed for the analysis) are also listed for comparison. For major elements, such as sulfur and lead, where systematics dominate over counting statistics, measurements were reproducible to about 1%.

To determine the fluctuations in aerosol concentration measurement caused by the samplers, three of the units were operated side by side for three 12 h

Table 1. Long-term performance of the pulsed XRF spectometer. Results of repetitive analysis of a sample over a 2 month period

ELEMENT	MEAN CONCENTRATION (ng/cm^2)	STANDARD DEVIATION (ng/cm^2)	MINIMUM DETECT LIMIT (ng/cm^2)
Al	226	150	153
Si	214	37	63
P	44	41	40
S	11675	121	34
Cl	224	24	28
K	222	8	12
Ca	196	41	62
Ti	26	19	31
V	11	6	21
Cr	10	6	16
Mn	16	9	13
Fe	337	9	11
Ni	6	2	6
Cu	91	21	7
Zn	495	7	6
Ga	0	0	4
As	3	4	3
Se	6	2	3
Br	84	2	3
Rb	1	1	3
Sr	1	1	4
Hg	0	0	7
Pb	632	3	10
Cd	24	4	4
Sn	30	8	5
Sb	2	2	5
Ba	5	6	22

sampling periods. These samplers came straight from the field with no special preparation. The mean concentration and the standard deviations calculated for the samples produced in this test are listed in Table 2 for the representative elements S, Pb and Fe in each of the size fractions analyzed. For example, the standard deviations between the three samples of fine-particle sulfur and lead are less than 0.5 and 1.3%, respectively, illustrating the uniformity and long-term stability of the samplers.

SULFUR ANALYSIS

The importance of the sulfur measurement requires that the limitations and systematic effects associated with both sampling and analysis be fully understood. Results of the XRF analysis must be examined in detail since the analyses of light elements, such as sulfur, are known to be susceptible to systematic bias arising from X-ray attenuation effects. We discuss here four areas of possible systematic error.

Calibration

As shown earlier, the precision or long-term reproducibility of the XRF spectrometer is better than 2%. The absolute accuracy of the method is well established in cases where X-ray attenuation is not significant. Calibration is normally performed either directly with gravimetrically weighed thin-film standards or indirectly through bootstrap calibrations based on laboratory generated aerosols of known elemental ratios (Giauque, 1977).

For light element calibrations, corrections must be made for X-ray attenuation. We have established the accuracy of our sulfur calibration by comparing two independent bootstrap procedures using a series of accurately measured thin-film Cu standards as the starting point. Figure 2 is an illustration of the two paths by which we arrived at our sulfur calibration. The $CuSO_4 \cdot 5H_2O$ and K_2SO_4 specimens used were in the form of 0.3 μm particles (generated at the Particle Technology Laboratory, University of Minnesota) deposited on the surface of 0.1 μm pore size Nuclepore filters. The results of the [Cu → S] and the [Cu → Cr → K → S] calibrations agreed to within 1%.

Table 2. Mean and standard deviations in analysis of samples collected simultaneously by 3 samplers for 3 sampling periods

ELEMENT	PART SIZE FRACT.	12 HR SAMPLING PERIOD NUMBER	MEAN CONCEN-TRATION (ng/m^3)	STAND DEVIA. (ng/m^3)
S	F	1	771	3
		2	1145	6
		3	1738	9
	C	1	109	11
		2	231	5
		3	269	7
Pb	F	1	632	6
		2	1716	22
		3	1114	15
	C	1	229	6
		2	501	11
		3	378	15
Fe	F	1	67	4
		2	230	5
		3	186	4
	C	1	1015	42
		2	2113	113
		3	1737	79

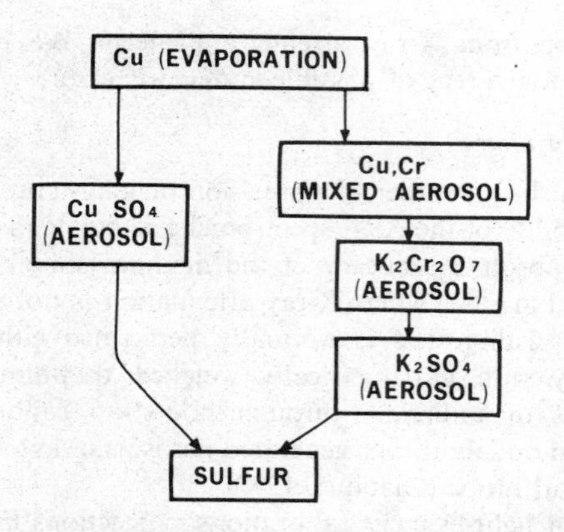

Fig. 2. Two independent sulfur XRF bootstrap calibrations from a copper primary standard.

X-ray attenuation corrections

A detailed treatment of this subject has previously been given (Loo *et al.*, 1977). Here we review a few of the results.

(1) Since almost 90% of the sulfur-carrying particulates are in the fine-particle fraction, the use of the dichotomous sampler, which removes coarse-particle interference, greatly reduces the uncertainty due to absorption in individual particles (~1% for 0.3 μm particles).

(2) The attenuation correction is typically about 5% for sulfur within a layer of fine ambient particles 200 $\mu g\,cm^{-2}$ thick. This is the maximum loading of fine particles that the Millipore filter can accept without clogging.

(3) Direct measurements show that particles in the range of 0.05 μm to 1 μm diameter deposition the filter surface and do not penetrate significantly into the filter medium. Under these circumstances, the correction for sulfur X-ray attenuation by filter substrate is 3 ± 3%.

Particle migration

Our sulfur X-ray attenuation studies have shown that aerosol particles generated both in the field and in the laboratory collect almost entirely on the surface of the filter without significant penetration. However, systematic examination of a large collection of field samples reveals that, on occasion, aerosol material does migrate into a filter substrate. Figure 3 is an illustration of this problem: when all the sulfur particles reside on the front surface of a filter, the ratio of S_F (the sulfur signal when the front side of the filter faces the detector) to S_B (the sulfur signal when the back side of the filter faces the detector) should be in the range of 6–8 (depending on the thickness of the filter blank).

Figure 3 shows the sulfur front-to-back ratio S_F/S_B (open circles) for 6 h samples at RAPS stations 103 and 105 during the summer of 1975. Also shown are the sulfur levels at stations 103 and SO_2 levels at the station 105, together with the relative humidity variations. Occasional low values of S_F/S_B (solid dots) occurred, indicating a significant penetration of sulfur into the filter. The observation that these rare events occur simultaneously at two sampling locations suggests that they are weather-related. However, the coarse-particle filters obtained at the same time as these anomalous samples do not have low values of S_F/S_B, which eliminates precipitation as the cause. The poor correlation of low S_F/S_B with SO_2 indicates

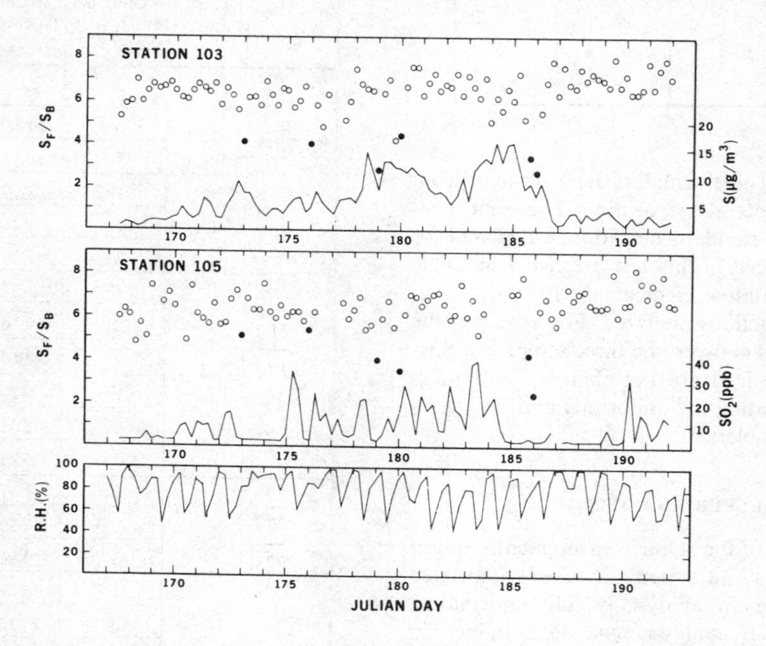

Fig. 3. Migration of aerosol material into the filter medium as evidenced by the ratios of the sulfur XRF yield for the front and back sides of the filters.

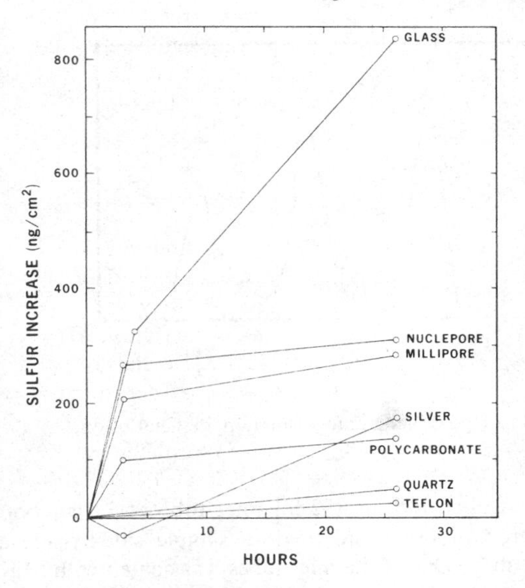

Fig. 4. Spurious sulfate formation on clean filters under extreme exposures (4000 ppm SO_2 and 100% r.h.).

that very little gaseous SO_2 conversion can be occurring on the filter substrate.

Low S_F/S_B does seem to correlate with a combination of high humidity and high sulfur loading. Subsequent laboratory tests have shown that typical fine aerosol samples loaded with $>10\ \mu g\ cm^{-2}$ of sulfur and exposed to 100% relative humidity exhibit a tendency for sulfur to migrate into the filter. Such migration is probably the result both of the hygroscopic nature of the particles and of the hydrophilic nature of the filter.

We have developed a simple correction technique for filter penetration based on measuring the front/back ratio of samples. This is now being applied to the St. Louis data.

The sulfur migration problem can be totally eliminated, if necessary, by using Teflon filters which are hydrophobic. Furthermore, the thinner Teflon filters ($1\ \mu g\ cm^{-2}$) also lowers the detection limits for the X-ray fluorescence analysis and improves the accuracy of mass determination by beta-gauging. However, Teflon filters become clogged with only a third of the mass loading that can be accepted by Millipore filters. In some applications this can be a major drawback.

Spurious sulfate formation

It is well known that gaseous SO_2 may be converted to form spurious sulfate on the filter when the filter material is alkaline, (Pierson, 1977) and that Millipore filters, being nonalkaline, are fairly immune to this process. Recently, however, it was reported that as much as 40% spurious sulfate (produced by SO_2 conversion) was found on ambient aerosols collected on Millipore filters (Lasko, 1977).

In collaboration with R. K. Stevens and T. G. Dzubay of the Environmental Protection Agency (EPA), we have reinvestigated the spurious sulfate formation. Clean filters of various kinds were exposed to extreme

Table 3. Results of spurious sulfate formation experiment in Charleston, West Virginia

START (MAY)		PERIOD		TUBE WITH MgO COATING				TUBE WITHOUT COATING		
DAY	HR.	HR.	MIN.	SAMPLER	TUBE NO.	S (ng/cm²)	Pb (ng/cm²)	SAMPLER	S (ng/cm²)	Pb (ng/cm²)
12	9 am	10	52	B	1	1980	409	A	1931	392
12	8 pm	11	57	A	1	2412	630	B	2482	631
13	8 am	11	39	A	2	3420	532	B	3300	518
13	8 pm	11	50	B	2	4371	802	A	4382	808
14	8 am	11	44	B	3	3893	377	A	4012	384
14	8 pm	12	03	A	3	4658	652	B	4541	624
15	8 am	11	33	A	4	1830	122	B	1782	119
15	8 pm	11	52	B	4	1164	437	A	1127	476
16	8 am	12	12	A	4	2955	716	B	2849	695
16	8 pm	11	38	B	4	2775	505	A	2778	508
17	8 am	23	49	B	3	9058	1073	A	9112	1111
18	8 am	23	45	A	6	9437	924	B	9173	877

$$\frac{\sum S/\sum Pb\ (\text{without MgO})}{\sum S/\sum Pb\ (\text{with MgO})} = \frac{47469/7143}{47953/7179} = 0.995 \qquad \frac{\sum Pb\ (A)}{\sum Pb\ (B)} = 1.027$$

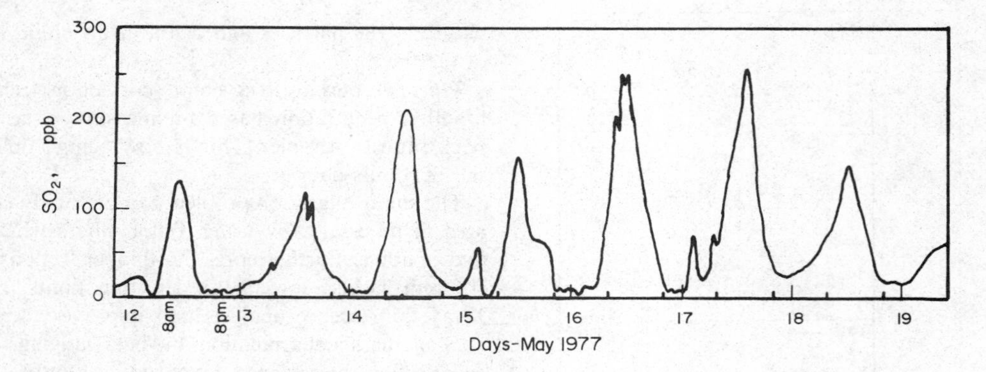

Fig. 5. Ambient SO_2 variations during the investigation of possible sulfate formation in Charleston, West Virginia.

environmental conditions (100% r.h. air at one atmosphere with 3 Torr partial pressure of SO_2). As seen in Fig. 4, the sulfur saturated on these at a relatively low level, except for the case of the glass fiber filter. It is clear that the saturation level of spurious sulfur formation on Millipore filters is less than $0.5\ \mu m\ cm^{-2}$. Similar extreme exposures of field samples from the RAPS network indicate that the sulfur increase due to SO_2 conversion is limited to about 1.5% of the normal fine-particle load and 4% of the coarse-particle load.

An experiment was recently carried out to test the spurious sulfate formation under field conditions. Two samplers, A and B, each equipped with a flow controller and an 152 cm inlet tube of 2.8 cm bore, were calibrated and used to sample side-by-side at $5\ l\ m^{-1}$. One of the inlet tubes was coated with MgO to remove SO_2 from the inlet airstream. The efficiency of one of these SO_2 denuder tubes has been measured by EPA to be 99.5% at 50% r.h. and at an inlet SO_2 level of 400 ppb. The inlet tubes were alternated between the two samplers after each sampling period of 12 or 24 h. Fresh denuder tubes were installed after one or two days of sampling to eliminate any possible saturation effects. The parallel samples, collected on $1\ \mu m$ pore size Teflon filters (Ghia Corporation, Pleasanton, California), were then compared by XRF analysis for sulfur and lead concentrations.

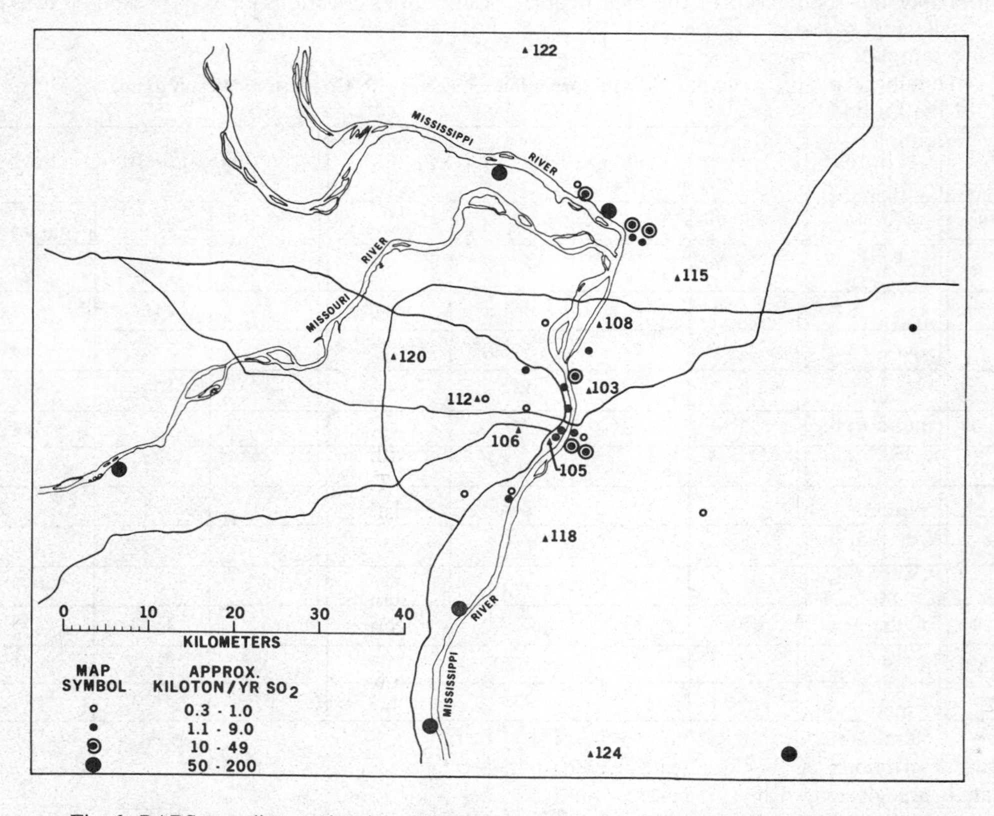

Fig. 6. RAPS sampling station locations (numbered) and major SO_2 emission point sources.

Table 4. Brief descriptions of the aerosol sampling locations in the RAPS network

STATION	LOCAL ENVIRONMENT
103	National City, Illinois. Amidst heavy industries, railroad, army depot and vehicular traffic. Open area 0.5 km around site. Power plant 2.2 km to the NW.
105	South central St. Louis. Among trucking and warehousing operations. Close to chemical plants. Heavy industry 3 km to the E.
106	Within botanical garden. Central residential district. Freeway 0.1 km to the N. Steel company 0.8 km to the W.
108	6 km NW of Granite City, Illinois, among farmland. Steel operations 4.5 km to the SE. Cement plant 5.8 km to the W.
112	On athletic field of Washington University. Highway 2 km to the S.
115	23 km NE of St. Louis. Argicultural area. Petrochemical plant 5.8 km to the NW.
118	16 km S of St. Louis on farmland. Petrochemical plants 7 km to the NE.
120	24.5 km W of St. Louis. Light industrial park. Heavy traffic 1.8 km to W and 0.2 km to the S.
122	45 km N of St. Louis. Agricultural area. No major pollution source nearby.
124	38.5 km S of St. Louis. Agricultural area. No major pollution source within several km.

The results of this experiment are summarized in Table 3, while Fig. 5 shows the ambient SO_2 level during the sampling period. By periodically interchanging the tubes and by normalizing the results in terms of the lead concentrations, any asymmetries between the samplers were eliminated. No detectable spurious sulfate formation was observed in these specimens. We, therefore, have no explanation for the reported SO_2 conversion except that it may have occurred under a unique atmospheric condition.

RESULTS AND DISCUSSIONS

We are now completing measurements on the RAPS aerosol samples and applying the recently-derived systematic corrections retroactively to data already in the EPA data bank. However, since substantial corrections, such as that for the sulfur migration effect, are required only for a small number of samples, we feel that some preliminary conclusions can be drawn from the present data.

Figure 6 is a map of the St. Louis area showing the location of the ten sampling sites (stations 103, 105, 106, 108, 112, 115, 118, 120, 122 and 124) that were equipped with automated dichotomous samplers. The metropolitan area is roughly encompassed by a 15 km radius circle centered about station 112. Brief descriptions of the local environment of each station are given in Table 4. Included in Fig. 6 are the major fixed SO_2 emission sources categorized according to their SO_2 output in kilotons per year (Lippman, 1977).

Temporal and spatial variations

The short-term variations of fine-particle sulfur, lead, bromine and zinc for 2 h samples collected at Site 103 are contrasted in Fig. 7. Salient features of the results are: (1) The time variations of sulfur are much slower than those of lead or bromine, which generally follow diurnal traffic patterns. Sulfur also exhibits little correlation with other elements. (2) A constant lead-to-bromine ratio indicates automobile emission as the primary source of these elements. In general, our data corresponds to this situation but variations in the Pb/Br ratio occur on occasion. (3) The peaks due to non-automotive lead were well correlated with the zinc peaks on Julian (J) days 196 and 197. Trace elements such as zinc and copper characteristically show extreme short-term fluctuations which indicates highly varying source activities, short residence time in the air, or the proximity of the sampling site to a source (making it sensitive to wind direction). (4) During the sulfur episode on J days 210–212, which were plagued by ground fog and haze, zinc was nearly absent.

The intimate relationship between fine-particle sulfur and total particulate mass is illustrated in Fig. 8 where their temporal variations at station 120 are delineated in parallel; Fig. 9 is a scatter plot of the same data. The correlation coefficient between sulfur

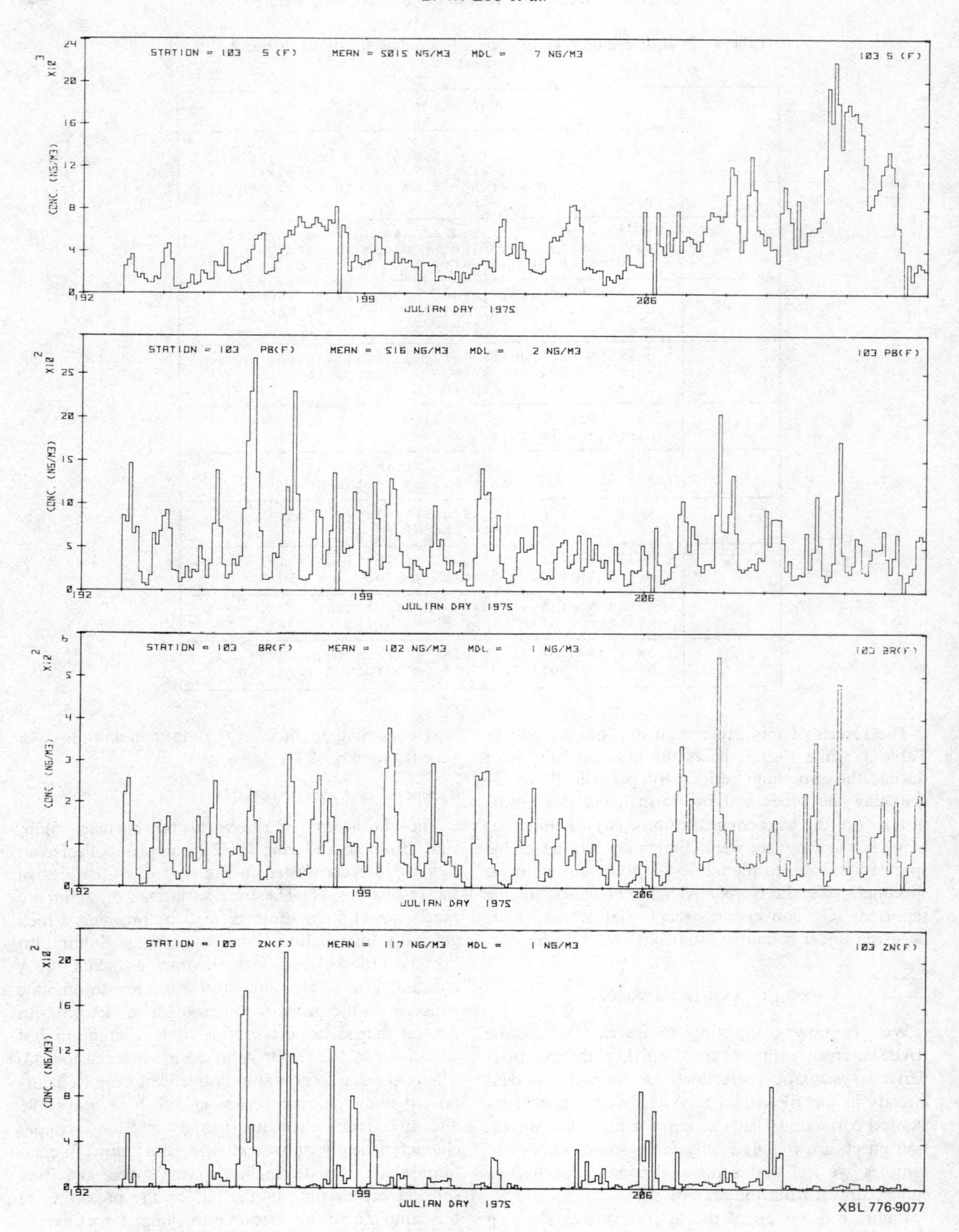

Fig. 7. Short-term temporal variations of fine particulate S, Pb, Br and Zn.

and mass was calculated to be 0.92 for these samples.

The spatial distributions of fine-particle sulfur and total aerosol mass are shown in Table 5, where the monthly averages for all ten stations during the latter half of 1975 are tabulated. The seasonal trends are apparent. Ratios of fine-particle sulfur to mass S(F)/M(F) are seen to be fairly constant over all sampling stations, with the average ratio for the entire region being 0.117. Typically, the sulfate mass might be three times that of elemental sulfur; therefore, sul-

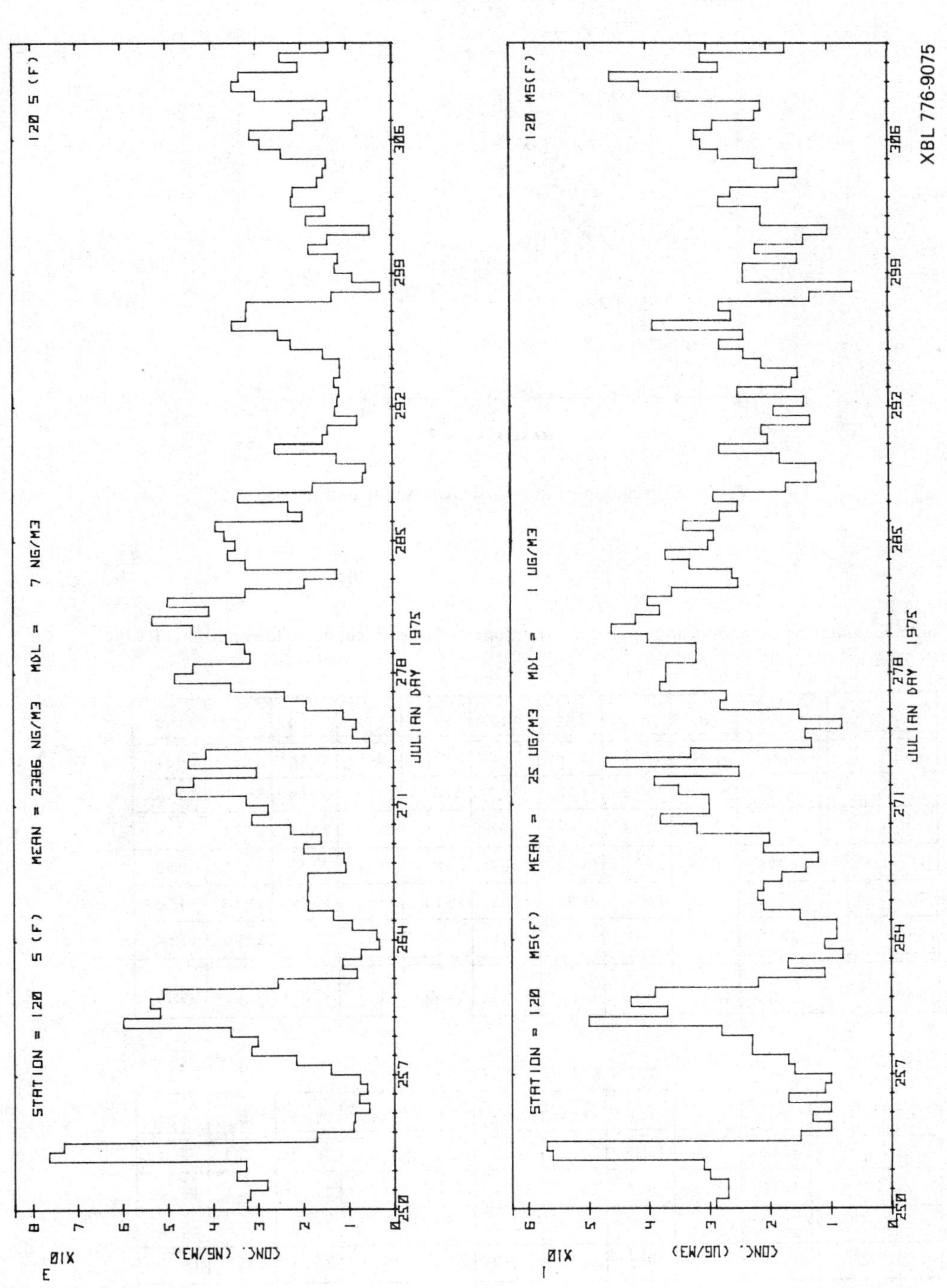

Fig. 8. Temporal variation of fine-particulate sulfur and mass.

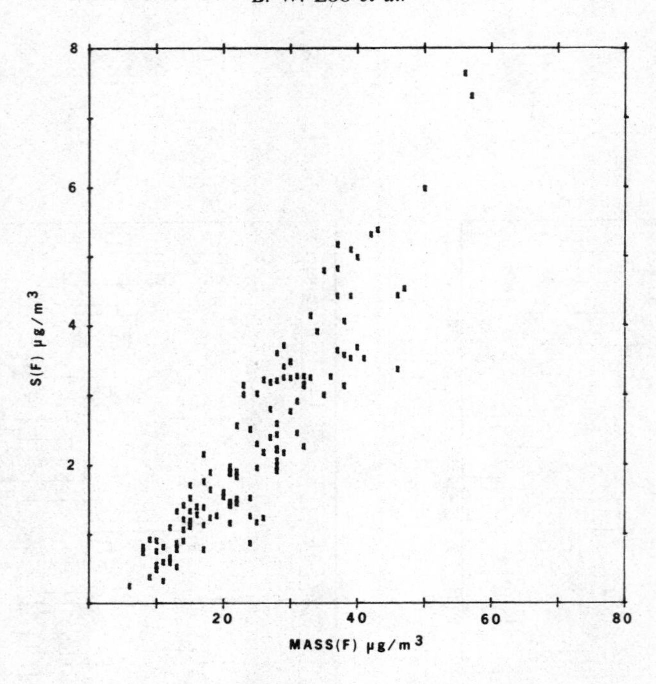

Fig. 9. Correlation of fine-particulate sulfur and mass.

Table 5. Long-term temporal and spatial distributions of fine-particulate sulfur and mass (1975)

SULFUR (FINE) (ng/m³)

STATION	103	105	106	108	112	115	118	120	122	124
JULY	5663	5786	5427	4726	5277	5531	4940	4646	3954	3967
AUG.	4900	4455	4142	3698	3719	3760	2818	3619	3607	3045
SEPT.	3776	3793	3158	2761	3037	2626	2627	2684	2181	2916
OCT.	2986	2818	2705	2196	2584	2301	2100	2205	1721	1972
NOV.	2352	2228	2128	2093	2055	1911	1496	1789	1843	1470
DEC.	2818	2748	2589	2576	2716	2492	1961	2397	2196	1584
SIX MO. AVERAGE	3733	3638	3358	3008	3231	3104	2657	2890	2584	2492

MASS (FINE) (µg/m³)

STATION	103	105	106	108	112	115	118	120	122	124
JULY	55.6	50.0	48.6	36.3	50.0	42.3	34.5	36.2	33.0	27.4
AUG.	46.8	31.2	29.4	28.1	29.4	25.6	18.3	21.2	27.1	18.7
SEPT.	31.7	28.9	27.2	23.7	22.4	19.5	17.7	25.6	20.3	20.6
OCT.	43.4	37.2	25.7	25.5	24.2	20.0	15.1	24.5	23.2	16.5
NOV.	20.8	27.0	20.1	25.3	22.1	17.3	12.8	23.6	16.9	17.3
DEC.	24.1	26.6	20.6	23.2	21.3	19.7	15.6	21.5	17.5	13.5
SIX MO. AVERAGE	37.1	33.5	28.6	27.0	28.2	24.1	19.1	25.4	23.0	19.0
S(F)/M(F)	0.10	0.11	0.12	0.11	0.11	0.13	0.14	0.11	0.11	0.13

Table 6. Wind directional sulfur distribution averaged over a four-month period

SULFUR (FINE)

STATION	NORTH		EAST		SOUTH		WEST	
	ng/m^3	ratio	ng/m^3	ratio	ng/m^3	ratio	ng/m^3	ratio
103	2819	1.70	3466	1.50	2622	1.29	3251	2.21
105	2949	1.78	3149	1.36	2450	1.20	2636	1.80
106	3440	2.08	2266	0.98	2464	1.21	2556	1.74
112	2935	1.77	2399	1.04	2267	1.11	2300	1.57
115	2396	1.45	2614	1.13	2233	1.10	2298	1.57
120	1841	1.11	3625	1.56	2349	1.15	1468	1.00
122	1657	1.00	2317	1.00	2129	1.05	2044	1.39
124	1862	1.12	2719	1.17	2037	1.00	1735	1.18

fates constitute ~35% of the fine aerosol mass. During a smog episode in 1975 (J day 178–186), this rose to as much as 41%. The relative constancy of the sulfur-to-mass ratio can be used as a useful quick cross-check between the two independent measurements (i.e. sulfur determined by X-ray fluorescence and mass by beta-gauging).

Wind directional spatial variations

To better assess the impact of SO_2 sources on local particulate sulfur levels, the sulfur data from the last four months of 1975 have been sorted into four groups according to the predominant wind directions: north, east, south, or west. (National Weather Service Office, 1975) The values are listed in Table 6.

For each of the four groups, the average fine-sulfur levels at various sites were compared with the level at an upwind reference stations and the ratios were calculated. Stations 108 and 118 were omitted in this comparison because they were sampled less frequently. Referring to Fig. 6, we see that stations 122, 124 and 120, respectively, can be used as background stations when the winds are from the north, south, or west. Since there is no station located far upwind in the east, the remote station 122 was selected as a substitute background station.

The fact that minor SO_2 sources are not included in the analysis and that wind directions are divided only roughly into four quadrants makes the interpretation of the results in Table 6 rather general. However, the following features seem to emerge: (1) The downwind stations always exhibit higher sulfur levels than the background stations showing that the city contributes a significant burden to the atmosphere. (2) As a whole, the background air masses arriving from west and north of St. Louis have about 30% lower particulate sulfur than those from the east and south. This result probably reflects the long-distance transport of sulfur from sources in the industrial east coast areas of the United States. (3) Considering the orientation of the sampling stations with respect to the major emission sources, we find the relative amount of sulfur increase to be consistent with the observed wind directions. For example, station 120 had the lowest sulfur level of all stations when the wind was from the west, the largest increase (56% when the wind was from the east, and moderate increases of 11 and 15% when the winds were from the north and south, respectively. These crosswind comparisons are only qualitative because of the uncertainties in comparing data sets taken for different time periods and under different conditions. (4) The remote stations show a smaller sulfur increase than the urban stations, even though the remote stations are downwind from the city. For instance, station 124 shows only a 12% sulfur increase when the wind is from the north and station 122 shows only a 5% increase when the wind is from the south, while the urban stations experience increases of more than 50% in both cases. This suggests that the effect of the emission sources (in terms of paticulate sulfate) is either short-ranged or highly directional. To address this question, we examined some short-term sulfur distributions, where the effect on a distant station is less diluted by changes in wind direction that occurs in long-term averages. Table 7 shows examples of sulfur distributions averaged over six-hour periods during which wind direction and speed were known to be quite steady. The distant downwind stations now have sulfur increases comparable to those of the city

Table 7. Short-term (6 h average) wind directional sulfur distribution

SULFUR (FINE)

JULIAN DAY (1975)	218.25		211.25		222.0		208.25	
WIND DIRECTION	15°		85°		190°		255°	
WIND SPEED (KNOT)	8.5		3.5		7.0		6.0	
RELATIVE HUMIDITY (%)	98		87		90		88	
STATION	ng/m^3	ratio	ng/m^3	ratio	ng/m^3	ratio	ng/m^3	ratio
103	4706	5.91	14605	1.55	7123	1.52	5245	1.17
105	5399	6.78	13306	1.41	5859	1.25	4735	1.06
106	4398	5.53	11184	1.19	5663	1.20	4311	0.96
112	3102	3.90	14011	1.48	6478	1.38	4319	0.96
115	2302	2.89	10201	1.08	5207	1.11	5642	1.26
120	3410	4.28	10093	1.07	6624	1.41	4481	1.00
122	796	1.00	9437	1.00	5919	1.26	4452	0.99
124	3774	4.74	8226	.87	4700	1.00	4252	0.95

stations. We therefore conclude that the emission sources have long-range (at least 40 km) and rather directional impact on their downwind environment. Note that the air mass from the north was quite clean on J day 218 (1975); thus, even though the sulfur levels downwind were not excessive on that date, the percentage increase was exceptionally high.

Aerosol composition

Table 8 shows results for a number of elements for four representative stations throughout the region. Both size fractions are represented. The results given are mean concentrations over about one-month period in June/July 1975. Although the aerosol mass is almost evenly divided between the two size-frac-

Table 8. Aerosol compositions over four representative stations

STATION	103	112	118	122
Mass (C) μg/m^3	38	34	25	22
Mass (F)	51 (57%)	35 (51%)	30 (55%)	27 (55%)
S (C) μg/m^3	701	816	579	354
S (F) μg/m^3	5620 (89%)	3933 (83%)	3736 (87%)	3387 (91%)
Fe (C) μg/m^3	1421 (88%)	1261 (88%)	671 (87%)	528 (87%)
Fe (F) μg/m^3	200	180	98	80
Zn (C) μg/m^3	73	40	12	15
Zn (F) μg/m^3	133 (65%)	64 (62%)	30 (71%)	45 (75%)
Pb (C) μg/m^3	128	192	60	23
Pb (F) μg/m^3	573 (82%)	705 (79%)	213 (78%)	153 (87%)
Br (C) μg/m^3	24	52	8	3
Br (F) μg/m^3	108 (82%)	133 (72%)	26 (76%)	16 (84%)
Pb (F)/Br (F)	5.3	5.3	8.2	9.6

tions, nearly 90% of the sulfur is found in the fine fraction. Iron (generally a soil component) is found largely in the coarse-particle fraction. Similarly for other elements, while the actual concentrations show wide spatial variations, the percentage in each of the size-fractions is quite constant.

It is of interest to note that lead to bromine ratio in the fine-particle fraction, an indicator of the age of the automobile emission aerosol (Robbins, 1972), is significantly higher at the rural sites than at the urban stations. However, the actual Pb/Br ratios shown in Table 6 are not typical. Long-term data (4 month averages) reveal that the range of such ratios between urban and rural sites are from 3.1 to 6.1.

CONCLUSION

The sampling and analysis system used in this study represents an integral approach to large-scale aerosol monitoring, including particle sizing, mass analysis and elemental concentration measurements. We feel that the inherent advantages of the XRF technique have been fully demonstrated in the study and that the analytical corrections required for sulfur analysis are sufficiently well defined to make the technique accurate and reproducible.

In the course of this study, we have advanced from utilizing a large computer center for complex data sorting to using a dedicated data system with a large (40M word) disc. More complete analysis of the experimental data obtained in large-scale studies will require the systematic treatment of very large data sets including visibility, meteorological and gaseous data. The disc-based system will vastly improve the capacity to handle large volumes of data and of our ability to examine the correlation patterns.

Acknowledgements—A large-scale study of this nature of necessity involves many contributors. We thank all our collaborators at the Lawrence Berkeley Laboratory including R. Adachi, R. Fisher, R. Giauque, B. Jarrett, D. Landis, N. Madden, J. Meng, A. Ramponi and W. Searles. We also appreciate the work of the staff of the Rockwell International Science Center in St. Louis and particularly that of D. Hern, A. Jones, L. Myers and E. Nelson.

REFERENCES

Appel B. R., Kothny E. L., Hiffer E. M., Buell G. C., Wall S. M. and Wesolowski J. J. (1977) A comparative study of wet chemical and instrument methods for sulfate determination. Symposium Preprint, Division of Environmental Chemistry, ACS, 173rd National Meeting, New Orleans, Louisiana, March 20–27, pp. 117–120.

Camp D. C., VanLehn A. L., Rhodes J. R. and Pradzynski A. H. (1975) Intercomparison of trace element determinations in simulated and real air particulate samples. *X-ray Spectrometry*, **4**, 123–137.

Dzubay T. G. and Lamothe P. J. (1977) Polymer film as calibration standards for X-ray fluorescence analysis. *Adv. X-ray Anal.* **20**, 411–421.

Giauque R. D., Garrett R. B. and Goda L. Y. (1977) Calibration of energy-dispersive X-ray spectrometers for analysis of thin environmental samples. In *X-ray Fluorescence Analysis of Environmental Samples* (edited by T. G. Dzubay) pp. 153–164. An Arbor Science.

Goulding F. S., Jaklevic J. M. and Loo B. W. (1975) Fabrication of aerosol monitoring system for determining mass and composition as a function of time. Environmental Protection Agency Report No. EPA-650/2-75-048, EPA Research Triangle Park, North Carolina.

Jaklevic J. M., Loo B. W. and Goulding F. S. (1977a) Photon-induced X-ray fluorescence analysis using energy-dispersive detector and dichotomous sampler. In *X-ray Fluorescence Analysis of Environmental Samples* (edited by T. G. Dzubay) pp. 3–18, Ann Arbor Science.

Jaklevic J. M. and Walter R. L. (1977b) Comparison of minimum detectable limits among X-ray spectrometers. In *X-ray Fluorescence Analysis of Environmental Samples* (edited by T. G. Dzubay) pp. 63–75, Ann Arbor Science.

Lasko L. J., Washeleski M. C., Noll K. E. and Allen H. E. (1977) Continuous sulfate monitoring program in a large urban area, Symposium Preprint, Division of Environmental Chemistry, ACS, 173rd National Meeting, New Orleans, Louisiana, March 20–27, pp. 175–176.

Lippman F. (1977) Major SO_2 sources information provided by Mr. Lippman, Rockwell International Science Center from the point source emission inventory of the St. Louis Regional Air Pollution Study.

Liu B. Y. H. and Kuhlmey G. A. (1977) Efficiency of air sampling filter media. In *X-ray Fluorescence Analysis of Environmental Samples* (edited by T. G. Dzubay) pp. 107–119. An Arbor Science.

Loo B. W., Jaklevic J. M. and Goulding F. S. (1976) Dichotomous virtual impactors for large-scale monitoring of airborne particulate matter. In *Fine Particles: Aerosol Generation, Measurement, Sampling and Analysis* (edited by B. Y. H. Liu) pp. 311–350. Academic Press, New York.

Loo B. W., Gatti R. C., Liu B. Y. H., Kim C. S. and Dzubay T. G. (1977) Absorption corrections for submicron sulfur collected in filters. In *X-ray Fluorescence Analysis of Environmental Samples* (edited by T. G. Dzubay) pp. 187–202. Ann Arbor Science.

NWSO (1975) Wind data taken from Local Climatological Data, National Weather Service Office, St. Louis International Airport.

Pierson W. R. (1977) Spurious sulfate in aerosol sampling: a review. Symposium Preprint, Division of Environmental Chemistry, ACS, 173rd National Meeting, New Orleans, Louisiana, March 20–27, pp. 165–167.

Robbins J. A. and Snitz F. L. (1972) Bromide and chlorine loss from lead halide automobile exhaust particulates. *Environ. Sci. Tech.* **6**, 164–169.

Stevens R. K., Dzubay T. G., Russwurm G. and Rickle D. (1978) Sampling and analysis of atmospheric sulfate and related species. *Atmospheric Environment* **12**, 55–68.

Whitby K. T., Husar R. B. and Liu B. Y. H. (1972) The aerosol size distribution of Los Angeles smog. In *Aerosols and Atmospheric Chemistry* (edited by G. M. Hidy) pp. 137–264. Academic Press, New York.

Whitby K. T. (1973) On the multimodal nature of atmospheric aerosol size distribution. Particle Technology Lab Publication No. 218. University of Minnesota.

Atmospheric Environment Vol. 12, pp. 773–777. Pergamon Press 1978. Printed in Great Britain.

CHARACTERISTICS OF SULFUR AEROSOL IN FLORIDA AS DETERMINED BY PIXE ANALYSIS

M. S. AHLBERG, A. C. D. LESLIE and J. W. WINCHESTER

Department of Oceanography, Florida State University, Tallahassee, FL 32306, U.S.A.

Abstract—The particle size distribution of sulfur has been determined in Florida during two seasons, July–August and December 1976. Approximately 120 cascade impactor samples from 10 widely distributed urban and nonurban sites in the state, each sample consisting of 6 separate particle size fractions from $<0.25\,\mu m$ to $>4\,\mu m$ aerodynamic diameter, were analyzed by proton induced X-ray emission for elemental constituents in over 800 individual specimens, including blanks. Most of the sulfur occurred in a fine particle mode, $<2\,\mu m$ diameter, with lesser amounts in a coarse mode, $>2\,\mu m$. The mass median aerodynamic diameter, MMAD, of sulfur in the fine mode was found to be greater for samples collected under higher average relative humidity, r.h., conditions than samples from lower humidities. The trend of MMAD with r.h. indicates that ammonium sulfate, rather than sulfuric acid, is more likely to be the principal chemical form of sulfur in the fine mode.

INTRODUCTION

The atmospheric concentration of sulfur in particles of different size can be indicative of the principal natural and pollution sources, atmospheric transport and removal, chemical speciation and potential biological impacts of importance to man. Because sulfur is generally a reactive element under natural conditions, however, its concentration in discrete particle size ranges may not be as indicative of these aspects as may be the relationships between concentrations in different particle size ranges under various environmental conditions. In this paper (paper 1) we explore some of these relationships using as a data base a series of samples collected in Florida during two periods in middle and late 1976. In a separate paper (paper 2) (Leslie *et al.*, 1978) we examine certain geographic trends in the sulfur concentrations.

EXPERIMENTAL

Cascade impactors were operated during the periods 28 July–7 August and 5–18 December 1976, at 10 widely separated nonurban and urban sites for six 24 h intervals in each period. The sites were located as shown in the index map, Fig. 1. A discussion of the field program for sample collection and the specific sampling sites has been

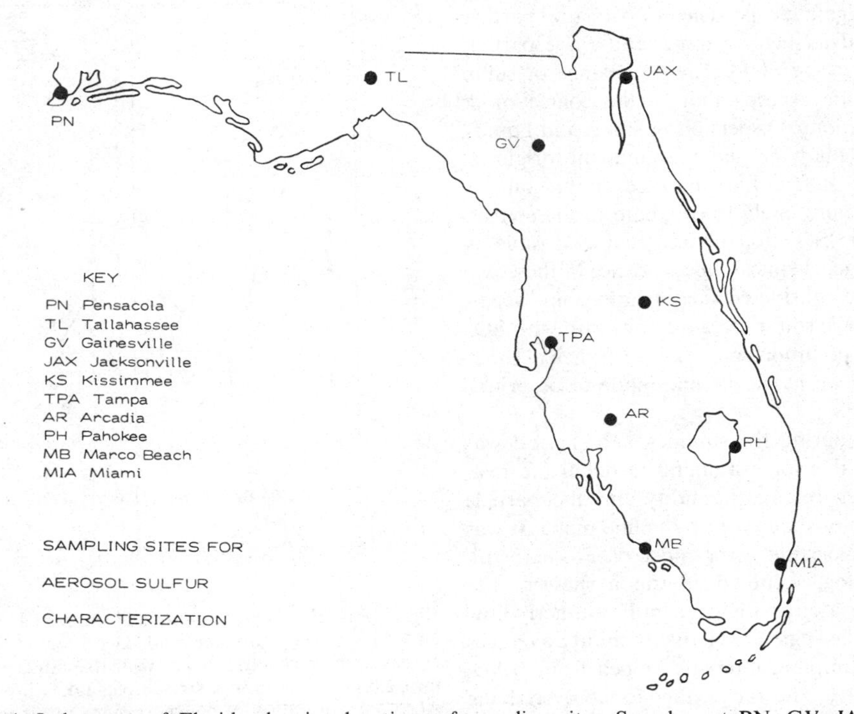

Fig. 1. Index map of Florida showing locations of sampling sites. Samplers at PN, GV, JAX, KS, AR, PH and MB were located at small airports and at TL, TPA and MIA within the cities.

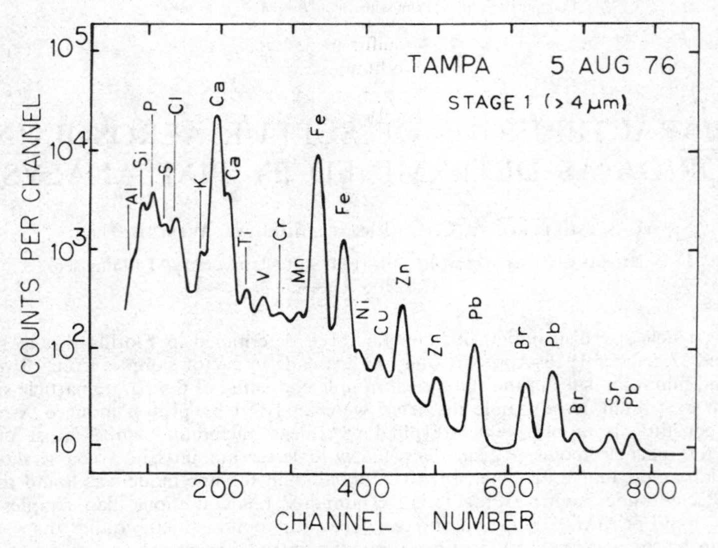

Fig. 2. Example of X-ray spectrum from a PIXE analysis showing characteristic X-ray peaks of the elements resolved from background radiation.

given by Leslie *et al.* (1977). Six particle size fractions, from $>4\ \mu m$ to $<0.25\ \mu m$ dia. were analyzed for S as well as for Cl, K, Ca, Ti, V, Fe, Zn, Br and Pb using proton induced X-ray emission, PIXE (Johansson *et al.*, 1975). The X-ray spectra, exemplified by Fig. 2, were resolved by computer (Kaufmann *et al.*, 1976).

RESULTS

Typical of the results of this investigation are the geometric mean particle size distributions for sulfur shown in Fig. 3 for Pensacola, Tampa and Marco Beach in the summer and December 1976 periods. The trends suggest the presence of a coarse particle mode, $\gtrsim 2\ \mu m$ dia. (stages 1 and 2) and a fine particle mode, $\lesssim 2\ \mu m$ (stages 3–6). The abundance of sulfur in the fine mode exceeds that in the coarse mode at all three locations. Dispersion of solids and liquids, e.g. sea spray, may be the predominant origin of coarse particle sulfur. Fine particle sulfur, on the other hand, is more likely to originate in the conversion of gaseous SO_2 to sulfuric acid and sulfates. Because of the geochemical independence of these two modes, we have considered their relationships separately. Fine particle sulfur, because of its link with SO_2 emissions from pollution sources, is of principal interest. Empirically we judge the fine mode to be principally stages 3–6.

The weather during the summer 1976 period was especially favorable for examining in detail the relationship between relative humidity and the particle size distribution of sulfur within the fine mode. It was typical for summertime over most of Florida with early morning fog, scattered afternoon showers and thundershowers, along with a generally southerly flow as a rule over the region. Relative humidity averages over the 24 h sampling intervals ranged from below 70% to over 90%. The range, therefore, covered the region on both sides of the critical humidity of 81% for the conversion of crystalline ammonium sulfate to aqueous solution droplets.

In the discussion which follows it is important to note some details concerning a frontal passage or instability lines which developed during the sampling period. From 28 to 31 July the Bermuda high was centered well off the east coast of Florida with a narrow ridge extending westward through the central and south regions. Surface winds were southwesterly over the north and northwest, were variable over cen-

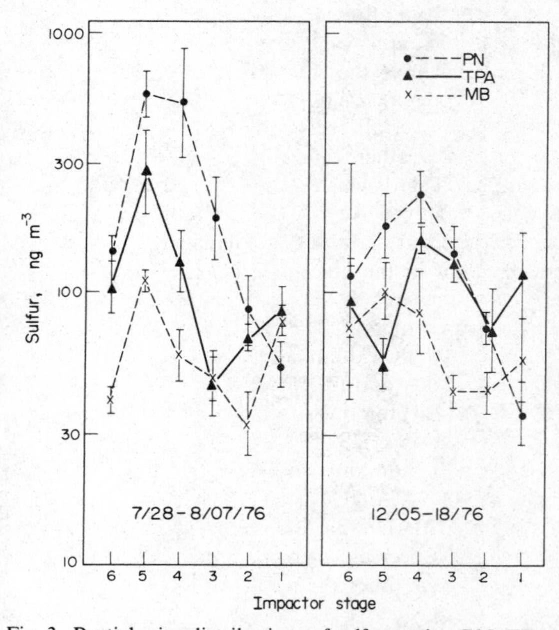

Fig. 3. Particle size distributions of sulfur at sites PN, TPA and MB during summer and December 1976 periods, expressed as geometric mean, with its standard deviation, for each group of approx six samples taken in each period. Absolute calibration of concentrations in air is significant to a factor of 2.

Table 1. Mass median aerodynamic diameter, in μm, for sulfur in the small mode $\leq 2\ \mu m$, and corresponding averaged relative humidity.

Sampling site	Date, July–August 1976											
	28–29	29–30	30–31	31–1	1–2	2–3	3–4	4–5	5–6	6–7	7–8	8–9
PN			0.39	0.41	0.40		0.58	0.65				
			74.8	78.9	75.6	81.3	74.1	80.1				
TL		0.46	0.43		0.45	0.68		0.74	0.56			
		79.6	82.4		80.2	93.0		85.4	78.8			
GV			0.42	0.43	0.47	0.69	0.56	0.57				
			77.9	80.5	76.2	80.7	88.7	84.1				
JAX	0.55	0.70			0.39	0.60	0.83					
	82.5	75.6			70.8	83.5	91.1					
KS	0.41	0.42		0.36	0.31	0.45	0.56					
	79.8	75.5		74.9	72.7	76.4	86.8					
TPA				0.32	0.38	0.40	0.55	0.38	0.39			
				73.5	71.3	74.6	83.0	71.8	70.5			
AR		0.32	0.44		0.58	0.39	0.51	0.44				
		72.8	72.1		74.3	68.8	75.2	75.1				
PH	0.27	0.45		0.43	0.45	0.43	0.55					
	75.1	78.1		78.0	82.0	73.6	81.3					
MB		0.40	0.42	0.50	0.50	0.36	0.67					
		72.5	72.1	72.4	70.2	69.6	75.8					
MIA							0.50	0.63	0.60	0.54	0.57	0.58
							68.5	73.8	77.5	75.6	70.1	73.1

tral Florida, and were southeasterly in the south with some local sea breezes. Intermittent nighttime fog and haze and afternoon to evening thunderstorms occurred throughout the state. From 1 to 4 August unsettled weather was observed over all but the extreme south of Florida. A weak Canadian cool front passed through the north and northwest regions and caused a wind shift from southwest to north and northeast. Instability lines formed over the north and moved south as far as Miami, accompanied by heavy thunderstorms and a switch to northerly air flow and heavy morning fog. From 5 to 10 August Hurricane Belle was forming to the east of Miami, and a weak Pacific cool front affected the northwest region of the state. Whereas on 4 August a generally easterly flow was prevalent over the entire state, by 6 August the tropical depression had caused enhanced north-easterly flow over all except the northwest region.

With these weather features in mind, let us examine the data of Table 1 which covers the period of frontal passage, 28 July–5 August. In the table we present the mass median aerodynamic diameter, MMAD in μm for sulfur in the fine mode (stages 3–6) together with the average relative humidity in per cent for each 24 h sampling interval. The north Florida sites PN, TL, GV and JAX, exhibit a distinct increase in both relative humidity and fine mode sulfur MMAD from about 0.4 to 0.6 μm on 2 August, but the central and south Florida sites, KS, TPA, AR, PH and MB show an increase in humidity and MMAD delayed until 3 August. On 4 August at TPA and AR a return to smaller MMAD was observed, apparently corresponding to a decrease in relative humidity. We also noted the exact times of frontal passage across our north Florida sites: PN, 03Z on 8/03; TL, 18Z on 8/03; JAX, 00Z on 8/04; and GV, 04Z on 8/04. (In local EDT the times are 4 h earlier). Apparently the gradual passage of the front and its attendant humidity changes are associated with the MMAD increases we observe.

DISCUSSION

It is of some interest to compare aerosol sulfur MMAD with that expected for its principal chemical compounds under the relative humidity conditions during sampling. On Fig. 4 we have plotted the fine mode sulfur MMAD vs average per cent relative humidity over the corresponding 24 h sampling intervals based on hourly airport weather data. In all cases, (except sites AR, PH and MB where observations at Sarasota, West Palm Beach and Fort Myers were used) the weather data were taken at or very near the sampling sites.

With very few exceptions, before frontal passage humidities were well below 81% and MMAD 0.3–0.5 μm. After frontal passage MMAD were $>0.5\ \mu m$ and 24 h average humidities ranged both above and below 81%. On Fig. 4 we also show the theoretical variation of MMAD with per cent relative humidity for H_2SO_4 and $(NH_4)_2SO_4$, based on Hänel (1976), with the H_2SO_4 and dry $(NH_4)_2SO_4$ curves fit to the $<0.5\ \mu m$ points before frontal passage. The phase change at 81% from solid to aqueous $(NH_4)_2SO_4$ is experienced with rising humidity, but with falling humidity supersaturated solutions may persist well below this value. Consequently, the supersaturated curve is also shown. For H_2SO_4, which is liquid at all humidities, no phase change occurs. The data points appear to conform to $(NH_4)_2SO_4$, rather than H_2SO_4, as the principal chemical form of sulfur. It seems significant that there are no points near the H_2SO_4 curve at high humidities. Of course, in view of the long 24 h sampling times, in which considerable diurnal humidity variation occurred, our ability to distinguish between the two forms is not as precise as if sampling were performed under constant humidity conditions. If sampling was restricted to shorter time intervals, with precise humidity recording at the sampler location or by using samplers activated by humidity sensing devices, greater precision in dis-

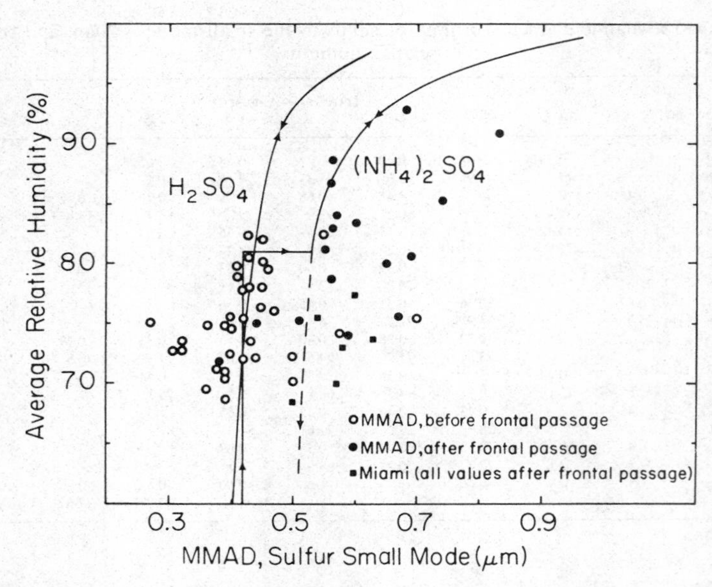

Fig. 4. Plot of mass median aerodynamic diameter of sulfur in $<2\ \mu m$ particles (impactor stages 3–6) against r.h. averaged over the 24 h sampling periods.

tinguishing these two chemical forms may be achieved. Alternatively, the humidity of the air before sampling could be controlled.

The MMAD values were calculated using concentrations from all the impactor stages 3–6, in which sulfur is most abundant in stages 4 and 5 (Fig. 3). The method, therefore, does not reveal possible differences in the chemical form of sulfur which may exist

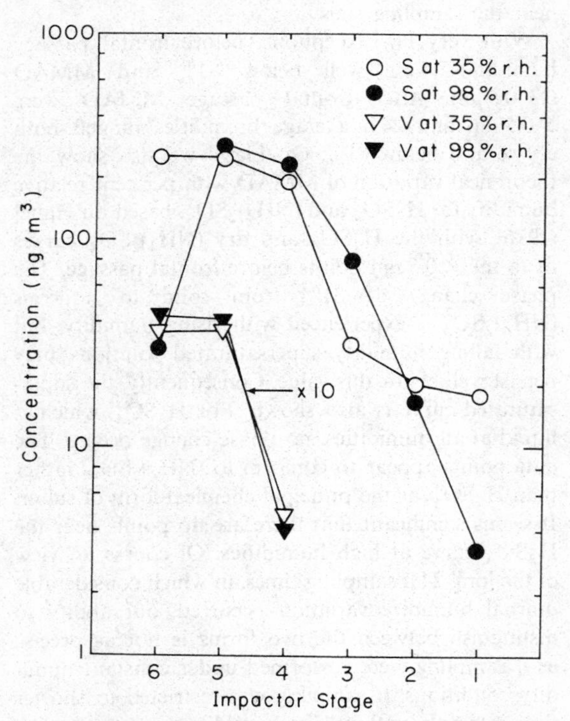

Fig. 5. Particle size distributions of sulfur and vanadium in Tallahassee laboratory air before and after humidification pretreatment.

in the different fine mode particle size ranges. With more extensive data of this kind, such as mentioned above, an extension of the method may provide evidence for such differences.

Figure 5 reports laboratory measurements of aerosol-humidity relationships for ambient sulfur and vanadium (Ahlberg *et al.*, 1977). Two impactors, one equipped with a horizontal humidification tube half-filled with water simultaneously sampled air in the laboratory. One sample represented the particle size distributions of the elements in air at its ambient relative humidity of 35%, and the other sample represented distributions at a relative humidity of over 95% because of passage of the air over water in the tube. Sulfur showed substantial growth of stage 6 particles ($<0.25\ \mu m$) whereas vanadium did not, indicating a relatively greater hygroscopic character for S in finest particles. The calculated MMAD values for S were 0.34 μm before and 0.58 μm after humidification, in agreement with the data of Fig. 4. It should be noted that this evidence also indicates that S and V must be constituents of different particles; the humidification-controlled sampling technique, therefore, appears to be useful in distinguishing trace element associations with different populations of particles.

Acknowledgements—We are indebted to Stephen L. Cohn and Scott Rheingrover for assistance with meteorological analysis. The study was supported in part by the Florida Sulfur Oxides Study, Inc., by the National Institutes of Health, by the Federal Interagency Energy/Environment Research and Development Program through EPA grant R803887, and by the National Science Foundation for accelerator facilities support. We also acknowledge technical support of William E. Wilson, ESRL, EPA, Director of Project MISTT. The conclusions reached do not necessarily represent endorsement by FSOS, Inc.

REFERENCES

Ahlberg M. S., Leslie A. C. D. and Winchester J. W. (1977) Environmental and occupational health analyses using proton induced X-ray emission. In Proceedings of Symposium on Electron Microscopy and X-Ray Applications to Environmental and Occupational Health Analyses, Denver, Colorado. Ann Arbor Science Publishers, Ann Arbor, Michigan.

Hänel G. (1976) The properties of atmospheric aerosol particles as functions of the relative humidity at thermodynamic equilibrium with the surrounding moist air. *Adv. Geophys.* **19**, 73–188.

Johansson T. B., Van Grieken R. E., Nelson J. W. and Winchester J. W. (1975) Elemental trace analysis of small samples by proton induced X-ray emission. *Analyt. Chem.* **47**, 855–860.

Kaufmann H. C., Akselsson K. R. and Courtney W. J. (1976) REX: a computer program for PIXE spectrum resolution of aerosols. Advances in X-Ray Analysis. (Edited by R. W. Gould, C. S. Barrett, J. B. Newkirk and C. O. Ruud) Vol. 19, pp. 355–366. Kendall Hunt, Dubuque, Iowa.

Leslie A. C. D., Ahlberg M. S., Winchester J. W. and Nelson J. W. (1978) Aerosol characterization for sulfur oxide health effects assessment. *Atmospheric Environment* **12**, 729–733.

Leslie A. C. D., Ahlberg M. S., Winchester J. W. and Nelson J. W. (1977) Aerosol characterization for sulfate health effects assessment in Florida. In *Trace Substances in Environmental Health-XI*. (Edited by D. D. Hemphill) University of Missouri Press, Columbia (in press).

Atmospheric Environment Vol. 12, pp. 779–784. Pergamon Press 1978. Printed in Great Britain.

THE SAME-DAY IMPACT OF POWER PLANT EMISSIONS ON SULFATE LEVELS IN THE LOS ANGELES AIR BASIN

W. H. WHITE
1180 N. Chester Avenue Pasadena, CA 91104, U.S.A.
S. L. HEISLER, R. C. HENRY and G. M. HIDY
Environmental Research & Technology, Inc. Westlake Village, CA 91361, U.S.A.
and
I. STRAUGHAN
Southern California Edison Co. Rosemead, Cal. U.S.A.

(*First received* 14 *June* 1977 *and in final form* 27 *September* 1977)

Abstract—On the basis of emissions inventories, power plants are estimated to contribute about 45% of the SO_2 released annually in the Los Angeles Basin. The day-to-day variability of SO_2 emissions from power plants is comparable to the day-to-day variability of the basin's ambient sulfate concentrations. This paper examines the statistical relationship between the daily SO_2 emissions of Los Angeles power plants and the 24-h average sulfate concentration monitored at West Covina, a 'receptor' site characterized by high sulfate levels. Little correlation between emissions and ambient levels is found, even after much of the influence of meteorology has been factored out. The lack of correlation is consistent with the hypothesis that sulfate production in the Los Angeles Basin is limited by factors other than sulfur emissions. It is also consistent with the hypothesis that sulfate production is a linear function of sulfur emissions, if only a small fraction of the average sulfate concentration at West Covina is contributed by the same-day emissions of Los Angeles power plants.

INTRODUCTION

Although the Los Angeles basin is best known for its oxidant concentrations, it frequently experiences high sulfate loadings as well. Sulfates are thought to be responsible for much of the reduction in visibility associated with photochemical smog (White, 1976; Cass, 1976; White and Roberts, 1977) and there is concern that their effects on the respiratory system may be potentiated by the oxidizing atmosphere (Lee and Duffield, 1977).

A substantial fraction of the SO_2 released in the Los Angeles basin comes from power plants. Due to the sensitivity of power plant emissions to the availability of low-sulfur fuels, their relationship to ambient sulfate levels is currently a topic of considerable interest. The contribution of power plants to ambient levels cannot simply be proportioned on the basis of emissions, because their emissions are released at significantly higher altitudes than those of other sources. On the other hand, the direct measurement of an individual plant's impact is made difficult by the multitude of sources, complicated airflow, and generally high pollutant levels characteristic of the basin (Smith *et al.*, 1975; Richards *et al.*, 1976).

The demand for electric power and the availability of low-sulfur fuels often fluctuate sharply from day to day in Los Angeles, and the resulting variations in SO_2 emissions can be viewed as a serendipitous experimental test of the impact of power plants on ambient concentrations. This approach has previously been taken by Thomas (1962), who found little or no correlation between daily power plant emissions and ambient visibilities, and by Cass (1975), who found no correlation between annual power plant emissions and ambient SO_2 concentrations. In this paper we consider the statistical relationship of daily power plant emissions to ambient sulfate concentrations; it is not surprising, in view of the foregoing results, that we should find little correlation between the two. The novel contribution of the present study is the derivation, from this observed lack of correlation and the assumption of a linear relationship, of a quantitative upper bound for the fraction of the ambient sulfate concentration attributable to recent emissions by power plants.

DATA BASE

The variables considered in our analysis were routinely monitored by a variety of agencies during the May–October 'smog season' of 1975. This data base is summarized in Table 1.

The South Coast Air Quality Management District (SCAQMD)* calculates, from records of fuel use, the daily SO_2 emissions from each power plant in Los

* Incorporating the former Los Angeles County Air Pollution Control District.

Table 1. Data set for May–October 1975

Symbol	Description	Source
$[SO_4^{2-}]_{WC}$ ($\mu g\ m^{-3}$)	Average sulfate concentration at West Covina over 24-h period beginning 11 a.m.	CHAMP
$[O_3]_{AZ}$ (pphm)	Maximum 1-h average oxidant concentration at Azusa (SCAQMD).	California ARB
$r.h._{LB}$ (%)	Average daytime relative humidity at Long Beach Airport (NWS).	SCAQMD
T_{850} (°C)	Temperature at 850 mb, from 6 a.m. sounding at Los Angeles International Airport (LAX).	SCAQMD
T_{LA} (°C)	Maximum surface temperature at downtown Los Angeles (NWS)	U.S. Dept. of Commerce
P_{AB} (tons)	Midnight to midnight SO_2 emissions from Alamitos (SCE) and Haynes (LADWP) power plants (both at Alamitos Bay).	SCAQMD
P_{LA} (tons)	Midnight to midnight SO_2 emissions from all Los Angeles County power plants	SCAQMD

Angeles County. These plants account for over 80% of the steam generating capacity in the air basin (which also includes portions of Orange, San Bernardino, and Riverside counties). All but one of the basin's major plants (≥600 MW) are sited along the coast, the two largest side by side at Alamitos Bay (Fig. 1).

The U.S. Environmental Protection Agency operates six Community Health Air Monitoring Program (CHAMP) stations in the Los Angeles basin. Daily 24-h high-volume filter samples from each station are analyzed for sulfates by wet chemistry. During the 6-month period covered by this study, the highest sulfate concentrations recorded by the CHAMP network were measured at West Covina, near the eastern boundary of Los Angeles (Fig. 1). West Covina lies downwind of the major source areas of Los Angeles and Orange-counties during the afternoon seabreeze regime, and limited time-resolved measurements show sulfate concentrations in West Covina peaking at this time (Hidy *et al.*, 1974).

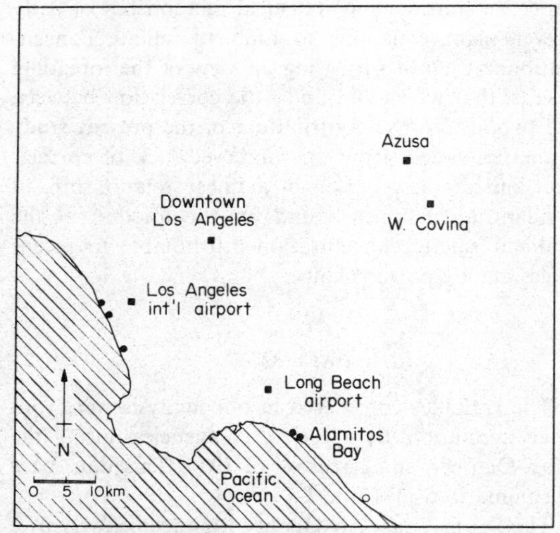

● power plants included in study ○ other power plants
Fig. 1. Map of Los Angeles.

Power plant emissions are calculated for the calendar day, midnight to midnight, while CHAMP filters are switched at approx 11 a.m. In the statistical analysis which follows, the emissions of power plants on a given day are compared with the sulfate concentration measured over the 24-h period beginning at 11 a.m. on that day. The 11-h phase lag is an advantage in the present study; the early morning emissions of coastal sources do not reach West Covina in strength much before 11 a.m., since the sea breeze does not develop until mid-morning.

Relative humidity and the strength of the persistent temperature inversion have been identified as important influences on sulfate concentrations in downtown Los Angeles (Cass, 1975; Zeldin *et al.*, 1976). These meteorological factors are also associated with much of the variability in the sulfate concentration at West Covina. During the 184-day study period, sulfate concentrations in excess of 20 $\mu g\ m^{-3}$ were measured at West Covina on 42 days. All but 5 of these days were among the 81 characterized by high humidities near the coast ($r.h._{LB} \geq 60\%$), high temperatures aloft ($T_{850} \geq 16°C$), and a strong inversion ($T_{850} \geq T_{LA} - 4°C$). (The 850 mb level is approx 1600 m above mean sea level, where under neutral or unstable conditions the air would be at least 14°C colder than that at the surface.) This is shown in Fig. 2.

Since the relationship of emissions to ambient levels is of greatest interest when the meteorological potential for pollution is high, days of low humidities or weak inversions were eliminated from the statistical analysis. The effects of this stratification on the data are summarized in Table 2. Average concentrations of both sulfates and ozone were significantly higher on days of high meteorological potential for sulfate, while power plant emissions were significantly lower. The association of lower emissions with days of high pollution potential confounds the search for a relationship between emissions and ambient levels if meteorological influences are not taken into account.

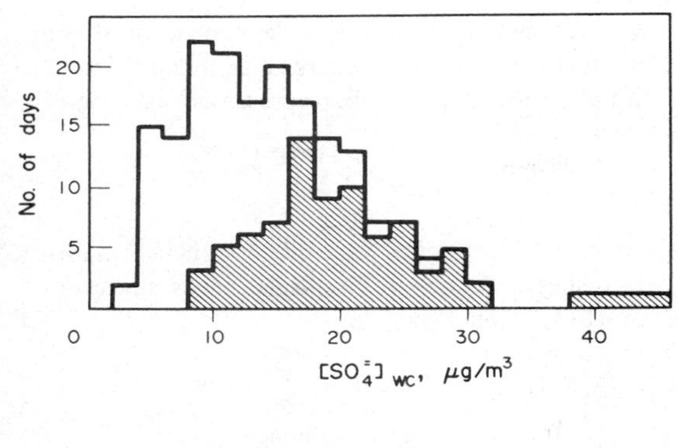

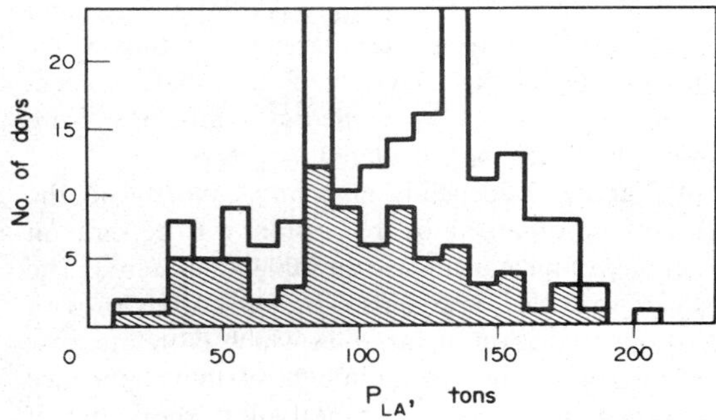

Fig. 2. Frequency distribution of power plant SO_2 emissions and ambient sulfate concentrations during May–October 1975. Shaded portion shows distribution on days of high sulfate potential.

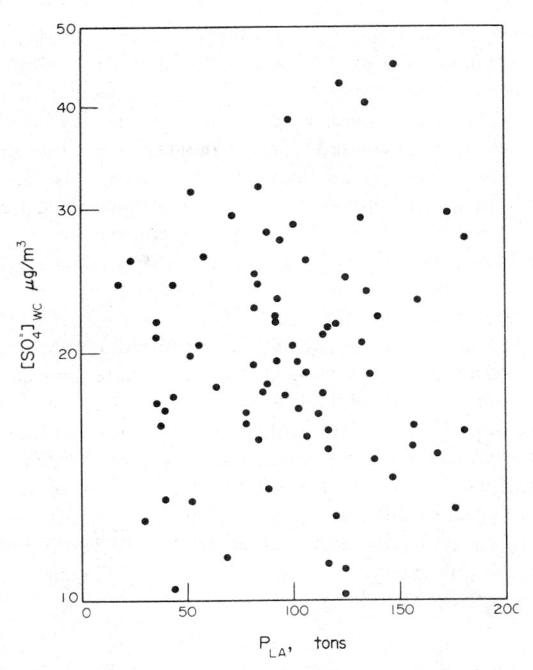

Fig. 3. Relationship of ambient sulfate concentration to power plant SO_2 emissions on days of high sulfate potential during May–October 1975.

ANALYSIS

Sulfate concentrations at West Covina did not correlate with the emissions of Los Angeles power plants on days of high sulfate potential in May–October 1975 (Fig. 3). This observation can be interpreted in terms of the linear roll-back model to given an upper bound for the average same-day contribution of Los Angeles power plant emissions to West Covina sulfate concentrations.

Our analysis will be based on the assumption that, for given conditions of meteorology and air chemistry, the sulfate concentration attributable to SO_2 emissions from power plants is proportional to the magnitude of those emissions. In mathematical terms,

$$[SO_4^{2-}] = AP + B,$$

Table 2. Statistical summary, May–October 1975

	Means								
	Geometric							Arithmetic	
	N	$[SO_4^=]_{WC}$	$[O_3]_{AZ}$	RH_{LB}	T_{LA}	P_{AB}	P_{LA}	P_{AB}	P_{LA}
All days:	184	13.2	13.8	63.1	22.5	66.4	99.7	72.1	109.6
Days of high sulfate potential:*	81	19.2	18.3	69.4	22.1	60.9	88.3	66.9	97.9
	Standard Deviations								
	Geometric							Arithmetic	
	N	$[SO_4^=]_{WC}$	$[O_3]_{AZ}$	RH_{LB}	T_{LA}	P_{AB}	P_{LA}	P_{AB}	P_{LA}
All days:	184	1.70	1.78	1.28	1.13	1.56	1.62	25.6	41.0
Days of high sulfate potential:*	81	1.41	1.47	1.07	1.08	1.60	1.64	25.9	39.7

*Days of high sulfate potential are those for which $RH_{LB} \geq 60\%$ and $T_{850} \geq \max(16^{\circ}C, T_{LA} - 4^{\circ}C)$.

where P and $[SO_4^{2-}]$ are, respectively, a given day's SO_2 emissions from power plants and average sulfate concentration. (Recall that the 'day' over which $[SO_4^{2-}]$ is measured starts 11 h after the day over which P is calculated.) The 'same-day' contribution of power plant emissions to ambient sulfate concentrations is AP, the coefficient A being determined by meteorology and air chemistry. The concentration of sulfate contributed by the same-day emissions of other sources, or carried over from previous days, is B.

It will be convenient also to assume that most of the meteorological variance in the sulfate concentration on days of high sulfate potential is due to factors which tend to scale the contributions of all sources equally. Ventilation is an example of such a 'universal' scaling factor, while wind direction is not. The effect of this assumption on the equation for sulfate concentration is to allow day-to-day variations in the influence of meteorology and air chemistry to be factored as follows:

$$[SO_4^{2-}] \cong X\,(aP + b). \qquad (1)$$

In Equation (1), X is a variable determined by 24-h average meteorology and air chemistry, a is a constant determined by long-term average meteorology and air chemistry, and b is a constant determined by non-power-plant emissions, which vary only slightly from day to day, and by long-term-average meteorology and air chemistry.

In logarithmic form, Equation (1) becomes

$$\log [SO_4^{2-}] \cong \log X + \log (aP + b).$$

Given the observed lack of correlation between $[SO_4^{2-}]$ and P, it is reasonable to suppose that same-day emissions from power plants contribute a relatively small fraction of the total sulfate concentration. In this case, the term in P can be linearized; $\log (aP + b) = \log b + \log (1 + aP/b) \cong \log b + aP/b$, so that

$$\log [SO_4^{2-}] \cong \log X + \log b + aP/b. \qquad (2)$$

If emissions (P) are independent of meteorology (X), or if the data set is stratified by meteorology to the point where correlations between X and P on the days of interest are unimportant, then Equation (2) furnishes a model for linear regression of $\log [SO_4^{2-}]$ on P (Crow *et al.*, 1960). Writing $c_0 = \log b + \log \bar{X}$, $c_1 = a/b$, and $\epsilon = \log (X/\bar{X})$, we have

$$\log [SO_4^{2-}] = c_0 + c_1 P + \epsilon.$$

The parameter $a/b = c_1$, which determines the relative importance of power plant emissions in Equation (1), can thus be estimated by regression. The average fraction, F, of the total sulfate concentration contributed by same-day emissions from power plants is then given by

$$F = \frac{a\bar{P}}{a\bar{P} + b} = \frac{(a/b)\bar{P}}{(a/b)\bar{P} + 1} = \frac{c_1\bar{P}}{c_1\bar{P} + 1}.$$

More generally, X is not independent of P, but is of the form $X = YZ$, where Y is independent of P and Z is not. If Z can be represented as a power function

$$Z = \prod_{i=2}^{n} W_i^{e_i}$$

of known meteorological variables W_i, then Equation (2) furnishes a model for multiple linear regression of $\log [SO_4^{2-}]$ on P and $\log W_i$:

$$\log[SO_4^{2-}] = c_0 + c_1 P + \sum_{i=2}^{n} c_i \log W_i + \epsilon,$$

where $c_0 = \log b + \log \bar{Y}$, $c_1 = a/b$, $c_i = e_i$, and $\epsilon = \log (Y/\bar{Y})$. Given the regression estimate for $a/b = c_1$, the average fraction of the sulfate concentration contributed by same-day emissions from power plants can be calculated as before.

In practice, X depends on many factors, and the value of c_1 obtained by regression will depend on the choice of meteorological variables considered. The critical variables are those which correlate with emissions; omission of variables which correlate only with the sulfate concentration expands the confidence interval about c_1, but does not affect the value of c_1 itself. This can be seen in the simple case where $Z = W^e$ is a function of one variable. The regression coefficient c_1 can then be written in terms of standard deviations (s) and correlation coefficients (r) as follows:

$$c_1 = \frac{s_{\log [SO_4^{2-}]}}{s_P} \times \left(\frac{r_{P,\log [SO_4^{2-}]} - r_{\log W,\log [SO_4^{2-}]}\, r_{P,\log W}}{1 - r_{P,\log W}^2} \right). \qquad (3)$$

If $r_{P,\log W} = 0$, then this expression simplifies to that for the regression coefficient obtained by regression of $\log [SO_4^{2-}]$ on P alone.

The relationship between X and P cannot be determined directly, because X is known only as the ratio $[SO_4^{2-}]/P$ and the relationship between $[SO_4^{2-}]$ and P is not known. It must instead be assessed indirectly by identifying, on physical grounds, the likely influences of meteorology and air chemistry on the two factors which determine P: the sulfur content and quantity of fuel burned in power plants. The type of fuel burned is determined, on a day to day basis, solely by the availability and cost of alternative fuels. The quantity of fuel burned, on the other hand, is affected by the demand for space heating and cooling, which is a strong function of temperature and a weaker function of r.h. and (as an index of photochemical smog levels) the ozone concentration. Of these, only temperature showed significant correlation with emissions on days of high sulfate potential (Table 3).

The above analysis suggests that P_{LA} was linked to X on days of high meteorological potential for sulfate principally through ambient temperature, T_{LA}. Regression of $\log [SO_4^{2-}]_{WC}$ on P_{LA} and $\log T_{LA}$ over

Table 3. Correlation coefficients for days of high sulfate potential during May–October 1975

	$\text{Log}[SO_4^=]_{WC}$	$\text{Log}[O_3]_{AZ}$	$\text{Log } RH_{LB}$	$\text{Log } T_{LA}$	P_{AB}	P_{LA}
$\text{Log } [SO_4^=]_{WC}$	1.00	.05	.35	-.23	.00	.03
$\text{Log } [O_3]_{AZ}$		1.00	-.34	.27	-.04	-.08
$\text{Log } RH_{LB}$			1.00	-.11	.00	.03
$\text{Log } T_{LA}$				1.00	.28	.28
P_{AB}					1.00	.94
P_{LA}						1.00

the 81 days of high sulfate potential yields the following relationship, with a multiple correlation coefficient of 0.25:

$$\log [SO_4^{2-}]_{WC} = 6.2 + 0.00089\, P_{LA} - 1.1 \log T_{LA}. \quad (4)$$

The coefficient $c_1 \cong a/b$ of P_{LA} corresponds to an estimate of

$$F = \frac{a\bar{P}_{LA}}{a\bar{P}_{LA} + b} \cong 8\%$$

for the average fraction of the West Covina sulfate concentration attributable to same-day emissions by Los Angeles power plants. (It may be noted here that $a/b\bar{P}_{LA} = 0.09$, justifying the approximation $\log(1 + a/bP) \cong a/bP$ made earlier.)

It is possible that there are additional linkages between meteorology and power plant emissions that have not been considered here. The sensitivity of the estimate for F to such potential confounding influences can be calculated from Equation (3). In order to obtain a value of $F = 45\%$, for example, it would be necessary to attribute about half of the variance of both emissions and ambient concentrations to common meteorological factors. Considering that local power plant operations are affected by a variety of factors, including the economics and availability of alternate modes of generation over a large regional grid, this is an improbably high degree of correlation.

The relative uncertainties in the above partial regression coefficients are large, due to the degree of scatter in the regression relationship. The actual values of the coefficients are therefore not as meaningful as their confidence intervals. At the 90% confidence level, the upper bound for the coefficient of P_{LA} in Equation (4) is 0.0021, corresponding to an estimate that

$$F < 17\%.$$

DISCUSSION

The statistical analysis of the preceding section indicated that, on days of high sulfate potential during the 1975 smog season, the 24-h average sulfate concentration at West Covina was relatively insensitive to the emissions of Los Angeles power plants. Under the assumptions of linear roll-back, it was estimated that an average day's power plant emissions accounted for less than 17% of the average sulfate concentration at West Covina. This figure is much smaller than the fraction of local SO_2 emissions attributed to power plants; power plants contributed 45% of the inventoried SO_2 emissions in Los Angeles County during May–October 1975 (Cass, 1975), and the corresponding figure for the basin is similar (White and Roberts, 1977). The discrepancy may reflect the spatial distribution of individual source plumes in the basin, or the differing chemical environments into which power plants and low-level sources release their effluents.

It must be emphasized that our estimate for the impact of power plants does not include contributions from emissions on previous days. Our analysis indicates that scattered one-day cutbacks in power plant emissions would yield disappointing reductions in ambient sulfate levels; in the absence of data on basin residence times, it can offer little guidance on the expected effect of an extended cutback. The characteristic residence time of air in the Los Angeles Basin is not known, and is difficult to judge from our data because the successive-day autocorrelation of both emissions and ambient concentrations is relatively high.

Our estimate for the impact of power plants was derived on the assumption that sulfates are produced in proportion to SO_2 emissions. An alternative hypothesis, consistent with the low correlation observed between SO_2 emissions and sulfate concentrations, would be that the atmosphere is to some degree 'saturated' with SO_2 (cf. Trijonis *et al.*, 1975), in the sense that sulfate formation is limited by other factors, such as acidities in the liquid phase. In this case, Equation (1) and our estimate for a/b would still have meaning as a description of the response, under existing conditions, of sulfate levels to incremental changes in power plant emissions. The overall share of the sulfate concentration attributable to power plants would not be well defined, however, because the effects of individual emissions would no longer be additive.

Acknowledgements—The authors are grateful to P. T. Roberts, J. P. Lodge, and the referees for helpful criticism of the manuscript.

REFERENCES

Cass G. R. (1975) Dimensions of the Los Angeles SO_2/Sulfate Problem. Memorandum No. 15, Environmental Quality Laboratory, California Institute of Technology.

Cass G. R. (1976) The relationship between sulfate air quality and visibility at Los Angeles. Memorandum No. 18, Environmental Quality Laboratory, California Institute of Technology.

Crow E. L., Davies F. A. and Maxfield M. W. (1960) *Statistics Manual*, Chap. 6, Dover, New York.

Hidy G. M. *et al.* (1974) Characterization of aerosols in California (ACHEX). Final Report to California Air Resources Board by Rockwell International Science Center.

Lee R. E. and Duffield F. V. (1977) EPA's catalyst research program: environmental impact of sulfuric acid emissions. *J. Air Pollut. Control Ass.* **27**, 631–635.

Richards L. W. (1976) The chemistry, dispersion and transport of air pollutants emitted from fossil fuel power plants in California: ground level pollutant measurements and analysis. Final Report to California Air Resources Board by Rockwell Air Monitoring Center.

Smith T. B., White W. H., Anderson J. A. and Marsh S. L. (1975) The chemistry, dispersion, and transport of air pollutants from fossil fuel power plants in California: airborne pollutant measurement and analysis. Final Report to California Air Resources Board by Meteorology Research, Inc.

Thomas M. D. (1962) Sulfur dioxide, sulfuric acid aerosol and visibility in Los Angeles. *Int. J. Air. Wat. Pollut.* **6**, 443–454.

Trijonis J. *et al.* (1975) An implementation plan for suspended particulate matter in the Los Angeles region. Final Report to U.S. Environmental Protection Agency by TRW Transportation and Environmental Operations.

White W. H. (1976) Reduction of visibility by sulfates in photochemical smog. *Nature* **264**, 735–736.

White W. H. and Roberts P. T. (1977) On the nature and origins of visibility-reducing aerosols in the Los Angeles air basin. *Atmospheric Environment* **11**, 803–812.

Zeldin M. D., Davidson A., Brunelle M. F. and Dickinson J. E. (1976) A meteorological assessment of ozone and sulfate concentrations in Southern California. Evaluation and Planning Division Report 76-1, Southern California Air Quality Management District.

Atmospheric Environment Vol. 12, pp. 785–790. Pergamon Press 1978. Printed in Great Britain.

SULPHATES AND SULPHURIC ACID IN THE ATMOSPHERE IN THE YEARS 1971–1976 IN THE NETHERLANDS

A. J. ELSHOUT, J. W. VILJEER and H. VAN DUUREN
N.V. Kema, Utrechtseweg 310, Arnhem, The Netherlands

(*First received* 6 *June* 1977 *and in final form* 8 *August* 1977)

Abstract—In 1971 an orientating study was started concerning the ground-level concentrations of sulphate and sulphuric acid relative to sulphur dioxide. In order to achieve a more selective method for the determination of sulphuric acid, the separation of sulphuric acid from sulphates by isopropanol was used. After separation sulphuric acid and sulphate are determined spectrophotometrically with barium chloranilate. Air samples were taken with air samplers using 0.8 μm pore cellulose triacetate filters. In the winter 1971–1972 at Arnhem the ratio sulphate/total sulphur correlated significantly with the relative humidity and decreased with increasing SO_2 concentration. The mean value of the ratio was 12.5% at a mean SO_2 concentration of 71 $\mu g\,m^{-3}$. The mean ratio H_2SO_4/total sulphur was 0.34%. In summer 1972, with a mean ratio sulphate/total sulphur of 36.8% at a mean SO_2 level of 8 $\mu g\,m^{-3}$, there was a correlation between sulphate and NH_3 on days with a relative humidity >80%; this suggests an influence on the oxidation rate by NH_3. The amounts of NH_4^+, SO_4^{2-}, and NO_3^- give rise to the supposition that the greater part of sulphate and nitrate is present as ammonium salts. The measurements were continued at Arnhem and Rotterdam (Rijnmond area) in 1973, 1975 and 1976. In these years there was a strong decrease in SO_2 concentration with a less stronger decrease in sulphate concentration. This together with the uniform distribution of sulphates in The Netherlands indicates that the sulphate level is caused mainly by long range transport. There was no increased reactivity in the formation of sulphates in the industrial Rijnmond area as compared to the eastern part of The Netherlands. Increased H_2SO_4 concentrations occurred as a result of transport of aerosols over sea by western winds. In summer 1976 there were significant correlations between both sulphate and nitrate with ozone at all the measuring stations, from which the existence of photochemical reactions over large areas can be concluded.

INTRODUCTION

The objective of the study was to get an impression of the amounts of sulphates and sulphuric acid in relation to the amount of SO_2 in the atmosphere over The Netherlands and to analyze the results in terms of correlations. In the first phase of the investigation (1971, 1972) measurements were carried out at Arnhem, a city in the eastern part of The Netherlands and partially simultaneously at Amsterdam in the western region. In the second phase (1973–1976) measurements were made at Arnhem and simultaneously at two points in the urbanized and industrialized area in and near Rotterdam.

The measuring period 1971–1976 was characterized in The Netherlands by a decrease in the SO_2 emission due to the large scale introduction of natural gas as fuel. The total emission of SO_2 decreased from 680 Gg in 1970 to 390 Gg in 1975 (Hartogensis, 1977). This resulted in a significant reduction of the ground-level SO_2 concentration, which was particularly noticeable in towns where coal and oil were replaced by natural gas as domestic fuel and where the average ground-level concentration dropped by a factor 2–3 (Ministry of Health and Environmental Protection, 1976; Evendijk and Post van der Burg, 1977).

SAMPLING AND ANALYSIS

Air samples are taken with air samplers, with a capacity of 4 $m^3\,h^{-1}$ when applying 0.8 μm pore cellulose triacetate filters (Gelman metrical GA-4). These filters are pretreated by rinsing them with water and drying again.

In order to achieve a more selective method for the determination of sulphuric acid it was decided to separate free sulphuric acid from sulphates by isopropanol. This principle, already applied to separate SO_3/H_2SO_4 from SO_2 in stack gas examinations, is recommended by Barton (1970) for use in air sample analysis too. In the resulting solutions sulphate is determined spectrophotometrically with barium chlorinate as described by Bertolacini and Barney (1957, 1958) and further developed by Schafer (1967).

Further evaluation of this method showed that it is necessary to centrifuge carefully after the reaction has finished; this results in considerably lower blanks. It was also necessary to use barium chloranilate prepared in the laboratory from chloranilic acid and barium chloride which, after having been dried over silica gel, showed much lower blank extinctions than did the formerly applied commercial product.

Under these precautions a calibration curve was established with sulphuric acid in 80% iso-propanol. Extinctions were measured at 332 nm in a 5 cm cell. These measurements showed per 100 ml initial solution: (a) an extinction of 0.011 for each μg SO_4^{2-}; (b) a detection limit of 1.5 μg SO_4^{2-}; (c) a precision of 0.36 μg SO_4^{2-}.

Aerosol collected on filters was analyzed according to Barton (1970) in such a way, that first the free

Table 1. Results of measurements in the period 1971–1976

Period	Time	Station	SO_2 ($\mu g\,m^{-3}$)			SO_4^{2-} ($\mu g\,m^{-3}$)			H_2SO_4 ($\mu g\,m^{-3}$)			NH_3 ($\mu g\,m^{-3}$)		
			n	x	s	n	x	s	n	x	s	n	x	s
13 Dec. 1971–31 March 1972	00h00–24h00*	Arnhem	296	71	49	302	11	6	302	0.3	0.3	79†	12	5
1 August–25 Sept. 1972	00h00–24h00*	Arnhem	102	8	6	102	5	5	101	0.1	0.1	92	4	3
17 Sept.–9 Nov. 1973	11h00–15h00	Arnhem	39	43	57	40	12	4	39	0.6	0.3			
		Rotterdam (centre)				29	10	4	29	0.9	0.7			
		Rotterdam (Waalhaven)	28	70	48	30	10	3	30	0.8	0.7			
1 Sept.–10 Oct. 1975	11h30–15h30	Arnhem	29	8	10	29	5	4	29	2.2	3.0			
		Rotterdam (Waalhaven)	30	48	29	27	3	2	27	1.6	1.7			
		Maasvlakte	29	8	10	29	4	3	29	1.7	1.1			
14 June–6 August 1976	11h00–15h00	Arnhem	33	4	8	40	6	6	40	0.9	0.8			
		Rotterdam (Waalhaven)	38	40	18	38	10	11	38	0.9	0.8			
		Maasvlakte	40	29	32	40	7	6	40	0.8	0.9			

* Four periods of 6 hours.
† March only.
n = number of samples, x = mean value, s = standard deviation.

sulphuric acid was extracted with 100% iso-propanol ($H_2O < 0.03\%$) which after extraction was diluted to 80%. After that, the filter was treated with new 80% iso-propanol to dissolve the residual sulphates. In both solutions other materials may be present showing some absorption at 332 nm. For this effect a correction is made in subtracting the extinction at 332 nm of the 80% iso-propanol solution before the barium chloranilate addition from the final extinction measured after this addition. The method of separation used for sulphuric acid and sulphates was studied in more detail. For the extraction with 100% iso-propanol a recovery was found of 95–105% for H_2SO_4 and 70% for NH_4HSO_4. $(NH_4)_2SO_4$, Na_2SO_4, K_2SO_4, $MgSO_4$ and $CaSO_4$ were not detectable in the extract after extraction of filters, loaded with these salts, with 100% iso-propanol. For these sulphates a recovery was found >95% after extraction with 80% iso-propanol. These results are in agreement with the results of Barton (1970) and Leahy *et al.* (1975). Leahy *et al.* (1972) reported recoveries of 40–100% for extraction of NH_4HSO_4 with 100% iso-propanol. So NH_4HSO_4 is partly or fully determined as sulphuric acid by this method. Besides, only phosphoric acid is known to be detected in small part as sulphuric acid (Barton, 1970). Hence it appears that the described method is specific to sulphuric acid and acid sulphates, allowing a sharper separation than does the also used microdiffusion method of Dubois *et al.* (1969).

On some filters NH_4^+ and NO_3^- were determined after extraction of a part of the filter with distilled water. NH_4^+ was analyzed using the indophenol-blue method (Tetlow and Wilson, 1964). NO_3^- was determined after reduction to NO_2^- (Hendriksen and Selmer-Olson, 1970). The concentration of NH_3 in the atmosphere was measured by absorption in 0.01 n H_2SO_4 after filtering the aerosol. NH_3 was determined as NH_4 with indophenol-blue (Tetlow and Wilson, 1964). SO_2 was determined using the TCM method (Scaringelli *et al.*, 1969).

RESULTS

(a) *First phase* (*1971–1972*)

Samples have been collected during successive 6 h periods in Arnhem from December 1971 until March 1972 (winter) and from August 1972 until September 1972 (summer). Sulphur dioxide content and relative humidity were recorded. Meteorological observations of the Weather Station Deelen near Arnhem were used for information about weather conditions, visibility, wind direction and wind speed. The results for the winter and summer periods are summarized in Table 1. The mean value of the ratio sulphate/total sulphur was 12.5% in winter 1971–1972 and 36.8% in summer 1972. For the ratio H_2SO_4/total sulphur the mean value was 0.34% in winter 1971–1972 and 1.05% in summer 1972.

A relationship was established between sulphate and sulphur dioxide in the winter period. For the total of measurements the linear regression equation $SO_4^{2-} = a \cdot SO_2 + b$ yielded values for a and b of 0.06 and 6.62 respectively. The coefficient of correlation of first order was 0.47 with a significance level >99%. The correlations divided in ranges of relative humidity showed significant differences in the value of a.

From these results a ratio sulphate:total sulphur is obtained, which ratio decreases with increasing sulphur dioxide concentrations as given in Fig. 1. This tendency is in agreement with the results of Thomas (1962) and Altshuller (1976), which implies that the average concentrations of sulphate and sulphuric acid aerosol, expressed in per cent of the total sulphur are mostly lower than 10% for SO_2 concentrations above 100 $\mu g\,m^{-3}$.

The measured sulphuric acid concentrations are relatively low and, expressed in per cent of total sul-

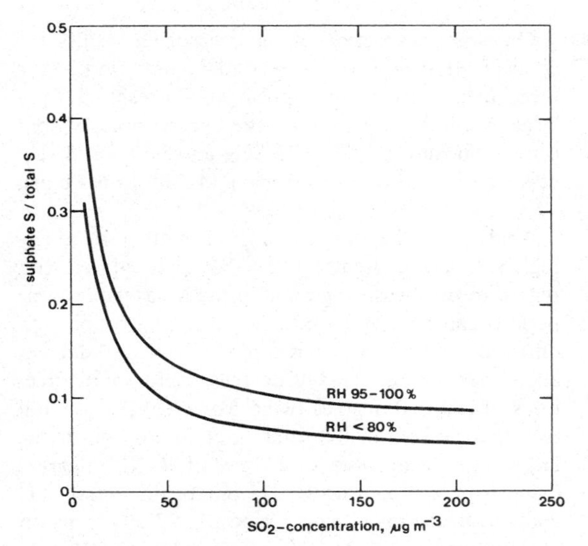

Fig. 1. The relation between the relative content of sulphate sulphur and sulphur dioxide at Arnhem in the winter period 1971–1972. (r.h. = relative humidity.)

phur, in the same order as found by Persson (1967). Persson determined sulphuric acid by titration to pH 5 of a part of the washing solution giving 'strong acids'. A comparison of the results of simultaneous measurements made on twenty days in Amsterdam and Arnhem showed that on these days the percentage ratio sulphate:total sulphur was higher in Amsterdam than in Arnhem: 18.5% compared with 8.8%. On the twenty days concerned, the most frequent wind direction was between east and south, with a number of days with a wind direction of 130–140 degrees. On those days the wind direction was equal to the direction of the line Ruhr-area (Western Germany)–Arnhem–Amsterdam. In view of the situation of the measuring stations at Arnhem and Amsterdam and with regard to the local urbanized and industrial areas in both cities, it can be concluded that at a wind direction of 130–140 degrees the measuring stations are influenced only for a relatively small extent by SO_2 emissions from their local surroundings. Between Arnhem and Amsterdam (distance 80 km) there are no important sources of SO_2, therefore the measured concentrations are mainly due to SO_2 emissions from the Ruhr-area in Western Germany.

On the days with a wind direction of 130–140 degrees a decrease in SO_2 concentration and an increase in sulphate ($SO_4^{2-} + H_2SO_4$) concentration was found in the transport direction, as given in Table 2. The pollutants arriving at Arnhem, at a distance of about 100 km from the centre of the Ruhr area, can be assumed to be already mixed in the available mixing layer. The horizontal dispersion over this relatively wide front of transported pollutants can also be considered to have no important influence on the concentration by further transport to Amsterdam, and therefore the oxidation rate of SO_2 can be approximated. Using the method of estimation of the transformation rate of SO_2 to sulphate from atmospheric concentration data as given by Alkezweeny and Powell (1977), including their assumptions, transformation rates were calculated for the days given in Table 2.

Assuming a difference between the deposition velocities of SO_2 and sulphate of 0.7 cm s^{-1} (grass and wood in the winter period) and a representative value of about 500 m for the depth of the mixing layer, the calculated transformation rates are 1.0, 4.4, 0.6 and 0.7% h^{-1} respectively. Moreover the possible influence of atmospheric NH_3 on the amount of sulphates was considered. During the winter period there appeared to be no distinct correlation. During the summer period a positive correlation $r = 0.81$ (significance level $>99\%$) was found for the linear regression between SO_4^{2-} and NH_3 (in $\mu g\,m^{-3}$) when grouping the results of days with a visibility <10 km, relative humidity $>80\%$ and SO_2 concentrations in the range of 0–25 $\mu g\,m^{-3}$. The regression equation is: $[SO_4^{2-}] = 0.92[NH_3] + 0.91$.

Results of measurements of the amounts of NH_4^+, NO_3^- and SO_4^{2-} of the sampled aerosols carried out in March 1973 at Arnhem give rise to the supposition that most of the sampled sulphate and nitrate particles consist of $(NH_4)_2SO_4$ and NH_4NO_3. Figure 2 gives the correlation between the amounts of NH_4^+ and $NO_3^- + SO_4^{2-}$ in micro-equivalents. The significance level of $r = 0.83$ is $>99\%$.

(b) *Second phase (1973–1976)*

Sampling of aerosols present in the atmosphere was carried out in this period at the previously mentioned measuring point at Arnhem and at two measuring points at Rotterdam, one of which was located in

Table 2. Concentrations of SO_2 and sulphate ($SO_4^{2-} + H_2SO_4$) at the stations Arnhem and Amsterdam on days in 1972 with a wind direction of 130–140 degrees

		Arnhem			Amsterdam		
Date	Mean wind velocity m s^{-1}	SO_2 $\mu g\,m^{-3}$	SO_4 $\mu g\,m^{-3}$	Sulphate S / total S %	SO_2 $\mu g\,m^{-3}$	SO_4 $\mu g\,m^{-3}$	Sulphate S / total S %
14 January	4	124	8	4.1	96	19	11.5
3 February	7	138	17	7.0	53	22	21.8
28 February	2	68	13	11.2	36	19	25.5
13 March	5	103	9	5.3	85	14	9.7

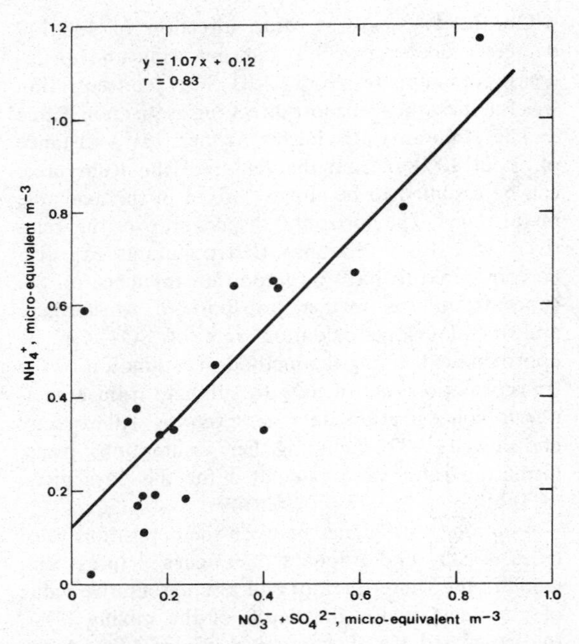

Fig. 2. The relation between NH_4^+ and $NO_3^- + SO_4^{2-}$ in the aerosol sampled at Arnhem in March 1973. (r = coefficient of correlation.)

Rotterdam itself, the other at the Maasvlakte situated directly on the sea almost 30 km west of Rotterdam.

The measurements were first made particularly in September and October in view of the air pollution episodes which had occurred in corresponding periods in 1970 and 1971 in the Rijnmond area (IG-TNO, 1971). The pollution in these episodes had in various quarters been referred to as photochemical smog formation, the component sulphuric acid being mentioned as a possible cause of the discomfort, a statement for which no corroborating evidence in the form of measuring results were available.

In Table 1 the results are summarized for the various periods in the three years that measurements were made in concert with the Gemeente-Energiebedrijf Rotterdam. From this table it can be seen that in these annual periods the average sulphuric acid content in the atmosphere was relatively low.

On closer inspection of the measuring results the concentration rates of the secondary pollution at the three measuring points appear to correspond to a large extent. Figure 3 shows the average weekly concentration rates of SO_4^{2-}, H_2SO_4 and NO_3^- in 8 successive weeks in the period from 14 July to 6 August 1976.

Analysis of the periods in 1975 with a relatively high sulphate content reveals that particularly with regard to the measuring point at the Maasvlakte these periods can be divided into two distinct groups with different wind directions: directions of 90–180 degrees and directions of 230–240 degrees. With wind directions of 230–240 degrees (wind from sea) the fraction of sulphuric acid in the total sulphate appears to be larger than with wind directions of 90–180 degrees (wind from the continent) at a practically equal SO_2 content. In the measuring period of 1975 the fraction of sulphuric acid in the total sulphate (H_2SO_4 + SO_4^{2-}) at this measuring point averaged 53% at wind directions of 230–240 degrees and 22% at wind directions of 90–180 degrees.

In the measuring period of 1975 a marked concentration peak of sulphuric acid was observed at all measuring points in the second week as can be seen from Fig. 4. A sea wind was blowing practically constant throughout this week. This observation is similar to that of Brosset *et al.* (1975) in their study of the nature and possible origin of acid particles observed at the Swedish West Coast. On a few occasions they found high-acid particles (type 2 particles). In these situations the air masses had moved over the North Sea for some days from the south-west. According to Brosset *et al.* the origin of these particles is a photochemical oxidation of SO_2. In slightly humid air the reaction results in a large number of particles consisting of hydrated SO_3. If the reaction takes place in an air mass with a low NH_3 concentration (sea air) a large amount of H_2SO_4-drops are obtained which in turn are neutralized by the present NH_3 to only a rather small extent.

Considering that photochemical reactions may occur in summer, samples were taken in the summer months of 1976 (14 July–6 August). In this part of

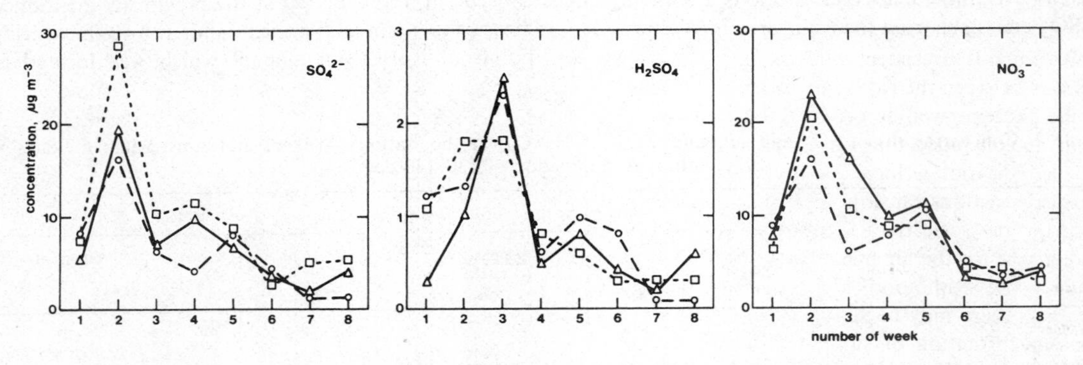

Fig. 3. The weekly mean concentrations of sulphate, sulphuric acid and nitrate at three stations in the period 14 June–6 August 1976. ○ Arnhem, □ Rotterdam (Waalhaven), △ Maasvlakte.

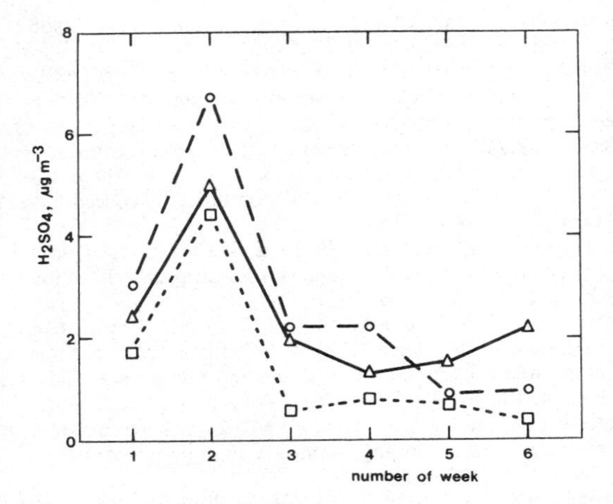

Fig. 4. The weekly mean concentration of sulphuric acid at three stations in the period 1 September–10 October 1975. ○ Arnhem, □ Rotterdam (Waalhaven), △ Maasvlakte.

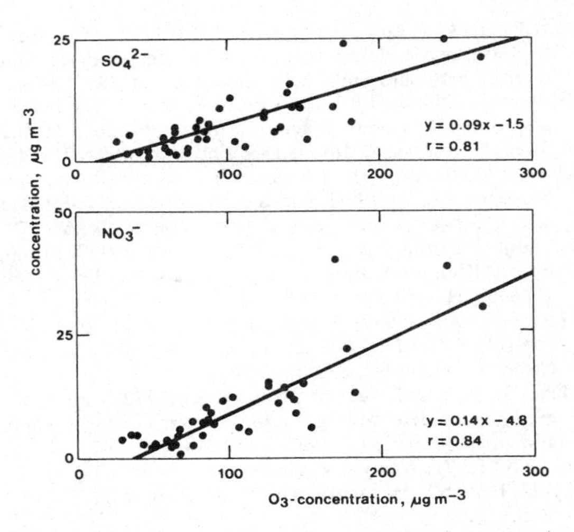

Fig. 5. The relations between sulphate (SO_4^{2-} + H_2SO_4)–ozone and nitrate–ozone at Maasvlakte in the period 14 June–6 August 1976. (r = coefficient of correlation.)

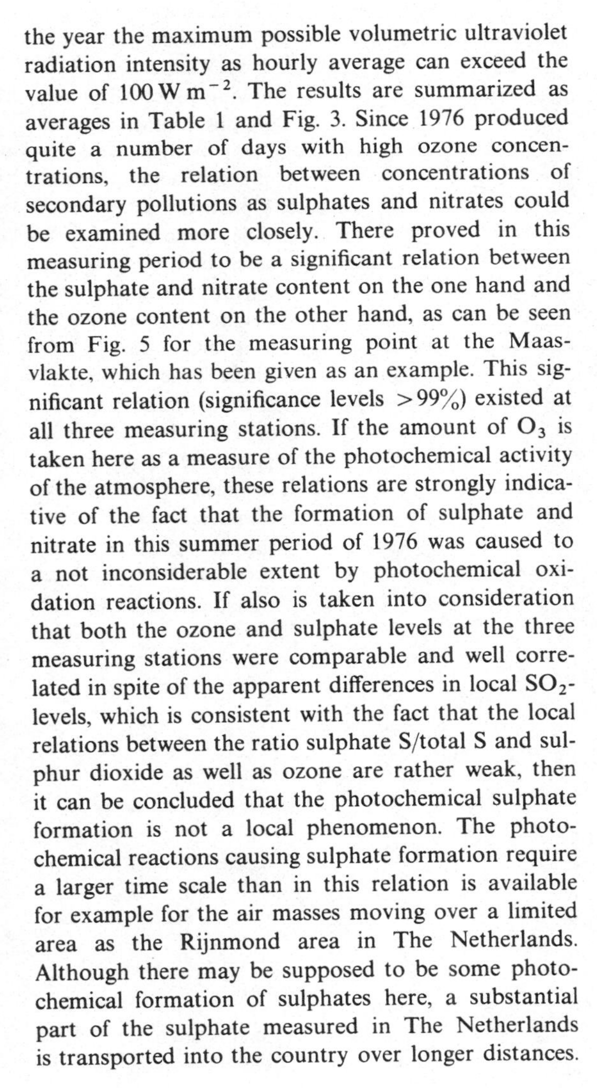

the year the maximum possible volumetric ultraviolet radiation intensity as hourly average can exceed the value of 100 W m^{-2}. The results are summarized as averages in Table 1 and Fig. 3. Since 1976 produced quite a number of days with high ozone concentrations, the relation between concentrations of secondary pollutions as sulphates and nitrates could be examined more closely. There proved in this measuring period to be a significant relation between the sulphate and nitrate content on the one hand and the ozone content on the other hand, as can be seen from Fig. 5 for the measuring point at the Maasvlakte, which has been given as an example. This significant relation (significance levels $>99\%$) existed at all three measuring stations. If the amount of O_3 is taken here as a measure of the photochemical activity of the atmosphere, these relations are strongly indicative of the fact that the formation of sulphate and nitrate in this summer period of 1976 was caused to a not inconsiderable extent by photochemical oxidation reactions. If also is taken into consideration that both the ozone and sulphate levels at the three measuring stations were comparable and well correlated in spite of the apparent differences in local SO_2-levels, which is consistent with the fact that the local relations between the ratio sulphate S/total S and sulphur dioxide as well as ozone are rather weak, then it can be concluded that the photochemical sulphate formation is not a local phenomenon. The photochemical reactions causing sulphate formation require a larger time scale than in this relation is available for example for the air masses moving over a limited area as the Rijnmond area in The Netherlands. Although there may be supposed to be some photochemical formation of sulphates here, a substantial part of the sulphate measured in The Netherlands is transported into the country over longer distances.

SUMMARY AND CONCLUSIONS

The levels of total sulphate measured in The Netherlands, which in various periods and places average between 5 and 12 μg m^{-3}, are caused to an important extent by transport over longer distances. These sulphates largely consist of ammonium compounds. Relatively high concentrations of sulphuric acid (including acid ammonium sulphates) of 5 μg m^{-3} and more are occasionally measured when the wind is blowing from the sea. In case of continental wind with pollution transport from Germany and Belgium the fraction of sulphuric acid in the total sulphur is lower than in the case of wind from the sea.

The average sulphuric acid level of 1 μg m^{-3} in The Netherlands is relatively low. In none of the measuring areas were indications found of higher locally formed sulphuric acid concentrations, neither in Arnhem nor in the Rijnmond area. There are clear indications that part of the sulphate present in summer is formed by photochemical oxidation reactions; however these reactions take place in a comparatively long time and a comparatively large space.

REFERENCES

Alkezweeny A. J. and Powell D. C. (1977) Estimation of transformation rate of SO_2 to SO_4 for atmospheric concentration data. *Atmospheric Environment* **11,** 179–182.

Altshuller A. P. (1976) Regional transport and transformation of sulfur dioxide to sulfates in the United States. *J. Air Pollut. Control Ass.* **36,** 318–324.

Barton S. C. (1970) A specific method for the automatic determination of ambient H_2SO_4-aerosol. Proceedings 2nd International Clean Air Congress, paper CP-7D, Washington.

Bertolacini R. J. and Barney J. E. (1957) Colorometric determination of sulfate with barium chloranilate. *Analyt. Chem.* **29,** 281–283.

Bertolacini R. J. and Barney J. E. (1958) Ultraviolet spectrophotometric determination of sulfate, chloride and fluoride with chloranilic acid. *Analyt. Chem.* **30**, 202–205.

Brosset C., Andreasson K. and Ferm M. (1975) The nature and possible origin of acid particles observed at the Swedish west coast. *Atmospheric Environment* **9**, 631–642.

Dubois L., Baker C. J., Teichman T., Zdrowjeski A. and Monkman J. L. (1969) The determination of sulphuric acid in air: a specific method. *Microchim. Acta* 269–279.

Evendijk J. E. and Post van der Burg P. A. R. (1977) Monitoring 'Rijnmond smog' for alert conditions. *Envir. Sci. Technol.* **11**, 450–455.

Hartogenesis F. (1977) A general policy for air pollution control in The Netherlands. Proc. Fourth International Clean Air Congress, 873–874, Tokyo.

Hendriksen A. and Selmer-Olson A. R. (1970) Automatic methods for determining the nitrate and nitrite in water and soil and extracts. *Analyst* **95**, 514–518.

IG-TNO (1971) Analyse van de smogsituatie in de Randstad Holland. Werkrapport G500, Delft, The Netherlands.

Leahy D., Siegel R., Klotz P. and Newmann L. (1965) The separation and characterization of sulfate aerosol. *Atmospheric Environment* **9**, 219–229.

Ministry of Health and Environmental Protection (1976) Bestrijding van de luchtverontreiniging, 56 Verslag Adviezen Rapporten.

Persson G. (1967) A study of the ratio sulphate:total sulphur and sulphuric acid:total sulphur in Gothenburg during different meteorological conditions. Proceedings OECD DAS/DSI/67.99.

Scaringelli F. P., Saltzman B. E. and Frey S. A. (1967) Spectrophotometric determination of atmospheric sulfur dioxide. *Analyt. Chem.* **39**, 1709–1719.

Schafer H. N. S. (1967) An improved spectrophotometric method for the determination of sulfate with barium chloranilate as applied to coal ash and related materials. *Analyt. Chem.* **39**, 1719–1726.

Tetlow J. A. and Wilson A. L. (1964) An absorptiometric method for determining ammonia in boiler feed-water. *Analyst* **89**, 453–465.

Thomas M. D. (1962) Sulfur dioxide, sulfuric acid aerosol and visibility in Los Angeles. *Int. J. Air Wat. Pollut.* **6**, 443–454.

Atmospheric Environment Vol. 12, pp. 791–794. Pergamon Press 1978. Printed in Great Britain.

VISIBILITY IN LONDON AND THE LONG DISTANCE TRANSPORT OF ATMOSPHERIC SULPHUR

R. A. BARNES

'Grendon',* Brackendale, Potters Bar, Hertfordshire, Great Britain

and

D. O. LEE

Department of Geography, Birkbeck College, Gresse Street, London W1, Great Britain

(*First received* 10 *June* 1977 *and in final form* 26 *August* 1977)

Abstract—The long distance transport of air pollutants and summer visibility in London is related to sulphate concentration and wind direction. The results suggest that a non-linear, approximately inverse relationship exists between visibility and aerosol sulphate concentration. The worst mean visibility (6.4 km) and the highest daily mean sulphate concentration (16.0 $\mu g\ m^{-3}$) both occur in airflows between east and south. The most likely source of the sulphate on these occasions would be emissions on the continental mainland.

1. INTRODUCTION

The transport of any pollutant in the atmosphere is governed by windspeed, wind direction and atmospheric stability. These variables are related to weather type. Thus, particularly when cleaning by precipitation is taken into account, the transport features of certain air flows can be identified over a period of time. These characteristics are obviously important in the context of air quality management.

Atkins *et al.* (1972) reporting on photochemical ozone and sulphuric acid aerosol formation over southern England, found that sulphate concentrations during days of easterly winds were very much higher (as were ozone levels) than during periods of westerlies. Because sulphate is formed only slowly in the atmosphere and is removed almost solely by precipitation, these authors conclude that the origin of the sulphate observed in central southern England on days of east winds probably originated over Northern France or the Low Countries. Cox *et al.* (1975) found that the rise of ozone concentrations in easterly winds was a regional, rather than a local phenomenon, and only loosely related, if at all, to indigenous emissions of precursor compounds. It therefore seems probable that the rise of ozone—and sulphate—concentrations reported by Atkins *et al.* (1972) was also of regional extent. Such a conclusion is confirmed by the work of Lee *et al.* (1974) who studied total suspended particulate (including sulphate) data obtained at six sites in England and Scotland over a six month period. However, the most remarkable finding of these authors was that at sites within 25 and 30 km of Manchester and London respectively, lower concentrations were to be observed in airflows from the conurbations (north-west and south-west, respectively) than in south-east winds.

Barnes (1976), in a study of data collected at 21 rural sites in England and Wales, examined the average decay rate of sulphur dioxide concentrations from a common source (London) in north-east and in south-west winds. He found that the mean distance required for concentrations to fall to a given level was about two-thirds greater in north easterlies than in south-westerlies. In a subsequent paper, Barnes and Eggleton (1977) show that mean sulphate (and nitrate) concentrations monitored at two remote sites, one on the English east coast and the other on the south-east coast are at their highest with airflows from the continent. The influence of foreign sulphur dioxide is also quite apparent at these two sites, although it is less marked than for the longer lived sulphate species.

The results of Barnes and Eggleton have recently been confirmed on a continental scale by NILU (1977). Their sector analysis is based on data obtained in 1973 and 1974 from 76 non-urban sites in 11 West European countries. It shows that mean aerosol sulphate concentrations are generally at their highest in winds from between north-east and south, the precise sector of highest concentration at a given site seemingly being determined by the site's orientation to the industrial heartlands of Central Europe. Again, sulphur dioxide shows a similar, but less distinct pattern.

To summarise, it seems that in north-west Europe winds from the north-east, east, south-east and, to a lesser extent, south, tend to produce higher concentrations at a given distance from an emission area

* Present address: Esso Research Centre, Abingdon, Oxfordshire, Great Britain.

than do winds from other directions, and that the higher concentrations are observed on a regional rather than a local scale. The higher concentrations are probably related to characteristics of these airstreams, such as stagnation over source areas, low mixing height, stability and general lack of precipitation. In some cases the effect of source strength, even when emissions are relatively local, can be swamped by differences in the concentration decay rate from one wind direction to another. In the case of the British Isles, which lie on the western margin of the European industrial complex, both wind direction, climatology and emission field combine in adversity. It is the purpose of this paper to show one facet of the environmental impact which such polluted airstreams bring to south-east England.

2. EXPERIMENTAL

Starting in October 1975, visibility observations have been made daily at 10.00 and 14.00 from the top floor of a nineteen storey office block in central London. The method of observation was based on that described in the Meteorological Office Observer's Handbook (1969). The closest standard object was the Big Ben clocktower (Palace of Westminster), 0.6 km distant, and the furthest was the northern tip of the Epping Ridge, 25 km away. All the standard objects lay between north-west and north-north-east of the observer, with the exception of one, 10.3 km away, to the south-south-east. Caution was exercised when reporting the visibility using the latter, since silhouetting was often experienced. Unfortunately, no easily identifiable object existed at a similar distance to the north which could be used as an alternative.

Aerosol sulphate data are obtained on a daily basis at 20 sites throughout the U.K. by Warren Spring Laboratory. Sampling began at most sites in March 1976. The sample is collected on a Whatman 40 filter paper from midnight to midnight and later analysed by X-ray fluorescence for sulphur. Full details of site locations, instrumentation and analytical technique are given by McInnes (1977). Results from the site at Gravesend, a Thameside town 30 km east of Central London, have kindly been supplied to the authors by Warren Spring Laboratory. The latter have themselves recently correlated sulphate concentrations with six meteorological variables, including visibility, but not wind direction (McInnes, in preparation).

Wind direction data were obtained from the British Meterological Office Daily Weather Report for Heathrow, a well exposed airport site about 20 km west of Central London. Observations at Heathrow are usually taken to be representative of most of south-east England and there is no reason to suppose they would be atypical of London or Gravesend.

From the data analysis point of view it was fortunate that both visibility and sulphate observations covered the summer of 1976. During winter, visibility is reduced by natural processes (fog, mist, etc.) and pollutants emitted locally by space heating requirements (especially open coal fires) in addition to aerosols transported from distant sources. The significance of the latter would therefore be difficult to isolate under such circumstances.

During a typical summer, 24-hour mean sulphate concentrations might not have sufficient temporal resolution to be used successfully in conjunction with visibility observations made only 4 h apart. Rainfall is a very effective atmospheric cleansing mechanism. Thus, if the visibility observations were made in the absence of precipitation, but within a 24-h sulphate sampling period during which precipitation had occurred, the reported sulphate concentration would probably not be representative of that at the time when the visibility observations were made. However, 1976 was an exceptionally dry summer over much of Europe. In the London area there was only a quarter of the average rainfall in the months of April–August and in the middle of the summer a period of three consecutive months saw precipitation on only ten days with consistently low relative humidity. It was therefore considered reasonable to attempt to determine any relationship between visibility, wind direction and sulphate concentration for the summer of 1976 using the customary 6 month period of April–September.

3. ANALYTICAL

The data available does not lend itself to detailed analysis for the reasons already discussed. However, a very simple technique was used to investigate if visibility is affected by sulphate concentration and whether the latter, in turn, is influenced by wind direction.

For the period April–September 1976 the mean of the two daily visibility observations was calculated

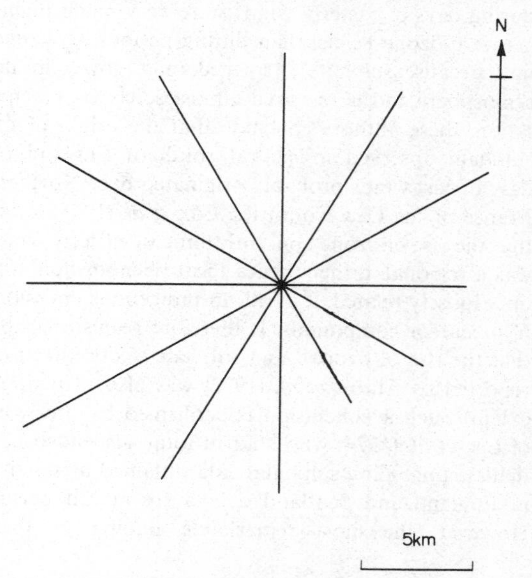

Fig. 1. Visibility according to wind direction London, summer 1976.

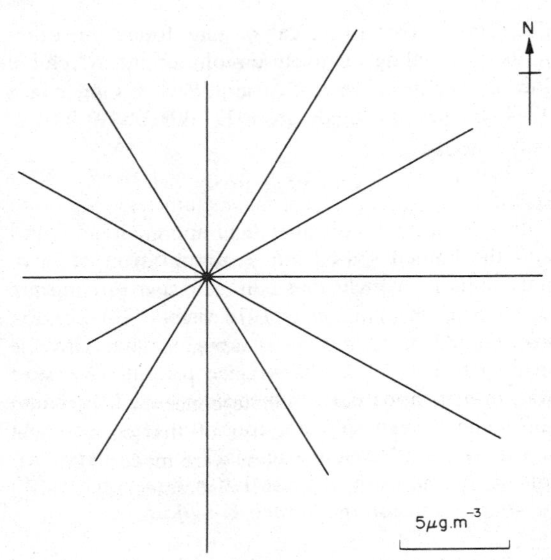

Fig. 2. Aerosol sulphate concentration according to wind direction London, summer 1976.

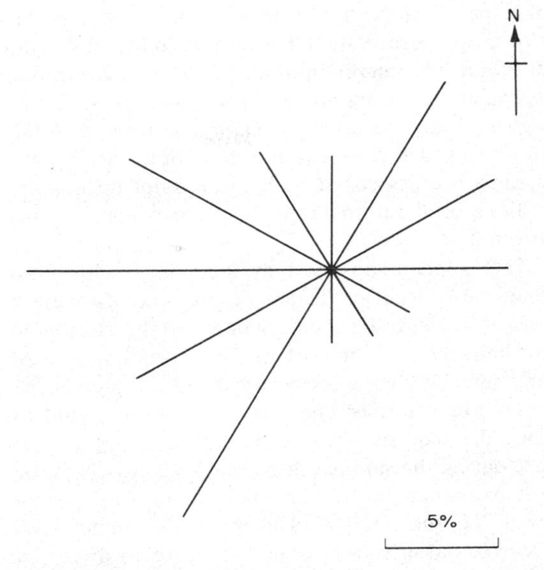

Fig. 3. Wind direction frequency London (Heathrow) 12.00 h, summer 1976.

and assigned to one of 12 sectors according to the wind direction at noon. Days on which one, instead of the normal two observations was made were omitted from the study. The mean visibility for each sector was calculated and the results used to construct a rose diagram (Fig. 1). Using a similar technique, a rose of mean aerosol sulphate concentration according to wind direction was also prepared (Fig. 2). For comparison purposes a wind direction frequency diagram for the period of study is included as Fig. 3. Finally, visibility was plotted against sulphate concentration, irrespective of wind direction, the results being shown in Fig. 4.

4. DISCUSSION

Figure 1 shows that in east and south-east winds visibility in London, during summer 1976, was below the typical 10–12 km in other wind directions. The worst mean visibility, 6.4 km, occurred in south-south-east winds. At the other end of the spectrum, south-westerlies were apparently the clearest airstreams with a mean visibility of 13.7 km.

As would be expected from the use of a single observation of wind direction to group 24-h mean sulphate concentrations, Fig. 2 does not show such a clear pattern as Fig. 1. However, the relationship

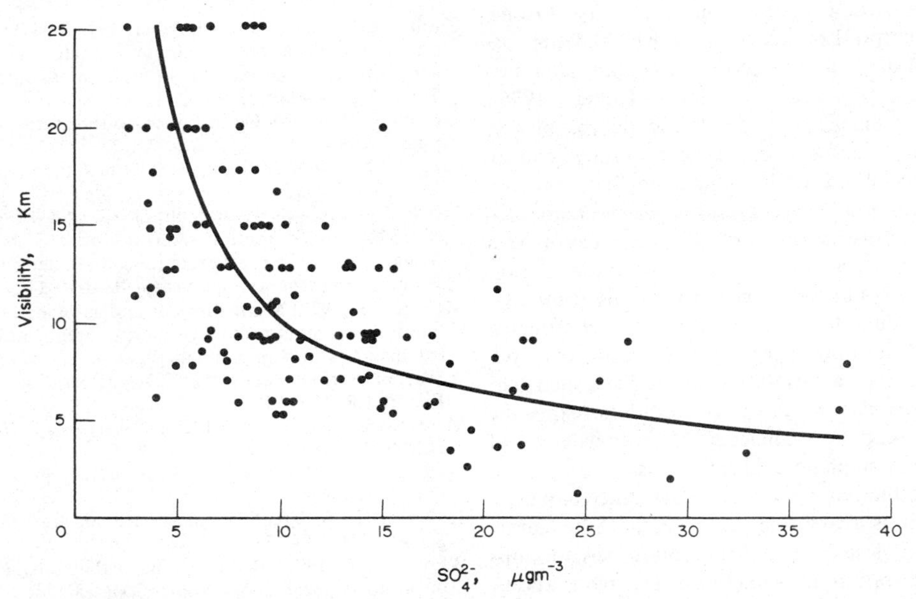

Fig. 4. Relationship between visibility and aerosol sulphate concentration London, summer 1976.

of sulphate concentration to wind direction is generally the inverse of that between visibility and wind direction. The minimum mean sulphate concentration ($6.4\,\mu g\,m^{-3}$) occurs in south-west winds with the highest means occurring in directions between south and north-east. Both easterly and south-easterly airstreams are associated with mean sulphate concentrations of about $16.0\,\mu g\,m^{-3}$, the maximum of any direction.

The sulphate observed in these latter directions could be of local or distant origin. Since Gravesend lies to the east of London, emissions in the Metropolis are unlikely to be important. Local sulphur dioxide emissions may have been oxidised in the presence of metal ion catalysts but this seems rather unlikely since the requisite relative humidities ($>70\%$) were rare during the summer in question. Alternatively, direct oxidation due to the uncommonly high ozone levels of the period (Tinker, 1976) could have occurred but in view of other work already discussed, sulphate produced in this way would seem to be less important than that coming from continental sources.

Figure 4 suggests that a non-linear inverse relationship probably exists between visibility and aerosol sulphate concentration but the spread of points prevents a more precise description from being made. The curve shown was fitted by eye.

In addition to the poor temporal compatibility of the data, the spread of points in Fig. 4 is probably explained by a number of secondary factors: (a) The finite limit of 25 km imposed on the visibility observations by the choice of standard objects would obviously introduce a bias since the true visibility would almost certainly have been greater on some occasions. (b) While visibility reduction due to natural causes or to locally produced sulphate would be unimportant in the long hot dry summer of 1976, other aerosols may have adversely affected visibility on days experiencing strong photochemical activity. Ozone levels in Central London rose above 20 pphm on several occasions and exceeded 25 pphm at times elsewhere in south-east England (Tinker, 1976). However, too much emphasis should not be placed on non-sulphate aerosols during these photochemical episodes since White (1976) concluded that they are much less important in reducing visibility than sulphate aerosols, even when representing the major part of the aerosol mass. (c) The size distribution of sulphate aerosol could well vary considerably from day to day according to the meteorological conditions prevailing at the time it was formed, its history and other factors. Such a variation in aerosol size distribution from one day to the next would be expected to alter the scattering efficiency of air masses with similar aerosol sulphate concentrations.

In view of the relatively crude data and analytical procedures it is reassuring to note that the curve shown in Fig. 4 suggests a background aerosol sulphate concentration of something less than about $2.0\,\mu g\,m^{-3}$. This is typical of that found by other authors sampling relatively unpolluted air at remote locations (Junge, 1960; Georgii, 1970; Cuong *et al.*, 1974; Barnes and Eggleton, 1977; NILU, 1977).

5. CONCLUSIONS

In view of the simple analytical approach employed and the limited spatial and temporal span of data, only tentative conclusions can be drawn. It appears that during the summer of 1976 visibility in London was related in a non-linear, approximately inverse manner to aerosol sulphate concentration. The latter was, in turn, apparently influenced by wind direction, the highest mean sulphate concentrations occurring in east and south-east airstreams. The most likely origin of the sulphate in these latter directions would be sources outside the United Kingdom.

Acknowledgements—The authors are indebted to Warren Spring Laboratory for supplying sulphate data. In addition, they wish to thank the Cartography Unit of Birkbeck College London for preparing the diagrams at very short notice.

REFERENCES

Atkins D. H. F., Cox R. A. and Eggleton A. E. J. (1972) Photochemical ozone and sulphuric acid aerosol formation in the atmosphere over southern England. *Nature* **235**, 372–376.

Barnes R. A. (1976) Long-term mean concentrations of atmospheric smoke and sulphur dioxide in country areas of England and Wales. *Atmospheric Environment* **10**, 619–631.

Barnes R. A. and Eggleton A. E. J. (1977) The transport of atmospheric pollutants across the North Sea and English Channel. *Atmospheric Environment* **11**, 879–892.

Cox R. A., Eggleton A. E. J., Derwent R. G., Lovelock J. E. and Pack D. H. (1975) Long-range transport of photochemical ozone in north-west Europe. *Nature* **255**, 118–121.

Cuong N., Bonsang B. and Lambert G. (1974) The atmospheric concentration of sulphur dioxide and sulphate aerosol over Antarctic, Subantarctic areas and oceans. *Tellus* **26**, 241–249.

Georgii H.-W. (1970) Contribution to the atmospheric sulphur budget. *J. geophys. Res.* **75**, 2365–2371.

Junge C. E. (1960) Sulphur in the atmosphere. *J. geophys. Res.* **65**, 227–237.

Lee R. E., Caldwell J., Akland G. G. and Frankhauser R. (1974) The distribution and transport of airborne particulate matter and inorganic components in Great Britain. *Atmospheric Environment* **8**, 1095–1109.

McInnes G. (1977) Multi-Element and sulphate in particulate surveys: monitoring locations, sampling and analytical methods and preliminary results reporting system. Warren Spring Laboratory, Department of Industry, Report LR 247 (AP).

Meteorological Office (1969) *Observer's Handbook.* H.M.S.O., London.

N.I.L.U. (1977) *Long Range Transport of Air Pollutants.* O.E.C.D., Paris.

Tinker J. (1976) Treating smog with air freshener. *New Scientist* **73**, 530–531.

White W. H. (1976) Reduction in visibility by sulphates in photochemical smog. *Nature* **264**, 737–736.

Atmospheric Environment Vol. 12, pp. 795–796. Pergamon Press 1978. Printed in Great Britain.

SHORT COMMUNICATION

EXPERIMENTAL RESULTS ON THE SO_2 TRANSFER IN THE MEDITERRANEAN OBTAINED WITH REMOTE SENSING DEVICES

(*First received September* 1977 *and in final form* 20 *October* 1977)

Experimental results have been recently obtained which concern the emission of SO_2 from anthropogenic sources and volcanoes in the Mediterranean. We used an automated moving air quality mapping laboratory (AMAQML). The concept was originated by L. Langan, of Environmental Measurement Inc. (EMI) at San Francisco, U.S.A.

The real time mapping system includes three operating units:

(a) an air quality data collection unit,
(b) a time and location data collection unit,
(c) a real time data acquisition and processing system.

A Barringer correlation spectrometer was used to measure overhead burdens of SO_2 along the traversing path of the moving laboratory.

The ground level concentrations of SO_2 and the overhead burden of SO_2 are expressed in map form. Other types of processing, like computing pollutant flow are also available.

The wind velocity at the plume level which is needed for computing the pollutant flow was obtained by teleanemometry, with an accuracy of 10–15% by means of a camera, a video recorder and geodimetric equipment.

The camera views the plume on an axis located by points on the azimuth and on site. The image received by the camera is reproduced on a screen on which the average apparent speed of the plume is easy to measure. Knowing the camera image multiplication factor, we calculate the real angular speed of the movement of the plume from the point of origin. The distance of the source being known by geodimetry and the direction of the speed of the wind being determined from AMAQML data, one can calculate the value of the absolute speed of the wind at the level of the plume.

RESULTS

(1) The Mount Etna volcano is an SO_2 source of prime importance in the western Mediterranean. As shown in Table 1, which combines results obtained from four experimental periods in 1975, 1976 and 1977, mass flow rate varies between 1100 T/day (low activity period) and 12,400 T/day (Punta Lucia eruption in 1976).

The SO_2 mass flow rate is therefore an indicator of the intensity of the volcanic activity and should be routinely monitored.

The uncertainties which accompany the measurement method of mass fluxes of SO_2 by correlation spectroscopy are examined in detail by Hamilton *et al.* (1978).

In ideal conditions (good luminosity and absence of particles and condensed water vapour) the precision is in the order of 30% (we have always profited from good luminosity).

We have taken samples on a number of occasions of the gas emitted by the Central crater (3200 m asl). Sampling was carried out on the lip of the crater and in general there was no condensed water vapour.

The SO_2 concentration was of the order of a hundred mg m^{-3} and the particle concentration inferior to 10 mg m^{-3}. Knowing the cross-section of the plume at the sampling level and having measured the wind speed by an anemometer, we determined the mass flux values. These values were found compatible with the values obtained by correlation spectrometry.

In the case of violent eruptions, the quantity of material in suspension is more important and although the measurements are made at a distance of about 10 km, the particles could be a cause of overestimation, without the order of magnitude being suspected.

The observed values in June 1975 correspond to the normal activity of Etna with a slow effusion of lava on the northern flank. Those obtained in January 1976 were correlated with a weak activity in the Punta Lucia cone which appeared two months before. In June 1976, the opening of a new summit vent on the 17th on the north-west flank of the north-east crater was preceded by two days of relatively slow SO_2 emission. Similarly in May 1977 the observed values were low—about seven weeks after a new eruption was observed in the north-east crater zone.

Year	Month	Day	SO_2 emission ($T\,d^{-1}$) at various times during the day
1975	June	14	4800 3800 3200 2700 3900 3600 4200 3700
		15	3300 4200 3800 4000
		18	4100 3700 3600 3600 3400
1976	January	31*	5400 12400 7400
	June	14	1600 1300 2300 2200
		15	2900 3700
		16	6100 6500
1977	May	22	1200 1100
		23	1100

* During Punta Lucia eruption.

(2) The plume of Mount Etna was identified at a distance up to 250 km, as the SO_2 polluted air was pushed above the Italian coast by southerly winds after crossing the Tyrrenian Sea. On 26 May 1977, the plume was observed from Reggio di Calabria up to Cosenza, the equivalent thickness being in the range of 4000–5000 m, and the width being 25 km at 50 km from the crater. On 27 May 1977, the plume was again observed above the Golfo di Policastro region. The equivalent thickness of the plume was in the range of 2000 m and the width was 20 km at 100 km from the crater. The trajectories of the polluted air are certainly influenced by the Apeninnes.

(3) During the same two days the excess of SO_2 at the ground level which could be attributed to the Etna emissions was in the range of 10–40 $\mu g\,m^{-3}$, the mass flow rate during this low activity period being only in the range

of 1000 T d^{-1}. This means that the transfer coefficient is about 10^{-9} s m^{-3} for distances of several hundred kilometers. During the periods of high activity, high concentrations of SO_2 should be observed downstream of the wind in regions which are fairly far away from Mount Etna. Studying the SO_2 plume during normal activity periods of Mount Etna should provide good experimental data for checking large distance transfer models.

CONCLUSIONS

The presented data are not a result of a systematic survey and therefore are not suitable for theoretical application. The main interest in these results is to draw attention to the importance of Etna as a major source of SO_2 in the Mediterranean region.

Commissariat à l'Energie Atomique
Dept. de Protection, B.P.6,
92260 *Fontenay-aux-Roses, France*

PIERRE ZETTWOOG
ROBERT HAULET

REFERENCES

Hamilton P. M., Varey R. H. and Millan M. M. (1978) Remote sensing of sulfur dioxide. *Atmospheric Environment* **12,** 127–133.

Haulet R., Zettwoog P. and Sabroux J. C. (1977) Sulphur dioxide discharge from Mount Etna. *Nature* **268,** 715–717.

AUTHOR INDEX

SUBJECT INDEX